FUNDAMENTALS OF

AQUATIC TOXICOLOGY

Second Edition

FUNDAMENTALS OF

AQUATIC TOXICOLOGY

Second Edition

Effects, Environmental Fate, and Risk Assessment

Edited by
Gary M. Rand, Ph.D.
Ecological Services Inc
North Palm Beach, Florida

CRC PRESS

Boca Raton London New York Washington, D.C.

Library of Congress Cataloging-in-Publication Data

Fundamentals of aquatic toxicology: effects, environmental fate, and
risk assessment / edited by Gary M. Rand.
p. cm.
Includes bibliographical references and index.
ISBN 1-56032-091-5 PB ISBN 1-56032-090-7 HB
1. Water quality bioassay. 2. Toxicity testing. 3. Waterl
pollution—Toxicity. 4. Aquatic organisms—Effect of water
pollution on. I. Rand, Gary M., date.
QH90.95.B5F86 1995
574.5′263—dc20 95-6381

International Standard Book Number 156032-091-5
Library of Congress Card Number 95-6381
Printed in the United States of America 3 4 5 6 7 8 9 0
Printed on acid-free paper

TABLE OF CONTENTS

Part II
ENVIRONMENTAL FATE

Part III
ASSESSMENT
Exposure/Effects, Environmental Legislation, and Hazard/Risk

Exposure/Effects

Environmental Legislation

CONTRIBUTORS

D. P. Anderson, Ph.D.
U.S. Fish and Wildlife Service
National Fish Health Research Laboratory
Box 700
Kearneysville, WV 25430

F. Balk
BKH Consulting Engineers
P.O. Box 5094
2600 GB, Delft,
The Netherlands

J. A. Bantle, Ph.D.
Department of Zoology
Oklahoma State University
201 Life Sciences East
Stillwater, OK 74078

A. Baril
Environment Canada
Wildlife Toxicology Division
National Wildlife Research Center
Ottawa, Ontario K1A 0H3
Canada

W. H. Benson, Ph.D.
University of Mississippi
Environmental Toxicology Research Program
School of Pharmacy
University, MS 38677

J. Blok, Ph.D.
BKH Consulting Engineers
P.O. Box 5094
2600 GB, Delft,
The Netherlands

L. A. Burns, Ph.D.
U.S. EPA
Environmental Research Laboratory
College Station Road
Athens, GA 30613

J. Cairns, Jr., Ph.D.
Virginia Polytechnic Institute and State University
University Center for Environmental and Hazardous Materials Studies
Derring Hall, Room 1020
Blacksburg, VA 24061-0415

G. A. Chapman, Ph.D.
U.S. EPA
2111 S.E. Marine Science Drive
Newport, OR 97365-5260

J. R. Clark, Ph.D.
Exxon Biomedical Sciences, Inc.
Research & Environmental Health Division
Mettlers Road CN 2350
East Millstone, NJ 08875-2350

J. D. Cooney, Ph.D.
New England Bioassay
77 Batson Road
Manchester, CT 06040

N. DiGiulio
Toxikon Corporation
225 Wildwood Avenue
Woburn, MA 01801

R. T. Di Giulio, Ph.D.
Duke University
School of Forestry and Environmental Studies
Ecotoxicology Laboratory
Durham, NC 27706

P. B. Dorn, Ph.D.
Shell Development Company
3333 Highway 6 South
Houston, TX 77251-1380

M. R. Ellersieck, Ph.D.
University of Missouri
Agricultural Experimental Station
105 Math Sciences Building
Columbia, MO 65211

J. P. Giesy, Ph.D.
Michigan State University
Department of Fisheries and Wildlife
#13 Natural Resources Building
East Lansing, MI 48824-1222

N. J. Grandy, Ph.D.
Organization for Economic Co-operation and Development
Chemicals Division
2, rue André-Pascal
75775 Paris Cedex 16, France

R. L. Graney, Ph.D.
Miles Inc.
17745 South Metcalf
Stilwell, KS 66085

W. E. Hawkins, Ph.D.
Gulf Coast Research Laboratory
P.O. Box 7000
703 East Beach Drive
Ocean Springs, MS 39564-7000

J. F. Hobson, Ph.D.
Technology Service Group, Inc.
1101 17th Street, N.W.
Suite 500
Washington, DC 20036

G. W. Hudiburgh, Jr., Ph.D.
U.S. EPA
Office of Wastewater Enforcement and Compliance
401 M Street, S.W.
Washington, DC 20460

C. G. Ingersoll, Ph.D.
Midwest Science Center
National Biological Service
4200 New Haven Road
Columbia, MO 65201

K. D. Jenkins, Ph.D.
California State University - Long Beach
Molecular Ecology Institute
1250 Bellflower Blvd.
Long Beach, CA 90840

T. W. La Point, Ph.D.
Clemson University
TIWET
One Tiwet Drive
Box 709
Pendleton, SC 29670

R. J. Larson, Ph.D.
The Procter and Gamble Co.
Ivorydale Technical Center
5299 Spring Grove Avenue
Cincinnati, OH 45217

C. R. Lee, Ph.D.
U.S. Army Corps of Engineers
3909 Halls Ferry Road
Vicksburg, MS 39180-0631

M. A. Lewis, Ph.D.
U.S. EPA
Sabine Island
Gulf Breeze, FL 32561

R. L. Lipnick, Ph.D.
U.S. EPA
Office of Pollution Prevention and Toxics
401 M Street, S.W.
Washington, DC 20460

W. J. Lyman, Ph.D.
Camp Dresser and McKee, Inc.
100 Cambridge Center
Cambridge, MA 02142

D. Mackay, Ph.D.
University of Toronto
Chemical Engineering Department
Toronto, Ontario M5S 1A4
Canada

A. V. Malloy
FMC Corporation
Box 8
Princeton, NJ 08540

R. K. Markarian, Ph.D.
Entrix
200 Bellevue Parkway
Suite 200
Wilmington, DE 19809

L. S. McCarty, Ph.D.
L.S. McCarty Scientific Research and Consulting
280 Glen Oak Drive
Oakville, Ontario L6K 2J2
Canada

P. Mineau, Ph.D.
Environment Canada
Wildlife Toxicology Division
National Wildlife Research Center
Ottawa, Ontario K1A 0H3
Canada

B. R. Niederlehner
Virginia Polytechnic Institute and State University
University Center for Environmental and Hazardous Materials Studies
Derring Hall, Room 1020
Blacksburg, VA 24061-0415

R. M. Overstreet, Ph.D.
Gulf Coast Research Laboratory
P.O. Box 7000
703 East Beach Drive
Ocean Springs, MS 39564-7000

D. J. Paustenbach, Ph.D.
McLaren/Hart
ChemRisk Division
1135 Atlantic Avenue
Alameda, CA 94501

G. M. Rand, Ph.D.
Ecological Services Inc.
North Palm Beach, FL 33408

B. M. Sanders, Ph.D.
California State University - Long Beach
Molecular Ecology Institute
1250 Bellflower Blvd.
Long Beach, CA 90840

P. J. Sheehan, Ph.D.
McLaren/Hart
ChemRisk Division
1135 Atlantic Avenue
Alameda, CA 94501

L. R. Shugart, Ph.D.
Oak Ridge National Laboratory
Environmental Sciences Division
P.O. Box 2008
Oak Ridge, TN 37831-6036

E. P. Smith, Ph.D.
Virginia Polytechnic Institute and State University
Department of Statistics
University Center for Environmental and Hazardous Materials Studies
Blacksburg, VA 24061

A. Spacie, Ph.D.
Purdue University
Department of Forestry and Natural Resources
West Lafayette, IN 47907

R. L. Steele, Ph.D.
U.S. EPA
2111 S.E. Marine Science Drive
Newport, OR 97365-5260

G. W. Suter II, Ph.D.
Oak Ridge National Laboratory
Environmental Sciences Division
P.O. Box 2008-MS 6038
Oak Ridge, TN 37831

G. B. Thursby, Ph.D.
U.S. EPA
Environmental Research Laboratory
27 Tarzwell Drive
Narragansett, RI 02882

L. W. Touart, Ph.D.
U.S. EPA
Ecological Effects Branch
401 M Street, S.W.
Washington, DC 20460

R. van Compernolle, Ph.D.
Shell Development Company
3333 Highway 6 South
Houston, Texas 77251-1380

P. A. Van Veld, Ph.D.
The College of William and Mary
Virginia Institute of Marine Science
P.O. Box 1346
Gloucester Point, VA 23062

P. G. Vincent, Ph.D.
U.S. FDA
Office of Center Director, HFD-004
5600 Fishers Lane
Rockville, MD 20857

J. D. Walker, Ph.D.
U.S. EPA
Office of Pollution Prevention and Toxics
Interagency Testing Committee
401 M. Street, SW (TS-792)
Washington, DC 20460

W. W. Walker, Ph.D.
Gulf Coast Research Laboratory
P.O. Box 7000
703 East Beach Drive
Ocean Springs, MS 39564-7000

G. S. Ward
Toxikon Environmental Sciences
106 Coastal Way
Jupiter, FL 33477

P. G. Wells, Ph.D.
Dalhousie University
School for Resource and Environmental Studies
1312 Robie Street
Halifax, Nova Scotia B3H 3E2
Canada

D. M. Woltering, Ph.D.
Environ Corporation
4350 N. Fairfax Drive
Arlington, VA 22314

M. G. Zeeman, Ph.D.
U.S. EPA
Office of Pollution Prevention and Toxics
401 M Street, S.W.
Washington, DC 20460

PREFACE

The intent and philosophy of the second edition of this textbook remain similar to those delineated in the preface to the first edition. It is designed primarily as a text for graduate courses in aquatic and environmental toxicology that deal with general principles and concepts, and the evaluation of toxicity data with emphasis on methodologies used in aquatic toxicology. The aspects of the prior edition that have made it useful as a reference to students and scientists have been retained. The second edition has been further expanded to include an update on current testing procedures, data evaluation and interpretation, fate, environmental legislation, and risk assessment. New authors have been added to broaden the input and provide expanded coverage of the changing field of aquatic toxicology. The text still provides the only comprehensive source of information in aquatic toxicology for graduate students and practitioners in the environmental sciences.

The introduction to the book reviews basic concepts and principles, examines concentration-response relationships, and presents an update on new findings for interpreting aquatic toxicity data. The following 32 chapters are divided into three parts. Part I, Effects—Toxicity Testing, describes basic toxicological concepts and methodologies used in aquatic toxicity testing, including the philosophies underlying testing strategies now required to meet and support regulatory standards. Each method is discussed from a state-of-the-art approach so that a student can perform the test and obtain insight into its rationale. Included in Part I are general chapters on freshwater and marine tests in addition to chapters on special tests on microalgae and vascular plants, macroalgae, sea urchin, frog embryos, sediment, and field impact. The quantification of data and the use of good laboratory practices are also reviewed. Part I concludes with chapters on toxicity testing for specific effects on the immune system (immunotoxicity), genetic makeup (genotoxicity), and tissues (carcinogenicity). The latter topics have received considerable attention as potential "biomarkers" of exposure and effects of chemicals. Fate, Part II, discusses various factors that affect the transport, transformation, and ultimate distribution of chemicals in the aquatic environment and thus influence the concentrations to which aquatic organisms may be exposed. Part II includes chapters on how the aquatic environment and the organism (uptake, distribution, biotransformation, elimination) affect chemical fate. Biochemical biomarkers are also discussed. Part II concludes with a chapter on fate modeling. Part III, Assessment, discusses types of functional effects (or endpoints) evaluated in field studies and the use of structure-activity relationships in aquatic toxicology to predict biological activity and physicochemical properties of a chemical. Part III also contains an extensive background of environmental legislation in the United States and within the European Community, and an introduction to hazard/risk assessment with case studies. The literature citations and supplemental reading sections at the end of each chapter will carry the reader beyond the methods and techniques presented in this text.

Because aquatic toxicology is a specialized discipline, with its own terminology, a glossary of the most commonly used terms is included at the end of the book. An appendix with four chapters from the first edition will also provide the reader with additional references on basic information in aquatic toxicology. The editor and publisher include them here in the second edition as classic chapters in aquatic toxicology that should be kept in print and available for reference by more recent works.

This second edition, like the first edition, provides a broad perspective on the subject of aquatic toxicology. The editor is grateful to the contributors from academia, industry and the government whose combined expertise make a volume with this breadth of coverage possible.

Gary M. Rand, Ph.D.

PREFACE TO THE FIRST EDITION

Fundamentals of Aquatic Toxicology is designed to fill the need for a single, comprehensive source of information concerning aquatic toxicology. It presents a definitive description of basic concepts and test methods employed in aquatic toxicology studies as well as examples of typical data and their interpretation.

This volume is designed to be used as a textbook for courses in aquatic and environmental toxicology. In addition, it should be a useful reference for those whose responsibilities include managing industrial plant operations and interacting with agencies charged with regulating chemical impacts on aquatic ecosystems.

The contributors are scientists and managers from academic, government, and industrial institutions. In addition to their extensive knowledge of the theoretical aspects of the individual topics, each contributor is experienced in the practical application of these theories to actual environmental situations.

The 23 chapters are divided into five parts. The first part–Toxicity Testing–describes the basic concepts and methodologies used in aquatic toxicity testing. Sublethal Effects, the second part, presents information on sublethal effects testing and its utility in evaluating the less obvious effects of chemical exposure on aquatic organisms. The third part–Specific Chemical Effects–summarizes the available literature on the toxicity of generic types of chemicals (such as pesticides and metals) to aquatic organisms. Chemical Distribution/Fate, the fourth part, discusses the various factors that affect the distribution and fate of chemicals in the aquatic environment and thus influence the chemical concentrations to which aquatic organisms may be exposed. The concluding fifth part–Hazard Evaluation–discusses the manner in which environmental fate and biological effects data are integrated to provide an assessment of the potential hazard posed by the use or discharge of chemicals in the aquatic environment. It also identifies the specific laws that provide regulatory agencies with enforcement powers to control discharges into the aquatic environment.

Since aquatic toxicology is a specialized discipline, with its own terminology, a glossary of the most commonly used terms is included.

This book represents an attempt to provide a broad perspective on the subject of aquatic toxicology. We are grateful to the contributors, whose combined theoretical knowledge and practical experience make a volume of this scope possible. We will consider that our objective has been met if this book proves pertinent to the needs of its readers and if it provides a stimulus for further advancement in this evolving science.

Gary M. Rand, Ph.D.
Sam R. Petrocelli, Ph.D.

INTRODUCTION

Chapter 1 sets the "stage" for the book. It provides background information on the characteristics and composition of water and sediment and it presents general concepts, principles, and terminology basic to aquatic toxicology including a description of general types of toxic agents and their effects. The chapter further defines the criteria and approaches to toxicity testing, general experimental design and methodology for single- and multispecies toxicity studies, the importance of good laboratory practices (GLPs), and considerations for interpreting aquatic toxicity data. A comprehensive supplemental reading section at the end of the chapter (as in many other chapters) contains additional references on selected topics that will be useful for teaching and practitioners.

The reader is encouraged to read Chapter 1 and return to it after reading the other chapters to obtain an understanding of important concepts and principles relevant to the science of aquatic toxicology.

Chapter 1

INTRODUCTION TO AQUATIC TOXICOLOGY

G. M. Rand, P. G. Wells, and *L. S. McCarty*

AQUATIC TOXICOLOGY

Aquatic toxicology is the study of the effects of manufactured chemicals and other anthropogenic and natural materials and activities (collectively termed toxic agents or substances) on aquatic organisms at various levels of organization, from subcellular through individual organisms to communities and ecosystems. Effects can cause both positive and negative deviations from previously existing circumstances, but aquatic toxicology focuses primarily on the deviations that are considered to be adverse in nature and on recovery processes in biota that may occur when exposures diminish. Adverse effects at the organismal level include both short-term and long-term lethality (expressed as mortality or survival) and sublethal effects such as changes in behavior, growth, development, reproduction, uptake and detoxification activity, and tissue structure. Adverse effects at the suborganismal level include induction or inhibition of enzymes and/or enzyme systems and their associated functions. At the supraorganismal level adverse effects include changes in species genotype and/or phenotype, as well as changes in the number, relative abundance, and physiological condition of species typically found in a given community type.

Effects at the different levels of organization may be quantified by a variety of criteria, such as number of organisms killed (or surviving), reproductive success (from egg production and hatchability to year-class recruitment), whole-body (length and weight) or organ condition factors, number of teratogenic abnormalities or incidence of tumors, induction or inhibition of enzyme activity, and number and abundance of species in an ecological community. Because exposure to toxic agents may be via the water, sediment, and food in the aquatic environment, the quantities, concentrations, and bioavailability of toxic agents in these compartments are of primary concern. Thus, the study of the sources, transport, distribution, inorganic and organic transformation, and ultimate fate of toxic agents in the aquatic environment is a vital component of aquatic toxicology.

Aquatic toxicology is a branch of the science of ecotoxicology that is multidisciplinary in scope and interdisciplinary in practice. *Ecotoxicology* was defined by Truhaut (1975, 1977) and later by Butler (1978) as the branch of toxicology that studies the toxic effects of natural or artificial substances on living organisms (e.g., fish, birds, plants), whether animal or vegetable, terrestrial or aquatic, that constitute the biosphere. It also includes the interaction of these substances with the physical environment in which these organisms live. Ecotoxicology is a young science compared to mammalian toxicology, whose long history evolved from pharmacology. The differences between classical mammalian toxicology and ecotoxicology are listed in Table 1. Ecotoxicology is sometimes used synonymously with *environmental toxicology*; however, the latter also encompasses the effects of environmental chemicals and other agents on humans.

Aquatic toxicology has evolved as a field of study, borrowing freely from several other basic sciences (Figure 1). It is necessary to understand the chemical (e.g., hydrolysis, oxidation, and photolysis), physical (e.g., molecular structure, solubility, volatility, and sorption), and biological (e.g., biotransformation) factors that affect environmental concentrations of chemicals, to determine how potentially toxic agents act in the environment and how the environment acts on these agents and to estimate the potential exposure of aquatic organisms. A working knowledge of aquatic ecology, one or more biological subdisciplines such as physiology, biochemistry, histology, and behavior, and environmental chemistry is required to understand the effects of toxic agents on aquatic organisms.

Structure-activity relationships are important as they enable biological activities (e.g., toxicity, bioaccumulation, biodegradation) and environmental fate to be explained, and subsequently predicted, from the structure of chemicals. Statistical analysis and

Table 1. Mammalian toxicology and ecotoxicology differ in many respects

Mammalian toxicology	Ecotoxicology[a]
Objective: to protect humans from exposure to toxic substances and materials at concentrations which are or may be associated with adverse effects	Objective: to protect populations and communities of many diverse species from exposure to toxic substances and materials at concentrations which are or may be associated with adverse effects
Must almost always rely on animal models (e.g., rat, mouse, guinea pig, rabbit) since experimentation with humans is not feasible	Can experiment directly on species of concern (although there may be uncertainty on whether the most appropriate "indicator" or "sensitive" species is used)
Species of interest (man) is known; thus degree of extrapolation is more certain	Not able to identify and test all species of concern; thus, degree of extrapolation is uncertain. Organism responses and toxicity may be different in more complex natural systems because of bioavailability of chemical, organic matter concentrations and other environmental interactions
Test organisms are homeothermic or warm-blooded (body temperature is relatively uniform and nearly independent of environmental temperature); thus, toxicity is predictable	Test organisms (aquatic) live in a variable environment and most are poikilothermic or cold-blooded (body temperature varies with the environmental temperature), birds and aquatic mammals being the exception; thus toxicity may not be sufficiently predictable
The dose of a test chemical usually can be measured directly and accurately, and may be administered by a number of routes. However, unless "absorbed dose" measurements are made via tissue dosimetry, the typical LD50 (e.g., oral bolus) estimate is an external or exposure dose	The external or exposure "dose" is known in terms of the chemical's concentration in a medium (typically water, but also sediment and/or food) and the length of exposure to it; the actual "absorbed dose" is often determined now experimentally using bioconcentration/bioaccumulation and metabolism studies
Extensive "basic" research has been conducted; emphasis has been on understanding mechanisms of toxic action	Much less "basic" research has been conducted, as emphasis has been on measuring toxic effects and generating media-based threshold concentration data, with an eye toward regulatory needs. More recently, emphasis has been on mechanisms of action and structure-activity relationships
Test methods are well developed, their usefulness and limits well understood	Many commonly used test methods are relatively new and some are formalized (standardized). However, their usefulness in many cases at predicting field impacts and protecting natural ecosystems is often uncertain

[a]Organisms can include aquatic and terrestrial species including plants, invertebrates, fish, birds, and mammalian wildlife.
Adapted from Rand, 1991.

mathematical modeling are used to quantitate and predict biological effects and determine the probability of their occúrrence under different conditions. Finally, it is essential that the practitioner comprehends sophisticated analytical chemistry methodologies and instrumentation well enough to ensure accurate identification and quantification of extremely low concentrations of chemicals in various aquatic environmental compartments (e.g., sediment, water, and tissues).

A SHORT HISTORY OF AQUATIC TOXICOLOGY

There are various perspectives on the origin, development, and landmark events of aquatic toxicology. Many of its current practitioners have come from many other disciplines; only in the past 25 years have university graduate programs been available for training aquatic toxicologists and the field's historical roots are intertwined with those of classical toxicology and environmental chemistry. This section gives an overview of some of the major developments and events contributing to the science of aquatic toxicology, circa 1994. It also acts as a useful chronological framework illustrating advances in understanding of impacts of pollutants on aquatic biota.

One account of early history is given by Manahan (1989): "The origins of modern toxicology can be traced to M.J.B. Orfila (1787–1853), a Spaniard born on the Island of Minorca. In 1815, Orfila published a classic book, the first ever devoted to the harmful effects of chemicals on organisms. This work discussed many aspects of toxicology recognized as valid today. Included are the relationships between the demonstrated presence of a chemical in the body and observed symptoms of poisoning, mechanisms by which chemicals are eliminated from the body, and treatment of poisoning with antidotes."

Likewise, Rodricks (1992) attributed modern, mostly mammalian toxicology as coming from activities in clinical toxicology, pharmacology, and occupational medicine, together with nutritional science, radiation biology, and pesticide science. He states that "the systematic study of toxic effects in

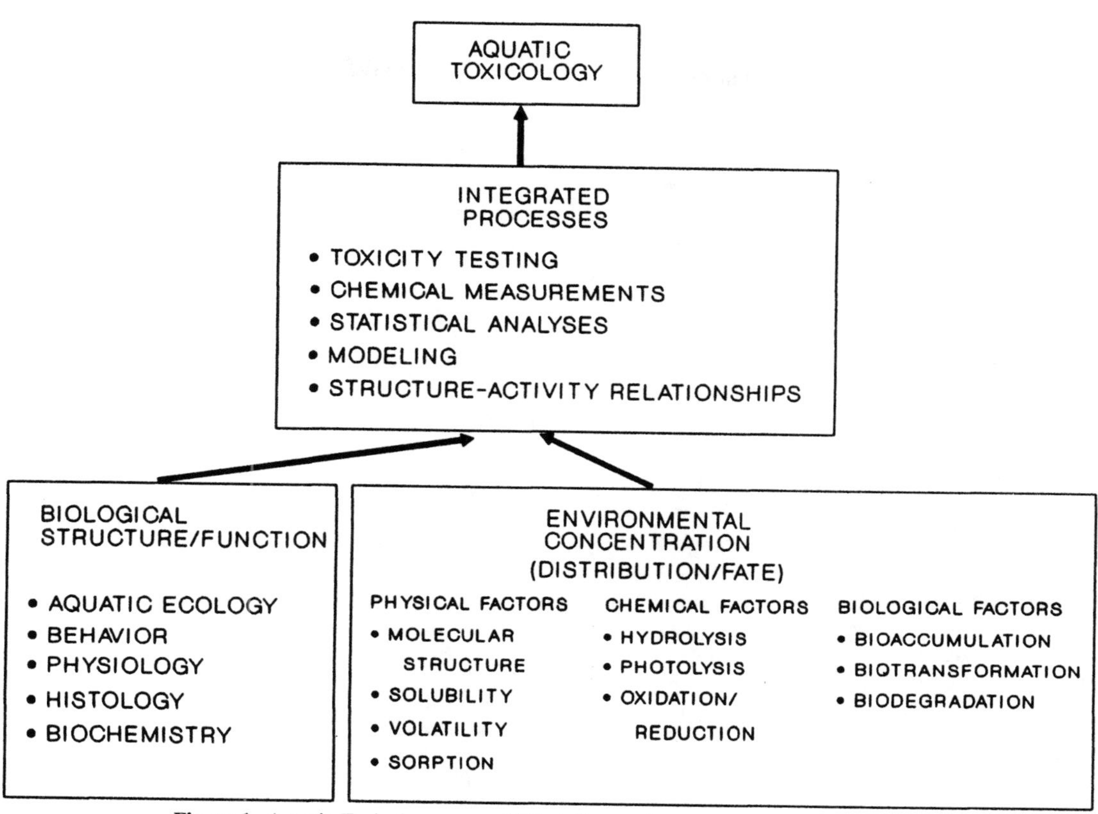

Figure 1. Aquatic Toxicology—a multidisciplinary science.

laboratory animals began in the 1920's, in response to concerns about the unwanted side effects of food additives, drugs, and pesticides." He acknowledges the major impact of Rachel Carsons classic book *Silent Spring* (Carson, 1962), which "almost single-handedly created modern society's fears about synthetic chemicals in the environment and, among other things, fostered renewed interest in the science of toxicology." This new public and governmental interest, as well as the ongoing research in university, private, and industrial laboratories, enhanced toxicology in the 1960s and 1970s and stimulated the new area of environmental toxicology and ecotoxicology.

According to Buikema et al. (1982), "the formal discipline of toxicology arose in the early 1880's in response to the development of organic chemistry (Zapp, 1980)." Basic toxicological research and practice were started in a number of countries in the 1800s, with work on rodents, fish, and insects, and major developments in pharmacology, toxicology, and applied entomology occurred from that time onward. A comprehensive overview of the history and scope of toxicology is also given by Gallo and Doull (1991).

Early environmental work included pesticide testing with insects and water quality testing with cladocerans or water fleas. However, with few exceptions, such as Anderson's classic early work (Anderson, 1944), it was not until the 1940s and 1950s that concerns about the effects of wastes and chemicals on nonhuman biota became a public issue. In the meantime, some basic acute aquatic toxicity tests, especially with fish, had been developed and applied (Hunn, 1989).

The early years prior to World War II saw investigations on the toxicity of various materials and metals to fish, early attempts at method descriptions, and suggestions for appropriate test organisms (e.g., cladocerans, goldfish, salmonids). Cladocerans and fish were used in water pollution investigations in both the laboratory and field (Anderson, 1944, 1980; Hunn, 1989).

The years after World War II saw efforts to standardize techniques for acute toxicity testing. Some major laboratories were established in the United Kingdom, Canada, and the United States, and large numbers of chemicals either found in wastes or possibly useful as piscicides were screened. It was the era of the "pickle-jar bioassays" (Doudoroff, 1976) in which toxicity tests with aquatic organisms were conducted in 1- and 5-gallon jars. Classic papers included those by Doudoroff and Katz (1950, 1953), Hart et al. (1945), Doudoroff et al. (1951), Cairns (1957), Henderson (1957), and Tarzwell (1958). Hart et al. (1945) introduced the single-species fish bioassay and Doudoroff et al. (1951) described and approved it (Cairns and Pratt, 1989). These workers, as well as T. W. Beak and co-workers in Canada from the F. E. J. Fry "school of comparative physiology" at the University of Toronto, demonstrated the usefulness of exposing fish and other aquatic organisms to chemicals and wastes in toxicity tests (Sprague, 1992). Such fish tests were also used to study the effects of various environmental factors on toxic responses, as a way of predicting field impacts. Work by this group of scientists established the acute toxicity test as the cornerstone of methodologies to study and monitor pollution effects (Buikema et al., 1982), a position it still retains. Also in the 1950s, the idea evolved "that picking representative organisms from different levels of the food chain would more faithfully display the range of response to toxicants than fish alone" (Cairns, 1956a,b), thus introducing the concept of using a range of organisms in aquatic toxicity testing.

The fish toxicity test was the basis of the first protocol of the American Society for Testing and Materials (ASTM) for aquatic toxicity testing (ASTM, 1954). In 1960, the American Public Health Association (APHA) included fish toxicity test procedures in its well-established "Standard Methods"; this section (Part 8000, Toxicity Test Methods for Aquatic Organisms) has expanded to cover many other species in a very detailed format and is a standard reference to current methods (see Chapters 2 and 3).

By the 1960s, biological and aquatic toxicological research on a wide range of pollution problems was being conducted in Canada, the United States, the United Kingdom, and various European countries to acquire data on the acute toxicities of effluents and chemicals from industries and municipalities and to determine effects on biota and their habitats for each water body receiving wastes (Wells, 1989). Considerable work was being published by J. Cairns Jr., W. A. Brungs, P. Doudoroff, D. I. Mount, Q. H. Pickering, and C. M. Tarzwell in the United States; J. B. Sprague, D. McLeese, and M. Waldichuk in Canada; and J. S. Alabaster, V. M. Brown, K. Wilson, and R. Lloyd in the United Kingdom, to name a few. The impacts of industrial effluents on receiving streams and other waters was given particular attention (Sprague, 1971, 1992).

The 1960s produced advances in static and flow-through exposure techniques, for example, continuous flow diluters, by Mount and Brungs (1967); measurement of chemical exposure concentrations; multifactorial experimental designs; and precise control of water quality. There was great concern about chronic sublethal effects of low-level chemical exposures, brought on by knowledge of the distribution of organochlorine compounds (as pesticides) and their effects on fish and wildlife (Carson, 1962) and the discovery of the wide occurrence of polychlorinated biphenyls (PCBs). In this era multifactorial designs were introduced in toxicity studies to study the sources of toxicity, the various chemical interactions in mixtures, and the effects of water quality and various physicochemical factors (particularly pH, temperature, and salinity) and modifications of methods on toxicity thresholds. Experimentation proceeded with young life stages and the measurement of various sublethal end points (e.g., growth, development, and behavior), life cycle assays (e.g., flagfish), and studies on a range of species from microorganisms to fish.

In the 1960s, APHA Standard Methods continued its section on bioassays. The ASTM started to expand its coverage of other organisms (e.g., diatoms) (ASTM, 1964). Sprague's water research trilogy on aquatic toxicology (Sprague, 1969, 1970, 1971) was published, establishing a sound conceptual, methodological, and scientific framework for practitioners and setting the stage for much of the renewed effort in aquatic toxicology in North America in the 1970s. Journals such as Water Research, Marine Pollution Bulletin, and Environmental Pollution started during this time.

The 1970s heralded a new period of public, scientific, and governmental awareness and concern about water pollution. This was reflected by an increase in the numbers of laboratories employing toxicity procedures and conducting marine and freshwater pollution research, including U.S. Environmental Protection Agency laboratories and Fisheries and Environment laboratories in Canada (Johnstone, 1977). More emphasis was placed on studying sublethal effects (Cole, 1979), use of a wider range of species in toxicity testing (Kenaga, 1978; Buikema and Cairns, 1980), and the description and computerization of advanced statistical approaches for toxicity threshold estimation (Stephan et al., U.S. EPA Duluth Laboratory). Also, more attention was paid to the chemistry of external waterborne exposures and internal tissue and organ residues and the use of multispecies tests (microcosms) as research tools, with applications in specific fields such as pesticide evaluation and oil spill impact assessment.

The 1970s also saw acute fish "toxicity" being accepted as a valid parameter for governmental regulations and guidelines for water pollution control. The rainbow trout test was used in Canada (Pessah and Cornwall, 1980; Wells and Moyse, 1981), a brown trout test in Britain (V. M. Brown, personal communication), and the goldfish (Smith et al., 1974) and other fish (mummichogs, fathead minnows, bluegill, trout, silversides) in the United States. "The number, variety and complexity of aquatic bioassays steadily increased during the 1970s. Bioassays were used to assess acute toxicity, sublethal effects, plant nutrient status, mutagenicity, bioconcentration potential, and a myriad of other pollutant effects on aquatic organisms, as well as drinking water for human consumption" (Maciorowski et al., 1980). During this decade aquatic toxicologists organized various symposia (ASTM, Canadian Annual Aquatic Toxicity Workshops, Pellston

Workshops, and the Marine Pollution and Physiology series in the United States). There was continued effort on standardized methods and interlaboratory calibration. The field expanded well beyond acute testing and was rapidly becoming a science of its own (Stephan, 1982). There was tremendous growth in the literature. By the late 1970s, research interests had expanded to hazard evaluation and risk analysis (see various ASTM, Pellston, and American Fisheries Society symposia volumes as references).

The 1980s saw the startup of the journal Aquatic Toxicology, the establishment of the Society of Environmental Toxicology and Chemistry (SETAC) in North America, and an increase in pollution and toxicology symposia. Aquatic toxicology was integrated into the hazard assessment process (Cairns and Pratt, 1987). Scientifically, aquatic toxicology matured rapidly, with advances especially being made in quantitative structure-activity relationships (QSARs), sediment evaluation, biochemical toxicology, the identification of suitable "biomarkers" for monitoring, microscale toxicity assays, and the analysis of effluents (through the use of toxicity in the Toxicity Reduction/Toxicity Identification Evaluation approach) and other chemical mixtures.

The role of aquatic toxicology in the context of toxic substances management was given more considered attention, at meetings such as SETAC (Maki, 1983; Wells, 1985; Wells and Côté, 1988), Alliston, Ontario (Day et al., 1989), and Lancaster, England (Munawar et al., 1989). It was recognized that the science of aquatic toxicology needed to move away from a reliance on single-species bioassays, even though they have remained attractive to regulators, and toward use of both multispecies approaches and full field ecosystem end points for verification of predicted impacts of substances (Cairns and Pratt, 1989; Macek, 1982).

During the 1980s, many books were published, notably the Belgium MARTOX volumes (Persoone et al., 1984), Concepts in Marine Pollution Measurements (White, 1984), Fundamentals of Aquatic Toxicology (Rand and Petrocelli, 1985), Ecotoxicology: The Study of Pollutants in Ecosystems (Moriarity, 1988), Environmental Bioassay Techniques and Their Application (Munawar et al., 1989), Ecotoxicology: Problems and Approaches (Levin et al., 1989), the continued ASTM (USA), SETAC (USA) and Aquatic Toxicity Workshop Proceedings (Canada), and many more specific texts (see Literature Cited and the Supplemental Reading list) too numerous to mention here. Much progress was made in developing more suitable procedures for effects monitoring in receiving waters, as evidenced by recent volumes on biomarkers (e.g., McCarthy and Shugart, 1990; Huggett et al., 1992; Peakall, 1992) and the United Nations International Oceanographic Commission Group of Experts on Environmental Pollution (GEEP) Practical Workshops on Marine Biomonitors (e.g., Bayne et al., 1988).

The 1990s have seen continued growth and activity in the field. In particular, there have been significant advances on organismal responses (biomarkers) incorporating procedures from human health studies, the choice of alternative testing systems (Goldberg and Frazier, 1989; Blaise, 1991), automation of various techniques for measuring acute toxicity, measurement of genotoxic and mutagenic effects, and new syntheses contributing to the conceptual development of aquatic toxicology in both the laboratory and the field (e.g., Bartell et al., 1992; Burton, 1992; Munkittrick and Dixon, 1989; McCarty and Mackay, 1993; Suter, 1993).

Over the past few decades, there has been an internationalization of aquatic toxicology. To deal with wastewater discharges, environmental protection agencies in a number of countries, particularly in Europe, North America, Japan, and Australia have recognized and restated the value of applying aquatic hazard assessment principles and procedures to effluents and their component chemicals and properties. Examples of such applications are described in the MARTOX Conference Proceedings (Persoone et al., 1984), the 1984 OECD-sponsored Workshop on the Biological Testing of Effluents (Environment Canada, 1984), the OECD guidelines document of effluents (OECD, 1987), the SETAC/Pellston Meeting and proceedings on the hazard assessment of effluents (Bergman et al., 1986), and in Mackay et al. (1989), who describe the U.K. experience with bioassay techniques. The growing application of biological tests for environmental protection, which is rapidly occurring in Europe, North America, South America, Japan, Southeast Asia, and Australia–New Zealand is an outcome of key policy decisions regarding their application and of their increasing versatility, predictability, cost-effectiveness, and interpretive value.

Testing of effluents via "end-of-pipe" and "receiving water" approaches has served a number of countries well in their efforts to curb aquatic pollution (e.g., McLeay, 1991). There is a better understanding of the applications, advantages, and limitations of the various laboratory and field approaches. It is commonly accepted now that biology/ecotoxicology (effects identification) and chemistry (cause identification) are inseparable components of the various hazard assessment approaches being developed or employed (Blaise et al., 1988).

Biological procedures of many kinds, including toxicity tests, are being built into current policies, frameworks, strategies, guidelines, regulations, and codes of good practice in many countries, on most continents. The reason is simple—biological tests and toxicological end points offer rapid, reliable and unequivocal, integrated measures of environmental effects or their potential. Examples of freshwater concerns include the industrial effluent testing framework of West Germany (P. Hansen, personal communication); the U.S. EPA Clean Water Act NPDES Program, in which both freshwater and marine acute and chronic tests are being employed for site-specific regulation of wastewaters; the OECD toxic chemical protocols for toxicity testing; the OECD industrial effluent management plan with components for biological procedures; and the Great Lakes Water Quality Program under the Great Lakes Water Quality Agreement (1978, revised in 1983; IJC, 1989).

Various marine pollution frameworks encompass ecotoxicological testing (Wells and Côté, 1988; GESAMP, 1991). As described by Reish (1988) and Parker et al. (1991), international organizations and a range of countries are recognizing and adopting marine biological tests in their assault on land-based sources of marine pollution and other marine pollution problems through environmental research. For example, the OECD Water Management Policy Group and the Expert Group on Biological Testing of Effluents concluded that "toxicity testing can be used successfully for measuring the acute and chronic toxic effects of effluents on the biota of receiving waters, and to generate the quantitative data on effluent toxicity which provide the regulatory basis for the control of toxic substances" (OECD, 1987). In addition, the European Economic Community (EEC)* established an ecotoxicology advisory group in the 1980s, in part to control industrial discharges (G. Persoone, personal communication). There are also aggressive programs in some European countries to apply toxicity tests in regulatory schemes for industrial effluents (P. Hansen, personal communication).

The successful international SETAC Congress in Portugal (1993) is testimony to the growing acceptance of aquatic toxicity procedures in the regulatory sphere worldwide and testimony to the challenges ahead to ensure a sound conceptual and practical footing for aquatic toxicology. However, the field must get beyond the data rich/principle poor stage (Macek, 1982; Cairns and Pratt, 1989; Hodson, 1994). It must "start to account for ecosystem dynamics in the methods of ecotoxicological science, for predicting effects of stressors such as chemicals on system health" (Cairns and Pratt, 1989); and it must consider shifting from a dose-response paradigm or approach to a stress-response approach in aquatic toxicology (Cairns, 1993). "If environmental, including aquatic, toxicology is to come of age, it must begin to ask more searching questions, develop broader hypotheses involving natural systems, and develop models that are validated in landscapes, not laboratory jars" (Cairns, 1992). Finally, "aquatic toxicologists and ecologists should interact more frequently, because the goal of aquatic toxicology or ecotoxicology is to protect natural ecosystems and their inherent organisms" (Cairns and Mount, 1990).

THE AQUATIC ENVIRONMENT

The aquatic environment is highly complex and diverse. It includes several distinct ecosystem types—freshwater streams, lakes, ponds and rivers; estuaries; and marine coastal and deep ocean waters—that have many different biotic and abiotic components and unique characteristics. The *biotic* or living components consist of many combinations of plants, animals, and microorganisms that inhabit specific ecological niches in each ecosystem. The *abiotic* or nonliving components include the physical environment (e.g., water, substrate-sediment, and suspended particulate material) within the boundaries of the ecosystem. Each aquatic ecosystem is thus a dynamic product of complex interactions of living and nonliving components, with both constant and changing features in time and space.

The physical and chemical properties of aquatic ecosystems can have a profound effect on the biological activity and impact of chemicals and other xenobiotics. The vulnerability of the aquatic environment to chemical insult depends on several factors, including (1) physical and chemical properties of the chemical and its transformation products; (2) concentrations and total loading of the chemical entering the ecosystem; (3) duration and type of inputs (acute or chronic, intermittent spill or continuous discharge); (4) properties of the ecosystem that enable it to resist changes that could result from the presence of the chemical (e.g., pH buffering capacity of seawater or dissolved organic matter concentrations) or return it to its original state after the chemical is removed from the system (e.g., flushing of water from estuaries by tidal action); and (5) location of the ecosystem in relation to the release site of the chemical.

Because aquatic ecosystems involve complex interactions of physical, chemical, and biological factors, it is difficult to understand the response of a system to a chemical unless the relationships among components of the system are well defined. Assessment is further complicated by the ability of the biotic components to adapt, the diversity of species present in the ecosystem (which normally changes over time), and the differences in structural and functional responses among the biological components. Furthermore, similar ecosystems are not necessarily affected similarly by addition of the same chemical. Even minor differences in the chemical and physical environment and in species composition can result in differences in the fate of a chemical and different effects on the systems. Therefore, site-specific conditions must be considered in precisely evaluating potential chemical hazards.

Before a discussion of fundamental concepts in aquatic toxicology is initiated, a background of the basic properties and composition of water and sediment will be presented. This will provide the reader with an appreciation of the complexities involved when a foreign chemical enters and interacts with the components of a natural aquatic system. For further information on these topics, refer to the Supplemental Reading section at the end of the chapter.

Water—Structure and Properties

Water is the medium common to all aquatic ecosystems and maintains the integrity of the aquatic or-

*European Community (EC) is now referred to as European Union (EU).

ganisms that live, grow, and reproduce in it. The unique properties of water derive from the structure and bonding of a water molecule, which is formed by joining two atoms of hydrogen with one atom of oxygen. The higher electronegativity of the oxygen atom and its two pairs of unshared electrons produce a net negative charge at the oxygen end of the molecule. The hydrogen ends have small net positive charges. Because of the arrangement of hydrogen atoms and the unequal charge distribution, a *polar covalent bond* is formed within the water molecule. This leads to *hydrogen bonding* of water molecules. This is perhaps the most important characteristic of the molecule, because it makes water an excellent solvent for ionic (i.e., electrically charged) substances, as well as making it highly reactive with such substances.

Table 2 summarizes the properties of water. The *specific heat* is the amount of heat in calories required to raise the temperature of a unit mass (i.e., 1 g) of a substance by 1°C. Water holds significant thermal energy (or heat) with a relatively small change in temperature. Water has a specific heat of 1 and is exceeded only by a few substances including liquid ammonia, liquid hydrogen, and lithium at high temperatures. Water thus holds significant thermal energy (or heat) with a relatively small change in temperature. This enables water to serve as a buffer, ameliorating sudden large fluctuations in temperature in aquatic systems and preventing shock to aquatic organisms. Aquatic organisms in the water column are thus subjected to narrow temperature ranges. The slow rate of cooling and warming in lakes, ponds, and streams is linked to the high specific heat of water. Dissolved substances in natural water lower its specific heat.

The high latent heat of vaporization of water also stabilizes the temperature of aquatic systems and land masses. The *latent heat of vaporization* is the amount of heat required for the vaporization of 1 g of water. Water thus vaporizes slowly when heated. This influences heat and water vapor transfer between the atmosphere and bodies of water. As much heat is needed to vaporize 1 g of water as to raise the temperature of 540 g of water by 1°C. The *latent heat of fusion* is the heat absorbed in the change of water (1 g) from a solid state to a liquid state with no change in temperature. About 79.7 calories of heat are required to melt 1 g of ice at 0°C.

Density is a regulating factor in the physical and chemical dynamics of freshwater systems. Density of fresh water reaches a maximum of 1.00 (g/cm^3) at a temperature close to 4°C. As the water cools beyond this point, density decreases at a progressively increasing rate. The density characteristics of water are the reasons why lakes and ponds freeze from the surface downward and only in the coldest regions do natural bodies of water freeze solidly. Furthermore, vertical circulation processes in lakes are largely controlled by temperature-density relationships in

Table 2. Important properties of liquid water

Property	Comparison with other substances	Importance in aquatic environment
Specific Heat Capacity	Higher than all solids and other liquids except liquid ammonia, liquid hydrogen and lithium	Prevents wide fluctuations in temperature in water and stabilization of body temperatures in organisms
Latent Heat of Fusion	Higher except for ammonia	Temperature stabilization at freezing point of water due to absorption or release of latent heat
Latent Heat of Vaporization	Higher than any other substance	Determines transfer of heat and water molecules between atmosphere and aquatic systems
Density (or Thermal Expansion)	Maximum density for pure fresh water is at 4°C and for sea water it is near its freezing point (−1.9°C)	Freshwater and dilute seawater reach maximum density at temperatures above freezing; controls temperature distribution and vertical circulation in stratified lakes
Surface Tension	Higher than any liquid except mercury	Controlling factor in cell physiology; surface phenomena and drop formation
Dissolving (Solvent) Power	Dissolves more substances and in greater amounts than any other naturally occurring liquid	Facilitate chemical reactions (e.g., hydrolysis) and transport of nutrients and by-products in biological processes
Dielectric Constant	Pure freshwater higher than all liquids except hydrogen peroxide and hydrogen cyanide	High solubility of inorganic substances because of ionization
Transparency	Absorption of radiant energy is high for infrared and longer wave lengths of ultraviolet light; little selective absorption in visible portion	Allows light for photosynthesis and photolysis to occur at significant depths

Sources: Drever (1988), Libes (1992), Pytkowicz (1983), Reid and Wood (1976).

water. Fine organic materials like detritus (organic matter from nonliving and decomposing organisms), clay, and silt increase the density of water in lakes and streams. Density also increases with increasing concentrations of dissolved salts. Dissolved salts also depress the freezing point of water. For seawater with a salinity (or salt content) of 35‰ (in parts per thousand), the freezing point is −1.91°C. As seawater cools toward its freezing point it becomes progressively more dense. Seawater is thus heaviest (density = 1.02 g/cm^3) near freezing and at this time is about 6° cooler than fresh water. Oceans do not freeze because of their sheer volume, movement (from wind and tides), and convective currents, not because of their density properties.

Viscosity is a measure of how much a fluid will resist changing its form when a force is applied to it. It is a measure of mobility and flow. Compared to many liquids, water is resistant to flow because of the energy contained in the hydrogen bonds of the molecule. Water viscosity decreases as temperature increases and is affected little by the salinity of water. The high viscosity of water affects energy expenditure of organisms because it offers great frictional resistance to a moving organism. The sinking rates and distribution of sedimenting particles and planktonic algae are affected by density-related changes in viscosity. Viscosity obviously is of great importance for smaller organisms because they possess little momentum. The ratio of momentum to viscosity is expressed in the *Reynolds number*. Small organisms have low Reynolds numbers, illustrating that apparently simple activities like locomotion are complicated because the organisms must expend considerable energy to move about as well as overcome changes in viscosity they encounter.

Within the interior of water, molecules are hydrogen bonded to neighboring molecules on all sides with an average force, but surface molecules are not attached on one side. Therefore, at the surface of a water mass (air-water interface) there is an unbalanced inward molecular attractive force to the liquid phase below. The result is an interface surface film under tension. The water surface thus has a higher potential energy than the interior. Water has a high *surface tension*, which is important in sorption of chemical substances to particulate solids suspended in water. The substances that lower surface tension between water and particulate matter become positively sorbed to the particulate. The local concentration of a substance can be affected by the interfacial tensions between the substance and water. Surface tension decreases with increasing water temperature, increases only slightly with dissolved salts, and is reduced markedly with organic substances.

Water is a good *solvent* because it can dissolve a little of most substances. Traces of most chemical elements can be found in natural waters, including lakes, estuaries, or oceans. Water as a solvent coupled with its high dielectric constant (i.e., tendency to ionize dissolved substances) enables it to serve both as a medium to facilitate chemical reactions (e.g., hydrolysis) and as a conduit in the transport of nutrients and waste products. Ionic solids, or salts, are very soluble in water because of the ability of water to hydrate the cations and anions that compose the salts.

Water is *transparent*; sunlight can penetrate into water at considerable depths. The absorption of light at different wavelengths is a function of transparency. The major absorption of light in pure water is at the red end of the spectrum and the minimum absorption is at the blue range. About 90% of the radiation above 750 nm is absorbed by the time sunlight has penetrated to a depth of 1 m in water. Relationships between sunlight and the transparency of water are complex but are nonetheless of great relevance in plant-animal interactions in natural aquatic ecosystems. Transparency of water is also an important property to consider for chemicals in water such as oil-derived hydrocarbons and pesticides that are susceptible to degradation by sunlight (i.e., photolysis).

Water has a good capacity to maintain substances in solution and to enter into chemical reactions. All natural water thus typically contains background substances in solution, which include dissolved gases and inorganic and organic substances. "Pure water" truly does not occur in natural systems. The substances typically found in natural waters greatly influence their characteristics.

Water—Composition

Most natural aquatic systems are open and dynamic and thus vary in chemical composition. The most abundant constituents (ions) in seawater are the *conservative components*, including chlorine, sodium, calcium, potassium, magnesium, and sulfate. They constitute over 99% of the mass of solutes dissolved in seawater. Sodium and chlorine account for about 86% alone. The ratios of the conservative components to chloride concentration are the same everywhere. Alternatively, the *nonconservative components*, which include dissolved gases (e.g., O_2, CO_2, H_2S, nitrogen at nM to mM concentrations), nutrients (e.g., phosphate, nitrate at μM concentrations), dissolved silica, dissolved organic compounds (e.g., amino acids, humic substances at ng/L to mg/L concentrations), trace metals (e.g., nickel, lithium, iron at concentrations <0.05 μM), colloids (e.g., matter from ≈ 0.001 to 0.45 μm at concentrations ≤mg/L), and particulate matter (e.g., sand, clay, nonliving tissues, excreta, and organisms at concentrations ≤mg/L), vary significantly from place to place in the oceans as a result of reactions that occur in the ocean and sediment. Although most substances in seawater are nonconservative, they compose only a small fraction of the total mass of the ocean.

Most seawater in the ocean has a salinity between 33 and 37‰ with an average of 35‰, which is equivalent to a 3.5% salt solution. Salt waters may be classified by the Venice system, which categorizes waters according to salinity zones. In this system the salinities of marine (or ocean) waters, estuaries, and fresh water are >30‰, 0.5–30‰, and <0.5‰, respectively.

As in seawater, oxygen and carbon dioxide are the two most abundant gases in fresh water. Oxygen is important as a regulator of organismal and community metabolic processes. The solubility of oxygen in water is dependent on temperature, partial pressure of oxygen in the atmosphere, and the salt content (i.e., salinity) of water. For example, as the temperature of fresh water falls from 25°C to freezing, the solubility of oxygen increases by more than 40%. As the concentration of dissolved salts increases in water, the solubility and consequent oxygen concentration decrease. At 0°C, fresh water at saturation contains approximately 2 mg/L more oxygen than does seawater (35‰). Most oxygen in water comes from the atmosphere, and therefore it is also the primary means of reoxygenation. If oxygen-consuming processes are occurring in water at a rapid rate, with little replenishment, such as in microbial degradation (i.e., biodegradation) of chemicals, adverse effects on fish and other biota may occur. The rate at which atmospheric oxygen passes across the air-water surface and is dissolved in water depends on a number of factors. Increased wave action results in more of the gas in solution, and the greater the difference in partial pressure between air and water, the greater the movement into solution. Oxygen in natural water may also be produced by the photosynthetic action of algae, but this is not an efficient means of reoxygenating water because some oxygen produced during the daylight hours may be used by the algae at night.

Carbon dioxide serves as a "buffer" to prevent rapid shifts in acidity or alkalinity and it regulates certain biological processes (e.g., plant growth, animal respiration) in aquatic communities. Carbon dioxide is highly soluble in natural waters. The solubility of carbon dioxide varies in natural water because it normally exists in a delicate equilibrium with bicarbonate (HCO_3^-) and carbonate (CO_3^{2-}) ions and the predominant species is dependent on pH. Un-ionized CO_2 predominates below pH 6, HCO_3^- ion in the pH range 6–10 (i.e., natural water), and CO_3^{2-} in alkaline water above pH 10. Carbon dioxide in natural water is derived from bacterial decomposition of organic matter and respiration by animals and plants. Within natural water, certain reactions between acids and carbonate compounds release carbon dioxide. The atmosphere also supplies some of the carbon dioxide to natural water.

The most important and abundant dissolved solids in freshwater systems are the compounds of calcium and magnesium, sodium and potassium, phosphorus, iron, sulfur, and silicon. Calcium is typically the most abundant cation, followed by magnesium. Carbonates are usually the most abundant salts in fresh water.

Freshwater systems, like saltwater systems, contain natural organic compounds that are produced by autotrophic organisms. In fresh water, a portion of the total organic compounds may also come from the terrestrial watershed. Organic compounds fall into various classes of biomolecules and include hydrocarbons, carbohydrates, lipids, fatty acids, amino acids, and nucleic acids. Most of the organics are composed of elements such as carbon, oxygen, and hydrogen and some contain nitrogen, sulfur, and phosphorus but usually in smaller quantities. Organic matter in the aqueous environment is an important source of electrons that drives most biologically mediated redox reactions of heterotrophs, which require these compounds as their source of carbon and chemical energy. The distribution of organic compounds is therefore controlled by the activities of organisms and is the result of kinetic rather than thermodynamic controls, because equilibrium is rarely attained in biologically mediated reactions. Furthermore, the biogeochemical cycle of organic matter has a significant influence on the cycle of the elements, especially for nitrogen and phosphorus because of their low levels in water.

In natural waters it is evident that there are organic and inorganic dissolved and nondissolved fractions of substances. Dissolved and nondissolved substances are divided into true solutes (≤0.001 μm), colloids (≈0.001 to 0.45 μm), and solids (≈ ≥0.45 μm) based on theoretical particle size criteria. In practice, the concentration of substances can be determined in dissolved and nondissolved fractions by filtering water through a membrane filter with a pore size of 0.45 μm (Figure 2). The particles or substances retained by the filter (> 0.45 μm) are termed solids or *particulate matter* (e.g., sand, clay, nonliving tissues, excreta, organisms). The substances that pass through (< 0.45 μm) are referred to as *dissolved matter* and include *colloids*. Colloids are composed of inorganic and/or organic substances. The common organic substances that contribute to the colloidal particle load in an aquatic system include humic acid polymers (see below) and acidic polysaccharides. Particulate matter, including some colloids, may be retained by the filter because of clogging during filtration. This is an operational definition because the separation is dependent on the filtration conditions. The quantities of organic compounds are therefore determined as operationally defined fractions of either particulate organic matter (POM) or dissolved organic matter (DOM). DOM may also adsorb to particles and form aggregates and therefore constitutes part of the operationally defined POM fraction. Chemical analyses of operationally defined fractions have provided information on the biogeochemistry of organic matter in natural waters. Analysis has included measurement of percent carbon and nitrogen, as well as concentrations of particulate organic carbon (POC) and dissolved organic carbon (DOC).

DOC quantifies the chemically reactive fraction and provides the mass of organic carbon dissolved in water. It is a measure of the organic molecules making up the dissolved organic load. POC is the

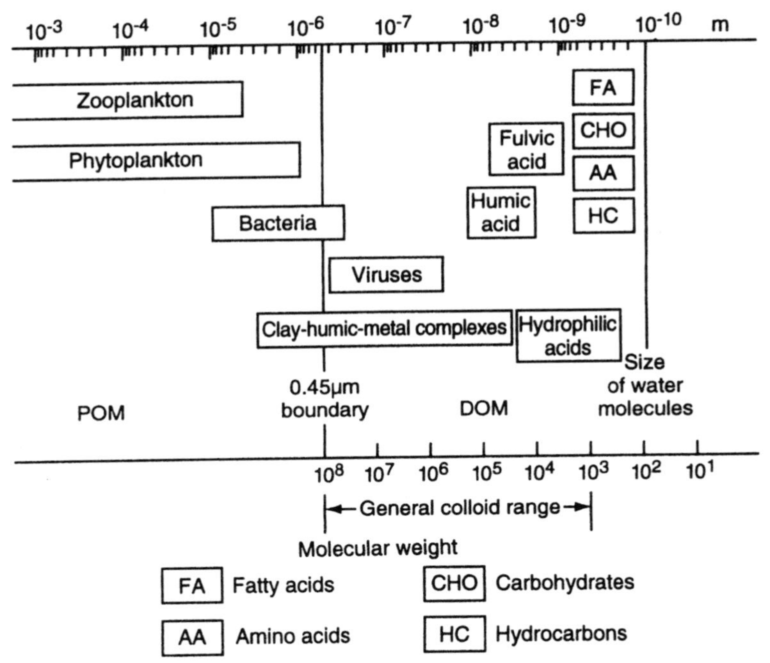

Figure 2. Continuum of particulate and dissolved organic matter in natural water. (Modified from Thurman, 1985, reprinted by permission of Kluwer Academic Publishers.)

same as the suspended organic carbon which is also retained by the 0.45-μm filter. It is composed of plant and animal organic matter and the organic coatings on silt and clay. Total organic carbon (TOC) is the sum of DOC, POC, and colloids. Dissolved, particulate, and total organic matter (DOM, POM and TOM) are analogous to DOC, POC, and TOC (Thurman, 1985). However, organic matter refers to the whole organic molecule, including other elements such as oxygen and hydrogen, and is therefore difficult to quantify. Measures of organic carbon are typically used. Measurements of DOM, POM, and TOM are often equal to two times the DOC, POC, and TOC, respectively.

The average concentrations of DOC and POC vary with the type of water. Seawater has the lowest concentration of DOC with a mean of 1.0 mg/L for surface water (deep water has 0.5 mg/L), and the concentration of POC is much less, ranging from 0.01 to 0.1 mg/L. The greater amount of DOC in surface waters is produced by the photosynthetic activity of phytoplankton, which is also the source of POC. Coastal waters have considerably higher concentrations of dissolved organic carbon (1–3 mg/L) because of increased rates of primary production and inputs of organic matter from rivers. The DOC of large rivers ranges from 2 to 10 mg/L and the POC may range from 10% of the DOC to equal to the DOC. POC is strongly related to sediment loading and changing discharges in rivers (and streams). In rivers, POC generally makes up 2–3% of the sediment as coatings on mineral grains and as discrete organic detritus. Estuaries have concentrations of DOC that range between the DOC values of rivers and seawater. DOC behaves conservatively in estuaries; there is a linear relationship between salinity and DOC. The concentrations of POC in estuaries range between those of rivers and seawater and POC may be another source of DOC in the estuary. Most of the organic carbon in lakes is dissolved, and POC is only about 10% of the dissolved organic carbon. DOC in lakes increases with productivity and therefore with trophic status. Oligotrophic and mesotrophic lakes have the lowest concentration of DOC (1–4 mg/L) because there is little algal productivity. Eutrophic lakes have greater concentrations of DOC (3–34 mg/L). Dystrophic lakes that receive organic matter from bogs and marshes have the greatest concentrations (20–50 mg/L) of DOC because of the buildup of fulvic acids (see below). In interstitial waters of sediment, DOC concentrations are high when oxygen concentrations are low. Generally, concentrations of DOC in interstitial waters are much larger than in surface waters. Interstitial waters do not contain POC.

A large part (30–50%, but > 50% for marshes, bogs, and swamps) of the DOC of natural water is composed of a class of high-molecular-weight (500–5000) organic compounds that are collectively referred to as *humic substances* and are believed to be resistant to microbial degradation (Baker, 1991a,c). Their elemental composition is 50% carbon, 4–5% hydrogen, 35–40% oxygen, 1–2% nitrogen, and <1% sulfur. The major oxygen-containing functional groups include carboxylic acid, phenolic hydroxyl, carbonyl, and alcoholic hydroxyl groups. Humic substances are formed in abiotic chemical reactions (i.e., condensation, polymerization) that link together fragments of biomolecules or biopolymers (e.g., proteins, carbohydrates) generated during the microbial degradation of organic matter.

Humic substances occur naturally as deposits of solids on sediment particles (and soil). The lowest concentrations of humic substances are in seawater at a range of 0.06–0.60 mg carbon/L. Streams, rivers, and lakes contain from 0.5 to 5.0 mg carbon/L (includes oligotrophic, mesotrophic, and eutrophic lakes); marshes, bogs, and swamps vary from 10 to 30 mg carbon/L (includes dystrophic lakes) as humic substances.

Humic substances are classified by solubility. Those that are insoluble at acidic pH and soluble at basic pH are termed *humic acid* and those that are soluble at acidic and basic pHs are termed *fulvic acid*. Most of the humic substance of DOC in natural waters is usually in the form of fulvic acid followed by humic acid. The rest of the DOC is hydrophilic acid (30%) and simple organic compounds (20%). *Humin* is the fraction of humic substances that is nonextractable because it is insoluble in aqueous solutions at any pH. Humin is typically the largest fraction of organic matter in recent freshwater sediments. In fact, humin organic carbon may contribute as much as 80% of the total organic carbon in a sediment. Humin, humic acid, and fulvic acid are not single compounds but refer to a wide range of compounds with similar characteristics.

Humic substances serve as natural sorbants and chelators, affecting the distribution and fate of metals and organics through chemical bonding. Fulvic acid exerts its effect as the soluble species and provides binding sites for metals. Humin and humic acid are relatively insoluble in water and effectively exchange cations and organic substances with water. Humin and humic acid provide a relatively nonpolar environment into which a hydrophobic chemical can escape. A portion of the humics may precipitate, thus providing a mechanism for sedimentary concentration of organics and metals, which could include foreign chemicals.

As shown in Figure 2, biomolecules, colloids, humic substances, some bacteria, and viruses are included in the operational DOM fraction of natural water. Macroscopic particulate organic matter, which includes zooplankton, phytoplankton, bacteria, and detrital organic matter, can be removed by filtration. In the ocean most organic matter is present in dissolved or colloidal form (Figure 3). Most of the POC is nonliving detritus; less than one-quarter of the POC is alive and is mainly composed of phytoplankton and bacteria. DOM serves as a large reservoir of carbon and a major source of food for heterotrophs.

In summary, organic matter in freshwater systems (e.g., lakes) is *autochthonous* (i.e., derived in the lake from primary producers like phytoplankton and macroalgae in photosynthesis or other autotrophic processes) and *allochthonous* (i.e., plant and soil organic matter transported from the terrestrial watershed) in origin. However, in most small lakes or ponds (including streams) with a high proportion of surface area as littoral zone much of the organic matter is terrestrial in origin. The larger the lake, the more important autochthonous inputs of organic matter become. Most marine organic matter is generated in situ by phytoplankton; however, in coastal areas and estuaries organic matter is a mixture from terrestrial and marine sources.

The organic composition of sediment (see below) and the chemical characteristics of the humic substances and dissolved organic carbon in a body of water are influenced by whether autochthonous and/or allochthonous source input is involved and by all the biogeochemical processes involved in carbon cycling. Allochthonous particulate matter may be rich in plant structural and cell wall material (e.g., lignins and cellulose). These materials are largely biologically recalcitrant, so turnover following heterotrophic uptake is slow. For autochthonous particulate material, which originates from algae-dominated systems, lipids, pigments, proteins, and polysaccharides

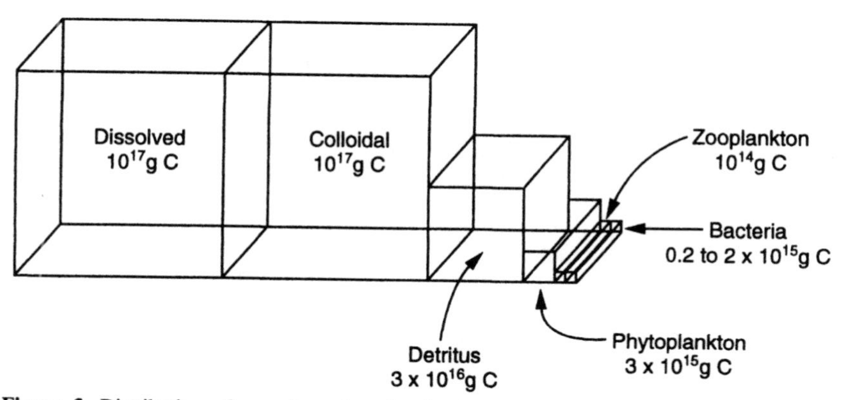

Figure 3. Distribution of organic carbon in the oceans. (From Cauwet, 1978, used with permission.)

are the most abundant components of the particulate organic matter. During settling through the water column autochthonous POC generally settles more slowly and undergoes greater microbial degradation than allochthonous POC. Degradation continues when the POC reaches the sediments.

Sediment-Water Interactions

Organic matter consumed by heterotrophs (e.g., fungi, protozoans, bacteria, animals) is passed along the food chain to drive biologically mediated reactions. A substantial amount of the nonliving or detrital fraction of the particulate organic matter (i.e., large biopolymers like proteins) is decomposed by the respiratory activities of heterotrophs, which releases DOM and returns nitrogen, phosphorus, and carbon to their soluble forms, greatly enhancing nutrient concentrations. This process regenerates nutrients and enhances nutrient recycling rates and is called *remineralization.*

Much of the POM is remineralized after the death of organisms or excretion. However, small amounts of POM in the water column sink and reach the sediment before becoming remineralized. Because respiration occurs at all depths, less than a few percent of the POM resulting from primary production survives the trip to the sediment. The sediment-water interface is the site of much biological activity, so that POM that reaches the sediment is thus subject to extensive degradation by the benthos prior to burial. Remineralization is significant through about the top 10 cm of sediment.

Particulate (including colloidal) matter can bind to foreign chemicals, which can be suspended in the water column or deposited on or in the sediment. ''Free'' and/or dissolved chemicals are often more available for uptake by organisms and may be potentially toxic. Conversely, reduced bioavailability is associated in the aquatic environment with chemicals that are bound. Furthermore, microorganisms like bacteria adhere to abiotic particles, including colloids, forming aggregates (or ''bioflocs'') and thus enhance the production of larger aggregate matrices, which in turn increases the binding of metals and organic chemicals. Bacterial densities, particle size, and the concentration and spatial distribution of inorganic and organic particles are key factors that influence the adsorption of foreign chemicals onto suspended particles.

Sedimentation rate of bound chemicals is a function of the size and density of particles and hydrographic features. Studies of the transport of suspended sediment demonstrate that particles that are routinely suspended in the water column tend to range from 0.1 to 200 μm in diameter (e.g., includes fine sands and silt and clay within this size range).

Aquatic sediments are typically heterogenous in composition and geographically variable in distribution. They typically consist of mixtures of particles such as clay, silt, sand, viable organic biota, and nonviable organic substances or coatings (e.g., mixtures of proteins, carbohydrates, lipids, humic substances; see Chapter 8). Although sediments often consist of an inorganic matrix of silica, alumina, and carbonates coated with organic matter and iron and manganese oxides, they can vary from pure mineral to pure organic matter. In slow-moving waters sediments tend to be fine-grained (diameter < 64 μm), whereas in fast-flowing waters sediments are composed of cobble and gravel. As a result of a large surface-area-to-volume ratio and other factors (e.g., surface electrical charge), fine-grained sediments typically have relatively large adsorptive capacities for chemicals compared to coarse-grained sediments. For this reason, fine-grained sediments are responsible for much of the transport of chemicals in lakes, estuaries, and the nearshore areas of oceans. Furthermore, small-grained sediment environments are often more complex than large-grained sediments because they have more microhabitats and interrelating redox gradients. Chemicals may be transformed in the sediment by biological degradation (biodegradation) or abiotic processes (e.g., photolysis, hydrolysis, pH change, various salt effects). Some of the processes are discussed in Chapter 15.

The largest volume of surface sediments is occupied by the fluid that is found in open spaces between the particles of sediment and is termed *pore* or *interstitial water* (Libes, 1992). Conditions may vary considerably between the overlying water and the pore water as a result of the interaction between particulate matter and water. Because sediment is the site of chemical reactions such as sulfate reduction as well as mineral precipitation and dissolution, which can change ion ratios, the natural composition of chemicals in pore water may be quite different from that of the overlying water. Furthermore, the movement of particle-sorbed foreign chemicals to the sediment is an additional source of chemicals for pore water and is the main route for clearance of these chemicals from the water column. Consequently, concentrations of both natural and foreign chemicals in pore water may be many times higher than in the overlying water. In turn, the overlying water of the sediment-water interface may also have higher concentrations of chemicals because of expulsion of pore water as result of compaction from particles settling on sediment, bioturbation, or other activities.

Three physical activities (processes) transport chemicals vertically from the sediment into the water column: advection, diffusion, and convection. These activities result in the bulk transport of contaminated interstitial water including colloids and particles from sediment to the overlying water. Micro- and macrobenthic processes can also affect these physical activities. Microbenthic processes include microbial activity that results in changes in the sediment chemical environment through the cycling of elements and compounds. Of particular concern here is the metabolism of complexes that include foreign chemicals and organic carbon, which can result in mobilization of chemicals. Benthic bacteria can also induce changes in the oxidation-reduction and pH conditions of sediment. This is illustrated by the release of metals as a result of a reduction in pH when bacteria metabolize metal sulfides. Conversely, bacteria can convert metals into metal organic complexes that may be toxic and mobile (e.g., methyl mercury).

The activities (bioturbation) of macrobenthic infaunal and epifaunal organisms alter physical processes that mediate chemical transport more so than do microbenthic organisms. Macrobenthic effects are primarily by the construction and ventilation of burrows and the swimming of larger benthic organisms, which intensify advective and turbulent diffusive events and result in contaminated fluid and sediment particle resuspension in the water column. The transport of chemicals between sediment and water is significantly greater in the presence of benthic organisms than with purely abiotic physical processes alone. This effect may be more pronounced in saltwater systems, where the structure of the benthic community is dominated by active crustaceans and infaunal polychaetes and mollusks, unlike freshwater benthic communities, where epibenthic arthropods, such as insect larvae, tend to dominate. Bioturbation may also change sediment particle size distribution (e.g., deposit feeders gluing detritus into fecal pellets) and increase sediment water content, which in turn increases porosity and increases diffusional and advective fluxes. Bioturbation thus affects the depth of the redox layer by enhancing the rate of oxygen supply to the pore waters. Therefore, the oxic zone of sediments (upper 10 cm) is characterized by relatively homogeneous chemical distributions and the anoxic sediments are more likely to have well-defined concentration gradients.

Sediment-associated organisms may be exposed to chemicals whether dissolved in the pore water or overlying water or through ingestion of contaminated particles. The latter route of exposure may play the largest role in bioaccumulation of sediment-associated contaminants. Organisms (infaunal and microbes) can also increase the concentration of DOM through excretion or by in situ decomposition, thus affecting interstitial chemicals. For example, decomposing diatoms can release humic and fulvic acids but none are found in living organisms (Lee, 1991). Although infaunal organisms can also reduce DOM by assimilating dissolved carbohydrates and amino acids, it is not likely that those organisms would substantially reduce interstitial water DOM concentrations.

Particulate and dissolved organic matter and colloids are important in water chemistry and the binding and transport of metals and organic chemicals. Sorption interactions are discussed in Baker (1991a,b), Förstner (1990), Förstner and Wittman (1979), Stumm and Morgan (1981), Whitfield and Turner (1987), and Newman and McIntosh (1991). However, a mechanistic understanding of the sorption interaction is generally not well defined. Currently the most well-modeled and described chemicals are nonionic organics and metals. The dominant mechanism of interaction with particle surfaces for nonionic, nonpolar organic chemicals is partitioning by a solubility mechanism into organic matter coatings. The most significant characteristic affecting the partitioning behavior of many nonionic organics is the fraction of organic carbon associated with the sediments and interstitial waters. Moderately polar organic chemicals are sorbed through a combination of hydrogen bonding, cation exchange, and nonpolar partitioning interactions. Both polar and nonpolar chemicals can also form a variety of bound conjugates with organic matter through various biochemical processes.

Metals interact with particles through physical-chemical sorption effects as a result of nonspecific forces of attraction (van der Waals forces) acting between charges present on the surface of the particle and the dissolved metal species. Strictly chemical processes can include ion exchange, precipitation, coprecipitation, solid-state diffusion, and isomorphic substitution. In comparison to organic chemicals, the interaction of metals with sediments and interstitial and overlying waters is very complex. The factors that control the form of the metal in sediment/water are not clearly understood. However, it is believed that four chemical processes may cause metals to move out of the sediment: elevation of salinity, changes in oxidation-reduction state of the sediments toward reduced conditions, reduction in pH, and the presence of complexing agents. The first three processes release mobile free metal; the fourth process releases organically bound mobile metal. Also, in anoxic sediments when the concentration of metal exceeds the binding of the acid volatile sulfide pool metals are more available.

Modifications of the organic material composition of sediment and the chemical characteristics of DOC in a body of water are important in the transport, transformation, availability, and potential accumulation and toxicity of chemicals in aquatic ecosystems. Changes in their concentration should be considered when assessing toxic effects of chemicals in sediment-water systems.

Reactions in Water

Most chemical reactions in natural aquatic systems can be categorized into one of the following phenomena:

- Acid-base
- Complexation
- Dissolution-precipitation
- Oxidation-reduction
- Sorption-desorption
- Hydrolysis
- Photolysis

The importance of these reactions in aquatic chemistry is presented in Chapter 15, and in Morel and Hering (1993), Stumm (1987; 1990), Pagenkopf

(1978), Pankow (1991), and Manahan (1991). Most aquatic reactions are discussed from the thermodynamic equilibrium approach, which is adequate for describing natural water chemistry; however, the chemical kinetics or rates of chemical reactions should also be considered. Biological processes, like algal photosynthesis and biodegradation, play a key role in aquatic chemistry, and their effects on reactions should not be underestimated.

FACTORS THAT AFFECT THE ENVIRONMENTAL CONCENTRATION OF CHEMICALS

In the aquatic environment the concentration, transport, transformation, and disposition (collectively called *fate* or environmental fate) of a chemical or material are primarily controlled by the following:

- The physical and chemical properties of the compound
- The physical, chemical, and biological properties of the ecosystem
- The sources and rate(s) of input of the chemical into the environment.

Transport and transformation processes and mechanisms in the aquatic environment are discussed in Chapters 15 and 17.

The physical and chemical properties of the compound that are important include molecular structure, solubility in water, and vapor pressure. Rate constants for hydrolysis, photolysis, biological degradation, evaporation, sorption, uptake, and depuration by organisms and partition coefficients (air-water, sediment-water, octanol-water) also provide significant information.

Some of the properties of aquatic environments that may affect the fate of a chemical are surface area–volume relationships, temperature, salinity, pH, flow, depth, amount of suspended material, sediment particle size, and organic carbon content in sediment.

Knowledge of the average rates of input and the occurrence of single large "slugs" of chemicals entering the aquatic environment is important for predicting environmental concentrations. However, the average rates of entry and the short-term higher inputs associated with manufacture, use, and disposal are variable and are difficult to estimate; hence they can only be crudely approximated. Information on background concentrations of chemicals and on transformation products is also important for predicting concentrations in the aquatic environment.

The types of data outlined above are used not only to predict environmental concentrations of chemicals but also to determine the following:

- The mobility of a chemical and the parts of the aquatic environment in which it most likely would be distributed
- The kinds of chemical and biological reactions that take place during transport and after deposition
- The eventual chemical form (i.e., parent substance or metabolite) found
- The persistence of each chemical form

Knowledge of the physical and chemical properties of a compound permits an estimate of the part of the aquatic environment likely to receive the greatest exposure and the ability of the compound to move within and through the environment (Mackay, 1991). For example, organic chemicals with high vapor pressure and low water solubility tend to dissipate from the water into the atmosphere (volatilize). Chemicals with low vapor pressure and low water solubility tend to become associated with sediment and particulate matter, and chemicals with high water solubility tend to remain in water. Factors that increase water solubility are ionic functional groups (e.g., carboxyl) and hydroxy and amine functional groups. Thus, chemicals substituted with these polar functional groups do not volatilize rapidly from solution. Chemicals with low molecular weight, and which do not contain functional groups, have greater vapor pressures. In the aquatic environment water-soluble chemicals also tend to be more widely and homogeneously distributed than insoluble chemicals.

To summarize, in water, a chemical can exist in three different basic forms that affect its availability to organisms (*bioavailability*): (1) dissolved, (2) sorbed to biotic or abiotic components and suspended in the water column or deposited on the bottom, and (3) incorporated (accumulated) into organisms. Dissolved (i.e., water soluble or hydrophilic) chemicals are readily available to organisms in the water column. Hydrophobic (i.e., not liking water but fat soluble or lipophilic) chemicals may be sorbed on sediments, large suspended organic colloids, or other particulates and may be irreversibly bound and generally unavailable. However, some of the bound chemicals may be available to benthic organisms through ingestion or direct uptake from interstitial water. Sediment-bound chemicals may also be covered and unavailable until the sediment is disturbed. Hydrophobic chemicals can be accumulated by organisms in various tissues, biotransformed (metabolized), and excreted back into the water. Bioaccumulation/bioavailability and biotransformation are discussed in Chapters 16 and 17.

Bioavailability is a critical concept in ecotoxicology because toxic agents that are not bioavailable are, by definition, unlikely to produce any adverse effects as organisms are not exposed. As scientists with a variety of backgrounds and objectives conduct ecotoxicology investigations, there have been some incompatible uses of the term. The following definitions have been proposed to clarify the situation (Landrum et al., 1994).

Environmental availability is the portion of the total chemical or material present in all or a specified part of the environment that can be involved in a particular process or processes and is subject to all physical, chemical, and biological modifying influences. For example, it might be said that only 40% of chemical X is environmentally available for various abiotic chemical reactions because 60% is effectively isolated by being buried deep in marine and freshwater sediment. Broadly speaking, this term defines the total amount of material potentially available to organisms. Bioavailability can be viewed as a subset of environmental availability where organisms, as indicated by the prefix "bio," actually encounter some of the material in question. Thus, bioavailability incorporates not only the characteristics of the chemical and environmental speciation but also the behavior and physiology of the organism.

Environmental bioavailability is the fraction of the environmentally available material that an organism actually accumulates when processing or encountering a given environmental medium (e.g., fraction of a chemical extracted by the gills from the total amount in water passing through the respiratory cavity or fraction absorbed from the total encountered in moving through and/or consuming sediment). It is the ratio of the uptake clearance rate (amount removed or cleared from the medium by the activity of the organism) to the rate at which an organism encounters a given contaminant in a medium (e.g., water, sediment, food) being processed by the organism. This is a measure of an organism's general extraction efficiency, via respiratory, dietary, and surface absorption processes, from the environmentally available portion of a material.

Toxicological bioavailability is the fraction of the exposure dose or concentration which is adsorbed and/or absorbed by an organism, distributed by the systemic circulation, and ultimately presented to the receptors or sites of toxic action (i.e., target). The fraction of the exposure dose (or concentration) that actually reaches the eventual target(s) is the net result of such internal processes as absorption (uptake), distribution, biotransformation (metabolism), and excretion (elimination). This is the classical pharmacological/toxicological definition of bioavailability. It works well for chemicals with systemic toxicity, but it is of less value for topically active materials. Environmental bioavailability and toxicological bioavailability are similar in that both are estimates of an extraction efficiency, but environmental bioavailability addresses the total amount adsorbed and/or absorbed by the organism whereas toxicological bioavailability addresses the amount that reaches the site(s) of toxic action.

The average amount of time a chemical spends in a particular compartment (e.g., water, sediment, biota) or reservoir before it is removed through some transport process(es) is called its *residence time*. Steady-state residence time is the ratio of the total amount of a chemical (dissolved and particulate) in a reservoir to the rate of supply or removal of the chemical to or from the reservoir per unit time. Despite transport mechanisms, the mobility of chemicals is affected by partitioning at the air-water and sediment-water interfaces. Transport mechanisms and mobility of chemicals are discussed in Chapter 15.

Water-soluble chemicals may persist and retain their physical and chemical characteristics while being transported and distributed in the aquatic environment. Chemicals that are persistent (not significantly degraded) can accumulate in the environment to toxic levels. The persistence of a chemical in a compartment may be expressed in terms of its *half-life*, which is defined as the time required to reduce the initial concentration of the chemical by one-half. The use of half-life implies that all fate processes can be described as first-order rate constants. First-order kinetics are those whose rates are proportional to the concentration of the chemical substrate. Although a useful simplifying procedure, this assumption is not typically true, and caution must be used when moving beyond very general discussions about persistence.

Chemicals may also be converted to other forms as a result of biotic and abiotic transformations. The predominant abiotic transformation reactions in water are hydrolysis, oxidation, and photolysis. These reactions may render a chemical more or less available for biotic transformation (biotransformation). Fish, invertebrates, microorganisms, and plants transform chemicals to various metabolites after their uptake and absorption. These biotransformations are enzymatically mediated, unlike the purely nonmetabolic chemical and photochemical reactions that occur in water itself.

Biotransformations mediated by plants, animals, and higher organisms affect the environmental concentrations of chemicals. However, for most organic compounds in the aquatic environment, these effects are insignificant in comparison to those of microbial transformations. *Biodegradation* refers to those transformations of chemicals by microorganisms whereby parts of the chemical are incorporated into cellular material or used as an energy source for the organisms and the remainder is converted to simple inorganic molecules. The complete biodegradation of a substance to CO_2, H_2O, and simple organic and inorganic end products is called *mineralization*. Generally, biotransformation tends to degrade a chemical to a more polar and water-soluble form of lower toxicity. However, this is not always the case, and the transformation products may also be toxic. Biotic transformations by microorganisms are discussed in Chapter 15 and those by fish and invertebrates in Chapter 17.

Mathematical modeling can be used to integrate data on complex mechanisms such as transport, degradation and transformation processes, transfer between environmental media (e.g., air, water, and sediment), and biological uptake into a mathematical picture of the behavior of a chemical. The model may be used to predict the concentrations of the chemical in the aquatic environment, the potential exposure areas, and the fate of the chemical. The use of mathematical models is described in Chapter 18. Other useful references are Calamari (1993), James

(1993), Jorgensen (1984, 1986, 1990), Mackay (1991), Samiullah (1990), Schnoor et al. (1987), and Sheehan et al. (1985).

BASIC TOXICOLOGICAL CONCEPTS AND PRINCIPLES

Concepts

A *toxicant* is an agent that can produce an adverse response (effect) in a biological system, seriously damaging its structure or function or producing death. The adverse response may be defined in terms of a measurement that is outside the "normal" range for healthy organisms. A toxicant or foreign substance (i.e., xenobiotic) may be introduced deliberately or accidentally into the aquatic ecosystem, impairing the quality of the water and making it unfavorable for aquatic life. Toxicants enter aquatic ecosystems from (1) nonpoint (diffuse) sources or from more than one point, such as agricultural runoff from land, contaminated ground water and bottom sediments, urban runoff, dredged sediment disposal, and atmospheric fallout, and (2) point (discrete, localized) sources such as discharges (effluents) from processing and manufacturing plants, hazardous waste disposal sites, and municipal wastewater treatment plants. A chemical or oil spill on or near the surface of a body of water is also considered a point source. The effects of contamination on the aquatic environment from diffuse sources may be detrimental but are often less obvious than those from point sources, because often there is no adjacent uncontaminated area with which comparisons can be made. Point sources can be characterized because the amount and location of the discharge can be measured more accurately than with nonpoint sources, which cover large areas or are a composite of many point sources.

Pollutant or *toxic agent* is used occasionally in this text to indicate an agent that can produce adverse effects. However, these terms have a broader meaning than toxicant as they encompass adverse abiotic changes, such as extremes in temperature and pH, decreases in dissolved oxygen, and increases in suspended solids or sedimentation. Furthermore, there are formal, internationally accepted definitions of pollution (e.g., GESAMP, 1993), important for international laws, and often a distinction is made between a contaminant (a chemical or physical change is present) and a pollutant (a chemical or physical change is present and causing adverse effects on biota, on their habitat, or on ecosystem structure and function).

Pollution means the introduction by humans, directly or indirectly, of substances or energy into the aquatic environment resulting in such deleterious effects as harm to living resources, hazards to human health, hindrance to aquatic activities including fishing, impairment of quality for use of water, and reduction of amenities (modified from marine pollution definition by GESAMP, 1993). *Effluents* may be classified as water pollutants because they may be toxic and also change the temperature, pH, and salinity of a receiving water in which they enter. However, effluents usually contain one or more chemical toxicants that vary in composition over time (see Chapter 33 for a discussion of effluents).

This book focuses on the major categories of anthropogenic toxic materials, especially those known to cause significant biological and ecological problems in the aquatic environment. These include synthetic organic pesticides, other industrial chemicals from chemical manufacturing, petroleum hydrocarbons, metal salts, other inorganic compounds, and liquid industrial waste discharges (effluents). The sources and characteristics of some of these toxic materials are briefly summarized below. Although some of these chemicals (e.g., certain metal salts or other inorganics) may normally be found in minimal nutritive background concentrations, most of those discussed here are foreign, nonnutritive, and not normally present in significant background concentrations in water bodies.

Toxicity is a relative property reflecting a chemical's potential to have a harmful effect on a living organism. It is a function of the concentration and composition/properties of the chemical to which the organism is exposed (i.e., externally *and* internally) and the duration of exposure. Traditionally, toxicity data have been used in comparing chemical substances or the sensitivities of different species to the same substance. Information about the biological mechanisms affected and the conditions under which the toxicant is harmful is also important for this comparison. *Toxicity tests* are therefore used to evaluate the adverse effects of a chemical on living organisms under standardized, reproducible conditions that permit comparison with other chemicals or species tested and comparison of similar data from different laboratories.

Toxicity can be divided into the broad categories *direct* and *indirect*. *Direct toxicity* results from the toxic agent acting more or less directly at sites of action in and/or on organisms; *indirect toxicity* occurs as a result of the influence of changes in the chemical, physical, and/or biological environment (e.g., changes in the quality and/or quantity of food organisms or habitat changes and/or losses). Direct toxicity is largely the result of internal biochemical changes, whereas indirect toxicity is more a function of changes in general organism viability produced by factors external to the organism. Although many indirect toxic effects on a population or community may be traced back to direct toxicity in a particular group or species (e.g., a decrease in a food species resulting from a direct effect may have an indirect effect in predator species as their viability is decreased by poor nutrition), this is not always the case. Laboratory-oriented aquatic toxicity studies tend to focus on examination of direct toxic effects, and field-oriented studies may more often include consideration of indirect effects and changes in communities and populations. Most experimental aquatic toxicology studies to date have been concerned with direct toxicity to individual species. The direct toxicological information gained is then used to estimate indirect effects or interpret site-specific situations, although experimental examination of indirect toxicity is increasing. Futher discussion related to effects is presented in the section on types of toxic effects later in this chapter.

In general, toxic effects may be manifested *immediately* during exposure or after termination of exposure to a chemical, or they may be *delayed* until some time after exposure. This is determined by the properties of the chemical, its mode of toxic action, and the ability of the organism to metabolize (or biotransform) the chemical. For example, chemicals that are susceptible to biotransformation (e.g., low-molecular-weight hydrocarbons) have a short half-life in the organisms and are therefore excreted rapidly. Chemicals with a short half-life should not produce delayed effects. In the natural aquatic environment, chemicals may have a short half-life in the water and sediment because they are transformed and degraded by chemical and physical processes as well as biological processes.

Some toxic effects are *reversible* and others are *irreversible*. They may be reversible by normal repair mechanisms, such as by regeneration of damaged or lost tissue and recovery from narcosis. In many cases effects are reversible only if the organism can escape the toxic medium and find a toxicant-free environment. Serious damage or injury to an organism may be irreversible and may eventually result in death. In the laboratory, the reversibility or irreversibility of chemical effects may be studied by transferring organisms from a medium containing the toxicant to a medium that is free of the toxicant (e.g., a recovery study). In the field, this may be studied by monitoring affected species or populations for a period of time after input of the chemical is stopped.

A distinction can be made on the basis of the general site of action of a toxicant. *Local effects* occur at the primary site of contact. An example of a local effect is a skin or gill reaction (e.g., discoloration, inflammation, or erosion) in fish exposed to various organic and inorganic compounds. For a chemical to have an effect at a site other than the primary site of contact, it must gain access to the internal environment of the organism, generally by transport through the circulatory system. *Systemic effects* are those that require absorption and distribution of the chemical to a site distant from the original contact or entry site. Effects on the nervous system or various organs are classified as systemic effects.

The toxicity of certain chemicals is *nonselective* in that they may have an undesirable effect on the membranes of numerous cells and tissues of aquatic organisms. Nonselective toxicity is also referred to as *narcosis* or anaesthesia and is a result of a general disruption of cell membrane function, the details of which are poorly understood (Abernethy et al., 1988). Many industrial organics, especially solvents, are nonselective toxicants.

In contrast, a chemical may be so *selective* (e.g., organophosphate insecticides) that it adversely affects only one type of cell or tissue or process (e.g., nerve transmission) without harming others (Albert, 1965, 1973). The affected and nonaffected cells, tissues, or processes may be in the same or in different organisms. If they are in different organisms, the chemical is *species specific* in its selective activity. If they are in the same organism, the chemical may or may not be species specific, but it will be selectively active on certain types of cells, tissues, or processes within a species. *Selective activity* or *toxicity* results from biological diversity and variability in the response of cells and tissues to chemicals. This diversity may prevent the toxicologist from predicting the effects of a chemical in one species from results of experiments performed in another species. Further details on types of toxic action are presented later in the section on interpreting toxicity test data.

There appear to be two significant mechanisms for the selective action of a chemical. The first mechanism involves the presence or absence of specific target or receptor sites (see Chapter 20) in the exposed cell system, because selectivity indicates that the chemical reacts only with specific normal components (or targets) of the cell. The target may be vital to the cell and may be altered by the chemical so that it can no longer carry on its function; as a result, its viability is altered. Or the target may be a protein or lipid that is not vital to the function of the cell, so that the reaction between the chemical and the target does not produce a direct alteration in cell function. The second mechanism involves factors that are responsible for the distribution and alteration of the concentration of the chemical at a specific cell or tissue site. This is usually the result of such processes as selective absorption, translocation, biotransformation, and excretion.

Small changes in the structure of a chemical agent may alter its biological activity. This is because chemical-biological effects are the result of a physical-chemical reaction or interaction between chemicals and organisms. Examination of the relationships between chemical structure (and related properties) and biological responses is referred to as the study of *structure-activity relationships*. Such studies are used to define as precisely as possible the limits of variation in chemical structure that are consistent with the production of a specific biological response (e.g., death). If enough of these studies are performed, a hypothesis may be developed concerning the most likely targets or receptors in a reaction with a chemical and/or the critical processes involved in the chemical reaching those target sites. These studies also enable the investigator to synthesize analogs of the chemical that may be more active in producing the same biological effect or that may be inactive in the biological system under investigation. Such information is also important for the development of a hypothesis concerning the structure of the target and predicting bioaccumulation, toxicity, and biodegradation. Structure-activity relationships are discussed in Chapter 20.

Principles

The underlying assumptions of toxicology are critical to its conduct and to the understanding, interpretation, and application of toxicity data, whether they are from laboratory tests or from field observations. Perhaps the primary goal of all toxicological investigations is "to know the relative toxicity of the substance with a view to determining its position among other toxic substances whose toxicities are already known" (Filov et al., 1979). In other words, it is important to determine whether a substance is more or less toxic relative to other known toxic agents. The fundamental assumptions of toxicology are as integral and applicable for ecotoxicology as they are for human or mammalian toxicology, where they were initially developed (Filov et al., 1979; Amdur et al., 1991). Although basic toxicological principles are often summarized in the simple expressions "dose-response" or "concentration-response," as discussed below, there are a number of vital assumptions:

1. A cause-effect relationship exists. The effect or response in question is clearly a direct or indirect result of the exposure of the organism(s) to the toxic agent(s) being examined.
2. A dose-response or concentration-response relationship exists.
   a. The effect or response in question is a result of the toxic agent(s) reaching and interacting with the site(s) of toxic action in or on the organism.
   b. The amount of toxic agent reaching the site(s) of toxic action is some function of the exposure of the organism to the toxic agent or, in the case in which a metabolite is the agent, its parent compound.
   c. Above a real or statistically based effect threshold, the magnitude of the effect or response is proportional to the amount of toxicant reaching the site(s) of toxic action.
3. Effects can be quantified. Observed effects or responses of toxic action can be measured and quantified in a reproducible way that is relevant to the toxic processes under examination.

Causality is a critical toxicological principle as, for quantitative purposes, it must be reasonably certain that there is a causal relationship between the observed effect or response and the presence of the toxic agent. However, there may be some doubt about the identity of the chemical, which may have changed during the exposure, the actual exposure concentration in the water, and the specificity of the response, because aquatic organisms may respond similarly to a variety of stresses. Thus, until proved, it is only a reasonable working presumption that the effect or response being observed is a result of exposure to the known concentration of chemical.

Apparently innocuous chemical substances can have undesirable or distinctly harmful effects when taken up by an organism in sufficient amounts over short time periods. In contrast, the uptake of minute quantities of toxic chemicals may result in no apparent adverse effects. No chemical is completely safe and no chemical is completely harmful. One factor that determines whether a chemical agent is potentially harmful or safe is the relationship between the concentration (quantity) of the chemical to which an organism is exposed and the duration of the exposure. In mammalian toxicology the key principle that measures the severity of the response or effect resulting from such exposure is termed the *dose-response relationship* as toxicant is usually introduced orally as a bolus or dose (i.e., by gavage with a feeding needle or by gastric intubation) into the body of the organism. In aquatic toxicology, where exposure is usually via a waterborne concentration of the toxicant, this basic principle is referred to as the *concentration-response relationship.* For any toxic agent, contact with a biological membrane or system may not produce a measurable adverse effect if the concentration of the chemical is below a minimally effective (threshold) level. This principle also implies that all chemicals are capable of producing a deleterious effect if a high enough concentration of the chemical comes into contact with a biological membrane or system.

The concept of exposure dose or exposure concentration is vital to toxicology and warrants further examination. For convenience the exposure amount or concentration in the exposure medium is usually termed the toxic dose or toxic concentration. However, the amount reaching the site(s) of toxicant action in the organism is the correct and most accurate definition of dose. Any other measure of toxicant, such as the concentration of toxicant in the exposure medium (e.g., water), must be considered a surrogate for the actual concentration of toxicant reaching the site(s) of toxic action. If a surrogate for the target dose is to be employed, it is crucial that the relationship between the target dose and the surrogate be understood. Although this may appear obvious, it is a primary source of confusion. Franks and Lieb (1982) state the situation clearly: "One must be very careful, when comparing potencies, not to get confused between observed values and the potency at the site itself."

In other words, estimates of toxicological *potency*, which is a measure of the relative effectiveness of inducing a toxic response of a given type and magnitude, may vary substantially depending on whether the exposure concentrations in environmental media, such as water, or the concentrations reaching the site(s) of toxic action in the body of the organism are used. For example, the acutely lethal water concentration is simply a surrogate for the amount of toxicant in the exposed organism which has reached the site(s) of toxic action and is producing the effects that result in death. If, for a given chemical, the relationship between the amount in the water and the amount in the organism is not known, then the primary toxicological objective of determining relative potency cannot be reliably satisfied, because it is uncertain if the water-based surrogate in this case is equivalent to water-based surrogates whose relationship with the organism or target dose has been validated. Figure 10, which appears much later in this chapter, illustrates apparent differences that can be observed when using either exposure water concentrations or whole-body concentrations for potency comparisons. The importance of this concept to the accurate comparison of species sensitivities to chemicals cannot be overstated.

The general phenomena of toxicity can be placed into perspective, and into a useful practical framework, by dividing toxicological processes into three broad, basic aspects: the exposure phase, the toxicokinetic phase, and the toxicodynamic phase (see Figure 4). The *exposure phase* is concerned with duration of the period an organism is exposed and the environmental bioavailability of a chemical toxicant during that exposure. The *toxicokinetic phase* comprises the uptake, distribution, metabolism, and elimination of the bioavailable portion of the toxic agent by the organism. The *toxicodynamic phase* involves the time course of biological response resulting from the agent reaching the site(s) of toxic action in the organism and interacting with receptors at the site(s) to produce its effect(s). As basic to toxicological theory and, in the tradition of the "three Rs" of education, these could be referred to as the "three Ps" of toxicology: exPosure, Partitioning, and Potency (McCarty and Mackay, 1993).

Figure 4 presents a summary of the factors controlling toxicity divided into the major phases and indicating which occur outside and inside exposed organisms. The phases associated with both bioaccumulation and toxicity of a chemical are also indicated by arrows at the bottom of the figure. When making comparisons between the toxic effects of two or more chemicals, one chemical is said to have greater *potency* than the other if it requires a lesser amount to induce the same degree of biological response. Although potency is often referenced to the exposure dose or concentration (i.e., LD50 or LC50), it is used in Figure 4 with reference to the amount of chemical in the organism that is associated with the adverse effect, a more toxicologically appropriate definition.

FACTORS THAT INFLUENCE TOXICITY

The general term that describes various abiotic and biotic factors that modulate toxicity is *modifying factors*. A brief examination appears below; a more detailed review is presented in Appendix Chapter C.

Factors Related to Exposure

For a chemical, its metabolites, or its conversion products to elicit an adverse response or have a toxic effect on aquatic organisms, the chemical must come into contact and react at an appropriate target site(s) on or in each organism at a high enough concentration and for a sufficient length of time. The concentration and time required to produce an adverse effect vary with the chemical, species of organism, and severity of the effect. This contact-reaction between the organism and the chemical is called *exposure*. In the assessment of toxicity the most significant factors related to exposure are the kind, duration, and fre-

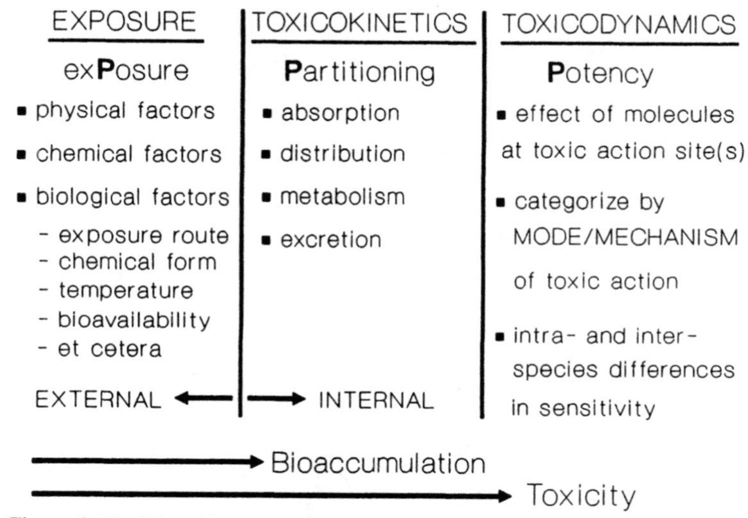

Figure 4. The Three Phases of Toxic Action. (Modified from McCarty, 1990, used with permission.)

quency of exposure and the concentration of the chemical. Exposure assessment is discussed in the context of ecological risk assessment in Chapters 28 through 33.

Aquatic organisms may be exposed to chemicals present in water, sediment, or food, and the air (for aquatic birds, mammals, amphibians, and reptiles). Water-soluble (hydrophilic) chemicals are more readily available to organisms than water-insoluble (hydrophobic) chemicals that are tightly adsorbed or otherwise bound to suspended particles, organic matter, or biological systems. Water-soluble chemicals may enter an organism through the general body surface (i.e., dermal exposure) and respiratory tissue such as gills or mouth. Chemicals in food may be ingested (i.e., oral exposure) and absorbed through the gastrointestinal tract. Adsorbed chemicals may also enter an organism through the body surface and respiratory surfaces (e.g., gills) as they gradually dissociate from particles to the water in immediate contact with these areas. Oral exposure provides direct venous blood input of the chemical, which is subject to first-pass elimination in liver and gills, whereas chemicals absorbed dermally are subject to first-pass elimination in the kidney and gills. Each elimination pathway may produce a reduction in the amount of parent chemical and/or metabolite that reaches the systemic arterial circulation. In contrast, chemicals absorbed across the gills enter the systemic circulation directly with no possibility of first-pass elimination and associated mitigation. Thus, the route(s) of exposure may affect kinetic factors such as absorption, distribution, biotransformation, and excretion and may ultimately determine the toxicity of a chemical. Only a fraction of the chemical absorbed may reach the target (i.e., receptor tissue) site of action.

Adverse or toxic effects can be produced in the laboratory or in the natural environment by acute (short-term) or chronic (long-term) exposure to chemicals or other potentially toxic agents. In *acute exposure*, organisms come in contact with the chemical delivered either in a single event or in multiple events that occur within a short period of time, generally hours to days. Acute exposures to chemicals that are rapidly absorbed generally produce immediate effects, but they may also produce delayed effects similar to those caused by chronic exposure. During *chronic exposure*, organisms are exposed to low concentrations of a chemical delivered either continuously or at some other periodic frequency over a long period of time (weeks, months, or years), measured in relation to the organism's life cycle. Chronic exposure to chemicals may induce rapid, immediate effects similar to acute effects, in addition to effects that develop slowly. Acute and chronic exposure and toxicity in freshwater and saltwater systems are discussed in Chapters 2 and 3, respectively.

In general, an acute exposure involves a short period of time compared to the life cycle of an organism, whereas a chronic exposure may involve an entire generation time or reproductive life cycle (i.e., egg to egg) or several of them. Exposures that are intermediate in duration (a month to several months) may be less than a complete reproductive life cycle and include exposure during sensitive early stages of development. These are referred to as subchronic exposures. In aquatic toxicology, studies involving subchronic exposure are generally called early life stage, critical life stage, embryo-larval, or egg-fry tests. These tests are discussed in Chapters 2 and 3.

The frequency of exposure may also affect toxicity. For example, one acute exposure to a single concentration of a chemical may have an immediate adverse effect on an organism, whereas two successive exposures cumulatively equal to the single acute exposure may have little or no effect. This may occur as a result of metabolism (detoxification) of the chemical between exposures (e.g., phenol) or acclimation of the organism to the chemical (e.g., copper). However, if there is minimal metabolism (e.g., some PCBs), the chemical may not be easily transformed and excreted and may remain in the organism, eventually producing a chronic toxic effect.

External environmental factors may influence the toxicity of a chemical by altering the form of the chemical or the nature of the exposure. Factors associated with the bioavailability of the chemical in the water medium include dissolved oxygen, pH, temperature, and dissolved solids. These factors are discussed by Abel (1989), Alabaster and Lloyd (1980), and Buikema et al. (1982), as well as in Appendix Chapter C.

Factors Related to the Organism

Species differ in susceptibility to chemicals. This may be due to differences in accessibility, with certain species effectively excluding a toxic medium for short periods of time (e.g., clams can close and maintain anaerobic metabolism). In addition, rates and patterns of metabolism and excretion can substantially affect susceptibility. Differences in susceptibility to chemical agents among fish of different strains also result from genetic factors. Evidence for genetic selection in the natural environment has been observed in mosquito fish after exposure to high levels of insecticides and in some crustaceans exposed to metals.

Dietary factors also influence toxicity, by producing changes in body composition, physiological and biochemical functions, and nutritional status of the organism.

Immature or young neonatal organisms often appear to be more susceptible to chemical agents than are adult organisms. This may be due to differences in degree of development of detoxification mechanisms between young and adult organisms. Differences in rates of excretion of toxic chemicals may also be involved in age-dependent toxicity effects and the influence of differences in body size on toxicokinetics in general must be considered (see following section on body size). However, the reverse is also true. Embryos may be less sensitive than adults because, at particular stages, they may have protective or impermeable membranes.

The toxicity of a particular chemical agent is traditionally evaluated on the basis of tests carried out with robust or healthy organisms. However, test organisms that are in poor health or are stressed in some other manner, such as by previous or concurrent exposure to other toxicants or being in particular growth phases, are likely to be more susceptible to a toxic chemical.

Factors Related to the Chemical

The toxicity of a chemical agent can be influenced by its composition. Impurities or contaminants that are considerably more toxic than the chemical itself may be present. Impurities may vary from one batch of the chemical to another, so that the results obtained with a particular batch may not be reproducible. Therefore, the identity and purity of chemicals are important in toxicity testing. However, toxicity tests conducted with highly purified samples of a chemical agent may not accurately predict the aquatic hazard (see Chapter 28) associated with exposure to the chemical in its less purified form when it is discharged into the environment. Other factors that are directly related to the chemical are its physical and chemical properties such as solubility, vapor pressure, pH, and lipophilicity. These factors affect the persistence, transformation, bioavailability, and ultimate fate of the chemical in water.

TOXIC AGENTS AND THEIR EFFECTS

Toxic Agents

Toxic agents are classified in a number of ways depending on the needs of the practitioner. For example, as the field of aquatic toxicology gains in importance and matures as a science, more research is being conducted on mechanisms of action and target sites (e.g., liver, kidney) of the agents. These are two ways in which toxic agents can be classified. In addition, toxic agents may be classified according to their effects (e.g., immune system, mutation, cancer; see Chapters 12, 13, and 14), use (e.g., pesticide), physical state (e.g., liquid), chemistry (e.g., hydrocarbon), toxicity potential (e.g., extremely toxic, very toxic), or impact on aquatic resources (e.g., tainting). No single classification system is appropriate for the wide diversity and variety of toxic agents. Classification systems that consider the properties of the toxicant (e.g., toxicity, bioaccumulation, persistence) and the anticipated exposure conditions have proved to be most useful for regulatory purposes.

Exposures in the field may be to mixtures of a variety of chemical and physical toxic agents resulting from loadings due to the discharge of complex industrial and domestic effluents or from combinations of pesticides and other chemicals found in agricultural runoff in an area. As each such local area is subject to a more or less unique mixture of toxic agents, each thus represents a site-specific problem. The major concern therefore is to determine the contribution of each component to the overall toxicity. Toxic components (e.g., ammonia, cyanide, phenols) of complex industrial effluents are discussed in Abel (1989), Alabaster and Lloyd (1980), Hellawell (1986), and Russo (1985). Pesticides are discussed in the following.

The focus of this book is on clearly identifiable chemical toxicants with the objective of understanding the toxic processes associated with each general classification. With adequate knowledge about the fate and toxicity of the component chemicals, evaluating various site-specific situations becomes easier, although this problem is still daunting. A general, chemistry-based classification system is used in the following to facilitate discussion, but it must be remembered that there is no guarantee that toxicants grouped on the basis of general chemical characteristics will have all or any common toxicological characteristics in every situation.

Metals and Metalloids

Heavy metals are a group of metallic elements with atomic weights greater than 40 and are characterized by similar electronic distribution in their external shell. These exclude alkali earth metals (e.g., calcium, magnesium), alkali metals (e.g., sodium, potassium), lanthanides, and actinides (e.g., uranium). The broader definition of *trace metals* includes both heavy metals and the latter metals. *Metalloids* are nonmetallic elements, such as silicon and arsenic, that have many properties similar to those of true metals. In aquatic systems, the heavy metals of greatest concern are copper, zinc, cadmium, mercury, and lead. These elements are toxic to organisms above specific threshold concentrations but many of them (e.g., copper and zinc) are essential for metabolism at lower concentrations. Lead and cadmium have no known biological function. Other elements of concern are aluminum, chromium, selenium, silver, arsenic, and antimony, which have contributed to serious problems in freshwater, estuarine, and coastal ecosystems.

Metal contamination in the aquatic environment arises from industrial processes (e.g., mining, smelting, finishing and plating of metals, paint and dye manufacture) and from pipes and tanks in domestic systems. The toxicity of metals varies with aquatic species and environmental conditions; water quality (e.g., hardness) greatly affects the chemical speciation of metals. The metabolic processes of an organism play an important role in its susceptibility to metal toxicity. For example, metal-binding proteins (i.e., metallothioneins; see Chapter 17), which typically control elevated levels of trace metals, may also sequester heavy metals, modifying the toxicological bioavailability of accumulated metals and subsequently reducing toxicity.

The toxicity of heavy metals is discussed in Alabaster and Lloyd (1980), Depledge et al. (1994), Furness and Rainbow (1990), Hellawell (1986), Kennish (1992), Laws (1981), Leland and Kuwabara (1985), Mance (1987), Newman and McIntosh (1991), and Whitton and Say (1975).

Inorganics

Inorganic, nonmetallic toxic agents include a variety of elements, such as chloride, chlorine, nitrogen, phosphorus, and boron, as well as ammonia, arsenic, nitrites, nitrates, and sulfides. The chemical and toxicological properties of the inorganics differ and, as with the metals, may vary with aquatic species and conditions (e.g., pH, temperature). The toxicity of inorganics is discussed in Alabaster and Lloyd (1980), Furness and Rainbow (1990), Hellawell (1986), Kennish (1992), Laws (1981), Leland and Kuwabara (1985), Mance (1987), Newman and McIntosh (1991), Russo (1985), and Whitton and Say (1975).

Organics

The twentieth century has seen widespread exploitation of the developments in organic chemistry that occurred in the latter part of the nineteenth century. As a result of large-scale use of manufactured organic chemicals in industrial and domestic applications and products, there has been a tremendous increase in both direct and indirect discharges to the general environment. Chemicals of concern include the polychlorinated biphenyls, 2,3,7,8-tetrachlorodibenzo-*p*-dioxin (TCDD) and related dioxin and furan congeners, polycyclic aromatic hydrocarbons (PAHs), and various organic solvents. There is tremendous variability in the fate, bioavailability, and toxicity characteristics, both within and between various groups of organic chemicals.

Polychlorinated biphenyls (*PCBs*) are a group (209 congeners/isomers) of organic chemicals, based on various substitutions of chlorine atoms on a basic biphenyl molecule. These manufactured chemicals have been widely used in various processes and products because of the extreme stability of many of the isomers, especially those with five or more chlorines. As a result of this stability and their general hydrophobic nature, PCBs released to the environment have dispersed widely throughout the ecosystem. Some tend to accumulate in living organisms, where a variety of adverse effects have been reported. As a result, most countries have banned the production of PCBs and eliminated or severely reduced their use.

Chlorinated dioxins, commonly referred to as *dioxins*, consist of a group of 75 congeners/isomers based on substitution of chlorine atoms on a basic dibenzo-*p*-dioxin (or PCDD) molecule. Chlorinated furans, commonly referred to as *furans*, consist of a group of 135 congeners/isomers based on substitution of chlorine atoms on a basic dibenzofuran (or PCDF) molecule. Dioxins and furans, which are widespread in the environment, are formed naturally, mainly through natural combustion processes such as forest fires, and anthropogenically, through various combustion processes (waste incineration, fossil-fuel combustion, wood burning) and as a by-product of certain manufacturing activities (e.g., production of PCBs and chlorophenols, pulp and paper production). Of all of the dioxins and furans the most studied is the most toxic isomer *2,3,7,8-tetrachlorodibenzo-*p*-dioxin (TCDD)*. A variety of adverse environmental effects have been reported to result from elevated levels of dioxins and furans in organisms.

Organic solvents include a wide variety of aliphatic and aromatic hydrocarbons, both halogenated and nonhalogenated (e.g., benzene and related aromatic chemicals). Information on some groups of organic chemicals can be found in Eisler (1986a,b, 1989) and Kennish (1992). Niimi (1994) presents a summary on PCBs, PCDDs, and PCDFs.

Polycyclic aromatic hydrocarbons (PAHs), also known as polynuclear aromatic hydrocarbons, encompass a wide range of naturally occurring (direct biosynthesis by microbes and plants, diagenesis of sediment organic matter) and non-naturally occurring chemicals. However, the largest fraction is attributable to human activities. Sources of PAHs include municipal and industrial effluents, petroleum spills, creosote oil, combustion of fossil fuels, brush fires, and atmospheric deposition. Wastewater released from refineries and offshore oil rigs contribute volumes of PAH contaminants which may exceed those from major accidents. Some PAHs are potent mutagens (see Chapter 13), carcinogens (see Chapter 14), and teratogens (see Chapter 7). The general toxicity of PAHs varies as well. Many aquatic organisms can metabolize and detoxify certain PAHs, but some of these compounds become carcinogenic or mutagenic or both when activated through metabolism (see Chapters 14 and 17).

Carcinogenic metabolites have taken on increasing environmental importance because of potential ecological and human health impacts associated with their occurrence in natural systems. The situation is further exacerbated because PAHs tend to concentrate in sediments as a result of their low solubility in water and strong sorption to particulate matter. They also disperse in the water column, concentrate in aquatic biota, and undergo chemical oxidation and biodegradation. Most PAHs in water do not occur in dissolved form but rather are associated with particulate matter. The substantial quantities of PAHs in the aquatic environment and the potential threat they pose to humans and aquatic organisms have triggered numerous studies on their fate and toxicity. Additional information is given in Eisler (1987), GESAMP (1993), Kennish (1992), MacKay et al. (1992), Mix (1984), and Neff (1979, 1981, 1985).

Synthetic detergents, like the alkylbenzene sulfonates, replaced soap as domestic and industrial cleaning agents because of their efficiency as surfactants and because they did not precipitate calcium salts in hard water. However, they were not readily degraded in sewage treatment processes and were toxic to biota in receiving waters. As a result, detergent manufacturers changed the detergent manufacturing process to produce linear alkylbenzene sulfonate (LAS) detergents. These detergents are biodegraded in conventional waste treatment plants but they may still be toxic at low concentrations to aquatic organisms. Other surfactants compositionally different from common detergents are also used as components of oil-spill dispersants so that exposure of coastal marine habitats to such compounds is likely. Detergents are discussed in Chapters 27 and 31. More information on detergents is given in Abel (1989) and Hennes-Morgan and DeOude (1994). For more information on oil-spill dispersants refer to NRC (1989) and GESAMP (1993).

Pesticides

Pesticides are a diverse group of widely varying chemical structures ranging from simple inorganic substances to complex organic molecules. Of the latter, some are natural derivatives of plants (e.g., pyrethrins) and others are synthetic derivatives of natural products or completely synthetic substances produced in chemical manufacturing facilities. Unlike most toxic agents, pesticides (insecticides, herbicides, and fungicides) have been selected and synthesized for their biocidal properties and are applied to kill or control organisms. Thus they are all toxic to some forms of life. Pesticides may be introduced into natural aquatic systems by various means: incidentally during manufacture, during their application (i.e., through aerial spray drift), and through surface water runoff from agricultural land after application. In addition, some pesticides are deliberately introduced into aquatic systems to kill undesirable pests such as weeds, algae, and vectors of human disease. In many countries, the significance of pesticides as aquatic contaminants has led to strict regulatory controls on their manufacture, transport, use, and disposal (see Chapters 21 and 27).

A number of generalizations can be made about pesticides. First, effective pesticides are designed to be selective in their effects; they are extremely toxic to some forms of life and relatively harmless to others. Few are absolutely specific to their target organisms, so other related and unrelated species may be affected. Second, the mode of application of pesticides varies according to the circumstance. Third, in stagnant lentic (i.e., nonflowing) aquatic systems draining agricultural areas, certain pesticides are more likely to be present at low but persistent concentrations. Such pesticides may be resistant to abiotic and biotic degradation and cause sublethal effects in a wide range of species, including wildlife.

As a result of their ubiquitous use in the environment and their potential exposure to aquatic organisms, as well as regulatory requirements for least hazardous formulations, the literature on pesticides is enormous. Useful references include Hellawell (1986), Hill (1985), Laws (1981), Mayer and Ellersieck (1986), Muirhead-Thomson (1987), Nimmo (1985), Nimmo and McEwen (1994), and Nimmo et al. (1987).

Radionuclides

Atoms of the same chemical element whose nuclei contain different numbers of neutrons are known as *isotopes* of the element. Isotopes are collectively referred to as *nuclides*. There is more than one isotope for all known elements. Some isotopes are unstable and seek stability by giving off particles or electromagnetic rays (i.e., *radioisotopes* or *radionuclides*). In effect, the nucleus of the parent radionuclide spontaneously disintegrates or decays and ultimately leads to the production of a stable daughter nuclide. During decay the unstable nuclide may emit one or more of four types of radiation: alpha particles, beta particles, gamma rays, and neutron radiation (produced anthropogenically in nuclear reactors). The emission of gamma photons commonly accompanies the release of alpha and beta particles during the decay process. Radionuclide decay is a probabilistic phenomenon in that the radionuclide should decay with a certain time period. Because of radioactive decay the radioactivity of a material declines with time, each radionuclide having a specific half-life, and eventually becomes stable.

There are both natural and anthropogenic sources of ionizing radiation. The environment is naturally radioactive with radiation coming from outer space (cosmic or galactic radiation) and primordial radionuclides in the earth's crust. Anthropogenic sources include nuclear power plants and the nuclear weapons industry. In recent years the contribution from nuclear detonations has decreased because of nuclear test bans.

The use of radioactive materials (nuclear armament production and power plants) produces wastes that must be disposed of and that pose a potential risk to aquatic organisms. Radionuclides enter biota of freshwater and saltwater systems by uptake from water, sediment, and food. Radionuclides accumulate in bottom sediments and enter the organic detritus reservoir after the death of the plants and animals that sequestered radionuclides in their tissues when still alive. Solubilization or resuspension of nuclides in detritus enables them to reenter food chains or remobilize to other areas of the system. Aquatic organisms bioaccumulate radionuclides and body levels higher than that typical of background may be found in areas where there are high natural levels (e.g., natural radioactive deposits) or near areas where there have been substantial losses during use or processing of radioactive materials (e.g., various facilities). Large amounts of radionuclides were released into the environment at Oak Ridge, Tennessee and Hanford, Washington during processing associated with nuclear armament production, and an accident at the Chernobyl nuclear power plant in the Ukraine resulted in considerable loss of radionuclides from this civilian facility.

Acute and chronic effects of radiation exposure have been noted in laboratory testing but only a few of the many studies conducted have demonstrated appreciable effects in the field. For more information on radioactive materials see Kennish (1992), Laws (1981), and Mason (1991).

Types of Toxic Effects

Just as a distinction is made between acute and chronic exposures, one can be made between acute and chronic effects. *Acute effects* are those that occur rapidly as a result of short-term exposure to a chemical. In fish and other aquatic organisms, effects that occur within a few hours, days, or weeks are considered acute. Generally, acute effects are relatively severe. The most common one measured in aquatic organisms is lethality or mortality. A chemical is considered acutely toxic if by its direct action it kills 50% or more of the exposed population of test organisms in a relatively short period of time, such as 96 h to 14 d.

Chronic or *subchronic toxic effects* may occur when the chemical produces deleterious effects as a result of a single exposure (e.g., to a strong acid), but more often they are a consequence of repeated or long-term exposures to low levels of persistent chemicals, alone or in combination. There may be a relatively long latency (time to occurrence) period for the expression of these effects, particularly if the exposure concentration is very low. *Chronic* effects also may be lethal or sublethal. An example of a lethal chronic effect is failure of the chronically exposed organisms to produce viable offspring.

Some confusion is created by the use of the terms acute and chronic to describe both the duration of the exposure and the nature of the toxic effect elicited. Clarity may be improved by using terms that unequivocally separate the two. For example, duration can be described as short-term or long-term. Rand (1980) has suggested that short-term be defined as any exposure duration which is $\leq 10\%$ of the expected life span of the organism. As well, effects can be divided into the broad categories of lethal and nonlethal or sublethal. The latter doesn't require that mortality be absent; rather it indicates that death is not the primary toxic end point being examined. This allows clearer discussion of some newer toxicity tests, which often combine brief exposures with nonlethal end points and prolonged exposure with mortality or survival as the toxic end point.

The most common *sublethal effects* are behavioral (e.g., swimming, feeding, attraction-avoidance, and prey-predator interactions), physiological (e.g., growth, reproduction, and development), biochemical (e.g., blood enzyme and ion levels), and histological changes (Sheehan et al., 1984). Some sublethal effects may indirectly result in mortality. For example, certain behavioral changes (e.g., swimming or olfactory) may diminish the ability of aquatic organisms to find food or to escape from predators and may ultimately lead to death. Some sublethal effects may have little or no effect on the organism because they are rapidly reversible or diminish or cease with time (e.g., growth may be reduced at high concentrations early in a toxicity study but not be significantly different from that in controls by the end of the study). In laboratory studies, sublethal effects may be unnoticed in acute tests. A way to study sublethal toxicity in the laboratory is by using longer term exposures and by measuring multiple biological end points, structural and functional. Subchronic and chronic laboratory exposure studies and the sublethal effects that may result from such exposures are discussed throughout this book.

Most of the preceding discussion of effects is relevant to laboratory and field chemical exposure studies. However, laboratory studies are typically based on single-species exposure, whereas in field studies the effects of a toxicant on a myriad of species (or communities) are often evaluated in either experimental model (e.g., microcosm/mesocosm) or natural (e.g., ponds, streams, coastal) ecosystems (see Chapter 9). The quantification of toxicant-induced effects in such systems has involved changes or effects in community (or ecosystem) *structure* and/or *function* (Cairns and Pratt, 1986). A biological community consists of numerous species, each of which is represented by a number of individuals. The diversity that exists in the community is a function of numbers of individuals and species. To assess changes in *ecosystem structure* the composition of different taxonomic groups of the biological community is quantified as to numbers and kinds of species at a point in time in contaminated and noncontaminated areas. This may include biomass, life histories, the quantity and distribution of abiotic materials (nutrients), and the range or gradient of conditions of existence, such as temperature and light (Matthews et al., 1982; Odum, 1962).

Common quantitative measures (or variables) of effects on ecosystem structure that may be monitored are the number of distinct taxonomic units (e.g., species, also called species richness) in a community, the species composition, and how the species abundances or the numbers of individuals are distributed among the species (also called species evenness or equitability). Estimates of the relative abundance (number of individuals in one species relative to the total number of individuals in the community) or species richness of periphyton, plankton, benthic, and fish communities can yield information on the degree of disturbance relative to a baseline condition, perhaps due to contamination in an aquatic ecosystem. Species composition and richness are important because the loss of a specific species in an ecosystem can be critical when that species plays a crucial role in community or ecosystem function, such as in predation, grazing, or competition (LaPoint and Fairchild, 1992; Paine, 1969; Giesy, 1980; Matczak and MacKay, 1990).

Using a community structure approach, various numerical indices derived from direct quantitative measures of taxa presence have been developed to assess stress of toxicants (Hellawell, 1986; LaPoint and Fairchild, 1992). Indices such as evenness, relative abundance, similarity (comparing likeness of community composition between two different sites) and diversity (combine both species richness and evenness into a single calculated value) are helpful in data reduction, but they limit data interpretation because information on the fate of individual species may be overlooked. Effects on ecosystem structure under natural conditions to assess the effects of toxicants, nonetheless, continue to be most widely used. Reviews on measures of changes on ecosystem (or community) structure and the use of indices relative to aquatic systems can be found in Abel (1989), Green (1979), LaPoint and Fairchild (1992), Ludwig and Reynolds (1988), Rosenberg and Resh (1993), and Spellerberg (1991).

Ecosystem function denotes the set of ecological processes that is vital to the maintenance of each ecosystem and the continued evolution of its species. It includes measures of the dynamics or rate changes in the system over time (Cairns, 1992). Three classes of ecological processes are important from a scientific viewpoint (Landres, 1992). These are processes that affect the rate and quantity of energy flow (i.e., primary/secondary production, production/respiration or production/biomass ratios), the rate and total quantity of nutrient cycling (i.e., decomposition, nutrient and nutrient mobilization/immobilization), and ecosystem services (e.g., biodegradation of pollutants, pollination of food crops, plant nutrients released from decomposition) important to humans.

Examples of measures of functional effects in aquatic ecosystems are rates of production, respiration, mineralization, and nutrient regeneration. Effects on ecosystem function are not used extensively because the database on methodology for these end points is limited, many factors influence such end points, and the effects on ecosystem function are often difficult to interpret. Functional assessment of perturbed aquatic ecosystems is discussed in Chapter 19.

There is much debate about whether structural or functional components are most appropriate for assessing change (Levin et al., 1989). It is evident that structural and functional components are linked together in different ways in different ecosystems and that both structural and functional observations are needed to characterize a system. This was recognized by Odum (1962) and summarized by Schindler (1987): "After 18 years of manipulating whole [lake] ecosystems, I find that changes in ecosystem function, such as production, decomposition or nutrient cycling, cannot be properly interpreted without analogous information on the organization and structure of the biotic communities which perform the function." Theoretically, ecosystems are so complex that no single observation or index could adequately summarize this complexity.

A framework that combines examination of both ecosystem structure and function was proposed by Munkittrick and Dixon (1989) and Munkittrick (1992). The objective was to provide *top-down* ecotoxicological evaluation, based on determination of actual adverse effects at an aquatic community level, to complement the traditional *bottom-up* approach, which is based on single-species laboratory testing. The essence of the approach is to use conventional fisheries data, which contains information on the structure and function of selected fish populations, to determine the presence and nature of overt stresses being placed on an aquatic community. Once a stress is identified and classified, its nature regarding the entire aquatic community can be obtained by characterizing the community response in terms of changes in physiology, nutrition, reproduction, food chain structure, and habitat alteration. The principles can also be applied to other circumstances, such as to sediment-dwelling invertebrates, where sufficient population information is available (Power et al., 1991). In addition to providing an ecologically valid regulatory tool, these and other such approaches are a useful guide to research on community-level stresses and responses (Munawar et al., 1989).

The top-down versus bottom-up debate represents a long-standing difference of opinion about how best to determine environmentally significant adverse biological effects. Ecological/biological systems comprise a series of levels of organization ranging from the suborganismal levels (biochemical, cell, and tissue/organ) through the organismal level to the supraorganismal levels (population, community, and ecosystem). Field ecologists tend to prefer a top-down approach, looking for effects at the higher levels of organization such as ecosystem or community, whereas laboratory investigators tend toward the bottom-up approach, examining lower levels of organization, such as the organism and below. As a rule, investigators working at any particular level look at levels above for the significance of the adverse response being examined and at levels below for the cause and mechanism. Most investigations in aquatic toxicology are "middle-out," as laboratory and field investigators tend to emphasize data at the organismal level and look down for mechanisms and up for ecological significance (McCarty and Mackay, 1993). Clearly, to understand completely the nature of adverse ecological effects, the processes at each level of organization and all of the interrelationships between the levels should be known.

Mixtures

In assessing chemically induced effects (responses), it is important to consider that in the natural aquatic environment organisms may be exposed not to a single chemical but rather to a myriad or mixture of different substances at the same or nearly the same time. Exposure to mixtures may result in toxicological interactions (Alabaster and Lloyd, 1980; Calabrese, 1991). A *toxicological interaction* is one in which exposure to two or more chemicals results in a biological response quantitatively or qualitatively different from that expected from the action of each of the chemicals alone. Interaction between chemicals can occur by mechanisms such as alteration in absorption, protein binding, and biotransformation or excretion of one or more of the interacting chemicals. The multiple chemical exposures may be sequential or simultaneous in time and the altered response may be greater or smaller in magnitude.

Exposure to two chemicals simultaneously may produce a response that is simply additive of the individual responses or one that is greater or less than expected from addition of their individual responses. Several terms have been used to describe toxicological interaction. An *additive effect* occurs when the combined effect of two chemicals is equal to the sum of the effects of the individual chemicals applied alone (e.g., 1 + 1 = 2). The additive effect is most commonly observed when two chemicals are applied together. A *synergistic effect* occurs when the combined effect of two chemicals is much greater than the sum of effects of the individual chemicals applied alone (e.g., 1 + 1 = 5). *Potentiation* occurs when one chemical has a toxic effect only when applied with another chemical (e.g., 0 + 1 = 4). *Antagonism* occurs when two chemicals, applied together, interfere with each other's action or one interferes with the other chemical (e.g., 2 + 3 = 4, 3 + 0 = 1). It should be noted that interactions may be different, even opposite, depending on the magnitude of the exposure (e.g., additive at low doses and antagonistic at high doses), so discussion and especially prediction of interactions require clear information about the magnitude of the exposures in question.

There are four types of antagonism: functional, chemical, dispositional, and receptor antagonism. *Functional antagonism* occurs when two chemicals counterbalance one another by eliciting opposite effects on the same physiologic function. *Chemical antagonism* is a chemical reaction between two chemicals to produce a less toxic product. *Dispositional antagonism* occurs when the absorption, biotransformation, distribution, or excretion of a chemical is changed so that the concentration and/or duration of the chemical at the target site is decreased. *Receptor antagonism* occurs when two chemicals that bind to the same receptor site produce less of an effect when given together than the sum of their individual effects (e.g., 3 + 4 = 6) or when one chemical antagonizes the effect of the second chemical (e.g., 0 + 2 = 1).

There are many examples in the aquatic toxicology literature of both additive (e.g., copper and zinc; Sprague and Ramsey, 1965) and more than additive (e.g., organophosphate insecticides; Marking, 1977) toxicity. However, examples of less than additive toxicity (i.e., antagonism) are not common in aquatic organisms. Antagonism is found, however, when mammalian organisms are exposed to two chemicals. Antagonistic effects of chemicals are desirable in humans and are the basis of many antidotes.

More detailed reviews of mixture toxicity can be found in Marking (1985), Christensen and Chen (1989), and Broderius (1991).

EXAMINATION OF CONCENTRATION-RESPONSE RELATIONSHIPS

The concentration-response relationship is the fundamental concept in aquatic toxicology. Although the basic assumptions were reviewed earlier, there are some practical aspects that must be considered. Substantial differences may exist among individual organisms in a supposedly homogeneous population. These differences become evident when the organisms are challenged by exposure to a chemical or potentially toxic stress. For example, not all of the organisms would respond in a quantitatively identical manner to the same concentration of a toxicant. The effects of such an exposure might vary from very intense in some organisms to minimal or none in other organisms. That is, some organisms may die and others survive with apparently minimal adverse effects. These differences in response are due to natural biological variation, reflecting the genetic makeup of the population and the condition of the individuals. The variation is normally small for organisms of the same species and similar age and health and generally greater between species.

In measuring the toxicity of a chemical, the objective is to estimate as precisely as possible the range of chemical concentrations that produce some selected, readily observable, quantifiable response in groups of the same test species under controlled laboratory conditions. The results of the exposure are plotted on a graph that relates the percentage of organisms in test groups exhibiting the defined response (dependent variable, Y axis) to the exposure concentration of the test chemical (independent variable, X axis). The concentration-response relationship is a graded relationship between the concentration of the test chemical to which the organisms are exposed and the severity of the response elicited.

Generally, within certain limits, the greater the exposure concentration of the test chemical, the more severe the response. The curve drawn to represent this relationship will generally be asymptotic to the Y axis, because at all concentrations below some minimum (threshold) value no measurable adverse response will be elicited, whereas at all concentrations above some maximum value most or all of the test group will be adversely affected. The steeper the slope of the central portion of the curve, the sharper the threshold of effect—that is, the more intense the response over a narrow range of concentrations. Figure 5 shows the typical sigmoid form of the concentration-response curve. For a statement about the concentration-response relationship to have a precise meaning, the duration of exposure must also be defined.

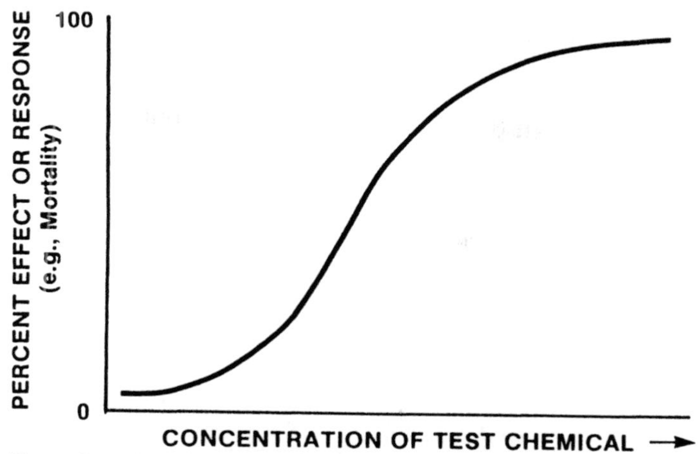

Figure 5. Typical form of the concentration-response curve.

In aquatic toxicity testing, the test organism is exposed to the chemical indirectly by mixing the chemical into the water in which the animal lives, thus producing a test *concentration*. The concentration of a test chemical in water is usually expressed in parts per million (ppm), or units of test chemical (usually expressed as mass) per 10^6 units of untreated dilution water (diluent). The commonly used ratio is milligrams of test chemical per liter of water (1 ppm approximately equals 1 mg/L in fresh water and in salt water). The concentration may also be expressed in parts per billion (ppb), or units of test chemical per 10^9 units of diluent water. Here the commonly used ratio is micrograms of test chemical per liter of water (1 ppb approximately equals 1 μg/L in fresh water and in salt water).

If the organisms are exposed to test solutions of liquid industrial wastes of an effluent, the concentration of effluent in water is usually expressed as a volume percent [100 × (volume of effluent)/(volume of effluent plus volume of dilution water)]. For example, a 10% dilution equals 1 volume of effluent in 9 volumes of dilution water. Corresponding expressions of chemical concentrations in solid media such as tissues and sediments are in parts per million or milligrams of test chemical per kilogram of tissue or sediment (1 ppm = 1 mg/kg), which is equivalent to micrograms of test chemical per gram of tissue or sediment. Likewise, parts per billion correspond to micrograms of test chemical per kilogram of tissue or sediment or nanograms per gram. Weight-specific concentrations in tissues and sediments may be expressed as either wet or dry weight. Where dry weight is used, sufficient information should be available to estimate wet weight concentrations.

Despite the long-standing use of mass-based units for reporting doses or exposure concentrations, more appropriate units should be used whenever possible. The approved SI (Système International) metric unit for reporting amounts of chemicals whose relative molecular mass is known is the *mole*. Thus, rather than reporting in units of mg/L or μg/g, toxicity test data obtained with specific chemicals or chemical mixtures should be reported in molar units such as mmol/L or μmol/g. The use of molar concentrations is very important from a toxicological point of view, as toxic effects resulting from exposure to chemicals are usually a function of the number of molecules present at the target sites, not the mass of those molecules. Wherever possible results should be expressed in molar units or, at the very least, molecular weight information for the chemicals being examined should be reported.

Criteria for Effects and LC50

In evaluating the safety of chemical substances, it is necessary to have a precise means of expressing the toxicity and a quantitative method of measuring it. Various *criteria for effects* or end points of toxicity could be used to compare chemically exposed organisms with unexposed organisms. The ideal criteria are those closely associated with the molecular events that result from exposure to the chemical. However, in aquatic toxicology this is difficult to achieve because the molecular events are usually unknown. Alternatively, one may select a measure of toxicity that is unequivocal, clearly relevant, readily observable, describable, measurable, biologically significant, and reproducible. For the initial test measurement in toxicological evaluation, it is customary to use lethality (or mortality) as an index.

Measurement of lethality is precise, important (i.e., biologically interpretable), unequivocal, and therefore useful for estimating the concentration and potency of a chemical. It provides a means of comparing substances whose mechanisms of action may be quite different and indicates whether further toxicity studies should be conducted. Mortality and survival over a specific period of time are typical effect criteria in short-term (acute) exposure tests. Data from lethality tests are *quantal*; that is, the animals live or die (an all-or-none response). However, it is important to have sublethal effect criteria that indicate toxic stress at a stage before death, so that early observation will permit rapid action to prevent mortalities. Growth (length and weight), number of normal embryos, morphological anomalies (e.g., double-headedness and deformed spines), and number of offspring are typical sublethal effect criteria in long-term (chronic) exposure tests. All of these responses are *quantitative* (or graded); they are measured not in terms of incidence but in some unit of measured response (milligrams, centimeters) that can be used to compare chemically treated test organisms with unexposed controls and determine whether the differences between them are statistically significant. Because there are usually many test organisms, a series of several graded measurements is generated for each concentration. Whereas in a quantal test it is sufficient merely to determine for every organism whether or not the selected response (effect) has occurred, a quantitative test requires measuring the extent of the response shown by each organism.

Whatever the effect or response selected for measurement, the relationship between the degree of response of the organisms and the quantity (or concentration) of the chemical almost always assumes a classic concentration-response form. In Figure 6 the ordinates represent percent mortality and the abscissas represent concentration of the test chemical; both measures increase with increasing distance from the origin. The graphs represent the results of tests in which groups of test organisms of the same species were exposed to various concentrations of a chemical for a specific length of time. In Figure 6a the mean percent mortality for each test group has been plotted on ordinary graph paper (arithmetic scale) against the concentration producing that mortality. The responses of test organisms to the different chemical concentrations yield a characteristic S-shaped (sigmoid) curve. Each point on the curve represents an average cumulative response to the specific concentration, and each average has an associated variation due to different responses of individual organisms. The least variability in the curve is at the 50% level of response. The concentration at which 50% of the individuals react (the median) after a specified length of exposure (e.g., 24 or 48 h) is therefore used as a measure of the activity or toxicity of the chemical agent.

In determining the relative toxicity of a new chemical to aquatic organisms, an acute toxicity test is first conducted to estimate the median lethal concentration (LC50) of the chemical in the water to which test organisms are exposed. The *LC50* is the concentration estimated to produce mortality in 50% of a test population over a specific time period. The length of exposure is usually 24 to 96 h, depending on the species (see Chapters 2–8). When effects other than mortality are measured, the expression EC50 is used. The *EC50* (median effective concentration) is the concentration of a chemical estimated to produce a specific effect (e.g., behavioral or physiological) in 50% of a population of test species after a specified length of exposure (e.g., 24 or 48 h). Typical effect criteria include immobility, a developmental abnormality or deformity, loss of equilibrium, failure to respond to an external stimulus, and abnormal behavior.

The LC50 can be interpolated from the curve in Figure 6a by drawing a horizontal line from the 50% mortality point on the ordinate to the concentration-response curve and then drawing a vertical line from the point of intersection with the curve to the abscissa. The vertical line intersects the abscissa at the LC50 value, which is then read off the graph. The LC50 is an *estimate* of the concentration of a chemical that would produce a specific effect (mortality) in 50% of an infinitely large population of the test species under the stated conditions.

The normally distributed sigmoid curve in Figure 6a approaches a mortality of 0% as the concentration is decreased and approaches 100% as the concentration is increased, but theoretically never passes through 0 and 100%. The middle portion of the curve, in the region between 16 and 84%, is linear. These values represent the limits of 1 standard deviation (SD) of the mean (and the median) in a normally distributed population of organisms. In a normally distributed population, the mean ± 1 SD represents 68.3% of the test population, the mean ± 2 SD 95.5% of the test population, and the mean ± 3 SD 99.7% of the test population.

Figure 6b is a plot of the data in Figure 6a, but with the concentration shown on a logarithmic scale (transformed data; see Chapter 10). The sigmoid shape is again evident but the curve approaches a straight line. Figure 6c represents another transformation of the same data, with the logarithm of concentration plotted against percent mortality expressed as probits. The probit transformation adjusts mortality data to an assumed normal population distribution, which results in a straight line being the best-fit curve. The LC50 is obtained by drawing horizontal and vertical lines in the manner described for Figure 6a.

The logarithmic conversion was introduced by Krogh and Henningsen (1928) and later by Gaddum (1933). Because quantal concentration-response data are normally distributed, the percent response was converted to units of deviation from the mean or normal equivalent deviates (NED) by Gaddum (1933). The NED for a 50% response is zero and that for an 84.1% response is +1. Bliss (1934a,b) later suggested that 5 be added to the NED to eliminate negative numbers. The converted units of NED plus 5 were called *probits*. A 50% response with this transformation becomes a probit of 5 (NED = 0), and an 84.1% response or +1 deviation becomes a probit of 6. The probit transformation adjusts mortality or other quantal data to an assumed normal population distribution that yields a straight line. The log-probit

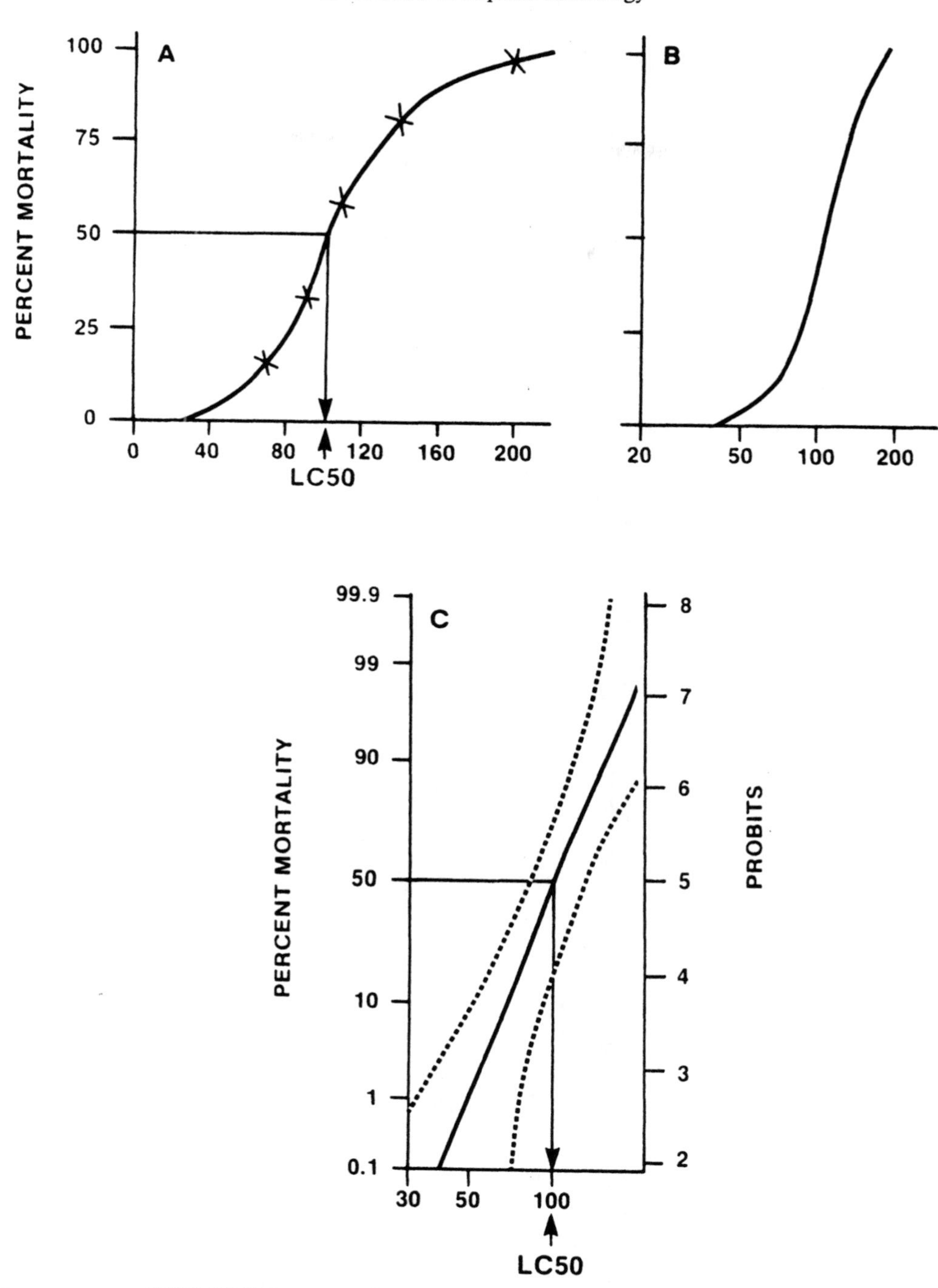

Figure 6. Mortality in a fish population exposed to a range of concentrations of a chemical in water. (a) Percent mortality versus concentration plotted on an arithmetic scale. (b) The same data as in "(a)", but with mortality on an arithmetic scale and concentration on a logarithmic scale. (c) The same data as in (a), but with mortality expressed as probits versus concentration on a logarithmic scale. The dotted lines on each side of the curve represent the 95% confidence limits.

conversion is valuable for description and statistical analysis (see Chapters 2 and 10). The LC50 estimated by graphical interpolation is accurate and usually similar to the LC50 obtained by formal, statistical, computer-assisted analysis.

Confidence Limits

The degree of scatter of observed values may be evaluated by calculation and expressed as *confidence limits*. Confidence limits are shown by dotted lines on both sides of the solid line in Figure 6c. The limits indicate the area or range within which the concentration-response line would be expected to fall in replicate tests in 19 of 20 samples (95% confidence limits) taken at random from the same test population under the same conditions. A series of such curves would be well correlated with each other at the 50% mortality level but probably not well correlated as the mortality approaches 0 or 100%. This indicates that the LC50 may be estimated more precisely than greater or lesser effects (e.g., LC99 or LC1).

Slope

Figure 7 shows concentration-response curves (log-probit scale) for two chemicals, X and Z. The LC50 values for the chemicals are the same but the slopes of the concentration-mortality curves are different. The curve corresponding to Z represents a set of data of greater variability, as reflected by the larger confidence limits. The flat slope of line Z indicates that mortality increases by small increments as the concentration of the chemical to which the organisms are exposed increases. Conversely, the steep slope of line X indicates that large increases in mortality are associated with relatively small increases in the concentration to which the test organisms are exposed. The *slope* is thus an index of the range of sensitivity to the chemical within the test sample of organisms (e.g., fish).

A flat slope, as for chemical Z, may also be indicative of slow absorption, rapid excretion or detoxification, or delayed toxification. A steep slope, as for chemical X, usually indicates rapid absorption and rapid onset of effects; for some organics, the chemical's fate in the test solution is reflected. Although the slope alone is not a reliable indicator of toxicological mechanisms, it is a useful parameter and should always be reported along with the confidence limits.

More discussion of the statistical procedures used to determine the LC50, the slope, and the confidence limits may be found in Chapters 2 and 10. Chapter 10 also discusses statistical procedures used to evaluate quantitative responses in long-term exposure tests.

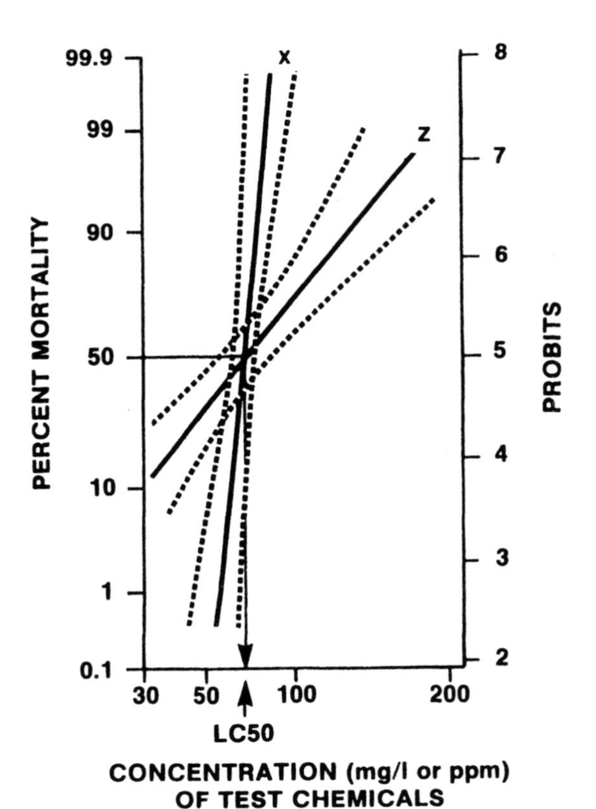

Figure 7. Hypothetical concentration-response curves for two chemicals, X and Z, demonstrating differences in slope.

Toxicity Curves

In acute toxicity tests mortality or survival information is collected for each exposure concentration at various exposure durations. Although Sprague (1973) recommended a logarithmic time series for the first day, followed by daily observations, in practice observations are often made only at 24, 48, 72, and 96 h and sometimes even less frequently. The number of exposure concentrations employed in the test design is dependent on the nature of the substance being tested, but again Sprague (1973) recommends a minimum series of four or five logarithmically distributed exposure concentrations spanning the expected toxicity range.

The matrix of information collected—mortality in the exposure concentrations at various exposure durations—can be plotted in two ways. The first way, illustrated in Figure 8a, is to plot percent mortality (probit transformed) versus the logarithm of time for each exposure concentration used in the test. As shown in the figure by the dotted line, the LT50s (median lethal times) or ET50s (median effective times) can be estimated for each concentration (usually following Litchfield, 1949) by interpolating the time at which the 50% mortality line intersects each concentration line. The second way, illustrated in Figure 8b, is to plot percent mortality versus the logarithms of exposure concentrations employed for each observation period. As shown in the figure by the dotted line, the LC50s (median lethal concentra-

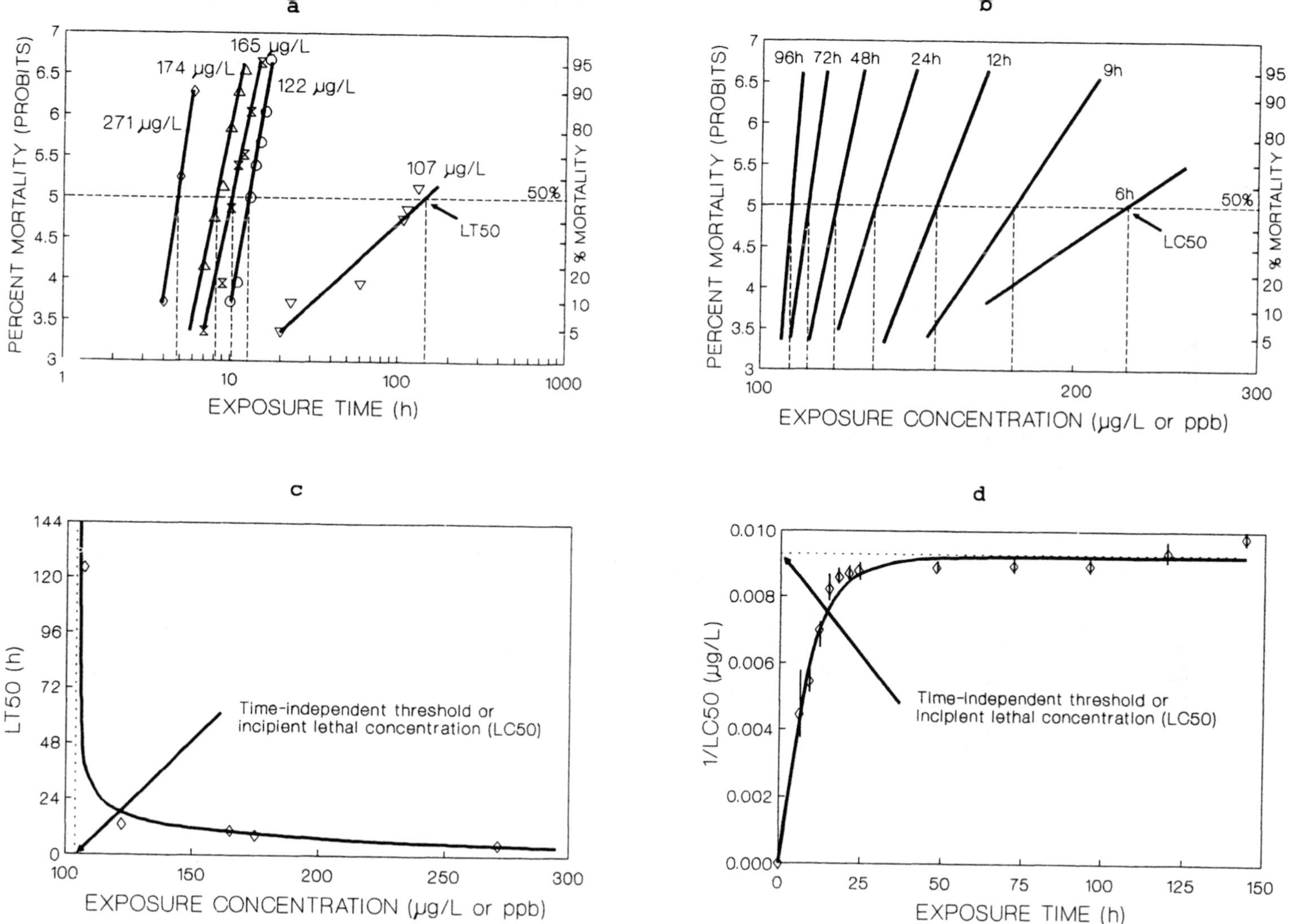

Figure 8. Cumulative percentage mortality for a fish population exposed to a range of concentrations of a chemical in water. (a) Plot of exposure time versus mortality at various exposure concentrations (drawn with data for low fat fish exposed to pentachlorophenol, from van den Heuvel et al., 1991). (b) Plot of exposure concentration versus mortality at various exposure times (idealized pentachlorophenol results).

Toxicity curves of time and toxicity. (c) Plot of LT50 (median time to death) versus exposure concentration (LT50s from Figure 8a). (d) Plot of the inverse of LC50 versus exposure time (LC50s for low fat fish exposed to pentachlorophenol, from van den Heuvel et al., 1991). (Both figures redrawn from van den Heuvel et al., 1991,

tions) or EC50s (median effective concentrations) can be estimated for each exposure duration by interpolating the exposure concentration at which the 50% mortality line intersects each exposure duration line.

Figure 8c presents a toxicity curve based on the exposure concentration and median time to death (LT50; ET50 may also be used for nonlethal end points) data presented in Figure 8a. This toxicity curve uses logarithmic scales for both time to median effect (the response) and exposure concentration (the dose). Figure 8d presents an alternative toxicity curve based on the exposure time and effective exposure concentration (LC50 or EC50) data presented in Figure 8b. The toxicity curve in Figure 8d uses the inverse of the LC50 data, which both provides a normalizing statistical transformation (the inverse transformation) and places the start of the curve at the origin (zero). The latter facilitates statistical curve fitting as discussed below. If untransformed or log-transformed LC50 or EC50 data were used, the curve would be the inverse of that presented and be similar in general appearance to that in Figure 8c.

Such curves give the investigator an idea of how the test is progressing and may indicate when acute lethality has ceased. Mortality has stopped where the curve becomes asymptotic to the time axis, as noted in Figure 8c and d. The concentration at which this occurs is called the *threshold* or *incipient LC50* (also called the asymptotic LC50, incipient lethal level or ILL, ultimate median tolerance limit, or lethal threshold concentration). The incipient LC50 is the concentration at which 50% of the test population can live for an indefinite time, or the lethal concentration for 50% of the test organisms in a long-term exposure. It is the most important feature of the toxicity curve, as it allows a true comparison of different substances' toxicities (see Figure 9), avoiding possible errors in comparing substances' toxicities using LC50s for arbitrarily chosen time periods, a point often forgotten by some of today's practitioners.

Figure 9 presents toxicity curves for two hypothetical chemicals A and B; Figure 9a is plotted using LT50 data and 9b is plotted using LC50 data. If the toxicity observations were taken only at the exposure duration marked as time X on the graphs, chemical B would appear to be more toxic (a lower exposure concentration is required to produce death in 50% of exposed organisms) than chemical A. However, at a longer exposure duration, approaching the time-independent incipient lethal exposure, it is clear that chemical A is the more toxic of the two.

The toxicity curve may assume a variety of shapes (Sprague, 1969; Abel, 1989) and may be fitted to the points by eye or more precisely by computer-assisted curve-fitting statistical programs. The shape of the curve may provide information about the mode of toxic action of the chemical or metabolic degradation or activation and may indicate the presence of more than one chemical agent in the water. As discussed by McCarty et al. (1992a) and van den Heuvel et al. (1991), toxicokinetic information may be obtained from analysis of toxicity curves. This entails fitting a *one-compartment first-order kinetics* (1CFOK) model, via nonlinear curve fitting, to time and toxicity data. The appropriate, statistically valid method for this approach (Kooijman, 1981; Chew and Hamilton, 1985) employs time-to-death data, such as those presented in Figure 8c. Unfortunately, this method cannot always be used, as many toxicity test data sets are incompletely reported, often presenting only LC50 estimates for one or a few exposure durations (e.g., 24-h or 96-h LC50). In these cases, the method proposed by Matida (1960) must be used. This method fits the 1CFOK model to time and effective exposure concentration data, such as those presented in Figure 8d.

Although Matida's method is statistically invalid because of pseudoreplication (lack of independence) between the LC50 estimates (see Chapters 10 and 19), it does allow estimation of simple rate constants, similar to those obtained with the time-to-death methodology (van den Heuvel et al., 1991), when nothing else can be done. Such kinetic information can be valuable for toxicity test interpretation and is vital for both simple and sophisticated toxicological modeling (see the following section on toxicological modeling plus Chapters 16 and 18).

TOXICITY TESTING

Aquatic toxicology has been defined as the study of the effects of chemicals and other toxic agents on aquatic organisms with special emphasis on adverse or harmful effects. *Toxicity tests* are used to evaluate the concentrations of the chemical and the duration of exposure required to produce the criterion effect(s). Such tests represent standardized methodologies for implementation of the toxicological principles outlined earlier. As the basic paradigm of aquatic toxicology has been the concentration-response relationship, most of the toxicity testing protocols have been designed to exploit it to generate useful data while meeting toxicological and statistical requirements to ensure validity.

The effects of a chemical may be of such minor significance that the aquatic organism is able to carry on its functions in a normal manner and that only under conditions of additional stress (e.g., changes in pH, low dissolved oxygen, and high temperature) can a chemically induced effect be detected. The influence of these factors on chemical toxicity is discussed in Abel (1989) and in Appendix Chapter C. Effects may also result from the interaction (often solely addition) of small amounts of some chemicals and larger amounts of other chemicals without these additional stresses.

Aquatic toxicity tests are used to detect and evaluate the potential toxicological effects of chemicals on aquatic organisms. Since these effects are not necessarily harmful, a principal function of the tests is to identify chemicals that can have adverse effects on aquatic organisms at relatively low exposure con-

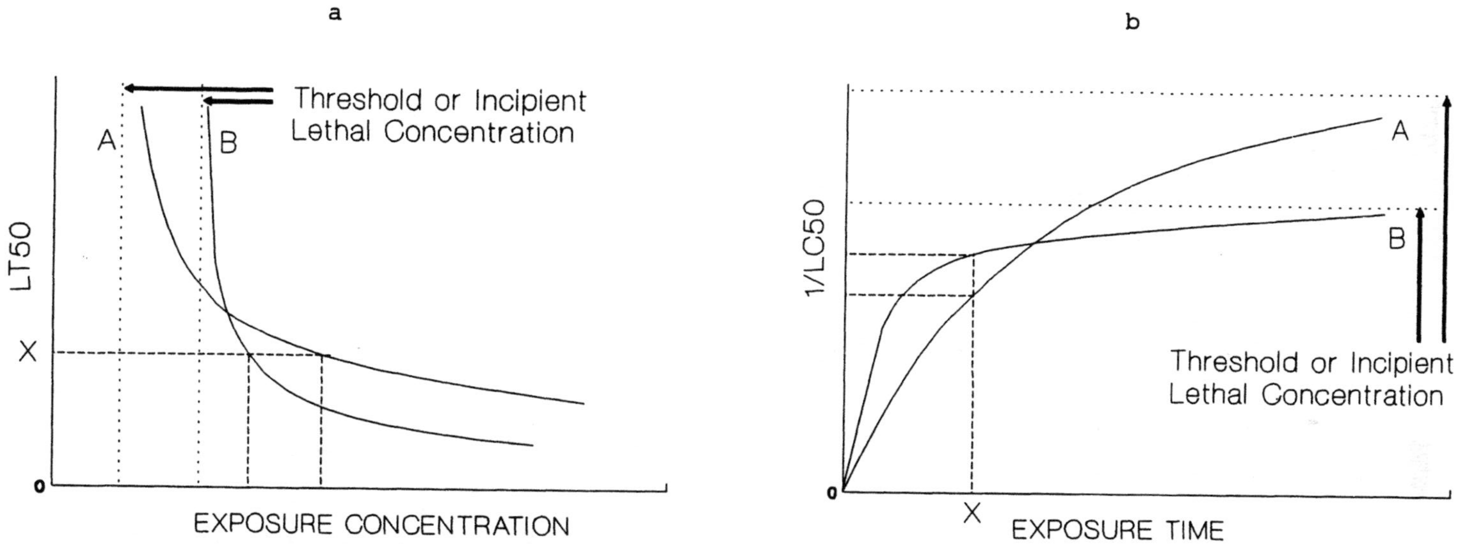

Figure 9. Two hypothetical toxicity curves using the same data; 9a using LT50 estimates and 9b using LC50 estimates. Substance A is clearly more toxic than substance B, since the lethal threshold of A is smaller. However, had the test been discontinued at time X, B would have appeared more toxic than A (9A redrawn from Abel, 1989, used with permission from Ellis Hørwood).

centrations or body residues. These tests provide a database that can be used to assess the risk associated with a situation in which the chemical agent, the organism, and the exposure conditions are defined.

An aquatic toxicity test is frequently called a *bioassay*. This would be erroneous if the narrow pharmaceutical definition of a bioassay is used: "a test to evaluate the relative potency of a chemical by comparing its effect on a living organism with that of a standard preparation." Here a bioassay refers to specific procedures performed to determine the strength of the chemical from the degree of response elicited in the test organisms, not to estimate the concentration of the chemical that is toxic to those organisms. Such procedures are frequently used in the pharmaceutical industry to evaluate the potency of vitamins and other pharmacologically active compounds. On the other hand, an aquatic toxicity test is performed to measure the degree of response produced by a specific level of stimulus (test chemical concentration). The above notwithstanding, the two terms (bioassay and toxicity test) are used synonymously by practitioners of aquatic toxicology.

According to Finney (1978): "A biological assay (or bioassay) is an experiment for estimating the nature, constitution, or potency of a material (or of a process), by means of the reaction that follows its application to living matter." A typical bioassay involves a *stimulus* of a measured external exposure or absorbed *dose* that is applied to a *subject* whose *response* to the stimulus is estimated in the change of some biological characteristic or state of the subject. This broad definition of bioassay excludes neither the pharmaceutical nor the ecotoxicological usage.

Aquatic bioassays fall into two general categories: the *toxicity test*, where there are various implementations employing lethal and nonlethal response end points with diverse exposure durations, and the *bioconcentration test*, where the accumulation kinetics of a chemical are studied without adverse effects being elicited. Finney (1978) characterizes bioassays as either direct or indirect. (Note: direct and indirect are used here in a statistical sense, in contrast to the toxicological definitions presented earlier in the section on basic toxicological concepts and principles). Finney classifies a direct bioassay as one in which the potency of an unknown preparation is determined as the ratio of exposure doses of an unknown and a standard preparation where each elicits the same specified response (e.g., if 10 μl of an unknown preparation produces the same biological response as 1 μl of a standard preparation, then the potency of the unknown is $\frac{1}{10}$ = 0.1 of the standard). This methodology is more often seen in the older pharmaceutical literature. An indirect bioassay is the more familiar method in which potency is estimated from the quantal or quantitative responses of multiple subjects and multiple dose or exposure levels. In this classification system acute (short-term), lethal tests should be categorized as *indirect quantal bioassays* and chronic (longer-term), nonlethal tests should be considered *indirect, quantitative bioassays*. If a change in body concentration can be considered a biological response (e.g., a water/body concentration ratio of 1 representing the standard response) then a bioconcentration test could be considered a *direct, quantitative bioassay*.

The confusion about toxicity bioassays and toxicity tests is in part related to the objectives for which they are employed and the use of the information gained. The primary objective of carrying out a toxicity bioassay is to determine the potency of the toxicant being examined relative to the potency of those that are already known (Bliss, 1957; Filov et al., 1979). Although the ultimate goal is increasing toxicological knowledge, there may be more immediate uses of toxicity test data (see Toxicity Data and Environmental Regulations section), such as regulatory compliance monitoring. Over the past decade or so the streamlining of many existing tests and the introduction of faster, less data-intensive tests have reached the point where some toxicity tests generate only a "number" and little useful general toxicological information. Furthermore, validation of the fundamental toxicological assumptions of the bioassay design may not be possible as insufficient information has been collected. The point has been made directly and succinctly by Calow (1989), who noted "too many tests and not enough experiments."

Although the existing toxicity testing protocols (see Chapter 11) have served the science well, it is clear that some modifications would be useful to enhance the nature and quality of the toxicological information obtained (McCarty, 1991). Change for the sake of change is neither useful nor effective, yet it is clear that improvements in existing protocols and promulgation of new investigative procedures are not just desirable but necessary if aquatic toxicology is to continue to develop and expand.

Data obtained in toxicity tests exhibit variability (see Chapter 2). Variability in toxicity tests can be described in terms of two types of precision: "within" or intralaboratory precision and "between" or interlaboratory precision. *Intralaboratory precision* is a reflection of the ability of trained personnel to obtain consistent results repeatedly when performing the same test on the same species using the same chemical or effluent. *Interlaboratory precision* is a measure of how reproducible a method is when conducted by a large number of laboratories using the same method, species, and chemical or effluent. Generally, intralaboratory results are less variable than interlaboratory results. Typically interlaboratory tests (i.e., round-robin or ring tests) are used to determine the reproducibility of a new test method so that it can become a standardized method.

Criteria and Approaches to Testing

Before discussing general approaches to toxicity testing, the criteria used to determine the suitability of a standard or model test procedure should be established. These criteria may include some of the following:

- The test should be widely accepted by the scientific community.
- The test procedures should have a sound statistical basis and should be repeatable and generate similar results in different laboratories. This requires that tests be standardized and carried out according to defined protocols.
- The data should include effects of a range of concentrations within realistic durations of exposure. They should also be quantifiable through graphical interpolation, statistical analysis, or other accepted methods of quantitative evaluation.
- The test should have some field predictive capability for similar organisms.
- The data should be useful for risk assessment.
- The test should be economical and easy to conduct.
- The test should be sensitive and as realistic as possible in design to detect and measure the effect.

Current test methods have been designed predominantly to examine the responses of a few individuals within a species (single-species tests), notwithstanding methods with bacteria, protozoa, and phytoplankton. Considerably less attention has been given to the impact of chemicals on species interactions and on structure and function within different aquatic ecosystems. Tests on single species, species interactions (multispecies tests), and ecosystems are all important for evaluating the potential impact of a chemical on an aquatic ecosystem.

Three general approaches have been used to conduct these tests, and each has advantages and limitations.

1. Effects can be studied in a controlled laboratory experiment with a limited number of variables.
2. Effects can be studied in an experimental model ecosystem (i.e., simulated indoor or outdoor).
3. Effects can be studied in a natural ecosystem (in situ).

Most single-species tests are conducted in the laboratory. These tests can provide a great deal of information on the external and internal concentrations of chemicals (McCarty and Mackay, 1993) and duration of exposure that produce changes in mortality, growth, reproduction, pathology, behavior, physiology, and biochemistry of organisms within species. However, the results can seldom be used to assess the chemical impact above this level of biological organization.

Cause-and-effect relationships can easily be established from single-species tests because of the degree of control over laboratory conditions. These tests are also straightforward and relatively simple to conduct, and many are standardized and can be replicated. Current single-species tests are conducted with individual species that are considered representative of broad classes of organisms (e.g., fish or invertebrates), so that the results provide information on the toxicity of specific chemicals to different types of organisms under given conditions. The utility of single-species laboratory tests is, however, a function of the criteria used to select the test organisms (discussed below). The main limitation of these studies is that effects observed in the laboratory may not occur in the same way or to the same degree at similar concentrations in the natural environment. Single-species laboratory tests generally do not account for the adaptive ability of natural populations of organisms.

Bioavailability will also be different in laboratory tests because of the use of "standard" laboratory waters, which typically do not contain realistic environmental concentrations of dissolved and particulate organic matter. Thus, effects observed in laboratory tests may appear more severe than those seen in the field. Furthermore, laboratory studies are unrealistic in that they cannot simulate the many complex species interactions and environmental influences and changes that continually occur in natural systems. Given the complexity of aquatic ecosystems, it is unlikely that a single-species test approach could serve as an index of the structure and function of an entire community or ecosystem.

Multispecies tests and small ecosystem tests may also be conducted in the laboratory. These studies often involve "laboratory microcosms" or "model ecosystems." Laboratory microcosms for toxicological studies are small-scale enclosures (plastic or glass) containing samples from the natural ecosystem (water, sediment, invertebrates, and plants). Their advantage is that effects beyond the level of a single species can be identified, providing information more predictive of ecological consequences of the chemical's release. In principle, if conditions are uniform, these tests should be easy to replicate and standardize for different chemical substances; however, in practice this is not always the case. Because the environmental influences are controlled, cause-and-effect relationships are more easily analyzed than in natural systems. Microcosms have several limitations, however, as they are simple simulations of natural systems. For example, the impact of the physical environment (e.g., temperature and seasonal changes) on natural aquatic ecosystems can be very different from that represented by a microcosm. In addition, significant biotic components of the natural system (e.g., invaders) may be absent. Laboratory aquatic microcosm studies are described in Boyle (1985), Cairns (1985, 1986), Cairns and Cherry (1993), Cairns and Pratt (1989), Giesy (1980), Taub (1993), and in Chapter 9.

Single-species, multispecies, and ecosystem tests are most realistically conducted in the field or natural ecosystem. The ecosystem may be a pond or a stream, a lake, or an estuary, and it is usually selected because of the possibility of chemical exposure. The ecosystem may also be an outdoor experimental model microcosm or mesocosm, such as the "limnocorrals" in freshwater ponds. The distinction between microcosms and mesocosms is discussed in Chapter 9. Some influences and interactions of biotic and abiotic components that are not present in laboratory studies can be identified in such field studies. Field studies can also be used to validate laboratory studies and mathematical ecosystem models and to estimate environmental concentrations of chemicals. They may be more useful for predicting the potential fate and effects of chemicals on the aquatic environment.

There are, however, limitations of field studies. Environmental variables are not stable, so that field work is difficult to monitor and impossible to replicate. Results are often equivocal because they can be attributed to many factors. Cause-and-effect relationships are often difficult to establish and interpret because there are too many variables that are not adequately controlled and many are undefined. Natural and experimental model (e.g., microcosms, mesocosms) aquatic ecosystem field studies are discussed in Chapter 9.

Ideally, data on priority chemicals and mixtures should be obtained from a combination of laboratory and field studies before widespread use. In designing these studies, consideration should be given to site-specific factors appropriate to the particular ecosystem. No single test can provide enough information to make adequate predictions of the impact of a chemical on the aquatic environment. Tests should be conducted to detect effects on single species, species interactions, and the ecosystem. However, the development and standardization of tests beyond the single-species level are difficult because of the complexity of both species interactions and natural aquatic ecosystems.

Because most aquatic toxicity tests to date (circa 1994) have been laboratory studies with single species or suites of single species, the discussion throughout most of this book concerns these test systems. Many single-species tests have yielded results that are well correlated with the observed effects of chemicals under natural conditions (e.g., pesticides, organotins, organometals). Mathematical models are used to link the data obtained from laboratory toxicity tests with predicted field effects (Reed et al., 1984). The combined data can provide more complete information about the potential impact of chemicals on the aquatic environment.

General Test Design

Toxicity testing in the laboratory usually follows a stepwise tier approach, progressing from simple short-term tests to more complex and sophisticated longer-term tests based on the results of previous tests. Although the details of the protocol for each test may differ, the general test design is similar; it requires careful control of conditions such as pH, temperature, dissolved oxygen concentration, and photoperiod. Test organisms are exposed in test chambers (e.g., glass tanks) to various concentrations of the test material (e.g., pesticides or industrial effluents) in water solutions. The criteria for effects (e.g., mortality, growth, and reproduction), which are established before testing, are then evaluated by comparing the chemically exposed (treated) organisms with the untreated organisms (controls).

All toxicity tests should include a concurrent control to ensure that the effects observed are associated with or attributable to exposure to the test material. There are three basic types of controls:

1. The untreated (negative) water control consists of a group of organisms with the same dilution water (without test material or solvent added) and the same conditions and procedures. The organisms are from the same source as those used in the remainder of the treatments (exposed to the test material concentrations). This type of control is used to determine the inherent background effects in the test, such as effects related to the health of the organisms and the quality of the dilution water. It provides a baseline and a point of correction for interpreting the test results.
2. In some tests with water-insoluble or poorly soluble test materials, an organic solvent or carrier may be used to prepare stock solutions of the test material. When this occurs, a solvent or vehicle control should be included. A solvent is an organic chemical that is miscible with water and in which the test material is more soluble than it is in the diluent water. The solvent control is essentially an untreated control, except that the maximum volume of solvent used to prepare a test material concentration is added. This control also provides a baseline, taking into account effects of the solvent on the test organisms. The toxicity of the solvent to the test species should be relatively low compared with that of the test material and, at worst, the solvent should only interact additively with the test material. Typical solvents are acetone, dimethyl formamide (DMF), dimethyl sulfoxide (DMSO), and triethylene glycol (TEG). The reported toxicities (96-h LC50s) of these solvents to fish are acetone, 9,100 mg/L; DMF, 10,410 mg/L; DMSO, 33,500 mg/L; and TEG 92,500 mg/L (C. E. Stephan, personal communication; Willford, 1968).
3. A positive (reference) control is a material known from previous experience to produce a defined effect on the test organisms. An ideal reference toxicant is toxic at low concentrations, rapidly lethal, stable, nonselective, and detectable by known analytical techniques. It is used to determine the health and sensitivity of the organisms, to compare the relative toxicities of substances by using the control as an internal standard, to perform interlaboratory calibrations, and to evaluate the reproducibility (precision) of test data with time. A positive control is optional if an untreated or solvent control is maintained but can be highly instructive if routinely included as an intralaboratory quality control measure. Compounds used or evaluated as reference toxicants include sodium

pentachlorophenate (Adelman and Smith, 1976; Davis and Hoos, 1975), dehydroabietic acid (Davis and Hoos, 1975), phenol and sodium azide (Klaverkamp et al., 1975), *p,p'*-DDT (Marking, 1966), the surfactant dodecyl sodium sulfate (DSS) (LaRoche et al., 1970; Pessah et al., 1975; Wells et al., 1982), antimycin (Hunn et al., 1968), and sodium chloride (Adelman and Smith, 1976). Based on the characteristics of an ideal reference toxicant, sodium pentachlorophenate has been recommended with phenol as an alternative (Lee, 1980).

More recently, four organics (4-chlorophenol, dodecyl sodium sulfate, sodium pentachlorophenate, and phenol) and nine inorganics (cadmium chloride, copper sulfate, potassium dicromate, potassium chromate, potassium chloride, sodium chloride, silver nitrate, zinc chloride, and zinc sulfate) were evaluated to determine their suitability as reference toxicants in the control of toxicity test precision for fish, invertebrates, and algae in acute and chronic tests (Environment Canada, 1990). Results indicate that the most suitable reference toxicants for a wide variety of tests are two organics (sodium pentachlorophenate, phenol) and two inorganics (hexavalent chromium, zinc). It was also evident from this review that little research has been conducted to verify that a reference toxicant can consistently reflect poor health of organisms or genetically different stocks of organisms.

Canadian and European scientists have embraced the concept of using reference controls as one component of intra- and interlaboratory quality assurance/quality control (QA/QC) (see Chapter 11).

Whenever possible, chemical analysis should be performed to measure the concentrations to which test organisms are exposed. In addition, it is valuable to measure chemical residues in tissues of exposed organisms. Methods and instrumentation for the analysis of chemical residues in water, sediment, and tissues of aquatic organisms are described in Gilbert and Kakareka (1985), Hunt and Wilson (1986), and Purser and Hume (1994).

Test Organisms

Criteria for Selection

In order to extrapolate meaningful, relevant, and ecologically significant results from aquatic toxicity tests, not only appropriate tests but also appropriate organisms should be used. Several criteria should be considered in selecting organisms for toxicity testing:

1. Because sensitivities vary among species, a group of species representing a broad range of sensitivities should be used whenever possible.
2. Widely available and abundant species should be considered.
3. Whenever possible, species should be studied that are indigenous to or representative of the ecosystem that may receive the impact.
4. Species that are recreationally, commercially, or ecologically important should be included.
5. Species should be amenable to routine maintenance in the laboratory and techniques should be available for culturing and rearing them in the laboratory so as to facilitate both acute and chronic toxicity tests.
6. If there is adequate background information on a species (i.e., its physiology, genetics, and behavior), the data from a test may be more easily interpreted.

The species selected for testing may differ from ecosystem to ecosystem and the selection will often be based on site-specific considerations. The objective(s) of the testing should always influence the choice of test species (Sprague and Fogels, 1977). For example, when assessing the potential impact of a chemical on cold-water streams, a salmonid such as a trout would be selected, while for warm-water streams, a centrachid such as a sunfish would be chosen. There is no standard test species for all questions and all ecosystems. The kinds and number of species will depend on the complexity of the ecosystem. However, criteria 1, 5, and 6 above are so important at times that several standard aquatic organisms have been used. These include the water flea (*Daphnia* sp.), fathead minnow (*Pimephales promelas*), bluegill (*Lepomis macrochirus*), rainbow trout (*Oncorhynchus mykiss*; formerly *Salmo gairdneri*), mysid shrimp (e.g., *Mysidopsis* sp.), and sheepshead minnow (*Cyprinodon variegatus*).

Because of the variation between species in sensitivity to waterborne chemicals, a range of different effects and degrees of effect may be expected when different species are exposed to the same concentrations of the same chemical. It is therefore important to conduct tests with several species, perhaps from different taxonomic groups, to get some indication of the natural variability in levels causing an effect. The selection of test species should be based on as many of the foregoing criteria as possible.

Species Types

Toxicity tests have traditionally been performed with a variety of freshwater and saltwater test species representing algae, fish, and invertebrates. These are described in Chapters 2–8. These species are recommended for use in toxicity tests because a substantial amount of toxicity information is available for many of them, they are sensitive, and they are ecologically important. If one of these species is not available, organisms from the recommended genus should be considered. Test species can be collected from wild populations in relatively unpolluted areas, purchased from commercial suppliers, or cultured in the laboratory. Species should not be collected by use of electroshock or chemicals because of the physiological stress induced. All species in a particular test should be from the same source.

It should also be noted that a range of nontraditional toxicity tests are now being developed and utilized, employing fish cell lines, bacteria, yeasts, protozoa, and other microscopic groups or life stages (Blaise, 1991; Mayfield, 1993; Persoone and Janssen, 1993; Tarazona et al., 1993; Wells, 1993).

It has been questioned whether fish acute toxicity data should be mandatory for testing new industrial chemicals and whether alternatives should be considered (Fentem and Balls, 1993; Weideman, 1993). Alternatives have a special meaning in animal experimentation and include procedures that completely replace the need for animal experiments, reduce the number of animals required, or diminish any pain or suffering that may occur. Replacement alternatives include computer modeling and QSAR techniques, use of lower organisms (e.g., bacteria, protozoa, fungi), and in vitro cell culture tests. These nontraditional tests have been suggested as alternatives for fish acute tests because the number of acute tests conducted each year has been increasing throughout the world.

Exposure Systems

In aquatic toxicity tests the controls and treated organisms can be exposed to the dilution water or the test water solutions containing the chemical toxicant (or effluent) by four different techniques.

1. In a *static test* the organisms are exposed in still water. The test material is added to the dilution water to produce the desired test concentrations. The control and test organisms are then placed in test chambers and there is no change of water for the duration of the test.
2. A *recirculation test* is similar to a static test except that the test solutions and control water are pumped through an apparatus, such as a filter, to maintain water quality but not reduce the concentration of test material. The water is returned to the test chamber. This type of test is not routinely used because it is expensive to set up and maintain and because of uncertainty about the effect of the apparatus (aerator, filter, sterilizer) on the test material.
3. A *renewal test* is similar to a static test because it is conducted in still water, but the test solutions and control water are renewed periodically (usually 24-h intervals) by transferring the test organisms to chambers with freshly prepared material or by removing and replacing the material in the original containers.
4. In a *flow-through test*, the test solutions and control water flow into and out of the chambers in which the test organisms are maintained. The once-through flow may be intermittent or continuous. A stock solution of test material to be mixed with dilution water to prepare the test concentrations may be prepared once at the beginning of the test, or fresh stock solutions may be prepared daily. In either case, metering pumps or diluters control the flow (and thus the volume) of dilution water/test material stock solution so that the proper proportions of each will be mixed. Chapter 2 illustrates the flow-through technique.

All four techniques are discussed in detail in Appendix Chapter A. The static and flow-through techniques are the most widely used means of exposure. Exposure concentrations in flow-through tests are usually measured several times during the exposure and reported as a mean concentration. Exposure concentrations in static tests are usually reported as calculated nominal values, although many investigators now provide measured values for some types of non-flow-through tests, especially recirculation and renewal tests, as illustrated by the oil pollution and PAH literature. Despite being uncomplicated and inexpensive, the static technique has disadvantages:

- The test material may be degraded and volatilized, adsorbed onto the test chambers or test organisms, or otherwise changed. The concentrations to which the organisms are exposed thus decrease and change as the test progresses.
- The test material may have a high biochemical oxygen demand (BOD) so that the toxicity is masked by depletion of dissolved oxygen from the test solution.
- The metabolic products of the test organism build up and may react with the material being tested, producing a toxic response different from the one that would have been produced had the excretory products been removed.

If any of these conditions exist to a significant extent, the results of the static test may not accurately represent the effects of the material being tested. Therefore, the flow-through procedure should be used wherever possible. This results in more uniform and stable test conditions, with the test material concentration, dissolved oxygen concentration, and other water quality characteristics remaining relatively constant and optimal while the waste products are removed. Although flow-through tests may involve a more complex delivery system, they yield the best and most accurate estimate of toxicity. In some cases, unique miniaturized flow-through systems have been developed to enable testing of sensitive young life stages (Tjeerdema et al., 1991).

Most of the toxicity tests discussed in this book involve static or flow-through techniques. Both are used for acute tests, but for subchronic and chronic tests the flow-through technique is preferred because there is a greater degree of assurance that the concentrations of test material and conditions to which the organisms are exposed during the test period have remained constant (i.e., within desired ranges). Cause-and-effect relationships can thus be more easily and accurately established.

Standard Procedures

A variety of test methods have been developed by the American Public Health Association (APHA), U.S. Environmental Protection Agency (U.S. EPA), American Society for Testing and Materials (ASTM), International Standardization Organization (ISO), Environment Canada, and Organization for Economic Cooperation and Development (OECD) to evaluate the potential toxicity and hazard of materials to aquatic organisms. A background on some of the different test method standardization groups is presented in Chapter 2 (Freshwater Tests) and Chapter 26 (OECD). Doudoroff et al. (1951) first recognized the need to develop uniform, standardized test procedures to maximize the comparability of data from tests. The advantages of using standardized test procedures were summarized by Davis (1977) as follows:

- Allows selection of one or more uniform and useful tests by a variety of laboratories.
- Facilitates comparison of data and results and thus increases usefulness of published data.
- Increases accuracy of the data.
- Allows replication of the test.
- Allows the test to be easily initiated and conducted by a variety of personnel (if the procedure is well documented).
- Legal advantage if procedures are accepted by the courts.
- Useful for routine monitoring purposes.

Davis (1977) stated that the initial step to standardization is a thorough knowledge of the various chemical, physical, and biological factors that affect toxicity test results. Standardization can then be achieved by:

- Adoption of detailed test protocols that minimize or standardize disturbing effects.
- Use of a standard test species.
- Selection of specific test types designed to meet specific objectives.
- Use of reference toxicants or "disease-free" certified test animals.

A standardized method (protocol) with reference toxicants and standard test species theoretically maximizes comparability, replicability, and reliability and thus is essential for answering questions on relative toxicity and sensitivity or replicability of tests (Buikema et al., 1982). The effectiveness of a test method can be judged according to the five Rs—relevance, reproducibility, reliability, robustness, and repeatability (Calow, 1993).

Although there is an obvious need for standardized toxicity test protocols, there is also a danger in overstandardization because it "may stifle innovative and creative work," especially if regulatory agencies "do not recognize work that does not coincide with their own priorities" (Davis, 1977). For example, a standardized toxicity test may not be appropriate for answering specific questions about a particular body of water. For assessing the hazard of a particular chemical to an indigenous fish community in a specified body of water, it may not be useful to employ standard test species that are not normally present in that body of water. Furthermore, it is desirable to conduct the tests under the physical and chemical conditions characteristic of that body of water.

Toxicity test protocols typically specify the exposure of test organisms to fixed concentrations of chemical compounds for a defined time period. However, chemicals rarely, if ever, enter the environment at a constant concentration and rate (Cairns and Thompson, 1980). Most chemicals enter the aquatic environment sporadically in "pulses," intermittently as "slugs," or as a one-time "spill," creating episodic exposures. Therefore, the toxicant is present in high concentrations for a relatively short period of time and is usually diluted to lower concentrations with time. In these cases mortality is a function of a changing toxicant concentration and the length of exposure. Generally, although there are marked exceptions (e.g., acids), organisms can tolerate high concentrations for short periods of time and lower concentrations for longer periods of time (Buikema et al., 1982). A typical standard test, in which all test conditions are kept constant or maintained optimally, may be inappropriate for predicting responses in a changing natural system. In that case, site-specific studies should be conducted.

The objectives of a study will determine which toxicity test and which experimental design are most applicable. If the results are to be used to compare the toxicity of one chemical to another, rigid standardization is necessary. However, if the data are to be used to describe or predict the behavior of the chemical in a specific water system, there is a danger in overstandardization and there are many advantages to "customizing" the choice of test(s).

Some methods that have been widely accepted and are now standardized or in the process of being standardized are listed in Chapters 2–8.

Description of Test Methods

Aquatic toxicity test methods may be categorized according to length of exposure, test situation, criteria of effects to be evaluated, and organisms to be tested. The data generated in these tests may enable the researcher to determine the *no observed effect concentration* (*NOEC*) or *no-effect concentration*, which is the maximum concentration of the test material that produces no statistically significant harmful effect on test organisms compared to controls in a specific test. The *lowest observed effect concentration* (*LOEC*) or *minimum threshold concentration* (*MTC*) may also be obtained. This is the lowest concentration that has a statistically significant deleterious effect on test organisms compared to controls in a specific test. The effects evaluated are biological end points selected because they are based on processes important to the survival, growth, behavior, and perpetuation of a species. These end points differ depending on the type of toxicity test being conducted and the species used. The statistical approach also changes with the type of toxicity test conducted. Some of the commonly used tests and the end points (effect criteria) that they measure are briefly described in the following.

Acute Toxicity Tests

These are tests designed to evaluate the relative toxicity of a chemical to selected aquatic organisms upon short-term exposure to various concentrations of the test chemical. Common effect criteria for fish are mortality; for invertebrates, immobility and loss of equilibrium; and for algae, growth. These tests may be conducted for a predetermined length of time (*time-dependent test*) to estimate the 24- or 96-h LC50 or the 48- or 96-h EC50. An acute toxicity test may also have a duration that is not predetermined, in which case it is referred to as a *time-independent* (*TI*) test. In a TI test, exposure of the test organisms continues until the toxic response manifested has ceased or economic or other practical considerations dictate that the test be terminated. For example, the acute TI test should be allowed to continue until acute toxicity (mortality or a defined sublethal effect) has ceased or nearly ceased and the toxicity curve indicates that a threshold or incipient effect concentration can be estimated. With most test materials this point is reached in 7–14 d, but it may not be reached within 21 d.

In the early development of acute toxicity tests, data were expressed as the *median tolerance limit* (TLm or TL50)—the test material concentration at which 50% of the test organisms survive for a specified exposure time (usually 24–96 h). This term has been replaced by median lethal concentration (LC50) and median effective concentration (EC50). Acute toxicity tests, lethal and sublethal, are discussed in Chapters 2 (Freshwater), 3 (Saltwater), 4 (Algae and Vascular Plants), 5 (Seaweed), 6 (Sea Urchin), 7 (Frog Embyro), and 8 (Sediment). For more information on acute toxicity tests see Appendix Chapter A.

Chronic Toxicity Tests

The fact that a chemical does not have adverse effects on aquatic organisms in acute toxicity tests does not necessarily indicate that it is not toxic to these species. Chronic toxicity tests permit evaluation of the possible adverse effects of the chemical under conditions of long-term exposure at sublethal concentrations. In a full chronic toxicity test, the test organism is exposed for an entire reproductive life cycle (e.g., egg to egg) to at least five concentrations of the test material. Partial life cycle (or partial chronic) toxicity tests involve only several sensitive life stages; these include reproduction and growth during the first year but do not include exposure of very early juvenile stages. In full chronic toxicity tests exposure is generally initiated with an egg or zygote and continues through development and hatching of the embryo, growth and development of the young organism, attainment of sexual maturity, and reproduction to produce a second-generation organism. Tests may also begin with the exposed adult and continue through egg, fry, juvenile, and adult to egg. With fish, for example, exposure begins with fertilized eggs and criteria for effect include growth, reproduction, development of gametes, maturation, spawning success, hatching success, survival of larvae or fry, growth and survival of different life stages, and behavior. The duration of a chronic toxicity test varies with the species tested; it is approximately 21 d for the water flea *Daphnia magna* and can be 275–300 d for the fathead minnow, *Pimephales promelas*.

From the data obtained in partial life cycle and complete life cycle tests the *maximum acceptable toxicant concentration* (*MATC*) can be estimated. This is the estimated threshold concentration of a chemical within a range defined by the highest concentration tested at which no significant deleterious effect was observed (NOEC) and the lowest concentration tested at which some significant deleterious effect was observed (LOEC) (Mount and Stephan, 1967). Because it is not possible to test an unlimited number of intermediate concentrations, an MATC is generally reported as being greater than the NOEC and less than the LOEC (NOEC < MATC < LOEC; e.g., 0.5 ppm < MATC < 1.0 ppm). For regulatory purposes, the MATC is sometimes calculated as the geometric mean of the LOEC and NOEC, so it can be used as a point estimate.

Life cycle tests are also used to determine application factors (AFs) to predict "safe" concentrations. Hart et al. (1945) first suggested that a factor be applied to mortality data for a chemical to obtain a long-term safe concentration for aquatic organisms. At this time data on the chronic effects of chemicals was limited, so "arbitrary" (e.g., 10, 100, 1000) safety factors, which depended on the nature, quality, and quantity of available data, were proposed as a first approach to protect aquatic life. Historically (Warren and Doudoroff, 1958; Mount and Stephan, 1967), the AF was first empirically derived as the numerical value of the ratio of the MATC to the time-independent or incipient LC50 estimated in a dynamic acute toxicity test, if possible. If the TI LC50 was not available, a time-dependent LC50 (e.g., 96-h LC50) was used; that is, AF = MATC/LC50. The AF was intended to provide an estimate of the relationship between a test material's chronic and acute toxicity, which could then be applied to aquatic organisms for which an MATC could not be easily derived. In order to obtain an MATC for these organisms, the assumption was made that the AF for a given toxicant is relatively constant over a range of test species. The AF derived for species 1 and the LC50 for species 2 could then be used to estimate the MATC for species 2; that is $AF_1 \times LC50_2 = MATC_2$.

Although the AF has value as a crude estimator, there are cases in which it does not hold true (Mount, 1976). Kenaga (1979) showed that the AF could range from 0.91 to 0.00009, or inversely (1/AF) from 1.1 to 11,100—four orders of magnitude difference. It is more widely accepted to use whole numbers, i.e., the inverse of the AF (Kenaga, 1982). The *acute-to-chronic toxicity ratio* (*ACR*) is the inverse of AF and is also used to estimate an MATC for species for which only acute toxicity data exist. Both the AF and ACR approaches have limitations conceptually as they are based on empirical observations. As will be seen later in the section on interpreting toxicity test data, it appears that there is a toxicological explanation for many of the observed acute to chronic relationships. With a clearer understanding of the underlying toxicological processes, the utility of this methodology may be substantially expanded. Chronic toxicity tests are discussed in Chapters 2, 3, and 4.

Short-Term Sublethal Tests

To evaluate the toxicity of effluents to aquatic organisms, the EPA has developed short-term sublethal tests (i.e., 7–9 d or less; often misleadingly called short-term "chronic" tests) by focusing on the most sensitive life stages. The short-term sublethal toxicity end points include changes in growth, reproduction, and survival. The NOEC, LOEC, EC50, and/or LC50 are reported in these tests. Short-term sublethal test methodology is discussed in Chapters 2 and 3 and their application in Chapter 33.

Early Life Stage Tests

These tests include continuous exposure of the early life stages (e.g., egg, embryo, larva, and fry) of aquatic organisms to various concentrations of a chemical for 1–2 mo, depending on the species. Although these tests do not provide total life cycle exposure and lack a full assessment of reproduction, they do include exposure during the sensitive life stages. They have been used to predict accurately MATC values estimated in the fish life cycle tests. Early life stage tests with fish are discussed in Chapters 2 and 3. For more information on fish early life stage (ELS) tests, see Appendix Chapter B and von Westernhagen (1988); for invertebrates see Buikema and Cairns (1980) and Persoone et al. (1984).

Bioaccumulation Tests

Chemicals with low solubility in water usually have an affinity for fatty tissues and thus can be stored and concentrated in tissues with a high lipid content. Such hydrophobic chemicals may persist in water and demonstrate cumulative toxicity to organisms. Chemicals with these characteristics are usually considered for bioconcentration tests, which are designed to determine or predict the *bioconcentration factor* (*BCF*). The BCF is the ratio of the average concentration of test chemical accumulated in the tissues of the test organisms under steady-state conditions to the average measured concentration in the water to which the organisms are exposed.

Bioconcentration is the process by which chemicals from the water enter organisms, through gills or epithelial tissue, and are accumulated. *Bioaccumulation* is a broader term and includes not only bioconcentration but also accumulation of chemicals through consumption of food. *Biomagnification* refers to the total process, including bioconcentration and bioaccumulation, by which tissue concentrations of accumulated chemicals increase as the chemical passes through several trophic levels. Bioaccumulation and bioavailability are discussed in Chapter 16. For more information on bioconcentration, bioaccumulation, and biomagnification in aquatic systems see Connell (1990), Nagel and Loskill (1991), Phillips (1993), and Appendix Chapter D.

Other Sublethal Effects Tests

In the aquatic environment organisms are not usually exposed to high, acutely toxic concentrations of chemicals unless they are restricted to the vicinity of a chemical release site or spill area. Beyond the initial impact site, dilution and dispersion occur, decreasing these acute concentrations to lower, potentially sublethal levels. In general, a greater biomass is exposed to sublethal concentrations of chemicals than to acutely toxic lethal concentrations. The lower concentrations may not produce death, but they may have a profound effect on the future survival of the organisms.

Sublethal effects may be studied in the laboratory by a variety of procedures. Such effects generally are divided into three classes: biochemical and physiological, behavioral, and histological (Sheehan et al., 1984). Biochemical and physiological effects tests include studies of proteins (e.g., enzyme inhibition, stress proteins), clinical chemistry, hematology, and respiration. Behavior represents an integrated response corresponding to complex biochemical and physiological functions, so chemically induced behavioral changes may reflect effects on internal homeostasis. Behavioral end points may thus be sensitive indicators of sublethal effects. Behavioral effects that have received considerable attention in aquatic organisms are locomotion and swimming, attraction-avoidance, prey-predator relationships, aggression and territoriality, and learning. These are all ecologically significant behaviors. They are discussed in Rand (1985) and SETAC (1990).

Histological studies are also useful because changes in histological structure and the occurrence of pathologies, sometimes caused by chemicals, may often significantly modify the function of tissues and organs (Couch and Fournie, 1993). Together, behavioral, biochemical and physiological, and histological tests are useful for evaluating the environmental hazard of a chemical, and they may provide important information on its mode of action. The literature (e.g., WHO/UNEP, 1989) also indicates that chemicals in aquatic organisms may affect the immune system (i.e., immunotoxic) and produce alterations in DNA (i.e., genotoxic) and tissue changes including cancer (i.e., carcinogenic). Chapters 12, 13, 14, and 16 discuss immunotoxic, genotoxic, histologic (carcinogenicity), and biochemical end points, all which may be used as biomarkers of effects and/or exposure to chemicals (Huggett et al., 1992). More information may be found on biochemical and physiological end points in Mehrle and Mayer (1985) and on histopathological end points in Meyers and Hendricks (1985) and Couch and Fournie (1993).

Good Laboratory Practices

Important in all phases of aquatic toxicity testing, but especially critical with respect to regulatory and litigious matters, are quality assurance (QA) in the laboratory and adherence to good laboratory practices (GLPs) to promote the development of quality test data. Good laboratory practice is concerned with the conditions under which laboratory (includes field) studies are planned, performed, monitored, recorded, and reported. Aquatic toxicity testing programs are classified as nonclinical laboratory studies, the GLP regulations published by the U.S. Department of Health, Education and Welfare, Food and Drug Administration (1978), and the U.S. Environmental Protection Agency (1980) describe the requirements for an acceptable study.

These GLP requirements deal with all related aspects of testing, including *personnel*—their qualifications, responsibilities, management, and QA function; *facilities*—test animal maintenance and handling, chemical handling and storage, laboratory testing areas, specimen storage, and data handling and storage; *equipment*—design, maintenance, and calibration; *laboratory operations*—standard operating procedures, methods of animal care, and standard reagents and solutions; *test chemicals*—handling and storage, characterization, and use of solvents or carriers; *protocols*—test methods, data collection, and data handling; *reports*—data storage, retrieval, verification, use in reports, and retention; and *disqualification of testing facilities*—grounds for disqualification and public notice of such.

These comprehensive regulations must be closely adhered to and the adherence must be supported by documentation capable of passing government inspection before a study is considered to meet the minimum test of adequacy. Failure to meet the major provisions of the GLP requirements can result in a ruling by the FDA or U.S. EPA that a study is unacceptable and must be repeated. A testing facility that has failed to comply with the GLP regulations can also be disqualified. Furthermore, these regulations apply to any facilities, including academic and industrial laboratories, that prepare data for submission in support of regulated products. Good laboratory practices are discussed in Chapter 11.

INTERPRETING TOXICITY TEST DATA

In the previous sections the theory and practice of toxicity testing have been outlined. When data are obtained by following recommended testing protocols it is imperative to remember that *toxicity test data must be interpreted.* Unlike the physical-chemical properties of chemicals, which are often constants (although not always simple to measure!), toxicity is a relative measure and a function of the interaction of numerous influencing factors related to both the toxic agent and the organism. For example, it is not sufficient to know only that the LC50 for two or more chemicals is the same. Since the LC50 end point is a product of three types of factors—*exposure duration, accumulation kinetics*, and *chemical potency*—LC50 test results for various chemicals and organisms can have the same numerical value. However, this does not mean that the toxic processes are the same in each circumstance or that the toxicological significance of each is comparable. Similarities or differences can be judged only when each set of experimental data is interpreted by quantifying the influence of parameters in each of the three component categories. Toxicity test results themselves (i.e., the numerical estimates of effect and no-effect concentrations) are not the ultimate toxicological objective, rather they are the basis for gaining toxicological insights which can be used to make comparisons with other toxic agents and/or organisms.

In employing the concentration-response relationship in toxicity bioassays, several operational assumptions are employed in addition to, or in extension of, the fundamental ones outlined earlier. It is assumed that sometime during the test a steady state is reached between the external chemical concentration and the concentrations of chemical at target sites in the organism. Steady state in this situation implies that a constant concentration of chemical is reached at the target sites during continuous exposure. In the ideal situation, at the threshold exposure concentration where toxic effects have reached the incipient stage, the exposure water concentration should be a valid surrogate for the unknown amount of toxicant in the organism that is actually causing the effect. Once a threshold relationship is established, the potency and kinetic information obtained can be used to interpolate the time course prior to steady state. Steady state is not required for a toxic effect to occur; a response can happen at any time that target site concentrations reach a critical level. Furthermore, steady state is not required for interpretation of the indirect bioassays used in aquatic toxicity testing. However, attempting to interpret such surrogate-based toxicity test data without knowledge of steady-state conditions is a very complicated undertaking.

The art of the toxicity bioassay involves creating a situation in which a steady state occurs, interpreting the data by means of the basic principles and assumptions, and using understanding to focus on the many circumstances in the laboratory and field where steady states are rare. Unfortunately, Sprague's admonishment to get steady-state toxicity test data by obtaining and reporting time-independent bioassay results, such as incipient or threshold LC50s (Sprague, 1969), has been largely ignored.

The purpose of reported time-independent data is to determine if and when steady state is reached so that the nature of the surrogate-target relationship can be established (also see the earlier section on toxicity curves).

Although it is not possible to obtain steady-state toxicity data with some toxic agents and organisms, that does not prejudice use of the rule. Lack of steady state simply means that more detailed toxicological investigation is required because the underlying assumptions of the toxicity test design have been violated. Such test results have no firm meaning and cannot be compared legitimately with results obtained where steady state is or can be reached. Indiscriminate use of improperly executed or interpreted toxicity test results will result in misleading conclusions.

Toxicological Models

The use of models appears to be a specialized aspect of toxicology, but in fact it is an integral part of both toxicity testing and interpretation. A model is simply an operational framework that allows practical application of basic principles and concepts (see Chapter 18). A model may be purely conceptual in nature, a detailed mathematical algorithm, or something in between. Much of the practical application of aquatic toxicology testing and interpretation is based on application of the one-compartment, first-order kinetics (1CFOK) model (see Chapter 16 and Appendix Chapter D; McCarty, 1990). This assumes that organisms are a single homogeneous, well-mixed compartment and all uptake, transformation, and elimination processes associated with toxicant accumulation are first order; that is, rates are proportional to differences in toxicant concentrations between phases.

Although viewing organisms as a 1CFOK model is clearly a dramatic oversimplification, it has and continues to be of great utility. Living organisms are better described by complex, multicompartment, *physiologically based pharmacokinetic (PB-PK) models* (also called PB-TK models; physiologically-based toxicokinetic) that take into account the various organs and tissues present in real organisms, the nature and magnitude of the blood perfusion and toxicant transport between compartments, the actual nature of exposure, uptake and elimination via respiration and diet, and the multiple types of kinetics that are involved (O'Flaherty, 1981). Much of the basic biological and physiological information needed to apply such sophisticated PB-PK models is unavailable for many aquatic organisms. Barron et al. (1990) and McKim and Nichols (1994) have reviewed the situation for aquatic organisms and Nichols et al. (1990) and McKim and Nichols (1994) have provided detailed examples of PB-PK (and PB-TK) models for fish.

Models and associated assumptions are not exclusive to toxicokinetics. For example, variability in toxicity test results can result from the variability in the time course of accumulation and amount delivered to the sites of toxic action (toxicokinetics) and/or the variability in the time course of the response of individual organisms to the target dose (toxicodynamics). As it is usually impractical to separate the two, interpretation is usually facilitated by assuming all variability is associated with one or the other factors. Depending on the subsequent use of the interpreted data, substantial errors may result from either assumption. It is clear that, irrespective of whether conceptual or detailed mathematical models are employed, it is vital to consider the assumptions on which they are based to ensure that experimental design and interpretation are properly constrained. Model assumptions must be clearly stated so that subsequent investigators are not misled by either the approach or the conclusions drawn.

Influence of Modifying Factors

As previously discussed, water quality factors (e.g., temperature, dissolved oxygen) may affect the bioavailability of a chemical during a toxicity test and thus produce differences in an organism's response (e.g., LC50) to a chemical (Appendix Chapter C). In toxicity tests the exposure regime itself is an important modifying factor, as many testing protocols effectively manipulate exposure water concentration and exposure duration to overcome toxicokinetic factors and produce a given concentration of toxic agent in the organism(s) that is associated with the adverse toxicological response being studied. Even when a relationship between the surrogate water concentration and the organism concentration is established, other factors may confound interpretation and subsequent comparisons with other organisms and chemicals. These factors are related to physiological and biochemical differences between species that may change the kinetics of uptake, distribution, biotransformation, and elimination. The possible influences of several of these important modifying factors often present in aquatic toxicity testing are reviewed below.

Toxic Agents in Nontarget Compartments

Much of the discussion has been on using whole-body residues as an improved surrogate for the concentration of toxic agent reaching the site(s) of toxic action. Although this simplification is useful it is clearly incorrect, because the organisms are not homogeneous single compartments. Nontarget phases or compartments in organisms can have a significant impact in toxicological comparisons where organisms have substantially different ratios of target to nontarget compartments. For hydrophobic organic chemicals, storage lipids (i.e., fat, adipose tissue) in the body represent one of the major nontarget sites in organisms. Although metallothioneins play a similar role for metals, metallothioneins are actively produced by the organism to sequester metals, whereas organics are sequestered in lipid largely through their partitioning behavior, so little or no additional metabolic cost is required.

Lipid content is a well-known modifying factor for organic chemicals. In fact, the phenomenon has been described by Lassiter and Hallam (1990) as "survival of the fattest." Bioconcentration studies often provide lipid-corrected results in an attempt to overcome this. Unfortunately, the situation for toxicity is more complicated, especially because hydrophobic target sites may not actually be lipid in nature. Nevertheless, the ratio of the lipid at the site of toxic action to nontarget lipid can vary widely and the amount of nontarget lipid in organisms, consisting largely of storage lipids throughout the body, is affected by a variety of factors and varies widely (Henderson and Tocher, 1987; Halver, 1989). Geyer et al. (1990) provide an excellent example of the importance and magnitude of the influence of nontarget lipids on toxicity test results. Toxicity test results for mammals of various species, strains, body weight, sex, and age exposed to 2,3,7,8-tetrachlorodibenzo-*p*-dioxin (TCDD) vary widely, with oral LD50 estimates ranging from 1 μmol/kg for male guinea pigs (4.5% lipid) to 5000 μmol/kg for male golden Syrian hamsters (17–20% lipid). However, the apparent differences can be explained mainly by differences in total body fat content (log LD50 = 5.3 log total body fat − 3.22, $r^2 = 0.834$, $n = 20$).

More basic physiology-biochemistry data and detailed information on body composition must be acquired for the aquatic test species usually employed. With the additional information, multicompartment models can be used to investigate the effect of nontarget compartments on the accumulation of toxicant in the target compartment(s). In the meantime, critical body residue comparisons between organisms of substantially different lipid contents should be considered preliminary.

Metallothioneins are specialized proteins produced by many organisms that form complexes with metals absorbed into the body. The bound metals are thus sequestered away from sites where adverse effects might otherwise be initiated. This active process protects the organism from toxic effects in much the same way that nontarget lipid pools provide protection from some of the organic chemicals accumulated. See Chapter 17 for more information on metallothioneins.

Metabolism

The relationship between exposure and target site concentrations can change considerably when substantial metabolic breakdown of the toxic chemical occurs, because more toxicant is needed in the exposure phase to overcome the loss due to metabolism in the organism, where the biological response is elicited. Unless the effect of metabolism is accounted for, the exposure surrogate will be higher by an amount related to the metabolic degradation rate. If such data were compared with data obtained in a similar situation where no metabolic degradation was occurring, differences might be erroneously attributed to toxicity-related factors. To avoid such interpretational errors it should be confirmed that the metabolic transformation of the chemicals being compared is similar in each case.

Work published by de Bruijn et al. (1991) and de Wolf et al. (1992) clearly illustrates the effect of metabolic breakdown of organic chemicals as a modifying factor in chemical accumulation and toxicity. Perhaps the most dramatic example was provided by Sijm and Opperhuizen (1988). They postulated that the considerable difference in the toxicity of various dioxin (TCDD) congeners and isomers to fish is largely related to differences in metabolic breakdown by the fish. For example, 2,8-DCDD (2,8-dichlorodibenzo-*p*-dioxin, a congener of TCDD) does not cause toxic effects in exposed fish and is undetectable in tissue analysis. However, when metabolism of 2,8-DCDD is inhibited by blocking the activity of the enzyme that biotransforms it, 2,8-DCDD becomes quite toxic to fish. In fact, the body residue associated with lethality is within about two orders of magnitude of the lethal residue of 2,3,7,8-TCDD itself.

It is possible to compensate for the effects of differences in metabolism on accumulation and toxicity; however, this is not usually done because metabolic transformation rates are usually unavailable for aquatic organisms typically used in tests. Until more routine estimation of metabolic transformation rates are available, Mackay et al. (1992) have suggested a method for obtaining rough estimates using existing toxicity data and relationships.

Body Size

Body size is considered to be an important modifying factor that is strictly controlled in recommended bioassay protocols. This ensures that the bioassay produces reasonable results within the species and test regime used. However, size is often neglected when interspecies comparisons are made. This is particularly relevant to the debate concerning use of the most sensitive species in both laboratory and field situations. Body size affects kinetic rate constants due, among other things, to the dramatic differences in surface area/volume ratios between organisms of different sizes. Much of what has been attributed to sensitivity differences may be, in fact, artifacts of the bioassay process caused by size differences affecting kinetics of accumulation. Organisms may be equally sensitive in terms of the toxic potency of the chemical, but the interaction of body size and the kinetics of accumulation and the use of the exposure water concentration as the surrogate, result in apparent differences that may be very misleading. An example is presented in Table 3 and discussed below.

Allometry is the general term for the study of the influence of organism body size on organism characteristics (e.g., metabolic rate, ingestion rate, reproduction rate) and *allometric relationships* express such relationships in terms of equations. The sim-

Table 3. Estimates of influence of body size on time to steady state in an idealized organism

Relative time	Relative organism body sizes					
	1	10	100	1000	10,000	100,000
Time to steady state: scaling factor 0.67	1	2.2	4.6	10	22	44.7
Time to steady state: scaling factor 0.75	1	1.8	3.2	5.6	10	17.8

plest allometric relationship is the exponential or power equation $Y = aW^b$, where Y is an animal characteristic, W is animal size, a is a constant, and b is the scaling factor (Peters, 1983).

A simple illustration is offered using this simple allometric equation to scale the time required for idealized organisms of various sizes to reach a steady state with a constant concentration of a chemical in the exposure water. Allometric scaling factors of 0.67 and 0.75 will be employed. Although 0.67 is correct for a spherical object, observed scaling factors range between about 0.6 to 0.8, and 0.75 is often used as an empirical estimate based on metabolism–body size relationships (Filov et al., 1979; Peters, 1983). It is assumed that the transport across the surface of an organism is constant per unit exchange surface area and transport occurs across the entire surface area. Mass and volume are considered interchangeable as most aquatic organisms have densities close to 1. Only the external surface area and volume are scaled allometrically.

Table 3 illustrates allometric scaling by modeling the effect of body size on the time to reach steady state for organisms of size 1 through 100,000. This size range is easily found in a lake or river, where some larval fish and adult invertebrates are in the range of 0.01 g and large fish are 10,000 g or larger. The effect should be readily evident when large size differences occur. For example, organisms that are 1000 times larger should take 5.6–10 times longer to reach steady state. Although using scaling factors of 0.67 or 0.75 affects absolute values, the effect of body size on relative time to steady state is consistent and substantial in both cases.

The body size effect is important in understanding a toxicity test and when making comparisons between the results of toxicity tests. For example, when exposed to a constant lethal water concentration of a chemical a small fish usually dies much sooner than a large fish. A simple explanation is that both sizes of fish die when a similar concentration of chemical is reached in the body, but because of its smaller size and the resultant size-related differences in kinetics of accumulation, the small fish reaches the lethal body residue much sooner. If observations are extended until the time-independent incipient lethal concentration is reached for both sizes of fish, the residues and any associated responses should be similar in both cases. However, the time taken to reach the respective incipient lethal concentration is a function of the body size differences (see Table 3). Thus, it is not always necessary to invoke a difference in toxicant sensitivity to explain the difference beween the toxicity bioassay results for large and small organisms. Obviously, body size effects on accumulation kinetics are not always a valid explanation for such a situation, but fundamental similarities between living organisms are often ignored and complex explanations offered where simple ones may suffice.

If this simple model often approximately describes the influence of size on kinetics, there are important implications when comparing toxicity test data for organisms with substantially different body sizes. As seen earlier in the section on toxicity curves, toxicity results can vary through time until a time-independent threshold is reached. For example, within an individual toxicity test, LC50 estimates at 24 h are often higher than at 96 h or longer. Because body size can influence the time taken to reach an incipient toxicity level, the fact that one size of organism has reached an incipient level in 24 h is no guarantee that another size of organism has reached threshold in 24 or 96 h or any other fixed exposure time. Meaningful toxicological comparisons start with the use of time-independent threshold or incipient toxicity data for all organisms and chemicals being compared.

Examples of the effects of body size on bioaccumulation kinetics and/or toxicity in aquatic organisms have been provided by Matida (1960), Murphy (1971), Anderson and Weber (1975), Bradley and Sprague (1985), and Tarr et al. (1990).

Quantitative Structure-Activity Relationships

Quantitative structure-activity relationships (QSARs; see Chapter 20) are based on a well-established theory with the following assumptions:

1. Biological activity is a function of the structure of the organic chemicals in question.
2. Structure implies both global (for the whole molecule) and local (at certain sites on the molecule) properties, such as lipophilicity, charge, and shape, which can be quantified by molecular descriptors.
3. A statistical model that relates changes in biological response to changes in global and local properties of the toxicants does exist.

There are several approaches to QSAR analysis, but the additivity or de novo model, proposed by Free and Wilson (1964), and the linear free energy–related (LFER) model, formalized by Hansch and Fujita (1964) from pioneering work by Ferguson

(1939) and others, are the most commonly used. Although the two are closely related and theoretically equivalent, the latter is the most widely used method. The general form of the LFER approach is

$$BA = a*(\text{hydrophobic}) + b*(\text{electronic}) + c*(\text{steric}) + \text{constant}$$

Here BA is the biological activity end point, usually log $1/C$, where C is a measure of biological activity in terms of the applied external molar concentration of the chemical producing a standard response such as LC50 or LD50. The hydrophobic, electronic, and steric factors are represented by molecular descriptors that describe the characteristics of the toxicant molecules and are each associated with a constant. In this approach the biological activity–molecular descriptor relationship is usually assumed to be a log-linear function (see Chapter 20).

A variety of molecular descriptors, such as molecular weight, lipophilicity (usually octanol-water partition coefficient, K_{ow} or P), Hammett constants, molecular connectivity indexes, pKa (i.e., acid dissassociation constant of a chemical), molar volume, molar refractivity, water solubility, and solvatochromic parameters, can be employed. In practice, the approach is to regress bioassay test results for a given test organism to a group of organic chemicals against one or more molecular descriptors. This indicates the proportion of the variance attributable to the molecular descriptors(s) and thus quantifies the ability of the model to explain the biological activity. In some cases, especially involving chemicals causing narcosis, the electronic and steric terms are not always required to obtain reasonable QSARs.

An assortment of end points from various biological tests and organisms have been successfully employed in QSAR analysis (see review by Hermens, 1989; Chapter 20). Despite the initial promise of QSARs as predictors of toxicity, they have had more success as interpretive tools.

Figure 10 (and Chapter 20) presents examples of both bioconcentration and acute toxicity QSARs for some poorly metabolized, nonpolar, narcotic organic chemicals and fish using hydrophobicity (log K_{ow} or log P) as the molecular descriptor. Above a log K_{ow} of about 1.5 the bioconcentration relationship presented is the linear one proposed by Mackay (1982). Below a log K_{ow} of about 1.5 the relationship becomes nonlinear because chemicals in this log K_{ow} range are hydrophilic rather than hydrophobic. Although the amount in the lipid phase of the organism may continue to follow the relationship described in the previous equation, the whole-body concentration is now dominated by amounts in the water phase. The acute toxicity QSAR is linear throughout with a negative slope.

Most of the effort in QSAR research has been focused on the chemical parameters; however, the overall importance of the biological data should not be underestimated. When generating or comparing QSARs it is vital that biological effects be comparable both within and between the data sets. In many cases variability and uncertainty may be traced to the use of raw, uninterpreted toxicity test data. Even when steady state has been confirmed, which is not commonly carried out in current QSAR work, modifying factors can obscure relationships that might otherwise be found or suggest false relationships (McCarty et al., 1993).

Although most QSAR work has been carried out with organic chemicals, the approach is not limited to organics. However, the difficulty with metals and inorganics is the determination of appropriate molecular descriptors for both partitioning behavior and induction of toxic effects. Hoeschele et al. (1991) review the current status of this area of research.

Critical Body Residues

With recent technological advances in analytical chemistry it has become feasible to examine whole-body or tissue concentrations of many chemicals routinely. This makes it possible to move from surrogates based on exposure media (water, sediment, soil, air, and food) to those based on amounts in the body of the exposed organisms. In mammalian toxicology this has been termed *dosimetry*, and in ecotoxicology the general approach has been termed the *critical body residue (CBR)* approach (McCarty, 1990; McCarty et al. 1992a; Mackay et al., 1992; McCarty and Mackay, 1993). The use of lethal body burdens (LBBs) in toxicity investigations is a specific example of this approach.

Although neither new nor radical, the CBR approach does represent a more complete exploitation of existing information using the fundamental toxicological principles outlined earlier. It has several advantages over the traditional exposure-based concentration-response approach:

- Bioavailability is explicitly considered.
- Accumulation kinetics are considered, reducing the confounding effect of organism exposure duration when interpreting results.
- Uptake from food (as distinct from water) is explicitly considered.
- Toxic potencies are expressed in a less ambiguous manner, facilitating identification and investigation of different modes of toxic action.
- Effects of metabolism on accumulation are considered.
- Mixture toxicity may be more readily assessed.
- Experimental verification can be readily sought in the laboratory and the field.

An illustration of the CBR approach for neutral, poorly metabolized organic chemicals producing mortality by narcosis is presented in Figure 10.

Both bioconcentration (log bioconcentration factor, BCF) and toxicity (in this case log LC50) data are plotted against the octanol-water partition coefficient (log K_{ow}) for the chemicals in question. Although the relationships shown in Figure 10 are idealized (slopes have been forced to unity), they

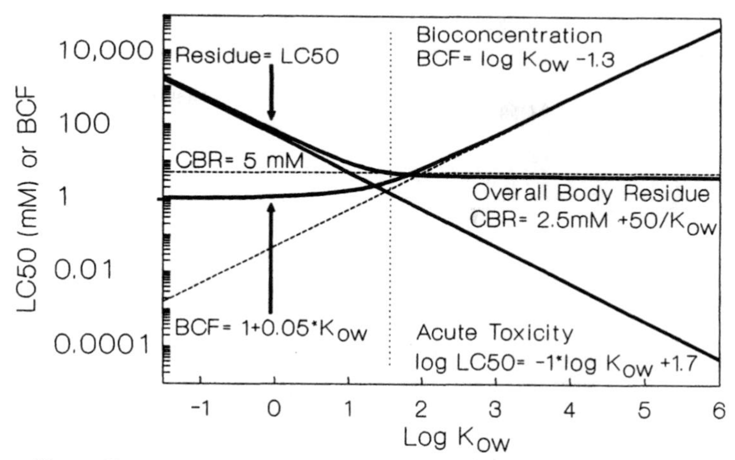

Figure 10. Bioconcentration, acute and chronic toxicity, and estimated critical body residue relationships as related to octanol-water partition coefficient for narcotic organic chemicals and small freshwater fish. (From McCarty et al. 1992a, reprinted with permission. Copyright 1992 SETAC.)

closely reflect actual relationships seen in toxicity and estimated residue data (McCarty et al., 1992a). This example reinforces the critical importance of interpretation of toxicity test results.

For organic chemicals with a log K_{ow} above 1.5, combination of the appropriate bioconcentration and toxicity equations indicates that, to a first approximation, narcotic toxicity results from a near-constant body residue.

$$\begin{aligned}\log \text{CBR (mM)} &= \log \text{BCF} + \log \text{LC50} \\ &= (\log K_{ow} - 1.3) \\ &\quad + (-\log K_{ow} + 1.7) \\ &= 0.4\end{aligned}$$

The millionfold difference in exposure-based bioassay results for narcotic organics is due largely to differences in partitioning behavior (i.e., toxicokinetics). Although the LC50 estimates are different, all of these chemicals exhibit similar potency when organism concentrations are considered (i.e., the LC50 occurs at about 2.5 mmol/kg of fish, the antilog of 0.4), assuming that the whole-body residues are an adequate surrogate for chemical concentrations at the site(s) of toxic action. This estimate is close to the CBR value of approximately 4.4 mmol/kg obtained with an actual fish data set (McCarty, 1990; McCarty et al., 1992a). Narcotic CBRs can also be related to molar volumes rather than molar concentrations. The CBR estimate above corresponds to volume fractions of about 0.0003–0.0012 (i.e., the chemical reaches a level in the organism of 0.03–0.12% by volume). See Abernethy et al. (1988) and Mackay et al. (1992) for further discussion of this approach.

When the very water-soluble chemicals (log K_{ow} < 1.5) are included in the process the toxic residue in the hydrophobic phase remains the same but the whole-body residue becomes dominated by the concentration of chemical in the water phase of the organism (i.e., it should be approximately equal to LC50). CBR estimates over the log K_{ow} range of about −2 to 6 can be described as follows using linear versions of the appropriate toxicity and bioconcentration QSARs:

$$\begin{aligned}\text{CBR (mM)} &= \text{BCF} * \text{TOX} \\ &= (1 + 0.05K_{ow}) * (50/K_{ow}) \\ &= 2.5\text{mM} + 50/K_{ow}\end{aligned}$$

This is a well-recognized relationship elucidated in the classic work of Ferguson (1939). He hypothesized that, for narcotics, it is the activity of the number of molecules of a toxicant in an organism, not their actual number, which is most closely related to the biological response. The relationship between concentration and activity is similar to that encountered with hydrogen ion concentration and pH, where a fixed number of molecules will yield different measures of activity (i.e., pH) with changes in factors such as temperature and ionic strength. Thus, the whole-body concentration associated with the biological response, incorporating adjustments for known modifying factors, can be taken as a reasonable approximation of the total internal toxicant dose, which itself is an approximation for the activity of the effective dose at the site(s) of toxic action.

Investigations reporting on the significance, utility, and methods of estimating toxicant residue levels associated with biological response end points have been published several times in the past (e.g., Ferguson, 1939; McGowan, 1952, 1963; Nightingale, 1971; McGowan and Mellors, 1986) and Kooijman (1983) has provided statistical methods for incorporating consideration of body residues into standard toxicity test protocols, but the concept has been largely ignored (Friant and Henry, 1985). This may be related to the confusion arising from the conflict between the theoretical toxicology approach, which focuses on toxicant concentration in the organism, and the applied toxicology approach, which treats the exposure concentration or dose as the critical factor.

Other similar residue-effect relationships for acute toxicity, chronic toxicity, and other effect end points obtained with various organisms can be found in the literature. Furthermore, although the CBR associated with lethal narcosis may be somewhat different for different species, much of this may be due to disparity in body character and composition rather than differences in target site concentrations. Toxicity differences are probably not as dramatic as suggested by the water concentration data derived from exposure-based bioassays. This is borne out by various studies, such as those of Hodson (1985), who reported that good correlations existed between rat or mouse oral LD50s and fish intraperitoneal LD50s for some organic chemicals, and Kaiser and Palabrica (1991), who reported good correlations between toxicity of organics to fish or invertebrates and *Photobacterium phosphoreum* toxicity.

Although the previous discussion is focused on organic chemicals, CBR principles generally apply to metals and inorganics, but the relationship between the whole-body residue and the site of target action may be less well defined than for organics. Residue-effect relationships for metals have been reviewed by Foulkes (1990) and McCarty and Mackay (1993).

Modes of Toxic Action

When applying the CBR approach, it is vitally important to recognize the existence of various modes of toxic action. In addition to the general, nonspecific mode of toxic action known as narcosis that was discussed earlier, there are other more specific modes of action. *Mode of action* can be defined as a common set of physiological and behavioral signs that characterize a type of adverse biological response. Pioneering work on modes of action in aquatic toxicology was carried out by investigators at the U.S. EPA-Duluth laboratory. They employed a combination of behavioral, physiological, and biochemical responses, which they termed *fish acute toxicity syndromes (FATS)*, to investigate and classify common modes of toxic action (Bradbury et al., 1989, 1991; McKim et al., 1987a,b). As noted by Drummond and Russom (1990), more than one specific biochemical mechanism may be associated with response of whole organisms in a common mode of toxic action. FATS have been used to identify seven major modes of toxic action: nonpolar narcosis, polar narcosis, uncouplers of oxidative phosphorylation, acetylcholinesterase (AChE) inhibitors, irritants, a general group of central nervous system seizure agents, and respiratory blockers. Dioxin (2,3,7,8-tetrachlorodibenzo-*p*-dioxin, TCDD) has not been investigated with the FATS method but it is well recognized as having a different mode of toxic action (Eisler, 1986b).

Modes of toxic action can be subdivided into two general categories: *narcosis* and *specific-action*. Narcosis is a generalized depression in biological activity due to the presence of toxicant molecules in the organism. Although narcosis is exploited daily by humans in a variety of ways, ranging from alcohol intoxication to general anaesthesia for surgery, the site(s) and mechanism of action remain unclear. The current theories are the ''critical-volume'' hypothesis and the ''protein-binding'' hypothesis (Abernethy et al., 1988). The critical-volume theory suggests that changes in the lipid component of cell membranes caused by an increase in volume (i.e., not concentration), due to toxicant dissolved in this phase, substantially modifies the characteristics of the membrane, bringing on narcosis. The protein-binding theory hypothesizes that narcosis results from the toxicant binding to receptor sites of specific dimensions that are located in hydrophobic regions of proteins found in cell membranes. Although this action is thought to occur in nerve cells in higher animals, narcosis is a phenomenon found in all organisms, including bacteria and plants. Therefore, the effect is likely generalized to all cell membranes. Any organic chemical should cause narcosis, if it is able to reach the site(s) of action, as narcosis appears to be largely a result of the presence of foreign molecules rather than their absolute character. Narcosis is reversible by stopping exposure before death and allowing the organism to recover by eliminating the chemical.

Specifically-acting toxicants are not a single category but rather a group composed of a variety of non-narcotic modes of action. Although it appears that virtually all organic chemicals are ultimately narcotics (hence the term ''baseline toxicity'' refers to narcosis; see Chapter 20), specific action occurs at lower concentrations; in essence, fewer molecules are required to cause the specific toxic response than is the case for narcosis. If the chemical has specific activity, it is the mode of action that is expressed first. A chemical can be a specific toxicant in one test where the organism has a susceptible system, yet be a narcotic in another where the test organism does not. Specifically acting chemicals work by binding to a site on a specific molecule and modifying or inhibiting some biological process. Binding to the site and the subsequent effect may be readily reversible depending on the chemical and the situation. Although polar narcosis appears to be a separate mode of action based on FATS analysis, there is some question of whether this is a real mode-of-action group or a mixture of narcotics and specific toxicants whose toxicity evaluation is complicated by their polar properties (McCarty et al., 1993).

Available estimated and measured aquatic acute and chronic residue data for small aquatic organisms exposed to toxic agents in the seven major FATS-identified groups plus TCDD are presented in Table 4 and summarized graphically in Figure 11. As noted in Table 4, the nonpolar narcotics, which represent ''baseline toxicity,'' exhibit the highest residues

Table 4. Summary of modes of toxic action and associated critical body residue estimates in fish[a]

Chemical and effect	Estimated residue (mmol/kg)	Source
1. Narcosis		
Acute (various chemicals)	2—8	McCarty et al. (1992a)
Chronic (various chemicals)	0.2—0.8	McCarty et al. (1992a)
Acute (octanol, MS222)	1.68 or 6.32[b]	McKim and Schmieder (1991)
2. Polar Narcosis		
Acute (various chemicals)	0.6—1.9	McCarty et al. (1993)
Acute (2,3,4,5-tetrachloroaniline)	0.7—1.8	de Wolf et al. (1991)
Chronic (various chemicals)	0.2—0.7 (Chronic/acute = 0.1–0.3)	Kenaga (1982); McCarty (1987); Hodson et al. (1991)
Chronic (2,4,5-trichlorophenol)	0.2	Arthur (1991)
Acute (aniline, phenol, 2-chloroaniline, 2,4-dimethylphenol)	0.68 or 1.76	McKim and Schmieder (1991)
3. Respiratory Uncoupler		
Acute (pentachlorophenol)	0.3	Hickie (1990); McCarty, (1990); van den Heuvel et al. (1991)
Acute (2,4-dinitrophenol)	0.0015 or 0.2	McCarty (1990)
Chronic (pentachlorophenol, 2,4-dinitrophenol)	0.09—0.00015 (chronic/acute = 0.1–0.3)	Kenaga (1982); McCarty (1987); Hodson et al. (1991)
Chronic (pentachlorophenol)	0.094	Spehar et al. (1985)
Chronic (pentachlorophenol)	0.08	Arthur (1991)
Acute (pentachlorophenol, 2,4-dinitrophenol)	0.11 or 0.20	McKim and Schmieder (1991)
4. AChE Inhibitor		
Acute (malathion and carbaryl, chlorpyrifos)	0.5 & 2.7	McCarty (1990)
Acute (chlorpyrifos)	2.2	Hansen et al. (1986)
Acute (aminocarb)	0.05 & 2	Doe et al. (1988); Richardson and Quadri (1986)
Acute (parathion in blood)	0.13—0.2	Mount and Boyle (1969)
Chronic (chlorpyrifos)	0.003	Hansen et al. (1986)
Acute (malathion, carbaryl)	0.16 or 0.38	McKim and Schmieder (1991)
5. Membrane Irritant		
Acute (benzaldehyde)	0.16	(Estimated with BCF = 1.6, acute fathead LC50 = 0.1 mmol/L)
Acute (benzaldehyde)	2.1 or 13.2	McKim and Schmieder (1991)
Acute (acrolein)	0.0014 or 0.94	McKim and Schmieder (1991)

Table 4. Summary of modes of toxic action and associated critical body residue estimates in fish[a] (*Continued*)

Chemical and effect	Estimated residue (mmol/kg)	Source
6. CNS Convulsant[c]		
Acute (fenvalerate, permethrin, cypermethrin)	0.002—0.017	Haya (1989)
Acute (fenvalerate, permethrin, cypermethrin)	0.000048—0.0013	McLeese et al. (1980)
Acute (endrin in blood)	0.0007	Mount et al. (1966)
Acute (endrin)	0.0018—0.0026	Argyle et al. (1973)
Acute (endrin)	0.005	Anderson and Defoe (1980)
Chronic (fenvalerate & permethrin)	0.0005 & 0.015	Hansen et al. (1983)
7. Respiratory Blockers		
Acute (rotenone)	0.0006—0.003	Rach and Gingerich (1986)
Acute (rotenone)	0.008	Mackay et al. (1992)
Acute (rotenone)	0.0009 or 0.0028	McKim and Schmieder (1991)
8. Dioxin (TCDD)-like		
Lethal (TCDD)	0.000003—0.00004	Yochim et al. (1978); Adams et al. (1986) Lothenbach et al. (1988); Mehrle et al. (1988); Cook et al. (1991)
Growth/Survival (TCDD)	0.0000003—0.000008	Branson et al. (1985); Mehrle et al. (1988)
Early life stages, lethal (TCDD)	0.00000015—0.0000014	Wisk and Cooper (1990); Cook et al. (1991); Walker and Peterson (1991)
Early life stages, NOAEL[d] (TCDD)	0.0000001—0.0000002	Cook et al. (1991)

[a]the rainbow trout employed by McKim and Schmieder (1991) weighed 600–1000 g while the other data presented are mostly for small fish, sometimes early life stages, that were typically less than 1 g in weight. Most estimates have been converted from mass-based data.
[b]the two values represent residues estimated by two different methods.
[c]includes three subgroups characterized by strychnine; fenvalerate and cypermethrin; endosulfan and endrin (Bradbury et al., 1991).
[d]NOAEL; no observed adverse effect level.

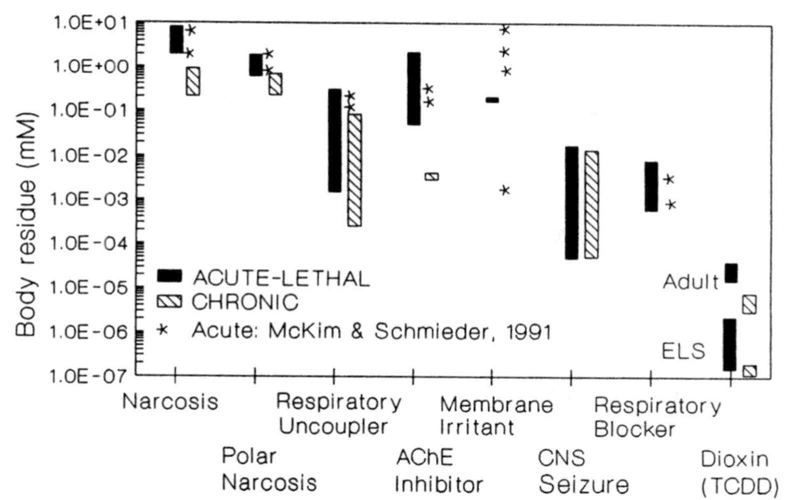

Figure 11. Estimated body residue levels associated with acute and chronic toxicity endpoints for fish exposed to eight categories of organic chemicals. (From McCarty and Mackay, 1993, used with permission from the American Chemical Society. Copyright 1993 American Chemical Society.) Data summarized from Table 4.

measured at death. Tissue concentrations of all other mode-of-action groups were generally lower, indicating a more toxic and more specific mode of action. Despite the noisiness in some places, the dearth of data for some modes of action, and the diversity of species of organisms employed, different modes of toxic action producing lethal effects appear to be associated with differing ranges of body residues. Although fewer data exist for chronic toxicity, a similar type of residue-toxicity relationship is apparent. Overall, this information is supportive of whole-body residues of many organic chemicals being reasonable first approximations of the amount of chemical reaching the site(s) of toxic action which is responsible for the adverse response observed. As noted earlier, using whole-body residues as surrogate for target tissue residues in the organism has a number of shortcomings, many which are shared with the external exposure approach. As with exposure-based interpretation, a number of these potential difficulties can be minimized by proper experimental design.

In addition to improving the ability to interpret toxicity test results for single chemicals, body residues and modes of toxic action may provide a way to better interpret and explain mixture toxicity (McCarty et al., 1992b). Although the mixture interactions outlined earlier may be present, approximate additivity is common in real-world situations with many organic chemicals. In a variety of mixture studies with equitoxic contributions from 3 to 50 organic chemicals, Hermens and co-workers reported that mixture toxicity was generally additive (Hermens et al., 1984, 1985; Deneer et al., 1988). They found that closer approximations of concentration addition were achieved with larger numbers of chemicals in a mixture. This observation held when either acute or chronic end points were examined and was unaffected by the presence of chemicals with various specific modes of toxic action in the mixture. McCarty and Mackay (1993) have suggested a likely explanation for this phenomenon. When nonnarcotic organic chemicals are present below their threshold for specific toxic action (i.e., below about 0.3 to 0.02 of their threshold LC50), they do not express a specific toxic action; rather they contribute to narcotic activity. Thus, simple addition of the narcotic toxicity of the components of the mixture, rather than any interaction between specific modes of toxic action, produces the biological response. Although this is but one example, it is obvious that toxicological comprehension can be improved substantially by knowledge of modes of toxic action and associated internal dose concentrations.

Relationships between Acute and Chronic Toxicity End Points

For much of the history of aquatic toxicology data the bulk of the data has been generated by acute testing, based on short exposures and using mortality as the response end point. It was recognized that, for evaluation of ecological significance, information about longer exposure durations and nonlethal response end points was needed. This need was particularly pressing for the development of water quality and aqueous effluent regulations (Hart et al., 1945). As previously discussed, safety factors or application factors were often used to generate water quality criteria regulations from laboratory-based acute toxicity data (see previous section on chronic toxicity tests). Kenaga (1979, 1982) used the inverse of the application factor, the acute-to-chronic toxicity ratio (ACR), and found that the ACR for a variety of organisms and chemicals was typically of the order of about 10, averaging 12 for organics (Kenaga, 1982). This relationship has also been seen in QSARs for some groups of organic chemicals for which acute and chronic regressions are about an order of magnitude apart (Call et al., 1985; McCarty, 1986; Abernethy et al., 1988). For example, if the appropriate chronic toxicity data were included in Figure 10 they would form a toxicity QSAR that is roughly parallel to the acute one but about an order of magnitude lower.

The explanation for this phenomenon can be found in work by Mayer et al. (1986, 1992) in which a probit-based method was developed which employed acute toxicity test data to predict chronic toxicity. It was found that chronic end points used in early life stage tests with fish were similar to 0.01% acute lethality estimates (i.e., LC0.01) obtained from probit-based extrapolation of the acute toxicity concentration-response curve for the same species and chemical. This held for chronic tests in which growth or survival was used as the response end point but was less reliable when reproduction was used as the response end point. Thus, many chronic toxicity test results appear to be estimates of a different magnitude of response (0.01% rather than 50%) within the same mode of action as for acute toxicity, rather than resulting from a different mode of toxic action. The 50% response level (i.e., LC50) was chosen for optimal statistical efficiency, but lower percentage responses can be employed, with a corresponding loss of statistical efficiency.

Once the nature of the toxicity tolerance curve is established, it is possible not only to interpolate to LC0.01, LC1, LC10, or other response levels of interest but also to estimate the associated body residues. As can be seen in Figure 12, body residues (CBRs) associated with acute effects at the conventional 50% response level (i.e., LC50) can be estimated by calculation of the product of toxicity and bioconcentration factor. To estimate residues for other response levels, simply substitute the appropriate toxicity estimate. In the example in Figure 12 the CBR associated with the LC0.01% response is calculated as the LC0.01 estimate, 1 μmol/L, multiplied by the bioconcentration factor, 1000. The CBR for 0.01% mortality is thus calculated to be 1 mmol/L of fish or 1 mmol/kg of fish if the fish has a density of 1.0.

In an alternative approach Mackay et al. (1992) prepared a model that incorporated a constant lethal body residue for the mode of toxic action, the log K_{ow} for the chemical, a Weibull distribution factor, and, where necessary, a factor to account for metabolic breakdown of the toxicant. The response surface generated (i.e., a three-dimensional plot of exposure times, exposure concentrations, and biological responses) follows the time course of a toxicity bioassay, from initiation to the point where the time-independent incipient or threshold point is reached. The results were in reasonable agreement with both the 50% response level and responses on either side of this midpoint (10%, 30%, 70%, 90%, etc.) when compared with experimental toxicity data. It is apparent that considerable toxicological insight can be gained from a thorough analysis of a properly executed toxicity test exploiting basic toxicological principles.

BIOMONITORING

Monitoring is the systematic measurement of variables and processes over time related to a specific problem (Spellerberg, 1991). Monitoring of the en-

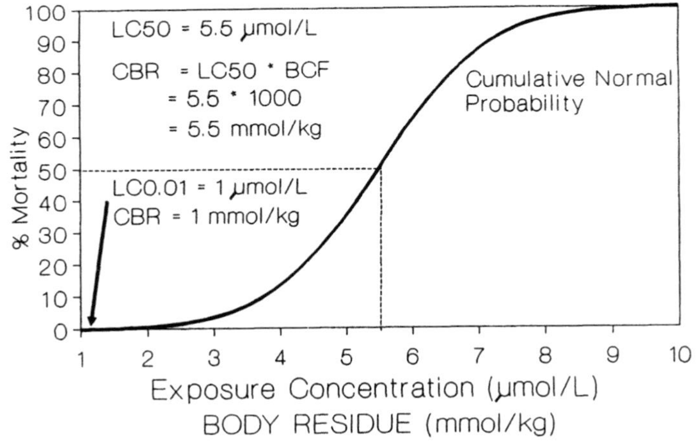

Figure 12. Relationship between acute and chronic toxicity on the basis of both exposure concentrations and critical body residues. (From McCarty and Mackay, 1993, used with permission from the American Chemical Society. Copyright 1993 American Chemical Society.)

vironment includes observations and measurements of biological, chemical, and physical parameters according to prearranged schedules and comparable methodologies. The objective of monitoring is to collect sufficient data to assess the quality of the environment (i.e., air, water, soil). Central to aquatic toxicology is *biological monitoring* (or *biomonitoring*) or regular, systematic use of living organisms to evaluate changes in environmental or water quality (Cairns and van der Schalie, 1982). Repetitive measurements are involved and these are taken within the framework of a statistical design (McBride, 1985; Ward et al., 1986). The definition thus includes analysis of the state of the environment by monitoring individuals, species, populations, and communities to understand changes that may occur as a result of chemical exposure over short or long periods of time. Biomonitoring is important because biological responses (effects) may be elicited at chemical concentrations below analytical detection limits or after chemical exposure has ceased. It is still very important to conduct chemical monitoring in conjunction with biomonitoring to determine what the organisms are being exposed to and their concentrations.

The concept of a *biological indicator* (or bioindicator) or *indicator species* is fundamental to biomonitoring. A species cannot survive in an environment that does not provide its physical, chemical, and nutritional requirements. Once the requirements are defined, the presence of a particular species in a habitat indicates that its requirements are being met. However, absence of this species does not mean the converse—because one species may be competitively excluded by another from a habitat or because of species turnover, some short-lived organisms may be absent. The presence, absence, or relative abundance of a species (or species assemblage), either suddenly or gradually, may be used as an *indicator* of changes in environmental quality or conditions.

Biotic indices are an approach for defining water pollution making use of the indicator organism concept. They use values for sensitive and tolerant species to calculate an index to classify the degree of pollution. A biotic index is specific for one or two types of pollution because indicator organisms are not equally sensitive to all pollutants. Biotic indices are pollution (usually organic) and geography specific. Biotic indices should never be compared to diversity indices as indicators of water quality because one measures community structure, which may change with pollution, whereas the other is a specific measure of pollution (generally organic matter causing decreases in oxygen concentrations) based on the reaction of physiologically sensitive organisms.

Indicator organisms can include any member of the flora or fauna of an aquatic habitat. The following groups (i.e., whole organisms) have been used as indicators to assess water quality changes: fish, birds, macroinvertebrates, protozoa, macrophytes, algae, yeasts, fungi, bacteria, and viruses. All members of a community can be considered potential indicators of water quality; however, the group most commonly used as indicators is macroinvertebrates (Rosenberg and Resh, 1993). The advantages and disadvantages of the different groups as indicators are discussed in Abel (1989), Boyle (1987), Cairns (1974, 1979), Haslam (1982), Hellawell (1977, 1986), McIntyre and Pearce (1980), NRC (1991), Shubert (1984), and Whitton (1979).

The indicator concept may be expanded to go beyond simply noting the presence or absence of a species. It can include organisms that continue to survive in a contaminated environment and act as "biosensors" because they elicit one or more physiological, biochemical, or histological changes which may be manifested in diminished rate of growth and development, impaired reproductive capacity, or modified behavior. The term *biomarker* has been adopted to refer to the use of physiological, biochemical, and histological changes as "indicators" of exposure and/or effects of xenobiotics at the suborganismal or organismal level (Huggett et al., 1992).

The actual indicator or biomarker of these changes may be biological material ranging from biomolecules such as nucleic acids and proteins (e.g., enzymes) to organelles, cells, tissues, organs, and whole organisms. Chapters 12, 13, 14, and 17 address changes in the immune system, genetic makeup, and tissues and biotransformation (biochemical indicators) that are all viable biomarkers of responses to stressors including chemicals. Although their utility has been extensively investigated in the laboratory, their application in actual field assessment has not yet been adequately validated. Biomarkers are discussed in Huggett et al. (1992), Peakall (1992), McCarthy and Shugart (1990), and Peakall and Shugart (1993).

From the foregoing discussion on indicator species, it is evident that they have different uses (Spellerberg, 1991), which include:

- Sensitive organisms as an early warning system at a point source discharge site (i.e., effluent).
- Resident indigenous species as "detectors" in an area of potential interest, which may show responses to environmental changes.
- Species whose presence indicates probable chemical contamination or pollution. Very few organisms are known whose presence specifically indicates that the water is contaminated with a chemical.
- Organisms that accumulate measurable quantities of chemicals in their tissues.
- Organisms used to rank chemicals to determine their relative toxicity.

The use of indicator species in biomonitoring has led to its main objectives, namely:

- To act as means of *surveillance*. This may include surveys before and after a chemical spill or effluent release. Surveillance can also be used to determine the effectiveness of an effluent or water treatment process (see Chapter 33).

- To ensure *compliance* either to meet state or federal requirements or to control long-term water quality. For example, organisms may be used to assess the toxicity of effluents and to verify receiving-water quality standards (see Chapter 33).

Most of the objectives can be met with on-site (in situ or in-plant) biomonitoring programs (Cairns, 1982). However, one of the most widely used procedures in biomonitoring of water quality is toxicity testing.

TOXICITY DATA AND ENVIRONMENTAL REGULATIONS

Aquatic toxicity test data have a variety of applications:

- Corporate industrial decisions on product development, manufacture, and commercialization.
- Registration of products to satisfy regulatory requirements.
- Permitting for the discharge of municipal and industrial wastes.
- Environmental (or ecological) hazard-risk assessments.
- Prosecution and defense of chemical-related activities in environmental litigation.
- Derivation of numerical water and sediment quality criteria for protection of aquatic organisms.

Such applications have been internationally recognized (e.g., Environment Canada, 1984; Wells and Côté, 1988).

In developing a chemical as an intermediate or a final product it is necessary to consider the potential impact of that chemical on human health and on the terrestrial and aquatic environments. Corporate industrial decisions to proceed with the development of a chemical may include consideration of the potential effects on the aquatic environment resulting from its manufacture, transport, use, and disposal. In the United States, government authorization for the manufacture of new chemicals is regulated under the Toxic Substances Control Act (TSCA), which requires that a premanufacturing notification (PMN) be submitted for U.S. EPA review before full-scale manufacture of the chemical. For chemicals expected to be used, transported, or otherwise released into the aquatic environment, the PMN should include information on their potential toxicity to aquatic organisms. Many industrial chemicals do not survive the initial stages of development because they are toxic to aquatic organisms or are highly persistent and readily bioaccumulated in the aquatic environment.

For biocides such as insecticides, rodenticides, fungicides, and herbicides, the U.S. EPA requires that specific aquatic toxicity tests be conducted to evaluate the potential hazard to nontarget aquatic species and that the resulting data be submitted in support of registration permits for sale and use of these chemicals. In addition, biocide manufacturers frequently conduct aquatic toxicity tests to evaluate the efficacy of chemicals proposed to control pests (piscicides, molluscicides, algicides) in the aquatic environment. For pharmaceuticals, the Food and Drug Administration also requires that aquatic toxicity data be submitted for some New Drug Applications (NDAs).

Aquatic toxicity test data are also used to evaluate the potential hazards and risks resulting from the discharge of municipal and industrial wastes (effluents) into water bodies. Industrial discharges are regulated by the U.S. EPA under the National Pollutant Discharge Elimination System (NPDES) permit program, as authorized by the Clean Water Act (CWA). Permits, in fact, may require biomonitoring of industrial discharges. The biomonitoring required in the permits is a water quality surveillance-compliance program in which aquatic toxicity tests are conducted periodically (e.g., once every month or every 3–4 mo) to ascertain whether aquatic life may be endangered by discharge of the wastes. Similar programs are conducted in other countries (e.g., Canada, Germany, United Kingdom, etc.).

Aquatic toxicity tests with indigenous species and site-specific conditions can be used to evaluate the potential environmental hazards and risks of an unanticipated, accidental release of a chemical into the aquatic environment. This evaluation includes the potential human health hazard posed by contamination of commercially important aquatic species with a chemical. For example, bioconcentration studies indicate the extent to which a chemical present at some measurable concentration in the aquatic environment can be accumulated in the tissues of aquatic organisms, to be consumed by humans.

With the trend in the past two decades toward use of litigation to resolve concerns about environmental pollutants, particularly point-source industrial effluents, aquatic toxicity tests have been used to evaluate the effects of long-term chemical releases under controlled conditions and accidental spills, which permit the evaluation of causal relationships. That is, toxicity tests can be used to confirm or refute a relationship between observed effects and the concentrations of a chemical shown to exist or to have existed in the aquatic environment.

Numerical water and sediment quality criteria, guidelines, and objectives for chemicals are produced for fresh water and salt water to protect populations and communities of aquatic organisms and other water uses from unacceptable chemical effects. Such effects may be due to exposures to high concentrations for short periods of time, lower concentrations for longer periods of time, or combinations of the two. Toxicity data for acute and chronic effects to plants and animals and estimates of bioconcentration and bioaccumulation rates are used as criteria to develop such water quality guidelines (Stephan et al., 1985). Similar work is proceeding rapidly for sediment quality guidelines and objectives for persistent toxic substances and compound classes (e.g., Di Toro et al., 1991; see Chapter 8).

Some of these applications and uses of aquatic toxicity data are discussed in Chapters 21 to 25 and

27 on environmental legislation and in Chapters 28 through to 33 on risk assessment. It is clear that aquatic toxicologists are actively contributing to more realistic environmental guidelines and improved legislation over time. Bioassay data are a requirement of a growing number of national and international conventions and regulations (e.g., MARPOL 73/78; the London Dumping Convention, 1972; the revised Great Lakes Water Quality Agreement of 1987; Canada's Environmental Protection Act (CEPA), of 1988; directives and guidelines on chemicals of the OECD). Practicing and future aquatic toxicologists have a responsibility to ensure the continued development and standardization of appropriate methods and the use of bioassay and other aquatic toxicology data for the protection of aquatic environments and their resources.

SUMMARY

Aquatic toxicology is a relatively new and still evolving discipline, originating from concern for the safety, conservation, and protection of aquatic environments. Scientists from academic, industrial, consulting/private, and government institutions have made and are making significant contributions to this multidisciplinary science and its many applications in managing toxic substances and complex wastes.

This chapter highlights a number of basic principles of aquatic toxicology, as well as some of the advances in approach and understanding since the mid-1980s (Rand and Petrocelli, 1985). The chapter is not all-inclusive. More could have been written about the important role of nutrition in organism health and sensitivity to toxicants (Lanno et al., 1989), as well as abiotic factors influencing toxicity. The toxicity of mixtures, or multiple interactions of chemicals, is only briefly discussed; it has been given extensive attention elsewhere due to considerable concern about cumulative effects and long-term, low-level multiple exposures from chemicals (Calabrese, 1991; Howells et al., 1990). There is a wide range of plant and animal models for toxicology, and only a selection is covered comprehensively; some groups such as bacteria and protozoa are being studied extensively and are described elsewhere (Mayfield, 1993; Munawar et al., 1989; Personne and Janssen, 1993). Of necessity, the discussion is brief for pesticides, metals, and PAHs; these are extensively discussed in other recent literature. The reader is encouraged to pursue these and other topics, guided by the supplemental readings.

The chapters in this volume were written by scientists with expertise in their respective subdisciplines of aquatic toxicology. They represent a comprehensive overview of the general concepts and state of the art in techniques, methodologies, research, and assessment of aquatic toxicology. Above all, the collective contributions that make up this book not only illustrate the current state of the science but also convey the excitement and commitment of the practitioners to the dynamic field called aquatic toxicology.

LITERATURE CITED

Abel, P. D.: Water Pollution Biology. New York: Ellis Horwood Ltd., Halsted Press (Wiley), 1989.

Abernethy, S. G.; Mackay, D.; and McCarty, L. S.: "Volume fraction" correlation for narcosis in aquatic organisms: the key role of partitioning. *Environ. Toxicol. Chem.*, 7:469–481, 1988.

Adams, W. J.; DeGraeve, G. M.; Sabourin, T. D.; Cooney, J. D.; and Mosher, G. M.: Toxicity and bioconcentration of 2,3,7,8-TCDD to fathead minnows (*Pimephales promelas*). *Chemosphere*, 9–12:1503–1511, 1986.

Adelman, I. R., and Smith, L. L., Jr.: Fathead minnow (*Pimephales promelas*) and goldfish (*Carassius auratus*) as a standard fish in bioassays and their reaction to potential reference toxicants. *J. Fish. Res. Bd. Can.*, 33:209–214, 1976.

Alabaster, J. S., and Lloyd, R. (eds.): Water Quality Criteria for Freshwater Fish. London: Butterworths for the Food and Agriculture Organization, 1980.

Albert, A.: Fundamental aspects of selective toxicity. *Ann. N.Y. Acad. Sci.*, 123:5–18, 1965.

Albert, A.: Selective Toxicity. London: Chapman & Hall, 1973.

Amdur, M. O.; Doull, J.; and Klaassen, C. D.: Casarett and Doull's Toxicology the Basic Science of Poisons, 4th ed. New York: Pergamon, 1991.

American Society for Testing and Materials (ASTM) D1345-45T: Standard Test Methods for Evaluating Acute Toxicity of Water to Fresh-Water Fishes. Annual Book of Standards—Water. Philadelphia: American Society for Testing and Materials, 1954.

American Society for Testing and Materials (ASTM) D2037-64T: Standard Test Method for Evaluating Inhibitory Toxicity of Waters to Diatoms. Annual Book of Standards—Water. Philadelphia: American Society for Testing and Materials, 1964.

Anderson, B. C.: Toxicity thresholds of various substances found in industrial wastes as determined by the use of *Daphnia magna*. *Sewage Works J.*, 16:1156–1165, 1944.

Anderson, B. C.: Chapter One. Aquatic invertebrates in tolerance investigations from Aristotle to Naumann. Aquatic Invertebrate Bioassays, edited by A. L. Buikema, Jr., and J. Cairns, Jr., pp. 3–35. ASTM STP 715. Philadelphia: American Society for Testing and Materials, 1980.

Anderson, P. D., and Weber, L. J.: Toxic response as a quantitative function of body size. *Toxicol. Appl. Pharmacol.*, 33:471–483, 1975.

Anderson, R. L., and DeFoe, D. L.: Toxicity and bioaccumulation of endrin and methoxychlor in aquatic invertebrates and fish. *Environ. Pollut.*, 22A:111–121, 1980.

APHA, AWWA, WPCF: Standard Methods for the Examination of Water and Wastewater, 15th ed. Washington, DC: American Public Health Association, American Water Works Association, and Water Pollution Control Federation, 1981.

Argyle, R. L.; Williams, G. C.; and Dupree, H. K.: Endrin uptake and release by fingerling channel catfish (*Ictalurus punctatus*). *J. Fish. Res. Bd. Can.*, 30:1743–1744, 1973.

Arthur, A. D.: Verification studies of a body-residue-based model for predicting the sublethal toxicity of time-variable exposures to organic chemicals in small fish. M. Sc. thesis, Department of Biology, Univ. of Waterloo, Waterloo, Ontario, 1991.

Baker, R. A. (ed.): Organic Substances and Sediments in Water, Vol. 1, Humics and Soils. Boca Raton, FL: Lewis, 1991a.

Baker, R. A. (ed.): Organic Substances and Sediments in Water, Vol. 2, Process and Analysis. Boca Raton, FL: Lewis, 1991b.

Baker, R. A. (ed.): Organic Substances and Sediments in Water, Vol. 3, Biological. Boca Raton, FL: Lewis, 1991c.

Barnes, R. S. K., and Mann, K. H.: Fundamentals of Aquatic Ecology. Oxford: Blackwell Scientific, 1991.

Barron, M. G.; Stehly, G. R.; and Hayton, W. L.: Pharmacokinetic modeling in aquatic animals. I. models and concepts. *Aquat. Toxicol.*, 18:61–86, 1990.

Bartell, S. M.; Gardner, R. H.; and O'Neill, R. V.: Ecological Risk Estimation. Chelsea, MI: Lewis, 1992.

Bayne, B. L.; Clarke, K. L.; and Gray, J. S.: Biological effects of pollutants: results of a practical workshop. *Mar. Ecol. Progr. Ser.*, 46:1–278, 1988.

Bergman, H. L.; Kimerle, R. A.; and Maki, A. W. (eds.): Environmental Hazard Assessment of Effluents. New York: Pergamon, 1986.

Blaise, C.: Microbiotests in aquatic ecotoxicology: characteristics, utility, and prospects. *Environ. Toxicol. Water Qual.*, 6:145–155, 1991.

Blaise, C.; Sergy, G.; Wells, P. G.; Bermingham, N.; and Van Coillie, R.: Biological testing—development, application and trends in Canadian environmental protection laboratories. *Toxicity Assessment: An International Journal*: 3:385–406, 1988.

Bliss, C. I.: The method of probits. *Science*, 79:38–39, 1934a.

Bliss, C. I.: The method of probits—a correction. *Science*, 79:409–410, 1934b.

Bliss, C. I.: Some principles of bioassay. *Am. Sci.*, 45:449–466, 1957.

Boyle, T. P. (ed.): Validation and Predictability of Laboratory Methods for Assessing the Fate and Effects of Contaminants in Aquatic Ecosystems. ASTM STP 865. Philadelphia: American Society for Testing and Materials, 1985.

Boyle, T. P. (ed.): New Approaches to Monitoring Aquatic Ecosystems. ASTM STP 940. Philadelphia: American Society for Testing and Materials, 1987.

Bradbury, S. P.; Henry, T. R.; Niemi, G. J.; Carlson, R. W.; and Snarski, V. M.: Use of respiratory-cardiovascular responses of rainbow trout (*Salmo gairdneri*) in identifying acute toxicity syndromes in fish. Part 3. Polar narcotics. *Environ. Toxicol. Chem.*, 8:247–261, 1989.

Bradbury, S. P.; Carlson, R. W.; Niemi, G. J.; and Henry, T. R.: Use of respiratory cardiovascular responses of rainbow trout (*Oncorhynchus mykiss*) in identifying acute toxicity syndromes in fish. Part 4. Central nervous system seizure agents. *Environ. Toxicol. Chem.*, 10:115–131, 1991.

Bradley, R. W., and Sprague, J. B.: Accumulation of zinc by rainbow trout as influenced by pH, water hardness and fish size. *Environ. Toxicol. Chem.*, 4:685–694, 1985.

Branson, D. R.; Takahashi, I. T.; Parker, W. M.; and Blau, G. E.: Bioconcentration kinetics of 2,3,7,8-tetrachlorodibenzo-*p*-dioxin in rainbow trout. *Environ. Toxicol. Chem.*, 4:779–788, 1985.

Broderius, S. J.: Modeling the joint toxicity of xenobiotics to aquatic organisms: basic concepts and approaches. Aquatic Toxicology and Risk Assessment: Fourteenth Volume, edited by M. A. Mayes, M. G. Barron, pp. 107–127. ASTM STP 1124. Philadelphia: American Society for Testing and Materials, 1991.

Buikema, A. L., Jr., and Cairns, J., Jr. (eds.): Aquatic Invertebrate Bioassays. ASTM STP 715. Philadelphia: American Society for Testing and Materials, 1980.

Buikema, A. L., Jr.; Niederlehner, B. R.; and Cairns, J., Jr.: Biological monitoring. Part IV. Toxicity Testing. *Water Res.*, 16:239–262, 1982.

Burton, G. A. (ed.): Sediment Toxicity Assessment. Boca Raton, FL: Lewis, 1992.

Butler, G. C. (ed.): Principles of Ecotoxicology. SCOPE 12. New York: Wiley, 1978.

Cairns, J., Jr.: Effect of heat on fish. *Ind. Wastes*, 1:180–183, 1956a.

Cairns, J., Jr.: Effects of increased temperatures on aquatic organisms. *Ind. Wastes*, 1:150–152, 1956b.

Cairns, J., Jr.: Environment and time in fish toxicity. *Ind. Wastes*, 2:1–5, 1957.

Cairns, J., Jr.: Protozoans. Pollution Ecology of Freshwater Invertebrates, edited by C. W. Hart, S. H. Fuller, pp. 1–25. New York: Academic Press, 1974.

Cairns, J., Jr.: A strategy for the use of Protozoans in the evaluation of hazardous substances. Biological Indicators of Water Quality, edited by A. James, L. Evison, pp. 6-1–6-17. Chichester: Wiley, 1979.

Cairns, J., Jr. (ed.): Biological Monitoring in Water Pollution. New York: Pergamon, 1982.

Cairns, J., Jr. (ed.): Multispecies Toxicity Testing. New York: Pergamon, 1985.

Cairns, J., Jr.: Overview. Community Toxicity Testing, edited by J. Cairns, Jr., pp. 1–5. ASTM STP 920. Philadelphia: American Society for Testing and Materials, 1986.

Cairns, J., Jr.: Paradigms flossed: the coming of age of environmental toxicology. *Environ. Toxicol. Chem.*, 11: 285–287, 1992.

Cairns, J., Jr.: Environmental science and resource management in the 21st century: scientific perspective. *Environ. Toxicol. Chem.*, 12:1321–1329, 1993.

Cairns, J., Jr., and Cherry, D. S.: Freshwater multi-species test systems. Handbook of Ecotoxicology, Vol. 1, edited by P. Calow, pp. 101–116. Boston: Blackwell Scientific, 1993.

Cairns, J., Jr., and Mount, D. I.: Aquatic toxicology (part 2 of a four-part series). *Environ. Sci. Technol.*, 24(2): 154–161, 1990.

Cairns, J., Jr., and Pratt, J. R.: On the relation between structural and functional analyses of ecosystems. *Environ. Toxicol. Chem.*, 5:785–786, 1986.

Cairns, J., Jr., and Pratt, J. R.: Ecotoxicological effect indices: a rapidly evolving system. *Water Sci. Tech.*, 19(11):1–12, 1987.

Cairns, J., Jr., and Pratt, J. R.: The scientific basis of bioassays. *Hydrobiologia* 188/189:5–20, 1989.

Cairns, J., Jr., and Thompson, R. W.: A computer interfaced toxicity testing system for simulating variable effluent loading. Second Symposium on Process Measurements for Environmental Assessment, edited by P. L. Levin, J. C. Harris, and K. D. Drewitz, pp. 183–198. Cambridge, MA: Arthur D. Little, 1980.

Cairns, J., Jr., and van der Schalie, W. H.: Review paper. Biological monitoring. Part I—early warning systems. *Water Res.*, 14:1179–1196, 1980.

Cairns, J., Jr.; Neiderlehner, B. R.; and Smith, E. P.: The emergence of functional attributes as endpoints in ecotoxicology. Sediment Toxicity Assessment, edited by G. A. Burton, Jr., pp. 111–128. Boca Raton, FL: Lewis, 1992.

Calabrese, E. J.: Multiple Chemical Interactions. Chelsea, MI: Lewis, 1991.

Calamari, D. (ed.): Chemical Exposure Predictions. Boca Raton, FL: Lewis, 1993.

Call, D. J.; Brooke, L. T.; Knuth, M. L.; Poirier, S. H.; and Hoglund, M. D.: Fish subchronic toxicity prediction model for industrial organic chemicals that produce narcosis. *Environ. Toxicol. Chem.*, 4:335–341, 1985.

Calow, P.: The choice and implementation of environmental bioassays. *Hydrobiologia* 188/189:61–64, 1989.

Calow, P.: General principles and overview. Handbook of Ecotoxicology, Vol. 1, edited by P. Calow, pp. 1–5. Cambridge, MA: Blackwell Scientific, 1993.

Carson, R.: Silent Spring. Boston: Houghton Mifflin, 1962.

Cauwet, G.: Organic chemistry of seawater particulates: concepts and development. *Oceanol. Acta.*, 1:99–105, 1978.

Chew, R. D., and Hamilton, M. A.: Toxicity curve estimation: fitting a compartment model to median survival times. *Trans. Am. Fish. Soc.*, 114:403–412, 1985.

Christensen, E. R., and Chen, C. Y.: Modeling of combined toxic effects of chemicals. Hazard Assessment of Chemicals, Vol. 6, edited by J. Saxena, pp. 125–186. New York: Hemisphere, 1989.

Cole, H. A. (ed.): The Assessment of Sublethal Effects of Pollutants in the Sea. London: The Royal Society, 1979.

Connell, D. W.: Bioaccumulation of Xenobiotic Compounds. Boca Raton, FL: CRC Press, 1990.

Cook, P. M.; Kuehl, D. W.; Walker, M. K.; and Peterson, R. E.: Bioaccumulation and toxicity of TCDD and related compounds in aquatic ecosystems. Biological Basis for Risk Assessment of Dioxins and Related Compounds, edited by M. A. Gallo; R. J. Scheuplein; and K. A. van der Heijden, pp. 143–165. Banbury Report 35. Plainview, NY: Cold Spring Harbor Laboratory Press, 1991.

Couch, J. A., and Fournie, J. W. (eds.): Advances in Fisheries Science. Pathobiology of Marine and Estuarine Organisms. Boca Raton, FL: CRC Press, 1993.

Davis, J. C.: Standardization and protocols of bioassays—their role and significance for monitoring, research and regulatory usage. Proceedings of the 3rd Aquatic Toxicity Workshop, Halifax, Nova Scotia, November 2–3, 1976, edited by W. R. Parker, E. Pessah, P. G. Wells, G. F. Westlake, pp. 1–14. EPS-5-AR-77-1, 1977.

Davis, J. C., and Hoos, R. A. W.: Use of sodium pentachlorophenate and dehydroabietic acid as reference toxicants for salmonid bioassays. *J. Fish. Res. Bd. Can.*, 32: 411–416, 1975.

Day, K. E.; Ongley, E. D.; Scroggins, R.; and Eisenhauer, H. R. (eds.): Biology in the New Regulatory Framework for Aquatic Protection. Proceedings of the Alliston Workshop, April 1988. Burlington, Ont: Environment Canada, National Water Research Institute, 1989.

de Bruijn, J. H. M.; Yedema, E.; Seinen, W.; and Hermens, J. L. M.: Lethal body burdens of four organophosphorous pesticides in the guppy (*Poecilia reticulata*). *Aquat. Toxicol.*, 20:111–122, 1991.

de Wolf, W.; de Bruijn, J. H. M.; Seinen, W.; and Hermens, J. L. M.: Influence of biotransformation on the relationship between bioconcentration factors and octanol-water partition coefficients. *Environ. Sci. Technol.*, 26:1197–1201, 1992.

de Wolf, W.; Opperhuizen, A.; Seinen, W.; and Hermens, J. L. M.: Influence of survival time on the lethal body burden of 2,3,4,5-tetrachloroaniline in the guppy, *Poecilia reticulata. Sci. Total. Environ.*, 109:457–459, 1991.

Deneer, J. W.; Sinnige, T. L.; Seinen, W.; and Hermens, J. L. M.: The joint acute toxicty to *Daphnia magna* of industrial organic chemicals at low concentrations. *Aquat, Toxicol,*, 12:33–38, 1988.

Depledge, M. H.; Weeks, J. M.; and Bjerregaard, P.: Heavy metals. Handbook of Ecotoxicology, Vol. 2, edited by P. Calow, pp. 79–105. Cambridge, MA: Blackwell Scientific, 1994.

Di Toro, D. M.; Zarba, C. S.; Hansen, D. J.; Berry, W. J.; Swartz, R. C.; Cowan, C. E.; Pavlou, S. P.; Allen, H. E.; Thomas, N. A.; and Paquin, P. R.: Technical basis for establishing sediment quality criteria for nonionic organic chemicals using equilibrium partitioning. *Environ. Toxicol. Chem.*, 10:1541–1583, 1991.

Doe, K. G.; Ernst, W. R.; Parker, W. R.; Julien, G. R. J.; and Hennigar, P. H.: Influence of pH on the acute lethality of fenitrothion, 2,4-D, and aminocarb and pH-altered sublethal effects of aminocarb on rainbow trout (*Salmo gairdneri*). *Can. J. Fish Aquat. Sci.*, 45:287–293, 1988.

Doudoroff, P.: Keynote address: reflections on "pickle-jar" ecology. Biological Monitoring of Water and Effluent Quality, edited by J. Cairns, Jr.; K. L. Dickson; and G. F. Westlake, pp. 3–19. ASTM STP 607. Philadelphia: American Society for Testing and Materials, 1976.

Doudoroff, P., and Katz, M.: Critical review of literature on the toxicity of industrial wastes and their components to fish. I. Alkalies, acids, and inorganic gases. *Sewage Ind. Wastes*, 22:1432–1458, 1950.

Doudoroff, P., and Katz, M.: Critical review of literature on the toxicity of industrial wastes and their components to fish. II. The metals, as salts. *Sewage Ind. Wastes*, 25: 802–839, 1953.

Doudoroff, P.; Anderson, B. G.; Burdick, G. E.; Galtsoff, P. S.; Hart, W. B.; Pattrick, R.; Stronge, E. R.; Surber, E. W.; and Van Horn, W. M.: Bio-assay for the evaluation of acute toxicity of industrial wastes to fish. *Sewage Ind. Wastes*, 23:1380–1397, 1951.

Drever, J. I.: The Geochemistry of Natural Waters, 2nd ed. Englewood Cliffs, NJ: Prentice-Hall, 1988.

Drummond, R. A., and Russom, L. C.: Behavioral toxicity syndromes: a promising tool for assessing toxicity mechanisms in juvenile fathead minnows. *Environ. Toxicol. Chem.*, 9:37–46, 1990.

Eisler, R.: Polychlorinated Biphenyl Hazards to Fish, Wildlife, and Invertebrates: A Synoptic Review. Biological Report 85(1.7). Laurel, MD: U.S. Fish and Wildlife Service, 1986a.

Eisler, R.: Dioxin Hazards to Fish, Wildlife, and Invertebrates: A Synoptic Review. Biological Report 85(1.8). Laurel, MD: U.S. Fish and Wildlife Service, 1986b.

Eisler, R.: Polycyclic Aromatic Hydrocarbon Hazards to Fish, Wildlife and Invertebrates: A Synoptic Review, Biol. Rep. 85(1.11). Washington, DC: U.S. Fish and Wildlife Service, 1987.

Eisler, R.: Pentachlorophenol Hazards to Fish, Wildlife, and Invertebrates: A Synoptic Review. Biological Report 85(1.17). Laurel, MD: U.S. Fish and Wildlife Service, 1989.

Environment Canada: Proceedings of the OECD Workshop on the Biological Testing of Effluents (and Related Receiving Waters), Duluth, MN, September 1984. Ottawa: Environment Canada, 1984.

Environment Canada: Guidance Document on Control of Toxicity Test Precision using Reference Toxicants. Report EPS1/RM/12, August 1990. Ottawa, Ontario, 1990.

Fentem, J., and Balls, M.: Replacement of fish in ecotoxicology testing: use of bacteria, other lower organisms and fish cells in vitro. Ecotoxicology Monitoring, edited by M. Richardson, pp. 71–81. New York: VCH Publishers, 1993.

Ferguson, J.: The use of chemical potentials as indices of toxicity. *Proc. R. Soc. Lond.*, 127B:387–404, 1939.

Filov, V. A.; Golubev, A. A.; Liublina, E. I.; and Tolokontsev, N. A.: Quantitative Toxicology: Selected Topics. New York: Wiley, 1979.

Finney, D. J.: Statistical Method in Biological Assay, 3rd ed. London: Griffin, 1978.

Förstner, U.: Inorganic sediment chemistry and elemental speciation. Sediments: Chemistry and Toxicity of In-Place Pollutants, edited by R. Baudo, J. Giesy, and H. Muntau, p. 61–106. Boca Raton, FL: Lewis, 1990.

Förstner U: Wittman G: Metal Pollution in the Aquatic Environment. Berlin: Springer-Verlag, 1979.

Foulkes, E. C.: The concept of critical levels of toxic heavy metals in target tissues. *Crit. Rev. Toxicol.*, 20:327–339, 1990.

Franks, N. P., and Lieb, W. R.: Molecular mechanisms of general anaesthesia. *Nature*, 300:487–493, 1982.

Free, S. M., Jr., and Wilson, J. W.: A mathematical contribution to structure-activity studies. *J. Med. Chem.*, 7: 395–399, 1964.

Friant, S. L., and Henry, L.: Relationship between toxicity of certain organic compounds and their concentrations in tissues of aquatic organisms: a perspective. *Chemosphere* 14:1897–1907, 1985.

Furness, R. W., and Rainbow, P. S.: Heavy Metals in the Marine Environment. Boca Raton, FL: CRC Press, 1990.

Gaddum, J. H.: Reports on biological standards. III. Methods of biological assay depending on quantal response. Medical Research Council Special Report Series 183. London: HMSO, 1933.

Gallo, M. A., and Doull, J.: Chapter 1. History and scope of toxicology. Toxicology. The Basic Science of Poisons, 4th ed., edited by M. O. Amdur, J. Doull, and C. D. Klaassen, pp. 3–11. New York: Pergamon, 1991.

GESAMP (IMO/FAO/UNESCO/WMO/WHO/IAEA/UN/UNEP Joint Group of Experts on the Scientific Aspects of Marine Pollution): Global Strategies for Marine Environmental Protection. GESAMP Reports and Studies No. 45. London: International Maritime Organization, 1991.

GESAMP (IMO/FAO/UNESCO/WMO/WHO/IAEA/UN/UNEP Joint Group of Experts on the Scientific Aspects of Marine Pollution): Impact of Oil and Related Chemicals and Wastes on the Marine Environment. GESAMP Reports and Studies No. 50. London: International Maritime Organization, 1993.

Geyer, H. J.; Scheunert, I.; Rapp, K.; Kettrup, A.; Korte, G.; Greim, H.; and Rozman, K.: Correlation between acute toxicity of 2,3,7,8-tetrachlorodibenzo-*p*-dioxin (TCDD) and total body fat content in mammals. *Toxicology*, 65:97–107, 1990.

Giesy, J. P. (ed.): Microcosms in Ecological Research. DOE COMF-781101. Springfield, VA: NTIS, 1980.

Gilbert, T. R., and Kakareka, J. P.: Analytical chemistry. Fundamentals of Aquatic Toxicology, edited by G. M. Rand, S. R. Petrocelli, pp. 475–494. New York: Hemisphere, 1985.

Goldberg, A. M., and Frazier, J. M.: Alternatives to animals in toxicity testing. *Sci. Am.*, 261(2):24–30, 1989.

Green, R. H.: Sampling Design and Statistical Methods for Environmental Biologists. New York: Wiley, 1979.

Halver, J. E.: Fish Nutrition, 2nd ed. San Diego: Academic Press, 1989.

Hansch, C., and Fujita, T.: Rho-theta-pi analysis. A method for the correlation of biological activity and chemical structure. *J. Am. Chem. Soc.* 86:1616–1626, 1964.

Hansen, D. J.; Goodman, L. R.; Moore, J. C.; and Higdon, P. K.: Effects of the synthetic pyrethroids AC 222,705, permethrin and fenvalerate on sheepshead minnows in early life stage toxicity tests. *Environ. Toxicol. Chem.*, 2: 251–258, 1983.

Hansen, D. J.; Goodman, L. R.; Cripe, G. M.; and Macauley, S. F.: Early life-stage toxicity test methods for gulf toadfish *Opsanus beta* and results using chlorpyrifos. *Ecotox. Environ. Safety*, 11:15–22, 1986.

Hart, W. B.; Doudoroff, P.; and Greenbank, J.: The Evaluation of the Toxicity of Industrial Wastes, Chemicals and Other Substances to Fresh Water Fishes. Philadelphia: Waste Control Laboratory, Atlantic Refining Co., 1945.

Haslam, S. M.: A proposed method for monitoring river pollution using macrophytes. *Env. Technol. Lett.*, 3:19–34, 1982.

Haya, K.: Toxicity of pyrethroid insecticides to fish. *Environ. Toxicol. Chem.*, 8:381–391, 1989.

Hellawell, J. M.: Biological surveillance and water quality monitoring. Biological Monitoring of Inland Fisheries, edited by J. S. Alabaster, pp. 69–88. London: Applied Science, 1977.

Hellawell, J. M.: Biological Indicators of Freshwater Pollution and Environmental Management. London: Elsevier Applied Science, 1986.

Henderson, C.: Application factors to be applied to bioassays for the safe disposal of toxic wastes. Biological Problems in Water Pollution, pp. 31–37. Washington, DC: U.S. Department of Health, Education and Welfare, Public Health Service, 1957.

Henderson, R. J., and Tocher, D. R.: The lipid composition and biochemistry of freshwater fish. *Prog. Lipid Res.*, 25: 281–347, 1987.

Hennes-Morgan, E. C., and De Oude, N. T.: Detergents. Handbook of Ecotoxicology, Vol. 2, edited by P. Calow, pp. 130–154. Cambridge, MA: Blackwell Scientific, 1994.

Hermens, J. L. M.: Quantitative structure-activity relationships of environmental pollutants. The Handbook of Environmental Chemistry, Vol. 2, Part E, edited by O. Hutzinger, pp. 111–162. New York: Springer-Verlag, 1989.

Hermens, J. L. M.; Canton, H.; Janssen, P.; and de Jong, R.: Quantitative structure-activity relationships and toxicity studies of mixtures of chemicals with anaesthetic potency: acute lethal and sublethal toxicity to *Daphnia magna*. *Aquat. Toxicol.*, 5:143–154, 1984.

Hermens, J. L. M.; Konemann, H.; Leeuwangh, P.; and Musch, A.: Quantitative structure-activity relationships in aquatic toxicity studies of chemicals and complex mixtures of chemicals. *Environ. Toxicol. Chem.*, 4:273–279, 1985.

Hickie, B. E.: Development and testing of a toxicokinetic-based model for pulse exposure toxicity of organic contaminants. Ph.D. thesis, Univ. of Waterloo, Waterloo, Canada, 1990.

Hill, I. R.: Effects on non-target organisms in terrestrial and aquatic environments. The Pyrethroid Insecticides, edited by J. P. Leahey, pp. 151–262. Philadelphia: Taylor & Francis, 1985.

Hodson, P. V.: A comparison of the acute toxicity of chemicals to fish, rats and mice. *J. Appl. Toxicol.*, 5:220–226, 1985.

Hodson, P. V.: Editorial. As our science matures, so should our journal. *Environ. Toxicol. Chem.*, 13:1, 1994.

Hodson, P. V.; Parisella, R.; Blunt, B.; Gray, B.; and Kaiser, K. L. E.: Quantitative structure-activity relationships for chronic toxicity of phenol, *p*-chlorophenol, 2,4-dichlorophenol, pentachlorophenol, *p*-nitrophenol, and 1,2,4-trichlorobenzene to early life stages of rainbow trout (*Oncorhynchus mykiss*). Canadian Technical Report of Fisheries and Aquatic Sciences 1784. Mont-Joli, Quebec: Fisheries and Oceans Canada, 1991.

Hoeschele, J. D.; Turner, J. E.; and England, M. W.: Inorganic concepts relevant to metal binding, activity, and toxicity in a biological system. *Sci. Total. Environ.*, 109: 477–492, 1991.

Howells, G.; Calamari, D.; Gray, J.; and Wells, P. G.: An analytical approach to assessment of long-term effects of low levels of contaminants in the marine environment. *Mar. Pollut. Bull.*, 21(8):371–375, 1990.

Huggett, R. J.; Kimerle, R. A.; Mehrle, P. M.; and Bergman, H. L. (eds.): Biomarkers: Biochemical, Physiological and Histological Markers of Anthropogenic Stress. Boca Raton, FL: Lewis, 1992.

Hunn, J. B.; Schoettger, R. A.; and Wealdon, E.: Observations on the handling and maintenance of bioassay fish. *Prog. Fish Cult.*, 30:164–167, 1968.

Hunn, J. B.: History of Acute Toxicity Tests with Fish, 1863–1987. Investigations in Fish Control 98. LaCrosse, WI: U.S. Department of the Interior, Fish and Wildlife Service, 1989.

Hunt, D. T. E., and Wilson, A. L.: The Chemical Analysis of Water—General Principles and Techniques, 2nd ed. The Royal Society of Chemistry, Burlington House, London, England. Oxford: Alden Press, 1986.

International Joint Commission (IJC): Revised Great Lakes Water Quality Agreement of 1978 as Amended by Protocol Signed October 16, 1983. International Joint Commission, Windsor, Ontario, 1989.

James A. (ed.): An Introduction to Water Quality Modeling, 2nd ed. New York: Wiley, 1993.

Johnstone, K.: The Aquatic Explorers: A History of the Fisheries Research Board of Canada. Toronto: University of Toronto Press, 1977.

Jorgensen, S. E. (ed.): Modelling the Fate and Effect of Toxic Substances in the Environment. Developments in Environmental Modelling, 6. New York: Elsevier, 1984.

Jorgensen, S. E. (ed.): Fundamentals of Ecological Modelling. Developments in Environmental Modelling, 9. New York: Elsevier, 1986.

Jorgensen, S. E. (ed.): Modelling in Ecotoxicology. Developments in Environmental Modelling, 16. New York: Elsevier, 1990.

Kaiser, K. L. E., and Palabrica, V. S.: *Photobacterium phosporeum* toxicity data index. *Water Pollut. Res. J. Can.*, 26:361–431, 1991.

Kenaga, E. E.: Test organisms and methods useful for early assessment of acute toxicity of chemicals. *Environ. Sci. Technol.*, 12:1322–1329, 1978.

Kenaga, E. E.: Aquatic test organisms and methods useful for assessment of chronic toxicity of chemicals. Analysing the Hazard Evaluation Process, edited by K. L. Dickson; A. W. Maki; and J. Cairns, Jr., pp. 101–111. Bethesda, MD: American Fisheries Society, 1979.

Kenaga, E. E.: Predictability of chronic toxicity from acute toxicity of chemicals in fish and aquatic invertebrates. *Environ. Toxicol. Chem.*, 1:347–358, 1982.

Kennish, M. J.: Ecology of Estuaries: Anthropogenic Effects. Boca Raton, FL: CRC Press, 1992.

Klaverkamp, J. F.; Kenney, A.; Harrison, S. E.; and Danell, R.: An evaluation of phenol and sodium azide as reference toxicants in rainbow trout. Second Annual Aquatic Toxicity Workshop 1975 Proceedings, Toronto, Ontario, November 4–5, 1975, edited by G. R. Craig, pp. 73–92. Toronto, Ontario: Ontario Ministry of the Environment, 1975.

Kooijman, S. A. L. M.: Parametric analysis of mortality rates in bioassays. *Water Res.*, 15:107–119, 1981.

Kooijman, S. A. L. M.: Statistical aspects of the determination of mortality rates in bioassays. *Water Res.*, 17: 749–759, 1983.

Krogh, A., and Hemmingsen, A. M.: The assay of insulin on rabbits and mice. Det Kgl Danske Videnskebernes Selskab Biol VIII, 1928.

Landres, P. B.: Ecological indicators: panacea or liability. Ecological Indicators, Vol. 2, edited by D. H. McKenzie, D. Eric Hyatt, and V. J. McDonald, pp. 1295–1318. New York: Elsevier Applied Science, 1992.

Landrum, P. L.; Hayton, L. W.; Lee, H., II; McCarty, L. S; Mackay, D.; and McKim, J. M.: Chapter 3: Synopsis of discussion session on the kinetics behind environmental bioavailability. Bioavailability: Physical, Chemical, and Biological Interactions, edited by J. L. Hamelink; P. F. Landrum; H. L. Bergman; and W. H. Benson, pp. 203–219. SETAC Pellston Workshop Series. Boca Raton, FL: Lewis, 1994.

Lanno, R. P.; Hickie, B. E.; and Dixon, D. G.: Feeding and nutritional considerations in aquatic toxicology. Hydrobiologia 188/189:525–531, 1989.

LaPoint, T. W., and Fairchild, J. F.: Evaluation of sediment contaminant toxicity: the use of freshwater community structure. Sediment Toxicity Assessment, edited by G. A. Burton, Jr., pp. 87–110. Boca Raton, FL: Lewis, 1992.

LaRoche, E.; Eisler, R.; and Tarzwell, C. M.: Bioassay procedures for oil and oil dispersant toxicity evaluation. *J. Water Pollut. Control Fed.*, 42:1982–1989, 1970.

Lassiter, R. R., and Hallam, T. G.: Survival of the fattest—implications for acute effects of lipophilic chemicals on aquatic populations. *Environ. Toxicol. Chem.*, 9:585–595, 1990.

Laws, E. A.: Aquatic Pollution. New York: Wiley, 1981.

Lee, D. R.: Reference toxicants in quality control of aquatic bioassays. Aquatic Invertebrate Bioassays, edited by A. L. Buikema, Jr., and J. R. Cairns, Jr., pp. 188–199. ASTM STP 715. Philadelphia: American Society for Testing and Materials, 1980.

Lee, H.: A clam's eye view of the bioavailability of sediment-associated pollutants. Organic Substances and Sediments in Water, Vol. 3, Biological, edited by R. A. Baker, pp. 73–93. Boca Raton, FL: Lewis, 1991.

Leland, H. V., and Kuwabara, J. S.: Trace metals. Fundamentals of Aquatic Toxicology, edited by G. M. Rand, and S. R. Petrocelli, pp. 374–415. New York: Hemisphere, 1985.

Levin, S. A.; Harwell, M. A.; Kelly, J. R.; and Kimball, K. D.: Ecotoxicology: Problems and Approaches. New York: Springer-Verlag, 1989.

Libes, S. M.: An Introduction to Marine Biogeochemistry. New York: Wiley, 1992.

Litchfield, J. T.: A method for rapid graphic solution of time-percent effect curves. *J. Pharmacol. Exp. Ther.*, 97: 399–408, 1949.

Lothenback, D. B.; Henry, T. R.; and Johnson, R. D.: 2,3,7,8-TCDD toxicity in medaka. SETAC Ninth Annual Meeting, Poster Abstracts. Society of Environmental Toxicology and Chemistry, Washington, DC, p. 108, 1988.

Ludwig, J. A., and Reynolds, J. F.: Statistical Ecology—A Primer on Methods of Computing. New York: Wiley, 1988.

Macek, K. J.: Aquatic toxicology: anarchy or democracy? Aquatic Toxicology and Hazard Assessment, 5th Conference, edited by J. G. Pearson; R. B. Foster; and W. E. Bishop, pp. 3–8. ASTM STP 766, Philadelphia: American Society for Testing and Materials, 1982.

Maciorowski, A. F.; Little, L. W.; and Sims, J.: Bioassays—procedures and results. *Water Pollut. Control Fed.*, 52: 1630–1656, 1980.

Mackay, D.: Correlation of bioconcentration factors. *Environ. Sci. Technol.*, 16:274–278, 1982.

Mackay, D.: Multimedia Environmental Models. The Fugacity Approach. Boca Raton, FL: Lewis, 1991.
Mackay, D. W.; Holmes, P. G.; and Redshaw, C. J.: The application of bioassay techniques to water pollution problems—the United Kingdom experience. *Hydrobiologia* 188/189:77–86, 1989.
Mackay, D.; Puig, H.; and McCarty, L. S.: An equation describing the time course and variability in uptake and toxicity of narcotic chemicals to fish. *Environ. Toxicol. Chem.*, 11:941–951, 1992.
Mackay, D.; Shiu, W. Y.; and Ma, K. C.: Illustrated Handbook of Physical-Chemical Properties and Environmental Fate for Organic Chemicals. Vol. II. Polynuclear Aromatic Hydrocarbons, Polychlorinated Dioxins, and Dibenzofurans. Boca Raton, FL: Lewis, 1992.
Maki, A. W.: Editorial: Ecotoxicology—critical needs and credibility. *Environ. Toxicol. Chem.*, 2:259–260, 1983.
Manahan, S. E.: Toxicological Chemistry. A Guide to Toxic Substances in Chemistry. Chelsea, MI: Lewis, 1989.
Manahan, S. E.: Environmental Chemistry, 5th ed. Boca Raton, FL: Lewis, 1991.
Mance, G.: Pollution Threat of Heavy Metals in Aquatic Environments. New York: Elsevier, 1987.
Marking, L. L.: Evaluation of *p,p'*-DDT as a reference toxicant in bioassays. Investigations in fish control. U.S. Department of Interior Fish and Wildlife Service Resource Publication 14, 1966.
Marking, L. L.: Methods for assessing additive toxicity of chemical mixtures. Aquatic Toxicology and Hazard Evaluation, edited by F. L. Mayer, J. L. Hamelink, pp. 99–108. ASTM STP 634. Philadelphia: American Society for Testing and Materials, 1977.
Marking, L. L.: Toxicity of chemical mixtures. Fundamentals of Aquatic Toxicology, edited by G. M. Rand and S. R. Petrocelli, pp. 164–176. New York: Hemisphere, 1985.
Mason, C. F.: Biology of Freshwater Pollution, 2nd ed. London: Longman Scientific and Technical, 1991.
Matczak, T. Z., and MacKay, R. J.: Territoriality in filter-feeding caddisfly larvae: laboratory experiments. *J. North Am. Benthol. Soc.*, 9:26–34, 1990.
Matida, Y.: Study on the toxicity of agricultural control chemicals in relation to fisheries management. No. 3: A method of estimating the threshold value and a kinetic analysis of the toxicity curve. *Bull. Freshwater Fish Res. Lab. Tokyo*, 9:1–12, 1960.
Matthews, R. A.; Buikema, A. L., Jr.; Cairns, J., Jr.; and Rodgers, J. H., Jr.: Biological monitoring. Part IIa: Receiving system functional methods, relationships, and indices. *Water Res.*, 16:129–139, 1982.
Mayer, F. L., Jr., and Ellersieck, M. R.: Manual of acute toxicity: interpretation and database for 410 chemicals and 66 species for freshwater animals. U.S. Fish and Wildlife Service Resource Publication 160, 1986.
Mayer, F. L., Jr.; Mayer, K. S.; and Ellersieck, M. R.: Relation of survival to other endpoints in chronic toxicity tests with fish. *Environ. Toxicol. Chem.*, 5:737–748, 1986.
Mayer, F. L.; Krause, G. F.; Ellersieck, M. R.; and Lee, G.: Statistical Approach to Predicting Chronic Toxicity of Chemicals to Fishes from Acute Toxicity Test Data. EPA/600/R-92-091. Gulf Breeze, FL: U.S. Environmental Protection Agency, 1992.
Mayfield, C. I.: Microbial systems. Handbook of Ecotoxicology, Vol. 1, edited by P. Calow, pp. 9–27. Cambridge, MA: Blackwell Scientific, 1993.
McBride, G. B.: The Role of Monitoring in the Management of Water Resources. Biological Monitoring in Freshwaters: Proceedings of a Seminar, Hamilton, November 21–22, 1984, Part 1, edited by R. D. Pridmore, and A. B. Cooper, pp. 7–16.
McCarthy, J. F., and Shugart, L. R.: Biomarkers of Environmental Contamination. Boca Raton, FL: Lewis, 1990.
McCarty, L. S.: The relationship between aquatic toxicity QSARs and bioconcentration for some organic chemicals. *Environ. Toxicol. Chem.*, 5:1071–1080, 1986.
McCarty, L. S.: Relationship between toxicity and bioconcentration for some organic chemicals. I. Examination of the relationship. II. Application of the relationship. QSAR in Environmental Toxicology—II, edited by K. L. E. Kaiser, pp. 207–220 and 221–230. Dordrecht: Reidel, 1987.
McCarty, L. S.: A kinetics-based analysis of quantitative structure-activity relationships in aquatic toxicity and bioconcentration bioassays with organic chemicals. Ph.D. thesis, Univ. of Waterloo, Waterloo, Canada, 1990.
McCarty, L. S.: Toxicant body residues: implications for aquatic bioassays with some organic chemicals. Aquatic Toxicology and Risk Assessment: Fourteenth Volume, edited by M. A. Mayes, and M. G. Barron, pp. 183–192. ASTM STP 1124. Philadelphia: American Society for Testing and Materials, 1991.
McCarty, L. S.; Mackay, D.; Smith, A. D.; Ozburn, G. W.; and Dixon, D. G.: Residue-based interpretation of toxicity and bioconcentration QSARs from aquatic bioassays: neutral narcotic organics. *Environ. Toxicol. Chem.*, 11:917–930, 1992a.
McCarty, L. S.; Ozburn, G. W.; Smith, A. D.; and Dixon, D. G.: Toxicokinetic modeling of mixtures of organic chemicals. *Environ. Toxicol. Chem.*, 11:1037–1047, 1992b.
McCarty, L. S., and Mackay, D.: Enhancing ecotoxicological modeling and assessment: body residues and modes of toxic action. *Environ. Sci. Technol.*, 27:1719–1728, 1993.
McCarty, L. S.; Mackay, D.; Smith, A. D.; Ozburn, G. W.; and Dixon, D. G.: Residue-based interpretation of toxicity and bioconcentration QSARs from aquatic bioassays: polar narcotic organics. *Ecotoxicol. Environ. Safety*, 24:253–270, 1993.
McGowan, J. C.: The physical toxicity of chemicals. III. A systematic treatment of physical toxicity in aqueous solutions. *J. Appl. Chem.*, 2:651–658, 1952.
McGowan, J. C.: Partition coefficients and biological activities. Nature 200:1317, 1963.
McGowan, J. C., and Mellors, A.: Molecular Volumes in Chemistry and Biology: Applications Including Partitioning and Toxicity. Chichester: Ellis Horwood, 1986.
McIntyre, A. D., and Pearce, J. B. (eds.): Biological Effects of Marine Pollution and the Problems of Monitoring. Proceedings from ICES Workshop held in Beauford, NC, 26 February–2 March 1979. Rapp. Reun. Cons. Int. Explor. Mer. 179, 1980.
McKim, 1985. See Appendix Chapter B.
McKim, J. M., and Nichols, J. W.: Use of physiologically based toxicokinetic models in a mechanistic approach to aquatic toxicology. Aquatic Toxicology—Molecular, Biochemical and Cellular Perspectives, edited by D. C. Malins, and G. K. Ostrander, pp. 469–519. Boca Raton, FL: Lewis, 1994.
McKim, J. M.; Schmieder, P. K.; Carlson, R. W.; Hunt, E. P.; and Niemi, G. J.: Use of respiratory-cardiovascular responses of rainbow trout *Salmo gairdneri* in identifying acute toxicity syndromes in fish: Part 1. Pentachlorophenol, 2,4-dinitrophenol, tricaine methanesulfonate and 1-octanol. *Environ. Toxicol. Chem.*, 6:295–312, 1987a.

McKim, J. M.; Schmieder, P. K.; Niemi, G. J.; Carlson, R. W.; and Henry, T. R.: Use of respiratory-cardiovascular responses of rainbow trout *Salmo gairdneri* in identifying acute toxicity syndromes in fish: Part 2. Malathion, carbaryl, acrolein and benzaldehyde. *Environ. Toxicol. Chem.*, 6:313–328, 1987b.

McKim, J. M., and Schmieder, P. K.: Bioaccumulation: does it reflect toxicity. Bioaccumulation in Aquatic Systems: Contributions to Assessment. Proceedings of an International Workshop, Berlin 1990, cited by R. Nagel, and R. Loskill, pp. 161–183. Weinheim: VCH Verlagsgesellschaft, 1991.

McLeay, D. J.: Canadian approaches and test methods for determining and monitoring the aquatic toxic effects of BKME. Proceedings of the International Conference on Bleached Kraft Pulp Mills—Technical and Environmental Issues, pp. 281–311, Melbourne, Australia, February 4–7, 1991.

McLeese, D. W.; Metcalfe, C. D.; and Zitko, V.: Lethality of permethrin, cypermethrin, and fenvalerate to salmon, lobster, and shrimp. *Bull. Environ. Contam. Toxicol.*, 25: 950–955, 1980.

Mehrle, P. M., and Mayer, F. L.: Biochemistry/physiology. Fundamentals of Aquatic Toxicology, edited by G. M. Rand, and S. R. Petrocelli, pp. 264–282. New York: Hemisphere, 1985.

Mehrle, P. M.; Buckler, D. R.; Little, E. E.; Smith, L. M.; Petty, J. D.; Peterman, P. H.; Stalling, D. L.; DeGraeve, G. M.; Coyle, J.; and Adams, W. J.: Toxicity and bioconcentration of 2,3,7,8-tetrachlorodibenzodioxin and 2,3,7,8-tetrachlorodibenzofuran in rainbow trout. *Environ. Toxicol. Chem.*, 7:47–62, 1988.

Meyers, T. R., and Hendricks, J. D.: Histopathology. Fundamentals of Aquatic Toxicology, edited by G. M. Rand, and S. R. Petrocelli, pp. 283–331. New York: Hemisphere, 1985.

Miller, W. E.; Greene, J. C.; and Shiroyama, J.: *Selenastrum capricornutum* Printz algal assay bottle test: experimental design, application, and data interpretation protocol. EPA-600/9-78-018. Corvallis, OR: U.S. EPA, 1978.

Mix M. C.: Polycyclic aromatic hydrocarbons in the aquatic environment: occurrence and biological monitoring. Reviews in Environmental Toxicology, Vol. 1, edited by E. Hodgson, pp. 58–102. Amsterdam: Elsevier, 1984.

Morel, F. M. M., and Hering, J. G.: Principles and Applications of Aquatic Chemistry. New York: Wiley, 1993.

Moriarity, F.: Ecotoxicology—The Study of Pollutants in Ecosystems, 2nd ed. New York: Academic Press, 1988.

Mount, D. I.: Quarterly report. U.S. Environmental Research Laboratory, Duluth, MN, 1976.

Mount, D. I., and Boyle, H. W.: Parathion—use of blood concentration to diagnose mortality of fish. *Environ. Sci. Technol.*, 3:1183–1185, 1969.

Mount, D. I., and Brungs, W. A.: A simplified dosing apparatus for fish toxicology studies. *Water Res.*, 1:21–29, 1967.

Mount, D. I., and Stephen, C. E.: A method for establishing acceptable limits for fish—malathion and the butoxyethanol ester of 2,4-D. *Trans. Am. Fish Soc.*, 96:185–193, 1967.

Mount, D. I.; Vigor, L. W.; and Schafer, M. L.: Endrin: Use of concentration in blood to diagnose acute toxicity to fish. *Science* 152:1388–1390, 1966.

Muirhead-Thomson, R. C.: Pesticide Impact on Stream Fauna with Special Reference to Macroinvertebrates. Cambridge, UK: Cambridge Univ. Press, 1987.

Munawar, M.; Dixon, D. G.; Mayfield, C. I.; Reynoldson, T.; and Sadar, M. H. (eds.): Environmental Bioassay Techniques and Their Application. Proceedings of the 1st International Conference held in Lancaster, England, July 11–14, 1988. *Hydrobiologia* 188/189:1–680, 1989a.

Munawar, M.; Munawar, I. F.; Mayfield, C. I.; and McCarthy, L. H.: Probing ecosystem health: a multi-disciplinary and multi-trophic assay strategy. *Hydrobiologia* 188/189:93–116, 1989b.

Munkittrick, K. R.: A review and evaluation of study design and evaluation for site-specifically assessing the health of fish populations. *J. Aquat. Eco. Health*, 1:283–293, 1992.

Munkittrick, K. R., and Dixon, D. G.: Use of white sucker *Catostomus commersoni* populations to assess the health of aquatic ecosystems exposed to low-level contaminant stress. *Can. J. Fish Aquat. Sci.*, 46:1455–1462, 1989.

Murphy, P. G.: The effect of size upon the uptake of DDT from water by fish. *Bull. Environ. Contam. Toxicol.*, 6: 20–23, 1971.

Nagel, R., and Loskill, R. (eds.): Bioaccumulation in Aquatic Systems. Proceedings of an International Workshop, Berlin, 1990. New York: VCH, 1991.

National Research Council (NRC): Using Oil Spill Dispersants on the Sea. Marine Board, National Research Council. Washington, DC: National Academy Press, 1989.

National Research Council (NRC): Animals as Sentinels of Environmental Health Hazards. Washington, DC: National Academy Press, 1991.

Neff, J. M.: Polycyclic aromatic hydrocarbons. Fundamentals of Aquatic Toxicology, edited by G. M. Rand and S. R. Petrocelli, pp. 416–454. New York: Hemisphere, 1985.

Neff, J. M.: Polycyclic Aromatic Hydrocarbons in the Aquatic Environment. Sources, Fates and Biological Effects. London: Applied Science, 1979.

Neff, J. M., and Anderson, J. W.: Responses of Marine Animals to Petroleum and Specific Petroleum Hydrocarbons. London: Applied Science, 1981.

Newman, M. C., and McIntosh, A. W. (eds.): Metal Ecotoxicology. Concepts and Applications. Chelsea, MI: Lewis, 1991.

Nichols, J. W.; McKim, J. M.; Andersen, M. E.; Gargas, M. L.; Clewell, H. J., III; and Erickson, R. J.: A physiologically based toxicokinetic model for the uptake and disposition of waterborne organic chemicals in fish. *Toxicol. Appl. Pharmacol.*, 106:433–447, 1990.

Niemi, G. J.; Bradbury, S. P.; and McKim, J. M.: The use of fish physiology literature for predicting fish acute toxicity syndromes. Aquatic Toxicology and Risk Assessment: Fourteenth Volume, edited by M. A. Mayes, and M. G. Barron, pp. 245–260. ASTM STP 1124. Philadelphia: American Society for Testing and Materials, 1991.

Nightingale, C. H.: Drug concentration in goldfish at the pharmacologic endpoint. *J. Pharm. Sci.*, 60:1762, 1971.

Nimmi, A. J.: PCBs, PCDDs and PCDFs. Handbook of Ecotoxicology, Vol. 2, edited by P. Calow, pp. 204–243. Cambridge, MA: Blackwell Scientific, 1994.

Nimmo, D. R.: Pesticides. Fundamentals of Aquatic Toxicology, edited by G. M. Rand, and S. R. Petrocelli, pp. 335–373. New York: Hemisphere, 1985.

Nimmo, D. R.; Coppage, D. L.; Pickering, Q. H.; and Hansen, D. J.: Assessing the toxicity of pesticides to aquatic organisms. Silent Spring Revisited, edited by G. J. Marco; R. M. Hollingworth; and W. Durham, pp. 49–70. Washington, DC: American Chemical Society, 1987.

Nimmo, D. R., and McEwen, L. C.: Pesticides. Handbook of Ecotoxicology, Vol. 2, edited by P. Calow, pp. 155–203. Cambridge, MA: Blackwell Scientific, 1994.

Odum, E. P.: Relationship between structure and function in ecosystems. *Jpn. J. Ecol.*, 12:108–118, 1962.
O'Flaherty, E. J.: Toxicants and Drugs: Kinetics and Dynamics. New York: Wiley, 1981.
Organization for Economic Co-operation and Development (OECD): OECD Guidelines for Testing of Chemicals. Paris: OECD, 1981.
Organization for Economic Co-operation and Development (OECD): The use of biological tests for water pollution assessment and control. OECD Environmental Monograph No. 11, October 1987.
Pagenkopf, J. F.: Introduction to Natural Water Chemistry. New York: Marcel Dekker, 1978.
Paine, R. T.: A note on trophic complexity and community stability. *Am. Nat.*, 103:91–93, 1969.
Pankow, J. F. (ed.): Aquatic Chemical Concepts. Boca Raton, FL: Lewis, 1991.
Parker, W. R.; Doe, K. G.; Wells, P. G.: Application of marine biological tests in assessing and controlling the acute and chronic toxicities of effluents. Proceedings of the 17th Annual Aquatic Toxicity Workshop: November 5–7, Vancouver, BC, Vol. 1, edited by P. Chapman, F. Bishay; E. Power; K. Hall; L. Harding; D. McLeay; M. Nassichuk; and W. Knapp, pp. 135–151. Can. Tech. Rep. Fish. Aquat. Sci. No. 1774 (Vol 1). Department of Fisheries and Oceans, Ottawa, 1991.
Parrish, 1985. See Appendix Chapter A.
Peakall, D. B.: Animal Biomarkers as Pollution Indicators. London: Chapman & Hall, 1992.
Peakall, D. B., and Shugart, L. R. (eds.): Biomarkers—Research and Application in the Assessment of Environmental Health. NATO ASI Series. Berlin: Springer-Verlag, 1993.
Peltier, W.: Methods for measuring the acute toxicity of effluents to aquatic organisms. EPA 600/4-78-012. U.S. EPA, Washington, DC, 1978.
Persoone, G., and Janssen, C. R.: Freshwater Invertebrate Toxicity Tests. Handbook of Ecotoxicology, Vol. 1, edited by P. Calow, pp. 51–63. Cambridge, MA: Blackwell Scientific, 1993.
Persoone, G.; Jaspers, E.; and Claus, C.: Ecotoxicological Testing for the Marine Environment. State University of Ghent, Belgium and Institute for Marine Scientific Research, Bredene, Belgium. Vols. I and II, 1984.
Pessah, E., and Cornwall, G. M.: Use of toxicity tests in regulating the quality of industrial wastes in Canada. Aquatic Toxicology, edited by J. G. Eaton; P. R. Parrish; and A. C. Hendricks, pp. 130–141. ASTM STP 707. Philadelphia: American Society for Testing and Materials, 1980.
Pessah, E.; Wells, P. G.; and Schneider, J. R.: Dodecyl sodium sulfate (DDS) as an intralaboratory reference toxicant in fish bioassays. Second Annual Aquatic Toxicity Workshop 1975 Proceedings, Toronto, Ontario, November 4–5, 1975, edited by G. R. Craig, pp. 93–121. Toronto, Ontario: Ontario Ministry of the Environment, 1975.
Peters, R. H.: The Ecological Implications of Body Size. Cambridge: Cambridge Univ. Press, 1983.
Phillips, D. J. H.: Quantitative Aquatic Biological Indicators: Their Use to Monitor Trace Metal and Organochlorine Pollution. London: Applied Science, 1980.
Phillips, D. J. H.: Bioaccumulation. Handbook of Ecotoxicology, Vol. 1, edited by P. Calow, pp. 378–396. Cambridge, MA: Blackwell Scientific, 1993.
Power, E. A.; Munkittrick, K. R.; and Chapman, P. M.: An ecological impact assessment framework for decision-making related to sediment quality. Aquatic Toxicology and Risk Assessment: Fourteenth Volume, edited by M. A. Mayes, and M. G. Barron, pp. 48–64. ASTM STP 1124. Philadelphia: American Society for Testing and Materials, 1991.
Purser, C. A., and Hume, A. S.: Detection—analytical. Basic Environmental Toxicology, edited by L. G. Cockerham, B. S. Shane, pp. 501–528. Boca Raton, FL: Lewis, 1994.
Pytkowicz, R. M.: Equilibria, Nonequilibria and Natural Waters, Vol. 1. New York: Wiley, 1983.
Rach, J. J., and Gingerich, W. H.: Distribution and accumulation of rotenone in tissues of warmwater fishes. *Trans. Am. Fish Soc.*, 115:214–219, 1986.
Rand, G. M.: Bioassays. Introduction to Environmental Toxicology, edited by F. E. Guthrie, and J. J. Perry, pp. 390–403. New York: Elsevier, 1980.
Rand, G. M.: Behavior. Fundamentals of Aquatic Toxicology, edited by G. M. Rand, S. R. Petrocelli, pp. 221–263. New York: Hemisphere, 1985.
Rand, G. M.: Basic toxicological considerations. Controlling Chemical Hazards/Fundamentals of the Management of Toxic Chemicals, edited by R. T. Côté, P. G. Wells, pp. 47–78. London: Unwin Hyman, 1991.
Rand, G. M., and Petrocelli, S. R. (eds.): Fundamentals of Aquatic Toxicology. Methods and Applications. Washington, DC: Hemisphere, 1985.
Reed, M.; Spaulding, M. L.; Lourda, E.; Walker, H.; and Saila, S. B.: Oil spill fishery impact assessment modeling the fisheries recruitment problem. *Estuarine Coastal Shelf Sci.*, 19(6):591–610, 1984.
Reid, G. K., and Wood, R. D.: Ecology of Inland Waters and Estuaries, 2nd ed. New York: Van Nostrand, 1976.
Reish, D. J.: Chapter 10. The use of toxicity testing in marine environmental research. Marine Organisms as Indicators, edited by D. V. Soule, and G. S. Kleppel, pp. 231–245. New York: Springer-Verlag, 1988.
Richardson, G. M, and Qadri, S. U.: Tissue distribution of ^{14}C-labeled residues of aminocarb in brown bullhead (*Ictalurus nebulosus* Le Sueur) following acute exposure. *Ecotoxicol. Environ. Safety*, 12:180–186, 1986.
Rodricks, J. V.: Calculated Risks. Understanding the Toxicity and Human Health Risks of Chemicals in Our Environment. Cambridge: Cambridge Univ. Press, 1992.
Rosenberg, D. M., and Resh, V. H. (eds.): Freshwater Biomonitoring and Benthic Macroinvertebrates. New York: Chapman & Hall, 1993.
Russo, R. C.: Ammonia, nitrite and nitrate. Fundamentals of Aquatic Toxicology, edited by G. M. Rand, and S. R. Petrocelli, pp. 455–471. New York: Hemisphere, 1985.
Samiullah, Y.: Prediction of the Environmental Fate of Chemicals. New York: Elsevier Applied Science, 1990.
Schindler, D. W.: Detecting ecosystem responses to anthropogenic stress. *Can. J. Fish Aquat. Sci.*, 44(Suppl 1):6–25, 1987.
Schnoor, J. L.; Sato, C.; McKechnie, D.; and Sahoo, D.: Processes, coefficients, and models for simulating toxic organics and heavy metals in surface waters. EPA/600/3-87/015. U.S. EPA Environmental Research Laboratory, Athens, GA, June 1987.
Sheehan, P.; Korte, F.; Klein, W.; and Bourdeau, P.: Appraisal of Tests to Predict the Environmental Behaviour of Chemicals. SCOPE 25. Chichester: Wiley, 1985.
Sheehan, P. J.; Miller, D. R.; Butler, G. C.; and Bourdeau, P.: Effects of Pollutants at the Ecosystem Level. SCOPE 22. Chichester: Wiley, 1984.
Shubert, E.: Algae as Ecological Indicators. New York: Academic Press, 1984.
Sijm, D. T. H. M., Opperhuizen A: Biotransformation, bioaccumulation, and lethality of 2,8-dichlorodibenzo-*p*-

dioxin: a proposal to explain the biotic fate of PCDD's and PCDF's. *Chemosphere* 17:83–89, 1988.

Smith, L. L.; Auerback, S. I.; Cairns, J., Jr.; Mount, D. I., Rohlich, G. A.; Sprague, J. B.; and Kleim, W. L.: ORSANCO 24-hour bioassay. Ohio River Valley Water Sanitation Commission, 1974.

Society of Environmental Toxicology and Chemistry (SETAC): Special Issue—Symposium on behaviorial toxicology convened by E. E. Little, and M. G. Henry, *Environ. Toxicol. Chem.*, 9:1–119, 1990.

Spacie and Hamelink, 1985. See Appendix Chapter D.

Spehar, R. L.; Nelson, H. P.; Swanson, M. J.; and Renoos, J. W.: Pentachlorophenol toxicity to amphipods and fathead minnows at different test pH. *Environ. Toxicol. Chem.*, 4:389–398, 1985.

Spellerberg, I. F.: Monitoring Ecological Change. New York: Cambridge Univ. Press, 1991.

Sprague, J. B.: Review paper. Measurement of pollutant toxicity to fish. I. Bioassay methods for acute toxicity. *Water Res.*, 3:793–821, 1969.

Sprague, J. B.: Review paper. Measurement of pollutant toxicity to fish. II. Utilizing and applying bioassay results. *Water Res.*, 4:3–32, 1970.

Sprague, J. B.: Review paper. Measurement of pollutant toxicity to fish—III. Sublethal effects and "safe" concentrations. *Water Res.*, 5:245–266, 1971.

Sprague, J. B.: The ABCs of pollutant bioassay with fish. J. Cairns, Jr., and K. L. Dickson (eds.), Biological Methods for the Assessment of Water Quality, edited by J. Cairns, Jr., and K. L. Dickson, pp. 6–30. STP 528. Philadelphia: American Society for Testing and Materials, 1973.

Sprague, 1985. See Appendix Chapter C.

Sprague, J. B.: Perspective on a career. Changing approaches to water pollution evaluation. *Mar. Pollut. Bull.*, 25(1-4): 6–13 1992.

Sprague, J. B., and Fogels, A.: Watch the Y in bioassay. Proceedings 3rd Aquatic Toxicity Workshop, Halifax, NS, November 2–3, 1976. Environmental Protection Sevice Technical Report No. EPS-5-AR-77-1, pp. 107–114, Halifax, NS, 1977.

Sprague, J. B., and Ramsay, B. A.: Lethal effects of mixed copper and zinc solutions for juvenile salmon. *J. Fish Res. Bd. Can.*, 22:425–432, 1965.

Stephan, C. E.: Increasing the usefulness of acute toxicity tests. Aquatic Toxicology and Hazard Assessment: Fifth Conference, edited by J. G. Pearson; R. B. Foster; and W. E. Bishop, pp. 69–81. ASTM STP 766. Philadelphia: American Society for Testing and Materials, 1982.

Stephan, C. E.; Mount, D. I.; Hansen, D. J.; Gentile, J. H.; Chapman, G. A.; and Brungs, W. A.: Guidelines for deriving national water quality criteria for the protection of aquatic organisms and their uses. NTIS (PB 85-227049). Washington, DC: U.S. EPA, 1985.

Stumm, W., and Morgan, J. J.: Aquatic Chemistry, 2nd ed. New York: Wiley-Interscience, 1981.

Stumm, W. (ed.): Aquatic Surface Chemistry. New York: Wiley, 1987.

Stumm, W. (ed.): Aquatic Chemical Kinetics—Reaction Rates of Processes in Natural Waters. New York: Wiley, 1990.

Suter, G. W. II (ed.): Ecological Risk Assessment. Boca Raton, FL: Lewis, 1993.

Tarazona, J.V.; Cebrain, M.; and Castaño, A.: Development of in vitro cytotoxicology tests using fish cell lines. Progress in Standardization of Aquatic Toxicity Tests, edited by M. V. M. Soares, and P. Calow, pp. 119–128. SETAC Special Publication Series. Boca Raton, FL: Lewis, 1993.

Tarr, B. D.; Barron, M. G.; and Hayton, W. L.: Effect of body size on the uptake and bioconcentration of di-2-ethylhexyl phthalate in rainbow trout. *Environ. Toxicol. Chem.* 9:989–995, 1990.

Tarzwell, C. M.: The use of bioassays in the safe disposal of electroplating wastes. American Electroplaters Society, 44th Annual Technical Proceedings, 1958.

Taub, F. B.: Standardizing an aquatic microsom test. Progress in Standardization of Aquatic Toxicity Tests, edited by M. V. M. Soares, and P. Calow, pp. 159–188. SETAC Special Publication Series. Boca Raton, FL: Lewis, 1993.

Thurman, E. M.: Organic Geochemistry of Natural Waters. Dordrecht: Kluwer, 1985.

Tjeerdema, R. S.; Singer, M.M.; and Smalheer, D.L.: Continuous-flow toxicity tests using the microscopic life stages of various marine organisms. Proceedings 17th Annual Aquatic Toxicity Workshop, November 5–7, 1990, Vancouver, BC, Vol I, pp. 348–354. Can. Tech. Rept. Fish. Aquat. Sci. No. 1774, Feb. 1991. Fisheries and Oceans, Ottawa, 1991.

Truhaut, R.: Ecotoxicology—a new branch of toxicology: a general survey of its aims, methods, and prospects. Ecological Toxicology Research, edited by AD McIntyre, CF Mills, pp. 3–24. New York: Plenum, 1975.

Truhaut, R.: Ecotoxicology: objectives, principles and perspectives. *Ecotoxicol Environ. Safety* 1:151–173, 1977.

U.S. Department of Health, Education, and Welfare, Food and Drug Administration: Nonclinical laboratory studies. Good laboratory practice regulations. *Fed. Regist.* 43: 59986–60025, 1978.

U.S. Environmental Protection Agency: Proposed environmental standards and proposed good laboratory practice standards for physical, chemicals, persistence, and ecological effects testing. *Fed. Regist.* 45:77332–77365, 1980.

van den Heuvel, M. R.; McCarty, L. S.; Lanno, R.P.; Hickie, D.E.; and Dixon, D. G.: Effect of total body lipid on the toxicity and toxicokinetics of pentachlorophenol in rainbow trout (*Oncorhynchus mykiss*). *Aquat. Toxicol.* 20:235–252, 1991.

von Westernhagen, H.: Sublethal effects of pollutants on fish eggs and larvae. Fish Physiology, Vol XI, The Physiology of Developing Fish, Part A, Eggs and Larvae, edited by W. S. Hoar, and D. J. Randall, pp. 253–347. New York: Academic Press, 1988.

Walker, M. K., and Peterson, R. E.: Potencies of polychlorinated dibenzo-*para*-dioxin, dibenzofuran, and biphenyl congeners, relative to 2,3,7,8-tetrachlorodibenzo-*para*-dioxin, for producing early life stage mortality in rainbow trout (*Oncorhynchus mykiss*). *Aquat. Toxicol.* 21: 219–238, 1991.

Walker, M. K.; Spitsbergen, J. M.; Olsen, J. R.; and Peterson, R. E.: 2,3,7,8-Tetrachlorodibenzo-*p*-dioxin toxicity in early life stage development of lake trout (*Salvelinus namaycush*). *Can. J. Fish. Aquat. Sci.* 48:875–883, 1991.

Ward, R. C.; Loftis, J. C.; and McBride, G.B.: The data-rich but information-poor syndrome in water quality monitoring. *Environ. Manag.* 10:291–297, 1986.

Warren, C. E., and Doudoroff, P.: The development of methods for using bioassays in the control of pulp mill disposal. *Tappi* 41:211A–216A, 1958.

Weideman, M.: Toxicity tests in animals: historical perspectives and new opportunities. *Environ. Health Perspect.* 101:222–225, 1993.

Wells, P. G.: Ecotoxicology—its role in the management of toxic substances. SETAC Sixth Annual Meeting Abstracts, New Perspectives in Environmental Toxicology and Chemistry, p. 35. November 1985, SETAC, Washington, DC (now Pensacola, FL), 1985.

Wells, P. G.: History and practice of biological effects assessment for aquatic protection in Canada: a synopsis. Biology in the New Regulatory Framework for Aquatic Protection. Proceedings of the Alliston Workshop, April 1988, edited by K. E. Day; E. D. Ongley; R.P. Scroggins; and H.R. Eisenhauer, pp. 25–29. Burlington, Ont: National Water Research Institute, 1989.

Wells, P. G.: Assessing and protecting the health of coastal ecosystems: the role and future of microscale marine toxicity tests. Paper given at the First SETAC World Congress, Ecotoxicology and Environmental Chemistry—a Global Perspective, Lisbon, Portugal, March 1993. Unpublished, 1993.

Wells, P. G., and Côté, R. P.: Protecting marine environmental quality from land-based pollutants: the strategic role of ecotoxicology. *Mar. Policy* 12:9–21, 1988.

Wells, P. G., and Moyse, C: A selected bibliography on the biology of *Salmo gairdneri* Richardson (rainbow, steelhead, Kamloops trout), with particular reference to studies with aquatic toxicants, 2nd ed. Environment Canada, Environmental Protection Service, Economic and Technical Review, Report EPS-3-AR-81-1, 1981.

Wells, P. G.; Abernethy, S. G.; and Mackay, D.: Study of oil-water partitioning of a chemical dispersant using an acute bioassay with marine crustaceans. *Chemosphere* 11:1071–1086, 1982.

White, H. H. (ed.): Concepts in Marine Pollution Measurements. College Park, MD: Maryland Sea Grant College, 1984.

Whitfield, M., and Turner, D. R.: The role of particles in regulating the composition of seawater. Aquatic Surface Chemistry, edited by W. Stumm, pp. 457–494. New York: Wiley, 1987.

Whitton, B. A.: plants as indicators of river water quality. Biological Indicators of Water Quality, edited by A. James, and L. Evison, pp. 5.1–5.35. Chichester: Wiley, 1979.

Whitton, B. A., and Say, P. J.: Heavy metals. River Ecology, edited by B. A. Whitton, pp. 268–311. Oxford: Blackwell, 1975.

Willford, W. A.: Toxicity of dimethyl sulfoxide (DMSO) to fish. *Bur. Sport Fish Wildl. Invest. Fish Control* 20:3–8, 1968.

Wisk, J. D., and Cooper, K. R.: The stage specific toxicity of 2,3,7,8-tetrachlorodibenzo-*p*-dioxin in embryos of the Japanese medaka (*Oryzias latipes*). *Environ. Toxicol. Chem.* 9:1159–1169, 1990.

World Health Organization (WHO) and United Nations Environment Programme (UNEP): Advances in Applied Biotechnology Series, Vol 5, Carcinogenic, Mutagenic, and Teratogenic Marine Pollutants: Impact on Human Health and the Environment. Houston: Gulf, 1989.

Yockim, R. S.; Isensee, A.R.; and Jones, G.E.: Distribution and toxicity of TCDD and 2,4,5-T in an aquatic model ecosystem. *Chemosphere* 7:215–220, 1978.

Zapp, J. A., Jr.: Historical consideration of interspecies relationships in toxicity assessment. Aquatic Toxicology, edited by J. G. Eaton; P. R. Parrish; and A. C. Hendricks, pp. 2–10. ASTM STP 707. Philadelphia: American Society for Testing and Materials, 1980.

SUPPLEMENTAL READING

The following are a group of books recommended for additional reading. Books are categorized into different subject areas.

Aquatic Ecology

Barnes, R. S. K., and Mann, K. H. (eds.): Fundamentals of Aquatic Ecology. Cambridge, MA: Blackwell Scientific, 1991.

Day, J. W., Jr.; Hall, C. A. S.; Kemp, W. M.; and Yáñez-Arancibia A.: Estuarine Ecology. New York: Wiley, 1989.

Jeffries, M., and Mills, D.: Freshwater Ecology—Principles and Applications. London: Bellhaven, 1993.

Sommer, U. (ed.): Plankton Ecology. Succession in Plankton Communities. New York: Springer-Verlag, 1989.

Aquatic/Environmental Chemistry (Fate)

Brown, J.; Colling, A.; Park, D.; Phillips, J.; Rothery, D.; and Wright, J.: Seawater: Its Composition, Properties and Behavior. Prepared by Open University Course Team. New York: Pergamon, 1992.

Drever, J. I.: The Geochemistry of Natural Waters, 2nd ed. Englewood Cliffs, NJ: Prentice-Hall, 1988.

Gobas, A. P. C., and McCorquodale, J. A. (eds.): Chemical Dynamics in Fresh Water Ecosystems. Boca Raton, FL: Lewis, 1992.

Hemond, H. F., and Fechner, E. J.: Chemical Fate and Transport in the Environment. New York: Academic Press, 1994.

Libes, S. M.: An Introduction to Marine Biogeochemistry. New York: Wiley, 1992.

Lyman, W. J.; Rheel, W. F.; and Rosenblatt, D. H.: Handbook of Chemical Estimation Methods. New York: McGraw-Hill, 1982.

Mackay, D.: Multimedia Environmental Models. The Fugacity Approach. Boca Raton, FL: Lewis, 1991.

Mackay, D.; Shiu, W. Y.; and Ma, K. C.: Illustrated Handbook of Physical-Chemical Properties and Environmental Fate for Organic Chemicals, Vol. I, Monoaromatic Hydrocarbons, Chlorobenzenes, and PCBs. Boca Raton, FL: Lewis, 1991.

Mackay, D.; Shiu, W. Y.; and Ma, K.C.: Illustrated Handbook of Physical-Chemical Properties and Environmental Fate for Organic Chemicals, Vol. II, Polynuclear Aromatic Hydrocarbons, Polychlorinated Dioxins, Dibenzofurans. Boca Raton, FL: Lewis, 1992.

Mackay, D.; Shiu, W. Y.; and Ma, K.C.: Illustrated Handbook of Physical-Chemical Properties and Environmental Fate for Organic Chemicals, Vol. III, Volatile Organic Chemicals. Boca Raton, FL: Lewis, 1993.

Manahan, S. E.: Environmental Chemistry, 5th ed. Chelsea, MI: Lewis, 1991.

Morel, F. M. M., and Hering, J. G.: Principles and Applications of Aquatic Chemistry. New York: Wiley, 1993.

Neely, W. B., and Blau, G. (eds.): Environmental Exposure from Chemicals, Vol. I and II. Chelsea, MI: CRC Press, 1985.

Pankow, J. F.: Aquatic Chemistry Concepts. Chelsea, MI: Lewis, 1991.

Schnoor, J. L. (ed.): Fate of Pesticides and Chemicals in the Environment. New York: Wiley, 1992.

Schwarzenbach, R. P.; Gschwend, P. M.; and Imboden, D. M.: Environmental Organic Chemistry. New York: Wiley, 1993.

Stumm, W., and Morgan, J. J.: Aquatic Chemistry. New York: Wiley, 1981.

Stumm, W. (ed.): Aquatic Surface Chemistry. New York: Wiley, 1987.

Stumm, W. (ed.): Aquatic Chemical Kinetics. New York: Wiley, 1990.

Stumm, W.: Chemistry of the Solid-Water Interface. New York: Wiley, 1992.

Thurman, E. M.: Organic Geochemistry of Natural Waters. Dordrecht: Martinus Nijhoff/Kluwer Academic, 1985.

Aquatic Microbiology

Ford, T. E. (ed.): Aquatic Microbiology—An Ecological Approach. Boston: Blackwell Scientific, 1993.
Rheinheimer, G.: Aquatic Microbiology, 4th ed. New York: Wiley, 1992.

Aquatic Pollution/Toxicology

Abel, P. D.: Water Pollution Biology. New York: Ellis Horwood, Halsted, 1989.
Boudou, A., and Ribeyre, F. (eds): Aquatic Ecotoxicology: Fundamental Concepts and Methodologies, Vols. I and II. Boca Raton, FL: CRC Press, 1989.
deKruijf, H. A. M.; deZwart, D.; Ray, P. K.; and Viswanathan, P. N. (eds.): Manual on Aquatic Ecotoxicology. Cambridge, MA: Kluwer Academic, 1988.
Heath, A. G.: Water Pollution and Fish Physiology. Boca Raton, FL: CRC Press, 1987.
Hellawell, J. M.: Biological Indicators of Freshwater Pollution and Environmental Management. New York: Elsevier Applied Science, 1986.
Kennish, M. J.: Ecology of Estuaries: Anthropogenic Effects. Boca Raton, FL: CRC Press, 1992.
Laws, E. A.: Aquatic Pollution, 2nd ed. New York: Wiley, 1993.
Malins, D. C., and Ostrander, G. K. (eds.): Aquatic Toxicology—Molecular, Biochemical and Cellular Prospectives. Boca Raton, FL: Lewis, 1994.
Mason, C. F.: Biology of Freshwater Pollution, 2nd ed. New York: Longman Scientific and Technical, 1991.
Munawar, M.; Dixon, G.; Mayfield, C. I.; Reynoldson, T.; and Sadar, M. H. (eds.): Environmental Bioassay Techniques and Their Application. Hydrobiologia 188/189. Dordrecht: Kluwer Academic, 1989.
Rand, G. M., and Petrocelli, S. R. (eds.): Fundamentals of Aquatic Toxicology. New York: Hemisphere, 1985.

Bioaccumulation

Connell, D. W.: Bioaccumulation of Zenobiotic Compounds. Boca Raton, FL: CRC Press, 1990.
Nagel, R., and Loskill, R. (eds.): Bioaccumulation in Aquatic Systems. New York: VCH, 1991.
Walker, C. H., and Livingstone, D. R. (eds.): Persistent Pollutants in Marine Ecosystems. Oxford: Pergamon, 1992.

Biomarkers

Huggett, R. J.; Kimerle, R. A.; Mehrle, P.M.; and Bergman, H. L. (eds.): Biomarkers: Biochemical, Physiological and Histological Markers of Anthropogenic Stress. Boca Raton, FL: Lewis, 1992.
McCarthy, J. F., and Shugart, L. R. (eds.): Biomarkers and Environmental Contamination. Boca Raton, FL: Lewis, 1990.
Peakall, D. B.: Animal Biomarkers as Pollution Indicators. New York: Chapman & Hall, 1992.
Peakall, D. B., and Shugart, L. R. (eds.): Biomarkers-Research and Application in the Assessment of Environmental Health. NATO ASI Series. Berlin: Springer-Verlag, 1993.

Biomonitoring/Water Quality

Bergman, H. L.; Kimerle, R. A.; and Maki, A.W. (eds.): Environmental Hazard Assessment of Effluents. New York: Pergamon, 1986.
Cairns, J., Jr., (ed.): Biological Monitoring in Water Pollution. New York: Pergamon, 1982.
Chapman, D. (ed.): Water Quality Assessments. A Guide to the Use of Biota, Sediments and Water in Environmental Monitoring. New York: Chapman & Hall, 1992.
Rosenberg, D. M., and Resh, V. H. (eds.): Freshwater Biomonitoring and Benthic Macroinvertebrates. New York: Chapman & Hall, 1993.
Spellerberg, I. F.: Monitoring Ecological Change. New York: Cambridge Univ. Press, 1991.

Ecotoxicology (General)

Calo, P. (ed.): Handbook of Ecotoxicology, Vols. 1, 2. Cambridge, MA: Blackwell Scientific, 1993, 1994.
Forbes, V. E., and Forbes, T. L.: Ecotoxicology in Theory and Practice. Ecotoxicology Series 2. New York: Chapman & Hall, 1994.
Levin, S. A.; Harwell, M. A.; Kelly, J. R.; and Kimball, K. D. (eds.): Ecotoxicology: Problems and Approaches. New York: Springer-Verlag, 1989.
McKenzie, D. H.; Hyatt, D. E.; and McDonald, V. J. (eds.): Ecological Indicators, Vols. 1 and 2. New York: Elsevier Applied Science, 1992.
Moriarity, F.: Ecotoxicology—The Study of Pollutants in Ecosystems, 2nd ed. New York: Academic Press, 1988.

Multispecies Toxicity Studies (Includes Laboratory Microcosms and Outdoor Microcosms/Mesocosms)

Cairns, J. Jr., (ed.): Multispecies Toxicity Testing. New York: Pergamon, 1985.
Cairns, J. Jr., and Niederlehner, B. R. (eds.): Ecological Toxicity Testing. Boca Raton, FL: Lewis, 1995.
Graney, R. L.; Kennedy, J. H.; and Rodgers, J. H., Jr. (eds.): Aquatic Mesocosm Studies in Ecological Risk Assessment. SETAC Special Publication Series. Boca Raton, FL: Lewis, 1993.
Grice, G. D., and Reeve, M. R. (eds.): Marine Mesocosms—Biological and Chemical Research in Experimental Ecosystems. New York: Springer-Verlag, 1982.
Lalli, C. M. (ed.): Enclosed Experimental Marine Ecosystems: A Review and Recommendations. New York: Springer-Verlag, 1990.

Risk Assessment

Bartell, S. M.; Gardner, R. H.; and O'Neill, R. V.: Ecological Risk Estimation. Chelsea, MI: Lewis, 1992.
Maughan, J. T. (ed.): Ecological Assessment of Hazardous Waste Sites. New York: Van Nostrand Reinhold, 1993.
Suter, G. W., II (ed.): Ecological Risk Assessment. Boca Raton, FL: Lewis, 1993.

Sediment

Baker, R. A. (ed.): Organic Substances and Sediments in Water, Vol. 1, Humics and Soils. Boca Raton, FL: Lewis, 1991.
Baker, R. A. (ed.): Organic Substances and Sediments in Water, Vol 2, Process and Analysis. Boca Raton, FL: Lewis, 1991.
Baker, R. A. (ed.): Organic Substances and Sediments in Water, Vol 3, Biological. Boca Raton, FL: Lewis, 1991.
Baudo, R.; Giesy, J. P.; and Muntau, H. (eds.): Sediments: Chemistry and Toxicity of In-Place Pollutants. Chelsea, MI: Lewis, 1990.
Burton, G. A., Jr., (ed.): Sediment Toxicity Assessment. Boca Raton, FL: Lewis, 1992.
Horowitz, A. J.: Sediment-Trace Element Chemistry, 2nd ed. Chelsea, MI: Lewis, 1991.

PART I

EFFECTS—TOXICITY TESTING

The first part of the book describes in detail single-species laboratory toxicity tests and both natural and simulated multispecies field studies to evaluate the effects of toxic agents on aquatic organisms. Chapter 2 reviews the design and procedures for laboratory tests on freshwater vertebrates and invertebrates with single chemicals and effluents, along with examples of test data and general statistical methods for analysis of data. General test designs presented in Chapters 1 and 2 are applicable for laboratory tests discussed throughout the book. Chapter 3 presents saltwater laboratory test procedures for vertebrates and invertebrates, and Chapter 4 is devoted to algae (micro) and vascular plant tests in fresh water and salt water. Chapters 5, 6, and 7 review procedures for single-species tests with macroalgae (or seaweeds), sea urchin sperm cells, and frog embryos. Chapter 8 focuses on tests with sediment, because this is an important habitat and a potential reservoir for chemicals with low water solubility.

Multispecies field studies in simulated (e.g., microcosm, mesocosm) and natural (e.g., ponds) systems are described in Chapter 9. Although statistical procedures are discussed in many of the chapters on single-species tests and in the chapter on field studies, Chapter 10 is included to provide the reader with a more comprehensive background on the design of experiments and analysis of data in aquatic toxicology studies. Because of the importance of developing reliable and defensible quality data, good laboratory practices (GLPs) are summarized in Chapter 11. Finally, Chapters 12, 13, and 14 review several new tests on changes in immune function, DNA, and tissues (carcinogenicity) as a result of chemical exposure. Such changes may be used as "biomarkers" (see Chapter 1) or indicators of either exposure to or effects of a chemical(s) (see also Chapter 17).

Chapter 2

FRESHWATER TESTS

J. D. Cooney

INTRODUCTION

Background

Natural fresh waters (i.e., ground water, surface waters) are the ultimate recipients of most toxic substances generated by industrial, agricultural, and domestic activities and released into the environment (Anderson and D'Apollonia, 1978). Management of toxic substances entering natural waters is difficult because contaminants often enter an aquatic system from multiple or diffuse sources. Although aquatic ecosystems are adaptable with a variety of physical, chemical, and biological mechanisms by which toxic substances may be assimilated without serious implications for endemic biota, when chemical contaminants reach levels in excess of the assimilative capacity of the receiving waters they may affect survival, development, growth, reproduction, or behavior (movement) of the organisms (Anderson and D'Apollonia, 1978).

Adverse effects on aquatic ecosystems may result from toxicant exposure that directly causes the death of an organism (acute effects) or produces sublethal effects on the organism's ability to develop, grow, and reproduce in the ecosystem (chronic effects). Economic, recreational, and agricultural use of aquatic ecosystems may also be impaired by toxic chemicals; human health may be adversely affected through uptake of contaminants from drinking water and/or consumption of flesh from contaminated organisms.

To identify and control toxic inputs to aquatic systems requires that laboratory toxicity tests be performed and that pollutant levels are extrapolated from these laboratory-derived data that can be deemed "safe" for aquatic ecosystems (Anderson and D'Apollonia, 1978). Based on results of acute and chronic toxicity tests, it then becomes necessary to decide what level of impairment observed under laboratory conditions would constitute a detrimental effect in natural aquatic systems. Thus, toxicity tests are the initial "tools" that provide the qualitative and quantitative data on adverse or toxic effects of chemicals and other toxicants on aquatic organisms that can then be used to assess the potential for, or degree of, damage to an aquatic ecosystem. Toxic effects in the laboratory may include both lethality (mortality) and sublethal effects such as changes in development, growth, reproduction, metabolism, physiology, and behavior; the effects measured during the tests may be expressed by quantifiable criteria such as number of organisms killed, number of eggs produced (fecundity), percent of eggs hatched, number of neonates produced, changes in length and/or weight, and rates of oxygen consumption or other metabolic parameters and qualitatively by changes in behavior of the organisms (e.g., avoidance response in fish).

This chapter summarizes general methodology and techniques used in freshwater laboratory toxicity testing; however, most of the discussion is also relevant for saltwater organisms (see Chapter 3). Emphasis is also on new effluent (versus single chemical) toxicity methodology, because general aquatic toxicity testing techniques for single chemicals have not changed significantly from those in Rand and Petrocelli (1985).

Regulatory History

Until recently, most pollution abatement efforts were aimed at controlling sewage and wastewater discharges into receiving waters. These discharges not only were highly visible and potentially dangerous to public health but also could produce odorous anaerobic water, fish kills, and aesthetic impairment of receiving waters, especially during critical low-flow periods.

The emphasis on water pollution control activities in the United States and the increase in the use of freshwater and saltwater aquatic toxicity tests have been driven primarily by various federal regulations for protection of humans (human health) and, at a later date, protection of aquatic life. During the past 25 to 30 yr, emphasis has changed from protection of human health to protection of aquatic life; this change is important because water quality requirements of fish and other aquatic organisms may often be more restrictive than the requirements for human health. For example, a "safe" level of copper in drinking water for humans is 1 mg/L (Federal Register, 1989) compared with 0.018 mg/L for acute aquatic life criteria (U.S. EPA, 1986a).

With the formation of the U.S. Environmental Protection Agency (EPA) in 1970, the regulatory authority in the United States and basis for water pollution control activities, pesticide risk assessment, and toxic substance control have rested with the EPA. The regulatory authority comes mainly from three acts:

- Federal Insecticide, Fungicide, and Rodenticide Act (FIFRA)
- Toxic Substances Control Act (TSCA)
- Clean Water Act (CWA)

The Federal Insecticide, Fungicide, and Rodenticide Act (FIFRA), passed in 1947 and reauthorized in 1972, requires that any pesticide distributed or sold in the United States be registered with the EPA (see Chapter 21 for detailed information on FIFRA). The passage of the Federal Toxic Substances Control Act (TSCA) of 1976 gave the government broad authority to control the production, distribution, and use of all potentially hazardous industrial chemicals; TSCA represents an attempt to establish a mechanism whereby the hazard to human health and the environment of a chemical substance can be assessed before it is introduced into the environment (see Chapters 22 and 23 for detailed information on TSCA). Under these acts, EPA must regulate pesticides and chemicals so as to protect the environment and yet reap the benefits that derive from their use (Moore, 1987). Acute and chronic aquatic toxicity tests are a significant portion of the testing performed under FIFRA (U.S. EPA, 1982a; 1985a, b; 1986b,c,d) and TSCA (U.S. EPA, 1982b; Federal Register, 1985) to assess the degree of risk to the aquatic environment.

In 1984 EPA issued a national policy under the Clean Water Act (CWA) for development of water quality-based permit limitations for toxic pollutants (Federal Register, 1984) that addressed control of toxic pollutants in effluent discharges beyond technology-based requirements to meet water quality standards (see Chapter 24 for detailed information on CWA). The national policy and subsequently the Technical Support Document for water quality-based toxics control (U.S. EPA, 1991a) emphasize an integrated approach using chemical-specific analyses and acute and chronic aquatic toxicity tests (U.S. EPA, 1989a, 1991b) to evaluate effects of effluent discharges and to establish permit limitations under the National Pollutant Discharge Elimination System (NPDES) program.

Other international organizations are following the lead of the United States by increasing their focus on controlling toxicants in aquatic systems. The Organization of Economic Cooperation and Development (OECD), comprising an international group of nations (mainly in Europe), was formed to ensure better health conditions as new products reach the international marketplace. In addition, the founders of OECD hoped that some standardization of toxicological testing methodology would enable regulatory agencies of different countries to work in concert. Many of the specific testing procedures required by EPA under TSCA are the same as or similar to guidelines issued by the OECD (OECD, 1987; see Chapter 26 for detailed information on OECD). Similar water quality-based toxics control activities are being performed in Canada, through activities within Environment Canada, with the development of numerous biological test methods to protect aquatic life (Environment Canada, 1990a,b,c,d,e,f; 1992a,b,c,d). More recently, the Food and Drug Administration has implemented the use of acute freshwater toxicity tests in their Environmental Assessment for all new Drug Applications (see Chapter 25).

The enactment of FIFRA, TSCA, and CWA (and their amendments) in the United States has been a powerful stimulus to the development of standardized toxicity testing procedures to evaluate acute and chronic hazards to human health and the environment associated with potentially toxic substances. However, effective implementation of these acts depends on the development and use of an effective toxicity evaluation process. No single test method or test organism can be expected to satisfy a comprehensive approach to environmental conservation and protection; therefore, an evaluation process must be developed that is not only cost-effective, but also truly protective and does not depend on use of only a few organisms and end points as indicators of effects on health of humans, animals, and plants.

FRESHWATER TOXICITY TEST PROCEDURES

During the past three decades, the state of the art of freshwater toxicological testing has advanced to meet or exceed most expectations. However, regulatory requirements for biological testing have escalated during the same period. The EPA alone has been mandated under TSCA (i.e., TSCA Inventory List) to assess the hazard potential of 50,000 to 70,000 chemicals already in production in the United States. EPA adds about 1000 new chemicals to the TSCA Inventory List each year.

The quantification of toxicant stress on aquatic biota relies principally on traditional acute and chronic test methods. As previously described in detail in Chapter 1, the terms acute and chronic are sometimes interpreted differently by various investigators; this lack of precision and accuracy has led to much ambiguity in terminology in aquatic toxicology (Chapman, 1989). Acute toxicity tests are usually defined by their short duration (usually 2 to 4 d) and simple experimental design; the end points used most commonly in acute tests are lethality or immobilization of juvenile organisms. Because of these factors, acute tests tend to be less costly to perform than chronic tests; consequently, the majority of the toxicological database is skewed in favor of acute responses (Birge et al., 1985). Historically, data from acute tests were often used to extrapolate to chronic effects through the use of "safety" or "application" factors (Stephan and Mount, 1973; Mount, 1977; Anderson and D'Apollonia, 1978; Laws, 1981; Cairns, 1991; see Chapter 1). Acute aquatic toxicity tests are discussed in detail in Appendix Chapter A.

However, the need to assess more accurately the potential hazards of toxicants to aquatic biota required more empirical methods for studying the effects of a toxicant through an entire reproductive period (i.e., embryo, larva, juvenile, adult, and reproduction). Thus, during the mid-1960s to mid-1970s, the use of full life cycle chronic toxicity tests with fish and invertebrates increased as methods for culturing indigenous fish and invertebrates improved (Appendix Chapter B). The first life cycle test with fish was performed using fathead minnows in 1967 by Mount and Stephan. Within the next 5 to 6 yr, life cycle tests were performed with other freshwater fish including bluegill (Eaton, 1970, 1973), brook trout (McKim and Benoit, 1971, 1974), and flagfish (Smith, 1973).

Invertebrate life cycle chronic tests were also being performed during this time with cladocerans, *Daphnia magna*, (Biesinger and Christensen, 1972), amphipods (Arthur, 1980), gastropods (Arthur, 1970), and midges (Derr and Zabik, 1972a,b; Nebeker, 1973). Benthic organisms used in freshwater toxicity tests are discussed more fully in Chapter 8 on sediment toxicity tests.

Although the use of chronic full life cycle tests provided more accurate information on potential toxicant effects by exposing the organisms from embryo to embryo, the time and costs associated with these tests were much greater than with acute tests. Full life cycle tests with fathead minnows and brook trout may take 9 months and 30 mo, respectively, to complete and cost between 100 and 200 thousand dollars (Appendix Chapter B). The effort to perform these tests successfully also requires strict standardization of the methods and closely controlled experimental conditions during the entire test period (Petrocelli, 1985). Although life cycle tests with invertebrates were significantly longer than acute tests (3 wk to 5 mo for chronic tests versus 1 to 4 d for acute tests), the costs, in terms of manpower and dollars, were much less than for fish life cycle tests. Chronic full life cycle tests in invertebrates and fish are discussed in detail by Petrocelli (1985).

Therefore, investigators focused on developing partial life cycle and early life stage (ELS) chronic tests with more diverse fish species that could be used to determine relative sensitivities to toxicant stress but would be shorter than the 1 to 3 yr to perform full life cycle tests. Partial life cycle tests still expose the critical life stages (i.e., embryos, larvae, early juveniles) but usually start with juvenile fish (F_0 generation) approaching maturity (about 1 yr old for bluegills and brook trout); the tests begin before active gonad development and end 30 to 60 d after the hatching of the next generation (F_1 generation). ELS tests range from 28 to 32 d for warmwater fish (such as fathead minnows and bluegills) to 60 d posthatch for salmonids; the tests begin with ova fertilization and extend through embryonic, larval, and early juvenile development. See Appendix Chapter B for a complete description of these tests.

The use of early life stage, partial life cycle, and full life cycle tests was emphasized in the mid-1970s in the evaluation of sublethal effects of toxicants to fish. Although life cycle tests with invertebrates were substantially shorter than chronic tests with fish, most invertebrate chronic toxicity tests are still life cycle (i.e., embryo to embryo) tests. Partial life cycle and ELS fish tests required less cost, time, and effort to perform, however, economic constraints and time required to perform these tests still limit the number of tests that can be done.

The duration of exposure and, in many tests, dependence on flow-through test conditions limit the use of these chronic methods for broad-based toxicological testing and biomonitoring studies of effluent discharges. Thus, the traditional full life cycle, partial life cycle, and ELS chronic tests with fish are not practical for routine toxicity evaluations. In response to the regulatory need for rapid screening of effluents for chronic toxicity, EPA investigators and others (Mount and Norberg, 1984; Norberg and Mount, 1985; Birge et al., 1985; U.S. EPA, 1989a) developed four freshwater short-term tests ($\leq$7 d in duration) as substitutes for the traditional chronic tests which attempt to estimate chronic effects of effluents within a short time frame. Saltwater short-term chronic tests are discussed in Chapter 3. The freshwater short-term chronic tests were:

- 7-d static-renewal larval survival and growth test using the fathead minnow, *Pimephales promelas*
- 7-d static-renewal embryo-larval survival and teratogenicity test using the fathead minnow, *P. promelas*
- 7-d static-renewal survival and reproduction test using the cladoceran, *Ceriodaphnia dubia*
- 4-d static growth test using the green alga, *Selenastrum capricornutum*

Two of the four short-term chronic tests use early life stages (eggs or larvae) of fathead minnows; one test is a modification of the full life cycle (i.e., 21–28-d tests) test with *D. magna*, but uses another daphnid, *C. dubia*, that can be exposed from embryo-to-embryo in 7 to 9 d. The fourth test is an algal test originally developed to assess nutrient enrichment of natural surface waters (U.S. EPA, 1971) that has been modified for use in effluent toxicity assessments; toxicity tests with algae and other aquatic plants are discussed in more detail in Chapters 4 and 5. For *C. dubia*, in some instances, the 7-d test has been reduced to 4 d using reproductively mature animals instead of animals less than 24 h old (Oris et al., 1991; Masters et al., 1991). Other investigators have developed shorter acute and chronic life cycle test with various freshwater rotifers, *Brachionus* spp. (Snell and Persoone, 1989a,b; Snell et al., 1991a,b; Snell and Moffat, 1992; ASTM, 1992l). In these tests, rotifers are exposed to various concentrations of the test substance for 1 d for acute tests and for a minimum of 2 d for chronic toxicity tests; the acute end point is lethality and the chronic end point is measured by using the intrinsic rate of population increase [$r = \ln N_t$ (number of rotifers at end of study) $- \ln N_0$ (number of rotifers at beginning of study) divided by t (time)].

Toxicologists are now faced with an increased demand on resources. Because of the increase in toxicity testing requirements, it is not feasible or practical to perform even acute toxicity tests, let alone full life cycle toxicity tests with all the vast number of species and chemicals of concern. To that extent, the degree of extrapolation in the application of aquatic toxicity data is frequently much greater than that associated with human health assessment. To address in a timely and cost-effective manner the pressing need for determining "safe levels" of toxicants in aquatic systems, it is necessary to expand into the realm of predictive aquatic toxicology using short-term chronic tests. We have only recently developed an empirical database on chronic toxicity of chemicals to a variety of aquatic forms against which to measure the predictive capabilities of short-term tests or to extrapolate chronic data from one aquatic species to another.

Currently, water quality criteria (see Chapter 24) for the protection of aquatic life (U.S. EPA, 1986a) are derived from the maximum acceptable toxicant concentrations (MATCs) as the geometric mean (i.e., point estimate) of the no observed effects and lowest observed effects concentrations (NOECs and LOECs) from the most sensitive biological end points from the chronic toxicity tests. However, with the effort to reduce the duration of chronic tests, Suter (1990) has pointed out that the validity of the assumption that MATCs developed from short-term chronic tests are thresholds for chronic toxicity is questionable because of reductions in the duration of chronic tests and disagreement about the appropriate expression of results. The disagreement results in part from the existence of four distinct interpretations of the term "chronic" effects among aquatic toxicologists:

- Chronic effects result from long-term exposures.
- Chronic effects include exposure of all life stages and processes.
- Chronic effects are not severe.
- Chronic effects occur at lower chemical concentrations than other effects.

Outside aquatic toxicology, the term "chronic" still refers to the duration of the test and is formally defined as an exposure of greater than 10% of the organism's life span (Rand, 1980). An implication of this is that either the toxicant or its effects accumulate in the organism over time, resulting in effects that are not seen with shorter exposures. The 7-d short-term effluent toxicity test with *C. dubia* is a sensitive test that fulfills the chronic requirement of exposure to all life stages (i.e., embryo to embryo). However, a 7-d fish test (i.e., one that includes only one life stage and has a duration of less than 1% of the life span of the organism) is not a chronic test, and these tests are predictive of chronic toxic effects in the sense that a 96-h LC50 is predictive of chronic toxic effects: they are correlated, but error in correlation may be large. Accurate predictions of chronic toxicity still require chronic tests (Suter, 1990; Norberg-King, 1989).

The 7-d short-term chronic test with fathead minnows is sensitive because fish larvae are usually among the most vulnerable stages of the entire life cycle (Woltering, 1984; Norberg and Mount, 1985; Suter et al., 1987; Norberg-King, 1989; see also Appendix Chapter B). In general, the 7-d test may be expected to predict the toxicity in a 30-d exposure of early life stages of fathead minnows very closely in some cases (from NOECs reported by Norberg-King, 1989) and within a factor of 2 in most cases, but they may underpredict by an order of magnitude. The 7-d fish test may underpredict the sublethal toxicity in a complete life cycle test of fathead minnows by factors of 2 to 3 in many cases, but sometimes by factors of 25 to 45 or more. Thus, the 7-d larval test does not replace the results from chronic toxicity tests, but it comes closer to such chronic results than a conventional 96-h acute lethality test with juvenile fish. For practical and economic reasons, the short-term (4 to 7 d) chronic tests may be the best tests of effluent toxicity. However, for testing of single chemicals under TSCA and FIFRA, the longer-term tests that expose the critical life stages (development, growth, and reproduction) are more appropriate.

METHODOLOGY

For the past 90 yr, the methods manual most commonly used by investigators for the examination of water and wastewaters has been Standard Methods for the Examination of Water and Wastewater, published jointly by the American Public Health Association (APHA), the American Water Works Association (AWWA), and the Water Pollution Control Federation (WPCF)*. The first edition of the Standard Methods of Water Analysis appeared in 1905 (APHA, 1905) and included techniques suitable for the physical, chemical, microscopic, and bacteriological examination of water. Each subsequent edition presented improvements of methodology and an enlarged scope to include techniques suitable for examination of many types of samples encountered in the assessment and control of water quality and water pollution.

*Water Pollution Control Federation (WPCF) is now called Water Environment Federation (WEF).

By 1971 the 13th edition of Standard Methods (APHA, 1971) still contained only one method for acute toxicity tests with fish. In 1973 EPA published a manual to complement the acute laboratory methods in the 13th edition of Standard Methods and included two procedures for chronic full life cycle toxicity tests with fathead minnows (*Pimephales promelas*) and brook trout (*Salvelinus fontinalis*) (U.S. EPA, 1973). The trend toward standardization of toxicity test methods was further promoted by the Committee on Methods for Toxicity Tests with Aquatic Organisms, which resulted in the publication in 1975 of EPA's first guidance manual on conducting acute toxicity tests (U.S. EPA, 1975). By 1985, acute toxicity test methods for aquatic organisms accounted for approximately 10% of the text of the 16th edition of the Standard Methods (Part 800) (APHA, 1985) (Table 1). The most recent edition (18th edition) of Standard Methods (1992) phylogenetically reorganized the various test methods but did not add any substantially new freshwater methods. The subsections are currently arranged by type of test organism as follows: Algae (Part 8110), Biostimulation (Algal Productivity; Part 8111), Duckweed (Part 8211), Aquatic Plants (Part 8220; Proposed), Ciliated Protozoa (Part 8310), Annelids (Part 8510), Mollusks (Part 8610), Microcrustacea (Part 8710), Macrocrustacea (Part 8720), Aquatic Insects (Part 8750), and Fish (Part 8910). Algae and vascular plants are discussed in Chapter 4.

Another consensus organization, the American Society for Testing and Materials (ASTM), has also been in the forefront of methods standardization for aquatic toxicity testing; Committee E-47 on Biological Effects and Environmental Fate has provided guidance for standardization of acute and chronic test methods with freshwater organisms. In the 1992 ASTM Annual Book of Standards, 16 standard guides or practices for freshwater organisms were listed (Table 1).

The aquatic toxicity test procedures described in ASTM, APHA Standard Methods, EPA, and other test procedure manuals (Table 1) are typically classified according to duration (i.e., short-term, intermediate, or long-term), method of adding test solutions (i.e., flow conditions), and purpose (i.e., toxicity end points). Toxicity end points (i.e., criteria for effects) are often related to duration of exposure to the test solution and life stage(s) of the exposed organisms.

Four types of exposure conditions (i.e., static, static-renewal, recirculation, and flow-through) are commonly used in both acute and chronic freshwater and saltwater toxicity tests (ASTM, 1992a,b,c,d,e,f,g,h,i,j,k,l,m,n,o,p; APHA, 1992; U.S. EPA, 1991b). The four exposure types are defined in Chapter 1. Descriptions of flow-through systems are presented in this chapter (see *Daphnia* chronic case study) and in Appendix Chapter A. The determination of which exposure condition to use in acute and chronic toxicity tests usually depends on test substance characteristics, test duration, and regulatory requirements.

Dilution Water

The source of water for culturing of test organisms and of dilution water used in acute and chronic toxicity tests depends on the objectives of the study. If the objective of the study is to estimate the inherent toxicity of the test substance, a standard laboratory-prepared water is usually used. If the objective of the study is to estimate the toxicity of the test substance in an uncontaminated surface water, which is often the case with many toxicity tests with effluents, the water sample should be representative of the receiving water. It should be collected from an uncontaminated source outside the zone of influence of the effluent and be used as the test dilution and control water.

The dilution water should be available in adequate supply, be acceptable to the test organisms, be of uniform quality, and not unnecessarily affect results of the test. The minimal requirement for an acceptable dilution water is that healthy test organisms survive in it through acclimation and testing without showing signs of stress, such as discoloration, unusual behavior, or death (ASTM, 1992a,b).

A good freshwater supply, such as from a spring, well, or surface water, should be uncontaminated, constant in quality, and not contain more than the designated amounts of the following (Federal Register, 1985):

Parameter	Concentration
Total suspended solids	20 mg/L
Total organic carbon	2 mg/L
Chemical oxygen demand	5 mg/L
Un-ionized ammonia	1 μg/L
Total residual chlorine	<3 μg/L
Total organophosphorus pesticides	50 ng/L
Total organochlorine pesticides plus PCBs	50 ng/L
Organic chlorine	25 ng/L

Laboratory-prepared water is usually practical only for culturing small organisms; it is usually not cost-effective to use reconstituted water for large-scale rearing or for flow-through tests with fish. Where laboratory-prepared dilution water is required for a test, the water is generally considered acceptable if test organisms show the required survival, growth, and reproduction in the controls during the test. Standard fresh water is prepared by adding reagent-grade chemicals or mineral water to glass-distilled or deionized water to make up water to various chemical compositions (APHA, 1992; ASTM, 1992a; U.S. EPA, 1989a, 1991b) (Table 2). Deionized water may be obtained from a Millipore Milli-Q system or equivalent.

Tap water may be used as a source of dilution water if properly treated. However, tap water may contain high levels of metals, such as copper (from copper piping) and/or zinc (from corrosion control), or total residual chlorine (from disinfection) that are toxic to

Table 1. Freshwater toxicity test guidelines

Designation	Title
	Standard Methods, Part 800 (APHA et al., 1985)
802[a] (8111)[b]	Biostimulation (Algal Productivity)
803 (8112)	Toxicity Testing with Phytoplankton
804 A (8310)	Toxicity Test Procedures for Ciliated Protozoa
804 B (8710)	Toxicity Test Procedures for *Daphnia*
806 (8510)	Toxicity Test Procedures for Annelids
807 (8720)	Toxicity Test Procedures for Crustaceans
808 (8750)	Toxicity Test Procedures for Aquatic Insects
810 A (8910)	Toxicity Test Procedures for Fish
	ASTM (ASTM, 1992a–p)
E-729-88a	Guide for Conducting Acute Toxicity Tests with Fishes, Macroinvertebrates, and Amphibians
E-1192-88	Guide for Conducting Acute Toxicity Tests on Aqueous Effluents with Fishes, Macroinvertebrates, and Amphibians
E-1193-87	Guide for Conducting Renewal Life-Cycle Toxicity Tests with *Daphnia magna*
E-1241-92	Guide for Conducting Early Life-Stage Toxicity Tests with Fishes
E-1295-89	Guide for Conducting Three-Brood, Renewal Toxicity Tests with *Ceriodaphnia dubia*
E-1023-84	Guide for Assessing the Hazard of a Material to Aquatic Organisms and Their Uses (Reapproved in 1988)
E-1383-92	Guide for Conducting Sediment Toxicity Tests with Freshwater Invertebrates
E-1440-91	Guide for Acute Toxicity Test with the Rotifer, *Brachionus*
D-4229-84	Practice for Conducting Static Acute Toxicity Tests on Wastewaters with *Daphnia* (Discontinued in 1990)
E-1022-84	Practice for Conducting Bioconcentration Tests with Fishes and Saltwater Bivalve Molluscs (Reapproved in 1988)
D-3696-89	Practice for Evaluating an Effluent for Flavor Impairment to Fish Flesh
E-1203-87	Practice for Using Brine Shrimp Nauplii as Food for Test Animals in Aquatic Toxicology
E-1366-91	Practice for Standardized Aquatic Microcosm: Fresh Water
E-1218-90	Guide for Conducting Static 96-h Toxicity Tests with Microalgae (see Chapter 4)
E-1415-91	Guide for Conducting Static Toxicity Tests with *Lemma gibba* G3 (see Chapter 4)
D-3978-80	Practice for Algal Growth Potential Testing with *Selenastrum capricornutum* (Reapproved in 1987; see Chapter 4)
E-1439-91	Guide for Conducting the Frog Embryo Teratogenesis Assay-*Xenopus* (FETAX) (see Chapter 7)
	EPA Acute Effluent Tests (U.S. EPA, 1991b)
—	Acute Effluent Toxicity Tests with *Ceriodaphnia dubia*
—	Acute Effluent Toxicity Tests with *Daphnia pulex* and *D. magna*
—	Acute Effluent Toxicity Tests with Fathead Minnows (*Pimephales promelas*)
—	Acute Effluent Toxicity Tests with Rainbow Trout (*Oncorhynchus mykiss*) and Brook Trout (*Salvelinus fontinalis*)
	EPA Chronic Effluent Tests (U.S. EPA, 1989a)
1000.0	Fathead minnow (*Pimephales promelas*) Larval Survival and Growth Effluent Toxicity Test
1001.0	Fathead minnow (*Pimephales promelas*) Embryo-Larval Survival and Teratogenicity Effluent Toxicity Test
1002.0	Cladoceran (*Ceriodaphnia dubia*) Survival and Reproduction Effluent Toxicity Test
1003.0	Algal (*Selenastrum capricornutum*) Growth Test
	EPA TSCA Test Guidelines—Subpart B (U.S. EPA, 1982b; Federal Register, 1985)
797.1050	Algal Acute Toxicity Test (see Chapter 4)
797.1060	Freshwater Algae Acute Toxicity Test (see Chapter 4)
797.1160	*Lemna* Acute Toxicity Test (see Chapter 4)
797.1300	Daphnid Acute Toxicity Test
797.1330	Daphnid Chronic Toxicity Test
797.1350	Daphnid Chronic Toxicity Test
797.1400	Fish Acute Toxicity Test
797.1440	Fish Acute Toxicity Test
797.1520	Fish Bioconcentration Test
797.1560	Fish Bioconcentration Test
797.1600	Fish Early Life-Stage Toxicity Test

	EPA-FIFRA Test Guidelines (U.S. EPA, 1982a, 1985a,b; 1986b–e)
—	SEP[c]: Acute Toxicity Test for Freshwater Invertebrates (EPA-540/9-85-005; 1985)
—	SEP: Acute Toxicity Test for Freshwater Fish (EPA-540/9-85-006; 1985)
—	SEP: Fish Life-Cycle Toxicity Tests (EPA-540/9-86-137; 1986)
—	SEP: Fish Early Life-Stage Test (EPA-540/9-86-138; 1986)
—	SEP: *Daphnia magna* Life-Cycle (21-Day Renewal) Chronic Toxicity Test (EPA-540/9-86-141; 1986)
—	SEP: Nontarget Plants: Growth and Reproduction of Aquatic Plants-Tiers 1 and 2 (EPA-540/9-86-134; 1986) (see Chapter 4)
	OECD Test Guidelines (OECD, 1987)
201	Alga, Growth Inhibition Test
202	*Daphnia* sp. Acute Immobilisation Test and Reproduction Test
203	Fish, Acute Toxicity Test
204	Fish, Prolonged Toxicity Test: 14-day Study
	Environment Canada (Environment Canada, 1990a–e; 1992a–c)
—	Acute Lethality Test Using Rainbow Trout (1990a)
—	Acute Lethality Test Using Threespine Stickleback (1990b)
—	Acute Lethality Test Using *Daphnia* spp. (1990c)
—	Test of Reproduction and Survival Using the Cladoceran, *C. dubia* (1992a)
—	Test of Larval Growth and Survival Using Fathead Minnows (1992b)
—	Growth Inhibition Test Using the Freshwater Alga (*S. capricornutum*) (1992c)
—	Early Life-Stage Toxicity Tests Using Salmonid Fishes (1992d)
—	Guidance Document on Control of Toxicity Test Precision Using Reference Toxicants (1990d)
—	Reference Method for Determining Acute Lethality of Effluents to Rainbow Trout (1990e)
—	Reference Method for Determining the Acute Lethality of Effluents to *Daphnia magna* (1990f)
	FDA—Environmental (Aquatic) Assessment (FDA, 1987)
4.01	Algal Assay
4.08	*Daphnia* Acute Toxicity
4.09	*Daphnia* Chronic Toxicity
4.10	*Hyalella azteca* Acute Toxicity
4.11	Freshwater Fish Acute Toxicity

[a]In APHA 1985; [b]In APHA 1992; [c]SEP = Standard Evaluation Procedure

aquatic life. Dechlorination of tap water should be used only as a last resort because dechlorination is often incomplete, resulting in toxic levels of residual chlorine and chlorine-produced oxidants. Dechlorination can be accomplished by aerating for 24 h, carbon filtration, or the use of sodium bisulfite, sodium sulfite, or sodium thiosulfate (1.0 mg anhydrous sodium thiosulfate/L reduces 1.5 mg chlorine/L).

Test Organisms

Whenever possible, freshwater tests should be performed with species listed in Table 3. These species are routinely selected on the basis of availability; commercial, recreational, and ecological importance; past successful use; ease of handling in the laboratory; and regulatory use (ASTM, 1992a,b). Their use is encouraged to increase comparability of results and availability of much information about a few species rather than a little information about many species (ASTM, 1992a). However, because of the diversity of toxicological responses, the progress of hazard assessment may be constrained by using only one or a few surrogate species (Moore, 1987). It is important to be familiar with the life history and general requirements of the test organisms to ensure that extraneous variability is not added to the study because of lack of control (i.e., standardization) of a

Table 2. Recommended composition for reconstituted fresh water

	Salts required (mg/L)				Water quality		
Water type	$NaHCO_3$	$CaSO_4$	$MgSO_4 \cdot 2H_2O$	KCl	pH (SU)	Hardness (mg/L as $CaCO_3$)	Alkalinity (mg/L as $CaCO_3$)
Very Soft	12	7.5	7.5	0.5	6.4–6.9	10–13	10–13
Soft	48	30	30	2.0	7.2–7.6	40–48	30–35
Hard	192	120	120	8.0	7.6–8.0	160–180	110–120
Very Hard	384	240	240	16.0	8.0–8.4	280–320	225–245

From U.S. EPA, 1991b.

Table 3. Test species commonly used for freshwater toxicity tests

Vertebrates
Brook trout, *Salvelinus fontinalis*
Coho salmon, *Oncorhynchus kisutch*
Chinook salmon, *Oncorhynchus tshawytscha*[a]
Rainbow trout, *Oncorhynchus mykiss* (formerly *Salmo gairdneri*)
Goldfish, *Carassius auratus*
Common carp, *Cyprinus carpio*
Fathead minnow, *Pimephales promelas*
White sucker, *Catostomus commersoni*
Channel catfish, *Ictalurus punctatus*
Bluegill, *Lepomis macrochirus*
Green sunfish, *Lepomis cyanellus*
Northern pike, *Esox lucius*[a]
Threespine stickleback—*Gasterosteus aculeatus*[a]
Zebra fish, *Brachydanio rerio*—tropical fish[a]
Guppy, *Poecilia reticulata*—tropical fish[a]
Invertebrates
Daphnids, *Daphnia magna, D. pulex, D. pulicaria, Ceriodaphnia dubia*[a]
Amphipods, *Gammarus lacustris, G. fasciatus, G. pseudolimnaeus*
Crayfish, *Orconectes* sp., *Cambarus* sp., *Procambarus* sp., *Pacifastacus leniusculus*
Stoneflies, *Pteronarcys* sp.
Mayflies, *Baetis* sp., *Ephemerella* sp., *Hexagenia limbata, H. bilineata*
Midges, *Chironomus tentans, C. riparius*
Snails, *Physa integra, P. heterostropha, Amnicola limnosa*
Planaria, *Dugesia tigrina*
Rotifers, *Brachionus calyciflorus, B. rubens, B. plicatilis* (brackish water)[a]

[a]Added by author.

critical parameter. For this reason, the use of standard test species is recommended because their testing requirements are well known. Of the group of freshwater species listed in Table 3, the majority of acute and chronic tests are performed with three species of invertebrates (*Daphnia magna*, *Daphnia pulex*, and *Ceriodaphnia dubia*) and four species of fish (fathead minnows, bluegill, rainbow trout, and brook trout). With the increase in the use of freshwater benthic organisms to assess the toxicity of potentially contaminated sediments, there has been a great deal of activity in developing and validating test methods with amphipods, aquatic insect larvae, and oligochaetes. For more detail on sediment toxicity testing, see Chapter 8.

Under the Clean Water Act for acute effluent testing *C. dubia*, *D. magna*, and *D. pulex* are the preferred invertebrate species and fathead minnows are the preferred vertebrate species (U.S. EPA, 1991b). For chronic effluent tests, *C. dubia* and fathead minnows are the recommended test species (U.S. EPA, 1989a). The specific test species are listed in the state-specific NPDES permit. Some states (e.g., Maine) that have a large cold-water fishery may substitute a trout species (either rainbow trout, *O. mykiss*, or brook trout, *Salvelinus fontinalis*) for fathead minnows in the NPDES permit.

For FIFRA, the preferred acute and chronic invertebrate test species is *D. magna* (U.S. EPA, 1985a, 1986d); the preferred vertebrate species for acute tests are bluegill (warm-water fish) and rainbow trout (cold-water fish) (U.S. EPA, 1986b,c). However, because of the shorter life cycle of fathead minnows, they are typically used for chronic tests in FIFRA. Both fathead minnows and rainbow trout are used for ELS tests in FIFRA. For TSCA, the preferred acute and chronic invertebrate test species are *D. magna* and *D. pulex*; the preferred vertebrate test species for acute tests are bluegill and fathead minnows for warm water and rainbow trout for cold water (U.S. EPA, 1982b; Federal Register, 1985). The recommended freshwater vertebrate species for ELS tests under TSCA are fathead minnows, brook trout, and rainbow trout (U.S. EPA, 1982b; Federal Register, 1985). For FDA, the preferred acute invertebrate test species is *D. magna* or *D. pulex* and the preferred vertebrate species are bluegill, fathead minnow, and trout.

Brief life histories of these commonly used invertebrate and vertebrate species are summarized in the following sections.

Daphnids

Daphnids are freshwater microcrustaceans, commonly referred to as water fleas, belonging to the class Crustacea, order Cladocera (Figure 1). Cladocerans from the family Daphniidae, which include *Daphnia* spp. (e.g., *D. magna* and *D. pulex*) and *Ceriodaphnia* spp., are ubiquitous in temperate fresh wa-

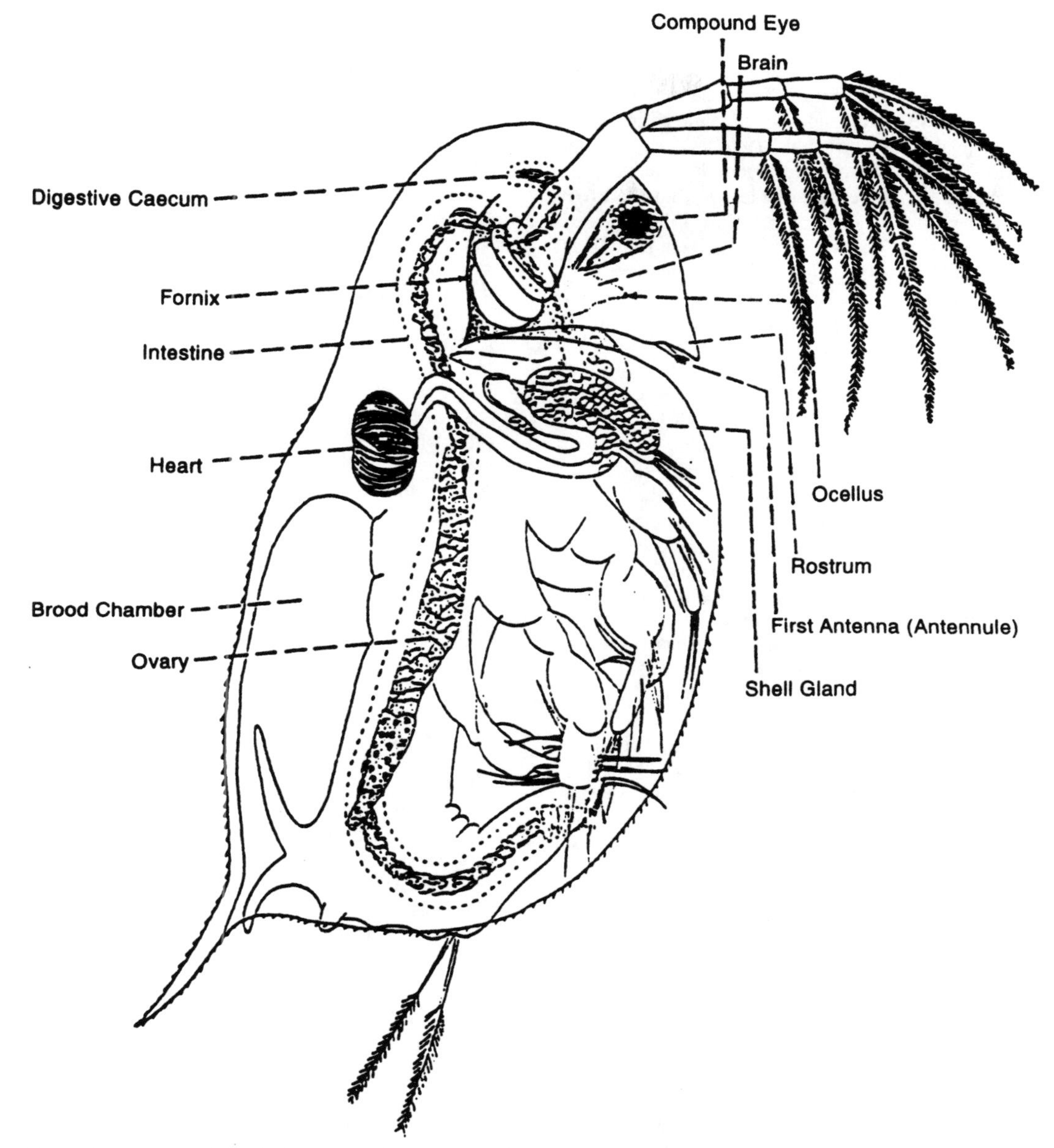

Figure 1. Anatomy of female *Daphnia pulex*. (From Pennak R. W.: Freshwater Invertebrates of the United States, 2nd ed., p. 351. Copyright © 1978 John Wiley & Sons, Inc. Reprinted by permission of John Wiley & Sons, Inc.)

ters (Berner, 1986). Both genera are abundant in lakes, ponds, and quiescent sections of streams and rivers throughout North America (Pennak, 1978). Within these habitats, cladocerans are ecologically important species because they convert phytoplankton and bacteria into animal protein and form a significant portion of the diet of numerous fish species. Thus their importance in aquatic communities makes them logical choices for inclusion among species that need protection.

During most of the year, natural populations of daphnids consist almost entirely of females; the males are usually abundant only in the spring or autumn (Barker and Hebert, 1986). Production of males apears to be induced principally by low temperatures or high densities and subsequent accumulation of excretory products and/or a decrease in available food (U.S. EPA, 1991b). These adverse conditions may induce the appearance of a sexual (resting) egg (embryo) in a case called an ephippium (pl. ephippia), which is cast off with the next molt. *D. magna* and *C. dubia* reproduce only by cyclic parthenogenesis, in which males contribute to the genetic makeup of the young during the sexual stage of reproduction, whereas *D. pulex* may reproduce either by cyclic or obligate parthenogenesis, in which zygotes develop within the ephippium by ameoitic parthenogenesis with no genetic contribution from the males.

Four distinct periods may be recognized in the life history of daphnids (Pennak, 1978): (1) egg, (2) juvenile, (3) adolescence, and (4) adult. The life span of daphnids, from release of the egg into the brood chamber until the death of the adult, is highly variable, depending on the species and environmental conditions (Pennak, 1978); life span generally increases as temperature decreases, because of lowered metabolic activity. The average life span of *D. magna* is about 40 d at 25°C and 56 d at 20°C; the average life span of *D. pulex* is about 50 d at 20°C. The average life span of *C. dubia* is about 30 d at 25°C and 50 d at 20°C (U.S. EPA, 1991b).

The eggs hatch in the brood chamber and the juveniles, which are already similar in form to the adults, are released in approximately 2 d when the female molts (casts off her exoskeleton or carapace). *D. magna* and *D. pulex* generally mature in 6 to 10 d and have three to five juvenile instars; each instar stage is terminated by a molt. *C. dubia* matures in 3 to 5 d. Four events take place in a matter of minutes at the end of each adult instar: (1) release of young from the brood chamber to the outside, (2) molting, (3) increase in size, and (4) release of a new clutch of eggs into the brood chamber (Pennak, 1978). Growth occurs immediately after each molt while the new carapace is still elastic. The adolescent stage consists of one instar and it is in this stage that the first clutch of eggs reaches full development in the ovary; eggs are generally deposited in the brood chamber within minutes after molting, and the young that develop are released just before the next molt. An instar generally lasts about 2 d for *D. magna* and *D. pulex* and 1 d for *C. dubia* under favorable conditions, but when conditions are unfavorable, it may last longer (up to a week in *D. magna*).

Daphnia magna and *D. pulex* are extensively used for acute and chronic aquatic toxicity testing because they are readily available, adaptable to laboratory conditions, require little space, and are frequently among the more sensitive aquatic animals to chemicals. Either *D. magna* or *D. pulex* may be used in acute or full life cycle chronic toxicity tests and the choice of species is somewhat dependent on the hardness of the control/dilution water (Environment Canada, 1990c,f). *D. pulex* may be used at any hardness, and this species is recommended if the hardness of the control and dilution water is <80 mg/L. *D. magna* may be used as the test species if the hardness of the control/dilution water is ≥80 mg/L as $CaCO_3$. The neonates of *D. magna* are larger and easier to observe in the test solutions. However, *D. magna* is found naturally only in hard waters (>150 mg/L) (Pennak, 1978) and the use of *D. magna* in soft-water solutions may lead to mortality caused by osmotic stress (Greene et al., 1988). Sublethal stress resulting from low hardness might affect resistance to the substance being tested. As *D. pulex* resides naturally in both hard and soft waters, the use of this species for test solutions with hardness values <80 mg/L is recommended.

The microcrustacean cladoceran *C. dubia* is used for acute tests and in the three-brood, short-term chronic tests, primarily with effluents. *Ceriodaphnia* are closely related and morphologically similar to *Daphnia* but are smaller and have a shorter generation time (Berner, 1986). *C. dubia* is thought to be synonymous with *C. affinis*, and the designation *C. dubia* has taxonomic preference (Berner, 1986). *C. dubia* does not have a long history of routine use in toxicity testing; the first paper describing this organism's use in effluent toxicity evaluations was published in 1984 (Mount and Norberg, 1984). However, during the past 11 yr, the use of *C. dubia* as a test organism has greatly increased, mainly in acute and short-term chronic toxicity tests with effluents. *C. dubia* has been successfully cultured and tested in natural and synthetic waters of various hardnesses ranging from soft (20 to 25 mg/L) to hard synthetic water (160 to 180 mg/L) (U.S. EPA, 1989a, 1991b; Cooney et al., 1992a; Patterson et al., 1992).

To obtain organisms for testing, adult daphnids with young in their brood chambers are placed in isolation chambers containing the untreated control test water and food during the 24-h period preceding the test. All young hatched during this time in the isolation chambers will be <24 h old and can be used in either acute or chronic toxicity tests. For *C. dubia*, observations are made at more frequent intervals to decrease the age "window" (e.g., all daphnids produced within 8 or 12 h of each other) used in short-term chronic tests.

The selection of daphnids for routine use in toxicity tests is appropriate for a number of reasons (Environment Canada, 1990c,f; 1992a):

- Daphnids are broadly distributed in freshwater bodies and are present throughout a wide range of habitats.
- Daphnids are important links in many important food chains and a significant source of food for small fish.
- Daphnids have a relatively short life cycle and are relatively easy to culture in the laboratory.
- Daphnids are sensitive to a broad range of aquatic contaminants and are widely used as test organisms for evaluating acute and chronic toxicity of chemicals or effluents.
- The small size of daphnids requires only small volumes of test and dilution water.

Fish (Fathead Minnows, Rainbow Trout, Brook Trout, and Bluegill)

The fathead minnow (*Pimephales promelas*) is currently the most commonly used warm-water species for acute and chronic (full life cycle, early life stage, and short-term) toxicity tests (Braughn and Schoettger, 1975; Benoit, 1982; Denny, 1987). Fathead minnows belong to the family Cyprinidae, the carps and minnows, the dominant freshwater fish family in terms of number of species. The fathead minnow thrives in ponds, lakes, ditches, and slow muddy streams. It is an omnivore, feeding opportunistically on anything from living invertebrates to detritus, but is well suited to a diet high in vegetable matter. It has a widespread distribution throughout North America, is an important forage fish in the food chain, and can be readily cultured in the laboratory. It is a popular bait fish, and the ease with which it is propagated has led to its widespread introduction both within and outside the native range of the species. As a result of the early interest in raising the fish for bait, the natural history and spawning behavior of the fathead minnow are well known. Fathead minnows have been used for life cycle tests in the United States since the 1960s (Mount and Stephan, 1967) and it is now one of the standard species for tests of both acute lethality and sublethal or chronic effects (U.S. EPA, 1989a, 1991b).

Fathead minnows readily spawn under laboratory conditions in 10- to 20-gallon aquaria. Breeding males are territorial and select sites for spawning on the underside of objects. In the laboratory, the most common spawning substrates are half-cylinders of schedule 40 PVC pipe (3 × 3 inches). The female releases adhesive eggs (1.1 to 1.5 mm in size; 100 to 150 eggs per spawn) under the spawning substrate, where they attach to the underside of the substrate; the eggs are immediately fertilized by the male and the female is driven off. Once the eggs are deposited on the substrate, the male becomes aggressive and drives off all intruding fish. The incubation time to hatch depends on the temperature and is about 5 d (range 4.5 to 6 d) at 25°C (Table 4).

The methods used to obtain fathead minnows for testing depend on the life stage and age requirements for the specific test protocol. For life cycle chronic tests, fertilized eggs (≤24 h old) are manually removed from the spawning substrates by gently rolling the water-hardened eggs off the substrate with a finger; the embryos are then randomly distributed to incubation cups. For the 7-d short-term chronic effluent tests, the eggs on several spawning substrates are placed together in an isolation chamber; the chambers are checked every 24 h for hatching fry. All fry hatched within the 24 h preceding the test in the isolation chambers may be used in short-term chronic effluent tests. If embryos and fry are not used in testing, they are "grown out" in glass aquaria to whatever age is required for other test protocols (e.g., for acute effluent tests, the required age is 1 to 14 d, with all fry hatched within 24 h; EPA, 1991b).

If fathead minnow fry are maintained at 25°C and a constant 16-h light photoperiod, the fry mature in 5 to 6 mo; daily feeding should include unrestricted quantities of live brine shrimp and frozen adult brine shrimp or other suitable food (Benoit, 1982; U.S. EPA, 1991b). With proper care and maintenance, fathead minnows consistently produce large numbers of embryos for 6 to 8 mo.

The rainbow trout (*Oncorhynchus mykiss*, formerly *Salmo gairdneri*) and brook trout (*Salvelinus fontinalis*) of the family Salmonidae are widely distributed salmonid species in North America. Both species have great commercial value as sport fish. Rainbow trout thrive in most cool, fresh water bodies (lakes, streams, and rivers) and have become the world standard cold-water fish for freshwater toxicity studies and research in aquatic toxicology (Braughn and Schoettger, 1975; Environment Canada, 1990a,e, 1992d). Brook trout is more widespread in northeastern North America east of the Mississippi River. Culturing of rainbow and brook trout is well established.

In nature, rainbow trout are basically spring spawners but can spawn at the beginning of summer or early winter, depending on climate, elevation, and genetic strain. Females deposit demersal eggs (3 to 5 mm in size) in a spawning pit or depression (redd) in the gravel and sand substrate; the eggs are immediately fertilized by the male and fall into the

Table 4. Average time-to-hatch for representative freshwater fish

Species	Temperature (°C)	Days to hatch
Brook trout, *Salvelinus fontinalis*	6.1	75
	10	44–50
Coho salmon, *Oncorhynchus kisutch*	10	31
Chinook salmon, *O. tshawytscha*	10	56
Rainbow trout, *O. mykiss*	7	48
	10	31
Lake trout, *Salvelinus namaycush*	7	72
Brown trout, *Salmo trutta*	10	41
Fathead minnow, *Pimephales promelas*	25	5
Channel catfish, *Ictalurus punctatus*	26	6–7
Bluegill, *Lepomis macrochirus*	28	6
Northern pike, *Esox lucius*	15	6

From U.S. EPA, 1986b (modified by author).

spaces between the gravel and are then covered with loose gravel and sand by the female. Eggs usually hatch in 4 to 7 wk, depending on the water temperature (at 7°C, eggs hatch in about 48 d; at 10°C, eggs hatch in about 31 d; see Table 4). The newly hatched fish are called alevins and have a yolk sac that is absorbed within 3 to 7 d. After the yolk sac is absorbed, the young are called fry.

Brook trout spawn in late summer or autumn, the date varying with latitude and temperature, usually from late October to January. Spawning time may be regulated by artificial control of the photoperiod or by injection of pituitary hormones. Some females may spawn when they are a year old. The adhesive eggs (3.5 to 5.0 mm) are laid in a redd, where they are fertilized by the male and then adhere to the gravel. The eggs are covered with gravel and sand and the redd is then deserted; eggs hatch in 75 d at 6.1°C and approximately 44 to 50 d at 10°C (Table 4). After the eggs hatch, the sac fry remain in the gravel of the redd until the yolk is absorbed. Depending on the water temperature, it may take 1 to 3 mo to absorb the yolk sac.

Salmonid embryos and fry can be obtained in several ways, such as by collecting ripe wild fish at the time of spawning, by procuring gametes/embryos (or fry) from hatcheries, or by providing fish with the proper conditions for spawning in the laboratory. One common method is to strip gametes from ripe males and females and artificially fertilize the eggs in the laboratory. The eggs and milt (semen) are individually stripped into a container and then gently mixed. After about 10 min, the eggs are rinsed to remove the milt. The fertilized eggs are then set aside for about 30 min to harden before they are transferred to oscillating incubation chambers. Full life cycle, partial life cycle, and ELS chronic tests should begin within 96 h of fertilization; the younger the embryo, the more sensitive the test. Test protocols used by Environment Canada (1992d) require test initiation within 30 min of egg fertilization to ensure that the eggs are exposed to the test substance before water hardening. For acute tests, fish of the correct age and size are usually obtained from hatcheries and then acclimated to the test conditions for a minimum of 2 to 3 d before testing.

The bluegill, *Lepomis machrochirus*, is one of the largest and most common sunfish (family Centrarchidae) with a native range restricted to eastern and central North America. The family includes some of most highly colored and attractive North American fish and several species (e.g., smallmouth and largemouth bass) are important sport fish (Scott and Crossman, 1973). The bluegill (average adult size 7 to 8 inches long) inhabits the shallow, weedy, warm water of large and small lakes, ponds, and heavily vegetated slowly moving areas of small rivers and large creeks (Scott and Crossman, 1973). In the wild, spawning takes place in late spring to early summer. Males prepare a nest (circular depression 18 to 24 inches in diameter) in the shallow water, which they defend aggressively. Females lay small demersal eggs, which are immediately fertilized by the male; the male then chases the female away and protects the eggs in the nest. Hatching takes place in about 6 d at 28°C (Table 4). The breeding period is long and spawning may take place more than once during a season. Breeding and cultivation of bluegills may be carried out in a variety of ponds and tanks by controlled natural reproduction (APHA, 1992). Techniques for consistent, successful spawning of bluegill in the laboratory are not presently available (ASTM, 1992d). However, some natural methods (Eaton, 1970, 1973a) and artificial methods (Banner and Van Arman, 1973) are available. For the artificial method, fish are field collected and isolated by sex in large aquaria kept at 26° to 28°C and a 16-h light photoperiod. To obtain embryos for testing, females are injected at set time intervals intraperitoneally with carp pituitary hormone until ripe eggs are produced; the production of ripe gametes may require up to 3 mo. Eggs and milt (sperm) are then stripped into a shallow pan where the gametes are mixed together, similar to the methods used for salmonids. Chronic tests should begin with 2-h- to 24-h-old embryos. For acute tests, bluegill of the correct age and size are usually obtained from hatcheries and then acclimated to the test conditions for a minimum of 2 to 3 d before testing.

A toxicological data bank of appreciable magnitude has been assembled for fathead minnows, bluegill, rainbow trout, and brook trout. The use of fathead minnows and bluegill as warm-water species and rainbow and brook trout as cold-water species for standard toxicity tests in fresh waters is supported by:

- Extensive toxicological data base
- Proven sensitivity to aquatic toxicants
- Commercial value (rainbow and brook trout)
- Widespread availability

GENERAL PROCEDURES

Although ASTM, Standard Methods, and EPA are the focal point for general method standardization, the specific procedures used to conduct either acute or chronic toxicity tests depend on the objectives of the study, the regulatory requirements, and the test species selected (Table 1). This is also the case for saltwater toxicity tests (see Chapter 3). For pesticide testing under FIFRA (see Chapter 21), pesticide assessment guidelines and standard evaluation procedures (SEPs) for acute and chronic toxicity tests with freshwater fish and invertebrates are available (U.S. EPA, 1982a, 1985a,b,c; 1986b,c). The individual SEPs are discussed in more detail in Chapter 11. For testing of industrial chemicals under TSCA (see Chapters 22 and 23), a different set of environmental effects test guidelines is employed for acute and chronic tests with freshwater organisms (U.S. EPA, 1982b; Federal Register, 1985). For effluent testing under the Clean Water Act's NPDES permitting program acute test methods are provided in U.S. EPA (1991b); chronic methods for freshwater organisms are provided in an EPA guidance document (U.S. EPA, 1989a). Regardless of the specific test requirements, some general characteristics of toxicity tests are listed in the following:

- Source and age of test organisms
- Preparation of test solutions and design
- Experimental test conditions
- Test observations and measurements
- Test end points and calculations

Source and Age of Test Organisms

All organisms in a test should be from the same source because organisms of the same species from different sources may have different sensitivities to the toxicants, thus increasing extraneous variability in test results. During holding and acclimation, animals should not be unnecessarily stressed by excessive handling. To maintain aquatic animals in good condition and avoid unnecessary stress, they should not be crowded or subjected to rapid changes in temperature or water quality.

For acute tests, immature organisms should be used whenever possible, because they are often more sensitive than older organisms of the same species. A group of organisms should not be used for a test if more than 5% of the individuals show signs of disease or stress, such as discoloration, unusual behavior, or death during the 48-h period immediately preceding the test (ASTM, 1992a,b). Fish embryos may be obtained using the artificial methods previously described.

For *Daphnia* sp., neonates ≤24 h old are used as the test organisms for both acute and chronic tests. Neonates of *C. dubia* (≤24 h old) are used to start acute tests; however, an additional stipulation for 7-d short-term chronic tests is that all organisms are produced within the same window of time (e.g., 6, 8, or 12 h of the same age). For 4-d tests with *C. dubia* or 7-d tests with *D. magna*, the organisms are obtained similarly; however, neonates are held in clean water for 3 d for *C. dubia* and 4 d for *D. magna* before being introduced into the test solutions. Amphipods, mayflies, and stoneflies should be tested in an early instar; midges should be tested in the second or third instar (ASTM, 1992a). Rotifers are usually hatched from commercially available cysts by placing the cysts in synthetic fresh water at a temperature of 25°C and light intensity of 2000 to 4000 lux about 16 to 22 h before test initiation (ASTM, 1992h; Snell and Moffat, 1992).

For acute tests with fish, fish weighing between 0.1 and 0.5 g (wet weight) each are desirable. For acute testing under FIFRA (U.S. EPA, 1985b) and TSCA (U.S. EPA, 1982b; Federal Register, 1985), the recommended weight of the fish should be between 0.5 and 5 g. A standard requirement for acute fish tests is that the length of the largest fish should not be more than twice that of the smallest fish in the same test. Acute tests should be conducted with juvenile fish, that is postlarval or older and actively feeding, but not sexually mature, spawning, or spent. For acute tests with effluents, the age depends on the test species; for fathead minnows, brook trout, and rainbow trout, the age requirements are 1 to 14 d, 15 to 30 d (after yolk absorption), and 30 to 60 d, respectively. For short-term chronic effluent tests with fathead minnows, larvae that are ≤24 h old are used in the tests.

The grams of organism (wet weight) per liter of solution in the test chambers (i.e., test chamber loading) should not be so high as to affect the results of the test. Therefore, the loading of the organisms into the test chambers should be limited to ensure that the concentrations of dissolved oxygen and test substance do not fall below acceptable levels, concentrations of metabolic products do not exceed acceptable levels, and the test organisms are not stressed because of aggression or crowding. For static or static-renewal tests with fish, the loading should not exceed 0.5 to 0.8 g/L of test solution depending on the test temperature (higher temperatures should have reduced loading); for flow-through tests, the loading should not exceed 0.5 to 1.0 g/L/d.

Test Solution Preparation and Design

For most tests that are intended to estimate an acute end point (i.e., LC50) or a chronic end point (NOEC or LOEC), at least five concentrations of the test substance plus a control solution (100% dilution water) are usually prepared. If a solvent or carrier (e.g., dimethyl formamide, triethylene glycol, methanol, acetone, and ethanol) is used because of low solubility of the test substance, a solvent control is also included. The solvent control is prepared by adding solvent to laboratory-prepared water at the same concentration of solvent used to dissolve the test substance in the test water. An appropriate geometric dilution series should be used, in which each successive concentration is about 0.5 (U.S. EPA, 1989a) or 0.6 (ASTM, 1992a,b) of the previous concentration (e.g., at a dilution factor of 0.5: 100, 50, 25, 12.5, and 6.25). Test concentrations may also be selected from other appropriate logarithmic dilution series (APHA, 1992; ASTM, 1992a).

The test begins when the test organisms are first placed in test chambers containing test solutions. An equal number of replicates should be used for each concentration including the control(s). Test chambers are defined as the smallest physical units between which there are no water connections; the size, shape, and construction of the test chambers depend on the size of the individuals and the allowable loading. The numbers of organisms added to the test chambers at the beginning of a test should be equal; the number of organisms per chamber is dependent on the type of test and the species. For an acute fish test, 10 fish are commonly used per replicate with two replicates per test concentration; for an acute daphnid test, either 5 or 10 organisms are used per replicate with two or four replicates per concentration (20 organisms per concentration). For short-term chronic effluent tests with fathead minnows, a minimum of three replicates (four replicates are recommended based on statistical considerations) is used with 10 animals per replicate; for daphnid chronic effluent tests, one organism is used per replicate with 10 replicates per concentration. For full life cycle chronic daphnid single-chemical tests 10 organisms are used per replicate, with four replicates per concentration including controls (e.g., untreated and solvent, if necessary). Table 5 describes the design of a full life cycle chronic test with fathead minnows.

Experimental Test Conditions

Acute test durations are usually 48 h for daphnids and midge larvae and 48 to 96 h for fish and other invertebrates. For short-term chronic tests with *C. dubia* and fathead minnows, the test duration is 7 d; for *C. dubia*, the test may be extended to a maximum of 9 d to ensure the production of three broods in the control organisms. The duration of full chronic tests is dependent on the species. For full life cycle tests with *D. magna*, the recommended duration is 21 to 28 d (ASTM, 1992c). A life cycle test with rotifers can be completed within 2 d with the chronic test end point being the intrinsic rate of population increase (Snell and Moffat, 1992). For full life cycle, partial life cycle, and ELS chronic tests with fish, the range in duration can be 12 to 30 mo, 12 to 15 mo, and 30 to 60 d posthatch, respectively (ASTM, 1992d; U.S. EPA, 1986b,c).

The test temperature for acute and chronic tests is dependent on the specific regulatory requirements. When possible, test temperatures should be selected from the series 7, 12, 17, 22, 27, and 32°C (ASTM, 1992a,b); this series was selected because it provides temperatures that are better suited to more species and is usually more convenient for use by investigators than any other series. However, for daphnids and warm-water fish, such as fathead minnows or bluegills, the most common test temperatures are usually 20°C and 25°C; for cold-water species, such as rainbow and brook trout, the most common test temperatures are 12°C and 15°C. The time-weighted average measured temperature at the end of the test should be within ±1°C of the selected test temperature with no instantaneous readings > 3°C measured in any test chamber above or below the selected temperature (ASTM, 1992a). Lighting should be < 500 lux at the water surface and cool white fluorescent lights are suitable. For most species, a photoperiod of 16 h light and 8 h dark is acceptable.

In addition to temperature, commonly measured water quality parameters are dissolved oxygen and pH. Test solutions are usually not aerated during the test; however, during the test, dissolved oxygen should not fall below 40% saturation for warm-water

Table 5. Description of design and approximate timing for milestones during the conduct of a full life-cycle chronic toxicity test with fathead minnows

Time period	Task
Month 1	Solubility trials, solvent selection, flow-through system preparation and calibration, preliminary egg and/or larvae exposure toxicity screen.
Month 2	Finish preliminary exposure. Selection and setup of chronic exposure concentrations (five), controls (solvent[a] and water), analysis of water samples to demonstrate stability of chemical in test system. Total of seven treatments (five concentrations and two controls).
Month 3	Days 0–30 of chronic exposure, egg incubation (100 embryos/replicate, 2 replicates/treatment), larvae selection and 30-d exposure with photographs for total length and survival determination of F_0 generation. Survival of embryos, time required to complete hatch, and hatching success are recorded. After completion of hatch, the total number of larvae including dead or deformed, are counted.
Month 4	Days 30–60 of chronic exposure, larvae feeding, 60-d photographs for total length and thinning to 25 juvenile fish/aquarium (100 fish/treatment).
Month 5	Days 60–90 of chronic exposure, maintain unit and test fish. (Statistical evaluation of hatch, 30- and 60-d survival and growth of F_0 fish). Reduce to 50 fish/treatment. Length and weight of fish discarded are recorded.
Month 6	Days 90–120 of chronic exposure, maintain unit and test fish.
Month 7	Days 120–150 of chronic exposure, maintain unit and test fish, prepare for spawning and embryo incubation.
Month 8	Days 150–180 of chronic exposure, establish spawning groups, begin checking for spawns, incubate embryos, start F_1 larvae groups. Embryos are counted in spawning tanks. A total of 100 embryos/treatment are used for F_1.
Month 9	Days 180–210 of chronic exposure, continue to check spawns, incubate embryos and start larval group.
Months 10/11	Days 210–270 of chronic exposure, complete spawning data and survival and growth of F_1 larval groups.
Month 12	Complete parental exposure, terminate fish. Statistical analysis of survival, length-weight data, F_0 reproduction, and F_1 embryo hatchability.

[a]Solvent is used with chemicals that are insoluble in water.
From Springborn Laboratories, Wareham, MA, 1993.

species or 60% saturation for cold-water species. If the dissolved oxygen concentration falls below these values, the stress from low dissolved oxygen could interact with that from the toxic substance under investigation and make the latter appear more toxic. When the dissolved oxygen falls below acceptable levels in a test, gentle aeration is used in all test chambers, including controls, to maintain adequate dissolved oxygen levels in the test solutions and consistent treatment of all test chambers.

Toxicity tests are typically conducted without adjustment of pH. In instances where the chemical or wastewater causes the pH of the test solutions to be outside the range of 6 to 9 SU, it is usually desirable to adjust the pH of the samples to a neutral value ($\approx$ 7 SU) before the addition of the test organisms, because the objective of most chemical tests is to evaluate the chemical (or effluent) and not the pH.

Test organisms are not fed during acute tests; during chronic tests, food type, ration, and frequency of feeding are fixed by the specific protocol (ASTM, 1992c,d,e,j; U.S. EPA, 1986b,c,d; 1989a). Nutritional state of the animals before and during the test and the diets used in testing are important because these factors may affect test results (Appendix Chapter C). Daphnids are usually fed green algae or a mixed diet of trout chow food, yeast, and dried cereal grasses (U.S. EPA, 1989a); type and quantity of food provided to daphnids and other crustaceans can dramatically affect nutrition and the sensitivity of the test organisms to toxicants (Cowgill et al., 1985; Winner, 1989; Keating and Dagbusan, 1986; Knight and Waller, 1992, Patterson et al., 1992; Cooney et al., 1992a).

For many toxicity tests with larval fish, the food of choice is often live brine shrimp because of their availability, ease of use, and nutritional value (Bengtson et al., 1985; ASTM, 1992l); for older fish, starter-grade commercial salmon or trout food or other dried flake food is commonly used (ASTM, 1992a,b). The primary requirement for the acceptability of food used in toxicity tests is that it is the appropriate size for the test organisms, has adequate nutritional value, and does not contain excessive concentrations of contaminants (ASTM, 1992a,b; U.S. EPA, 1989a).

Acute tests are not valid if mortality in the control water(s) exceeds 10%. Chronic tests are usually invalid if mortality exceeds 20% (30% in some fish ELS tests) and minimum growth or reproduction criteria are not achieved. For short-term chronic tests with *C. dubia*, 60% of the control organisms must have produced their third broods and produced an average of 15 neonates per female (ASTM, 1992e; EPA, 1989a); for fathead minnows, average dry weights should be $\geq$0.25 mg per fish (U.S. EPA, 1989a).

Test Observations and Measurements

Test organisms should be observed daily for the duration of both acute and chronic tests. For daphnids, the test duration for acute tests is routinely 48 h, which allows all daphnids to proceed through one molt, at which time they are usually most sensitive to toxicant stresses, but does not prolong the test to the point that starvation becomes a major factor; midge larvae are also exposed for 48 h. For fish and other invertebrates, the acute test duration is usually 48 to 96 h. For acute tests with aquatic organisms, death and immobilization are the adverse effects most often used. The criteria for death are usually lack of movement, especially absence of respiratory movements in fish, and lack of reaction to gentle prodding. The criterion for immobilization is lack of movement except for minor spontaneous, random activity of the appendages. Because death of some invertebrates is not easily distinguished from immobilization, an EC50 is usually determined rather than an LC50. For daphnids and midge larvae, the EC50 is usually based on death plus immobilization (ASTM, 1992a).

For chronic toxicity tests with fish and invertebrates (such as *Daphnia*) (ASTM, 1992c,d), end points, other than survival of the various life stages, are routinely measured during the test (e.g., daphnid reproduction) or at the end of the test (weight or length). For daphnid tests, the number of live young in each test chamber is recorded daily throughout the test; time to first brood is also usually recorded. Dry weight (dried at 60°C to constant weight) is often used as a test end point for both daphnid and ELS tests with fish. Length at the end of the test is determined as distance from apex of helmet to base of spine in daphnids or total length in fish. Test organisms are also observed regularly during the test for abnormal development and aberrant behavior. End points typically measured in a chronic fish test are listed in Table 5.

The frequency of water quality measurements such as hardness, alkalinity, specific conductivity, pH, temperature, dissolved oxygen, and test material concentration depends on the type of test material (e.g., single chemical or effluent), the test duration, and the flow conditions (i.e., static, static-renewal or flow-through conditions). For static and static-renewal tests (acute and chronic) with invertebrates such as *Daphnia* (ASTM, 1992c), hardness, alkalinity, specific conductivity, dissolved oxygen, and pH are usually measured at the beginning and end of a test. In addition, for static-renewal chronic tests, these parameters should be measured at least weekly on a fresh test solution (i.e., newly prepared solution prior to changing test solutions) from the control treatment(s); and at least once on an old test solution (i.e., sample collected from the individual test chambers) from the control treatment. Alkalinity and pH should also be measured at the highest test concentration to determine whether they are affected by the test material at least once in new and old test solutions. Dissolved oxygen should also be measured in the old test solutions from the low, medium, and high concentrations of test material near the beginning, middle, and end of the test. For effluent tests, dissolved oxygen should be measured daily in the low, middle, and high test concentrations. Temperature should be monitored at least hourly in one test chamber.

For flow-through tests such as ELS tests with fish (ASTM, 1992d), hardness, alkalinity, specific conductivity, dissolved oxygen, and pH are usually measured at the beginning and end of a test and at least weekly in chronic tests in the control treatments. Because the solutions in the test chambers are changed continuously, there is no distinction between new and old solutions in a flow-through test. Alkalinity and pH should also be measured in the highest test concentration to determine whether they are affected by the test material at least once during the test. Dissolved oxygen should be measured in at least one test chamber in each treatment containing live organisms at the beginning and end of the test and at least weekly during the test. Dissolved oxygen is also measured at any time there is an interruption of flow of test solutions to the chambers or if the behavior of the organisms (e.g., fish swimming near the surface and gulping air) indicates potential low dissolved oxygen concentrations. Temperature should be monitored at least hourly in one test chamber. For static, static-renewal and flow-through acute tests (i.e., with single chemicals) the test material concentration in each treatment must be measured at the beginning and end of the test and for chronic tests it must be measured frequently (at least weekly) during the test to establish average concentration and variability during the test. For flow-through tests measurements should also be made if there are any malfunctions in the metering system of the diluter.

Test End Points and Calculations

Acute toxicity tests are generally used to determine the concentration of test material that produces a specific adverse effect on a specified percentage of test organisms during a short exposure (see Chapter 1). Because death and immobilization are obviously important adverse effects and are easily detected for most species, the most common end points in acute toxicity tests are lethality and immobilization (invertebrates primarily). In acute toxicity tests, groups of aquatic animals are exposed to progressively increasing concentrations of a toxicant in an attempt to elicit a quantal response (i.e., counts of the number of organisms in two mutually exclusive categories such as dead or alive). The primary purpose of the test is to estimate the concentration of the test substance that is lethal to or immobilizes 50% of the animals of a given species within a specific length of time (usually 24, 48, 72, or 96 h). Experimentally, an effect on 50% of a group of test organisms is the most reproducible and easily determined measure of toxicity (see Chapter 1). The exposure concentration that kills 50% of the organisms is referred to as the median lethal concentration (LC50); the exposure concentration that immobilizes 50% of the organisms is referred to as the median effective concentration (EC50). The LC50 is a statistically or graphically estimated concentration that is expected to be lethal to 50% of a group of organisms under specified conditions; the EC50 is a statistically or graphically estimated concentration that is expected to cause one or more specified effects (e.g., immobilization or loss of equilibrium) in 50% of a group of organisms under specified conditions.

A variety of methods can be used to calculate estimates of LC50 or EC50 values. For example, TOXDAT, a computer program cited in the EPA acute manual (U.S. EPA, 1985; 1991b), estimates an LC50 using one of three methods: binomial, moving-average, or probit method. The method selected is based on the shape of the concentration-response curve and the number of concentrations with partial mortalities (mortality greater than 0% but less than 100%).

The probit and moving-average methods estimate the LC50 with 95% confidence limits. The bounds placed on the LC50 by using the binomial test are not 95% confidence limits but can be used as statistically sound conservative bounds that are always above 95% when the animal sample size per concentration is large enough ($N \geq 6$) (Stephan, 1977). Hamilton et al. (1977) recommended use of the trimmed Spearman-Karber method, a nonparametric technique, which produces a fail-safe LC50 estimate (Gelber et al., 1985).

In full life-cycle, partial life-cycle, and short-term chronic tests, the objective is to determine the effects of a test substance on the viability of a species for an extended period of time (see Chapter 1). Another objective of chronic aquatic toxicity tests with effluents and single compounds is to estimate the highest "safe" or "no-effect" concentration of these substances. A maximum acceptable toxicant concentration (MATC) is often calculated from chronic data. The upper end of the MATC range is represented by the lowest test concentration that has a statistically significant effect (lowest observed effect concentration, LOEC). The lower end of the MATC range is represented by the highest test concentration for which there is no statistically significant difference from the control (highest no observed effect concentration, NOEC). Two less frequently used terms to describe similar chronic effects are the no observed effect level (NOEL) and lowest observed effect level (LOEL) values, synonymous with NOEC and LOEC.

NOEC and LOEC values may be derived statistically by the hypothesis-testing approach (see Chapter 10). However, Gelber et al. (1985) pointed out that NOEC and LOEC values have the undesirable property that the likelihood of observing an effect at a given concentration is as much a function of experimental design as of pollutant toxicity. The MATC is generally obtained by calculating the geometric mean of the NOEC and LOEC. The MATC represents an arbitrary estimate of an effect threshold that might lie anywhere in the range between the LOEC and NOEC (Suter, 1990). The calculated value of the mean is governed by whatever concentrations the investigator happened to select for the test; the selection of the dilution factor (0.5 or 0.6) determines, to some degree, the MATC. In addition, no confidence limits can be estimated for the MATC, NOEC, and LOEC values. For effluent tests, the geometric mean of the NOEC and LOEC is sometimes called the chronic or subchronic value (Norberg-King, 1991), but this can be misleading for short-term larval fish exposures (7 d), which represent only 1 to 2% of a fathead minnow's lifetime and are therefore not chronic tests (see Chapter 1).

Another chronic end point is the percentage inhibition concentration (IC_p), which can be calculated as a point estimate of the concentration that causes a specified degree of effect and can be useful as an additional end point (Norberg-King, 1989; U.S. EPA, 1989a). The IC_p is defined as a point estimate of the toxicant concentration that would cause a given percentage reduction (e.g., IC25) in a sublethal biological measurement of the test organisms, such as reproduction, growth, or fertilization (U.S. EPA, 1991a). The percentage is usually selected by the investigator and is customarily 25 to 50% reduction in growth or reproductive success or decrease in survival compared with the control. The IC_p is a useful measure of effect and, in particular, confidence limits may be calculated, allowing statistical comparisons with other values. Point estimates tend to avoid some of the statistical anomalies cited by Gelber et al. (1985) and Suter (1990) in estimating safe levels of toxicants.

The end points for chronic toxicity tests using *C. dubia* and *Daphnia* sp. are based on mortality of daphnids and their success in reproduction. For some chronic tests with *Daphnia* spp., carapace length is used as an additional test end point (Geiger et al., 1980; Winner, 1981; Dunbar et al., 1983; Knight and Waller, 1987). For fish short-term chronic tests, the end points are adverse effect on growth, measured as mean dry weights of the groups of fish from the test vessels, and increased mortality. For full life cycle, partial life cycle, and ELS tests with fish, end points, such as percent egg hatchability, length, weight, and reproductive success (e.g., number of eggs produced) are measured. The adverse effect is assessed by statistical comparison with data derived from the controls. If there is complete mortality in all replicates at a given concentration, that concentration is usually excluded from the analysis.

Chronic toxicity data can be analyzed by using a variety of statistical software packages. For example, TOXSTAT (Release 3.4) was developed by Gulley et al. (1989) of the University of Wyoming (Laramie) to address specifically the statistical requirements for analysis of chronic effluent toxicity data (U.S. EPA, 1989a) (Figure 2). Response end points from chronic tests (i.e., effluent and single chemical) include percent survival, percent egg hatchability, length, weight, and reproductive success. These effects can also be analyzed by using either parametric or nonparametric tests. The chronic data are analyzed by comparing either survival, egg hatchability, growth (length or weight), or reproductive success of the test organisms in the test concentrations with the selected parameter mean value in control water. NOEC and LOEC values for end points other than survival are calculated by using only the concentrations in which survivorship is not significantly reduced compared with the controls.

If the chronic data are not normally distributed (using Shapiro-Wilks or Chi-Square tests) or homogeneous (using Bartlett's, Hartley's, or Levene's test), chronic data are analyzed by using nonparametric techniques (e.g., Steel's many-one rank test for equal replicate size or Wilcoxon's rank sum test for unequal replicate size) (U.S. EPA, 1989a). If chronic data are normally distributed and homogeneous, they can be analyzed by using a parametric ANOVA followed by either Dunnett's test or Bonferroni's *T*-test (if the ANOVA indicates significant differences). For an in-depth discussion of statistical procedures used to evaluate acute and chronic aquatic toxicity data, see Chapter 10. The following sections provide some individual case studies of acute and short-term chronic effluent toxicity test results. Chapter 33 also illustrates the use of effluent toxicity tests in various case studies. One case study is included below on the use of a full life cycle *Daphnia magna* test to assess the effects of a 21-d continuous exposure to a single chemical.

Acute Effluent Toxicity Tests

Under the Clean Water Act NPDES permit program, an industrial discharger is required to monitor the quality of its effluent monthly using acute toxicity tests (U.S. EPA, 1991b). The NPDES permit compliance limit for this industrial discharge is an LC50 $\geq$ 100% effluent. Each month, static-acute multiconcentration toxicity tests using *C. dubia* and fathead minnows (*P. promelas*) are performed with a single 24-h composite sample of final effluent collected from the industrial facility. Immature *C. dubia* ($\leq$ 24 h old at test initiation) and fathead minnows (1 to 14 d old at test initiation; all fish produced within 24 h) are continuously exposed for 48 h under static conditions to five concentrations of the final effluent using a standard 0.5 dilution series (6.25, 12.5, 25, 50, and 100% effluent) and a dilution water control. Because the receiving stream for the industrial discharge was previously shown to be unacceptable as dilution water (i.e., survival < 90% at test completion), laboratory-prepared dilute mineral water is used as dilution/control water for the acute tests. *C. dubia* are exposed in groups of five animals in 50-ml beakers containing 30 ml of test solution or control water. Four replicates are used for each test concentration and control (20 animals per concentration). Fathead minnows are exposed in groups of 10 animals in 1000-ml beakers containing 900 ml of test solution or control water; two replicates are used for each test concentration and control (20 animals per concentration). Test

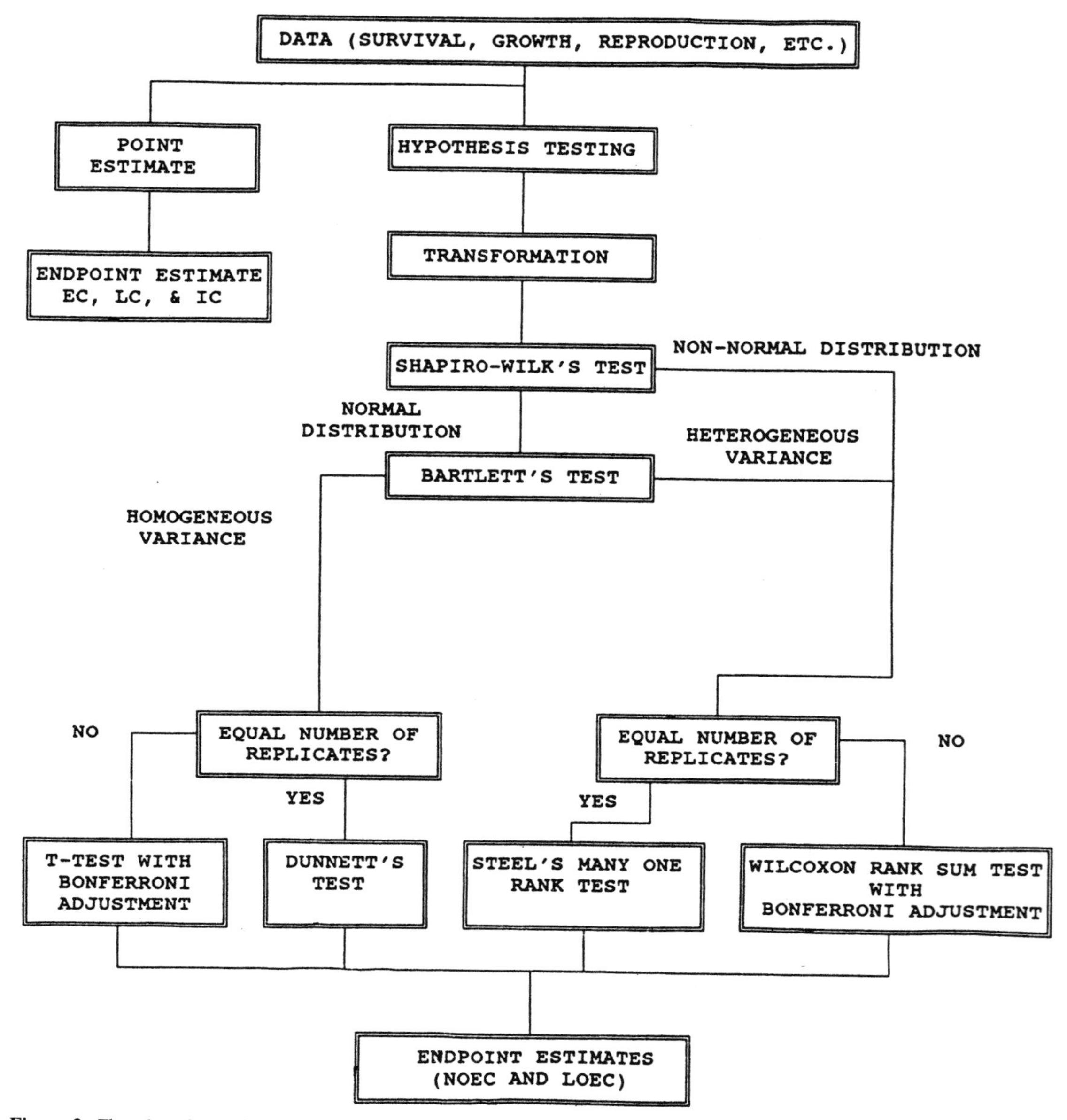

Figure 2. Flowchart for statistical analysis of chronic effluent test data. (From EPA, 1989a.)

chambers are maintained under the specified test conditions (mean temperature 25 ± 2°C and individual temperature observations 25 ± 2°C; photoperiod 16 h light and 8 h dark; ambient laboratory illumination). Test organisms are not fed during the 48-h test. Temperature, dissolved oxygen, pH, specific conductivity, hardness, and alkalinity are measured during the tests. Observations on the number of live and dead (or immobilized) animals are made at 24 and 48 h (Table 6). Data from the acute toxicity tests with *C. dubia* and fathead minnows are used to estimate median lethal concentrations (LC50) (Table 7). For this effluent sample, the 48-h LC50 values for *C. dubia* and fathead minnows were 15% and 71% effluent, respectively (Tables 7 and 8); the test results indicated that for this specific effluent sample, the industrial discharger is in violation of its NPDES permit compliance limit of LC50 ≥ 100% effluent.

Short-Term Chronic Effluent Toxicity Tests

Under the Clean Water Act NPDES permit program (see Chapter 24) a municipal discharger is required to monitor the quality of its effluent quarterly using short-term chronic toxicity tests (U.S. EPA, 1989a). The NPDES permit compliance limit for this munic-

Table 6. Results of 48-h *Ceriodaphnia dubia* and fathead minnow (*Pimephales promelas*) static-acute toxicity tests with an industrial effluent sample using dilute mineral water (DMW) as dilution water

Test concentration (% effluent)	Survival (%) 24 h	Survival (%) 48 h	pH (SU)	D.O. (mg/L)	Temp. (°C)	Cond. (μmho/cm)
Ceriodaphnia dubia						
DMW control	100	100	7.2 7.0–7.5	7.6 7.4–7.8	24.6 24.4–24.7	118
6.25	100	100	7.2 7.0–7.4	7.6 7.6	24.6 24.4–24.7	117
12.5	100	80	7.3 7.3	7.6 7.6	24.6 24.6–24.7	118
25	5	0	6.8 6.4–7.2	7.8 7.4–8.2	24.6 24.5–24.8	120
50	0	0	6.8 6.6–6.9	7.3 6.8–7.8	24.6 24.5–24.6	126
100	0	0	6.6 6.1–7.0	8.1 8.0–8.2	24.8 24.6–24.9	137
Fathead minnows (*Pimephales promelas*)						
DMW control	100	95	7.4 7.0–7.7	7.1 6.4–7.4	25.2 24.7–25.6	118
6.25	100	100	7.5 7.4–7.7	7.2 6.7–7.6	25.0 24.7–25.3	117
12.5	100	100	7.4 7.3–7.7	7.1 6.5–7.6	25.0 24.7–25.3	118
25	100	100	7.4 7.2–7.6	7.1 6.6–7.5	25.0 24.8–25.1	120
50	60	60	7.1 6.6–7.5	7.2 6.5–7.8	24.8 24.4–25.2	126
100	40	40	6.4 6.1–6.8	7.2 6.6–8.2	24.8 24.1–25.3	137

ipal discharge is a chronic NOEC (C-NOEC) of 14% effluent. Thus, to be in compliance with its NPDES permit, the municipal discharge must not be chronically toxic [i.e., no significant effects on survival, growth (fish only), or reproduction (daphnids only)] to either test species at an effluent dilution of 14% (i.e., 14% effluent and 86% test dilution water). Each quarter, static-renewal multiconcentration 7-d short-term chronic toxicity tests are performed with *C. dubia* (≤ 24 h old at test initiation) and fathead minnows (≤ 24 h old at test initiation). Organisms are continuously exposed for 7 d to concentrations of effluent from a municipal wastewater treatment facility (WWTF) using a standard 0.5 dilution series

Table 7. Median lethal concentrations (LC50 values) for 48-h static-acute toxicity tests conducted with an industrial effluent sample

Test species	Control[a] survival (%)	LC50[b] (% effluent)		95% Confidence limits LCL[c]	UCL[c]
C. dubia	100	24 h:	18[d]	17	20
	100	48 h:	15[d]	14	17
P. promelas	100	24 h:	71[e]	56	>100[f]
	95	48 h:	71[e]	56	>100

[a]Control and dilution water was 12% dilute mineral water.
[b]The test duration was 48 h.
[c]LCL, lower confidence limit; UCL, upper confidence limit.
[d]LC50 and confidence limits estimated using the trimmed Spearman-Karber method.
[e]LC50 and confidence limits estimated using the moving-average method.
[f]estimate of UCL is greater than highest test concentration (undiluted 100% effluent).

Table 8. LC50 summary printout for static-acute effluent toxicity tests

Binomial, Moving-Average, Probit, and Spearman Methods

<u>*Ceriodaphnia dubia* Survival Data</u>

Spearman-Karber

Trim:	.00%
LC50:	15.389
95% Lower confidence:	13.595
95% Upper confidence:	17.421

Conc. %	Number exposed	Number dead	Percent dead	Binomial prob. (%)
6.25	20.	0.	.00	.9537D−04
12.50	20.	4.	20.00	.5909D+00
25.00	20.	20.	100.00	.9537D−04
50.00	20.	20.	100.00	.9537D−04
100.00	20.	20.	100.00	.9537D−04

The binomial test shows that 12.50 and 25.00 can be used as statistically sound conservative 95 percent confidence limits since the actual confidence level associated with these limits is 99.4090 percent. An approximate LC50 for this data set is 15.506

When there are less than two concentrations at which the percent dead is between 0 and 100, neither the moving average nor the probit method can give any statistically sound results.

Date: xx-xx-xx
Sample: Industrial Effluent
Method LC50

Test Number: 93-xxxx Duration: 48 hours
Species: *Ceriodaphnia dubia*

		Confidence Limits		
		Lower	Upper	Span
Binomial	15.506	12.500	25.000	12.500
Moving-Average	********	********	********	********
Probit	********	********	********	********
Spearman	15.389	13.595	17.421	3.826

**** = Limit does not exist

<u>Fathead Minnow Survival Data</u>

Spearman-Karber

Trim:	40.00%
LC50:	70.711
95% Lower confidence:	41.334
95% Upper confidence:	120.966

Conc. %	Number exposed	Number dead	Percent dead	Binomial prob. (%)
6.25	20.	0.	.00	.9537D−04
12.50	20.	0.	.00	.9537D−04
25.00	20.	0.	.00	.9537D−04
50.00	20.	8.	40.00	.2517D+02
100.00	20.	12.	60.00	.2517D+02

The binomial test shows that 25.00 and +infinity can be used as statistically sound conservative 95 percent confidence limits since the actual confidence level associated with these limits is 99.9999 percent. An approximate LC50 for this data set is 70.711

Results using moving average

Span	G[a]	LC50	95% Confidence limit	
2	.157	70.71	55.52	102.51

Results calculated by probit method

Iterations	G	H[b]	Goodness of fit
6	.200	1.00	.35

Slope = 3.55 95% Confidence limits: 1.97 and 5.14
LC50 = 74.98 95% Confidence limits: 58.89 and 107.68
LC1 = 16.61 95% Confidence limits: 5.69 and 25.81

[a]G; if G is >0.2 caution should be exercised in using the LC50 and confidence limits. As G approaches 1, the LC50 and confidence limits become meaningless.
[b]H = Heterogeniety factor.

Table 8. LC50 summary printout for static-acute effluent toxicity tests (*Continued*)

Binomial, Moving-Average, Probit, and Spearman Methods

Date: xx-xx-xx
Sample: Industrial effluent

Test Number: 93-xxxx Duration: 48 hours
Species: *Pimephales promelas*

Method	LC50	Confidence Limits Lower	Upper	Span
Binomial	70.711	25.000	*******	*******
Moving-Average	70.711	55.522	102.512	46.990
Probit	74.983	58.888	107.682	48.794
Spearman	70.711	41.334	120.966	79.632

**** = Limit does not exist

(6.25, 12.5, 25, 50, and 100% effluent). Because the NPDES permit compliance concentration for this municipality is 14% effluent, an additional effluent concentration is added to the test. Three 24-h composite effluent samples are collected during a 7-d period from the WWTF by using an automatic sampler. Receiving water collected upstream of the effluent discharge is used as the test dilution water. Because receiving water is being used as the test dilution and control water, an additional control consisting of laboratory-prepared water is added to the test series to assess the general health of the test organisms and the adequacy of the test dilution water.

For the *C. dubia* test, daphnids are individually exposed in 30-ml test beakers containing 15 ml of test solution or control water with 10 replicate beakers per concentration (10 animals per concentration). For the fathead minnow test, fish are exposed in groups of 10 animals in 1000-ml beakers containing 500 ml of test solution or control water, with four replicate beakers per concentration (40 animals per concentration). Test animals are fed daily [brine shrimp for fathead minnows and the alga, *S. capricornutum*, and a yeast–trout chow–Cerophyl suspension for *C. dubia* (U.S. EPA, 1989a)]. Test beakers are maintained under the specified conditions (mean temperature 25 ± 1°C and individual temperature 25 ± 2°C; photoperiod 16 h light and 8 h dark) for the 7-d tests.

Temperature, dissolved oxygen, pH, and specific conductivity are measured daily on newly prepared solutions and on 24-h-old test solutions at each concentration. Observations of the number of live and dead (or immobilized) fathead minnows and *C. dubia* are made daily. Reproduction of *C. dubia* is also monitored daily by counting the number of live and dead young per female. New test solutions are prepared daily using the most recently collected effluent and receiving-water samples. For *C. dubia*, the adults are transferred to new test solutions each day and the young produced during the previous 24-h period are counted and then discarded. *C. dubia* tests are completed when ≥ 60% of the diluent control animals produce their third brood. This reproductive criterion commonly is met by test day 7; however, the test may be ended after 6d or extended to 8 or 9 d to ensure that this reproductive criterion is met.

For fathead minnows, after survival observations and analytical measurements are made each day, the 24-h-old test solutions are removed from the test chambers by siphoning and then replaced with newly prepared test solutions; about 75% of the 24-h old solution is removed from the test chambers during each renewal and replaced with freshly prepared solutions. At the end of 7 d when the fathead minnow test is terminated, all live fry within a replicate are placed in preweighed drying pans and kept in a drying oven overnight; the weigh pans containing the dried fry are then weighed. The total dry-fry weight per replicate is then divided by the number of fry weighed to obtain the average dry-fry weight per replicate.

Survival at test completion, fish growth, and daphnid reproduction are analyzed using the statistical analyses outlined in Figure 2 (Table 9). These analyses are used to estimate C-NOEC and C-LOEC values. The short-term chronic test results showed no significant effects on survival of either *C. dubia* (Fisher's exact test; $P > 0.05$) or fathead minnows (Steel's many-one rank test; $P > 0.05$). Growth of fathead minnows also was not significantly affected during the 7-d test (ANOVA; $P > 0.05$). However, *C. dubia* reproduction was significantly reduced in the 100% effluent concentration (ANOVA and Dunnett's multiple comparisons test; $P < 0.05$) but not in the lower effluent concentrations (6.25 to 50% effluent). Thus, the C-NOEC values for the fathead minnow and *C. dubia* short-term chronic tests were 100% and 50% effluent, respectively. Although some chronic toxicity was measured in the *C. dubia* test, the C-NOEC value of 50% effluent was > 14% effluent. Thus, the municipal discharge was in compliance with its NPDES permit limit.

Life Cycle Daphnid Chronic Toxicity Test

As part of the Federal Insecticide, Fungicide, and Rodenticide Act (FIFRA), certain pesticidal substances must be evaluated in acute and chronic toxicity tests with freshwater organisms before commercial manufacturing, distribution, and use in the United States (see tier testing flowchart in Chapter 21 on FIFRA; also includes saltwater organisms). The full life-cycle chronic toxicity test with the

Table 9. Results of 7-day *Ceriodaphnia dubia* and fathead minnow (*Pimephales promelas*) chronic toxicity tests with effluent samples collected from a municipal wastewater treatment facility using receiving water collected upstream of the effluent discharge as dilution water

	Ceriodaphnia dubia										
Test concentration (% effluent)	Daily survival (%)								Total young[a] per female		
	1	2	3	4	5	6	7	Sig.	*N*	Mean	Sig.
Dilution water control	100	100	100	100	100	100	100	—	10	20.2	—
Laboratory water control	100	100	100	100	100	100	100	—	10	23.0	—
6.25	100	100	100	100	100	100	100	NS	10	15.3	NS
12.5	100	100	100	100	100	90	90	NS	10	21.4	NS
14	100	100	100	100	100	100	100	NS	10	16.7	NS
25	100	100	100	100	100	100	100	NS	10	19.8	NS
50	100	100	100	100	100	100	100	NS	10	23.6	NS
100	100	100	80	80	80	80	80	NS	10	8.8	[b]

	Fathead minnows (*Pimephales promelas*)									
Test concentration (% effluent)	Daily survival (%)								Fish weight[c] (mg)	
	1	2	3	4	5	6	7	Sig.	Mean	Sig.
Dilution water control	100	100	100	100	100	100	100	—	0.38	—
Laboratory water control	100	100	100	98	98	98	98	—	0.61	—
6.25	100	100	100	100	100	100	100	NS	0.41	NS
12.5	100	100	100	100	100	100	100	NS	0.43	NS
14	100	100	100	100	100	98	98	NS	0.47	NS
25	100	100	100	100	100	100	100	NS	0.50	NS
50	100	100	100	100	100	100	100	NS	0.47	NS
100	100	100	98	92	92	90	90	NS	0.43	NS

[a]*N*, number of females at start of test; mean, average number of young per female.
[b]Significantly reduced ($P < 0.05$) compared with reproduction in the dilution-water control using ANOVA and Dunnett's multiple comparisons test.
[c]Mean, average dry fish weight per replicate.

daphnid *D. magna* is one of a suite of tests generally required under FIFRA to assess potential hazards of pesticides to aquatic life (U.S. EPA, 1986d). For the test substance, a flow-through 21-d life-cycle chronic toxicity test was performed with *D. magna* (< 24 h old at test initiation); daphnids were continuously exposed to concentrations of the test substance using a dilution series based on a 96-h static acute range-finding toxicity test. Based on the lethal-nonlethal range estimated in the acute test, the nominal test concentrations, using a 0.5 dilution factor, were 12.5, 25, 50, 100, 200, and 400 μg test substance/L dilution water. A proportional vacuum-siphon diluter (Figure 3) was used to deliver the test concentrations to the chambers (Mount and Brungs, 1967) with approximately 17 volume additions every 24 h. The dilution water was a moderately hard fresh water (56 to 70 mg/L as $CaCO_3$). Because a solvent carrier (dimethyl formamide) was used to dissolve the test substance, a carrier control was used in addition to the freshwater control to assess the potential impact on the organisms of the solvent.

D. magna were exposed in four 100-mm-diameter crystallizing dishes containing 300 ml of test solution or control water with 10 animals per replicate (40 animals per concentration). Test animals were fed continuously with the green alga *Selenastrum capricornutum*, metered into the dilution water during each diluter cycle. A yeast–trout chow–Cerophyl suspension was also added to each test chamber. Beakers were maintained under the specified conditions (mean test temperatures 20 ± 1°C and individual test temperatures 20 ± 2°C; photoperiod 16 h light and 8 h dark) for the 21-d test.

Water quality measurements, including temperature (hourly), dissolved oxygen (alternate days), pH (weekly), specific conductivity (weekly), hardness (weekly), and alkalinity (weekly), were measured periodically during the test. Test substance concentrations were measured at 3- to 7-d intervals. Observations of the number of live and dead (or immobilized) *D. magna* were made daily. Reproduction of *D. magna* was monitored daily by counting the number of live and dead young per female. At the end of the 21-d test, all surviving first-generation daphnids were fixed in ethanol for subsequent determination of growth; the lengths (apex of the helmet to the base of the spine) were measured using a compound microscope equipped with an eyepiece micrometer.

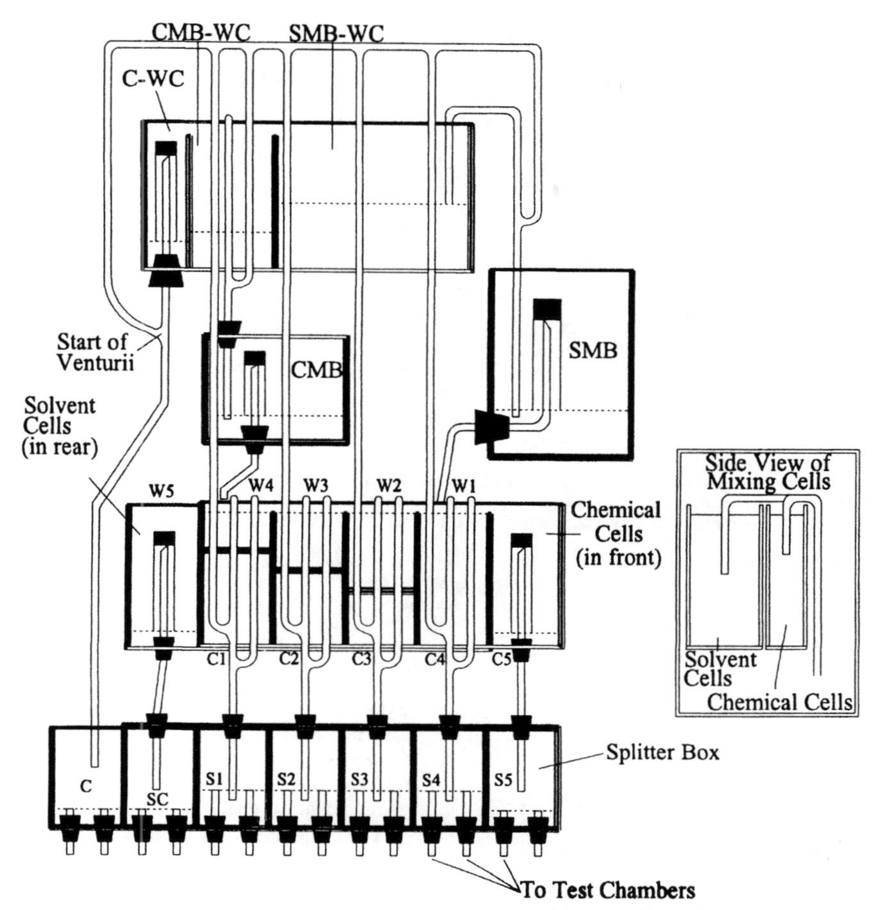

Figure 3. Diagrammatic representation of an intermittent diluter flow-through system designed to deliver five different concentrations of test substance in dilution water and two different controls (dilution water and solvent/carrier) to replicate test chambers. C-WC, control water cell; CMB-WC, chemical mixing box water cell; SMB-WC, solvent mixing box water cell; CMB, chemical mixing box; SMB, solvent mixing box; W1–W5, water cells; C1–C5, chemical cells; C, control; SC, solvent control; S1–S5, flow splitting cells. (From Toxikon Environmental Sciences, 1993.)

The survival, reproduction, and growth results for the 21-d chronic test are summarized in Table 10. Analysis of survival data was performed using Fisher's exact test and the Chi-Square test of independence and was confirmed by using a Student's *t*-test on transformed % survival data. The survival data for the two controls were not significantly different ($P > 0.05$). Therefore, the data for the two control groups were pooled before the statistical analysis. A significant survival effect ($P < 0.05$) was measured in the two highest test concentrations (measured concentrations of 99.3 and 298 μg/L) compared with survival of the pooled control. All daphnids were dead in the highest concentration by test day 10 and survival in the 99.3 μg/L concentration was only 75% after 21 d.

Because a significant survival effect was measured in the two highest concentrations, these two concentrations were excluded from the analysis of the growth and reproductive data. There was a significant increase ($P < 0.05$; Student's *t*-test) in both reproduction and growth of daphnids in the solvent control compared with the laboratory water control; therefore the data were not pooled and all comparisons were made using the solvent control. For both growth and reproduction, significant reductions ($P < 0.05$) were measured (ANOVA and Dunnett's multiple comparison tests) in all test concentrations compared with the growth or reproduction in the solvent control.

The C-NOEC and C-LOEC values for survival were 49.9 and 99.3 μg/L, respectively. An MATC is typically derived for this type of analysis; an MATC of 70.4 μg/L (geometric mean of the C-NOEC and C-LOEC for survival) was estimated for survival. However, no MATC values could be estimated for either the reproduction or length data because significant reductions in both reproduction and length were measured in all test concentrations compared with control data.

Table 10. Analysis of toxicity data for a 21-d flow-through chronic test with a single test chemical using *Daphnia magna*

Mean measured concentration[a] (μg/L)	Survival (%)[b]		Reproduction[c]		Length[d]	
	Mean	Sig.[e]	Mean	Sig.[f]	Mean	Sig.[f]
Control	97.5	—	9.4	—	4.41	—
Solvent control	92.5	—	13.3	—	4.76	—
Pooled control	94.9	—	11.4	—	4.58	—
7.59	97.6	NS	9.4	*	4.53	*
16.4	87.5	NS	11.2	*	4.73	*
29.5	97.5	NS	9.7	*	4.72	*
49.9	97.5	NS	9.7	*	4.64	*
99.3	75.0	*	5.7	NE[g]	4.02	NE[g]
298	0.0[h]	*	—	—	—	—

[a]Nominal test concentrations = 12.5, 25, 50, 100, 200, 400 μg/L.
[b]Percent survival of initial animals after 21 d.
[c]Reproduction expressed as number of offspring per adult per reproductive day.
[d]Length expressed in millimeters (measured from tip of helmet to base of spine).
[e]NS, not significant at $P > 0.05$; *significantly different from the pooled control at $P < 0.05$ using a one-tailed chi-square test and Fisher's exact test.
[f]NS, not significant at $P > 0.05$; *significantly different from the solvent control at $P < 0.05$ using a one-tailed parametric Dunnett's mean comparisons procedure. However, no test concentrations were significantly different ($P > 0.05$) from the pooled control using the same procedure.
[g]Concentration not included in analysis of reproduction or growth because of significant survival effects.
[h]All daphnids were dead by day 10.

The data indicate that the MATC values for these two end points are at some test substance concentration less than the lowest test concentration evaluated (7.59 μg/L).

FACTORS THAT MODIFY TOXICITY

Many factors may affect the results of toxicity tests with aquatic organisms (see Chapter 1). These factors may either be characteristics of the water or experimental design or biological traits associated with the test species (Appendix Chapter C). Therefore, it is essential, when performing acute and chronic toxicity tests, to use standardized test procedures to minimize extraneous variability in the test end point(s). The goal of standardization is to eliminate and/or control as many extraneous factors in the test as possible to reduce variability of the test results and improve their precision and reproducibility.

Normal criteria for acceptability of test results generally include minimum requirements for control survival (e.g., ≥ 90% survival for acute tests, ≥ 80% survival for chronic tests, ≥ 70% for some ELS fish tests) and minimum levels of reproduction or growth in controls. However, other abiotic and biotic factors that may modify test results should also be considered in evaluating the validity of test results.

Biotic factors, such as life stage, size, age, disease, and nutritional status of the test organisms, may have a modifying effect on the test results. Therefore, it is important to ensure that all organisms used in a test come from the same healthy cohort and are impartially or randomly assigned to the various test chambers. Abiotic factors such as temperature and dissolved oxygen are routinely monitored throughout a test because significant changes in these factors can adversely affect test results. For this reason, ASTM (1990a) has set maximum allowable limits for temperature and dissolved oxygen deviations that can occur within a test.

The pH of a test solution can also affect the toxicity of the test substance (U.S. EPA, 1991b,c; Schubauer-Berigan et al., 1993); however, adjustments in pH are usually made only at the beginning of a test and the pH of the test solutions is monitored only during the test. Because many toxicants can ionize under the influence of pH (e.g., ammonia, cyanide, metals), there has been more interest in trying to control pH fluctuations during a test through the use of buffers (U.S. EPA, 1991b; Mount and Mount, 1992).

The concentration of dissolved cations such as calcium and magnesium (i.e., cations primarily responsible for water hardness) can also affect the toxicity of various pollutants, especially metals; therefore, it is an important factor affecting test results that should be controlled by the investigator.

The EPA guidance document titled Quality Criteria for Water (U.S. EPA, 1986a), commonly called the "Gold Book," provides toxicity data on acute and chronic effects for many toxicants. For many metals, water hardness has a distinct effect on toxicity. To protect freshwater aquatic organisms, EPA provides, when sufficient data are available, both acute and chronic water quality criteria based on hardness. For example, the copper national criteria for acute and chronic effects are given by the following formulas:

$$\text{Acute:} \quad \exp_e (0.9422*[\ln (\text{hardness})] - 1.464).$$

$$\text{Chronic:} \quad \exp_e (0.8545*[\ln (\text{hardness})] - 1.465).$$

Based on these formulas, acute and chronic criteria for hardnesses of 50, 100, and 200 mg/L as $CaCO_3$ are

	Hardness (mg/L as $CaCO_3$)		
Criterion	50	100	200
Acute (μg/L)	9.2	18	34
Chronic (μg/L)	6.5	12	21

Thus, when designing a toxicity test or interpreting toxicity test results, it is important to evaluate the potentially modifying influence of hardness on toxicity.

As an outgrowth of EPA's whole-effluent approach to toxics control (Federal Register, 1984; U.S. EPA, 1991a) under the NPDES permitting program (see Chapter 24), EPA has developed toxicity identification evaluation (TIE) methods as a tool to isolate and characterize the physical-chemical nature of toxicants in complex effluents based on a knowledge of how factors, such as pH, modify toxicity (Burkhard and Ankley, 1989; U.S. EPA, 1991b,c, 1989b, 1989c). The goal of a toxicity-based approach to TIE is to separate the toxicants from the nontoxic components in the effluent. Toxicants are "tracked" through all sample manipulations using the most relevant detector available, the test organism (Burkhard and Ankley, 1989). Once the toxicants are characterized, instrumental analyses are performed. In this way, there is a direct relationship between the toxicants and measured analytical data.

Many of the effluent manipulations used in TIEs make use of the knowledge that abiotic factors, like pH and hardness, can alter the nature of the toxicants. It is possible through interpretation of the toxicity test results and subsequent chemical analyses of the fractions following the TIE manipulations to isolate, identify, and confirm many of the sources of toxicity in complex effluents.

A TIE is divided into three phases. In Phase I (U.S. EPA, 1991b,c), the physical-chemical nature of the toxic effluent constituents is characterized. During Phase II (U.S. EPA, 1993a), the specific toxicants are identified using analytical techniques specific for the effluent characteristics determined in Phase I. In Phase III (U.S. EPA, 1993b), the suspected toxicants that were characterized in Phase I and identified in Phase II are confirmed through additional chemical and toxicological testing. The Phase I methods focus on defining a consistent set of characteristics of the effluent toxicant(s) using effluent manipulations and aquatic toxicity tests. Each characterization is designed to alter or render biologically unavailable a group of toxicants, such as oxidants, cationic metals, volatile compounds, nonpolar organic compounds, or chelatable metals (U.S. EPA, 1991b). Aquatic toxicity tests, performed before and after the individual characterization treatment, indicate the effectiveness of the treatment and provide information on the nature of the toxicants. Although the TIE Phase I methodology was developed using the freshwater cladoceran *C. dubia*, EPA and other investigators have used these techniques with *D. magna*, *D. pulex*, and fathead minnows.

For one complete Phase I set of characterizations, nine categories of tests are conducted (Figure 4) (U.S. EPA, 1991b):

- Initial toxicity test (day 1)
- Baseline toxicity test; toxicity persistence (day 2)
- EDTA addition (chelation) test (day 2)
- Sodium thiosulfate addition (oxidant reduction) test (day 2)
- pH adjustment tests (day 2)
- pH adjustment/C18 solid-phase extraction tests (day 2)
- pH adjustment/aeration tests (day 2)
- pH adjustment/filtration tests (day 2)
- Graduated pH tests (pH 6, pH 7, and pH 8) (day 2)

The initial toxicity test is performed upon receipt of the effluent sample (day 1) to assess the level of toxicity in the fresh effluent sample. A 24-h LC50 is estimated using the selected test species. On the day following sample receipt (day 2), a second effluent toxicity test is performed to assess persistence of effluent toxicity. The *baseline toxicity test* is performed concurrently with the effluent manipulation tests.

The pH of a sample has a substantial effect on the toxicity of many compounds. Changes in pH can affect solubility, polarity, volatility, stability, and speciation of a compound, thereby affecting its bioavailability as well as its toxicity. Thus, several of the manipulations (e.g., aeration, filtration, and C18 resin extraction) are performed not only at the initial pH (pHi) of the effluent sample, but also at an acidic pH (3 SU) and a basic pH (11 SU). After the TIE manipulations, the pH is readjusted to the original pH of the sample (pHi).

The *pH adjustment/filtration tests* provide information on effluent toxicants associated with filterable materials. Three aliquots of the effluent are manipulated. The first effluent aliquot is pressure filtered through a 1-μm glass fiber filter; the second and third aliquots are first pH adjusted to 3 SU and 11 SU using HCl or NaOH (ACS grade in high-purity water) before filtration. After filtration, the pH-adjusted effluent samples are returned to the initial pH of the effluent (pHi) by adding either NaOH or HCl. By filtering pH-adjusted aliquots of effluent, the compounds in solution at unadjusted pH but insoluble or associated with particles to a greater extent at extreme pH values (i.e., 3 or 11 SU) are removed. When the toxicant-contaminated particles or precipitated compounds are removed before readjustment of the sample to pHi, these toxicants are no longer available for dissolution in the effluent. The pH change may also destroy or dissolve particles, thereby removing the sorption surfaces or driving the dissolved/sorbed equilibrium in the opposite direc-

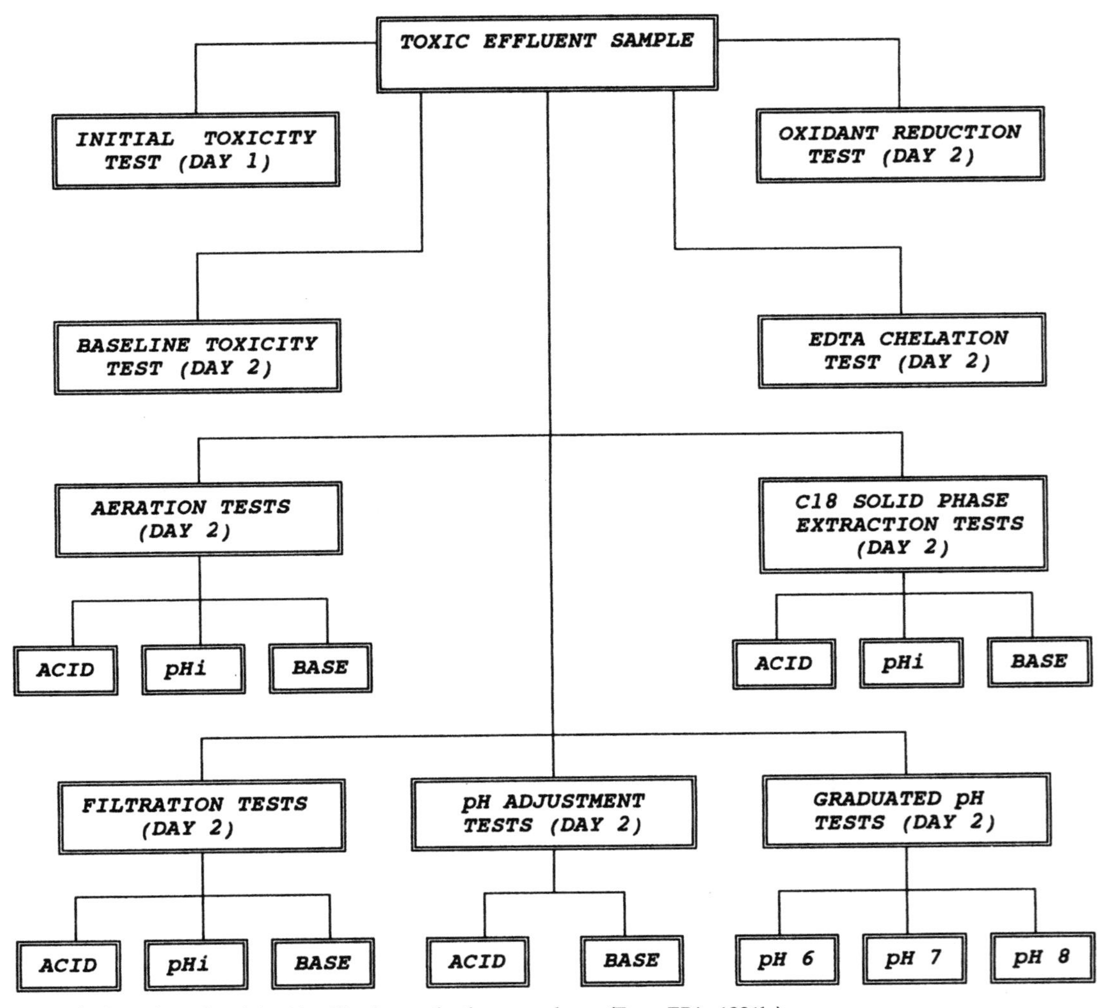

Figure 4. Overview of toxicity identification evaluation procedures. (From EPA, 1991b.)

tion. After filtration and pH readjustment, aquatic toxicity tests are performed.

The *pH adjustment/aeration tests* are designed to determine how much toxicity can be attributed to volatile, sublatable, or oxidizible compounds. Air is used for sparging so that oxidation is included. Nitrogen gas can also be used for sublatable compounds. The pHi, pH3, and pH11 effluent aliquots are placed in tall glass graduated cylinders or burets and the solutions are aerated for 60 min. The pH is checked frequently and readjusted to the target pH if it drifts significantly. After aeration, the samples of aerated effluent are readjusted to the initial pH and then evaluated for toxicity.

The *pH adjustment/solid phase extraction (SPE) tests* using C18 resin columns are designed to determine the extent of effluent toxicity caused by the organic compounds and metal chelates that are relatively nonpolar. The pHi, pH3, and pH9 (column degradation occurs at pH 11; therefore, the pH 11 aliquot is adjusted to pH 9 before use) effluent aliquots are passed through a small column packed with an octadecyl (C18) sorbent. Compounds in the effluent interact through solubility and polarity with the sorbent and are extracted from the effluent onto the sorbent. If the toxicity is reduced by the C18 column, the column may be eluted with 100% methanol to see if the toxic constituents removed by the column can be recovered.

The *oxidant reduction test* is designed to determine the extent to which toxic effluent constituents are reduced by the addition of sodium thiosulfate ($Na_2S_2O_3$). Chlorine, a commonly used biocide and oxidant, is frequently found at toxic levels in municipal effluents. Other potential toxicants used in disinfection, such as ozone and chlorine dioxide, chemicals formed during chlorination (such as mono- and dichloramines), bromine, iodine, manganous ions, and electrophilic organic chemicals are also neutralized by this test. Although the thiosulfate addition test was initially designed to determine whether oxidants are responsible for effluent toxicity, sodium thiosulfate can also be a chelating agent for some cationic metals, such as cadmium, copper, silver, and mercury (Smith and Martell, 1981).

The *chelation test* is designed to determine the extent to which toxicity is caused by certain cationic metals. This is determined by adding a chelating agent (EDTA; ethylenediaminetetraacetate ligand) to the effluent. EDTA is a strong chelating agent, and its addition to water solutions produces relatively nontoxic complexes with many metals (e.g., aluminum, barium, cadmium, cobalt, copper, iron, lead, manganese, nickel, strontium, and zinc).

The *graduated pH test* is designed to determine whether effluent toxicity can be attributed to compounds whose toxicity is pH dependent (such as ammonia, hydrogen sulfide, cyanide, and some organic compounds, e.g., pentachlorophenol). The pH of the effluent is usually adjusted to ± 1 SU of the pHi. The greatest challenge in the graduated pH test is to maintain a constant pH in the test solution. The pH may be held constant by the addition of a CO_2 blanket or buffers (U.S. EPA, 1991b; Mount and Mount, 1992).

Throughout the TIE manipulations, procedural blank controls (i.e., fresh water manipulated in the same manner as the effluent) are prepared and evaluated for toxicity as a check on the general health of the test organisms, dilution water quality, test conditions, and any artifactual toxicity associated with the manipulation.

After a suite of Phase I tests are performed, the results usually show that some manipulations increase toxicity, some decrease toxicity, and others have no measurable effect on effluent toxicity (U.S. EPA, 1991b). The investigator must carefully evaluate the data to determine why the manipulations affected the effluent toxicity by understanding the underlying chemistry of the manipulations. For a TIE to be successful, it must be able to resolve all of the analytical and toxicological problems posed by the effluent. Successful TIEs require a complete understanding of the concentration-response curve for the toxicants, including the influences of the effluent matrix on the toxicants, synergistic and antagonistic interactions among toxicants, and effects of modifying factors on toxicity (Burkhard and Ankley, 1989). Thus, a good knowledge of the factors that modify toxicity is useful for identifying toxicants in complex mixtures such as effluents (see Chapter 33 for detailed case studies).

VARIABILITY OF TEST RESULTS

Toxicity tests, like all measurements, exhibit variability (U.S. EPA, 1991a). Variability of test results can be measured in two ways—the ranges of the NOEC values for survival, growth, or reproductive success obtained by hypothesis testing, and the coefficients of variation (CVs) associated with point estimates such as the LC50 and IC25 values. Ranges of the NOEC values can be simply expressed as the frequency of occurrence of NOEC values and the number of concentrations spanned by the observations, ignoring the actual exposure levels. The range of the NOEC values allows the investigator to observe variability simply by inspection (i.e., wider ranges of NOEC values indicate greater variability in individual NOEC values), whereas CVs associated with point estimates such as LC50 and IC25 values are a quantitative expression of variability (i.e., the standard deviation expressed as a percentage of the mean LC50 or IC25 value).

Another way (and the only quantitative way) to assess variability in test results is by comparing CVs for LC50 values and IC25 values with intralaboratory CVs for other procedures used in regulatory settings, such as other toxicity tests and analytical chemistry procedures. CVs are an accepted way to measure variability for both toxicity testing procedures (Grothe and Kimerle, 1985; Dorn et al., 1987; DeGraeve et al., 1991, 1992; see Chapter 10) and analytical measurements (U.S. EPA, 1986f,g).

Toxicity test variability can be described in terms of two types of precision—"within" or intralaboratory precision and "between" or interlaboratory precision (U.S. EPA, 1991a; see Chapter 1). Intralaboratory precision refers to the ability of trained laboratory personnel to obtain consistent results repeatedly when performing the same test on the same species using the same toxicant. Interlaboratory variability is a measure of how reproducible a method is when conducted by a large number of laboratories using the same methods, species, and toxicant (U.S. EPA, 1991a).

Rue et al. (1988) reviewed intra- and interlaboratory variability in acute toxicity test results for a variety of effluents. Using the data evaluated by Rue (excluding CVs of 0.0 where the LC50 was ≥ 100%), the average intralaboratory CV for LC50 values was 15.8% for the tests examined. The data reviewed by Rue et al. (1988) were all based on analyses of acute freshwater and saltwater toxicity data from tests with *Daphnia* sp., *Ceriodaphnia*, *Mysidopsis bahia*, *Pimephales promelas*, *Oncorhynchus mykiss*, or *Photobacterium phosphoreum*. Obtaining precise data in effluent toxicity tests is more difficult than in pure chemical (e.g., single pesticide) testing because of the continuously changing nature of the effluent. For example, a freshly collected effluent sample tested onsite may be more toxic than a sample shipped off site for testing if the effluent sample contained a nonpersistent toxicant (e.g., chlorine or volatile organics). Also, because of the day-to-day variability of the effluent characteristics, the results obtained with one set of effluent samples cannot be directly compared with those for a second set of effluent samples collected at a different time (i.e., an effluent toxicity test cannot be repeated like a toxicity test with a pure chemical).

Longer test periods used in chronic toxicity tests offer greater opportunities for random physical, chemical, and biological factors to affect test results.

Thus, results from longer-term tests might be expected to be more variable (larger CVs) than results from shorter-term tests. However, ranges of intralaboratory CVs for survival (LC50 values), growth, and reproductive success (IC25 values) for 7-d short-term chronic toxicity tests are similar to those found by Rue et al. (1988). For two freshwater short-term chronic tests, the fathead minnow and *C. dubia* 7-d tests (DeGraeve et al., 1991, 1992), intralaboratory CVs averaged 19.5% and 24.6% for 7-d LC50 values, respectively. For the chronic end points (growth in the fathead minnow test and reproductive success in the *Ceriodaphnia* test), variability for fathead minnows averaged 19.8% and for *Ceriodaphnia* averaged 32.1%. Because of the high costs and associated manpower requirements, intra- and interlaboratory variability data are generally lacking on longer chronic toxicity tests. However, an EPA interlaboratory validation study of a *D. magna* 21-d life cycle test with 11 laboratories found that for four test substances, the interlaboratory variability was typically two- to threefold for acute EC50 values and two- to fourfold for chronic MATC values (U.S. EPA, 1986e).

Analytical chemistry procedures have been used extensively for regulatory compliance assessments, and the intra- and interlaboratory variabilities of many of those procedures have been thoroughly evaluated and published by EPA (1986f,g). DeGraeve et al. (1991) summarized intralaboratory and interlaboratory precision data for a large number of analytical methods and found that they were similar to those found with acute and short-term chronic effluent toxicity tests. For example, the CVs for phthalate esters ranged from as low as 1% to as high as 80%. The higher variabilities were generally associated with analyses performed near the analytical detection limits, whereas lower variabilities were associated with midrange or higher analyses. Other examples of intralaboratory CVs for analytical measurements are:

- Semivolatile compounds: 13 to 90%, mean CV of 36.5%
- Pesticides and PCB mixtures: 20 to 61%, mean CV of 30%
- PAHs: 20 to 50%, mean CV of 36%.

Thus, acute and chronic tests, when performed according to standardized procedures using healthy organisms, are as precise and reproducible as many analytical chemistry methods used today.

LITERATURE CITED

American Public Health Association (APHA): Standard Methods of Water Analysis, 1st ed. Washington, DC: Committee on Standard Methods of Water Analysis, 1905.

American Public Health Association (APHA), American Water Works Association (AWWA), and Water Pollution Control Federation (WPCF): Standard Methods for the Examination of Water and Wastewater, 13th ed. Washington, DC: APHA, 1971.

APHA, AWWA, and WPCF: Standard Methods for the Examination of Water and Wastewater, 16th ed. Washington, DC: APHA, 1985.

APHA, AWWA, and WPCF: Standard Methods for the Examination of Water and Wastewater, 18th ed. Washington, DC: APHA, 1992.

American Society for Testing and Materials (ASTM): Standard guide for conducting acute toxicity tests with fishes, macroinvertebrates, and amphibians. Designation E 729-88a. Annual Book of ASTM Standards, Vol. 11.04:403–422, 1992a.

American Society for Testing and Materials: Standard guide for conducting acute toxicity tests on aqueous effluents with fishes, macroinvertebrates, and amphibians. Designation E 1192-88. Annual Book of ASTM Standards, Vol. 11.04:799–811, 1992b.

American Society for Testing and Materials: Standard guide for conducting renewal life-cycle toxicity tests with *Daphnia magna*. Designation E 1193-87. Annual Book of ASTM Standards, Vol. 11.04:812–828, 1992c.

American Society for Testing and Materials: Standard guide for conducting early life-stage toxicity tests with fishes. Designation E 1241-92. Annual Book of ASTM Standards, Vol. 11.04:886–913, 1992d.

American Society for Testing and Materials: Standard guide for conducting three-brood renewal toxicity tests with *Ceriodaphnia dubia*. Designation E 1295-89. Annual Book of ASTM Standards, Vol. 11.04:973–991, 1992e.

American Society for Testing and Materials: Standard guide for assessing the hazard of a material to aquatic organisms. Designation E 1023-84 (Reapproved 1988) Annual Book of ASTM Standards, Vol. 11.04:669–684, 1992f.

American Society for Testing and Materials: Standard guide for conducting sediment toxicity tests with freshwater invertebrates. Designation E-1383-92. Annual Book of ASTM Standards, Vol. 11.04:1116–1138, 1992g.

American Society for Testing and Materials: Standard guide for acute toxicity test with the rotifer, *Brachionus*. Designation E-1440-91. Annual Book of ASTM Standards, Vol. 11.04:1210–1216, 1992h.

American Society for Testing and Materials: Standard practice for conducting static acute toxicity tests with *Daphnia*. Designation E 4229-84 (Discontinued 1990). Annual Book of ASTM Standards, Vol. 11.04:47, 1992i.

American Society for Testing and Materials: Standard practice for conducting bioconcentration tests with fishes and saltwater bivalve molluscs. Designation E 1022-84 (Reapproved 1988). Annual Book of ASTM Standards, Vol. 11.04:652–668, 1992j.

American Society for Testing and Materials: Standard practice for evaluating an effluent for flavor impairment to fish flesh. Designation D 3696-89. Annual Book of ASTM Standards, Vol. 11.04:22–27, 1992k.

American Society for Testing and Materials: Standard practice for using brine shrimp nauplii as food for test animals in aquatic toxicology. Designation E 1203-87. Annual Book of ASTM Standards, Vol. 11.04:869–873, 1992l.

American Society for Testing and Materials: Standard practice for standardization aquatic microcosm: fresh water. Designation E-1366-91. Annual Book of ASTM Standards, Vol. 11.04:1048–1082, 1992m.

American Society for Testing and Materials: Standard guide for conducting static 96-h tests with microalgae. Designation E-1218-90. Annual Book of ASTM Standards, Vol. 11.04:874–885, 1992n.

American Society for Testing and Materials: Standard guide for conducting static toxicity tests with *Lemma gibba* G3. Designation E-1415-91. Annual Book of ASTM Standards, Vol. 11.04:1171–1180, 1992o.

American Society for Testing and Materials: Standard practice for algal growth potential testing with *Selenastrum capricornutum*. Designation D 3978-80 (Reapproved 1987). Annual Book of ASTM Standards, Vol. 11.04: 32–36, 1992p.

American Society for Testing and Materials: Standard guide for conducting the frog embryo teratogenesis assay-*Xenopus* (FETAX). Designation E1439.91. Annual Book of ASTM Standards, Vol. 11.04: 1199–1209, 1992q.

Anderson, P. D., and D'Apollonia, S.: Aquatic animals. Principles of Ecotoxicology (Scope 12), edited by G. C. Butler, ed., p. 187–221. Toronto: Wiley, 1978.

Arthur, J. W.: Chronic effects of linear alkylate sulfonate detergent on *Gammarus pseudolimnaeus, Campeloma decisum*, and *Physa integra*. *Wat. Res.*, 4:251–257, 1970.

Arthur, J. W.: Review of freshwater bioassay procedure for selected amphipods. Aquatic Invertebrate Bioassays, edited by A. L. Buikema, Jr., and J. Cairns, Jr., pp. 98–108. ASTM STP 715. Philadelphia: ASTM, 1980.

Banner, A., and Van Arman, J. A.: Thermal effects on eggs, larvae, and juveniles of bluegill sunfish. EPA-R-373-041. U.S. Environmental Protection Agency, 1973.

Barker, D. M., and Hebert, P. D. N.: Secondary sex ratio of the cyclic parthenogen *Daphnia magna* (Crustacea: Cladocera) in the Canadian arctic. *Can. J. Zool.*, 64: 1137–1143, 1986.

Bengtson, D. A.; Beck, A. D.; and Simpson, D. L.: Standardization of the nutrition of fish in aquatic toxicological testing. Cowey, C. B.; A. M. Mackie; and J. G. Bell (eds.), Nutrition and Feeding of Fish, edited by C. B. A. M. Cowey, and J. G. Bell, pp. 431–446. London: Academic Press, 1985.

Benoit, D. A.: User's guide for conducting life-cycle chronic toxicity tests with fathead minnows (*Pimephales promelas*). EPA-600/8-81-011. Duluth, MN: Environmental Research Laboratory, U.S. Environmental Protection Agency, 1982.

Berner, D. B.: Taxonomy of *Ceriodaphnia* (Crustacea: Cladocera) in U.S. Environmental Protection Agency cultures. EPA/640/4-86/032. Cincinnati: U.S. Environmental Protection Agency, 1986.

Biesinger, K. E., and Christensen, G. M.: Effects of various metals on survival, growth, reproduction, and metabolism of *Daphnia magna*. *J. Fish. Res. Bd. Can.*, 29: 1691–1700, 1972.

Birge, W. J.; Black, J. A.; and Westerman, A. G.: Short-term fish and amphibian tests for determining the effects of toxicant stress on early life stages and estimating chronic values for single compounds and complex effluents. *Environ. Toxicol. Chem.*, 49:807–821, 1985.

Braughn, J. D., and Schoettger, R. A.: Acquisition and culture of research fish: rainbow trout, fathead minnows, channel catfish, and bluegill. U.S. Environmental Protection Agency, Ecol. Res. Ser. EPA-660/3-75-011, 1975.

Burkhard, L. P., and Ankley, G. T.: Identifying toxicants: NETAC's toxicity-based approach. *Environ. Sci. Technol.*, 23:1438–1443, 1989.

Cairns, J., Jr.: Application factors and ecosystem elasticity: the missing connection. *Environ. Toxicol. Chem.*, 10: 1235–1236, 1991.

Chapman, P. M.: A bioassay by any other name might not smell the same. *Environ. Toxicol. Chem.*, 8:551, 1989.

Cooney, J. D.; DeGraeve, G. M.; Moore, E. L.; Palmer, W. D.; and Pollock, T. L.: Effects of food and water quality on culturing of *Ceriodaphnia dubia*. *Environ. Toxicol. Chem.*, 11:823–837, 1992a.

Cooney, J. D.; DeGraeve, G. M.; Moore, E. L.; Lenoble, B. J.; Pollock, T. L.; and Smith, G. J.: Effects of environmental and experimental design factors on culturing and toxicity testing of *Ceriodaphnia dubia*. *Environ. Toxicol. Chem.*, 11:839–850, 1992b.

Cowgill, U. M.; Keating, K. I.; and Takahashi, I. T.: Fecundity and longevity of *Ceriodaphnia dubia/affinis* in relation to diet at two different temperatures. *J. Crustacean Biol.*, 5:420–429, 1985.

DeGraeve, G. M., and Cooney, J. D.: *Ceriodaphnia*: an update on effluent toxicity testing and research needs. *Environ. Toxicol. Chem.*, 6:331–333, 1987.

DeGraeve, G. M.; Cooney, J. D.; McIntyre, D. O.; Pollock, T. L.; Reichenbach, N. G.; Dean, J. H.; and Marcus, M. D.: Variability in the performance of the 7-d fathead minnow (*Pimephales promelas*) larval survival and growth test: an intra- and interlaboratory study. *Environ. Toxicol. Chem.*, 10:1189–1203, 1991.

DeGraeve, G. M.; Cooney, J. D.; Marsh, B. H.; Pollock, T. L.; and Reichenback, N. G.: Variability in the performance of the 7-d *Ceriodaphnia dubia* survival and reproduction test: an intra- and interlaboratory study. *Environ. Toxicol. Chem.*, 11:851–866, 1992.

Denny, J. S.: Guidelines for the culture of fathead minnows (*Pimephales promelas*) for use in toxicity tests. EPA-600/3-87-001. U.S. Environmental Protection Agency, Environmental Research Laboratory, Duluth, MN, 1987.

Derr, S. K., and Zabik, M. J.: Biologically active compounds in the aquatic environment: the effect of DDE on the egg viability of *Chironomus tentans*. *Bull. Environ. Contam. Toxicol.*, 7:366–368, 1972a.

Derr, S. K., and Zabik, M. J.: Biologically active compounds in the aquatic environment: the uptake and distribution of [1,1-dichloro-2,2-bis(*p*-chlorophenol) ethylene], DDE by *Chironomus tentans* Fabricus (Diptera: Chironomidae). *Trans. Am. Fish. Soc.*, 101:323–329, 1972b.

Dorn, P. B.; Rodgers, J. H., Jr.; Jop, K. M.; Raia, J. C.; and Dickson, K. L.: Hexavalent chromium as a reference toxicant in effluent toxicity tests. *Environ. Toxicol. Chem.*, 6:435–444, 1987.

Dunbar, A. M.; Lazorchak, J. M.; and Waller, W. T.: Acute and chronic toxicity of sodium selenate to *Daphnia magna*. *Environ. Toxicol. Chem.*, 2:239–244, 1983.

Eaton, J. G.: Chronic malathion toxicity to the bluegill (*Lepomis macrochirus* Rafinesque). *Water Res.*, 4:673–684, 1970.

Eaton, J. G.: Chronic cadmium toxicity to the bluegill (*Lepomis macrochirus* Rafinesque). *Trans. Am. Fish. Soc.*, 103:729–735, 1973a.

Eaton, J. G.: Chronic toxicity of copper, cadmium and zinc mixture to the fathead minnow (*Pimephales promelas*). *Water Res.*, 7:1723–1736, 1973b.

Environment Canada: Biological Test Methods: Acute Lethality Test Using Rainbow Trout. Report EPS 1/RM/9. Environment Canada, Conservation and Protection, Ottawa, Ontario, 1990a.

Environment Canada: Biological Test Methods: Acute Lethality Test Using Threespine Stickleback (*Gasterosteus aculeatus*). Report EPS 1/RM/10. Environment Canada, Conservation and Protection, Ottawa, Canada, 1990b.

Environment Canada: Biological Test Methods: Acute Lethality Test Using the *Daphnia* spp. Report EPS 1/RM/11. Environment Canada, Conservation and Protection, Ottawa, Ontario, 1990c.

Environment Canada: Guidance Document on Control of Toxicity Test Precision Using Reference Toxicants. Report EPS 1/RM/12. Environment Canada, Conservation and Protection, Ottawa, Ontario, 1990d.

Environment Canada: Biological Test Methods: Reference Method for Determining Acute Lethality of Effluents to Rainbow Trout. Report EPS 1/RM/13. Environment Canada, Conservation and Protection, Ottawa, Ontario, 1990e.

Environment Canada: Biological Test Methods: Reference Method for Determining Acute Lethality of Effluents to *Daphnia magna*. Report EPS 1/RM/14. Environment Canada, Conservation and Protection, Ottawa, Ontario, 1990f.

Environment Canada: Biological Test Methods: Test of Reproduction and Survival Using the Cladoceran *Ceriodaphnia dubia*. Report EPS 1/RM/21. Environment Canada, Conservation and Protection, Ottawa, Ontario, 1992a.

Environment Canada: Biological Test Methods: Test of Larval Growth and Survival Using Fathead Minnows. Report EPS 1/RM/22. Environment Canada, Conservation and Protection, Ottawa, Ontario, 1992b.

Environment Canada: Biological Test Methods: Growth Inhibition Test Using the Freshwater Alga *Selenastrum Capricornutum*. Report EPS 1/RM/25. Environment Canada, Conservation and Protection, Ottawa, Ontario, 1992c.

Environment Canada: Biological Test Methods: Toxicity Tests Using Early Life Stages of Salmonid Fish (Rainbow trout, Coho Salmon, or Atlantic Salmon). Report EPS 1/RM/28. Environment Canada, Conservation and Protection, Ottawa, Ontario, 1992d.

Food and Drug Administration (FDA): Environmental Assessment Technical Assistance Handbook, U.S. Food and Drug Administration. From National Technical Information Service, Springfield, VA. PB87-175345, 1987.

Federal Register: Development of Water-Quality-Based Permit Limitations for Toxic Pollutants: National Policy. *Fed. Regist.*, 49:9016–9019, 1984.

Federal Register: Toxic Substances Control Act Test Guidelines; Final Rules. *Fed. Regist.*, 50(188):39252–39516, 1985.

Federal Register: National Primary and Secondary Drinking Water Regulations. *Fed. Regist.*, 54(97):22062–22160, 1989.

Geiger, J. G.; Buikema, A. L., Jr.; and Cairns, J., Jr.: A tentative seven-day test for predicting effects of stress on populations of *Daphnia pulex*. J. G. Eaton, P. R. Parrish, and A. C. Hendricks (eds.), pp 13–26. ASTM STP 707. Philadelphia: American Society for Testing and Materials, 1980.

Gelber, R. D.; Lavin, P. T.; Mehta, C. R.; and Schoenfeld, D. A.: Statistical analysis. Fundamentals of Aquatic Toxicology, edited by G. M. Rand, and S. R. Petrocelli, pp. 110–123. Washington, DC: Hemisphere, 1985.

Greene, J. C.; Bartels, C. L.; Warren-Hicks, W. J.; Parkhurst, B. P.; and Miller, W. E.: Protocols for Short-term toxicity Screening of Hazardous Waste Sites. U.S. Environmental Protection Agency, 1988.

Grothe, D. R., and Kimerle, R. A.: Inter- and intralaboratory variability in *Daphnia magna* effluent toxicity test results. *Environ. Toxicol. Chem.*, 4:189–192, 1985.

Gulley, D. D.; Boelter, A. M.; and Bergman, H. L.: TOXSTAT Release 3.2. Laramie: University of Wyoming, 1989.

Hamilton, M. A.; Russo, R.; and Thurston, R. V.: Trimmed Spearman-Karber method for estimating median lethal concentrations in toxicity bioassays. *Environ. Sci. Technol.*, 11:714–718, 1977.

Keating, K. I., and Dagbusan, B. C.: Diatoms in daphnid culture and bioassay. *Environ. Toxicol. Chem.*, 5:299–307, 1986.

Knight, J. T., and Waller, W. T.: Incorporating *Daphnia magna* into the seven-day *Ceriodaphnia* effluent toxicity test method. *Environ. Toxicol. Chem.*, 6:635–645, 1987.

Knight, J. T., and Waller, W. T.: Influence of the addition of Cerophyl on the *Selenastrum capricornutum* diet of the cladoceran, *Ceriodaphnia dubia*. *Environ. Toxicol. Chem.*, 11:521–534, 1992.

Krenkel, P. A., and Novotny, V.: Water Quality Management. New York: Academic Press, 1980.

Laws, E. A.: Aquatic Pollution. New York: Wiley, 1981.

Macek, K. J.: Aquatic toxicology: fact or fiction. *Environ. Health Perspect.*, 34:159–163, 1980.

Macek, K. J., and Sleight, B. H.: Utility of toxicity tests with embryos and fry of fish in evaluating hazards associated with the chronic toxicity of chemicals to fishes. Aquatic Toxicology and Hazard Evaluation, edited by F. L. Mayer, and J. L. Hamelink, pp. 137–146. ASTM STP 634. Philadelphia: ASTM, 1977.

Masters, J. A.; Lewis, M. A.; and Davidson, D. H.: Validation of a four-day *Ceriodaphnia* toxicity test and statistical considerations in data analysis. *Environ. Toxicol. Chem.*, 10:47–55, 1991.

McKim, J. M.: Evaluation of tests with early life stages of fish for predicting long-term toxicity. *J. Fish. Res. Bd. Can.*, 34:1148–1154, 1977.

McKim, 1985. See Appendix Chapter B.

McKim, J. M., and Benoit, D. A.: Effects of long-term exposures to copper on the survival, growth, and reproduction of brook trout. *J. Fish. Res. Bd. Can.*, 28:655–662, 1971.

McKim, J. M., and Benoit, D. A.: Duration of toxicity tests for establishing "no effect" concentrations for copper with brook trout. *J. Fish. Res. Bd. Can.*, 31:449–452, 1974.

Moore, J. A.: Environmental effects: rhetoric versus reason. *Environ. Toxicol. Chem.*, 6:727–729, 1987.

Mount, D. I.: An assessment of application factors in aquatic toxicology. EPA-600/3-77-085. Corvallis, OR: U.S. Environmental Protection Agency, 1977.

Mount, D. I., and Brungs, W.: A simplified dosing apparatus for fish toxicology studies. *Water Res.*, 1:21–29, 1967.

Mount, D. R., and Mount, D. I.: A simple method of pH control for static and static-renewal aquatic toxicity tests. *Environ. Toxicol. Chem.*, 11:609–614, 1992.

Mount, D. I., and Norberg, T. J.: A seven-day life-cycle cladoceran toxicity test. *Environ. Toxicol. Chem.*, 3:425–434, 1984.

Mount, D. I., and Stephan, C. E.: A method for establishing acceptable toxicant limits for fish—malathion and butoxyethanol ester of 2,4-D. *Trans. Am. Fish. Soc.*, 96: 185–193, 1967.

Nebeker, A. V.: Temperature requirements and life cycle of the midge *Tanytarsus dissimilis*. *J. Kans. Entomol. Soc.*, 46:160–165, 1973.

Norberg-King, T. J.: An evaluation of the fathead minnow seven day subchronic test for estimating chronic toxicity. *Environ. Toxicol. Chem.*, 8:1075–1089, 1989.

Norberg, T. J., and Mount, D. I.: A new fathead minnow (*Pimephales promelas*) subchronic toxicity test. *Environ. Toxicol. Chem.*, 4:711–718, 1985.

Organization for Economic Cooperation and Development (OECD): Guidelines for Testing of Chemicals. Paris, 1987.

Oris, J. T.; Winner, R. W.; and Moore, M. V.: A four-day survival and reproduction toxicity test for *Ceriodaphnia dubia*. *Environ. Toxicol. Chem.*, 10:217–224, 1991.

Parrish, 1985. See Appendix Chapter A.

Patterson, P. W.; Dickson, K. L.; Waller, W. T.; and Rodgers, J. H., Jr.: The effects of nine diet and water combinations on the culture health of *Ceriodaphnia dubia*. *Environ. Toxicol. Chem.*, 11:1023–1035, 1992.

Pennak, R. W.: Freshwater Invertebrates of the United States. 2nd ed. New York: Wiley, 1978.

Petrocelli, S. R.: Chronic toxicity tests. Fundamentals of Aquatic Toxicology, edited by G. M. Rand, S. R. Petrocelli, pp. 96–109. Washington, DC: Hemisphere, 1985.

Rand, G. M.: Detection: bioassay. Introduction to Environmental Toxicology, edited by F. E. Guthrie, J. J. Perry, pp. 390–403. New York: Elsevier, 1980.

Rand, G. M., and Petrocelli, S. R.: Fundamentals of Aquatic Toxicology. Washington, DC: Hemisphere, 1985.

Rue, W. J.; Fava, J. A.; and Grothe, D. R.: A review of inter- and intralaboratory effluent toxicity test method variability. Aquatic Toxicology and Hazard Assessment: Tenth Symposium, edited by W. J. Adams; G. A. Chapman; and W. G. Landis, pp. 190–203. ASTM STP 971. Philadelphia: American Society for Testing and Materials, 1988.

Schubauer-Berigan, M. K.; Dierkes, J. R.; Monson, P. D.; and Ankley, G. T.: pH-dependent toxicity of Cd, Cu, Ni, Pb, and Zn to *Ceriodaphnia dubia*, *Pimephales promelas*, *Hyalella azteca*, and *Lumbriculus variegatus*. *Environ. Toxicol. Chem.*, 12:1261–1266, 1993.

Scott, W. B., and Crossman, E. J.: Freshwater Fishes of Canada. Bulletin 184. Ottawa: Fisheries Research Board of Canada, 1973.

Smith, R. M., and Martell, A. E.: Critical Stability Constants, Vol. 4: Inorganic Complexes, p. 87. New York: Plenum, 1981.

Smith, W. E.: A cyprinodontid fish, *Jordanella floridae*, as reference animals for rapid chronic bioassays. *J. Fish. Res. Bd. Can.*, 39:329–330, 1973.

Snell, T. W., and Moffat, B. D.: A 2-d life-cycle test with the rotifer *Brachionus calyciflorus*. *Environ. Toxicol. Chem.*, 11:1249–1257, 1992.

Snell, T. W., and Persoone, G.: Acute toxicity bioassays using rotifers. I. A test for brackish and marine environments with *Brachionus plicatilis*. *Aquat. Toxicol.*, 14: 65–80, 1989a.

Snell, T. W., and Persoone, G.: Acute toxicity bioassays using rotifers. II. A freshwater test with *Brachionus rubens*. *Aquat. Toxicol.*, 14:81–92, 1989b.

Snell, T. W.; Moffat, B. D.; Janssen, C.; and Persoone, G.: Acute toxicity bioassays using rotifers. III. Effects of temperature, strain, and exposure time on the sensitivity of *Brachionus plicatilis*. *Environ. Toxicol. Water Qual.*, 6:63–75, 1991a.

Snell, T. W.; Moffat, B. D.; Janssen, C.; and Persoone, G.: Acute toxicity bioassays using rotifers. IV. Effects of cyst age, temperature, and salinity on the sensitivity of *Brachionus calyciflorus*. *Ecotoxicol. Environ. Safety*, 21: 308–317, 1991b.

Sprague, 1985. See Appendix Chapter C.

Stephan, C. E.: Methods for calculating an LC50. Aquatic Toxicology and Hazard Evaluation. Edited by F. L. Mayer, and J. L. Hamelink, pp. 65–84. ASTM STP 634. Philadelphia: American Society for Testing and Materials, 1977.

Stephan, C. E., and Mount, D. I.: Use of toxicity tests with fish in water pollution control. Biological Methods for the Assessment of Water Quality, pp. 164–177. ASTM STP 528. Philadelphia: American Society for Testing and Materials, 1973.

Suter, G. W., II: Seven-day tests and chronic tests. *Environ. Toxicol. Chem.*, 9:1435–1436, 1990.

Suter, G. W., II; Rosen, A. E.; Linder, E.; and Parkhurst, D. F.: Endpoints for responses of fish to chronic toxic exposures. *Environ. Toxicol. Chem.*, 6:793–809, 1987.

U.S. Environmental Protection Agency: Algal assay procedures: bottle test. National Eutrophication Program, Environmental Research Laboratory, Corvallis, OR, 1971.

U.S. Environmental Protection Agency: Water quality criteria. EPA-R3-73-033. Washington, DC, 1972.

U.S. Environmental Protection Agency: Biological Field and Laboratory Methods for Measuring the Quality of Surface Waters and Effluents, edited by C. I. Weber. EPA-670/4-73-001. Office of Research and Development, Cincinnati, OH, 1973.

U.S. Environmental Protection Agency: Methods for Acute Toxicity Tests with Fish, Macroinvertebrates, and Amphibians. EPA-660/3-75-009. Committee on Methods of Acute Toxicity Tests with Aquatic Organisms, National Environmental Research Center, Corvallis, OR, 1975.

U.S. Environmental Protection Agency: Pesticide Assessment Guidelines. EPA 560/9-82-018 through 028. Office of Pesticide Programs, Washington, DC, 1982a.

U.S. Environmental Protection Agency: Toxic Substances Test Guidelines. EPA 560/6-82-01 through 003. Office of Toxic Substances, Washington, DC, 1982b.

U.S. Environmental Protection Agency: Standard Evaluation Procedure: Acute Toxicity Test for Freshwater Invertebrates. Hazard Evaluation Division, EPA-540/9-85-005, 1985a.

U.S. Environmental Protection Agency: Standard Evaluation Procedure: Acute Toxicity Test for Freshwater Fish. Hazard Evaluation Division, EPA-540/9-85-006, 1985b.

U.S. Environmental Protection Agency: Methods for Measuring the Acute Toxicity of Effluents to Freshwater and Marine Organisms, 3rd ed., Edited by W. H. Peltier, and C. I. Weber. EPA-600/4-85-013. Environmental Monitoring and Support Laboratory, Cincinnati, OH, 1985c.

U.S. Environmental Protection Agency: Quality Criteria for Water. EPA 440/5-86-001. Washington, DC, 1986a.

U.S. Environmental Protection Agency: Standard Evaluation Procedure: Fish Life-Cycle Toxicity Tests. Hazard Evaluation Division, EPA-540/9-86-137, 1986b.

U.S. Environmental Protection Agency: Standard Evaluation Procedure: Fish Early Life-Stage Test. Hazard Evaluation Division, EPA-540/9-86-138, 1986c.

U.S. Environmental Protection Agency: Standard Evaluation Procedure: *Daphnia magna* Life-Cycle (21-day Renewal) Chronic Toxicity Tests. Hazard Evaluation Division, EPA-540/9-86-141, 1986d.

U.S. Environmental Protection Agency: Standard Evaluation Procedure: Non-Target Plants: Growth and Reproduction of Aquatic Plants-Tiers 1 and 2. Hazard Evaluation Division, EPA-540/9-86-134, 1986e.

U.S. Environmental Protection Agency: Collaborative study of *Daphnia magna* static-renewal assays. EPA/600/x-86-115. Environmental Monitoring and Systems Laboratory, Las Vegas, NV. 1986e.

U.S. Environmental Protection Agency: Test Methods for Evaluating Solid Waste, Vol. 1B: Laboratory Manual—Physical/Chemical Methods, 3rd ed. Office of Solid Waste and Emergency Response, Washington, DC, 1986f.

U.S. Environmental Protection Agency: Test Methods for Evaluating Solid Waste. Vol. 1A: Laboratory Manual—

Physical/Chemical Methods, 3rd ed. Office of Solid Waste and Emergency Response, Washington, DC, 1986g.

U.S. Environmental Protection Agency: Methods for estimating the chronic toxicity of effluents and receiving waters to freshwater organisms. EPA-600/4-89-001. Environmental Monitoring Systems Laboratory, Cincinnati, OH, 1989a.

U.S. Environmental Protection Agency: Technical Support Document for Water Quality-based Toxics Control. EPA 505/2-90-001. U.S. Environmental Protection Agency, Office of Water, Washington, DC, 1991a.

U.S. Environmental Protection Agency: Methods for Measuring the Acute Toxicity of Effluents and Receiving Waters to Freshwater and Marine Organisms, 4th ed., edited by C. I. Weber. EPA-600/4-90/027, 1991b.

U.S. Environmental Protection Agency: Methods for aquatic toxicity identification evaluations: Phase I toxicity characterization procedures, 2nd ed., edited by T. J. Norberg-King; D. I. Mount; E. Durhan; G. Ankley; and L. Burkhard. EPA-600/6-91/003, 1991c.

U.S. Environmental Protection Agency: Toxicity identification evaluation: Characterization of chronically toxic effluents, Phase I, edited by T. Norberg-King; D. I. Mount; J. R. Amato; D. A. Jensen; and J. A. Thompson. EPA-600/6-91/005, 1991d.

U.S. Environmental Protection Agency: Sediment toxicity identification evaluation: Phase I (Characterization), Phase II (Identification), and Phase III (Confirmation) Modifications of Effluent Procedures, edited by G. T. Ankley; M. K. Schubauer-Berigan; J. R. Dierkes; and M. T. Lukasewycz. EPA-600/6-91/007, 1991e.

U.S. Environmental Protection Agency: Methods for Aquatic Toxicity Identification Evaluations: Phase II Toxicity Identification Procedures. For Samples Exhibiting Acute and Chronic Toxicity. EPA-600/R92/080, 1993a.

U.S. Environmental Protection Agency: Methods for Aquatic Toxicity Identification Evaluations: Phase III Toxicity Confirmation Procedures. For Samples Exhibiting Acute and Chronic Toxicity. EPA-600/R92/081, 1993b.

Weir, P. A., and Hine, C. H.: Effects of various metals on behavior of conditioned goldfish. *Arch. Environ. Health*, 20:45–51, 1970.

Winner, R. W.: A comparison of body length, brood size, and longevity as indices of chronic copper and zinc stresses in *Daphnia magna*. *Environ. Poll.*, (Ser. A) 26: 33–37, 1981.

Winner, R. W.: Evaluation of the relative sensitivities of 7-d *Daphnia magna* and *Ceriodaphnia dubia* toxicity tests for cadmium and sodium pentachlorophenol. *Environ. Toxicol. Chem.*, 7:153–159, 1988.

Winner, R. W.: Multigeneration life-span tests of the nutritional adequacy of several diets and culture waters for *Ceriodaphnia dubia*. *Environ. Toxicol. Chem.*, 8:513–520, 1989.

Woltering, D. M.: The growth response in fish chronic and early life-stage toxicity tests: a critical review. *Aquat. Toxicol.*, 5:1–21, 1984.

Chapter 3

SALTWATER TESTS

G. S. Ward

INTRODUCTION

Most chemical substances either directly or indirectly reach saltwater ecosystems. With salt water covering over 70% of the earth's surface and the vast diversity of plant and animal life in the estuaries and open sea, toxicological evaluations of effects of chemicals on saltwater organisms are essential to humankind.

Saltwater fishes, invertebrates, and plants are currently tested under several U.S. and international environmental regulations. In the United States, the four major environmental laws utilizing saltwater testing are the Federal Insecticide, Fungicide, and Rodenticide Act (FIFRA), the Clean Water Act (CWA), the Marine Protection, Research, and Sanctuaries Act (MPRSA; see Foster, 1985), and the Toxic Substances Control Act (TSCA). FIFRA, TSCA, and CWA are discussed in Chapters 21, 22, 23, and 24.

As a result of the great diversity of marine plant and animal life, hundreds of species have been used in saltwater toxicity tests. However, relatively few tests have been developed and considered standard for incorporation into the requirements for various environmental legislation (see Chapters 21, 22, 23, and 24). In fact, based on the limited life history, dietary, and disease information available on most of the currently utilized saltwater test species, few of the tests would probably be considered standard. What follows is an overview of some of the most frequently utilized acute and chronic (including both early life stage and full life cycle) toxicity tests with saltwater plants and animals that are either commonly cultured or easily collected. References for more detailed information on each of the tests are provided, as well as a discussion of some of the problems encountered in the performance of these tests.

In addition to single-species toxicity tests, multispecies procedures have been developed for evaluating chemical effects on saltwater communities. These procedures allow a greater diversity of species to be tested, may identify sensitive species, and present a more real-world evaluation of toxicity because sediment is included and there are assemblages of organisms. An example of a saltwater multispecies test is included in the chapter. Chapter 9 on Field Studies also provides examples of the use of natural and simulated experimental systems to assess chemical stress in salt water using a multispecies approach.

GENERAL SALTWATER TOXICITY TEST METHODOLOGY/PROCEDURES

As with freshwater species, standard toxicity testing is limited to acute, short-term tests of 4 d or less with fish and invertebrates to determine the median lethal concentration (LC50) or median effective concentration (EC50) of a test material (i.e., single chemical or effluent). Longer-duration early life stage (ELS) tests can be conducted for 28 d or more with several saltwater fish to evaluate potential chemical effects from the egg stage through early development. Tests measure embryo survival and hatchability and juvenile survival and growth. As in freshwater tests, the LOEC, NOEC, and MATC can be determined for each biological end point. These values can also be determined using a full life cycle chronic test that measures the effect of a material on survival, growth, and reproduction. Full life cycle chronic tests with saltwater invertebrates are 2 to 4 wk and a minimum of 6 mo for fish. Only a few fish and invertebrate species can be routinely tested in full life cycle toxicity tests. In the past 10 yr in response to regulatory need to screen effluents for chronic toxicity, several saltwater short-term chronic tests (≤7–9 d) were developed to estimate chronic effects of effluents within a short time frame.

Standardized methods used to conduct saltwater tests are listed in Table 1. ASTM, APHA/AWWA/WPCF (Standard Methods) and U.S. EPA list procedures according to test organism, duration of ex-

Table 1. Saltwater toxicity test guidelines

Designation	Title
	Standard Methods, Part 8000 (APHA et al., 1992)
8111	Biostimulation (Algal Productivity)
8112	Toxicity Test Procedures for Phytoplankton
8310	Toxicity Test Procedures for Ciliated Protozoa
8410	Toxicity Test Procedures for Scleractinian Coral
8510	Toxicity Test Procedures for Annelids
8610	Toxicity Test Procedures for Mollusks
8710	Toxicity Test Procedures for Microcrustacea
8712	Toxicity Test Procedures for Acartia
8720	Toxicity Test Procedures for Macrocrustaceans
8910	Toxicity Test Procedures for Fish
	ASTM (ASTM, 1993a-j; two *proposed* guides for annelids not included-ASTM, 1991a, b)
E-1191-90	Guide for Conducting Life-Cycle Toxicity Tests with Mysids
E-729-88a	Guide for Conducting Acute Toxicity Tests with Fishes, Macroinvertebrates, and Amphibians
E-1241-92	Guide for Conducting Early Life-Stage Toxicity Tests with Fishes
E-1440-91	Guide for Acute Toxicity Test with the Rotifer, *Brachionus* (and Estuarine and Marine Rotifers)
E-724-89	Guide for Conducting Static Acute Toxicity Tests Starting with Embryos of Four Species of Saltwater Bivalve Molluscs
E-1463-92	Guide for Conducting Static and Flow-Through Acute Toxicity Tests with Mysids from the West Coast of the United States
E-1367-92	Guide for Conducting 10-Day Static Sediment Toxicity Tests with Marine and Estuarine Amphipods (see Chapter 8)
E-1192-88	Guide for Conducting Acute Toxicity Tests on Aqueous Effluents with Fishes, Macroinvertebrates, and Amphibians
E-1498-92	Guide for Conducting Sexual Reproduction Tests with Seaweeds (see Chapter 5)
E-1218-90	Guide for Conducting Static 96-Hour Toxicity Tests with Microalgae (see Chapter 4)
	EPA Acute Effluent Tests (U.S. EPA, 1991)
________	Acute Effluent Toxicity Tests with *Mysidopsis bahia*
________	Acute Effluent Toxicity Tests with *Cyprinodon variegatus*
________	Acute Effluent Toxicity Tests with *Menidia* sp.
	EPA Chronic Effluent Tests (U.S. EPA, 1988)
1004	Sheepshead Minnow (*Cyprinodon variegatus*) Larval Survival and Growth Test Method
1005	Sheepshead Minnow (*Cyprinodon variegatus*) Embryo-Larval Survival and Teratogenicity Test Method
1006	Inland Silverside (*Menidia beryllina*) Larval Survival and Growth Method
1007	Mysid (*Mysidopsis bahia*) Survival, Growth, and Fecundity Test Method
1008	Sea Urchin (*Arbacia punctulata*) Fertilization Test Method (see Chapter 6)
1009	Algal (*Champia parvula*) Sexual Reproduction Test Method (see Chapter 5)
	EPA TSCA Test Guidelines (Code of Federal Regulations, 1990)
797.1050	Algal Acute Toxicity Test (see Chapter 4)
797.1075	Freshwater and Marine Algae Acute Toxicity Test (see Chapter 4)
797.1400	Fish Acute Toxicity Test
797.1440	Fish Acute Toxicity Test
797.1520	Fish Bioconcentration Test
797.1560	Fish Bioconcentration Test
797.1600	Fish Early Life-Stage Toxicity Test
797.1800	Oyster Acute Toxicity Test
797.1930	Mysid Shrimp Acute Toxicity Test
797.1950	Mysid Shrimp Chronic Toxicity Test
797.1970	Penaeid Shrimp Acute Toxicity Test

Table 1. Saltwater toxicity test guidelines (*Continued*)

Designation	Title
	EPA-FIFRA Test Guidelines (U.S. EPA, 1985a–d; 1986a–c)
———	SEP[a]: Acute Toxicity Test for Estuarine and Marine Organisms (Estuarine Fish 96-Hour Acute Toxicity) (EPA-540/9-85-009, 1985)
———	SEP: Acute Toxicity Test for Estuarine and Marine Organisms (Shrimp 96-Hour Acute Toxicity Test) (EPA-540/9-85-010, 1985)
———	SEP: Acute Toxicity Test for Estuarine and Marine Organisms (Mollusc 96-Hour Flow-Through Shell Deposition Study) (EPA-540/9-85-011, 1985)
———	SEP: Acute Toxicity Test for Estuarine and Marine Organisms (Mollusc 48-Hour Embryo Larvae Study) (EPA-540/1-85-012, 1985)
———	SEP: Fish Early Life-Stage (EPA-540/9-86-138, 1986)
———	SEP: Fish Life-Cycle Toxicity Tests (EPA-540/9-86-137, 1986)
———	SEP: Nontarget Plants: Growth and Reproduction of Aquatic Plants—Tiers 1 and 2 (EPA-540/9-86-134, 1986) (see Chapter 4)
	OECD Test Guidelines (OECD)
———	No specific saltwater toxicity test guidelines.
	Environment Canada (Environment Canada 1992a,b)
———	Acute Test for Sediment Toxicity Using Marine or Estuarine Amphipods (1992a) (see Chapter 8)
———	Fertilization Assay with Echinoids (Sea Urchins and Sand Dollars) (1992b) (see Chapter 6)

[a]SEP = Standard Evaluation Procedure

posure, exposure condition (or method—static, static-renewal, or flow-through), or biological (or toxicological) end point.

Procedures described in Chapter 2 for freshwater testing also apply to saltwater testing for acute and chronic exposure studies. This includes test design [i.e., five or more test material concentrations (or effluent percent) and appropriate control(s) for a total of six or seven (with solvent and untreated water controls) treatments], monitoring of biological end points, water quality, and chemical analyses (i.e., if necessary for single chemicals). The only differences in saltwater tests are in the test organisms used (see Table 2), dilution water (i.e., salt water; see ASTM E729-88a and APHA et al., 1992), and monitoring of salinity during tests.

The preferred saltwater invertebrate and vertebrate test species under FIFRA are mysid (*Mysidopsis bahia*) and sheepshead minnow (*Cyprinodon variegatus*), respectively, for acute and chronic exposure studies. Oysters (e.g., *Crassostrea virginica*) are also suggested for sublethal tests (EC50). For TSCA, the preferred invertebrate for acute and chronic tests is the mysid and secondarily the penaeid shrimp (*Penaeus* spp.) for acute tests. Oyster acute sublethal tests may also be conducted under TSCA. The preferred saltwater vertebrate species for ELS tests under TSCA are silverside (*Menidia* sp.) and sheepshead minnow. Mysid is the preferred invertebrate saltwater species for acute effluent testing under the Clean Water Act; sheepshead minnows and silversides are the preferred saltwater vertebrate species. For chronic (short-term) effluent tests, mysids, sheepshead minnows, and silversides are the preferred species.

Table 2. Test species commonly used for saltwater toxicity tests

Vertebrates
Sheepshead minnow, *Cyprinodon variegatus*
Mummichog, *Fundulus heteroclitus*
Longnose killifish, *Fundulus similis*
Silverside, *Menidia* sp.
Threespine stickleback, *Gasterosteus aculeatus*
Pinfish, *Lagodon rhomboides*
Spot, *Leiostomus xanthurus*
Shiner perch, *Cymatogaster aggregata*
Tidepool sculpin, *Oligocottus maculosus*
Sanddab, *Citarichthys stigmaeus*
Flounder, *Paralichthys dentatus, P. lethostigma*
Starry flounder, *Platichthys stellatus*
English sole, *Parophrys vetulus*
Herring, *Clupea harengus*
Invertebrates
Copepods, *Acartia clausi, Acartia tonsa*
Shrimp, *Penaeus setiferus, P. duorarum, P. aztecus*
Grass shrimp, *Palaemonetes pugio, P. intermedius, P. vulgaris*
Sand shrimp, *Crangon septemspinosa*
Shrimp, *Pandalus jordani, P. danae*
Bay shrimp, *Crangon nigricauda*
Mysid, *Mysidopsis bahia, M. bigelowi, M. almyra*
Blue crab, *Callinectes sapidus*
Shore crab, *Hemigrapsus* sp., *Pachygrapsus* sp.
Green crab, *Carcinus maenas*
Fiddler crab, *Uca* sp.
Oyster, *Crassostrea virginica, C. gigas*
Polychaete, *Capitella capitata*

Source: ASTM, 1992a. Copyright ASTM. Reprinted with permission.

Micro- and macroalgae (including vascular plants) and sea urchin tests may also be required under some of these regulations (see below).

General acute, early life stage, and full life cycle chronic toxicity testing procedures (including bioaccumulation) were described in detail in Rand and Petrocelli (1985; see also Appendix Chapters A, B, and D).

Algae and Macrophytes

Many species of marine algae are currently cultured and tested throughout the world. The most commonly utilized unicellular species for testing under the current environmental regulations include several species of marine diatoms (*Skeletonema costatum* and *Nitzschia punctata*) and the dinoflagellate *Dunaliella tertiolecta*. Procedures for microalgal and vascular plant testing are described in Chapter 4.

Only a few saltwater macroalgae are tested and all tests are relatively new. These include the species of red and brown macroalgae. Procedures for testing the red alga, *Champia parvula*, and the brown alga, *Laminaria* sp., are described in Chapter 5.

Cnidarians

The phylum Cnidaria includes the hydras, jellyfish, sea anemones, and corals. Stebbing and Brown (1984) reviewed the literature on ecotoxicological tests utilizing Cnidaria and found that current testing is primarily involved with hydroids (colonial hydrozoans) and corals, although they described one test with a scyphozoan (jellyfish) and some work with anemones. Although hydroids are easily cultured and can be sensitive to chemicals, they have not been recommended for marine pollution monitoring programs because testing and culture are labor-intensive and they are not of particular ecological or economic importance. However, Karbe et al. (1984) described a standard toxicity test with the marine hydroids *Laomedea flexuosa* (Figure 1) and *Eirene viridula*, based on the effects on colony growth rate and morphological changes of the hydranths. The vast majority of hydroids are colonial. In hydras, buds form on the stalk as evaginations of the body wall. The distal end of the bud forms a mouth and tentacles; then the whole bud drops off to form a new individual. In the development of colonial forms like *L. flexuosa* (Figure 1), the buds remain attached; these in turn produce buds so that each polyp is connected to the others. Such a collection of polyps is known as a hydroid colony. The term hydranth refers to the oral end of the polyp bearing the mouth and tentacles. The colony is anchored by a horizontal root-like stolon.

In the standard procedure described by Karbe et al. (1984), hydroid colonies are induced to grow on glass plates prior to testing (one colony per plate). Culture and test water is 0.15-μm filtered offshore natural seawater, although artificial seawater can also

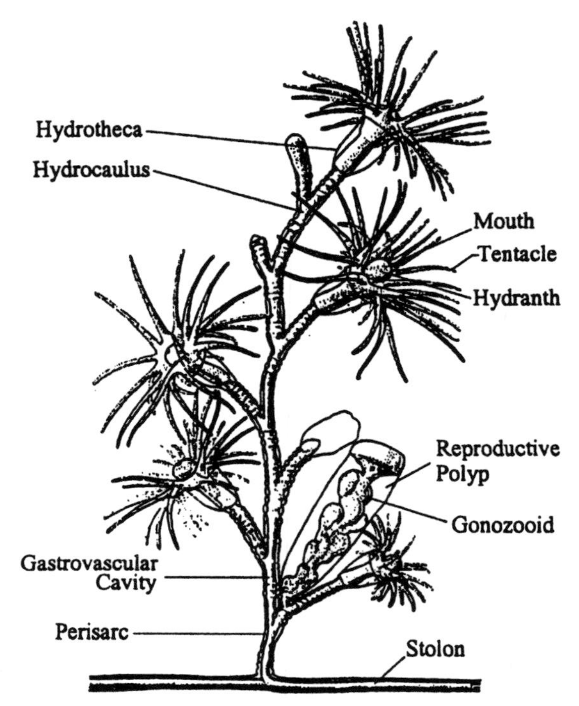

Figure 1. A single upright of a *Laomedea flexuosa* colony showing hydranths, a gonozooid, and the stolon. The upright is approximately 5 mm high. Hydroids are surrounded by a perisarc (protein-chitin envelope) that extends upward to enclose the hydranth in a casing known as the hydrotheca or the perisarc may be confined to the hydrocaulus. The hydrotheca is bell shaped in this species. Hydroids with a hydrotheca surrounding the polyp are said to be thecate. (Modified from Fig. 1, Stebbing ARD: An experimental approach to the determinants of biological water quality. *Philos. Trans. R. Soc. Lond. Ser. B*, 286:465–481, 1979.)

be used. At the beginning of the test, the number of individual polyps per colony is reduced to approximately 10 by removing excess uprights with a scalpel or razor blade. Six glass plates (each with one attached colony of 10 individual polyps) are then placed at random in the test containers. Each test container is one treatment and at least five treatments and control are tested. The test solutions are aerated and maintained at 20°C. During the test (minimum duration of 2 wk), *E. viridula* are fed live 3-d-old *Artemia salina* twice a week and *L. flexuosa* are fed every day. One hour after feeding, the plates with hydroids are transferred to freshly prepared test treatment solutions.

The colony size is determined at the start of the test, on feeding days, or at the end of each week, at a minimum. The number of hydranths with fully developed tentacles or visible parts of tentacles is determined. In tests with *L. flexuosa*, both hydranths and well-developed polyps called gonozooids are counted; with *E. viridula*, morphological effects may also be evaluated.

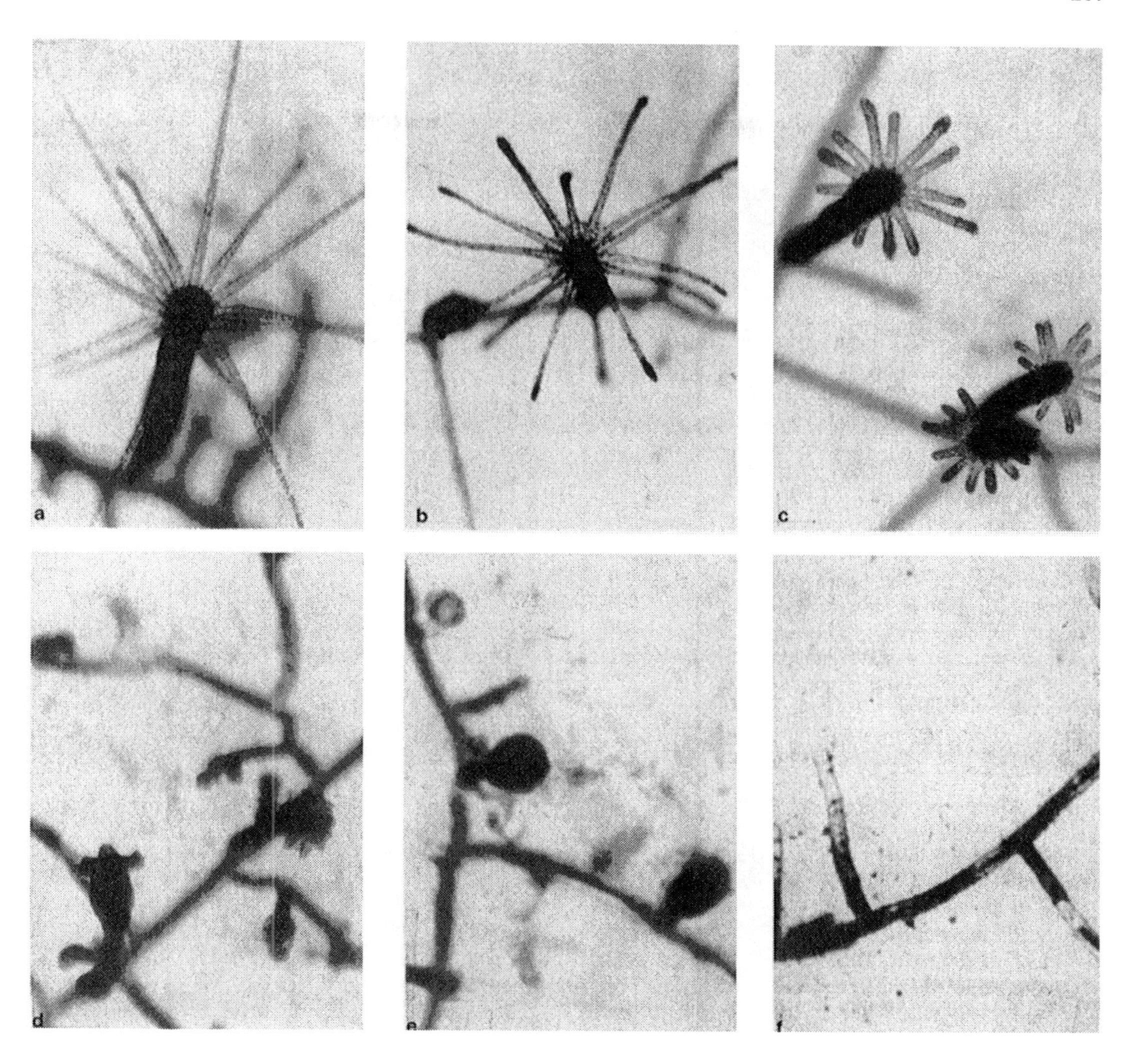

Figure 2. The six stages of redifferentiation in *Eirene viridula*. (From Karbe, 1972. Reprinted with permission of Springer-Verlag New York, Inc.)

The effects on colony growth rate are examined through the calculation of budding rates utilizing the formula:

$$k = \frac{\ln n_2 - \ln n_1}{t_2 - t_1}$$

with k = budding coefficient
n_2 = number of hydranths (+ gonozoids) at t_2
n_1 = number of hydranths (+ gonozoids) at t_1
$t_2 - t_1$ = difference in time between beginning and end of the respective period of growth (in days)

For each experimental set of six glass plates (one chemical treatment), the arthmetic mean and standard deviation of the k values is determined. These values are recalculated as a percentage of the mean budding rate of the control. From these data, statistically significant effective concentration (EC) values can be calculated.

In addition to budding rates, distinct morphological alterations of the hydranths can be observed in *E. viridula* because its hydranths are not protected by a theca in which the polyp is fixed in a casing. Six levels of redifferentiation can be described as stages of increasing damaging effects (Figure 2):

1. Polyps are of nearly normal appearance with only slight damage of the tentacles (swellings, clefts, coalescing). Occasionally, stolon ends become twisted.

2. Sequential redifferentiation of the tentacles, which appear rigid and frequently swollen.
3. Complete redifferentiation of the tentacles, mostly accompanied by a rounding of the polyp heads from which tentacles are absent.
4. Polyp heads drop off, or their contents are retracted into the base of the uprights.
5. Complete redifferentiation of the polyp colony with the exception of the stolonial parts.
6. Stolon tissue is damaged as well.

Hydranths can frequently recover from stages 1 to 5 if transferred to uncontaminated seawater, by budding of new hydranths or by regeneration of partially affected polyps. An effect is considered significant when more than 20% of the colony is affected. The threshold concentrations for significant morphological effects on different levels of redifferentiation are specified as $EC20_1$, $EC20_2$, $EC20_3$, etc. Because classification of redifferentiation is difficult to interpret, Stebbing (1980) developed a test based on the production of gonozooids rather than hydranths by *L. flexuosa*. Gonozooid production increases when stressed, an apparent adaptive response that results in the parent colony releasing planktonic medusae. Increases in gonozooid production have been observed in exposures to cyanide, copper, cadmium, mercury, zinc, and tributyltin fluoride, as well as reduced salinity (Karbe et al., 1984). Gonozooid production increases until a point at which colony growth is inhibited (Figure 3). The increase in gonozooid frequency appears to be a generalized response to unfavorable conditions that allows the sessile colonies to produce planktonic medusae to escape to areas of more favorable conditions.

In addition to hydroids, corals (anthozoans) have been tested extensively because of the importance of coral species to reef ecosystems. Although corals are difficult to maintain in the laboratory for long periods and their size and sessile nature make them good organisms for in situ studies, laboratory test procedures for corals have been developed and are presented in Standard Methods (APHA, AWWA, and WPCA, 1992). A list of recommended species is presented in Table 3. *Acropora cervicornis* is recommended as the primary test species for the Atlantic because it is widely distributed, very sensitive, and similar to the important Indo-Pacific species, *A. formosa*.

The procedure outlined in Standard Methods incorporates at least 20 coral colonies for each test substance concentration and control. Colonies must be uniform in size; colonies of 10 g wet weight are recommended. The dilution water should be a tropical open-ocean surface water away from strong terrestrial influence with a salinity of 33 to 35 parts per thousand (‰). Tests should be conducted under flow-through conditions with the tank volume replaced hourly and a flow rate that varies by no more than ±5%. Temperature should be at 27 ± 1°C, the dissolved oxygen concentration within 10% of saturation, and the pH between 8.1 and 8.4. The top of the coral colonies should be covered with at least 2 cm of water and the test system illuminated with a 12-h light and 12-h dark photoperiod with appropriate twilight periods.

Lethal and sublethal effects on corals are monitored. Although each coral colony consists of different numbers of polyps, each colony is generally considered as an individual because a lethal concentration generally kills all polyps. Death of a colony is defined when polyps become insensitive to stimulation and opaque in appearance. Corals begin to disintegrate within a few hours after death. Several sublethal responses may be observed, including extrusion of zooxanthellae (symbiotic dinoflagellates) that leave the polyps transparent, destruction of the thin tissue covering the septa or cenosarc making the skeleton visible, production of large amounts of mucus, and contraction of the polyps. The mortality of coral colonies may be utilized to calculate LC50 values. Control colony survival must be ≥90% for a valid test.

Rotifers

Tests with rotifers have been developed because they provide a relatively easy and economical means of testing invertebrates. Snell and Persoone (1989a) point out that because of rotifers' capacity for producing cysts (or dormant eggs), which allows test animals to remain on a shelf until needed, stock cultures can be eliminated with their inherent variability. In addition, test animals can be hatched synchronously with the resulting neonates emerging in a uniform physiological condition. Cyst production in two species of rotifers in the genus *Brachionus* has been developed. One species, *B. rubens*, has been evaluated for freshwater toxicity testing (Snell and Persoone, 1989b; Halbach et al., 1983); another, *B. plicatilis*, has been developed for saltwater testing (Snell and Persoone, 1989a). *B. plicatilis* is a cosmopolitan estuarine rotifer.

B. plicatilis cyst hatching is initiated approximately 24 h prior to the initiation of a toxicity test. The cysts are hatched by placing them in 15‰ salt water and incubating at 25°C in light of 1000 to 3000 lux. Hatching begins in about 22 h and approximately 50% of the cysts hatch within 24 h. Animals between 0 and 2 h of hatching should be collected for testing, because animals are not fed during the test and because mortality as a result of starvation may begin after about 80 h at 25°C (ASTM, 1993a). Rotifers are small (approximately 250 μm in length), which necessitates the use of a stereomicroscope with a dark field and a micropipet for transferring animals.

Acute toxicity tests with rotifers should be conducted in sterile, 24-well polystyrene tissue culture plates, which are used once and discarded. One milliliter of test solution is placed in each well and 10 neonate rotifers are introduced by micropipet. Each treatment is in triplicate, resulting in a total of 30 animals per test treatment; generally five test substance concentrations and appropriate controls are

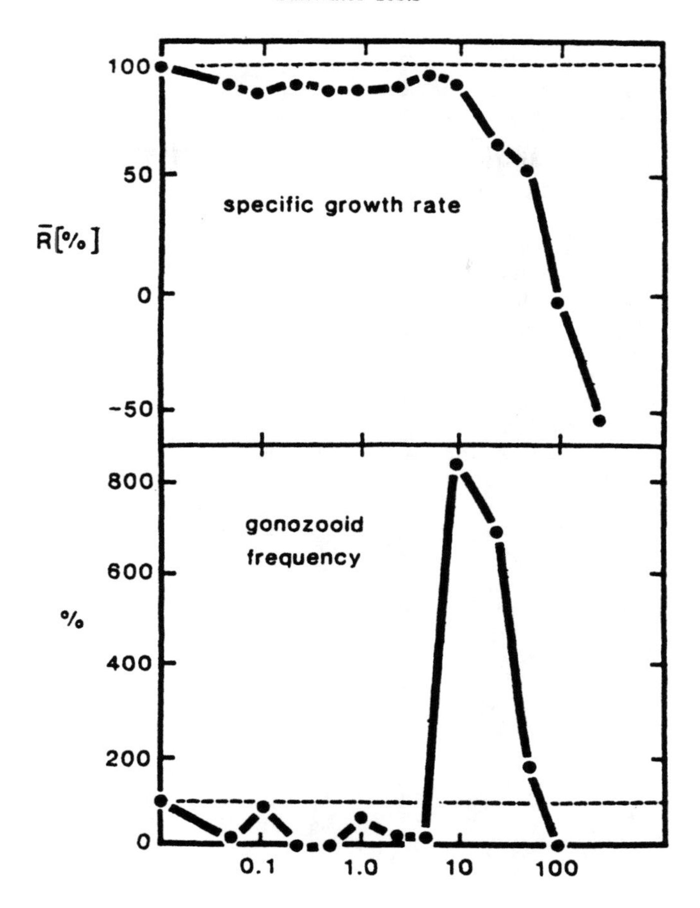

Figure 3. Concentration-response curve for the effects of cyanide on mean specific colonial growth rates as percentages (R %) and gonozooid frequency of *Laomedea flexuosa*. Gonozooid frequency is the number of gonozooids as a proportion of the total number of colony members expressed as a percentage of the frequency in control colonies. (From Stebbing and Brown, 1984.)

tested. The tests should be conducted using salt water of either 15‰ or 30‰ salinity and maintained in darkness at a temperature of 25°C. The animals are not fed during the test. A piece of parafilm should be stretched across the top of the plate and the cover placed on tightly. The test should be conducted for 24 h. After the 24-h test period, mortality is recorded and an LC50 calculated. A rotifer is considered dead if it does not exhibit any external or internal movement within a 5- to 10-s observation period; dead rotifers also change from translucent to opaque. For a valid test, control mortality must be 10% or less. Control survival for 24 h is reported by Snell and Persoone (1989b) to be 100% in most cases. No chronic toxicity test procedures have been developed for *B. plicatilis*.

Table 3. Test species of scleractinian corals

Indo-Pacific area	Atlantic area
1) Branching *Acropora*	1) Branching *Acropora*
A. formosa	*A. cervicornis*
2) Branching *Pocillopora*	2) Branching *Porites*
P. damicornis	*P. porites*
3) Branching *Porites*	3) Other widely used
P. compressa	*Meandrina meandrites*
	Montastrea annularis
	Montastrea cavernosa
4) Hermatypic coral	4) Hermatypic coral
Fungia scutaria	*Scolymia lacera*
5) Ahermatypic coral	5) Ahermatypic coral
Tubastrea aurea	*Tubastrea aurea*

Copepods

Copepods are a desired marine test species because of their wide distribution, because they are representative of zooplankton, and because they have a short life cycle (Lee, 1977). Both calanoid and harpacticoid copepods have been utilized for testing (Wells, 1984).

The calanoid copepod *Acartia tonsa* (Figure 4a) was selected for use because it is one of the most abundant species in coastal waters during the summertime and it has a life cycle of approximately 3 wk at 25°C. Because of its marine distribution and abundance, *A. tonsa* is considered an acceptable test species for evaluation of the effects of dredged materials (U.S. EPA/COE, 1978), ocean disposal of wastes (U.S. EPA, 1978), and oils and oil dispersants (U.S. EPA, 1990). Both acute testing and chronic testing are possible with *A. tonsa*.

Methods for conducting acute tests with *A. tonsa* were described by Gentile and Sosnowski (1978). The procedure recommended is a static test requiring 15 adult copepods to be tested in each of three replicate test chambers per treatment and control—a total of 45 animals per treatment and control. Test chambers are flat-bottom borosilicate glass crystallizing dishes containing 100 ml of salt water. The depth of the dilution water must be ≥2.0 cm. The animals are monitored daily for 96 h. The numbers of dead, moribund, and living copepods are recorded. Results are reported in terms of the 96-h LC50 (or 96-h EC50). No feeding is required during the exposure. Testing is acceptable when mortality is ≤15% in controls. Although most copepod tests are conducted under static conditions, flow-through tests are possible (Heinle and Beaven, 1980).

Although copepods can be cultured in the laboratory (Gentile and Sosnowski, 1978; Heinle, 1969;

a.

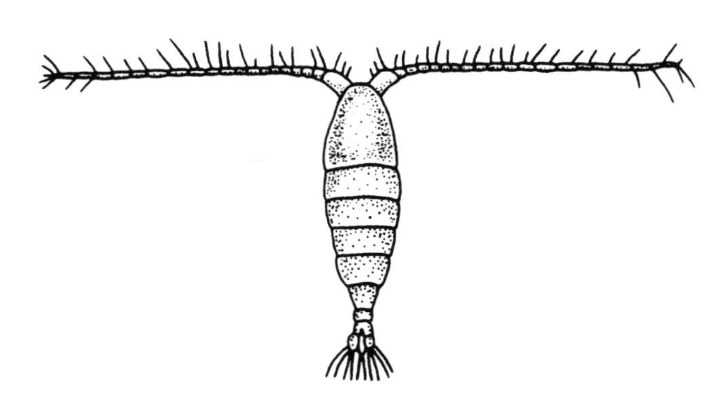

b.

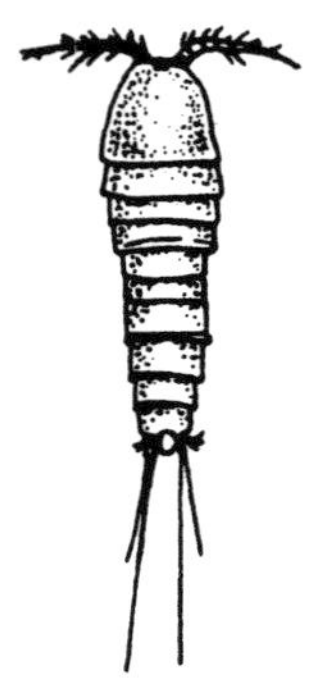

Figure 4. Typical calanoid and harpacticoid copepods. a. Calanoid copepod, *Acartia tonsa*, b. Harpacticoid copepod.

APHA et al., 1992; Zillioux, 1969; Zillioux and Wilson, 1966; Zillioux and Lackie, 1970) chronic testing is limited. Heinle and Beaven (1980) conducted one chronic test with *Eurytemora affinis*, but many other chronic attempts failed for various reasons, including high control mortalities.

Ward et al. (1979) described the results of five chronic tests conducted with *A. tonsa* on ocean-disposed material. The 30-d chronic test consisted of three phases: two 5-d static test phases (Phases 1 and 3) and one 20-d flow-through test phase (Phase 2). Phase 1, the production of F_1 organisms, was conducted under static conditions because of the small size and fragility of the eggs. Forty adult *A. tonsa* per test treatment were placed in a breeding cage immersed in a 2-L culture dish containing test substance solution. The adults were randomly selected from a reproducing culture under the assumption that each group had the same ratio of males to females. The adults were fed a diet of four algal species (*Skeletonema costatum, Thalassiosira pseudonana, Isochrysis galbana*, and *Chroomonas salina*) at the rate of 5.0×10^7 total cells/L/d. The chambers were maintained at 20°C with a 14-h light, 10-h dark photoperiod. After 4 d of exposure, all adults were removed and the eggs allowed 24 h to hatch. The number of nauplii in each chamber was determined by direct microscopic count of triplicate samples using a Sedgewick-Rafter counting cell. During Phase 2, the F_1 animals developed to sexual maturity in the chemical treatments. Phase 2 was conducted using a flow-through test system presented in Figure 5, with all treatments conducted in triplicate. A total of 100 nauplii from each breeding cage was added to each of three 2-L culture dishes containing 1 L of the appropriate test solution. A metering pump was used to replace the test solution in each culture dish at the rate of 1 L per 12 h. Test solution exited the culture dishes through thistle tubes covered with plankton netting. Animals were fed the four-species algal diet every 2 d at the rate of 2.0×10^6 total cells/L for the first 5 d and 2.0×10^7 total cells/L for the remainder of the test. At the end of 20 d, all surviving adult animals were counted. During Phase 3, the reproductive success of the copepods was examined under static conditions. From the surviving adult animals of Phase 2, 40 adults were selected for each treatment and placed in the breeding chambers containing the appropriate test solution. The animals were maintained for 4 d, during which they were fed the previously specified algal diet at the rate of 5.0×10^7 total cells/L/d. After 4 d, the adults were removed and the eggs allowed 24 h to hatch. All nauplii were counted as described previously.

Typical control responses for each of the three phases of the test are provided in Table 4. Based on the results of these five tests, average control survival for this test should be $\geq$70%. Results are used to calculate the NOEC and LOEC values based on the most sensitive biological indicator (i.e., survival or reproduction) and reported as an MATC.

Harpacticoid copepods (see Figure 4b) are considered to be less difficult to rear than the larger calanoids and therefore have also been developed for toxicity testing (Bengtsson, 1978; Lassus et al., 1984). The common euryhaline species *Tigriopus brevicornis*, which is distributed in tidal pools, has been developed as a standardized test species by Lassus et al. (1984).

The recommended procedure for *T. brevicornis* is to isolate 30 ovigerous (egg-carrying) females in 15-ml glass containers. Test solutions are prepared with

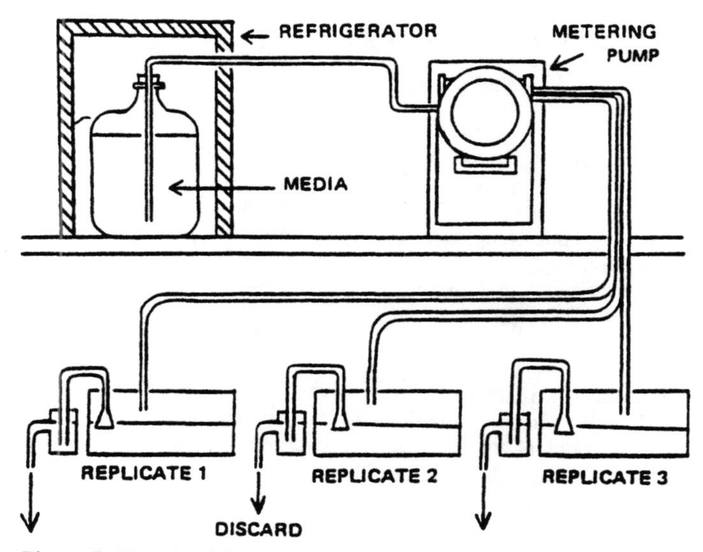

Figure 5. Flow-through test system design for flow-through portion of the chronic toxicity test with *Acartia tonsa*. (From Ward et al., 1979. Copyright ASTM. Reprinted with permission.)

Table 4. Control responses during a three-phase chronic toxicity test with *Acartia tonsa*[a]

Test Number	Phase 1: Average no. viable F_1 young	Phase 2: Average no. live F_1 animals	Phase 3: Average no. viable F_2 young
1	1180.0	71.3	—
2	54.0	83.0	22.3
3	41.0	80.7	37.0
4	79.7	79.7	54.3
5	81.0	79.7	49.3

[a]All values given in the table represent the means of the triplicate control chambers at each test phase.

Excerpted from Ward et al., 1979. Copyright ASTM. Reprinted with permission.

30‰ membrane-filtered seawater. The test is maintained at 20 ± 1°C. Test solutions are inoculated with a concentrated algal culture of the flagellate *Tetraselmis suecica* to achieve an average cell concentration of at least 1×10^6 cells/ml. Daily, each group of adults are monitored for mortality, frequency of newly produced egg sacs, and survival of hatched nauplii larvae. Egg sacs are produced approximately every 3 d. The test is terminated after 7 to 15 d. At that time, the produced nauplii and copepodites (eggs typically hatch as nauplius larvae and then pass through five or six naupliar instars and five copepodite instars) are fixated in formalin and their total number per concentration determined.

Survival of harpacticoid copepods for standard 96-h test periods is within the normal limit of 90%. During three 7- to 12-d exposures, Lassus et al. (1984) determined that the total number of successfully hatching egg sacs was 52 during the 7-d exposure, 63 during the 10-d exposure, and 101 during the 12-d test. Larval production (total number of nauplii and copepodites) during the same tests ranged from approximately 600 during the 7-d exposure to 1500–2,300 during 12-d exposures.

Annelids

Polychaetous annelids, as pointed out by Reish (1980), are the most neglected major group of marine invertebrates in aquatic toxicological work. However, because polychaetes constitute over 40% of both the number of species and the specimens in subtidal soft-bottom ecosystems and are sensitive indicators of marine pollution, polychaete testing should be an important part of any chemical evaluation program in the marine environment. Several species of polychaetes have been utilized in toxicity tests, including *Capitella capitata, Ctenodrilus serratus, Nereis virens, Neanthes arenaceodentata, Ophryotrocha diadema, Dinophilus gyrociliatus*, and *Arenicola cristata* (ASTM, 1991a; Reish, 1980; Carr et al., 1986; Walsh et al., 1986). *N. arenaceodentata* is widely distributed throughout the world, is easily cultured in the laboratory, and has been one of the most tested polychaetes. Procedures for testing *N. arenaceodentata* follow; procedures for other polychaetes are given in Reish (1980, 1984), ASTM (1991a), and Walsh et al. (1986).

Culture procedures for *N. arenaceodentata* are also described in a currently developing document from ASTM (1991a) and various stages of its life cycle are pictured in Figure 6. *N. arenaceodentata* may be tested with or without sediment depending on the purpose of the toxicity test. Aqueous testing procedures are described below, and sediment testing procedures are described in an ASTM draft document (ASTM, 1991b; see also Chapter 8).

In acute (96-h) tests, 2- to 3-mo-old worms (age determined from the time of emergence from the parent's tube or egg capsule) are collected from a laboratory culture (worms freed from their tubes and placed in clean water should be kept isolated or with only four to five in a petri dish) and single worms placed in a glass test container with 100 ml of test solution. All worms should be examined under a dissecting microscope prior to use in a test to check for injuries and females with developing eggs in their coelom. Injured and gravid animals must not be used. A minimum of 10 worms should be used per treatment. For cannibalistic species, such as *N. arenaceodentata*, one worm is generally tested per test container. For other species, 2 to 10 worms are placed per container, depending on biomass and container size. Dilution water salinity is dependent on species tested. The test is conducted at 20 ± 1°C; photoperiod and light intensity do not appear to be a factor in polychaete tests. Worms are not fed during this test. Test containers should be examined daily for dead worms. Stressed worms generally leave their tubes prior to death. Dead animals are characterized by lack of movement, everted proboscis, and a pale appearance. As with most acute toxicity tests, survival of polychaetes must not exceed 10% in the controls for the test to be considered acceptable. The test results are used to calculate a 96-h LC50.

Life cycle reproductive tests begin with juvenile worms, approximately 1 mo old. The following procedure describes a static or static-renewal exposure; the test design can be modified to conduct the test under flow-through conditions. Four worms (sex undetermined) are placed in 2500 ml of test solution in a 3.8-L glass jar. The dilution salt water should have a salinity between 30 and 35‰. Ten replicate test jars are set up per control and test concentration. The test should be conducted at 20 ± 2°C under a 16-h light and 8-h dark photoperiod. Using a glass pipet inserted into each test solution, very gentle aeration should be supplied to maintain acceptable dissolved oxygen concentrations. Tetramarin should be fed every other day at a rate of 8 mg (dry weight) per animal. In order to prevent a deterioration in water quality, no food should be added if worms fail to eat during the previous feeding. Mortality should be monitored daily during the test. Beginning at 15 d of exposure, worms should be examined for the presence of eggs in the coelom. Observations should be

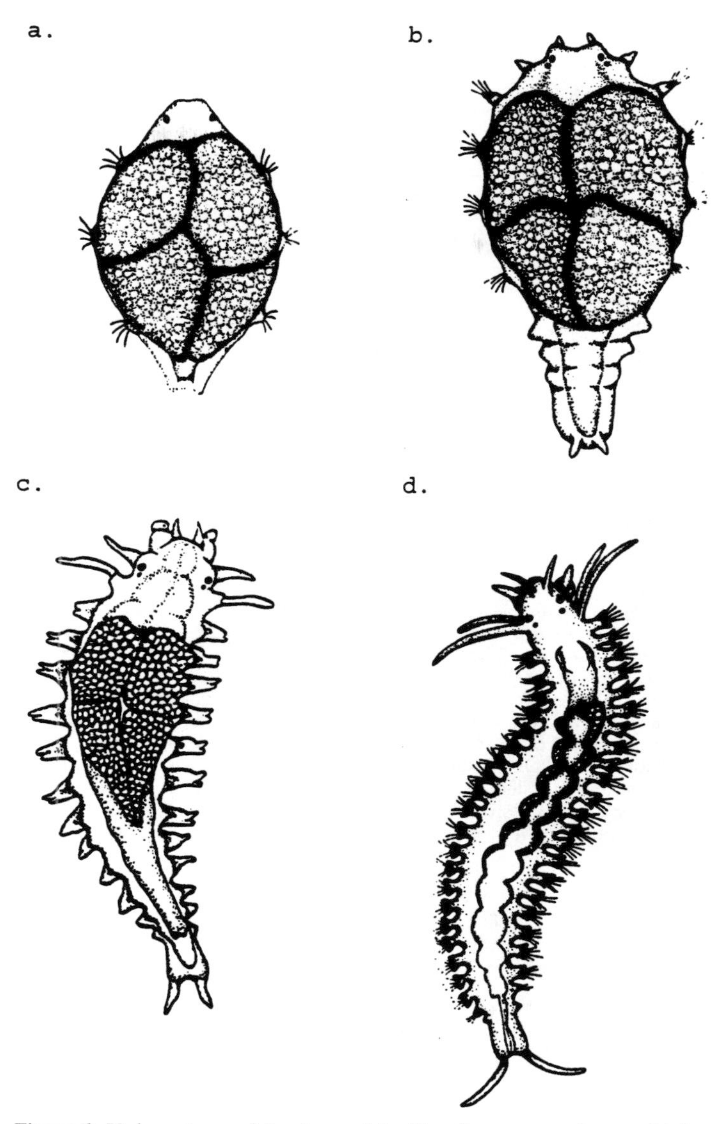

Figure 6. Various stages of development for *Neanthes arenaceodentata*: (a) 3-segmented larval stage; (b) 4-segmented larval stage; (c) 12-segmented stage; (d) juvenile with 21 segments. (From ASTM, 1991b. Copyright ASTM. Reprinted with permission.)

made at 5-d intervals until eggs are observed. Once eggs are observed, the test containers should be examined every 2 to 3 d for the deposition of these eggs. Egg masses should be carefully removed when observed and the number of eggs present enumerated. Eggs should be discarded after counting. Test containers should be maintained if worms are still surviving. If only two worms are alive and they are both the same sex (worms of the same sex fight, but worms of the opposite sex will lie side by side) and there is another test container within the same concentration series in which the two remaining worms are of the opposite sex, they may be paired. The test may be terminated after eggs are produced in all control test containers or after a set period of time (e.g., 3 mo).

A life cycle test is considered invalid if animals fail to produce eggs in the control or if mortality of the first-generation animals exceeds 10%. Egg production is temperature dependent. Reish and Gerlinger (1984) observed between 524 and 668 eggs laid per reproducing female at test temperatures of 15 and 20°C, respectively.

Pesch et al. (1991) reported results of chronic tests of contaminated sediments with *N. arenaceodentata*. The researchers used a flow-through test system

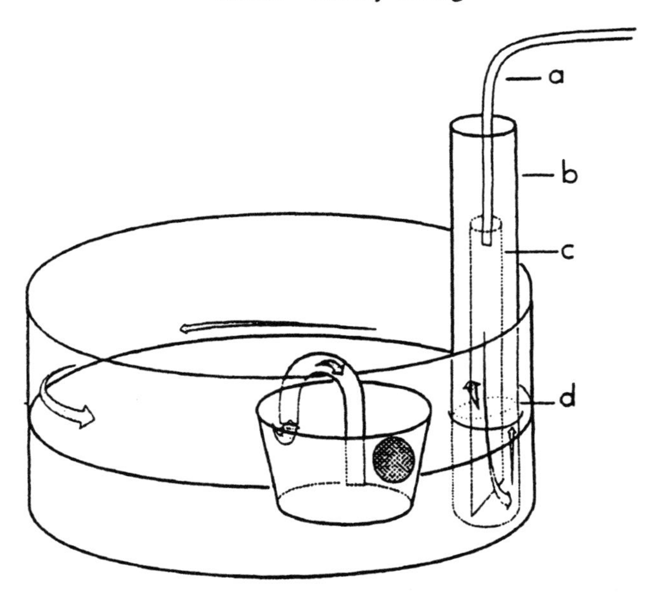

Figure 7. Flow-through exposure chamber for flow-through tests with polychaetes. The exposure chamber is a glass crystallizing dish with an inflow of water over the sediment surface. Arrows show flow of water into the test tube (b) through silicone tubing (a), which has a piece of glass tubing (c) attached at the bottom, then through an elliptical opening (d) cut in the side of the test tube and into the dish just above the sediment surface. Water circulates around the dish and leaves through a siphon and catch cup. (Reprinted with permission from Pesch, C. E., Munns, W. R. Jr., Gutjahr-Gobell, R.: Effects of a contaminated sediment on life history traits and population growth rate of *Neanthes arenaceodentata* (Polychaeta: Nereidae) in the laboratory. Environmental Toxicology and Chemistry 10(6):805–815. Copyright 1991, SETAC.)

composed of 190-mm-diameter by 100-mm-high glass crystallizing dishes as the experimental chambers (Figure 7). Water delivery into each chamber was maintained at a rate of 33 ± 4 ml/min delivered in regular pulses from a water distribution system consisting of self-priming siphons and splitter boxes. The test chambers were maintained at a test temperature of 19.7 ± 1°C under a controlled photoperiod of 10-h light and 14-h darkness. The experiment was initiated with approximately 30-d-old worms distributed in groups of 28 to each test chamber. Each treatment was initiated with 20 replicates. These replicates were destructively sampled six times during the 153-d test (days 58, 75, 95, 110, 131, and 153). On each sampling day, adults were counted, sexed, and sized (dry weight and number of segments), broods of eggs and larvae were counted, juveniles were counted, and a subsample of juveniles was taken to determine size (dry weight and number of setigerous segments). The following reproductive end points were determined: estimated time to egg laying, age-specific fecundity, number of broods, number of eggs or larvae per brood, and number of juveniles. Test chambers without sediment were also used and supplemented with 850 mg of dried, powdered, *Enteromorpha* sp. to provide material for worms to construct tubes. Worms were fed powdered (sieved to ≤0.335 mm) prawn flakes three times weekly until day 49 and then every weekday for the rest of the test. The amount of food added was the same for all treatments and was increased in increments throughout the test.

Representative values for the life history traits evaluated in this chronic test are presented in Table 5. Because of density differences relative to the pres-

Table 5. Representative values for life history traits of *Neanthes arenaceodentata*

Life history trait	Mean and standard deviation			
	No sediment		Sediment	
Total no. of broods of eggs	11		13	
Total no. of broods of larvae	10		9	
No. of eggs per brood[a]	1018	(635)	1556	(591)
No. of larvae per brood[a]	512	(489)	850	(379)
No. of juveniles[b]	236	(480)	663	(641)
No. of segments, juveniles[c]	21.8	(3.8)	33.4	(3.3)
Dry weight (mg), juveniles[c]	0.05	(0.01)	0.38	(0.22)

[a]Third, fourth, fifth, and sixth sampling days.
[b]Fifth and sixth sampling days.
[c]Sixth sampling day.
Data compiled from Pesch et al., 1991.

ence or absence of sediment, the responses of some life history traits of worms tested with no sediment differed from those of worms tested with sediment. When no sediment was present, adult males and females were smaller and there were generally fewer juveniles, which were smaller at test termination.

Molluscs

Both gastropods and bivalve molluscs have been utilized in acute toxicity tests, but because of their commercial importance and sessile nature, testing has concentrated on bivalves, which include oysters, clams, and mussels. Two testing procedures are utilized for regulatory purposes (i.e., FIFRA). One procedure examines the effects of test materials on the embryonic development of bivalve larvae; the other procedure examines the growth of new shell formation.

Bivalve Embryo/Larval Test

Several species of bivalves have been successfully spawned and gametes fertilized in the laboratory. These include the hardshell clam, *Mercenaria mercenaria*; the Eastern oyster, *Crassostrea virginica*; the Pacific oyster, *Crassostrea gigas*; and the blue mussel, *Mytilus edulis*. The test is initiated with newly fertilized embryos and is terminated 48 h later after the embryos have developed through a trochophore stage and metamorphosized into fully hinged veliger larvae (Figures 8 and 9). The test evaluates

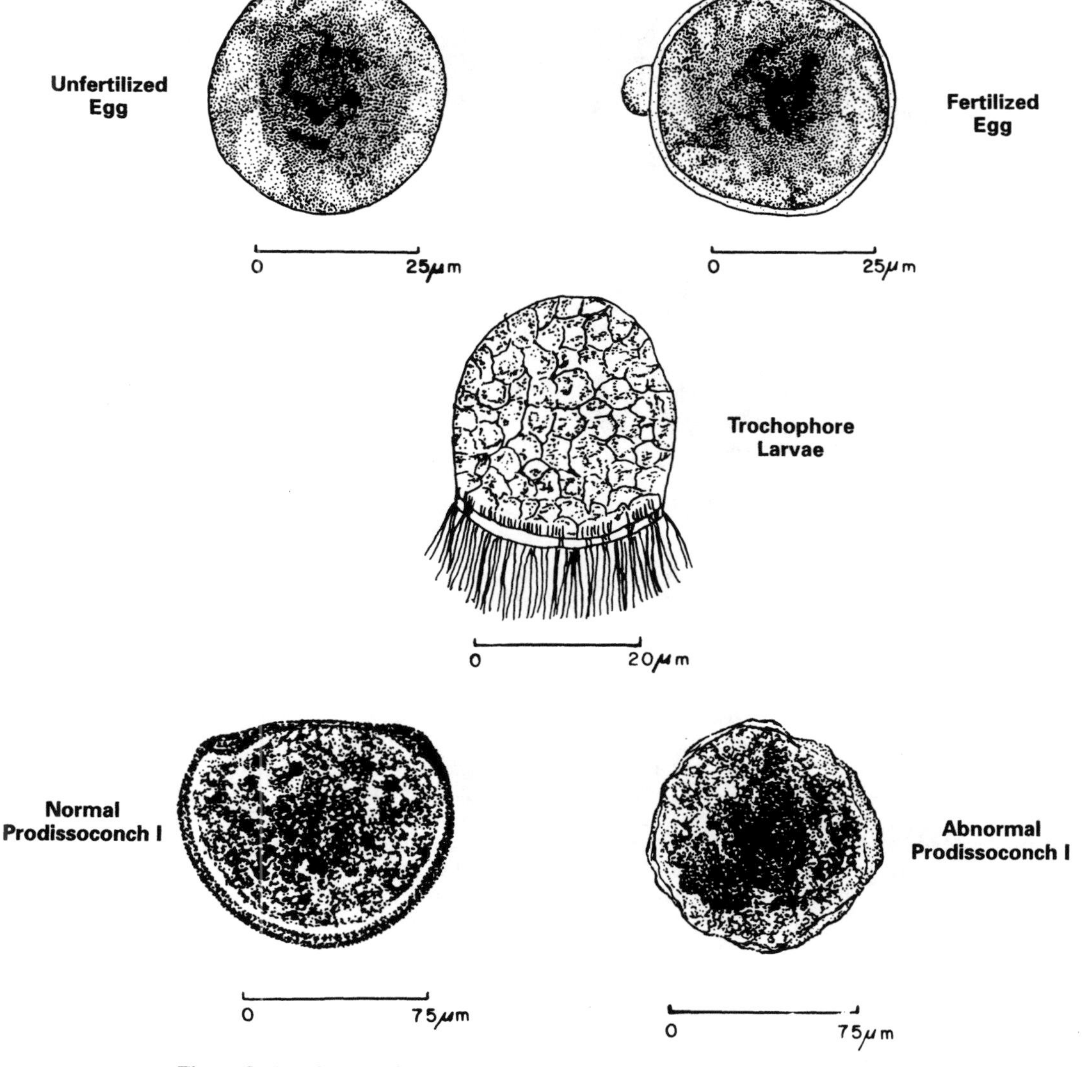

Figure 8. Developmental stages of bivalve larvae during first 48 h of development. (ASTM, 1991d. Copyright ASTM. Reprinted with permission.)

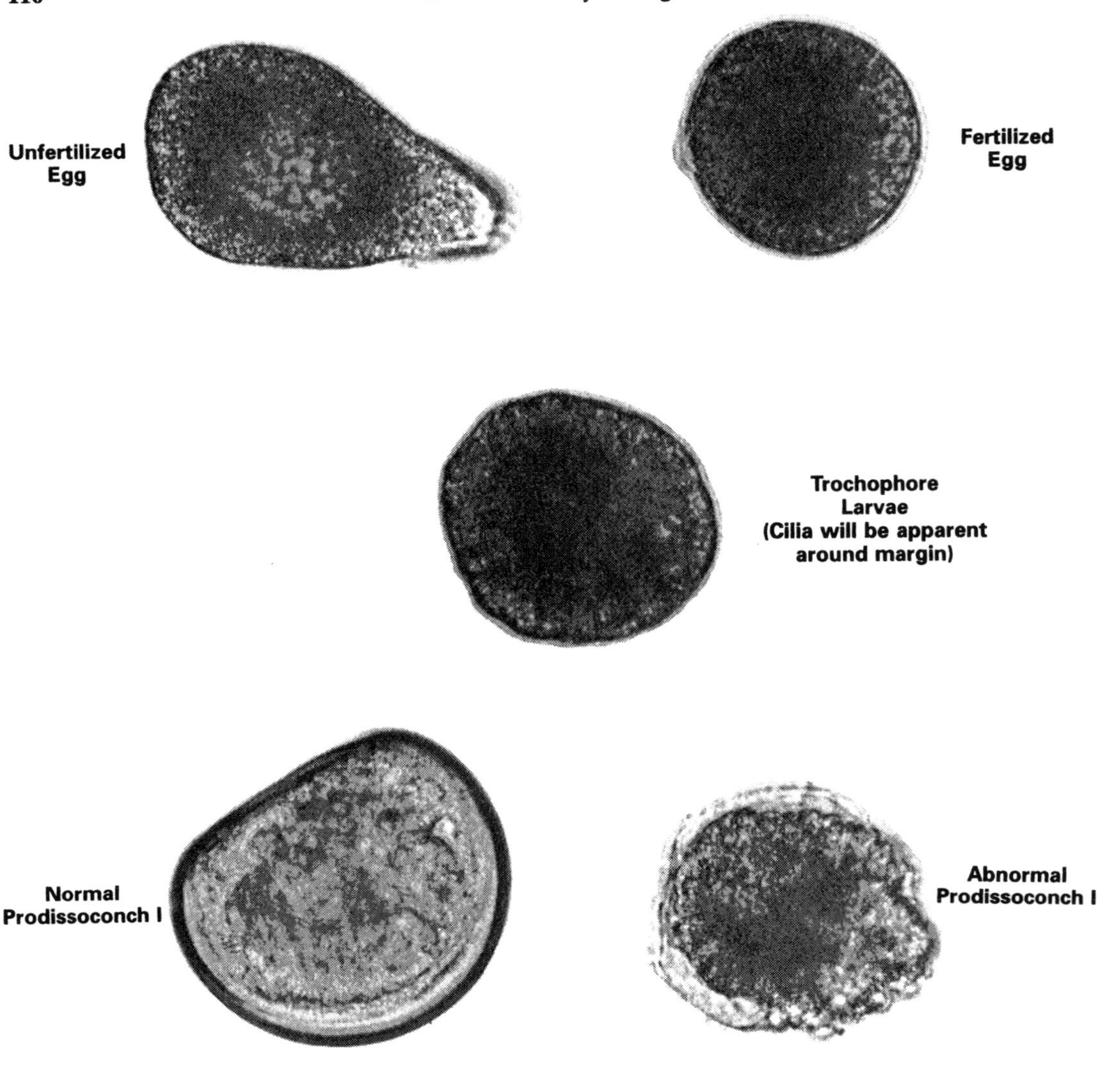

Figure 9. Photomicrographs of developmental stages of bivalve larvae during the first 48 h of development. (ASTM, 1991d. Copyright ASTM. Reprinted with permission.)

the success of the developmental process by comparing the total number of live larvae with completely developed shells in the control(s) with the numbers of live, completely developed larvae within each chemical treatment solution (ASTM, 1993e).

Adult bivalves possessing mature gonads can easily be induced to spawn during certain times of the year (generally spring through summer). Spawning can be induced with a variety of biological, chemical, and physical stimuli, and until spawning is desired these stimuli must be minimized. If adults with ripe gonads cannot be obtained, animals may be conditioned properly for an appropriate duration to promote gametogenesis and production of mature gametes. Temperature ranges for holding, conditioning, and induction of spawning for the four species are presented in Table 6.

The ripeness of the brood stock can be determined by sacrificing several animals and examining the gonads. A sample of the gametes can be obtained from the gonads and examined microscopically. Gametes from a ripe male are small and become highly active when placed in salt water; those from a ripe female are large and teardrop shaped. Eggs rapidly become spherical when placed in salt water (usually less than 30 min).

To spawn bivalves, adults should be placed individually into containers with a sufficient volume of clean test water to keep them covered. After allowing sufficient time for the animals to begin pumping water, the water temperature should be rapidly raised from 5 to 10°C above the conditioning temperature (see Table 6). Additional stimuli may be applied during thermal stimulation, including heat-killed sperm

Table 6. Recommended temperatures (°C) for holding, conditioning, and testing bivalves

Species	Holding	Conditioning	Induction	Test
Crassostrea gigas (Pacific oyster)	14–15	20	25–32	20
Crassostrea virginica (eastern oyster)	14–15	20–25	25–32	25
Mercenaria mercenaria (quahogs or hard clam)	14–15	20–25	25–32	25
Mytilus edulis (bay mussel)	8	12–14	15–20	16

Source: ASTM, 1993e. Copyright ASTM. Reprinted with permission.

(preferably from a naturally spawning male, but may be from a sacrificed animal), and for bay mussels injection of potassium chloride into the posterior adductor muscle may be used. Other stimuli may be used but are not always reliable for producing high-quality embryos (ASTM, 1993e). If no animals spawn within 30 to 60 min, the animals should be placed into fresh saltwater at holding temperature and then the stimulation process repeated. Difficulty in spawning might be caused by insufficient conditioning or by unacceptable water quality (salinity too high or too low for the particular species). Spawning can be recognized by the appearance of a whitish stream of gametes emerging from the animal. A sample of released gametes should be collected while spawning is occurring and examined to determine whether they are eggs or sperm and their condition. Males are generally easier to stimulate and normally spawn first.

Fertilization is generally more successful if sperm are obtained from a natural spawn, although sperm stripped from a male may be used. Use of stripped eggs for fertilization is not recommended because it often results in poorly developed and malformed embryos (ASTM, 1993e). Following release of eggs, the eggs should be rinsed through a 75-μm screen into untreated clean test water to remove tissue and other debris associated with the spawning process. A 1-ml aliquot of the solution should then be counted to determine the concentration of eggs. Unrounded eggs should not be counted because they are unlikely to fertilize. The total number of eggs available should be calculated to determine whether there is a sufficient number for testing. The egg density should be adjusted to between 20 and 50 eggs/ml by adding additional clean test water. Once the density of fertilizable eggs has been established, sufficient sperm suspension should be added to yield a final sperm concentration in the egg suspension of 10^5 to 10^7 sperm/ml. High concentrations of sperm should be avoided to reduce the possibility of polyspermy (Stephano and Gould, 1988; Staeger and Horton, 1976). Polyspermy results in aberrant development of larvae and failure of larvae to metamorphize properly. Once fertilization is confirmed microscopically, the embryo suspension should be poured through a 54-μm screen to remove debris and then onto a 22-μm screen and rinsed with clean test water to remove excess sperm, protozoa, and bacteria. The embryos are then backwashed into clean test water and the density of well-developed embryos (those that have developed to a two-cell stage or beyond) is then determined.

The test is generally conducted with at least three replicates for each control and test material treatment; at least five test treatments are utilized. A test is initiated within 4 h of fertilization, that is, when equal volumes of the homogeneously mixed embryo suspension are added by automatic pipet to each triplicate test container that already contains test solution. The inoculation volume of embryo suspension should be calculated to yield a concentration of embryos in each test solution between 15 and 30 embryos/ml. The initial number of embryos added to each test container should be confirmed by directly counting samples from three or more test containers.

The test is terminated after 48 h. At that time, an aliquot of each control and test solution is collected and either counted immediately or preserved in 5% buffered formalin. Because embryos and larvae sink following preservation, most of each sample can be discarded before collecting the residual volume containing the test organisms for counting. The residual volume is counted using a Sedgewick-Rafter cell and enumerating all normal larvae (those with completely developed shells and containing meat). Empty shells should be counted as dead because the larvae were not alive at the end of the test. Larvae with misshapened shells are counted as normal because a malformed shell will not necessarily reduce survival; however, numbers of malformed embryos may be enumerated to assess the proportion of live larvae with malformed shells. Larvae with incompletely developed shells are counted as dead because retarded development is considered likely to reduce survival.

Mollusc embyro-larval tests are difficult to conduct because of the many problems that affect the success of the test. Of primary importance is the condition of the adult bivalves. As stated previously, during certain times of the year adult bivalves possess ripe gametes and can be spawned easily. However, at other times, bivalves must be held for several weeks to months in order to condition them (i.e., provide proper conditions for a time period sufficient to allow gametogenesis and production of mature gametes to occur). During conditioning, the adults must be furnished with an acceptable food supply, salinity, and temperature to promote gametogenesis but not stimulate spawning. Conditioning temperatures are presented in Table 6. If the natural phytoplankton concentration is insufficient to support good survival and growth, saltwater algae should be added to the water. Three recommended algal species are *Monochrysis lutheri, Isochrysis galbana*, and *Tetraselmis suecica*.

Once ripe bivalves are obtained, problems may still occur in the spawning process. The spawning water should be similar in salinity to the holding water; improper salinity may prevent bivalves from spawning. In addition, small collections of adults from the same area may be almost entirely males or females, thus preventing the collection of both mature eggs and sperm for fertilization (Galtsoff, 1964). Depending on the ripeness and conditioning of adults, the induction of spawning may require several attempts. Animals should be maintained at the recommended induction temperature (Table 6) for only 60 min (30 min if pumping water extremely actively). If spawning does not occur within this time frame, the water temperature should be lowered back to the holding temperature and the stimulation process repeated. Low fertilization success may result if the gametes are immature or of poor quality. In addition, polyspermy resulting from a high concentration of sperm and/or the collection of many sperm around each egg may result in aberrant embryos and large losses of larvae. Proper sperm concentrations and constant agitation during fertilization will reduce problems associated with polyspermy. Once fertilization is accomplished, it is important to reduce the concentration of embryos to the desired exposure concentration (i.e., 15 to 30 embyros/ml) and to transfer the embryos into clean water as quickly as possible. Concentrations >30 embryos/ml will result in abnormally developing embryos for *C. virginica, M. mercenaria*, and *M. edulis. C. gigas* embryo development has been shown not to be impaired by embryo concentrations up to 100 embryos/ml (Staeger and Horton, 1976). Poor embryo development will also occur in waters with high bacterial populations. Therefore, proper rinsing of debris from the embryos prior to transferring them into test solutions and use of filtered water are essential to maintaining acceptable water quality throughout the test.

The criterion for test acceptability is based on the percentage of embryos that achieve completely developed shells by test termination. A test is considered unacceptable if less than 70% of oyster embryos or 60% of hard clam embryos introduced into a required control treatment results in live larvae with completely developed shells. A 48-h EC50 is calculated based on a reduction in the number of live larvae with completely developed shells in exposed treatments as compared to the number of live larvae in the control.

Oyster Shell Deposition Test

The oyster shell deposition test was first described by Butler (1965) and presented later as a standard test procedure (Butler and Lowe, 1978). The test measures the deposition of new shell growth by juvenile eastern oysters (*C. virginica*) during a 96-h exposure period. The degree of inhibition in shell deposition can be directly related to the amount of stress induced by exposure to chemicals.

The test is based on two principles: (1) peripheral deposition of new shell may be maximized by removing all new or thin shell and (2) noxious chemicals will directly reduce pumping activity, which in turn will reduce the rate of shell deposition. Butler (1965) found that shell growth in the oyster does not occur at a uniform linear rate but that there is only an initial increase in the terminal deposition of shell followed by a period during which the new shell is thickened by internal deposits. During the thickening period, there may be no increase in length. By removing new or thin shell, the oyster occupies all of the available space and is forced to deposit shell at the posterior end. Under average conditions, Butler (1965) determined that shell deposits are made at the rate of approximately 4 mm/wk (approximately 0.57 mm/d). Together with observations of Butler et al. (1960), it was determined that many chemicals (e.g., agricultural pesticides) reduced the number of hours per day that an oyster was open and actively pumping. The oyster shell growth test is thus useful for determining the concentrations at which certain chemicals affect physiological functions in oysters.

The test utilizes juvenile oysters, approximately 25 to 50 mm in height, which are cleaned of epizootic organisms. Approximately 3 to 5 mm of the shell periphery is removed by grinding or filing. This is to provide a smooth rounded blunt profile from which to measure any new shell growth and to force shell growth to occur at the posterior end of the oyster.

Twenty ground oysters should be positioned on their left valve (the left valve is larger and cuplike) in each test chamber approximately equidistant from one another with the anterior hinged ends oriented away from the incoming water flow. Generally five test substance treatments and appropriate control treatments are utilized.

The test system should provide a flow of unfiltered salt water at a rate sufficient to obtain adequate shell growth (mean new shell growth >1.0 mm, although mean shell growth >2.0 mm is desirable). Butler and Lowe (1978) stated that oysters feed and grow readily at a flow rate of 5 L of water per hour per oyster. Rates as low as 1 L water/h/oyster have also been satisfactorily used to achieve growth greater than 2.0 mm (G. S. Ward, personal observations).

The test should be conducted at a test temperature between 15 and 30°C in unfiltered salt water with a salinity between 10 and 30‰. Growth may be reduced by exposures with temperatures and/or salinities exceeding these ranges (Galtsoff, 1964).

Following exposure, all oysters are removed from the test chambers and new shell growth is measured using calipers. Because shell deposition is not uniform on the periphery, the length of the longest "finger" of new shell is measured to the nearest 0.1 mm. Percentage reduction in shell growth of exposed oysters is calculated relative to the mean new shell growth of the control oysters. EC50 values are then obtained by statistical analysis of the percent reduction values at the different test concentrations.

Mysids

Since first presented as a life cycle toxicity test by Nimmo et al. (1977), the saltwater mysid, *Mysidopsis bahia*, has become the model invertebrate of choice for evaluation of the effects of chemical exposure in saltwater systems. Their ecological significance, short life cycle, and documented sensitivity to pollutants (Nimmo and Hamaker, 1982) are some of the reasons for their rapid incorporation into environmental regulations. Culture procedures for mysids are described in Lussier et al. (1988).

Procedures for acute toxicity tests with mysids are similar to those used for most invertebrates. Tests are generally initiated with young animals, from less than 24 h to 5 or 6 d. Because animals may produce young within 12 d of birth, animals older than 6 d should be avoided. Mysids from 1 to 2 d old may be collected using a pipet, and older mysids are easily netted on a nylon screen. Although mysids can survive a 96-h test period without food, feeding with live brine shrimp nauplii is recommended on a daily basis during the test because of their cannibalistic nature.

With the exception of a life cycle test developed by Tyler-Schroeder (1979) with the palaemonid shrimp *Palaemonetes pugio*, the test with the mysid is the only life cycle test currently available for saltwater invertebrates. The length and difficulties associated with the grass shrimp test, however, have prevented it from developing into a routinely used standard test procedure.

The mysid chronic is generally conducted for 28 d, which necessitates that the test be conducted in a flow-through system. The test begins with newly released postlarval (<24-h-old) animals, continues through development to adulthood at approximately 10 to 12 d (Rodgers et al., 1986), and terminates following evaluation of reproductive success and the effect on F_1 generation survival.

Postlarval mysids can be collected from recirculating culture systems (Reitsema and Neff, 1980) or from an isolated population of gravid females. Following collection, between 10 and 15 mysids are distributed to each of four Nitex screen (approximately 350-μm mesh size) chambers per test treatment. Generally two test chambers are positioned in each duplicate test tank for ease in counting. The mysids are fed live brine shrimp (*Artemia salina*) at least once per day, although multiple feedings are preferable. The brine shrimp nauplii should be monitored for acceptable concentrations of essential 20:5ω3 fatty acid (Leger et al., 1985) or the newly hatched nauplii should be exposed to a fatty acid supplement prior to use as a mysid food (Friedman et al., 1989). Lack of or low concentrations of 20:5ω3 fatty acids in the diet of mysids have been shown to result in poor survival and growth and low production of young (Friedman et al., 1989; Leger et al., 1985). Mysids should attain adulthood between days 10 and 12. At that time, brood pouches begin to form on females. As eggs are deposited in the marsupium (brood pouch) and the larvae develop, the pouch grows in size and becomes darker in color. It may be desirable at this stage to pair ovigerous females with males (i.e., adults exhibiting no sign of brood pouch formation) to evaluate reproductive success of individual females. When young are released into the screen chamber, they are enumerated. The effect of the chemical on the second generation of mysids may be determined by collecting a representative group from each treatment (e.g., 20) and placing them in another screen chamber for examination of survival for at least 96 h. Young may be monitored through their reproductive stage to provide further information. Excess young produced can be simply enumerated and discarded. The test ends after 28 d of exposure of the F_0 mysids, unless reproduction data are desired on the F_1 generation. Growth of first-generation mysids can be evaluated by measuring total length and/or dry weight at test termination. Because differences in size between males and females exist, all mysids should be sexed at test termination for possible evaluation of growth by sex.

Generally accepted biological criteria for chronic test acceptance are (1) ≥70% survival of first-generation control mysids, (2) ≥75% of paired first-generation females in the control(s) produce young, and (3) the average number of first-generation females in the control(s) is ≥3 (ASTM, 1993a).

A review of existing literature on 28-d mysid chronics shows that survival of greater than 80% is normally obtainable. Table 7 presents typical ranges for chronic end points for a 28-d mysid chronic. Total length (anterior margin of the carapace to the end of the uropods, not including setae), averages from 7.24 to 8.38, with females being slightly larger than males. On a dry weight basis, mysids average between 0.752 and 1.187 mg. Reproductive success can be reported either as the number of young produced per female during the test or as the number of young

Table 7. Typical values for chronic end points of mysid life cycle tests

Chronic end point	Range
Survival, %	78–93
Growth	
Length (male), mm	7.24–7.86[a]
Length (female), mm	7.28–8.38[a]
Dry Weight (male), mg	0.752–0.860[a]
Dry Weight (female), mg	0.967–1.187[a]
Reproductive Success	
Young/female	8.25–18.00[a]
Young/female reproductive day	0.28–1.11[b]

[a]From Breteler et al., 1982.
[b]Lussier et al., 1985.

produced per female reproductive day. The latter presentation normalizes the data to account for deaths of females during the exposure and for differences in maturation of mysids from chamber to chamber. This calculation also allows comparison of control data from tests of various durations. Average brood sizes for *M. bahia* range from 8 to 12 (Rodgers et al., 1986).

Problems that can occur during the performance of mysid chronics include losses of adult animals by jumping out of the water onto the sides of the screen chambers and poor survival and poor reproduction due to a dietary deficiency of essential fatty acids. Jumping may be caused by several factors, including low dissolved oxygen concentrations (i.e., less than 3 mg/L) and physical disturbances (e.g., lights switching on or off). Animals that jump are lost because they adhere to the side of the test container and desiccate. Jumping may be reduced by using a transition period between periods of light and darkness and by shielding the test from external activities. Maintenance of acceptable dissolved oxygen concentrations is essential, as mysids will tend to jump in an attempt to escape from low-oxygen water.

In 1987, a 7-d short chronic test was developed for estimating the chronic toxicity of effluents and receiving waters (Lussier et al., 1987). Although this test could easily be adapted for pure chemical testing, it has been utilized principally for the evaluation of chronic effects of effluents discharged into estuarine and marine environments. As one of the goals of this test is to determine the effects of exposure on fecundity, the test is initiated with animals that are 7 d old and will attain sexual maturity during the 7-d exposure period. Fecundity is measured as the percentage of female mysids observed with eggs in the oviduct and/or brood pouch at test termination. Survival and growth (dry weight) of the mysids are also measured during this short chronic test.

In preparation for a 7-d short chronic, newly released (<24 h old) mysids must be collected and maintained in dilution water at 26 to 27°C for 7 d. During the holding period, the mysids are fed newly hatched brine shrimp nauplii twice a day in sufficient quantity to keep some food constantly present (generally at least 150 nauplii per mysid per day). A static renewal procedure for this test is described in "Short-term methods for estimating the chronic toxicity of effluents and receiving waters to marine and estuarine organisms" (U.S. EPA, 1988). Following this procedure, five 7-d-old mysids are placed in each of eight replicate test chambers per treatment and tested for 7 d. The mysids are placed in 150 ml of test solution in appropriate-sized test vessels and maintained at 26 to 27°C. The mysids are fed live brine shrimp nauplii twice a day and test solutions renewed on a daily basis. Mysid survival is observed and recorded daily and dead animals are removed along with excess food prior to removing old test solutions. At test termination, the live animals are examined for eggs and the sex of each animal is determined. Using a stereomicroscope at 240×, the number of immature animals, the sex of mature animals, and the presence or absence of eggs in the oviducts or brood sacs of the females are determined (Figures 10, 11, and 12). Observations of eggs must be made while the animals are alive because they turn opaque upon dying. After sexing and establishing the presence or absence of eggs, the mysids in each test chamber are rinsed in deionized water and then placed on tared pieces of aluminum foil for dry weight analyses. The animals are dried at 60°C for 24 h or 105°C for at least 6 h.

Decapods

Of the crustaceans, decapod crustaceans are primarily used in toxicity testing because of their commercial importance. However, like the crustaceans mentioned previously, decapods are also sensitive to chemicals and ecologically important. Table 8 summarizes some of the marine decapod crustaceans most frequently utilized in ecotoxicological testing. Within the decapod crustaceans, several developmental stages are commonly used for toxicological testing. Of the four life stages (larva, megalops, juvenile, and adult), Gentile et al. (1984) reported that adult and juvenile stages have been utilized the most and the larval and megalops stages the least. This trend reflects the difficulties associated with utilizing the early life stages (larva and megalops) of decapods. Juvenile and adult decapod crustaceans may be acutely tested following procedures set forth in Standard Methods (APHA et al., 1992) and ASTM (1993b). Special procedures are generally required for handling and testing crustaceans, especially larval and megalops stages. Acute tests are generally performed using five or six chemical test concentrations along with appropriate controls. A minimum of 10 animals per treatment is tested. When testing adult shrimp and crabs, containers must be covered to prevent escape by jumping or crawling. Cannibalism is another problem when testing some species of decapods. With some species, the chelipeds may be removed or the animals isolated in separate test containers or in screened chambers within the main test chambers. Problems associated with testing crustacean larvae usually involve handling techniques and special diets to ensure proper survival, growth, and development.

Of the decapod crustaceans listed in Table 8, the grass shrimp (*Palaemonetes* spp.) is the only genus for which both acute and life cycle toxicity test procedures have been developed. Acute and chronic testing procedures for grass shrimp are described in detail.

Grass Shrimp

Because grass shrimp normally exhibit variability in molting and developmental rates that make it unfeasible to produce sufficient larvae of individual stages for testing, Tyler-Schroeder (1978a) recommends that acute testing of early life stages of grass

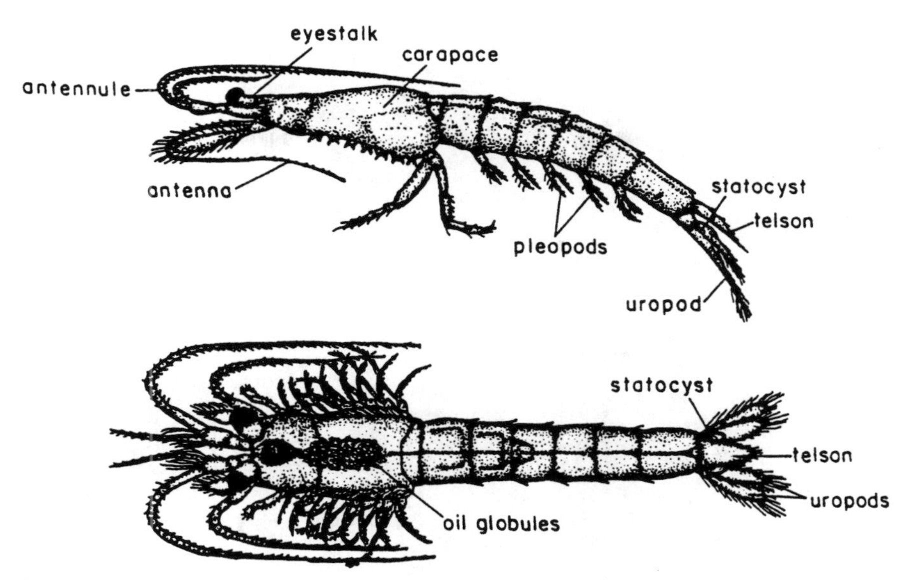

Figure 10. Immature *M. bahia*. Above: lateral view. Below: dorsal view. (From Lussier et al., 1987.)

shrimp be conducted with larvae of specified ages and not individual life stages. Most grass shrimp larvae metamorphose to postlarvae between 18 and 21 d after release at 25°C, so Tyler-Schroeder recommends that both 1-d-old and 18-d-old animals be tested. Larvae are obtained by isolating three ovigerous female shrimp per 8-inch glass culture bowl in 1 L of filtered salt water (salinity of 15 to 25‰). At least 17 to 25 bowls are required per acute test. The chelipeds are removed with fine surgical scissors to prevent removal of the eggs by the females. In the definitive test procedure for conducting a 96-h static test with grass shrimp larvae, Tyler-Schroeder proposes that three replicates per treatment of 30 larvae per replicate be tested to allow for the inherent variability of each age group of larvae. Each replicate consists of a 20.3-cm glass culture dish containing 1 L of filtered salt water. The larvae are fed an excess of live *Artemia* nauplii daily throughout the test. The test solutions should be renewed daily and *Artemia* added with each change of test solution. The test should be conducted at 25°C in total darkness or on

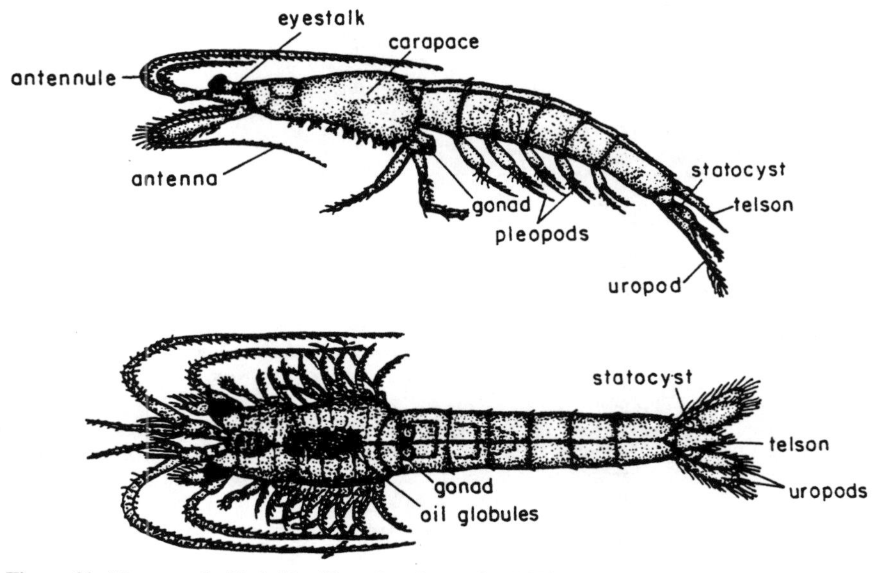

Figure 11. Mature male *M. bahia*. (From Lussier et al., 1987.)

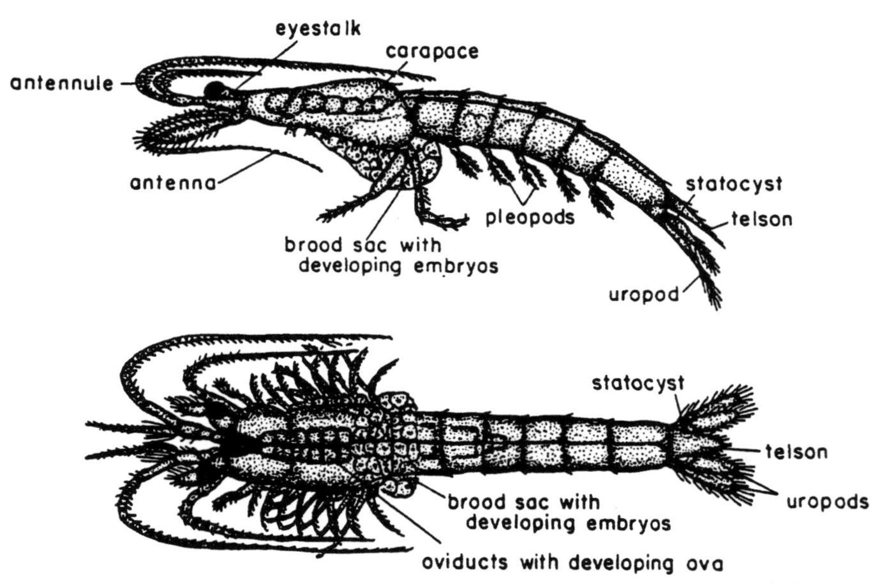

Figure 12. Mature female *M. bahia* with eggs in oviducts and developing embryos in the brood sac. Above: lateral view. Below: dorsal view. (From Lussier et al., 1987.)

a 12-h light and 12-h dark photoperiod. Control mortality should not exceed 10% for test acceptability.

Tyler-Schroeder (1978b, 1979) also describes a life cycle (145-d) toxicity test with *Palaemonetes*. Although described below, this test procedure has been of limited use since the introduction of the life cycle test with *M. bahia*, which is shorter and easier to conduct. The grass shrimp test is conducted under flow-through conditions. A flow-through diluter system may be utilized with a modified overflow drain to prevent escape and entrapment of larvae (Figure 13). The dilution water should have a salinity of approximately 20‰ (25‰ optimal for larval development) and be maintained at 25 ± 1°C. The life cycle test is initiated with 100 juvenile shrimp (less than 15 mm rostrum-telson length) per test treatment concentration, including controls. Counts of individual shrimp should be made every 4 to 6 wk during the test. Lengths (rostrum-telson) of 30 shrimp per concentration are measured at test initiation and every 4 wk until test termination. During the 2 to 3 wk required for the shrimp to attain sexual maturity, the photoperiod is maintained at a constant 8 h light and 16 h dark using 15-W incandescent bulbs to prevent premature induction of gonad development and spawning. After the 2- to 3-wk period, shrimp should range from 20 to 25 mm in length and spawning may be induced. This is accomplished by using 100-W incandescent lights and switching to a 10-h light and 14-h dark photoperiod. The photoperiod is increased in 47-min increments every 2 wk to a maximum of 15-h, 29-min light and 8-h, 31-min darkness. Ovigerous females should be separated from the rest of the population to assess the number of females spawning per day.

Reproductive effects evaluated include number of ovigerous females produced, egg production, and embryo hatching success. The number of ovigerous females produced in each exposure concentration is recorded daily. Egg production is determined by counting eggs from at least 10 females per concentration. Note that the rostrum-telson length of each female stripped of eggs for counting purposes must be measured because egg production is proportional to length (Buikema et al., 1980). The effects on hatching success of embryos are determined by isolating ovigerous females with well-developed eggs in a special hatching apparatus (Figure 14). Water from each diluter cycle flushes newly released larvae

Table 8. Summary of marine decapod crustacea most frequently utilized in ecotoxicological testing

Family	Species
Palaemonidae	*Palaemon adspersus*
	P. macrodactylus
	Palaemonetes pugio
	P. vulgaris
Penaeidae	*Penaeus aztecus*
	P. duorarum
	P. setiferus
	P. stylirostris
Caridea	*Crangon* sp.
Canceridae	*Cancer irroratus*
	C. magister
	C. productus
Portunidae	*Callinectes sapidus*
	Carcinus maenas
Nephropidae	*Homarus americanus*

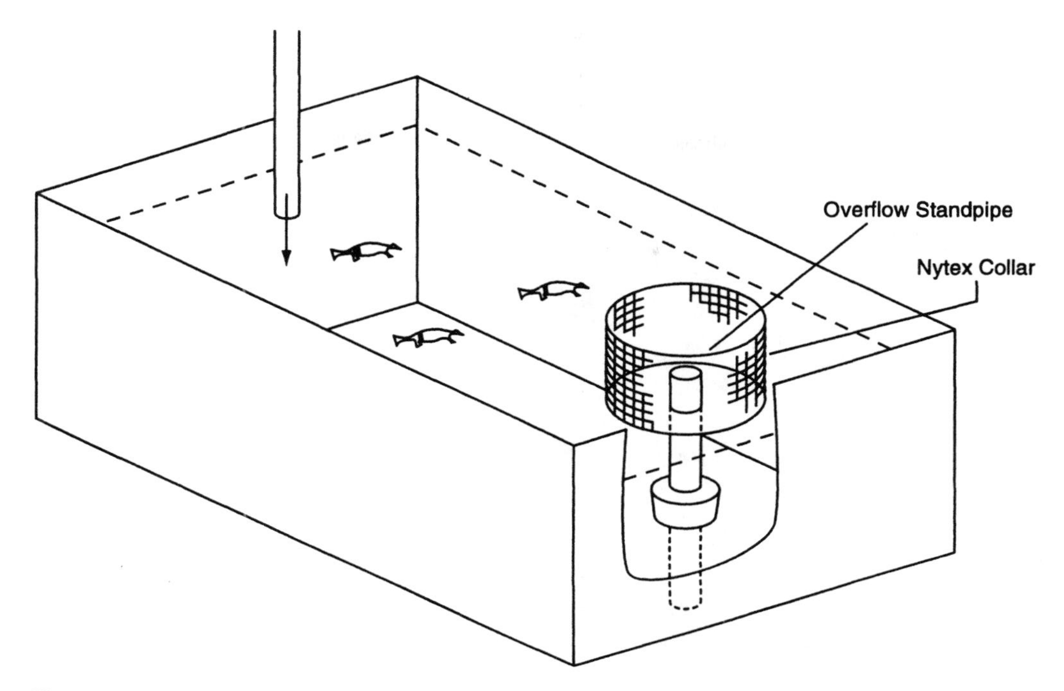

Figure 13. Flow-through grass shrimp test tank with modified overflow drain. (Modified from Tyler-Schroeder, 1978c.)

from each hatching apparatus into a screened egg chamber [100-mm-diameter petri dish fitted with a 363-μm nylon mesh collar, 13 mm high]. Larvae from at least 10 females in each exposure concentration are collected and counted.

Hatched larvae are maintained in these screened egg chambers and the chambers are positioned in rearing trays (small elevated tanks located within the main test tank and equipped with a self-starting siphon to promote water exchange through the screened chambers). Larvae from several females are collected from the screened egg chambers and placed in large-diameter (150-mm-diameter) screen chambers to observe the effects on larval development. A total of 20 larvae per screen chamber and four to five screen chambers per treatment are used including the controls. Survival of larvae should be monitored daily and observations of metamorphosis recorded.

The rostrum-telson length of all postlarvae are measured 35 d after hatching to evaluate the potential effects of the test material on growth. Lengths of 30 shrimp per concentration are recorded weekly until test termination.

The test can be terminated after the F_1 larvae metamorphose to postlarvae and grow to 15-mm-long juveniles or extended to evaluate the effects of the test material on F_1 reproduction and F_2 hatching success, larval development, and growth. An MATC is estimated based on determination of the NOEC and LOEC values, which are statistically calculated from the most sensitive biological indicator (i.e., survival, reproduction, and growth).

Echinoderms

Echinoderms, such as the sea urchin, can be tested in acute tests following standard toxicity testing procedures. Recently, using a process long utilized in the classroom for obtaining gametes from sea urchins, a test was developed to examine the potential toxic effects of chemicals on sea urchin fertilization. Although utilized for the evaluation of effluent discharges into estuarine and marine environments, this test can be used for all types of toxicants. This test is described in Chapter 6.

Fish

Many saltwater fish species are available for acute toxicity testing and procedures are well described (APHA et al., 1982; ASTM, 1993b). However, the most commonly tested fish species in the United States are the sheepshead minnow, *Cyprinodon variegatus*, and the Atlantic or inland silverside, *Menidia menidia* or *M. beryllina*. Sheepshead minnows are generally considered hardier and can also be utilized for full life cycle chronic toxicity tests (Hansen, 1978). Although early life stage procedures are available for *Menidia* spp. (Goodman et al., 1985), young are especially sensitive to handling and spawning requirements make full chronic testing difficult. No full life cycle chronic tests have been conducted with *Menidia* spp. to date. General procedures for the conduct of early life stage fish tests and full life cycle

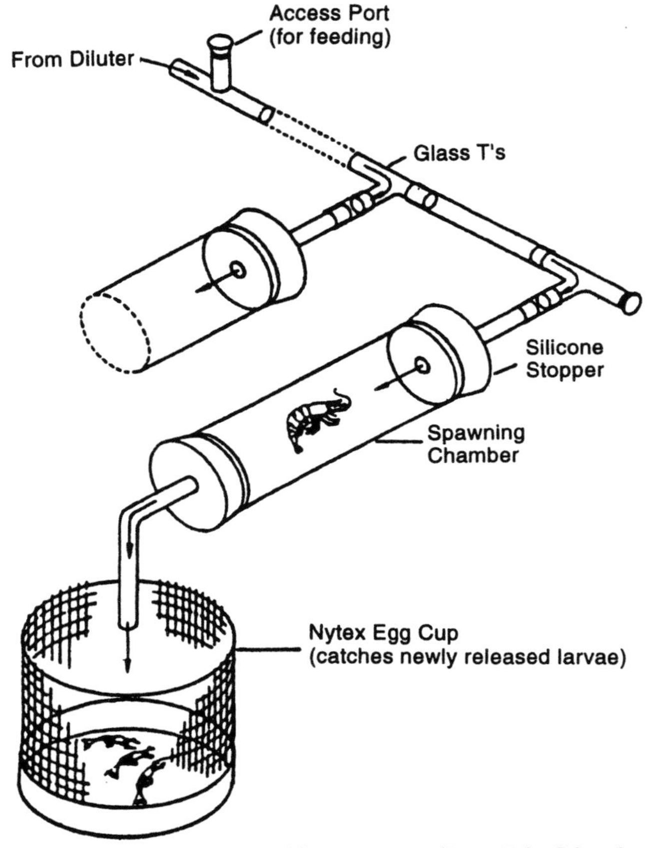

Figure 14. Grass shrimp hatching apparatus. (From Tyler-Schroeder, 1979. Reprinted with permission. Copyright ASTM.)

chronic fish tests are described in Rand and Petrocelli (1985) and Appendix Chapter B.

Short-term chronic tests have been developed with sheepshead minnow and the inland silverside for testing of effluents to determine effects on survival and growth (U.S. EPA, 1988). The test with both fish species begins with larval fish of the same age and exposes the fish for 7 d under static renewal conditions. Sheepshead minnow larvae are generally less than 24 h old at test initiation, and inland silversides are between 7 and 11 d old. During the 7-d exposure period, fish larvae are fed live brine shrimp nauplii daily. The recommended feeding regime is 0.10 g wet weight of live brine shrimp nauplii per replicate test chamber from days 0 to 2 and 0.15 g wet weight brine shrimp per replicate chamber from days 3 to 6. Three to four replicate chambers per test concentration are used with 10 to 15 larvae tested per replicate test chamber. The exposures are conducted at 25 ± 2°C in salt water with a salinity between 20 and 32‰ for sheepshead minnows and between 5 and 32‰ for silversides. All test solutions are renewed daily. Dead brine shrimp, fecal matter, and other debris are removed while renewing test solutions to maintain acceptable water quality. Test containers should be 300-ml to 1-L beakers or the equivalent and contain between 250 and 750 ml of dilution water. Survival of larvae is monitored daily and at test termination all surviving larvae in each test chamber are dried at 60°C for 24 h or at 105°C for a minimum of 6 h. Dried larvae are cooled in a desiccator and dry weights determined.

The criteria for acceptability of these tests include (1) the average survival of control larvae equals or exceeds 80% and (2) the average dry weight of unpreserved control sheepshead minnow larvae is ≥0.60 mg or ≥0.50 mg for unpreserved inland silversides.

Sheepshead minnow eggs may be obtained naturally from in-house cultures following procedures described by Hansen (1978) or by stimulation of ovulation artificially through injection of females with human chorionic gonadotropic hormone. The eggs are stripped from the females and fertilized with sperm obtained from sacrificed males (Hansen, 1978). Fertilized eggs may be hatched either by placing the embryos in screened chambers under flowing salt water or by aerating in static salt water (Hansen et al., 1978; Hansen, 1978). Embryos hatch in 4 to 6 days at temperatures of 25 to 30°C. Sheepshead minnows less than 24 h old are used to initiate the short-term chronic tests.

It is preferable to obtain Inland silversides from in-house cultures. Acceptable culture practices for *M. beryllina* are described by Middaugh et al. (1987). Embryos hatch in 6 to 7 d when incubated at 25°C. Unlike sheepshead minnows, inland silversides are too small to ingest brine shrimp nauplii immediately after hatching and must be fed rotifers (*Brachionus plicatilis*) for at least 5 days before *M. beryllina* will begin feeding on newly hatched brine shrimp nauplii. inland silversides should be 7 to 11 d old when short-term chronic tests are initiated because of their sensitivity to handling during the first week after hatching.

The 7-d short-term chronic test procedures, although developed for estimating the chronic toxicity of effluents in marine waters, may be also used as screening tests for other xenobiotics.

The short-term chronic test with sheepshead minnow may also be modified to screen effluents for teratogens (U.S. EPA, 1988). This test is initiated with sheepshead minnow embryos less than 24 h old obtained as described above. The exposure is performed as described for the 7-d survival and growth test. At 25°C, the embryos hatch in approximately 6 d. The test is terminated after 9 d or 4 d after hatching, whichever comes first. Test results are based on mortality and frequency of gross morphological deformities (terata). Deformed larvae are those with gross morphological abnormalities such as curved spines, lack of appendages, lack of fusiform shape, lack of mobility, abnormal swimming, or any other characteristic that would preclude survival. Normal morphological development of the sheepshead minnow is illustrated in Figure 15. At least 80% survival must occur throughout the study in the controls for the test results to be considered acceptable.

Although procedures for conducting full life cycle chronic and early life stage toxicity tests were generally described in Rand and Petrocelli (1985), specific procedures for conducting early life stage and full life cycle chronic toxicity tests with the sheepshead minnow and historical values for chronic end points for these tests are presented below.

The early life stage and full life cycle tests with sheepshead minnow are conducted under flow-through conditions and are initiated similarly. Fertilized eggs are obtained either naturally from spawning groups or by artificial means as described previously. Generally, two groups of 40 to 50 fertilized eggs are placed in duplicate test chambers, with 80 to 100 eggs per treatment. At least five chemical treatments are tested along with appropriate controls. The eggs are transparent and the development of the embryos can be monitored daily until hatching. Following hatching and determination of hatching success, fry are fed live brine shrimp nauplii two or three times per day ad libitum. Fry survival and development are monitored for the duration of the test. The early life stage test is terminated after 28 d posthatch, at which time all fish are sacrificed and length and wet and/or dry weights measured. During the life-cycle test, fry are simply measured photographically at 28 to 30 d and the exposure continued. After 60 d posthatch, all fry are photographically measured and then thinned to 25 per replicate chamber, 50 per treatment. The discarded fry are weighed and measured and the tissues analyzed for the test substance. The remaining fry are allowed to grow until sexual maturity is attained (approximately 3 to 4 months posthatch). When fish reach sexual maturity (approximately 26 mm in length and exhibiting sexual dimorphism), egg productivity is measured with three 2-wk spawning periods. For each spawning period, a different spawning group (three females and two males) is maintained. Each spawning group is held in a spawning chamber designed to allow eggs extruded by female fish and fertilized by male fish to drop through a large mesh screen onto a finer mesh screen (Figure 16). The adult fish must not have access to the eggs or they will consume the eggs. The eggs are collected and enumerated daily throughout each 2-wk spawning period. During at least one of the three spawning periods, 50 eggs per replicate should be collected and placed in incubation chambers for evaluation of hatching success and fry survival of the second generation of fish. Typical values for chronic end points during early life stage and life-cycle toxicity tests are presented in Table 9. The LOEC and NOEC are typically reported for each end point, if possible.

Several problems are associated with the conduct of sheepshead minnow early life stage and full life cycle chronic toxicity tests. One problem is the type of solvent used and the solvent concentration. Although typical organic solvents (i.e., triethylene glycol, dimethyl formamide, and acetone) are routinely utilized at concentrations up to 100 mg/L, sheepshead minnow egg hatchability can be significantly reduced by these solvents at these concentrations. It is believed that bacteria utilizing the organic solvent as a carbon source increase significantly in numbers, coating the outside of the egg. This coating of bacteria either directly or indirectly reduces the penetration of oxygen through the chorion to the embryo and suffocates the embryo. Therefore, solvent concentrations should be as low as possible; a maximum concentration of 20 mg/L is recommended by the author. In addition, triethylene glycol is recommended over dimethyl formamide and acetone as the solvent of choice because it appears to result in less bacterial growth.

Another problem that affects sheepshead minnow fry following hatching is an external parasite (believed to be a trematode). This parasite appears to be transferred from the adult fish during spawning to the eggs. Once the fish hatch, the parasites infest the fry, resulting in mortalities during the first 14 d. Formalin dips (20 mg/L) of the adults can reduce or

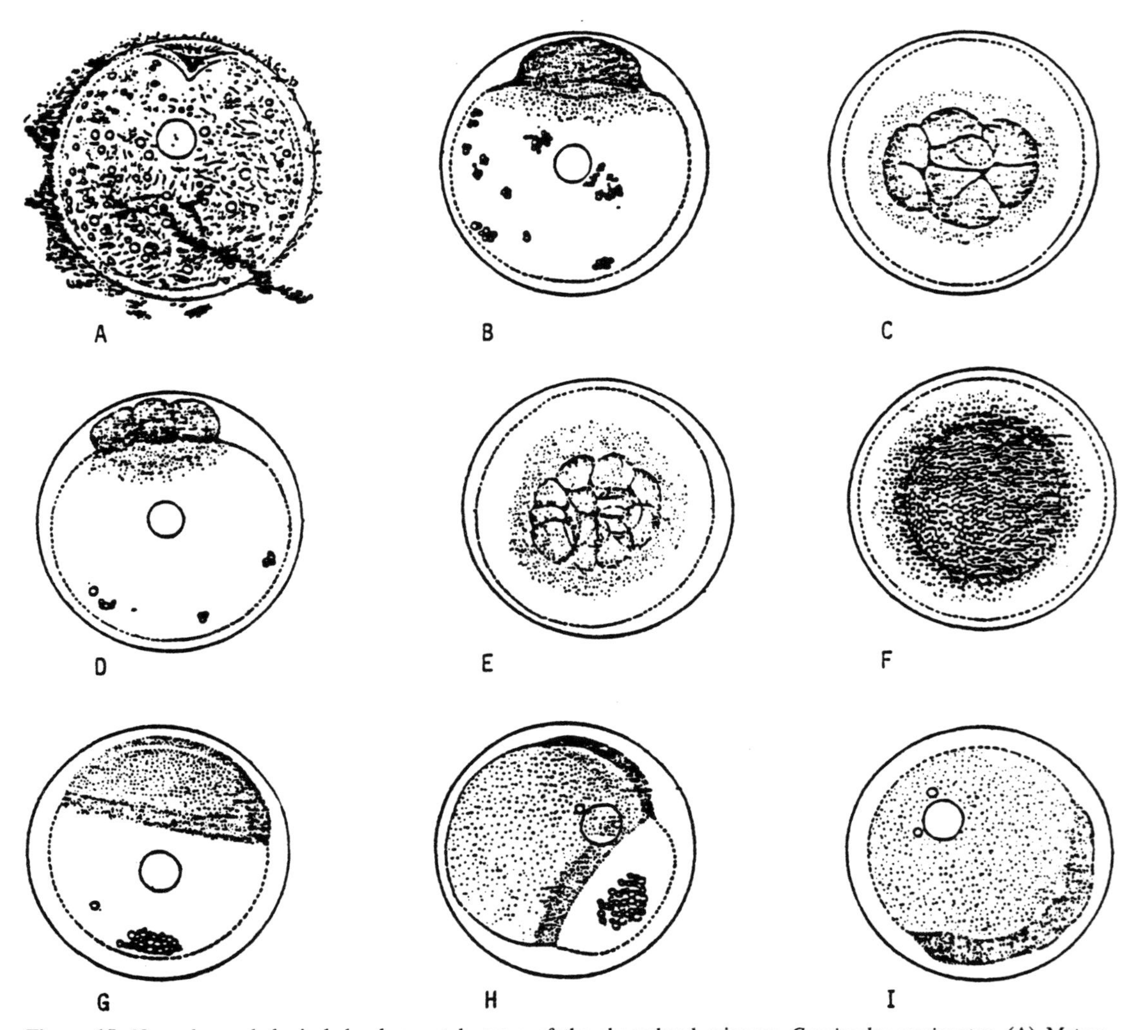

Figure 15. Normal morphological developmental stages of the sheepshead minnow, *Cyprinodon variegatus.* (A) Mature unfertilized egg, showing attachment filaments and micropyle; (B) blastodisc fully developed; (C and D) blastodisc, 8 cells; (E) blastoderm, 16 cells; (F) blastoderm, late cleavage stage; (G) blastoderm with germ ring formed, embryonic shield developing; (H) blastoderm covers over three-quarters of surface of yolk, yolk noticeably constricted; (I) early embryo.

eliminate this parasite, allowing acceptable fry survival. In some cases, however, the infestation cannot be entirely eliminated and it has been found that redipping the adults in a formalin solution immediately prior to stripping eggs and fertilizing will improve the chances of obtaining good fry survival in the test.

DEVELOPING COMMUNITY TESTS

Developing community tests were designed to assess the impacts of chemical substances on developing communities (i.e., multispecies) of benthic estuarine organisms. These tests allow testing of a wide variety of nonstandard organisms during their early and presumably most sensitive stages of development. Developing community tests have been conducted on a polychlorinated biphenyl (PCB), several insecticides and biocides, a herbicide, a drilling mud, and a drilling mud component, barite (Hansen, 1974; Hansen and Tagatz, 1980; Tagatz et al., 1977, 1978a, 1978a,b; Tagatz and Tobia, 1978; Tagatz et al., 1979a,b; Tagatz and Deans, 1983; Parrish et al., 1981).

Basically, the community development tests consisted of clean substratum exposed for a period of 7 to 16 wk to a series of treatments including control and three to five test substance treatments. At the end of each exposure period, the substratum was sieved through a 1-mm mesh sieve and all benthic macroinvertebrates were identified and enumerated. Treatments and controls were then compared to evaluate changes in community structure including changes

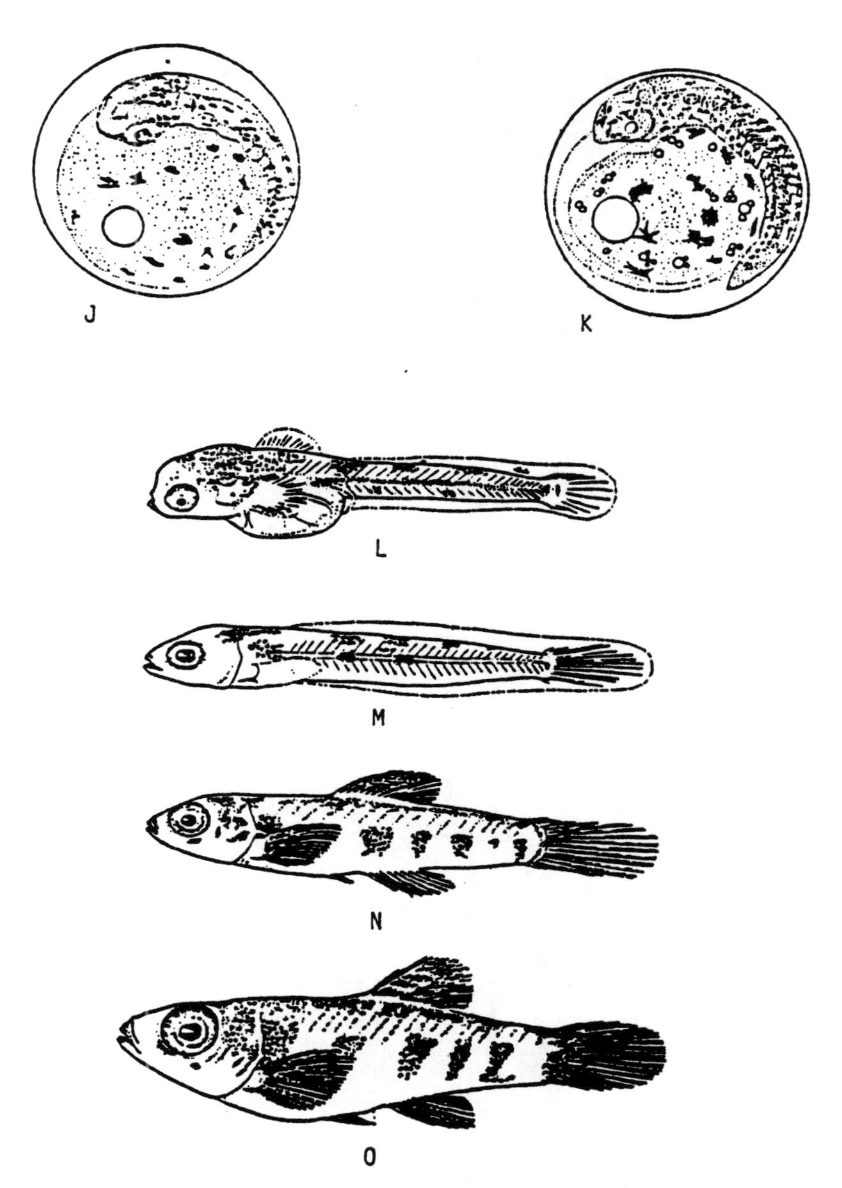

Figure 15. Normal morphological developmental stages of the sheepshead minnow, *Cyprinodon variegatus* (*Continued*). (J) embryo 48 h after fertilization, now segmented throughout, pigment on yolk sac and body, otoliths formed; (K) posterior portion of embryo free from yolk and moving freely within egg membrane, 72 h after fertilization; (L) newly hatched fish, actual length 4 mm; (M) larval fish 5 d after hatching, actual length 5 mm; (N) young fish 9 mm in length; (O) young fish 12 mm in length. (From U.S. EPA, 1988.)

in relative abundance of animals by species and phylum.

The early community development tests were conducted using a system of test tanks, each containing 10 adjacent chambers (each chamber was 56 cm long × 9 cm wide × 12 cm high) (Figure 17). Each chamber was filled with clean sand to a depth of 6 cm and provided a water flow of 200 ml/min by siphon from a head box. The water level was maintained at 9 cm (3 cm above the substratum). Water was unfiltered seawater. The two end chambers of each test aquarium were not used, to eliminate any potential "end of row" differences. The testing system was redesigned to eliminate potential distribution effects. The new system consisted of a central constant head box from which eight test chambers (40 cm long × 10 cm wide × 12 cm high) radiated (Figure 18). Each test chamber was filled with 5 cm

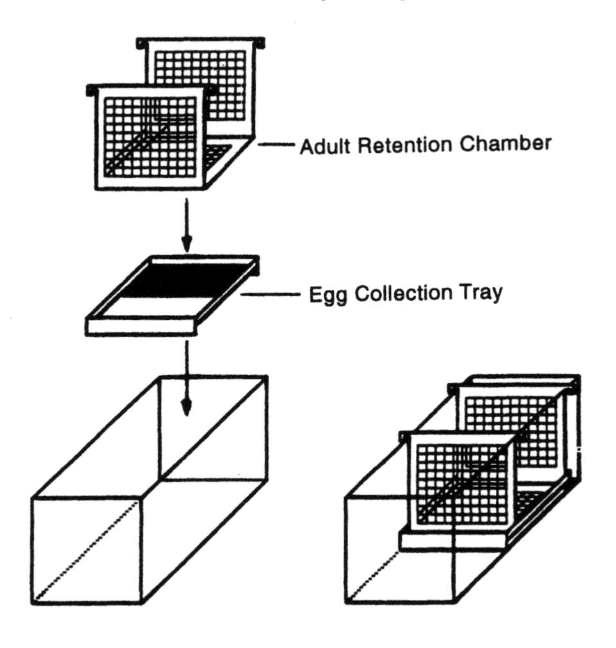

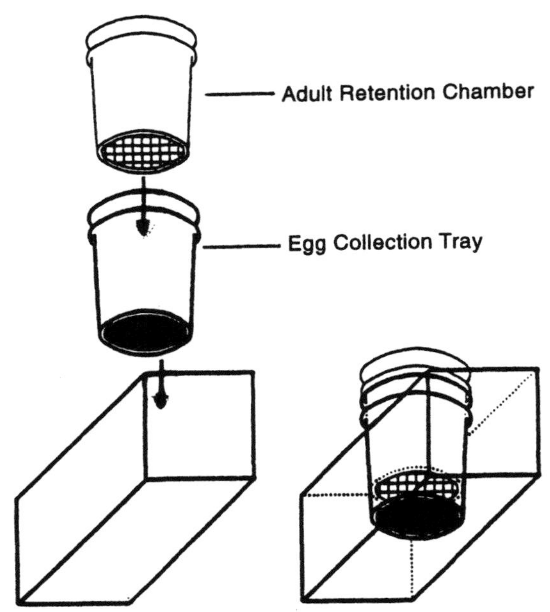

Figure 16. Two types of spawning chambers for collection of sheepshead minnow eggs. (From Toxikon Environmental Sciences.)

Table 9. Typical values for chronic end points during life cycle toxicity tests with the sheepshead minnow

Chronic end point	Range of values
Embryo hatchability[a]	77–99
Survival[b]	
Day 28–30 posthatch	74–96
Growth	
Day 28–30 posthatch	10–14 mm 40–70 mg wet weight
Day 103 posthatch	31–36 mm 99–190 mg wet weight

Fecundity (eggs/d/female)[c]	Mean ± SD	Range of replicates
Test A	17 ± 5	11–26
Test B	17 ± 9	9–30
Test C	10 ± 4	4–16
Test D	25 ± 9	14–37

[a]Percentage of embryos that hatched. Embryos hatch in 4 to 7 d at test temperatures between 25 and 30°C.
[b]Percentage of fry surviving of embryos that successfully hatched.
[c]Tests A and B from Parrish et al., 1977. Tests C and D from Parrish et al., 1978.

of clean sand and provided with unfiltered seawater at a flow rate of 200 ml/min. The water depth was maintained at 9 cm or 3 cm above the surface of the substratum.

Tagatz and Deans (1983) reported that the benthic communities that developed in the community development toxicity tests were structurally similar to the benthic communities in natural aquatic systems adjacent to the laboratory. Depending on the season and annual and geographic differences in the reproductive cycles of animals with pelagic larvae, the communities that developed were very diverse and variable. In the six tests conducted at the U.S. EPA Gulf Breeze laboratory, an average of 7 phyla, 52 species, and 4367 individuals were collected per test (Table 10). In the one test conducted at another laboratory in the same vicinity, 8 phyla, 43 species, and 1049 individuals were collected. Annelids, arthropods, chordates, coelenterates, or molluscs were dominant numerically or by species in these tests.

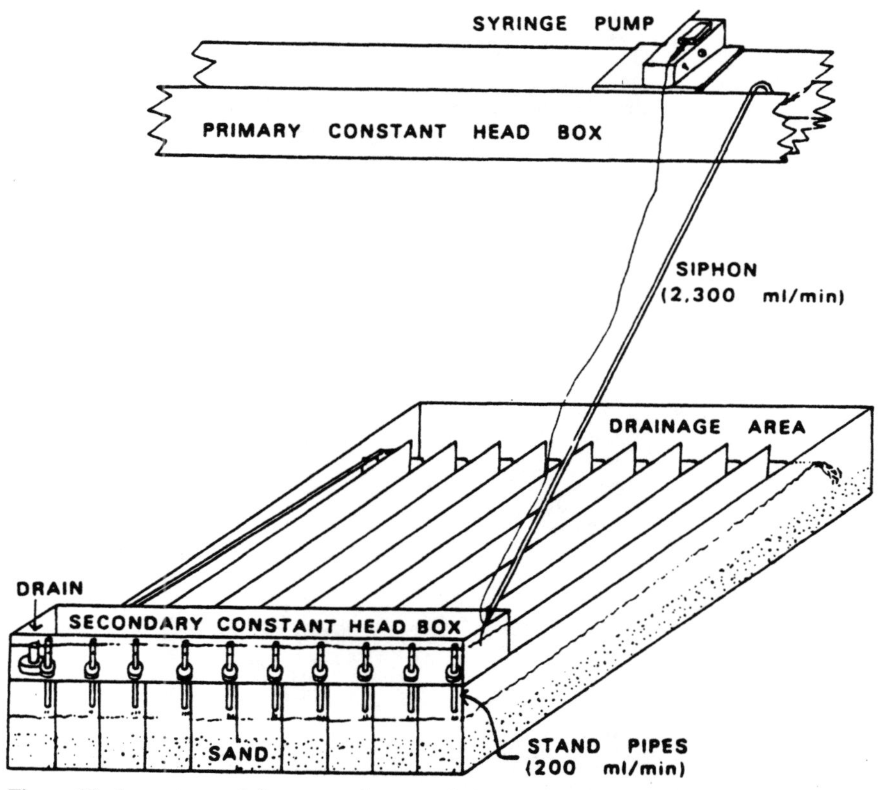

Figure 17. Apparatus used for community tests. Only one of four identical apparatuses is shown. (From Hansen: Aroclor® 1254, 1974. Reprinted with permission. Copyright ASTM.)

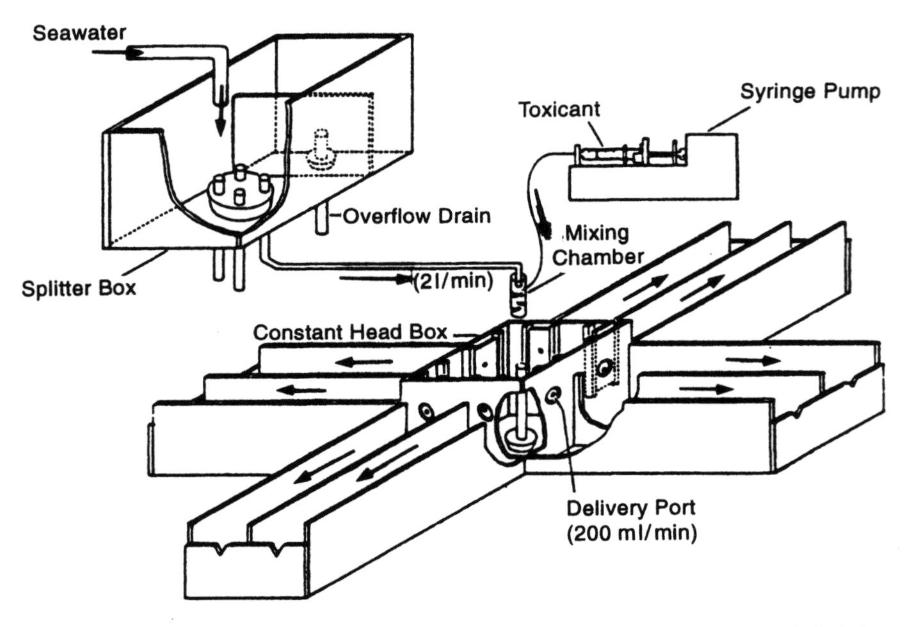

Figure 18. Revised developing community development apparatus. (From Tagatz, Ivey & Oglesby, 1979a.)

Table 10. Summary of estuarine community development studies designed to determine the effects of test substances on development of communities of benthic estuarine organisms

Test substance	Individuals	Species	Phyla	Exposure (wk)	Abundance (%) by phyla
Aroclor 1254	5,897	67	9	18	Arthropods (65) Chordates (22) Annelids (8) Molluscs (4)
Toxaphene	12,129	81	8	12	Molluscs (58) Annelids (26) Arthropods (15)
Pentachlorophenol	267	28	5	9	Annelids (69) Molluscs (15) Arthropods (15)
Dowicide G-ST	5,066	32	9	13	Molluscs (59) Arthropods (20) Annelids (20)
Barite	3,020	59	8	10	Molluscs (56) Annelids (23) Chordates (14)
Drilling mud	1,025	45	8	8	Annelids (62) Molluscs (16) Arthropods (12)
Atrazine	1,049	43	8	8	Arthropods (53) Molluscs (34) Annelids (8)
Sevin®	7,844	29	7	10	Molluscs (75) Arthropods (18) Annelids (6)

LITERATURE CITED

American Public Health Association, American Water Works Association, and Water Pollution Control Federation (APHA, AWWA, and WEF): *Standard Methods for the Examination of Water and Wastewater*, 18th ed. Washington, DC: APHA, 1992.

American Society for Testing and Materials (ASTM): Standard guide for conducting life-cycle toxicity test with saltwater mysids. E1191-90, pp. 736–751, 1990. Annual Book of ASTM Standards, Vol. 11.04:837–852, 1993a.

American Society for Testing and Materials (ASTM): Standard guide for conducting toxicity tests with fishes, macroinvertebrates, and amphibians, E729-88a, pp. 360–379, 1990. Annual Book of ASTM Standards, Vol. 11.04:456–475, 1993b.

American Society for Testing and Materials (ASTM): Standard guide for conducting early life-stage toxicity tests with fishes. E1241-88, pp. 827–852, 1990. Annual Book of ASTM Standards, Vol 11.04:941–968, 1993c.

American Society for Testing and Materials (ASTM): Standard guide for acute toxicity test with the rotifer *Brachionus*, E1440-91, pp. 1210–1216, 1991. Annual Book of ASTM Standards, Vol. 11.04:1271–1277, 1993d.

American Society for Testing and Materials (ASTM): Standard guide for conducting static acute toxicity tests starting with embryos of four species of saltwater bivalve molluscs, E724-89. Annual Book of ASTM Standards, Vol. 11.04:430–447, 1993e.

American Society for Testing and Materials (ASTM): Standard guide for conducting static and flow-through acute toxicity tests with mysids from the west coast of the United States, E1463-92. Annual Book of ASTM Standards, Vol. 11.04:1278–1299, 1993f.

American Society for Testing and Materials (ASTM): Standard guide for conducting 10-day static sediment toxicity tests with marine and estuarine amphipods. E1367-92. Annual Book of ASTM Standards, Vol. 11.04:1138–1163, 1993g.

American Society for Testing and Materials (ASTM): Standard guide for conducting acute toxicity tests on aqueous effluents with fishes, macroinvertebrates, and amphibians. E1192-88. Annual Book of ASTM Standards, Vol. 11.04:853–865, 1993h.

American Society for Testing and Materials (ASTM): Standard guide for conducting sexual reproduction tests with seaweeds. E1498-92. Annual Book of ASTM Standards, Vol. 11.04:1318–1328, 1993i.

American Society for Testing and Materials (ASTM): Standard guide for conducting static 96-h toxicity tests with microalgae. E1218-90. Annual Book of ASTM Standards, Vol. 11.04:929–940, 1993j.

American Society for Testing and Materials (ASTM): Proposed guide for conducting acute, chronic, and life-cycle aquatic toxicity tests with polychaetous annelids. Subcommittee E47.01, 1991a.

American Society for Testing and Materials (ASTM): Proposed guide for conducting acute and chronic sediment toxicity tests with the marine polychaetous annelid *Neanthes arenaceodentata*. Subcommittee E47.01, 1991b.

Bengtsson, B.-E.: Use of a harpacticoid copepod in toxicity tests. *Mar. Pollut. Bull.*, 9:238–241, 1978.

Breteler, R. J.; Williams, J. W.; and Buhl, R. L.: Measurement of chronic toxicity using the opossum shrimp *Mysidopsis bahia*. *Hydrobiologia* 93:189–194, 1982.

Buikema, A. L., Jr.; Niederlehner, B. R.; and Cairns, J., Jr.: Toxicant effects on reproduction and distribution of the egg-length relationship in grass shrimp. *Bull. Environ. Contam. Toxicol.*, 24:31–36, 1980.

Butler, P. A.: Reaction of some estuarine mollusks to environmental factors. Biological Problems In Water Pollution—Third Seminar—1962. Washington, DC: U.S. Department of Health, Education, and Welfare, Public Health Service Publication No. 999-WP-25, June 1965.

Butler, P. A., and Lowe, J. I.: Flowing sea water toxicity test using oysters (*Crassostrea virginica*). Bioassay Procedures for the Ocean Disposal Permit Program. EPA-600/9-78-010, pp. 25–27, 1978.

Butler, P. A.; Wilson, A. J., Jr.; and Rick, A. J.: Effect of pesticides on oysters. *Proc. Natl. Shellfish Assoc.*, 51: 23–32, 1960.

Carr, R. S.; Curran, M. D.; and Mazurkiewicz, M.: Evaluation of archiannelid *Dinophilus gyrociliatus* for use in short-term life-cycle toxicity tests. *Environ. Toxicol. Chem.*, 5:703–712, 1986.

Code of Federal Regulations: Environmental Effects Testing Guidelines (TSCA). Part 797—Environmental Effects Testing Guidelines, Subpart B—Aquatic Guideline. 40 CFR: 298–392, 1990.

Environment Canada: Biological Test Method: Acute Test for Sediment Toxicity Using Marine or Estuarine Amphipods. Report EPS1/RM/26. Environment Canada, Conservation and Protection, Ottawa, Ontario, 1992a.

Environment Canada: Biological Test Method: Fertilization Assay Using Echinoids (Sea Urchins and Sand Dollars). Report EPS1/RM/27. Environment Canada, Conservation and Protection, Ottawa, Ontario, 1992b.

Foster, R. B.: Environmental legislation. Fundamentals of Aquatic Toxicology, edited by G. M. Rand, S. R. Petrocelli, pp. 587–600. New York: Hemisphere, 1985.

Friedman, S. D.; Deans, C. H.; Lytle, J. S.; and Parrish, P. R.: Food supplements and their effects on broodstock nutrition and reproduction of mysids (*Mysidopsis bahia*). Presented at the 119th Annual Meeting of the American Fisheries Society, Anchorage, Alaska, September 4–8, 1989.

Galtsoff, P. S.: The American oyster, *Crassostrea virginica* Gmelin. Fishery Bulletin of the Fish and Wildlife Service, Vol. 64. Washington, DC: Government Printing Office, 1964.

Gentile, J. H., and Sosnowski, S. L.: Methods for the culture and short term bioassay of the calanoid copepod (*Acartia tonsa*). Bioassay Procedures for the Ocean Disposal Permit Program, pp. 28–45. EPA–600/9-78-010. U.S. Environmental Protection Agency. 1978.

Gentile, J. H.; Johns, D. M.; Cardin, J. A.; and Heltshe, J. F.: Marine ecotoxicological testing with crustaceans. *Ecotoxicological Testing for the Marine Environment* edited by G. Persoone; E. Jaspers; and C. Claus, Vol. 1 State Univ. of Ghent and Institute for Marine Scientific Research, Bredene, Belgium, 1984.

Goodman, L. R.; Hansen, D. J.; Middaugh, D. P.; Cripe, G. M.; and Moore, J. C.: Method for early life-stage toxicity tests using three atherinid fishes and results with chloropyrifos. Aquatic Toxicology and Hazard Evaluation, edited by R. D. Cardwell; R. Purdy; and R. C. Bahner, pp. 145–154. ASTM STP 854. Philadelphia: American Society for Testing and Materials, 1985.

Halbach, U.; Siebert, M.; Westermayer, M.; and Wissel, C.: Population ecology of rotifers as a bioassay tool for ecotoxicological tests in aquatic environments. *Ecotoxicol. Environ. Safety*, 7:484–513, 1983.

Hansen, D. J.: Aroclor® 1254: effect on composition of developing esturarine animal communities in the laboratory. *Contrib. Mar. Sci.*, 18:19–33, 1974.

Hansen, D. J.: Laboratory culture of sheepshead minnows (*Cyprinodon variegatus*). Bioassay Procedures for the

Ocean Disposal Permit Program, pp. 107–108. EPA-600/9-78-010. 1978.

Hansen, D. J., and Tagatz, M. E.: A laboratory test for assessing impacts of substances on developing communities of benthic estuarine organisms. Aquatic Toxicology, edited by J. G. Eaton; P. R. Parrish; and A. C. Hendricks, pp. 40–57. ASTM STP 707. Philadelphia: American Society for Testing and Materials, 1980.

Hansen, D. J.; Parrish, P. R.; Schimmel, S. C.; and Goodman, L. R.: Life-cycle toxicity test using sheepshead minnows (*Cyprinodon variegatus*). Bioassay Procedures for the Ocean Disposal Permit Program, pp. 109–117. EPA-600/9-78-010. 1978.

Heinle, D. R.: Culture of calanoid copepods in synthetic sea water. *J. Fish. Res. Bd. Can.*, 26(1):150–153, 1969.

Heinle, D. R., and Beaven, M. S.: Toxicity of chlorine-produced oxidants to estuarine copepods. Aquatic Invertebrate Bioassays, edited by A. L. Buikema, Jr., and J. Cairns, Jr., pp. 109–130. ASTM STP 715. Philadelphia: American Society for Testing and Materials, 1980.

Karbe, L.: Marine Hydroiden als Testorganismen zur Prufung der Toxizitat von Abwasserstoffen (Marine hydroids as test organisms for assessing the toxicity of water pollutants). *Mar. Biol.*, 12:316–328, 1972.

Karbe, L.; Borchardt, T.; Dannenberg, R.; and Meyer, E.: Ten years of experience using marine and freshwater hydroid bioassays. Ecotoxicological Testing for the Marine Environment, edited by G. Persoone; E. Jaspers; and C. Claus, Vol. 2. State University Ghent and Inst. Mar. Scient. Res., Bredene, Belgium. 1984.

Lassus, P.; Le Baut, C.; Le Dean, L.; Bardouil, M.; Truquet, P.; and Bocquene, G.: Marine ecotoxicological tests with zooplankton. Ecotoxicological Testing for the Marine Environment, edited by G. Persoone; E. Jaspers; and C. Claus, Vol. 2. State Univ. of Ghent and Institute for Marine Scientific Research, Bredene, Belgium, 1984.

Lee, W. Y.: Some laboratory cultured crustaceans for marine pollution studies. *Mar. Pollut. Bull.*, 8(11):258–259, 1977.

Leger, P.; Sorgeloos, P.; Millamena, O. M.; and Simpson, K. L.: International study on *Artemia*. XXV. Factors determining the nutritional effectiveness of *Artemia*: the relative impact of chlorinated hydrocarbons and essential fatty acids in San Francisco Bay and San Pablo Bay *Artemia*. *J. Exp. Mar. Biol. Ecol.*, 93:71–82, 1985.

Lussier, S. M.; Gentile, J. H.; and Walker, J.: Acute and chronic effects of heavy metals and cyanide on *Mysidopsis bahia* (Crustacea: Mysidacea). *Aquat. Toxicol.*, 7: 25–35, 1985.

Lussier, S. M.; Kuhn, A.; and Sewall, J.: Guidance manual for conducting seven day mysid survival/growth/reproduction study using the estuarine mysid, *Mysidopsis bahia*. Contribution No. X106. Users Guide to the Conduct and Interpretation of Complex Effluent Toxicity Tests at Estuarine/Marine Sites, edited by S. C. Schimmel. Environmental Research Laboratory, U.S. Environmental Protection Agency, Narragansett, RI, Contribution No. 796, 1987.

Lussier, S. M.; Kuhn, A.; Chammas, M. J.; and Sewall, J.: Techniques for the laboratory culture of *Mysidopsis* species (Crustacea: Mysidacea). *Environ. Toxicol. Chem.*, 7: 969–977, 1988.

Middaugh, D. P.; Hemmer, M. J.; and Goodman, L. R.: Methods for spawning, culturing and conducting toxicity-tests with early life stages of four antherinid fishes: the inland silverside, *Menidia beryllina*, Atlantic silverside, *M. menidia*, tidewater silverside, *M. peninsulae*, and California grunion, *Leuresthes tenuis*. EPA/600/8-87/004. Washington, DC: Office of Research and Development, U.S. Environmental Protection Agency, 1987.

Nimmo, D. R.; Bahner, L. H.; Rigby, R. A.; Sheppard, J. M.; and Wilson, A. J., Jr.: *Mysidopsis bahia*: an estuarine species suitable for life-cycle toxicity tests to determine the effects of a pollutant. Aquatic Toxicology and Hazard Evaluation, edited by F. L. Mayer, and J. L. Hamelink, pp. 109–116. ASTM STP 634. Philadelphia: American Society for 1977.

Nimmo, D. R., and Hamaker, T. L.: Mysids in toxicity testing—a review. *Hydrobiologia* 93:171–178, 1982.

Parrish, P. R.; Dyar, E. E.; Lindberg, M. A.; Shanika, C. M.; and Enos, J. M.: Chronic toxicity of methoxychlor, malathion, and carbofuran to sheepshead minnows (*Cyprinodon variegatus*). Ecological Research Series, EPA-600/3-77-059, 1977.

Parrish, P. R.; Dyar, E. E.; Enos, J. M.; and Wilson, W. G.: Chronic toxicity of chlordane, trifluralin, and pentachlorophenol to sheepshead minnows (*Cyprinodon variegatus*). Ecological Research Series, EPA-600/3-78-010, 1978.

Parrish, P. R.; Heitmuller, P. T.; Ward, G. S.; and Ballantine, L. G.: Chemical effects on esturine communities. Presented at the First Annual SETAC Symposium, 1981.

Pennak, R. W.: Fresh-Water Invertebrates of the United States: Protozoa to Mollusca, 3rd ed. New York: Wiley, 1989.

Pesch, C. E.; Munns, W. R., Jr.; and Gutjahr-Gobell, R.: Effects of a contaminated sediment on life history traits and population growth rate of *Neanthes arenaceodentata* (Polychaeta: Nereidae) in the laboratory. *Environ. Toxicol. Chem.*, 10:805–815, 1991.

Rand, G. M., and Petrocelli, S. R. (eds.): Fundamentals of Aquatic Toxicology. Washington, DC: Hemisphere, 1985.

Reish, D. J.: Use of polychaetous annelids as test organisms for marine bioassay experiments. Aquatic Invertebrate Bioassays, edited by A. L. Buikema, Jr., J. Cairns, Jr., pp. 140–154. ASTM STP 715. Philadelphia: American Society for Testing and Materials, 1980.

Reish, D. J.: Marine ecotoxicological tests with polychaetous annelids. Ecotoxicological Testing For The Marine Environment, edited by G. Persoone, E. Jaspers, C. Claus, Vol. 1. State Univ. of Ghent and Institute for Marine Scientific Research, Bredene, Belgium, 1984.

Reish, D. J., and Gerlinger, T. V.: The effects of cadmium, lead, and zinc on survival reproduction in the polychaetous annelid *Neanthes arenaceodentata* (F. Nereididae). Proceedings of the First International Polychaete Conference, Sydney, edited by P. A. Hutchings, pp. 383–389. Linnean Society of New South Wales, 1984.

Reitsema, L. A., and Neff, J. M.: A recirculating artificial seawater system for the laboratory culture of *Mysidopsis almyra* (Crustacea; Pericaridea). *Estuaries* 3(4):321–323, 1980.

Rodgers, J. H.; Dorn, P. B.; Duke, T.; and Parrish, R.: *Mysidopsis* sp.: life history and culture, a report from a workshop held in Gulf Breeze, FL on October 15–16, 1986.

Snell, T. W., and Persoone, G.: Acute toxicity bioassays using rotifers. I. A test for brackish and marine environments with *Brachionus plicatilis*. *Aquat. Toxicol.*, 14:65–80, 1989a.

Snell, T. W., and Persoone, G.: Acute toxicity bioassays using rotifers. II. A freshwater test with *Brachionus rubens*. *Aquat. Toxicol.*, 14:81–92, 1989b.

Staeger, W. H., and Horton, H. F.: Fertilization method quanitfying gamete concentrations and maximizing larvae production in *Crassostrea gigas*. *Fish Bull.*, 74(3): 698–701, 1976.

Stebbing, A. R. D.: An experimental approach to the determinants of biological water quality. *Philos. Trans. R. Soc. Lond.*, Ser B 286:465–481, 1979.

Stebbing, A. R. D.: Increase in gonozooid frequency as an adaptive response to stress in *Campanularia flexuosa*. Developmental and Cellular Biology of Coelenterates, edited by P. Tardent, and R. Tardent, pp. 27–32. Amsterdam: Elsevier, 1980.

Stebbing, A. R. D., and Brown, B. E.: Marine ecotoxicological test with coelenterates. Ecotoxicological Testing for the Marine Environment, edited by G. Persoone; E. Jaspers; and C. Claus, Vol 1. State Univ. of Ghent and Institute for Marine Scientific Research, Bredene, Belgium, 1984.

Stephano, J. L., and Gould, M.: Avoiding polyspermy in the oyster (*Crassostrea gigas*). *Aquaculture* 73:295–307, 1988.

Tagatz, M. E., and Deans, C. H.: Comparison of field- and laboratory-developed esturarine benthic communities for toxicant-exposure studies. *Water Air Soil Pollut.*, 20: 199–209, 1983.

Tagatz, M. E., and Tobia, M.: Effect of barite ($BaSO_4$) on development of estuarine communities. *Estuarine Coastal Mar. Sci.*, 7:401–407, 1978.

Tagatz, M. E.; Ivey, J. M.; Moore, J. C.; and Tobia, M.: Effects of pentachlorophenol on the development of estuarine communities. *Toxicol. Environ. Health*, 3:501–506, 1977.

Tagatz, M. E.; Ivey, J. M.; Lehman, H. K.; and Oglesby, J. L.: Effects of a lignosulfonate-type drilling mud on development of experimental estuarine macrobenthic communities. *Northeast Gulf Sci.* 2(1):35–42, 1978a.

Tagatz, M. E.; Ivey, J. M.; and Tobia, M.: Effects of Dowicide® G-ST on development of experimental estuarine macrobenthic communities. Pentochlorophenol, edited by K. Ranga Rao, pp. 157–163. New York: Plenum, 1978b.

Tagatz, M. E.; Ivey, J. M.; and Oglesby, J. L.: Toxicity of drilling-mud biocides to developing estuarine macrobenthic communities. *Northeast Gulf Sci.* 3(2):88–95, 1979a.

Tagatz, M. E.; Lehman, H. K.; and Oglesby, J. L.: Effects of sevin on development of experimental estuarine communities. *Toxicol. Environ. Health*, 5:643–651, 1979b.

Tyler-Schroeder, D. B.: Static bioassay procedure using grass shrimp (*Palaemonetes* sp.) larvae. Bioassay Procedures for the Ocean Disposal Permit Program, pp. 73–82. EPA-600/9-78-010, 1978a.

Tyler-Schroeder, D. B.: Entire life-cycle toxicity test using grass shrimp (*Palaemonetes pugio* Holthuis). Bioassay Procedures for the Ocean Disposal Permit Program, pp. 83–88. EPA-600/9-78-010, 1978b.

Tyler-Schroeder, D. B.: Culture of the grass shrimp (*Palaemonetes pugio*) in the laboratory. Bioassay Procedures for the Ocean Disposal Permit Program, pp. 69–72. EPA-600/9-78-010, 1978c.

Tyler-Schroeder, D. B.: Use of the grass shrimp (*Palaemonetes pugio*) in a life-cycle toxicity test. Aquatic Toxicology, edited by L. L. Marking, R. A. Kimerle, pp. 153–170. ASTM STP 667. Philadelphia: American Society for Testing and Materials, 1979.

U.S. Environmental Protection Agency (U.S. EPA): Bioassay Procedures for the Ocean Disposal Permit Program. EPA-600/9-78-010, 1978.

U.S. EPA: Short-term methods for estimating the chronic toxicity of effluents and receiving waters to marine and estuarine organisms. EPA-600/4-87/028, 1988.

U.S. EPA/Corps of Engineers (COE): Ecological evaluation of proposed discharge of dredged material into ocean waters, 1978.

U.S. EPA: Standard evaluation procedure: acute toxicity test for estuarine and marine organisms (estuarine fish 96-hour acute toxicity). EPA-540/9-85-009. 1985a.

U.S. EPA: Standard evaluation procedure: acute toxicity test for estuarine and marine organisms (shrimp 96-hour acute toxicity test). EPA-540/9-85-010. Hazard Evaluation Division, 1985b.

U.S. EPA: Standard evaluation procedure: acute toxicity test for estuarine and marine organisms (mollusk 96-hour flow-through shell deposition study). EPA-540/9-85-011. Hazard Evaluation Division, 1985c.

U.S. EPA: Standard evaluation procedure: acute toxicity test for estuarine and marine organisms (mollusc 48-hour embryo larvae study). EPA-540/9-85-012. Hazard Evaluation Division, 1985d.

U.S. EPA: Standard evaluation procedure: fish early life-stage. EPA-540/9-86-138. Hazard Evaluation Division, 1986a.

U.S. EPA: Standard evaluation procedure: fish life-cycle toxicity tests. EPA-540/9-86-137. Hazard Evaluation Division, 1986b.

U.S. EPA: Standard evaluation procedure: non-target plants: growth and reproduction of aquatic plants—tiers 1 and 2. EPA-540/9-86-134. Hazard Evaluation Division, 1986c.

U.S. EPA: Methods for measuring the acute toxicity of effluents and receiving waters to freshwater and marine organisms, 4th ed., edited by CI Weber. EPA-600/4-90-027, 1991.

Ward, T. J.; Rider, E. D.; and Drozdowski, D. A.: A chronic toxicity test with the marine copepod *Acartia tonsa*, Aquatic Toxicology, edited by L. L. Marking, R. A. Kimerle, pp. 148–158. ASTM STP 667. Philadelphia: American Society for Testing and Materials, 1979.

Walsh, G. E.; Louie, M. K.; McLaughlin, L. L.; and Lores, E. M.: Lugworm (*Arenicola cristata*) larvae in toxicity tests: survival and development when exposed to organotins. *Environ. Toxicol. Chem.*, 5:749–754, 1986.

Wells, P. G.: Marine ecotoxicological tests with zooplankton. Ecotoxicological Testing for the Marine Environment, edited by G. Persoone, E. Jaspers, C. Claus, Vol. 1. State Univ. of Ghent and Institute for Marine Scientific Research, Bredene, Belgium, 1984.

Zillioux, E. J.: A continuous recirculating culture system for planktonic copepods. *Mar. Biol.*, 4(3):215–218, 1969.

Zillioux, E. J., and Lackie, N. F.: Advances in the continuous culture of planktonic copepods. *Helgolander Wiss Meeresunters* 20:325–332, 1970.

Zillioux, E. J., and Wilson, D. F.: Culture of a planktonic calanoid copepod through multiple generations. *Science* 151:996–998, 1966.

Chapter 4

ALGAE AND VASCULAR PLANT TESTS

M. A. Lewis

INTRODUCTION

Plant communities are important to the functioning of freshwater, estuarine, and marine ecosystems. Algae associated with the plankton (i.e., phytoplankton) and periphyton form the base of most food chains, produce oxygen, and are important in the cycling of nutrients. Vascular plants or macrophytes provide habitat and shelter for many forms of animal life in near-coastal and freshwater environments and are more important in photosynthetic activity and constitute a larger biomass in rivers than planktonic algae. Submerged and emergent macrophytes serve as food and shelter for aquatic life and wildlife and also are important in nutrient cycling and erosion control (Wolverton et al., 1975). The relationship of aquatic plants to fish and macroinvertebrates has been previously discussed (Peets et al., 1994; Kilgore et al., 1993).

Plants also play a role in controlling the physical and chemical quality of natural waters. They can affect light intensity, temperature, pH, hardness, and dissolved oxygen in the water column. Algae can mobilize metals from sediment, which may reduce water quality (Laube et al., 1979), but they are able to biodegrade some potential toxicants such as pesticides (O'Kelley and Deason, 1976; Boyle, 1984). Algae and vascular plants have been used for 35 years to remove nutrients, metals, coliforms, and other contaminants from wastewater (Lavoie and de la Noüe, 1985; Tripathi and Shukla, 1991).

Many aquatic plants have seasonal cycles (i.e., structure of community) that are influenced by biotic and abiotic factors which can be affected by changes in water quality due to the presence of nutrients and toxicants. Plant growth can be inhibited or stimulated by toxicants entering the ecosystem and both effects can have a detrimental impact. However, nuisance plant growths are more visible to the public and historically, this impact has received considerable attention. Nuisance algal blooms and dense stands of vascular plants can affect the recreational use of water bodies and the quality of drinking water. It has been estimated that over 12,000 lakes in 38 states and 61% of the drinking water in the United States was adversely affected by aquatic plants in the 1980s (AWWA, 1987; Welch and Lindell, 1992). In addition, dense blooms of several marine and freshwater algae produce toxins that can affect other algae, fish, wildlife, domestic animals, and humans (Palmer, 1977; Graneli et al., 1990). Excessive growths of algae can also cause corrosion in concrete and metal pipes, clog filters and screens, cause odor problems, and be parasitic to fish. The impact of excessive plant growths, management control techniques, and plant-non-plant interactions have been summarized in an excellent overview by the Aquatic Plant Management Society (1993).

Because of their ecological importance, freshwater and saltwater plants have been used for many years as biomonitors (see Chapter 1) of environmental change. Their use as indicator species (see Chapter 1) has been discussed by Palmer (1977), Whitton (1979), Haslam (1982), and Shubert (1984). The presence of some phytoplankton indicates clean water (e.g., *Melosira islandica*, *Cyclotella ocellata*) and others may indicate polluted conditions (e.g., *Microcystis aeruginosa*, *Aphanizomenon flos-aquae*). The latter two species have been associated with noxious blooms and anoxic conditions creating offensive tastes and odors. Because phytoplankton have short life cycles, they respond quickly to environmental change and their diversity and density can indicate the quality of their habitat. Floristic surveys are an integral part of most environmental monitoring programs and several reviews have summarized this use (Weitzel, 1979; Dennis and Isom, 1983; Shubert, 1984; Farmer, 1990). In addition, Palmer (1969, 1977) and Patrick (1973), among others, have described the pollution tolerance of algal species and the various techniques used to monitor their distribution.

Although the importance of plants has been recognized by ecologists, a similar attitude has not been as widespread among ecotoxicologists. In contrast to aquatic vertebrates and invertebrates (see Chapters 2, 3, 6, and 7), freshwater and saltwater plants have not been used routinely as test species in toxicity studies. Only recently has this changed, a positive trend attributable in part to environmental regulations (Table 1). For an in-depth discussion of some of these regulations, see Chapters 21, 22, 23, 24, and 25. More specifically, phytotoxicity data are considered or required in the United States for the calculation of national and site-specific water quality criteria (Stephan et al., 1985; Carlson et al., 1985), premanufacturer notices (PMNs) for new chemicals (U.S. EPA, 1985a) and the registration of pesticides (Holst and Ellwanger, 1982) and to assess the environmental impact of human and animal drugs (FDA, 1987), effluents (Weber et al., 1988, 1989), and oil spill bioremediation products (NETAC, 1991). In addition, algae and vascular plants are used increasingly to monitor the toxicities of contaminated sediments and hazardous waste leachates (Thomas et al., 1986; Giesy and Hoke, 1989; Cheung et al., 1993).

Phytotoxicity data are considered for chemical registrations and effluent evaluations in other countries as well as the United States (OECD, 1984a; EEC, 1987; Freemark et al., 1990; Vasseur et al., 1991). See Chapters 26 (OECD) and 27 (EEC).

Despite the applications mentioned, the importance of phytotoxicity data in the regulatory process has not been as significant as that of toxicity results for invertebrates and fish, which currently predominate the decision-making process. For example, of the 12,403 PMNs submitted for new chemicals under the Toxic Substances Control Act (TSCA) during the past 10 years, only 149 contained phytotoxicity data (Benenati, 1990) and only 7% of the research projects conducted by the U.S. Environmental Protection Agency (EPA) regions in 1989 contained algal data (Smith, 1991). Based on the aquatic toxicity information retrieval database (AQUIRE, 1987), a green alga was thirteenth on the list of most frequently used species. This limited consideration and unavailability of data will change in the future, however, as the scientific community gains more experience in conducting phytotoxicity tests and also realizes that freshwater and saltwater plants can be sensitive test species. The objective of this chapter is to describe the current use of freshwater plants and, to a lesser extent, saltwater species in ecotoxicology. The chapter includes discussions of the types of tests currently conducted, which are in order of decreasing frequency: algal population growth, duckweed, algal photosynthesis, rooted whole plant macrophytes, and macrophytic seeds and microcosms/mesocosms. Chapter 5 also describes the use of seaweeds (macroalgae) in saltwater toxicity tests. Because of the broad scope of this topic and limitations on the length of this chapter, only introductory information is provided. Additional and more specific details are included in the literature cited/supplemental reading and should be consulted before any phototoxicity test is conducted.

Table 1. Several federal regulations that require consideration of phytotoxicity data

- Clean Water Act
- Comprehensive Environmental Response, Compensation and Liability Act
- Federal Food, Drug and Cosmetic Act
- Federal Insecticide, Fungicide and Rodenticide Act
- Marine Protection, Research and Sanctuaries Act
- Resource Conservation and Recovery Act
- Toxic Substances Control Act

TERMINOLOGY

Algae. A group of plants that are single-celled, colonial, or multicelled containing chlorophyll and lack roots, stems, and leaves.

Algicidal concentration. A concentration of a chemical (e.g., pesticide) or effluent that is lethal to the plant population.

Algistatic concentration. A concentration of a chemical or effluent that completely inhibits growth but allows the test species to resume growth when resuspended in test medium not containing the test substance.

Axenic culture. A pure culture containing only one strain or species of algae and no other organisms.

Batch culture. A culture in which the supply of nutrients is not renewed and continually decreases.

Biomass. Weight of algal cells per volume of nutrient medium for a given time period.

Chlorosis. Decrease in pigment content as seen as yellowing of a plant.

Chemostat. An apparatus used for the continuous culture of algal populations.

Continuous culture. A culture in which the supply of nutrients is continuous and can be regulated so that the algal growth rate and population development can be controlled.

ECx. An estimated concentration of a chemical or effluent that reduces the measured parameter (e.g., growth) x percent relative to the control population.

Growth rate. Increase in biomass or cells/ml in a specified period of time.

Macrophytes. Large plants that can be seen without magnification.

Maximum standing crop. The biomass at test termination.

Necrosis. Death of the algal cell.

Periphyton. The community of microorganisms attached to or living on submerged surfaces.

Phytoplankton. Plant part of the plankton. Microscopic algae that may occur as unicellular, colonial, or filamentous forms.

Phytotoxicity. Toxicity of environmental pollutants determined on the growth and survival of plants.

Plankton. The community of microscopic animal and plants suspended or freely floating in the water column.

Primary productivity. Rate at which light energy is accumulated by the photosynthetic and chemosynthetic activity of plants in the form of organic material.

SC20. The test concentration that stimulates algal growth 20% above that of the control algal population.

Unialgal culture. A culture containing one species of algae and also possibly containing bacteria, protozoa, and fungi.

THE NEED FOR PHYTOTOXICITY DATA

The need for aquatic phytotoxicity data in the environmental risk assessment process has been debated for years. Toxicity results for invertebrates and fish are considered by many in the regulatory and scientific communities to be representative of those for algae and vascular plants. This perception of general insensitivity has been based on several published reports, such as that of Kenaga and Moolenar (1979). Several aquatic vascular plants and a green alga, *Chlorella* sp., were less sensitive than animal species in this report to many chemical compounds. Morgan (1972) also reported a similar trend. In contrast, algae have been found more sensitive than animals to several heavy metals, pulp mill and industrial effluents, pesticides, cationic surfactants, soil elutriates, and hazardous waste leachates (Table 2) as well as to other chemicals (Walsh, 1982; Hartman and Martin, 1984, 1985; LeBlanc, 1984; Winner et al., 1990). *Selenastrum capricornutum*, a freshwater green alga, was found more sensitive than daphnids and fish in 78 of the 155 PMNs submitted during the past 10 yr that contained aquatic phytotoxicity data (Benenati, 1990). Algal sensitivities in an additional 15 PMNs were comparable to those for animal test species. The pattern of sensitivity observed for freshwater microalgae is also true for the macroalgae or seaweeds, which have been reported to be more sensitive than marine animals in several cases (Thursby et al., 1993; see Chapter 5).

Overall, the relative sensitivities of plants and animals have been found to be species and chemical specific and, therefore, unpredictable. The concept of a universally sensitive group of test species, plants or animals, is not realistic and the assumption that animal toxicity data can serve as a surrogate for freshwater and saltwater plants is not technically valid without supporting data. Consequently, phytotoxicity tests are necessary to evaluate the impact of potential contaminants that may enter the environment. Microalgae have been used more frequently for this purpose and their use is discussed first.

TEST METHODS: ALGAE

History

Algae were used as early as 1910 in toxicity tests (Allen and Nelson, 1910), primarily to evaluate the effectiveness of algicides (e.g., Palmer and Maloney, 1955). The first consensus algal toxicity test method in the United States was developed and evaluated in the mid-1960s for freshwater algae and published as the Algal Assay Procedure Bottle Test (AAP) (U.S. EPA, 1971a). A similar method for marine algae was developed shortly thereafter by representatives of industry, academia, and government (U.S. EPA, 1974). The focus of the first early standard methods was to determine the limiting algal nutrients, algal growth potential, and algal productivity of natural waters. They were not specifically designed for monitoring the toxic effects of chemicals, which limited their routine use for this purpose. A modified version of the AAP method (Miller et al., 1978), however, has served as the basis for many of the current test methods that have been published by international societies, regulatory agencies, and standard-writing organizations (Table 3).

Test Method Description

Most algal toxicity tests are chronic tests because effects are assessed over several generations during the usual 3- to 4-d exposure period. Freshwater green algae have a rapid growth rate under the standard test conditions. For example, the density of *S. captricornutum* is expected to increase from 1×10^4 cells/ml to 3.5×10^6 cells/ml after 96 h (U.S. EPA, 1985a). Growth rates of $1.5-1.9\ d^{-1}$ can be attained for other species (ISO, 1987), and with constant exponential growth, algal biomass theoretically can increase 37-fold after 48 h and as much as 1339 times after 96 h (Nyholm and Källqvist, 1989).

Most of the tests listed in Table 3 are similar in technique. Static tests are used to monitor the effects of a toxicant on an algal population growing exponentially for 72 to 96 h in a nutrient-enriched medium. A control and five test substance concentrations are used in at least triplicate on a variable-speed rotary or oscillatory shaker at usually 100 (±10%) oscillations per minute. The test flasks can be hand-shaken once or twice daily if a shaker is not available. The tests are conducted under conditions of controlled temperature, light, and initial pH. Biomass is determined or estimated several times during the test to provide information on the effects of the test substance on algal growth. Several methods in Table 3 can be used with municipal and industrial effluents. These tests vary in technique from those with commercial chemicals and are discussed later in greater detail.

Toxicity tests conducted with marine microalgae and chemicals are similar to those with freshwater microalgae; static tests are conducted with rapidly growing and easily cultured species. Several methods have been reported for marine algae in addition to those in Table 3 and include those of ISO (1988), Anderson and Hunt (1988), and Blanck and Bjornsater (1989). Effluent toxicity tests with marine micro- and microalgae have been described. The salinity of the effluent needs adjustment in these tests because most effluents discharged into marine and

Table 2. Comparative sensitivities of plant and animal test species reported in selected studies

Test substance	Test species	Most sensitive	Reference
Surfactant Sodium salts	Alga Snail Fish	Alga	Patrick et al. (1968)
PCB, DDT	Algae Protozoan Daphnid Ostracod Fish	Daphnid	Morgan (1972)
27,781 chemicals	Algae Macrophytes Daphnid Fish	Substance specific	Kenaga and Moolenar (1979)
Cationic surfactant	Alga Fish	Alga	Shehata and Nawar (1979)
Textile effluents	Algae Crustaceans Fish	Algae	Walsh et al. (1980)
Organic and inorganic compounds	Alga Bacteria Protozoa	Substance specific	Bringmann and Kühn (1980)
Surfactants Copper Diquat	Duckweed Daphnid Fish	Substance specific	Bishop and Perry (1981)
Sodium lauryl sulfate Cadmium Pesticide	Marine algae Marine invertebrates Marine copepods Marine fish	Substance specific	Roberts et al. (1982)
Industrial effluents	Freshwater and Marine algae Invertebrates	Algae	Walsh et al. (1982)
Fluorene	Algae Invertebrates	Fish and invertebrates	Finger et al. (1985)
Herbicide	Alga Daphnid Fish	Alga	Meyerhoff et al. (1985)
Copper Acridine	Algae Daphnids Fish	Substance specific	Blaylock et al. (1985)
Zinc Copper 2,4-D Soil elutriates Chemical waste site water	Alga Daphnid Earthworms Microtox	Alga	Thomas et al. (1986)
Surfactant Atrazine PCP 4-Nitrophenol	Algae Macrophytes Invertebrate	Substance specific	Huber et al. (1987)
Atrazine	Algae Duckweed	Algae	Hughes et al. (1988)
13 chlorophenols	Algae Fish	Substance specific	Shigeoka et al. (1988)
Phenol	Insects Crustaceans Worms Bivalves Fish Algae	Insects	Walker (1988)
Chromium(VI)	Marine algae Invertebrates Fish	Algae	Jop (1989)

Table 2. Comparative sensitivities of plant and animal test species reported in selected studies (*Continued*)

Test substance	Test species	Most sensitive	Reference
Ammonia and paper mill effluent	Marine algae Shrimp Fish	Algae	Schimmel et al. (1989)
Copper sulfate Sodium dodecyl sulfate	Marine alga Marine invertebrates Marine fish	Alga	Morrison et al. (1989)
Refinery and herbicide manufacturing plant effluents	Duckweed Daphnid Fish	Duckweed	Taraldsen and Norberg-King (1990)
Chlorinated benzenes and biphenyls	Fish Macrophyte	Substance specific	Gobas et al. (1991)
River water extracts	Alga Daphnid	Alga	Galassi et al. (1992)

estuarine environments are fresh water. The recommended test method using a marine macroalga (Weber et al., 1988) differs from that with microalgae. The macroalga test consists of exposing male and female plants to the effluent for 2 d, followed by a 5- to 7-d recovery period in control medium to determine cystocarp maturation. The maximum concentration in this test is 50% effluent.

A comparative review of the experimental conditions in the more widely used algal test methods appears in Table 4 and several of the more important experimental factors are discussed briefly below. Additional reviews have been presented for freshwater algae (Nyholm and Källqvist, 1989) and marine algae (Walsh, 1982, 1988; Fletcher, 1991; Thursby et al., 1993; and Chapter 5).

Table 3. Several published phytotoxicity test methods

Method	Algae		Reference
	Freshwater	Saltwater	
Algal assay procedure bottle test	•		U.S. EPA (1971a)
Marine algal assay procedure bottle test		•	U.S. EPA (1974)
A method for measuring algal toxicity and its application to safety assessment of new chemicals	•	•	Payne and Hall (1979)
Marine algal bioassay method with pesticides and industrial wastes		•	Walsh and Alexander (1980)
Alga, growth inhibition test	•		OECD (1984)[a]
Algal acute toxicity test	•	•	U.S. EPA (1985a)
Water quality–algal growth inhibition test	•		ISO (1987)[b]
Algal assay	•		FDA (1987)[c]
Short-term methods for estimating chronic toxicity of effluents and receiving waters to marine and estuarine organisms		•	Weber et al. (1988)
Short-term methods for estimating chronic toxicity of effluents and receiving waters to freshwater organisms	•		Weber et al. (1989)
Toxicity testing with phytoplankton	•	•	APHA et al. (1989a)[d]
Standard guide for conducting static 96-h toxicity tests with microalgae	•	•	ASTM (1990a)[e]
Standard practice for conducting static chronic 96-h toxicity tests on hazardous chemical wastes using freshwater alga, *Selenastrum capricornutum*		•	ASTM (1990c)
Oil spill bioremediation products testing protocol methods manual	•		NETAC (1991)

[a]OECD, Organization for Economic Cooperation and Development.
[b]ISO, International Organization for Standardization.
[c]FDA, Food and Drug Administration.
[d]APHA et al., American Public Health Association, American Water Works Association, Water Pollution Control Federation.
[e]ASTM, American Society for Testing and Materials.
The methods by Walsh and Alexander (1980), Weber et al. (1988, 1989), and ASTM (1990a) are suitable for effluents.

Table 4. Comparison of selected variables recommended in several standard toxicity test procedures for micro- and macroalgae

Experimental variable	ASTM (1990a)	Weber et al. (1989)	Weber et al. (1988)[a]	ISO (1987)	U.S. EPA (1985a)	OECD (1985)	Holst and Ellwanger (1982)
Initial cell density (cells/ml)	2×10^4; 5×10^4	1×10^4	5 female branches and 1 male branch	1×10^4	1×10^4; 7.7×10^4	1×10^4	NS[b]
Test duration	4 d	4 d	2-d exposure, 5–7-d recovery	3 d	4 d	3 d	5 d
pH	7.5–8.0	7.5 (± 0.1)	NS	8.3 (± 0.2)	7.5 (± 0.2); 8.0 (± 0.1)	8.0	7.5–8.5
Illumination ($\mu E\ m^{-2}s^{-1}$)	30–90 (±15)	86 (± 8.6)	100	120 (± 20%)	300 (± 25)	120 (± 20)	40–85
Temperature (°C)	24 (± 2); 20 (± 2)	25 (± 1)	23 (± 1)	23 (± 2)	20 ± 2; 24 (± 1)	21–25 (± 2)	24 (± 2)
Calculations	IC50, NOEC	LOEC, NOEC	LOEC, NOEC	EC50, NOEC	EC0, EC10, EC50, EC90	EC50, NOEC	EC10, EC50, EC90
Number of replicates	3	3	3–4	3; 6 in controls	3	3; 6 in controls	3
Test acceptability	>1 × 10^5 cells/ml after 96 h in controls	2 × 10^5 cells/ml in control: > 20% replicate variability in controls	Control plants should average 10 or more cystocarps	Cell density in controls should increase 16× after 72 h	3.5 × 10^6 cells/ml after 96 h in control for freshwater species, 1.5 × 10^6 cells/ml for marine species	Cell density in controls should increase 16 × after 72 h	NS

[a]Test is for a marine macroalga.
[b]NS, not stated.

Test and Culture Media

Algae are cultured and exposed to the test substance in a nutrient-enriched medium. A variety of culture and test media have been described in the published test methods. Millington et al. (1988), among others, have compared the ingredients in several media for freshwater algae. Freshwater growth medium is a mixture of micronutrients, macronutrients and distilled, filtered, or deionized water. Various amounts of the nutrients are added to the dilution water, which is then pH adjusted between 7.5 and 8.5. The medium is then filter sterilized (0.22 μm) or autoclaved before use. It is recommended in tests using effluents that the nutrient stock solutions be added directly to the effluent, which is then diluted with medium prepared with the dilution water. This compensates for the possibility that the effluent dilutions may be nutrient poor and have an inhibitory effect independent of that due to the potential toxicant. Marine algal growth media can be natural seawater or synthetic seawater to which salts, vitamins, and trace metals are added. Numerous descriptions and references for suitable marine and freshwater algal media and culture techniques can be found in Guillard and Ryther (1962), Provasoli (1968), Droop (1969), Stein (1973), Ukeles (1976), Harrison et al. (1980), and Walsh (1988).

The composition of the nutrient medium is not representative of that found in natural waters. This affects the environmental realism of most phytotoxicity results. In addition, the composition of the nutrient medium has been found to affect toxicity significantly in some cases (Smith et al., 1987; Walsh et al., 1987a; Millington et al., 1988; Fernandez-Piñas et al., 1991) but not in others (Vasseur and Pandard, 1988). The effect of one constituent, EDTA, has been investigated more than that of most others. The low concentration of EDTA that is needed to complex trace metal nutrients in the culture medium can reduce the toxicity of metal test substances if used in the test medium. In contrast, if chelators are omitted, algal growth can be reduced significantly during the test and log growth may not be attained or delayed. EDTA is not recommended for use in the test media described in U.S. EPA methods for effluents (Weber et al. 1988, 1989), chemicals (U.S. EPA, 1985a), and pesticides (Holst and Ellwanger, 1982) but it is recommended in others (OECD, 1984a; ISO, 1987; ASTM, 1990a). Greene et al. (1991) provides more information on the relevance of EDTA in phytotoxicity tests and additional examples of its impact have been reported by Walsh and Alexander (1980) and Wren and McCarroll (1990). In addition to EDTA, the nitrogen and phosphorus content in the media can have a significant impact on toxicity (Hall et al., 1989a).

Inoculum—Initial Cell Density

An algal toxicity test begins when the test medium is inoculated with exponentially growing algae obtained from a laboratory culture. Most test methods describe the batch culture techniques needed to provide the algae required for testing. Chemostat cultures are not used frequently but they have been recommended (Hall et al., 1989b; Peterson, 1991). A minimum inoculum level is recommended in most standard test methods to reduce the possibility of a lag phase and to increase the likelihood of achieving log growth during the usual 3- to 4-d test. This is important because one of the criteria for a "successful" toxicity test in many methods is that the control population must attain a certain density at test termination (Table 4). The inoculum is usually made within 3 to 4 h of the test initiation using, for example, a green alga collected from a 7- to 10-d-old culture in the exponential growth phase. Longer culture periods may be necessary to provide sufficient inocula for blue-green algae and diatoms. The inocula should contain enough algae to result in the usually recommended initial densities of 1×10^4 to 2×10^4 cells/ml. The correct amount of inoculum can be determined using guidance provided in most published methods.

Temperature, Light, and pH

Temperature and light need to be controlled during algal toxicity tests because they can significantly affect growth. This has been shown by a variety of reports, more recently by Gaur and Singh (1991). The recommended temperatures are between 20 and 25°C and are species specific (Table 4). For example, tests with freshwater green and blue-green algae are conducted at 24 (±2°)C and those with marine species and diatoms at 20 (±2°)C (ASTM, 1990a). The recommended light intensity is usually between 80 to 120 $\mu E\ m^{-2}\ s^{-1}$ and is less for freshwater blue-green algae such as *Microcystis aeruginosa* and marine algae than for freshwater green species. The toxicity tests are conducted under constant light for freshwater green species but 14-h light and 10-h dark periods are recommended for several marine species.

The initial pH is an additional factor that needs control and guidelines are provided in the published methods. Even with control, the pH may shift by two units (Nyholm and Källqvist, 1989) during a test, which can alter the toxicity of some test substances such as metals (Skowronski et al., 1991).

Test Duration

Laboratory algal toxicity tests have ranged in duration from less than 1 h to 3 wk but the typical exposure period in the traditional population growth studies is 3 to 4 d. Test periods less than these are preferred by some to lessen the chance of any reduction in the test concentrations due to adsorption and biodegradation. Tests longer than 4 d are used to determine the toxic effects of degradation products, stimulatory effects, and algistatic concentrations. Examples of these longer tests (range = 5 to

21 d) have been described by Payne and Hall (1979) and the Food and Drug Administration (FDA, 1987).

Test Species

Several recommended test species are listed in Table 5 and appear in Figure 1. Many of these were chosen because of their availability and ease of culture. Ecological relevance and sensitivity to toxicants were factors also considered but to a lesser degree. Of the species in Table 5, the freshwater green alga *S. capricornutum* and the marine diatom *Skeletonema costatum* are the most frequently used. *S. capricornutum* was the test species of choice in 131 of 149 PMNs containing phytotoxicity data (Benenati, 1990). The predominance of *Selenastrum* and, to a lesser extent, *Skeletonema* as test species can be attributed in part to their use in the AAP bottle tests (U.S. EPA, 1971a,b, 1974). A bibliography describing the use of *Selenastrum* in toxicity tests, although dated, is available (Leischman et al., 1979). The OECD (1984a) and ISO (1987) methods recommend the use of the green algae *Scenedesmus subspicatus* and *S. capricornutum*. Freshwater blue-green and diatom species are used infrequently in toxicity tests because of the difficulty in maintaining long-term viable cultures, lack of commercial sources, and difficulty in obtaining sufficient growth during the usual 3- to 4-day test. However, these species need to be used because of the reported interspecific variation in response of algae to toxicants.

Table 5. Marine and freshwater microalgae recommended for use in various published toxicity test methods

Species	Method
Green algae	
Freshwater	
Selenastrum capricornutum Printz	ASTM (1990a), APHA et al. (1989a), FDA (1987), ISO (1987), Weber et al. (1989), U.S. EPA (1985a), OECD (1984), Holst and Ellwanger (1982), U.S. EPA (1971a)
Scenedesmus subspicatus Chodat	OECD (1984a), ISO (1987), ASTM (1990a)
Scenedesmus quadricauda (Turp.) Breb.	U.S. EPA (1985a)
Chlorella vulgaris (Beij.	U.S. EPA (1985a) ASTM (1990a), OECD (1984a), U.S. EPA (1985a)
Marine	
Dunaliella tertiolecta Butcher	ASTM (1990a), APHA et al (1989a), U.S. EPA (1974)
Blue-green algae	
Freshwater	
Microcystis aeruginosa Kutzing	ASTM (1990a), APHA et al. (1989a), FDA (1987), U.S. EPA (1971a)
Anabaena flos-aquae (Lyng) Breb	ASTM (1990a), APHA et al. (1989a), Holst and Ellwanger (1982), U.S. EPA (1971a)
Diatoms	
Freshwater	
Cyclotella spp.	APHA et al. (1989a)
Navicula pelliculosa Grun.	ASTM (1990a)
Nitzschia sp.	APHA et al. (1989a)
Synedra sp.	APHA et al. (1989a)
Marine	
Skeletonema costatum (Grev.) Clevel.	ASTM (1990a), APHA et al. (1989a), U.S. EPA (1985a), Holst and Ellwanger (1982), U.S. EPA (1974)
Thalassiosira pseudonana Halse and Heimdal	ASTM (1990a), APHA et al. (1989a), U.S. EPA (1985a), U.S. EPA (1974)
Red algae	
Marine	
Champia parvula (C. Agardh) Harvey	Weber et al. (1988)

Several of the recommended freshwater test species are widely distributed in nature. Greeson (1982) reports that 500 genera of freshwater algae have been identified in the United States and, of these, 58 genera were considered dominant. These included *Cyclotella*, *Navicula*, *Scenedesmus*, and *Anabaena*. *Selenastrum*, although not considered dominant, was considered "commonly occurring."

Walsh and Alexander (1980), Missimer et al. (1989), and Cowgill et al. (1989), among others, have evaluated the use of *S. costatum*. It is considered a sensitive test species (Walsh, 1988) and it is more sensitive than *S. capricornutum* to several chemicals (LeBlanc, 1984). Toxicity tests with marine macroalgae have been conducted less frequently than those with microalgae. *Champia parvula*, a red macroalga, has been recommended for determining the toxicity of effluents discharged to marine environments (Weber et al., 1988) but it is not frequently used. *C. parvula* occurs commonly in temperate and tropical coastal waters worldwide and its culture requirements have been described (Thursby and Steele, 1984; Steele and Thursby, 1988). A variety of marine macroalgae have been used other than the "standard" species and several of these are given in Table 6. For example, kelp, such as *Macrocystis pyrifera*, has been used in effluent testing. Reproduction and spore germination have been the primary end points in these tests, which are discussed in more detail by Fletcher (1991) and Thursby et al. (1993) and also in Chapter 5.

Additional information on culture availability and the types of freshwater and marine algae used in toxicity tests can be obtained from the reviews of Swanson et al. (1991) and Thursby et al. (1993). Swanson et al. (1990) list the limitations of a variety of test species based on their availability and culture requirements. Cultures of algae are available from several commercial sources, such as the following:

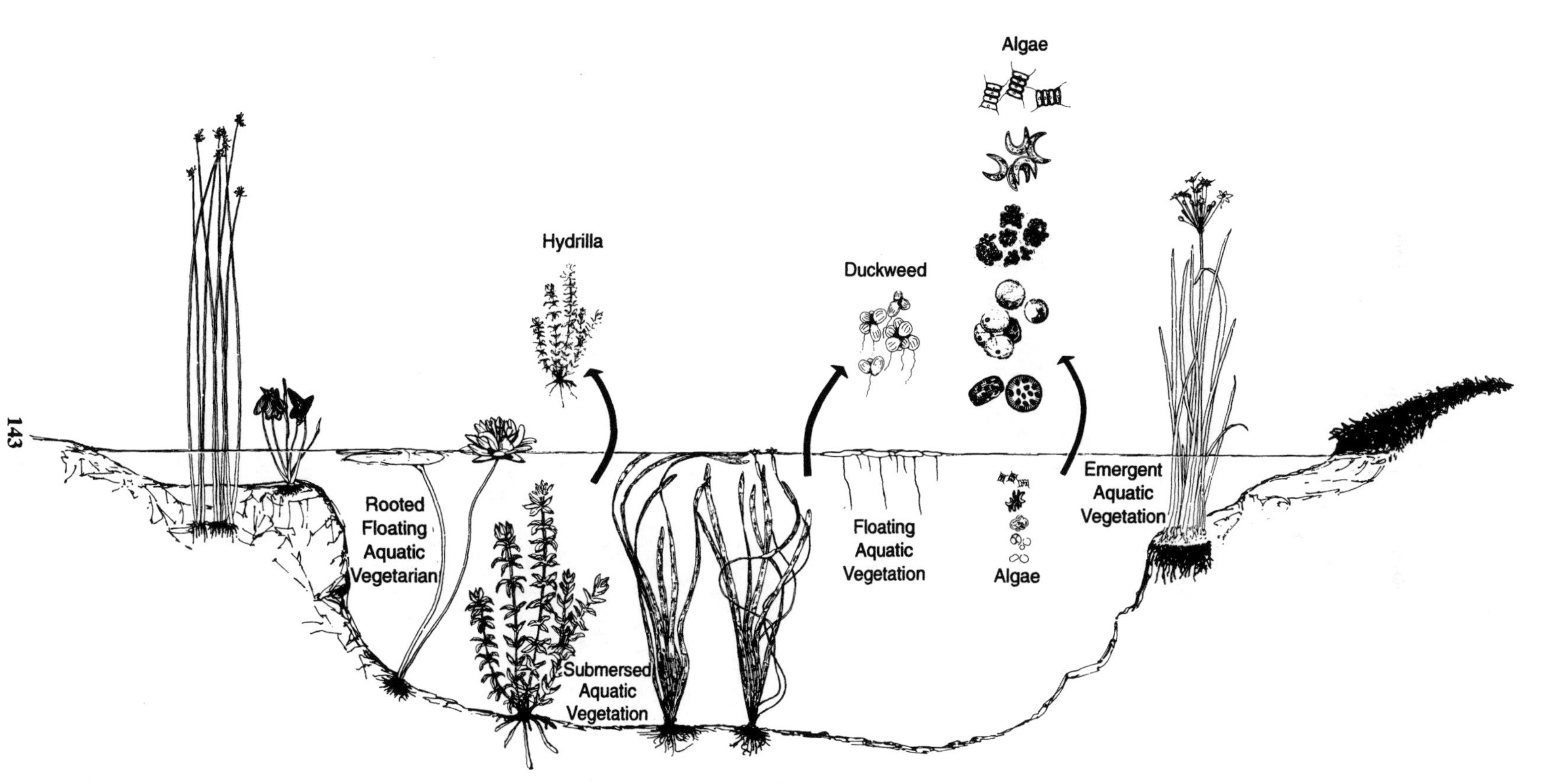

Figure 1. Types of aquatic plants and representative test species used in freshwater phytotoxicity tests.

Table 6. Several species of marine macroalgae used occasionally in toxicity tests

Chlorophyta
Enteromorpha compressa
Enteromorpha intestinalis
Enteromorpha prolifera
Ulva lactuca
Urospora penicilliformis
Fucophyceae
Ascophyllum nodosum
Cytoseua osmundacea
Eisenia arborea
Ectocarpus siliculosus
Fucus edentatus
Fucus serratus
Fucus spiralis
Laminaria dentigera
Laminaria ephemera
Laminaria farlowii
Laminaria saccharina
Macrocystis pyrifera
Sargassum agardhianum
Saragassum muticum
Rhodophyta
Antithamnion plumula
Callithamnion hookeri
Callithamnion tetricum
Ceramium flabelligerum
Ceramium pedicellatum
Ceramium rubrum
Champia parvula
Plumaria elegans
Polysiphonia brodiaei
Polysiphonia nigrescens
Polysiphonia lanosa
Spermothamnion repens

Source: Adapted from Fletcher RL: Marine microalgae as bioassay test organisms, Ecotoxicology and the Marine Environment, edited by PD Abel, V Axiak, 111–131. New York: Ellis Horwood, 1991.

- American Type Culture Collection, 12301 Parklawn Drive, Rockville, MD 10852.
- Culture Collection of Algae, Botany Department, University of Texas, Austin, TX 78712.

Test Species Sensitivity

A key consideration in algal toxicity tests is the choice of the test species, as this decision can influence the test results (Bringmann and Kühn, 1978; Wängberg and Blanck, 1988; Swanson et al., 1990). Most standard test methods have been developed for freshwater green algae; other species are listed in these methods but they are seldom used. The importance of these "other" species cannot be overstated, because the interspecific variation in response of algae to chemicals can vary several orders of magnitude (Table 7). For example, the difference in EC100 values for 13 algal species and disodium hydrogen arsenate was 200-fold (Blanck et al., 1984). The variation is not always this great, but one to two orders of magnitude difference are not uncommon. In addition to interspecific variation, the tolerances of different algal clones and geographic races can differ (Fisher et al., 1973; Blanck et al., 1984; Millie and Hersh, 1987; Hersh and Crumpton, 1989) as well as the sensitivity of one species in the presence of another (Walsh and Alexander, 1980).

A consistently sensitive algal species has not been identified. *S. capricornutum*, the most common test species, has been reported to be relatively sensitive to a wide range of substances (Palmer, 1969; Blanck et al., 1984) including effluents and dyes (U.S. EPA, 1985a) but not to anionic surfactants (Lewis and Hamm, 1986), streptomycin (Harrass et al., 1985), and several insecticides (Aronson, 1973). This inconsistent trend suggests that several algae representing different taxonomic groups (e.g., green, blue-green, and diatom species) should be used to screen the toxicities of potential toxicants. This "battery" approach has been recommended for toxicity tests with pesticides (Holst and Ellwanger, 1982), and 56 freshwater and 7 marine candidate species have been evaluated for this use (Swanson et al., 1990).

Effect Parameters and End Points

Algae exhibit several responses to toxicants, including growth inhibition and stimulation (Figure 2) and morphological and physiological changes. The effect most frequently reported is growth inhibition based on changes in biomass. Biomass can be determined by direct and indirect methods, which have been described in detail (Brezonik et al., 1975; APHA et al., 1989b). Whole-cell measurements include dry weight calculations and cell counts. Dry weight is an accurate direct measurement but it is very time consuming and is not frequently used. Several microscopic techniques are available to count cells, and if they are used morphological changes in cell shape, color, size, and unusual algal aggregations can also be observed. Algal viability can be differentiated using mortal stains such as Evans Blue (U.S. EPA, 1985a) or using a test method specifically designed to determine algicidal effects (Payne and Hall, 1979). When many samples are to be analyzed, an automatic particle counter is often used, such as a Coulter counter. The use of a particle counter is rapid and relatively sensitive but it has limitations that should be considered (Rehnberg et al., 1982). For example, counters should not be used for colonial algae and sonification is needed for filamentous forms such as *A. flos-aquae*. In addition, effluents may contain a high concentration of particles, which can interfere with the accuracy of the results, and prior filtration may be required.

Biomass can also be determined by monitoring cellular constituents such as chlorophyll *a*. Chlorophyll *a* can be estimated by several spectrophotometric and fluorometric techniques (ASTM, 1979; APHA et al., 1989b). Measurements based on fluorescence are considered rapid and sensitive (U.S. EPA 1976). Other techniques utilized less frequently

Table 7. Examples of interspecific variation in response of several algae exposed to the same test substance[a]

Test substance	Test species	Toxicity difference	Reference
Copper sulfate	*Microcystis aeruginosa* *Chlorella pyrenoidosa*	20	Fitzgerald and Faust (1963)
Atrazine Neburon Ametryne Diuron	18 marine species	7–17	Hollister and Walsh (1973)
Sodium dichromate[b]	*Microcystis aeruginosa* *Scenedesmus quadricauda*	300	Bringmann and Kühn (1978)
Aniline		52	
Lindane		6	
Atrazine		10	
Phenol		2	
Pyridine		4	
Copper	*Chlorella pyrenoidosa* *Scenedesmus quadricauda* *Selenastrum capricornutum* *Chlamydomonas sp.*	80–150×	U.S. EPA (1980)
10 industrial wastes	*Selenastrum capricornutum* *Skeletonema costatum*	1200×	Walsh et al. (1982)
Sodium lauryl sulfate	*Pseudoisochrysis paradoxa* *Skeletonema costatum*	<2	Roberts et al. (1982)
Cadmium	*Prorocentrum minimum*	57	
pesticide		<2	
Glyphosate[c]	13 freshwater species	8	Blanck et al. (1984)
Copper sulfate		6	
Hydroquinone		258	
Aliquat		4	
Tributyltin chloride		32	
Sodium lauryl sulfate		129	
Textile wastewater		>8	
Streptomycin	15 freshwater species	733	Harrass et al. (1985)
Atrazine	8 freshwater species	17	Larsen et al. (1986)
Anionic surfactants	*Selenastrum capricornutum*	32–83	Lewis and Hamm (1986)
Nonionic surfactants	*Microcystis aeruginosa*	7–35	
Cationic surfactants	*Navicula pelliculosa*	1–3	
LAS	*Scendesmus communis*	20	Huber et al. (1987)
PCP	*Oocystis lacustris*	0	
4-Nitrophenol		<2	
Chlordane	*Skeletonema costatum* *Isochrysis galbana* *Dunaliella tertiolecta* *Porphyridium cruentum*	1–10	Walsh et al. (1987b)
Organotin compounds	*Skeletonema costatum* *Thalassiosira pseudonana*	1–5	
13 chlorophenols	*Selenastrum capricornutum* *Chlorella vulgaris*	1–25	Shigeoka et al. (1988)
Chromium(VI)	10 estuarine species	50	Riedel (1989)
Landfill Leachates	*Chlorella pyrenoidosa* *Chlorella vulgaris* *Dunaliella tertiolecta* *Scenedesmus* sp.	1–5	Cheung et al. (1993)

[a]Values represent difference in effect concentrations, usually the EC50 values, determined for the different test species. Additional and more specific information is available in the corresponding references.
[b]Several of the 180 substances tested.
[c]Several of the 18 substances tested.

to estimate biomass utilize DNA and ATP. The microplate technique described by Blaise et al. (1986) is based on ATP measurements.

Whether phytotoxic effects should be based on changes in the growth rate or standing crop has been discussed. The relative merits of these effect parameters have been discussed in several of the published test methods and by Payne and Hall (1979), Nyholm (1985, 1990), and Nyholm and Källqvist (1989). Although this issue is not resolved, the use

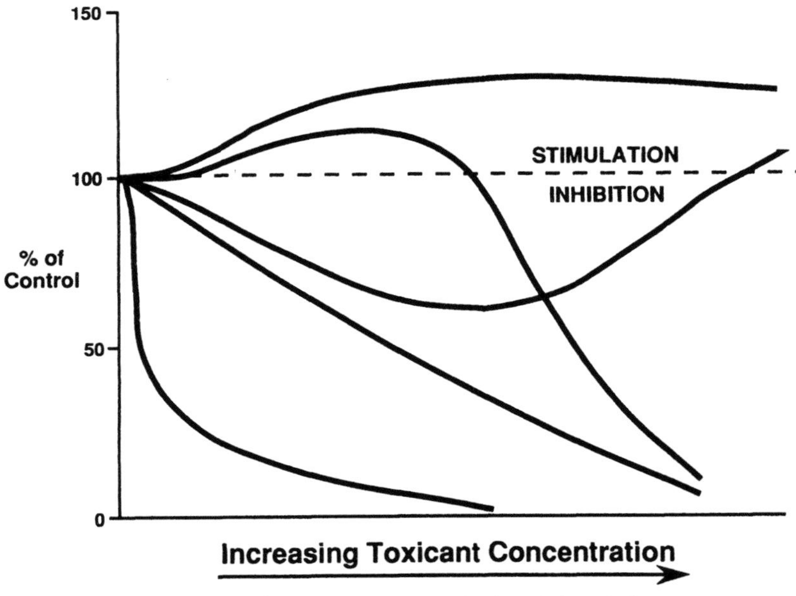

Figure 2. Several representative response patterns of cultured algae during exposure to toxicants in laboratory toxicity tests.

of either parameter is acceptable, realizing that the variation in results of algal toxicity tests is probably affected more by the composition of the nutrient medium and choice of the test species.

The area under the growth curve is a calculation recommended for consideration by the OECD (1984a), ISO (1987), and American Society for Testing and Materials (ASTM, 1990a) methods. Although the area under the growth curve was shown to be a sensitive parameter for several organic chemicals (Adams et al., 1985), it is not commonly used in the United States.

Algae grow rapidly and can therefore recover quickly if not killed by a short-term exposure during transitory episodic events, i.e., seasonal herbicide applications. Therefore, algistatic and algicidal effects may be more environmentally meaningful in some cases than the usually reported calculations such as the EC50 value which represents a partial reduction in growth rate and standing crop. Payne and Hall (1979) recommended the use of these parameters and stated that reductions such as 50% in the maximum standing crop are not meaningful because in a laboratory culture growth remains logarithmic and the results are time dependent. Furthermore, they stated that reductions in growth rate do not correspondingly reduce the final standing crop, and its calculation requires frequent analysis of biomass. Hughes et al. (1988) also concluded that algistatic and algicidal effects are the more environmentally realistic responses in algal toxicity tests. Nevertheless, these effect levels are not usually reported in the literature and in regulatory data submissions.

Several end points have been used to express the results of algal toxicity tests (Figure 3). Most published methods recommend reporting the EC50 and NOEC values (Table 4). The EC50 value is more commonly reported; it was used in 127 of the 149 PMNs containing algal toxicity data (Benenati, 1990). The test methods for effluents described by Weber et al. (1988, 1989) recommend reporting the NOEC and the LOEC.

A variety of statistical techniques have been used to analyze algal toxicity data. Probit analysis, the binomial method, the moving-average method, and regression analysis have been used to determine the EC50 value. Parametric (Dunnett's procedure) and nonparametric methods (Steel's many-one rank test) have been used for the determination of the NOEC and LOEC. The algistatic concentration can be determined using linear regression analysis and "inverse estimation" (Payne and Hall, 1979). Bruce and Versteeg (1992) described a nonlinear regression estimation procedure and presented several examples of its use with algal toxicity data. More information on appropriate statistical methods can be obtained from Kooijman et al. (1983), Weber et al. (1988, 1989), and Chapters 2 and 10.

Graphical interpolation is a technique that can be used to determine the degree of inhibition and stimulation at any test concentration. Walsh et al. (1978b) compared the EC50 values determined by graphical interpolation and the moving-average, probit, and binomial methods for 21 pesticides, 18 tin compounds, and 5 species of marine unicellular algae. The EC50 values were very similar regardless of the statistical technique and it was concluded that graphical interpolation was a suitable method although no confidence intervals could be reported.

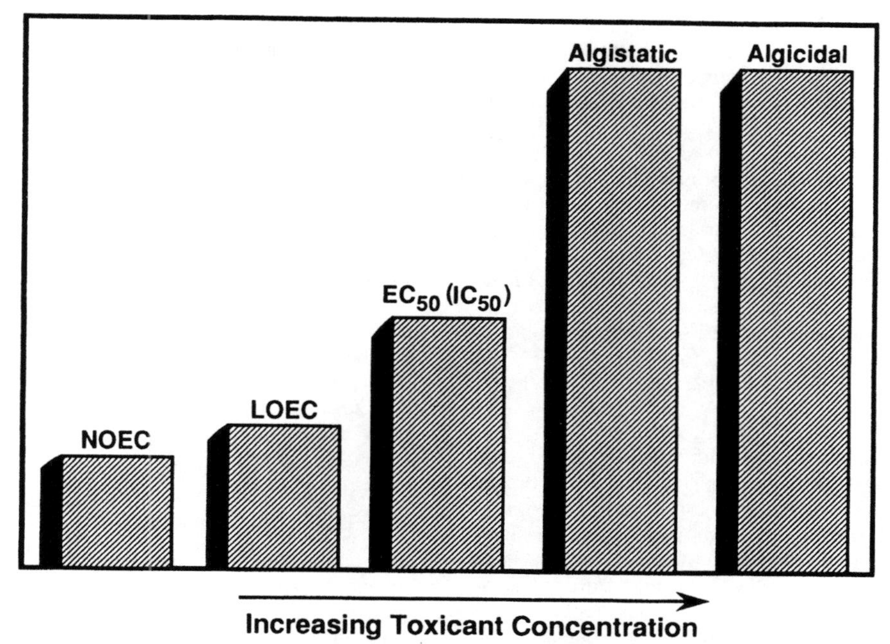

Figure 3. Relationship of various end points reported from phytotoxicity tests. See terminology section for term description.

Growth inhibition is the usual response of algae exposed to chemicals, but stimulation also occurs, particularly in tests with effluents (Claesson, 1984). Stimulation is considered an important effect in several of the published test methods but few offer guidelines for its calculation and interpretation. The test method of Weber et al. (1989) describes the calculation of the SC20 concentration, which was first derived by Walsh et al. (1980). The *SC20* value represents the concentration that increases growth 20% above that of the control population. It can be determined by graphical interpolation and by the moving-average method. This calculation has not been widely used and a consensus on the degree of stimulation considered biologically relevant in laboratory phytotoxicity tests has not been attained. Considering the ecological consequence of excessive plant growth as a result of stimulation, clarification of its significance in laboratory phytotoxicity tests is needed.

Physiological changes in algal cells have been monitored in response to chemical exposure but typically not using the traditional test methods. Changes in cellular composition, pigment content, and amino acid composition have been observed after exposure to surfactants, metals, DDT, and effluent (Goulding et al., 1984, Nyberg, 1985; Lustigman, 1986; Shaw et al., 1989; Maita and Kawaguchi, 1989). The relationship of these effects to those on population growth and community structure of natural algal assemblages has not been established. Consequently, physiological and morphological effect concentrations have not played a significant role in phytotoxicity evaluations.

Other Experimental Considerations

The precision of algal toxicity tests conducted either in the same laboratory or at different laboratories has been reported (U.S. EPA, 1971b; OECD, 1984a; ISO, 1987; Weber et al., 1988, 1989; Morrison et al., 1989). The coefficient of variation for multiple tests conducted at one laboratory with three reference toxicants and *S. capricornutum* ranged from 47 to 83% (Weber et al., 1989). The NOEC values in another study varied less than twofold in multiple toxicity tests conducted with copper and *Champia parvula* (Weber et al., 1988). In an interlaboratory comparison (i.e., round-robin or ring test), the EC50 values for potassium dichromate and freshwater algae varied by a factor of less than 4 (ISO, 1987). Nyhold and Källqvist (1989) discuss the results of several other interlaboratory comparisons of algal toxicity test results.

There is no single preferred reference substance (see Chapter 1) in algal toxicity tests; several have been used, such as cadmium chloride, sodium dodecyl sulfate, sodium pentachlorophenate, copper, potassium dichromate, and sodium chloride. Likewise, no single solvent (see Chapter 1) has been commonly utilized in algal toxicity tests. The phytotoxicities of several solvents have been evaluated (Hughes and Vilkas, 1983; Adams et al., 1986; Stratton and Smith, 1988).

The EC50 values for a green alga and acetone, ethanol, methanol, hexane, dimethyl sulfoxide (DMSO), and *N,N*-dimethyl formamide (DMF) ranged between 0.94 and 3.6 mg/l (Stratton and Smith, 1988).

The analytical verification of the toxicant concentrations is not usually performed in laboratory phytotoxicity tests despite recommendations to do so. Verification is important because biodegradation, adsorption, and changes in pH due to the increasing algal biomass may reduce the bioavailabiltiy of the test substance. Despite these factors, most toxicity results for algae are based on nominal toxicant concentrations.

Utility of Test Results

Laboratory algal toxicity studies are conducted often as screening toxicity tests but in many cases the results are used as estimates of ecotoxic concentrations. Their use to screen for toxicity is more justified than to predict environmental impact. The environmental significance of results from the standard population growth studies has been debated (Kooijman et al., 1983; Wängberg and Blanck, 1988; Freemark et al., 1990), and overall the results have uncertain applicability and are generally considered less relevant than those for animal test species (Lewis, 1990a). This opinion is reflected in Macek et al. (1978), where the usefulness of a variety of aquatic toxicity tests was ranked for several factors. The results of algal toxicity tests were judged to be of lesser value in several categories than those of studies conducted with freshwater animal test species (Table 8). This view of lesser significance contrasts to some degree with that of Giesy and Hoke (1989), who ranked the utility of several types of toxicity tests used to determine the effects of contaminated sediments. In this evaluation, at least one type of algal test was judged equivalent to those using freshwater invertebrates and fish.

The environmental relevance of the results from the laboratory algal toxicity tests in Table 3 is unclear based on method guidelines. The ASTM (1990a) guideline, for example, states that the data can be used for determining algal species sensitivity, the toxicities of different chemicals, and the impact of environmental modifying factors on toxicity. Algal test results submitted for TSCA "provide an indication of effect on algal populations" (U.S. EPA, 1985a), and the ISO (1987) and OECD test guidelines (OECD, 1984a) recommend that the NOEC and EC50 values be used not to predict environmental impact but only to provide the likelihood of effect or no effect. In contrast, the U.S. EPA guidelines for effluent toxicity tests with *S. capricornutum* and *C. parvula* (Weber et al., 1988, 1989) suggest that "safe levels" be established with the laboratory-derived data.

Overall, the uncertain relevance of algal toxicity data expressed by many in the scientific community is understandable. It is largely attributable to the considerable interspecific variation in response of algae and the unrealistic experimental conditions in the traditional laboratory tests.

Complexity

Algal toxicity tests are considered by some to be simple to conduct and relatively inexpensive (Elnabarawy and Welter, 1984; Nyholm and Källqvist, 1989; Thursby et al., 1993). Blaise et al. (1986) reported that 12 standard tests could be maintained during a 7-h working day. Realistically, however, these laboratory tests are not as simple as conducting acute animal toxicity tests and even chronic studies, particularly if the test species is not a green algae. The technical expertise needed to conduct algal toxicity tests and to culture the test species is different from that for animal test species and it is not always available.

Method Improvements

Efforts are continually being made to improve the precision, reproducibility, and efficiency of the current standard methodologies. The effects of several experimental factors, such as test solution volumes, temperature, pH, test durations, statistical methodologies, and test species, have been evaluated (Hughes and Vilkas, 1983; Lukavsky, 1983; Lundy et al., 1984; Wong et al., 1984; Adams et al., 1985; Stratton and Smith, 1988; Stratton and Giles, 1990). Some researchers have developed and, in some cases, validated test designs that use small test volumes, volatile test compounds, and flow cytometric techniques (Blaise et al., 1986; Blanck and Bjornsater, 1989; Herman et al., 1990; Wren and McCarroll, 1990; Blaise, 1991; St. Laurent et al., 1992).

TEST METHODS: VASCULAR PLANTS

Vascular plants or macrophytes are multicellular and are larger than most algae. They inhabit a variety of niches (Figure 1) and serve many useful purposes in freshwater and near-coastal ecosystems. For this reason, the depletion of macrophytic vegetation by pollution can negatively affect aquatic ecosystems, such as the Chesapeake Bay (Jones et al., 1986). In addition, excessive growths of rooted emergent and submersed plants in freshwater environments can be detrimental, and this issue has been of more concern to the public than their depletion. As a result, the development of eradication procedures for these species has received more attention in the past (Nichols, 1991) than the development of toxicity tests to determine the effects of pollutants on their survival.

Toxicity data for freshwater algae have been used as a surrogate for vascular plants, aquatic and terrestrial (Garten and Frank, 1984). The technical support for this practice is not established but algae, in one review, were less sensitive than vascular plants in 20% of the comparisons (Fletcher, 1990). These data extrapolations will continue, however, because of the limited database available for aquatic vascular plants.

Table 8. Comparative scientific value of several toxicity tests in decreasing order.

Ecological significance	Predictive utility	Scientific and legal defensibility	Present relative utility	Availability of routine methods
Reproduction	Reproduction	Acute lethality	Acute lethality	Acute lethality
Acute lethality	Bioaccumulation	Bioaccumulation	Embryo/larval	*Algal assay*
Embryo/larval	Acute lethality	Embryo/larval	Reproduction	Embryo/larval
Algal assay	Embryo/larval	Reproduction	Bioaccumulation	Bioaccumulation
Bioaccumulation	*Algal assay*	*Algal assay*	*Algal assay*	Reproduction

Source: Macek et al, 1978. Copyright ASTM. Reprinted with permission.

Basic information concerning reliable culture techniques, test method development, and identification of sensitive test species is needed before this information will become more available. Nevertheless, a few macrophytic species have been used in toxicity tests and a brief description of this use follows. For more information concerning this topic see Sortkjaer (1984), Guilizzoni (1991), and Swanson et al. (1991).

Duckweed

Duckweeds are floating plants that lack stems and true leaves (Figure 1). They are the most commonly used vascular plants in phytotoxicity tests because of their small size, rapid growth, and structural simplicity. Their use is suggested or required for TSCA and FIFRA (Holst and Ellwanger, 1982; U.S. EPA, 1985b). *Lemna minor* and *L. gibba* are the more commonly used test species, but others such as *Spirodella oligorhiza* and *L. perpusilla* have also been used. *L. minor* is one of the smaller forms (2–4 mm across) and has oval fronds and a single root. Hillman (1961) describes in greater detail the life history characteristics of *L. minor* as well as those of other species. Cultures of *Lemna* spp. are available from several sources including:

- Carolina Biological Supply Company, Burlington, NC 27215 (*Lemna minor*).
- Dr. Charles F. Cleland, U.S. Department of Agriculture, Aerospace Bldg., Rm 323, 901 D Street SW, Washington, DC 20051 (*Lemna gibba* G3).

Wang (1990) reviewed the use of duckweed in toxicity tests and has reported the results for several metals (Wang, 1986, 1987 a,b). Furthermore, a review of critical issues associated with these tests has been discussed (Huebert and Shay, 1993). Tests with duckweed can be of a static, static-renewal, or flow-through design and several methodologies and culture techniques have been reported (Wallbridge, 1977; Bowker et al., 1980; Davis, 1981; Holst and Ellwanger, 1982; U.S. EPA, 1985b; Cowgill and Milazzo, 1989; APHA et al. 1989d; ASTM, 1990b; Huebert et al., 1990). Typically, the tests are conducted under continuous light at intensities ranging from 1615 to 6700 lux. Duckweed plants with two to four fronds are usually used in a test. Twelve to 16 fronds are added to each test chamber, which can be a glass beaker, flask, jar, culture dish, or test tube. The tests are conducted in a nutrient-enriched medium that may be the same as that used for algae but increased in strength as much as 20-fold (ASTM, 1990b). Other media used include Hoagland's E medium (Hillman, 1961) and several of its variations (Holst and Ellwanger 1982). EDTA is not included in most recommended test media but, as with algae, this issue is a matter of continuing debate.

Typically, the EC50 value and the NOEC are calculated based on changes in the number of fronds, although effects on biomass, root number, plant number, root length, and chlorophyll content have also been determined. The phytostatic, phytocidal, EC10, and EC90 values have also been reported in the literature but less frequently (Holst and Ellwanger, 1982; Hughes et al., 1988).

The comparative sensitivities of the various species of duckweed are not well understood. The few available reports such as those by King and Coley (1985) and Cowgill et al. (1991) have shown no consistent trend. The comparative sensitivity of duckweed and other aquatic plants and animals has been investigated to a greater extend and the results indicate chemical-specific trends. For example, duckweed was not as sensitive as some algae to chromium(VI), solid waste leachates, chlorinated phenolic compounds, and coal ash waste (Mangi et al., 1978, Rodgers et al., 1978; Rowe et al., 1982; Klaine, 1985) but was more sensitive than fish to Cd, Cr, Pb, Ni and Se (Wang, 1990). Walker and Evans (1978) and Hughes et al. (1988) reported the similar sensitivities of duckweed and freshwater and marine algae to cationic surfactants and atrazine. Duckweed was more sensitive than three species of macrophytes to a herbicide (Forney and Davis, 1981). Bishop and Perry (1981) reported that *L. minor* was more sensitive in a few cases to diquat, copper, and cationic surfactants, but overall its sensitivity was either comparable to or less than that of bluegills and daphnids. Cowgill et al. (1991) reported that the sensitivities of freshwater and marine algae, two daphnids, and five clones of *Lemna* to eight chemicals were within one order of magnitude.

Most toxicity tests with duckweed are conducted to meet the requirements of FIFRA. However, these tests have also been recommended to evaluate effluent toxicity because of their cost-effectiveness and reliability (Wang, 1990; Wang and Williams, 1990; Taraldsen and Norberg-King, 1990). In addition, the photosynthetic activity of duckweed is unaffected by colored and turbid effluents, unlike algae, and successful tests can be conducted under these conditions. Nevertheless, the use of duckweed to determine effluent toxicity is uncommon and it plays almost no role in the NPDES permitting process in the United States (U. S. EPA, 1986).

In summary, the duckweeds are the "representative" macrophytic species currently used in most phytotoxicity tests but their use is considerably less than that of algae. This is due primarily to the lack of information concerning their sensitivities relative to other aquatic plants and animal species. The duckweeds are sensitive to some chemicals and effluents but their use in wastewater treatment as bioremediative agents does not suggest high sensitivity to others. Therefore, additional information and experience is needed before they will become more widely used and accepted by the scientific and regulatory communities.

Rooted Vascular Plants

These species are used less in toxicity tests than duckweeds because they are more difficult to work with due to their large size, slow growth, and culture difficulty. In contrast, they are more fully exposed to contaminants in the water column and sediments (roots) than duckweed. For this latter reason, rooted plants are better suited to determining the phytotoxic effects of contaminated sediment than floating and planktonic species.

A standard toxicity test method designed specifically for rooted macrophytes does not exist although a draft of the American Society for Testing and Materials guideline is available (ASTM, 1993a). The availability of "non-standard" methods is greater particularly for use in field tests and with contaminated sediments (Walsh et al., 1991a; Crossland et al., 1992). Due to the lack of standardized techniques, a variety of experimental techniques have been used. Culture techniques, test chamber size, test species, number of replicates, and light intensity are some experimental factors that have varied considerably in the reported studies. For example, test durations have ranged from 1 h to 6 wk and the light intensities have been as high as 170 $\mu E\ m^{-2}\ s^{-1}$ (Forney and Davis, 1981; Mhatre and Chaphekar, 1985). The test species are usually collected from natural sources as either whole plants or cuttings. Several species of *Spartina*, *Juncus*, *Sagittaria*, *Scirpus*, and *Nuphar* are available commerically (Environmental Concern Inc., St. Michaels, MD and Kester's Wild Game Food Nurseries, Omro, WI) and exposed in a nutrient-enriched solution such as Hoagland's medium. See Smart and Barko (1985) for a discussion of the culture techniques used for submerged macrophytes. The test chambers have been an assortment of aquaria and jars that may contain sediment, either natural, prepared, or spiked with the toxicant. Several specific experimental designs follow.

Forney and Davis (1981) described the effects of several herbicides on macrophytes such as *Elodea canadensis* Michx. (waterweed), *Myriophyllum spicatum* L. (Eurasian water milfoil), and *Vallisneria americana* Michx. (wild celery). These species were exposed in 3.8-L glass jars and 19-L plastic buckets containing either well-rinsed sand, sandy soil, or natural sediments. Four to five cuttings of the test species were exposed in a nutrient-enriched solution for 3 to 6 wk. Kay et al. (1984) exposed water hyacinths to lead, cadmium, and copper in containers lined with plastic bags. Tap water, commercial liquid nutrient, chelator, Hoagland's solution, and a metal were added to each test chamber that contained five plants. Effects on chlorosis, necrosis, and growth were observed after 6 wk of exposure. Hinman and Klaine (1992) exposed *Hydrilla verticillata* R. to copper and several insecticides. The whole-plant and root exposures were for 14 d in glass jars containing the test compounds. Shoot length, root length, chlorophyll *a*, and changes in plant dehydrogenase activity were determined. Root growth was the most sensitive end point.

Toxicity tests with rooted species have been conducted with test compounds and species other than those described above. These have inlcuded surfactants, chlorine, heavy metals, and pesticides such as atrazine, glyphosate, lindane, chlordane, simazine, and terbutyan (Clearly and Coleman, 1973; Stanley, 1974; Brown and Rattigan, 1979; Correll and Wu, 1982; Watkins and Hammerschlag, 1984; Jones et al., 1986; Thomas et al., 1986; Huber et al., 1987; Mitchell, 1987; Gurney and Robinson, 1989; Byl and Klaine, 1991; Garg and Chandra, 1990). The test species in these studies, among others, were *Najas quadulepensis* (bushy pondweed), *Potomogeton perfoliatus* (pondweed), and *Ceratophyllum demersum* L. (coontail). See Swanson et al., (1991) for more information on the types of macrophytic test species used in studies with pesticides.

The sensitivities of rooted vascular plants relative to algae, duckweed, and animal test species are largely unknown. Studies by Forney and Davis (1981), Hartman and Martin (1985), Larsen et al. (1986), Huber et al. (1987), and Hinman and Klaine (1992) have provided some insight but the data are too few and inconsistent to indicate a trend.

The use of rooted macrophytes to detect effluent toxicity has been almost nonexistent. Walsh et al. (1991a) and Watkins and Hammerschlag (1984) have reported test methods and discussed their significance for this purpose. Overall, the routine use of rooted plants to assess the phytotoxicities of effluents as well as commercial chemicals will not be common in the foreseeable future. Duckweed is the more promising macrophytic species for this purpose. The more likely role for rooted vascular plants will be their use for detection of contaminated sediments, particularly in relation to wetland preservation, for dredge spoil evaluation and for derivation of sediment quality criteria in the United States (see Chapter 8 on Sediment Toxicity Tests).

Seed Germination and Seedling Growth Toxicity Tests

Toxicity tests have been developed that assess the effects of toxicants on seed germination and early seedling growth. These tests are currently more developed for terrestrial plant species than aquatic plant species (U. S. EPA, 1982; OECD, 1984b). Freshwater and marine wetland plants have been used in seed germination tests only recently (Wang, 1992) and include such species as millet (*Echinochloa crusgalli*), rice (*Oryza sativa*), watercress (*Zizania aquatica*), and a marine cordgrass, *Spartina alterniflora*.

There are few standard guidelines for seed germination tests using aquatic vascular plants. A method has been proposed (APHA et al., 1989c) that can be used with chemicals, effluents, contaminated sediment extracts, and leachates. The seeds usually 10 to 15 are exposed to the toxicant for 4 to 5 d in culture dishes either suspended in the liquid test solution or on filter paper soaked with the test solution. Effects on germination, root elongation, and shoot and root biomass are determined. Specific experimental factors such as temperature and light are species specific.

Toxicity tests with vascular plant seeds, like those using whole plants, are not widely used and are in the initial stages of development. To some in the scientific community these types of tests are considered to be less complex than those with algae since there is no need for time-consuming cultures and the use of nutrient media. Although the comparative sensitivity of algae and macrophytic seeds has been evaluated in only a few cases (Garten and Fran, 1984; Thomas et al., 1986), some results show that seed germination is more sensitive than algal population growth. Nevertheless, more time and information are needed to better judge the utility of these types of tests.

BIOCONCENTRATION

Many vascular aquatic plants and algae bioconcentrate (see Chapter 16) contaminants to a greater extent than aquatic animals. This ability has been widely reported and has resulted in their use as bioindicators or "sentinel" species (Chapter 1) of water quality and as bioremediative agents in the treatment of various types of wastewaters (Reed, 1990; Nichols, 1991). In contrast to these beneficial roles, their bioconcentration capacity is a potential threat to the upper trophic levels in the ecosystem.

A significant bioconcentration database exists for freshwater and saltwater plants. Detailed information concerning the test methods, application of the data, and uptake rates can be obtained from Davies (1978), Rai et al. (1981), Kelly (1988), Boyle (1984), Whitton (1984), and Guilizzoni (1991). The degree of bioconcentration has depended on factors such as the pH, temperature, contaminant type, test concentration, type of tissue, tissue age, growth rate, and test species. Metals and pesticides have been the focus of most studies. For example, the uptake of metals such as lead, nickel, cadmium, copper, cobalt, and chromium has been reported for several algae, such as *Cladophera*, *Scenedesmus*, and *Dunaliella* spp., duckweed, *Lemna paucicostata*, and the rooted macrophytic species *Ceratophyllum demersum*, *Eichhornia crassipes*, and *Myriophyllum exalbescens* (Stokes, 1975; Nasu et al., 1983; Kay et al., 1984; Lustigman et al., 1985; Garg and CHandra, 1990; Vymazal, 199)) It can be seen in Figure 4 that plants can accumulate relatively high levels of some metals. Bioconcentration or enrichment factors for zinc have ranged between 254 and 163,750 and for copper between 36 to 691,250 (Kelly, 1988). One of the highest bioconcentration factors reported is 1,200,000 for lead (Satake et al., 1989).

As stated earlier, the bioconcentration ability of algae and macrophytes has led to their use in the bioremediation of a variety of wastewaters. Natural and constructed wetlands are used to remove nutrients and contaminants from municipal and industrial effluents, landfill and agricultural runoff, urbasn stormwater, and acid mine wastes (Hadden, 1994; Lekven et al., 1994). There are 300 constructed wetlands currently in use to treat municipal wastes and 140 to treat acid mine drainage (Knight, 1994; Wieder, 1989). The removal rates of plants in wetland treatment systems are affected by shading, turbidity, organic content and sediment oxygen. Uptake is usually greater in the roots than in other tissues. The impact of these factors and overviews on the general principles, case histories, design criteria and appropriate plant species and their removal efficiencies have been provided by Hammer (1989), Adamus et al, (1991), Dortch (1992), Moshiri (1993) and Knight (1994).

PHOTOSYNTHESIS

Photosynthetic acitivty has been a common effect parameter monitored in laboratory toxicity tests with cultured algae and in situ with natural phytoplankton and periphyton communities. The more common methods used for its analysis have been $^{14}CO_2$ uptake and oxygen evolution. The effects of oil (Gaur and Singh, 1990), metals (Turbak et al., 1986), effluents (Delistraty, 1986), herbicides (Hollister and Walsh, 1973; Jones and Winchell, 1984; Yoshida et al., 1986; Millie and Hersh, 1987; Goldsborough and Brown, 1988), coal-derived oil (Giddings et al., 1984), and surfactants (Lewis and Hamm, 1986; Huber et al., 1987) have been reported using these techniques.

The primary advantage of photosynthesis tests is their short duration, which is usually 2 to 4 h, but exposure times have also ranged form 30 min to 24 h (Kuivasniemi et al., 1985; Versteeg, 1990). Several investigators have compared the results of these short-term tests with those of the 3- to 4-day algal population growth studies. Photosynthetic acitivty has been the less sensitive indicator of the effect in most cases. This has been observed for zinc, copper, cadmium, glyphosate, and simazine (Turbak et al., 1986) as well as for oil shale process waters, metals, pesticides, surfactants, phenolic compounds, and effluents (Kuivasniemi et al., 1985; Delistraty, 1986; Lewis and Hamm, 1986; Versteeg, 1990).

MULTISPECIES TEST METHODS

Natural communities of phytoplankton and periphyton contain many species, most of which are not used in the traditional laboratory phytotoxicity tests. The sensitivities of these naturally occurring species are almost unknown. Furthermore, these species exist in dynamic communities whose composition and balance are affected by seasonal climatic factors. Because these conditions are not like those in the traditional laboratory tests, alternative test designs have been used to increase the environmental realism of the test results. Several of these are briefly discussed below.

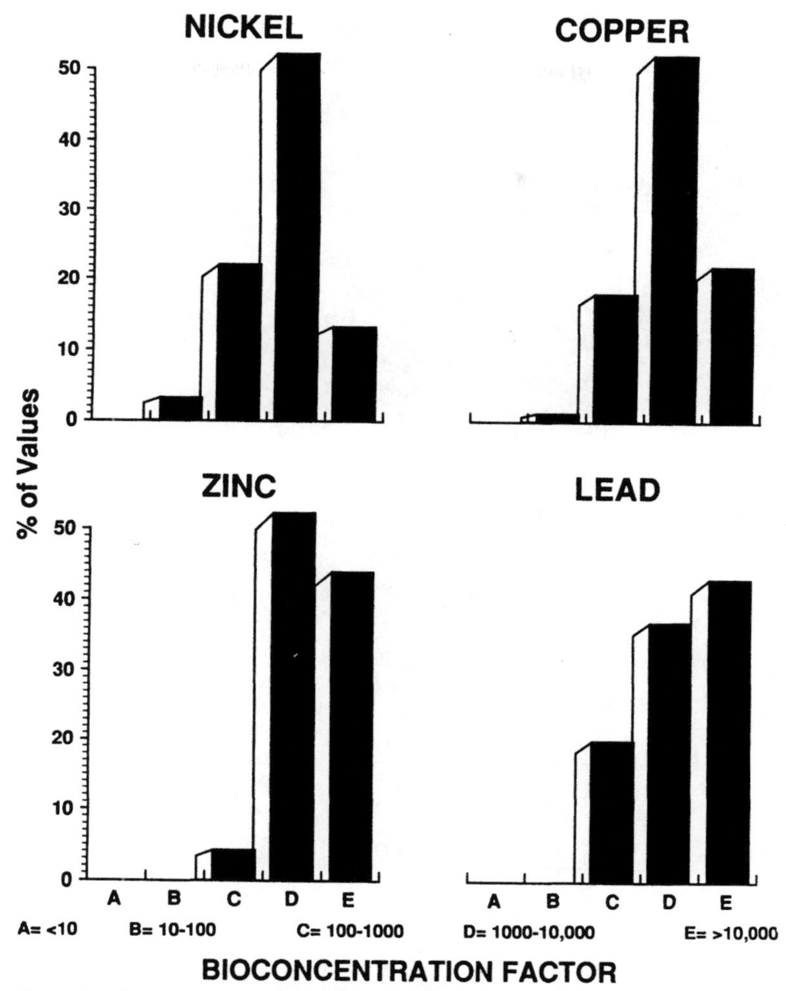

Figure 4. Bioconcentration factors reported for several heavy metals and various species of aquatic plants. (Values from Kelly, 1988.)

Laboratory Methodologies

Several types of laboratory toxicity studies have been conducted with freshwater and saltwater plants which incorporate more realistic experimental conditions (Bates and Weber, 1981; Grice and Reeve, 1982; Cairns, 1985; 1986; Boyle, 1985; Clark, 1989; LaPoint et al., 1989). These "adaptations" include the use of river water for dilution (Wang 1985a, 1987c), the simultaneous exposure of several algal species (Lundy et al., 1984; Claesson, 1984), and the use of "transplanted" natural communities (Niederlehner and Cairns, 1990). The incorporation of these conditions has led to several findings. For example, the presence of multiple test species has been shown to affect the test results. Mosser et al. (1972) showed that *Thalassiosira pseudonana* was more sensitive to polychlorinated biphenyl (PCB) and DDT when in the presence of other marine species, and Walsh and Alexander (1980) reported that the presence of "resistant" algal species reduced the toxicity of a herbicide to the "sensitive" species.

Laboratory microcosms (see Chapter 1) have been used successfully to evaluate the phytotoxic effects of a variety of chemicals, and a review of the methodologies and their advantages and disadvantages has been presented by Draggan (1980) and Taub (1984). A laboratory microcosm is a small-scale model "ecosystem" contained in test chambers to which the test chemical is added. An example appears in Figure 5. There are no widely accepted test procedures, although several have been proposed and evaluated. Taub et al. (1989) describe a design referred to as the standardized aquatic microcosm (SAM) that utilizes a defined biotic community containing, among other biota, 10 species of freshwater algae. This method formed the basis for a standard-

Figure 5. A small laboratory microcosm containing daphnids, fish, and an aquatic macrophyte, *Elodea canadensis*.

ized test guideline (ASTM, 1993b). Another well-developed design is that of Leffler (1984), which is often referred to as the mixed flask culture (MFC) technique. This laboratory method utilizes several species of algae, zooplankton, and microorganisms, which develop from a natural source such as lakes (Stay et al., 1989). Other laboratory microcosm designs have also been reported (Franco et al., 1984; Sheehan et al., 1986).

Field Methodologies

Outdoor ponds (see Chapter 9) have been used to determine the effects of several contaminants on phytoplankton and vascular plant communities. The interest in pond mesocosms and their utilization in the United States are due primarily to regulatory requirements for pesticides (Holst and Ellwanger, 1982; U.S. EPA, 1988; Hobson et al., 1989). Despite attempts to standardize the methodology, the experimental techniques have varied considerably, e.g., in the pond size and test duration. For example, ponds 2 m deep and 470 m^3 in volume were utilized to determine the effects of atrazine on photosynthesis, respiration, and algal biomass for up to 3 yr (Larsen et al., 1986). The 15-m^3 ponds used by Giddings et al. (1984) were 1 m deep and were dosed with coal-derived oil for 8 wk, and effects were observed for an additional 52 wk. Muir et al. (1982) utilized three ponds 0.5 m deep to determine the effects of phosphate esters on duckweed and cattails.

Natural periphyton communities have been exposed to toxicants in experimental streams, in situ floating exposure units, and enclosures. Reviews of these methods have been presented by Patrick (1973), Rodgers et al. (1979), Weitzel (1979), and Kosinski (1989). Experimental streams of various designs have been utilized to determine the effects of effluents and a variety of chemicals on periphyton communities. For example, Bothwell and Stockner (1979) monitored the seasonal effects of a kraft mill effluent on colonized periphyton in experimental troughs of 3.6 m × 19 cm located on the bank of the Mackenzie River. Yount and Richter (1986) used eight outdoor 520-m channels to investigate the effects of PCP on periphyton. In another example, indigenous biota were added to ten 6.1-m indoor experimental streams and allowed to develop for 3 mo before the effects of diflubenzuron were assessed on periphyton community structure for 5 mo (Hansen and Garton, 1982). Recirculating 2.4 m long and 12.5 cm wide artificial streams have been used to evaluate the effects of several herbicides on colonized periphyton (Kosinski and Merkle, 1984; Moorhead and Kosinski, 1986).

A flow-through test has been used in situ to determine the effects of added nutrients and effluents on periphyton (Peterson et al., 1983; Tease et al., 1983). This procedure was modified by Lewis et al. (1986, 1993) to determine the toxicities of several detergent surfactants on colonized periphyton in river areas above and below a municipal effluent outfall (Figure 6). The colonized periphyton were exposed to four test concentrations for 21 d. The test design proved useful in providing realistic toxicity estimates but was labor-intensive.

Figure 6. In situ toxicity test design in which periphyton colonized on plexiglass plates were exposed to detergent surfactants for 21 d.

Enclosures (see Chapter 9) have been used to determine the phytotoxicities of contaminants and the methodologies, advantages, and disadvantages have been discussed (Grice and Reeve, 1982; Taub, 1984; Schelske, 1984; Solomon and Liber, 1988). These enclosures, either totally closed or open, isolate a portion of the phytoplankton community to which a chemical can be applied without contaminating the surrounding ecosystem. Historically, studies of this type have been conducted more frequently in salt water than fresh water, as in the *Controlled Ecosystem Pollution Experiment* (CEPEX) program (Grice and Reeve, 1982). Several considerations related to the use of enclosures include the size, shape, method of test compound application, replication, sampling frequency, and impact of the enclosure or "bottle" effects on the enclosed biotic communities. The volumes of enclosures have ranged from 41 to 18,000 m^3 and test durations have ranged from 4 d to 22 mo. Brazner et al (1989) reported a cost-effective 5 × 10 m enclosure that includes the zones of rooted emergent and submergent plants in littoral areas. Twelve of these were used to evaluate the effects of chlorpyrifos in a Minnesota pond.

Small enclosures have been used to determine the effects of metals and surfactants on phytoplankton communities (Figure 7). For example, the effect of cadmium on lake plankton enclosed in 8-L carboys was determined for 3 wk in Lake Michigan by Marshall and Mellinger (1980). Lewis (1986) conducted surfactant studies in 21-L translucent carboys with phytoplankton for 10 to 21 d, and Winner et al. (1990) and Moore and Winner (1989) utilized 100-L polyethylene bags to expose pond phytoplankton to copper. Dialysis bags have also been used to evaluate the effects of contaminants and effluents on algae (Walsh, 1982).

Figure 7. Small polycarbonate enclosures, 21L, used to determine the phytotoxic effects of detergent surfactants on phytoplankton.

Biological surveys and stream and lake dosing are additional methods that have been used to monitor the effects of a variety of potential phytotoxic compounds on phytoplankton and periphyton communities. These techniques are not routinely used, but several studies have been reported (Geckler et al., 1976; Scorgie, 1980; Schindler et al., 1985; Leland and Carter, 1986).

A relatively new multispecies test measuring pollution-induced community tolerance (PICF) has been proposed (Blanck et al., 1988). This test compares the tolerance levels of exposed and nonexposed biotic communities to a toxicant. The test has been used with freshwater and marine algae and can be conducted in the laboratory and the field. The usefulness of this test design is discussed in Blanck et al. (1988) and it has been evaluated using a microcosm and mesocosm (see Chapter 9) with a polychlorinated phenolic compound, tributyltin, and diuron (Molander et al., 1990, 1992).

Collection and Analysis

Phytoplankton and periphyton are the flora utilized in most multispecies toxicity tests. The methods for the collection and analysis of these communities have been described in detail by Weber (1973), APHA et al. (1989b), and ASTM (1993c). Phytoplankton are usually collected using mechanical equipment such as Juday, Kemmerer, and Van Dorn samplers. Net collection may also be used. After preservation, qualitative and quantitative analysis is conducted using a compound binocular microscope and "specialized" counting chambers such as Sedgwick-Rafter counting chambers, Palmer-Maloney cells, and hemocytometers. Prior to analysis and if the density is low, the phytoplankton may be concentrated using filtration, sedimentation, and centrifugation.

Periphyton are collected in many cases using floating artificial substrates that are usually glass or Plexiglas slides (Figure 6). These substrates are suspended in the water column using a periphytometer. The analysis and preservation techniques for periphyton are similar to those for phytoplankton.

End Points and Effect Parameters

The end points reported from multispecies toxicity studies conducted either in the laboratory or field are usually the NOEC and LOEC, which are based on the most sensitive effect parameter. The more commonly monitored effects are those on community composition (e.g., diversity and similarity), community respiration, biomass, population abundance, chlorophyll *a*, phaeophytin, and primary productivity. The use of diversity and similarity indices to indicate a change in community composition (i.e., structure) has been discussed by Brock (1977), Perkins (1983), Washington (1984), and Boyle et al. (1990). Measurements for biomass, chlorophyll *a*, and photosynthetic activity have been discussed earlier in this chapter. The phaeophytin content, a degradation product of chlorophyll, can may be determined using the method described by Weber (1973). The presence of large concentrations may indicate a water quality impact.

Laboratory–Field Data Comparisons

The results of several multispecies toxicity tests have been compared with those of the traditional single species tests conducted in the laboratory with the same toxicant. The results of these comparisons have been compound specific and too unpredictable for valid extrapolation from one chemical to another (Table 9). The comparative database is largely for pesticides such as atrazine, to which the laboratory test species have similar or greater sensitivity than natural plant communities. This was observed by Stay et al. (1989) for atrazine and four natural plankton communities using Leffler microcosms and natural ponds. Plumley and Davis (1980) and Larsen et al. (1986) reported similar trend in comparing results from experimental ponds, a laboratory microcosm, and single-species toxicity tests conducted with freshwater and salt marsh algae. A somewhat different finding was reported by deNoyelles and Kettle (1985), in which the photosynthetic response of phytoplankton in laboratory tests was predictive of the immediate but not long-term photosynthetic effects of phytoplankton in experimental ponds.

Data comparisons also exist for substances other than pesticides. In several cases, the single species are less sensitive than natural plant communities. This was observed for several detergent surfactants, with which the results from multispecies tests were up to eight times more sensitive than those from single-species tests (Guhl and Gode, 1989). Giddings et al. (1984) reported that periphyton community composition in ponds was altered at 5% of the 4-h EC50 value based on the photosynthetic response of *S. capricornutum* to a synthetic oil. Laboratory bioconcentration tests with duckweed and flame retardants underestimated uptake by duckweed in ponds (Muir et al., 1982). Huber et al. (1987) found that algal and macrophyte communities in experimental ponds were more sensitive than single species to several compounds. In contrast to this trend, Boyle et al. (1985), Lewis and Hamm (1986), and Kuivasniemi et al. (1985) reported that single algal species were more sensitive than natural phytoplankton to other chemicals. More information concerning laboratory–field data evaluations can be found in Boyle (1985).

Regulatory Use

Harrass and Sayre (1989) and several papers in Cairns (1985) discuss the advantages and disadvantages of multispecies toxicity tests in the regulatory process. These studies can be a valuable tool, but their worth is dependent on the regulatory agency's experience and the ability of the test design to provide meaningful and useful information. Multi-

Table 9. Examples of the relative sensitivity of single test species and multiple species exposed to the same test compound

Test substance	Single species	Multiple species	Test design	Single species were:	Reference
Atrazine	Two algal species	Salt marsh algae	Microcosm	More sensitive	Plumley and Davis (1980)
Phosphate ester flame retardants	Duckweed	Duckweed	Experimental pond	Less sensitive[a]	Muir et al. (1982)
Diflubenzuron	*Selenastrum capricornutum*	Algae	Experimental streams	Less sensitive	Hansen and Garton (1982)
Synthetic coal-derived oil	*Selenastrum capricornutum*	Periphyton	Experimental pond	Less sensitive	Giddings et al. (1984)
Fluorene	*Selenastrum capricornutum*	Phytoplankton	Experimental pond	More sensitive	Boyle et al. (1985)
Atrazine	Phytoplankton	Phytoplankton	Experimental pond	Similar[b]	deNoyelles and Kettle (1985)
Chlorinated phenolic compounds	*Selenastrum capricornutum*	Phytoplankton	Laboratory tests	More sensitive	Kuivasniemi et al. (1985)
Atrazine	Eight algal species	Algae	Experimental pond, laboratory microcosm	Similar	Larsen et al. (1986)
Surfactants	Three algal species	Phytoplankton	Enclosures	More sensitive	Lewis (1986)
Surfactants	Three algal species	Periphyton	Floating bioassay	More sensitive	Lewis et al. (1986)
Surfactants	Three algal species	Phytoplankton	Enclosures	More sensitive	Lewis and Hamm (1986)
LAS 4-Nitrophenol PCP Atrazine	Two algae, three vascular plants	Phytoplankton, vascular plants	Experimental ponds	Less sensitive	Huber et al. (1987)
Atrazine	Two algal species	Algae	Laboratory microcosm	Similar	Stay et al. (1989a)
Fifteen surfactants	Algae	Algae	Laboratory microcosm	Less sensitive	Guhl and Gode (1989)
Industrial effluent Pulp/paper mill effluent Copper	Two freshwater algal species	Mixed phytoplankton	Enclosure	Similar	Ludyanskiy and Pasichny (1992)
Anionic surfactant	Three freshwater algal species	Periphyton	Floating bioassay	Less sensitive	Lewis et al. (1993)

[a]Duckweed in laboratory bioconcentrated less of the test substance.
[b]Outcome was dependent on duration of laboratory tests monitoring ^{14}C uptake of natural phytoplankton.

species phytotoxicity results are not common because of the economic resources and time needed for their development. Nevertheless, the realization that these data are more meaningful than those for single species ensures their continuing development, particularly for commercially important chemicals found highly phytotoxic in the traditional laboratory tests.

REPRESENTATIVE PHYTOTOXICITY DATA

Some of the first phytotoxicity tests were conducted to determine the algicidal properties of cationic surfactants and metals such as copper (Williams et al., 1952; Fitzgerald et al., 1952). The first studies with copper sulfate were conducted in the early 1900s and investigations with this compound still continue (Hawkins and Griffiths, 1987). Consequently, the phytotoxicity database for these compounds is more extensive than those for most other chemicals. Other databases have been reviewed for surfactants by Lewis (1990b) and for metals by Whitton (1970, 1984), Davies (1978), Rai et al. (1981), Stratton (1987), and Goudey (1987). More recently, for obvious reasons, herbicides have been the focus of many phytotoxicity tests and reviews of the results have been presented by O'Kelley and Deason (1976), Larsen et al. (1986), Stratton (1987), Stay et al. (1989), and Swanson et al. (1990). Comprehensive listings of phytotoxicity data for a variety of other chemicals have been presented by Bringmann and Kühn (1978) and Blanck et al. (1984) and several data retrieval systems are available for additional information, e.g., the Aquatic Plant Information Retrieval System (University of Florida, Gainesville 32606). The relative phytotoxicities of several types of contaminants are shown in Figure 8.

The large-scale effects of acid precipitation, acid mine drainage, thermal effluents, and oil pollution on aquatic plants have been reviewed (Kelly, 1988; Stokes et al., 1989; Shales et al., 1989; Langford 1987).

SUMMARY

The use of algae and vascular plants as bioindicators of environmental quality has been common for many years. In contrast, their use at test species in toxicity

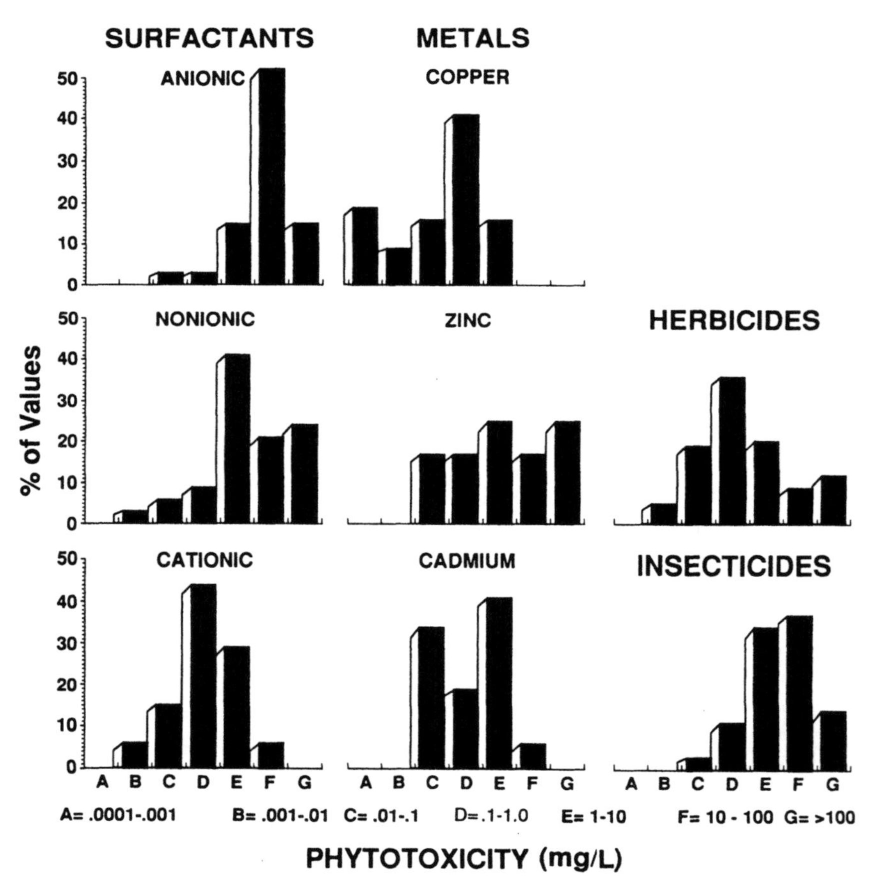

Figure 8. Comparative phytotoxicities of several contaminants. (Data from Stratton, 1987; Kelly, 1988; Lewis, 1990b.)

tests has a realtively brief history and their value is a subject of continuing discussion. Historically, toxicity data for animal species were used as a surrogate for freshwater and saltwater vegetation. However, as the comparative toxicity database has increased, there is considerable evidence that this data extrapolation is not technically valid. This realization and environmental regulations have resulted in the increased development of phytotoxicity information. Nevertheless, these data are still relatively uncommon and their importance in the regulatory decision-making process continues to be less than that of invertebrates and fish. This trend is attributable to the absence of proven test methods and culture techniques for some species, the insensitivity of the one test species recommended for use when reliable test methodologies are available, and the uncertain environmental relevance of the test results.

The following types of phytotoxicity tests are used by ecotoxicologists, in order of decreasing frequency.

- Algal population growth: green algae
- Duckweed
- Algal population growth: blue-green algae and diatoms
- Short-term algal photosynthesis
- Seed germination/early seedling growth
- Multispecies toxicity i.e., microcosms and mesocosms

It can be seen that the current scientific understanding of the phytotoxic effects of potential toxicants is usually based on the results derived for a few species of freshwater green algae. These data have been used as a surrogate for other algae, for vascular aquatic plants, and in some cases for terrestrial species (Fletcher, 1990). This "extended" use of the data should be viewed with skepticism in view of the interspecific variation in response of cultured algae, which can range several orders of magnitude. Consequently, it has been recommended that a species battery approach be used to better determine the potential phytotoxic effects of toxicants and that data extrapolations be limited until supporting data are available.

The environmental significance of laboratory-derived phytotoxicity results for algae has been questioned not only because of the interspecific variation but also because of the unrealistic experimental conditions of the standard laboratory tests. More realistic designs have been used that include laboratory microcosms, experimental ponds, experimental streams, and in situ enclosures. Results of these multispecies tests have been compared in a few cases to those of the more conventional single-species studies. The results of these comparisons have been substance specific and usually unpredictable.

Studies to identify effective eradication procedures and to determine the bioconcentration ability of floating and rooted vascular plants have been more numerous than those designed to evaluate the effects of environmental toxicants. Only the duckweeds (*Lemna* spp.) have been used to any significant extent for this purpose and their use is less frequent than that of algae.

Phytotoxicity data have been derived largely to support the registrations of chemicals and pesticides. They have played almost no role in the derivation of water quality criteria and their use in effluent toxicity evaluations to support the NPDES permitting process (see Chapter 24) is uncommon in the United States (U.S. EPA, 1986; Parkhurst et al., 1992). This trend persists despite the presence of effluent test methodologies (Weber et al., 1988, 1989; see Chapters 2, 3 and 33) and numerous recommendations and examples of their suitability for this purpose (Walsh and Alexander, 1980; Walsh, 1982; Walsh and Merrill, 1984; Walsh and Garnas, 1983; Claesson, 1984; Taraldsen and Norberg-King, 1990). Currently effluent toxicity is regulated with minimal understanding of the potential phytotoxic effects.

The use of algae and vascular plants to assess the toxicities of contaminated soils, sediments (see Chapter 8), and dredged materials has been limited despite the availability of several methods and recommendtions that they would be useful test species (Munawar and Munawar, 1987; Giesy and Hoke, 1989; U.S. COE, 1989; U.S. EPA-U.S. COE, 1991; Folsom and Price, 1991; Ireland et al., 1991). Algae have been used to determine the toxicity of hazardous chemical wastes (Greene et al., 1988) and efforts are in progress by ASTM (1990c) to develop a standard test method for this purpose.

Overall, phytotoxicity tests should be considered an integral component of ecotoxicology. Phytotoxicity data are necessary for the complete hazard evaluation of contaminants entering freshwater and saltwater environments and will be viewed as such by regulatory agencies if useful data are provided by the scientific community. With this in mind, the primary needs are to improve the quality and quantity of test data as derived from the current standard test methods by focusing future research in the following areas:

- Additional freshwater and saltwater algal test species representing a wide taxonomic range need to be identified and used in the current laboratory test methods (i.e., compensate for interspecific variation).
- The amount of algal stimulation considered environmentally meaningful in laboratory toxicity tests needs to be determined and its relevance to inhibitiory effects and data interpretation evaluated.
- The relationship of the effects of toxicants on algal population growth observed in the laboratory tests to the structural and functional effects on natural plant communities needs to be determined (i.e., field validation).
- Reliable test methods and culture techniques are needed for macrophytic plants and their seeds.
- Sensitive test species need to be identified and their sensitivities compared to those of algae and

animal test species that are commonly used in the current standard toxocitiy tests.

LITERATURE CITED

Adams, N.; Goulding, K. H.; and Dobbs, A. J.: Toxicity of eight water soluble organic chemicals to *Selenastrum capricornutum*: a study of methods for calculating toxic values using different growth parameters. *Arch. Environ. Contam. Toxicol.*, 14:333–345, 1985.

Adams, N.; Goulding, K. H.; and Dobbs, A. J.: Effect of acetone on the toxicity of four chemicals to *Selenastrum capricornutum. Bull. Environ. Contam. Toxicol.*, 36:254–259, 1986.

Adamus, P. R.; Stockwell, L. T.; Ellis, J.; Clarian, E. J.; Morrow, M. E.; Rozas, L. P.; and Smith, R. D.: Wetland Evaluation Technique Volume 1: Literature Review and Evaluation Rationale. U. S. Army Corps of Engineers Report WRP-DE-Z, 1991.

Allen, E. J., and Nelson, E. W.: On the artificial culture of marine plankton organisms. *J. Mar. Biol. Assoc.*, 8:421–474, 1910.

American Public Health Association, American Water Works Association, and Water Pollution Control Federation: Toxicity testing with phytoplankton. Standard Methods for Examination of Water and Wastewater, 17th ed. Washington, DC: American Public Health Association, 1989a.

American Public Health Association, American Water Works Association, and Water Pollution Control Federation: Standard Methods for Examination of Water and Wastewater, 17th ed, Suppl. Washington, DC: APHA, 1989b.

American Public Health Association, American Water Works Association and Water Pollution Control Federation: Toxicity test procedures using aquatic vascular plants (Proposed) 8220. Standard Methods for Examination of Water and Wastewater, Washington, DC: APHA, 1989c.

American Public Health Association and Water Pollution Control Federation: Duckweed (Proposed) 8211. Standard Methods for Examination of Water and Wastewater. Washington, DC: APHA, 1989d.

American Society for Testing and Materials: Standard practices for measurement of chlorophyll content of algae in surface waters. D3731-79. Philadelphia: American Society for Testing and Materials, 1979.

American Society for Testing and Materials: Standard guide for conducting static 96-h toxicity tests with microalgae. E1218-90. Philadelphia: American Society for Testing and Materials, 1990a.

American Society for Testing and Materials: New standard guide for conducting static toxicity tests with *Lemna gibba*. E1415-91. Philadelphia: American Society for Testing and Materials, 1990b.

American Society for Testing and Materials: Standard practice for conducting static chronic 96-h toxicity tests on hazardous chemical wastes using the freshwater green alga, *Selenastrum capricornutum* (Draft). Philadelphia: American Society for Testing and Materials, 1990c.

American Society for Testing and Materials: Standard guide for conducting a toxicity test with rooted aquatic emergent macrophytes (draft). Philadelphia: American Society for Testing and Materials, , 1993a.

American Society for Testing and Materials: Standard practice for standardized aquatic microcosms. Freshwater. E1366-91. Philadelphia: American Society for Testing and Materials, 1993b.

American Society for Testing and Materials: Annual Book of Standards. Section II Water and Environmental Technology. Philadelphia: American Society for Testing and Materials, 1993c.

American Water Works Association: Current methodology for the control of algae in surface reservoirs. Denver: American Water Works Association Research Foundation, 1987.

Anderson, B. S., and Hunt, J. W.: Bioassay methods for evaluating the toxicity of heavy metals, biocides and sewage effluent using microscopic stages of giant kelp *Marcocytis pyrifera* (Agardh); A preliminary report. *Mar. Environ. Res.*, 26:113–134.

AQUIRE: Aquatic toxicity information retrieval data base. Duluth, MN: U.S. EPA, 1987.

Aronson, J. G.: The effect of some common insecticides upon C-14 uptake in phytoplankton. M.S. thesis, Univ. of Nebraska, Lincoln, p. 96, 1973.

Bates, J. M., and Weber, C. I. (eds.): Ecological Assessements of Effluent Impacts on Communities of Indigenous Aquatic Organisms. ASTM STP 730, p. 333. Philadelphia: American Society for Testing and Materials, 1981.

Benenati, F.: Keynote address: plants—keystone to risk assessment. Plants for Toxicity Assessment, edited by W. Wang, J. W. Gorsuch, W. R. Lower, pp. 5–13. ASTM STP 1091. Philadelphia: American Society for Testing and Materials, 1990.

Bishop, W. E., and Perry, R. L.: Development and evaluation of a flow-through growth inhibition test with duckweed (*Lemna minor*). In D. R. Branson and K. L. Dickson eds., Aquatic Toxicology and Hazard Assessment: Fourth Conference, edited by D. R. Branson, K. L. Dickson, pp. 421–435. STP 737. Philadelphia: American Society for Testing and Materials, 1981.

Blaise, C. B.: Microbiotests in aquatic ecotoxicology: characteristics, utility and prospects. *Environ. Toxicol. Water Qual.*, 6:145–155, 1991.

Blaise, C. B.; Legault, R.; Bermingham, N.; and Coillie, R. V., Vasseur, P.: A simple algal assay technique for aquatic toxicity assessment. *Toxicity Assessment*, 1:261–281, 1986.

Blanck, H., and Bjornsater, B.: The algal microtest battery—a manual for routine tests of growth inhibition. KEMI Report No. 3198. Goteborg: The Swedish National Chemicals Inspectorate, 1989.

Blanck, H.; Wallin, G.; and Wängberg, S.: Species-dependent variation in algal sensitivity to compounds. *Ecotoxicol. Environ. Safety*, 8:339–351, 1984.

Blanck, H.; Wängberg, S.; and Molander, S.: Pollution induced community tolerance—a new ecotoxicological tool. Functional Testing of Aquatic Biota for Estimating Hazard of Chemicals, edited by J. Cairns, J. R. Pratt, pp. 219–230. ASTM STP 988. Philadelphia: American Society for Testing and Materials, 1988.

Blaylock, B. G.; Frank, M. L.; and McCarthy, J. F.: Comparative toxicity of copper and acridine to fish, *Daphnia* and algae. *Environ. Toxicol. Chem.*, 4:63–71, 1985.

Boston, H. L.; Hill, W. R.; and Stewart, A. J.: Evaluating direct toxicity and food-chain effects in aquatic systems using natural periphyton communities. Plants for Toxicity Assessment: Second Volume, edited by J. W. Gorsuch, W. R. Lower, W. Wang, M. A. Lewis, pp. 126–145. ASTM STP 1115. Philadelphia: American Society for Testing and Materials, 1991.

Bothwell, M. L. K., and Stockner, J. G.: Influence of secondarily treated Kraft mill effluent on the accumulation rate of attached algae in experimental continuous-flow troughs. *Can. J. Fish. Aquat. Sci.*, 37:248–254, 1979.

Bowker, D. W.; Duffuld, A. N.; and Deny, P.: Methods for the isolation, sterilization and cultivation of Lemnaceae. *Freshwater Biol.*, 10:385–388, 1980.

Boyle, T. P.: The effect of environmental contaminants on aquatic algae. Algae as Ecological Indicators, edited by L. E. Shubert, pp. 237–256. New York: Academic Press, 1984.

Boyle, T. P. (ed.): Validation and Predictability of Laboratory Methods for Assessing the Fate and Effects of Contaminants in Aquatic Ecosystems. ASTM Special Publication. STP 865. Philadelphia: American Society for Testing and Materials, 1985.

Boyle, T. P.; Finger, S. E.; Paulson, R. L.; and Rabeni, C. F.: Comparison of laboratory and field assessment of fluorene—Part II: Effects on the ecological structure and function of experimental pond ecosystems. Validation and Predictability of Laboratory Methods for Assessing the Fate and Effects of Contaminants in Aquatic Ecosystems, edited by T. P. Boyle, pp 134–151. ASTM STP 865. Philadelphia: American Society for Testing and Materials, 1985.

Boyle, T. P.; Smillie, G. M.; Anderson, J. C.; and Beeson, D. R.: A sensitivity analysis of nine diversity and seven similarity indices. *Res. J. Water Pollut. Control Fed.*, 62: 749–762, 1990.

Brazner, J. C.; Heinis, L. J.; and Jensen, D. A.: A littoral enclosure for replicated field experiments. *Environ. Toxicol. Chem.*, 8:1209–1216, 1989.

Brezonik, P. L.; Browne, F. X.; and Fox, J. L.: Application of ATP to plankton biomass and bioassay studies. *Water Res.*, 9:155–162, 1975.

Bringmann, G., and Kühn, R.: Testing of substances for their toxicity threshold: model organisms *Microcystis aeruginosa* and *Scenedesmus quadricauda*. *Mitt. Verein. Limnol.*, 21:275–284, 1978.

Bringmann, G., and Kühn, R.: Comparison of toxicity thresholds of water pollutants to bacteria, algae, and protozoa in the cell multiplication test. *Water Res.*, 14:231–241, 1980.

Brock, D. A.: Comparison of community similarity indexes. *J. Water Pollut. Control Fed.*, 49:2488–2495, 1977.

Brown, W. E., and Rattigan, B. M.: Toxicity of soluble copper and other metal ions to *Elodea canadensis*. *Environ. Pollut.*, 20:303–315, 1979.

Bruce, R. D., and Versteeg, D. J.: A statistical procedure for modeling continuous toxicity data. *Environ. Toxicol. Chem.*, 11:1485–1494, 1992.

Byl, T. D., and Klaine, S. J.: Peroxidase activity as an indicator of sublethal stress in the aquatic plant *Hydrilla verticillata*. Plants for Toxicity Assessment: Second Volume, edited by J. W. Gorsuch, W. R. Lower, M. A. Lewis, W. Wang, pp. 101–106. Philadelphia: American Society for Testing and Materials, 1991.

Cairns, J. (ed.): Multispecies Toxicity Testing. Society of Environmental Toxicology and Chemistry, Special Publication Series. New York: Pergamon, 1985.

Cairns, J. (ed.): Community Toxicity Testing. ASTM-STP 920. Philadelphia: American Society for Testing and Materials, 1986.

Carlson, R.; Brungs, W.; Chapman, G.; and Hansen, D.: Guidelines for deriving numerical aquatic site-specific water quality criteria by modifying national criteria. Duluth, MN: U.S. Environmental Protection Agency, 1985.

Cheung, K. C.; Chu, L. M.; and Wong, M. H.: Toxic effect of landfill leachate on microalgae. *Water, Air, Soil Pollut.*, 69:337–349, 1993.

Claesson, H.: Use of a mixed algal culture to characterize industrial wastewater. *Ecotoxicol. Environ. Safety*, 8:80–96, 1984.

Clark, J. R.: Field studies in estuarine ecosystems: a review of approaches for assessing contaminant effects. Aquatic Toxicology and Hazard Assessment: 12th Volume edited by U. M. Cowgill Philadelphia: L. R. Williams, pp. 120–133. ASTM STP 1027. Philadelphia: American Society for Testing and Materials, 1989.

Clearley, J. E., and Coleman, R. L.: Cadmium toxicity and accumulation in southern naiad. *Bull. Environ. Contam. Toxicol.*, 9:100–102, 1973.

Correll, D. L., and Wu, T. L.: Atrazine toxicity to submersed vascular plants in simulated estuarine microcosms. *Aquat. Bot.*, 12:151–158, 1982.

Cowgill, U. M., and Milazzo, D. P.: The culturing and testing of two species of duckweed. Aquatic Toxicology and Hazard Assessment: 12th volume, edited by U. M. Cowgill, L. R. Williams pp. 379–391. ASTM STP 1027. Philadelphia: American Society for Testing and Materials, 1989.

Cowgill, U. M.; Milazzo, D. P.; and Landesburger, B. D.: The sensitivity of *Lemna gibba* G-3 and four clones of *Lemna minor* to eight common chemicals using a 7-day test. *Res. J. Water Pollut. Control Fed.*, 63:991–998, 1991.

Cowgill, U. M.; Milazzo, D. P.; and Landesburger, B. D.: Toxicity of nine benchmark chemicals to *Skeletonema costatum*, a marine diatom. *Environ. Toxicol. Chem.*, 8: 451–455, 1989.

Crossland, N. O.; Heimbach, F.; Hill, I. R.; Boudou, A.; Leeuwangh, P.; and Persoone, G.: Summary and recommendations of the European Workshop on freshwater field tests. Potsdam, June 25-26, 1992. (Copies available from authors.)

Davies, A. G.: Pollution studies with marine plankton, II. Heavy metals. Advances in Marine Biology, Vol. 15, edited by F. S., Russell, M. Young, pp. 382–508. London: Academic Press, 1978.

Davis, J. A.: Comparison of static-replacement and flow-through bioassays using duckweed, *Lemna gibba* G-3. EPA 560/6-81-003, Washington, DC: U.S. EPA, 1981.

Delistraty, D.: Growth and photosynthetic response of a freshwater alga, *Selenastrum capricornutum* to an oil shale by-product water. *Bull. Environ. Contam. Toxicol.*, 36:114–121, 1986.

Dennis, W. M., and Isom, B. G.: Ecological Assessment of Macrophyton: Collection, Use and Meaning of Data, edited by W. M. Dennis, B. G. Isom. ASTM STP 843. Philadelphia: American Society for Testing and Materials, 1983.

Dorth, M. S.: Literature Analysis of The Functional Ability to Wetlands to Improve Vicksburg, MS: U.S. Army Corps of Engineers, December, 1992.

deNoyelles, F., Jr., and Kettle, W. D.: Experimental ponds for evaluating bioassay predictions. Validation and Predictability of Laboratory Methods for the Assessment of Fate and Effects of Environmental Contaminants in Aquatic Ecosystems, edited by T. P. Boyle, pp. 91–103. ASTM STP 865. Philadelphia: American Society for Testing and Materials, 1985.

Draggan, S. (ed.): The microcosm: Biological model of the ecosystem. Report No. 19, Monitoring and Assessment Research Center, Chelsea College, University of London, 1980.

Droop, M. R.: Algae. Methods in Microbiology 3B, edited by J. R. Norris, W. D. Ribbons, pp. 269–313. New York: Academic Press, 1969.

Elnabarawy, M. T., and Welter, A. N.: Utilization of algal cultures and assays by industry, Algae as Ecological Indicators, edited by L. E. Shubert, pp. 317–328. New York: Academic Press, 1984.

European Economic Community (EEC): Methods for the determination of ecotoxicity: algal inhibition test. EEC Directive 79/831, Annex V, Part C, 1987.

Farmer, A. M.: The effects of lake acidification on aquatic macrophytes—a review. *Environ. Pollut.*, 65: 219–240, 1990.

Fernandez-Piñas, F.; Mateo, P.; and Bonilla, I.: Binding of cadmium by cyano bacterial growth media: free ion concentration as a toxicity index to the cyanobacterium *Nostoc* VAM 208. *Arch. Environ. Contam. Toxicol.*, 21:425–431, 1991.

Finger, S. E.; Little, E. F.; Henry, M. G.; Fairchild, J. F.; and Boyle, T. P.: Comparison of laboratory and field assessment of fluorene—Part I: Effects of fluorene on the survival, growth, reproduction, and behavior of aquatic organisms in laboratory tests. Validation and Predictability of Laboratory Methods for Assessing the Fate and Effects of Contaminants in Aquatic Ecosystems, edited by T. P. Boyle, pp. 120–133. ASTM STP 865. Philadelphia: American Society for Testing and Materials, 1985.

Fisher, N. S.; Graham, L. B.; Carpenter, E. J.; and Wurster, C. F.: Geographic differences in phytoplankton sensitivity to PCBs. *Nature*, 241:548549, 1973.

Fitzgerald, G. P., and Faust, S. L.: Bioassay for algicidal vs algistatic chemicals. Water Sewage Works August: 296–298, 1963.

Fitzgerald, G. P.; Gerloff, G. C.; and Skoog, F.: Studies on chemicals with selective toxicity to blue-green algae. *Sewage Indust. Wastes*, 24:888–896, 1952.

Fleming, W. J.; Ailstock, M. S.; Momot, J. J.; and Norman, C. M.: Response of sago pondweed, a submerged aquatic macrophyte, to herbicides in three laboratory culture systems. Plants for Toxicity Assessment: Second Volume, edited by J. W. Gorsuch, W. R. Lower, W. Wang, M. A. Lewis, pp. 267–275. ASTM STP 1115. Philadelphia: American Society for Testing and Materials, 1991.

Fletcher, J. S.: Use of algae versus vascular plants to test for chemical toxicity. Plants for Toxicity Assessment, edited by W. Wang, J. W. Gorsuch, W. R. Lower, pp. 33–39. ASTM STP 1091. Philadelphia: American Society for Testing and Materials, 1990.

Fletcher, R. L.: Marine microalgae as bioassay test organisms. Ecotoxicology and the Marine Environment, edited by P. D. Abel, V. Aviak, pp. 111–131. New York: Ellis Horwood, 1991.

Folsom, B. L., Jr., and Price, R. A.: A plant bioassay for assessing plant uptake of contaminants from freshwater soils or dredged material. Plants for Toxicity Assessment: Second Volume, edited by J. W. Gorsuch, W. R. Lower, W. Wang, M. A. Lewis, pp. 172–177. ASTM STP 1115. Philadelphia: American Society for Testing and Materials, 1990.

Food and Drug Administration: Algal assay test. Environmental Assessment Technical Guide. Washington, DC: Bureau of Veterinary Medicine and Bureau of Foods, 1987.

Forney, D. R., and Davis, D. E.: Effects of low concentrations of herbicides on submersed aquatic plants. *Weed Sci.*, 29:677–685, 1981.

Franco, P. J.; Giddings, J. M.; Herbes, S. E.; Hook, L. A.; Newbold, J. D.; Roy, W. K.; Southworth, G. R.; and Stewart, A. J.: Effects of chronic exposure to coal-derived oil on freshwater ecosystems: I. Microcosms. *Environ. Toxicol. Chem.*, 3:447–463, 1984.

Freemark, K.; MacQuarrie, P.; Swanson, S.; and Peterson, H.: Development of guidelines for testing pesticide toxicity to nontarget plants for Canada. Plants for Toxicity Assessment, edited by W. Wang, J. W. Gorsuch, W. R. Lower, pp. 14–29. Philadelphia: American Society for Testing and Materials, 1990.

Galassi, S.; Guzzela, L.; Mingazzini, M.; Vigano, L.; Capri, S.; and Sora, S.: Toxicological and chemical characterization of organic micropollutants in River Po water (Italy). *Water Res.*, 26:19–27, 1992.

Garg, P., and Chandra, P.: Toxicity and accumulation of chromium in *Ceratophyllum demersum L. Bull. Environ. Contam. Toxicol.*, 44:473–478, 1990.

Garten, C. T., and Frank, M. L.: Comparison of toxicity to terrestrial plants with algal growth inhibition by herbicides. ORNL/TM-9177. Oak Ridge, TN: Oak Ridge National Laboratory, 1984.

Gaur, J. P., and Singh, A. K.: Growth, photosynthesis and nitrogen fixation of *Anabaena doliolum* to Assam crude extract. *Bull. Environ. Contam. Toxicol.*, 44:494–500, 1990.

Gaur, J. P., and Singh, A. K.: Regulatory influence of light and temperature on petroleum toxicity to *Anabaena doliolum. Environ. Toxicol. Water Qual.*, 6:342–359, 1991.

Geckler, J. R.; Horning, W. B.; Neiheisel, T. M.; Pickering, Q. H.; Robinson, E. L.; and Stephan, C. E.: Validity of laboratory tests for predicting copper toxicity in streams. EPA 600/3-76-116, pp. 1–192. Duluth, MN: U.S. EPA. 1976.

Giddings, J. M.; Franco, P. J.; Cushman, R. M.; Hook, L. A.; Southworth, G. R.; and Stewart, A. J.: Effects of chronic exposure to coal-derived oil on freshwater ecosystems: II. Experimental ponds. *Environ. Toxicol. Chem.*, 3:465–488, 1984.

Giesy, J. P., and Hoke, R. A.: Freshwater sediment quality criteria: toxicity bioassessment. Sediments: Chemistry and Toxicity of In-place Pollutants, edited by R. Baudo, J. Giesy, H. Muntau, pp. 265–348. Chelsea, MI: Lewis, 1989.

Gobas, F. A. P. C.; Lovett-Doust, L.; and Haffner, G. D.: A comparative study of the bioconcentration and toxicity of chlorinated hydrocarbons in aquatic macrophytes and fish. Plants for Toxicity Assessment: Second Volume, edited by J. W. Gorsuch, W. R. Lower, W. Wang, M. A. Lewis, pp. 178–193. ASTM STP 1115. Philadelphia: American Society for Testing and Materials, 1991.

Goldsborough, L. G., and Brown, D. J.: Effect of glyphosate (Roundup formulation) on periphytic algal photosynthesis. *Bull. Environ. Contam. Toxicol.*, 41:253–260, 1988.

Goudey, J. S.: Modeling the inhibitory effects of metals on phytoplankton growth. *Aquat. Toxicol.*, 10:265–278, 1987.

Goulding, K. H.; Connolly, A. K.; Ellis, S. W.; Freeman, P.; and Kill, R. C.: The effect of DDT (1,1,1-trichloro-2,2 bis(*p*-chlorophenyl ethane) on the cellular composition of *Chlorella fusca. Environ. Pollut.* (Ser A) 34:23–35, 1984.

Graneli, E.; Sundström, B.; Edler, L.; and Anderson, D. M.: Toxic Marine Phytoplankton. New York Elsevier, 1990.

Greene, J. C.; Miller, W. E.; Debacon, M.; Long, M. A.; and Bartels, C. L.: Use of *Selenastrum capricornutum* to assess the toxicity potential of surface and ground water contamination caused by chromium waste. *Environ. Toxicol. Chem.*, 7:35–39, 1988.

Greene, J. C.; Peterson, S. A.; Parrish, L.; and Nimmo, D.: Zinc Sensitivity of *Selenastrum capricornutum* in algae assay medium with various EDTA concentrations. *Can. Tech. Rep. Fish Aquat. Sci.*, 1774:252–254, 1991.

Greeson, P. E.: An annotated key to the identification of commonly occurring and dominant genera of algae observed in the phytoplankton of the U.S. Geological Survey Water Supply, Paper 2079. Washington, DC: U.S. Government Printing Office, 1982.

Grice, G. D., and Reeve, M. R.: Marine Mesocosms: Biological and Chemical Research in Experimental Ecosystems. New York: Springer-Verlag, 1982.

Guhl, W., and Gode, P.: Correlations between lethal and chronic/biocenotic effect concentrations of surfactants. *Tenside Surf. Det.*, 26:282–287, 1989.

Guilizzoni, P.: The role of heavy metals and toxic materials in the physiological ecology of submersed macrophytes. *Aquat. Bot.*, 41:87–109, 1991.

Guillard, R. R. L., and Ryther, J. H.: Studies on marine planktonic diatoms. I. *Cyclotella nana* Hustedt and *Detonula confervacaceae* (Cleve) Gran. *Can. J. Microbiol.*, 8:229–239, 1962.

Gurney, S. E., and Robinson, G. G. C.: The influence of two triazine herbicides on the productivity biomass and community composition of freshwater marsh periphyton. *Aquat. Bot.*, 36:1–22, 1989.

Hadden, D. A.: A new slant on wetlands. *Water Environ. Tech.*, p. 46, February, 1994.

Hall, J.; Healy, F. P.; and Robinson, G. G. C.: The interaction of chronic copper toxicity with nutrient limitation in two chlorophytes in batch culture. *Aquat. Toxicol.*, 14: 1–14. 1989a.

Hall, J.; Healey, F. P.; and Robinson, G. G. C.: The interaction of chronic copper toxicity with nutrient limitation in chemostat cultures of *Chlorella. Aquat. Toxicol.*, 14: 15–26. 1989b.

Hammer, D. A. (ed.): Constructed Wetlands for Wastewater Treatment. Chelsea, MI: Lewis, 1989.

Hansen, S. R., and Garton, R. R.: Ability of standard toxicity tests to predict the effects of the insecticide deflubenzuron on laboratory stream communities. *Can. J. Fish Aquat. Sci.*, 39:1273–1288, 1982.

Harrass, M. C., and Sayre, P. G.: Use of microcosm data for regulatory decisions. Aquatic Toxicology and Hazard Assessment: 12th Volume, edited by U. M. Cowgill, L. R. Williams, pp. 204–223. ASTM STP 1027. Philadelphia: American Society for Testing and Materials, 1989.

Harrass, M. C.; Kindig, A. C.; and Taub, F. B.: Responses of blue-green and green algae to streptomycin in unialgal and paired culture. *Aquat. Toxicol.*, 6:1–11, 1985.

Harrison, P. J.; Waters, R. E.; and Taylor, F. J. R.: A broad spectrum artificial seawater medium for coastal and open water phytoplankton. *J. Phycol.*, 16:28–35, 1980.

Hartman, W. A., and Martin, D. B.: Effect of suspended bentonite clay on the acute toxicity of glyphosate to *Daphnia pulex* and *Lemna minor*. *Bull. Environ. Contam. Toxicol.*, 33:355–361, 1984.

Hartman, W. A., and Martin, D. B.: Effects of four agricultural pesticides on *Daphnia pulex*, *Lemna minor* and *Potamogeton pectinatus*. *Bull. Environ. Contam. Toxicol.*, 35:646–651, 1985.

Haslam, S. M.: A proposed method for monitoring river pollution using macrophytes. *Environ. Technol. Lett.*, 3: 19–34. 1982.

Hawkins, P. R., and Griffiths, D. J.: Copper as an algicide in a tropical reservoir. *Water Res.*, 4:475–480, 1987.

Herman, D. C.; Inniss, W. E.; and Mayfield, C. I.: Impact of volatile aromatic hydrocarbons, alone and in combination on growth of the freshwater alga *Selenastrum capricornutum*. *Aquat. Toxicol.*, 18:87–100, 1990.

Hersh, C. M., and Crumpton, W. G.: Atrazine tolerance of algae isolated from two agricultural streams. *Environ. Toxicol. Chem.*, 8:327–332, 1989.

Hillman, W. S.: The lemnaceae or duckweeds: a review of the descriptive and experimental literature. *Bot. Rev.*, 27: 221–287, 1961.

Hinman, M. L., and Klaine, S. J.: Uptake and translocation of selected organic pesticides by the rooted aquatic plant *Hydrilla verticillata* Royle. *Environ. Toxicol. Chem.*, 26: 609–613, 1992.

Hobson, J. F.; Sherman, J. W.; and Palmieri, M. A.: Evolution of the farm pond study. Aquatic Toxicology and Hazard Assessment: 12th Volume, edited by U. M. Cowgill, L. R. Williams, pp. 113–119. ASTM STP 1027. Philadelphia: American Society for Testing and Materials, 1989.

Hollister, T., and Walsh, G. E.: Differential responses of marine phytoplankton to herbicides: oxygen evolution. *Bull. Environ. Contam. Toxicol.*, 9:281–295, 1973.

Holst, R. W., and Ellwanger, T. C.: Pesticide Assessment Guidelines, Subdivision J, Hazard Evaluation: Nontarget Plants. EPA-54019-82-020. Washington DC: U.S. EPA, 1982.

Huber, W.; Zieris, F. J.; Feind, D.; and Neugebaur, K.: Ecotoxicological evaluation of environmental chemicals by means of aquatic model. Research Report 03-7314-0, Bonn, 1987.

Huebert, D. B., and J. M. Shay: Considerations in the assessment of toxicity using duckweeds. *Environ. Toxicol. Chem.*, 12:481–483, 1993.

Heubert, D. B.; McIlraith, A. L.; Shay, J. M.; and Robinson, G. G. C.: Axenic culture of *Lemna trisulca* L. *Aquat. Bot.*, 38:295–301, 1990.

Hughes, J. S., and Vilkas, A. G.: Toxicity of *N,N*-dimethylformamide used as a solvent in toxicity tests with the green alga, *Selenastrum capricornutum*. *Bull. Environ. Contam. Toxicol.*, 31:98–104, 1983.

Hughes, J. S.; Alexander, M. M.; and Balu, K.: An evaluation of appropriate expressions of toxicity in aquatic plant bioassays as demonstrated by the effects of atrazine on algae and duckweed. Aquatic Toxicology and Hazard Assessment: 10th Volume, edited by W. J. Adams, G. A. Chapman, W. G. Landis, pp. 531–547. ASTM STP 971. Philadelphia: American Society for Testing and Materials, 1988.

International Organization for Standardization (ISO): Water Quality—algal growth inhibition test. No. 8692. Paris, 1987.

International Organization for Standardization (ISO): Water quality: marine algal growth inhibition test with *Skeletonema costatum* and *Phaeodactylum tricornutum*. Kalfjeslaan, Netherlands, 1988.

Ireland, F. A.; Judy, B. M.; Lower, W. R.; Thomas, N. W.; Krause, G. F.; Asfaw, A.; and Sutton, W. W.: Characterization of eight soil types using the *Selenastrum capricornutum* bioassay. Plants for Toxicity Assessment: Second Volume, edited by J. W. Gorsuch, W. R. Lower, W. Wang, M. A. Lewis, pp. 217–229. ASTM STP 1115. Philadelphia: American Society for Testing and Materials, 1991.

Jones, T. W., and Winchell, L.: Uptake and photosynthetic inhibition by atrazine and its degradation products on four species of submerged vascular plants. *J. Environ. Qual.*, 13:243–247, 1984.

Jones, T. W.; Kemp, W. M.; Estes, P. S.; and Stevenon, J. C.: Atrazine uptake, photosynthetic inhibition and short-term recovery for the submersed vascular plants, *Potamogeton perfoliatus* L. *Arch. Environ. Contam. Toxicol.*, 15:277–283, 1986.

Jop, K. M.: Acute and rapid-chronic toxicity of hexavalent chromium to five marine species. Aquatic Toxicology and Hazard Assessment: 12th volume, edited by U. M. Cowgill, L. R. Williams, pp. 251–260. ASTM STP 1027. Philadelphia: American Society for Testing and Materials, 1989.

Kay, S. H.; Haller, W. T.; and Garrard, L. A.: Effects of heavy metals on water hyacinths (*Eichhornia crassipes* (Mart.) Solms). *Aquat. Toxicol.*, 5:117–128, 1984.

Kelly, M.: Mining and Its Freshwater Environment. New York: Elsevier Applied Science, 1988.

Kilgore, K. J.; Dibble, E. D.; and Hoover, J. J.: Relationship between fish and aquatic plants: a plan of study. Miscellaneous paper A-93-1. Vicksburg, MS: U.S. Army Corps of Engineers, 1993.

Kenaga, E., and Moolenar, R.: Fish and *Daphnia* toxicity as surrogates for aquatic and vascular plants and algae. *Environ. Sci. Technol.*, 13:1479–1480, 1979.

King, J. M., and Coley, K. S.: Toxicity of aqueous extracts of natural and synthetic oils to three species of *Lemna*, pp. 302–309. ASTM STP 891. Philadelphia: American Society Testing and Materials, 1985.

Klaine, S. J.: Toxicity of coal gasifier solid waste to the aquatic plants *Selenastrum capricornutum* and *Spirodela oligorhiza*. *Bull. Environ. Contam. Toxicol.*, 35:551–555, 1985.

Knight, R. L.: Treatment wetlands database now available. *Water Environ. Tech.*, p. 31, February, 1994.

Kooijman, S. A. L. M.; Hanstviet, A. O.; and Oldersma, H.: Parametric analyses of population growth in bioassays. *Water Res.*, 17:527–538, 1983.

Kosinski, R.: Artificial streams in ecotoxicological research. Aquatic Toxicology: Fundamental Concepts and Methodologies, edited by A. Boudou, F. Ribeyre, pp. 297–316. Boca Raton, FL: CRC Press, 1989.

Kosinski, R. J., and Merkle, M. G.: The effect of terrestrial herbicides on the community structure of stream periphyton. *Environ. Pollut.*, (Ser A) 36:165–189, 1984.

Kuivasnieme, K.; Eloranta, V.; and Knuutinen, J.: Acute toxicity of some chlorinated phenolic compounds to *Selenastrum capricornutum* and phytoplankton. *Arch. Environ. Toxicol. Chem.*, 14:43–49, 1985.

Langford, T. E. L.: Ecological Effects of Thermal Discharges, New York: Elsevier Applied Science, 1987.

LaPoint, T. W.; Fairchild, J. F.; Little, E. E.; and Finger, S. E.: Laboratory and field techniques in ecotoxicological research: strengths and limitations. Aquatic Ecotoxicology and Fundamental Concepts and Methodologies, edited by A. Boudou, F. Ribeyre, Vol. II. Boca Raton, CRC Preess, FL: 1989

Larsen, D. P.; deNoyelles, D. P. F.; Stay, F.; and Shiroyama, T.: Comparisons of single species, microcosm and experimental pond responses to atrazine exposure. *Environ. Toxicol. Chem.*, 5:179–190, 1986.

Larson, L. J.: The influence of test length and bacteria on the results of algal bioassays with monophenolic acids. Plants for Toxicity Assessment: Second Volume, edited by J. W. Gorsuch, W. R. Lower, W. Wang, and M. A. Lewis, pp. 230–239. ASTM STP 1115. Philadelphia: American Society for Testing and Materials, 1991.

Laube, V. S.; Ramamoorthy, S.; and Kushner, D. J.: Mobilization and accumulation of sediment bound heavy metals by algae. *Bull. Environ. Contam. Toxicol.*, 21: 763–770, 1979.

Lavoie, A., and de la Nolauue, J.: Hyperconcentrated cultures of *Secenedesmus obliques*: a new approach for wastewater biological tertiary treatment. *Water Res.*, 19: 1437–1442, 1985.

LeBlanc, G. A.: Interspecies relationships in acute toxicity of chemicals to aquatic organisms. *Environ. Toxicol. Chem.*, 3:37–50, 1984.

Leffler, J. W.: The use of self-selected, generic aquatic microcosms for pollution effects assessment. Concepts in Marine Pollution Meausurements, edited by H. H. White, pp 139–158, College Park, MD: Univ. of Maryland, 1984.

LeKven, G. C.; William, C. R.; Charney, R. D.; and Crites, R. W.: Wetlands put to the test. *Water Environ. Technol.*, p. 40, February, 1994.

Leischman, A. A.; Greene, J. C.; and Miller, W. E.: Bibliography of literature pertaining to the genus *Selenastrum*. EPA-600/9-79-021, U.S. EPA, Corvallis, OR: 1979.

Leland, H. V., and Carter, J. L.: Use of detrended correspondence analysis in evaluating factors controlling species composition of periphyton. Rationale for Sampling and Interpretation of Ecological Data in the Assessment of Freshwater Ecosystems edited by B. G. Isom, ASTM STP 894, Philadelphia: American Society for Testing and Materials, 1986.

Lewis, M. A.: Are laboroatory-derived toxicity data for freshwater algae worth the effort? *Environ. Toxicol. Chem.*, 9:1279–1284, 1990a.

Lewis, M. A.; Pittinger, C. A.; Davidson, D. H.; and Ritchie, C. J.: In-situ response of natural periphyton to an anionic surfactant and an environmental safety assessment of phytotoxic effects. *Environ. Toxicol. Chem.*, 12: 1803–1812, 1993.

Lewis, M. A.: Chronic toxicities of surfactants and detergent builders to algae: a review and risk assessment. *Ecotoxicol. Environ. Safety*, 20:1279–1284, 1990b.

Lewis, M. A.: Comparison of the effects of surfactants on freshwater phytoplankton communites in experimental enclosures and on algal population growth in the laboratory. *Environ. Toxicol. Chem.*, 5:319–332, 1986.

Lewis, M. A., and Hamm, B. G.: Environmental modification of the photosynthetic response of lake plankton to surfactants and significance to a laboratory-field comparison. *Water Res.*, 12:1575–1582, 1986.

Lewis, M. A.; Taylor, M. J.; and Larson, R. J.: Structural and functional repsonse of natural phytoplankton and periphyton communities to a cationic surfactant with considerations on environmental fate. Community Toxicity Testing, edited by J. Cairns, pp. 241–268. STP 920 Philadelphia: American Society for Testing and Materials, 1986.

Ludyanskiy, M. L., and Pasichny, A. P.: A system for water toxicity estimation. *Water Res.*, 26:689–694, 1992.

Lukavsky, J.: The evaluation of algal growth potential by cultivation on solid media. *Water Res.*, 17:549–558, 1983.

Lundy, P.; Wurster, C. F.; and Rowland, R. F.: A two-species marine algal bioassay for detecting aquatic toxicity of chemical pollutants. *Water Res.* 18:187–194, 1984.

Lustigman, B.; Korky, J.; Zabady, A.; and McCormick, J. M.: Absorption of Cu^{++} by long-term cultures of *Dunaliella saline*, *D. tertiolecta* and *D. viridis*. *Bull. Environ. Contam. Toxicol.*, 35:362–367.

Lustigman, B. K.: Enhancement of pigment concentrations in *Dunaliella tertiolecta* as a result of copper toxicity. *Bull. Environ. Contam. Toxicol.*, 37:710–713, 1986.

Macek, K.; Birge, W.; Mayer, F.; Buikema, A.; and Maki: Discussions session synopsis. Estimating the Hazard of Chemical Substances to Aquatic Life, edited by J. Cairns, K. L. Dickson, and A. W. Maki, pp. 27–32. ASTM STP 657. Philadelphia: American Society for Testing and Materials, 1978.

Maita, Y., and Kawaguchi, S.: Amino acid composition of cadmium-binding protein induced in a marine diatom, *Phaeodactylum tricornutum*. *Bull. Environ. Contam. Toxicol.*, 43:394–301, 1989.

Mangi, J.; Schmidt, K.; Pankow, J.; Gaines, L.; and Turner, P.: Effects of chromium on some aquatic plants. *Environ. Pollut.*, 16:285–291, 1978.

Marshall, J. S., and Mellinger, D. L.: An in situ experimental method for toxicological studies on natural plankton communities. Aquatic Toxicology, edited by J. G. Eaton, P. R. Parrish, A. C. Hendricks, pp. 27–39. ASTM STP 707. Philadelphia: American Society for Testing and Materials, 1980.

Meyerhoff, R. D.; Grothe, D. W.; Sauter, S.; and Dorulla, G.: Chronic toxicity of tebuthiuron to an alga (*Selenastrum capricornutum*) a cladoceran (*Daphnia magna*) and the fathead minnow (*Pimephales promelas*). *Environ. Toxicol. Chem.*, 4:695–701, 1985.

Mhatre, G. N., and Chaphekar, S. B.: The effects of mercury on some aquatic plants. *Environ. Pollut.*, (Ser A) 39:207–216, 1985.

Miller, W. E.; Greene, J. C.; and Shiroyama, T.: The *Selenastrum capricornutum* Printz algal assay bottle test: experimental design, application, and data interpretation protocol EPA-600/9-78-018, pp. 1–126. Corvallis, OR: U.S. EPA, 1978.

Millie, D. F., and Hersch, C. M.: Statistical characterizations of the atrazine induced photosynthetic inhibition of *Cyclotella menghiniana* (Bacillariophyta). *Aquat. Toxicol.*, 10:239–249, 1987.

Millington, L. A.; Goulding, K. H.; and Adams, N.: The influence of growth medium composition on the toxicity of chemicals to algae. *Water Res.*, 22:1593–1597, 1988.

Missimer, C. L.; Lemarie, D. P.; and Rue, W. J.: Evaluation of a chronic estimation toxicity test using *Skeletonema costatum*. Aquatic Toxicology and Hazard Assessment: 12th Volume, edited by U. M. Cowgill, L. R. Williams, pp. 345–354. ASTM STP 1027. Philadelphia: American Society for Testing and Materials, 1989.

Mitchell, C.: Growth of *Halodule wrightii* in culture and the effects of cropping, light, salinity and atrazine. *Aquat. Bot.*, 28:25–37, 1987.

Molander, S.; Blanck, H.; and Söderström, M.: Toxicity assessment by pollution-induced community tolerance (PICT), and identification of metabolites in periphyton communities after exposure to 4,5,6-trichloroguaiacol. *Aquat. Toxicol.* 18:115–136, 1990.

Molander, S.; Dahl, B.; Blanck, H.; Jonsson, J.; and Sjostrom, M.: Combined effects of tributly tin (TBN) and diuron on marine periphyton communities detected as pollution-induced community tolerance. *Arch. Environ. Contam. Toxicol.*, 22:419–427, 1992.

Moore, M. W., and Winner, R. W.: Relative sensitivity of *Ceriodaphnia dubia* laboratory tests and pond communities of zooplankton and benthos to chronic copper stress. *Aquat. Toxicol.*, 15:311–330, 1989.

Moorhead, D. L., and Kosinski, R. J.: Effect of atrazine on the productivity of artificial stream algal communities. *Bull. Environ. Contam. Toxicol.*, 37:330–336, 1986.

Morgan, J. A.: Effects of Aroclor 1242 and DDT on cultures of an alga, protozoan, daphnid, ostracod and guppy. *Bull. Environ. Contam. Toxicol.*, 8:129–137, 1972.

Morrison, G.; Torello, E.; Comeleo, R.; Walsh, R.; Kuhn, A.; Burgess, R.; Tagliabue, M.; and Greene, W.: Interlaboratory precision of saltwater short-term chronic toxicity tests. *Res. J. Water Pollut. Control Fed.*, 51:1708–1710, 1989.

Moshiri, G. A. (ed.): Constructed Wetlands for Water Quality Improvement. Chelsea, MI: Lewis, 1993.

Mosser, J. L.; Fisher, N. S.; and Wurster, C. F.: Polychlorinated biphenyls and DDT alter species composition in mixed cultures of algae. *Science*, 176:633–635, 1972.

Muir, D. C. G.; Grift, N. P.; and Lockhart, W. L.: Comparison of laboratory and field results for prediction of the environmental behavior of phosphate esters. *Environ. Toxicol. Chem.*, 1:113–119, 1982.

Munawar, M., and Munawar, I. F.: Phytoplankton bioassays for evaluating toxicity of in-situ sediment contaminants. *Hydrobiologia*, 149:87–105, 1987.

Nasu, Y.; Kugimoto, M.; Tanaka, O.; and Takimoto, A.: Comparative studies on the absorption of cadmium and copper in *Lemna paucicostata*. *Environ. Poll.*, (Ser A) 32:201–209, 1983.

National Environmental Technology Applications Corporation (NETAC): Oil spill bioremediation products testing protocol methods manual. Univ. of Pittsburgh Applied Research Center, Pittsburgh, 1991.

Nichols, S. A.: The interaction between biology and the management of aquatic macrophytes. *Aquat. Bot.*, 41: 225–252, 1991.

Niederlehner, B. R., and Cairns, Jr. J.: Effects of ammonia on periphytic communities. *Environ. Pollut.*, 55:207–221, 1990.

Nyberg, H.: Physiological effects of four detergents on the algae *Nitzschia actinastroides* and *Porphyridium purpureum*. Publication No. 12. Department of Botany, Univ. of Helsinki, Finland, 1985.

Nyholm, N.: Response variable in algal growth inhibition tests—biomass or growth rate. *Water Res.*, 19:273–279, 1985.

Nyholm, N.: Expression of results from growth inhibition toxicity tests with algae. *Arch. Environ. Contam. Toxicol.*, 19:518–522, 1990.

Nyholm, N., and Källqvist, T.: Methods for growth inhibition toxicity tests with freshwater algae. *Environ. Toxicol. Chem.*, 8:6899–703, 1989.

O'Kelley, J. C., and Deason, T. R.: Degradation of pesticides by algae. Office of Research and Development. ERPA-60013-76-022. Athens, GA: U.S. Environmental Protection Agency, 1976.

Organization for Economic Cooperation and Development (OECD): Alga growth inhibition test. Test Guidline No. 201. OECD Guidelines for Testing of Chemicals. Paris, 1984a.

Organization for Economic Cooperation and Development (OECD): Terrestrial plants: growth test. OECD Guidelines for Testing of Chemicals No. 208. Paris, 1984b.

Palmer, C. M.: A composite rating of algae tolerating organic pollution. *J. Phycol.*, 5:78–82, 1969.

Palmer, C. M.: Algae and Water Pollution. EPA/600/14. Cincinnati, OH: Office of Research and Development, U.S. EPA, 1977.

Palmer, C. M., and Maloney, T. E.: Preliminary screening for potential algicides. *Ohio J. Sci.*, 1:1–8, 1955.

Parkhurst, B. R.; Hicks, W.; and Noel, L. E.: Performance characteristics of effluent toxicity tests: summarization and evaluation of data. *Environ. Toxicol. Chem.*, 11:771–791, 1992.

Patrick, R.: Use of algae, especially diatoms, in the assessment of water quality. Biological Methods for the Assessment of Water Quality, edited by J. C. Cairns, K. L. Dickson, pp. 76–95. ASTM STP 528. Philadelphia: American Society for Testing and Materials, 1973.

Patrick, R.; Cairns, J., Jr.; and Scheir, A.: The relative sensitivity of diatoms, snails and fish to twenty common constituents of industrial wastes. *Prog. Fish Culturist*, July: 137–140, 1968.

Payne, A. G., and Hall, R. H.: A method for measuring algal toxicity and its application to the safety assessment of new chemicals. Aquatic Toxicology, edited by L. L.

Marking, R. A. Kimerle, pp. 171–180. ASTM STP 667. Philadelphia: American Society for Testing and Materials, 1979.

Peets, R.; Miller, A. C.; and Beckett, D. C.: Effects of three species of aquatic plants on macroinvertebrates in Lake Seminole, Georgia. Technical Report A-94-5. Vicksburg, MS: U.S. Army Corps. of Engineers, 1994.

Perkins, J. L.: Bioassay evaluation of diversity and community comparison indices. *J. Water Pollut. Control Fed.*, 55:522–530, 1983.

Peterson, B. J.; Hobbie, J. E.; Corliss, T. L.; and Kreit, K.: A continuous-flow periphyton bioassay: tests of nutrient limitation in a tundra stream. 28:583–591, 1983.

Peterson, H. G.: Toxicity testing using a chemostat-grown green alga, *Selenastrum capricornutum*. Plants for Toxicity Assessment: Second Volume, edited by J. S. Gorsuch, W. R. Lower, W. Wang, M. A. Lewis, pp. 107–117. ASTM STP 1115. Philadelphia: American Society for Testing and Materials, 1991.

Plumley, F. G., and Davis, D. E.: The effects of a photosynthetic inhibitor atrazine on salt marsh edaphic algae, in culture, microecosystems and in the field. *Estuaries*, 3:271–277, 1980.

Provasoli, L.: Media and prospects for the cultivation of marine algae. Cultures and Collections of Algae, edited by A. Watanabe, A. Hattori, p. 63 Proceedings U.S.-Japan Conference, Habone, Japanese Society of Plant Physiology, p. 63, 1968.

Rai, L. C.; Gaur, M. J. P.; and Kumar, H. D.: Phycology and heavy-metal pollution. *Biol. Rev.*, 56:99–151, 1981.

Reed, S. (ed.): Natural systems for wastewater treatment. Manual of Practice FD-15. Washington, DC: Water Pollution Control Federation, 1990.

Rehnberg, B. G.; Schultz, D. A.; and Raschke, R. L.: Limitations of electronic particle counting in reference to algal assays. *J. Water Pollut. Control Fed.*, 54:181–186, 1982.

Riedel, G. F.: Interspecific and geographical variation of the chromium sensitivity of algae. Aquatic Toxicology and Environmental Fate: Eleventh Volume, edited by G. W. Suter, II, M. A. Lewis, pp. 537–548. ASTM STP 1007. Philadelphia: American Society for Testing and Materials, 1989.

Roberts, M. H.; Warinner, J. E.; Tsai, C. F.; Wright, D.; and Cronin, L. E.: Comparison of estuarine species sensitivities to three toxicants. *Arch. Environ. Contam. Toxicol.*, 11:681–692, 1982.

Rodgers, J. H.; Cherry, D. S.; and Guthrie, R. K.: Cycling of elements in duckweed (*Lemna perpusilla*) in an ash settling basin and swamp drainage system. *Water Res.*, 12:765–770, 1978.

Rodgers, J. H.; Dickson, K. L.; and Cairn, J., Jr.: A review and analysis of some methods used to measure functional aspects of periphyton. Methods and Measurements of Periphyton Communities: A review, edited by R. L. Weitzel, pp. 1432–167. Philadelphia: American Society for Testing and Materials, 1979.

Rowe, E. L.; Ziobro, R. J.; Wang, C. J.; and Dence, C. W.: The use of an alga *Chlorella pyrenoidosa* and a duckweed *Lemna perpusilla* as test organisms for toxicity bioassays of spent bleaching liquors and their compounds. *Environ. Pollut.*, (Ser A) 27:289–296, 1982.

Satake, K.; Takamatsu, T.; Soma, M.; Shibata, K.; Nishikawa, M.; Say, P. J.; and Whitton, B. A.: Lead accumulation and location in the shoots of the aquatic liverwort *Scapania undulata* Dum. in stream water at Greenside mine, England. *Aquat. Bot.*, 33:111–122, 1989.

St. Laurent, D.; Blaise, C.; MacQuarrie, P.; Scroggins, R.; and Trottier, B.: Comparative assessment of herbicide phytotoxicity to *Selenastrum capricornutum* using microplate and flask bioassay procedures. *Environ. Toxicol. Water Qual.*, 7:35–48, 1992.

Schelske, C. L.: In-situ and natural phytoplankton assemblage bioassyays. Algae As Ecological Indicators, edited by L. E. Shubert, pp. 15–47. New York: Academic Press, 1984.

Schimmel, S. C.; Thursby, G. B.; Herber, M. A.; and Chammas, M. J.: Case study of a marine discharge: comparison of effluent and receiving water toxicity. Aquatic Toxicology and Environmental Fate: Eleventh Volume, edited by G. W. Suter, II, M. A. Lewis, p. 159–173. ASTM STP 1007. Philadelphia: American Society for Testing and Materials, 1989.

Schindler, D. W.; Mills, K. H.; Malley, D. F.; Findlay, D. L.; Shearer, J. A.; Davies, I. J.; Turner, M. A.; Linsey, G. A.; and Cruikshank, D. R.: Long-term ecosystem stress. The effects of years of experimental acidification on a small lake. *Science*, 228:1395–1401, 1985.

Scorgie, H. R. A.: Ecological effects of the aquatic herbicide cyanatryn on a drainage channel. *J. Appl. Ecol.*, 17: 2076–2254, 1980.

Sebastian, S., and Nair, K. V. K.: Total removal of coliforms and *E. coli* from domestic sewage by high-rate pond mass culture of *Scendesmus obliques*. *Environ. Pollut.*, (Ser A) 34:197–206. 1984.

Shales, S.; Thake, B. A.; Frankland, B.; Khan, D. H.; Hutchinson, J. D.; and Mason, C. F.: Biological and ecological effects of oils, p81–173. The Fate and Effects of Oil in Freshwater, edited by J. Green, M. W. Treet pp. 81–173, New York: Elsevier Applied Science, 1989.

Shaw, B. P.; Shau, A.; and Panigrahi, A. K.: Effect of the effluent from a chlor-alkali factory on a blue-green alga: changes in the pigment content. *Bull. Envrion. Contam. Toxicol.*, 43:618–626, 1989.

Sheehan, P. J.; Axler, R. P.; and Newhook, R. C.: Evaluation of simple generic aquatic ecosystem tests to screen the ecological impacts of pesticides. Community Toxicity Testing, edited by J. Cairns, Jr., pp. 158–179. ASTM STP 920. Philadelphia: American Society for Testing and Materials, 1986.

Shehata, S. A., and Nawar, S. S.: Toxicity effect of antigerm 50 to algae and fish. *Z. Wasser Abwasser Forsch.*, 12;226–229, 1979.

Shiegoka, T.; Sato, Y.; Takeda, Y.; Yoshida, K.; and Yamauchi, F.: Acute toxicity of chlorphenols to green algae, *Selenastrum capricornutum* and *Chlorella vulgaris* and quantitative structure-activity relationships. *Environ. Toxicol. Chem.*, 7:847–854, 1988.

Shubert, L. E. (ed.): Algae as Ecological Indicators. London: Academic Press, 1984.

Skowronski, T.; Szubinska, S.; Pawlik, B.; and Jakubowski, M.: The influence of pH on cadmium toxicity to the green alga *Stichococcus bacillaris* and on the cadmium forms present in the culture medium. *Environ. Pollut.*, 74:89–100, 1991.

Smart, R. M., and Barko, J. W.: Laboratory culture of submersed freshwater macrophytes on natural sediments. *Aquat. Bot.*, 21:251–263, 1985.

Smith, B. M.: An inter- and intra-agency survey of the use of plants for toxicity assessment. Plants for Toxicity Assessment: Second Volume, edited by J. W. Gorsuch, W. R. Lower, W. Wang, M. A. Lewis pp. 41–59. ASTM STP 1115. Philadelphia: American Society for Testing and Materials, 1991.

Smith, P. D.; Brockway, D. L.; and Stancil, F. E.: Effects of hardness, alkalinity and pH on the toxicity of pentachlorophenol to *Selenastrum carpicornutum* (Printz). *Environ. Toxicol. Chem.*, 6:891–900, 1987.

Solomon, K. R., and Liber, K.: Fate of pesticides in aquatic mesocosm studies—an overview of methodology. Proceedings of the Brighton Crop Protection Conference, Brighton, England, pp. 139–148, 1988.

Sortkjaer, O.: Macrophytes and macrophyte communities as test systems in ecotoxicological studies of aquatic systems. *Ecol. Bull. Stockh.*, 36:75–80, 1984.

Stanley, R. A.: Toxicity of heavy metals and salts to Eurasian watermilfoil (*Myriophyllum spicatum L.*) *Arch. Environ. Contam. Toxicol.*, 2:331–342, 1974.

Stay, F. S.; Katko, A.; Rohm, M.; Fix, M. A.; and Larsen, D. P.: The effects of atrazine on microcosms developed from four natural plankton communities. *Arch. Environ. Contam. Toxicol.*, 18:866–875, 1989.

Steele, R., and Thursby, G. B.: Laboratory culture of gametophytic stages of the marine macroalgae *Champia parvula* (Rhodophyta) and *Laminaria saccharina* (Phaeophyta). *Environ. Toxicol. Chem.*, 7:997–1002, 1988.

Stein, J. R. (ed.): Handbook of Phycological Methods Cambridge: Cambridge Univ Press, 1973.

Stephan, C.; Mount, D.; Hansen, D.; Gentile, J.; Chapman, G.; and Grungs, W.: Guidelines for Deriving National Water Quality Criteria for the Protection of Aquatic Organisms and Their Uses. PB85-227049, U.S. Environmental Protection Agency, Office of Water Regulations and Standards. Springfield, VA: U.S. Government Printing Office, 1985.

Stokes, P. M.; Howell, E. T.; and Krantzberg, G.: Effects of acid precipitation on the biota of freshwater lakes. Acid Precipitation, Vol. 2, Biological Effects, pp. 273–305. New York: Springer-Verlag, 1989.

Stokes, P.: Uptake and accumulation of copper and nickel by metal-tolerance strains of *Scenedesmus*. *Verh. Int. Verein. Limnol.*, 19:2128–2127, 1975.

Stratton, G. W.: The effects of pesticides and heavy metals towards phototrophic microorganisms. Reviews in Environmental Toxicology 3, edited by E. Hodgson, pp. 71–147. New York: Elsevier, 1987.

Stratton, G. W., and Giles, K.: Importance of bioassay volume in toxicity tests using algae and aquatic invertebrates. *Bull. Environ. Contam. Toxicol.*, 44:420–427, 1990.

Stratton, G. W., and Smith, T. M.: Interaction of organic solvents with the green alga *Chlorella pyrenoidosa*. *Bull. Environ. Contam. Toxicol.*, 40:736–742, 1988.

Swanson, S. M.; Rickard, C.P.; Freemark, K. E.; and MacQuarrie, P.: Testing for pesticide toxicity to aquatic plants: recommendation for test species. Plants for Toxicity Assessment: Second Volume, edited by J. W. Gorsuch, W. R. Lower, W. Wang, M. A. Lewis, pp. 77–97. ASTM STP 1115. Philadelphia: American Society for Testing and Materials, 1990.

Taraldsen, J. E., and Norberg-King, T. J.: New method for determining effluent toxicity using duckweed (*Lemna minor*). *Environ. Toxicol. Chem.*, 9:761–767, 1990.

Taub, F. B.: Synthetic microcosms as biological models of algal commuinities. Algae as Ecological Indicators, edited by L. E. Shubert, pp. 363–394. New York: Academic Press, 1984.

Taub, F. B.; Kindig, A. C.; Conquest, L. L.; and Meador, J. P.: Results of interlaboratory testing of the standardized aquatic microcosm protocol. Aquatic Toxicology and Environmental Fate: Eleventh Volume, edited by G. W. Suter II and M. A. Lewis, pp. 368–394. ASTM STP 1007. Philadelphia: American Society for Testing and Materials, 1989.

Tease, B.; Hartman, E.; and Coler, R. A.: An in-situ method to compare the potential for periphyton productivity of lotic habitats. *Water Res.*, 17:589–591, 1983.

Thomas, J. M.; Skalski, J. R.; Cline, J. F.; McShane, M. C.; Miller, W. E.; Peterson, S. A.; Callahan, C. A.; and Greene, J. C.: Characterization of chemical waste site contamination and determination of its extent using bioassays. *Environ. Toxicol. Chem.*, 5:487–501, 1986.

Thursby, G. B., and Steele, R. L.: Toxicity of arsenite and arsenate to the marine macroalga *Champia parvula* (Rhodophyta). *Environ. Toxicol. Chem.*, 3:391–397, 1984.

Thursby, G. B.; Anderson, F. B.; Walshe, G. E.; and Steele, R. L.: A review of the current status of marine algal testing in the United States. First Symposium on Environmental Toxicology and Risk Assessment: Aquatic, Plant and Terrestrial, edited by Landis, W. G., Hughes, J. S., Lewis, M. A., pp. 362–377. ASTM STP 1179. Philadelphia: American Society for Testing and Materials, 1993.

Tripathi, B. D., and Shukla, S. C.: Biological treatment of wastewater by selected aquatic plants. *Environ. Pollut.*, 69:69–78, 1991.

Turbak, S. C.; Olson, S. B.; and McFeters, G. A.: Comparison of algal systems for detecting water-borne herbicides and metals. *Water Res.*, 20:981–96, 1986.

Ukeles, R.: Cultivation of plants: unicellular plants. *Marine Ecology*, Vol. III, edited by O. Kinne, pp. 367–529. New York: Wiley, 1976.

U.S. Army Corps of Engineers (COE): A plant bioassay for assessing plant uptake of heavy metals from contaminated freshwater dredged matieral. EEDP-04-11. Vicksburg, MS: U.S. Army Engineer Waterways Experiments U.S. Environmental Protection Agency: Algal assay: Bottle test. Corvallis, OR: National Eutrophication Research Program, Pacific Northwest Environmental Research Laboratory, 1971a.

U.S. Environmental Protection Agency: Inter-laboratory precision test—an eight laboratory evaluation of the Provisional Algal Assay Procedure Bottle Test. Corvallis, OR: National Eutrophication Research Program, Pacific Northwest Environmental Research Laboratory, 1971b.

U.S. Environmental Protection Agency: Marine algal assay procedure bottle test. Corvallis, OR: Eutrophication and Lake Restoration Branch National Environmental Research Center, 1974.

U.S. Environmental Protection Agency: Bioassay Procedures for the Ocean Disposal Permit Program. EPA-60019-76-010. Gulf Breeze, FL: U.S. EPA, 1976.

U.S. Environmental Protection Agency: Seed germination/root elongation toxicity test. EG-12. Washington, DC: Office of Toxic Substances, 1982.

U.S. Environmental Protection Agency: Toxic substances control act test guidelines; final rules. Fed Regist 50: 797.1050, 797.1075, and 797.1060, 1985a. (Also Technical Support Document.)

U.S. Environmental Protection Agency: *Lemna* acute toxicity test. Fed. Regist. 50:39331–39333, 1985b.

U.S. Environmental Protection Agency: Aquatic mesocosm tests to support pesticide registrations. U.S. EPA-EEB/HED/OPP, 540/09-88-035. Washington, DC: U.S. EPA, 1988.

U.S. Environmental Protection Agency: Program survey—biological toxicity testing in the NPDES permits program. Washington, DC: Permits Division EN-336, Office of Water Enforcement and Permits, 1986.

U.S. Environmental Protection Agency: Ambient water quality criteria for copper. EPA 440/5-80-036. Springfield, VA: National Technical Information Service, 1980.

U.S. Environmental Protection Agency-U.S. Army Corps of Engineers: Evaluation of dredged materials proposed for ocean disposal. Testing manual. EPA-503/9-01/001. Washington, DC, 1991.

Vasseur, P., and Pandard, P.: Influence of some experimental factors on metal toxicity to *Selenastrum capricornutum*. *Toxicity Assessment*, 3:331–343, 1988.

Vasseur, P.; Ferard, J. F.; and Babut, M.: The biological aspects of the regulatory control of industrial effluents in France. *Chemosphere* 22:626–633, 1991.

Versteeg, D. J.: Comparison of short-and long-term toxicity test results for the green alga, *Selenastrum capricornutum*. Plants for Toxicity Assessment, edited by W. Wang, J. W. Gorsuch, W. R. Lower, pp. 4–48. ASTM STP 1091. Philadelphia: American Society for Testing and Materials, 1990.

Vymazal, J.: Uptake of lead, chromium, cadmium and cobalt by *Cladophera glomerata*. *Bull. Environ. Contam. Toxicol.*, 44:468–472, 1990.

Walker, J., and Evans, S.: Effect of quaternary ammonium compounds on some aquatic plants. *Mar. Pollut. Bull.*, 9: 136–137, 1978.

Walker, J. D.: Relative sensitivity of algae, bacteria, invertebrates and fish to phenol: analysis of 234 tests conducted for more than 149 species. *Toxicity Assessment*, 3:415–447. 1988.

Wallbridge, C. T.: A flow through testing procedure with duckweed (*Lemna minor*). EPA-600/3-77-108, Duluth, MN: U.S. Environmental Protection Agency, 1977.

Walsh, G. E.: Algal bioassay of industrial and energy process effluents. EPA-600/D-82-141, Gulf Breeze, FL, 1982.

Walsh, G. E.: Principles of toxicity testing with marine unicellular algae. *Environ. Toxicol. Chem.*, 7:979–987, 1988.

Walsh, G. E., and Alexander, S. V.: A marine algal bioassay method: results with pesticides and industrial wastes. *Water Air Soil Pollut.*, 13:45–55, 1980.

Walsh, G. E., and Garnas, R. L.: Determination of bioactivity of chemical fractions of liquid wastes using freshwater and saltwater algae and crustaceans. *Environ. Sci. Technol.*, 17:180–182, 1983.

Walsh, G. E., and Merrill, R. G.: Algal bioassays of industrial and energy process effluents. Algae as Ecological Indicators, edited by L. E. Shubert, pp. 329–360. New York: Academic Press, 1984.

Walsh, G. E.; Bahner, L. H.; and Horning, W. B.: Toxicity of textile mill effluents to freshwater and estuarine algae, crustaceans, and fishes. *Environ. Pollut.*, (Ser A) 21: 169–179, 1980.

Walsh, G. E.; Duke, K. M.; and Foster, R. B.: Algae and crustaceans as indicators of bioactivity of industrial wastes. *Water Res.*, 16:879–883, 1982.

Walsh, G. E.; Yoder, M. J.; McLaughlin, L. L.; and Lores, E. M.: Responses of marine unicellular algae to brominated organic compounds in six growth media. *Ecotoxicol. Environ. Safety*, 14:215–222, 1987a.

Walsh, G. E.; Deans, C. H.; and McLaughlin, L. L.: Comparison of the EC_{50}s of algal toxicity tests calculated by four methods. *Environ. Toxicol. Chem.*, 6:767–770, 1987b.

Walsh, G. E.; Weber, D. E.; Simon, T. L.; and Brashers, L. K.: Toxicity tests of effluents with marsh plants in water and sediment. *Environ. Toxicol. Chem.*, 10:517–525, 1991b.

Walsh, G. E.; Weber, D. E.; Simon, T. L.; Brashers, L. K.; and Moore, J. C.: Use of marsh plants for toxicity testing of water and sediment. Plants for Toxicity Assessment: Second Volume, edited by J. W. Gorsuch, W. R. Lower, W. Wang, M. A. Lewis, pp. 341–354. ASTM STP 1115. Philadelphia: American Society for Testing and Materials, 1991a

Wang, W.: Toxicity tests of aquatic pollutants by using common duckweed. *Environ. Pollut.*, (Ser B) 11:1–14, 1986.

Wang, W.: Chromate ion as a reference toxicant for aquatic phytotoxicity tests. *Environ. Toxicol. Chem.*, 6:953–960, 1987a.

Wang, W.: Toxicity of nickel to common duckweed (*Lemna minor*). *Environ. Toxicol. Chem.*, 6:961–967, 1987b.

Wang, W.: The effect of river water on phytotoxicity of Ba, Cd, and Cr. *Environ. Pollut.*, (Ser B) 11:193–204, 1987c.

Wang, W.: Review: Literature review on duckweed toxicity testing. *Environ. Res.*, 52:7–22, 1990.

Wang, W.: Use of plants for the assessment of environmental contaminants. *Rev. Environ. Contam. Toxicol.*, 126:87–127, 1992.

Wang, W., and Williams, J. M.: The use of phytotoxicity tests (common duckweed, cabbage and millet) for determining effluent toxicity. *Environment Monitor Assess.*, 14:45–58, 1990.

Wängberg, S., and Blanck, H.: Multivariate patterns of algal sensitivity to chemicals in relation to phylogeny. *Ecotoxicol. Environ. Safety*, 16:72–82, 1988.

Washington, H. G.: Diversity, biotic and similarity indices: a review with special reference to aquatic ecosystems. *Water Res.*, 18:653–694, 1984.

Watkins, C. H., and Hammerschlag, R. S.: The toxicity of chlorine to a common vascular aquatic plant. *Water Res.*, 19:1037–1043, 1984.

Weber, C. I. (ed.): Biological field and laboratory methods for measuring the quality of surface waters and effluents. Program Element 1BA 027, EPA/670/4-783-001. Cincinnati, OH: U.S. EPA, 1973.

Weber, C.; Horning, W. B.; Klemm, D. J.; Neiheisel, T. W.; Lewis, P. A.; Robinson, E.; Menkendick, J. R.; and Kessler, F. A.: Short-term methods for estimating the chronic toxicity of effluents and receiving waters to marine and estuarine organisms. EPA 600/4-876/028. Cincinnati, OH: Environmental Monitoring Support Laboratory, 1988.

Weber, C. I.; Peltier, W. H.; Norberg-King, T. J.; Horning, W. B.; Kessler, F. A.; Menkedick, J. R.; Neiheisel, T. W.; Lewis, P. A.; Klemm, D. J.; Pickering, Q. H.; Robinson, E. L.; Lazorchak, J. M.; Wymer, L. J.; and Freyberg, R. W.: Short-term methods for estimating the chronic toxicity of effluents and receiving waters to freshwater organisms. EPA 600/4-89/001. Cincinnati, OH: Environmental Monitoring Systems Laboratory, 1989.

Weider, R. K.: A survey of constructed wetlands for acid-coal mine drainage treatment in the eastern United States. *Wetlands*, 9:299–315, 1989.

Weitzel, R. L. (ed.): Methods and Measurement of Periphyton Communities. ASTM STP 690. Philadelphia: American Society for Testing and Materials, 1979.

Welch, E. B., and Lindell, T.: Ecological Effects of Wastewater. New York: Ellis Horwood, 1992.

Whitton, B. A.: Toxicity of heavy metals to freshwater algae: a review. *Phykos.*, 9:116–25, 1970.

Whitton, B. A.: Plants as indicators of river water quality. Biological Indicators of Water Quality, edited by A. James, L. Evison, p. 5.1–5.35. Chicester: Wiley, 1979.

Whitton, B. A.: Algae as monitors of heavy metals. Algae as Ecological Indicators edited by L. E. Shubert, pp. 257–280. New York: Academic Press, 1984.

Weider, R. K.: A survey of constructed wetlands for acid-coal mine drainage treatment in the eastern United States. *Wetlands*, 9:299–315, 1989.

Williams, O. B.; Groniger, C. R.; and Albritton, N. F.: The algicidal effect of certain quaternary ammonium compounds. Producers Monthly June: 14–15, 1952.

Winner, R. W.; Owen, H. A.; and Moore, M.: Seasonal variability in the sensitivity of freshwater lentic communities to a chronic copper stress. *Aquat. Toxicol.*, 17: 75–92, 1990.

Wolverton, B. C.; Barlow, R. M.; and McDonald, R. C.: Application of vascular aquatic plants for pollution removal, energy and food production in a biological system. NASA Tech. Memo. No. TMX-72726. Bay St. Louis, MS: National Space Technology Laboratory, 1975.

Wong, M. H.; Chu, L. M.; and Chan, W. C.: The effects of heavy metals and ammonia in sewage sludge and animal manure on the growth of *Chlorella pyrenoidosa*. *Environ. Pollut.*, (Ser A) 34:55–71, 1984.

Wren, M. J., and McCarroll, D.: A simple and sensitive bioassay for the detection of toxic materials using a unicellular green algae. *Environ. Pollut.*, 64:87–91, 1990.

Yoshida, T.; Maruyama, T.; Imamaura, H.; Allahpichay, I.; and Mori, S.: Evaluation of the effect of chemicals on aquatic ecosystem by observing the photosynthetic activity of a macrophyte, *Porphyra yezoensis*. *Aquat. Toxicol.*, 9:207–214, 1986.

Yount, D. J., and Richter, J. E.: Effects of pentachlorophenol on periphyton communities in outdoor experimental streams. *Arch. Environ. Contam. Toxicol.*, 15:51–60, 1986.

SUPPLEMENTAL READING

Bates, J. M., and Weber, C. I. (eds.): Ecological Assessments of Effluent Impacts on Communities of Indigenous Aquatic Organisms. ASTM 730. Philadelphia: American Society for Testing and Materials, 1981.

Fletcher, J., and Ratsch, H.: Plant tier testing: a workshop to evaluate nontarget plant testing in subdivision J pesticide guidelines. EPA/600/9-91/0941. Corvallis, Orgeon: U.S. EPA, 1991.

Gorsuch, J. W.; Lower, W. R.; Lewis, M. A.; and Wang, W. (eds.): Plants for Toxicity Assessment: Second Volume. ASTM STP 1115. Philadelphia: American Society for Testing and Materials, 1991.

Lewis, M. A.: Freshwater primary producers. Handbook of Ecotoxicology, Volume 1 edited by P. Calow, pp. 28–50. Cambridge: Blackwell Scientific, 1993.

Riemer D. N. (ed.): Introduction to Freshwater Vegetation. Westport CT: AVI, 1984.

Shubert, L. E. (ed.): Algae as Ecological Indicators. Orlando, FL: Academic Press, 1984.

Wang, W.; Gorsuch, J. W.; and Lower, W. R. (eds.): Plants for Toxicity Assessment. ASTM STP 1091. Philadelphia: American Society for Testing and Materials, 1990.

Weitzel, R. L. (ed.): Methods and Measurements of Periphyton Communities: A Review. ASTM STP 690. Philadelphia: American Society for Testing and Materials, 1979.

Chapter 5

SEXUAL REPRODUCTION TESTS WITH MARINE SEAWEEDS (MACROALGAE)

G. B. Thursby and *R. L. Steele*

INTRODUCTION

New and exotic chemicals, as well as many known compounds, are released into the air and water daily. Most of these compounds find their way into the marine environment and need to be tested as potential toxicants to different trophic levels of marine organisms, including algae. Algal toxicity testing is not a new field, but only within the past few years have standards for allowable contamination considered algal data (see Chapter 4). The field of aquatic toxicology started with acute (short-term, lethal) testing on fish and other aquatic animals. Aquatic plant testing was later added, with an emphasis on eutrophication studies using growth of freshwater phytoplankton (similar tests were eventually developed using marine species of phytoplankton). Standardized test procedures using seaweeds, however, have only recently begun to receive attention.

Seaweeds are saltwater macroalgae belonging to the divisions Chlorophyta, Phaeophyta, and Rhodophyta, commonly known as green, brown, or red algae, respectively. The life histories of seaweeds can be among the most complex in the plant kingdom, with sexual reproduction for many species being the key to perennial survival. Payne and Hall (1979) proposed an algistatic response as the best end point for toxicity tests with microalgae. This end point, however, would not be useful with seaweeds because asexual and sexual reproduction can be impaired at toxicant concentrations at which growth does not cease (Steele and Thursby, 1983; Steele and Hanisak, 1978; Anderson et al., 1990). Work with red and brown algae has incorporated sexual reproduction as the primary end point for deciding toxicity. Sexual reproduction tests using green seaweeds have not been developed to date, as it is much more difficult to find a definitive reproductive end point with green algae; many gametes play a dual role as zoospores.

Seaweeds represent a different ecological niche from microalgae. They are generally sessile and represent a different food source from microalgae. Phytoplankton are grazed by filter-feeders, whereas it is generally held that macroalgae contribute primarily to detrital food chains (Mann, 1972). Many coastal macrophytes provide significant habitat and community structure for marine animals (e.g., kelp forests), and several species of macrophytic algae in North America are economically important as well. The kelps *Macrocystis* and *Laminaria* are a source of alginic acid, and the red alga *Chondrus crispus* is a source of carrageenan. Any assessment of potential effects of chemicals on the marine environment must include data from toxicity tests with this environmentally and economically important component. Currently, the primary algal toxicity tests used routinely employ planktonic microalgae (see Chapter 4). The level of expertise required to use seaweeds in reproductive tests is approximately equal to that necessary to run microalgal toxicity tests.

Most seaweeds are sessile and many are of sufficient biomass to be useful for bioaccumulation studies (integrating an exposure response to toxicants over time; see Chapter 16) to monitor pollution with heavy metals and organic chemicals (Levine, 1984, and references therein). This type of study requires the use of tolerant species or life history stages because the goals is to have an alga that will accumulate a chemical but not die because of the exposure (Phillips, 1977). The opposite is true in toxicological studies. One selects a sensitive species or life history stage; if sensitive species are protected, tolerant species should be protected as well. Historically, seaweeds have been considered less useful for toxicity testing than microalgae (Jensen, 1984), and microalgae have often been considered less sensitive than aquatic animals (Kenaga and Moolenaar, 1976; Kenaga, 1982, see Chapter 4). Jensen (1984) said that seaweeds "seem to be rather insensitive to many chemicals and will probably survive pollution better than many other organisms in the marine environment." There was also a presumption (based on early tests with insensitive species of algae) that "results of toxicity tests with plants usually indicated that criteria which adequately protect aquatic animals and their uses will probably also protect aquatic plants and their uses" (Stephan et al., 1985).

These previous decisions concerning seaweed insensitivity to toxicants were based on data for only a few hardy species and based generally on vegetative growth of adult stages as the primary end point. The conclusion that seaweeds are essentially insensitive was based partly on vegetative growth studies with species of the rockweeds *Fucus* and *Ascophylum*. It is not surprising that these algae are insensitive because they are primarily perennial intertidal species. As such, they must tolerate extremes of exposure to the heat of summer and the freezing of winter. Seaweed sensitivity to toxicants, however, increases dramatically when effects on sexual reproduction are assessed. This has been shown for the red alga *Champia parvula* (Thursby et al., 1985) and the brown algae *Fucus edentatus* and *Laminaria saccharina* (Steele and Hanisak, 1978) and *Macrocystis pyrifera* (Anderson et al., 1990). Often seaweeds are more sensitive to toxicants than the most sensitive aquatic animals (Thursby et al., 1985).

A reoccurring question with macroalgal tests is whether they produce acute or chronic data. It is recommended that the data be referred to as neither, substituting instead a concept of whether the data are vegetative or reproductive. This chapter focuses on three macroalgal test protocols that are currently being used to varying degrees to assess the effects of chemicals on sexual reproduction. The first method uses *C. parvula*, a subtidal red alga in the order Rhodymeniales. It is the only one of the three methods that has an "official" Environmental Protection Agency (EPA) test protocol (U.S. EPA, 1988), and an ASTM version has recently been published for testing with this species (ASTM, 1994). *C. parvula* is one of the test species that can be required for NPDES permitting (Chapter 24) of marine effluents. The other two test methods both use species of brown algae. Test procedures are being completed by EPA for *L. saccharina*, a kelp in the order Laminariales. Sexual reproduction data for this species have also been used to propose National Water Quality Criteria for thallium and acenaphthene. The final test procedure described is for several species of *Fucus*, rockweeds in the order Fucales that grow both subtidally and intertidally. Sexual reproduction data for *Fucus* spp. have been used to assess damage and recovery from oil spills.

TERMINOLOGY

Antheridium (*antheridia*). The structure that produces sperm in oogamous sexual reproduction.

Conceptacle. A nearly spherical invagination or cavity containing reproductive structures in certain brown and red algae.

Cystocarp. A structure produced by female gametophytes in response to fertilization.

Gametophyte. The sexual, gamete-producing phase in the life history of a plant.

Parthenogenetic. Production of a new individual from a single, unfertilized gamete, often an egg.

Receptacle. A fertile area on which gametangia or sporangia arise.

Sorus (*sori*). A group or cluster of reproductive organs.

Spermatium (*spermatia*). Male gamete in red algae, nonmotile and colorless; released from spermatangium.

Sporophyte. The diploid (usually) spore-producing plant or phase in a life cycle.

Thallus (*thalli*). A plant body not differentiated into roots, stems, and leaves.

Vegetative. Not associated with reproduction, as in vegetative growth.

CHAMPIA PARVULA

The toxicity test procedure using *C. parvula* was originally developed as a 2-wk test to assess chronic effects of pollutants to marine seaweeds and was evaluated with heavy metals and cyanide (Steele and Thursby, 1983). The original procedure also tested the effects of arsenite and arsenate (Thursby and Steele, 1984) and 10 different organic compounds (Thursby et al., 1985) on *C. parvula*. This method showed that sexual reproduction was generally a sensitive and practical end point to use for *Champia* (Table 1). The current, 2-d sexual reproduction test is as sensitive to single chemicals as the 2-wk procedure (Thursby and Steele, 1986) and is of short enough duration to fit time constraints imposed on tests used in effluent evaluations. Effluents cannot be easily tested using the 2-wk test procedure, because it requires that the test chambers remain unialgal during the test period. Any microalgae introduced with the effluent would compete with *Champia* for light and nutrients, thus influencing *Champia*'s growth rate. Fine filtering or autoclaving, which would be necessary to eliminate unwanted microalgae, could change the character of the effluent. However, if sexual reproduction is the primary end point, then plants need be exposed for only a few days (long enough to show effects on fertilization). Any effects of other organisms on the growth rate of *Champia* should not be serious because the variable of interest is whether sexual reproduction takes place.

Distribution and Life History

Champia parvula (C. Agardh) Harvey is a common species in many parts of the world (Taylor, 1957; Abbott and Hollenberg, 1976; Lewis, 1973; Reedman and Womersley, 1976). It grows in Mexico and southern California on the West Coast of North America, and from the Caribbean to Massachusetts on the East Coast. It has also been reported in Brazil, France, Spain, Korea, and southern Australia.

Table 1. Effect concentrations for *Champia parvula* based on statistically significant differences from controls for various chemical compounds[a]

	Females		Tetrasporophytes	
Compound	Vegetative growth	Cystocarp production	Vegetative growth	Production of tetrasporangia
Toxaphene	39	39	14	65
Naphthalene	2,210	2,210	2,210	1,160
2,4,5-Trichlorophenol	1,290	1,290	2,040	7,780
Pentachlorophenol	4,700	4,700	7,900	1,680
Phenol	21,600	7,800	7,800	7,800
Benzene	57,100	57,100	34,260	34,260
Isophorone	83,070	49,840	83,070	49,840
Silver	2.6	1.6[b]	2.6	2.6
Copper	4.3	6.0[b]	4.1	10.6
Lead	15.8	15.8[b]	16.2	16.2
Cadmium	11.4	11.4[b]	12.5	>189
Cyanide	5.5	5.5[b]	13.5	20.5

[a] Exposure durations were 11 d for tetrasporophytes and 14 d for females (and males). Values are expressed in μg/L.
[b] Effect based on total absence of cystocarps.
Data are from Steele and Thursby, 1983 and Thursby et al., 1985.

The plants of *C. parvula* are bushy and 5 to 10 cm tall in the field. The main axis and branches are cylindrical, hollow, and septate. Healthy plants have many colorless, sterile hairs throughout the surface of the thallus. The exact function of these hairs is not known, but they are thought to aid in nutrient uptake. In addition, on the female, these hairs may help to "capture" spermatia from the water column and then pass them on to the trichogynes (the reproductive hairs). *Champia*'s life history is an alternation of isomorphic generations (Figure 1). Fertilization of the female gametophyte by the spermatia from the male gametophyte results in formation of the microscopic carposporophyte, which is "parasitic" on the female and housed within the cystocarp. Each carposporophyte can produce many diploid spores, which in turn germinate into another life history stage, the tetrasporophyte. Meiosis occurs

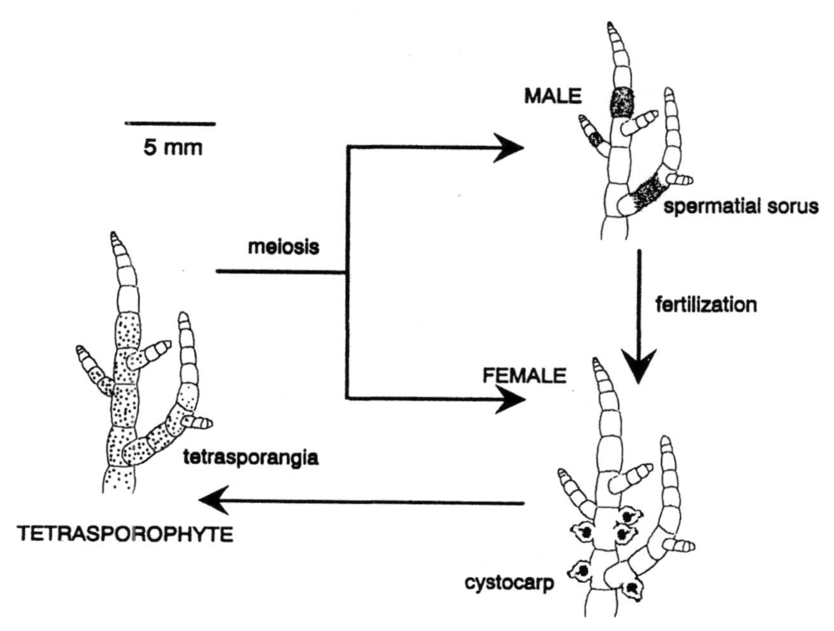

Figure 1. Life history of *Champia parvula*. All three macroscopic stages superficially look the same. Only the male and female gametophyte are used in toxicity testing. The end point measured for toxic effects is the production of cystocarps (evidence of sexual reproduction). One cystocarp is produced for each successful sexual fusion.

within tetrasporangia of this stage and the resultant haploid tetraspores give rise to new female and male gametophytes.

Culture

Although the expertise required to culture *Champia* in the laboratory is not greater than that required to rear phytoplankton, some laboratories that have successfully cultured phytoplankton have had problems culturing *Champia*. Many phytoplankton species that are cultured in toxicology laboratories are hardy species and success is judged by vegetative growth. Success with *Champia*, on the other hand, is determined by ability to produce gametes. Seemingly slight differences in laboratory conditions can make a big difference in the ability to culture *Champia* for successful gamete production. These differences, however, are really no greater than those between conditions to maintain healthy fish or invertebrate populations and conditions to culture reproducing populations.

Male and female plants of *Champia* can easily be maintained in separate unialgal cultures in the laboratory, providing clonal test plants with a similar history. To guard against microalgal contamination sterile technique is used, including autoclaving all stock solutions and culture vessels and flaming all tools before cutting or transferring plants. New cultures can be started from excised branches, making it possible to keep clonal material indefinitely with no special preconditioning required to induce reproduction. Thus, plant material can be available for testing any time of the year.

Natural seawater is the medium of choice for *Champia* and should be filtered to at least 0.45 μm and sterilized either by further filtration or by autoclaving. Additional nutrients are added for proper growth (Steele and Thursby, 1988). Almost any nutrient recipe can be used for *Champia* cultured in natural seawater. Medium f (Guillard and Ryther, 1962), Provasoli's medium (Provasoli, 1971), GP2 (Spotte et al., 1984), and the commercially available AlgoGro (Carolina Biological Supply) have all yielded successful results. Healthy, actively growing and reproducing plants are the goal, not a standard culture recipe. Many commercial artificial seawaters or seawater made from recipes with reagent-grade salts will support good growth of *Champia* as long as enough chelation is present in the nutrient medium. Chelation must be reduced or eliminated, however, for toxicity testing, especially if metals are tested. In addition, although *Champia* grows well in these artificial seawaters, it does not always exhibit its correct morphology or reproduce well. Although *Champia* cannot be cultured well in artificial seawater, artificial seawater can be used during the exposure portion of toxicity testing.

Cultures are gently aerated and illuminated with 40 to 75 $\mu E\ m^{-2}\ s^{-1}$ of cool-white fluorescent light on a 16:8, light-dark cycle. A dark period is essential for good gamete development. The best water temperature is 22 to 24°C with a salinity of 28 to 30 ppt. Media are changed once a week, although longer periods between medium changes may be acceptable, depending on the biomass-to-volume ratio.

Preparation of Plants for a Test

Stock cultures should be checked for their readiness for use in toxicity tests. The color of the tissue should be a uniform reddish to reddish brown color. The exact color varies with culture conditions, especially irradiance level; the brighter the light, the paler the plants. Under conditions of stress the tips of the branches turn pink and the older tissue is generally much paler. Under conditions of nutrient deficiency the entire plant turns pale yellow. If the stress is severe enough the older tissue (main axes) or occasionally the branch tips turn white (evidence of necrotic tissue). Healthy female gametophytes have several trichogynes (reproductive hairs to which the spermatia attach) easily seen near the apex (Figure 2-1). Mature spermatial sori can be identified by looking at the edge of the male thallus (Figure 2-2), usually 1 to 2 cm back from the tip in healthy plants. On healthy male gametophytes the sorus area is generally thicker and lighter in color than the rest of the plant body. Fertilization takes place on the female when a nonmotile spermatum attaches to a trichogyne (Figure 2-3) and the male nucleus migrates down the trichogyne to the egg nucleus. The resulting fusion starts the process for producing a cytocarp (Figure 2-4), the structure counted at the end of a toxicity test. Once cultures are determined to be healthy enough for toxicity testing (i.e., sufficient gametes are present), branch tips should be cut into their final size. For females, 7- to 10-mm branch tips are used, generally five per treatment chamber. Only one male branch, 2.5 to 3 cm long, is used per treatment chamber, but it should have two or more spermatial sori present.

Test Conditions

The most common test volume is 100 ml, although larger volumes can be used. Testing can be performed in polystyrene "party" cups or 125-ml Erlenmeyer flasks. The use of either of these is generally a matter of preference, although some consideration of the expected toxicant may be necessary (e.g., metals adsorb more readily to glass than to plastic). If glass test chambers are used, then before reuse they should be acid-stripped in 10 to 15% HCl (v/v) and rinsed thoroughly with deionized water after washing. The acid removes potentially toxic residues left by the detergent as well as heavy metals. If organic compounds have been previously tested in the glassware, a rinse with acetone prior to washing is recommended and baking in a muffle oven may be necessary.

The test exposure duration to toxicants is 2 d, followed by a 5- to 7-d development period without toxicants for maturation of cystocarps. The exposure

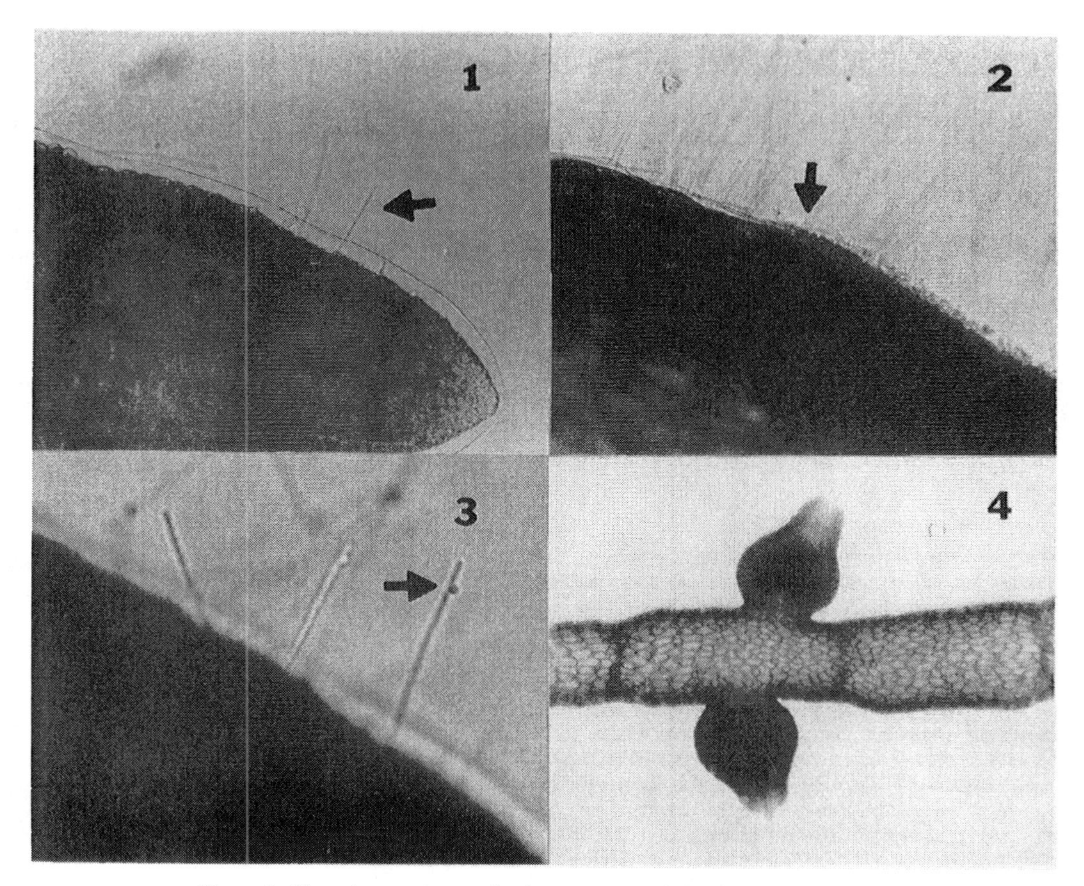

Figure 2. *Champia parvula* reproductive structures: (1) tip of branch on a female showing receptive hairs, called trichogynes (~ 200 μm long), which are attached to the egg cell (larger hairs in background are sterile; (2) portion of a male plant showing the spermatial areas on the surface; (3) single spermatia (~ 10 μm in diameter) attached to a trichogyne; (4) cystocarps.

water temperature is between 22 and 24°C, with a test salinity of 28 to 32 ppt. For testing receiving waters, salinity will often be below the desired range and must be adjusted with artificial sea salts or hypersaline brine. The photoperiod should be a 16:8 light-dark cycle of 40 to 75 μE m^{-2} s^{-1} of cool-white fluorescent light. It is not necessary for the recovery conditions to be the same as the exposure conditions; however, the more optimal they are, the faster cystocarps develop.

Plants are not aerated during the exposure period. Chambers are either shaken at 75 to 100 rpm on a rotary shaker or briefly hand-swirled twice a day (even when shaken, a daily hand-swirling helps ensure spermatia suspension). Spermatia are not motile, so some water motion is essential, but attachment of spermatia to trichogynes is a rapid event. In an early experiment, female branch tips were placed into seawater containing spermatia for different amounts of time (unpublished results). The results were striking: all time exposures from 1 s (a quick dip) to 4 h produced essentially the same number of cystocarps after a week of recovery (Table 2). Aeration, however, enhances the growth rate of plants in the recovery bottles but is not essential. Shaking can be substituted, although cystocarp maturation may be delayed a day or two.

The nutrients added to the exposure medium are not as many or as concentrated as those used for culture (Steele and Thursby, 1988). The recipe is modified for the exposure medium during toxicity testing, including reduction or elimination of EDTA, omission of trace metals, and a general reduction in the concentration of the rest of the nutrients. Although a standard culture medium may not be necessary, a standard exposure nutrient medium is recommended because it helps to make exposure conditions more consistent.

Females are added to the treatment chambers first, followed by the male plants, so that any reproduction must take place in the presence of the toxicant. After 48 h the females are removed and placed in control

Table 2. Effect of time of exposure to seawater containing spermatia on the production of cystocarps[a]

Time exposed to spermatia	Cystocarps per plant (mean ± SD)
Control (not exposed)	0
1 s	16 ± 5
15 s	11 ± 6
60 s	12 ± 5
5 min	12 ± 5
15 min	13 ± 4
60 min	15 ± 5
2 h	12 ± 3
4 h	16 ± 2

[a] Values represent the mean number of cystocarps per five branch tips with the corresponding standard deviation. Cystocarps were counted after a 1-wk development period.

medium to allow the development of any cystocarps. Although sexual reproduction is the primary end point, notes on any necrotic tissue within any of the replicates can be useful in later interpretation of results. Necrosis generally first appears on the male spermatial sori or on the cut ends of the branches of both sexes. It is probably no coincidence that these areas of the plants are also the only places where a clear cuticle does not cover the plant surface. Counts of cystocarps are made with the aid of a stereomicroscope at the end of a 5- to 7-d development period. One advantage of this test procedure is that if there is uncertainty about the identification of an immature cystocarp, the plants can be held a little longer in the recovery vessels. No new cystocarps will form because the males have been removed; plants will only get bigger.

The effect of concentrations of chemicals or effluents on cystocarp production can be determined by either analysis of variance (ANOVA) followed by a mean separation test (e.g., Dunnett's) or by calculating an IC50 value (see Chapter 10 on Statistical Analysis). The *IC50* is the concentration of a chemical that causes a 50% inhibition in cystocarp production by treated plants compared with control plants.

Criteria for Acceptability of a Test

A test should not be accepted if there is any control mortality. Plants from the controls and the lowest exposure concentration should be in good physical condition. The branches should not be fragmented or necrotic. This could mean that the plants were unhealthy from the beginning of the test. A test is not acceptable if the controls average fewer than 10 cystocarps per plant. Ideally, cystocarps should be clustered in groups on the controls. The results from the replicate control chambers should be identical, and all replicates from the affected concentration chambers should show an effect.

Test Performance

Precision

One area of concern in any toxicity test is the repeatability of the results. This concern is usually addressed by estimating the precision of the procedure with repeated tests of the same toxicant. Copper and sodium dodecyl sulfate (SDS) have been used in several laboratories as reference toxicants with *C. parvula* (Table 3). The coefficients of variation for these reference toxicants are within the range seen for other test species with the same two toxicants (Morrison et al., 1989). There is also good agreement among the IC50 values calculated from tests conducted at different laboratories. Of perhaps even more practical importance, tests using complex whole effluents instead of single chemical compounds have shown the same degree of repeatability (Table 4).

Artificial Seawater

Although an artificial seawater medium would allow more consistency in water quality among different

Table 3. Single laboratory precision test series for copper and sodium dodecyl sulfate (U.S. EPA, 1991)

Laboratory	Compound	Number of tests	Mean IC50[b] (μg/L)	Coefficient of variation (%)
One	Copper	6	1.5	44
	SDS	9	360	37
Two	Copper	4	3.4	34
	SDS	4	750	23

[a] Data from laboratory one were collected by four different investigators over approximately 4 yrs. Data from laboratory two were collected by a single investigator over approximately 1 yr.
[b] 50% inhibition concentration.

Table 4. Single laboratory precision with complex effluents[a]

Effluent	Number of tests	Mean IC50[b] (% effluent)	Coefficient of variation (%)
Rhode Island STP[b]			
December 1985	6	2.1	71
July 1986	7	4.2	17
September 1986	5	2.9	34
Puerto Rico			
STP 1	4	2.8	21
STP 2	4	3.3	18
STP 3	4	3.0	53
Naval Air Station	3	6.5	12

[a] All tests were run at the U.S. EPA's ERL-Narragansett. All tests for a given effluent were run within a 7-d period.
[b] 50% inhibition concentration.

Table 5. Comparison of the number of cystocarps produced by females of the marine red alga, *Champia parvula*, when exposed to fertile male plants for 2 d in either natural of artificial seawater[a]

Date	Natural seawater (mean ± SD)	Artificial seawater[b] (mean ± SD)
November 1985	21.5 ± 5.0	19.5 ± 0.7
November 1985	13.0 ± 5.7	13.5 ± 0.7
May 1988	25.5 ± 4.3	20.7 ± 2.9
May 1988	18.1 + 5.0	18.4 ± 4.9
April 1990	17.6 ± 3.2	10.8 ± 0.6
April 1990	17.6 ± 5.2	8.4 ± 4.9
May 1990	17.0 ± 3.6	11.9 ± 2.9
May 1990	24.5 ± 3.5	22.9 ± 1.2
July 1990	11.9 ± 3.8	12.1 ± 1.9

[a] After the exposure, all plants were recovered in natural seawater.
[b] November 1985 artificial seawater was straight GPS (Spotte et al., 1984). The artificial seawater medium used for all other comparisons was a dilution of 3× GP2 brine (salinity of 90 ppt).

tests and among different laboratories, attempts to culture *C. parvula* in artificial media have not been completely successful. However, as stated earlier, artificial seawater can be used during the exposure portion of the test procedure. In side-by-side tests with both natural and artificial seawater the response of control plants was the same as long as the plants were cultured and recovered in more optimal media (Table 5). Test results with copper and silver have also been similar in both natural and artificial seawaters, again because the artificial seawater plants were cultured and recovered in more optimal media (Figure 3).

Salinity Adjustment

Because most effluents are fresh water and many ambient samples that require toxicity testing are of low salinity, a way to adjust salinity and still yield good control responses is essential if tests are to be performed with *C. parvula*. One convenient way to adjust salinity is with hypersaline brine. Table 6 shows the results of adjusting the salinity to 30 ppt of samples of various initial salinities. Three types of brine were used: natural seawater brine prepared by evaporating seawater, artificial brine prepared with reagent-grade salts, and brine made by mixing reagent-grade salts into natural seawater. In all cases the numbers of cystocarps produced were similar, demonstrating that either of these methods can be used to adjust salinity of test samples.

Success Rate

Figure 4 shows a frequency distribution for the control data from 337 *Champia* tests performed from 1985 through 1990. Included are effluent, receiving water, and single chemical compound tests. The success rate (meeting the criterion of an average of 10 cystocarps or greater) was 83.4%. The mean control response for acceptable tests was 18.7 cystocarps per plants with a coefficient of variation of 36%.

Minimum Detectable Difference

For each of 109 successful effluent tests performed between 1985 and 1990, the minimum detectable sta-

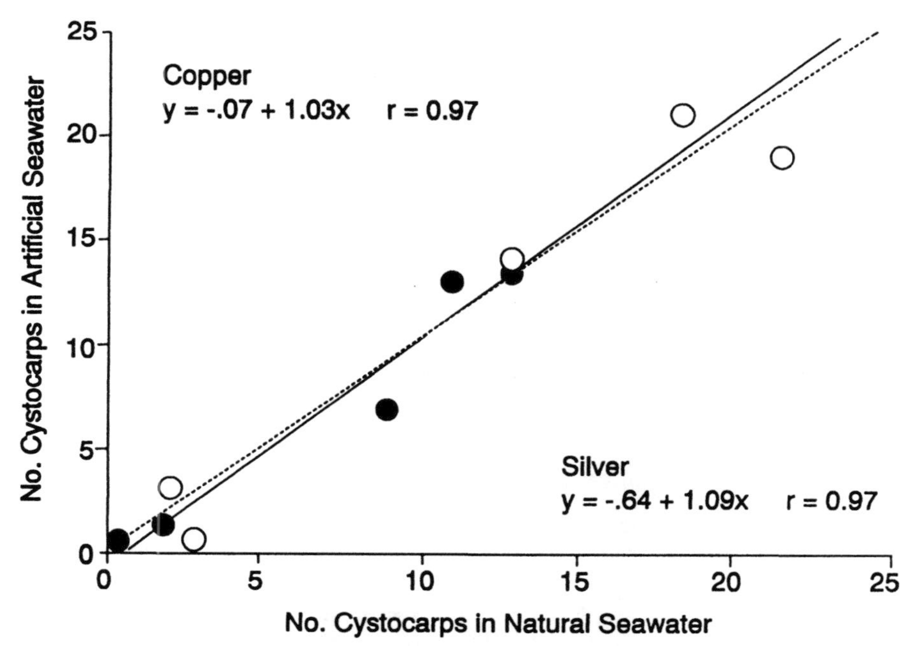

Figure 3. Comparison of the number of cystocarps produced by *Champia parvula* when exposed to different concentrations of copper or silver in either natural or artificial seawater.

Table 6. Cystocarp production by the marine red alga *Champia parvula*[a]

Initial salinity (ppt)	3× GP2 brine[b] (90 ppt)	2× GP2 in NSW[c] (90 ppt)	NSW brine[d] (100 ppt)
Control (30 ppt)	19.7 + 4.2	19.3 + 4.2	19.0 ± 5.1
0	16.0 ± 5.9	17.3 ± 6.0	16.2 ± 8.0
5	17.4 ± 5.4	18.8 ± 3.7	17.6 ± 8.3
10	15.8 ± 6.3	18.6 ± 5.4	19.3 ± 12.8
15	18.1 ± 6.4	18.0 ± 5.7	17.3 ± 10.4

[a]A comparison of different methods of adjusting low-salinity water back to 30 ppt. Initial salinities were prepared with combinations of different amounts of natural seawater from Narragansett Bay, RI and either freshwater from a well or deionized water. Controls were straight seawater at 30 ppt. Each control value represents the mean of three separate experiments of three replicates with five plants per replicate. Each salinity treatment represents the mean of six such experiments.

[b]Triple-strength GP2 (Spotte et al. 1984) made up with deionized water.

[c]Double-strength GP2 made up with 30 ppt seawater.

[d]Brine made by evaporation of natural seawater.

tistical difference was calculated. The statistics performed for each test consisted of one-way analysis of variance followed by Dunnett's mean separation test. The detectable difference was expressed as a percentage of the control response and the results are shown in the frequency distribution in Figure 5. The mean detectable difference was 38.2% of the control, with a range of 13.2 to 47.9%. The test was able to detect statistically a 50% difference (a common cutoff point in many toxicity test procedures) 80 to 90% of the time.

Comparison with Other Isolates and Red Algal Species

Toxicity testing completed to date using *C. parvula* demonstrates its usefulness as a test species and its sensitivity to a variety of toxicants. However, most of the testing with this species has been performed with a single clone isolated in 1979 from Rhode Island waters. *C. parvula* can undergo spontaneous somatic mutation in the laboratory (Steele et al., 1986); therefore it is important to establish that the isolate that is being selected for routine use in toxicity testing is not an unusually sensitive mutant. Testing with two other isolates of *C. parvula*, as well as two other species of red algae, was used to address this concern. The two additional *Champia* isolates were collected from Rhode Island in 1981 and from Pusan, South Korea. In addition, isolates of *Agardhiella subulata* (C. Agardh) Kraft and Wynne and *Grinnellia americana* (C. Agardh) Harvey were cultured from plants collected from Rhode Island. The responses of these isolates and species to three organic compounds (phenol, toxaphene, and isophorone) were compared with that of the standard isolate. All tests lasted 2 to 3 wk (depending on species) so that both growth (increase in dry weight) and reproduction could be compared. Data from these tests showed that the standard isolate is not unusually sensitive relative to the other plants tested (Table 7).

Mechanism of Toxicity

Inhibition of cystocarp production by female gametophytes can be due to toxic effects at one or more of a variety of locations. Spermatia can fail to form

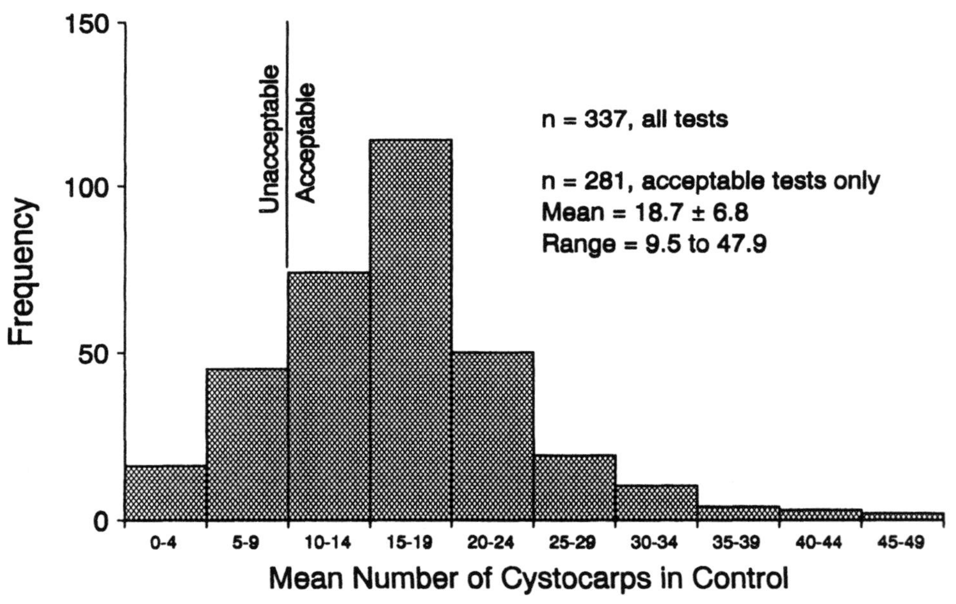

Figure 4. Frequency diagram summarizing the control response for cystocarp production by *Champia parvula* from toxicity tests performed from 1985 through 1990.

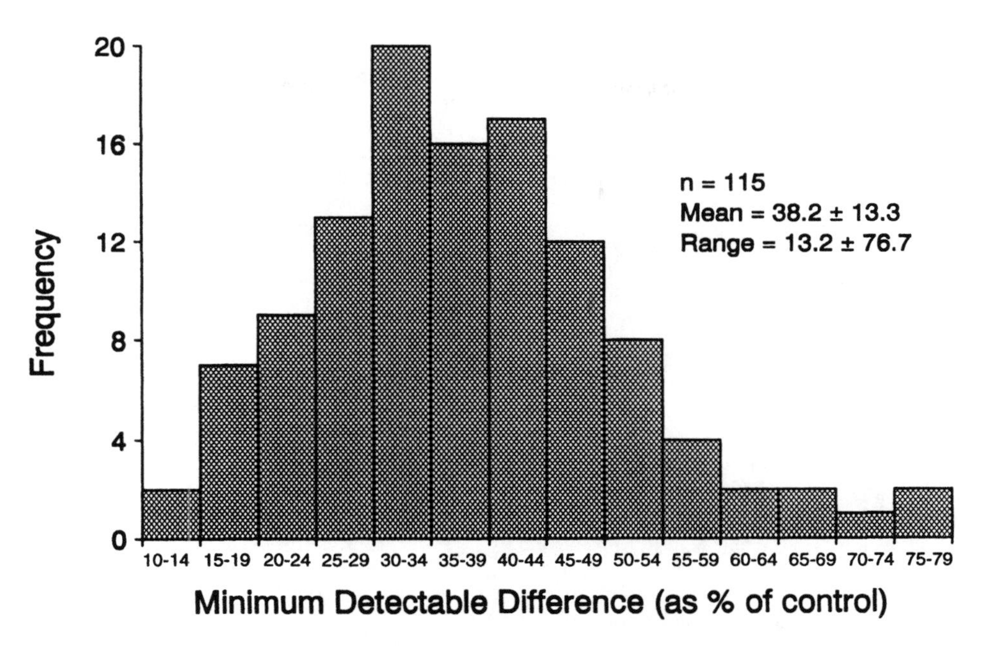

Figure 5. Frequency diagram summarizing the ability to detect statistical differences in effluent toxicity tests performed from 1985 through 1990.

on or be released from male plants. There can be a failure of sperm to attach to trichogynes or, once attached, a failure to fuse with the nucleus of the egg cell. Chemicals can also affect reproduction by interfering with trichogyne formation on female plants. Finally, there can be toxic effects on postfertilization development events. All of these can cause a failure of female plants to produce cystocarps, but toxic effects on cystocarp formation often seem due to effects on male gametophytes. Cystocarp formation is

Table 7. Comparison of the sensitivity of the standard isolate of the marine red alga *Champia parvula* with two other isolates and two other species of red algae[a]

		Concentration that affects cystocarp production (μg/L)	
Species	Compound	Statistical difference[b]	Absence[c]
Champia parvula 1979 isolate	Phenol	7,800	36,100
Champia parvula 1981 isolate		7,800	36,100
Champia parvula Korean isolate		7,800	13,000
Agardhiella subulata		7,800	21,700
Grinnellia americana		—	13,000
Champia parvula 1979 isolate	Toxaphene	39	108
Champia parvula 1981 isolate		108	108
Champia parvula Korean isolate		65	108
Agardhiella subulata		180	>180
Grinnellia americana		108	108
Champia parvula 1979 isolate	Isophorone	83,000	83,000
Champia parvula 1981 isolate		49,800	83,000
Champia parvula Korean isolate		29,500	83,000
Agardhiella subulata		49,800	138,000
Grinnellia americana		49,800	49,800

[a] All tests with *Champia* and *Grinnellia* lasted 2 wk. Tests with *Agardhiella* lasted 3 wk. All values represent the geometric mean of the effect and no-effect concentrations.
[b] Concentration causing a statistically significant decrease relative to controls (ANOVA followed by Dunnett's mean separation test).
[c] Concentration causing a total absence of cystocarp production.

Table 8. Effect of pre- and postfertilization exposure of gametophytes of *Champia parvula* to a toxic industrial effluent on the production of cystocarps[a]

% effluent	Number of cystocarps per plant	
	Prefertilization exposure	Postfertilization exposure
Control	36 ± 11	26 ± 10
0.6	24 ± 8	23 ± 6
1.2	24 ± 5	20 ± 7
2.4	18 ± 10	20 ± 6
4.8	0	12 ± 5
7.0	0	10 ± 4

[a] Values are the mean ± SD of three replicates.

often more sensitive in prefertilization exposures than in exposures that are made after females have been fertilized (Table 8). Exposing male plants to various copper concentrations in seawater for 2 d and then adding them to clean seawater with females for several hours had similar effects on cystocarp formation as exposing males and females together for the 2 d (Figure 6). The response was due to effects on the male plant (or on spermatia formation and release), because exposure of spermatia to copper without the male plants had little effect on cystocarp formation when females were later added for 1 h.

Formation of spermatia takes place in several steps on male gametophytes of *C. parvula*. First there is a proliferation of small cells on the surface of the area of sorus development, with subsequent formation of small upright chains of cells. Next, the tips of these chains begin producing the single-celled spermatia. Finally, the cuticle breaks down, releasing the spermatia into the water column. Plants have been observed that exhibit effects at all stages of spermatia development, depending on the toxicant. Obviously, at high toxicant concentrations the effect can be cell death of the spermatia, but it is at the lower, moderately toxic concentrations that more interesting things happen. In some males the process of spermatia formation is inhibited at one of several levels of development, but in others the cuticle fails to break down. In the latter case the spermatia appear to form "naturally" but cannot be released.

LAMINARIA SACCHARINA

Kelp and other brown algae were used in toxicity testing before *C. parvula*. Some earlier tests were concerned with effects of discharges on the giant kelp, *Macrocystis pyrifera*, in California (Clendenning, 1958, 1959, 1960; North, 1964). Kelps are normally found in colder waters and thus make good complementary algal test species to the warm-water *Champia*. As with *Champia*, sexual reproduction is used as the end point for assessing the effects of toxicants on *Laminaria*. The test differs from that for *Champia* in that it requires an initial period for gamete induction to perform a test. In addition, a compound microscope equipped with ultraviolet (UV) fluorescence is necessary to distinguish the products of sexual reproduction from those that may be produced parthenogenetically. Some parthenogensis oc-

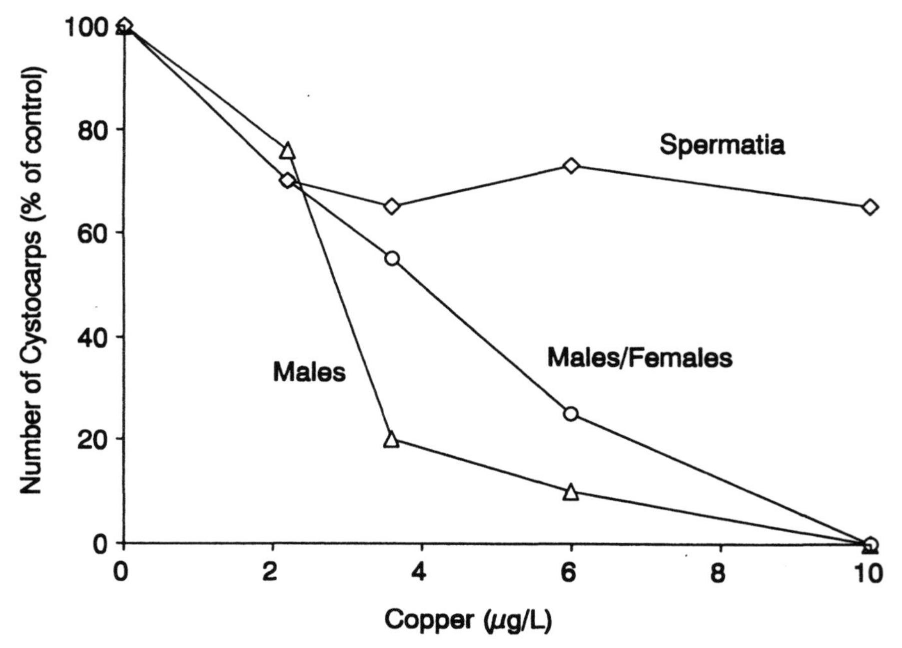

Figure 6. Effect of copper exposure on cystocarp production by *Champia parvula* males only, males and females simultaneously, and spermatia only.

curs in all species of kelp that have been tested to date. Usually these parthenogenetically derived "sporophytes" do not progress as far as sexually derived sporophytes and they often become misshapen after a few days' growth. However, they are normal enough to be potentially confused with sexually produced sporophytes.

The current kelp sexual reproduction test is based on techniques that were first used with oil studies several years ago (Steele and Hanisak, 1978). Those tests used gametophytes derived from field-collected sporophytes for each test, whereas the present proposed test method relies on cultured gametophytic material. The current procedure was developed with an East Coast isolate of *L. saccharina* but should be applicable to most kelp species with minor, if any, modification.

The life history of kelp species consists of an alternation between a macroscopic diploid stage (the sporophyte) and microscopic haploid gametophytes; toxicity tests have been developed to examine several end points within the life history. Two tests using *M. pyrifera* were developed by researchers for the State of California (Anderson and Hunt, 1988; Anderson et al., 1990): a short-term 48-h test and a 15- to 20-d test. The 48-h test has two end points: germination of the haploid kelp zoospores and initial growth of the germ tube of the developing gametophyte. The longer test focuses on sporophyte production (via sexual reproduction), requiring time for zoospore germination, gametophyte maturation, and subsequent fertilization. The 48-h toxicity test is more practical for use in routine testing, as the long-term test is time-consuming and more susceptible to microalgal contamination when testing complex effluents.

An alternative kelp test protocol, using cultured gametophyte clones, has been developed with an East Coast isolate of *L. saccharina*. The procedure was developed as a 48-h exposure, using fragmented male and female gametophytes that have undergone gametogenesis. This allows sexual reproduction to be used as an end point in effluent toxicity tests short enough to eliminate the problem with microalgal contamination experienced using *M. pyrifera*. The test method can probably be adapted to most kelp species and has been used to test the toxicity of a variety of single chemical substances.

Distribution and Life History

Laminaria saccharina (L) Lamour belongs to a large group of brown algae (Phaeophyta) known as kelps that are widely distributed throughout the temperate zones of both the northern and southern hemispheres. *L. saccharina* is common in the North Atlantic and North Pacific (Taylor, 1957; Bold and Wynne, 1985; Scagel, 1978) and is ecologically important in establishing the foundational structure for certain coastal communities. Many kelps do so throughout temperate latitudes (Mann, 1973).

Laminaria's life history is an alternation of microscopic gametophytes with a large diploid blade-like thallus (Figure 7). Meiosis occurs in the central region of the adult sporophyte, releasing haploid, biflagellated zoospores. In theory, half of the zoospores germinate into branched, uniseriate male gametophytes and the other half into female gametophytes

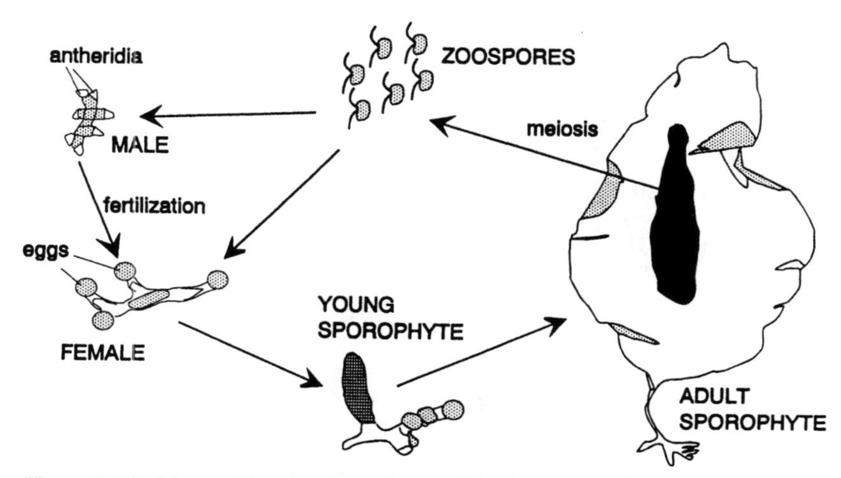

Figure 7. Life history of *Laminaria saccharina*. There is an alternation between haploid microscopic male and female plants and a diploid macroscopic blade (as much as several meters long). When sexually mature, eggs release a collection of organic compounds known as pheromones. These compounds cause the release of mature sperm from the male gametophyte and subsequent attraction of the sperm to the egg for fertilization. The young sporophyte begins attached to, and eventually overgrows, the female gametophyte. The end point for toxicity tests is the production of young sporophytes. One sporophyte is produced for each successful sexual fusion.

(also uniseriate). The female produces eggs that remain attached to the female, and male produce antheridia, each containing one biflagellated sperm cell. Sperm are released from the male plants and attracted to eggs in response to a collection of pheromones produced by the egg. After fertilization, the developing sporophyte (blade) remains attached to and eventually overgrows the female.

Culture

Male and female kelp gametophytes can be cultured easily in the laboratory from field-collected zoospores. All glassware and media are autoclaved to minimize the possibility of microalgal contamination. For maintenance cultures, a nutrient medium without added iron is used to inhibit gametogensis (Motomura and Sakai, 1984) but allow vegetative growth in white light. Although cultures can be maintained in red light to inhibit gamete formation (Luning and Dring, 1975), it is easier to control gamete formation through iron nutrition. Some preconditioning with added iron is required to induce reproduction, and under the conditions listed below, male and female gametophytes produce gametes in about 6 d. Thus, plant material can be made available any time for testing.

Unlike the case of *Champia*, the culture medium for *L. saccharina* can be artificial or natural seawater, to which additional nutrients are added. Artificial seawater is preferred for maintaining plants in a vegetative condition because it allows better control over iron concentration in the medium, which in turn allows better control over gametogensis. In addition, salinity adjustments are necessary in testing most complex effluents and receiving waters. These adjustments are easily made with artificial sea salts; GP2 (Spotte et al., 1984) has been used successfully with *L. saccharina*.

To change the medium or start new cultures, old culture medium is poured off and the tissue is poured into a food blender and blended on high speed for about 30 s. The resulting "soup" is filtered through a fine Nytex screen (45 to 74 μm porosity). The portion that passes through the screen is discarded (or used for testing; see below). That remaining on the screen is used for starting new cultures. These cells may be washed with sterile seawater to remove cell debris before being rinsed into a clean flask. Filters and blending containers are cleaned by rinsing all surfaces with at least 70% ethanol and then rinsing thoroughly with sterile water. Each sex or clone should have its own filter apparatus and blending container to lessen the possibility of cross-contamination. With this procedure performed every 10 to 14 d, actively growing plants are continuously available.

Preparation of Plants for a Test

To be usable for toxicity testing, plants are blended in about 100 ml of chilled (12°C) seawater using a commercial food blender at the highest speed for about 30 s. The resulting suspension is used as is for males or filtered through a 45 μm (30–60 μm) nylon screen for females, and the portion that passes through the filter should be used for testing. These cells can then be dispensed into the test chambers after dilution. They should be diluted to yield two to three fragments per field of view under a compound microscope at 200×. Most testing to date has been performed in 20 ml of medium using 60 × 25 mm polystyrene petri dishes. Males are settled onto the dishes, and females are settled onto glass coverslips on the bottom of larger, 10 cm, petri dishes. The blending and settling is done 6 d prior to test initiation. A nutrient medium containing a high iron concentration is used at this time to stimulate gametogenesis. During gametogenesis plants are placed at 12 to 14°C in about 75 μE m^{-2} s^{-1} of cool-white fluorescent light on a 16:8 light-dark cycle.

After 5 d, males and females are checked for the presence of pregametic cells and an occasional released egg. Cell contents appear different from those of vegetative cells, usually being darker with indistinct chromatophores, and the area of the nucleus is apparent in some cells (females). Pre-egg cells appear as side branches perpendicular to the original filament of cells. Male gametes begin as small triangular side branches. Eventually, on day 5 or 6, male gametophytes are composed mostly of these triangular cells with very pale chromatophores, each containing a single sperm cell.

Test Conditions

The duration of test exposure to the chemical (or effluent) is 2 d, allowing time for fertilization to occur. The exposure water temperature should be 12 ± 1°C, the salinity 28 to 32 ppt, and the pH between 7.8 and 8.2. For receiving waters and effluents, salinity will often be below the desired range and must be adjusted upward with artificial sea salts. The photoperiod should be a 16:8 light-dark cycle of about 75 μE m^{-2} s^{-1} of cool-white fluorescent light. The nutrient receipe used for toxicity tests is the same as that used for *Champia*.

On day 6 after preconditioning, water in the male dishes is replaced with control or treatment water with the appropriate nutrients added. The coverslips with female plants are briefly rinsed in sterile seawater and one is placed into each male dish. After 2 d, these coverslips are removed, rinsed briefly with sterile seawater, and place into a calcofluor-white solution. Calcofluor white is a vital stain specific for cellulose and is commonly used to stain cell walls. Zygotes and young sporophytes stain much more brightly than gametophytic tissue, unfertilized eggs, or parthenogenetically produced structures. After at least 10 min (or up to several hours) the coverslips with the females are observed using a UV fluorescent microscope. Fifty females should be examined and scored as to whether or not they have any sexual

Table 9. Effects of no. 2 fuel oil on vegetative growth and production of sporophytes of *Laminaria saccharina*

Added oil	Measured[a]	Diameter of gametophyte after 21 days (μm)	% female gametophytes with at least one sporophyte
Control	ND	190	85
2 ppb	ND	310	2
20 ppb	ND	220	5
200 ppb	ND	250	5
2 ppm	0–30 ppb	245	5
20 ppm	1–3 ppm	104	0
200 ppm	18–28 ppm	0	0
2000 ppm	45–50 ppm	0	0

[a]Total extractable hydrocarbons as measured by infrared spectrophotometry; ND, not detected.
Data are from Steele and Hanisak, 1978.

structures. Alternatively, the total number of sexual structures can be counted.

Sample Data

Formation of sporophytes by gametophytes of *L. saccharina* (sexual reproduction) is a sensitive indicator of petroleum hydrocarbon pollution. The difference between the sensitivity of vegetative growth and that of sexual reproduction can be dramatic. Table 9 shows some sample data from Steele and Hanisak (1978) on the effects of No. 2 fuel oil on gametophytes of *L. saccharina*. The most severe effect on vegetative growth (death) occurred at added oil concentrations of 200 ppm. However, sexual reproduction was almost eliminated at the lowest concentration of added oil, 2 ppb. This is a difference in effect levels of five orders of magnitude. The sexual reproduction end point is particularly sensitive to petroleum hydrocarbons, presumably because of interference with the sperm's ability to detect the presence of pheromones (''sex attractants'') produced by the egg. Petroleum hydrocarbons have been shown to interfere with the pheromone system in at least two species of brown algae (Derenbach and Gereck, 1980; Derenbach et al., 1980). Because pheromones are effective at very low concentrations, it does not take much hydrocarbon to mask their presence and thus prevent the sperm from finding the egg.

L. saccharina sexual reproduction has shown sensitivity to several single chemicals. Data for two of these, thallium and acenaphthene (Thursby and Steele, 1988b; Thursby et al., 1989b), have been proposed to be used to set the chronic limits for the National Water Quality Criteria (U.S. EPA, 1990; Thursby, 1990). The effect concentration for *L. saccharina* gametophytes exposed for 2 d to thallium was 8.8 times lower than that for *Mysidopsis bahia* exposed during a 28-d life cycle test (Table 10). The effect concentration for acenaphthene was 3.6 times lower for *L. saccharina*. The use of these data represents the first attempt to set national criteria based on plant data.

FUCUS

Besides kelp, several other species of brown algae have been used successfully to measure sublethal effects of chemicals using sexual reproduction as the end point. For example, the prefertilization stages of species of *Fucus* are very sensitive to a variety of petroleum products (Steele and Hanisak, 1978; Steele, 1977; Johnston, 1977). Zygotes are considerably less sensitive, suggesting there is interference with the pheromone system of attracting sperm to egg (Derenbach and Gereck, 1980; Derenbach et al., 1980). Sexual reproduction in several species of *Fucus* was used successfully to assess damage and recovery in intertidal areas in Narragansett Bay, Rhode Island following an oil spill (Thursby et al., 1990).

Distribution and Life History

Species of *Fucus* are common in North America in the intertidal and subtidal regions of the temperature shoreline (Taylor, 1957; Abbott and Hollenberg, 1976). The life history of this genus (Figure 8) is simpler than that of either *Laminaria* or *Champia*. The adult plant (sporophyte) is diploid. Swollen reproductive tips called receptacles contain numerous conceptacles where meiosis takes place. Nonmotile eggs and biflagellated sperm are released from their respective conceptacles, generally on tidal cues. The egg releases a pheromone that attracts the sperm. After fertilization the zygote settles to the substrate and

Table 10. Comparison of the sensitivity of sexual reproduction in *Laminaria saccharina* to that of *Mysidopsis bahia* in a 28-d life cycle test[a]

Compound	*Mysidopsis* 28-d	*Laminaria* 2-d	Ratio
Thallium[b]	272	31	8.8
Acenaphthene[c]	64	18	3.6

[a]Values represent the geometric mean of the effect and no effect concentrations express in μg/L.
[b]Data from Thursby and Steele, 1988b; Thursby and Berry, 1988.
[c]Data from Thursby et al., 1989.

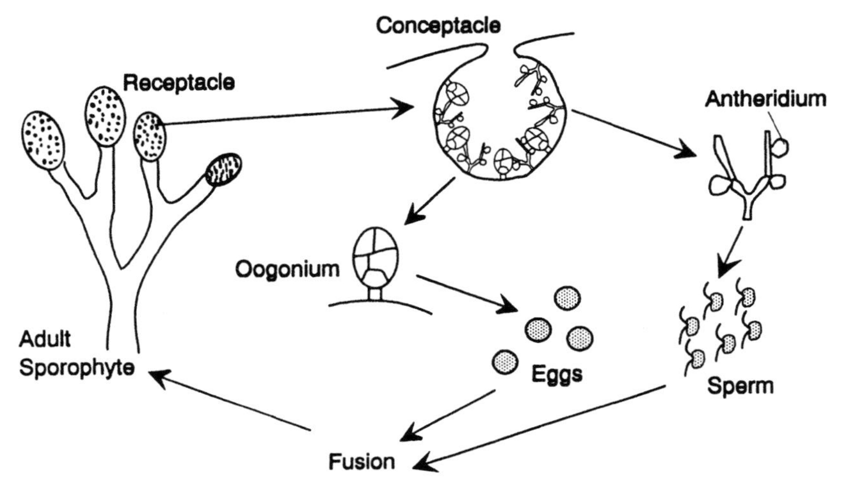

Figure 8. Life history of *Fucus spiralis.* There is no alternation of generations; gametes are released into the environment directly from the adult plant and zygotes geminate directly back into new adults. The end point for toxicity tests is the germination of zygotes into bipolar, two-celled embryos.

germinates and develops directly into a new adult plant.

Test Procedure

Fucus plants are collected from the shoreline prior to testing; they are not cultured in the laboratory. Although reproductive plants can be found year-round, they are more difficult to find during the summer months. Ripe material can be held in the laboratory and refrigerated for several days. Plants are prepared for use in toxicity tests by removing the ripe receptacles and briefly rinsing them in deionized water. The tips are then placed overnight in a damp petri dish in a temperature control chamber at about 12°C (although a refrigerator can work fine).

The next day the plant tips are dropped into chemical test treatments, and sperm and eggs begin to release in a few minutes. The release process is caused by the physical rewetting of the thallus. Tests are performed at 12–16°C and a salinity of 28–36 ppt. The release process is allowed to proceed under chemical exposure for 1 h, when the receptacles are removed from the exposure vessels. Eggs (or presumably zygotes at this time) are examined with a stereomicroscope and transferred with a pipet to control media for 24 to 48 h, when they are reexamined and the percent germination counted as an indication of the original rate of fertilization. Germinated zygotes are easily identified by their bipolar division (Figure 9).

Two types of tests can be performed using the sexual reproduction end point with *Fucus*. First, plants can be collected from suspect polluted areas and their fertilization rate in control medium compared with that of plants collected from clean sites. Second, plants can be collected from clean sites and used as

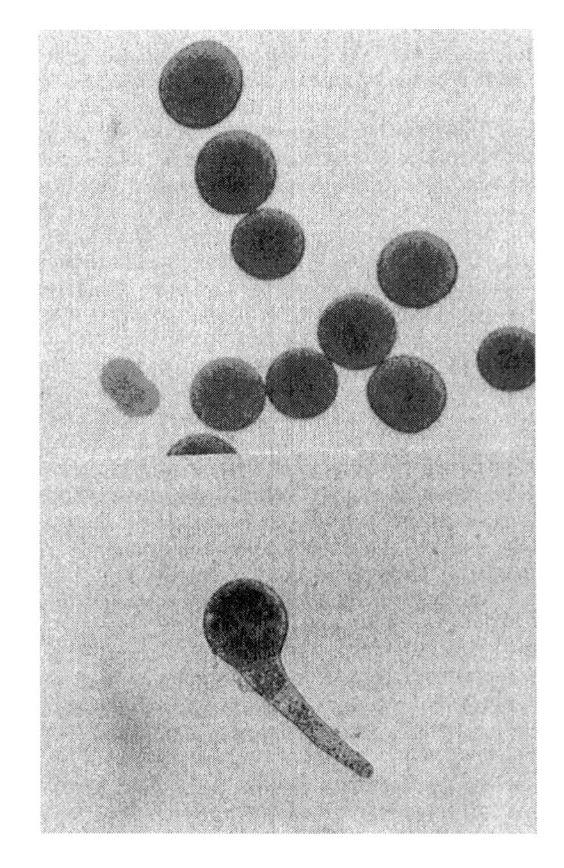

Figure 9. Eggs (top) and a 48-h-old embryo (bottom) of *Fucus*. The bipolar division makes it easy to distinguish young embryos. The eggs are approximately 50 μm in diameter.

Table 11. Comparison of effects of various petroleum products on growth of *Fucus edentatus* zygotes when exposure was either before or after fertilization[a]

Added oil	Measured[b]	No. 2 fuel oil		JP-4 jet fuel		Willamar crude oil	
		Post[c]	Pre	Post	Pre	Post	Pre
Control	ND	694	699	704	701	776	710
2 ppb	ND	702	0	712	0	764	0
20 ppb	ND	748	0	732	0	748	0
200 ppb	ND	696	0	687	0	750	0
2 ppm	0–30 ppb	671	0	635	0	756	0
20 ppm	1–3 ppm	0	0	0	0	734	0
200 ppm	18–28 ppm	0	0	0	0	399	0
2,000 ppm	45–50 ppm	0	0	0	0	0	0

[a] Values represent the length of young plants after 12 days in μm.
[b] Total extractable hydrocarbons as measured by infrared spectrophotometry; ND, not detected.
[c] Post, postfertilization exposure; pre, prefertilization exposure.
Data are from Steele, 1977.

in other methods to test the effects of ambient waters, effluents, or single chemicals.

Sample Data

The procedure with *Fucus* species has been used to test the effects of petroleum products (Steele, 1977; Steele and Hanisak, 1978; Thursby et al., 1990). As with *Laminaria*, the difference in the effect concentrations between exposing receptacles (prefertilization exposure so that fertilization has to take place in the presence of the oil product) and exposing zygotes (oil exposure after fertilization) is several orders of magnitude (Table 11). *Fucus* was also shown to be a sensitive indicator of the toxicity of water samples and of the status of intertidal communities after a spill of no. 2 fuel oil into the mouth of Narragansett Bay, Rhode Island (Thursby et al., 1990). The oil spill occurred on June 23, 1989, and a shoreline survey of the intertidal and upper subtidal macroalgae was begun on June 24. There was little evidence of damage to vegetative tissue among the

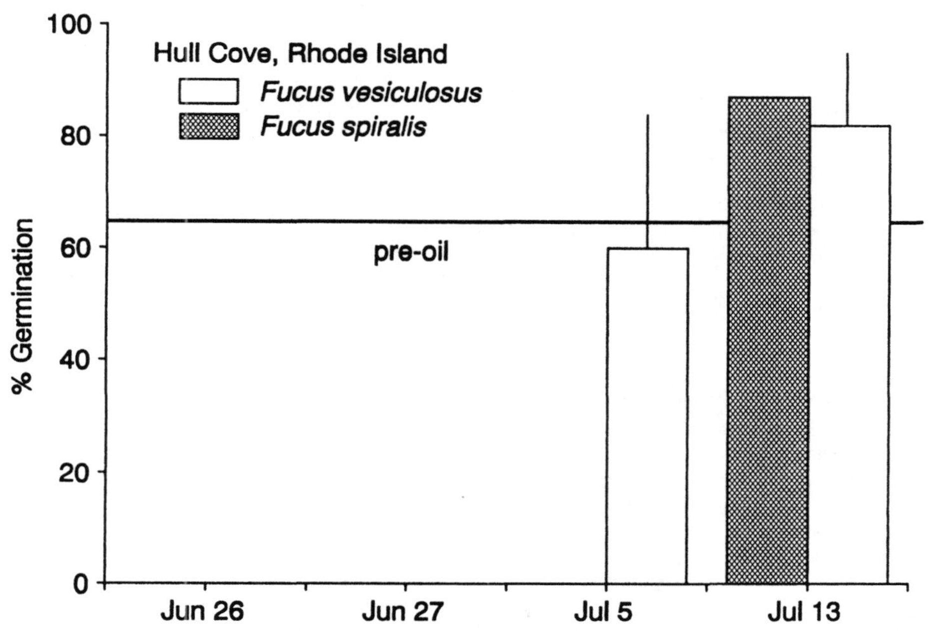

Figure 10. Percent germination for embryos of *Fucus vesiculosus* and *Fucus spiralis* collected from Hull Cove. "Pre-oil" line is germination rate for embryos of *F. spiralis* collected from Narragansett Pier on June 24, prior to the arrival of oil. The low pre-oil rate (63%) is due to the fact that summer is not the optimal time for reproduction by this species.

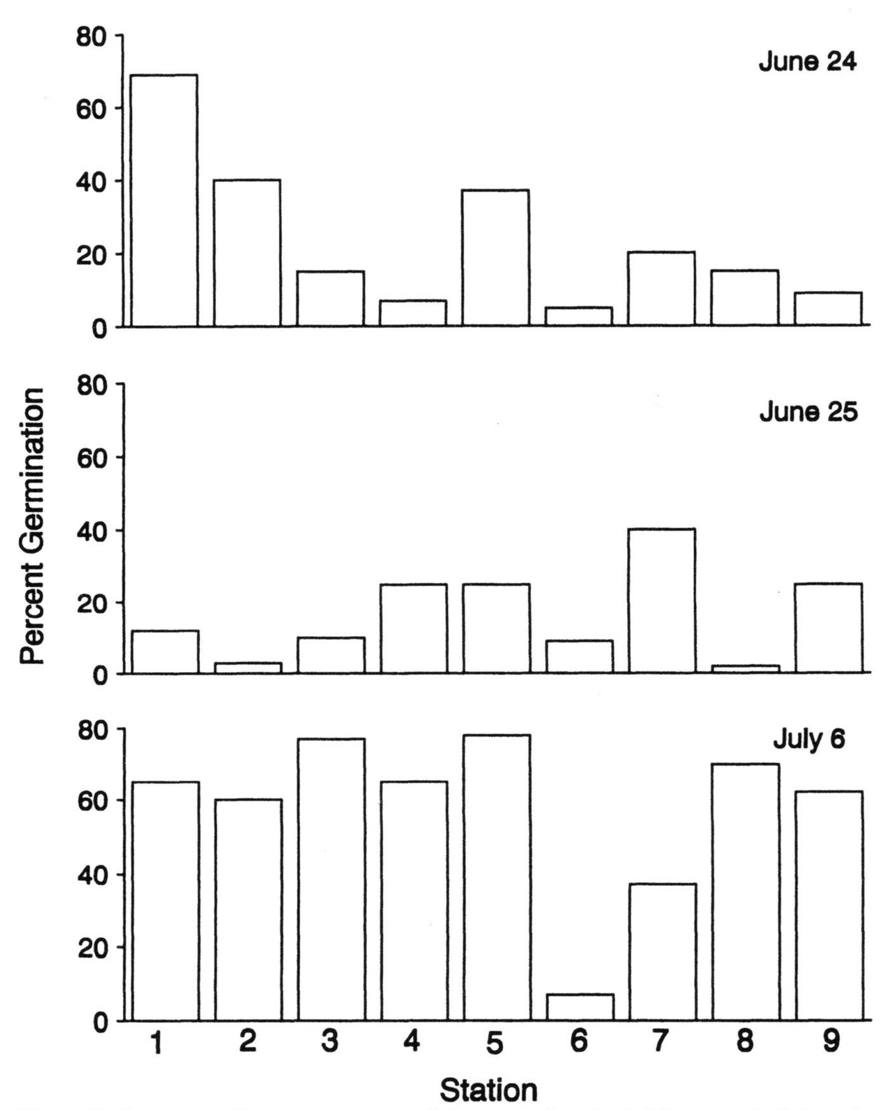

Figure 11. Percent germination for embryos of *Fucus spiralis* collected from a control site and exposed to water collected from stations at various distances from the site of the *World Prodigy* oil spill. The spill occurred on June 23. Stations 7, 8, and 9 surround the site of the spill. Station 1 is the farthest from the spill.

attached plants at most of the sites visited. Several species of *Fucus*, however, showed inhibition of sexual reproduction. Through the use of sexual reproduction tests with *Fucus* it was shown that even at one of the most heavily contaminated oil sites, Hull Cove, germination rates for both *Fucus vesiculosus* and *Fucus spiralis* were similar to rates before the oil spill for that time of the year (Figure 10). In addition, tests using *F. spiralis* plants collected from a clean site were used to monitor the changes in water column toxicity at nine stations various distances from the spill site. The toxicity observed on June 24 and 25 had all but disappeared by July 6 (Figure 11).

SUMMARY

Sexual reproduction is a sensitive indicator of toxic effects of chemicals to marine algae. *Champia* is an annual plant and inhibition or absence of sexual reproduction reduces or eliminates the next stage in its life history, giving ecological significance to the end point. Kelp and fucoid species are perennial; however, tagging experiments have shown that individual plants for at least some species may survive only 9 mo (Brady-Campbell et al., 1984). Sexual reproduction is the only mechanism in nature for reestablishing individuals in a population. Toxicity tests using sexual reproduction as an end point have great relevance to field populations of these species. Attitudes concerning the necessity of testing with algae for risk assessment are changing with the use of macroalgae and the emergence of more sensitive end points.

LITERATURE CITED

Abbott, I. A., and Hollenberg, C. J.: Marine Algae of California. Stanford, CA: Stanford University Press, 1976.

American Society for Testing and Materials (ASTM): Standard guide for conducting sexual reproduction tests with seaweeds, E1498-92. Annual Book of ASTM Standards, Vol. 11.04:1344–1354, 1994.

Anderson, B. S., and Hunt, J. W.: Bioassasy methods for evaluating the toxicity of heavy metals, biocides, and sewage effluent using microscopic stages of giant kelp *Macrocystis pyrifera* (Agardh): a preliminary report. *Mar. Environ. Res.*, 26:113–134, 1988.

Anderson, B. S.; Hunt, J. W.; Turpen, S. L.; Coulan, A. R.; and Martin, M.: Copper toxicity to microscopic stages of giant kelp *Macrocystsis pyrifera*: interpopulation comparisons and temporal variability. *Mar. Ecol. Prog. Ser.*, 68:147–156, 1990.

Bold, H. C., and Wynne, M. J.: Introduction to the Algae, 2nd ed. Englewood Cliffs, NJ: Prentice-Hall, 1985.

Brady-Campbell, M. M.; Campbell, D. B.; and Harlin, M. M.: Productivity of kelp (*Laminaria* spp.) near the southern limit in the northwestern Atlantic Ocean. *Mar. Ecol. Prog. Ser.*, 18:79–88, 1984.

Buikema, A. L., Jr.; Niederlehner, B. R.; and Cairns, J., Jr.: Biological monitoring: Part IV—Toxicity testing. *Water Res.*, 16:239–262, 1982.

Clendenning, K. A.: The effects of waste discharge on kelp. Water Pollution Control Board: Annual Progress Report. IMR reference 58-11:27–38, 1958.

Clendenning, K. A.: The effects of waste discharge on kelp. Water Pollution Control Board: Quarterly Progress Report. IMR reference 59-4:1–13, 1959.

Clendenning, K. A.: The effects of waste discharge on kelp. Water Pollution Control Board: Quarterly Progress Report. IMR reference 60-4:45–47, 1960.

Derenbach, J. B., and Gereck, M. V.: Interference of petroleum hydrocarbon with the sex phermone reaction of *Fucus vesiculosus* (L). *J. Exp. Mar. Biol. Ecol.*, 44:61–65, 1980.

Derenbach, J. B.; Boland, W.; Folster, E.; and Muller, D. G.: Interference tests with the phermone system of the brown alga *Cutleria multfida*. *Mar. Ecol. Prog. Ser.*, 3: 357–361, 1980.

Guillard, R. R. L., and Ryther, J. H.: Studies on marine planktonic diatoms. I. *Cyclotella nana* Hustedt and *Detonula confervaceae* (Cleve) Gran. *Can. J. Microbiol.*, 8: 229–239, 1962.

Jensen, A.: Marine ecotoxicological tests with seaweeds. Ecotoxicological Testing for the Marine Environment, Vol. 1, edited by G. Persoone, E. Jaspers, and C. Claus. Bredene, Belgium: State University of Ghent and Institute for Marine Scientific Research, 1984.

Johnston, C. S.: The sub-lethal effects of water-soluble extracts of crude oil on the fertilisation and development of *Fucus serratus* (L). *Rapp. P. Reun. Cons. Perm. Int. Explor. Mer.*, 171:184–185, 1977.

Kenaga, E. E.: The use of environmental toxicology and chemistry data in hazard assessment; progress, needs, challenges. *Environ. Toxicol. Chem.*, 1:69–79, 1982.

Kenaga, E. E., and Moolenaar, R. J.: Fish and *Daphnia* toxicity as surrogates for aquatic vascular plants and algae. *Environ. Sci. Technol.*, 13:1479–1480, 1976.

Levine, H. G.: The use of seaweeds for monitoring coastal waters. Algae as Ecological Indicators, edited by L. E. Shubert, pp. 189–210. New York: Academic Press, 1984.

Lewis, E. J.: The protein, peptide and free amino acid composition in species of *Champia* from Saurashtra Coast, India. *Bot. Mar.*, 16:145–147, 1973.

Luning, K., and Dring, M. J.: Reproduction, growth and photosynthesis of gametophytes of *Laminaria saccharina* grown in blue and red light. *Mar. Biol.*, 29:195–200, 1975.

Mann, K. H.: Introductory Remarks. *Mem. Ist. Ital. Idrobiol.*, 29 (Suppl):353–383, 1972.

Mann, K. H.: Seaweeds: their productivity and strategy for growth. *Science*, 182:975–981, 1973.

Morrison, G.; Torello, E.; Camello, R.; Walsh, R.; Kuhn, A.; Burgess, R.; Tagliabue, M.; and Greene, W.: Intralaboratory precision of saltwater short-term chronic toxicity tests. *Res. Water Pollut. Control Fed.*, 61:1707–1710, 1989.

Motomura, T., and Sakai, Y.: Regulation of gametogensis of *Laminaria* and *Desmarestia* (Phaeophyta) by iron and boron. *Jpn. J. Phycol.*, 32:209–215, 1984.

North, W. J.: Ecology of rocky nearshore environment in Southern California and possible influence of discharged wastes. *Adv. Water Pollut. Res.*, 3:247–262, 1964.

Payne, A. G., and Hall, R. H.: A method of measuring algal toxicity and its application to the safety assessment of new chemicals. Aquatic Toxicology, edited by L. L. Marking, and R. A. Kimerle, pp. 171–180. Philadelphia: American Society for Testing and Materials, 1979.

Phillips, D. J. H.: The use of biological indicator organisms to monitor trace metal pollution in marine and estuarine environments—a review. *Environ. Pollut.*, 13:281–317, 1977.

Provasoli, L.: Media and prospects for the cultivation of algae. Selected Paper in Phycology, edited by S. R. Rosowski, B. C. Parker, pp. 599–604. Lincoln: Department of Botany, University of Nebraska, 1971.

Reedman, D. J., and Wormersley, H. B. S.: Southern Australian species of *Champia* and *Chylocladia* (Rhodomeniales: Rhodophyta). *Trans. R. Soc. South Aust.*, 100:75–104, 1976.

Scagel, R. F.: Guide to Common Seaweeds of British Columbia. Handbook No. 27. Victoria, Canada: British Columbia Provincial Museum, 1978.

Spotte, S.; Adams, G.; and Bubucis, P. M.: GP2 medium is an artificial seawater for culture or maintenance of marine organisms. *Zoo Biol.*, 3:229–240, 1984.

Steele, R. L.: Effects of certain petroleum products on reproduction and growth of zygotes and juvenile stages of the alga *Fucus edentatus* (Phaeophyta). pp. 138–142: Fate and Effects of Petroleum Hydrocarbons in Marine Ecosystems and Organisms, edited by D. A. Wolfe, pp. 138–142. Oxford: Pergamon, 1977.

Steele, R. L., and Hanisak, M. D.: Sensitivity of some brown algal reproductive stages to oil pollution. Proceedings of the Ninth International Seaweed Symposium, edited by A. Jensen, and J. R. Stein. Princeton: Science Press, 1978.

Steele, R. L., and Thursby, G. B.: A toxicity test using life stages of *Champia parvula* (Rhodophyta). Aquatic Toxicology and Hazard Assessment: Sixth Symposium, edited by W. E. Bishop; R. D. Cardwell; and B. B. Heidolph, pp. 73–89. ASTM STP 802. Philadelphia: American Society for Testing and Materials, 1983.

Steele, R. L., and Thursby, G. B.: Laboratory culture of gametophytic stages of the marine macroalgae *Champia parvula* (Rhodophyta) and *Laminaria saccharina* (Phaeophyta). *Environ. Toxicol. Chem.*, 7:997–1002, 1988.

Steele, R. L.; Thursby, G. B.; and van der Meer, J. P.: Genetics of *Champia parvula* (Rhodymeniales, Rhodophyta): Mendelian inheritance of spontaneous mutants. *J. Phycol.*, 22:538–542, 1986.

Stephan, C. E.; Mount, D. I.; Hansen, D. J.; Gentile, J. H.; Chapman, G. A.; and Brungs, W. A.: Guidelines for Deriving Numerical National Water Quality Criteria for the Protection of Aquatic Organisms and Their Uses. NTIS PB85-227049. Springfield, VA: National Technical Information Service, 1985.

Taylor, W. R.: Marine Algae of the Northeastern Coast of North America. Ann Arbor: Univ. of Michigan Press, 1957.

Thursby, G. B.: Saltwater section of the ambient aquatic life water quality criteria for acenaphthene. Report submitted to D. J. Hansen. U.S. EPA, Environmental Research Laboratory-Narragansett, 1990.

Thursby, G. B., and Berry, W. J.: Acute and chronic toxicity of thallium to *Mysidopsis bahia*: flow-through. Univ. of Rhode Island (Memorandum to D. J. Hansen, U.S. EPA and K. J. Scott, SAIC, Narragansett), May 20, 1988.

Thursby, G. B., and Steele, R. L.: Toxicity of arsenite and arsenate to the marine macroalga *Champia parvula* (Rhodophyta). *Environ. Toxicol. Chem.*, 3:391–397, 1984.

Thursby, G. B., and Steele, R. L.: Comparison of short- and long-term sexual reproduction tests with the marine red alga *Champia parvula*. *Environ. Toxicol. Chem.*, 5:1013–1018, 1986.

Thursby, G. B., and Steele, R. L.: Laboratory culture of gametophytic stages of the marine macroalgae *Champia parvula* (Rhodophyta) and *Laminaria saccharina* (Phaeophyte). *Environ. Toxicol. Chem.*, 7:997–1002, 1988a.

Thursby, G. B., and Steele, R. L.: Toxicity of thallium to the kelp, *Laminaria saccharina*. Univ. of Rhode Island and U.S. EPA (memorandum to D.J. Hansen, EPA and K. J. Scott, SAIC, Narragansett), June 6, 1988b.

Thursby, G. B.; Steele, R. L.; and Kane, M. E.: Effects of organic chemicals on growth and reproduction in the marine red alga, *Champia parvula*. *Environ. Toxicol. Chem.*, 4:797–805, 1985.

Thursby, G. B.; Berry, W. J.; and Champlin, D.: Flow-through acute and chronic tests with acenaphthene using *Mysidopsis bahia*. (Memorandum to D. J. Hansen. U.S. EPA, Narragansett), September 19, 1989a.

Thursby, G. B.; Tagliabue, M.; and Sheehan, C.: Toxicity of acenaphthene to seaweeds. (Memorandum to D. J. Hansen, U.S. EPA, Narragansett) December 20, 1989b.

Thursby, G.; Steele, R.; Tagliabue, M.; and Sheehan, C.: Sexual reproduction in species of the brown seaweed, *Fucus*, to assess damage and recovery from the *World Prodigy* oil spill. Proceedings of the Conference on Oil Spills: Management and Legislative Implications, edited by M. L. Spaulding, and M. Reed, pp. 291–301. American Society of Civil Engineers, 1990.

U.S. Environmental Protection Agency: Short-term methods for estimating the chronic toxicity of effluents and receiving waters to marine and estuarine organisms. EPA Report 600/4-87/028, 1988.

U.S. Environmental Protection Agency: Ambient aquatic life water quality criteria for thallium. Environmental Research Laboratory, Duluth, MN and Narragansett, RI, 1990.

U.S. Environmental Protection Agency: Technical support document for water quality-based toxics control. EPA Report 505/2-90-001, 1991.

Chapter 6

SEA URCHIN SPERM CELL TEST

G. A. Chapman

INTRODUCTION

Aquatic toxicity testing is necessarily conducted with only a few species intended to serve as surrogates for the broad array of aquatic species that may be threatened with toxic effects of specific chemicals or effluents. In most cases, the exposure duration and test material concentrations in the toxicity test attempt to mimic exposure conditions that might be experienced by organisms in the real world. Occasionally, a test is devised with organisms or conditions that do not necessarily simulate real-world exposures but that do provide a convenient indicator of toxic materials and their probable toxic concentrations. Familiar examples are the Ames mutagenicity test and the Microtox test, which utilize surrogate test organisms (bacteria) and often use artificial exposure conditions (an extract or concentrate) (Bulich et al., 1981; Vasseur et al., 1984; Allen et al., 1983).

Sea urchin sperm cell tests are usually surrogate tests in both of these respects: (1) they are intended to provide toxicity information for the protection of other marine organisms and (2) the typical 60-min duration of the sperm test exposure probably exceeds real-world exposure conditions for sea urchin sperm. The latter surrogacy has been a point of some misunderstanding and contention in the regulatory application of this test as one of several marine toxicity tests stipulated for use in effluent discharge permits (U.S. EPA, 1988).

The primary impediment to utilizing surrogate tests, such as the sperm test, is uncertainty that concentrations causing toxic effects in the surrogate tests will also produce toxic effects on other organisms of concern. The only practical way to reduce this uncertainty is to generate data comparing the sensitivity of the surrogate test to the sensitivities of tests with other organisms. Data available to date indicate that the sperm cell test will tend to indicate no toxicity in the presence of conditions toxic (i.e., false negative) to at least one common test organism much more frequently than to indicate toxicity in the presence of conditions not toxic to any other common test organisms. The latter case will occur primarily with materials that are particularly toxic to sperm compared with other life stages and species (e.g., 2,4-pentanedione: Nacci et al., 1986).This situation argues for the use of the surrogate sperm cell test in conjunction with other tests. It cannot be emphasized too strongly that this situation also applies to extrapolation from any standard toxicity test; *no single test is consistently so sensitive as never to produce false-negative results.*

The chief advantages of the sea urchin sperm cell test are its rapidity, small test volume, and great potential for using many test treatments and multiple replications of each treatment. The test is completed in 1 d and uses only a few milliliters of test solution. Most other tests providing similar toxicity estimates require several days to a week to conduct and utilize a significantly greater volume of test material.

The primary disadvantage of the sea urchin sperm cell test is the often limited seasonal availability of urchins with mature gametes. This disadvantage has been minimized in three different ways: (1) switching among test species to take advantage of various seasonal maturation periods, (2) shipping sea urchins from localities where the maturation period differs from that of local populations, and (3) holding sea urchins under controlled laboratory conditions to extend normal maturation or induce asynchronous maturation cycles (Leahy et al., 1978). The latter procedure is the most desirable, but, at least with sea urchins from intertidal populations, the necessary methods have been neither well documented nor widely applied.

In practice, the standardization of the sea urchin sperm cell test has seen a number of variations on the basic design. Although apparently minor, these variations can influence test results through both obvious and subtle mechanisms. The major difference is related to the time elapsed between the sperms' metabolic activation, which is triggered by pH and ionic changes following dilution of semen with sea-

This chapter is ERL-Narragansett contribution number N-134.

water, and the introduction of eggs for fertilization. In various methods, this has been produced by different toxicant exposure periods ranging from 5 to 60 min and by diffent sperm ages (time elapsed between semen dilution/activation and egg introduction) of 10 to 150 min. One significance of exposure duration is obvious; the longer the exposure to toxicant, the longer the opportunity for toxic action and the more sensitive the test. A much more subtle and controversial aspect is whether toxicity test results based on egg fertilization by sperm that have been active for an hour or more can have ecological significance. Regardless of the arguable degree of fidelity to sea urchin fertilization in nature, experience has shown that the toxicity data obtained from these tests provide good estimates of results from other standard marine toxicity tests.

This chapter discusses primarily methods for tests with four species of North American echinoids including three species of sea urchins and the closely related sand dollar of the Pacific Coast. These species include:

Purple urchin (Atlantic): *Arbacia punctulata*
Purple urchin (Pacific): *Strongylocentrotus purpuratus*
Green urchin (circumpolar): *Strongylocentrotus droebachiensis*
Sand dollar (Pacific): *Dendraster excentricus*

For convenience, these species will be referred to in the following text as *Arbacia*, the purple urchin, the green urchin, and the sand dollar, respectively. Some of the references are related to work performed with other species of sea urchins common to Europe or Asia. Unless considered important for comparative toxicological purposes, the identification of these test organisms as to genus and species has been avoided.

BASIC STUDIES OF SEA URCHIN GAMETES

A review of sea urchin spermatozoa (Rothschild, 1951) noted that "fertilization has probably been studied more in the echinoderms, and particularly in sea-urchins, than in any other animal or plant phylum." Indeed, for many years in the early 1900s, such journals as the Biological Bulletin of Wood's Hole and the American Journal of Physiology seldom contained an issue that did not contain one or more articles on sea urchin gametes, fertilization, or embryology. An excellent review of the early literature is contained in Harvey's (1956) The American *Arbacia* and Other Sea Urchins. Another excellent source of information on general sea urchin reproductive physiology is Boolotian's The Physiology of Echinodermata, especially Chapter 25 (Reproductive Physiology, Boolotian, 1966), and Chapters 27 and 28 (The Gametes; Some Procedures and Properties, Tyler and Tyler, 1966a; and Physiology of Fertilization and Early Development, Tyler and Tyler, 1966b).

A generalized description of fertilization of sea urchin eggs provides a basis for understanding this test. Sea urchin gametes in nature are shed into the water from male and female sea urchins. When the semen is diluted by the seawater, mature sperm are activated and show intense swimming activity. When sperm make contact with eggs, the sperm undergo a reaction (the acrosome reaction) in which a short gelatinous filament appears at the head of the sperm, enabling the sperm to adhere to the egg and probably to release an enzyme that dissolves a portion of the egg's vitelline membrane, allowing sperm penetration. Immediately after fertilization the egg undergoes water uptake and a resultant rapid elevation of the outer (vitelline) membrane (then termed the fertilization membrane) preventing multiple sperm penetration (polyspermy). Absence of this fertilization membrane is the indicator of toxicity in the sperm cell test; i.e., chemical concentrations that result in sperm populations that elicit this response in a smaller percentage of eggs, relative to control treatments, are deemed toxic.

The toxicity test involves the exposure of sea urchin sperm to the test solutions (usually for 60 min) followed by addition of eggs. The relative percentage of unfertilized eggs in each treatment is compared to that in the control; an increase in the percentage of unfertile eggs is an indication of toxicity.

Prior to introduction, the eggs are rinsed and allowed to settle several times to reduce the amount of egg jelly (Figure 1a) around the eggs and in solution. Egg jelly can interfere with fertilization of the eggs by inducing the acrosome reaction before sperm-egg contact (Vacquier, 1979) and components of the jelly significantly increase sperm respiration (Gray, 1927c; Ohtake, 1976b). After a brief period for fertilization (usually 20 min), the test is stopped by the addition to the test tubes of a fixative such as formaldehyde or glutaraldehyde. Eggs are subsequently pipetted from each test tube and at least 100 eggs from each tube are examined for presence or absence of a fertilization membrane.

The life span of a sea urchin sperm cell is relatively short, with loss of viability occurring over a period of minutes to hours. Measurements of sperm viability have included respiration (Gray, 1927a,b,c; Ohtake, 1976a,b), ATP level (Christen et al., 1983), swimming activity (Timourian and Watchmaker, 1970; Grave 1934), and ability to fertilize eggs.

Using purple urchin sperm, Timourian and Watchmaker (1970) looked at sperm motility measured spectrophotometrically at 18–20°C. They found total loss of motility in 9 to 20 h. Motility dropped linearly from an index value of 17 in freshly activated sperm to an index value of about 3 after 5 h. Thereafter, motility was stable until about 12 h, when the final drop to zero occurred over some period of hours.

In an early study of sperm motility, Grave (1934) observed eggs of *Arbacia* isolated at one end of glass tubes of various lengths. Sperm were introduced at

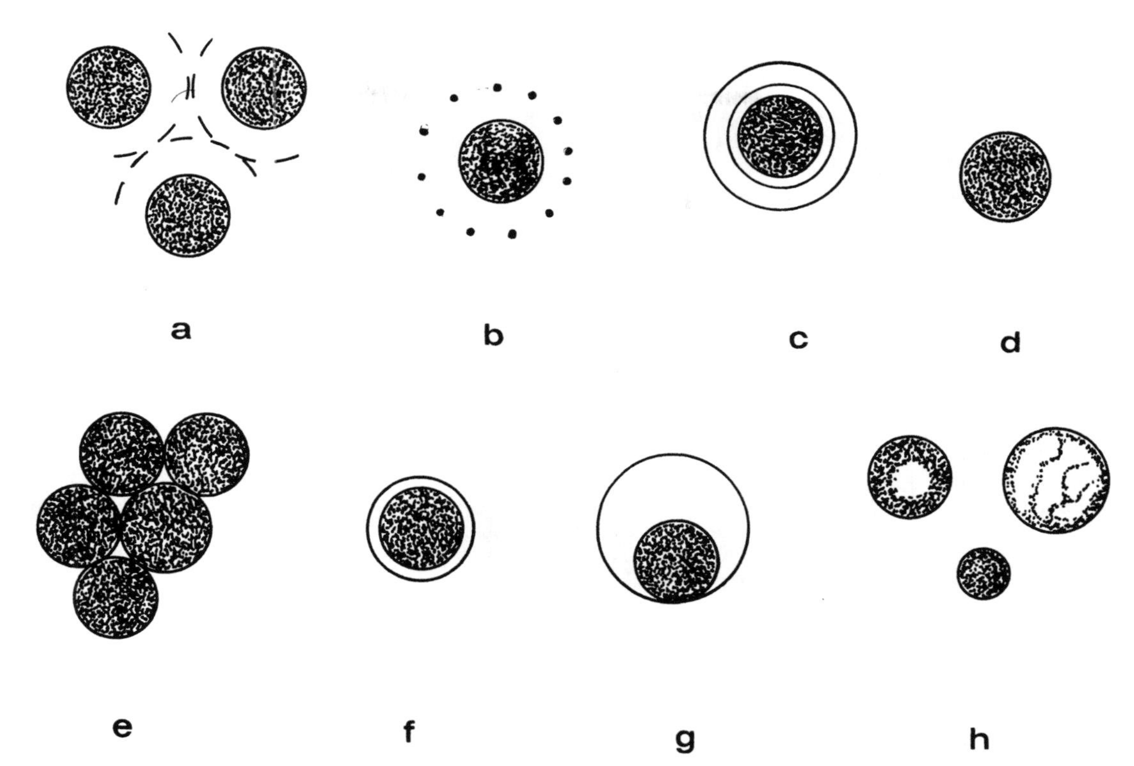

Figure 1. Typical appearance of purple sea urchin eggs: (a) newly shed eggs encased in a usually invisible jelly coating (shown with dashed lines); (c) a newly fertilized egg with outer fertilization membrane and inner hyalin membrane; (d) an unfertilized egg; (e) adhesion of unfertilized eggs into clumps (sometimes seen after preservation); (f and g) fertilized eggs often seen as artifacts in preserved samples of eggs, even those initially containing 100% normally fertile eggs (as c); (h) poor quality eggs include those with vacuoles (almost never fertilize), very small eggs with otherwise normal appearance, and very large, poorly pigmented eggs (the latter two may be fertilized at high sperm density)—all should be excluded from initial and final egg counts. Sand dollar eggs usually have similar appearances except that newly shed eggs (b) have a jelly coating that contains about 130 purple pigment cells per egg (Chia and Atwood, 1982).

the opposite ends, carefully avoiding disturbance of the water in the tube. The time of appearance of fertile eggs was noted and, along with tube length, provided an indication of swimming rate as a function of time. Assuming a direct sperm path to the eggs (highly improbable), the sperm showed an effective swimming rate of about 7.5 cm/h for several hours and a marked decrease in swimming rate thereafter (Table 1). Indeed, the true swimming speed of newly activated sea urchin sperm is probably much greater, being in the vicinity of 0.1 mm/s (Levitan, et al. 1991).

According to Grave, "The sharply marked reduction in rate after 2 hours of swimming . . . is significant. It probably means that although spermatozoa live and function for much longer periods, their vitality and energy are practically spent within 2.5 hours." However, he also noted that "spermatozoa which have been stimulated to great activity and weakened by hours of swimming may become quiescent and lie almost motionless. If these spent spermatozoa are now brought into the presence of eggs, or of sea water which has contained eggs for some time, they often revive and again swim actively."

A considerable body of literature exists about longevity of sea urchin sperm and various factors, including degree of dilution in seawater (Gray, 1927a,b; Hayashi, 1945), the presence of chelating agents (Tyler, 1953; Johnson and Epel, 1983), and the effects of pH (Ohtake, 1976a; Christen et al.,

Table 1. *Arbacia punctulata* sperm swimming rate and egg fertilization time (at 24.5°C) based on observations of time to egg fertilization in tubes of various lengths with eggs confined at one end and sperm introduced at the other

Tube length (cm)	Fertilization time (h)	Theoretical swimming rate (cm/h)	
		Overall	Last interval
7	0.9	7.6	—
12	1.6	7.5	(7.5)
15	2.5	6.0	(3.3)
18	8.0	2.3	(0.5)

Calculations based upon data presented in Grave, 1934.

1983), potassium (Gati and Christen, 1985), and metals (Lillie, 1921; Hoadley, 1923). In general, greater dilution, more alkaline pH (>8), and metal (e.g., copper) presence decrease the life span of the sperm. Conversely, lesser dilution, more acidic pH (<8), increases in potassium content, and the presence of metal chelators increase the life span of sperm. Many of these conditions are linked to a complex web of factors, including the internal sperm pH, the pH of the medium, ATP levels and ATPase activities, mitochondrial respiration, the presence of egg jelly (Ohtake, 1976b), and sperm activity.

It appears that the pH of the medium may be a very critical aspect of sea urchin sperm longevity. Christen et al. (1983) showed that ATPase activity decreased linearly by a factor of 10 from pH 8.1 to 7.3, with concomitant higher levels of ATP, and lower sperm activity and respiration. Sperm respiration decreased linearly with pH, with an 80% reduction between pH 8.2 and 6.2 (Ohtake, 1976a).

A detailed discussion of these factors, both as indicators of normal sperm physiology and as possible factors in the sea urchin sperm cell test, is beyond the scope of this chapter. In general, the more sperm are diluted, the more alkaline the pH, and the greater the water temperature, the more rapidly they lose their capability to fertilize eggs. With this in mind, any study of sperm viability must be regarded with special attention to these factors.

As a practical matter, the sea urchin sperm cell toxicity test should be run under conditions that allow sperm in the control (uncontaminated seawater) solutions to achieve a high degree of egg fertilization. Conditions of spawning, handling, storage, and exposure of gametes that result in an acceptably high percentage of egg fertilization in control seawater are generally considered adequate for the sperm cell toxicity test. However, in toxicity testing, especially for regulatory purposes, quality control and precision considerations dictate that the conditions and mechanics of the test be suitably standardized.

SPERM CELL TEST DEVELOPMENT

Early Toxicity Tests

Most of the early data on effects of chemicals on sea urchin sperm, eggs, and fertilization came from research into the physiology of sea urchin gametes and fertilization. These very short-term tests often used high chemical exposure concentrations in order to elicit effects (Lillie, 1921; Hoadley, 1923). These early data have little relevance to today's sea urchin test results or to applications in aquatic toxicology. Only since 1970 have effluent-related toxicity tests been conducted with sea urchin sperm.

Effects of petroleum fractions on fertilization and early development of purple sea urchins were reported by Allen (1971). Sperm were exposed to the test solutions for 5 min before the eggs were added, and eggs were examined for fertilization after 2 additional minutes. Examination for cell division was made 2 and 4 h after fertilization. No description was given of the sperm and egg densities used in the test or of the sperm/egg ratio used. The authors concluded that under these test conditions, cell division was a much more sensitive indicator of petroleum toxicity than fertilization.

Kobayashi (1980) conducted sperm toxicity tests with a Japanese sand dollar (*Peronella japonica*) using just 3-min sperm exposure time to copper, alkyl benzene sulfate (ABS), and ammonia. Data from these tests provided EC50 values of about 37 μg/L for copper, 3 mg/L for ABS, and 9 mg/L for ammonia. A large number of test methods and results were summarized by Kobayashi (1984), including tests with fertilized eggs and various embryonic and larval stages. In general, tests including exposure of gametes were the most sensitive.

The possible effects of chlorination of wastewater on purple urchin egg fertilization were investigated by Muchmore and Epel (1973). They used a 5-min exposure of sperm prior to the introduction of eggs for fertilization, and they showed that the concentration of sperm can be very important to the sensitivity of egg fertilization to chlorine toxicity. With a lower sperm density (1.6 million/ml), a 0.5% concentration of chlorinated sewage limited fertilization to only 1% of the eggs, but at a 10-fold higher sperm density (16 million/ml) fertilization was 100% at the same sewage concentration. Similar results were obtained with hypochlorite (64 μg/L) with 100% and 2% fertilization at high and low sperm densities, respectively. According to the authors, poor fertilization was clearly due to effects on sperm and not on eggs, and thiosulfate-neutralized wastewater had much less toxicity, indicating that sperm were almost certainly being killed or immobilized by chlorine.

The key point regarding these early tests is that eggs were introduced into the test solutions prior to, or shortly after, the sperm exposure began. Thus, they attempted to simulate presumably natural exposure conditions; current surrogate test methodologies use much longer sperm exposure times prior to introduction of eggs.

Development of a Standard Test Procedure

Availability of Gametes

In the 1970s, development of a standardized sea urchin sperm cell test was undertaken by Dinnel, Stober, and co-workers at the University of Washington. Working with green sea urchins, they first determined the seasonal inducibility to laboratory spawning using the KCl injection technique of Tyler (1949). This commonly used technique involves an injection of 0.5 to 1.0 ml of 0.5 M KCl into the coelomic cavity of the urchin through the peristomal membrane surrounding the mouth of the urchin (Figure 2).

Stober et al. (1979) used two populations of green sea urchins held at ambient temperature and about 2°C below ambient temperature. The spawning season at the cooler temperature was delayed by several

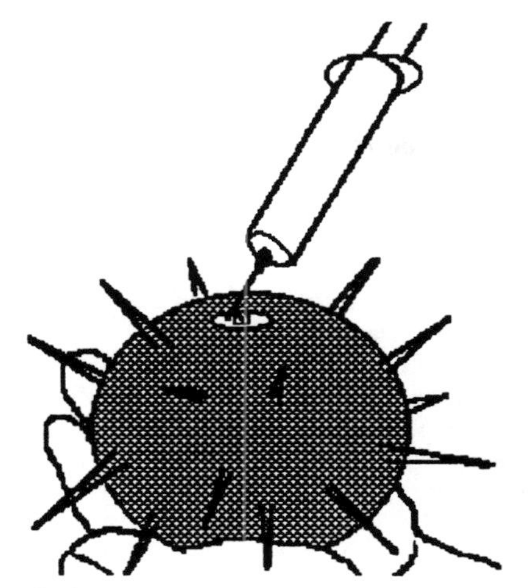

Figure 2. Spawning technique showing injection of KCl through the peristomal membrane and into the coelem.

weeks and extended by several months. Combined, the spawning at the two temperatures lasted from January through June. This is about the same as the spawning season observed in laboratory-held purple sea urchins collected from an intertidal population on the Oregon coast and held at the laboratory of the U.S. Environmental Protection Agency (EPA), Newport (G. A. Chapman, personal communication). Purple sea urchins at the Bodega Marine Laboratory north of San Francisco can be spawned over a period approaching 10 mo (G. N. Cherr, personal communication), and the Southern California Coastal Water Research Project (SCCWRP) laboratory in Long Beach, California, reports successful year-round spawning of laboratory populations of the purple sea urchin held at about 13°C (S. Bay, personal communication). The conditions responsible for this discrepancy are not known, but the laboratories reporting year-round spawning success also routinely held temperature conditions constant.

Over the past 2 yr, all three spontaneous spawnings of purple sea urchins observed in holding tanks at the U.S. EPA laboratory in Newport, Oregon coincided with significant late-winter and late-spring plankton blooms; no plankton blooms were noted during this period other than those associated with the spawns. This suggests that temperature history and larval food availability may interact to trigger many instances of natural spawning.

Annual reproductive cycles appear common in the purple urchin on the Pacific Coast (Boolootian, 1966). Populations north of about Pt. Conception, California show regular, pronounced peaks of gonadal development during late fall and winter, and those farther south show weaker, less regular cycles.

As a result of the seasonal green sea urchin spawning period, Stober et al. (1979) also investigated the spawning and testing of the sand dollar (*Dendraster excentricus*), a summer spawning echinoid. This species is found intertidally in the Puget Sound area of Washington state, but nearly all of the more southerly populations are subtidal. Initial work indicated that the sand dollar could be spawned from July through September. This spawning period extended the potential sperm testing season to include all but the fall season, with October and November notably lacking in spawning echinoids. Significant extension of the spawning season of sand dollars has been reported from the National Council for Air and Stream Improvement (NCASI) laboratory in Anacortes, Washington (T. Hall, personal communication). They utilize no special holding conditions other than temperature control (10–12°C) to prolong summer spawning and to induce winter spawning of recently collected organisms.

The current problems of a year-round population of Pacific Coast echinoids for sperm testing are in contrast to the experience on the East Coast with *Arbacia*. This species has been held in the laboratory (15°C) and spawned throughout the year. It is interesting to note that although this species can be spawned by KCl injection, a more practical method involving electrical stimulation is used. This technique may be characterized by fewer instances of complete spawnout of urchins, because release of gametes usually ceases upon cessation of electrical stimulation. Use of electrical stimulation has generally failed with *Strongylocentrotus* and *Dendraster*, but it is possible that a systematic investigation of both AC and DC voltages and amperages would discover a successful combination for spawning these species. Indeed, Harvey (1956) cites Iwata (1950) as having developed an electrical spawning method for an unspecified *Mytilus* mussel and the Japanese sea urchin, *Heliocidaris crasspina*; Harvey also mentions that the *Arbacia* technique was successful with a sand dollar (not identified).

Typically, *Arbacia* populations show a natural spawning season much like that of West Coast echinoids. Harvey (1956) reports normal spawning seasons for *Arbacia* as June–August at Woods Hole, earlier ripening (by May) at Beaufort, North Carolina, and ripe from January to June at Alligator Harbor, Florida. It thus appears that more southern populations of *Arbacia* spawn earlier than their more northerly counterparts.

Although it is possible to extend the period of gamete availability by taking advantage of geographical differences in spawning season, it is probable that year-round use of any of the test species will require spawning and holding techniques that (1) allow repeat spawning of organisms after a period of renewed gonad development or after only a partial spawning, (2) provide adequate dietary and thermal conditioning, and (3) prevent mass spawning by environmental triggers. Repeat spawning has been achieved regularly with most species. With both *Arbacia* and *Strongylocentrotus*, repeat spawning is aided by incomplete initial spawning. As mentioned earlier, this is achieved by use of electrical stimulation of *Arbacia*, which allows taking only the quantity of gametes needed at any one time. With *Strongylocentrotus*, the KCl injection technique frequently stimulates release of gametes from only a few of the five gonopores, possibly because the KCl does not distribute evenly in the coelomic activity. Even following apparently complete spawning of purple urchins (especially early in the spawning season), a second successful spawning has been regularly achieved at the U.S. EPA laboratory following several months of holding under ambient conditions and laboratory feeding (G. A. Chapman, personal communication).

Most species of sea urchins are routinely fed with seasonally available kelps, including *Laminaria* (on both coasts) and *Alaria*, *Hedophyllum*, and *Nereocystis* among many acceptable West Coast kelps. During periods when kelp is unavailable, urchins have consumed other foods, such as romaine lettuce, shrimp, and carrots. It is doubtful that any sea urchins, being generally omnivorous, would starve under conditions of laboratory culture, but obtaining proper nutrition for the generation of viable gametes almost certainly can require more than simply consumption of an offered foodstuff.

Laboratory cultures of sand dollars are usually held in an aquarium or tray containing a layer of sand. The sand dollars feed on a variety of small organisms including diatoms and other plant and animal species found in the substrate. They also feed on planktonic organisms in the water column (De Ridder and Lawrence, 1982). Indeed, during periods of plankton blooms and spontaneous spawning of sea urchins at the U.S. EPA laboratory at Newport, Oregon, the sand dollars are observed to be standing upright, with about three-fourths of their body above the sand, presumably to feed on the abundant food entrained in the incoming water. In a crowded sand dollar culture, the amount of food provided by the substrate and the incoming water could be insufficient to provide adequate caloric or nutrient input for suitable gonadal development.

Determining Suitable Sperm and Egg Densities for Tests

Following their determination of the spawning period for green sea urchins and sand dollars, Stober et al. (1979) looked at the effect of various sperm and egg densities on the percent fertilization of eggs. In their initial tests, they used 5000 eggs in test tubes containing 25 ml of seawater and sperm densities ranging from 50,000 to 25,000,000 per test tube. The resulting S:E (sperm-to-egg ratios), which ranged from 10:1 to 5000:1, produced egg fertilization of about 10% at the lowest sperm density and 100% at the highest sperm density (Figure 3). These data also showed that the longer the sperm were activated prior to egg introduction, the lower the percent egg fertilization at a given S:E. Similar qualitative results were obtained with sand dollar sperm and eggs, although >90% fertility was obtained at all S:E values of 125:1 or greater, even with sperm activation times as long as 120 min. Later refinements in the sea urchin sperm test (Dinnel et al., 1982, 1983, 1987) led to decreasing the test volume to 10 ml but maintained the standard 200 eggs per ml of test solution.

Throughout this period of test method development, the observation that greater S:E results in higher percent egg fertilization has remained qualitatively valid. This is an important point, because if, for example, a test solution contained twice the num-

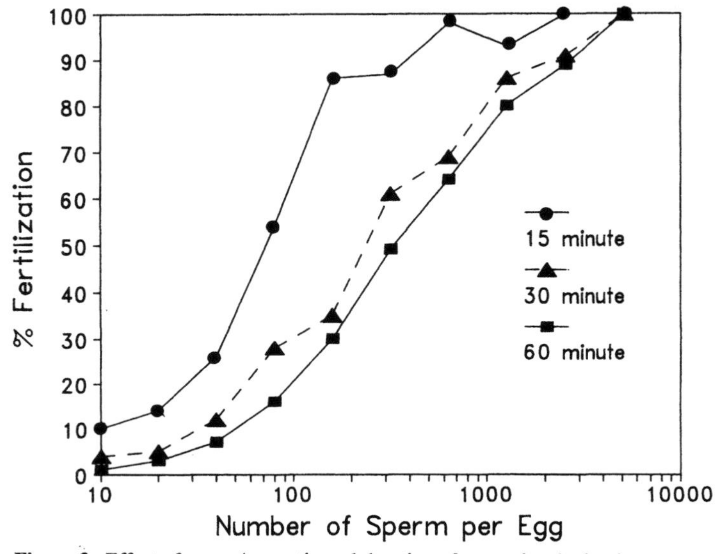

Figure 3. Effect of sperm/egg ratio and duration of sperm incubation in seawater on the fertilization of eggs of green sea urchin (*Strongylocentrotus droebachiensis*). (From Stober et al., 1979.)

ber of sperm needed to achieve 100% fertilization of the eggs, then a 50% debilitation of the fertilization capacity of the sperm population could go unnoticed as the remaining sperm would be sufficient to fertilize all the eggs.

Experience has also shown that the optimum S:E is species dependent and, most important, that the optimum ratio can vary from spawn to spawn within a single population of any of the common test species. Generally, at the peak of the spawning season the required S:E is lower than either early or late in the spawning season.

Recommended egg density is 200 per ml for the commonly used sperm cell test protocols in the United States (Dinnel et al., 1987; U.S. EPA, 1988). This density has been chosen so that a single milliliter of test solution provides a convenient egg density for evaluating fertilization in at least 100 eggs in a small (usually 1 ml) sample of test solution. Recommended S:E for the four most common test species are 2500:1 for *A. punctulata* (U.S. EPA, 1988), 200:1 for *S. purpuratus*, 1200:1 for *D. excentricus*, and 2000:1 for *S. droebachiensis* (Dinnel et al., 1987).

Effect of S:E and Percent Control Fertilization on Test Sensitivity

As pointed out above, the presence of more sperm than are necessary to fertilize all eggs in a test solution could decrease the sensitivity of the sea urchin sperm test. The magnitude of this effect was demonstrated in a series of silver toxicity tests with sand dollar sperm (Dinnel et al., 1987). In this study, a generally linear effect was observed, with an S:E range of 900:1 to 10,500:1 producing EC50s increasing from 22.9 to 37.0 μg/L (Figure 4). The results were expressible as

$$\text{EC50} = 13.0(\log \text{S:E}) - 13.5$$

Based on their analysis of sperm and egg density variability produced by dilution and counting vagaries, these authors concluded that "minor variations in the sperm:egg ratios between toxicity tests should not lead to results of significant error." Two iterations of the U.S. EPA test protocol for *A. punctulata* provided a tentative second evaluation of the effect of S:E on sensitivity of that test (Nacci et al. 1987; U.S. EPA, 1988). Initially using an S:E of 250:1 in precision testing with copper provided a mean EC50 for copper of 12.05 μg/L; subsequent test modification to more regularly achieve a high percentage of control egg fertilization led to the use of a 2500:1 S:E, which produced a mean EC50 of 29.9 μg/L. Analysis of just these two data points yields the following relationship:

$$\text{EC50} = 16.22(\log \text{S:E}) - 25.27$$

The similar slope functions of the two data sets (13.0 and 16.22) suggest that these results may be generally applicable (Figure 4). However, the 12.05 μg/L EC50 for *Arbacia* was based on nominal copper concentrations (D. Nacci, personal communication); use of measured copper concentrations could change the slope function to some unknown extent.

Research at the U.S. EPA laboratory in Newport, Oregon has provided additional information on the

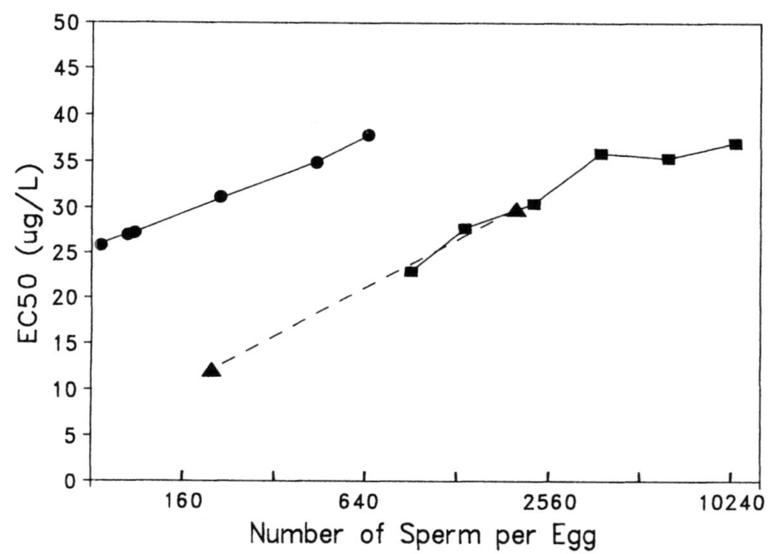

Figure 4. Effect of S:E on the sensitivity of the sea urchin sperm test: (●) copper—Pacific Coast method (20-min exposure) using purple sea urchin (*Strongylocentrotus purpuratus*) sperm (unpublished data, U.S. EPA, Newport, Oregon); (▲) copper—U.S. EPA method using purple sea urchin (*Arbacia punctulata*) sperm (Nacci et al., 1987; U.S. EPA, 1988); (■) silver—Dinnel method using green sea urchin (*Strongylocentrotus droebachiensis*) sperm. (From Dinnel et al., 1983.)

sensitivity of the sperm cell test as a function of S:E and percent control egg fertilization. Using purple urchin sperm and copper, a continuous linear effect on EC50 values was found (Figure 4) over a range of S:E from 0.20 to 1.53 of the S:E (439:1) that was just sufficient to fertilize 100% of control eggs (Figure 5). The control fertilization ranged from 49% at the lowest S:E to 100% at the two highest S:E ratios. The relationship between S:E and EC50 was expressed by the following regression:

$$EC50 = 13.24(\log S{:}E) + 0.28$$

It was also determined that seasonal differences in gamete viability produced seasonal differences in optimum S:E. Early and late in the spawning season, ratios in excess of 1000:1 were sometimes necessary to produce 100% control fertilization. During the middle of the spawning season an S:E of 100:1 was sufficient. Finally, as the optimum S:E increased from 100:1 to 1000:1, the sensitivity of the test decreased at a given percent control fertilization. The cause of this shift in sensitivity is not known, but it may be due to the increased opportunity for sperm and egg contact in test solutions with greater sperm densities. Measurement of dissolved copper concentrations in the test solutions indicated that the sperm were not detoxifying the solution by reducing the amount of dissolved copper (an alternative hypothesis for the decreased sensitivity of the test at high S:E).

Test sensitivity can vary with two interacting factors: (1) sperm density (equivalent to the S:E because egg density is constant) and (2) percent control fertilization (Figure 5). With any given batch of gametes, the sperm density and percent control fertilization are codependent and their relative influence cannot be separated. As the sperm concentration required to achieve a particular percent control fertilization changes over the season or among different batches of gametes, the test sensitivity also can change, even at a single, targeted percent control fertilization (because the sperm concentration has been changed). Conversely, at a constant sperm concentration, the percent control fertilization can change with the season or among different batches of gametes, and this can also change test sensitivity. The critical issue regarding these effects is how they influence the precision of the test.

Precision of the Sperm Cell Test

The U.S. EPA laboratory in Narragansett, Rhode Island has conducted several series of intralaboratory precision tests with variations of the *Arbacia* sperm

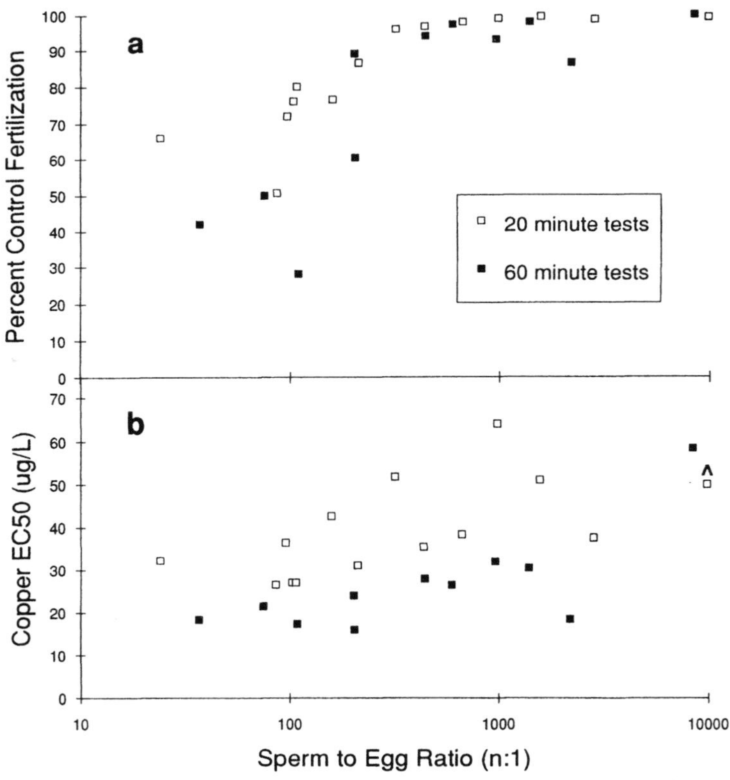

Figure 5. Effect of sperm-to-egg ratio and sperm exposure time to copper (20 vs. 60 min) on (**a**) percent control fertilization and (**b**) sperm sensitivity to copper as measured by the copper concentration causing a 50% reduction in egg fertilization. (Unpublished data, U.S. EPA, Newport, OR, using the purple sea urchin.)

cell test protocol. The largest test series, with copper as the reference toxicant, yielded coefficients of variation [CV = 100(standard deviation of the EC50s/ mean EC50)] ranging from 22.6 to 48% in five different precision checks. Similar U.S. EPA precision studies with the reference toxicant sodium dodecyl sulfate (SDS) provided a range of CVs of 10 to 35.7% in intralaboratory tests.

There is considerable discussion regarding what constitutes an acceptable coefficient of variation in various toxicity tests. Certainly the precision of a given test method can be compared with other methods for the same test organism as well as with precision estimates for other common test organisms. Morrison et al. (1989) published results of inter- and intralaboratory tests with common marine and freshwater test species. For marine short-term chronic estimator tests (including the *Arbacia* sperm cell test), the CVs for copper and SDS ranged from 1.8 to 46.4% with a median value of 24.1%. For marine acute tests the CVs ranged from 22 to 104% with a median value of 44%. Finally, for freshwater acute tests, the CVs ranged from 33 to 71% with a median value of 47%.

Despite the fact that a few major variables are known which can lead to loss of precision in the sperm cell test, the CVs observed with the sperm cell tests are within the range of other tests that are in common use. Test variation is often due to changes in condition of test organisms (e.g., in response to size, diet changes, reproductive cycle phase)—changes that merely mimic changes that occur constantly in the natural populations of organisms that the tests are designed to protect. Chronic toxicity tests, by definition, are working at levels of toxicity that are just above the detection limits. The sperm cell test EC50 values are usually only two or three times the toxicity test detection limit; state-of-the-art chemical analytical methods seldom yield better CVs so near the threshold of detection.

Standard Sperm Test Methods

Although many test methods have been described in the literature since 1971, a few presently being used in toxicological evaluations for regulatory programs are the most widely used and evaluated. For other toxicological purposes, such as screening for toxicity or generally comparing toxicity of various materials or waste fractions, any of the existing methods can provide valid comparative information.

The most widespread use of sea urchin sperm cell test methods is in regulatory programs dealing with evaluations of the toxicity of effluents discharged into marine and estuarine environments (U.S. EPA, 1988). The goal of many of these regulatory programs is to use toxicity testing to prevent acute and chronic toxicity in the receiving water. This presents an enigma to those involved in test development, test conduct, and test evaluation: how to determine concentrations that can produce chronic toxicity using tests that are sufficiently short and sensitive to be practical to employ?

During development of a sperm test method for sea urchins (*S. droebachiensis*, *S. purpuratus*, and *S. franciscanus*) and sand dollars (*D. excentricus*), the sperm cell tests were compared in sensitivity with tests with other marine organisms, including sea urchin and sand dollar embryos, crab zoeae (*Cancer magister*), and squid (*Loligo opalescens*) and cabezon (*Scorpaenichthys marmoratus*) larvae (Dinnel et al., 1983; Dinnel, 1989). Test materials included five metals and four pesticides. Larvae and embryos were most sensitive to metals, and larvae and adults were most sensitive to pesticides. The sperm cell test was a reasonably good toxicity estimator for other sensitive species and life stages with copper, silver, zinc, dieldrin, and DDT. Because of the short duration and ease of the sperm cell test, its slightly less sensitive response was not considered a major impediment to its utility as a toxicity indicator, although sperm cell tests were not particularly good surrogates for more sensitive species with chemicals such as cadmium, lead, Endrin, and endosulfan. The great convenience and the potential sensitivity of the method described by Dinnel et al. (1982) led to development of a similar test with *A. punctulata* at the U.S. EPA laboratory in Narragansett, Rhode Island (Nacci et al., 1987).

Despite the fact that the sperm cell test can be relatively insensitive to certain single chemicals, its use in the monitoring of effluent toxicity indicates that it is frequently a good surrogate for other short-term chronic estimator tests including those with fish (inland silverside, *Menidia beryllina*; fathead minnow, *Pimephales promelas*; sheepshead minnow, *Cyprinodon variegatus*; mysids, *Mysidopsis bahia*; and plants, *Champia parvula* [Schimmel et al., 1989]).

Most regulatory applications of sea urchin sperm tests have utilized the U.S. EPA method with *Arbacia* (U.S. EPA, 1988) or the Dinnel method with purple urchins or sand dollars (Dinnel et al., 1987). On the Pacific Coast, use of the Dinnel method is commonplace, but many individual modifications that are in common use can alter the sensitivity of the test.

Differences in Sperm Test Procedures

Variations in sperm cell test procedures derive from a number of motives, including the desire for a test that is more or less sensitive (based on a desire for tighter or looser regulations), a test that is less of a surrogate and more of a true simulation of sea urchin spawning, a test that is even easier to conduct, a test that is more precise, and a test with a failure rate (usually due to poor control fertilization) that approaches 0%.

Stober and Dinnel (Stober et al., 1979; Dinnel et al., 1982) have emphasized switching from species to species to provide test coverage over the greater portion of the year on the Pacific Coast (particularly in Puget Sound). Work with *Arbacia* on the Atlantic Coast has generally overcome the problem of a period without adequate gametes for tests, but more work is needed to extend this type of success to the Pacific Coast in general.

A primary modification of the Dinnel procedure was made in the development of the *Arbacia* test procedure. Instead of allowing males to spawn directly into seawater (the "wet" method) and the sperm to be activated by dilution in seawater, *Arbacia* sperm are collected "dry" (Figure 6). Most recent Pacific Coast modifications of this test also use the dry collection technique (Anderson et al., 1990; Cherr et al., 1987; Chapman, 1992). The dry collection technique has several effects: it allows the sperm to be held longer without losing viability; it decreases the sensitivity of a test conducted at a fixed S:E because the sperm are more viable at the beginning of the exposure period; the increased viability decreases the frequency of test failure due to poor control fertilization during periods of marginal gamete viability; and it is sometimes more difficult to get an adequate supply of sperm because sea urchins and (especially) sand dollars tend to spawn a little more readily when the gonopores are in water and because semen is lost into the surface irregularities of the urchin's body.

In order to minimize instances of test failure due to poor control fertilization, the *Arbacia* test procedure was later modified to use a ratio of 2500:1 rather than the initial 250:1. As noted in an earlier section of this chapter, this change in S:E may have increased the EC50 for copper from 12.05 to 29.9 μg/L in the *Arbacia* sperm test.

Test protocol modifications on the Pacific Coast (Anderson et al., 1990; Cherr et al., 1987) have utilized the dry sperm collection method, a very short period of preexposure activation of sperm, and shorter exposure periods (usually 10 or 20 min). All of these modifications probably tend to reduce the sensitivity of the test by varying degrees. During periods of good gamete viability, this reduction is probably due solely to the shorter exposure duration, and the magnitude of the effect ranges from essentially undetectable to perhaps as much as a threefold difference. During periods of marginal gamete viability, the use of these modifications may increase the chance of completing a successful test.

Retaining sperm viability does not always require special procedures such as collecting sperm by the dry technique or holding sperm on ice. During early development of the sperm cell test, Stober et al. (1979) were able to hold diluted sperm of green sea urchins at 7.7°C for over 24 h with little loss of viability. Indeed, experience at the U.S. EPA laboratory in Newport, Oregon, shows that sperm of sand dollars appear to retain viability longer if diluted slightly with seawater and stored at 12°C than if collected by the dry technique and stored on ice.

Use of shorter exposure periods has several logistic advantages. Shorter tests are more convenient to repeat and more tests can be conducted during a given period. However, exposures that are shorter than 10 to 20 min can make large-scale tests hectic because of the time required to complete introduction of sperm to all test vessels in time to begin the introduction of eggs; this depends somewhat on the pipetting method utilized. As a practical matter, tests much shorter than 20 min provide little advantage other than potentially greater environmental realism

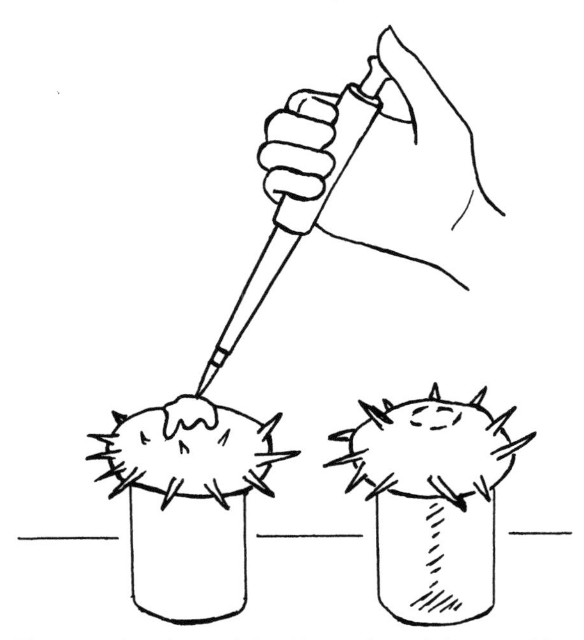

Figure 6. The "dry" and "wet" spawning techniques used for the purple sea urchin (*Strongylocentrotus purpuratus*). Semen is collected directly using a pipet (dry) or shed directly into a beaker of seawater (wet). Eggs are usually collected by the wet method.

for sea urchin egg fertilization per se, and the decreased test sensitivity weakens the value of the test result as a surrogate for more sensitive marine and estuarine species.

With a general understanding of how collection, storage, sperm activation time, and exposure period of sperm can influence test sensitivity and test success, differences in the various sperm cell methods can be more fully appreciated. Table 2 shows characteristics of three test methods, including those of Dinnel et al. (1987) and U.S. EPA (1988) for *Arbacia* and a more recent U.S. EPA method for use on the Pacific Coast (Chapman, 1992). The latter 60-min exposure test is similar to procedures used in 20-min toxicity tests in California (Anderson et al., 1990).

It is clear that the primary difference among these methods is the time during which the sperm are activated prior to egg introduction. This duration ranges from 65 min for the 60-min Pacific Coast method to 120 min for the U.S. EPA *Arbacia* method and 180 min for the Dinnel procedure. These differences should presumably make the longer-duration activation tests slightly more sensitive at a common S:E because of a potential loss of sperm viability, which may vary significantly from spawn to spawn. However, in practice these methods often use different species and different S:E values, and direct comparison of the effect of activation period alone cannot be based on historical data. The comparative toxicity data that are available indicate generally similar results between species and test methods (Table 3).

Sperm Cell Test Mechanics

Sperm Counts and Dilution

Sea urchin semen is diluted about 10,000-fold and the sperm are counted on a hemacytometer. At this density, the full field of a Neubauer hemacytometer yields a count of several hundred sperm. The sperm concentration in semen of the purple urchin *S. purpuratus* is usually about 4×10^{10}/ml (40,000,000,000/ml). This is the concentration in semen collected by the dry method. A ripe male can easily shed 1 ml of semen. Using the wet method and collecting the semen in 100 ml of seawater will dilute 1 ml of sperm about 100-fold (about 400,000,000 sperm/ml). Prior to counting, a subsample of the sperm is diluted and the sperm are immobilized with seawater containing 10% acetic acid. Indirect sperm density estimates might be acceptable with proper calibration; these include turbidimetric, photometric, and particle counting methods.

Dinnel Method

Sperm from individual males are collected in 100-ml beakers of seawater. A 1 in 100 dilution of the sperm solution is prepared for counting. Sperm solutions from several males could be pooled. Based on count, sperm stock is diluted to desired sperm density for introduction of 0.1 ml of sperm into 10 ml of test solution. Tests are usually run at 12°C.

U.S. EPA Method

From 0.5 to 1 ml of semen from each of four male *Arbacia* are collected dry in a test tube (on ice). Serial dilutions are prepared, the last two being made into 10% acetic acid in seawater to kill sperm prior to counting. Based on sperm count, the desired sperm density for testing is achieved by further dilution of the closest more concentrated member of the series. Tests begin with the addition of 0.1 ml of sperm solution into 5 ml of test solution. Tests are run at 20°C.

Pacific Coast Method

Semen from at least three males is collected dry in separate test tubes (on ice). A tiny quantity (about 1 μl) of semen from each male is placed into a drop of seawater and viewed microscopically to check motility (batches of semen showing nonmotile or poorly motile sperm are not used). Equal volumes of semen from good males are pooled into a composite of at least 0.15 ml and held on ice. Twenty-five microliters of semen is diluted into 179 ml of seawater, killed with acetic acid, and a sperm count performed. Once the dilution of semen necessary to provide a 500:1 sperm/egg ratio is determined, a second 25-μl aliquot of semen is diluted accordingly. This test also uses 0.1 ml of sperm solution in 5 ml of test solution. Tests are run at 12°C. The sperm density of the final sperm solution is counted to provide data for reporting the actual S:E.

Egg Preparation

Eggs (Figure 1) are shed into seawater, rinsed several times, and subsampled and counted to determine the concentration of eggs in the stock. Egg concentration in the final egg stock is checked and readjusted if necessary.

Dinnel Method

Eggs from each female are shed into 100-ml beakers filled with seawater. Eggs from different females may be pooled. Eggs are decanted into a large beaker with more seawater and allowed to settle. Decanting, rinsing, and settling are repeated two or three times. Egg concentration is determined by counting two or three subsamples under a dissecting scope. A final egg stock solution of 2000 eggs/ml is prepared. After sperm have been exposed in test solutions for 60 min, 1 ml of egg suspension (2000 eggs) is added to each test tube.

U.S. EPA Method

Eggs are shed from females slightly submerged in shallow trays. Eggs are collected with a syringe equipped with a large-bore needle and transferred to seawater in a conical centrifuge tube. The pooled eggs are rinsed twice by gentle centrifugation ($50 \times g$ for 3 min) with resuspension in clean seawater each time. Egg stock is brought to a volume of 200 ml, subsampled (1 ml of a 1:10 dilution is counted in a Sedgewick-Rafter chamber), and density adjusted to 2000 eggs/ml. Final egg stock density is

Table 2. Characteristics of several current sea urchin sperm test methods

Spawning period	
Dinnel et al. (1987)	30 min
U.S. EPA (1988)	Brief period of electrical stimulation
Chapman (1992)	Not to exceed 30 min
Egg rinsing period	
Dinnel et al.	During activation and exposure periods
U.S. EPA	During activation and exposure periods
Chapman	30 min; before sperm activation
Sperm activation period (before test exposure)	
Dinnel et al.	Spawning period plus exactly 90 min
U.S. EPA	Exactly 60 min
Chapman	Not specified; <5 min in practice
Exposure period	
Dinnel et al.	60 min
U.S. EPA	60 min
Chapman	60 min
Fertilization period	
Dinnel et al.	20 min
U.S. EPA	20 min
Chapman	20 min
Total sperm activation period prior to egg introduction	
Dinnel et al.	180 min
U.S. EPA	120 min
Chapman	65 min
Overall test duration	
Dinnel et al.	200 min including spawning period
U.S. EPA	140 min plus spawning period
Chapman	85 min plus spawning period

Table 3. EC50 values (mg/L) from the three standard sperm cell test procedures discussed in the text

	Dinnel		U.S. EPA	Pacific Coast
Chemical	Purple urchin	Sand dollar	*Arbacia*	Purple urchin
Copper	0.025	0.026	0.030	0.026
Silver	0.115	0.054		
Cadmium	18.	8.		
Zinc	0.262	0.028		
Lead	8.2	17.		
SDS			2.98	3.65
Endrin	0.342	0.441		
Endosulfan	0.081	0.352		
Dieldrin	0.013–0.025	0.088		
DDT	0.0005–0.001			
Pentachlorophenol			0.9	
Hexachloroethane			29.1	
2,4-Pentanedione			0.9	
2,2,2-Trichloroethanol			291.9	
2-Chloroethanol			550.4	
2-Methyl-1-propanol			718.3	
2-Methyl-2,4-pentanedione			6.8	
2-(2-Ethoxyethoxy) ethanol			3370.	

Data for the Dinnel method from Dinnel et al., 1989; data for the U.S. EPA method from Nacci et al., 1986, and U.S. EPA, 1988; data for the Pacific Coast method from G. A. Chapman, personal communication.

confirmed by a second count. Eggs are stored at room temperature. After sperm have been exposed to test solutions for 60 min, 1 ml of egg suspension (2000 eggs) is added to each tube. Egg mixing for sampling is achieved with air bubbling.

Pacific Coast Method

Eggs are shed into seawater in 100-ml beakers. Eggs from each female are examined under a microscope to determine their suitability (Figure 1a and h). Trial fertilization on a microscope slide is recommended using a small amount of sperm solution. Eggs from the good females are pooled, brought to 600 ml in a 1-L beaker, and allowed to settle for 15 min at 12°C; overlying water is siphoned off and the rinse repeated for a second 15-min period. Eggs are brought to a known volume (100 or 250 ml depending on their numbers), and either a 1 to 10 or a 1 to 100 subsample is counted in a 1-ml Sedgewick-Rafter chamber. Eggs are diluted to a final target density of 2240 eggs/ml. Eggs are kept at 12°C and, after sperm exposure is completed, 0.5 ml of egg stock (1120 eggs) is added to each test tube. During egg introduction, the stock is mixed in a 150-ml beaker by gentle agitation with a perforated plunger.

Test Conclusion and Egg Fertilization Counts

All three test methods allow 20 min for egg fertilization before the test is stopped by the addition of a fixative. Egg samples from each tube are usually examined for fertilization within a few days. The presence or absence of a fertilization membrane is used as the end point (Figure 1). Sampling bias can occur in tubes with large numbers of unfertilized eggs (at least following preservation with glutaraldehyde) because they tend to become adhesive and may form clumps (Figure 1*d*) or stick to the bottom of the test tube. Fertilized eggs show no such tendency and so may be obtained more easily in samples pipetted from settled eggs. Samples should be especially well mixed prior to sampling if this phenomenon is observed.

Dinnel Method

Tests are stopped by the addition of 1.0 ml of concentrated (37%) formaldehyde. After eggs settle, the sample is concentrated by pipetting off most of the overlying solution. A subsample of the concentrated eggs are placed onto a counting slide (unspecified) and 100 to 200 eggs are evaluated for fertilization.

U.S. EPA Method

Tests are stopped by the addition of 2 ml of 10% buffered formalin. After eggs settle, 1 ml of eggs is pipetted from the bottom of a tube, transferred to a Sedgewick-Rafter chamber, and about 100 eggs evaluated for fertilization.

Pacific Coast Method

Tests are stopped by addition of either formaldehyde or glutaraldehyde (no volumes or concentrations are specified; the U.S. EPA laboratory, Newport, Oregon uses 0.5 ml of 1.0% glutaraldehyde). In order to check the actual egg density in the tubes, five tubes per test are selected and the eggs resuspended and mixed with an automatic pipet. One milliliter of sample is transferred to a Sedgewick-Rafter chamber, 100 eggs are scored for fertilization, and all eggs in the subsample are counted. The other samples are concentrated by pipetting off most of the overlying solution, resuspending and mixing the eggs, and examining 100 eggs for fertilization.

Data Analysis

Most investigators report the results of sea urchin sperm tests as EC50 values (estimates of the concentration of test material reducing egg fertilization to 50% of control values). When control fertilization is below 99% (unfertile >1%) the percent unfertile data are normalized to control values by using the ubiquitous Abbott's formula (Abbott, 1925):

$$\text{Adjusted \% test response} = \frac{\text{\% test response} - \text{\% control response}}{100 - \text{\% control response}} \times 100$$

For estimating chronic toxicity, it has been common to use hypothesis testing (see Chapter 10) to determine treatments that were statistically significantly different from controls. However, the ability to detect effects statistically can vary so greatly between tests that apparent increases in infertility of 35 and 5% (for example) can be identified statistically as insignificant and significant, respectively. Because of this phenomenon, it is likely that sea urchin sperm cell test data that are used to estimate chronic toxicity will be analyzed using regression techniques. This approach allows calculation of a specific end point (e.g., the EC20 estimate of the concentration of test material causing a 20% increase in infertility) along with its confidence interval.

One problem with the use of Abbott's formula and having significantly less than 100% control fertilization is that low concentrations of some chemicals stimulate fertilization to over 100% of control values. As a result, normalized percent fertilization can exceed 100% at these low chemical concentrations. This causes slight but potentially significant shifts in results of probit and some hypothesis tests. Low concentrations of a number of chemicals can cause this type of stimulation, including seven of nine chemicals tested by Dinnel et al. (1983) (copper, silver, lead, endosulfan, endrin, dieldrin, and DDT). These results are one argument for establishing an S:E that allows essentially 100% control egg fertilization unless detection of such stimulation is considered to be an important aspect of the toxicity evaluation.

DISCUSSION

Regulatory Applicability

In regulatory application, test methods are generally selected based on sensitivity, repeatability, cost-effectiveness, environmental applicability, and need. All sea urchin sperm cell test methods generally fulfill these requirements, but some are better than others in each respect. Insensitive tests are out of favor because they do not provide adequate protection; sensitive tests are often impugned because of perceived lack of precision, difficulty in conducting the tests, or environmental inapplicability (but often only because they drive regulatory actions). Selection of a test method therefore relies on a firm foundation of established precision, ease of conducting the test, and environmental applicability.

The U.S. EPA *Arbacia* method, of the test methods discussed above, has the best-documented precision; this does not say that it necessarily has the best precision, but it has been subjected to thorough intralaboratory precision testing. All three sea urchin sperm cell test methods have environmental applicability based largely on comparative tests against other test species (they are sometimes excellent surrogates for other sensitive species, especially for complex effluents, and are not unrepresentatively sensitive).

Each toxicity test species and method is a tool for the evaluation of toxicity. The sea urchin sperm cell test is often used as an estimator of chronic toxicity because it is a life stage sensitive to many toxic materials. In the U.S. EPA's NPDES programs (see Chapter 24) to eliminate the discharge of toxic materials in toxic amounts, estimations of chronically toxic effluent concentrations rely on small batteries of short-term tests of freshwater and marine organisms. Few, if any, of these tests are as sensitive as a full life cycle chronic test. Yet the use of short-term tests allows practical direct measurement of the toxicity of chemicals whose toxicological properties may be only poorly known and of mixtures of toxic chemicals whose interactions are entirely unknown.

Ideally, short-term toxicity tests such as the sea urchin sperm test would be able to identify chronic toxicity to the same level of resolution as the battery of acute and chronic toxicity tests required for a national water quality criterion (see Chapter 24) for a specific chemical. In reality this is highly unlikely; an examination of the capability of the sea urchin sperm test to estimate safe concentrations of copper and lead demonstrates this fact (Table 4). An estimate of the safe, or acceptable, concentration of these metals is provided by the U.S. EPA water quality criteria documents (U.S. EPA, 1985a,b).

Sea urchin sperm cell tests are very sensitive to copper. The most sensitive sea urchin sperm test (the 200:1 S:E *Arbacia* test) produced an EC50 of 12.05 μg/L, compared to the water quality criteria value of 2.916 μg/L. In fact, representatives of the battery of sensitive life stage tests often used in monitoring the toxicity of marine effluents (sea urchin sperm tests and mussel and oyster embryo tests) almost always fall within a factor of 10 of the chronic criterion for copper. A shortfall of this magnitude, however, causes little consternation in a monitoring program, primarily because effluent discharge limits using direct toxicity measurements are applied on undiluted effluents or are applied at the edge of a small zone of initial dilution in the receiving water (see Chapters 24 and 33). This is different from a water quality criterion that may be interpreted as describing allowable levels over an entire body of water. Thus, the utility of the sea urchin sperm test (and other toxicity monitoring tests) derives from the tenet, based on databases such as that shown in Table 4, that one or more of the commonly used tests will have sufficient sensitivity to be a close surrogate for the more sensitive organisms in the receiving water.

The need to use an array of monitoring tests is illustrated by comparing the sensitivity of sea urchin sperm and *Mysidopsis* in tests with copper and lead (Table 4). As discussed previously, sea urchin sperm tests with copper produce EC50 values that are reasonably good estimates of the chronic water quality criteria value. Chronic tests with mysids were only about half as sensitive as the sea urchin sperm tests.

Data from tests with lead show that sea urchin sperm tests provide EC50 values that are very poor estimators of chronic lead toxicity, being about 1000 times higher than the marine chronic toxicity value of 5.6 μg/L. Mysid chronic tests, on the other hand, provided an excellent estimator for the chronic water quality criteria value (within a factor of 4).

These examples illustrate two basic principles of using toxicity tests as monitoring tools: (1) no single test is sufficiently sensitive that its use alone will always provide a reasonable indication of toxicity to all organisms in a receiving water and (2) rarely will a small battery of monitoring tests provide a toxicity estimate as sensitive as a much larger number of acute and chronic tests. These principles lead to the practical conclusions that the use of convenient toxicity-monitoring tests should include a reasonable number of test species and that the results should be applied with awareness that some degree of chronic toxicity might occur at the higher chemical or effluent concentrations that produced no effect in the monitoring tests.

The Future of the Sea Urchin Sperm Test

The sea urchin sperm test will continue to be a valuable tool in the monitoring of effluent toxicity. It is being considered as a test for monitoring sediment toxicity (see Chapter 8), at least with sediment elutriates and pore water samples. The very short duration and logistical simplicity of the test may make it extremely useful in identifying the toxic components of effluents and in evaluating the efficacy of waste treatment alternatives. The period between egg fertilization and test termination may be lengthened by one to several hours, allowing one or two cell divisions, making the fertilization end point easier to determine even after longer periods of preservation, and including potentially toxicant-sensitive cell division in the test. The sperm test may become a component of a broader environmental monitoring "sea

Table 4. The theoretically acceptable concentrations (final chronic value, FCV; final acute value, FAV) and copper or lead to marine animals compared to acute toxicity test results (ranked by increasing EC50 values by test organism genus), mysid chronic test results, and sea urchin sperm test results

Rank	Genus/test	Copper concentration (μg/L)	Table value/ final chronic value
	Marine FCV	2.916	
1.	*Mytilus* (embryo)	5.8	2.0
	Marine FAV	5.832	
	Arbacia SPERM[a]	12.05	4.1
2.	*Paralichthys*	13.93	
3.	*Crassostrea* (embryo)	14.9	5.1
	Purple urchin sperm	25	8.6
	Sand dollar sperm	26	8.9
	Arbacia sperm[b]	29.9	10
4.	*Mya*	39	
5.	*Acartia*	39.97	
6.	*Cancer*	49	
	Chronic—*Mysidopsis* sp.	54.09	19
	Green urchin sperm	59	20
7.	*Haliotis*	65.6	
8.	*Homarus*	69.28	
9.	*Phyllodoce*	120	
10.	*Pseudopleuronectes*	128.9	
		Lead concentration (μg/L)	
	Marine FCV	5.603	
	Chronic—Mysidopsis	21.08	3.8
	Marine FAV	287.4	51
1.	*Fundulus*	315	
2.	*Mytilus* (embryo)	476	85
3.	*Ampelisca*	547	
4.	*Cancer*	575	
5.	*Acartia*	668	
6.	*Mercenaria* (embryo)	780	139
7.	*Crassostrea* (embryo)	1,363	243
8.	*Mysidopsis*	3,130	
9.	*Cyprinodon*	>3,140	
	Arbacia SPERM[c]	5,400	963
10.	*Menidia*	>5,604	
	Purple urchin sperm	8,200	1463
	Sand dollar sperm	13,000	2320
	Green urchin sperm	19,000	3391

[a] 200:1 S:E ratio.
[b] 2500:1 S:E ratio.
[c] 200:1 S:E ratio.
Sperm cell test results are from Dinnel (1989), U.S. EPA (1988), and Nacci et al. (1986); all other values are from U.S. EPA water quality criteria documents (U.S. EPA 1985a,b).

urchin test system'' proposed by Dinnel et al. (1988) and including all sea urchin life stages and a variety of toxicological responses including acute toxicity, behavior, bioaccumulation, cytotoxicity, fertilization, development, and genotoxicity.

The methods currently being evaluated at the U.S. EPA laboratory in Newport, Oregon use several quality control features that may be included in future test methods, at least in modified form. These include the use of ''egg controls'' (to which no sperm are added) to check for the presence of unintentionally fertilized eggs, which would bias the test data, and the use of ''toxicant egg controls'' (to which no sperm are added) to determine whether the material being tested might induce an artificial elevation of the fertilization membrane.

The use of regression analysis to measure toxicity thresholds (e.g., an EC10 or EC20) may make the test more sensitive and allow the use of fewer replicates of each treatment. Using automated procedures may make sperm counts, egg counts, and fertilization determinations more rapid. Turbidimeters have been used for rapidly adjusting sand dollar sperm density (T. Hall, personal communication). Image analyzers, particle counters, and blood cell analyzers may be useful in egg counts and counting fertilized and unfertilized eggs.

Laboratory culture of sea urchins (and sand dollars) will be improved, and development of a kelp-based prepared diet may improve year-round nutrition. Sperm and egg collection and storage procedures will be improved, thus reducing major sources of test variability. A ceiling on acceptable S:E may be instituted to prevent testing with gametes or conditions that may produce spurious results. The value of reference toxicants in quality control will increase with a better understanding of the uses, abuses, and selection of reference toxicants.

Finally, tests with shorter sperm exposure periods (10–20 min) may be developed to provide acceptably sensitive results compared to current 60-min standard tests. If this occurs, the use of shorter sperm cell test exposures could avoid the surrogacy problem inherent in the use of arguably unnatural sperm exposure periods. Indeed, an echinoid sperm test procedure using a 10-min sperm exposure period is likely to be adopted by Environment Canada (J. Sprague, personal communication).

LITERATURE CITED

Abbott, W. S.: A method of computing the effectiveness of an insecticide. *J. Econ. Entomol.*, 18:265–267, 1925. Cited in Finney, D. J.: Probit Analysis, 3rd ed., Chapter 7. Cambridge: University Press, 1980.

Allen, H.: Effects of petroleum fractions on the early development of a sea urchin. *Mar. Pollut. Bull.*, 2:138–140, 1971.

Allen, H. E.; Noll, K. E.; and Nelson, R. E.: Methodology for assessment of potential mutagenicity of dredged materials. *Environ. Lett.*, 4:101–106, 1983.

Anderson, S.; Hoffman, E.; Steward, D.; and Harte, J.: Ambient toxicity characterization of San Francisco Bay and adjacent wetland ecosystems. Report LBL-29579, Lawrence Berkeley Laboratory, Univ. of California, pp. 1–85, 1990.

Boolootian, R. A.: Reproductive physiology. Physiology of Echinodermata, edited by R. A. Boolootian, Chapter 25. New York: Interscience, Wiley, 1966.

Bulich, A. A.; Greene, M. W.; and Isenberg, D. L.: Reliability of the bacterial luminescence assay for determination of the toxicity of pure compounds and complex effluents. Aquatic Toxicology and Hazard Assessment, edited by D. R. Branson, and K. L. Dickson, pp. 338–347. ASTM STP 737. Philadelphia: ASTM, 1981.

Chapman, G. A.: Sea urchin (*Strongylocentrotus purpuratus*) fertilization test method, pp. 1–35. Newport, OR: U.S. EPA, ERL-Narragansett, Pacific Ecosystems Branch, 1992.

Cherr, G. N.; Shenker, J. M.; Lundmark, C.; and Turner, K. O.: Toxic effects of selected bleached kraft mill effluent constituents on the sea urchin sperm cell. *Environ. Toxicol. Chem.*, 6:561–569, 1987.

Chia, F. S., and Atwood, D. G.: Pigment cells in the jelly coat of sand dollar eggs. International Echinoderms Conference, Tampa Bay, edited by J. M. Lawrence, pp. 481–484. Rotterdam: Balkema, 1982.

Christen, R; Schackmann, R. W.; and Shapiro, B. M.: Metabolism of sea urchin sperm—interrelationships between intracellular pH, ATPase activity and mitochondrial respiration. *J. Biol. Chem.*, 258:5392–5399, 1983.

De Ridder, C., and Lawrence, J. M.: Food and feeding mechanisms: Echinoidea. Echinoderm Nutrition, edited by M. Jangoux, and J. M. Lawrence, pp. 57–115. Rotterdam: Balkema, 1982.

Dinnel, P. A.: Comparative sensitivity of sea urchin sperm bioassays to metals and pesticides. *Arch. Environ. Contam. Toxicol.*, 18:748–755, 1989.

Dinnel, P. A.; Stober, Q. J.; Crumley, S. C.; and Nakatani, R. E.: Development of a sperm cell toxicity test for marine waters. Aquatic Toxicology and Hazard Assessment; Fifth Conference, edited by J. G. Pearson; R. B. Foster; and W. E. Bishop, pp. 982–989. ASTM STP 766. Philadelphia: American Society for Testing and Materials, 1982.

Dinnel, P. A.; Stober, Q. J.; Link, J. M.; Letourneau, M. W.; Roberts, W. E.; Felton, S. P.; and Nakatani, R. E.: Methodology and validation of a sperm cell toxicity test for testing toxic substances in marine waters. Final Report, Grant R/TOX-1, Univ. of Washington Sea Grant Program in cooperation with the U.S. EPA, 1983.

Dinnel, P. A.; Link, J. M.; and Stober, Q. J.: Improved methodology for a sea urchin sperm cell bioassay for marine waters. *Arch. Environ. Contam. Toxicol.*, 16:23–32, 1987.

Dinnel, P. A.; Pagano, G. G.; and Oshida, P. S.: A sea urchin test system for marine environmental monitoring. Echinoderm Biology, edited by R. D. Burke; P. V. Mladenov; P. Lamber; and R. L. Parsley, pp. 611–619. Rotterdam: Balkema, 1988.

Gatti, J., and Christen, R.: Regulation of internal pH of sea urchin sperm—a role for the Na/K pump. *J. Biol. Chem.*, 260:7599–7602, 1985.

Grave, B. H.: Further studies on the longevity and swimming ability of spermatozoa. *Biol. Bull.*, 67:513–518, 1934.

Gray, J.: The effect of dilution on the activity of spermatozoa. *Br. J. Exp. Biol.*, 5:337–344, 1927a.

Gray, J.: The senescence of spermatozoa. *Br. J. Exp. Biol.*, 5:345–361, 1927b.

Gray, J.: The effect of egg-secretions on the activity of spermatozoa. *Br. J. Exp. Biol.*, 5:362–363, 1927c.

Harvey, E. B.: The American *Arbacia* and other sea urchins. Princeton, NJ: Princeton Univ. Press, 1956.

Hayashi, T.: Dilution medium and survival of the spermatozoa of *Arbacia punctulata*. I. Effect of the medium on fertilizing power. *Biol. Bull.*, 89:161–179, 1945.

Hoadley, L.: Certain effects of the salts of heavy metals on the fertilization reaction in *Arbacia punctulata*. *Biol. Bull.*, 44:255–280, 1923.

Iwata, K.: A method of determining the sex of sea urchins and of obtaining eggs by electric stimulation. *Annot. Zool. Jpn.*, 23:39–42, 1950.

Johnson, C., and Epel, D.: Heavy metal chelators prolong motility and viability of sea urchin sperm by inhibiting spontaneous acrosome reactions. *J. Exp. Zool.*, 226:431–440, 1983.

Kobayashi, N.: Comparative sensitivity of various developmental stages of sea urchins to some chemicals. *Mar. Biol.*, 58:163–171, 1980.

Kobayashi, N.: Marine ecotoxicological testing with echinoderms. Ecotoxicological Testing for the Marine Environment. Proceedings of the International Symposium on Ecotoxicological Testing for the Marine Environment, edited by B. Persoone; E. Jaspers; and C. Claus, pp. 341–405. Ghent, Belgium, September 12–14, 1983. Bredene, Belgium: State Univ. of Ghent, Belgium and Institute for Marine Scientific Research, 1984.

Leahy, P.; Tutschulte, T.; Britten, R.; and Davidson, E.: A large-scale laboratory maintenance system for gravid purple sea urchins (*Strongylocentrotus purpuratus*). *J. Exp. Zool.*, 204:369–380, 1978.

Levitan, D. R.; Sewell, M. A.; and Chia, F.: Kinetics of fertilization in the sea urchin *Strongylocentrotus franciscanus*: interaction of gamete dilution, age, and contact time. *Biol. Bull.*, 181:371–378, 1991.

Lillie, F. R.: Studies of fertilization: X. The effects of copper salts on the fertilization reaction in *Arbacia* and a comparison of mercury effects. *Biol. Bull.*, 41:125–143, 1921.

Morrison, G.; Torello, E.; Comeleo, R.; Walsh, R.; Kuhn, A.; Burgess, R.; Tagliabue, M.; and Greene, W.: Intralaboratory precision of saltwater short-term chronic toxicity tests. *J. Water Poll. Control Fed.*, 61:1707–1710, 1989.

Muchmore, D., and Epel, D.: The effects of chlorination of wastewater on fertilization in some marine invertebrates. *Mar. Biol.*, 19:93–95, 1973.

Nacci, D.; Jackim, E.; and Walsh, R.: Comparative evaluation of three rapid marine toxicity tests: sea urchin early embryo growth test, sea urchin sperm cell toxicity test and Microtox. *Environ. Toxicol. Chem.*, 5:521–525, 1986.

Nacci, D.; Walsh, R.; and Jackim, E.: Guidance manual for conducting sperm cell tests with the sea urchin, *Arbacia punctulata*, for use in testing complex effluents. Narragansett, RI: U.S. EPA, 1987.

Ohtake, H.: Respiratory behaviour of sea-urchin spermatozoa. I. Effect of pH and egg water on the respiratory rate. *J. Exp. Zool.*, 198:303–312, 1976a.

Ohtake, H.: Respiratory behaviour of sea-urchin spermatozoa. II. Sperm-activating substance obtained from jelly coat of sea-urchin eggs. *J. Exp. Zool.*, 198:313–322, 1976b.

Rothschild, Lord: Sea urchin spermatozoa. *Biol. Rev.*, 26: 1–27, 1951.

Schimmel, S.; Morrison, G.; and Heber, M.: Marine complex effluent toxicity program: test sensitivity, repeatability and relevance to receiving water toxicity. *Environ. Toxicol. Chem.*, 8:739–746, 1989.

Stober, Q. J.; Dinnel, P. A.; and Crumley, S. C.: Development of the echinoderm fertilization bioassay for testing toxic substances. Fisheries Research Institute report to EPA for grant R805839-91-1, Univ. of Washington, Seattle, 1979.

Timourian, H., and Watchmaker, G.: Determination of spermatozoan motility. *Dev. Biol.*, 21:62–72, 1970.

Tyler, A.: A simple, non-injurious, method for inducing repeated spawning of sea urchins and sand dollars. *Collect Net* 19:19–20, 1949.

Tyler, A.: Prolongation of life-span of sea urchin spermatozoa and improvement of the fertilization-reaction by treatment of spermatozoa and eggs with metal-chelating agents (amino acids, versene, DEDTC, oxine, cupron). *Biol. Bull.*, 104:224–239, 1953.

Tyler, A., and Tyler, B. S.: The gametes; some procedures and properties. Physiology of Echinodermata, edited by R. A. Boolootian, Chapter 27. New York: Interscience, Wiley, 1966a.

Tyler, A., and Tyler, B. S.: Physiology of fertilization and early development. Physiology of Echinodermata, edited by R. A. Boolootian, Chapter 28. New York: Interscience, Wiley, 1966b.

U.S. Environmental Protection Agency: Ambient water quality criteria for copper—1984. EPA 440/5-84-031. Washington, DC: U.S. EPA, 1985a.

U.S. Environmental Protection Agency: Ambient water quality criteria for lead—1984. EPA 440/5-84-027. Washington, DC: U.S. EPA, 1985b.

U.S. Environmental Protection Agency: Short-term methods for estimating the chronic toxicity of effluents and receiving waters to marine and estuarine organisms, edited by C. Weber et al. EPA-600/4-87-028. Cincinnati, OH: U.S. EPA, 1988.

Vacquier, V.: The fertilizing capacity of sea urchin sperm rapidly decreases after induction of the acrosome reaction. *Develop., Growth and Differ.*, 21:61–69, 1979.

Vasseur, P.; Ferard, J. F.; Rast, C.; and Larbaigt, G.: Luminescent marine bacteria in acute toxicity testing. Ecotoxicological Testing for the Marine Environment, edited by G. Persoone; E. Jaspers; and C. Claus, pp. 381–396. Ghent, Belgium, September 12–14, 1983. Bredene, Belgium: State Univ. of Ghent, Belgium and Institute for Marine Scientific Research, 1984.

SUPPLEMENTAL READING

Biology and Physiology

Boolootian, R. A.: Reproductive physiology. Physiology of Echinodermata, edited by R. A. Boolootian, Chapter 25. New York: Interscience, Wiley, 1966.

Harvey, E. B.: The American *Arbacia* and other sea urchins. Princeton, NJ: Princeton Univ. Press, 1956.

Levitan, D. R.; Sewell, M. A.; and Chia, F.: Kinetics of fertilization in the sea urchin *Strongylocentrotus franciscanus*: interaction of gamete dilution, age, and contact time. *Biol. Bull.*, 181:371–378, 1991.

Tyler, A., and Tyler, B. S.: The gametes; some procedures and properties. Physiology of Echinodermata, edited by R. A. Boolootian, Chapter 27. New York: Interscience, Wiley, 1966a.

Tyler, A., and Tyler, B. S.: Physiology of fertilization and early development. Physiology of Echinodermata, edited by R. A. Boolootian, Chapter 28. New York: Interscience, Wiley, 1966b.

Toxicology

Cherr, G. N.; Shenker, J. M.; Lundmark, C.; and Turner, K. O.: Toxic effects of selected bleached Kraft mill effluent constituents on the sea urchin sperm cell. *Environ. Toxicol. Chem.*, 6:561–569, 1987.

Dinnel, P. A.: Comparative sensitivity of sea urchin sperm bioassays to metals and pesticides. *Arch. Environ. Contam. Toxicol.*, 18:748–755, 1989.

Dinnel, P. A.; Link, J. M.; and Stober, Q. J.: Improved methodology for a sea urchin sperm cell bioassay for marine waters. *Arch. Environ. Contam. Toxicol.*, 16:23–32, 1987.

Kobayashi, N: Marine ecotoxicological testing with echinoderms. Ecotoxicological Testing for the Marine Environment. Proceedings of the International Symposium on Ecotoxicological Testing for the Marine Environment, edited by B. Persoone; E. Jaspers; and C. Claus, pp. 341–405. Ghent, Belgium, September 12–14, 1983. Bredene, Belgium: State Univ. of Ghent, Belgium and Institute for Marine Scientific Research, 1984.

Nacci, D.; Jackim, E.; and Walsh, R.: Comparative evaluation of three rapid marine toxicity tests: sea urchin early embryo growth test, sea urchin sperm cell toxicity test and Microtox. *Environ. Toxicol. Chem.*, 5:521–525, 1986.

Chapter 7

FETAX—A DEVELOPMENTAL TOXICITY ASSAY USING FROG EMBRYOS

J. A. Bantle

INTRODUCTION

The Need for Developmental Toxicity Testing

Developmental toxicity tests are designed to detect xenobiotic agents that affect embryonic development. Embryonic development can be considered a "weak link" in the life cycle of an organism. During this period unique cellular and molecular processes operate to generate a complex multicellular organism from a zygote. These processes are sensitive and easily perturbed by many chemicals. Developmental toxicants are chemicals that can exert their effects at concentrations lower than that required to affect adults or cause general cellular toxicity. For example, semicarbazide causes malformation in frog embryos at 1/3000 the concentration required to kill embryos and affects embryonic growth at even lower concentrations (Schultz et al., 1988). Chronic full life cycle tests account for all significant life stages and usually take longer to run for vertebrates than the 4-d developmental toxicity test presented here. Short-term developmental toxicity tests can then be considered acute or subchronic tests that may predict chronic effects in far less time and cost.

FETAX (frog embryo teratogenesis assay—*Xenopus*) is a 4-d whole-embryo developmental toxicity test that utilizes the embryos of the South African clawed frog, *Xenopus laevis*. FETAX was initially designed as an indicator of potential human developmental health hazards. The assay is well suited for complex mixtures (e.g., industrial effluents) testing and has been validated using single chemicals with known mammalian developmental toxicity (see Validation Study Results). The assay is also applicable to aquatic toxicology assessments. This chapter emphasizes the latter use.

Uses of FETAX

FETAX data can be extrapolated to other species, because an evolutionarily conserved genetic program controls embryonic development. This program may be thought of as a series of consecutively expressed genes that guide the formation of basic embryonic structures. If differences such as metabolic activation and placentation are taken into account, it is even possible to extrapolate the data to mammals. However, there are some features of the amphibian egg that make it unique. These features allow the use of this assay to find developmental toxicants that affect amphibians. This amphibian developmental toxicity test may help in studies designed to discover the reasons for the reported worldwide disappearance of amphibians even in pristine locations (Wake and Morowitz, 1990; Wake, 1991). This decline may be due in part to normal population fluctuations caused by climatological factors or by anthropogenic factors (Pechmann et al., 1991). In at least one case, frog eggs failed to develop in pond water but developed normally when moved to the laboratory (Science Briefings, 1991). It is, therefore, possible that some decline may be due to chemical pollution and FETAX may be used to investigate the extent and causes of the decline.

When FETAX is used for aquatic toxicology evaluation, it must be remembered that stunted and malformed embryos would be swiftly removed from the normal population through the inability to feed or by predation. This means that species survival can be compromised by developmental toxicants. For humans, developmental abnormalities persist in live offspring with attendant social and health costs.

Developmental Toxicity Assay Design

In designing a developmental toxicity assay, it is imperative to account for the normal molecular and cellular mechanisms that guide embryonic development. As mentioned earlier, a genetic program guides development and it entails the sequential expression and repression of genes. Many of these genes are expressed for a short period only during a specific stage of embryonic development. Therefore, genotoxic agents (see Chapter 13) are often developmentally toxic as well. Five cellular mechanisms operate in concert during development and each is critical in embryogenesis. These mechanisms are cell division, interaction (induction), migration, differentiation, and selective cell death. Division is an obvious mechanism, as a single-celled zygote cleaves to form a hollow ball of cells followed by normal mitosis during embryogenesis. This continues until a complex multicellular organism forms. Cell interaction is an important mechanism by which cells signal one another via short- and long-range mechanisms. The signaling is informational and target cells respond by changing developmental pathways. The induction of the lens of the eye by underlying neural ectoderm is a classic example of this process. Cells also migrate in the embryo either as individual cells, as tissues, or as entire organs. Thus, primordial germ cells (future spermatozoa and ova) wander from their point of origin throughout the body until they find the presumptive gonad. The kidneys of mammals form in one location and move posteriorly as development progresses. After cells arrive in their final locations they differentiate and acquire their cell-specific functions. Neurons form by cell enlargement and then elaborate complex cytoplasmic processes that differentiate to become axons and dendrites. Perhaps that most difficult of cell processes for the student to comprehend is cell death. Programmed cell death is an important feature in the generation of normal embryonic structure. Whole embryonic systems form only to be modified or removed during development. The eradications of the pronephros and mesonephros in mammals are typical examples. Interruption of any of these mechanisms may cause abnormal development or even death. This means that end points of any developmental toxicity assay must consider all of these mechanisms. FETAX end points are the 96-h LC50 (embryo death), the 96-h EC50 (malformation), and the minimum concentration to inhibit growth (MCIG). These end points account for all important cell and molecular mechanisms because the assay is based on the whole embryo and not on embryo parts or cultured cells.

Some developmental toxicants affect only certain stages of embryonic development. The drug thalidomide exerts its devastating effect of abnormal limb development and stunting only during a very short period. Treatment with thalidomide before or after this period has little or no effect. Therefore, exposure conditions in any assay must be designed to ensure that exposure occurs during all the sensitive developmental stages. Exposure is continuous in FETAX throughout the 4-d period of primary organogenesis, ensuring that all sensitive stages are affected. During the 4-d exposure period, the embryo proceeds from a hollow blastula stage of a few hundred cells (Figure 1) to a free-swimming larva (Figure 2) that is ready to feed. All primary organogenesis is complete, although limbs have not yet formed.

Lastly, it is important to note two other fundamental concepts in developmental toxicology. Karnofsky's law (Karnofsky, 1965) states that any material can be teratogenic when administered at concentrations approaching general cell toxicity. This feature will be seen later in the description of the teratogenic index (TI) concept. The second concept is that insult to early stages is far more deleterious than damage to later stages of development. Early injury to a primordium (i.e., a group of undifferentiated cells destined to produce an organ) can result in damage to whole organ systems, whereas

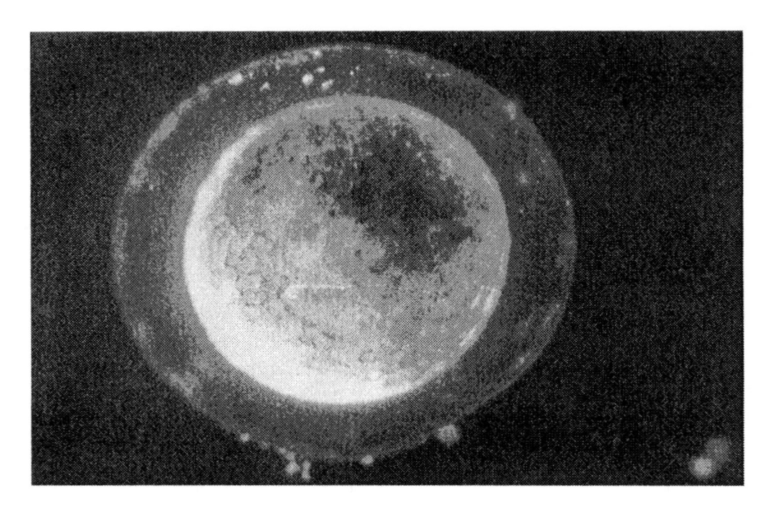

Figure 1. Dorsal view of a typical *Xenopus* stage 8 blastula as it might be collected from the bottom of the breeding aquarium. The jelly coat has not been removed. It is quite sticky and attempts to manipulate the embryo with a spatula prove time consuming. (Photo by M. A. Hull.)

damage later may affect only a single organ. Damage to an early group of cells is transmitted to succeeding generations of daughter cells and the initial fault is magnified throughout the embryo. Therefore, a rule is that the earlier damage occurs, the worse and more widespread the damage.

FETAX LITERATURE REVIEW

Amphibian embryos and larvae have been exposed experimentally to a wide variety of toxic chemicals. Herbicides (Anderson and Prahlad, 1976), fungicides (Bancroft and Prahlad, 1973), insecticides (Cabejszed and Wojcik, 1968), metals (Abbasi and Soni, 1984; Chang et al., 1974), and many other chemicals and mixtures have been tested in a variety of species (Cooke, 1972; Ghate and Mulherkar, 1980; Ghate, 1983, 1985a,b; Green, 1954). The results of this work convinced early researchers that amphibian embryos were sensitive indicators of water quality. However, factors such as species used, exposure time, diluent water differences, temperature, and number of embryos per dish varied greatly from one experiment to another. Comparison of results became impossible and none of the experiments set good criteria for determining whether a chemical was a developmental toxicant or simply caused malformations at or near the concentration of general cellular toxicity. Birge's group did extensive work in comparing the relative sensitivities of several different anuran species, although assay conditions often varied (Birge and Black, 1979; Birge et al., 1979). In the mid-1970s Dial (1976) and Greenhouse (1976a,b, 1977) began the process of standardization by simply publishing the results of several studies that used the same basic methodology. It was then possible to begin comparing the developmental toxicity of several chemicals. Greenhouse's use of *Xenopus laevis* paved the way for FETAX because he recognized the many advantages of using this imported frog (see Test Organism).

It was Dumont and his co-workers who named the assay, defined the basic protocol, fixed the end points, and proposed the concept of the teratogenic index (TI) that helped to differentiate developmental toxicants from other chemicals (Dumont et al., 1979, 1983a). FETAX was first used in studying the developmental toxicity of selenium (Browne and Dumont, 1979, 1980), aromatic amines (Davis et al., 1981), and complex shale oil mixtures (Dumont and Schultz, 1980). Work soon turned to solidifying the protocol and initial efforts were made to validate the assay for use in screening for human developmental toxicants (Dumont et al., 1983a). The laboratories of Sabourin and Bantle followed Dumont's work by first comparing FETAX to other developmental toxicity assays (Sabourin et al., 1985) and then continuing the validation process (Sabourin and Faulk, 1987; Bantle and Courchesne, 1985; Bantle and Dawson, 1988; Bantle et al., 1989a,b, 1990a; Courchesne and Bantle, 1985; Dawson and Bantle, 1987a; Dawson et al., 1988a, 1989; DeYoung et al., 1991; Fort et al., 1989; Rayburn et al., 1991a,b). As the assay began to prove to be both sensitive and specific, the validation process demonstrated certain weaknesses in the protocol. An American Society for Testing and Materials (ASTM) E.47 Aquatic Toxicology Taskforce was formed to define the best protocol and write a New Standard Guide, which has just been accepted for publication (Bantle and Sabourin, 1991). This guide recommends a basic FETAX protocol to be used for all testing. An Atlas of Abnormalities—A Guide for the Performance of FETAX (Bantle et al., 1990b) was written as a companion to the guide. This atlas facilitates learning the assay and helps in staging the embryos and identifying malformations. Many investigators wish to add end points or alter exposure conditions. Dumpert and

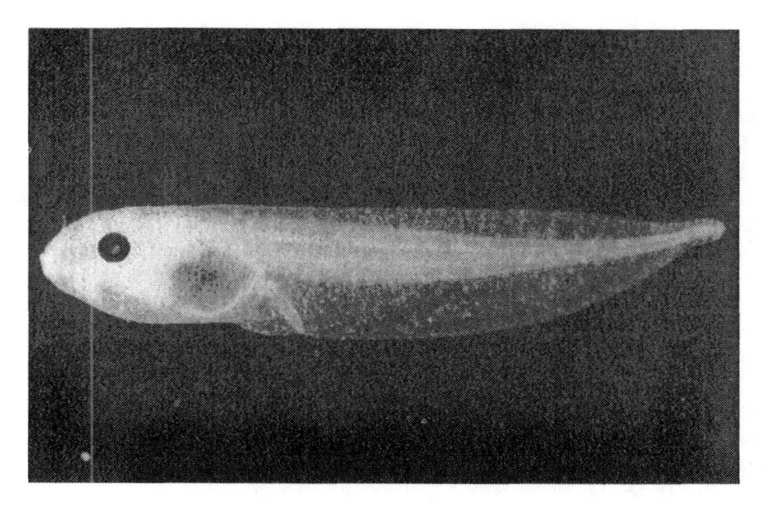

Figure 2. The normal stage 46 (96-h) larva, a full-length view. Refer to Figure 7 for structures. (Photo by M. A. Hull.)

co-workers have extended exposure periods and have waited extended time periods before recording results (Dumpert and Zeitz, 1984; Dumpert, 1986, 1987). This practice is not discouraged so long as the investigator indicates that a deviation from the standard FETAX has been made so that comparisons are not made erroneously. FETAX serves as a base from which other work can proceed, so it is recommended that standard FETAX be performed and modifications made later.

Bantle and co-workers realized early that a metabolic activation system (MAS) was required if the predictive accuracy of FETAX was to be high enough to be useful as a developmental toxicity screen for mammals. Early experimentation showed that *Xenopus* embryos lacked a competent metabolic activation system like that of mammals. Cultured hepatocytes were initially tried, but the system was costly and awkward (J. A. Bantle, unpublished data). S9 supernatant proved to be too toxic for the frog embryos (J. A. Bantle, unpublished data). S9 supernatant is a postmitochondrial supernatant produced by the high-speed centrifugation of rat liver homogenate. Residual Aroclor 1254 was probably the toxic component. The best system proved to be a mixture of microsomes derived from Aroclor 1254– and isoniazid-induced rat livers (Bantle and Dawson, 1988; Bantle et al., 1989a, 1990b; Dawson et al., 1988; Fort et al., 1988, 1989, 1991; Fort and Bantle, 1990a; Fort and Bantle, 1990b).

Work during this time period also proved that FETAX could be used to assess the developmental toxicity of surface water (Dawson et al., 1984), ground water (Bantle et al., 1989b), and sediment extracts (Dawson et al., 1988b).

Schultz and co-workers and Dawson have showed that FETAX can be used in structure-activity studies (Schultz et al., 1980, 1988; Schultz and Ranney, 1988; Dawson et al., 1990a,b, 1991a,b). These studies have also progressed to complex mixture analysis, which is currently ongoing (Dawson and Bantle, 1987b; Dawson and Wilke, 1991a,b). R. A. Finch (unpublished) has worked with FETAX using flow-through exposure and is developing methods for the use of the assay in a biomonitoring (see Chapter 1) trailer. Linder et al. (1990) have also shown that the assay is flexible enough to be used in situ for on-site biomonitoring. Linder has devised a simple inexpensive plastic mesh cage that confines the embryos securely. This is staked to a stainless steel pole and marked with a bobber.

Currently, an interlaboratory study is being conducted by laboratories in academia, private industry and government to assess the repeatability and reliability of FETAX. Attempts are being made to discover useful biomarkers in *Xenopus* and to couple the effects of toxicants at the molecular level to whole-embryo effects. This will allow FETAX to become an important model system with which to study molecular mechanisms of developmental toxicity.

TERMINOLOGY

For general terminology the reader is referred to the glossary in this book. A *developmental toxicant* is a chemical material that affects any developmental process. Developmental toxicants exert their effects (embryo mortality, malformation, growth inhibition, etc.) at concentrations lower than that required to cause general cellular toxicity or adult effects. Various criteria for effects or end points of developmental toxicity could be used to compare chemically exposed organisms with unexposed organisms. Embryo death, malformation, and growth are typically criteria that account for most of the potentially adverse effects that may occur in developmental processes. These effects are used in FETAX end points to define the 96-h LC50 (embryo death), 96-h EC50 (malformation), and MCIG. Rarely used end points such as pigmentation and swimming ability can be used to account for neural damage (Courchesne and Bantle, 1985). Other end points are possible but entail added labor. It is important to differentiate between a developmental toxicant and a teratogen. *Teratogens* cause malformation at concentrations lower than those required to cause general cellular toxicity or adult effects. Thus, teratogens are a subset of developmental toxicants. The *teratogenic index or TI* is a measure of developmental hazard (Dumont et al., 1983a; Dawson and Bantle, 1987a). The TI is defined as the 96-h LC50 divided by the 96-h EC50 (malformation). TI values higher than 1.5 signify a larger separation of the concentration ranges that produce mortality and malformation and, therefore, a greater potential for all embryos to be malformed in the absence of significant embryo mortality. TI values have ranged from 1 to 3000 depending on the nature of the test material. The *minimum concentration to inhibit growth (MCIG)* is the lowest concentration of test material that significantly inhibits growth as measured by head-tail length. A significant difference in growth may be determined by the *t*-test for group observations at the $P = 0.05$ level.

An in vitro developmental toxicity assay is defined here as any assay that does not use whole mammalian embryos in vivo. FETAX is classified as an in vitro test even though whole embryos are used. An exogenous *metabolic activation system (MAS)* is added to an in vitro test system to provide conditions for the metabolism of the test material as if it were in a whole-animal system. FETAX uses a 1:1 combination of Aroclor 1254– and isoniazid-induced rat liver microsomes plus generator system (Bantle et al., 1990b). A MAS is incorporated when FETAX is used for predicting developmental toxicity in mammals.

FETAX REFERENCE SOURCES

The ASTM new standard guide for the conduct of FETAX gives the basic protocol for FETAX without the MAS (Bantle and Sabourin, 1991). Different options for exposure are given, as well as the possible utilization of alternative anuran species when it is necessary to use endemic species. The Atlas of Abnormalities (Bantle et al., 1990b) is a companion manual to the ASTM guide. It covers topics that could not be included in the guide, such as adult care, identification of stages and malformations, and the preparation of the MAS. It also contains standard data forms for use in conducting the assay.

Two other reference sources are available which aid in animal husbandry and breeding. Dawson et al. (1992) provide information on continuous flow tanks that simplify care and alternative methods of inducing mating. The second source is edited by Kay and Peng (1992). Besides care and breeding information, other uses of *Xenopus* are given that show the tremendous flexibility of the *Xenopus* model and further prove that FETAX is a high-connectivity system.

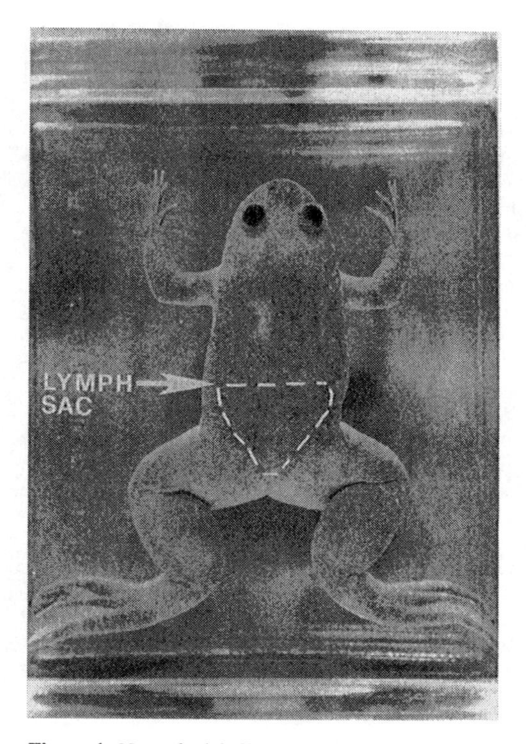

Figure 4. Normal adult *Xenopus* male. Males are generally smaller than females and do not possess cloacal lips. They have dark forearm pads (nuptial pads) on the ventral surface of the hand and forearm which darken even further after HCG injection. The pale marks on the back of this male are the result of liquid nitrogen branding for identification purposes. (Photo by M. A. Hull.)

TEST ORGANISM

The selection of a test organism is critical to the success of an assay. It is a critical factor in the specificity, sensitivity, repeatability, and reliability of the assay. It also plays a role in determining assay costs. *X. laevis*, South African clawed frogs, were selected over North American species for the following reasons: they can be raised from birth to death in the laboratory, they eat dead food as opposed to the live food required by most North American species, they can be kept in aquaria like fish because the adults are totally aquatic, the adults are resistant to most diseases, and they can be bred a number of times before replacement.

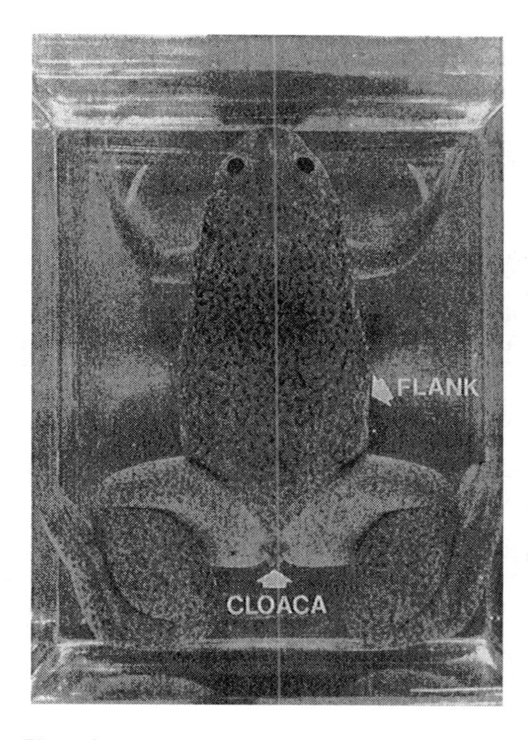

Figure 3. Normal adult *Xenopus* female. Females are identified by their fleshy cloacal lips (arrow), which increase in size after injection with HCG. Gravid females show bulging flanks. This female had been bred recently so it does not show the enlarged flanks. (Photo by M. A. Hull.)

Breeding is accomplished any time of the year by a simple injection of commercially available human chorionic gonadotropin into the dorsal lymph sacs of the male and female (Figures 3–5). Amplexus ensues (i.e., the male clasps the female and both release gametes) (Figure 6) and several thousand eggs are available the next day. Dejellied normally developing eggs must be selected from abnormal or overripe eggs but this is not time-consuming. This is important because selection of abnormal or overripe eggs would lead to high mortality and malformation in controls.

In contrast, most North American frogs require injection of pituitary gland extracts to induce ovulation and the eggs must be stripped from the female by pressing firmly on her flanks. Fertilization must be accomplished by artificial semination because the frogs will not go into amplexus. Artificial semination involves mincing testes in buffer and applying the sperm suspension over the freshly stripped eggs. When donor frogs are taken into account, several frogs are killed for every breeding. In addition to this costly breeding scheme, it is possible to carry out breeding only during the spring. As native species are in decline in most parts of the globe, it makes little sense to use the remaining frogs as test organisms when a readily available substitute is available.

Another important reason for the selection of *Xenopus* is that it is a high-connectivity model (NRC, 1985). The reason for this high connectivity is that *Xenopus* has been used for many years in biology. A great deal is already known about its normal development (Nieuwkoop and Faber, 1975; Deuchar, 1972), biology, and biochemistry (Deuchar, 1975). Consequently, test results from FETAX may be more readily explainable in light of this knowledge. This makes it much easier to understand and interpret the results of mechanistic studies and can help in the development of biomarkers in amphibians.

It is important to note that the larvae of *Xenopus* are transparent (Figure 2). This makes it easy to observe internal malformations as a number of internal organs are easily visible from the outside (Figure 7). This obviates the need for dissection or histological sectioning when assessing abnormal development.

In cases in which endemic species must be tested, the ASTM guide lists alternative species that Birge and co-workers (Birge and Black 1979; Birge et al., 1979) found to give suitable numbers of eggs per year. When native species are used, the FETAX protocol can be employed, but the results cannot be compared with typical FETAX results.

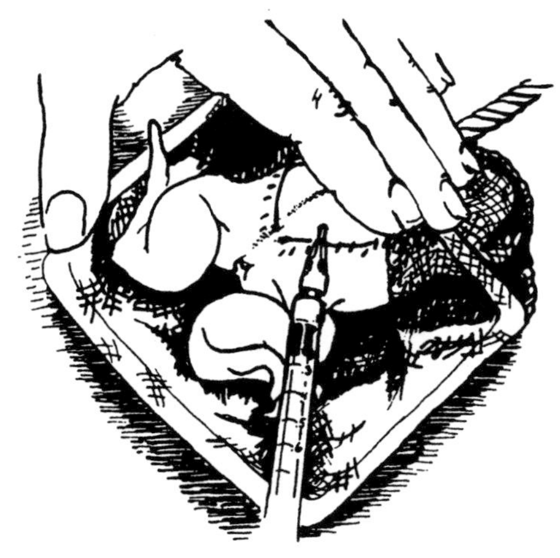

Figure 5. Method of injecting human chorionic gonadotropin. It is easy to inject *Xenopus* by carefully immobilizing the adult in an ordinary aquarium net. Use a tuberculin syringe with a 1/2 inch long, 26-gauge needle to inject the HCG into the dorsal lymph sac. The lymph sac is bounded by the lateral line that runs along the side of the frog and appears as "stitching" on the skin. The dorsal lymph sac is surrounded by a dashed white line in Figure 4. Injections should be inside (centrad) the lateral line. Note that one-fourth of the needle tip enters the skin at a shallow (10–15°) angle. When injecting the frog, wrinkle the skin so that the injection can be administered subcutaneously. Keep the point of the needle well away from the spinal cord. (Sketch by D. J. DeYoung.)

Although *Xenopus* generally exhibits good breeding characteristics, breeding is generally less successful in the fall and it may be necessary to breed as many as three pair of frogs simultaneously to ensure that at least one breeding pair produces acceptable eggs. Even if eggs are produced, there can be problems with the fertilization rate. Problems with egg quality can be ameliorated by breeding the adults every 3 mo and conditioning the frogs prior to breeding. Conditioning starts by selecting healthy, sexually mature adults and feeding them every day for 3 wk prior to mating. Daily water changes (or use of a flow-through system) will help the conditioning of the frogs. As with all toxicity tests, consistent success is directly dependent on the quality of the animal husbandry. It is better to maintain a small colony of well cared for frogs than a large colony that is maintained indifferently.

FETAX END POINTS AND EXPOSURE REGIMENS

FETAX End Points and Assay Data

FETAX has three standard end points and a TI ratio calculated from two of the end points. Embryo death is measured by the 96-h LC50 and embryo malformations by the 96-h EC50 (malformation). Only malformations in live embryos are recorded. Standard concentration-response experiments are performed and mortality and malformation curves constructed using probit analysis. The probit analysis results in the appropriate LC50 or EC50 values with 95% confidence limits. Both concentration-response curves are usually placed on the same plot to demonstrate the separation between the two curves. The TI is found by dividing the 96-h LC50 by the 96-h EC50 (malformation). The ASTM task force agreed that values less than 1.5 indicate little developmental hazard and values greater than 1.5 indicate increasing developmental hazard. It is very rare that TI values exceed 1000 and developmental toxicants more commonly have TIs in the range 10–200.

The MCIG, the minimum concentration to inhibit growth, is calculated by measuring the head-tail length of each embryo following the contour of the embryo. The embryos are fixed in 3.0% formalin prior to this procedure and fixation does not seem to alter embryo length. Length data from each concentration set are compared to control length data using the *t*-test for grouped observations. The lowest concentration set that inhibits growth at the $P = 0.05$ level of significance is the MCIG. The data are usually plotted as the percentage of control versus the percentage of the 96-h LC50. This makes it possible to compare results from different test materials.

Figure 6. Amplexus in the mating tank. A male and female frog are shown here in amplexus. A 10-gallon glass aquarium has been used a mating chamber. A plastic screen made from a fluorescent light diffuser grate (available from most hardware stores) has been used as a grate to support the adults. Mating is carried out in the dark at a temperature of 23°C. FETAX solution (ASTM Standard Guide) is used as the medium and there should 2.5 inches of solution above the grate in the mating chamber. The eggs fall to the bottom, where they can be scraped off into plastic petri dishes. Another useful grate material is a 1-cm plastic mesh (Cat. no. XV-0350) manufactured by InterNet Inc. [2730 Nevada Ave. North, Minneapolis, MN 55427; phone (612) 541-9690]. However, this must be purchased in large quantities. Heavy-duty aluminum foil is used as a top and a bubbler is used for aeration. If a tank with shorter sides is used, a weighted lid may be necessary to preclude escape. (Photo by M. A. Hull.)

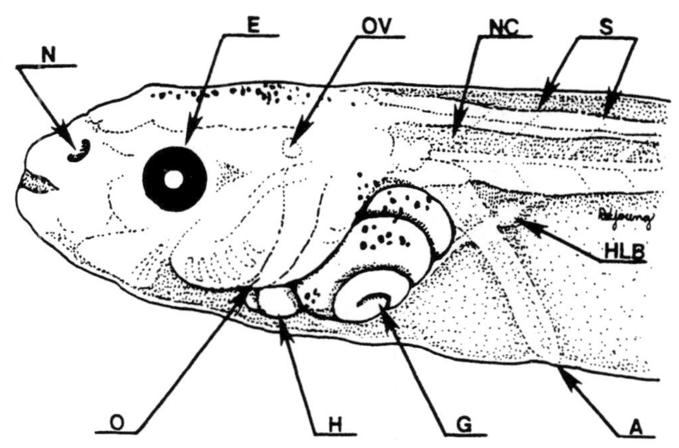

Figure 7. Diagram of the head region of the stage 46 (96-h) larva. A stage 46 larva is recognized by the appearance of the hindlimb bud, the coiling of the gut, and the shape of the operculum covering the gills. The best indicator that the larva has attained stage 46 is the appearance of the hindlimb bud. Gut coiling is also easily observed. N=nares, E=eye, OV=otic vesicle, S=somite, O=operculum, H=heart, G=gut, HLB=hindlimb bud, A=anus. (Diagram by D. J. DeYoung.)

Many other end points are possible in FETAX as long as the required end points are performed. Additional end points such as pigmentation, locomotion, and hatchability are described in the ASTM New Standard Guide (Bantle and Sabourin, 1991). As EC50 can usually be obtained for all three of the latter end points. However, pigmentation and locomotion are very subjective and much work needs to be done to make data collection objective.

Exposure Regimens

Dumont initially used a 4-d static exposure in all the early work performed in his laboratory (Dumont et al., 1979; Dumont and Schlutz, 1980). The advantage of this exposure regimen is that it is very gentle on the embryos. Test material is added only at the beginning of the test. Dead embryos are removed every 24 h to prevent bacteria from multiplying and killing other embryos. However, decomposition of the test material is possible and the ASTM task force decided that the static-renewal (also called renewal) procedure was the best compromise for standard FETAX. Under this exposure regimen, test material is renewed every 24 h. Dead embryos are removed at these times. This also helps reduce the number of bacteria and fungi in the dish. When rat liver microsomes for metabolic activation are used in the test, they are also renewed at this time along with the generator system.

It is possible to use continuous exposure with FETAX (Bantle and Sabourin, 1991). One simple method of exposing the embryos is to cut the bottom off 25-ml beakers and glue 50-μm monofilament mesh to the open bottom using a small amount of silicon sealant. The embryos are placed inside and the beaker clipped to the sidewall of an exposure tank. In this manner FETAX can be conducted simultaneously with other exposure assays such as acute fish toxicity tests. Because of the limited availability of diluters for most laboratories and their attendant costs, the 24-h renewal procedure was chosen as the best exposure compromise.

Linder et al. (1990) developed an exposure chamber for in situ exposure of *Xenopus* or other amphibian embryos. This chamber is constructed of a 2.5-cm-wide, 7.0-cm-diameter circlet of large plastic polyethylene mesh (1-mm mesh size). This material can be purchased from hobby shops. The circlet is held together by means of a small nylon bolt and nut. Two large 7.5-cm disks of the same mesh are pressed to each side of the circlet and held by a centrally located nylon bolt and wing nut. The inside of the large mesh is lined with Teflon 50–100-μm mesh to keep the embryos from passing through the larger mesh. The assembled chamber with embryos is lowered into the site just below the surface of the water and anchored in place using a stainless steel stake. A common fishing bobber is used to locate the chamber and each chamber is checked daily.

FETAX POSITIVE AND NEGATIVE CONTROLS

For FETAX data to be valid, it is necessary to use appropriate controls. The standard negative control is four dishes of 25 embryos each in either a standard buffer solution (FETAX solution) or diluent water. With 100 embryos in the control group, a single death or malformation is only a 1% response. Mortality must be $< 10\%$ and malformation $< 7\%$ for the test to be valid. Generally, mortality and malformation rates can be held to 5% or less. When carrier solvents are to be used to solubilize water-insoluble materials, appropriate solvent controls must be performed to ensure that the solvent is not causing mortality or malformations. Even with these controls it is still difficult to discount the possibility that the carrier solvent is not interacting with the test material to cause synergism or antagonism.

When a positive control is required, 6-aminonicotinamide is used to cause a predicted level of mortality or malformation (Bantle and Sabourin, in press). Concentrations of 5.5 and 2500 mg/L are used. At 5.5 mg/L, malformation should be between 40 and 60%, and at 2500 mg/L mortality should be between 40 and 60%. 6-Aminonicotinamide was selected as a reference toxicant because it has a TI of 455, representing good separation between mortality and malformation concentration-response curves. It is relatively stable and solution concentrations can be checked spectrophotometrically by absorbance at 257 nm. The compound has an extinction coefficient of 13.7 mM at a pH of 1.8. Growth inhibition is not considered with the 6-aminonicotinamide reference toxicant.

Cyclophosphamide is used as a reference toxicant for the metabolic activation system. At 4.0 mg/ml cyclophosphamide and no MAS, there should be no mortality. However, if MAS and a generator system are included, 100% mortality can be expected. If there is less than this level of mortality at 96 h, the MAS did not bioactivate properly and the test must be repeated.

As expected, the controls are critical in establishing the reliability of the data. Every effort must be made to set up controls accurately. If there are enough embryos during test setup, it is advisable to set up at least two plastic dishes as a control for proper washing of the glass petri dishes. If mortality, malformation, and growth inhibition are greater in the glass dishes than the plastic dishes, it is likely that the glassware was not cleaned properly or that residual detergent was present in the dishes. If mortality and malformation in the controls remain high and the solution becomes slightly turbid with time, it is possible that bacterial contamination is causing these effects. In subsequent experiments, at least two dishes of 25 embryos each should be made up with 100 U/ml penicillin G and 100 U/ml streptomycin sulfate. If this improves the mortality and/or malformation rates relative to standard controls, bacterial contamination is the cause of the problem. The experiments can be repeated, using penicillin and streptomycin in all the dishes, although a caution will have to be added when presenting the data. Although there are no data indicating antibiotic–test material interactions, the possibility of interactions cannot be eliminated.

FETAX METHODOLOGY

Introduction

Both the ASTM New Standard Guide (Bantle and Sabourin, 1991) and the Atlas of Abnormalities (Bantle et al., 1990b) adequately cover basic FETAX methodology. This section emphasizes rationale and methods not included previously that make it easier to perform the assay and understand the data. The strict format of ASTM does not allow an explanation of why procedures are carried out, and the Atlas of Abnormalities provides photographic information not allowed in the ASTM guide.

Preparation

Animals

Animal husbandry is the greatest single determinant of the success of FETAX. *Xenopus* can be purchased from a number of commercial sources, but many supply very small adults that will breed but not provide the 700 plus embryos required for most tests. Seven hundred are usually required for the controls and a concentration series. Proven breeders should be purchased and animals should not be bred out or bred within the preceding 60 d. Animals that have been bred too often and not conditioned prior to breeding generally degrade over time and are termed "bred out." About 60 d between breedings are required for the frogs to produce an adequate number of gametes. Males should be 7.5 to 10 cm in crown-rump length and females 10–12.5 cm. In nature, it is likely that the female will continue to increase clutch size as she ages. Some females can produce 4000 plus eggs per mating. However, for a variety of reasons, the literature suggests that animals kept in laboratory situations progressively decline in their reproductive capabilities. Every attempt should be made to provide quality care for the adults. This process is known as conditioning and entails feeding the frogs daily and changing their water each day (if the adults are kept in static tanks) at least 3 wk prior to breeding. Because their food is typically beef liver and/or lung, considerable accumulation of debris may build up on the side of the tank. An inexpensive toilet cleaning brush may accompany each aquarium and be used to clean the sidewalls prior to water changes. Having a brush for each aquarium helps to keep down the spread of disease.

Although very hardy and able to spend long periods in foul water without contracting disease, *Xenopus* can still become diseased, most notably with redleg and nematode infections. The Atlas covers the recognition and treatment of both diseases. However, proper cleaning and care may be more effective than any treatment regimen in reducing mortality in the colony. Sick frogs stop eating even before they manifest any outward sign of disease. As frogs sicken, their skin color darkens and skin subsequently sloughs off. Care must be exercised here as *Xenopus* have the ability to change skin color with their surroundings. In the advanced stages of any disease, the frogs waste away, a process that takes several days. Diseased frogs should be removed from the colony as soon as possible to reduce the probability that healthy frogs will subsequently be infected.

Another problem that must be overcome is cyclical breeding. Despite housing the adults in a light-proof room, placing them on a constant 12-h day-night light cycle, and rigidly controlling the water temperature (23 ± 3°C), *Xenopus* still seem to display seasonality in their breeding. When breeding success seems to decline in the fall despite the best animal husbandry, it is time to increase the number of breeding pairs in order to increase the chances of success. In spring, two pair should suffice, but in fall it may be necessary to breed three pairs. However, the embryos from separate breeding pairs must never be mixed. This caution stems from the observation that some animals produce abnormal offspring even though the early embryos are initially acceptable. Mixing embryos may invalidate a test that would otherwise be successful. If embryo numbers are few, the test should be scaled back or converted to a small range-finding experiment.

Another caution is to make sure that the temperature of the animal room housing the adults does not fall below 21°C. Older literature suggests that reducing the temperature below this level prevents the females from shedding their eggs. Temperatures above 26°C can be harmful to embryonic development. This means that the animal room temperature should be precisely controlled even if it becomes necessary to install electric heaters or air conditioning. Precise data are unavailable, but it seems that the animals can sense impending weather fronts and that this may affect mating and egg deposition. Good animal husbandry remains as much an art as it is a science, and experience and patience play important roles in the success of any test employing live animals.

Glassware

Because FETAX is a sensitive developmental toxicity test, any residual contaminant or detergent on glassware following the wash procedure given in the ASTM guide will cause increased mortality, malformation, and growth inhibition in controls. Only soap that is intended for cell culture glassware should be used for cleaning and only minimal amounts of this detergent should be used. It is wise to separate FETAX glassware from other laboratory glassware and have investigators wash their own glassware to assure cleanliness. Running controls in plastic petri dishes concurrently with glass petri dishes helps to identify glassware washing problems. When the problem of cleaning glassware proves intractable, glassware may be taken to a glassblower who is equipped with a high-temperature bake oven in which glassware can be heated to 500°F. Most organics are removed by high-temperature treatment such as this and the glassware becomes nontoxic.

The ASTM guide gives the criteria for selection of glass or plastic dishes for FETAX. Glass dishes should be chosen when the nature of the test material is unknown. Generally, the decision is made when it is known that the test material will bind to either plastic or glass. Dishes made of the nonbinding material are then selected. Covered 60-mm petri dishes should be used as test chambers. Limited capacity is the only problem when the 60-mm plastic petri dishes are used. Generally, they are able to accommodate only 20 embryos in 8 ml of test solution, instead of the standard 25 embryos in 10 ml of test solution in glass dishes. However, plastic dishes are the best choice for experiments employing a metabolic activation system because they can be purchased sterile and can be disposed of following the test.

Apparatus and Equipment

FETAX was designed to be compatible with common fish tests in terms of equipment and glassware. For handling test materials, safety equipment such as hoods, respirator masks, gloves, acid buckets, and bench matting is required along with safety guides and material safety data sheets. Analytical and pan balances are required for weighing test materials and weighing salts for balanced salt solutions. Standard weights should be used prior to each weighing. A pH meter is required for the test. Disposable pipets and adjustable pipettors are also required for precisely handling liquids. Adjustable pipettors should be calibrated prior to the start of each test. This is easily done by repeatedly filling disposable micropipets of appropriate volumes available from science supply houses. Glassware should include volumetric flasks, graduated cylinders, and a variety of beakers and Erlenmeyer flasks. The 60-mm glass and plastic petri dishes can be purchased by the case. Dishes must remain covered throughout the test.

A high-quality binocular dissection microscope and scope light are essential for staging embryos and recording death and malformations. Fine camel hair artist brushes and watchmakers forceps are handy for manipulating the embryos and larvae in the dish.

Animal culture equipment is covered in the ASTM guide, but considerable flexibility is allowed. Fiberglass raceways, fiberglass bathtubs, vegetable crispers (restaurant sizes), and standard fish aquaria may be used to house adult frogs. Sidewalls must be at least 1 foot high to prevent adults from escaping and hardware cloth must never be used for covers or mating platforms because of the potential for metal ion contamination. Cost seems to be the determining factor in the selection of animal husbandry equipment. Heaters may be required for the animal room and, if other than well water is used to culture the adults, carboys and carbon filters are required for water treatment. Some investigators have successfully used reverse osmosis water to culture the adults with no apparent adverse effects on results. Only nontoxic tubing is to be used to pipe water to the tanks. Standard Tygon tubing is toxic. A variety of commercially available valves may be attached to frog holding tanks to allow drainage from tanks. Some investigators use simple siphoning to empty tanks. Small water pumps are available from hardware stores to help drain water from the tank.

One of the best systems for culturing adults is to use a continuous-flow system. Investigators with high-quality well water can frequently afford to keep slowly flowing water going through the tanks. A simple 5-cm standpipe with a plastic mesh covering over the drain converts a tank to a flow-through system. In flow-through systems, it is still necessary to brush sidewalls. Because *Xenopus* are wholly aquatic there is no need to provide "dry land" for the adults.

Perhaps the single most important piece of equipment for FETAX is the incubator or environmental chamber that holds the embryos throughout the 4-d test. Incubators are now available that can regulate $\pm^1/_4$°C, and these are the most desirable type. A recording thermometer is also important to ensure that temperature maxima and minima are not exceeded. The ASTM guide for randomization of dishes is probably unnecessary for modern circulating-air incubators. If there is any question about temperature variation within the incubator, dish randomization must be carried out.

A personal computer is essential for FETAX data analysis. An inexpensive darkroom enlarger is used to enlarge the images of the embryos two to three times on a digitizing pad for length measurements. When selecting a darkroom enlarger, a transparent ruler should initially be projected in several different orientations and the length of the projected image measured. It should be the same regardless of orientation. Head-tail length measurements, which once took many hours by hand using a map measurer, now take only 2 h per test by computer digitizing. Software for data analysis includes Toxstat, available from the University of Montana, and EPA probit analysis. The Manual of Pharmacologic Calculations by Tallarida and Murray (1987) is commercially available in software for IBM-compatible computers. Sigmascan (Jandell Scientific, Corte Madera, CA) software is used in analyzing head-tail length and for performing the *t*-test for grouped observations. Sigmaplot (Jandell Scientific) is used in graphing results from concentration-response curves for mortality, malformation, and growth inhibition.

Water

FETAX solution was developed as a standard balanced salt solution and its formulation is listed in the ASTM guide (Bantle et al., 1991). Only a standard solution would allow FETAX to yield data that are repeatable, reliable, and comparable. In designing FETAX solution, Dawson and Bantle (1987a) found that relatively high Mg^{2+} concentrations were required for normal growth and development of frog embryos. This may affect results in metal toxicity studies. It is difficult to dissolve lead salts in FETAX solution (unpublished), and even if it gets into solution the Mg^{2+} concentration may affect toxicity. This must be taken into account when toxicity tests with metals are performed and it may be desirable to alter FETAX solution in certain cases. This and any other alteration must be reported as a variation from standard procedure and additional controls may need to be performed. Another problem area with FETAX solution is buffering capacity. A number of zwitterionic buffers such as HEPES and MOPS were initially tried (unpublished). All caused unacceptably high levels of malformation and it was decided that no buffer should be included. Consequently, the buffering capacity of FETAX solution is small and the test material may change the pH. FETAX should be performed between pH 6.5 and 9 and it is best if the pH is between 7.0 and 7.8. However, there are situations in which it is inadvisable to adjust the pH after the addition of test material beyond pH 6.5 or 9.0. In some cases toxicity is altered and in others the test material will not stay in solution. This shows the importance of the requirement in the FETAX guide for daily measurement of pH in the control and the highest test concentration. These measurements are made in the dishes after each 24-h of exposure. Bacteria can also change the pH of the test solution and this situation must be monitored. The ASTM guide allows the use of other diluent water in continuous-flow experiments, but pH, metals content, and other chemical factors must be monitored as specified in the guide. The best indicator of diluent water quality is the embryo. If mortality is less than 10%, malformation less than 7%, and embryos grow to at least 0.9 cm by stage 46 (see Selection of Normal Embryos for Staging), the water is usable as diluent so long as the pH is between 6.5 and 9.0.

Beginning the Test

Breeding

One of the advantages of using *X. laevis* is that breeding may be induced by injection of commercially available human chorionic gonadotropin (HCG) into the dorsal lymph sac of the male and female (Figure 4). Advice has already been given on preparing the adults for mating by conditioning and adjusting the number of breeding pairs. Additional modifications can be made to ensure an acceptable clutch of embryos for FETAX. The standard scheme of injecting males with 400 international units (IU) of HCG into the dorsal lymph sac and females with 1000 units is given in the ASTM guide, and the Atlas of Abnormalities shows pictures of the injection procedure. The animals are then placed in a breeding tank together and left to breed overnight. The next day eggs that have been laid and fertilized have fallen through the mesh suspended 3 cm off the bottom of the aquarium. The eggs and embryos are then deposited on the bottom of the aquarium and away from the adults.

In the injection of the HCG, it is important to use a 26-gauge needle because the wound in the frog's skin does not close as readily as that in mammalian skin. A larger needle would allow release of HCG from the lymph sac. If several breedings are unsuccessful, a 2-d injection scheme may be initiated to allow the HCG more time to act. In this scheme, the male and female are injected with 200 and 400 units, respectively, of HCG 48 h prior to mating. On the night before mating, inject the male with 400 units and the female with 800 units and then place them into the breeding tank as usual. It might be necessary to monitor breeding success by observing egg deposition around 11:30 PM on the night of mating. If the female has begun to deposit eggs, the test should start very early the next morning. When breeding *Xenopus*, try to minimize disturbances that may interfere with amplexus. If the sidewalls of the breeding tank are glass, they should be painted in order not to disturb the frogs. If a large colony is established and individual frogs marked by liquid N_2 branding, then consistently unsuccessful breeders can be eliminated from the colony.

Collection and Staging

The embryos that are found on the tank bottom following mating are covered in a viscous, sticky jelly coat. It was decided that this jelly coat should be removed (dejellying) for FETAX because it slowed production and several experiments suggested its presence did not alter toxicity. Two methods can be employed to remove the embryos from the tank. The first method is simply to invert a petri dish and place it on the bottom of the tank. Move the dish slowly into the eggs with an even motion. The embryos will pile up on the leading edge of the dish. When a sufficient number accumulate, slowly invert the dish, raising the edge with the embryos first. With practice, the embryos will fall into the open dish after it is inverted. A spatula can then be used to scrap the embryos into a 125-ml bottle or Erlenmeyer flask for dejellying. The second method is easier for the novice but requires more dejellying solution. If a small breeding aquarium is used and the distance from the support mesh to the bottom is kept about 2.5 cm, it is possible to dejelly the embryos in the breeding aquarium. This also helps recover embryos stuck to the support mesh and the embryos can simply be poured out into a large beaker. The greatest danger of this technique is continuing the dejellying process too long. If this occurs, the embryos may be irreversibly damaged and will not be usable in FETAX. Perhaps the best approach is to pour the embryos into the flask as soon as they come off the bottom and continue dejellying in the flask.

Dejellying Embryos

Embryos are dejellied in a *fresh* 2% w/v cysteine solution at pH 8.1. This efficiently strips the jelly coat from the embryo. Treatment should not last more than 3 min or the embryos could be irreversibly damaged, resulting in death or malformations. This can occur even though the embryos appear normal after treatment. In practice, dejellying is not a difficult process when properly performed. A stopwatch should be started upon addition of the 2% cysteine. The solution is gently swirled and watched. As the jelly is stripped, the solution progressively turns cloudy and gray. The embryos then begin to roll freely on the bottom. At this point the cysteine solution should be poured off and the embryos allowed to pile up on the bottom of the flask. Two to three additions of fresh FETAX solution are made immediately with gentle swirling to wash away all of the cysteine solution. Underdejellied embryos stick to glassware and overdejellied embryos die or are malformed. If high mortality or malformation rates are observed in both controls in glass and plastic, the possibility of overdejellying should be considered. Large-bore blood bank Pasteur pipets or regular pipets that have had the ends snapped off and have been fire polished are used to transfer the 1.5- to 2-mm embryos to large 100-mm petri dishes. The large dishes prevent overcrowding and allow large amounts of FETAX solution to bathe the embryos. If some embryos and unfertilized eggs disintegrate and release degradative enzymes into the medium, the stress on the normal embryos will be minimized by the large volume.

Selection of Normal Embryos

With the exception of judging malformation, no other procedure is more important in FETAX than selecting normally developing stage 8 to 11 embryos (Nieuwkoop and Faber, 1975). The pace of embryonic development is dependent on a number of factors, the most important of which is temperature. Staging is a process of accounting for the speed of embryonic development by noting the appearance of certain morphological markers. Stage 8 embryos are blastulae (Figure 8). A *blastula* is an embryo composed of a hollow ball of cells just approaching primary organogenesis. Selecting embryos prior to this stage is risky because apparently normal cleavage stage embryos can quickly become abnormal. By waiting until stage 8, the chance of selecting embryos that will grow and develop normally is greatly enhanced. It must be remembered that the female uses egg deposition to shed normal, abnormal and overripe eggs. This is a normal process. Because FETAX measures what goes wrong during development, it is imperative that only normal embryos be used at the start of the test. Stage 11 embryos are at the gastrula stage (Figure 9). At this stage, the formation of the embryonic gut has commenced and events are leading to the formation of the neural tube. Waiting past this stage risks missing a sensitive stage of development. This is important because damage done during early developmental stages causes extensive damage to a number of organ systems. A double selection process materially helps the proper selection of normal embryos between stages 8 and 11. After the embryos are placed in large petri dishes, an attempt should be made to quickly eliminate large white necrotic eggs before they begin to disintegrate. The number of these varies with the mating but they are usually more prevalent in females that have not been bred for long periods of time. Most technicians put too many embryos in a single dish prior to sorting. It is better to provide

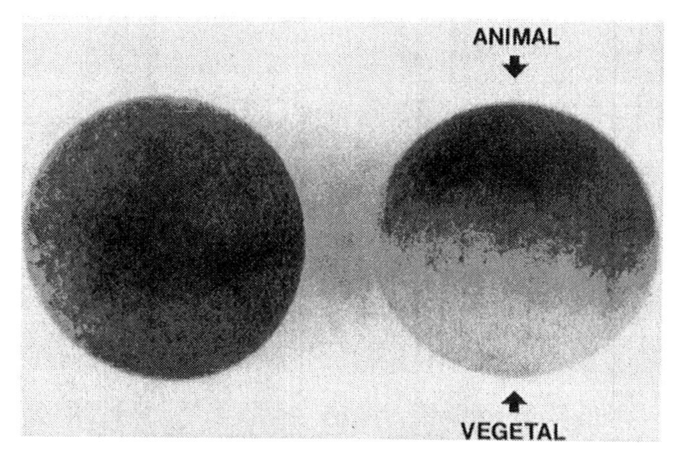

Figure 8. Normal stage 8 blastulae. These are medium cell blastulae. On the left is a dorsal view and on the right a lateral view. Notice that there is a gradual reduction in size of the cells in the animal hemisphere area compared with the vegetal hemisphere. The lateral view of the blastula shows the gradual reduction in cell size and the progressive movement of the pigmented animal hemisphere cells down over the larger white vegetal hemisphere cells. (Photo by M. A. Hull.)

ample volume for the embryos. If the FETAX solution becomes clouded, the solution should be quickly changed and the selection process accelerated. After the necrotic eggs and bad embryos are removed, normally pigmented embryos are selected; the Atlas of Abnormalities should be consulted for examples of normal stage 8–11 embryos. In order to select only normally cleaving embryos, a double selection process is used. Double selection means that embryos are sorted quickly a first time and then a second, more careful, selection is performed.

During their training period, new technicians should breed frogs, select normal stage 8–11 embryos, and culture them in FETAX solution for 96 h. If mortality is $< 10\%$ and malformation $< 7\%$, actual experiments may be conducted.

Adding Test Material

The best method of preparing text material dilutions is to pipet the stock solution into volumetric flasks (TC). Volumetric flasks either deliver a prescribed volume (marked TD: to deliver) or contain a prescribed volume (marked TC:to contain). FETAX solution is then added to the line. Once the solutions are prepared, add the embryos to the empty dish, quickly remove the solution with a Pasteur pipet, and then immediately add the test material from the volumetric flask. The concentration of the stock solution should be adjusted so that reasonable amounts of stock are added to the flasks in the dilution series, as it is fairly difficult to add small volumes.

Another approach to making test solutions is to calculate the number of milliliters of stock solution and the number of milliliters of FETAX solution required to prepare a specific concentration of the test material. Embryos are added to the dish and all excess FETAX solution removed using a Pasteur pipet. The calculated amount of FETAX solution is added first and then the test material is added with gentle swirling. This avoids the need for volumetric flasks and subsequent washing problems. It allows the use of disposable pipets. The technique yields adequate concentration-response curves assuming that the proper size pipet is used and the pipetting is done very carefully. Great care should be taken not to apply the concentrated stock solution directly on the embryos. This latter method is not as accurate as using volumetric flasks. The standard FETAX test follows the renewal procedure, which entails fresh replacement of test material every 24 h during the test. Renewal is accomplished by removing test solution with a Pasteur pipet then adding 10 ml of fresh solution to each dish.

Assignment to the Incubator (or Constant-Temperature Chamber)

When all test and control dishes have been prepared, the dishes are placed on an ordinary cafeteria tray lined with absorbent paper. It is possible to fit all of the dishes on a single tray and thus use only a single shelf of the incubator. This allows several tests to be performed concurrently. In addition, locating the dishes on a single shelf reduces the chance for temperature variation. If the dishes are randomized, a simple random number generator can be used to assign the dishes to tray positions. Incubator temperature must be between 23 and 24°C and it is best to monitor the temperature continuously throughout the test using a recording thermometer. When a recording thermometer is not available, temperature measurements should be made at least twice daily. Many electronics stores sell digital thermometers that report the temperature and keep track of the high and low events. These are ideal for FETAX as long as they are calibrated before use. If the concentration of test material is to be analyzed to ensure that the dilutions were made properly, additional dilution dishes free of embryos should be prepared and analyzed. Any analysis samples taken should be placed in vials with minimum head space and tightly sealed. If the analysis cannot be performed immediately, the samples should be frozen. Some head space is warranted so that the vials will not crack if frozen. If

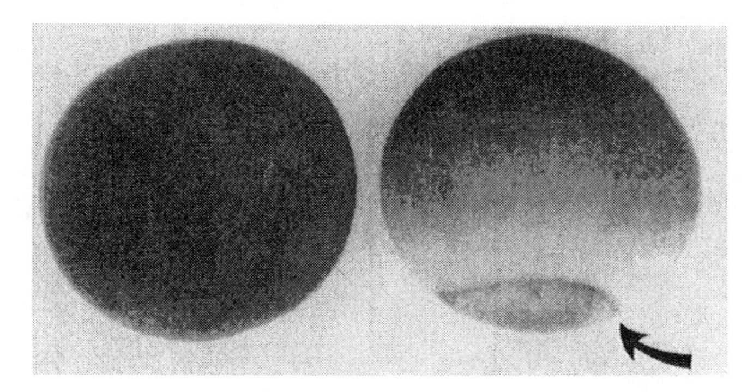

Figure 9. Normal stage 11 gastrulae. On the left is a dorsal view and on the right a lateral view. The blastopore (arrow) now encircles the lower part of the embryo and the white circle of yolk cells is now referred to as a yolk plug. The lateral view shows the extent which the animal hemisphere cells have now enveloped the embryo. (Photo by M. A. Hull.)

shipment is required, they should be shipped on dry ice to prevent test material degradation.

Conducting the Test

Temperature and pH Measurements

Incubator temperature plays a primary role in determining whether the control embryos reach stage 46 by 96 h. Over 90% of the controls should be at stage 46 at the end of the test if the temperature averages 23.5°C. Too low a temperature retards development and too high a temperature leads to high rates of malformation in the controls. This is why temperature must be carefully controlled and monitored. When mortality is low but rates of malformation exceed 7%, the temperature record of the incubator should be reviewed.

The ASTM guide (Bantle and Sabourin, 1991) requires the measurement of pH in the diluent control and the highest test concentration every 24 h of the test. This measurement should be made just before changing the test solutions and at the end of the test (see below). A semimicro pH probe can be used by tilting the dish so that the electrode bulb and reference wick are fully immersed. The pH should be recorded on the standard forms. If the pH in the highest test concentration is low, it may be a sign of bacterial growth. If the solution appears cloudy, this is another sign that bacterial growth has occurred. All previous attempts to perform FETAX gnotobiotically have failed. Some test materials even serve as a substrate for bacterial growth. In these cases it might prove necessary to conduct FETAX in the presence of 100 U/ml penicillin and 100 U/ml streptomycin and report it as a deviation from standard FETAX procedure. It should be noted that antibiotics may interact with the test material and alter toxicity.

Changing Test Solutions

A large-bore fire-polished Pasteur pipet is used to remove test solutions from dishes. By starting with the lowest concentrations and proceeding to higher concentrations, it is possible to use the same pipet to remove the test material. The procedure should be accomplished quickly to avoid drying the embryos and care should be taken to count and remove dead embryos from the dish. If the embryo has disintegrated, all remaining yolk material and tissue should be removed or it will serve as a substrate for bacterial growth. New test material solution is quickly renewed, as is any MAS and generator system. The embryos have usually hatched from the fertilization membrane and are quite fragile. Great care must be taken to avoid damaging them in the pipet. The tops of the Pasteur pipet can be broken off and fire polished to produce larger bores to accommodate larger embryos without damage in case they are accidently picked up. Depending on the toxicity of test material, this operation is performed in a safety hood or while using a respirator mask.

Ending the Test and Biological Data

General Procedure

If the temperature has been held constant at 24 ± 2°C, 90% of the controls should attain stage 46 by the end of 96 h. It may be acceptable to delay ending the test for an hour or two until the controls reach stage 46 (Figure 2). The Atlas of Abnormalities shows the morphological markers indicative of stage 46 (Figures 2 and 7). Hindlimb buds are the easiest markers for the novice to recognize (see arrow on top larva in Figure 10). Staining procedures covered

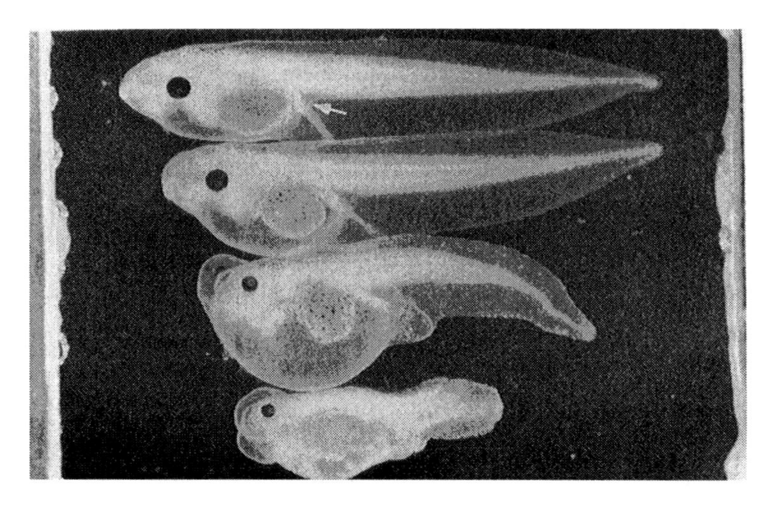

Figure 10. Increasing concentrations of a developmental toxicant. The effects of increasing concentration of the teratogen hydroxyurea. Note that as the concentration increases, the malformations become more severe. Concentrations from top to bottom are control, 0.2, 0.4, and 0.7 mg/ml. (Photo by M. A. Hull.)

in the Atlas make it easy to recognize the hindlimb bud during training.

After determining that a sufficient number of embryos have reached stage 46 by 96 h, the next step is to remove the dead larvae. The absence of heartbeat is an unambiguous sign of death that can be used at 48 (stage 35), 72 (stage 42), and 96 h (stage 46) (see heart labeled "H" in Figures 7 and 11). At 24 h, live embryos will move if gently irritated. Necrosis is easily observable in dead embryos. Death at 24 h is ascertained by skin pigmentation, structural integrity, and irritability. Once the dead larvae are removed, the total number dead at 96 h is recorded on the standard form (Figure 12) and MS-222 anesthetic is added followed by 3% formalin to fix the remaining live embryos. Some laboratories observe the anesthetized embryos for cardiac malformations prior to fixation. It is much easier to see heart structure while the heart is slowly pumping red blood. Gross congenital malformations are now recorded with particular reference to severity, type, and number in each category (Figure 13). Some test materials cause only slight malformations whereas others may cause extremely severe malformations. Increasing concentrations of test materials usually increase the severity of the malformations (Figure 10). It is important to refer to the Atlas of Abnormalities in order to identify and categorize the various malformations. During the technician training period, it is advisable to have a second dissection microscope present with control larvae for direct comparison with abnormal larvae. Larvae should be moved and rolled over during inspection and the scope light can be moved to place light on different parts of the embryo. After several tests, data collection becomes routine, as most malformations are severe enough to be easily observable. It takes time to observe minor malformations such as improper gut coiling. Despite the subjective nature of the malformation end point, the ongoing interlaboratory validation study of FETAX indicates that the data are slightly more variable than the objective mortality end point.

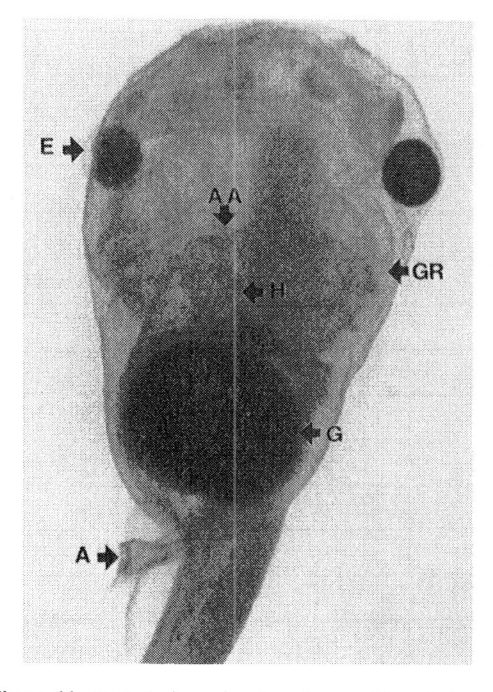

Figure 11. Ventral view of a stained stage 46 control embryo. A ventral view of the heart of a normal 96-h *Xenopus laevis* embryo. The transparency of the embryos allows a technician to easily identify malformations. Since the embryos are transparent, this specimen was stained for clarity in the photo. H, heart; AA, aortic arches; E, eye; GR, gill rakers; G, gut; A, anus. (Photo by M. A. Hull.)

Figure 11 shows a ventral view of a control larva with a tightly coiled gut. This figure illustrates the transparency of the larvae at stage 46, which makes it unnecessary to section the embryo to observe internal malformations. Gills and aortic arches are easily seen. Figure 14 shows a severely malformed larva. All organs are malformed in this larva and the malformations are easily observed. Most technicians have trouble detecting facial malformations, and Figure 15 shows a control larva on top and a larva with obvious facial malformation on the bottom. Once again, staining helps in identification during the training phase until competency is attained. In this figure, the tails of the larvae are inserted into fiberglass insect screen suspended in a petri dish.

Some developmental toxicants cause very specific malformations. For example, one ground water sample caused all larvae to be completely devoid of pigmentation and have incomplete gut coiling (Bantle et al., 1989b). Another study with plant glycoalkaloids caused anencephaly (headless larvae) in most embryos at high concentrations of toxicant (Friedman et al., 1990). These abnormalities give important information as to the target organ(s) of the toxicant.

Growth inhibition data are collected by placing each petri dish between the condenser and lens in the negative carrier compartment of a standard darkroom enlarger. A glass plate can substitute for the negative carrier to hold the dishes. Before purchasing the enlarger, the opening of the negative carrier should be measured to make sure that the petri dish will fit and that the image can be enlarged and focused on the digitizing pad. Enlargement of 2.5 to 3 diameters greatly increases the resolution of each measurement. The enlargement should be great enough to measure length accurately but not too great to image all of the larvae on the pad at once.

The Sigmascan software is loaded and the length program calibrated by magnifying a transparent ruler onto the pad and measuring a 1-cm length to serve as a standard reference. This is the approximate length of a stage 46 larva. Once calibrated, each petri dish in ascending concentration series is enlarged and each larva's head-tail length measured. Sigmascan allows the user to follow the contours of larvae that have been misshapen by the toxicant. The *t*-test for grouped observations is part of the Sigmascan software package, so data analysis can proceed immediately after collection. Each concentration is compared with the diluent control. In cases in which a cosolvent is required, the concentrations must be compared with the solvent control. Summary data can be sent to Sigmaplot software for graphic analysis along with concentration-response data for mortality and malformation. There is teratogenic hazard when growth is significantly affected (P = 0.05) at concentrations below 30% of the LC50 (see Data Analysis and Interpretation).

Frequently Encountered Problems

With care, FETAX can be routinely performed with excellent results. When problems occur they are likely to be in the following areas: breeding, selection of good embryos, identification of malformations, and controls not falling within acceptable limits. Breeding problems have already been discussed but many problems surface when the investigator attempts to stretch an experiment by selecting embryos that are marginal in quality. It is better to scale back the experiment and reduce high and low concentrations than choose poorly pigmented or abnormally cleaving blastulae. It is important to remember that the stage 8 embryos are

FETAX MORTALITY DATA

Test Material:	Investigator
Test No. Date: / /	*Stock Concentration:

Concentration	Stock (ml)*	FETAX (ml)	24-hr	48-hr	72-hr	96-hr	Total	No.	%/dish	T.M.
Controls (1&2)	0.00	10.00								X
Controls (3&4)	0.00	10.00								

[1]T.M. stands for total mortality at that concentration.

Figure 12. FETAX standard mortality form. This form is used to record the number of dead embryos in each dish at each 24-h period. This procedure is performed at the same time as the solution renewal for each dish.

very fragile and must be handled with great care. If the double selection process is performed carefully and the glassware to be used in the assay is properly cleaned, the solvent or FETAX solution controls will fall within acceptable limits. When they are not within acceptable limits, each phase of the experiment needs to be carefully analyzed starting with embryo selection and proceeding to glassware washing, bacterial contamination, and pH changes.

Another area of difficulty is in identifying malformations in the stage 46 larvae. This problem usually disappears with experience but it is important to remember that developmental toxicants may delay development in a manner similar to cold temperatures. Ordinarily, this is obvious as the embryo appears normal but simply at an earlier stage of development. Sometimes there is confusion as to whether the gut is abnormally coiled or simply delayed in its development. If there are doubts, the surviving embryos should be placed in clean FETAX solution and cultured past the 96-h time period. If the gut finally coils and the embryo is normal in appearance, developmental delay is indicated. This should not be recorded as a malformation but it should be noted as developmental delay in the report. Pericardial or abdominal edema may be observed during the test. Many times, the edema disappears and all embryos

FETAX MALFORMATION DATA

Test Material: Test No. Date / / Investigator:

CONCENTRATION:	Controls																												
MALFORMATION ▾	1	2	3	4	1	2	1	2	1	2	1	2	1	2	1	2	1	2	1	2	1	2	1	2	1	2	1	2	TOTAL
Severe																													
Gut																													
Edema - optic																													
- cardiac																													
-abdominal																													
- facial																													
- cephalic																													
- blisters																													
Tail																													
Notochord																													
Fin																													
Face																													
Eye																													
Brain																													
Hemorrhage																													
Cardiac																													
Other-specify*																													
No. Malformed																													
Total No.																													

*Comments:

Figure 13. FETAX standard malformation form. This form is used to record the number of different types of malformations seen in each individual dish. The number of malformed embryos is recorded at the end of the test. The judging of malformations entails comparing the exposed embryos in the different concentrations to the control embryos. The control embryos are staged by comparing them to pictures of normal stage 46 embryos in the Atlas of Abnormalities.

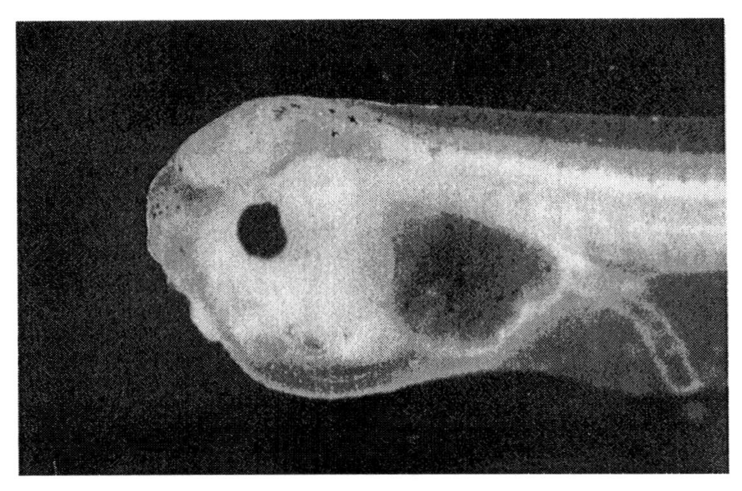

Figure 14. Embryo exposed to a severe teratogen. This embryo displays severe abnormalities. Its face, brain, eye and gills are damaged. It was exposed to 5 ng/ml 13-*cis*-retinoic acid. (Photo by M. A. Hull.)

appear normal at the end of 96 h. This should be recorded but not as a malformation.

DATA ANALYSIS AND INTERPRETATION

FETAX gives concentration-response data for mortality, malformation, and growth. These data can be compared with similar data on a molar basis using other pure chemicals to yield relative developmental toxicity. For example, saccharin has an LC50 of 18.4 mg/ml and hydroxyurea has an LC50 of 1.8 mg/ml. Once corrected for molarity, hydroxyurea will be at least 10 times more toxic.

For assessing the developmental toxicity of complex mixtures, it must be remembered that any significant difference between the 100% concentration and controls represents some developmental toxicity and associated hazard. It is more difficult to assign the teratogenic hazard to a single test material or pure compound. Recall that a teratogen is any agent that causes malformation in living offspring. In order to rank compounds according to their teratogenic hazard, Dumont and co-workers (1983a) developed the teratogenic index. This value is found by dividing the 96-h LC50 by the 96-h EC50 (malformation). Values less than 1.5 indicate little or no teratogenic hazard, and values greater than 1.5 indicate increasing hazard. The TI is similar to the commonly used therapeutic index in pharmacology in that a specific effect (malformation) is compared to general toxicity (mortality). The LC50 and EC50 are used in the determination of TI because they are automatically calculated by probit analysis programs and the least variation in confidence intervals is at the 50% response level (see Chapter 1). There is some danger the TI

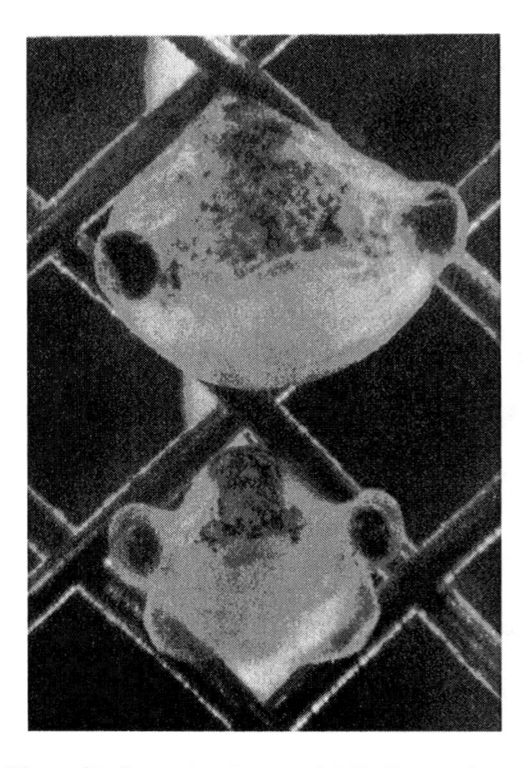

Figure 15. Contrasting the normal 96-h *Xenopus laevis* embryo with one that has facial malformations. Facial abnormalities are also easy to distinguish in the *Xenopus* embryo. At the top is a 96-h control, and below lies a severely malformed embryo. By using the control as a comparison, the malformations in the lower embryo are easily detected. The lower embryo was exposed to 0.07 mg/ml of the teratogen coumarin. (Photo by M. A. Hull.)

will not correctly identify the teratogenic hazard if the slopes of the mortality and malformation curves are different. However, this has not occurred very often. Most of the concentration curves obtained during the validation phase of FETAX demonstrated similar slopes for mortality and malformation (Figure 16). There is no reason to assume that the TI in FETAX will be indicative of the TI generated by another species. However, it is likely that these values will be fairly close, given that the predictive accuracy of FETAX is nearly 90% when compared to chemicals of known mammalian and human developmental toxicity (see Validation Study Results).

Figure 16 shows several concentration-response curves that illustrate how decisions are made in assessing developmental toxicity. It is important to remember that most probit analysis programs do not run when 0 and 100% response data points are entered. Generally, only partial response concentrations are used in establishing the curve. Therefore, there are usually many other data points collected than appear in Figure 16. α-Chaconine (Figure 16, top) is a plant glycoalkaloid that is quite toxic to embryos (Friedman et al., 1990). Normally a TI of only 1.03 would allow a determination of low teratogenic hazard. However, this compound causes extremely severe head abnormalities in most surviving embryos. Nonteratogens such as cycloheximide and puromycin (Courchesne and Bantle, 1985) have low TIs like that of α-chaconine but cause only slight abnormalities even at high concentrations. Figure 17 shows the growth inhibition curves for α-chaconine, hydroxyurea, and isoniazid. The data are graphed as the percentage of mean control head-tail length (ordi-

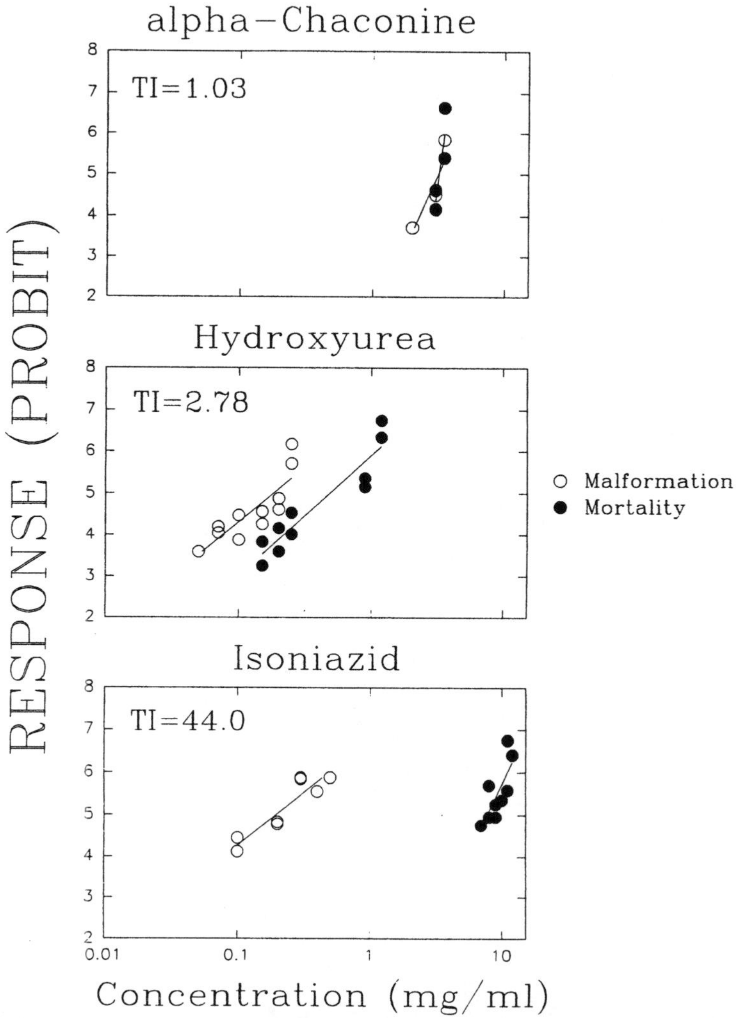

Figure 16. Probit analysis graphs of chemicals tested with FETAX. Representative concentration-response curves and respective teratogenic index values for three compounds tested with FETAX.

nate) versus the percentage of the LC50 concentration (abscissa). Note that stronger teratogens such as hydroxyurea and isoniazid have significant effects on growth at much lower concentrations than α-chaconine and have steeper slopes. If the MCIG is less than 30% of the 96-h LC50, the teratogenic hazard of a compound is considered significant. α-Chaconine has a significant effect on growth only at much higher concentrations. Because of these differences, α-chaconine can be considered a low-hazard teratogen qualifying as a teratogen only on the basis of the severity of the malformations caused.

Note that in Figure 16 (middle) the mortality and malformation concentration-response curves for hydroxyurea are farther apart than for α-chaconine, resulting in a TI of 2.78. The curves are parallel and the correspondence of the data points to the lines is good. The growth-inhibition curve shows a significant effect near 30% of the LC50 level, indicating some teratogenic hazard (Figure 17). Note that at the highest concentrations embryos are only 40% of control lengths. Hydroxyurea, a DNA synthesis inhibitor, also causes severe malformations. Thus, hydroxyurea poses a significant teratogenic hazard by all three criteria.

Isoniazid, an antibiotic, is a false positive in FETAX (Figure 16, bottom). In mammals it is normally metabolized before it can affect development but it is a severe teratogen in FETAX unless isoniazid-induced microsomes are added. Note that the mortality

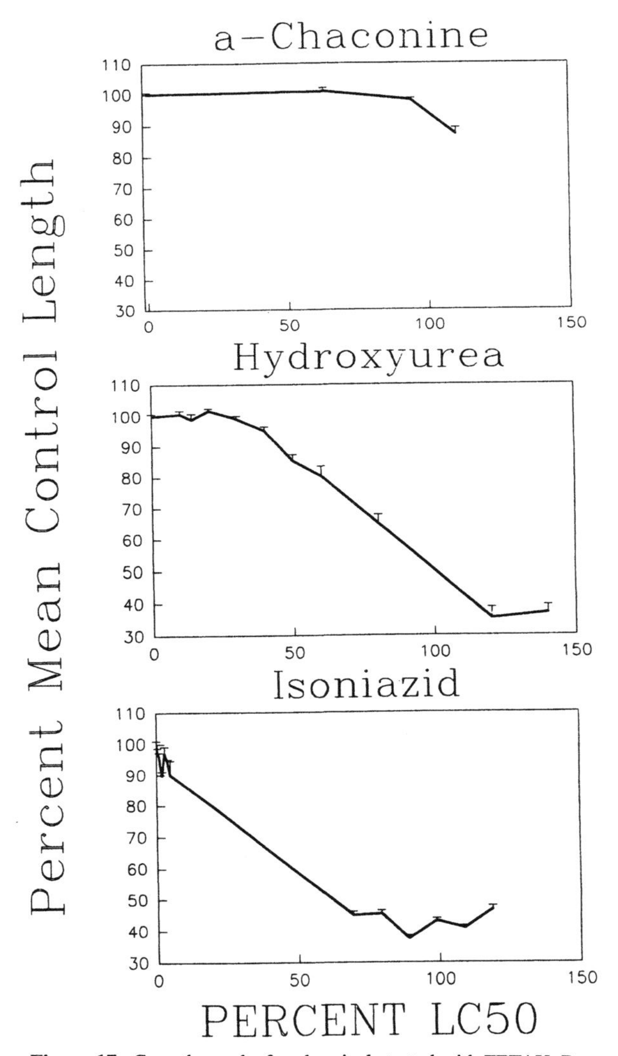

Figure 17. Growth graphs for chemicals tested with FETAX. Representative embryo growth curves for the three compounds. Concentrations are expressed as percentages of the respective compound LC50. Growth is expressed as a percentage of mean FETAX solution control length.

Table 1. Results of FETAX validation studies

Substance	Units	96-h LC50	96-h EC50	TI	MCIG	MAS LC50	MAS EC50	MAS TI	MAS MCIG
Acetaminophen	(mg/ml)	0.158	0.140	1.1	0.105				
Acetone	(% v/v)	2.190	1.287	1.7	1.430				
Acetylaminofluorine(2-)[a]	(mg/L)	88.50	7.15	12.4		42.50	2.60	16.35	
Acetylhydrazide	(mg/ml)	12.42	0.050	248.4	0.050	11.14	0.06	185.6	0.04
Actinomycin D(**FN**)[c]	(mg/ml)	0.019	0.217	0.1	0.016	1.80	0.12	15.00	0.02[b]
Amaranth	(mg/ml)	3.386	3.500	1.0	>4.0				
Aminonicotinamide(6-)	(mg/ml)	3.070	0.006	511.7	0.100				
Ascorbic acid	(mg/ml)	19.700	12.130	1.6	<10.00				
Aspartame	(mg/ml)	13.920	13.140	1.1	7.000				
Azacytidine(5–)	(mg/ml)	0.516	0.044	11.7	0.090				
Benzo[*a*]pyrene[a]	(mg/L)	>10.0	11.0	0.9		>10.0	1.65	6.11	none
Busulfan	(mg/ml)	>0.20	0.160	>1.25					
Caffeine	(mg/L)	0.257	0.128	2.0	0.100				
Chaconine( )	(mg/L)	1.880	1.730	1.1	<4.00				
Chaconine(β1)	(mg/L)	2.520	1.750	1.4	4.00				
Chaconine(β2)	(mg/L)	>10.0	6.500	>1.5	6.000				
Chaconine(gamma)	(mg/L)	>10.0	7.820	>1.28	4.000				
Copper sulfate	(mg/L)	1.190	0.130	9.2	0.050				
Cotinine	(mg/ml)	4.340	0.720	6.0	0.325				
Coumarin[a] (**FP**)[d]	(mg/ml)	0.129	0.040	3.2					
Cycloheximide	(mg/ml)	0.159	0.119	1.3	0.056				
Cyclophosphamide	(mg/ml)	1.370	0.390	3.5		>10.0	1.65	>6.06	
Cytochalasin D	(mg/L)	461.5	121.0	3.8		805.5	546.0	1.48	
Cytosine arabinoside	(mg/ml)	5.410	0.760	7.1	0.7				
Diazapam	(mg/ml)	0.032	0.020	1.6	0.013				
Dimethyl sulfoxide	(% v/v)	1.183	1.310	0.9	1.500				
Diphenhydramine HCl	(mg/ml)	0.031	0.003	10.7	0.003				
Diphenylhydantoin[a]	(mg/L)	74.450	32.000	2.3		126.35	69.15	1.83	
Doxlyamine succinate	(mg/ml)	0.22	0.04	4.8	0.035				
Ethanol (L)	(% v/v)	1.440	1.010	1.4	1.000				
Ethidium bromide	(mg/ml)	0.050	0.035	1.4	0.050				
Ethyl(*N*-)-*N*-nitrosourea	(mg/ml)	0.258	0.051	5.0	4.870	3.70	7.30	5.07	
Flourouracil(5-)	(mg/ml)	1.620	0.137	11.8	0.120				
Furazolidone	(mg/ml)	0.014	0.007	2.0	0.008				
Hydroxydilantin(M-)[a]	(mg/L)	>150.0	>150.0	1.0		>150	>150		
Hydroxydilantin(P-)[a]	(mg/L)	>150.0	>150.0	1.0		>150	>150		
Hydroxyurea	(mg/ml)	1.820	0.430	4.2	0.3				
Isoniazid (**FP**)[d]	(mg/ml)	9.893	0.270	36.6	0.170	8.90	0.29	30.17	0.25
						6.34	0.41	15.28	0.13[b]
Isonicotinic acid	(mg/ml)	3.180	1.260	2.5	1.250	3.29	1.51	2.18	1.13
						3.29	153.0	0.02	1.27[b]
Methotrexate	(mg/ml)	0.504	0.026	19.4	0.019				
Methylmercury chloride	(mg/L)	0.08	0.02	3.6	0.038				
Naphthalene	(mg/L)	6.540	5.750	1.1					
Nicotine	(mg/L)	136.50	0.41	332.9	0.463	20.35	5.85	3.48	
Nitriloacetic acid	(mg/ml)	0.570	0.52	1.1	0.375				
Nitrosodimethylamine(N-)	(mg/ml)	3.38	2.295	1.5	1.250				
Procarbazine	(mg/ml)	3.170	1.310	2.4	1.000				
Propylene glycol	(% v/v)	2.697	1.740	1.5	1.375				
Pseudoephedrine (**FP**)[d]	(mg/ml)	0.414	0.237	1.7	0.200				
Retinoic acid(-*cis*)	(mg/ml)	30.42	2.49	12.2	9.000				
Retinoic acid(-*trans*)	(mg/L)	0.375	0.034	11.0	0.070				
Rifampicin[a]	(mg/ml)	>2.00	>2.000	1.0	1.38	0.50	2.75		
Saccharin	(mg/ml)	18.370	19.340	0.9	17.000				
Serotonin	(mg/ml)	3.005	0.370	8.1	0.425				
Sodium acetate	(mg/ml)	4.200	3.300	1.3	2.500				
Sodium cyclamate	(mg/ml)	16.135	14.845	1.1	12.000				
Sodium salicylate	(mg/ml)	2.320	1.450	1.6	1.250				
Sodium selenate	(mg/ml)	0.021	0.007	3.0	0.009				
Solanine (a)	(mg/L)	10.900	8.800	1.2	12.000				
Solasonine	(mg/L)	5.600	5.110	1.1					
Trichloroethylene	(% v/v)	0.030	0.008	3.9	0.020				
Triethylene glycol	(% v/v)	2.450	2.150	1.1	1.430				
Urethane	(mg/ml)	5.645	1.730	3.3	1.250				
Zinc sulfate (**FP**)[d]	(mg/L)	34.40	2.69	12.8		36.70	2.88	12.72	

[a]These chemicals were used with DMSO as the solvent.
[b]These values were obtained using isoniazid-induced microsomes. All other MAS data were generated using Aroclor-induced microsomes.
[c]This chemical gave false-negative results.
[d]These chemicals gave false-positive results.

and malformation concentration-response curves are widely separated, resulting in a TI of 44.0. Between concentrations of 0.1 and 6.0 mg/ml, all embryos survive but are all malformed. This is indicative of a highly hazardous teratogen. This hazard is also demonstrated in the growth inhibition curve presented in Figure 17, as a significant effect is observed within 10% of the 96-h LC50. Once again, the embryos are reduced in size to as low as 40% of controls at 65% of the LC50. Isoniazid causes severe malformations in surviving embryos.

Concerning the foregoing sections, it is important to realize that all three criteria (mortality, malformation, and growth inhibition) play a role in assessing relative teratogenic hazard. Each criterion should be considered on its own merits and reported. Only then can teratogenicity be adequately assessed.

VALIDATION STUDY RESULTS

Table 1 shows the results of validation studies carried out in the author's laboratory using 65 compounds (see also Dumont et al., 1983b). Attempts were made to select compounds from several different chemical classes and to choose nearly as many nonteratogens as teratogens. "Blind" testing was used for many of the compounds. In cases in which a metabolic activation system was known to play a role in developmental toxicity in mammals, the MAS was added to FETAX. For the purposes of routine testing for hazard to humans, MAS must be routinely added to all samples. Although the validation study was of more use for human developmental toxicity testing purposes, it showed that FETAX gives predictable results with a wide variety of known developmental toxicants. The confidence intervals are narrow in most cases and the results are repeatable. In this list only actinomycin D (false negative, FN), isoniazid (false positive, FP), coumarin (FP), and zinc (FP) (Table 1) gave incorrect results. It must be remembered that considerable controversy surrounds the classification of many of these compounds as mammalian and human teratogens. This makes it difficult to carry out this type of study.

SUMMARY AND CONCLUSIONS

FETAX is a useful developmental toxicity test that has already been used in assaying environmental mixtures in a laboratory, in a biomonitoring trailer, and in situ. FETAX can be used in relation to ecotoxicological risk and human health. The validation study and the in vitro metabolic activation system support the latter use of FETAX. An ongoing interlaboratory validation study suggests that FETAX will prove to be repeatable and reliable. Once the basic assay is performed, FETAX is flexible enough to allow a number of variations in terms of exposure, end points, and species used. This can provide additional valuable data. The greatest drawback to the assay is occasional breeding problems. When good embryos are available, however, FETAX is straightforward and provides excellent data.

The use of different species and different end points is needed to safeguard the environment adequately. Acute toxicity tests alone may not suffice. It is likely that reproductive and developmental toxicity tests will be needed along with other testing (see Chapters 12, 13, and 14) to evaluate adequately the hazards posed by environmental toxicants. An effort is already being made to replace FETAX with even simpler molecular-based tests. These tests involve special dyes that fluoresce quantitatively when excited by light. The dyes can give a specific indication of cell pH, membrane function, DNA content, etc. (unpublished). FETAX will indicate the predictive value of these new tests because results of the new cellular tests will be compared with FETAX results. Because a consistent protocol was followed in FETAX it will be possible to compare results. Additional efforts must be expended to ensure that developmentally relevant, rapid, and inexpensive tests are available to the scientific community.

ACKNOWLEDGMENTS

This work was supported by U.S. Army Medical Research and Development Contract DAMD17-91-C-1048. The opinions, interpretations, conclusions, and recommendations are those of the author and not necessarily endorsed by the U.S. Army. In conducting research using animals, the investigator adhered to the Guide for Care and Use of Laboratory Animals, prepared by the committee on Care and Use of Laboratory Animals of the Institute of Laboratory Resources, National Research council (NIH Publication No. 86-23, Revised 1985). Special thanks go to Jim Rayburn and Mendi Hull for their assistance in the laboratory and in preparation of the manuscript.

LITERATURE CITED

Abbasi, S. A., and Soni, R.: Teratogenic effects of chromium (VI) in the environment as evidenced by the impact of larvae of amphibian *Rana tigrina*: implications in the environmental management of chromium. *Int. J. Environ. Stud.* 23(2):131–138, 1984.

Anderson, R. J., and Prahlad, K. V.: The deleterious effects of fungicides and herbicides on *Xenopus laevis* embryos. *Arch. Environ. Contam. Toxicol.* 4:312–323, 1976.

Bancroft, R., and Prahlad, K. V.: Effect of ethylenebis[dithiocarbamic acid] disodium salt (nabam) and ethylenebis[dithio carbamato] manganese (maneb) on *Xenopus laevis* development. *Terat.* 7:143–150, 1973.

Bantle, J. A., and Courchesne, C. L.: The combined use of genotoxicity and whole embryo teratogenicity screening assays in predicting teratogenic risks. *Proc. 2nd Int. Conf. Ground-Water Qual. Res.*, Oklahoma, Stillwater, pp. 175–176, 1985.

Bantle, J. A., and Dawson, D. A.: Uninduced rat liver microsomes as an in vitro metabolic activation system for the Frog Embryo Teratogenesis Assay—*Xenopus* (FETAX). Aquatic Toxicology and Hazard Assessment: Tenth Volume, edited by W. G. Landis, and W. H. van der Schalie, pp. 316–326. ASTM STP 971. Philadelphia: ASTM, 1988.

Bantle, J. A., and Sabourin, T. D.: Standard Guide for Conducting the Frog Embryo Teratogenesis Assay—*Xenopus* (FETAX). ASTM E1489. Philadelphia: ASTM, in press.

Bantle, J. A.; Fort, D. J.; and Dawson, D. A.: Bridging the gap from short-term teratogenesis assays to human health hazard assessment by understanding common modes of teratogenic action. Aquatic Toxicology and Hazard Assessment: Twelfth Volume, edited by W. G. Landis, and W. H. van der Schalie, pp. 46–58. ASTM STP 1096. Philadelphia: ASTM, 1989a.

Bantle, J. A.; Fort, D. J.; and James, B. L.: Identification of developmental toxicants using the frog embryo teratogenesis assay—*Xenopus* (FETAX). *Hydrobiology* 188/189:577–585, 1989b.

Bantle, J. A.; Fort, D. J.; Rayburn, J. R.; DeYoung, D. J.; and Bush, S. J.: Further validation of FETAX: evaluation of the developmental toxicity of five known mammalian teratogens and non-teratogens. *Drug Chem. Toxicol.* 13(4):267–282, 1990a.

Bantle, J. A.; Dumont, J. N.; Finch, R. A.; and Linder, G.: Atlas of Abnormalities: A Guide for the Performance of FETAX. Stillwater: Oklahoma State Publications Department, 1990b.

Birge, W. J., and Black, J. A.: Copper in the Environment, Part II, Health Effects. New York: Wiley, 1979.

Birge, W. J.; Black, J. A.; Westerman, A.G.; and Hudson, J.E.: The Biogeochemistry of Mercury in the Environment. Amsterdam: Elsevier/North-Holland Biomedical, 1979.

Browne, C. L., and Dumont, J. N.: Toxicity of selenium to developing *Xenopus laevis* embryos. *J. Toxicol. Environ. Health* 5:699–709, 1979.

Browne, C. L., and Dumont, J. N.: Cytotoxic effects of sodium selenite on *Xenopus laevis* tadpoles. *Arch. Environ. Contam. Toxicol.* 9:181–191, 1980.

Cabejszed, I., and Wojcik, J.: Trial application of tadpoles of *Xenopus laevis* for estimation of harmfulness of pesticides in water. *Rocz. Panstq. Zakl. Hig.* 19:499–506, 1968.

Chang, L. W.; Reuhl, K. R.; and Dudley, A. W.: Effects of methylmercury chloride on *Rana pipiens* tadpoles. *Environ. Res.* 8:82–91, 1974.

Cooke, A. S.: The effects of DDT, dieldrin, and 2,4-D on amphibian spawn and tadpoles. *Environ. Pollut.* 3:51–68, 1972.

Courchesne, C. L., and Bantle, J. A.: Analysis of the activity of DNA, RNA, and protein synthesis inhibitors on *Xenopus* embryo development. *Teratogenesis Carcinog. Mutagen.* 5:177–193, 1985.

Davis, K. R.; Schultz, T. W.; and Dumont, J. N.: Toxic and teratogenic effects of selected aromatic amines on embryos of the amphibian *Xenopus laevis*. *Arch. Environ. Contam. Toxicol.* 10:371–391, 1981.

Dawson, D. A., and Bantle, J. A.: Development of a reconstituted water medium and initial validation of FETAX. *J. Appl. Toxicol.* 7:237–244, 1987a.

Dawson, D. A., and Bantle, J. A.: Co-administration of methylxanthines and inhibitor compounds potentiates teratogenicity in *Xenopus* embryos. *Teratology* 35:221–227, 1987b.

Dawson, D. A., and Wilke, T. S.: Initial evaluation of developmental malformation as an end point in mixture toxicity hazard assessment for aquatic vertebrates. *Ecotoxicol. Environ. Safety* 21:215–226, 1991a.

Dawson, D. A., and Wilke, T. S.: Evaluation of the Frog Embryo Teratogenesis Assay: *Xenopus* (FETAX) as a model system for mixture toxicity hazard assessment. *Environ. Toxicol. Chem.* 10:941–948, 1991b.

Dawson, D. A.; McCormick, C. A.; and Bantle, J. A.: Detection of teratogenic substances in acidic mine water samples using the Frog Embryo Teratogenesis Assay—*Xenopus* (FETAX). *J. Appl. Toxicol.* 5:234–244, 1984.

Dawson, D. A.; Fort, D. J.; Smith, G. L.; Newell, D. L.; and Bantle, J. A.: Comparative evaluation of the developmental toxicity to nicotine and cotinine with FETAX. *Teratogenesis Carcinog. Mutagen* 8:329–338, 1988a.

Dawson, D. A,; Stebler, E.; Burks, S. A.; and Bantle, J. A.: Teratogenicity of metal contaminated sediment extracts to frog (*Xenopus laevis*) and fathead minnow (*Pimephales promelas*) embryos. *Environ. Toxicol. Chem.* 7: 27–34, 1988b.

Dawson, D. A.; Fort, D. J.; Newell, D. L.; and Bantle, J. A.: Developmental toxicity testing with FETAX: evaluation of five validation compounds. *Drug Chem. Toxicol.* 12(1):67–75, 1989.

Dawson, D. A.; Schultz, T. W.; Baker, L. L.; and Wilke, T. S.: Comparative developmental toxicity of acetylenic alcohols on embryos and larvae of *Xenopus laevis*. Aquatic Toxicology and Risk Assessment: Thirteenth Volume, edited by W. G. Landis, and W. H. van der Schalie, pp. 267–277. ASTM STP 1096. Philadelphia: ASTM, 1990a.

Dawson, D. A.; Schultz, T. W.; Baker, L. L.; and Mannar, A.: Structure-activity relationships for oesteolathyrism. III. Substituted thiosemicarbazides. *J. Appl. Toxicol.* 10: 59–64, 1990b.

Dawson, D. A.; Schultz, T. W.; and Baker, L. L.: Structure-activity relationships for osteolathyrism: IV. *Para*-substituted benzoic acid hydrazides and alkyl carbazates. *Environ. Toxicol. Chem.* 10:455–461, 1991a.

Dawson, D. A.; Wilke, T. S.; and Schultz, T. W.: Joint action of binary mixtures of osteolathyrogens at malformation-inducing concentrations for *Xenopus* embryos. *J. Appl. Toxicol.* 11:415–421, 1991b.

Dawson, D. A.; Schultz, T. W.; and Shroeder, E. C.: Laboratory care and breeding of the African clawed frog. *Lab. Anim.* 21:3136, 1992.

Deuchar, E. M.: *Xenopus laevis* and developmental biology. *Biol. Rev.* 47:37–112, 1972.

Deuchar, E. M.; *Xenopus*: The South African Clawed Frog. New York: Wiley, 1975.

DeYoung, D. J.; Bantle, J. A.; and Fort, D. J.: Assessment of the developmental toxicity of ascorbic acid, sodium selenate, coumarin, serotonin, and 13-*cis* retinoic acid using FETAX. *Drug Chem. Toxicol.* 14(1&2):127–141, 1991.

Dial, N.: Methylmercury: Teratogenic and lethal effects in frog embryos. *Teratology* 13:327–334, 1976.

Dumont, J. N., and Schultz, T. W.: Effects of coal-gasification sour water on *Xenopus laevis* embryos. *J. Environ. Sci. Health* A15:127–138, 1980.

Dumont, J. N., Schultz, T. W., Jones, R. D.: Toxicity and teratogenicity of aromatic amines to *Xenopus laevis*. *Bull. Environ. Contam. Toxicol.* 22:159–166, 1979.

Dumont, J. N.; Schultz, T. W.; Buchanan, M. V.; and Kao, G. L.: Frog embryo teratogenesis assay: *Xenopus* (FETAX)—a short-term assay applicable to complex environmental mixtures, Symposium on the Application of Short-Term Bioassays in the Analysis of Complex Environmental Mixtures, edited by M. Waters et al., pp. 393–405. New York: Plenum, 1983a.

Dumont, J. N.; Schultz, T. W.; and Epler, R. G.: The response of the FETAX to mammalian teratogens. *Teratology* 27(2):A39–A40, 1983b.

Dumpert, K.: Tests with the South African clawed toad (*Xenopus laevis*) for detecting chemical causes of the decrease of amphibians. *Chemosphere* 15:807–811, 1986.

Dumpert, K.: Embryotoxic effects of environmental chemicals: tests with the South African clawed toad (*Xenopus laevis*). *Ecotoxicol. Environ. Safety* 13:324–338, 1987.

Dumpert, K., and Zietz, E.: Platanna (*Xenopus laevis*) as a test organism for determining the embryotoxic effects of environmental chemicals. *Ecotoxicol. Environ. Safety* 8: 55–74, 1984.

Fort, D. J., and Bantle, J. A.: Use of frog embryo teratogenesis assay—*Xenopus* (FETAX) and an exogenous metabolic activation system to evaluate the developmental toxicity of diphenylhydantoin. *J. Fund. Appl. Toxicol.* 14:420–733, 1990a.

Fort, D. J., and Bantle, J. A.: Analysis of the developmental toxicity of isoniazid using exogenous metabolic activation systems with frog embryo teratogenesis assay—*Xenopus* (FETAX). *Teratogenesis Carcinog. Mutagen.* 10:463–476, 1990b.

Fort, D. J.; Dawson, D. A.; and Bantle, J. A.: Development of a metabolic activation system for the frog embryo teratogenesis assay: *Xenopus* (FETAX). *Teratogenesis Carcinog. Mutagen.* 8:251–263, 1988.

Fort, D. J.; James, B. L.; and Bantle, J. A.: Evaluation of the developmental toxicity of five compounds with the frog embryo teratogenesis assay: *Xenopus* (FETAX). *J. Appl. Toxicol.* 9:377–388, 1989.

Fort, D. J.; Rayburn, J. R.; DeYoung, D. J.; and Bantle, J. A.: Developmental toxicity testing with frog embryo teratogenesis assay—*Xenopus* (FETAX): efficacy of Aroclor 1254–induced exogenous metabolic activation system. *Drug Chem. Toxicol.* 14(1&2):143–160, 1991.

Friedman, M.; Rayburn, J. R.; and Bantle, J. A.: Developmental toxicology of potato alkaloids in the frog embryo teratogenesis assay—*Xenopus* (FETAX). *Food Chem. Toxicol.* 29:537–547, 1990.

Ghate, H. V.: Notochordal anomaly in frog embryos exposed to tetraethythiuram monosulphide and tetraethylthiuram disulphide. *Toxicol. Lett.* 19:253, 1983.

Ghate, H. V.: Toxicity and teratogenic effects of carbon disulphide in frog embryo. *Riv. Biol.* 78(1):129–132, 1985a.

Ghate, H. V.: Dithiocarbamate induced teratogenesis in frog embryo. *Riv. Biol.* 78(2):288–291, 1985b.

Ghate, H. V., and Mulherkar, L.: Effect of sodium diethyldithiocarbamate on developing embryos of the frog *Microhyla ornata*. *Indian J. Exp. Biol.* 18:1040–1042, 1980.

Green, E. V.: Effects of hydrocarbon-protein conjugates on frog embryos. I. Arrest of development by conjugates of 9,10-dimethyl-1,2-benzanthracene. *Cancer Res.* 14:591–598, 1954.

Greenhouse, G. A.: Evaluation of the teratogenic effects of hydrazine, methylhydrazine, and dimethylhydrazine on embryos of *Xenopus laevis*, the South African clawed toad. *Teratology* 13:167–178, 1976a.

Greenhouse, G. A.: The evaluation of toxic effects of chemicals in freshwater by using frog embryos and larvae. *Environ. Pollut.* 11:303–315, 1976b.

Greenhouse, G. A.: Toxicity of *N*-phenyl-α-naphthylamine and hydrazine to *Xenopus laevis* embryos and larvae. *Bull. Environ. Contam. Toxicol.* 18:503–511, 1977.

Karnofsky, D. A.: Mechanisms of action of certain growth inhibiting drugs. *Teratology*: Principles and Techniques, edited by J. G. Wilson, and J. Warkany, pp. 185–213, 1965.

Kay, B. K., and Peng, H. B. (ed.): Methods in Cell Biology Volume 36. New York: Academic Press, 1992.

Linder, G.; Barbitta, J.; and Kwaiser, T.: Short-term amphibian toxicity tests and paraquat toxicity assessment. Aquatic Toxicology and Risk Assessment: Thirteenth volume, edited by W. G. Landis, and W. H. van der Schalie, pp. 189–198. ASTM STP 1096. Philadelphia: ASTM, 1990.

NRC. Models for Biomedical Research: A New Perspective. Washington, DC: Committee on Models for Biomedical Research, Board on Basic Biology, Commission of Life Sciences, National Research Council, National Academy Press, 1985.

Nieuwkoop, P. D., and Faber, J.: Normal Table of *Xenopus laevis* (Daudin), 2nd ed. Amsterdam: North Holland, 1975.

Pechmann, J. H. K.; Scott, D. E.; Semlitsch, R. D.; Caldwell, J. P.; Vitt, L. J.; and Gibbons, J. W.: Declining amphibian populations: the problem of separating human impacts from natural fluctuations. *Science* 253:892–895, 1991.

Rayburn, J. R.; Fort, D. J.; McNew, R.; and Bantle, J.A.: Synergism and antagonism induced by three carrier solvents with *t*-retinoic acid and 6-aminonicotinamide. *Bull. Environ. Contam. Toxicol.* 46:625–632, 1991a.

Rayburn, J. R.; Fort, D. J.; DeYoung, D. J.; McNew, R.; and Bantle, J. A.: Altered developmental toxicity caused by three carrier solvents. *J. Appl. Toxicol.* 11:253–260, 1991b.

Sabourin, T. D., and Faulk, R. T.: Comparative evaluation of a short-term test for developmental effects using frog embryos. Banbury Report 26: Developmental Toxicology: Mechanisms and Risk, pp. 203–223. Cold Spring Harbor, New York: Cold Spring Harbor Laboratory, 1987.

Sabourin, T. D.; Faulk, R. T.; and Goss, L. B.: The efficacy of three non-mammalian test systems in the identification of chemical teratogens. *J. Appl. Toxicol.* 5:227–233, 1985.

Schultz, T. W., and Ranney, T. S.: Structure-activity relationships for osteolathyrism: II. Effects of alkyl-substituted acid hydrazides. *Toxicology* 53:147–159, 1988.

Schultz, T. W.; Cajina-Quezada, M.; and Dumont, J. N.: Structure-toxicity relationships of selected nitrogenous heterocyclic compounds. *Arch. Environ. Contam. Toxicol.* 9:591–598, 1980.

Schultz, T. W.; Ranney, T. S.; Riggin, G. W.; and Cajina-Quesada, M.: Structure-activity relationships for osteolathyrism. I. Effects of altering the semicarbazide structure. *Trans. Am. Microsc. Soc.* 107(2):113–126, 1988.

Science Briefings: New task force on declining amphibians. *Science* 253:509, 1991.

Tallarida, R. J., and Murray, R. B.: Manual of Pharmacologic Calculations with Computer Programs, 2nd ed. New York: Springer-Verlag, 1987.

Wake, D. B.: Declining amphibian populations. *Science* 253:860, 1991.

Wake, D. B., and Morowitz, H.J.: Declining amphibian populations—a global phenomenon. Report of a workshop sponsored by the Board on Biology, National Research Council, held at Irvine, CA, February 19–20, 1990.

Chapter 8

SEDIMENT TESTS

C. G. Ingersoll

INTRODUCTION

Sediment provides habitat for many aquatic organisms and is a major repository for many of the more persistent chemicals that are introduced into surface waters. In the aquatic environment, most anthropogenic chemicals and waste materials including toxic organic and inorganic chemicals eventually accumulate in sediment. Mounting evidence exists of environmental degradation in areas where U.S. Environmental Protection Agency (EPA) water quality criteria (WQC) are not exceeded, yet organisms in or near sediments are adversely affected (Chapman, 1989; U.S. EPA, 1994a). The WQC were developed to protect organisms in the water column and were not intended to protect organisms in sediment. Concentrations of contaminants in sediment may be several orders of magnitude higher than in the overlying water; however, bulk sediment concentrations are not highly correlated to bioavailability (Burton, 1991). Partitioning or sorption of a compound between water and sediment depends on many factors, including aqueous solubility, pH, redox, affinity for sediment organic carbon and dissolved organic carbon, grain size of the sediment, sediment mineral constituents (oxides of iron, manganese, and aluminum), and the quantity of acid volatile sulfides in sediment (Di Toro et al., 1990, 1991; U.S. EPA, 1992). Although certain chemicals are highly sorbed to sediment, these compounds may still be available to the biota. Contaminated sediments may be directly toxic to aquatic life (Swartz et al., 1985; ASTM, 1992, 1995a) or can be a source of contaminants for bioaccumulation in the food chain (Mac et al., 1984; U.S. EPA, 1994a).

In an overview of sediment quality in the United States, Lyman et al. (1987) concluded that very few areas have been well characterized because of a lack of national guidelines or standard methods for evaluating toxic concentrations of contaminants in sediment. All surface waters receiving input from anthropogenic sources contained some in-place contaminants; however, the extent or significance of the contamination has not been evaluated. The United States has 39.4 million acres of lakes, 1.8 million miles of rivers, 32 thousand square miles of estuaries, 23 thousand ocean coastline miles (excluding Alaska), and hundreds of thousands of square miles of nearshore and continental shelf habitat that receive anthropogenic contaminants. A wide variety of contaminants including heavy metals, pesticides, industrial chemicals, and energy-related hydrocarbons accumulate in sediment (Lyman et al., 1987). Contaminated sediment may be located in small, isolated hot spots or distributed over large areas. The locations of contaminated sediments have not been comprehensively assessed on a national scale; however, industrial harbors have been most severely contaminated (NRC, 1989). The U.S. EPA is currently conducting a National Sediment Inventory with a variety of existing databases to better evaluate the extent of sediment contamination in the United States (B. Southerland, U.S. EPA, Washington, DC, personal communication).

Historically, the assessment of sediment quality has often been limited to chemical characterizations. However, quantifying contaminant concentrations alone cannot provide enough information to evaluate adequately the potential adverse effects, interactions among chemicals, or time-dependent availability of these materials to aquatic organisms (Rodgers et al., 1984; Dillon and Gibson, 1986). Because relationships between concentrations of contaminants in sediment and their bioavailability are poorly understood, determination of contaminated sediment effects on aquatic organisms requires controlled toxicity and bioaccumulation tests (Seelye and Mac, 1984; Burton and Ingersoll, 1994).

TERMINOLOGY

Contaminated sediment. Sediment containing chemical substances at concentrations that pose a known or suspected threat to environmental or human health (NRC, 1989).

Control sediment. A sediment that is essentially free of contaminants and is used routinely to assess the acceptability of a test (ASTM, 1995a; U.S. EPA-U.S. COE, 1991, 1994). Any contaminants in control sediment may originate from the global spread of pollutants and do not reflect any substantial input from local or nonpoint sources (ASTM, 1995b). Comparing test sediments to control sediments gives a measure of the toxicity of the test sediment beyond inevitable background contamination (ASTM, 1995b).

Elutriate. The water collected from vigorous shaking (15 to 30 min) of one part sediment with four parts water. This mixture is allowed to settle and the liquid phase is centrifuged and might be filtered (U.S. EPA-U.S. COE, 1977). The elutriate test was developed for evaluating the potential effects of disposing dredged material in open water. Analyses of elutriate samples measure the water-soluble constituents potentially released from sediment to the water column during dredge disposal operations (Shuba et al., 1978).

Interstitial water or pore water. Water occupying space between sediment or soil particles (ASTM, 1994c; see Chapter 1).

Overlying water. The water placed over whole sediment in a test chamber (ASTM, 1995a).

Reference sediment. A whole sediment near an area of concern used to assess sediment conditions exclusive of material(s) of interest (ASTM, 1994c; U.S. EPA-U.S. COE, 1991, 1994). The reference sediment may be used as an indicator of localized sediment conditions exclusive of the specific pollutant input of concern. Such sediment would be collected near the site of concern and would represent the background conditions resulting from any localized pollutant inputs as well as global pollutant input (ASTM, 1995b). This is the manner in which reference sediment is used in dredge material evaluations (U.S. EPA-U.S. COE, 1991, 1994).

Reference toxicity test. A test conducted in conjunction with sediment tests to determine possible changes in condition of the test species (Lee, 1980). Deviations from an established normal range indicate a change in the condition of the test organism population. Reference toxicant tests are most often acute lethality tests performed in the absence of sediment (U.S. EPA-U.S. COE, 1991; U.S. EPA, 1994a). Sediment spiked with a toxicant might also be included as a positive control for the sediment toxicity test.

Sediment. Particulate material normally lying below water, or for experimental purposes a sediment could be formulated from particulate material (ASTM, 1994c).

Spiked sediment. A sediment to which a material has been added for experimental purposes (ASTM, 1994c).

Whole sediment. Sediment and associated pore water that have had minimal manipulation (ASTM, 1994c).

CHARACTERISTICS OF SEDIMENT

Sediment has been defined as particulate material normally lying below water, or for experimental purposes a sediment could be formulated from particulate material (ASTM, 1994c). Whole sediment is typically heterogeneous in physical, chemical, and biological characteristics and may include organic matter in various stages of decomposition or synthesis, particulate mineral matter, and inorganic matter of biogenic origin (Knezovich et al., 1987; see Chapter 1). Power and Chapman (1992) describe four main components of sediment. The largest component is interstitial water, which surrounds sediment particles and is typically 50% by volume of sediment. The second largest component consists of inorganic phases include rock fragments and minerals. The inorganic phases control the bioavailability of many divalent metals (Di Toro et al., 1990). The third component is organic matter, which occupies a smaller proportion of the sediment. Organic matter is an important component because it controls the sorption and bioavailability of many nonionic organic contaminants (Di Toro et al., 1991). The fourth component is anthropogenically derived materials including contaminants.

Grain size of sediment is generally classified in the following categories: clay (<2 μm), silt (2 to <50 μm), sand (50 to 2000 μm), and cobble and gravel (>2000 μm; U.S. Department of Agriculture classification scheme in Foth et al., 1982). The coarse fractions are composed primarily of stable, inorganic materials and are not generally associated with chemical contamination. Problem contaminants are most often associated with fine sediment because this fraction consists of particles with relatively large ratios of surface area to volume, which increase the sorptive capacity for contaminants (Power and Chapman, 1992; see Chapter 1).

GENERAL PRINCIPLES OF SEDIMENT TOXICITY TESTS

Historically, the evaluation of contaminant effects has emphasized surface waters, not sediments. Most assessments of water quality focus on water-soluble substances, and sediment is considered to be a safe repository of sorbed contaminants (Maki et al., 1984). This approach emphasizes testing organisms in the water column without considering the fate of chemicals in sediment. Investigations of sediment toxicity have been limited by a lack of understanding of the factors that control contaminant availability in sediment. A lack of standardization has also limited the use of toxicity tests in assessments of sediment contamination (Giesy and Hoke, 1989).

Sediment toxicity tests can provide rapid information on the potential toxicity of contaminants to benthic organisms. The primary incentive for sediment testing has been dredge material permitting. In 1977, the U.S. EPA and the U.S. Army Corps of Engineers (U.S. COE) recommended a series of 10-

d toxicity and bioaccumulation tests with clams, polychaetes, shrimps, and fishes to evaluate proposed discharge of dredged material into ocean waters (U.S. EPA-U.S. COE, 1977). Sediment tests have also been recommended in combination with other methods such as evaluations of benthic community structure to assess the toxicity of contaminated sediment (Chapman, 1986; Barrick et al., 1988; Canfield et al., 1994; Swartz et al., 1994). The International Joint Commission (IJC, 1988a) suggested that surveys of the benthic community should be used first to evaluate areas of the Great Lakes with suspected sediment contaminant problems. If no effects were demonstrated in an initial survey, IJC (1988a) recommended no further assessment. However, the absence of benthic organisms in sediment does not necessarily indicate that sediment toxicity caused the effect. Benthic invertebrate distributions may exhibit high spatial or temporal variability. Furthermore, short-term chemical (e.g., ammonia, oxygen) or physical (e.g., temperature, abrasion) effects can influence benthic invertebrate distributions in the absence of measurable contaminants in sediment. Therefore, the distribution of benthic invertebrates is not always indicative of the sediment toxicity (Giesy and Hoke, 1989).

The objective of a sediment toxicity test is to determine whether sediment is harmful to benthic organisms. The tests measure interactive toxic effects of complex contaminant mixtures in sediment. Furthermore, the tests do not require knowledge of specific pathways of interactions among sediment and test organisms (Kemp and Swartz, 1988). Toxicity testing of sediment can be used to (1) determine the relationship between toxic effects and bioavailability, (2) investigate interactions among contaminants, (3) determine spatial and temporal distribution of contamination, (4) evaluate hazards of dredge material, (5) rank areas for cleanup, and (6) estimate the effectiveness of remediation and management. Furthermore, sediments spiked with known concentrations of contaminants can be used to establish cause-and-effect relationships between chemicals and biological responses.

Various methods have been developed to evaluate sediment toxicity. These procedures range in complexity from short-term lethality tests that measure effects of individual contaminants on single species to long-term tests that determine the effects of chemical mixtures on the structure and function of communities. The evaluated sediment phase may include whole sediment, suspended sediment, elutriates, or sediment extracts. The amount of sediment tested can range from a few milliliters to over 800 L (Lamberson and Swartz, 1988). The test organisms include algae, macrophytes, fishes, and benthic, epibenthic, and pelagic invertebrates (Burton, 1992).

Most methods used for measuring the toxicity of contaminants in sediments rely on relatively short-term (≤10 d) exposures (U.S. EPA-U.S. COE, 1977, 1991, 1994; Shuba et al., 1978; Nebeker et al., 1984a; Dillon and Gibson, 1986). These test procedures have a number of limitations: (1) only lethal responses are measured, (2) the experimental design is usually limited to static tests, (3) the commonly tested species may not be the most sensitive and these tests do not include the most sensitive life stages, (4) elutriate and suspended sediment phases are often tested instead of the whole sediment (often referred to as the solid phase), and (5) when the whole sediment is tested, the species are often not in direct contact with the sediment (Anderson et al., 1984).

Ideally, a sediment toxicity test should be rapid, simple, and inexpensive if the objective of the study is to screen a large number of samples (Giesy and Hoke, 1989). Acutely lethal sediment can be identified with existing methods. However, concentrations of contaminants in sediments may not be lethal but may interfere with the ability of an animal to develop, grow, or reproduce (Chapman et al., 1985; Samoiloff, 1989). A better understanding of the sublethal effects of chemicals in sediment is needed to identify areas with moderate contamination and evaluate chemicals that do not elicit acutely lethal responses. Most estimations of sublethal effects of contaminated sediment have been based on 7- to 14-d exposures with the midges *Chironomus riparius* and *C. tentans*; the amphipods *Rhepoxynius abronius, Hyalella azteca*, and *Diporeia* spp. (formerly *Pontoporeia hoyi*); or the cladoceran *Daphnia magna* (Adams et al., 1985; Nebeker et al., 1984a; Swartz et al., 1985; Giesy et al., 1988; Landrum et al., 1989; ASTM, 1992, 1995a). However, these partial life cycle exposures may not always include the most sensitive life stage(s) of the test species. Testing sensitive life stages in long-term exposures may provide a more subtle measure of chemical toxicity (Scott and Redmond, 1989; Breteler et al., 1989; Ingersoll and Nelson, 1990; Kemble et al., 1994; see Chapters 1, 2, and 3).

SEDIMENT QUALITY ASSESSMENT

The U.S. EPA has authority under a variety of statutes to manage contaminated sediment (Table 1). Until recently, the U.S. EPA has not addressed sediment quality except in relation to disposal of material removed during navigational dredging. Southerland et al. (1992) outlined four goals of a U.S. EPA management strategy for contaminated sediments: (1) in-place sediment should be protected from contamination to ensure beneficial uses of surface waters, (2) protection of in-place sediment should be achieved through pollution prevention and source control, (3) in-place remediation should be limited to locations where natural recovery will not occur in an acceptable period of time, and (4) consistent methods should be used to trigger regulatory decisions.

The Clean Water Act (CWA) is the single most important law dealing with environmental quality of surface waters in the United States (Table 1; see Chapter 24). The goal of the CWA is to restore and maintain physical, chemical, and biological integrity

Table 1. Statutory needs for sediment quality assessment

Law[a]	Area of need
CERCLA	Assess need for remedial action with contaminated sediments; assess degree of cleanup required, disposition of sediments
CWA	NPDES permitting, especially under BAT (best available technology) in water quality–limited water
	Section 403(c) criteria for ocean discharges; mandatory additional requirements to protect marine environment
	Section 301(g) waivers for POTWs (publically owned treatment works) discharging to marine waters
	Section 404 permits for dredge and fill activities (administered by the Corps of Engineers)
EPA/ORD	Evaluation of sediment quality for research or screening purposes
FIFRA	Review uses of new and existing chemicals
	Pesticide labeling and registration
MPRSA	Permits for ocean dumping
NEPA	Preparation of environmental impact statements for projects with surface water discharges
TSCA	Section 5: Premanufacture notice reviews
	Section 6: Reviews for existing chemicals
RCRA	Assess suitability (and permit) on-land disposal or beneficial use of contaminated sediments considered "hazardous"

[a] CERCLA = Comprehensive Environmental Response, Compensation and Liability Act ("Superfund"); CWA = Clean Water Act; EPA/ORD = U.S. Environmental Protection Agency, Office of Research and Development; TSCA = Toxic Substances Control Act; FIFRA = Federal Insecticide, Fungicide, and Rodenticide Act; MPRSA = Marine Protection, Resources and Sanctuary Act; NEPA = National Environmental Policy Act; RCRA = Resources Conservation and Recovery Act.
Modified from Dickson et al. (1984) and Southerland et al. (1992)

of the nation's waters (Southerland et al., 1992). Federal and state monitoring programs have traditionally focused on evaluating water column problems caused by point-source dischargers. No national directive requires states to monitor sediments for contamination; however, during the next few years, the U.S. EPA is currently developing a National Sediment Inventory of contaminated sediment sites. In addition, consistent sediment testing methods have been developed by the U.S. EPA for preventing, remediating, and managing contaminated sediment (Southerland et al., 1992; U.S. EPA, 1994a,b).

Table 2 lists several approaches the U.S. EPA is currently considering for the assessment of sediment quality and the development of sediment quality criteria (SQC). These approaches include (1) reference area comparisons, (2) equilibrium partitioning, (3) tissue residues, (4) interstitial water toxicity, (5) whole-sediment toxicity and sediment-spiking tests, (6) benthic community structure, and (7) Sediment Quality Triad (see Chapman (1989) and U.S. EPA (1989, 1990a,b) for a critique of these methods). The sediment assessment approaches listed in Table 2 can be classified as numeric (e.g., equilibrium partitioning), descriptive (e.g., whole-sediment toxicity tests), or a combination of numeric and descriptive approaches (e.g., Effect Range Median; U.S. EPA, 1992). Numeric methods can be used to derive chemical-specific SQC. Descriptive methods such as toxicity tests with field-collected sediment cannot be used alone to generate numerical SQC for individual chemicals. Although each approach can be used to make site-specific decisions, no single approach can adequately address sediment quality or be used to develop numeric criteria. An integration of several methods using the weight of evidence is needed to assess the effects of contaminants associated with sediment. Hazard evaluations integrating data from laboratory exposures, chemical analyses, and benthic community assessments (the sediment quality triad) provide strong complementary evidence of the degree of pollution-induced degradation in aquatic communities (Chapman, 1986; Burton, 1991).

SEDIMENT QUALITY CRITERIA AND BIOAVAILABILITY

The U.S. EPA has published draft Sediment Quality Criteria for five nonionic organic compounds based on the equilibrium partitioning approach (Table 3) (U.S. EPA, 1991a–e). These criteria are currently under development and are cited in this chapter only to illustrate the approach. The SQC are numerical concentrations of an individual chemical intended to be predictive of biological effects on benthic organisms and are to be applicable to a range of natural sediments in lakes, streams, estuaries, and near-coast marine waters (U.S. EPA, 1991a). The SQC are to be used in a similar way as WQC (Stephan et al., 1985).

The following section summarizes the approach described in Di Toro et al. (1991) for establishing SQC using the equilibrium partitioning approach. Developing numerical SQC requires knowledge of factors controlling the bioavailability of sediment-associated chemicals. Similar concentrations of a chemical in units of mass of chemical per mass of sediment dry weight often exhibit a range in toxicity in different sediments. Effect concentrations of chemicals in sediment have been correlated with interstitial water concentrations, and effect concentra-

Table 2. Sediment quality assessment procedures

Method	Type			Approach
	Numeric	Descriptive	Combination	
Equilibrium partitioning	*			A sediment quality value for a given contaminant is determined by calculating the sediment concentration of the contaminant that corresponds to an interstitial water concentration equivalent to the U.S. EPA water quality criterion for the contaminant.
Tissue residues	*			Safe sediment concentrations of specific chemicals are established by determining the sediment chemical concentration that results in acceptable tissue residues.
Interstitial water toxicity	*			Toxicity of interstitial water is quantified and identification evaluation procedures are applied to identify and quantify chemical components responsible for sediment toxicity.
Benthic community structure		*		Environmental degradation is measured by evaluating alterations in benthic community structure.
Whole-sediment toxicity and sediment spiking	*	*	*	Test organisms are exposed to sediments that may contain known or unknown quantities of potentially toxic chemicals. At the end of a specified time period, the response of the test organisms is examined in relation to a specified end point. Dose-response relationships can be established by exposing test organisms to sediments that have been spiked with known amounts of chemicals or mixtures of chemicals.
Sediment quality triad	*	*	*	Sediment chemical contamination, sediment toxicity, and benthic community structure are measured on the same sediment sample. Correspondence between sediment chemistry, toxicity, and field effects is used to determine sediment concentrations that discriminate conditions of minimal, uncertain, and major biological effects.
Apparent effects threshold	*	*	*	The sediment concentration of a contaminant above which statistically significant biological effects (e.g., sediment toxicity) are always expected. AET values are empirically derived from paired field data for sediment chemistry and a range of biological effects indicators.

Modified from U.S. EPA (1992).

tions in interstitial water are often similar to effect concentrations in water-only exposures. Effect concentrations for nonionic organic compounds are often inversely correlated with the organic carbon concentration of the sediment. Whatever the route of exposure, these correlations of effect concentrations with interstitial water concentrations indicate predicted or measured concentrations in interstitial water and can be used to quantify the exposure concentration to an organism. Therefore, information on partitioning of chemicals between solid and liquid phases of sediment is useful for establishing SQC.

The SQC (μg contaminant/g sediment, dry weight) can be computed using the partitioning coefficient, K_p (L/g) between sediment and interstitial water:

$$SQC = K_p * WQC \tag{1}$$

where the WQC is the final chronic value calculated using the WQC Guidelines (Stephan et al., 1985).

On a sediment organic carbon bases the SQC_{oc} ($\mu g/g_{oc}$) would be

$$SQC_{oc} = K_{oc} * WQC \tag{2}$$

where K_{oc} is the partitioning coefficient of a chemical to organic carbon and is used to describe the distribution of chemical between the organic fraction of sediment and the interstitial water (Table 3). Organic carbon appears to be the predominant sorption phase for nonionic organic chemicals in naturally occurring sediments (U.S. EPA, 1991a–e).

Assumptions of the equilibrium approach for establishing SQC include the following: (1) partitioning of a chemical between sediment organic carbon, interstitial water, and the organism is at equilibrium (not kinetically limited); (2) concentrations in sediment or interstitial water can be measured or predicted using appropriate partitioning coefficients; (3) organisms receive equivalent exposure from water-only exposures or from any sediment phase that is in equilibrium; (4) for nonionic organic chemicals, effect concentrations in sediment on an organic carbon basis can be predicted using the organic carbon partitioning coefficient (K_{oc}) and effect concentrations in water; and (5) the WQC are appropriate effect concentrations (U.S. EPA, 1990a, 1991a–e).

The SQC for the nonionic organic chemicals listed in Table 3 are about 20- to 1250-fold higher in sediment (assuming a sediment with 1% organic carbon) than water-only WQC concentrations. Elevated concentrations of chemicals can accumulate in sediments and not cause a biological effect. However, a portion of the chemical is still potentially bioavailable to organisms inhabiting the sediment. Organic carbon is a predominant sorption phase for nonionic organic chemicals in naturally occurring sediments (U.S. EPA, 1991a–e). However, other sediment characteristics may also control sorption of nonionic organic chemicals (e.g., grain size, mineralogy, dissolved organic carbon, and the source of organic carbon). Current research efforts are evaluating factors controlling the bioavailability of polar organic chemicals, ionic organic chemicals, and metals in sediment (Di Toro et al., 1990, 1991; Ankley et al., 1991a,b). Additional research is needed to better evaluate the assumptions of the equilibrium approach. Furthermore, the equilibrium approach for establishing SQC has not yet been extensively field validated, was developed using acute toxicity tests with single compounds, and the uncertainty in the approach has not been adequately evaluated (U.S. EPA, 1990a, 1991a, Di Toro et al., 1991). Additional data are needed to evaluate the utility of SQC using chronic sediment toxicity tests, benthic colonization experiments, in situ exposures, and mixtures of chemicals.

REMEDIATION OF CONTAMINATED SEDIMENT

Several remediation options are available (Francinques et al., 1985; IJC, 1988b) once sediment has been identified as containing chemicals at concentrations that pose a problem (e.g., U.S. EPA-U.S. COE, 1991, 1994). The sediment can either be removed, stabilized, capped, or treated in situ, or "no action" may be taken. Removal involves dredging and disposal of sediment in open water or placing the material in upland, shoreline, or in-water confined disposal facilities. Stabilization involves injection of a grout, cement, or sealant to eliminate free water from the sediment. Capping involves placing a layer of inert material such as clay over contaminated sediment outside the navigation channels. In situ treatment includes either chemical (e.g., solvent extraction) or biological (e.g., microbial degradation) treatment of sediment. The no-action alternative leaves sediment in place undisturbed. Taking no action may be appropriate in areas where (1) contaminants do not pose an immediate problem, (2) natural processes may bury the contaminated sediment in a short time, (3) biodegradation may detoxify the contaminants, (4) remediation is too costly, or (5) remediation might cause more of a problem than doing nothing (Grigalunas and Opaluch, 1989). The remediation procedure or combination of procedures chosen is site specific and depends on ecological, chemical, physical, engineering, economic, human health, and political considerations (Lyman et al.,

Table 3. Water quality criteria (WQC) and sediment quality criteria (SQC) reported in draft EPA documents (U.S. EPA, 1991a–e).

Compound	log K_{oc} (L/kg$_{oc}$)	WQC (μg/L)	SQC$_{oc}$ (μg/g$_{oc}$)	SQC at 1% OC (μg/g)
Dieldrin				
Freshwater	5.16	0.0625	9.03[a]	0.0903
Salt water	5.16	0.1147	16.6	0.166
Endrin				
Freshwater	4.82	0.061	4.03	0.0403
Salt water	4.82	0.011	0.73	0.0073
Acenaphthene				
Freshwater	3.78	23.0	138	1.38
Salt water	3.78	40.4	243	2.43
Fluoranthene				
Freshwater	5.10	8.12	1022	10.22
Salt water	5.10	10.65	1341	13.41
Phenanthrene				
Freshwater	4.36	6.32	123	1.23
Salt water	4.36	8.26	161	1.61

[a] $SQC_{oc} = K_{oc} * WQC$
$= (10^{5.16}\ L/kg_{oc}) * (10^{-3}\ kg_{oc}/g_{oc})\ (0.0625\ \mu g\ dieldrin/L)$
$= 9.03\ \mu g\ dieldrin/g_{oc}$.

1987; Grigalunas and Opaluch, 1989). Furthermore, source control and continued monitoring must be included with any remediation effort to avoid creation of new problem areas.

METHODOLOGY

The American Society for Testing and Materials (ASTM) through a consensus balloting process has recently developed five standard guides for testing sediment toxicity: (1) ASTM 1995a: E 1706-95a "Standard guide for conducting sediment toxicity tests with freshwater invertebrates"; (2) ASTM (1992): E 1367-92 "Standard guide for conducting 10-d static sediment toxicity tests with estuarine and marine amphipods"; (3) ASTM (1994d): E 1525-94 "Standard guide for designing biological tests with sediments"; (4) ASTM (1994e): E 1611-94 "Standard guide for conducting sediment toxicity tests with marine and estuarine polychaetous annelids"; and (5) ASTM (1994a): E 1391-94 "Standard guide for collection, storage, characterization, and manipulation of sediment for toxicological testing." The following section summarizes the methods described in these five guides for conducting whole-sediment toxicity tests in the laboratory. In addition to these five approved guides, additional guides are currently being balloted by ASTM (ASTM, 1993, 1994b; Ingersoll, 1991).

ASTM initially developed guides rather than test methods for assessing the bioavailability of contaminants in sediments because most sediment testing procedures have been recently developed. A guide is defined by ASTM as "a series of options or instructions that does not recommend a specific course of action." A test method is defined by ASTM as "a definitive procedure for identification, measurement, and evaluation of one or more qualities, characteristics, or properties of a material, product, system, or service that produces a test result" (ASTM, 1986). The U.S. EPA (1994a,b) and ASTM (1995a) are currently developing test methods for selected species described in the ASTM guides.

Sediment Collection and Storage

Contaminants in field-collected sediment may include carcinogens, mutagens, and other potentially toxic compounds. Inasmuch as toxicity testing of sediments is often started before chemical analyses can be completed, contact of workers with sediment needs to be minimized by using gloves, laboratory coats, safety glasses, and respirators; manipulating sediments under a ventilated hood or in an enclosed glove box; and enclosing and ventilating the exposure system (Ingersoll and Nelson, 1990).

Sediments exhibit high spatial and temporal variability (Stemmer et al., 1990a). Replicate samples should be collected to determine variance in sediment characteristics. Sediment should be collected with as little disruption as possible; however, subsampling, compositing, or homogenization of sediment samples may be necessary for some experimental designs. Sampling may cause loss of sediment integrity, change in chemical speciation, or disruption of chemical equilibrium (ASTM, 1994a). A benthic grab or core should be used rather than a dredge to minimize disruption of the sediment sample. If the sediment is obtained with a grab, collection of sample from the upper 2 cm is preferred because this is the layer most benthic organisms inhabit (ASTM, 1994a). However, the depth of the sediment sampled depends on the objective of the study (U.S. EPA, 1994a). Exposure to direct sunlight during collection should be minimized if the sediment contains photolytic compounds (Oris and Giesy, 1985). Sediment samples should be cooled to 4°C in the field (ASTM, 1995a).

Given that the contaminants of concern and the influencing sediment characteristics are not always known *a priori*, it is desirable to hold sediments in the dark at 4°C and to start tests soon after collection from the field (U.S. EPA, 1994a). Recommended sediment holding time ranges from less than two weeks (Shuba et al., 1978; ASTM, 1992) to less than eight weeks (U.S. EPA-U.S. COE, 1994). If whole-sediment toxicity tests cannot be started within this time period, it is desirable to conduct an interstitial water toxicity test with the sediment sample soon after sediment collection. The interstitial water toxicity test should then be repeated at the beginning of the whole-sediment test in order to evaluate potential changes in toxicity with storage of sediment (ASTM, 1995a). Freezing and longer storage might further change sediment properties such as grain size or partitioning and should be avoided (ASTM, 1994a; Day et al., 1992). In general, sediment samples are not sieved to remove indigenous organisms, but large indigenous organisms and large debris can be removed physically (e.g., using forceps; U.S. EPA, 1994a). If sediments are sieved to remove indigenous organisms (Reynoldson et al., 1994), the influence of sieving on the subsequent concentrations of contaminants should be documented.

Control and Reference Sediment

Sediment tests generally include a control and a reference sediment. A control sediment is a sediment that is essentially free of contaminants, is used routinely to assess the acceptability of a test (ASTM, 1994c), and is not necessarily collected near the site of concern (U.S. EPA-U.S. COE, 1991, 1994). Any contaminants in control sediment may originate from the global spread of pollutants and does not reflect any substantial input from local or nonpoint sources (ASTM, 1994b). Comparing test sediments to control sediments is a measure of the toxicity of the test sediment beyond inevitable background contamination (ASTM, 1994b). A control sediment provides a measure of test acceptability, evidence of test organism health, and a basis for interpreting data obtained from the test sediments. A reference sediment is collected near an area of concern and is used to assess sediment conditions exclusive of the material(s) of interest (ASTM, 1994c; U.S. EPA-U.S. COE, 1991, 1994). Testing a reference sediment provides a site-specific basis for evaluating toxicity. If the geochemical characteristics of the test sediment exceed the tolerance range of the test organism, a reference or control sediment encompassing these characteristics needs to be evaluated (DeWitt et al., 1988; Ankley et al., 1994).

Laboratory-Spiked Sediment

Test sediment can be prepared by manipulating the properties of a control or reference sediment. The source of sediment toxicity and the interactive effects of contaminants can be determined by spiking a sediment with chemicals or complex waste mixtures (Lamberson and Swartz, 1992). Data from studies using sediments spiked with known concentrations of contaminants can be used to establish concentration-response and cause-effect relationships between chemicals and adverse biological responses. The influence of sediment physicochemical characteristics on chemical toxicity can also be determined with sediment-spiking studies (Adams et al., 1985).

Sediments spiked with a range of concentrations can be used to generate either LC50 data or a minimum concentration at which effects are observed (lowest observable effect concentration, LOEC; see Chapters 2 and 10 on statistical procedures). Mixing time of spiked sediment should be limited to a few hours and temperature should be kept low to minimize potential changes in the physicochemical and microbial characteristics of the sediment (ASTM, 1994a). Duration of contact between the chemical and sediment can affect partitioning and bioavailability (Word et al., 1987; Lamberson and Swartz, 1992; ASTM, 1994a). U.S. EPA (1994a) recommends that spiked sediment be aged at least one month before starting a test; however, equilibration for some chemicals may not be achieved for even longer periods. Some chemicals such as pesticides may enter sediment in a pulse; hence, equilibration of the spiked sediment may not be appropriate. See ASTM (1994a) for additional detail regarding spiking methods.

Use of organic carrier solvents should be avoided because of potential changes to the dissolved organic carbon concentration in pore water (U.S. EPA, 1994a). If a solvent other than water must be used, a solvent control sediment in addition to a negative control sediment should be tested. If statistically significant differences in the response of the test organisms are detected between the negative and solvent controls, only the solvent control can be used for determining test acceptability or for calculation of results. However, the negative control might provide additional information on the general health of the test organisms. If no statistically significant difference is detected, the data from both controls can be used for calculating results for the test (ASTM, 1995a).

Triethylene glycol is often a good organic solvent for preparing stock solutions because of its low toxicity to aquatic organisms, low volatility, and ability to dissolve many organic chemicals. Other water-miscible organic solvents (methanol, ethanol, or acetone) can be used, but these compounds might be toxic, affect organic carbon concentration, or stimulate undesirable growth of microorganisms. Use of acetone may be preferable because it is highly volatile and might leave the system more readily than methanol or ethanol (ASTM, 1992). Addition of chemicals to sediment has been done by a variety of procedures including rolling mills, feed mixers, and hand mixing (ASTM, 1994a). Care should be taken to ensure that the chemical is thoroughly and evenly distributed in the sediment. Analyses of sediment subsamples is advisable to determine the degree of mixing homogeneity (Ditsworth et al., 1990).

Sediment Characterization

Characterization of sediment should include factors known to control the availability of contaminants in sediment because bulk chemical concentrations alone cannot be used to evaluate bioavailability (Di Toro et al., 1991). These measures should always include pH, organic carbon, inorganic carbon, acid volatile sulfides, percent water, and grain size (e.g., percent sand, silt, and clay). Depending on the experimental design, other analyses might include biological oxygen demand (BOD), chemical oxygen demand (COD), dissolved organic carbon (DOC), cation exchange capacity (CEC), redox (eH), total volatile solids, total ammonia, metals, organosilicones, synthetic organic compounds, oil and grease, petroleum hydrocarbons, and chemical analysis of interstitial water (ASTM, 1995a; U.S. EPA, 1994a). Macrobenthos may also be determined by subsampling sediment collected from the field.

Experimental Procedures for Elutriate and Interstitial Water Sediment Toxicity Tests

Methods have been developed for preparing elutriates (U.S. EPA-U.S. COE, 1977) or for isolating sediment interstitial water (Ankley et al., 1990, 1991a,b) for toxicity testing. The elutriate test was developed for evaluating the potential short-term effects (hours or days) of open-water disposal of dredged material. Tests with elutriate samples measure the potential release of water-soluble constituents from sediment to the water column during disposal of dredge material (Shuba et al., 1978). The elutriate water sample is typically produced by vigorously shaking (15 to 30 min) one part sediment with four parts water. The mixture is allowed to settle and the liquid phase is centrifuged and might be filtered (U.S. EPA-U.S. COE, 1977). Toxicity testing of the elutriates with water column organisms generally indicates that little acute lethality is associated with the discharge material (Lamberson and Swartz, 1988). However, some elutriate samples are reportedly more toxic than interstitial water samples (Giesy and Hoke, 1989).

Toxicity tests of sediment interstitial water were developed for evaluating the potential in situ effects of contaminated sediment on aquatic organisms (Ankley et al., 1991a,b). For many benthic invertebrates, the toxicity and bioaccumulation of sediment-associated contaminants such as metals and nonionic organic contaminants have been correlated with concentrations of these chemicals in interstitial water (U.S. EPA, 1992). Methods for sampling interstitial water have not been standardized. Sediment interstitial water is isolated by several methods including centrifugation, squeezing, suction, and dialysis (Pittinger et al., 1988; ASTM, 1994a; Schults et al., 1992). If large volumes of water are needed, sediment centrifugation or squeezing is the most practical option. Filtration or dialysis may alter the concentration of contaminants in the interstitial water sample. Centrifugation may be adequate for isolating metals in interstitial water; however, studies with hydrophobic organic compounds should be conducted using suction or squeezing techniques that will not be as likely to affect sorption (Knezovich et al., 1987). If sediments are anoxic, sample processing needs to minimize oxidation of reduced species (limit contact of sediment with air). Interstitial water should be isolated and tested as soon as possible because chemical changes might occur when samples are stored for even short periods of time (ASTM, 1994a; Kemble et al., 1994).

Once an interstitial water or an elutriate sample has been isolated from the whole sediment, the testing procedures are similar to effluent toxicity testing procedures with nonbenthic species described in Chapters 2 and 3. Benthic species are not routinely used to test interstitial water or elutriate samples, because they may be stressed by the absence of sediment (Lamberson and Swartz, 1988). U.S. EPA (1994a) recommends placing inert materials (e.g., sand or plastic screen) in water-only tests with the *Hyalella azteca* and the midge, *Chironomus tentans*.

Experimental Procedures for Whole-Sediment Toxicity Tests

Experimental Design

The experimental design depends on the purpose of the study. Variables that need to be considered include the number and type of control and reference sediments, the number of treatments and replicates, and water quality characteristics. For instance, the purpose of the study might be to determine a specific end point such as an LC50 and may include a negative control sediment, a solvent control, and several concentrations of sediment spiked with a chemical. If the purpose of the study is to conduct a reconnaissance field survey to identify toxic sites for further investigation, experimental design might include only one sample from each site to allow maximum spatial coverage. The lack of replication usually precludes statistical comparisons, but these surveys can be used to identify toxic sites for further study. In other instances, the purpose of the study might be to conduct a quantitative sediment survey to determine statistically significant differences between effects among control, reference, and test sediments from several sites. The number of field replicates per site should be based on the need for sensitivity or power (see Chapter 10). In a quantitative survey, field replicates (separate samples from different grabs collected at the same site) would need to be taken at each site. Separate subsamples might be used to determine within-sample variability or for comparisons of test procedures (e.g., comparative sensitivity among test species), but these subsamples cannot be considered to be true field replicates for statistical comparisons among sites (ASTM, 1995a; U.S. EPA, 1994a).

Test Chambers

Test chambers are the smallest physical units without water connections and overlying water cannot flow from one chamber to another (ASTM, 1995a). Test chambers can be constructed in several ways using various materials depending on the experimental design and the contaminants of interest. Test chambers and all equipment that come in contact with sediment and overlying water should not contain substances that leach or dissolve in amounts that adversely affect the test organisms. In addition, equipment that touches sediment or water should be chosen to minimize sorption of chemical from water. Glass, type 316 stainless steel, nylon, fluorocarbon (Teflon) plastics, and high-density polyethylene or polycarbonate should be used to minimize leaching, dissolution, and sorption (ASTM, 1995a).

The interaction between sediment and overlying water in the test chambers may influence the availability of the contaminants (Gunnison et al., 1989). For example, the ratio of sediment to overlying water in the test chambers may modify bioavailability and toxic effects. Stemmer et al. (1990b) investigated the influence of sediment volume and surface area on the toxicity of selenium-spiked sediment to *Daphnia magna*. Varying surface area within a constant 1:4 ratio of sediment to water did not alter survival; however, a decrease in the sediment/water ratio (1:8) and increased sediment surface area decreased survival. These results indicate that test conditions deviating substantially from the conventional method of testing 1:4 sediment to water may affect contaminant availability and aqueous concentrations. Therefore, standardizing sediment/water ratios may be necessary to make comparisons between species and methods. Tests may also need to be conducted with a range of expected environmental conditions in water over the sediment. For example, overlying water hardness or pH may influence sediment toxicity (Anderson et al., 1984).

Test Organisms

Species commonly used for toxicity testing of whole sediment are listed in Table 4. The choice of the test organism has a major influence on the ecological

Table 4. Commonly used species for whole-sediment toxicity testing

Organism	End point[m]	Test duration (d)[n]	Habitat	Feeding habit
Freshwater				
Hyalella azteca (amphipod)[a]	S, G, R	28	Burrow, epibenthic	Deposit feeder
Diporeia sp. (amphipod)[b]	S	28	Burrow, infaunal	Deposit feeder
Chironomus riparius (midge)[a]	S, G, E	14	Tube dweller	Suspension and deposit feeder
Chironomus tentans (midge)[a]	S, G	10	Tube dweller	Suspension and deposit feeder
Hexagenia limbata (mayfly)[c]	S, G, M	10	Tube dweller	Suspension and deposit feeder
Ceriodaphnia dubia (cladoceran)[a,d]	S, R	7	Water column	Suspension feeder
Daphnia magna (cladoceran)[c]	S, G, R	10	Water column	Suspension feeder
Lumbriculus variegatus[e]	S, G, R	28	Burrow, infaunal/ epibenthic	Deposit feeder
Tubifex tubifex[f]	S	28	Burrow, infaunal/ epibenthic	Deposit feeder
Salt water				
Rhepoxynius abronius (amphipod)[g]	S	10	Burrow, infaunal	Deposit feeder, predator
Eohaustorius estaurius (amphipod)[g]	S	10	Burrow, infaunal	Deposit feeder
Ampelisca abdita (amphipod)[g]	S, G, R	20	Tube dweller	Suspension and deposit feeder
Grandidierella japonica (amphipod)[g]	S, G	10	Tube dweller	Deposit feeder
Hyalella azteca (amphipod)[h]	S, G, R	28	Burrow, epibenthic	Deposit feeder
Leptocheirus plumulosus (amphipod)[i]	S, G, R	28	Burrow, infaunal	Deposit feeder
Neanthes sp. (polychaete)[j]	S, G, R	85	Tube dweller	Deposit feeder
Capitella capitata (polychaete)[k]	S, G	35	Tube dweller	Deposit feeder
Nereis virens (polychaete)[l]	S	12	Tube dweller	Deposit feeder

[a] ASTM (1995a); Ingersoll and Nelson (1990).
[b] Landrum (1989); Landrum et al. (1989); ASTM (1995a); formerly *Pontoporeia hoyi*.
[c] Nebeker et al. (1984a); ASTM (1995a); Bahnick et al. (1981).
[d] Burton et al. (1989); ASTM (1995a).
[e] Phipps et al. (1992); ASTM (1995a).
[f] Reynoldson et al. (1991); ASTM (1995a).
[g] ASTM (1992), Swartz et al. (1985).
[h] Nebeker and Miller (1988b); U.S. EPA (1994a).
[i] ASTM (1992), Schlekat et al. (1991).
[j] Johns and Ginn (1990), Pesch (1979), ASTM (1994e).
[k] Chapman and Fink (1984).
[l] McLeese et al. (1982); ASTM (1994e).
[m] S = survival, G = growth, R = reproduction, M = molting frequency, E = adult emergence.
[n] Maximum duration of tests.

relevance, success, and interpretation of the test. Furthermore, no one species is best suited for all conditions over the wide range of sediment characteristics. The species should be selected based on sensitivity to contaminant behavior in sediment and feeding habit, ecological relevance, geographic distribution, taxonomic relation to indigenous animals, acceptability for use in toxicity assessment (e.g., a standardized method), availability, and tolerance to natural geochemical sediment characteristics (e.g., grain size; U.S. EPA, 1994a). Many species that might be appropriate for sediment testing do not meet these selection criteria because historically little emphasis has been placed on developing testing procedures for benthic organisms. Organisms can be obtained from laboratory cultures or they can be collected from natural populations. Organisms should not be collected near a disposal site, where they may have developed a resistance to contaminants.

Sensitivity of an organism is related to the degree of contact with sediment and biochemical sensitivity to a particular contaminant. Sediment-dwelling organisms can receive a dose from three primary sources: interstitial water, whole sediment, and overlying water. Feeding habit, including the type of food and feeding rate, will control the dose of contaminant from sediment (Adams, 1984). Benthic invertebrates often selectively consume particles with a higher organic carbon concentration, which, as previously discussed, may have higher contaminant concentrations. Detrital feeders may receive most of their body burden directly from sediment ingestion. In amphipods (Landrum, 1989) and clams (Boese et al., 1990) uptake by the gut can exceed uptake across the gill for certain higher K_{ow} compounds (see Chapter 16). Organisms in direct contact with sediment may also accumulate contaminants by direct adsorption to the body wall or by absorption through the integument (Knezovich et al., 1987). Further, estimates of bioavailability are more complex for epibenthic animals that inhabit both the sediment and the water column.

Even with the complexities in estimating the dose that an animal receives from sediment, the toxicity and bioaccumulation of many contaminants in sediment such as kepone, fluoranthene, organochlorines, and metals have been correlated with the concentration of these chemicals in the interstitial water (U.S. EPA, 1992; Di Toro et al., 1990, 1991). The relative importance of whole sediment and interstitial water routes of exposure depends on the test organism and the specific contaminant (Knezovich et al., 1987). Because benthic communities contain a diversity of species, many combinations of exposure routes may be important. Therefore, the behavior and feeding habits of an animal can influence its ability to accumulate contaminants from sediment and must be considered when selecting animals for sediment toxicity testing.

Relative species sensitivity frequently varies among contaminants; consequently, a battery of tests including species representing different trophic levels is needed to assess sediment quality (Craig, 1984; Williams et al., 1986a; Burton et al., 1992; Long et al., 1990; Ingersoll et al., 1990; Burton and Ingersoll, 1994). For example, Reish (1988) reported the relative toxicity of six metals (As, Cd, Cr, Cu, Hg, and Zn) to crustaceans, polychaetes, pelecypods, and fishes and concluded that no one species or group of animals was the most sensitive to all of the metals.

Various organisms and methods (Table 4) are commonly used to test the toxicity of whole sediments. The amphipod *Hyalella azteca* has been routinely used in freshwater testing of sediment. Tests with *H. azteca* generally start with animals 7 to 14 days old and can be conducted for up to 4 wk through reproductive maturation (ASTM, 1995a). End points in tests with *H. azteca* include survival, growth, and reproductive maturation. Although a direct measurement of amphipod reproduction is appealing, the quantitative isolation of young amphipods from sediment is difficult because of their small size (<1 mm). Indirect measures of reproduction, such as time to reproductive maturation or the number of young in the marsupium, are more easily quantified than the number of young produced. Moreover, the total number of young produced during the exposure not only may reflect a direct effect on reproduction but also may be affected by a reduction in adult survival (Borgmann and Munawar, 1989; Ingersoll and Nelson, 1990; Kemble et al., 1994).

Sediment toxicity tests with marine and estuarine amphipods (*Rhepoxynius abronius, Eohaustorius estuarius, Ampelisca abdita, Grandidierella japonica Leptocheirus plumulosus*; Table 4) are generally started with either immature or adult animals collected from the field and acclimated in the laboratory before the start of a test. The length of the test is generally 10 d and the monitored response includes survival and behavior (ASTM, 1992; U.S. EPA, 1994b). Although 10-d tests with either freshwater, estuarine, or marine amphipods have been routinely used to measure lethality, the duration of these exposures is often too short to determine contaminant effects on growth or reproduction (Scott and Redmond, 1989; Ingersoll and Nelson, 1990). Methods are currently being developed to conduct long-term (20 to 60 d) sediment exposures with *A. abdita, L. plumulosus* and *H. azteca* (U.S. COE, 1990; Schlekat et al., 1991; ASTM, 1992; U.S. EPA, 1994a,b).

Midges are perceived to be relatively insensitive organisms in toxicity assessments. This conclusion is based on the practice of conducting short-term tests with fourth instar larvae in water-only exposures, a procedure that may underestimate the sensitivity of midges to toxicants. The first and second instars are more sensitive to contaminants than the third or fourth instars. For example, first instar *C. tentans* larvae were 6 to 27 times more sensitive than fourth instar larvae to acute copper exposure (Nebeker et al., 1984b; Gauss et al., 1985) and first instar *C. riparius* larvae were 127 times more sensitive than second instar larvae to acute cadmium exposure (Williams et al., 1986b). In chronic tests with first instar animals, midges were often as sensitive as daphnids to inorganic and organic compounds (Ingersoll et al., 1990).

Sediment tests with the midge *Chironomus tentans* typically start with second instar to third instar larvae (10–14 d old), continue for 10 to 17 d until the fourth instar, and measure effects on larval survival or growth (ASTM, 1995a; U.S. EPA, 1994a). Exposures of *C. tentans* that start with first instar larvae or that measure adult emergence have met with only limited success (Nebeker et al., 1984b, 1988a). Whole-sediment testing procedures with the midge *Chironomus riparius* are started with 1- to 3-d-old larvae and continue through pupation and adult emergence (ASTM, 1995a). Ingersoll and Nelson (1990) observed growth of mold and bacteria on the surface of sediment in 28-d adult emergence studies with *C. riparius*. When feeding levels were reduced enough to eliminate visible mold-bacterial growth on the sediment, the survival of larvae was not affected, but the emergence of adults was delayed beyond 28 d. Because of the dependence of adult emergence on feeding and the problem of mold-bacterial growth at higher feeding levels, tests with *C. riparius* were limited to 14 d with renewal of overlying water. In 14 d at 20°C, the first instar larvae develop to the fourth instar and larval survival and growth can be monitored as toxicity end points. Methods are currently being developed to conduct long-term (20–42 d) sediment exposures with *C. tentans* (D. A. Benoit, U.S. EPA, Duluth, Minnesota, personal communication).

Sediment toxicity tests with mayflies and cladocerans are generally conducted for up to 10 d (Bahnick et al., 1981; Nebeker et al., 1984b; Burton et al., 1989; Giesy et al., 1990; ASTM, 1995a). Survival, growth, or molting frequency are the toxicity end points measured in the mayfly tests; survival, growth, or reproduction are measured in the cladoceran tests. Although cladocerans are not in direct contact with sediment, they are frequently in contact with the sediment surface and are probably exposed to both water-soluble contaminants in the overlying water and particulate-bound contaminants at the sediment surface (ASTM, 1995a). Cladocerans are also one of the more sensitive groups of organisms used in toxicity testing (Mayer and Ellersieck, 1986).

Most oligochaetes are relatively tolerant of many classes of contaminants; however, this tolerance may be a positive attribute for assessing bioaccumulation or the toxicity of severely contaminated sites (Phipps et al., 1993; U.S. EPA, 1994a). Oligochaetes have complex life cycles and, therefore, culturing requirements has prohibited their use in routine toxicity testing (Dillon and Gibson, 1986). The most frequently described sediment testing methods for oligochaetes are acute toxicity tests (Keilty et al., 1988), although Wiederholm et al. (1987) described methods for conducting 500-d oligochaete exposures that measure effects of sediment on growth and reproduction. Reynoldson et al. (1991) and ASTM (1995a) described a 28-d test starting with sexually mature *Tubifex tubifex*. In this shorter test, effects on growth and reproduction are monitored and the duration of the exposure makes the test more useful for routine assessments of sediment toxicity. Phipps et al. (1993) and U.S. EPA (1994a) outlined testing methods for *Lumbriculus variegatus* to assess lethal and sublethal toxicity and bioaccumulation of sediment contaminants in 28-d exposures.

Tests for determining sublethal sediment toxicity with larval, juvenile, or adult life stages of marine polychaetes monitor effects on survival or growth in 12-d (McLeese et al., 1982) to 85-d (Pesch, 1979) exposures. Methods have been described for evaluating effects of marine and estuarine sediments on polychaetes in 4-d lethality and 28-d growth tests (ASTM, 1994e).

Exposure Conditions

Water for culturing organisms and testing should be acceptable to the test organisms and uniform in quality. Acceptable water allows satisfactory survival, growth, and behavior of test organisms. Natural overlying water should be uncontaminated and of constant quality as specified by ASTM (1988) and U.S. EPA (1985, 1991f). For certain applications, the experimental design might require water from the same site as the sediment.

Sediment Homogenization

The day before the test starts, sediment is generally mixed in the storage container and a subsample of the whole sediment is added to each test chamber. Sediment depth in the chamber is dependent on experimental design and the test species (see previous section on test chambers). Overlying water is then gently poured along the side of the chambers to minimize the resuspension of sediment. Gentle aeration is started and the chambers are left to equilibrate overnight in a water bath (ASTM, 1995a).

Static and Water Renewal Tests

The operation of the exposure system (e.g., aeration, temperature) should be checked daily and be adjusted as necessary. The pH, alkalinity, hardness, dissolved oxygen, conductivity, and dissolved organic carbon of the overlying water samples should be measured at the beginning and end of the test and at least weekly during the test in each sediment treatment. To monitor changes in interstitial water or whole sediment during the test, separate chambers should be set up and destructively sampled during the exposure (ASTM, 1995a).

In static tests, the volume of overlying water sampled for water quality determinations should be minimized and replaced with fresh overlying water. In addition, the overlying water may have to be aerated in static tests and evaporated water should be replaced at least weekly with deionized water.

In water renewal tests (e.g., flow-through) with one to four volume additions of overlying water per day, water quality characteristics generally remain similar to the inflowing water (Ingersoll and Nelson, 1990; Ankley et al., 1993); however, in static tests, water quality may change profoundly during the exposure (Shuba et al., 1978). For example, in static whole-sediment tests, the alkalinity, hardness, and conductivity of overlying water more than doubled in several treatments during a 4-wk exposure (Ingersoll and Nelson, 1990). Concentrations of metabolic products (e.g., ammonia) may also increase during static exposures and these compounds can be directly toxic to the test organisms or may contribute synergistically to the toxicity of the contaminants in the sediment (Wuhrmann and Woker, 1948). Furthermore, changes in water quality characteristics such as hardness may influence the toxicity of many inorganic (Gauss et al., 1985) and organic (Mayer and Ellersieck, 1986) contaminants. Although contaminant concentrations are reduced in the overlying water in water renewal tests, organisms in direct contact with sediment (Table 4) generally receive a substantial proportion of a contaminant dose directly from either the whole sediment or from the interstitial water. The water delivery system used in water renewal tests should be able to deliver similar volumes of overlying water (±5%) to each test chamber. Mount and Brungs (1967) diluters have been modified for sediment testing and other diluter systems have also been used (Maki, 1977; Ingersoll and Nelson, 1990; Benoit et al., 1993; Zumwalt et al., 1994).

Conducting the Test

Before the start of a test, the animals may need to be acclimated to the appropriate overlying water, temperature, lighting, and photoperiod (ASTM, 1995a). Animals should be handled as little as possible and should be introduced into the overlying water below the air-water interface. During the test, all chambers need to be checked daily and observations can be made to assess test organism behavior such as sediment avoidance or reproductive behavior. Monitoring effects on burrowing test organisms is difficult because the animals are not normally visible during the exposure.

Survival of organisms in short-term exposures (≤10 d) is the most commonly monitored response in sediment toxicity tests (Table 4). In these short-term exposures, test organisms are generally not fed. Methods have been developed to determine sublethal effects of contaminated whole sediment on growth, maturation, reproduction, or behavior of test organisms (Scott and Redmond, 1989; Ingersoll and Nelson, 1990; Kemble et al., 1994; Reynoldson et al., 1994). In these longer exposures, supplemental food is often added to the test chambers. Without addition of food, the test organisms may starve during longer exposures. On the other hand, the addition of the food may alter the availability of the contaminants in the sediment (Wiederholm et al., 1987). Furthermore, if too much food is added to the test chamber or if the mortality of test organisms is high, mold-bacterial growth may develop on the sediment surface. If test organisms are fed in whole-sediment tests, the amount of food should be kept to a minimum. If food accumulates on the sediment or if a mold or bacterial growth is observed on the surface of the sediment, feeding should be suspended for one or more days. Detailed records of feeding rates and the appearance of the sediment surface should be made daily (ASTM, 1995a).

At the end of an exposure, test organisms are typically removed from the chambers by wet-sieving the sediment. Smaller test organisms such as *H. azteca* may be difficult to isolate from the sieved material. Ingersoll and Nelson (1990) recommend preserving (Haney and Hall, 1973) all material collected after sieving for later examination. Animals not found when the sieved material is examined are presumed to have died during the exposure. Dead animals generally decompose during the exposure or break apart during sieving (ASTM, 1995a; Tomasovic et al., 1995).

Quality Control

Before a toxicity test is conducted in a new facility, a "noncontaminant" test should be conducted in which all test chambers contain a control sediment and overlying water. This information is used to demonstrate that the facility, control sediment, water, and handling procedures provide acceptable species-specific responses. The within- and between-replicate variance should be determined and the statistical precision of the test should also be evaluated in relation to sample size (ASTM, 1995a). Performance-based criteria have been recommended for use in judging the quality of the culture and the test (U.S. EPA, 1994a). For example, different culturing procedures would be acceptable if consistent organisms are produced for testing. Performance could be evaluated using criteria such as control survival, reference toxicants, or percent lipid in the organisms.

Laboratories should also demonstrate that their personnel are able to recover an average of at least 90% of the organisms from whole sediment (U.S. EPA, 1994a). For example, test organisms could be added to control or test sediment and recovery could be evaluated after 1 h. Results of these recovery evaluations of test organisms could be used to separate effects on survival from the inability to recover small organisms such as *H. azteca* from sediment. Moreover, recovery of preserved test organisms from an externally supplied "organism-spiked" sediment could be an additional measure of quality control for sediment tests (Tomasovic et al., 1994).

Reference toxicity tests should be conducted in conjunction with sediment tests to determine possible changes in condition of the test species (Lee, 1980; U.S. EPA, 1994a). Deviations outside an established normal range indicate a change in the condition of the test organism population. Results of reference toxicant tests also enable interlaboratory comparisons of test organism sensitivity. Reference toxicant tests are most often acute lethality tests performed in the absence of sediment (U.S. EPA-U.S. COE, 1991, 1994; U.S. EPA, 1994a). Sediment spiked with a toxicant might also be included as a positive control for the sediment toxicity test. Many chemicals have been used as reference toxicants, including sodium chloride, potassium chloride, cadmium, copper, chromium, sodium lauryl sulfate, and phenol. No one reference toxicant can be used to measure the condition of test organisms in respect to another toxicant with a different mode of action (Lee, 1980). However, it is unrealistic to test more than one or two reference toxicants routinely.

INTERPRETATION AND SIGNIFICANCE OF RESULTS

Data Analysis

Chemical and biological data are compared with control or reference data to determine toxic effects, calculate an LC50, or estimate a no observed effect concentration (NOEC). All calculations should be based on measured initial or average concentration of the test material(s). The statistical comparisons and interpretation of the results have to be appropriate to the experimental design. No statistical procedure should be used without consideration of the advantages and disadvantages of various approaches and appropriate preliminary statistical tests. For further detail see Chapter 10 on Statistical Analysis.

For calculations of an LC50 or EC50, either probit, binomial, or moving-average methods should be used as appropriate (Gelber et al., 1985; Stephan, 1977; U.S. EPA, 1994a; see Chapter 2). When sediment samples are independently replicated, effects can be statistically compared with control or reference sediments by *t*-tests, analysis of variance (ANOVA), or regression analysis. An ANOVA is used to determine differences among treatments. The null hypothesis states that no differences exist between all treatments and the control (or reference) sediment. If the *F*-test is not statistically significant, it can be concluded that the observed effects were not large enough to be detected as statistically significant by the experimental design and hypothesis test used. In this instance, the NOEC would be the highest concentration tested. The lowest concentration for which there is a statistically significant difference is the lowest observed effect concentration (LOEC). All effects can be compared by mean separation procedures such as Fisher's, Dunnett's, or Williams' method (ASTM, 1995a; U.S. EPA, 1994a).

Swartz (1984) and Mac and Schmitt (1992), and U.S. EPA (1994a) discuss statistical hypothesis testing for making decisions about the discharge of contaminated sediments into the aquatic environment. If a null hypothesis of no difference in means between control and test treatments is evaluated at the 95% significance level (α = 0.05), only a 5% probability exists for rejecting a true null hypothesis (type I error; a nontoxic sediment is classified as toxic). However, the probability of rejecting a false null hypothesis (type II error; a toxic sediment is classified as nontoxic) may be as large as 95%. In this instance, normal statistical procedures allow the discharge of test sediments more toxic than the control (or reference) sediment. The environmental consequences of a type II error are potentially more serious than those of a type I error. A type II error allows unrestricted disposal of toxic sediment that could adversely affect aquatic communities or human health. A type I error requires restricted disposal of nontoxic sediment (e.g., confined disposal or capping; Mac and Schmitt, 1992). The probability of committing a type II error and discharging toxic sediment could be reduced if the null hypothesis of no difference between control and test treatment means is rejected at an α value exceeding 0.5 (e.g., the probability of a type I error is greater than 50%; Swartz, 1984). Type II error can also be reduced by increasing the power (sample size) of the test or by setting β at 0.2 or 0.1. For additional detail, see Swartz (1984), Mac and Schmitt (1992), and U.S. EPA (1994a).

Advantages and Limitations of Sediment Toxicity Tests

Advantages and limitations of sediment toxicity tests are outlined in Table 5. Because of the heterogeneity of natural sediments, extrapolation from laboratory studies to in situ conditions is difficult (Burton, 1991). Sediment collection, handling, and storage may alter bioavailability and concentration by changing the physical, chemical, or biological characteristics of the sediment. Maintaining the integrity of a field-collected sediment during removal, transport, mixing, storage, and testing is extremely difficult and may complicate the interpretation of effects. Direct comparisons of animals exposed in the laboratory and in the field are needed to verify laboratory results. Furthermore, spiked sediment may not be representative of contaminated sediment in the field. The duration and contact between the sediment and the added material may affect partitioning (Word et al., 1987). Most studies with spiked sediment have been started a few days after the chemical has been added to the sediment.

Depletion of aqueous and sediment-sorbed contaminants resulting from uptake by an organism may also influence availability. In most cases, the organism is a minor sink for contaminants relative to the sediment. However, within the burrow of an organism, sediment desorption kinetics may limit uptake

Table 5. Advantages and limitations of sediment toxicity tests

Advantages
- Provide a direct measure of benthic effects.
- Limited special equipment is required.
- Methods are rapid and inexpensive.
- Legal and scientific precedence exist for use; ASTM standards are available.
- Tests with spiked chemicals provide data on cause-effect relationships.
- Sediment toxicity tests can be applied to all chemicals of concern.
- Tests applied to field samples reflect cumulative effects of all contaminants and contaminant interactions.
- Toxicity tests are amenable to field validation.

Limitations
- Sediment collection, handling, and storage may alter bioavailability.
- Spiked sediment may not be representative of field-contaminated sediment.
- Natural geochemical characteristics of sediment may affect the response of test organisms.
- Indigenous animals may be present in field-collected sediments.
- Route of exposure may be uncertain and data generated in sediment toxicity tests may be difficult to interpret if factors controlling the bioavailability of contaminants in sediment are unknown.
- Tests applied to field samples cannot discriminate effects of individual chemicals.
- Few comparisons have been made of methods or species.
- Only a few chronic methods for measuring sublethal effects have been developed or extensively evaluated.
- Laboratory tests have inherent limitations in predicting ecological effects.
- Tests do not directly address human health effects.

rates. Within minutes to hours, a major portion of the total sorbed chemical may be inaccessible to the organisms because of depletion of available residues. The desorption of a particular compound from sediment may range from easily reversible (labile; within minutes) to irreversible (nonlabile; within days or months) (Karickhoff and Morris, 1985). Interparticle diffusion or advection and the quality and quantity of sediment organic carbon can also affect sorption kinetics. Two assumptions of sediment toxicity tests that must be better evaluated are that (1) uptake of contaminants from sediment is rate limited by the organisms and not by sediment and (2) the behavior of chemicals in field-collected and spiked sediment is similar to that in sediment in situ.

Natural geochemical properties such as sediment texture may influence the response of animals (DeWitt et al., 1988; Ankley et al., 1994). The geochemical properties of test sediment need to be within the tolerance limits of the test species. Ideally, the limits of the test species should be determined in advance; however, controls for factors including grain size and organic carbon can be evaluated if the limits are exceeded in the test sediment. The effects of sediment characteristics can also be addressed with regression equations (DeWitt et al., 1988; Ankley et al., 1994). Formulated sediment has also been used to evaluate geochemical properties of sediment on test organisms (Walsh et al., 1991; Suedel and Rodgers, 1994; U.S. EPA, 1994a). Future research is needed to determine the influence of "noncontaminant" factors such as sediment moisture, type of organic content, and interstitial water quality on the response of test animals. This information is needed to distinguish responses to contaminants from responses to natural sediment characteristics.

Indigenous animals may be present in field-collected sediments. An abundance of the same species or species taxonomically similar to the test species in the sediment sample may make interpretation of treatment effects difficult or impossible (Reynoldson et al., 1994). Previous investigators have inhibited the biological activity of sediment with heat, mercuric chloride, antibiotics, or gamma irradiation (ASTM, 1994a; Day et al., 1992). Gamma irradiation is probably the most desirable method because it would cause the least alteration in either the physical or chemical characteristics of the sediment. Further research is needed to determine the effects on contaminant bioavailability from treating sediment to remove or destroy indigenous organisms.

The route of exposure may be uncertain and data generated in sediment toxicity tests may be difficult to interpret if factors controlling the bioavailability of contaminants in sediment are unknown. Bulk sediment chemical concentrations have to be normalized to factors other than dry weight. For example, concentrations of nonionic organic compounds might be normalized to sediment organic carbon content (U.S. EPA, 1992) and certain metals normalized to acid volatile sulfides (Di Toro et al., 1990). Even with the appropriate normalizing factors, determination of toxic effects from ingestion of sediment or from dissolved chemicals in the interstitial water can still be difficult (Lamberson and Swartz, 1988).

Laboratory sediment toxicity testing with field-collected sediments are useful in estimating cumulative effects and interactions of multiple contaminants in a sample. Tests with field samples cannot usually discriminate between effects of individual chemicals. However, most sediment samples contain a complex matrix of inorganic contaminants, organic contaminants, and numerous unidentified compounds. Laboratory toxicity studies that test single chemical compounds spiked into the sediment are required to determine more directly the specific contaminants causing a toxic response (Adams et al., 1985). Sediment toxicity tests with spiked chemicals provide evidence of causal relationships and can be applied to many chemicals of concern. Sediment spiking can also be used to investigate additive, antagonistic, or synergic effects of specific contaminant mixtures in a sediment sample (Swartz et al., 1988). Confirmation of sediment toxicity or the interactive effects of sediment toxicants can also be determined using Toxicity Identification Evaluations (TIE; U.S. EPA, 1992; see Chapters 2 and 33). Once the cause of sediment toxicity has been identified, better decisions can be made regarding remediation options.

Methods that measure sublethal chronic effects are either not available or have not been routinely used to evaluate sediment toxicity (Craig, 1984; Dillon and Gibson, 1986; Ingersoll and Nelson, 1990). Most assessments of contaminated sediment rely on acute lethality testing methods (U.S. EPA-U.S. COE, 1991, 1994; Swartz et al., 1994). Acute lethality tests are useful in identifying "hot spots" of sediment contamination, but these tests cannot be used to evaluate moderately contaminated areas. Additional method development is needed for culturing and chronic sediment testing procedures for infaunal species with a variety of feeding habits including suspension and deposit feeders. Sediment quality assessments of sublethal responses of benthic animals provide data to measure long-term effects on growth and reproduction of organisms and can be used to evaluate structure and function of benthic communities in moderately contaminated areas (Scott, 1989).

Existing sediment testing methods provide a rapid and direct measure of effects of contaminants on benthic communities. Laboratory tests with field-collected sediment can also be used to determine temporal, horizontal, or vertical distribution in sediment toxicity (Swartz et al., 1984). Most tests can be completed within 2 to 4 wk and require limited specialized equipment. Legal and scientific precedence exists for toxicity tests in regulatory decision making (e.g., U.S. EPA, 1986). Furthermore, sediment toxicity tests with complex contaminant mixtures are important tools for making decisions about the extent of remedial action for contaminated aquatic sites and for evaluating the success of remediation activities.

CASE STUDIES

Effect of DDT- and Endrin-Spiked Sediment on the Amphipod *Hyalella azteca*

Sediment total organic carbon (TOC) concentration has a major influence on the bioavailability and toxicity of nonionic organic contaminants in sediment. Nebeker et al. (1989) investigated the influence of sediment organic carbon on the bioavailability of two nonionic contaminants, DDT and endrin, to *H. azteca* in 10-d static exposures. Sediments with higher concentrations of TOC have a greater capacity to sorb nonionic organic compounds, which reduces the concentration of these compounds in the interstitial water in contact with the benthic organisms. The toxicity and bioaccumulation of many contaminants associated with sediment have been correlated to interstitial water concentrations (Di Toro et al., 1991).

Toxicity tests were used by Nebeker et al. (1989) to determine the LC50 for DDT or endrin in sediments with three different TOC concentrations (Figures 1 and 2). LC50 concentrations were calculated for DDT or endrin in (1) interstitial (or pore) water, (2) whole sediment normalized to dry weight, and (3) whole sediment normalized to TOC and dry weight. All three sediments were spiked with five different concentrations of pesticide by shaking sediment in glass flasks coated with DDT or endrin for 7 d. The sediment was then transferred into glass flasks and kept on a shaker table for 6 to 8 wk until use. The amount of pesticide in the sediment that elicited the required response was determined with range-finding tests.

Tests were started with 20 juvenile animals added to each replicate 1-L beaker containing 200 ml of sediment and 800 ml of overlying water. Each beaker was aerated and received 40 mg of rabbit pellets on days 3 and 7 of the test. Five sediment concentrations and a control were tested in triplicate and survival was recorded after a 10-d exposure. A second set of 18 beakers also contained water, sediment, and test organisms. Six of these beakers were destructively sampled on days 0, 5, and 10 for chemical analyses of whole sediment and interstitial water. Interstitial water was isolated by centrifugation. For a complete description of the testing procedures see Nebeker et al. (1989).

The toxicity of DDT (μg/g dry weight of sediment) decreased with increasing sediment TOC (Figure 1). When dry weight LC50s were divided by concentrations of TOC (μg DDT/g carbon in sediment), there was still a tendency for toxicity of DDT to decrease with increasing sediment concentration (Figure 1). However, the range in toxicity was less (< twofold) when concentrations were expressed relative to organic carbon normalization compared to dry weight alone (> fourfold). Interstitial water LC50s for DDT indicated the lowest toxicity with sediment at intermediate TOC concentrations (Figure 1).

The toxicity of endrin (μg/g dry weight of sediment) did not substantially differ across the range of sediment TOC tested (Figure 2). The LC50s expressed relative to organic carbon (μg endrin/g carbon in sediment) increased with increasing sediment TOC concentrations (Figure 2). Interstitial water LC50s for endrin were similar across the range of sediment TOC tested (Figure 2).

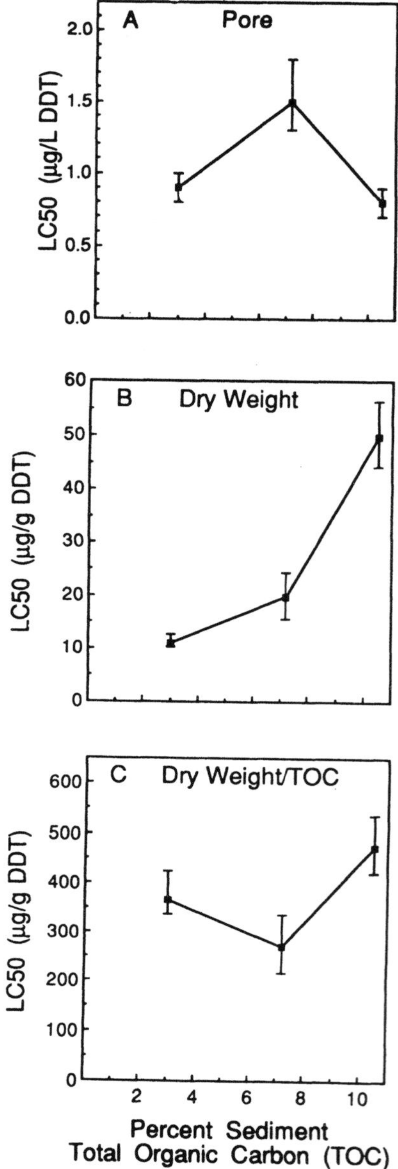

Figure 1. DDT LC50 concentrations and 95% confidence intervals at three sediment total organic carbon concentrations for *Hyalella azteca*. Modified from Nebeker et al. (1989). (Reprinted with permission from Nebeker, A. V., Schuytema, G. S., Griffis, W. L., Barbitta, J. A., Carey, L. A.: Effect of sediment organic carbon on survival of *Hyalella azteca* exposed to DDT and endrin, Environmental Toxicology and Chemistry, 8(8):705–718. Copyright 1989 SETAC).

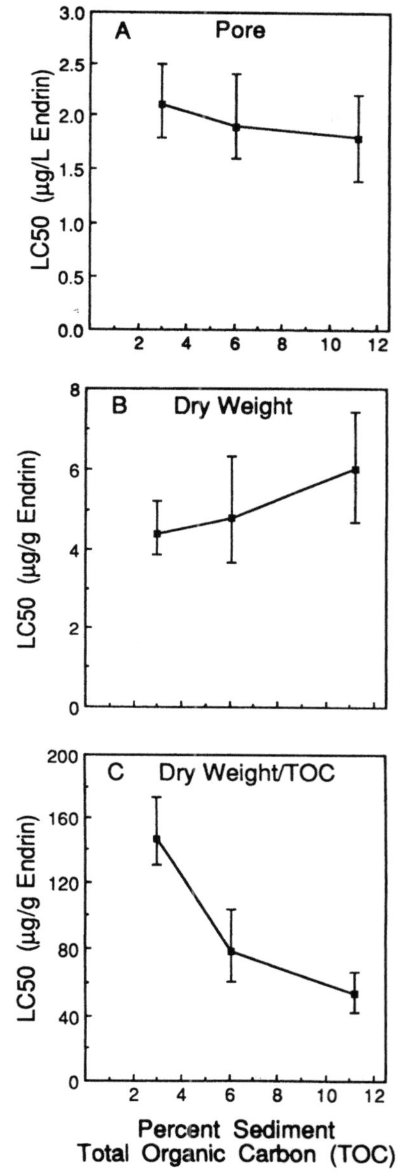

Figure 2. Endrin LC50 concentrations and 95% confidence intervals at three sediment total organic carbon concentrations for *Hyalella azteca*. Modified from Nebeker et al. (1989). (Reprinted with permission from Nebeker, A. V., Schuytema, G. S., Griffis, W. L., Barbitta, J. A., Carey, L. A.: Effect of sediment organic carbon on survival of *Hyalella azteca* exposed to DDT and endrin, *Environmental Toxicology and Chemistry*, 8(8):705–718. Copyright 1989 SETAC).

The DDT data generally support the carbon normalization theory. Increasing sediment organic carbon decreased whole-sediment toxicity and the toxicity of DDT in the interstitial water remained constant. However, the endrin data did not support the carbon normalization theory. Increasing sediment TOC did not decrease the toxicity of endrin normalized to dry weight. Moreover, when LC50s were normalized for TOC, toxicity increased with increasing TOC. Nebeker et al. (1989) suggested that physical differences between the two chemicals might account for the differences in toxicity and TOC relationships. The sorptive capacity of organic carbon (organic carbon partitioning coefficient, K_{oc}) is lower for endrin ($K_{oc} = 1.7 \times 10^3$) than for DDT ($K_{oc} = 4 \times 10^6$). Perhaps the amphipods received a dose of endrin from ingestion of sediment particles in addition to receiving a dose of endrin from the interstitial water. Compounds like endrin with a lower K_{oc} are less tightly associated with the organic carbon in the sediment and may be more rapidly released in the gut during ingestion. The different sources of organic carbon in the sediments tested by Nebeker et al. (1989) may have differentially controlled partitioning of the DDT and endrin (Word et al., 1987). The influence of sediment TOC on the availability of nonionic organic contaminants may depend on both carbon partitioning and the specific properties of the compound.

Sediment Toxicity and the Distribution of Amphipods in Commencement Bay

Swartz et al. (1982) evaluated sediment toxicity, contamination, and macrobenthic community structure in Commencement Bay, Washington. Commencement Bay has higher concentrations of organic and inorganic contaminants in water, suspended material, sediments, and biota than other urban embayments of the Puget Sound. The sound receives organic and inorganic contaminants from a variety of sources including chemical manufacturing plants, petroleum refineries, pulp mills, and marinas. The objective of the study was to determine the relationship between the survival of the amphipod *Rhepoxynius abronius* in laboratory exposures with field-collected sediment and ecological conditions in Commencement Bay. To achieve this objective, Swartz et al. (1982) determined the degree and spatial extent of sediment toxicity in Commencement Bay and correlated toxicity patterns with the distribution of amphipods in the bay.

Toxicity tests were conducted on a total of 175 sediment samples from Commencement Bay and its waterways. Control sediment was collected from Yaquina Bay, Oregon. Most samples (125) were used in a reconnaissance survey and were not replicated in order to provide maximum spatial coverage of the area. A quantitative survey included collecting five replicate sediment samples at each of 10 stations for toxicity testing and benthic community analyses. Sediment was sampled with a modified 0.1-m^2 van Veen grab sampler. After removal of a sediment subsample from the upper 2 cm of each grab for the toxicity testing, the remaining sample was sieved through a 1.0-mm screen. Animals retained on the screen were identified as to species and counted. A

2-cm-deep layer of each test sediment was placed in a 1-L beaker and covered with 800 ml of seawater (25 ppt, 15°C). *R. abronius* were collected from Yaquina Bay within 4 d of starting the static tests. Aeration was provided and the beakers were covered with watch glasses. Twenty animals were placed in each beaker and after 10 d the contents of the beakers were sieved and the number of surviving animals determined. For a complete description of the testing procedures see Swartz et al. (1982, 1985) and ASTM (1992).

Amphipod survival was higher in the offshore deeper portions of the bay and lower in some parts of the waterways (Figures 3 to 5, Tables 6 and 7). Survival was significantly different between the Yaquina Bay control and the samples from the Hylebos lower turning basin (HI), the City Waterway opposite Slip No. 1 (CII) and Browns Point (B). No significant difference in survival was observed between the control and the five stations on a transect between the deep and shallow stations in the deeper portion of Commencement Bay, and station SI in the Sitcum waterway and LI in the Blair waterway (Table 6).

In all waterways, there was a patchy distribution of sediment toxicity. Within each of the waterways and the nearshore areas of Commencement Bay, the range of survival was tremendous (Figures 4 and 5). At some of the stations in the waterway, most amphipods died and at other stations most or all amphipods survived. No single station was representative of an entire waterway. Habitat difference, sedimentation rates, proximity to contaminant sources and sinks, disruption of the bottom sediment by ship traffic, and dredging could all contribute to the patchiness of the sediment toxicity.

Amphipods were present at all but one station (CII) in the city waterway. Species richness and density of amphipods were higher in deeper portions of the bay and at Browns Point than in the waterways (Table 6). The same trends were also evident in the family Phoxocephalidae (the family of *R. abronius*). Phoxocephalid amphipods were not collected at five stations, but survival of amphipods in the laboratory

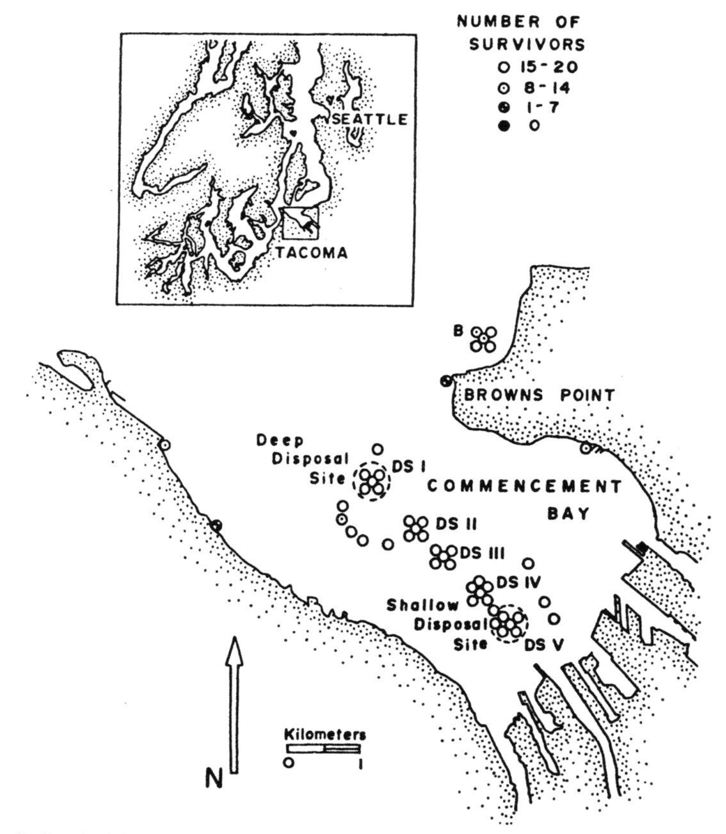

Figure 3. Survival of the the infaunal amphipod, *Rhepoxynius abronius* in sediment from Commencement Bay, Washington. (Reprinted from Marine *Pollution Bulletin* 13: Swartz, R. C., Deben, W. A., Sercu, K. A., Lamberson, J. O., Sediment toxicity and the distribution of amphipods in Commencement Bay, Washington, USA, pp. 359–364, Copyright 1982, with permission from Pergamon Press Ltd, Headington Hill Hall, Oxford OX3 OBW, UK.)

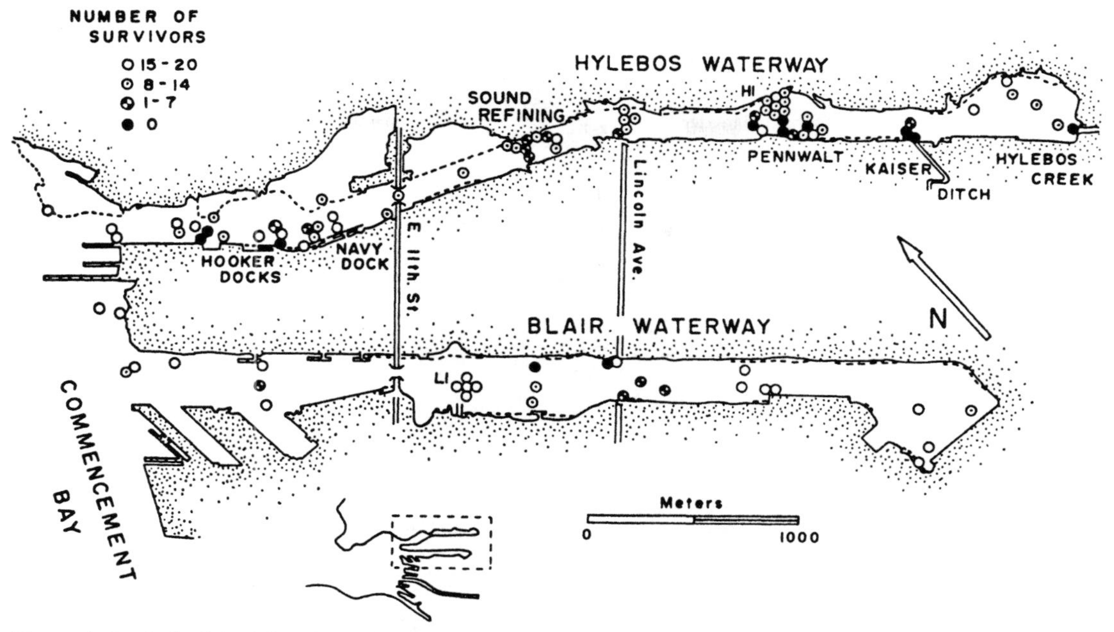

Figure 4. Survival of the infaunal amphipod, *Rhepoxynius abronius* in sediment from the Hylebos and Blair Waterways adjacent to the Commencement Bay, Washington. (Reprinted from Marine *Pollution Bulletin* 13: Swartz, R. C., Deben, W. A., Sercu, K. A, Lamberson, J. O., Sediment toxicity and the distribution of amphipods in Commencement Bay, Washington, USA, pp. 359–364, Copyright 1982, with permission from Pergamon Press Ltd, Headington Hill Hall, Oxford OX3 OBW, UK.)

was significantly reduced with exposure to sediment from only three of these five stations. This discrepancy may reflect the inability of 10-d toxicity tests to measure the sublethal effects of exposure in the field (Swartz et al., 1994). The distribution of amphipods was similar to the pattern of sediment toxicity and contamination in Commencement Bay. Although certain areas in the waterways were not toxic or contaminated, the structure of the amphipod community and the toxicity results reflected a stressed community. This relationship between laboratory and field results supports the ecological relevance of the sediment toxicity test (Swartz et al., 1994). The tests can be used to identify locations of greatest concern where more intensive assessments could be applied. However, Swartz et al. (1982) cautioned that *R. abronius* sediment toxicity survey should not be the only measurement in an assessment of contaminated sediment. The absence of significant toxicity in the laboratory does not demonstrate the absence of adverse environmental effects.

SUMMARY

Sediment toxicity tests can provide rapid information on the potential toxicity of contaminants to benthic organisms. Laboratory toxicity tests of sediment can be used to determine the relationship between toxicity and bioavailability, investigate interactions among contaminants, determine the spatial and temporal distribution of contamination, evaluate hazards of dredge material, rank areas for cleanup, estimate the effectiveness of management and remedial options, identify areas of contaminated sediment for further investigation, and establish cause-and-effect relationships between chemicals and biological responses by spiking sediments with known concentrations of contaminants. Most tests can be completed within 2 to 4 wk and require limited specialized equipment. Legal and scientific precedence exist for use of toxicity tests in regulatory decision making.

Sediment toxicity tests have limitations because collection, handling, and storage of sediment may alter bioavailability; depletion of sediment-sorbed contaminants and kinetics may reduce bioavailability; natural geochemical characteristics of sediment and indigenous animals may influence the response of test organisms; and chronic testing methods that measure sublethal effects are generally not available for evaluating contaminated sediments. Although sediment toxicity tests can be used to make site-specific decisions, no one single approach can adequately address sediment quality or be used to develop numeric criteria. An integration of several methods (Table 2) is needed to evaluate the problems resulting from contaminants associated with sediment.

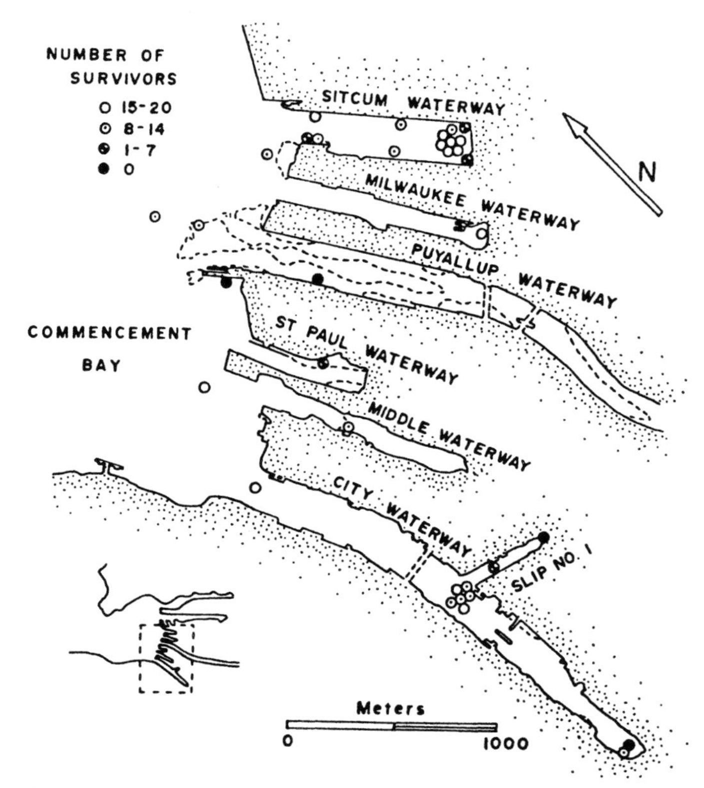

Figure 5. Survival of the infaunal amphipod, *Rhepoxynius abronius* in sediment from waterways adjacent to Commencement Bay, Washington. (Reprinted from Marine *Pollution Bulletin* 13: Swartz, R. C., Deben, W. A., Sercu, K. A., Lamberson, J. O., Sediment toxicity and the distribution of amphipods in Commencement Bay, Washington, USA, pp. 359–364, Copyright 1982, with permission from Pergamon Press Ltd, Headington Hill Hall, Oxford OX3 OBW, UK.)

Table 6. Mean survival of *Rhepoxynius abronius* and distribution of amphipods in Commencement Bay, Washington and adjacent waterways[a]

Station	Survival (%)	All amphipods		Phoxocephalid amphipods	
		S	*N*	*S*	*N*
Commencement Bay Transect					
Deep disposal site (DSI)	86 A	6.4 A	15.4 A	1.2	3.0
Transect station (DSII)	89 A	5.4 A	12.0 AB	2.4	6.2
Transect station (DSIII)	88 A	5.6 A	18.6 A	1.6	9.6
Transect station (DSIV)	89 A	2.8 AB	6.2 BC	0.8	2.2
Shallow disposal site (DSV)	86 A	2.2 B	3.0 CD	0.5	0.8
Browns Point (B)	76 B	5.0 A	11.4 AB	0	0
Hylebos Waterway (HI)	61 B	1.4 B	2.6 D	0	0
Sitcum Waterway (SI)	89 A	1.2 B	1.6 D	0	0
Blair Waterway (LI)	85 A	1.4 B	1.4 D	0	0
City Waterway (CII)	71 B	0	0	0	0

[a] Density data are mean number of individuals (*N*) and mean number of species (*S*) in the 0.1-m^2 grab sample. Survival of amphipods in the Yaquina Bay control was 91%. Means sharing a common letter in each column are not significantly different.
Modified from Swartz et al., 1982.

Table 7. Habitat distribution of samples with high toxicity (less than 40% survival of amphipods) in Commencement Bay, Washington waterways

Habitat	High-toxicity samples	Total samples	Percent of samples with high toxicity
Midchannel	7	54	13
Nearshore subtidal	15	50	30
Intertidal	12	25	48
Total	34	129	26

Modified from Swartz et al., 1982.

LITERATURE CITED

Adams, W. J.: Bioavailability of neutral lipophilic organic chemicals contained on sediments: a review. Fate and Effects of Sediment-Bound Chemicals in Aquatic Systems, edited by K. L. Dickson; A. W. Maki; and W. A. Brungs, pp. 219–244. New York: Pergamon, 1984.

Adams, W. J.; Kimerle, R. A.; and Mosher, R. G.: An approach for assessing the environmental safety of chemicals sorbed to sediments. Aquatic Toxicology and Hazard Evaluation: Seventh Symposium, edited by R. D. Cardwell; R. Purdy; and R. C. Bahner, pp. 429–453. ASTM STP 854. Philadelphia: ASTM, 1985.

American Society for Testing and Materials: Form and Style for ASTM Standards, Seventh edition. ASTM 13-000001-86. Philadelphia: ASTM, 1986.

American Society for Testing and Materials: Standard guide for conducting 10-d static sediment toxicity tests with estuarine and marine amphipods. ASTM E 1367-92. Philadelphia: ASTM, 1992.

American Society for Testing and Materials: Standard guide for determination of the bioaccumulation of sediment-associated contaminants by fish. ASTM subcommittee E47.03 on Sediment Toxicology. Philadelphia: ASTM, 1993.

American Society for Testing and Materials: Standard guide for conducting sediment toxicity tests with oysters and echinoderm embryos and larvae. ASTM subcommittee E47.03 on Sediment Toxicology. Philadelphia: ASTM, 1994b.

American Society for Testing and Materials: Terminology. ASTM E 943-94, Philadelphia: ASTM, 1994c.

American Society for Testing and Materials: Standard guide for designing biological tests with sediments. ASTM E 1525-94, Philadelphia: ASTM, 1994d.

American Society for Testing and Materials: Standard guide for conducting sediment toxicity tests with marine and estuarine polychaetous annelids. ASTM E 1611-94, Philadelphia: ASTM, 1994e.

American Society for Testing and Materials: Standard guide for collection, storage, characterization, and manipulation of sediments for toxicological testing. ASTM E 1391-94, Philadelphia: ASTM, 1994a.

American Society for Testing and Materials: Standard test methods for measuring the toxicity of sediment-associated contaminants with freshwater invertebrates. ASTM E 1706-95a, Philadelphia: ASTM 1995a.

American Society for Testing and Materials: Standard guide for determination of the bioaccumulation of sediment-associated contaminants by benthic invertebrates. ASTM E 1688-95, Philadelphia: ASTM, 1995b.

Anderson, J.; Birge, W.; Gentile, J.; Lake, J.; Rodgers, J., Jr.; and Swartz, R.: Biological effects, bioaccumulation, and ecotoxicology of sediment-associated sediments. Fate and Effects of Sediment-Bound Chemicals in Aquatic Systems, edited by K. L. Dickson; A. W. Maki; and W. A. Brungs, pp. 267–297. New York: Pergamon, 1984.

Ankley, G. T.; Benoit, D. A.; Balough, J. C.; Reynoldson, T. B.; Day, K. E.; and Hoke, R. A.: Evaluation of potential confounding factors in sediment toxicity tests with three freshwater benthic invertebrates. *Environ. Toxicol. Chem.*, 13:627–635, 1994.

Ankley, G. T.; Katko, A.; and Arthur, J. W.: Identification of ammonia as an important sediment-associated toxicant in the Lower Fox River and Green Bay, Wisconsin. *Environ. Toxicol. Chem.*, 9:313–332, 1990.

Ankley, G. T.; Schubauer-Berigan, M. K.; Dierkes, J. R.; and Lukasewycz, M. T.: Sediment toxicity identification evaluation: Phase I (Characterization), Phase II (Identification) and Phase III (Confirmation). Modifications of effluent procedures. EPA 600/6-91/007. Duluth, MN: U.S. EPA, 1991a.

Ankley, G. T.; Phipps, G. L.; Leonard, E. N.; Benoit, D. A.; Mattson, V. R.; Kosian, P. A.; Cotter, A. M.; Dierkes, J. R.; Hansen, D. J.; and Mahony, J. D.: Acid-volatile sulfide as a factor mediating cadmium and nickel bioavailability in contaminated sediment. *Environ. Toxicol. Chem.*, 10:1299–1307, 1991b.

Ankley, G. T.; Hoke, R. A.; Benoit, D. A.; Leonard, E. N.; West, C. W.; Phipps, G. L.; Mattson, V. R.; and Anderson, L. A.: Development and evaluation of test methods for benthic invertebrates and sediments: effects of flow rate and feeding on water quality and exposure conditions. *Arch. Environ. Contam. Toxicol.*, 25:12–19, 1993.

Bahnick, D. A.; Swenson, W. A.; Markee, T. P.; Call, D. J.; Anderson, C. A.; and Morris, R. T.: Development of bioassay procedures for defining pollution of harbor sediments. Part I. EPA-600/3-81-025. Duluth, MN: U.S. EPA, 1981.

Barrick, R.; Becker, S.; Brown, L.; Beller, H.; and Pastorok, R.: Sediment quality values refinement: 1988 update and evaluation of Puget Sound AET, Vol. I. Bellevue, WA: PTI, 1988.

Benoit, D. A.; Phipps, G. L.; and Ankley, G. T.: A sediment testing intermittent renewal system for the automated renewal of overlying water in toxicity tests with contaminated sediment. *Water Res.*, 27:1403–1412, 1993.

Boese, B. L.; Lee, H.; Specht, D. T.; Randall, R. C.; and Winsor, M. H.: Comparison of aqueous and solid-phase uptake for hexachlorobenzene in the tellinid clam *Macoma nasuta* (Conrad): a mass balance approach. *Environ. Toxicol. Chem.*, 9:221–231, 1990.

Borgmann, U., and Munawar, M.: A new standardized bioassay protocol using the amphipod *Hyalella azteca*. *Hydrobiologia*, 188/189:425–431, 1989.

Breteler, R. J.; Scott, K. J.; and Shepherd, S. P.: Application of a new sediment toxicity test using the marine amphipod *Ampelisca abdita* to San Francisco Bay sediments. Aquatic Toxicology and Hazard Assessment: Twelfth Symposium, edited by U. M. Cowgill, and L. R. Williams, pp. 304–314. ASTM STP 1027. Philadelphia: ASTM, 1989.

Burton, G. A., Jr.: Assessment of freshwater sediment toxicity. *Environ. Toxicol. Chem.*, 10:1585–1627, 1991.

Burton, G. A., Jr.: Sediment Toxicity Assessment. Chelsea, MI: Lewis, 1992.

Burton, G. A., Jr., and Ingersoll C. G.: Evaluating the toxicity of sediments. The ARCS Assessment Guidance Document. EPA/905-B94/002, Chicago, IL: U.S. EPA, 1994.

Burton, G. A., Jr.; Stemmer, B. L.; Winks, K. L.; Ross, P. E.; and Burnett, L. C.: A multitrophic level evaluation of sediment toxicity in Waukegan and Indiana Harbors. *Environ. Toxicol. Chem.*, 8:1057–1066, 1989.

Burton, G. A., Jr.; Nelson, M. K.; and Ingersoll, C. G.: Freshwater benthic toxicity tests. Sediment Toxicity Assessment, edited by G. A. Burton, Jr., pp. 213–240. Chelsea, MI: Lewis, 1992.

Canfield, T. J.; Kemble, N. E.; Brumbaugh, W. G.; Dwyer, F. J.; Ingersoll, C. G.; and Fairchild, J. F.: Use of benthic invertebrate community structure and the sediment quality triad to evaluate metal-contaminated sediment in the upper Clark Fork River, MT. *Environ. Toxicol. Chem.*, 13: 1999–2012, 1994.

Chapman, P. M.: Sediment quality criteria for the sediment quality triad—an example. *Environ. Toxicol. Chem.*, 5: 957–964, 1986.

Chapman, P. M.: Current approaches to developing sediment quality criteria. *Environ. Toxicol. Chem.*, 8:589–599, 1989.

Chapman, P. M., and Fink, R.: Effects of Puget Sound sediments and their elutriates on the life cycle of *Capitella capitata*. *Bull. Environ. Contam. Toxicol.*, 33:451–459, 1984.

Chapman, P. M.; Dexter, R. N.; Kocan, R. M.; and Long, E. R.: An overview of biological effects testing in Puget Sound, Washington: methods, results, and implications. Aquatic Toxicology and Hazard Assessment: Seventh Symposium, edited by R. D. Cardwell; R. Purdy; and R. C. Bahner, pp. 344–363. ASTM STP 854. Philadelphia: ASTM, 1985.

Craig, G. R.: Bioassessment of sediments: review of previous methods and recommendations for future test protocols. Mississauga, Ontario: IEC Beak Consultants Ltd, 1984.

Day, K. E.; Kirby, R. S.; and Reynoldson, T. B.: The effects of sediment manipulations on chronic sediment bioassays with three species of benthic invertebrates. Presented at the 13th Annual Society of Environmental Toxicology Meeting, Cincinnati, November 8–12, 1992.

DeWitt, T. H.; Ditsworth, G. R.; and Swartz, R. C.: Effects of natural sediment features on the phoxocephalid amphipod, *Rhepoxynius abronius*: implications for sediment toxicity bioassays. *Mar. Environ. Res.*, 25:99–124, 1988.

Dickson, K. L.; Maki, A. W.; and Brungs, W. A.: Fate and Effects of Sediment-Bound Chemicals in Aquatic Systems. New York: Pergamon, 1984.

Dillon, T. M., and Gibson, A. B.: Bioassessment methodologies for the regulatory testing of freshwater dredged material. Miscellaneous Paper EL-86-6, U.S. Army Engineer Waterways Experiment Station. Vicksburg, MS: U.S. COE, 1986.

Di Toro, D. M.; Mahony, J. H.; Hansen, D. J.; Scott, K. J.; Hicks, M. B.; Mayr, S. M.; and Redmond, M.: Toxicity of cadmium in sediments: the role of acid volatile sulfides. *Environ. Toxicol. Chem.*, 9:1487–1502, 1990.

Di Toro, D. M.; Zarba, C. S.; Hansen, D. J.; Berry, W. J.; Swartz, R. C.; Cowan, C. E.; Pavlou, S. P.; Allen, H. E.; Thomas, N. A.; and Paquin, P. R.: Technical basis for establishing sediment quality criteria for nonionic organic chemicals using equilibrium partitioning. *Environ. Toxicol. Chem.*, 10:1541–1583, 1991.

Ditsworth, G. R.; Schults, D. W.; and Jones, J. K. P.: Preparation of benthic substrates for sediment toxicity testing. *Environ. Toxicol. Chem.*, 9:1523–1529, 1990.

Foth, H. D.; Withee, L. V.; Jacobs, H. S.; and Thien, S. J.: Laboratory Manual for Introductory Soil Science, pp. 13–26. Dubuque, IA: Brown, 1982.

Francinques, N. R., Jr.; Palermo, M. R.; Lee, C. R.; and Peddicord, R. K.: Management strategy for disposal of dredged material: contaminant testing and controls. Miscellaneous Paper D-85-1, U.S. Army Engineer Waterways Experiment Station. Vicksburg, MS: U.S. COE, 1985.

Gauss, J. D.; Woods, P. E.; Winner, R. W.; and Skillings, J. H.: Acute toxicity of copper to three life stages of *Chironomus tentans* as affected by water hardness-alkalinity. *Environ. Pollut.*, (Ser A) 37:149–157, 1985.

Gelber, R. D.; Lavin, P. T.; Mehta, C. R.; and Schoenfeld, D. A.: Statistical analysis. Fundamentals of Aquatic Toxicology, edited by G. M. Rand, S. R. Petrocelli, pp. 110–123. Washington, DC: Hemisphere, 1985.

Giesy, J. P., and Hoke, R. A.: Freshwater sediment toxicity bioassessment: rationale for species selection. *J. Great Lakes Res.*, 15:539–569, 1989.

Giesy, J. P.; Graney, R. L.; Newstead, J. L.; Rosiu, C. J.; Benda, A.; Kreis, R. G., Jr.; and Horvath, F. J.: Comparison of three sediment bioassay methods using Detroit river sediments. *Environ. Toxicol. Chem.*, 7:483–498, 1988.

Giesy, J. P.; Rosiu, C. J.; Graney, R. L.; and Henry, M. G.: Benthic invertebrate bioassays with toxic sediment and pore water. *Environ. Toxicol. Chem.*, 9:233–248, 1990.

Grigalunas, T. A., and Opaluch, J. J.: Economic considerations of managing contaminated marine sediments. Contaminated Marine Sediments—Assessment and Remediation, pp. 291–310. Washington, DC: National Research Council, National Academy Press, 1989.

Gunnison, D.; Brannon, J. M.; Mills, A. L.; and Blum, L. K.: Sediment-water interactions and contaminants in Corps of Engineers reservoirs. Technical Report E-89-2, U.S. Army Engineer Waterways Experiment Station. Vicksburg, MS: U.S. COE, 1989.

Haney, J. F., and Hall, D. J.: Sugar-coated *Daphnia*: a preservation technique for Cladocera. *Limnol. Oceanogr.*, 18: 331–333, 1973.

Ingersoll, C. G.: Sediment toxicity and bioaccumulation testing. Standardization News 19:28–33, 1991.

Ingersoll, C. G., and Nelson, M. K.: Testing sediment toxicity with *Hyalella azteca* (Amphipoda) and *Chironomus riparius* (Diptera). Aquatic Toxicology and Risk Assessment: Thirteenth Volume, edited by W. G. Landis, W. H. van der Schalie, pp. 93–109. ASTM STP 1096. Philadelphia: ASTM, 1990.

Ingersoll, C. G.; Dwyer, F. J.; and May, T. W.: Toxicity of inorganic and organic selenium to *Daphnia magna* (Cladocera) and *Chironomus riparius* (Diptera). *Environ. Toxicol. Chem.*, 9:1171–1181, 1990.

International Joint Commission: Procedures for the assessment of contaminated sediment problems in the Great Lakes. Sediment Subcommittee and its Assessment Work Group to the Great Lakes Water Quality Board Report to the International Joint Commission, Dec. 1988a.

International Joint Commission: Options for the remediation of contaminated sediments in the Great Lakes: Sediment Subcommittee and its Remedial Options Work Group to the Great Lakes Water Quality Board Report to the International Joint Commission, Oct. 1988b.

Johns, D. M., and Ginn, T. C.: The relationship between juvenile growth and reproductive success in the marine polychaete, *Neanthes arenaceodentata*. Abstract from the 11th annual meeting of the Society of Environ. Toxicol. Chem., Arlington, VA, No. 214, p. 53, 1990.

Karickhoff, S. W., and Morris, K. R.: Sorption dynamics of hydrophobic pollutants in sediment suspensions. *Environ. Toxicol. Chem.*, 4:469–479, 1985.

Keilty, T. J.; White, D. S.; and Landrum, P. F.: Short-term lethality and sediment avoidance assays with endrin-contaminated sediment and two oligochaetes from Lake Michigan. *Arch. Environ. Contam. Toxicol.*, 17:95–101, 1988.

Kemble, N. E., Brumbaugh, W. G., Brunson, E. L., Dwyer, F. J., Ingersoll, C. G., Monda, D. P., and Woodward, D. F. Toxicity of metal-contaminated sediments from the upper Clark Fork River, MT to aquatic invertebrates in laboratory exposures. *Environ. Toxicol. Chem.*, 13: 1985–1997, 1994.

Kemp, P. F., and Swartz, R. C.: Acute toxicity of interstitial and particle-bound cadmium to a marine infaunal amphipod. *Mar. Environ. Res.*, 26:135–153, 1988.

Knezovich, J. P.; Harrison, F. L.; and Wilhelm, R. G.: The bioavailability of sediment-sorbed organic chemicals: a review. *Water Air Soil Pollut.*, 32:233–245, 1987.

Lamberson, J. O., and Swartz, R. C.: Use of bioassays in determining the toxicity of sediment to benthic organisms. Toxic Contaminants and Ecosystem Health: A Great Lakes Focus, edited by M. S. Evans, pp. 257–279. New York: Wiley, 1988.

Lamberson, J. O., and Swartz, R. C.: Spiked-sediment toxicity test approach. Sediment Classification Methods Compendium. U.S. EPA 823-R-92-006. Washington, DC: U.S. EPA, 1992.

Landrum, P. F.: Bioavailability and toxicokinetics of polycyclic aromatic hydrocarbons sorbed to sediments for the amphipod, *Pontoporeia hoyi*. *Environ. Sci. Technol.*, 23: 588–595, 1989.

Landrum, P. F.; Faust, W. R.; and Eadie, B. J.: Bioavailability and toxicity of a mixture of sediment associated chlorinated hydrocarbons to the amphipod, *Pontoporeia hoyi*. Aquatic Toxicology and Hazard Assessment: Twelfth Symposium, edited by U. M. Cowgill, L. R. Williams, pp. 315–329. ASTM STP 1027. Philadelphia: ASTM, 1989.

Lee, D. R.: Reference toxicants in quality control of aquatic bioassays. Aquatic Invertebrate Bioassays, edited by A. L. Buikema, Jr., J. Cairns, Jr., pp. 188–199. ASTM STP 715. Philadelphia: ASTM, 1980.

Long, E. R.; Buchman, M. F.; Bay, S. M.; Breteler, R. J.; Carr, R. S.; Chapman, P. M.; Hose, J. E.; Lissner, A. L.; Scott, J.; and Wolfe, D. A.: Comparative evaluation of five toxicity tests with sediments from San Francisco and Tomales Bay, California. *Environ. Toxicol. Chem.*, 9: 1193–1214, 1990.

Lyman, W. J.; Glazer, A. E.; Ong, J. H.; and Coons, S. F.: An overview of sediment quality in the United States. EPA-905/9-88-002. Washington, DC: U.S. EPA, 1987.

Mac, M. J., and Schmitt, C. J.: Sediment bioaccumulation testing with fish. Sediment Toxicity Assessment, edited by G. A. Burton, pp. 295–311, Boca Raton, FL: Lewis, 1992.

Mac, M. J.; Edsall, C. C.; Hesselberg, R. J.; and Sayers, R. E., Jr.: Flow through bioassay for measuring bioaccumulation of toxic substances from sediment. EPA-905/3-84-007. Washington, DC: U.S. EPA, 1984.

Maki, A. W.: Modifications of continuous flow test methods for small aquatic organisms. *Prog. Fish Cult.*, 39:172–174, 1977.

Maki, A. W.; Dickson, K. L.; and Brungs, W. A.: Introduction. Fate and Effects of Sediment-Bound Chemicals in Aquatic Systems, edited by K. L. Dickson; A. W. Maki; and W. A. Brungs, pp. xv–xxi. New York: Pergamon, 1984.

Mayer, F. L., Jr., and Ellersieck, M. R.: Manual of acute toxicity: Interpretation and data base for 410 chemicals and 66 species of freshwater animals. U.S. Fish and Wildlife Service Resource Publication 160. Washington, DC: U.S. FWS, 1986.

McLeese, D. W.; Burridge, L. E.; and Van Dinter, J.: Toxicities of five organochlorine compounds in water and sediment to *Nereis virens*. *Bull. Environ. Contam. Toxicol.*, 28:216–220, 1982.

Mount, D. I., and Brungs, W. A.: A simplified dosing apparatus for fish toxicology studies. *Water Res.*, 1:21–29, 1967.

National Research Council: Contaminated Marine Sediments—Assessment and Remediation. Washington, DC: National Research Council, National Academy Press, 1989.

Nebeker, A. V.; Cairns, M. A.; Gakstatter, J. H.; Malueg, K. W.; Schuytema, G. S.; and Krawczyk, D. F.: Biological methods for determining toxicity of contaminated freshwater sediments to invertebrates. *Environ. Toxicol. Chem.*, 3:617–630, 1984a.

Nebeker, A. V.; Cairns, M. A.; and Wise, C. M.: Relative sensitivity of *Chironomus tentans* life stages to copper. *Environ. Toxicol. Chem.*, 3:151–158, 1984b.

Nebeker, A. V.; Onjukka, S. T.; and Cairns, M. A.: Chronic effects of contaminated sediment on *Daphnia magna* and *Chironomus tentans*. *Bull. Environ. Contam. Toxicol.*, 41: 574–581, 1988a.

Nebeker, A. V., and Miller, C. E.: Use of the amphipod crustacean *Hyalella azteca* in freshwater and estuarine sediment toxicity tests. *Environ. Toxicol. Chem.*, 7:1027–1033, 1988b.

Nebeker, A. V.; Schuytema, G. S.; Griffis, W. L.; Barbitta, J. A.; and Carey, L. A.: Effect of sediment organic carbon on survival of *Hyalella azteca* exposed to DDT and endrin. *Environ. Toxicol. Chem.*, 8:705–718, 1989.

Oris, J. T., and Giesy, J. P.: The photoenhanced toxicity of anthracene to juvenile sunfish (*Lepomis* spp.). *Aquat. Toxicol.*, 6:133–146, 1985.

Pesch, C. E.: Influence of three sediment types on copper toxicity to the polychaete *Neanthes arenaceodentata*. *Mar. Biol.*, 52:237–245, 1979.

Phipps, G. L.; Ankley, G. T.; Benoit, D. A.; and Mattson, V. R.: Use of the aquatic oligochaete *Lumbriculus variegatus* for assessing the toxicity and bioaccumulation of sediment-associated contaminants. *Environ. Toxicol. Chem.*, 12:269–274, 1993.

Pittinger, C. A.; Hand, V. C.; Masters, J. A.; and Davidson, L. F.: Interstitial water sampling in ecotoxicological testing: partitioning of a cationic surfactant. Aquatic Toxicology and Hazard Assessment: Tenth Volume, edited by W. J. Adams, G. A. Chapman, W. G. Landis, pp. 138–148. ASTM STP 971. Philadelphia: ASTM, 1988.

Power, E. A., and Chapman, P. M.: Assessing sediment quality. Sediment Toxicity Assessment, edited by G. A. Burton, Jr., pp. 1–18. Chelsea, MI: Lewis, 1992.

Reish, D. J.: The use of toxicity testing in marine environmental research. Chapter 10. Marine Organisms as Indicators, edited by D. F. Soule, G. S. Kleppel, pp. 231–245. New York: Springer-Verlag, 1988.

Reynoldson, T. B.; Thompson, S. P.; and Bamsey, J. L.: A sediment bioassay using the tubificid oligochaete worm *Tubifex tubifex*. *Environ. Toxicol. Chem.*, 10:1061–1072, 1991.

Reynoldson, T. B.; Day, K. E.; Clarke, C.; and Milani, D.: Effect of indigenous animals on chronic endpoints in freshwater sediment toxicity tests. *Environ. Contam. Toxicol.*, 13:973-977, 1994.

Rodgers, J. H., Jr.; Dickson, K. L.; Saleh, F. Y.; and Staples, C. A.: Bioavailability of sediment-bound chemicals to aquatic organisms—some theory, evidence, and research needs. Fate and Effects of Sediment-Bound Chemicals in Aquatic Systems, edited by K. L. Dickson; A. W. Maki; and W. A. Brungs, pp. 245–266. New York: Pergamon, 1984.

Samoiloff, M. R.: Toxicity testing of sediments: problems, trends, and solutions. Aquatic Toxicology and Water Quality Management, edited by J. O. Nriagu, and J. S. S. Lakshminarayana, pp. 143–152. New York: Wiley, 1989.

Schlekat, C. E.; McGee, B. L.; and Reinharz, E.: Testing sediment toxicity in Chesapeake Bay using the amphipod *Leptocheirus plumulosus*: an evaluation. *Environ. Toxicol. Chem.*, 11:225–236, 1991.

Schults, D. W.; Ferraro, S. P.; Smith, L. M.; Roberts, F. A.; and Poindexter, C. K.: A comparison of methods for collecting interstitial water for trace organic compounds and metal analyses. *Water Res.*, 26:989–995, 1992.

Scott, K. J.: Effects of contaminated sediments on marine benthic biota and communities. Contaminated Marine Sediments—Assessment and Remediation, pp. 132–154. Washington, DC: National Research Council, National Academy Press, 1989.

Scott, K. J., and Redmond, M. S.: The effects of a contaminated dredged material on laboratory populations of the tubicolous amphipod, *Ampelisca abdita*. Aquatic Toxicology and Hazard Assessment: Twelfth Symposium, edited by U. M. Cowgill, and L. R. Williams, pp. 289–303. ASTM STP 1027. Philadelphia: ASTM, 1989.

Seelye, J. G., and Mac, M. J.: Bioaccumulation of toxic substances associated with dredging and dredged material disposal. A literature review. EPA-905/3-84-005. Washington, DC: U.S. EPA, 1984.

Shuba, P. J.; Tatem, H. E.; and Carroll, J. H.: Biological assessment methods to predict the impact of open-water disposal of dredged material. Technical Report D-78-5Q. Washington, DC: U.S. COE, 1978.

Southerland, E.; Kravitz, M.; and Wall, T.: Management framework for contaminated sediments (the U.S. EPA sediment management strategy). Sediment Toxicity Assessment, edited by G. A. Burton, Jr., pp. 341–370. Chelsea, MI: Lewis, 1992.

Stemmer, B. L.; Burton, G. A., Jr.; and Sasson-Brickson, G.: Effect of sediment spatial variance and collection method on cladoceran toxicity and indigenous microbial activity determinations. *Environ. Toxicol. Chem.*, 9: 1035–1044, 1990a.

Stemmer, B. L.; Burton, G. A., Jr.; and Leibfritz-Frederick, S.: Effect of sediment test variables on selenium toxicity to *Daphnia magna*. *Environ. Toxicol. Chem.*, 9:381–389, 1990b.

Stephan, C. E.: Methods for calculating an LC_{50}. Aquatic Toxicology and Hazard Evaluation, edited by F. L. Mayer, and J. L. Hamelink, pp. 65–84. ASTM STP 634. Philadelphia: ASTM, 1977.

Stephan, C. E.; Mount, D. I.; Hansen, D. J.; Gentile, J. H.; Chapman G. A.; and Brungs, W. A.: Guidelines for deriving numerical national water quality criteria for the protection of aquatic organisms and their uses. PB85-227049, National Technical Information Service. Springfield, VA: U.S. EPA, 1985.

Suedel, B. C., and Rodgers, J. H., Jr.: Development of formulated reference sediments for freshwater and estuarine sediment toxicity testing. *Environ. Toxicol. Chem.*, 13: 1163–1176, 1993.

Swartz, R. C.: Toxicological methods for determining the effects of contaminated sediment on marine organisms. Fate and Effects of Sediment-Bound Chemicals in Aquatic Systems, edited by K. L. Dickson; A. W. Maki; and W. A. Brungs, pp. 183–198. New York: Pergamon, 1984.

Swartz, R. C.: Marine sediment toxicity tests. Contaminated Marine Sediments—Assessment and Remediation, pp. 115–129. Washington, DC: National Research Council, National Academy Press, 1989.

Swartz, R. C.; Cole, F. A.; Lamberson, J. O.; Ferraro, S. P.; Schults, D. W.; DeBen, W. A.; Lee, H., II; and Ozretich, J. R.: Sediment toxicity, contamination, and amphipod abundance at a DDT and dieldrin-contaminated site in San Francisco Bay. *Environ. Toxicol. Chem.*, 13:949–962, 1994.

Swartz, R. C.; De ben, W. A.; Sercu, K. A.; and Lamberson, J. O.: Sediment toxicity and the distribution of amphipods in Commencement Bay, Washington, USA. *Mar. Pollut. Bull.*, 13:359–364, 1982.

Swartz, R. C.; Schults, D. W.; Ditsworth, G. R.; and DeBen, W. A.: Toxicity of sewage sludge to *Rhepoxynius abronius*, a marine amphipod. *Arch. Environ. Contam. Toxicol.*, 13:207–216, 1984.

Swartz, R. C.; DeBen, W. A.; Jones, J. K. P.; Lamberson, J. O.; and Cole, F. A.: Phoxocephalid amphipod bioassay for marine sediment toxicity. Aquatic Toxicology and Hazard Assessment: Seventh Symposium, edited by R. D. Cardwell; R. Purdy; and R. C. Bahner, pp. 284–307. ASTM STP 854. Philadelphia: ASTM, 1985.

Swartz, R. C.; Kemp, P. F.; Schultz, D. W.; and Lamberson, J. O.: Effects of mixtures of sediment contaminants on the marine infaunal amphipod, *Rhepoxynius abronius*. *Environ. Toxicol. Chem.*, 7:1013–1020, 1988.

Tomasovic, M.; Dwyer, F. J.; Greer, I. E.; and Ingersoll, C. G.: Recovery of *Hyalella azteca* from sediment toxicity tests. *Environ. Toxicol. Chem.*: In press, 1995.

U.S. Army Corps of Engineer Waterways Experiment Station: Chronic sublethal sediment bioassays for the regulatory evaluation of marine and estuarine dredged material: proceedings of a workshop. EEDP-01-22. Vicksburg, MS: U.S. COE, 1990.

U.S. Environmental Protection Agency: Short-term methods for estimating the chronic toxicity of effluents and receiving waters to freshwater organisms. EPA-600/4-85/014. Cincinnati, OH: U.S. EPA, 1985.

U.S. Environmental Protection Agency: Quality criteria for water. EPA-440/5-86-001. Washington, DC: U.S. EPA, 1986.

U.S. Environmental Protection Agency: Evaluation of the apparent effects threshold (AET) approach for assessing sediment quality. Report of the Sediment Criteria Subcommittee. SAB-EETFC-89-027. Washington, DC: U.S. EPA, 1989.

U.S. Environmental Protection Agency: Evaluation of the equilibrium partitioning (EqP) approach for assessing sediment quality. Report of the Sediment Criteria Subcommittee of the Ecological Processes and Effects Committee. EPA-SAB-EPEC-90-006. Washington, DC: U.S. EPA, 1990a.

U.S. Environmental Protection Agency: Evaluation of the sediment classification methods compendium. Report of the Sediment Criteria Subcommittee of the Ecological Processes and Effects Committee. EPA-SAB-EPEC-90-018. Washington, DC: U.S. EPA, 1990b.

U.S. Environmental Protection Agency: Proposed sediment quality criteria for the protection benthic organisms: dieldrin. EPA draft report dated November 1991. Washington, DC: U.S. EPA, 1991a.

U.S. Environmental Protection Agency: Proposed sediment quality criteria for the protection benthic organisms: endrin. EPA draft report dated November 1991. Washington, DC: U.S. EPA, 1991b.

U.S. Environmental Protection Agency: Proposed sediment quality criteria for the protection benthic organisms: Acenaphthene. EPA draft report dated November 1991. Washington, DC: U.S. EPA, 1991c.

U.S. Environmental Protection Agency: Proposed sediment quality criteria for the protection benthic organisms: fluoranthene. EPA draft report dated November 1991. Washington, DC: U.S. EPA, 1991d.

U.S. Environmental Protection Agency: Proposed sediment quality criteria for the protection benthic organisms: phenanthrene. EPA draft report dated November 1991: Washington, DC: U.S. EPA, 1991e.

U.S. Environmental Protection Agency: Methods for measuring the acute toxicity of effluents and receiving waters to freshwater and marine organisms. EPA-600/4-90/027. Washington, DC: U.S. EPA, 1991f.

U.S. Environmental Protection Agency: An SAB report: review of sediment criteria development methodology for non-ionic organic contaminants. Report of the Sediment Criteria Subcommittee of the Ecological Processes and Effects Committee. EPA-SAB-EPEC-93-002. Washington, DC: U.S. EPA, 1991g.

U.S. Environmental Protection Agency: Sediment Classification Methods Compendium. U.S. EPA 823-R-92-006. Washington, DC: U.S. EPA, 1992.

U.S. Environmental Protection Agency: Methods for measuring the toxicity and bioaccumulation of sediment-associated contaminants with freshwater invertebrates. EPA 600/R-94/024. Duluth, MN: Washington, DC: U.S. EPA, 1994a.

U.S. Environmental Protection Agency: Methods for measuring the toxicity and bioaccumulation of sediment-associated contaminants with marine invertebrates. EPA 600/R-94/025, Duluth, MN: Washington, DC: U.S. EPA, 1994b.

U.S. Environmental Protection Agency and U.S. Corps of Engineers: Technical Committee on Criteria for Dredged and Fill Material. Ecological Evaluation of Proposed Discharge of Dredged Material into Ocean Waters; Implementation Manual for Section 103 of Public Law 92-532. Vicksburg, MS: Environmental Effects Laboratory, U.S. Army Engineer Waterways Experimental Station, 1977.

U.S. Environmental Protection Agency and U.S. Army Corps of Engineers: Evaluation of dredge material proposed for ocean disposal. EPA-503/8-91/001. Washington, DC: U.S. EPA-U.S. COE, 1991.

U.S. Environmental Protection Agency and U.S. Army Corps of Engineers: Evaluation of dredged material proposed for discharge in inland and near coastal waters (Draft). EPA-823-B-94-002, Washington, DC. In press: U.S. EPA-U.S. COE, 1994.

Walsh, G. E., Weber, D. E., Simon, T. L., Brashers, L. K.: Toxicity tests of effluents with marsh plants in water and sediment. *Environ. Toxicol. Chem.*, 10:517–525, 1991.

Wiederholm, T.; Wiederholm, A. M.; and Goran, M.: Freshwater oligochaetes. *Water Air Soil Pollut.*, 36:131–154, 1987.

Williams, L. G.; Chapman, P. M.; and Ginn, T. C.: A comparative evaluation of marine sediment toxicity using bacterial luminescence, oyster embryo and amphipod sediment bioassays. *Mar. Environ. Res.*, 19:225–249, 1986a.

Williams, K. A.; Green, D. W. J.; Pascoe, D.; and Gower, D. E.: The acute toxicity of cadmium to different larval stages of *Chironomus riparius* (Diptera:Chironomidae) and its ecological significance for pollution regulation. *Oecologia* 70:362–366, 1986b.

Word, J. Q.; Ward, J. A.; Franklin, L. M.; Cullinan, V. I.; and Kiesser, S. L.: Evaluation of equilibrium partitioning theory for estimating the toxicity of the nonpolar organic compound DDT to the sediment dwelling amphipod *Rhepoxynius abronius*. U.S. Environmental Protection Agency Criteria and Standards Division. SCD #11, Washington, DC: U.S. EPA, 1987.

Wuhrman, D., and Woker, H.: Beitrage zur toxikologie der fische. II. Experimentelle untersuchungen uber die ammoniak- und blausaurevergiftung. *Schweiz. Z. Hydrol.*, 11:210–244, 1948.

Zumwalt, D. C.; Dwyer, F. J.; Greer, I. E.; and Ingersoll, C. G.:A water-renewal system that accurately delivers small volumes of water to exposure chamber. *Environ. Toxicol. Chem.*, 13:1311–1314, 1994.

SUPPLEMENTAL READING

Adams, W. J.; Kimerle, R. A.; and Mosher, R. G.: An approach for assessing the environmental safety of chemicals sorbed to sediments. Aquatic Toxicology and Hazard Evaluation: Seventh Symposium, edited by R. D. Cardwell; R. Purdy; and R. C. Bahner, pp. 429–453. ASTM STP 854. Philadelphia: ASTM, 1985.

Burton, G. A., Jr.: Assessment of freshwater sediment toxicity. *Environ. Toxicol. Chem.*, 10:1585–1627, 1991.

Chapman, P. M.: Sediment quality criteria for the sediment quality triad—An example. *Environ. Toxicol. Chem.*, 5: 957–964, 1986.

Di Toro, D. M.; Zarba, C. S.; Hansen, D. J.; Berry, W. J.; Swartz, R. C.; Cowan, C. E.; Pavlou, S. P.; Allen, H. E.; Thomas, N. A.; and Paquin, P. R.: Technical basis for establishing sediment quality criteria for nonionic organic chemicals using equilibrium partitioning. *Environ. Toxicol. Chem.*, 10:1541–1583, 1991.

Lyman, W. J.; Glazer, A. E.; Ong, J. H.; and Coons, S. F.: An overview of sediment quality in the United States. EPA-905/9-88-002. Washington, DC: U.S. EPA, 1987.

Nebeker, A. V.; Cairns, M. A.; Gakstatter, J. H.; Malueg, K. W.; Schuytema, G. S.; and Krawczyk, D. F.: Biological methods for determining toxicity of contaminated freshwater sediments to invertebrates. *Environ. Toxicol. Chem.*, 3:617–630, 1984a.

Swartz, R. C.; DeBen, W. A.; Jones, J. K. P.; Lamberson, J. O.; and Cole, F. A.: Phoxocephalid amphipod bioassay for marine sediment toxicity. Aquatic Toxicology and Hazard Assessment: Seventh Symposium, edited by R. D. Cardwell; R. Purdy; and R. C. Bahner, pp. 284–307. ASTM STP 854. Philadelphia: ASTM, 1985.

Chapter 9

FIELD STUDIES

R. L. Graney, J. P. Giesy, and *J. R. Clark*

INTRODUCTION

Classical evaluations of the biological effects of chemicals on aquatic organisms have been based on the results of single-species laboratory toxicity tests developed over the past decade. Whether such studies are protective of aquatic ecosystems is currently under debate (Cairns, 1989; National Academy of Sciences, 1981). Standardization of these laboratory test procedures has provided an efficient and cost-effective means for assessing the potential adverse effects of chemicals (Rand and Petrocelli, 1985) (see Chapters 2 to 8). Concentration-response information (see Chapter 1) developed under controlled laboratory conditions can be compared with predicted or measured concentrations in the environment and estimates of relative risk can be made (see Chapters 28 to 33). The accuracy of such predictions is based on the reliability of the exposure estimates and the extrapolatability of the effects data to the "real world." Laboratory toxicity tests are designed to provide consistent reproducible results; however, they are not designed to predict effects on single species that may occur under natural conditions. In the laboratory, water quality conditions (e.g., low total organic carbon, high oxygen) and organism health are optimized and the exposure regime does not represent real-world exposure conditions.

Few studies have been conducted in such a way that the results of single-species laboratory tests can be compared with the results of multispecies tests, such as those involving micro- or mesocosms, which are enclosed simulated experimental aquatic ecosystems (see Terminology) (Cairns, 1989). Responses of communities to the effects of chemicals are dependent to some extent on community structure (Burnett and Liss, 1990). The predicted threshold for effects would be expected to be different in a complex multispecies community and in a single-species toxicity test. One way to evaluate this is to compare the thresholds for no observed adverse effect levels (NOAEL values), which have been determined from the use of a battery of sensitive species from several families (calculation of final chronic values, FCVs) (La Point et al., 1989). When this was done by Giesy (1985), he concluded that there is no evidence to support the contention that multispecies toxicity tests inherently allowed more accurate or sensitive predictions about a chemical's toxicity than those made from the use of simple laboratory tests. This conclusion assumes that the exposures in the laboratory and field are the same. Simple laboratory tests generally overestimate the magnitude and duration of exposure relative to that observed in more natural field systems (Pratt et al., 1990). A primary reason to conduct outdoor field tests, especially in the evaluation of agricultural chemicals (i.e., pesticides), is to provide a more realistic exposure regime, not to provide more realistic concentration-response information than that typically generated from laboratory tests. Similarly, in analysis of the utility of microcosms and mesocosms to predict the possible adverse impacts of pesticides (and other chemicals) on aquatic life, a panel of experts concluded that the effects of nonpersistent, acutely toxic compounds in field tests can accurately be predicted from laboratory-scale studies (SETAC RESOLVE, 1992). If the exposure concentration of chemicals is rapidly changed, single-species tests may over estimate the potential effects of chemicals in the field. On the other hand, if the chemicals require metabolic activation or are only chronically toxic and need to be biomagnified to express their effects, laboratory-scale toxicity tests would tend to underestimate the potential adverse effects.

Although the effects of most chemicals are less under field conditions because of reduced persistence and bioavailability, some chemicals are actually more toxic under field conditions. One example of this phenomenon is the photoenhanced toxicity of some polycyclic aromatic hydrocarbons (PAHs) such as anthracene. Under laboratory conditions anthracene is not acutely toxic to plants or animals at concentrations less than the water solubility. However, in the presence of ecologically relevant intensities of ultraviolet (UV) light (285–400 nm) anthracene is approximately 50,000-fold more toxic than would be predicted from laboratory tests, which exclude the ecologically relevant UV light (Allred and Giesy, 1985).

It has been assumed that by testing the most sensitive species in the laboratory and applying appropriate safety factors, the structural and functional components of the ecosystem would be protected (Cairns, 1989; see Chapters 1 and 19). Although there is some evidence that this approach can work and afford adequate environmental protection (Hansen, 1989), many questions remain to be addressed through the use of complex multispecies ecosystem-level tests such as microcosms, mesocosms, or full field studies. The multitude of environmental factors, the dynamic aspects of short- and long-term biological cycles, and the critical importance of organism interactions in determining the response of ecosystems to a stressor can create considerable uncertainty in the degree to which results of single-species tests can be extrapolated to field situations. Single-species tests do not provide direct information on the indirect effects that may occur in complex multispecies ecosystems. In a general sense, the types of indirect effects that may occur are predictable. However, the details and resolution associated with the specific changes are difficult to predict with any degree of accuracy. The specific fate (see Chapter 15) and toxicological profile of the chemical will determine the level of this uncertainty and thus the ability to predict the concentrations at which adverse environmental impacts will occur.

The strategy of protecting target species or monitoring indicator species (i.e., a sensitive organism that is sampled and representative of a community or population; see Chapter 1) has been implemented because of the relative ease with which single species can be tested in the laboratory and single populations can be monitored in the environment. This has resulted in the use of quantal, acute, toxicity tests (see Chapters 1, 2, 3). Furthermore, concentration-response tests have been favored over time-to-occurrence methods because they are more rapid, less expensive, and more reproducible. This does not mean that concentration-response tests have good predictability for the effects that chemical substances will have in an aquatic ecosystem. The standard LC50 is more of a comparative tool than a predictive one. For predictive purposes, all of the biotic and abiotic accessory factors that will mediate responses of organisms in a complex ecosystem must be considered, as well as the intensity and duration of the concentration of chemical. This is not to say that the acute toxicity testing methodology has not been useful as a point of departure. In fact, for chemicals that exhibit a great degree of acute toxicity and a small acute-to-chronic ratio (i.e., little to no chronicity; see Chapter 1) the results of these types of tests can be useful in protecting ecosystems.

Many of the important biotic and abiotic interactions between a population of organisms and the surrounding environment are eliminated in single-species toxicity testing; therefore, the National Academy of Sciences (1981) has concluded that single-species tests alone are inadequate for assessing ecological field impacts. In addition, the U.S. Environmental Protection Agency (EPA) has initiated a program to develop multispecies tests. Cairns et al. (1972) state that a complex community with many interlocking recriprocal cause-effect linkages cannot be preserved by protecting one or a few target species and that monitoring the effects of stressors should consider more important aspects of ecosystem structure and function (Cairns et al., 1972; Levin et al., 1989) (see Chapters 1 and 19).

Field testing may be needed to address the uncertainty associated with extrapolating from the results of single-species toxicity tests to the real world. Field testing can serve a number of basic functions, including (1) validation of the risk assessment process (see Chapter 28), which currently is based on single-species laboratory toxicity tests; (2) evaluation of the potential fate and effects of a specific chemical; and (3) basic research on the influence of stressors on the structure and function of aquatic ecosystems and the development and validation of fate and effects models (see Chapter 18).

An important reason for conducting large-scale, field-type, multispecies tests is to investigate higher-order structural and functional properties of ecosystems. The structural properties are the absolute and relative numbers of organisms. In addition to simple enumeration of individuals and taxa, various indices and multivariate statistics can be used as more information-rich, and thus more sensitive and predictive, end points. Effects of chemical exposure investigated by comparing the absolute numbers of individuals and taxa can be as sensitive as or more sensitive than indices of species diversity and evenness.

Integrative measures of ecosystem function have been advocated as a more appropriate measure for determining an ecologically relevant no-effect concentration of a chemical (see Chapter 19). These measures, such as community primary production and respiration (i.e., community energetics) or the ratio of the two, or nutrient cycling, are seldom more sensitive than the no-effect concentrations that are calculated from measures of community structure (Giesy, 1985; La Point et al., 1989). In fact, quantified changes in structural components of ecosystems are generally more sensitive indicators of the effects of chemical stressors than are functional measures (Schindler, 1987). In an extensive review of the responses of lakes to eutrophication and acidification, functions such as primary production, organic matter decomposition, and nutrient cycling were affected only after dramatic changes in the structure of the planktonic communities had occurred (Schindler, 1987). In addition, the latter functional changes contradicted results of other studies. For example, laboratory studies generally overestimated the direct effects of acidification on aquatic systems due to rapid pH shock common in artificial systems. Small-scale studies also underestimated the "functional redundancy" of an ecosystem, which allowed ecosystem function (e.g., primary production) in an acidified lake to continue unaffected due to rapid replacement of sensitive species by more resistant species.

To have perfect predictability of the response of an ecosystem to a given intensity and duration of exposure to a chemical substance, one would literally need to know all of the information embodied in the ecosystem. This type of knowledge is impossible to obtain (Maguire et al., 1980). An ecosystem cannot be reduced into its individual components and then reassembled component by component. The study of ecosystems as such has been actively pursued because the study of all of the components has not allowed understanding of ecosystem structure and function. Most environmental assessments have been focused at either the population or the community level of organization; detailed information on the structural and functional properties of ecosystems is seldom collected. Ecosystems have emergent properties such that the whole system is greater than the sum of its parts (National Academy of Sciences, 1981). The many nonadditive interactions observed in ecological studies are evidence of this. Thus, studies of ecosystem components, such as populations or individuals, will not adequately describe or allow an understanding of the effects of perturbations on the entire system or even on individuals or populations. Individuals and populations in ecosystems do not exist in isolation from the surrounding biotic and abiotic components. They are intimately associated as they actively exchange energy and materials. These exchanges determine the fates of the individuals and populations. Thus an important reason for using multispecies tests in predicting the effects of contaminants on ecosystems, rather than using single-species toxicity tests, is the perceived need to be able to investigate ecosystem-level properties. Ecosystem science and hence aquatic toxicology are built on the previously mentioned concept that ecosystems have emergent properties such that the whole of the ecosystem is greater than the simple sum of its parts. For this reason, predicted effects on whole ecosystems from a "reduced system" (i.e., single-species tests) are simplifications and may thus be inaccurate.

Multispecies toxicity tests such as mesocosms or microcosms are simplifications or models of natural ecosystem behavior and are intermediates between the entire field ecosystem and single-species toxicity tests conducted under laboratory conditions. It should be noted that no mesocosm or microcosm could behave identically to the natural ecosystem because the natural system is usually patchy, spatially and temporally variable, and possibly chaotic (Pilson, 1990). Multispecies tests, to be useful, must be in some way structurally or functionally analogous to natural ecosystems or some identifiable subset of an ecosystem to provide information in a more efficient way. Two powerful aspects of multispecies field testing are the ability to design a system to investigate the potential effects of a chemical before release into the environment and to reduce unexplained variability within and among tests.

Field tests allow researchers to investigate interacting and/or compensatory mechanisms, which when studied individually may provide misleading or uninterpretable results. Examples of these types of interactions that would not be manifested in single-species tests are predator-prey couplets or competition that may involve compensatory growth or reproduction or density-dependent behavioral changes. One may also observe changes at suborganismal levels of organization, such as enzyme inhibition or changes in behavior to which ecological significance cannot be attributed. For example, when copper was added to limnocorrals in the CEPEX or MELIMEX studies (Grice and Reeve, 1982; McAllister et al., 1961; Menzel and Case, 1977), the population densities of phytoplankton increased. This effect was due to depressed grazing by zooplankton and would not have been predicted from single-species tests used to evaluate the effects of metals on phytoplankton.

This chapter addresses the use of field studies (i.e., natural and simulated) to assess the potential effects of chemicals on aquatic organisms. It presents an overview of the experimental design, procedures, and advantages and disadvantages of different "outdoor" aquatic field studies. Although much of the aquatic toxicology literature focuses on freshwater ecosystems, marine ecosystems will also be considered.

TERMINOLOGY

Numerous types of aquatic field studies have been conducted. The tremendous variation in the design of test systems (e.g., physical enclosures to represent lentic or lotic aquatic systems or natural ponds or streams) and objectives of these studies makes it difficult to categorize them logically for discussion purposes. As indicated by the chapter title, the studies will be broadly classified to include *natural field studies* and *simulated field studies* and will primarily address experimentation outdoors.

Natural field studies will be defined as those in which both the test system (i.e., lentic or lotic) and exposure to the stressor are "naturally" derived. Examples of such studies include a pond in an agricultural setting (i.e., adjacent to a crop on a field) in which the influence of pesticide application(s) to the surrounding watershed is evaluated, an evaluation of the impact of acid rain on acidified and nonacidified lakes, and a lotic field survey upstream and downstream of a power plant or chemical plant effluent. Also considered in this section will be "natural" systems to which chemicals are specifically applied and effects evaluated. Such studies have the advantage of studying "undisturbed" ecosystems under "controlled" application or exposure scenarios. All natural field studies are site specific and designed to evaluate the specific impact of chemicals from identified sources on specific ecosystems. These studies are not designed to be predictive. Thus extrapolation of the results to other sites or situations can be difficult. These types of studies have used a variety of field procedures for evaluating effects (i.e., community and population sampling, in situ toxicity tests involving, e.g., caged fish, functional measurements, rapid bioassessment protocols).

Simulated field studies are composed of either an isolated subsection of the natural environment or a man-made physical model of a lotic or lentic ecosystem. In these studies, the test systems are manually treated with the test chemical at predetermined test concentrations. Numerous test systems and designs utilized by various researchers (Table 1) will be discussed, including the following:

Mesocosms. Artificially constructed earthen ponds ranging in size from 0.01 to 0.1 ha and in volume from 100 to 1000 m^3. Ponds are allowed to colonize for a predetermined period and fish are stocked prior to treatment. These systems have been required by the Environmental Protection Agency for the registration of specific pesticides that present potential ecological risk (see Chapter 21).

Outdoor "Microcosms". Fabricated tanks large enough to be representative of lentic ecosystems (e.g., ponds) and not greatly influenced by ambient environmental conditions (e.g., temperature, light, wind). These are generally 2000 to 15,000 L in volume and differ from mesocosms primarily in their ability to evaluate reproductive effects for certain fish, although reproduction of some small fish (e.g., minnows) can be studied in these systems. The influence of scaling in these test systems is critical.

Although not discussed in this chapter, numerous small laboratory microcosm systems have been used to study the effects of chemicals on aquatic organisms. Such studies can be important as a first-step screening tool prior to investigating effects in the field. However, these small systems are limited in size and number of organisms that can be maintained and sampled and they are typically held under laboratory conditions of light, temperature, etc. so that extrapolation to the field is difficult. One major advantage of these smaller systems is that radiolabeling studies can be conducted economically.

Limnocorrals. Artificial enclosures placed in the pelagic region of ponds, lakes, or marine environments. These systems vary in size from as little as 2 L to over 2.5 million liters; however, most of these systems have a volume between 1000 and 10,000 L. These systems may or may not be in contact with the profundal region. Fish are generally excluded from these test systems.

Littoral enclosures. Plastic dividers are used in this test system to isolate the littoral region of ponds. These tests systems generally have a volume of 1000 to 50,000 L and a maximum depth of 2 m.

Lotic mesocosms or microcosms. Streams of various sizes have been used to evaluate the effects of chemicals. Unlike lentic systems, no standardized design has been developed for flowing-water test systems.

In general, simulated field studies can also be defined by the term "experimental ecosystem" which, by design, is a complete ecosystem approach which should be:

> physically confined; self-maintaining; multitrophic; have a duration time exceeding the generation time of the penultimate trophic level present; and of size sufficient to enable pertinent sampling and measurements to be made without seriously influencing the structure and dynamics of the system. (Lalli, 1990)

Mesocosms, microcosms, limnocorrals and littoral enclosures may be considered experimental ecosystems.

Throughout this chapter, there is reference to the concepts of *direct effects* and *indirect effects.* Numerous definitions have been provided to explain these terms (see Chapter 1), and for purposes of this chapter the definitions provided by the EPA in their Framework For Ecological Risk Assessment will be used (Risk Assessment Forum, 1992). Specifically:

Direct effect. An effect where the stressor acts on the ecological component of interest itself, not through effects on other components of the ecosystem.

Indirect effect. An effect where the stressor acts on the supporting components of the ecosystem, which in turn have an effect on the ecological component of interest.

The most obvious example of an indirect effect is depletion of the food source for one component via direct effects on another.

Natural Field Studies

Field studies can be an important component for evaluating and understanding the biological and/or ecological effects of chemicals under natural conditions. Field studies are most applicable to site-specific projects that are extremely well defined. A major strength of such studies is the incorporation of realistic exposure regimens, which directly determine the effects observed. Exposure in field studies is the result of both stochastically determined transport (e.g., drift, runoff) of the chemical to the aquatic environment and, once it has entered the aquatic system, the mitigative influences of the environment on the chemical's fate and effects. These mitigative factors directly influence the exposure regimen for aquatic organisms. The other obvious strength to such study designs is the use of naturally occurring ecosystems. These systems are the real world, results obtained are real results, and thus extrapolation to the ecosystem of concern is unnecessary.

These two strengths also reflect some of the weaknesses of natural field studies. First, the stochastic mechanisms controlling field exposure conditions

Table 1. Some representative outdoor simulated field studies

Reference	Volume (L)	Description	Studies
McLaren, 1969	—	Plastic tubes under ice	Zooplankton
Hellebust, et al., 1975	—	Metal cylinders (6)	Oil
Miller et al., 1978	—	Artic ponds	Oil
Snow and Scott, 1975	—	Lakes divided with plastic sheeting	Oil
Marshall and Mellinger, 1980	8	Carboys in Lake Michigan	Effects of cadmium
Giddings, 1980	700	Pond enclosures	Arsenic
Giddings and Eddleman, 1978			
Jackson et al., 1980	860	Cylinders (5)	Fate of metals at different pH
Weinberger et al., 1982	1,000	Lake enclosure, with sediment	1 mo, pesticides
Brunskill et al., 1980	1,000	Cylinders (8), with sediments	Effects of sucrose and arsenic
Bender and Jordan, 1970	2,000	Plastic bags in lake	Primary production
Goldman, 1962	3,000	Plastic tube	
Hellebust et al., 1975	5,700	Plastic cylinders (6)	3 mo, oil
Goldman, 1962	11,000	Plastic tubes	
Boyd, 1981	42,000	Steel tower	Effects of cold and pressure on fish turbulence
Hesslein and Quay, 1973	74,000	Triangular, plastic limnocorral	
Bower and McCorkle, 1980	77,000	Plastic cylinders	Primary production
Dutka and Kwan, 1984	81,000	Ponds (5), Lake Huron	Fate and effects of oil and dispersant
Nagy et al., 1984		Huron	
Scott and Glooschenko, 1984			
Scott et al., 1984a			
Scott et al., 1984b			
Sherry, 1984			
Kaushik et al., 1985	125,000	Enclosures (3), with sediments	Effects of pesticides
Solomon et al., 1980			
Solomon et al., 1985			
Stephenson et al., 1984			
Rudd and Turner, 1983a	130,000	Enclosures (5), some with sediment	Effects of mercury and selenium
Rudd and Turner, 1983b			
Salki et al., 1984			
Turner and Rudd, 1983			
Jackson et al., 1980	150,000	Plastic cylinders (4) in sediments	Fate and effects of metals and acidification
Marshall and Mellinger, 1980			
Muller, 1980			
Schindler et al., 1980			
Hall et al., 1970	650,000	20 ponds	Effects of nutrients and predators
Stephenson et al., 1984	1,000,000	Enclosures with sediments (3)	Effects of size
Jones, 1973	18,000,000	Rubber limnocorrals	Effects of enclosure, nutrients, and mixing
Jones, 1975		UK, Lund tubes	
Jones, 1976			
Lack and Lund, 1974			
Lund, 1972			
Lund, 1978			
Lund and Reynolds, 1982			
Reynolds, 1983			
Reynolds, 1983			
Reynolds et al., 1983			
Smyly, 1976			
Thompson et al., 1982			
Baccini et al., 1979	—	Limnocorral	Heavy metals
Imboden et al., 1979			
Lang and Lang-Dobler, 1979		MELIMEX	

Source: Gearing et al., 1989. Reprinted with permission from Springer-Verlag, New York, Inc.

make planning and organizing such studies difficult. If specific weather conditions are required to create exposure (i.e., rainfall to produce agricultural surface runoff), the success of the study can be dependent on uncontrollable factors. This problem can be avoided by treating the system directly; however, with this approach the advantage of evaluating effects generated via natural exposure is lost. In addition, it is difficult, and of questionable ethical practice, to treat directly large natural ecosystems with potentially toxic chemicals.

The site-specific nature of these studies can make extrapolation and predictability more difficult. This is especially true given the importance of the exposure regimen in determining the response of the system and the tremendous variation in exposure that may occur at different sites or regions. In agricultural systems, regional differences in climate and soil types may alter the exposure potential for a particular chemical used in each region. Because field studies usually do not include multiple treatment concentrations and are often not replicated, it is impossible to develop a concentration-response relationship. Furthermore, the inter- and intrasystem variability of natural systems makes designing statistically powerful (see Chapter 10) studies difficult. Variability among systems within the same year and within single systems among different years makes it difficult to identify any treatment-related effects with any statistical confidence. These factors interfere with the ability to extrapolate among sites. The costs of using an adequate number of test systems to address this problem can be prohibitive, although EPA has developed procedures that provide states with practical and cost-effective biological assessment tools for lotic systems (Plafkin et al., 1989). In addition, in order to account for impacts and recovery over ecologically relevant time frames, field studies must be conducted over several years, also affecting experimental costs.

Simulated Field Studies

By definition, simulated field studies (SFSs) are artificially bounded systems that are simplifications of specific ecosystems. The "art" of conducting a valid SFS is one of being able to make simplifications on bound ecosystems in such a way as to reduce unexplained variability or to isolate mechanisms without invalidating the conclusions or predictions that can be made. Thus, both spatial scaling and temporal scaling are important in determining the appropriateness of SFSs. To make ecosystem-specific predictions, SFSs must be appropriately scaled to mimic closely the system of interest. An attractive attribute of SFSs is their replicability, which is amenable to statistical testing of the impacts of manipulation. However, many variables cannot be well controlled to make these tests completely analogous to natural systems. The question of realism of SFSs is critical in evaluating their utility in risk assessment schemes. It is impossible to separate completely the discussion of realism versus replicability. Conceptually, realism and replicability (i.e., spatially and temporally) can be linked in such a way that a completely realistic (i.e., natural) system is unique in space and time and by definition is thus not replicable. However, given the objectives of most studies and the resolution in the end points being evaluated, test systems established in the same manner can be considered replicates at least for a specified time period. Divergence of the systems will occur over time, at which point classification of the systems as "replicates" may become questionable.

An underlying assumption of all SFSs is that the test system represents either a single function or a complex set of functions expressed by larger uncontrollable natural ecosystems, regardless of simplifications. Thus, all simplifications made must be evaluated to determine the validity of the results obtained. It is important that SFSs have sufficient complexity in their functional attributes so that there is competitive pressure on all major functional groups. If this is not true, erroneous results can be obtained because of biased compensatory behavior of the various populations. One can state, for example, that the primary interest of a study is the periphyton community. However, any study of the effects of a stressor on this component must be conducted in the context of all other controlling influences (e.g., macroinvertebrate consumers and microbial decomposers). Patrick (1949) found that some chemical contaminant loading could reduce the numbers of "higher" organisms such as fish while there was a concomitant increase in the number of species and density of "lower" organisms. Thus, if only the responses of the lower trophic level are monitored, an inaccurate interpretation of ecosystem health will result. To ensure balance in the test system, it may be appropriate to impose an artificial cropping on herbivores that are not preyed upon or on predator populations that are not under selective tension. Therefore, multispecies tests should contain higher trophic levels to express the controlling influence they have on lower trophic levels (Bowling et al., 1980). Although a dynamic equilibrium among all trophic levels that is reproducible in space or time would be desirable for purposes of screening toxic substances, this condition is seldom achieved in multispecies tests or natural ecosystems.

An important advantage in using SFSs is the ability to reduce specific variance components and unexplained residual variability. Natural physical and biological assemblages are often variable and undergo seasonal succession. In SFSs, controlling the variability in space and time can result in decreased similarity to natural systems. The magnitude and timing of stochastic events (i.e., high-flow conditions in streams) and the within-component variability may be the most salient features determining the observed structure and function of a community, and reducing this variability may result in artifacts in information from multispecies tests. For example, population densities, as well as diversity of stream benthic organisms, are greatly influenced by colonization from upstream areas through invertebrate drift. Thus, artificial bounding that restricts this type of variability will drastically influence the results and interpretations of a system of this type. When it is suspected that these events are important in controlling the structure or regulating the function of an ecosystem, SFSs will have low predictability for the behavior of that system.

Some of the original multispecies tests were conducted in laboratory microcosms, which were small systems enclosed in flasks or aquariums (Giesy, 1980). Although useful information was obtained from these studies, the systems were very small and thus had a number of limitations. For example, they did not allow the inclusion of larger predatory animals such as fish, nor could they be easily sampled over time without greatly affecting the balance of the system. Because laboratory microcosms had unrepresentatively large surface-to-volume ratios, there were artifacts caused by the interactions between the walls and chemical tested, as well as with the organisms in the test system. The chemical and physical properties of the laboratory microcosms were much different from those in nature and thus, although these were multispecies tests, they did not provide a great deal of reality relative to exposure conditions. In an attempt to obtain greater realism and predictive power, larger outdoor systems, open to the atmosphere and subject to ambient temperature and lighting conditions, were developed. The advantages and disadvantages of these outdoor systems for laboratory single-species and microcosm tests will be discussed later.

To summarize, numerous advantages are associated with conducting SFSs (Table 2). These multispecies tests generate additional information on a plethora of species that have not been tested or are not amenable to testing under laboratory conditions. Such tests provide a database for evaluating the relative sensitivity of laboratory test organisms and are useful in determining whether certain laboratory test species have an unusual sensitivity to a particular chemical or class of chemicals. In addition to providing direct toxicity information, the test systems are generally complex functioning ecosystems that can provide information on the indirect effects resulting from direct toxicity to one or more components of the ecosystem. For example, for phytoplankton, changes in biomass or species composition due to alterations in the zooplankton community can occur (Siefert et al., 1989; Hughes et al., 1980; Papst and Boyer, 1980; Crossland and Wolff, 1985; Crossland, 1984). In addition, more realistic exposure regimens are provided in natural or simulated field studies. Laboratory studies are normally conducted under continuous exposure regimens, which, except for more persistent chemicals or for point sources with a constant input, do not occur in the field. Mitigating water quality conditions that can alter the exposure

Table 2. Generic advantages and limitations of simulated mesocosm field studies

Advantages	Limitations
The test system contains a complete, functioning ecosystem in roughly the appropriate proportions.	Difficult to establish realistic fish community and associated predator-prey relationships.
The system can be characterized and maintained constant for a defined period of time.	Expensive to construct and maintain.
The conditions within the system can be monitored before and after the addition of the chemical of interest.	Divergence of the replicated systems occurs with time.
The system can be replicated in an appropriate statistical design.	Scaling factors must be carefully considered.
Certain components of the ecosystem can be isolated, removed, or augmented to determine the effects of contaminants on these components.	
Environmental conditions and thus exposure are more realistic than those in laboratory studies.	
The effects of chemicals on a large number of species with different sensitivities can be investigated simultaneously.	
Ecosystem-level effects and interactions among species can be investigated.	

regime are also accounted for in SFSs. For example, the suspended solids or organic carbon content of the water column can reduce the availability of hydrophobic organic compounds to organisms (Chapter 1). This reduction in bioavailability, which can result in reduced toxicity, is not generally accounted for in laboratory toxicity tests.

The disadvantages of SFSs have already been alluded to. Ecosystems are by nature variable and somewhat uncontrollable. The identification of ecosystem-level effects requires the ability to separate treatment- or chemical-induced changes in the system from natural or background variability. In addition, although field studies or SFSs provide a test system that evaluates ecosystem-level effects, interpretation of the ecological and regulatory significance of such effects can be difficult. A final limitation of field or SFSs can be cost. Single-species tests have been the mainstay of ecological risk assessment (see Chapter 28) and have provided a cost-effective means of regulating toxic chemicals. For the purpose of evaluating the potential effect of toxic chemicals, the environment may be better served by evaluating several different chemicals in single-species tests or in simple microcosm systems than by conducting a large field study for a single chemical. Resources should be allocated to maximize the amount of useful information obtained without adversely interfering with continued scientific and economic progress in society.

"NATURAL" FIELD STUDIES

Numerous "natural" field studies in both lentic and lotic systems have been conducted. It is not possible, or desirable, to attempt to discuss and review all of these studies in this chapter. In an attempt to provide an overview, generalizations will be made which may conflict with the results or conclusions of some studies. The extremely broad subject of aquatic field studies makes it impossible for such conflicts to be avoided.

Experimental Procedures and Methodology

The first task in the design and conduct of a field study is an accurate statement of the study objectives and establishment of the hypothesis to be tested (see Chapter 10). This is basic to any scientific experiment and will determine, a priori, the details of the experimental design (see Chapter 10). Unfortunately, the circumstances associated with the practical aspects of conducting field studies often limit our ability to address the stated objective rigorously. Specifically, the exact design of the study is often influenced by one's ability to identify a field site or location that meets the criteria for the study. The inability to find appropriate sites, along with the high cost associated with multiple sites or field station studies, often results in poor statistical power for field studies. This restriction can inadvertently lead to pseudoreplication within field studies or in some cases a complete lack of replication (see Chapters 10 and 19).

Hurlbert (1984) defined pseudoreplication as "the use of inferential statistics to test for treatment effects with data from experiments where either treatments are not replicated (though samples may be) or replicates are not statistically independent." Numerous instances of pseudoreplication in ecological field studies are cited by Hurlbert (1984). His paper provides a good overview of the factors that should be considered when designing a statistically valid field study. Unfortunately, few options exist when it comes to replicating large ecosystems. The simple lack of replication does not invalidate a study, nor does it mean that useful information cannot be attained from such a study. It simply means that the researchers must interpret the data within the limits of the experimental design and need to be careful if they attempt to extrapolate their results to other systems.

After the broad objectives of the study are defined, it is often necessary to initiate site selection procedures before the design can be finalized. With these objectives in mind, criteria should be developed for the selection of field study sites. For example, Fischer (1993) conducted a pond study designed to evaluate the effects of the insecticide endosulfan on aquatic habitats adjacent to tomato fields treated with the chemical. Criteria established for selecting the site included consideration of the use pattern of endosulfan, the ability of the site to produce tomatoes, cooperation of the farmer in applying the desired chemicals, use of a watershed that drained into the pond, viable invertebrate and vertebrate test populations in the ponds, and easy access to the site. Over 500 ponds were evaluated before 4 ponds were identified as meeting all of these criteria. Once the sites were identified, it was possible for the researchers to finalize the specific aspects of the study design. However, although more than a single pond was used in the study, these ponds could not be considered true statistical replicates.

Although end points to be measured are often implicit in the stated objectives (i.e., determine the influence of contaminant on fish reproduction in the receiving system), this is not always the case. Before embarking on large and costly field projects, all laboratory acute and chronic effects data should be considered in relation to end points to be measured. This can provide information on the effects most likely to be observed and can save wasted effort in areas where effects may be minor or unimportant. Broad ecological effects surveys or "shotgun" approaches for evaluating the impact of chemicals on aquatic systems are expensive and can dilute efforts so that an end point that should be of primary concern may not be adequately replicated or sampled. Studies that have the stated objective of understanding the ecological pathways through which biological effects are manifested can require extensive sampling at multiple trophic levels. Such studies are best conducted under controlled conditions such as a laboratory microcosm or in a simulated field study.

Site-specific aspects of the study design include the identification of specific collection stations (if appropriate), setting up sampling schedules, establishing the number of samples per sampling interval, identifying and validating all sampling procedures, and ensuring that an adequate quality assurance plan (see Chapter 11) is developed. Collection stations should be chosen randomly from within a defined sampling zone. For studies in lentic systems, zones can be established for the littoral and pelagic regions and locations within those zones randomly chosen for each sampling interval. For studies in lotic systems, identification of zones within a sampling station can also be based on habitat characteristics such as riffle areas or pools. Due to tidal fluxes, it can be especially difficult to identify and establish appropriate sampling stations in estuarine systems (Clark and Noles, 1993). Independent of system type, sampling schedules are typically established based on a fixed time interval; usually weekly or biweekly. Sampling may not be on a fixed interval if there is specific interest in examining effects immediately after a stochastic exposure event (e.g., random effluent discharges, pesticide runoff after specific storm events), although the logistical aspects of sampling on such a schedule can be extremely difficult.

The number of samples to be taken at each interval is dependent on the within-system variability and, if replicate test systems are available, the between-treatment variability for each end point (see Chapters 10 and 19). If the between-treatment variability can be estimated before the test, then an estimate of the number of replicates required to show a specific change can be calculated. Generally, in natural field studies, the between-treatment variability is high and quantifying statistically significant changes is difficult. At higher levels of taxonomic identification (i.e., family versus species), variability is generally less and if broad effects are observed these can often be identified statistically. The appropriate level of taxonomic identification is dependent on the objectives of the study.

The problem of field variability often raises the question of whether the expense associated with having control or untreated sites is worth the effort. Data also can be analyzed by comparing the specific end points during pretreatment with posttreatment responses. This may be confounded by seasonal changes in the structural and functional characteristics of ecosystems; however, if seasonal or historical information is available for the test system it may be possible to account for such fluctuations.

Many of the limitations and difficulties discussed above were taken into consideration when EPA developed rapid bioassessment protocols (RBPs) (Plafkin et al., 1989). These protocols incorporate different levels of biosurvey effort for evaluating the relative impairment of specific aquatic habitats. Specific survey techniques were formalized for evaluating effects on invertebrates and fish. Several community, population, and functional parameters or metrics are analyzed and collated into a single measure of biological condition. These metrics are presented relative to reference site conditions and the relative condition of the system is reported. Identification of reference sites that have similar habitat characteristics is thus a critical component of the evaluation. Obviously the relative "health" of a particular system will depend on the status of the chosen reference site or the established reference baseline. Because of the type of biological indices used, classical statistical techniques (i.e., hypothesis testing) are not used for evaluating the data. Using the RBP approach, if an area of concern is identified, additional in situ biological and/or chemical evaluations are usually required to identify the causative agent.

Field procedures for sampling aquatic habitats are fairly well established. Standard equipment is available for sampling benthic biota and emerging aquatic insects (Downing and Rigler, 1984; Plafkin et al., 1989; Weber, 1973). Artificial substrates are often used for sampling specific types of benthic macroinvertebrates and periphyton and offer an easy, reproducible sampling method. Procedures for sampling of the water column for zooplankton and phytoplankton are also well developed (Downing and Rigler, 1984). For any invertebrate sampling procedure, it must be remembered that any sampling device will, in some manner, be biased toward the collection of certain types of organisms (Peckarsky, 1984). Methods for sampling and assessing the fish population of ponds, lakes, and rivers are also extremely well established (Bagenal, 1978). Procedures for measurement of water quality parameters such as dissolved oxygen, hardness, and alkalinity are well developed and generally available (APHA et al., 1989). Procedures are also available for measurement of the composition and physical and chemical characteristics of sediment.

Case Histories

Freshwater Field Studies

Freshwater field studies have been conducted to evaluate the effects of specific stressors on various structural and functional components of both lentic and lotic ecosystems. Many of the problems and difficulties already discussed have been encountered in these studies. The specific objectives have varied from general biological surveys in contaminated and uncontaminated sites to more controlled experiments evaluating the effect of a specific chemical on a specific ecosystem component.

The effects of pesticides on aquatic ecosystems have received particular attention because these chemicals are intentionally released in the environment and their toxicological properties have been well characterized. Effects resulting from the application of diflubenzuron to citrus were evaluated by Ali et al. (1988). Many of the problems discussed above relative to natural field studies were realized in this study. No statistically significant effects of the chemical were observed; however the data were variable and the potential for exposure was not well documented. In a similar study, the effects of phenthoate on fish after application to citrus were evaluated (Nigg et al., 1984). In this study, mortality of caged fish was observed and related directly to aerial drift onto the pond surface after application to the citrus. The study by Fischer (1993) previously discussed provides another good example of the types of natural field studies conducted for pesticides. In this study, separating the potential effects of the chemical from variability within and among the systems was difficult. The effects of aerial application of cypermethrin on aquatic organisms in drainage ditches adjacent to application sites were evaluated (Shires and Bennett, 1985). Although residues in the ditches were noted, no significant biological effects were observed. Aquatic field studies were discussed in a comprehensive technical review on the impacts of pesticides on wildlife in the Prairie Pothole Region of Canada (Sheehan et al., 1987). Studies reviewed were useful in demonstrating the direct and indirect effects of pesticides on aquatic biota in the region. See Chapter 30 for more information.

As discussed previously, in non-pesticide-related studies, Schindler et al. (1985) conducted the largest and most extensive series of investigations on the effects of acidification on the structural and functional properties of lake ecosystems. In these studies, entire lakes were acidified and effects on fish, invertebrate, phytoplankton, and water chemistry were measured. Specific effects noted included declines in sensitive invertebrate species, shifts in the size and species composition of invertebrates, and alterations in fish recruitment. No effects on primary productivity were observed.

The effects of effluents on receiving aquatic systems have also been investigated. Macroinvertebrate communities located near a sewage outfall were evaluated by the classical survey approach of establishing sampling stations at different distances from the point-source discharge (Swartz et al., 1985). Combined with sediment toxicity tests (see Chapter 8), the results indicated an inverse relationship between relative contamination and macroinvertebrate abundance. Field studies have also been conducted on the effects of complex effluents on primary productivity. Effluents released into the River Raisin and ultimately into Lake Huron directly inhibited phytoplankton primary productivity, as measured by carbon assimilation (Bridgham et al., 1988).

Field studies have been conducted on the effects of pesticides on lotic systems. Specific pesticides have been applied directly to natural streams and parameters such as insect drift, community composition, and primary productivity have been measured. Methoxychlor significantly increased insect drift and decreased macroinvertebrate abundance when applied to a headwater stream at 300 μg/L (Sebastien and Brust, 1989). The greatest impact was observed for univoltine species, which had a low propensity to drift. For these organisms, effects were more persistent because recolonization was slow. Methoxychlor has also been shown to alter significantly chironomid secondary production, macroinvertebrate assemblages, and leaf litter processing rates in headwater streams (Wallace et al., 1986; Lugthart et al., 1990). In a study that evaluated more realistic exposure routes, the effects of aerial applications of fenitrothion were investigated in mountain streams (Hatakeyama et al., 1990). These applications resulted in a significant increase in the number of drifting insects.

In an effort to determine whether the laboratory-based estimates of adverse effect levels for copper were actually observed in the field, an extensive field trial was conducted via the continuous dosing of a natural stream ecosystem (Geckler et al., 1976). Shayler Run (Ohio) was continuously dosed with copper at 120 μg/L for 33 mo. Extensive evaluation of effects on fish and benthic macroinvertebrates were monitored before (26 mo in duration), during, and after the exposure period. Significant impacts on both fish and invertebrates were observed. Given a comprehensive set of laboratory acute and chronic tests on a variety of species for copper, water quality criteria could be established, suggesting that these tests could estimate safe concentrations for copper that appeared to be protective of the aquatic habitat. However, acute or chronic test results based on one or a few species were not adequate. The authors also indicated that chemical factors such as alkalinity, hardness, and other potentially detoxifying properties must be factored into laboratory tests to estimate accurately safe copper concentrations. Laboratory tests were not capable of predicting the observed avoidance of copper by fishes in Shayler Run.

Various types of in-stream biological surveys have been used by states and during the development of rapid bioassessment protocols to assess water quality status (Karr et al., 1986, 1987; Ohio EPA, 1988; Barbour et al., 1992). Suter (1992) has reviewed some of the advantages and disadvantages of such survey approaches.

Marine Field Studies

Their immense spatial scale, their heterogeneous distributions, and their variable composition over time and space pose the most difficult challenges when conducting field studies in marine ecosystems. Hines et al. (1987) present results of several years of invertebrate and fish population monitoring and characterize the tremendous degree of temporal and spatial variation in their distributions. Studies by Livingston et al. (1978) linked a variety of estuarine physicochemical functions with a number of seasonal, annual, and long-term biological cycles to explain contaminant dynamics in an estuarine system.

Studies of contaminant-induced effects in offshore habitats have been conducted following oil spills (Jarnelov and Linden, 1981) or disposal of sewage or contaminated sediments at offshore sites (Phelps et al., 1987). Short-term or localized effects on algae and macroinvertebrate plankton have been demonstrated in these systems and have generally been consistent with laboratory-derived exposure-response relationships for single species. Studies of effects on organisms in the water column have demonstrated how metals or bioaccumulative contaminants such as polychlorinated biphenyls (PCBs) are cycled in planktonic food webs. This has furthered our understanding of how contaminants in vast, open ocean systems can still pose a threat to fishery resources and human health. In offshore habitats, benthic organisms may endure contaminant exposures of relatively longer duration. Studies of deep-water benthic communities have also shown how contaminants are cycled or degraded and the extent to which benthic systems contribute to contamination of resources of concern. Assessment of open-ocean ecosystems remains an active area of research as contamination problems are assessed on global scales.

Studies of the effects of contaminants in near-shore, marine ecosystems are much more common than studies in open-water ecosystems, in part because of easier problem and study site definition, along with the perception that the problem is more severe. National monitoring programs have been implemented to characterize changes in near-coastal-ecosystems as water quality improvement plans are implemented and as coastal development continues (Phelps et al., 1987, Pearce and Despresj-Patanjo, 1988; Turgeon et al., 1992). Problems caused by contaminants and nutrient enrichment have received focused attention in the Chesapeake Bay system, where anthropogenic stresses from a variety of land, water, and atmospheric sources affect the marine ecosystem (Mountford and Mackiernam, 1987). Algal blooms, anaerobic water masses, loss of seagrass beds, and changes in the abundance of sport and commercial fishery resources have been linked to increased nutrient inputs into the system (Flemer et al., 1987). Research activities on ecological effects of anthropogenic stresses in the Chesapeake Bay system have been divided into smaller, focused, exposure-response studies of particular contaminants and specific resources. Examples of herbicide-seagrass studies, wastewater–algal bloom studies, and industrial effluent–fishery resource studies (Macalaster et al., 1982) show how large-scale systems can be broken down into more manageable assessments. However, the ultimate evaluation for understanding the ecological effects of contaminant inputs to Chesapeake Bay must come from a synthesis of these studies of ecosystem components (Flemer et al., 1987).

Field studies to assess impacts of contaminated sediments are common occurrences in marine habitats, in part because estuaries receive and process contaminants that are sorbed to particulates that have been transported to coastal areas by river systems. In addition, agricultural chemicals used near coastal waterways and industrial development near major harbor areas contribute significantly to inputs of contaminants into coastal habitats. Estuarine field studies assessing pesticide impacts have been reviewed by Clark and Noles (1993). Assessment of the impacts of contaminants in sediments requires the use of a combination of field survey data of benthic communities, chemical analyses of sediments, and laboratory toxicity tests with sediment-dwelling species (see Chapter 8) to develop a comprehensive assessment database, commonly referred to as the "triad approach" (Chapman, 1986; Chapman et al., 1987). When this approach was used to evaluate sediments from 15 sampling sites along the West Coast, researchers concluded that no single species or chemical analysis was most useful or reliable for all the sites; comprehensive evaluations of large marine systems require information from multiple assessment measurements (Long et al., 1990). This also was true of major sediment contamination studies along the East Coast (Rogerson et al., 1985; Folsom et al., 1988).

Many concerns regarding wide-scale effects of contaminated sediments on marine biota have been raised because of links between a multitude of sublethal effects and diseases (including neoplasms) among bottom-dwelling species (see Chapter 14). Malins et al. (1984) summarized a number of studies on relationships between sediment contaminants and diseases of fishery resources and presented conclusive evidence of impacts on the health of benthic fishes resulting from a variety of sediment-associated contaminants (including polycyclic aromatic hydrocarbons; PAHs) in areas of Puget Sound, Washington. Their approach has been incorporated into a number of site-specific studies and national monitoring programs (Turgeon et al., 1992). National sediment quality criteria (see Chapter 8) are being developed by the U.S. EPA in part as a response to these concerns.

SIMULATED FIELD STUDIES

The various types and designs of SFSs were previously described. Factors relevant to the experimental design of such studies are somewhat independent of whether a pond mesocosm or microcosm, limnocorral, littoral enclosure, or lotic test system is chosen. Many of the more theoretical aspects of study design will be discussed prior to the specific sections dealing with individual test systems. The choice of the most appropriate system is dependent on the individual researcher's objectives.

Overview of Simulated Field Study Concepts

Because many SFSs have been conducted for the registration or reregistration of pesticides under the U.S. Federal Fungicide, Rodenticide and Insecticide Act (FIFRA; see Chapter 21), this framework will be used to provide generic background information on the use of such studies in aquatic risk assessment (see Chapter 28). Relative to these studies, the EPA has provided a generic guidance document on how to conduct aquatic mesocosm tests (Touart, 1988). This document identifies two regulatory objectives for SFSs. The first objective is to "provide a pesticide manufacturer supportable means for negating presumptions of unacceptable risk." SFSs will not be required in the registration process unless laboratory studies of toxicity and probable environmental fate indicate imminent hazard to aquatic organisms. The second objective is to "provide risk managers descriptive information on the extent of adverse impacts, both in duration and magnitude, likely to occur in aquatic systems which can thus be evaluated in risk-benefit analysis." The document goes on to discuss the need for addressing both structural and functional properties of ecosystems and states that "risk managers must know how expected exposures of a potentially hazardous pesticide impact population or community structure or ecosystem functions in a representative aquatic system before making regulatory decisions." An additional objective, which may or may not be included within the first objective, is to identify a threshhold or no-effect or no observed effect concentration (NOEC).

Relative to the first objective, it is important to provide a mechanism to refute a presumption of risk or to evaluate the appropriateness of the laboratory-based risk assessment, because of the *uncertainty* associated with the current risk assessment procedures and the fact that laboratory-derived estimates of toxicity may not be indicative of effects that may occur in the field. Although a conclusion of "unacceptable risk" may be appropriate based on laboratory toxicity and estimates of the potential for exposure under field conditions, this conclusion is based on an "unvalidated" risk assessment procedure containing numerous sources of uncertainty. This uncertainty can be categorized based on the two primary components of risk assessment: effects and exposure (see Chapter 28). Understanding the relative uncertainty associated with each component will provide guidance in designing the most appropriate test system.

Relative to the effects component, collecting ecosystem-level descriptive information in lieu of the known chemical and toxicological properties of a chemical may not greatly reduce the uncertainty associated with the risk assessment. The level of uncertainty is dependent on the fate of the chemical and its toxicological characteristics. Because of the tiered approach (see Chapter 21) of the current FIFRA risk assessment procedure, it is not likely that SFSs would demonstrate greater sensitivity or unexpected positive interactions that would not, on a broad basis, be predictable from laboratory studies (assuming the laboratory and field studies had similar exposure profiles). For example, if cladocerans are sensitive to the test chemical and expected to be affected in the field, it is reasonable to assume that such impacts on cladocerans may cause competitive induced alterations in other zooplankters, such as increases in rotifiers due to reduced competition for food or changes in the species compositions of the prey (i.e., algae). However, the exact magnitude and nature of such an indirect effect will vary among SFSs and may be as much a function of the structural and functional properties of the particular test systems being studied or the timing of the treatment of the test chemical (Hanazato and Yasuno, 1990) as they are a function of the toxicological properties of the compound. Given this, although "new" or different effects may be observed in SFSs, predictability relative to performing a risk assessment may not be greatly enhanced. The very nature of the tiered system is such that SFSs are required only if the laboratory-scale tests have led one to expect adverse environmental effects at an estimated (expected) environmental concentration (EEC). The null hypothesis to be tested in such studies is that there is no difference between NOECs predicted from laboratory toxicity studies and those observed in a field study. That is to say, there is no environmental mitigation of exposure to the test chemical. The SFSs would never be conducted if the laboratory-scale tests indicated that adverse effects would not be expected at the EEC. Therefore, since the structure of the regulatory tiered testing is not designed for the SFS to uncover additional adverse effects, the SFSs should not be designed in that way. These studies should be designed to further investigate end points (i.e., direct effects on fish reproduction) identified to be of toxicological importance in less complex, controlled experiments. If community interactions are of primary interest, these may be investigated more easily in controlled laboratory multispecies microcosm experiments than in large, complex field studies, where exposure concentrations cannot be accurately controlled and biological/chemical variability is high.

Within any risk assessment process, a major portion of the uncertainty is often associated with determining the exposure component (EEC). Relative to evaluating the utility of simulated field studies in reducing this uncertainty, exposure will be discussed as consisting of two components: system loading (i.e., chemical input) such as effluents or pesticide runoff or spray drift and actual exposure of an organism within an aquatic system. In SFSs conducted with pesticides, test system chemical loading (i.e., mass) is calculated using computer simulation models (see Chapter 18) such as the Simulator for Water Resources in Rural Basins (SWRRB) (Computer Sciences Corporation, 1980; Arnold et al., 1990) or the Pesticide Root Zone Model (PRZM) (Carsel et al., 1984) (see review later in chapter for discussion of simulation models and loading). Simulated field studies themselves do not provide any information concerning real-world loading to aquatic ecosystems. In this context, these studies are strictly "effects" studies that do not provide realistic information on the potential for exposure and, therefore, do not address a major component of the uncertainty associated with the original risk assessment. Given this, the most useful information SFSs can provide is not whether there are any observed effects at a specific exposure concentration (EEC), because the exposure concentration may vary, but establishment of a concentration-response relationship. Once the concentration-response relationship has been determined, "sensitivity analyses" can be performed by evaluating the EEC associated with different chemical use scenarios (or real-world loading data) and determining how far above or below the exposure is relative to some effects threshold value.

Experimental Design Strategies

The overall experimental design (see Chapters 10 and 19) for a simulated field study should be chosen based on the objectives of the study and the method to be used for the final risk assessment. The design will directly influence how the results of the study can be interpreted. In addition, the design will influence how the treatment concentrations are established and their importance relative to the risk assessment process. The specific end points to be measured will be determined primarily by the objectives of the study and will not necessarily be affected by the experimental design.

Potential design approaches can be clearly divided into two categories, hypothesis testing or point estimates and concentration-response relationships. In reality, the objectives of the different approaches are often combined in an attempt to attain the positive attributes of each approach (Touart, 1993). For discussion purposes, however, separating the design provides the most effective way to illustrate the rationale behind the different approaches.

Hypothesis Testing

The testing of specific hypotheses is a frequent application of statistics in biological/chemical research. A vast array of statistical techniques have been established for determining the probability that a given set of data is within the expected distribution of a larger population, given a certain variability and error associated with the data. Depending on the amount of acceptable type I (alpha) and type II (beta) errors established, the concentration identified as causing an adverse effect in toxicological studies can be inferred (Giesy and Allred, 1985; Kirk, 1968; Shaw et al., 1993).

A test of significance can be defined as a method of analyzing data so as to discriminate between two hypotheses (see Chapter 10). The first hypothesis is the null hypothesis; the second hypothesis is called the alternative hypothesis. Once the appropriate hypotheses have been stated, they are evaluated by selecting the appropriate statistical test (e.g., analysis of variance, ANOVA) with the greatest power and robustness. The selection of the appropriate test is based on the needs of the investigator, design of the experiment, explicit and implicit assumptions, and sample sizes (Regal and Lozano, 1993; Kirk, 1968). As the choice of statistical design will obviously determine or influence how the results are interpreted, it is essential that the rationale for interpretation of each approach is understood prior to designing the study.

The currently recommended design for simulated field studies conducted for pesticide registration in the United States is four treatment concentrations (three test substance concentrations and an untreated control) exposed in triplicate (Touart, 1988). The treatment concentrations are chosen based on specific estimates of loading to aquatic ecosystems. The generic null hypothesis for this design and a given parameter such as fish growth would be that there is no difference between the control and treated populations. An ANOVA and specific multiple range tests could be used to identify treatment concentrations that are statistically different from the control (see Chapter 10).

Hypothesis testing and subsequent statistical analyses, such as ANOVA, test for a difference(s) among populations based on the variability within and among treatments. With this approach the ability to identify treatment-related effects and probability of error depends on this variability and on the number of replicates. In SFSs, variability between or among systems may fluctuate regionally and temporally and may be influenced by such factors as climate, time for colonization, and length of experiment. Under these circumstances it is possible that two separate SFSs treated at the same level and with the same chemical, but conducted in different years or geographic regions, may result in different conclusions because the variabilities of the studies may be different enough to influence the statistical analysis. As stated by Stephan and Rogers (1985) in their paper comparing regression analysis with hypothesis testing in chronic toxicity tests, "an investigator who uses a large number of experimental units and achieves better reproducibility between replicate experimental units will tend to make test materials appear more toxic than an investigator who uses few experimental units and achieves less reproducibility."

The hypothesis-testing approach is most appropriate when a well-defined estimate of exposure is available. Under these circumstances, the uncertainty associated with exposure is low and the experimental design can focus on identifying effects observed at the established exposure concentrations. Depending on the probability of making a type I (alpha) error, one simply determines whether the response of a particular parameter (e.g., phytoplankton density) to preselected treatment concentrations is significantly different from a control and other treatment concentrations. With the hypothesis-testing approach, the question of ecological relevance must always be considered and should not be lost in the regulatory need to focus on statistical significance. A statistically significant effect could be identified that is very small and transient, such as a 5% decrease in fish growth, which may result in unnecessary regulatory judgments. Alternatively, one could fail to demonstrate statistically an effect such as a 50% reduction in fish egg production, which could have long-lasting effects on populations.

The hypothesis testing contains the implicit assumption that the chemical loading rate can be established prior to test initiation. The test concentrations are then set based on calculations of loading to aquatic systems, and after test completion the risk assessment is performed by statistically comparing responses of different treatment concentrations to responses of systems receiving no chemical (i.e., untreated controls). Therefore, after the estimated loading rates are determined and the treatment concentrations established, the fate or exposure component is factored out of the risk assessment process. This approach is acceptable if the only objective of the study is to negate the presumption of risk identified for a specific chemical use pattern and if the estimate of the EEC is accurate. The risk assessment procedure resulting from a hypothesis test design is such that decisions are based on the effects observed, and no posttreatment comparison of effects and exposure is necessary. This fact highlights the importance of setting the most appropriate treatment concentrations when using this design.

Relative to the classical risk assessment procedure of comparing exposure and effects, one needs to evaluate the usefulness of the hypothesis-testing approach for conducting simulated field studies. Remember, with the hypothesis-testing approach, estimates of chemical loading are used to set the treatment concentrations, and the risk assessment will be based solely on effects at the treatment concentrations. Future research and subsequent simulation model improvements in loading estimates may result in EECs considerably different from those tested in the simulated field study. In the case of pesticides, this will make it very difficult to predict how changes in the pesticide label, such as reductions in crop application rates, elimination or restrictions of use, or expansion of use to new geographic areas for new crops, will influence potential risk. The difficulty arises primarily because the hypothesis test design is not conducive to analyzing retroactively the effects data relative to new exposure information. Because of the expense of conducting SFSs (e.g., mesocosm tests) and the difficulty in accurately determining the EEC, flexibility is gained if the design of the mesocosm test is not dependent on the establishment of an EEC, but rather independent of the EEC.

As an alternative approach, SFSs can be designed to obtain "concentration-response" information for predefined end points and these relationships compared to a given EEC. In this way, if further study allows one to make more accurate predictions of an EEC from improved model simulations or field observations, one can compare the EEC to the concentration-response relationship without the need to conduct the entire simulated field study again.

Concentration-Response Approach

For the concentration-response approach, multiple exposure concentrations (treatments) are established with the objective of obtaining a concentration-response relationship and an NOEC. Treatment concentrations may or may not include the EEC. The primary criterion for this approach is that a series of concentrations, possibly arranged arithmetically, is established to span the range from no effect to complete system disruption, thereby providing greater predictive power.

The concentration-response approach enables researchers or regulators to identify significant adverse effect and no-effect concentrations. From either regression equations or simplistic evaluation of the concentration-response relationship, one can estimate the concentration of pesticide required to produce a specific proportional change in an end point of interest. The specific regression model used to analyze the data depends on the type of data collected. In contrast to the hypothesis test design, replicate sampling is not essential to test the equality of system responses, because the components of variance (concentration-to-concentration and random sample variability) can be ascertained from the differences between all samples and their associated theoretical responses. Although some replication, particularly in the control response, is desirable, a greater number of test concentrations can be investigated than with the hypothesis approach without compromising the ability to identify specific effect concentrations.

Assuming concentration-response data can be developed, concentrations causing specific effects, such as 20% reduction in fish biomass, can be identified and acceptable end points adjusted to protect the environment. The selected end points can be based on an understanding of the biology of the system instead of arbitrarily determining whether effects are observed at concentrations that are equivalent to a specific EEC. Appropriate no-effect concentrations can be identified by comparing the relative effects on all end points of interest across a wide range of test concentrations.

Although the ability to identify uncertainty and establish confidence intervals makes the concentration-response approach appear more flexible than the hypothesis test design, there still exists a need to establish statistical significance. Regulatory criteria are normally based on statistically significant effects such that any experimental design is strengthened if there is a method of identifying statistical significance. Different procedures for analyzing data via regression analysis are discussed by Liber et al. (1992) and Thompson et al. (1993).

Regardless of statistical considerations, the use of concentration-response information is related more directly to the fundamental concept of how risk assessments should be conducted. After the concentration-response curve is established, effects that may occur at any given exposure concentration can be predicted with a given probability of occurrence. Under this scenario, the exposure estimates can be either model-generated EEC values (see Chapter 18) or actual field monitoring data. For pesticides, exposure concentrations resulting from different crop uses and application rates can be compared to effect concentrations, and in cases in which exposure estimates are close to effects thresholds, regulatory restrictions can be implemented to mitigate or reduce exposure below the level of concern.

Even though, for given circumstances, the concentration-response approach may be more flexible than the hypothesis test approach, it is still limited by the fact that it is difficult to establish what level of effect is biologically significant and then to develop appropriate regulatory criteria. For example, what percent reduction in fish growth or fish production is biologically/ecologically significant such that regulatory criteria are defensible? Is a 1% reduction in fish production at the estimated environmental concentration considered an unacceptable adverse impact? At present, the data do not exist to answer such questions categorically; however, that does not mean that regulatory end points cannot be rationally defined.

Currently, studies based on hypothesis testing do not directly consider biological significance, so any attempt to incorporate biological significance in the risk assessment process would be an improvement. In addition, it is hoped that future trends in risk assessment will utilize probabilities of occurrence instead of point estimates. In this scenario, the probability of biological effects occurring at different exposure concentrations would be known and would be considered concurrently with an analysis of biological significance of the observed effects. For example, a 1% probability of a 25% decrease in egg production may be acceptable, whereas a 75% probability of a 10% decrease may not be acceptable. Therefore greater use of a probabilistic approach to risk assessment in the future will restrict the utility of data generated from experiments using the hypothesis-testing design. A similar problem currently exists with laboratory early life stage and chronic toxicity tests designed and analyzed via hypothesis testing (see Chapters 2 and 3). Development and improvement of risk assessment as a more probabilistic science are partially hindered by the fact that much of the toxicological data generated thus far and data that are continuing to be generated are not amenable to this type of analysis (Suter, 1992).

Another difficulty associated with the regression approach is the possibility of nonlinearity of effects caused by the varying sensitivities of different species and by the occurrence of indirect effects (Thompson et al., 1993; Shaw et al., 1993). This factor makes it difficult to apply simple regression analysis to the data set; it requires that each end point be evaluated individually and will probably require different regression techniques for different parameters.

System Loading or Treatment

Much of the controversy in the design and conduct of SFSs to set ecologically acceptable exposures has focused on the treatment (i.e., dosing) regime. The experimental design will determine the relative importance of establishing a firm loading rate prior to initiating a study. Currently there are two basic experimental designs. The first design is based solely on hypothesis testing and requires a priori identification of the loading rate. The second approach sets the concentrations for the SFS based on the chemical's toxicological properties, as it may or may not have been tested in the same range as the estimated exposure concentration. Also, a number of "hybrid" designs have been proposed which attempt to incorporate the different objectives of these basic designs into a single study. Hybrid designs, as discussed by Touart (1993), attempt to incorporate the advantages of the hypothesis approach and the concentration-response approach into a single experimental design. Depending on the approach, loading for the simulated field study may be calculated using different assumptions or procedures.

It should be noted that the term EEC (estimated or expected environmental concentration) will not be used in this section, as it refers to the concentration in a body of water and is volume dependent. The present discussion refers primarily to the test system treatment or dosing rate, which is only a function of the specific loading or transport of the chemical to the receiving aquatic system.

As an example of the rationale for determining specific treatment rates, typical approaches for establishing the loadings for a SFS conducted for pesticide regulatory purposes will be described. The two primary transportation routes for pesticides to the aquatic environment are aerial spray drift and surface runoff. For test system treatment, these routes can be addressed separately or combined into a single mass treatment, depending on the experimental design and or the physical-chemical properties of the chemical. For the treatment of a test system by either of these routes, three important factors to be considered are (1) the number of applications to be made, (2) the interval between the applications, and (3) the mass loading for each application. Although not discussed, it is also possible to dose a test system continually, although logistically this can be difficult for large SFSs. Some of the basic considerations are discussed in the following. A more detailed discussion of the topic is provided by Hill et al. (1993).

Surface Runoff Loading for Pesticides

If one is testing a hypothesis concerning the presence or absence of an effect at a specific pesticide loading rate, it is extremely important to identify that rate prior to initiating the study. As previously mentioned, pesticide surface runoff simulation models (e.g., SWRRB, PRZM) are generally used for estimating the mass of material being transported from an agricultural field. Although a considerable amount of validation work has been conducted on some of these models, there is still considerable uncertainty. As with all models, the output is only as valid as the data input values, so realistic model parameterization is imperative. For example, the day of pesticide application used in the simulation can drastically influence runoff because the proximity of the application date to a rainfall event will determine the amount of pesticide available for runoff. This is especially important for nonpersistent, water-soluble chemicals. Also, many of the important fate processes (e.g., sorption, volatilization) associated with pesticide runoff are not well understood to be adequately incorporated in currently used models. For example, the fraction of a pesticide residue that would be bound to either plants or soil and is "resistant" to desorption is not well known. Di Toro (1985) reported that a significant portion of hydrophobic chemicals is essentially "nondesorbable." Similarly, it is generally not known what fraction of foliar pesticides can be washed off during a rainfall event. These and many other unknowns (e.g, degradation rate constants) associated with simulation models make the use of EECs for setting test concentrations an especially critical and inexact design component.

Currently, the U.S. EPA uses the Pesticide Root Zone Model (PRZM) or the Simulator for Water Resources in Rural Basins (SWRRB) model to calculate the annual loading for a given scenario. Other models that can also be used include the Chemicals, Runoff, and Erosion from Agricultural Management Systems (CREAMS) model (Knisel, 1980) and the Erosion-Productivity Impact Calculator (EPIC) model (Williams et al., 1990). The runoff predicted by one of these models is then used to determine the mass to be added to each test system. The exposure scenario to be used in the model is based on the use pattern and practices (i.e., number of applications, minimum interval between applications), as stated on the pesticide label.

Because the models provide chemical outputs for numerous years, a difficult decision is which year is most representative of typical or "worst-case" exposure. The runoff predicted for the median year is often used. As an alternative approach, a specific probability of occurrence function can be used to determine the most appropriate loading rate. For example, the median treatment rate for a simulated field study could be based on a 10% probability of occurrence or a 1 in 10 yr storm event for the geographical region of interest. Before such an approach can be implemented, however, progress is needed in the area of probabilistic modeling. Such progress includes the standardization of procedures and scenarios, which is required for regulatory implementation. In addition, such an approach requires an agreement on what probability of exposure is appropriate for regulatory action. This may be the most difficult aspect of developing a probabilistic approach to treatment of mesocosms and other SFSs.

Relative to pesticides, a difficult question concerning the calculation of runoff loading is the most appropriate "runoff interval" for loading—that is, how frequently does a runoff event occur and thus on what time interval should the test system be dosed? Due to the temporal limitations associated with SFSs, it is difficult to treat the test system over the longer time periods that may be predicted by the model. For more persistent chemicals, the model may predict a large portion of the annual runoff during the fall or winter months when precipitation may be greatest. Because most simulated field studies are conducted during the summer months (time of greatest biological activity, etc.), is it appropriate to include this portion (fall-winter runoff) of the runoff in the loading estimates? Use of the annual loading predicted by the simulation model represents the most conservative approach; however, the appropriateness of concentrating all the loadings from a full year into a short treatment interval during the summer is questionable. As alluded to already, the importance of this question is primarily dependent on the chemical's persistence. For a chemical (i.e., pesticide) that is applied in the spring and degrades quickly, the majority of the exposure will occur immediately after application.

The output of simulation models can be used to calculate the number of applications and the interval between applications. In theory, the most appropriate pesticide mesocosm loading sequence should be guided by models containing accurate and reliable weather data for the agronomic regions of interest (i.e., crops where pesticides are used). In using a simulation model for this purpose, it is important to identify the number of significant runoff events that occurred during the year. "Significant" can be based on a number of factors, including specific mass transport per event (i.e., any event transporting more than 10 g is considered significant), specific water yield per event, or events resulting in the receiving water containing concentrations of toxicological concern (i.e., any runoff event resulting in instantaneous water concentrations greater than 1 ppb). The most important parameters determining the number of significant events include the time of application (spring vs. summer), the number of applications, and the chemical's persistence.

Unfortunately, the tremendous variation in the number and timing of runoff events and the subjective nature of defining a significant event make it difficult to standardize procedures for determining treatment number and intervals for different chemicals. Depending on the scenario, the number of significant runoff events per year can vary from 2–3 to as many as 40–50 and the interval between applications can be as short as 1–2 d or as long as 2–3 mo. Given this, along with the relatively short time frame of most SFSs, it is often not practical to simulate the timing of "real" runoff events. Often, the number of applications and interval between applications are based on logistical considerations. A currently used scenario for treating test systems with pesticides having use labels that permit multiple applications to a single crop involves five runoff applications at 14-d intervals. The advantages of standardizing such a scenario include easier comparison among chemicals with similar toxicological properties, easier definition of cause-effect relationships, and simplification of logistical aspects of study. The major disadvantage, as with all attempts at standardization, is that the flexibility in designing a study to reflect the specific properties of a specific chemical can be compromised. In addition, realistic extrapolation may be limited as the base data may not be representative of actual use patterns.

For experimental designs based on toxicity testing or obtaining a concentration-response relationship, the question of mass loading is not as important; however, a decision still must be made on the number and interval of applications. At this point, the best approach would be a scenario similar to that outlined above. The number of treatments, as discussed above, will depend on the number of applications permitted by the label and the persistence of the chemical. For a short-lived chemical with only a few applications per year permitted by the use label, two or three applications may be appropriate, whereas more treatments would be required for more persistent chemicals or for chemicals having labels permitting many (e.g., 10) applications. Simulation modeling would be helpful in determining the appropriate number of applications, although a standardized 7–14-d interval may be most appropriate.

Procedures for simulating runoff for SFSs have varied tremendously. As it is impossible to simulate exactly a real runoff event, the amount of effort expended in attempting to simulate an actual event is dependent on the objectives of the study. If the study is designed with only a single concentration representing the regulatory loading rate, considerable effort may be justified in making the loading procedure as real as possible. This may entail having two to three specific entry points in a mesocosm (or other SFS test system) for a runoff "slurry" containing a chemical-soil mixture. Such a slurry mixture has been prepared in a variety of ways (Johnson et al., 1993; Howick et al., 1993; Hill et al., 1993). The critical factor in this approach is that the particle size distribution for the slurry should be, as much as possible, representative of the particle sizes that may be transported from the agricultural field of interest. When using such a point-source approach, direct use of topsoil is inappropriate. During erosional runoff events, the smaller particles transported in runoff water do not settle rapidly on entering a pond and are thus usually distributed throughout the pond. If slurries containing larger particles (i.e., sand) are used as point-source inputs into test systems, rapid settling will occur at the point of entry, thus minimizing the transport and exposure throughout the test system. These types of problems have resulted in the development of treatment procedures that evenly distribute the test chemical over the entire test system (Johnson et al., 1993; Hill et al., 1993) (Figure 1). Although less realistic, these systems ensure an even and uniform exposure throughout the system. Such uniformity should help to reduce within-system variability in the response of the test organisms, thus increasing the power of the test to identify effects.

Spray Drift Loading for Pesticides

The treatment of test systems with a spray drift simulation in addition to a runoff simulation is justified when application of the pesticide product could result in off-target drift. In general, only aerial application and airblast treatment of tree crops may result in significant off-target drift. Under these circum-

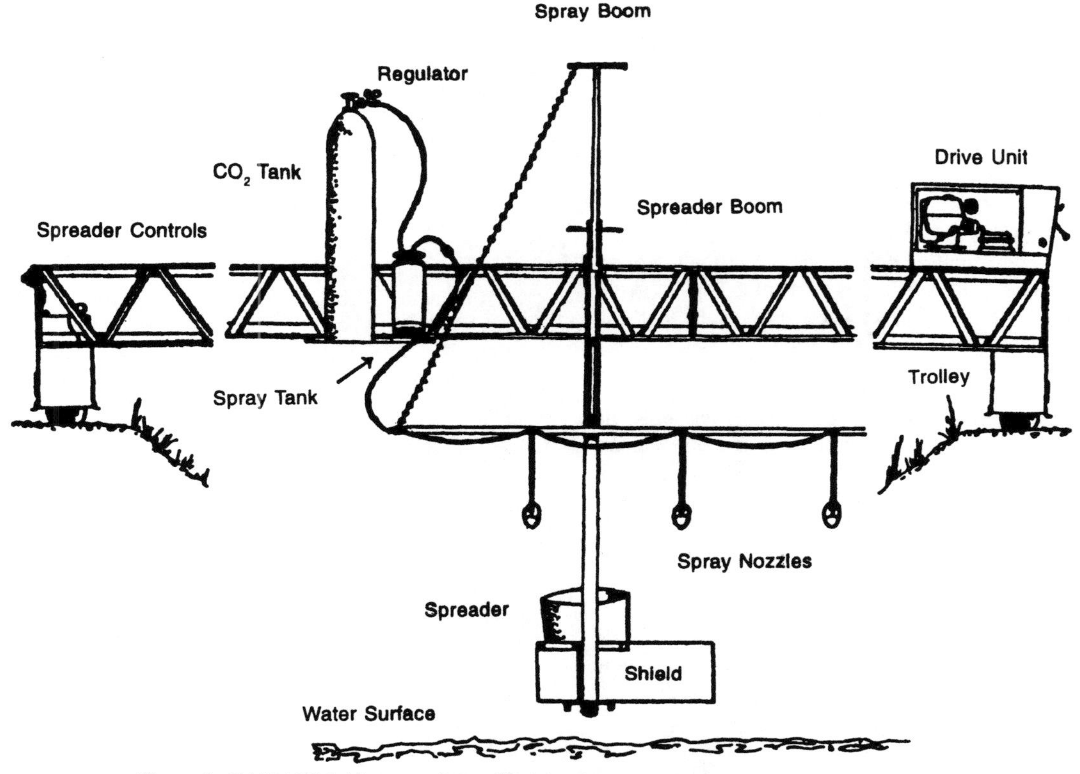

Figure 1. GAMACO bridge spanner modified for the application of test chemicals in either soil slurries, via a spreader, or in solution, via spray nozzles. The spanner extends the entire width of the mesocosm pond and moves across the surface via a hydraulic drive lift. (From University of North Texas.)

stances, the number of treatments, the interval between treatments, and the mass loading must all be determined. Probability values are not presently available to quantify the relative frequency of spray drift entry to an individual body of water, across regions with different use patterns, climates, etc. In addition, at present no aerial drift models are adequately developed to be useful in setting concentration levels. A review of the literature indicates that, depending on a variety of chemical properties such as tank additives, meteorology (i.e., wind speed, humidity), and spray application (i.e., ground vs. air) conditions, spray drift can vary from 0.1 to 20.0% of the field application rate (Yates et al., 1978; Goering, 1973; Adair et al., 1971). This large range of potential exposures makes it difficult to establish realistic treatment concentrations.

Under EPA's current pesticide testing scheme for determining the number and interval for dosing, the maximum number of applications and the minimum interval between applications, as specified on the product label, represent the treatment regime. For standardization purposes, the spray drift mass loading is often set at 5% of the maximum per acre application rate specified on the label. This mass loading is kept constant for all spray drift applications. Considering that few farmers apply all the applications allowed at the most frequent interval and that it is unlikely that drift will deposit on adjacent bodies of water after every application, this percentage has been discussed by Hill et al., (1993) as not representative of typical use. As more information becomes available and improved techniques in probabilistic drift modeling are developed, it may be possible to develop treatment approaches that more accurately reflect the exposure regimens under actual use conditions.

Simulating spray drift to the test systems is simpler than simulating runoff. Boom systems have been developed with which the chemical is applied across the entire pond surface (Hill et al., 1993; Johnson et al., 1993) (Figure 1). Less complex systems using backpack sprayers and boats have been equally successful (Howick et al., 1993). The key problem associated with spray drift applications has been the minimization of the potential for cross-pond contamination via drift during application. This problem can be minimized in a number of ways, including making applications only during low winds, making subsurface applications, and erecting barriers or berms between simulated systems.

Residue Analysis

Measuring the concentration of the test substances in SFSs is essential. This provides information on the actual exposure concentration for the organisms in the test system. Such data can be useful in the interpretation of the results. In addition, if follow-up laboratory or field monitoring studies are conducted, information on the actual exposure concentration that caused specific effects is important. However, for pesticide studies, the regulatory triggers are based primarily on the toxicological properties of the chemical. Exposure is obviously a component of this triggering process; however, for most chemicals the uncertainty in this exposure component is associated primarily with the transport of the chemical (i.e., runoff or drift), not the chemical's dissipation once in the aquatic environment. The design of the residue analysis program for an SFS should be based on the objectives of the study and what is known about the chemical fate of the compound. Extensive chemical analysis is not essential unless this is a defined objective of the study.

As discussed previously, SFSs can be designed to test a specific hypothesis based on a priori assumptions concerning loading, or they can be designed to obtain a concentration-response relationship, in which case risk is evaluated after the test is completed by comparing a specific loading estimate or EEC value for a specific use to the concentration-response curve. When the mass loading for regulatory decisions is made prior to the SFS, extensive measurement of test concentrations during the study is not essential. In this case, a specific hypothesis can be tested with limited information or knowledge concerning the actual water or sediment concentrations. The effects observed are assumed to be associated with a specific loading rate and fate of the chemical. Obviously, treatment rates should be verified, and any additional information on residues may provide data useful or supportive in the study's interpretation, but such information is not essential. When the design involves developing a concentration-response relationship, the measurement of exposure concentrations is more important. In this case, interpretation of the study is dependent on measuring the actual concentration required to produce explicit responses.

Regardless of the design, it is useful to measure the dissipation half-life of the chemical in the test system. (Note: The dissipation of the compound refers to the total loss rate and is not equivalent to degradation. For example, dissipation from the water column may be very rapid for a hydrophobic compound; however, the main dissipation route may be adsorption and sedimentation.) This information can be used to confirm the fate predictions, which are normally based on data from laboratory studies (e.g., photolysis, hydrolysis, aerobic and anaerobic sediment–aquatic metabolism; see Chapter 15). In addition, water column and sediment dissipation rates of the chemical will provide information on the exposure profile for the organisms. Pulsed dose and continuous exposure (normally conducted in laboratory toxicity tests) of organisms can produce different toxicological responses, and having dissipation information may be helpful in explaining the effects observed in the field. However, such fate data are generally concentration independent, so the half-life or dissipation measurements are required only at one test concentration and after one or two applications. It is generally best to measure dissipation after the first application, where interference from metabolites formed after the previous applications is not a problem, and after the last application, to get a final dissipation rate, as adapted microbial communities might increase biodegradation rates over the course of the study.

The matrices (water, sediment, and biological) selected for residue analysis should be based on the physical-chemical properties of the compound. For example, hydrophobic chemicals that partition into sediments or bioaccumulate in fish should be measured in all matrices. For water-soluble compounds, which have laboratory fate and bioaccumulation data to indicate they do not partition into sediment or fish, the measurement of residues in the water phase should be the major focus of the analytical resources.

Measurement of spatial distribution in the system (i.e., vertical distribution in the water column) should be required only if the data to be obtained fit within the objectives of the study and provide information necessary for interpretation of the data. As already stated, most simulated field studies with pesticides are effects studies, not detailed environmental fate studies. Information on the concentration of chemical in specific zones is useful only if information on the effects in those zones is also collected and evaluated in light of the exposure concentrations in those zones. For example, in most studies, integrated water column samples are collected for evaluating effects on zooplankton. Collecting information on chemical concentrations in specific horizontal layers is inconsistent with this study design and will provide information of limited usefulness in assessing the biological effects observed. In addition, the test systems are artificial and are not representative of typical systems in nature. Given the importance of wind mixing in the vertical distribution of a chemical, extrapolating any information on the vertical distribution in mesocosms (or microcosms) to the real world is not possible.

The need to measure metabolites in simulated field studies is totally dependent on the test substance and the objectives of the study (i.e., hypothesis test vs. concentration-response, as discussed earlier). If a test substance has a metabolite (or metabolites) of toxicological significance, it is important to include these in the residue analysis. If the toxicity of the metabolite is considerably different from that of the parent substance, or if the fate is considerably different, it may also be advantageous to ensure that the residue method is capable of separating the parent from the metabolite(s). If the toxicities are the same, a combined residue method may be acceptable.

Lentic Freshwater Mesocosm Studies

According to Odum (1984), "the term mesocosm seems appropriate for such middle sized worlds falling between laboratory microcosms and the large, complex, real world macrocosms." For the most part, this reference is appropriate for this discussion, with the qualifier that the following discussion refers specifically to lentic systems representative of small ponds.

Experimental Procedures

Mesocosms are typically man-made, earthen (Giddings et al., 1985, 1993; Howick et al., 1993; Fairchild et al., 1992; Weber et al., 1992; Hill et al., 1993; Johnson et al., 1993; Mayasich et al., 1993) or concrete ponds (Crossland, 1982; Crossland and Bennett, 1989) constructed for aquacultural or ecological research purposes. They have ranged in size from a surface area of 0.01 acres up to 0.1 ha. Depths are typically 1–2 m and most of the systems have incorporated both a pelagic and littoral zone. Clay liners are normally used for water retention, although systems have also been completely lined with plastic. Topsoil that has been allowed to age and colonize naturally can be used as a substratum in the systems (Johnson et al., 1993), although it is generally more desirable to line the systems with sediment removed from a natural pond (Howick et al., 1993). All systems have allowed specific colonization times for the development of appropriate plant and invertebrate communities. Systems that use sediment from natural ponds have developed macroinvertebrate populations representative of local natural communities (Ferrington et al., 1993). Depending on region and previous pond history, it is generally desirable to allow the ponds to colonize a minimum of 6–12 mo before chemical treatment. To ensure the similarity of water chemistry and plankton communities, water recirculation systems have been developed at some research facilities. Figure 2 is an aerial photograph of the Water Research Field Station at the University of North Texas, and Figure 3 is a schematic of the pond design, layout, and water recirculation system.

The requirement for fish as part of the test system should be determined based on the objectives of the study, which are based partially on the toxicological properties of the chemical. Fish are usually stocked in the system after the invertebrate communities have become established. For most regions of the United States, fish are stocked in March to April of the treatment year. For example, ponds may be filled with water in August and allowed to colonize and overwinter; fish are then stocked in March of the following year and treatment may occur in May–September, depending on the chemical and objectives of the study. If possible, colonization of the systems for longer periods of time prior to treatment (i.e., 1–2 yr) is desirable.

Typically mesocosm ponds can be stocked with reproductively mature bluegill (*Lepomis macrochirus*) at a rate of 25–35 adults per tenth-acre surface area. Some studies have attempted to stock multispecies fish communities representing various feeding groups or habitats (deNoyelles et al., 1989). Having such a predator-prey system can help to stabilize the fish communities and control the overpopulation that can occur when a single-species "community"

Figure 2. Aerial photograph of lentic mesocosm ponds at the Water Research Field Station of the University of North Texas, Denton.

is established. This is a problem in some currently designed mesocosms in which the adult bluegills can produce 10,000–20,000 young per tenth-acre pond during the experimental period (Johnson et al., 1993; Hill et al., 1993). Such a large bluegill population can drastically influence the invertebrate food base and interfere with the sensitivity of the test system to identify treatment-related effects. It has been suggested that largemouth bass (*Micropterus salmoides*) be added to the test system as a predator for the control of the bluegill (Deutsch et al., 1992). However, the unknowns in stocking rates and the potential influence of predator mortality on the variability in bluegill populations have prevented this option from being exercised. Even though such unknowns exist relative to stocking bass, continued evaluation of a means of controlling bluegill overproduction in these test systems is essential (Anon, 1992).

End Points in Freshwater Mesocosm Studies

The end points to be measured in the mesocosm study are dependent on the objectives of the study, which are partially determined by the toxicological properties of the chemical and the size of the test systems. A research program with the objective of understanding the influence of a herbicide on nutrient cycling within pond ecosystems will certainly focus on different end points than a study addressing a chemical that is nontoxic to plants and invertebrates but has the potential for reproductive effects in fish. Broad ecological field surveys should not be conducted independent of the known toxicological profile of the chemical.

It is important to differentiate between studies being conducted for risk assessment purposes and studies being conducted for basic research purposes. The majority of the mesocosm studies that have been conducted thus far for pesticide registration (i.e., compliance under FIFRA; see Chapter 21) have been for insecticides, which are typically very toxic to fish and/or invertebrates. For discussion purposes, consider a chemical that has been determined in the laboratory to be very toxic to fish (LC50 = 1.0 ppb), toxic to invertebrates (LC50 = 20.0 ppb), but nontoxic to algae or aquatic plants (EC50 = ppm range). For such chemicals, the overall objective should be to determine the effect of the insecticide on the growth and/or reproduction of the fish and invertebrates present in the mesocosm. Effects on algae and macrophytes are secondary responses and information on them is not necessary for making an informed risk assessment decision, even though the secondary effects may be of interest from a research perspective. Given that fish are the most sensitive species, followed by invertebrates, it would be

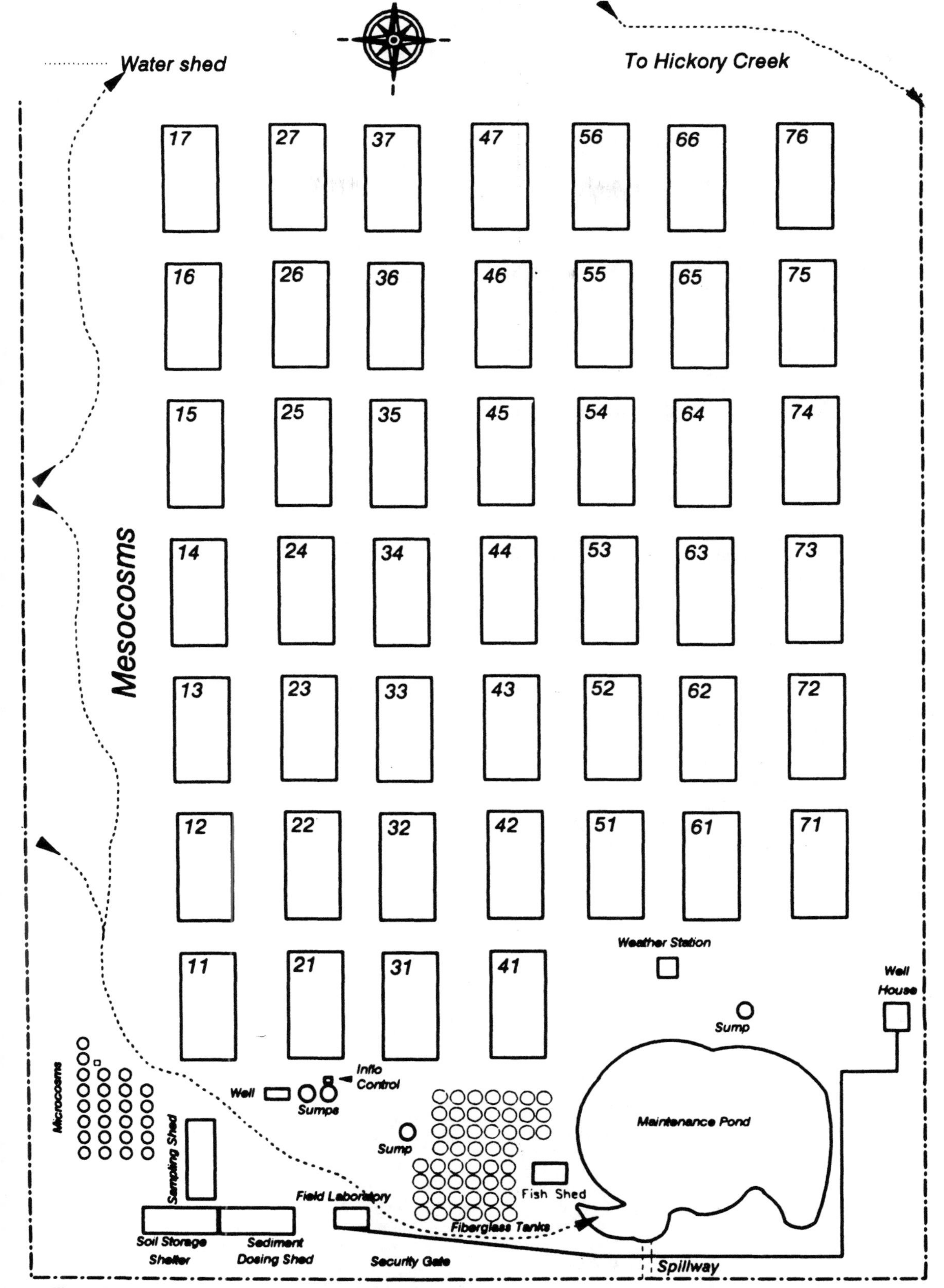

Figure 3. Schematic diagram of the lentic mesocosm ponds at the Water Research Field Station of the University of North Texas. Smaller circles represent the location of fiberglass microcosm test systems.

highly unlikely that any indirect effects on primary producers would influence the risk assessment. Quantitative or qualitative changes that may occur in the algae would be the result of changes in the invertebrates and/or fish populations. Because direct measurements of effects of the chemical on fish or invertebrates would be made, monitoring changes in the primary producers would not be necessary for a risk assessment decision. That is not to say that effects on algae or macrophytes would not occur or would not be of interest, but for a regulatory decision, measurement of such changes would not be required to assess the potential for unacceptable adverse effects of this specific chemical.

A number of end points have been measured in mesocosm studies (Table 3). These can be divided into water quality and biological end points. The primary water quality variables include temperature, pH, dissolved oxygen (DO), alkalinity, conductivity, hardness, and total suspended solids. Other end points that can be measured include nutrients (total nitrogen and phosphorus levels) and dissolved organic carbon. These analyses can be expensive and time-consuming, however, and one should be sure that they are necessary to meet the identified objectives of the study before expending the effort to measure them.

Of these end points, pH and DO are likely the most critical and can reflect ecosystem metabolism or energetics (e.g., productivity, community respiration; see Chapter 19). Diurnal measurements of DO (dawn, dusk and dawn) can provide information on system respiration and primary production. Increases in the DO values during the day provide an estimate of primary production and decreases in DO during the night reflect system respiration (Lind, 1979). Similarly, due to shifts in the carbonate-bicarbonate system via the photosynthetic removal of carbon dioxide, increases in pH can also be indicative of high primary productivity.

Table 3. End points often measured in simulated field studies

Water quality	Biological
pH	Phytoplankton species composition
Dissolved oxygen	Phytoplankton biomass
Temperature	Zooplankton species composition
Alkalinity	Zooplankton biomass
Total nitrogen	Benthic macroinvertebrates
Total phosphorus	Insect emergence
Conductivity	Species diversity
Dissolved organic carbon	Ecosystem metabolism
Particulate organic carbon	Leaf litter decomposition
	Fish survival, growth, and reproduction

Phytoplankton, zooplankton, macroinvertebrates, and emerging aquatic insects are often critical components of the test system and should be measured when the objectives warrant inclusion. Standard methods exist for sampling these compartments. These methods are basically the same as those outlined in the natural field studies section.

Harvesting of fish at the end of the study provides estimates of total fish production and size class distribution. Fish are generally not sampled during the study, except possibly for residue analysis. Adult fish individually tagged prior to stocking will allow individual growth measurements. Numerous methods for tagging exist; however, the most useful approach has been the intramuscular implantation of microtransponders (Johnson et al., 1993).

Case Histories

Mesocosm studies have been conducted with a variety of test substances. Many of the studies have focused on the direct and indirect effects of pesticides, although studies have also been conducted for heavy metals (Kettle and deNoyelles, 1986) and phenols (Giddings et al., 1985). Studies have focused on pesticides because specific regulatory requirements mandate the conduct of such studies, the toxicological profile of pesticides makes them ideal test substances for evaluating the response of ecosystems to stress, and small pond ecosystems are often a primary receiving system when off-target movement of pesticides occurs, thus making pond mesocosms an appropriate test system.

Because of the great toxicity of pyrethroid and organophosphate insecticides to fish and/or invertebrates (Hill, 1985; Mayer and Ellersieck, 1986), a number of lentic mesocosm studies have been performed with these classes of pesticides. Since organophosphates are generally nonpersistent, the majority of studies have focused primarily on acute effects. The influence of carbaryl (a carbamate insecticide) on zooplankton dominance and species composition was evaluated in concrete tanks lined with polyethylene film (Hanazato and Yasuno, 1987; Hanazato et al., 1989). Although these authors observed recovery from the reductions measured in zooplankton species, the extent of recovery was dependent on the time of application (Hanazato and Yasuno, 1990). A common indirect effect observed in mesocosm studies is the competition-induced increase in microzooplankton (Hurlbert et al., 1970, 1972; Papst and Boyer, 1980). In these studies, rotifers and/or copepods have been shown to increase when the larger cladocerans are directly affected by insecticide treatment. In addition, the reduction in larger zooplankton grazing pressure can produce an increase in phytoplankton populations. Increases in copepods were also noted in ponds treated with azinphos-methyl, an organophosphate insecticide (Giddings et al., 1993). The latter authors speculated that this increase was the result of decreased predatory pressure due to insecticide-induced mortality of fish in the test system. Similar secondary effects were observed in ponds treated with methyl parathion (Crossland, 1984). In this test, however, indirect effects on fish growth were also noted. This effect was attributed to the decreased invertebrate food supply caused by methyl parathion treatment. Secondary effects on fish growth have also been observed in ponds treated with the organophosphate insecticide sulprofos (Howick et al., 1993). The growth effect observed in this study was attributed primarily to the reduction in cladocerans in the test system.

Under laboratory conditions, pyrethroid insecticides are extremely toxic to insects and fish (Coats and O'Donnell-Jeffrey, 1979; Mayer and Ellersieck, 1986). However, unlike organophosphates, these compounds are extremely hydrophobic and water insoluble. This property has created considerable debate concerning their actual availability and consequent toxicity to aquatic organisms under field conditions (Hill, 1985). Mesocosm studies indicate that pyrethroid toxicity can be mitigated under field conditions (Crossland, 1982; Hill, 1985), although adverse effects on sensitive aquatic invertebrates still occur. In a study conducted by Crossland (1982), zooplankton and mayflies were severely affected by cypermethrin. Esfenvalerate was found to have direct and indirect effects on zooplankton (Fairchild et al., 1992); however, because of the chemical's short half-life, recovery of the affected populations was relatively rapid. The recovery of populations from esfenvalerate exposure was discussed relative to the trophic structure and life history characteristics of the affected organisms (LaPoint and Fairchild, 1993). Other pyrethroid mesocosm studies include evaluations of the effects of lambda-cyhalothrin (Hill et al., 1993), tralomethrin (Mayasich et al., 1993), cyfluthrin (Johnson et al., 1993), and esfenvalerate (Weber et al., 1992). For all these studies, multiple treatment concentrations were chosen to cover the estimated field exposure concentrations. Effects were observed on specific zooplankton and macroinvertebrate species, although significant effects on fish at relevant environmental concentrations were not observed.

Given the different toxicological characteristics of most herbicides, the effects observed in lentic ecosystems are considerably different from those reported for insecticides. Most impacts are the result of direct effects on plant production and biomass, followed by habitat-related indirect effects on invertebrates and possibly fish. The most extensive research has been performed at the University of Kansas on the influence of atrazine (triazine herbicide) on pond ecosystems (Dewey, 1986; deNoyelles et al., 1982, 1989; Kettle et al., 1986; Larsen et al., 1986). In these studies, direct effects on phytoplankton and macrophytes were observed. Phytoplankton production and biomass recovered fairly rapidly, although species composition was altered and resistant or tolerant algal species emerged. Because of their longer generation time, resistance in macrophytes did not develop and affected macrophytes did not recover quickly. The loss of macrophyte habitat produced secondary effects on invertebrates that were dependent on the macrophytes for either food or refugia. This impact on invertebrates also resulted in a decline in bluegill populations due to the loss of their primary food source. The impact of 2,4-D (phenoxy herbicide) on pond ecosystems was not as severe as those reported for atrazine, although effects on macrophytes and macroinvertebrates were noted (Boyle, 1980; Stephenson and Mackie, 1986). In contrast to the effects observed for atrazine, metamitron (triazinone herbicide) was studied in outdoor ponds and shown to have no effects on phytoplankton, macrophytes, macroinvertebrates, or fish (Heimbach et al., 1992, 1993). The lack of effects was attributed to the extremely short half-life of metamitron.

Lentic mesocosms have also been used to evaluate the impact of nonpesticide chemicals. The effects of multiple treatment levels of fluorene were investigated in 0.1-ha mesocosms (Boyle et al., 1985). The half-life of fluorene in the ponds was 6 d. There were no effects on emergent insects; however, zooplankton were affected at the highest treatment concentration. Bluegill and largemouth bass stocked in the ponds were severely affected (i.e., survival and biomass yield) and were found to be more sensitive in the field than in laboratory toxicity tests. This was thought to be the result of greater stress for the fish under field conditions because of actual stress factors such as higher water temperatures, lower dissolved oxygen, and greater intraspecific competition for food and disease. However, laboratory tests on phytoplankton, zooplankton, macroinvertebrates, and insects overestimated potential field effects. Cadmium (Kettle and deNoyelles, 1986) and phenols (Giddings et al., 1985) have also been investigated in mesocosms. Direct and indirect effects were observed in both studies. In the phenol experiments, the sensitivity of individual ecosystem components was similar to that measured in the laboratory. Indirect effects of phenol exposure, including changes in water quality, were observed at all trophic levels.

A number of studies have been conducted that emphasize the importance of chemical fate in determining the toxicological response of the test system. As discussed, the fate of the chemical and its subsequent influence on exposure within the pond system represent a critical attribute of the mesocosm study. Laboratory-observed toxicity values do not represent the actual exposure under field conditions and may thus lead to inaccurate conclusions regarding the actual risk of a chemical.

LaPoint and Fairchild (1993) discussed the importance of exposure relative to the trophic status of the system and the life history of the organisms. Understanding the influence of trophic status and other ecosystem properties on bioavailability is complex. Under eutrophic conditions the water column organic content (i.e., phytoplankton) is usually great. For hydrophobic compounds that bind to organic matter, bioavailability and thus toxicity may be reduced under these conditions. LaPoint and Fairchild review and discuss field data that tend to support this hypothesis. Other factors are also modified under eutrophic conditions, so bioavailability is not the only mechanism that may alter the exposure profile with ecosystem trophic status. For example, microbial concentrations may be greater in eutrophic waters so that degradation rates are greater. Conversely, under the same conditions, light penetration may be greatly reduced so that photolysis is minimized relative to that observed in oligotrophic conditions. In any case, the specific environmental fate properties of the chemical will determine the influences of the trophic status on exposure and thus effects in aquatic systems.

Another example of the importance of chemical fate and the resultant exposure profile on the toxicity observed under field conditions is provided by Liber et al. (1993). Limnocorrals (see page 285) were treated with Diatox, a photolabile tetrachlorophenol, in the morning and in the evening. Because of the immediate exposure to sunlight, the morning treatments resulted in a shorter exposure time (i.e., faster degradation). After the evening application, the water residues were more persistent and the observed effects on zooplankton were considerably greater. A factor as simple as the time of day when dosing was initiated had considerable impact on the effects observed. Given the size and complexity of some pond mesocosms, it often takes the entire day to treat all the test systems. If a photolabile compound is being investigated, this factor may add considerable variability to the results obtained.

The papers by LaPoint and Fairchild (1993) and Liber et al. (1993) serve to emphasize the complexity of and importance associated with treatment and exposure in SFS test systems. The effect observed can be influenced by the procedures used for treatment, the trophic status and water quality of the systems, the life stage of the organisms, and the time of day when the systems are dosed. Given the importance of all of these factors, it becomes difficult for a single study to provide predictive information regarding the likelihood of adverse effects occurring when an ecosystem is exposed to a specific chemical. It is impossible to know and understand all possible consequences associated with any action. Mesocosms are only surrogate ecosystems designed to provide additional information on the response of natural communities to a "typical" exposure profile that may occur. They cannot, and should not be expected to, address all possible factors or answer all possible questions. By understanding the processes that control the chemical's fate (hydrolysis/photolysis studies are typically conducted in the laboratory; see Chapter 15) and by measuring the effects that occur for defined exposure profiles, reasonable interpretations can be made regarding the potential biological effects that may occur in the field. Such interpretations can be greatly enhanced by understanding how the effects change with changing chemical concentrations.

Lentic Marine Mesocosm Studies

Marine systems are not especially amenable to mesocosm-type tests as we have defined them in this chapter. Marine ecosystems are especially difficult to study because of their large scale, high degree of spatial and temporal heterogeneity, and open tidal exchange. The distribution of biota and the system process rates may be made more spatially and temporally uniform in a mesocosm by controlling the energy and material exchange of these normally open ecosystems. Nevertheless, marine mesocosms still require substantial and regular inputs of energy and materials in order to incorporate the energy and mass associated with tidal exchange and to sustain system-level functions that realistically model natural marine systems. These more uniform conditions are more amenable to manipulation and study, enabling us to enhance our understanding of the biotic and abiotic interactions that sustain marine ecosystems. Several excellent articles or books have reviewed the various designs and applications of mesocosm-type systems for studying marine ecosystems (Giesy, 1980; Grice and Reeve, 1982; Gearing, 1989; Lalli, 1990).

Compared to enclosure devices used to study marine systems, land-based mesocosm systems offer greater ease of engineering control and operation as well as enhanced access for sampling. The biological components of these systems are derived from colonization; thus systems exert selective pressure for ecosystem components that can adapt to the specific conditions designed into the mesocosm system. Perhaps the most widely known example of such a system is from the Marine Ecosystem Research Laboratory (MERL) at the University of Rhode Island (Figure 4). This land-based exposure system consists of 14 tanks (5.4 m deep and 1.8 m in diameter with a volume of 13 m^3) (Nixon et al., 1980). The tanks contain about 0.7 m^3 of sediment, the type of which can be selected by the investigator. Water is pumped from an estuary and treated as necessary for the experiment, by adding toxicants, for instance. The water in the tanks is mixed vertically and the water in the tanks has an average residence time of 27 d. The MERL system is very flexible and allows one to conduct many types of studies that would be much more difficult if not impossible to conduct with in situ enclosures. These MERL-type systems have been well studied and characterized relative to processes occurring in the larger bodies of water, which they are used to simulate.

The fate and effects of chemicals including metals and the effects of a number of parameters, such as concentrations of nutrients, mixing rate, temperature, and physical stressors, have been tested using MERL-type systems. Results of these tests have been widely published and reviewed (Santschi, 1985; Pilson, 1990). The numerous studies conducted dem-

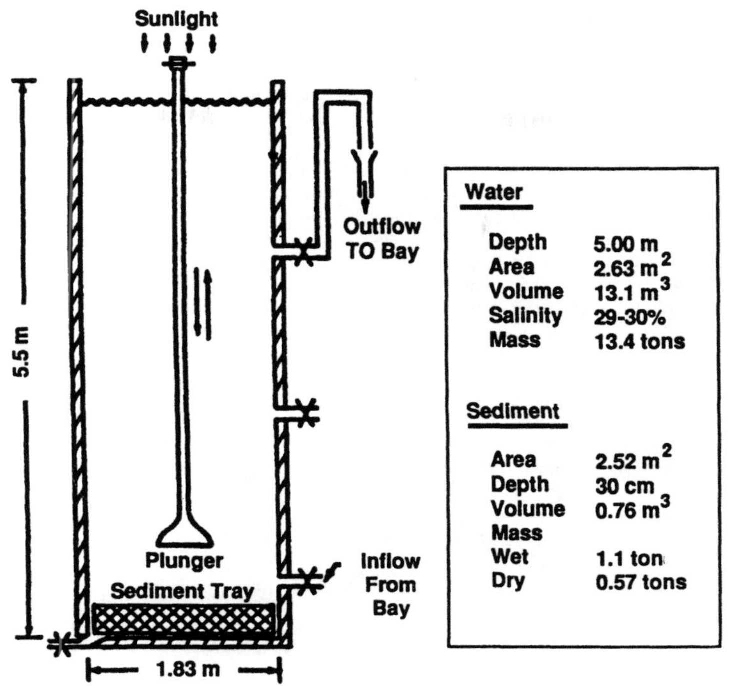

Figure 4. Diagram of MERL mesocosm tank.

onstrate the utility of this system to address a variety of scientific interests in marine ecology and to focus on system components at various trophic levels or test ecosystem function as a whole. de Wilde and Kuipers (1977) describe a land-based mesocosm system that allows focused testing of benthic communities.

Outdoor Freshwater Lentic Microcosms

Outdoor microcosms have been previously defined. The primary difference between these systems and mesocosms is scale. The importance of scaling cannot be minimized and is critical in deciding whether an outdoor microcosm study can adequately address the stated objectives of the study.

A workshop on Aquatic Microcosms for Ecological Assessment of Pesticides was held at Wintergreen, Virginia, February 6–11, 1991 (SETAC-RESOLVE, 1992). This workshop produced a generic guidance document on how to conduct microcosm studies and, through the evaluation of five hypothetical pesticides, developed chemical-specific protocols to address specific questions. Conclusions from the workshop include the following:

1. Outdoor microcosms can be useful tools for ecological assessment of pesticides. Results of microcosm studies may resolve uncertainties in laboratory-based assessments and may preclude the need for larger mesocosm studies. If further uncertainties remain, mesocosms or other aquatic studies can be designed to address those specific issues.
2. Multiple exposure regimens are critical to the interpretation of microcosm results, at least for the case-specific applications considered in the workshop.

Other workshops have also been held on the use of outdoor microcosms in the ecological assessment of pesticides (SETAC-EUROPE, 1992; Crossland et al., 1992). The conclusions of these workshops were very similar to the conclusions of the Wintergreen Workshop. The workshop participants proposed that "test procedures should remain flexible enough to cope with the variety of situations that may be encountered" (SETAC-EUROPE, 1992). In other words, "no single microcosm or mesocosm method can be designed which is applicable to represent the myriad of aquatic scenarios that may prevail following application of a pesticide (or any chemical) in a field situation (e.g., application to different crops)."

Some of the specific advantages and limitations of microcosm studies are outlined in Table 4.

Experimental Procedures

Guidance documents have been developed for the performance of outdoor microcosm studies for pesticide regulatory purposes (SETAC RESOLVE, 1992; SETAC EUROPE, 1992). These documents describe microcosm design and construction, biological and water chemistry measurements, test chemical treatment, and residue analysis. Briefly, outdoor

Table 4. Advantages and limitations of microcosms, as identified by the aquatic effects dialogue group (Anon, 1992)

Advantages

1. Microcosm studies can be performed more quickly than mesocosm studies. A microcosm study can be completed within 2 mo of initiation, although most studies have lasted for 4 to 6 mo.
2. Microcosms can be replicated more economically than mesocosms, because they are smaller and easier to construct. Because microcosm studies can be designed with more experimental units, a single study can address a wide range of conditions or treatment regimes. Greater replication also confers statistical power to an experimental design.
3. Microcosms can address a wider range of ecological issues than laboratory test systems, because they contain many of the elements of natural ecosystems (such as sediments, rooted vegetation, algae, micro- and macroinvertebrates, and fish).
4. Microcosms provide a more realistic physical environment than laboratory test systems, because they incorporate sediments, solar radiation, wind, and precipitation.
5. Microcosms allow studies to be conducted using radioactively labeled test materials. Such studies would be difficult to conduct in mesocosms because large quantities of radiolabeled materials would be required, and containment and disposal of these materials would be more difficult to ensure.

Limitations

1. If bluegill is the focal species, microcosms cannot readily be applied to studies of fish reproduction or recruitment. Adult fish are generally not included in microcosm studies.
2. The confined physical system of a microcosm may distort chemical exposure regimes. For example, disproportionate amounts of pesticide may be removed from circulation by becoming adsorbed to the walls of the microcosm container.
3. The number of samples that can be taken from a microcosm is limited if destructive sampling techniques are used.
4. There is limited potential for populations of organisms in a microcosm to find refugia from which ecological recovery could occur after the chemical has dissipated.
5. Microcosms are more vulnerable to catastrophe (such as destruction by waterfowl sediment resuspension during storms) than larger systems.
6. Recolonization or recovery of microcosms may be more variable than that of mesocosms.

microcosm tanks are seeded with sediment, water, and organisms from an established pond community. After a defined invertebrate colonization period (e.g, several weeks to months), which will vary with geographic region, the tanks can be stocked with juvenile bluegill sunfish or another appropriate fish species. Fish need to be included only if they are necessary to meet the specific objectives of the study. Six to eight weeks after the fish are introduced, the tanks may be treated with the test chemical. Test concentrations are typically chosen to bracket the expected environmental exposure concentration, although the discussion of treatment levels and experimental design for mesocosms is, for the most part, applicable to microcosms. Treatment procedures can be developed to mimic the expected exposure route (e.g., surface runoff, aerial drift) for the chemical into the aquatic environment. The treatment concentrations and the number of replicates per treatment depend on the specific objectives and goals of the study. In general, the experimental design is similar to the mesocosm studies. Photographs of representative microcosm test systems are provided in Figures 5 and 6. In one system, tanks have a roof cover (i.e., to control rain) and are buried in the ground and surrounded by soil to control temperature fluctuations. In the second system, tanks are submerged in a pond to control temperature fluctuations.

Ecological end points are basically the same as those discussed for mesocosm studies and depend on the specific toxicological properties of the chemical. The only end point not easily addressed in microcosm studies is effects on fish reproduction and recruitment, especially if bluegill or large mouth bass are the test species. However, if small minnows (e.g., fathead minnow, *Pimephales promelas*) are used the effects of the chemical on fish reproduction can be evaluated. The growth of fish can be addressed if fish are measured at the beginning of the study, at an intermediate time, and again at the end of the study (i.e., after 4 to 6 mo). Most other end points, including water quality, abundance and composition of insects, zooplankton, phytoplankton, and periphyton, and insect emergence, can be assessed in these test systems.

Case Histories

Outdoor microcosms have been used on a number of occasions to evaluate the potential for ecosystem-level effects. Cypermethrin was applied at seven different concentrations to 1000-L tanks located in a small outdoor pond (Shires, 1983). Residues and fish mortality were monitored and water removed from the test systems was used for laboratory toxicity tests. Experimental ponds with a volume of 11,000 –14,000 L were used to evaluate the impacts of synthetic crude oil on pond benthic insects (Cushman and Goyert, 1984). Both direct and indirect effects were noted and partial system recovery was observed after 3 mo.

Scaling is a critical component in microcosm testing. Relative to risk assessment, a critical aspect of scaling is whether the same results are obtained in smaller, outdoor microcosms as are obtained in larger mesocosm studies or in the real world. This question has been addressed by a number of re-

Figure 5. Photograph of outdoor microcosm test systems (20,000-L tanks) located at Toxikon Environmental Sciences in Jupiter, FL. These tanks have been used to evaluate the impact of pesticides on aquatic communities.

Figure 6. Photograph of outdoor microcosm test systems (10,000-L tanks) located at the University of Kansas. These fiberglass tanks have been used by Springborn Laboratories, Wareham, MA. to evaluate the impact of pesticides on aquatic communities.

searchers. Overall, research has demonstrated that outdoor microcosms can be representative of freshwater lentic systems and respond similarly to larger mesocosm systems when exposed to specific toxicants (Heimbach et al., 1992, 1993; Howick et al., 1993; Morris et al., 1993).

During the conduct of a mesocosm study (0.1 acre) on sulprophos, outdoor microcosms (10,000 L) were established and treated with sulprophos at the same time (Howick et al., 1993). The effects observed on fish and invertebrates in the microcosm study were similar to the results of the mesocosm study. Specific differences in effects were noted, although such differences were generally explained based on differences in exposure, scaling, or preexposure test conditions. For example:

- There were greater effects on chironomids in the microcosm study. These effects were correlated with higher sediment residues, which were the result of differences in treatment procedures.
- Differences in water quality factors, such as dissolved oxygen, were noted (i.e., microcosms had higher DO levels). These differences reflect the greater predominance of macrophytes in the microcosm.
- Overall, zooplankton population densities in the microcosms were greater and the species composition more diverse. This probably reflected differences in fish predation and macrophyte abundance. This, in many ways, can be seen as an advantage of the microcosms and a disadvantage of mesocosms. The overproduction of young bluegill in the mesocosms can cause serious problems relative to the predatory pressures exerted by the young fish on invertebrate food supplies. This was not a problem in the microcosm studies.
- Although similar trends were seen relative to bluegill growth, the microcosm growth effects were not as severe as those observed in the mesocosm study.

A microcosm/mesocosm comparative study was also conducted with the pyrethroid cyfluthrin (Morris et al., 1993). This study found results very similar to those of the sulprophos comparative study. Observations that were similar in the sulprophos and cyfluthrin comparative studies include the following:

- Zooplankton populations were much more diverse in the microcosms than in mesocosms. Again, this can be attributed to differences in fish predation.
- The macrophytes were more dominant in the microcosm systems, and this resulted in slight differences in water quality (i.e., DO).
- The effects of cyfluthrin and sulprophos on invertebrate communities in the microcosms were similar to the results observed for both compounds in the larger mesocosm test systems.

The main difference between the two comparative studies is that in the cyfluthrin study the fish appeared to be more sensitive to indirect growth effects in the microcosm study than in the mesocosm study. This is contrary to the sulprophos comparative study,

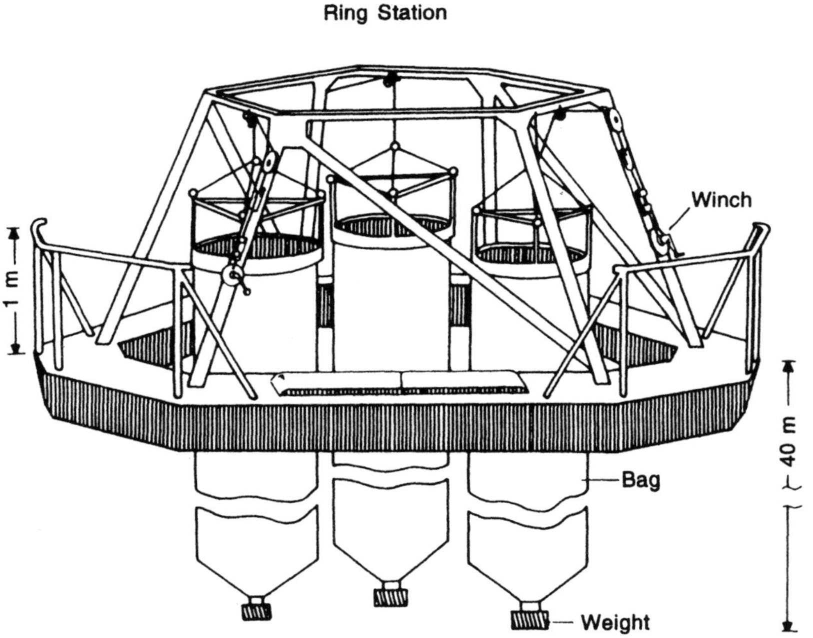

Figure 7. Schematic of Hamburg enclosure. (From Grice and Reeve, 1982. Reprinted with permission from Springer-Verlag, New York, Inc.)

in which the microcosm appeared less sensitive to identifying indirect effects on fish.

In a separate study, stainless steel tanks were also used to evaluate the effects of cyfluthrin and the impacts observed in these systems were compared to those in studies performed in larger pond systems (Heimbach et al., 1992). The authors concluded that the smaller test system (4500 L) allowed detection and better resolution of pesticide-induced effects on invertebrate species. In general, the effects observed in the smaller test system were similar to those observed in the larger system.

Limnocorrals

Limnocorrals are enclosures that are placed in natural aquatic systems (i.e., both freshwater and marine) so that a large volume, but less than the entire ecosystem, can be manipulated (Grice and Reeve, 1982; Gearing, 1989). These test systems are generally cylindrical and made of clear sheets of flexible plastic (Figures 7–10). These enclosures are suspended in the water column from floating platforms and tethered in place with drogue lines to anchors on the bottom. The cylindrical shape is often main-

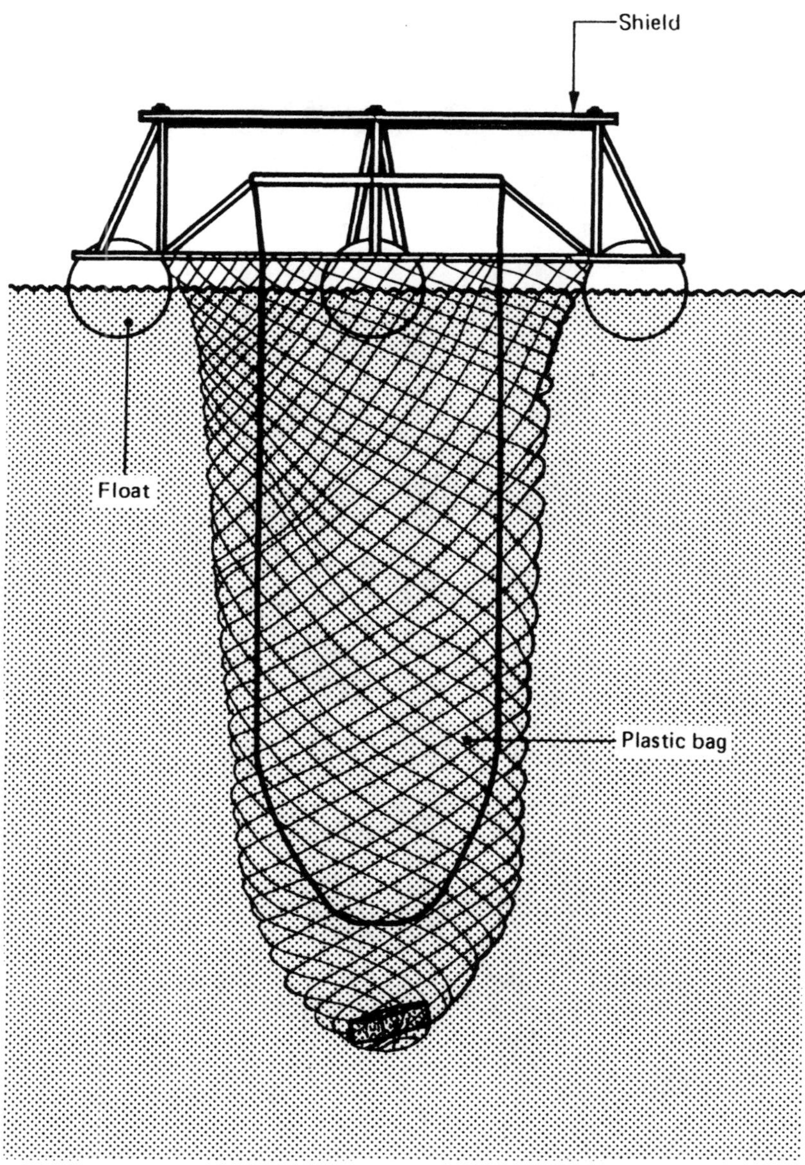

Figure 8. Diagram of suspended plastic bag. (Reprinted with permission from Kuiper, 1977b.)

tained with rigid rings or hoops, which are attached to the sides at the top and bottom and possibly along the sides. The limnocorral can be suspended above the bottom or the enclosed water can be allowed to be in contact with the sediment. The top of the cylinder is generally open to allow gas exchange and light penetration. The top is high enough to preclude possible exchange of water with the ambient water column. The size of limnocorrals used in studies of toxicants has ranged from as little as a few liters to over a million liters, but generally they range from 1000 to 1,000,000 L (Table 5).

Limnocorrals have been made of many materials such as rubber, nylon, polyethylene, or polyvinyl chloride (PVC) (Table 5). In some cases, enclosures have been rigid and constructed of aluminum, steel, Plexiglas, or glass. The materials used to construct the enclosure need to be appropriate to the study objectives and type of chemical being studied. The materials must resist deterioration for a duration appropriate for the study, as well as resisting physical stressors in the environment. Also, the material should be inert and not contribute any toxic materials to the enclosed water or to interact with the chemical of interest. As examples, metals can leach from painted surfaces in sufficient quantities to have effects on the enclosed community and organic toxicants can be sorbed to organic materials used in construction (i.e., polyethylene or PVC) and, in some cases, chemicals can actually diffuse through the walls into the surrounding water. Such interactions can be exacerbated and cause treatment-specific effects if the chemical of interest is toxic to the aufwuchs community, which will colonize the surfaces of the walls of the enclosure.

Limnocorrals are designed to capture a portion of the natural water column, including its assemblage of biota and other seston, and to hold it, isolated in the water column. When filling the corrals, it is important to obtain a representative sample of the entire water column. It is not possible to capture the intact biological assemblage of the water column, so it is advised to fill the enclosure with water from several depths. This should ensure that a representative sample of biota is included in the enclosure, regardless of the time of day when the samples are taken. Alternately, during filling one could leave the bottom open and pump water from several depths

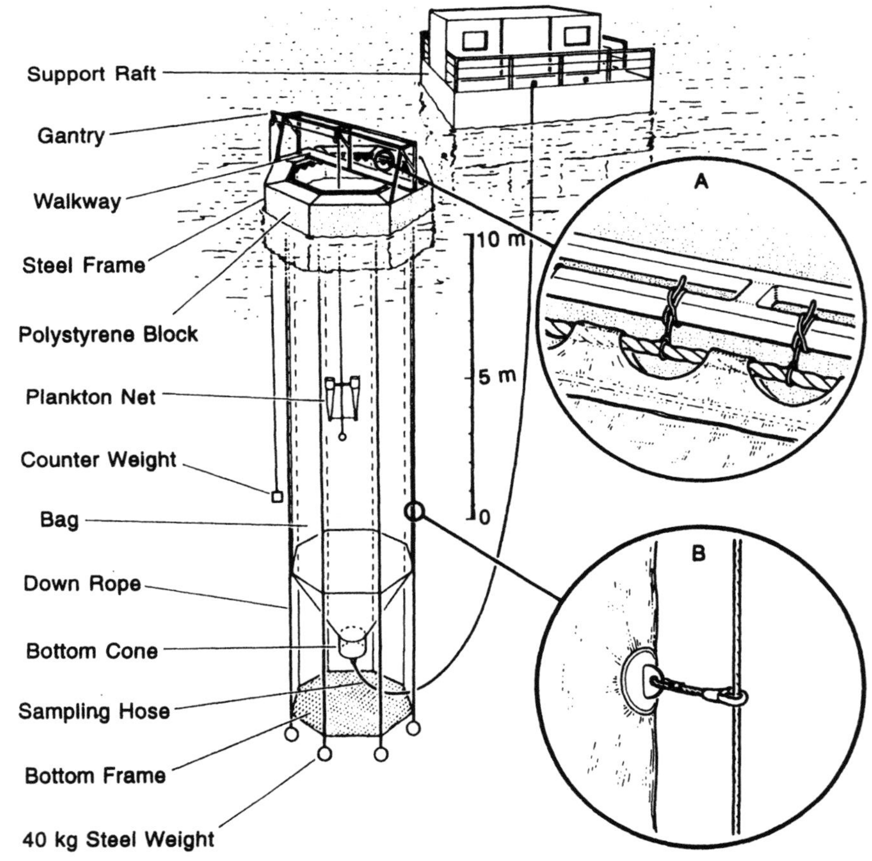

Figure 9. Schematic of Loch Ewe Enclosure. (From Grice and Reeve, 1982. Reprinted with permission from Springer-Verlag, New York, Inc.)

through the enclosure to mix the assemblage. The assemblage will be, of course, a function of the time and season of filling. The community in the enclosure may or may not reflect changes in the ambient community. Regardless of how the enclosure is filled, it should be sampled to ascertain the similarity with the surrounding water column. Because of the patchy nature of pelagic communities, it is very difficult to enclose a community that is similar to the entire pelagic community. For this reason the initial conditions in otherwise identical limnocorrals can be very different, which will affect the ability to draw statistically significant conclusions about the effects of added test chemicals. One way to minimize these initial differences and subsequent diversion of the biological communities among enclosures is to pump water from one enclosure to another for several days before adding the test chemical.

There are a number of reasons why the biological communities contained within enclosures will differ from that in the surrounding water column. The primary determining factors are a decrease in water currents and the effects of periphyton growth on the walls. Nutrient depletion inside the enclosure may lead to very different responses inside and outside the enclosure. Also, it is difficult to include fish in enclosure studies, because the relatively small volume will not provide sufficient production to sustain their growth. For this reason, it is important to be sure that all fish are excluded from the enclosures (Steele and Gamble, 1982). The inadvertent enclosure of fish in some limnocorrals, but not others, may cause divergent results in the enclosures that are independent of the treatments (Liber et al., 1993). These types of artifacts are the greatest limitations in the use of enclosures.

Limnocorrals, like other simulated aquatic field studies, have been used for basic ecological research and applied toxicological evaluations of the impact of chemicals on the structure and function of freshwater and marine communities. A variety of end points have been measured using these test systems (Table 6). The interactions of predator-prey communities have been studied in detail and the importance of planktivorous fish density on the composition and abundance of zooplankton was studied in two lakes in central Ontario (Mazumder et al., 1990). These factors were related to nutrient dynamics in the systems. Basic zooplankton dynamics have been studied in other systems (Smyly, 1976; Stephenson et al., 1984).

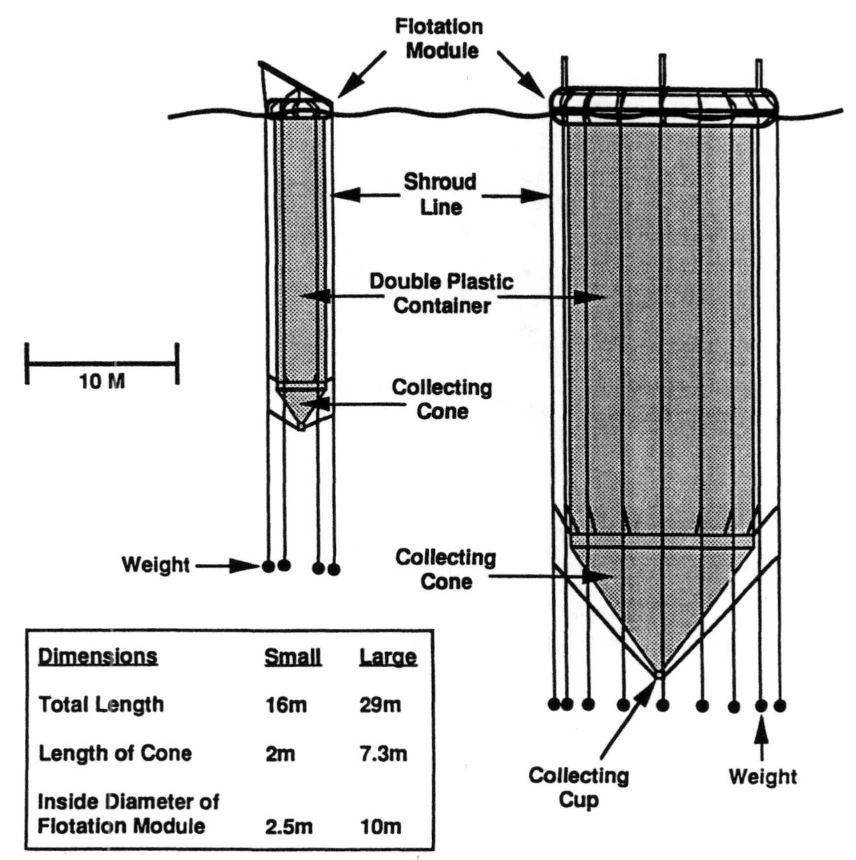

Figure 10. Diagram of CEPEX.

Table 5. Sizes and materials used in constructing limnocorrals

Enclosure	Diameter (m)	Length (m)	Volume (l)	Materials
POSER Brockman et al., 1974	1	40	3.0×10^4	Polyethylene inside, polyamide outside
Kiel plankton tower von Bodungen et al., 1976	2	16.5	1.0×10^5	Nylon-coated polyethylene
Den Helder Kuiper, 1977a	20	3.5	1.6×10^4	Nylon-coated polyethylene
Loch Ewe Gamble et al., 1977	4.7	17	2.5×10^5	Vinyl-reinforced polyethylene
Blelham Tarn Lack and Lund, 1974	45.5	16	2.3×10^6	Butyl rubber
MELIMEX Gachter, 1979	12	12.5	4.7×10^5	Rubber/foil polyethylene
Pallanza				
Ravera and Annoni, 1980	1	3	9.4×10^4	Nylon mesh–reinforced polyethylene
Havens and DeCosta, 1985	1	2	6.3×10^3	Polyethylene
Salki et al., 1984	10	4	1.3×10^5	Polyethylene
Goldman, 1962	0.6	10	1.8×10^4	Polyethylene
McAllister et al., 1961	6	1.1	6.6	Polyethylene
Blin et al., 1977	0.6	3.8	6.5	Nylon-reinforced polyethylene
Kirstiz, 1978	0.8	1	2.5×10^3	Aluminum
Devek, 1976	6	12	7.2×10^6	Split fiber nylon
Uehlinger et al., 1984	3	11	1.0×10^5	Polyethylene
CEPEX				
Grice et al., 1980	9.5	23.5	1.6×10^6	Nylon-reinforced polyethylene
Gamble et al., 1977	3	17	1.6×10^5	Nylon-reinforced PVC
Kuiper, 1977a	0.75	0.6	1.4×10^3	PVC inside, polyamide outside
Shapiro, 1973	1	1.5	4.7×10^3	Polyethylene

Most of the limnocorral research in freshwater systems has addressed aquatic toxicity questions. Solomon and co-workers at the University of Guelph have studied the impact of a variety of chemicals on aquatic pelagic communities (Solomon et al., 1986; Stephenson et al., 1986; Kaushik et al., 1985; Hamilton et al., 1989; Herman et al., 1986; Liber et al., 1992). Most of the studies have been conducted in Lake St. George in southern Ontario. The plastic walls of the corrals were generally anchored in the sediments and fish were often excluded from the test systems, although this was not always possible. Chemicals evaluated in these systems included the herbicide atrazine and the insecticides permethrin and methoxyclor. Other researchers have used plastic enclosures to study the effects of acid and aluminum (Havens and Heath, 1989; Havens and DeCosta, 1985).

Relative to marine systems, limnocorrals that enclose and isolate a portion of a natural system have been used to study plankton dynamics, benthic community processes, and system-level effects of pollution stresses such as oil spills, nutrient enrichment, and additions of metals (Gearing, 1989; Pilson,

Table 6. Toxicants and end points studied in limnocorrals

Chemical	Reference	End point
Mercury	Koeller and Wallace, 1977 Koeller and Parsons, 1977	Growth of chum salmon
Mercury	Beers et al., 1977	Zooplankton community structure
Napthalene	Lee and Anderson, 1977	Phytoplankton and zooplankton
Mercury	Kuiper, 1977b	Phytoplankton and zooplankton
Mercury	Blin et al., 1977	Primary Productivity
Emulsifiers	Lacaze, 1971	Plankton
2,4-Dichlorophenol	Kuiper and Hanstveit, 1984a	Plankton
2,4-Dichloroaniline	Kuiper and Hanstveit, 1984b	Plankton
Copper	Vaccaro et al., 1977	Bacteria, phytoplankton, zooplankton

1990). An international consortium of institutions implemented the Controlled Ecosystem Pollution Experiment (CEPEX) to look at long-term effects of low levels of pollution stress on marine, planktonic ecosystems (Menzel and Case, 1977). The test system for these studies was a moored flotation collar with polyethylene sheeting configured to form a large cylindrical enclosure with a cone-shaped bottom that facilitated collection of settled particulates (Figure 10). Systems containing 68 or 1335 m^3 of seawater were utilized for a number of experiments (Menzel and Case, 1977). Testing focused on pollutant effects on primary and secondary producers, with several studies providing particular attention to effects on the food web interactions of phytoplankton, zooplankton, and larval or juvenile fishes (Grice and Menzel, 1978; Thomas et al., 1977; Beers et al., 1977a,b; Koeller and Parsons, 1977; Koeller and Wallace, 1977). These systems maintained a number of trophic levels over an extended period of time and were designed to allow zooplankton to migrate in relation to vertical phytoplankton distribution. Selective nutrient additions and different nutrient regimens were studied in the 1335-m^3 enclosures to test the hypothesis that the structure of the phytoplankton community determines the type and number of links in the food chain and controls the transfer efficiency from primary production to upper trophic levels such as fish (Greve and Parsons, 1977; Steele and Frost, 1977). CEPEX-type experiments have been listed and reviewed by Gearing (1989). Other marine systems (Figures 7, 8, and 9) have employed different designs to allow study of plankton and sediment interactions, to utilize various mooring systems adapted to site characteristics, or to incorporate design features that enhance operational access (Lacaze, 1974; Grice and Reeve, 1982; Lalli, 1990).

There are a number of advantages and disadvantages of the use of limnocorrals to study the effects of chemicals on pelagic aquatic communities (Giesy and Odum, 1980; Davies and Gamble, 1979; Verduin, 1969) (Table 7). However, often the greatest advantage of these systems is also their greatest limitation. For example, by bounding a water mass in limnocorrals so that it can be studied, there can be no exchange (in a natural way) with surrounding water. This can have a serious affect on natural mixing. It would be impossible to make the limnocorral large enough to mimic the patchy, dynamic nature of the surrounding water. In addition, currents and light penetration can be altered by the presence of the enclosure and the walls can have a direct effect by acting as a substrate for growth and by excluding predators. All of these factors have a dominant role in controlling the structure of planktonic communities that develop within the limnocorral.

Littoral Enclosures

A modification of the limnocorral concept has been developed cooperatively by researchers from the U.S. EPA Environmental Research Laboratory in Duluth, Minnesota, and the University of Wisconsin-Superior (Brazner et al., 1989). In an attempt to evaluate a different component of aquatic ecosystems, these researchers studied the littoral community of freshwater ponds by isolating these areas via plastic enclosures (Figure 11). These enclosures have a natural shoreline and three plastic walls constructed to isolate them from the natural pond. The littoral enclosures are approximately 10 m long, 5 m wide, and 1 m deep and have a total volume of approximately 30,000 L. The systems were developed primarily to study the effects of pesticides on freshwater ecosystems.

Construction of the enclosures is fairly simple and cost-effective. Wooden walkways are first constructed around the perimeter of each enclosure area. This provides a support structure for isolating the littoral region and a platform for chemical and biological sampling of the enclosed area. Walls of plastic (0.20 μm, woven-filament, polyolefin plastic) and slatted fencing are attached to the support structure. The plastic is firmly pushed into the sediment to ensure a "seal" of the enclosure. During construction, it is also critical to exclude fish species from the enclosure or, if this is not possible, ensure that a similar number is included in each enclosure. Failure to do this will introduce variability in the test systems, thus impeding the sensitivity of the systems to identify adverse effects.

Table 7. Advantages and limitations of limnocorrals

Advantages
1. Generic advantages outlined in Table 2 (page 263).
2. The water contained within the corral is natural and of known origin.
3. The body of enclosed water can be manipulated without treating the entire system.

Limitations
1. The enclosures are generally not large enough to contain ecologically relevant populations of large carnivores such as fish.
2. The systems are not large enough to simulate the "patchy" nature of naturally occurring pelagic ecosystems.
3. Enclosures are expensive to construct and maintain.
4. It is difficult to uniformly close the enclosure in an ecologically relevant way and maintain the desired concentration of test chemical.
5. The "wall" effects result in a very different system than a natural pelagic community and can affect the chemical dynamics of nutrients and test chemicals.
6. The enclosures restrict vertical and horizontal diffusions and advection which are important in structuring communities and affecting chemical dynamics.
7. The chemical and physical properties of the body of water enclosed in limnocorrals changes relative to the adjacent water as a function of time.
8. The chemical and biological properties of the water bodies in different, replicate limnocorrals can diverge due to different critical conditions or replicate-specific perturbations.

Greater detail on the construction procedures is provided by Brazner et al. (1989).

Littoral enclosures provide a unique tool for studying the effects of pesticides and other chemicals on a critical component of aquatic ecosystems. An advantage of such systems is that they are isolated, naturally derived components that can be replicated in a rather simple and cost-effective manner. Variability should be rather low in such systems because they simply enclose the natural shoreline, water column, and benthic sediment of the same pond. Most of the biological and water quality end points described for mesocosms can be evaluated in these test systems.

Littoral enclosure test systems have a number of limitations. Relative to end points, the main limitation is the degree to which fish reproductive effects can be evaluated. Although successful studies have been conducted which evaluated bluegill reproduction, fish population studies are difficult to conduct because of the currently defined size of the test systems. In addition, as with the microcosm test systems, care must be taken not to oversample the populations, especially the univoltine invertebrates. Another potential disadvantage of these test systems is the ability to find appropriate test sites. Fluctuations in water levels in many areas of the United States, due to rainfall and/or evaporation, may limit

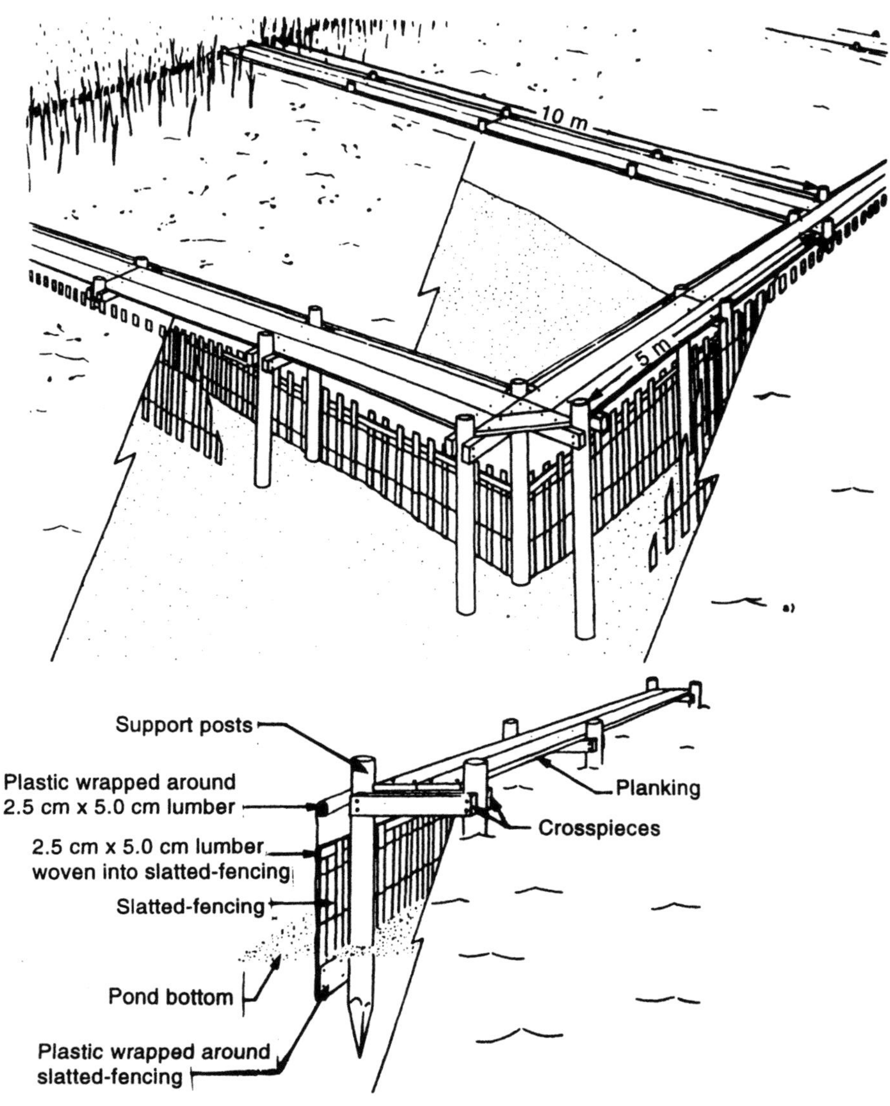

Figure 11. Schematic diagram of a littoral enclosure (Reprinted with permission from *Environmental Toxicology and Chemistry* 8(12): 1209–1216, Brazner, J. C., Heinis, L. J., Jensen, D. A. A littoral enclosure for replicated field experiments, Copyright 1989 SETAC.)

broad applicability of this approach. In addition, it is difficult to find truly uniform shorelines for construction of multiple enclosures. Variations in shoreline development in freshwater ponds can be substantial and placement of the test systems in areas that can be considered "replicates" may be difficult.

The test systems developed at Duluth, Minnesota have been used to conduct extensive evaluations of two agricultural chemicals: chlorpyrifos and esfenvalerate. The first series of tests, conducted during the development of the littoral enclosure methodologies, were performed with the organophosphate insecticide chlorpyrifos (Siefert et al., 1989; Brazner et al., 1988, 1989). Detailed information on the environmental fate and ecological effects of chlorpyrifos was obtained. Overall, water residue concentrations were reproducible among replicates and the water column half-life was found to be relatively short. Surface water concentrations were higher than subsurface and mixing within the enclosures was slow. Considerable partitioning of chlorpyrifos to other compartments such as sediment, plants, and sidewalls occurred. Both primary and secondary effects on fish and invertebrates were observed. Overall, effects were observed at concentrations lower than those predicted by laboratory toxicity tests.

Esfenvalerate is a pyrethroid insecticide highly toxic to both fish and invertebrates. Tests performed in littoral enclosures confirmed the sensitivity of aquatic invertebrates (Lozano et al., 1992). Effects on certain species were observed at concentrations as low as 0.08 μg/L. The results of the residue analysis indicate an extremely short water half-life and a relatively short sediment half-life. The adsorptive properties of esfenvalerate, as with all pyrethroids, have been proposed to mitigate effects predicted to occur from laboratory toxicity tests (Hill, 1985). In this study, the potential for mitigation was not evaluated because the water column was low in suspended solids and organic matter.

Marine Coastal Enclosures

Enclosures designed for studying benthic communities in intertidal or nearshore subtidal habitats must overcome the erosional and depositional forces of tidal exchange along the shoreline (de Wilde, 1990). Benthic mesocosm enclosures can exclude large, mobile predators such as fish and crabs and control material inputs and biota colonization, in an attempt to yield a more uniform distribution of biota within the substrate. Benthic test systems can range from complex, engineered systems designed to control multiple variables (Farke et al., 1984) to relatively simple nylon mesh, predator exclusion-inclusion screens installed at selected sites within a mudflat (Livingston, 1988).

The Bremmerhaven caisson (Farke et al., 1984) provide an example of large-scale engineering and construction that can be committed to study shallow-water, benthic community structure and function by using controlled, in situ enclosures for extended periods of time. This open-top and open-bottom test system is floated to the study site (Figure 12). When lowered and anchored into position, it encloses a 5.6 $\times$ 2.3 m area with walls extending 2 m above the sediment and plates extending some variable depth into the sediment. The system thus controls exchange of overlying and interstitial pore water as well as immigration and emigration of macrobiota. Access for sampling the substrate is provided by a movable bridge system suspended above the sediment. This system has been used to study effects of lead and chromium (Schulz-Baldes et al., 1983) on the pro-

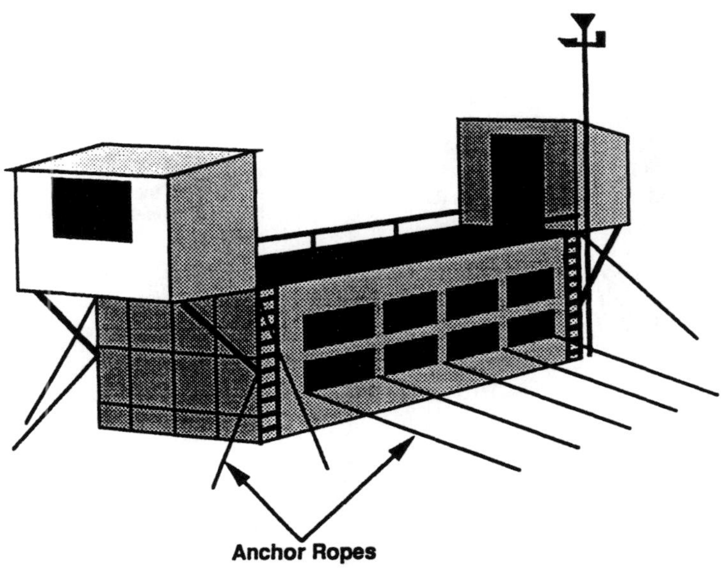

Figure 12. Bremmerhaven caisson.

Figure 13. Artificial stream system of the Water, Soil and Air Hygiene Office, Marienfelde, Germany. Facility has been used to study the effects of sewage, nutrients, and detergents on stream ecosystem. Large building in the center is a pilot sewage treatment plant which contains automated sampling equipment. (Photo by P. D. Hansen.)

ductivity and food web interactions within the benthic community.

Enclosures that couple planktonic and benthic compounds of marine ecosystems have been used to study fate and effects of contaminant releases and have application to understanding basic ecological processes (de Wilde, 1990). Systems discussed by Lacaze (1974), von Bodungen et al. (1976), and Smetacek et al. (1976) were deployed in nearshore environments and allowed study of interactions between planktonic and subtidal benthic communities. Outdoor model tidal flats (MOTIFs) are presently being used at the Netherlands Institute for Sea Research (NIOZ) to study the effects of oil spills and contaminated harbor sludge on coastal and intertidal ecosystems. These studies provide an example of approaches to complex fate and effects studies necessary to fully understand impacts of contaminants on marine ecosystems.

Lotic Systems

Artificial channels or artificial streams have been used to study basic ecological principles of lotic ecosystems as well as the effects of toxic chemicals on specific components of the ecosystems and the interactions among and between species and their en-

Table 8. Advantages and limitations of artificial streams

Advantages

1. Generic advantages outlined in Table 2 (page 263).
2. Represents unique properties of flowing water systems not present and evaluated in lentic systems.

Limitations

1. Most artificial streams are not large enough to support a completely natural fish fauna.
2. The nature of water movement or currents are generally not the same as the system being simulated.
3. The substratum, including refugia, is generally not similar to the entire system one is trying to simulate.
4. It is difficult to maintain a source of colonizing organisms, which are present in natural systems.
5. It is difficult to obtain a uniform, ecologically relevant concentration in the channels.
6. The chemical and physical properties of channels tend to diverge due to relative positioning and initiation.
7. Artificial stream facilities are expensive and time-consuming to operate.
8. It is difficult to treat or dispose of the large volumes of contaminated water during the study and sediments after the study has been completed.
9. It is difficult to maintain ecologically relevant riparian zones to simulate "external" controls important to most stream communities.

Table 9. Representative artificial stream studies with contaminants (reprinted with permission from Kosinski, 1989)

Toxicant	Tr.	Ch.	Dimensions	Cur.	R&P	Circ.	Time	Exp	Biota	I/G/O	Design	Reference
Anthracene	2	2					36	C	P	O		Bowling et al., 1984
Acid, Al	3	3	6 × 0.215 × 0.24		N	FT	41 + 85	C	P + I	O	NRT	Allard and Moreau, 1985
Heat	4	12	114 × 4.3 × 2	0.23P 0.93R	Y	FT	67 + 177	C	P	O	RT	Armitage, 1980
Zn	3	3			N	R	21 + 30	C	I	I + O	NRT	Belinger et al., 1986
Heat	2	2	20 × 1.3 × 0.8	60	Y			C	F	O	NRT	Bisson and Davis, 1976
NTA[a]	5	5	0.56 × 0.23		N	R	10	C	P	G	NRT	Bott et al., 1977
NTA[a] + Cu	4	4	7.3 × 0.6 × 0.18	30–42	Y	FT or R	730			G	NRT	
Acid + Al	3	3	2.1 × 0.6 × 0.56			R	28 + 28	C	I	I	NRT	Burton and Allan, 1986
Description		6	4 m long	9	N	R	14		P	O		Clark et al., 1980
p-Cresol	2	2	1040 × 3.9 × 0.8	12.2P 25.0R	Y	R R		I	I + F	O	NRT	Cooper and Stout, 1984
Sediments	6	12	6.6 m long			PR	180 + 90	P	I + F	G	RT	Crouse et al., 1981
Zn	1	1	1 m long		N	FT or R			P	I	NRT	Cushing and Rose, 1971
Heat	3	3	0.86 × 0.27 × 0.1		N	R		C	I	I	NRT	de Kozlowski and Bunting, 1981
Chlorpyrifos	3	3	245 × 3.9 × 0.8		Y	FT	100	C + I	I + F	O	NRT	Eaton et al., 1985
Treated sewage	4	4	200 × 0.35 × 0.2	15–25	N	FT	365 + 730	C	P	O	NRT	Eichenburger, 1967
Co, Cu, Zn	5	8	75 × 0.2 × 0.2	8–12				C	P	O	NRT	Eichenburger et al., 1981
Coal leachate	2	5	2.3 × 0.1	10–30	N	PR		C	P	I	RT	Gerhart, 1977
Volcanic ash			3.2 × 0.5 × 0.15	18	N	R		P	I	I		Gersich and Brusven, 1982
Cd	3	6	91.5 × 0.61 × 0.31	1.3	Y	FT	365	C	B,Fn,P,M,I,F	O	RT	Giesy et al., 1979
Cd	8	4	2 × 0.15 × 0.15		N	FT or R	14	C	I	I	NRT	Graney et al., 1984
Cd, Cu, Zn	4	6	3.5 × 0.7 × 0.3		N		30	C	I	I	NRT	Graney et al., 1983
Diflubenzuron	5	10	12.2 m long	30	Y	PR	90 + 150	C	B,Fn,P,I	I	RT	Hansen and Garton, 1982
Description		6	6.7 × 0.3 or 0.6		Y	FT		C	P,I,M	G		Harvey, 1973
PCP	4	4	152 × 3.9 × 0.8	1.4	Y	FT	84	C	B,P,I,F	O	NRT	Hedtke and Arthur, 1985
NH_4Cl	12	12	6 × 0.25 × 0.2	24	N	PR		C	F	I	NRT	Hedtke and Norris, 1980
Sewage		12	5 × 0.15 × 0.2	40–74	N	FT	23	C	P	O		Hoffman and Horne, 1980
Cu	4	4	3.5 × 0.7 × 0.3	8	N		7 + 18	C + I	P	I	NRT	Kaufman, 1982
Atrazine, trifluralin, MSMA,[b] paraquat	8	32	2.4 × 0.12 × 0.1	3	N	PR	28 + 21	P	P	O	RT	Kosinski, 1984; Kosinski and Merkle, 1984
Cl, chloramines	4	12	6.6 × 0.66 × 0.25	24	Y	PR	240 + 54	C	P + I	G	RT	Larson et al., 1978
Atrazine, HCBP[c]	4	6	4 × 0.58 × 0.27		N	FT	365 + 90	C	P,I,M,F	I	NRT	Lynch et al., 1985
TFM	2	6	8 × 0.6 × 0.25	2.4	Y	FT	300 + 3	P	P,I	I	RT	Maki and Johnson, 1976
TFM			4.86 × 0.29 × 0.19	15	N	R	1 + 14	P	M	I		Maki and Johnson, 1977
N + P	6	6	3.46 × 0.29	26	N	PR			P	O	NRT	Manuel and Minshall, 1980
N + P	4	4	20 × 0.5	45	N	FT	10	C	P	O	NRT	
Atrazine	4	16	2.4 × 0.12 × 0.1	3	N	PR	35 + 7	P	P	O	RT	Moorhead and Kosinski, 1986
Permethrin, temephos			0.25 × 0.04 × 0.005	30	N	FT	1	C	I	I		Muirhead-Thompson, 1977
Heat	2	2	520 × 3.9 × 0.8	1.4	Y	FT	365	C	I	O	NRT	Nordlie and Arthur, 1981
PCP	4	4	488 × 3.9 × 0.8	1.4	Y	FT	84 or 107	C	B	O	NRT	Pignatello et al., 1983

Table 9. Representative artificial stream studies with contaminants (reprinted with permission from Kosinski, 1989) (*Continued*)

Toxicant	Tr.	Ch.	Dimensions	Cur.	R&P	Circ.	Time	Exp	Biota	I/G/O	Design	Reference
PCP	3	3	180 × 3.4 × 0.8	1.4	Y	FT	88	C	B	O	NRT	Pignatello et al., 1985
Heat	4	12	112 × 4.3 × 2	0.23P 0.93R	Y	FT		C	I	O	RT	Rodgers, 1980
Cu, Cr, $Ca(OCl)_2$		6	4 × 0.35	9	N	FT	20	C	P	I		Rodgers et al., 1980
Dieldrin	3	6	6 × 0.3 × 0.18	28 or 0		PR	60 + 120	C	P	G		Rose and McIntre, 1970
	4	4										
	4	6										
Temephos, chlorpyifos		1	4 × 0.36	9 or 15	N	R			I	I		Ruber and Kocor, 1976
Kraft mill effluent	3	6	6.6 × 0.66 × 0.25	10	Y	PR		C	I,F	G	RT	Seim et al., 1977
Description		5	6 × 0.22 × 0.24		N	FT		C	P,I	O		Serodes et al., 1984
Hg	3	6	91.5 × 0.61 × 0.31	1.3	Y	FT	150 + 365	C	P,I	O	RT	Sigmon et al., 1977
p-Cresol	2	2	1000 × 3.9 × 0.8	12.2	Y	R		C + I	I	O	NRT	Stout and Cooper, 1983
p-Cresol	2	2	1000 × 3.9 × 0.8	12.2	Y	R			P	O	NRT	Stout and Kilham, 1983
Sucrose	4	1	152 × 1.5		N	FT		P	I,F	O	NRT	Warren et al., 1964
Treated sewage	3	3	300 × 1.25 × 0.4	10R 43P	Y	FT	620	C	I	O	NRT	Watton and Hawkes, 1984
Heat	5	6	6.7 × 0.3 × 0.6	8.2	Y	FT	1095 + 365	C	P	G	NRT	Wilde and Tilly, 1981
Zn	4	4	6.1 × 0.3 × 0.2	0.056	N	FT	98	C	P + I	O	NRT	Williams and Mount, 1965
Description		4			N	FT		C	I	O		Wilton and Travis, 1965
Heat	2	2	10 × 0.15	30	N	FT		C	I	O	NRT	Wojtalik and Waters, 1970
Heat	4	12	112 × 4.3 × 2	0.23P 0.93R	Y	FT	1095 + 185	C	P,I,Z,F	O	RT	Wrenn et al., 1979
Heat	4	12	112 × 4.3 × 2	0.23P 0.93R	Y	FT	1095 + 185	C	F	O	RT	Wrenn and Grannemann, 1980
Temephos, chlorphoxim	2	2	100 × 0.23	12	N	FT		I	P,I	O	NRT	Yasuno et al., 1985
PCP	4	4	520 × 3.9 × 0.8		Y	FT	7 + 20	C	P	O		Yount and Richter, 1986
Acid	3	3	122 × 3.9 × 0.8		Y	FT	119	C	I + F	O	NRT	Zischke et al., 1982

[a] Nitrilotriacetic acid.
[b] Monosodium methanarsonate.
[c] Hexachlorobiphenyl.

Toxicant — Test chemical studied; if "description" is listed for toxicant, it means that the paper described the system rather than reporting results.
Tr. — Refers to number of treatments.
Ch. — Refers to number of channels.
Dimensions — Stream dimensions in meters.
Cur. — Current speed in cm/s, with pool and riffle speeds indicated by P and R, respectively.
R&P — Refers to presence (*Y*) or absence (*N*) of both riffles and pools.
Circ. — FT = flow-through, R = recirculating, PR = partially recirculating.
Time — Length of experiment in days.
Exp. — C = continuous, I = intermittent, P = pulsed, exposure (Exp.).
Biota — Organisms studied; B = bacteria, Fn = fungi, P = periphyton, Z = zooplankton, M = macrophytes, I = macroinvertebrates, F = fish.
I/G/O — Studies conducted: I = indoor, G = greenhouse, O = outdoor.
Design — NRT = nonreplicated treatments; RT = replicated treatments.

Source: Reprinted with permission from Kosinski RJ: Artificial streams in ecotoxicological research, in Aquatic Ecotoxicology: Fundamental Concepts and Methodologies, Vol. I, edited by A. Boudou, F. Ribeyre, Chapter 11. Copyright 1989. CRC Press, Boca Raton, FL.

vironment (Giesy, 1980; Kosinski, 1989). The primary purposes for the use of artificial streams are the same as those for other multispecies tests, which have already been given. Their special feature is that they contain flowing water, which is an important structuring determinant of riverine ecosystems. Artificial streams have been small (2 × 10 feet) or large ($^1/_2$ mile long) and conducted under either field or laboratory conditions (Giesy, 1980; Kosinski, 1989) (Figure 13). The realism gained by the use of artificial streams must be balanced against the cost, which restricts the number of replicate channels that can be used (Kosinski, 1989). The advantages and limitations of artificial streams are outlined in Table 8.

Outdoor artificial channels can be constructed of many materials. These include soil, either natural or selected, concrete, concrete covered with a persistent coating such as epoxy paint, and concrete or soil covered with removable liners, such as PVC. Some channels have been made of metal, but these are generally not really artificial streams or channels but rather flumes.

A number of different artificial stream designs have been used to study the effects of a number of chemicals. In his review of studies of toxicants and other stressors in artificial streams, Kosinski (1989) identified 57 studies in which artificial streams or channels were used to study the effects of contaminants on aquatic systems and summarized the information, which is reproduced here as Table 9.

The artificial stream system located at the Shell Research facility in Sittingbourne, UK, is illustrated in Figure 14. This is a partially recirculating system that contains both pools and riffle areas. Studies of changes in the structural and functional components of stream communities have been conducted using this system for petrochemical effluents (Crossland and Mitchell, 1992) and lindane (Crossland et al., 1991).

CONCLUSIONS

An extensive amount of field research has been conducted on the impact of a variety of stressors on the structural and functional aspects of aquatic communities. Aquatic field toxicology research is expensive and difficult to conduct in a statistically reliable fashion so that definitive conclusions can be drawn. The trend in the area of scientific and regulatory science is towards controlled SFSs designed to answer specific questions. It is hoped that the days of ''shotgun'' aquatic toxicology field studies are in the past. Not all possible end points can be measured in all studies. Extensive research is needed on the development of better, more defined studies. Decisions need to be made on the ability to extrapolate results to the real world, and regulators need to better define the specific end points on which regulatory decisions will be based.

Figure 14. Photograph of recirculating artificial stream system developed by Shell Research (England).

LITERATURE CITED

Adair, H. M.; Harris, F. A.; Kennedy, M. V.; Laster, M. L.; and Threadgill, E. D.: Drift of methyl parathion aerially applied low volume and ultra low volume, *J. Econ. Entomol.*, 64:718–721, 1971.

Ali, A.; Nigg, N.; Stamper, H.; Kok-Yokomi, M. L.; and Weaver, M.: Diflubenzuron application to citrus and its impact on invertebrates in an adjacent pond. *Bull. Environ. Contam. Toxicol.*, 41:781–790, 1988.

Allard, M., and Moreau, G.: Short-term effect on the metabolism of lotic benthic communities following experimental acidification. *Can. J. Fish. Aquat. Sci.*, 42:1676–1680, 1985.

Allred, P. M., and Giesy, J. P.: Solar radiation induced toxicity of anthracene to *Daphnia pulex*. *Environ. Toxicol. Chem.*, 4:210–226, 1985.

APHA, AWWA, and WPCF: Standard Methods for the Examination of Water and Wastewater, 17th ed. Washington, DC: APHA, 1989.

Anonymous: Improving Aquatic Risk Assessment under FIFRA. Report of the Aquatic Effects Dialogue Group. Resolve. World Wildlife Fund, Washington, DC, 1992.

Armitage B: Effects of temperature on periphyton biomass and community composition in the Browns Ferry Experimental Channels. Microcosms in Ecological Research, edited by J. P. Giesy, pp. 668–683. DOE Symposium Series 52, 1980.

Arnold, J. G.; Williams, J. R.; Nicks, A. D.; and Sammons, N. B.: SWRRB: A Basin Scale Simulation Model for Soil and Water Resources Management. College Station, TX: Texas A & M Univ. Press, 1990.

Baccini, P.; Ruchti, J.; Warner, O.; and Grieder, E.: MELIMEX, an experimental heavy metal pollution study: regulation of trace metal concentrations in limnocorrals. *Schweiz. Z. Hydrol.*, 41:202–227, 1979.

Bagenal, T.: Methods for Assessment of Fish Production in Fresh Waters. IBP Handbook Number 3. Boston, MA: Blackwell Scientific Publications, 1978.

Barbour, M. T.; Pflakin, J. L.; Bradley, B. P.; Graves, C. G.; and Wisseman, R. W.: Evaluation of EPA's rapid bioassessment benthic metrics: metric redundancy and variability among reference stream sites. *Environ. Toxicol. Chem.*, 11:437–449, 1992.

Barnthouse, L. W.; Suter, G. W., II; Rosen, A. A.; and Beauchamp, J. J.: Estimating responses of fish populations to toxic contaminants. *Environ. Toxicol. Chem.*, 6: 811–824, 1987.

Beers, J. R.; Reeve, M. R.; and Grice, G. D.: Controlled ecosystem pollution experiment: effect of mercury on enclosed water columns. IV. Zooplankton population dynamics and production. *Mar. Sci. Comm.*, 3:355–394, 1977a.

Beers, J. R.; Steward, G. L.; and Hopkins, K. D.: Dynamics of microzooplankton populations treated with copper: controlled ecosystem pollution experiment. *Bull. Mar. Sci.*, 27:66–79, 1977b.

Belanger, S. E.; Farris, J. L.; Cherry, D. S.; and Cairns, J., Jr.: Growth of Asiatic clams during and after long-term zinc exposure in field-located and laboratory artificial streams. *Arch Environ. Contam. Toxicol.*, 15:427–434, 1986.

Bender, M. E., and Jordon, R. E.: Plastic enclosure versus open lake productivity measurements. *Trans. Am. Fish. Soc.*, 99:607–610, 1970.

Bisson, P. A., and Davis, G. E.: Production of juvenile chinook salmon, *Oncorhynchys tshawyncha*, in a heated model stream. *Fish Bull.*, 74:763–774, 1976.

Blinn, D. W.; Tompkins, T.; and Zaleski, L.: Mercury inhibition on primary productivity using large volume plastic chambers "in situ." *J. Phycol.*, 13:58–61, 1977.

Bott, T. L.; Preslan, J.; Finlay, J.; and Brunker, R.: The use of flowing-water streams to study microbial degradation of leaf litter and nitrolotriacetic acid (NTA). *Dev. Ind. Microbiol.*, 18:171–184, 1977.

Bower, P., and McCorkle, D.: Gas exchange, photosynthetic uptake and carbon budget for a radiocarbon addition to a small enclosure in a stratified lake. *Can. J. Fish. Aquat. Sci.*, 37:464–471, 1980.

Bowling, J. W.; Giesy, J. P., Jr.; Kania, H. J.; and Knight, R. L.: Large scale microcosms for assessing fates and effects of trace contaminants. Microcosms in Ecological Research, edited by J. P. Giesy, Jr., pp. 224–247. DOE Symposium Series 52. Springfield, VA: T National Technical Information Center, 1980.

Bowling, J. W.; Haddcock, J. D.; and Allred, P. M.: Dissipation of Anthracene in Water and Aufwuches Matrices of a Large, Outdoor Channel Microcosm: A Data Set for Mathematical Simulation Models. EPA 600/S3-84-036, 1984.

Boyd, C. D.: Microcosms and experimental planktonic food chains. Analysis of Marine Ecosystems, edited by A. R. Lonhurst, pp. 627–649. New York: Academic Press, 1981.

Boyle, T. P.: Effects of the aquatic herbicide 2,4-D DMA on the ecology of experimental ponds. *Environ. Pollut. A.*, 21:35–49, 1980.

Boyle, T. P.; Finger, S. E.; Paulson, R. L.; and Rabeni, C. F.: Comparison of laboratory and field assessment of fluorene. Part II: Effects on the ecological structure and function of experimental pond ecosystems. Validation and Predictability of Laboratory Methods for Assessing the Fate and Effects of Contaminants in Aquatic Ecosystems, edited by T. P. Boyle, pp. 134–151. ASTM STP 865. Philadelphia: ASTM, 1985.

Brazner, J. C.; Lozano, S. J.; Knuth, M. L.; Bertelsen, S. L.; Heinis, L. J.; Jensen, D. A.; Kline, E. R.; O'Halloran, S. L.; Sargent, K. W.; Tanner, D. K.; and Siefert, R. E.: The Effects of Chloropyriphos on a Natural Aquatic System: A Research Design for Littoral Enclosure Studies and Final Research Report. CERL-600/X-88-150. Duluth, MN: U.S. EPA, 1988.

Brazner, J. C.; Heinis, L. J.; and Jensen, D. A.: A littoral enclosure for replicated field experiments. *Environ. Toxicol. Chem.*, 8:1209–1216, 1989.

Bridgham, S. D.; McNaught, D. C.; and Meadows, C.: Effects of complex effluents on photosynthesis in Lake Erie and Lake Huron. Functional Testing of Aquatic Biota for Estimating Hazards of Chemicals, edited by J. Cairns, Jr., and J. R. Pratt, pp. 74–85. ASTM STP 988. Philadelphia: ASTM, 1988.

Brockman, U. H.; Ederlein, K.; Junge, H. D.; Trageser, M.; and Trahms, K. J.: Einfach Folien-tanks zur Planktonuntersuchung in situ. *Mar. Biol.*, 24:163–166, 1974.

Brunskill, G. J.; Graham, B. W.; and Rudd, J. W. M.: Experimental studies on the effect of arsenic on microbial degradation of organic matter and algal growth. *Can. J. Fish. Aquat. Sci.*, 37:415–423, 1980.

Burnett, K. M., and Liss, W. J.: Multi-steady-state toxicant fate and effect in laboratory aquatic ecosystems. *Environ. Toxicol. Chem.*, 9:637–647, 1990.

Burton, T. M., and Allan, J. W.: Influence of pH, aluminum and organic matter on stream invertebrates. *Can. J. Fish. Aquat. Sci.*, 43:1285–1289, 1986.

Cairns, J., Jr.: Applied Ecotoxicology and Methodology. Aquatic Ecotoxicology: Fundamental Concepts and

Methodologies, edited by A. Boudou, and F. Ribeyre, pp. 275–290. Boca Raton, FL: CRC Press, 1989.

Cairns, J.; Lanza, G. R.; and Parker, B. G.: Pollution related structural and functional changes in aquatic communities with emphasis on freshwater algae and protozoa. *Proc. Acad. Nat. Sci. Phila.*, 124:70–127, 1972.

Carsel, R. F.; Smith, C. N.; Mulkey, L. A.; Dean, J. D.; and Jowise, P.: Users Manual for the Pesticide Root Zone Model (PRZM). EPA-600/3-84-109, 1984.

Chapman, P. M.: Sediment quality criteria from the sediment quality triad: an example. *Environ. Toxicol. Chem.*, 5:957–964, 1986.

Chapman, P. M.; Dexter, R. N.; and Long, E. R.: Synoptic measures of sediment contamination, toxicity and infaunal community composition (the sediment quality triad) in San Francisco Bay. *Mar. Ecol. Prog. Ser.*, 37: 75–96, 1987.

Clark, J. R.: Field studies in estuarine ecosystems: a review of approaches for assessing contaminant effects. Aquatic Toxicology and Hazard Assessment; 12th Volume, edited by U. M. Cowgill, and L. R. Williams, pp. 120–133. ASTM STP 1027. Philadelphia: American Society for Testing and Materials, 1989.

Clark, J. R., and Noles, J. L.: Pesticide effects in marine/esturaine systems: field studies and scaled simulations. Aquatic Mesocosm Studies in Ecological Risk Assessment, edited by R. L. Graney; J. H. Kennedy; and J. H. Rodgers, pp. 47–60. Boca Raton, FL: Lewis, 1993.

Clark, J.; Rodgers, J. H., Jr.; Dickson, K. L.; and Cairns, J., Jr.: Using artificial streams to evaluate perturbation effects on aufwuches structure and function. *Water Res. Bull.*, 16:100–104, 1980.

Coats, J. R., and O'Donnell-Jeffery, N. L.: Toxicity of four synthetic pyrethroid insecticides to rainbow trout. *Bull. Environ. Contam. Toxicol.*, 23:250–255, 1979.

Computer Sciences Corporation: Pesticide Runoff Simulator (SWRRB)—User's Manual. Falls Church, VA: U.S. EPA, 1980.

Cooper, W. E., and Stout, R. J.: Assessment of transport and fate of toxic materials in an experimental stream system. Modeling the Fate of Chemicals in the Aquatic Environment, edited by K. L. Dickson; A. W. Maki; and J. Cairns, Jr., pp. 347–378. Woburn, MA: Ann Arbor Science, 1982.

Crossland, N. O.: Aquatic toxicology of cypermethrin. II. Fate and biological effects in pond experiments. *Aquat. Toxicol.*, 2:205–222, 1982.

Crossland, N. O.: Fate and biological effects of methyl parathion in outdoor ponds and laboratory aquaria. II: Effects. *Ecotoxicol. Environ. Safety*, 8:482–495, 1984.

Crossland, N. O.: A method for evaluating effects of toxic chemicals on the productivity of freshwater ecosystems. *Ecotoxicol. Environ. Safety*, 16:279–292, 1988.

Crossland, N. O., and Bennett, D.: Outdoor ponds: their use to evaluate the hazards of organic chemicals in aquatic environments. Aquatic Ecotoxicology: Fundamental Concepts and Methodologies, edited by A. Boudou, F. Ribeyre, pp. 273–296. Boca Raton, FL: CRC Press, 1989.

Crossland, N. O.; Heimbach, F.; Hill, I. R.; Boudou, A.; Leeuwangh, P.; Matthiessen, P.; and Persoone, G.: Summary and recommendations of the European workshop on freshwater field tests. Potsdam, Germany: SETAC, 1992.

Crossland, N. O.; Mitchell, G. C.; and Dorn, P. B.: Use of outdoor artificial streams to determine threshold toxicity concentrations for a petrochemical effluent. *Environ. Toxicol. Chem.*, 11:95–59, 1992.

Crossland, N. O.; Mitchell, G. C.; Bennett, D.; and Maxted, S.: An outdoor artificial stream system designed for ecotoxicological studies. *Ecotoxicol. Environ. Safety*, 22: 175–183, 1991.

Crossland, N. O., and Wolff, C. J.: Fate and biological effects of pentachlorophenol in outdoor ponds. *Environ. Toxicol. Chem.*, 4:73–86, 1985.

Crouse, M. R.; Callahan, C. A.; Malueg, K. W.; and Domingeuz, S. E.: Effects of fine sediment on the growth of juvenile coho salmon in laboratory streams. *Trans. Am. Fish. Soc.*, 110:281–286, 1981.

Cushing, C. E., and Rose, F. L.: Cycling of zinc-65 by Columbia River periphyton in a closed lotic microcosm. *Limnol. Oceanogr.*, 15:762–767, 1971.

Cushman, R. M., and Goyert, J. C.: Effects of a synthetic crude oil on pond benthic insects. *Environ. Pollut. A.*, 33:163–186, 1984.

Davies, J. M., and Gamble, J. C.: Experiments with large enclosed ecosystems. *Philos. Trans. R. Soc. Lond. B. Biol. Sci.*, 286:523–544, 1979.

deNoyelles, F., Jr.; Kettle, W. D.; Fromm, C. H.; Moffett, M. F.; and Dewey, S. L.: Use of experimental ponds to assess the effects of a pesticide on the aquatic environment. Using Mesocosms to Assess the Aquatic Ecological Risk of Pesticides: Theory and Practice, edited by J. R. Voshell, Jr., *Misc. Publ. Entomol. Soc. Am.*, 75:41–56, 1989.

deNoyelles, F.; Kettle, W. D.; and Sinn, D. E.: The responses of plankton communities in experimental ponds to atrazine, the most heavily used pesticide in the United States. *Ecology*, 63(5):1285–1293, 1982.

Deutsch, W. G.; Webber, E. C.; Bayne, D. R.; and Reed, C. W.: Effects of largemouth bass stocking on fish populations in aquatic mesocosms used for pesticide research. *Environ. Toxicol. Chem.*, 11:5–10, 1992.

Devek, O.: Marine greenhouse systems: the entrainment of large water masses. *Environ. Sci. Res.*, 8:113–141, 1976.

Dewey, S. L.: Effects of the herbicide atrazine on aquatic insect community structure and emergence. *Ecology*, 67(1):148–162, 1986.

de Wilde, P. A. W. J.: Benthic mesocosms: I. Basic research in soft-bottom benthic mesocosms. Enclosed Experimental Marine Ecosystems: A Review and Recommendations, edited by C. M. Lalli, pp. 110–121. New York: Springer-Verlag, 1990.

de Wilde, P. A. W. J., and Kuipers, B. R.: A large indoor tidal mud-flat ecosystem. *Helgol. Wiss. Meeresunters.*, 30:334–342, 1977.

Di Toro, D. M.: A particle interactive model of reversible organic chemical sorption. *Chemosphere*, 14:1503–1538, 1985.

Downing, J. A., and Rigler, F. H.: A Manual on Methods for the Assessment of Secondary Productivity in Freshwaters. IBP Handbook Number 17. New York: Blackwell Scientific, 1984.

Dutka, B. J., and Kwan, K. K.: Study of long term effects of oil and oil-dispersant mixtures on freshwater microbial populations in man made ponds. *Sci. Total. Environ.*, 35:135–148, 1984.

Eaton, J.; Arthur, J.; Hermanutz, R.; Kiefer, R.; Mueller, L.; Anderson, R.; Erickson, R.; Nordling, B.; Rogers, J.; and Pritchard, H.: Biological effects of continuous and intermittent dosing of outdoor experimental streams with chloropyrifos. Aquatic Toxicology and Hazard Assessment, 8th Symposium, edited by R. C. Bahner, and D. J. Hansen, pp. 85–118. Philadelphia: American Society for Testing and Materials, 1985.

Eichenbruger, E.: Okologische Untersuchungen an modelfleissengewassern. I. Die Jahreszeitliche Verteilung der bestandesbildenden pflanzlichen Organismen bei verscheider Abwasserbelastung. *Schweitz. Z. Hydrol.*, 29:1–31, 1967.

Eichenburger, E.; Schlatter, F.; Weilenmann, H.; and Wuhrmann, K.: Toxic and eutrophying effects of cobalt, copper and zinc on algal benthic communities in rivers. *Verh. Int. Ver. Limnol.*, 21:1131–1134, 1981.

Fairchild, J. F.; La Point, T. E.; Zajicek, J. L.; Nelson, M. K.; Dwyer, F. J.; and Lovely, P. A.: Population-, community- and ecosystem-level responses of aquatic mesocosms to pulsed doses of pyrethroid insecticide. *Environ. Toxicol. Chem.*, 11:115–129, 1992.

Farke, H.; Schultz-Blades, M.; Ohm, K.; and Gerlach, S. A.: Bremerhaven caisson for intertidal field studies. *Mar. Ecol. Prog. Ser.*, 16:193–197, 1984.

Ferrington, L. C., Jr.; Blackwood, M. A.; Wright, C. A.; and Goldhammer, D. S.: Sediment transfers and representativeness of mesocosm test fauna. Aquatic Mesocosm Studies in Ecological Risk Assessment, edited by R. L. Graney; J. H. Kennedy; and J. H. Rodgers, pp. 179–200. Boca Raton, FL: Lewis, 1993.

Fischer, R.: Simulated or actual field testing? A comparison. Aquatic Mesocosm Studies in Ecological Risk Assessment, edited by R. L. Graney; J. H. Kennedy; and J. H. Rodgers, pp. 35–46. Boca Raton, FL: Lewis, 1993.

Flemer, D. A.; Biggs, R. B.; Tippie, V. K.; Nelsen, W.; Mackierman, G. B.; and Price, K. S.: Characterizing the Chesapeake Bay ecosystem and lessons learned. Proceedings of the Tenth National Conference of the Coastal Society, pp. 153–177. Washington, DC: The Coastal Society, 1987.

Folsom, B. L.; Skogerboe, J. G.; Palermo, N. R.; Simmers, J. W.; and Pranger, S. A.: Synthesis of the results of the field verification program upland disposal alternative. U.S. Army Corps of Engineer Waterways Experiment Station Technical Report D-88-7, 1988.

Gachter, R.: MELIMEX, an experimental heavy metal pollution study: goals, experimental design and major findings. *Schweiz. Z. Hydrol.*, 41:169–176, 1979.

Gamble, J. C.; Davis, J. M.; and Steele, J. H.: Loch Ewe bag experiment, 1974. *Bull. Mar. Sci.*, 27:146–175, 1977.

Gaufin, A. R.: Use of aquatic invertebrates in the assessment of water quality. Biological Methods for the Assessment of Water Quality, edited by J. Cairns, Jr., and K. L. Dickson, pp. 96–116. ASTM STP 528. Philadelphia: American Society for Testing and Materials, 1973.

Gearing, J. N.: The role of aquatic microcosms in ecotoxicologic research as illustrated by large marine systems. Ecotoxicology: Problems and Approaches, edited by S. A. Levin; M. A. Harwell; J. R. Kelley; and K. D. Kimball, pp. 411–470. New York: Springer-Verlag, 1989.

Geckler, J. R.; Horning, W. B.; Neiheisel, T. M.; Pickering, Q. H.; Robinson, E. L.; and Stephan, C. E.: Validity of laboratory tests for predicting copper toxicity in streams. EPA-600/3-76-116, 1976.

Gerhart, D. E.; Anderson, S. M.; and Richter, J.: Toxicity bioassays with periphyton communities: design of experimental streams. *Water Res.*, 11:567–570, 1977.

Gersich, F. M., and Brusven, M. A.: Volcanic accumulation and ash-voiding mechanisms of aquatic insects. *J. Kans. Entomol. Soc.*, 55:290–296, 1982.

Giddings, J. M.: Types of aquatic microcosms and their research application. Microcosms in Ecological Research, edited by J. P. Giesy, Jr., pp. 248–266. Symposium Series 52 (Conf-781101). Washington, DC: U.S. Department of Energy, 1980.

Giddings, J. M., and Eddleman, G. K.: Photosynthesis/respiration ratios in aquatic microcosms under arsenic stress. *Water Air Soil Pollut.*, 9:207–212, 1978.

Giddings, J. M.; Franco, P. J.; Bartell, S. M.; Cushman, R. M.; Herbes, S. E.; Hook, L. A.; Newbold, J. D.; Southworth, G. R.; and Stewart, A. J.: Effects of Contaminants on Aquatic Ecosystems: Experiments with Microcosms and Outdoor Ponds. ORNL Environmental Sciences Division, Publication No. 2381.

Giddings, J. M.; Biever, R. C.; Helm, R. L.; Howich, G. L.; and deNoyelles, F. J.: The fate and effects of Guthion (azinphos methyl) in mesocosms. Aquatic Mesocosm Studies in Ecological Risk Assessment, edited by R. L. Graney; J. H. Kennedy; and J. H. Rodgers, pp. 469–496. Boca Raton, FL: Lewis, 1993.

Giesy, J. P., Jr.; Kania, H. J.; Bowling, J. W.; Knight, R. L.; Mashburn, S.; and Clarkin, S.: Fate and Biological Effect of Cadmium Introduced into Channel Microcosms. EPA-600/3-79-039. Athens, GA: U.S. EPA, 1979.

Giesy, J. P.: Microcosms in Ecological Research. DOE Symposium Series 52. CONF-781101. Springfield, VA: Technical Information Center, U.S. Department of Energy, 1980.

Giesy, J. P.: Multispecies tests: research needs to assess the effects of chemicals on aquatic life. Aquatic Toxicology and Hazard Assessment: Eighth Symposium, edited by R. C. Bahner, and D. J. Hansen, pp. 67–77. ASTM 891. Philadelphia: American Society for Testing Materials, 1985.

Giesy, J. P., and Allred, P. M.: Replicability of aquatic multispecies test systems. Multispecies Toxicity Testing, edited by J. Cairns, Jr., pp. 187–247. New York: Pergamon, 1985.

Giesy, J. P., and Odum, E. P.: Microcosmology: introductory comments. Microcosms in Ecological Research, edited by J. P. Giesy, pp. 1–13. CONF-781101. Springfield, VA: Technical Information Center, U.S. Department of Energy, 1980.

Goering, C. E.: Paired field studies of herbicide drift. *Trans. Am. Soc. Agric. Eng.*, 73-1575:27–34, 1973.

Goldman, C. R.: A method studying nutrient limiting factors "in situ" in water columns isolated by polyethylene film. *Limnol. Oceanogr.*, 7:99–101, 1962.

Greve, W., and Parsons, T. R.: Photosynthesis and fish production: The possible effects of climatic change and pollution. *Helgol. Wiss. Meeresunters.*, 30:666–672, 1977.

Grice, G.; Harris, R. P.; Reeve, M. R.; Heinbokel, J. F.; and Davis, C. O.: Large scale enclosed water column ecosystem. An overview of Foodweb I, the final CEPEX experiment. *J. Mar. Biol. Assoc. U.K.*, 60:401–414, 1980.

Grice, G. D., and Menzel, D. W.: Controlled ecosystem pollution experiment: effect of mercury on enclosed water columns. VIII. Summary of results. *Mar. Sci. Comm.*, 4:23–31, 1978.

Grice, G. D., and Reeve, M. R. (eds.): Marine Mesocosms: Biological and Chemical Research in Experimental Ecosystems. New York: Springer-Verlag, 1982.

Graney, R. L.; Cherry, D. S.; and Cairns, J., Jr.: Heavy metal indicator potential of the Asiatic clam (*Corbicula fluminea*) in artificial stream systems. *Hydrobiologia* 102:81–88, 1983.

Graney, R. L.; Cherry, D. S.; and Cairns, J., Jr.: The influence of substrate, pH, diet and temperature upon cadmium accumulation in the Asiatic clam (*Corbicula fluminea*) in laboratory artificial streams. *Water Res.*, 18: 833–842, 1984.

Graney, R. L.; Giesy, J. P., Jr.; and DiToro, D.: Mesocosm experimental design strategies: advantages and disadvantages in ecological risk assessment. Using Mesocosms to Assess the Aquatic Ecological Risk of Pesticides: Theory and Practice, edited by J. R. Voshell, Jr. *Misc. Publ. Entomol. Soc. Am.*, 75:74–88, 1989.

Hall, D. J.; Cooper, W. E.; and Werner, E. E.: An experimental approach to the production dynamics and structure of freshwater animal communities. *Limnol. Oceanogr.*, 15:839–928, 1970.

Hamilton, P. B.; Lean, D. R. S.; Jackson, G. S.; Kaushik, N. K.; and Solomon, K. R.: The effect of two applications of atrazine on the water quality of freshwater enclosures. *Environ. Pollut.*, 60:291–304, 1989.

Hanazato, T., and Yasuno, M.: Effects of a carbamate insecticide, carbaryl, on the summer phyto- and zooplankton communities in ponds. *Environ. Pollut.*, 48:145–159, 1987.

Hanazato, T., and Yasuno, M.: Effects of carbaryl on the spring zooplankton communities in ponds. *Environ. Pollut.*, 56:1–10, 1989.

Hanazato, T., and Yasuno, M.: Influence of time of an application of an insecticide on recovery patterns of a zooplankton community in experimental ponds. *Arch. Environ. Contam. Toxicol.*, 19:77–83, 1990.

Hanazato, T.; Iwakuma, T.; Yasuno, M.; and Sakamoto, M.: Effects of temephos on zooplankton communities in a shallow eutrophic lake. *Environ. Pollut.*, 59:305–314, 1989.

Hansen, D. J.: Status and development of water quality criteria and advisories. Water Quality Standards for the 21st Century, pp. 163–169. Washington, DC: U.S. EPA, Office of Water, 1989.

Hansen, S. R., and Garton, R. R.: Ability of standard toxicity tests to predict the effect of the insecticide, diflubenzuron, on a laboratory stream communities. *Can. J. Fish. Aquat. Sci.*, 39:1273–1288, 1982.

Harte, J.; Terry, D.; Rees, J.; and Saegebarth, E.: Making microcosms an effective tool. Microcosms in Ecological Research, edited by J. P. Giesy, Jr., pp. 105–137. Springfield, VA: Technical Information Center, 1980.

Harvey, R. S.: A flowing stream laboratory for studying the effects of water temperature on the ecology of stream organisms. *Assoc. Southeast Biol. Bull.*, 20:3–7, 1973.

Hatakeyema, S.; Shiraishi, H.; and Kobayashi, N.: Effect of aerial spraying of insecticides on nontarget macrobenthos in a mountain stream. *Ecotoxicol. Environ. Safety*, 19:254–270, 1990.

Havens, K., and De Costa, J.: The effect of acidification in enclosures on the biomass and population size structure of Bosmina longirostris. *Hydrobiologia,* 133:153–158, 1985.

Havens, K. E., and Heath, R. T.: Acid and aluminum effects on freshwater zooplankton: an in situ mesocosm study. *Environ. Pollut.*, 62:195–211, 1989.

Hedtke, S. F., and Arthur, J. W.: Evaluation of a Site-Specific Criterion Using Outdoor Experimental Streams. EPA-600/D-85-041. Monticello, MN: U.S. EPA, 1985.

Hedtke, J. L., and Norris, L. A.: Effect of ammonium chloride on predatory consumption rates of brook trout (*Salvelinus fontinalis*) on juvenile chinook salmon (*Oncorhynchus tshawytscha*) in laboratory streams. *Bull. Environ. Contam. Toxicol.*, 24:81–89, 1980.

Heimbach, F.; Pflueger, W.; and Ratte, H.: Use of small artificial ponds for assessment of hazards to aquatic ecosystems. *Environ. Toxicol. Chem.*, 11:27–34, 1992.

Heimbach, F.; Berndt, J.; and Pflueger, W.: Fate and biological effects of a herbicide on two artificial pond ecosystems of different sizes. Aquatic Mesocosm Studies in Ecological Risk Assessment, edited by R. L. Graney; J. H. Kennedy; and J. H. Rodgers, pp. 303–320. Boca Raton, FL: Lewis, 1993.

Hellebust, J. A.; Hanna, B.; Sheath, R. G.; Gergis, M.; and Hutchinson, T. C.: Experimental crude oil spills on a small subartic lake in the Mackenzie Valley, N.W.T.: effects on phytoplankton, periphyton and attached aquatic vegetation. 1975 Conference on Prevention and Control of Oil Pollution, pp. 509–515. Washington, DC: American Petroleum Institute, 1975.

Herman, D.; Kaushik, N. K.; and Solomon, K. R.: Impact of atrazine on periphyton in freshwater enclosures and some ecological consequences. *Can. J. Fish. Aquat. Sci.*, 43:1917–1925, 1986.

Hesslein, R. H., and Quay, P.: Vertical eddy diffusion studies in the thermocline of a stratified lake. *J. Fish. Res. Bd. Can.*, 30:1491–1500, 1973.

Hill, I. R.: Effects on non-target organisms in terrestrial and aquatic environments. The Pyrethroid Insecticides, edited by J. P. Leahey, pp. 151–262, London: Taylor and Francis, 1985.

Hill, I. R.; Travis, K. Z.; and Ekoniak, P.: Spray-drift and run-off simulations of foliar-applied pyrethriods to aquatic mesocosms: rates, frequencies and methods. Aquatic Mesocosm Studies in Ecological Risk Assessment, edited by R. L. Graney; J. H. Kennedy; and J. H. Rodgers, pp. 201–240. Boca Raton, FL: Lewis, 1993.

Hines, A. H.; Haddon, P. J.; Miklas, J. J.; Wiechert, L. A.; and Haddon, A. M.: Estuarine invertebrates and fish: sampling design and constraints for long-term measurements of population dynamics. New Approaches to Monitoring Aquatic Ecosystems, edited by T. P. Boyle, pp. 140–154. ASTM STP 940. Philadelphia, American Society for Testing and Materials, 1987.

Hoffman, R. W., and Horne, A. J.: On-site flume studies for assessment of effluent impacts on stream aufwuchs communities. Microcosms in Ecological Research, edited by J. P. Giesy, pp. 610–624. DOE Symposium Series 52. Oak Ridge, TN: Technical Information Center, U.S. Department of Energy, 1980.

Howick, G. L.; deNoyelles, F., Jr.; Giddings, J. M.; and Graney, R. L.: Earthen ponds versus fiberglass tanks as venues for assessing the impact of pesticides on aquatic environments: a parallel study with sulprofos. Aquatic Mesocosm Studies in Ecological Risk Assessment, edited by R. L. Graney; J. H. Kennedy; and J. H. Rodgers, pp. 321–336. Boca Raton, FL: Lewis, 1993.

Hughes, D. N.; Boyer, M. G.; Papst, M. H.; Fowle, C. D.; Reese, G. A.; and Baula, P.: Persistence of three organophosphorus insecticides in artificial ponds and some biological implications. *Arch. Environ. Contam. Toxicol.*, 9:269–279, 1980.

Hurd, L. E.; Mellinger, M. V.; Wolf, L. L.; and McNaughton, J.: Stability and diversity at three trophic levels in terrestrial ecosystems. *Science*, 173:1134–1136, 1971.

Hurlbert, S. H.: Pseudoreplication and the design of ecological field experiments. *Ecol. Monogr.*, 54(2):187–211, 1984.

Hurlbert, S. H.; Mulla, S.; Keith, J. O.; Westlake, W. E.; and Dusch, M. E.: Biological effects and persistence of dursban in freshwater ponds. *J. Econ. Entomol.*, 63(1): 43–52, 1970.

Hurlbert, S. H.; Mulla, S.; and Willson, H. R.: Effects of an organophosphorous insecticide on the phytoplankton, zooplankton and insect populations of fresh-water ponds. *Ecol. Monogr.*, 42(2): 269–299, 1972.

Imboden, D. M.; Eid, B. S. F.; Joller, T.; Schurter, M.; and Wetzel, J.: MELIMEX, an experimental heavy metal pollution study: vertical mixing in a large limnocorral. *Schweiz. Z. Hydrol.*, 41:177–189, 1979.

Jackson, T. A.; Kipphut, G.; Hesslien, R. H.; and Schindler, D. W.: Experimental study of trace metal chemistry in soft-water lakes at different pH levels. *Can. J. Fish. Aquat. Sci.*, 37:387–402, 1980.

Jarnelov, A., and Linden, O.: IXTOC I: a case study of the world's largest oil spill. *Ambio.*, 6:299–306, 1981.

Johnson, P. C.; Kennedy, J. H.; Morris, R. G.; and Hambleton, F. E.: Fate and effects of Cyfluthrin (pyrethroid insecticide) in pond mesocosms and concrete microcosms. Aquatic Mesocosm Studies in Ecological Risk Assessment, edited by R. L. Graney; J. H. Kennedy; and J. H. Rodgers, pp. 337–372. Boca Raton, FL: Lewis, 1993.

Jones, J. G.: Studies on freshwater bacteria: the effect of enclosure in large experimental tubes. *J. Appl. Bacteriol.*, 36:445–456, 1973.

Jones, J. G.: Some observation on the occurrence of the iron bacterium *Leptoothrix ochracea* in fresh water, including reference to large experimental enclosures. *J. Appl. Bacteriol.*, 39:63–72, 1975.

Jones, J. G.: The microbiology and decomposition of seston in open water and experimental enclosures in a productive lake. *J. Ecol.*, 65:241–278, 1976.

Karr, J. R.; Fausch, K. D.; Angermeier, P. L.; Yant, P. R.; and Schlosser, I. J.: Assessing biological integrity in running waters: a method and its rationale. Special Publication 5. Champaign, IL: Illinois Natural History Survey, 1986.

Karr, J. R.; Yany, P. R.; and Fausch, K. D.: Spatial and temporal variability of the index of biotic integrity in three midwestern streams. *Trans. Am. Fish. Soc.*, 116:1–11, 1987.

Kaufman, L. H.: Stream aufwuchs accumulation: disturbance frequency and stress resistance and reliance. *Oecologia.*, (Berl) 52:57–63, 1982.

Kaushik, N. K.; Stephenson, G. L.; Solomon, K. R.; and Day, K. E.: Impact of permethrin on zooplankton communities in limnocorrals. *Can. J. Fish. Aquat. Sci.*, 42: 77–85, 1985.

Kelley, J. R.: Ecotoxicology beyond sensitivity: a case study involving "unreasonableness" of environmental change. Ecotoxicology: Problems and Approaches, edited by S. A. Levin; M. A. Harwell; J. R. Kelley; and K. D. Kimball, pp. 473–496. New York: Springer-Verlag, 1989.

Kettle, W. D., and deNoyelles, F., Jr.: Effects of cadium stress on the plankton communities of experimental ponds. *J. Freshwater Ecol.*, 3:433–443, 1986.

Kettle, W. D.; deNoyelles, F., Jr.; Heacock, B. D.; and Kadoum, A. M.: Diet and reproduction success of bluegill recovered from experimental ponds treated with atrazine. *Bull. Environ. Contam. Toxicol.*, 38:47–52, 1986.

Kirk, R. E.: Experimental design: procedures for the behavioral sciences. Monterey, CA: Brooks/Cole, 1968.

Kirstiz, R. U.: Recycling of nutrients in an enclosed aquatic community of decomposing macrophytes (*Myriophyllum spicatum*). *Oikos* 30:561–569, 1978.

Klaassen, H. E., and Kadoum, A. M.: Distribution and retention of atrazine and carbofuran in farm pond ecosystem. *Arch. Environ. Contam. Toxicol.*, 8:345–353, 1979.

Knisel, W. G. (ed.): CREAMS, A field scale model for chemicals, runoff and erosion from agricultural management systems. USDA Conserv. Res. Rep., No. 26, 1980.

Koeller, P., and Parsons, T. R.: The growth of young salmonids (*Onchorhynchus keta*): controlled ecosystem pollution experiment. *Bull. Mar. Sci.*, 27:114–118, 1977.

Koeller, P. A., and Wallace, G. T.: Controlled ecosystem pollution experiment: effect of mercury on enclosed water columns. V. Growth of juvenile chum salmon (*Orcorhynchus keta*). *Mar. Sci. Commun.*, 3:395–406, 1977.

Kosinski, R. J.: The effect of terrestrial herbicides on the community structure of stream periphyton. *Environ. Pollut., Ser. A.*, 36:165–189, 1984.

Kosinski, R. J.: Artificial streams in ecotoxicological research. Chapter 11. Aquatic Ecotoxicology: Fundamental Concepts and Methodologies, edited by A. Boudou, F. Ribeyre, Vol. 1, pp. 297–316. Boca Raton, FL: CRC Press, 1989.

Kosinski, R. J., and Merkle, M. G.: The effect of four herbicides on the productivity of artificial stream algal communities. *J. Environ. Qual.*, 13:75, 1984.

de Kozlowski, S. J., and Bunting, D. L., II: A laboratory study on the thermal tolerances of four southeastern stream insect species (Trichopetra, Ephemeroptera). *Hydrobiology*, 79:141–145, 1981.

Kuiper, J.: Development of North Sea coastal plankton communities in separate plastic bags under identical conditions. *Mar. Biol.*, 44:97–107, 1977a.

Kuiper, J.: An experimental approach in studying the influence of mercury on a North Sea coastal plankton community. *Helgol Wiss. Meeresunters*, 30:652–665, 1977b.

Kuiper, J., and Hanstveit, A. O.: Fate and effects of 4-chlorophenol and 2,4-dichlorophenol in marine plankton communities in experimental enclosures. *Ecotoxicol. Environ. Safety*, 8:15–33, 1984a.

Kuiper, J., and Hanstveit, A. O.: Fate and effects of 3,4-dichloroaniline (DCA) in marine plankton communities in experimental enclosures. *Ecotoxicol. Environ. Safety*, 8:34–54, 1984b.

La Point, T. W., and Fairchild, J. F.: Use of mesocosm data to predict effects in aquatic ecosystems: limits to interpretation. Aquatic Mesocosm Studies in Ecological Risk Assessment, edited by R. L. Graney; J. H. Kennedy; and J. H. Rodgers, pp. 241–256. Boca Raton, FL: Lewis, 1993.

La Point, T. W.; Fairchild, J. F.; Little, E. E.; and Finger, S. E.: Laboratory and field techniques in ecotoxicological research: strengths and limitations. Aquatic Ecotoxicology: Fundamental Concepts and Methodologies, Vol. II, edited by A. Boudou, and F. Ribeyre, pp. 239–255. Boca Raton, FL: CRC Press, 1989.

Lacaze, J.: Ecotoxicology of crude oils and the use of experimental marine ecosystems. *Mar. Pollut. Bull.*, 5: 153–156, 1974.

Lack, T. J., and Lund, J. W.: Observations and experiments on the phytoplankton of Blelham Tarn, English Lake District. I. The experimental tubes. *Freshwater Biol.*, 4: 399–415, 1974.

Lalli, C. M.: Enclosed Experimental Marine Ecosystems: A Review and Recommendations. New York: Springer-Verlag, 1990.

Lang, C., and Lang-Dobler, B.: MELIMEX, an experimental heavy metal pollution study: oligochaetes and chironomid larvae in heavy metal loaded and control limnocorrals. *Schweiz. Z. Hydol.*, 41:271–276, 1979.

Larsen, D. P.; deNoyelles, F., Jr.; and Stay, F.: Comparison of single-species, microcosm and experimental pond responses to atrazine exposure. *Environ. Toxicol. Chem.*, 5:179–190, 1986.

Larson, G. L.; Warren, C. E.; Hutchins, F. E.; Lamperti, L. P.; Schlesinger, D. A.; and Seim, W. K.: Toxicity of Residual Chlorine Compounds to Aquatic Organisms. EPA-60/3-78-023. Washington, D.C.: U.S. EPA, 1978.

Lee, R. F., Anderson, J. W.: Fate and effect of naphthalenes: controlled pollution experiment. *Bull. Mar. Sci.*, 27:127–134, 1977.

Levin, S. A.; Harwell, M. A.; Kelley, J. R.; and Kimball, K. D.: Ecotoxicology: Problems and Approaches. New York: Springer-Verlag, 1989.

Liber, K.; Kaushik, N. K.; Solomon, K. R.; and Carey, J. H.: Experimental designs for aquatic mesocosm studies: a comparison of the "Anova" and "regression" design for assessing the impact of tetrachlorophenol on zooplankton populations in limnocorrals. *Environ. Toxicol. Chem.*, 11:61–77, 1992.

Liber, K.; Solomon, K. R.; Kaushik, N. K.; and Carey, J. H.: Impact of Diatox, a commercial 2,3,4,6-tetrachlorophenol formulation, on plankton communities in limnocorrals. Aquatic Mesocosm Studies in Ecological Risk Assessment, edited by R. L. Graney; J. H. Kennedy; and J. H. Rodgers, pp. 257–294. Boca Raton, FL: Lewis, 1993.

Lind, O. T.: Handbook of Common Methods in Limnology, pp. 147–149. St. Louis: Mosby, 1979.

Livingston, R. J.: Field verification of multispecies microcosms of marine macroinvertebrates. Aquatic Toxicology and Hazard Assessment: 10th Volume, edited by W. J. Adams; G. A. Chapman; and W. G. Landis, pp. 369–383. ASTM STP 971. Philadelphia: ASTM, 1988.

Livingston, R. J.; Thompson, N. P.; and Meeter, D. A.: Long-term variation of organochlorine resides and assemblages of epibenthic organisms in a shallow north Florida (USA) estuary. *Mar. Biol.*, 46:355–372, 1978.

Long, E. R.; Buchman, M. F.; Bay, S. M.; Breteler, R. J.; Carr, R. S.; Chapman, P. M.; Hose, J. E.; Lissner, A. L.; Scott, J.; and Wolfe, D. A.: Comparative evaluation of five toxicity tests with sediments from San Francisco Bay and Tomales Bay, California. *Environ. Toxicol. Chem.*, 9: 1193–1214, 1990.

Lozano, S. J.; O'Halloran, S. L.; Sargent, K. W.; and Brazner J. C.: Effects of esfenvalerate on aquatic organisms in littoral enclosures. *Environ. Toxicol. Chem.*, 11:35–47, 1992.

Lugthart, G. J.; Wallace, J. B.; and Huryn, A. D.: Secondary production of chronomid communities in insecticide-treated and untreated headwater streams. *Freshwater Biol.*, 24:417–427, 1990.

Lund, J. W. G.: Preliminary observation on the use of large experimental tubes in lakes. *Verh. Int. Ver. Limnol.*, 18: 71–77, 1972.

Lund, J. W. G.: Experiments with lake phytoplankton in large enclosures. *Rep. Freshwater Biol. Assoc.*, 46:32–39, 1978.

Lund, J. W. G., and Reynolds, C. S.: The development and operation of large limnetic enclosures in Blelham Tarn, English Lake District, and their contribution of phytoplankton ecology. *Prog. Phycol. Res.*, 1:1–65, 1982.

Lynch, T. R.; Johnson, H. E.; and Adams, W. J.: Impact of atrazine and hexachlorobiphenyl on the structure and function of model stream ecosytems. *Environ. Toxicol. Chem.*, 4:399–413, 1985.

McAllister, C. D.; Parsons, T. R.; Stephes, K.; and Strickland, J. D. H.: Measurements of primary production in coastal sea water using a large volume plastic sphere. *Limnol. Oceanogr.*, 6:237–258, 1961.

McLaren, I. A.: Population and production ecology of zooplankton in Ogac Lake, a landlocked fiord on Baffin Island. *J. Fish. Res. Bd. Can.*, 26:1485–1559, 1969.

Macalaster, E. G.; Barker, D. A.; and Kasper, N. E. (eds.): Chesapeake Bay Program Technical Studies: a synthesis. Washington, D.C.: U.S. EPA, 1982.

Maguire, B. L.; Slobokin, L. B.; Morowitz, H. J.; More, B.; and Botkin, D. B.: A new paradigm for the examination of closed ecosystems. Microcosms in Ecological Research, edited by J. P. Giesy, Jr., pp. 30–68. Springfield, VA: Technical Information Center, 1980.

Maki, A. W., and Johnson, H. E.: Evaluation of a toxicant on the metabolism of model stream communities. *J. Fish. Res. Bd. Can.*, 33:2740–2746, 1976.

Maki, A. W., and Johnson, H. E.: The influence of larval lampricide (TFM: 3-trifluoromethyl-4-nitrophenol) on growth and production of two species of aquatic macrophytes, *Elodea canadensis* (Michx.) Planchon and *Myriophyllum spicatum* (L.). *Bull. Environ. Contam. Toxicol.*, 17:57–65, 1977.

Malins, D. C.; McCain, B. B.; Brown, D. W.; Chan, S. L.; Myers, M. S.; Landah, J. T.; Prohaska, P. G.; Friedman, A. J.; Rhodes, L. D.; Burrows, D. G.; Gronlund, W. D.; and Hodgins, H. O.: Chemical pollutants in sediments and diseases of bottom-dwelling fish in Puget Sound, Washington. *Environ. Sci. Technol.*, 18:705–713, 1984.

Manuel, C. Y., and Minshall, G. W.: Limitations on the use of microcosms for predicting algal response to nutrient enrichment in lotic systems. Microcosms in Ecological Research, edited by J. P. Giesy, Jr., pp. 645–667. Oak Ridge, TN: Technical Information Center, U.S. Department of Energy, 1980.

Marshall, J. S., and Mellinger, D. L.: Dynamics of cadmium-stressed plankton communities. *Can. J. Fish. Aquat. Sci.*, 37:403–414, 1980.

Mayasich, J. M.; Kennedy, J. H.; and O'Grodnick, J. S.: Evaluation of the effects of the pyrethroid tralomethrin utilizing an experimental pond system. Aquatic Mesocosm Studies in Ecological Risk Assessment, edited by R. L. Graney; J. H. Kennedy; and J. H. Rodgers, pp. 497–516. Boca Raton, FL: Lewis, 1993.

Mayer, F. L., Jr., and Ellersieck, M. R.: Manual of Acute Toxicity: Interpretation and Data Base Species for 410 Chemicals and 66 Species of Freshwater Animals. U.S. Fish and Wildlife Service, Resource Publication 160, 1986.

Mazumder, A.; Taylor, W. D.; McQueen, D. J.; Lean, D. R. S.; and Lafontaine, N. R.: A comparison of lakes and lake enclosures with contrasting abundances of planktivorous fish. *J. Plankton Res.*, 12:109–124, 1990.

Menzel, D. W., and Case, J.: Concept and design: controlled ecosystem pollution experiment. *Bull. Mar. Sci.*, 27:1–7, 1977.

Miller, M. C.; Alexander, V.; and Barsdate, R. J.: The effects of oil spills on phytoplankton in an Artic lake and ponds. *Artic.*, 31:192–198, 1978.

Moorhead, D. L., and Kosinski, R. J.: Effect of atrazine on the productivity of artificial stream algal communities. *Bull. Environ. Toxicol.*, 37:330–336, 1986.

Morris, R. G.; Kennedy, J. H.; Johnson, P. C.; and Hambleton, F. E.: Pyrethroid insecticide effects on bluegill sunfish in microcosms and mesocosms and bluegill impact on microcosm invertebrate fauna. Aquatic Mesocosm Studies in Ecological Risk Assessment, edited by R. L. Graney; J. H. Kennedy; and J. H. Rodger, pp. 373–396. Boca Raton, FL: 1993.

Mountford, K., and Mackiernam, G. B.: A multidecade trend-monitoring program for Chesapeake Bay, a temperate east coast estuary. New Approaches to Monitoring Aquatic Ecosystems, edited by T. P. Boyle, pp. 91–106. ASTM STP 940. Philadelphia: ASTM, 1987.

Muirhead-Thomson, R. C.: Comparative tolerance levels of blackfly (Simulium) larvae to permethrin (NRDC 143) and Temephos. *Mosq. News.*, 37:172–179, 1977.

Muller, P.: Effects of artificial acidification on the growth of periphyton. *Can. J. Aquat. Sci.*, 37:355–363, 1980.

Nagy, E.; Scott, B. F.; and Hart, J.: The fate of oil and oil-dispersant mixtures in freshwater ponds. *Sci. Total. Environ.*, 35:115–133, 1984.

National Academy of Sciences: Testing for effects of Chemicals on Ecosystems. Washington, D.C.: National Academy Press, 1981.

Nigg, H. N.; Stamper, J. H.; Queen, R. M.; and Knapp, J. L.: Fish mortality following application of phenthioate to Florida citrus. *Bull. Environ. Contam. Toxicol.*, 32:587–596, 1984.

Nixon, S. W.; Alonso, D.; Pilson, M. E. Q.; and Buckley, B. A.: Turbulent mixing in aquatic microcosms. Microcosms in Ecological Research, edited by J. P. Giesy, pp. 818–849. Springfield, VA: Technical Information Center, U.S. Department of Energy, 1980.

Nordlie, K. J., and Arthur, J. W.: Effects of elevated temperature on insect emergence in outdoor experimental channels. *Environ. Pollut. Ser. A.*, 25:53–65, 1981.

Odum, E. P.: The mesocosm. *BioScience* 34:558–562, 1984.

Ohio EPA: Biological Criteria for the Protection of Aquatic Life: Vol. 1: the role of biological data in water quality assessment. Division of Water Quality Planning and Assessment. Columbus, OH: U.S. EPA, 1988.

Papst, M. H., and Boyer, M. G.: Effects of two organophosphorus insecticides on the chlorophyll *a* and pheopigment concentration of standing ponds. *Hydrobiologia*, 69(3):245–250, 1980.

Patrick, R.: A proposed biological measure of stream conditions based on a survey of the Conestoga Basin, Lancaster Co., PA. *Proc. Acad. Nat. Sci. Phila.*, 101:277–341, 1949.

Pearce, J. B., and Despres-Patanjo, L.: A review of monitoring strategies and assessments of estuarine pollution. *Aquat. Toxicol.*, 11:323–343, 1988.

Peckarsky, B. L.: Sampling the Stream Benthos. A Manual on Methods for the Secondary Productivity in Fresh Waters, edited by J. A. Downing, and H. Rigler, pp. 131–160. Boston, MA: Blackwell Scientific Publications, 1984.

Phelps, D. K.; Datz, C. H.; Scott, K. J.; and Reynolds, B. H.: Coastal monitoring: evaluation of monitoring methods in Narragansett Bay, Long Island Sound and New York Bight, and a general monitoring strategy. New Approaches to Monitoring Aquatic Ecosystems, edited by T. P. Boyle, pp. 107–124. ASTM STP 940. Philadelphia: ASTM, 1987.

Phillips, D. J. H.: Quantitative Aquatic Biological Indicators. London: Applied Science, 1980.

Pignatello, J. J.; Martinson, M. M.; Steiert, J. G.; Carlson, R. E.; and Crawford, R. L.: Biodegradation and photolysis of pentachlorophenol in artificial freshwater streams. *Appl. Environ. Microbiol.*, 46:1024–1031, 1983.

Pignatello, J. J.; Johnson, K. L.; Martinson, M. M.; Carlson, R. E.; and Crawford, R. L.: Response of the microflora in outdoor experimental streams to pentachlorophenol: compartmental contributions. *Appl. Environ. Microbiol.*, 50:127–132, 1985.

Pilson, M. E. Q.: Application of mesocosms for solving problems in pollution research. Enclosed Experimental Marine Ecosystems: A Review and Recommendations, edited by C. M. Lalli, pp. 155–168. New York: Springer-Verlag, 1990.

Plafkin, J. L.; Barbour, M. T.; Porter, K. D.; Gross, S. K.; and Hughes, R. M.: Rapid Bioassessment Protocols for Use in Streams and Rivers: Benthic Macroinvertebrates and Fish. EPA/444/4-89-001. U.S. EPA Office of Water, 1989.

Pratt, J. R., and Bowers, N. J.: Effect of selenium on microbial communities in laboratory mesocosms and outdoor streams. *Toxic. Assess.*, 5:293–307, 1990.

Pratt, J. R.; Bowers, N. J.; and Cairns, J., Jr.: Effect of sediment on estimates of diquat toxicity in laboratory microcosms. *Water Res.*, 24:51–57, 1990.

Rand, G. M., and Petrocelli, S. R.: Fundamentals of Aquatic Toxicology. New York: Hemisphere, 1985.

Ravera, O.: The enclosure method: Concepts, Technology, and some Examples of Experiments with trace metals. Aquatic Ecotoxicology: Fundamental Concepts and Methodologies, edited by A. Boudou, and F. Ribeyre, pp. 249–272. Boca Raton, FL: CRC Press, 1989.

Ravera, O., and Annoni, D.: Effecti ecologici del rame studiati per mezzo di un ecosistema sperimentale. Atti 3 Congr. Associazione Italiana Oceanologia e Limnologia, edited by R. de Bernardi, p. 417. Pallanza, Italy: Istituto Italiano di Idrobiologia, 1980.

Regal, R. R., and Lozano, S. J.: Optimal design of field studies. Aquatic Mesocosm Studies in Ecological Risk Assessment, edited by R. L. Graney; J. H. Kennedy; and J. H. Rodgers, pp. 157–172. Boca Raton, FL: Lewis, 1993.

Reynolds, C. S.: Growth-rate responses of *Volvox aureus* Ehrenb (Chlorophyta, Volvcales) to variability in the physical environment. *Phycol. J.*, 18:422–433, 1983.

Reynolds, C. S.; Wiseman, S. W.; Godfrey, B. M.; and Butterwick, C.: Some effects of artificial mixing on the dynamic of phytoplankton populations in large limnetic enclosures. *J. Plankton Res.*, 5:203–234, 1983.

Reynolds, C. S.; Graham, G. P.; Harris, P.; and Gouldney, D. N.: Comparison of carbon specific growth rates and rates of cellular increase in phytoplankton in large limnetic enclosures. *J. Plankton Res.*, 7:791–820, 1985.

Risk Assessment Forum: Framework for Ecological Risk Assessment. EPA/630/8-92/001. U.S. EPA, 1992.

Rodgers, E. B.: Effects of elevated temperatures on macroinvertebrate populations in the Browns Ferry experimental ecosystems. Microcosms in Ecological Research, edited by J. P. Giesy, Jr., pp. 684–702. Oak Ridge, TN: Technical Information Center, U.S. Department of Energy, 1980.

Rodgers, J. H., Jr.; Clark, J. R.; Dickson, K. L.; and Cairns, J., Jr.: Nontaxonomic analysis of structure and function of aufwuches communities in lotic microcosms. Microcosms in Ecological Research, edited by J. P. Giesy, Jr., pp. 625–644. Oak Ridge, TN: Technical Information Center, U.S. Department of Energy, 1980.

Rogerson, P. F.; Schimmel, S. C.; and Hoffman, G.: Chemical and biological characterization of Black Rock Harbor dredged material. Field verification program. Technical Report D-85-9. Narragansett, RI: U.S. EPA, 1985.

Rose, F. H., and McIntire, C. D.: Accumulation of dieldrin by benthic algae in laboratory streams. *Hydrobiologia*, 35:481–493, 1970.

Ruber, E., and Kocor, R.: The measurement of upstream migration in a laboratory stream as an index of the potential side-effects of Temephos and Chlorpyrifos on *Gammarus fasciatus* (Amphipoda: Crustacea). *Mosq. News*, 36:424–429, 1976.

Rudd, J. W. M., and Turner, M. A.: The English-Wabigoon river system: II. Suppression of mercury and selenium bioaccumulation by suspended and bottom sediments. *Can. J. Fish. Aquat. Sci.*, 40:2218–2227, 1983a.

Rudd, J. W. M., and Turner, M. A.: The English-Wabigoon river system: V. Mercury and selenium bioaccumulation as a function of aquatic primary productivity. *Can. J. Aquat. Sci.*, 40:2251–2259, 1983b.

Salki, A.; Turner, M.; Patalas, K.; Rudd, J.; and Findlay, D.: The influence of fish-zooplankton-phytoplankton interaction on the results of selenium toxicity experiments within large enclosures. *Can. J. Fish. Aquat. Sci.*, 42: 1132–1143, 1984.

Santschi, P. H.: The MERL mesocosm approach for studying sediment-water interactions and ecotoxicology. *Environ. Tech. Lett.*, 6:335–350, 1985.

Schindler, D. W.: Detecting ecosystem responses to anthropogenic stress. *Can. J. Fish. Aquat. Sci.*, 44(Suppl. 1): 6–25, 1987.

Schindler, D. W., and Hesslien, R. H.: Effects of acidification on mobilization of heavy metals and radionuclides from the sediment of a freshwater lake. *Can. J. Fish. Aquat. Sci.*, 37:373–377, 1980.

Schindler, D. W.; Mills, K. H.; Malley, D. F.; Finday, D. L.; Shearer, J. A.; Davies, I. J.; Turner, M. A.; Linsey, G. A.; and Cruikshank, D. R.: Long-term ecosystem stress: the effects of years of experimental acidification on a small lake. *Science*, 228:1395–1401, 1985.

Schulz-Baldes; M., Rehm, E.; and Farke, H.: Field experiments on the fate of lead and chromium in an intertidal benthic mesocosm, the Bremerhaven caisson. *Mar. Biol.*, 75:307–318, 1983.

Scott, B. F., and Glooschenko, V.: Impact of oil and oil-disperant mixtures on flora and water chemistry parameters in freshwater ponds. *Sci. Total. Environ.*, 35: 169–190, 1984.

Scott, B. F.; Wade, P. J.; and Taylor, W. D.: Impact of oil and oil-dispersant mixtuers on the fauna of freshwater ponds. *Sci. Total. Environ.*, 35:191–206, 1984a.

Scott, B. F.; Nagy, E.; Dutka, B. J.; Sherry, J. P.; Hart, J.; and Taylor, W. D.: The fate and impact of oil and oil-dispersant mixtures in freshwater pond ecosystems: introduction. *Sci. Total. Environ.*, 35:105–113, 1984b.

Sebastien, R. J., and Brust, R. A.: Impact of Methoxychlor on selected nontarget organisms in a riffle of the Souris River, Manitoba. *Can. J. Fish. Aquat. Sci.*, 46:1047–1061, 1989.

Seim, W. K.; Lichatowich, J. A.; Ellis, R. H.; and Davis, G. E.: Effects of kraft mill effluents on juvenile salmon production in laboratory streams. *Water Res.*, 11:189–196, 1977.

Sérodes, J. B.; Moreau, G.; and Allard, M.: Dispositif experimental pour l'étude de divers impact sur la faune benthique d'un cours d'eau. *Water Res.*, 18:95–101, 1984.

SETAC-EUROPE: Guidance document on testing procedures for pesticides in freshwater mesocosms. From the Workshop: A meeting of experts on guidelines for static field mesocosm tests. Monks Wood Experimental Station, U.K., July, 1991. Society of Environmental Toxicology and Chemistry, Europe, 1992.

SETAC-RESOLVE: Proceedings of a workshop on aquatic microcosms for ecological assessment of pesticides (Wintergreen, Virginia, October 1991). SETAC Foundation for Environmental Education and the RESOLVE Program of the World Wildlife Fund, 1992.

Shapiro, J.: Blue-green algae: why they become dormant. *Science*, 179:382–384, 1973.

Shaw, J. L.; Moore, M.; Kennedy, J. H.; and Hill, I. R.: Design and statistical analysis of field mesocosm studies. Aquatic Mesocosm Studies in Ecological Risk Assessment, edited by R. L. Graney; J. H. Kennedy; and J. H. Rodgers, pp. 85–104. Boca Raton, FL: Lewis, 1993.

Sheehan, P. J.; Baril, A.; Smith, D. J. K.; Harenfenist, A.; and Marshall, W. K.: The impact of pesticides on the ecology of pair nesting ducks. Technical Report Services No. 19, Canadian Wildlife Service, Headquarters, 1987.

Sherry, J. P.: The impact of oil and oil-dispersant mixtures on fungi in freshwater ponds. *Sci. Total. Environ.*, 35: 149–167, 1984.

Shires, S. W.: The use of small enclosures to assess the toxic effects of cypermethrin in fish under field conditions. *Pestic. Sci.*, 14:475–480, 1983.

Shires, S. W., and Bennett, D.: Contamination and effects in freshwater ditches resulting from an aerial application of Cypermethrin. *Ecotoxicol. Environ. Safety*, 9:145–158, 1985.

Siefert, R. E.; Lozano, S. J.; Brazner, J. C.; and Knuth, M. L.: Littoral Enclosures for Aquatic Field Testing of Pesticides: Effects of Chlorpyrifos on a Natural System. Using Mesocosms to Assess the Aquatic Ecological Risk of Pesticides: Theory and Practice. *Misc. Publ. Entom Soc. Am.*, 75:57–73, 1989.

Sigmon, C. F.; Kania, H. J.; and Beyers, R. J.: Reductions in biomass and diversity resulting from exposure to mercury in artificial streams. *J. Fish. Res. Bd. Can.*, 34:493–500, 1977.

Smetacek, V.; von Bodungen, B.; von Brockel, K.; and Aietzschel, B.: The plankton tower. II. Release of nutrients from sediments due to changes in the density of bottom water. *Mar. Biol.*, 34:373–378, 1976.

Smyly, W. P. J.: Some effects of enclosures on the zooplankton in a small lake. *Freshwater Biol.*, 6:241–251, 1976.

Snow, N. B., and Scott, B. F.: The effect and fate of crude oil spilt on two lakes. 1975 Conference on Prevention and Control of Oil Spills, pp. 527–534. Washington, D.C.: American Petroleum Institute, 1975.

Solomon, K. R.; Smith, K.; Guest, G.; Yoo, J. Y.; and Kaushik, N. K.: Use of limnocorrals in studying the effects of pesticides in the aquatic ecosystem. *Can. Tech. Rep. Fish. Aquat. Sci.*, 975:1–9, 1980.

Solomon, K. R.; Yoo, J. Y.; Lean, D.; Kaushik, N. K.; Day, K. E.; and Stephenson, G. L.: Dissipation of permethrin in limnocorral. *Can. J. Fish. Aquat. Sci.*, 42:70–76, 1985.

Solomon, K. R.; Yoo, J. Y.; Lean, D.; Kaushik, N. K.; Day, K. E.; and Stephenson, G. L.: Methoxychlor distribution, dissipation, and effects in freshwater limnocorrals. *Environ. Toxicol. Chem.*, 5:577–586, 1986.

Steele, J. H., and Frost, B. W.: The structure of plankton communities. *Philos. Trans. R. Soc. Lond. B. Biol. Sci.*, 280:485–534, 1977.

Steele, J. H., and Gamble, J. C.: Predator control in enclosures, Marine Mesocosms: Biological and Chemical Research in Experimental Ecosystems, edited by G. D. Grice, and M. R. Reeve, pp. 227–237. New York: Springer-Verlag, 1982.

Stephan, C. E., and Rogers, J. W.: Advantages of using regression analysis to calculate results of chronic toxicity tests. Aquatic Toxicology and Hazard Assessment: 8th Symposium, edited by R. C. Bahner, and D. J. Hansen, pp. 328–339. ASTM STP 891. Philadelphia: American Society for Testing and Material, 1985.

Stephenson, G. L.; Hamilton, P., Kaushik; N. K., Robinson, J. B.; and Solomon, K. R.: Spatial distribution of plankton in enclosures of three sizes. *Can. J. Fish. Aquat. Sci.*, 41:1048–1054, 1984.

Stephenson, G. L.; Kaushik, N. K.; Solomon, K. R.; and Day, K.: Impact of methoxychlor on freshwater communities of plankton in limnocorrals. *Environ. Toxicol. Chem.*, 5:587–603, 1986.

Stephenson, M., and Mackie, G. L.: Effects of 2,4-D treatment on natural benthic macroinvertebrate communities in replicate artificial ponds. *Aquat. Toxicol.*, 9:243–251, 1986.

Stout, R. J., and Cooper, W. E.: Effect of *p*-cresol on leaf decomposition and invertebrate colonization in experimental outdoor streams. *Can. J. Fish. Aquat. Sci.*, 40: 1647–1657, 1983.

Stout, R. J., and Kilham, S. S.: Effects of *p*-cresol on photosynthesis and respiratory rates of a filamentous green alga (Spirogyra). *Bull. Environ. Contam. Toxicol.*, 30:1–5, 1983.

Suter, G. W. II (ed.): Ecological Risk Assessment. Chelsea, MI: Lewis, 1992.

Swartz, S. C.; Schults, D. W.; Ditworth, G. R.; DeBen, W. A.; and Cole, F. A.: Sediment toxicity, contamination and macrobenthic communities near a large sewage outfall. Validation and Predictability of Laboratory Methods for Assessing the Fate and Effects of Contaminants in Aquatic Ecosystems, edited by T. P. Boyle, pp. 152–190. ASTM STP 865. Philadelphia: ASTM, 1985.

Thomas, W. H.; Seibert, D. L. R.; and Takahashi, M.: Controlled ecosystem pollution experiment: Effect of mercury on enclosed water columns. III. Phytoplankton population dynamics and production. *Mar. Sci. Comm.*, 3: 331–354, 1977.

Thompson, D. G.; Holmes, S. B.; Solomon, K. R.; and Wainio-Keizer, K. L.: Applying concentration-response theory to aquatic mesocosm studies. Aquatic Mesocosm Studies in Ecological Risk Assessment, edited by R. L. Graney; J. H. Kennedy; and J. H. Rodgers, pp. 129–156. Boca Raton, FL: Lewis, 1993.

Thompson, J. M.; Ferguson, A. D. J.; and Reynolds, C. S.: Natural filtration rates of zooplankton in a closed system: the derivation of a community grazing index. *J. Plankton Res.*, 4:545–560, 1982.

Touart, L. W.: Aquatic Mesocosm Test to Support Pesticide Registrations. Hazard Evaluation Division Tech. Guidance Document. EPA 540/09-88-035. Washington, D.C.: U.S. EPA, 1988.

Touart, L. W.: Regulatory endpoints and the experimental design of aquatic mesocosm tests. Aquatic Mesocosm Studies in Ecological Risk Assessment, edited by R. L. Graney; J. H. Kennedy; and J. H. Rodgers, pp. 25–34. Boca Raton, FL: Lewis, 1993.

Turgeon, D. D.; Bricker, S. B.; and O'Connor, T. P.: National status and trends program: chemical and biological monitoring of U.S. coastal waters. Ecological Indicators, Vol. 1, edited by D. E. Hyatt; V. J. McDonald; and D. H. McKenzie, pp. 425–458. Essex, England: Elsevier Applied Science, 1992.

Turner, M. A., and Rudd, J. W. M.: The English–Wagigoon River System: III. Its geochemistry, bioaccumulation and ability to reduce mercury bioaccumulation. *Can. J. Aquat. Sci.*, 40:2228–2240, 1983.

Uehlinger, J.; Bossard, P.; Bloesch, J.; Burgi, H. R.; and Solomon, K. R.: Ecological experiments in limnocorrals: methodological problems and quantification of the epilimnetic phosphorus and carbon cycle. *Verh. Int. Ver. Limnol.*, 22:163–171, 1984.

Vaccaro, R. F.; Azam, A.; and Hodson, R. E.: Response of natural marine bacterial populations to copper: controlled ecosystem pollution experiment. *Bull. Mar. Sci.*, 27:17–22, 1977.

Verduin, J.: Critique of research methods involving plastic bags in aquatic environments. *Trans. Am. Fish. Soc.*, 98: 335–336, 1969.

von Bodungen, B.; von Brockel, V.; Smetacek, V.; and Zietzschel, B: The plankton tower. I. A structure to study water/sediment interactions in enclosed water columns. *Mar. Biol.*, 34:369–372, 1976.

Wallace, J. B.; Vogel, D. S.; and Cuffney, T. F.: Recovery of a headwater stream from an insecticide-induced community disturbance. *J. North Am. Benthol. Soc.*, 3(2): 115–120, 1986.

Warren, C. E.; Wates, J. H.; Davis, G. E.; and Doudoroff, P.: Trout production in an experimental stream enriched with sucrose. *J. Wildl. Manage.*, 28:617–660, 1964.

Watton, A. J., and Hawkes, H. A.: Studies on the effects of sewage effluent on a gastropod population in experimental streams. *Water Res.*, 18:1235–1247, 1984.

Webber, C. E.; Deutsch, W. G.; Bayne, D. R.; and Seesock, W. C.: Ecosystem-level testing of a synthetic pyrethroid insecticide in aquatic mesocosms. *Environ. Toxicol. Chem.*, 11:87–105, 1992.

Weber, C. I.: Biological Field and Laboratory Methods for Measuring the Quality of Surface Waters and Effluents. EPA-670/4-73-001. Washington, D.C.: U.S. EPA, 1973.

Weinberger, P.; Greenhalgh, R.; Moody, R. P.; and Boulton, B.: Fate of fenitrothion in aquatic microcosms and the role of aquatic plants. *Environ. Sci. Technol.*, 16:470–473, 1982.

Wilde, E. W., and Tilly, L. J.: Structural characteristics of algal communities in thermally altered artificial streams. *Hydrobiologia*, 76:57–63, 1981.

Williams, J. R.; Dyke, P. T.; Fuchs, W. W.; Benson, V. W.; Rice, O. W.; and Taylor, E. D.: EPIC—Erosion/Productivity Impact Calculator: 2. User Manuel, edited by A. N. Sharpley, and J. R. Williams. U.S. Department of Agriculture Technical Bulletin No. 1768, 1990.

Williams, L. G., and Mount, D. L.: Influence of zinc on periphyton communities. *Am. J. Bot.*, 52:26–34, 1965.

Wilton, D. P. and Travis, B. V.: An improved method for simulated stream tests of blackfly larvae. *Mosq. News*, 25:118–123, 1965.

Wojtalik, T. A., and Waters, T. F.: Some effects of heated water on the drift of two species of stream invertebrates. *Trans. Am. Fish. Soc.*, 99:782–788, 1970.

Wrenn, W. B., and Grannemann, K. L.: Effect of temperature on bluegill reproduction and young-of-the-year standing stocks in experimental ecosystems. Microcosms in Ecological Research, edited by J. P. Giesy, Jr., pp. 703–714. Oak Ridge, TN: Technical Information Center, U.S. Department of Energy, 1980.

Wrenn, W. B.; Armitage, B. J.; Rodgers, E. B.; and Forsythe, T. D.: Browns Ferry Biothermal Res. Ser. II. Effects of Temperature on Bluegill and Walleye, and Periphyton, Macroinvertebrates and Zooplankton Communities in Experimental Ecosystems. EPA-600/3-79-092. Duluth, MN: U.S. EPA, 1979.

Yasuno, M.; Sugaya, Y.; and Iwakuma, T.: Effects of insecticides on the benthic community in a model stream. *Environ. Pollut. Ser. A.*, 38:31–43, 1985.

Yates, W. E.; Akseson, N. B.; and Bayer, D. E.: Drift of glyphosphate sprays applied with aerial and ground equipment. *Weed Sci.*, 26:497–604, 1978.

Yount, J. D., and Richter, J.: Effects of pentachlorophenol on periphyton communities in outdoor experimental streams. *Arch. Environ. Contam. Toxicol.*, 15:51–61, 1986.

Zischke, J. A.; Athur, J.; Nordlie, K. J.; Hermanutz, R. O.; Standen, D. A.; and Henry, T. P.: Acidification effects on macroinvertebrates and fathead minnows (*Pimephales promelas*) in outdoor experimental channels. *Water Res.*, 17:47–63, 1982.

SUPPLEMENTAL READING

Cairns, J., Jr., (ed.): Multispecies Toxicity Testing. New York: Pergamon, 1985.

Cairns, J., Jr., and Niederlehner, B. R. (eds.): Ecological Toxicity Testing. Boca Raton, FL: Lewis, 1995.

Crossland, N. O., and La Point, T. W.: Symposium on Aquatic Mesocosms in Ecotoxicology. *Environ. Toxicol. Chem.*, 11(1):1–130, 1992.

Graney, R. L.; Kennedy, J. H.; and Rodgers, J. H., Jr. (eds.): Aquatic Mesocosm Studies in Ecological Risk Assessment. SETAC Special Publication Series. Boca Raton, FL: Lewis, 1993.

Grice, G. D., and Reeve, M. R. (eds.): Marine Mesocosms—Biological and Chemical Research in Experimental Ecosystems. New York: Springer-Verlag, 1982.

Hill, I. R.; Heimbach, F.; Leeuwangh, P.; and Matthiessen, P. (eds.): Freshwater Field Tests for Hazard Assessment of Chemicals. Boca Raton, FL: Lewis, 1994.

Lalli, C. M. (ed.): Enclosed Experimental Marine Ecosystems: A Review and Recommendations. New York: Springer-Verlag, 1990.

Chapter 10

STATISTICAL ANALYSIS

M. R. Ellersieck and *T. W. La Point*

INTRODUCTION

Background

During the past 15 years, the field of aquatic toxicology has become more detailed and complex in theory and practice. Studies are designed to generate more data on a variety of controversial problems to address and resolve various areas of environmental concern. The change in aquatic toxicology stems from the evolution of the field. Like other sciences, aquatic toxicology started as a descriptive science. Aquatic organisms were exposed to a toxic chemical or a change in water quality (e.g., temperature, dissolved oxygen, pH) in the laboratory or field and the adverse effects quantified. As the descriptive database accumulated, it became possible to study mechanisms of action and make inferences about why some adverse effects occur. Aquatic toxicology thus entered the mechanistic stage of development. This has become increasingly apparent at annual symposia at which papers and forums are presented on the biochemical (e.g., biomarkers) and physiological mechanisms of toxicant-induced stress. In fact, many studies today encompass both descriptive and mechanistic aspects. In the evolution of this science, it is also evident that an understanding of effects observed in the laboratory must be validated and interpreted with field experimentation to define the "biological reality" and implications of a laboratory-measured effect and its meaning for organisms in the environment. As with other sciences, aquatic toxicology has entered its most significant stage of development, the ecological-reality stage, in which active contributions link descriptive and mechanistic results to better understand the fate and effects of chemicals in the aquatic environment.

As studies today are designed and conducted to generate increasing amounts of data, resulting problems in data analyses have become more complex. Hence, aquatic toxicology has had to become more proficient in its use of *statistics*. This chapter introduces the reader to statistical analyses routinely used by researchers and to aspects of experimental design that should be considered prior to collecting laboratory and field data. It is not intended to substitute for textbooks in statistics (see Literature Cited); rather, it directs the reader to specific tests and problems in aquatic toxicology.

This chapter discusses two main subjects. The first is a general background in statistics and includes basic concepts, types of data, decision theory, and sampling. The second deals with statistical tests for significance, transformations, and experimental design. Several statistical tests are discussed and illustrated.

What Is Statistics?

Statistics is a science of making decisions with incomplete data. It is a branch of mathematics dealing with the collection, analysis, and interpretation of population subsamples. The collection, analysis, and interpretation of data are conducted according to certain rules and statistical designs. A common question in statistics is, "Are two populations alike or are they different?" This question can be answered by making particular measurements on a population, such as the probability of a success, mean, and variance, and comparing these measurements between or among populations. If the measurements are alike, the populations are considered to be the same or equal; if they are different, at least two populations exist (i.e., there is inequality). There is a problem in that there is no uniform method for testing equality or inequality among populations. If there were, the many statistical books and articles could be summarized in one manuscript. Many methods have been developed, primarily from mathematical theory, for testing differences between or among populations. The methods may require different rules for conducting an experiment. Conclusions that are based on inappropriate methods may be misleading or incorrect.

Statistical methods may perform three roles. The one most familiar to toxicologists is hypothesis testing—that is, determining whether two or more groups of data (from populations) differ from each other at a predetermined level of confidence. Another role of statistics is to construct and use models to predict the potential outcome of chemical-biological interactions. In model fitting, one's interest is in relating one variable (treatment or independent variable) to another variable (the effect or dependent variable) or in determining whether a pattern of effect is related to a pattern of treatment. Finally, statistics may simplify the understanding of data, either visually or numerically, by reducing the number of variables (not information!) to more easily resolve or understand a problem. Examples of the latter techniques include multidimensional scaling, time series, and cluster analysis. Most of this chapter concerns hypothesis testing.

The Experiment

A simple definition of "experiment" (or "study") is that it is a description of some action to be taken. Experiments can range from very simple, such as flipping a coin, to very sophisticated, such as determining the interaction of pH and aluminum in describing the effect of acid rain. An experiment should describe the statistical design (see section on experimental design), how the experimental units are randomized, how treatments are allocated, and how, when, and where samples are to be taken.

CONSIDERATIONS FOR DESIGNING EXPERIMENTS

The goal of experimental design is statistical efficiency and limiting the outlay of economic resources. The design may also be looked at in a logical manner by doing a stepwise analysis of the problem. This approach should be taken every time a study is done. The following should be considered when designing an experiment:

- What are the objectives and hypotheses of the study and what are the questions to be answered?
- What statistical model should be used in order to maximize the chance of detecting a statistical difference?
- Have the experimental units been properly randomized?

Randomization ensures that every treatment has the potential to contain both extreme high and low data values. It allows the toxicologist to proceed as if there were no known systematic bias in how the data are collected (i.e., fits the assumption of sample independence).

- How is the variable to be measured?
- What is the nature of the data generated by the measure?
- Are there interactions between measurements (can they be separated and identified)?
- Is the *n* (sample size) sufficient?
- Is the control appropriate?

Comparisons between treatment and control groups should be done between experimental units (i.e., organisms) that originated from the same defined population. Organisms used in the "control" group(s) should also come from the same source, lot, etc. as the chemically treated test group(s). Except for the actual test substance or chemical treatment being evaluated, treatment and control organisms should be maintained and handled in the same manner.

- Are confounding variables inadvertently being added?
- Are other undesirable variables present (e.g., different technicians observing different groups of organisms)?
- How large an effect will be considered biologically significant?
- What are the possible results of the study?
- How will the results be used?
- Do the results offer a reasonable expectation of achieving the objectives that prompted the initiation of the study?
- What new questions may be proposed based on the results? Can the study be redesigned before it is started, so that these "new" questions may be answered in the original study?

For further exploration of experimental design, detailed texts are available that include extensive treatment of statistical aspects of experimental design. Those recommended are Cochran and Cox (1957), Diamond (1981), Federer (1955), Hicks (1982), Gad and Weil (1988), and Meyer (1975).

Efficiency in Conducting Toxicological Experiments

Toxicological studies generally have a twofold purpose. The first is to determine whether a chemical produces an effect on a biological system. The second is to determine how much of an effect is present. The cost of conducting research to answer these questions and the value that society places on the results of such efforts have rapidly increased. Because costs will not lessen in the future, it seems reasonable to assume that every experiment or study should yield as much information as possible to answer the questions. Furthermore, the results should be as unequivocal as possible.

Toxicologists typically prepare extensively detailed protocols for an experiment or study prior to its initiation. Protocols usually include a detailed plan for statistical analysis of the data generated by the study. Selecting the appropriate statistical methodology is a significant part of protocol development because it structures the experiment (experimental design) and maximizes the possibility of success.

Prior selection of statistical methodologies is essential to determine the number of organisms per group, sampling intervals, etc. The aquatic toxicologist should, therefore, have a good understanding of the system under investigation and the general principles of experimental design. The analysis of data is dependent on the manner in which the data were obtained. Data are often collected without an appropriate experimental design or with insufficient sample size. In such cases, the person who is asked to analyze the data must become a magician rather than a statistician. However, often a population is not known and thus no logical experimental design can be planned. In this case, a pilot study should be performed to obtain some initial statistical information and to develop appropriate sampling techniques.

Designing an experiment gives the researcher a higher probability of a project being scientifically acceptable. However, in many cases experimental design is not easy to determine. One must not only choose appropriate methodology for statistical analysis but also determine the sample size needed to maintain a statistical power level. If the experimental units are inexpensive, the experimental facility extensive, and data determinations not cost prohibitive, experimental design can be easy.

Sample Size

The main question in designing an experiment is, "Given the response variable, what should the sample size be for each treatment?" Without more information about the population, a statistician cannot answer this question. However, additional questions may be asked to estimate an appropriate sample size. The questions are: How large a difference is one trying to detect? and How large a variance exists in the population? If these questions cannot be answered, a pilot study should be conducted to estimate these parameters. If a pilot study cannot be conducted, perhaps the parameters may exist in published literature. Let us examine the questions in closer detail.

How Large Must the Sample Size Be?

The power of a statistical test is simply 1 − the probability (Pr) of a type II error. The type II error is discussed later. If an investigator sets Pr (type II error) = β = 0.2, the power of the statistical test is $1 - 0.2 = 0.8$, or the power of the test is 80%. Power, in this sense, is the probability of detecting a significant difference if a true difference exists. If the power of a statistical test is 30%, the researcher has only a 30% chance of detecting a true difference. Obviously, this would not provide a high degree of resolution among mean differences. In designing an experiment, one should strive for 80% power or above.

Table 1 presents equations for estimating sample sizes required for increased statistical power for paired and unpaired t-tests and for one- or two-tailed tests.

For example, one may want to test whether a difference exists between the population means of two randomized groups with a power setting of 80%, and an estimate of the population variance equal to 200. The ability of estimating a minimal difference between two population means of 10, with $\alpha = 0.05$, and a two-tailed test, results in a sample size of 32 experimental units from each population using the formula in Table 1:

$$n = \frac{(\text{t-value} + t\text{-value})^2 \times 2}{\beta = 0.2 \quad \alpha/2 = 0.025} \times \frac{\sigma^2}{\Delta^2}$$

$$= (0.8416 + 1.96)^2 \times 2 \times \frac{200}{100}$$

$$= 31.40$$

$$= 32/\text{treatment}$$

(Snedecor and Cochran, 1989). The variance estimate may come from variances reported in the literature for similar experiments or from previous pilot studies.

In most field studies, the method of calculating the appropriate number of samples (n) is dependent on the distribution of the population (Elliott, 1977; Green, 1979). If it can be assumed that the data came from a normal distribution, then the allowable error of the population estimate (e.g., 10%, 20%) and the desired confidence level (0.95 confidence limit, or a 1-in-20 chance of being wrong, or $\alpha = 0.05$) should be determined. This information, along with an estimate of the variance (s^2) from the pilot study, may

Table 1. Equations for calculating sample size for testing differences between two treatments

	One-tailed test	Two-tailed test
Paired t-test	$\frac{(t\text{-value}^a \text{ at } \beta + t\text{-value at } \alpha)^2 \sigma^2}{\Delta^2}$	$\frac{(t\text{-value at } \beta + t\text{-value at } \alpha/2)^2 \sigma^2}{\Delta^2}$
Two-group t-test	$(t\text{-value at } \beta + t\text{-value at } \alpha)^2 \, 2\frac{\sigma^2}{\Delta^2}$	$(t\text{-value at } \beta + t\text{-value at } \alpha/2)^2 \, 2\frac{\sigma^2}{\Delta^2}$

[a] t-values are obtained from t tables at df = ∞ from most statistics books. t value is based on df = infinity from the Student t-table book values. β = Pr(of a type II ERROR); α = Pr(of a type I ERROR); σ^2 = population variance; Δ = minimal difference one is trying to detect.

be used to calculate n. The formula for n is given by:

$$\text{Sample size } (n) = \frac{t^2 s^2}{L^2} \quad (1)$$

where t is the value of Student's t-distribution for the degrees of freedom $(n - 1)$ associated with the estimate of variance and the desired confidence level, and L is the allowable error in the sample mean. For example, assume 10 random samples were collected in the pilot study, the mean $(\bar{X}) = 20$, and the variance $(s^2) = 5$. If the 95% confidence level is chosen, $t = 2.262$ $(P = \frac{\alpha}{2} = 0.025$, df $= n - 1 = 9)$ and if 20% of the mean is the allowable error, then the number of samples that would be appropriate is

$$n = \frac{(2.262)^2(5)}{(4)^2} = \frac{25.58}{16} = 1.598 \text{ or } (2)$$

If you decrease the allowable error to 10%, the n would quadruple in the above example to 6.395. Note: Sample size (n) must always be greater than 1.

If the data do not approximate the normal distribution but rather approximate a Poisson distribution, the variance may be either equal to or greater than the mean. Sampling intensity should increase in these cases. Elliott (1977) proposed a method for estimating the number of samples when the distribution is unknown. The formula for n when the distribution is unknown is

$$n = \frac{s^2}{D^2\bar{X}^2} \quad (2)$$

Decide on D, the allowable size of the ratio of the standard error to the mean (e.g., 0.1 or 10%). Determine the estimate of the population mean $\bar{X}$ and variance s^2 from a pilot study. As the estimated ratio of the variance to the mean increases, so does the number of samples required to keep the same ratio of standard error to mean.

These formulas are handy but may not show the total picture. The following example shows the possible complexity in determining sample size. Assume a researcher conducting a field study wants to determine (by measuring copper concentration or invertebrate counts) whether the water toxicity in stream 1 is equal to that in stream 2. How should each stream be sampled? One may look at different sites within a stream (see caution below on *pseudoreplication*), take multiple samples at each site, and make multiple determinations for each sample. Where should the emphasis be? The objective is to minimize the stream population standard error. The formula for the stream standard error (SE) from the example is

Population SE of each stream

$$= \sqrt{\frac{\sigma_D^2}{abc} + \frac{\sigma_S^2}{ab} + \frac{\sigma_{\text{site}}^2}{a}} \quad (3)$$

where σ_D^2 = variation among determinations within samples and sites
σ_S^2 = variation among samples within site
σ_{site}^2 = variation among sites
a = number of sites
b = number of samples per site
c = number of determinations within samples per site

In order to understand Eq. (3), it is important to understand the different variation terms. In this example, multiple determinations per sample were performed to get a good estimate for each sample. If chemical determinations or invertebrate counts are not repeatable due to instrument, field technician, or other causes, this type of variation increases (σ_D^2). This type of variation is usually referred to as laboratory (or field) error. If a technician is familiar with an instrument and/or technique, this type of variation is reduced.

Multiple samples per site are needed to get a good estimate for individual sites. If one is sampling a site for invertebrates, one sample per site may not be sufficient. Invertebrates often congregate and counts will be patchy. One sample from a site may have a count of 0, whereas another sample may have a count in the thousands. This is a true source of variation (σ_S^2).

Because in this example one is testing stream 1 versus stream 2, multiple sites per stream are needed. The site is the experimental unit and the variation among sites is given the symbol σ_{site}^2 in Eq. (3). All of the subsamples (pseudoreplications) are instrumental in obtaining a good estimate for each site.

Sampling emphasis depends on which variance component in Eq. (3) is largest. If the measured component "laboratory error" is quite variable (as indicated by σ_D^2), a larger number of determinations should be made in this category (c). If the number of samples per site is the most variable component (σ_S^2), more samples per site should be taken (b). Notice that if more sites are sampled (a), it affects all variance components. This demonstrates that simply increasing the total number of values per stream will not necessarily minimize the standard error. In this example, site is the experimental unit. Another subsampling (pseudoreplication) is needed to obtain an accurate measure of site, and all observations are instrumental in minimizing the SE.

The Experimental Unit

Sample size (n) describes how many experimental units are needed per treatment. Experimental units are typically called replicates. Care must be taken when replicating an experiment. Sometimes replicates are not true replicates. For instance, when several samples are taken in close conjunction in a field program, there is often a desire to call them "replicate" samples. However, the samples may not be independent and they are usually not replicates. For example, typically, a fish is the experimental unit. If three determinations of tissue residue are evaluated for each fish, the determinations are not replicates; rather, multiple fish per treatment are the replication. Subsampling of experimental units is sometimes called pseudoreplication. For further reading, see Hurlbert (1984).

For replication to be successful, the first step is to determine the experimental unit. This is sometimes difficult. For example, consider an experiment with six different tanks, each holding the same number of fish, with the random variable "fish weight." Treatment 1 is given to tank one, treatment 2 to tank two, . . . , treatment 6 to tank six. What is the experimental unit? In this case, if one uses fish as the experimental unit, treatment and tank are said to be *completely confounded.* That is, if one tests a difference between treatments, one is also testing a difference due to tank. The effect of tank and treatment cannot be separated. The experimenter can say that there is no tank difference and all the variation is due to treatment, but this cannot be proved. In this example the tank is the experimental unit, not the fish. This means that to have a legitimate analysis the experimenter needs at least two tanks, not one, per treatment. Confounding treatment with the experimental unit is one of the major errors in statistical design (Millikan and Johnson, 1989).

The Random Variable (Data)

To perform experiments properly and analyze results successfully, one must understand the different types of data and the usage and value of such data. In toxicological research, optimal design and interpretation of experiments require that the researcher understand the system under study and the data generated. In an aquatic toxicity study, data (variables) are collected on an effect measured in the experimental subjects (i.e., fish, invertebrates, algae, microorganisms) as a result of their association with a chemical treatment. These are, therefore, called treatment variables, which are independent (can be controlled by humans or nature), and effect variables, which are dependent (e.g., growth, mortality) on the chemical treatment under investigation.

The measurements of a variable for an entire population cannot be done—for example, one cannot measure the reproduction of all bluegill sunfish that exist. Therefore, in statistics one deals with a representative group or sample of data. If a sample is properly collected and is of adequate size, it will serve to provide an estimate of the parent population from which it was taken.

An experiment produces certain random variables (data). A random variable is what one measures as a result of an experiment. For example, in flipping a coin, the random variable is which side lands up. If the experiment describes different concentrations of toxicant exposure to fish, the random variable may be survival, weight, or tissue concentration of the toxicant. Protocols of some experiments are established to measure not one random variable but many. For instance, in looking at different sites within a stream, e.g., above and below waste expulsion (this describes the experiment), the random variables may be tissue concentration of metals, pesticides, and growth parameters in different species of organisms.

Realizations

The random variable has realizations. The realizations are the possible outcomes of the random variable. For example, if the experiment is flipping a coin, the random variable is which side is up and the realizations are heads or tails. If one looks at the lethality of a toxicant to biological units, the random variable is the percentage dead and the possible realizations range from 0 to 100%.

The nature of the data collected is determined by several considerations. These include the system under study, the instrumentation and methodology used to make the measurements, and the experimental design. The investigator has some degree of control over each of these. The least control is over the biological system and the most is over the design of the study.

There exist two main types of random variables (value imparted to a variable is called data) (see Table 2). The first type of random variable is called the continuous random variable and the second is called the discrete random variable, sometimes called categorical data. The underlying assumption for many parametric statistical procedures (e.g., analyses of variance versus nonparametric procedure) is that data be continuous. This implies, theoretically, that the data associated with a variable can assume any of a number of values between two fixed points on a real number line (e.g., line measurements of body weight between 1.0 and 2.0 mg). Instead of being placed in a category such as small, medium, or large, an actual measurement is taken. Most variables studied in toxicology are, in fact, continuous. Examples are length, weight, concentration, temperature, exposure duration, and percent lethality. For these types of continuous variables, the sample may be described with measures of central tendency and dispersion (see section on descriptive statistics). Data (measurements) collected on continuous variables often have mound- or bell-shaped frequency distributions. A continuous

Table 2. Types of random data variable (data) and an example of each type

Classification	Type	Example
Continuous	Scalar	Body weight, length, % hatchability, % survival
	Ranked	Severity of lesion
Discrete	Attribute	Colors of fish
	Quantal	Dead/alive or present/absent

variable that is normally distributed (the bell-shaped, normal curve) provides a good model for these types of data.

The normally distributed variable also plays a very important role in statistical inference. The normal probability distribution, used for continuous variables, is symmetrical about the population mean. The breadth of the curve is determined by the variance (σ^2). The normal curve works well for measurements (length, weight, etc.) when the population is not skewed toward large or small individuals. Even though some measurements in toxicology do not conform to the normal distribution, a population of means from the measurements will be normally distributed as the number of observations per mean approaches infinity. This is known as the central limit theorem. It is also known that if a measurement is selected at random from a population of measurements that possesses a normal distribution, the probability is approximately 0.68 that the measurement will be within one standard deviation (SD) of the mean (i.e., 68% of all values will be within 1 SD of the mean). Similarly, the probability is approximately 0.95 that a value will be within two standard deviations of the mean (i.e., 95% of all values will be within 2 SD of the mean). Essentially, all measurements (0.99) in a normal distribution will be within three standard deviations of the mean (the standard deviation is described in the section on descriptive statistics).

Other frequency distributions are the binomial, Poisson, and chi-square. Methods for calculating the different distributions are covered in most statistics books. It should be noted that as the sample size increases, the distribution of a population of means approaches normality even when the raw data are not normally distributed. In addition, it may be possible to "transform" the data to a normal distribution (see section on transformation).

In contrast to continuous data, there are discrete (categorical) data, which can assume only certain fixed numerical values. No intermediate values are possible between the "fixed points" in discrete data. Examples of such data are listed in Table 2. The realization of this type of random variable is the number or frequency that exists in each category (e.g., 5 animals alive, 15 animals dead).

Thus, for different types of random variables, different statistical methodologies are required to describe the population. Therefore, both the description of the experiment and the type of random variable measured are very important when choosing the appropriate statistical method.

Statistical Decision Theory

As stated earlier, a primary question in statistics is whether two populations are alike or different. Thus, there are two possible conclusions one can make. Statistical methods are used only to make one of these two conclusions. Statistical methods cannot explain why populations do or do not differ. Differences measured in a variable among two or more biological populations, although statistically significant, may not be biologically significant (Skalski, 1981). A statistical difference may exist but this difference may not be associated with a biological difference.

Using statistical methods to make a decision has a slight drawback. There is a small probability of making a wrong decision. However, researchers can set their own probability of making a wrong decision. The kinds of errors that can be made are referred to as type I and type II errors.

To demonstrate how these types of errors may occur, suppose a random set of 100 random numbers is drawn from a statistics book. This means that the numbers 0, 1, 2, 3, 4, 5, 6, 7, 8, and 9 are equally likely to be chosen (probability=0.1). Also, assume that if the number 0, 1, 2, 3, or 4 is chosen, this will be considered a success. Now, if 100 numbers are investigated one would expect 50 successes. In this example, the experiment is to investigate 100 random numbers. The random variable is the number of successes in 100 trials. The realizations range from 0 to 100 successes. One would expect to get 50. For this experiment, two hypotheses are possible:

1. Observed population = 50%
2. Observed population ≠ 50%

Statisticians refer to the first hypothesis as the null hypothesis (H_0) and to the second as the alternative hypothesis (H_A) or research hypothesis. The research hypothesis is the actual prediction one wants to test. The research hypothesis is the logical opposite of the null hypothesis, so that if the null hypothesis is found to be improbable (according to statistical criteria) then, by implication, the research hypothesis is considered acceptable. The example deals with a known population for which the null hypothesis is true; but through statistical methodology, could H_A be considered the acceptable conclusion? The realizations range from 0 successes to 100 successes. Certainly, the probability of getting only numbers 0 through 4 out of 100, or only numbers 5 through 9, is very low, but it could happen. These are possible realizations. If this occurs, the statistical methodology would instruct the experimenter to conclude H_A when H_0 was really true. If this was the case, one would have committed a statistical error (type I error). This type of error was made because a rare sample was obtained.

To illustrate this, Table 3 shows how these errors can occur. A truth does exist. The observed distribution either equals or does not equal that of the true population. The problem is that the experimenter must decide what is truth. This is accomplished by statistical methodology. Bear in mind that a probability level is selected as a way of rejecting the null hypothesis. That is, if the probability calculated for the null hypothesis is at this level or less, the null hypothesis will be rejected and the alternative accepted. The probability (or significance) level is the

Table 3. Description of type I and type II errors[a]

		The experimenter's decision	
		HO accepted	HA accepted
Truth	HO	Correct	Type I ERROR
	HA	Type II ERROR	Correct

[a]HO: observed population = hypothesis; HA: observed population ≠ hypothesis.

reasonable doubt about the null hypothesis. An actual probability value is calculated, and if it is equal to or less than 5% (0.05) or 1% (0.01), the null hypothesis is rejected. If the experimenter concludes that H_A is true and, in fact, the truth is that the observed population does not equal the true population, the experimenter has made the correct decision. If the conclusion is made that the observed and hypothesis populations are equal and, in fact, they are, a correct decision is also concluded.

However, if the experimenter calls the observed and true populations different but the truth is equality, the experimenter has made an error. This error is called a type I error, rejecting a null hypothesis when it is true. How could this type of error happen using our random number example? The expected frequency of success was 50. What would happen if the experimenter obtained a rare realization of 0 successes or 100 successes? If these realizations came about, the experimenter would have to conclude that the population does not equal 50. This conclusion is in error because the experimenter obtained a rare sample. Investigators set the probability of making this type of error (due to rare sampling) low, usually 5 or 1%. These probabilities are given the symbol α (alpha). Probability levels of 5% and 1% are standard in most scientific journals.

Another error can be committed if the experimenter makes the decision that the observed population equals the true population when the truth is inequality. This type of error is referred to as a type II error (accepting the null hypothesis when the alternative is true) and is given the symbol β (beta). The level of β should be set prior to initiating the experiment because it is instrumental in determining the sample size of an experiment. As sample size increases, β decreases (smaller type II error). The power of a statistical test is 1 minus the probability of a type II error, $1 - \beta$. The *power* of the test is defined as the probability of detecting a statistical difference or effect if a true difference or effect exists. It is desirable to have the power high.

How to Make a Statistical Decision

Making a statistical decision is called statistical hypothesis testing. There are basically five steps in making a statistical decision.

1. Formulate the hypothesis. Either the population conforms to what the experimenter expected it to be (H_0) or it does not conform (H_A). One of these conclusions must exist.
2. Set the probability of concluding nonconformity (H_A) when conformity is true to a very low probability (α = type I error). The experimenter can select any probability for a type I error; however, most journals use an α level of 5 or 1%.
3. Randomly sample a population and make certain measurements. These measurements are usually point estimates such as a mean, regression coefficient, probability, and a measure of dispersion, which is the variance of the point estimates. The point estimates are a function of the random variable; thus, the point estimate itself is a random variable and not a constant. This means the point estimates have a variance. These measurements, sometimes called moments, are unique identifiers of a population. The objective is to compare the unique identifiers between or among populations to determine equality or inequality of the populations.
4. Calculate a test statistic. Because the test statistic is a function of the random variable, the test statistic itself is a random variable. The distribution of the test statistic is known and tabulated. Most statistics books have the various test statistics tabulated, usually in an appendix. Some common tables are called chi-square, *t*-value, and *F*-value.

   How does a test statistic work? Statistics is a study of how far observed values deviate from what is expected. As these deviations increase, the value of the test statistic increases. If the deviation of an observed value from its expected value becomes large, there is a point at which the experimenter concludes that the observed values are different from what was expected. For example, assume a population of animals has a sex ratio of 1:1. If a random sample of 10 animals were evaluated for gender, there is a small probability that all sampled animals would be exclusively male or exclusively female. If this occurs, the investigator would obtain a very large test statistic because the observed value would deviate significantly from what was expected.
5. Make the decision. The tabulated values of the test statistic are calculated assuming H_0 is true. These values of the test statistic at the level of α are known and tabulated and are generally called critical values. Generally, the rule for making a decision is to reject the null hypothesis (statement of equality) if the calculated test statistic is greater than or equal to the critical value of the test statistic at the probability level of alpha.

Outliers

Data outliers are a critical consideration that must be evaluated in both laboratory and field toxicity studies. Outliers are high or low numerical values that are extreme compared with the remainder of the data in a study. They may arise from a faulty measurement of instrument, or they may be real. Outliers can be detected either by visually reviewing the data, by quantifying the dispersion in the data via standard deviation, or through histograms (described later).

If an outlier can be linked to some faulty equipment or human error, it can be eliminated from the analysis. However, if the outlier cannot be linked to one of those errors, the outlier could be the result of a real effect or it could be a sampling error. Outliers in a study may actually reflect a biologically significant effect. Outliers increase variability within a sample and, therefore, decrease sensitivity of the statistical tests used. Outliers may eliminate any statistically significant effects (Barnett and Lewis, 1984; Beckman and Cook, 1983). Outlier data generated by obvious instrument or technician errors should be eliminated and rejected from the rest of the data. The question arises of whether these rejection outliers can be easily detected.

Several techniques can be used for outlier rejection. The use of these techniques depends on the reviewer's familiarity and experience with the distribution of the data under investigation. Chauvenet's criterion (Meyer, 1975) is one technique that can be used with normally distributed data with a single extreme value. This method is good for samples sizes of $n \leq 20$. Other references that describe detection of outliers are Snedecor and Cochran (1989) and Neter et al. (1985).

Sampling

Sampling is the collection of data points in a study. In aquatic toxicology, most laboratory tests (for example, acute toxicity determinations) sample and analyze the same data points from all samples collected. In some aquatic laboratory tests, however, only certain organisms (and water) are collected for specific analyses. For example, in bioconcentration tests, during uptake and depuration phases, water and exposed organisms are sampled at different times for chemical analysis to determine accumulation (and elimination) of test substance from water. Another example is aquatic field studies, in which samples of biota (e.g., fish, phytoplankton, zooplankton, periphyton), sediment, and water are collected (see Chapter 9). Typically, several assumptions about sampling must be used when statistical analysis techniques are to be followed. Samples should be collected without bias and each member of a sample should be collected independently of the other. Eliminating bias means that, at the time the sample is taken from a population, each portion of that population has an equal chance of being selected. Independent implies that the selection of any portion of the sample is not affected by and does not affect the selection of any other portion of the sample.

There are several major types of sampling methods:

Simple random sampling occurs when the researcher has a source of experimental units that are uniform, such as fish of the same age, size, and so forth. If this situation exists, experimental units can be allotted to experimental treatments at random. This is referred to as a simple random sample. If experimental units are not uniform, other sampling techniques are needed.

Stratified random sampling is performed by dividing the population of *n* units into subpopulations (or strata), followed by a simple random sampling from each stratum. These subpopulations are nonoverlapping and together constitute the whole population. This method may be employed when the environment sampled has a number of homogeneous subsets but there may be large differences among individual subsets and smaller differences within a subset. Stratified random sampling can add precision to the population estimate with minimum effort. It may be possible to divide a heterogeneous population into subpopulations, each of which is internally homogeneous. This is suggested by the name stratum, with its implication of division into layers. If each stratum is homogeneous, in that the measurements vary little form one unit to another, a precise estimate of any stratum mean can be obtained from a small sample in that stratum. These estimates can then be combined into a precise estimate for the whole population. When properly used, stratification nearly always results in a smaller variance for the estimated mean or total than is given by a comparable simple random sample. It is not true, however, that stratified random sampling always gives a smaller variance than a simple random sample.

In a farm pond of mesocosm field study (see Chapter 9), the pond area may be divided into a number of subpopulations or strata. Samples may then be collected individually in each stratum. The sampling strategy (in this technique) is to collect a number of samples in each stratum proportional to the product of the weight and the standard deviation of the stratum, thus reducing the standard error. However, in field sampling, one must be aware of pseudoreplication. Statistical techniques used in stratified random sampling are described by Cochran (1963).

Systematic sampling is conducted when samples are taken at predetermined intervals and are evenly spaced throughout a designated sampling area. The initial sampling point is chosen randomly. For example, a random point may be chosen in a pond and samples then collected at that point and at 10 points 50 m apart in four directions. Such systematic sampling (compared with random sampling) is often easier to conduct and may be more precise and easier to execute without mistakes. This is often an advantage in field sampling. If the sampled area varies in a "regular" fashion, systematic sampling is about as precise as stratified random sampling. However, because systematic samples may be typically spread more evenly over the population (for instance, in the case of zooplankton in a pond), systematic sampling is sometimes more precise than stratified random sampling.

Cluster sampling is used when the pool of samples is divided into several separate groups and then several elements from each set are selected. The result is a group or cluster of measures from the elements. This method is used when the effort and expense of physically collecting a small group of units are significant.

Composite samples are often taken by aquatic biologists in the laboratory or field by pooling several samples into a composite sample to estimate one parameter (e.g., collecting three water samples and combining the water for analysis of chemical residues). This process is economical; however, information on spatial variation is lost. Sampling programs using composite samples should take into account natural variation in the biological communities. Also, if chemical residues in sediment and water are being composited, natural spatial variations should be considered. Replicate composite samples should be taken whenever possible to estimate population variances.

For further reading on sampling techniques, one should consult Cochran (1976).

Computer Programs

Computer programs provide useful support for analyzing data. Numerous software packages for use on large mainframe computers and microcomputers are available for most procedures discussed or suggested in this chapter. The most commonly used statistical software includes BMDP (Dixon, 1985), SAS Institute (1989), and SPSS (1988). Various specialized statistical analysis programs are also available for microcomputers.

Statistical Methodology

The type of statistical method used to analyze data depends on what type of random variable is measured. As it is impossible to describe all the possible methods available in one chapter, this section describes the more commonly used procedures. Perhaps the best way to discuss statistical methodology is to describe methods used for continuous and discrete random variables.

METHODS COMMONLY USED FOR CONTINUOUS RANDOM VARIABLES

Descriptive Statistics

As explained earlier, a continuous random variable is one that is actually measured rather than placed in a category. For example, fish weight can usually be measured with a scale rather than simply classifying the fish as small, medium, or large. A number of different measures of location and dispersion can be made. Two of the more important are the mean, a measure of location (index of central tendency), and the variance, a measure of dispersion. Many statistical methods estimate and test differences of these measures.

As mentioned previously, statistics is a science concerned with how far an experimental unit deviates from what is expected. This is true for both types of random variables (continuous and discrete). For example, examine the continuous random variable of fish lengths (cm) in Table 4. The mean for fish length is calculated by adding all fish lengths (equal to 60) and dividing by the sample size (10):

$$\text{Mean } (\overline{X}) = \frac{\Sigma X_i}{n} = 6 \quad (4)$$

The statistic $\overline{X}$ (sample mean) is an estimate of the parameter μ, the true population mean. $\overline{X}$ should lie close to μ if the sample is large enough and not unusual. The mean (μ) is referred to as the expected value, $E(x)$. The mean or average is the measure of central tendency most used in statistical analyses.

The mean is the value one would expect to get if an infinite number of samples were taken. The symbol X_i represents the actual random variable for the ith individual. The deviations are calculated by determining how much the observed data deviate from what is expected. Thus, these data may be represented by the following equation or linear additive model (see below), which states that an observation consists of a mean plus an error.

$$\text{Linear additive model: } X_i = \mu + e_i \quad (5)$$

where X_i = observed value for the ith individual
μ = expected value (= mean)
e_i = deviation of the ith individual from the mean (error term)

Such a model is applicable to the problem of estimating or making inferences about population means and variances. If one has the mean and the deviation from the mean, the actual random variable can be generated. For example, if one looks at fish number 5, $X_5 = 6 - 3 = 3$.

The sample variance (s^2) is the squared sum of the deviations of the observed data minus the mean, divided by the sample number minus 1 (i.e., df, the

Table 4. Example of fish lengths (cm)

Fish number	Length (X_i)	Mean	Deviation from mean
1	7	6	1
2	6	6	0
3	9	6	3
4	4	6	−2
5	3	6	−3
6	5	6	−1
7	5	6	−1
8	8	6	2
9	6	6	0
10	7	6	1

degrees of freedom). The sample variance is an estimate of "true" variance, σ^2.

$$\text{Variance } (s^2) = \frac{\Sigma (X_i - \overline{X})^2}{n - 1} \quad (6)$$

The degrees of freedom may be thought of as the number of deviations from the mean that need to be directly estimated. The deviation of the ith individual from the mean is given by x_i (referred to as "little" x_i), thus $x_i = (X_i - \overline{X})$. For example, if one sums all the deviations from the mean of a sample, the result is zero. Thus, $\Sigma x_i = 0$. This is always true. Hence, if one estimates all the deviations (x_i) within the sample except one, the last deviation can be calculated indirectly by subtraction. As an example, add all the deviations from the mean of fish numbers 1 through 9 in Table 4. The result is -1. Thus, the value of fish number 10 must be 1 to make $\Sigma x_i = 0$. Therefore, degrees of freedom are determined by $n - 1$.

An alternative was to calculate the variance is as follows:

$$\text{Variance } (s^2) = \frac{\Sigma x^2}{n - 1} = \frac{\Sigma X_i^2 - (\Sigma X_i)^2/n}{n - 1} \quad (7)$$

This is the calculation formula. The square root of the variance (s^2) is referred to as the standard deviation. Variance and standard deviation are powerful measures of dispersion. The range of a sample (i.e., distance between maximum and minimum values in the sample set) is also used to describe sample variation; however, it is only a rough estimate of dispersion because it is easily affected by outliers and sample size. As mentioned earlier, for a normally distributed population, approximately 68% of points fall between $\pm$ 1 SD of the mean and approximately 95% of points fall between $\pm$ 2 SD of the mean.

We may also calculate a standard deviation of a population of means. To explain, the sample mean ($\overline{X}$) is itself a random variable because it is calculated from a random variable. The standard deviation of a population of means is commonly called the standard error (SE) and is calculated as

$$\text{Standard error (SE)} = \frac{\text{SD}}{\sqrt{n}} = \sqrt{\frac{s^2}{n}} \quad (8)$$

Note that, as the sample size becomes very large, the standard error becomes small, despite a fixed standard deviation for the population. A mean $\pm$ 1 SD represents approximately 68% of data points and the mean $\pm$ 1 SE represents approximately 68% of a population of means.

The coefficient of variation, CV, is a relative measure of dispersion in a sample. It is the ratio of the standard deviation to the mean:

$$\text{Coefficient of variation (CV)} = \frac{\text{SD}}{\overline{X}} \quad (9)$$

The CV is a descriptive statistic that enables one to compare relative amounts of variation between two variables. For example, the biomass of one species (e.g., brook trout) of fish from a stream may average 0.80 kg with a standard deviation of 0.40 (see below), and the average biomass of a second species (e.g., brown trout) from a stream may be 3.0 kg with a standard deviation of 1.0. Notice that the standard deviation is larger for the second species. However, a comparison of the CV's indicates that brook trout have proportionately more variation in average biomass. For example,

$$\text{Brook trout CV} = \frac{0.4}{0.8} = 0.5$$

$$\text{Brown trout CV} = \frac{1}{3.0} = 0.3$$

Statistical Testing

A primary question in statistics is, "Are two populations alike or are they different?" The type of random variable, the number of treatments, and the experimental design (discussed later) determine the analytical method to be used to answer that question. The following sections describe methods used for continuous random variables.

Standard (or Student's) t*-Test*

Three types of standard or Student's t-test are commonly used in the field of toxicology: one-sample t-tests, paired t-tests, and two-group t-tests (unpaired). Each of these tests may have a one-sided (one-tailed test) or two-sided (two-tailed test) alternate. One-tailed and two-tailed tests will be discussed with the one-sample t-test.

One-Sample t*-Test*

To test whether a population is equal to a known or standard value, a one-sample t-test is used. As an example, assume the adult biomass of bluegill sunfish to be about 0.25 kg, given normal environmental conditions (this is, of course, unknowable in nature!). A researcher may want to determine whether the mean adult biomass of a population of a bluegill in a newly discovered lake is equal to 0.25 kg ("standard").

The steps in completing this test are as follows (data are in Table 5):

Table 5. t-test example for fish biomass (kg)

Fish number	Biomass (X_i)	X_i^2
1	0.3	0.09
2	0.2	0.04
3	0.4	0.16
4	0.5	0.25
5	0.3	0.09
	$\Sigma X_i = 1.7$	$\Sigma X_i^2 = 0.63$

1. Set the null and alternative hypothesis.

   H_0: Mean adult biomass of bluegill = 0.25 kg

   H_A: Mean adult biomass of bluegill ≠ 0.25 kg

2. Set the probability of a type I error, $\alpha = 0.05$.
3. Take a random sample of size n and measure biomass; then calculate the mean, variance, and standard error, using Eqs. (4), (7), and (8).

$$\text{Mean } (\bar{X}) = \frac{\Sigma X_i}{n} = \frac{1.7}{5} = 0.34 \quad (4)$$

$$\text{Variance } (s^2) = \frac{\Sigma x^2}{n-1} = \frac{\Sigma X^2 - (\Sigma X_i)^2/n}{n-1} = \frac{63 - (1.7^2/5)}{4} = 0.013 \quad (7)$$

$$\text{Standard error (SE)} = \sqrt{\frac{s^2}{n}} = \sqrt{\frac{.013}{5}} = 0.051 \quad (8)$$

4. Calculate the t-test statistic:

$$t = \frac{\bar{X} - 0.25}{\text{SE}} = \frac{0.34 - 0.25}{0.051} = 1.76 \quad (10)$$

5. Choose the appropriate hypothesis based on the following rule: Reject H_0 if the absolute value of the calculated test statistic is greater than or equal to the tabulated t-value with df = $n - 1$ and $\alpha/2$.

In this example, df = 4 and $\alpha/2 = 0.025$, so the t-value = 2.776, and 1.76 (calculated value) ≥ 2.776 (tabulated value). H_0 cannot be rejected. Thus, the researcher would conclude no difference between biomass of the new pond and the ''standard'' biomass value (0.25 kg).

This procedure is a simple t-test comparing a sample population to a known or hypothesized value. The specific example was a two-sided alternate or a two-tailed test. A two-tailed test has equality signs in the hypothesis statement. However, often a researcher does not want to know if a population is equal to a known value; rather, one wants to determine whether a population is greater than or less than a known value. This produces an inequality in the hypothesis and is referred to as a one-sided alternative or one-tailed test. An example of a one-tailed test follows.

Use the same bluegill data as before; however, the question now is whether the bluegill biomass in the new body of water is greater than 0.25 kg. The way this statement is cast determines how one sets up the alternative hypothesis. Using the bluegill data (Table 5), the steps in the analysis are:

1. H_0: Bluegill population mean biomass ≤ 0.25 kg
   H_A: Bluegill population mean biomass > 0.25 kg

   Because the hypothesis has inequalities, the test is called a one-sided alternative or one-tailed test. In this example, the researcher is testing in only one direction, not two.

Steps 2, 3, and 4 are the same as for the two-tailed test.

5. Choose the appropriate hypothesis based on the following rules:
   Reject H_0 if both the following conditions are true:

   (a) The test statistic must be positive (1.76); and
   (b) The absolute value of the test statistic is greater than or equal to a t-value with df = $n - 1$ and α (not $\alpha/2$). Hence, 1.76 (calculated ≥ 2.132 (tabulated). Both of these statements must be true to reject H_0. Because only one is a true statement, H_0 cannot be rejected.

The same rules apply when testing whether a population mean is less than a hypothesized value. For example,

1. H_0: Population ≥ hypothesized value
   H_A: Population < hypothesized value

Steps 2, 3, and 4 follow the same as for the two-tailed test.

5. Reject H_0 if both the following conditions are true:

   (a) The test statistic is negative; and
   (b) The absolute test statistic ≥ t value at df $n - 1, \alpha$.

Decision for a One- or Two-Tailed Test

Often a researcher has no idea if a population is less than or greater than a hypothesized value. In these cases, a two-sided alternative should be used. If the question specifically tests a mean less than (or greater than) a value, a one-sided alternative should be used. Notice that one uses $\alpha/2$ and α for the two-tailed and one-tailed tests, respectively, to find the critical value in a t-table.

Paired t*-Test*

The t-test discussed above compares one population to a known or hypothesized value. The more common t-test determines whether a difference exists between two population means. There are two principal two-population tests, the paired t-test and the two-group (unpaired) t-test.

The paired t-test deals with a phenomenon called natural pairs. Examples of natural pairs would be leaf halves, eyes, identical twins, split-homogenetic, or use of an experimental unit for self-control (example: blood sample before and after treatment). When one tests whether population I is equal to population II, the null and alternative hypotheses are:

H_0: $\mu_1 = \mu_2$

H_A: $\mu_1 \neq \mu_2$

However, in the paired t-test the hypotheses are:

H_0: $\mu_D = 0$ where μ_D = average difference

H_A: $\mu_D \neq 0$

Example

If a fish is homogenized and one-half of the homogenate is sent to one laboratory and the other half to a different laboratory, and if the resulting chemical analyses are identical between laboratories, then the mean of homogenate half 1 minus the mean of homogenate half 2 should equal zero. thus, in a paired t-test, differences between units have to be calculated and all statistics performed on the differences. After the differences have been determined, the method is the same as testing a population versus a known or hypothesized value (as already discussed). However, in this case, the hypothesized value = 0.

Two-Group t-*Test*

If unpaired data are taken from two populations, a two-group t-test should be performed. The appropriate two-tailed hypotheses are:

H_0: $\mu_1 = \mu_2$

H_A: $\mu_1 \neq \mu_2$

The two-group test for differences between means can take on two different forms, depending on the presence of equal or unequal variances between the two populations. The test to determine equal or unequal variances is done by dividing the larger variance by the smaller variance. (The ratio of the two variances is distributed as an F and is referred to as an F statistic.) The hypotheses and rejection criterion for H_0 are as follows:

Hypotheses H_0: $\sigma_1^2 = \sigma_2^2$

H_A: $\sigma_1^2 \neq \sigma_2^2$

Reject H_0 if:

$$\frac{\text{Larger variance}}{\text{Smaller variance}} \geq \begin{array}{l}\text{tabulated } F \text{ value at df of}\\ \text{numerator, df of denominator,}\\ \text{and } \alpha/2\end{array}$$

The correct F-value is determined by the numerator df (larger variance df), the denominator df (smaller variance df), and $\alpha/2$. These parameters determine where in the F-table one looks up the critical value of F. The $\alpha/2$ is the probability associated with a two-tailed test parameter. This F-test is a two-tailed test because which population variance is larger is not known beforehand.

If the two population variances are equal and both populations are assumed to be normally distributed and independent of each other, then a two-group t-test can be performed. If the variances of the two populations are unequal, then an approximation to a t-test can be performed. Alternatively, a transformation of the data can be performed to equalize the variances of the transformed data. Data transformations are discussed in a later section.

The following example data set will be used to demonstrate a t-test when sample variances are equal and an approximate t-test when sample variances are unequal. Normally, the decision of which t-test to use is made prior to the actual analysis; however, both types of t-test can be demonstrated with one example. Given the following example of fish lengths from two different ponds (Table 6):

1. Hypothesis H_0: $\mu_1 = \mu_2$
   H_A: $\mu_1 \neq \mu_2$
2. Probability of committing a type I error, $\alpha = 0.05$.
3. The data are as shown in Table 6.
4. (a) The test statistic used if the variances are equal is the two-group t-test statistic (equal variance):

$$t = \frac{\bar{X}_1 - \bar{X}_2}{\sqrt{\frac{\Sigma x_1^2 + \Sigma x_2^2}{(n_1 - 1) + (n_2 - 1)}\left(\frac{1}{n_1} + \frac{1}{n_2}\right)}} \tag{11}$$

or pooled variance*

$$s_P^2 = \frac{\Sigma x_1^2 + \Sigma x_2^2}{(n_1 - 1) + (n_2 - 1)} \tag{12}$$

(b) The test statistic used if sample variances are unequal is the two-group t-test statistic (unequal variances)

$$t' = \frac{\bar{X}_1 + \bar{X}_2}{\sqrt{(s_1^2/n_1) + (s_2^2/n_2)}} \tag{13}$$

When the example is worked out, the result is as follows:

$$t' = \frac{56.21 - 61.25}{\sqrt{1.288 + .8835}} = -3.42$$

*Note that S_P^2 in Eq. (12) is referred to as the pooled variance or the mean square error. Equation (11) is the same for equal or unequal sample sizes. For the example data:

$$t = \frac{56.21 - 61.25}{\sqrt{\frac{54.09 + 26.50}{6 + 5}\left(\frac{1}{7} + \frac{1}{6}\right)}} = -3.35$$

Table 6. Two randomized group t-test

		Length in cm	
		Pond 1	Pond 2
Fish lengths	1	57.8	64.2
	2	56.2	58.7
	3	61.9	63.1
	4	54.4	62.5
	5	53.6	59.8
	6	56.4	59.2
	7	53.2	
ΣX		393.5	367.5
ΣX^2		22,174.4	22,535.87
$\bar{X}$		56.21	61.25
Σx^2 [a]		54.09	26.50

[a] The Σx^2 is the numerator in Eq. (7).

5. Rejecting H_0:

(a) If $\sigma_1^2 = \sigma_2^2$, then reject H_0 if the absolute value of the calculated t-value $\geq$ tabulated t-value at df = $n_1 + n_2 - 2 = df_1 + df_2$, and $\alpha/2$. In this example $3.35 \geq 2.201$. H_0 is rejected.

(b) If $\sigma_1^1 \neq \sigma_2^2$, then reject H_0 if the absolute value of the calculated $t' \geq t'$-test where:

t'-test

$$= \frac{(s_1^2/n_1)\text{ (tabulated } t \text{ at df} = n_1 - 1, \alpha/2) + (s_2^2/n^2)\text{ (tabulated } t \text{ at df} = n_2 - 1, \alpha/2)}{(s_1^2/n_1) + (s_2^2/n_2)} \tag{14}$$

For the example data:

$$t'\text{-test} = \frac{(1.288)(2.447) + (.884)(2.571)}{1.288 + .8835} = 2.50$$

Since $3.42 \geq 2.50$, H_0 is rejected.

Which test should have been used: the two-group t when $\sigma_1^2 = \sigma_2^2$ or the approximation to a two-group t when $\sigma_1^2 \neq \sigma_2^2$? To answer this, test the two variances using the F-test described earlier:

$$F = \frac{\text{larger } s^2}{\text{smaller } s^2} = \frac{9.015}{5.3} = 1.7$$

Reject H_0 if F-value $\geq$ tabulated F at df of numerator, df of denominator, $\alpha/2$; in this case, $1.7 \geq 6.98$.

Because the above statement is not true, the H_0 of equal variances cannot be rejected. Hence, the appropriate t-test is a two-group t and not an approximation to a t-test.

A standard t-test assumes that the sample variances are homogeneous. Homogeneity of variances means a "closeness" in magnitude of sample variances. The alternate situation, in which sample variances are not similar, is referred to as "heterogeneity." If the assumption of homogeneous sample variances does not hold, the investigator can either transform the variables to reduce heterogeneity or modify the t-test to compare the means. The standard t-test will not be accurate for heterogeneous sample variances because the null hypothesis will be rejected fewer times then expected if the larger sample has the larger variance. Furthermore, the null hypothesis will be rejected many more times than expected if the larger sample has the smaller variance. If heterogeneity of variances is not corrected by transformation of the data, the t-test must be modified so the t-value approximates a t-distribution (Steel and Torrie, 1980). A weighted variance is not used to calculate the standard error of the mean $S_{\bar{x}1-\bar{x}2}$ because the samples from the two populations have unequal variances. See Cochran and Cox (1957), Snedecor and Cochran (1989), and Steel and Torrie (1980) for further details of the use of different t-tests based on differences in variance and sample size.

LINEAR STATISTICAL MODEL AND THE ANALYSIS OF VARIANCE

One of the primary statistical methods used to describe differences between or among populations is the analysis of variance (ANOVA). This test is used to compare three or more groups (means) of continuous data when the variances are homogeneous and the data are independent and normally distributed. If two groups exist, the ANOVA will give the same results as a two-group t-test (Sokal and Rohlf, 1981).

Analysis of variance can be expressed as an extension of Eq. (5). One type of ANOVA may be represented by the following equation [note the similarity to Eq. (5)]: linear additive model for a one-way ANOVA:

$$X_{ij} = \mu + t_i + e_{ij} \tag{15}$$

where X_{ij} = the random variable
μ = population mean
t_i = effect of the ith treatment or population; $\Sigma\, t_i = 0$
e_{ij} = deviation from the mean of individuals given the ith treatment

This equation is also referred to as a linear statistical model (LSM) or the general linear model (GLM). It is referred to as linear because all the parameters are additive. There exist many LSMs because there exists a large number of statistical designs. The statistical design represented by Eq. (15) is called a completely randomized design. Different designs are discussed in the section on experimental designs.

The name "analysis of variance" is appropriate for testing means in that it determines whether different treatments produce a significant amount of variation in a population. It is an arithmetic process for partitioning the total sum of squares into components associated with recognized sources of variation. For example, if one group of fish is placed in clean water and another placed in water with a toxicant, one would expect normal growth in clean water and retarded growth in contaminated water. This difference would presumably cause variation in growth between the two groups of fish. Hence, the sample variances are analyzed to determine the amount of variance caused by treatment and the amount of variance not caused by treatment (also referred to as the residual error or random error).

How Does the Linear Statistical Model Represent the Population?

Suppose an ANOVA is completed for the results of an experiment with fish in clean versus contaminated water. The given statistical model may look like this:

$$X_{ij} = \mu + t_i + e_{ij}$$

$$X_{ij} = 6 + 2 + e_{ij} \quad \text{for clean water}$$

$$X_{ij} = 6 - 2 + e_{ij} \quad \text{for contaminated water}$$

t_i is referred to as a treatment effect. In this example, if fish are in clean water, the value +2 is added to the average growth of 6. Thus the average for fish growth in control water is 6 + 2 = 8. If fish are in contaminated water, they get a −2 value, so the average of their growth is 6 − 2 = 4. Consequently, the effect of the treatment on sample means can be calculated by adding the total population means to the treatment effect. The calculation of means using LSM is called least-square means.

What does an ANOVA estimate? An ANOVA partitions the total variance into the amount of variance attributed to differences among treatments and to the variation within treatments. A simple ANOVA table presents both the source of variation and the amount associated with each treatment or source. For example, Table 7 presents a simple one-way ANOVA with the formula representing each partitioned variance component (referred to as expectation).

σ_e^2 is the pooled within-variance component (sometimes called the residual mean square or error mean square). Note that σ_e^2 is also a component of the among-treatment estimates. Also, the among-treatment estimate contains a constant k and the term $\Sigma\, t_i^2$. The t_i were discussed earlier and represent treatment effects. If there is no treatment effect, the t_i will all equal 0 and $k\,\Sigma\, t_i^2 = 0$. An F-value is the ratio of among-group variance (treatment mean square − independent estimate of the variance when the null hypothesis is true) divided by the within-group variance (error mean square − generalized error term, an average of the components contributed by the several populations or treatments). If all treatment effects are zero, then all t_i are zero, leaving $k\,\Sigma\, t_i^2 = 0$, and the F-ratio will theoretically be 1. As treatment effects become large, $k\,\Sigma\, t_i^2$ becomes large, thus producing a large F-value. There is a point at which the F-value becomes "significantly" large, which indicates that at least two t_i are not zero or at least two treatment means differ.

The following example illustrates the calculation of a one-way ANOVA, completely randomized design. An experiment is conducted to determine the effect of a chemical exposure on fish length. Three treatments (chemical concentrations) are used: 0 mg/L of chemical (control), 25 mg/L, and 50 mg/L. Mathematically, the basic calculations performed for a two-group t-test could also be computed for this example. However, since there are three rather than two treatments, it is more powerful to analyze all the experimental data at the same time, which is the purpose of the analysis of variance.

The data and basic calculations are in Table 8. A SAS program for analyzing this data is presented in Appendix 1.

Table 7. One-way analysis of variance table

Source	Variance components (expectations)
Among treatments	$\sigma_e^2 + k\,\Sigma\, t_i^2$
Within treatments	σ_e^2

Table 8. Data for calculating analysis of variance

	Fish length (cm)		
	0 mg/L	25 mg/L	50 mg/L
	14.8	12.2	9.5
	18.1	11.4	10.8
	27.6	15.6	11.9
	19.0	18.0	10.6
	18.2	14.2	8.6
	22.9	18.0	13.1
	30.7	17.0	10.4
ΣX_i	151.3	106.4	74.9
ΣX_i^2	3467.55	1660.8	814.59
$(\Sigma X_i)^2/n$	3270.24	1617.28	801.43
$\Sigma\, x^{2a}$	197.31	43.52	13.16
$\bar{X}$	21.61	15.2	10.70

[a] The $\Sigma\, x^2$ is the numerator in Eq. (7).

After the basic statistics are complete, the ANOVA is calculated (Table 9). Table 10 is the condensed version of the ANOVA presented in Table 9.

If the among-treatment mean square is divided by the within-treatment mean square and if this fraction is not significantly greater than the tabulated F-value (Fisher tables) at numerator degrees of freedom (df) = 2 and denominator df = 18, then all the t_i are 0. In this example, the among-treatment mean square divided by the within-treatment mean square is equal to 15 (210.6/14) (Table 10). For this F-value not to be significant, it would have to be less than 3.55 (from the F-table). Obviously, this is not the case. This significant F-value implies the $\Sigma\, t_i^2$ is not equal to 0; thus, at least one of the treatments (chemical concentration) produced an average response in fish length significantly different from the total population mean of 15.84.

An ANOVA is normally thought of as a pretest to determine whether treatments differ from each other. If a significant F-value is obtained, at least two treatments may be assumed to differ from each other. The ANOVA does not specifically indicate which ones differ, just that at least one differs from the others. A follow-up (referred to as a posteriori) test is usually conducted subsequent to an ANOVA to indicate specific treatment mean differences.

Mean Separation Techniques

If an experiment has only two treatments, the test for significance is the two-group t-test previously discussed. However, if more than two treatments are tested and a significant variance among treatments is indicated, which treatments are different? To test where the differences occur, mean separation techniques are used.

To understand mean separation techniques, one needs to recognize exactly what is being tested. One may ask, is treatment 1 equal to treatment 2? To answer this, the first step is to subtract $\bar{X}_1 - \bar{X}_2$ and determine a difference. Recall that $\bar{X} = \mu +$

Table 9. Formula for calculating a one-way ANOVA

Source of variation	Degrees of freedom (df)	Sum of squares (Ss)	Mean square
Among treatments	$a - 1$ Number of treatments (a) minus 1 = 2	$(\Sigma X_1)^2/n_1 + (\Sigma X_2)^2/n_2 + (\Sigma X_3)^2/n_3 - CF^a$ For each treatment: sum all observations, then square the sum, which is then divided by the n for that treatment. After computing this value for each treatment, add all values together. This sum is subtracted from the correction factor described below. This value is called the treatment sum of squares. From the example: 3270.24 + 1617.28 + 801.43 − 6267.75 = 421.2	$\frac{\text{Among treatment SS}}{\text{Among treatment df}}$ $\frac{421.2}{2} = 210.6$
Within treatment (residual, error)	$a(n - 1)$ The number of treatments times the number of observations per treatment (n) minus 1, providing n is the same for all treatments 18	Total sum of square minus treatment sum of squares or $\Sigma x_1^2 + \Sigma x_2^2 + \Sigma x_3^2$ 675.19 − 421.2 = 254 197.30 + 43.52 + 13.67 = 254	$\frac{\text{Within treatment sum of squares}}{\text{Within treatment df}} = \frac{254}{18} = 14$
Total	$an - 1$ Total number of experimental units minus 1 20	Sum of the squares of all experimental units minus CF. $\Sigma X^2 - CF$ 5942.94 − 5267.75 = 675.19	

[a] Correction factor = CF = sum all observations, then square the sum, then divide this value by the total number of observations (an). CF = $(\Sigma X_1 + \Sigma X_2 + \Sigma X_3)^2/(n_1 + n_2 + n_3)$ = 110,622.76/21 = 5267.75.

t_1 and $\overline{X}_2 = \mu + t_2$. Thus, subtracting the two means produces

$$\overline{X}_1 - \overline{X}_2 = (\mu - t_1) - (\mu - t_2)$$
$$\overline{X}_1 - \overline{X}_2 = t_2 - t_1$$

This indicates that the test for differences between two treatment means is really a test for differences between treatment effects.

Mean separation may deal with pairwise comparisons in which each treatment is compared with every other treatment. This technique is used mainly if the number of treatments is minimal. Another type of mean separation with more power to differentiate among means is referred to as a contrast. The contrast method allows researchers to pool a group of means and test it against another group of means or a single mean. For example, one may want to test a control treatment against all other treatments pooled together. Only a certain number of independent comparisons can be made using contrasts. If two contrasts are independent of each other, they are called orthogonal contrasts.

Table 10. Condensed version of table 9

Source	df	Mean square	Expectation of mean square
Among treatment	2	210.6	$\sigma e^2 + k\, \Sigma\, t_i^2$
Within treatment	18	14	σe^2
Total	20		

Pairwise Comparisons

Many pairwise comparison techniques exist. One may begin to look at methods for treatment mean separation by describing whether treatments are qualitative or quantitative and how treatments are arranged. There are a number of references detailing how and when to use certain mean separation techniques (Chew, 1977; Day and Quinn, 1989). Qualitative treatments are treatments that are not quantitatively distinct. For example, a qualitative treatment may be "pesticide 1 versus pesticide 2." There is no quantitative measure associated with the grouping "pesticide 1 or pesticide 2." A quantitative treatment would be represented if different concentrations of the pesticide were included in the analysis, for example, 0 ppm, 2 ppm, . . . , 10 ppm, etc. Some analyses may have both qualitative and quantitative treatments. For example, both pesticides might have several concentrations. In this case, the pesticides are qualitative treatments and concentrations of each pesticide are quantitative. This is referred to as a factorial arrangement of treatments and is discussed later.

When to use a particular pairwise mean separation technique is difficult to decide; however, there are some basic rules. Multiple range tests such as Duncan's multiple range test (Duncan, 1955) or Newman-Keuls multiple range test (Chew, 1977) are acceptable for qualitative treatments with equal n per treatment. They are not the best choice if treatments are quantitative, have unequal observations per treatment, or are factorially arranged.

If treatments are qualitative, a multiple range test can be performed. Most multiple range tests do not need a prior significant F-statistic from an ANOVA. For example, assume nine salmonid species are tested at sublethal concentrations of a pesticide. The steps needed to perform a Duncan's multiple range test are:

1. Order the population means and identify them.
2. No F-test is required; however, one must calculate the pooled standard error (SE).

$$(\text{SE}) = \sqrt{\frac{\text{EMS}}{n}} \tag{16}$$

where EMS = error mean square or within-mean square from the ANOVA table
n = number of observations per treatment

3. Tabulated values are obtained from Duncan's tables and are determined by two criteria: the within mean square degrees of freedom and by how many positions means are apart, inclusive of self. For any two adjacent means, their position number is 2. The lowest and the highest mean will have a position number equal to the number of treatments (Chew, 1977).
4. Multiply the tabulated value by the pooled SE. This results in critical values based on position application.
5. The test begins by comparing the largest mean to the smallest mean. If $|X_{\text{largest}} - X_{\text{smallest}}|$ is $\geq$ the largest-position critical value, reject H_0.
6. Proceed by comparing the largest mean to the second smallest using the new position critical value. Once a nonsignificant difference is obtained, draw a line from the largest mean to the mean where the nonsignificant difference occurred. Do not test means within this line.
7. Proceed to the second largest mean and repeat steps 1 through 6.
8. Proceed until all means are assigned to a line.

Worked Example: Duncan's Multiple Range Test

1. Nine sample means of fish length are ranked from smallest to largest and identified:

ID	$\bar{X}$
11	119.7
31	127.4
13	125.6
33	130.8
21	132.6
12	134.6
32	137.2
22	147.3
23	149.8

2. Pooled SE = 6.26
3. From Duncan's tables: $\alpha = 0.05$, df = 18, Chew (1977).

Position no.	2	3	4	5	6	7	8	9
Tabulated value	2.97	3.12	3.21	3.27	3.32	3.35	3.37	3.39

4. (Tabulated value) * (pooled SE)

Position no.	2	3	4	5	6	7	8	9
Tabulated value	2.97	3.12	3.21	3.27	3.32	3.35	3.37	3.39
Tabulated value * Pooled SE	18.59	19.53	20.09	20.47	20.78	20.97	21.10	21.22

5. Compare the largest mean to the smallest:

Largest − smallest mean = 149.8 − 119.7 = 30.1

This difference is compared to the position 9 critical value of (21.21) because the smallest mean versus the largest is 9 positions apart. Results indicate these two means are statistically different.

6. Continue by comparing the largest mean to the second smallest:

$|149.8 - 127.4| \geq$ position 8 CV = 21.09.

As one proceeds, the first mean in which the largest mean does not differ significantly is ID number 33. At this time, a line should be drawn from the largest mean to 130.83.

7 and 8. Continue the same procedure with the second largest mean. The results produce the following three lines:

ID	11	31	13	33	21	12	32	22	23
Mean	119.7	127.4	125.6	130.8	132.6	134.5	137.2	147.3	149.8
Line 1									
2									
3									

Most multiple range tests follow the same procedure except that the tabulated values are different. Appendix 2 lists a SAS program which calculates Duncan's multiple range test and other mean separation techniques. Finally, we note that if treatments are quantitative, the response effects should be orthogonally partitioned into linear, quadratic, and possibly further polynomial components. This method is described by Snedecor and Cochran (1989) and Chew (1977).

Another type of a posteriori pairwise comparison is Fisher's least significant difference (LSD). It may be used if the numbers of observations among treatment groups are equal or unequal, the overall F-

value is significant, and the number of comparisons made is equal to or less than the number of treatments. However, this method is questionable if the overall F-value is not significant. If one uses the data from the ANOVA table calculated earlier (Table 9), the LSD value is as follows:

$$\text{LSD} = t - \text{value}\sqrt{\text{EMS}\left(\frac{1}{n_i} + \frac{1}{n_j}\right)} \quad (17)$$

where t − value = the table t value for degrees of freedom of the EMS and $\alpha/2$
EMS = error mean square
n_i and n_j = number of observations for the ith and jth means

The completed example shows:

$$\text{LSD} = 2.101\sqrt{254\left(\frac{1}{7} + \frac{1}{7}\right)} = 17.9$$

If $|\overline{X}_i - \overline{X}_j| \geq$ LSD, reject H_0.

If the main hypothesis is to compare all treatments versus a control treatment, Dunnett's method (Dunnett, 1955), Gupta and Sobel's method (Gupta and Sobel, 1958), or Williams' method (Williams, 1971) should be investigated (see Appendix 2).

Contrasts

Contrasts can be calculated whether treatments are qualitative or quantitative. Qualitative treatment contrasts can be used to compare one group of treatments versus another. For example, if the ANOVA has eight pesticide treatments and four of these treatments have one mode of action and the other four a different mode of action, a contrast may be used to compare one set of treatments versus the other. This type of contrast is called a pooled mean contrast. If treatments are quantitative, a pooled mean contrast can be made, but a more meaningful contrast is to test whether a significant response occurs. This is the same as testing a linear, quadratic, cubic, etc. response. This type of contrast is called a polynomial contrast. Both types of contrasts are calculated in the same way.

Orthogonal contrasts are contrasts that are independent of each other. The number of possible orthogonal contrasts depends on the number of treatments in the ANOVA. The number of orthogonal contrasts is limited to the treatment degrees of freedom. For example, if three treatments are present, there would be two possible orthogonal contrasts. The following sequence of equations demonstrates how to calculate contrast and test for significance. This is followed by a method to determine if the contrasts are orthogonal. The method is followed by an example of the two types of contrasts using the data from the ANOVA presented in Table 9.

$$\text{Contrast } (C) = \lambda_1 T_1 + \lambda_2 T_2 + \cdots + \lambda_a T_a = \Sigma\, \lambda_i T_i \quad (18)$$

where T_i = treatment sum based on equal number of observations n
λ = multipliers, $\lambda_1 + \lambda_2 + \cdots \lambda_a = 0$
$\Sigma\, \lambda_i = 0$
a = number of treatments
n = number of observations per treatment, given equal n per treatment

Sum of squares (Ss) for a contrast:

$$\text{Ss} = \frac{C^2}{n\, \Sigma\, \lambda_i^2} = \frac{(\Sigma\, \lambda_i T_i)^2}{n\, \Sigma\, \lambda_i^2} \quad (19)$$

This sum of squares has 1 df; thus the sum of squares is the mean square (Ss/df − MS).

The average difference of a contrast:

$$\hat{\mu}_D = \frac{C}{n\, \Sigma \text{ positive } \lambda_i} \quad (20)$$

To determine whether two contrasts are orthogonal, proceed as follows: If a different contrast (H) has a different set of multipliers (ϵ), then

$$H = \epsilon_1 T_1 + \epsilon_2 T_2 + \epsilon_a T_a \quad \text{and} \quad \Sigma\, \epsilon_i = 0$$

is also a contrast. The two contrasts (C and H) are orthogonal if:

$$\lambda_1\epsilon_1 + \lambda_2\epsilon_2 + \cdots + \lambda_a\epsilon_a = 0$$

There are $(a - 1)$ orthogonal contrasts possible among treatments (set of orthogonal contrasts).

Pairwise independence among all comparisons in a set of orthogonal contrasts ensures mutual independence. If the $a - 1$ contrasts are all independent of each other, then the sum of squares for the contrast, $\text{Ss}_1 + \text{Ss}_2 + \cdots \text{Ss}_{(a-1)}$ is equal to the treatment sum of squares.

The hypothesis test for significant contrast is as follows:

H_0: $\mu_D = 0$

H_A: $\mu_D \neq 0$

Reject H_0 if Eq. (21) is a true statement. Test for significant contrast (1 df F-test):

$$\frac{\text{Ss of contrast}}{\text{Within mean square}} \geq F\text{-tabulated at 1, EMS df} \quad (21)$$

Example of Pooled Mean Contrast and Polynomial Contrast

The ANOVA example (Table 9) will be used to calculate contrasts. For example, to test whether the control dose response is equal to the 25 and 50 mg/L doses, the contrast is performed in the following manner:

1. λ_i	−2	1	1
T_i	151.3	106.4	74.9

$$C = \lambda_1 T_1 + \lambda_2 T_2 + \lambda_3 T_3$$
$$C = -2(151.30) + (106.40) + (74.9)$$
$$C = -302.6 + 106.4 + 74.9$$
$$C = -121.3$$
$$\Sigma \lambda_i = 0$$
$$\Sigma \lambda_i^2 = 6$$

2. Equation (19) is used to calculate the sums of squares:

$$\text{Ss} = \frac{C^2}{n \Sigma \lambda^2} = \frac{121.3^2}{7(6)} = 350.33$$

3. The mean (= average difference) is calculated following Eq. (20):

$$\hat{\mu}_D = \frac{121.3}{7(2)} = 8.7$$

The contrast compares the control against the 25 and 50 mg/L concentrations and is referred to as a pooled mean contrast. It is important to understand how the multipliers (λ) work in a pooled mean contrast. If a researcher is testing treatment mean 1 versus treatment mean 2, mean 1 is subtracted from mean 2. What is actually being performed is that treatment mean 1 is multiplied by +1 and treatment mean 2 is multiplied by −1 and then they are added together. Essentially, this is being done in a pooled mean contrast except that the groups have to have equal weights. In the previous example, the control group is multiplied by 2 because it is being compared to two treatments, 25 mg/L and 50 mg/L. It does not matter which group gets the plus sign or the negative sign as long as all the multipliers, when added, equal 0 ($\Sigma \lambda_i = 0$).

4. Another contrast could be the control and 25 mg/L concentration against the 50 mg/L concentration. Both are pooled mean contrasts. Are these two contrasts independent (orthogonal)?

Multipliers for	0 mg/L	25 mg/L	50 mg/L
0 vs. 25 and 50	−2	+1	+1
0 and 25 vs. 50	−1	−1	+2

Both are legitimate contrasts because $\Sigma \lambda_i = 0$ in both situations. To determine if the contrasts are orthogonal, multiply the multiplier coefficients between the contrasts and then add the products over all treatments. For example,

$$(-2 * -1) + (1 * -1) + (1 * +2) = 3$$

Because the multipliers do not sum to 0, these contrasts are not orthogonal. Therefore, at least some information in one contrast is also in the other. The polynomial contrasts below are orthogonal.

Multipliers for	0 mg/L	25 mg/L	50 mg/L
Linear	−1	0	1
Quadratic	−1	2	−1

Note that $(-1 * -1) + (0 * 2) + (1 * -1) = 0$.

5 and 6. The linear and quadratic polynomial contrasts are a set of independent comparisons. No other comparisons will be independent of these two.

7. The example ANOVA (Table 9) has two degrees of freedom; thus, there cannot be more than two orthogonal contrasts. Each orthogonal contrast has one degree of freedom and a sum of squares. The sum of the orthogonal sum of squares will equal the among-treatment sum of squares in the ANOVA. Using the linear and quadratic contrasts,

Sum	151.3	106.4	74.9
Linear	−1	0	1

$$C = 151.7 + 0 + 74.9 = -75.4$$
Ss linear = 416.93 [from Eq. (19)]

Quadratic	−1	2	−1

$$C = 151.3 + 212.8 - 74.9 = -13.4$$
Ss quadratic = 4.27 [from Eq. (19)]

The sum of these two orthogonal contrasts add to the treatment sum of squares = 416.93 + 4.27 = 421.20.

8. Each contrast can be tested for significance:

H_0: $\mu_i = 0$

H_A: $\mu_i \neq 0$

Reject H_0 if Eq. (21) is a true statement. For the example data:

$$\frac{\text{Ss of contrast}}{\text{Within mean square}} \geq F\text{-Table 1, within DF, } \alpha = 4.41$$

$$\text{Linear contrast } \frac{416.93}{14} = 29.78 \geq 4.41$$

$$\text{Quadratic contrast } \frac{4.27}{14} = 0.305 \geq 4.41$$

In this example, the linear contrast is statistically significant but not the quadratic contrast. Also, the treatments are quantitative and the contrasts are polynomial. If the multipliers for the linear (−1, 0, 1) are plotted, it can be seen that a linear function is being tested. Plotting the quadratic contrast multipliers (−1, 2, −1) shows that a curvature is being tested. Interpretation of these data would say the response is linear with no curvature (Chew, 1977). This conclusion is based on a significant F-value for the linear but not for the quadratic contrast. The multipliers for polynomial contrasts are presented in Snedecor and Cochran (1989).

Parametric Assumptions

There are assumptions that need to be satisfied when performing an ANOVA or other parametric tests. The basic assumptions of all parametric analyses are:

1. Treatment observations are taken at random and are independent of each other.

2. The data follow a normal distribution.
3. Variances among treatments are homogeneous.

A normal distribution is illustrated by a bell-shaped curve (approximated in Figure 1a). The area underneath the curve represents 100% of a statistical population. Sometimes experimental units are not normally distributed. In such cases, the experimenter may handle the problem of nonnormality by increasing sample size. The central limit theorem states that in selecting random samples of size n from a population with a mean (μ) and standard deviation (σ), the sample distribution of $\bar{X}$ becomes normally distributed with mean (μ) and standard deviation ($\sigma/\sqrt{n}$) as the sample size approaches ∞. In actuality, convergence to normality is quick. For practical purposes, an n of 30 is indistinguishable from an infinitely large sample size.

Distribution of Data—Histograms

The use of appropriate statistical tests and the effects of small sample size on the selection of statistical techniques are two major points made in this chapter. Preliminary examination of the nature and distribution of data collected from a study can suggest unexpected patterns for which additional or alternative statistical procedures should be used. If data sets contain 25 or more values, then kurtosis (flatness of the curve) or skewness (distributions greater or less than expected with a normal curve) of the population can be computed to determine whether the population follows a normal distribution. A χ^2 goodness-of-fit test (see below) for normality can be used for this purpose. However, if each group of data contains fewer than 25 values (typically the case in aquatic toxicology), measures or tests such as kurtosis, skewness, and χ^2 goodness of fit are not accurate indicators of normality. In these cases, a frequency histogram may be prepared to determine visually the location and approximate distribution of data to estimate whether the data are normally distributed. The horizontal scale or abscissa should be the same scale as the data values and it should be divided to cover all the observed values. Histograms reveal information about the association between two or more variables, the existence of outliers, and data clustering (Chambers et al., 1983).

For example, consider the shell length data for two different groups of oysters (n = 16 per group) (Figure 1a and b). The range of each group is divided into

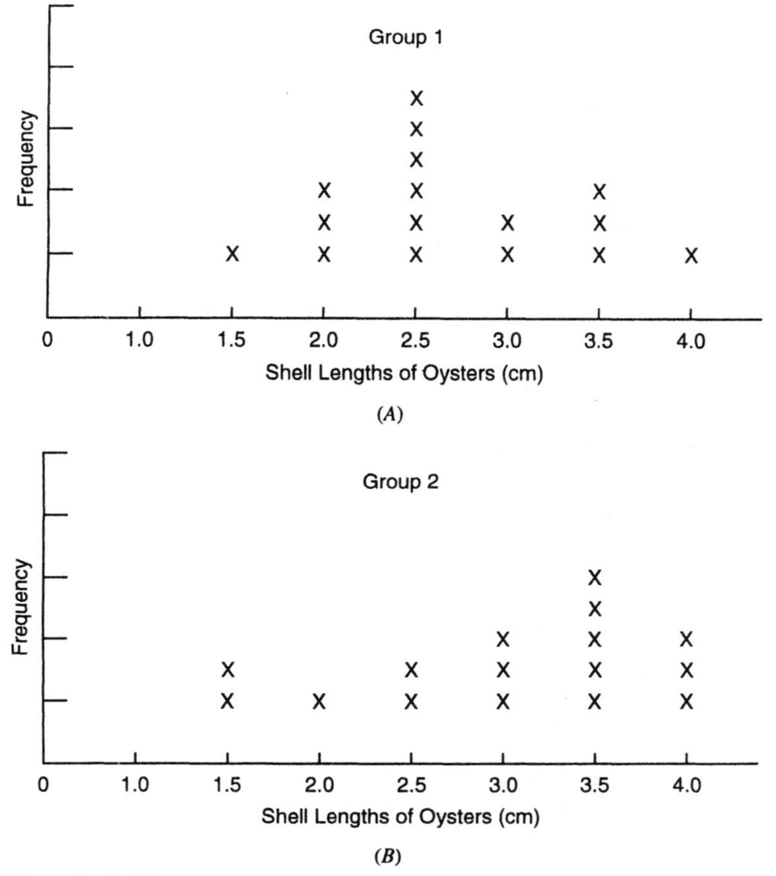

Figure 1. Shell length data for two groups of oysters.

equal size classes (seven classes of 0.5 cm each) and the observed values and frequencies (number within each categorical range) used to draw a frequency histogram. Figure 1a represents a group with an approximately normal distribution (bell-shaped) and one in which parametric tests can most probably be used. However, the data in Figure 1b are clearly not normally distributed and the data from this group should be quantified using nonparametric techniques.

Alternatively, a graph based on cumulative frequency distribution can be drawn using arithmetic probability paper. On such graph paper, the abscissa is linear and the ordinate is the cumulative percentage rising from 0.01 to 9.99. Relatively speaking, the ordinate is "stretched" at both ends so that when the percentages of observations less than or equal to each of the upper class limits are plotted, the points lie on a straight line if the distribution is normal. In such cases, the data may be divided into size classes, the frequency calculated for each class, and the cumulative frequency plotted against the upper class limit on probability paper. A straight line is drawn through the points and the most weight is given to the points lying between cumulative frequencies of 25% to 75%. This method is good for large sample sizes ($n > 60$). For small sample sizes, the Rankit method is a suitable graphical test for normality and is discussed in Sokal and Rohlf (1981).

Another approach for large samples to determine whether a sample follows a normal distribution is to calculate the theoretical or expected distribution and compare it with the observed distribution. There are five steps in calculating the expected normal frequency distribution:

1. A Z value (standard deviation unit) is calculated for the upper limit on each size class (category):

$$Z = \frac{X_u - \overline{X}}{s} \tag{22}$$

where X_u = upper limit
$\overline{X}$ = sample mean
s = standard deviation

2. The area under the normal curve (A) is then read from a table of the cumulative normal frequency distribution, which is based on the value of Z and is available in most statistics books.
3. The cumulative area under the curve is calculated as follows: for negative Z, use $(0.5 - A)$; for positive Z, use $(0.5 + A)$. For example, when $Z = 2.5$, $A = (0.5 + 0.4938)$.
4. The expected class probabilities are calculated by subtracting successive cumulative probabilities.
5. To obtain the expected frequencies, class probabilities are multiplied by the sample size.

To test whether the observed frequency distribution fits or conforms with the theoretical or expected frequency distribution, a χ^2 goodness-of-fit test is typically applied. It is also used to test whether two or more sets of discrete data have similar distributions. When the sample size is small, a Kolmogorov-Smirnov test for goodness of fit may be applied because it is more powerful than the χ^2 test (Sokal and Rohlf, 1981).

Bartlett's Test for Homogeneity of Variances

Another important assumption for parametric analyses is equal variances among treatments. If variances among treatments are unequal, a data transformation may be used. One test for equality of variance, the F-test, has been discussed. This test is carried out by dividing the larger variance by the smaller variance to calculate an F-value. If a number of treatments exist, this is a rough estimate of equality of variance. However, for situations with more than two treatments, a common method for testing homogeneous variances is Bartlett's test.

Bartlett's test is used to compare the variability of data among three or more groups of continuous data (e.g., body weight, length, organ weight). These types of data are typically analyzed with parametric statistical methods (see below). Bartlett's test is based on calculating a corrected χ^2 value with the formula

$$\chi^2_{corr} = 2.3026 \frac{\Sigma\, df(\log_{10})\left(\frac{\Sigma\,[df(s^2)]}{\Sigma\, df}\right) - \Sigma\,[df(\log_{10} s^2)]}{1 + \frac{1}{3(K-1)}\left[\Sigma \frac{1}{df} - \frac{1}{\Sigma\, df}\right]} \tag{23}$$

where s^2 = variance for each treatment
Σ = sum over all treatments
N = number of data points within a group
K = number of groups being compared
df = degrees of freedom for each group = $(N - 1)$
2.3026 = natural log of 10

The χ^2 value generated by this calculation is compared with the values listed in a χ^2 table according to number of degrees of freedom and α (see a statistics book such as Snedecor and Cochran, 1989; Steele and Torrie, 1980). If the calculated χ^2 value is smaller than the tabulated value (at the selected α level like 0.05), the groups are assumed to be homogeneous and an analysis via ANOVA is appropriate. However, if the calculated χ^2 is greater than the tabulated value, the groups are heterogeneous and a data transformation or other tests may be necessary (see Table 11).

Example

The lengths of fish (cm) in three different treatment groups at the end of a chronic toxicity test are as shown in Table 11. Sample variances are calculated as follows:

$$s_1^2 = \frac{136 - [(20)^2/3]}{3 - 1} = 5.33$$

$$S_2^2 = \frac{113 - [(21)^2/4]}{4 - 1} = 0.92$$

$$S_3^2 = \frac{366 - [(38)^2/4]}{4 - 1} = 1.67$$

Table 12 should be created.

Table 11. Data used for Bartlett's test

(200 μg/L)		(100 μg/L)		(0 μg/L)	
(X_1)	$(X_1)^2$	(X_2)	$(X_2)^2$	(X_3)	$(X_3)^2$
8	64	6	36	11	121
6	36	6	36	10	100
6	36	5	25	9	81
		4	16	8	64
$\Sigma X_1 = 20$	$\Sigma X_1^2 = 136$	$\Sigma X_2 = 21$	$\Sigma X_2^2 = 113$	$\Sigma X_3 = 38$	$\Sigma X_3^2 = 366$

Now substitute into Eq. (23) for corrected χ^2:

$$\chi^2 = 2.3026 \frac{8[\log_{10}(10.43/8)] - 0.8073}{1 + [1/(3(3-1))](1.1667 - 1/8)} -$$

$$= 2.3026 \frac{8(0.1148) - 0.8073}{1 + 0.1667(1.0416)}$$

$$= 2.3026 \frac{0.1111}{1.1736} = 0.218$$

The critical χ^2 value with two degrees of freedom (= number of groups minus 1) at the 0.05 level is 5.99. Because the calculated value (0.218) is less than this, the χ^2 is not significant. The treatment group variances are accepted as homogeneous. Thus, parametric methods (e.g., ANOVA) may be used for further quantitative comparisons.

Bartlett's test for homogeneity of variances is valid for three or more samples (groups), and the standard *F*-test is a two-sample (control and treatment group) test for homogeneity of variance. The *F*-test is sensitive to data that depart from normality; such departures may be produced by outliers, especially in studies with small sample sizes. Bartlett's test is more efficient than the *F*-test.

Transformations

A number of transformations may be used when variances are unequal. Transformations most often used in conjunction with ANOVA are square root, log, and arcsin–square root. When should these transformations be used? The arcsin–square root transformation should be used when binomial proportions are to be analyzed by ANOVA. This transformation is called an angular transformation. For example, a binomial random variable estimate may be calculated as "the proportion dead in a number tested" (*p*). The variance of this point estimate is *p* (1 − *p*)/number tested. As can be seen, as *p* approaches 0 or 1 the variance becomes 0. This generally produces unequal variances among treatments. The arcsin–square root transformation attempts to control this problem. It stretches out the values at the tails of a distribution of percentages or proportions and compresses the middle of the distribution.

To test a random variable distributed as a binomial, it is important that the number of observations per treatment be equal. The transformation of the binomial is accomplished using the arcsin–square root of the proportion. However, statistical methods have been developed to analyze the binomial using discrete random variable methods. These methods are discussed later.

The square root and the logarithm transformations are also used to stabilize the variance. The square root transformation is applied to data that follow a Poisson distribution, in which the variance is proportional to the mean. The logarithmic transformation is commonly used for biological data. Biological

Table 12. Basic calculations from Table 11

Concentration (μg/L)	N	df = $(N-1)$	s^2	(df)(s^2)	log s^2
200	3	2	1.33	2.67	0.1239
100	4	3	0.92	2.76	−0.0362
0	4	3	1.67	5.00	0.2227
Sums (Σ)	11	8		10.43	

Concentration (μg/L)	(df)(log s^2)	1/df
200	0.2478	0.5000
100	−0.1086	0.3333
0	0.6681	0.3333
Sums (Σ)	0.8073	1.1666

data (examples would be mass, population growth, metabolic rate) typically respond in logarithmic fashion to changes in environmental conditions. Such data may be transformed as $Z = \log(X + 1)$, which allows zero values to be used. Also, $\log(X + 1)$ is a better transformation for small values than $\log X$. The log transformation is typically applied when the mean is positively correlated with the standard deviation. Applying this transformation may make the variance independent of the mean.

An ANOVA assumes that, as the means among treatments increase, the variance associated with each sample mean remains the same. In some cases, this does not happen. One can test for the relationship between the mean and variance by plotting treatment means against the variances. As an example, plot treatment means along the X axis and treatment variances along the Y axis and visually draw a line among all points (Figure 2). If the line is flat, no transformation is needed. This indicates that, as treatment means increase, treatment variances remain constant. If the line increases at a linear rate, a square root transformation may be needed. If the line increases at an increasing rate (nonlinear), a logarithmic transformation may be needed. Figure 2 demonstrates the types of transformation that may be needed.

Of course, once data have been transformed, the data should be rechecked to see whether the transformation has corrected the problem. Several other methods help to stabilize treatment variance, such as the Box and Cox (1964) method for normal data (Sokal and Rohlf, 1981; Box et al., 1978; Bartlett, 1947), Taylor's (1961) method for variance stabilization, and Tukey's method for nonadditivity (Snedecor and Cochran, 1989). If treatments have equal or near-equal observations per treatment, the assumptions of normality and equal variance are not as critical for testing treatment mean differences. On the other hand, these assumptions of normality and equal variances are critical for unequal sample sizes (Scheffe, 1959; Milliken and Johnson, 1984). As an exercise, examine the data used for analysis in Table 8. A closer look indicates nonequal variances; therefore a transformation may be needed. We suggest that the reader transform the data using the log and recalculate the results.

NONPARAMETRIC TECHNIQUES

Analysis of Rank Data

Sometime the assumption of normality or homogeneous variances cannot be met. When this is true, the data can still be analyzed using nonparametric techniques. In cases in which the sample distribution is unknown and the data cannot be transformed to approximate the normal distribution, a nonparametric alternative is recommended. Nonparametric tests make fewer assumptions about data distribution than do standard parametric tests. For example, for *distribution-free* statistics, there are no assumptions about the distribution of data or homogeneity of variance. However, nonparametric tests tend to be less powerful than their parametric alternatives, particularly when sample size is small. Two tests are presented below for evaluating statistical significance in rank data: the Wilcoxon rank sum test and the Mann-Whitney U-test. Both tests are nonparametric alternatives to the standard t-test.

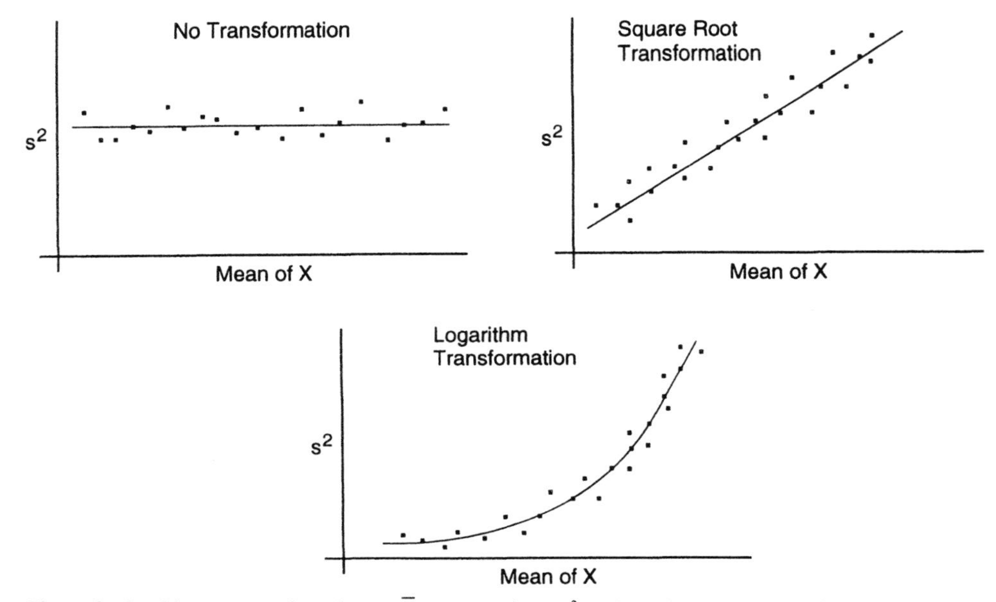

Figure 2. Graphic representation of mean $\overline{X}$ versus variance s^2 to determine possible transformations.

Wilcoxon Rank Sum Test

This test compares two groups of nonparametric data and assumes no set distribution in the data. Data in both groups are ranked from smallest to largest, without regard to sign. Each number is then assigned a rank value, with the smallest number receiving a rank of 1. If there are duplicate numbers (or "ties"), each number of equal size receives the median rank for the entire group. For example, if the lowest number appears twice, both values receive the rank of 1.5. This indicates that the ranks of 1.0 and 2.0 have been used and the next highest number has a rank of 3.0. If the lowest appears three times, then each is ranked as 2.0 and the next rank is 4.0. Each tied number gets a median rank. After ranking the numbers in the two columns (one column for each group), the ranks in each group are totaled to give a "sum of ranks" for each group. To check this, calculate the value $[N(N + 1)]/2$, where N is the total number of data values in both groups. This calculation should equal the sum of ranks for both groups. The sum of rank values is compared to table values to determine the degree of significant difference, if any (see Beyer, 1976). The table includes an upper and a lower value for each significance level. If the number of data points in each group is the same and the calculated sum of ranks for both groups is the same, then the calculated sum of ranks for both groups should be between the two limit values. When this is true, the two groups are considered similar and not significantly different. If one or both sums of ranks are equal to or fall outside the table limits, the groups are statistically different. If the two groups have different numbers of data points, the smaller sum of ranks is compared to the table limits to determine significance.

Example

The times to hatch (in days) of fish exposed to 10 mg/L (group A) and 5 mg/L (group B) of a test substance are compared. The exposures might yield the data in Table 13.

From a probability table, for $N_1 = 8$, the limit values for $\alpha = 0.05$ are 49 and 87. Because one of the numbers falls outside the limits, the two groups of fish differ significantly in days to hatch. This example has a high percentage (>10%) of ties in the samples. This high incidence of ties causes the test to overestimate α and, therefore, creates a high false-positive rate (Berry and Lindgren, 1990). This test should not be used in cases with a high percentage of ties. It should be noted that the Wilcoxon signed-ranks test is used for paired observations.

Table 13. Ranked days to hatch

Days to hatch		Ranks	
Group A	Group B	Group A	Group B
1	1	2.0	2.0
1	2	2.0	5.0
2	3	5.0	7.5
2	4	5.0	10.5
3	4	7.5	10.5
4	4	10.5	10.5
5	5	13.5	13.5
18	23	45.5	59.5

Sum of sums = 45.5 + 59.5 = 105
Check = [14 * 15]/2 = 105

Mann-Whitney *U*-Test

Mann-Whitney U is another nonparametric test in which data from two groups are ordered from lowest to highest values, after which the entire set of values is combined and ranked. Average ranks are assigned to tied values. Ranks are then summed for each group and U is determined by the following equations. Mann-Whitney U for treated group:

$$U_t = n_c n_t + \frac{n_t(n_t = 1)}{2} - R_t \tag{24}$$

Mann-Whitney U for control group:

$$U_c = n_c n_t + \frac{n_c(n_c + 1)}{2} - R_c \tag{25}$$

where n_c and n_t are the sample sizes for the control and treatment groups and R_c and R_t are the sums of ranks for the two groups. To compare the two groups, the larger value of U_c or U_t is used and compared with critical table values (Siegel, 1956). The Mann-Whitney U has the same problem as the Wilcoxon test with a high percentage of ties. This test should *not* be used for paired observations.

Other Nonparametric Tests (Three or More Groups)

The Kruskal-Wallis nonparametric one-way analysis is used for a study with three or more groups of data when the assumptions of ANOVA fail. The data will commonly be rank-type evaluation data (Pollard, 1977). Steel's multiple treatment comparison rank sum test (Steel, 1959) can also be used (see Chapter 2). See Hollander and Wolfe (1973) for other nonparametric tests.

THE MAIN KINDS OF EXPERIMENTAL DESIGNS

Many research papers state "an ANOVA will be performed on the data." This statement, in most cases, is very general. Analyses of variance are based on experimental design. The number of observations is important; however, if the experimental units are not applied to an appropriate experimental design, the experiment may fail. As an example, let us say three treatments are to be tested and a large number of fish (the experimental unit) are to be exposed in each treatment. The random variable is weight after 90 d. What if treatment 1 is applied to 1-cm fish, treatment

2 to 3-cm fish, and treatment 3 to 5-cm fish? If an ANOVA is computed, a significant *F*-statistic will surely be present. Is the ANOVA testing treatment differences or initial size differences? Because each treatment was applied to a different size of fish, the question cannot be answered. Even if the number of experimental units were sufficient, the experimental design has failed. Designing an experiment is important to the success of the experiment. This section discusses five of the major experimental designs.

Complete Randomized Design (CRD)

Assume that all of the experimental units are homogeneous (i.e., same sex, weight, length, etc.) and that treatments are randomly assigned to each experimental unit. This design is called a completely randomized design. The ANOVA will contain the effects of treatment and random error.

Randomized Complete Block Design (RCBD)

In some cases, the experimental units are heterogeneous but can be partitioned into homogeneous parts. If such is the situation, a randomized complete block design should be employed. In the preceding example in which different treatments were given to three different sizes of fish, a proper design could have been performed. Treatments 1 through 3 could have been randomly assigned within "small" unit, "medium" unit, and "large" unit. The effect of size could have been explained and removed so that a treatment effect could be demonstrated with no interference from size. The experiment in this case should have been blocked according to size, with all treatments occurring within a size. Whenever a heterogeneous group of experimental units exists, one should evaluate how these units may be placed into homogeneous group and treatments applied within groups. The title randomized complete block indicates that treatments are randomly assigned to experimental units within a block and that combinations of all treatments exist within a block. The ANOVA will contain the effect of block, treatment, and random error. If all treatments cannot exist within a block, a lattice or balanced incomplete block should be investigated.

The Latin Square

Often facilities are limited or the experimental units are expensive. In these cases, a Latin square design may be appropriate. Assume that four ultraviolet light chambers exist and that treatments are four different levels of light. The random variable is percentage survival of fish in each chamber. In this case, if the experiment is performed only once, the investigator cannot perform a statistical analysis, because there is only one observation per treatment. Also, let us assume there is no assumption of equality of chamber effects. How should the experiment be mapped? The following is a possible map of the experiment.

		Chamber number			
		1	2	3	4
Period	1	A	B	C	D
	2	B	C	D	A
	3	C	D	A	B
	4	D	A	B	C

The letters represent treatments, the columns represent chambers, and the basic experiment must be conducted four times (period or rows). Notice that all treatments appear in a row or column combination only once. The ANOVA will contain row, column, treatment, and random error. The number of treatments dictates the dimensions of a Latin square design. If six treatments are proposed, the dimension of the Latin square is six rows by six columns.

There are variations of the Latin square. Latin squares can be computed with a missing row or missing column, but this is not recommended. Also, if the treatments are arranged so that each follows another treatment the same number of times, residual effects can be tested. These designs are commonly used to test carryover effects of herbicide and pesticide rotations on field crops. If a researcher has only two treatments, a 2 by 2 Latin square is not valid, because no degrees of freedom are left for the error term. The solution is to design an experiment with multiple 2 by 2 latin squares. This type of variation of the Latin square is called a crossover design.

Many other designs and variations of these basic designs exist:

Nesting

Many times multiple samples are taken from an experimental unit and multiple determinations are made within each sample, etc. The repeated sampling and subsampling gives rise to a nested design. These types of designs are important in determining sources of variation. This was mentioned earlier in the section on the difficulty of selecting an appropriate sample size (i.e., each treatment had multiple sites, each site had multiple samples, and each sample had multiple determinations, see equation 3). Care must be taken in the nested design when computing the *F*-statistic. A good knowledge of the expectation of the mean square is needed to calculate correct *F*-values and variance components.

Split Plot

Split plots are designs allowing an investigator to produce an ANOVA for one factor but also apply a second factor within each experimental unit. For example, assume an experiment is conducted to test five treatments (say, five different pesticides) with 10 fish per treatment. After a period of time, the investigator removes different organs from each fish and makes measurements on each organ. In this case, the main plot is treatment and the subplot is organ. Researchers often make repeated measurements on the same experimental units given certain treatments (multiple weights, length, blood chemistries). In this case, treatment is the main plot effect and time is the subplot.

Factorial Arrangement of Treatments

A factorial arrangement is not a design; rather, it describes how treatments are arranged. All designs mentioned above may have treatments arranged factorially. For example, an experiment may be conducted to evaluate different levels of pH on different levels of aluminum. Assume five levels of pH (5.0, 5.5, 6.0, 6.5, and 7) and, at each pH, four concentrations of aluminum. This is called a 5 by 4 factorial arrangement of treatments. In this example, 20 separate combinations or treatments exist. The factorial arrangement is important in that it allows the researcher to determine synergistic and antagonistic effects.

Unequal Sample Size

Most experiments are designed to have equal experimental units per treatment. However, it is inevitable that experimental units are eliminated for one reason or another. Most computer packages have no problem in analyzing data with unequal numbers of observations per treatment. Some experimenters believe that more experimental units should be placed in a toxic treatment to compensate for the death loss. If this is done, this treatment may be biased upward, leaving only the hardier experimental units, while the control has a full range of hardy to nonhardy units. Increasing the sample size for certain treatments is not recommended and, if it is done, care should be taken when interpreting data. For more reading on experiment designs, see Cochran and Cox (1957).

LINEAR AND MULTIPLE REGRESSION

Regression analysis is similar to an ANOVA in that it partitions the amount of variation in a trend into two main parts, explained and not explained. Regression is used to estimate cause-and-effect types of experiments (dependence of one variable on another variable), such as the effect of increasing toxicant concentration on weight change in experimental units. Often the experimenter cannot control the independent variable as in an ANOVA. For example, a study may look at the growth of a biological unit in relation to tissue pesticide concentration in the same experimental unit. In this type of analysis, the experimenter does not control the independent variable of pesticide level; thus a regression analysis can be performed. If the experimenter controls the concentrations of a pesticide, an ANOVA would be performed first, followed by a regression analysis to determine trends or rates. This type of analysis is referred to as polynomial orthogonal contrast. This indicates the similarity between regression and ANOVA as subsets of general linear models. (See section on contrasts.) A regression analysis cannot be used if treatments are qualitative (toxicant versus toxicant), but it can be used if treatments in the ANOVA are quantitative (dose levels, time levels, etc.). The simplest regression analysis is to estimate the linear effect of X on Y. The regression coefficient in this case is simply the slope of a straight line.

The following is an example of linear regression calculations and a simple t-test for testing significant relationships between X (independent variable) and Y (dependent variable). The data (Table 14) come from the previous example used for the ANOVA calculations (Table 8).

1. The calculations in Table 14 use capital letters to indicate individual variates and sums of squares are represented by small letters.

   Sum of squares for X:

$$\Sigma x^2 = \Sigma X^2 - \frac{(\Sigma X)^2}{n} \tag{26}$$

   For the example data:

$$\Sigma x^2 = 21875 - \frac{(525)^2}{21} = 8750$$

   Sum of squares for Y:

$$\Sigma y^2 = \Sigma Y^2 - \frac{(\Sigma Y)^2}{n} \tag{27}$$

   For the example data:

$$\Sigma y^2 = 5942.94 - \frac{(332.6)^2}{21} = 675.19$$

Table 14. Example of fish lengths from Table 8

X (mg/L)	Y fish length (cm)	XY cross-products
0	14.8	0
0	18.1	0
0	27.6	0
0	19.0	0
0	18.2	0
0	22.9	0
0	30.7	0
25	12.2	305
25	11.4	285
25	15.6	390
25	18.0	450
25	14.2	355
25	18.0	450
25	17.0	425
50	9.5	475
50	10.8	540
50	11.9	595
50	10.6	530
50	8.6	430
50	13.1	655
50	10.4	520
$\Sigma X = 525$	$\Sigma Y = 332.6$	$\Sigma XY = 6405$
$\Sigma X^2 = 21875$	$\Sigma Y^2 = 5942.94$	

Cross product sum of squares:

$$\Sigma\, xy = \Sigma\, XY - \frac{(\Sigma\, X)\,(\Sigma\, Y)}{n} \tag{28}$$

For the example data:

$$\Sigma\, xy = 6405 - \frac{(525)(332.6)}{21} = -1910$$

2. Regression coefficient (slope of a straight line) is calculated according to Eq. (29).

Regression coefficient:

$$b = \frac{\Sigma\, xy}{\Sigma\, x^2} \tag{29}$$

For the example data:

$$b = \frac{-1910}{8750} = -0.218$$

Predicted values:

$$\hat{\mu}_{y.x} = \overline{Y} + b(X - \overline{X}) = a + bX \tag{30}$$

where $\hat{\mu}_{y.\bar{x}}$ = predicted value of Y at a specific value of X
$\overline{Y}$ = average value of Y
$\overline{X}$ = average value of X
b = regression coefficient
a = y intercept

For the example data calculate the predicted value of Y when $X = 18$. $\overline{X} = \frac{525}{21}$, $\overline{Y} = \frac{332.6}{21}$, $b = -0.218$.

$$17.37 = 15.84 + (-0.218)(18 - 25)$$

3. Partitioning $\Sigma\, y^2$ into sum of squares explained by regression and into residual, or not explained by regression is represented by Eqs. (31) and (32).

Sum of squares explained by regression:

$$\frac{(\Sigma\, xy)^2}{\Sigma\, x^2} = b\,\Sigma\, xy \tag{31}$$

For the example data:

$$b\,\Sigma\, xy = -0.218(-1910) = 416.38$$

Sum of squares not explained by regression:

$$\Sigma\, d^2_{y.x} = \Sigma(Y_i - \hat{\mu}_{y.x})^2 = \Sigma\, y^2 - b\,\Sigma xy \tag{32}$$

Notice that the total sum of squares in Y equals the sum of the two partitions:

$$\Sigma\, y^2 = b\,\Sigma\, xy + \Sigma\, dx.y^2$$

For the example data:

$$\Sigma\, d^2_{y.x} = 675.19 - 416.38 = 258.81$$

4. The coefficient of determination (R^2), represented by Eq. (33), describes the amount of variation in the dependent variable explained by variation in the independent variable. It is usually a percentage.

$$R^2 = \frac{b\,\Sigma\, x.y}{\Sigma\, y^2} * 100 \tag{33}$$

For the example data:

$$R^2 = \frac{416.38}{675.19} * 100 = 61.67$$

5. Standard error of regression coefficient is represented by Eq. (34).

Standard error of b:

$$\mathrm{SE}_b = \sqrt{\frac{(\Sigma\, d^2_{y.x})/(n-2)}{\Sigma\, x^2}} \tag{34}$$

For the example data:

$$\mathrm{SE}_b = \sqrt{\frac{258.81/19}{8750}} = .039$$

Test for a Significant Relationship Between X and Y

X may cause a significant amount of variation in Y. This can be tested in two different ways, the F-test or the t-test. The regression computations can be put into an ANOVA table:

Source	df	Sum of squares	Mean square
Explained by linear regression	1	$b\,\Sigma\, xy = 416.38$	$b\,\Sigma\, xy = 416.38$
Not explained by linear regression (error)	$n - 2 = 19$	$\Sigma\, d^2_{y.x} = 258.81$	$\frac{\Sigma\, d^2_{y.x}}{n-2} = 13.62$
Total	$n - 1 = 20$	$\Sigma\, y^2 = 675.19$	

The following two statistical methods (F-test and t-test) are used to determine whether a significant relationship exists between X and Y.

F-test statistic:

$$F = \frac{b\,\Sigma\, xy}{(\Sigma\, d^2_{y.x})/(n-2)} \tag{35}$$

A 1-df F-test is the same as a t-test.

t-test statistic:

$$t = \left|\frac{b}{\mathrm{SE}_b}\right| \tag{36}$$

The hypotheses for both the F-test and t-test are as follows:

H_0: true slope = 0

H_A: true slope ≠ 0

In order to reject H_0:

F-test statistic ≥ F-value at 1 and EMS df and α

or

t-test statistic ≥ t-value at EMS df and $\alpha/2$

For the example data:

Reject H_0 if

F-test $416.38/13.62 \geq 4.38$

or

t-test $|-.218/.039| \geq 2.093$

The data indicate that as the amount of pesticide increases 1 mg/L, the fish length decreases by a factor of 0.218. How much of the variation in fish length is affected by increased levels of pesticide? The answer is that 61.67% of the variation in fish length is explained by pesticide concentration. Is this statistically significant? Yes, both the F-test and the t-test indicate significance. Also, the linear sum of squares calculated in this regression example equals (within round-off error) the sum of squares calculated for the linear polynomial orthogonal contrast calculated in the ANOVA section [Eqs. (19)–(22)]. Note that this data set, used to calculate both the ANOVA and regression analyses, has slightly different error terms. This is because of a lack-of-fit term. The ANOVA had three treatments with 2 df. These 2 df can be orthogonally partitioned into a linear and a quadratic portion. The linear portion was estimated for both the ANOVA and regression analyses. The quadratic was estimated only for the ANOVA and not for the regression analyses. Normally, this quadratic is placed into a component of regression analyses called lack of fit rather than being placed in the error term.

Regression analysis can be used for many experiments—for example, to estimate the rate of uptake or elimination of a substance. If a toxicologist measures the amount of residue in an experimental unit at fixed dose levels and specific points in time, a regression analysis may examine the effects of dose at each time, or time and dose may be entered together in the linear statistical model.

If more than one independent variable is entered into a regression model, the analysis is called a multiple regression, because there is more than one regression coefficient. These coefficients are called partial regression coefficients. For example, the following equation represents the residue of a pesticide found in an organism given different doses and measured at different times.

$$\begin{aligned}\text{Pesticide residue} &\\ = 1.9666 &- 0.073 * \text{TM} + 0.0014 * \text{TM}^2\\ &+ 0.0625 * \text{TOX} + 0.0048 * \text{TOX}^2\\ &+ 0.0230 * \text{TM} * \text{TOX}\\ &- 0.001 * \text{TM} * \text{TOX}^2\\ &- 0.0007 * \text{TM}^2 * \text{TOX}\\ &+ 0.00002 * \text{TM}^2 * \text{TOX}^2\end{aligned}$$

where TM = exposure time
TOX = pesticide concentration

Often, multiple regression coefficients can be plotted by generating the X, Y, and Z coordinates. The equation above may be plotted to represent response in a figure, called a surface response (Figure 3). This surface response may give the experimenter insight into how treatments interact. An SAS program for calculating and plotting the surface response can be found in Appendices 3A and 3B.

Another type of regression analysis is referred to as nonlinear regression. In this type of regression, the individual partial regression coefficients are not additive as in the previous multiple regression model. In most cases, knowledge of calculus and use of computer programs are needed to compute these types of regression models.

THE ANALYSIS OF COVARIANCE

Analysis of covariance is a combination of ANOVA and regression analysis. The ANOVA includes treatments controlled by the experimenter. The regression analysis sometimes uses variables not controlled by the experimenter (i.e., weight, length, or age). When these two types of independent variables are combined for analysis, the approach is called analysis of covariance. The uncontrolled independent variable is called the covariate. The procedure performs two main functions. The first function determines whether regression coefficients among treatments are equal. For example, in an experiment in which fish weights (dependent variable) are measured, the treatments (controlled independent variable) represent different toxicants, and the uncontrolled independent variable (the covariate) represents age, the experimenter may want to know if the growth rate is similar among treatments. This is done by comparing regression coefficients.

The second function of the analysis of covariance is to adjust all data to a constant covariate. For example, if treatments represent different toxicants and the covariate represents age of the experimental units, analysis of covariance adjusts all treatments to a constant age. This is especially valuable if the average age in one treatment is different from that in another treatment. Treatment differences and age differences in this case are said to be confounded. Analysis of covariance can remove the effect of age to test the effect of toxicant without interference from age (Snedecor and Cochran, 1989; Neter et al., 1985).

MULTIVARIATE TECHNIQUES

All the preceding methods have dealt with univariate approaches. If a statistical analysis is performed to look at a single dependent variable, the analysis is univariate. If a statistical method is performed to look at more than one dependent variable, it is called multivariate. In many cases, a person describes toxicity by evaluating more than one variable. For example, if species numbers are sampled at different sites, toxicity differences among sites may be measured not by looking at a specific species but rather by looking at the pattern of species present at each

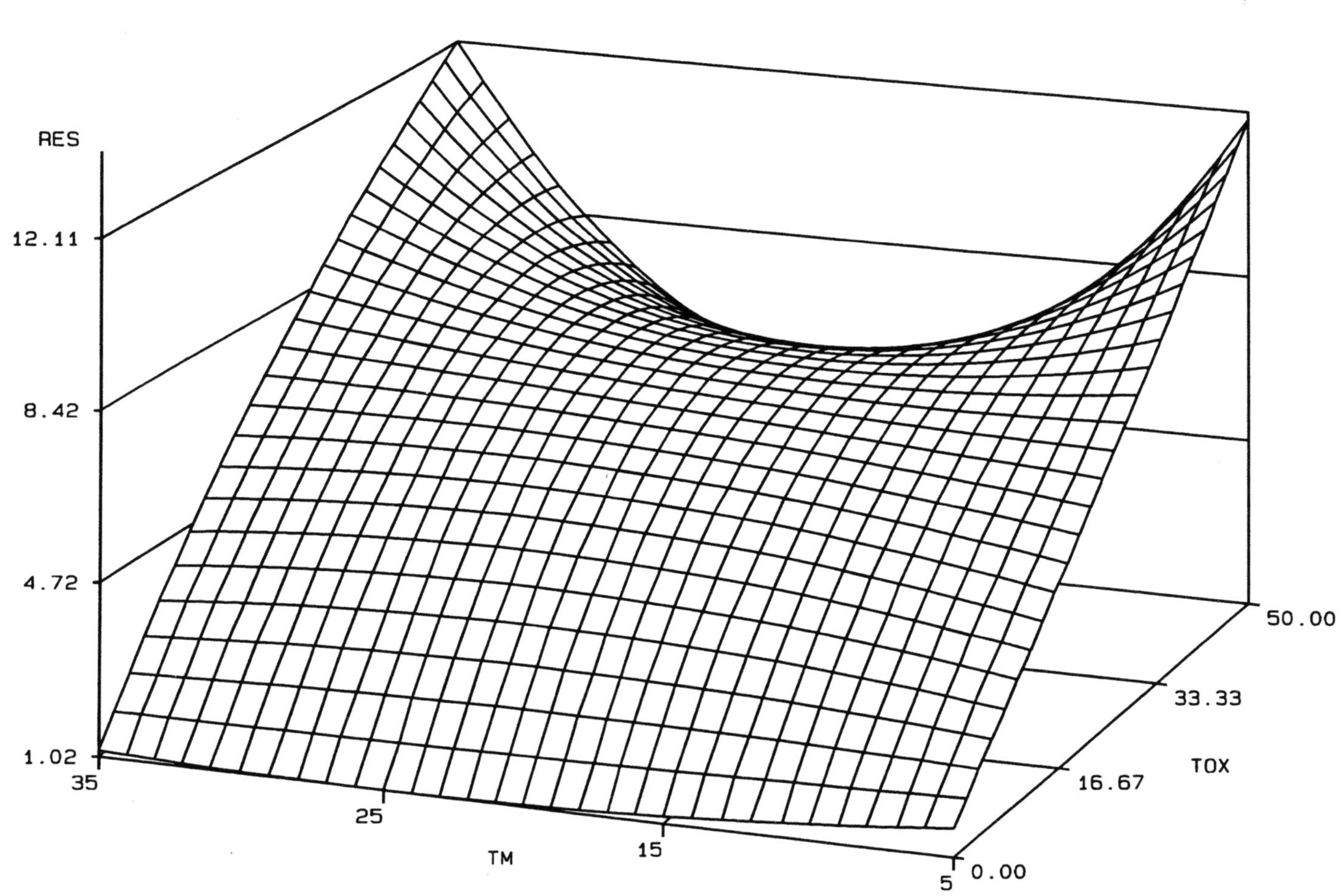

Figure 3. Example of surface response. RES, pesticide residue; TM, exposure time; TOX, pesticide concentration.

site. Looking at one specific species would be a univariate analysis; looking at the pattern of more than one species would be a multivariate analysis. In a univariate case, the analysis is called ANOVA; in a multivariate case, the analysis is called a multivariate analysis of variance (MANOVA).

Multivariate techniques require a good understanding of matrix algebra. Even with this knowledge, interpretation of data is difficult. If a multivariate technique demonstrates significance, univariate techniques can be used to describe the data further. In the example given, if a difference exists in the pattern of species between sites (multivariate), a researcher will ask, which is the most important species (univariate)? There are many techniques for analyzing multiple random variables. It is important to have a general knowledge of multivariate techniques before planning an experiment that requires multivariate analysis (Morrison, 1967; Anderson, 1958).

METHODS COMMONLY USED FOR THE DISCRETE RANDOM VARIABLE

Discrete random variable (sometimes called categorical data) methods are also important to the toxicologist. Often, data are put into categories, such as live versus dead. The discrete random variable is generated by counting the number of individuals that fall into each category. Categories are sometimes expressed in terms of proportions or percentages. Tests are presented in this section for data (variables) that cannot be measured but are expressed qualitatively as a frequency (i.e., categorical data).

Categorical Data

Methods involving discrete random variables, as with continuous random variables, test how far observed values deviate from what is expected. For example, in higher animals the sex ratio is 1:1, meaning 50% female and 50% male. This is what is expected. If the ratio is reported in terms of frequency based on 500 animals, $0.5 \times 500 = 250$ will be female and $0.5 \times 500 = 250$ will be male. This simple example is the expectation of the binomial probability function, which is

$$\text{Expectation} = nP$$

where n = sample size
P = probability of a specific category

In toxicological terms, 50% is an important number. A standard point estimate is LC50, which refers to the concentration of a toxicant that will kill 50% of a specific population (LC50 = lethal concentration for 50% of a population). Sometimes toxicologists are not interested in the lethality of a chemical but rather in the chemical's capability of affecting some biological function such as growth or reproduction. This point estimate is referred to as the EC50, or the effective concentration to produce a specific sublethal effect in population.

Goodness of Fit

The concentration associated with the LC50 is the expected concentration to produce mortality in 50% of a population of organisms. To see if a concentration is equal to an LC50, one would use a simple statistical method for testing "observed kill" versus "expected kill," called a goodness-of-fit χ^2 (chi-square) test. Consider the following example.

An experimenter exposes 500 cutthroat trout to a hypothesized LC50 value of 280 μg/L malathion for 96 h (Mayer and Ellersieck, 1986). As this is the hypothesized LC50, we may expect 50% of the trout to die (expected dead 500×0.50). But, in actuality, 200 die (observed value). Does the observed value deviate significantly from what is expected? To determine this, perform a goodness-of-fit χ^2 test. Goodness-of-fit χ^2 test statistic:

$$\sum_{i=1}^{2} = \frac{(O_i - E_i)^2}{E_i} \tag{37}$$

where O_i = observed value of the ith class
E_i = expected value of the ith class

In this experiment, there are two classes, alive and dead. This example produces a χ^2 value of 20. This calculated χ^2 value is greater than a tabulated χ^2 value ($k - 1$ degree of freedom or $2 - 1$); thus, the null hypothesis that this concentration of malathion kills 50% of the cutthroat trout (or that the observed frequency distribution is similar to the hypothesized distribution) is rejected.

Row by Column (R × C) χ^2 Test

The goodness-of-fit χ^2 determines whether an observed distribution of values conforms to a theoretical or expected distribution of values. This is used for discrete variables. The row by column (R × C) χ^2 (sometimes referred to as contingency table) tests for differences between or among observed values from different treatments. The R × C test is used to determine whether two or more sets of discrete data have similar distributions. For example, an experimenter wants to know the effectiveness of two insecticides (pyrethrin and Lethane) in killing fathead minnows. Five hundred minnows are in one tank and 500 in another tank. One tank is dosed with pyrethrin and the other with Lethane. The hypothetical results are as in Figure 4. The results indicate that Lethane killed 450 out of 500 and pyrethrin killed 430 out of 500 minnows.

If the two insecticides have equal killing power, the two populations' ratios should be the same. The expected ratios are determined by multiplying the row frequency by the column frequency and dividing the product by the total sample size. This is done for each cell. For example, the expected frequency for the cell Lethane, Dead = $(500 \times 880)/1000 = 440$. After computing the expected frequency for each cell represented in the upper right-hand portion of each cell in Figure 4, an inspection of the expected ratios for these two insecticides shows equal ratios (both

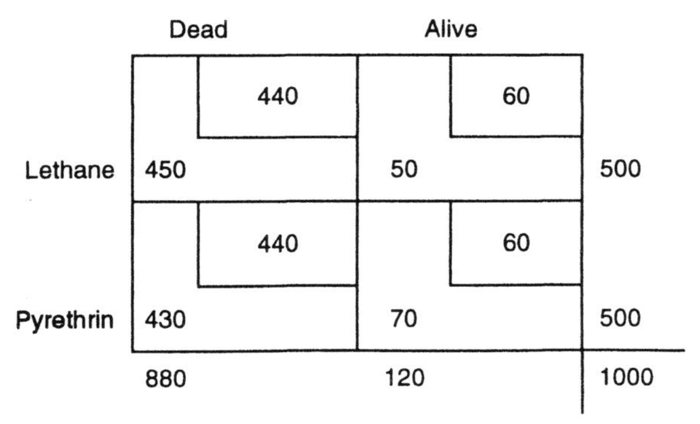

Figure 4. Row by column χ^2 example.

are 440:60). The object now is to determine whether the observed alive/dead ratio deviates from the expected alive/dead ratio. This is accomplished by using the following χ^2 formula.

Row-by-column χ^2 test statistic:

$$\sum_{i=1}^{r} \sum_{j=1}^{c} \frac{(O_{ij} - E_{ij})^2}{E_{ij}} \tag{38}$$

where O_{ij} = observed frequency of the ith row, jth column
E_{ij} = expected frequency of the ith row, jth column
r = number of rows
c = number of columns

This formula gives instructions to square the observed minus the expected frequency deviation and then divide by the expected frequency, for each cell. If this is computed for this data set, the answer is 3.788, which, when compared with a tabulated χ^2 value at $\alpha = 0.05$ and degrees of freedom = $(r - 1)(c - 1) = 1$ ($\chi^2 = 3.84$), indicates no difference between insecticides. This method works for simple experiments. Other more elaborate statistical methods have been developed for more complex statistical designs. These designs are sometimes called categorical models (see below). Many times some of the expected frequencies are small. In these cases, Fisher's exact test should be used.

Fisher's Exact Test

This test can be used to compare two sets of discrete, quantal data. These include frequencies of incidence of mortality, survival, hatchability, and so forth. The data involve numbers of responses denoted as ratios and set up in a contingency table. To conduct the analysis, a 2 × 2 contingency table is established (as in the R × C χ^2 test) to summarize the discrete (yes-or-no) response data. Fisher's exact test is performed below:

	Yes	No	Total
Test group 1	A	B	$A + B$
Test group 2	C	D	$C + D$
Totals	$A + C$	$B + D$	$A + B + C + D$ = total X

The test uses a probability (P) value to determine the relationship between the two variables (i.e., yes or no, such as hatched and not hatched) and therefore tests whether the two groups differ significantly at a certain probability level, such as 0.05. Tables are available for individual probabilities for small sample sizes (Zar, 1984). The formula for P follows:

$$P = \frac{(A + B)\,!\,(C + D)\,!\,(A + C)\,!\,(B + D)!}{X\,!\,A\,!\,B\,!\,C\,!\,D} \tag{39}$$

[A! is a factorial. For example, 3 ! means (3)(2)(1) = 6.] This is a powerful test with small sample sizes ($N < 50$) or when an expected value is less than 5.

If an aquatic laboratory toxicity test has been conducted with several types of controls (e.g., untreated or solvent) and several treatment concentrations of a chemical with a fish or invertebrate species, a Fisher's exact test can be conducted first on the control groups (if solvent and untreated controls are present) to determine whether the control groups should be pooled. Controls are pooled and used for comparison with treatment groups when there is no difference between the solvent controls and untreated controls. After this comparison test, a Fisher's exact test (two-tailed) can be conducted on each treatment concentration group (using pooled data in tanks) versus the control. If the calculated value of P is less than the tabulated book value at designated degrees of freedom (rows minus 1 × columns minus 1) and significance level = 0.05, then there is no association between control and treatment. However, if P is greater than this tabulated value, there is a statistically significant difference between the control and treatment.

Appendix 4 contains a SAS program that calculates row by column χ^2 and Fisher's exact test.

Categorical Models

A methodology similar to the analysis of variance for the continuous random variable has been developed for the discrete random variable (Grizzle et al., 1969; Forthofer et al., 1971). This analysis allows the researcher to analyze data from multidimensional contingency tables. This means that a general linear model can be analyzed when the data are discrete. For example, a researcher can analyze four levels of pH, each having four different levels of aluminum, when the response (dependent variable) is live or dead. Thus, the effect of the interaction of pH and aluminum can be investigated. This methodology is valuable when testing whether two toxicants have antagonistic or synergistic effects. A procedure in SAS (1989), CATMOD, can analyze this type of experiment. Not only ANOVA-type procedures but also multiple regressions can be performed using discrete data. The multiple regression type of analysis is sometimes called a logistic regression. The SAS procedure called LOGISTIC is used for logistic regression. This type of analysis can generate a prediction equation when the response is categorical. For example, a researcher collects fish at a specific location and determines if the experimental unit has a cancerous tumor. At the same time, a water sample is taken and analyzed for certain pesticides and heavy metals. A logistic multiple regression can be performed to determine whether a relationship exists between cancer or no cancer (dependent variable) and pesticide and heavy metal combinations (independent variable). Procedures such as CATMOD and LOGISTIC are being used more frequently in toxicology.

Testing for Acute and Chronic Quantal Responses

As discussed previously, the lethal concentration or lethal dose to kill 50% of a population, LC50 or LD50, is widely used by toxicologists as the standard point estimate for describing the potency of a chemical toward an organism. Many statistical methodologies have been proposed for obtaining the point estimate of LC50. Some of the methods used in the past were essentially "eyeball" methods using graph paper and a calculator to obtain an LC50 value. In modern times, computers and programs are readily available, so some of these methods are now ignored except for teaching purposes. Many procedures today use the iterative-maximum-likelihood method to calculate LC50 values.

The general use of statistics in acute toxicity testing involves estimating the concentration sufficient to kill a certain percentage of organisms. An implicit assumption is that the organisms represent a valid sample from the population at large and that their responses reflect population responses. A further assumption, typically, is that the toxicity responses follow a monotonic nondecreasing function, that is, that increasing concentrations of a chemical will elicit a toxic response in an increasing number of the test organisms. As stated above, several techniques developed to measure the median response were derived because the lack of computational power forced investigators to rely on visual examination of cumulative probability plots of regression analyses of transformed toxicity response variables. The techniques commonly used include the probit, logit, Spearman-Karber, and others that are less used, such as the binomial and moving average (see Chapter 2).

The probit (**prob**ability **unit**) was originally defined by Bliss (1934). The term derives from generating a parametric normalized distribution of the percentage of organisms responding to a chemical concentration (Bliss, 1967; see Chapter 1). The standard normal deviate (also referred to in the literature as a normal equivalent deviate, NED) is defined as 1 standard deviation from the normal distribution of quantal responses. Thus, $X = (Y - \mu)/\delta$, where X is the standard normal deviate, μ is the parametric mean of the normal probability distribution of tolerance to the concentration, and δ is the associated parametric standard deviation (Bliss, 1967).

The standard normal deviate for a 50% response (e.g., death) is zero; hence, Bliss proposed adding 5 to each term to eliminate dealing with negative numbers. The graphical technique for estimating the LC50 has been to plot the probit (probability)–transformed percentage response data (death or probit scale) against the logarithmic-transformed concentrations (see Chapter 1). This typically results in a straight line, from which the 50th percentile response is read on a horizontal line from the probit value of 5 to the concentration-response slope and from there down to the log concentration. The linear form of the slope is $Y = 5 + (\log \text{dose} - \mu)/\delta$; the probit is a linear function of the log concentration and has a slope of $1/\delta$. Current practice calls for using such programs as the SAS PROC PROBIT procedure (SAS Institute, 1989), in which a maximum-likelihood regression estimation procedure is used for probit and logit models.

The logit (**log**istic un**it**) model is similar to the probit model (Hamilton et al., 1977) in that the distribution of tolerances follows a parametric logistic model, where the logit is defined as $P(x) = \{1 + \exp[B(x - \mu]\}^{-1}$, where x is the log of concentration, $P(x)$ is the proportion of organisms responding, μ is the log(LC50), and β is a measure of the dose-response slope (Sanathanan et al., 1987). The differences in predicting an EC50 are slight using either the logit or probit models; for toxicity responses ranging from 1 to 99%, Cox (1970) found the two distributions varied by only 1.3%.

Difficulties in estimating specific response concentrations (e.g., EC70 or EC20) arise when the toxicity response data do not fit the assumptions underlying the parametric models (Hamilton, 1979; Debanne and Haller, 1985; Sanathanan et al., 1987). If the data do not conform to the parametric distributions, the maximum-likelihood estimates can be very unstable. Also, if the estimate of the median response is obtained, heterogeneity may result in such wide confidence limits as to negate the value of the estimate (Miller, 1973).

Generally, lack of fit to the predicted model stems from outliers in one or both of the tails of the response distribution (Sanathanan et al., 1987). One solution has been to "trim" the tails mathematically by a given proportion (Miller, 1973; Hamilton et al., 1977). This restricts the distribution to the more central, approximately linear, portion of concentration responses. The most common approach is to use the Spearman-Karber estimator, described in detail by Finney (1964) and modified (trimming the tails of the observed response distributions) by Hamilton et al. (1977, 1978).

The Spearman-Karber is a distribution-free (nonparametric) estimator, requiring only symmetry of the tolerance distribution, a concentration producing 100% mortality, and a concentration producing no mortality, hence a monotonic, nondecreasing function (Hamilton et al., 1977, 1978). For the trimmed method, α is the percentage of extreme values excluded from each tail of the response distribution prior to calculating the LC50. By trimming the extreme values, the procedure relies on the toxicity responses plotted along the central, linear portion of the concentration-response curve. The portion of the distribution in either tail to be trimmed, α, is $\leq 50\%$ of the outermost values. The α chosen is typically 5 or 10%. Hand calculations for the Spearman-Karber method are in Gelber et al. (1985) and are not repeated here.

As already stated, the Spearman-Karber technique is nonparametric and requires only a monotonic nondecreasing sequence of mortalities and symmetry about the LC50 estimate. If the data fit a parametric distribution, the tests for those have power to discern the difference among means. However, there is no computational technique considered best for all types of toxicity responses. It should be kept in mind that the objective in using these calculations is to estimate concentrations expected to produce a toxicological response in a defined proportion of the tested population. Small sample sizes, lack of partial kills (e.g., exposures resulting in only 0 and 100% mortalities), and incidents wherein the monotonic nondecreasing requirement is violated lead to estimates of LC50s that are less robust. For instance, the binomial method is appropriate when there are no partial kills (Bliss, 1967). This technique uses the arithmetic mean of the two concentrations in which none and 100% of the organisms were killed. However, do not rely too closely on such an estimate of the LC50; a third concentration providing a partial kill could substantially alter the estimate of the quantal response. See Appendix 5 for an example SAS program to calculate a LC50.

The methods outlined above can be used to calculate effect concentrations under either acute or chronic conditions. Acute tests are defined as tests that are short-lived. Typically, acute tests are performed for 24, 48, 72, or 96 h, depending on the organism (see Chapter 1). Chronic tests are performed over an extended period of time (see Chapter 1). Chronic toxicity refers to the long-term effect of a chemical. Chronic tests are important if chemicals persist or are frequently added to the environment. The levels may not be high enough to produce a short-lived effect (acute) but may be high enough to produce a long-term effect (chronic). Furthermore, in chronic testing it is often desirable to incorporate an estimate of a concentration below which no further toxicity is observed. These procedures and estimates are discussed in the following section.

New methods are being developed that can use data from more than one time period. Many acute studies measure the response of lethality at different times (see Chapter 1). Instead of calculating an LC percent at a single time period, models are being developed which incorporate all time periods and all dose levels into a more accurate measurement (Lee et al., 1991).

Chronic Test and No Observable Effect Concentration (NOEC)

Chronic tests are time consuming and expensive. Sometimes chronic tests with fish run for 9 or more months. Consequently, chronic tests are rarely done. Several attempts have been made to estimate chronic toxicity point estimates from acute data. If this could be accomplished, long-time exposure effects could be estimated without the time and expense of testing.

In the past, long-time exposure toxicity was determined by calculating an application factor or an acute-chronic ratio for a limited number of organisms and then applying these factors to other species (Mount and Stephan, 1967; Kenaga, 1979, 1982; Buikema et al., 1982; see Chapter 1). Another method, proposed by Mayer et al. (1992, 1994), estimates lethal concentration values by regressing the probit percent dead (dependent variable) with toxicant concentration (independent variable) at 24, 48, 72, and 96 acute time values. A second analysis regresses the concentration at an LC percent as the dependent variable and the reciprocal of time (1/time) as the independent variable. The value at the Y intercept is where time equals infinity, and the estimate at this intercept is the concentration of a toxicant that will kill a percentage of the population at time infinity. This method can be taken a step further. If the first regression analysis for each acute time estimate uses LC0.01 (lethal concentration at 0.01%) and these concentrations are used for the second regression, the resulting intercept indicates the concentration of a chemical that can exist in an environment with essentially no effects on aquatic life at extended periods of time. This estimate of a chemical is called the no observable effect concentration (NOEC; see Chapter 1).

Another technique has been developed called multifactor probit analysis (MPA) (Lee et al., 1991, 1995) that will predict long-time exposure effect toxicity based on acute data. This procedure uses the iterative reweighed least-squares procedure to estimate the parameters of the probit surface. The independent variables are time of exposure and concentration of the toxicant, which are used in the model simultaneously. The independent variable is the probit of the proportion responding to a concentration. MPA allows the user to predict the concentration of a toxicant at any time for a percent mortality. The MPA program calculates a point estimate and measure of dispersion (95% approximate confidence limits). Also, methods in life testing and accelerated life testing are being developed for toxicological purposes. These methods were originally used for mechanical devices placed under stress (e.g., a generator running constantly at full power and at high temperature). The measured variable was time to failure. These methods are now being applied to organisms under stress (i.e., exposed to toxic chemical) and the variable that is measured is time to failure or death (Sun et al., 1994).

SUMMARY

This chapter has described the underlying assumptions from which inferences about differences in biological variables are drawn. There is increasing sophistication among toxicologists in using general linear methods and transformations to analyze the effects of chemicals on aquatic populations. With the present ease of use of computer programs to analyze sets of quantal data, it is imperative that the investigator be aware of the parametric justification (e.g., type of variable, correct H_0 model, parametric distribution of the statistic) of the analyses. Verifying the experimental design before the experiment is conducted can help ensure that the data can address the question asked in the hypothesis.

ACKNOWLEDGMENTS

The authors are indebted to Pam Haverland for her critical review and suggestions pertaining to the content of this chapter and to Karen J. Ellersieck for her invaluable editorial review.

LITERATURE CITED

Anderson, T. W.: An Introduction to Multivariate Statistical Analysis. New York: Wiley, 1958.

Barnett, V., and Lewis, T.: Outliers in Statistical Design, 2nd ed. New York: Wiley, 1984.

Bartlett, M. S.: The use of transformation. *Biometrics*, 3: 39–52, 1947.

Beckman, R. J., and Cook, R. D.: Outliers. *Technometrics*, 25:119–163, 1983.

Berry, D. A., and Lindgren, B. W.: Statistics Theory and Methods. Belmont, CA: Brooks/Cole, 1990.

Beyer, W. H.: Handbook of Tables for Probability and Statistics. Boca Raton, FL: CRC, 1976.

Bliss, C. I.: The method of probits. *Science*, 79:38–39, 1934.

Bliss, C. I.: Statistics in Biology, pp. 92–124. New York: McGraw-Hill, 1967.

Box, G. E. P., and Cox, D. R.: An analysis of transformation. *J. R. Stat. Soc.*, B26:211–252, 1964.

Box, G. E. P.; Hunter, W. G.; and Hunter, I. S.: Statistics for Experimenters: An Introduction to Design, Data Analysis and Model Building. New York: Wiley, 1978.

Buikema, A. L. Jr.; Niederlehner, B. R.; and Cairns, J. Jr.: Biological monitoring. Part IV—toxicity testing. *Water Res.* 16:239–262, 1982.

Chamber, J. M.; Cleveland, W. S.; Kleiner, B.; and Tukey, P.A.: Graphical Methods for Data Analysis. Boston: Duxbury, 1983.

Chew, V.: Comparisons Among Treatment Means in and Analysis of Variance. Report No. O-280-931/SEA-5, Gainesville, FL: U.S. Department of Agriculture, 1977.

Cochran, W. G.: Sampling Techniques, 2nd ed. New York: Wiley, 1963.

Cochran, W. G., and Cox, G. M.: Experimental Designs, 2nd ed. New York: Wiley, 1957.

Cox, D. R.: Analysis of Binary Data. London: Methuen, 1970.

Day, R. W., and Quinn, G. P.: Comparisons of treatments after an analysis of variance in ecology. *Ecol. Monogr.* 59(4):433–463, 1989.

Debanne, S. M., and Haller, H. S.: Evaluation of statistical methodologies for estimation of median effective dose. *Toxicol. Appl. Pharmacol.*, 79:274–282, 1985.

Diamond, W. J.: Practical Experimental Designs. California: Lifetime Learning Publications, 1981.

Dixon, W. J.: BMDP Statistical Software. Berkeley: Univ. of California Press, 1985.

Duncan, D. B.: Multiple range and multiple *F* tests. *Biometrics*, 13:164–176, 1955.

Dunnett, C. W.: A multiple comparisons procedure for comparing several treatments with a control. *J. Am. Stat. Assoc.*, 50:1096–1121, 1955.

Elliott, J. M.: Some Methods for the Statistical Analysis of Samples of Benthic Invertebrates. *Sci. Publ. No. 25.* Ferry House, U.K.: Freshwater Biological Association, 1977.

Federer, W. T.: Experimental Design. New York: Macmillan, 1955.

Finney, D. J.: Statistical Method in Biological Assay. 2nd ed. London: Griffin, 1964.

Forthofer, R. N.; Starmer, C. F.; and Grizzle, J. F.: A program for the analysis of categorical data by linear models. *J. Biomed. Syst.*, 2:3–48, 1971.

Gad, S. C., and Weil, C. S.: Statistics and Experimental Design for Toxicologists. Caldwell, N.J.: Telford, 1988.

Gelber, R. D.; Lavin, P. T.; Mehta, C. R.; and Schoenfeld, D. A.: Statistical analysis. Fundamentals of Aquatic Toxicology, edited by G. M. Rand and S. R. Petrocelli, pp. 112–123. Washington, D.C.: Hemisphere, 1985.

Green, R. J.: Sampling Design and Statistical Methods for Environmental Biologists. New York: Wiley-Interscience, 1979.

Grizzle, J. E.; Starmer, C. F.; and Koch, G. G.: Analysis of categorical data by linear models. *Biometrics*, 25:489–504, 1969.

Gupta, S. S., and Sobel, M.: On selecting a subset which contains all populations better than a standard. *Ann. Math. Stat.*, 29:235–244, 1958.

Hamilton, M. A.: Robust estimates of the ED50. *J. Am. Stat. Assoc.*, 74:344–354, 1979.

Hamilton, M. A.; Russo, R. C.; and Thurston, R. V.: Trimmed Spearman-Karber method for estimating median lethal concentrations in toxicity bioassays. *Environ. Sci. Technol.*, 7:714–719, 1977.

Hamilton, M. A.; Russo, R. C.; and Thurston, R. V.: *Correction. Environ. Sci. Technol.*, 12:417, 1978.

Hicks, C. R.: Fundamental Concepts in the Design of Experiments. New York: Holt, Rinehart & Winston, 1982.

Hollander, M., and Wolfe, D. A.: Nonparametric Statistical Methods. New York: Wiley, 1973.

Hurlbert, S. H.: Pseudoreplication and the design of ecological field experiments. *Ecol. Monogr.*, 54(2):187–211, 1984.

Kenaga, E. E.: Aquatic test organisms and methods useful for assessment of chronic toxicity of chemicals. Analyzing the Hazard Evaluation Process, edited by K. L. Dickson, A. W. Maki, and J. Cairns Jr., pp. 101–111. Washington, D.C.: American Fisheries Society, 1979.

Kenaga, E. E.: Predictability of chronic toxicity from acute toxicity of chemicals in fish and aquatic invertebrates. *Environ. Toxicol. Chem.*, 1:347–358, 1982.

Lee, G.; Ellersieck, M.; and Krause, G.: Multifactor Probit Analysis. Report EPA/600/X-91/101. Gulf Breeze, FL: U.S. Environmental Protection Agency, 1991.

Lee, G.; Ellersieck, M. R.; Mayer, F. L.; and Krause, G. F.: Predicting chronic lethality of chemicals to fishes from acute toxicity data: Multifactor probit analysis. *Environ. Toxicol. Chem.*, 14:345–349, 1995.

Mayer, F. L., and Ellersieck, M. R.: Manual of Acute Toxicity: Interpretation and Data Base for 410 Chemicals and 66 Species of Freshwater Animals. Resource Publication 160. U.S. Department of the Interior, Fish and Wildlife Service, 1986.

Mayer, F. L.; Krause, G. F.; Buckler, D. R.; Ellersieck, M. R.; and Lee, G.: Predicting chronic lethality of chemicals to fishes from acute toxicity data: concepts and linear regression. *Environ. Toxicol. Chem.*, 13:671–678, 1994.

Mayer, F. L.; Krause, G. F.; Ellersieck, M. R.; and Lee, G.: Statistical Approach to Predicting Chronic Toxicity of Chemicals to Fishes from Acute Toxicity Test Data. PB92-503119/AS. Springfield, VA: NTIS, 1992.

Meyer, S. L.: Data Analysis for Scientists and Engineers, pp. 17–18. New York: Wiley, 1975.

Milliken, G. A., and Johnson, D. A.: Analysis of Messy Data, Vol. I, Designed Experiments. New York: Van Nostrand Reinhold, 1984.

Milliken, G. A., and Johnson, D. E.: Analysis of Messy Data, Vol. II, Nonreplicated Experiments. New York: Van Nostrand Reinhold, 1989.

Miller, R. G.: Nonparametric estimators of the mean tolerance in bioassay. *Biometrika*, 60:535–542, 1993.

Morrison, D. F.: Multivariate Statistical Methods. New York: McGraw-Hill, 1967.

Mount, D. I., and Stephan, C. E.: A method for establishing acceptable limits for fish—malathion and the butoxyethanol ester of 2,4-D. *Trans. Am. Fish. Soc.*, 96:185–193, 1967.

Neter, J.; Wasserman, W.; and Kutner, M. H.: Applied Linear Statistical Models, 2nd ed. Homewood, IL: Irwin, 1985.

Pollard, J. H.: Numerical and Statistical Techniques. New York: Cambridge Univ. Press, 1977.

Sanathanan, L. P.; Gade, E. T.; and Shipkowitz, N. L.: Trimmed logit method for estimating the ED50 in quantal bioassay. *Biometrics*, 43:825–832, 1987.

SAS Institute Inc.: SAS/Stat User's Guide. Version 6, 4th ed. Carey, N.C.: SAS Institute, 1989.

Scheffe, H.: The Analysis of Variance. New York: Wiley, 1959.

Shapiro, M.; Preisler, H. K.; and Robertson, J. L.: Enhancement of baculovirus activity on gypsy moth (Lepidoptera: Lymantriidae) by chitinase. *J. Econ. Entomol.*, 806: 1113–1116, 1987.

Skalski, J. R.: Statistical inconsistencies in the use of no-observed effect levels in toxicity testing. Aquatic Toxicology and Hazard Assessment, Proceedings of the 4th Annual Symposium on Aquatic Toxicology, edited by P. R. Branson and K. L. Dickson, pp. 277–287. ASTM STP 237. Philadelphia: American Society for Testing and Materials, 1981.

Siegel, S.: Nonparametric Statistics for the Behavioral Sciences. New York: McGraw-Hill, 1956.

Snedecor, G. W., and Cochran, W. G.: Statistical Methods, 8th ed. Ames: Iowa State Univ. Press, 1989.

Sokal, R. R., and Rohlf, F. J.: Biometry. San Francisco: Freeman, 1981.

SPSS: SPSS-X User's Guide, 3rd ed. Chicago: SPSS, 1988.

Steel, R. G. D.: A multiple comparison rank sum test: treatments versus control. *Biometrics*, 15:560–572, 1959.

Steel, R. G. D., and Torrie, J. H.: Principles and Procedures of Statistics, 2nd ed. New York: McGraw-Hill, 1980.

Sun, K.; Krause, G. F.; Mayer, F. L.; Ellersieck, M. R.; and Basu, A. P.: Predicting chronic toxicity based on the theory of accelerated life testing. EPA/600/R94-058. Gulf Breeze, FL: U.S. Environmental Protection Agency, 1994.

Taylor, L. R.: Aggregation, variance and the mean. *Nature*, 189:732–735, 1961.

William, D. A.: A test for differences between treatment means when several dose levels are compared with a zero dose control. *Biometrics*, 27:103–117, 222, 1971.

Zar, J. H.: Biostatistical Analysis, 2nd ed. New Jersey: Prentice Hall, 1984.

SUPPLEMENTAL READING

Newan, M. C.: Quantitative Methods in Aquatic Ecotoxicology. Boca Raton, FL.: Lewis, 1995.

```
***************APPENDIX 1*********************************;

*************ANOVA AND REGRESSION*************************;

********************PC SAS********************************;

OPTIONS LS=132 PS=60;
DATA ONE;
INPUT TRT LENGTH @@;
CARDS;
0 14.8 0 18.1 0 27.6 0 19.0 0 18.2 0 22.9 0 30.7
25 12.2 25 11.4 25 15.6 25 18.0 25 14.2 25 18.0 25 17.0
50  9.5 50 10.8 50 11.9 50 10.6 50  8.6 50 13.1 50 10.4
PROC PRINT;
*************ANALYSIS OF VARIANCE***************************;
PROC GLM; CLASSES TRT;
MODEL LENGTH=TRT;
CONTRAST '0 VS 25+50' TRT 2 -1 -1;
CONTRAST 'LINEAR' TRT -1 0 1;
CONTRAST 'QUADRATIC' TRT -1 2 -1;
MEANS TRT/LSD;
*************REGRESSION ANALYSIS***************************;
PROC GLM;
MODEL LENGTH=TRT;
RUN;
```

```
****************APPENDIX 2************************************/

/* EXAMPLE OF PCSAS MEAN SEPARATION TECHNIQUES INCLUDING DUNCAN'S
MULTIPLE RANGE TEST*/
/****************************;
DATA ONE;
 INPUT TRT WEIGHT @@;
CARDS;
11  111.4  11  133.0  11  114.7  12  117.1  12  150.4  12  136.4
13  121.5  13  134.2  13  127.0  21  136.6  21  136.8  12  124.4
22  141.0  22  151.0  22  147.0  23  153.9  23  152.2  23  143.3
31  123.4  31  142.2  31  116.8  32  135.8  32  149.8  32  126.1
33  125.6  33  146.0  33  120.9
 PROC PRINT;
PROC SORT; BY TRT;
PROC MEANS; BY TRT;
PROC GML;CLASSES TRT;
MODEL WEIGHT=TRT;
MEANS TRT/LSD DUNCAN SCHEFFE SNK TUKEY WALLER;
MEANS TRT/DUNNETT ('CONTROL');
**** THE FIRST ALPHA NUMERIC TREATMENT MUST BE THE CONTROL;
RUN;
```

```
***************APPENDIX 3A**********************************;

***********ANOVA FOR SURFACE RESPONSE*********************;
**********************PCSAS********************************;
OPTIONS LS=132 PS=60;

DATA ONE;
INPUT TOX TM RES1 RES2;
RES=RES1; OUTPUT;
RES=RES2; OUTPUT;
DROP RES1 RES2;
CARDS;
0 5 2 1
0 15 1 1.5
0 25 0 2
0 35 1 1
20 5 5 7
20 15 5 8
20 25 6 7
20 35 7 8
30 5 8 7
30 15 7 7
30 25 7 8
30 35 7 9
40 5 10 9
40 15 7 8
40 25 9 7
40 35 10 11
50 5 12 12
50 15 6 7
50 25 6 5
50 35 12 13
PROC PRINT;
*****************ANOVA***********;
PROC GLM; CLASSES TM TOX;
MODEL RES=TM|TOX;
MEANS TM|TOX;
*************************MULTIPLR REGRESSION************;
PROC GLM;
MODEL RES=TM TM*TM TOX TOX*TOX TM*TOX TM*TOX*TOX TM*TM*TOX
TM*TM*TOX*TOX; RUN;
```

```
*************APPENDIX 3B***************************************;

********PCSAS PLOT OF SURFACE RESPONSE*************************;

GOPTIONS RESET=GLOBAL GUNIT=PCT FTEXT=SWISSB HTITLE=6 HTEXT=3

DEVICE=HPLJS2;
TITLE1 'SURFACE RESPONSE';
DATA ONE;
DO TM=5 TO 35 BY 1;
DO TOX=0 TO 50 BY 2;
RES=1.95581-.0734499*TM+.0014479*TM*TM+.062487*TOX+.0048*TOX*TOX
+.0298299*TM*TOX-.001036377*TM*TOX*TOX-.0006521*TM*TM*TOX
+.000024245*TM*TM*TOX*TOX;
OUTPUT; END;END;
PROC G3D;
PLOT TM*TOX=RES/GRID;
RUN;


*******************APPENDIX 4 **********************************;

****EXAMPLE OF ROW BY COLUMN CHI-SQUARE ******************;
*************AND FISHER'S EXACT TEST*****************;
OPTIONS LS=132 PS=60;
DATA ONE;
INPUT ROW COL X;
CARDS;
1 1 450
1 2 50
2 1 470
2 2 70
PROC PRINT;
PROC FREQ;
WEIGHT X;
TABLES ROW*COL/EXPECTED CHISQ;
RUN;


*****************APPENDIX 5***************************;

****************PROBIT ANALYSIS USING PCSAS*********;
DATA ONE;
INPUT DOSE N RESPONSE;
CARDS;
2 20 0
3.6 20 4
7.1 20 9
14 20 15
27 20 20
PROC PROBIT LOG10;
VAR DOSE N RESPONSE;
RUN;
```

Chapter 11

GOOD LABORATORY PRACTICES

N. DiGiulio and *A. V. Malloy*

INTRODUCTION

Scientific measurements in tests (i.e., experimental studies) are often the basis for critical decisions regarding human health and protection of the environment. Without proper control of these measurements, test data may not be reliable and any findings based on these data may be indefensible. Because of this, the Food and Drug Administration (FDA), the United States Environmental Protection Agency (EPA), and international organizations such as the Organization for Economic Cooperation and Development (OECD) have finalized guidelines for tests to ensure the integrity of data submitted to support product registrations and labeling requirements.

These guidelines are referred to as good laboratory practice regulations (GLPs) and have been in use since the late 1970s. In general, the GLP regulations can be described as a series of guidelines or conformance specifications that are utilized during the process of conducting a study. These regulations not only assist in maintaining data quality but also serve as the foundation for the advancement toward a multilateral standard for the conduct of any scientific study.

The GLP regulations define the organization, personnel, facilities, equipment, procedures, controls, and records that are required for regulated studies conducted in the laboratory (FDA, 1992). More specifically, they provide the framework upon which the individual scientist or organization can build to enhance the generation of precise, accurate, and reproducible data.

With technological advances that have enabled analytical detection limits of chemical substances down to parts per trillion (ppt) levels, quality assurance (qualitative) and quality control (quantitative) assessments of data have been forced to keep pace in controlling and verifying these measurements. These terms are further defined in the terminology section. GLP regulations have also moved outside the laboratory environment to encompass both aquatic and terrestrial field studies. These studies generate large volumes of data with multiple end points, often in environmental conditions that are less than optimum. Studies of this type further increase the complexity of ensuring data quality.

In this chapter, quality assurance and quality control applications are examined that govern regulated studies related to health and environmental effects of chemicals and pesticides. GLP guidelines used by the EPA to assess product registration (FIFRA) are reviewed in detail, including further discussion of the trends for future organization and control of scientific data.

BACKGROUND

Good laboratory practice regulations have been evolving since the original mandate of the Federal Food, Drug, and Cosmetic Act of 1938 (Lucas, 1978). Requirements under this law obligated companies to ensure proof of safety for new products, more specifically related to drugs. This Act authorized the FDA to promulgate regulations for the efficient enforcement of the Act, which included expanded labeling requirements, unannounced inspections by the FDA, and injunctions by federal courts. The Act was designed to provide reliable data for investigational new drugs (INDs) and new drug applications (NDAs).

Later, in 1962, the Defauver-Harris Drug Amendments were passed. These laws called for an increase in regulatory action by further requiring proof of effectiveness of drugs before marketing, compliance with good manufacturing practices (GMPs), specific approval requirements from the FDA for NDAs, informed consent of clinical research subjects, and reporting of adverse reactions. These amendments also required the FDA to inspect each pharmaceutical establishment at least once every 2 years.

In an effort to comply with ever-increasing regulatory demands and to safeguard against inconsistent, unreliable, and fraudulent data submissions, the first

GLP regulations were promulgated by the FDA in 1976 (U.S. FDA, 1976). The final rules were published in the Code of Federal Regulations (CFR) Section 21, Part 58 in December 1978 (U.S. FDA, 1978). They became effective on June 20, 1979. The most recent amendments to the FDA GLP regulations were finalized on September 4, 1987 (U.S. FDA, 1987b) and July, 1991 (U.S. FDA, 1992) (see Chapter 25 for discussion of FDA).

The FDA GLP regulations are the basis on which all subsequent GLP regulations have been written. These regulations describe good laboratory practices for conducting nonclinical laboratory studies that support or are intended to support applications for research or marketing permits for products regulated by the Food and Drug Administration. This encompasses food and color additives, animal food additives, human and animal drugs, medical devices for human use, biological products, and electronic products.

In Europe, chemical control laws were passed in Organization for Economic Cooperation and Development (OECD) member countries in the 1970s and 1980s which called for the testing of chemicals with emphasis on GLP standards (see Chapters 26 and 27 for discussion of OECD and EEC). The overall goal was to ensure that the assessment of chemicals and any associated hazards would be based on reliable test data of assured quality. OECD member countries include Australia, Austria, Belgium, Canada, Denmark, Finland, France, Germany, Greece, Iceland, Ireland, Italy, Japan, Luxembourg, Mexico, the Netherlands, New Zealand, Norway, Portugal, Spain, Sweden, Switzerland, Turkey, the United Kingdom, and the United States.

Obviously, the OECD GLP regulations have broad international scope. If data are produced by one member country under the same set of technical and quality standards utilized in another member country, countries may rely on each other's studies. Moreover, common principles and procedures for GLPs would facilitate the exchange of scientific information among member countries while enhancing environmental and health protection.

On May 12, 1981, the OECD Council formally adopted a decision that data generated in the testing of chemicals in an OECD member country in accordance with OECD technical guidelines and OECD Principles of Good Laboratory Practice should be accepted by another member country (OECD, 1982). This was a major breakthrough for the acceptability of data that could be considered reliable and utilized to assess the protection of human health and the environment.

At the same time, the EPA was working on its own GLP standards. In 1979, under the Toxic Substances Control Act (TSCA), the EPA proposed GLP regulations for studies that generated health effects data (see Chapters 22 and 23 for discussion of TSCA). In 1980, guidelines were proposed to expand the scope of the GLP regulations to include environmental (i.e., ecological) effects and chemical fate data. The TSCA GLP regulations were finalized November 29, 1983 and became effective December 29, 1983 (U.S. EPA, 1983a).

A second set of EPA GLP regulations was drafted concurrently with the TSCA GLP regulations (U.S. EPA, 1983b). In 1980, under the Federal Insecticide, Fungicide and Rodenticide Act (FIFRA), the EPA proposed guidelines for the conduct of studies in support of pesticide registration (see Chapter 21 for discussion of FIFRA). The TSCA and FIFRA GLP regulations are similar in scope, and they were both amended in 1989 to include field testing; tests for ecological effects, chemical fate, and residue chemistry; and in certain cases efficacy testing.

The GLP regulations are strongly enforced today, by both the EPA and the FDA. They have interagency agreements that allow them to conduct study audits and facility audits for one another and share common information as required. The Food and Drug Administration has policies and procedures to enforce the National Environmental Policy Act (NEPA) under 21 CFR, Part 25. This now allows the FDA to request information related to the environmental (i.e., ecological) impact of any product regulated by the FDA, a regulation no longer exclusive to EPA registrations.

Today, GLP regulations also govern clinical laboratory studies under standards referred to as good clinical practices (GCPs), demonstrating their far-reaching applications. However, because this book specifically concerns aquatic toxicology, which falls mainly under the purview of FIFRA and TSCA, the major focus includes the details and interpretation of the GLP regulations by the EPA.

TERMINOLOGY

Many terms that are used in this chapter and may be mentioned in other chapters are defined below.

Code of Federal Regulations (CFR). A series of books published by the U.S. Government Printing Office on an annual basis which summarize all of the general and permanent rules of the executive departments and agencies of the U.S. federal government.

Data gaps (DG). Omissions in the information supplied by a registrant (i.e., chemical or drug company) to a reviewing regulatory agency.

Enforcement response policy (ERP). Document describing the procedures and criteria used to determine appropriate enforcement response for violations of EPA GLP regulations (FIFRA).

European Community (EC). A unique political entity/supranational organization composed of sovereign member states whose goal is to establish a political, economical, and monetary union. See Chapter 27.

Good laboratory practice (GLP) Regulations. A set of standards set forth in the Code of Federal Regulations (CFR) governing the experimental design and conduct of toxicological and analytical studies.

Investigational new drug (IND). The term FDA assigns to unregistered drugs that have undergone complete safety and efficacy testing (preclinical testing) and are approved for experimental use in humans (clinical testing).

New drug application (NDA). The term FDA assigns to a submission by a drug/pharmaceutical company to request marketing approval for a drug that has undergone complete preclinical and clinical testing requirements.

Organization for Economic Cooperation and Development (OECD). A group of member countries promoting policies designed to achieve sustainable economic growth and stability, and to contribute to the development of the world economy.

Protocol. A written document detailing the objectives and methods that are designed to govern the conduct of a GLP study.

Quality assurance (QA). Standardized processes or qualitative measures designed to ensure consistent and reliable output from laboratory or field test systems.

Quality assurance unit (QAU). Any person or organized group designated by either testing facility management or the study sponsor to perform quality-related activities for a study as defined in the GLP regulations.

Quality control (QC). Standardized specifications or quantitative measures (e.g., control limits, tolerances) designed to interface with QA processes to ensure conformance with stated output objectives.

Sponsor/registrant. A person or organization that initiates the conduct of a GLP study through financial provision or submits a study as part of an application to the EPA in support of a product registration, amended registration, reregistration, experimental use permit, exemption, petition for modification, or any other application, petition, or submission to the EPA.

Standard evaluation procedure (SEP). Technically specific documents written and utilized by the EPA as a guideline to evaluate laboratory toxicity and field study data.

Standard operating procedures (SOPs). Written documents that define the elements of laboratory or field operations as well as the general and discrete specifications for the conduct of a GLP study.

Study director. Scientist or technically qualified and trained professional who is designated by testing facility management to oversee and manage the conduct of a GLP study.

Testing facility. A person(s) or company that actually conducts a study, i.e., uses the test substance in a test system. "Testing facility" encompasses only operational units that are being or have been used to conduct studies (U.S. EPA, 1992b).

Test system. Any animal, plant, microorganism, chemical, or physical matrix including but not limited to soil and water, or subparts thereof, to which the test, control, or reference substance is administered or added for study (U.S. EPA, 1992b).

Total quality management (TQM). Program that encompasses all the concepts of quality assurance and quality control and promotes continuous improvement (CI) in systems, services, and products.

SCOPE AND GENERAL DESCRIPTION OF THE GLP REGULATIONS

Standard GLP regulations are published by country or organization as itemized in Table 1 (Broad and Dent, 1990). However, quality assurance regulations and activities may vary significantly among these regulating bodies.

For this reason, the current emphasis has been on "harmonization" of quality assurance and GLP principles and practices, especially in the European Community (EC) (see Chapter 27) and the United States. GLP principles in Europe are those accepted by the OECD and translated into EC Directives. This will allow greater "transportability," or acceptability of test data from one agency or group to another. The acceptability or transportability of data can be greatly enhanced if the data quality objectives, quality assurance approaches, and data reporting requirements are harmonized among the producing organizations (OECD, 1981).

Currently, there are extensive efforts to work together on global standards for data acceptability. The formation of multinational societies and organizations has assisted in creating an atmosphere for more "generic" or globally recognized quality standards to ease the regulatory burden. In the meantime, it is still necessary to be cognizant of the similarities and differences in good laboratory principles and practices and their adaptability for future changes. Copies of GLP regulations can be obtained directly through these regulating agencies and are discussed at length in many books and periodicals.

Although some GLP regulations are nearly identical to others, subtle differences in and noncompliance with even minor details of the GLPs can result in study rejection by a reviewing agency. It is important to remember that studies conducted according to high scientific standards that are not in compliance with GLPs may be readily rejected by a regulatory agency. Therefore, understanding the regulations and the degree of flexibility or interpretation allowable within the GLPs is critical in experimental design, data generation, and finally study acceptability.

The original GLPs issued by the EPA required only compliance for health effects testing for data review purposes. Today, the EPA requires compliance with the FIFRA GLPs (40 CFR, Part 160) for most data submitted to them that are intended to support pesticide product research or marketing permits. Studies requiring compliance with FIFRA GLP regulations are described in product performance data requirements (40 CFR, Part 158; see Chapter 21), including selected chemical analyses.

Even though some studies may be exempt from the GLP regulations (such as basic chemical characterization, e.g., color, odor, density), it is wise to utilize GLPs whenever data collection occurs. This is especially true because the GLPs continue to ex-

Table 1. Evolution of GLPs

Year	Country	Proposal
1976	USA	FDA Proposals 19 November 1976 Federal Register 41-51206.
1978	USA	FDA Final rule 22 December 1978 Federal Register 43-59986.
1979	USA	EPA Proposals (TOSCA) 9 May 1979 (Health effects). Federal Register 44-27362.
1980	USA	EPA Proposals (FIFRA) 18 April 1980. Federal Register 45-26373.
	USA	EPA Proposals (TOSCA) 21 November 1980. (Environmental effects and chemical fate) Federal Register 44-77357.
1980	USA	FDA Amendment to GLP. Federal Register 11 April 1980.
1981	OECD	Guidelines for testing of chemicals. (1981 and continuing series) OECD Paris. ISBN 92-64-12221-4
	OECD	Guidelines for national GLP Inspections and Study Audits, Paris 1981.
1982	OECD	Good Laboratory Practice in the testing of Chemicals, Paris 1982, ISBN 92-64-12367-9.
1982	UK	Health and Safety Commission, Approved Code of Practice, Principles of Good Laboratory Practice, ISBN 0 11 883658 7.
1983	USA	EPA Final rule. TOSCA, Health and Environmental Effects and Fate of Chemical Substances and Mixtures. 29 November 1983. Federal Register 48-53922. 40 CFR, Part 792.
	USA	EPA Final rule. FIFRA. Health safety of regulated products and pesticides. 29 November 1983. Federal Register 48-53946 40 CFR, Part 160.
1984	USA	FDA Amendments to GLP (Proposals) 19 October 1984. Federal Register 49-43530.
1984	Japan	MOHW and MITI, Japanese Directive of 31 March 1984 (Kanpogyo No. 39. Environmental Agency Yakuhatsu No. 229, MOHW, Kikyoku No. 85 MITI).
1984	Japan	J.MAFF 59 NohSan, Notification No. 3850, Agricultural Production Bureau. 10 August 1984.
1986	UK	DHSS GLP Compliance Program, December 1986
1986	EC	Official Journal of the European Communities No. C 13/5. Proposal for a Council Directive on the Inspection and verification of the organizational processes and conditions under which laboratory studies are planned, performed, recorded and reported for the non-clinical testing of chemicals (Good Laboratory Practice), 17 January 1987.
1986	EC	Council Directive of 18 December 1986 (87/18/EEC). No. L15/29. The harmonization of laws, regulations and administrative provisions relating to the applications of principles of Good Laboratory Practice and the verification of their applications for tests on chemical substances.
1987	USA	FDA Amendment Final Rule. 4 September 1987 Federal Register 52-33768.
	USA	EPA (Proposals) FIFRA. Ecological effects, chemical fate, residue chemistry and product performance in efficacy testing 28 December 1987. Federal Register 52-48920.
	USA	EPA Amendment (Proposals) FIFRA. Health effects. 28 December 1987. Federal Register 52-48920.
	USA	EPA Amendment (Proposals) TOSCA. Health and environment effects, chemical fate and field studies. 28 December 1987. Federal Register 52-48933.
1988	OECD	Environment Monographs No. 15. Final report of the working group on mutual recognition of Compliance with Good Laboratory Practice March 1988. W.5902K 10650.
1988	EC	Council Directive of 9 June 1988 on the inspection and verification of Good Laboratory Practice (GLP) No. L 145/35 (88/320/EEC)
1989	USA	EPA Final Rule. TOSCA. Health and Environmental Effects and Fate of Chemical Substances and Mixtures, 17 August 1989. Ref 40 CFR Part 792.
1989	USA	EPA Final Rule FIFRA, 17 August 1989. Ref 40 CFR Part 160.
1989	UK	DoH GLP Compliance Program, December 1989.
1990	EC	EEC Directive 90/18/EEC.

Notes: DHSS/DoH, Department of Health and Social Security; EC, European Community; EPA, Environmental Protection Agency; FDA, Food and Drug Administration; FIFRA, Federal Insecticide, Fungicide and Rodenticide Act, J-MAFF, Japanese Ministry of Agriculture, Forestry and Fisheries; MITI, Ministry of International Trade and Industry; MOHW, Ministry of Health and Welfare; OECD, Organization for Economic Co-operation and Development; TOSCA, Toxic Substance Control Act.

From Broad RD, Dent NJ: An introduction to good laboratory practice (GLP). In: Good Laboratory and Clinical Practices, edited by PA Carson, NJ Dent. Oxford, UK: Heinemann Newnes (Butterworth Heinemann Ltd.), 1990.

pand in scope and data collection that does not presently require GLP compliance may become regulated some time in the future.

As an example, since 1989 the GLP regulations have moved outside the laboratory and now involve all aspects of field testing, including ground water and soil studies, air monitoring studies, terrestrial and aquatic field studies, plant metabolism studies, and other types of biologically or chemically mediated environmental studies. From a regulatory standpoint, it may be strategic to anticipate GLP requirements for long-term data acceptability.

The basic outline of the GLP regulations under FIFRA, 40 CFR, Part 160, Good Laboratory Practice Standards is summarized below. Highlights of these regulations are discussed further to provide a better understanding of GLP applications and interpretation.

Subpart A—General Provisions

Section
160.1 Scope
160.3 Definitions
160.10 Applicability to studies performed under grants and contracts
160.12 Statement of compliance or non-compliance
160.15 Inspection of a testing facility
160.17 Effects of non-compliance

Subpart B—Organization and Personnel

Section
160.29 Personnel
160.31 Testing facility management
160.33 Study director
160.35 Quality assurance unit (QAU)

Subpart C—Facilities

Section
160.41 General
160.43 Test system care facilities
160.45 Test system supply facilities
160.47 Facilities for handling test, control, and reference substances
160.49 Laboratory operation areas
160.51 Specimen and data storage facilities

Subpart D—Equipment

Section
160.61 Equipment design
160.63 Maintenance and calibration of equipment

Subpart E—Testing Facilities Operation

Section
160.81 Standard operating procedures
160.83 Reagents and solutions
160.90 Animal and other test system care

Subpart F—Test, Control, and Reference Substances

Section
160.105 Test, control, and reference substance characterization
160.107 Test, control, and reference substance handling
160.113 Mixtures of substances with carriers

Subpart G—Protocol for and Conduct of a Study

Section
160.120 Protocol
160.130 Conduct of a study
160.135 Physical and chemical characterization studies

Subpart J—Records and Reports

Section
160.185 Reporting of study results
160.190 Storage and retrieval of records and data
160.195 Retention of records

Before reviewing these GLP subparts in greater detail, it is important to understand how a GLP study is generally organized and executed.

GLP Project Management

The basic outline of GLP study planning before study implementation is illustrated in the GLP project management flowchart (Figure 1). There is a data requirement that a sponsor/registrant produce information relative to their product (see terminology section for definitions).

Either the sponsor or the testing facility may develop the study protocol, which is the document outlining the scope of the testing procedure (see Figure 4 for required elements of a study protocol). Often, the sponsor collaborates with the testing facility in developing the study protocol or may provide the testing facility with a protocol to use. Either way, the protocol must be approved by both the sponsor and the testing facility's study director.

After protocol approval, the study is conducted once the test, control, and reference substances are made available to the testing facility and the testing facility has prepared the required test system. A sponsor is usually not directly involved in the actual study conduct if the study has been subcontracted to an independent laboratory. However, many sponsors/registrants have their own testing facilities and will develop the study protocol and conduct the study within their own organization.

Once the study is completed, a final report is generated and the data are organized into an appropriate format for submission to a reviewing agency. The agency may request more information from the sponsor or clarification of data provided before granting any product approvals, permits, etc.

GLP Regulations/General Provisions

An important part of general provisions (Subpart A), beyond the description of studies that require the application of GLP standards, is a list of definitions and terminology (Section 160.3). If a question arises concerning the situations under which the GLPs should

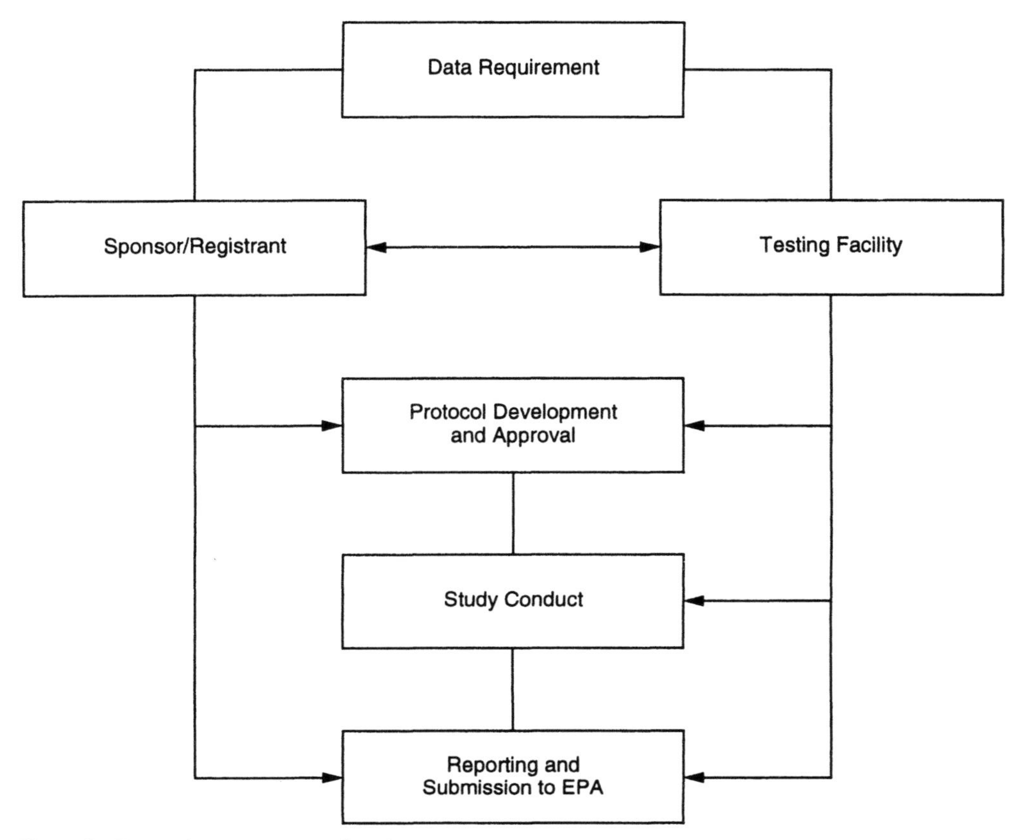

Figure 1. GLP project management flowchart.

be followed, one should refer to the definition of an application for research or marketing permit, which describes various scenarios and applicability related to the pesticide registration process (Section 160.10). In general, GLPs should be applied to any studies that would be included in an application, petition, or submission to the EPA.

Section 160.3 also describes the definition of dates that are required to be included in the study protocol. These are the experimental start and termination dates, which refer to application (i.e., treatment) of the test chemical and data collection. This section also defines the study start and completion dates, which are the dates on which the study director signs the protocol and the final report, respectively.

It is important to understand what constitutes raw data. This was easier to discern when GLPs applied only to health effects testing that generally occurred in laboratories. Now that the GLPs apply to both laboratory and environmental field testing, it is important to understand that raw data means any worksheets, records, memoranda, or notes that are the result of original observations and activity of a study. These records are required for the reconstruction and evaluation of a study. For example, data written on a "paper towel" in the field to record an observation when the technician did not have his or her notebook would be considered raw data. Nevertheless, this type of practice is not recommended and proper data recording forms should always be available before the conduct of a GLP study.

Also of interest are the definitions of control substance and test substance. A *control substance* means any chemical substance or mixture that is administered to the test system in the course of a study for the purpose of establishing a basis for comparison with the test substance. The *test substance* is defined as the material administered or applied to the test system (i.e., organism).

The regulations also define reference substances, carriers, batches, and other standardized notations used in study design per GLP regulations. *Reference substances* are similar to control substances in that they are used as the basis for comparison with the test substance. *Carriers* are defined as any material with which the test substance is combined for administration to the test system. *Batches* refer to a specific quantity or lot of test, control, or reference substance that has been characterized for identity, strength, purity, composition, or other defining characteristics.

Compliance with the GLPs is the ultimate responsibility of the *sponsor*. The *sponsor* is defined as the person (i.e., representative from a company) who initiates and supports the GLP study or who submits a study to the EPA in support of an application for research or marketing permit as previously described. A *contract testing or research facility* must be notified by the sponsor that the study is to be conducted in compliance with GLP regulations. A contract testing facility must conform to GLP regulations if it agrees to perform a GLP study for a sponsor. Because several sites or contract facilities may be participating in a field study (for biological analysis, chemical analysis, etc.), it is important to ensure that all of the necessary facilities are aware of the compliance requirement.

The EPA has issued enforcement policies to further mandate compliance with the provisions of the GLP regulations. Testing facilities (e.g., contract laboratories) must allow EPA inspectors to audit their GLP compliance programs as well as individual studies. Based on the enforcement policy of the EPA, violations of the FIFRA GLP standards will be charged as an unlawful act of FIFRA under FIFRA section 12(a)(2)(B)(i), 12(a)(2)(M), 12(a)(2)(Q), or 12(a)(2)(R). The determination of the appropriate unlawful act with which a violator is charged will depend on the specific facts of the case.

Violations of the FIFRA GLPs may affect reliability or scientific merits of test data, the ability of EPA to validate or reconstruct test results, and the ability of the EPA to make sound and timely regulatory decisions regarding a pesticide (U.S. EPA, 1991). It may also affect the overall administration of the GLP inspection and enforcement program and ultimately human health and the environment.

Statements of compliance or noncompliance must be submitted to the EPA with any study used in the application for a research or marketing permit. These statements are included in the study report verifying whether the study was or was not conducted according to GLP regulations. The statement must be signed by the applicant, the sponsor, and the study director. If a statement of compliance is considered to be false by the EPA, it may serve as the basis for cancellation, suspension, or modification of a research or marketing permit or denial or disapproval of such an application (U.S. EPA, 1992b).

Organization and Personnel

Three main roles are described in this section: the testing facility management, the study director, and the quality assurance unit (QAU). Section 160.29 is generic to all three roles, but perhaps most important are the requirements for sufficient numbers of personnel and documentation that each person has the education, training, and/or experience to perform his or her assigned tasks. This becomes especially important for studies in which the technology is constantly evolving (e.g., aquatic field studies).

Demonstrating technical competence and proficiency has become increasingly stringent. A requirement of the GLPs is that the testing facility maintain a current training summary (e.g., outside courses and in-house training on equipment and technical procedures) and job description for each person either engaged in or supervising a study. The training summaries should be signed and dated by the trainer at the conclusion of each training session. Training records must be updated on an ongoing basis and these documents must be readily available at the time of an inspection by either an in-house QAU or a government agency.

Section 160.31 specifies seven responsibilities of the testing facility management. These include assigning the study director; ensuring that there is a QAU; ensuring that the test, control, and reference substances have been characterized properly; ensuring that there are sufficient resources to conduct the work; ensuring that all study participants understand their assigned functions; and most important, ensuring that any deviations noted in the study by the QAU are communicated to the study director and corrective action is taken and documented.

Since the addition of field studies to the EPA GLPs, perhaps the most controversial section has been 160.33, describing the role of the study director. According to this section, every study must have a study director who has the overall responsibility for the technical conduct of the study as well as the documentation, analysis, interpretation, and reporting of results. The study director is the single point of study control.

Before the publication of the revised GLPs, the role of the study director did not present a problem because studies were generally conducted at a single site. This, however, is not the case with aquatic and terrestrial field studies (e.g., aquatic mesocosm studies; see Chapter 9), in which the test substance may be applied at one field site and the analysis of soil, water, tissue, and sediment may occur at another site. Several subcontract testing groups may be involved in this process as well.

It is possible to accommodate a field study as a single study under the GLPs even though several sites are involved. For example, the testing facility would encompass all the organizational units involved in the actual work, and the testing facility management duties would be assumed by the sponsor facility or by a lead contracting facility. Nonetheless, certain overall responsibilities, such as those of the study director, must be centralized and cannot be delegated.

Although testing facility management is responsible for designating the study director and making appropriate assurance cited in Section 160.31, the GLPs do not state that the testing facility management actually perform the duties it is providing assurance for under this section. There is considerable flexibility for contractors (i.e., the sponsor) to provide testing facilities with their own standard operating procedures (SOPs), QAUs, or other requirements provided that the overall testing facility management can ensure that there was compliance with the good laboratory practice regulations. In some instances, the study sponsor will even designate its own scientific staff member to act as the study director for a study contracted to a testing facility.

The last major personnel classification is the quality assurance unit. As with the study director, it is a requirement of the testing facility management to ensure that there is a QAU which is responsible for monitoring each study. The primary purpose of a QAU is to assure management that the facilities, equipment, personnel, methods, practices, records, and controls are in conformance with the regulations. An important requirement of the QAU is that it be entirely separate from and independent of the personnel engaged in the direction and conduct of that study. This is to assure management and the EPA or other regulatory agencies that the study was conducted according to GLPs and that the final report reflects the data collected for the study.

Quality assurance personnel must be involved in every critical phase of a study, including protocol preparation, application of the test substance, collection and analysis of samples, and verification of all raw data and final report. The QAU must also conduct phase audits of ongoing studies and ascertain that all SOPs are in place and accurately reflect the procedures being followed in the planning and implementation of the study. The QAU also reviews equipment calibration and maintenance procedures and practices and ensures that all appropriate documentation is complete, verifiable, and accurate for each GLP study.

The quality assurance unit may comprise one or more individuals, and the EPA does allow a person technically involved in one study to act in QA for another study with which he or she is not associated in any other way. This allows personnel to be utilized efficiently, especially for small organizations, which often do not have the resources to support a separate QA staff. Most of the larger organizations have a discrete QA unit that fulfills these requirements on a full-time basis.

The complexity of large-scale field studies can be quite challenging for the quality assurance unit. Because these studies can encompass many sites but are still considered one study, a person cannot act in QA for a portion of a field study if that person is involved in the same study, even if the sites involved are geographically separated. For this reason, the QAU must be highly organized and coordinated in order to satisfy all the requirements of the GLPs. Communication lines must be closely maintained between the QAU, study director, and field personnel in order to make the most efficient and cost-effective use of QAU personnel.

The remaining duties of the QAU are relatively simple and are defined in Section 160.35 (b) 1–7. These include maintaining a master schedule (i.e., index of all ongoing GLP studies), retaining protocols, ensuring that deviations from protocols were authorized and documented, submitting status reports to management, reviewing final reports and raw data for completeness and accuracy, and including a compliance statement in the final report which specifies the dates of inspection and the dates when the findings of those inspections were reported to management and the study director (see Figure 2). Figure 3 is an example of a standard internal quality assurance reporting and documentation format.

It should be noted that during an agency audit, the EPA may request copies of the master schedule, SOPs, and other records maintained by the QAU. They are not, however, entitled to copies of actual inspection reports or audit findings from the testing facility's QAU. The agency may request proof that inspections and auditing have taken place and that findings have been reported to management and the study director. Therefore, the records maintained by the QAU should include sufficient evidence of inspection dates (i.e., inspection logs), which will, in some specific cases, be verified by status reports.

The master schedule or list of studies is usually reviewed during an EPA audit to determine the study "burden" of a testing facility. From this document, it can be determined whether the facility is adequate for the number and types of studies being conducted. In addition, the QAU is required to maintain standard operating procedures related to each responsibility of the QAU just as SOPs are required for study methods. The SOPs must detail each step taken to ensure that appropriate quality activities are taking place and how they are documented. The standard operating procedure requirements under the revised GLPs are described in Section 160.81.

Up to this point, the background and scope of the GLPs have been discussed, including the important players: facility management, study director, and QAU. The remainder of the GLPs deal with things rather than people, such as facilities; equipment; testing facilities operations; test, control, and reference substances; and "documents" such as protocols, reports, and other records. These document subparts describe the requirements regarding the content of a protocol and study conduct as well as the reporting of the study results and the storage, retrieval, and retention of records related to the study.

It is these last subparts (C–J) that provide the framework for the actions of the QAU. Each subpart can be catalogued into checklists, which become the basic tool by which QAU monitors a study. For example, Figure 4 illustrates the required items of the study protocol that can easily be reviewed utilizing this format. Similar types of checklists can be prepared and utilized in the review of study reports and other itemized GLP requirements.

Facilities

Subpart C of the GLP regulations discusses facilities, which includes laboratory areas used for animal studies and other test system areas such as those used for field studies. The primary focus of this subpart is to prevent cross-contamination of different chemicals used in different tests. For example, in a me-

QUALITY ASSURANCE STATEMENT

TEST: Acute Toxicity to the Mysid, *Mysidopsis bahia*, Under Flow-through Conditions

PROJECT NUMBER: J9302003a

TEST SUBSTANCE: Chemical A

Based on review of this study report, protocol and standard operating procedures utilized, and inspection of study events listed below, this test conformed to the current FIFRA GLP (Good Laboratory Practice) regulations, 40 CFR, Part 160.

The Sponsor was responsible for all test and control substance purity, stability, and characterization data as specified in 40 CFR, Part 160.105.

Date of Inspection	Phase Inspected	Date Reported to Management	Date Reported to Study Director
03/02/93	Treatment	03/02/93	03/02/93
04/15/93	Raw Data & Report Draft	04/17/93	04/17/93
04/20/93	Final Report	04/20/93	04/20/93

Jean J. Auditor, M.S. Date
Quality Assurance

Figure 2. Quality assurance statement (example).

socosm or other aquatic field study (see Chapter 9), the locations of the ponds or outdoor tanks (i.e., for microcosm studies) are very important, not only geographically but in relation to each other. For instance, during chemical treatment (application), test tanks or ponds receiving the test chemical should not be situated in such a way that their chemical treatment causes cross-contamination of control or untreated tanks. Facilities are also required to have provisions to regulate environmental conditions such as temperature, humidity, and photoperiod for studies or parts of studies conducted in a laboratory setting.

There should also be provisions for collection and disposal of contaminated water, soil or sediment, or other spent materials. Any facility utilized for GLP studies must be operated to minimize vermin infestation, odors, disease hazards, and environmental contamination. Proper storage areas are also required for supplies and equipment. These areas, particularly those of a temporary nature used in field studies, must be protected against infestation or contamination. Perishable supplies must be preserved by appropriate means.

Most important, there must be separate storage areas for test, control, and/or reference substances and mixtures. These areas must be away from areas housing the test systems and should be adequate to preserve the identity, strength, purity, and stability of these substances and mixtures.

For studies requiring laboratory activities, separate space is required for the performance of routine toxicity testing and specialized procedures (e.g., histological assessment of tissue; see Chapter 14). Space is also required for archives, limited to access by authorized personnel only, for the storage and retrieval of all raw data, reports, and specimens from completed studies. These archives can be as simple as a fireproof, lockable filing cabinet or as complex as an entire room, protected by a fire suppression system and alarm. Wherever the archives are located,

QA PHASE AUDIT AND STATUS REPORT

Project Number:
Sponsor: Company Name
Study: Microcosm
Study Director: GMR

Protocol Approval Date:
Experimental Start Date:
Inspection (Phase): Water Quality Sampling
Date of Inspection:

ITEM Standard Operating Procedures (SOPS):	**Yes**	**No**	**N/A**
1) Are SOPs relative to the study immediately available?	____	____	____
2) Are applicable SOP's adhered to?	____	____	____
3) Are deviations documented and authorized by Study Director?	____	____	____
ITEM Protocol:			
4) Is a written approved protocol immediately available?	____	____	____
5) Is the protocol adhered to?	____	____	____
6) Are protocol deviations properly authorized and documented?	____	____	____

FINDINGS/PROBLEMS: (Reference ITEM # if applicable)

____

____

____

____

____

Inspector: ______________________________ **Date:** ____________

REVIEWED BY:

Study Director: ______________________________ **Date:** ____________

Management: ______________________________ **Date:** ____________

MANAGEMENT/STUDY DIRECTOR RESPONSE/RESOLUTION:

____ No Corrective Action Required

____ Corrective Action Is Required. Estimated Completion Date:

Action To Be Taken: ______________________________

____

QAU FINAL REVIEW: QAU Signature/Title: ______________________________ Date: ____________

____ Acceptable

____ Unacceptable

Corrective Action Completed On: ______________________________

Study Director: ______________________________ Date: ____________

Management: ______________________________ Date: ____________

Figure 3. QA phase audit and status report (example).

Testing Facility:

Form No:
Revision No:
Effective Date:

Laboratory Project ID Number:____________________

Each study shall have an approved written protocol that clearly indicates the objectives and all methods for the conduct of the study. The protocol shall contain but not be limited to the following information:

1.	A descriptive title and statement of the purpose of the study.		9.	A. Where applicable, a description and/or identification of the:	
2.	Identification of the:			Diet used in the study	
	Test substance			Solvents	
	Control			Emulsifiers and/or materials used to solubilize or suspend Test	
	Reference substance			Control	
	CAS #/Code#			Reference substances before mixing with the carrier	
3.	Name/address sponsor			B. Description shall include specifications for acceptable levels of contaminants that are reasonably expected to be present in the dietary materials.	
	Name/address of testing facility at which study is being conducted		10.	A. The route of administration	
4.	Proposed experimental start date		11.	B. The reason for its choice.	
	Termination date		12.	Each dosage level in milligrams per kilogram of body or test system weight or:	
5.	Justification for selection of test system			Other appropriate units of the test sub.	
6.	Where applicable:			Control	
	Number			Reference substance to be administered	
	Body weight range			Method	
	Sex			Frequency of administration	
	Source of supply		13.	Type and frequency of tests, analyses and measurements to be made.	
	Species		14.	The records to be maintained.	
	Strain		15.	The date of approval of protocol by sponsor and dated signature of the study director.	
	Substrain		16.	A statement of the proposed statistical method to be used.	
	Age of test system		17.	All changes in or revisions of an approved protocol and the reasons therefore shall be documented, signed by the study director, dated, and maintained with the protocol.	
7.	The procedure for identification of the test system			Conclusions:	
8.	A description of:				
	Experimental design				
	Including methods for the control bias				

Key: Y = Yes
N = No
C = Comments

Auditor:____________________ Date:__________

Figure 4. FIFRA GLP compliance checklist/study protocol (example from Toxikon Environmental Sciences).

it is most important to have them secure from fire and unauthorized access.

Equipment

Subpart D concerns equipment and its maintenance and calibration. Adequate procedures must be in place for each piece of equipment for inspection, cleaning, and calibration. This includes any equipment used in the generation, measurement, or assessment of data and equipment used for facility environmental control. All equipment must be of appropriate design and adequate capacity to function according to the protocol and suitably located for operation, inspection, cleaning, and maintenance.

Standard operating procedures are required and should describe in sufficient detail the methods, materials, and schedules to be used in the routine inspection, cleaning, maintenance, testing, calibration, and/or standardization of the equipment. They must also specify, when appropriate, remedial action to be taken in the event of failure or malfunction of the equipment. The SOPs must indicate the person responsible for the performance of each of these operations.

Written records are to be maintained for all inspection, maintenance, testing, calibrating, and/or standardizing operations. These records should include the dates of these operations and indicate whether the maintenance operations were routine and followed the written standard operating procedures. Written records are also required for nonroutine repairs performed on equipment as a result of failure and malfunction. These records should document the nature of the defect, how and when the defect was discovered, and any remedial action taken in response to the defect.

Testing Facilities Operation

The SOPs previously described are the subject of Subpart E—Testing Facilities Operation. This subpart requires that the SOPs be in writing, setting forth study methods that management is satisfied are adequate to ensure the quality and integrity of the data generated in the course of a study (EPA, 1992b). Deviations from SOPs must be documented and authorized by the study director in the raw data. Significant changes in SOPs must be authorized in writing by management. An SOP numbering system should be devised whereby revisions from the original SOP can be identified. Maintenance of a historical SOP file, which includes all revisions and revision dates, is also required.

The GLPs specify 12 areas for which SOPs are required. This list is not all-inclusive and may be expanded. The 12 areas include test system area preparation; test system care; test material preparation including receipt, identification, storage, handling, mixing, and method of sampling of the test, control, and reference substances; test system observation; laboratory or other tests; handling of test systems found moribund or dead during the study; necropsy or postmortem examination of test systems; collection and identification of specimens; histopathology; data handling, storage, and retrieval; maintenance and calibration of equipment; and transfer, proper placement, and identification of test systems.

An important requirement of this subpart is that SOP manuals be available to the personnel performing the testing and in close proximity to where the actual procedure or testing is performed. This can present some problems when conducting field studies, because environmental conditions often make it difficult to store SOP manuals in areas where they are required. Waterproof paper, plastic sleeves, and other barrier techniques should be considered under these conditions. In some cases, published literature may be used as a supplement to a standard operating procedure but may not be used in place of one. In these cases, literature is appended to a written and approved SOP for reference.

Another requirement of this subpart is that all reagents and solutions in the laboratory areas be labeled to indicate identity, titer or concentration, storage requirements, and expiration date. Deteriorated and outdated reagents and solutions should be discarded and not used.

Section 160.90 of this subpart refers to animal and other test system care. This section is appropriate for laboratory animals as well as aquatic and avian test systems. It requires SOPs for the housing, feeding, handling, and care of animals and other test systems. As animals are received, they should be isolated until their health status or appropriateness for the study is evaluated in accordance with acceptable veterinary medical practice or scientific methods. If a test system is found to have a disease or condition that might interfere with the conduct of the study, the test system should be isolated, if necessary.

Complete documentation of the diagnosis and treatment of disease should be retained in the study records. In addition, if an animal or other test system requires removal from its housing unit in order to be observed, the test system needs to be identified in some way. The identification information must appear on the outside of the housing unit and on the animal itself (i.e., for small mammals or birds), whenever possible. Test systems should be housed separately to avoid potential confusion between studies or inadvertent exposure to test, control, or reference substances from other studies.

Housing units such as cages, racks, pens, enclosures, aquaria, holding tanks, ponds, growth chambers, and other holding, rearing, and breeding areas and accessory equipment must be cleaned and sanitized at appropriate levels. Feed, soil or sediment, and water used for the test systems must be analyzed periodically to ensure that contaminants known to be capable of interfering with the study and reasonably expected to be present in such feed, soil, sediment, or water are not present at levels above allowable tolerances as stated in the protocol. Documentation of these analyses must be maintained in the raw data.

The last three items in this section require that bedding (i.e., for small mammals or birds) used in housing units such as cages or pens be changed as often as necessary and that bedding not interfere with the purpose or conduct of the study. This requirement generally refers to animal studies for health effects testing in which direct bedding is used. Pest control materials are allowed to be used, but their use must be fully documented and they must contain no substances that would interfere with the study. The last item requires test systems to be acclimatized to the environmental conditions of the test prior to their use in a study. This is important for laboratory animals as well as avian and aquatic test systems.

Test, Control, and Reference Substances

Test, control, and reference substances are the topic of Subpart F, which includes GLP requirements of characterization, handling, and mixture with carriers. The characterization section (160.105) requires complete identification of test, control, and reference substances. Methods of synthesis, fabrication, or derivation of these substances must be documented by the sponsor or testing facility. The location of this documentation must be specified.

The GLPs are very specific about the timing of test, control, and reference substance characterizations. When relevant to the conduct of the study, the solubility of these substances should be determined by the testing facility or the sponsor before the experimental start date. The stability should also be determined prior to the experimental start date, or concurrently according to written SOPs, which provide for periodic analysis of each batch.

The GLPs also require each storage container for a test, control, or reference substance to be labeled by name, chemical abstracts service number (CAS) or code number, batch number, expiration date, if any, and where appropriate to maintain the identity, strength, purity and composition of the substances. Storage containers are to be assigned to a particular test substance for the duration of the study. In addition, for studies with an experimental duration longer than 4 weeks, reserve samples of each batch of test, control, or reference substance must be retained for a specific amount of time as specified in Part 160.195. The stability of each test, control, and reference substance under storage conditions is required to be known for all studies. This requirement could be particularly complicated for avian and aquatic field studies, in which drums of test materials may be stored in a holding area near the field or pond site. The stability under these conditions must be determined and documented.

The requirements for test, control, and reference substance handling are described in Sections 160.107 and 160.113. The first section includes substances in their natural state and the second section concerns the substance when mixed with carriers. These sections require procedures for receipt, storage, identification, and distribution of substances and are designed to preclude the possibility of contamination, deterioration, or damage to the substances.

It is most important that an accurate audit trail of test or reference chemical be maintained. The results of an expensive and time-consuming study are seriously compromised if the documentation of chemical usage cannot be substantiated. Analytical testing must be performed in order to verify homogeneity of test substances mixed with carriers or vehicles and to verify appropriate concentrations or dosages. If a mixture is to be used, the stability of the test substance in the carrier or vehicle must be determined either before the experimental start date or concurrently according to standard operating procedures. In essence, complete characterization data and other physical/chemical parameters that may be necessary for proper study conduct must be procured prior to or during the initial stages of a GLP study for test, control, and reference substances.

Protocol for and Conduct of a Study

Subpart G of the regulations discusses the study protocol and its necessary components (i.e., elements) in order to conduct a GLP study. As previously mentioned, this section of the regulations lends itself well to a QA checklist type of format (see Figure 4) to ensure that all required elements are included. The protocol is the document that provides study-specific information in order to conduct the experiment. Every study must have an approved written protocol describing the objectives of the study and the methodology.

An important requirement for study protocols is that all changes or revisions to an approved protocol and the reasons for the changes be documented, signed by the study director, dated, and maintained with the protocol. There must be a historical documented account of each and every protocol event, similar to an audit trail of the test substance and its usage in the study. Depending on the complexity of a study protocol, it may be unrealistic to expect a study always to follow a protocol precisely. Changes and revisions are common and acceptable to GLP study protocols provided they are approved by the study director. If a contract testing facility is performing the study, protocol amendments should likewise be approved by the study sponsor.

The required elements of a study protocol include a descriptive title and statement of study purpose (or objective); identification of test, control, and reference substance by name, CAS number or code; name and address of the sponsor and testing facility; proposed experimental start and termination dates; justification for selection of the test system; number, body weight, range, sex, source of supply, species, strain, substrain, and age of the test system (where applicable); procedures for identification of the test system; description of the experimental design and methods for the control of bias; description of diet used and other solvent or materials used to suspend the test, control, or reference substances; route of test substance administration and the reason for its choice; dose level (i.e., for rats or mice) or concentrations (i.e., for aquatic organisms) to be employed expressed in appropriate units (e.g., mg/kg body weight for rats or mice or mg/L for fish) and the methods and frequency of administration; type and frequency of tests, analyses, and measurements made; records to be maintained; date of protocol approval by the sponsor and study director; and a statement of the proposed statistical methods that will be utilized. Other information may also be included in the study protocol at the discretion of the sponsor and/or study director.

There are fewer specifications described in the regulations (Section 160.130) for the conduct of the study than for the protocol requirements. The primary concern is that the study is conducted and monitored according to the protocol. A major requirement, particularly in field studies or microcosm/mesocosm studies in which numerous specimens are collected, is to have each specimen identified by test system, study, nature [test parameters(s) or type of analysis], and date of collection. This information must accompany the specimen in a way that precludes any confusion in the recording or storage of data, and it should be entered onto a chain of custody record or sample transmittal form (Figure 5) that accompanies the samples to the laboratory and/or respository.

A convenient way to ensure that the correct information is collected is to use a computer-generated label that contains the items to be recorded. Especially for field studies, it is often necessary to enter information with a waterproof pen. When specimens are being collected in a field study or microcosm/mesocosm study, nonwaterproof ink may invalidate some data because of smearing of the label. This is not a specific requirement of the GLPs but makes good scientific sense considering the audit trails required.

The GLPs do require that data be recorded promptly, directly, and legibly in ink. All entries must be dated on the date of entry and signed or initialed by the person entering the data. Most important, if a mistake is made, the entry must not be obliterated. A single line should be drawn through the error and the correct entry made as close to the error as possible. The reason for the change and the date and initials of the person making the change are also required. Many automated data collection systems allow changes to data entries as well but maintain an automatic audit trail that saves the original entry in a separate file. No data should ever be discarded or written over during the conduct of a study.

The last section of this subpart (160.135) itemizes physical and chemical characterization studies that may be required as part of product registration. GLP requirements now apply to these studies and they have been added to the scope of GLP regulations. These characterization studies include stability, solubility, octanol-water partition coefficient, volatility, and studies to determine persistence such as biodegradation, photodegradation, and chemical degradation studies of test, control, or reference substances.

Records and Reports

The final subpart (J) of the GLP standards is related to records and reports, including the storage and retrieval of records and data. As with the study protocol requirements, the minimum standards for study report requirements are itemized in the regulations. They include name and address of testing facility and dates of initiation, completion, termination, or discontinuance; objectives and procedures and any changes from the original protocol; statistical methods utilized; test, control, and reference substance information; stability and often solubility information related to test, control, and reference substances; methods used; description of test system used and procedures for identification thereof; description of dosage, dosage regimen, route of administration, and duration; description of circumstances that may have affected data quality or integrity; name of the study director and other professionals or supervisory personnel involved in the study; description of transformation, calculations or data manipulation, data analysis, and summary of conclusions; signed and dated reports of individual scientists or professionals involved in the study; location where specimens, raw data, and final report are to be stored; and a quality assurance statement.

In addition to these GLP requirements for the final report, all FIFRA-type GLP studies must be specifically formatted according to the requirements designated in Pesticide Registration Notice 86-5 (U.S. EPA, 1986b). The final report should also meet the reporting requirements of the data reporting guidelines (DRGs), the addenda to the pesticide assessment guidelines, and the pesticide registration (PR) notices (Hochman and Garner, 1992).

Quality assurance review of final reports may also lend itself to a checklist system, similar to the protocol checklist, because the guidelines are relatively clear on required elements. However, there are often logistical problems when signatures of professionals at different locations are required. Because of this, the final report can often present some unique coordination problems. Once the study director signs the final report, it is considered completed as defined earlier in the definitions section. Therefore, every aspect of the report must be finalized prior to the study director's final signature.

After a study report is completed, there are standard evaluation procedures (SEPs) published by the EPA that are designed to ensure comprehensive and consistent study data review. They also provide interpretive policy guidance where appropriate. Many of the quality assurance provisions for each type of study performed are often defined in SEPs. They should be utilized in tandem with GLP regulations in designing, performing, and reviewing studies.

Testing Facility:

Form No:
Effective:

CHAIN OF CUSTODY RECORD (Sample Transmittal Form)		
Project Number:	**Site (Lab or Field):**	**Date:**
Packed By:	**Shipped to:** ______	
Shipped By:	**Contact:** ______	

Upon receipt of these samples, send signed copy of form to: Quality Assurance Officer Testing Facility and Address	Sample Type (Matrix): 1. Water 2. Sediment 3. Fish Tissue 4. Soil 5. Plant Tissue 6. Other ______

Sample Number	Sample Location	Matrix	Collection Date	Storage Condition (see codes below)

Storage Codes: RT = Room temp.; FR = Refrigerate; F = Frozen (-40°); DI = Dry Ice

Received by:	**Date:** **Page** ______ **of** ______

Figure 5. Chain of custody record (sample transmittal form from Toxikon Environmental Sciences).

This is discussed in greater detail in the section that follows on QA/QC program designs.

If any corrections or changes are needed after the study director has signed the report, an amendment is required. The amendment should identify the part of the final report that is being corrected or changed, with appropriate explanations. The amendment must be signed by the study director. All amendments, along with the final report and all supporting raw data and specimens, must be maintained by the sponsor and/or testing facility.

Once the study is completed, a system is required for storing and retrieving data and test materials. The regulations clearly state that all raw data, documentation, records, protocols, specimens, and final reports generated as a result of a study shall be retained. Specimens obtained from mutagenicity tests; specimens of soil, water, and plants; and wet specimens of blood, urine, feces, and biological fluids do not need to be retained after quality assurance verifications. Other documents related to the study, including telephone logs, memoranda, and other communicated material, should be archived with the study materials.

Conditions of storage must minimize deterioration of the documents or specimens in accordance with the retention requirements of the GLP regulations. These time periods are described in Section 160.195. The archives may be an in-house facility, or the testing facility may contract with commercial archivists to provide a repository for all materials to be retained.

As mentioned earlier, the archives may consist of a fireproof, lockable cabinet or a large room with shelves and cabinets and a required fire-suppression system. A person must be identified as responsible for archives, and all of these activities are generally described in SOPs for archives description, operations, and maintenance procedures.

For in-house archives, the facilities must be restricted to authorized personnel. Most important, the archival system must be indexed so that items are easily retrievable. Archive procedures and indices may be subject to audit by the EPA or other auditing agencies.

The GLP regulations also require the retention of master schedules and copies of protocols, records of quality inspections, summaries of training and experience of laboratory/field staff, and job descriptions. Other records include maintenance, calibration, and inspection of equipment and all other ancillary information and documentation pertaining to each GLP study. All of these records must be retained as either original documents or certified copies such as microfilm, microfiche, or other accurate reproductions of the original record.

QA/QC PROGRAM DESIGNS

For every test performed in aquatic toxicology and other testing disciplines, a variety of technical requirements are part of the overall study design. These specifications are incorporated in the study protocol. Good laboratory practice requirements are likewise included in the study protocol and often become an integral part of the technical requirements beyond the generic GLP requirements already discussed. Because of this, it can become difficult to differentiate between good laboratory practices and the technical "quality" requirements when designing a study or developing the protocol. Therefore, in order to standardize the review process for study auditors, the EPA has developed a series of standard evaluation procedures, which were previously described.

These procedures are utilized by the agency as a guidance document and have been developed for specific studies. To fully understand the scope and complexity of GLP regulations in health effect and environmental study design, it is helpful to become familiar with inherent quality issues required within each study type. Technical study review may be distinct from a GLP audit of the same study, but quality and technical review inevitably interface during the review process.

Furthermore, the manner in which the GLPs are incorporated in the overall study design may also be subject to some degree of interpretation by the study director working together with the QAU. At all times, the study director and quality assurance staff should anticipate potential data gaps and concurrent GLP or quality assurance deficiencies in the study design prior to finalizing study protocols. Likewise, any other study specific documentation (SOPs, forms, etc.) should be carefully reviewed.

Highlights of requisite study elements that require quality review are described in the following sections. These sections describe the testing requirements for laboratory studies through a description of field testing requirements and data analysis. It is important to remember that the majority of quality assurance and quality control requirements described for laboratory testing also apply to field studies. Field study quality management practices are merely an expanded application of good laboratory practices and principles.

Analytical Methods

Many environmental and human health effects studies require analytical chemistry. Chemical analysis is needed either to characterize the test or control substances used in biologically mediated studies or as part of the overall study objective. The analytical portion of any study must be fully described and performed according to detailed specifications, which include an initial method validation study. The overall quality of chemical data may be judged on the basis of two aspects: the accuracy of the parameter measured and the numerical accuracy (Taylor, 1987).

Elements of method validation to demonstrate that the assay is appropriate for its intended purpose include resolution, linearity, accuracy, precision, sensitivity, and specificity. Other aspects of quality assurance/quality control for analytical studies include system check and performance samples (e.g., blanks, fortification/spike samples, initial calibration verification standards, continuing calibration standards, duplicate analyses, and midpoint standard checks). Additional types of QA/QC analyses may be performed during study conduct depending on protocol design and specifications.

Constant advances in technology and instrumentation allow for greater sensitivity and specificity in the analysis of chemicals and their degradates of interest. Because of these developments and the importance of chemical data, the scope of the FIFRA GLPs was expanded in 1989 to include analytical studies. However, studies have shown that analytical chemistry data (residue data) are commonly rejected because of an inadequately described or validated methodology (U.S. EPA, 1992a).

Likewise, a standard evaluation procedure was prepared to aid the Dietary Exposure Branch data reviewers in their evaluation of the analytical method(s) submitted by petitioners/registrants for pesticides and their metabolites that together constitute the ''total toxic residue'' in raw and processed commodities (U.S. EPA, 1989a). Chemical analysis is a requirement for pesticide residue studies and is also required in any study design that contains an analytical component. Proper project planning and a cooperative effort must be undertaken by the study director, sampling personnel, and analytical chemists to define the levels of quality that shall be required for the data (U.S. EPA, 1984).

Analytical study review includes:

- Full and reproducible description of the methodology; equipment and conditions; reagents and standards.
- Principles of the analytical method utilized.
- Specific information regarding preparation or extraction procedures and efficiency; cleanup procedures and derivitization procedures.
- Extensive data requirements including logbooks; standard and sample preparation logs; column logs; instrument run logs; printouts from instrumentation (chromatograms, spectra); all raw data and calculations; statistical treatment of data and verification; and any other entries or notations made by the technician and study director.

For chemistry studies in general, accurate methodology, adequate calibration, and systematic quality control (QC) procedures are required to provide reliable analytical data (Taylor, 1986).

Fate Studies

Soil Metabolism

These studies are required by 40 CFR Part 158.130 in support of pesticide registration of an end-use product intended for terrestrial use or forestry use and to support registration of a product for manufacturing use that may also be legally used to formulate such an end-use product (U.S. EPA, 1985a).

Soil metabolism studies should provide information on the rate of degradation in soil, identification of soil degradation products, rate of formation and decline of degradation products, and material balance. These data are important because information obtained is used in designing and conducting field studies, considering both aerobic and anaerobic conditions. Accordingly, it is important to plan the study carefully because it affects more involved and costly field programs.

Study review includes:

- Complete information on the test substance; soil characterization; methods of inoculation; sampling frequency; analytical methodology sensitivity and specificity; and replication and frequency specifications and tolerances.
- Other information on compound purity (as well as site of radiolabeling); complete characterization of the soils; treatment rate and method of treatment of the soil; the methods for quantitative and qualitative analyses of the degraded compounds; and formation and decline of degradation products.

Other questions that should be considered include whether the soil was maintained in a manner conducive to maintaining microbial populations and whether quality procedures were maintained during sampling, storage, and analysis of soil samples (U.S. EPA, 1989b). Aquatic metabolism studies are similarly reviewed, utilizing an aqueous test system.

Soil Column Leaching

Leaching and sorption/desorption studies provide information on the movement of pesticides and their degradates to and dispersion in aquatic sites. These analyses provide mobility data on the test substance and aged residues (U.S. EPA, 1985f). Many of the same quality elements required in the soil metabolism studies are required for soil column leaching studies.

Other specific quality review includes:

- Appropriate use of textured soils with organic matter content at sufficient levels; whether study treatment is representative of the use rate; and ensuring test chemical and soil degradates are mobile in soil under the conditions of the test.

Photolysis Studies

Photolysis studies are required for each active ingredient intended for any terrestrial, aquatic, or forestry use. Data on rates of photolysis in water of the parent chemical and its photodegradates are needed to establish the importance of this transformation process and the persistence characteristics of photoproducts formed (U.S. EPA, 1985d).

As with all studies, the data provided must be applicable to support the proposed use. If the data are not relevant to the proposed use application, in most cases the study will not be reviewed. Photolysis studies, like many other environmental studies, require critical review for all of the quality components expected to maintain data integrity.

Key elements for photolysis study review include:

- Appropriate identification of photoproducts present at >10%; test conditions (temperature, sterility, pH, and darkness); sunlight source documentation (latitude, time of year, atmospheric cover, and any other major variations that could affect incident light). For artificial light sources, emission wavelength spectrum and intensity should be reviewed.
- Use of high-quality solvents and glassware; verification of photolysis rate calculations and derivations; explanation of identification products; appropriate test chemical and buffer concentrations; radiolabeling location in a stable portion of the chemical; testing duration (at least one half-life or 30 days); identification of any impurities that might interfere with the analytical method; and the proper use of control samples.

Hydrolysis Studies

Knowledge of the hydrolytic fate of pesticides is critical to understanding the overall fate of pesticides in the environment (U.S. EPA, 1985e). Because hydrolysis data are needed for all outdoor use pesticides, the studies are reviewed in detail. Hydrolysis is one of the most common reactions occurring in the environment and therefore represents one of the most potentially important degradation pathways for many classes of chemicals (U.S. FDA, 1987a).

Data quality review includes some of the major topics already discussed for other fate/distribution studies, including site of radiolabeling, solubility information, and other standard technical quality requirements.

Data review specific to hydrolysis studies includes:

- Potential effects from microbial degradation; appropriate initial concentrations of the test compound (above or below water solubility); other potential confounding factors (effects of photolysis also contributing to degradation, other cosolvents, buffer catalysis effects); appropriate mass balance and sampling frequency to establish kinetics; and test compound stability under appropriate storage conditions and accountability.

Toxicity Testing

Acute Freshwater Studies

The review of acute toxicity laboratory testing for freshwater fish and invertebrates corresponds to the ecological effects data specified in 40 CFR, Part 158.145. For the most part, these types of studies have an established technique for assessing toxicity of a chemical to freshwater species and for data analysis (see Chapter 2). The criteria have been outlined in published literature (U.S. EPA, 1985c). Whether the studies are static or flow-through designs, basic study elements must be provided by the sponsor and testing facility for data acceptability. Data recording forms (see Figures 6, 7, and 8 for examples) for these and other studies should provide for all study-specific information.

Study review includes *pretest criteria*:

- Species acceptability, including sex, age, and physical condition (length, weight, and general health); source and acclimatization conditions; source and conditions of dilution water; temperature; appropriate test vessels or housing conditions; regulated photoperiod; appropriate loading (test organism mass per volume of test solution); and monitoring of test and control systems for pH, dissolved oxygen (DO), and temperature on a continuous basis.

Analytical component:

- Conditions of analysis (aerated vs. nonaerated); determination of volatility or insolubility of the test chemical; adsorption properties of the test chemical to the test container structure or materials; verification of test chemical concentration in test system from automated chemical application (diluter system); and minimization of solvents used (less than 0.5 ml/L in static systems and 0.1 ml/L in flow-through systems).

Test conditions:

- Use of appropriate concentrations (dose levels) established from range-finding study; number of test animals per concentration group; use of concurrent controls per test level; use of solvent control if necessary; use of the technical grade of the test chemical, including source, batch, and exact purity; complete characterization of dilution water (hardness, alkalinity, DO, conductivity); preparation of stock and test solutions; complete physical and behavioral observations; and calculation of LC50 and statistical analysis verification.

It is important to remember that the data derived from acute toxicity testing are used in ecological risk assessment. Therefore, data accuracy with respect to LC50 calculations in toxicity tests is critical to limit the propagation of error. For example, a high degree of uncertainty is introduced when extrapolating LC50 values to exposure and consequently calculating risk. There are other methods such as the quotient method that correlate the toxicity end point (e.g, LC50) directly with the exposure value (Bascietto et al, 1990; see Chapter 28). Because calculation of risk is directly dependent on toxicity data, the importance of precision and accuracy of toxicity data cannot be overstated.

Marine Studies

Marine acute studies are necessary to support the application of any chemical intended for direct application to estuarine or marine environments. The quality elements of freshwater studies generally apply to marine toxicity tests (see Chapter 3), with a few exceptions. The seawater utilized in these

Testing Facility:

Page:
Form No:
Effective Date:

STATIC TOXICITY TEST -- TEST DESIGN | Data By:

Sponsor: | Project Number: | Date:

Test Substance

Test Substance: ______
See Page R______ of ______
Receipt Log for Test Substance Information.

Test Conditions

[ ] Preliminary
[ ] Definitive
[ ] Screening

[ ] Static [ ] Renewal

Duration :

Test Animal History

Species : ______
Batch/Lot Number : ______
Age / Life Stage : ______
Date Acclimation / Maintenance Began : ______
See Page ______ of ______ Log
for raw data.
Mortality (%) 48 Hrs prior to testing: ______ %

Test Area Used	Temperature	Hard/Sal
	± °C	

Protocol Followed:

Lighting

Light Intensity Range: ______ to ______ uE/m^2/s
Dawn/Dusk Transition Period: ______ Minutes
Photoperiod: ______ hr Light : ______ hr Dark

Dilution Water:

Test Chamber Dimensions : ___L x ___ W x ___ H cm
or: ___D x ___ H cm
Test Solution Height : ______ cm
Test Chamber : [ ] Open [ ] Covered
Test Chamber Volume : ______ Liters
Dilution Volume : ______ Liters

Reps/Conc.: | # Animals/Rep.:

Time Test Substance Added:

Time Test Animals Added :

Solvent Used : ______
Solvent CTL Concentration : ______
Amount Solvent Added : ______

Concentrations Based on:
[ ] A. I. [ ] W. M.

Test Concentrations: (Units =):	Control							
Amount Dilution Water Added ( ):								
Amount Test Substance Added ( ):	N/A							
Stock Solution Used (e.g. 1°, 2°, 3°):	N/A							

Figure 6. Static toxicity test—test design; data record form (example from Toxikon Environmental Sciences).

Testing Facility:

Page:
Form No:
Effective Date:

FLOW-THROUGH TOXICITY TEST -- TEST DESIGN | **Data By:**

Sponsor: | Project Number: | Date:

Test Substance

Test Substance: ______
See Page R______ of ______
Receipt Log for Test Substance Information.

Test Conditions

[ ] Preliminary
[ ] Definitive
[ ] Screening

[ ] Acute [ ] Chronic

Duration :

Test Animal History

Species : ______
Batch/Lot Number : ______
Age / Life Stage : ______
Date Acclimation / Maintenance Began : ______
See Page ______ of ______ Log
for raw data.
Mortality (%) 48 Hrs prior to testing: ______ %

Test Area Used	Reps / Conc	# Animals / Rep

Lighting

Photoperiod : ______hr Light : ______hr Dark
Dawn/Dusk Transition Period: ______Minutes
Light Intensity Range : ______ to ______ uE/m^2/s

Dilution Water :

Test Chamber Dimensions : ____L x ____W x ____H cm
or : ____D x ____H cm
Test Solution Height : ______cm
Test Chamber Volume : ______Liters
Dilution Volume : ______Liters

Protocol Followed:

Solvent Used : ______
Amount Solvent Added : ______
Solvent CTL Concentration: ______

Conc. as: [ ] A.I. [ ] W.M. | Time Test Animals Added: | Temp: ± °C | Hard/Sal:

Test Concentrations: (Units =):	Control							

See pages WC-____ and SB-____ of Diluter #___ Calibration and Maintenance Log for diluter calibration.

Was test aerated? [ ] Yes [ ] No. If Yes, give date and time initiated:

Describe retention chambers, if any were used:

Describe method of chemical delivery (include SN's):

Figure 7. Flow-through toxicity test—test design; data record form (example from Toxikon Environmental Sciences).

Testing Facility:

Page:
Form No:
Effective Date:

ACUTE TOXICITY TEST -- 0, 24, 48-HOUR DATA AND MORTALITY RECORD [1]

Test Substance:								Project No:							Species:							Final Results			
Nominal Concentration	Day:			Time: 0-Hour				Day:			Time: 24-Hour				Day:			Time: 48-Hour							
	Date:			By:				Date:			By:				Date:			By:							
	Rep.	# Alive	Obs.	Temp	Sal.	DO	pH	Rep.	# Alive	Obs.	Temp	Sal.	DO	pH	Rep.	# Alive	Obs.	Temp	Sal.	DO	pH	Rep.	# Dead	Total Dead	% Dead
	A							A							A							A			
	B							B							B							B			
	A							A							A							A			
	B							B							B							B			
	A							A							A							A			
	B							B							B							B			
	A							A							A							A			
	B							B							B							B			
	A							A							A							A			
	B							B							B							B			
	A							A							A							A			
	B							B							B							B			
	A							A							A							A			
	B							B							B							B			
	Meter ID #:							Meter ID #:							Meter ID #:										

Observation Codes:

N	Nothing out of the ordinary was observed	ERR	Erratic	CLDY	Cloudy solution	FF	Flake food
AS	At the surface	GYR	Gyrating	PRE	Precipitate on bottom of vessel	YCT	Yeast/Cerophyll/Trout Chow
MSP	Muscle spasms	IMM	Immobile	FOS	Film on surface	EE	Entry error
PLE	Partial loss of equilibrium	DRK	Dark coloration/pigmentation	UC	Undissolved chemical	N/A	Not applicable
CLE	Complete loss of equilibrium	HEM	Hemorrhagic	NF	Not found	AI/WM	Active ingredient/Whole Material
LETH	Lethargic	RAR	Rapid respiration	D	Dead		
HYP	Hyperactive	GUL	Gulping	BS	Brine shrimp		

1 For 96-hour test add additional sheet for 72 and 96-hour data.

Figure 8. Acute toxicity test data and mortality record (example from Toxikon Environmental Sciences).

studies is usually fresh unfiltered seawater with a salinity of approximately 10 to 30 parts per thousand (ppt) depending on species. Controls should be available at the laboratory to monitor the seawater parameters and to ensure that the monthly pH range does not deviate by more than 0.8 unit (U.S. EPA, 1985b). Similar data recording formats utilized for freshwater studies are employed for marine studies.

The use of seawater can also affect the chemical analysis of a chemical. In these cases, the analytical methodology appropriate for a test chemical in fresh water may have to be revalidated for marine studies. Various interferences due to the presence of salt and other naturally occurring constituents of seawater often necessitate reevaluation of the analytical methodology.

Chronic Toxicity and Life Cycle Tests

To study long-term effects of chemical exposure in the aquatic environment, short-term subchronic (e.g., fish early life stage tests), chronic (e.g., *Daphnia*, mysid), and full life cycle (i.e., fish) tests are conducted with freshwater and marine organisms. Some of the quality parameters that are considered in acute testing are also taken into account in these studies. Additional quality considerations include the following:

Chronic Toxicity

Documentation of health conditions of the test animals and pretest diet; renewal schedules to count live and dead animals; control of test temperature, DO, and pH tolerances; analysis of test water for pesticides, metals, and other contaminants; test vessel specifications; randomization procedures for assigning test organisms to test and control groups; and physical characteristics of water at the end of the study.

Life Cycle

Specific quarantine conditions; appropriate procedures and documentation for fertilization (procuring eggs from adult fish); feeding schedules; embryo removal; embryo exposure; larval-juvenile exposure; juvenile-adult exposure; second-generation embryo exposure; and second-generation larval-juvenile exposure. Appropriate aeration and flow rates; methodology to distinguish developmental stages; measurements of growth and other parameters such as locomotion and behavior; records of all observations; and statistical analysis methods and references.

Plant Toxicity (Phytotoxicity)

Growth and reproduction studies of aquatic plants are designed to provide data on pesticide phytotoxicity (U.S. EPA, 1986a). The basic principles of quality and GLP practices for animal studies also apply to plant studies. Because these studies are conducted in growth chambers, complete and accurate descriptions of the laboratory facilities, treatments, and procedures are required. Other information that should be incorporated as part of the overall quality review includes:

- Sampling data and the basis or explanation of the phytotoxicity rating system utilized; storage information of the samples; chemical analysis and chemical content (when applicable); and data reporting including statistical analysis.
- Control measures/precautions followed to ensure fidelity of phytotoxicity determinations; recordkeeping procedures and availability of logbooks; skill of the laboratory personnel; equipment status at the laboratory (e.g., growth chambers); adherence to good agricultural practices to maintain healthy plants; and any other information to provide a complete and thorough description of the test procedures and results such as climatological data (i.e., if study is conducted outdoors), trial identification number, and appropriate replication data.

Other Toxicity Tests

There are numerous other types of biological effects tests, including sediment, seaweed, frog embryo, sea urchin, and effluent studies, which are discussed in detail in earlier chapters. These studies are subject to the same general GLP and quality practices in their design and review as described for toxicity testing. As with any test, study-specific technical components should be integrated as part of the quality review process to ensure data integrity.

Field Studies

The conduct of a field study entails a well-organized, coordinated effort between the technical staff and the quality assurance unit. Because of the large scale of both aquatic and terrestrial field studies, quality becomes more difficult to control than for studies conducted entirely in a laboratory setting. Also, because of the outdoor location of field studies, there are often other factors that are difficult or even impossible to anticipate or control. Examples include severe weather conditions and animal predation on the test systems. Because of these factors, effective responses to developments in the field may be delayed or not taken, causing such a study to be compromised (U.K. Department of Health, 1990). Nonetheless, these studies are regulated under GLPs, and specific areas that an agency will examine for study acceptability should be understood.

Aquatic Field Testing

Aquatic field testing in general is designed to provide toxicity data in support of registration of a pesticide that either directly will be applied to an aquatic system (e.g., aquatic herbicide) or indirectly (i.e., via surface runoff or drift) may be transported and find residence in an aquatic system.

These tests are conducted only after laboratory tests have indicated that the chemical is potentially hazardous. Field testing is important in the ecological risk assessment process (see Chapters 9 and 28). Field testing may include assessments of effects on plants, fish, and invertebrates.

Phytotoxicity

Field tests are conducted on pesticides (e.g., herbicides) that may adversely affect aquatic plants. Many of the elements described in relation to laboratory phytotoxicity testing also apply to aquatic field testing. However, the outdoor component of the study necessitates further examination, and documentation includes:

- Site-specific information (aquaria, lake, swamp); geographic region; physical environment characteristics (water turbidity, flow rates, salinity, degree of exposure); substrate characterization (types of growth media, soil types); appropriate selection and use of aquatic plant species; and adequate and comprehensive data reporting.

Fish/Invertebrate Toxicity

Mesocosm and microcosm (experimental pond and tank) studies are also conducted in the field, and numerous end points (e.g., biological, chemical; see Chapter 9) and interactive effects of an ecosystem are studied. Because these types of studies are usually conducted on chemicals that are presumed hazardous to aquatic organisms on the basis of laboratory data, the data are critically reviewed. The registrant typically wants to demonstrate that these complex systems will mitigate the exposure or toxicity indicated by laboratory data, and the regulators need information on ecosystem-level responses to evaluate risk-benefit analysis (U.S. EPA, 1988). For these studies to be scientifically credible and reproducible so the data are useful in ecological risk assessment, the following information should be considered.

Pretest criteria:

- Acceptable study design (physical, chemical, biological, and mechanical); sediment type and classification (percent clay, silt, sand, organic carbon, organic nitrogen, ion exchange capacity); water characteristics (pH, DO, temperature, dissolved solids, total organic carbon, dissolved organic carbon, total phosphorus, NO_2, NO_3, etc.); appropriate acclimatization or seasoning of the ponds/tanks; appropriate biotic loading and pretreatment chemical and biologic monitoring. Background metal and pesticide screens are also conducted on water and sediment.

Application and postapplication phase:

- Appropriate dose levels (based on exposure modeling or residue monitoring data); treatment design and duration; monitoring data for chemical, biological, and physical properties; and test chemical residue analysis (sediment and water for parent compound and metabolites/degradates); collection and identification techniques and complete taxonomic evaluation of biological samples; equipment usage and calibration (application equipment, scales, balances, weather station, spectrophotometers, freezers, refrigerators, and other necessary equipment including the water treatment system for replenishing water).

Interpretation of results:

- Complete data auditing for fish biomass, growth, fecundity, survival, residues in tissues, zooplankton and benthic community structure; phytoplankton and periphyton (analysis of chlorophyll); community metabolism parameters; residue data; water chemistry data; climatological data; and statistical analysis.

Terrestrial Field Studies

Terrestrial field studies are utilized to study the dissipation and movement of chemical residues in the field. The main focus of these studies is the conduct of analytical chemistry procedures to measure either the parent chemical or degradates. Additional quality parameters that should be reviewed include:

- Climatic data including precipitation and temperature; irrigation amounts; depth to water table; seasonal variability; slope of test plots; soil temperature data; techniques and times of planting and harvesting; stages of crop and test development; application rates; sampling (soil core extraction through chemical analysis); appropriate sampling intervals; and complete chemical analysis.

Although it can be difficult to monitor and control a terrestrial field study (or any field study), lack of attention to every detail can invalidate an entire study. For example, experience has demonstrated that if some test plots are not properly irrigated, the analytical data for concentrations of test substances sampled from these areas will be inconsistent with the data for irrigated test plots.

Protocol changes are common in field studies in order to accommodate weather conditions and often equipment availability. Consequently, changes in application and sampling schedules can occur frequently. In addition, storage stability samples (sampling to analysis time or extraction through analysis time) should adequately reflect the storage conditions relative to the study design. Stability data from time sequences that do not adequately represent the conditions of the study may be rejected.

The study director is required to document any planned or unplanned protocol event, any inconsistencies in data collection, and any deviations from SOPs in the raw data. The study director is the single point of control for these large-scale studies. Although it may seem a daunting task, the study director must be constantly aware of all activities and their impact on study quality and integrity.

Statistical Analysis

The proper statistical treatment (see Chapters 2, 10 and 19) of study data is as important as the data themselves. If data are not properly measured and analyzed, false-positive or false-negative conclusions may result, invalidating a study despite all the GLP precautionary measures previously described. Qualitative identification, quantitative accuracy, representativeness, completeness, and comparability of what is being measured must be known with confidence approaching certainty if the data are to have any use whatsoever (Taylor, 1990).

Although it is not the direct responsibility of the quality assurance staff to ensure proper statistical treatment, some knowledge of key points in statistical design and analysis is useful. For example, there are some basic questions a quality assurance professional might ask when reviewing a study that utilized regression analysis:

- Power—has a sufficient amount of data been collected to attain a desired level of precision or confidence?
- Assumptions—have the assumptions for the particular statistical treatment of data been met? For example, have the basic assumptions about existence, independence, linearity, homoscedasticity, and normality been met?
- Transformations—if assumptions have not been met, have the data been transformed in an attempt to meet the assumptions by normalizing, stabilizing variance, or performing diagnostic (residual) or outlier tests on the data?
- Analysis—have anticipated problems with the statistical treatment been considered, such as lack of fit, collinearity, significance testing of large residuals, use of leverage statistics and influence statistics, analysis of covariance, or use of weighted least squares (Kleinbaum et al., 1988)?
- Conclusions—based on the statistical evaluation, do the conclusions drawn accurately and confidently reflect the data and meet the objectives stated in the study protocol?

The same general exercise can be utilized for other statistical reviews, such as analysis of variance (ANOVA). Whatever statistical treatment is used, the effects of confounding variables should always be taken into consideration and controlled, wherever possible.

During protocol development for large-scale GLP studies or more complex laboratory studies, it may be useful to consult one or more statisticians so that appropriate design and power are taken into consideration. These professionals should also be involved in data testing and analysis during and at the conclusion of a study. Nonetheless, it remains the responsibility of the study director, technical staff, and the quality assurance unit to maintain integrity throughout the entire data collection and analysis phase of a GLP study. From a regulatory standpoint, GLP studies in their entirety are the responsibility of the study director.

The amount of documentation specified in the regulations and the practical application thereof facilitate organized and systematic auditing of a study. An independent auditor should always be able to reconstruct a GLP study from beginning to end with the available documentation. As mentioned earlier in this chapter, any possible questions from a reviewing agency should be anticipated during GLP project organization and protocol development.

Even after the study has been initiated, the study director must constantly monitor its progress in case the study needs to be amended or modified while in progress. Whatever the course of study events, data and record keeping must be extremely well documented and reproducible. It is expected that GLP regulations and quality practices with respect to toxicological data submissions will most likely increase, not diminish, in the foreseeable future.

FUTURE FOCUS

There are many ways to accomplish similar data quality goals or objectives, and agencies and organizations occasionally enter into conflict because the approaches of one agency are either nonstandard or foreign to the other and are therefore not recognized (Williams, 1989). These situations have initiated the development of many domestic and international organizations to harmonize quality objectives for study management.

For example, in the United States there has been an effort to reach a consensus on measurement methodology by such standardization organizations as the American Society for Testing Materials (ASTM) and the Association of Official Analytical Chemists (AOAC). However, as with any guideline, the implementation of standardized quality practices has been left up to individual organizations, similar to implementation of GLP objectives within each organization.

As quality assurance programs continue to grow domestically and abroad, many companies have begun to restructure management to incorporate quality assurance positions on a much higher level. In many corporations, management positions for quality assurance professionals have noticeably increased over the last few years.

In addition to elevation of quality assurance positions to senior management status, quality improvement seems to be emerging from the education sectors. Many colleges and universities are implementing quality objectives in their administrations and offering quality-related courses and degrees to full-time students (Axland, 1992). Education in quality practices is only one step toward a planned process to meet all of the GLP objectives. For full and consistent compliance with the regulations, total quality management (TQM) processes should be implemented by testing sponsors and facilities as described by Johnson (1993). TQM is essentially a process of system and performance review and continual improvement of laboratory and testing operations (see Terminology).

Societies directly involved in quality issues related to toxicological data quality continue to make their presence known. Within the United States there are many societies including the Regulatory Affairs Professional Society (RAPS) and the Society of Quality Assurance (SQA). The International Society of Quality Assurance (ISQA), a more recently established organization, focuses on addressing international harmonization of GLP practices. These and other organizations are extremely helpful in facilitating the exchange of quality practices and new ideas.

Bilateral and multilateral agreements, or MOUS as they are commonly called (memorandum of understanding), continue to be a mechanism by which countries can mutually agree to accept data in support of product registration (Snyder, 1992). As mentioned previously in this chapter, the EC-GLP directives are based on OECD GLP principles and practices, and although one internationally recognized set of GLP standards is not currently available, most international GLP programs, including that of Japan, are quite similar to the U.S. programs. The major problems facing harmonization are mainly regulatory or administrative in nature, rather than being an issue of data quality or acceptability. Because of this, continued emphasis and cooperation are needed for multilateral GLP agreement.

Technological improvements such as electronic data systems to expedite data submissions while simultaneously improving the quality of review have already been discussed at the FDA. More advances in computerized systems within both the EPA and FDA are likely future considerations for improving the review process and GLP enforcement policy. More sophisticated computerization will also expedite products through the regulatory process. These data systems must be accessible, well managed and presented, and easily analyzed.

Since January 1993, it has become increasingly difficult to market new products and services and protect current markets without having recognized certified quality systems in place. With international implications, ISO 9000 is a process whereby a company or organization can register its quality systems to conform to a standard of worldwide recognition. A principal factor in the performance of an organization is the quality of its products or services (ISO, 1992). Registrants and testing laboratories should consider pursuing ISO 9000 or other internationally recognized certification and accreditation programs (e.g., EN45000) to compete effectively in a global economy.

Elements of these accreditations and certifications include, but are not limited to, a quality statement, organizational responsibilities, quality policies, customer objectives, work procedures (process steps), skill analysis and training, quality measures, analysis, reporting, and corrective action. Most of these elements are already included in the GLP requirements.

It is clear that good laboratory practice regulations continue to evolve toward greater sophistication and refinement together with technological advances in study design and testing methodologies. Constant advances in quality activities through cooperative efforts, automation, and statistical control are necessary to produce data that present and future generations can rely upon to safeguard their biological and ecological systems.

LITERATURE CITED

Axland, S.: A higher degree of quality. *Quality Prog.*, 25: 41–61, 1992.

Bascietto, J.; Hinkley, D.; Plafkin, J.; and Slimak, M.: Ecotoxicity and ecological risk assessment: regulatory applications at EPA. *Environ. Sci. Technol.*, 24:10–15, 1990.

Broad, R. D., and Dent, N. J.: An introduction to good laboratory practice. *Good Laboratory and Clinical Practices* (GLP), edited by P. A. Carson, and N. J. Dent, pp. 3–15. Oxford, UK: Heinemann Newnes (Butterworth Heinemann Ltd.), 1990.

Carson, P. A., and Dent, N. J. (eds.).: *Good Laboratory and Clinical Practices.* Oxford, UK: Heinemann Newnes (Butterworth Heinemann Ltd.), 1990.

Hochman, J. H., and Garner, W. F.: Reports study results. *Good Laboratory Practice Standards, Applications for Field and Laboratory Studies*, edited by W. Y. Garner, M. S. Barge, J. P. Ussary, pp. 127–139. Washington, DC: American Chemical Society, 1992.

ISO 9000: International Standards for Quality Management. ISBN 92-67-10172-2, Geneva: ISO, 1992.

Johnson, R. S.: TQM: leadership for the quality transformation. *Quality Prog.*, 26:73–75, 1993.

Kleinbaum, D. G.; Kupper, L. L.; and Muller, K. E.: *Applied Regression Analysis and Other Multivariable Methods.* Boston: PWS-Kent, 1988.

Lucas, S.: The FDA. Millbrae, CA: Celestial Arts, 1978.

OECD: Good Laboratory Practice in the Testing of Chemicals, A Final Report of the Group of Experts on Good Laboratory Practice. ISBN 92-64-12367-9. Paris: Director of Information, 1982.

OECD: Issues for Agreement: OECD Test Guidelines, Principles for Laboratory Practice and Mutual Acceptance of Data. OECD Survey Report, 1981.

Snyder, F. G.: Harmonization and prospects for the future. *Good Laboratory Practice Standards, Applications for Field and Laboratory Studies*, edited by W. Y. Garner, M. S. Barge, and J. P. Ussary, pp. 419–430. Washington, DC: American Chemical Society, 1992.

Taylor, J. K.: *A Collection of Abstracts of Selected Publications Related to Quality Assurance of Chemical Measurements.* NBSIR 86-3352. Gaithersburg, MD: U.S. Department of Standards, National Bureau of Standards, 1986.

Taylor, J. K.: *Quality Assurance of Chemical Measurements.* Chelsea, MI: Lewis, 1987.

Taylor, J. K.: Statistical Treatment for Data Analysis. Chelsea, MI: Lewis, 1990.

U.K. Department of Health: Good Laboratory Practice; Application of GLP Principles to Field Studies. Advisory Leaflet Number 2. United Kingdom Compliance Programme, London, 1990.

U.S. EPA: Toxic substance control; good laboratory practice standards. *Fed. Regist.*, 48:53922–53944, 1983a.

U.S. EPA: Pesticide programs; good laboratory practice standards. *Fed. Regist.*, 48:53946–53969, 1983b.

U.S. EPA: Guidance Document for Preparation of Combined Work/Quality Assurance Project Plans for Environmental Monitoring. OWRS QA-1, EPA 600/4-83-004, Washington, DC: Office of Water Regulations and Standards, 1984.

U.S. EPA: Hazard Evaluation Division Standard Evaluation Procedure, Aerobic Soil Metabolism Studies. Office of Pesticide Programs, EPA-540/9-85-015, 1985a.

U.S. EPA: Hazard Evaluation Division Standard Evaluation Procedure, Acute Toxicity Test for Estuarine and Marine Organisms. Office of Pesticide Programs, EPA-540/9-85-011, 1985b.

U.S. EPA: 23. Hazard Evaluation Division Standard Evaluation Procedure, Acute Toxicity for Freshwater Fish. Office of Pesticide Programs, EPA-540/9-85-006, 1985c.

U.S. EPA: Hazard Evaluation Division Standard Evaluation Procedure, Aqueous Photolysis Studies. Office of Pesticide Programs, EPA-540/9-85-014, 1985d.

U.S. EPA: Hazard Evaluation Division Standard Evaluation Procedure, Hydrolysis Studies. Office of Pesticide Programs, EPA-540/9-85-013, 1985e.

U.S. EPA: Hazard Evaluation Division Standard Evaluation Procedure, Soil Column Leaching. Office of Pesticide Programs, EPA-540/9-85-017, 1985f.

U.S. EPA: Hazard Evaluation Division Standard Evaluation Procedure, Non-Target Plants: Growth and Reproduction of Aquatic Plants, Tiers I and II. Office of Pesticide Programs, EPA 540/9-86-134, 1986a.

U.S. EPA: "PR Notice 86-5, Notice to Producers, Formulators, Distributors, and Registrants," U.S. Environmental Protection Agency. Washington, DC: Government Printing Office, July 29, 1986b.

U.S. EPA: Hazard Evaluation Division Technical Guidance Document, Aquatic Mesocosm Tests to Support Pesticide Registrations. OPP/HED (TS-769C), EPA 540/09-88/035, 1988.

U.S. EPA: Hazard Evaluation Division Standard Evaluation Procedure, Analytical Methods. Office of Pesticide Programs, EPA-540/09-89-062, 1989a.

U.S. EPA: Hazard Evaluation Division Standard Evaluation Procedure, Anaerobic Soil Metabolism Studies. Office of Pesticide Programs, EPA-540/09-88-104, 1989b.

U.S. EPA: Enforcement Response Policy for the Federal Insecticide, Fungicide, and Rodenticide Act, Good Laboratory Practice (GLP) Regulations. Office of Compliance Monitoring, 1991.

U.S. EPA: Pesticide Reregistration/Rejection Rate Analysis/Residue Chemistry. Pesticides and Toxic Substances (H-7508W), EPA 738-R-92-001, 1992a.

U.S. EPA: Good Laboratory Practice Standards. 40 CFR, Part 160, 1992b edition.

U.S. FDA: Good laboratory practice regulations, proposed rule. *Fed. Regist.*, 41:51206, 1976.

U.S. FDA: Good laboratory practice regulations, final rule. *Fed. Regist.*, 43:59986, 1978.

U.S. FDA: Environmental Assessment Technical Assistance Handbook. Report No. FDA/CFSAN-87/30, PB87 175345, 1987a.

U.S. FDA: Good Laboratory Practice Regulations, final rule. *Fed. Regist.*, 52:33762–33768, 1987b.

U.S. FDA: Good Laboratory Practice for Nonclinical Laboratory Studies. 21 CFR, Part 58, 1992 edition.

Williams, L. R.: Harmonization of quality assurance—an interagency perspective. Aquatic Toxicology and Hazard Assessment, 12th Volume, ASTM STP 1027, edited by U. M. Cowgill and L. R. Williams, pp. 11–18. Philadelphia: American Society for Testing and Materials, 1989.

SUPPLEMENTAL READING

Advisory/Policies

U.S. FDA: Requirements of Laws and Regulations Enforced by the U.S. Food and Drug Administration, U.S. Department of Health and Human Services Superintendent of Documents, U.S. Government Printing Office, Washington, DC; GPO 1989 0-236-682; DHHS Publication Number (FDA) 89-115.

U.S. EPA: FIFRA Good Laboratory Practice Standards (GLPs) Regulations Questions and Answers Document. Office of Compliance Monitoring: Office of Prevention, Pesticides, and Toxic Substances, 1992.

U.S. EPA: GLP Regulations Advisories, 1–52, issued by the EPA Policy and Grants Division. Laboratory Data Integrity Assurance Division, U.S. Environmental Protection Agency, 1992.

Articles

Chirsitan, M. C.: Study-planning with science and compliance as the driving forces. *J. Am. Coll. Toxicol.*, 10(3): 381–386, 1991.

Goeke, J. E.: Improving the study process: reviewing problems and making improvements. *J. Am. Coll. Toxicol.*, 10(3):397–402, 1991.

Siconolfi, R. M.: Applications of the principles of total quality will improve science and compliance. *J. Am. Coll. Toxicol.*, 10(3):371–376, 1991.

Snyder, F. G., and King, H. M.: Science versus compliance equals less than total quality. *J. Am. Coll. Toxicol.*, 10(3): 403–405, 1991.

Books

Goeke, J. E.: Quality assurance for academic toxicology research. Principles of Research Data Audit, edited by A. E. Shamoo, chapter 12. New York: Breach Science Publishers, 1989.

News Articles

Hamilton, D. P.: In the trenches, doubts about scientific integrity. *Science*, 255:1636, 1992.

Kauffman, J. M., and Weiler, E. D.: Application of ISO 9002 and FDA's good manufacturing practices to general chemical manufacturing. Quality Assurance: Good Practice, Regulation, and Law 1(3):213–216, 1992.

U.S. EPA: FIFRA good laboratory practice complaint settled for $132,000. *Pesticide and Chemical News Weekly*, January 29:3, 1992.

Chapter 12

IMMUNOTOXICOLOGY IN FISH

D. P. Anderson and *M. G. Zeeman*

INTRODUCTION—JUSTIFICATIONS AND INVESTIGATIONAL PROCEDURES

In fish as in other animals, the immune system functions in resistance against diseases and in protection against neoplastic cells. The cells in the immune system providing these functions are leukocytes arising from stem cells within the anterior kidney, spleen, or other immunopoietic sites, where they proliferate and differentiate before giving rise to functionally mature cells. The majority of leukocytes are divided into granulocytes, lymphocytes, monocytes, and macrophages. The interaction of environmental contaminants (e.g., pesticides) or drugs (e.g., antibiotics) with lymphoid tissue may alter the functions and balances of the immune system and result in such undesirable effects as (1) immunosuppression, (2) uncontrolled cell proliferation, (3) alterations of host defense mechanisms against pathogens, and (4) neoplasia.

Recent methods for toxicologic assessment have implicated the immune system as a target of toxicant insult after exposure to some chemicals and drugs. Blood and tissue samples from fish exposed to sublethal concentrations of toxic agents show aberrations in hematopoietic and immunopoietic parameters and indicate impaired health of fish (Anderson 1990; Weeks et al. 1992). Indications of potential immune alterations may include changes in lymphoid organ weights; changes in leukocyte differential counts; increased susceptibility of fish to infections by opportunistic organisms, including latent viral and bacterial infections; and increased risk of tumors and cancer. These effects can occur at contaminant concentrations below those that produce overt toxicity such as gross morphological aberrations or death.

The immune system is an important target for studying the toxicant-induced effects of chemical exposure for the following reasons: immunocompetent cells are required for host resistance and therefore exposure to immunotoxicants can result in increased susceptibility to disease, and incompetent cells require continued proliferation/differentiation for renewal and are thus sensitive to chemical agents that affect proliferation. The biology of the immune system of most animals is well understood and thus mechanisms by which toxicants are immunoalterative can be elucidated. In addition, functional assessment or enumeration of leukocytes can often be achieved with nonlethal sampling using small volumes of blood or lymphoid tissue.

Fish live in an aqueous environment that is unfortunately conducive to the maintenance of some disease-causing organisms. Fish diseases can cause large losses of animals, compromise results of aquatic toxicity tests, and waste time, money, and materials. When fish are held in close proximity to each other or in great numbers, as in modern fish culture, the probability of outbreaks of diseases increases. Indeed, the causes of this situation are complicated because many fish carry living pathogens or potential pathogens in their tissues and organs throughout their lives. The balances between parasites, bacteria, and viruses and the fish themselves are also dependent on the characteristics of the environmental confinements. For example, culturists know that rainbow trout survive best at temperatures of 8–14°C in hatchery ponds. In most hatcheries, as the temperature approaches 17°C, the numbers of the skin parasite *Ichthyophthirius* sp. increase, overwhelming the host's resistance. This overload of protozoan parasites can eventually kill the host.

The immune responses of warm-blooded animals are extremely important in disease prevention, in maintaining good health, and even in the normal surveillance for destroying cancerous cells and tumors. The immune systems in fish perform much the same functions for these aquatic, cold-blooded animals. It is also well known that natural waters and sediment where fish live can be significantly contaminated by pesticides and other industrial chemicals, affecting fish immune systems.

Demonstrating the effects of a toxicant on a fish population is complicated by the diversity of the fish species and individual adaptations to their habitats. Immunological assays that show the health of the fish or the influence of contaminants have to be modified to the particular specialization and ecological niche of each fish. For the most part, the individual species live intimately with the environment, their physiology being regulated by the ambient temperatures, light cycles, food availability, and predators. Even seasonality must be considered when sampling for the effect of toxicants on the immune response of fish (Zapata et al., 1992). Tana and Nikunen (1986), working with pulp mill effluents, showed that the time of the year strongly influences the hematological parameters of the immune response in rainbow trout. During the time of smolt, when salmon are migrating to the sea, the immune response may be depressed, as noted by changes in the specific immune response as indicated by the reduction of plaque-forming cells (PFCs) in the spleen (Maule et al., 1987).

Aquaculturists and fish farmers use chemicals and antibiotics to maintain healthy animals and reduce disease losses (Stoskopf, 1993). Although there is a great benefit to fish culture productivity, these additives may be toxic and immunotoxic in cumulative doses (Grondel et al., 1987). The consequences of a fish immune system compromised by chemical contaminants and even therapeutics in fish culture can be substantial. These consequences have been discussed in several reviews (Zeeman and Brindley, 1981; Walczak, 1985; Fries, 1986). Figure 1 shows schematically some of the substances that may affect fish health. Treatments of fish with antibiotics, vaccines, or chemicals that kill or rid the animals of pathogenic microorganisms can be beneficial, if the agents are administered at appropriate doses. Even low levels of contaminants may harmfully affect some physiological aspects of the animal; when the immune protection is suppressed, opportunities for the entry of disease-causing microorganisms are opened.

The literature on the subject of aquatic immunotoxicology is rapidly expanding as governmental, academic, and private organizations recognize the importance of supporting programs for clean, habitable environments (Zeeman and Brindley, 1981; Adams, 1990). Fish immunology texts and laboratory manuals are becoming available (Anderson, 1974; Anderson and Hennessen, 1981; van Muiswinkel and Cooper, 1982; Manning and Tatner, 1985; Stolen et al., 1986; Austin and Austin, 1989; Stolen et al., 1990; Stolen et al., 1992). New instrumentation and methods for sophisticated immunological assays and microscopic examination enable presentation of results with greater validity. References in the following tables were selected in this chapter to emphasize the possibilities and adaptations of present techniques. Readers are advised to refer to current journals, review articles, and literature searches for recent developments.

TERMINOLOGY

The reader may be helped by the following current definitions of functional components and products of the hematopoietic (blood-originating) and immunopoietic (defense-originating) organs and cells in fish. Many of the terms are similar to and have been taken from those used in veterinary and medical mammalian models.

Antibody. A specific protein molecule produced by the plasma cells of fish in response to an antigen. Humoral or circulatory antibody in the blood of fish forms complexes with antigens as a first step in inactivating invasive pathogens.

Antigen. A complex chemical or particle that is foreign to the body of the fish and can stimulate the production of antibodies and combine specifically with them. A virus or bacterium is composed of many antigens.

Background, natural or nonspecific antibodies. Complement-dependent antibody-like molecules involved in defense. Sera from normal fish often show low levels of destructive activity toward bacterial agents and other cells such as heterologous red blood cells.

Bacterin. Killed bacterial cells or products that are used to immunize an animal against a specific disease. Often the words bacterin and vaccine are used synonymously; however, in the strict sense "to vaccinate" means to expose the animal to vaccinia, the virus causing cowpox. "Vaccine" is sometimes limited to describing viral immunogens.

Complement (C'). A multiple group of proteins in fish plasma that are triggered in a cascade reaction with antibody-antigen complexes or other activators to lyse bacteria and other foreign cells or toxic products.

Immune. Derived from the Latin *immunire*, meaning to defend or strengthen. Fish fortify and protect themselves against disease-causing pathogens by the specific immune response.

Immunogen. An antigen that induces a specific immune response resulting in protection against a disease. This is a narrowing definition for an antigen.

Immunosuppression. A reduction in an animal's defense mechanism against disease. The immune response may be suppressed by cortisol, drugs, or environmental contaminants, leaving the animal more susceptible to pathogenic microorganisms.

Nonspecific defense mechanisms (NSDMs). Lines of defense against microorganisms that are already in place in fish, without stimulation of an antigen. The mucous layer, hardened scales, blood lysozyme, neutrophils, and phagocytic cells are part of the NSDM, as is natural antibody. This system can be elevated in its activity by antigen exposure, but it remains generalized and the responses are not directed toward specific agents or antigens.

Opsonin. Factors in blood that assist in phagocytosis of microorganisms. Some complement molecules may opsonize or increase engulfment rates by phagocytic cells and destruction of pathogenic agents. Antibody can also act as an opsonin.

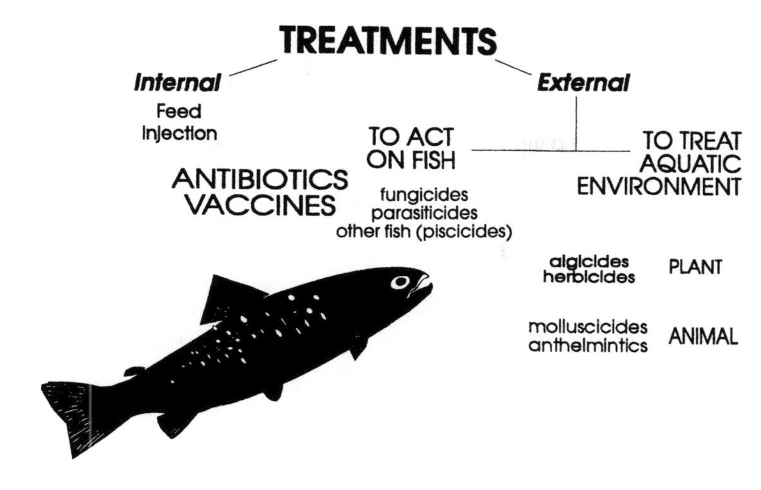

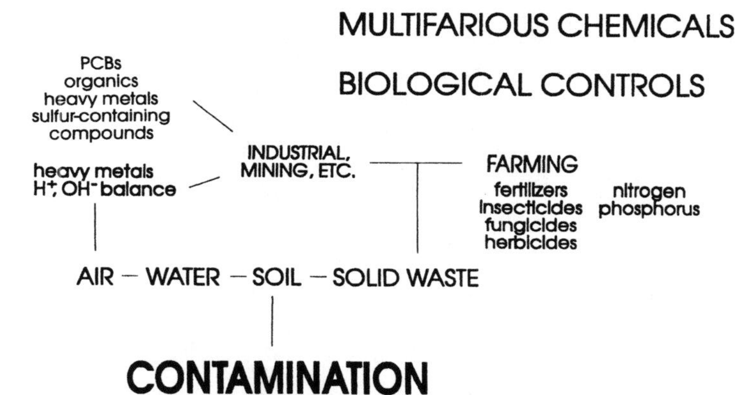

Figure 1. In culture or in the field fish are exposed to many potential immunomodulators. In hatcheries, treatments with drugs or chemicals to prevent diseases may adversely suppress the specific immune response if given in high doses. Vaccines are immunostimulating by their nature and can be used successfully to prevent diseases, increasing survival after immunization. Contaminants such as metals from mining operations or from industrial waste are known to cause immunosuppression. Interactions of the agents and changes in the environment may potentiate toxicities and make fish more susceptible to diseases. Acid rain changes the pH of lakes, increasing the toxicity of aluminum ions to fish.

Phagocytic cells. Blood monocytes, tissue macrophages, and most neutrophils. These cells have the ability to engulf foreign particles and destroy or break them up in cytoplasmic vacuoles by digestive enzymes. Derived from Latin, the word means "eating cell."

Pinocytosis. Engulfment and ingestion of soluble fluid-containing vesicles by phagocytic cells.

Specific immune response (SIR). A reaction by the fish to an individual antigen, resulting in the generation of plasma cells producing antibody, and the activation of cells of the immune system, such as neutrophils, phagocytes, and cytotoxic killer cells to destroy pathogenic agents. The SIR includes other molecules of cell activation and defense in the immunoglobulin supergene family, for instance, cytokine molecules that enable communication between cells and receptors on T lymphocytes that recognize antigens.

The following descriptions of fish cells pertain to rainbow trout (*Oncorhynchus mykiss*). Cells from other fish species may have some variation in size and function.

Erythrocytes (red blood cells, RBCs). Circulatory cells containing hemoglobin that carries oxygen to body tissues. Oval in shape, 7×12 μm, and nucleated, the RBCs are good gauges for size comparisons with other circulatory cells in microscopic studies.

Granulocytes. Large mononuclear leukocytes with cytoplasmic granules. In rainbow trout granulocytes, most cytoplasmic particles stain blue, indicating a basophilic nature (basophils). Warm-water fish (e.g., bluegill *Lepomis macrochirus*) often have red-staining granulocytes (eosinophils) in the kidney.

Leukocytes (white blood cells, WBCs). A diverse group of colorless, circulatory cells including lymphocytes, granulocytes, monocytes, and their precursors. Most are involved in defensive mechanisms.

Lymphocytes. Round leukocytes, usually 8 μm in diameter. Subpopulations include B lymphocytes that mature into plasma cells, producing specific antibody and T lymphocytes that regulate the antibody-producing B lymphocytes and other immune functions. T lymphocytes may be further divided into T helper (Th) and T suppressor (Ts) lymphocytes. The narrow band of the lymphocytes' cytoplasm usually stains dark blue with Wright's stain.

Macrophages. Large, slow-moving, or mostly stationary, mononuclear, phagocytic, leukocytes whose activity includes the engulfment and destruction of foreign particles. Another main activity of these phagocytes is to break down foreign particles into fragments to present information to lymphocytes for the initiation of a specific immune response. The term antigen-presenting cell (APC) is used for this characteristic. The phagocytic cells include the following:

Kupffer cells: Large, stationary phagocytic cells in the liver.

Dendritic cells: Fibrous phagocytic cells found in the spleen and other tissues that trap foreign particles.

Langerhans cells: Highly branched phagocytic cells of the epidermal and subdermal areas.

Monocytes: Circulatory cells with phagocytic and APC activities. These circular size cells are about 9–25 μm in diameter.

Natural cytotoxic cells or killer cells. Lymphocyte-like immunopoietic and circulatory cells that attack and destroy foreign cells, particularly tumor cells.

Neutrophils. Circulatory granulocytes with a diameter of 13 μm and, usually, a clear cytoplasm. Granules in neutrophils require special staining or detection by electron microscopy as they are often not visible with commonly used blood stains. The nucleus of a neutrophil is often segmented, hence neutrophils are termed polymorphonuclear leukocytes (PMNs). The cell carries destructive enzymes and reactive oxygen species (O_2^-, H_2O_2, OH^-) that can lyse bacteria. Neutrophils are sometimes phagocytic but rarely function as APCs.

Plaque-forming cell (PFC). An antibody-producing cell that can be identified in the Jerne or hemolytic plaque assay by the clear plaque that forms around it as the diffusing antibody lyses indicator cells. This plasma cell is a mature B cell.

Thrombocytes. Small circulatory cells that produce clotting factors. Usually the fragile thrombocytes appear spindle shaped when freshly stained in a blood slide. The nuclei resemble those of erythrocytes.

A flow diagram of the derivation and interrelations of the blood cells from a rainbow trout is shown in Figure 2. These pathways are somewhat similar to the mammalian models, when the anterior kidney is substituted for bone marrow as the hematopoietic and immunopoietic tissue.

The immune system is composed of many cell types and subpopulations (e.g., granulocytes, monocytes, macrophages, lymphocytes, killer cells) that are found in circulation or in organized lymphoid tissues of the kidney, spleen, and thymus. The system is in a constant state of renewal and maturation. It is a regulated network of cells that must discriminate self from nonself and react to nonself with rapid defense responses.

FISH IMMUNOLOGY—DEFENSIVE RESPONSES

The defensive system in fish functions can be loosely divided into an "in place" or immediate response system and a delayed, refined system. A *nonspecific defense mechanism* does not require prior contact with the inducing agent to elicit a response and lacks specificity, and a *specific immune response* is directed against a specific agent to which the organism has been previously sensitized. Penetrance of the first defense barriers by invading microorganisms may result in nonspecific reactions involving such actions as activating mononuclear phagocytes (i.e., monocytes and tissue macrophages) and granulocytes. This activation may be followed by specific immune responses involving antibody production and the induction of effector lymphocytes, responding through mediators to seek out and destroy the agent. Both antibody-producing lymphocyte B cells and lymphocyte T cells can be triggered by mechanisms in which the foreign antigen is recognized, digested, and presented by monocytes and macrophages (i.e., APCs). Following antigen-induced activation, B cells proliferate and differentiate into plasma cells that produce specific antibodies (e.g., immunoglobulins, Igs). The antibodies enter the circulation, where they bind foreign material and either neutralize, lyse, or facilitate phagocytosis of the agent.

Two types of phagocytic leukocytes—the polymorphonuclear phagocyte (PMN) or granulocyte and the mononuclear phagocyte, a macrophage—are involved in nonspecific mechanisms of resistance. In rainbow trout, both cell types originate from progenitor cells in the anterior kidney and enter the bloodstream. PMNs traverse blood vessels and provide the first line of defense against infectious agents. Both PMNs and macrophages exhibit phagocytic activity toward foreign material, especially in the presence of specific opsonic antibodies and complement, and can destroy most microorganisms. Monocytes and macrophages can be enhanced to an activated state of bactericidal or tumoricidal activity by soluble products (opsonins, cytokines, and lymphokines) produced by lymphocytes that are sensitized to a specific microbial agent.

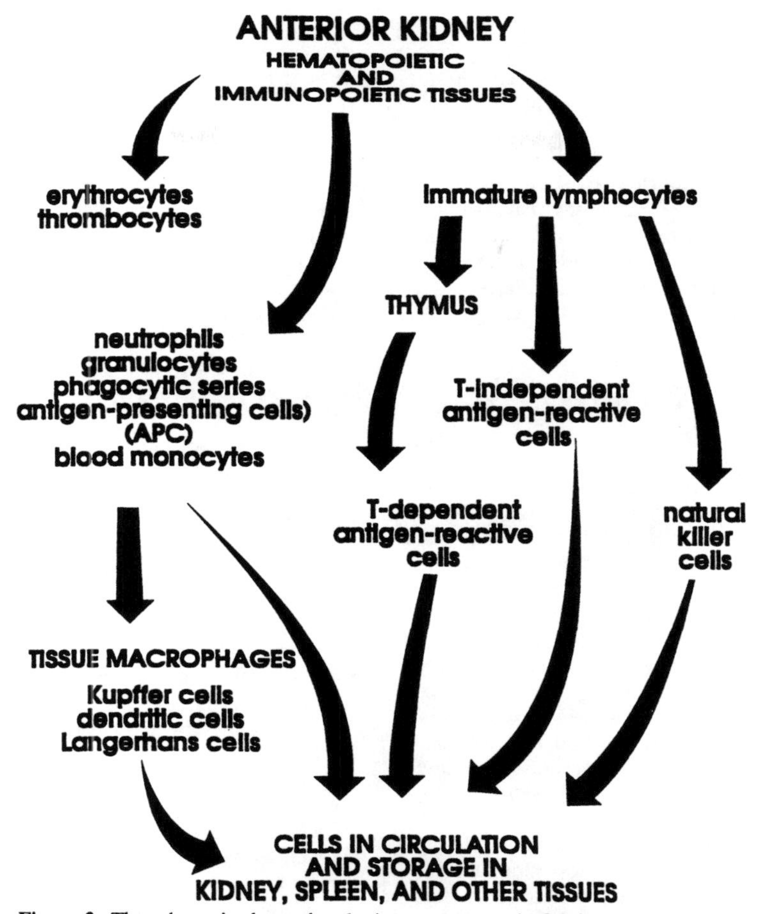

Figure 2. The schematic shows that the immune system in fish is a dynamic mechanism. The cells migrate and tissues adapt to the physiological and protective needs of the animal. After vaccination, rapid cell division occurs, generating effector cells such as antibody-producing cells. Precursor and immature cells generated in the kidney mature and migrate to the spleen and other temporary storage sites until the cells are employed in fending off pathogens, recycled or destroyed. Fish exposed to contaminants may show reduction of some effector mechanisms; for example, the presence of polynuclear aromatic hydrocarbons may reduce macrophage activity.

Pluripotent stem cells in the trout kidney are a unique group of cells that have renewal capacity. Indeed, these stem cells are found in the yolk of the embryo and later in the anterior kidney of trout. They differentiate along different pathways, giving rise to the myeloid series cells (e.g., granulocytes) or the lymphoid series. Cell maturation occurs mostly in the anterior kidney and spleen.

Another group of mononuclear cells present in the blood are the killer cells (K cells). These may be similar in morphology to the lymphocytes. Natural killer cells (NK cells) act spontaneously to destroy invasive microorganisms or neoplasia; cytotoxic killer cells may possess a receptor for a region of the immunoglobulin molecule. Antibody, complement, or other opsonins may assist the cytotoxic killer cells in lysing their target. Their functions may involve direct contact with and destruction of offending agents or aberrant tumor cells by specific mechanisms (Evans et al., 1987).

Whether an antigen induces a strong cellular response or is more centered on antibody production may depend on the physical and chemical nature of the antigen, the mode of presentation of the antigen to lymphocytes, the localization pattern of the antigen within lymphoid tissue, and the configuration of the antigen. Antigens that produce strong cellular responses include tissue-associated antigens, chemical agents, and drugs that bind covalently to proteins and antigenic determinants or persistent intracellular microorganisms. Antigens such as bacterial lipopolysaccharides may be more active in stimulating the production of specific antibody; the T cells are not so involved.

Induction of strong cellular response proceeds when immature cells differentiate and ultimately give rise to cells that are responsible for effector functions. Cells can further differentiate into effector cells endowed with cytotoxic potential, helper cells that facilitate antibody responses by B lymphocytes and aid in some T lymphocyte responses, capable of inhibiting both T and B cell responses. The humoral factors or cytokines elaborated by these cells that regulate interactively are termed lymphokines; they include interferons, chemotactic factors, macrophage activation factors, interleukins, and B cell growth factors.

Nonspecific Defense Mechanisms

Fish live more intimately with their environment than terrestrial mammals. The epidermal surfaces of fish are composed of living cells rather than hardened, dead corneal layers of skin, as is the case in humans and most other terrestrial animals. The aquatic surroundings influence how fish defend themselves against invasive microorganisms. The specific immune response of most fish is believed not to be as highly evolved as that of mammals. Part of the reason may be that most fish are relatively short-lived, and tissue specialization does not have time to develop. Another reason for a slower defensive response is the physiological limits of low temperatures in cold-water fish. Microorganisms coming in contact with a fish may first be entrapped by the viscous mucus, which contains lytic enzymes. As the invaders are destroyed by enzymatic digestion in the mucous layer, the debris is harmlessly sloughed off in the water. The next barrier is the cells of the epidermis. As microscopic investigations show, a thin layer of living cells often covers the scales. If the invader enters through the skin at a point where there is a scale, unless there is lateral movement by the invader, this plate-like layer is almost impenetrable.

Fish are vulnerable in many surface areas of the body that have no scales or a thin mucus. Fortunately, there are numerous other nonspecific defense mechanisms prepared to fight the invasive agents. Tissue macrophages in the subdermal layers phagocytize invading bacteria into their intracellular vacuoles or lysosomes and release destructive enzymes for digestion of the pathogen. Neutrophils in the circulation and residing in tissues are capable of moving into contact with offending organisms and directionally releasing reactive oxygen species to destroy foreign cells. In addition, these defensive cells set off an inflammatory response that draws more phagocytic cells, including neutrophils, into the area of invasion. The inflammation response in mammals is characterized by heat, pain, swelling, and redness. These signs are less distinct in most fish; however, a 3°C rise in temperature has been described in sunfish (*L. macrochirus*) after injection of killed *Aeromonas hydrophila* (Reynolds, 1977).

Soluble signals and messengers of inflammatory responses are stimulated by the initial cells involved in contact with the invader. Mammalian immunologists have demonstrated complex schemes showing cellular interactions of numerous cytokines. Soluble messengers are released from macrophages or other leukocytes to signal to other cells. For example, neutrophils in the area of an injury can signal by release of cytokines to other cells to migrate to the infraction and act in the animal's defense. Counterparts have been described for fish (Smith and Braun-Nesje, 1982). The cytokine interleukin-2 (IL-2), which is known to stimulate T lymphocyte activity, was putatively described in fish (Graham and Secombes, 1990). Medical immunologists describe at least 11 other interleukins that carry specific intercellular messages; presumably many of these also occur in fish.

Interferons (IFNs) are also soluble cell messenger molecules, produced most specifically in the defense of animals against viral infections. They have been described in fish (in vivo) and cell culture experiments (in vitro). De Kinkelin et al. (1982) showed that rainbow trout (*O. mykiss*) and carp (*Cyprinus carpio*) produce IFN after injection with rhabdoviruses that cause diseases. Different kinds of IFN may be produced by many types of cells such as leukocytes or fibroblasts in the mammalian system; presumably this is also true in cultivation of tissues from fish. The infected tissue cells signal alarms by IFN molecules to adjacent cells and help the uninfected cells prepare a barrier against the spread of a virus. The induced protection is nonspecific; it protects the stimulated cells against the inducing virus as well as heterologous viruses. IFN may cause interference problems when scientists use fish cell lines to grow viruses. As a purposeful viral infection stimulates IFN in tissue culture, there is a feedback and suppression of further viral proliferation.

C-reactive protein (CRP) and heat shock protein (HSP) (see Chapter 17 for discussion of stress proteins) are among other soluble factors that have been described in fish sera. The level of these glycoproteins usually rises during inflammation; therefore, they are considered for use as health indicators (Fletcher, 1981). C-reactive protein was named when it was first described in mammals for its binding affinity with pneumomococcus C-polysaccharide. In rainbow trout, CRP has two subunit molecules that combine to give a molecular mass of 81,400 daltons (Murai et al., 1990). Levels of CRP increased threefold after a bacterial infection. HSPs are smaller molecules occurring throughout the animal kingdom. They were first reported in response to temperature changes. The HSPs may also indicate tissue damage and have potential as diagnostic indicators (Welch et al., 1991).

Nonspecific defense mechanisms are eminently important in fish, because fish have fewer complex specific immune capabilities than higher vertebrates to delay and prevent the entrance of a violating microorganism. These nonspecific defenses must respond relatively quickly, even at low temperatures.

Indeed, fish place a greater emphasis on physiological availability or energy spent on the nonspecific defense mechanisms than on the more elaborate specific defense response used by mammals.

Sometimes distinctions between nonspecific defense mechanisms and specific immune responses are difficult to make. Major factors, such as complement, specific antibodies, or other molecules of the specific response, act as antigen receptors and augment and increase the activity and direction of the nonspecific mechanisms after stresses or antigen stimulation. For example, if a fish already possesses circulating antibody to *Yersinia ruckeri* (the causative agent of the disease enteric redmouth) from a previous exposure to the disease or from immunization, the fish is better prepared to fight the pathogen when it reappears. The combination of specific antibody molecule with the receptors on defense cells such as macrophages reacts with antigen to opsonize or enhance phagocytic and destructive abilities (Griffin, 1983).

The concept of antibody as an opsonin in fish immunology is important because the blood of many adult fish contains natural, background, nonspecific antibodies or antibody-like molecules. These factors are difficult to characterize. Some of these serum proteins have characteristics similar to those of an antibody, as they require complement to lyse bacterial or sheep red blood cells (SRBCs). Their molecular weight is also in the range of that of a small divalent fish antibody molecule. Most of the time, however, these factors seemingly are not induced by any previous known specific stimulation.

Simple experiments to show the levels of the antibody-like action in sera of fish can be done by making titer dilutions of the sera, adding SRBCs, and demonstrating that this natural lytic action exists in fish that have never knowingly been exposed to SRBCs. Attempts have been made to explain this action. Some fish immunologists believe that *natural antibody* may arise as it is encoded and translated from the genome and generated as the fish matures. This hypothesis speculates that it is produced to fight microorganisms that evolved in parallel with the fish. The animals are preprogrammed to produce the specific antibodies, which incidentally also lyse SRBCs. *Background antibody* may come from stimulation through interactions with environment pathogens or other microorganisms as the fish matures at a particular site. For instance, specific antibody is stimulated and identical antigens are found in low amounts in other bacteria or on SRBCs. *Nonspecific antibody* may be stimulated by antigens that are naturally present in the environment and widely cross-reactive with somewhat similar antigens on other microorganisms. Alternatively, the antibody molecules themselves may not be so specifically directed. The presence of low levels of nonspecific antibody sometimes clouds diagnostic results in fish disease analysis. An example is seen in the case of specific immunization against furunculosis diseases in trout caused by *Aeromonas salmonicida*. When one set of rainbow trout was injected with *A. salmonicida* growth medium supernatants and another injected with phosphate-buffered saline (PBS), sera from the latter group also showed activity against the *A. salmonicida* in agglutination assays (Michel et al., 1990). Concomitantly, protection after pathogen challenge was also observed in PBS-injected fish, indicating how important these nonspecific defense mechanisms are in this function. The phenomenon of nonspecific protection against furunculosis has also been observed when only Freund's complete adjuvant was injected (Cipriano and Pyle, 1985). This mineral oil–based formula evidently stimulates many of the factors of the nonspecific defense mechanisms. These examples illustrate that when fish immunologists measure specific immune responses in fish, the baseline for nonspecific defense mechanisms may also have been stimulated to a higher level of activity by the insult. When fish vaccines are tested for efficacy and potency, the inducible background nonspecific factors have to be accounted for by comparing immunized fish with sham-immunized fish and with undisturbed negative controls. Likewise, when testing for immunosuppression caused by the presence of contaminants, normal baseline responses should be known.

Specific Immune Response

The specific immune response is a unique system in the fish's physiology, because of production of molecules (antibodies) that are specifically directed against antigens. Almost all other cellular proteins produced by animals are constructed directly from the genome exactly as encoded. In contrast, the production and assembly of antibody arising from stable genes may be randomized by genetic segmentation and somatic mutations, giving rise to unique molecular sequences that are able to complex with any of the multitude of antigens a fish may meet. Low levels of specific antibody to many antigens may be present in the animal all its life; however, increased production against a specific antigen is stimulated mainly in two ways: by exposure to the disease agent itself or by active immunization.

Fish culturists use the specific immune response to protect their fish against diseases. Commercially produced bacterins are given by injection, bath, shower, or flush exposure to large numbers of young salmonids to prevent enteric redmouth disease, caused by *Y. ruckeri*, vibriosis, caused by *Vibrio* sp., and furunculosis, caused by *A. salmonicida* (Figure 3). The topical application of these bacterins shows another unique feature of fish: the bacterins do not need to be injected to be effective at inducing protection. Fish have the ability to recognize antigens from potentially harmful pathogens and produce a specific response when the antigens or bacterins are given through the water (Amend and Fender, 1976). The bacterins are taken in by gills and other skin

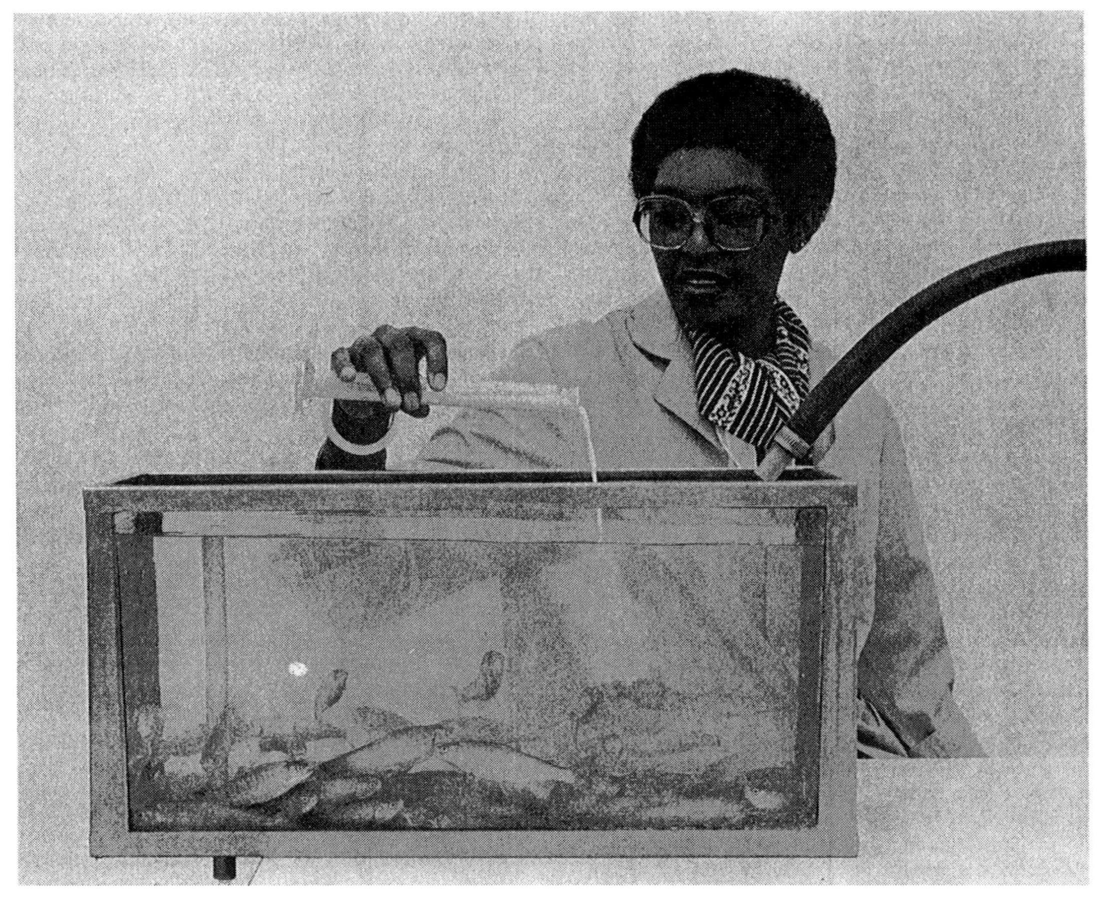

Figure 3. Fish can be immunized by bath or flush exposure as demonstrated in this photograph showing immunization of rainbow trout by pouring the vaccine solution in the water. The animals take up the vaccine through the gills and after several days antigens are found deposited in the spleen and kidney. Environmental conditions such as pH or the presence of contaminants can inhibit the uptake of the vaccine from the water by fish, resulting in reduced effectiveness to resist diseases.

surfaces. They are then internalized and transported to immunopoietic organs.

The uniqueness of that first step is the recognition of antigen from ambient water by the fish. Immunologists speculate how fish are able to pick out the particular bacterin from a background of other detritus for uptake, in spite of the fact that levels of natural and background organic materials are often much higher than the concentrations of the immunizing bacterins.

Laboratory experiments on kinetics of bacterin uptake have been done by immunofluorescent antibody techniques and electron microscopy. Particles can be shown entering epidermal gill pavement cells and transferring through this external cellular barrier into the circulatory system (Zapata et al., 1987). Electron photomicrographs show that particles first adhere to epidermal surfaces of the pavement cell; then invagination occurs and particles are seen in vacuoles. Other portals of entry for vaccines into fish have been implicated, including the lateral line, pseudobranch, gut, and general skin surface (Figure 4). Blood monocytes have been observed carrying antigen in circulation. The antigen seems to accumulate in macrophage aggregates of the spleen, occurring in the kidney and liver. For trout, the anterior kidney area is also the main location for generation of the specific immune response. Splenic tissues of trout, striped bass, and other teleosts have immunopoietic functions with more specialization evident than in the more highly evolved species. The spleen in most fish, as in mammals, is the major filter of blood-borne antigens (Bodammer et al., 1990). It is also a major site for removal of damaged blood cells.

Trout have a distinct thymus tissue located directly under the upper half of the gill opercula. This whitish tissue is covered with a thin layer of skin and often a lateral nerve can be seen passing directly between the transparent skin and the thymus. The thymus in trout is much less differentiated than that of mam-

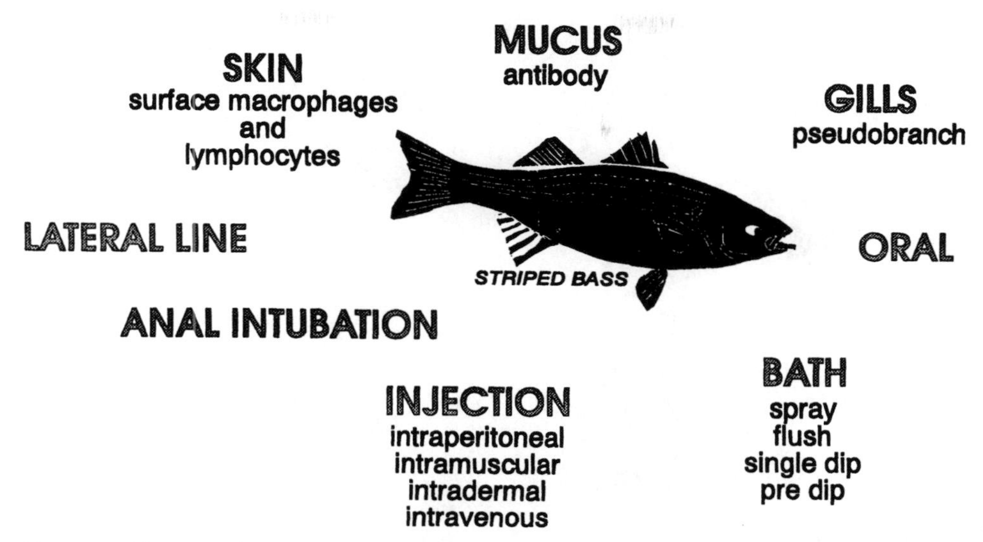

Figure 4. Fish can be given vaccines by several routes, including injection, bath, or feed. When vaccines are given by bath, the main portal may be the gills. Microscopic examination of the gill shows that the particles become lodged in the gill pavement cells after fish have been bathed in a bacterin. Exposing the fish to low concentrations of phenol before bath immunization reduces the immune response.

mals, appearing to have a somewhat loose organization of immature lymphocytes in the center (medulla) and smaller, more developed lymphocytes in the outer portion (cortex). In mammals, migrating undifferentiated lymphocytes are processed and signaled for maturation into their specific actions in the thymus. Fish have subpopulations of lymphocytes with different functions in the specific immune response; distinctions among these lymphocytes can be observed by microscope with specific antiserum that is labeled with visual markers such as those used in immunofluorescence studies. The techniques involve isolating specific receptors or surface proteins on the cells and making polyclonal or monoclonal antibodies against the distinguishing proteins. The subpopulations of B lymphocytes can be characterized by labeling with anti-antibody because these cells secrete copious amounts of antibody, especially in the plasma cell stage. T lymphocytes are more difficult to identify in fish. The cytotoxic killer cells are also morphologically similar to lymphocytes.

The spleen of trout contains more mature cells of the immunopoietic series and is an easy organ to sample (Figure 5). Generally, this spindle-shaped organ is described as having filtering functions that destroy older cells and collect debris. Fish bathed or injected with bacterins collect antigens in splenic macrophage aggregates. A large number of antibody-producing plasma cells are also stored in the spleen. The functioning of the spleen as an immunopoietic organ depends on the fish species (Figure 6).

What has been described up to this point in this section is the afferent immune system. This is the part of the immune response that takes up, processes, and presents information to the efferent immune system. The latter half of the specific immune pathway is the productive phase. The efferent immune system has been signaled by the afferent system to produce products that act against the invaders, the production of specific antibody being one of the most important functions.

Antibodies and the Antibody-Antigen Complex

Antibody molecules produced by plasma cells are classified as immunoglobulins. Mammalian models of Ig are IgM (macroglobulin, a large, early-appearing antibody molecule), IgG (gamma, the most common, lighter antibody), IgA (secretory antibody), IgE (cell-associated antibody, often implicated in allergies), and IgD (on membranes of B lymphocytes; may be important in recognition). The first two classes, IgG and IgM, are major components in the mammalian blood.

The basic molecular construction of antibody molecules in fish is essentially similar to that in mammalian models. Figure 7 shows schematically examples of catfish (*Ictalurus punctatus*) antibody molecules. Two L (light) and H (heavy) protein chains, held together by disulfide bonds, compose a subunit (Ghaffari and Lobb, 1989). Whereas mammals have a pentameric IgM, however, most fish

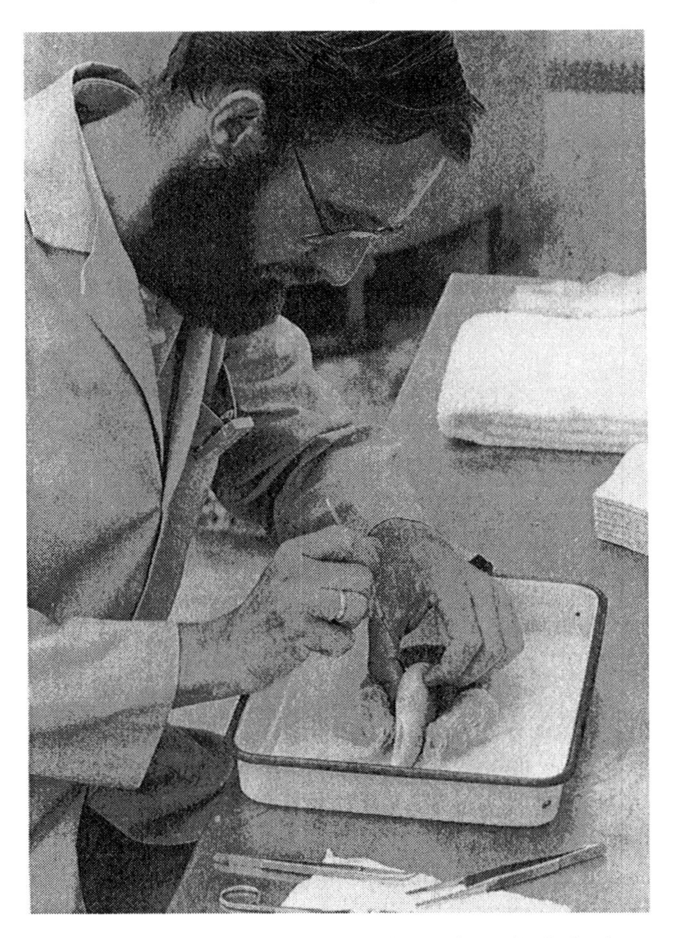

Figure 5. To determine the changes in the numbers of splenic plaque-forming cells after contaminant exposure and immunization experiments a section of the spleen is excised from a rainbow trout. The section is then drawn over a wire mesh to make a cell suspension that can be used in the immunological assays. With proper technique and use of antibiotics, the fish survives the procedure.

have a tetrameric IgM molecule as the main class of antibody. The subunit molecules (H and L) may be held together by a J (joining) molecule. Each heavy and light chain is subdivided into a variable region that determines the molecular specificity for antigen and a constant region that is responsible for many biological activities of the molecule. After an immunization, mammals generally show a switch from IgM to IgG, the more efficient molecule. Similar Ig switches have been described in fish, and although a memory response is present, it may not be as strong as described for mammals. With a memory response, sometimes called an anamnestic response, repeated exposures result in more rapid and higher production of antibody. A memory develops so that upon secondary immunization, a more rapid immune response occurs. That memory is most often held in the T lymphocytes (Figure 8).

Antibody molecules described for rainbow trout occur in two sizes: a large tetrameric molecule weighing about 700 kDa which, upon centrifugation in a cesium chloride gradient, concentrates in a band of 17S, and a smaller dimeric molecule of about 95 kDa which bands at 7S (Hodgins et al., 1973; Cisar and Fryer, 1974; Sanchez et al., 1989). The larger, designated an IgM-like globulin, is known to increase in quantity in the blood when the fish are actively immunized. Some of the smaller molecules are lytic for SRBCs and require complement for that action. These smaller IgM-like molecules may be involved in some of the background and nonspecific antibody peculiarities of fish and, indeed, have been speculated to be disjoined IgM molecules. Antibody in other higher fish shows variations on the rainbow trout models. Tilapia (*Oreochromis aureus*) IgM has a molecular mass between 780 and 788 kD (Smith

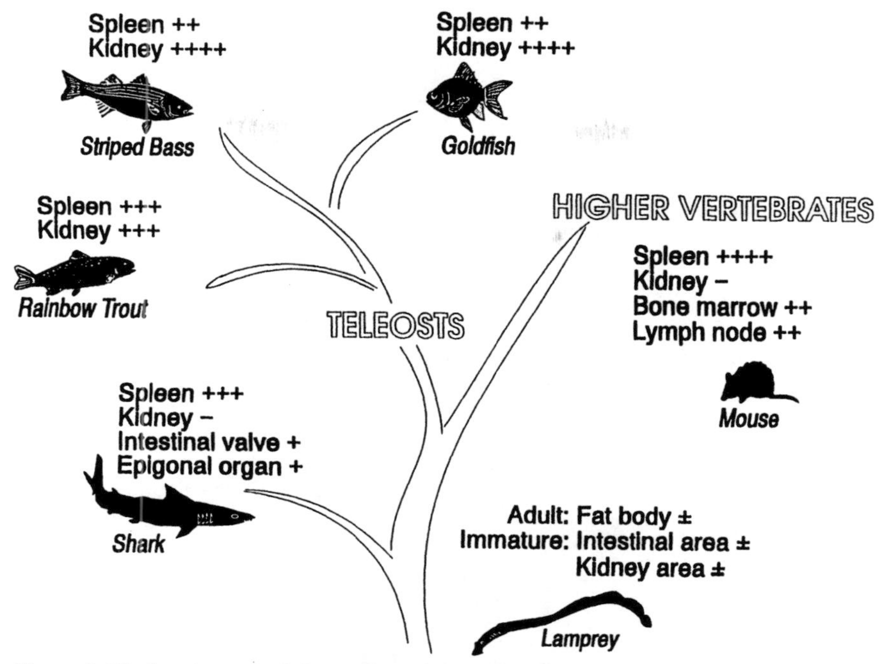

Figure 6. The location and relative number of plaque-forming cells in spleen and kidney demonstrate the diversity of the immune response among different species of fish. In higher teleosts such as the goldfish, the kidney is the most active immunopoietic organ (+ + + +); whereas, the kidney of a shark shows little or no activity.

et al., 1993). Research on protein sequencing is characterizing antibody assembly in fish and showing the phylogenetic relationships of different species. The H chains in *Raja erinacea* show isotypes, leading to clues to how Ig classes and the switch from IgM to IgG may have evolved (Harding et al., 1990). Nevertheless, basic antibody molecules among fish and different animals are remarkably similar and can be used to show phylogenetic relationships at the genomic level. Indeed, the antibody Igs are also grouped with cell receptor and signal molecules involved in the immune response. Some T lymphocyte receptor molecules and macrophage-derived cytokines, for instance, are also members of the Ig supergene family.

Antibody-antigen complexes form when a pathogen invades an immunocompetent host. Antigens from the invader complex with host antibodies and the complexes set off a complement (C′) cascade that is capable of lysing invasive cells such as bacteria, protozoans, or heterologous erythrocytes such as SRBCs (the latter are used as a diagnostic indicator). Higher fish have two C′ systems, the classical C′ pathway, usually associated with the specific immune response and antibody, and the alternative C′ pathway, whose destructive abilities can be activated against some gram-negative bacteria containing endotoxin without involving specific antibody. Jenkins and Ourth (1990) show that the alternative C′ pathway of catfish increased the permeability of the cytoplasmic membrane of *Escherichia coli*. This C′ lysis is caused by C′ molecules physically opening holes in the cells. C′ is functionally generic or species specific. For instance, in diagnostic assays with rainbow trout in which the presence of C′ is used as an indicator to lyse target cells, C′ from goldfish cannot be used (the C′ from one genus may not be substituted for another). There is a complete interchange of functions, however, among human, mouse, and guinea pig, perhaps demonstrating that these animals are more closely related! This phenomenon of species specificity further illustrates the diversity of fish species.

Immunosuppression

The nonspecific defense mechanisms and specific immune response act in concert to protect fish from invasive microorganisms and diseases. Problems arise when toxicants cause immunosuppression. If any disease-causing agents are present during immunosuppression, the animals may succumb to disease. Furthermore, tumor immunologists (oncologists) believe that the immune response controls or suppresses some cancers. It is theorized that some tumors are caused by naturally occurring, subtle mutations in genetic and tissue structure in rapidly proliferating cells. These changes are recognized as foreign and rejected through actions of the immune system. Exposure to specific viruses is known to

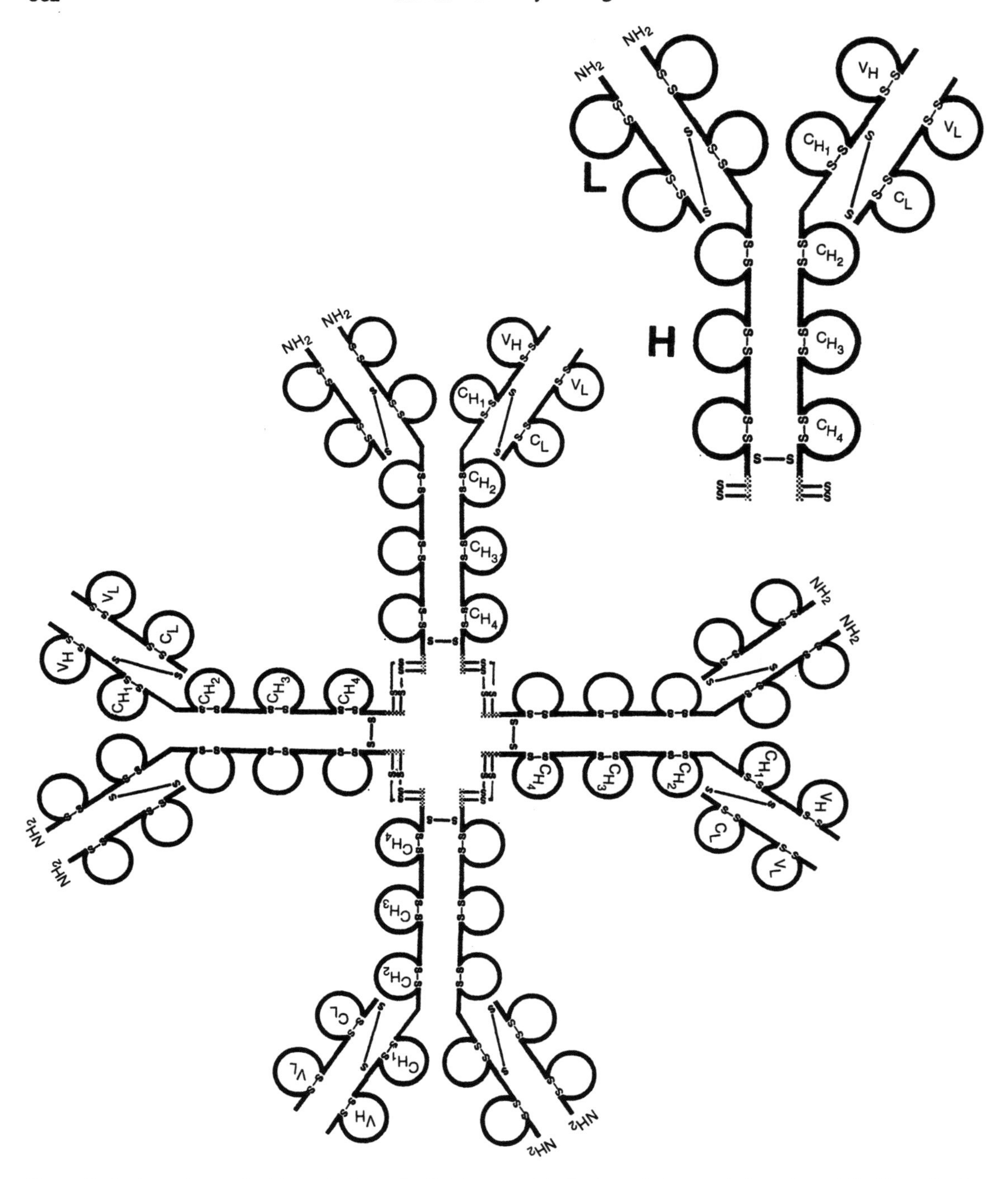

Figure 7. A schematic model of the major type of catfish immunoglobulin shows its tetrameric form (lower left). The antibody molecule has eight heavy (H) and light (L) chains. The specificity of the eight antigen combining sites in the terminal regions is determined by the variable regions of the H and L chains. Disulfide bonds (–S–S–) occur within the chains and at points near the bases holding the chains together. A monomeric form is shown upper right. (Models derived from Ghaffari, S. H., and Lobb, S. J., 1989, *Journal of Immunology*. Copyright 1989, *The Journal of Immunology* 143:2730–2739.)

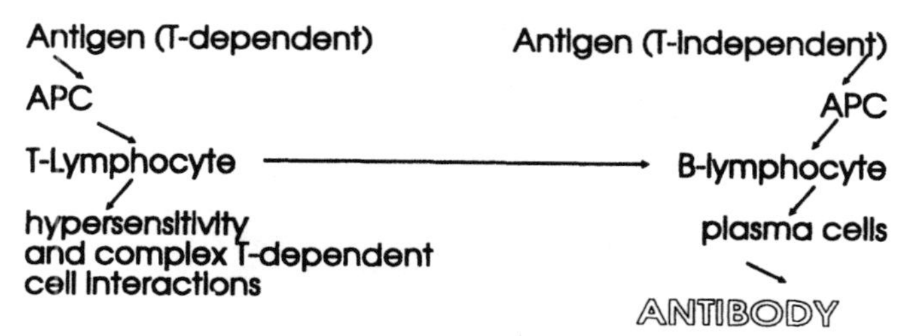

Figure 8. T-dependent antigens are often proteinaceous materials and require processing through the antigen-producing cells (APCs) to stimulate and guide the production of antibody by the plasma cells. T-independent antigens, such as bacterial lipopolysaccharides, have a more direct route for antibody stimulation.

cause tumors in fish (Mulcahy and O'Leary, 1970). Other tumors in fish are induced by certain chemicals such as the aflatoxins, responsible for hepatomas in rainbow trout (Sinnhuber et al., 1968). In some cases, the tumors themselves may generate immunosuppressive or immune evasive factors that suppress actions against the tumor.

Scientists correlate incidences of tumors with populations of fish under suppressive agents. Winter flounders (*Pseudopleuronectes americanus*) in Boston Harbor, Massachusetts, have a high incidence of hepatic vacuolated cell foci developing into neoplasia; flathead soles (*Hippoglossoides elassodon*) showed a high incidence of epidermal papillomas in the area of the San Juan Islands, Washington (Wellings et al., 1968; Murchelano and Wolke, 1985). Fish are increasingly used for predicting carcinogenicity in human and fish health (Metcalfe, 1989). Tumor immunology in fish will become an important part of immunotoxicology as observational surveys are correlated with information on how the infectious agents and chemicals induce cancers.

ASSAYS AND APPLICATIONS—IMMUNOLOGICAL METHODOLOGIES

In the past decade, advances have been tremendous in bringing hematological, serological, and immunological assays to practical use in fish culture (Wedemeyer and McLeay, 1981; Stolen et al., 1990, 1992). These assays are used extensively in fish disease diagnosis and in following the effectiveness of administering vaccines to fish. Fish disease agents, such as the bacterium *Renibacterium salmoninarum* found in kidney tissues, can be easily identified by immunological assays using specific antiserum that is made in goats or rabbits by injection of formalin-killed *R. salmoninarum*. Likewise, fish injected with *R. salmoninarum* bacterins generate specific antibody which may indicate that the fish is now protected against this disease. Research is also focusing on using these immunological tests to determine genetic strains and differences in fish populations (e.g., different populations may have different antigens), investigate cell structure and tissue functions by using labeled antibodies, and develop forensic tests. In the latter instance, detectives can confirm the presence of fish products in meats sold in markets or the identification of fish blood samples taken from fishermen's clothes. This has applications in cases in which endangered species are illegally fished and, because the animal has been discarded or hidden, only blood samples remain as evidence. Specific antigen occurring in Atlantic salmon blood, for example, would be determined in immunological assays, to distinguish salmon blood from carp blood.

Use of immunological assays in studying and demonstrating the effects of toxicants on fish is a relatively new development. In the following, assay techniques are discussed along with examples of their application (see Table 1).

Hematocrits

One of the most important and simplest assays a hematologist uses is the hematocrit, which shows the ratio of centrifuged, packed erythrocytes to total volume of blood. Nonlethal blood samples can be taken by heart puncture or from the caudal artery. Likewise, samples can easily be taken from fish by severing the caudal peduncle. The cut should be clean and the first few drops of blood discarded. A heparinized capillary tube is then placed near the flowing artery to load two thirds full. The tube is plugged at one end with clay or plastic putty and centrifuged in a special hematocrit centrifuge for 2 min. For an easy measurement, the tube is then placed over a gradient scale that shows the percentage of erythrocytes. In healthy coho salmon (*Oncorhynchus kisutch*), a batch of fish samples may give erythrocyte readings that range from 30 to 60% (Smith, 1967). Information on hidden infections, progress of diseases, nutritional regimens, presence of contaminants, and reproductive cycles is indicated by hematocrit readings. For example, stress in handling, such as during transport, has been shown to reduce hematocrit values in catfish (Ellsaesser and Clem, 1986).

On the other hand, strenuously exercised trout showed a higher hematocrit reading, because of erythrocytes enlarged to optimize oxygen-carrying

Table 1. Some hematological and immunological assays considered for use in fish immunotoxicology investigations

Assay	What it measures
Hematocrit	Levels of blood erythrocytes
Leukocrit	Levels of blood leukocytes
Cell differential count	Blood cell populations
Phagocytic percent	Percent of phagocytic cells with engulfed materials
Phagocytic index	Amount of phagocytized material
Pinocytosis	Phagocytosis of fluids
Adherence	Particle adherence to leukocytes
Rosette-forming cells	SRBC-surrounding leukocytes
Nitroblue tetrazolium test (NBT test)	Neutrophil oxidative radical NBT activity by dye color change
Chemiluminescence	Oxidative radical release generating photoburst
Blastogenesis or mitogenesis, lymphocyte proliferation	Leukocyte activity
Trypan blue	Cell viability
Chromium release	Killer cell activity
Scale rejection (transplantation)	Individual similarities (tissue compatibilities), rapidity of rejection
Delayed hypersensitivity	Allergic reactions (cellular) (topical sensitivity)
Macrophage aggregates, melanomacrophage centers	Immunopoietic accumulation of antigens
Passive hemolytic plaque assay	Antibody-producing cells (plaque-forming cells, PFC)
Complement fixation	Complement levels or circulating antibody
Hemagglutination	Circulating antibody
Agglutination	Circulating antibody
Ouchterlony precipitin gel, agar gel	Circulating antibody
Immunoelectrophoresis, counterimmunoelectrophoresis	Circulating antibody
Immunoblot	Circulating antibody
Enzyme-linked immunosorbent assay (ELISA)	Circulating antibody
Fluorescent antibody	Circulating antibody
Neutralization	Circulating antibody (usually with viruses)
Radioimmunoassay	Circulating antibody
Disease challenges	Protection against disease

capacity of the cells (Wells and Weber, 1991). Investigators thought that this swelling was due to the fact that hemoglobins are more efficient oxygen carriers when in dilute solutions; the erythrocytes had enlarged to enable more oxygen transport. Coho salmon held in the presence of pulp mill effluents also had elevated hematocrits (McLeay and Gordon, 1977). The major disadvantage of hematocrit readings is the wide variation among individuals in the same population (Blaxhall, 1972). Reliable sources of control fish for comparison readings may be difficult to find, and standards may be different for various locations and fish populations. Some hatchery biologists use data for samples from previous years for comparison. In spite of these drawbacks, hematocrit readings are simple to obtain and, used consistently, are helpful in supporting data obtained with other indicators.

Leukocrits

On top of the erythrocyte pack in a hematocrit sample appears a layer of leukocytes or white blood cells (WBCs). In healthy rainbow trout, the whitish layer of cells is about 1–2% of the total blood volume. This narrow band or buffy coat of leukocytes is often difficult to read or estimate, and because of the small percentage, this is a somewhat unreliable gauging figure. In a new technique a buoyant bead is placed in the capillary tube and, upon loading and centrifugation, enlarges the buffy coat area to expand the accuracy of the reading (Levine and Wardlaw, 1988). Perhaps the leukocrit's greatest importance is showing whether the cell layer is absent, possibly indicating leukopenia (depression of WBC numbers), or greatly increased above normal, showing, for example, that the animal may be fighting a viral infection. The leukocrit may give a better indication than the hematocrit of the status or health of the immunopoietic system in fish and their resistance to diseases (Wedemeyer et al., 1983; Pickering and Pottinger, 1987; 1988) because it more specifically indicates the quantity of cells directly responsible for protection and immunity. McLeay and Gordon (1977) noted that reduction in leukocrit values corresponded to lower hematocrits when salmonids were exposed to pulp mill effluents.

Occasionally, a lymphoma or myeloma condition is detected in a fish by the leukocrit sample. A higher leukocrit results as cells are released from the tumor and enter the circulatory system. Uncontrolled growth of a lymphosarcoma was described by Dunbar (1969) in brook trout (*Salvelinus fontinalis*) and splake (brook trout × lake trout, *Salvelinus namaycush*). Tumor nodules in the thymus implicated the source, and evidently these lymphocytes were accumulating in the blood. Leukemia has been also described in cutthroat trout (*Salmo clarki*) by Smith (1971). Nodules of undifferentiated blastoid cells were found in the kidney, spleen, and liver. These cells were also observed in the circulatory system. Warr et al. (1984) described a malignant lymphoma of probably thymic origin. At this time, attempts were made to obtain a continuous cell line for possible use in making hybridomas in order to make monoclonal antibodies from fish cell lines. The lymphoma cell line would be fused with temporal plasma cells producing specific antibody, thus creating hybridoma cells that would produce antibody and be cultured continuously. Cells were taken from the blood and organs and placed in culture medium, but the lymphoma cells died after 36 d in culture. Herman (1969) noted that lymphosarcomas are observed occasionally by hatchery biologists but causes are unknown.

Cell Differentials

A blood drop placed on a microscope slide and smeared quickly with another slide to form a thin film layer can be used to examine blood cell morphology. This blood smear can also be used to count and differentiate cell populations. A cell count can be calculated to give a ratio—for instance, numbers of lymphocytes relative to erythrocytes—or simply to note numbers of specific cells in a certain field or area and characteristics of those cells. Fish biologists commonly use Wright or Leishman-Giemsa stains to examine blood cells. Erythrocytes, thrombocytes, and lymphocytes are easily identifiable, but views differ on classifications and subpopulations of neutrophils and granulocytes for different fish. For instance, Doggett and Harris (1989) described three types of granulocytes in tilapia (*Oreochromis mossambicus*), based on cytoplasmic granule characteristics and the ability to engulf colloidal carbon particles. When biologists carefully make their own standards and note differences among populations of fishes, cell differential counts can be a valuable tool, albeit tedious to perform. Differentials can also be obtained with organ smears or blots. Peters and Schwarzer (1985) demonstrated changes in rainbow trout spleen and head kidney after handling and social stress, including increases in macrophage-like cells and reductions of hemoblasts and lymphocytes.

Cell Counts

Vital stains (stains that are excluded by living cells) or gross observations without staining of cell suspensions on a hemacytometer can give important information about materials affecting cell viability. The most commonly used vital stain is trypan blue, often used with fresh blood cell suspensions. The percentage of live lymphocytes is easily determined as dead cells take up the blue stain.

The hemacytometer is a significant tool in immunology for the calculation of mean corpuscular volume (MCV) and mean corpuscular hemoglobin (MCH) of erythrocytes. The MCV is related to the size or volume of erythrocytes and was used by McCleay and Gordon (1977) and Tana and Nikunen (1986), who reported erythrocyte swelling in fish exposed to chemicals. The calculation of MCH requires a measurement of hemoglobin concentration, usually done by lysing erythrocytes and obtaining spectrophotometric readings to determine amounts of soluble hemoglobin. Both assays require the calculation of number of erythrocytes/ml. Smith (1967) used both assays to show that folic acid–deficient diets in coho salmon caused higher MCV and MCH values.

Phagocytic Assays

Various methods are used for measuring uptake of particulate or soluble materials by phagocytic cells in investigating the health of fish. An example of a phagocytic cell isolated from striped bass and incubated with formalin-killed bacteria to show engulfment is shown in Figure 9. Nonantigenic particles such as fluorescently labeled latex beads injected into fish accumulate in macrophage aggregates of the spleen, kidney, and liver (Ziegenfuss and Wolke, 1991). These groups of cells, sometimes designated as melanomacrophage centers when melanin is apparent, can be observed in situ after histological sectioning of affected tissues and examined for changes in enzymatic activity or the uptake of beads or carbon particles (Herraez and Zapata, 1986). For instance, tilapia (*Oreochromis aureus*) stressed through starvation or injection with *A. hydrophila* bacterins showed increased alkaline phosphatase activity (Agius and Couchman, 1986). The activity was linked to increased catabolic tissue breakdown in melanomacrophage centers (Agius and Roberts, 1981).

Phagocytic cells can be isolated and assayed in functional tests in vitro. Weeks and Warinner (1984) isolated macrophages from the kidney of spot (*Leiostomus xanthurus*) and hogchokers (*Trinectes maculatus*) to study the toxic effects on fish in the Elizabeth River, Virginia. They separated the macrophages from the kidneys by discontinuous density gradient centrifugation in isosmotic Percoll and let the cells settle on coverslips. The adherent cells were then incubated with formalin-killed *E. coli* bacteria. Numbers of macrophages containing intracellular bacteria were counted by microscope and the percentage of active cells was derived. The macrophages from these fish were compared with macrophages from spot that were residents of an unpolluted river to show the decreased activity associated with the presence of contaminants (Figure 10). This research group obtained similar data by using automated assays that

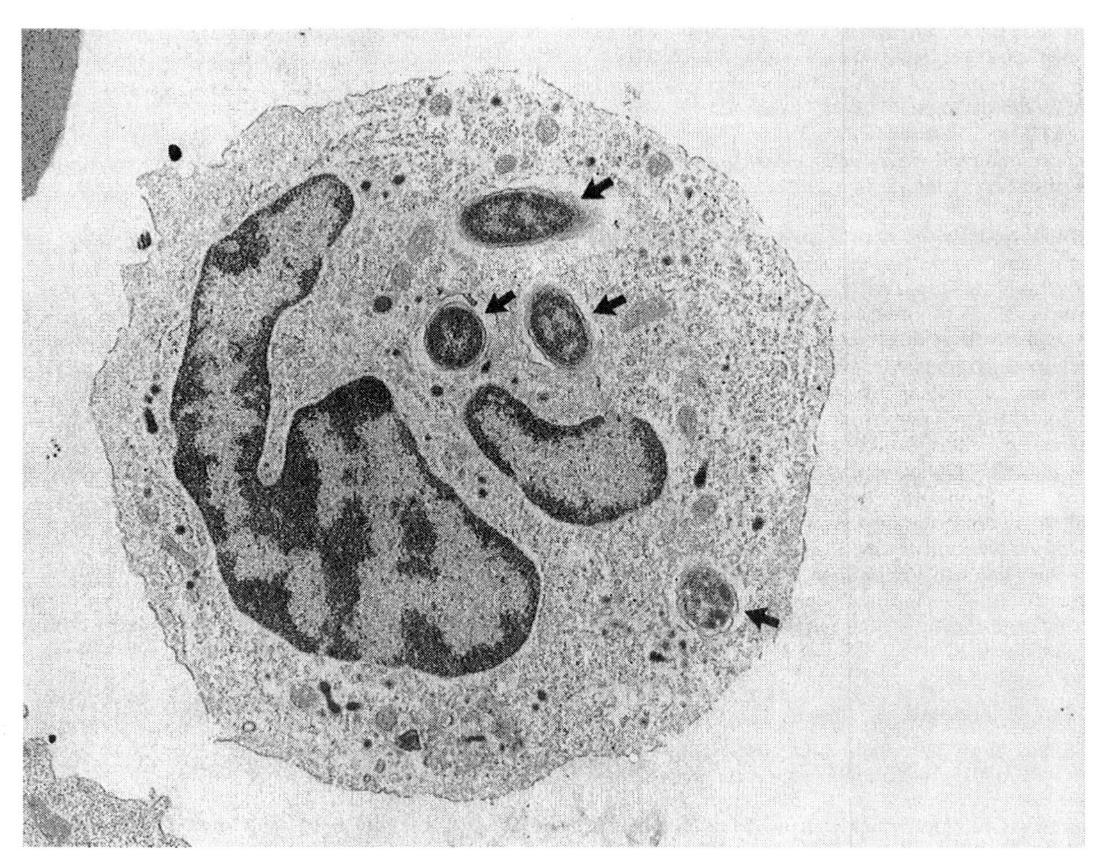

Figure 9. Electron microphotograph of a phagocytic cell isolated from the spleen of a striped bass shows engulfed bacterial cells. The fish spleen cells were incubated in vitro with a suspension of a *Yersinia ruckeri* bacterin. Vacuoles enclose the particles; lysozyme, peroxidase, and other enzymes may result in further destruction. (Photograph by J. E. Bodammer.)

rely on the uptake of neutral red vesicles and then lysing the cells for a spectrophotometric determination (Weeks et al., 1987; Seeley and Weeks-Perkins, 1991). The pinocytosis of the dye serves as a biomarker of environmental degradation.

With the use of spleen cell suspensions, Anderson et al. (1989) demonstrated the immunostimulation of phagocytic cells by the injection of levamisole alone or in conjunction with bacterins. They used splenic cell suspensions mixed with formalin-killed *Staphylococcus aureus* bacterin. After an appropriate incubation period, the suspension was smeared on a microscope slide. Leishman-Giemsa staining showed increased numbers of engulfed bacteria in macrophages from the levamisole-treated fish. Another way to use direct observation by microscopy to determine phagocytosis is by using glutaraldehyde-fixed SRBCs. Being much larger than bacteria, SRBCs are easier to visualize when inside phagocytic cells (Sövényi and Kusuda, 1987).

Several methods have been developed for eliminating the subjectivity from phagocytic cell assays, including use of a chemiluminescence assay. The technique is based on the emission of oxygen radical species from macrophages and neutrophils when stimulated with various agents such as yeast cell extracts or bacterial preparations. Minute amounts of light are emitted during this reaction and can be measured with a photometer or luminometer. This chemiluminescence assay is a good health indicator and has been useful for examining the effect of heavy metals on pronephros cells of striped bass (Stave et al., 1983).

Neutrophil Assays

During the inflammatory process, circulating neutrophils first adhere to blood vessel epithelium and then migrate in a process of diapedesis into tissues by passing between arterial or venule walls. In addition, activated neutrophils, similar to many phagocytic cells as discussed in the previous section, release reactive oxygen species to destroy invasive microorganisms. The activated cells give a positive nitroblue tetrazolium (NBT) reaction (Figure 11). The advantage of NBT assays is that they are remarkably sim-

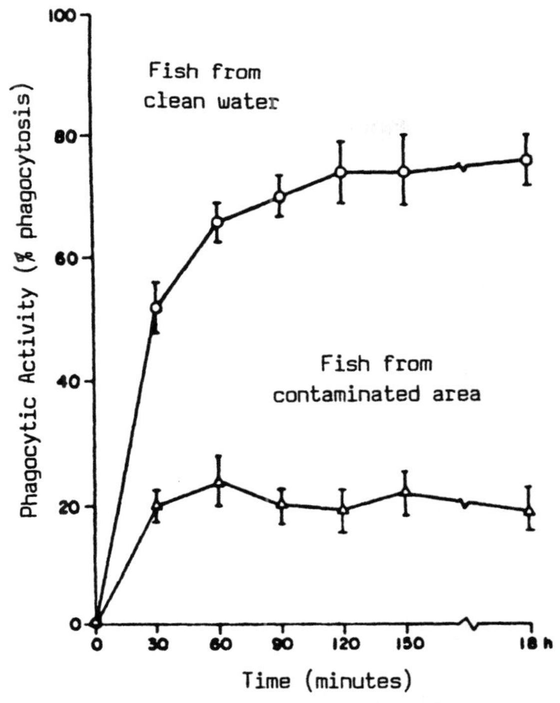

Figure 10. Relative activity of macrophage phagocytic activity in spots from the Ware (o) and Elizabeth (△) rivers in Virginia. Each value represents the mean ± SEM of four experiments with 5–10 fish/experiment. The Elizabeth River group differed significantly ($P < .05$) from the fish taken from the contaminated site at each time interval. (Reprinted from *Marine Environmental Research* 14, Weeks, B. A., and Warinner, J. E. Effects of toxic chemicals on macrophage phagocytosis in two estuarine fishes, pp. 327–335, Copyright 1984, with permission from Elsevier Science Ltd., The Boulevard, Langford Lane, Kidlington OX5, 1GB, UK.)

ple and easily used by field biologists. A drop of blood from a fish is placed directly on a glass coverslip and allowed to incubate for 30 min for the cells to stick to the glass. Nonadherent cells are rinsed off with saline, and the coverslip is then placed on top of a few drops of the yellow NBT solution on a glass slide for staining. A blue halo of formazan develops around the neutrophils that stick to the glass and emit reactive oxygen species.

NBT tests were used to show potential immunosuppressive effects of drugs used in treating fish bacterial agents (Siwicki et al., 1990a). Oxytetracycline and other drugs are sometimes used in conjunction when fish are immunized, especially when aquaculturists are dealing with disease problems in large net pens. Suppressive effects should be considered in drug treatment regimens. Reduced NBT responses were observed when fish were exposed to 10 mg/kg oxytetracycline in conjunction with a *Y. ruckeri* bacterin but not when oxolinic acid, another drug for treating fish bacterial diseases, was used (Siwicki et al., 1989). These immunomodulatory effects can also be shown by in vitro tests. Rainbow trout spleen sections were incubated in vitro with levamisole to demonstrate the kinetics of immunomodulation (stimulation or suppression) of this T lymphocyte stimulator. Although neutrophils lose some of their reactive power upon prolonged incubation in vitro, the immunosuppression was shown at higher doses of levamisole.

Modifications of NBT protocols for macrophage and neutrophil activities have been suggested by Secombes (1990) and Anderson et al. (1992). These simple and rapid assays should be adaptable for general diagnostic field use. For example, rainbow trout held at 12°C and injected with a bacterin show a neutrophil activation curve giving evidence within 2 d after injection that this nonspecific defense mechanism is taking place (Figure 12).

Adherence Assays

When leukocytes are activated from a resting state, new and increased numbers of receptors or other cell membrane molecules become available on their surfaces. Some receptors promote stickiness of leukocytes to a variety of substances including bacteria, antigen-labeled latex particles, glass, and heterologous red blood cells (RBCs). For example, injection or bath exposure of sockeye salmon (*Oncorhynchus nerka*) to *Vibrio anguillarum* bacterins resulted in stimulation of antigen-binding cells in the peripheral blood (Gosting et al., 1981). This assay involves a nonspecific response; the technician can use any indicator particle that shows adherence to the leukocytes. Two days after injection with a *Y. ruckeri* bacterin, those antigen-binding properties are evident in the heterologous *A. salmonicida* formalin-killed cells and with the injected antigen. The formations of cells or particles circling around a center fish cell should not be confused with *rosettes* formed by SRBCs or other RBCs around T lymphocytes. This latter phenomenon is used for enumerating subpopulations of lymphocytes (Yoshida and Kitao, 1990).

Adherence of bacteria to leukocytes of rainbow trout was used as an assay by Siwicki et al. (1989) to show the immunostimulating effects of levamisole given with or without the *Y. ruckeri* bacterin. Blood samples taken 10 d after immunization showed that the leukocytes had increased adherence capacities for *S. aureus* formalin-killed cells and that the fish given the bacteria with levamisole had the highest adherent cell capacities.

Lymphocyte Proliferation Assays

After immunization, the efferent immune response begins the productive phase by initiating proliferation or increases in numbers of lymphocytes. This phenomenon can also be induced by mitogens, agents that induce mitosis in cells. This rapid cell division can be detected by simply observing and

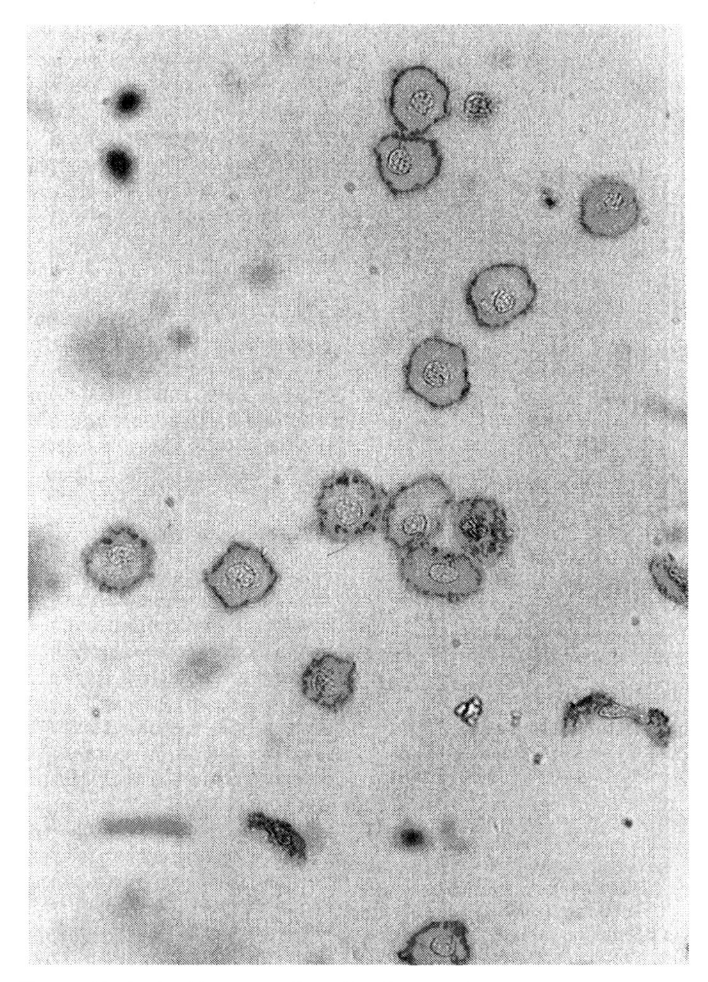

Figure 11. A nitroblue tetrazolium (NBT) assay demonstrates the release of oxygen radicals from neutrophils as the yellow dye is reduced from blue-colored formazan by activated cells. Two days after immunization, rainbow trout were bled and a drop of blood was placed on a glass slide. The activated neutrophils become glass adherent and show a blue halo of stain.

counting increases in numbers of cells with a hemocytometer or by sophisticated radioimmunoassays detecting rapid uptake of radiolabeled elements. Spitsbergen et al. (1986) used a radiolabeled chromium release assay to show that TCCD-dioxin decreased the mitogenic response of rainbow trout lymphocytes after antigen stimulation (Table 2). This was also found when they used pokeweed mitogen (PWM), which preferentially stimulated B lymphocytes. However, when the cells were exposed to concanavalin A, a T lymphocyte stimulator, no changes were found.

Grondel and Boesten (1982) showed that oxytetracycline suppressed the mitogenic response in carp by exposing lymphocytes to mitogens, phytohemagglutinin (PHA), or lipopolysaccharide (LPS). Their studies indicated that the antibiotic interfered with the B lymphocyte proliferation.

Plaque-Forming Cell Assays

In vivo, antibody-producing cells (plasma cells) release antibody into the circulating (humoral) blood. In Jerne assays or hemolytic plaque assays, plasma cells can be detected because antibody forms complexes with antigen on SRBCs and C′ (complement) acts secondarily to lyse the SRBCs. The direct hemolytic plaque assay can be done by simply immunizing fish with SRBCs, resulting in the creation of plaque-forming cells (PFCs) against the SRBC antigens. This direct Jerne assay method was reported by Sharma and Zeeman (1980) to demonstrate effects of the pes-

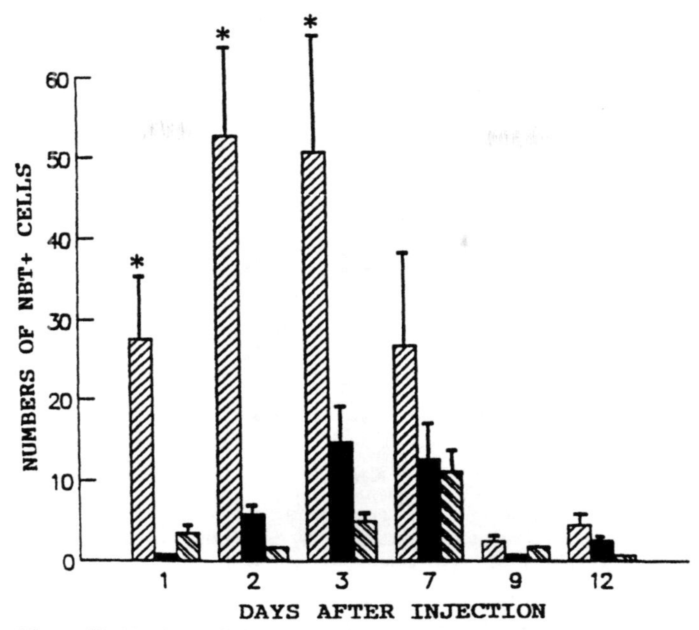

Figure 12. Numbers of glass-adherent, NBT-positive blood neutrophils from rainbow trout given an injection of 10 μg of *Yersinia ruckeri* O-antigen (▨), a PBS injection (■), or no injection (▧) and sampled at various times post injection. Error bars represent SE (n = 10). *Significant differences between fish injected with bacterin doses compared to controls injected with PBS (P < .05).

ticide DDT (p, p'-dichlorodiphenyltrichloromethane) on the immune response in goldfish. In fish injected with DDT and later with SRBCs, the PFC numbers were reduced, and therefore serum antibody titers were below those of the control groups.

Antigens can be labeled on the SRBC for an indirect Jerne assay test. For example, *Y. ruckeri* O antigen is attached easily onto the SRBC by incubation at 37°C. When spleen cell suspensions from fish are immunized with this antigen and are mixed with the *Y. ruckeri*–SRBC, the antibody released from isolated antibody-producing cells lyses the SRBC and a clear plaque is formed around the plasma cell.

As with other immunological assays in fish, the time or kinetics of appearance of the PFC is dependent on ambient temperatures. The optimal time for sampling goldfish after injection with a bacterin, for example, PFC, is 7–9 d when fish are held at 25°C, whereas for mice the maximum time for sampling PFCs after immunization is 3–5 d. That time is further extended in rainbow trout held at 12°C to 14–21 d (Figure 13).

Table 2. Effects of TCDD on lymphocyte blastogenesis in rainbow trout[a]

		Dose of TCDD (μg/kg)[c]				
		0				
Mitogen	Organ[b]	(Sham)	(Vehicle)	0.1	1.0	10
Con A	Spleen	1834 ± 437	1303 ± 343	2366 ± 766	1467 ± 358	643 ± 207
	Thymus	1467 ± 447	1248 ± 417	757 ± 218	980 ± 326	1454 ± 321
PWM	Spleen	1209 ± 269	955 ± 242	1200 ± 253	1039 ± 277	366 ± 89*

[a]Ten fish sampled per treatment group, two each day on d 16–20 after receiving TCDD.

[b]Mean cpm ± SEM for unstimulated splenic and thymic lymphocyte cultures were 247 ± 25 and 212 ± 12, respectively. No statistically significant differences were evident between treatment groups in mean cpm of unstimulated splenic or thymic cultures.

[c]Triplicate mean specific cpm ± SEM. Data reflect the reference of individual fish lymphocyte cultures at the optimal concentration of mitogen (50 or 100 μg/ml for Con A and 25 or 250 μg/ml for PWM).

*Significantly different from vehicle control (P < .05).

Reprinted with permission from Spitsbergen et al. 1986.

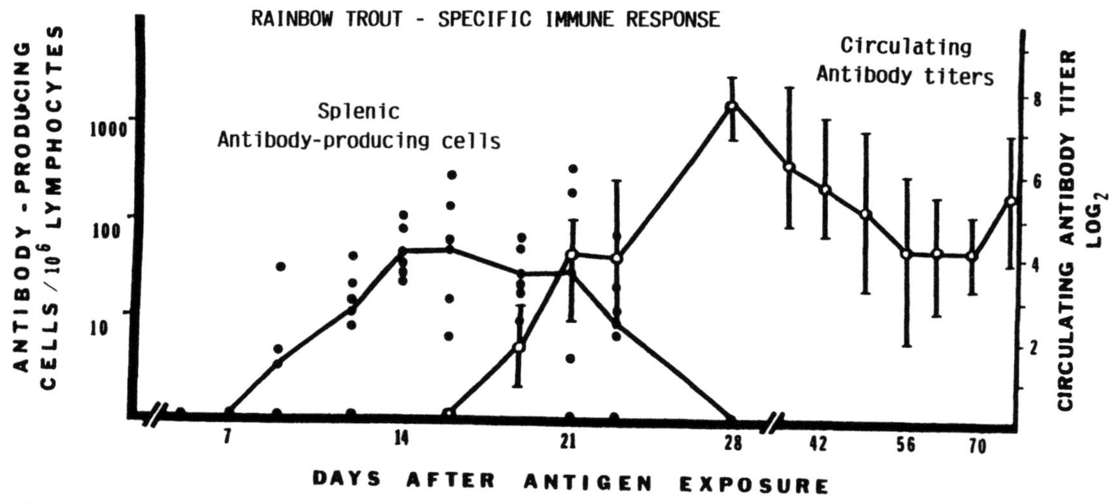

Figure 13. The immune kinetics of a response of rainbow trout held at 12°C and given a bath in a *Yersinia ruckeri* O-antigen bacterin show the appearance of splenic plaque-forming cells on day 10 after the immunization bath, a 1-wk plateau, then a reduction by d 28. The circulatory antibody titers were first detected around d 19 and continued beyond the 60-d experiment.

Reduction of PFCs can implicate immunosuppression caused by the presence of contaminants. Figure 14 shows results of experiments in which trout were bathed in low concentrations of phenol and then transferred to a bacterin bath. The phenol-bathed fish showed reduced numbers of PFCs 14 d later compared with control groups (Anderson et al., 1990). The phenol prebath may have inhibited antigen uptake by the gill cells. In demonstrating the effects of copper on suppressing numbers of PFCs in spleen sections in vitro, Anderson et al. (1989) maintained the immunized fish spleen sections at 15°C and sampled on d 10. In sections cultured with the high copper concentrations (100 μg/ml), all cells died; at concentrations of 0.1–10 μl, leukocytes remained viable but fewer PFCs were present than in sections cultured without the presence of copper.

Antibody Assays

Circulating antibody released by plasma cells into serum can be detected by agglutination of particulates; precipitation by complexing with antigen, usually in a gel; activation of complement—complement fixation; fluorescent antibody techniques; enzyme immunosorbent assays; or radioimmunoassays. All of these methods have been used for fish disease diagnosis and may be used to detect the influence of contaminants in circulating antibody titers. Variations on each serological or immunological assay method exist. For example, most bacterial and viral fish pathogens are composed of multiple antigenic fractions and not all of these may stimulate an antibody response in the same manner. This influences the type of antibody assay selected for use. Antigens on large particles or carriers would not be used in gel precipitin assays that depend on the antigen moving through a small-mesh matrix.

Methods used to determine levels of circulating antibody are the most commonly used assays in immunology, partly because taking a blood sample does not require sacrificing the animal (Figure 15). In addition, agglutination and precipitin assays are simple and fast and require less equipment and technical training than most other immunological assays. One of the disadvantages, however, is that they measure a late time in the immune response, thus samplings are delayed after antigen exposure. Also, in some cases, antibody assays are less sensitive and there may be interference from nonspecific reactions. Some researchers like to use the antibody assays to confirm other immunological assays such as phagocytic assays; thus a humoral and cellular response is measured. For example, trout exposed to the pesticide endrin showed a suppressed antibody-induced macrophage inhibition factor (MIF) and PFC numbers, but the phagocytic ability was not effected. The same *Y. ruckeri* antigen-labeled SRBCs as applied in the passive hemolytic plaque assay for determining PFCs were used to demonstrate that the circulating antibody titers were reduced (Bennett and Wolke 1987a). As noted previously, SRBCs spontaneously lyse when mixed with fresh serum because of the presence of nonspecific antibody and complement; therefore, for these experiments, the serum was heated to destroy the complement.

Disease Challenges

A simple method for determining immunomodulatory chemicals or drugs is to challenge the exposed fish with an infectious agent and compare mortality rates among exposed and control groups of fish. The procedure provides a way to determine whether the substance interferes with host resistance to patho-

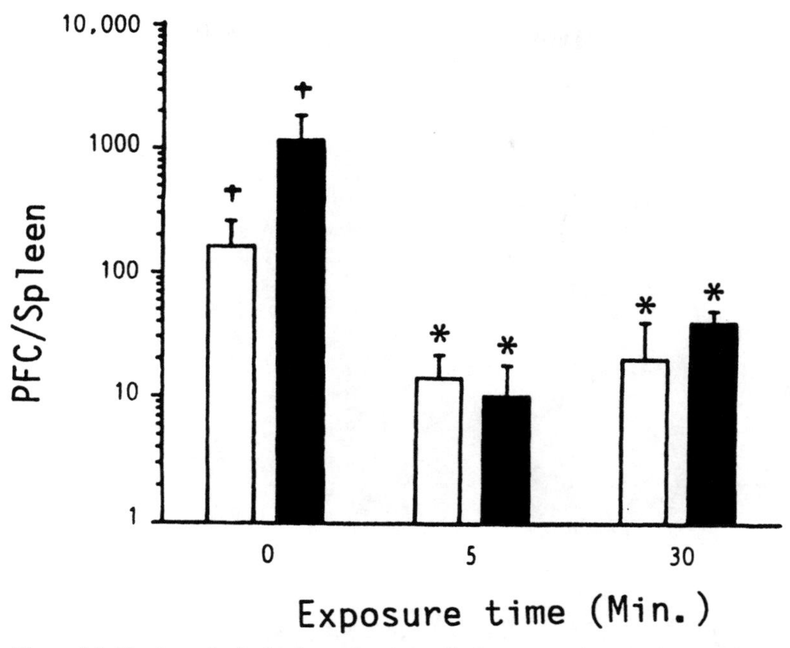

Figure 14. Number of splenic plaque-forming cells from trout given 5 min, 30 min, or no treatment of 10 ppm phenol before immunization by bath. (Open bars) 10 μg/ml dose; (filled bars) 100 μg/ml dose. No background PFCs were evident in these experiments when no immunization was given. Error bars represent SEM ($n = 5$). † denotes the difference between immunogen doses: * denotes significant difference of phenol treatment compared to fish without treatment ($P < 0.01$) (From Anderson et al. 1990. Copyright ASTM. Reprinted with permission.)

gens. Common problems arise when considering disease challenges to determine level of resistance from immunization or immunomodulations from pollutants or drugs. By injecting fish with virulent pathogen doses, exact numbers of pathogens can be calculated, but fish are rarely exposed to pathogens in this way in the environment. When water-borne exposures are used, such as pouring pathogens in the water or dipping the fish into a pathogen suspension, a natural exposure is simulated; however, the exact pathogen dosage to individual fishes is not under control. As an alternative to these artificial methods, Fryer et al. (1972) used an environmental challenge in order to test the bacterin's effectiveness. They transferred immunized and control fish from the pathogen-free area where they had been raised and immunized with a vibriosis bacterin to an estuary where the *Vibrio* sp. pathogens were endemic. Unfortunately, such environmental situations are not available for most immunotoxicological tests. In all cases disease challenges have to be standardized in the individual laboratory, following guidelines for immunological assays as suggested by Zeeman (1986).

LD50 values ("lethal doses" of a pathogen required for 50% mortality of the animals tested) are used to determine resistance after immunization, analogous to the LC50s used for ascertaining lethal exposure concentrations of toxicants. These doses are usually administered by injection. Inconsistent results may arise because when fish age they may become more resistant, and there are variances in resistance among fish from different sources. Nevertheless, the immunological questions remain. Can fish resist infections after contaminant treatments? Can that be demonstrated in a meaningful manner?

IMMUNOTOXICOLOGY—EFFECTS OF CLASSES OF TOXICANTS

Immunotoxicology of fish is a new field of study that is expanding as pathologists acquire more sophisticated methodologies and sensitive equipment to measure subtle changes in the hematopoietic and immune systems. Publications from previous decades describe changes from toxicant exposure in fish in hematopoietic systems in an observational manner. For instance, alterations in hematocrit and leukocrit, MCH, and MCV due to stress and the presence of chemicals were noted. As more sensitive and diverse assays became available, functional changes in the immune system due to contaminant exposure have been increasingly reported in fish. Because many of the contaminants affecting protective immune systems in the environment are nonlethal, the problem is to show if chronic exposure to toxicants makes

Figure 15. Blood can be taken from the caudal artery of anesthetized trout. This procedure can be repeated several times from the same fish to show the changes in the kinetics of an immune response due to the presence of a contaminant.

fish more susceptible to diseases. In the following section, examples from the literature are summarized to demonstrate effects of metals, pesticides, drugs, polychlorinated biphenyls (PCBs), and other toxicants on fish protective mechanisms and disease susceptibility.

Metals

The presence of metals, including copper, aluminum, cadmium, mercury, arsenic, and selenium, in the aquatic environment may adversely affect fish health (Table 3). In particular, the mechanisms that protect fish against diseases are affected (Hetrick et al., 1979; Zeeman and Brindley, 1981; Sorensen and Bauer, 1983; MacFarlane et al., 1986; Banerjee and Kumari, 1988; Goss and Wood, 1988; Thuvander, 1989).

Direct injections of lead or cadmium in brown trout (*Salmo trutta*) immunized with an MS2 bacteriophage reduced the resultant specific circulating antibody response (O'Neill, 1981b). The fish were given high doses of the metal by an unnatural route. Thuvander (1989) showed that when rainbow trout immunized with a vaccine against *V. anguillarum* were exposed to cadmium in doses of 0.7 or 3.6 μg/L in the water for 12 wk, no clinical or histological changes were observed; however, the cellular response as indicated by exposure of lymphocytes to mitogens was reduced. Conversely, fish exposed to cadmium had higher humoral antibody levels, but no differences in protection against the disease were observed between the cadmium-exposed and controls in the immunized fish.

In high concentrations, most metals seem to affect the hematopoietic system, as observed by the decrease in circulatory erythrocytes or change in the number of blast cells in the kidney. Dogfish (*Scyliorhinus canicula*) exposed to cadmium had lower hematocrits (Tort and Torres, 1988), as did rainbow trout exposed to arsenic contamination (Oladimeji et al., 1984). Under some conditions, the presence of low concentrations of metals like cadmium can elevate immune parameters and protection against the disease caused by *Cytophaga columnaris* (MacFarlane et al., 1986). This phenomenon is not yet

Table 3. Effects of metals on fish indicated by changes in immune response parameters[a]

Metal	Fish species	Immunizing and (indicator) antigens	Immune[b] parameter affected	Concentration and route	Reference
Aluminum	Rainbow trout	None (multiple[c])	Reduced CL	10 μg/ml (in vitro)	Elsasser et al. (1986)
Arsenic	Rainbow trout	None (latex beads)	Phagocytosis elevated at low levels, reduced at high levels	1–100 μg (in vitro)	Thuvander et al. (1987)
Cadmium	Rainbow trout	*V. anguillarum*	Enhanced SA	0.7 or 3.6 μg/L in water, 12 wk	Thuvander (1989)
	Goldfish	None (cell numbers followed)	Lymphocyte number and mitogenic response reduced, granulocytes increased	90 or 445 μg/g in water, 3 wk	Murad and Houston (1988)
	Rainbow trout	None (multiple[c])	Reduced CL after	1 μg/ml (in vitro)	Elsasser et al. (1986)
	Cunners	*Bacillus cereus* (same)	Heightened SA	12 μg/ml in water, 96 h	Robohm (1986)
	Striped bass	*Bacillus cereus* (same)	Lowered SA	10 μg/ml in water, 96 h	Robohm (1986)
Chromium	Brown trout, carp	MS2 bacteriophage (same)	SA greatly reduced	1.01 μg/L in water, 38 wk	O'Neill (1981a)
Copper	Rainbow trout	None (multiple[c])	Reduced CL	10 μg/ml (in vitro)	Elsasser et al. (1986)
	Brown trout and carp	MS2 bacteriophage (same)	SA greatly reduced (by neutralization)	0.29 μg/L in water, 38 wk	O'Neill (1981a)
	Rainbow trout	*Yersinia ruckeri* (same)	PFC reduced	0.1, 1.0, 10, or 100 μg/ml (in vitro)	Anderson et al. (1989)
	Eel	None (*Vibrio anguillarum*)	Increased susceptibility to *V. anguillarum*	30 to 60 μg/L in water for 150+ d	Rodsaether et al. (1977)
	Rainbow trout	None (IHNV[d])	Increased susceptibility to IHNV	2.1, 3.9, 5.1, or 9.5 μg/L in water, 9 d	Hetrick et al. (1979)
Lead	Brown trout	MS2 bacteriophage (same)	SA reduced (by neutralization)	0.01 to 0.3 μg/L injection	O'Neill (1981b)
Mercury	*Barbus conchonius*	None (cell numbers followed)	Lymphocyte number reduced	181 μg/ml (mercurin chloride) in water, 96 h	Gill and Pant (1985)
	Blue gourami	IPNV (same)	Increased susceptibility to IPNV	9 ppb in water, 3 wk	Roales and Perlmutter (1977)

[a] Experiments often involve immunizing the fish with an antigen, subjecting the fish to contaminant exposure, then using that antigen (indicator antigen, in parentheses) in an immunological assay to test for changes in the response. In some cases the fish were not immunized before the contaminant exposure, and just general changes in the response measured. Likewise, some of the immune assays didn't involve the use of the homologous antigen; general responses or responses to multiple antigens were followed.

[b] PFC = plaque-forming cells; SA = serum antibody; CL = chemiluminescence, phagocytosis.

[c] *Staphylococcus aureaus, Aeromonas hydrophila, Yersinia ruckeri.*

[d] IHNV = Infectious hematopoietic necrosis virus; IPNV = Infectious pancreatic necrosis virus.

completely understood, but it may be, as some scientists suggest, that metals act on suppressing actions of the pathogen rather than directly on the fish. Other hypotheses to explain this phenomenon include the concept of hormesis: minimally stressed animals sometimes show higher physiological responses than animals at rest or without stresses. Low levels of metal stress may be beneficial!

Copper is toxic to fish; fish culturists avoid the use of copper in tubing and pipes in hatcheries. Demonstrations of the effects of copper on the immune response in fish are numerous. In vitro experiments using copper as the contaminant in tissue culture media holding immunopoietic cells from the anterior kidney of rainbow trout demonstrated that copper causes a decreased phagocytic response when measured by chemiluminescence assays (Elsasser et al., 1986). Numbers of antibody-producing cells were reduced after incubating immunized spleen sections in vitro in tissue culture media containing higher concentrations of copper (Anderson et al., 1989). The Jerne assay was conducted 10 d after in vitro immunization with *Y. ruckeri* to detect the activity of the antibody-producing cells in lysing antigen-labeled sheep red blood cells.

Hetrick et al. (1979) showed in laboratory-controlled studies that salmonids held in low dilutions of copper and exposed to infectious hematopoietic necrosis virus became more susceptible to the disease. Knittel (1981) demonstrated that copper makes rainbow trout more susceptible to *Y. ruckeri* bacterial infections. Although these researchers could not pinpoint defensive systems that were most affected, they noted that copper also affects other physiological systems besides the immune response in fish. In a similar study with columnaris for disease challenges, MacFarlane et al. (1986) showed that moderate concentrations (4–10 times the average environmental background) of arsenic made striped bass (*Morone saxatilis*) more susceptible to columnaris disease, whereas fish treated at these levels with copper, cadmium, lead, and selenium actually showed increased disease resistance. Thuvander et al. (1987) using fluorescently stained latex beads, showed increased uptake of these particles by phagocytic cells from rainbow trout held in low concentrations of arsenic (1 μg/ml); higher concentrations decreased the activity.

Selenium, like vitamins C and E, seems to be a micronutrient necessary for a healthy immune system in most animals. When this heavy metal is deficient in feed, fish may become immunosuppressed by having a weaker peroxidase defense, so their own cells may be damaged during oxygen radical release. Comparisons between hatchery-reared and wild coho salmon showed the survival rates were higher in the wild groups with the higher selenium levels (Felton et al., 1990).

Biologists would like to use the immunological parameters discussed here as monitoring tools or indicators (i.e., biomarkers) to detect metals that fish have been exposed to in the wild. If fish are found to have some changes in immunological parameters, there may be some substantiation for this idea that metals are contaminating or have contaminated their environment. In most cases, this information might be coupled to changes in other physiological parameters. A good comparison test might be to measure the levels of metallothionein in the liver, as this metal-detoxifying enzyme is elevated during cadmium stress in brook trout livers (Hamilton et al., 1987).

Acid Rain

Acidic effluents discharged into the water and air by industry have changed the aquatic environment for fish indirectly by destroying the phytoplankton and zooplankton food base and directly by acting destructively on the animals' epidermal surfaces. Acidic water conditions may also exacerbate influences of other contaminants on fish. Aluminum, cadmium, copper, and mercury become potentially more stressful to striped bass and rainbow trout in acidified waters (Hall et al., 1988; Akielaszek and Haines, 1981). Similar signs are evident when carbamate pesticides are present in high concentrations (Mayer and Ellersieck, 1986). Fish under the stress of acidification showed plasma cortisol raised to levels that could be harmful to immune response capabilities (Barton et al., 1985). A bibliographic report with abstracts details over 200 publications by the U.S. Fish and Wildlife Service on acid rain (Villella, 1989). Salmonids often inhabit rivers and lakes influenced by acid rain. These fish survive best in the pH range from 5 to 9 (Mayer et al., 1984). Rainbow trout had progressively higher hematocrits and hemoglobin concentrations in parallel with heightened acidification (Giles et al., 1984).

Pesticides

There is great concern about the increased use of pesticides for agriculture, silviculture, and public health protection (FAO, 1986). Sublethal levels of pesticides may accumulate in fish from the water or food. By their nature, these diverse groups of toxic chemicals are often persistent in the environment and leach into aquatic ecosystems. Effects of pesticides and their breakdown products on fish need to be thoroughly investigated; information on immunotoxicology of pesticides is summarized in Table 4. Pesticides may act strongly on fish populations by indirectly affecting food chains.

Bennett and Wolke (1987a) studied sublethal effects of the organochlorine compound endrin on rainbow trout. Exposure for over 60 d had no effect on the phagocytic ability of peritoneal macrophages; however, the migration inhibition factor, numbers of PFCs, and serum agglutination titers were all reduced compared with values for untreated controls. In a follow-up study, they confirmed that the elevation of serum cortisol had a central role in repression

Table 4. Effects of pesticides on fish indicated by changes in immune response parameters[a]

Pesticide	Fish species	Immunizing and (indicator) antigens	Immune[b] parameter affected	Concentration and route	Reference
Nickel	Brown trout, carp	MS2 bacteriophage (same)	SA reduced (by neutralization)	0.75 μg/L in water, 38 wk	O'Neill (1981a)
Zinc	Brown trout, carp	MS2 bacteriophage (same)	SA reduced (by neutralization)	1.06 μg/L in water, 38 wk	O'Neill (1981a)
	Zebra fish	IPNV (*Proteus vulgaris* and IPNV)	Increased susceptibility to *P. vulgaris* but not immunizing antigen	0.3 ppm in water, 5 wk	Sarot and Perlmutter (1976)
Endrin	Rainbow trout	*Yersinia ruckeri* O-antigen (same)	No effect on phagocytosis, MIF, PFC SA reduced	0.12–0.15 μg/L in water, 30 d and longer	Bennett and Wolke (1987a)
Malathion	Channel catfish	(Lymphocyte) number	Reduced lymphocyte count	4.5 mg/L in water, 96 h	Areechon and Plumb (1990)
Methyl bromide	Medaka	None (thymic tissue)	Thymic necrosis	1.8 mg/L in water, 1–3 mo	Wester et al. (1988)
Trichlorphon	Carp	*Y. ruckeri* (same)	No effect	10–20,000 ppm in water	Cossarini-Dunier et al. (1990)
	Carp	None (*Pseudomonas alcalignes*	Reduced phagocytic neutrophilic activity and lysozyme	10,000 ppm in water	Siwicki et al. (1990b)
DDT	Goldfish	SRBC (same)	Reduced PFC, reduced SA	10–50 mg/kg injection	Sharma and Zeeman (1980)
Lindane	Carp	*Y. ruckeri* (skin graft rejection)	No effect	1000 mg/kg of food for 1 mo	Cossarini-Dunier et al. (1987)
Atrazine	Carp	*Y. ruckeri* (skin graft rejection)	No effect	1,000 mg/kg of food for 1 mo	Cossarini-Dunier et al. (1987)
Bayluscide	African catfish	SRBC (same)	Reduced SA, numbers of Stannius corpuscles increased	0.1–0.3 ppm in water for 6 mo	Faisal et al. (1988)
Tributyltin	Oyster toadfish Hogchoker Croaker	None (*E. coli*)	Reduced CL	4–400 μg/L in vitro Brief exposure	Wishkovsky et al. (1989)
	Toadfish	None (mitogens)	Reduced CL	0.25–25 mg/kg	Rice and Weeks (1990)

[a] Experiments often involve immunizing the fish with an antigen, subjecting the fish to contaminant exposure, then using that antigen (indicator antigen, in parentheses) in an immunological assay to test for changes in the response. In some cases the fish were not immunized before the contaminant exposure, and just general changes in the response measured. Likewise, some of the immune assays didn't involve the use of the homologous antigen; general responses or responses to multiple antigens were followed.

[b] MIF = macrophage inhibition factor; PFC = plaque-forming cells; SA = serum antibody; CL = chemiluminescence, phagocytosis.

of the immune response (Bennett and Wolke, 1987b). Lindane (γ-hexachlorocyclohexane) given in the feed up to concentrations of 1000 ppm had no effect on hematocrit levels or antibody production in fish immunized with a *Y. ruckeri* bacterin (Cossarini-Dunier et al., 1987).

Some pesticides can be present in remarkably high concentrations, yet show little toxicity or lethality to fish populations. Methyl bromide, a fumigant used against insects, was found to cause thymic necrosis when Japanese medaka (*Oryzias latipes*) were held in concentrations of 1.8 mg/L for over a month; guppies (*Poecilia reticula*) showed no observed lethal exposure concentration value in 10,000 mg/L after 3 mo with sodium bromide (Wester et al., 1988).

Carbamyl analogues such as Sevin generally have been reported to be innocuous to fish (Woodward and Mauck, 1980).

Sublethal concentrations of malathion, an organophosphate insecticide, caused hematological and histopathological damage to channel catfish (*Ictalurus punctatus*) (Areechon and Plumb, 1990). Deformities were found in the vertebrae and blood samples showed increases in numbers of erythrocytes and decreases in numbers of leukocytes. Jeney and Jeney (1986) also reported a rise in hematocrit levels after high-dose (2500 mg/L) treatment with trichlorphon, an organophosphate used in fisheries for control of planktonic invertebrates and prevention of certain parasite infections. Tributyltin, used as an antifouling agent in boat paints, has been shown to have a suppressive effect on chemiluminescence of oyster toadfish (*Opsanus tau*) macrophages. Rice and Weeks (1990) administered doses of tributyltin of 0.25 to 2.5 mg/kg in vivo weekly for 6 wk and found dose-dependent reductions in chemiluminescence. They proposed that calcium flux across the macrophage membrane may be suppressed, impairing function.

Direct effects of herbicides on fish can be detected in laboratory trials. However, indirect effects of herbicides on fish may be more devastating by depleting phyto- and zooplankton communities.

The most commonly used herbicides, Roundup, a glyophosate, and Bolero, a thiobencarbamate, have low to intermediate acute toxicities for rainbow trout (Finlayson and Faggella, 1986; and Mitchell et al., 1987). Concentrations of Roundup at 10 times the levels encountered in the environment after aerial spraying did not affect hematocrit readings of coho salmon (*O. kisutch*) smolts (Mitchell et al., 1987).

The herbicide atrazine was placed in diets fed to carp (*Cyprinus carpio*) for 84 d at 1000 ppm (Cossarini-Dunier et al., 1988). Changes were seen in immunological parameters of splenic index and numbers of PFCs in these carp, which were subsequently immunized with a *Y. ruckeri* bacterin.

Polychlorinated Biphenyls and Other Aromatics

The widespread use of PCBs for many years in electrical transformers and their environmental persistence make them dangerous aquatic contaminants. Aroclor 1254 and Aroclor 1232 are PCBs which have been discussed in the previous section. PCBs, polychlorinated dibenzodioxins (PCDDs), and polychlorinated dibenzofurans (PCDFs) are often lipophilic and are found as trace contaminants in many commercial products (Mehrle and Petty, 1987). Results of immunotoxicology data on PCBs and other aromatics are shown in Table 5.

Aroclor 1254 added to diets of rainbow trout for up to 12 mo had no effect on a PFC response (Cleland et al., 1988). Jones et al. (1979) injected the Aroclor 1232 intraperitoneally into channel catfish (*I. punctatus*) immunized with an *Aeromonas hydrophila* bacterin. Although none of the immune parameters such as circulating antibody titers were diminished, the cellular function of phagocytosis was reduced. After fish were challenged, all of the PCB-exposed fish died; in contrast, there were no mortalities in the controls. Stolen (1985), using Aroclor 1254 in much the same immunization regimen, found a slightly suppressed delay in the humoral immune response in fish immunized with *E. coli.*

Disease challenges were given to rainbow trout after exposure to waste transformer oil and PCBs, Aroclors 1254 and 1260 (Mayer et al., 1985). Exposure to the PCBs increased disease resistance when the *Y. ruckeri* disease agent was given by flush (external) challenge, but when the pathogen was administered by intraperitoneal injection contaminated fish died more rapidly than the controls. Spitsbergen et al. (1988), in contrast, found that even high doses of Aroclor 1254 did not make rainbow trout more susceptible to infectious hematopoietic necrosis viral infections.

In a study of effects of the transfer of halogenated aromatic hydrocarbons (HAHs) through food chains, Cleland et al. (1989) fed coho salmon (*O. kisutch*) that had naturally bioaccumulated HAHs to mice and checked for immunosuppressive effects in the mice by immunizing mice with SRBCs. Although no changes appeared in total numbers of spleen lymphocytes (total T lymphocytes or T lymphocyte subsets as determined by flow cytometry), the humoral response was suppressed in proportion to elevated HAH levels.

The effect of environmental polyaromatic hydrocarbons (PAHs) on spot (*Leiostomus xanthurus*) and hogchoker (*Trinectes maculatus*) was investigated by Weeks and Warinner (1986). Peritoneal macrophages were isolated from fish and placed in Boyden chambers, with *E. coli* as a chemotactic stimulus. This apparatus demonstrates active cellular migration by placing cells on one side of a filter and the attractant on the other. As the macrophages are stimulated, they migrate through the filter, similar to what might be found in an inflammatory response in vivo. Lower macrophage migrations were found in fish sampled from PAH-contaminated wild environments of the Elizabeth River in Virginia, showing loss of this activity due to the presence of the contaminant.

Phenol and related cyclohexanes are common aquatic contaminants from paper processing and paint industries. Fish have a natural avoidance response to phenol levels as low as 3.2 mg/L (DeGraeve, 1982). Exposures as short as 2 h at 6.5 mg/L can cause inflammatory gill damage, leaving only naked pillar cells (Mitrovic et al., 1968). McLeay and Brown (1979), working with bleached draft pulp mill effluents, showed that fish had reduced leukocrits after sublethal exposures. In a study of the effects of phenol on the specific immune response, Anderson et al. (1990) showed that splenic antibody-producing cells were reduced in rainbow trout when they were exposed to phenol at 10 ppm for 5 min before an immunization bath in *Y. ruckeri* O antigen (Figure 14). They speculated that the gills may have been temporarily damaged; thus the bacterin was not taken up in their functional assay.

Table 5. Effects of some environmental contaminants on fish immune response[a]

Chemical	Fish species	Immunizing and (indicator) antigens	Immune[b] parameter affected	Concentration and route	Reference
Phenol	Rainbow trout	*Yersinia ruckeri* O-antigen (same)	PFC depressed	10 ppm in water, 5 min	Anderson et al. (1990)
Benzidine		None	Nonspecific rise in agglutinins	1 ppm in water	Middlebrooks and Meador (1984)
Polychlorinated biphenyls (PCBs)					
Aroclor 1254	Rainbow trout	SRBC (same)	No effect on PFC	Diet	Cleland et al. (1988)
Aroclor 1254	Coho salmon	SRBC (same)	Depressed PFC	Contaminated food	Cleland et al. (1989)
Aroclor 1254	Summer flounder	*Escherichia coli* bacterin (same)	Slight suppression of SA	5μg/g injection	Stolen (1985)
Aroclor 1232	Channel catfish	*Aeromonas hydrophila* (same)	Increased disease susceptibility, decreased phagocytosis	Injection 70 mg/kg	Jones et al. (1979)
Aroclor 1254/1260	Rainbow trout	None (*Yersinia ruckeri*)	Disease susceptibility increased (by injection), decreased (by flush exposure)	0.38 to 6.0 μg/L in water for 90 d	Mayer et al. (1985)
Chlorinated dioxin (TCDD)	Rainbow trout	SRBC (same)	Mitogenic response partially suppressed	0.1–10 μg/kg injection	Spitsbergen et al. (1986)
Polynuclear aromatic hydrocarbons (PAHs)	Spot	None (*E. coli*)	Macrophage response suppressed	Unknown (fish in rivers)	Weeks and Warner (1986)
	Hogchoker	None (*E. coli*)	Macrophage response suppressed	Unknown (fish in rivers)	Weeks and Warner (1986)
	Flounder	None (MMC) numbers	Reduced MMC numbers	25–50 ppm, 4 mo on sediments	Payne and Fancey (1989)

[a] Experiments often involve immunizing the fish with an antigen, subjecting the fish to contaminant exposure, then using that antigen (indicator antigen, in parentheses) in an immunological assay to test for changes in the response. In some cases the fish were not immunized before the contaminant exposure, and just general changes in the response measured. Likewise, some of the immune assays didn't involve the use of the homologous antigen; general responses or responses to multiple antigens were followed.

[b] PFC = plaque-forming cells, MMC = melanomacrophage center; SA = serum antibody.

Crude Oil and Dispersion Chemicals

It is difficult to make specific comments about the toxicity of crude oil because the mixture of volatile and long-chain hydrocarbons often contains traces of heavy metals. In addition, environmental damage caused by oil spills is dependent on temperature, water current source and composition, and cleanup efforts. Hodgins et al. (1977) reviewed the literature on the potential effects of petroleum in marine fish, gathering information prior to the Alaskan oil exploration. Much of their review concerned the effects of heavy metals and hydrocarbons, as discussed here previously. In addition, they pointed out that higher occurrences of tumors were apparent in fish in oil-contaminated sites. They implied that immunosuppression may result from the presence of these compounds. In some cases the mortalities were exacerbated by latent viral infections.

Detergents (see Chapter 1) are often used in controlling dispersion of crude oil spills. Discretionary use is advised, and little information about detergents is available in fish immunotoxicity. In one study, a detergent wash of rainbow trout before immunization resulted in fewer PFCs (Anderson et al., 1984). However, it is doubtful that the high concentrations used in these experiments would ever be found in point spills.

Drugs

Drugs, chemicals, and antibiotics are used extensively for treatment of fish diseases and may be

Table 6. Effects of antibiotics, drugs, and other chemicals on fish indicated by changes in immune response parameters[a]

Chemical	Fish species	Immunizing and (indicator) antigens	Immune[b] parameter affected	Concentration and route	Reference
Oxytetracycline	Carp (*C. carpio*)	None (PHA, LPS)[b]	Mitogenic response of leukocytes	10 μg/ml (in vitro)	Grondel et al. (1982)
	Rainbow trout	*Y. ruckeri* O-antigen (same)	Reduced PFC[c] in feed	3333 ppm in feed, 27 d	van Muiswinkel et al. (1985)
	Rainbow trout	*Y. ruckeri* O-antigen (same)	Reduced PFC	10 mg/kg injection	Siwicki et al. (1989)
Aflatoxin B_1	Rainbow trout	TNP-KLH (same)	Loss of B cell memory	0.5 ppm for 1/2 to 1 h as eggs	Arkoosh and Kaattari (1987)
Cortisol/ Kenalog-40	Rainbow trout	*Y. ruckeri* O-antigen (same)	Reduced PFC	2 to 200 mg/kg injection	Anderson et al. (1982)
Hydrocortisone	Striped bass	None (*Aeromonas hydrophila*)	Reduced phagocytosis (chemiluminescent) (in vitro)	25 to 280 μg/ ml (in vitro)	Stave and Roberson (1985)

[a] Experiments often involve immunizing the fish with an antigen, subjecting the fish to contaminant exposure, then using that antigen (indicator antigen, in parentheses) in an immunological assay to test for changes in the response. In some cases the fish were not immunized before the contaminant exposure, and just general changes in the response measured. Likewise, some of the immune assays didn't involve the use of the homologous antigen; general responses or responses to multiple antigens were followed.
[b] PHA = phytohemagglutinin; LPS = lipopolysaccharide.
[c] PFC = plaque-forming cells; TNP-KLH = trinitrophenyl–keyhole limpet hemocyanin.

responsible for immunomodulation (Table 6). The five chemicals and drugs presently approved by the U.S. Food and Drug Administration for therapeutic treatment of salmonid diseases are oxytetracycline, sulfadimethoxine/ormetroprim, formaldehyde, salt (NaCl), and acetic acid. Some drugs to combat fish diseases have been withdrawn because of heavy metal accumulation in tissues. Pyridyl mercuric acid and ethyl mercury phosphate, for example, are effective in controlling gill diseases in trout. However, the accumulation of mercury from the drug in tissues is dangerous to consumers (Amend, 1970). Another agent, malachite green, effective against fungal infections, was withdrawn because of potential carcinogenic properties (Meyer and Jorgenson, 1983).

Some drugs are known immunosuppressants in fish. Cortisol and hydrocortisone have been used in several experiments showing a diminished immune response as assayed by a lower chemiluminescent response of anterior kidney cells in striped bass (Stave and Roberson, 1985) or by reduced numbers of plaque-forming cells in salmonids (Anderson et al., 1982). High doses of Kenalog-40, a synthetic corticosteroid, depleted circulatory lymphocyte populations in rainbow trout; however, if the fish were healthy and not carrying any pathogens, the lymphocyte numbers returned to normal within 4 wk (Anderson et al., 1982).

Oxytetracycline is immunosuppressive in reducing numbers of plaque-forming cells in several studies (Grondel et al., 1987; van Muiswinkel et al., 1985; Siwicki et al. 1989). Although use of the antibiotic for the prevention of bacterial diseases is still recommended, fish culturists should observe discretion in use and be aware that drug treatments may interfere with immunization regimens. A promising new drug for treating fish bacterial agents is oxolinic acid. It was found not to have immunosuppressive effects by the Jerne assay when injected with bacterins (Siwicki et al., 1989).

Radiation

Zeeman and Brindley (1981) extensively reviewed the scientific literature on the effects of radiation on fish immune structures and functions. They showed that the effects of radiation on fish are similar to those on mammals. Further information was contributed by Nakanishi (1986), who showed that when the thymus was excised from rockfish (*Sebasticus marmoratus*) the animals' immune system did not recover from X-irradiation. Likewise, Chilmonczyk and Oui (1988) demonstrated that rainbow trout subjected to gamma-irradiation were more susceptible to infection with fish pathogens.

SUMMARY, CONCLUSIONS, AND FUTURE RESEARCH

The organizations and individuals that are or should be interested in the immunotoxicology of fish are diverse. Universities, state and federal fish hatcheries, contract laboratories, pet stores, distributors of live fish for human consumption, and others need ready access to healthy stocks of fish.

Fish live intimately with their environment and the health of the immune system reflects their surroundings. Hematopoietic and immunopoietic assays were discussed as a means of monitoring the functioning of the immune system. These assays are now ready for use as diagnostic tools in immunotoxicology. This review has provided additional evidence that fish may be exposed to chemicals that are capable of compromising the immune response(s). Several heavy metals, pesticides, drugs used in aquaculture, PCBs, and other aromatic contaminants (such as TCDD) have been identified as frequent environmental contaminants that can interfere with normal fish immune responses.

The immune response of fish plays a key role in the control of fish diseases and in the normal surveillance of fish tumor cells. Nonspecific and specific immune mechanisms of fish were identified that can be useful in preventing or controlling disease outbreaks. The extensive review presented the types of immune test assays and methodologies that may be useful to fish culturists. In the past, these assays were all done in vivo; however developments of in vitro tests promise to give immunotoxicologists more reliable information because of better-controlled laboratory conditions. In addition, fewer test animals are used (Figure 16). The effect of normal biological rhythms in fish immunology on the results of fish toxicity tests was also discussed.

Aquatic toxicologists frequently use fish as their experimental animals and are justifiably concerned when these animals are not received in a healthy condition or cannot be kept healthy in an environment that is conducive to disease. For more consistency in the evaluation of immunology-toxicology studies of fish, suggestions are made for greater control of experiments. Zeeman (1986) includes the following suggestions:

1. Where reasonable and feasible, experimental fish in immunology (and disease) studies need to be suitably stabilized and maintained in a defined light-dark cycle.
2. The basic circadian (seasonal or other) rhythms in fish hematology and immunology need to be determined in representative fish species (e.g., patterns of minimum and maximum sensitivity to immunizing agents or to immunomodulators, numbers of circulating leukocytes).
3. Fish immunology researchers using the in vitro techniques of cell, tissue, and organ culture need to know that biological rhythms are likely to persist in experimental organisms.
4. When in vivo and in vitro background immunology information is unavailable for an experiment, researchers must determine whether each specific experimental procedure should be done at the same time every day [in actuality, at the same

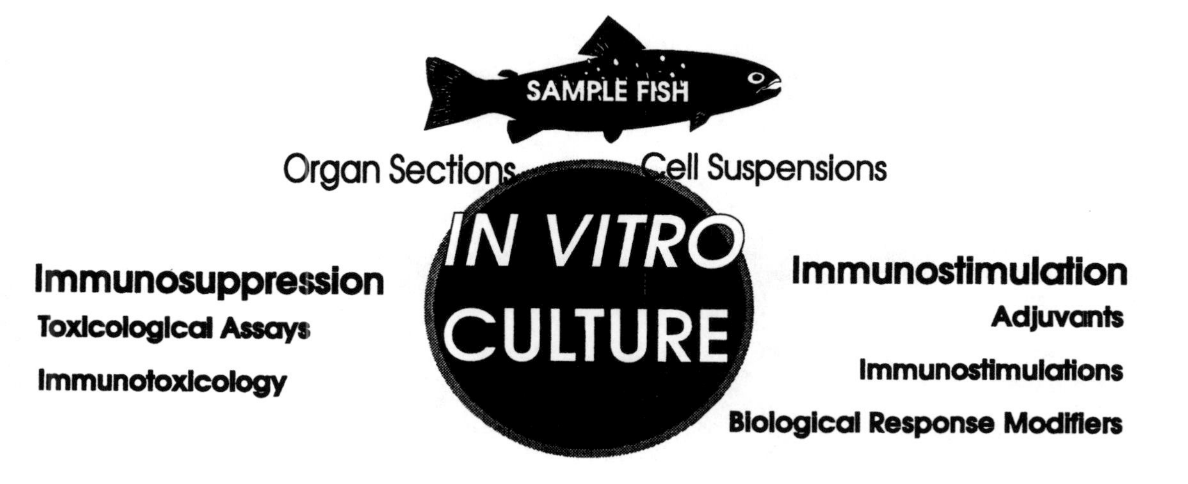

Figure 16. In vitro culture of trout cell suspensions or organ sections and exposing the immunopoietic cells to contaminants or immunostimulants have some advantages over in vivo methods where fish are held in flowing water. For example, in vitro cultures can be incubated in a precisely temperature-controlled, incubator environment; fewer fish are necessary for obtaining multiple samples, damping the problems of variability among individual fish; and cells or organs can be shipped from sample site to laboratory more easily than live fish.

period(s) during the light or dark phase of the photoperiod]. Procedures that have to be controlled are sampling of blood and other immunological tissues, immunization and exposure to other antigens, and dosing of fish or their cells or tissues with compounds or agents that modify their immune response.
5. Nonspecific items, such as stress, can seriously affect immune systems. Time and duration of such occurrences have to be recorded. The same stressor may have a different influence at different times of day.
6. Reviewers of research articles in fish immunology and funding agencies have to consider the significance of fish circadian and seasonal rhythms. Journals and books on fish immunology studies must be encouraged to acknowledge the importance of documenting such factors as environment and age of fish.

In an ideal situation, damage by a single chemical would leave a specific deficiency or lesion in the immune pathway that is distinctive for that chemical. The chemical could then be traced by the damage to the immune parameter(s). Unfortunately, exposure to chemicals is not that distinctive because, in addition to the immune system, other tissues and cells are usually damaged. Nevertheless, if coupled with other physiological assays, gross observations, and environmental records, these immunological assays can contribute much to informing us about a toxicant that affected a fish species or about the chronic effects of continuous exposure to chemicals. Fish culturists need to be aware of the variety of nonspecific defense mechanisms and specific immune response actions and their interactions with chemical agents that could adversely affect the normal functioning of fish health.

ACKNOWLEDGMENTS

We thank Mrs. Robin B. Owens for manuscript preparation, Ms. Joyce A. Mann for assisting in literature searches, and Dr. Joanne S. Stolen for preparing graphics.

LITERATURE CITED

Adams, S. M. (ed.): Biological Indicators of Stress in Fish. Bethesda, MD: American Fisheries Society Symposium 8, 1990.

Agius, C., and Couchman, W.: Induction of enhanced alkaline phosphatase activity in the melano-macrophage centres of *Oreochromis aureus* (Steindachner) through starvation and vaccination. *J. Fish Biol.*, 28:87–92, 1986.

Agius, C., and Roberts, R. J.: Effects of starvation on the melano-macrophage centres of fish. *J. Fish Biol.*, 19: 161–169, 1981.

Akielaszek, J. J., and Haines, T. A.: Mercury in the muscle tissue of fish from three northern Maine lakes. *Bull. Environ. Contam. Toxicol.*, 27:201–208, 1981.

Amend, D. F.: Retention of mercury by salmon. *Prog. Fish Cult.*, 32:192–194, 1970.

Amend, D. F., and Fender, D. C.: Uptake of bovine serum albumin by rainbow trout from hyperosmotic solutions: a model for vaccinating fish. *Science*, 192:793–794, 1976.

Anderson, D. P.: Fish immunology, Book 4. *Diseases of Fishes*, edited by S. F. Snieszko, and H. R. Axelrod, Neptune, NJ: T.F.H. Publications, 1974.

Anderson, D. P.: Immunological indicators: effects of environmental stress on immune protection and disease outbreaks. *Biological Indicators of Stress in Fish*, edited by S. M. Adams, pp. 38–50. Bethesda, MD: American Fisheries Society Symposium 8, 1990.

Anderson, D. P., and Hennessen, W., (eds.): 1981. Fish Biologics: Serodiagnostics and Vaccines. *Dev. Biol. Standard*, Vol. 49. Basel, Switzerland: Karger.

Anderson, D. P.; Roberson, B. S.; and Dixon, O. W.: Immunosuppression induced by a corticosteroid or an alkylating agent in rainbow trout (*Salmo gairdneri*) administered a *Yersinia ruckeri* bacterin. *Dev. Comp. Immunol.*, Suppl. 2:197–204, 1982.

Anderson, D. P.; Dixon, O. W.; and van Ginkel, F. W.: Suppression of bath immunization in rainbow trout by contaminant bath pretreatments. *Chemical Regulation of Immunity of Veterinary Medicine*, edited by M. Kende; J. Gainer; and M. Chirigos, pp. 289–293. New York: Alan R. Liss, 1984.

Anderson, D. P.; Dixon, O. W.; Bodammer, J. E.; and Lizzio, E. F.: Suppression of antibody-producing cells in rainbow trout spleen sections exposed to copper in vitro. *J. Aquat. Anim. Health*, 1:57–61, 1989.

Anderson, D. P.; Dixon, O. W.; and van Muiswinkel, W. B.: Reduction in the numbers of antibody-producing cells in rainbow trout, *Oncorhynchus mykiss*, exposed to sublethal doses of phenol before bath immunization. Aquatic Toxicology and Risk Assessment, Vol. 13, ASTM STP 1096, edited by W. G. Landis, and W. A. H. van der Schalie, pp. 331–337. Philadelphia: American Society for Testing and Materials, 1990.

Anderson, D. P.; Moritomo, T.; and de Grooth, R.: Neutrophil, glass-adherent nitroblue tetrazolium assay gives early indication of immunization effectiveness in rainbow trout. *Vet. Immunol. Immunopathol.*, 30:419–429, 1992.

Areechon, N., and Plumb, J. A.: Sublethal effects of malathion on channel catfish, *Ictalurus punctatus*. *Bull. Environ. Contam. Toxicol.*, 44:435–442, 1990.

Arkoosh, M. R., and Kaattari, S. L.: Effect of early aflatoxin B_1 exposure on in vivo and in vitro antibody responses in rainbow trout, *Salmo gairdneri*. *J. Fish Biol.*, 31 (suppl A):19–22, 1987.

Austin, B., Austin, D.A. (eds.): Methods for the microbiological examination of fish and shellfish. Chichester: Ellis Horwood, 1989.

Banerjee, V., and Kumari, K.: Effect of zinc, mercury and cadmium on erythrocyte and related parameters in the fish *Anabas testudineus*. *Environ. Ecol.*, 6:737–739, 1988.

Barton, B. A.; Weiner, G. S.; and Schreck, C. B.: Effect of prior acid exposure on physiological responses of juvenile rainbow trout (*Salmo gairdneri*) to acute handling stress. *Can. J. Fish Aquat. Sci.*, 42:710–717, 1985.

Bennett, R. O., and Wolke, R. E.: The effect of sublethal endrin exposure on rainbow trout, *Salmo gairdneri* Richardson. I. Evaluation of serum cortisol concentrations and immune responsiveness. *J. Fish Biol.*, 31:375–385, 1987a.

Bennett, R. O., and Wolke, R. E.: The effect of sublethal endrin exposure on rainbow trout, *Salmo gairdneri* Richardson. II. The effect of altering serum cortisol concentrations on the immune response. *J. Fish Biol.*, 31:387–394, 1987b.

Blaxhall, P. C.: The hematological assessment of the health of freshwater fish. A review of selected literature. *J. Fish Biol.*, 4:593–604, 1972.

Bodammer, J. E.; Anderson, D. P.; and Dixon, O. W.: Ultrastructure of the spleen and head kidney of striped bass. *J. Aquat. Anim. Health*, 2:182–193, 1990.

Chilmonczyk, S., and Oui, E.: The effects of gamma irradiation on the lymphoid organs of rainbow trout and subsequent susceptibility to fish pathogens. *Vet. Immunol. Immunopathol.*, 18:173–180, 1988.

Cipriano, R. C., and Pyle, S. W.: Adjuvant-dependent immunity and the agglutinin response of fishes against *Aeromonas salmonicida*, cause of furunculosis. *Can. J. Fish Aquat. Sci.*, 42:1290–1295, 1985.

Cisar, J. O., and Fryer, J. L.: Characterization of anti-*Aeromonas salmonicida* antibodies from coho salmon. *Infect Immun* 9:236–243, 1974.

Cleland, G. B.; McElroy, P. J.; and Sonstegard, R. A.: The effect of dietary exposure to Aroclor 1254 and/or mirex on humoral immune expression of rainbow trout (*Salmo gairdneri*). *Aquat. Toxicol.*, 12:141–146, 1988.

Cleland, G. B.; McElroy, P. J.; and Sonstegard, R. A.: Immunomodulation in C57B1/6 mice following consumption of halogenated aromatic hydrocarbon–contaminated coho salmo (*Oncorhynchus kisutch*) from Lake Ontario, Canada. *J. Toxicol. Environ. Health*, 27:477–486, 1989.

Cossarini-Dunier, M.; Monod, G.; Damael, A.; and Lepot, D.: Effects of γ-hexachlorocyclohexane (lindane) on carp (*Cyprinus carpio*): 1. Effect of chronic intoxication on humoral immunity in relation to tissue pollutant levels. *Ecotoxicol. Environ. Safety*, 13:339–345, 1987.

Cossarini-Dunier, M.; Demael, A.; Riviere, J. L.; and Lepot, D.: Effects of oral doses of the herbicide atrazine on carp (*Cyprinus carpio*). *Ambio.*, 17:401–405, 1988.

Cossarini-Dunier, M.; Damael, A.; and Siwicki, A. K.: In vivo effect of the organophosphorus insecticide trichlorphon on the immune response of carp (*Cyprinus carpio*): effect of contamination on antibody production in relation to residue level in organs. *Ecotoxicol. Environ. Safety*, 19:93–98, 1990.

DeGraeve, G. M.: Avoidance response of rainbow trout to phenol. *Prog. Fish Cult.*, 44:82–87, 1982.

de Kinkelin, P.; Dorson, M.; and Hattenberger-Baudouy, A. M.: Interferon synthesis in trout and carp after viral infection. *Dev. Comp. Immunol. Suppl.*, 2:167–174, 1982.

Doggett, T. A., and Harris, J. E.: Ultrastructure of the peripheral blood leukocytes of *Oreochromis mossambicus*. *J. Fish Biol.*, 34:747–756, 1989.

Dunbar, C. E.: Lymphosarcoma of possible thymic origin in salmonid fishes. *Natl. Cancer Inst. Monog.*, 31:167–171, 1969.

Ellsaesser, C. F., and Clem, L. W.: Haematological and immunological changes in channel catfish stressed by handling and transport. *J. Fish Biol.*, 28:511–521, 1986.

Elsasser, M. S.; Roberson, B. S.; and Hetrick, F. M.: Effects of metals on the chemiluminescent response of rainbow trout (*Salmo gairdneri*) phagocytes. *Vet. Immunol. Immunopathol.*, 12:243–250, 1986.

Evans, D. L.; Smith, E. E.; and Brown, F. E.: Nonspecific cytotoxic cells in fish (*Ictalurus punctatus*). VI. Flow cytometric analysis. *Dev. Comp. Immunol.*, 11:95–104, 1987.

Faisal, M.; Cooper, E. L.; El-Mofty, M.; and Sayed, M. A.: Immunosuppression of *Clarias lazera* (Pisces) by a molluscicide. *Dev. Comp. Immunol.*, 12:85–97, 1988.

Felton, S. P.; Wenjuan, J.; and Mathews, S. B.: Selenious concentrations in coho salmon outmigrant smolts and returning adults: a comparison of wild versus hatchery-reared fish. *Dis. Aquat. Org.*, 9:157–161, 1990.

Finlayson, B. J., and Faggella, G. A.: Comparisons of laboratory and field observations of fish exposed to the herbicides molinate and thiobencarb. *Trans. Am. Fish. Soc.* 115:882–891, 1986.

Fletcher, T. C.: The identification of non-specific humoral factors in the plaice (*Pleuronectes platessa* L). *Dev. Biol. Stand.*, 49:321–327, 1981.

Food and Agriculture Organization of the United Nations: International Code of Conduct on the Distribution and Use of Pesticides. Rome, 1986.

Fries, C. R.: Effects of environmental stressors and immunosuppressants on immunity in *Fundulus heteroclitus*. *Am. Zool.*, 26:271–282, 1986.

Fryer, J. L.; Nelson, J. S.; and Garrison, R. L.: Vibriosis in fish. *Prog. Fish Food Sci.*, 5:129–133, 1972.

Ghaffari, S. H., and Lobb, C. J.: Nucleotide sequence of channel catfish heavy chain cDNA and genomic blot analyses. *J. Immunol.*, 143:2730–2739, 1989.

Giles, M. A.; Majewski, H. S.; and Hobden, B.: Osmoregulatory and hematological responses of rainbow trout (*Salmo gairdneri*) to extended environmental acidification. *Can. J. Fish Aquat. Sci.*, 41:1686–1694, 1984.

Gill, T. S., and Pant, J. C.: Mercury-induced blood anomalies in the fresh water teleose *Barbus conchonius*. *Water Air Soil Pollut.*, 24:165–171, 1985.

Goss, G. G., and Wood, C. M.: The effects of acid and acid/aluminum exposure on circulating plasma cortisol levels and other blood parameters in rainbow trout, *Salmo gairdneri*. *J. Fish Biol.*, 32:63–76, 1988.

Gosting, L.; Mirando, D. M.; and Gould, R. W.: Antigen-binding cells in the peripheral blood of sockeye salmon, *Oncorhynchus nerka* Walbaum, induced by immersion or intraperitoneal injection of *Vibrio anguillarum* bacterin. *J. Fish Biol.*, 19:83–86, 1981.

Graham, S., and Secombes, C. J.: Do fish lymphocytes secrete interferon-γ? *J. Fish Biol.*, 36:563–573, 1990.

Griffin, B. R.: Opsonic effect of rainbow trout (*Salmo gairdneri*) antibody on phagocytosis of *Yersinia ruckeri* by trout leukocytes. *Dev. Comp. Immunol.*, 7:253–259, 1983.

Grondel, J. L., and Boesten, H. J. A. M.: The influence of antibiotics on the immune system. I. Inhibition of the mitogenic leukocyte response in vitro by oxytetracycline. *Dev. Comp. Immunol. Suppl.*, 2:211–216, 1982.

Grondel, J. L.; Nouws, J. F. M.; DeJong, M.; Schutte, A. R.; and Driessens, F.: Pharmacokinetics and tissue distribution of oxytetracycline in carp *Cyprinus carpio* L., following different routes of administration. *J. Fish Dis.*, 10:153–163, 1987.

Hall, L. W., Jr.; Bushong, S. J.; Ziegenfuss, M. C.; Hall, W. S.; and Herman, R. L.: Concurrent mobile on-site and in situ striped bass contaminant and water quality studies in the Choptank River and upper Chesapeake Bay. *Environ. Toxicol. Chem.*, 7:815–830, 1988.

Hamilton, S. J.; Mehrle, P. M.; and Jones, J. R.: Cadmium-saturation technique for measuring metallothionine in brook trout. *Trans. Am. Fish Soc.*, 116:541–550, 1987.

Harding, F. A.; Amemiya, C. T.; Litman, R. T.; Cohen, N.; and Litman, G. W.: Two distinct immunoglobulin heavy chain isotypes in a primitive, cartilaginous fish, *Raja erinacea*. *Nucleic Acids Res.*, 18:6369–6376, 1990.

Herman, R. L.: Lymphosarcoma of the thymus of salmonids. Comparative Leukemia Research, Bibliotheca Haematologica 36, edited by RM Dutcher, p. 646. Basel: Karger, 1969.

Herraez, M. P., and Zapata, A. G.: Structure and function of the melano-macrophage centres of the goldfish *Carassius auratus*. *Vet. Immunol. Immunopathol.*, 12:117–126, 1986.

Hetrick, F. M.; Knittel, M. D.; and Fryer, J. L.: Increased susceptibility of rainbow trout to infectious hematopoietic necrosis virus after exposure to copper. *Appl. Environ. Microbiol.*, 37:198–201, 1979.

Hodgins, H. O.; Wendling, F. L.; Braaten, B. A.; and Weiser, R. S.: Two molecular species of agglutinins in rainbow trout (*Salmo gairdneri*) serum and their relation to antigenic exposure. *Comp. Biochem. Physiol.*, 45B: 975–977, 1973.

Hodgins, H. O.; McCain, B. B.; and Hawkes, J. W.: Marine fish and invertebrate diseases, host disease resistance, and pathological effects of petroleum. Effects of Petroleum on Artic and Subartic Marine Environments and Organisms, Vol. II, pp. 95–173. New York: Academic Press, 1977.

Jeney, Z., and Jeney, G.: Studies on the effect of trichlorphon on different biochemical and physiological parameters of common carp (*Cyprinus carpio* L) *Aquacult. Hung.*, 5:79–89, 1986.

Jenkins, J. A., and Ourth, D. D.: Membrane damage to *Escherichia coli* and bactericidal kinetics by the alternative complement pathway of channel catfish. *Comp. Biochem. Physiol.*, 79B:477–481, 1990.

Jones, D. H.; Lewis, D. H.; Eurell, T. W.; and Cannon, M. S.: Alteration of the immune response of channel catfish (*Ictalurus punctatus*) by polychlorinated biphenyls. Animals as monitors of environmental pollutants. Symposium on Pathobiology of Environmental Pollutants: Animal Models and Wildlife as Monitors, pp. 385–386, 1979.

Knittel, M. D.: Susceptibility of steelhead trout (*Salmo gairdneri* Richardson) to redmouth infection *Yersinia ruckeri* following exposure to copper. *J. Fish Dis.*, 4: 33–40, 1981.

Levine, R. A., and Wardlaw, S. C.: A new technique for examining blood. *Am. Sci.*, 76:592–598, 1988.

MacFarlane, R. D.; Bullock, G. L.; and McLaughlin, J. J. A.: Effects of five metals on susceptibility of striped bass to *Flexibacter columnaris*. *Trans. Am. Fish Soc.*, 115: 227–231, 1986.

Manning, M. J., and Tatner, M. F. (eds.): Fish Immunology. London: Academic Press, 1985.

Maule, A. G.; Schreck, C. B.; and Kaattari, S. L.: Changes in the immune system of coho salmon (*Oncorhynchus kisutch*) during the parr-to-smolt transformation and after implantation of cortisol. *Can. J. Fish Aquat. Sci.*, 44: 161–166, 1987.

Mayer, F. L., and Ellersieck, M. R.: Manual of acute toxicity: interpretation and data base for 410 chemical and 66 species of freshwater animals. Washington, DC: U.S. Fish and Wildlife Service Resource Pub. 160, 1986.

Mayer, K. S.; Multer, E.-P.; and Schreiber, R. K.: Acid rain: effects on fish and wildlife. Washington, DC: U.S. Fish and Wildlife Leaflet 1, 1984.

Mayer, K. S.; Mayer, F. L.; and Witt, A., Jr.: Waste transformer oil and PCB toxicity to rainbow trout. *Trans. Am. Fish Soc.*, 114:869–886, 1985.

McLeay, D. J., and Brown, D. A.: Stress and chronic effects of untreated and treated bleached draft pulpmill effluent on the biochemistry and stamina of juvenile coho salmon (*Oncorhynchus kisutch*). *J. Fish Res. Board Can.*, 36: 1049–1059, 1979.

McLeay, D. J., and Gordon, M. R.: Leucocrit: a simple hematological technique for measuring acute stress in salmonid fish, including stressful concentrations of pulpmill effluent. *J. Fish. Res. Board Can.*, 34:2164–2175, 1977.

Mehrle, P., and Petty, J.: Dioxin (TCDD) and furan (TCDF) toxicity to rainbow trout. U.S. Fish and Wildlife Service, Research Information Bulletin 87-55, 1987.

Metcalfe, C. D.: Tests for predicting carcinogenicity in fish. *Can. J. Fish Aquat. Sci.*, 1:111–119, 1989.

Meyer, F. P., and Jorgenson, T. A.: Teratological and other effects of malachite green on development of rainbow trout and rabbits. *Trans. Am. Fish Soc.*, 112:818–824, 1983.

Michel, C.; Gonzalez, R.; Bonjour, E.; and Avrameas, S.: A concurrent increasing of natural antibodies and enhancement of resistance to furunculosis in rainbow trout. *Ann. Rech. Vet.*, 21:211–218, 1990.

Middlebrooks, B. L., and Meador, C. B.: Effects of benzidine exposure on the immune response of an estuarine fish (*Cyprinodon variegatus*). Abstracts of the Annual Meeting of the American Society for Microbiology, p. 82, 1984.

Mitchell, D. G.; Chapman, P. M.; and Long, T. J.: Seawater challenge testing of coho salmon smolts following exposure to Roundup herbicide. *Environ. Toxicol. Chem.*, 6:875–878, 1987.

Mitrovic, V. V.; Brown, V. M.; Shurben, D. G.; and Berryman, M. H.: Some pathological effects of sub-acute and acute poisoning of rainbow trout by phenol in hard water. *Water Res.*, 2:249–254, 1968.

Mulcahy, M. F., and O'Leary, A.: Cell-free transmission of lymphosarcoma in the northern pike *Esox lucius*, L. (Pisces: Esocidae). *Experientia*, 26:891, 1970.

Murad, A., and Houston, A. H.: Leukocytes and leucopoietic capacity on goldfish, *Carassius auratus*, exposed to sublethal levels of cadmium. *Aquat. Toxicol.*, 13:141–154, 1988.

Murai, T.; Kodama, H.; Naiki, M.; Mikami, T.; and Izawa, H.: Isolation and characterization of rainbow trout C-reactive protein. *Dev. Comp. Immunol.*, 14:49–58, 1990.

Murchelano, R. A., and Wolke, R. E.: Epizootic carcinoma in the winter flounder, *Pseudopleuronectes americanus*. *Science*, 228:587–589, 1985.

Nakanishi, T.: Effects of X-irradiation and thymectomy on the immune response of the marine teleost, *Sebastiscus marmoratus*. *Dev. Comp. Immunol.*, 10:519–527, 1986.

Oladimeji, A. A.; Qadri, S. U.; and deFreitas, A. S. W.: Long-term effects of arsenic accumulation in rainbow trout, *Salmo gairdneri*. *Bull. Environ. Contam. Toxicol.*, 32:732–741, 1984.

O'Neill, J. G.: The humoral immune response of *Salmo trutta* L. and *Cyprinus carpio* L. exposed to heavy metals. *J. Fish Biol.*, 19:297–306, 1981a.

O'Neill, J. G.: Effects of intraperitoneal lead and cadmium on the humoral immune response of *Salmo trutta*. *Bull. Environ. Contam. Toxicol.*, 27:42–48, 1981b.

Payne, J. F., and Fancey, L. F.: Effect of polycyclic aromatic hydrocarbons on immune responses in fish: change in melanomacrophage centers in flounder (*Pseudopleuronectes americanus*) exposed to hydrocarbon-contaminated sediments. *Mar. Environ. Res.*, 28:431–435, 1989.

Peters, G., and Schwarzer, R.: Changes in hemopoietic tissue of rainbow trout under influence of stress. *Dis. Aquat. Organisms*, 1:1–10, 1985.

Pickering, A. D., and Pottinger, T. G.: Crowding causes prolonged leucopenia in salmonid fish, despite interrenal acclimation. *J. Fish Biol.*, 30:701–712, 1987.

Pickering, A. D., and Pottinger, T. G.: Lymphocytopenia and the overwinter survival of Atlantic salmon parr, *Salmo salar* L. *J. Fish Biol.*, 32:689–697, 1988.

Reynolds, W. W.: Fever and antipyresis in the bluegill sunfish, *Lepomis macrochirus*. *Comp. Biochem. Physiol.*, 57C:165–167, 1977.

Rice, C. D., and Weeks, B. A.: The influence of in vitro exposure to tributyltin on reactive oxygen formation in oyster toadfish macrophages. *Arch. Environ. Contam. Toxicol.* 19:854–857, 1990.

Roales, R. R., and Perlmutter, A.: The effects of sub-lethal doses of methylmercury and copper, applied singly and jointly, on the immune response of the blue gourami, (*Trichogaster trichopterus*) to viral and bacterial antigens. *Arch. Environ. Contam. Toxicol.*, 5:325–331, 1977.

Robohm, R. A.: Paradoxical effects of cadmium exposure on antibacterial antibody responses in two fish species: inhibition in cunners (*Tautogolabrus adspersus*) and enhancement in striped bass (*Morone saxatilis*). *Vet. Immunol. Immunopathol.*, 12:251–262, 1986.

Rodsaether, M. C.; Olafsen, J.; Raa, J.; Myhre, K.; and Steen, J. B.: Copper as an initiating factor in vibriosis (*Vibrio anguillarum*) in eel (*Anguilla anguilla*). *J. Fish Biol.*, 10:17–21, 1977.

Sanchez, C.; Dominguez, J.; and Coll, J.: Immunoglobulin heterogeneity in the rainbow trout, *Salmo gairdneri* Richardson. *J. Fish Dis.*, 12:459–465, 1989.

Sarot, D. A., and Perlmutter, A.: The toxicity of zinc to the immune response of the zebrafish, (*Brachydanio rerio*), injected with viral and bacterial antigens. *Trans. Am. Fish Soc.*, 105:456–459, 1976.

Secombes, C. J.: Isolation of salmonid macrophages and analysis of their killing activity. Techniques in Fish Immunology (Fish Immunology Technical Communications No. 1), edited by J. S. Stolen; T. C. Fletcher; D. P. Anderson; B. S. Roberson; and W. B. van Muiswinkel, pp. 137–154. Fair Haven, NJ: SOS Publications, 1990.

Seeley, K. R., and Weeks-Perkins, B. A.: Altered phagocytic activity of macrophages in oyster toadfish (*Opsanus tau*) from a highly polluted subestuary. *J. Aquat. Anim. Health*, 3:224–227, 1991.

Sharma, R. P., and Zeeman, M. G.: Immunologic alterations by environmental chemicals: relevance of studying mechanisms versus effects. *J. Immunopharmacol.*, 2: 285–307, 1980.

Sinnhuber, R. O.; Wales, J. H.; Ayres, J. L.; Engebrecht, R. H.; and Amend, D. L.: Dietary factors and hepatoma in rainbow trout (*Salmo gairdneri*). I. Aflatoxins in vegetable protein feedstuffs. *J. Natl. Cancer Inst.*, 41:711–718, 1968.

Siwicki, A. K.; Anderson, D. P.; and Dixon, O. W.: Comparisons of nonspecific and specific immunomodulation by oxolinic acid, oxytetracycline and levamisole in salmonids. *Vet. Immunol. Immunopathol.*, 23:195–200, 1989.

Siwicki, A. K.; Anderson, D. P.; and Dixon, O. W.: In vitro immunostimulation of rainbow trout (*Oncorhynchus mykiss*) spleen cells with levamisole. *Dev. Comp. Immunol.*, 14:231–237, 1990a.

Siwicki, A. K.; Cossarini-Dunier, M.; Studnicka, M.; and Damael, A.: In vivo effect of the organophosphorus insecticide trichlorphon on immune response of carp (*Cyprinus carpio*): effect of high doses of trichlorphon on nonspecific immune response. *Ecotoxicol. Environ. Saf.*, 19:99–105, 1990b.

Smith, C. E.: Hematological changes in coho salmon fed a folic acid deficient diet. *J. Fish Res. Board Can.*, 25: 151–156, 1967.

Smith, C. E.: An undifferentiated hematopoietic neoplasm with histologic manifestations of leukemia in cutthroat trout (*Salmo clarki*). *J. Fish Res. Board Can.*, 28:112–113, 1971.

Smith, P. D., and Braun-Nesje, R.: Cell mediated immunity in the salmon: lymphocyte and macrophage stimulation, lymphocyte/macrophage interactions, and the production of lymphokine-like factors by stimulated lymphocytes. *Dev. Comp. Immunol. Suppl.*, 2:233–238, 1982.

Smith, S. A.; Gebhard, D. H.; Housman, J. M.; Levy, M. G.; and Noga, E. J.: Isolation, purification, and molecular weight determination of serum immunoglobulin from tilapia (*Oreochromis aureus*). *J. Aquat. Anim. Health*, 5: 23–35, 1993.

Sorensen, E. M. B., and Bauer, T. L.: Hematological dyscrasia in teleosts chronically exposed to selenium-laden effluent. *Environ. Contam. Toxicol.*, 12:135–141, 1983.

Sövényi, J. F., and Kusuda, R.: Kinetics of in vitro phagocytosis by cells from head-kidney of common carp, *Cyprinus carpio*. *Fish Pathol.*, 22:83–92, 1987.

Spitsbergen, J. M.; Schat, K. A.; Kleeman, J. M.; and Peterson, R. E.: Interactions of 2,3,7,8-tetrachlorodibenzo-*p*-dioxin (TCDD) with immune responses of rainbow trout. *Vet. Immunol. Immunopathol.*, 12:263–280, 1986.

Spitsbergen, J. M.; Schat, K. A.; Kleeman, J. M.; and Peterson, R. E.: Effects of 2,3,7,8-tetrachlorodibenzo-*p*-dioxin (TCCD) or Aroclor 1254 on the resistance of rainbow trout, *Salmo gairdneri* Richardson, to infectious haematopoietic necrosis virus. *J. Fish Dis.*, 11:73–83, 1988.

Stave, J. W., and Roberson, B. S.: Hydrocortisone suppresses the chemiluminescent response of striped bass phagocytes. *Dev. Comp. Immunol.*, 9:77–84, 1985.

Stave, J. W.; Roberson, B. S.; and Hetrick, F. M.: Chemiluminescence of phagocytic cells isolated from the pronephros of striped bass. *Dev. Comp. Immunol.*, 7:269–276, 1983.

Stolen, J.: The effect of the PCB, Aroclor 1254, and ethanol on the humoral immune response of a marine teleost to a sludge bacterial isolate of *E. coli*. *Marine Pollution and Physiology: Recent Advances*, edited by F. J. Vernberg; F. P. Thurberg; A. Calabrese; and W. B. Vernberg, pp. 419–426. New York: Academic Press, 1985.

Stolen, J. S.; Anderson, D. P.; and van Muiswinkel, W. B. (eds.): Fish immunology. New York: Elsevier, 1986.

Stolen, J. S.; Fletcher, T. C.; Anderson, D. P.; Roberson, B. S.; and van Muiswinkel, W. B. (eds.): Techniques in Fish Immunology (Fish Immunology Technical Communications No. 1). Fair Haven, NJ: SOS Publications, 1990.

Stolen, J. S.; Fletcher, T. C.; Anderson, D. P.; Kaattari, S.L.; and Rowley, A. F. (eds.): Techniques in Fish Immunology—II. Fair Haven, NJ: SOS Publications, 1992.

Stoskopf, M. K. (ed.): Fish medicine. Philadelphia: WB Saunders, 1993.

Tana, J., and Nikunen, E.: Physiological responses of rainbow trout in a pulp and paper mill recipient during four seasons. *Ecotoxicol. Environ. Safety*, 12:22–34, 1986.

Thuvander, A.: Cadmium exposure of rainbow trout, *Salmo gairdneri* Richardson: effects on immune functions. *J. Fish Biol.*, 35:521–529, 1989.

Thuvander, A.; Norrgren, L.; and Fossum, C.: Phagocytic cells in blood from rainbow trout, *Salmo giardneri* (Richardson), characterized by flow cytometry and electron microscopy. *J. Fish Biol.*, 31:197–208, 1987.

Tort, L., and Torres, P.: The effects of sublethal concentrations of cadmium on hematological parameters in the dogfish, *Scyliorhinus canicula*. *J. Fish Biol.*, 32:277–282, 1988.

van Muiswinkel, W. B.; Anderson, D. P.; Lamers, C. H. J.; Egberts, E.; van Loon, J. J. A.; and Ijssel, J. P.: Fish immunology and fish health. Fish Immunology, edited by M. J. Manning, and M. F. Tatner, pp. 1–8. London: Academic Press, 1985.

van Muiswinkel, W. B., and Cooper, E. L. (eds.): Immunology and Immunization of fish. *Dev. Comp. Immunol. Suppl. 2.*. New York: Pergamon, 1982.

Villella, R. F.: Acid rain publications by the U.S. Fish and Wildlife Service, 1979–1989. U.S. Fish and Wildlife Service Biological Report 80(40.28). Washington, DC, 1989.

Walczak, B. Z.: Immune capability of fish: a literature review. *Can. Tech. Rep. Fish Aquat. Sci.*, No. 1334, 1985.

Warr, G. W.; Griffin, B. R.; Anderson, D. P.; McAllister, P. E.; Lidgerding, B.; and Smith, C. E.: A lymphosarcoma of thymic origin in the rainbow trout, *Salmo gairdneri* Richardson. *J. Fish Dis.*, 7:73–82, 1984.

Wedemeyer, G. A., and McLeay, D. J.: Methods for determining the tolerance of fishes to environmental stress. Stress and Fish, edited by A. D. Pickering, pp. 247–275. New York: Academic Press, 1981.

Wedemeyer, G. A.; Gould, R. W.; and Yasutake, W. T.: Some potentials and limits of the leucocrit test as a fish health assessment method. *J. Fish Biol.*, 23:711–716, 1983.

Weeks, B. A., and Warinner, J. E.: Effects of toxic chemicals on macrophage phagocytosis in two estuarine fishes. *Mar. Environ. Res.*, 14:327–335, 1984.

Weeks, B. A., and Warinner, J. E.: Functional evaluation of macrophages in fish from a polluted estuary. *Vet. Immunol. Immunopathol.*, 12:313–320, 1986.

Weeks, B. A.; Keisler, A. S.; Warinner, J. E.; and Mathews, E. S.: Preliminary evaluation of macrophage pinocytosis as a technique to monitor fish health. *Mar. Environ. Res.*, 22:205–213, 1987.

Weeks, B. A.; Anderson, D. P.; DuFour, A. P.; Fairbrother, A.; Goven, A. J.; Lahavis, G. P.; and Peters, G.: Immunological biomarkers to assess environmental stress. Biomarkers: Biochemical, Physiological, and Histological Markers of Anthropogenic Stress, edited by R. J. Huggett; R. A. Kimerle; P. M. Mehrle, Jr.; and H. L. Bergman, pp. 211–234. Boca Raton, FL: Lewis, 1992.

Welch, W. J.; Kang, H. S.; Beckmann, R. P.; and Am Mizzen, L.: Response of mammalian cells to metabolic stress: changes in cell physiology and structure/function of stress proteins. *Curr. Top. Microbiol. Immunol.*, 167: 31–55, 1991.

Wellings, S. R.; Chuinard, R. G.; Gouley, R. T.; and Cooper, R. A.: Epidermal papillomas in the flathead sole, *Hippoglossoides elassodon*, with notes on the occurrence of similar neoplasms in other pleuronectids. *J. Natl. Cancer Inst.*, 33:991–1004, 1968.

Wells, R. M. G., and Weber, R. E.: Is there an optimal haematocrit for rainbow trout, *Oncorhynchus mykiss* (Walbaum)? An interpretation of recent data based on blood viscosity measurements. *J. Fish Biol.*, 38:53–65, 1991.

Wester, P. W.; Canton, J. H.; and Dormans, J. A. M. A.: Pathological effects in freshwater fish *Poecilia reticulata* (guppy) and *Oryzias latipes* (medaka) following methyl bromide and sodium bromide exposure. *Aquat. Toxicol.*, 12:323–344, 1988.

Wishkovsky, A.; Mathews, E. S.; and Weeks, B. A.: Effect of tributyltin on the chemiluminescent response of phagocytes from three species of estuarine fish. *Arch. Environ. Contam. Toxicol.*, 18:826–831, 1989.

Woodward, D. F., and Mauck, W. L.: Toxicity of five forest insecticides to cutthroat trout and two species of aquatic invertebrates. *Bull. Environ. Contam. Toxicol.*, 25:846–853, 1980.

Yoshida, T., and Kitao, T.: Spontaneous rosette formation of carp peripheral leukocytes with mammalian red blood cells. *Nippon Suisan Gakkaishi*, 56:1579–1585, 1990.

Zapata, A. G.; Torroba, M.; Alvarez, F.; Anderson, D. P.; Dixon, O. W.; and Wisniewski, W.: Electron microscopic examination of antigen uptake by salmonid gill cells after bath immunization with a bacterin. *J. Fish Biol.*, 31(suppl A):209–217, 1987.

Zapata, A. G.; Varas, A.; and Torroba, M.: Seasonal variations in the immune system of lower vertebrates. *Immunol. Today*, 13:142–147, 1992.

Zeeman, M.: Modulation of the immune response in fish. *Vet. Immunol. Immunopathol.* 12:235–241, 1986.

Zeeman, M. G., and Brindley, W. A.: Effects of toxic agents upon fish immune systems: a review. Immunologic Considerations in Toxicology, Vol. 2, edited by R. P. Sharma, pp. 1–60. Boca Raton, FL: CRC Press, 1981.

Ziegenfuss, M. C., and Wolke, R. E.: The use of fluorescent microspheres in the study of piscine macrophage aggregate kinetics. *Dev. Comp. Immunol.*, 15:165–171, 1991.

SUPPLEMENTAL READING

Adams, S. M. (ed.): Biological Indicators of Stress in Fish. Bethesda, MD: American Fisheries Society Symposium 8, 1990.

Anderson, D. P.: Fish immunology, Book 4. *Diseases of Fishes*, edited by S. F. Snieszko, and H. R. Axelrod. Neptune, NJ: TFH Publications, 1974.

Anderson, D. P., and Hennessen, W. (eds.): Fish Biologics: Serodiagnostics and Vaccines. *Dev. Biol. Stand.*, Vol. 49. Basel, Switzerland: Karger, 1981.

Austin, B., and Austin, D. A. (eds.): *Methods for the Microbiological Examination of Fish and Shellfish.* Chichester: Ellis Horwood, 1989.

Hunt, T. C., and Margetts, A. R. (eds.): Immunology and Disease Control Mechanisms of Fish. *J. Fish Biol.*, 31, Suppl A. London: The Fisheries Society of the British Island, 1987.

Kende, M.; Gainer, J.; and Chirigos, M.: Chemical regulation of immunity in veterinary medicine. *Progress in Clinical and Biological Research*, Vol. 161. New York: Alan R. Liss, 1984.

Landis, W. G., and van der Schalie, W. H. (eds.): Aquatic Toxicology and Risk Assessment, Vol. 13. Philadelphia: American Society for Testing and Materials, 1990.

Manning, M. J., and Tatner, M. F. (eds.): Fish Immunology. London: Academic Press, 1985.

Meyer, F. P., and Barclay, L. A. (eds.): Field manual for the investigation of fish kills. U.S. Fish and Wildlife Service, Resource Publication 177. Washington, DC, 1990.

Post, G.: Textbook of Fish Health, 2nd ed. Neptune City, NJ: TFH Publications, 1987.

Stolen, J. S.; Anderson, D. P.; and van Muiswinkel, W. B. (eds.): Fish Immunology. New York: Elsevier, 1986.

Stolen, J. S.; Fletcher, T. C.; Anderson, D. P.; Roberson, B. S.; and van Muiswinkel, W. B.: Techniques in Fish Immunology: Fish Immunology Technical Communications—1. Fair Haven, NJ: SOS Publications, 1980.

Stolen, J. S.; Fletcher, T. C.; Anderson, D. P.; Kaattari, S. L.; and Rowley, A. F. (eds.): Techniques in Fish Immunology—II. Fair Haven, NJ: SOS Publications, 1992.

van Muiswinkel, W. B, and Cooper, E. L. (eds.): Immunology and Immunization of Fish. *Dev. Comp. Immunol. Suppl.*, 2. New York: Pergamon, 1982.

Zeeman, M. G., and Brindley, W. A.: Effects of toxic agents upon fish immune systems: a review. Immunologic Considerations in Toxicology, Vol. 2, edited by R. P. Sharma, pp. 1–60. Boca Raton, FL: CRC Press, 1981.

Zelikoff, J. T.: Metal pollution-induced immunomodulation in fish. *Ann. Rev. Fish Dis.*, 2:305–325, 1993.

Chapter 13

ENVIRONMENTAL GENOTOXICOLOGY

L. R. Shugart

INTRODUCTION

Background

Genetic toxicology is an area of science in which the interaction of DNA-damaging agents with genetic material is studied in relation to the subsequent effect(s) on the health of the organism. Several agents damage genetic material at concentrations that also produce acute, nonspecific cytotoxicity and death. However, the objective of the genetic toxicologist is to detect at sublethal (or subtoxic) concentrations the potential for agents specific for interacting with nucleic acids to produce genetic alterations. Agents that produce alterations in nucleic acids at sublethal exposure concentrations, resulting in changes in hereditary characteristics or DNA inactivation, are classified as genotoxic. Genotoxic agents have common chemical and physical properties that enable them to interact with nucleic acids. The term genotoxic is generally used to describe and distinguish chemical (and physical) agents that have an affinity for direct DNA interaction from those that do not interact. At this point, the reader is reminded that much of the discussion to follow cuts across many disciplines (cell biology, molecular biology, biochemistry, biophysics, etc.) and for this reason terminology becomes an important consideration. Therefore, it is suggested that access to a dictionary on biological terms and/or a current textbook on cell biology is highly recommended (also see Terminology in this chapter).

Genetic toxicology has as its origin the somatic mutation theory of carcinogenesis (for a review see Knudson, 1985), which states that the tumorigenic state is a heritable property of affected cells. Subsequent hypotheses dealing with chemical carcinogenesis (see Chapter 14) suggest that the process involves a series of biochemical and cellular events that commence with damage to DNA and culminate in the tumorigenic state (Weinstein, 1978). Genetic toxicology has thus been applied as a screen for potential human carcinogens. It is in the context of human health studies that these assumptions have been extensively investigated and the various cellular events documented, thus establishing the field of genetic toxicology as an important component of medical toxicology.

Pollution of the environment through the accelerated release of synthetic chemicals has become a major concern of society. Perhaps one of the more serious concerns associated with pollution in the environment is the potential for exposure to substances that are genotoxic. This concern arises because some of the pollutants known to be present have been shown through corresponding laboratory studies to be chemical carcinogens and mutagens with the capacity to affect both the structural and biological integrity of DNA (Wogan and Gorelick, 1985). In fact, changes in the integrity of DNA have been proposed as useful end points for assessing the hazardous nature of environmental pollutants toward human health (Kohn, 1983; CBMNRC, 1987) or biota therein (Shugart, 1990a; Shugart et al., 1992a,b). A range of assays have been developed that either quantify the extent of exposure to known genotoxins or monitor exposed individuals for evidence that a genotoxic agent has interacted with and/or modified the DNA of somatic tissue (Ashby, 1988). Thus it has become possible to extrapolate backward from the type of DNA damage event observed to identify the potential agent(s) responsible for the insult or, conversely, to extrapolate foward and predict the consequence(s).

Environmental Genotoxicology

The major emphasis of this chapter is on the application of the principles and methodologies of genetic toxicology to the assessment of environmental pollution. In particular, an in situ approach is described in which aquatic species present in a contaminated environment are examined for the effect of genotoxins (Shugart et al., 1989; Shugart, 1990a). This is in contrast to the approach in which gentoxicity is studied in a laboratory setting using bioassays that range, for example, from the microbial mutation test of Ames (Ames et al., 1973) to the eukaryotic cell test for anaphase aberrations (Kocan and Powell, 1985a). This is not to imply that the latter approach should not be used to obtain scientifically relevant data that supplement and complement the former approach, or vice versa. In fact, cytogenetic studies with fish and other aquatic organisms are currently a major area of intense scientific interest. Two approaches are used in genotoxicological research: (1) survey of organs and tissues from fish to assess the effect of agents in the natural environment or under controlled laboratory exposures and (2) experimental manipulation of in vitro models for research purposes. For example, Hinton et al. (1988) described cytological changes during progression of experimentally induced hepatic neoplasia in fish in an effort to define and characterize responses to carcinogens. These responses may be useful indicators of exposure to carcinogens in field screening investigations (Adams et al., 1990). The application of cell and tissue culture systems to the study of genotoxic substances is based in part on the finding that species differences are retained in the in vitro systems. Both mutations and visible chromosomal damage can be observed in these systems and serve as predictors of potential long-term genetic damage (Kocan et al., 1985b; Baksi and Frazier, 1990). Short-term genetic bioassays are now being used to screen both individual chemicals and complex mixtures for mutagenic and potential carcinogenic activity. These bioassays are often employed in tiered or phased approaches to prioritize substances for more expensive chronic animal testing (Waters et al., 1984).

Environmental genotoxicology is an area of genetic toxicology that has received serious attention relatively recently. Early attempts at assessing the effects of genotoxins on species present in the environment most often involved direct observation, such as the visual occurrence of neoplasms and chromosomal aberrations in various plants, wild terrestrial mammals, and aquatic vertebrates (Sandhu and Lower, 1989). Public awareness of increasing contamination of streams, rivers, and oceans and the potential for associated health risks has sparked renewed efforts, and reports are appearing that suggest a correlation between environmental factors, including pollution, and cancer-like lesions in fish (Harshbarger and Clark, 1990). Lines of evidence supporting this conclusion for fish liver neoplasms include:

1. Historical—new discoveries of epizootic occurrences parallel exponential growth of industries producing synthetic chemicals.
2. Experimental—spontaneous neoplasms are rare; however, pure chemicals and contaminated sediments are capable of producing them.
3. Physiological—fish livers have the enzymatic capacity to activate indirect carcinogens.

TERMINOLOGY

Adduct. Structural modification to DNA resulting from the covalent attachment of a chemical or its metabolite usually via a base moiety.

Aneuploidy. Loss or gain of one or more individual chromosomes during meiosis or mitosis. Mutagens increase the frequency of this cellular event.

Chemical carcinogenesis. Generation of cancer as a result, initially, of a chemically induced change in the nucleotide sequence of the DNA molecule.

Chromosome. Cellular organelle containing hereditable information in the form of deoxyribonucleic acid (DNA) complexed with proteins and ribonucleic acids (RNA).

Chromosomal aberration. A change in normal complement of chromosomes involving either number or structure.

Duplex DNA. DNA structure consisting of two linear polymers (chains), antiparallel with respect to the two sugar-phosphate backbones, twisted in a helix and held together by hydrogen bonds between complementary base pairs on opposite chains.

Gene translocation. Erratic rejoining of chromosome fragments that somehow have become broken in more than one place.

Gene amplification. Gain of gene copies due to unequal crossing-over between sister chromosomes containing the gene.

Genotoxic. Capable of causing damage to chromosomes.

Micronuclei. Small, secondary nuclei formed following chromosome breakage.

Mutation. An alteration in the genetic material (DNA) that may be induced by factors external to an organism such as chemicals or radiation or that may appear accidentally during replication. Mutations occur when one base is substituted for another or when one or more bases are inserted or deleted from the DNA.

Oncogene. Alteration of a normal gene, whose function is to stimulate cell proliferation, to an abnormal, hyperactive mutated gene.

BASIC MECHANISMS OF DNA DAMAGE

Gene Structure

Informational molecules of all living organisms are composed of DNA, except for RNA viruses. Some of the basic characteristics of DNA are listed in Table 1. Storage of information and gene expression are similar in mechanism in all organisms. The simplest functional unit in a DNA molecule is the *gene*. Information on the structure and function of the gene has been acquired from studies with bacteria or bacteriophages. Genes of prokaryotic organisms (those without a defined nucleus, i.e., bacteria) and eukaryotic organisms (higher organisms with a nucleus—plants, animal cells, etc.) differ in number, location on the chromosome (complex of DNA, RNA, and protein), and mechanisms of gene regulation. In bacteria there is a single chromosome entity with little to no differentiation along the DNA molecule in terms of function. Eukaryotic cells, in contrast, have DNA with repetitive, nonfunctional gene sequences. These cells have regions of noncoding DNA called *introns* between coding sequences called *exons*. The function of repetitive DNA sequences and introns is unknown. Nucleotide composition and the mechanisms by which encoded information in a gene is transformed to gene products are universal. Gene expression in eukaryotic cell types is similar. Information is maintained in coding of sequences (exons) in the DNA located in the nucleus. Information is transcribed into messenger RNA (mRNA), which is transported to the ribosomes in the cytoplasm for translation into gene product. Intron regions are not present in prokaryotic cells, and the gene is read in one sequence. Somatic cells constitute the major portion of vertebrate organisms. The genomes of somatic cells are diploid, and damage to these cells is not transmissible to future generations. Germ cells undergo meiotic division to haploid gametes. Damage to germ cells has potential for transmission to the next generation.

Table 1. Basic biochemical characteristics of DNA

1. Inheritable properties reside in the DNA of chromosomes as information-containing elements called genes.
2. DNA composed of two, unbranched, complementary, antiparallel polymers twisted into a double helix structure.
3. DNA polymer composed of repeating units of deoxyribonucleotides. Each unit consists of a heterocyclic base and a sugar linked together by covalent phosphodiester bonds via the 5′ and 3′ carbons of the deoxyribose.
4. Hydrogen bonding between bases (two purines, guanine and adenine; and two pyrimidines, cytosine and thymine) on opposite strands result in unique paing (G with C and A with T) that hold the double helix together.
5. DNA replicates by semiconservative method in which the two strands separate and each is used as a template for the synthesis of a new complementary strand.
6. DNA content of cell varies (1.8×10^9 daltons for a bacterium to 1.9×10^{11} for mammal).

DNA Alterations

The cellular metabolism of genotoxic chemicals, once these agents become bioavailable, can be a relatively complex phenomenon and, in some cases, the lack of complete detoxication leads to the formation of highly reactive electrophilic metabolites (Phillips and Sims, 1979; Harvey, 1982; Phillips, 1983). These intermediates can undergo attack by nucleophilic centers in macromolecules such as lipids, proteins, DNA, and RNA, which often results in cellular toxicity. Interaction with DNA is manifested primarily by structural alterations in the DNA molecule and can take the form of adducts (where the chemical or its metabolite becomes covalently attached to the DNA), of strand breakage, or of chemically altered bases.

Conceptually, the preceding discussion of structural alterations by genotoxic agents can be depicted (Figure 1) as a model that reflects events and cellular processes related to DNA integrity (Shugart, 1990a; Shugart et al., 1992b). DNA is present in cells as a functionally stable, double-stranded polymer without discontinuity (strand breaks) or abnormal structural modifications (adducts or chemically altered bases) that is complexed with proteins in a chromosomal structure. As such, it is considered to have high integrity. Rigid maintenance of this integrity is important for survival and is reflected in the low mutation rate observed in living organisms, which has been estimated to be on the order of one mutation per average gene per 200,000 years (Alberts et al., 1989). DNA damage can occur as a result of wear and tear by normal cellular events such as metabolism and random thermal collisions (pathway 1), interaction with physical agents such as ultraviolet light and ionizing radiation (pathway 2), and interaction with chemical agents (pathway 3). These various processes give rise to structural alterations, which are usually rapidly repaired but in the process produce a transient population of DNA with an increase in strand breakage and thus a polymer of low integrity. Some chemicals exert their effect via a free radical mechanism and cause strand breaks directly, whereas other genotoxic agents can interfere with the fidelity of DNA repair or normal modifications to DNA (pathway 6). Spontaneous losses of bases from the DNA molecule (abasic sites) are frequent lesions that occur as a result of random thermal collisions or breakdown of chemically unstable adducts. Even the normal cellular process of replication (pathway 5) produces DNA with strand breaks. Therefore, at any one time, a background level of DNA with low integrity (DNA with various types of structural alterations) may exist in the cell. Fortunately, most cells have DNA repair mechanisms (Sancar and Sancar, 1988) that under normal circumstances efficiently eliminate DNA of low integrity (pathway 4). Several types of enzymatic DNA repair processes have developed during evolution to maintain the fidelity and integrity of genetic information. DNA is the only molecule with the capacity for self-repair. The common feature of repair is the ability to remove and replace damaged segments of DNA. Therefore, if a DNA lesion induced by a mutagen can be repaired before fixation, there may be no effect of the DNA insult. This is especially true after low-level exposure, where excision repair enzymes are not saturated by significant numbers of damaged DNA sites. Because all normal organisms are capable of some type of repair following chemical insult, a significant increase in the level of repair after exposure to sublethal concentrations of a chemical is a good indicator of DNA-directed toxicity.

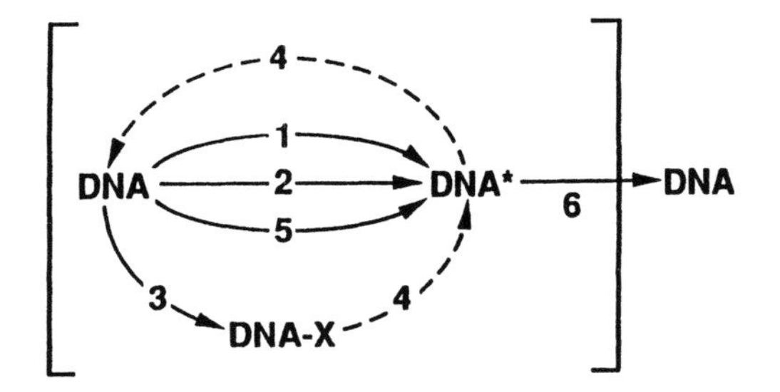

INSULT	REPAIR	SYNTHESIS
1. Normal wear and tear	4. Incision	5. Replication
2. UV and ionizing radiation	excision	6. Postreplication
(γ and X irradiation)	resynthesis	modification
3. Chemical	ligation	

DNA: Normal double-stranded DNA with no strand breaks.
DNA-X: Chemically modified DNA.
DNA*: DNA with strand breaks.

Figure 1. Schematic representation of the status of DNA in relation to insults that disrupt DNA integrity and cellular processes that maintain DNA integrity. (Reprinted with permission from Shugart L. R., Biological monitoring: Testing for genotoxicity. *Biological Markers of Environmental Contaminants*, edited by J. F. McCarthy, and L. R. Shugart, pp. 205–216, 1990a. Copyright Lewis Publishers, a subsidiary of CRC Press, Boca Raton, Florida.)

Irreversible Events

The consequence of the structural perturbations just discussed can be innocuous because of repair of the damage or death of the cells containing the damaged DNA. However, lesions that are not properly repaired may result in alterations (i.e., mutations) that become fixed and are eventually transmitted to daughter cells. It has been postulated that the reaction of chemicals with DNA and changes that result may cause deleterious, pathological conditions (Harvey, 1982). Affected cells often exhibit altered cellular functions indicative of subclinical manifestation of genotoxic disease. For example, structural alterations of DNA caused by genotoxic chemicals may potentiate irreversible changes in the DNA molecule and result in the expression of cellular responses such as chromosomal aberrations and oncogene activation. Such responses have been associated with the latent events of chemical carcinogenesis (Weinstein, 1978). Because of the fundamental role genes play in all living organisms, other types of toxicological end points may be affected, such as teratogenesis and reproductive capacity.

From the preceding discussion it is evident that upon exposure to genotoxic chemicals the integrity of an organism's DNA may be subject to damage. If the damage is not repaired or if it persists, the investigator is provided the opportunity to test for genotoxicity (Shugart, 1990a; Jones and Parry, 1992).

ANALYTICAL TECHNIQUES

Background

Exposure of an organism to a genotoxic chemical may result in the induction of a cascade of cellular events, starting with an initial insult to the DNA and culminating in the appearance of an overt pathological disease. The rationale for the selection of a given change in DNA integrity to assess this exposure or the consequence thereof is dictated and limited by two important considerations: the present state of our knowledge concerning the cellular mechanisms involved in the genotoxic response and the analytical technology available to define the response.

The success of current in situ efforts in the field of environmental genotoxicology is due mainly to the incorporation of new analytical techniques that are extremely selective and/or sensitive for the detection of changes in DNA integrity. Several select methods that utilize these techniques are detailed in this section in relation to the end point (i.e., change in DNA integrity) under consideration. Attention is given to a description of the analytical technique, its applicability, and the basic principles that establish its sensitivity and selectivity.

DNA Isolation, Purity, and Assay Conditions

In most of the methods to be described, DNA is subjected to analysis after it has been isolated from a suitable organ or tissue of an organism. Although it may seem a trivial point, the interpretation of data generated by a particular method will require knowledge about the status of the DNA in the preparation being analyzed. The degree of purity and the form of the DNA needed for analysis may be as complex as the intact chromatin structure residing inside the nucleus, which is suitable for measuring DNA content of cells by flow cytometeric analysis. A semipurified form with some ribonucleic acids and proteins present but not small cellular molecules or cell debris can be used for the detection of strand breaks, whereas a highly purified DNA is often required for adduct measurements. An investigator should never fall into the trap of assuming that if a given protocol is followed, the DNA obtained from a particular organism is in the form necessary for analysis. It is prudent to verify the status of the DNA preparation under study by other means.

The analytical techniques and assay conditions discussed in the following represent a concise summary of information available from the scientific literature. Therefore, before attempting to perform the assay in a laboratory setting, the original literature should be consulted for a more detailed description of the assay conditions.

Structural Damage

Structural modifications of DNA are probably the best understood genotoxic event (see discussion in the previous section) for which analytical techniques with the appropriate selectivity and sensitivity are available (Wogan and Gorelick, 1985; Bartsch et al., 1988; Santella, 1988; Shugart, 1988). For this reason, many investigations have as their main focus the analysis of the DNA molecule for structural alteration.

DNA Adduct Detection

Organisms have the metabolic capacity to modify enzymatically substances to which they are exposed, and shortly afterward the presence of interactive products may be detected in the form of a covalently bound DNA adduct in the target tissue. As an example, exposure to the highly carcinogenic chemical 7,12-dimethylbenz[*a*]anthracene (DMBA) results in metabolic transformation to several hydroxymethylated intermediates (12-HMBA, 7-HMBA, and DHBA) as depicted in Figure 2. Subsequent selective enzymatic activation may proceed to a sulfate derivative (7-HMBA sulfate) that can be inactivated via conjugation with glutathione or become attached to cellular macromolecules such as protein and DNA. Note that in the example given, the adduct is a modified guanine base in the DNA molecule and occurred as a result of the covalent bonding of a metabolite of the original chemical to the N^2 nitrogen of the base. Adduct detection and quantification are not simple tasks because techniques (methodologies and instrumentation) are limited in their sensitivity or specificity. These and other difficulties that may arise are discussed by Shugart et al. (1992b).

Methods of varying sensitivity have been devised to measure DNA adducts, including physiochemical (Rahn et al., 1982), immunological (Santella et al., 1987), and postlabeling (Randerath et al., 1981) methods. The methods most commonly employed are ^{32}P postlabeling and immunoassays using adduct-specific antibodies.

Principle of the ^{32}P Postlabeling Adduct Assay

DNA isolated from a suitable sample is enzymatically hydrolyzed with micrococcal nuclease and spleen phosphodiesterase. This results in the formation of 3′-monophosphates of both normal and adducted nucleosides. The products are then ^{32}P labeled by T_4 polynucleotide kinase–catalyzed phosphorylation with [γ-^{32}P]ATP, leading to 5′-^{32}P–labeled 3′,5′-bisphosphate nucleosides.

^{32}P-labeled adducts are resolved from normal nucleotides, ^{32}Pi, unused [γ-^{32}P]ATP, and unknown contaminants by anion exchange thin-layer chromatography on polyethyleneimine-cellulose using a multidirectional developmental scheme (Gupta et al., 1982; Gupta and Randerath, 1988). Dried chromatograms are placed in contact with X-ray film and exposed at −80°C. Adducts are detected by autoradiography and quantitated by scintillation counting. Carcinogen-related adduct patterns are obtained from carcinogen-exposed DNA.

The sensitivity of the assay for adducts can be enhanced by several orders of magnitude, which is most often accomplished by enrichment of aromatic or hydrophobic nucleosides prior to labeling. Extracting in 1-butanol in the presence of a phase transfer agent is one method (Gupta, 1985). An alternative procedure involves selective enzyme hydrolysis of the 3′-nucleotides with nuclease P1 (Reddy and Randerath, 1986) or use of with reverse-phase high-performance liquid chromatography (HPLC) (Dunn and San, 1988).

Levels of detection of adducts can be as low as one in 10^{10} normal nucleotides with this technique; however, ^{32}P-labeled spots of undetermined origin also occur and may interfere with the interpretation of adduct patterns (Randerath et al., 1986). Nevertheless, with the exception of methylating agents and mycotoxins, numerous known carcinogens have been shown to produce carcinogen-specific fingerprints by this method (Gupta and Randerath, 1988).

Principle of the Immunologic Detection of Adducts

Highly specific polyclonal and monoclonal antibodies have been developed to detect chemical adducts of DNA (Santella et al., 1987; Santella, 1988). The two types of immunogens used to generate antibodies are modified DNA electrostatically complexed with protein and the deoxyribonucleoside adduct covalently coupled with protein.

Competitive forms of radioimmunoassay (RIA), enzyme-linked immunosorbent assay (ELISA), or ul-

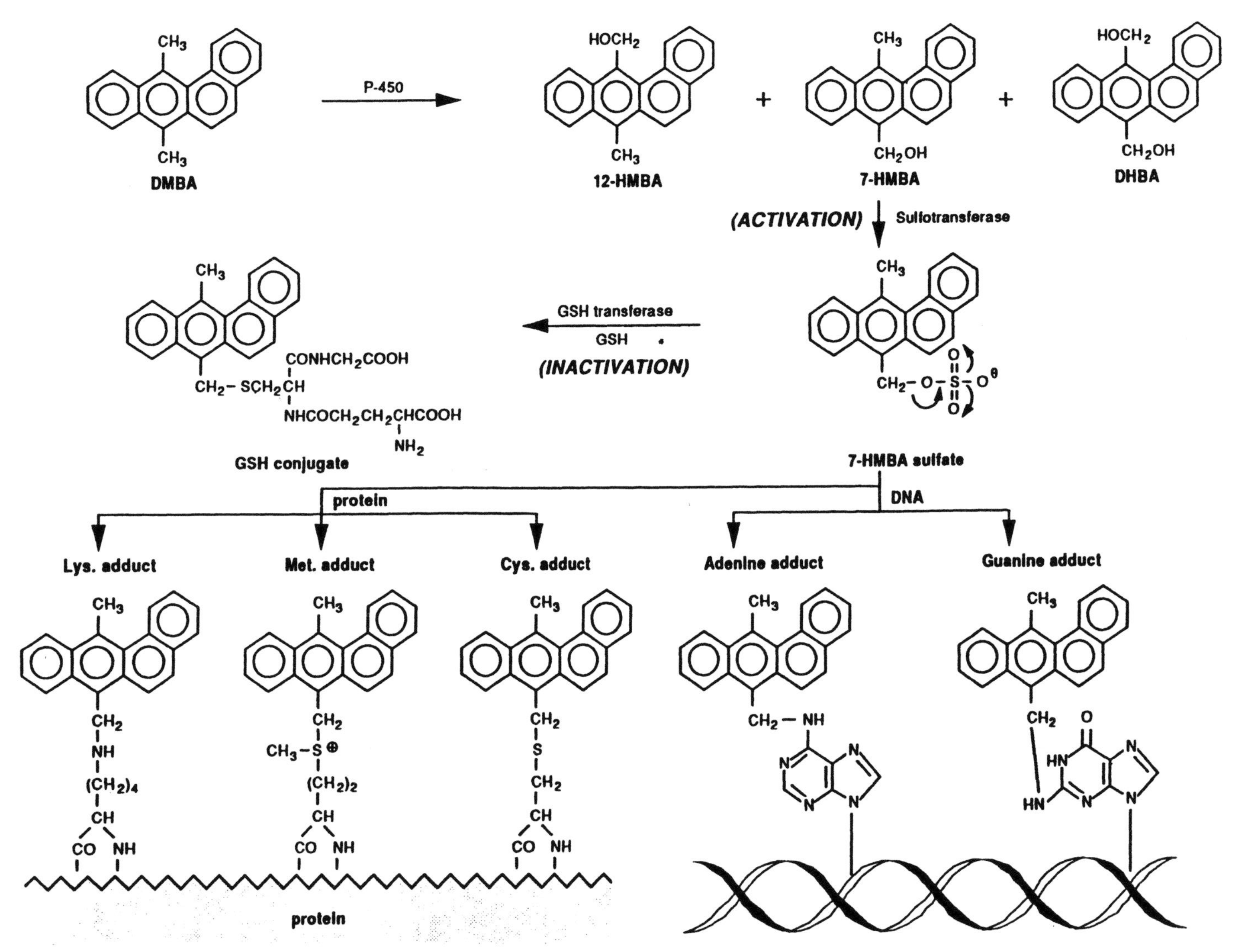

Figure 2. Metabolic conversion of chemical carcinogen 7,12-dimethy-benz[*a*]anthracene to activated products and subsequent adduction of cellular macromolecules (DNA and protein). Lys = Lysine, Met = Methionine, Cys = Cysteine.

trasensitive enzymatic radioimmunassay (USERIA) have been used for quantitation of adduct levels. In RIAs a radiolabeled antigen is mixed with antibody and increasing concentrations of the nonlabeled antigen are used to generate a standard curve of inhibition of antibody binding to the tracer. Bound tracer is determined by precipitation of antibody-antigen complex. An unknown sample is checked for its ability to inhibit antibody binding. In ELISA, the antigen is coated onto a surface and a standard curve of inhibition of antibody binding to the plate is constructed by serially diluting the antigen and mixing with antibody before addition to the coated surface. Binding of the antibody to the surface is quantitated with a second "enzyme-tagged" antibody. Substrates for the enzyme generate colored or fluorescent products, which are measured. The USERIA is similar to the ELISA except that the substrate is radiolabeled.

In general, antibody sensitivity is greatest for the original antigen; however, there can be cross-reactivity with the surrounding DNA structure and other forms of the adduct. Assay sensitivity may also depend on the assay used. In a competitive ELISA, for example, one adduct in 10^8 normal nucleotides can theoretically be obtained with an antibody to a monoadduct.

The antibody approach to adduct detection is limited by the need to develop specific antibodies for each DNA adduct of interest. In addition, because trace contaminants may be capable of inducing an antibody response, complete knowledge of adduct structure is needed to provide assurance that the antibody is recognizing the adduct of interest.

Other DNA Structural Aberrations

Exposure to genotoxic agents may cause, in addition to or concomitant with adduct formation, other types of damage to the DNA molecule. These structural aberrations can manifest themselves in various ways, including strand breaks in the polymer or chemically modified bases (other than adducts).

Principle of the Alkaline Unwinding Assay for the Detection of Strand Breaks

DNA strand breaks can be estimated in DNA by the alkaline unwinding assay (Daniel et al., 1985). In this assay the rate of transition of double-stranded DNA to single-stranded DNA during alkaline denaturation is determined under conditions in which the rate of unwinding is proportional to the number of breaks in the phosphodiester backbone. This technique has been modified (Shugart, 1988) to allow the estimation of strand breaks in the DNA from tissue taken from intact organisms. Highly polymerized DNA is released from the tissue by homogenization in 1 N NH_4OH, 0.2% Triton X-100. Subsequent to differential extraction with chloroform, isoamyl alcohol, phenol (24:1:25 v/v), the DNA is further purified by passage through a molecular sieve column (Sephadex G50). Following a controlled, time-dependent partial alkaline unwinding of the isolated DNA, F values (duplex DNA/total DNA) are calculated from measurements of the amount of DNA that did not unwind and is still in the double-stranded form (duplex DNA). Under conditions of the assay, DNA unwinding takes place at single-strand breaks within the molecule; therefore, the amount of double-stranded DNA remaining after a given period of alkaline unwinding will be inversely proportional to the number of strand breaks present at the initiation of the alkaline exposure, provided renaturation is prevented. The amounts of these two types of DNA (single-stranded and double-stranded) are quantified by measuring the fluorescence that results with bisbenzimidazole (Hoechst dye 33258), a specific DNA-binding dye that fluoresces with double-stranded DNA with about twice as much intensity as it does with single-stranded DNA (Kanter and Schwartz, 1982).

The theoretical background for estimating strand breaks in DNA by alkaline unwinding has been established (Rydberg, 1975) and is summarized by the equation:

$$\ln F = -\left(\frac{K}{M}\right)(t^b)$$

where K is a constant, t is time, M is the number average molecular weight between two breaks, and b is a constant less than 1 that is influenced by the conditions for alkaline unwinding.

The ln F values are used to determine the relative number of strand breaks (N value) between two DNA samples and is expressed as follows:

$$N = (\ln F_s/\ln F_r) - 1$$

where F_s and F_r are the mean F values of DNA from a sampled and a reference preparation, respectively. N values greater than zero indicate that the sampled DNA has more strand breaks than the reference DNA.

Assays for Determining Free Radical Damage to DNA

The microsomal metabolism (see Chapter 17) of some hydrocarbons produces free radicals, including oxyradicals, that are thought to play important roles in chemical carcinogenesis (see Chapter 14). The biochemical mechanisms (see Chapter 17) whereby fluxes of these radicals are produced in aquatic animals have been reviewed (DiGiulio et al., 1989). Both the superoxide radical and the hydroxyl radical are known to damage DNA directly through strand scission and oxidation of the bases of DNA, in particular guanine.

Free radical damage to DNA can result in the hydroxylation of guanosine to form 8-hydroxydeoxyguanosine (8-OH-dGuo), which is released from intact DNA when it is hydrolyzed enzymatically to the deoxynucleoside level. The 8-OH-dGuo is separated from other deoxynucleosides by reverse-phase HPLC and is identified with an electrochemical detector system (Floyd et al., 1986). The level of sensitivity is in the order of one 8-OH-dGuo residue per 10^5 normal dGuo residues.

Free radical damage to the guanine base may also result in the formation of the 2,6-diamino-4-hydroxy-5-formamidopyrimidine moiety (FapyGua), in which ring opening of the heterocyclic base occurs. The FapyGua lesion is removed from DNA by acid hydrolysis, trimethylsilylated, and subjected to gas chromatography–mass spectrometry with single-ion monitoring. Detection of 0.01 nmol of this compound in 1 mg of DNA has been reported (Malins et al., 1990).

Irreversible Events

Structural damage to DNA by genotoxins, if allowed to persist or not properly repaired, can lead to various other anomalies in genetic material, such as chromosomal aberrations, sister chromatid exchanges, gene mutation, and cellular effects (e.g., appearance of tumors, cell differentiation). In relation to the initiating damage, these other events are irreversible and latent. Some anomalies do occur shortly (relatively speaking) after the structural damage to DNA has become established, and several methods that measure or detect these events are discussed.

Principle of the Flow Cytometric Assay to Assess Clastogenic Events

Flow cytometric analysis is a technique that can measure several cellular variables in suspended cells. These may include levels of DNA, RNA, protein, specific chemicals (e.g., immunofluorescent probes), as well as numerous morphological attributes that affect time-of-flight and light-scatter parameters (Shapiro, 1988). The clastogenic action of mutagenic chemicals and ionizing radiation can be detected by flow cytometric analysis as an increase in the variability of the nuclear or chromosomal DNA content of affected cells (Bickham, 1990).

Cell suspensions, prepared from appropriate animal tissues, are stained with a stoichiometric DNA-binding fluorochrome (e.g., 4,6-diamidino-2-phenylindole) and passed through a flow system that carries them in a single stream through the excitation beam. The beam excites the dye and causes emission of visible light from the cell, which is measured by a photometer and processed by a pulse height analyzer. DNA flow histograms are generated and analyzed for information on cell cycle dynamics.

Principle of Oncogene Activation as an Assay for Gene Mutations

Proto-oncogenes are cellular genes that are expressed during normal growth and development processes. They are highly conserved and have been detected in species as divergent as yeast and humans. These genes can be activated to cancer-causing oncogenes by point mutations or by gross DNA rearrangement (e.g., chromosomal translocation or gene amplification). Studies of animal tumor models suggest that chemicals and radiation may play a role in the activation of oncogenes by point mutation (Bartsch et al., 1988), and that this activation creates a variant gene that eventually leads to the emergence of transformed cells.

Oncogenes can be detected by the DNA cotransfection assay in which the NIH/3T3 mouse fibroblast accepts and expresses genes from donor DNA of carcinogen-transformed cells and a selected marker gene. Selected cells are injected subcutaneously into the immunocompromised mouse and tumors that develop are analyzed to characterize the transforming gene (Fasano et al., 1984).

In Situ ENVIRONMENTAL GENOTOXICOLOGY STUDIES WITH AQUATIC SPECIES

Strategies

Genotoxic agents in environmental samples (e.g., sediment, food, or water) in all probability occur as complex mixtures. Even if the chemical composition of these mixtures were known, it would be impossible to predict the biological consequences of exposure. This is true for several reasons. First, toxicological properties of chemicals in mixtures do not necessarily follow the properties of individual components, and second, the compensatory and repair processes of organisms vary considerably. Meaningful information concerning the potential for genotoxicity can be obtained by examining the integrity of DNA of these organisms. This is because the organism functions as an integrator of exposure, accounting for abiotic and physiological factors that modulate the dose of toxicant taken up from the environment. The subsequent magnitude of the genotoxic response can then be used to estimate the severity of exposure, hopefully in time to take preventive or remedial measures (Shugart, 1990a). Although the preceding statement implies a dose-response relationship, it may not always be possible to demonstrate such a relationship in natural populations (those found outside the laboratory) for numerous reasons that tend to confound the issue. For example, organisms in the wild are typically exposed to complex mixtures of pollutants rather than single chemicals and to other stresses (e.g., temperature, disease, nutritional, predator-prey) that may modulate the magnitude of the genotoxic response. This subject is covered in more detail by Depledge et al. (1993).

A suite of markers for genotoxicity that could be used in a program designed to test for the influence of genotoxins on environmental species is given in Table 2. This list is derived from the discussion in the section on basic mechanisms of DNA damage that described the types of DNA damage that genotoxic agents cause as well as some of the consequences of the damage. Each marker represents a unique type of change in the integrity of the DNA molecule corresponding, to some extent, to the pro-

Table 2. Markers of genotoxicity

Marker type	Genotoxic response	Temporal occurrence[a]
Structural	Adducts	Early
	Strand breaks	Early
	Base change	Early
Irreversible	Abnormal DNA	Middle
	Mutation	Late
Other	DNA repair	Early

[a] Temporal occurrence of genotoxic response subsequent to exposure to genotoxin: early, hours to days; middle, days to weeks or months; late, weeks or months to years.

gressive nature of carcinogenesis. Furthermore, existing analytical capabilities have been described that have the sensitivity necessary to detect these genotoxic responses, except for DNA repair, in aquatic species in the wild.

Recent Studies

It is generally recognized that contamination of aquatic environments has led to increased occurrences of neoplasia in aquatic species (Harshbarger and Clark, 1990). Concerns have been expressed that not only ecosystems but also human health are being affected by these contaminants. Studies designed to address these specific issues have been conducted in certain geographic areas, particularly in estuarine and coastal but also in freshwater environments. Several field studies are described in the following that address environmental problems where genotoxins are suspected of being a principal factor.

Puget Sound, Washington

Study Area

As part of the National Benthic Surveillance Project of the National Oceanic and Atmospheric Administration, indicators of contaminant exposure and contaminant-induced stress are being studied in wild fish populations in Puget Sound, Washington. Results from field studies with English sole (*Parophrys vetulus*) taken from Puget Sound suggest a strong correlation between the levels of chemical contaminants, particularly polycyclic aromatic hydrocarbons (PAHs) and polychlorinated biphenyls (PCBs), in the sediments and the prevalence of liver neoplasms (Malins et al., 1984). Uptake and metabolism of these compounds by this species were demonstrated (Varanasi et al., 1987), and laboratory investigations showed that dosing with benzo[*a*]pyrene (BaP), a potent carcinogen of the PAH family of chemicals, resulted in the formation of BaP-DNA adducts and precancerous lesions (Schiewe et al., 1988).

Approach

To further document exposure to genotoxic chemicals, the ^{32}P-postlabeling assay was used to examine DNA for adducts in fish sampled immediately after capture from contaminated and reference areas, as well as those exposed in the laboratory to complex mixtures of xenobiotic compounds extracted from contaminated sediments (Varanasi et al., 1989).

Results

Results obtained indicated that hepatic DNA of fish from sites contaminated with a broad spectrum of chemicals contained a suite of adducts indicative of exposure to genotoxic aromatic compounds, while those sampled from a site of very low chemical contamination did not contain similar hepatic DNA adducts. Also, autoradiograms suggested that some of the adducts detected in wild fish from contaminated sites were due to the adduction of PAH metabolites to DNA, and the mean level of DNA adducts generally agreed with the level of contamination of the sediments. Furthermore, adduct analyses on DNA of the fish exposed to contaminated sediment had profiles qualitatively similar to those of fish exposed in the wild. Overall, it was concluded that the ^{32}P-postlabeling assay could distinguish between fish sampled from areas of high and low contamination and that levels of DNA adducts may be more indicative of cumulative exposure of fish to genotoxic compounds.

Additional Studies

Malins et al. (1990) have shown that FapyGua DNA lesions also occur in the hepatic DNA of English sole environmentally exposed to carcinogens (see Chapter 14). The occurrence of this damage is indicative of insult by a free radical mechanism of genotoxicity. Initially it was reported that the lesion was intimately associated with neoplastic tissue and not with normal tissue. Subsequent studies (Malins and Haimanot, 1991) showed that this type of DNA lesion did occur in histologically normal livers of fish.

Boston Harbor, Massachusetts

Study Area

Winter flounder (*Pseudopleuronectes americanus*) from Boston Harbor, in contrast to fish from adjacent, less polluted coastal areas, show a high incidence of liver lesions that include hepatocellular and cholangiocellular carcinomas. Sediments in the harbor are heavily contaminated with PAHs and other polychlorinated hydrocarbons, suggesting a possible chemical etiology (McMahon et al., 1990).

Approach

DNA structural damage in the form of adducts was assessed by the ^{32}P-postlabeling assay.

Results

The patterns of adducts that were detected in winter flounder from Boston Harbor by the ^{32}P-postlabeling assay (Varanasi et al., 1989) were similar to those found in the English sole from contaminated areas.

Adduct analyses were not performed on the winter flounder from noncontaminated areas.

Additional Studies

Other studies were performed to determine whether oncogene activation was also occurring in this species of fish (McMahon et al., 1990). The DNA from the livers of 13 winter flounder obtained from Boston Harbor and exhibiting gross liver lesions was transfected into NIH/3T3 mouse fibroblasts and assayed for the potential to induce subcutaneous fibrosarcomas in athymic nude mice. DNA from 7 of the 13 diseased livers was effective in eliciting tumor formation, whereas DNA from the remaining lesion-bearing fish and 5 control fish gave negative results. It was concluded that the finding of c-K-*raz* gene activation in most of the diseased flounder livers represented a strong correlation between environmental exposure to genotoxic chemicals and a deleterious disease process.

San Pedro Bay, California

Study Area

San Pedro Bay is located off the coast of southern California near Los Angeles. The sediment is heavily contaminated with PAHs and chlorinated hydrocarbons including PCBs. Important local sportfish, white croaker (*Genyonemus lineatus*) and kelp bass (*Paralabrax clathratus*), that occupy increasing trophic levels in the marine food web are found here. Because they have high tissue concentrations of PAHs and PCBs, human consumption warnings have been posted.

Approach

White croaker and kelp bass were obtained from contaminated areas and from less contaminated sites. Body burdens of chlorinated hydrocarbons were determined along with circulating erythrocyte micronuclei (Hose et al., 1987). Micronuclei are smaller, secondary nuclei formed following chromosomal breakage. Although they may arise spontaneously, induction occurs after exposure to genotoxic agents.

Results

In this study, micronuclei frequencies in fish from contaminated sites were higher than in fish from reference sites and were related to environmental concentrations of chlorinated hydrocarbons.

East Fork Poplar Creek, Tennessee

Study Area

East Fork Poplar Creek (EFPC) is the receiving stream for industrial effluent from the Y-12 plant, a facility in Oak Ridge, Tennessee managed by Martin Marietta Energy Systems for the U.S. Department of Energy. It also receives some urban and agricultural runoff. Water and sediments in EFPC downstream of the Y-12 plant contain metals, organic chemicals, and radionuclides discharged over many years of operation. Sediment samples contained 10 priority pollutant organics (including 7 polycyclic aromatic hydrocarbons, polychlorinated biphenyls, and total phenols) and 7 metals (As, Cd, Pb, Hg, Ni, Ag, and Zr) at concentrations above background. Monitoring of the discharge into EFPC identified potential toxicity problems due to copper, ammonia, residual chlorine, perchloroethylene, total nitrogen, and oil and grease (Loar, 1988). An extensive remedial action program has been implemented to reduce discharge of industrial pollution into the stream, and a Biological Monitoring and Abatement Program (BMAP), as stipulated under an interim National Pollutant Discharge Elimination System permit (see Chapter 24 for general discussion) issued to the Oak Ridge Y-12 Plant, was established.

Approach

Biological markers (Adams et al., 1990) are being examined in fish of EFPC as part of the BMAP, including the detection of strand breaks in the DNA of bluegill (*Lepomis macrochirus*) and redbreast sunfish (*Lepomis auritus*) as a biological monitor for environmental genotoxicity. Strand break analysis was tried for two important reasons. First, it is easy to perform (Shugart, 1988) and therefore cost effective; and second, it is indicative of nonspecific types of DNA damage such as might result from any of a wide range of genotoxic agents that have the potential to cause transient DNA strand breakage.

Results

Results obtained over a 14-mo period (Shugart, 1990b) indicate that the incidence of DNA strand breakage in sunfish at several EFPC sampling stations was significantly higher than in fish from the reference streams at the beginning of the study; however, the temporal and spatial decrease in DNA strand breakage in the fish population suggested a decrease in genotoxic stress indicative of an improving aquatic environment. These results assume even greater importance when integrated into a much larger set of data obtained through the BMAP. Analyses of these data have been used to identify early warning signs of impaired health in fish, to suggest causative relationships between contaminant exposure and effects at the organism level, and to document effects of contaminant stress on biota (Adams et al., 1990).

Savannah River Site, South Carolina

Study Area

A nuclear materials production facility of the U.S. Department of Energy near Aiken, South Carolina has catchment basins containing elevated concentrations of radionuclides such as cesium and strontium (Scott et al., 1986). The resident population of slider turtles (*Trachemys scripta*) have acquired relatively high burdens of both radionuclides. However, a preliminary study (Bickham et al., 1988) of the effect on these species of exposure to low-level radiation has been hampered by the presence of chemical pollutants in these basins.

Approach

Genetic damage in the form of aneuploidy and chromosomal deletions and duplications was evaluated by measuring the DNA content of cells from the tissues of turtles inhabiting reservoirs where exposure was limited to radiation.

Results

Flow cytometric assays of red blood cell nuclei demonstrated significantly greater variation in DNA content of the reservoir turtles than of turtles from a nearby, nonradioactive site (Lamb et al., 1991). Furthermore, two of the reservoir turtles had flow cytometric profiles that were indicative of aneuploid mosaicism.

Additional Studies

A settling basin for low-level radioactivity is located on the Department of Energy reservation in Oak Ridge, Tennessee. Two species of turtles, *T. scripta* (yellow-bellied slider) and *Chelydra serpentian* (snapping turtle), that occupy this basin have significantly higher concentrations of strontium-90, cesium-137, cobalt-60, and mercury in their tissues than do turtles from an uncontaminated reference site. In addition, significantly greater amounts of DNA damage, as measured by strand breakage, were detected in the turtles from the contaminated basin, suggesting exposure to genotoxic agents (Meyers-Schone et al., 1990).

Critique

The in situ studies just discussed were selected in order to emphasize strategies, approaches, and application of techniques. The salient features are summarized in Table 3 and consist of the following:

1. Both marine and freshwater environments were investigated.
2. Various changes in the integrity of DNA were measured that included both structural damage and irreversible events.
3. Sensitive and selective analytical methodologies were employed to detect these changes in DNA integrity.

It should be recognized that many other investigations, similar in approach to those just described, have been initiated to assess the environmental impact of chemical contaminants.

No array of genotoxic responses can be proposed that will provide a comprehensive approach for monitoring genotoxins in the environment. In this regard it should be noted that numerous factors represent barriers to the assessment of exposure and subsequent risk from that exposure. Some factors are intrinsic (e.g., age, sex, health, and nutritional status of the organism) and others are extrinsic (e.g., dose, duration, route of exposure to the contaminant, and the presence of a complex array of chemicals). Furthermore, the levels of DNA alterations that induce subsequent cellular responses can vary with the genotoxic agent. Background levels of DNA adducts, strand breaks, mutations, and other DNA alterations occur as a result of natural phenomena, such as ionizing radiation and dietary components. These levels can vary among species and among tissues within a single species. The ability to measure contamination-induced DNA alterations is directly dependent on an accurate measurement of the background levels of such alterations (Shugart et al., 1992a; Depledge et al., 1993; Peakall and Shugart, 1993).

Therefore, successful exploitation of these markers in an environmental genotoxicity program requires the development of appropriate strategies for their selection and use with the prime goal being the acquisition of data that will allow positive predictions. This is a complex subject and is covered in more detail in several articles on the application of the biological monitoring approach to the assessment of the effect of contaminants on biota in the environment (see Supplemental Reading list at end of this

Table 3. In situ environmental genotoxicology studies with aquatic species

Environment and species	Genotoxic response	Methodology
Marine		
English sole	Adduct	^{32}P postlabeling
	Modified base	Gas chromatography–mass spectrometry
Winter flounder	Adduct	^{32}P postlabeling
	Mutation	Oncogene
Croaker and bass	Chromosomal Aberration	Micronuclei
Freshwater		
Sunfish	Strand break	Alkaline unwinding
Turtles	Chromosomal Aberrations	Flow cytometry
	Strand break	Alkaline unwinding

chapter). Nevertheless, in the context of interpretation, the selection and prioritization of genotoxic biomarkers (see Chapter 1) before initiating an investigation cannot be overemphasized.

One last important point needs to be made. Environmental genotoxicology as a field of scientific investigation is in its infancy. The full potential of this concept and approach will be realized only if continued effort and emphasis are placed on research concerned with the basic cellular mechanisms by which chemicals elicit genotoxic responses; adaptation and/or development of new and sensitive analytical techniques; and field validation studies, particularly with respect to species specificity of individual tests and the selection of appropriate sentinel species (Shugart et al., 1989; Peakall and Shugart, 1993). With regard to the first point, several current areas of scientific endeavor seem appropriate, including genetic polymorphism, programmed cell death, and preferential DNA repair. However, there is an obvious need to correlate DNA damage with genotoxic responses and end points other than cancer (e.g., genetic diversity, reproductive capacity). This need bears on the importance of defining the significance of environmental genotoxin effects on living organisms in terms of biological and ecological risks. Organismal responses have a genetic basis; therefore understanding change at the genetic level should help define the more complex changes at the ecosystem level (Shugart et al., 1992a; Wurgler and Kramers, 1992).

POTENTIAL VALUE OF ENVIRONMENTAL GENOTOXICOLOGY

The field of environmental genotoxicology provides many tools, strategies, and approaches that may be applied to other important areas of regulatory concern. Two of these are risk assessment (see Chapters 28–33) and toxicity testing (see Chapters 2–8).

Environmental Genotoxicity and Risk Assessment

As a result of the many varied practices of our society, numerous substances and materials are constantly entering the environment. Many of these pose a long-term threat to public health and indigenous wildlife because they are nonbiodegradable. Some of the more persistent and widespread environmental chemicals can affect the structural and biological integrity of DNA, inducing genetic perturbations and potentiating subsequent deleterious effects. Traditional physicochemical methods work well for evaluating concentrations of exogenous chemical constituents in tissues or in a well-defined subset environmental compartment, but they do not detect biological effects. Environmental genotoxicology can offer a sensitive, practical, and cost-effective approach for monitoring the occurrence and effects of genotoxins in the environment and would greatly aid in assessing ecosystem health and delineating the effects of anthropogenic pollutants (Lyne et al., 1992; Shugart et al., 1992a; Wurgler and Kramers, 1992).

Genotoxic End Points in Toxicity Testing

Introduction

Toxicity tests are either short (e.g., acute, subchronic) in duration or extended, long-term (chronic) studies. In the former test systems, for example, early life stages of a species are examined for the effect(s) of exposure to toxic substances and the end points of concern are survival, growth, and development. With the latter tests, latent events such as neoplasia and reproductive capacity of survivors are measured. The development of neoplasia, tumors, or other pathological responses of carcinogenesis is a slow process dependent on numerous unknown and ill-defined factors. Therefore, these end points are not amenable to examination in short-term toxicity testing but require expensive long-term exposures for evaluation. The predictive power of toxicity testing in this area could be greatly increased if a suite of rapidly responding markers for genotoxicity is included as an additional component for analysis.

Genotoxicity Testing in Fish

The primary end point for most small fish carcinogenesis studies (see Chapter 14) is the histopathogenic identification of a neoplastic lesion. Such lesions occur long after exposure and are found mainly in the liver, although other tissues in which lesions have been observed include nervous tissue, kidney, mesenchymal tissue, and skeletal and smooth muscle. The histogenesis of liver neoplasms in fish is similar to that in rodents. Following exposure to chemical carcinogens, preneoplastic lesions appear as eosinophilic foci, or basophilic foci, followed by tumor formation (Hawkins et al., 1988). This is identical to the process that appears to occur in rodents and other mammals (Farber, 1987).

The most promising marker for detecting early exposure to carcinogenic chemicals are those that measure the interaction of these compounds with the cellular macromolecule, DNA. Thus, by definition, they are markers of genotoxicity. Various methods have been described for detecting and measuring damage to DNA, and their use provides the basis for assessing exposure to genotoxic agents. This biological end point of genotoxicity assumes even greater significance because the interactions that cause damage to DNA may be the critical events leading eventually to adverse effects (neoplasia) in the exposed organism (Harvey, 1982).

Application

In toxicity testing an evaluation is made on the basis of the suspected presence of some deleterious chemical(s) or contaminant(s), as in water quality testing (U.S. Environmental Protection Agency), sediment toxicity testing (U.S. Corps of Engineers), and animal bioassays for carcinogenesis (National Institutes of Health). Markers of genotoxicity would be useful in these tests because they could signal the presence of an unsuspected chemical with genotoxic properties that would normally go unnoticed with the conventional end points of the tests used as in the first two examples (e.g., measurement of strand breaks or adducts in DNA in fathead minnow larvae at the termination of a standard early life stage toxicity test). Or, these markers of genotoxicity could be a short-term predictor of potential long-term effects, as in the tests for carcinogenesis (e.g., correlation of early genotoxic responses with subsequent latent events such as the occurrence of aneuploidy or transformation of initiated cells). An often overlooked ancillary benefit derived from the application of biomarkers is the savings in cost, especially in tests involving expensive animal models of carcinogenesis and mutagenesis.

SUMMARY

The exposure of an organism to genotoxic chemicals may include a cascade of DNA-damaging events. Initially, structural alterations of DNA are formed. Next, the DNA damage is processed and subsequently expressed in mutant gene products. Finally, diseases result from the genetic damage. The detection and quantitation of the various events in this sequence may be used to establish exposure and to assess the effect of that exposure. Numerous techniques and methodologies are available to detect the various genotoxic responses that may be elicited in an exposed organisms. Adducts are a type of structural change involving covalent attachment of a chemical or its metabolite to DNA. Several analytical procedures are currently used to monitor the formation of DNA adducts directly; the most prominent of these procedures is the ^{32}P-postlabeling technique. A more general procedure that detects a nonspecific type of DNA damage is the measurement of strand breaks, a type of damage produced either directly by genotoxic chemicals or by the processing of structural damage. Detection of this type of damage is facilitated by the alkaline unwinding assay. The current methods available for monitoring the consequences of DNA damage are based on changes in chromosomes, such as aberrations, formation of micronuclei, detection of abnormal distribution of DNA within cells, and activation of proto-oncogenes. At present it is difficult to demonstrate the direct effect of environmental pollution at the ecosystem level. Future research must address mechanistic questions about how an organism is related to other organisms and in turn to its physical environment.

ACKNOWLEDGMENTS

The author is a senior research staff member of the Oak Ridge National Laboratory which is managed by Martin Marietta Energy Systems Inc., under contract DE-AC05-84OR21400 with the U.S. Department of Energy. This is Environmental Sciences Division publication no. 3994.

LITERATURE CITED

Adams, S. M.; Shugart, L. R.; Southworth, G. R.; and Hinton, D. E.: Application of bioindicators in assessing the health of fish populations experiencing contaminant stress. *Biological Markers of Environmental Contaminants*, edited by J. F. McCarthy, and L. R. Shugart, pp. 333–353. Boca Raton, FL: Lewis, 1990.

Alberts, B.; Bray, D.; Lewis, J.; Raff, M.; Roberts, K.; and Watson, J. D.: *Molecular Biology of the Cell*, 2nd ed, pp. 220–227. New York: Garland, 1989.

Ames, B. N.; Durston, W. E.; Yamasaki, E.; and Lee, R. D.: Carcinogens are mutagens: a simple test system combining liver homogenates for activation and bacterial for detection. *Proc. Natl. Acad. Sci. USA*, 70:2281–2285, 1973.

Ashby, J.: Comparison of techniques for monitoring human exposure to genotoxic chemicals. *Mutat. Res.*, 204:543–551, 1988.

Baksi, S. M., and Frazier, J. M.: Isolated fish hepatocytes—model systems for toxicology research. *Aquat. Toxicol.*, 16:229–256, 1990.

Bartsch, H.; Hemminki, K.; and O'Neill, I. K. (eds.): Methods for Detecting DNA Damaging Agents in Humans: Application in Cancer Epidemiology and Prevention. *IARC Scientific Publication No. 89*, IARC, Lyon, France, 1988.

Bickham, J. W.: Flow cytometry as a technique to monitor the effects of environmental genotoxins on wildlife populations. In Situ Evaluation of Biological Hazards of Environmental Pollutants, edited by S. S. Sandhu; W. R. Lower; F. J. deSerres; W. A. Suk; and R. R. Tice, pp. 97–108. New York: Plenum, 1990. *Environmental Science Research*, Vol. 38.

Bickham, J. W.; Hanks, B. G.; Smolen, J. J.; Lamb, T.; and Gibbons, J. W.: Flow cytometric analysis of the effects of low-level radiation exposure on natural populations of slider turtles (*Pseudemys scripta*). *Arch. Environ. Contam. Toxicol.*, 17:837–841, 1988.

CBMNRC/Committee on Biological Markers of the National Research Council: Biological markers in environmental health research. *Environ. Health Perspect.*, 74: 3–9, 1987.

Daniel, F. B.; Haas, D. L.; and Pyle, S. M.: Quantitation of chemically induced DNA strand breaks in human cells via an alkaline unwinding assay. *Anal. Biochem.* 144: 390–402, 1985.

Depledge, M. H.; Amaral-Mendes, J. J.; Daniel, B.; Halbrook, R. S.; Kloepper-Sams, P.; Moore, M. N.; and Peakall, D. B.: The conceptual basis of the biomarker approach. *Biomarkers: Research and Application in the Assessment of Environmental Health*, edited by D. B. Peakall, and L. R. Shugart, pp. 15–29. Heidelberg: Springer-Verlag, 1993.

DiGiulio, R. T.; Washburn, P. C.; Wenning, R. J.; Winston, G. W.; and Jewell, C. S.: Biochemical responses in aquatic animals: a review of determinants of oxidative stress. *Environ. Toxicol. Chem.*, 8:1103–1123, 1989.

Dunn, B. P., and San, R. H. C.: HPLC enrichment of hydrophobic DNA-carcinogen adducts for enhanced sensitivity of ^{32}P-postlabeling analysis. *Carcinogenesis*, 9: 1055–1060, 1988.

Farber, E.: Possible etiologic mechanisms in chemical carcinogenesis. *Environ. Health Perspect.*, 75:65–70, 1987.

Fasano, O.; Birnbaum, D.; Edlund, L.; Fogh, J.; and Wigler, M.: New human transforming genes detected by a tumorigenicity assay. *Mol. Cell Biol.*, 4:1695–1705, 1984.

Floyd, R. A.; Watson, J. J.; Wong, P. K.; Altmiller, D. H.; and Rickard, R. C.: Hydroxyl free radical adduct of deoxyguanosine: sensitive detection and mechanisms of formation. *Free Radical Res. Commun.*, 1:163–172, 1986.

Gupta, R. C.: Enhanced sensitivity of ^{32}P-postlabeling analysis of aromatic carcinogen-DNA adducts. *Cancer Res.*, 45:5656–5662, 1985.

Gupta, R. C., and Randerath, K.: Analysis of DNA adducts by ^{32}P labeling and thin layer chromatography. *DNA Repair*, Vol 3, edited by E. C. Friedberg, and P. C. Hanawalt, pp. 399–418. New York: Marcel Dekker, 1988.

Gupta, R. C.; Reddy, M. C.; and Randerath, K.: ^{32}P-labeling analysis of nonradioactive aromatic carcinogen-DNA adducts. *Carcinogenesis*, 3:1081–1092, 1982.

Harshbarger, J. C., and Clark, J. B.: Epizootiology of neoplasms in bony fish of North America. *Sci. Total Environ.*, 94:1–32, 1990.

Harvey, R. C.: Polycyclic hydrocarbons and cancer. *Am. Sci.*, 70:386–393, 1982.

Hawkins, W. E.; Overstreet, R. M.; and Walker, W. W.: Carcinogenicity tests with small fish species. *Aquat. Toxicol.*, 11:113–128, 1988.

Hinton, D. E.; Couch, J. A.; Teh, S. J.; and Courtney, L. A.: Cytological changes during progression of neoplasia in selected fish species. *Aquat. Toxicol.*, 11:77–112, 1988.

Hose, J. E.; Cross, J. N.; Smith, S. G.; and Diehl, D.: Elevated circulating erythrocyte micronuclei in fishes from contaminated sites off southern California. *Mar. Environ. Res.*, 22:167–176, 1987.

Jones, N. J., and Parry, J. M.: The detection of DNA adducts, DNA base changes and chromosome damage for the assessment of exposure to genotoxic pollutants. *Aquat. Toxicol.*, 22:323–344, 1992.

Kanter, P. M., and Schwartz, H. S.: A fluorescence enhancement assay for cellular DNA damage. *Mol. Pharmacol.*, 22:145–151, 1982.

Knudson, A. G., Jr.: Hereditary cancer, oncogenes, and antioncogenes. *Cancer Res.*, 45:1437–1443, 1985.

Kocan, R. M., and Powell, D. B.: Anaphase aberrations: an in vitro test for assessing the genotoxicity of individual compounds and complex mixtures. *Short-Term Genetic Bioassays in the Analysis of Complex Environmental Mixtures IV*, edited by M. D. Water; S. S. Sandhu; J. Lewtas; L. Claxton; G. Strauss; and S. Nesnow, pp. 75–86. New York: Plenum, 1985a.

Kocan, R. M.; Sabo, K. M.; and Landolt, M. L.: Cytotoxicity/genotoxicity: the application of cell culture techniques to the measurement of marine sediment pollution. *Aquat. Toxicol.*, 6:165–177, 1985b.

Kohn, H. W.: The significance of DNA-damaging assays in toxicity and carcinogenicity assessment. *Ann. N.Y. Acad. Sci.*, 407:106–118, 1983.

Lamb, T.; Bickham, J. W.; Gibbons, J. W.; Smolen, M. J.; and McDowell, S.: Genetic damage in a population of slider turtles (*Trachemys scripta*) inhabiting a radioactive reservoir. *Arch. Environ. Contam. Toxicol.*, 20:138–142, 1991.

Loar, J. M. (ed.): *First Annual Report on the Y-12 Plant Biological Monitoring and Abatement Program*, Draft ORNL/TM. Oak Ridge National Laboratory, Oak Ridge, TN, 1988.

Lyne, T. B.; Bickham, J. W.; Lamb, T.; and Gibbons, J. W.: The application of bioassays in risk assessment of environmental pollution. *J. Risk Anal.*, 12:361–364, 1992.

Malins, D. C., and Haimanot, R.: The etiology of cancer: hydroxyl radical–induced DNA lesions in histologically normal livers of fish from a population with liver tumors. *Aquat. Toxicol.*, 20:123–130, 1991.

Malins, D. C.; McCain, B. B.; Brown, D. W.; Chan, S. L.; Myers, M. S.; Landahl, J. T.; Prohaska, P. G.; Friedman, A. J.; Rhodes, L. D.; Burrows, D. G.; Gronlund, W. D.; and Hodgins, H. O.: Chemical pollutants in sediments and diseases in bottom-dwelling fish in Puget Sound, Washington. *Environ. Sci. Technol.*, 18:705–713, 1984.

Malins, D. C.; Ostrander, G. K.; Haimanot, R.; and Williams, P.: A novel DNA lesion in neoplastic livers of feral fish: 2,6-diamino-4-hydroxy-5-formamidopyrimidine. *Carcinogenesis*, 11:1045–1047, 1990.

McMahon, G.; Huber, L. J.; Moore, M. J.; Stegeman, J. J.; Wogan, G. N.: c-K-*ras* oncogenes: prevalence in livers of winter flounder from Boston harbor. *Biological Markers of Environmental Contaminants*, edited by J. F. McCarthy, and L. R. Shugart, pp. 229–235. Boca Raton, FL: Lewis, 1990.

Meyers-Schone, L.; Shugart, L. R.; and Walton, B. T.: *Comparison of two freshwater turtle species as monitors of environmental contamination*. ORNL/TM-1146. Oak Ridge National Laboratory, Oak Ridge, TN, 1990.

Peakall, D. B., and Shugart, L. R. (eds.): *Biomarkers: Research and Application in the Assessment of Environmental Health*. Heidelberg: Springer-Verlag, 1993.

Phillips, D.: Fifty years of benzo[*a*]pyrene. *Nature*, 303: 478–472, 1983.

Phillips, D., and Sims, P.: PAH metabolites: their reaction with nucleic acids. *Chemical Carcinogens and DNA*, Vol. 2, edited by P. L. Grover, pp. 9–57. Boca Raton, FL: CRC Press, 1979.

Rahn, R. O.; Chang, S. S.; Holland, J. M.; and Shugart, L. R.: A fluorometric HPLC assay for quantitating the binding of benzo[*a*]pyrene metabolites to DNA. *Biochem. Biophys. Res. Commun.*, 109:262–268, 1982.

Randerath, K.; Reddy, M. V.; and Gupta, R. C.: ^{32}P-labeling test for DNA damage. *Proc. Natl. Acad. Sci. USA*, 78: 6126–6129, 1981.

Randerath, K.; Reddy, M. C.; and Disher, R. M.: Age- and tissue-related DNA modifications in untreated rates: detection by ^{32}P-postlabeling assay and possible significance for spontaneous tumor induction and aging. *Carcinogenesis*, 7:1615–1617, 1986.

Reddy, M. V., and Randerath, K.: Nuclease P1–mediated enhancement of sensitivity of ^{32}P-postlabeling test for structurally diverse DNA adducts. *Carcinogenesis*, 7: 1543–1551, 1986.

Rydberg, B.: The rate of strand separation in alkali of DNA of irradiated mammalian cells. *Radiat. Res.*, 61:274–285, 1975.

Sancar, A., and Sancar, G. B.: DNA repair enzymes. *Annu. Rev. Biochem.*, 57:29–67, 1988.

Sandhu S. S., Lower W. R.: In situ assessment of genotoxic hazards of environmental pollution. *Toxicol. Ind. Health*, 5:73–83, 1989.

Santella, R. M.; Gasparo, F.; and Hsieh, L.: Quantitation of carcinogen-DNA adducts with monoclonal antibodies. *Prog. Exp. Tumor Res.*, 31:63–75, 1987.

Santella, R. M.: Application of new techniques for the detection of carcinogen adducts to human population monitoring. *Mutat. Res.*, 205:271–282, 1988.

Schiewe, M. H.; Landahl, J. T.; Meyers, M. S.; Plesha, P. D.; Jacques, F. J.; Stein, J. E.; McCain, B. B.; Weber, D. D.; Chan, S. L.; and Varanasi, U.: Relating field and laboratory studies: cause-and-effect research. *Proceedings of the 1st Annual Meeting of the Puget Sound Water Quality Authority*, Seattle, pp. 577–584, 1988.

Scott, D. E.; Whicker, F. W.; and Gibbons, J. W.: Effect of season on the retention of ^{137}Cs and ^{90}Sr by the yellow-bellied slider turtle (*Pseudemys scripta*). *Can. J. Zool.*, 64:2850–2853, 1986.

Shapiro, H. M.: *Practical Flow Cytometry*, 2nd ed. New York: Alan R. Liss, 1988.

Shugart, L. R.: Quantitation of chemically induced damage to DNA of aquatic organisms by alkaline unwinding assay. *Aquat. Toxicol.*, 13:43–52, 1988.

Shugart, L. R.: Biological monitoring: testing for genotoxicity. *Biological Markers of Environmental Contaminants*, edited by J. F. McCarthy, and L. R. Shugart, pp. 205–216. Boca Raton, FL: Lewis, 1990a.

Shugart, L. R.: DNA damage as an indicator of pollutant-induced genotoxicity. *13th Symposium on Aquatic Toxicology and Risk Assessment: Sublethal Indicators of Toxic Stress*, edited by W. G. Landis, W. H. van der Schalie, pp. 348–355. Philadelphia: ASTM, 1990b.

Shugart, L. R.; Adams, S. M.; Jimenez, B. D.; Talmage, S. S.; and McCarthy, J. F.: Biological markers to study exposure in animals and bioavailability of environmental contaminants. *Biological Monitoring for Pesticide Exposure: Measurement, Estimation, and Risk Reduction*, edited by R. G. M. Wang; C. A. Franklin; R. C. Honeycutt; and J. C. Reinert, pp. 86–97. Washington, DC: American Chemical Society, 1989. ACS Symposium Series 382.

Shugart, L. R.; McCarthy, J. F.; and Halbrook, R. S.: Biological markers of environmental and ecological contamination: an overview. *Risk Anal.*, 12:353–360, 1992a.

Shugart, L. R.; Bickham, J.; Jackim, G.; McMahon, G.; Ridley, W.; Stein, J.; and Steinert, S.: DNA alterations. *Biomarkers: Biochemical, Physiological, and Histological Markers of Anthropogenic Stress*, edited by R. J. Huggett; R. A. Kimerle; P. M. Mehrle; and H. L. Bergman, pp. 125–153. Boca Raton, FL: Lewis, 1992b.

Varanasi, U.; Stein, J. E.; Nishimoto, M.; Reichert, W. L.; and Collier, T. K.: Chemical carcinogenesis in feral fish: uptake, activation, and detoxification of organic xenobiotics. *Environ. Health Perspect.*, 71:155–170, 1987.

Varanasi, U.; Reichert, W. L.; and Stein, J. E.: ^{32}P-postlabeling analysis of DNA adducts in liver of wild english sole (*Paraophyrs vetulus*) and winter flounder (*Pseudopleuronectes americanus*). *Cancer Res.*, 49:1171–1177, 1989.

Waters, M. D.; Sandhu, S. S.; Lewtas, J.; Claxton, L.; Strauss, G.; and Nesnow, S. (eds.): *Short-Term Bioassays in the Analysis of Complex Environmental Mixtures IV*. New York: Plenum, 1984.

Weinstein, I. B.: Current concepts on mechanism of chemical carcinogenesis. *Bull. N. Y. Acad. Med.*, 54:336–383, 1978.

Wogan, G. N., and Gorelick, N. J.: Chemical and biochemical dosimetry to exposure to genotoxic chemicals. *Environ. Health Perspect.*, 62:5–18, 1985.

Wurgler, F. E., and Kramers, P. G. N.: Environmental effects of genotoxins (eco-genotoxicology). *Mutagenesis*, 7:321–327, 1992.

SUPPLEMENTAL READING

Adams, S. M. (ed.): *Biological Indicators of Stress in Fish*. Bethesda, MD: American Fisheries Society, 1990. American Fisheries Symposium 8.

Bartsch, H.; Hemminki, K.; and O'Neill, I. K. (eds.): Methods for Detecting DNA Damaging Agents in Humans: Application in Cancer Epidemiology and Prevention. Lyon, France: IARC, 1988. *IARC Scientific Publication No. 89*.

Huggett, R. J.; Kimerle, R. A.; Mehrle, P. M.; and Bergman, H. L. (eds.): *Biomarkers: Biochemical, Physiological, and Histological Markers of Anthropogenic Stress*. Boca Raton, FL: Lewis, 1992.

Maccubbin, A. E.: DNA adduct analysis in fish: laboratory and field studies. *Aquatic Toxicology-Molecular, Biochemical and Cellular Perspectives*, edited by D. C. Malins, and G. K. Ostrander, pp. 267–294. Boca Raton, FL: Lewis, 1994.

McCarthy, J. F., and Shugart, L. R. (eds.): *Biological Markers of Environmental Contamination*. Boca Raton, FL: Lewis, 1990.

Peakall, D. B., and Shugart, L. R. (eds.): *Biomarkers: Research and Application in the Assessment of Environmental Health*. Heidelberg: Springer-Verlag, 1993.

Pritchard, J. B. (ed.): Mechanisms of pollutant action in aquatic organisms. *Environ. Health Perspect.*, 71:1–193, 1987.

Varanasi, U. (ed.): *Metabolism of Polycyclic Aromatic Hydrocarbons in the Aquatic Environment*. Boca Raton, FL: CRC Press, 1989.

Chapter 14

CARCINOGENICITY TESTS USING AQUARIUM FISH

W. E. Hawkins, W. W. Walker, and *R. M. Overstreet*

INTRODUCTION

Environmental factors contribute to the occurrence of cancer in humans (Swenberg et al., 1991). Based on epidemiological evidence, Doll and Peto (1981) estimated that environmental factors including exposure to chemical carcinogens account for almost 80% of all human cancers in the United States. Experimentally, the link between exposure to certain chemicals and cancer has been established in whole-animal chronic bioassays (i.e., carcinogenicity tests) using inbred strains of rats and mice. These tests are conducted under highly standardized laboratory conditions. Typically, groups of 50 male and 50 female specimens of each species are exposed to two or more dose levels of the test substance over a substantial part of the life span of the test organism. Dose levels generally include the maximum tolerated dose (MTD) of the test substance and one or more fractions of the MTD (Weisburger and Williams, 1981). Sex- and age-matched animals constitute the control (untreated) group. The MTD is predicted from subchronic toxicity studies (i.e., 90 d) as the dose that causes no more than 10% weight loss compared with the control group and does not produce mortality, clinical signs of toxicity, or noncancerous pathologic lesions that would shorten the animal's natural life span. The MTD is not necessarily a nontoxic dose but is expected to elicit some level of toxicity. Chronic carcinogenicity bioassays usually last for 1 or 2 yr, after which time the test animals are examined pathologically for the occurrence of neoplasms, or cancerous lesions, and the incidence of neoplasms in the exposure groups is compared statistically with that in the control specimens. Generally, a *neoplasm* represents a pathological disturbance of growth characterized by an excessive and uncontrolled proliferation of cells. Although the rodent carcinogenicity test has been criticized because of its cost, the length of time required to establish effects, and insensitivity because of the necessity of using the MTD to ensure against false-negative results, the whole-rodent bioassay remains the method of choice for identifying chemicals that pose a carcinogenic threat to humans.

Mammals, however, are not the only species at risk of developing cancer from exposure to chemical carcinogens. Studies have shown that environmental chemicals cause cancer in fishes as well. Harshbarger and Clark (1990) documented 41 geographic regions in North America in which clusters, or *epizootics*, of cancer in wild fish have occurred. They defined an epizootic as a situation in which three or more cases of a neoplasm originating from a specific cell lineage or type have been found in a geographically defined area. Although some of the tumor types such as those involving hemic, neural, pigment cell, connective tissue, and gonadal neoplasms do not appear to be closely associated with environmental contamination, the occurrence of neoplasms involving epithelial tissues such as the liver, pancreas, and gastrointestinal tract and some epidermal neoplasms appear strongly correlated with environmental contamination, that is, exposure to chemical carcinogens. Reports also suggest that some epizootics of neoplasms in invertebrates are also the result of chemical contamination (Gardner, 1993).

It follows that certain fish species should be suitable test organisms for assessing a chemical's potential carcinogenicity, at least to that fish species. Cancer in fish has not been confined strictly to wild populations. In the 1960s wild stocked rainbow trout (*Onchorhynchus mykiss*) were found to have liver tumors (Halver, 1967; Sinnhuber et al., 1977). Laboratory studies later showed that aflatoxin, a mycotoxin produced in moldy feeds, was the cause of liver neoplasia in the trout. Aflatoxin and its analogues have since been shown to be potent liver carcinogens in a wide variety of species including mammals. The fact that many fish species appear to be sensitive to the carcinogenic effects of certain chemicals supports the use of fish in carcinogenicity tests as alternatives or supplements to rodent chronic bioassays and also provides opportunities to address some fundamental issues in cancer biology (Bailey et al., 1987; Ishikawa and Takayama, 1979; Masahito et al., 1988). This chapter examines the use of small fish species in carcinogenicity tests and describes some practical approaches to using small aquarium fish to detect environmental carcinogens.

FUNDAMENTAL CONCEPTS IN CARCINOGENESIS

Terminology Used in Carcinogenesis and Carcinogenicity Testing

As in any field of science, cancer biologists have developed their own lexicon. Recognizing that individuals wishing to conduct small-fish carcinogenicity bioassays or to use data derived from them might come from diverse backgrounds, a brief overview of current terms and concepts in carcinogenesis and carcinogen testing is provided. Because authorities sometimes disagree, we have limited the reference sources of our information to a popular treatise on cancer, Cancer: The Misguided Cell (Prescott and Flexer, 1986); a standard medical textbook, Pathological Basis of Disease (Robbins et al., 1984); and a monograph, The Pathobiology of Neoplasia (Sirica, 1989a).

The objective of the carcinogenicity test is to determine whether the test agent induces neoplastic lesions in the test organism. Neoplastic lesions, or neoplasms, are usually detected by histopathological techniques. Neoplasms have been shown to occur in organisms from numerous phyla including platyhelminthes, arthropods, mollusks, and chordates and can arise from many cell types representing nearly all types of tissues (Harshbarger et al., 1981). A neoplastic lesion, or neoplasm, is one that incorporates heritable changes that are not controlled by normal cellular processes that regulate growth. The change is passed on through successive (clonal) cell generations. Growth of a neoplasm is autonomous, exceeds that of the surrounding tissue, and continues after cessation of the original stimulus (for example, chemical exposure) that caused it (Sirica, 1989b). The term neoplasm is preferred to *tumor*, which means a swelling, or to *cancer*, which generally refers to malignant neoplasm (Sirica, 1989b). Two other terms, hyperplasia and dysplasia, are often associated with histopathological analysis in carcinogenicity tests and tend to cause confusion in both rodent (Eustis, 1989) and fish (Harshbarger, 1984) studies. Morphologically, *hyperplasia* refers to an increase in the number of normal cells that constitute a given tissue, whereas *dysplasia* refers to an alteration in adult cells that is characterized by variations in size, shape, and organization (Robbins et al., 1984). Hyperplasia, however, may play a role in the early stages of neoplasia.

Although there are exceptions, *benign lesions* tend to grow slowly, do not invade adjacent normal tissue (i.e., remain localized), and do not metastasize or spread to distant sites (Eustis, 1989). Cells of benign neoplastic lesions usually resemble their normal counterparts and are organized somewhat like the tissue from which they originated. Benign neoplasms are often encapsulated by fibrous connective tissue. *Malignant lesions*, on the other hand, tend to grow rapidly, invade adjacent normal tissues, and metastasize. Usually, malignant neoplasms are not encapsulated. Malignant cells often exhibit degrees of anaplasia. *Anaplasia* denotes a radical departure from the normal appearance and organization of the cells and from the tissue of origin. An anaplastic lesion is said to be poorly differentiated, and often the cells of the lesion are primitive or even embryonic in form and organization. The terms benign and malignant are often difficult to apply because they require the pathologist to predict the biological behavior of the lesions. Some statistical methods, however, now call for the pathologist to anticipate the lethality or malignancy of lesions, giving malignant ones greater statistical weight than benign ones.

Nomenclature of Neoplasms

Histopathological evaluation involves detecting neoplastic lesions and accurately diagnosing, or naming, them. Neoplasm nomenclature is complex, with few rules and many exceptions. Nomenclature sometimes varies among species and usage among investigators. Although it is often difficult to determine whether lesions are benign or malignant, those terms are entrenched in neoplasm nomenclature. Some common terms including suffixes, prefixes, and specific neoplasms, particularly as they relate to fish, are summarized in Table 1. The nomenclature of neoplastic lesions in fish attempts to follow the same guidelines as those in mammals (Mawdesley-Thomas, 1972, 1975). A review analyzes the pathobiology of the principal fish neoplasms and compares them with their counterparts in mammalian species (Dawe et al., in press).

Mechanistic Classification of Carcinogens

There are several ways to classify chemical carcinogens and no single classification is absolute. A chemical agent can be described in different ways depending on the classification system. One way is to identify a carcinogen with a particular class of chemicals that includes individual ones that have been proved to be carcinogenic when tested in animal models. Table 2 lists some well-known classes of carcinogens and provides some examples of carcinogenic members of the classes. As discussed below, carcinogenicity depends not only on chemical factors but also on many biological variables such as species, strain, sex, age, and hormonal state.

Another classification system is based on the metabolism and reactivity of the agent with cellular nucleophiles and includes (1) direct-acting versus indirect-acting carcinogens, (2) genotoxic versus epigenetic carcinogens, and (3) initiators versus pro-

Table 1. Some terms used to name neoplastic lesions

Term	Meaning
Suffix *-oma*	Benign neoplasm of a particular tissue
Carcinoma	Malignant neoplasm of epithelial origin
Sarcoma	Malignant neoplasm of mesenchymal origin
Prefix *-adeno*	Refers to a gland
Suffix *-blastoma*	Malignant neoplasm that resembles embryonic tissue
Cholangioma	Benign neoplasm of bile ducts
Cholangiocarcinoma	Malignant neoplasm of bile ducts
Adenoma	Benign neoplasm of a gland
Adenocarcinoma	Malignant neoplasm of a gland
Rhabdomyoma	Benign neoplasm of striated muscle
Rhabdomyosarcoma	Malignant neoplasm of striated muscle
Lymphoma	Malignant neoplasm of lymphoid tissue
Hepatocellular carcinoma	Malignant neoplasm of hepatocytes; preferred over *hepatoma*
Granuloma	Nonneoplastic lesion consisting mainly of macrophages usually associated with bacterial infection

Source: Sirica, 1989b.

moters (Weisburger, 1989) (Table 3). How a potential carcinogen acts within an organism can be dependent on many factors, such as those mentioned above, as well as the mechanism of action of the carcinogen. Although the database is smaller for fish than for mammals, several small aquarium fish species have been shown to be sensitive to members of each of these classes of carcinogenic agents (Hatanaka et al., 1982; Hawkins et al., 1988b).

Regulatory Classification of Carcinogens

Carcinogens are also classified according to the hazard or risk potential they pose to exposed populations (Matula and Somers, 1989). To protect humans from hazardous exposures to carcinogens, agencies and institutions such as the International Agency for Research on Cancer (IARC), the U.S. Environmental Protection Agency (EPA), and the National Toxicology Program (NTP) classify carcinogens on the basis of evidence derived from epidemiologic and laboratory studies. Each of these bodies has its specific criteria for carcinogenicity. For example, IARC considers evidence for carcinogenicity from animal studies sufficient if the studies showed an increased incidence of neoplasms in (1) two or more species, (2) two or more independent studies using the same species, or (3) one species demonstrating an exception-

Table 2. Some well-known chemical classes and examples of carcinogenic compounds in mammals

Chemical class	Examples
Polycyclic aromatic hydrocarbons	Benzo[*a*]pyrene
	7,12-Dimethylbenz[*a*]anthracene
Nitrosamines	*N*-Nitrosodiethylamine
	N-Nitrosodimethylamine
	N-Nitrosomorpholine
	N-Methyl-*N*′-nitronitrosoguanidine
Aromatic amines	Benzidine
	2-Acetylaminofluorene
	4-Aminobiphenyl
Hydrazines and related compounds	Hydrazine
	1,2-Dimethylhydrazine
	Azoxymethane
	Methylazoxymethanol acetate
Mycotoxins	Aflatoxin B_1
Halogenated aliphatic compounds	Chloroform
	Carbon tetrachloride
	Vinyl chloride
	Trichloroethylene
	1,2-Dibromoethane
	1,2-Dibromo-3-chloropropane

Source: Weisburger, 1989.

Table 3. Classification of chemical carcinogens based on reactivity of metabolites

Carcinogen classification	Characteristics	Some examples
Direct-acting	Metabolism not required to react with cellular macromolecules	Methylazoxymethanol acetate β-Propiolactone Bis(chloromethyl)ether *N*-methyl-*N*′-nitronitrosoguanidine
Indirect-acting	Metabolism required to react with cellular macromolecules; also called pre- or procarcinogens	2-Acetylaminofluorene Benzo[*a*]pyrene *N*-Nitrosodiethylamine Aflatoxin B_1
Genotoxic	Interacts with genetic material of the cell, especially DNA	Includes both direct- and indirect-acting carcinogens
Epigenetic	Does not react with DNA, nonmutagenic	Immunosuppressants Hormones Solid-state materials (asbestos)
Initiator	Capacity to cause irreversible alteration in cell genome	Presumably any of the above
Promoter	Increase carcinogenic response when applied after a carcinogen; non- or weakly carcinogenic themselves; not mutagenic and do not bind to DNA	Phorbol esters Dodecane Anthralin Mezerein Teleocidin

Source: Weisburger, 1989.

ally high neoplasm incidence. However, extrapolation of the results of animal tests to determine human risk is difficult. To aid in extrapolation, potency indices are being developed such as the TD50 (TD = tumorigenic dose) index of Peto et al. (1984), which indicates the carcinogen dose rate in mg/kg body wt/d that will result in half of the test animals remaining tumor free through the standard life span of the species. Potency indices aid in comparing carcinogens with one another and in estimating realistic exposure scenarios for humans (Goodman and Wilson, 1991).

Multistage Nature of Carcinogenesis

Inherent in interpreting the results of a carcinogenicity bioassay is an understanding that carcinogenesis occurs in stages that are usually described as initiation, promotion, and progression (Pitot, 1989; Sirica, 1989b). These stages, which are illustrated schematically in Figure 1, are associated with the development of some specific histopathologic lesions (Goodman et al., 1991) that are discussed later. Multistage carcinogenesis has been demonstrated to occur in studies in which neoplasm development was enhanced by sequential or simultaneous administration of two or more chemical compounds having different carcinogenic properties. Some specific types of neoplasm enhancement have been described as syncarcinogenesis and cocarcinogenesis (Peraino and Jones, 1989). *Syncarcinogenesis* occurs when the simultaneous or sequential administration of two carcinogens, individually having little carcinogenic activity, enhances neoplasm formation in the test organism. *Cocarcinogenesis* results from the simultaneous administration of a carcinogen and a cocarcinogen, a compound having no carcinogenic activity of its own.

The concept of initiation and promotion in carcinogenesis involves the administration of a subcarcinogenic dose of an initiating carcinogen (i.e., the initiator) followed by exposure to a promoter that has little or no carcinogenic activity of its own. Exposure to the *initiator* presumably results in mutations in DNA in cells of the target organ and makes them susceptible to the promoting stimulus. The *promoter* acts on the cell population that carries this mutation and stimulates clonal expansion, or hyperplasia, of initiated cells, resulting in a "precancerous" lesion. This condition often leads to *progression*, a third stage of carcinogenesis in which some of the promoted (preneoplastic) lesions develop further into neoplasms (Sirica, 1989b). Preneoplastic lesions are considered to be reversible. That is, they regress, or disappear, when the promoting stimulus is removed. *Complete carcinogens* do not require promotion but provide their own promoting stimulus. Potentially, certain progressor agents might also induce progression without having other carcinogenic properties.

Initiation-promotion models have been used widely in cancer research to investigate the stages of carcinogenesis. Multistage carcinogenesis has been shown to occur in all organs in which it has been studied: skin, urinary bladder, liver, respiratory tract, intestine, mammary gland, thyroid, exocrine pancreas, and stomach (see Peraino and Jones, 1989). Some compounds that have been used as initiators and promoters are listed in Table 4. Note that, in addition to chemical compounds, the list includes several environmental and biologic factors and conditions that can initiate and promote neoplasia. Sen-

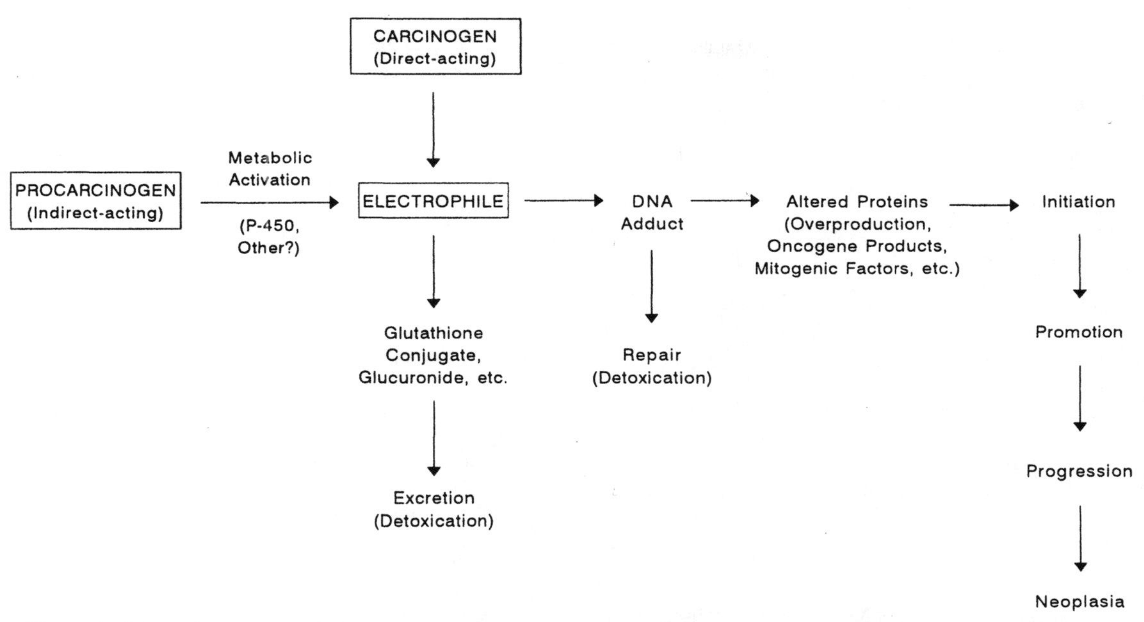

Figure 1. Schematic summary of the potential events and pathways in bioactivation of chemical carcinogens. (Reprinted from *Comp. Biochem. Physiol.*, 89C, Guengerich, F. P. Cytochromes P-450, pp. 1–4, Copyright 1988, with kind permission from Elsevier Scient Ltd., The Boulevard, Langford Lane, Kidlington, OX5 1GB, U.K.)

sitivity to carcinogenic effects of initiators and promoters, however, is influenced by other factors such as age and species (Diwan and Rice, 1989). The susceptibility of an organism to the carcinogenic effects of a chemical (i.e., species dependence) is often related to the ability of the organism to metabolize the carcinogen to its active chemical form. Age dependence appears to be related to cell proliferation associated with growth in which cells in S phase, the period of the cell cycle that is characterized by enhanced DNA replication, are thought to be especially vulnerable to the action of carcinogens (Diwan and Rice, 1989). Cell proliferation following hepatectomy appears to enhance the effects of subsequent carcinogen exposure and to promote the effects of previous carcinogen exposure (Berenblum, 1985).

Modification of carcinogenicity includes inhibition as well as enhancement of neoplastic development. The process of *anticarcinogenesis*, however, can be very complicated, with inhibition occurring in one

Table 4. Some chemical compounds used in initiation and promotion tests

Tissue or organ	Initiators	Promoters
Skin	Dimethylbenzanthracene	Physical wounding
	Benzo[*a*]pyrene	Oleic acid
		Croton oil (phorbol esters)
		Phenobarbital
Urinary bladder	Methylnitrosourea	Physical wounding
		Saccharin
		Cyclamates
Liver	2-Acetylaminofluorene	DDT
		Polychlorinated biphenyls
		2,3,7,8-Tetrachlorodibenzo-*p*-dioxin
		Some contraceptive steroids
		Choline-deficient diet
		Peroxisome proliferators
Other	Dimethylnitrosamine	Dietary fat (mammary gland and intestine)
	N-Methyl-*N*′-nitro-*N*-nitrosoguanidine	Various hormones (mammary gland)
		Tobacco use (respiratory tract)

Source: Peraino and Jones, 1989.

organ while enhancement occurs in another (Williams, 1984). Furthermore, inhibition may be dependent on timing of exposure, so that a chemical compound may be an inhibitor when administered after carcinogen exposure but a promoter when administered before the exposure.

Multistage carcinogenesis can also be described in terms of mutational events with at least two mutations being necessary for development of a malignant neoplasm (Moolgavkar, 1989). Presumably, in this multihit paradigm, an initiating agent causes specific mutations in a target with the number of mutations dependent on the dose and length of exposure. If exposure duration or concentration is sufficient, the mutation required to cause malignant neoplasia occurs and the agent is said to be a *complete carcinogen.* Promoting agents applied following subcarcinogenic exposures probably serve to increase the probability of subsequent mutations by increasing the number of initiated cells that, when stimulated, undergo clonal expansion.

Cell Proliferation in Multistage Carcinogenesis

A great deal of attention is being given to cell proliferation to help explain mechanisms of carcinogenesis, to determine relevant test doses in carcinogenicity tests, and to establish risk estimates (Butterworth, 1991). It appears that cell proliferation plays an important role in each of the three stages of multistage carcinogenesis. Cell proliferation is critical for the initiation stage because DNA synthesis in the presence of the carcinogen is required for the carcinogen-induced mutation to become "fixed" (Pitot et al., 1991). In promotion, the reversible stage of neoplasia, cell proliferation is required for the clonal expansion of initiated cells. The role of cell proliferation in progression, which is characterized by the development of malignant neoplastic lesions, is thought to be necessary but is not well understood at this time.

Biotransformation of Chemical Carcinogens in Fishes

Knowledge of the uptake, distribution, and metabolism of test agents is critically important in interpreting the results of a carcinogenicity bioassay. Although much information exists on various aspects and details of carcinogen metabolism in mammals, relatively little is known about carcinogen bioactivation in fish (Malins and Ostrander, 1991). However, the literature indicates that, as is the case in mammals (Weisburger, 1990) biotransformation in fish takes place mainly in the liver during phase I (bioactivation) and phase II (conjugation and excretion) reactions (Lech and Vodicnik, 1984; Stegeman and Lech, 1991; James, 1987; Kleinow et al., 1987; see Chapter 17).

As in mammals, activation of procarcinogens in fishes frequently depends on oxidative metabolism catalyzed mostly by microsomal cytochrome P-450–dependent mixed-function oxidases (MFOs) (Stegeman et al., 1990; see Chapter 17). Figure 1 summarizes the potential events and pathways in bioactivation of chemical carcinogens (Guengerich, 1988) and reviews some of the concepts already discussed. In mammals, many chemical carcinogens including representatives of most of the principal chemical classes are activated by MFOs (see Stegeman and Lech, 1991). These compounds include benzo[*a*]pyrene (B*a*P), benzanthracene, diethylnitrosamine (DEN), aflatoxin B_1, and 2-acetylaminofluorene (AAF). Likewise, procarcinogen metabolic activation by P-450–related enzymes is well documented in various fish species for B*a*P and other polycyclic aromatic hydrocarbons (PAHs) (Buhler and Williams, 1989), various aflatoxins (Loveland et al., 1988), and DEN (Kaplan et al., 1991). Correspondingly, representatives of each of these groups of compounds have been demonstrated to induce neoplastic lesions in fish. A principal difference between mammals and fish is that a phenobarbital-like P-450 metabolizing system does not appear to be active in fish (Elskus and Stegeman, 1989).

Other metabolic pathways might also be involved in activation of carcinogens in fishes. Some halogenated organic compounds, mainly dihaloalkanes such as ethylene dibromide, have an uncommon carcinogenic mechanism that utilizes what is normally a detoxification pathway, conjugation with glutathione (see Chapter 17), to form electrophilic intermediates that damage DNA and appear to initiate carcinogenesis in rats (Guengerich et al., 1987). Ethylene dibromide is carcinogenic to the medaka (*Oryzias latipes*), a small aquarium fish, possibly by a similar mechanism (unpublished).

Some Aspects of Molecular Mechanisms of Carcinogenesis in Fish

Researchers generally agree that chemical carcinogens induce neoplastic lesions by causing a permanent heritable change in genetic material and thus altering growth control mechanisms. As shown in Figure 1, carcinogen exposure results in the development of electrophilic intermediates that can bind covalently with DNA, forming adducts (Beland and Poirier, 1989; see Chapter 13). The extent of adduction correlates with the mutagenic and carcinogenic properties of the carcinogen. There is considerable evidence that the damaged DNA causes alterations in the expression of certain genes (oncogenes or suppressor genes) that appear to be involved in normal cellular growth and differentiation (Anderson and Reynolds, 1989). These genetic alterations set in motion the subsequent events that lead to the development of neoplastic lesions.

Many fish species can metabolize a wide variety of environmental carcinogens to electrophilic intermediates that can then form covalent DNA or protein adducts (Dunn, 1991). The relationship between exposure to environmental chemical carcinogens and occurrence of neoplastic lesions in exposed fish has been shown for both wild fish living in contaminated environments and fish exposed to carcinogens under laboratory conditions. For example, English sole (*Parophrys vetulus*) from Puget Sound, Washington, collected from sites with contaminated sediments, have a high incidence of hepatic neoplastic lesions (Myers et al., 1990) that are associated with high levels of DNA adducts (Stein et al., 1990; Malins et al., 1990). DNA adducts of fish are often species specific and their presence may not always indicate environmental contamination. Kurelec et al. (1989) reported that natural populations of several fish species had similar adducts whether the specimens were taken from known contaminated or presumably noncontaminated waters.

Because of the fundamental importance of their normal functions, oncogenes are highly conserved and occur in most organisms, even those at the lowest phylogenetic levels. It is not surprising, then, that the relatively few studies that have been conducted on fish oncogenes have revealed many parallels with those of mammals (Van Beneden et al., 1988, 1990). One of the first oncogenes from any organism was predicted and later confirmed in melanomas that develop in platyfish-swordtail (*Xiphophorus* sp.) hybrids (Anders et al., 1984). The platyfish-swordtail system has been widely used to examine the hereditary nature of cancer.

Some of the oncogene nucleotide sequences that have been reported in fish are homologous to those of oncogenes identified in mammals. These include goldfish *ras* (Nemoto et al., 1986) and rainbow trout c-*myc* (Van Beneden et al., 1986) from normal liver tissues. Also, oncogene activation has been associated with chemical carcinogenesis in fish. Activation of K-*ras* was reported from liver tumors that were probably induced by chemical contaminants in winter flounder from Boston Harbor (McMahon et al., 1988) and in Atlantic tomcod (*Microgadus tomcod*) from the Hudson River (Wirgin et al., 1989). This corresponds to activation of members of the *ras* oncogene family that has been observed in chemically induced neoplasia in rodents (Beer and Pitot, 1989). Oncogene activation has not been widely studied in aquarium fish carcinogenesis models, although activation of an unidentified oncogene was found in a chemically induced neoplasm in the medaka (RJ Van Beneden, personal communication).

Nongenotoxic (Nonmutagenic) Carcinogens

There appear to be many carcinogens that are not mutagenic in bacterial mutagenesis (Ames) assays and do not otherwise damage DNA (Tennant et al., 1987). There is a great deal of interest in these so-called nongenotoxic or epigenetic carcinogens because of their widespread occurrence and potential for extensive human exposure (Lijinsky, 1990). Little is known about how these compounds exert their effects because they do not interact with genetic material in the same ways as do the more widely studied genotoxic carcinogens.

Table 5. Some nongenotoxic (nonmutagenic) carcinogens

Carbon tetrachloride	Nitrosodiethanolamine
Chloroform	Polychlorinated biphenyls
Dieldrin	Sodium saccharin
Diethylstilbestrol	Safrole
Lindane	Tetrachlorodibenzodioxin
Methapyrilene	Urethane

Source: Reprinted from Lijinsky, 1990, pp. 45–87, by courtesy of Marcel Dekker, Inc.

Table 5 lists some specific chemical compounds that are suspected to have nongenotoxic mechanisms of carcinogenicity. The list includes many compounds that have already appeared on other lists in this chapter that have classified carcinogens by other means, adding to the confusion regarding carcinogen classification. Note that the list includes many commonly used compounds such as chlorinated solvents, pesticides, and drugs. The mechanisms by which nongenotoxic carcinogens induce neoplasia are poorly understood and the compounds appear to have few biologic properties in common. Induction of cell proliferation following toxic injury by such compounds as carbon tetrachloride and chloroform appears to be a feature of many nongenotoxic carcinogens. Other agents such as trichloroethylene and phthalate esters cause proliferation of peroxisomes (see Chapter 17), whereas the environmental carcinogen 2,3,7,8-tetrachlorodibenzo-*p*-dioxin (TCDD) acts at very low doses through specific cellular receptors (see Klaunig and Ruch, 1990). Many genotoxic carcinogens stimulate cell proliferation subsequent to the inhibition of intercellular communication through membrane modifications called gap junctions (Klaunig and Ruch, 1990; Trosko and Chang, 1984).

USE OF SMALL FISH TO DETECT CHEMICAL CARCINOGENS

The use of small fish in carcinogenesis research has received considerable attention (e.g., Black, 1984; Couch and Harshbarger, 1985; Dawe et al., 1981; Hawkins et al., 1988a,b; Hendricks, 1982; Hoover, 1984; Masahito et al., 1988; Metcalfe 1989; Mix, 1986; Powers, 1989), mainly because there are many advantages to using them in cancer research. For example, many species are available that are easily bred and reared, requiring little space and attention. They can be exposed for varying periods of time to test compounds in enclosed systems and they tend to develop neoplastic lesions rapidly.

Small-fish carcinogenesis models are being developed with several related applications in mind. One application of small fish is to examine the carcinogenicity of water-borne and sediment-bound compounds implicated in cancer of the liver and other organs in wild fish (e.g., Fabacher et al., 1991). Another application of small-fish carcinogenesis models concerns their potential for identifying and predicting human health effects. Reviews of Hendricks

(1982), Black (1984), and Metcalfe (1989) specifically dealt with tests to identify carcinogens. Because of factors such as economy, rapid response, and ability to work with large numbers of organisms, small fish species are being considered as supplements or in some cases replacements for rodent models in carcinogenicity bioasssays. Simon and Lapis (1984) illustrated this potential when they used bioassays with guppies (*Poecilia reticulata*) to identify carcinogenic isomers of a series of chemotherapeutics that had nitrosamine moieties. In unpublished studies at the Gulf Coast Research Laboratory, several halogenated hydrocarbons were tested individually and in mixtures with medaka (*Oryzias latipes*), guppy (*P. reticulata*), and sheepshead minnow (*Cyprinodon variegatus*). The protocols used for these tests involved chronic exposures at stable concentrations (Walker et al., 1985), which, in some respects, appear to be more practical than rodent chronic bioassays in which volatile compounds such as halogenated hydrocarbons are given by gavage or in drinking water.

There is a growing number of examples in which small fish models can enhance our understanding of the basic mechanisms of cancer (Masahito et al., 1988). For example, there are few whole-animal experimental models for ocular cancers. The medaka, however, develops retinal medulloepithelioma, a primitive type of ocular neoplasm, after a single, brief exposure to methylazoxymethanol acetate (MAM) (Hawkins et al., 1986). Similarly, MAM-induced exocrine pancreatic neoplasia in the guppy (Fournie et al., 1987) provides an additional model for studying the causes and progression of that disease. The use of the platy-swordtail melanoma system (Anders et al., 1984) to investigate cancer genetics has already been mentioned. The sensitivity of the medaka to many carcinogens, the large numbers that can easily be studied, and the ability to control many extraneous factors make this species a good model for studying the biology of hepatic neoplasia (Hinton et al., 1988), oncogene activation (Van Beneden, 1990), and DNA repair (Ishikawa et al., 1984). Furthermore, the medaka and other small-fish models can be used to study aspects of cancer in which large numbers of experimental organisms are required, such as in low-dose risk assessment, as well as to examine factors that only slightly increase cancer risk (Walker et al., 1992).

Environmental and Biomedical Regulatory Applications

Small-fish systems could have extensive environmental and biomedical applications in identifying carcinogens. If one considers a carcinogen to be any chemical or compound that causes cancer of any type in any exposed organism at any dose level, the application of small-fish carcinogenesis tests that are sensitive, rapid, efficient, and economical can provide sound carcinogenic data on many more substances than can now be tested with the expensive and lengthy rodent carcinogenicity bioassays. With regard to identifying environmental carcinogens, it can be anticipated that small-fish carcinogenicity tests will find broad application and eventual acceptance as a definitive means of assessing the genotoxicity (see Chapter 13) of aquatic sediments, effluents, and potential contaminants.

Ecological and Environmental Importance of Cancer in Fish

Several cancer epizootics have been identified in wild fish such as English sole (*Parophrys vetulus*) from Puget Sound, Washington; winter flounder (*Pseudopleuronectes americanus*) from Boston Harbor; and brown bullhead catfish (*Ictalurus nebulosus*), white suckers (*Catostomus commersoni*), and several other freshwater species from inland waterways, particularly in the Great Lakes area. Scientists are beginning to understand some of the carcinogenic mechanisms involved in the development of the cancers. A brief review of some of those epizootics is presented because one of the potential uses of small-fish carcinogenicity bioassays is to study the causes of neoplasia in wild fish that might not be easily cultured.

English sole from contaminated areas of the Puget Sound have high incidences of hepatic lesions that range from degenerative ones to neoplasms (Myers et al., 1991). Numerous detailed studies (e.g., Malins et al., 1984a,b, 1985, 1987, 1988) have established statistically significant associations between the presence of PAHs in the sediments and the incidence of liver lesions. Further studies have focused on the metabolism of PAHs in sole liver, including the distribution of PAH metabolites in bile and tissues and the covalent binding of genotoxic metabolites of PAH to sole hepatic DNA (Varanasi et al., 1989a,b). Malins et al. (1990) identified a novel DNA lesion, 2,6-diamino-4-hydroxy-5-formamidopyrimidine, in neoplastic livers of English sole from carcinogen-affected areas of the Puget Sound.

The hepatic cancer epizootic in English sole from the Puget Sound area is mirrored on the East Coast by epizootic hepatic neoplasia in winter flounder from Boston Harbor, Massachusetts (Murchelano and Wolke, 1985, 1991). As in the case of the Puget Sound sole but not as rigidly established, the hepatic lesions in the winter flounder were highly correlated with anthropogenic chemical contamination.

Although numerous incidences of cancer epizootics have occurred in fish from fresh water (Black and Baumann, 1991), none have been as well studied as the English sole and winter flounder epizootics. Neoplasm epizootics in populations of brown bullhead catfish (*I. nebulosus*) and Atlantic tomcod (*Microgadus tomcod*) are prominent. Sediments rich in PAH have generally been considered the principal causes of skin and liver neoplasia in brown bullhead in the contaminated Black River (Ohio), a tributary of Lake Erie (Baumann, 1989; Baumann et al., 1987; 1990). In laboratory tests, medaka exposed to extracts and fractions of PAH-contaminated sediments from tributaries of the Great Lakes, including the Black River, developed hepatic neoplastic lesions (Fabacher et al., 1991). Similarly, epizootics of hepatic neoplasia have been reported in Atlantic tomcod from the Hudson River (Smith et al., 1979; Cormier et al., 1989). These liver neoplasms have been associated with elevated tissue levels of polychlorinated biphenyls (PCBs) (Klauda et al., 1981). Enhanced cytochrome P450IA gene expression (Kreamer et al., 1991) with activation of K-*ras* oncogene characterizes these lesions (Wirgin et al., 1989).

White suckers from industrially polluted areas of Lake Ontario exhibited increased incidences of hepatic and skin neoplasia (Sonstegard, 1977; Hayes et al., 1990). As in other epizootics, the neoplasms have been associated with PAH contamination. Stalker et al. (1991) showed that the progression of hepatocellular and bile duct neoplasms in the white sucker is accompanied by a loss of immunoreactive glutathione *S*-transferases, which usually catalyze a major detoxification pathway.

Few studies of cancer epizootics have dealt with small fish species. However, studies suggest that some wild small fish species are also sensitive to the carcinogenic effects of environmental carcinogens. Histologic examination of mummichog (*Fundulus heteroclitus*) from a creosote-contaminated site in the Elizabeth River, Virginia, revealed high incidences of hepatic neoplastic lesions (Vogelbein et al., 1990) along with neoplasms with several other cellular origins (WK Vogelbein and JW Fournie, personal communication). Many of the attributes that small fish have for laboratory studies also apply to field studies, especially the use of large numbers of specimens, the option to conduct whole-body histological examination, and the ease of collection and processing.

Laboratory Studies on the Induction of Neoplasia in Fish

The database on the response of small fish to exposure to carcinogenic compounds is small but growing. In general, it appears that compounds that cause cancer in rodents also cause cancer in fish. The variability in sensitivity to certain compounds that is evident among strains and species of rodents also occurs among fish species. For thorough accounts of small-fish carcinogenicity data regarding compounds and species, see the reviews of Couch and Harshbarger (1985) and Metcalfe (1989). Following is a brief review of carcinogenic responses of small fish to representative chemicals from several carcinogen classes, including nitroso compounds, polynuclear aromatic hydrocarbons, aromatic amines, and halogenated hydrocarbons.

Nitroso Compounds

Nitroso compounds have been the most widely used class of carcinogens to test for and examine mechanisms of carcinogenesis in small fish species (Metcalfe, 1989). Khudoley (1984) demonstrated that several nitrosamines including dimethylnitrosamine, diethylnitrosamine (DEN) and nitrosomorpholine are hepatocarcinogenic to zebra fish and guppies. DEN has been the most widely used nitroso compound and has caused liver neoplasia in every species in which it has been tested and in some cases esophageal and intestinal neoplasia as well. In the topminnows, *Poeciliopsis* spp., tumorigenic sensitivity, as indicated by type of hepatic neoplasm and by incidence, is related to species and strain (Schultz and Schultz, 1982, 1988). This differential induction of tumorigenesis among genotypes was shown to be related in part to the ability of sensitive species and strains to take the initial steps toward metabolic activation of DEN to its carcinogenic metabolites (Kaplan et al., 1991).

Hepatic neoplasia induced by DEN in medaka has been widely studied and appears to progress in stages (Bunton, 1990; Lauren et al., 1990; Hinton et al., 1988) similar to those that characterize rodent hepatic neoplasia. Protocols for DEN hepatocarcinogenicity vary but generally involve continuous exposure to 15–100 ppm DEN in ambient water for about 8 wk, followed by an additional 8-wk or longer "grow-out" period in clean water for tumor development (see Hinton et al., 1985). Hepatic neoplasia induced by DEN in medaka can be enhanced by several factors, including partial hepatectomy preceding exposure (Kyono-Hamaguchi 1984), increased environmental temperature during exposure and grow-out phases (Kyono-Hamaguchi, 1984), and increased carcinogen exposure concentration, length of exposure, and length of grow-out period in clean water (Ishikawa and Takayama, 1979). Exposure concentration of DEN also affects the type of hepatic neoplasm induced. The modifying conditions also apply for other nitrosamines in other small fish species (Khudoley, 1984). Usually, DEN causes a spectrum of hepatocellular neoplastic lesions including foci of cellular alteration, adenoma, and carcinoma. Cholangiocellular lesions sometimes occur frequently, especially in genotypes of *Poeciliopsis* (Schultz and Schultz, 1988), but usually less frequently than hepatocellular neoplasms. High exposure concentrations and extended grow-out periods often result in the development of various kinds of sarcomas in medaka liver (Bunton, 1990). The ability of DEN-induced hepatic neoplasms, hepatocellular carcinoma, and hemangiopericytoma to grow independently when transplanted into non-tumor-bearing specimens of the same species was demonstrated in *Poeciliopsis* spp. (Schultz and Schultz, 1985).

Polynuclear Aromatic Hydrocarbons

The carcinogenicity of PAHs to fish is of interest because of associations that have been established between neoplasia in fish and PAH contamination and because of the potential effects of PAH exposure to humans (Hendricks et al., 1985). As discussed earlier, many fish species appear metabolically equipped to metabolize PAHs to their carcinogenic intermediates. At least two PAHs, benzo[*a*]pyrene (B*a*P) and

7,12-dimethylbenzanthracene (DMBA), known to be carcinogenic in mammals mainly to skin following topical application, are also carcinogenic to small fish species. Like DEN, these compounds typically cause liver neoplasia in exposed fish, but DMBA also appears to cause other types of neoplasia. Furthermore, these PAHs cause carcinogenic effects at much lower concentrations than does DEN. Hepatocellular neoplastic lesions occurred in guppy, medaka, and sheepshead minnow following two brief (6-h) exposures to B*a*P at measured concentrations of about 250 ppb, which were nontoxic levels (Hawkins et al., 1988c, 1990). The lesions developed by 24 wk after the initial exposure in these species. Medaka were more susceptible than guppies to the carcinogenic effects of B*a*P based on incidence (36% at 36 wk in medaka versus 19% in guppy) and the apparent progression of induced lesions, in that some developed earlier and progressed to advanced stages sooner. Several fish species appear to be sensitive to the hepatocarcinogenic effects of DMBA, including guppy, medaka, and sheepshead minnow (Hawkins et al., 1989, 1991b, unpublished data). Genotypes of *Poeciliopsis* develop liver neoplasia, both hepatocellular neoplasms and hemangiopericytoma, after being exposed to DMBA given as a 5 ppm aqueous suspension up to five times in exposures lasting up to 20 h each. Hepatocellular neoplasms developed in about 50% of the exposed fish 6 to 9 mo after the initial exposure (Schultz and Schultz, 1982, 1984). The DMBA-induced tumors were transplantable, as were the DEN-induced ones (Schultz and Schultz, 1985). Toxic effects of DMBA lead to cell proliferation, which appears to play a role in liver neoplasia in *Poeciliopsis* (Schultz et al., 1989).

There are species-related differences in DMBA carcinogenic sensitivity among medaka, guppy, and sheepshead minnow. Based on the types of neoplasms induced, progression of induced lesions, and neoplasm incidence, DMBA is much more carcinogenic than B*a*P, causing higher percentages of tumors with lower exposure concentrations in each species. Hepatic neoplasms occurred in medaka within several months after two 6-h exposures of 30–60 ppb DMBA and in the guppy after four 6-h exposures of 15–30 ppb (Hawkins et al., 1989). In a concentration-response study of hepatic neoplasia in medaka, exposure to 25–30 ppb DMBA for only 6 h and then removal to clean water for grow-out resulted in hepatic neoplastic lesions in specimens examined at 6 mo (WE Hawkins et al., unpublished). Furthermore, DMBA induced neoplasms in several organs in addition to the liver in both medaka and guppy but not in the sheepshead minnow. Both B*a*P and DMBA appear to be far more carcinogenic in small-fish models than in rats, in which DMBA induces liver carcinogenesis only after it is applied following partial hepatectomy (Marquardt et al., 1970).

Methylazoxymethanol Acetate

This compound, a component of cycasin and a metabolite of dimethylhydrazine (Zedeck, 1984), is hepatocarcinogenic in numerous small fish species after brief 1- to 2-h exposures in the low ppm range and it also induces extrahepatic neoplasia in several species (Couch and Harshbarger, 1985; Harada et al., 1989; Hawkins et al. 1985; Hinton et al., 1984; Metcalfe, 1989). Each of six small fish species including medaka, guppy, sheepshead minnow, gulf killifish (*Fundulus grandis*), inland silverside (*Menidia berylina*), and fathead minnow (*Pimephales promelas*) tested with MAM under similar conditions developed hepatic neoplastic lesions, with neoplasm incidence and latency (time to tumor) period varying from species to species (Hawkins et al., 1985). The compound is also neurocarcinogenic to medaka, causing a primitive type of retinal neoplasm called medulloepithelioma, which is closely related to retinoblastoma (Hawkins et al., 1986). In addition to hepatic neoplasia, MAM exposure in the guppy results in exocrine pancreatic neoplasia (Fournie et al., 1987). Carcinogenesis by MAM is modulated by many of the factors that affect DEN and DMBA carcinogenesis. Carcinogenesis induced by MAM is a function of exposure concentration and time, is accompanied by cell proliferation, is marginally enhanced by posttreatment with caffeine, and is inhibited by pretreatment with X-irradiation (Aoki and Matsudaira, 1981, 1984). MAM is considered a direct-acting carcinogen, but no studies have dealt with the carcinogen-related metabolism of MAM in fish.

Aromatic Amines

Compared with DEN, MAM, and PAHs, few aromatic amines, a class of chemicals that includes the carcinogens 2-acetylaminofluorene (AAF), benzidine, and aniline, have been tested for carcinogenicity in small fish species. From studies that are available, aromatic amines appear to be weakly to moderately carcinogenic to small fish. Aminoazotoluene and AAF induced liver neoplasms in medaka (Hatanaka et al., 1982) and guppies (Sato et al., 1973).

As already discussed, AAF has been used as a model carcinogen to study mechanisms of initiation and promotion in rodents. For AAF to be carcinogenic, it must be N-hydroxylated by a cytochrome P-450–dependent, microsome-bound enzyme (Weisburger, 1990). Ring hydroxylation by another P-450 enzyme appears to be a detoxification step. In studies of the metabolism of AAF in medaka and guppy, James et al. (1992; 1994) found that the guppy was able to produce more of the carcinogenic metabolite of AAF than the medaka. Correspondingly, in parallel carcinogenicity studies of AAF in the two species, the guppy developed hepatic neoplastic lesions sooner and in higher incidences than did the medaka.

Halogenated Hydrocarbons

The carcinogenicity of many short-chain halogenated hydrocarbons is of interest because of their widespread occurrence in the workplace and the environment. Because of their volatility and poor solubility in water they are difficult to work with in aquatic applications, but these technicalities can be overcome using flow-through exposure methodology (Walker et al., 1985). Some studies indicate that small fish species may be good models for examining the carcinogenicity of halogenated hydrocarbons. In 90-d flow-through studies, ethylene dibromide proved to be carcinogenic to medaka. In a similar study, vinylidene chloride was carcinogenic to both medaka and guppy (WE Hawkins et al., unpublished results). Some other halogenated hydrocarbons, tetrachloroethane, and chlorodibromomethane were not carcinogenic to either medaka or guppy in similarly conducted studies.

CONSIDERATIONS IN THE DESIGN OF CARCINOGENICITY TESTS USING SMALL FISH

The design of rodent carcinogenicity tests takes into account many factors, such as number and size of the experimental groups, selection of dose, route of administration, randomization, husbandry, diet, duration of exposure, purity of test chemicals, consistency of environmental conditions, and observation and sampling strategies (Gart et al., 1986; Robens et al., 1989). Each of these factors can be adequately addressed in small-fish carcinogenicity bioassays.

Number of Animals and Statistical Design

Statistical analysis is an important component of the interpretation of carcinogenicity tests and much effort has gone into developing sophisticated ways to conduct and analyze the tests from a statistical point of view (Gart et al., 1986). Carcinogenic activity of a test substance is usually demonstrated by an increased rate of occurrence of an otherwise spontaneously occurring neoplastic lesion or by the appearance of a neoplasm that usually does not appear in control organisms. The latter situation seems to be the more valid indicator of carcinogenesis. Another indicator of carcinogenic activity is a decrease in latency period, which is the time required for the development of a neoplastic lesion. A chemical carcinogen should cause a statistically significant increase in a specific type of tumor in treated versus nontreated groups. Guidelines for statistical analyses of mammalian carcinogenicity data have been established that allow pooling of certain types of lesions based on their cells of origin (McConnell et al. 1986). Furthermore, the tumor incidence is expected to increase with increasing dose of carcinogen (Gart et al., 1986).

The low rate of spontaneous neoplasia in fish species is a distinct advantage that these models have over rodent carcinogenesis models. In practical terms, the lower the spontaneous rate of neoplasia, the fewer specimens needed to achieve statistical significance. For example, Krewski et al. (1989) calculated that to validate statistically a 10% tumor incidence when the spontaneous rate is 1% would require 161 test animals with half assigned to the test group and half assigned to the control group (Table 6). The number of test animals needed more than doubles, however, if the spontaneous tumor rate in the test animal is 10% or if a tumor rate of 5% over the spontaneous rate is sought.

Dose or Exposure Concentrations

Dose or exposure concentrations vary with study objectives. The low cost of small-fish carcinogen bioassays compared with rodent tests affords flexibility in organizing tests with regard to numbers of exposure groups and sampling schedules. Dose selection is an important issue in rodent carcinogenicity tests (Haseman, 1985). As discussed earlier, exposure concentrations in rodent chronic carcinogenicity tests are based on an estimated maximum tolerated dose (MTD), for example, the dose that causes no more than a 10% weight loss in a 90-d preliminary study. The test groups include a control, one or more exposures at fractions of the MTD, and the MTD (Haseman, 1985; Weisburger and Williams, 1981). It is difficult for rodent carcinogenicity tests to identify weak or even moderately strong carcinogens because of the small number of dose groups and the limitations of using small sample sizes because of economic restraints. Small-fish tests allow multiple dose or exposure groups with large sample sizes for each group at lower costs.

Fish carcinogenicity tests should also utilize an MTD, probably one based on lethality in a short-term (e.g., 14- or 28-d) toxicity test. For the long-term test, one approach could have a group with a high exposure concentration that is an established percentage (e.g., 75%) of the lethally toxic concentration (LC50), a low exposure concentration at or near environmentally realistic levels (EEC, estimated environmental concentration), and several intermediate concentrations (Table 7). The idea behind having multiple exposure groups is not only to investi-

Table 6. Total sample sizes to ensure a false-negative rate of 20% in a carcinogen bioassay with a spontaneous rate of 1%[a]

Excess over spontaneous rate[b] (%)	Number of animals required[c]
0.1	260,784
1	3,952
5	376
10	161
15	101
20	74
25	58

[a] Data compiled from Krewski et al., 1989.
[b] Response rate in exposed group equal to control response rate plus excess over response rate.
[c] Based on Fisher's exact test at 5% level of significance; half of the specimens assigned to a control group, the other half to the exposed group.

Table 7. Suggested test groups in a small-fish bioassay

Control
Carrier or solvent control
Low concentration (e.g., estimated environmental concentration)
Several intermediate concentrations based on MTD
High concentration based on MTD

gate concentration-response effects but also to help ensure against a false-negative outcome. Appropriate control groups would include an untreated group, a sham-treated group if necessary, and a group treated with the solvent or other type of carrier that may be used to administer the test compound.

Under some circumstances, it might be advisable to include a positive control group. A positive control group in a carcinogenicity test is one in which the experimental organisms are exposed to a test substance already known to induce neoplastic lesions in the test model. For example, Simon and Lapis (1984) used DEN as a positive control to help evaluate responses in guppies to several industrial chemicals and pharmaceuticals and Fabacher et al. (1991) used MAM acetate to help assess responses of medaka to toxic sediments. The use of a positive control treatment group has both positive and negative attributes. On the negative side, carcinogens operate by such specific mechanisms that a response to a given test substance might mean that the organism is sensitive only to that compound (Peck, 1973). On the other hand, standardization of the use of positive controls can help establish the potency of the test substance by comparing the response it induces with that of a proven carcinogen (Weisburger, 1973) and with the results of prior tests. Furthermore, a positive control group could help indicate whether conditions of a given test differ from those of prior tests.

Test Duration

Typically, the duration of small-fish tests is about 1 yr. Although there is concern that this does not represent a "life span" test, it is probably a practical length from economic and biologic standpoints. In the small fish species, it appears that if a carcinogenic response is to occur it occurs within 1 yr. Furthermore, longer studies may be compromised by the development of spontaneous neoplasms, which typically increases with age. Studies longer than a year, however, can be useful for examining the properties of weak carcinogens, which might require a long latency period, or for characterizing the development of neoplasms in secondary target organs following exposure to strong carcinogens. Depending on the potency of the carcinogen or other factors, tests may be terminated at scheduled times of 4 to 6 mo or earlier than the projected duration.

Determination of histopathological sampling schedules depends on several factors such as the expected carcinogenic potency of the test compound and its target organ. It is often useful in a chronic test to include examination of some specimens for acute toxicological effects in order to estimate toxic exposures and, in some cases, to predict onset of carcinogenesis. All moribund and dead specimens, if they are not too autolyzed, should be examined histologically to determine cause of death. In interim samples, it is sometimes feasible to examine all high-concentration specimens histologically and then examine intermediate- and low-concentration specimens and controls only when positive responses are observed in the high-concentration specimens. A major advantage of the small-fish carcinogenicity test is that it facilitates histopathology because whole specimens can be processed and examined for carcinogenic effects on a few glass slides. Depending on the sectioning protocol (i.e., number of histologic levels examined), this allows the pathologist to examine simultaneously multiple tissues maintained in their normal relationships. This contrasts with histopathological examination of rodent tissues, in which individual tissues are usually examined separately.

Exposure Systems

Small-fish carcinogenesis tests are highly versatile and offer a wide range of approaches to testing suspect cancer-causing substances (Hawkins et al., 1988a). The three principal types of exposure systems are flow-through, static-renewal, and static (see Chapter 1). Other exposure methods, including dietary exposures, embryo injection, and intraperitoneal injection, are applicable under certain conditions. Decisions on the exposure methodology are based on the availability of the test substance, toxicity to test organisms, expected carcinogenicity, solubility, ease of chemical disposal, and possibly other factors.

Results

Daily observations of abnormal behavior are recorded. Fish found dead should be necropsied as soon as possible so that autolysis is minimized. However, because dead specimens autolyze rapidly in warm water, it is preferable to sacrifice moribund fish as soon as they are recognized. Body weight and length measurements can be recorded at the end of the study for all specimens and for those found dead or sampled during the study. Gross examination of organs and tissues of treated and untreated specimens may reveal specific target organ responses. For specimens of larger species, individual organ weights can be obtained and organ weight/body weight calculations made. Histopathological observations can range broadly depending on many factors, such as test species, test substance, specimen size and age, and histologic sectioning protocol (Hinton et al., 1992). Alterations of target and nontarget tissues can be recorded and treated and untreated groups compared. The incidences of toxic lesions or tumors in a particular organ can be determined and subjected to statistical tests.

Statistical Analyses

Properly designed, small-fish carcinogenicity tests offer good opportunities to analyze study results statistically. If in-test mortality is minimal, the Fisher-Irwin exact test is usually used to analyze neoplasm incidence for individual groups and the Cochran-Armitage test for linear trend is used if more than one test group is available (Robens et al., 1989). The Cochran-Armitage test determines whether the slope of the concentration-response curve is different from zero. In addition to neoplasm incidence, other components that can be analyzed include neoplasm-associated mortality, patterns of lesion co-occurrence (see Myers et al., 1987), tumor lethality and regression, comparison of the induction of malignant versus benign tumors, time of tumor onset, and dose or concentration response (Gart et al., 1986). Although it is usually advisable to use data from concurrent controls, it can be useful to consider historical control data in the case of occurrence of rare tumors (Haseman et al., 1984).

Personnel Safety and Disposal of Test Materials

Worker safety and the proper disposal of hazardous materials generated in the laboratory are important in carcinogenicity tests as they are in any laboratory activity. Any human exposure to carcinogenic hazards is considered unacceptable and appropriate guidelines for storing, handling, and disposing of carcinogens must always be followed.

ANALYSIS OF CARCINOGEN-INDUCED LESIONS IN SMALL-FISH CARCINOGENICITY STUDIES

Accurate diagnosis of carcinogen-induced lesions is critical to the analysis of carcinogenicity tests. Rodent tests are more highly standardized than fish tests as to pathological protocols, quality assurance (Boorman et al., 1985), and assessment procedures for regulatory purposes (Dua and Jackson, 1988). Criteria for neoplastic lesions in principal carcinogen target organs such as the liver have been established for rats (Maronpot et al., 1987; Rinde et al., 1986) and mice (Maronpot et al., 1986). In addition, there is a large historical database on the occurrence of different types of neoplasms in rats and mice. Similar approaches for fish tests are being considered and standards are being established.

Because the liver is the principal target of carcinogens in fish as well as in rodents, much of this section will be devoted to hepatocellular and biliary lesions. That the liver is a principal target is based on the fact that in teleosts it is the major site of cytochrome P-450 (see Chapter 17), which inactivates some chemicals and activates others. Furthermore, nutrients derived from gastrointestinal absorption are stored in hepatocytes and released for further metabolism by other tissues (Moon et al., 1985), bile synthesized by hepatocytes aids in the digestion of fatty acids (Boyer et al., 1976) and carries conjugated metabolites of toxicants (Gingerich, 1982) into the intestine for excretion or enterohepatic recirculation, and the yolk protein vitellogenin is synthesized within the liver (Vaillant et al., 1988). Some of the principal extrahepatic neoplastic lesions reported in small fish are discussed in the following because studies from several laboratories have shown that carcinogenic potency is sometimes reflected by the occurrence of extrahepatic primary neoplasms. As mentioned earlier, whole small fish specimens can be mounted on histologic slides, allowing relatively easy analysis of extrahepatic neoplasia.

Progression of Hepatic Neoplasia in Small Fish

In rodents, the principal hepatocellular proliferative lesions that result from carcinogen exposure are foci of cellular alteration, adenoma, and hepatocellular carcinoma (Maronpot et al., 1986, 1987) (Table 8). Carcinogen-induced hepatocellular proliferative lesions in rainbow trout (Hendricks et al., 1984) and in several species of small fish (e.g., Bunton, 1990; Couch and Courtney, 1987; Hawkins et al., 1989; Hinton et al., 1988) appear to be counterparts of those in rodents. Some of the typical lesions are illustrated here. Figures 2 and 3 show an altered focus in the liver of a guppy exposed to benzo[*a*]pyrene. An adenoma is illustrated in Figure 4 and hepatocellular carcinomas in Figures 5, 6, and 7. These lesions have been considered to represent stages of carcinogenesis in rodents (Williams, 1980) as well as in fish (Hinton et al., 1992). Statistical associations of incidence among these lesions in carcinoma-bearing wild fish also argue for their developmental relationships (Myers et al., 1987). Figure 8 shows a hypothetical scheme for sequences of neoplasm development in liver and reviews some of the concepts of neoplastic progression discussed earlier (Williams, 1990).

Spontaneous and Rarely Occurring Neoplastic Lesions

An understanding of the spontaneous neoplastic lesions known to occur in the experimental model is necessary in order to assess the carcinogenic effects of a test substance. The low frequency of spontaneous neoplasms in small-fish models, especially in principal carcinogen target organs such as the liver, is an important attribute. Data on the occurrence of hepatic neoplastic lesions in control specimens of medaka and guppy at 24-, 36-, and 52-wk sampling periods for carcinogen screening assays show low incidences for hepatocellular lesions (Table 9). Cholangiocellular neoplastic lesions, which are also usually associated with exposure to chemical carcino-

Table 8. Characteristics of hepatocellular proliferative lesions in rodents[a]

Lesion	Characteristics
Foci of cellular alteration	Common in rats but not in mice Exhibit altered staining and enzymatic patterns Hepatocytes of foci merge with surrounding normal cells Do not compress adjacent tissues Include foci subtypes: clear, cell, acidophilic, basophilic, and mixed
Adenoma	Discrete lesions Compress adjacent normal cells Composed of well-differentiated cells Cells may be eosinophilic, basophilic, or vacuolated Cells may exhibit uneven growth patterns within adenoma May have increased mitotic index
Hepatocellular carcinoma	May have distinct trabecular or adenoid pattern Cells may be poorly differentiated or anaplastic Lesion may be locally invasive or metastatic Some appear to arise within adenomas

[a] Data compiled from Maronpot et al., 1986, 1987.

gens in fish, were not observed in those samples. The occurrence of hepatocellular neoplasms generally appears age related. Masahito et al. (1989) examined nearly 1000 medaka from 1 to 5 yr of age and observed that liver tumors were rare in 1-yr-old specimens, but the incidence rose to 7.1% in 5-yr-old females.

Although it appears that the occurrence of liver neoplasms in small fish is highly correlated with carcinogen exposure, not enough data are available on the etiology and toxicologic pathology of nonhepatic neoplastic lesions to distinguish between spontaneous and carcinogen-induced ones. Some nonhepatic lesions are clearly inducible by carcinogens, for example, retinal medulloepithelioma in medaka (Hawkins et al., 1986) and exocrine pancreatic neoplasia in the guppy by methylazoxymethanol acetate (Fournie et al., 1987). Several other nonhepatic lesions appear to be spontaneous and not carcinogen inducible, at least not with the carcinogens tested in Gulf Coast Research Laboratory. For the medaka, these include pancreatic acinar carcinoma (Hawkins et al., 1991), thymic lymphosarcoma (Battalora et al., 1990; Harada et al., 1990; Okihiro and Hinton, 1989), and germ cell neoplasms that resemble spermatocytic seminomas (Hawkins et al., unpublished) or dysgerminoma (Harada et al., 1991). In life span examinations of medaka, spontaneous tumors were rare in extrahepatic organs but included squamous cell carcinomas, melanomas, and lymphosarcomas, none of which occurred in relation to sex or age (Masahito et al., 1989). Few extrahepatic neoplasms in the guppy have been reported. A single case of retinal adenocarcinoma reported in a guppy was considered to be spontaneous (Fournie et al., 1992).

In interpreting the results of a small-fish carcinogenicity bioassay, decisions on the significance of the

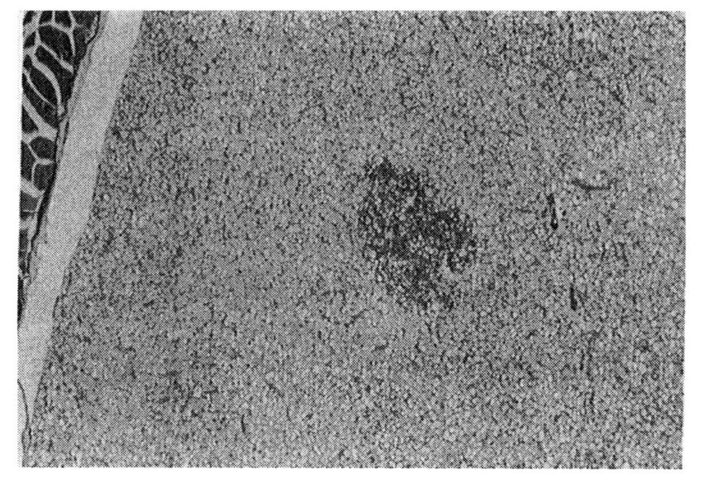

Figure 2. Altered focus in liver of a guppy exposed to benzo[*a*]pyrene and sampled at 24 wk after exposure. (From Hawkins et al., 1988c.)

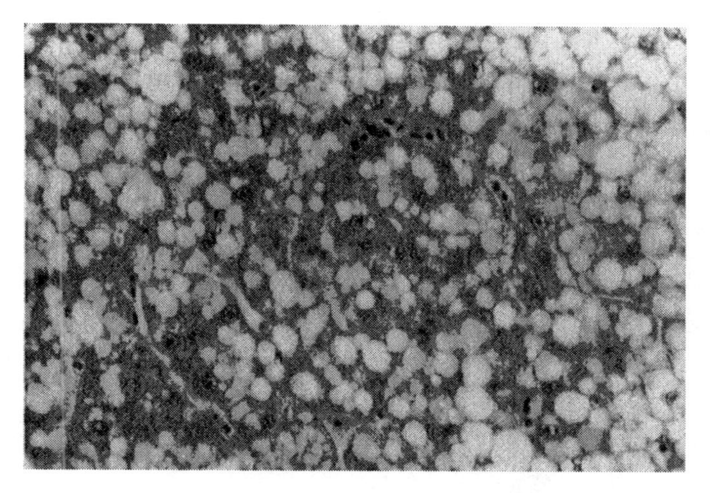

Figure 3. Higher magnification of altered focus in Figure 2, showing an indistinct border between focus and normal tissue. (From Hawkins et al., 1988c.)

occurrence of rare neoplasms should be made cautiously and both historical and concurrent cases of neoplasia in controls should be considered in evaluating the carcinogenic response. On the other hand, the occurrence of such lesions could be highly relevant in assessing carcinogenicity of a test substance even in the absence of statistical significance. Still, without an extensive database of spontaneous lesions occurring in control specimens, this situation could result in a false-positive error.

In rodent tests, pathologists usually attempt to distinguish between benign and malignant lesions. Malignant lesions are considered those that ultimately kill the organism and the lesions have certain behavioral and cytologic characteristics discussed earlier. Most of the more common carcinogen-induced lesions in small fish do not kill over test periods of about a year. Histologically, however, many of the lesions have the hallmarks of malignancy or appear to have the capability of progressing to malignancy. It may be advisable at this stage of development of the small-fish carcinogenicity bioassay that carcinogen-induced lesions are not distinguished according to whether they are benign or malignant until considerably more information concerning the biology of neoplasia in small fish is known.

Confidence in the results of small-fish carcinogenicity bioassays will increase as agreement between the results of small-fish tests and rodent tests occurs. It is also important that the results be interpretable

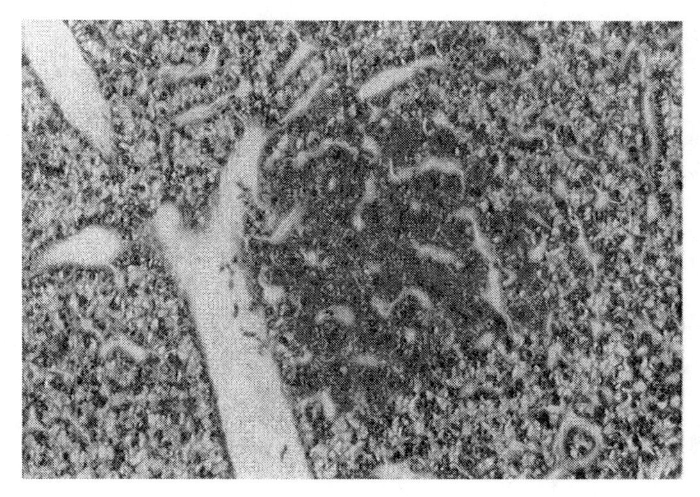

Figure 4. Adenoma in the liver of a medaka showing a distinct border and well-differentiated cells. Specimen exposed to benzo[*a*]pyrene and sampled at 24 wk. (From Hawkins et al., 1988c.)

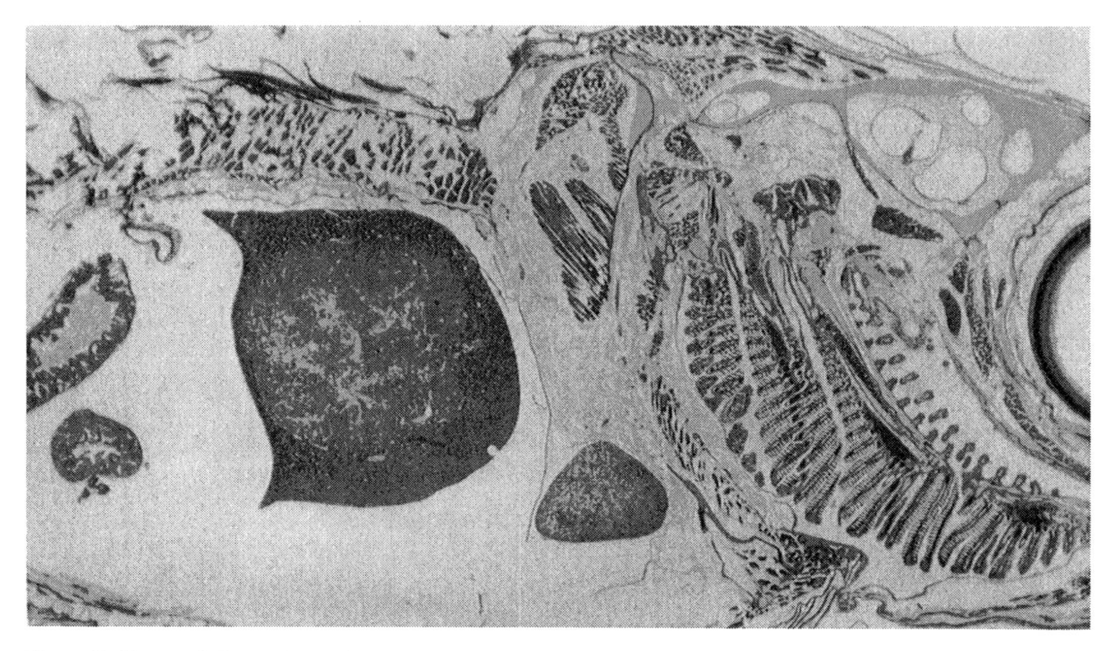

Figure 5. Hepatocellular carcinoma in center of liver from medaka exposed to 10 mg/L diethylnitrosamine and sampled at 6 mo.

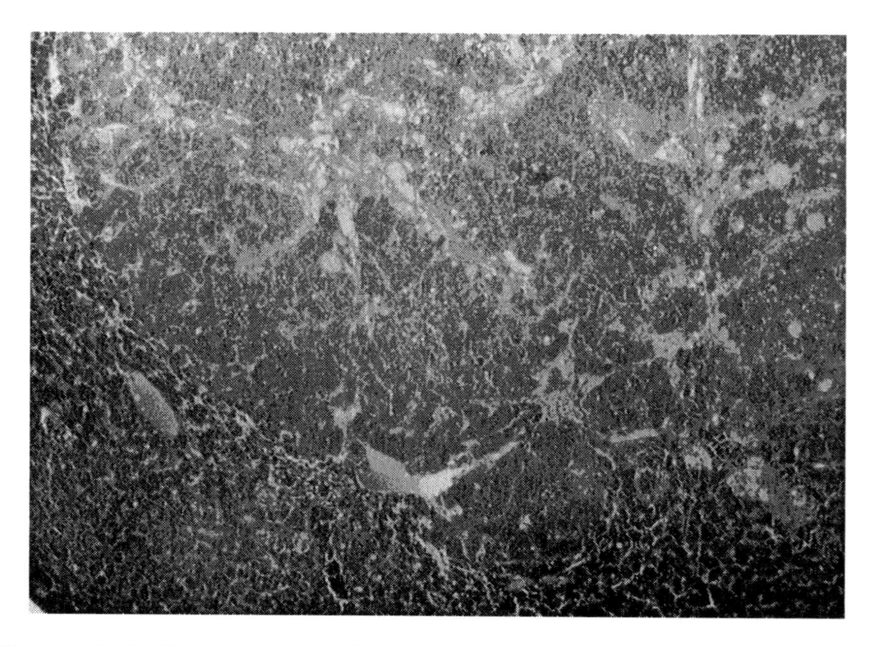

Figure 6. Poorly differentiated hepatocellular carcinoma showing cellular pleomorphism and an uneven and invasive border. From a medaka exposed to 10 mg/L diethylnitrosamine and sampled at 6 mo.

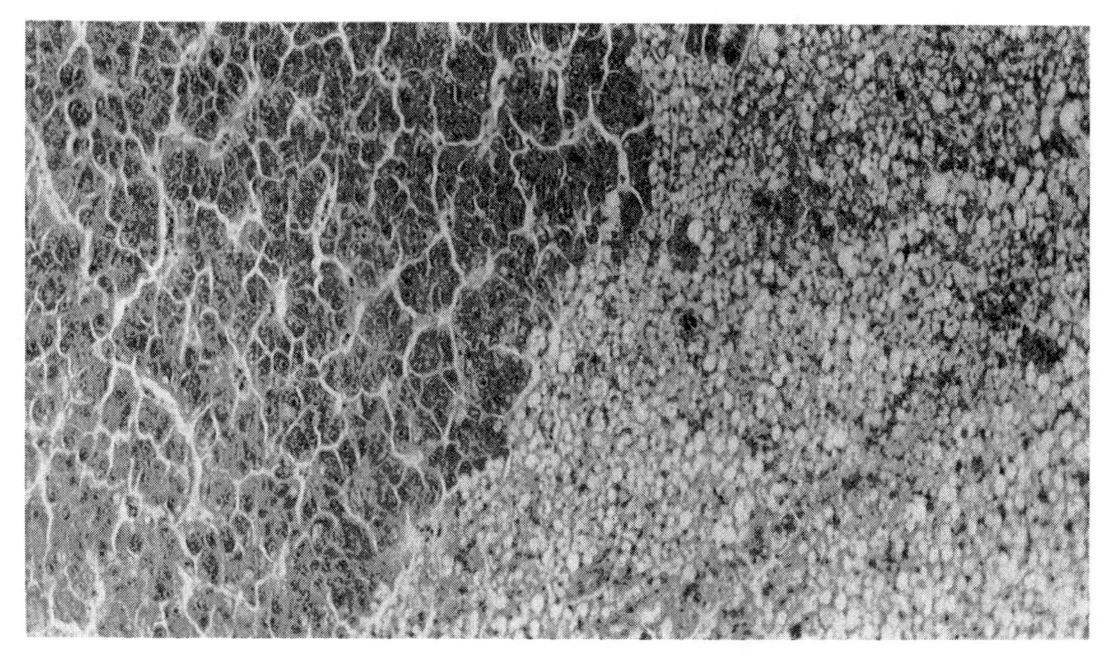

Figure 7. Well-differentiated hepatocellular carcinoma. Note uniformity of cells in carcinoma and evenness of border between lesion and normal liver. From a guppy exposed to acetylaminofluorene and sampled at 24 mo. (From James et al., 1994.)

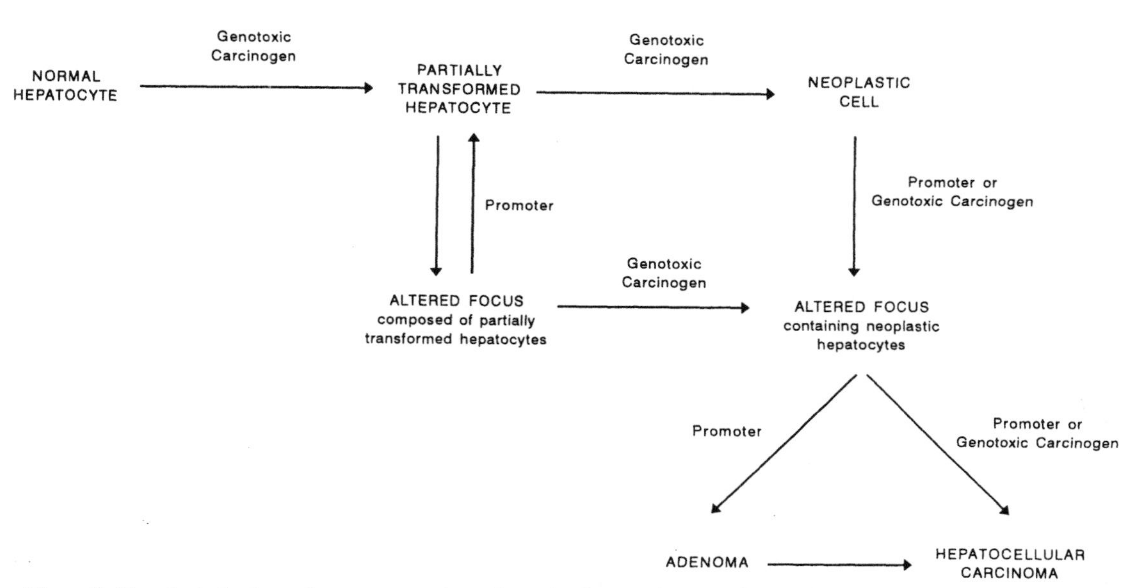

Figure 8. Hypothetical scheme for sequences of neoplasm development in liver. (From Williams, G. M.: Epigenetic mechanisms of liver tumor promotion. *Mouse Liver Carcinogenesis: Mechanisms and Species Comparisons*, edited by D. E. Stevenson; R. M. McClain; J. A. Popp; T. J. Slaga; J. M. Ward; and H. C. Pitot, Copyright © 1990. Alan R. Liss, Inc. Reprinted by permission of John Wiley & Sons, Inc.).

Table 9. Spontaneous hepatocellular proliferative lesions in control medaka (*Oryzias latipes*) and guppy (*Poecilia reticulata*)

	24-wk		36-wk		52-wk	
Lesion	Medaka (1815)[a]	Guppy (1035)	Medaka (2193)	Guppy (1240)	Medaka (939)	Guppy (555)
Altered focus	2	0	0	3	0	4
Adenoma	0	0	4	2	2	1
Carcinoma	0	0	3	1	1	2
Total incidence	2 (0.1%)	0 (0.0%)	7 (0.3%)	6 (0.5%)	3 (0.3%)	7 (1.3%)

[a]Number of specimens examined in parentheses.
From Hawkins et al. (unpublshed).

in light of what is known about the metabolic mechanisms by which test organisms handle carcinogens. A further objective should be the incorporation of ancillary in vivo tests into fish bioassays. This approach has been discussed for mammalian bioassays by Weisburger and Williams (1991) and involves a battery of tests that includes the Ames test (a reverse mutation assay in prokaryotic *Salmonella typhimurium*), assay for DNA repair in explanted liver cells, determination of DNA adducts, DNA breakage, chromosome aberrations, and sister chromatid exchange.

Confounding Lesions

Numerous pitfalls are associated with the histopathological evaluation of a carcinogenic response. Several toxic lesions appear to be associated with carcinogenic responses and target organ toxicity may be a critical stage in the process of the development of neoplastic lesions. Additional lesions seem to reflect noncarcinogenic toxicity or, at least, toxicity that is not part of the neoplastic process. Some inflammatory or reactive lesions in fishes also mimic neoplastic lesions (Harshbarger et al., 1984).

METHODOLOGY IN SMALL-FISH CARCINOGENICITY BIOASSAYS

Experimental Design

With only a few exceptions, small-fish carcinogenicity tests can be designed similarly to rodent tests. Factors such as numbers of test organisms, dose or concentration levels, lengths of exposure and total test periods, disease surveillance, and histopathological end points can be arranged to parallel those in rodent tests. Fish tests usually can offer many other attributes such as (1) genetically consistent test organisms; (2) uniformity of exposure to test substances; (3) uniformity of test conditions such as temperature, metabolite disposal, and lighting; (4) rapid onset of tumor development; (5) a short exposure period; and (6) low cost, at least compared to rodent carcinogenicity bioassays. In this section, the organization of tests with respect to appropriate control groups is discussed, including the use of positive control groups and experimental groups.

A few conditions that are unacceptable in rodent tests often must be accepted in most small-fish carcinogenesis tests. Because many of the test species reach sexual maturity at only a few weeks or months of age, it is typically necessary to hold breeding males and females together in test systems. Progeny should be removed periodically from the test system. To separate the sexes might place the organisms in undue endocrine stress, as many fish species are capable of changing from one sex to the other to take advantage of opportunities for dominance in social hierarchies and for establishing breeding pairs.

Another accepted condition is the environment in which the small fish are maintained. Algae and other microscopic organisms populate grow-out (maintenance) aquaria, but they do not appear to affect tests. Also, these organisms probably contribute favorably to the nutrition of the fish and may help stabilize the water quality of the aquarium.

Selection of Species and Culture Requirements

In this section, some of the core methods involved in small-fish carcinogenesis tests are described, including fish culture and maintenance, exposure methodology, pathological analyses, and statistical analyses.

Numerous small fish species are available as test models. Unfortunately, relatively few species have been tested for carcinogen sensitivity. Selection of test species is typically based on factors other than carcinogen sensitivity, such as availability, economy, ease of rearing, and fecundity. Because most carcinogen screening tests with small fish have been conducted with the medaka (*Oryzias latipes*) and guppy (*Poecilia reticulata*), culture requirements and handling of these species are described in detail. The saltwater species sheepshead minnow (*Cyprinodon variegatus*) and Gulf killifish (*Fundulus grandis*) also present opportunities that will be available with the development of saltwater models.

Fish Culture and Maintenance Techniques

Medaka (Oryzias latipes)

Some manuals and publications are available that deal with the culture of medaka (Denny et al., 1991; Kirchen and West, 1976; Yamamoto, 1975). Procedures described here are in use at the Gulf Coast Research Laboratory (GCRL) and incorporate information from all those sources. Medaka brood stock are maintained at a 3:2 female/male ratio with a density of about five fish per liter in 38-L or larger aquaria containing aged well water. Flow-through 380-L fiberglass raceways that undergo three volume exchanges daily are also used. Each contains 400 to 600 fish and these are kept as broodstock for about 5.5 mo.

Brood tanks are held at 27°C and provided a daily regimen of 16 h light and 8 h dark. Diets vary from laboratory to laboratory. At GCRL, fish are fed dry flake food (e.g., Prime Flakes, Zeigler Bros.) sparingly three times daily and are provided live brine shrimp (*Artemia franciscana*) nauplii once daily. Feces and other debris are removed from aquaria twice each week. Aeration and continual water filtration are provided by biological sponge filters.

Eggs are collected by siphoning them from the tank bottom, teasing them away from a netted female, or removing them from biological sponge filters following egg deposition by the female, which becomes sexually mature about 6 wk after hatching. Medaka usually produce 10 to 50 eggs per spawn. When eggs are first released, they remain attached near the vent until the fish attaches them to vegetation in the wild or some object in an aquarium. An effective way to collect eggs is to add extra sponge filters to the tank to provide a substrate for the fish to attach their eggs. If eggs are removed from these at about 0900 to 1000 h (2.5 h after initiation of light cycle), the chorion is hardened, few eggs are eaten by the fish, and over 90% of the days production can be collected. Individual eggs are placed in hatching solution at 24°C in a 3.8-L aerated jar. Hatching solution contains 100 mg NaCl, 3 mg KCl, 4 mg $CaCl_2 \cdot 2H_2O$, and 16.3 mg $MgSO_4 \cdot 7H_2O$ per 100 ml of glass-distilled water. Dead or diseased embryos are discarded.

As fry begin to hatch (about 10 d following fertilization at 24°C), they are transferred to finger bowls containing approximately 1400 ml of water with an average of 50 fry per bowl. The fry are provided cultured ciliates (*Paramecium caudatum*) for 3 d and cultured nematodes (*Pangrilius silusae*) for 6 or more days. Under those conditions, about 80 to 90% hatch, with 95% of those surviving. Of those that survive, usually fewer than 2% are abnormally large, small, or weak. Bowls with fry are maintained in a 27°C water bath. Feces and debris are removed daily with a concomitant 50% water change to ensure good water quality. After 3 d, newly hatched brine shrimp nauplii are added as a food source. After 6 to 10 d fry are fed dry flake food and brine shrimp. Fry are either utilized for testing at or about age d 6, maintained for future testing, or transferred to maintenance aquaria to provide future brood stock.

Guppy (Poecilia reticulata)

Procedures for rearing and maintaining guppies are similar to those for medaka with several important exceptions that are mainly related to the guppy being a live bearer rather than an egg layer. Brood guppies are maintained in either static, 75- to 115-L glass aquaria or flow-through, 380-L fiberglass raceway tanks filled with 300 L of water. A loading density of approximately 2.5 fish per liter and a 2:3 male/female ratio are desirable, even though one male will fertilize many females. Tanks are maintained at ambient temperature, generally 25 ± 2°C, but to increase fry production the temperature may be elevated to 25–27°C using submersible heaters. Feces and debris are removed twice weekly from each tank in conjunction with a minimum 20% water change. Holding tanks contain aged well water and are equipped with sponge filters. Fluorescent lighting is timer controlled to provide a photoperiod of 16 h light to 8 h dark. Brood fish are fed dry flake food three times daily and live brine shrimp nauplii once daily.

With appropriate conditions of temperature, light, diet, and water quality, female guppies begin to bear young when they are about 3 mo old. The gestation period is about 28 d. Unlike medaka, which consistently produce large numbers of eggs per female, the guppy has less predictable fry production that usually extends over several days. When production is desired, feeding is increased, debris is removed more often, and floating vegetation such as hornwort is added. The latter utilizes nitrogenous wastes from the fish and decreases carbon dioxide. Both factors stimulate production. Most important, the vegetation provides a means for the fry to escape predation by the adult fish.

Guppy fry require cover for protection from predation. Live floating aquatic plants such as hornwort work best, but artificial refuges such as plastic vegetation, plastic filaments, or nylon screen may also be used. Fry are best collected when adults are being fed. Floating feeding rings can be placed in the tanks to prevent the dispersal of food into the areas of fry cover. Placement of food into the tanks, with or without the feeding rings, distracts adults from the collection efforts. Within a few hours of being born, fry swim near the surface. The vegetation or other cover is moved aside, and fry are removed with a fine-mesh dip net or beaker and placed in a finger bowl or 38-L aquarium. To obtain fry of known age, all fry are continually removed from all brood tanks for 1 or 2 d before the initiation of collection of fry. Fry collection then commences the following morning and continues until the desired number has been isolated. Although production of guppies is somewhat erratic and unpredictable, at GCRL each 500 adults produce about 300 fry daily. This production rate lasts for about a week after production is stimulated. The required number of fry dictates the approximate number of broodstock to stimulate.

Recently collected fry are segregated into holding tanks on a daily basis so that each tank contains only fry collected on the same day. Tanks are static, supplied with biological sponge filters, and stocked with a loading density of about five per liter. About 25 to 30% of the water is exchanged every other day. Aquaria containing fry are maintained under feeding, lighting, filtration, and temperature regimens identical to those in brood aquaria. Fry are fed dry flakes three times daily and brine shrimp nauplii once each day. Generally, fry are used for testing or added to brood stock tanks 24–48 h or 6 d after birth.

Several weeks after birth, the anal fin of the male guppy develops as an elongated gonopodium that is used to deliver sperm to the female. This character can be used to determine an individual's sex and to gauge maturation. After being fertilized, the female develops a "gravid spot" in the shape of a triangle near the vent. The spot becomes darker and gradually enlarges and serves as a gauge of maturation as embryonic development progresses until young are born. Young guppies produce as few as two or three fry and older ones can have over 100 in one brood.

Exposure Techniques

Static, Single-Pulse Exposures

With medaka, static, single-pulse exposures are conducted in a commercial carcinogen glove box (Labconco 50350), laminar flow hood, or other adequately ventilated chamber using aged well water. Exposures are routinely conducted in the dark at a temperature of 26 ± 1.0°C maintained by a water bath. Duration of exposure varies among chemicals from 2 to 24 h. Individual specimens selected for testing are typically 6 d old at the initiation of a test and are healthy and uniform in size. During the exposure, they are housed in Nytex mesh chambers within glass beakers or small aquaria and are not fed. Following exposure, fish are rinsed with toxicant-free water and transferred to 38-L grow-out aquaria maintained at 27°C under a 16:8 h light/dark regime. Fish in grow-out are fed dry food three times and brine shrimp nauplii once daily. Exposed fish are observed at least once daily during the entire study for indications of stress or aberrant behavior and are sampled at predetermined times depending on the objectives of the study. Specimens are histologically processed for light microscopic examination and, in special cases, for electron microscopic examination. Moribund fish are sampled when observed.

Intermittent Multiple-Pulse Exposures

These exposure tests are conducted under conditions similar to those described above for the static single-pulse exposures. Exposures are performed at intervals, usually weekly through a maximum of 6 wk. Following each exposure, fish are rinsed and maintained within their meshed chambers in toxicant-free water at 26 ± 1°C under a 16:8 h light/dark regime and fed as described above. Dead, dying, and diseased fish are removed upon observation. Upon conclusion of the final exposure in the series, fish are rinsed and transferred to grow-out aquaria.

Flow-Through Exposures

Flow-through exposures of variable duration can be conducted in specially designed flow-through glove boxes (Walker et al., 1985). Untreated (chemical-free) control groups accompany all exposure tests. To maintain consistent concentrations of test chemicals throughout an extended exposure period, a reservoir consisting of two or three serially connected, sealed, 45-Liter Pyrex carboys is utilized (Walker et al., 1985). Test chemicals and test water are added to each carboy, and the contents are magnetically stirred until the concentration of test chemical stabilizes at or near its saturation limit in all the test groups.

When a test is initiated, chemical-laden water is withdrawn from the nearest, or dispensing, carboy in the series by precision liquid dispensing syringe pumps (PLD-II, Hamilton Company, Reno, NV) and delivered through microbore tubing to one of six mixing chambers. Chemical-free diluent water enters the system by gravity flow from an elevated head box through a solenoid-controlled valve, filling a seven-compartment water partitioner similar to that described by Schimmel et al. (1974). Float switches within the water partitioner activate a programmable laboratory controller (Idec PLE-30R, Industrial Electric Supply Co., Birmingham, AL), which in turn activates the series of PLD injectors. All injectors draw from the dispensing carboy but receive different instructions from the controller regarding number of injections per cycle. The flow of diluent water into the water partitioner is controllable in order to provide a range of cycling times. Syringe size, number of syringes, distance of plunger withdrawal, and number of injections can be varied, thereby facilitating introduction of a wide variety of chemical masses and hence test concentrations. Typically, for chronic exposure tests (3–6 mo), 48-L aquaria are used with 250–400 fish per aquarium. In this configuration, the exposure system can handle three chemical concentrations and a control. Fish can be contained in meshed chambers (10-cm inner diameter petri dishes, each with a 9-cm-high nylon mesh collar) or allowed to swim freely in treatment aquaria. Each treatment aquarium fills to a depth of 8 cm, at which time chemical-laden water discharges through activated carbon (Filtersorb 400, Calgon Corp., Houston, TX). Mixing chambers, splitter boxes, and treatment aquaria, all constructed of glass and silicone cement, are housed within a resin-coated plywood exposure chamber (341.6 cm long by 92.7 cm wide by 53.3 cm high) covered with a pitched top (34.3 cm high along its center). Access to the interior of the box is through capped ports, and manipulations within the chamber are conducted through three or four gloved ports along each of the two long sides of the chamber. Treatment aquaria are housed within a central water bath maintained at 26 ± 1°C in a 16:8 h light/dark regimen. The exposure chamber is maintained at a slightly negative pressure by exhaust fans, which also serve to draw incoming air and remove gaseous toxicants through carbon filters (BPL activated carbon, 12 × 30 mesh, Calgon Corp.). Usually, fish are fed flake food and brine shrimp nauplii throughout the exposure period. Fish are observed periodically each day during exposure and dead fish are removed and recorded upon discovery. Chemical concentrations are monitored once or twice each week throughout each exposure period by appropriate gas-liquid chromatographic methods. Upon termination of the exposure test, fish are transferred to grow-out aquaria, maintained at 27°C, and sampled as described above.

Grow-Out Techniques

Following exposure, fish are placed in 38-L aquaria (maximum 100 fish/aquarium) containing clean, aged well water. Aquaria are held in circulating water baths at 26 ± 1°C under a 16:8 h light/dark regimen. Aeration and continuous water filtration are provided by biological sponge filters. Feces and debris in the tanks are removed three times weekly, which results in a 20% water change. Temperature, pH, salinity, and ammonia are monitored monthly on a revolving schedule to ensure optimum water quality. Fish are fed dry flake food three times daily and live brine shrimp nauplii once daily.

Histological Procedures

Histological procedures are generally designed to allow the examination of large numbers of specimens and to survey most internal organs of each individual specimen. Usually, whole adult fish specimens are processed. Fish are narcotized in ice water or MS-222 (tricaine methanesulfonate), the bellies of larger specimens are slit open, and whole fish are placed into Lillie's fixative (10% formalin, 85% saturated aqueous solution of picric acid, 5% formic acid). This solution decalcifies as well as fixes. Fixation lasts 24 h to 1 wk depending on the size and relative amount of bone. Extended fixation times ensure decalcification of the larger specimens. Specimens are then dehydrated in ethanol, cleared in xylene or a xylene substitute, and embedded in paraffin. Usually, fish are embedded on one side or the other depending on the side on which the liver lies. To survey most major organs, sections are taken from one or more planes, mounted on glass slides, and routinely stained with hematoxylin and eosin. In some cases, specific tumors are sectioned selectively. For example, eye tumors are probably best studied in transverse sections so that both eyes appear in the same section, whereas neurogenic tumors that often arise from spinal nerves require midline sections through the spinal column to observe early stages and origins of those tumors. When appropriate, special stains or immunocytochemical tests are used to identify cellular components, secretory products, or tissues. Each tumor is described and measured along with nonneoplastic lesions.

Electron Microscopy

For electron microscopic studies, tissues are dissected, immersed in 3% glutaraldehyde in 0.1 M phosphate buffer, rinsed 2 h to overnight in buffer, and postfixed in 1% osmium tetroxide in phosphate buffer. Tissues are dehydrated in ethanol, soaked in propylene oxide, and embedded in one of several available epoxy resins. To take advantage of the improved resolution of plastic-embedded tissues and to choose areas for thin sectioning, 2-μm-thick sections are cut, stained with toluidine blue, and examined with a light microscope. Thin sections are cut on an ultramicrotome, mounted on copper grids, stained with uranyl acetate and lead citrate, and examined with an electron microscope.

SUMMARY

In this chapter, the rationale, guidelines, and approaches for conducting small-fish carcinogenicity tests were presented. Goals were to demonstrate that the tests are relatively easy to conduct, that they require little in equipment or expertise in addition to that already available in a modern aquatic toxicology laboratory, and that the tests can approach the power and relevance of rodent carcinogenicity bioassays. Because the development of neoplasms following exposure to a carcinogenic substance is a latent or chronic response, a whole-animal carcinogenicity test using a fish model can be considered an elaboration or extension of routine aquatic toxicity tests that might otherwise have mortality or physiological perturbations as the principal end points.

ACKNOWLEDGMENTS

We gratefully acknowledge the contributions of Sue Barnes and Don Barnes in the development of the fish culture technique, Alex Schesny and Steve Manning for the exposure techniques, and Robert Allen, Retha Edwards and Rena Krol for the histological techniques used in our studies. We also thank Dr. Gary K. Ostrander, Oklahoma State University, and Dr. Marilyn Wolfe, Experimental Pathology Laboratories, for critically reviewing early drafts of the manuscript. This study was supported in part by the U.S. Army Biomedical Research and Development Command (contract DAMD17-88-C-8050).

LITERATURE CITED

Anders F.; Schartl M.; Barnekow A.; and Anders A.: *Xiphophorus* as an in vivo model for studies on normal and defective control of oncogenes. *Adv. Cancer Res.*, 42: 191–275, 1984.

Anderson, M. W., and Reynolds, S. H.: Activation of oncogenes by chemical carcinogens. The Pathobiology of Neoplasia, edited by A. E. Sirica, pp. 291–304. New York: Plenum, 1989.

Aoki, K., and Matsudaira, H.: Factors influencing tumorigenesis in the liver after treatment with methylazoxymethanol acetate in a teleost, *Oryzias latipes*. Phyletic Approaches to Cancer, edited by C. J. Dawe et al., pp. 205–216. Tokyo: Japan Scientific Societies Press, 1981.

Aoki, K., and Matsudaira, H.: Factors influencing methylazoxymethanol acetate initiation of liver tumors in *Oryzias latipes*: carcinogen dosage and time of exposure. *Natl. Cancer Inst. Monogr.*, 65:345–351, 1984.

Bailey, G.; Selivonchick, D.; and Hendricks, J.: Initiation, promotion, and inhibition of carcinogenesis in rainbow trout. *Environ. Health Perspect.*, 71:147–153, 1987.

Battalora, M. St. J.; Hawkins, W. E.; Walker, W. W.; and Overstreet, R. M.: Occurrence of thymic lymphoma in carcinogenesis bioassay specimens of the Japanese medaka (*Oryzias latipes*). *Cancer Res.*, [Suppl.]50:5675–5678, 1991.

Baumann, P. C.: PAH, metabolites, and neoplasia in feral fish populations. Metabolism of Polycyclic Aromatic Hydrocarbons in the Aquatic Environment, edited by U. Varanasi, pp. 269–290, Boca Raton, FL: CRC Press, 1989.

Baumann, P. C.; Smith, W. D.; and Parland, W. K.: Tumor frequencies and contaminant concentrations in brown bullheads from an industrialized river and a recreational lake. *Trans. Am. Fish. Soc.*, 116:79–86, 1987.

Baumann, P. C.; Harshbarger, J. C.; and Hartman, K. J.: Relationship between liver tumors and age in brown bullhead populations from two Lake Erie tributaries. *Sci. Total Environ.*, 94:71–87, 1990.

Beer, D. G., and Pitot, H. C.: Proto-oncogene activation during chemically induced hepatocarcinogenesis in rodents. *Mutat. Res.*, 220:1–10, 1989.

Beland, F. A., and Poirier, M. C.: DNA adducts and carcinogenesis. Sirica, A. E. (ed.): *The Pathobiology of Neoplasia*. Plenum, New York, 1989, pp. 57–80.

Berenblum, I.: Challenging problems in cocarcinogenesis. *Cancer Res.*, 45:1917–1921, 1985.

Black, J. J.: Aquatic animal neoplasia as an indicator for carcinogenic hazards to man. Hazard Assessment of Chemicals: Current Developments, Vol. 3, edited by J. Saxena, pp. 181–232. New York: Academic Press, 1984.

Black, J. J., and Baumann, P. C.: Carcinogens and cancers in freshwater fishes. *Environ. Health Perspect.*, 90:27–33, 1991.

Boorman, G. A.; Montgomery, C. A.; Eustis, S. L.; Wolfe, M. J.; McConnell, E. E.; and Hardisty, J. F.: Quality assurance in pathology for rodent carcinogenicity studies. Handbook of Carcinogenesis Testing, edited by H. Milman, and E. Weisberger, pp. 345–357. Park Ridge, NJ: Noyes, 1985.

Boyer, J. L.; Swartz, J.; and Smith, N.: Biliary secretion in elasmobranchs. II. Hepatic uptake and biliary excretion of organic anions. *Am. J. Physiol.*, 230:974–981, 1976.

Buhler, D. R., and Williams, D. E.: Enzymes involved in metabolism of PAH by fishes and other aquatic animals: oxidative enzymes (or phase I enzymes). Metabolism of Polycyclic Aromatic Hydrocarbons in the Aquatic Environment, edited by U. Varanasi, pp. 151–184. Boca Raton, FL: CRC Press, 1989.

Bunton, T. E.: Hepatopathology of diethylnitrosamine in the medaka (*Oryzias latipes*) following short-term exposure. *Toxicol. Pathol.*, 18:313–323, 1990.

Butterworth, B. E.: Chemically induced cell proliferation as a predictive assay for potential carcinogenicity. Chemically Induced Cell Proliferation: Implications for Risk Assessment, edited by B. E. Butterworth; T. J. Slaga; W. Farland; and M. McClain, pp. 457–468. New York: Wiley-Liss, 1991.

Cormier, S. M.; Racine, R. N.; Smith, C. E.; Dey, W. P.; and Peck, T. H.: Hepatocellular carcinoma and fatty infiltration in the Atlantic tomcod, *Microgadus tomcod* (Walbaum). *J. Fish Dis.*, 12:105–116, 1989.

Couch, J. A., and Courtney, L. A.: *N*-Nitrosodiethylamine–induced hepatocarcinogenesis in estuarine sheepshead minnow (*Cyprinodon variegatus*), neoplasms and related lesions compared with mammalian lesions. *J. Natl. Cancer Inst.*, 79:297–321, 1987.

Couch, J. A., and Harshbarger, J. C.: Effects of carcinogenic agents on aquatic animals: an environmental and experimental overview. *Environ. Carcinogen Rev.*, 3: 63–105, 1985.

Dawe, C. J.; Harshbarger, J. C.; Kondo, S.; Sugimura, T.; and Takayama, S.: Phyletic Approaches to Cancer. Tokyo: Japan Scientific Societies Press, 1981.

Dawe, C. J.; Harshbarger, J. C.; Wellings, S. R.; and Stranberg, J. D. (eds.): Pathobiology of spontaneous and induced neoplasms in fishes: comparative characterization, nomenclature and literature. San Diego: Academic Press, in press.

Denny, J. S.; Spehar, R. L.; Mead, K. E.; and Yousuff, S. C.: Guidelines for culturing the Japanese medaka (*Oryzias latipes*). EPA/600/3-91/064, 1991.

Diwan, B. A., and Rice, J. M.: Organ and species specificity in chemical carcinogenesis and tumor promotion. The Pathobiology of Neoplasia, edited by A. E. Sirica, pp. 149–171. New York: Plenum, 1989.

Doll, R., and Peto, R.: The causes of cancer: quantitative estimates of avoidable risks of cancer in the United States today. *J. Natl. Cancer Inst.*, 66:1191–1308, 1981.

Dua, R. N., and Jackson, B. A.: Review of pathology data for regulatory purposes. *Toxicol. Pathol.*, 16:443–450, 1988.

Dunn, B. P.: Carcinogen adducts as an indicator for the public health risks of consuming carcinogen-exposed fish and shellfish. *Environ. Health Perspect.*, 90:111–116, 1991.

Elskus, A. A., and Stegeman, J. J.: Further consideration of phenobarbital effects on cytochrome P-450 activity in the killifish, *Fundulus heteroclitus. Comp. Biochem. Physiol.*, 92C:223–230, 1989.

Eustis, S. L.: The sequential development of cancer: a morphological perspective. *Toxicol. Lett.*, 49:267–281, 1989.

Fabacher, D. L.; Besser, J. M.; Schmitt, C. J.; Harshbarger, J. C.; Peterman, P. H.; and Lebo, J. A.: Contaminated sediments from tributaries of the Great Lakes: chemical characterization and carcinogenic effects in medaka (*Oryzias latipes*). *Arch. Environ. Contam. Toxicol.*, 20: 17–34, 1991.

Fournie, J. W.; Hawkins, W. E.; Overstreet, R. M.; and Walker, W. W.: Exocrine pancreatic neoplasms induced by methylazoxymethanol acetate in the guppy (*Poecilia reticulata*). *J. Natl. Cancer Inst.*, 78:715–725, 1987.

Fournie, J. W.; Hawkins, W. E.; and Walker, W. W.: Adenocarcinoma of the retinal pigment epithelium in the guppy *Poecilia reticulata* Peters. *J. Comp. Pathol.*, 106: 429–434, 1992.

Gardner, G. R.: Chemically induced histopathology in aquatic invertebrates. Pathobiology of Marine and Estuarine Organisms, edited by J. A. Couch, and J. W. Fournie, pp. 359–391. Boca Raton, FL: CRC Press, 1993.

Gart, J. J.; Krewski, D.; Lee, P. N.; Tarone, R. E.; and Wahrendorf, J.: Statistical Methods in Cancer Research.

Vol. III. The Design and Analysis of Long-Term Animal Experiments. Lyon, France: IARC, 1986. IARC Scientific Publication No. 79.

Gingerich, W. H.: Hepatic toxicology in fishes. *Aquatic Toxicology*, edited by L. Weber, pp. 55–105. New York: Raven Press, 1982.

Goodman, G., and Wilson, R.: Predicting the carcinogenicity of chemicals in humans from rodent bioassay data. *Environ. Health Perspect.*, 94:195–218, 1991.

Goodman, J. I.; Ward, J. M.; Popp, J. A.; Klaunig, J. E.; and Rice, T. R.: Mouse liver carcinogenesis: mechanisms and relevance. *Fundam. Appl. Toxicol.*, 17:651–665, 1991.

Guengerich, F. P.: Cytochromes P-450. *Comp. Biochem. Physiol.*, 89C:1–4, 1988.

Guengerich, F. P.; Peterson, L. A.; Cmarik, J. L.; Koga, N.; and Inskeep, P. B.: Activation of dihaloalkanes by glutathione conjugation and formation of DNA adducts. *Environ. Health Perspect.*, 76:15–18, 1987.

Halver, J. E.: Crystalline aflatoxin and other vectors of trout hepatoma. Trout Hepatoma Research Conference Papers, edited by J. E. Halver, and I. A. Mitchell, pp. 78–102. Washington, DC: Bureau of Sports Fisheries and Wildlife, Research Report 70, 1967.

Harada, T.; Hatanaka, J.; and Enomoto, M.: Liver cell carcinomas in medaka (*Oryzias latipes*) induced by methylazoxymethanol-acetate. *J. Comp. Pathol.*, 98:441–452, 1989.

Harada, T.; Hatanaka, J.; Kubota, S.S.; and Enomoto, M.: Lymphoblastic lymphoma in medaka, *Oryzias latipes*. *J. Fish Dis.*, 13:169–173, 1990.

Harada, T.; Okazaki, N.; Kubota, S. S.; Hatanaka, J.; and Enomoto, M.: Spontaneous ovarian tumour in a medaka (*Oryzias latipes*). *J. Comp. Pathol.*, 104:187–193, 1991.

Harshbarger, J. C.: Pseudoneoplasms in ectothermic animals. *Natl. Cancer Inst. Monogr.*, 65:251–273, 1984.

Harshbarger, J. C., and Clark, J. B.: Epizootiology of neoplasms in bony fish of North America. *Sci. Total Environ.*, 94:1–32, 1990.

Harshbarger, J. C.; Charles, A. M.; and Spero, P. M.: Collection and analysis of neoplasms in sub-homeothermic animals from a phyletic point of view. Phyletic Approaches to Cancer, edited by C. J. Dawe et al., pp. 357–384. Tokyo: Japan Scientific Societies Press, 1981.

Haseman, J. K.: Issues in carcinogenicity testing: dose selection. *Fundam. Appl. Toxicol.*, 5:66–78, 1985.

Haseman, J. K.; Huff, J.; and Boorman, G. A.: Use of historical control data in carcinogenicity studies in rodents. *Toxicol. Pathol.*, 12:126–135, 1984.

Hatanaka, J.; Doke, N.; Harada, T.; Aikawa, T.; and Enomoto, M.: Usefulness and rapidity of screening for the toxicity and carcinogenicity of chemicals in medaka, *Oryzias latipes*. *Jpn. J. Exp. Med.*, 52:243–253, 1982.

Hawkins, W. E.; Overstreet, R. M.; Fournie, J. W.; and Walker, W. W.: Development of aquarium fish models for environmental carcinogenesis: tumor induction in seven species. *J. Appl. Toxicol.*, 5:261–264, 1985.

Hawkins, W. E.; Fournie, J. W.; Overstreet, R. M.; and Walker, W. W.: Intraocular neoplasms induced by methylazoxymethanol acetate in Japanese medaka (*Oryzias latipes*). *J. Natl. Cancer Inst.*, 76:453–465, 1986.

Hawkins, W. E.; Overstreet, R. M.; and Walker, W. W.: Small fish models for identifying carcinogens in the aqueous environment. *Water Resource Bull.*, 24:941–949, 1988a.

Hawkins, W. E.; Overstreet, R. M.; and Walker, W. W.: Carcinogenicity tests with small fish species. *Aquat. Toxicol.*, 11:113–128, 1988b.

Hawkins, W. E.; Walker, W. W.; Overstreet, R. M.; Lytle, J. S.; and Lytle, T. F.: Carcinogenic effects of 7,12-dimethylbenz[*a*]anthracene on the guppy *Poecilia reticulata*. *Aquat. Toxicol.*, 15:63–82, 1989.

Hawkins, W. E.; Walker, W. W.; Overstreet, R. M.; Lytle, J. S.; and Lytle, T. F.: Carcinogenic effects of some polynuclear aromatic hydrocarbons on the Japanese medaka and guppy in waterborne exposures. *Sci. Total Environ.*, 94:155–167, 1990.

Hawkins, W. E.; Walker, W. W.; Overstreet, R. M.; Lytle, T. F.; and Lytle, J. S.: Dose-related carcinogenic effects of waterborne benzo[*a*]pyrene on livers of two small fish species. *J. Ecotox. Environ. Safety.*, 16:219–231, 1988c.

Hawkins, W. E.; Fournie, J. W.; Battalora, M. St. J.; and Walker, W. W.: Spontaneous acinar cell carcinoma of the exocrine pancreas in Japanese medaka (*Oryzias latipes*) from carcinogenesis bioassays. *J. Aquat. Anim. Health*, 3:213–220, 1991a.

Hawkins, W. E.; Walker, W. W.; Lytle, T. F.; Lytle, J. S.; and Overstreet, R. M.: Studies on the carcinogenic effects of benzo(*a*)pyrene and 7,12-dimethylbenz(*a*)anthracene on the sheepshead minnow (*Cyprinodon variegatus*). Aquatic Toxicology and Risk Assessment: Fourteenth Volume, edited by M. A. Mayes, and M. G. Barron, pp. 97–104. Philadelphia: American Society for Testing and Materials, 1991b. ASTM STP 1124.

Hayes, M. A.; Smith, I. R.; Rushmore, T. H.; Crane, T. L.; Thorn, C.; Kocal, T. E.; and Ferguson, H. W.: Pathogenesis of skin and liver neoplasms from industrially polluted areas in Lake Ontario. *Sci. Total Environ.*, 94:105–123, 1990.

Hendricks, J. D.: Chemical carcinogenesis in fish. Aquatic Toxicology, edited by L. J. Weber, pp. 149–211, New York: Raven Press, 1982.

Hendricks, J. D.; Meyers, T. R.; and Shelton, D. W.: Histological progression of hepatic neoplasia in rainbow trout (*Salmo gairdneri*). *Natl. Cancer Inst. Monogr.*, 65:321–336, 1984.

Hendricks, J. D.; Meyers, T. R.; Shelton, D. W.; Casteel, J. L.; and Bailey, G. S.: Hepatocarcinogenicity of benzo(*a*)pyrene to rainbow trout by dietary exposure and intraperitoneal injection. *J. Natl. Cancer Inst.*, 74:839–851, 1985.

Hinton, D. E.; Lantz, R. C.; and Hampton, J. A.: Effect of age and exposure to a carcinogen on the structure of the medaka liver: a morphometric study. *Natl. Cancer Inst. Monogr.*, 65:239–249, 1984.

Hinton, D. E.; Hampton, J. A.; and McCuskey, P. A.: Japanese medaka liver tumor model: review of literature and new findings. Water Chlorination: Chemistry, Environmental Impact, and Health Effects, edited by R. Jolley; R. Bull; W. Davis; S. Katz; M. Roberts; and V. Jacobs, Vol. 5, pp. 439–450. Chelsea, MI: Lewis, 1985.

Hinton, D. E.; Couch, J. A.; Teh, S. J.; and Courtney, L. A.: Cytological changes during progression of neoplasia in selected fish species. *Aquat. Toxicol.*, 11:77–112, 1988.

Hinton, D. E.; Baumann, P. C.; Gardner, G.; Hawkins, W. E.; Hendricks, J. D.; Murchelano, R.; and Okihiro, M. S.: Histopathologic biomarkers. Biomarkers: Biochemical, Physiological, and Histological Markers of Anthropogenic Stress, edited by R. J. Huggett; R. A. Kimerle; P. M. Mehrle; and H. L. Bergman, pp. 155–209. Chelsea, MI: Lewis, 1992.

Hoover, K. L. (ed.): Use of small fish species in carcinogenicity testing. *Natl. Cancer Inst. Monogr.* 65:1984.

Ishikawa, T., and Takayama, S.: Importance of hepatic neoplasms in lower vertebrate animals as a tool in cancer research. *J. Toxicol. Environ. Health*, 5:537–550, 1979.

Ishikawa, T.; Masahito, P.; and Takayama, S.: Usefulness of the medaka, *Oryzias latipes* as a test animal: DNA repair processes in medaka exposed to carcinogens. *Natl. Cancer Inst. Monogr.*, 65:35–43, 1984.

James, M. O.: Conjugation of organic pollutants in aquatic species. *Environ. Health Perspect.*, 71:97–104, 1987.

James, M. O.; Hawkins, W. E.; and Walker, W. W.: Acetylaminofluorene (AAF) metabolism in two small fish species, medaka and guppy. *Toxicologist*, 767, 1992.

James, M. O.; Hawkins, W. E.; and Walker, W. W.: Phase 1 and phase 2 biotransformation and carcinogenicity of 2-acetylaminofluorene in medaka and guppy. *Aquat. Toxicol.*, 28:79–95, 1994.

Kaplan, L. A. E.; Schultz, M. E.; Schultz, R. J.; and Crivello, J. F.: Nitrosodiethylamine metabolism in the viviparous fish *Poeciliopsis*: evidence for the existence of liver P450pj activity and expression. *Carcinogenesis* 12:647–652, 1991.

Khudoley, V. V.: Use of aquarium fish, *Danio rerio* and *Poecilia reticulata*, as test species for evaluation of nitrosamine carcinogenicity. *Natl. Cancer Inst. Monogr.*, 65:65–70, 1984.

Kirchen, R. V., and West, W. R.: The Japanese Medaka, Its Care and Development. Burlington, NC: Carolina Biological Supply Company, 1976.

Klauda, R. J.; Peck, T. H.; and Rice, G. K.: Accumulation of polychlorinated biphenyls in Atlantic tomcod (*Microgadus tomcod*) collected from the Hudson River estuary, New York. *Bull. Environ. Contam. Toxicol.*, 27:829–835, 1981.

Klaunig, J. E., and Ruch, R. J.: Role of inhibition of intercellular communication in carcinogenesis. *Lab. Invest.*, 62: 135–146, 1990.

Kleinow, K. M.; Melancon, M. J.; and Lech, J. J.: Biotransformation and induction: implications for toxicity, bioaccumulation and monitoring of environmental xenobiotics in fish. *Environ. Health Perspect.*, 71:105–119, 1987.

Kreamer, G.-L.; Squibb, K.; Gioeli, D.; Garte, S. J.; and Wirgin, I.: Cytochrome P450IA mRNA expression in feral Hudson River tomcod. *Environ. Res.*, 55:64–78, 1991.

Krewski, D.; Goddard, M. J.; and Murdoch, D.: Statistical considerations in the interpretation of negative carcinogenicity data. *Reg. Toxicol. Pharmacol.*, 9:5–22, 1989.

Kurelec, B.; Garg, A.; Krca, S.; Chacko, M.; and Gupta, R. C.: Natural environment surpasses polluted environment in inducing DNA damage in fish. *Carcinogenesis*, 10: 1337–1339, 1989.

Kyono-Hamaguchi, Y.: Effects of temperature and partial hepatectomy on the induction of liver tumors in *Oryzias latipes*. *Natl. Cancer Inst. Monogr.*, 65:337–344, 1984.

Lauren, D. J.; Teh, S. J.; and Hinton, D. E.: Cytotoxicity phase of diethylnitrosamine-induced hepatic neoplasia in medaka. *Cancer Res.*, 50:5504–5514, 1990.

Lech, J. J., and Vodicnik, M. J.: Biotransformation of chemicals by fish: an overview. *Natl. Cancer Inst. Monogr.*, 65: 355–358, 1984.

Lijinsky, W.: Non-genotoxic environmental carcinogens. *Environ. Carcinog. Rev.*, 8:45–87, 1990.

Loveland, P. M.; Wilcox, J. S.; Hendricks, J. D.; and Bailey, G. S.: Comparative metabolism and DNA binding of aflatoxin B_1, aflatoxin M_1, aflatoxicol and aflatoxicol-M_1 in hepatocytes from rainbow trout (*Salmo gairdneri*). *Carcinogenesis*, 9:441–446, 1988.

Malins, D. C., and Ostrander, G. K.: Perspectives in aquatic toxicology. *Annu. Rev. Pharmacol. Toxicol.*, 31:371–399, 1991.

Malins, D. C.; McCain, B. B.; Brown, D. W.; Chan, S. L.; Myers, M. S.; Landahl, J. T.; Prohaska, P. G.; Frideman, A. J.; Rhodes, L. D.; Burrows, D. G.; Gronlund, W. G.; and Hodgins, H. O.: Chemical pollutants in bottom-dwelling fish in Puget Sound, Washington. *Environ. Sci. Technol.*, 18:705–713, 1984a.

Malins, D. C.; Krahn, M. M.; Brown, D. W.; Rhodes, L. D.; Myers, M. S.; McCain, B. B.; and Chan, S. L.: Toxic chemicals in marine sediment and biota from Mukilteo, Washington: relationships with hepatic neoplasms and other hepatic lesions in English sole (*Parophrys vetulus*). *J. Natl. Cancer Inst.*, 74:487–494, 1984b.

Malins, D. C.; Krahn, M. M.; Myers, M. S.; Rhodes, L. D.; Brown, D. W.; Krone, C. A.; McCain, B. B.; and Chan, S. L.: Toxic chemicals in sediments and biota from a creosote-polluted harbor: relationships with hepatic neoplasms and other hepatic lesions in English sole (*Parophrys vetulus*). *Carcinogenesis*, 6:1463–1469, 1985.

Malins, D. C.; McCain, B. B.; Myers, M. S.; Brown, D. W.; Krahn, M. M.; Roubal, W. T.; Schiewe, M. H.; Landahl, J. T.; and Chan, S. L.: Field and laboratory studies of the etiology of liver neoplasms in marine fish from Puget Sound. *Environ. Health Perspect.*, 71:5–16, 1987.

Malins, D. C.; McCain, B. B.; Landahl, J. T.; Myers, M. S.; Krahn, M. M.; Brown, D. W.; Chan, S. L.; and Roubal, W. T.: Neoplastic and other diseases in fish in relation to toxic chemicals: an overview. *Aquat. Toxicol.*, 11:43–67, 1988.

Malins, D. C.; Ostrander, G. K.; Haimanot, R.; and Williams, P.: A novel DNA lesion in neoplastic livers of feral fish: 2,6-diamino-4-hydroxy-5-formamidopyrimidine. *Carcinogenesis*, 11:1045–1047, 1990.

Maronpot, R. R.; Montgomery, C. A., Jr.; Boorman, G. A.; McConnell, E. E.: National toxicology program nomenclature for hepatoproliferative lesions of rats. *Toxicol. Pathol.*, 14:263–273, 1986.

Maronpot, R. R.; Haseman, J. K.; Boorman, G. A.; Eustis, S. E.; Rao, G. N.; and Huff, J. E.: Liver lesions in B6C3F1 mice: the National Toxicology Program, experience and position. *Arch. Toxicol. Suppl.*, 10:10–26, 1987.

Marquardt, H.; Sternbert, S. S.; and Philips, F. S.: 7,12-Dimethylbenz(*a*)anthracene and hepatic neoplasia in regenerating rat liver. *Chem.-Biol. Interact.*, 2:401–403, 1970.

Masahito, P.; Ishikawa, T.; and Sugano, H.: Fish tumors and their importance in cancer research. *Jpn. J. Cancer Res.*, 79:545–555, 1988.

Masahito, P.; Aoki, K.; Egami, N.; Ishikawa, T.; and Sugano, H.: Life-span studies on spontaneous tumor development in the medaka (*Oryzias latipes*). *Jpn. J. Cancer Res.*, 80:1058–1065, 1989.

Matula, T.I., and Somers, E.: The classfication of carcinogens. *Reg. Toxicol. Pharmacol.*, 10:174–182, 1989.

Mawdesley-Thomas, L. E.: Some tumors of fish. *Symp. Zool. Soc. Lond.*, 30:191–284, 1972.

Mawdesley-Thomas, L. E.: Neoplasia in fish. The Pathology of Fishes, edited by W. E. Ribelin, and G. Migaki, pp. 805–870. Madison, WI: Univ. of Wisconsin Press, 1975.

McConnell, E. E.; Solleveld, H. A.; Swenberg, J. A.; and Boorman, G. A.: Guidelines for combining neoplasms for evaluation of rodent carcinogenesis studies. *J. Natl. Cancer Inst.*, 76:283–289, 1986.

McMahon, G.; Huber, L. J.; Stegeman, J. J.; and Wogan, G. N.: Indentification of a c-Ki-*ras* oncogene in a neoplasm isolated from winter flounder. *Mar. Environ. Res.*, 24:345–350, 1988.

Metcalfe, C. D.: Tests for predicting carcinogenicity in fish. *CRC Rev. Aquat. Sci.*, 1:111–129, 1989.

Mix, M. C.: Cancerous diseases in aquatic animals and their association with environmental pollutants: a critical review of the literature. *Mar. Environ. Res.*, 20:1–141, 1986.

Moolgavkar, S. H.: A two-stage carcinogenesis model for risk assessment. *Cell. Biol. Toxicol.*, 5:445–460, 1989.

Moon, T. W.; Walsh, P. J.; and Mommsen, T. P.: Fish hepatocytes: a model metabolic system. *Can. J. Fish. Aquat. Sci.*, 42:1772–1782, 1985.

Murchelano, R. A., and Wolke, R. E.: Epizootic carcinoma in winter flounder, *Pseudopleuronectes americanus*. *Science*, 228:587–589, 1985.

Murchelano, R. A., and Wolke, R. E.: Neoplasms and nonneoplastic lesions in winter flounder, *Pseudopleuronectes americanus*, from Boston Harbor, Massachusetts. *Environ. Health Perspect.*, 90:17–26, 1991.

Myers, M. S.; Rhodes, L. D.; and McCain, B. B.: Pathologic anatomy and patterns of occurrence of hepatic neoplasms, putative preneoplastic lesions, and other idiopathic hepatic conditions in English sole (*Parophrys vetulus*) from Puget Sound, Washington. *J. Natl. Cancer Inst.*, 78:333–363, 1987.

Myers, M. S.; Landahl, J. T.; Krahn, M. M.; Johnson, L. L.; and McCain, B. B.: Overview of studies on liver carcinogenesis in English sole from Puget Sound: evidence for a xenobiotic chemical etiology. I: Pathology and epizootiology. *Sci. Total Environ.*, 94:33–50, 1990.

Myers, M. S.; Landahl, J. T.; Krahn, M. M.; and McCain, B. B.: Relationships between hepatic neoplasms and related lesions and exposure to toxic chemicals in marine fish from the U.S. West Coast. *Environ. Health Perspect.*, 90:7–15, 1991.

Nemoto, N.; Kodama, K.; Tazawa, A.; Masahito, P.; Ishikawa, T.: Extensive sequence homology of the goldfish *ras* gene to mammalian *ras* genes. *Differentiation* 32: 17–23, 1986.

Okihiro, M. S., and Hinton, D. E.: Lymphoma in the Japanese medaka (*Oryzias latipes*). *Dis. Aquat. Org.*, 7:79–87, 1989.

Peck, H. M.: Design of experiments to detect carcinogenic effects of drugs. Carcinogenesis Testing of Chemicals, edited by L. Goldberg, pp. 1–13. Cleveland, OH: CRC Press, 1973.

Peto, R.; Pike, M. C.; Berstein, L.; Gold, L. S.; and Ames, B. N.: The TD50: a proposed general convention for the numerical description of the carcinogenic potency of chemicals in chronic-exposure animal experiments. *Environ. Health Perspect.*, 58:1–8, 1984.

Peraino, C., and Jones, C. A.: The multistage concept of carcinogenesis. The Pathobiology of Neoplasia, edited by A. E. Sirica, pp. 131–148, New York: Plenum, 1989.

Pitot, H. C.: Progression: the terminal stage in carcinogenesis. *Jpn. J. Cancer Res.*, 80:599–607, 1989.

Pitot, H. C.; Dragan, Y. P.; Neveu, M. J.; Rizvi, T. A.; Hully, J. R.; and Campbell, H. A.: Chemicals, cell proliferation, risk estimation, and multistage carcinogenesis. Chemically Induced Cell Proliferation: Implications for Risk Assessment, edited by B. E. Butterworth; T. J. Slaga; W. Farland; and M. McClain, pp. 517–532. New York: Wiley-Liss, 1991.

Prescott, D. M., and Flexer, A. S.: Cancer: The Misguided Cell, 2nd ed. Sunderland, MA: Sinauer Associates, 1986.

Powers, D. A.: Fish as model systems. *Science*, 246:352–358, 1989.

Rinde, E.; Chiu, A.; Hill, R.; and Haberman, B.: Proliferative hepatocellular lesions of the rat: review and future use in risk assessment. EPA/625/3-86/011, 1986.

Robbins, S. L.; Cotran, R. S.; and Kumar, V.: Pathologic Basis of Disease, 3rd ed. Philadelphia: Saunders, 1984.

Robens, J. F.; Piegorsch, W. W.; and Schueler, R. L.: Methods of testing for carcinogenicity. Principles and Methods of Toxicology, edited by A. W. Hayes, pp. 251–273. New York: Raven Press, 1989.

Sato, S.; Matsushima, T.; Tanaka, N.; Sugimura, T.; Takashima, F.: Hepatic tumors in the guppy (*Lebistes reticulatus*) induced by aflatoxin B_1, dimethylnitrosamine, and 2-acetylaminofluorene. *J. Natl. Cancer Inst.*, 50: 765–778, 1973.

Schimmel, S. C.; Hansen, D. J.; and Forrester, J.: Effects of Aroclor 1254 on laboratory reared embryos and fry of the sheepshead minnow (*Cyprinodon variegatus*). *Trans. Am. Fish. Soc.*, 103:582–586, 1974.

Schultz, M. E., and Schultz, R. J.: Induction of hepatic tumors with 7,12-dimethylbenz[*a*]anthracene in two species of viviparous fish (genus *Poeciliopsis*). *Environ. Res.*, 27:337–351, 1982a.

Schultz, M. E., and Schultz, R. J.: Diethylnitrosamine-induced hepatic tumors in wild vs. inbred strains of a viviparous fish. *J. Hered.* 73:43–48, 1982b.

Schultz, M. E., and Schultz, R. J.: Transplantable chemically-induced liver tumors in the viviparous fish *Poeciliopsis*. *Exp. Mol. Pathol.*, 42:320–330, 1985.

Schultz, M. E., and Schultz, R. J.: Differences in response to a chemical carcinogen within species and clones of the livebearing fish, *Poeciliopsis*. *Carcinogenesis*, 9: 1029–1032, 1988.

Schultz, M. E.; Kaplan, L. A. E.; and Schultz, R. J.: Initiation of cell proliferation in livers of the viviparous fish *Poeciliopsis lucida* with 7,12-dimethylbenz(*a*)anthracene. *Environ. Res.*, 48:248–254, 1989.

Schultz, R. J., and Schultz, M. E.: Characteristics of a fish colony of *Poeciliopsis* and its use in carcinogenicity studies with 7,12-dimethylbenz[*a*]anthracene and diethylnitrosamine. *Natl. Cancer Inst. Monogr.*, 65:5–13, 1984.

Simon, K., and Lapis, K.: Carcinogenesis studies on guppies. *Natl. Cancer Inst. Monogr.*, 64:71–81, 1984.

Sinnhuber, R. O.; Hendricks, J. D.; Wales, J. H.; and Putnam, G. B.: Neoplasm in rainbow trout, a sensitive animal model for environmental carcinogenesis. *Ann. N. Y. Acad. Sci.*, 298:389–408, 1977.

Sirica, A. E. (ed.): *The Pathobiology of Neoplasia*. New York: Plenum, 1989a.

Sirica, A. E.: Classification of neoplasms. The Pathobiology of Neoplasia, edited by A. E. Sirica, pp. 25–38. New York: Plenum, 1989b.

Smith, C. E.; Peck, T. H.; Klauda, R. J.; and McLaren, J. B.: Hepatomas in Atlantic tomcod *Microgadus tomcod* (Walbaum) collected in the Hudson River estuary in New York. *J. Fish Dis.*, 2:313–319, 1979.

Sonstegard, R. A.: Environmental carcinogenesis studies in fishes of the Great Lakes of North America. *Ann. N. Y. Acad. Sci.*, 298:261–269, 1977.

Stalker, M. J.; Kirby, G. M.; Kocal, T. E.; Smith, I. R.; and Hayes, M. A.: Loss of glutathione *s*-transferases in pollution-associated liver neoplasms in white suckers (*Catastomus commersoni*) from Lake Ontario. *Carcinogenesis*, 12:2221–2226, 1991.

Stegeman, J. J., and Lech, J. J.: Cytochrome P-450 monooxygenase systems in aquatic species: carcinogen metabolism and biomarkers for carcinogen and pollutant exposure. *Environ. Health Perspect.*, 90:101–109, 1991.

Stegeman, J. J.; Woodin, B. R.; and Smolowitz, R. M.: Structure, function and regulation of cytochrome P450 forms in fish. *Biochem. Soc. Trans.*, 18:19–21, 1990.

Stein, J. E.; Reichert, W. I.; Nishimoto, M.; and Varanasi, U.: Overview of studies on liver carcinogenesis in English sole from Puget Sound: evidence for xenobiotic chemical etiology. II: Biochemical studies. *Sci. Total Environ.*, 94:51–69, 1990.

Swenberg, J. A.; Hoel, D. G.; and Magee, P. N.: Mechanistic and statistical insight into the large carcinogenesis bioassays on *N*-nitrosodiethylamine and *N*-nitrosodimethylamine. *Cancer Res.*, 51:6409–6414, 1991.

Tennant, R. W.; Margolin, B. M.; Shelley, M. D.; Zieger, E.; Haseman, J. K.; Spalding, J.; Caspar, W.; Resnick, M.; Stasiewicz, S.; Anderson, B.; and Minor, R.: Production of chemical carcinogenicity in rodents from in vitro genetic toxicity assays. *Science*, 236:933–941, 1987.

Trosko, J. E., and Chang, C. C.: Role of intercellular communication in tumor promotion. Mechanisms of Tumor Promotion, Vol. IV, edited by T. J. Slaga, pp. 119–145. Boca Raton, FL: CRC Press, 1984.

Vaillant, C.; LeGuellec, C.; and Podkel, F.: Vitellogenin gene expression in primary culture of rainbow trout hepatocytes. *Gen. Comp. Endocrinol.*, 70:284–290, 1988.

Van Beneden, R. J.; Watson, D. K.; Chen, T. T.; Lautenberger, J. A.; and Papas, T. S.: Cellular *myc* (c-*myc*) in fish (rainbow trout): its relationship to other vertebrate *myc* genes and to the transforming genes of the MC29 family of viruses. *Proc. Natl. Acad. Sci. USA*, 83:3698–3702, 1986.

Van Beneden, R. J.; Watson, D. K.; Chen, T. T.; Lautenberger, J. A.; and Papas, T. S.: Teleost oncogenes: evolutionary comparison to other vertebrate oncogenes and possible roles in teleost neoplasms. *Mar. Environ. Res.*, 24:339–343, 1988.

Van Beneden, R. J.; Henderson, K. W.; Blair, D. G.; Papas, T. S.; and Gardner, H. S.: Oncogenes in hematopoietic and hepatic fish neoplasms. *Cancer Res.*, [Suppl]50: 5671s–5674s, 1990.

Varanasi, U.; Stein, J. E.; and Nishimoto, M.: Biotransformation and disposition of PAH in fish. Metabolism of Polycyclic Aromatic Hydrocarbons in the Aquatic Environment, edited by U. Varanasi, pp. 93–150. Boca Raton, FL: CRC Press, 1989a.

Varanasi, U.; Reichert, W. L.; and Stein, J. E.: ^{32}P-postlabeling analysis of DNA adducts in liver of wild English sole (*Parophyrs vetulus*) and winter flounder (*Pseudopleuronectes americanus*). *Cancer Res.*, 49: 1171–1177, 1989b.

Vogelbein, W. K.; Fournie, J. W.; Van Veld, P. A.; and Huggett, R. J.: Hepatic neoplasms in the mummichog *Fundulus heteroclitus* from a creosote-contaminated site. *Cancer Res.*, 50:5978–5986, 1990.

Walker, W. W.; Manning, C. S.; Overstreet, R. M.; and Hawkins, W. E.: Development of aquarium fish models for environmental carcinogenesis: an intermittent-flow exposure system for volatile, hydrophobic chemicals. *J. Appl. Toxicol.*, 5:255–260, 1985.

Walker, W. W.; Hawkins, W. E.; Overstreet, R. M.; and Friedman, M. A.: A small fish model for assessing cancer risk at low carcinogen concentrations. *Toxicologist*, 302, 1992.

Weisburger, E. K.: Chemical carcinogenesis in experimental animals and humans. The Pathobiology of Neoplasia, edited by A. E. Sirica, pp. 39–56. New York, Plenum, 1989.

Weisburger, E. K.: Mechanistic considerations in chemical carcinogenesis. *Reg. Toxicol. Pharmacol.*, 12:41–52, 1990.

Weisburger, J. H.: Inclusion of positive control compounds. Carciongenesis Testing of Chemical, edited by L. Goldberg, pp. 29–34. Cleveland, OH: CRC Press, 1973.

Weisburger, J. H., and Williams, G. M.: Carcinogen testing: current problems and new approaches. *Science*, 214, 401–407, 1981.

Weisburger, J. H., and Williams, G. M.: Critical methods to detect genotoxic carcinogens and neoplasm-promoting agents. *Environ. Health Perspect.*, 90:121–126, 1991.

Williams, G. M.: The pathogenesis of rat liver cancer caused by chemical carcinogens. *Biochim. Biophys. Acta.*, 605:167–189, 1980.

Williams, G. M.: Modulation of chemical carcinogenesis by xenobiotics. *Fundam. Appl. Toxicol.*, 4:325–344, 1984.

Williams, G. M.: Epigenetic mechanisms of liver tumor promotion. Mouse Liver Carcinogenesis: Mechanisms and Species Comparisons, edited by D. E. Stevenson; R. M. McClain; J. A. Popp; T. J. Slaga; J. M. Ward; and H. C. Pitot, pp. 131–145. New York: Alan R. Liss, 1990.

Wirgin, I.; Currie, D.; and Garte, S. J.: Activation of the K-*ras* oncogene in liver tumors of Hudson River tomcod. *Carcinogenesis* 10:2311–2315, 1989.

Yamamoto, T.-O.: Medaka (Killifish): Series of Stock Culture in Biological Field. Tokyo: Keigaku Publishing Company, 1975.

Zedeck, M. S.: Hydrazine derivatives, azo and azoxy compounds, and methylazoxymethanol and cycasin. Chemical Carcinogens, 2nd ed., edited by C. E. Searle, Vol. 2, pp. 915–944. Washington, D.C.: American Chemical Society, ACS Monograph 182, 1984.

SUPPLEMENTAL READING

Couch, J. A., and Fournie, J. W. (eds.): *Pathobiology of Marine and Estuarine Organisms*. Boca Raton, FL: CRC Press, 1993.

Moore, M. M., and Myers, M. S.: Pathobiology of chemical-associated neoplasia in fish. Aquatic Toxicology-Molecular, Biochemical, and Cellular Perspectives edited by D. C. Malins, and G. Ostrander, pp. 327–386. Boca Raton, FL: Lewis, 1994.

PART II

ENVIRONMENTAL FATE

The second part of the book begins with Chapter 15, on phase transfer/transport and transformation of chemicals in natural surface waters by abiotic processes, with a small section on microorganisms. Bioavailability of chemicals and chemical accumulation following absorption (uptake) and distribution in aquatic organisms are described in Chapter 16. Mechanisms for biotransformation (metabolism) of foreign chemicals and biochemical changes that may occur in response to chemical exposure are reviewed in Chapter 17. Various biochemical end points (e.g., changes in enzymes and other proteins) in Chapter 17 may also be used as biomarkers of exposure and/or effects of a chemical(s). Chapter 18 is a summary of the fundamentals, applications, and use of mathematical fate modeling.

Chapter 15

TRANSPORT AND TRANSFORMATION PROCESSES

W. J. Lyman

INTRODUCTION

If a chemical is introduced into the environment, there is a certain probability that it may move from the point where it is released and eventually be distributed over a broad geographic area as the original parent substance including different degradates (or metabolites). The different environmental pathways are summarized in Figure 1. A chemical can move between and within compartments (i.e., phase transfer and transport) and may ultimately change in form (transformation). For example, a chemical introduced into the aquatic environment can move with the water whether it is dissolved in solution or adsorbed on a particle. Movement would be defined by the appropriate hydrological parameters and transformation by the properties of the chemical. The chemical may also go directly or indirectly (i.e., from the aquatic environment) to the atmosphere and be transported by winds, so that meteorological processes will determine the rate and direction of movement. The chemical may also move back to the aquatic environment from the atmosphere or into the soil, where movement is accomplished primarily by a diffusion or mass transport process. In the atmosphere and in the aquatic environment, movement of the chemical is more a function of the characteristic transport processes and the effect of the properties of the chemical being transported is minimal. For soil it is completely opposite. In addition, the tendency for a chemical to move between compartments is more dependent on the properties of the chemical.

Studies of phase transfer/transport and transformation are used to determine the environmental fate of the chemical and thus yield the following information:

- Environment compartment(s) (i.e., soil, water, air matrices) in which the chemical will be distributed
- Form or chemical species (i.e., parent and/or metabolite) and concentrations in which the chemical will be distributed
- Residence time of the chemical
- Whether the chemical will be accumulated or exported from the compartment
- Major and minor routes and rates of transport and transformation
- Data (e.g., quantitative measurement of rates) for mathematical fate models (see Chapter 18)

The fate of a chemical in the environment is ultimately determined by its molecular structure and consequent reactivity in a variety of chemical, physical, and biologically mediated processes and the rate of its physical transport. Fate is a key component in exposure assessment (see Chapter 28) for organisms in the environment. All studies and assessments in aquatic toxicology should thus address the environmental fate or phase transfer/transport and transformation processes important for the toxicants of concern. This is especially important for studies that attempt to elucidate some basic toxicological response that may be applied and extrapolated to a variety of environmental situations. To do this requires an understanding of the environmental chemistry of the chemical. Properties of the chemical that are important in evaluating its environmental chemistry are listed in Table 1. Properties of the aquatic environment that may be important to evaluate are listed in Table 2.

Failure to address the appropriate environmental chemistry variables could lead to a misinterpretation of your own data or to significant limitations on the utility of your data to other scientists and/or regulators. The different U.S. and international regulations (see Chapters 21–25 and 27) require environmental chemistry testing for product development and safety assessment. Despite intensive research today, we have only a cursory knowledge of how chemicals move between the atmosphere, soil/sediment, and water and what transformations they undergo during transport. However, scientists have made progress in predicting rates of reactions (e.g., chemical hydrolysis, photolysis, volatilization, sorption, bioconcentration) and partitioning between

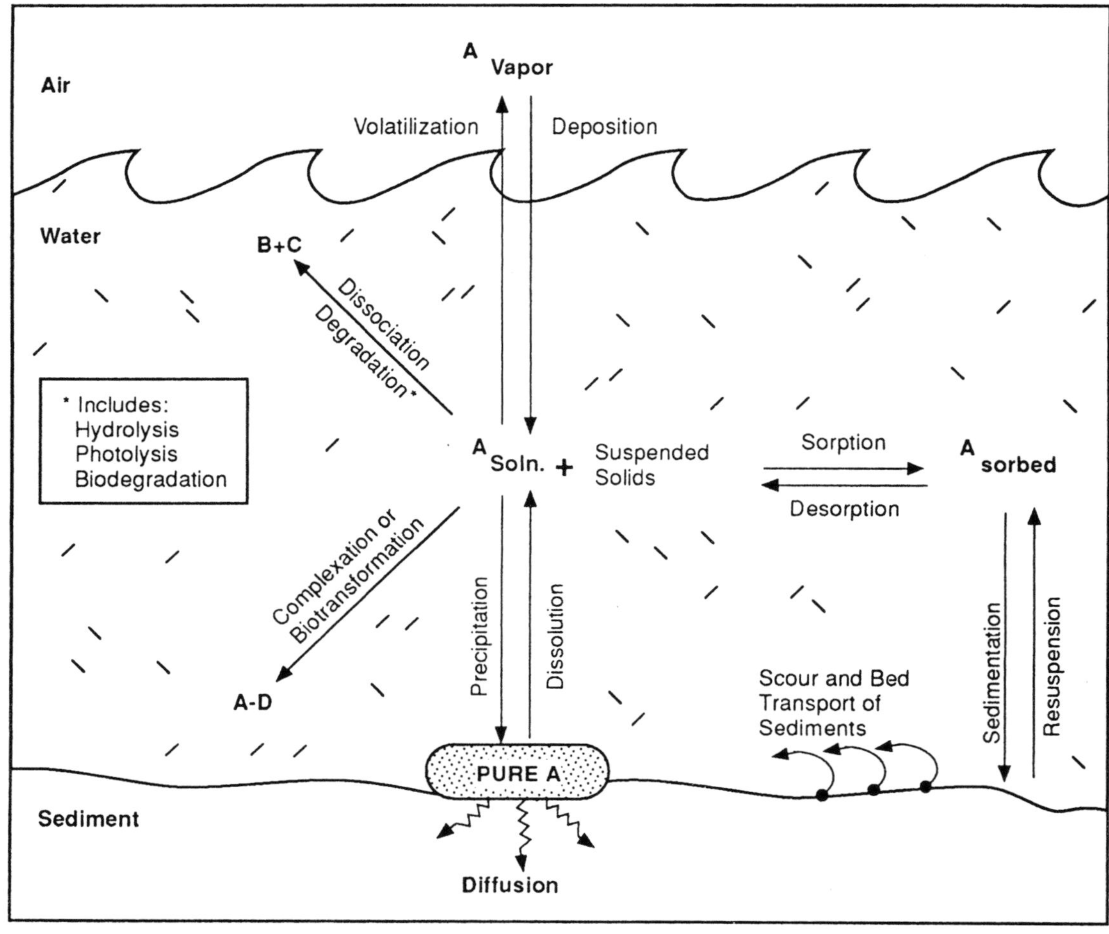

Note: Symbols B and C represent degradation products of chemical A, and symbol D represents a ligand that complexes with (or adds to) A.

Figure 1. Schematic representation of some transport and transformation phenomena important for a chemical in an aquatic environment.

compartments but little progress on biological transformation (biotransformation or biodegradation; see discussion later and Chapters 17 and 31).

The coverage of transport and transformation processes in toxicity tests will, in many cases, have to go well beyond simple measurement of the water quality parameters that control many of them, such as temperature, pH, redox potential, suspended solids, dissolved organic matter, ionic strength, and hardness. It will require additional work—which will frequently involve fairly simple tests and calculations—to consider more specifically the influence of specific processes such as:

- Partitioning between phases (e.g., sorbed to suspended sediment vs. dissolved in water)
- Losses to other environmental compartments (e.g., volatilization to air or sedimentation to bottom sediments)
- Chemical reactions (e.g., hydrolysis, oxidation, photolysis) or biological degradation (i.e., biodegradation)
- Speciation (e.g., dissociation, complexation, and polymerization to form new chemical species).

Figure 1 provided an overview of these processes and their interconnection within an aquatic system.

It is the purpose of this chapter to describe briefly most of the environmental transport and transformation processes that are important in aquatic toxicology. The coverage is subdivided first based on the major chemical classes:

1. Organic chemicals
2. Inorganic chemicals
3. Organometallic chemicals

The coverage of organometallic chemicals herein is limited for several reasons: (1) there are only a

Table 1. Properties of a chemical important in environmental chemistry

Property
Molecular structure
Molecular weight
Water solubility and precipitation[a]
Vapor pressure
Henry's law constant
Octanol-water partition coefficient[b]
Sorption constants for soils, sediments, or atmospheric particles[a]
Acid or base dissociation constants[a]
Activity coefficient[c]
Complexation constants[a]
Electron transfer (redox) constants[a,c]
Polymerization constants[a]
Diffusion coefficients
Light absorption spectrum and quantums yield
Bioconcentration factor[a,b]
Biodegradation or biotransformation constants[b]
Hydrolysis constants[b]
Particle size (for solids)

[a] Both equilibrium and rate constants are important.
[b] Primarily used for organic chemicals.
[c] Primarily used for inorganic chemicals.

few such chemicals of current interest; (2) relatively little is known about their environmental behavior; (3) many organometallics can be crudely classified (and evaluated) as either organic or inorganic in nature and, thus, evaluated on the basis of the process described for those two major classes. This classification scheme is outlined later.

Because of the breadth of coverage in this chapter, the depth of coverage for each process is necessarily limited. Each subsection does briefly define the process and its potential importance to toxicological studies and describes the key parameters (e.g., equilibrium or rate constant) and methods used for quantitative evaluations. References to supplemental reading material and data are also provided.

Table 2. Properties of the aquatic environment important in predicting the fate and transport of a chemical

Property
Physical properties
Surface area
Depth
Flow, extent of mixing, and bottom scouring
Sedimentation rate
Solar irradiation (at surface) and irradiance as function of wavelength and water body depth
Chemical properties
Temperature
pH
Eh[a] (for several redox couples, including oxygen)
Suspended solids (nature and concentration)
Hardness, salinity, ionic strength
Concentration of major ions
Concentration of dissolved organic matter
Bottom sediments (nature, including organic carbon content and redox status)
Biological properties
Microbiological populations and activity
Trophic status
Nutrient concentrations

[a] Eh = value which represents potential of redox reactions.

TERMINOLOGY

The terminology used in this chapter is defined in the following under two subheadings: (1) processes and properties and (2) chemical terminology.

Processes and Properties

Absorption. A process in which a solute becomes physically associated with a porous sorbent by a process equivalent to the sorbate's dissolution into the sorbent. (See also adsorption and sorption.)

Activity. (Used in this chapter only for inorganic chemicals.) An altered measure of a chemical's concentration in water obtained from the product of an activity coefficient (values less than 1.0 representing deviation from ideal behavior) and the actual chemical concentration.

Adsorption. A process in which a solute becomes physically associated with a solid sorbent by a process equivalent to the sorbate's condensation onto the surface(s) of the sorbent.

Advection. A process that transports a chemical from one location to another by virtue of the fact that the chemical is a component of a moving physical system (e.g., wind, flowing river, sediment transport).

Atmospheric deposition. A process that transfers a chemical from the atmosphere to the earth's surface (land or water) by either dry impingement or transport in rain or snow (wet deposition).

Bulk transport. Same as advection.

Biodegradation. A microbiologically mediated process (e.g., due to the action of bacteria, yeasts, and fungi) that chemically alters the structure of a chemical, the common result being the breakup of the chemical into smaller components (ultimately CO_2 and H_2O for aerobic biodegradation of hydrocarbons).

Biotransformation. A microbiologically mediated process that chemically alters the structure of a chemical, but does not necessarily degrade it (e.g., methylation of metals). This process can also refer to a biologically mediated process that is determined by fish or invertebrates (see Chapter 17).

Chemisorption. A process in which a solute becomes chemically associated with a solid sorbent (e.g., via ion exchange or more permanent chemical bonding).

Complexation. A process in which two or more solutes (e.g., a metal ion and various ligands) join chemically to form a new chemical complex.

Diffusion. The movement of a chemical in almost any form (gas, liquid, solute) through a host medium (e.g., air, water, soil) solely as a result of the chemical's concentration gradient. Movement is from regions of high to regions of low concentration.

Dispersion. The mixing of a chemical in a host fluid medium (e.g., air, water) as a result of the movement and turbulence in the medium. The mixing results in dilution of the chemical's concentration.

Dissociation. The breaking apart of two or more components of a chemical (typically in solution) as a result of some chemical reaction (e.g., acid-base, complex dissociation, photolysis).

Equilibrium constant. A constant representing the ratio of concentrations of the chemicals on two sides of a reversible chemical reaction when the rates of the forward and reverse reactions are equal. Such equilibrium conditions are primarily controlled by chemical thermodynamics.

Half-life. The time required for 50% of a chemical to degrade or to be removed from the local environment by some physical process. Typically associated with so-called first-order reactions and processes.

Hydrolysis. A chemical reaction in which a solute reacts with water to yield a new chemical, typically incorporating one or more OH groups. Reactions involving the photochemically produced OH radical (both in the atmosphere and in water) may also be considered a form of hydrolysis.

Mass transfer coefficient. A parameter that is usually a constant of proportionality between a chemical driving force (e.g., a concentration gradient) and the mass movement or flux of the chemical that results from that driving force.

Oxidation. A chemical reaction between two chemicals (typically solutes) in which one component, the reducing agent (reductant), loses electrons and is thereby "oxidized." The second solute, the oxidizing agent (oxidant), gains electrons and is thereby "reduced." The electron transfer occurs by direct interaction of the two reactants. The reaction may be microbiologically mediated.

Polymerization. A chemical reaction in which two or more identical molecules or ions are chemically joined to form a larger molecule or complex.

Precipitation. A chemical reaction that results in a solute, having exceeded its solubility limit, forming a solid phase composed of the chemical. The precipitation may be caused by an increase in the solute's concentration, evaporation of the solvent, chemical reaction, change in salinity, or change in temperature.

Photolysis. A chemical reaction, initiated directly or indirectly by sunlight, that typically results in the breakup of the initial molecule. In aqueous systems, oxygen or other naturally occurring elements or compounds may become incorporated into the parent chemical or resulting fragments. When the reaction follows the absorption of sunlight energy (photons) by the degrading chemical, it is called direct photolysis. When an intermediary absorbs the sunlight and then transfers it to the degrading chemical, it is called indirect photolysis.

Partition coefficient. A parameter defining the ratio of a chemical's concentration in two phases that are in contact and between which the chemical has been allowed to partition, although not necessarily to equilibrium conditions (see Chapter 16).

Reduction. See Oxidation.

Sedimentation. A process in which suspended solids in a surface water are deposited on the bottom of the water body. The process will also involve transport of chemicals to the bottom if they are sorbed to the suspended sediments.

Sorption. A process by which a solute becomes physically or chemically associated with a solid sorbent regardless of the mechanism (e.g., chemisorption, adsorption, absorption, see Chapter 16).

Speciation. A term referring to the existence or formation of a variety of chemical complexes generally related to a central element of interest.

Stability constant. A form of equilibrium constant used to represent acid-base or complexation reactions in which multiple ligand addition occurs on a central ion. The stability constants can reflect various steps in the ligand addition or the formation of a multiligand complex relative to the uncomplexed central ion (overall stability constant).

Volatilization. The transport of a solute from the aqueous phase (surface water) to the gas phase (atmosphere) as a result of natural conditions including the solute's physicochemical properties and relative gas- and liquid-phase concentrations.

Chemical Terminology

Complex. A chemical unit (specie) comprising two or more components (ions or neutral solutes) that have been chemically bound. Often comprises a central ion and one or more ligands attached to it.

Chelate. A water-soluble chemical, usually organic, containing two or more functional groups (e.g., carboxylic acid) that can act as ligands and bind to a ligand position on a central ion.

Element. A unique chemical defined only by the number of protons in its atomic nucleus (e.g., iron, nickel, hydrogen, chlorine).

Heavy metal. Metals with atomic numbers between 21 (scandium) and 92 (uranium).

Inorganic chemical. Any element or chemical not based on carbon as the main structural element.

Ligand. A solute, typically an ion, that binds with a central ion to form a complex.

Metal. A metal is typically distinguished from a nonmetal by its lustre, malleability, conductivity (of heat and electricity), and usual ability to form positive ions.

Metalloid. An element midway between metals and nonmetals with regard to its physical and chemical properties (e.g., arsenic).

Organic chemical. Any chemical based on carbon as the main structural element.

Solute. Any chemical dissolved in a liquid phase (i.e., water for this chapter).

Sorbate. Any chemical (typically a solute) that may be, or has become, sorbed to a solid sorbent.

Sorbent. Any solid or semisolid material to which solutes may adhere via a variety of chemical or physical processes. Includes suspended and bottom sediments, soils, sludges, minerals, and so forth.

Specie(s). A particular chemical, usually in solution, that represents one form in which a particular element may exist. Species include free elements (in neutral or ionized form) as well as various complexes they may form with different ligands.

ORGANIC CHEMICALS

Introduction

An organic chemical at low concentrations in natural waters (i.e., below its solubility limit) can exist in either or both of two forms—a dissolved phase and a sorbed phase. The most important transport and transformation processes acting on organic chemicals in the aquatic environment include the following:

Phase Transfer and Transport	Transformation
Phase Transfer	Abiotic processes
Dissolution	Hydrolysis
Sorption to solids	Photolysis
Volatilization to air	Dissociation (acid/base)
Atmospheric deposition	Oxidation/reduction
Transport	Biotic
Bulk transport and dispersion	Aerobic biodegradation
Sedimentation	Anaerobic biodegradation
Diffusion	
Entrainment (with sediment) from bottom deposits (scour)	

Most of these processes are sensitive to site-specific conditions such as temperature, pH, suspended solids concentration, and microbiological population (and activity). The following discussions generally indicate their influence on the different processes of interest.

Discussions of the behavior of organic chemicals in water usually need to be qualified with regard to their applicability to neutral, nonionizing molecules versus charged, ionizing molecules. This is due to the significant differences in the aquatic behavior of neutral and ionized molecules. Unless otherwise stated, the following discussions generally presume neutral, nonionizing molecules. The discussion of dissociation by organic acids and bases provides a basis for evaluating when the presence of charged or ionizing molecules must be considered.

Phase Transfer and Transport

Phase Transfer

One of the simplest phenomena that should always be considered for organic chemicals in aquatic systems is movement between the different phases that may be present. This section covers the following phase transfer processes:

- Dissolution (liquid or solid organic to aqueous solution)
- Sorption (aqueous solution to sorption on, or in, a solid)
- Volatilization [aqueous solution to atmosphere (vapor phase)]
- Atmospheric deposition [vapor phase (or sorbed) from atmosphere to earth's surface]

Dissolution

The process of dissolution—for organics in water—is probably important only when bulk quantities of the chemical are released into the water, such as from a spill or improper waste disposal. In such cases, the organic material may initially reside in one of three locations:

- On the water surface (e.g., gasoline, fuel oil and other lighter-than-water materials)
- In the water column, dispersed in small drops or colloid-sized particles
- On the bottom of the water body and/or interspersed in the bottom sediments

The rate of dissolution depends on a number of factors including:

- The local water/chemical volume ratio
- The degree of mixing (turbulence) in the water
- The temperature and, to a lesser extent, other water quality parameters (e.g., ionic strength and dissolved organic matter content)
- The solubility of the chemical in water or—for organic mixtures—the pattern coefficient of the constituent of concern between the organic and aqueous phases (Herbes et al, 1983; Southworth et al., 1983).
- The presence of other organic chemicals, primarily very soluble ones, that might act (as cosolvents) to enhance the solubility of other organic chemicals of lesser solubility.

Because, in open surface water systems, the volume ratio of water to organic chemical is typically very large, dissolution is seldom restricted by the solubility limit of the organic. Nevertheless, data on the solubility limit of organic chemicals under study should be obtained because the values will provide some insight into the anticipated rate of dissolution and the upper limit of aqueous concentrations that should be found. (It is not uncommon to have laboratory results indicating an aqueous solute concentration above a reported solubility limit; this could be due, for example, to the presence of fine particulates with high concentrations of the chemical sorbed to them. This should always trigger a reevaluation of one or both data.)

No organic chemical, except for some polymers, is completely insoluble in water. At the low end, solubilities below 1 ppb have been measured (e.g., 0.026 ppb for benzo[*g,h,i*]perylene). The solubilities of most common organic chemicals are in the range of 1 to 100,000 ppm at ambient temperatures, but several are higher and some compounds (e.g., ethyl alcohol) are infinitely soluble, that is, miscible with water in all proportions. An overall range of at least nine orders of magnitude is thus involved for organic chemicals (Lyman et al., 1981).

Compilations of water solubility data that are both comprehensive and reliable are almost nonexistent. Compilations that may be helpful include those provided by Mackay et al. (1991a), Howard (1989–1991), Keith and Walters (1991), and Verschueren (1993). A variety of empirical estimation methods for water solubility are also available, some based on property-property correlations (e.g., solubility and octanol-water partition coefficient) and others based on structure alone (Lyman et al., 1981). Between these available compilations and estimation methods, it is usually possible to obtain reasonably accurate data on water solubilities for neutral organic molecules.

The best predictions of water solubility (S) are generally achieved using correlations with the octanol-water partition coefficient (K_{OW}). Three examples are provided below.

For liquids and solids (modified from Yalkowsky, 1993):

$$\log S = -1.165 \log K_{OW} + 0.288 - 0.0095(t_m - 25), \quad n = 41,\ r^2 = 0.97,\ s = 0.386 \qquad (1)$$

For liquids (Muller and Klein, 1992):

$$\log S = -1.16 \log K_{OW} + 0.79, \quad n = 156,\ r^2 = 0.95,\ s = 0.298 \qquad (2)$$

For liquids (Mailhot and Peters, 1988):

$$\log S = -1.45 \log K_{OW} + 1.36, \quad n = 258,\ r^2 = 0.91 \qquad (3)$$

where S = water solubility (at about 25°C) (mol/L)
K_{OW} = octanol water partition coefficient
t_m = melting point (°C)
n = number of chemicals in test set used to derive regression equation
r^2 = square of correlation coefficient
s = standard deviation of correlation equation

The solubility data and estimation methods mentioned above are for pure water. In salt water, solubilities are reduced in proportion to the ionic strength of the solution (I). Hashimoto et al. (1984) give the following relationship for estimating solubility in sodium chloride solutions, which have been shown to approximate natural salt waters:

$$\log S_s = (0.0298I + 1)\log S_0 - 0.114I \qquad (4)$$

where S_s = solubility (mol/L) in salt water (NaCl solution) at 20°C
S_0 = solubility (mol/L) in pure water (at ~20°C)
I = ionic strength (mol/L) = $\frac{1}{2}(\Sigma\, C_i z_i^2)$
C_i = concentration of ion i in salt water (mol/L)
z_i = valence of ion i in salt water

At a typical seawater salinity of 35 ppt, I = 0.599, and Eq. (4) reduces to

$$\log S_s = 1.018 \log S_0 - 0.0683 \qquad (5)$$

In this case, for example, anthracene's solubility in pure water, 1.8×10^{-7} mol/L, is reduced to 1.2×10^{-7} mol/L in salt water.

It is often useful to be able to estimate the solubility limit of a constituent of an organic mixture (e.g., benzene in gasoline) in water. For common petroleum fuels, most studies have indicated that it is acceptable to assume ideal solution behavior in the organic phase (i.e., all activity coefficients = 1.0) (Mackay et al., 1991b; Cline et al., 1991; Groves, 1988). If one also assumes a reasonably large gasoline/water volume ratio, then the following relationship between a solute's concentration in the fuel and in water (in contact with the fuel) can be derived:

$$C_w = X_g S_w \qquad (6)$$

where C_w = concentration of solute in aqueous phase (mg/L)
X_g = concentration of solute in hydrocarbon fuel (mole fraction)
S_w = solubility limit of solute in pure water (mg/L)

Dilution of such a system, via increase of the water/fuel volume ratio, lowers all aqueous phase concentrations. However, sequential contact of the fuel—for example, water passing over a fuel spill with equilibrium partitioning present at all times—can lead to the temporary increase in solute concentrations for some solutes (Mackay et al., 1991b). Several studies have concluded that the presence of cosolutes with fairly high solubility (e.g., the oxygenated gasoline additive methyl-*t*-butyl ether, MTBE) does not significantly affect the water-fuel partitioning of other solutes such as benzene (Cline et al., 1991; Groves, 1988; Pinal et al., 1991).

Sorption

The process that involves the transfer of a chemical between a dissolved state and one in which it is attached to a solid (e.g., soil, sediment, biota)—the sorbed state—is called sorption. In some instances, the term "sorption" may refer only to the forward reaction [see Eq. (7)] in which a chemical becomes bound to a solid and the term "desorption" refers to the reverse reaction. In other cases it is clear that a discussion of sorption covers both reactions. Sorption of organic solutes onto suspended sediments or settled solids (bed sediments) in a surface water is, perhaps, the most important transfer process affecting the bioavailability and toxicity of the chemical. The transfer of a chemical, A, from the liquid to the solid phase may be expressed as

$$A_{(soln)} + \text{solid} \rightleftharpoons \text{A-solid} \qquad (7)$$

An understanding of sorption requires that the interaction of a solute (chemical that sorbs or the sorbate) with the surface of a solid (sorbent, or sorbing surface) be characterized in terms of the physical and chemical properties of the solute, solvent (water), and sorbent (Sigg, 1987; Whitfield and Turner, 1987). Chemical reactions of primary importance are those of the solute with the surface and those of the solute with the solvent. The interaction of the solvent with the surface is of less importance in understanding aquatic surface chemistry under normal circumstances when water is the only solvent. In the environment, water is usually the solvent that sorbs onto the solid surface, thus typically saturating it. The variation in the sorption process due to this factor is not considered unless the solvent is variable. The interactions and reactions that might be considered important in aquatic surface chemistry are illustrated in Table 3 (some are relevant for inorganics; see discussion later). Hydrogen bonding is weaker than the specific chemical reactions cited but it is generally important in the sorption of many polar organic molecules to natural organic and oxidic surfaces. Chemical interactions are described in terms of the surface and the solute; similar interactions occur between solvent and solutes and between different solutes. The distribution of species in solution that results from these reactions is referred to as chemical speciation. Solution speciation is determined independently of sorption reactions, and the latter reactions are interpreted in terms of the particular species in solution. One exception to the dominance of the usual solute-solvent and solute-surface interactions in sorption is the case of the hydrophobic effect. This effect can be viewed as the result of the attraction not of the solute to the surface or the solvent, but of the solvent for itself, which restricts the entry of the hydrophobic solute into the aqueous phase (Tanford, 1980).

Sorption reactions generally influence three processes in the aquatic environment: (1) the distribution between the aqueous phase and particulate matter, which affects the chemical's transport through water; (2) the electrostatic properties of suspended particles, which affect their aggregation and transport; and (3) reactivity at surfaces, including dissolution, precipitation, and surface-catalyzed reactions of solutes (Westall, 1987). The net effect of sorption is often to slow the overall rate of loss of a chemical through reactions in water; in such cases, sorption leads to longer half-lives.

Table 3. Interactions of importance in aquatic surface chemistry

Interactions
Chemical reactions with surfaces
Surface hydrolysis
Surface complexation
Surface ligand formation
Hydrogen bond formation
Electrical interactions at surfaces
Electrostatic interactions
Polarization
Interactions with solvents
Hydrophobic expulsion

Soil (sediment) is the major surface of significance in sorption. The study of soil is beyond the scope of this chapter; however, because the interest here is sorption of a chemical to a soil surface, the important consideration is the nature of the binding site on the soil surface, which will determine the sorption process. Soil consists of a mineral fraction, an organic fraction, and pore water.

The inorganic mineral fractions of soil are repositories of water and plant nutrients. The mineral fraction is composed primarily of layer silicates (occasionally aluminum), metal hydroxides, and oxygen. Layers of silicon and aluminum interact in various combination to give the characteristic layered structure of clay minerals. The layer silicates have a large surface area with negative charge that is pH dependent and results from the ionization of hydroxyl hydrogens. In sediments, sorption to clay minerals occurs through cation exchange of organic cations (and metal ions) with surface clay anions. Unlike partitioning to the organic fraction in sediment or soil, cation exchange is site specific and limited by the available sites and by competition among all cations. Sorption can also occur on silica and the metal oxides but the mechanisms are not well understood.

The organic fraction of soil is composed of polysaccharides, amino sugars, nucleotides, organic sulfur, and phosphorus compounds. Humus, the water-insoluble material, makes up the bulk of soil organic matter (up to 5% by weight) and is formed during decomposition of vegetation. Humus is composed of a base-soluble fraction called humic and fulvic acids and an insoluble fraction called humin (see Chapter 1). Humic material in soils strongly sorbs many solutes (including organics) and is an important complexing agent for heavy polyvalent cations and organic compounds with low water solubility.

There is a strong interaction between the organic and inorganic fractions of soil, especially between clay and humic (fulvic) acid compounds. Sediments and soils may contain up to 50% by weight of organic matter, over 50% of which may be complexed with clay. It is evident that the two soil fractions and their interaction (complexation) play a major role in determining the properties of soil (sediment) and its sorption.

As a chemical (sorbate) sorbs on a surface (soil), the characteristics of the surface change and approach that of the sorbate. This occurs through a relatively nonselective sorption process.

The pore water (or interstitial water) fraction of a soil (sediment) is a critical component related to sorption. The composition of this pore water may be significantly different from that of an overlying stream or lake. For example, it might contain higher concentrations of colloidal (nonsettling) particles, dissolved organic and inorganic chemicals, and a lower redox potential. Temperature and pH would also be different. Consequently, the sorption of a chemical from such pore water onto a soil or sediment may vary significantly from that (for the same chemical) represented by laboratory measurements using tap water and dilute soil-water slurries. In particular, sorption may be weaker in such pore waters. Additional discussion is provided by Brownawell and Farrington (1986), Chin and Gschwend (1992), and Di Toro et al. (1991).

The physicochemical basis for sorption can vary depending on the nature of the chemicals and soils. For neutral chemicals, two processes, neither involving chemical reactions, dominate:

- For low organic carbon soils or sediments ($f_{oc} < 0.001$*), physical *adsorption* dominates. Here, the solutes appear to precipitate (or condense) out of solution due to physical attraction to the soil's surface. Soil surface area is thus a key factor in determining sorption capacity. Soils with a high surface area per unit weight, e.g., clay and silts, absorb more of a contaminant than those with low surface areas, e.g., sand.
- For high organic carbon soils and sediments ($f_{oc} > 0.001$), chemical *absorption* dominates. Here, the solutes appear to partition into the spongelike organic-water matrix surrounding the soil mineral particles. The organic carbon content of the sediments is thus the key factor in determining sorption capacity.

In some surface water bodies, both types of sorption may take place because of the mix of solid particles present. Table 4 shows, for example, that the particle types contributing to suspended materials in Onondaga Lake, New York, and its tributaries include some fractions that are predominantly organic and others that are predominantly inorganic. In soils, the top layers (varying in depth from a few centimeters to a few meters) typically have sufficient humin to favor absorption. Deeper soils (also called unconsolidated sediments) are deficient in naturally occurring organic matter, favoring adsorption. In either case (i.e., absorption or adsorption), there is the presumption that the bioavailability—and thus the toxicity—of the sorbed fraction has been significantly reduced, if not eliminated.

*f_{oc} is the weight fraction of organic carbon in the sediment or soil.

Although the reversibility of Eq. (4) has been called into question by a number of experiments, for purposes of many environmental assessments it is appropriate to assume that equilibrium conditions for sorption reactions will usually be achieved within short time frames (hours to days) and that a sediment-specific equilibrium sorption constant, K_d, can be obtained:

$$K_d = \frac{C_{\text{sed}}}{C_{\text{water}}} \tag{8}$$

where K_d = sorption constant (kg/L)
C_{sed} = equilibrium concentration of solute on sediment (mg/kg)
C_{water} = equilibrium concentration of solute in water (mg/L)

K_d values are thus simple measures of the distribution of chemicals between aqueous and soil/sediment phases; values range from about 10^5 for strongly sorbed chemicals (e.g., DDT and polychlorinated biphenyls) to 1 or less for chemicals that are weakly sorbed and miscible in water (e.g., ethanol, acetone).

Equation (8) is equivalent to a Freundlich sorption isotherm with an isotherm constant, n, equal to 1.0 [see Eq. (42)]. Numerous studies have shown that, for the high organic carbon sediments and soils, the amount of a chemical sorbed is directly proportional to the organic carbon content of the soil. This has led to the common practice of converting sediment-specific sorption constants, K_d, into sediment-independent (chemical-specific) sorption constants, K_{oc}:

$$K_{oc} = \frac{K_d}{f_{oc}} \tag{9}$$

where both K_{oc} and K_d have units of kg/L and f_{oc} is the (dimensionless) weight fraction of organic carbon

Table 4. Particle types contributing to suspended materials in Onondaga Lake, New York

Particle type	% of total particle cross-sectional area[a] Onandaga Lake	Tributaries
"Organic" detritus[b]	39 ± 22	9 ± 11
$CaCO_3$ precipitate	22 ± 13	20 ± 15
Quartz plus diatoms	23 ± 15	14 ± 5
"Clay particles"	3 ± 2	20 ± 12
Calcium precipitate on clay silica "nucleus"	8 ± 6	27 ± 14
Anything else	5 ± 6	10 ± 16

[a] Mean and standard deviation given for each entry.
[b] This category may also contain some low average atomic weight inorganic materials as well as true "organic" residues.
Source: Reprinted with permission from Johnson et al., 1991. Copyright 1991 American Chemical Society.

in the sediment. Values of K_{oc} range from 1 to 10,000,000. Selected values exemplifying this range are shown in Table 5.

The primary utility of K_{oc} values is that they are easily predicted via correlations with other properties such as water solubility and octanol-water partition coefficients. Over 90 such correlations have been published (Lyman et al., 1981; Lyman and Loreti, 1987) enabling the prediction of K_{oc} values typically within a factor of 2 to 3. For example, Lyman and Loreti (1987) present the following correlation:

$$\log K_{oc} = 0.779 \log K_{ow} + 0.460, \quad n = 52,\ r^2 = 0.873 \tag{10}$$

which was derived from a data set screened to eliminate measurements that may have been affected by high solids concentrations. Examples of correlations based on water solubility (S) include Eqs. (11) (Lyman and Loreti, 1987) and (12) (Mill et al., 1982):

$$\log K_{oc} = -0.602 \log S(\text{mol/L}) + 0.656, \quad n = 110,\ r^2 = 0.77 \tag{11}$$

$$\log K_{oc} = -0.7882 \log S(\text{mol/L}) - 0.27, \quad n = 100 \text{ (approx.)},\ r^2 = \text{not available} \tag{12}$$

Compilations of K_{oc} values that are both reliable and comprehensive do not exist; however, many values are available in Howard (1989–1991) and Mackay et al. (1991a). With measured or estimated values of K_{oc} and f_{oc}, a site-specific K_d may be obtained with Eq. (9). The fraction of the solute associated with the solids in the system may then be obtained by application of Eqs. (8) and (44) (see later). Such calculations should be routinely carried out for aquatic assessments.

Table 5. Measured values of soil or sediment sorption constants for selected organic chemicals

Nonaromatic chemicals		Aromatic chemicals			
Chemical	log K_{oc}	Chemical	log K_{oc}	Chemical	log K_{oc}
Acrolein	0.70	Acetophenone	1.54	Glyphosate	3.42
Aldrin	4.30	Acridine	4.11	Hexachlorobenzene	3.59
Bromacil	1.86	Alachlor	2.28	Leptophos	3.97
Chlordane	5.15	Ametryn	2.59	Linuron	2.91
Crotoxyphos	2.23	2-Aminoanthracene	4.45	Methazole	3.42
Cycloate	2.54	6-Aminochrysene	5.16	Methoxychlor	4.90
2,4-D	1.23	Anthracene	4.78 (4.41)	2-Methoxy-3,5,6-trichloropyridine	2.96
DBCP	2.11	Aroclor 1254	6.25	9-Methylanthracene	4.81
Diallate	3.28	Asulam	2.48	3-Methylcholanthrene	6.09
cis-1,3-Dichloropropene	1.36	Atrazine	2.17	2-Methylnaphthalene	3.93
trans-1,3-Dichloropropene	1.41	Benzene	1.92	Methyl parathion	3.33
Dieldrin	3.82	Benzo[*a*]anthracene	6.24	Metobromuron	1.78
Dinoseb	2.09	Benzo[*a*]pyrene	6.74	Metribuzin	1.98
Disulfoton	3.25	2,2′-Biquinoline	4.02	Monolinuron	2.30
Endrin	3.56	Carbaryl	2.36	Monuron	2.00
EPTC	2.38	Carbofuran	1.45	Naphthalene	3.55 (3.11)
Ethion	4.19	Carbophenothion	4.66	1-Naphthol	2.72
Ethylene dibromide	1.64	Chlorbromuron	2.66	Napropamide	2.83
Heptachlor	4.00	Chloroneb	3.06	Neburon	3.36
Ipazine	3.22	Chloroxuron	3.51	Nitrapyrin	2.62
Lindane	2.96 (3.03)	Chlorpyrifos	4.13	Norfluorazon	3.28
Malathion	3.17	Chlorpyrifos-methyl	3.52	Oxadiazon	3.51
Methomyl	2.20	Chlorthiamid	2.03	2,2′,4,4′,5,5′-PCB	6.08
Methyl isothiocyanate	0.78	Chrysene	5.77	2,2′,4,5,5′-PCB	4.63
Monuron	2.20	Cyanazine	2.30	Parathion	3.93
Pebulate	2.80	DDD	5.38	Phenanthrene	4.36
Pentachlorophenol	2.95	DDE	5.17	Phenol	1.43
Phorate	3.51	DDT	5.48 (5.38)	Prometon	2.54
Picloram	1.17	13*H*-Dibenzo[*a,i*]carbazole	6.02	Prometryn	2.91
2,4,5-T	1.87	Dibenzothiophene	4.05	Pronamide	2.30
Tebuthiuron	2.79	Dibenzo[*a,h*]anthracene	6.23	Propachlor	2.42
Terbacil	1.71 (1.58)	Dicamba	0.27	Propazine	2.20
Toxaphene	3.00	Dichlobenil	2.27	Propham	1.71
Triallate	3.35	Diflubenzuron	3.83	Pyrazon	2.08
		7,12-Dimethylbenz[*a*]anthracene	5.35	Pyrene	4.92
		Dipropetryn	3.07	Pyroxychlor	3.48
		Diuron	2.60 (2.47)	Simazine	2.01
		Fenuron	1.43	Terbutryn	2.85
		Fluometuron	2.24	Tetracene	5.81
		Fluoranthene	5.31	3,5,6-Trichloro-2-pyridinol	2.11
		Fluorene	4.01	Trietazine	2.78
				Trifluralin	4.14

Source: Lyman and Loreti, 1987 (values compiled from several sources).

Part of the difficulty in the preparation of "reliable" compilations of K_{oc} values is that significant amounts of variability may be associated with laboratory measurement protocols as well as with a host of chemical and environmental parameters. Thus, K_{oc} should be considered not as a true constant but as a convenient parameter that may understandably range over a factor of 10 for significantly different soils and sediments.

The chemical and environmental parameters that can affect the value of K_{oc} include:

- Variability in the sorption capacity of organic matter, especially between that in soils and that in aquatic sediments
- Presence of nonsettling (and nonfilterable) microparticles in apparent solution in the aqueous phase
- Presence of organic cosolvents (in true solution) in the aqueous phase
- Salinity or ionic strength
- pH (primarily for organic acids and bases)
- Temperature
- Kinetics (i.e., the time required for solutes to diffuse into or out of the deeper layers of the organic matter and to achieve equilibrium)
- Chemical class of the solutes

One or more of these factors are covered in the works of Fu et al. (1983), Kadeg et al. (1986), Karickhoff (1983, 1984), La Poe (1985), Lyman and Loreti (1987), Voice and Weber (1983), and Weber et al. (1983).

The sorption of organic acids and bases is fundamentally different from that of neutral chemicals and thus deserves special attention. Organic acids include carboxylic acids and some substituted phenols (e.g., pentachlorophenol). Organic bases include most nitrogen-containing compounds (e.g., amines). Descriptions of such chemicals and ways to estimate their acid or base dissociation constants are given by Lyman et al. (1981) and Perrin et al. (1981). Further discussion is also provided in the following.

For acids, AH, and bases, B, we can represent the deprotonation reactions and acid dissociation constants (K_a) as follows:

$$AH \rightleftharpoons A^- + H^+, \qquad K_a = \frac{[A^-][H^+]}{[AH]} \tag{13}$$

$$BH^+ \rightleftharpoons B + H^+, \qquad K_a = \frac{[B][H^+]}{[BH^+]} \tag{14}$$

where the brackets, [ ], represent molar concentrations.

Because of pH effects on the extent of dissociation (i.e., loss of H^+) or association (e.g., acquisition of H^+), it is not uncommon for sorption to change by as much as a factor of 5 over a pH range of 2 to 5 units. For *acids*, the general rule is that the neutral species, AH, sorbs more strongly than the anion, A^-, and that the sorption (of AH) will become appreciable when the pH is 1.0 to 1.5 units above the pK_a ($pK_a = -\log K_a$) of the acid. For *most bases*, it appears that the protonated species, BH^+, dominates in the sorption process and that sorption is primarily via a cation exchange process (Lyman and Loreti, 1987).

Thus, for acids and bases it becomes important to estimate the fraction that is dissociated or protonated. From Eqs. (13) and (14), the concentrations ratio of neutral to ionized species can be related to the pK_a and solution pH as follows:

$$\log\left(\frac{[AH]}{[A^-]}\right) = pK_a - pH \tag{15}$$

$$\log\left(\frac{[B]}{[BH^+]}\right) = pH - pK_a \tag{16}$$

Similarly, the fraction (f) of total acid that is in the neutral form (AH) is obtained as

$$f_{AH} = \frac{[AH]}{[AH] + [A^-]} = \frac{1}{1 + K_a/[H^+]} \tag{17}$$

and for bases (B):

$$f_B = \frac{[B]}{[B] + [BH^+]} = \frac{1}{1 + [H^+]/K_a} \tag{18}$$

Note from Eqs. (15) and (16) that when pH = pK_a the concentrations of neutral and ionized species are the same. Equations (17) and (18) can thus be used to estimate the fractional concentrations of acids and bases that are more or less likely to be sorbed.

Volatilization

Many organic chemicals can easily be transported out of a surface water, across the air-water interface, and into the overlying air. The process is simply referred to as volatilization. Volatilization is a transfer process; it does not result in the breakdown of a substance, only its movement from the liquid (or solid) phase to the gas phase. Like other environmental transport and transformation phenomena, its importance is strongly linked to both environmental and chemical properties. Specifically, volatilization will be most important:

- For shallow, well-mixed water bodies with sufficient air movement over the water's surface
- For chemicals with a high air-water partition coefficient, referred to as Henry's law constant

Under the most favorable conditions, chemicals may volatilize with half-lives on the order of hours, thus nearly eliminating them as contaminants of concern in a surface water (except in locations near the source). On the other hand, under unfavorable conditions, volatilization may play no significant role in the aquatic fate of a contaminant.

Transfer of organic vapors at the air-water interface is often predicted from a two-film diffusion model (Liss and Slater, 1974; Mackay and Yuen, 1983; Liss, 1983). In this model, mass transfer of gas is governed by molecular diffusion through a stagnant liquid and gas film and is driven by the concentration gradient between the equilibrium concentrations at the interface and bulk reservoirs. Mass thus moves from areas of high to low concentration and transfer is limited at the gas or liquid-film level. For example, oxygen is controlled by the liquid-film resistance but nitrogen, although more abundant in the atmosphere than oxygen, has a greater liquid-film resistance.

Volatilization, as described by a two-film theory, is a function of Henry's law constant, the gas-film resistance, and the liquid-film resistance. The film resistance depends on diffusion and mixing.

The key chemical-specific parameter related to the importance of volatilization is Henry's law constant, *H*, defined as the *equilibrium* ratio of a chemical's air- and water-phase concentrations:

$$H \text{ (atm m}^3\text{/mol)} = \frac{C_{air} \text{ (atm)}}{C_{water} \text{ (mol/m}^3\text{)}} \tag{19}$$

Equation (19) shows H in units of atm m^3/mol. H is also commonly reported in "nondimensional" units obtained, for example, when C_{air} is in mg/L (air) and C_{water} is in mg/L (water). Interconversion of this nondimensional Henry's law constant, H', and values of H from Eq. (19) is effected by

$$H' = \frac{H}{RT} \tag{20}$$

where H' = Henry's law constant (nondimensional)
H = Henry's law constant (atm m^3/mol)
R = gas constant = 8.205×10^{-5} atm m^3/mol K
T = temperature (K)

Limited tabulations of H values are available in the following publications: Howard (1989–1991), Mackay et al. (1991a), Hine and Mookerjee (1975), Mackay and Shiu (1981), Eklund et al. (1991), Gossett (1987), Burkhard et al. (1985), Yaws et al. (1991), Fedinger and Glotfelty (1990), and Howe et al. (1987).

When reliable measured values of H are not available, it is appropriate to estimate values of H from the ratio of a chemical's pure component vapor pressure to its water solubility:

$$H = \frac{P}{S} \tag{21}$$

where P = pure component vapor pressure (atm)
S = solubility in water (mol/m^3)
H = Henry's law constant (atm m^3/mol)

Note that values of P and S should be for the temperature at which H is desired and should be for the same physical state (solid or liquid). Note also that, since P increases nearly exponentially with increasing temperature, H also increases significantly with increasing temperature. Equation (21) is quite accurate (as long as the P and S values are accurate) for chemicals with S above about 1 mol/L (1000 mol/m^3). Acceptable, but less accurate, estimates may be obtained for chemicals with higher solubilities. Other estimation methods, including some applicable to chemicals of infinite water solubility, are described by Lyman (1985) and Meylan and Howard (1991). As an example, Figure 2 shows values of P, S, and H for selected gasoline constituents, including additives.

As indicated by Figure 2, values of H can range over many orders of magnitude. The following general rules can help in a qualitative assessment of a chemical's volatilization potential (in reasonably mixed surface waters):

H (atm m^3/mol)	Volatilization potential
$< 3 \times 10^{-7}$	Chemical is less volatile than water.
3×10^{-7} to 10^{-5}	Chemical volatilizes slowly. Volatilization may be significant in shallow waters.
$> 10^{-5}$	Volatilization is significant in all (well-mixed) surface waters.
$> 10^{-3}$	Volatilization is rapid.

The value of H can also be used to develop simplifying assumptions for modeling volatilization. If either the liquid or gas film resistance controls, the lesser resistance can be neglected. The threshold of Henry's constant for gas or liquid film control is approximately 0.1 for dimensionless H, or 2.2×10^{-3} atm at m^3/mol. Above this threshold value the chemical is liquid film controlled, and below it it is gas film controlled.

Actual volatilization rates are seldom measured in the open environment, although laboratory measurements are more common. Such data have been used to derive a combination of theoretical and empirical equations by which volatilization rates may be estimated with specified conditions of temperature, water body depth, water current, air speed, water surface roughness, and other parameters (Lyman et al., 1981; Mackay, 1985; Mackay et al., 1982; Ince and Inel, 1991). Estimates derived from such equations have a significant amount of uncertainty. Lyman et al. (1981) suggest that it is advisable to assume that such estimated volatilization rates may be high by a factor of 10 at most and low by a smaller factor of possibly 3.

Perhaps more important than an estimate of a volatilization rate (which typically assumes an air concentration of zero for the contaminant) is an initial estimate of whether volatilization is expected at all, considering measured values of C_{air} and C_{water} at the site of interest. Three possible conditions may be inferred from such measurement:

Value of C_{air}/C_{water}	Condition
Greater than H	C_{air} is above equilibrium concentration and, thus, net flux of chemical is from air to water.
Approximately equal to H	System is in approximate equilibrium, and no net volatilization losses are expected.
Less than H	C_{air} is less than equilibrium amount and volatilization (net water-to-air transfer) will take place.

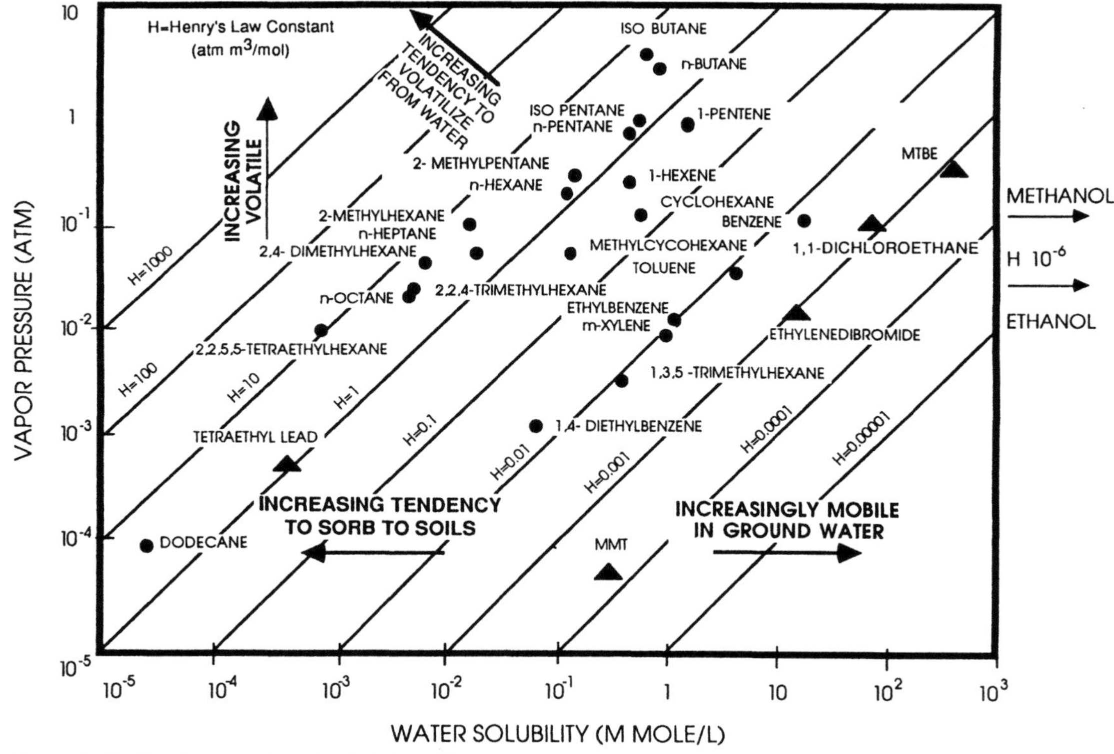

Figure 2. Partitioning properties of selected gasoline constituents.

Atmospheric Deposition

In many instances, especially for large-area water bodies, a significant fraction of the flux of toxic chemicals entering the water comes directly from the atmosphere through a process called atmospheric deposition. This process is considered to be made up of dry deposition and wet deposition. In dry deposition, the chemicals—in the form of gases, aerosols, or particulates—are transferred to the earth's surface (land, vegetation, and water) without the aid of precipitation. Wet deposition involves facilitated transport to the earth's surface in conjunction with falling rain, snow, or fog with which the chemicals have become associated (Hanna and Hosker, 1980; Hicks, 1984; Schroeder and Lane, 1988).

For dry deposition, mechanisms such as diffusion, impaction, and sedimentation are important. The rate of atmospheric deposition (e.g., in units of mass per unit area per unit time) depend on a large number of chemical-specific, surface-specific, and atmospheric properties. For water bodies, important surface-specific properties would include surface roughness and temperature.

Dry deposition can be expressed by the following equation:

$$D_d = V_d C_a \tag{22}$$

where D_d = dry deposition rate (g/cm^2 s)
V_d = deposition velocity (cm/s)
C_a = chemical concentrations in air (g/cm^3)

Values of the deposition velocity vary widely even for similar conditions and for gases span four orders of magnitude, from 2×10^{-3} to 26 cm/s. The range for particles is greater, from 10^{-3} to 180 cm/s (Schroeder and Lane, 1988). Some data are available in publications by Sehmel (1984), McMahon and Denison (1979), and Eisenreich et al. (1980).

For wet deposition, processes such as Brownian capture, nucleation, dissolution, and impaction play major roles in combining chemicals with water in the atmosphere. An empirical approach to representing the atmospheric chemical removal resulting from wet deposition uses the following equation:

$$W = \frac{C_r \rho_a}{C_a} \tag{23}$$

where W = washout ratio (dimensionless)
C_r = chemical concentration in precipitation (μg/g)
C_a = chemical concentration in air (μg/m^3)
ρ_a = density of air (1200 g/m^3 at NTP)

Values of W are typically less than 2×10^5 (Schroeder and Lane, 1984). The flux of a chemical carried to the earth's surface in a precipitation event may be estimated from

$$D_w = \frac{V_r C_r}{tA} \quad (24)$$

where D_w = wet deposition rate (g/cm^2 s)
V_r = volume of precipitation (L)
C_r = chemical concentration in precipitation (g/L)
t = duration of precipitation event (s)
A = area for which flux is being calculated (cm^2)

Transport

We briefly describe here three processes that can transport a chemical from one location to another in an aquatic environment:

- Bulk transport with dispersion
- Sedimentation
- Diffusion

Bulk Transport with Dispersion

The movement of chemicals within a water column due to movement of the water itself is called bulk transport or advection. The movement of the water creates turbulence and mixing, which, in turn, lead to dispersion and dilution of the chemicals within an ever-increasing volume of water. Although qualitatively these phenomena are easy to understand, the quantitative treatment is quite complicated. The reader is referred to other works for more detailed information on this subject (Hemond and Fechner, 1994; Lick, 1987; van Leeuwen and Buffle, 1992, 1993).

Transport with Sediments

As previously discussed, a chemical partitions between a dissolved and a particulate sorbed phase based on its sediment-to-water partition coefficient (Karickhoff et al., 1979). Particulate and dissolved concentrations can be calculated for the water column or bed sediment by knowing the concentration of suspended solids in water or in the bed. The suspended load of solids in a river or stream is defined as a flow rate times the concentration of suspended solids (e.g., kg/d) and is greatly influenced by peak flows. These flows cause large inputs of foreign material from erosion and runoff as well as increases in scour and resuspension of bed and bank sediment.

Several equations have been reported to calculate the rate of sediment movement near the bottom, but they have been developed for rivers and noncohesive sediments (i.e., fine to coarse sands and gravel). It should be noted that it is not sand but rather silts and clays to which most chemicals sorb. Therefore, the equations are of limited value. Bed load transport is a fraction of the total sediment transport (which equals suspended plus bed load). Although bed load transport of silts and clays may be important in estuaries, there are few predictive equations for such applications. Bed load consists of particles that creep, flow, or saltate near the bottom.

Suspended sediment particles and sorbed chemicals are transported downstream at nearly the mean current velocity. In addition, they are transported vertically downward by their mean sedimentation velocity. Generally, silt and clay particles settle according to Stokes' law, in proportion to the square of the particle diameter and the difference between sediment and water densities:

$$W = 8.64\left(\frac{g}{18\mu}\right)(\rho_s - \rho_w)\, d_s^2 \quad (25)$$

where W = particle fall velocity, ft/s
ρ_s = density of sediment particle, g/m^3
ρ_w = density of water, 1 g/m^3
g = gravitational constant, 981 cm/s^2
d_s = sediment particle diameter, mm
μ = absolute viscosity of water, 0.01 poise (g/cm s) at 20°C

Generally, the washload (i.e., fine silt and clay-size particles) carries most of the mass of sorbed chemical. These materials have very small fall velocities, on the order of 0.3–1.0 m/d for clays of 2–4 μm nominal diameter and 3–30 m/d for silts of 10–20 μm nominal diameter.

When a particle reaches the bed, there is a certain probability that it can be scoured from the bed sediment and resuspended. The difference between sedimentation and resuspension represents net sedimentation.

Sedimentation, the movement of suspended sediment particles and sorbed pollutants from the water column to the bottom sediments, is an important loss mechanism for many chemicals. If a chemical is found to undergo significant sorption to suspended solids, sedimentation should always be evaluated, even in fast-moving water bodies where net sedimentation may not be taking place. (In the latter situation there could be a continual exchange of contaminated and uncontaminated sediments, thus modifying both water column and bottom sediment concentrations.)

Diffusion

True molecular diffusion involves the movement of molecules, under the influence of a concentration gradient, from areas of high concentration to areas of low concentration. Diffusion in water is a relatively slow process and will be important in porous media (e.g., the pores of bottom sediments) only if the flow of water there is itself less than 2.5×10^{-4} cm/s; diffusion can probably be ignored if the water flow is greater than 2×10^{-3} cm/s (Tucker and Nelken, 1982). Diffusion of inorganic chemicals through solids or semisolids (e.g., the humin or other naturally occurring organic matter that frequently exists in surface waters) may also be important for neutral species as it is for neutral organic chemicals; however, it will probably not be important for electrolytes (charged species) because of the difficulties in moving ions through such materials. In any case, diffusion through solids is not covered in this text.

Limited information is provided by Lyman et al. (1992).

Diffusion is usually defined in terms of Fick's law (Robinson and Stokes, 1959):

$$F = -D\frac{dC}{dx} \quad (26)$$

In this equation, F is the mass flux or rate of transfer of the species per unit cross-sectional area (mol/m^2 s), dC/dx is the concentration gradient (mol/m^3/m), and D is the diffusion coefficient of the species in water (m^2/s). Although simple in concept, rigorous estimation of fluxes of electrolytes due to diffusion can become exceedingly complex because one must usually consider the species concentration, the simultaneous diffusion of other electrolytes and their interactions, the diffusion of water into the spaces left by the solvated electrolytes, and environmental variables such as ionic strength and temperature (Bodek et al., 1988).

Overall, however, diffusion coefficients in water are found to vary relatively little. The coefficients of many simple salts generally do not change by more than a factor of 2 over a range of concentrations and temperatures, and most values are within the range 1×10^{-5} to 5×10^{-5} cm/s at 25°C (Bodek et al., 1988). Values of D are tabulated by Bodek et al. (1988) and Robinson and Stokes (1959). Estimation methods are also provided by Bodek et al. (1988) for electrolytes and by Tucker and Nelken (1982) for nonelectrolytes.

Transformation

This second major section on organic chemicals describes a number of chemical (abiotic) and biological processes that can alter the identity of the original parent chemical. The transformation mechanisms that are covered are:

Abiotic processes	Biotic processes
Hydrolysis	Aerobic degradation
Photolysis	Anaerobic degradation
Dissociation	
Oxidation-reduction	

Abiotic Processes

Hydrolysis

Hydrolysis is a chemical transformation process in which an organic molecule, RX, reacts with water, forming a new carbon-oxygen bond and cleaving a carbon-X bond in the original molecule. (Here, R symbolizes an organic molecule and X represents any leaving group on the molecule, e.g., a chlorine atom.) The net reaction is typically a direct replacement of X by OH:

$$R{-}X + H_2O \rightarrow R{-}OH + X^- + H^+ \quad (27)$$

Unlike organic acid-base and hydration reactions, hydrolysis is not treated as a reversible reaction. Reactions closely related to hydrolysis include addition, in which H_2O adds a hydrogen (H) and a hydroxyl (OH) to a double bond, and elimination, in which a double bond is created by the elimination of a bonded atom or group from a neighboring pair of carbons.

Some examples of hydrolysis include conversion of alkyl halides to alcohols, esters to acids, and epoxides to diols. Hydrolysis is important for exposure assessment (and modeling) because hydrolytic products like alcohols and acids are typically more water soluble and less likely to be involved in biouptake or volatilization reactions than their parent substances. Hydrolysis reactions depend on the susceptibility of the chemical to attack by nucleophilic reagents such as the water molecule or hydroxide ion. Acid catalysis enhances nucleophilic attack by influencing the charge distribution in the molecule. Molecules that are susceptible to hydrolysis are molecules in which the electron distribution gives some charge separation, facilitating nucleophilic attack. Hydrolysis then is important in any chemical where alkyl carbon, carbonyl carbon, and imino carbon are bonded to halogen, oxygen, or nitrogen atoms (or groups) through σ-bonds.

It is generally observed that hydrolysis reactions are first-order in the concentration of the organic species, i.e.,

$$\frac{-d[RX]}{dt} = k[RX] \quad (28)$$

where [RX] = aqueous concentration of RX
d[RX]/dt = rate of change of concentration of RX per unit time, t
k = hydrolysis rate constant (reciprocal time units)

The rate constant, k, is related to the half-life, $t_{1/2}$, the time for 50% of RX to be transformed, by

$$t_{1/2} = 0.693/k \quad (29)$$

Since k has units of reciprocal time (d^{-1}), $t_{1/2}$ has units of time (d). Hydrolysis half-lives can range over many orders of magnitude. However, for purposes of an environmental assessment the interest is only in chemicals that will hydrolyze over the time frames of the assessment, typically hours to years or decades. Figure 3 provides examples of several chemical classes that have hydrolysis half-lives in this range.

Hydrolysis reaction rates are affected by a number of environmental variables including pH, the presence of catalysts, temperature, sorption onto solids, and ionic strength. Changes in pH are primarily important for chemicals that are susceptible to acid (H^+)-catalyzed or base (OH^-)-catalyzed hydrolysis, or both. In the latter case, Eq. (28) is expanded to a more general kinetic expression for hydrolysis:

$$\frac{-d[RX]}{dt} = k_a[H^+][RX] + k_n[RX] + k_b[OH^-[RX] \quad (30)$$

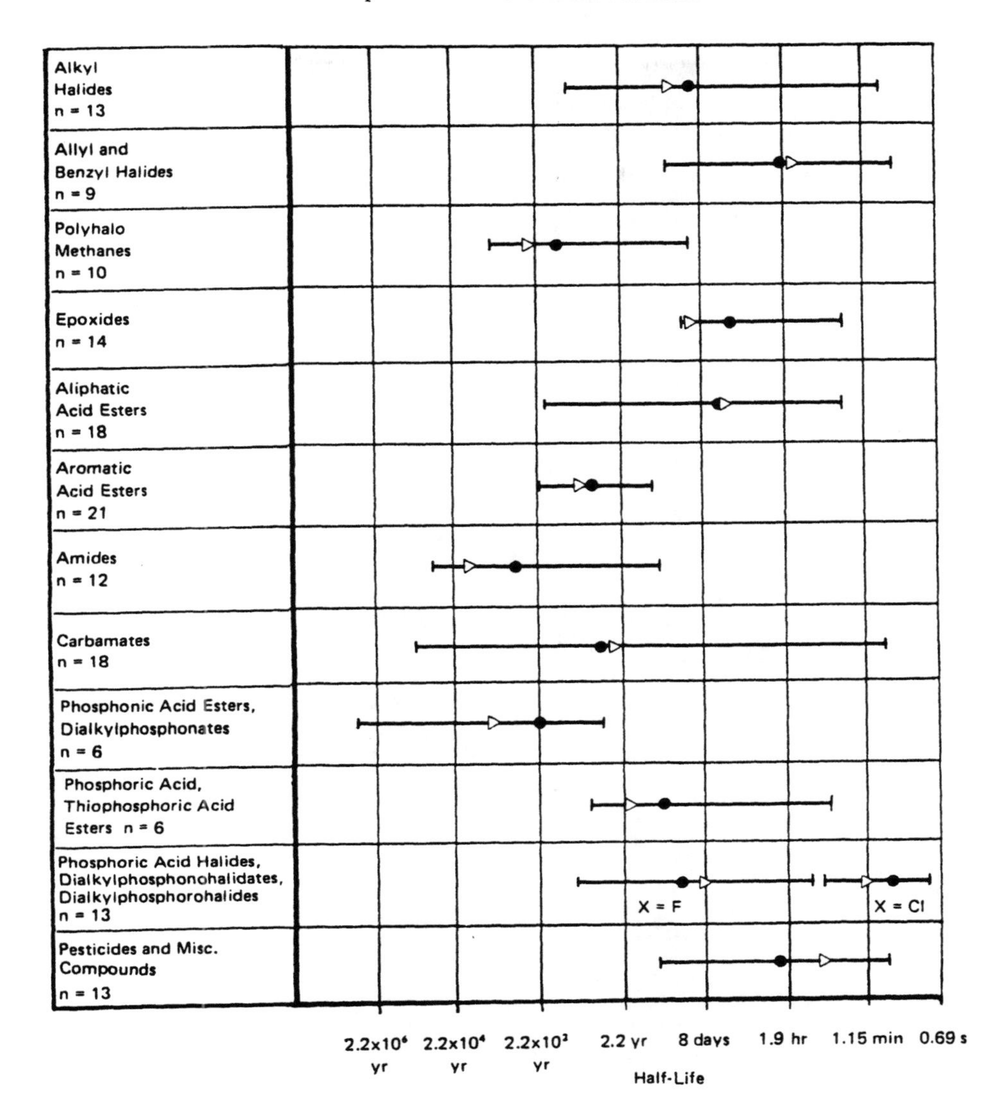

Key:
- • Average
- ▶ Median
- n No. of Compounds Represented

Figure 3. Examples of the range of hydrolysis half-lives for various types of organic compounds in water at pH 7 and 25°C. [Adapted by Lyman et al. (1981) from data of Mabey and Mill (1978)].

where k_a, k_n, k_b = rate constants for acid, neutral, and base promoted reactions (L/mol s for k_a and k_b; s^{-1} for k_n)

$[H^+]$ = hydrogen ion concentration (mol/L)

$[OH^-]$ = hydroxyl ion concentration (mol/L)

Examples of values of k_a, k_n, and k_b are provided in Table 6. From this equation it is easy to see how a lowering of pH would promote the acid-catalyzed hydrolysis of compounds such as warfarin and brucine.

The rate of hydrolysis of organic compounds increases with temperature in an exponential relationship. Some useful rules of thumb for temperatures in the range 0–50°C are that (Lyman et al., 1981):

For a temperature increase of	The hydrolysis rate constant increases about
1°C	10%
10°C	factor of 2.5
25°C	factor of 10

Table 6. Examples of hydrolysis rate constants (data are for 25°C)

Chemical	Rate constants: Acid (M^{-1} h^{-1})	Neutral (h^{-1})	Base (M^{-1} h^{-1})
Warfarin	1.4×10^{-4}	4.9×10^{-6}	0.026
Aldrin		$(3.8 \pm 2.3) \times 10^{-5}$	
Brucine	5.9×10^{-3}		0.21
Dieldrin		$(7.5 \pm 3.3) \times 10^{-6}$	
Disulfoton		$(2.8 \pm 0.4) \times 10^{-4}$	5.99
Endosulfan I	$(8.1 \pm 2.7) \times 10^{-3}$	$(3.2 \pm 2.0) \times 10^{-3}$	$(1.0 \pm 0.7) \times 10^{4}$
Endosulfan II	$(7.4 \pm 3.9) \times 10^{-3}$	$(3.7 \pm 2.0) \times 10^{-3}$	$(1.5 \pm 0.9) \times 10^{4}$
Fluoroacetic acid, sodium salt		$<1.7 \times 10^{-6}$	
2-Methyllactonitrile		4.47	
Nitroglycerine			77 ± 11
Famphur		$(2.5 \pm 9) \times 10^{-4}$	5.0
Acrylamine	$<3.6 \times 10^{-2}$	$<(2.1 \pm 2.1) \times 10^{-6}$	
Acrylonitrile[a]	$(4.2 \pm 0.3) \times 10^{-2}$		$(6.1 \pm 6.5) \times 10^{-1}$
Mitomycin C		3.7×10^{-4}	3.0 ± 1.7
Chloromethyl methyl ether		21	
1,2-Dibromo-3-chloropropane			20.6
Ethylene Dibromide		9.9×10^{-6}	
cis-1,4-Dichloro-2-butene		$(9.1 \pm 1.1) \times 10^{-3}$	
trans-1,4-Dichloro-2-butene		$(9.0 \pm 0.5) \times 10^{-3}$	
4,4-Methylene-bis(2-chloroaniline)	$(2.9 \pm 3) \times 10^{-4}$	$<9 \times 10^{-8}$	
Pentachloronitrobenzene		$(2.8 \pm 0.7) \times 10^{-5}$	
Pronamide	4.3×10^{-3}	$<1.5 \times 10^{-5}$	7.4×10^{-2}
Reserpine	0.82	$(4.5 \pm 1.8) \times 10^{-5}$	9.8 ± 10.9
Thiourea		$<5.3 \times 10^{-7}$	
Uracil mustard		0.57 ± 0.08	$(2.05 \pm 0.2) \times 10^{5}$
Ethyl carbamate		$<2.6 \times 10^{-7}$	1.1×10^{-1}
2,3-Dichloro-1-propanol		$(5.3 \pm 0.8) \times 10^{-5}$	20.6 ± 2.2
1,3-Dichloro-2-propanol		$(3.1 \pm 0.2) \times 10^{-3}$	854 ± 87
1,2,3-Trichlorpropane		$(1.8 \pm 0.6) \times 10^{-6}$	9.9×10^{-4}
1,2,3-Trichlorobenzene		$(1.6 \pm 1.3) \times 10^{-5}$	
1,2,4-Trichlorobenzene		$(2.3 \pm 9) \times 10^{-5}$	

[a] Calculated from alkaline second-order rate constant assuming zero neutral contribution.
Source: Ellington et al., 1987.

Mill and Mabey (1988) have shown that for neutral hydrolyses of alkyl halides there is a fourfold change in rate for each 10°C change near 25°C and acid-catalyzed hydrolysis of epoxides had a threefold change for the same temperature change. Amides fall in the same range for both acid- and base-promoted reactions, and carbamates have a wider range of rate changes.

Hydrolysis rates may also be catalyzed by certain alkaline earth and heavy metal ions (e.g., Mg, Mu, Cu), although the role in environmental hydrolysis is not considered to be significant (Mabey and Mill, 1978). Similarly, while ionic strength changes can lead to hydrolysis rate acceleration or retardation, the effect in fresh waters (which usually have ionic strength well below 0.01 M) is not likely to be significant.

The effect of sorption on hydrolysis rates has not been extensively investigated. Macalady and Wolfe (1984) reviewed the literature for pesticides and concluded that the rate constants of pH-independent ("neutral") hydrolyses for sorbed pesticides are the same within experimental uncertainties as the corresponding rate constants for dissolved (aqueous-phase) pesticides. Base-catalyzed rates, however, were substantially retarded by sorption and acid-catalyzed rates were substantially enhanced. Thus, it should be clear that sorption effects must be considered for hydrolyzable chemicals.

In addition to the fairly extensive compilation of hydrolysis rate constants given by Mabey and Mill (1978), smaller compilations are provided by Lyman et al. (1981), Ellington (1989), Ellington et al. (1987), and Howard (1989–1991). A few estimation methods are available (Lyman et al., 1981; Karickhoff et al., 1991), but they have limited applicability and accuracy. Table 7 provides a listing of chemical classes generally considered resistant and susceptible to hydrolysis. To go beyond such qualitative assessments, or estimates of uncertain reliability, usually requires laboratory testing, which has the added benefit of allowing the identification of various reaction products and their stability.

Laboratory methods for measuring hydrolysis rate constants are described in detail in the Environmen-

Table 7. Organic chemical classes and their susceptibility to hydrolysis

Generally resistant to hydrolysis[a]	Potentially susceptible to hydrolysis
Alkanes	Alkyl halides
Alkenes	Amides
Alkynes	Amines
Benzenes/biphenyls	Carbamates
Polycyclic aromatic hydrocarbons	Carboxylic acid esters
Heterocyclic polycyclic aromatic hydrocarbons	Epoxides
Halogenated aromatics/PCBs	Nitriles
Dieldrin/aldrin and related halogenated hydrocarbon pesticides	Phosphonic acid esters
Aromatic nitro compounds	Phosphoric acid esters
Aromatic amines	Sulfonic acid esters
Alcohols	Sulfuric acid esters
Phenols	
Glycols	
Ethers	
Aldehydes	
Ketones	
Carboxylic acids	
Sulfonic acids	

[a] Multifunctional organic compounds in these categories may, of course, be hydrolytically reactive if they contain a hydrolyzable functional group in addition to the alcohol, acid, etc. functionality.
Source: Lyman et al., 1981.

tal Protection Agency protocols (EPA, 1985). The disappearance of chemical is followed in the dark at constant pH and temperature, thus eliminating other processes. Environmental chemists have also used structure-activity relationships (SARs) to predict rate constants for hydrolysis of alkyl halides, epoxides, esters, carbamates, and phosphorus esters (see Chapter 20).

Photolysis

Sunlight photolysis (i.e., photoreaction) of organic chemicals occurs in surface waters, on soil, in the atmosphere, and on many surfaces (e.g., plant canopy). Photoreactions may be the dominant loss process for many compounds in the air and clear water and they often create oxidation products that are more water soluble, less volatile, and less subject to biouptake than their parent substances.

Electromagnetic radiation in the near-ultraviolet and visible range (240–700 nm) interacts with the electrons in a molecule and is probably most significant from an environmental point of view. Different classes of molecular structures absorb sunlight in different wavelength regions and with different intensities, but typically below 400 nm. In some cases the energy absorbed can produce molecular changes. In contrast, infrared radiation produces insignificant changes in chemical species.

Photolysis, the light-initiated transformation reaction, is a function of the incident energy (photon absorption) on the chemical and the quantum yield (process efficiency) in air or water at wavelengths above 300 nm in the solar spectrum. It is not necessary that the chemical of interest absorbs a solar photon to undergo photolysis (see indirect photolysis).

When light strikes a molecule and is absorbed at certain wavelengths, several photochemical changes occur: the energy content of the molecule is increased and it reaches an excited electron state, or the excited state is unstable and the molecule reaches a lower energy level by loss of the extra energy through emission (i.e., fluorescence or phosphorescence) or is converted to a different molecule through the new electron distribution that existed in the excited state.

In direct photolysis, the chemical itself absorbs the solar radiation, whereas with indirect (or sensitized) photolysis an intermediary chemical becomes energized and then energizes the chemical of interest.

The end result of the photolysis may include a variety of reaction types such as dissociation or fragmentation, rearrangement or isomerization, cyclization, photoreduction by hydrogen ion extraction from other molecules, dimerization and related addition reactions, photoionization, and electron transfer reactions (Lyman et al., 1981).

Both the reaction rates and distribution of reaction products depend on a number of chemical and environmental factors, including the following:

Chemical specific	Environment specific
Wavelengths absorbed	Intensity and spectrum of solar radiation
Molar absorptivity[a]	Presence of sensitizers, quenchers
Quantum yield[a]	Incident angle of sunlight
	Depth in water body
	Attenuation of light in waters

[a] As a function of wavelength

With data on all of these factors, it is possible to estimate—for direct photolysis—a photolysis transformation rate constant, k, which presumes a first-order reaction:

$$\frac{-d[\mathrm{R}]}{dt} = k_a\phi[\mathrm{R}] = k[\mathrm{R}] \qquad (31)$$

where $[\mathrm{R}]$ = concentration of chemical R (mol/L)
k = photolytic first-order rate constant (s^{-1})
k_a = rate constant for absorption of light by chemical (s^{-1})
ϕ = quantum yield or efficiency with which absorbed light causes a photoreaction (dimensionless)
t = time (s)

These estimation methods, which are best suited for computerized calculations, are discussed by Lyman et al. (1981), Leifer (1988), Zika (1987), and Karickhoff et al. (1991). These and other references (e.g., Zafiriou et al., 1984; Ollis et al., 1991) also provide good general discussions of the process.

Examples of some experimental and estimated photolytic half-lives are provided in Table 8. Because the conditions of these measurements and calculations are not specified, caution should be used in applying the data to other sites and conditions.

Although numerous measurements of photolytic rate constants appear in the literature, the environmental scientist should carefully examine the experimental conditions before applying the same rate to a different environment. Because of the ever-present potential for indirect (sensitized) photolysis, it is difficult to list, as with hydrolysis, the chemicals that are resistant to photolysis. It is known, however, that the presence of one or more chromophores (functional groups that absorb ultraviolet or visible light) in a molecule will promote direct photolysis. Examples of such chromophores include C=O (aldehydes, ketones), >C=S, −N=N−, $-NO_2$, conjugated double bonds, and aromatic rings. Generally, nitro, azo, and polyhalogenated benzenes; naphthalene derivatives; polycyclic aromatics; aromatic amines; and all ketones and aldehydes absorb sunlight in the range 300 to 500 nm. Some polycyclic and azoaromatics absorb light beyond 700 nm.

Table 8. Half-life for disappearance via direct photolysis in aqueous media

Compound	λ[a] (nm)	$t_{1/2}$
Pesticides		
Carbaryl	[b]	50 h
2,4-D, butoxyethyl ester	[b]	12 d
2,4-D, methyl ester	[b]	62 d
DDE	[b]	22 h (calc)
Malathion	[b]	15 h
Methoxychlor	[b]	29 d
Methyl parathion	[b]	30 d
Mirex	[b]	1 y
N-Nitrosoatrazine	[b]	0.22 h (calc)
Parathion	[b]	10 d (calc)
	[b]	9.2 d
Sevin	[b]	11 d
Trifluralin	[b]	0.94 h (calc)
Polycyclic aromatic hydrocarbons (PAH)		
Anthracene	366	0.75 h
Benz[*a*]anthracene	[b]	3.3 h
Benzo[*a*]pyrene	[b]	1 h
Chrysene	313	4.4 h
9,10-Dimethylanthracene	366	0.35 h
Fluoranthene	313	21 h
Naphthalene	313	70 h
Phenanthrene	313	8.4 h
Pyrene	313,366	0.68 h
Miscellaneous		
Benzo[*f*]quinoline	[b]	1 h
9*H*-Carbazole	[b]	3 h
p-Cresol	[b]	35 d
9*H*-Dibenzocarbazole	[b]	0.3 h
Dibenzothiophene	[b]	4–8 h
Quinoline	[b]	5–21 d

[a] Wavelength(s) at which photolysis rate was measured.
[b] Sunlight.
Source: Lyman et al., 1981 (data compiled from several sources).

Indirect photolysis occurs in water or soil when a photon is absorbed by dissolved organic matter to form selective oxidants such as singlet oxygen or peroxy radicals. Consequently, only phenols, furans, aromatic amines, sulfides, and nitro aromatics undergo indirect photolysis in water (Mill and Mabey, 1985). However, some of these compounds also undergo direct photolysis. In contrast, the OH radical dominates atmospheric photooxidations producing rapid oxidation of even alkanes, olefins, alcohols, and simple aromatics (Atkinson, 1986; Atkinson, 1987). In water, no one oxidant is dominant.

Several methods are used to measure direct photolysis rate constants and ϕ in water (Dulin and Mill, 1982; EPA, 1985). Atmospheric photolysis constants can be treated similarly to those in water, but the methodology is limited to organics with a vapor pressure >0.1 torr (about 5×10^{-6} M). Indirect photoreaction constants can be estimated with structure-activity relationships (Mill, 1989).

Dissociation

As indicated in the discussion of sorption, a number of organic chemicals may act as acids or bases and thus undergo dissociation reactions [see Eqs. (13) and (14)]. Organic acids, AH, dissociate to produce anions, A^-, and organic bases, B, typically combine with hydrogen ions (H^+) to produce cations, BH^+. This deviation from neutrality significantly affects all other environmental processes and properties such as sorption, bioconcentration, and toxicity. Thus, it is important to identify such acids and bases, evaluate their extent of dissociation, and consider the resulting effects on other processes.

Organic acids of interest are typically those containing a carboxylic acid group (−COOH), or a substituted alcohol or phenol (e.g., pentachlorophenol). Organic bases include most nitrogen-containing compounds (e.g., amines, anilines, pyridines). (The dissociation of organometallics is covered below.)

As indicated by Eqs. (13) and (14), acid and base dissociation is a reversible process. The extent of dissociation is determined by the acid dissociation constant, K_a, and the hydrogen ion concentration, $[H^+]$, or pH, as further indicated by Eqs. (15)–(18).

Compilations of pK_a values for organic chemicals are provided by Perrin (1981) and Lyman et al. (1981) and references cited therein. These authors and Karickhoff et al. (1991) discuss a variety of pK_a estimation methods, which can usually provide values within ±0.3–0.5 pK_a unit of the true value. Larger errors may be associated with aliphatic compounds and compounds containing two or more acid or base groups. Examples of several dissociation constants (K_a) are provided in Table 9. Values of pK_a are only modestly affected by changes in temperature and ionic strength. In general, K_a decreases with increasing temperatures, and increasing ionic strength tends to favor the ionic form of the conjugate acid-base pair (i.e., increases K_a for acids like AH and decreases K_a for bases like BH^+).

Other Reactions

In addition to the hydrolysis, photolysis, and dissociation reactions already described, some organic chemicals are susceptible to other abiotic transformation reactions variously referred to as nucleophilic substitution, addition, elimination (or dehalogenation if Cl^-, Br^-, F^-, or I^- is removed), oxidation-reduction, and complexation. Our understanding of the nature and importance of these transformations in natural environments is, at present, quite limited. In some cases (e.g., Jafvert and Wolfe, 1987), the presence of an effective abiotic reaction is inferred only from the continuing loss or degradation of a chemical in a sterilized sample. However, although sterilization may kill the microorganisms, it may not deactivate microbial exudates, which may enzymatically alter the chemical. In other cases, the degradation products are specifically identified (e.g., Weber and Wolfe, 1987; Cooper et al., 1987). Both homogeneous (solute-solute) and heterogeneous (solute-solid) reactions may be involved.

In one study by Haag and Mill (1988), the importance of various naturally occurring nucleophiles, particularly sulfur compounds, in transformation of haloalkanes under environmental conditions was evaluated. They found that the bisulfide ion (HS^-) was generally the most important nucleophile, considering naturally occurring concentrations, although polysulfides (S_4^{2-} and S_5^{2-}) were inherently more reactive. Other nucleophiles that were evaluated by Haag and Mill (1988) include sulfite (SO_3^{2-}), thiosulfate ($S_2O_3^{2-}$), bicarbonate (HCO_3^-), and chloride (Cl^-). Schwarzenbach et al. (1985) also showed that the bisulfide ion reacts fairly easily with a variety of bromo- and chloroalkanes. Both studies showed that, for some chemicals, bisulfide reactions will be more important than hydrolysis. The end products of such reactions include a variety of mercaptans (RSH) and dialkyl sulfides (RSR′) via the following types of reactions:

$$RX + HS^- \longrightarrow RSH + X^- \quad (32)$$

$$RSH \longrightarrow RS^- + H^+ \quad (33)$$

$$RS^- + R'X \longrightarrow RSR' + X^- \quad (34)$$

In wastewater effluents treated for pathogen control with oxidants (e.g., chlorine or ozone), a variety of oxidation reactions should certainly be expected for the organic chemicals present. Data on the nature of these reactions are fairly abundant and are found in a variety of references (Hoigne, 1982, 1988; Hoigne and Bader, 1983a, 1983b; Hoigne et al., 1985; Jolley et al., 1978–1990; Rice and Cotruvo, 1978; Yao and Haag, 1991). A detailed discussion is beyond the scope of this chapter.

Some organic chemicals can apparently form fairly strong complexes with inorganic ions. Betterton et al. (1988), for example, report high stability constants (K) for complexes formed by the reaction of acetaldehyde and hydroxyacetaldehyde with the bisulfite ion (HSO_3^-). Specifically, for

$$K = \frac{[RCH(OH)SO_3^-]}{[RCHO][HSO_3^-]} \quad (35)$$

When RCHO was	K =
Acetaldehyde	$(6.90 \pm 0.54) \times 10^5$ M^{-1}
Hydroxyacetaldehyde	$(2.0 \pm 0.5) \times 10^6$ M^{-1}

The importance of such complexation reactions in the natural aquatic environment is not known.

Biodegradation

Biodegradation (i.e., microbiologically mediated degradation) is frequently one of the most important transformation processes for organic chemicals in water and soil; but at the same time, because of the numerous environmental variables, it is one of the hardest to evaluate quantitatively.

Biodegradation may occur in water or soil under aerobic or anaerobic conditions by microorganisms, bacteria, fungi, and algae. In fact, the majority of chemical reactions occurring in water, particularly those involving organic matter, occur through microbial intermediates. In aerobic biodegradation, organisms utilize oxygen as the ultimate electron acceptor in metabolism, whereas in anaerobic biodegradation organisms use electrophilic substrates like sulfate and nitrate and some organisms can use both oxygen and inorganic salts as oxidants. Aerobic processes are typically faster than anaerobic processes in surface waters because there are greater numbers of aerobic microorganisms with rapid growth. In general, biodegradation is the least understood of environmental transformations because microorganisms have complex, rapid life cycles and they can affect many chemical reactions.

Fungi and bacteria (except photosynthetic) are classified as reducers. Reducers break down chemical compounds to simple species and extract energy needed for growth and metabolism. Algae are primary producers of organic matter because they utilize light energy and store it as chemical energy for their metabolic requirements. Aquatic microorganisms regulate and modify the chemical composition of their environment through production and consumption of organic matter, their involvement in nutrient cycling (see Chapter 19), and their scavenging of trace metals and contaminants, and they mediate specific reactions through extracellular enzymatic reactions. The enzymes enable the microbes to obtain what they need for growth (major nutrients from

Table 9. Examples of acid dissociation constants for several aromatic and aliphatic compounds

Compound	K_a	pK_a
Aromatic		
p-Aminobenzoic acid		
K_1 (NH_3^+ group)	5.13×10^{-3}	2.29
K_2 (COOH group)	1.37×10^{-5}	4.86
m-Aminobenzoic acid		
K_1 (NH_3^+ group)	8.51×10^{-4}	3.07
K_2 (COOH group)	1.86×10^{-5}	4.73
p-Methoxybenzoic acid	3.38×10^{-5}	4.47
m-Phenoxybenzoic acid	1.12×10^{-4}	3.95
m-Methylsulfonylbenzoic acid	3.02×10^{-4}	3.52
p-Tolylacetic acid	4.27×10^{-5}	4.37
p-Nitrophenylarsonic acid, K_1	1.27×10^{-3}	2.90
p-Cyanophenol	1.12×10^{-8}	7.95
Tetralol-2	3.31×10^{-11}	10.48
1,3,5-Trihydroxybenzene, K_1	3.56×10^{-9}	8.45
m-Aminophenol, K_1 (NH_3^+ group)	6.76×10^{-5}	4.17
m-Aminophenol, K_2 (OH group)	1.35×10^{-10}	9.87
3-Bromo-4-methoxyl anilinium ion	8.32×10^{-5}	4.08
4-Chloro-3-nitroanilinium ion	1.26×10^{-2}	1.90
Aliphatic		
Bromoacetic acid	1.25×10^{-3}	2.90
Dichloroacetic acid	5.53×10^{-2}	1.26
Trifluoroacetic acid	0.59	0.23
Cyanoacetic acid	3.36×10^{-3}	2.47
But-3-enoic acid	4.62×10^{-5}	4.34
Chloromethylphosphonic acid	3.98×10^{-2}	1.40
Hydroxymethylphosphonic acid	1.23×10^{-2}	1.91
Glycine		
K_1	4.47×10^{-3}	2.35
K_2	1.66×10^{-10}	9.78
Aminocyanomethane	4.57×10^{-6}	5.34

Source: Lyman et al., 1981 (using data from various sources).

macromolecules and trace elements from unavailable redox species or complexes).

Microorganisms typically degrade many natural organic chemicals because they serve as a carbon source for microbial growth and maintenance. Synthetic organic chemicals may not be subject to microbial degradation unless the structures of these chemicals are similar to naturally occurring organic structures. In addition, some chemicals can be degraded only by cometabolism, which refers to the transformation of a chemical that can be degraded only in the presence of other chemicals or nutrients that can serve as the sole carbon or energy source.

In considering biodegradation, one should first determine whether *primary biodegradation* (any biologically induced structural change in the parent compound), *ultimate biodegradation* (complete conversion to inorganic compounds such as CO_2, H_2O, and N_2; also methane if anaerobic microorganisms are involved), or some in-between stage is of interest. In most aquatic toxicology studies, consideration of primary biodegradation may suffice. In other cases, however, it will be necessary to consider the formation of intermediates—which in a few cases are more toxic than the parent chemical—and the subsequent biodegradation of the intermediates.

In most cases the identification of the initial degradation products, not to mention all intermediates, will be difficult because (1) there are many different types of biodegradation reactions, yielding different products, and (2) the importance of these reactions changes with changing environmental conditions. Examples of biodegradation reactions include (Lyman et al., 1981):

β-Oxidation	Ester hydrolysis
Oxidative dealkylation	Amide hydrolysis
Thioether oxidation	Phosphorus ester hydrolysis
Decarboxylation	Nitrile hydrolysis
Epoxidation	Hydrolytic dehalogenation
Aromatic hydroxylation	Halogen migration
Aromatic ring cleavage	Reductive dehalogenation
Nitro reduction	Dehydrohalogenation

Biodegradation should be considered for all organic chemicals because it has been found to be a viable process for nearly all of the manufactured chemicals. Biodegradation may be of minimal importance only for such classes as insoluble solids (e.g., creosote tars and some polymers), completely or highly chlorinated chemicals (e.g., carbon tetrachloride, hexachlorobenzene, decachlorobiphenyl, mirex, octachlorodibenzodioxin), and large molecules that do not easily pass through cell membranes (e.g., some dioxins, polycyclic aromatic hydrocarbons with more than three rings, and petroleum constituents with high carbon numbers). Over long time scales (years, decades, or centuries), however, even these chemicals may undergo significant biodegradation. In many cases, susceptibility to biodegradation correlates with water solubility, hydrolysis rate constants, and the octanol-water partition coefficient, so some general guidance may be derived from an assessment of these chemical properties. Other indications of the relative susceptibility to biodegradation may be derived from the presence of certain functional groups and structural features. Tables 10 and 11 provide two examples of such correlations. Discussions of such correlations between biodegradability and chemical properties and/or structures are provided by Lyman et al. (1981), Desai et al. (1990), Geating (1981), Banerjee et al. (1984), Paris et al. (1982, 1983, 1984), and others (see Chapter 20).

The more important variables potentially affecting the rate of biodegradation in an aquatic system are listed in Table 12. A large volume of literature exists that describes, in detail, the effects of these variables. The reader is referred to the following for additional information: Lyman et al. (1981), Klecka (1985), Battersby (1990), Alexander (1985), and Hutzinger (1980–1988).

Of the many environment-related variables, oxygen availability ranks high in importance. Both aerobic (oxygen-rich) and anaerobic (oxygen-depleted) zones exist within most surface waters and sediments, and the nature and rate of biodegradation will usually be quite different in these two zones. Historically, aerobic biodegradation was considered to be more important, not only because of the larger volumes (or times) it was present in such environments but also because of generally higher degradation rates. More data have become available showing that anaerobic degradation plays a significant role for a wide variety of chemicals including chlorinated hydrocarbons and even plain hydrocarbons. Figure 4 shows, for example, that anaerobic biodegradation plays a significant role in the transformations of the common two-carbon chlorinated solvents that are frequently found as pollutants in both ground waters and surface waters. Although oxygen may not be

Table 10. Example of correlation between biodegradation rate constant and presence of functional groups[a]

Group	Group constant, α
Methyl (CH_3)	−1.3667
Methylene (CH_2)	−0.0438
Hydroxy (OH)	−1.7088
Acid (COOH)	−1.3133
Ketone (CO)	−0.5073
Amine (NH_2)	−1.4654
Aromatic CH (ACH)	−0.5016
Aromatic carbon (AC)	1.0659

[a] The relationship between the first-order biodegradation rate constant, k (1/h), and the group constants, α, is $\ln k = \Sigma N_i \alpha_i$, where N_i is the number of groups of type i in the compound and the sum covers all group types. The database of rate constants used to derive these group constants was associated with the following experimental conditions: temperature 20°C, pH 7, sludge concentration 309 mg/L, and compound concentration 100 mg/L.

Source: Reprinted from *Environ. Toxicol. Chem.* 9: Desai SM, Govind R, Tabak HH, Development of quantitative structure-activity relationships for predicting biodegradation kinetics, pp. 473–477, Copyright (1990), with the kind permission of Pergamon Press Ltd., Headington Hill Hall, Oxford 0X3 OBW, UK.

Table 11. Molecular structures and groups used by Geating (1981) to evaluate biodegradability

Description	Coefficient[a]
Single occurrence of sulfur in a ring	−13.9
More than two carbocyclic rings	−10.5
Alkyl chain (CH_2) or $CH_3(CH_2)_{n-1}$ where n = 10 or more (chain fragment)	5.03
One benzene ring	3.94
More than one —N= or HN= group (substituent fragment)	−12.1
One —C=O group (sub. fragment)	4.71
Atoms other than C, H, O, N, S, or halogen	5.01
One —OH group (sub. fragment)	3.03
Substituent hydroxylamine	−16.4
Single occurrence of carbonyl in a ring	6.16
Substituent primary amide	−11.0
Presence of suffix	−4.80
More than one —O— group (chain fragment)	−5.44
True bridge indicator (ring linkage)	12.8
Pyrimidine analog	−11.1
Branching terminal nitro-group-NO_2 outside ring	−3.19
Ethyl/ethylene group (chain fragment)	2.57
More than one —C(=O)—OH (acid) group (chain fragment)	3.85
Halogenated aromatic	−8.45
Aromatic six-member ring(s)	−2.36
More than two heteroatoms in one ring	−7.35
One three-branch carbon atom outside ring	−4.30
More than one three-branch carbon atom (extension)	−3.37
Alkyl/chain $(CH_2)_n$ or $CH_3(CH_2)_{n-1}$ where n = 3–9 (chain fragment)	2.29
One —C(=O)—OH (acid) group (sub. fragment)	2.11
One —C=O group (chain fragment)	2.73
Barbiturate (sub. fragment)	12.5
Chain phosphonyl	−6.61
One heterocyclic ring	2.34
Chain tertiary amide	−6.20
One —O— group (chain fragment)	−2.98
One carbo/carbo fusion	−3.40
More than one —OH group (chain fragment)	2.78
Carbocyclic five-member ring(s)	−5.09
Molecular weight	−.005
Triple bond outside a ring	−8.25
More than one —NH— group (sub. fragment)	4.83
More than one —OH group (sub. fragment)	1.60
More than one methyl/methylene group (chain fragment)	−2.46
Bilinkage	−2.10
Constant	3.42

[a] A chemical's degradability score is obtained from the sum of the constant (3.42) plus the coefficient value for all other structures/groups present. The more positive the final score, the more likely the chemical is to be biodegradable.

Source: Geating, 1981.

Table 12. Variables potentially affecting the rate of biodegradation

Substrate-related
Physicochemical properties
Concentration (possible toxic effects at high concentration)
Organism-related
Species composition and population
Spatial distribution (in water column or sediments)
Population density (concentration)
Previous history (adaptation)
Inter- and intraspecies interactions
Enzymatic makeup and activity
Environment-related
Temperature
pH
Oxygen availability
Presence of sorbing solids (and their nature)
Salinity
Other substances (including nutrients and electron acceptors)

Source: Adapted from Lyman et al., 1981.

present in the anaerobic zones, microbiologically mediated oxidation of organic matter can still take place with nitrate, sulfate, or iron acting as the electron acceptor (see, e.g., Lovley, 1991; Hutchins et al., 1991; and Haag et al. 1991).

There have been numerous measurements of biodegradation rates in real or simulated aquatic environments. In general, most data have been found to fit either a generalized power rate law or a second-order rate expression.

Power rate law:

$$\frac{-d[\mathrm{R}]}{dt} = k[\mathrm{R}]^n \tag{36}$$

Second-order rate expression:

$$\frac{-d[\mathrm{R}]}{dt} = k[\mathrm{R}][\mathrm{B}] \tag{37}$$

where [R] = solute concentration
$-d[\mathrm{R}]/dt$ = solute degradation rate
k = degradation rate constant
n = order of the reaction (first order if n = 1)
[B] = microbial population concentration

Other, more complicated rate expressions are available that account for other variables including microorganism growth. Examples of several biodegradation rate constants, or associated half-lives, for organic chemicals in aquatic systems are shown in Table 13. Because these rates are very situation specific, extreme caution should be used in extrapolating the value to other situations.

Because of the situation-specific variability, biodegradation rate constants are not commonly tabulated as are other physiochemical and toxicological properties of organic chemicals. The best data sources are often reviews for individual chemicals

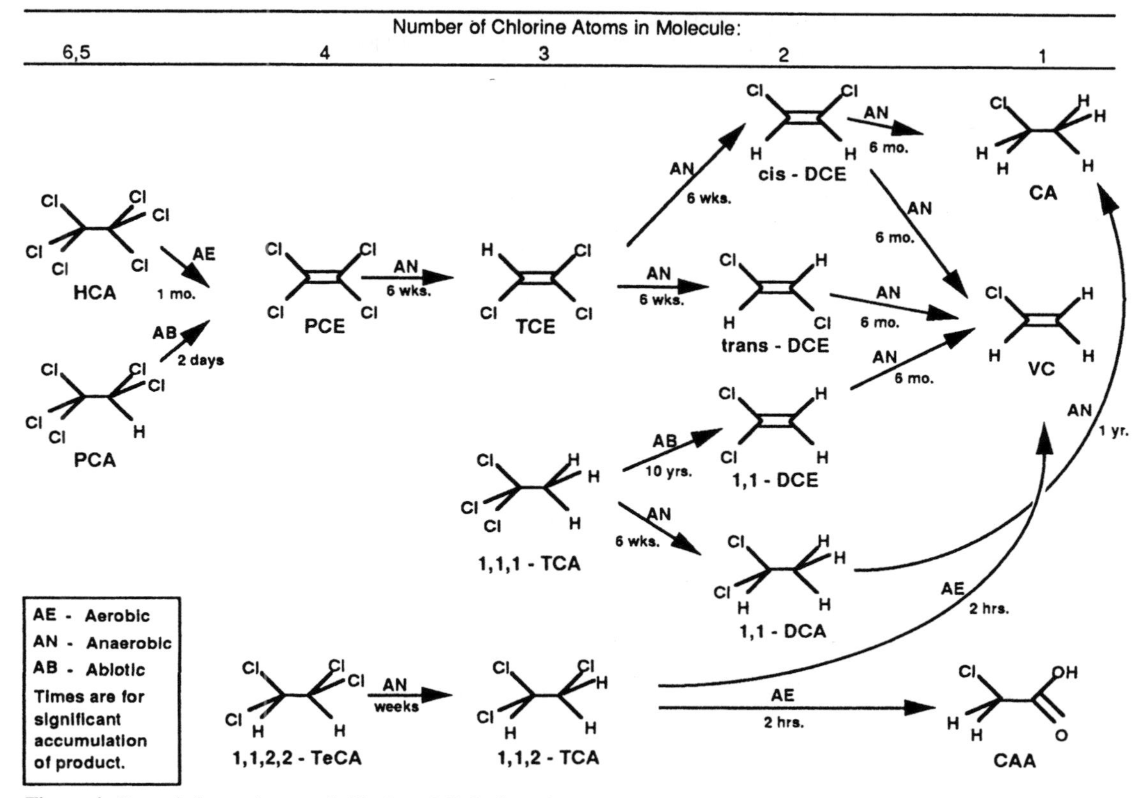

Figure 4. Degradation pathways of chlorinated C_2 hydrocarbons.

(e.g., Howard, 1989–1991) and related computerized databases (e.g., CHEMFATE issued by Syracuse Research Corp.) or the primary literature. There have been numerous attempts to develop estimation methods for biodegradation rate constants (and for related parameters like biological oxygen demand) but no current method has any significant breadth of applicability (to different chemical classes) or accuracy. Some further discussion of this subject is provided by Lyman et al. (1981), Enslein et al. (1984), and Geating (1981).

INORGANIC CHEMICALS

Introduction

Inorganic chemicals comprise all the elements and compounds that do not have carbon as the basic structural element. There is no unique or definitive listing of classes of inorganic chemicals, but common groups include metals (with subclasses for heavy metals and trace metals), metalloids, salts, acids, and bases. In addition, whole classes of inorganics are associated with the variety of compounds one particular element may form (e.g., oxides for oxygen and hydrides for hydrogen) or one particular group of elements may form (e.g., cyanides for CN, hypochlorites for OCl, and peroxides for O_2). Further classification of inorganic groups may be derived from common properties as reflected in their positioning within a column of the periodic chart of the elements (e.g., halides, rare gases).

Many transport and transformation processes can dramatically change the aquatic chemistry of inorganic chemicals. Thus, it should be clear that no assessment of the aquatic toxicology of an inorganic chemical should be undertaken without a simultaneous assessment of the chemical and biochemical processes acting on the chemical and of the resulting speciation. In aquatic chemistry, speciation refers to the formation, through covalent or ionic bonds, of different chemical compounds (species), which may be neutral or charged, dissolved, or precipitated. For example, the predominant calcium species in natural waters are Ca^{2+}, $CaHCO_3^+$, $CaCO_3$, and $CaSO_4$. For many inorganic chemicals, the assessment can be complex because of the multiplicity of species or complexes the inorganic element can form. Often the situation is so complicated, and reliable data are so scarce, that even with simplifying assumptions and state-of-the-art computerized speciation models there will still be significant uncertainty over the precise speciation and importance of individual transport and

Table 13. Biodegradation rate constants for organic compounds in aquatic systems

Compound	Rate constant[a]	See note
Anthracene	0.007–0.055 d^{-1}	b
Atrazine (N-phosphorylated)	$t_{1/2}$ 3.21 d	c
Benzo[*a*]anthracene	0	
Benzene	0.11 d^{-1}	
Benzo[*a*]pyrene	0	
Benzo[*f*]quinoline	8.6×10^{-7} $ml(cell)^{-1}$ d^{-1}	
Bis(2-ethylhexyl)phthalate	1.0×10^{-9} $ml(cell)^{-1}$ d^{-1}	
Carbaryl	2.4×10^{-10} $ml(cell)^{-1}$ d^{-1}	
Carbazole (9H)	1.2×10^{1} $ml(cell)^{-1}$ d^{-1}	
Chlorobenzene	0.0045 d^{-1}	
	0.0092 d^{-1}	d
Chlorodiphenyl oxide	7.2×10^{-3} $ml(gVS)^{-1}$ d^{-1}	
p-Chlorophenol	0.23 d^{-1}	d
Chlorpropham	$1.6–1.8 \times 10^{-8}$ $ml(cell)^{-1}$ d^{-1}	e
	$3.6–6.7 \times 10^{-10}$ $ml(cell)^{-1}$ d^{-1}	e,f
Crotoxyphos	$t_{1/2}$ = 7.5 d (pH 9)	
	= 22.5 d (pH 2)	
2,4-D (butoxyethyl ester)	$6.24–24.0 \times 10^{-6}$ $ml(cell)^{-1}$ d^{-1}	e,f
	6.2×10^{-5} $ml(cell)^{-1}$ d^{-1}	g
	9.6×10^{-7} $ml(cell)^{-1}$ d^{-1}	h
p,p'-DDE	0.0006 d^{-1}	d
Diazinon	$t_{1/2}$ = 4.91 d (pH 3.1)	
	= 185 d (pH 7.4)	
Diazoxon	$t_{1/2}$ = 0.016 d (pH 3.1)	
	= 27.9 d (pH 7.4)	
Dibenzo[*c,g*]carbazole	0	
Dibenzothiophene	1.27×10^{-5} $ml(cell)^{-1}$ d^{-1}	
Dimethyl phthalate	1.2×10^{-4} $ml(cell)^{-1}$ d^{-1}	
Di-*n*-butyl phthalate	7.4×10^{-7} $ml(cell)^{-1}$ d^{-1}	
Di-*n*-octyl phthalate	7.4×10^{-9} $ml(cell)^{-1}$ d^{-1}	
Galactose	$1.2–10 \times 10^{3}$ mg(g $bacteria)^{-1}$ d^{-1}	i
	1.4×10^{3} mg(g $bacteria)^{-1}$ d^{-1}	j
Glucose	0.24 d^{-1}	i
	$1.1–1.6 \times 10^{4}$ mg(g $bacteria)^{-1}$ d^{-1}	
	5.2×10^{3} mg(g $bacteria)^{-1}$ d^{-1}	j
Hexachlorophene	0.0024 d^{-1}	d
Malathion	$2.6–16.1 \times 10^{-7}$ $ml(cell)^{-1}$ d^{-1}	e,f
	6.2×10^{-8} $ml(cell)^{-1}$ d^{-1}	
	5.0×10^{-8} $ml(cell)^{-1}$ d^{-1}	
	1.9×10^{-1} mg (g $fungi)^{-1}$ d^{-1}	
Methyl anisate	1.3×10^{-8} $ml(cell)^{-1}$ d^{-1}	
Methyl benzoate	1.7×10^{-8} $ml(cell)^{-1}$ d^{-1}	
Mirex	0	
Nitrilotriacetate (NTA)	0.05–0.23 d^{-1}	k
Parathion	$t_{1/2}$ = >4250 d	
Paraoxon	$t_{1/2}$ = >4250 d	
p-Cresol	1.24×10^{-5} $ml(cell)^{-1}$ d^{-1}	
Phenol	0.079 d^{-1}	
Propham (IPC)	0.003–2.1 mg(g $bacteria)^{-1}$ d^{-1}	
Quinoline	7.4×10^{-5} $ml(cell)^{-1}$ d^{-1}	
Triallate	$t_{1/2} \approx$ 680 d (pH 6.8)	
	$\approx$ 1170 d (pH 7)	
2,4,5-T	0.001 d^{-1}	
	0.01–0.03 d^{-1}	d,l
1,4,5-Trichlorophenoxyacetic acid	0.0005 d^{-1}	
	0.0012–0.012 d^{-1}	d,l

[a] All tests assumed to be river die-away.
[b] First value is mean for d 0–15; second is for d 20–65.
[c] First-order half-life in aqueous solution.
[d] In sediment (slurry).
[e] Range due to measurement in different samples of river water.
[f] Rate constant does not account for lag period.
[g] Degradation by yeast culture (*Rhodotorula glutinis*).
[h] Degradation by bacterial culture (*Bacillus subtilus*).
[i] First value from unacclimated microbial population, second from acclimated population.
[j] River water bacterial culture.
[k] Dissolved concentrations ranging from 0.2 mg/L to saturation.
[l] Temperature range 9–21°C.
Source: Lyman et al., 1981 (using data from various sources).

transformation processes. Nevertheless, the assessment should be undertaken. The days of measuring and using "total" element concentrations in water to assess potential toxicological impacts are gone.

The most important transport and transformation processes acting on inorganic chemicals in water include the following:

Phase Transfer and Transport	Transformation
Phase transfer	Abiotic processes
Dissolution*	Acid-base equilibria (hydrolysis)
Precipitation*	Complexation
Sorption to solids*	Electron transfer (oxidation-reduction)
Volatilization to air	Polymerization
	Photolysis
Transport	Biotic
Bulk transport and dispersion	Oxidative-reduction
Sedimentation	Sulfide precipitation
Diffusion	Methylation-demethylation
	Dealkylation

Many, if not most, of these processes are sensitive to a number of environmental parameters, including:

- Temperature
- pH (negative log of the hydrogen ion concentration $[H^+]$)
- Oxidation-reduction potential (based on concentration of oxygen and relative concentrations of other redox couples, e.g., Fe^{2+}/Fe^{3+} and SO_4^{2-}/HS^-)
- Ionic strength
- Presence and concentration of complexing species (e.g., for silver, important complexing species are sulfur and halide ions such as S^{2-}, HS^-, Cl^-, and Br^-)
- Presence of solids on which sorption may take place
- Presence of microorganisms that may absorb and transform the chemical

Information allowing a quantitative assessment of the impact of these environmental parameters on the processes listed above is not abundant, and when available it may be difficult to apply rigorously because of uncertain applicability. Nevertheless, failure to consider the impact of these parameters could lead to erroneous conclusions.

Arsenic is a good example of a chemical that has some rather complicated aqueous chemistry. First, like many other metals and metalloids, it is widespread in the environment, being contained in over 100 minerals and ores and naturally present at up to 40 mg/kg in some soils (Bowen, 1979). Human activities, mostly involving the use of pesticides, have added to these levels in many areas. In water, arsenic is usually found as an oxyanion (e.g., $H_2AsO_4^-$, $HAsO_4^{2-}$, AsO_4^{3-}) or neutral compound (e.g., H_3AsO_3, H_3AsO_4). Methylated forms of arsenic are also formed and could be due to either discharge of pesticides (some of which are methylated arsenic compounds) or natural formation by methylating bacteria in the water, typically under reducing conditions (Anderson and Bruland, 1991). Altogether, some 5 to 10 arsenic species may be present in the water column of a pond or lake. Because these species can have significantly different environmental behaviors and toxicities, an analysis for "total" arsenic would tell extremely little about the speciation and potential aquatic toxicity.

The intent of this section on inorganic chemicals is to give the reader a brief primer on the transport and transformation processes listed above. It should be sufficient to alert you to the potential importance of one or more of these processes. For detailed assessments, more comprehensive works should be consulted (see Supplemental Reading).

*These processes often involve a chemical transformation as a part of the phase transfer; e.g., the dissolution of table salt (NaCl) results in the release of separate sodium (Na^+) and chloride (Cl^-) ions.

Phase Transfer and Transport

Phase Transfer

Although it should be understood that there may be many solid phases in contact with an aqueous solution, this section provides simple descriptions of the following fundamental phase transfer processes:

- Dissolution (solid to solution)
- Precipitation (solution to solid)
- Sorption (solution to sorption on, or in, a solid)
- Volatilization (solution to vapor phase)

Dissolution and precipitation refer to the two directions of the same reaction and are thus discussed together.

Dissolution and Precipitation

In assessments of water quality, the question often arises of whether a solid will dissolve in water and, if so, how much will dissolve and how fast. Alternatively, the assessment may start with the known presence of dissolved ions; the question is then whether these ions may combine to form a solid precipitate and, again, to what extent and how fast. Such questions can often be resolved through consideration of the chemical equilibria between solids and aqueous solutions. Answers regarding the preferred direction of the reaction—dissolution versus precipitation—are relatively easy to obtain relative to the reaction rate. In many cases, a solution or precipitation reaction may be favored by thermodynamic conditions but significantly hindered (slowed) by other factors.

It is easier to discuss dissolution-precipitation with the aid of a simple example, say of a metal salt, M_pA_q, that dissolves in water yielding a metal ion (M) and an anionic ligand (A):

$$\underset{\text{solid}}{M_pA_q} \underset{\text{Precipitation}}{\overset{\text{Dissolution}}{\rightleftharpoons}} \underset{\text{solution}}{pM^{z+} + qA^{z-}} \quad (38)$$

where p and q are stoichiometric coefficients, and z^+ and z^- are the positive and negative electrical charges on the ion. For all such reactions, a solubility product constant (K_{sp}) is defined as follows:

$$K_{sp} = [M]^p[A]^q \quad (39)$$

where the brackets, [ ], denote solution concentrations in units of moles per liter.

Values of K_{sp} vary widely, covering over 50 orders of magnitude (e.g., from over 10 for highly soluble solids to less than 10^{-50} for very insoluble solids). For example, the K_{sp} for FeS at 25°C is $10^{-18.1}$. As with hydrogen ion concentrations and pH, solubility product constants are often written in the negative log form (e.g., the pK_{sp} for FeS is 18.1). Tabulations of K_{sp} values, along with values for a related equilibrium constant K that considers ion activities, may be found in standard water chemistry texts (e.g., Morel, 1983; Stumm and Morgan, 1981) and chemistry handbooks (e.g., Weast, 1990) or in several specialized publications (e.g., Butler, 1964; Garrels and Christ, 1965; International Critical Tables, 1926–1930; Kotrly and Sucha, 1985; Linke, 1958, 1965; Sillen and Martell, 1964, 1971; Smith and Martell, 1976; Stephen and Stephen, 1963). A few selected values of K_{sp} are provided in Table 14.

Equation (39) holds true only when equilibrium conditions prevail [i.e., when the rates of the forward (dissolution) and backward (precipitation) reactions of Eq. (38) are equal]. If, however, the ion pair product ($[M]^p[A]^q$) is greater than K_{sp}, the solution is considered to be supersaturated and precipitation will be favored. Conversely, if the ion pair product is less than K_{sp}, the solution is undersaturated and dissolution of the solid, M_pA_q, will be favored.

Thus, with appropriate values of K_{sp}, an investigator should be able to make a qualitative assessment of the extent of undersaturation or oversaturation with respect to selected solid compounds. The extent of under- or oversaturation is also one factor that will

Table 14. Selected solubility product constants (at 25°C and zero ionic strength)

Solid	log K_{sp}	Solid	log K_{sp}
Hydroxides		Carbonates	
$Fe(OH)_2$	−15.1	$FeCO_3$	−10.68
$Fe(OH)_3$	−38.8	$CaCO_3$	−8.22
$Cd(OH)_2$	−14.39	$CuCO_3$	−9.63
$Cu(OH)_2$	−19.32	$ZnCO_3$	−10.00
$Mn(OH)_4$	~−56.0	$NiCO_3$	−6.87
$Hg(OH)_2$	−54.77		
$Ni(OH)_2$	−15.2	Sulfates	
$Tl(OH)_3$	−45.2	Ag_2SO_4	−4.83
$Zn(OH)_2$	−15.47	$CaSO_4$	−4.62
		$HgSO_4$	−6.13
Chlorides		$PbSO_4$	−7.79
AgCl	−9.752	$ZnSO_4$	3.41
Hg_2Cl_2	−17.91		
KCl	0.932	Sulfites	
$PbCl_2$	−4.78	Ag_2SO_3	−13.82
		$CaSO_3$	−6.5
Fluorides			
BaF_2	−5.76	Sulfides	
CaF_2	−10.50	Ag_2S	−50.1
CrF_3	−10.18	CdS	−27.0
LiF	−2.77	CuS	−36.1
MgF_2	−8.18	Cu_2S	−48.5
PbF_2	−7.44	FeS	−18.1
		HgS (black)	−52.7
Phosphates		α-NiS	−19.4
$AlPO_4$	−20.6	β-NiS	−24.9
$Ca(H_2PO_4)_2$	−1.14	γ-NiS	−26.6
$CaHPO_4$	−6.6	PbS	−27.5
$FePO_4$	−21.9	α-ZnS	−24.7
$Fe_3(PO_4)_2$	~−32.0	β-ZnS	−22.5
$Pb_3(PO_4)_2$	−43.5		
Chromates			
$BaCrO_4$	−9.67		
$PbCrO_4$	−13.75		
$AgCrO_4$	−11.92		

Source: Reprinted from Bodek et al. (eds.), *Environmental Inorganic Chemistry: Properties, Processes, and Estimation Methods*, Copyright 1988, 1280 pp, with permission from Elsevier Science Ltd., The Boulevard, Langford Lane, Kidlington 0X5 1GB, U.K.

determine the rate of dissolution or precipitation; other factors include temperature, solution volume and solids concentration, solid surface area, and the chemical composition of both the solid and solution phases. A detailed discussion of dissolution-precipitation rates is provided by Bodek et al. (1988).

Values for the solubility product constant are not constant but vary with temperature, pH, ionic strength, pressure, presence of complexing ligands, and the nature (e.g., particle size) of the solid phase formed. The effect of temperature may be either positive or negative, depending on the standard enthalpy of the reaction. For most chemicals, the change in K_{sp} is expected to be within about a factor of 2 (from the original value) for a temperature swing of up to 40°C (Bodek et al., 1988). The effect of pressure on K_{sp} values is much smaller than the effect of temperature, for the normal range of environmental pressures. Typically, values of K_{sp} increase slightly with increasing pressure. Decreasing particle size, for very small particles (typically with diameters less than 10^{-6} cm or 0.01 μm), leads to increases in K_{sp}. Changes in K_{sp} with ionic strength, pH, and the presence of other complexing ligands can be difficult to predict and are best addressed with one of the sophisticated computerized speciation models that are generally available (e.g., GEOCHEM, REDEQL, MINTEQ, and PHREEQE).

Sorption

The term sorption in environmental chemistry refers broadly to a variety of physical or chemical processes that result in the transfer of a chemical (A) from the solution phase to an existing solid phase:

$$A_{(soln)} + \text{solid} \rightleftharpoons \text{A-solid} \qquad (40)$$

Within the two broad categories of mechanisms that result in sorption (i.e., physical and chemical) are a number of contributing processes that may act alone or in combination. These include simple physical adsorption and several processes involving chemical interaction: ion exchange, organic complexation, precipitation and coprecipitation, solid-state diffusion, and isomorphic substitution. This multiplicity of potential mechanisms is one reason that sorption for inorganic chemicals is so difficult to characterize and predict.

From the viewpoint of aquatic toxicology, the primary importance of sorption derives from the fact that a portion of a chemical (A) of concern is taken out of solution—in which it is presumably bioavailable—and rendered partly inert or unavailable by attachment to a solid. Thus, analyses for "total" A in a water sample should be supplemented by analyses for "dissolved" A (i.e., via use of appropriate filtration techniques) to obtain a more toxicologically significant concentration.

As indicated by Eq. (40), the sorption reaction is usually considered reversible, although the equilibrium can lie so far to one side or the other that some chemicals may be considered to be (essentially) irreversibly sorbed or, at the other end of the spectrum, not sorbed at all. Also, as with precipitation-dissolution reactions, the rate of the (sorption) reaction can also be a confounding issue, with environmental systems often appearing to be far from presumed (or, perhaps, laboratory-measured) equilibrium conditions because of slow sorption or desorption reactions. Desorption from solids can be an especially slow process if it involves sorbate movement via molecular diffusion out of the inner pores of the solid to which it has "sorbed."

Because of the apparent existence of equilibrium sorption reactions, many environmental scientists and engineers seek a simple, unique sorption constant, as has been found for neutral organic chemicals (i.e., the K_{oc} parameter). Unfortunately, the sorption of inorganic species is highly influenced by the environmental conditions of the system under consideration, with different sorption mechanisms affected in different ways or to different degrees. Thus, unlike the case for most organic chemicals, *no species-unique sorption constant can be applied to a broad range of soils and sediments.* Depending on such factors as soil (or sediment) texture and chemistry, pH, redox potential, and solute and ligand concentrations in the water, a particular inorganic species may be strongly sorbed in one environment and weakly sorbed in another. Thus, only with great caution should a literature value of a sorption constant be applied to a different environment.

This confusing situation has not stopped researchers from seeking theoretical and empirical models to represent known sorption behaviors. From a practical point of view, the nonexpert should be aware of at least three forms of empirical equations used to obtain sorption constants and related parameters from experimental data presumed to represent equilibrium conditions. They are as follows.

Simple distribution coefficient:

$$S = K_d C \qquad (41)$$

Freundlich equation:

$$S = K_F C^{1/n} \qquad (42)$$

Langmuir equation:

$$S = \frac{K_L A_m C}{1 + K_L C} \qquad (43)$$

where S = mass sorbed at equilibrium per mass of sorbent
C = sorbate concentration in solution at equilibrium
K_d = distribution coefficient
K_F = Freundlich isotherm constant
n = Freundlich isotherm constant ($n \geq 1$)
K_L = Langmuir sorption constant
A_m = maximum sorption capacity of sorbent

A valuable compilation of empirical sorption constants for metals is provided by Rai et al. (1984). Values for a few selected metals are shown in Tables 15 and 16. Other values, also with a focus on soils rather than sediments, are provided by Bodek et al.

Table 15. Distribution coefficients for some metals[a]

Metal	Sorbent	Sorbate conc. (M)	Electrolyte Identity	Electrolyte Conc. (M)	pH	K_d (ml/g)
Ba	River sediment	$10^{-5.9}$	Seawater	≈0.7	8	530
	River sediment	$10^{-6.4}$	River water	—	—	2,800
Cd	Montmorillonite (Na form)	$10^{-6}-10^{-7}$	$NaNO_3$ + NaOAc	1.0,0.01	5.0	8
				1.0,0.01	6.5	100
				0.01,0.01	5.0	210
				0.01,0.01	6.5	900
Cu	Bentonite (Ca form)	$0-10^{-5.3}$	Seawater	≈0.7	8	43
	Kaolinite	Trace	Humic acid added	0 μg/L	6.4	43
				0.5 μg/L	6.4	3.2
				1.0 μg/L	6.4	2.5
				71.5 μg/L	6.4	2.2
	$Fe_2O_3 \cdot H_2O$ (am)	$10^{-6.5}$	Seawater	≈0.7	8	7,000
	MnO_2 (hydrous)	$0-10^{-5.5}$	Seawater	≈0.7	8	7,300
Mn	Fe oxide (hydrous)	$10^{-7.8}-10^{-5.8}$	Seawater	≈0.7	8	20,000
Hg	$Fe_2O_3 \cdot H_2O$ (am)	2.5×10^{-5}	$NaNO_3$	1.0	4.5	59,200
					5.95	2,550,000
	Bentonite	10^{-6}	$Ca(NO_3)_2$	0.01	6.7	408,000
					7.9	179,000
					8.9	119,000
Ni	Montmorillonite	$10^{-6.5}$	Seawater	≈0.7	8	200
	Fe oxide (hydrous)	$10^{-6.5}$	Seawater	≈0.7	8	100,000
Zn	δ-MnO_2	$10^{-8}-10^{-6}$	Seawater	≈0.7	8	800,000

[a]Definitions: $S = K_d C$, where: S = mass sorbed at equilibrium per mass of sorbent (μmol/g); K_d = distribution coefficient (ml/g), C = sorbate concentration in solution at equilibrium (μmol/ml).

Source: Reprinted from Bodek et al. (eds.), *Environmental Inorganic Chemistry: Properties, Processes, and Estimation Methods,* Copyright 1988, 1280 pp., with permission from Elsevier Science Ltd., The Boulevard, Langford Lane, Kidlington 0X5 1GB, U.K.

(1988). Large data compilations related to aquatic systems (i.e., where the sorbing material is either suspended or settled sediments) are not available. For those willing to invest time, effective K values can sometimes be derived from water quality databases (e.g., EPA's STORET), which often contain paired values of "total" and "dissolved" analyses for common heavy metals. However, care must be exercised in the use of highly averaged water quality data, as the composition of suspended solids (which may be of primary interest in some assessments) can vary significantly with location and time. Table 4 showed, for example, significant differences in the particle types making up the suspended material in a lake and in its tributaries. Significant temporal variations in the particle makeup were also seen, the major factor being large storm water runoff events.

It should be clear from the discussion of sorption variability (and complexity), that the extent of sorption of an inorganic chemical on a particular soil or sediment cannot be estimated, ab initio, with the degree of reliability usually desired for quantitative (or even semiquantitative) modeling and toxicological assessments. If literature values representing very similar systems cannot be found, then assessments should proceed with the use of reasonable (sometimes reasonable worst-case) assumptions, or laboratory measurements should be made to obtain the necessary data.

A simple calculation that can be made when assessing the importance of sorption is to determine the fraction (F) of the chemical that is associated with the solids in a system. For example, in a surface water with X g/L of suspended solids and dissolved and sorbed concentrations of chemical A given by $[A]_d$ (mg/L) and $[A]_s$ (mg/g), respectively, the fraction is given by

$$F = \frac{X[A]_s}{X[A]_s + [A]_d} \quad (44)$$

Volatilization

In this text, volatilization refers to the transport of aqueous solutes across the air-water interface and into the air. In many cases, limited essentially to neutral solutes, it can be an important loss mechanism, removing chemicals from the water and transferring them to the air. In a few other cases, the reverse reaction (vapor dissolution*) can be quite important; examples include sulfur and nitrogen oxides—precursors of acid rain—and mercury, a globally transported air pollutant.

For all chemicals that do exchange across the air-water interface, there is a point at which the volatilization and dissolution rates will be equal. The ratio of gas- and liquid-phase concentrations under such

*The transfer of gases from the air to water can occur by a number of mechanisms including simple dissolution (may also be called dry deposition or impingeness) and wet deposition, where the gases are first dissolved in falling rain or snow.

Table 16. Langmuir constants for some metals[a]

Metal	Sorbent	Sorbate conc. (M)	Electrolyte Identity	Electrolyte Conc. (M)	pH	Constants A_m (μmol/g)	Constants K_L (log M^{-1})
As	Hydroxide, $Al(OH)_3$ (am)	As(V) $10^{-4.2}-10^{-3.1}$	—	—	5	1,600	5.08
					6	1,487	5.17
					7	1,179	5.24
					8	538	5.12
					8.5	681	4.85
					9	501	4.82
	Oxyhydroxide, $Fe_2O_3 \cdot H_2O$ (am)	As(III) as AsO_2^- $10^{-7}-10^{-5}$	$NaNO_3$	1.0	4.0	457	5.98
					5.7	490	6.26
					7.0	513	6.36
					8.8	417	6.18
	Montmorillonite	$10^{-4}-10^{-3}$ As(V) + $10^{-4}-10^{-3}$ As(III)	Leachate	—	5	9.9[b]	3.57[b]
Ba	MnO_2 (hydrous)	$10^{-5}-10^{-3}$	$NaClO_4$	0.01	5	2,050	4.6
Cd	Montmorillonite (Na form)	$10^{-6.9}-10^{-6}$	$NaClO_4$	0.01	6.5–7	0.5	7.7
				0.03	6.5–7	0.4	7.2
				0.05	6.5–7	0.4	6.8
				0.17	6.5–7	0.2	6.9
				1.0	6.5–7	0.3	6.7
	River sediment	$10^{-6.7}-10^{-3.8}$	—	—	~7.5	10–173	4.4–6.0
Cr	Kaolinite	$10^{-4}-10^{-2.2}$ Cr(VI)	Leachate	—	3	3.64	—
					4	2.5	—
					7	0.98	—
		$10^{-3.3}-10^{-1.8}$ Cr(III)	Leachate	—	3	96.3	—
					4	283	—
Cu	Kaolinite	$10^{-5.8}-10^{-3.8}$	$CaCl_2$	0.05	5.5	1.9	4.5
	Peat	$0-10^{-3.8}$	$CaCl_2$	0.05	5.5	184	5.0
	$Fe_2O_3 \cdot H_2O$ (am)	$0-10^{-4.5}$	Seawater	~0.7	8	1,120	5.3
Pb	Montmorillonite	$10^{-4.3}-10^{-2.7}$	Landfill Leachate	~0.1	5.0	8.8 (A_{m1}) 5.38 (A_{m2})	4.3 (K_{L1}) 2.1 (K_{L2})
	Goethite	$10^{-3}-10^{-2.2}$	KNO_3	0.1	5.0	85	2.9
	Mn oxide (hydrous)	$10^{-3.5}-10^{-2.8}$	—	—	6	7,000	3.7
	Various soils	$10^{-4.6}-10^{-3.3}$	KCl	0.1	7	7.0–23	3.7–4.9
Mn	Montmorillonite	0.005–0.05	—	—	5–6	370,610	2.7,2.8
	Cryptomelane (α-MnO_2 as $K_2Mn_8O_{16}$)	$10^{-3}-10^{-2.2}$	KNO_3	0.1	5	1,130	3.5
	Clay loam	$0-10^{-2.9}$	Na_2SO_4	0.01	5.4	14.7,19.3	3.4
Hg	Sediments	$10^{-6}-10^{-4}$	—	—	—	11.9–237	5.5–6.5
Ni	δ-MnO^2	$10^{-3}-10^{-2.2}$	KNO_3	0.1	5	690	3.6
Se	Kaolinite	$10^{-3}-10^{-4}$ Se(IV)	Landfill Leachate	—	3	4.20	3.23
					5	2.93	3.37
					7	2.40	3.26
Zn	Montmorillonite, Ca-saturated	$10^{-3.6}-10^{-2.6}$	Landfill Leachate	~0.1	5	29	3.0
					6	35	3.1
	$Fe(OH)_3$ (am)	$10^{-6.7}-10^{-5}$	—	—	6	25	5.4
					7	170	5.9
					8	440	6.7
	Calcite	$10^{-7}-10^{-5.3}$	—	—	8.4	59	5.2
	Clay loam	$0-10^{-5.5}$	$CaCl_2$	0.016	7.7[c]	9.4	5.5

[a]Definitions: $S = K_L A_m C/[1 + K_L C]$, where S = mass sorbed at equilibrium per mass of sorbent (μmol/g), A_m = maximum sorption capacity of sorbent (μmol/g), K_L = Langmuir constant related to binding energy of sorbate (L/μmol), given above as log [L/mol], C = sorbate concentration in solution at equilibrium (μmol/L), A_{m1},K_{L1},A_{m2},K_{L2} refer to two sorption sites.

Source: Reprinted from Bodek et al. (eds.), *Environmental Inorganic Chemistry: Properties, Processes, and Estimation Methods*, Copyright 1988, 1280 pp., with permission from Elsevier Science Ltd., The Boulevard, Langford Lane, Kidlington 0X5 1GB, U.K.

equilibrium conditions is called the Henry's law constant (H) (discussed earlier):

$$H = \frac{[A]_{air}}{[A]_{water}} \qquad (45)$$

where H = Henry's law constant
$[A]$ = equilibrium concentration of chemical A in specified phase

With values of $[A]_{air}$ and $[A]_{water}$ in atmospheres and moles per liter, respectively, H is obtained in units of atm L/mol. Other units are also commonly used. Values of H for selected gases are shown in Table 17. Also shown are the products of reaction with water.

Values of H depend on a number of environmental parameters, especially temperature. Values of H increase with increasing temperature in amounts that are usually nearly proportional to the increase of the chemical's vapor pressure over the same temperature interval. For sparingly soluble chemicals, H can be estimated as the ratio of the vapor pressure (P) to the water solubility (S) of the chemical (i.e., $H = P/S$).

For many inorganic chemicals, the importance of volatilization depends as much on the aqueous speciation and complexation of the chemical as on the value of H. For example, the volatilization of hydrogen sulfide (H_2S) from water depends heavily on how far to the right the dissociation reaction [Eq. (46)] is forced:

$$\begin{array}{l} H_2S \rightleftharpoons H^+ + HS^- \\ \qquad\qquad\quad \updownarrow \\ \qquad\qquad H^+ + S^{2-} \end{array} \qquad (46)$$

A high pH, as well as the presence of sulfide (S^{2-}) complexing agents, tends to move this reaction to the right. This, in turn, lowers the concentration of H_2S and thus lowers volatilization losses of this chemical. As implied above, charged species in water (e.g., HS^- and S^{2-}) have little or no tendency to volatilize from water because of their high solubility.

In addition to the chemicals listed in Table 17, there are many other chemicals for which volatilization may be an important transport mechanism. This includes many hydrides (e.g., SiH_4, GeH_4, AsH_3, SbH_3 and SeH_2), carbonyls (e.g., $COCl_2$), methylated metals [e.g., $(CH_3)_2Hg$ and $(CH_3)_4Pb$], and elemental mercury (Hg^o). The value of H for mercury is reported to be 8.5 atm L/mol (Stumm and Morgan, 1981), and the ratio of the mass transfer coefficient of mercury to that of oxygen has been reported as 0.94 ± 0.08 between 0°C and 30°C. The latter ratio implies that, with an equivalent concentration gradient across the air-water interface, mercury and oxygen would tend to volatilize at about the same rate.

Mathematical models are available with which to estimate the rates of volatilization of chemicals from water. However, even under the best of circumstances, uncertainties of a factor of 5 are usually still present in the estimates. The simpler expressions show the volatilization flux, N (units are mass/area × time), being proportional to the concentration gradient across the interface, ΔC, and to another term, k, that is related to Henry's law constant, H, and to liquid- and gas-phase mass transfer coefficients:

$$N = k\,\Delta C \qquad (47)$$

The mass transfer coefficients reflect both chemical and environmental conditions, the former including molecular diffusion coefficients and the latter including such items as wind and current velocity. Useful descriptions of volatilization models are given by Thomas (1982).

In many environmental assessments it may suffice to determine qualitatively whether volatilization is important. This may be done by taking field measurements of the air and water concentrations for the chemical of interest and comparing the ratio with measured or estimated values of H (which represent

Table 17. Henry's law constant and aqueous reaction products for some gases

Gas	Henry's law constant, H, at 20°C (mm Hg/mole fraction)	Products of reaction with water
CO	4.07×10^7	—
CO_2	1.08×10^6	H_2CO_3, HCO_3, CO_3^{-2}, H^+
Cl_2	5.72×10^5[a]	HOCl, Cl^-, H^+
H_2S	3.67×10^5	HS^-, S^{-2}, H^+
NH_3	2.10×10^3	NH_4^+, OH^-
N_2	6.11×10^7	—
N_2O	1.50×10^6	—
NO	2.01×10^7	Converted to NO_2 on contact with O_2
O_2	3.04×10^7	—
O_3	2.86×10^6	—
SO_2	2.62×10^4	$SO_2 \cdot H_2O$, HSO_3^-, SO_3^{2-}, H^+

[a]Henry's law constant for molecular chlorine, determined at 1 atm pressure of pure chlorine.

Source: Reprinted from Bodek et al. (eds.), *Environmental Inorganic Chemistry: Properties, Processes, and Estimation Methods*, Copyright 1988, 1280 pp., with kind permission from Elsevier Science Ltd., The Boulevard, Langford Lane, Kidlington 0X5 1GB, U.K.

equilibrium conditions). If the field-determined ratio is significantly smaller than H, then the aqueous phase concentration is higher than the equilibrium amount and the chemical will volatilize into the air. If the field-determined ratio is larger than H, the direction of net transport will be the other way, from the air to the water.

Transport

The reader is referred to the previous subsection on transport, which covers both inorganic and organic chemicals.

Transformation

This second major section on inorganic chemicals describes a number of abiotic and biotic processes that can change the speciation of an inorganic element present. The change mechanisms, or transformations, that are covered are as follows:

Abiotic processes	Biotic processes
Acid-base equilibria	Oxidation-reduction
Complexation	Sulfide precipitation
Oxidation-reduction	Methylation-demethylation
Polymerization	Dealkylation
Photolysis	

Abiotic Processes

In an aqueous environment a metal ion is usually written as M^{n+}, which indicates a simple metal ion $M(H_2O)_x^{n+}$. Metal ions do not exist as a separate bare entity in water because they try to reach a state of maximum stability with their outer electron shells through chemical reactions in which the ions are bonded (coordinated) to water molecules or other stronger bases (i.e., electron-donor partners). Acid-base, precipitation, complexation, and oxidation-reduction reactions provide means through which metal ions in water are transformed to more stable forms.

Acid-Base Equilibria

One of the more important aqueous transformations involves the interaction of the chemical with water. The reaction is sometimes referred to as hydrolysis, but it is also called acid dissociation or base dissociation. As with other reactions, a forward reaction and a reverse reaction are always present with equilibrium representing equal forward and reverse rates. According to one simple definition an acid (e.g., AH) is a species that can donate a proton (H^+):

$$AH \rightleftharpoons A^- + H^+ \tag{48}$$

and a base (e.g., A^- or BOH) is a species that can accept a proton (H^+) or donate a hydroxyl ion (OH^-):

$$A^- + H^+ \rightleftharpoons HA \tag{49}$$

$$B{-}OH \rightleftharpoons B^+ + OH^- \tag{50}$$

For each such reaction there is a parameter, the acid dissociation constant, K_a, that relates the equilibrium concentrations of species on the right and left sides of each reaction. For HA, K_a would be given by

$$k_a = \frac{[A^-][H^+]}{[HA]} \tag{51}$$

where the brackets [ ], strictly represent activities but can be taken as molar concentrations at low ionic strengths. Equivalent base dissociation constants, K_b, are associated with reactions such as those shown in Eqs. (49) and (50).

Where an acid can donate more than one proton (e.g., two for H_2SO_4 and three for H_3PO_4), a series of dissociation reactions exist, each with its own value of K_a. Even aqueous species that are not directly bonded to hydrogen atoms can act as acids or bases. For example, Fe^{3+}, because it is hexacoordinated with water molecules, can have stepwise dissociation reactions as follows:

$$Fe(OH_2)_6^{3+} \rightleftharpoons Fe(OH_2)_5(OH)^{2+} + H^+, \quad K_1 \tag{52}$$

$$Fe(OH)_5(OH)^{2+} \rightleftharpoons Fe(OH_2)_4(OH)_2^+ + H^+, \quad K_2 \tag{53}$$

$$Fe(OH_2)_4(OH)_2^+ \rightleftharpoons Fe(OH_2)_3(OH)_3^\circ + H^+, \quad K_3 \tag{54}$$

The reader is directed to other works for a detailed discussion of these systems (e.g., Stumm and Morgan, 1981; Bodek et al., 1988). The important point to note for all such acids (or bases) is that there will be dissociation in water and that the pH of the system, which defines the value of $[H^+]$, will strongly influence the equilibrium concentration of the dissociated and undissociated species.

When evaluating the aquatic chemistry of a chemical that can act as an acid or base, it is helpful to calculate the theoretically anticipated concentrations of all species in the reaction chains or the fraction present in a certain form. For example, the fraction (f) of total A in a solution, containing HA and A^- according to Eq. (18), that is in the form of A^- is given (at equilibrium) by

$$f = \left(1 + \frac{[H^+]}{K_a}\right)^{-1} \tag{55}$$

Note that the value of $[H^+]$ would be obtained from the solution pH as 10^{-pH}. Equations, similar to Eq. (55), for more complicated systems are given by Bodek et al. (1988). Selected values of acid dissociation constants are provided in Table 18 in the negative log form (i.e., $pK_1 = -\log K_1$). Note that they range over 13 orders of magnitude. Useful compendia of acid dissociation constants and related thermodynamic parameters are given by Sillen and Martell (1964), Martell and Smith (1977, 1982), Smith and Martell (1982), and Perrin (1981). For complicated systems, the speciation calculations can become very complicated and one would be strongly advised to use one of the many computerized spe-

Table 18. Values of acid dissociation constants for a variety of inorganic species

Species[a,b]	pK_1	pK_2	pK_3	pK_4
Ag^+	11.9	11.9	—	—
Al^{3+}	5.4	4.6	5.7	7.9
H_3AsO_4	2.1	6.7	11.2	—
H_3AsO_3	9.1	12.1	13.4	—
Ba^{2+}	13.2	—	—	—
Be^{2+}	5.7	8.2	9.9	14.0
H_3BO_3	9.1	12.7	13.8	—
HCN	9.0	—	—	—
H_2CO_3	6.2	10.0	—	—
H_2CrO_4	0.74	6.5	—	—
Cr^{3+}	4.2	6.2	8.3	9.1
Cd^{2+}	10.3	10.3	13.2	13.1
Co^{2+}	9.9	8.9	12.7	—
Cu^{2+}	8.2	9.3	10.3	11.3
HF	2.9	—	—	—
Fe^{2+}	9.7	11.1	10.2	15.0
Fe^{3+}	2.6	3.6	3.8	11.9
Hg^{2+}	3.8	2.4	14.9	—
Mn^{2+}	10.8	11.6	12.4	13.1
Ni^{2+}	10.2	9.0	10.8	14.0
H_3PO_4	2.0	6.7	11.7	—
Pb^{2+}	7.9	9.4	10.7	—
H_2S	7.0	12.0	—	—
H_2SO_3	1.8	6.8	—	—
Sn^{2+}	3.6	3.7	9.3	—
Tl^{3+}	0.94	1.2	1.7	—
V^{3+}	2.7	3.8	7.0	—
Zn^{2+}	9.2	7.9	11.3	12.3

[a] The terminology assumes an aq subscript on each species, which refers to the fact that the ion is in aqueous solution. The acid dissociation constants thus refer to loss of H^+ from successive water molecules, e.g.,

$$Fe(H_2O)_6^{3+} \rightleftharpoons Fe(H_2O)_5\,OH^{2+} + H^+ \quad K_1$$
$$Fe(H_2O)_5\,OH^{2+} \rightleftharpoons Fe(H_2O)_4\,(OH)_2 + H^+ \quad K_2$$

Values are given as $pK_i = -\log K_i$. Temperature is 20–25°C and ionic strength is 0.1 M.

[b] Data obtained from Kragten (1978) was used to calculate values as follows: $K_1 = {}^*\beta_1$, $K_2 = {}^*\beta_2/{}^*\beta_1$, $K_3 = {}^*\beta_3/({}^*\beta_1 K_2)$, and $K_4 = {}^*\beta_4/({}^*\beta_1 K_2 K_3)$.

Source: Reprinted from Bodek et al. (eds.): *Environmental Inorganic Chemistry: Properties, Processes and Estimation Methods*, Copyright 1988, 1280 pp., with permission from Elsevier Science Ltd., The Boulevard, Langford Lane, Kidlington 0X5 1GB, UK.

ciation programs available (e.g., GEOCHEM, REDEQL, MINTEQ, and PHREEQE).

Acid dissociation constants may increase or decrease with increasing temperature, although the changes are usually not large over the temperature ranges in most aquatic environments. In comparison with other aqueous environmental inorganic reactions, acid dissociation reactions are usually considered instantaneous. Thus there is no practical reason to consider reaction rates.

Complexation

Most inorganic ions in aqueous solution can undergo a variety of complexation reactions with other ligands (i.e., complexing agents) that can result in strongly bound complexes or loose ion pairs. Overall, the complexation of a metal (M) and a ligand (L), with charges of $x+$ and $y-$ (respectively), can be written as

$$M^{x+} + L^{y-} \rightleftharpoons ML^{x-y} \quad (56)$$

When equilibrium conditions exist for this reaction, a complexation equilibrium constant or stability constant, K_1, is defined as follows:

$$K_1 = \frac{[ML^{x-y}]}{[M^{x+}][L^{y-}]} \quad (57)$$

In reality, the simple reaction described above may proceed through a two-step process involving (1) substitution of the ligand (L) for a water molecule held in an outer coordination sphere around the central metal ion (M) and (2) substitution of L for a water molecule held in an inner coordination sphere. For many ions, the exchange between the inner and outer coordination ligands is so rapid that it is not possible to distinguish the inner- and outer-sphere complexes.

Also, as with acid-base reactions, complexation reactions may proceed in a stepwise manner, each step replacing a coordinated water molecule:

$$M + L \rightleftharpoons ML, \quad K_1 \quad (58)$$
$$ML + L \rightleftharpoons ML_2, \quad K_2 \quad (59)$$
$$ML_2 + L \rightleftharpoons ML_3, \quad K_3 \quad (60)$$

For such a complexation series, an overall stability constant, β, may be defined for

$$M + 3L \rightleftharpoons ML_3, \quad \beta = K_1K_2K_3 \quad (61)$$

The environmental importance of complexation derives from the fact that the complexed metal (ML) will behave differently from the uncomplexed metal (M), as well as differently from other complexes. Chemical properties such as solubility, attenuation behavior on soils, bioconcentration factors, and toxicity are modified through complexation.

The total number of ligands possible for any metal is difficult to define and can greatly exceed the number of ligand positions in the inner coordination sphere (typically six or four).

Complexation is most commonly considered for metal ions in water (e.g., for Fe^{3+}, Al^{3+}, Co^{2+}). Common inorganic anion ligands that are found in most natural waters include OH^-, Cl^-, SO_4^{2-}, CO_3^{2-}, and PO_4^{3-}. Metal cations also complex easily with a variety of natural and synthetic organic chemicals such as humic acids, acetic acid, and nitrilotriacetic acid. Some of these ligands may have more than one donor atom that can form a bond to the metal ion. These are called chelates. For example nitrilotriacetic acid, $N(CH_2COOH)_3$, has three acetic acid units, each of which can bind to a ligand position on a metal ion. The resulting chelate complex ion can have markedly different physicochemical and toxicological properties from the uncomplexed metal.

Selected values for equilibrium constants (K) and stability constants (β) for metals with inorganic and

Table 19. Logarithms of complexation equilibrium constants (K) and overall stability constants (β_2) for metals with commonly occurring ligands (at 25°C and zero ionic strength, unless noted)

Metal	M-Cl[a]	M-Cl$_2$[b]	M-SO$_4$[a]	M-OH[a]	M(OH)$_2$[b]
Na^+	−1.85	—	1.06	−0.2	—
K^+	−0.5	—	0.96	−0.5	—
Ca^{2+}	—	—	2.31	1.15	—
Mg^{2+}	—	—	2.36	2.56	—
Ba^{2+}	−0.13[c]	—	2.7	0.5	—
Cr^{3+}	0.6	—	3.0	10.0	18.3
Al^{3+}	—	—	—	9.0	18.7
Fe^{3+}	1.5	2.1	4.0	11.8	22.3
Mn^{2+}	0.6	—	2.3	3.4	5.8
Fe^{2+}	0.36[d]	0.40[c]	2.2	4.5	7.4
Co^{2+}	0.5	—	2.2	4.3	5.1
Ni^{2+}	0.6	—	2.3	4.1	9.0
Cu^{2+}	0.4	−0.4	2.4	6.5	11.8
Zn^{2+}	0.4	0.0	2.1	5.0	11.1
Pb^{2+}	1.6	1.8	2.8	6.3	10.9
Hg^{2+}	7.2	14.0	2.5	10.6	21.8
Cd^{2+}	2.0	2.6	2.3	3.9	7.6
Ag^+	3.3	5.3	1.3	2.0	4.0

[a] Refers to K values for the reaction $M + L \rightleftharpoons ML$ where $L = Cl^-$, OH^- or SO_4^{2-}.
[b] Refers to β_2 values for the reaction $M + 2L \rightleftharpoons ML_2$ where $L = Cl^-$ (or OH^-).
[c] 18°C.
[d] 2 *M* ionic strength.
Source: Data from Kotrlý and Šůcha (1985); Morel (1983); Sillén and Martell (1964).

organic ligands are shown in Tables 19 and 20, respectively. As implied by the notes in these tables, complexation constants are affected by such properties as temperature and ionic strength. For many metal-organic complexes, pH also affects the extent of complexation since H^+ competes for the basic coordination site. As with precipitation-dissolution and acid-base reactions, calculations of the equilibrium concentrations of complexed and uncomplexed metals are best done with computerized speciation models unless only a few reacting species are present.

Complexation reaction rates vary widely, over 15 orders of magnitude, and are affected by a variety of variables, including (1) the identity and oxidation

Table 20. Logarithms of complexation equilibrium constants (K) for metals with sequestering agents of commercial interest (1:1 complex formation at 25°C and 0.10 *M* ionic strength, unless noted)

	Sequestering Agent[a]					
Metal	NTA	HIDA	EDTA	HEDTA	TPP	CIT[b]
Mg^{2+}	5.47	3.46	8.83	7.0	5.8	3.40
Ca^{2+}	6.39	4.77	10.61	8.2	5.2	3.55
Mn^{2+}	7.46	5.56	13.81	10.8	7.0	3.7
Fe^{2+}	8.82	6.77	14.27	12.2	—	4.4
Fe^{3+}	15.9	11.6	25.0	19.8	—	11.40
Co^{2+}	10.38	8.02	16.26	14.5	6.9	5.00
Ni^{2+}	11.50	9.33	18.52	17.1	6.7	5.40
Cu^{2+}	12.94	11.72	18.70	17.5	8.1	5.90
Zn^{2+}	10.66	8.45	16.44	14.6	7.3	4.98
Cd^{2+}	9.78	7.24	16.36	13.1	6.5	3.75
Hg^{2+}	14.6	5.4	21.5	20.0	—	—
Pb^{2+}	11.34	9.5	17.88	15.5	—	4.08[c]

[a] NTA = nitrilotriacetic acid, $N(CH_2COOH)_3$
HIDA = ethanolaminediacetic acid, $(HOCH_2CH_2)N(CH_2COOH)_2$
EDTA = ethylenediaminetetraacetic acid, $(HOOCCH_2)_2NCH_2CH_2N(CH_2COOH)_2$
HEDTA = *N*-(2-hydroxyethyl)-ethylene-dinitrilotriacetic acid, $(HOCH_2CH_2)(HOOCCH_2)NCH_2CH_2N(CH_2COOH)_2$
TPP = tripolyphosphoric acid, $(HO)_2OPOPO_2(OH)PO(OH)_2$
CIT = citric acid, $(HOOCCH_2)(OH)(COOH)C(CH_2COOH)$.
[b] 20°C.
[c] 2.0 *M* ionic strength.
Source: Martell, 1975.

state of the central metal ion; (2) the identity of the ligand(s); and (3) the chemical composition, temperature, and ionic strength of the solution. The subject is described in detail by Bodek et al. (1988).

Oxidation-Reduction

An important class of aqueous inorganic and organic reactions involves the transfer of electrons between two species—the oxidizing agent (oxidant) and the reducing agent (reductant). In such oxidation-reduction reactions (also called redox reactions), the oxidant gains electrons and the reductant loses them. For example, the oxidation of a metal (M) by oxygen (O_2) from a 2+ to a 3+ oxidation state is represented as

$$4M^{2+} + O_2 + 4H^+ \rightleftharpoons 4M^{3+} + 2H_2O \quad (62)$$

Here M^{2+} loses an electron as it is oxidized to M^{3+} and oxygen is reduced from O_2 to the 2− oxidation state (H_2O) by gaining an electron. As with other reaction types already described, redox reactions have an equilibrium point at which the rates of the forward and reverse reactions are equal. Hypothetical equilibrium constants can be assigned at least to each half-reaction*, but it is more common to define the equilibrium in terms of electrode potentials ($E°$, in volts) or theoretical electron concentrations, *pe* ($= -\log[e^-]$), at equilibrium. Values of $E°$ for redox couples vary from about −2 to +3 V. The values of *pe* are related to the values of equilibrium constants and can vary widely. Selected values of $E°$ are provided in Table 21 as examples.

A measure of the redox potential in a system is given by E_h, which can be measured with a platinum electrode. Values of E_h and $E°$ are related by the Nernst equation. For a half-reaction with a couple potential of E°;

$$E_h = E° + \frac{2.3RT}{nF} \log \frac{[\text{oxid}]}{[\text{red}]} \quad (63)$$

where E_h = potential measured at platinum electrode (V)
$E°$ = electrode potential of half-reaction (V)
R = gas constant
T = temperature (K)
n = number of electrons transferred
F = Faraday constant (23.06 kcal/V mol equiv)
[oxid] = concentration of oxidant
[red] = concentration of reductant

The range of E_h in aqueous systems is approximately −0.8 to +1.2 V. Thus, knowledge of E_h (e.g., a field-measured value) and $E°$ allows an estimate to be made of the relative concentration of the oxidized and reduced species to be expected in a solution at equilibrium.

Oxidation-reduction reactions, like acid-base, complexation, and precipitation-dissolution reactions,

*The half reactions for Eq. (62) would be
$M^{2+} \rightarrow M^{3+} + e^-$
$O_2 + 4H^+ + 4e^- \rightarrow 2H_2O$

Table 21. Electrode potentials of redox half-reactions[a]

Oxidation-reduction reaction	$E°$ (V)
$Ag^{2+} + e^- \longrightarrow Ag^+$	+2.00
$Ag^{2+} + e^- \longrightarrow Ag(s)$	+0.799
$Al^{3+} + 3e^- \longrightarrow Al(s)$	−1.66
$Am^{4+} + e^- \longrightarrow Am^{3+}$	+2.40
$H_3AsO_4 + 2H^+ + 2e^- \longrightarrow HAsO_2 + H_2O$	+0.56
$As(s) + 3H^+ + 3e^- \longrightarrow AsH_3(g)$	−0.61
$BrO_3^- + 3H_2O + 6e^- \longrightarrow Br^- + 6OH^-$	+0.61
$\frac{1}{2}Br_2(aq) + e^- \longrightarrow Br^-$	+1.08
$Ce^{4+} + e^- \longrightarrow Ce^{3+}$	+1.74
$ClO_3^- + 2H^+ + e^- \longrightarrow ClO_2(g) + H_2O$	+1.15
$ClO_2 + e^- \longrightarrow ClO_2^-$	+0.93
$\frac{1}{2}Cl_2(aq) + e^- \longrightarrow Cl^-$	+1.39
$Co^{3+} + e^- \longrightarrow Co^{2+}$	+1.95
$HCrO_4^- + 7H^+ + 3e^- \longrightarrow Cr^{3+} + 4H_2O$	+1.20
$Cr^{3+} + e^- \longrightarrow Cr^{2+}$	−0.41
$Cu^{3+} + e^- \longrightarrow Cu^{2+}$	+2.3
$Cu^{2+} + e^- \longrightarrow Cu^+$	+0.17
$Cu^+ + e^- \longrightarrow Cu(s)$	−0.52
$\frac{1}{2}F_2(g) + e^- \longrightarrow F^-$	+2.87
$Fe^{3+} + e^- \longrightarrow Fe^{2+}$	+0.771
$2Hg^{2+} + 2e^- \longrightarrow Hg_2^{2+}$	+0.907
$Hg_2^{2+} + 2e^- \longrightarrow 2Hg(l)$	+0.792
$\frac{1}{2}I_2(aq) + e^- \longrightarrow I^-$	+0.621
$I_3^- + 2e^- \longrightarrow 3I^-$	+0.536
$MnO_4^- + e^- \longrightarrow MnO_4^{2-}$	+0.57
$MnO_4^- + 4H^+ + 3e^- \longrightarrow MnO_2(s) + 2H_2O$	+1.68
$MnO_2(s) + 4H^+ + 2e^- \longrightarrow Mn^{2+} + 2H_2O$	+1.23
$Mn^{3+} + e^- \longrightarrow Mn^{2+}$	+1.488
$Mn^{2+} + 2e^- \longrightarrow Mn(s)$	−1.17
$Mo(VI) + e^- \longrightarrow Mo(V)$	+0.53
$NO_3^- + 3H^+ + 2e^- \longrightarrow HNO_2 + H_2O$	+0.94
$Ni^{2+} + 2e^- \longrightarrow Ni(s)$	−0.25
$O_2(g) + 4H^+ + 4e^- \longrightarrow 2H_2O$	+1.229
$O_2(g) + 2H^+ + 2e^- \longrightarrow H_2O_2$	+0.69
$H_3PO_3 + 2H^+ + 2e^- \longrightarrow H_3PO_2 + H_2O$	−0.50
$Pb(IV) + 2e^- \longrightarrow Pb^{2+}$	+1.655
$Pb^{2+} + 2e^- \longrightarrow Pb(s)$	−0.126
$Pu^{4+} + e^- \longrightarrow Pu^{3+}$	+0.967
$Ru^{3+} + e^- \longrightarrow Ru^{2+}$	+0.249
$SO_4^{2-} + 4H^+ + 2e^- \longrightarrow H_2SO_3 + H_2O$	+0.17
$S(s) + 2H^+ + 2e^- \longrightarrow H_2S$	+0.141
S (s, rhombic) $+ 2e^- \longrightarrow S^{2-}$	−0.48
$2S(s) + 2e^- \longrightarrow S_2^{2-}$	−0.49
$3S(s) + 2e^- \longrightarrow S_3^{2-}$	−0.45
$4S(s) + 2e^- \longrightarrow S_4^{2-}$	−0.36
$5S(s) + 2e^- \longrightarrow S_5^{2-}$	−0.34
$6S(s) + 2e^- \longrightarrow S_6^{2-}$	−0.36
$Sb(V) + 2e^- \longrightarrow Sb(III)$	+0.75
$SeO_4^{2-} + 4H^+ + 2e^- \longrightarrow H_2SeO_3 + H_2O$	+1.15
$H_2SeO_3 + 4H^+ + 4e^- \longrightarrow Se(s) + 3H_2O$	+0.74
$Se(s) + 2H^+ + 2e^- \longrightarrow H_2Se(g)$	−0.37
$Sn(IV) + 2e^- \longrightarrow Sn(II)$	+0.144
$Ti(IV) + e^- \longrightarrow Ti(III)$	+0.130
$Ti^{3+} + 2e^- \longrightarrow Tl^+$	+1.26
$U^{4+} + e^- \longrightarrow U^{3+}$	−0.609
$VO_2^+ + 2H^+ + e^- \longrightarrow V^{3+} + H_2O$	+0.34

[a]Gaseous concentrations are in atmospheres.
Source: Kotrlý and Šůcha, 1985.

change the speciation of the basic element of concern. The newly formed species will, similarly, have different physicochemical and toxicological properties from the original materials. Perhaps the best-known example in this regard is chromium, whose oxidized [Cr(VI)] and reduced [Cr(III)] forms are fairly easily interconverted in nature, and one form [Cr(VI)] is much more toxic to many aquatic biota. Thus, information should always be sought on the actual or expected oxidation state of a dissolved species before assessing its probable environmental fate and aquatic toxicity. Figure 5 summarizes the elements that are capable of undergoing redox reactions in the environment and their oxidation (or valence) states.

Environmentally important redox reactions are not limited to the interaction of species with oxygen but may involve any pair of species—including an inorganic-pair—that are capable of such reactions. Examples include the following:

$$Fe^{2+} + SO_4^{2-} \rightleftharpoons FeSO_4^{+} + e^- \quad (64)$$

$$Fe^{2+} + Cl^- \rightleftharpoons FeCl^{2+} + e^- \quad (65)$$

$$Fe^{2+} + 2HS^- \rightleftharpoons FeS_2 + 2H^+ + 2e^- \quad (66)$$

The electrons (e^-) released in these iron oxidation reactions would be consumed in the simultaneous reduction of oxygen or some other species. Redox cou-

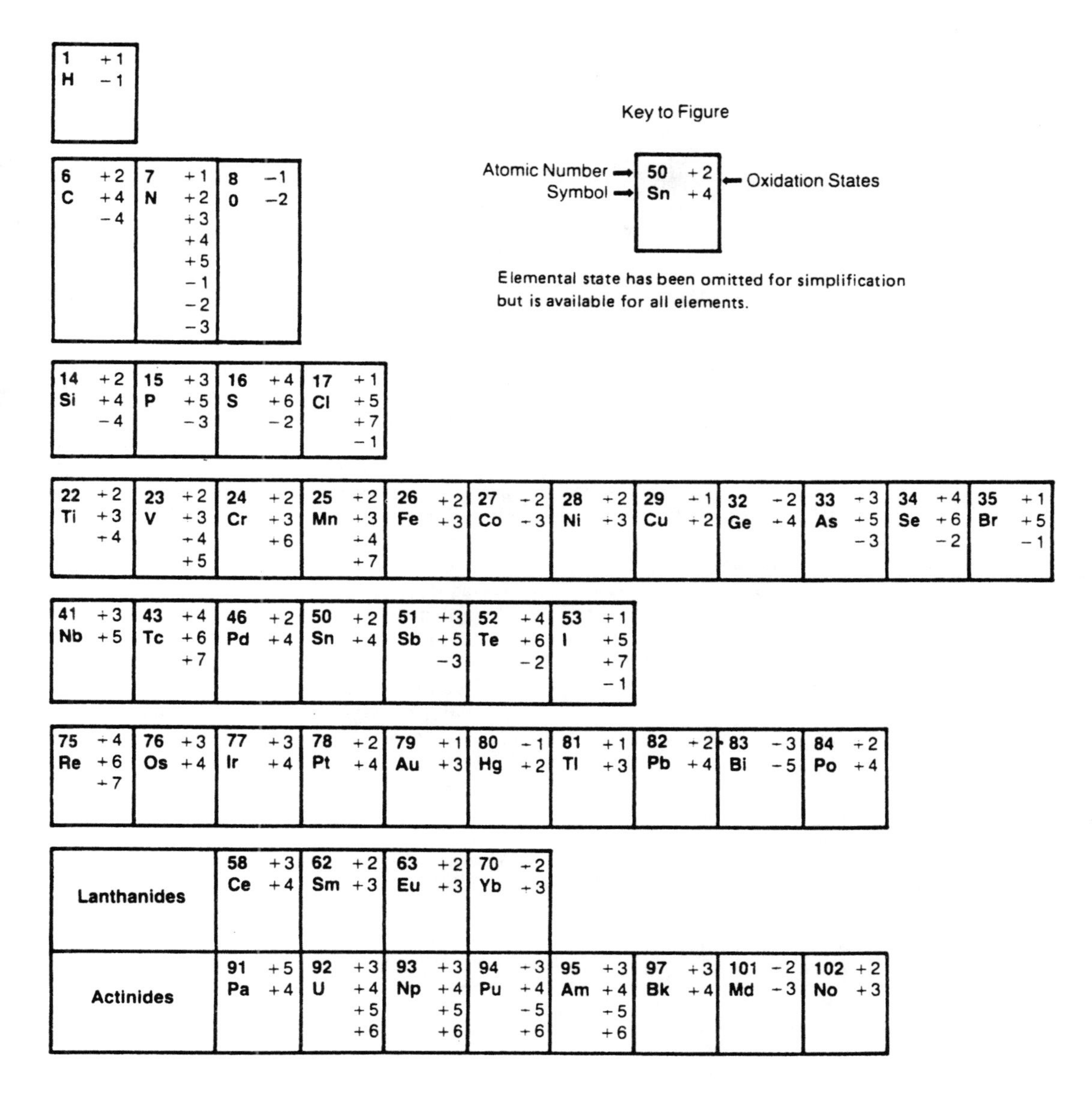

Source: Lyman et al. (1987)

Figure 5. Summary of elements that can undergo redox reactions and their valence states.

ples (pairs of oxidized and reduced species of an element) that often determine the overall redox state or potential of an aquatic system include O_2/H_2O, Fe^{3+}/Fe^{2+}, $Fe^{2+}/Fe(OH)_{3(s)}$, HS^-/SO_4^{2-}, $HS^-/S_{rhombic}$, NO_2^-/NO_3^-, NH_4^+/NO_3^-, NH_4^+/NO_2^-, $NH_4^+/N_{2(aq)}$, $CH_{4(aq)}/HCO_3^-$, and various Mn couples. Knowledge of the relative abundance of these species will help provide valuable information on the oxidation potential of a system. It is however, dangerous to assume that such couples have reached their equilibrium concentrations, as redox reactions can be significantly hindered by kinetic limitations and local heterogeneities. In practice, significant concentrations of both couples may be found even though the underlying oxidation potential may indicate that only one should be present in significant quantities. Many redox reactions may be catalyzed by a variety of chemicals and some are microbiologically mediated.

Information on the speciation of elements under various redox (E_h) and pH conditions is often provided in the form of stability diagrams (e.g., Figure 6 for the $Fe/H_2O/SO_4/CO_2$ system). The specific location of the lines in these diagrams, which represent borders where the concentrations (or activities) of species on either side are equal, depend on the conditions of the system, which should always be specified. Many such diagrams are available in water chemistry texts and can be helpful evaluating the redox behavior of an element. If the environmental conditions in the system being assessed are significantly different from those assumed for any existing diagram, one of the available computerized speciation models should be used to obtain more precise estimates of the speciation.

*Polymerization**

A limited number of inorganic species are capable of reacting with themselves to form larger molecules, or polymers. These polymers may be linear, linear with cross-linkages, or nonlinear. Many can be formed spontaneously in aqueous systems, and others that are manufactured are of environmental interest because of their degradation rather than their formation.

Examples of inorganic polymer species include:

- Polymeric metal hydroxo complexes, $M[OH]_t M]_{x-1}^{xz-xt+t}$
- Polysulfides, S_n^{2-}
- Polyphosphates, $PO_4(PO_3)_n^-$
- Polyborates, $B_3O_3(OH)_4^-$
- Paracyanogen, $(CN)_n$

The formation of such species should be taken into account when evaluating the likely speciation of polymer-forming units. Each of these polymers is briefly described below.

Polymeric metal hydroxo complexes are formed by reaction of metal ions and water. Some of the complexes formed are mononuclear (include only one metal ion), while others, referred to as polynuclear hydroxo complexes, involve several metal ions. Formation of the latter has been reported for most metals, although only a few form larger polynuclear species. Many of these polynuclear complexes are not present under equilibrium conditions; they are often formed only under conditions of oversaturation with respect to the metal hydroxide or oxide and are kinetic intermediates in the transition from free metal ions to a solid precipitate. However, they can persist as metastable species for years and thus can be significant in natural water systems.

Polysulfide ions (S_n^{2-}) may be formed in natural waters by the interaction of aqueous sulfide with elemental sulfur, the latter being formed initially either by bacterial oxidation of sulfide or by partial oxidation of sulfide by dissolved oxygen. These reactions are intermediate steps in the chemical oxidation of sulfide (e.g., to thiosulfate, $S_2O_3^{2-}$). An equilibrium is eventually reached among the various S_n^{2-} species. The dominant species are thought to be those with $n \geq 4$, especially the $n = 5$ and $n = 6$ species. Polysulfide species are not stable under anaerobic conditions. However, they are sometimes found in the natural environment, especially in hydrothermal or heavily polluted waters. They may also be formed in industrial sulfide precipitation reactions used to remove metals from industrial effluents.

Polyphosphoric acids and polyphosphates, consisting of multiple PO_4 units, are formed by the condensation of two or more molecules of phosphoric acid, H_3PO_4. These chemicals are manufactured in large quantities for use in fertilizers, detergents, water-softening agents, and other industrial formulations. They are thus released to the environment both directly (e.g., as fertilizers) and in industrial wastes. In the aqueous environment, polyphosphates are gradually hydrolyzed back to phosphoric acid, the reaction being influenced by several factors including pH, temperature, metal ion catalysis, and enzymatic catalysis.

Polyborates are formed by the reaction of monoborate ions, $B(OH)_4^-$, with free boric acid molecules, $B(OH)_3$. The most abundant product in the pH range 5 to 11 is $B_3O_3(OH)_4^-$, but other significant species include $B_3O_3(OH)_5^{2-}$, $B_4O_5(OH)_4^{2-}$, and $B_5O_6(OH)_4^-$. The formation of these species is most favored around pH 9.

Cyanide (CN) can polymerize at high temperatures (300–500°C) to form *paracyanogen,* $(CN)_4$, a dark brown, water-insoluble solid. It is thus not likely to form in natural waters but could be present in industrial wastes, especially those that have been heat treated.

Photolysis (in Water)

Many inorganic compounds and organometallic complexes are susceptible to decomposition, dissociation, or chemical change upon exposure to sunlight. Table 22 provides several examples resulting

*Much of the information in this subsection is from Bodek et al. (1988).

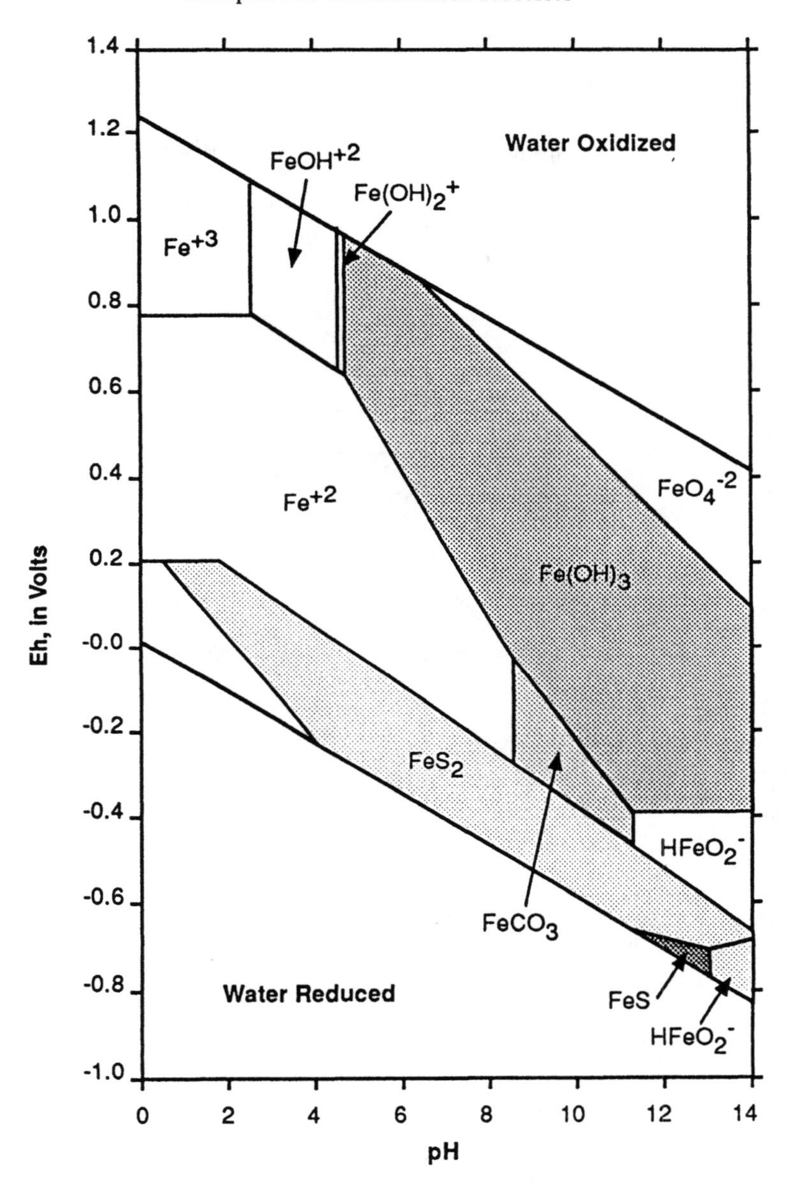

Shaded areas are stability regions for solids; predominant dissolved species are shown in unshaded areas. Activity of sulfur species = 96 mg/l as SO_4^{-2}. CO_2 species = 1,000 mg/l as HCO_3^-, and total dissolved iron activity = 0.0056 mg/l.

Source: Hem (1970)

Figure 6. Fields of stability for solid and dissolved forms of iron as a function of E_h and pH at 25°C and 1 atmosphere. (Courtesy of Hem, 1970).

from *direct photolysis*, in which the species absorbs solar radiation, is promoted to an excited state, and either decomposes or undergoes secondary reactions. Inorganic chemicals may also be susceptible to *indirect photolysis*, in which energy or electrons are transferred from some other photolytically sensitized species, which may be dissolved or in the form of a solid suspension.

There are no clear rules for assessing the potential importance of photolysis in the speciation of inorganic chemicals in an aquatic system. Literature data on photolytic reactions or light absorption must usu-

Table 22. Examples of direct photolysis involving inorganic or organometallic complexes

Species	Products	Probable mechanism
NO_2^-	NO + OH	Decomposition
MnO_2	Mn(II)	Electron transfer to metal
Cu(II)	Cu(I)	Reduction; dissociation
Cu(II)-organic	Cu(I)	Reduction; dissociation
Fe(III)-organic	Fe(II) + CO_2	Oxidation of organics; reduction of O_2; dissociation
Fe(III)-organic	Fe(II), CO_2, amine	Electron transfers to metal; decomposition
Organic mercurials	Elemental Hg, Hg salts	Decomposition
$Fe(CN)_6^{4-}$	$Fe(CN)_5^{3-}$ + CN^-	Reduction; decomposition
$Pb(CH_2CH_3)_4$	Ethane and eventually, inorganic Pb salts	Decomposition
BrO^-	Br + O^-; Br^- + O	Decomposition
ClO_2	ClO + O; or Cl + O_2	Decomposition; cage recombination of photolysis products an intermediate
ClO^-	Cl^- + O; Cl + O^-	Decomposition
ClO_2^-	ClO^- + O; ClO + O^-	Decomposition

[a] References in original source.

Source: Reprinted from Bodek et al. (eds.), *Environmental Inorganic Chemistry: Properties, Processes and Estimation Methods*, Copyright 1988, 1280 pp., with permission from Elsevier Science Ltd., The Boulevard, Langford Lane, Kidlington 0X5 1GB, UK.

ally be sought and evaluated. The possible importance of indirect photolysis, including the role of catalysts (e.g., TiO_2), is even more difficult to ascertain. The photolytic rate will typically be:

- Directly proportional to:
  - the concentration of the chemical
  - the total sunlight intensity
  - the amount of sunlight absorbed (e.g., the molar absorption)
  - the quantum yield for the reaction
- Indirectly proportional to:
  - the depth in the water body
  - the attenuation coefficient that increases with decreasing water clarity

When sufficient data are available on these, quantitative calculations of the rate of photolysis may be made (Bodek et al., 1988).

*Microbial Transformations**

Microorganisms such as bacteria, yeasts, and fungi may catalyze the modification of inorganics in the environment. These transformations can affect the mobility and toxicity of these substances. The high metabolic activity and versatility of diverse populations of microorganisms, combined with their rapid reproduction and mutation rates, enable them to adapt and develop enzyme systems that detoxify a variety of substances. Table 23 lists some of the processes known to be facilitated by microorganisms and the elements known to be affected by them.

*Much of the information in this section is from Bodek et al. (1988).

Table 23. Elements affected by microbial transformation processes

Process	Elements known to be affected
Oxidation-reduction	Hg, As, Fe, Mn, Sb, Te, Se, S
Sulfide precipitation	Cu, Pb, Zn, Ag, others
Methylation	Hg, As, Se, Pb, Cd, Sn, Te, Pu, Tl(?), Pt(?), Sb(?)
Dealkylation	Hg, Sn, As

Source: Reprinted from Bodek et al. (eds.), *Environmental Inorganic Chemistry: Properties, Processes, and Estimation Methods*, Copyright 1988, 1280 pp., with permission from Elsevier Science Ltd., The Boulevard, Langford Lane, Kidlington 0X5 1GB, UK.

Examples of oxidation-reduction reactions are (1) the reduction of $HgCl_2$ to Hg° by various bacteria and yeast, (2) the oxidation of As(III) to As(V) by pseudomonads, (3) the reduction of As(V) to As(III) by both bacteria and yeast, and (4) the oxidation of Fe(II) to Fe(III) by bacteria.

Sulfide minerals of metals such as Cu, Pb, Zn, and Ag are produced by the action of *Desulphovibrio desulphuricans* and other sulfate-reducing bacteria, which are found in mud, swamps, and drained pools. The reactions catalyzed by these microorganisms are inhibited under aerobic and acidic conditions.

Microorganisms can catalyze the methylation of metals and metalloids. This may be a detoxification mechanism for the organism, but the methylated products formed can be more toxic to higher organisms. Mercury, for example, can be methylated, by aerobic and anaerobic organisms, from Hg(II) to either monomethyl mercury, CH_3Hg^+, or dimethylmercury, CH_3HgCH_3. Monomethyl mercury is the predominant form found in fish. Dimethylmercury is volatile and may be released to the atmosphere. The relative amounts of these two forms produced are sensitive to several environmental parameters including temperature, pH, and redox potential. Mercury can also be demethylated by a variety of aerobic, anaerobic, and facultative microorganisms. Dealkylation of phenylmercuric acetate and tributylin has also been reported.

Arsenic can also be methylated by bacteria and fungi to produce a variety of methylated compounds. Methylation is thought to be a detoxification mechanism and is important in the transfer of arsenic from sediments to the water and atmosphere. Starting with arsenate, $AsO(OH)_3^{5+}$, or arsenite, $AsO(OH)^{3+}$, the methylation can yield methylarsonic acid, $CH_3AsO(OH)_2^{3+}$, dimethylarsenic acid, $(CH_3)_2ASO(OH)^+$, dimethylarsine, $(CH_3)_2AsH$, and trimethylarsine, $(CH_3)_3As$.

ORGANOMETALLICS*

In assessing the environmental fate of an organometallic, it is important to know what happens to the chemical when it dissolves in water. Does it remain essentially unchanged in solution, permitting it to be treated (for property estimation and fate assessment purposes) like a neutral organic chemical? Or does it react with water to yield a hydrated metal ion and an organic ion, each of which can be considered separately in an environmental assessment?

Need for Classification

The term "organometallic" has been applied to almost any substance containing a metal or metalloid and carbon atoms, but it most commonly refers to compounds containing metal-carbon bonds. Because of the wide variety of substances that contain both a metal or metalloid and a carbon atom [e.g., chelate complexes, metalorganic salts, and metal/metalloid-to-R bonds, where the bond to the organic group (R) may be to carbon, oxygen, nitrogen, or sulfur atoms], it is useful to classify organometallics on the basis of the processes that affect their environmental behavior. These processes include dissolution, complexation, and breaking of particular metal- or metalloid-to-carbon or other atom bonds. This method of classification characterizes an organometallic in terms of its initial process of environmental speciation, which is of primary concern in environmental assessment. It also helps to associate the compound with particular processes and estimation methods described in other chapters.

*Most of this section has been excerpted from Bodek et al. (1988).

Classes and Environmental Behavior

Organometallics have been grouped below into three categories based on their expected environmental behavior and the coverage of relevant properties in this chapter. The three categories reflect the compounds that are on various regulatory lists as well as those that may have major uses (Table 24).

- Category I: When dissolved in water, the compound yields an initial species with the same metal-to-organic bonds for the metal of interest, as shown in Eq. (67) for a metal or metalloid (M) and an organic group (R). Subsequent environmental reactions are based on the reactivity of the particular M—R framework bonds. This category can thus include salts of a metal-organic species ([M—R][X]) where, although the original compound separates into the respective ions in solution, all the metal-organic bonds in the ion are retained upon dissolution, as shown in Eq. (68) (charges on the ions are omitted).

$$M{-}R + H_2O \rightarrow M{-}R_{aq} \qquad (67)$$

or

$$[M{-}R][X] + H_2O \rightarrow M{-}R_{aq} + X_{aq} \qquad (68)$$

where R = carbon-containing organic fragment
M = metal or metalloid

Table 24. Categorization of some organometallics of environmental importance[a]

Category I	Category II	Category III
Ethyl silicate	Sodium fluoroacetate	Zinc stearate
Alkyl mercury compds	Lead acetate	Ferric dextran
Alkyl lead compds	Thallium(I) acetate	Mercury fulminate
Alkyl tin compds	Calcium naphthenate	Cobalt naphthenate
Nickel carbonyl	Sodium salt of naphthalenedisulfonic acid	Metal polyisobutenylsuccinate
Iron pentacarbonyl	Calcium carbide	
Diethylarsine	Sodium propionate	
Phenylmercuric acetate	Calcium propionate	
Dibutyltin bis(isooctylmaleate)		
Alkoxymethylpolysilazanes		
Lead alkyls		
Aluminum tris-*o*-ethyl phosphonate		
Disodium methylarsonate[b]		
Triphenyltin hydroxide		
Triphenyltin acetate		

[a] See text for explanation of categories.
[b] Although upon dissolution this compound yields separate dissolved ions, one of these ions (methylarsonate) retains the metalloid-organic bond of the original compound and is thus placed in category I.
Source: Reprinted from Bodek et al. (eds.), *Environmental Inorganic Chemistry: Properties, Processes, and Estimation Methods*, Copyright 1988, 1280 pp., with permission from Elsevier Science Ltd., The Boulevard, Langford Lane, Kidlington 0X5 1GB, UK.

X = counterion of salt (e.g., Cl^-, Na^+)
aq = designation of aqueous species

- Category II: Salts of organic anions with labile metals ([M][R]) which, upon dissolution in water, rapidly yield solutions containing predominantly the separated components of the salts, that is, the hydrated metal ions and the organic portions, as shown in Eq. (69) (charges omitted).

$$[M][R] + H_2O \rightarrow M_{aq} + R_{aq} \quad (69)$$

- Category III: Metal-organic complexes (e.g., chelates), $[M—R_n]$, of relatively nonlabile metals with strongly complexing ligands that, upon dissolution in water, yield metal-organic complexes for which reversible ligand exchange reactions are of importance, as shown in Eq. (70) (with charges omitted), along with any associated counterions.

$$\begin{aligned} M—R_n + H_2O \rightarrow\ & [M—R_n]_{aq} \\ & [M—R_{n-1}]_{aq} + R_{aq} \quad (70) \\ & [M—R_{n-2}]_{aq} + R_{aq} \\ & \text{etc.} \end{aligned}$$

Table 24 listed some compounds of regulatory and environmental importance in each of these categories. These categories are not mutually exclusive, as particular compounds can behave in a manner consistent with one or more categories; a gradation occurs within each category as compounds begin to exhibit properties associated with the next category. However, this method of classification provides a useful indication of the processes that are most important in environmental speciation.

Examples of category I compounds are tetraethyl lead, $Pb(CH_2CH_3)_4$, and diethylarsine, $(CH_3CH_2)_2AsH$, both of which can be considered to initially produce a dissolved molecular species of the same composition in water. Similarly, the sodium salt of methylarsonate, $Na_2CH_3AsO_3$, produces a methylarsonate anion in solution that retains the metalloid-organic bond. For all these compounds the chemical stability of the metal- or metalloid-to-carbon bond is of paramount importance in estimating persistence and environmental fate. (Acid-base equilibria should also be considered, if they modify the species in question.)

Examples of compounds in category II are sodium acetate, $NaC_2H_3O_2$, and calcium propionate, $Ca(C_3H_5COO)_2$; upon dissolution in water, these yield the labile aquo metal ions of sodium and calcium and the counter-organic anions of acetate and propionate, respectively. For this category, aqueous environmental chemistry is determined by the properties of the respective ions and not the original molecule.

Examples of category III compounds include transition metal cyanides, which form distinct coordination complexes in solution and whose subsequent environmental speciation can be based on reactions such as reversible complexation equilibria. The distinction between all these categories, especially for category II and III compounds, is very dependent on the kinetic stability or lability of the metal-to-organic ligand bonds.

Bond Character Classification

Of particular interest are the compounds in category I, whose behavior is strongly influenced by the specific metal or metalloid and its bonding to the elements that are linked to the organic framework (R).

Bodek et al. (1988) provide a means, using atomic electronegativity values, to assign category I compounds to subcategories Ia, Ib, and Ic and thus to form the basis for a preliminary assessment of the probable behavior in water. The protocol calls for the calculation of the difference in literature-provided elemental electronegativities for the metal (or metalloid) and the various elements bound directly to it. The absolute values of the differences in electronegativity are interpreted as follows:

Difference in electronegativity	Bond character	Compound subcategory
<1.0	Covalent	Ia
1.0 to 1.5	Intermediate	Ib
>1.5	Ionic	Ic

LITERATURE CITED

Alexander, M.: Biodegradation of organic chemicals. *Environ. Sci. Technol.*, 19:106–111, 1985.

Anderson, L. C. D., and Bruland, K. W.: Biogeochemistry of arsenic in natural waters. The importance of methylated species. *Environ. Sci. Technol.*, 25:420–427, 1991.

Atkinson, R.: Kinetics and mechanisms of gas phase reactions of hydroxyl radical with organic compounds. *Chem. Rev.*, 86:69–201, 1986.

Atkinson, R.: A structure-activity relationship for the estimation of rate constants for the gas-phase reactions of OH radicals with organic compounds. *Int. J. Chem. Kinet.*, 19:799–828, 1987.

Banerjee, S.; Howard, P. S.; Rosenberg, A. M.; Dombrowski, A. E.; Sikka, H.; and Tullis, D. L.: Development of a general kinetic model for biodegradation and its application to chlorophenols and related compounds. *Environ. Sci. Technol.*, 18:416–422, 1984.

Battersby, N. S.: A review of biodegradation kinetics in the aquatic environment. *Chemosphere*, 21:1243–1284, 1990.

Betterton, E. A.; Erel, Y.; and Hoffman, M. R.: Aldehyde-bisulfite adducts: prediction of some thermodynamic and kinetic properties. *Environ. Sci. Technol.*, 22:92–99, 1988.

Bodek, I.; Lyman, W. J.; Reehl, W. F.; and Rosenblatt, D. H. (eds.): Environmental Inorganic Chemistry: Properties, Processes, and Estimation Methods. New York: Pergamon, 1988.

Boethling, R. S., and Sabljic, A.: Screening Level Model for Aerobic Biodegradability Based on a Survey of Expert Knowledge. *Environ. Sci. Technol.*, 23:672–679, 1989.

Bowen, H. J. M.: Environmental Chemistry of the Elements. London: Academic Press, 1979.

Brownawell, B. J., and Farrington, J. W.: Biogeochemistry of PCBs in interstitital waters of a coastal marine sediment. *Geochim. Cosmochim. Acta.*, 50:157–169, 1986.

Burkhard, L. P.; Armstrong, D. E.; and Andreu, A. W.: Henry's law constants for the polychlorinated biphenyls. *Environ. Sci. Technol.*, 19:590–596, 1985.

Butler, J. N.: Ionic Equilibrium. Reading, MA: Addison-Wesley, 1964.

Chin, Y. P., and Gschwend, P. M.: Partitioning of polycyclic aromatic hydrocarbons to marine pore water organic colloids. *Environ. Sci. Technol.*, 26:1621–1626, 1992.

Cline, P. V.; Delfino, J. J.; and Rao, P. S. C.: Partitioning of aromatic constituents into water from gasoline and other complex solvent mixtures. *Environ. Sci. Technol.*, 25:914–920, 1991.

Cooper, W. J., Mehran, M., Riusech, D. J.; and Jones, J. A.: Abiotic transformations of halogenated organics. 1. Elimination reactions of 1,1,2,2-tetrachloroethane and formation of 1,1,2-trichloroethene. *Environ. Sci. Technol.*, 21:1112–1114, 1987.

Desai, S.M.; Govind, R.; and Tabak, H. H.: Development of quantitative structure-activity relationships for predicting biodegradation kinetics. *Environ. Toxicol. Chem.*, 9:473–477, 1990.

Di Toro, D. M.; Zarba, C. S.; Hansen, D. J.; Berry, W. J.; Swartz, R. C.; Cowan, C. E.; Pavlou, S. P.; Allen, H. E.; Thomas, N. A.; and Paquin, P. R.: Technical basis for establishing sediment quality criteria for nonionic organic chemicals using equilibrium partitioning. *Environ. Toxicol. Chem.*, 10:1541–1583, 1991.

Dulin, D., and Mill, T.: Development and evaluation of sunlight actinometers. *Environ. Sci. Technol.*, 16:815–820, 1982.

Eisenreich, S. J.; Looney, B. B.; and Thornton, J. D.: Appendix to the Great Lakes Science Advisory Board's 1980 Annual Report. Windsor (Ont.): Great Lakes Regional Office, The International Joint Commission, 1980.

Eklund, B.; Smith, S.; and Hunt, M.: Air/Superfund National Technical Guidance Study Series: Estimation of Air Impacts for Air Stripping of Contaminated Water. Report No. EPA/450/1-91/002. Washington, D.C.: U.S. Environmental Protection Agency, 1991.

Ellington, J. J.: Hydrolysis Rate Constants for Enhancing Property-Reactivity Relationships. Report No. EPA/600/3-89/063. Athens, GA: U.S. Environmental Protection Agency, 1989.

Ellington, J. J.; Stancil, F. E.; and Payne, W. D.: Measurement of Hydrolysis Rate Constants for Evaluation of Hazardous Waste Land Disposal: Volume 1. Data on 32 Chemicals. Report No. EPA/600/S3-86/043. Athens, GA: U.S. Environmental Protection Agency, 1987.

Enslein, K.; Tomb, M. E.; and Lander, T. R.: Structure-activity models of biological oxygen demand. *QSAR in Environmental Toxicology*, edited K. L. E. Kaiser, pp. 89–109. Dordrecht, Reidel, 1984.

EPA: Toxic Substances Control Act test guidelines, final rule. 40 CFR Parts 796–798. *Fed. Regist.*, 39252–39323, 1985.

Fendinger, N. J., and Glotfelty, D. E.: Henry's law constants for selected pesticides, PAHs and PCBs. *Environ. Toxicol. Chem.*, 9:731–735, 1990.

Fu, J.; Luthy, R. G.; and Dzombak, D. A.: Adsorption of Polycyclic Aromatic Hydrocarbon Compounds onto Soil and Transport of Naphthalene in Unsaturated Porous Media. Report No. DOE/PC/30246-1557. Morgantown, WV: U.S. Department of Energy, 1983.

Garrels, R. M., and Christ, C. L.: Solutions, Minerals, and Equilibria. New York: Harper & Row, 1965.

Geating, J.: Literature Study of the Biodegradability of Chemicals in Water. Vol. I: Biodegradability Prediction, Advances in Chemical Interferences with Wastewater Treatment. Report. No. EPA-600/2-81-175, Cincinnati: U.S. Environmental Protection Agency, 1981.

Gossett, J. M.: Measurement of Henry's law constants for C_1 and C_2 chlorinated hydrocarbons. *Environ. Sci. Technol.*, 21:202–208, 1987.

Groves, F. R.: Effect of cosolvents on the solubility of hydrocarbons in water. *Environ. Sci. Technol.*, 22:282–286, 1988.

Haag, F.; Reinhard, M.; and McCarty, P. L.: Degradation of toluene and *p*-xylene in anaerobic microcosms: Evidence of sulfate as a terminal electron acceptor. *Environ. Toxicol. Chem.*, 10:1379–1389, 1991.

Haag, W. R., and Mill, T.: Some reactions of naturally occurring nucleophiles with haloalakanes in water. *Environ. Toxicol. Chem.*, 7:917–924, 1988.

Hanna, S. R., and Hosker, R. P.: Atmospheric-Removal Processes for Toxic Chemicals. NOAA Technical Memorandum ERL ARL-102. Silver Spring, MD: Air Resources Laboratories, 1980.

Hashimoto, Y.; Tokura, K.; Kishi, H.; and Stracham, W. M. J.: Prediction of seawater solubility of aromatic compounds. *Chemosphere*, 13:881–888, 1984.

Hem, J. D.: Study and Interpretation of the Chemical Characteristics of Natural Water, 2nd ed. Water Supply Paper No. 1473. Washington, D.C.: U.S. Geological Survey, 1970.

Hemond, H. F., and Fechner, E. J.: Chemical Fate and Transport in the Environment, San Diego: Academic Press, 1994.

Herbs, S. E.; Southworth, G. R.; and Allen, C. P.: Rates of dissolution of constituent organic contaminants from coal liquefaction oil films into water. *Water Res.*, 17:1639–1646, 1983.

Hicks, B. B.: Deposition Both Wet and Dry. Stoneham, MA: Batterworth, 1984. Volume 4 of the Acid Precipitation Series.

Hine, J., and Mookerjee, P. K.: The intrinsic hydrophilic character of organic compounds. Correlations in terms of structural contributions. *J. Org. Chem..*, 40:292–298, 1975.

Hoigne, J.: Mechanisms, rates and selectivities of oxidations of organic compounds initiated by ozonation in water. Handbook of Ozone Technology and Applications, Vol. 1, pp. 341–379. Ann Arbor, MI: Ann Arbor Science Publishers, 1982.

Hoigne, J.: The chemistry of ozone in water. Process Technologies for Water Treatment, edited by S. Stucki, pp. 121–143. New York: Plenum, 1988.

Hoigne, J., and Bader, H.: Rate constants for reactions of ozone with organic and inorganic compounds in water —I. *Water Res.*, 17:173–183, 1983a.

Hoigne, J., and Bader, H.: Rate constants of reactions of ozone with organic and inorganic compounds in water —II. *Water Res.*, 17:184–194, 1983b.

Hoigne, J., Bader, H., Haag, W. R., and Staehelin, J.: Rate constants of reactions of ozone with organic and inorganic compounds in water—III. *Water Res.*, 19:993–1004, 1985.

Howard, P. H.: Handbook of Environmental Fate and Exposure Data for Organic Chemicals, Vol. 1, Large Production and Priority Pollutants (1989); Vol. 2, Solvents (1990); Vol. 3, Pesticides (1991). Chelsea, MI: Lewis, 1989–1991.

Howe, G. B.; Mullins, M. E.; and Rogers, T. N.: Evaluation and Prediction of Henry's Law Constants and Aqueous

Solubilities for Solvents and Hydrocarbon Fuel Components. Vol. I: Technical Discussion. Report No. ESL-TR-86-66, Tyndall AFB, FL; Air Force Engineering and Services Center, 1987.
Hutchins, S. R.; Sewell, G. W.; Kovacs, D. A.; and Smith, G. A.: Biodegradation of aromatic hydrocarbons by aquifer microorganisms under denitrifying conditions. *Environ. Sci. Technol.*, 25:68–76, 1991.
Hutzinger, O. (ed.): The Handbook of Environmental Chemistry, (12 volumes). New York: Springer-Verlag, 1980–1988.
Ince, N., and Inel, Y.: A semi-empirical approach to relate the volatilization rates for organic chemicals to their physical properties. *Water Res.*, 25:903–910, 1991.
International Critical Tables, 7 vols. New York: McGraw-Hill, 1926–1930.
Jafvert, C. T., and Wolfe, N. L.: Degradation of selected halogenated ethanes in anoxic sediment-water systems. *Environ. Toxicol. Chem.*, 6:827–837, 1987.
Johnson, D. L.; Jiao, J.; and DosSantos, S. G.: Individual particle analysis of suspended materials in Onondaga Lake, New York. *Environ. Sci. Technol.*, 25:736–744, 1991.
Jolley, R. L.; Condie, L. W.; Johnson, J. D.; Katz, S.; Mattice; J. S.; Minear, R. A.; and Jacobs, V. A.: Water Chlorination—Chemistry, Environmental Impact, and Health Effects, Vols. I–VI. Boca Raton, FL: CRC Press (Lewis Publishers), 1978–1990.
Kadeg, R. D.; Pavlou, S. P.; and Duxbury, A. S.: Elaboration of Sediment Normalization Theory for Nonpolar Hydrophobic Organic Chemicals. Report to U.S. Environmental Protection Agency, Office of Water Programs, Criteria and Standards Division, Washington, D.C., 1986.
Karickhoff, S. W.: Pollutant Sorption in Environmental Systems. Athens, GA: U.S. Environmental Protection Agency, Environmental Research Laboratory, 1983. (Available from NTIS as PB83-231381.
Karickhoff, S. W.: Organic pollutant sorption in aquatic systems. *J. Hydraulic Eng.*, 10:707–735, 1984.
Karickhoff, S. W.; Brown, D. S.; and Scott, T. A.: Sorption of hydrophobic pollutants on natural sediments. *Water Res.*, 13:241–248, 1979.
Karickhoff, S. W.; McDaniel, V. K.; Melton, C.; Vellino, A. N.; Nute, D. E.; and Carreira, L. A.: Predicting chemical reactivity by computer. *Environ. Toxicol. Chem.*, 10: 1405–1416, 1991.
Keith, L. H., and Walters, D. B.: National Toxicology Program's Chemical Solubility Compendium. Boca Raton, FL: Lewis, 1991.
Klecka, G. M.: Biodegradation. Environmental Exposure from Chemicals, edited by W. B. Neely, and G. E. Blau, Chapter 6, pp. 109–155. Boca Raton, FL: CRC Press, 1985.
Kragten, J.: Atlas of Metal-Ligand Equilibria in Aqueous Solutions, New York: Halstead Press (Wiley), 1978.
Kotrly, S., and Sucha, L.: Handbook of Chemical Equilibria in Analytical Chemistry. Chichester, UK: Ellis Horwood, 1985.
La Poe, R. G.: Sorption and Desorption of Volatile Chlorinated Aliphatic Compounds by Soils and Soil Components. Ph.D. dissertation. Report No. AFIT/CI/NR-85-99T Air Force Institute of Technology, Wright-Patterson AFB, OH, 1985. (Available from NTIS as AD-A158 096/8/XAB.)
Leifer, A.: The Kinetics of Environmental Aquatic Photochemistry: Theory and Practice. Washington, DC: American Chemical Society, 1988.
Lick, W.: The transport of sediments in aquatic systems. Fate and Effects of Sediment-Bound Chemicals in Aquatic Systems, edited by K. L. Dickson; A. W. Maki; and W. A. Brungs, pp. 61–82. New York: Pergamon Press, 1987.
Linke, W. F.: Solubilities—Inorganic and Metal-Organic Compounds, A–Ir, 4th ed, Vol I. New York: Van Nostrand, 1958.
Linke, W. F.: Solubilities—Inorganic and Metal-Organic Compounds, K–Z, 4th ed, Vol. II. Washington, DC: American Chemical Society, 1965.
Liss, P. S., and Slinn, W. G. N. (eds.): Air-Sea Exchange of Gases and Particles, NATO SAI Series. Reidel, Boston, 1983.
Liss, P. S., and Slater, P. G.: Flux of gases across the air-sea interface. *Nature*, 247:181–184, 1974.
Lovley, D. R.: Dissimilatory Fe(III) and Mn(IV) reduction. *Microbiol. Rev.*, 55:259–287, 1991.
Lyman, W. J.: Estimation of Physical Properties. Environmental Exposure from Chemicals, Vol. I, edited by W. B. Neeley, G. E. Blau, chapter 2, pp. 13–47. Boca Raton, FL: CRC Press, 1985.
Lyman, W. J., and Loreti, C. P.: Prediction of Soil and Sediment Sorption for Organic Compounds. Final Report on Task 15 of Contract 68-01-6951 to the U.S. Environmental Protection Agency, Office of Water Regulations and Standards, Washington, DC, 1987.
Lyman, W. J.; Reehl, W. F.; and Rosenblatt, D. H. (eds.): Research and Development of Methods for Estimating Physicochemical Properties of Organic Chemicals of Environmental Concern (Parts I and II). Report to the U.S. Army Medical Bioengineering Research and Development Laboratory, Fort Detrick, Frederick, MD, 1981. (Available from NTIS as AD A118754 and AD A119779. Also currently published by the American Chemical Society under the title Handbook of Chemical Property Estimation Methods. Previously published under this title by McGraw-Hill.)
Lyman, W. J.; Reidy, P. J.; and Levy, B.: Mobility and Degradation of Organic Contaminants in Subsurface Environments. Chelsea, MI: C. K. Smoley, Inc., 1992.
Mabey, W., and Mill, T.: Critical review of hydrolysis of organic compounds in water under environmental conditions. *J. Phys. Chem. Ref. Data*, 7:383–415, 1978.
Macalady, D. L., and Wolfe, N. L.: Abiotic Hydrolysis of Sorbed Pesticides. Report No. EPA-600/D-84-240. Athens, GA: U.S. Environmental Protection Agency, 1984.
Mackay, D.: Air water exchange coefficients. Neely, W. B.; and Blau, G. E. (eds.): *Environmental Exposure from Chemicals*, CRC Press, Boca Raton, FL, 1985.
Mackay, D., and Shiu, W. Y.: A critical review of Henry's law constants for chemicals of environmental interest. *J. Phys. Chem. Ref. Data*, 10:1175–1199, 1981.
Mackay, D., and Yuen, A. T. K.: Mass transfer coefficient correlations for volatilization of organic solutes from water. *Environ. Sci. Technol.*, 17:211–217, 1983.
Mackay, D.; Shiu, W. Y.; Bobra, A.; Billington, J.; Chau, E.; Yeun, A.; Ng, C.; and Szeto, F.: Volatilization of Organic Pollutants from Water. Report No. EPA-600/S3-82-019. Athens, GA: U.S. Environmental Protection Agency, 1982.
Mackay, D.; Shiu, W. Y.; and Ma, K. C.: Illustrated Handbook of Environmental Fate for Organic Chemicals. Boca Raton, FL: Lewis, 1991a.
Mackay, D.; Shiu, W. Y.; Maijanen, A.; and Feenstra, S.: Dissolution of non-aqeuous phase liquids in groundwater. *J. Contaminant Hydrol.*, 8:23–42, 1991b.
Mailhot, H., and Peters, R. H.: Empirical relationships between the 1-octanol/water partition coefficient and nine physicochemical properties. *Environ. Sci. Technol.*, 22: 1479–1488, 1988.

Martell, A. E.: The influence of natural and synthetic ligands on the transport and function of metal ions in the environment. *Pure Appl. Chem.*, 4:81, 1975.

Martell, A. E., and Smith, R. M.: Critical Stability Constants, Vol. 3: Other Organic Liquids. New York: Plenum, 1977.

Martell, A. E., and Smith, R. M.: Critical Stability Constants, Vol. 5: First Supplement. New York: Plenum, 1982.

McMahon, T. A., and Denison, P.: Empirical atmospheric deposition parameters: survey. *Atmos. Environ.*, 13:571–585, 1979.

Meylan, W. M., and Howard, P. H.: Bond contribution method for estimating Henry's law constants. *Environ. Toxicol. Chem.*, 10:1283–1293, 1991.

Mill, T.: Structure-activity relationships for photodegradation processes in the environment. *Environ. Toxicol. Chem.*, 8:31–43, 1989.

Mill, T., and Mabey, W. R.: Photodegradation in water. Environmental Exposure from Chemicals, edited by W. B. Neeley, G. E. Blau, pp. 175–216. Boca Raton, FL: CRC Press, 1985.

Mill, T.; Mabey, W. R.; Bomberger, D. C.; Chou, T.; Hendry, D. G.; and Smith, J. H.: Laboratory Protocols for Evaluating the Fate of Organic Chemicals in Air and Water. Report No. EPA-600/3-82-022. Athens, GA: U.S. Environmental Protection Agency, 1982.

Morel, F. M. M.: Principles of Aquatic Chemistry. New York: Wiley, 1983.

Mueller, M., and Klein, W.: Comparative evaluation of methods predicting water solubility for organic compounds. *Chemosphere*, 25:769–782, 1992.

Ollis, D. F.; Pelizzeti, E.; and Serpone, N.: Photocatalyzed destruction of water contaminants. *Environ. Sci. Technol.*, 9:1523–1529, 1991.

Paris, D. F.; Wolfe, N. L.; and Steen, W. C.: Structure-activity relationships in microbial transformations of phenols. *Appl. Environ. Microbiol.*, 44:153–158, 1982.

Paris, D. F.; Wolfe, N. L.; Steen, W. C.; and Banghman, G. L.: Effects of phenol molecular structure on bacterial transformation rate constants in pond and river samples. *Appl. Environ. Microbiol.*, 45:1153–1155, 1983.

Paris, D. F.; Wolfe, N. L.; and Steen, W. C.: Microbial transformation of esters of chlorinated carbocylic acids. *Appl. Environ. Microbiol.*, 47:7–11, 1984.

Perrin, D. D.: Ionization Constants of Inorganic Acids and Bases in Aqueous Solutions. New York: Pergamon, 1981. IUPAC Chemical Data Series, No. 29.

Perrin, D. D.; Dempsey, B.; and Serjeant, E. P.: pK_a prediction for organic acids and bases. New York: Chapman & Hall, 1981.

Pinal, R.; Lee, L. S.; and Rao, P. S. C.: Prediction of the solubility of hydrophobic compounds in nonideal solvent mixtures. *Chemosphere*, 22:939–951, 1991.

Rai, D.; Zachara, J.; Schwab, A.; Schmidt, R.; Girvin, D.; and Rogers, J.: Chemical Attenuation Rates, Coefficients and Constants in Leachate Migration, Vol. I: A Critical Review. Report EA-3356 to the Electric Power Research Institute (Palo Alto, CA) by Pacific Northwest Laboratories (Battelle Institute), Richland, WA, 1984.

Rice, R. G., and Cotruvo, J. A.: Ozone/Chlorine Dioxide Oxidation Products of Organic Materials. Cleveland: Ozone International Press, 1978.

Robinson, R. A., and Stokes, R. H.: Electrolyte Solutions, 2nd ed. New York: Academic Press, 1959.

Schroeder, W. H., and Lane, D. A.: The fate of toxic airborne pollutants. *Environ. Sci. Technol.*, 22:240–246, 1988.

Schwarzenbach, R. P.; Giger, W.; Schaffner, C.; and Wanner, O.: Groundwater contamination by volatile halogenated alkanes: abiotic formation of volatile sulfur compounds under anaerobic conditions. *Environ. Sci. Technol.*, 19:322–327, 1985.

Sehmel, G. A.: Atmospheric Science and Power Production, edited by D. Randerson, Washington, D.C.: U.S. Department of Energy, 1984.

Sigg, L.: Surface chemical aspects of the distribution and fate of metal ions in lakes. Aquatic Surface Chemistry, edited by W. Stumm, pp. 219–350. New York: Wiley Interscience, 1987.

Sillen, L. G., and Martell, A. E.: Stability Constants of Metal-Ion Complexes. London: The Chemical Society, 1964. Spec. Publ. No. 17.

Sillen, L. G., and Martell, A. E.: Stability Constants of Metal-Ion Complexes, Supplement No. 1. London: The Chemical Society, 1971. Spec. Publ. No. 25.

Smith, R. M., and Martell, A. E.: Critical Stability Constants, Vol. 4, Inorganic Ligands. New York: Plenum, 1976.

Smith, R. M., and Martell, A. E.: Critical Stability Constants, Vol. 4, Inorganic Complexes. New York: Plenum, 1982.

Southworth, G. R.; Herbes, S. E.; and Allen, C. P.: Evaluating a mass transfer model for the dissolution of organics from oil films into water. *Water Res.*, 17:1647–1651, 1983.

Stephen, H., and Stephen, T.: Solubilities of Inorganic and Organic Compounds, Vol. 1, Part 1. Oxford: Pergamon, 1963.

Stumm, W., and Morgan, J. J.: Aquatic Chemistry. New York: Wiley, 1981.

Tanford, C.: The Hydrophobic Effect. New York: Wiley, 1980.

Thomas, R. G.: Volatilization from Water. Handbook of Chemical Property Estimation Methods, chapter 15. New York: McGraw-Hill, 1982.

Tucker, W. A., and Nelken, L. H.: Diffusion coefficients in air and water. Handbook of Chemical Property Estimation Methods, chapter 17. New York: McGraw-Hill, 1982.

van Leeuwen, H. P., and Buffle, J. (eds.): Environmental Particles. Volume I and II. Boca Raton, FL: Lewis, 1992 (Vol. I), 1993 (Vol. II).

Verschueren, K.: Handbook of Environmental Data on Organic Chemicals, 2nd ed. New York: Van Nostrand Reinhold, 1993.

Voice, T. C., and Weber, W. J.: Sorption of hydropholic compounds by sediments, soils and suspended solids. I: Theory and background. *Water Res.*, 17:1433–1441, 1983.

Washburn, E. W. (ed.): International Critical Tables of Numerical Data, Physics, Chemistry and Technology, Vol. 3. New York: McGraw-Hill, 1928.

Weast, R.C. (ed.): Handbook of Chemistry and Physics. Boca Raton, FL: CRC Press, 1990.

Weber, E. J., and Wolfe, N. L.: Kinetic studies of the reduction of aromatic azo compounds in anaerobic sediment/water systems. *Environ. Toxicol. Chem.*, 6:911–919, 1987.

Weber, W. J.; Voice, T. C.; Pirbazari, M.; Hunt, G. E.; and Ulanoff, D. M.: Sorption of hydrophobic compounds by sediments, soils, and suspended solids. II: Sorbent evaluation studies. *Water Res.*, 17:1443–1452, 1983.

Westall, J.C.: Adsorption mechanisms in aquatic surface chemistry. Aquatic Surface Chemistry, edited by W. Stumm, pp. 3–32. New York: Wiley-Interscience, 1987.

Whitfield, M., and Turner, O. R.: Role of particles in regulating the composition of seawater. Aquatic Surface

Chemistry, edited by W. Stumm, pp. 457–494. New York: Wiley-Interscience, 1987.
Yalkowsky, S. H.: Estimation of the aqueous solubility of complex organic compounds. *Chemosphere*, 26:1239–1261, 1993.
Yao, C. C., and Haag, W. R.: Rate constants for direct reactions of ozone with several drinking water contaminants. *Water Res.*, 25, 761–773, 1991.
Yaws, C.; Yang, H. C.; and Pan, X.: Henry's law constants for 362 organic compounds in water. *Chem. Eng.*, November: 179–185, 1991.
Zafiriou, O. C.; Joussot-Dubien, J.; Zepp, R. G.; and Zika, R. G.: Photochemistry of natural waters. *Environ. Sci. Technol.*, 18:358A–371A, 1984.
Zika, R. G.: Photochemistry of Environmental Aquatic Systems. Washington, D.C.: American Chemical Society, 1987.

SUPPLEMENTAL READING

Bodek, I.; Lyman, W. J.; Reehl, W. F.; and Rosenblatt, D. H. (eds.): Environmental Inorganic Chemistry: Properties, Processes, and Estimation Methods. New York: Pergamon, 1988.
Drever, J. I.: The Geochemistry of Natural Waters. Englewood Cliffs, NJ: Prentice Hall, 1982.
Hemond, H. F., and Fechner, E. J.: Chemical Fate and Transport in the Environment. San Diego: Academic Press, 1994.
Howard, P. H.: Environmental Fate and Exposure for Organic Chemicals, Vols. I–IV. Boca Raton, FL: CRC Press (Lewis Publishers), 1989–1993.
Hutzinger, O. (ed.): The Handbook of Environmental Chemistry, Vols. I–III (in several parts). New York: Springer-Verlag, 1980–1984.
Lindsay, W. L.: Chemical Equilibrium in Soils. New York: Wiley, 1979.
Lyman, W. J.; Reehl, W. F.; and Rosenblatt, D. H. (eds.): Handbook of Chemical Property Estimation Methods, The Environmental Behavior of Organic Compounds, McGraw-Hill, New York, 1982. (Now published by the American Chemical Society.)
Mackay, D.; Shiu, W. Y.; and Ma, K. C.: Illustrated Handbook of Physical-Chemical Properties and Environmental Fate for Organic Chemicals, Vols. I–III. Chelsea, MI: Lewis, 1992–1993.
Morel, F. M.: Principles of Aquatic Chemistry. New York: Wiley, 1983.
Morel, F. M., and Hering, J. G.: Principles and Applications of Aquatic Chemistry. New York: Wiley, 1993.
Neely, W. B., and Blau, G. E.: Environmental Exposure from Chemicals, Vols. I and II. Boca Raton, FL: CRC Press, 1985.
Schwarzenbach, R. P.; Gschwend, P. M.; and Imboden, D. M.: Environmental Organic Chemistry. New York: Wiley, 1993.
Stumm, W., and Morgan, J. J.: Aquatic Chemistry. New York: Wiley, 1981.
Thibodeaux, L. J.: Chemodynamics—Environmental Movement of Chemicals in Air, Water, and Soil. New York: Wiley, 1979.

Chapter 16

BIOACCUMULATION AND BIOAVAILABILITY IN MULTIPHASE SYSTEMS

A. Spacie, L. S. McCarty, and *G. M. Rand*

INTRODUCTION

A life in water requires that gills and other body surfaces be designed for the efficient exchange of oxygen and other essential molecules. Unfortunately, the same physiological designs that make aquatic life so successful also lead to the efficient uptake of many other nonessential chemicals. Thus, aquatic organisms are well known for their ability to absorb a variety of chemicals from water. Some chemicals may be found only at low levels in various tissues, whereas others may build up to significant concentrations. Accumulation of chemicals by organisms in the aquatic environment is based on the interaction of a variety of physical, chemical, and biological characteristics and processes. The discussion in this chapter focuses largely on the characteristics and processes that influence the availability of chemicals at the environment-organism interface in multiphase systems, such as sediment, and the rate and magnitude of absorption of available chemical into the organism.

Bioaccumulation is the general term describing the net uptake of chemicals (usually nonessential ones) from the environment by any or all of the possible routes (i.e., respiration, diet, dermal) from any source in the aquatic environment where chemicals are present (i.e., water, dissolved, colloidal or particulate organic carbon, sediment, other organisms). *Bioconcentration* is a more specific term reserved for describing accumulation from water alone. Given the heterogeneous nature of aqueous phases in the environment, it is difficult to exclude the possibility that some uptake of chemical from that associated with dissolved, colloidal, and particulate organic carbon phases may also occur. Thus, the term bioconcentration should be used with caution and be employed in circumstances where information about the character of the aqueous phase is available.

Bioaccumulation is of concern both for its possible effect on the organism and for the contamination of higher trophic levels, including humans, that may occur. However, it must be emphasized that the presence or accumulation of a chemical or chemicals in an organism is not an adverse biological effect in and of itself; only the biological responses induced by the presence of the chemical(s) and/or its metabolite(s) are potential adverse effects. Because the type and nature of biological responses and the potency to induce such responses vary with organisms and chemicals, judgments about adverse effects must be referenced to known concentration-response relationships rather than based simply on the magnitude of the accumulation.

Thus, in terms of the basic phases of toxicological processes outlined in Chapter 1 (exposure, toxicokinetics, toxicodynamics), the study of bioaccumulation is directed toward the first two categories:

1. The nature of the exposure (including bioavailability of the chemical)
2. The nature and kinetics of processes that determine the rate, distribution, and magnitude of chemical accumulation in the organism.

Bioaccumulation investigations usually focus only on the nature of the exposure and toxicokinetic processes and degree of accumulation. The character and time course of any adverse effects that may result from the accumulation of a chemical (toxicodynamics) are not routinely considered.

Major advances in the understanding of the bioaccumulation process have occurred in the past decade. The process is now viewed essentially as a distribution or partitioning of chemical between the organism and its environment. The capacity of each phase, such as tissue or water, is determined by the thermodynamic behavior of the chemical—its solubility in each phase. Life is a nonequilibrium process and many of the most significant factors determining residue levels of chemicals in tissues are apt to change over time. For that reason, both the physicochemical characteristics of the chemical(s) and the physiology and biochemistry of the organism are necessary for a full understanding of the bioaccumulation process. A variety of quantitative structure-activity relationships (QSARs) have been developed for the major classes of organic chemicals in order to predict persistence, bioaccumulation, and biodegradation (see Chapter 20) and these are useful for making general predictions about bioaccumulation potential. At the same time, a much clearer understanding of the actual biological processes involved in accumulation has come mainly from the development of a wide range of kinetic models coupled with laboratory and field measurements (Appendix Chapter D, 1985; Connell, 1988, 1990; Barron, 1990).

One outcome of such research has been the realization that organisms vary widely in their bioaccumulation and biotransformation (Chapter 17) abilities. Even within genera, there are important species differences that may not be revealed in general modeling approaches. For that reason, the development of physiologically based pharmacokinetic (PB-PK) models has been a major advance, because a considerable amount of species-specific physiological and biochemical information is required (Barron et al., 1990; McKim and Nichols, 1994). PB-PK models are mathematical descriptions of the nature of various compartments in an organism, the processes that occur in these compartments, and the transport activities that occur between them. They usually consist of algorithms describing a group of tissue compartments that are arranged as parallel shunts between arterial and venous blood supplies. Estimates of physiological parameters are usually used rather than empirical rate constants obtained by curve fitting to bioassay data. Although in aquatic toxicology such models might be more correctly termed physiologically based, toxicokinetic models (PB-TK), the PB-PK terminology originating from pharmacology appears to be becoming generally accepted.

Much of the information on bioaccumulation comes from studies with certain chlorinated pesticides, polychlorinated biphenyls (PCBs), and other nonpolar, persistent organics, some of which accumulate to high levels in organisms. Because of this emphasis, the process of bioaccumulation has often been considered important only for relatively nontoxic, poorly metabolized chemicals. Actually, bioaccumulation is a precursor to all chemical toxicity. Without some degree of accumulation, however slight, toxic effects resulting from interactions at a target site(s) in organisms cannot take place. Much of the understanding that has come from research on the more persistent organochlorines may eventually lead to a clearer view of uptake processes for other groups of chemicals. Toxicokinetics, the uptake and tissue distribution process, is now seen as a key element in understanding the behavior of all classes of toxic chemical agents. As discussed here, there is a wealth of mathematical models and experimental techniques available for studying toxicokinetics.

Most aquatic toxicokinetic studies have focused on the transfer of chemical from the external medium (water) to the respiratory membranes, fluids, and organs of the target organism. As important as this process may be, it includes only part of the system that ultimately determines the delivery of toxicant to the organisms. Aquatic organisms live in complex, multiphase systems that include solid, colloidal, and perhaps gas phases as well. Myriad transfer and exchange processes occur in the organism's local environment that can affect the delivery of toxicant. These factors, considered under the general heading of *bioavailability*, can be modeled and measured by techniques that are somewhat analogous to toxicokinetic models (reviewed by Adams, 1987; Knezovich et al., 1987; Landrum and Robbins, 1990; Farrington, 1991). Advances in this area of contaminant dynamics now make it possible to consider both sides of the transfer process, both external and internal to the organism. As more and more attention is being given to contaminated sediments and to other complex systems such as wetlands, multimedia approaches are becoming essential.

The purpose of this chapter is to discuss the bioaccumulation process as it is affected by the distribution of organic contaminants in multiphase systems such as sediment. Basic principles of chemical partitioning are presented, as well as specific applications to aquatic organisms. The emphasis here is on the "supply side" of the process or the delivery of chemical up to and into the organism. Later steps in the internal bioaccumulation process, particularly biotransformation, are discussed in Chapter 17. Organochlorines and similar persistent organics are the only chemicals for which adequate distribution models are currently available. Therefore, ionizable organics and metals are not considered directly. Perhaps some of the approaches described for neutral organics will lead eventually to parallel approaches for other chemical classes as well.

TERMINOLOGY

Bioaccumulation factor (BAF). Ratio of tissue chemical residue to chemical concentration in an external environmental phase (i.e., water, sediment, or food). BAF is measured at steady state in situations in which organisms and food are exposed, unless noted otherwise. For equilibrium partitioning, at steady state, BAF may approximate K_b (see partition coefficient).

Biota-sediment accumulation factor (BSAF). Ratio of lipid-normalized tissue chemical residue to carbon-normalized sediment chemical concentration (units of g lipid/g organic carbon). Note that various other terms for this are used in the literature, including accumulation factor (AF).

Bioavailability. The portion of the total quantity or concentration of a chemical in the environment or a portion of it that is potentially available for biological action, such as uptake by an aquatic organism. More specifically, *environmental bioavailability* is the portion of chemical in an available form which the organism encounters that it actually absorbs. In pharmacology, the term "bioavailability" refers to the fraction of an administered dose that reaches target sites within an organism.

Bioconcentration factor (BCF). Ratio of tissue chemical residue to chemical concentration in water (unitless). BCF is measured at steady state where there is no food-chain exposure, unless stated otherwise. BCF is equivalent to K_b for equilibrium partitioning (see partition coefficient).

Biomagnification. The increase in tissue chemical residues at higher trophic levels, primarily as a result of dietary accumulation. If biomagnification occurs, the fugacity of the chemical(s) increases with trophic level and is greater than that of the chemical in water (see partition coefficient).

Body residue or burden. Total amount of a chemical in an individual organism.

Boundary layer. A relatively stagnant microlayer of water adjacent to a surface such as gill or sediment. The boundary layer becomes thinner as turbulent mixing of the water increases.

Depuration. Loss of accumulated chemical residues from an organism placed in clean water. See Elimination for a more general term.

DOC. Dissolved organic carbon (see Chapter 1).

Dose. Exposure dose is the amount or concentration an organism is exposed to, and absorbed dose is the total quantity of chemical absorbed by the organism. That proportion of the amount or concentration of the absorbed dose that reaches a target site(s) in specific organs or tissues and interacts there to elicit a toxic response is termed the target dose.

Elimination. General term for loss or disappearance of a substance from an organism by any active or passive transport mechanism, including diffusion and metabolic transformation (i.e., biotransformation).

Equilibrium partitioning approach (EqP). An approach for estimating the fate of chemicals (primarily organics) in the aquatic environment that is based on the assumption that a steady-state can, and usually is, achieved between the activity of chemicals (usually approximated as concentration) in the various component phases—water, sediment, organisms. The EqP approach is often exploited, for interpretation and extrapolation purposes, by normalizing chemical concentrations based on the lipid content of the aquatic organisms and the organic carbon content of the sediments. These normalized BSAF values are considered to be independent of particular sediments and species.

Exposure. Contact of an organism with a chemical.

First-order reaction kinetics. Reactions in which the rate of change of chemical flux (gain or loss) is directly proportional to chemical concentration.

Fugacity (f_i). The thermodynamic activity or "escaping tendency" of a solute (i.e., chemical) in a particular phase, given in units of partial pressure or pascals (Pa). It is a measure of the activity of the chemical in that phase and can be viewed as the partial pressure a chemical exerts as it attempts to escape from one phase and migrate to another. For example, if the fugacity of a chemical in water exceeds that in air, the chemical will evaporate until a new equilibrium is established. At equilibrium, f_i values in all phases are equal; however, common fugacities rarely correspond to common concentrations as the activity (i.e., fugacity) of a given number of molecules of a chemical varies depending on the phase they are in. Using fugacity instead of concentration immediately reveals the status of phases and the likely direction of diffusive transfer. Furthermore, the magnitude of the fugacity difference controls the rate of transfer (Mackay, 1991).

Hydrophobic chemical. A chemical of low water solubility and correspondingly high solubility in lipid or nonpolar solvents.

Partition coefficient (K_U). The distribution of a chemical solute between two phases at thermodynamic equilibrium or steady-state (denoted with the letter *K*), calculated as the ratio of its concentrations in the two phases. Important partition coefficients include K_{ow} (octanol-water), K_{oc} (organic carbon–water), K_{pw} (particle-water), K_{bl} (blood plasma proteins–water), K_b (organism-water).

Physiologically based pharmacokinetic (PB-PK) model. A kinetic model describing the uptake and internal distribution of chemicals that uses coefficients derived from measured anatomical/physiological characteristics of the organism and physicochemical properties of the chemical rather than simple rate constants obtained from empirical observations (i.e., curve fitting to bioassay data). Also called physiologically-based toxicokinetic (PB-TK) models.

POC. Particulate organic carbon. See Chapter 1.

Pore water (or interstitial water). The water between the particles of sediment (see Chapters 1 and 8).

Sorption. A general term used to describe adsorption to solid surfaces and absorption into liquid phases. Simple adsorption is relatively rare in the aquatic environment, occurring mainly on clay particles in carbon-poor environments such as ground water. For ionic and amphoteric contaminants, reversible adsorption can occur through ion exchange, hydrogen bonding, electrostatic binding, or weaker van der Waals forces. For weak acids and bases, the pH and ionic strength of the medium are important factors, both for sorption to natural particles and for uptake by organisms. For nonpolar organic chemicals, sorption occurs mainly through the organic-phase partitioning described earlier (see Chapters 1 and 15). For complex environmental particles, sorption is a continuum of binding processes covering a range of possible interactions.

Steady state. A condition in which exchange of matter or energy (flux) into and out of a system are equal, so that no net change in the system can be observed. Such a system may be in balance but still far from equilibrium.

TOC. Total organic carbon. This is the total amount in the water column and represents the sum of contributions from truly dissolved, colloidal, and suspended particulate organic carbon (POC).

Toxicokinetics. Uptake of chemical(s) from environmental media and subsequent distribution and elimination processes within an exposed organism.

Uptake clearance. The amount (real or hypothetical) of environmental medium (e.g., water or sediment) cleared of chemical by uptake into the organism per unit mass of organism per unit time, expressed as a flow rate (e.g., $L/g^{-1}\ h^{-1}$ or $kg\ g^{-1}\ h^{-1}$).

Uptake, assimilation, or absorption efficiency. The fraction of total available chemical in a medium (e.g., water or food) that is actually transferred to the organism, measured as the difference between the incoming (C_i) and outgoing (C_o) concentrations: Efficiency = $(C_i - C_o)/C_i$. For gill uptake, the inspired water and expired water are compared. For gut assimilation, the concentrations in food and feces are compared. For organic chemicals uptake efficiency is affected by chemical hydrophobicity and, although often assumed for modeling purposes to be constant for a particular species, it is known to be affected by respiration/consumption rate and medium characteristics (i.e., organic carbon content, lipid content).

Volume of distribution (V_d, V_{ss}, or apparent volume of distribution). The volume of a compartment that would account for the total quantity of drug or chemical in the body if it were distributed throughout the body in the same concentration as in blood plasma: $V_d = A_b/C_p$, where A_b is the total amount of drug or chemical in the body and C_p is the concentration of drug or chemical in plasma. V_d increases with increased lipophilicity and increased tissue binding and with decreased protein binding, factors that decrease C_p. Conversely, factors that keep the chemical in the plasma or increase C_p, such as low lipid solubility, increased plasma protein binding, or decreased tissue binding, reduce the apparent volume of distribution. Although it does not correspond to any real physical compartment in the body of an organism, it does provide some information about both the chemical and the organism being examined. In environmental work the term V_{ss} (apparent volume of distribution at steady state) is often used. $V_{ss} = A_b/C_w$, where C_w is the exposure water concentration.

MAJOR PROCESSES AFFECTING UPTAKE

Organisms accumulate chemicals by direct absorption across gills, skin, and other outer surfaces, or by ingestion of contaminated food and water (Appendix Chapter D). The feeding route is particularly complex in aquatic organisms because it may involve the ingestion of chemicals in live prey items, suspended particles, or sediments, depending on the specific feeding behavior of the organism. Several basic processes common to all routes of uptake are important to understand because they can profoundly affect the rate and extent of bioaccumulation. These are equilibrium partitioning, sorption kinetics, diffusion kinetics, membrane flux, and organism uptake efficiency. Some of the processes occurring on either side of the biological membrane are shown in Figure 1. Particulate burial or resuspension in sediment also occurs and may affect bioaccumulation.

Equilibrium Partitioning in Organic Phases

Many physical distribution processes in the environment can be treated as a thermodynamic equilibrium between phases. A simple example is the liquid-liquid partitioning of a nonpolar organic solute (i.e., chemical) between a solution of octanol and water. The solute is expected to distribute in a predictable way according to its solubilities in the two immiscible phases. At equilibrium the octanol-water partition coefficient (K_{ow}) is given by the ratio of concentrations (C_i) in the two phases:

$$K_{ow} = \frac{C_{\text{octanol}}}{C_{\text{water}}} \tag{1}$$

Also at equilibrium the "escaping tendency" or fugacity (units of pressure; Pa) of the solute in each phase will, by definition, be equal:

$$f_{\text{octanol}} = f_{\text{water}} \tag{2}$$

and a proportionality constant (Z; in mol/m^3 Pa), also known as a fugacity capacity factor, can be determined that relates fugacity (f) in each phase (i) to its concentration (C; mol/m^3) in that phase:

$$Z_i * f_i = C_i \tag{3}$$

This calculation emphasizes that a thermodynamic equilibrium occurs only when the chemical activities or fugacities in different phases are equal. Equality in chemical concentrations is not required, since equivalent fugacity is usually associated with different molar concentrations in disparate phases, due to the differences in chemical activity resulting from dissimilar solute-solvent interactions in each phase. Although concentrations of chemical in water and organism may be dramatically different in a bioconcentration study and reported as a bioconcentration factor, which is the ratio of these concentrations at the thermodynamic equilibrium, the activity of the chemical in water and organism is the same and a corresponding activity-based ratio would be 1.0. Z values can be established for each chemical in each phase so that the partition coefficient

$$K_{ow} = \frac{C_{\text{octanol}}}{C_{\text{water}}} = \frac{Z_{\text{octanol}}}{Z_{\text{water}}} \tag{4}$$

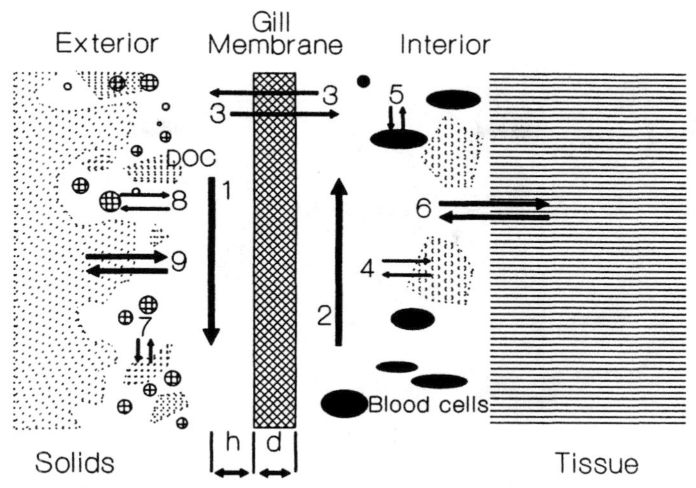

Figure 1. Conceptual view of major mass transfer processes affecting bioaccumulation in a multiphase system; in this case the gill is used as an illustrative example. Key: (1) flow of water along the membrane; (2) flow of blood in the organism; (3) flux of chemical across the membrane; (4) binding and release from blood serum proteins; (5) absorption to/desorption from blood cells; (6) transfer of chemical from blood to tissues via perfusion of major tissues; (7) complexation to/decomplexation from dissolved organic carbon (DOC); (8) adsorption to/desorption from particulate organic carbon (POC); (9) adsorption to/desorption from large particulate solids plus internal diffusion within these particulates. *d*, diffusion distance across the membrane; *h*, thickness of the stagnant water layer adjacent to membrane.

can be interpreted on the basis of well-defined thermodynamic properties that are calculable for various chemicals (Mackay and Paterson, 1981). If there are additional phases in the system such as particles, extra equations can be written that equate the fugacities in these various phases (Mackay, 1991). In principle, equilibrium distributions can also be written for ionizable organics and for metals, provided that the chemical equilibria relating the various chemical species to pH are also included, because in this approach un-ionized and various ionized forms of a chemical effectively represent a mixture of different chemicals with distinct characteristics. For ionized species chemical activities (estimated as an activity coefficient * C) are more suitable than fugacities for modeling purposes. Equilibrium partitioning models involving multiple chemical species have been proposed by Diamond et al. (1992).

Regardless of the approach chosen, *equilibrium partitioning* always assumes:

1. A closed system without external exchanges of energy or mass, at constant temperature and pressure
2. Reversibility
3. Sufficient time ($t \rightarrow \infty$) for the solute to distribute throughout the system

If these assumptions are reasonably approximated, equilibrium partitioning can be applied not only to organic solvents like octanol but also to the organic phases of natural materials in water (particulate organic carbon) and, as will be seen later, aquatic organisms themselves. The time required for equilibration of the solute between small organic particles and water may be seconds, days, or years, depending on particle size and K_{ow} of the solute (Figure 2). The extremely long equilibration times reflect the fact that the particle must randomly encounter and "extract" millions of units of the hydrophobic chemical from a very dilute solution. Although organic chemicals range from hydrophilic (miscible with water to K_{ow} of 10^1) to superhydrophobic ($K_{ow} > 10^6$), many industrially and/or environmentally important chemicals are found in the moderate to high K_{ow} range of 10^2 to 10^6 (Table 1) and sorb well to organic particles.

Methyl mercury is included in Table 1 as an example of an important contaminant that does not follow this ideal type of liquid-liquid partitioning. Although it accumulates in aquatic organisms to a high degree (bioaccumulation of the order of 10,000 to 100,000 times water concentrations), the small K_{ow} does not reflect this potential. Transfer from water to organic phases occurs mainly in the form of neutral ion pairs (CH_3HgCl^0 and CH_3HgOH^0) (Faust, 1992). However, once inside a biological matrix, the methyl mercuric ion (CH_3Hg^+) binds readily to the sulfhydryl groups of proteinaceous material (R–SH) of membranes and is effectively removed from solution. The binding violates the assumptions of equilibrium partitioning and shifts the partitioning process in favor of uptake by the organic phase. Organic chemi-

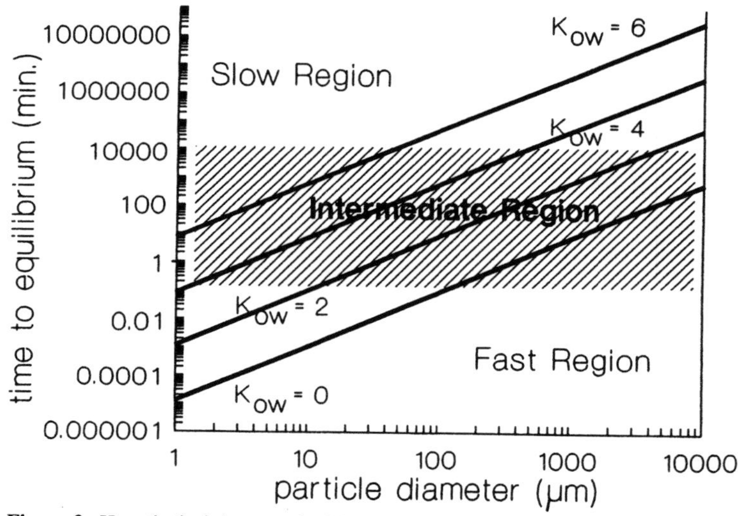

Figure 2. Hypothetical time required for equilibration of a chemical between water and particle, by diffusion only. The particle-water partition coefficient (K_{pw}) for these particles is assumed to be well-approximated by K_{ow}. In the region where equilibrium processes require intermediate time, kinetics may also be important. Stirring would decrease equilibrium time required by approximately 1 log unit. (Based on a concept of Mackay, 1992.)

cals with functional groups such as carbonyl (C=O) may also bind to biological material through hydrogen bonding or weaker van der Waals forces, causing "nonideal" behavior (Manahan, 1991).

Marine and freshwater environments possess a wide variety of particulate and colloidal material made up of minerals and natural organic substances (Table 2; see Chapter 1). The definition of these classes is operational, depending on the method of filtration used to fractionate the sample. Typical concentrations of particulate organic material in many water bodies vary between 10^{-6} and 10^{-7} kg/L (Gobas, 1992). Whether the matrix is mineral or biological in origin, virtually all aquatic particles carry an organic coating that can potentially sorb chemical contaminants (Stumm, 1992). In general, the solvent properties of natural coatings are similar to those of octanol, so an organic carbon–water (K_{oc}) partition coefficient can be written as

$$K_{oc} = \frac{C_{oc}}{C_w} \tag{5}$$

where C_{oc} is the concentration of chemical in the organic carbon phase and C_w is the concentration in water. The value of K_{oc} can vary with the type and age of organic matter but is usually nearly equal to K_{ow}, since the density of the organic material is often approximately 1 ($K_{oc} = K_{ow}/d_{oc}$, where d_{oc} is the density of the organic carbon material; Di Toro, 1985). For particles with a fraction by weight of organic carbon (f_{oc}) and a density of approximately 1, a particle-water partition coefficient K_{pw} (with typical units of L/kg) can then be calculated as

$$K_{pw} = \frac{C_s}{C_w} = f_{oc}K_{oc} \approx f_{oc}K_{ow} \tag{6}$$

where C_s is the concentration of chemical sorbed to suspended organic particles.

Organic carbon–based partitioning provides a good estimate of the amount of organic chemical sorbed to natural particles, even though the actual sorption mechanism may involve a variety of adsorption (surface binding) and partitioning processes. Sorption isotherms (see Chapter 15) can be developed to quantify particular sorbent-sorbate interac-

Table 1. Examples of octanol-water partition coefficient (K_{ow}) ranges for some organic chemicals of environmental importance

Chemical type	K_{ow}
Polycyclic aromatic hydrocarbons (PAHs)	10^3–10^7
Halogenated aliphatic hydrocarbons	10^1–10^3
Chlorinated ethers	10^3–10^4
Organic pesticides	10^0–10^7
Chlorinated benzenes and phenols	10^2–10^6
Polychlorinated biphenyls (PCBs)	10^4–10^{10}
Polychlorinated dioxins and dibenzofurans (PCDD, PCDF)	10^6–10^{10}
Methyl mercury (CH_3HgCl^0, CH_3HgOH^0)	2

Table 2. Classification scheme for particle sizes in water[a]

Type	Particle size (cm)	Example constituents
Solution	$<10^{-7}$	Inorganic simple and complex ions; molecules and polymeric species; polyelectrolytes; organic molecules; undissociated solutes; small aggregates
Colloidal dispersion (submicroscopic)	10^{-7}–10^{-5}	Mineral substances, hydrolysis and precipitation products; macromolecules; biopolymers; detritus
Fine particulate dispersion (visible under microscope)	10^{-5}–5×10^{-3}	Mineral substances, precipitated and coagulated particles; detritus; bacteria, plankton, and other microorganisms
Coarse dispersion (visible to the eye)	$> 5 \times 10^{-3}$	Mineral substances; precipitated and coagulated particles; detritus, macroplankton; organisms of all types including fish

[a]Particle sizes may differ slightly from those mentioned in Chapter 1.
Source: Horne RA: *Marine Chemistry*, Copyright 1969 Wiley Interscience. Reprinted by permission of John Wiley & Sons, Inc.

tions. However, at the very low concentrations typical of xenobiotics in aquatic environments, sorption relationships are usually linear and take the same form as a partition coefficient (K_{pw}).

An important application of Eq. (6) is to calculate the fraction of a chemical that remains freely dissolved in water (C_w), because it is C_w and not the total concentration ($C_{\text{total}} = C_w + C_{\text{sorbed}}$) that is available for uptake by organisms. The freely dissolved fraction can be expressed as

$$\frac{C_w}{C_{\text{total}}} = \frac{1}{1 + K_{pw}X} \tag{7}$$

where x is the volume fraction (kg/L) of particles (typically 10^{-5}). Substituting Eq. (6) into Eq. (7) gives the bioavailability of freely dissolved chemical in natural water:

$$C_w = \frac{C_{\text{total}}}{1 + f_{oc}K_{oc}} \tag{8}$$

Equation (8) can be used to examine a situation where a chemical with a K_{ow} value of 10^6 (~ log K_{oc} = 6) is present in aqueous solution with an organic carbon fraction of 10^{-6} kg/L (e.g., 1 mg/L of organic material found in a total suspended solids fraction of perhaps 10^{-5} kg/L or 10 mg/L, containing both organic and inorganic components). In this case 50% of the total water column chemical concentration is estimated to be freely dissolved (bioavailable) and 50% sorbed. Other examples can be interpolated

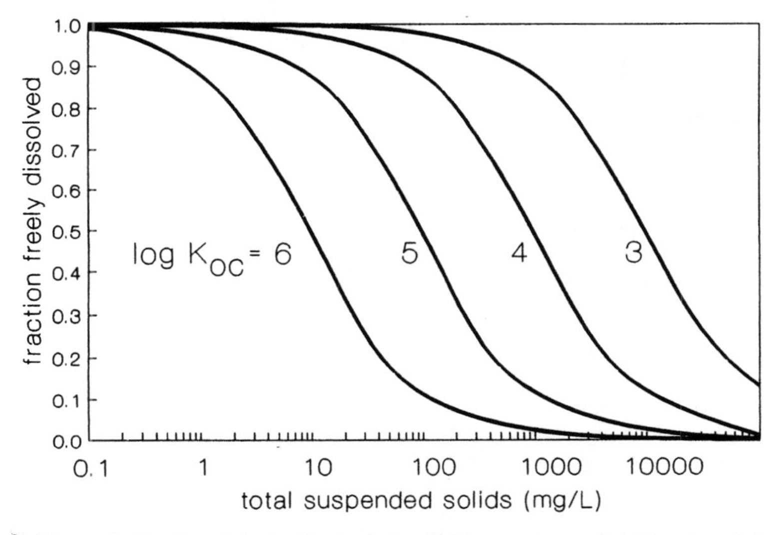

Figure 3. Fraction of freely dissolved chemical in a system containing suspended solids, where K_{oc} is the organic carbon-based partition coefficient between particle and water. Organic carbon content of suspended solids is assumed to be 10% of total (Adapted with permission from Pavlou and Dexter, 1979. Copyright 1979 American Chemical Society).

from Figure 3. The relationships show how the freely dissolved, biologically available concentration of chemical dissolved in water is reduced by sorption to particles and that the effect becomes greater with higher K_{ow} and with a greater mass of particles. In essence, the presence of a second phase (particles) reduces the fugacity of the chemical in the first phase.

Natural waters actually contain at least three phases—water, particles, colloidal/dissolved organic carbon—that affect chemical distributions (Baker et al., 1986). Because there is a broad continuum of sizes between particles and dissolved phases, the latter two phases have been defined in a number of classification schemes that are typically functional in nature and related to the subject under study, rather than being correct in any absolute sense (Horne, 1969). Equation (8) can therefore be modified to include sorption to dissolved as well as particulate matter. Accordingly,

$$C_w = \frac{C_{\text{total}}}{1 + f_{oc}K_{oc} + f_{doc}K_{doc}} \tag{9}$$

where f_{doc} is the fraction of dissolved organic carbon present and K_{doc} is dissolved organic carbon–water partition coefficient. Similarly, for i number of phases:

$$C_w = \frac{C_{\text{total}}}{1 + \Sigma f_i K_i} \tag{10}$$

where $\Sigma\ f_i K_i$ is the sum of the products of organic phase fraction and the phase-water partition coefficient for phases 1 through i.

Sediment pore waters can be rich in humics (see Chapter 1) and other dissolved organic carbon. As much as 50–90% of pore water DOC is colloidal (Krom and Sholkovitz, 1977) and organic chemicals partition to this DOC almost as if it were an octanol phase (Resendes et al., 1992). Brownawell and Farrington (1986) observed that the pore waters of marine sediments are best described as three-phase systems because of the presence of DOC. Examples of chemical concentration calculations based on this three-phase relationship are provided by Pankow and McKenzie (1991). The presence of a third phase may increase the total concentration of chemical in pore water relative to that found in the overlying water, where the DOC content may be much lower. Thus, higher chemical concentrations may be measured in pore water samples versus overlying water samples and measurements of whole sediment concentrations will increase in proportion to increasing pore water DOC content. Although the amount of DOC in the pore water is small relative to the total POC in sediment (mg/L versus percent) and may have only a small influence on the overall POC level, it dominates bioaccumulation processes from sediment as chemicals in the pore water DOC are more bioavailable than those in the general sediment POC.

These interactions make chemical analysis and interpretation of in situ sample results complicated. However, the uncomplexed, freely dissolved fraction of the chemical in water constitutes the most bioavailable portion to benthic organisms for respiratory and dermal absorption, and this concentration may not be dramatically different pore water and overlying water.

Mass Transfer Kinetics

If the distribution of a chemical between phases is not at thermodynamic equilibrium, there will be mass transfer down a gradient in the direction of that equilibrium state. For the simple octanol-water system, a nonpolar chemical dissolved in water diffuses into the organic phase until no more net change in either direction occurs. The rate of change of chemical concentration in octanol (dC_{octanol}/dt) as it approaches equilibrium is described by simple first-order kinetics:

$$\frac{dC_{\text{octanol}}}{dt} = k_1 C_{\text{water}} - k_2 C_{\text{octanol}} \tag{11}$$

where k_1 is the uptake rate constant (units of reciprocal time) and k_2 is the elimination or depuration rate constant (units of reciprocal time). At equilibrium, $dC_{\text{octanol}}/dt = dC_w/dt = 0$, so $k_1 C_{\text{water}} = k_2 C_{\text{octanol}}$ and

$$K_{ow} = \frac{C_{\text{octanol}}}{C_{\text{water}}} = \frac{k_1}{k_2} \tag{12}$$

Accumulation in an aquatic organism may be described by a similar simple model, sometimes referred to as a one-compartment, first-order kinetics (1CFOK) model. The organism is considered to be a single, homogeneous, membrane-bound compartment placed in water containing an infinite supply of the chemical in question at a given concentration. The accumulation process is described by the equation

$$\frac{dC_{\text{organism}}}{dt} = k_1 C_w - k_2 C_t \tag{13}$$

where dC_{organism}/dt is the rate of change of chemical concentration in the organism over time, C_w is the concentration of chemical in an infinite water source, and C_t is the concentration in an organism at time t. At steady state,

$$C_e = \frac{k_1 C_w}{k_2} = \frac{k_1}{k_2} C_w \tag{14}$$

where C_e is the chemical concentration in the organism at steady state. Combining the equations, and assuming k_1, k_2, and C_s remain constant, the integrated form of the model becomes

$$C_t = \frac{k_1}{k_2} C_w(1 - e^{-k2t}) = C_e(1 - e^{-k2t}) \quad \text{(integrated form)} \tag{15}$$

which indicates that the concentration in the whole body or some organ or tissue compartment increases over time at a rate that is maximal at the time of first exposure. The rate of accumulation then gradually decreases over time until some asymptotic concentration is attained.

Similar first-order kinetics models have been used to examine a wide variety of environmental fate and transport processes (Connell and Miller, 1984) and bioaccumulation processes (Spacie and Hamelink, 1982; Appendix Chapter D; Barron et al., 1990) that are reviewed elsewhere. Additional expressions may be added for the effects of temperature change, mixing, and so forth, but Eq. (11) represents the simplest example of a first-order kinetics model. In each case, k_1 is referred to as the apparent first-order uptake rate constant (units of L/kg^{-1} h^{-1} simplify to h^{-1} for the assumption L of water = kg of tissue), k_2 is the elimination rate constant, and K (in this case K_{ow}) is the partition or distribution coefficient. Models of this type can also be used for nonequilibrium processes in open systems that approach or achieve a steady state. There are two formulations for such 1CFOK models. The version outlined above can be termed the dual-rate-constant-based (DRC) version; the other, termed the clearance-volume-of-distribution (CVD) version, will be discussed later. Although very similar and closely related, the latter formulation has some particular advantages for investigations into chemical uptake by organisms.

Modifications of the 1CFOK model may incorporate multiple sources, multiple elimination components, growth, trophic efficiency, and internal exchange. However, enhanced realism with such models cannot be independent of the assumptions of the simple model: constant uptake rate, instantaneous mixing within the compartment(s), time- or age-dependent probability of transition between compartments, and a negative exponential depuration process for all compartments. Under field conditions the rate of contaminant intake may not be constant due to changes in behavior, food type, food availability, or other environmental factors. Violations of other assumptions are also likely to occur under field conditions. For example, elimination by poikilotherms will be linked to seasonal fluctuations in temperature.

Under free-living conditions, accumulation of contaminants may deviate significantly from the classical pattern and show an S-shaped or sigmoidal pattern of body residue buildup, characterized by an initial lag in the rate of chemical uptake compared with that described by Eq. (11). Such deviations from the classic model can be deduced from a qualitative assessment of the data set. Although sigmoidal patterns of accumulation have been noted (e.g., mercury accumulation in mosquito fish, *Gambusia affinis*), the ultimate accumulation has not been shown to deviate from the predictions of Eq. (11) (Brisbin et al., 1990).

One application of Eq. (11) in examining environmental fate processes is description of the first-order loss of a dissolved chemical from a system containing both sorbed and dissolved material. A specific example is one in which bioaccumulation, biotransformation, or microbial degradation processes remove only the dissolved form. The disappearance of chemical from a phase has the familiar form of logarithmic decay. In this example, where dissolved chemical is being lost from the total aqueous phase, C_w is the starting chemical concentration in the total aqueous phase and C_t is total chemical concentration in the aqueous phase at time t:

$$\frac{dC_t}{dt} = -k_2 C_w \tag{16}$$

$$C_t = C_w e^{-k2t} \quad \text{(integrated form)} \tag{17}$$

The time for one-half of the loss to occur (i.e., loss half-life) = $\ln 2/k_2 = 0.693/k_2$.

However if Eq. (16) is rewritten in terms of $C_w = C_{\text{total}} = (C_{\text{sorbed}} + C_{\text{dissolved}})$ using Eq. (7), the loss of dissolved chemical becomes

$$\frac{dC_t}{dt} = -k_2 \frac{C_{\text{total}}}{1 + f_{oc}K_{pw}} \tag{18}$$

$$C_t = \frac{C_{\text{total}}}{1 + f_{oc}K_{pw}} e^{-k2t} \quad \text{(integrated form)} \tag{19}$$

If sorption to particulate matter (K_{pw}) is negligible, then all of the material is subject to removal and the rate is maximal. Greater sorption slows the rate of loss from the bulk medium by the factor $1/(1 + f_{oc}K_{pw})$ because the dissolved concentration has been effectively reduced. Thus, the greater the hydrophobicity of the sorbed nonpolar organic, the more "resistance" is caused by sorption. For extremely hydrophobic substances, such as the highly substituted dioxins and PCBs (K_{ow} = 6–10), most of the chemical in an aqueous environment containing a significant sediment-colloid environment is sorbed (see Figure 3) and unavailable to processes that act on the chemical in dissolved form. In this case, the rate of desorption can strongly influence bioaccumulation, bacterial mineralization, and other biological processes because it may be the overall rate-limiting step.

Even chemicals with lower lipid solubility may have significant "lags" caused by slow desorption. Karickhoff and Morris (1985b), for example, have shown that desorption from natural particles occurs in both fast and slow stages. The fast stage may represent the release of surface-sorbed material, whereas the slower phase may involve the diffusion of chemical through the interior pores and channels of the particle. The desorption rate may also be dependent on the age of the particle and geochemical process causing its formation. Hydrocarbon-bearing particles created during pyrolysis, for example, may desorb polycyclic aromatic hydrocarbons (PAHs) more slowly than those that recently acquired PAHs from water (Farrington, 1991). Although both may have the same analytically determined PAH content, the former would be less readily available to organisms.

Sorption-desorption kinetics may be a major factor altering the seasonal availability of chemical contaminants, although the magnitude of the effect is K_{ow} dependent (Landrum et al., 1992a). Adsorption is a strongly exothermic process. Consequently, higher summer temperatures should cause less adsorption (K_{pw} reduced) and a greater quantity of chemical in the aqueous phase. Larsson (1986) found strong seasonal cycles for the bioaccumulation of PCBs in zooplankton. Greater desorption from sediments during the summer was cited as one of the likely causes for this pattern, although Landrum et al. (1992a) indicate that uptake clearance from sediment by organisms is probably temperature dependent and would also be involved in a seasonal bioaccumulation cycle.

Partitioning in Organisms

Partitioning of chemicals from the aquatic environment into organisms is similar in many respects to the processes described above for partitioning into nonliving organic materials. However, living organisms are both sufficiently different from the nonliving organic phase and of sufficient importance to require separate consideration. Aquatic organisms are sometimes, for the sake of simplifying both models and data interpretation, thought of as a membrane-delimited sac containing a mixture of lipid and water, where the lipid has properties similar to those of octanol. In fact, surrogate organisms composed of a semipermeable membrane surrounding a lipid (e.g., triolein) or nonpolar solvent (e.g., hexane) have been used as an exposure and bioconcentration monitoring tool in lieu of actual organisms (Sodergren, 1987; Huckins et al., 1993). According to this simplified view, the equilibrium partition coefficient K_b between organism and water is directly related to K_{ow}, provided that the chemical is nonreactive and not metabolized by the organism.

A large number of K_b-K_{ow} correlations (i.e., bioaccumulation QSARs; see Chapter 20) have been published in the literature (Esser, 1986; Connell and Hawker, 1988; Hawker and Connell, 1991) and are represented here by two regressions for small fish:

$$K_b = 1.00 \log k_{ow} - 1.32 \ (r^2 = 0.95) \quad \text{or}$$

$$K_b = 0.048 K_{ow} \quad \text{(Mackay, 1982)} \qquad (20)$$

$$\log K_b = 0.79 \log K_{ow} - 0.40 \ (r^2 = 0.86) \quad \text{(Veith and Kosian, 1982)} \qquad (21)$$

These particular regressions are suitable for small fish with a lipid content of approximately 7.6% and can be adjusted for other lipid values (L%) by multiplying the calculated K_b by (L%/7.6). Equivalent expressions based on water solubility rather than K_{ow} are also available (e.g., $\log K_b = -0.862 \log S_w + 2$, where S_w is water solubility in mol/m^3 and fish lipid content is 5%; Hawker and Connell, 1991).

Because the partitioning is assumed to involve the lipid portions of the organism, K_b values are often expressed on a lipid weight basis to reduce the variability caused by differences in lipid content among individuals. Metabolic transformation of a chemical acts to reduce the observed K_b, and other deviations in K_b are observed if the assumptions of equilibrium partitioning are violated. For example, extremely hydrophobic chemicals (i.e., high values of K_{ow}; see Table 1) may require long uptake times to reach steady state with respect to C_w (one or more years; see Figure 2) and many organisms with life spans of a few months or years may simply not live long enough to reach steady state. Another factor to consider is that rapidly growing organisms may appear to have constant or declining tissue residues (C_b) because of "growth dilution" (i.e., the increase in mass due to growth occurs at a higher rate than chemical uptake); thus, chemical concentration in the body declines even as the total amount in the body increases (Niimi and Cho, 1981; Sijm et al., 1992).

Simple partitioning theory indicates that bioconcentration (K_b) is related to k_1/k_2 (uptake rate constant/elimination rate constant), so the two first-order rate constants should also be expected to show some correlation to K_{ow}. Greater bioaccumulation of increasingly hydrophobic organic chemicals has been shown experimentally to be related largely to slower elimination kinetics rather than to increases in uptake kinetics. Various correlations to K_{ow} have been calculated that support an inverse relationship with k_2 and a direct, but weaker, relationship to k_1 (Gobas and Mackay, 1987; Hawker and Connell, 1988).

Most of the reported correlations of K_b, k_1, and k_2 with K_{ow} have found that the relationships break down both for very hydrophobic (superhydrophobic) substances ($\log K_{ow} > 6.0$), which appear to accumulate less than expected (Hawker and Connell, 1988, Gobas et al., 1989), and for hydrophilic substances ($\log K_{ow} < 1$), where chemical concentrations in the internal aqueous phase of organisms dominate K_b (McCarty et al., 1992). Although $\log K_b - \log K_{ow}$ relationships tend to be linear over the $\log K_{ow}$ range of about 2 to 6, such regressions should be used with caution because they tend to become nonlinear for extremely hydrophilic or hydrophobic chemicals (Figure 4). Furthermore, such relationships are valid only when there is little in the way of metabolic breakdown of the accumulated chemical by the organism.

Membrane Diffusion

In the simple view of chemical absorption from the aquatic environment, organisms can be seen as membrane-bound sacs of lipid and water and treated in a similar manner (with certain caveats) to nonliving organic material in the water column, as already discussed. However, in reality organisms are complex and a number of biological factors associated with the biological membrane, and organism activities and processes on either side of this barrier, can substantially affect the time associated with achieving a steady-state status as well as influence its magnitude.

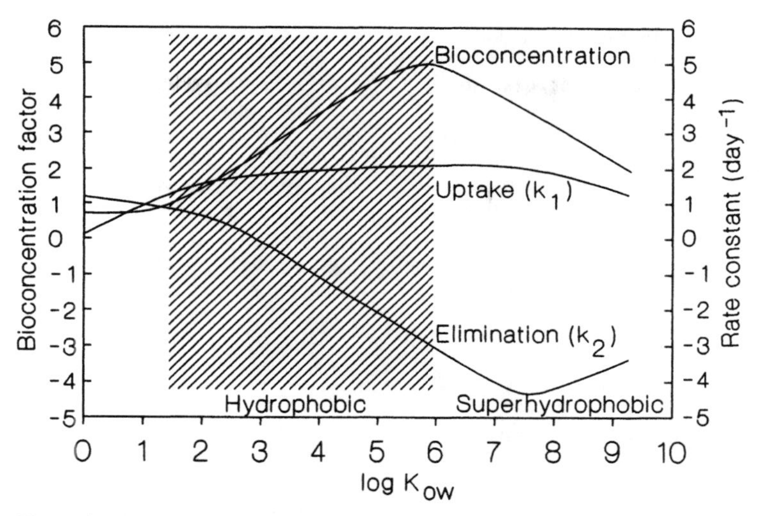

Figure 4. Idealized relationships between K_b, k_1, k_2, and K_{ow} for various nonreactive organic chemicals and small fish. Relationships illustrated for superhydrophobic chemicals are less well defined than for hydrophobic chemicals due to experimental difficulties. (Based on information from Gobas and Mackay, 1987; Connell and Hawker, 1988; McCarty et al., 1992.)

Transport of chemicals in the water column (see Chapter 15) is controlled by advection and eddy diffusion (i.e., dispersion), which are evaluated as part of a typical fate and transport model (see Chapter 18). Bioaccumulation studies generally do not deal with these factors except as they affect the "local" supply of chemical near the organism. Only recently has attention focused on the dynamics of this local environment. Uptake at the gill or other surface depletes the supply in the surrounding medium, creating a gradient that could affect other equilibria (see Figure 1). The conditions at such membranes are controlled by kinetic rather than equilibrium processes due to the various active metabolic processes (e.g., pumping of water and blood, osmoregulation, ionoregulation) that are carried out there specifically to counteract equilibrium tendencies, since life itself is a nonequilibrium process. The structure and function of the gills of aquatic organisms, especially fish, have been extensively studied and can be reviewed in most standard animal physiology textbooks. Hayton and Barron (1990) provide a useful succinct summary in their examination of the processes associated with the uptake of nonessential chemicals via gills, and Gobas et al. (1993) provide a similar review for uptake via the gut.

Although there are many details and nuances related to specific structures, functions, and circumstances, in general membranes at the interface between environment and organism are involved in transmembrane movement of water and chemicals and represent a series of barriers to diffusion that can potentially control the rate and extent of chemical uptake. Passage through the membrane itself depends on membrane thickness and other structural characteristics and on the molecular charge, size, and solubility of the chemical. Passage through the stagnant layer of water adjacent to the outer membrane surface depends on the gradient and flow of water.

For diffusion through the lipid portions of the membrane from water, the steady-state flux of chemical, where dX/dt is the rate of chemical transfer into the organism (mmol/min or mg/min), is given by (Hayton and Barron, 1990)

$$\frac{dX}{dt} = \frac{D_m A K_{mw}(C_o - C_i)}{d} \tag{22}$$

where D_m = diffusion coefficient of chemical in membrane (cm²/min)
A = membrane area participating in diffusion (cm²), or membrane absorbing surface area
K_{mw} = partition coefficient of chemical for membrane-water; log K_{mw} is generally proportional to log K_{ow}
C_o = concentration of chemical on outer membrane surface; it may be $\leq C_w$ for bulk water if there is an aqueous stagnant layer that has resistance; concentration is in appropriate molar or mass units per liter
C_i = concentration on the interior membrane surface, approximately equals blood plasma water concentration (in appropriate molar or mass units per liter)
d = membrane thickness (cm)

The expression $(D_m K_{mw}/d)$ is called the permeability factor, with dimensions of velocity (cm/h). The uptake of ionic species is probably limited by membrane permeability, but membrane permeability is likely to be large for hydrophobic organic chemicals. If membrane resistance is small, diffusion across the aqueous boundary layer next to the membrane may be the factor controlling the overall transport rate (Hayton and Barron, 1990). Such a stationary layer of water occurs whenever water is in contact with a surface and, assuming laminar flow, it varies in width as a function of the water flow rate over the surface in question. An expression for diffusion through an aqueous boundary layer, when it is considered to be a stagnant microlayer of thickness h, is:

$$\frac{dX}{dt} = \frac{D_a A(C_w - C_o)}{h} \quad (23)$$

$$P = \frac{D_a A}{h} \text{ (aqueous film diffusion control)} \quad (24)$$

where D_a is the diffusion coefficient of the solute in water, A is the effective surface area of the stagnant microlayer or film, and P is the uptake clearance. Molecular diffusion in either medium tends to decline with increasing molecular size and to increase with temperature. However, flux in the aqueous boundary phase is unrelated to K_{ow}, and flux through the membrane itself increases with lipophilicity. In aquatic organisms, membrane permeability may differ according to the salinity or hardness of the water, causing differences in uptake, even of un-ionized chemical (Part, 1989). The diffusive flux can also change with size of the organism, because gill surface area (A) is proportionally smaller in larger individuals. Allometric equations based on this function are typically included in uptake models (Barber et al., 1988).

The process of membrane diffusion and the broader mechanisms associated with absorption through gills or gut are central to understanding bioaccumulation. Extremely hydrophobic organic chemicals such as hexabromobenzene (log K_{ow} = 7.8) bioaccumulate less than predicted by a direct relationship with K_{ow} (see Figure 4). The reasons for this poor uptake are much debated. Gobas et al. (1986) suggested that uptake by fish gills for moderately lipophilic chemicals (up to log K_{ow} = 3 or 4) increases somewhat with K_{ow} and that poor diffusion through the aqueous boundary layer retards the uptake of chemicals with higher molecular weight. They further suggest that the membrane-water partition coefficient, K_{mw}, declines in proportion to K_{ow} for highly lipophilic solutes, perhaps because of increasing energy required to force large molecules into the structured phospholipid of the membrane (Gobas et al., 1988a). This was shown experimentally by partitioning chemicals of various K_{ow} and molar volume into phospholipid vesicles. Octanol-water partitioning was linear for log K_{ow} < 5.5 and for molar volumes <230 cm^3/mol. Opperhuizen (1986) found poor gill membrane permeability in guppies caused by steric hindrance for molecules of chain length > 4.3 nm or membrane cross sections > 0.95 nm. Similar but not identical results were found for trout, suggesting some interspecific differences (Sijm et al., 1993) for very lipophilic chemicals.

Uptake in vivo could be limited not only by membrane-boundary layer resistances but also by the flow of water (exterior) or blood (interior) (Hayton and Barron, 1990). The inward flux of chemical to the organism, dX/dt, is directly proportional to the concentration gradient between that bioavailable in external water (C_w) and blood plasma (C_i) (or average organism water phase concentration in a one-compartment model):

$$\frac{dX}{dt} = P(C_w - C_i) \quad (25)$$

$$X = V_d C_w(1 - e^{-(P/v)t}) \text{ (integrated form)} \quad (26)$$

where the apparent volume of distribution is related to the inward flux and the internal water-phase concentration by the equation $V_d = X/C_i$ and P, called the uptake or absorption clearance, which has units of flow (cm^3/min). Uptake clearance can be visualized as the flow of water across the gill that is completely cleared of an amount of chemical equal to ($C_w - C_f$) per unit time. This formulation of the 1CFOK model is the clearance-volume-of-distribution (CFD) version noted earlier (see Volume of Distribution and Uptake Clearance). Depending on the rate-limiting step in the overall uptake process, the clearance is determined by one of the following:

$$P = \frac{D_m A K_{mv}}{d} \text{ (membrane limitation)}$$

$$P = \frac{D_a A}{h} \text{ (boundary layer limitation)}$$

$$P = K_{bl} V'_{bl} \text{ (blood flow limitation)}$$

$$P = V'_w \text{ (ventilation limitation)}$$

where K_{bl} is the binding coefficient for blood plasma (i.e., blood-water partition coefficient) and V'_{bl} is the flow rate of the blood (cm^3/min). If the delivery of inspired water to the gill is limiting, the uptake clearance becomes $V'_{w'}$, the ventilation rate (cm^3/min). For very hydrophobic chemicals, the binding capacity of plasma proteins and lipids far exceeds the supply of chemical, so that blood flow is unlikely to be limiting. A series of direct measurements of uptake clearance in live trout (McKim and Goeden, 1982; McKim et al., 1985, 1986; Erickson and McKim, 1990a,b; Nichols et al., 1990; McKim and Erickson, 1991; Schmieder and Weber, 1992), coupled with physiologically based kinetic models of water and blood flow (Erickson and McKim, 1990b; Nichols et al., 1990), have shown that hydrophobic chemicals (log K_{ow} = 3–6) are indeed water flow limited. Hydrophilic chemicals (log K_{ow} < 1) partition poorly to gills and blood, so their uptake is blood flow limited. For chemicals of moderate lipid solubility (log K_{ow} = 1–3) uptake efficiency is proportional to K_{ow}. In a single pass across the gills the maximum extraction efficiency for the chemicals tested by McKim et al. (1985) was about 60%, similar to the uptake efficiency of oxygen from water by fish gills.

Rapid binding to plasma proteins and lipids has been shown by modeling to be necessary for the very efficient uptake observed experimentally (Streit et al., 1991). The same properties that lead to sorption to organic matter in water also contribute to binding internally (Figure 1). The fraction of freely dissolved chemical in plasma can also be calculated from a plasma or blood binding coefficient (K_{bind}) and known protein concentration, as in Eq. (8). Changes in the strength of the binding or in the concentration of proteins can thus affect the effective internal dose and, ultimately, the toxicity of the chemical. Small changes in these properties, which can vary with the health of the organism, are especially critical for hydrophobic chemicals with large binding capacities, where very little remains freely dissolved. This interaction is well known in pharmacology but has not been investigated extensively in aquatic toxicology. The few binding coefficients measured for trout blood show a direct correlation between log K_{bind} and K_{ow} over the range 1 to 4 log units (Schmieder and Weber, 1992). Apparently no equivalent information is available for the respiratory fluids of aquatic invertebrates.

Volume of Distribution and Uptake Clearance

As already discussed, various biological activities and processes associated with membrane diffusion in living organisms may exert considerable influence on chemical accumulation by organisms. Although specific experiments can be carried out to investigate the effect of each type of activity and process, as discussed below, a general idea of their nature and influence can be readily obtained by evaluation of uptake clearance and volume of distribution.

The K_b or BCF for an organism at steady state is expressed as

$$K_b = \frac{C_b}{C_w} = \frac{\text{mg chemical/g tissue}}{\text{mg chemical/ml water}} \qquad (27)$$

The ratio has units of ml/g (frequently omitted in the BCF literature) and can be thought of as the volume of water at steady state (V_{ss}) needed to supply the chemical found in each gram of organism (Figure 5). It is only an apparent value, not an actual physical volume, because it does not take into account any uptake inefficiencies that undoubtedly occur in the real system. Nevertheless, it is useful for conceptualizing various dynamic processes. For example, it is evident that any change in the partitioning behavior of the chemical caused by a change in sorbent phases in the water or by temperature change can greatly affect this volume. The whole-body V_{ss} described here is analogous to the volume of distribution V_d used in pharmacology with multicompartment models. In the example shown in Figure 5, the V_{ss} is referenced to clearance of an apparent volume of water at the chemical concentration of the exposure water, whereas V_d is usually referenced to clearance of chemical from an apparent volume of water that has the same concentration of chemical as blood plasma.

If V_{ss} is converted to a rate, determined by the organism's "pumping" ability, then the uptake clearance from water (CL_w) can be visualized as the volume of water the organism could completely clear of chemical per hour. This flow rate gives an indication of the minimum time required to achieve a particular body burden. The uptake clearance can be calculated as

$$CL_w = k_1 * V_{ss} \quad (\text{L/g}^{-1}\,\text{h}^{-1}) \qquad (28)$$

Here the proportionality constant k_1 is the familiar apparent first-order uptake rate coefficient (1/h).

The advantage of expressing the kinetic relationship using the CVD approach, as employed for Eqs. (25), (26), and (28), rather than the more typical DRC reaction kinetics [e.g., Eq. (11)] is that with the CVD approach the values CL and V_{ss} can clearly be seen to be independent properties of the system, arising, respectively, from the K_b for the particular chemical-tissue combination and the extraction ability of the organism. As discussed by Stehly et al. (1990) and Barron et al. (1990), both of the typical rate constants (k_1, k_2) are dependent variables of CL and V. The uptake rate constant k_1 and the uptake clearance flow P are equivalent for either approach. With the standard DRC approach it is not possible to distinguish the reason for changes in k_2 or its commonly employed inverse $T_{1/2}$ ($T_{1/2} = 0.693/k_2$). However, in the CVD approach it is obvious that $k_2 = P/V_{ss}$ and that k_2 varies with either P or V_{ss} or both. Furthermore, with the CVD approach it is possible to determine which combination of changes in clearance, volume of distribution, and metabolic degradation has affected the net uptake.

Clearly, a chemical with a small V_{ss} can reach steady state more rapidly than one with a large V_{ss}. This relationship for living organisms parallels that illustrated in Figure 2, where time to steady state is seen to increase as function of particle size. Applied to toxicity testing, an examination of V_{ss} for various chemicals can show those most likely to reach steady state during the test period and could be used to further examine the DRC-based kinetics relationships determined with existing toxicity tests (McCarty, 1991; McCarty et al., 1992; Mackay et al., 1992). Although both simple and complex CVD models are routinely employed in pharmacology, they were not widely adopted in environmental and aquatic investigations until the increased interest in CVD-based PBPK models for aquatic organisms. Schultz and Hayton (1991) provide a general guide to using CVD-based models in aquatic testing.

The concepts of clearance and volume of distribution can be adapted to a variety of bioaccumulation processes, such as uptake during sinking (algal cell wall), swimming (dermal uptake), respiration (gill uptake), or feeding (gut uptake). In Figure 5, for example, the dots could represent dissolved oxygen molecules or the caloric units of prey items (for pred-

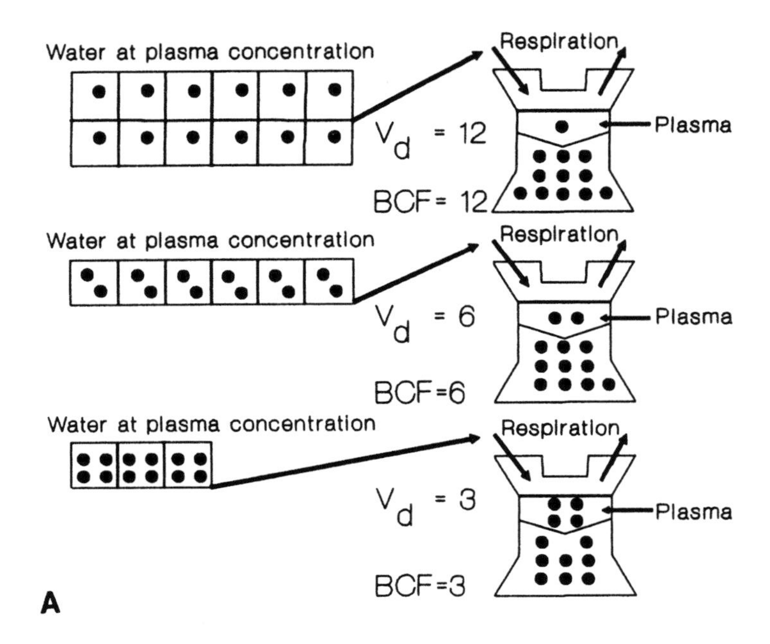

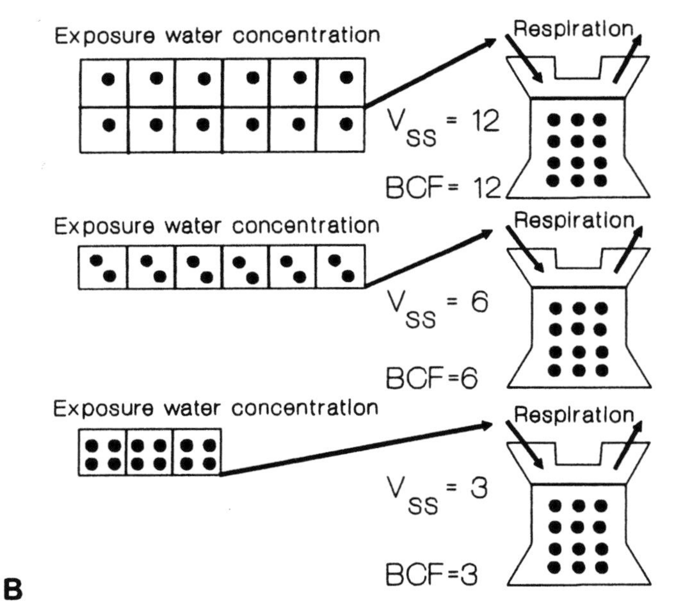

Figure 5. Illustration of volume of distribution in an organism for chemicals of varying hydrophobicity. (A) Model with two or more compartments (plasma and remainder of body). V_d is the apparent volume of water at the plasma chemical concentration which must be cleared completely to produce the known whole-body concentration. Factors that decrease plasma concentration increase V_d. Conversely, factors that increase plasma concentration decrease V_d. (B) One-compartment model. V_{ss} is the apparent volume of water at the exposure water concentration which must be cleared completely to produce the known whole-body concentration. In either situation in the upper example an organism must respire a large apparent volume of water to accumulate chemical which is present at low water concentrations. The chemical is of high hydrophobicity (i.e., high K_{ow}) and bioconcentration (BCF) is high. In the lower example the organism must respire much less water to obtain chemical which is present at higher water concentrations. The chemical is of low hydrophobicity (i.e., low K_{ow}) and BCF is low. The middle example is an intermediate case. (Adapted by permission from *Basic Clinical Pharmacokinetics*, third edition, edited by Michael E. Winter, published by Applied Therapeutics, Inc., Vancouver, Washington, © 1994.)

ators), suspended particulate matter (for filter feeders), or sediment (for benthic deposit feeders). For uptake from particles or sediments, clearance may be expressed on a dry weight basis (i.e., $CL_{sediment}$ units of $kg/g^{-1}\ h^{-1}$ based on dry weight concentrations in sediment) to conform to usual practice in sediment chemistry (Landrum and Robbins, 1990).

For the respiratory route it is useful to relate chemical uptake to oxygen uptake, because the physiological relationships for respiration are well known. By analogy to oxygen, effects of modifying factors such as temperature, body size, and stress can be incorporated into an exchange model. An example from Black and McCarthy (1990) shows the interrelationship of oxygen and chemical uptake for trout exposed to sublethal combinations of chlorine and PCBs. The chlorine exposure caused gill hyperplasia, which effectively increased the membrane diffusion distance (d) and decreased uptake efficiency of PCB and oxygen. Interestingly, there was no net change in PCB uptake rate because the fish compensated by increasing their ventilatory rate to maintain oxygen tension in the blood.

Uptake clearances of xenobiotic and oxygen have also been compared for several aquatic invertebrates. Zebra mussels (*Dreissena* sp.) in the Great Lakes show an uptake clearance of 800 $mlg^{-1}\ h^{-1}$ at 20°C for hexachlorobiphenyl (HCBP), implying a filtration rate of at least that much (Landrum et al., 1991). By comparison, oxygen uptake is 33.6 $mlg^{-1}\ h^{-1}$, or approximately 4% of the HCBP clearance rate. The large flow of HCBP from water, coupled with possible ingestion of HCBP-sorbed particles, shows the tremendous ability of *Dreissena* to bioaccumulate halogenated organics. Similar kinetic studies with the mayfly *Hexagenia limbata* and the amphipod *Diporeia* sp., suggest that contaminant uptake is significantly more efficient than oxygen uptake for these invertebrates as well (Stehly et al., 1990; Landrum and Stubblefield, 1991). On the other hand, in *Mysis relicta* oxygen and chemical uptake clearance rates are very similar (Landrum et al., 1992b).

Relationship of Bioaccumulation to Toxic Dose

As noted in the introduction to this chapter, bioaccumulation studies explicitly avoid consideration of adverse effects that might result from accumulation of chemicals. However, any exposure and toxicokinetic information obtained is of great use in toxicology. One important application of the partitioning principles discussed here is their use in the design and interpretation of standard toxicity tests, because they produce results, such as LC50s, which are a composite of toxicokinetic and toxicodynamic influences (see Chapter 1). For example, many common organic chemicals cause adverse biological effects primarily through narcosis, a general, nonspecific mode of toxic action (see Chapters 1 and 20). Abernethy et al. (1988) showed that the lethal "volume fraction" (volume toxicant/volume target tissue) is essentially constant for all general narcotics and that it correlates with K_{ow} and molar volume as well as total body residue. McCarty and co-workers (McCarty, 1986, 1987; McCarty et al., 1992) related acute and chronic lethality in fish to relatively constant whole-body concentrations (about 2–5 mmol/kg of fish for acute and about one-tenth of this for chronic); specifically, LC50 data ranged over a factor of a million but organism residues were within an order of magnitude.

Hydrophobicity, which determines partitioning behavior of narcotics, controls the expression of toxicity for such chemicals because lethality and other effects are related to the quantity of toxicant reaching internal target sites (probably the hydrophobic phase of cell membranes, especially cell membranes in nervous tissue in multicellular organisms). Both volume fraction and chemical concentration are closely related to chemical activity (i.e., fugacity), so given narcotic effects can be said to occur at relatively constant levels of activity in the organic phase of organisms. However, given the link between chemical activity and chemical concentration, it is often convenient to continue to use familiar concentration-based units. The toxicological principle being clearly illustrated is that the concentration of chemical in exposure water producing a biological response is a surrogate for the concentration of chemical in the body of the exposure organism, which is in turn a surrogate for the amount reaching the site(s) of toxic action where the effects are initiated.

Furthermore, McCarty and co-workers observed that since LC50 is related to K_b the two can be interconverted using K_{ow} and a simple first-order kinetics model (see Figure 10 in Chapter 1). Such interpretation requires that test organisms approach steady state during the toxicity test, as often (although not always!) appears to be the case for many chemicals tested with small invertebrates and fish.

The correlation between lethal dose and bioaccumulation is important because it allows interpretation of toxicity over a range of exposure times and concentrations, beyond those specifically used in testing protocols (Mackay et al., 1992). It also provides the link between toxicity predictions based on laboratory water exposures and the complex environmental processes that affect exposure and bioaccumulation in the field. Although organic chemicals can produce toxic effects by a variety of modes of action other than narcosis (McCarty and Mackay, 1993; see Chapter 1), knowledge of bioaccumulation processes is vital to understanding laboratory and field toxicity data by assisting in the separation of factors and effects related to toxicokinetics from those connected with toxicodynamics.

PREDICTIVE MODELS AND EXPERIMENTAL APPROACHES

Bioaccumulation research involves a combination of predictive modeling, experimentation, and field validation. Each approach has its advantages, and all three are needed ultimately. Current research on phytoplankton, microorganisms, invertebrates, and fish illustrate the application of these approaches.

Uptake by Phytoplankton and Microorganisms

Bioconcentration in phytoplankton has received much attention because of its obvious importance to pelagic food webs. Small size and proportionately large surface area suggest that plankton reach steady state relatively rapidly and that bioconcentration may be predicted directly from K_{ow} (Brown et al., 1982; Geyer et al., 1984). Based on reviews of experimental field data, most models have equated the lipid-normalized bioconcentration factor with K_{ow} for chemicals up to log K_{ow} of approximately 5 (Oliver and Niimi, 1988; Thomann, 1989; Connolly, 1991; Abernethy et al., 1986). However in the range of 5 to 8, bioconcentration factors appear to be independent of K_{ow}.

There are several possible explanations for this plateau. Swackhammer and Skoglund (1991) attribute reduced bioconcentration in phytoplankton to a high growth rate, perhaps because of increased removal processes associated with organic exudates produced by actively growing cells. Other explanations are possible, including analytical artifacts caused by partitioning to dissolved or colloidal organic matter in the aqueous phase. Mechanisms such as reduced membrane permeability have apparently not been examined.

A predictive model of PCB uptake by lake phytoplankton was developed by Richer and Peters (1993) from measurements of hexachlorobiphenyl uptake in laboratory cultures of *Selenastrum capricornutum*. Both initial uptake kinetics and bioconcentration factors were determined in relation to pH, dissolved organic carbon (DOC estimated from light absorbance at 440 nm), and algal biomass. Capacity factors (K') on High Performance Liquid Chromatography were also determined as a surrogate measure of K_{ow} (according to data from Mailhot, 1987: $\log K' = 0.444 \log K_{ow} - 1.019$, $n = 9$ chemicals, $r^2 = 0.982$). Results of model development showed that the initial uptake rate of PCB (%/min) was directly proportional to biomass and negatively related to DOC. The measured BCF values were predicted by

$$\begin{aligned}\log \text{BCF} = 4.11 &+ 0.86 \log K' \\ &- 0.87 \log \text{biomass} \\ &- 0.22 \log \text{Absorbance}\end{aligned} \tag{29}$$

where algal biomass was measured in mg/L. Good correlations were observed with field measurements of levels of PCBs in plankton from 11 lakes. There was essentially a simple inverse relationship between BCF and biomass, showing that large populations absorbed proportionately less of the available PCB per cell. The DOC content of the water acted as a strong retarding factor on instantaneous uptake and a weaker factor in reducing the final BCF values.

Microorganisms attached to sediment and suspended matter play key roles in contaminant fate processes and also serve as food sources for benthic invertebrates. Yet little is known about the bioavailability of sorbed xenobiotics to this important group. The few studies on biotransformation of sorbed organic contaminants in soil-water systems have reached diverse conclusions, depending on the type of sorbate, contaminant, and microorganism tested.

A study by Guerin and Boyd (1992) does much to clarify the situation by testing the hypothesis that sorbed fractions are unavailable for uptake. Two bacterial species (*Pseudomona putida* and a gram-negative soil isolate called NP-Alk) were exposed to naphthalene (log $K_{oc} = 2.74$) in two soil-free and soil-containing systems. Rates of mineralization of naphthalene by the two species were determined by first-order kinetics and by a three-parameter coupled desorption-degradation model. The bioavailable fraction was markedly different for the two species: for the NP-Alk strain both the rate and extent of mineralization were limited by the aqueous phase concentration. In contrast, *P. putida* was able to enhance the desorption of napthalene and achieve approximately sevenfold greater mineralization rates.

These results underscore the need for species-specific estimates of bioavailability and kinetics. The role of various microbial strains on assimilation efficiencies of deposit-feeding invertebrates would be a logical extension of this work. Conversely, there is some indication that the activities of benthic invertebrates can alter the mineralization rates of sediment-sorbed bacteria (Karickhoff and Morris, 1985a).

Uptake by Invertebrates

Zooplankton are expected to equilibrate with C_w (i.e., the fraction of chemical that remains freely dissolved in water) rapidly, not only because of their small size but also because swimming and ventilatory movements increase their rate of encountering chemicals in dilute solution. Experiments by McCarthy (1983) showed that live *Daphnia magna* accumulate benzo[*a*]pyrene about 10 times faster than heat-killed *Daphnia* given the same exposure. For chemicals of low to intermediate hydrophobicity (i.e.,log K_{ow} of about 1–4), uptake from water should be directly proportional to C_w. However, the contribution from chemical in food should become increasingly important as K_{ow} increases. The critical factor for zooplankton, as well as for other groups, is the exposure concentration, C_w, and its relationship to other phases encountered by the organism.

Experiments by McCarthy (1983) also clearly showed that the presence of suspended solids (yeast) in water acts to reduce bioaccumulation and that the effect is greater for chemicals with higher K_{ow} (Figure 6). Sorption to yeast was linear and proportional to K_{ow} for benzo[*a*]pyrene (log $K_{ow} = 6.0$) and anthracene (log $K_{ow} = 4.5$). The reduced bioavailability was evident even though *Daphnia* ingested particles;

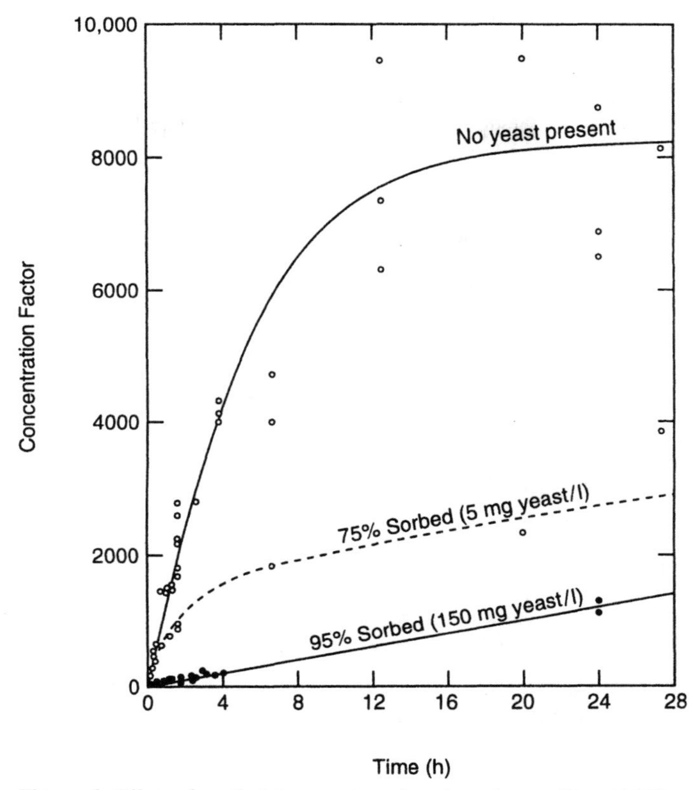

Figure 6. Effect of particulate organic carbon in water on bioavailability: Reduction of benzo[*a*]pyrene bioconcentration in *Daphnia* by varying levels of suspended organic material (yeast in this case). Particle-free water (open circles), 5 mg yeast/L (open squares), and 150 mg yeast/L (filled circles). (From McCarthy, 1983.)

ingestion of sorbed chemical represented only 3 to 15% of uptake from water. McCarthy (1983) also found predictable reductions in hydrocarbon uptake caused by binding to humic acids (i.e., DOC) in water. Humic-bound contaminants are also unavailable for uptake by fish (Black and McCarthy, 1988) and amphipods (Landrum et al., 1987). Desorption kinetics for particles and DOC are apparently too slow to supply significant amounts of xenobiotic to gill membranes during the brief passage along the respiratory membranes.

Benthic invertebrates are probably the most diverse and difficult group to model because of their wide variety of feeding behaviors and activity patterns. Accumulation in benthic organisms becomes increasingly important as contaminants are transported downward by sedimentation processes. Benthic organism exposure to persistent hydrophobic chemicals and their subsequent consumption by predators is an important pathway for food chain accumulation. For example, *Mysis relicta* migrates vertically and is preyed upon by pelagic fish in the Great Lakes. Klump et al. (1991) reported that mysids can accumulate sediment-sorbed hydrophobic organic contaminants efficiently (≈ 40–60% assimilation efficiency) and quickly. They reach whole-body levels similar to levels in the contaminated sediments (several orders of magnitude higher than the water concentration of chemical in the exposure system) in about 2 wk while on the bottom sediments and may contribute significantly to chemical residue levels in consumers at higher trophic levels. Similarly, the emergence of benthic midges and onshore movements of fiddler crabs both provide terrestrial transport routes (Larsson, 1984; Clark et al., 1986).

Benthic invertebrates are potentially exposed to chemicals through direct contact with sediment ventilation of overlying or pore water, or ingestion of sediment and/or food particles (see Chapter 1). Epifaunal species may have relatively little exposure to contaminated sediments if they respire overlying water and feed on suspended particulates. Freshwater mussels and some marine bivalves are exposed mainly through the aqueous and suspended solids routes (Ekelund et al., 1987). Thus, suspension feeders are suitable for monitoring effluents, ambient water quality, or resuspension of dredged sediments (Rice and White, 1987), but would not be the best choice for direct sediment toxicity (see Chapter 8) or bioaccumulation tests.

The use of infaunal species (Table 3) ensures that sediment ingestion contributes to exposure. Among the infauna, the subsurface deposit feeders (SSDF) probably provide the most direct test of sediment quality. Deposit feeders selectively feed on fine sediments that are high in organic content. The same fine particles tend to contain the highest concentrations of sorbed chemicals because of their TOC content and large surface area. Rubinstein et al. (1983) found higher concentrations of PCBs in *Nereis,* which forages in pore water and feeds on sediment of high organic content, than in the clam *Mercenaria mercenaria,* which burrows but is a suspension feeder, or the grass shrimp *Palaemonetes pugio,* a browser on plants. Similarly, Roesijadi et al. (1978) found the deposit feeders *Macoma inquinata* and *Phascolosoma agassizii* (sipunculan) were able to accumulate hydrocarbons to a greater extent than a suspension feeder (*Protothaca staminea*). A study of sediment-sorbed tributyltin (TBT) found consistently higher levels in the deposit-feeding clam *Scrobicularia plana* than in other benthic organisms (Langston and Burt, 1991).

A wide variety of sediment-water bioacumulation tests have been conducted with invertebrates, although the results are difficult to generalize because of the myriad chemical and biological factors affecting the outcome. Inevitably the test involves at least three phases (e.g., solids, colloids, dissolved forms) and the distribution among these phases may be affected by the density of organisms and their activity patterns. Test protocols are becoming more standardized and researchers are encouraged to consult the most current OECD (see Chapter 26) and ASTM guides (see Chapter 8), as well as those of the U.S. EPA (Boese and Lee, 1992) for specific recommendations on bioaccumulation-sediment toxicity test designs.

Key factors in any test are the choice of test species, sediment type (i.e., TOC, grain size), method of chemical application (i.e., field-collected contaminated sediment, spiked clean sediment, or chemical added to water), test dimensions (i.e., ratio of biota to water to sediment), and flow regime (i.e., flow-through or recirculating). Most bioaccumulation tests are performed with the assumption that tissue residues will reach steady state before the end of the test, but this must be verified with a series of chemical residue analyses. Steady state may be impractical for larger organisms and for chemicals of high bioaccumulation potential and a kinetic model may be needed to estimate steady-state concentrations in these situations (Landrum et al., 1992a).

Some tests have been designed to compare aqueous and sediment uptake routes by isolating organisms above or in contact with treated sediment. If such a system is at steady state, the fugacities, but not necessarily the free concentrations in overlying and pore water, will be equal and related to K_{ow}. If contaminated sediment is used in a flow-through or static-renewal test with clean water, the difference in concentrations between overlying and pore water may be large. Thus, under ideal circumstances bioaccumulation factors for many hydrophobic chemicals should be higher for organisms in direct contact with a given sediment than those held in cages above, because the latter are exposed primarily through respiration whereas the former also receive indirect and/or direct dietary exposure. However, experimental results are sometimes equivocal, reflecting technical and analytical difficulties, as well as the complex interaction of animal behavior and physiology, contaminant physicochemical properties, and sediment-water characteristics and chemistry that determines both bioavailability and organism uptake and accumulation (Muir et al., 1982, 1983, 1985; Landrum, 1989; Servos et al., 1992).

A general result of sediment-water tests has been the observation that higher sediment TOC values lead to less bioaccumulation and/or lower toxicity (Nebeker et al., 1989; Rubinstein et al., 1983; Muir et al., 1983, 1985). This is consistent with the idea that exposures are governed by the concentration of freely dissolved chemical in sediment pore water (C_{pore}), which is inversely related to fraction of organic carbon (f_{oc}) for a given sediment:

$$C_{pore} = \frac{C_s}{K_{oc} f_{oc}} \quad \text{(at equilibrium)} \tag{30}$$

where C_{pore} = chemical concentration in pore water (μg/L)
C_s = chemical concentration in sediment (μg/g dry weight of sediment)
K_{oc} = organic carbon-water partition coefficient (ml/g organic carbon)
f_{oc} = total organic C content in sediment expressed as a fraction mass (g organic carbon/g sediment dry weight)

The colloidal phase could also be included, as discussed above. If organic-based partitioning among phases controls exposure, then the fugacity in the lipids of the organisms should match the fugacity in the sediment organic phases. This is the basis for an equilibrium partitioning (EqP) approach to evaluate sediment contamination (Di Toro et al., 1991; see Chapter 8). Accordingly, a biota-sediment accumulation factor, BSAF (in units of g carbon/g lipid), can be defined as

$$\text{BSAF} = \frac{C_b/f_L}{C_s/f_{oc}} \tag{31}$$

or

$$\text{BSAF} = \frac{C_b f_{oc}}{f_L C_s} \approx \frac{K_L}{K_{ow}} \quad \text{when } K_{ow} \approx K_{oc} \tag{32}$$

where C_b = steady-state tissue residue (μg/g tissue wet weight)
C_s = chemical concentration in sediment (μg/g dry weight sediment)
f_L = concentration of lipid in organism expressed as a fraction mass (g lipid/g tissue)

Table 3. Invertebrate species used in sediment testing

Species	Type	Feeding type[a]	Source
Marine			
Abarenicola sp.	Polychaete	FUN	Augenfeld et al., 1982
Calliannassa sp.	Shrimp	SSDF	Ahsanullah et al., 1984
Macoma balthica	Bivalve	SDF/FF	Foster et al., 1987
Macoma nasuta	Bivalve	SDF/FF	Boese et al., 1990
Nephtys incisa	Polychaete	SSDF	Lake et al., 1990
Neanthes arenaceodentata	Polychaete	SDF/O	Pesch and Morgan, 1978
Nereis virens	Polychaete	SDF/O	Haya and Burridge, 1988; McElroy et al., 1990
Nereis diversicolor	Polychaete	SDF/O	Jennings and Fowler, 1980; Louma and Bryan, 1982
Palaemonetes pugio	Grass shrimp	SDF	Rubenstein et al., 1983; Clark et al., 1987
Yoldia limatula	Bivalve	SSDF	Lake et al., 1990
Freshwater			
Chironomus riparus	Dipteran insect	FF/SDF	Lohner and Collins, 1987; Pittinger et al., 1989
Chironomus tentans	Dipteran insect	FF/SDF	Swindoll and Applehans; 1987; Wood et al., 1987
Diporeia sp.	Amphipod	SSDF	Landrum, 1988, 1989
Hexagenia sp.	Mayfly, insect	COL	Stehly et al., 1990
Hyalella azteca	Amphipod	SSDF	Schuytema et al., 1990
Lubriculus variegatus	Oligochaete	SSDF	Ankley et al., 1992; Phipps et al., 1993
Lumbricus terrestris	Terrestrial earthworm	SSDF	Mac et al., 1990

[a] FUN, funnel feeder; SSDF, subsurface deposit feeder; SDF, surface deposit feeder; O, omnivore; FF, filter feeder; COL, collects surface particles.
Source: Adapted from Boese and Lee, 1992.

f_{oc} = total organic carbon in sediment expressed as a fraction mass (g carbon/g sediment)
K_L = lipid-water partition coefficient (L/kg lipid)

Such an accumulation factor should have a consistent relationship to K_{ow} regardless of sediment type or test species, provided that the contaminant is neutral, nonreactive, and at steady state. Bierman (1990) reported a BSAF range of 0.1 to 10 based on accumulation of PCBs in invertebrates in marine sediments. Ferraro et al. (1990) reported that mean BSAF values in burrowing clams (*Macoma nasuta*) exposed to 10 organic chemicals in laboratory experiments with various sediments collected in the field were usually less than 2 and less variable in highly polluted, organically enriched sediments. Analysis of PCBs in mollusks and polychaetes at field sites representing a range of TOC and contaminant concentrations also showed that BSAF calculations (i.e., normalization for lipid and organic carbon content) significantly reduce the variability in the raw tissue-sediment data (Lake et al., 1990). Nevertheless, there were significant differences among some PCB congeners, species, and sediment combinations. BSAF values were about twofold higher at sites of low PCB concentrations than at more contaminated sites. In a laboratory exposure of marine benthic amphipods, *Rhepoxynius abronius*, on sandy soil, Swartz et al. (1990) found excellent agreement between EqP calculations and the predicted toxicity of fluoranthene (log K_{oc} = 4.9). Subsequent experiments with fluoranthene toxicity to this species on sediments containing a variety of particulate organic matter showed that organic carbon type had little effect on the EqP predictions (DeWitt et al., 1992).

Meyer et al. (1993) tested the EqP approach using a mixture of chlorinated ethers including tetrachloropropyl ether (TCPE, log K_{oc} = 2.9) on sediments with a range of TOC. Toxicity tests were performed with *Daphnia magna*, *Hyalella azteca*, and *Chironomus tentans*. Good agreement between predicted and measured toxicity was found only for the cladoceran, *D. magna*, which remains in the overlying water. The responses of the two benthic species varied more widely with sediment TOC. The authors felt that sediment ingestion was not a factor contributing to the disagreement but that other sediment properties, such as grain size or food content, might have affected the sensitivity of the test organisms.

Relatively few field validations have been made as yet in support of the EqP approach for bioaccumulation. In one such study, Ankley et al. (1992)

Table 4. BSAF results for PCBs in laboratory and field-collected invertebrates and fish[a]

Homologue	*Lubriculus variegatus*[b]	Field oligochaetes[c]	*Pimephales promelas*[b]	*Ameiurus melas*[c]
Dichloro	0.21 (±0.17)[a]	2.59 (±4.69)[b]	0.04 (±0.03)[a]	0.30 (±0.37)[a]
Trichloro	0.51 (±0.33)[a,b]	0.71 (±0.51)[a]	0.11 (±0.07)[b]	0.83 (±0.68)[a]
Tetrachloro	1.35 (±0.71)[a]	1.10 (±0.56)[a]	0.50 (±0.25)[a]	2.92 (±3.14)[b]
Pentachloro	1.30 (±0.62)[a]	0.91 (±0.27)[a]	0.73 (±0.32)[a]	4.00 (±2.79)[b]
Hexachloro	1.03 (±0.46)[a]	1.42 (±0.53)[a]	0.48 (±0.18)[b]	3.37 (±0.59)[c]
Heptachloro	0.98 (±0.41)[a]	2.15 (±0.79)[b]	0.54 (±0.21)[a]	3.59 (±0.92)[c]
Octachloro	0.93 (±0.39)[a]	1.56 (±0.24)[b]	0.54 (±0.17)[a]	3.51 (±0.90)[c]
Nonachloro	0.82 (±0.34)[a,b]	1.25 (±0.38)[a]	0.58 (±0.25)[a,b]	2.54 (±1.03)[c]
Decachloro	0.57 (±0.20)[a]	0.79 (±0.36)[a]	0.27 (±0.11)[a]	2.26 (±1.36)[b]
Total PCBs	0.84 (±0.35)[a]	0.87 (±0.38)[a]	0.31 (±0.14)[a]	1.91 (±1.44)[b]

[a]Values with different superscript letters within a given row (homologue series) differed significantly at $P < 0.05$
[b]Results from laboratory testing
[c]Results from field collections
Source: Ankley et al., 1992.

compared BSAF values for PCBs in freshwater oligochaetes and two fish species for laboratory exposures and field collections (Table 4). The agreement for the infaunal oligochaetes was good, with many of the BSAF values being in the range of about 1 to 5 reported in many other laboratory studies. Analysis of the individual congeners showed species-specific differences in accumulation patterns. For example, fathead minnows (*Pimephales promelas*) studied in the laboratory exhibited much lower BSAF values for all chemicals tested, probably due to a combination of lack of dietary exposure and an exposure duration too brief for steady state to be reached.

The EqP approach has utility for evaluating the bioaccumulation or toxicity potential of in-place sediments that have equilibrated over a period of time. It also appears to be most suitable for deposit feeders and other organisms at the lower trophic levels. Conditions that would not favor the EqP approach include nonequilibrium distribution of contaminant in the sediments (i.e., recent inputs, dredging events, recent burial, rapid degradation, polar sorption), and non-steady state of the organisms (i.e., rapid growth, recent spawning event, migrations). Food chain accumulation could also lead to discrepancies with EqP predictions. For any of these situations a kinetic or bioenergetic modeling approach would be more appropriate (Landrum et al., 1992c).

For example, a kinetic model for BSAF that describes the change in tissue residues (normalized for lipid content) with time due to absorption from sediments (normalized for organic carbon content) could be based on the same information used in Eqs. (31) and (32) and written as

$$\frac{d(C_b/f_L)}{dt} = (k_1\ C_s/f_{oc}) - k_2C_b/L \quad (33)$$

$$\frac{C_b}{f_L} = \frac{k_1}{k_2}\frac{C_s}{f_{oc}}(1 - e^{-k2t}) \quad \text{(integrated form)} \quad (34)$$

Once steady state is reached, these equations are equivalent to Eqs. (31) and (32).

A simple mass transfer model can also be used to express the relative contributions of sediment and water sources of chemical to a benthic organism (Landrum et al., 1992a):

$$\frac{dC_a}{dt} = k_wC_w + k_sC_s - k_dC_a \quad (35)$$

$$C_a = \frac{k_wC_w + k_sC_s}{k_d}(1 - e^{-kdt}) \quad \text{(integrated form)} \quad (36)$$

where k_w = uptake clearance from overlying water (ml overlying water/g wet weight organism/h)
k_s = uptake clearance from sediment and/or food or other phase (g dry sediment/g wet weight organism/h)
k_d = elimination coefficient (1/h)
C_w = concentration of chemical in overlying water (ng chemical/ml water)
C_b = concentration of chemical in organism (ng chemical/g wet weight)
C_s = concentration of chemical in sediment, food, or other phase (ng chemical/g dry weight

The rate coefficients represented by the k values are empirical coefficients that represent the combined effects of chemical characteristics, organism characteristics, and environmental parameters on chemical uptake and elimination rates. K terms are affected more importantly by chemical hydrophobicity, organism mass, and environmental temperature. The model could be modified, depending on the expected routes of exposure, provided that the terms are distinguishable experimentally. Examples of experimental design to differentiate the exposure routes of a burrowing marine bivalve are given by Winsor et al. (1990) and Boese et al. (1990). Other modifications of the basic kinetic approach for uptake of PAHs by *Hexagenia* mayfly are described by Stehly et al. (1990). Landrum and Scavia (1983) used a kinetic approach to show that approximately 75% of the anthracene accumulated by *Hyalella azteca* is derived from pore water and sediment. Ratios of accumulated tissue residues to sediment concentrations were much lower than those predicted from ratios of tissue residues to C_w in the overlying water.

Extensive modeling of the freshwater amphipod *Diporeia* sp. (formerly classified as *Pontoporeia hoyi*) has shown the relative importance of sediment, pore water, and overlying water sources (Landrum and Robbins, 1990; Landrum et al., 1992a). Polycyclic aromatic hydrocarbons such as benzo[*a*]pyrene bound to sediments or to the DOC of pore water are not directly available for uptake by *Diporeia* (Landrum et al., 1987) and an inverse relationship exists between uptake clearance of ingested sediment and log K_{ow} (Landrum, 1989). However, desorption kinetics appear to control the bioavailability of PAH to the organisms. Bioavailability decreases with aging of the sediments, perhaps because contaminants diffuse farther into the interior of particles over time. Landrum and Robbins (1990) have also shown a strong seasonal fluctuation in residues that is apparently tied to feeding during the seasonal plankton blooms.

The importance of feeding to the bioaccumulation process was studied in the omnivorous invertebrate *Mysis relicta* by application of a kinetic model similar to Eq. (35) (Landrum et al., 1992b). Individuals were fed two types of food, diatom or daphnid, contaminated with either benzo[*a*]pyrene (BaP) or hexachlorobiphenyl (HCBP). Consumption of HCBP-contaminated food nearly doubled the tissue residues in comparison to water-only exposures. HCBP is poorly eliminated, and the additional uptake through the food route produced greater net bioaccumulation. In contrast, consumption of BaP in food had the opposite effect of decreasing body burdens. The hydrocarbon BaP is metabolized by *Mysis* to more water-soluble products that are eliminated in the feces. Gut passage of the food, especially diatoms, provided an elimination route for BaP.

Accurate predictions of tissue residues over full life cycles or seasonal cycles of organisms require a bioenergetics-based kinetics approach, because growth and related food consumption affect biomass, lipid storage, and chemical intake. However, few benthic invertebrates have been modeled in this way because of the species-specific parameters required (Moreno et al., 1992). Gut uptake efficiency from food is particularly difficult to measure in benthic invertebrates, because it varies with feeding rate and with contaminant concentration (Jumars and Self, 1986; Klump et al., 1987). Boese et al. (1990) modeled uptake in the deposit-feeding clam *Macoma nasuta* and determined that it derived approximately three-quarters of its hexachlorobenzene residues from sediment ingestion. About one-half of the remaining uptake came from respiration of pore water, and very little came from overlying water.

Uptake by Fish

The kinetics of uptake from water has been well characterized in a variety of marine and freshwater fish, and several models are available to calculate bioaccumulation of, primarily, organic chemicals (Barber et al., 1988; Barron et al., 1990). Such models, which may vary in their level of refinement, describe the thermodynamic process of bioconcentration considering both the biological characteristics of the fish and the physicochemical characteristics of the chemical. They address the kinetics of achieving the ultimate whole-body bioconcentration. More sophisticated physiologically based pharmacokinetic (toxicokinetic) models have been developed to describe chemical uptake and subsequent disposition within fish, a few examples of which are reported by Kulkarni and Karara (1990), Michel et al., (1990), Nichols et al., (1990) and Law et al., (1991).

PB-PK models treat major organ or tissue groups as separate compartments and use physiologically realistic compartment descriptions (i.e., volume, lipid content). Communication between compartments is via blood perfusion, so realistic blood flow information is required. Once a suitable physiologically based model of an organism is established, the physicochemical properties of the chemical and known physicochemical-biological relationships are incorporated and refinements are made by comparison with experimental data. Considerable physiological and biochemical information, some of which is not routinely available for many aquatic organisms, is required for successful use of a PB-PK model (e.g., blood flow rates in major blood vessels at various levels of activity and organism age or chemical biotransformation rates for a chemical at various exposure levels and nutritional states). The selection of a model type for a particular problem should be driven by the level of sophistication needed for the answer and the resources available. Good results from a simple bioconcentration model may often be more appropriate and obtained more quickly than poor or uncertain results from a PB-PK model.

Many of the chlorinated dioxins and furans, as well as many PCBs, are widely distributed in the aquatic environment. Since certain congeners and isomers are of considerable toxicological interest and, as a group, they are highly hydrophobic and environmentally persistent, they are the object of much scientific research. Some of the research that is pertinent to bioavailability and bioaccumulation in aquatic organisms is presented below. This is not to say that other organic chemicals are unimportant in this regard, only that dioxins and PCBs are both well studied and well suited to illustrating accumulation in aquatic organisms.

Both suspended solids and colloidal organic matter reduce the bioavailability of hydrophobic organics accumulated through respiration. Apparently desorption from these phases is slow relative to the brief contact time between water and gill surfaces (Eaton et al., 1983; McCarthy, 1983; Schrap and Opperhuizen, 1990). However, desorption and assimilation from ingested particles may be quite significant. Ingestion of sediment is a less common route of exposure for fish than for benthic invertebrates, although there are some types such as *Carpiodes* sp. (carp suckers) that feed specifically on organic-rich deposits. In an experimental exposure of *Cyprinus carpio* to suspended fly ash, Kuehl et al. (1987) found that carp accumulated various particle-sorbed dioxin (TCDD) and furan (TCDF) isomers differentially. A major route of uptake for these hydrophobic organics was ingestion and uptake through the gut.

Studies have shown that fish exposed to contaminated sediments are able to accumulate chemical residues (Jones et al., 1989). Both laboratory-held minnows (*Pimephales promelas*) and field collected catfish (*Ameiurus melas*) accumulated PCBs from sediments (Table 4). However, the difference in BSAF values for the two species is probably related to dietary intake. The minnows were fed uncontaminated food and held for a relatively short time, whereas the catfish fed on contaminated prey items during their full life span. Other studies on PCB-contaminated sediments include that of Califano et al. (1982), who found that larval striped bass, *Morone saxatilis*, accumulated PCBs initially from water but were eventually able to accumulate additional material that had desorbed from sediments or suspended solids. Rubinstein et al. (1984) found higher PCB residues in demersal fish (*Leiostomus xanthurus*) held in contact with contaminated sediment or fed PCB-containing food. In general, laboratory experiments with active fish held in close contact with contaminated sediments tend to produce higher residue accumulations than those with fish that are physically isolated from sediments, because of the contribution resulting from sediment-water mixing.

Because of the relatively minor role of sediment ingestion and the good predictability of respiratory uptake, most recent experimental and modeling approaches with fish have focused on dietary uptake from contaminated prey as an important factor in bioaccumulation. Research on dietary uptake has also been sparked by controversy regarding the mechanisms of food chain biomagnification. Although the nature and importance of this process have been debated for some time, studies confirm that organisms at higher trophic levels can acquire tissue residues in excess of those expected from aqueous partitioning alone (i.e., organism fugacity is higher than water fugacity) (Connolly and Pedersen, 1987; Connolly, 1991; Clark et al., 1988). Some cases of biomagnification are readily explained by the fact that such elevated fugacities are expected as they are the result of combined water-borne and dietary exposure and would be expected to be higher than aqueous exposure alone. However, in some cases the fugacity in upper trophic level organisms appears to be higher than even the fugacity of the diet. A possible explanation for this apparent nonsteady state is that the fugacity of contaminant in food increases during the digestion process, forcing a fugacity gradient from gut to fish (Gobas et al., 1988b; Gobas et al., 1993).

Alternatively, such levels of biomagnification can be viewed as the result of dietary chemical accumulation being "piggy-backed" on the active metabolic processes associated with food selection, digestion, and absorption in the gut and internal distribution and energy storage elsewhere in the body (i.e., adipose tissue, storage lipids). Once being effectively "pumped" into storage lipid depots, any outward losses would be largely passive (i.e., not associated with metabolic activities) and against a large chemical activity gradient, as such sites have a vast storage capacity for hydrophobic chemicals (i.e., very large Z capacity in fugacity terms). As well, losses due to metabolic breakdown would probably be low in such storage sites. These two biomagnification hypotheses are not necessarily mutually exclusive.

At steady state, for highly hydrophobic chemicals ($\log K_{ow} = 6$), there is a net inward flux of contaminant from the food to gut and tissue and a corresponding net outward flux of contaminant from the body to gill and water (Clark et al., 1990; Clark and Mackay, 1991). The actual bioaccumulation potential appears to be maximal at about $\log K_{ow} = 6$; beyond that, the environmental bioavailability of more hydrophobic chemicals is reduced by sorption processes. Uptake through the gut may also be poorer for very hydrophobic or large molecules. Assimilation efficiencies can also vary with the feeding regime. For example, assimilation efficiency for various chlorohydrocarbons increases proportionally with contaminant concentration in the food but less than proportionally with increased feeding frequency, which produces a reduced passage time through the gut and increased elimination due to faster egestion of fecal matter (Clark and Mackay, 1991).

Direct measurements of dietary assimilation have shown that efficiencies of 40–70% are typical for many halogenated hydrophobic organics. Assimilation efficiencies for young rainbow trout are about 50% for a dioxin (2,3,7,8-TCDD) (Kleeman et al., 1986) and 41–44% for 2,3,4,7,8-pentachlorodibenzofuran (PnCDF) (Muir et al., 1990). A TCDD assimilation efficiency of 38% for lake trout fed contaminated forage fish was determined by Cook et al. (1991). Opperhuizen and Schrap (1988) have proposed a figure of 50% for general modeling purposes, based on a review of numerous organochlorine studies. The uptake efficiency (E) for hydrophobic chemicals appears to be largely independent of $\log K_{ow}$ except above about 6, where efficiency declines ($1/\text{efficiency} = (5.3{*}E{-}8)K_{ow} + 2.3$; Gobas et al., 1989).

Many hydrophobic organics that might be expected to accumulate through the diet have very little net uptake because of metabolic and elimination processes. Gobas and Schrap (1990) found that the dietary assimilation of some polychlorinated dibenzo-*p*-dioxins and octachlorodibenzofuran in the guppy (*Poecilia reticulata*) was insignificant and bioaccumulation was far lower than for PCBs of the same hydrophobicity. Absorption efficiencies of penta- to octachlorinated dioxins and dibenzofurans (PCDDs and PCDFs) in guppies are low (1.7–21.7%) compared with those for other hydrophobic chemicals. They may be accumulated from food, but only slowly (Loonen et al., 1991). Dioxins and furans with four or more chlorine substituents, like octachlorodibenzo-*p*-dioxin, have very poor efficiencies. Low membrane permeability may be a factor (Opperhuizen and Sijm, 1990). Many of the lower chlorinated dioxins and furans are readily excreted and for some this rapid elimination may be associated with the substantial metabolic biotransformation capabilities of some organisms (Sijm and Opperhuizen, 1988). Dietary routes are also unimportant for the accumulation of some PAHs in trout due to a combination of poor absorption efficiencies, biotransformation, and rapid elimination (Niimi and Dookhran, 1989).

The steady-state model of Clark and Mackay (1991) illustrated the biomagnification expected for a nonreactive hydrophobic chemical (log K_{ow} = 6) in a simple four-step food chain: plankton are equilibrated with water (1× fugacity), small fish have a slight increase derived from consumption of plankton (1.32×), larger fish-eating fish have higher levels (2.37×), and the top predatory fish have highest levels (5.52×). Such "food chain multipliers" have been calculated for various models and vary as a function of the log K_{ow} of the chemical and the physiological and food chain parameters and relationships used in each. For example, Rasmussen et al. (1990) examined lake trout in a number of freshwater lakes of various sizes and found that much of the variability in tissue PCB levels could be explained by the length of the food chain in the lake; each trophic level contributing about a 3.5× multiplier factor. Evans et al. (1991) reported that multiplier factors for PCB, total DDT, and toxaphene ranged from 0.4 to 28.7 for successive trophic levels in Lake Michigan.

Models of the type originated by Thomann and Connolly (1984) have been used to fit the observed bioaccumulation data from several large water bodies. For example, Lake Michigan fish populations were found to contain approximately 3–5× increases in PCBs in omnivorous fish, 8× in small predators, and 14× in large lake trout (Connolly and Pedersen, 1988). Connolly (1991) further evaluated PCB distributions in lobster and winter flounder food chains from a New England harbor. The model gave an excellent fit for tri-, tetra-, penta-, and hexachlorobiphenyl concentrations observed at all levels of the food chain and across a concentration range of two orders of magnitude. Although most models have been for pelagic food chains, this one included a sediment component that proved to be a major source of PCBs to the fish and lobster. Dietary assimilation was highest for the trichlorobiphenyl and declined with increasing chlorine substitution.

Complete models of this type require extensive parameterization (i.e., realistic estimates have to be obtained for a wide variety of physical, chemical, and biological parameters and relationships before the model can be run and the degree of success is often related to how good such estimates are). Gobas (1992) has provided a simpler mass balance model that requires fewer coefficients and provides a good fit to PCB data from the Lake Ontario food chain studied by Oliver and Niimi (1988).

The chlorinated dioxin TCDD has also been successfully modeled for Lake Ontario, although the difficulties of parameterizing and field validating a dioxin model are much greater than for PCBs. Because of its low concentration and very high partition coefficient (log K_{ow} = 6.8), TCDD is virtually undetectable in lake waters and a BCF value cannot be directly calculated. Cook et al. (1991) exposed lake trout to various combinations of TCDD-contaminated Lake Ontario sediment and food fish (sculpin and smelt) for 120 d in order to measure the first-order kinetics of uptake from these routes. The estimated time to steady state was 300 d. Approximately 75% of the TCDD residues were derived from food; uptake from water was negligible. The food and sediment sources were essentially additive. Excellent agreement was found between the laboratory-derived BSAF values and those of Lake Ontario fish (BSAF value of 0.07 in each case). However, the values were far less than expected from estimates of dissolved TCDD concentration or from equilibrium partitioning calculations based on sediment organic carbon.

A likely reason for the discrepancy is that the sediments, which have a historic burden of TCDD residues, are not in equilibrium with the overlying water, which reflects more recent inputs to the lake system. Approximately one-half of the TCDD flux comes from suspended particles delivered by the Niagara River, and the remainder is associated with sediments (Whittle and Fitzsimons, 1983). The lake trout study shows the utility of laboratory-derived kinetic parameters as well as the difficulty of coupling ambient sediment and water concentrations that are not at steady state.

CONCLUSIONS AND FUTURE DIRECTIONS

The factors controlling bioaccumulation are important to understand, not only to predict tissue residues of the most persistent and hydrophobic chemicals but also to understand the uptake processes that occur with all chemicals. This chapter has shown the competing roles of binding and partitioning processes in membranes, tissues, and organic phases of the surrounding medium. Determining the rate-limiting processes of particle desorption and transfer at the gill or gut is critical to predicting the rate and extent of delivery of any toxic dose.

Correlations using the K_{ow} of organic chemical contaminants have been largely successful for predicting bioaccumulation in organisms exposed mainly through water. Such correlations are applicable only if the contaminant is not readily biotransformed and if K_{ow} is not too high. For example, PAHs are widely distributed in the marine environment and have physicochemical properties that make them suitable for bioaccumulation, but metabolic transformations at higher trophic levels prevent extensive biomagnification. In the case of extremely hydrophobic chemicals, sorption to environmental organic phases and poor membrane permeability can cause bioaccumulation to be substantially less than that found for chemicals with a lower hydrophobicity.

Nevertheless, a variety of persistent, hydrophobic xenobiotic chemicals, such as PCBs and polyhalogenated dioxins and furans, are characterized by relatively efficient uptake and slow depuration. During the past decade many of the major point sources of these chemicals have been reduced or eliminated. Still there remains a large reservoir of such materials and much of this resides in sediment. All aquatic organisms, especially benthic fauna, are potentially exposed to sediment-associated contaminants, even where the overlying waters are relatively clean. The contaminant burden varies both spatially and temporally and is most closely associated with fine organic-rich deposits. For these reasons, correlations based on sediment concentrations are now viewed as better predictors of tissue residues than predictions based only on water. The equilibrium partitioning approach shows promise as an efficient means of normalizing many of the variables associated with exposures through sediment.

It is unclear, as yet, whether the equilibrium partitioning approach will be successful in situations that are not near steady state. For example, receiving waters with variable inputs and high-energy environments such as estuaries, littoral zones, and rivers where sediment distributions change periodically may not be good candidates for an equilibrium approach. Much more needs to be learned about the specific physiological and behavioral patterns of benthic organisms before such an approach can be considered anything but a rapid screening tool. Sediment concentrations may also be suitable predictors for species at higher trophic levels, provided food chain transfer is included as a kinetic or bioenergetic modeling component. Since these predacious species are usually the ones destined for human consumption, the additional effort to develop and validate species- and habitat-specific models is clearly worthwhile.

In summary, it is clear that the focus of the study of bioaccumulation in aquatic systems is quickly shifting away from the simpler water-based exposure systems employed as the discipline developed. More complex multiphase approaches are mandatory if many real-world problems are to be adequately addressed, because the contribution to organism residues from sediment and food chains may be substantial, even dominant in some circumstances. The considerable additional effort required to employ multiphase approaches should ultimately produce a much better scientific understanding of the accumulation of chemicals by aquatic organisms and form the basis for advances in regulatory policies.

ACKNOWLEDGMENTS

Contributions to this chapter by the senior author were prepared during a visit to the U.S. Environmental Protection Agency Environmental Research Laboratory, Duluth, MN, under Cooperative Agreement CR820433 with the Lake Superior Research Institute, University of Wisconsin, Superior.

LITERATURE CITED

Abernethy S.; Bobra, A. M.; Shiu, W. Y.; and Mackay, D.: Acute lethal toxicity of hydrocarbons and chlorinated hydrocarbons to two planktonic crustaceans: the key role of organism-water partitioning. *Aquat. Toxicol.*, 8:163–174, 1986.

Abernethy, S. G.; Mackay, D.; and McCarty, L. S.: "Volume fraction" correlation for narcosis in aquatic organisms: the key role of partitioning. *Environ. Toxicol. Chem.*, 7:469–481, 1988.

Adams, W. J.: Bioavailability of neutral lipophilic organic chemicals contained on sediments: a review. Fate and Effects of Sediment-Bound Chemicals in Aquatic Systems, edited by K. L. Kickson; A. W. Maki; and W. A. Brungs, pp. 219–244. New York: Pergamon, 1987, SETAC Special Publication Series.

Ahsanullah, M.; Mobley, M.; and Negilski, D.: Accumulation of cadmium from contaminated water and sediment by the shrimp *Callianassa australiensis*. *Mar. Biol.*, 82:191–197, 1984.

Ankley, G. T.; Cook, P. M.; Carlson, A. R.; Call, D. J.; Swenson, J. A.; Corcoran, H. F.; and Hoke, R. A.: Bioaccumulation of PCBs from sediments by oligochaetes and fishes: comparison of laboratory and field studies. *Can. J. Fish. Aquat. Sci.*, 49:2080–2085, 1992.

Augenfeld, J. M.; Anderson, J. W.; Riley, R. G.; and Thomas, B. L.: The fate of polyaromatic hydrocarbons in an intertidal sediment exposure system: bioavailability to *Macoma inquinata* (Mollusca: pelecypoda) and *Abarenicola pacifica* (Annelida: polychaeta). *Mar. Environ. Res.*, 7:31–50, 1982.

Baker, J. E.; Capel, P. D.; and Eisenreich, S. J.: Influence of colloids on sediment-water partition coefficients of polychlorinated biphenyl congeners in natural waters. *Environ. Sci. Technol.*, 20:1136–1143, 1986.

Barber, M. C.; Suarez, L. A.; and Lassiter, R. R.: Modeling bioconcentration of nonpolar organic pollutants by fish. *Environ. Toxicol. Chem.*, 7:545–558, 1988.

Barron, M. G.: Bioconcentration. *Environ. Sci. Technol.*, 24:1612–1618, 1990.

Barron, M. G.; Stehly, G. R.; and Hayton, W. L.: Pharmacokinetic modeling in aquatic animals. 1. Models and concepts. *Aquat Toxicol.*, 17:187–212, 1990.

Bierman, V. J., Jr.: Equilibrium and biomagnification partitioning of organic chemicals in benthic animals. *Environ. Sci. Technol.*, 24:1407–1412, 1990.

Black, M. C., and McCarthy, J. F.: Dissolved organic macromolecules reduce the uptake of hydrophobic organic contaminants by the gills of the rainbow trout (*Salmo gairdneri*). *Environ. Toxicol. Chem.*, 7:593–600, 1988.

Black, M. C., and McCarthy, J. F.: Effects of sublethal exposure to chlorine on the uptake of polychlorinated biphenyl congeners by rainbow trout, *Salmo gairdneri* (Richardson). *Aquat. Toxicol.*, 17:275–290, 1990.

Boese, B. L.; Lee, H., II; Specht, D. T.; Randall, R. C.; and Winsor, M. H.: Comparison of aqueous and solid-phase uptake of hexachlorobenzene in the tellinid clam *Macoma nasuta* (Conrad): a mass balance approach. *Environ. Toxicol. Chem.*, 9:221–231, 1990.

Boese, B. L., and Lee, H., II: Synthesis of methods to predict bioaccumulation of sediment pollutants. Newport, OR: U.S. Environmental Protection Agency, ERL-N, Pacific Ecosystems Branch, 1992, ERL-N No N232.

Brisbin, I. L., Jr.; Newman, M. C.; McDowell, S. G.; and Peters, E. L.: Prediction of contaminant accumulation by free-living organisms: application of a sigmoidal model. *Environ. Toxicol. Chem.*, 9:151–157, 1990.

Brown, M. P.; McLaughlin, J. J. A.; O'Connor, J. M.; and Wyman, K.: A mathemathical model of PCB bioaccumulation in plankton. *Ecol. Model*, 15:29–47, 1982.

Brownawell, B. J., and Farrington, J. W.: Biogeochemistry of PCBs in interstitial waters of a coastal marine sediment. *Geochim. Cosmochim. Acta.*, 50:157–169, 1986.

Califano, R. J.; O'Connor, J. M.; and Hernandez, J. A.: Polychlorinated biphenyl dynamics in Hudson River striped bass. 1. Accumulation in early life history stages. *Aquat. Toxicol.*, 2:187–204, 1982.

Clark, J. R.; Patrick, J. M.; Moore, J. C.; and Forester, J.: Accumulation of sediment-bound PCBs by fiddler crabs. *Bull. Environ. Contam. Toxicol.*, 36:571–578, 1986.

Clark, J. R.; Patrick, J. M., Jr.; Moore, J. C.; and Lores, E. M.: Waterborne and sediment-source toxicities of six organic chemicals to grass shrimp (*Palaemonetes pugio*) and amphioxus (*Branchiostoma caribaeum*). *Arch. Environ. Contam. Toxicol.*, 16:401–407, 1987.

Clark, K. E., and Mackay, D.: Dietary uptake and biomagnification of four chlorinated hydrocarbons by guppies. *Environ. Toxicol. Chem.*, 10:1205–1217, 1991.

Clark, K. E.; Gobas, F. A. P. C.; and Mackay, D.: Model of organic chemical uptake and clearance by fish from food and water. *Environ. Sci. Technol.*, 24:1203–1213, 1990.

Clark, T.; Clark, K.; Paterson, S.; Mackay, D.; and Norstrom, R. J.: Wildlife monitoring, modeling, and fugacity. *Environ. Sci. Technol.*, 22:120–127, 1988.

Connell, D. W., and Miller, G. J.: Chemical and Ecotoxicology of Pollution. New York: Wiley-Interscience, 1984.

Connell, D. W.: Bioaccumulation behavior of persistent organic chemicals with aquatic organisms. *Rev. Environ. Contam. Toxicol.*, 101:121–154, 1988.

Connell, D. W.: Bioaccumulation of Xenobiotic Compounds. Boca Raton, FL: CRC Press, 1990.

Connell, D. W., and Hawker, D. W.: Use of polynomial expressions to describe the bioconcentration of hydrophobic chemicals by fish. *Ecotoxicol. Environ. Safety*, 16:242–257, 1988.

Connolly, J. P.: Application of a food chain model to polychlorinated biphenyl contamination of the lobster and winter flounder food chains in New Bedford Harbor. *Environ. Sci. Technol.*, 25:760–770, 1991.

Connolly, J. P., and Pedersen, C. J.: A thermodynamic-based evaluation of organic chemical accumulation in aquatic organisms. *Environ. Sci. Technol.*, 22:99–103, 1988.

Cook, P. M.; Kuehl, D. W.; Walker, M. K.; and Peterson, R. E.: Bioaccumulation and toxicity of 2,3,7,8-tetrachlorodibenzo-*p*-dioxins and related compounds in aquatic ecosystems. Banbury Report: Biological Basis for Risk Assessment of Dioxins and Related Compounds, edited by M. A. Gallo; R. J. Scheuplein; and C. A. Vande Heijde, pp. 143–167. Cold Spring Harbor, NY: Cold Spring Harbor Laboratory Press, 1991.

Cullen, M. C., and Connell, D. W.: Bioaccumulation of chlorohydrocarbon pesticides by fish in the natural environment. *Chemosphere*, 25:1579–1587, 1992.

DeWitt, T. H.; Ozretich, R. J.; Swartz, R. C.; Lamberson, J. O.; Schults, D. W.; Ditsworth, G. R.; Jones, J. K. P.; Hoselton, L.; and Smith, L. M.: The influence of organic matter quality on the toxicity and partitioning of sediment-associated fluoranthene. *Environ. Toxicol. Chem.*, 11:197–208, 1992.

Diamond, M. L.; Mackay, D.; and Welboum, P. M.: Models of multi-media partitioning of multi-species chemicals: the fugacity/equivalence approach. *Chemosphere*, 25: 1907–1921, 1992.

Di Toro, D. M.: A particle interaction model of reversible chemical sorption. *Chemosphere*, 14:1503–1538, 1985.

Di Toro, D. M.; Zarba, C. S.; Hansen, D. J.; Berry, W. J.; Swartz, R. C.; Cowan, C. E.; Pavlou, S. P.; Allen, H. E.; Thomas, N. A.; and Paquin, P. R.: Technical basis for establishing sediment quality criteria for nonionic organic chemicals using equilibrium partitioning. *Environ. Toxicol. Chem.*, 10:1541–1583, 1991.

Eaton, J. G.; Mattson, V. R.; Mueller, L. H.; and Tanner, D. K.: Effects of suspended clay on bioconcentration of Kelthane in fathead minnows. *Arch. Environ. Contam. Toxicol.*, 12:439–445, 1983.

Ekelund, R; Granmo, A.; Berggren, M.; Renberg, L.; and Wahlberg, C.: Influence of suspended solids on the bioavailability of hexachlorobenzene and lindane to the deposit feeding marine bivalve, *Abra nitida* (Müller). *Bull. Environ. Contam. Toxicol.*, 38:500–508, 1987.

Erickson, R. J., and McKim, J. M.: A model for the exchange of organic chemicals at fish gills: flow and diffusion limitations. *Aquat. Toxicol.*, 18:175–198, 1990a.

Erickson, R. J., and McKim, J. M.: A simple flow-limited model for exchange of organic chemicals at fish gills. *Environ. Toxicol. Chem.*, 9:159–165, 1990b.

Esser, H. O.: A review of the correlation between physicochemical properties and bioaccumulation. *Pestic. Sci.*, 17:265–276, 1986.

Evans, M. S.; Noguchi, G. E.; and Rice, C. P.: The biomagnification of polychlorinated biphenyls, toxaphene, and DDT compounds in a Lake Michigan offshore food web. *Arch. Environ. Contam. Toxicol.*, 20:87–93, 1991.

Farrington, J. W.: Biogeochemical processes governing exposure and uptake of organic pollutant compounds in aquatic organisms. *Environ. Health Perspect.*, 90:75–84, 1991.

Faust, B. C.: The octanol/water distribution coefficients of methylmercuric species: the role of aqueous-phase chemical speciation. *Environ. Toxicol. Chem.*, 11:1373–1376, 1992.

Ferraro, S. P.; Lee, H., II; Ozretich, R. J.; and Specht, D. T.: Predicting bioaccumulation potential: a test of a fugacity-based model. *Arch. Environ. Contam. Toxicol.*, 19:386–394, 1990.

Foster, G.; Maksi, S. M.; and Means, J. C.: Bioaccumulation of trace organic contaminants from sediment, Baltic clams (*Macoma abalthica*), and soft-shell clams (*Mya arenaria*). *Environ. Toxicol. Chem.*, 6:969–976, 1987.

Geyer H.; Politzki, G.; and Freitag, D.: Prediction of ecotoxicological behavior of chemicals: relationship between *n*-octanol/water partition coefficient and bioaccumulation of organic chemicals by alga *Chlorella*. *Chemosphere* 13:269–284, 1984.

Gobas, F. A. P. C.: Modeling the accumulation and toxicity of organic chemicals in aquatic food chains. Chemical Dynamics in Fresh Water Ecosystems, edited by F. A. P. C. Gobas, and J. A. McCorquodale, pp. 129–151. Boca Raton, FL: Lewis, 1992.

Gobas, F. A. P. C., and Mackay, D.: Dynamics of hydrophobic organic chemical bioconcentration in fish. *Environ. Toxicol. Chem.*, 6:495–504, 1987.

Gobas, F. A. P. C.; Opperhuizen, A.; and Hutzinger, O.: Bioconcentration of hydrophobic chemicals in fish: relationship with membrane permeation. *Environ. Toxicol. Chem.*, 5:637–646, 1986.

Gobas, F. A. P. C.; Lahittete, J. M.; Garofalo, G.; Shiu, W. Y.; and Mackay, D.: A novel method for measuring membrane-water partition coefficients of hydrophobic organic chemicals: comparison with 1-octanol–water partitioning. *J. Pharm. Sci.*, 77:265, 1988a.

Gobas, F. A. P. C.; Muir, D. C. G.; and Mackay, D.: Dynamics of dietary bioaccumulation and faecal elimination of hydrophobic organic chemicals in fish. *Chemosphere* 17:943-962, 1988b.

Gobas, F. A. P. C.; Clark, K. E.; Shiu, W. Y.; and Mackay, D.: Bioconcentration of polybrominated benzenes and biphenyls and related superhydrophobic chemicals in fish: role of bioavailability and elimination into the feces. *Environ. Toxicol. Chem.*, 8:231–245, 1989.

Gobas, F. A. P. C., and Schrap, S. M.: Bioaccumulation of some polychlorinated dibenzo-*p*-dioxins and octachlorodibenzofuran in the guppy (*Poecilia reticulata*). *Chemosphere*, 20:495–512, 1990.

Gobas, F. A. P. C.; McCorquodale, J. R.; and Haffner, G. D.: Intestinal absorption and biomagnification of organoclorines. *Environ. Toxicol Chem.*, 12:567–576, 1993.

Guerin W. F., and Boyd, S. A.: Differential bioavailability of soil-sorbed naphthalene to two bacterial species. *Appl. Environ. Microbiol.*, 58:1142–1152, 1992.

Hawker, D. W., and Connell, D. W.: Influence of partition coefficient of lipophilic compounds on bioconcentration kinetics with fish. *Water Res.*, 22:701–707, 1988.

Hawker, D., and Connell, D.: An evaluation of the relationship between bioconcentration factor and aqueous solubility. *Chemosphere*, 23:231–241, 1991.

Haya, K., and Burridge, L. E.: Uptake and excretion of organochlorine pesticides by *Nereis virens* under normoxic and hypoxic conditions. *Bull. Environ. Contam. Toxicol.*, 40:170–177, 1988.

Hayton, W. L., and Barron M. G.: Rate-limiting barriers to xenobiotic uptake by the gill. *Environ. Toxicol. Chem.*, 9:151–157, 1990.

Horne, R. A.: Marine Chemistry. New York: Wiley Interscience, 1969.

Huckins, J. N.; Manuweera, G. K.; Petty, J. D.; Mackay, D.; and Lebo, J. A.: Lipid-containing semipermeable membrane devices for monitoring organic contaminants in water. *Environ. Sci. Technol.*, 27:2489–2496, 1993.

Jennings, C. D., and Fowler, S. W.: Uptake of ^{55}Fe from contaminated sediments by the polychaete *Nereis diversicolor. Mar. Biol.*, 56:227–280, 1980.

Jones, P. A.; Sloan, R. J.; and Brown, M. P.: PCB congeners to monitor with caged juvenile fish in the upper Hudson River. *Environ. Toxicol. Chem.*, 8:793–803, 1989.

Jumars, P. A., and Self, R. F. L.: Gut-marker and gut-fullness methods for estimating field and laboratory effects of sediment transport on ingestion rates of deposit-feeders. *J. Exp. Mar. Biol. Ecol.*, 98:293–310, 1986.

Karickhoff, S. W., and Morris, K. R.: Impact of tubificid oligochaetes on pollutant transport in bottom sediments. *Environ. Sci. Technol.*, 19:51–56, 1985a.

Karickhoff, S. W., and Morris, K. R.: Sorption dynamics of hydrophobic pollutants in sediment suspensions. *Environ. Sci. Technol.*, 4:469–479, 1985b.

Kleeman, J. M.; Olson, J. R.; Chen, S. M.; and Peterson, R. E.: Metabolism and disposition of 2,3,7,8-tetrachlorodibenzo-*p*-dioxin in rainbow trout. *Toxicol. Appl. Pharmacol.*, 83:391–401, 1986.

Klump, J. V.; Krezoski, J. R.; Smith, M. E.; and Kaster, J. L.: Dual tracer studies of the assimilation of an organic contaminant from sediments by deposit feeding oligochaetes. *Can. J. Fish. Aquat. Sci.*, 44:1574–1583, 1987.

Klump, J. V.; Kaster, J. L.; and Sierszen, M. E.: *Mysis relicta* assimilation of hexachlorobiphenyl from sediments. *Can. J. Fish. Aquat. Sci.*, 48:284–289, 1991.

Knezovich, J. P.; Harrison, F. L.; and Wilhelm, R. G.: The bioavailability of sediment-sorbed organic chemicals: a review. *Water, Air, Soil Pollut.*, 32:233–245, 1987.

Krom, M. D., and Sholkovitz, E. R.: Nature and reactions of dissolved organic matter in the interstitial waters of marine sediments. *Geochim. Cosmochim. Acta.*, 41: 1565–1573, 1977.

Kuehl, D. W.; Cook, P. M.; Batterman, A. R.; and Butterworth, B. C.: Isomer dependent bioavailability of polychlorinated dibenzo-*p*-dioxins and dibenzofurans from municipal incinerator fly ash to carp. *Chemosphere*, 16: 657–666, 1987.

Kulkarni, M. G., and Karara, A. H.: A pharmacokinetic model for the disposition of polychlorinated biphenyls (PCBs) in channel catfish. *Aquat. Toxicol.*, 16:141–150, 1990.

Lake, J. L.; Rubinstein, N. I.; Lee, H., II; Lake, C. A.; Heltshe, J.; and Pavignano, S.: Equilibrium partitioning and bioaccumulation of sediment-associated contaminants by infaunal organisms. *Environ. Toxicol. Chem.*, 9: 1095–1106, 1990.

Landrum, P. F.: Toxicokinetics of organic xenobiotics in the amphipod, *Pontoporeia hoyi*: role of physiological and environmental variables. *Aquat. Toxicol.*, 12:245–271, 1988.

Landrum, P. F.: Bioavailability and toxicokinetics of polycyclic aromatic hydrocarbons sorbed to sediments for the amphipod *Pontoporeia hoyi. Environ. Sci. Technol.*, 23: 588–595, 1989.

Landrum, P. F., and Robbins, J. A.: Bioavailability of sediment-associated contaminants to benthic invertebrates. Sediments: Chemistry and Toxicity of In-Place Pollutants, edited by R. Baudo; J. P. Giesy; and H. Muntau, pp. 237–263. Chelsea, MI: Lewis, 1990.

Landrum, P. F., and Scavia, D.: Influence of sediment on anthracene uptake, depuration, and biotransformation by the amphipod *Hyalella azteca. Can. J. Fish. Aquat. Sci.*, 40:298–305, 1983.

Landrum, P. F., and Stubblefield, C. R.: Role of respiration in the accumulation of organic xenobiotics by the amphipod *Diporeia* sp. *Environ. Toxicol. Chem.*, 10:1019–1028, 1991.

Landrum, P. F.; Nihart, S. R.; Eadie, B. J.; and Herche, L. R.: Reduction in bioavailability of organic contaminants to the amphipod *Pontoporeia hoyi* by dissolved organic matter of sediment interstitial waters. *Environ. Toxicol. Chem.*, 6:11–20, 1987.

Landrum, P. F.; Gossiaux, D. C.; Fisher, S. W.; and Bruner, K. A.: The role of zebra mussels in contaminant cycling in the Great Lakes. *J. Shellfish. Res.*, 10:252–253, 1991.

Landrum, P. F.; Fontaine, T. D.; Faust, W. R.; Eadie, B. J.; and Land G. A.: Modeling the accumulation of polycyclic aromatic hydrocarbons by the amphipod *Diporeia* (spp.). Chemical Dynamics in Fresh Water Ecosystems, edited by F. A. P. C. Gobas, and J. A. McCorquodale, pp. 111–128. Boca Raton, FL: Lewis, 1992a.

Landrum, P. F.; Frez, W. A.; and Simmons, M. S.: Relationship of toxicokinetic parameters to respiration rates in *Mysis relicta*. *J. Great Lakes Res.*, 18:331–339, 1992b.

Landrum, P. F.; Lee, H., II; and Lydy, M. J.: Toxicokinetics in aquatic systems: model comparisons and use in hazard assessment. *Environ. Toxicol. Chem.*, 11:1709–1725, 1992c.

Langston, W. J., and Burt, G. R.: Bioavailability and effects of sediment-bound TBT in deposit-feeding clams, *Scrobicularia plana*. *Mar. Environ. Res.*, 32:61–77, 1991.

Larsson, P.: Transport of PCBs from aquatic to terrestrial environments by emerging chironomids. *Environ. Pollut. Ser. A* 34:283–289, 1984.

Larsson, P.: Zooplankton and fish accumulate chlorinated hydrocarbons from contaminated sediments. *Can. J. Fish Aquat. Sci.*, 43:1463–1466, 1986.

Law, F. C. P.; Abedini, S.; and Kennedy, C. J.: A biologically based toxicokinetic model for pyrene in rainbow trout. *Toxicol. Appl. Pharmacol.*, 110:390–402, 1991.

Lohner, T. W., and Collins, W. J.: Determination of uptake rate constants for six organochlorines in midge larvae. *Environ. Toxicol. Chem.*, 6:137–146, 1987.

Loonen, H.; Parsons, J. R.; and Govers, H. A. J.: Dietary accumulation of PCDDs and PCDFs in guppies. *Chemosphere*, 23:1349–1357, 1991.

Luoma, S. N., and Bryan, G. W.: A statistical study of environmental factors controlling concentrations of heavy metals in the burrowing bivalve *Scrombicularia plana* and the polychaete *Nereis diversicolor*. *Estuar. Coast Shelf Sci.*, 15:95–108, 1982.

Mac, M. J.; Noguchi, G. E.; Hesselberg, R. J.; Edsall, C. C.; Shoesmith, J. A.; and Bowker, J. D.: A bioaccumulation bioassay for freshwater sediments. *Environ. Toxicol. Chem.*, 9:1405–1414, 1990.

Mackay, D.: Correlation of bioconcentration factors. *Environ. Sci. Technol.* 16:274–278, 1982.

Mackay, D.: Multimedia Environmental Models. The Fugacity Approach. Chelsea, MI: Lewis, 1991.

Mackay, D., and Paterson, S.: Calculating fugacity. *Environ. Sci. Technol.*, 15:1006–1014, 1981.

Mackay, D.; Puig, H.; and McCarty, L. S.: An equation describing the time course and variability in uptake and toxicity of narcotic chemicals to fish. *Environ. Toxicol. Chem.*, 11:941–951, 1992.

Mailhot, H.: Predication of algal bioaccumulation and uptake rate of nine organic compounds by ten physicochemical properties. *Environ. Sci. Technol.*, 21:1009–1013, 1987.

Manahan, S. E.: Environmental Chemistry, 5th ed. Boca Raton, FL: Lewis, 1991.

McCarthy, J. F.: Role of particulate organic matter in decreasing accumulation of polynuclear aromatic hydrocarbons by *Daphnia magna*. *Arch. Environ. Contam. Toxicol.*, 12:559–568, 1983.

McCarty, L. S.: The relationship between aquatic toxicity QSARS and bioconcentration for some organic chemicals. *Environ. Toxicol. Chem.*, 5:1071–1080, 1986.

McCarty, L. S.: Relationship between toxicity and bioconcentration for some organic chemicals. II. Application of the relationship. QSAR in Environmental Toxicology—II, edited by K. L. E. Kaiser, pp. 221–230. Dordrecht, Holland: Reidel, 1987.

McCarty, L. S.: Toxicant body residues: implications for aquatic bioassays with some organic chemicals. Aquatic Toxicology and Risk Assessment, edited by M. A. Mayes, and M. G. Barron, pp. 183–192, Philadelphia: ASTM, 1991.

McCarty, L. S., and Mackay, D.: Enhancing ecotoxicological modeling and assessment: body residues and modes of toxic action. *Environ. Sci. Technol.*, 27:1719–1728, 1993.

McCarty, L. S.; Mackay, D.; Smith, A. D.; Ozburn, G. W.; and Dixon, D. G.: Residue-based interpretation of toxicity and bioconcentration QSARs from aquatic bioassays: neutral narcotic organics. *Environ. Toxicol. Chem.*, 11:917–930, 1992.

McElroy, A. E.; Farrington, J. W.; and Teal, J. M.: Influence of mode of exposure and the presence of a tubiculous polychaete on the fate of benz[*a*]anthracene in the benthos. *Environ. Sci. Technol.*, 24:1648–1655, 1990.

McKim, J. M., and Erickson, R. J.: Environmental impacts on the physiological mechanisms controlling xenobiotic transfer across fish gills. *Physiol. Zool.*, 64:39–67, 1991.

McKim, J. M., and Goeden, H. M.: A direct measure of the uptake efficiency of a xenobiotic chemical across the gills of brook trout. (*Salvelinus fontinalis*) under normoxic and hypoxic conditions. *Comp. Biochem. Physiol. Ser. C*, 72:65–74, 1982.

McKim, J. M., and Nichols, J. W.: Use of physiologically based toxicokinetic models in a mechanistic approach to aquatic toxicology. Aquatic Toxicology—Molecular, Biochemical, and Cellular Perspectives, edited by D. C. Malins, and G. K. Ostrander, pp. 469–519. Boca Raton: CRC Press, 1994.

McKim, J.; Schmieder, P.; and Veith, G.: Absorption dynamics of organic chemical transport across trout gills as related to octanol-water partition coefficient. *Toxicol. Appl. Pharmacol.*, 77:1–10, 1985.

McKim, J. M.; Schmieder, P. K.; and Erickson, R. J.: Toxicokinetic modeling of ^{14}C-pentachlorophenol in the rainbow trout (*Salmo gairdneri*). *Aquat. Toxicol.*, 9:59–80, 1986.

Meyer, C. L.; Suedel, B. C.; Rodgers, J. H., Jr.; and Dorn, P. B.: Bioavailability of sediment-sorbed chlorinated ethers. *Environ. Toxicol. Chem.*, 12:493–505, 1993.

Michel, C. M. F.; Squibb, K. S.; and O'Connor, J. M.: Pharmacokinetics of sulphadimethoxine in channel catfish (*Ictalurus punctatus*). *Xenobiotica* 20:1299–1309, 1990.

Moreno, M. D.; Cooper, K. R.; Brown, R. P.; and Georgopoulos, P.: A physiologically based pharmacokinetics model for *Mya arenaria*. *Mar. Environ. Res.*, 34:321–325, 1992.

Muir, D. C. G.; Grift, N. P.; Townsend, B. E.; Metner, D. A.; and Lockhart, W. L.: Comparison of the uptake and bioconcentration of fluridone and terbutryn by rainbow trout and *Chironomus tentans* in sediment and water systems. *Arch. Environ. Contam. Toxicol.*, 11:595–602, 1982.

Muir, D. C. G.; Townsend, B. E.; and Lockhart, W. L.: Bioavailability of six organic chemicals to *Chironomus tentans* larvae in sediment and water. *Environ. Toxicol. Chem.*, 2:269–281, 1983.

Muir, D. C. G.; Rawn, G. P.; Townsend, B. E.; Lockhart, W. L.; and Greenhalgh, R.: Bioconcentration of cypermethrin, deltamethrin, fenvalerate and permethrin by *Chironomus tentans* larvae in sediment and water. *Environ. Toxicol. Chem.*, 4:51–62, 1985.

Muir, D. C. G.; Yarechewski, A. L.; Metner, D. A.; Lockhart, W. L.; Webster, G. R. B.; and Friesen, K. J.: Dietary accumulation and sustained hepatic mixed function oxidase enzyme induction by 2,3,4,7,8-pentachlorodibenzofuran in rainbow trout. *Environ. Toxicol. Chem.*, 9: 1463–1472, 1990.

Nebeker, A. V.; Schuytema, G. S.; Griffis, W. L.; Barbitta, J. A.; and Carey, L. A.: Effect of sediment organic carbon on survival of *Hyalella azteca* exposed to DDT and endrin. *Water Res.*, 8:705–718, 1989.

Nichols, J. W.; McKim, J. M.; Andersen, M. E.; Gargas, M. L.; Clewell, H. J., III, and Erickson, R. J.: A physiologically based toxicokinetic model for the uptake and disposition of waterborne organic chemicals in fish. *Toxicol. Appl. Pharmacol.*, 106:433–447, 1990.

Niimi, A. J., and Cho, C. Y.: Elimination of hexachlorobenzene (HCB) by rainbow trout (*Salmo gairdneri*), and an examination of its kinetics in Lake Ontario salmonids. *Can. J. Fish. Aquat. Sci.*, 38:1350–1356, 1981.

Niimi, A. J., and Dookhran, G. P.: Dietary absorption efficiencies and elimination rates of polycyclic aromatic hydrocarbons (PAHs) in rainbow trout (*Salmo gairdneri*). *Environ. Toxicol. Chem.*, 8:719–722, 1989.

Oliver, B. G., and Niimi, A. J.: Trophodynamic analysis of polychlorinated biphenyl congeners and other chlorinated hydrocarbons in the Lake Ontario ecosystem. *Environ. Sci. Technol.*, 22:388–397, 1988.

Opperhuizen, A.: Bioconcentration of hyrophobic chemicals in fish. Aquatic toxicology and Environmental Fate, edited by T. M. Poston, and R. Purdy, pp. 304–315. Philadelphia: ASTM, 1986.

Opperhuizen, A., and Schrap, S. M.: Uptake efficiencies of two polychlorobiphenyls in fish after dietary exposure to five different concentrations. *Chemosphere*, 17:253–262, 1988.

Opperhuizen, A., and Sijm, D. T. H. M.: Bioaccumulation and biotransformation of polychlorinated dibenzo-*p*-dioxins and dibenzofurans in fish. *Environ. Toxicol. Chem.* 9:175–186, 1990.

Pankow, J. F., and McKenzie, S. W.: Parameterizing the equilibrium of chemicals between the dissolved, solid particulate matter, and colloidal matter compartments in aqueous systems. *Environ. Sci. Technol.*, 25:2046–2053, 1991.

Part, P.: Comparison of absorption rates of halogenated phenols across fish gills in fresh and marine water. *Mar. Environ. Res.*, 28:275–278, 1989.

Pavlou, S. B., and Dexter, R. M.: Distribution of polychlorinated biphenyls (PCB) in estuarine ecosystems. Testing the concept of equilibrium partitioning in the marine environment. *Environ. Sci. Technol.*, 13:65–71, 1979.

Pesch, C., and Morgan, D.: Influence of sediment in copper toxicity tests with the polychaete *Neanthes arenaceodentata*. *Water Res.*, 12:747–751, 1978.

Phipps, G. L.; Ankley, G. T.; Benoit, D. A.; and Mattson, V. R.: Use of the aquatic oligochaete *Lumbriculus varlegatus* for assessing the toxicity and bioaccumulation of sediment-associated contaminants. *Environ. Toxicol. Chem.*, 12:269–279, 1993.

Pittinger, C. A.; Woltering, D. M.; and Masters, J. A.: Bioavailability of sediment-sorbed and aqueous surfactants to *Chironomus riparius* (midge). *Environ. Toxicol. Chem.*, 8:1023–1033, 1989.

Rasmussen, J. B.; Rowan, D. J.; Lean, D. R. S.; and Carey, J. H.: Food chain structure in Ontario lakes determine PCB levels in lake trout (*Salvelinus namaycush*) and other pelagic fish. *Can. J. Fish Aquat. Sci.*, 47:2030–2038, 1990.

Resendes, J.; Shiu, W. Y.; and Mackay, D.: Sensing the fugacity of hydrophobic organic chemicals in aqueous systems. *Environ. Sci. Technol.*, 26:2381–2387, 1992.

Rice, C. P., and White, D. S.: PCB availability assessment of river dredging using caged clams and fish. *Environ. Toxicol. Chem.*, 6:259–274, 1987.

Richer, G., and Peters, R. H.: Determinants of the short-term dynamics of PCB uptake by plankton. *Environ. Toxicol. Chem.*, 12:207–218, 1993.

Roesijadi, G.; Anderson, J. W.; and Blaylock, J. W.: Uptake of hydrocarbons from marine sediments contaminated with Prudhoe Bay crude oil: influence of feeding type of test species and availability of polycyclic aromatic hydrocarbons. *Soil Sci. Soc. Am. J.*, 35:608–614, 1978.

Rubinstein, N. I.; Lores, E.; and Gregory, N. R.: Accumulation of PCBs, mercury and cadmium by *Nereis virens*, *Mercenaria mercenaria* and *Palaemonetes pugio* from contaminated harbor sediments. *Aquat. Toxicol.*, 3: 249–260, 1984.

Schmieder, P. K., and Weber, L. J.: Blood and water flow limitations on gill uptake of organic chemicals in the rainbow trout (*Oncorhynchus mykiss*). *Aquat. Toxicol.*, 24:103–122, 1992.

Schrap, S. M., and Opperhuizen, A.: Relationship between bioavailability and hydrophobicity: reduction of the uptake of organic chemicals by fish due to the sorption on particles. *Environ. Toxicol. Chem.*, 9:715–724, 1990.

Schultz, I. R., and Hayton, W. L.: Experimental design for pharmacokinetic data analysis using compartmental pharmacokinetic models. Aquatic Toxicology and Risk Assessment: Fourteenth Volume, edited by M. A. Mayes, and M. G. Barron, pp. 139–148, ASTM STP 1124. Philadelphia: American Society for Testing and Materials, 1991.

Schuytema, G. S.; Krawczyk, D. F.; Griffis, W. L.; Nebeker, A. V.; and Robideaux, M. L.: Hexachlorobenzene uptake by fathead minnows and macroinvertebrates in recirculating sediment/water systems. *Arch. Environ. Contam. Toxicol.*, 19:1–9, 1990.

Servos, M. R.; Muir, D. C. G.; and Barrie, W. G. R.: Bioavailability of polychlorinated dibenzo-*p*-dioxins in lake enclosures. *Can. J. Fish Aquat. Sci.*, 49:735–742, 1992.

Sijm, D. T. H. M., and Opperhuizen, A.: Biotransformation, bioaccumulation, and lethality of 2,8-dichlorodibenzo-*p*-dioxin: a proposal to explain the biotic fate of PCDD's and PCDF's. *Chemosphere*, 17:83–89, 1988.

Sijm, D. T. H. M.; Seinen, W.; and Opperhuizen, A.: Life-cycle biomagnification study in fish. *Environ. Sci. Technol.*, 26:2162–2174, 1992.

Sijm, D. T. H. M.; Part, P.; Opperhuizen, A.: The influence of temperature on the uptake rate constants of hydrophobic compounds determined by the isolated perfused gills of rainbow trout (*Oncorhynchus mykiss*). *Aquat. Toxicol.*, 25:1–14, 1993.

Sodergren, A.: Solvent-filled dialysis membranes simulate uptake of pollutants by fish. *Environ. Sci. Technol.*, 21: 859–863, 1987.

Spacie, A., and Hamelink, J. L.: Alternative models for describing the bioconcentration of organics in fish. *Environ. Toxicol. Chem.*, 1:309–320, 1982.

Spacie, A., and Hamelink, J. L.: Bioaccumulation. Fundamentals of Aquatic Toxicology, edited by G. M. Rand, and S. R. Petrocelli, pp. 495–525. New York: Hemisphere, 1985. [See Appendix D of this volume.]

Stehly, G. R.; Landrum, P. F.; Henry, M. G.; and Klemm, C.: Toxicokinetics of PAHs in Hexagenia. *Environ. Toxicol. Chem.*, 9:167–174, 1990.

Streit, B.; Sire, E.-O.; Kohlmaier, G. H.; Badeck, F. W.; and Winter, S.: Modeling ventilation efficiency of teleost fish gills for pollutants with high affinity to plasma proteins. *Ecol. Model*, 57:237–262, 1991.

Stumm, W.: Chemistry of the Solid-Water Interface. New York: Wiley Interscience, 1992.

Swackhammer, D. L., and Skoglund, R. S.: The role of phytoplankton in the partitioning of phydrophobic organic contaminants in water. Organic Substances and Sediments in Water, edited by R. A. Baker, pp. 91–105. Chelsea, MI: Lewis, 1991.

Swartz, R. C.; Schults, D. W.; DeWitt, T. H.; Ditsworth, G. R.; and Lamberson, J. O.: Toxicity of fluoranthene in sediment to marine amphipods: a test of the equilibrium partitioning approach to sediments quality criteria. *Environ. Toxicol. Chem.*, 9:1071–1080, 1990.

Swindoll, C. M., and Applehans, F. M.: Factors influencing the accumulation of sediment-sorted hexachlorobiphenyl by midge larvae. *Bull. Environ. Contam. Toxicol.*, 39: 1055–1062, 1987.

Thomann, R. V.: Bioaccumulation model of organic chemical distribution in aquatic food chains. *Environ. Sci. Technol.*, 23:699–707, 1989.

Thomann, P. V., and Connolly, J. P.: Model of PCB in the Lake Michigan lake trout food chain. *Environ. Sci. Technol.*, 18:65–71, 1984.

Veith, G. D., and Kosian, P.: Estimating bioconcentration potential from octanol/water partition coefficients. Physical Behaviour of PCBs in the Great Lakes, edited by D. Mackay; S. Paterson; S. Eisenreich, and Simmons, pp. 269–282. Ann Arbor, MI: Ann Arbor Science, 1983.

Whittle, D. M., and Fitzsimons, J. D.: The influence of the Niagara River on contaminant burdens of Lake Ontario biota. *J. Great Lakes Res.*, 9:295–302, 1983.

Winsor, M.; Boese, B. L.; Lee, H., II; Randall, R. C.; Specht, D. T.: A method of estimating the fluxes of interstitial and overlying water by the clam *Macoma nasuta. Environ. Toxicol. Chem.*, 9:209–213, 1990.

Winter, M. E.; Katcher, B. S.; and Koda-Kimble, M. A. (eds.): Basic Clinical Pharmacokinetics. Spokane, WA: Applied Therapeutics, 1980.

Wood, L. W.; Rhee, G.-Y.; Bush, B.; and Barnard, E.: Sediment desorption of PCB congeners and their bio-uptake by diptern larvae. *Water Res.* 21:875–884, 1987.

SUPPLEMENTAL READING

Barron, M. G.; Stehly, G. R.; and Hayton, W. L.: Pharmacokinetic modeling in aquatic animals. 1. Models and concepts. *Aquat. Toxicol.*, 17:187–212, 1990.

Hamelink, J. L.; Landrum, P. F.; Bergman, H. L.; and Benson, W. H. (eds.): Bioavailability—Physical, Chemical, and Biological Interactions. Boca Raton: Lewis, 1994.

Landrum, P. F., and Robbins, J. A.: Bioavailability of sediment-associated contaminants to benthic invertebrates. Sediments: Chemistry and Toxicity of In-Place Pollutants, edited by R. Baudo; J. P. Giesy; and H. Muntau, pp. 237–263. Chelsea, MI: Lewis, 1990.

Landrum, P. F.; Lee, H., II; and Lydy, M. J.: Toxicokinetics in aquatic systems: model comparisons and use in hazard assessment. *Environ. Toxicol. Chem.*, 11:1709–1725, 1992.

Lee, H., II; Winsor, M.; Pelletier, J.; Randall, R.; Bertling, J.; and Coleman, B.: Computerized Risk and Bioaccumulation System. Newport, OR: 1990, ERLN-N137. U. S. Environmental Protection Agency, Environmental Research Laboratory, Narragansett, Pacific Ecosystems Branch.

McKim, J. M., and Nichols, J. W.: Use of physiologically based toxicokinetic models in a mechanistic approach to aquatic toxicology. Aquatic Toxicology—Molecular, Biochemical, and Cellular Perspectives, edited by D. C. Malins, and G. K. Ostrander, pp. 469–519. Boca Raton: CRC Press, 1994.

Nagel, R., and Loskill, R. (eds.): Bioaccumulation in Aquatic Systems. Weinheim, Germany: VCH, 1991.

Chapter 17

BIOCHEMICAL MECHANISMS: METABOLISM, ADAPTATION, AND TOXICITY

R. T. Di Giulio, W. H. Benson, B. M. Sanders, and *P. A. Van Veld*

INTRODUCTION

In the realm of "traditional" (i.e., human health–oriented) toxicology, the study of biochemical phenomena occupies a central position. Elucidations of the biotransformation or metabolism of chemical compounds into more or less toxic products, of the underlying modes of toxic action, and of early biochemical responses have been accepted as critical to the informed diagnosis of human exposures to chemicals, as well as to the development of clinical treatments to counteract toxic effects. In aquatic toxicology, however, biochemical studies have played a lesser role. There are probably two reasons for this difference between these related fields. First, traditional toxicology is concerned ultimately with only one species—the human; other species are studied as models (e.g., rats and mice) for humans. The number of species of concern in aquatic toxicology, in contrast, is vast. The generation of information on even a representative cross section of these species in as much detail as that available for humans and key models is far beyond available resources and will remain so for the foreseeable future.

The other reason that biochemical studies have played a lesser role is more fundamental. In traditional toxicology, the biological level of chief concern is the *individual* organism, a person. In aquatic toxicology, on the other hand, concern is principally for higher levels of organization—populations, communities, and ecosystems. In the management of aquatic and wildlife resources, a few individuals of nonendangered species are generally considered expendable as long as these higher levels of organization remain unaffected. The difficulty in extrapolating effects of chemicals on these higher levels of organization from effects observed in individuals, including biochemical responses, has damped enthusiasm for the potential role of biochemistry in aquatic toxicology.

In recent years, however, biochemical studies have received greater prominence in aquatic toxicology. This has occurred for several reasons. First, there is a growing appreciation that basic research into fundamental processes can provide a more solid theoretical foundation that can be applied to the problem-solving endeavors of aquatic toxicology. Medical science provides an instructive model here; basic studies in fields such as molecular biology, physiology, and immunology have provided a solid theoretical underpinning upon which clinical decisions are often based. Aquatic toxicology is a relatively young science, and considerable work is required in fields from biochemistry to systems ecology in order to provide a solid theoretical basis in this discipline. Another reason for this increased focus on biochemical processes is the cross-fertilization between aquatic toxicology and medical toxicology resulting from a growing realization that the issues of ecosystem health and human health are highly related and that continued isolation between scientists in these areas is inefficient and counterproductive. Increasingly, environmental scientists are applying modern tools of molecular biology and biochemistry to explore problems posed by environmental pollution. Finally, interest has increased in the development of sensitive, biochemically based tools ("biomarkers"; see Chapter 1) for monitoring environmental quality (Stegeman et al., 1992). A key notion underlying biomarkers is that selected biochemical responses measured in feral organisms, for example, can provide sensitive indices, or early warning signals, for potential ecosystem degradation caused by contaminants. The development of biomarkers clearly illustrates a case in which basic research can yield practical tools that are the environmental equivalent of clinical diagnostics.

Although biomarkers do not directly provide information concerning the effects of chemicals at the higher levels of biological organization, they can provide an early warning of biological impact. Furthermore, studies suggest that there are predictable relationships between the physiological state of the organism and population-level parameters. As a consequence, biomarkers can play a significant role in ecological risk assessments (see Chapter 28).

In this chapter, several biochemical phenomena of particular interest in aquatic toxicology are discussed. These include (1) enzyme systems that metabolize, or biotransform, a broad array of organic contaminants; (2) protective and toxic responses associated with oxyradicals, the production of which is enhanced by many chemicals; (3) metal-binding proteins that play important roles in the regulation of both essential and nonessential trace metals; and (4) a heterogeneous group of proteins, collectively known as stress proteins, that are induced by a variety of environmental stressors, including chemical contaminants, and confer protection against environmentally induced damage to other proteins. The goals of this chapter are to provide the reader with a fundamental understanding of these systems, including specific nuances observed in aquatic organisms, to describe how these systems provide a degree of adaptation to chemicals, and to illustrate how responses associated with these systems may provide useful tools for problem solving in aquatic toxicology.

TERMINOLOGY

Activation. A process by which a foreign compound (i.e., xenobiotic) is made more toxic as a result of enzymatic biotransformation.

Apoenzyme. The protein component of an enzyme that requires a nonprotein component (cofactor) for catalytic activity.

Biotransformation. A process by which the structure of a compound is changed as a result of enzyme activity.

Detoxification. Process in which the toxicity of a foreign compound is reduced.

Endogenous. Naturally occurring in an organism.

Enzyme inhibition. Loss of catalytic activity of an enzyme due to various chemical interactions.

Exogenous. From a source external (foreign) to an organism.

Free radical. A molecule containing an unshared electron.

Hydrophilic. Relatively soluble in aqueous media; having an affinity for, or being soluble in, aqueous solutions.

Induction. The phenomenon of increased de novo synthesis of a protein; the process whereby the level of a protein is increased as a result of xenobiotic exposure. Induction can be regulated at the transcriptional and translational levels.

Lipophilic (hydrophobic). Relatively soluble in nonaqueous media (e.g., nonpolar, organic solvents); having an affinity for lipids (fats) and poorly soluble in aqueous solutions.

Metallothionein. A metallo derivative of the soluble, sulfur-rich protein, thionein.

Microsomes. Membrane vesicles derived from the smooth endoplasmic reticulum and obtained by differential centrifugation of tissue (e.g., liver) homogenates.

Oxidative stress. Cellular responses, including both adaptive responses (e.g., antioxidant enzyme inductions) and deleterious responses (e.g., lipid peroxidation) emanating from increased fluxes of oxyradicals.

Oxyradical. Reactive (or activated) species of O_2, including true oxygen free radicals ($O_2^{\cdot -}$, $\cdot OH$), and other, related species (H_2O_2, $^1O_2$).

Soluble enzymes, proteins, etc. Occurring in the aqueous (cytosolic) part of the cell (i.e., not associated with membranes).

Substrate. A chemical that undergoes biotransformation.

Xenobiotic. A foreign (exogenous) compound.

BIOTRANSFORMATION: Phase I and Phase II Reactions

Perhaps the most extensively studied biochemical systems in aquatic animals related to toxic chemicals are the enzymes involved in the metabolism or transformation of organic chemical contaminants. These systems were the subject of a chapter in the first volume of this text (Lech and Vodicinik, 1985), and the reader is referred to that excellent treatise for a detailed review of some aspects of this topic. Here, we provide an overview of these enzyme systems, including recent advances in basic research and applications to aquatic toxicology.

Although we generally associate chemical contamination of the aquatic environment with human activities, it is important to realize that organisms have had to deal with foreign compounds or xenobiotics since life began. For example, bacteria, fungi, algae, and plants all produce natural organic toxicants in order to deter competitors and predators. Marine organisms in particular produce a huge variety of toxic chemicals. In fact, the field of marine natural products chemistry arose from interest in isolating chemicals from marine organisms such as algae, seaweed, sponges, and coral for use in the biomedical field. In addition, polycyclic aromatic hydrocarbons (PAHs) are naturally produced during combustion of organic material (e.g., forest and grass fires) and can also enter the aquatic environment via natural seeps in the ocean floor.

These are just a few examples of the many sources and types of natural toxicants that aquatic organisms have had to deal with. It is likely that the enzyme systems that are discussed in this chapter evolved to reduce the toxicity and enhance the clearance of these natural compounds. Only recently have people begun to synthesize a vast number of foreign chemicals. Many of these are structurally similar to natural xenobiotics and can be readily biotransformed, whereas others (e.g., chlorinated organic pesticides, polychlorinated biphenyls or PCBs) are structurally dissimilar. The relative resistance of the latter compounds to enzymatic transformation may be a consequence of their recent appearance in the biosphere. In other words, enzymes capable of efficiently transforming these compounds have not had sufficient time to evolve.

Most organic contaminants of concern are lipophilic, a property that makes them readily absorbed across lipid membranes of gill, skin, and digestive tract. Following absorption, the fate and effects of these compounds are largely dependent on their susceptibility to biotransformation. Compounds that are recalcitrant to metabolism tend to accumulate in lipid-rich tissues such as triglyceride depots. Those that are subject to biotransformation can be converted into more water-soluble metabolites that are easier to eliminate than the parent compound. Although biotransformation is generally regarded as a detoxification process, there are many examples in which the toxicity (including carcinogenicity) of a compound is increased as a result of enzymatic transformation.

Biotransformation often consists of two phases. The first step (phase I) is frequently an oxidative step whereby a polar functional moiety, such as a hydroxyl group, is introduced into the lipophilic contaminant, making it more hydrophilic. In phase II, the metabolites produced in phase I are covalently linked (conjugated) to various naturally occurring endogenous compounds in the cell such as a sugar derivative, a peptide, or sulfate. These reactions generally result in very water-soluble products that are nontoxic, ionizable, and readily excreted. In some cases the parent xenobiotic may already possess an appropriate functional group and directly undergo phase II reactions. In the following sections, phase I and phase II reactions are discussed in more detail.

Phase I Reactions

Cytochrome P450 Monooxygenases

The primary oxidative enzymes (or monooxygenases) involved in phase I metabolism are a family of enzymes collectively referred to as cytochrome P450 (Black and Coon, 1987; Buhler and Williams, 1989; Stegeman, 1989). In excess of 100 P450 genes (or proteins) have been sequenced from prokaryotes and eukaryotes (Nebert et al., 1991). Cytochrome P450 (or P450) is a heme-containing protein involved in electron transport (i.e., a cytochrome). When the iron of its heme is in the reduced (Fe^{2+}) state, P450 can bind to carbon monoxide and absorb light maximally at a wavelength of 450 nm (Figure 1). This spectral property occurs only when P450 is intact and catalytically functional. The ''P'' in P450 comes from early studies of red ''pigments'' of the endoplasmic reticulum. P450 catalyzes the following general hy-

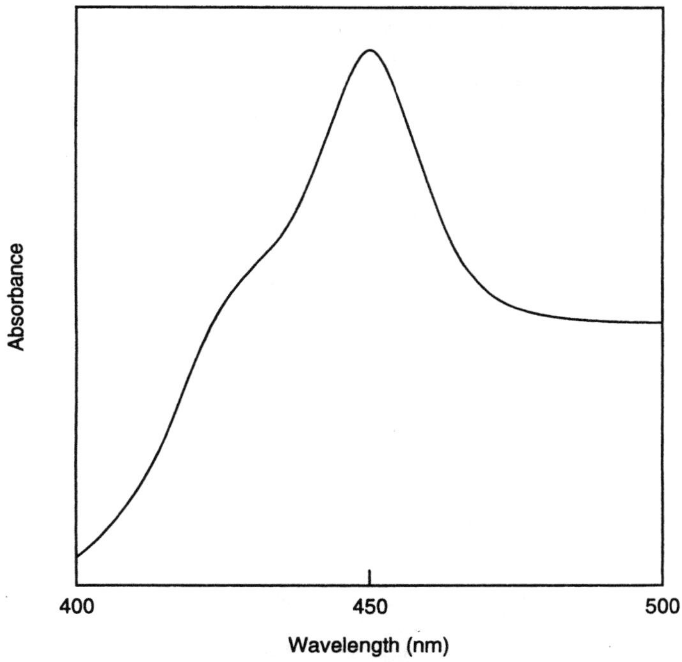

Figure 1. Typical CO-ligated cytochrome P450 difference spectrum for fish liver microsomes.

droxylation reaction, where R is a substrate such as a xenobiotic (e.g., pesticide or chemical):

$$RH + O_2 \xrightarrow{NADPH \rightarrow NADP^+} ROH + H_2O \quad (1)$$

chemical — oxidized chemical

In addition to hydroxylations, P450-dependent monooxygenases catalyze a variety of other reactions, including epoxidation, dealkylation, deamination, sulfoxidation, and desulfuration (Figure 2). This family of enzymes is involved not only in metabolism of foreign compounds but also in metabolism of many endogenous compounds, such as steroid hormones, prostaglandins, bile salts, and fatty acids (Table 1). Some forms catalyze a wide variety of reactions, whereas others are more specific as to the types of reactions performed and substrates transformed.

Several components are necessary for P450-catalyzed reactions; these components constitute the cytochrome P450–mediated monooxygenase system (Table 2). The system is also referred to as the mixed-function oxidase or mixed-function oxygenase (MFO) system. The monooxygenase system is composed of two primary enzymes: cytochrome P450 and NADPH-cytochrome P450 reductase. Cytochrome b_5 and NADH-cytochrome b_5 reductase may also participate in some monooxygenase reactions (Klotz et al., 1986). Other necessary components are NADPH, phospholipid, and O_2. The reductases function by transferring electrons from NADPH (or NADH in the case of NADH-cytochrome b_5 reductase) to cytochrome P450 during the oxidation of xenobiotics. Collectively, these components drive the monooxygenase cycle whereby one atom of molecular oxygen (O_2) is incorporated into a xenobiotic, rendering the compound more water soluble (Figure 3). The substrate specificities of the monooxygenase system are controlled by the distinct form of P450 involved.

P450 and other enzymes of the monooxygenase system are membrane-associated proteins located primarily in the smooth endoplasmic reticulum. During isolation procedures, the endoplasmic reticulum is obtained in the form of membrane vesicles called microsomes. Thus, we often see the term microsomal P450. Cytochrome P450 is present in virtually all species studied to date (Lee et al., 1981; Stegeman, 1989; James, 1989a,b; Livingstone et al., 1989). In vertebrates, P450 is often most abundant in the liver, although the enzymes are present in most extrahepatic organs as well (Stegeman et al., 1979). The presence of P450 in organs directly exposed to the environment such as gills (Miller et al., 1989) and intestine (Van Veld, 1990) may be particularly important, since these are also the organs involved in xenobiotic uptake. P450 is located primarily in the hepatopancreas of crustaceans (James, 1989b) and in the digestive gland of mollusks (Livingstone et al., 1989). In general, the concentration and activity of P450 in invertebrates are lower than in fish and are particularly low in mollusks (James 1989a,b; Livingstone et al., 1989).

Cytochrome P450 makes a chemical more or less toxic (see Significance of Biotransformation). The amounts of some types of P450 can also be induced by exposure to chemicals, including natural substrates (see Enzyme Inductions). Induction involves increases in the transcription and translation of messenger RNA coding for P450. As a result of induction, the rate of chemical transformation catalyzed by monooxygenases may be increased.

Flavin-Containing Monooxygenases

Although probably not as versatile and certainly not as extensively studied as P450, the flavin-containing monooxygenase systems also play a role in xenobiotic transformation in aquatic organisms (Kurelec, 1985; Kurelec, et al., 1986; Schlenk and Buhler, 1989, 1990, 1991). Within the cell, flavin-containing monooxygenases would be expected to compete with P450 systems for some nitrogen- and sulfur-containing molecules such as aromatic amines and organophosphate and carbamate insecticides (Figure 4). They convert tertiary amines to amine oxides, secondary amines to hydroxyl amines and nitrones, and primary amines to hydroxylamines and oximes. Flavin-containing monooxygenases use flavin as a coenzyme in oxidation reactions. Like P450 monooxygenases, this system resides in the smooth endoplasmic reticulum (microsomal fraction) and requires NADPH and O_2 (Ziegler, 1990). In some cases it is difficult to determine the relative contribution of P450 and flavin monooxygenase systems to a particular biotransformation. Kurelec and colleagues (1985, 1986) suggested that in mollusks, in which levels of P450 are low, flavin-containing monooxygenases may be the predominant phase I enzyme system.

Cooxidation During Prostaglandin Biosynthesis

Studies in mammalian toxicology have shown that during the synthesis of prostaglandins from polyunsaturated fatty acids (e.g., arachidonic acid), xenobiotics (e.g., PAH) can be cooxidized, resulting in biotransformation products similar to those produced by cytochrome P450 (Eling et al., 1990). Although this pathway has not received a great deal of attention in aquatic toxicology, prostaglandins have been detected in several marine species (Stanley-Samuelson, 1987) and their participation in xenobiotic cooxidation has been suggested (Buhler and Williams, 1989).

Reduction Reactions

Studies in the mammalian literature indicate that under conditions of low oxygen tension, cytochrome P450 can catalyze reduction reactions in which substrate rather than molecular oxygen accepts electrons and is reduced. Little progress has been made in this area in the aquatic field, but it is likely that azo, nitro, and halogenated compounds may undergo reductive metabolism in aquatic organisms.

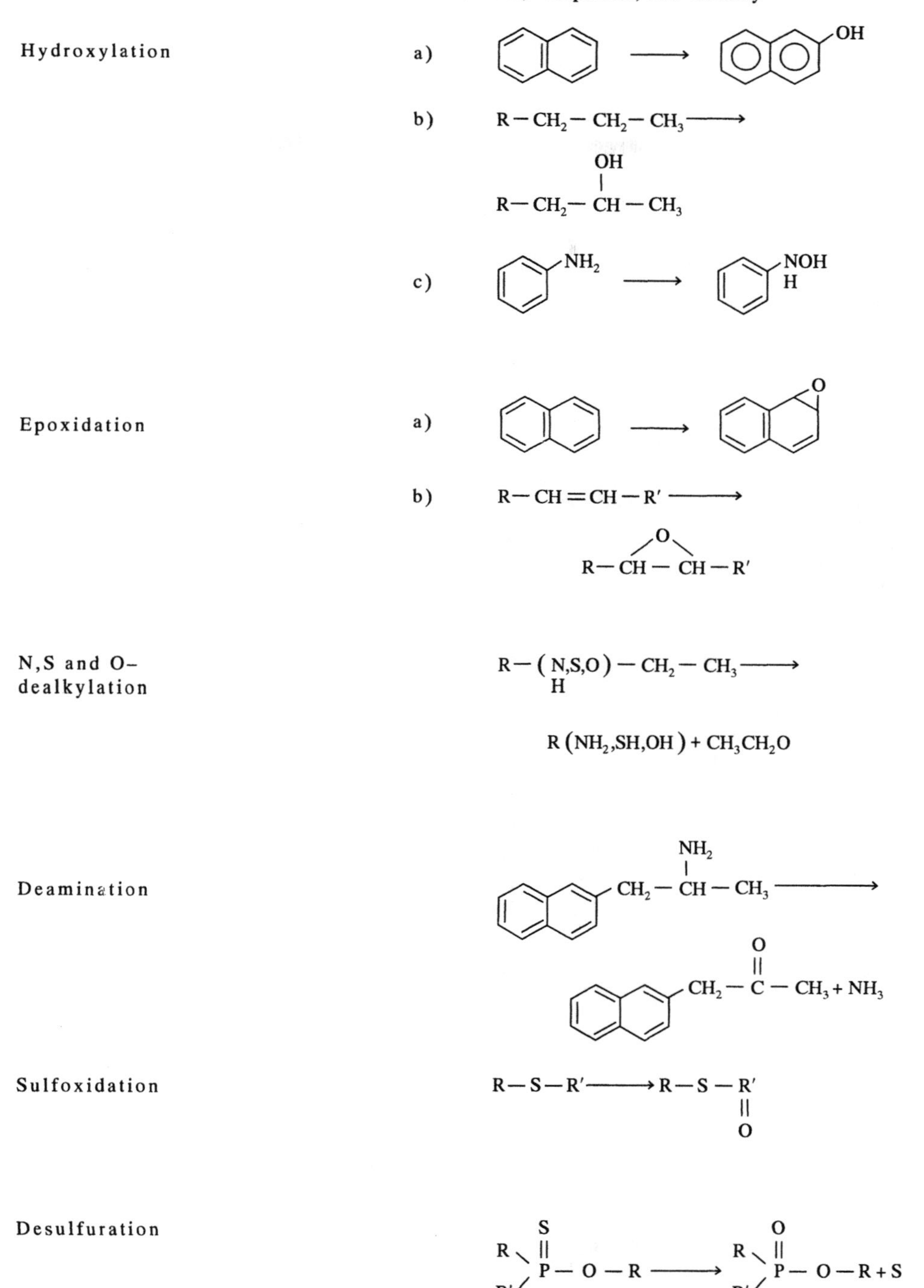

Figure 2. Examples of reactions catalyzed by the cytochrome P450 monooxygenase system.

Table 1. Examples of compounds metabolized by cytochrome P450–dependent monooxygenase systems

Polycyclic aromatic hydrocarbons
Aromatic amines
Aromatic amides
Pesticides (many)
Azo compounds
Flavonoids
Drugs and other chemicals
Endogenous substrates
Prostaglandins
Steroid hormones
Fatty acids
Bile salts

Table 2. Components of the monooxygenase system

Cytochrome P450[a]
NADPH-cytochrome P450 reductase
NADH-cytochrome b_5 reductase[b]
Phospholipid membrane
NADPH
NADH[b]
O_2

[a] Consists of a family of enzymes.
[b] May participate in some reactions.

Phase II Reactions

Phase II reactions serve to link metabolites produced during phase I to various water-soluble endogenous compounds naturally present in the cell at high concentrations (James, 1987; Foureman, 1989). Linkage (conjugation) of metabolites to sugar derivatives, amino acids, peptides, and sulfate produces metabolites with greatly increased water solubility and, in general, reduced toxicity. Because phase II reactions are biosynthetic, they require energy to drive the reactions. This is accomplished by activating the endogenous compound (e.g., sugar, sulfate) or the xenobiotic substrate to high-energy intermediates. The most widely studied and most important phase II enzymes are glutathione *S*-transferases, UDP-glucuronosyltransferases, and sulfotransferases. Although not always considered a phase II enzyme, epoxide hydrolase inserts a molecule of water into an epoxide group and is discussed in this section. Examples of some common phase II reactions are provided in Figure 5.

Glutathione S-Transferase

The glutathione *S*-transferases (GSTs) are a family of biotransformation enzymes involved in conjugating a variety of electrophilic metabolites, such as epoxides produced by P450, to glutathione (Armstrong, 1987; Clark, 1989; George and Buchanan, 1990). Glutathione (GSH) is an endogenous tripep-

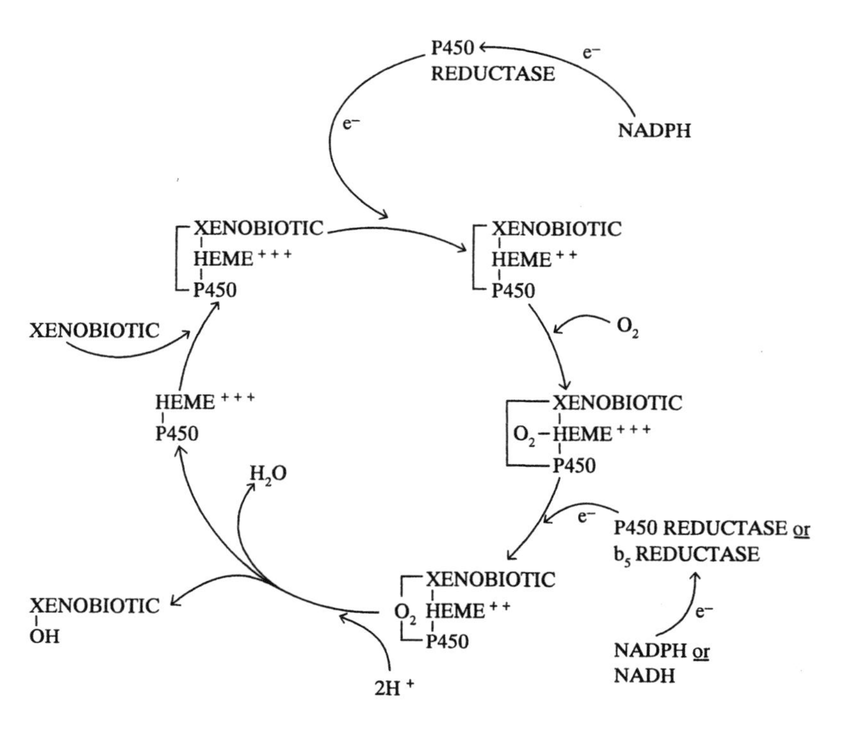

Figure 3. Cytochrome P450–mediated monooxygenation of a xenobiotic.

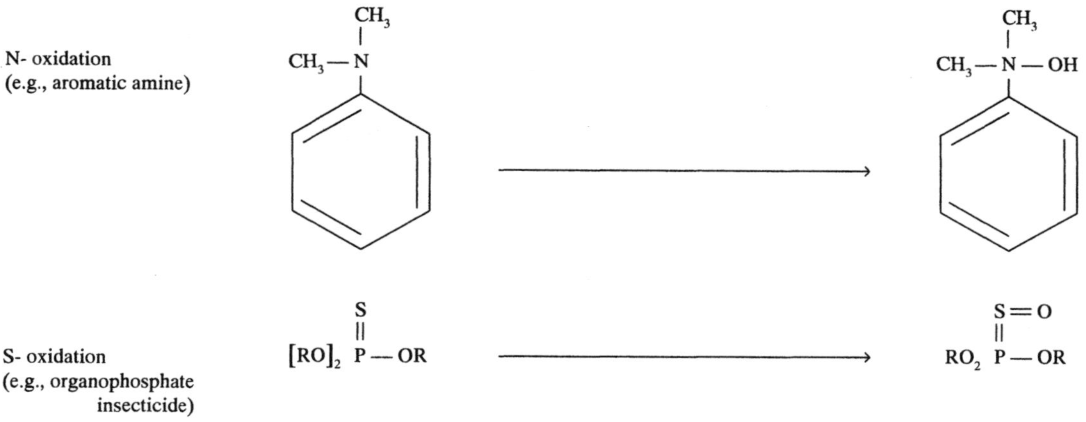

N- oxidation (e.g., aromatic amine)

$CH_3-N(CH_3)-C_6H_5 \longrightarrow CH_3-N(CH_3)(OH)-C_6H_5$

S- oxidation (e.g., organophosphate insecticide)

$[RO]_2\ P(=S)-OR \longrightarrow RO_2\ P(=S{=}O)-OR$

Figure 4. Typical reactions catalyzed by flavin-containing monooxygenases.

tide composed of the amino acids glutamate, cysteine, and glycine. In addition to their role in conjugations, GSTs play additional roles in detoxification (Jakoby, 1978). For example, these enzymes appear to be involved in the transport of endogenous and exogenous lipophilic compounds (xenobiotics) through the cytoplasm to sites of phase I metabolism (Hanson-Painton et al., 1983; Van Veld et al., 1987). GSTs are also capable of forming covalent bonds with activated epoxides produced by phase I enzymes (Schelin et al., 1983), thereby preventing binding of the activated species to DNA and other cellular macromolecules (Figure 6).

Several forms of GST appear to be present in fish; most are soluble (cytosolic), although microsomal forms exist as well (Nimmo, 1987; James, 1987; Lee et al., 1988; Foureman, 1989). The tissue distribution of GST appears to be similar to that of P450 and other biotransformation enzymes. Thus, high activities are found in the liver and hepatopancreas, with substantial activities occurring in extrahepatic organs as well (Clark, 1989). Multiple forms of GST exhibit what may be described as fairly broad, overlapping ranges of specificity toward different substrates. Typical reactions include aryl transferase, alkyl transferase, and epoxide transferase (Figure 5). The latter reaction is particularly important in detoxification of PAH epoxides produced by P450. Although fewer forms have been identified in invertebrates, GST levels and activities can be relatively high in these organisms (Keeran and Lee, 1987; Clark, 1989).

Epoxide Hydrolase

In addition to conjugation with glutathione, epoxides produced by P450 monooxygenases can be transformed by epoxide hydrolase (EH). EH functions by inserting a molecule of water into an epoxide group, resulting in the formation of a diol (Thomas and Oesch, 1988; Foureman, 1989). The corresponding diol is less electrophilic and therefore less chemically reactive than the epoxide. Because of the high specificity of EH toward epoxides, there are probably only one or two forms present in any tissue. The enzyme exists in all species studied to date. The liver of vertebrates and hepatopancreas of invertebrates are rich sources of EH, although the enzyme is present in other organs as well (James et al., 1979; Buhler and Williams, 1988). EH is primarily a microsomal enzyme, although in mammals soluble forms have been described as well (Thomas and Oesch, 1988).

Other Hydrolytic Enzymes

Although not fitting well into the overall phase I and phase II classification scheme for biotransformation enzymes, fish and other aquatic animals possess hydrolytic enzymes in addition to epoxide hydrolase that play important roles in the biotransformation of esters and amide groups, such as those present in organophosphate and carbamate insecticides.

UDP-Glucuronosyltransferase

UDP-glucuronosyltransferases (UDPGTs) are important phase II enzymes that catalyze the conjugation of a variety of substrates to glucuronic acid, a glucose derivative (Armstrong, 1987; Foureman, 1989). The formation of a high-energy intermediate, uridine diphosphate glucuronic acid (UDPGA), is required for these conjugations (Figure 7). The enzyme thus catalyzes the transfer of the glucuronic acid from UDPGA to the functional group on the acceptor molecule (the substrate). Compounds having hydroxyl and amino groups are particularly good substrates for UDPGT (Figure 5).

UDPGTs are located in the endoplasmic reticulum of most tissues studied. The location of these enzymes in the microsomal membrane may be important physiologically, since they may have direct access to the products formed by the microsomal P450 system. This would give rise to a very integrated system within the microsomal membrane that results in sequestration of lipophilic compounds, addition or

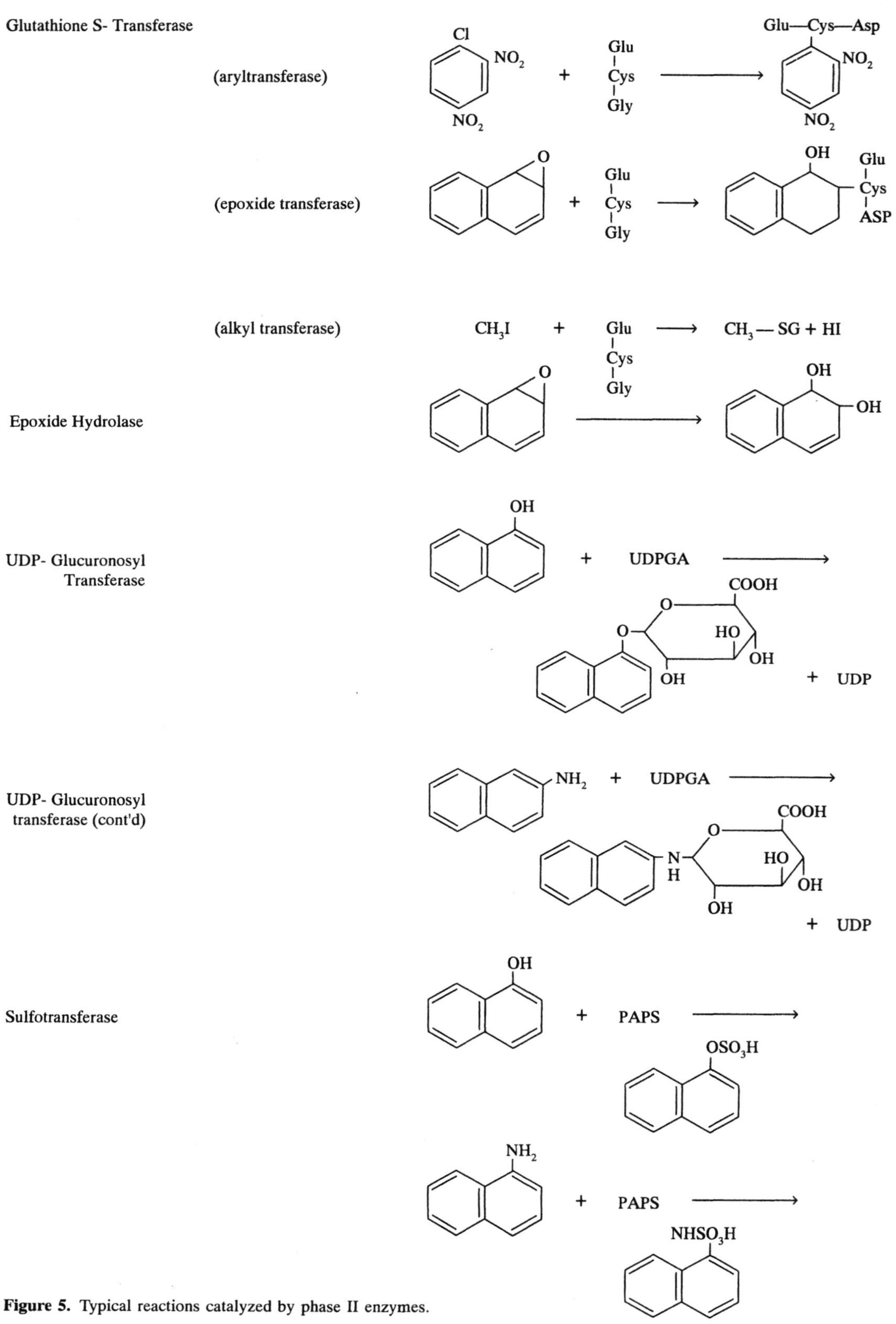

Glutathione S- Transferase

(aryltransferase)

Cl, NO_2, NO_2 + Glu–Cys–Gly ⟶ Glu—Cys—Asp, NO_2, NO_2

(epoxide transferase)

O + Glu–Cys–Gly ⟶ OH, Glu–Cys–ASP

(alkyl transferase)

CH_3I + Glu–Cys–Gly ⟶ CH_3— SG + HI

Epoxide Hydrolase

O ⟶ OH, OH

UDP- Glucuronosyl Transferase

OH + UDPGA ⟶ COOH, O, O, HO, OH, OH + UDP

UDP- Glucuronosyl transferase (cont'd)

NH_2 + UDPGA ⟶ COOH, O, N H, HO, OH, OH + UDP

Sulfotransferase

OH + PAPS ⟶ OSO_3H

NH_2 + PAPS ⟶ $NHSO_3H$

Figure 5. Typical reactions catalyzed by phase II enzymes.

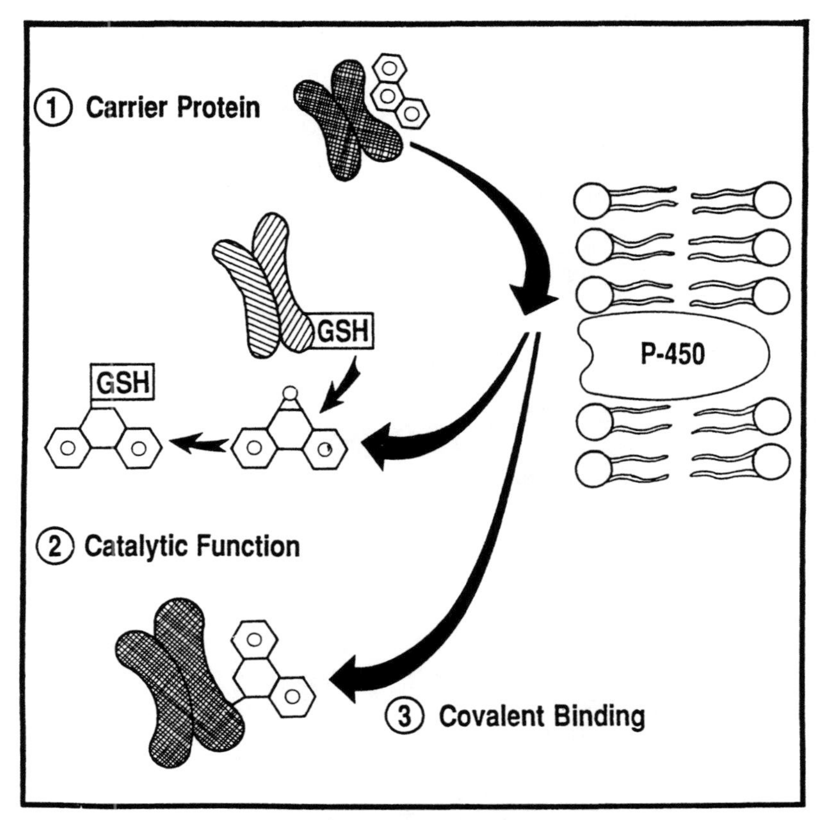

Figure 6. Triple role of glutathione *S*-transferase in xenobiotic metabolism. (1) Transport of lipophilic toxicant to site of phase I biotransformation. (2) Conjugation of phase I metabolite with glutathione. (3) Formation of covalent bond between activated metabolite and glutathione *S*-transferase.

unmasking of a functional group, and conjugation of this group with the polar glucuronic acid moiety. In fish, gill and intestinal activities often exceed those of liver (Koivusaari et al., 1981; Lindstrom-Seppa et al., 1981).

Sulfotransferases

Hydroxy or amine compounds can be conjugated with sulfate. Their action is catalyzed by the sulfotransferases (STs), cytosolic enzymes that compete with UDPGT for many of the same xenobiotics such as hydroxylated metabolites and amino compounds (Armstrong, 1987; James, 1987; Foureman, 1989) (Figure 5). STs conjugate these xenobiotic compounds with sulfate via the high-energy intermediate 3′-phosphoadenosine-5′-phosphosulfate (PAPS), which is the sulfate donor and cofactor synthesized from inorganic sulfate and ATP (Figure 7). ST activities have not been extensively studied in aquatic organisms but appear to be substantial in the vertebrate and invertebrate species examined thus far.

Factors Influencing Biotransformation

The levels and activities of biotransformation enzymes can be influenced by a number of factors (Table 3), which are considered in this section.

Enzyme Inductions

In the mammalian field, there is a great deal of information on enzyme induction or elevation of various biotransformation enzymes following exposure of animals to specific inducing agents (Lu, 1979; Whitlock, 1986). The enhanced activity results from an increase in the rate of protein synthesis of these enzymes. Thus, exposures of animals to carcinogens such as benzo[*a*]pyrene, drugs such as phenobarbital, and environmental contaminants such as PCBs, chlorinated dibenzodioxins, and chlorinated dibenzofurans can result in increased synthesis of specific phase I and phase II enzymes involed in xenobiotic metabolism. As with mammals, exposure of fish to certain PAHs, coplanar PCBs, chlorinated dibenzo-*p*-dioxins and chlorinated dibenzofurans results in induction of a specific form of P450 active in PAH metabolism (Stegeman, 1989). This form bears catalytic, struc-

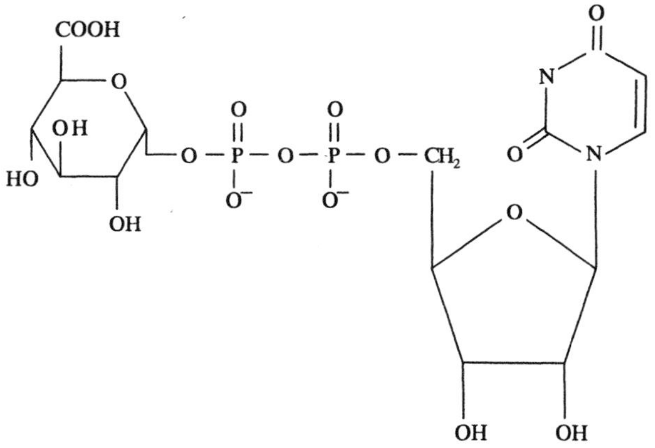

Uridine diphosphate glucuronic acid (UDPGA)

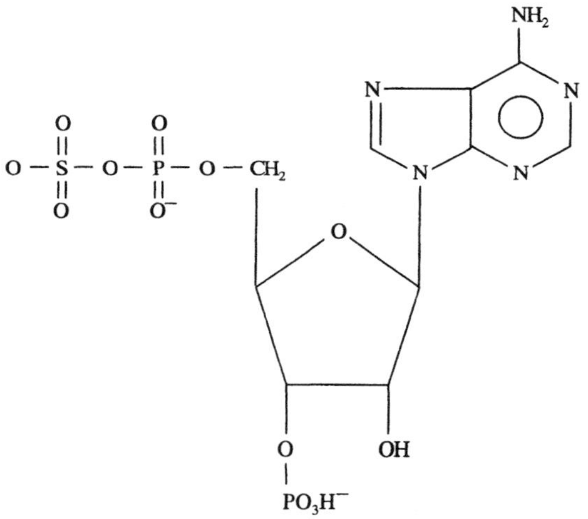

Phosphoadenosine phosphosulfate (PAPS)

Figure 7. High-energy cofactors of UDP-glucuronosyltransferase and sulfotransferase.

tural, and immunological similarities to cytochrome P4501A (CYP1A) proteins, the major PAH-inducible forms in mammals (Heilmann et al., 1988; Stegeman, 1989). Induction of CYP1A has been observed following treatment of fish with PAH-type inducers such as benzo[*a*]pyrene, 3-methylcholanthrene, and β-naphthoflavone. In addition, induction occurs following treatment with specific coplanar PCBs, chlorinated dibenzodioxins, and chlorinated dibenzofurans (Table 4).

In mammals, the mechanism for induction of CYP1A has been studied extensively (Whitlock, 1986). The same or a very similar mechanism probably operates in fish (Heilmann, 1988; Hahn et al., 1992, 1993). The response involves a soluble protein present in cells at low concentration and known as the Ah (aryl hydrocarbon) receptor. The Ah receptor binds strongly to incoming inducing agents (Figure 8). The receptor-inducer complex apparently binds to another protein (a translocating factor), which allows

Table 3. Factors influencing biotransformation rates of xenobiotics by aquatic organisms

Enzyme induction
Species differences
Temperature
Seasonal effects
Reproductive status
Diet
Exposure route
Age and development
Disease state (e.g., cancer)

Table 4. Examples of inducers of cytochrome CYP1A

Polycyclic aromatic hydrocarbons
(e.g., benzo[*a*]pyrene, 3-methylcholanthrene, dimethylbenz[*a*]anthracene)
Polychlorinated biphenyls
(e.g., 3,3′,4,4′-tetrachlorobiphenyl)
Polychlorinated dibenzo-*p*-dioxins
(e.g., 2,3,7,8-tetrachlorodibenzo-*p*-dioxin)
Polychlorinated dibenzofurans
(e.g., 2,3,7,8-tetrachlorodibenzofuran)
β-Napthoflavone

the complex to enter the nucleus. Once inside the nucleus, the complex attaches to specific sites on DNA, distorting the DNA chain and resulting in transcription of mRNA that codes for CYP1A. The mRNA is subsequently translated into new CYP1A, which is inserted into the endoplastic reticulum and supplied with a molecule of heme. Heme required at the active site of P450 is synthesized in the mitochondria.

The induction of CYP1A is rapid and can be detected within a few to several hours after exposure (Van Veld et al., 1988a; Haasch et al., 1989; Kloepper-Sams and Stegeman, 1989). Levels and activities of CYP1A in tissues of exposed animals can exceed those of uninduced animals by over 100-fold (Payne et al., 1987). This process can greatly influence the rates of metabolism of PAH entering the organism because of higher concentrations of enzyme present to perform biotransformation (Van Veld et al., 1988b).

In addition to affecting P450, exposures of mammals to various inducing agents can result in elevations of several phase II enzymes (Hammock and Ota, 1983; Burchell and Coughtrie, 1989). Studies with fish, however, have yielded conflicting results. Numerous studies report two- to threefold increases in hepatic GST activity following treatment of fish with PAH and PCB (Chatterjee and Bhattacharya,

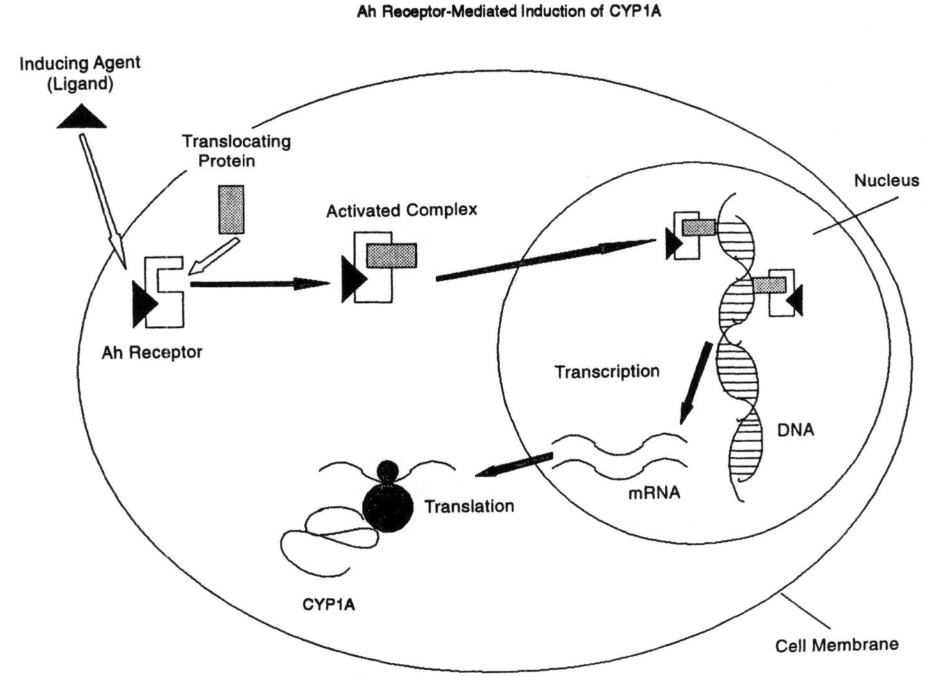

Figure 8. Induction of cytochrome CYP1A.

1984; Andersson et al., 1985; Gøksoyr et al., 1987; Van Veld et al., 1991). In other studies, however, treatment with similar agents did not result in a detectable response (James and Little, 1981; Ankley et al., 1986). Hepatic UDPGT induction (two- to three-fold) has been reported in a number of studies (Andersson et al., 1985; Pesonen et al., 1987). There is no strong evidence for induction of epoxide hydrolase or sulfotransferase in aquatic species.

Blue crab (*Callinectes sapidus*) GST was induced about two-fold after treatment with butylated hydroxyanisole (BHA) and butylated hydroxytoluene (BHT) (Lee et al., 1988). Studies with this species and other aquatic invertebrates suggest that the enzyme systems in these organisms are not sensitive to environmental inducing agents, however. Studies by Hahn and Stegeman (1992) indicate that the Ah receptor is lacking in many if not all marine invertebrates. Thus, treatment of mollusks and crustaceans with PAH and related compounds has yielded little or no detectable differences in activities of phase I or phase II enzymes (Foureman, 1989; James, 1989a,b; Livingstone et al., 1989).

Other Factors

In addition to enzyme induction, several other factors can influence the activity of biotransformation enzymes (Binder et al., 1984) (Table 3). For example, there can be species differences in basal levels of activity and in sensitivity to induction (James et al., 1979, 1988; Gøksoyr et al., 1987; Sternerson et al., 1987). Basal levels and induction of enzymes may be influenced by environmental factors such as temperature and season (Andersson and Koivusaari, 1985; Smolarek et al., 1988; George et al., 1990). Temperature changes in biotransformation may be the result of changes in the physicochemical properties of microsomal membranes or changes in hormonal regulation. Levels of P450 and perhaps other enzymes are influenced by circulating levels of steroid hormones that result in sexual differences during the reproductive season (Snowberger-Gray et al., 1991). Diet can have a profound effect on enzyme activities (Van Veld, 1988a). Much of this effect may be due to low levels of both natural inducing agents (or inhibitors) and contaminants present in the diet. Dietary fat can reduce the availability of lipophilic toxicants to biotransformation enzymes (Van Veld et al., 1987). Starvation can reduce enzyme levels and activities (Van Veld et al., 1988a). Age and developmental stage can affect biotransformation (Thomas et al., 1989). For example, basal aryl hydrocarbon hydroxylase activity in fish can be elevated several-fold immediately after hatching (Binder and Stegeman, 1984). In addition, levels and activities of enzymes can be altered in chemically induced tumors (Lorenzana et al., 1989; Van Veld et al., 1991, 1992).

Significance of Biotransformation

Studies of the biochemistry and function of biotransformation enzymes are central to our understanding of the toxic effects of many foreign compounds. The activities of biotransformation enzymes influence clearance rates and therefore the degree to which xenobiotics bioaccumulate in an organism (Lech and Bend, 1980; Spacie and Hamelink, 1982; Barron et al., 1990; also see Chapter 16). Bioaccumulation potential in turn affects the level of exposure required to elicit toxic effects. In addition, by changing the structure of a xenobiotic, the toxicity of that compound can be drastically affected (Lech and Bend, 1980; Armstrong, 1987; Varanasi et al., 1989). These points are addressed in the following discussion.

For bioaccumulation of a foreign compound to occur, the rate of uptake of that compound must exceed the rate of clearance (Spacie and Hamelink, 1982; Barron et al., 1990). Many lipid-soluble compounds would accumulate almost indefinitely if the organism did not have a means of transforming them into water-soluble metabolites capable of being excreted. A good example of the influence of biotransformation on bioaccumulation is the relative bioaccumulation potential of PAHs by fish and by mollusks (Lu et al., 1977). Fish have a very efficient set of phase I and phase II biotransformation enzymes located in the liver and other organs (Buhler and Williams, 1989; Foureman, 1989). These systems can rapidly metabolize PAHs into water-soluble compounds that are readily excreted. The half-life of PAHs in fish is on the order of days (Niimi and Palazzo, 1986). Thus, within a few days after exposure, detectable levels of PAHs are often found only in the bile in the form of water-soluble metabolites (Vetter et al., 1985). PAH levels in muscle and other tissues may be undetectable by this time. As with fish, PAHs and other lipophilic compounds are readily absorbed by mollusks. In contrast to the situation in fish, however, the activities of biotransformation enzymes in mollusks are low (James, 1989a; Livingstone et al., 1989). Clearance rates are therefore low and these compounds tend to accumulate to high levels in mollusks relative to accumulation in fish (Lu et al., 1977; James, 1989a). The importance of biotransformation to the bioaccumulation potential of organic toxicants is further demonstrated in studies of inhibitors of monooxygenase activity (Lu et al., 1977). Treatment of fish with monooxygenase inhibitors resulted in increased bioaccumulation of PAHs because of reduced rates of metabolism and clearance. The same inhibitors have little effect on accumulation in mollusks, in which monooxygenase activity is already low.

Biotransformation affects the toxicity of a foreign molecule by influencing not only the degree to which that molecule accumulates but also its reactivity with cellular macromolecules such as nucleic acids and proteins. Biotransformation is generally regarded as a protective mechanism to reduce the accumulation and toxicity of xenobiotics. However, there are numerous examples in which the product of detoxification enzymes is more toxic than the parent compound (Guengerich and Liebler, 1985). A well-studied example of enzymatic "activation" is biotransformation of the common environmental carcinogen benzo[*a*]pyrene (Gelboin, 1980; Varanasi et al., 1987, 1989; see Chapter 14). Benzo[*a*]pyrene is not a carcinogen itself but requires enzymatic activation. Metabolism of benzo[*a*]pyrene by the cytochrome P450 monooxygenase system produces a variety of hydroxylated and epoxide forms. Many of these are further metabolized to harmless water-soluble metabolites. Epoxides, for example, are good substrates for glutathione *S*-transferases and the resulting conjugates are considered detoxification products. However, epoxides are also substrates for epoxide hydrolase. Activation of benzo[*a*]pyrene into its carcinogenic form involves two sequential epoxidations at the 7,8 and 9,10 positions, with an intermediate hydrolysis of the 7,8-epoxide by epoxide hydrolase. This sequence of reactions yields a benzo[*a*]pyrene diol epoxide, a carcinogenic form of this PAH. This product can form covalent bonds with the guanosine moiety of DNA, resulting in DNA point mutations that can lead to such events as oncogene activation (Marshall et al., 1984; Barbacid, 1987; McMahon, 1990, see Chapter 14). These multiple pathways of detoxification/activation of benzo[*a*]pyrene are summarized in Figure 9. Benzo[*a*]pyrene is an example of a naturally occurring environmental compound that can become activated by an enzyme system that evolved for detoxification. There are also examples of synthetic compounds that must be enzymatically activated in order to exert their toxic effects. Organophosphate insecticides are examples of compounds that are activated by the monooxygenase system (Murty, 1986). Coadministration of monooxygenase inhibitors has been shown to decrease the toxicity of organophosphates by reducing activation by the monooxygenase system (Ludke et al., 1972).

Use of Biotransformation Enzymes in Environmental Monitoring

Because of the inducibility of some of the biotransformation enzymes, there has been considerable effort to determine the feasibility of using these enzymes in environmental monitoring (i.e., as biomarkers; see Chapter 1) of contaminant exposure and effects. Perhaps no other enzyme has received as much attention in this regard as CYP1A, the major hydrocarbon-inducible form of P450 in fish and other vertebrates. Here, important aspects of the use of CYP1A and other biotransformation enzymes as tools in environmental monitoring are highlighted. The reader is referred to reviews for more detailed treatments of this subject (Payne et al., 1987; Stegeman et al., 1992).

Evidence for induction of CYP1A has been obtained by three primary methods: catalytic assays (Dehnen et al., 1973; Burke and Mayer, 1974; Klotz et al., 1984), immunodetection of proteins (Williams and Buhler, 1984; Gøksoyr, 1985; Stegeman, 1989), and detection of messenger RNA synthesis using complementary DNA (cDNA) (Heilman et al., 1988; Haasch et al., 1989; Kloepper-Sams and Stegeman, 1989).

Catalytic assays are measurements of CYP1A activity (i.e., rate of transformation of a substrate). CYP1A is highly active toward PAH-type substrates, and the earliest method for detecting the activity of this enzyme involved measurement of the hydroxylation of benzo[*a*]pyrene. The assay is known as benzo[*a*]pyrene hydroxylase or more commonly as aryl hydrocarbon hydroxylase (AHH). Ethoxyresorufin *O*-deethylase (EROD) is another activity catalyzed by CYP1A. Although ethoxyresorufin, the substrate in this reaction, is not an environmental toxicant, this assay is highly specific for CYP1A. AHH and EROD activities occur at low, often undetectable levels in untreated organisms but are highly induced by treatment with hydrocarbon compounds. Other monooxygenase activities are not increased by this treatment, supporting selective induction of AHH and EROD.

Purification of CYP1A by several laboratories (Klotz et al., 1983; Williams and Buhler, 1984; Gøksoyr, 1985) has allowed production of polyclonal and monoclonal antibodies that can be used to measure directly levels of this enzyme using Western blot and enzyme-linked immunosorbent assay (ELISA) (Stegeman, 1989; Gøksoyr, 1991). Direct measurement of the enzyme itself by these immunochemical procedures offers advantages over catalytic assays in cases in which degradation of enzyme has occurred because of sample handling, storage, protease enzyme digestion, and so forth. Antibodies detect degraded as well as active enzyme, whereas catalytic measurements require that the enzyme is active. In addition, some inducers bind to the active site of CYP1A, "masking" catalytic activity (Gooch et al., 1989). In these cases, immunodetection would be a more sensitive and accurate indicator of induction than measurement of catalytic activity.

cDNA is a single-stranded DNA that is formed from an mRNA template by the enzyme reverse transcriptase. Among other applications, cDNA probes can be used to detect newly sysnthesized mRNA including that which codes for CYP1A in fish. Heilmann et al. (1988) cloned and sequenced a cDNA for fish PAH-inducible P450 from rainbow trout (*Oncorhynchus mykiss*) liver. This confirmed that the PAH-inducible P450 from teleosts is a member of the cytochrome CYP1A gene family. The cDNA hybridizes to rainbow trout genomic DNA and to mRNA from induced trout liver. cDNA probes have also been effective with other fish (Haasch et al., 1989; Kloepper-Sams and Stegeman, 1989).

The first suggestion of the utility of CYP1A in environmental monitoring came in the 1970s after the discovery of elevated AHH activities in livers of fish collected in a body of water with a history of hydrocarbon contamination (Payne and Penrose, 1975). Later, a similar AHH pattern was observed in liver and gill of fish collected near the site of a large oil refinery and a small boat marina (Payne, 1976).

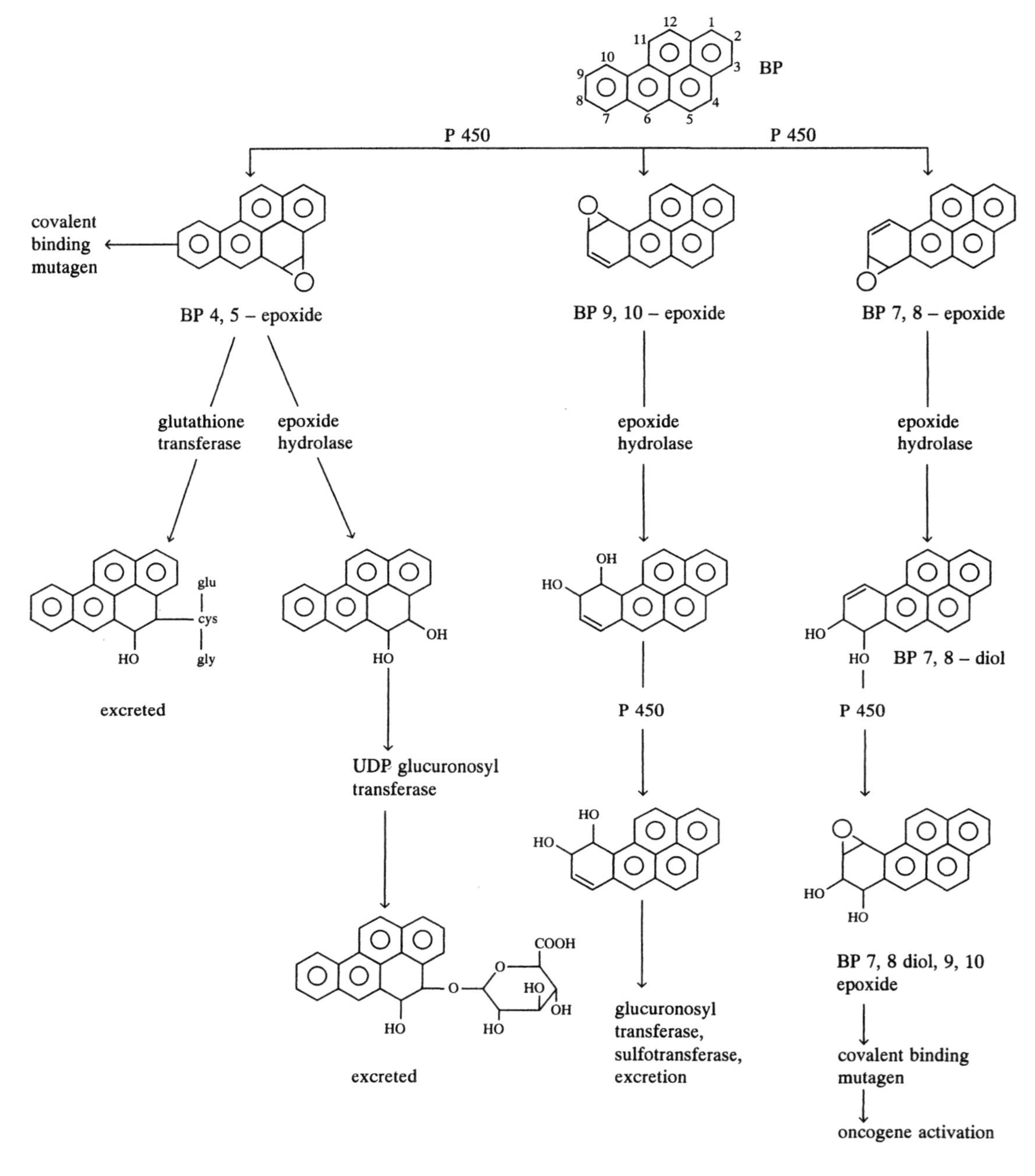

Figure 9. Pathways of activation and detoxification of benzo[*a*]pyrene.

Since then, numerous studies have provided evidence for induction of CYP1A in fish collected from various PAH- and PCB-contaminated sites (Elskus and Stegeman, 1989; Van Veld et al., 1990) as well as from waters receiving pulp mill effluents (Andersson et al., 1988; Mather-Mihaich and Di Giulio, 1991), oil spills, and domestic sewer outfalls (Spies et al., 1982) (Table 5). Although most studies of this type involve liver, sensitive induction has also been observed in extrahepatic tissues such as intestine (Van Veld, 1990) and gill (Payne, 1976) of fish collected from contaminated sites. Immunochemical analysis of extrahepatic tissues has been useful in identifying specific cell types involved in the induction response (Lorenzana et al., 1988; Smolowitz et al., 1992; Stegeman et al., 1991).

Phase II enzymes undoubtedly play important roles in biotransformation and clearance of foreign compounds. However, our present state of knowledge indicates that the response of these enzymes in

Table 5. Evidence for environmental induction of CYP1A and associated EROD and AHH activities in feral fish

Species	Tissue	Assay	Fold induction	Suspected contaminant	Reference
Brown trout (*Salmo trutta*)	Liver	AHH	15×	Petroleum hydrocarbons	Payne and Penrose, 1975
Cunner (*Tautogolabrus adspersus*)	Liver	AHH	3×	Petroleum hydrocarbons	Payne, 1976
	Gill	AHH	<27×	Petroleum hydrocarbons	Payne, 1976
Mummichog (*Fundulus heteroclitus*)	Liver	EROD, immunodetection	2–4×	PAH, PCB	Elskus and Stegeman, 1989
Spot (*Leiostomus xanthurus*)	Liver	EROD, immunodetection	8–23×	PAH, creosote	Van Veld et al., 1990
	Intestine	EROD, immunodetection	33–100×	PAH, creosote	Van Veld et al., 1990
Rattail (*Coryphaenoides armatus*)	Liver	EROD, AHH	10×	PCB	Stegeman et al., 1986
Flatfish (*Citharicthys sordidus*)	Liver	AHH	10×	Sewage outfall, PAH, PCB	Spies et al., 1982
Perch (*Perca fluviatilis*)	Liver	EROD	8×	Pulp mill effluent	Andersson et al., 1988

aquatic organisms to environmental toxicants is not sufficiently sensitive to warrant their use in environmental monitoring. Some field studies indicate that activities of GST and UDPGT are elevated up to threefold in fish collected from PAH- and halogenated hydrocarbon–contaminated sites (Andersson et al., 1988; Collier and Varanasi, 1991; Van Veld et al., 1992). Other studies indicate no significant differences in phase II activities between fish collected from control and contaminated sites (Lindstrom-Seppa and Oikari, 1988). Similarly, there is little evidence for induction of either monooxygenase activity or phase II enzymes activities by environmental exposures of invertebrates to foreign compounds.

Based on our current knowledge of the response of biotransformation enzymes to environmental contaminants, it appears that fish CYP1A is the most useful of these enzymes in environmental monitoring programs. This enzyme responds rapidly to environmental levels of foreign compounds in a manner that often allows identification of pollution gradients (Stegeman et al., 1992). Because of the role of this enzyme in detoxification and activation of foreign compounds including carcinogens, there is interest in establishing relationships between the activity of this enzyme and higher-order effects such as cancer in wild fish populations (Baumann, 1989; Vogelbein et al., 1990; see Chapter 14). In addition, although CYP1A is not one of the major steroid hydroxylating forms, there is interest in exploring the possible relationship between enzyme inductions, altered steroid hormone levels, and reproductive success (Forlin and Haux, 1985).

ACTIVATED OXYGEN METABOLISM AND OXIDATIVE STRESS

Oxyradicals

Molecular oxygen (O_2) is generally thought of as a benign compound; it is essential for aerobic organisms, which include all higher life forms. Its dominant role in eukaryotes is that of terminal electron acceptor in mitochondrial respiration, where it is ultimately reduced to water during the complex process of oxidative phosphorylation, the major source of ATP in aerobes. It is important to note, however, that the reduction of O_2 to water requires four electrons, and this reduction proceeds sequentially through the one-, two-, and three-electron products. These univalent reductions of O_2 to water are illustrated in Eq. (2) to (6):

$$O_2 + e^- \longrightarrow O_2^{\cdot -} \quad (2)$$

$$O_2^{\cdot -} + e^- \xrightarrow{2H+} H_2O_2 \quad (3)$$

$$H_2O_2 + e^- \xrightarrow{H+} \cdot OH + H_2O \quad (4)$$

$$\cdot OH + e^- \xrightarrow{H+} H_2O \quad (5)$$

$$\text{Sum} \quad O_2 + 4e^- \qquad 2H_2O \quad (6)$$

The one, two, and three reduction products displayed in Eq. (2) to (4) are the superoxide radical anion ($O_2^{\cdot -}$), hydrogen peroxide (H_2O_2), and the hydroxyl radical ($\cdot OH$). These activated species of oxygen, particularly $\cdot OH$, are very reactive and potentially deleterious to biological systems. Both $O_2^{\cdot -}$ and $\cdot OH$ are oxygen-based free radicals (*oxyradicals*). Although not a free radical (i.e., one possessing an unshared electron), H_2O_2 is also reactive and, via the Haber-Weiss reaction with $O_2^{\cdot -}$, it serves as an important precursor to $\cdot OH$ [Eq. (7)] (Haber and Weiss, 1934). Therefore, it is often treated as an oxyradical.

$$O_2^{\cdot -} + H_2O_2 \longrightarrow \cdot OH + OH^- + O_2 \quad (7)$$

While thermodynamically favorable, reaction (7) is kinetically quite slow. However, catalysis by transition metals such as iron and copper facilitates significant $\cdot OH$ production. In biological systems, metal (e.g., chelated iron)–catalyzed Haber-Weiss reactions are considered an important source of $\cdot OH$ [Eqs. (8) and (9)]. In this reaction $O_2^{\cdot -}$ serves as the reductant for the transition-metal oxidation-reduction catalyst [e.g., chelated iron, Eq. (8)]. The reduced metal then reacts with H_2O_2 to yield $\cdot OH$ [Eq. (9)].

$$O_2^{\cdot -} + Fe^{3+}\text{-chelate} \longrightarrow O_2 + Fe^{2+}\text{-chelate} \quad (8)$$

$$Fe^{2+}\text{-chelate} + H_2O_2 \longrightarrow \cdot OH + OH^- + Fe^{3+}\text{-chelate} \quad (9)$$

The net sum of these two reactions is Eq. (7), the Haber-Weiss reaction. Equation (9) is referred to as the Fenton reaction (Walling, 1975). Other noteworthy species of activated oxygen, treated here as oxyradicals, include singlet oxygen ($^1O_2$), alkoxy radicals ($RO\cdot$), and peroxy radicals ($ROO\cdot$).

Sources of Oxyradicals

In addition to mitochondrial electron transport, several other sources of endogenous cellular oxyradical production have been identified. These include the electron transport chains of microsomes (Winston and Cederbaum, 1983) and chloroplasts (Asada et al., 1974), the respiratory burst associated with active phagocytosis by leukocytes (Babior, 1984; see Chapter 12), and the activities of a number of enzymes such as xanthine oxidase, tryptophan dioxygenase, diamine oxidase, and prostaglandin synthase (Fridovich, 1978; Halliwell, 1978). Although oxyradicals are generally considered in light of their deleterious effects, the examples of prostaglandin synthesis and phagocytosis indicate they occasionally play important beneficial roles (i.e., at least for the host!).

In toxicological studies, however, oxyradicals are examined chiefly in relation to chemicals that enhance their production and the resulting damaging effects. Classes of compounds particularly noted for their ability to enhance the flux of oxyradicals include quinones and diols, bipyridyls, aromatic nitro compounds, aromatic hydroxylamines, aromatic azo dyes, and transition metal chelates (Table 6). A key

Table 6. Organic compounds that generate oxygen free radicals

Compound class	Examples
Quinones	Dichlone
	Menadione
	Benzo[*a*]pyrene quinones
	Adriamycin
Bipyridilium herbicides	Paraquat
	Diquat
Aromatic nitro compounds	Nitrofurantoin
	Nitrobenzoic acid
	Dinitrobenzene
	Nitropyrene
Aromatic hydroxylamines	Oxidation products of aromatic amines (e.g., *N*-OH-2-aminofluorene)
Azo dyes	Arsenazo III
	Sulfanazo III
Metal chelates	Fe-EDTA
	Cu-bleomycin

mechanism by which many of these compounds enhance cellular production of oxyradicals is referred to as redox cycling (Kappus, 1987). In this cycle, the compound is first reduced to its corresponding free radical, with reducing equivalents typically provided by NADH or NADPH. This reaction is generally catalyzed by NAD(P)H reductases such as cytochrome P450 reductase, xanthine oxidase, ferredoxin reductase, and NADH-ubiquinone oxidoreductase (Kappus and Sies, 1981; Cohen and d'Arcy Doherty, 1987). To complete the cycle, the unshared electron of the free radical is transferred to O_2, giving rise to $O_2^{\cdot -}$ and the parent compound; the latter can undergo continued "futile" cycling. Therefore, redox cycling is characterized by concomitant oxidations of NAD(P)H and catalytic yields of oxyradicals; both events can underlie the toxicities of redox-active compounds. The redox cycle, together with antioxidant defenses and toxic consequences (described in the following), is illustrated in Figure 10.

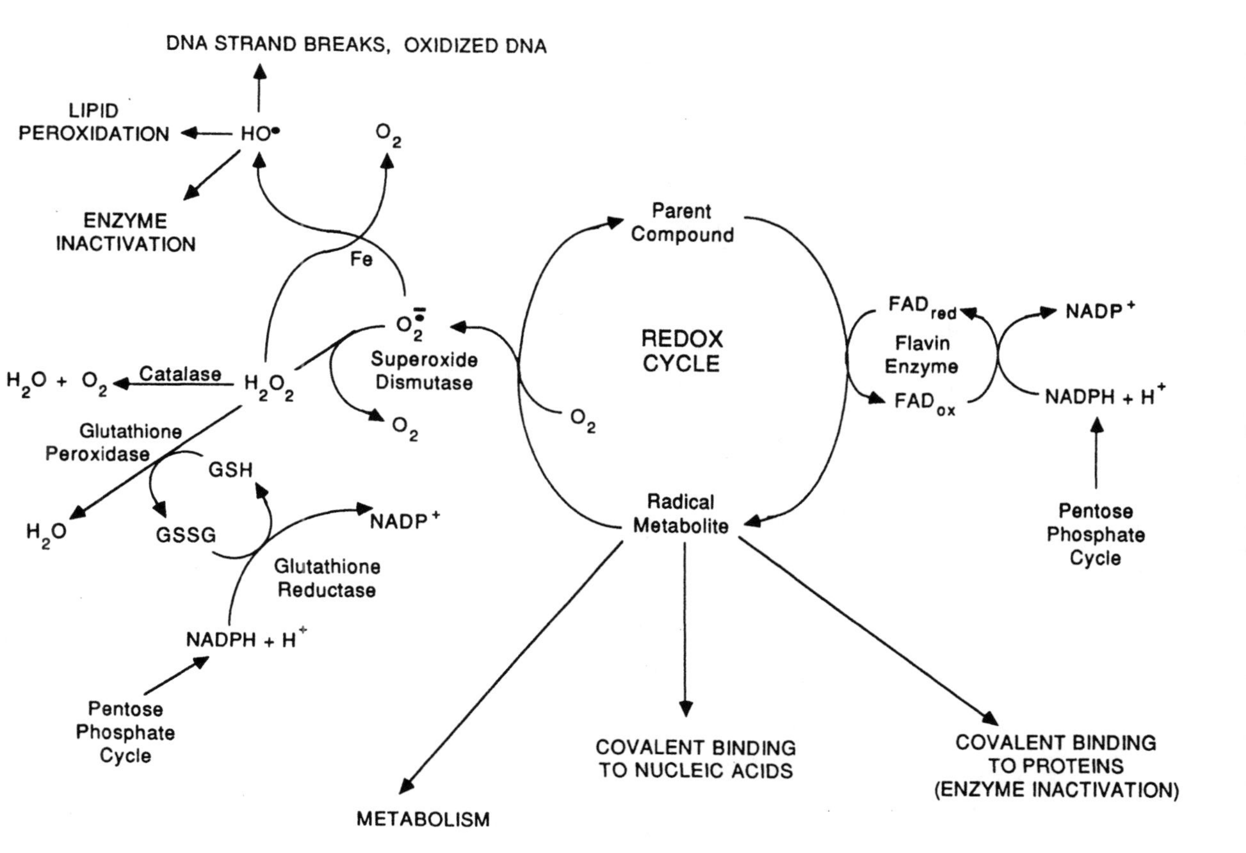

Figure 10. Overview of redox cycling, summarizing free radical production, antioxidant defenses, and toxicological consequences.

Antioxidant Defense Mechanisms

In order to counter the potentially deleterious effects of activated oxygen, aerobic organisms have evolved an extensive array of antioxidant defense mechanisms to detoxify oxyradicals. The development of these diverse antioxidant mechanisms can be viewed as the "cost" for the high energetic efficiency afforded by O_2 as a terminal electron acceptor. These defenses are in general highly modulated; that is, levels or activities of antioxidants are highly responsive to changes in the environmental milieu that affects the cellular production of activated oxygen species. Examples of modulating factors include fluctuations in ambient O_2 levels (e.g., dissolved oxygen), changes in solar ultraviolet (UV) irradiation (particularly important in chloroplasts), and exposures to redox-active chemicals. In short, antioxidant defenses provide a fundamental adaptation to both natural and anthropogenic stressors that influence the flux of oxyradicals in aerobic organisms.

Antioxidants can be classified as water-soluble reductants such as glutathione, uric acid, and ascorbate (vitamin C); lipid-soluble radical scavengers such as α-tocopherol (vitamin E), β-carotene (vitamin A), and various xanthophylls; and enzymes including superoxide dismutase (SOD), catalase (CAT), peroxidases (particularly glutathione and ascorbate peroxidase; GPx and AsPx), glutathione reductase (GR), and DT-diaphorase (DTD). Reduced glutathione (GSH) plays prominent roles in contaminant detoxifications (Dolphin et al., 1989; Reed, 1990). In addition to its role in phase II conjugations described earlier, GSH can scavenge radicals directly and provide reducing equivalents for the reduction of peroxides by GPx. Similarly, ascorbate in chloroplasts can scavenge oxyradicals such as ·OH, as well as serve as a cofactor for AsPx (Halliwell and Gutteridge, 1989). GSH and ascorbate are paradoxical in that they possess prooxidant as well as antioxidant activity; their autooxidations, particularly in the presence of transition metals, give rise to oxyradicals (Cohen et al., 1981). However, ascorbate can provide protection synergistically with α-tocopherol, the predominant membrane-associated free radical scavenger (Leung et al., 1981). β-Carotene primarily affords protection from $^1O_2$ (Kappus and Sies, 1981).

SOD is a group of metalloenzymes that disproportionates $O_2^{\cdot -}$ to H_2O_2 [Eq. (10)] (Fridovich, 1986):

$$2O_2^{\cdot -} + 2H^+ \xrightarrow{SOD} H_2O_2 + O_2 \qquad (10)$$

SOD is considered to play a pivotal antioxidant role; it occurs in all aerobic organisms examined and catalyzes the dismutation of $O_2^{\cdot -}$ at rates approximating diffusion limits, making it among the most active enzymes described. Specific isozymes are typically found in cytosol, mitochondria, and chloroplasts. Numerous studies have indicated induction of SOD in many organisms by factors associated with increased oxyradical production, such as elevated O_2 and exposure to redox-active contaminants.

The product of SOD activity, H_2O_2, can be removed by the activities of CAT [Eq. (11)] or peroxidases such as GPx in animals [Eq. (12)] or AsPx in plants (Frew and Jones, 1984; Hossain et al., 1984).

$$2H_2O_2 \xrightarrow{CAT} 2H_2O + O_2 \qquad (11)$$

$$H_2O_2 + 2GSH \xrightarrow{GPx} 2H_2O + GSSG \qquad (12)$$

CAT is associated primarily with peroxisomes, where it detoxifies H_2O_2 arising as a by-product of fatty acid oxidation (Fahimi and Sies, 1987). GPx is a cytosolic enzyme; in addition to reducing H_2O_2 to H_2O, it can reduce lipid peroxides (ROOH) to their corresponding alcohols (ROH) (Reed, 1990), an important reaction for quenching lipid-peroxidizing chain reactions (described in the following). In chloroplasts, AsPx similarly detoxifies H_2O_2, wherein ascorbate serves as the reductant (Alscher and Amthor, 1988). In plants, GSH plays a role by serving as reductant for the regeneration of ascorbate; GSH is concomitantly oxidized to GSSG (formed from two molecules of glutathione joined by a disulfide bond). In animals and plants, the enzyme glutathione reductase plays an important antioxidant role by catalyzing the reduction of GSSG to GSH at the expense of NADPH [Eq. (13)] (Reed, 1990).

$$GSSG + NADPH + H^+ \xrightarrow{GR} 2GSH + NADP^+ \qquad (13)$$

DT diaphorase [NAD(P)H: (quinone acceptor) oxidoreductase, DTD] catalyzes the two-electron reduction of quinones to the corresponding hydroquinone (Lind et al., 1982). This activity is generally considered to provide antioxidant protection by diverting quinone metabolism away from one-electron reduction to semiquinone radicals that can give rise to $O_2^{\cdot -}$ via autoxidation (Figure 11). However, in some instances hydroquinones can readily autoxidize to yield $O_2^{\cdot -}$; in such cases DTD may actually enhance $O_2^{\cdot -}$ generation by the parent quinone (Hasspieler and Di Giulio, 1992).

Oxidative Damage

Free radicals, including oxyradicals, can react with a large variety of biomolecules and are often nonspecific with respect to biochemical targets. This is particularly true of highly reactive radicals such as ·OH. Fundamental lesions associated with oxyradicals include oxidations of membrane lipids, proteins, and nucleic acids and altered cellular redox status. These perturbations are believed to underlie specific tissue injuries associated with redox-active contaminants and, more broadly, may be associated with aspects of chemical carcinogenesis and aging (Ames, 1989). However, our understanding of the progression of events from a primary biochemical effect (e.g., oxidative DNA damage) to an ultimate toxic expression (e.g., cancer) is in most cases incomplete. Summarized next are several examples of primary biochem-

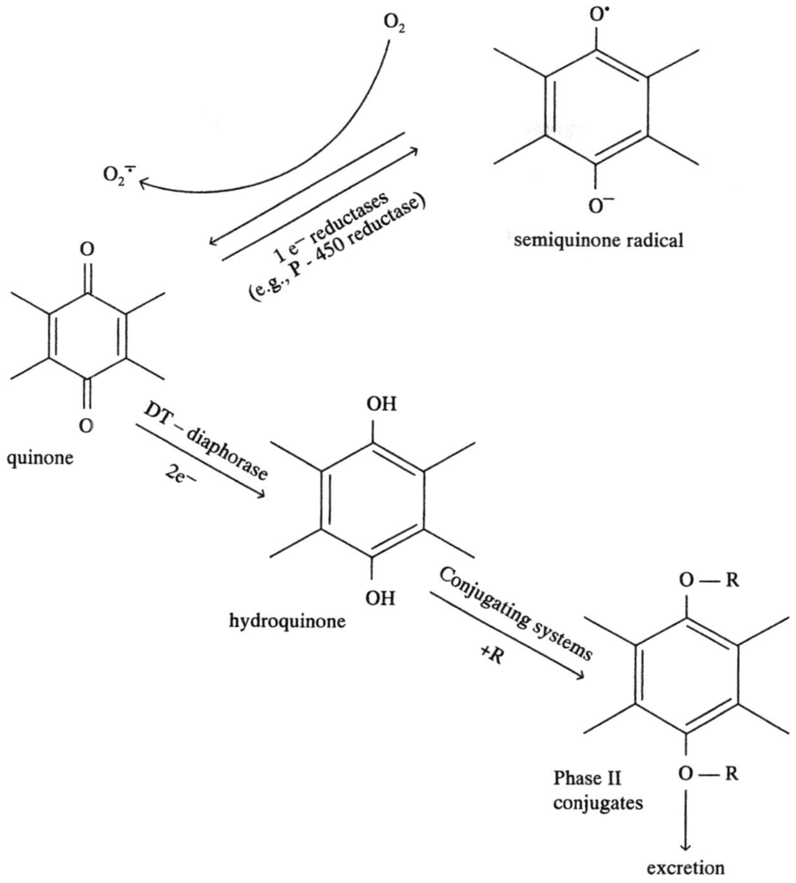

Figure 11. Quinone metabolism, including DT diaphorase catalysis of two-electron reductions of quinones to hydroquinones. This reaction can limit one-electron reductions that result in free radicals (i.e., the semiquinone and superoxide radicals). Also, DT-diaphorase–mediated reductions promote phase II conjugations and hence, excretion. R is likely to be glucuronic acid or sulfate.

ical and physiological events associated with fluxes of oxyradicals.

Lipid Peroxidation

Lipid peroxidation, the oxidation of polyunsaturated fatty acids (PUFAs), is an important consequence of oxidative stress and has been investigated extensively (Chan, 1987; Horton and Fairhurst, 1987). Lipid peroxidation (Figure 12) proceeds by a chain reaction and, as in the case of redox cycling, demonstrates the ability of a single radical species to propagate a number of deleterious biochemical reactions. Lipid peroxidation is initiated by the abstraction of a hydrogen atom from a methylene group (—CH_2—) of a PUFA (or "LH"); oxyradicals, particularly ·OH, can readily perform this abstraction, yielding the lipid radical, L·. This carbon-based radical (L·) tends to be stabilized by molecular rearrangement to a conjugated diene radical. The L· radical readily reacts with O_2 to produce the peroxy radical, LOO·. This radical can readily abstract a hydrogen from another LH, yielding a lipid hydroperoxide, LOOH, and a new L·, which can then continue the chain reaction, propagating additional LOOH and L·. LOOH is relatively stable in isolation but can react with transition metal complexes (including cytochrome P450) to yield alkoxyl radicals (LO·). The lipid peroxidation chain reaction can be terminated by two lipid radicals reacting to form a nonradical product or by quenching by a radical scavenger such as vitamin E.

Methemoglobinemia

In vertebrate erythrocytes, hemoglobin can also be a sensitive target for oxyradical attack. Under normal conditions, a small proportion (less than 1 or 2%) of hemoglobin exists as methemoglobin (MetHb). In MetHb, the iron centers of the heme moieties exist in the oxidized (Fe^{3+}) state; this prevents the molecules from serving their normal function of O_2 bind-

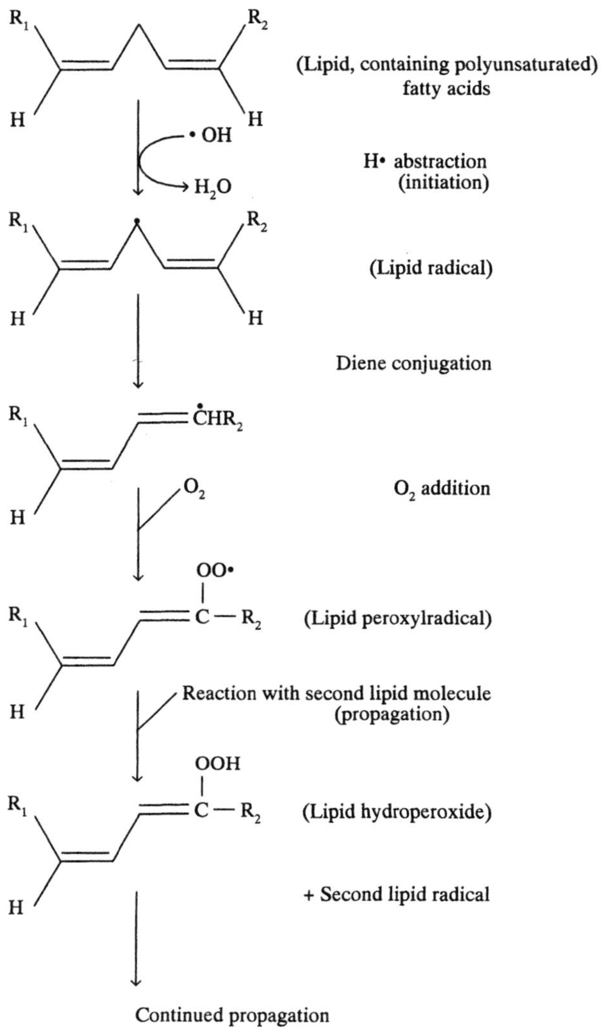

(Lipid, containing polyunsaturated) fatty acids

$\cdot OH$ → H_2O

H• abstraction (initiation)

(Lipid radical)

Diene conjugation

$\dot{C}HR_2$

O_2 addition

OO•

(Lipid peroxylradical)

Reaction with second lipid molecule (propagation)

OOH

(Lipid hydroperoxide)

+ Second lipid radical

Continued propagation

Figure 12. Hydroxyl radical (·OH)–mediated lipid peroxidation, a free radical chain reaction.

ing and transport (Stern, 1985). The enzyme methemoglobin reductase catalyzes the reduction of Fe^{3+} to Fe^{2+}, thereby regenerating functional oxyhemoglobin. Several compounds can enhance MetHb formation, including a number of aromatic hydrazines, quinones, nitrite, and some transition metals, particularly copper. The mechanism underlying MetHb formation by these compounds has been shown to proceed by the production of oxyradicals that facilitate the oxidation of heme-bound Fe^{2+}. Increased concentrations of methemoglobin may serve as a sensitive indicator of oxidative damage in red blood cells. In fish aquaculture, excess buildup of nitrite due to incomplete ammonia oxidation is a common problem that results in methemoglobinemia, or "brown blood" disease, so named due to the dark color of MetHb (Bowser et al., 1983).

DNA

The role that free radical intermediates, including oxyradicals, play in chemical carcinogenesis is a topic of intense research interest. The microsomal metabolism of some hydrocarbon procarcinogens (such as benzene and benzo[*a*]pyrene) produces organic free radical intermediates. These may play important roles in DNA alterations, including adduct formation, associated with these compounds (Irons and Sawahata, 1985; Cavalieri and Rogan, 1990; see Chapter 13). Furthermore, oxyradicals have been implicated in the metabolism of PAH procarcinogens into adduct-forming products (Georgillis et al., 1987). In addition to the roles played by oxyradicals in procarcinogen activation, these compounds can be genotoxic and directly damage DNA through the pro-

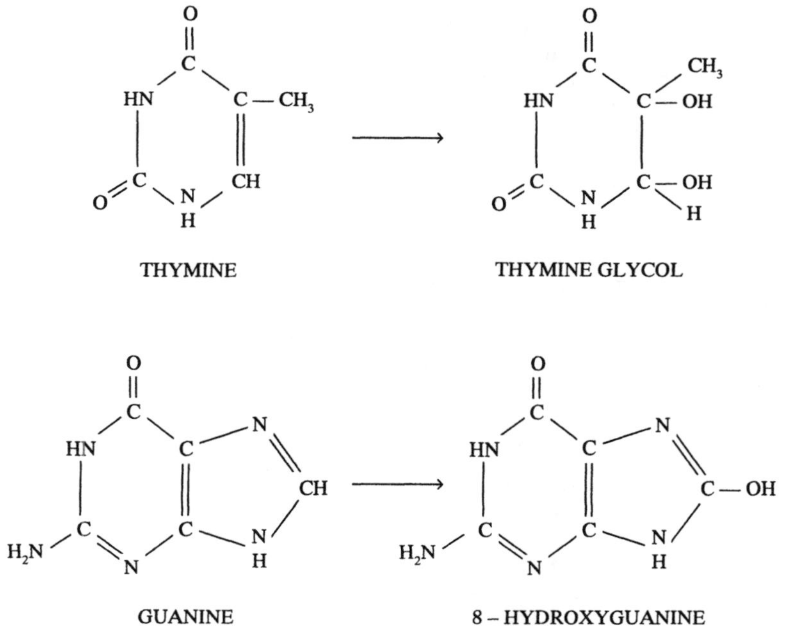

THYMINE → THYMINE GLYCOL

GUANINE → 8-HYDROXYGUANINE

Figure 13. Examples of DNA base oxidations resulting from hydroxyl radical (·OH) attack.

duction of strand scissions and chromosomal breakage (Imlay and Linn, 1988). These effects are not specific for free radicals and are discussed more completely in Chapter 13. A more specific genotoxic effect associated with oxyradicals (particularly ·OH) is the oxidation of DNA resulting in oxidized bases such as thymine glycols and 8-hydroxyguanine (Dizdaroglu and Bergtold, 1986) (Figure 13). Oxyradicals may also play important roles in tumor promotion (Cerutti, 1985).

Redox Status

Many of the effects just described—oxidations of lipids, proteins (i.e., hemoglobin), and DNA—represent direct consequences of oxyradical attack on biomolecules. Another fundamental biochemical response that can both modulate these oxidations and also broadly impair cellular metabolism is perturbed redox status. Normal cells typically regulate and maintain a critical pool of reducing equivalents, particularly as NADH and NADPH, that drive energy-requiring metabolic processes. GSH, described earlier, is another key reductant that serves various functions in protein synthesis and transport, in addition to its detoxification roles via phase II metabolism and oxyradical scavenging. Healthy cells typically maintain very high ratios (>10:1) of reduced to oxidized glutathione (GSH/GSSG) (Kosower and Kosower, 1978).

Free radical–generating compounds, however, can impose a drain on intracellular reducing equivalents with potentially profound consequences for a variety of metabolic processes (Stubberfield and Cohen, 1989). This loss of reducing equivalents can arise directly, for example, by electron abstractions from pyridine nucleotides by redox-active compounds; this phenomenon is facilitated by various cytochrome oxidases. The oxidation of GSH due to direct scavenging of oxyradicals or peroxide metabolism by GSH-dependent peroxidase activities represents another drain; NADPH is oxidized to regenerate GSH from GSSG by the activity of GR. More indirectly, oxidative stress can reduce the cell's reductant pool as a consequence of the energetic costs of mounting a defense against an increased flux of oxyradicals (e.g., biosynthesis of antioxidants).

Studies in Aquatic Organisms

Many fundamental aspects of oxyradical production, antioxidant defenses, and biochemical manifestations of oxidative injury are shared among biological systems. Studies of these phenomena in aquatic organisms are relatively recent, however. These studies have illustrated particular nuances regarding mechanisms of free radical generation and biochemical responses in these organisms, as well as indicating potential roles for oxidant-mediated responses as indices of environmental quality.

Oxyradical Production

Most studies of oxyradical generation in aquatic organisms have employed in vitro investigations of microsomal electron transport. For example, ·OH production by hepatopancreatic (in invertebrates) or hepatic (in fish) microsomal preparations has been investigated in red swamp crayfish (*Procambarus clarkii*) (Jewell and Winston, 1989), the common murine mussel (*Mytilus edulis*) (Winston et al., 1990), and three salmonid species of the genus *Oncorhynchus* (Kennish et al., 1989). These studies employed ·OH scavengers such as 2-keto-4-thiomethylbutyric acid (KMBA) that produce readily measured products (e.g., ethylene) upon oxidation by ·OH. These studies indicated that, in contrast to analogous mammalian systems, ·OH generation by these aquatic animals was greater when NADH was employed as a cofactor rather than NADPH. The reasons for and implications of this cofactor difference remain to be elucidated.

In addition to microsomal electron transport, active phagocytosis has been observed to be a source of oxyradicals in fish, as demonstrated by Secombes et al. (1988) for macrophages of rainbow trout that were stimulated by phorbol myristate acetate (PMA). In the respiratory burst, membrane stimulation of phagocytes (e.g., by PMA) activates NADPH oxidase; this enzyme (which is independent of the mitochondrial cytochrome system) employs the oxidation of NADPH to $NADP^+$ to drive the reduction of O_2 to $O_2^{\cdot -}$. The $O_2^{\cdot -}$ thus produced serves as the basis for various microbiocidal oxyradicals. The activation of NADPH oxidase is linked to activation of the hexose monophosphate (HMP) shunt, in which the conversion of glucose-6-phosphate to 6-phosphogluconate drives the generation of NADPH from $NADP^+$. The mechanism underlying the respiratory burst in fish phagocytes appears generally similar to that described for analogous mammalian white blood cells.

Xenobiotic-mediated increases in oxyradical production have also been investigated with in vitro studies in aquatic organisms. For example, several nitroaromatics were shown to generate $O_2^{\cdot -}$ in hepatic microsomes isolated from rainbow trout, channel catfish (*Ictalurus punctatus*), and largemouth bass (*Micropterus salmoides*) (Washburn and Di Giulio, 1989). Mechanisms underlying nitroaromatic-stimulated $O_2^{\cdot -}$ generation were found to be similar to those observed in mammals for both microsomes and cytosol of channel catfish (Washburn and Di Giulio, 1988). In similar studies with bivalves, the herbicide paraquat was found to enhance $O_2^{\cdot -}$ production in microsomes isolated from hepatopancreas of wedge clams (*Rangia cuneata*) and ribbed mussels (*Guekensia demissa*) (Wenning and Di Giulio, 1988). In contrast to observations with mammals and fish, these invertebrate studies demonstrated greater enhancements with NADH than with NADPH.

Antioxidant Responses

Numerous studies have investigated the activities of antioxidant enzymes and concentrations of oxyradical scavengers such as GSH and vitamins A, C, and E in aquatic animals. These studies have included observations suggesting adaptive responses of antioxidant system components in aquatic organisms exposed in vivo to conditions or xenobiotics likely to enhance oxyradical fluxes. For example, Dykens and Shick (1982) observed far greater activities of SOD in sea anemones (*Anthopleura elegantissima*) harboring symbiotic dinoflagellates than in anemones without symbionts. This difference was attributed to hyperoxic intracellular conditions produced by the photosynthetic activity of the dinoflagellates. In studies with bivalves, Wenning et al. (1988) observed increased activities of SOD and catalase in paraquat-exposed ribbed mussels, and Livingstone et al. (1990) observed increased catalase and DTD activities in blue mussels exposed to menadione, a naphthoquinone. Studies with fish, however, have generally failed to reveal consistent antioxidant responses.

Cellular Injury

Deleterious cellular effects associated with oxidative stress, such as lipid peroxidation, methemoglobinemia, and DNA oxidations, have also been investigated in aquatic animals. In this regard, lipid peroxidation has received the greatest attention. A wide variety of contaminants have been shown to be effective at enhancing lipid peroxidation in tissues of various organisms. These compounds include metals such as cadmium, copper, iron, and mercury (Viarengo et al., 1988; Wofford and Thomas, 1988; Bano and Hasan, 1989), redox-active organics such as paraquat and acetaminophen (Wenning et al., 1988; Wofford and Thomas, 1988), PCBs (Wofford and Thomas, 1988), and other organics such as carbon tetrachloride (CCl_4) and diethylmaleate (DEM) (Wofford and Thomas, 1988). Elevated lipid peroxidation was also observed in livers of channel catfish exposed to sediments enriched with PAHs, PCBs, and metals (Di Giulio et al., 1993). As discussed earlier, the mechanisms underlying lipid peroxidation are complex and, furthermore, variable among compounds. For example, the preceding list of chemicals includes ones likely to enhance lipid peroxidation via the direct generation of oxyradicals (e.g., redox-active organics and transition metals such as copper and iron). Others (cadmium, mercury, DEM) may act more indirectly by binding to GSH and thereby reducing its availability to combat lipid peroxidation. Enhanced microsomal electron transport may account in part for the responses observed with cytochrome P450 inducers such as PCBs and PAHs. From a comparative standpoint, there is evidence that aquatic animals may be more resistant to lipid peroxidation than mammals. Singh et al. (1992) found that lipids from rat microsomes incubated with an ·OH-generating system were about 30 times more extensively peroxidized than similarly incubated rainbow trout microsomes. Due to their poikilothermic nature, fish lipids are more highly unsaturated than those from homeotherms, which would seem to predispose them to lipid peroxidation. However, Singh et al. noted a 43-fold greater concentration of vitamin E, a principal lipid-soluble antioxidant, in trout microsomes versus rat microsomes. This large difference in vitamin E concentrations may have accounted for the greater resistance of trout microsomes to lipid peroxidation.

Methemoglobinemia is an important deleterious response of red blood cells to oxidative stress and, as described earlier, has been noted routinely as a consequence of nitrite accumulation in aquaculture systems (Bowser et al., 1983). However, relatively little attention has been paid to this response in fish as a consequence of exposures to other toxicants. Laboratory exposures of channel catfish to *n*-butyl mercaptan and to two naphthoquinones (menadione and the fungicide dichlone) resulted in dose-dependent increases in MetHb (Mather-Mihaich and Di Giulio, 1986; Andaya and Di Giulio, 1987). In addition, elevated concentrations of MetHb were observed in erythrocytes from perch (*Perca fluviatilis*) inhabiting portions of the Baltic Sea in Sweden receiving effluents from a bleached kraft mill (Andersson et al., 1988). The chemical factors underlying this response are unknown, although bleached kraft effluent contains a variety of phenolic compounds that may redox cycle in animal tissues.

Similarly, few studies have addressed oxidative DNA damage in aquatic organisms. Elevated concentrations of oxidized bases, including 8-hydroxydeoxyguanosine (8-OHdG), have been reported in livers of English sole (*Parophrys vetulus*) from contaminated regions of the Puget Sound, Washington (Malins et al., 1990; Malins and Haimanot, 1991). These lesions, believed to arise from ·OH attack (on the C-8 position of guanine in the case of 8-OHdG), were observed in both hepatic tumors and histologically normal liver tissue. Elevated concentrations of the oxidized base were not noted in livers of fish from a reference (noncontaminated) site. Further studies of these phenomena may elucidate roles of oxygen radicals in carcinogenesis, as well as indicate the contribution these measurements can make to the assessment of bioavailable genotoxins in aquatic systems.

In summary, studies of oxyradicals and oxidative stress in aquatic organisms lag behind investigations of other important biochemical processes such as phase I metabolism and metal-binding proteins (discussed below). However, the key roles free radical processes have in the effects of many contaminants and the complex armamentarium all aerobic organisms possess to defend themselves from radicals and related oxidants underscore the importance of this subject in aquatic biochemical toxicology.

METAL METABOLISM AND TOXICITY

Trace metals of concern in aquatic toxicology include both essential (e.g., Cu, Mn, Zn) and nonessential (e.g., Cd, Pb, Hg) elements. Elevated exposures to either essential or nonessential metals can produce toxic effects. These effects generally result from the nonspecific binding of reactive metal cations to biologically important macromolecules, causing modifications of molecular function. The regulation of the distribution of trace metals among the various pools of macromolecules is central to maintaining metal homeostasis and optimal cellular function. At the cellular level, metabolism involves the binding of metals to inducible metal-binding ligands such as metallothionein (MT) and phytochelatin and the sequestration of metals in membrane-bound vesicles such as secondary and tertiary lysosomes, inclusion bodies, and mineral concretions (Viarengo, 1989; Steffins, 1990; Roesijadi, 1992). These metal binding sites act to regulate the availability of both essential and nonessential metals within the cell. For example, they can provide high-affinity sinks for nonessential metals to reduce their interactions with macromolecules. They can also sequester essential metals to modulate their availability for metal-requiring apoproteins, while minimizing their nonspecific binding. Specific patterns of subcellular metal accumulation depend on such factors as route of exposure, tissue, species, and stage of the life cycle.

Figure 14 presents a schematic diagram of metabolism and toxicity of metals in eukaryotic cells. As depicted, metal uptake and metabolism are the result of interactions between the "free" metal ion, $\{Me^+\}$, and a diverse group of metal-binding ligands. At steady state, metal distribution is a function of both the number and relative binding affinity of each ligand. This distribution can be altered by changes in either $\{Me^+\}$ or ligand concentration. In the aquatic environment, $\{Me^+\}$ can interact with a series of abiotic metal-binding ligands (such as humic substances; see Chapter 1) in the sediment and water column that serve to reduce $\{Me^+\}$ (1). This is an important control of metal uptake by cells because there is a direct relationship between the free metal ion activity in the environment and metal transport into cells (e.g., of gills). Metals generally enter the cell through interaction of the $\{Me^+\}$ with transport proteins or ion channels in the plasma membrane (2). Once in the cytoplasm, the free metal ion interacts with high-molecular-weight ligands (HMW; e.g., metalloenzymes), low-molecular-weight ligands (LMW; e.g., glutathione, phytochelatin), and MT (3). The free ion may also bind specific MT transcription factors (TF_{MT}), which interact with metal-specific elements in the promoters of the MT gene known as metal-responsive elements (MREs), increasing transcription and consequently MT synthesis (3). Elevations in the $\{Me^+\}$ can also increase metal concentrations in other cellular compartments that are often sites of toxicity. For example, in the mitochondria they can bind to crucial enzymes and respiratory protein complexes, reducing energy conversion efficiency and, by uncoupling oxidative phosphorylation, causing oxidative damage (4). In some cell types metals are directed to tertiary lysosomes and unexported out of the cell by exocytosis (5). Finally, physiological

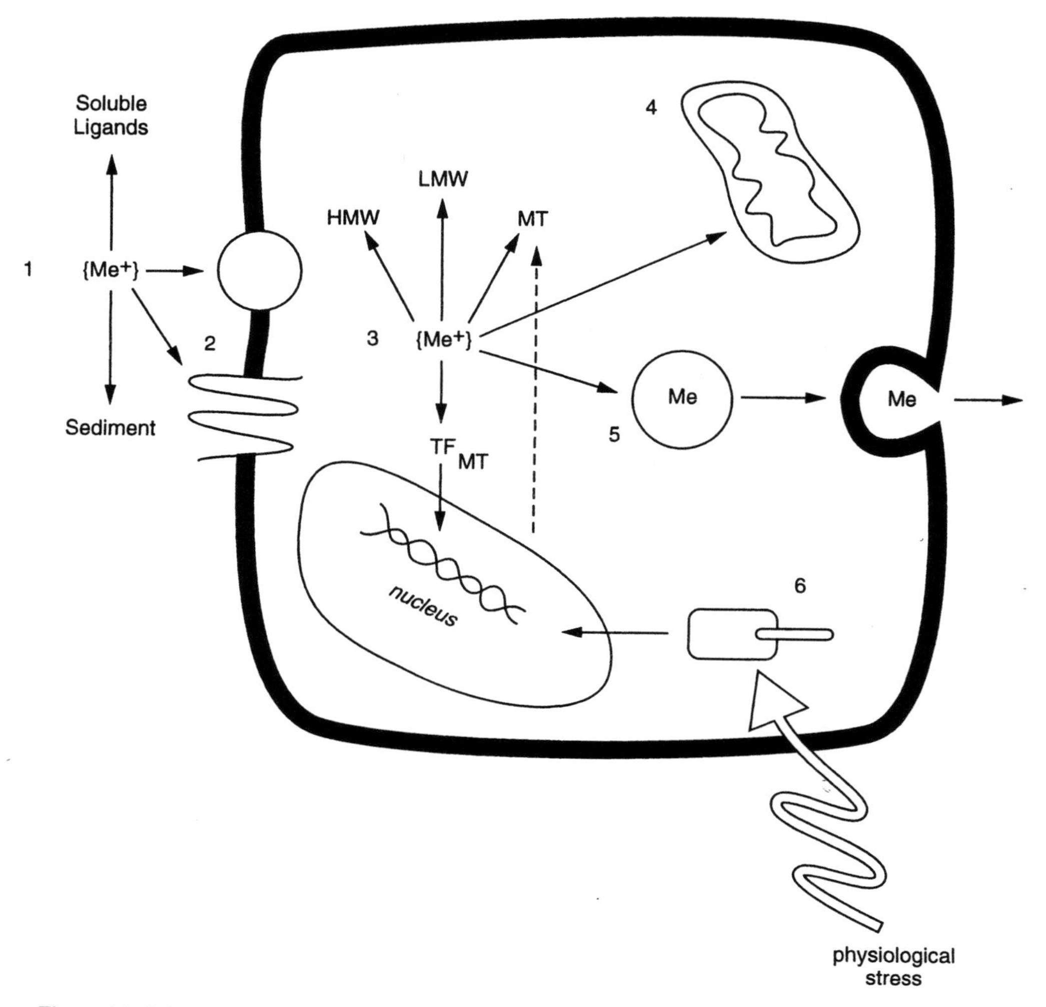

Figure 14. Schematic diagram of metabolism and toxicity of metals in eukaryotic cells. See text for detailed explanation.

stress can elevate steroid hormones such as glucocorticoids, which bind to their receptor and increase MT gene transcription and synthesis (6).

Metallothionein

Among metal-binding ligands, considerable attention has been focused on the role of MT in the biological activity of metals in aquatic organisms (Engel, 1987; Engel and Brouwer, 1987, 1991; Schlenk and Brouwer, 1991; Roesijadi, 1992). Metallothionein was first isolated from equine renal cortex as a soluble metalloprotein of low molecular weight that contained cadmium and zinc (Margoshes and Vallee, 1957). It was later purified, characterized, and designated metallothionein—a metallo derivative of the sulfur-rich protein thionein (Kägi and Vallee, 1960). The properties of MT, derived from many extensive isolation and characterization procedures, are given in Table 7. It is recommended that upon identification of a protein as MT, amino acid composition be documented. Less rigorously isolated and characterized proteins that are similar to MT but do not meet the full criteria of identification should be referred to as MT-like proteins.

Most vertebrate tissues contain two major isoforms of MT, designated as MT-I and MT-II based on their elution position during ion exchange chromatography (Hamer, 1986; Dunn et al., 1987). In many cases, each class actually consists of several different subforms usually designated as MT-I_A, MT-I_B, MT-I_C, etc. These terms apply to subforms of MT differing in genetically determined primary structure and not to differences in metal composition. The relative proportions of the different isoforms of MT (MT-I and MT-II) vary depending on species, tissue, physiological status, and exposure to metals (Dunn et al., 1987). Metals are bound to MT by thiolate

Table 7. Summary of the properties of metallothionein

Cytosolic localization in the cell
Low molecular mass, 6000 to 7000 daltons from amino acid composition and 15000 daltons from gel filtration chromatography
Single polypeptide chain with a unique amino acid composition (high cysteine, 23 to 33 mol %, and no aromatic amino acids or histidine)
High metal content (4 to 12 atoms/mole) bound exclusively by clusters of thiolate bonds (absorption maximum at 254 nm for cadmium-mercaptide, little to no absorption at 280 nm)
Heat stability (60°C for 5 min)
Induced synthesis of thionein by certain metals (Cd, Zn, Cu, Hg, Co, Ni, Bi, Ag)
Excellent sequence homology indicating a highly conserved primary structure

bonds to cysteine residues. Electron density mapping and ^{113}Cd nuclear magnetic resonance (NMR) studies have revealed two domains that form separate polynuclear metal-cysteine thiolate clusters for Cd,Zn MT-II) (Furey et al., 1986; Vašák, 1986). Cluster A, the carboxyl-terminal domain, is essentially a four-metal cluster with metals coordinated by six terminal and five bridging cysteine thiolates. Cluster B, the amino-terminal domain, is essentially a three-metal cluster with metals coordinated by six terminal and three bridging cysteine thiolates. The two clusters have different affinities for metals and can function independently of each other. For example, binding of metals in cluster B decreases in the order Cu > Zn > Cd. However, binding in cluster A decreases in the order Cd > Zn > Cu (Boulanger et al., 1983).

Metallothionein and MT-like proteins have been isolated from a diverse array of eukaryotic species (Hamer, 1986). Despite the fact that MT and the regulation of MT genes have been studied extensively, there is no unified understanding of its significance in biological systems. Possible biological functions of these metalloproteins include storage, transport or compartmentalization of essential metals and detoxification of heavy metals through chelation. Although chemical differences between metalloproteins from various species exist, most of these proteins appear to function as major sequestering mechanisms for metals in a manner analogous to that of mammalian MT. Upon cellular internalization of metals, these metalloproteins appear to occupy a central position in both essential (Zn, Cu) and nonessential (Cd, Hg) metal homeostasis. Shifts in metal composition of MT have been demonstrated following toxic metal exposure. It has been suggested that this disruption of normal essential metal homeostasis at the molecular level may be involved in toxicity and/or repair after cell injury by these agents. Therefore, sequestration of nonessential metal ions by MT may not be without consequence relative to metabolic homeostasis of essential metals. Also, although nonessential metal sequestration by MT is generally viewed as a detoxification process, this may not always be the case. For example, administration of the Cd-MT complex produced greater mammalian renal toxicity than free Cd (Suzuki and Cherian, 1989).

Use of Metallothionein in Environmental Monitoring

There is evidence of heavy metal–induced biosynthesis of MT in fish and other aquatic organisms. Roch et al. (1982) studied a river in Canada and correlated MT induction in rainbow trout with increased metal (Zn, Cu, Cd) concentrations in water. Similar increases in MT were observed following cadmium exposure in eel (Noël-Lambot et al., 1978), white suckers (Klaverkamp and Duncan, 1987), and rainbow trout (Olsson et al., 1988). Findings in rainbow trout were consistent with data for perch demonstrating that increased hepatic MT concentrations resulting from environmental exposure to cadmium were correlated with increased hepatic cadmium concentrations (Olsson and Haux, 1986). Also, Jenkins et al. (1982) reported hepatic metal burdens 1.3 to 3.5 times higher in white croaker from contaminated environments, and examination of cytosolic metal distribution revealed that excess metal in liver tissue was effectively sequestered by MT. Advances in recombinat DNA technology have allowed the measurement of MT transcripts (mRNA). This has increased the sensitivity of MT determinations and reduced the quantity of tissue required for MT analysis. Increases in oyster gill (Roesijadi, 1992) and bluecrab hepatopancreas (Schlenk and Brouwer, 1993) MT mRNA have been observed in animals treated with cadmium. In addition, significant inductions of MT transcripts in livers of fish taken from metal-contaminated environments have been observed (Kille et al., 1992).

Important to the assessment of the degree of toxicity is the determination of the amount of metal that is bound to MT. The fact that organisms have the capacity to synthesize metalloproteins that can sequester and subsequently detoxify heavy metals implies that an increased body burden of metals will not necessarily result in increased toxic effect. Pruell and Engelhardt (1980) used liver catalase activity as an indicator of toxic stress in killifish that were pretreated with low concentrations of cadmium and challenged with a subsequent higher cadmium concentration. Hepatic MT was apparently induced and correlated with adaptive responses in liver catalase activity. Increased hepatic cadmium levels were also observed. Kito et al. (1982) reported that carp exposed to sublethal cadmium or zinc concentrations were more resistant to cadmium toxicity and contained elevated cadmium concentrations in the MT fraction of kidney and liver but not gill. Benson and Birge (1985), however, demonstrated that tolerance to cadmium in fathead minnows was associated with gill MT concentrations and observed elevated metal content in the gills of cadmium-exposed organisms.

It has been suggested that measurement of MT may be useful in evaluating metal exposure as well

Table 8. Environmentally relevant inducers of stress protein synthesis

Anoxia and recovery	Elevated temperatures
Arsenate	Butanol
Arsenite	Ethanol
H_2O_2	Octanol
Transition metals	Propanol
Cd^{2+}	Tributyltin
Cu^{2+}	Carcinogens
Cr^{6+}	Teratogens
Hg^{2+}	Mutagens
Pb^{2+}	Nicotine
Zn^{2+}	Salinity changes
Organophosphate (diazinon)	UV irradiation
Organochlorine (lindane)	

as predicting metal toxicity. Although the significance of MT in biological systems is not fully understood, it appears that MT has an important protective function by sequestering metals and inhibiting interactions with sensitive cellular components. Cellular toxicity may result after the metal-binding capacity of MT has been exceeded. Therefore, it has been proposed that measurement of MT may provide information about potential health hazards of metals in exposed organisms. Variations in MT concentrations as a result of season, reproductive cycle, and stress may, however, hinder the use of this metalloprotein as a monitoring tool. In blue crab, Engel and Brouwer (1991) demonstrated that copper and zinc binding to MT vary inversely during ecdysis. Overnell et al. (1987) demonstrated increased amounts of hepatic MT with large variations during the breeding season. Olsson et al. (1987) observed fluctuations in zinc levels associated with MT in female trout during the annual reproductive cycle. Other factors such as stress response, cold, and hypoxia can also induce MT and MT-like proteins (Benson et al., 1990). Therefore, before MT can be used effectively in environmental monitoring and risk assessment, its physiological and biochemical function under normal conditions must be understood. In addition, potential toxicities associated with MT in mammals, described earlier, merit attention in aquatic organisms.

THE CELLULAR STRESS RESPONSE

The cellular stress response entails the rapid synthesis of a suite of proteins that are involved in protecting organisms from damage as a result of exposure to a wide variety of stressors of environmental concern, including heavy metals, organic compounds, and UV light (Table 8) (Lindquist and Craig, 1988; Sanders, 1993). A subset of these stress proteins are heat inducible and part of the classical heat shock response, whereas other stress proteins that are more stressor specific can be included in the broader "stress response."

The heat-inducible stress proteins have been studied most extensively (Nover, 1991). They are highly conserved over evolution and have been found in all organisms examined (Lindquist, 1986; Schlesinger, 1986). Four major heat shock protein families of 90, 70, 60, and 16–24 kDa are the most prominent and have been frequently referred to as hsp90, hsp70, hsp60 and the low-molecular-weight (LMW) hsps, respectively (Schlesinger et al., 1982). The 7-kDa protein, ubiquitin, is also included in this heat-inducible group (Schlesinger and Hershko, 1988). The term stress protein, instead of heat shock protein, is now more commonly used because the induction of these proteins occurs in response to many other types of environmental stressors (Sanders, 1993). A new convention has also been suggested to establish consistency and minimize the confusing terminology in the literature. The terms stress90 and stress70 are used for proteins in the 90-kDa and 70-kDa families, respectively, and the 60-kDa family is referred to as chaperonin 60 (cpn60) (Gething and Sambrook, 1992). The term stress protein is not meant to imply that stress proteins are present in cells only under stressful environmental conditions, because some members of the stress90, stress70, and chaperonin families are synthesized under normal conditions (Gething and Sambrook, 1992). These constitutively synthesized stress proteins, often referred to as cognates, play an important role in regulating protein homeostasis.

Molecular Chaperones

Stress90, stress70, and cpn60 are molecular "chaperones" that, under normal conditions, direct the folding and assembly of other cellular proteins (Rothman, 1989; Ellis, 1990; Martin et al., 1991; Gething and Sambrook, 1992; Hartl et al., 1992; Kelley and Georgopoulos, 1992; Seckler and Jaenicke, 1992). These stress proteins are also involved in regulating the kinetic partitioning between protein folding, translocation reactions, and protein aggregation. Under adverse environmental conditions, the synthesis of stress90, stress70, and cpn60 increases and they take on the additional roles of repairing denatured proteins and protecting cellular proteins from environmentally induced damage. Stress90, stress70, and chaperonin each have distinct cellular functions under both normal and stressful environmental conditions.

Figure 15. Schematic diagram of the functions of the major stress proteins, stress70 and cpn60, under normal conditions (a) and in response to environmental stress (b). See text for detailed explanation. L = lysosome; N = nucleus; ER = endoplasmic reticulum; M = mitochondria.

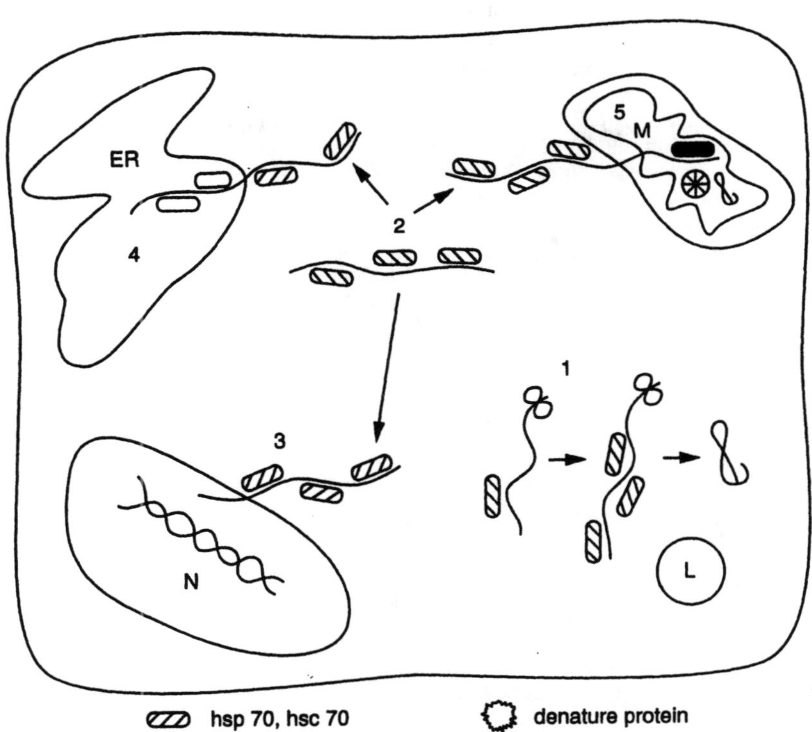

ER
4
2
5
M
3
N
1
L
hsp 70, hsc 70
hsp 75
bip (grp 78)
denature protein
folded protein
cpn 60

(a)

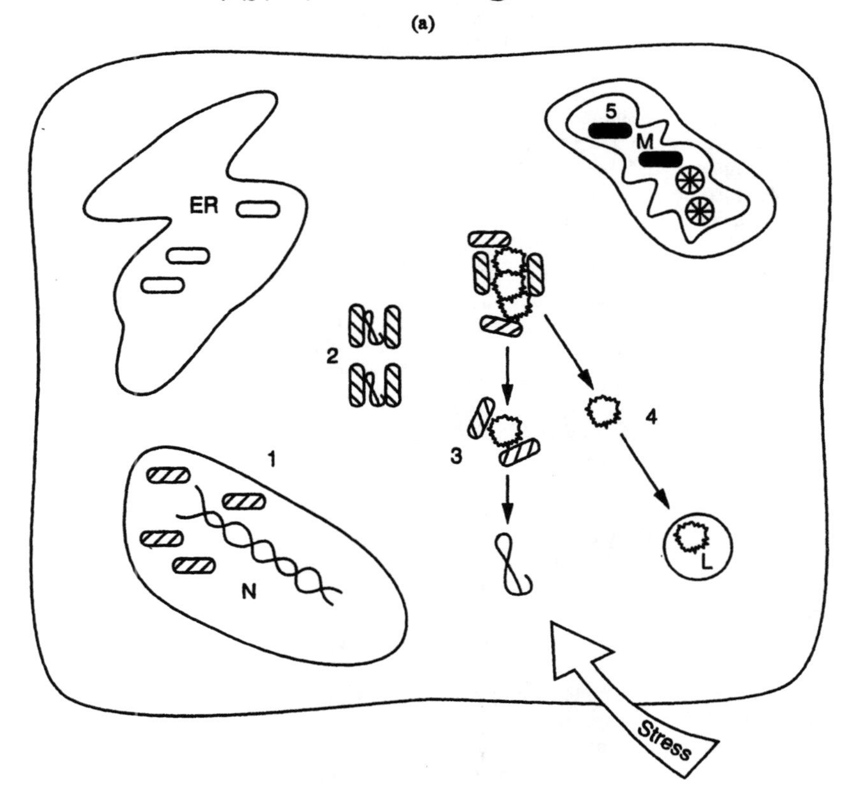

ER
5
M
2
1
3
4
N
L
Stress

(b)

Figure 15a illustrates the diverse functions of stress70 and cpn60 under normal conditions. Stress70 is part of a large multigene family whose members reside in a number of subcellular compartments including the cytoplasm, mitochondria, and endoplasmic reticulum (Pelham, 1990). Under normal conditions they bind to target proteins to modulate protein folding, transport, and repair. Cytoplasmic members of the stress70 family (e.g., hsp70 and hsc70) facilitate the correct folding of nascent peptides by stabilizing the peptide chains in a loosely folded state until synthesis has been completed, preventing the peptide from folding incorrectly (see 1 in Figure 15a; Beckmann et al., 1990). In all cases ATP hydrolysis appears necessary for dissociation of stress70 with the protein during the folding process. In addition, stress70 maintains some proteins in an unfolded intermediate configuration for targeting and translocation to cellular compartments, including the endoplasmic reticulum, chloroplast, and mitochondria (2) (Gething and Sambrook, 1992). Transport of proteins into the nucleus also requires stress70 (3). The stress70 member grp78 facilitates the subsequent folding of peptides after import into the endoplasmic reticulum (4) (Chirico et al., 1988; Craig, 1990). Another stress70, hsp75, facilitates folding after translocation to the mitochondria (5) (Hartl and Neupert, 1990).

Chaperonin 60 is found in the mitochondria and chloroplast of eukaryotes (McMullin and Hallberg, 1988; Ellis, 1990). Unlike stress70, which binds to target proteins as a monomer, chaperonin forms a 14-member oligomeric complex (5) (Saibil et al., 1993). Under normal conditions this complex binds incompletely folded proteins and directs the folding peptide to the correct conformation in a specific ATP-dependent manner (Cheng et al., 1989; Ostermann et al., 1989). The complex also prevents aggregation of incompletely folded proteins until they are competent for oligomer assembly (Martin et al., 1991).

Exposure to adverse physical and chemical conditions often cause protein denaturation through the weakening of polar bonds and exposure of hydrophobic groups, resulting in misfoldings and protein aggregation (Wedler, 1987). The term proteotoxicity, analogous to genotoxicity, is used to describe this primary mechanism of toxicity (Hightower, 1991). Under adverse environmental conditions the synthesis of stress70 and cpn60 increases and they function to protect the cell from proteotoxicity. In addition to an increased demand to perform their functions under normal conditions, they take on the roles of repairing denatured proteins and protecting cellular proteins from environmentally induced damage. Figure 15b illustrates the additional functions of these stress proteins under stressful conditions. Cytoplasmic stress70 migrates to the nucleus, where it binds to preribosomes and other protein complexes to help protect them from denaturation (1) (Lindquist, 1986; Gething and Sambrook, 1992). Stress70 also binds to vulnerable cytosolic proteins to protect them from stressor-induced damage and prevent the formation of insoluble aggregates, which are particularly damaging to the cell (2) (Pelham, 1990). It can also break up existing aggregates and refold the proteins back to their original conformation (3) (Skowyra et al., 1990; Gaitanaris et al., 1990; Ellis, 1991). Finally, stress70 vectors badly damaged proteins to the lysosomes for degradation (4) (Chiang et al., 1989). Chaperonin also takes on the roles of repairing damaged mitochondrial proteins and refolding them to their native conformation and protecting against protein denaturation and aggregation (5) (Martin et al., 1992). Unlike stress70, however, chaperonin is not able to break up existing aggregates (Buchner et al., 1991).

The Low-Molecular-Weight Stress Proteins

A more diverse group of stress proteins, referred to as the LMW group, are heat-inducible proteins in the 20–30 kDa range. The LMW stress proteins are more species specific and less highly conserved than the stress proteins previously discussed (Sanders, 1993). Unlike stress70 and cpn60, these stress proteins are not synthesized under normal conditions. Their synthesis is induced under adverse environmental conditions and is also developmentally regulated. The LMW stress proteins are particularly predominant on exposure to chemicals that are suspected carcinogens, mutagens, and teratogens (Buzin and Bournais-Vardiabasis, 1984; Bournais-Vardiabasis and Buzin, 1986).

Ubiquitin

Ubiquitin is a low-molecular-weight, highly conserved protein involved in the nonlysosomal degradation of intracellular proteins (Schlesinger and Hershko, 1988; Hochstrasser, 1992). It also participates in DNA repair and replication, transcription, rRNA processing, and maintaining ribosomal structure. Small multigene families code for both the constitutively expressed and inducible ubiquitin proteins. The synthesis of this protein increases on exposure to stress and increases the capacity for turnover of severely damaged proteins along the nonlysosomal pathway. This function of ubiquitin complements the role of stress70 in vectoring badly damaged peptides to the lysosome for degradation.

Induction of the Stress Response

Studies of the activation of the stress protein genes suggest an intriguing relationship between gene activation and function of these stress proteins. In eukaryotes, heat-inducible genes include a conserved sequence called the heat shock element (HSE). Activation of transcription involves the binding of a regulatory protein, the heat shock factor, to the HSE (Morimoto et al., 1990; 1992). The specific mechanism by which adverse environmental conditions activate the HSE is not known. The current working hypothesis is the "abnormal protein hypothesis," which states that heat shock or other stressors cause an increase in damaged proteins in the cell that in some manner activates the HSE (Hightower, 1980). Consequently, induction of the stress response involves a closed regulatory loop in which denatured proteins are both the signal that activates transcription of the stress protein genes and the substrate for the gene product. Therefore, the regulation of stress protein synthesis at the tissue level may be closely linked to the extent of environmentally induced protein damage.

Contaminant-Specific Stress Proteins

A number of studies have demonstrated the induction of other stress proteins that are not heat inducible but are instead induced by adverse exposure to chemicals. Many of these proteins participate in specific biochemical pathways involved in the metabolism of chemicals or detoxification of harmful byproducts rather than conferring protection against proteotoxicity. As a consequence, this class does not encompass a homologous group of proteins and is not related functionally or structurally to the heat shock proteins. Numerous inducible enzyme systems might fall within such a broad definition, including cytochrome P-450 and metallothionein, which are discussed in other sections of this chapter. Novel stressor-specific stress proteins are being discovered at a rapid rate. Most of these are currently identified only by their apparent molecular weight on a sodium dodecyl sulfate gel and we know little about their structure or function. Therefore we limit this discussion to two non-heat-inducible proteins, heme oxygenase and P-glycoprotein, which have recently been characterized.

Heme oxygenase was initially identified as a 32-kDa stress protein that is only slightly induced by heat shock but strongly induced by sodium arsenite, iodoacetamide, and heavy metals (Caltabiano et al., 1986). It is an enzyme that catalyzes the breakdown of heme to form biliverdin, which is subsequently reduced to bilirubin. It is most highly inducible by a variety of stressors that cause oxidative damage, such as UVA radiation, sodium arsenite, and hydrogen peroxide (Sanders, 1990; Nover, 1991). It has been suggested that because breakdown products of heme can readily react with peroxyl radicals, they may play an important role in protecting cells from oxidative damage as free radical scavengers in concert with glutathione. Furthermore, as heme is an important active component of cytochromes, in particular the cytochromes involved in the electron transport chain in the mitochondria, it may facilitate the turnover of these heme-containing proteins that are damaged by chemical exposure.

The *P-glycoprotein (P-gp)* multigene family comprises several isoforms of highly conserved proteins that are transport proteins in the plasma membrane (Abraham et al., 1993). The class I isoform is believed to serve as an energy-dependent transport mechanism for export of endogenous metabolites (Roninson, 1991). For example, in the adrenal gland it plays a role in normal physiological secretory processes. It is found in relatively high concentrations in kidney, liver, pancreas, small intestine, and adrenal gland.

Heat shock, sodium arsenite, methylcholanthrene, and dioxins can induce P-gp, although it appears able to transport only planar ring structures out of the cell (Gottesman and Pastan, 1988; Dellinger et al., 1992). It has been suggested that in addition to its "normal" function it plays a role in detoxification by transporting xenobiotics from the cell (Chan et al., 1992). P-gp overexpression has been shown to confer multidrug resistance in a host of organisms from *Plasmodium falciparum* to normal human tissue and mammalian tumor cell lines (Roninson, 1991). The P-glycoprotein gene has been sequenced in winter flounder and found to be homologous to the mammal gene (Chan et al., 1992). It has also been identified and characterized in marine sponges (Kurelec et al., 1992) and there is strong support for its presence in marine mussels (Kurelec and Pivcevic, 1991).

Environmentally Relevant Studies

A number of studies have characterized the stress response in aquatic organisms including mollusks, crustaceans, echinoderms, and fish. Some of these studies have examined the extent to which stress proteins are elevated under conditions that might occur naturally or as a result of anthropogenic activities in the environment. Although a thorough review of stress proteins in aquatic organisms is beyond the scope of this chapter, one can be found in Sanders (1993). The roles of stress proteins in acquired tolerance, adaptation, and stress physiology are major areas of emphasis in research on aquatic systems. Using the stress response as an environmental monitoring tool (i.e., biomarker) has also provided an important focus.

Acquired Tolerance

Evidence for a role of the stress response in conferring tolerance to environmental stressors is abundant in both aquatic and nonaquatic species. Stress proteins have been implicated in thermotolerance, a phenomenon in which a mild conditioning temperature confers tolerance, at a cellular and organismal level, to subsequent temperatures that would otherwise be lethal (Lindquist and Craig, 1988; Black and Subjeck, 1991). Boon-Niermeijer et al. (1986) correlated the synthesis of stress proteins with acquired tolerance in the mollusk *Lymnaea stagnalis*. Correlations between the ability to synthesize stress proteins and the expression of thermotolerance have also been observed in fish, echinoderm, amphibian, and mammalian embryonic systems (Heikkila et al., 1986; Mosser et al., 1987; Laszlo, 1988). Although heat shock has been the most thoroughly characterized inducer, other stressors that induce stress protein synthesis, such as arsenite, cadmium, and ethanol, also induce thermotolerance (Li, 1983; Nover, 1991).

Target Organ Toxicity

According to the abnormal protein hypothesis, regulation of stress protein synthesis is closely linked to the extent of protein damage. Therefore, it follows that the intensity of the stress response among the various tissues can help identify tissues that are most vulnerable to damage caused by a particular environmental stressor. Studies indicate that accumulation of stress70 and cpn60 is highly tissue specific in a manner that reflects tissue-level damage (Sanders, 1993). Furthermore, these tissue-level differences are stressor specific and are probably a function of the physical and chemical characteristics of the stressor.

In the fathead minnow the stress response is induced in a tissue-specific manner by elevated temperatures, arsenite, chromate, lindane, and diazinon (Dyer et al., 1991, 1993a,b). Moreover, the response reflects the mode of action for each chemical and was greatest in their respective target tissues. In *Mytilus* exposed to elevated copper in seawater, cpn60 in gill tissue is over 10 times greater than in mantle tissue (Sanders et al., 1994). In addition, a greater number of stress70 isoforms are present in gill tissue in response to Cu than in response to heat (Sanders et al., 1992). Induction of the stress response in rat kidney by mercuric chloride coincides with the early stages of nephrotoxicity (Goering et al., 1992). Also, stress70 accumulates in rat brain in patterns related to the histopathology of the neural injury (Gonzalez et al., 1989). These studies suggest that differences in the levels of accumulation of stress proteins among tissues may prove useful in identifying target tissues as well as routes of exposure. More research is needed on the induction and accumulations of stress proteins among tissues by contaminants with different mechanisms of action to determine whether this is a promising approach to increasing our understanding of the physiological mechanisms of toxicity.

Stress Proteins in Environmental Monitoring

For stress proteins to be useful as biomarkers of adverse biological effects for organisms exposed to contaminants in their environment, they should meet the following criteria: (1) their synthesis should be induced by a wide variety of stressors; (2) elevations in stress proteins should occur in organisms exposed to contaminants for long durations; and (3) the relationships between stress protein accumulation, contaminant exposure, and organismal stress should be linked in a predictable fashion (Mayer et al., 1992). A number of studies suggest that stress proteins may meet these criteria. The list of inducers of stress proteins that are of major environmental concern includes trace metals, pesticides, teratogens, and UV irradiation (Sanders, 1993). Induction of the response by these stressors appears to be regulated by the heat shock factor and to reflect stressor-induced protein damage (Nover, 1991).

Studies have also demonstrated that the stress response is induced in organisms under realistic environmental conditions and that it persists over time. In *Mytilus*, hsp60 and hsp70 remain elevated in tissues for at least 8 wk upon continuous exposure to an elevated temperature (Sanders et al., 1992). Stress proteins are elevated in mussels relative to laboratory controls that were collected from contaminated sites as part of the NOAA National Status and Trends Program (Sanders and Martin, 1993). Chemical analysis of sediments, water, and tissues from these sites indicated extensive chemical contamination. Changes in induction of stress proteins were also observed in mussels transplanted to contaminated sites in the North Sea (Veldhuizen-Tsoerkan et al., 1991).

The relationship between induction of the stress response and proteotoxicity at the tissue level suggests that induction of stress proteins in target tissues may provide an early warning of adverse effects at the organismal level. There is a concentration-response relationship between accumulation of hsp60 in mussels and the range of Cu concentrations to which they are exposed (Sanders et al., 1991). Moreover, significant increases in the accumulation of hsp60 are observed at Cu concentrations substantially lower than that which causes physiological impairment, as measured by a significant decrease in scope for growth. Studies of soil invertebrates (Köhler et al., 1992) suggest that stress protein accumulation examined in wild populations from contaminated sites or the induction of stress proteins following a laboratory challenge may be useful for quantifying the integrated "stress load" on an organism resulting from exposure to multiple stressors in the environment. Numerous studies suggest that stress proteins are involved in organismal adaptation to both anthropogenic and natural stressors. However, additional research is needed in a wide range of organisms to further delineate the relationships between induction of stress proteins in tissues and the physiological condition of the organism.

A number of precautions should be taken when using stress proteins as biomarkers of adverse biological effects (Stegeman et al., 1992; Sanders, 1993). First, it is important to characterize the regulation of the stress response thoroughly in each new species using a combination of metabolic labeling studies and immunological techniques (see Chapter 12). The specific members of the stress70 family that are recognized by an antibody should be determined by one- and two-dimensional Western blot analysis before it is used for quantifying stress70, since the antibody might be recognizing a constitutive and not inducible form. Also, the response should not be induced accidently by heat shock or by otherwise stressing the organism at the time of sample collection.

It is also important to note that the presence of stress70 and cpn60 does not necessarily signify that the organism has been exposed to environmental stress, as some members of these stress protein families are present under normal conditions. Also, organisms at certain stages of early development and females during their reproductive cycle may express high concentrations of some stress proteins as a consequence of normal physiological processes. Furthermore, many organisms such as those found in the intertidal zone use stress proteins to adapt to natural environmental extremes. As with many field studies, a careful experimental design (see Chapters 10 and 19) and appropriate controls will allow the distinction of elevations in stress proteins from natural variables from induction of stress proteins by contaminants.

SUMMARY

Significant advances have been made in our understanding of biochemical interactions of chemical contaminants with aquatic organisms. Four systems are described in this chapter which deal with biotransformations of organic compounds, oxidative stress, metal metabolism, and the cellular stress response. These systems play fundamental roles in these biochemical interactions and have received particular attention from aquatic toxicologists. It is important to bear in mind that these systems apparently evolved to function in routine aspects of cellular metabolism and homeostasis that are independent of exposures to anthropogenic chemicals. They also share the feature of providing organisms with mechanisms for coping, within strict limitations, with environmental stressors, including anthropogenic chemicals. But, as is generally the case in biology, exceptions occur. In some important instances (phase I activation of PAHs, redox cycling), these systems can actually exacerbate the toxicity of chemicals.

Biochemical and molecular studies, certainly not limited to the systems described herein, occupy an important position in aquatic toxicology. They contribute greatly to the theoretical basis for this young science, which ultimately serves as the foundation upon which practical applications, including regulatory and management decisions, depend. They also provide a key avenue for cross-fertilizations between environmental and human health scientists, the importance of which are becoming increasingly appreciated. An example of this is provided by the study of chemical carcinogenesis (see Chapter 14). Biochemical approaches adapted from biomedicine, including some of those described both in this chapter and in Chapter 13, are being used to explore this phenomenon in aquatic animals. Such studies will contribute to the understanding of this disease process in various organisms, including humans, as well as elucidate causative factors for the high rates of neoplasia observed in benthic fishes inhabiting some contaminated ecosystems. Finally, an understanding of biochemical mechanisms of contaminant metabolism, adaptation, and toxicity is useful for assessing environmental quality or aquatic ecosystem health. In this regard, such tools are particularly useful for gauging bioavailability (see Chapters 1 and 16) of metabolizable contaminants and for assessing the more subtle, sublethal effects often associated with environmental contamination.

LITERATURE CITED

Abraham, E. H.; Prat, A. G.; Gerwerck, L.; Senevereatine, T.; Arceci, R. J.; Kramer, R.; Guidotti, G.; and Cantiello, H. F.: The multidrug resistance (*mldl*) gene product functions as an ATP channel. *Proc. Natl. Acad. Sci. USA*, 90: 312–316, 1993.

Alscher R. G., and Amthor J. S.: The physiology of free radical scavenging: Maintenance and repair processes. In Air Pollution and Plant Metabolism, edited by S. Schulte-Hostede, N. M. Darall, L. W. Blank, and A. R. Welburn, pp. 95–115. New York: Elsevier Applied Science, 1988.

Ames, B. N.: Endogenous oxidative DNA damage, aging, and cancer. *Free Radical Res. Commun.*, 7:121–128, 1989.

Andaya, A. A., and Di Giulio, R. T.: Acute toxicities and hematological effects of two substituted naphthoquinones in channel catfish. *Arch. Environ. Contam. Toxicol.*, 16:233–238, 1987.

Andersson, T., and Koivusaari, U.: Influence of environmental temperature on the induction of xenobiotic metabolism by β-naphthoflavone in rainbow trout (*Salmo gairdneri*). *Toxicol. Appl. Pharmacol.*, 80:43–50, 1985.

Andersson, T.; Pesonen, M.; and Johansson, C.: Differential induction of cytochrome P-450 monooxygenase, epoxide hydrolase, glutathione transferase and UDP glucuronsyl transferase activities in the liver of rainbow trout by β-naphthoflavone or Clophen A50. *Biochem. Pharmacol.*, 34:3309–3314, 1985.

Andersson T.; Forlin, L.; Hardig, J.; and Larsson, A: Physiological disturbances in fish living in coastal waters polluted with bleached kraft pulp mill effluents. *Can. J. Fish Aquat. Sci.* 45:1525–1536, 1988.

Ankley, G. T.; Blazer, V. S.; Reinert, R. E.; and Agosin, M.: Effects of Aroclor 1254 on cytochrome P-450 dependent monooxygenase, glutathione *S*-transferase and UDP-glucuronosyl transferase in channel catfish liver. *Aquat. Toxicol.*, 9:91–103, 1986.

Armstrong, R. N.: Enzyme-catalyzed detoxication reactions: mechanisms and stereochemistry. *CRC Crit. Rev. Biochem.*, 22:39–88, 1987.

Asada, K.; Kiso, K.; and Yoshikawa, K.: Univalent reduction of molecular oxygen by spinach chloroplasts on illumination. *J. Biol. Chem.*, 249:2175–2181, 1974.

Babior, B. M.: The respiratory burst of phagocytes. *J. Clin. Invest.*, 73:599–601, 1984.

Bano, Y., and Hasan, M.: Mercury induced time-dependent alterations in lipid profiles and lipid peroxidation in different body organs of cat-fish *Heteropneustes fossilis*. *J. Environ. Sci. Health*, B24:145–166, 1989.

Barbacid, M.: *ras* genes. *Annu. Rev. Biochem.*, 56:779–827, 1987.

Barron, M. G.; Stehly, G. R.; and Hayton, W. L.: Pharmacokinetic modeling in aquatic animals. I. Models and concepts. *Aquat. Toxicol.*, 17:187–212, 1990.

Baumann, P. C.: PAH metabolites and neoplasia in feral fish populations. *Metabolism of Polycyclic Aromatic Hydrocarbons in the Marine Environment*, edited by U. Varanasi, pp. 269–290. Boca Raton, FL: CRC Press, 1989.

Beckmann, R. B.; Mizzen, L. A.; and Welch, W. J.: Interactions of hsp70 with newly synthesized proteins: implications for protein folding and assembly. *Science*, 248: 850–853, 1990.

Benson, W. H., and Birge, W. G.: Heavy metal tolerance and metallothionein induction in fathead minnow: results from field and laboratory investigations. *Environ. Toxicol. Chem.*, 4:209–217, 1985.

Benson, W. H.; Baer, K. N.; and Watson, C. F.: Metallothionein as a biomarker of environmental metal contamination: species-dependent effects. Biomarkers of Environmental Contamination, edited by J. F. McCarthy and L. R. Shugart, pp. 255–265. Boca Raton, FL: Lewis, 1990.

Berger, H. M., and Woodward, M. P.: Small heatshock proteins in *Drosophilia* may confer thermal tolerance. *Exp. Cell Res.*, 147:437–442, 1983.

Binder, R. L., and Stegeman, J. J.: Microsomal electron transport and xenobiotic monooxygenase activities during the embryonic period of development in the killifish (*Fundulus heteroclitus*). *Toxicol. Appl. Pharmacol.*, 73: 432–443, 1984.

Binder, R. L.; Melanchon, M. J.; and Lech, J. J.: Factors influencing the persistence and metabolism of chemicals in fish. *Drug Metab. Rev.*, 15:697–724, 1984.

Black, A. R., and Subjeck, J. R.: The biology and physiology of the heat shock and glucose-regulated stress protein systems. *Methods Achieve Exp. Pathol.*, 15:126–166, 1991.

Black, S. D., and Coon, M. J.: P-450 cytochromes: structure and function. *Adv. Enzymol. Relat. Areas Mol. Biol.*, 60: 35–87, 1987.

Boon-Niermeijer, E. K.; Tuyl, M.; and van der Scheur, H.: Evidence for two states of thermotolerance. *Hyperthermia*, 2:93–105, 1986.

Boulanger, Y.; Goodman, C. M.; Forte, P. C.; Fesik, S. W.; and Armitage, I. M.: Model for mammalian metallothionein structure. *Proc. Natl. Acad. Sci. USA*, 80:1501–1505, 1983.

Bournais-Vardiabasis, N., and Buzin, C. H.: Developmental effects of chemicals and the heat shock response in *Drosophila* cells. *Teratog. Carcinog. Mutag.*, 6:523–537, 1986.

Bowser, P. R.; Falls, W. W.; Van Zandt, J.; Collier, N.; and Phillips, J. D.: Methemoglobinemia in channel catfish: methods of prevention. *Prog. Fish Cult.*, 45:154–158, 1983.

Buchner, J.; Schmidt, M.; Fuchs, M.; Jaenicke, R.; Rudolph, R.; Schmid, F. X.; and Kiefhaber, T.: GroE facilitates refolding of citrate synthase by suppressing aggregation. *Biochemistry*, 30:1586–1591, 1991.

Buhler, D. R., and Williams, D. E.: The role of biotransformation in the toxicity of chemicals. *Aquat. Toxicol.*, 11:19–28, 1988.

Buhler, D. R., and Williams, D. E.: Enzymes involved in metabolism of PAH by fishes and other aquatic animals: oxidative enzymes (or phase I enzymes). Metabolism of Polycyclic Aromatic Hydrocarbons in the Marine Environment, edited by U. Varanasi, pp. 151–184. Boca Raton, FL: CRC Press, 1989.

Burchell, B., and Coughtrie, W. H.: UDP-glucuronsyltransferases. *Pharmacol. Rev.*, 43:261–289, 1989.

Burke, M. D., and Mayer, R. T.: Ethoxyresorufin: direct fluorometric assay of a microsomal dealkylation which is preferentially inducible by 3-methylcholanthrene. *Drug Metab. Dispos.*, 2:583–588, 1974.

Buzin, C. H., and Bournais-Vardiabasis, N.: Teratogens induce a subset of small heat shock proteins in *Drosophila* primary embryonic cell cultures. *Proc. Natl. Acad. Sci. USA*, 81:4075–4079, 1984.

Caltabiano, M. M.; Koestle, T. P.; Poste, G.; and Greig, R. G.: Induction of 32 and 34 kDa stress proteins by sodium arsenite, heavy metals, and thiol-reactive agents. *J. Biol. Chem.*, 261:13381–13386, 1986.

Cavalieri, E. L., and Rogan, E. G.: Radical cations in aromatic hydrocarbon carcinogenesis. *Free Rad. Res. Commun.*, 11:77–87, 1990.

Cerutti, P. A.: Prooxidant states and tumor production. *Science*, 227:375–381, 1985.

Chan, H. W.-S. (ed.): Autoxidation of Unsaturated Lipids. New York: Academic Press, 1987.

Chan, K. M.; Davies, P. L.; Childs, S.; Veinot, L.; and Ling, V.: P-glycoprotein genes in the winter flounder, *Pleuronectes americanus*: isolation of two types of genomic clones carrying 3′ terminal exons. *Biochim. Biophys. Acta.*, 1171:65–72, 1992.

Chatterjee, S., and Bhattacharya, S.: Detoxification of industrial pollutants by the glutathione *S*-transferase system in the liver of (*Anabas testrudineus*) (Bloch). *Toxicol. Lett.*, 22:187–198, 1984.

Cheng, M. Y.; Hartl, F.-U.; Martin, J.; Pollock, R. A.; Kalousek, F.; Neupert, W.; Hallberg, E. M.; Hallberg, R. L.; and Horwich, A. L.: Mitochondrial heat shock protein hsp60 is essential for assembly of proteins imported into yeast motochondria. *Nature*, 337:620–624, 1989.

Chiang, H.-L.; Terlecky, S. R.; Plant, C. P.; and Dice, J. F.: A role for a 70-kilodalton heat shock protein in lysosomal degradation of intercellular proteins. *Science*, 246: 382–385, 1989.

Chirico, W. J.; Waters, M. G.; and Blobel, G.: 70K heat shock related proteins stimulate protein translocation into microsomes. *Nature*, 333:805–810, 1988.

Clark, A. G.: The comparative enzymology of the glutathione *S*-transferases from non-vertebrate organisms. *Comp. Biochem. Physiol.*, 92B:419–446, 1989.

Cohen, G. D.; Lewis; and Sinet, P. M.: Oxygen consumption during the Fenton type reaction between hydrogen peroxide and ferrous chelate (Fe^{2+}-DTPA). *J. Inorg. Biochem.*, 15:143–151, 1981.

Cohen, G. M., and d'Arcy Doherty, M.: Free radical mediated cell toxicity by redox cycling chemicals. *Res. J. Cancer*, 55:46–52, 1987.

Collier, T. K., and Varanasi, U.: Hepatic activities of xenobiotic metabolizing enzymes and biliary levels of xenobiotics in English sole (*Parophrys vetulus*) exposed to environmental contaminants. *Arch. Environ. Contam. Toxicol.*, 20:462–473, 1991.

Craig, E. A.: Role of hsp70 in translocation of proteins across the membranes. Stress Proteins in Biology and Medicine, edited by R. I. Morimoto; A. Tissieres; and C. Georgopoulos, pp. 279–286. Cold Spring Harbor, NY: Cold Spring Harbor Laboratory Press, 1990.

Dehnen, W.; Tomingas, R.; and Roos, J.: A modified method for assay of benzo[*a*]pyrene hydroxylase. *Anal. Biochem.*, 53:373–383, 1973.

Dellinger, M.; Pressman, B. C.; Calderon-Higginson, C.; Savaraj, N.; Tapiero, H.; Kolonias, D.; and Lampidis, T. J.: Structural requirements of simple organic cations for recognition by multidrug-resistant cells. *Cancer Res.*, 52:6385–6389, 1992.

Di Giulio, R. T.; Habig, C.; and Gallagher, E. P.: Effects of Black Rock Harbor sediments on indices of biotransformation, oxidative stress, and DNA integrity in channel catfish. *Aquat. Toxicol.*, 26:1–22, 1993.

Dizdaroglu, M., and Bergtold, D. S.: Characterization of free radical-induced base damage in DNA at biologically relevant levels. *Anal. Biochem.*, 156:182–188, 1986.

Dolphin, D.; Avramoric, O.; and Poulson, R. (eds.): Glutathione: Chemical, Biochemical, and Medical Aspects. New York: Wiley, 1989.

Dunn, M. A.; Blalock, T. L.; and Cousins, R. J.: Metallothionein. *Proc. Soc. Exp. Biol. Med.*, 185:107–119, 1987.

Dyer, S. D.; Dickson, K. L.; Zimmerman, E. G.; and Sanders, B. M.: Tissue specific patterns of heat shock protein synthesis and thermal tolerance of the fathead minnow (*Pimephales promelas*). *Can. J. Zool.*, 69:2021–2027, 1991.

Dyer, S. D.; Brooks, G. L.; Dickson, K. L.; Sanders, B. A.; and Zimmerman, E. G.: Synthesis and accumulation of stress proteins in tissues of arsenite exposed fathead minnows (*Pimephales promelas*). *Environ. Toxicol. Chem.*, 12:1–12, 1993a.

Dyer, S. D.; Dickson, K. L.; and Zimmerman, E. G.: A laboratory evaluation of the use of stress proteins in fish to detect changes in water quality. Environmental Toxicology and Fish Assessment, edited by W. G. Landis; J. J. Huges; and M. A. Lewis, pp. 273–287. ASTM STP 1179. Philadelphia: ASTM Publishers, 1993b.

Dykens, J. A., and Shick, M.: Oxygen production by endosymbiotic algae control superoxide dismutase activity in their animal host. *Nature*, 297:579–580, 1982.

Eling, T. E.; Thompson, D. C.; Foureman, G. L.; Curtis, J. F.; and Hughes, M. F.: Prostaglandin H synthase and xenobiotic oxidation. *Annu. Rev. Pharmacol. Toxicol.*, 30:1–45, 1990.

Ellis, J.: The molecular chaperone concept. *Semin. Cell Biol.*, 1:1–9, 1990.

Ellis, R. J.: Chaperoning protein repair. *Curr. Biol.*, 1: 177–178, 1991.

Elskus, A. A., and Stegeman, J. J.: Induced cytochrome P-450 in (*Fundulus heteroclitus*) associated with environmental contamination by polychlorinated biphenyls and polynuclear aromatic hydrocarbons. *Mar. Environ. Res.*, 27:31–50, 1989.

Engel, D. W.: Metal regulation and molting in the blue crab, *Callinectes sapidus*: copper, zinc, and metallothionein. *Biol. Bull.*, 172:69–82, 1987.

Engel, D. W., and Brouwer, M.: Metal regulation and molting in the blue crab, *Callinectes sapidus*: metallothionein function in metal metabolism. *Biol. Bull.*, 173:239–251, 1987.

Engel, D. W., and Brouwer, M.: Short-term metallothionein and copper changes in blue crabs at ecdysis. *Biol. Bull.*, 180:447–452, 1991.

Fahimi, H. D., and Sies, H. (eds.): Peroxisomes in Biology and Medicine. New York: Springer-Verlag, 1987.

Forlin, L., and Haux, C.: Increased excretion in the bile of 17β-[^{3}H]estradiol–derived radioactivity in rainbow trout treated with β-naphthoflavone. *Aquat. Toxicol.*, 6:197–208, 1985.

Foureman, G. L.: Enzymes involved in metabolism of PAH by fishes and other aquatic animals: hydrolysis and conjugation enzymes or phase II enzymes. Metabolism of Polycyclic Aromatic Hydrocarbons in the Marine Environment, edited by U. Varanasi, pp. 185–202. Boca Raton, FL: CRC Press, 1989.

Frew, J. E., and Jones, P.: Structure and functional properties of peroxidases and catalases. Advances in Inorganic and Bioinorganic Mechanisms, Vol. 3, edited by A. G. Sykes, pp. 175–212. New York: Academic Press, 1984.

Fridovich, I: The biology of oxygen radicals. *Science*, 201: 875–880, 1978.

Fridovich, I.: Superoxide dismutases. *Adv. Enzymol.*, 58: 61–97, 1986.

Furey, W. F.; Robbins, A. H.; Clancy, L. L.; Winge, D. R.; Wang, B. C.; and Stout, C. D.: Crystal structure of Cd, Zn metallothionein. *Science*, 231:704–710, 1986.

Gaitanaris, G. A.; Papavassiliou, A. G.; Rubock, P.; Silverstein, S. J.; and Gottesman, M. E.: Renaturation of denatured gamma repressor requires heat shock proteins. *Cell*, 61:1013–1020, 1990.

Gelboin, H. V.: Benzo[*a*]pyrene metabolism, activation and carcinogenesis: role and regulation of mixed function oxidases and related enzymes. *Physiol. Rev.*, 60:1107–1166, 1980.

George, S. G., and Buchanan, G.: Isolation, properties and induction of plaice liver cytosolic glutathione *S*-transferases. *Fish Physiol. Biochem.*, 8:437–449, 1990.

George, S.; Young, P.; Leaver, M.; and Clarke, D.: Activities of pollutant metabolizing and detoxication systems in the liver of the plaice (*Pleuronectes platessa*): sex and seasonal variations in non-induced fish. *Comp. Biochem. Physiol.*, 96C:185–192, 1990.

Georgillis, A.; Montelius, J.; and Rydstrom, J.: Evidence for a free-radical-dependent metabolism of 7,12-dimethylbenz(*a*)anthracene in rat testis. *Toxicol. Appl. Pharmacol.*, 87:141–154, 1987.

Gething, M. J., and Sambrook, J.: Protein folding in the cell. *Nature*, 355:33–45, 1992.

Goering, P. L.; Fisher, B. R.; Chaudhary, P. P.; and Dick, C. A.: Relationship between stress protein induction in rat kidney by mercuric chloride and nephrotoxicity. *Toxicol. Appl. Pharmacol.*, 113:184–191, 1992.

Gøksoyr, A.: Purification of hepatic microsomal cytochromes P-450 from β-naphthoflavone–treated Atlantic cod (*Gadus morhua*), a marine teleost fish. *Biochim. Biophys. Acta.*, 840:409–417, 1985.

Gøksoyr, A.: A semi-quantitative cytochrome P450IA1 ELISA: a simple method for studying the monooxygenase response in environmental monitoring and ecotoxicological testing of fish. *Sci. Total Environ.*, 101:255–262, 1991.

Gøksoyr, A.; Andersson, T.; Hansson, T.; Klungsoyr, J.; Zhang, Y.; and Forlin, L.: Species characteristics of the hepatic xenobiotic and steroid biotransformation systems of two teleost fish Atlantic cod (*Gadus morhua*) and rainbow trout (*Salmo gairdneri*). *Toxicol. Appl. Pharmacol.*, 89:347–360, 1987.

Gonzalez, M. F.; Shiraishi, K.; Hisanaga, K.; Sagar, S. M.; Mandabach, M.; and Sharp, F. R.: Heat shock proteins as markers of neural injury. *Mol. Brain Res.*, 6:93–100, 1989.

Gooch, J. W.; Elskus, A. A.; Kloepper-Sams, P. J.; Hahn, M. E.; and Stegeman, J. J.: Effects of ortho and non-ortho substituted polychlorinated biphenyl congeners on the hepatic monooxygenase system in scup (*Stenotomus chrysops*). *Toxicol. Appl. Pharmacol.*, 98:422–433, 1989.

Gottesman, M. M., and Pastan, I.: The multidrug transporter, a double-edged sword. *J. Biol. Chem.*, 253: 12163–12166, 1988.

Guengerich, F. P., and Liebler, D. C.: Enzymatic activation of chemicals to toxic metabolites. *CRC Crit. Rev. Toxicol.*, 14:259–307, 1985.

Haasch, M. L.; Wejksnora, P. J.; Stegeman, J. J.; and Lech, J. J.: Cloned rainbow trout P_1 450 complementary DNA as a potential environmental monitor. *Toxicol. Appl. Pharmacol.*, 98:362–368, 1989.

Haber, F., and Weiss, J.: The catalytic decomposition of hydrogen peroxide by iron salts. *Proc. R. Soc. Lond. Ser. A*, 147:332–351, 1934.

Hahn, M. E., and Stegeman, J. J.: Phylogenetic distributional of the Ah receptor in non-mammalian species: implications for dioxin toxicity and Ah receptor evolution. *Chemosphere*, 25:931–937, 1992.

Hahn, M. E.; Poland, A.; Glover, E.; and Stegeman, J. J.: The Ah receptor in marine animals: phylogenic distribution and relationships to cytochrome P4501A inducibility. *Mar. Environ. Res.*, 34:87–92, 1992.

Hahn, M. E.; Lamb, T. M.; Schultz, M. E.; Smolowitz, R. M.; and Stegeman, J. J.: Cytochrome P450IA induction and inhibition by 3,3′,4′-tetrachlorobiphenyl in an Ah receptor–containing fish hepatoma cell line (PLHC-1). *Aquat. Toxicol.*, 26:185–208, 1993.

Halliwell, B.: Superoxide dependent formation of hydroxyl radicals in the presence of iron chelates. *FEBS Lett.*, 92: 321–326, 1978.

Halliwell, B., and Gutteridge, J. M. C.: Free radicals in Biology and Medicine, 2nd ed. Oxford: Clarendon, 1989.

Hamer, D. H.: Metallothionein. *Annu. Rev. Biochem.*, 55: 913–951, 1986.

Hammock, B. D., and Ota, K.: Differential induction of cytosolic epoxide hydrolase, microsomal epoxide hydrolase and glutathione *S*-transferase activities. *Toxicol. Appl. Pharmacol.*, 71:254–265, 1983.

Hanson-Painton, O.; Griffin, M. J.; and Tang, J.: Involvement of a cytosolic carrier protein in the microsomal metabolism of benzo[*a*]pyrene in rat liver. *Cancer Res.*, 43: 4198–4206, 1983.

Hartl, F.-U., and Neupert, W.: Protein sorting to mitochondria: evolutionary conservations of folding and assembly. *Science*, 247:930–938, 1990.

Hartl, F. U.; Martin, J.; and Neupert, W.: Protein folding in the cell: the role of molecular chaperones Hsp70 and Hsp60. *Annu. Rev. Biophys. Biomol. Struct.*, 21:293–322, 1992.

Hasspieler, B. M., and Di Giulio, R. T.: DT diaphorase [NAD(P)H: (quinone acceptor) oxidoreductase] facilitates redox cycling of menadione in channel catfish (*Ictalurus punctatus*) cytosol. *Toxicol. Appl. Pharmacol.*, 114:156–161, 1992.

Heikkila, J. J.; Browder, L. W.; Gedamu, L.; Nickells, R. W.; and Schultz, G. A.: Heat-shock gene expression in animal embryonic systems. *Can. J. Genet. Cytol.*, 28: 1093–1105, 1986.

Heilmann, L. J.; Sheen, Y.-Y.; Bigelow, S. W.; and Nebert, D. W.: Trout P450IA1: cDNA and deduced protein sequence, expression in liver and evolutionary significance. *DNA*, 7:379–387, 1988.

Hightower, L. E.: Cultured animal cells exposed to amino acid analogues or puromycin rapidly synthesize several polypeptides. *J. Cell Physiol.*, 102:407–427, 1980.

Hightower, L. E.: Heat shock, stress proteins, chaperons, and proteotoxicity. *Cell*, 66:1–20, 1991. Meeting review.

Hochstrasser, M.: Ubiqutin and intracellular degradation. *Curr. Biol.*, 4:1024–1031, 1992.

Horton, A. A., and Fairhurst, S.: Lipid peroxidation and mechanisms of toxicity. *CRC Crit. Rev. Toxicol.*, 18:27–79, 1987.

Hossain, M. A.; Nakano, Y.; and Asada, K.: Monodehydroascorbate reductase in spinach chloroplast and its participation in regeneration of ascorbate for scavenging hydrogen peroxide. *Plant Cell Physiol.*, 25:385–395, 1984.

Imlay, J. A., and Linn, S.: DNA damage and oxygen radical toxicity. *Science*, 240:1302–1309, 1988.

Trons, R. D., and Sawahata, T.: Phenols, catechols, and quinones. Bioactivation of Foreign Compounds, edited by M. W. Anders, pp. 259–281. Orlando, FL: Academic Press, 1985.

Jakoby, W. B.: The glutathione *S*-transferases: a group of multifunctional detoxification proteins. *Adv. Enzymol.*, 46:383–414, 1978.

James, M. O.: Conjugation of organic pollutants in aquatic species. *Environ. Health Perspect.*, 71:97–103, 1987.

James, M. O.: Biotransformation and disposition of PAH in aquatic invertebrates. Metabolism of Polycyclic Aromatic Hydrocarbons in the Marine Environment, edited by U. Varanasi, pp. 69–92. Boca Raton, FL: CRC Press, 1989a.

James, M. O.: Cytochrome P450 monooxygenase in crustaceans. *Xenobiotica*, 19:1063–1076, 1989b.

James, M. O., and Little, P. J.: Polyhalogenated biphenyls and phenobarbitol: evaluation as inducers of drug metabolizing enzymes in the sheepshead (*Archosargus probatocephalus*). *Chem. Biol. Interact.*, 36:229–248, 1981.

James, M. O.; Bowen, E. R., Dansette, P. M.; and Bend, J. R.: Epoxide hydrolase and glutathione *S*-transferase activities with selected alkene and arene oxides in several marine fish. *Chem. Biol. Interact.*, 25:321–324, 1979.

James, M. O.; Heard, C. S.; and Hawkins, W. E.: Effect of 3-methylcholanthrene on monooxygenase, epoxide hydrolase and glutathione *S*-transferase activities in small estuarine and freshwater fish. *Aquat. Toxicol.*, 12:1–15, 1988.

Jenkins, K. D.; Brown, D. A.; Hershelman, G. P.; and Meyer, W. C.: Contaminants in white croakers, *Genyonemus lineatus* (Ayers, 1855), from the southern California bight. I. Trace metal detoxification/toxification. Physiological Mechanisms of Marine Pollutant Toxicity, edited by W. B. Vernberg; A. Calabrese; F. P. Thurberg; and F. J. Vernberg, pp. 177–196. New York: Academic Press, 1982.

Jewell, C. S. E., and Winston, G. W.: Oxyradical production by hepatopancreas microsomes from the red swamp crayfish, *Procambarus clarkii. Aquat. Toxicol.*, 14:27–46, 1989.

Kägi, J. H. R., and Vallee, B. L.: Metallothionein: A cadmium- and zinc-containing protein from equine renal cortex. *J. Biol. Chem.*, 235:3460–3465, 1960.

Kappus, H.: Oxidative stress in chemical toxicity. *Arch. Toxicol.*, 60: 144–149, 1987.

Kappus, H., and Sies, H.: Toxic drug effects associated with oxygen metabolism: redox cycling and lipid peroxidation. *Experientia*, 37:1223–1241, 1981.

Kelley, W. L., and Georgopoulos, C.: Chaperones and protein folding. *Curr. Biol.*, 4:984–991, 1992.

Keeran, W. S., and Lee, R. F.: The purification and characterization of glutathione *S*-transferase from the hepatopancreas of the blue crab (*Callinectes sapidus*). *Arch. Biochem. Biophys.*, 255:233–243, 1987.

Kennish, J. M.; Russell, M. L.; and Netzel, S. J.: The role of iron chelates in the NAD(P)H-dependent oxidation of 2-keto-4-thiomethylbutyric acid (KMBA) by rainbow trout and Pacific salmon microsomal fractions. *Mar. Environ. Res.*, 28:87–91, 1989.

Kille, P.; Kay, J.; Leaver, M.; and George, S.: Induction of piscine metallothionein as a primary response to heavy metal pollutants: applicability of new sensitive molecular probes. *Aquat. Toxicol.*, 22:279–286, 1992.

Kito, H.; Ose, Y.; Mizuhira, V.; Sato, T.; Ishikawa, T.; and Tazawa, T.: Separation and purification of (Cd, Cu, and Zn)-metallothionein in carp hepato-pancreas. *Comp. Biochem. Physiol.*, 73C:121–127, 1982.

Klaverkamp, J. F., and Duncan, D. A.: Acclimation to cadmium toxicity by white suckers: cadmium binding capacity and metal distribution in gill and liver cytosol. *Environ. Toxicol. Chem.*, 6:275–289, 1987.

Kloepper-Sams, P. J., and Stegeman, J. J.: The temporal relationships between P450E protein content, catalytic activity and mRNA levels in the teleost (*Fundulus heteroclitus*) following treatment with β-naphthoflavone. *Arch. Biochem. Biophys.*, 268:525–535, 1989.

Klotz, A. V., Stegeman, J. J.; and Walsh, C.: An aryl hydrocarbon hydroxylating cytochrome P450 from the marine fish (*Stenotomus chrysops*). *Arch. Biochem. Biophys.*, 226:578–592, 1983.

Klotz, A. V.; Stegeman, J. J.; and Walsh, C.: An alternative 7-ethoxyresorufin *O*-deethylase assay: a continuous spectrophotometric method for measurement of cytochrome P-450 monooxygenase activity. *Anal. Biochem.*, 140: 138–145, 1984.

Klotz, A. V.; Stegeman, J. J.; Woodin, B. R.; Snowberger, E. S.; and Walsh, C.: Cytochrome P-450 isozymes from the marine teleost (*Stenotomus chrysops*): their roles in steroid hydroxylation and the influence of cytochrome b_5. *Arch. Biochem. Biophys.*, 249:326–338, 1986.

Köhler, H.-R.; Triebskorn, R.; Stöcker, W.; Kloetzel, P.-M.; and Alberti, G.: The 70 kD heat shock protein (hsp 70) in soil invertebrates: a possible tool for monitoring environmental toxicants. *Arch. Environ. Toxicol.*, 22:334–338, 1992.

Koivusaari, U.; Lindstrom-Seppa, P.; and Hanninen, O.: Xenobiotic metabolism in rainbow trout intestine. *Adv. Physiol. Sci.*, 29:433–440, 1981.

Kosower, N. S., and Kosower, E. M.: The glutathione status of cells. *Int. Rev. Cytol.*, 54:109–160, 1978.

Kurelec, B.: Exclusive activation of aromatic amines in the marine mussel (*Mytilus edulis*) by FAD containing monooxygenase. *Biochem. Biophys. Res. Commun.*, 127: 773–778, 1985.

Kurelec, B., and Pivcevic, B.: Evidence for a multixenobiotic resistance mechanism in the mussel *Mytilus galloprovincialis.*, 19:291–302, 1991.

Kurelec, B.; Britvik, S., Krca, S.; and Zahn, R. K.: Metabolic fate of aromatic amines in the mussel (*Mytilus galloprovincialis*). *Mar. Biol.*, 91:523–527, 1986.

Kurelec, B.; Krca, S.; Pivcevic, B.; Ugarkovic, D.; Bachmann, M.; Imsiecke, G.; and Muller, W. E. G: Expression of p-glycoprotein gene in marine sponges. Identification and characterization of the 125 kDa drug-binding glycoprotein. *Carcinogenesis*, 13:69–76, 1992.

Laszlo, A.: The relationship of heat-shock proteins, thermotolerance, and protein synthesis. *Exp. Cell. Res.*, 178: 401–414, 1988.

Lech, J. J., and Bend, J. R.: Relationship between biotransformation and the toxicity and fate of xenobiotic chemicals in fish. *Environ. Health Perspect.*, 34:115–131, 1980.

Lech, J. J., and Vodicinik, M. J.: Biotransformation. Fundamentals of Aquatic Toxicology, edited by G. M. Rand, and S. R. Petrocelli, pp. 526–557. New York: *Hemisphere*, 1985.

Lee, R. F.; Keeran,W. S.; and Pickwell, G. V.: Marine invertebrate glutathione *S*-transferases: purification, characterization and induction. *Mar. Environ. Res.*, 24:97–100, 1988.

Lee, R. F.; Singer, S. C.; and Page, D. S.: Responses of cytochrome P-450 systems in marine crabs and polychaetes to organic pollutants. *Aquat. Toxicol.*, 1:355–365, 1981.

Leung, H. W.; Vang, M. J.; and Mavis, R. D.: The cooperative interaction between vitamin E and vitamin C in suppression of peroxidation of membrane phospholipids. *Biochim. Biophys. Acta*, 664:266–272, 1981.

Li, G. C.: Induction of thermotolerance and enhanced heat shock protein in transiently thermotolerant Chinese hamster fibroblasts by sodium arsenite and ethanol. *J. Cell Physiol.*, 122:91–97, 1983.

Lind, C.; Hochstein, P.; and Ernster, L.: DT diaphorase as a quinone reductase: a cellular control device against semiquinone and superoxide radical formation. *Arch. Biochem. Biophys.*, 216:178–185, 1982.

Lindquist, S.: The heat shock response. *Annu. Rev. Biochem.*, 55:1151–1191, 1986.

Lindquist, S., and Craig, E. A.: The heat shock proteins. *Annu. Rev. Genet.*, 22:631–677, 1988.

Lindstrom-Seppa, P., and Oikari, A.: Hepatic biotransformation in fishes exposed to pulp mill effluents. *Water Sci. Tech.*, 20:167–170, 1988.

Lindstrom-Seppa, P.; Koivusaari, U.; and Hanninen, O.: Extrahepatic xenobiotic metabolism in north-European freshwater fish. *Comp. Biochem. Physiol.*, 69C:259–263, 1981.

Livingstone, D. R.; Kirchin, M. A.; and Wiseman, A.: Cytochrome P-450 and oxidative metabolism in molluscs. *Xenobiotica*, 19:1041–1062, 1989.

Livingstone, D. R.; Garcia Martinez, P.; Michel, X.; Narbonne, J. F.; O'Hara, S.; Ribera, D.; and Winston, G. W.: Oxyradical production as a pollutant-mediated mechanism of toxicity in the common mussel, *Mytilus edulis* L., and other molluscs. *Funct. Ecol.*, 4:415–424, 1990.

Lorenzana, R. M.; Hedstrom, O. R.; and Buhler, D. R.: Localization of cytochrome P450 in the head and trunk kidney of rainbow trout (*Salmo gairdneri*). *Toxicol. Appl. Pharmacol.*, 96:159–167, 1988.

Lorenzana, R. M.; Hedstrom, O. R.; Gallagher, J. A.; and Buhler, D. R.: Cytochrome P450 isozyme distribution in normal and tumor-bearing hepatic tissue from rainbow trout (*Salmo gairdneri*). *Exp. Mol. Pathol.*, 50:348–361, 1989.

Lu, A. Y. H.: Multiplicity of liver drug metabolizing enzymes. *Drug Metab. Rev.* 10:187–208, 1979.

Lu, P.; Metcalf, R. L.; Plummer, N.; and Mandel, D.: The environmental fate of three carcinogens: benzo[*a*]pyrene, benzidine, and vinyl chloride evaluated in laboratory model ecosystems. *Arch. Environ. Contam. Toxicol.*, 6: 129–142, 1977.

Ludke, J. L.; Gibson, J. R.; and Lusk, C. I.: Mixed function oxidase activity in freshwater fishes: aldrin epoxidation and parathion activation. *Toxicol. Appl. Pharmacol.*, 21: 89–97, 1972.

Malins, D. C., and Haimanot, R.: The etiology of cancer: hydroxyl radical–induced DNA lesions in histologically normal livers of fish from a population with liver tumors. *Aquat. Toxicol.*, 20:123–130, 1991.

Malins, D. C.; Ostrander, G. K.; Haimanot, R.; and Williams, P.: A novel DNA lesion in neoplastic livers of federal fish: 2,6-diamino-4-hydroxy-5-formamidopyrimidine. *Carcinogenesis*, 11:1045–1047, 1990.

Margoshes, M., and Vallee, B. L.: A cadmium protein from the equine kidney cortex. *J. Am. Chem. Soc.*, 79:4813–4814, 1957.

Marshall, C. J.; Vousden, K. H.; and Phillips, D. H.: Activation of c-Ha-*ras*-1 protooncogene by in vitro modification with a chemical carcinogen, benzo[*a*]pyrene diolepoxide. *Nature*, 310:586–589, 1984.

Martin, J.; Langeer, T.; Boteva, R.: Schramel, A.; Horwich, A. L.; and Hartl, F.-U.: Chaperonin-mediated protein folding at the surface of groEL through a 'molten globule'–like intermediate. *Nature*, 352:36–42, 1991.

Martin, J.; Horwich, A. L.; and Hartl, F.-U.: Prevention of protein denaturation under heat stress by the chaperonin hsp60. *Science*, 258:995–998, 1992.

Mather-Mihaich, E., and Di Giulio, R. T.: Antioxidant enzyme activities and malandialdehyde, glutathione and methemoglobin concentrations in channel catfish exposed to DEF and *n*-butyl mercaptan. *Comp. Biochem. Physiol.*, 85C:427–432, 1986.

Mather-Mihaich, E., and Di Giulio, R. T.: Oxidant, mixed function-oxidase and peroxisomal responses in channel catfish exposed to a bleached kraft mill effluent. *Arch. Environ. Contam. Toxicol.*, 20:391–397, 1991.

Mayer, F. L.; Versteeg, D. J.; McKee, M. J.; Folmar, L. C.; Graney, R. L.; McCume, D. C.; and Rattner, B. A.: Physiological and nonspecific biomarkers. Biomarkers: Biochemical, Physiological and Histological Markers of Anthropogenic Stress, edited by R. J. Huggett; R. A. Kimerle; P. M. Mehrle, Jr.; and H. L. Bergman, p. 5–86. Boca Raton, FL: Lewis, 1992.

McMahon, G.; Huber, L. J.; Moore, M. J.; Stegeman, J. J.; and Wogan, G. N.: Mutations in c-ki-*ras* oncogenes in diseased livers of winter flounder from Boston Harbor. *Proc. Natl. Acad. Sci. USA*, 87:841–845, 1990.

McMullin, T. W., and Hallberg, R. L.: A highly evolutionary conserved mitochondrial protein is structurally related to the protein encoded by the *Escherchia* gro EL gene. *Mol. Cell. Biol.*, 8:371–380, 1988.

Miller, M. R.; Hinton, D. E.; and Stegeman, J. J.: Cytochrome P450E induction and localization in gill pillar (endothelial) cells of scup and rainbow trout. *Aquat. Toxicol.*, 14:307–322, 1989.

Morimoto, R. I.; Tissieres, A.; and Georgopolis, C.: Stress Proteins in Biology and Medicine. Cold Spring Harbor, NY: Cold Spring Harbor Laboratory Press, 1990.

Morimoto, R. I.; Sarge, K.D.; and Abravaya, K.: Transcriptional regulation of heat shock genes. *J. Biol. Chem.*, 31: 21987–21990, 1992.

Mosser, D. D., van Oostrom, J.; and Bols, N. C.: Induction and decay of thermotolerance in rainbow trout fibroblasts. *J. Cell. Physiol.*, 132:155–160, 1987.

Murty, A. S.: Toxicity of Pesticides to Fish. Boca Raton, FL: CRC Press, 1986.

Nebert, D. W.; Nelson, D.; Coon, M.; Estabrook, R.; Feyereisen, R.; Fujii-Kuriyama, Y.; Gonzales, F.; Guengerich, F.; Gunsalus, I.; Johnson, E.; Loper, J.; Sato, R.; Waterman, M.; and Waxman, D.: The P450 superfamily: update on new sequences, gene mapping and recommended nomenclature. *DNA Cell Biol.*, 10:1–14, 1991.

Niimi, A. J., and Palazzo, V.: Biological half-lives of eight polycyclic aromatic hydrocarbons (PAHs) in rainbow trout (*Salmo gairdneri*). *Water Res.*, 20:503–507, 1986.

Nimmo, I. A.: The glutathione *S*-transferases of fish. *Fish Physiol. Biochem.*, 3:163–172, 1987.

Noël-Lambot, F.; Gerday, C.; and Disteche, A.: Distribution of Cd, Zn, and Cu in liver and gills of the eel, *Anguilla anguilla*, with special reference to metallothioneins. *Comp. Biochem. Physiol.*, 64C:177–187, 1978.

Nover, L.: Heat Shock Response. Boca Raton, FL: CRC Press, 1991.

Olsson, P.-E., and Haux, C.: Increased hepatic metallothionein content correlates to cadmium accumulation in environmentally exposed perch (*Perca fluviatilis*). *Aquat. Toxicol.*, 9:231–242, 1986.

Olsson, P.-E.; Haux, C.; and Forlin, L.: Variations in hepatic metallothionein, zinc and copper levels during an annual reproductive cycle in rainbow trout, *Salmo gairdneri*. *Fish Physiol. Biochem.*, 3:39–47, 1987.

Olsson, P.-E.; Larsson, A.; and Haux, C.: Metallothionein and heavy metal levels in rainbow trout (*Salmo gairdneri*) during exposure to cadmium in the water. *Mar. Environ. Res.*, 24:151–153, 1988.

Ostermann, J.; Horwich, A. L.; Neupert, W.; and Hartl, F. U.: Protein folding in mitochondria requires complex formation with hsp60 and ATP hydrolysis. *Nature*, 341: 125–130, 1989.

Overnell, J.; McIntosh, R.; and Fletcher, T. C.: The levels of liver metallothionein and zinc in plaice, *Pleuronectes platessa* L., during the breeding season, and the effect of oestradiol injection. *J. Fish Biol.*, 30:539–546, 1987.

Payne, J. F.: Field evaluation of benzopyrene hydroxylase induction as a monitor for marine pollution. *Science*, 191:945–946, 1976.

Payne, J. F.; and Penrose, W. R.: Induction of aryl hydrocarbon (benzo[*a*]pyrene) hydroxylase in fish by petroleum. *Bull. Environ. Contam. Toxicol.* 14:112–116, 1975.

Payne, J. F.; Fancey, L. L.; Rahimtula, A. D.; and Porter, E. L.: Review and perspective on the use of mixed-function oxygense enzymes in biological monitoring. *Comp. Biochem. Physiol.*, 86C:233–245, 1987.

Pelham, H. R. B.: Functions of the hsp70 protein family: an overview. Stress Proteins in Biology and Medicine, edited by R. I. Morimoto; A. Tissieres, and C. Georgopoulos, pp. 287–300. Cold Spring Harbor, NY: Cold Spring Harbor Laboratory Press, 1990.

Pesonen, M.; Celander, M.; Forlin, L.; and Andersson, T.: Comparison of xenobiotic biotransformation enzymes in kidney and liver of rainbow trout (*Salmo gairdneri*). *Toxicol. Appl. Pharmacol.*, 91:75–84, 1987.

Pruell, R. F., and Engelhardt, F. R.: Liver cadmium uptake, catalase inhibition, and cadmium thionein production in the killifish (*Fundulus heteroclitus*) induced by experimental cadmium exposure. *Mar. Environ. Res.*, 3:101–111, 1980.

Reed, D. J.: Glutathione: toxicological implications. *Annu. Rev. Pharmacol. Toxicol.*, 30:601–631, 1990.

Roch, M., and McCarter, J. A.: Chronic exposure of coho salmon to sublethal concentrations of copper. III. Kinetics of metabolism of metallothionein. *Comp. Biochem. Physiol.*, 77C:83–87, 1984.

Roch, M.; McCarter, J. A.; Matheson, A. T.; Clark, M. J. R.; and Olafson, R. W.: Hepatic metallothionein in rainbow trout (*Salmo gairdneri*) as an indicator of metal pollution in the Campbell River system. *Can. J. Fish Aquat. Sci.*, 39:1596–1601, 1982.

Roesijadi, G.: Metallothioneins in metal regulation and toxicity in aquatic animals. *Aquat. Toxicol.*, 22:81–114, 1992.

Roninson, I. B. (ed.): Molecular and Cellular Biology of Multidrug Resistance in Tumor Cells. New York: Plenum, 1991.

Rothman, J. E.: Polypeptide chain binding proteins: catalysts of protein folding and related processes in cells. *Cell*, 59:591–601, 1989.

Sailbil, H. R.; Zheng, A. M.; Roseman, A. M.; Hunter, A. S.; Watson, G. M. F.; Chen, S.; auf der Mauer, B. P.; O'Hara, B. P.; Wood, S. P.; Mann, N. H.; Barnett, L. K.; and Ellis, R. J.: ATP induces large quaternary rearrangements in a cage-like chaperonin structure. *Curr. Biol.*, 3:265–273, 1993.

Sanders, B. M.: Stress proteins: potential as multitiered biomarkers. Biomarkers of Environmental Contamination, edited by J. F. McCarthy and L. R. Shugart, pp. 165–191. Boca Raton, FL: Lewis, 1990.

Sanders, B. M.: Stress proteins in aquatic organisms: an environmental perspective. *Crit. Rev. Toxicol.*, 23:49–75, 1993.

Sanders, B. M., and Martin, L. S.: The use of stress proteins as biomarkers for environmental specimen banking. *Sci. Total Environ.*, 139/140:459–470, 1993.

Sanders, B. M.; Martin, L. S.; Nelson, W. G.; Phelps, D. K.; and Welch, W.: Relationships between accumulation of a 60 kDa stress protein and scope-for-growth in *Mytilus edulis* exposed to a range of copper concentrations. *Mar. Environ. Res.*, 31:811–897, 1991.

Sanders, B. M.; Pascoe, V. M.; Nakagawa, P. A.; and Martin, L. S.: Persistence of the heat-shock response over time in a common *Mytilus* mussel. *Mol. Mar. Biol. Biotech.*, 1:147–154, 1992.

Sanders, B. M.; Martin, L. S.; Howe, S. R.; Nelson, W. G.; Hegre, E. S.; and Phelps, D. K.: Tissue specific stress proteins in *Mytilus edulis* exposed to a range of copper concentrations. *Toxicol. Appl. Pharmacol.*, 125:206–213, 1994.

Schelin, C.; Tunek, A.; and Jergil, B.: Covalent binding of benzo[*a*]pyrene to rat liver cytosolic proteins and its effect on the binding to microsomal proteins. *Biochem. Pharmacol.*, 32:1501–1506, 1983.

Schlenk, D., and Brouwer, M.: Isolation of three copper metallothionein isoforms from the blue crab. (*Callinectes sapidus*). *Aquat. Toxicol.*, 20:25–34, 1991.

Schlenk, D., and Brouwer, M.: Induction of metallothionein mRNA in the blue crab (*Callinectes sapidus*) after treatment with cadmium. *Comp. Biochem. Physiol.*, 104C: 317–321, 1993.

Schlenk, D., and Buhler, D. R.: Xenobiotic biotransformation in the Pacific oyster. (*Crassostrea gigas*). *Comp. Biochem. Physiol.*, 94C:469–475, 1989.

Schlenk, D., and Buhler, D. R.: The in vitro biotransformation of 2-aminofluorene in the visceral mass of the Pacific oyster (*Crassostrea gigas*). *Xenobiotica*, 6:563–572, 1990.

Schlenk, D., and Buhler, D. R.: Flavin-containing monooxygenase activity in liver microsomes from the rainbow trout (*Oncorhynchus mykiss*). *Aquat. Toxicol.*, 20:13–24, 1991.

Schlesinger, M. J.: Heat shock proteins: the search for functions. *J. Cell Biol.*, 103:321–325, 1986.

Schlesinger, M., and Hershko, A.: The ubiqutin system. Cold Spring Harbor, NY: Cold Spring Harbor Laboratory Press, 1988.

Schlesinger, M. J.; Ashburner, M.; and Tissieres, A.: Heat shock from Bacteria to Man. Cold Spring Harbor, NY: Cold Spring Harbor Laboratory Press, 1982.

Seckler, R., and Jaenicke, R.: Protein folding and protein refolding. *FASEB J.*, 6:2545–2552, 1992.

Secombes, C. J.; Chang, S.; and Jefferies, A. H.: Superoxide anion production by rainbow trout macrophages detected by the reduction of ferricytochrome *c*. *Dev. Comp. Immunol.*, 12:201–206, 1988.

Singh, W.; Hall, G. L.; and Miller, M. G.: Species differences in membrane susceptibility to lipid peroxidation. *J. Biochem. Toxicol.*, 7:97–105, 1992.

Skowyra, D.; Georgopoulos, C.; and Zylicz, W.: The *E. coli dnaK* gene product, the hsp70 homolog, can reactivate heat-inactivated RNA polymerase in an ATP hydrolysis–dependent manner. *Cell*, 62:939–944, 1990.

Smolarek, T. A.; Morgan, S.; and Baird, W. M.: Temperature-induced alterations in the metabolic activation of benzo[*a*]pyrene in the bluegill fry cell line BF-2. *Aquat. Toxicol.*, 13:89–98, 1988.

Smolowitz, R. M.; Shultz, M. E.; and Stegeman, J. J.: Cytochrome P5450IA induction in tissues, including olfactory epithelium, of top minnows (*Poeceliopsis* spp.) by water borne benzo[*a*]pyrene. *Carcinogenesis*, 13:2395–2402, 1992.

Snowberger-Gray, E.; Woodin, B.; and Stegeman, J. J.: Sex differences in hepatic monooxygenases in winter flounder (*Pseudopleuronectes americanus*) and scup (*Stenotomus chrysops*) and regulation of P450 forms by estradiol. *J. Exp. Zool.*, 259:330–342, 1991.

Spacie, A., and Hamelink, J. L.: Alternative models for describing the bioconcentration of organics in fish. *Environ. Toxicol. Chem.*, 1:309–320, 1982.

Spies, R. B.; Felton, J. S.; and Dillard, L.: Hepatic mixed-function oxidases in California flatfishes are increased in contaminated areas and by oil and PCB ingestion. *Mar. Biol.*, 70:117–127, 1982.

Stanley-Samuelson, D. W.: Physiological roles of prostaglandins and other eicosanoids in invertebrates. *Biol. Bull.*, 173:92–109, 1987.

Steffins, J. C.: Heavy metal stress and the phytochelatin response. Stress Responses in Plants: Adaptation and Acclimation Mechanisms, edited by R. G. Alscher, and J. R. Cumming, pp. 377–394. New York: Wiley-Liss, 1990.

Stegeman, J. J.: Polynuclear aromatic hydrocarbons and their metabolism in the marine environment. Polycyclic Hydrocarbons and Cancer, Vol. 3, edited by H. Gelboin, and P. Tso, pp. 1–60. New York: Academic Press, 1981.

Stegeman, J. J.: Cytochrome P450 forms in fish: catalytic immunological and sequence similarities. *Xenobiotica*, 19:1093–1110, 1989.

Stegeman, J. J.; Binder, R. L.; and Orren, A.: Hepatic and extrahepatic microsomal electron transport components and mixed function oxygenases in the marine fish (*Stenotomus versicolor*). *Biochem. Pharmacol.*, 28:3431–3439, 1979.

Stegeman, J. J.; Smolowitz, R. M.; and Hahn, M. E.: Immunohistochemical localization of environmentally induced eytochrome P450IA1 in multiple organs of the marine teleost *Stenotomus chrysops* (scup). *Toxicol. Appl. Pharmacol.*, 110:486–504, 1991.

Stegeman, J. J.; Brouwer, M.; Di Giulio, R. T.; Forlin, L.; Fowler, B.; Sanders, B.; and Van Veld, P. A.: Molecular responses to environmental contamination: proteins and enzymes as indicators of contaminant exposure and effects. Biomarkers: Biochemical, Physiological and Histological Markers of Anthropogenic Stress, edited by R. J. Huggett; R. A. Kimerle; P. M. Mehrle, Jr.; and H. L. Bergman, pp. 235–335. Boca Raton, FL: Lewis, 1992.

Stegeman, J. J.; Kloepper-Sams, P. J.; and Farrington, J.: Monooxygenase inducton and chlorobiphenyls in the deep sea fish *Coryphaenoides armatus*. *Science*, 231: 1287–1289, 1986.

Stern, A.: Red cell oxidative damage. Oxidative Stress, edited by H. Sies, pp. 331–349. London: Academic Press, 1985.

Sternerson, J.; Kobro, S.; Bjerk, M.; and Arenal, U.: Glutathione transferases in aquatic and terrestrial animals from nine phyla. *Comp. Biochem. Physiol.* 86C:73–82, 1987.

Stubberfield, C. R., and Cohen, G. M.: Interconversion of NAD(H) to NADP(H). A cellular response to quinone-induced oxidative stress in isolated hepatocytes. *Biochem. Pharmacol.*, 38:2637, 1989.

Suzuki, C. A. M., and Cherian, M. G.: Renal glutathione and nephrotoxicity of cadmium-metallothionein in rats. *Toxicol. Appl. Pharmacol.*, 98:544–552, 1989.

Thomas, H., and Oesch, F.: Functions of epoxide hydrolase. *ISI Atlas Sci. Biochem.*, 1:287–291, 1988.

Thomas, R. E.; Rice, S. D.; Babcock, M. M.; and Moles, A.: Differences in hydrocarbon uptake and mixed function oxidase activity between juvenile and spawning adult coho salmon. (*Oncorhynchus kisutch*) exposed to Cook Inlet crude oil. *Comp. Biochem. Physiol.*, 93C: 155–159, 1989.

Van Veld, P. A.: Absorption and metabolism of dietary xenobiotics by the intestine of fish. *Rev. Aquat. Sci.*, 2: 185–203, 1990.

Van Veld, P. A.; Vetter, R. D.; Lee, R. F.; Patton, J. S.: Dietary fat inhibits the intestinal metabolism of the carcinogen benzo[*a*]pyrene in fish. *J. Lipid Res.*, 28:810–817, 1987.

Van Veld, P. A.; Stegeman, J. J.; Woodin, B. R.; Patton, J. S.; and Lee, R. F.: Induction of monooxygenase activity in the intestine of spot (*Leiostomus xanthurus*), a marine teleost, by dietary polycyclic aromatic hydrocarbons. *Drug Metab. Dispos.*, 16:659–665, 1988a.

Van Veld, P. A.; Patton, J. S.; and Lee, R. F.: Effect of preexposure to dietary benzo[*a*]pyrene (BP) on the first-pass metabolism of BP by the intestine of toadfish (*Opsanus tau*): in vivo studies using portal vein–catheterized fish. *Toxicol. Appl. Pharmacol.*, 92:255–262, 1988b.

Van Veld, P. A.; Westbrook, D. J.; Woodin, B. R.; Hale, R. C.; Smith, C. L.; Huggett, R. J.; and Stegeman, J. J.: Induced cytochrome P-450 in intestine and liver of spot (*Leiostomus xanthurus*) from a polycyclic aromatic hydrocarbon–contaminated environment. *Aquat. Toxicol.*, 17:119–132, 1990.

Van Veld, P. A.; Ko U.; Vogelbein, W. K.; and Westbrook, D. J.: Glutathione *S*-transferase in intestine liver and hepatic lesions of mummichog (*Fundulus heteroclitus*) from a creosote-contaminated environment. *Fish Physiol. Biochem.*, 9:369–376, 1991.

Varanasi, U.; Stein, J. E.; Nishimoto, M.; Reichert, W.; and Collier, T. K.: Chemical carcinogenesis in feral fish: uptake, activation and detoxication of organic xenobiotics. *Environ. Health Perspect.*, 71:155–170, 1987.

Varanasi, U.; Stein, J. E.; and Nishimoto, M.: Biotransformation and disposition of PAH in fish. Metabolism of Polycyclic Aromatic Hydrocarbons in the Marine Environment, edited by U. Varanasi, pp. 93–150. Boca Raton, FL: CRC Press, 1989.

Vasák, M.: Dynamic metal-thiolate cluster structure of metallothioneins. *Environ. Health Perspect.*, 65:193–197, 1986.

Veldhuizen-Tsoerkan, M. B.; Holwerda, D. A.; de Bont, A. M. T.; Smaal, A. C.; and Zandee, D. I.: A field study on stress indices in the sea mussel, *Mytilus edulis*: application of the "stress approach" in biomonitoring. *Arch. Environ. Contam. Toxicol.*, 21:497–504, 1991.

Vetter, R. D.; Carey, M. D.; and Patton, J. S.: Coassimilation of dietary fat and benzo[*a*]pyrene in the small intestine: an absorption model using the killifish. *J. Lipid Res.*, 26:428–434, 1985.

Viarengo, A.: Heavy metals in marine invertebrates: mechanisms of regulation and toxicity at the cellular level. *CRC Crit. Rev. Aquat. Sci.*, 1:295–317, 1989.

Viarengo, A.; Pertica, M., Canesi, L.; Biasi, F.; Cecchini, G.; Orunescu, M.: Effects of heavy metals on lipid peroxidation in mussel tissues. *Mar. Environ. Res.*, 24:354, 1988. Abstract.

Vogelbein, W. K.; Fournie, J. W.; Van Veld, P. A.; and Huggett, R. J.: Hepatic neoplasms in the mummichog (*Fundulus heteroclitus*) froma creosote-contaminated site. *Cancer Res.*, 50:5978–5986, 1990.

Walling, C.: Fenton's reagent revisited. *Accounts Chem. Res.*, 8:139–142, 1975.

Washburn, P. C., and Di Guilio, R. T.: Nitrofurantoin-stimulated superoxide production by channel catfish. (*Ictalurus punctatus*) hepatic microsomal and soluble fractions. *Toxicol. Appl. Pharmacol.*, 95:363–377, 1988.

Washburn, P. C., and Di Giulio, R. T.: The stimulation of superoxide production by nitrofurantoin, *p*-nitrobenzoic acid, and *m*-dinitrobenzene in hepatic microsomes of three species of freshwater fish. *Environ. Toxicol. Chem.*, 8:171–180, 1989.

Wedler, F. C.: Determinants of molecular heat stability. Thermotolerance, Vol. II, Mechanisms of Heat Resistance, edited by K. J. Henle, pp. 1–18. Boca Raton, FL: CRC Press, 1987.

Wenning, R. F., and Di Giulio, R. T.: Microsomal enzyme activities, superoxide production, and antioxidant defenses in ribbed mussels (*Gueskensia demissa*) and wedge clams (*Rangia cuneata*). *Comp. Biochem. Physiol.*, 90C;21–28, 1988.

Wenning, R. J.; Di Giulio, R. T.; and Gallagher, E. P.: Oxidant-mediated biochemical effects of paraquat in the ribbed mussel, *Gueskensia demissa*. *Aquat. Toxicol.*, 12: 157–170, 1988.

Whitlock, J.P.: The regulation of cytochrome P-450 gene expression. *Annu. Rev. Pharmacol. Toxicol.*, 26:333–369, 1986.

Williams, D. E., and Buhler, D. R.: Benzo[*a*]pyrene hydroxylase catalyzed by purified isozymes of cytochrome P450 from β-naphthoflavone–fed rainbow trout. *Biochem. Pharmacol.*, 33:3742–3753, 1984.

Winston, G. W., and Cederbaum, A. I.: Oxyradical production by purified components of the liver microsomal mixed-function oxidase system I. Oxidation of hydroxyl radical scavenging agents. *J. Biol. Chem.*, 258:1508–1513, 1983.

Winston, G. W.; Livingstone, D. R.; Lips, F.: Oxygen reduction metabolism by the digestive gland of the common marine mussel, *Mytilus edulis* L. *J. Exp. Zool.*, 255: 296–308, 1990.

Wofford, H. W., and Thomas, P: Effect of xenobiotics on peroxidation of hepatic microsomal lipids from striped mullet (*Mugil cephalus*) and Atlantic croaker (*Micropogonias undulatis*). *Mar. Environ. Res.*, 24:285–289, 1988.

Ziegler, D. M.: Flavin containing monooxygenases: enzymes adapted for multisubstrate specificity. *Trends Pharmacol. Sci.*, 11:321–324, 1990.

SUPPLEMENTAL READING

Anderson, R. S.: Modulation of blood cell-mediated oxyradical production in aquatic species: Implications and applications. Aquatic Toxicology—Molecular, Biochemical, and Cellular Perspectives, edited by D. C. Malins, and G. K. Ostrander, pp. 241–266. Boca Raton, FL: Lewis, 1994.

Armstrong, R. N.: Enzyme-catalyzed detoxication reactions: mechanisms and stereochemistry. *CRC Crit. Rev. Biochem.*, 22:39–88, 1987.

Darnell, J.; Lodish, H.; and Baltimore, D.: Molecular Cell Biology. New York: Scientific American Books, 1986.

Di Giulio, R. T.; Washburn, P. C.; Wenning, R. J.; Winston, G. W.; and Jewell, C. S.: Biochemical responses in aquatic animals: a review of determinants of oxidative stress. *Environ. Toxicol. Chem.*, 8:1103–1123, 1989.

Engel, D. W., and Brouwer, M.: Metallothionein and metallothionein-like proteins: physiological importance. Advances in Comparative and Environmental Physiology, Vol. 5, pp. 53–75. Berlin: Springer-Verlag, 1989.

George, S. G.: Enzymology and molecular biology of phase II xenobiotic-conjugating enzymes in fish. Aquatic Toxicology—Molecular, Biochemical, and Cellular Perspectives, edited by D. C. Malins and G. K. Ostrander, pp. 37–85, Boca Raton, FL: Lewis, 1994.

Halliwell, B., and Gutteridge, J. M. C.: Free Radicals in Biology and Medicine, 2nd ed. Oxford: Clarendon, 1989.

Huggett, R. J.; Kimerle, R. A.; Mehrle, P. M., Jr.; and Bergman, H. L. (eds.): Biomarkers—Biochemical, Physiological, and Histological Markers of Anthropogenic Stress. Boca Raton, FL: Lewis, 1992.

Mason, A. Z., and Jenkins, K. D.: Metal toxicity and detoxification in aquatic organisms. Interactions Between

Trace Metals and Organisms, edited by A. Tessier and D. Turner. International Union of Pure and Applied Chemistry. Boca Raton, FL: Lewis, in press.

Morimoto, R. I.; Tissieres, A.; and Georgopoulos, C. (eds.): Stress Proteins in Biology and Medicine. Cold Spring Harbor, NY: Cold Spring Harbor Laboratory Press, 1990.

Nover, L.: Heat Shock Response. Boca Raton, FL: CRC Press, 1991.

Peakall, D.: Animal Biomarkers as Pollution Indicators. London: Chapman & Hall, 1992.

Roesijadi, G: Metallothioneins in metal regulation and toxicity in aquatic animals. *Aquat. Toxicol.*, 22:81–114, 1992.

Roesijadi, G., and Robinson, W. E.: Metal regulation in aquatic animals: mechanisms of uptake, accumulation and release. Aquatic Toxicology—Molecular, Biochemical, and Cellular Perspectives, edited by D. C. Malins and G. K. Ostrander, pp. 387–420. Boca Raton, FL: Lewis, 1994.

Sanders, B. M.: Stress proteins in aquatic organisms: An environmental perspective. *CRC Crit. Rev. Toxicol.*, 23: 49–75, 1993.

Stegeman, J. J., and Hahn, M. E.: Biochemistry and molecular biology of monooxygenases: current perspectives on forms, functions, and regulation of cytochrome P450 in aquatic species. Aquatic Toxicology—Molecular, Biochemical, and Cellular Perspectives, edited by D. C. Malins and G. K. Ostrander, pp. 87–206, Boca Raton, FL: Lewis, 1994.

Varanasi, U. (ed.): Metabolism of Polycyclic Aromatic Hydrocarbon in the Marine Environment. Boca Raton, FL: CRC Press, 1989.

Winston, G. W., and Di Giulio, R. T.: Prooxidant and antioxidant mechanisms in aquatic organisms. *Aquat Toxicol.*, 19:137–161, 1991.

Chapter 18

FATE MODELING

D. Mackay, L. A. Burns, and *G. M. Rand*

INTRODUCTION

The primary concern of the aquatic toxicologist is to assess the effects of chemicals on the well-being of individual organisms and communities of organisms resident in the aquatic environment. The obvious assessment approach is to conduct laboratory toxicity studies under well-controlled, standardized conditions to determine the effects of chemicals singly, or in combination, on the organisms of interest over a range of chemical concentrations. It is hoped that concentrations of these chemicals encountered in the aquatic environment are (or can be regulated to be) well below those that cause adverse effects. Often, however, an element of judgment is necessary in this assessment because environmental concentrations of chemicals may not be known accurately and they may vary from place to place (i.e., spatially) and from time to time (i.e., temporally). There may also be episodes in the aquatic environment in which exposure to high concentrations of chemicals coincides with other stresses on organisms such as reduced oxygen concentration, high temperature, or general changes in water quality. These episodes may occur at times in the organism's life cycle when it is particularly vulnerable to stress.

The aquatic toxicologist is thus concerned not only with the effects of chemical exposure on the organism but also with the factors that determine exposure (expressed as magnitudes of concentration and duration). There are two general approaches for estimating such exposure—field monitoring (see Chapter 9) and modeling.

In a field monitoring program various components of the aquatic ecosystem, such as water, suspended and bottom sediments, and biota resident in the water column and benthic regions are sampled and analyzed. Such programs can be expensive, especially when the contaminant is present at low concentrations, difficult to analyze, and the characteristics of the ecosystem are such that it is necessary to take a large number of samples to account for spatial and temporal variations. In the event that a source of significant contamination and toxicity is identified and remedial measures are implemented (e.g., treat an effluent by biological oxidation), it may be necessary to repeat the monitoring program periodically to ensure that exposures are being reduced as planned. It may not be obvious how long it will take for a source reduction program to translate into reduced concentrations in the ecosystem. Assessment is also difficult if there are multiple sources of contaminants, such as various industrial and municipal sites, upstream sources, agricultural sources, and even atmospheric sources in the form of wet and dry deposition. Possibly most perplexing are situations in which there is an "in place" source of contamination from bottom sediments that have become appreciably contaminated by past discharges and are now slowly "bleeding" contaminant back to the water column.

A second approach to assessing exposure is to assemble a mathematical model of the system under investigation and use it to predict the environmental fate and likely future behavior of the chemical(s). The model is essentially an abstract representation of the aquatic ecosystem in the form of equations describing numerous transport and transformation processes (see Chapter 15). Inevitably, the model gives only an approximate representation of reality. Whether or not this representation is sufficiently accurate is a matter of judgment that must be made in light of local prevailing circumstances and requirements.

Mathematical models are most widely and successfully used to describe purely physical systems that tend to behave in a highly repeatable manner. For example, engineers routinely assemble mathematical models of bridges and aircraft to assess how close operating conditions are to those of failure. Agricultural chemists use models to predict pesticide behavior and persistence in soils following application. Models are also used to describe more complex hydrologic and atmospheric systems; these, however, are less reliable and therefore less successful. Generally, as systems become more complex, the models become less accurate and less satisfactory. In the last 20 yr considerable effort has been devoted to developing, improving, and validating models of aquatic systems. Models have been used successfully in various rivers, lakes, and estuarine situations to describe the behavior of organic chemicals, metals, radionuclides, and inorganic nutrients (phosphorus, nitrogen, etc.). Because aquatic systems are often very complex, there is an inherent limitation to the fidelity with which a model can describe the fate of chemicals in these systems and thus express accurately the concentrations that occur in various compartments of the ecosystem which in total define exposure to organisms.

Models engender several problems. They may require extensive data on the nature of the aquatic system that can be obtained only at considerable expense. The model may become so complex that only the modeler really understands its nature, its strengths, and its weaknesses. It may contain errors in that a rate constant for degradation or evaporation may be underestimated by a factor of 10. It may also contain a mistake as a result of the modeler forgetting to convert milligrams to micrograms. The task of model validation may be so difficult or expensive that it is not done well, or even not done at all. Frequently the modeler may be guilty of not conveying properly the expected accuracy or sensitivity of the model, thus leading to misinterpretation of findings. Consequently, the user of the model may be skeptical and distrustful, with the result that the model output is ignored.

On the positive side, a model can be invaluable in obtaining order of magnitude estimates of concentrations of substances in the environment. It also provides a mechanism by which important data such as the discharge rate of the chemical (e.g., 10 kg/d) and its biodegradation rate constant (e.g., a half-life of 10 d) can be translated into numerical estimates of concentration. In the absence of a model, the knowledge that a particular discharge rate applies but the chemical is subject to rapid biodegradation or slow evaporation can only be translated into weak, qualitative statements of likely environmental fate. The model thus attempts to be quantitative about these processes, rather than qualitative.

It is important to emphasize that while environmental scientists may vary in how they define exposure, the monitoring and modeling approaches are complementary and not competitive. There are numerous examples of vast amounts of monitoring data being gathered at great expense only to lie underinterpreted. There are probably also cases of bad decisions having been made on the results of erroneous environmental models. One of the most satisfying moments in environmental science occurs when there is a successful validation or reconciliation of model assertions with monitoring data. The implication is not only that the system is well characterized but also that it is well understood and the effects of interventions, for example, to reduce chemical discharges, can be predicted with fair reliability. Remedial measures to improve environmental quality can then be more readily justified, and the model can be used to design future monitoring efforts with greater economy and reliability.

The aquatic toxicologist who understands both the effect of exposure on the organism and the determinants of that exposure through familiarity with a well-constructed quantitative model is in a doubly powerful position to contribute to satisfactory management of the aquatic environment.

The purpose of this chapter is to describe the general fundamentals of aquatic fate modeling, how models are structured, and how they are used. In addition, a review of several models used in aquatic toxicology with case studies is presented.

FUNDAMENTALS OF MODELING

Types of Models

Statistical or Empirical Models

When there exists a substantial body of knowledge describing the system of interest, it is sometimes possible to construct a mathematical model to capture the observed system behaviors. These models have the advantage of being able to make use of standard statistical techniques to assess their fidelity to the database and to estimate the uncertainties of their predictions. Statistical or "empirical" models suffer, however, from an inability to make "predictions" outside the domain of prior observation. That is, an empirical model can be used only to interpolate between observations that have already been made; predictions outside this domain are likely to be invalid. Because the domain of ecotoxicological problems is multidimensional and often requires an assessment of new situations, empirical models have seen little use in the field. Still, some specialized subdomains of ecotoxicological statistical modeling have met with significant successes, as in the quantitative structure-activity relationships (QSARs) discussed elsewhere in this volume. Statistical models are also useful for identifying cause and effect; for example, a plot of algal concentration versus nutrient concentration may show a significant relationship suggesting (but not proving) that nutrients promote algal growth.

Simulation Models

Simulation models summarize the factual information relative to a particular system in the form of equations that express the laws of nature as we perceive them. Such models thus incorporate facts about chemistry, physics, biology, and toxicology in a single unified framework with the equations often being written as computer code. Not surprisingly, such models are easier to assemble when they describe purely physical and chemical phenomena. Biological systems are more difficult. Simulation models of ecotoxicological processes usually express mechanisms at a level more fundamental than that of the system under investigation. For example, the functioning of the immune system is best understood through the actions of the cells and tissues that make it up, and the dynamics of a population are best understood through the birth and death processes that add and subtract members from the population. Although often more difficult and expensive to construct than a statistical model, simulation models, once validated, can be used to predict over a wider range and can be used to assess new situations.

Qualitative and Semiquantitative Models

In some instances, ecotoxicological analyses are conducted through comparisons of very approximate estimates of exposures and effects, through a technique sometimes called the "quotient method" (see Chapter 28). In these techniques, limited ecotoxicological data are used to categorize chemicals as being, for example, of high, moderate, or low concern. This is often done in the context of a priority-setting exercise and is accomplished, for example, by estimating that toxic effects may occur in the field at a concentration 100 times less than an observed LC50, and then comparing this value with a presumed maximum exposure level. For example, in an agricultural setting the maximum exposure could be deduced by assuming that all the pesticide applied to a field is conveyed to an adjacent pond in runoff water. This concentration can then be divided by the LC50 to obtain a quotient. If the quotient is low, there is little risk, and as the quotient increases (because more pesticide is applied, or the chemical is more toxic and thus has a lower LC50) the risk increases. Although these procedures are obviously numerical, their approximate nature has resulted in them being characterized as "semiquantitative" ecotoxicological models.

In the balance of this chapter, we will be dealing almost exclusively with simulation models that use mathematical procedures to infer or predict ecotoxicological events. This focus arises from the fact that qualitative models are used for little more than screening out obviously safe chemicals. Statistical models are only marginally useful in full-scale fate and transport evaluations.

Quality

The quality of a model is highly dependent on the quality of elements used in the model. If the knowledge and quality of data are poor for a particular problem, a model is unlikely to provide the necessary information. Models that use data of high quality will generate new and useful information about the reactions, properties, and changes of a system in response to chemical substances. Because we frequently lack an adequate database on aquatic processes, much of the data (e.g., elements, parameters) used in a model must be estimated. Models never contain all the elements of a natural ("real") system. It is, however, important that the model contain the elements that are essential to the problem to be resolved or described. The most difficult aspect of modeling is to provide the necessary elements of knowledge and be able to estimate which components and processes to include in the model. Perhaps the art of modeling is to select the essential elements from the mass of other less-essential elements.

Recent Developments

Ecotoxicological modeling has developed rapidly over the past 5 yr because of several factors.

- Computer technology and aquatic model software have developed significantly so we can use and apply complex mathematical equations to biological problems. Many models are user-friendly and can be used in a manner analogous to statistical software without a full understanding of detailed coding or internal mathematical processes.
- Increased use of chemicals and development of new chemicals has led to potentially greater environmental exposure.
- Concern over sensitive and endangered species has increased.
- Accidental environmental catastrophes such as oil spills have led to ecological damage.
- There has been a considerable increase in our understanding of chemical fate in aquatic systems as a result of research into this aspect of ecotoxicology.
- Perhaps most important is an increase in public dissatisfaction with obviously degraded or contaminated ecosystems, which has led to demands for remediation or "cleanup."

Modeling Transport and Transformation Processes: An Example

Figure 1 is a sketch of an aquatic ecosystem showing various transport and transformation processes. The ecosystem in this case is Lake Ontario, the chemical is polychlorinated biphenyl (PCB), and the quantities are rates, in units of kg/yr, as of about 1965 (Mackay, 1989).

In this case there is direct discharge or emission of PCB to the lake from sources such as industrial facilities, spills, and sewage treatment plants. There is also inflow to the lake, in this case from the Niagara River, of PCB both in dissolved form and associated with, or sorbed to, particles. There is input from the atmosphere by dissolution of the chemical in rain and by deposition of chemical associated with aerosol particles either in rainfall (wet deposition) or by dry deposition, which is essentially dust fall. There is diffusive exchange between gaseous chemical in the air and dissolved chemical in the water (i.e., absorption and volatilization). In this example there is net exchange from the atmosphere to the water but often the net exchange is in the other direction.

There is outflow of the chemical from the lake in both dissolved and particle-associated forms. The chemical present in the water column partitions between dissolved and sorbed states to various mineral, organic, and biotic phases present in the water. Of particular importance are sediment-water exchange processes such as particle deposition, particle resuspension, and diffusion between the water column and the sediment pore water. Diffusion can take place both in dissolved form and when the chemical is associated with colloidal organic matter. The chemical is also subject to transformation processes such as hydrolysis, biolysis, and photolysis in the water column and in the sediment. Finally, the chemical present in the bottom sediment may be subject to burial; i.e., it is conveyed to depths in the sediment where it becomes essentially inaccessible.

Under normal environmental conditions, the concentration and hence the amount of chemical in the system change with time; thus, the inventories of chemical (in kilograms) in the water column and in the surface sediment change annually.

The aim of the fate model is to attach numbers, in this case kilograms per year or kilograms, to all these processes and compartments and thus assess the amount of chemical in the system and how fast this is changing. When a complete mass balance can be assembled as illustrated in Figure 1, the important fate processes can be readily identified and certain deductions can be made about how rapidly the system will respond to changes in quantities such as emission rates or air concentrations.

Compiling such a mass balance is a formidable task. It requires equilibrium partitioning information to describe how the chemical partitions between air and aerosols, between water and various solid phases in the water column, and between sediment pore water and sediment solids. It requires hydrologic information such as rates of water and suspended sediment inflow and outflow, rainfall rates, and rates at which suspended matter is deposited from the water column to the sediment and resuspended. Information is needed on rates of transformation by a variety of processes. Often these rates vary considerably from season to season as a result of temperature changes, the presence of ice cover, and changes in biological activity. An element of judgment is required for assessing the depth of the surface or accessible sediment layer. Typically, the modeler is faced with a situation in which some quantities such as river inflow or rainfall rate can be predicted with accuracies of about 10%, but other quantities such as sediment resuspension rates may only be known within a factor of 5.

The modeler's initial task is thus to assemble equations describing various equilibrium partitioning processes, transport processes, and transformation processes. Usually these equations express the rate or quantity as an algebraic function of the unknown concentrations of the substance in compartments of the system. The various expressions can then be combined into overall mass balance statements and the equations corresponding to these statements solved to estimate the concentrations in the water column, the various phases present in the water column, and the sediment. Many of these expressions have been described in Chapter 15 of this book. Convenient reviews can be found in the text by Lyman et al. (1982) and the report by Schnoor et al. (1987). Although not shown in Figure 1, it is often desirable to compile a complementary model describing the chemical fate in an aquatic food chain consisting of

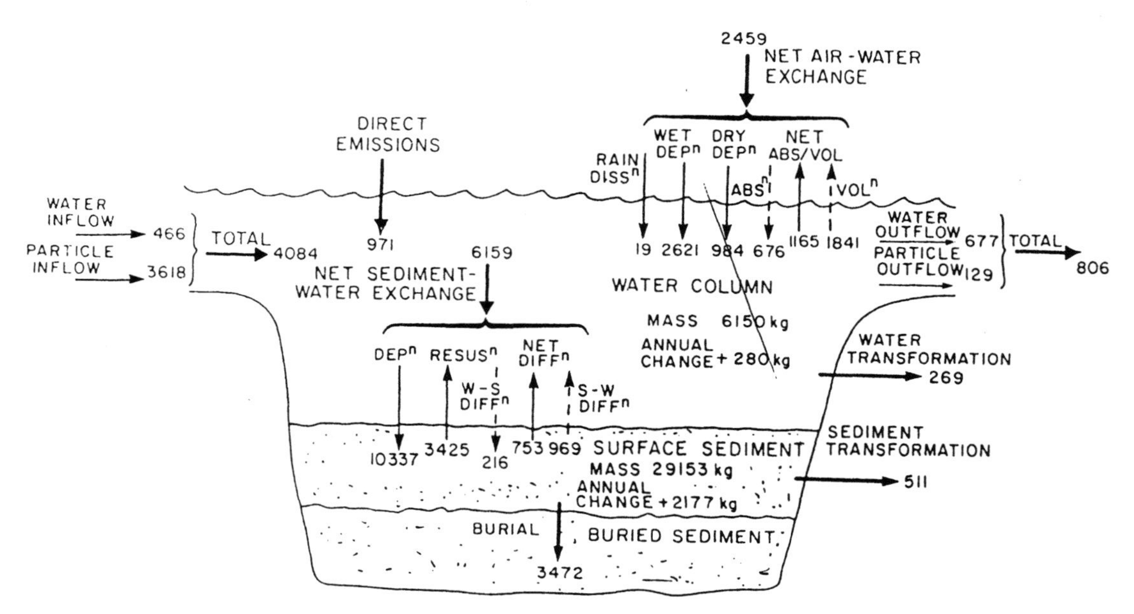

Figure 1. Results of a model showing estimated flows (kg/yr) of PCBs in Lake Ontario about 1965. (From Mackay, 1989.)

groups of organisms that take up chemical from the water or sediment and consume chemical contained in other food organisms.

The resulting abiotic and biotic concentrations can be compared with concentrations known (usually from laboratory experiments) to cause certain adverse effects on the organisms present in the aquatic ecosystem, and assertions can be made about the presence or absence of toxic effects. Obviously, the estimated concentrations should be compared with available monitoring data to validate the model.

Complete elimination of all chemical contamination from an aquatic system may be impossible. If the emissions can be related to their ecological implications for the aquatic environment under investigation, however, it may be possible to suggest or recommend how much the emissions must be reduced to eliminate adverse ecological effects. The use of the model in this case is to define the relationship between emission and ecological impact by synthesizing all the relevant data. The eventual result of the model is to recommend or suggest either emission limitations or a ban on the toxic chemical emissions that result in unacceptable adverse ecological effects.

Terminology

Forcing functions (i.e., driving forces, external variables) are variables of an external nature that affect the state of a system. A model is used to predict the potential changes that may occur in the system when forcing functions are varied with time. The forcing functions that can be controlled are called *control functions.* A control function in an ecotoxicological model is data on the input of chemical to the aquatic system, and other forcing functions of interest could be climatic variables (e.g., solar irradiance, rainfall) that affect biotic and abiotic processes.

State variables describe the state of the system. Selection of the correct state variable(s) is crucial for the structure of the model. For example, if the bioconcentration of a toxic chemical is to be modeled, the state variables would be the organisms in the most important food chains and the tissue concentrations of the toxic chemical. The values of the state variables predicted by changing the forcing functions can be considered as the results of a model.

Mathematical equations are used to represent the biological, chemical, and physical processes relative to a particular system. The equations define the relationship between forcing functions and state variables. The same types of processes may be used in different types of ecological systems, which indicates that the same equations can be used in different models. However, the same process may be formulated by different equations in different models because other factors may influence the process so that other equations may be a better fit for the process under investigation. In addition, the system or problem under investigation may be more complicated so that the model requires a great deal of detailed information.

Parameters are coefficients used in the mathematical expressions for the different processes. Coefficients may be constant for a part or the whole ecosystem. Sometimes parameters are not shown as constants but as ranges. Examples of parameters useful for ecotoxicological models include vapor pressure, water solubility, Henry's constant, biodegradability, and adsorption isotherms.

Universal constants, such as atomic and molecular weights and the gas constant, are also needed. Fortunately they are known exactly.

The Modeling Process

The most important steps of modeling are illustrated in Figure 2, which is adapted from Jorgensen (1988). The first step is the definition of the problem, i.e., how it is bounded by constituents of space, time, and subsystems. Complex models will contain more parameters and should account more accurately for the processes in the real system. However, with increased numbers of parameters in the model there will be more uncertainty, because more observations in the laboratory and field will have to be taken and some parameters may have to be estimated. Estimation of parameters is not error free and therefore may contribute to uncertainties in the results. Even if detailed information is available about a problem, it is unlikely that one can develop a model that is capable of accounting for the total input-output behavior of the real system and still be valid (Zeigler, 1976). Models should not be too complex, because it is then hard to obtain the data needed for calibration and validation.

To select the appropriate model complexity, it is important to remember not to make the model more complex than the data set available. The knowledge available about the processes, state variables, and the data set determine the model. If knowledge is inadequate, the model will have little detail and will have a high level of uncertainty. In contrast, if there is good knowledge of the problem that is to be modeled, a detailed model can be constructed with low uncertainty. The researcher must always present the shortcomings and uncertainties of the model and not present the model as a true picture of reality or nature with all its details. Given a certain set of data, addition of new state variables or parameters beyond a certain level of complexity rarely enhances the ability to model the ecosystem but only adds more uncertainty.

Referring to Figure 2, *data* should be available for all state variables. In only a few special cases should measurements of certain state variables be omitted, since the success of calibration and validation is closely linked to the *quality and quantity of data.*

After model complexity has been determined, it is important to develop a *conceptual process diagram* of the problem. The diagram (which could be similar

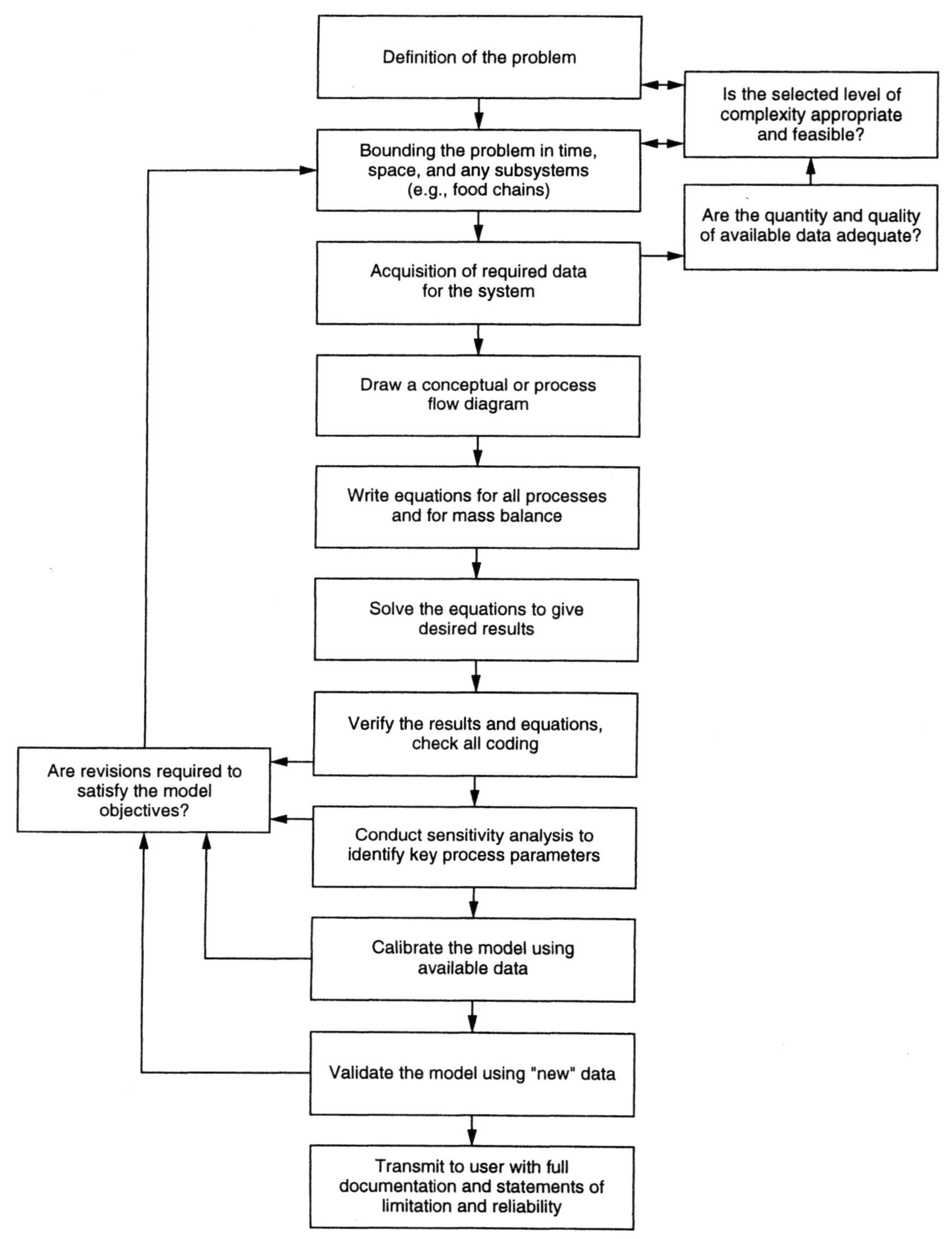

Figure 2. A tentative modeling procedure.

to Figure 1) should indicate the state variables, forcing functions, and processes needed in the model. The next step is the formulation of *mathematical equations* for the processes. Each process may have more than one equation. It is important that the most appropriate equation(s) be selected for the problem under consideration.

The equations are then *solved* by algebraic or numerical techniques to yield expressions for the state variables as a function of the forcing functions.

Verification can be conducted after the mathematical equations have been selected. This is a step that should not be deleted. At this step in the modeling procedure, several questions should be addressed.

If the model is run for a long period of time do the values of the state variables remain at approximately the same levels with variations in forcing functions?

Does the model respond the way it is supposed to?

At this stage in the modeling process, the modeler should apply the model to various conditions and varying parameters to become better acquainted with it and to understand how it responds.

Sensitivity analysis follows verification. In this procedure the modeler seeks an understanding of the sensitivity of the parameters, forcing functions, or submodel (process equations) to the state variables of greatest concern. If the modeler wants to simulate the concentration of a chemical in an omnivorous fish as a result of pesticide exposure, this will be the most important state variable in addition to the concentration of the chemical in insects, other invertebrates, and fish.

Sensitivity analysis is typically carried out by changing the parameter, the forcing function, or the submodel and observing the corresponding response of the state variables. Sensitivity analysis of submodels can also be done, in which the change in a state variable is documented when the equation of the submodel is either taken out of the model or changed to a different expression. These results are used to make structural changes in the model. For example, the sensitivity may show that a more detailed submodel is needed for the modeling. The model would thus be changed accordingly. The complexity and structure of the model should be closely aligned with the sensitivity analysis. In Figure 2 this is shown as a feedback from the sensitivity analysis. Furthermore, a sensitivity analysis of forcing functions provides an indication of the significance of various forcing functions and the accuracy needed of the forcing function data.

Calibration is used to improve the estimation of parameters when the model is being used to analyze a specific existing system, rather than in generalized assessments. Some parameters can be found in the literature as constants, approximate values, or intervals. Methods of parameter estimation are needed in ecotoxicological models because it is impossible to measure all parameters if we want to treat all possible combinations of toxic chemicals, their processes, and their interactions with living organisms. Approximate values for parameters are needed before the calibration process is started.

Calibration is basically a trial-and-error approach. The modeler tests the model by varying one or two parameters at a time and observing the response of the most crucial state variables. Models can be separated into submodels and calibrated independently. The set of parameters giving the closest accord between model output and measured values is selected.

The need for calibration is based on the following characteristics of ecotoxicological models and their parameters:

- In general, most ecotoxicological parameters are known as ranges, rather than as exact values. Most literature and estimated values have a degree of uncertainty.
- All ecotoxicological models are simplifications of nature. Many important components and processes may be incorporated in the model but its structure cannot account for all detail.
- In most ecotoxicological models one parameter represents the average value for several species. Since each species has its own parameter value, variation in species composition with time will give a corresponding variation in the average parameter used in the model.

Sometimes it may be difficult to calibrate the model. This may be due to the poor quality of the data set. It is also important that the observations reflect the dynamics of the system. If the purpose of the model is to describe one or a few state variables, it is imperative that the data show these dynamics. The frequency of data collection should reflect the dynamics of the state variables under consideration. It is suggested that the dynamics of all state variables be considered before the data collection program is determined in detail.

The following are suggestions for calibration of an ecotoxicological model:

- As many parameters as possible should be selected from the published literature.
- For parameters not found in the literature, estimation methods should be used or, when needed, parameters may be determined by laboratory or in situ field experiments.
- Sensitivity analysis should be used to determine which important parameters should be known with high certainty.
- An intense data collection program should be pursued for the most important state variables to provide better estimations for important parameters.
- Calibration should be done using data that have not been used in the model development. The most important parameters should be used for calibration. Calibration should first involve trial and error followed by an automatic calibration procedure.
- Results should then be used in a second sensitivity analysis.

- Another calibration should be done on parameters shown to be most important by the second sensitivity analysis. After this calibration the model can be considered calibrated and the validation step can begin.

In the *validation* step the model is tested against an independent data set and it is determined how well model simulations fit this data set. It should be emphasized that validation only confirms the behavior of the model under a given range of conditions represented by the available data. It is suggested that the model be validated using data from a period in which conditions prevailed other than those from the period of data collection for calibration. Even if an ideal validation cannot be done, it is still important to obtain a picture of reliability. The discipline of testing for reliability is a complex issue that is beyond the scope of this chapter.

A Taxonomy of Models in Ecotoxicology

Ecotoxicological mathematical models attempt to model the fate and/or effect of a toxic chemical in the ecosystem. They are typically different from general ecological models because they require more parameters, use more estimation methods, and may have an effect component to relate output exposure concentration to effect. Four classes of models are outlined below.

Steady-state models of the mass flows of toxic chemicals: Such models are used to describe and indicate the changes in a system if the input of a chemical is reduced or enlarged. These models assume that all variables and parameters are independent of time. They are based on the concept of mass balance (see next section).

Dynamic or unsteady-state models of toxic chemicals: These models express the time response of the system (abiotic and biotic) to changing inputs of chemical or to changes in other forcing functions. A simple example is given later. Usually the biotic community is defined and no allowance is made for the effect of the chemical on that community.

Food chain dynamics models: These models vary in complexity, can contain many state variables and parameters, and consider the movement of a chemical through the food chain consisting of one or more trophic levels. These models are typically used with more toxic, nonselective, and bioaccumulative chemicals that can affect the entire structure of an ecosystem.

Ecotoxicological models with effect components and population dynamics assessment: Ecotoxicological models with population assessment include the influence of a chemical substance on population dynamics parameters such as birth rate and mortality and interactions between different trophic levels. This type of model is a general population dynamics model with input on the relationship between concentrations of the chemical and the model parameters.

Included are relationships between the chemical and different types of sublethal and lethal biological effects. Effects may include any aberrant response at any level of biological organization from cells to ecosystems. Some of the questions that may be addressed on the effects are:

- To what extent does the chemical accumulate?
- What is the effect of chronic exposure?
- What is the effect of degradation products?

Obviously, a model that would be used to answer all of these questions would require an understanding of processes such as uptake, excretion rate, and biochemical transformation. A model of this type could become too complex in that there would be too many parameters to calibrate and more knowledge would be required than is available. Most models in this class consider only the accumulation of the chemical in the organism and its effect(s).

The static and dynamic models are simple and provide a realistic picture of the concentrations of chemicals in the ecosystem. Although many of the parameters are estimated, the application of safety factors makes them more acceptable. Models with effects components provide only a cursory understanding of what is happening because of our lack of basic knowledge about organisms and different levels of biological organization.

Validation

Mathematical models and computer simulation codes designed to aid in hazard assessment must be verified and validated before they can be used with confidence in a decision-making or priority-setting context. Operational validation, or full-scale testing via an ''appeal to nature'' in real-world situations, is usually the most ambiguous and least satisfactory part of the validation process. In most published studies, objective criteria of validity are lacking and evaluation of the models is subjective. This need not be the case, however, because acceptable model performance can usually be defined using relatively uncomplicated accuracy and precision criteria. Under these conditions, objective statistical tests can be formulated and executed to provide unambiguous proofs of validity in individual case studies or regulatory analyses. Such validations cannot, of course, demonstrate the global validity of a model; they merely provide a single instance of a failure to invalidate. The accumulation of a series of such validations, however, can give model users confidence in the general reliability and veracity of their decision tools.

The builders and users of ecotoxicological models are rightfully concerned that the models be valid, because decisions based on them have significant economic, public health, and ecological consequences. Substantial intellectual effort has been devoted to constructing a systematic framework for testing operational models (Mihram, 1972; Oren, 1981; Sargent, 1982). Despite the existence of these generalized guidelines, the specifics of model validation remain poorly developed and too frequently entirely subjective (Chapra and Reckhow, 1983). The problem of subjective validity criteria is especially acute in the case of models designed for hazard and risk analysis of toxic chemicals.

It must be recognized that objective test methods require objective criteria for judgment of model validity. The criteria usually must be based on considerations external to the model, that is, on the needs of the model user for accuracy and precision in model outputs. Chemical exposure models are usually used in a context of toxicological safety evaluations, in which a factor of 2 error is probably not excessive. We could thus formulate an objective test of validity of the following nature: Predicted values from the model must be within a factor of 2 of reality at least 95% of the time.

No one is in a better position to make such statements than the modeler, but a subtle "conflict of interest" may exist. The modeler naturally wants to present the work in a positive light, emphasizing the accomplishments. On the other hand, modesty and honesty dictate that the claims be reasonable and accurate.

THE MASS BALANCE CONCEPT

The concept of the mass balance, as illustrated in Figure 3, is fundamental to fate modeling. A volume of the environment is identified, and a mass balance equation is written for this volume. The volume may be that of a lake, a sediment, or an organism, but it must have defined physical boundaries and concentrations and preferably should be fairly homogeneous. Simple models normally treat one or two compartments but more complex models may have several water column layers and sediment layers, as well as compartments for organisms. As the number of compartments increases, the complexity of the system increases, the need for input data increases, and the model becomes more difficult to understand. But it is also likely to be more faithful to reality. The art of aquatic modeling can be viewed as the ability to select the "right" number of compartments and include only the processes that are most important using robust, simple expressions. This approach is perhaps best illustrated by a simple one-compartment model of an aquatic ecosystem as described in the next section.

From Conservation of Mass to Differential Equations

As illustrated in Figure 1, when a chemical is released into an aquatic ecosystem, it becomes subject to various transport and transformation processes. Transport from the point of entry into the bulk of the system takes place by advection and by turbulent dispersion (see Chapter 15). Transfers to sorbed forms and irreversible transformation processes take place simultaneously with transport. After a sufficient time, the chemical will be distributed throughout the system, with relatively smooth concentration gradients resulting from dilution, speciation, and transformation. The most efficient way to describe the parallel action of these processes is to combine them into a mathematical description of their total effect on the rate of change of chemical concentration in the system, through a set of coupled algebraic or differential equations.

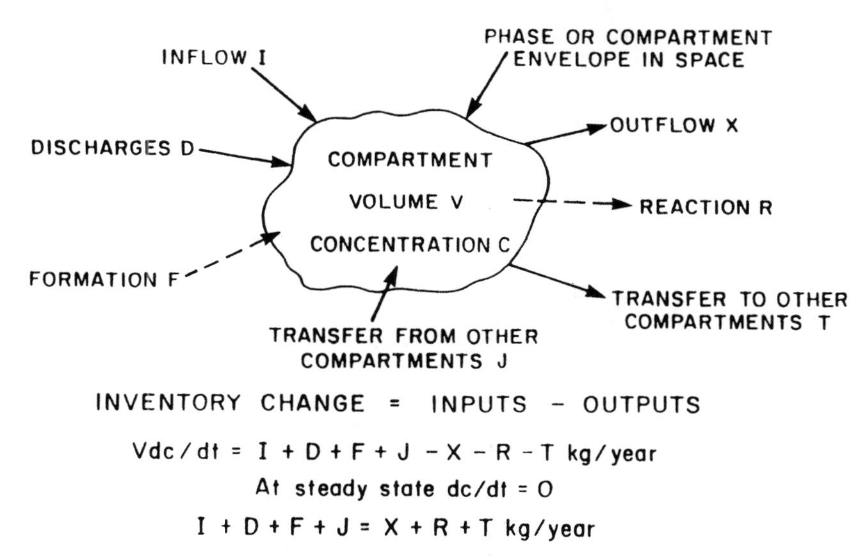

Figure 3. Illustration of the concept of mass balancing around a phase envelope control volume or compartment.

The simplest, and perhaps most important, principle used in constructing environmental models is the *law of conservation of mass*: matter can be neither created nor destroyed. Every molecule that enters a defined spatial zone (e.g., a laboratory tank or beaker, the epilimnion of a lake, the earth's atmosphere) must ultimately either leave the system, be transformed into another compound, or take up residence in the system. In other words, the inventory change equals the sum of the inputs less the sum of the outputs (Figure 3). Bankers use this to determine bank balances. The behavior of a chemical in an aquatic system thus can be rigorously circumscribed by using the conservation law as an accounting principle. Imagine, first, an invisible boundary drawn around some segment of the environment (Figure 3). This imaginary boundary encloses a *control volume*, or accounting unit to be used in a model. In real cases, boundaries are often chosen to correspond to actual physical discontinuities, such as riverbanks, the air-water interface, the benthic sediment–water column interface, or the depth of bioturbation of sediments. In many cases, however, the accuracy and detail of a model can be improved by including many relatively small accounting units bounded only by somewhat arbitrary lines drawn on a map or vertical profile of the water body.

Every chemical molecule entering the control volume now has three courses available to it: export (movement out of the control volume by outflow or transfer to other compartments), transformation, or residence. Environmental modeling begins by writing a mass balance around the control volume. This mass balance expresses the changes in concentration (increases and decreases in the number of chemical molecules per unit volume of the segment) that result from loadings (inputs), exports, and transformations. The product of concentration $[C]$ (mass/volume) and volume V is the accountable mass, so, using the usual notation of differential equations to denote a rate of change, one can write as shown in Figure 3

$$V\frac{d[C]}{dt} = I + D + F + J - X - R - T \qquad (1)$$

that is, rate of change of mass (the rate of accumulation or loss of chemical from the control volume) = discharges − export − transformations.

The terms in this equation must have consistent physical units for the equation to be valid. For example, when V is expressed in liters and $[C]$ in milligrams per liter, the input loadings and the exports must have units of milligrams per unit time (e.g., mg/s). If we consider waterborne exports only, X can be set equal to FC, where F is the flow rate of water leaving the system in liters per second. The transformation process R must also have the proper mass units (mg/s), usually as the product of volume (liters), concentration (mg/liter), and rate constant (s^{-1}).

Models of real ecosystems often consist of many such unit control volumes coupled by transport equations. An export from one unit then becomes an input or load to another. The loadings on each control volume can thus be the sum of many terms, including "allochthonous" (external) chemical loadings from industrial or agricultural discharges or contaminated rainfall plus "autochthonous" (internal) loadings due to hydrodynamic motions and other transport cycles within the system that move chemicals from one control volume to another. In models that include both parent compounds and progeny transformation products, the rate of transformation of the parent to the progeny compound is also an internal loading of the progeny compound.

Space and Time Scales

Synthetic chemicals enter the aquatic environment by many routes, including accidental spills, industrial effluents, surface runoff from agricultural fields, leaching from landfills and disposal sites, leaching of materials deposited on soils, from the atmosphere, and sewage treatment plant effluents. The rate, magnitude, and duration of releases by these pathways can be very different. The toxic impact of a substance and the rates of chemical transformations are, however, functions of the magnitude of current environmental concentrations, rather than the precise route of entry of the material. Even so, in some circumstances variable magnitudes and durations of chemical releases, in combination with variable transport regimes, result in highly variable exposures that can be evaluated only by detailed simulation models. In other cases, concentrations and exposure durations are relatively independent of short-term variations in releases, and very long-term or steady-state analyses are required. Consequently, it is not possible to specify unambiguously a single environmental model or computer program suitable for risk analysis or hazard evaluation. Models and programs are, of necessity, designed in the context of specific problems, and systematic risk assessments usually must employ a variety of tools in order to address the entire range of concerns at issue.

Chemical spills and, to a lesser degree, early post-application agricultural runoff, typically require an "event-oriented" model. One approach to this problem is to treat the "event" as a sudden injection of a given quantity of the chemical into the system. The time course of dilution and breakdown of the chemical can then be modeled to predict, among other things, the extent of contamination of downstream areas as a result of a chemical spill. For example, Neely et al. (1976) derived an analytical expression for transport and volatilization of soluble chemicals in flowing waters. They used the model to evaluate the consequences of a spill of chloroform in the Mississippi River. Thibodeaux (1979) summarized approaches needed to expand this analysis to account for situations such as the formation of pools of liquid chemical on the bottom of watercourses.

For many chemicals, relatively small but continuous or chronic releases are a more significant route of entry into aquatic systems than are sudden, massive accidents. Examples include runoff of persistent agricultural chemicals, industrial effluents from manufacturing plants, and dispersed consumer products entering aquatic systems from sewage treatment plants. In these situations, environmental evaluations must be projected over years or decades to assess the ultimate aquatic concentrations and fate of chemicals. These situations often require steady-state models, rather than the event-oriented models needed to evaluate the transient consequences of accidental spills.

Many significant fate processes occur so rapidly that their time-dependent behavior is almost never of interest in environmental situations. Ionization reactions of acids and bases typically reach equilibrium in a matter of seconds. Because the time scales of interest in toxicological investigations are seldom shorter than hours or days, ionization can usually be assumed to be in a continuous, constant equilibrium state. This is often true of sorption to nonliving sedimentary materials and of complexation reactions, although exceptions to this generalization have been discovered.

Models must be tailored to fit the problem-specific characteristics of the chemical and the timing and duration of chemical loadings. For example, a full evaluation of the possible risks posed by manufacture and use of a new chemical could begin from a detailed time series describing the expected releases of the chemical into aquatic systems over the projected duration of its manufacture. Given an equivalently detailed time series for environmental variables, machine integration would yield a detailed picture of estimated or expected environmental concentrations (EECs) in the receiving water body over this period. The model could also be used to forecast the time required for dissipation of the chemical residuals present in the system when manufacture is terminated. This approach is very costly, however, and errors accumulate in very large-scale simulation studies.

More realistically, then, risk analysis must begin from a general evaluation of the behavior of a chemical in the environment and the associated patterns of chemical loadings. Consider, for example, a chemical that is relatively short-lived in the environment. Suppose also that the chemical is released only sporadically and then only for limited periods. The manufacture of some low-volume specialty chemicals is one example fitting this description. In this case, transient lethal conditions (i.e., acute effects) are the primary frame of reference. Models patterned after classical analyses of dissolved oxygen problems in streams and rivers could be used to assess the risk associated with such a chemical.

Chemicals that are very persistent in the environment or are released at a steady rate over long periods fall at the opposite extreme. In this case, the primary toxicological frame of reference is usually that of chronic effects. For chemicals that are also extensively sorbed, the long-term impact on the benthos may be a major consideration in risk analysis. Models for analyzing these kinds of chemicals must include steady-state approaches; that is, the analysis must consider the long-term, ultimate residual concentrations to be expected in the environment.

Regardless of the time frame used, models that describe spatially heterogeneous systems and the associated transport processes lead to partial differential equations. These equations usually must be solved by numerical techniques with digital computers, rather than by explicit analytic solution of the differential equations. Standard numerical techniques in one way or another divide the system, which is continuously varying in space and time, into a set of discrete elements. Spatially discrete elements are often referred to as grid points, nodes, segments, or compartments. In a *compartmental* or *box* model, the physical space of the system is subdivided into a series of physically homogeneous segments (compartments) connected by advective and dispersive fluxes. Each compartment is a particular volume element of the system, containing water, sediments, biota, and dissolved and sorbed chemicals. Loadings and exports can then be represented as mass fluxes across the boundaries of the volume elements; reactive properties can be regarded as point processes centered within each compartment (Figure 3). Similarly, continuous time is often represented by fixing the system driving functions (e.g., rainfall or sunlight intensity) for a short interval, integrating over the interval, and then ''updating'' the forcing functions before evaluating the next time step.

When describing an aquatic system for simulation modeling, the number and sizes of the spatial segments must be prescribed. In most simulation techniques, each physical segment is sized so that it can be assumed to be *well mixed*, that is, so that the rate of chemical reactions in the segment does not depend on how rapidly materials circulate within the segment. In effect, the physical segments must not be so large that internal gradients in the forcing functions have a major effect on the calculated transformation rate constant of the chemical. Physical boundaries that can be used to demarcate system segments include the air-water interface, the thermocline, the benthic interface, and perhaps the depth of bioturbation of sediments. Some processes, however, are driven by environmental factors that occur as gradients in the system, and some processes are controlled by conditions at interfaces. For example, sunlight intensity decreases exponentially with depth in the water column, and volatilization occurs only at the air-water interface.

In well-mixed situations, gradients and processes at interfaces can be modeled easily. For example, the net rate of photochemical reaction in a well-mixed zone of a lake can be calculated by computing the average light intensity in the well-mixed zone. Similarly, volatilization models compute the flux of chemicals across the air-water interface by assuming that some depth of water below the interface zone continually renews the supply of chemical at the interface. The rate of resupply of chemical does not affect the rate of volatilization, because vertical turbulence moves the chemical into the interface zone faster than it is lost to the atmosphere. In very quiescent lakes, however, relatively fine-scale vertical segmentation may be needed in order to arrive at an accurate estimate of photolysis or volatilization rates. In these cases, the speed of vertical transport processes, which move the chemical to the interface or move it along a gradient in sunlight intensity, may control the overall speed of photochemical transformation or volatilization loss of chemical from the lake. Thus, although physical boundaries help define the appropriate segmentation for use in models, slow transport processes in physically homogeneous areas of the water body may require that each zone be subdivided, for computational purposes, into many additional segments (Plane et al., 1987).

Regardless of the temporal and spatial horizons involved, mathematical models designed to evaluate the behavior of chemicals must include descriptions of the physical, chemical, and biological processes governing the transport and transformation of chemicals in aquatic environments. These process models must be stated as a combination of the properties of the chemical and as functions of independently measurable environmental driving forces or coupling variables (e.g., pH, temperature). This approach makes it possible to study the fundamental chemistry and toxicology of compounds in the laboratory and then, based on independent studies of the kinetics and intensity of driving forces in aquatic systems, the probable behavior of the compound can be evaluated for systems that have never been exposed to it.

For example, the EXAMS program (Burns, 1990; Burns and Cline, 1985; Burns et al., 1982) as described later in this chapter is a modeling *system*, a framework for constructing very simple to very detailed models of aquatic ecosystems for ecotoxicological analyses, which includes three very different *modes* of operation to accommodate this large variety of problems. Mode 1 analysis carries long-term loadings to their ultimate steady state and then dynamically investigates persistence by examining the natural recovery of the system if the pollutant load is terminated. In mode 2, the analyst has direct access to the timing and sequencing of simulation segments and can examine detailed contamination and recovery scenarios. In mode 3, effluent discharges and event-driven non-point-source discharges (e.g., agricultural chemical surface runoff events) can be combined *ad libitum*, along with EXAMS databases of monthly average environmental conditions. Seasonal changes in climate and hydrology, agricultural practices, and sporadic industrial discharges thus all are accommodated within the EXAMS framework.

FATE AND EFFECT MODELS OF INTEREST IN AQUATIC TOXICOLOGY

There have been several reviews of "water quality" models and aspects of model formulation. Notable are those of Schnoor et al. (1987), Booty and Lam (1989), Swann and Eschenroeder (1983), Jorgensen (1984, 1988), Jorgensen and Gromiec (1989), and Hites and Eisenreich (1987). Some of the nutrient and toxic chemical models available to aquatic scientists are briefly reviewed below.

Nutrient Models

Much of the early modeling effort was devoted to establishing the relationship between discharge rates of phosphorus (P) and subsequent P concentrations in the water column, and the resulting enhancement of algal growth or "eutrophication." These models tackle a challenging problem because the presence of P controls the amount of algal biomass and hence the rates of P deposition. A wide range of models has been developed, varying in complexity from simple semiempirical expressions containing terms for loading, depth, water residence time, and sedimentation rate to very complex descriptions of the mass balances of P, oxygen, and carbon for multisegmented lakes. Examples are the studies by Vollenweider (1968, 1969), Lam and Schertzer (1987), Lam et al. (1982), Snodgrass (1987), and Schnoor and O'Connor (1980). Lake Erie has been the focus of much of this effort in North America.

Organic Chemical Models

Most models are elaborations of the simple model described earlier with additional compartments and improved or more detailed expressions for chemical fate. They are most frequently applied to organic chemicals that are hydrophobic (and thus bioaccumulate) or volatile (and thus evaporate) or particularly toxic, for example, pesticides. Muir (1991) has comprehensively reviewed the last class.

The EXAMS (exposure analysis modeling system) model was developed mainly by Burns, Cline, Zepp, Baughman, and Lassiter at the U.S. Environmental Protection Agency (EPA) Environmental Research Laboratory at Athens, Georgia (Baughman and Lassiter, 1978, Burns et al., 1982). EXAMS II is a more recent version suitable for use on personal computers (Burns and Cline, 1985; Burns, 1990). EXAMS is an interactive computer-based system for specifying and storing the properties of chemicals and ecosystems, modifying them using simple commands, and conducting rapid evaluations and error analyses of the probable aquatic fate of synthetic organic chemicals. EXAMS constructs simulation models by combining the loadings, transport, and transformations of a chemical into a set of differential equations, using the law of conservation of mass as an accounting principle. This is accomplished by computing the total mass of chemical entering and leaving each section of a body of water as the algebraic sum of external loadings, transport processes that distribute chemicals through the system and export them across its external boundaries, and transformation processes that convert chemicals to daughter products. The differential equations are then assembled and solved to give a picture of the behavior of chemicals in an aquatic ecosystem. The program produces output tables and simple graphics describing chemical

- Exposure: the expected or estimated environmental concentrations resulting from a particular pattern of chemical loadings
- Fate: the distribution of the chemical in the system and the fraction of the loadings consumed by each transport and transformation process
- Persistence: the time required for purification of the system (via export or transformation processes) should the chemical loadings cease

EXAMS includes separate mathematical models of the kinetics of the physical, chemical, and biological processes governing transport and transformations of chemicals—photolysis, hydrolysis, biolysis, oxidations, volatilization, and advective and dispersive transport. This set of unit process equations is the central core of EXAMS. EXAMS' "second-order" or "system-independent" approach makes it possible to study the fundamental chemistry of materials in the laboratory and then, based on independent studies of the levels of driving forces in aquatic systems, evaluate the potential fate of materials in systems that have not yet been exposed. EXAMS treats ionization and partitioning of the compound with sediments and biota as thermodynamic properties or purely local equilibria peculiar to each segment of the environmental model, as opposed to a treatment as system-wide "global" equilibria. In this way, EXAMS allows for the impact of spatial variation in sediment properties, pH, etc. on chemical reactivity. EXAMS computes the behavior of trivalent organic acids, bases, and ampholytes; each ionic species can have its own distinctive pattern of sorption and complexation with naturally occurring particulates and dissolved organic matter. Reaction pathways can be entered for the production of transformation products, whose further transport and transformations are then also simulated by EXAMS.

EXAMS computes the kinetics of transformations by direct photolysis, hydrolysis, biolysis, and oxidation reactions. The input chemical data for hydrolytic, biolytic, and oxidative reactions can be entered either as single-valued, second-order rate constants, or as pairs of values defining the rate constant as an Arrhenius function of the temperature in each segment of the water body. EXAMS has been designed to accept standard water-quality parameters and system characteristics that are commonly measured by limnologists throughout the world and chemical data sets conventionally measured or required by U.S. EPA regulatory procedures.

EXAMS is a computer-based system for building models of aquatic ecosystems and running simulation studies on the behavior of chemical contaminants. EXAMS' environmental models are maintained in a file composed of concise ("canonical") descriptions of aquatic systems, in which a body of water is described as a set of N segments or distinct zones in the system. By applying the principle of conservation of mass to the transport and transformation process equations, EXAMS compiles a differential equation for the net rate of change of chemical concentration in each segment. The resulting system of N differential equations describes the mass balance for the entire system. EXAMS includes a descriptor language that simplifies the specification of system geometry and connectedness.

EXAMS provides three operating "modes" of increasing complexity. In the simplest case (mode 1), EXAMS executes a direct steady-state solution of the dynamic system equations, thus generating a long-term analysis using a single set of environmental conditions (e.g., annual average driving forces). In mode 2, EXAMS makes available initial-value approaches that can be used to set initial conditions and introduce immediate "pulse" chemical loadings. To the extent that changes in hydrographic volumes (e.g., during spates) can be neglected, this mode can be used to evaluate shorter-term transport and transformation events by segmenting the input data sets and simulation intervals according to time slices under full user control. In mode 3, EXAMS uses a set of 12 monthly values of all environmental parameters, with input loads that can change monthly and can also include pulse events on individual dates, to compute the dynamics of chemical contamination over the course of one or more years. The outputs produced by the system are analogous for all modes of operation, although they differ in detail. For example, in mode 1, a summary table and sensitivity analyses of system fluxes are reported for steady-state conditions; in mode 2 the reports are generated for conditions at the close of each time slice; and in mode 3, the program reports annual (or interannual) average values and the size and location of exposure extrema. EXAMS produces 20 output tables; these include an echo of the input data and integrated analyses of the exposure, fate, and persistence of the chemical or chemicals under study.

A series of models has been developed by Connolly and Winfield (1984), Connolly and Thomann (1985), and O'Connor (1988a, 1988b) that have been named WASTOX, TOXIWASP, and related acronyms (Thomann and DiToro, 1983). TOXIWASP is another dynamic compartmentalized model that evaluates transport and transformation of organics (and heavy metals) in streams, stratified lakes and reservoirs, estuaries, and coastal waters. The mathematical formulation of TOXIWASP is based on the conservation law and was originally developed by combining the transport framework of WASP (Di Toro et al., 1982) with the kinetic structure of EXAMS (Burns et al., 1982) with the addition of a mass balance for solids and sediment. TOXIWASP is used to calculate sediment and chemical concentrations in the water and concentrations of dissolved and sorbed chemicals in the sediment bed. The program treats sediment as a conservative constituent that is advected and dispersed among water segments, that settles onto and erodes from benthic segments, and that moves between benthic segments through net sedimentation or erosion. Settling, deposition, resuspension, and burial velocities of sediments are not calculated but must be entered by the user. Reaction and transformation rates are based on addition of pseudo-first-order rate constants for hydrolysis, oxidation, biodegradation, volatilization, and photolysis of a toxic chemical dissolved in water or sorbed to sediments. Chemical transformation and reaction rates vary in time and space depending on the characteristics of the chemical and environmental conditions.

Input data requirements (physical, chemical, biological) for TOXIWASP are as intensive as they are for EXAMS II. TOXIWASP considers three sorption possibilities (dissolved, sediment sorbed, and biosorbed) for the forms of a chemical; it considers ionization of four species like EXAMS II. TOXIWASP handles both point and nonpoint source loads and can estimate time-varying chemical exposure resulting from chemical loadings delivered as slugs or pulses.

WASTOX is similar to TOXIWASP, differing mainly in its treatment of bioaccumulation. An empirically based food chain model is contained within WASTOX for calculating chemical concentrations in biota based on predicted aquatic concentrations (Connolly and Thomann, 1985). These models have been applied to the individual Great Lakes (especially Ontario and Michigan), to more contaminated sections of the Great Lakes such as Saginaw Bay, and to various riverine systems in the eastern United States. Most emphasis has been on persistent organic contaminants such as PCBs and kepone.

Halfon (1985) has developed an organic fate model for toxic substances in Lake Ontario (TOXFATE) and has applied it to persistent organochlorine chemicals such as mirex and chlorobenzenes. TOXFATE simulates time-varying concentrations in the water column, suspended and bottom sediments, and the percentage of loadings lost to the atmosphere. The equations can be parameterized to represent a variety of chemicals, and the model is formalized as a system of ordinary differential equations. The state variables (i.e., concentration of chemical in each part of the system) are chemical concentrations in suspended clay and colloidal matter, suspended silt, suspended sand, water, plankton and fish, bottom sediments, and benthos compartments. The model also includes a resuspension submodel. Process relationships are nonlinear to reflect movement of the chemical between sediments, water, and biota. The chemical is assumed to be immediately and completely mixed within each compartment. Two size classes of fish are also used in the model; small fish (about 5 g wet weight) and large fish (about 3.1 kg wet weight). Biouptake also includes an equation to describe excretion in fish.

HSPF, the hydrologic simulation program (FORTRAN) (Donigian et al., 1984, Johanson et al., 1984), is a comprehensive package for simulation of watershed hydrology and water quality for both conventional and toxic organic pollutants. HSPF incorporates the watershed-scale agriculture runoff management (ARM) model (Donigian and Davis, 1978) and non–point source (NPS) models (Donigian and Crawford, 1976) into a basin-scale analysis framework that includes fate and transport in one-dimensional stream channels. HSPF uses the SERATRA model (Onishi and Wise, 1982) for sediment transport, chemical decay, sediment-chemical partitioning, and risk assessment. It is a fully dynamic model that can simulate chemical behavior over an extended period of time and is applicable to nontidal rivers, streams, and narrow impoundments (one-dimensional reservoirs). HSPF requires a considerable amount of information, which includes time series data (e.g., precipitation, evapotranspiration data) and user-controlled inputs (e.g., characteristics of the land surface like soil types and agricultural practices). When channel processes are important, data like stream flow, channel geometry, and in-stream chemical concentrations are needed. The input, therefore, must represent spatial and temporal variations in flow rate and chemical input (loadings) that may result from combined meteorologic, hydrologic, chemical, and biological processes in the study area. The result of an HSPF simulation is a time history of runoff flow rate, sediment load, and nutrient and chemical concentrations, along with a time history of water quality and quantity in the watershed at different points. Underlying sediment and pore water are, however, not modeled. Effective use of HSPF requires a considerable amount of data and much more effort for data management than EXAMS II and TOXIWASP.

Models of Metallic Chemicals

The same general principles of partitioning, transport, and mass balance apply to metals as apply to organic chemicals. Obviously, degradation processes *per se* are negligible except in the case of radionuclides. The speciation of the metal becomes very important as influenced by pH, oxygen, and the presence of electrolytes and mineral surfaces. It is often necessary to treat separately the thermodynamic equilibrium existing between metal species with the aid of speciation models such as the MINTEQ series (Brown and Allison, 1987; Allison et al., 1991; Felmy et al., 1984a,b), which is supported by the U.S. EPA.

MINTEQ is a geochemical model that computes equilibrium aqueous speciation, adsorption-desorption, and the mass of metal transferred into and out of solution as a result of dissolution or precipitation of solid phases. The program requires thermodynamic (already contained in the MINTEQ database) and water quality (user-provided) data. Thermodynamic data include basic information (i.e., equilibrium constants) required to predict formation of each species or solid phase. Although the MINTEQ database is one of the most evaluated for any geochemical model, its main limitation is that it contains equilibrium constants for only a limited number of metals. In addition, the program considers every reaction as if it were at chemical equilibrium when in fact chemical reactions like precipitation-dissolution and oxidation-reduction are not and can be very slow. However, MINTEQ does not have the capability of computing kinetic, transfer, or transport processes. MINTEQ has been linked with EXAMS (MEXAMS) to assess fate and transport of heavy metals in aquatic systems.

An example of a model of metal fate in a lake is that of Dolan and Bierman (1982) for Saginaw Bay, Lake Huron. Another is the fate of lead in Lake Ontario as discussed by Mackay and Diamond (1989).

Multimedia Models

A family of models attempts to treat not only the aquatic environment but its interactions with the atmosphere and terrestrial environments. These *multimedia* models can become large and complex and can only afford to treat the aquatic environment in a simplistic manner. A review of developments in this area is that of Paterson and Mackay (1990).

Fugacity Models

The fugacity modeling system developed at the University of Toronto by Mackay and co-workers (Mackay and Paterson, 1981, 1982, 1988, 1990, 1991; Mackay, 1989, 1990, 1991) has been applied to organic chemical fate in lakes, in rivers (Holysh et al., 1986), and has been modified to treat nonvolatile chemicals such as metals (Mackay and Diamond, 1989). It is important to emphasize that fugacity and concentration approaches give exactly the same results, because ultimately the equations are algebraically identical. The advantage of the fugacity approach is that it is algebraically simpler and more elegant and can be more illuminating, especially when treating nonequilibrium situations. The disadvantage is that it requires the user to learn a new formalism. The concepts involved, however, transfer readily to other types of models, making this formalism particularly useful for general treatment of the subject.

The thermodynamic concept of fugacity, which was introduced in 1901 by Lewis, has been widely used in calculating phase equilibrium in the chemical processing industries. Only in the past decade has it been applied to calculations of the environmental fate of chemicals, and specifically to water quality models. In this section we introduce the concept of fugacity and describe how it may be applied to water quality modeling.

Fugacity, Concentration, and Z Values

Concentrations of contaminants are expressed in a diversity of units such as mg/L, μg/g, and ng/m^3, which creates some confusion when comparing numerical values for different media. It is first useful to convert all concentrations to a common unit such as g/m^3 or, as preferred here, mol/m^3. When a chemical such as DDT achieves equilibrium between phases such as air, water, and sediment, its concentrations in these media may differ by substantial ratios or partition coefficients—for example, a hundred or even a million. The criterion of equilibrium, which is common to DDT in all these phases at equilibrium, is its chemical potential, just as a common temperature is the criterion of thermal equilibrium. Unfortunately, chemical potential is logarithmically related to concentration, which causes a number of problems in calculation and interpretation. A more convenient criterion is the related quantity, fugacity, which is linearly or near-linearly related to concentration, especially at the low concentrations that generally apply in aquatic systems. The linear relationship is expressed for a chemical in solution in a phase such as air, water, or lipids as

$$C = Zf$$

where C is concentration (mol/m^3), f is fugacity (Pa), and Z is fugacity capacity (mol/m^3 Pa). Fugacity can be viewed as partial pressure or an escaping pressure or tendency. Indeed, it has units of pressure. Z is a measure of the capacity of a phase for a chemical, similar to a solubility. Z depends on the chemical nature of the phase (e.g., air or water), on the nature of the chemical, and on temperature.

If DDT is at equilibrium between, for example, air and water (subscripts A and W) with a partition coefficient K_{AW}, then

$$\frac{C_A}{C_W} = K_{AW} = \frac{Z_A f}{Z_W f} = \frac{Z_A}{Z_W}$$

It follows that a Z is essentially "half" a partition coefficient. When deducing values of Z the usual practice is first to calculate Z for the desired chemical in air and then, using reported partition coefficients such as K_{AW}, deduce Z for water and then other phases such as particles, fish, sediments, and soil. "Recipes" for deducing Z values are summarized in Figure 4 and are justified and discussed in a series of publications cited earlier in this section. The text by Mackay (1991) provides a comprehensive review of methods by which Z values can be deduced from chemical properties such as solubility, vapor pressure, and octanol-water partition coefficient as well as environmental parameters such as organic carbon content and lipid contents.

An Extended Example

The use of Z values is best illustrated by calculating the equilibrium distribution of a chemical such as hexachlorobenzene (HCB) in a hypothetical "unit world" consisting of defined volumes of air, water, soil bottom, and suspended sediment and fish. In this case the unit world has an area of 100,000 km^2 and has the volumes given in Table 1. A fuller explanation of the environmental properties is given in the

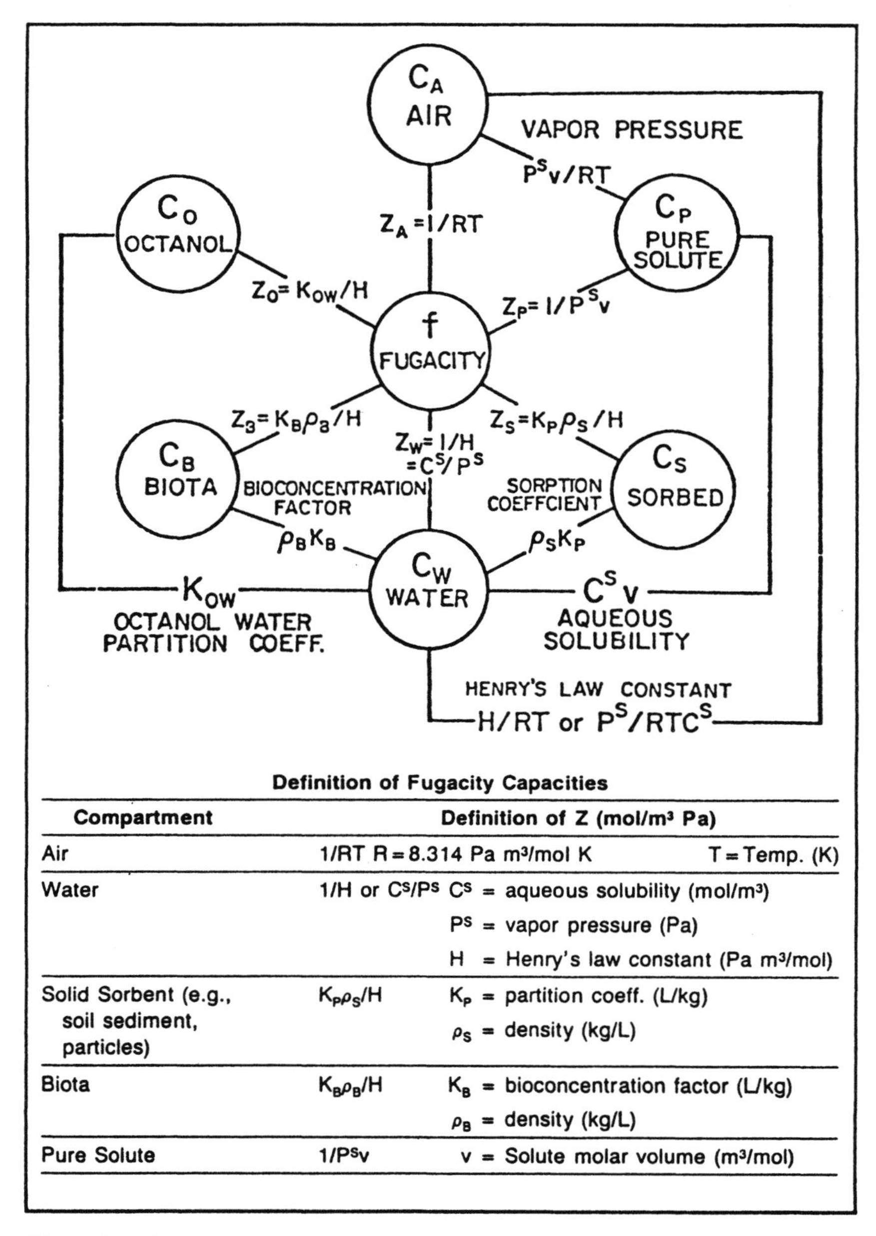

Definition of Fugacity Capacities

Compartment	Definition of Z (mol/m³ Pa)	
Air	1/RT	R = 8.314 Pa m³/mol K T = Temp. (K)
Water	1/H or C^S/P^S	C^S = aqueous solubility (mol/m³) P^S = vapor pressure (Pa) H = Henry's law constant (Pa m³/mol)
Solid Sorbent (e.g., soil sediment, particles)	$K_P \rho_S/H$	K_P = partition coeff. (L/kg) ρ_S = density (kg/L)
Biota	$K_B \rho_B/H$	K_B = bioconcentration factor (L/kg) ρ_B = density (kg/L)
Pure Solute	$1/P^S v$	v = Solute molar volume (m³/mol)

Figure 4. Definitions of Z values and their relationship to partition coefficients. (Reprinted with permission from Mackay D: *Multimedia Environmental Models: The Fugacity Approach*, 1991, Copyright Lewis Publishers, a subsidiary of CRC Press, Boca Raton, FL.)

handbook by Mackay et al. (1992). Using the stated properties of HCB, the Z values are calculated. If a total amount of HCB of say 100,000 kg or 351,000 mol is allowed to distribute between these phases at equilibrium (i.e., it achieves a common fugacity), it is easy to deduce this fugacity and hence the concentrations from the simple mass balance statement that the total amount M mol is the sum of the individual amounts m_i in the phases, which in turn are the concentration (C, mol/m³) volume (V, m³) products and thus the VZf products. Since f is common, its value can be deduced as follows

$$M = \Sigma\, m_i = \Sigma\, V_i Z_i f = f\, \Sigma\, V_i Z_i$$

$$f = \frac{M}{\Sigma\, V_i Z_i}$$

and each $C_i = Z_i f$ and each $m_i = C_i V_i$

This "level I" calculation enables the mass distribution to be deduced (for example, as percentages) and the relative concentrations to be determined. Table 1 illustrates this calculation and clearly shows the high tendency of HCB to bioconcentrate in fish by a factor of 15,800 over that in the water. This simple equilibrium partitioning calculation can obviously be applied to many phases with minimal increase in complexity. The results are illuminating by showing the relative concentrations and partitioning tendencies. It is noteworthy that where most of the chemical goes (in this case, soil) is not necessarily where the concentration is highest (in this case, fish). Despite its low vapor pressure, an appreciable 8.4% partitions into air and becomes subject to long-range transport.

Transport, Transformation, and **D** *Values*

In reality, chemicals are rarely at equilibrium. Accurate aquatic models thus require inclusion of expressions for transport between media (e.g., evaporation) or for transformation (e.g, biodegradation). This is accommodated by introducing D values such that the rate of the process N mol/h is given as the product of fugacity and a D value (mol/h Pa).

$$N = Df$$

D values are defined by rewriting the conventional concentration equation, replacing C by Zf, and then grouping all the terms except f as D. This can be done for a reaction with a first-order rate constant k (h^{-1}) as

$$N = VCk = (VZk)f = D_R f$$

For diffusion expressed by a mass transfer coefficient K (m/h) applied to an area A(m^2),

$$N = KAC = (KAZ)f = D_D f$$

In some cases K may be expressed as a ratio of diffusivity to a diffusion path length.

For advective or bulk flow in a phase of flow rate G m^3/h the expression is

$$N = GC = (GZ)f = D_A f$$

A key point is that when a chemical present in a phase is subject to a number of transport and transformation processes expressed by D values D_1, D_2, D_3, . . . , the total rate is obviously $f(D_1 + D_2 + D_3 + \cdots)$ and the relative importance of each process becomes immediately clear. A large D implies a fast and (usually) an important process. Again, methods of estimating D values are reviewed by Mackay (1991).

It becomes possible to assemble fairly complex mass balances for a number of phases into which a chemical is discharged and then becomes subject to a variety of intermedia transport and transformation processes. An example is shown in Figure 5, in which HCB is introduced into water at a steady rate of 1000 kg/h and then migrates to air, soil, and sediment, being subject to loss by reaction and advection loss. The steady-state solution illustrated in Figure 5 applies when input and output rates are equal for all four media and for the system as a whole. This example is described in more detail in the handbook by Mackay et al. (1992). In this case about half the HCB evaporates from the water to air, from which it is lost by advection. A large amount migrates to the sediment—indeed, of the total 5.168 million kg present in the system at steady state 90.6% is in sediment. The implications are that HCB will probably

- Evaporate and be transported to distant locations by the atmosphere
- Contaminate bottom sediments and their resident biota
- Reside in the system on the average for about 215 d

Aquatic Fugacity Models

Of interest in the context of this book are fugacity models of chemical fate in aquatic systems. An example of such an approach is the QWASI (quantitative water air sediment interaction) fugacity model, which has been applied to lakes (Mackay et al., 1983; Mackay, 1989, 1990) and to rivers (Holysh et

Table 1. Physical-chemical properties, Z values, and level I calculations for hexachlorobenzene at 25°C[a]

Compartment	Volume (m^3)	Z value	Amount (kg)	Concentrations: mol/m^3	Concentrations: Other units
Air	10^{14}	4.03×10^{-4}	8,420	2.9×10^{-10}	84 ng/m^3
Water	2×10^{11}	7.63×10^{-3}	320	5.6×10^{-9}	1.6 μg/m^3
Soil	9×10^9	47.5	89,200	3.5×10^{-5}	4.1 ng/g
Fish	2×10^6	121	5.0	8.8×10^{-5}	25 ng/g
Suspended sediment	10^7	297	62.0	2.2×10^{-4}	41 ng/g
Bottom sediment	10^8	95	1,980	7.0×10^{-5}	8.2 ng/g
Total			100,000		

[a] Properties of hexachlorobenzene: molecular mass, 284.8 g/mol; melting point, 230°C; vapor pressure (solid), 0.0023 Pa; solubility, 0.005 g/m^3; log K_{OW} (octanol-water partition coefficient), 5.5. Overall fugacity 7.33×10^{-7} Pa.
Data from Mackay et al., 1992; see pp. 298–299 for details.

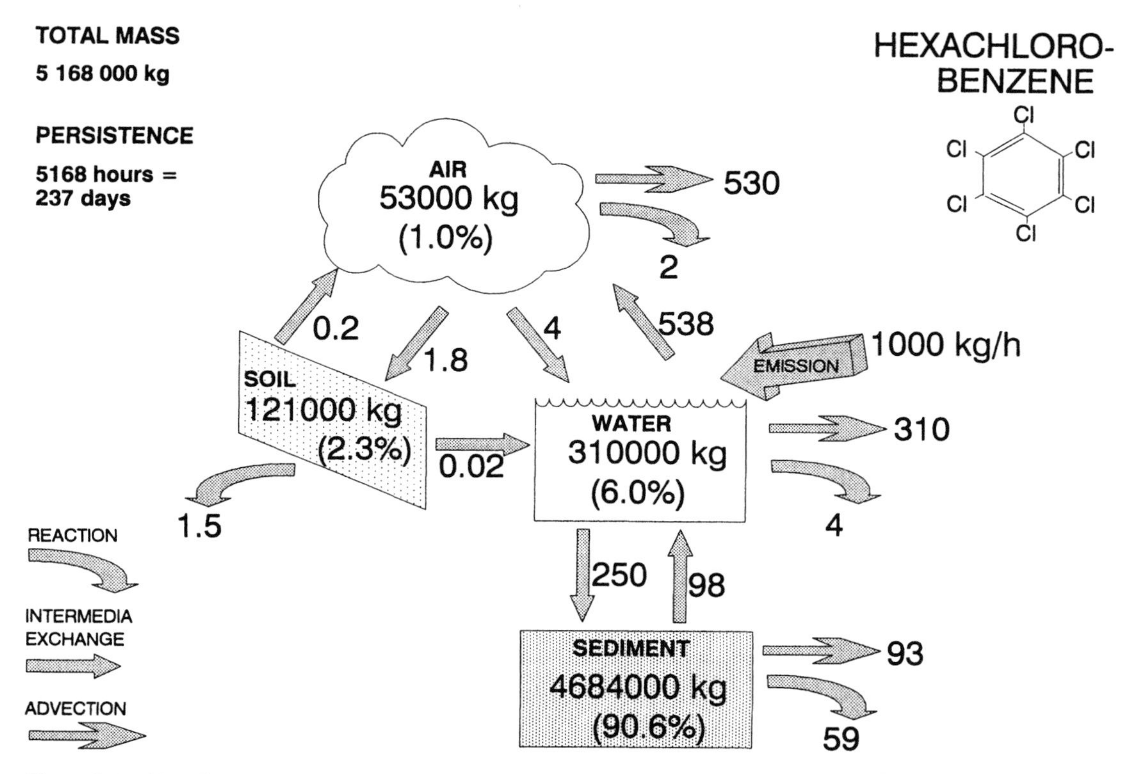

Figure 5. Multimedia mass balance diagram for hexachlorobenzene.

al., 1986). Obviously, as these models become more complex, there is a compelling incentive to perform the calculations on a computer, either as a program (as supplied on diskette with the text by Mackay, 1991) or as a spreadsheet as described by Southwood et al. (1989). This latter approach is particularly attractive because of the convenience of spreadsheet programs, permitting easy adjustment of parameters to explore sensitivities and rapid plotting of graphs showing the time dependence or response of concentrations to variations in input rates.

Figure 1 earlier showed a steady-state mass balance for PCB in Lake Ontario (Mackay, 1990) in which 22 chemical transport processes are included. For each arrow or process, a D value was calculated and the steady-state air, water, and sediment mass balance equations were solved simultaneously. A simpler, illustrative one-compartment fugacity calculation, as well as the equivalent concentration calculation, is presented later.

The primary advantage of fugacity calculations for aquatic systems is that a considerable volume of thermodynamic information is condensed in the Z values, and likewise a wide diversity of kinetic information is combined in the D values. The final mass balance equations become compact, elegant, and more easily interpreted and solved. The relative equilibrium status of chemical present in various media becomes immediately apparent. This is particularly useful when the technique is applied to uptake and clearance of chemicals by organisms such as fish. As Clark et al. (1990) have shown, the fugacity of a chemical in a fish can be more, less, or equal to that in the water or the food consumed by the fish, depending on factors such as bioavailability of the chemical in the water and the capacity of the fish to metabolize the chemical.

Fugacity calculations and interpretations can provide new insights into contaminant behavior in aquatic systems and can facilitate the quantitative expression of the diversity of phenomena that must be included in aquatic models.

AN ILLUSTRATIVE ONE-COMPARTMENT CASE STUDY

The aim of this section is to illustrate the mathematical processes involved in assembling a simple one-compartment model, by treating a situation in which a hypothetical chemical is introduced into a lake from several sources and is removed by various routes. The model can then be used to make certain deductions about the behavior of the system. Included in the model is a very simple bioconcentration model in which fish are in equilibrium with the water column. The reader is encouraged to formulate and solve a similar example for a chemical or lake of local interest.

Environmental Conditions

We examine the fate of a chemical in a lake subject to a direct chemical discharge and which is 10 m deep and has an area of 10^6 m^2, the volume of water then being 10^7 m^3. Water flows into and out of this pond at a rate of 1000 m^3/h, the residence time of the water thus being 10,000 h or approximately 14 mo. It is believed that there is inflow of 50 L/h of suspended sediment, there is deposition of 40 L/h to the bottom, and the remaining 10 L/h flows out of the system. The bottom (which we ignore here) consists of sediment. The chemical can react in the water column with a rate constant of 10^{-4} h^{-1}; that is, it has a half-life of 289 d. It evaporates with an overall water-side mass transfer coefficient of 0.0003 m/h.

The chemical (molecular mass 100 g/mol) has an air-water partition coefficient of 0.01, a particle-water partition coefficient K_{PW} of 20,000, and a biota-water partition coefficient K_{BW} or bioconcentration factor of 5000. These values are typical of a fairly hydrophobic chemical such as hexachlorobenzene. The concentration of suspended particles is 10 ppm and that of biota (including fish) is also 10 ppm, both on a volume/volume basis. There is a constant discharge of 0.02 mol/h or 2 g/h from one source with an additional 0.04 mol/h or 4 g/h from a second source. Chemical is also present in the inflowing water at a concentration of 10^{-5} mol/m^3, that is, 0.001 mg/L or 1 μg/L. The aim of the model is to calculate the steady-state or constant concentration in the system of water, particles, and fish and all the loss rates.

We can undertake this calculation in two ways—first a conventional concentration calculation, then a fugacity calculation.

Concentration Calculation

We let the total concentration of chemical in the bulk water (including chemical present in particles and fish) be an unknown C_W g/m^3. We then calculate the various process rates in terms of C_W, sum them, and equate this sum to the total input rate and solve for C_W. Finally, we deduce the various process rates. We use units of g/h for the mass balance calculation.

Input Rate

The discharges total 6 g/h. The inflow rate is the product of 1000 m^3/h and 0.001 g/m^3 (i.e., 0.001 mg/L) or 1.0 g/h. This gives a total input rate of 7 g/h.

Partitioning Between Water, Particles, and Fish

The total amount in the water is $V_W C_W$ g, where V_W is the volume of water (10^7 m^3). But this contains 10 ppm of particles (i.e., a volume of $10 \times 10^{-6} \times 10^7$ or 100 m^3) and similarly 10 ppm of biota or another 100 m^3. If the dissolved chemical concentration in water is C_D and we assume equilibrium to apply, the concentrations in the particles will be $K_{PW}C_D$ and in the biota $K_{BW}C_D$; thus the amounts are, respectively,

$10^7 C_D$	in water
$100K_{PW}C_D$	in particles
$100K_{BW}C_D$	in biota

From information about the chemical's partitioning properties, we know that K_{PW} is 20,000 and K_{BW} is 5000. Substituting gives

$$C_D\ (10 + 2 + 0.5) \times 10^6 = 12.5 \times 10^6 C_D$$

But this must equal $10^7 C_W$, thus C_D is $0.8C_W$ with 80% of the chemical being dissolved, 16% sorbed to particles, and 4% bioconcentrated in biota. The concentration on the particles is thus $K_{PW}C_D$ or $0.8K_{PW}C_W$ or $16{,}000C_W$. Note that we have used dimensionless partition coefficients K_{PW} and K_{BW}, that is, ratios of (mg/L)/(mg/L). Often K_{PW} is reported and used as a ratio of (mg/kg)/(mg/L) and thus has dimensions of L/kg.

We now calculate the unknown C_W from the mass balance equation.

Outflow

Because the outflow rate is 1000 m^3/h, the outflow rate of dissolved chemical must be $1000C_D$ or $800C_W$ g/h. In addition, there is outflow of 10 L/h (0.01 m^3/h) of particles containing 16,000 C_W g/m^3, giving $160C_W$ g/h. We assume the biota to remain in the lake, but they could be included if desired.

Reaction

The reaction rate is the product of water volume, concentration, and rate constant: $10^7 \times C_W \times 10^{-4}$, or $1000C_W$ g/h.

Deposition

The concentration on the particles is $16{,}000C_W$ g/m^3 particle. Since the particle deposition rate is 40 L/h or 0.04 m^3/h, the chemical deposition rate will be $16{,}000 \times 0.04C_W$ or $640C_W$ g/h.

Evaporation

The evaporation rate is the product of the mass transfer coefficient (0.0003 m/h), the water area (10^6 m^2), and the concentration of chemical dissolved in water. In this case, it is $0.0003 \times 10^6 \times 0.8C_W$, or $240C_W$. We assume that the air contains no chemical to create a ''back pressure'' or absorption from air to water. If this occurs, it could be included as another input term.

Removal Rate

Combining the process rates, we argue that at steady-state conditions they will equal the discharge rate of 7 g/h; thus adding water outflow, particle outflow, reaction, deposition, and evaporation gives

$$7 = 800C_W + 160C_W + 1000C_W + 640C_W + 240C_W = 2840C_W \text{ g/h}$$

Thus

$C_W = 7/2840 = 0.00246$ g/m^3 or mg/L,

or 2.46 μg/L

The dissolved concentration is thus 0.00197 g/m^3; that on the particles is 20,000 times this or 39.4 g/m^3 particle or if the particle density is 1.5 g/cm^3 approximately 26 mg/kg. This is also 0.00031 g/m^3 of water. The fish concentration will be 9.86 g/m^3 of fish or mg/kg fish and is equivalent to about 0.0001 g/m^3 water. The various process rates are thus

Outflow in water	1.97 g/h	or	28.2%
Outflow in particles	0.39 g/h		5.6%
Reaction	2.46 g/h		35.2%
Deposition	1.58 g/h		22.5%
Evaporation	0.59 g/h		8.5%

These total to the input of 7 g/h, thus satisfying the mass balance, the relative importance of the processes being immediately obvious. Figure 6 depicts this mass balance. There now emerges a clear picture of the chemical's fate. The rates of reaction and deposition are obviously important and may justify reexamination of the parameters controlling these rates.

Fugacity Calculation

As discussed earlier, in the fugacity approach, partitioning of chemical between phases is expressed in terms of the equilibrium criterion of fugacity. As in the concentration expressions, diffusive transfer rates between air and water, air and soil, and water and sediment are estimated from quantities such as mass transfer coefficients, interfacial areas, diffusivities, and path lengths. Nondiffusive processes are also treated, including air and water advection, wet and dry atmospheric deposition, leaching from ground water, sediment deposition, burial and resuspension, and surface runoff from soil, and are estimated from the appropriate flow rate of moving material (designated G). In each case the resulting rate parameter is expressed as a D value such that the rate is the product of D and fugacity. Degradation in each compartment or subcompartment is included by means of a rate constant, again expressed as a D value.

This calculation can be repeated in fugacity format by first calculating Z values, then D values, then equating input and output rates as before. It is preferable to use units of mol/h instead of g/h when doing fugacity calculations.

For air, Z is assumed to be $1/RT$ or 4.1×10^{-4} mol/m^3 Pa, where RT is the gas constant–absolute temperature group. Z for water is then calculated as Z_A/K_{AW} or 4.1×10^{-2}. For particles Z_P is $K_{PW}Z_W$ or 821 and for fish Z_B is $K_{BW}Z_W$ or 205. The total Z value for water, particles, and biota is the sum of these Z values, weighted in proportion of their volume fractions:

$$Z_{WT} = Z_W + 10 \times 10^{-6} Z_P + 10 \times 10^{-6} Z_B = 5.13 \times 10^{-2}$$

The D values are (in units of mol/Pa h)

Outflow in water	$D_1 = G_w Z_W = 1000 \times 4.1 \times 10^{-2} = 41.0$
Outflow in particles	$D_2 = G_P Z_P = 0.01 \times 821 = 8.2$
Reaction	$D_3 = VZ_{WT}k = 10^7 \times 5.13 \times 10^{-2} \times 10^{-4} = 51.3$
Deposition	$D_4 = G_D Z_P = 0.04 \times 821 = 32.8$
Evaporation	$D_5 = k_M A Z_W = 0.0003 \times 10^6 \times 4.1 \times 10^{-2} = 12.3$

The overall mass balance is then expressed in terms of the fugacity of the chemical in the water, f_W. The input rate is 0.07 mol/h.

$$\text{Input} = f_W D_1 + f_W D_2 + f_W D_3 + f_W D_4 + f_W D_5$$

or

$$0.07 = f_W \Sigma D_i = f_W 145.6$$

Thus

$f_W = 4.80 \times 10^{-4}$

$C_D = Z_W f_W = 1.97 \times 10^{-5}$ mol/m^3 or 0.00197 g/m^3

$C_P = Z_P f_W = 0.39$ mol/m^3 or 39 g/m^3 particle

$C_B = Z_B f_W = 0.098$ mol/m^3 fish or 9.8 g/m^3 or 9.8 mg/kg

The individual rates are Df and are identical to those calculated earlier. The D values give a useful, direct expression of the relative importance of each process. Once the concepts of Z values, D values, and fugacity are grasped, the algebra becomes simpler and more elegant. The numerical results are, of course, identical to those obtained using conventional concentration calculations.

Discussion

The model thus yields invaluable information about the status and fate of the chemical in this lake system. Obviously, it is desirable to measure concentrations in particles, water, and biota to determine whether the model assertions are correct. If (as is likely) discrepancies are observed, the model assumptions can be reexamined to test if some important process has been omitted or if the assumed parameters are correct in magnitude. When reconciliation is successful, the environmental scientist is in the satisfying position of being able to claim that the system is well understood. It is then possible to explore the effect of various changes in inputs to the system.

For example, if the discharge of 4 g/h was eliminated, the water, particle and biotic concentrations would eventually fall to three-sevenths or 43% of the values calculated above because the input would now be 3 g/h instead of 7 g/h. The biota would now contain 4.2 mg/kg instead of 9.8 mg/kg. To achieve a "target" of, say, 1.0 mg/kg would require reduction of the input rate to about 0.7 g/h. The nature and magnitude of regulatory measures necessary to achieve a desired environmental quality can thus be estimated.

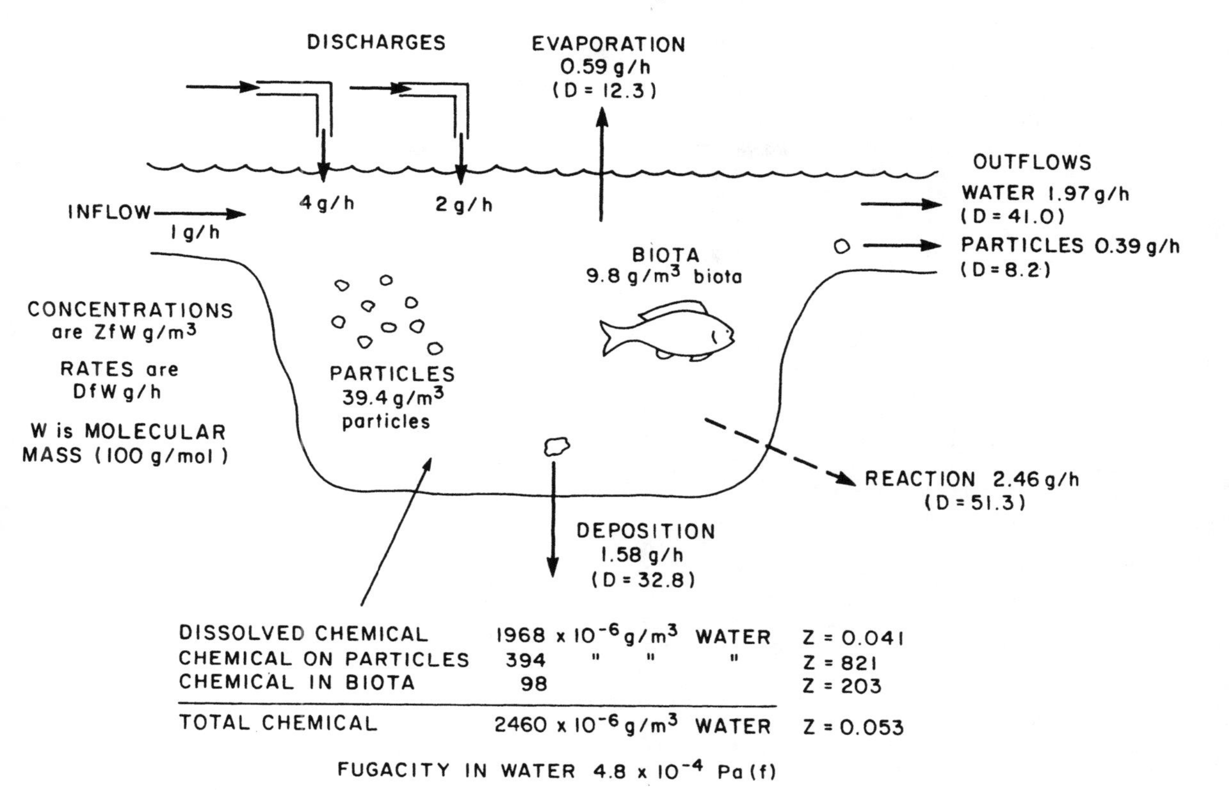

Figure 6. Mass balance diagram or model of chemical fate in the worked example.

A related issue is that of how long it will take for such measures to become effective. This requires solution of the differential mass balance equation in time t, which in this case is

$$V_W \frac{dC_W}{dt} = \text{inputs} - \text{outputs}$$

or

$$V_W \frac{dC_W}{dt} = 7 - 2840 C_W$$

or

$$\frac{dC_W}{dt} = 7 \times 10^{-7} - 2840 \times 10^{-7} C_W = A - BC_W$$

where $A = 7 \times 10^{-7}$ and $B = 2840 \times 10^{-7}$.

If the initial (time zero) water concentration is C_{W0}, for example, some high value like 8×10^{-3} g/m^3, at time zero this equation can be solved by separation of variables to give

$$C_W = C_{WF} - (C_{WF} - C_{W0}) \exp(-Bt)$$

where C_{WF} is the final value of C_W at a time of infinity and is A/B or 2.46×10^{-3} g/m^3 as before. C_W will thus change from C_{W0} to C_{WF} with a rate constant of B or a half-life of $0.693/B$ h.

Here B is 2840×10^{-7} h^{-1}; thus the half-life of 0.693/(rate constant) is 2440 h or 102 d. The water concentration would thus eventually fall from 8×10^{-3} g/m^3 to 2.46×10^{-3} g/m^3 with a half-life of 102 d.

In reality the biota would respond more slowly because of the delay in biouptake or release, but the important finding is that within a year the system would be well on its way to a new steady-state condition. In fugacity terms the equivalent value of B is $\Sigma D/V_W Z_{WT}$, and the individual processes contributing to this overall rate constant are $D/V_W Z_{WT}$. It is thus apparent that water outflow, sedimentation, and reaction have about equivalent influences on the rate or time of response, while evaporation and particle outflow are relatively unimportant. Such information can be of considerable regulatory value.

Clearly, the nature of these calculations is such that they are readily convertible into computer programs, either as a stand-alone program or on a spreadsheet such as Lotus 1-2-3. It is hoped that the enthusiastic reader will attempt to write such a program and explore the effects of changing the various parameters.

It is obviously possible to include a sediment compartment or another water compartment and expand the model to include more detail and thus achieve greater fidelity. If exposure to chemical through fish consumption is important, it may be necessary to develop a separate bioconcentration or food chain model [the FGETS model is one example (Barber et al., 1988, 1991)]. The nature and depth of the modeling activity should be tailored to the exposure assessment or to toxicological needs. The aquatic toxicologist who understands the nature of these models is well equipped to design and interpret such assessments.

CONCLUDING THOUGHTS

The primary purpose of a water quality model is to bring together available information on the partitioning tendencies of chemicals, along with transformation and transport rate data, to synthesize a coherent statement of the law of conservation of mass, as it applies to the single or multiple compartments of aquatic ecosystems. It is an attempt to be quantitative about chemical fate in the expectation that the resulting numbers will be useful for assessment of toxicity, fate, and remedial measures. It is thus merely a tool in a larger process; it is not an end in itself. Therefore, it must be carefully tailored to satisfy the requirements of the larger process and give sufficient accuracy, yet avoid excessive complexity, cost, and demands for data. Models have the potential to contribute significantly to aquatic toxicology, but their full potential is rarely realized. Perhaps there are two reasons.

First is failure of the modeler to integrate the model into the overall process and explain simply and fully the model's strengths, weaknesses, and sensitivities. Second, it must be recognized that a great deal of intellectual effort is contained in a model; thus it is a formidable task for the user to understand every line of computer coding. Regrettably, many potential users who could benefit from models are unable, or unwilling, to devote the time and effort necessary to become fully familiar with the model; and they remain skeptical of its utility. The simple example given in this chapter is an attempt to bridge that gap.

Considerable care must be exercised when using models developed by others. The most valuable are those that are fully documented and continuously supported in the form of a live, accessible expert who can provide advice on request, preferably by telephone. This can be a considerable financial burden on an agency but it is essential if the model is to be widely and correctly used.

A consensus is emerging that aquatic fate models can play an invaluable role in enhancing the applications of the science of aquatic toxicology to the solution of problems of environmental contamination. It is hoped that the account of modeling approaches and the simple illustration given in this chapter will encourage aquatic toxicologists to add model building, testing, and interpretation to their already considerable skills.

LITERATURE CITED

Allison, J. D.; Brown, D. S.; and Novo-Gradac, K. J.: MINTEQA2/PRODEFA2, A Geochemical Assessment Model for Environmental Systems: Version 3.0 User's Manual. Athens, GA: U.S. EPA Office of Research and Development, Athens Environmental Research Laboratory, 1991. EPA/600/3-91-021.

Barber, M. C.; Suárez, L. A.; and Lassiter, R. R.: Modeling bioconcentration of nonpolar organic pollutants by fish. *Environ. Toxicol. Chem.*, 7:545–558, 1988.

Barber, M. C.; Suárez, L. A.; and Lassiter, R. R.: Modelling bioaccumulation of organic pollutants in fish with an application to PCBs in Great Lakes salmonids. *Can. J. Fish Aquat. Sci.*, 48:318–337, 1991.

Baughman, G. L., and Lassiter, R. R.: Prediction of environmental pollutant concentration. Estimating the Hazard of Chemical Substances to Aquatic Life, edited by J. Cairns Jr.; K. L. Dickson; and A. W. Maki, pp. 35–54. ASTM STP 657. Philadelphia: American Society for Testing and Materials, 1978.

Booty, W. G., and Lam, D. C. L.: Freshwater Ecosystem Water Quality Modelling. Report NWRI 89–63. Burlington, Ontario: National Water Research Institute, 1989.

Brown, D. S., and Allison, J. D.: MINTEQA1, an Equilibrium Speciation Model: User's Manual. Athens, GA: U.S. Environmental Protection Agency, 1987. EPA/600/3-87/012.

Burns, L. A.: Exposure Analysis Modeling System: User's Guide for EXAMS II Version 2.94. Athens, GA: U.S. EPA Office of Research and Development, Athens Environmental Research Laboratory, 1990. EPA/600/3-89-084.

Burns, L. A., and Cline, D. M.: Exposure Analysis Modeling System: Reference Manual for EXAMS II. Athens, GA: U.S. EPA Office of Research and Development, Athens Environmental Research Laboratory, 1985. EPA-600/3-85-038.

Burns, L. A.; Cline, D. M.; and Lassiter, R. R.: Exposure Analysis Modeling System (EXAMS): User Manual and System Documentation. Athens, GA: U.S. EPA Office of Research and Development, Athens Environmental Research Laboratory, 1982. EPA-600/3-82-023.

Chapra, S. C., and Reckhow, K. H.: Engineering Approaches for Lake Management: Mechanistic Modeling. Woburn, MA: Butterworth, 1983.

Clark, K. E.; Gobas, F. A. P. C.; and Mackay, D.: Model of organic chemical uptake and clearance by fish from food and water. *Environ. Sci. Technol.*, 24:1203–1213, 1990.

Connolly, J. P., and Thomann, R. V.: WASTOX, a Framework for Modeling the Fate of Toxic Chemicals in Aquatic Environments. Part 2: Food Chain. Gulf Breeze, FL: U.S. EPA Office of Research and Development, Gulf Breeze Environmental Research Laboratory, and Duluth Minnesota, Duluth Environmental Research Laboratory, 1985.

Connolly, J. P., and Winfield, R. P.: A User's Guide for WASTOX, a Framework or Modeling the Fate of Toxic Chemicals in Aquatic Environments. Part I: Exposure Concentration. Gulf Breeze, FL: U.S. EPA Office of Research and Development, Gulf Breeze Environmental Research Laboratory, 1984. EPA-600/3-84-077.

Di Toro, D. M.; Fitzpatrick, J. J.; and Thomann, R. V.: Water Quality Analysis Simulation Program (WASP) and Model Verification Program (MVP)—Documentation. Hydroscience, Inc., Westwood, NJ. Prepared for U.S. Environmental Protection Agency, Duluth, MN, Contract No. 68-01-3872, 1982.

Dolan, D. M., and Bierman, V. J., Jr.: Mass balance modeling of heavy metals in Saginaw Bay, Lake Huron. *J. Great Lakes Res.*, 8:676–680, 1982.

Donigian, A. S. Jr., and Crawford, N. H.: Modeling Nonpoint Pollution from the Land Surface. Athens, GA: U.S. EPA Office of Research and Development, Athens Environmental Research Laboratory, 1976. EPA-600/3-76-083.

Donigian, A. S. Jr., and Davis, H. H. Jr.: User's Manual for Agricultural Runoff Management (ARM) Model. Athens, GA: U.S. EPA Office of Research and Development, Athens Environmental Research Laboratory, 1978. EPA-600/3-78-080.

Donigian, A. S. Jr.; Imhoff, J. C.; Bicknell, B. R.; and Kittle, J. L. Jr.: Application Guide for Hydrological Simulation Program—FORTRAN (HSPF). Athens, GA: U.S. EPA Office of Research and Development, Athens Environmental Research Laboratory, 1984. EPA-600/3-84-065.

Felmy, A. R.; Brown, S. M.; Onishi, Y.; Yabusaki, S. B.; Argo, R. S.; Girvin, D. C.; and Jenn, E. A.: Modeling the Transport, Speciation, and Fate of Heavy Metals in Aquatic Systems. Athens, GA: U.S. EPA Office of Research and Development, Athens Environmental Research Laboratory, 1984a. EPA-600/3-84-033.

Felmy, A. R.; Girvin, D. C.; and Jenne, E. A.: MINTEQ—A Computer Program for Calculating Aqueous Geochemical Equilibria. Athens, GA: U.S. EPA Office of Research and Development, Athens Environmental Research Laboratory, 1984b. EPA-600/3-84-032.

Halfon, E.: TOXFATE, a Model of Toxic Substances in Lake Ontario. NWRI Report, National Water Research Institute, Burlington, Ontario, 1985.

Hites, R. A., and Eisenreich, S. J. (eds.): Sources and Fates of Aquatic Pollutants. Washington, DC: American Chemical Society, 1987. ACS Symp Ser 216.

Holysh, M.; Paterson, S.; Mackay, D.; and Bandurraga, M. M.: Assessment of the environmental fate of linear alkybenzenesulphonates. Chemosphere 15:3–20, 1986.

Johanson, R. C.; Imhoff, J. C.; Kittle, J. L. Jr.; and Donigian, A. S. Jr.: Hydrological Simulation Program—FORTRAN (HSPF): User's Manual for Release 8.0. Athens, GA: U.S. EPA Office of Research and Development, Athens Environmental Research Laboratory, 1984. EPA-600/3-84-066.

Jorgensen, S. E. (ed.): Modelling the Fate and Effect of Toxic Substances in the Environment. Amsterdam: Elsevier, 1984.

Jorgensen, S. E.: Fundamentals of Ecological Modelling, 2nd ed. Amsterdam: Elsevier, 1988.

Jorgensen, S. E., and Gromiec, M. J. (eds.): Mathematical Submodels in Water Quality Analysis. Amsterdam: Elsevier, 1989.

Lam, D. C. L., and Schertzer, W. M.: Lake Erie thermocline model results: comparison with 1967–1982 data and relation to anoxic occurrences. *J. Great Lakes Res.*, 13: 757–769, 1987.

Lam, D. C. L.; Schertzer, W. M.; and Fraser, A. S.: Mass balance models of phosphorus in sediments and water. *J. Hydrobiol.*, 91:217–225, 1982.

Lyman, W. J.; Reehl, W. F.; and Rosenblatt, D. H.: Handbook of Chemical Property Estimation Methods. New York: McGraw-Hill, 1982.

Mackay, D.: An approach to modelling the long term behavior of an organic contaminant in a large lake: application to PCBs in Lake Ontario. *J. Great Lakes Res.*, 15: 283–297, 1989.

Mackay, D.: Atmospheric contributions to contamination of Lake Ontario. Long Range Transport of Pesticides, edited by D. A. Kurtz, chapter 21, pp. 317–328. Chelsea, MI: Lewis, 1990.

Mackay, D.: Multimedia Environmental Models: The Fugacity Approach. Chelsea, M. I.: Lewis, 1991.

Mackay, D., and Diamond, M. L.: Application of the QWASI (quantitative water air sediment interaction) fugacity model to the dynamics of organic and inorganic chemicals in lakes. *Chemosphere*, 18:1343–1365, 1989.

Mackay D, Paterson S: Calculating fugacity. *Environ. Sci. Technol.*, 15:1006–1014, 1981.

Mackay, D., and Paterson, S.: Fugacity revisited. *Environ. Sci. Technol.*, 16:654A–660A, 1982.

Mackay, D., and Paterson, S.: Partitioning models. Carcinogen Risk Assessment, edited by C. C. Travis, pp. 77–86. London: Plenum, 1988.

Mackay, D., and Paterson, S.: Fugacity models. Practical Applications of Quantitative Structure-Activity Relationships (QSAR) in Environmental Chemistry and Toxicology, edited by W. Karcher, and J. Devillers, pp. 433–460. Dordrecht: Kluwer Academic, 1990.

Mackay, D., and Paterson, S.: Evaluating the multimedia fate of organic chemicals: a level III fugacity model. *Environ. Sci. Technol.*, 25:427–436, 1991.

Mackay, D.; Joy, M.; and Paterson, S.: A quantitative water, air, sediment interaction (QWASI) fugacity model for describing the fate of chemicals in lakes. *Chemosphere*, 12: 981–997, 1983.

Mackay, D.; Shiu, W. Y.; and Ma, K. C.: Illustrated Handbook of Physical Chemical Properties and Environmental Fate for Organic Chemicals. Vol I, Monoaromatic Hydrocarbons, Chlorobenzenes and PCBs. Chelsea, MI: Lewis, 1992.

Mihram, G. A.: Some practical aspects of the verification and validation of simulation models. *Oper. Res. Q*, 23: 17–29, 1972.

Muir, D. C. G.: Dissipation and transformations in water and sediment. Environmental Chemistry of Herbicides, Vol II, edited by R. Grover, and A. J. Cessna, pp. 1–87. Boca Raton, FL: CRC Press, 1991.

Neely, W. B.; Blau, G. E.; and Alfrey, T. Jr.: Mathematical models predict concentration-time profiles resulting from chemical spill in a river. Environ Sci Technol 10:72–76, 1976.

O'Connor, D. J.: Models of sorptive toxic substances in freshwater systems, I: Basic equations. *J. Environ. Eng.*, 4:507–532, 1988a.

O'Connor, D. J.: Models of sorptive toxic substances in freshwater systems, II: Lakes and reservoirs. *J. Environ. Eng.*, 4:533–551, 1988b.

Onishi, Y., and Wise, S. E.: Mathematical Model, SERATRA, for Sediment-Contaminant Transport in Rivers and Its Application to Pesticide Transport in Four Mile and Wolf Creeks in Iowa. Athens, GA: U.S. EPA Office of Research and Development, Athens Environmental Research Laboratory, 1982. EPA-600/3-82-045.

Oren, T. I.: Concepts and criteria to assess acceptability of simulation studies: a frame of reference. Commun ACM 24:180–189, 1981.

Plane, J. M. C.; Zika, R. G.; Zepp, R. G.; and Burns, L. A.: Photochemical modeling applied to natural waters. Photochemistry of Environmental Aquatic Systems, edited by R. G. Zika, and W. J. Cooper, pp. 250–267. Washington, DC: American Chemical Society, 1987. ACS Symp Ser 27.

Paterson, S., and Mackay, D.: Models of environmental fate and human exposure to toxic chemicals. Reviews in Environmental Toxicology 4, edited by E. Hodgson, pp. 241–265. Raleigh, NC: Toxicology Communications, 1990.

Sargent, R. G.: Verification and validation of simulation models. Progress in Modelling and Simulation, edited by F. R. Cellier, pp. 159–169. London: Academic Press, 1982.

Schnoor, J. L.: Fate and transport of dieldrin in Coralville Reservoir: residues in fish and water following a pesticide ban. *Science*, 211:840–842, 1981.

Schnoor, J. L., and O'Connor, D. J.: A steady state eutrophication model for lakes. Water Res 14:1651–1665, 1980.

Schnoor, J. L.; Sato, C.; McKechnie, D.; and Sahoo, D.: Processess, Coefficients, and Models for Simulating Toxic Organics and Heavy Metals in Surface Waters. Athens, GA: U.S. EPA Office of Research and Development, Athens Environmental Research Laboratory, 1987. EPA/600/3-87/015.

Snodgrass, W. J.: Analysis of models and measurements for sediment oxygen demand in Lake Erie. *J. Great Lakes Res.*, 13:468–472, 1987.

Southwood, J. M.; Harris, R. C.; and Mackay, D.: Modeling the fate of chemicals in an aquatic environment: the use of computer spreadsheet and graphics software. *Environ. Toxicol. Chem.*, 8:987–996, 1989.

Swann, R. L., and Eschenroeder, A. (eds.): Fate of Chemicals in the Environment. Washington, DC: American Chemical Society, 1983. ACS Symp Ser 225.

Thibodeaux, L. J.: Chemodynamics: Environmental Movement of Chemicals in Air, Water and Soil. New York: Wiley, 1979.

Thomann, R. V., and Di Toro, D. M.: Physico-chemical model of toxic substances in the Great Lakes. *J. Great Lakes Res.*, 9:474–496, 1983.

Vollenweider, R. A.: Scientific fundamentals of the eutrophication of lakes and flowing waters, with particular reference to nitrogen and phosphorus as factors in eutrophication. OECD Tech Rep DAS/CSI/68.27, 1968.

Vollenweider, R. A.: Possibilities and limits of elementary models concerning the budget of substances in lakes. *Arch. Hydrobiol.*, 66:1–12, 1969.

Zeigler, B. P.: Theory of Modelling and Simulation. New York: Wiley, 1976.

SUPPLEMENTAL READING

Gobas, F. A. P. C., and McCorquodale, J. A.: Chemical Dynamics in Fresh Water Systems. Chelsea, MI: Lewis, 1992.

Hemond, H. F., and Fechner, E. J.: Chemical Fate and Transport in the Environment. New York: Academic Press, 1994.

James, A.: An Introduction to Water Quality Modeling. New York: Wiley, 1993.

Lerman, A.: Lakes—Chemistry, Geology, Physics. New York: Springer-Verlag, 1978.

Neely, W. B.: Chemicals in the Environment—Distribution, Transport, Fate, Analysis. New York: Dekker, 1980.

Schwarzenbach, R. P.; Gschwend, P. M.; and Imboden, D. M.: Environmental Organic Chemistry, New York: Wiley, 1993.

Stumm, W., and Morgan, J. J.: Aquatic Chemistry, 2nd ed. New York: Wiley, 1981.

Thibodeaux, L. J.: Chemodynamics—Environmental Movement of Chemicals in Air, Water and Soil. New York: Wiley, 1979.

Tinsley, I. J.: Chemical Concepts in Pollutant Behavior. New York: Wiley-Interscience, 1979.

Verschueren, K.: Handbook of Environmental Data on Organic Chemicals, 2nd ed. New York: Van Nostrand Reinhold, 1983.

PART III

ASSESSMENT

Most emphasis in toxicity studies is on structural end points (Chapter 1), although functional end points are an integral part of an aquatic toxicology program to evaluate the effects of chemicals. Part III, therefore, begins with Chapter 19 on the use of functional end points in aquatic ecosystems with a section on statistical issues. Chapter 20 provides a background with general concepts on structure-activity relationships and their use and application in exposure and effects assessment. Chapters 21–25 are extensive reviews of environmental legislation in the United States and their relevance in aquatic toxicology for regulation of specific pesticides (Chapter 21), industrial chemicals (Chapters 22 and 23), and drugs (Chapter 25) and for controlling discharges of toxic effluents (Chapter 24). Also included is a summary of environmental regulations in the European Community (Chapter 27). Although the Organization for Economic Cooperation and Development (OECD) is not a regulatory body, Chapter 26 is included because of the importance of the OECD in internationalization of test methods.

The final section in Part III on hazard and risk assessment begins with an introduction on general concepts and methods for aquatic effects (Chapter 28) followed by an overview of general aquatic assessments (Chapter 29). The remaining four chapters are case studies of procedures for hazard and risk assessments for insecticides (Chapter 30), an anionic surfactant (Chapter 31), a hazardous waste site (Chapter 32), and industrial effluents (Chapter 33).

Chapter 19

ECOSYSTEM EFFECTS: FUNCTIONAL END POINTS

J. Cairns, Jr., B. R. Niederlehner, and *E. P. Smith*

INTRODUCTION

The ecological consequences of environmental contamination are determined by a complex interplay of cause and effect. In the real world, a single stress does not affect single organisms. Instead, communities composed of thousands of different organisms experience changes in the quality of their physical habitat, food quality, harvesting pressures, and additions of enriching and toxic chemicals, all at the same time. In turn, the organisms themselves modify their environment by taking up, storing, and transforming chemicals and by eating or competing with other organisms. The complex nature of these interactions can confound our predictions of the environmental outcome of contamination.

Chemicals have unique effects at different levels of complexity and time-space scale, so the effects of contaminants identified will depend on what is measured, where it is measured, and how long it is measured. Table 1 presents one hierarchical scheme commonly used to describe different levels of complexity in natural systems. Each level of biological organization (e.g., Webster, 1979) is larger, slower, more complex, and composed of lower levels. In addition each level has its own characteristic structures and behaviors (Table 1). Tests of effects of contaminants on simple systems are generally easier, quicker, more replicable, and less expensive than those on more complex systems. For this reason, simple test systems are used to address problems that actually occur at larger, slower, and more complex scales. Although effects on complex systems can sometimes be predicted from effects on component parts, this requires a model of relevant processes at the next lower level (e.g., O'Neill et al., 1986; Webster, 1979). Developing such a model is clearly a daunting task given the complexity of most ecosystems. For example, a survey of the Flint River and Lake Blackshear in Georgia (United States) identified almost 1000 species of fish, macro- and microinvertebrates, and algae (Academy of Natural Sciences, 1984). This estimate does not include bacteria, and a study suggests that thousands of genomes would be present (Torsvik et al., 1990).

Even with a mechanistic model to predict ecosystem outcome from the dynamics of 5000 interacting components, the nature, importance, and acceptibility of an environmental change must be judged at a variety of scales because management goals are present at a variety of hierarchical levels. Not only is the protection of commercially important species necessary (a population-level goal), but also society has expressed a desire to protect biotic diversity and sustained ecosystem production (ecosystem-level goals). Whereas conventional toxicity tests assess direct effects on populations, ecosystem-level tests such as mesocosms (see Chapter 9) make it possible to assess the effects of chemical contamination on other properties of interest such as diversity, waste processing, and production that occur at higher levels of biological organization.

Two classes of responses are of interest: structural and functional. *Ecosystem structure* is studied through assessment of the numbers and kinds of species and other component parts at one point in time (see Chapter 1). Structure is what is present at one point in time; how many, how much, what kind, what condition. In contrast, *ecosystem function* is performance. Function describes the changes in the system over time or the lack of changes through homeostatic mechanisms. It is measured as rates of change: growth rates, production rates, transformation rates, transfer rates. Examples of structural and functional measures that have been used to assess the ecosystem-level effects of environmental contamination are presented in Table 2. Although the great majority of data collected to evaluate ecosystem effects has been structural (most commonly an inventory of species or the standing crop of a species that human society values), there is increasing appreciation that functions of ecosystems may be of great importance as well.

For the majority of the world's people, the most important argument for the protection and restoration

Table 1. Levels of biological organization and their characteristic structural and functional attributes

Individual—one whole organism. Individuals have size, shape, health, or condition. They grow, reproduce, and die over time.
Population—a group of individuals of the same species occupying the same area at a given time. Populations have abundance, biomass, size, and age class structures. They compete, exploit prey, and produce biomass.
Assemblage—a set of coexisting populations defined by phylogeny, location, or life-style. Assemblages are intermediate to populations and communities and have some of the properties of both groups, depending on how they are defined.
Community—the collection of all organisms that live in a specific region at a given time. Communities have biomass, diversity, evenness, richness, and a trophic structure. They produce biomass, process materials, and change through succession over time.
Ecosystem—all of the organisms in an area together with the physical environment with which they interact. Ecosystems have biological as well as physical and chemical structure. They move energy, materials, and nutrients.

of natural ecosystems is protection of the functional services that ecosystems provide to human society (Ehrlich and Ehrlich, 1991; Wilson, 1988). All ecosystems, whether pristine or modified, provide some functional services useful to society, including the production of oxygen; transformation of noxious waste materials to less harmful products and even useful materials; storage and transformation of nutrients; storage, purification, and movement of water; modification of the earth's climate at both micro and macro levels; and production of a variety of useful materials such as food, timber, and medicinal drugs or chemical models for medicinal drugs. Such functional services must be protected if society intends to maintain the environment in a robust, self-maintaining condition. If not, society will be required to restore many of these services at considerable cost or to accept diminished ecosystem services per capita. The most direct way to assure protection of these processes is by measuring the effects of stress on function. General changes in function could be useful in diagnosing ecosystem impairment in a manner analogous to the diagnosis of human disease in the practice of medicine. For example, in diagnosis of human disease, one rarely discusses the diversity of one's intestinal flora or the number of bacteria on one's teeth. It is more common to discuss heart rate, liver function, and other measures of the function of the human system and to assess health based on function.

Table 2. Examples of common structural and functional end points at the community and ecosystem level

Structural end points
- Species richness
- Species abundance
- Similarity
- Biomass

Functional end points
- Rate of biomass production
- Rate of primary productivity
- Rate of community respiration
- Rate of nutrient uptake/regeneration
- Rate of decomposition
- Rate of recovery after stress

This chapter describes some functional attributes of the ecosystems that may be affected by chemical contamination. Some of these, in combination with more commonly measured structural attributes, may be useful in assessing the effects of environmental contamination.

COMMONLY MEASURED FUNCTIONAL ATTRIBUTES

The most commonly measured functional end points in aquatic ecosystems include primary and secondary production rates, community respiration rates, decomposition rates, nutrient uptake and regeneration rates, and recovery rates of a structure or function after stress. These responses are important because they describe the basic machinery that supports life on earth.

Life depends on energy fixed from sunlight. If chemical contamination damages this ability to fix energy, the amount of energy available to support all kinds of life, including humans, will be diminished. The total energy fixed by all photosynthesis in an ecosystem is the *gross primary production* (GPP). GPP is a rate expressed as biomass produced per unit area over a length of time. After the organisms themselves have used the energy they need to stay alive, what is left is the *net primary production* (NPP). NPP is the energy that is available to all the other organisms that cannot capture energy from sunlight, but instead eat other organisms, their wastes, or their carrion—the *heterotrophs*. The energy contained within the heterotrophs is called *secondary production*. When we consider the entire ecosystem, both primary producers and heterotrophs, the energy represented is the *net community production* (NCP).

In aquatic ecosystems there are three spatially distinct groups that photosynthesize: phytoplankton in the water column, periphyton growing on surfaces, and aquatic macrophytes. Effects of contamination on primary production are often assessed by determining the *standing crop* of these groups, often measured as the dry weight (DW), ash-free dry weight (AFDW), or chlorophyll *a* content of photosynthesizers at any one time (American Public Health Association [APHA] et al., 1989). These are structural measurements. The related functional measurement, the rate of primary production, is obtained in several ways. Because photosynthesis takes carbon dioxide and water and uses energy obtained from sunlight to transform them to carbohydrates and oxygen, changes in dissolved oxygen content of overlying water and uptake of radiolabeled carbon dioxide are used to measure the rates of GPP (APHA et al., 1989). Changes in standing crop over time are used to measure NPP. Similarly, changes in standing stock of heterotrophs over time are used to measure secondary production. Changes in total biomass (B), both primary producers and heterotrophs, over time are used to measure NCP. Because all heterotrophs are ultimately dependent on primary production, estimates of GPP are sometimes used as an index of total energy available in a system. This is reasonable for some ecosystems but not for others, discussed below.

There are additional approaches that look at efficiency of energy exchanges rather than the abundance of production. In this way, effects of toxicants on the integrity of the process are isolated from effects on community structure. For example, a reduction in GPP in a lake might be due to a reduction in standing crop or a change in the rate at which the same standing crop fixes energy from sunlight. A reduction in GPP may be accompanied by a switch to use of energy fixed outside the system and then used by heterotrophic biomass. Or the loss in fixed energy from photosynthesis may not be compensated by other sources and the amount of living material supported in the system may be reduced. Some measures of energy exchange efficiency include the assimilation ratio (AR; Marker, 1976), the production-to-biomass ratio (P/B; Margalef, 1975), and biomass supported per unit of energy flow (B/E, where E = GPP + CR; Odum, 1967). AR is an assemblage-specific measurement. It describes changes in the efficiency of primary producers as the amount of photosynthesis per unit of plant or algal biomass (e.g., GPP/chlorophyll *a* biomass). In contrast, P/B (e.g., GPP/total biomass) is an ecosystem-level measure, looking at the total amount of living material supported by a given flow of energy. In general, one might expect toxic chemicals to reduce the efficiency with which plants and algae can fix energy (AR will decrease) and to decrease the amount of living material supported by the same amount of energy (P/B will increase).

But, in addition to primary production, many aquatic systems rely heavily on photosynthesis that occurred outside the water to support life within it. For this reason, descriptions of energy flux through a system cannot always rely entirely on estimates of primary production within the system. A common form of fixed energy that is imported into streams is leaves that fall from trees. *Decomposers* are organisms that live on the dead matter or wastes of other organisms, and decomposition of leaves is a major energy source for many low-order streams in temperate forests. But leaves are not the only source of fixed energy that is imported into aquatic systems. Most carbon-containing contaminants, for example, sewage or simple organic chemicals, can also provide food for decomposers. The rate at which decomposers break down these materials into smaller parts or more useful chemical constituents is sometimes measured. For example, the rate at which a pack of leaves loses mass can be tracked over time (e.g., Forbes and Magnuson, 1980). If chemical contamination reduces the ability of decomposers to use these imported sources of energy, the NCP of some systems will be adversely affected. Also, human societies depend on decomposers to break down wastes such as sewage. If contaminants interfere with this process, an essential ecosystem service is lost and must be replaced through human engineering.

Respiration is the energy it costs an organism just to stay alive. An organism must use energy to circulate essential materials such as nutrients and gases, to synthesis proteins, and to move around. Organisms use oxygen and expel carbon dioxide as they use energy to maintain their essential physiological activities and behaviors. Thus, *community respiration* (CR) is often measured as oxygen uptake from the water column. In order to eliminate the confounding effects of photosynthesis, in which oxygen is produced, determinations of CR are commonly made in the dark. CR can also be measured as enzyme activity of the dehydrogenases involved in energy usage (e.g., Burton et al., 1989). Carbon-containing contaminants like sewage often increase CR by encouraging the growth of heterotrophic biomass. If an aquatic system is contaminated with organic matter, oxygen consumption (respiration) exceeds oxygen production, resulting in oxygen depletion. Should it continue, anaerobic conditions would eliminate fish and other organisms. *Biological oxygen demand* (BOD; APHA et al., 1989) is the single most commonly performed functional test. It measures the uptake of oxygen by living organisms (mostly bacteria) as they use the organic matter present in a water sample. However, whereas respiration increases in response to organic enrichment, it will decrease in response to some levels and kinds of toxic chemicals. Toxic materials can reduce the metabolic activity of the organisms present; kill some organisms, reducing the total amount of biomass respiring; or select for resistant organisms. For contamination that is more toxic than enriching, Odum (1985) and others have predicted that the respiration to biomass (R/B) will increase with stress because organisms will be forced to spend energy to cope with physiological problems caused by contaminants. Thus, less energy will be available for production of biomass. This is another ecosystem-level measure of energy exchange efficiency when all compartments of the ecosystem are measured.

Other classification schemes characterize the bioenergetics of the ecosystem by defining whether the energy that runs the system comes from within or outside the system. The production-to-respiration ratio (P/R, usually calculated as GPP/CR) compares the amount of energy fixed to the maintenance costs of the entire community (Odum, 1956). If chemical contaminants set a community back to an earlier stage of succession, P/R is expected to move away from its mature value of 1 (Odum, 1985). For most toxicants this would mean an increase in P/R. But, in the case of organic enrichment, P/R may decrease. Net daily metabolism (NDM; Bott et al., 1985) is a similar measure. NDM is the net change in dissolved oxygen concentrations over a light-dark cycle (GPP−CR). If imported fixed carbon, as in leaves or sewage, is not important in the ecosystem, NDM is a reasonable index of NCP. An analogous structural measure is the trophic index (TI), which compares photosynthetic biomass to total biomass (Clark et al., 1979). The two factors that complicate the use of these measures as indicators of toxic stress have been mentioned. First, GPP is not always a good indication of the total fixed energy available to an aquatic ecosystem. Second, some chemical contaminants are themselves sources of fixed energy.

Nutrient cycling is another essential ecosystem-level process. Primary producers cannot fix sunlight if they are not provided with certain raw materials. The amount of primary production in aquatic systems is limited by the amount of nutrients such as nitrogen and phosphorus. If the supply of nutrients is inadequate or the ability of the organism to make effective use of them is impaired, primary production will decline and the energy available to support life will decline with it. Basic ecology courses describe the biogeochemical cycles by which these essential substances circulate repeatedly through organisms, being used, released, and used again. Nutrients are taken up and used to create biomass. The biomass is eaten. The nutrient is either excreted or used in creation of new biomass. Excreted nutrients are held in soil, sediment, leaf litter, or carrion until they are used again. Based on observations of terrestrial systems, it has been suggested that stress will disrupt the ability of communities to hold on to essential nutrients (Odum, 1985). Instead, the ecosystem becomes "leaky" and nutrients are exported rather than cycled tightly. With less essential nutrient available, the amount of life that can be supported within the system declines. Uptake and regeneration rates of essential nutrients by different assemblages have been measured (e.g., Makarewicz and Likens, 1979; Paul et al., 1989). In more holistic studies, the uptake and regeneration rates of entire stream lengths are integrated into a measure of the speed at which nutrients move downstream (e.g., DeAngelis, 1992). In unimpaired systems, nutrients are tightly cycled, reingested frequently, and move downstream slowly.

Another dynamic property of ecosystems is their *resilience*. Holling (1973) used the word *resiliency* to describe the ability to recover from natural disturbances. For the most part, impact studies (see Chapter 28) operationally define resilience as the inverse of the length of time required for some end point to return to a level within its preimpact operating range. This feature is important because stress is a fact of life. The ability of ecosystems to resist change and bounce back may be impaired by chemical contaminants making them more susceptible to pests, disease, fire, flood, and climatic changes.

In addition, judging the acceptability of an anthropogenic change necessitates knowing both the magnitude and the probable duration of an adverse effect. Short-lived, small changes in a redundant component on a local scale may be an acceptable consequence of development. However, if changes are irreversible (or reversible only over very long time spans) and they compromise unique resources, they may be judged unacceptable (see Chapter 28). An understanding of the ability of an ecosystem to recover its original structure and function, or some reasonable facsimile, provides one piece of information needed to judge the acceptability of risks from construction, operation, and ecoaccidents.

Table 3 summarizes some predictions that have been made about general effects of stress on both structural and functional attributes of ecosystems (Margalef, 1975; Odum, 1985; Pratt, 1990; Rapport et al., 1985; Resh et al., 1988; Schaeffer et al., 1988; Schindler, 1987; Woodwell, 1970).

OBSERVED EFFECTS OF CHEMICALS ON FUNCTION OF ECOSYSTEMS

There is ample evidence that severe stress can change ecosystem function. Some examples illustrate the nature of changes that have been observed. Examples are cited from both terrestrial and aquatic ecosystems because generalities and linkages between systems become important when addressing problems of more than local scale.

Whole Ecosystems

Cultural eutrophication is the increased input of nutrients to fresh waters resulting from human activities. For example, agricultural runoff and sewage contain phosphorus, which can increase algal productivity in lakes, streams, and bays. The overgrowth of algae blocks the penetration of sunlight into the water. Without light, aquatic macrophytes die. When the algae die, bacteria and fungi start to decompose them, but in the process can use up all the available dissolved oxygen in the water. Fish and other larger organisms can die as a result of lack of oxygen. Cultural eutrophication is a dramatic and well-documented example of the undesirable consequences of functional change in aquatic systems (e.g., Organization for Economic Cooperation and Development, 1982; Regier and Hartman, 1973; Vollenweider, 1968). In Lake Erie, increased phosphorus inputs resulted in a 20-fold increase in overall algal productivity (Regier and Hartman, 1973). Experimental additions of phosphorus, nitrogen, and carbon to whole lakes resulted in increased rates of production, decomposition, and nutrient cycling (Schindler, 1987). Mechanistic models have been developed to predict primary production, chlorophyll content, transparency, oxygen depletion, and fish yield from information about phosphorus loadings (Jones and Lee, 1988).

Functional changes have also been a consequence of other types of disturbances such as deforestation. One study at the Hubbard Brook Experimental Forest looked at the effects of deforestation and herbicide treatment on streams in the watershed (Likens et al., 1970). Trees were cut but left in place, and new growth was prevented by herbicide treatment. This disturbance resulted in gross nutrient export from the system, dramatic changes in hydrology, and disruption of the nitrogen cycle. After catastrophic disturbance from the Mount St. Helens eruption, small streams responded with increased algal productivity, respiration, nitrogen fixation, and denitrification (Sedell and Dahm, 1984). In addition, the trophic status (an index of eutrophication based on nitrogen and phosphorus levels, clarity of water, and algal biomass) of Spirit Lake was drastically changed (Wissmar et al., 1982).

In other whole-ecosystem studies, Woodwell (1970) found that forest growth rates were impaired by radiation even where oak-pine composition was maintained. Decomposition of leaf litter, nutrient pool size, and soil metabolic activity were impaired by smelter-derived metal contamination (Freedman and Hutchinson, 1980; O'Neill et al., 1977). Forbes and Magnuson (1980) found decreased rates of leaf decomposition in a coal-ash effluent–influenced stream. Hall et al. (1980) found a greater proportion of heterotrophs in periphyton after acidification of Norris Brook. In addition, nutrients were mobilized from sediments and exported. Effler et al. (1980) found that low levels (3–14 μg/L) of copper in Cazenovia Lake did not reduce algal standing crop but did transiently reduce AR. Although structure was maintained, functional integrity was affected for 5–6 d and then recovered.

Artificial Ecosystems

Effects of chemicals on function have also been observed in manipulative experiments using physical models of natural ecosystems, often with replication. For example, O'Neill et al. (1977) found loss of the nutrients calcium and nitrate a common consequence of stress in three sizes of terrestrial ecosystems, both natural and artificial. Maki and Johnson (1976) found that the lampricide TFM (3-trifluoromethyl-4-nitrophenol) applied at recommended concentrations for lamprey control significantly inhibited GPP and stimulated CR in artificial streams; consequently P/R decreased. Giesy et al. (1979) found that 5 and 10 μg/L cadmium additions to artificial streams reduced production, respiration, and leaf decomposition. Marshall and Mellinger (1980) found that cadmium additions of ≥0.2 μg/L to in situ microcosms in Lake Michigan depressed oxygen production. Brunskill et al. (1980) found that arsenic additions of 40 μmol/L depressed organic matter decomposition and planktonic uptake of phosphorus in enclosures of a lake in the Experimental Lakes Area of Canada. Primary production and decomposition were depressed by 190–200 mg/L by drilling fluids in seagrass microcosms (Kelly et al., 1987).

Clearly, functional impairments sometimes result from insults to ecosystems. However, questions about the relative efficiencies of structural and functional measures in detecting more subtle changes arise. Are functional end points unique or especially useful in assessing environmental harm?

RATIONALE FOR SELECTION OF FUNCTIONAL END POINTS

Among the characteristics commonly considered desirable in indicators of environmental health are generality, ease of measurement, ease of interpretation, sensitivity, and relevance (Hammons, 1981; Hunsaker and Carpenter, 1990; Kelly and Harwell, 1989; Kerr, 1990; Macek et al., 1978; Suter, 1990). The utility of functional indicators of ecosystem effects can be evaluated in these terms.

Generality

Functional end points can have an advantage over more commonly used indicators in being species neutral. Functional methods at community and ecosystem levels of organization do not depend on specific organisms or specific interactions and, instead, integrate the effects of an impact on many taxonomic subgroups and behaviors simultaneously. This increases the degree to which the same end points can be used in different systems and increases the odds for continuity in end points over the long term in a monitoring program. For example, lotic ecologists sometimes compare systems across continents by using functional guilds (groups of insects that feed in the same way) rather than species as their unit of study. In long-term monitoring programs, the disappearance of a single indicator species could create an uncorrectable breach in the data, making it impossible to assess trends across the gap. For example, lake trout populations are a useful indicator in the Great Lakes region (Ryder and Edwards, 1985). But if this one indicator species was eliminated by some unusual event (e.g., disease), there would be no way to look at trends over time in order to answer the question, are things getting better or worse? Once eliminated, the indicator species is unlikely to return. By using inclusive and integrative end points, it is harder to miss environmental changes entirely. As few studies can simultaneously monitor all taxonomic groups and interactions, end points that integrate the behavior of all groups and interactions may provide an indication of change when the "keystone" species or behavior is unknown at the outset.

However, this species neutrality is sometimes a disadvantage because humans clearly prefer some species to others. A kilogram of heterotrophic biomass in the form of trout is clearly preferable in human terms to the same kilogram in the form of lamprey. Similarly, integrative end points will not elucidate the mechanics of change, nor will they respond as early as constituent behaviors. Site-specific evaluations with a sufficient pilot study may focus more directly on specific interactions known to be important.

Another question is whether functional responses to stress will be so system specific as to be ungeneralizable from one system to any other. Clearly, the taxonomic units and interactions in two complex systems are never identical. However, because tolerance distributions and the mechanisms by which toxicity is expressed and repaired are similar, communities with distinctly different taxonomic compositions may respond similarly to stress. For example, Stay et al. (1989) found that microcosms derived from four different plankton communities responded similarly to atrazine stress. NPP, pH, and P/R were reduced at concentrations of 100–200 μg/L. However, functional changes in three pond mesocosm experiments with the pesticide esfenvalerate were less consistent. Fairchild et al. (1992) found no effects on GPP at 1.7 μg/L and Lozano et al. (1992) found no effects on P/R at 5.0 μg/L; however, Webber et al. (1992) found effects on GPP at concentrations an order of magnitude lower (0.18 μg/L).

Ease of Measurement

Cost-Effectiveness

Two distinct considerations must be used in evaluating the cost-effectiveness of any end point: (1) what is the cost to measure? and (2) what is the cost to act on the information obtained in that way?

Some functional measurements are easy and inexpensive to make, involving simple skills such as operation of analytical balances, spectrophotometers, or oxygen meters rather than requiring taxonomic expertise (Kelly et al., 1987). An example of a commonly used functional measure that is quite inexpensive is BOD. Other functional measures might require sophisticated biochemical techniques (e.g., enzyme induction in individuals). As the realism of the test and its complexity increase, sampling over time and space must increase, and costs for both structural and functional end points generally increase proportionally. Confusion sometimes arises between functional measures (which can be made at any level of biological organization) and ecosystem-level measures (which are limited to one scale; see Table 1). Although large, complex, long-term tests are more expensive than small, simple, short-term tests, functional end points are generally no more expensive to measure than structural end points at the same scale.

Costs of acting on information can result from underprotection when damage to the environment must be ameliorated (Cairns et al., 1977) or from overprotection when unnecessary additional treatment costs do not result in any biological benefit (Hall et al., 1992; Karr, 1991). Although evaluating environmental impact on larger and more realistic temporal-spatial scales is always more expensive regardless of whether the end points are structural or functional, the uncertainty involved in translating effects across scale is reduced. The cost of the information obtained may be higher, but the costs of acting on that information are likely to be lower.

Availability of Standard Methods

In general, there is less experience in the use of functional end points than of structural end points. Standard methods for analyses of ecosystem function have been published. However, there are unresolved problems, including the need to determine an appropriate scale for measurement, explicit guidance on sampling, and the elimination of insensitive end points in favor of more promising ones. APHA et al. (1989) provide methods for production estimates through standing crop or stock for plankton, periphyton, macrophyton, macroinvertebrates, and fish. Methods for assessing primary (Vollenweider, 1974) and secondary (Downing and Rigler, 1984) production have also been published. APHA et al. (1989) provide methods for metabolic rate measurements such as nitrogen fixation, production, and respiration of plankton and trophic indices such as the autotrophic index. Aloi (1990) reviews methods for measuring periphyton primary productivity. Methods for assessing nutrient dynamics in stream systems have also been published (Stream Solute Workshop, 1990; DeAngelis, 1992). A chamber method for determining the production and respiration of benthic communities in streams is undergoing peer review (American Society for Testing and Materials [ASTM], Committee E-47, personal communication). Methods for assessing cultural eutrophication through algal growth potential tests have been in use for decades (ASTM, 1990; National Eutrophication Research Program, 1971). Standard methods for microcosm and mesocosm toxicity tests incorporating functional parameters also exist (e.g., U.S. Environmental Protection Agency [U.S. EPA], 1987; Touart, 1988; SETAC Foundation, 1992; see also Chapter 9). All these methods will improve with increased experience.

Ease of Interpretation

Defined Normal Operating Range

Background variation in functional end points within and between unaffected natural systems and over successional stage (i.e., the normal operating range) is not well characterized for many ecosystem types. Yet, without knowledge about variation across time and space in the absence of anthropogenic disturbance, it is impossible to judge the acceptability of observed changes in both function and structure. For example, expected ranges for structural and functional attributes of fish and macroinvertebrates are being developed for different ecoregions to support the use of indices of *biotic integrity* in streams and rivers. Biotic integrity has been defined as the ability of an ecosystem to support and maintain "a balanced, integrated, adaptive community of organisms having a species composition, diversity, and functional organization comparable to that of natural habitat of the region" (Karr, 1991). The index of biotic integrity is a numerical measure of the health of the ecosystem based on a scoring system. Scores (usually on a scale of 1 to 5) are given for a number of criteria (e.g., number of fish species, number of omnivores) derived from "natural" and healthy systems and the scores are combined to form a single measure. Functional attributes are included (i.e., susceptibility to disease and functional group distributions).

Models of response to cultural eutrophication provide another predictive model for normal functioning in aquatic ecosystems (Jones and Lee, 1988). However, Carpenter and Kitchell (1987) point out that only 50% of the variability in production is explained by phosphorus load and that similar lakes vary over an order of magnitude.

Additional ecological data provide preliminary indications of functional variability. In streams, interbiome comparisons have been done in benthic community production and respiration and leaf decomposition (e.g., Bott et al., 1985; Minshall et al., 1983; Naiman, 1983). GPP generally increased as stream size increased. The four biomes studied varied over 10-fold. CR also increased with stream size with the four biomes varying approximately 10-fold. NDM changed from negative to positive with increasing stream size, from a mean of −0.73 to +0.50 g $O_2/m^2/d$.

Clearly, the normal operating ranges for functional end points in all ecosystem types need to be better characterized. There is more information on the normal operating characteristics for structural end points because they are more thoroughly studied. Until sufficient background data are available on functions of healthy ecosystems, structural measures will continue to be the primary approach to evaluation of ecosystem health.

Consistency

The degree to which functional responses to dissimilar stresses will be similar is also at issue. Few data exist for comparing functional responses across toxicant type. The responses listed in Table 3 have been suggested as general responses to stress, but experimental confirmation is needed. Cairns et al. (1988) evaluated the chronic response of microcosms to eight diverse toxicants. Functional end points included nutrient sequestering, alkaline phosphatase activity, NDM, P/R, CR, R/B, and spectral complexity of hydrogen ion cycling. They found NDM to be the most consistent of the functional end points examined. While structural changes were detected in response to low levels of all toxicants, significant differences in NDM were detected in response to seven of eight chemicals tested. Absolute nutrient pool sizes and alkaline phosphatase activity were quite sensitive in response to individual stresses but inconsistent over stress type.

Decisiveness

Decisiveness (i.e., the ability to make management decisions based on the information collected) usually depends on statistical determinations of difference or concentration-response models supplemented by subjective assessments of what level of impairment is unacceptable to society. This is true whether the end point is structural or functional. Some statistical methods appropriate to complex, large-scale tests are discussed below.

Sensitivity

Although there is little question that functional attributes are affected by severe stress, many questions

Table 3. Trends in function expected in stressed ecosystems

Energetics
1. Community respiration increases.
2. P/R ratio becomes unbalanced.
3. P/B and R/B ratios increase as energy is diverted from growth and reproduction into acclimation and compensation.
4. Importance of auxiliary energy increases (import becomes necessary).
5. Export of primary productivity increases.

Nutrient cycling
6. Nutrient turnover increases.
7. One-way transport increases and internal cycling decreases.
8. Nutrient loss increases.

Community dynamics
9. Life spans decrease, turnover of organisms increases.
10. Trophic dynamics shift, food chains shorten, functional diversity declines, proportion of energy flowing through grazer and decomposer food chains changes.

General system-level trends
11. Efficiency of resource use decreases.
12. Condition declines.
13. Mechanisms and capacity for damping undesirable oscillations change.

do exist about functional responses to more subtle, long-term stress. *Sensitivity* refers to signal-to-noise ratios. If an end point is to be useful in evaluating environmental impact, the response to an impact must be large enough to be distinguished from normal variations. Either the signal must be large or the normal variation must be small (Figure 1). In Figure 1A the variation or noise in the response is large relative to the change in the response. However, in Figure 1B the variation about the curve is relatively small and the large drop in the response when stress is applied is easily seen. Thus in part B the signal-to-noise (or response-to-variation) ratio is large while in part A it is small.

Several researchers contend that functional end points, especially those that are substrate limited, will always be less sensitive than structural measures because of *functional redundancy* in the community (Levine, 1989; Schindler, 1987). Functional redundancy means that any loss of functional capacity by one organism will be immediately compensated by increased activity by another organism that serves the same function. For example, primary production in aquatic systems is often limited by the amount of

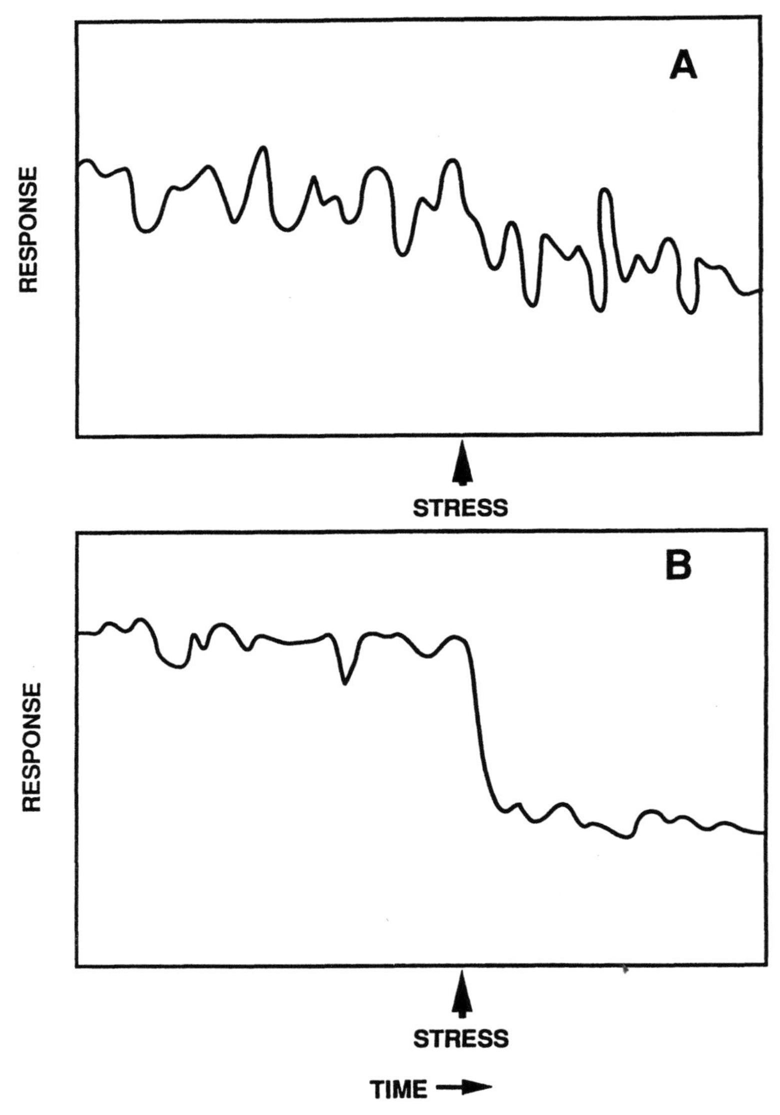

Figure 1. Differences in signal-to-noise ratios of end points in biomonitoring. (A) A low signal-to-noise ratio. The signal (i.e., the change in response after stress) is small but the response varies widely over time, so there is a large amount of noise. (B) A high signal-to-noise ratio. The change in response after stress is large and the response is less variable over time. In this case, it is easy to see the impact of the stress.

the nutrient phosphate that is available. If photosynthesis by one alga is impaired by some chemical contaminant, this will free up valuable phosphate that will immediately be put to use by other, more resistant algae. There would have to be concentrations of toxicant high enough to affect the majority of photosynthesizers adversely in order to exhaust this functional redundancy and result in overall changes in primary production. In this view, changes in function will be preceded by changes in taxonomic structure as resistant species replace sensitive ones in the same guild.

However, any comparison of the sensitivity of structure and function is heavily dependent on the hierarchical scale of the observations (e.g., Harris, 1980; Kolasa and Pickett, 1992). Structural and functional changes are expressed alternately as an impact is evaluated at increasing scale. One example of alternating structural and functional responses as the spatial and temporal scales increase is diagrammed in Figure 2. Atrazine concentrations of 20 μg/L in artificial ponds caused a reduction in phytoplankton biomass and primary production (DeNoyelles et al., 1989). But these changes were transient, and both biomass and primary production recovered within 3 wk. This recovery coincided with changes in the community composition. In general, the impairment of the function, in this example photosynthesis of specific populations, may be symptomatic of an inefficiency that is a competitive disadvantage. When invasion pressure exists, sensitive species are replaced by functionally similar, resistant species. As the community turns over, there is a structural change, and function is restored. The time it takes

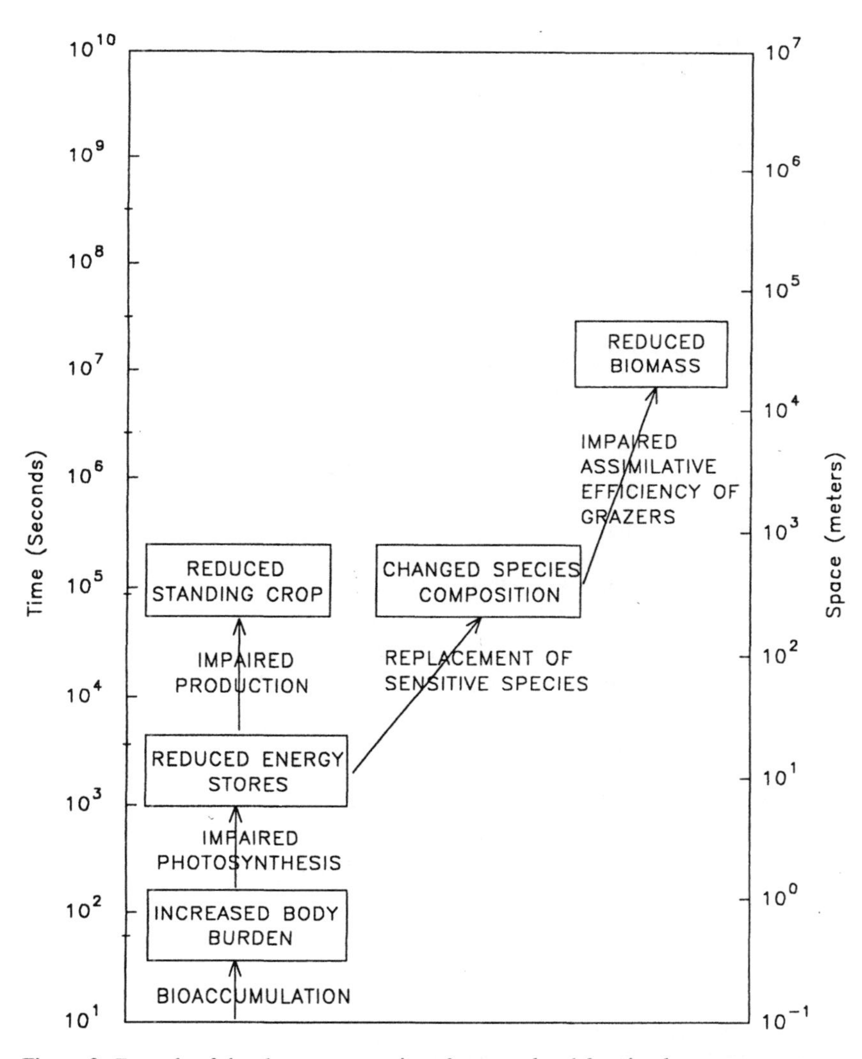

Figure 2. Example of the alternate expression of structural and functional responses to stress as time scale, spatial scale, and life spans of the observed organisms increase. Structural changes are shown within boxes. Arrows indicate functional changes.

the community to return to its preimpact normal operating range is a measure of resilience, another functional parameter. Restored function may end the sequence of events, or structural changes can affect function at the next level. For example, assimilative efficiencies of grazers feeding on phytoplankton might be reduced by the change to different algal taxa, and so on. Response at the lower level causes and explains. Response at the higher level integrates the initial functional change, the compensatory mechanisms, and the sum of the resulting structural changes. The relative sensitivity of structure and function depends on the scale at which each operates.

Consequently, some scientists find that function can be affected before compensatory mechanisms can operate. This is likely when short-term functions of long-lived organisms are being studied. In the atrazine experiments described earlier (DeNoyelles et al. 1989), the primary productivity of quickly changing communities of phytoplankton was not affected by 20 μg/L for periods of time much longer than it takes for a complete turnover of the assemblage, but the primary productivity of longer-lived aquatic macrophytes was permanently affected. Similarly, the primary productivity of longer-lived seagrass beds was affected by drilling fluids (Kelly, 1989), and the primary productivity of forests was affected by ionizing radiation (Woodwell, 1970) and air pollution (Bormann, 1983) before successional changes to impact-resistant species could compensate.

Because of this scale dependence, perhaps it is not surprising that a comparison of sensitivity of commonly employed structural and functional end points in ecosystem-level studies shows mixed results. Greater relative sensitivity of structural than of functional end points has been reported in whole-lake acidifications to pH 5.0, described by Schindler (1987). No significant changes in decomposition or nutrient concentrations were found, but species composition of phytoplankton was among the earliest indicators of change. Eaton et al. (1986) reported that of several functional end points monitored, only pondweed litter decomposition was affected in artificial streams treated with chlorpyrifos at levels causing structural changes (0.12–0.83 μg/L). There were no obvious changes in P/R, microbial biomass production and heterotrophic activity, organic matter mineralization, and total organic carbon production. Crossey and La Point (1988) found measures of GPP and CR to be more variable than macroinvertebrate taxonomic composition in a stream receiving mine drainage. Boyle et al. (1985) found that fluorene concentrations of 10 mg/L adversely affected survival of largemouth bass but did not affect primary production. Cairns et al. (1988) compared consistency and sensitivity of 25 end points, both structural and functional, over eight distinct stresses in microcosm toxicity tests. They found that species composition and species richness of protists were the most consistently sensitive end points. NDM was the most consistent functional end point, but its signal-to-noise ratio was still less than that for the structural end points.

In contrast, there are also many reports of adequate sensitivity of functional parameters. In terrestrial systems, functional measures have often been more sensitive than structural ones. Van Voris et al. (1980) found that functional changes such as export of the nutrient calcium was affected by cadmium contamination at concentrations that did not affect taxonomic measures (0.427 mg/cm^2). Woodwell (1970) found that functional attributes of forests such as growth rate were affected by radiation even where structural composition was maintained (1 roehtgenf/day). In aquatic systems, increasing phosphorus and nitrogen additions to whole lakes by 10-fold greatly increased algal production and enhanced decomposition by sufate-reducing bacteria (Schindler, 1987). Kelly (1989) found production and litter decomposition in seagrass communities as sensitive as structure to drilling fluids at concentrations of 190–200 ppm. Hedtke (1984) found primary production, dissolved organic carbon production, and macroalgal growth more sensitive than taxonomic structure to the addition of copper to microcosms (9.3 μg/L). Lewis et al. (1986) conducted in situ toxicity tests exposing phytoplankton and periphyton communities to the surfactant TMAC (lauryl trimethylammonium chloride), and found that photosynthesis was generally comparable in sensitivity to community similarity, responding at concentrations of 0.96 to 1.99 mg/L. Giddings and Franco (1985) found that changes in P/R, pH, and cladoceran densities were the earliest responses to synthetic crude oil treatment in pond mesocosms, responding at average phenol concentrations <0.05 mg/L. In a review of the literature, Levine (1989) found that of 100 articles describing the effects of chemical stresses on aquatic ecosystem functions, a majority reported finding significant changes. But, as she pointed out, published reports may be a biased sample as they are more likely to mention differences than failure to find differences.

Relative sensitivity of structural and functional end points can be expected to change with the selection of temporal scale, end points, measurement methods, and stress and the skill with which they are applied in individual studies. Because of longer experience and better developed methods for sampling, measuring, and analyzing structural end points, less methodological variation in these measurements can be expected. This factor alone will tend to increase statistical power and sensitivity for these measures. However, the importance of sensitivity in choosing a predictive end point is often overstated (Kelly, 1989). As pointed out by Mount (1987), "*More sensitive* doesn't mean *more accurate*!" The relevance of a functional parameter may dictate its inclusion.

Relevance

Clearly, some ecosystem services are worthy of protection. Humans depend on ecosystem services such as oxygen production, water purification, and production of foodstuffs. Specific examples of functional attributes of aquatic ecosystems that have been included in lists of end points for monitoring the state of the environment are presented in Table 4. Environmental management strategies can assure the protection of ecosystem function in two ways. Either tests that measure functional end points very similar to the ones of interest must be conducted, or models (see Chapter 18) must be developed to predict effects on function from data from commonly used tests on qualitatively different end points. Then these models must be validated. *Validation* involves comparing predictions based on the model to observed effects in natural systems in order to quantify the margins of error and establish whether the model is sufficiently accurate for management needs.

SCREENING FUNCTIONAL END POINTS FOR ASSESSMENT AND MONITORING

A review of ecotoxicological tests provides only a limited basis for screening functional end points for relative utility in assessing environmental health. Only a minority of ecotoxicological tests incorporate any functional end points. For the few functional end points that have been evaluated for more than one chemical, methodological variations and differences in the scale of test systems preclude conclusions about consistency. Many of the tests that have incorporated functional end points have not been designed to determine threshold concentration, so determinations of relative sensitivity are not possible. However, several functional end points seem more promising than others as general indicators of ecological health based on the data available and the desirable charcteristics for end points.

Production

Several factors complicate the use of production measures as a general indicator of chemical contamination. As previously mentioned, the response of production may depend on the time scale of observation compared with the length of the life cycle of the organisms. Whereas the production of long-lived producers may be affected by toxic contaminants, short-lived, easily dispersed organisms may show only transitory responses as sensitive species are replaced by resistant ones. A further contribution to confusion is that contaminants can stimulate as well as inhibit primary production. For example, a field study of metal mine drainage (Crossey and LaPoint, 1988) found GPP increased in affected stream reaches. In addition, primary production in natural systems may vary substantially due to differences in food web structure (Carpenter et al., 1987); that is, if there are many grazers, the standing crop of phytoplankton may be kept low. The wide range of naturally occurring levels of primary production makes it difficult to distinguish natural variation from that due to contaminants. For these reasons, expressions of primary production other than GPP may be more consistent from one site or toxicant to the next and more successful in identifying the effects of chemical contaminants.

When AR is used, expression of production is weighted to remove variance from differences in trophic structure of the community. Variance is reduced and sensitivity is improved. Several studies have found AR to be a more sensitive indicator of toxicant-induced effects than simultaneous measurements of GPP (Crossey and LaPoint, 1988; Effler et al., 1980; Pratt, 1990; Stay et al., 1985). The AR decreased in response to the majority of toxicants tested; 5 μg/L copper (Effler et al., 1980), 30 μg/L selenium (Pratt, 1990), acute zinc (10,078 μg/L; Niederlehner and Cairns, 1992), acute acid (pH 3.5; Niederlehner and Cairns, 1993), atrazine (337 μg/L and 100 μg/L; Pratt, 1990 and Stay et al., 1985, respectively), and a mixed heavy metal discharge (Crossey and LaPoint, 1988). However, there are contradictory reports. AR increased in response to sediment-associated copper (1000 mg/kg) but changes were transient (Scanferlato and Cairns, 1990). AR also increased in response to $\geq$84.9 μg/L selenium (Pratt and Bowers, 1990), but a successful gradient in AR was evident and may have interacted with toxicant stress (i.e., stressed communities were restricted to an earlier successional stage in which AR was higher).

A societal goal of sustainability or long-term productivity is often translated as a lack of impairment in secondary production (Bird and Rapport, 1986). It is most commonly assessed by monitoring change in standing stock of desirable fish communities through time. Because of its relevance, secondary production will remain an important monitoring end point (Bird and Rapport, 1986; Suter, 1990).

Table 4. Functional end points suggested for assessment and monitoring of aquatic systems

End points for regional ecological risk assessments (Suter, 1990):
1. Productivity
2. Nutrient export
3. Pollutant export
4. Susceptibility/contagion

USEPA Environmental Monitoring and Assessment Program (Hunsaker and Carpenter, 1990):
1. Trophic status
2. Production/biomass

State of the Environment Report for Canada (Bird and Rapport, 1986):
1. Production/biomass
2. Nutrient load

Respiration

Respiration has proven utility in delineating and modeling response to organic enrichment as evidenced by the widespread use of BOD. But respiration may increase or decrease in response to other kinds of contamination, and changes in overall respiration may indicate changes in biomass, changes in metabolic activity, or changes in species composition. Odum (1985) suggests that R/B should change more consistently, increasing under stressful conditions as a uniform indication of decreased efficiency in energy flow (Table 3). However, these energy costs may be apparent only late in the progression of impact when functional redundancy and the supply of resistant immigrants have been exhausted. In microcosm toxicity tests, R/B increased as predicted in response to an industrial effluent (Cairns et al., 1988), but R/B decreased in response to acute pH stress (pH 3.5; Niederlehner and Cairns, 1993). Schindler (1990) also found that R/B decreased in response to acidification in whole-lake experiments. R/B decreased in response to severe copper stress in a microcosm (1000 mg/kg sediment; Scanferlato and Cairns, 1990).

Community Metabolism

Although P/R is a commonly suggested functional measure of overall community metabolism, it is unlikely to be the most sensitive functional measure of adverse impact for several reasons. The combined measure has all of the drawbacks of its constituent parts, discussed previously, which make it difficult to predict a uniform, directional change that will occur over different toxicants and ecosystems. In addition, P/R has the statistical disadvantage of being a ratio. The distribution of P/R is skewed and widely variable when the values of the components (P and R) are not highly correlated. This results in low power. Log transformation of P/R helps some. But using other expressions of community metabolism such as NDM can eliminate the ratio and maintain units in a linear scale. By using a measure that is more statistically robust, sensitivity may be improved. Even more important is the development of explicit scale and sampling guidance for measures of community metabolism through pilot studies.

In studies comparing the responses of P/R and other functional end points to stress, the relative sensitivity of P/R varied. Giddings and Franco (1985) found that additions of coal liquid resulting in phenolic concentrations ≤0.05 mg/L significantly decreased P/R below a value of 1 and increased CR and that these were among the most sensitive responses monitored. However, Stay et al. (1985) found that atrazine concentrations of 500 μg/L were required to decrease P/R from 1, whereas AR was consistently decreased at 60 μg/L. Pratt and Bowers (1990) also found that AR was more sensitive than P/R to selenium stress. There were no differences in P/R at ≤160 μg/L selenium, but AR increased at ≥84.0 μg/L selenium. In stream reaches receiving metal mine drainage, P/R was reduced from ~3 in the control reach to ~1 in the affected reach (Crossey and LaPoint, 1988). AR was also depressed, while CR increased and GPP remained unchanged. In all these studies P/R decreased. But in one (Crossey and LaPoint, 1988), P/R moved closer to 1. This is inconsistent with the general prediction that stress will cause a shift away from 1 (Table 3).

Nutrient Cycling

Although nutrient export has been a significant and important effect of many studies with a terrestrial component (e.g., Likens et al., 1970; O'Neill et al., 1977; Van Voris et al., 1980), when the focus is strictly aquatic, the effects of contaminants on nutrient dynamics are less obvious. Schindler et al. (1985) found no disruptions in phosphorus cycling in Lake 223 when it was treated with acid to a pH of 5.0. Giesy et al. (1979) found no differences in phosphorus export in artificial streams treated with ≤10 μg/L cadmium. However, Wallace (1989) found increased particulate export in natural streams periodically treated with methoxychlor (10 mg/L for 4 h) and speculated that export of phosphate as part of the particulate matter may also be significant. Nutrient retention increased with addition of triphenyl phosphate–contaminated sediments (periodic additions of 50 to 1600 mg/L) in artificial streams (Fairchild et al., 1987), but it also increased with addition of uncontaminated sediments.

In contrast to these holistic determinations, enzyme activities associated with nutrient cycling, such as alkaline phosphatase, are more commonly measured and have been affected by many toxic stresses (e.g., Burton et al., 1989; Griffiths et al., 1982; Jones and Hood, 1980; Sayler et al., 1982). Because of the small spatial and temporal scales typically sampled and the patchiness of such activity, judging the acceptability of such impairments is difficult in an ecosystem context. As with primary production measures in phytoplankton communities, these are functional measures, but they are not ecosystem-level measures unless the sampling has been specifically designed to include all components of the system.

Decomposition

Disruption in decomposition is another reported result of contamination in terrestrial systems (e.g., Freedman and Hutchinson, 1980), but results in aquatic systems are mixed. For example, drilling fluids (190–200 mg/L; Kelly et al., 1987), cadmium (5 μg/L; Giesy et al., 1979), and coal ash effluent (Forbes and Magnuson, 1980) have been reported to decrease rates of decomposition, whereas eutrophication increased rates (Schindler, 1974). Leaf decomposition did not differ in other studies with copper (≤420 μg/L; Hedtke, 1984), cresol (8 mg/L; Cooper and Stout, 1985), and triphenyl phosphate–contaminated sediments (periodic additions of 50 to 1600 mg/L; Fairchild et al., 1987). The debate over the effects of acidification on decomposition, where contradictory results in a number of studies (see review by Baker and Christensen, 1991) must be reconciled, suggests that decomposition does not always respond simply and directly to toxicant addition. This is certainly a disadvantage for an end point selected for general monitoring. However, decomposition may still be an important end point in specific ecosystems.

As with nutrient cycling, activities of various enzymes, such as amylase, cellulase, and glucosidase (e.g., Burton et al. 1989), have also been used to monitor decomposition, and the same cautions about relative scale of measurement versus system of interest apply.

Resilience

Cairns et al. (1971) studied the recovery of riverine ecosystems from anthropogenic stresses approximately two decades ago and identified six factors contributing to ecosystem resilience: the existence of nearby species sources, the transportability of these species, the condition of the physical habitat after the stress, the persistence of residual toxiants, the condition of the water chemistry following the stress, and the potential for human activities to assist in remediation. Yount and Niemi (1990) provided a review of recovery in flowing fresh waters. In 150 studies reviewed, most riverine systems recovered structural integrity relatively rapidly after an acute insult, usually within 2 yr (Niemi et al., 1990). This information has been incorporated into management decisions as a limit to the permissible frequency of excursions outside chemical specific standards (U.S. EPA, 1991). However, there are few data on functional end points, chronic stresses, more subtle responses, or other ecosystem types. Because lotic ecosystems are characterized predominantly by rapid changes over time, they can be expected to recover more quickly than most ecosystem types. In forested watersheds that were disturbed, nutrient losses returned to normal within 3 yr at Hubbard Brook but elevated losses persisted for up to 20 yr at the Coweeta Hydrologic Laboratory (Vitousek et al., 1981).

STATISTICAL ISSUES IN ECOSYSTEM STUDIES

The ability to detect and measure change (recovery, effects of manufacturing plants, etc.) depends on the end point, either structural or functional), and its statistical properties. In ecosystem studies, these properties are controlled in part by nature and in part by the researcher and the measurement process. Reducing variation in ecosystem studies is important, and statistical understanding provides some aids for improvement of ecosystem studies. Of interest is the role of the study design in ecosystem studies and the use of statistical thinking to optimize functional measures (see Chapter 10).

Design of Experiments

Statistical design (see Chapter 10) is based on three basic concepts—*randomization*, *replication*, and *local control*. Randomization of experimental units to treatments provides a safeguard against potential sources of bias and confounding. Replication is needed to assess variation. Local control or blocking is a useful tool for removing unwanted variation and improving the sensitivity of designs. However, it is sometimes difficult to incorporate these features into ecosystem studies (further aspects are discussed in Chapter 9).

Randomization of experimental units to treatments is frequently not possible. For example, when the interest of the study is in assessing the impact of a power plant effluent discharge to a river, it is not possible to assign randomly half of the sites receiving effluent to locations above the plant and half to sites below the plant. Thus, the effect of the effluent is necessarily confounded with the location of the sampling sites. In a worst case, the sites below the discharge may have fast current and a cobble substrate, while sites above the plant may have pools and fine sediment substrate. Clearly, habitat may have more effect than the effluent, but the effects cannot be separated. The effect of the effluent is now confounded with the substrate; ecological differences between stations may be due to habitat rather than discharge.

Replication in aquatic ecosystem field studies is also difficult. A true statistical replicate in the power plant example would require a second power plant. Clearly, this is not feasible (would we build a second plant just to test for an effect?), and usually sites are viewed as replicates. One drawback to using the sampling sites as replicates is the problem of dependence. Sites that are close tend to be more similar biologically than sites farther apart. Hence, measurements on sites that are quite close can be expected to have similarity and would be spatially correlated. The spatial similarity would result in an estimate of variance that is lower than one based on using true replication. A consequence of the lack of replication in field studies is that no clear-cut assessment of within-treatment variability can be made and conclusions about the cause of any observed differences between sampling stations may be confounded with other factors. Hurlbert (1984) referred to this problem as the problem of *pseudoreplication*. As defined by Hurlbert (1984), pseudoreplication is the testing for treatment effects using an error term that is inappropriate for the hypothesis being considered (see Chapter 10).

Hurlbert (1984) provides numerous examples of pseudoreplication. For example, in a study of leaf decomposition in streams, a researcher may fill bags with leaves, place them in the stream, and measure the decomposition that occurs over time. However, because the leaf packs are really one group of packs, it is inappropriate to use the information to represent more than one observation. If two sets of leaf packs are placed at two different depths, it is inappropriate to view this as representing more than two observations. The problem is that the depth effect is confounded with a site effect and cannot be separated. A better approach is to select a number of sites with the same depth. Then the depth is replicated.

Another example of pseudoreplication occurs in toxicity testing. Suppose interest is in the effects of an effluent on fish. Put 15 fish in each of 10 jars (or test chambers), apply the effluent at different dilutions, and measure the rate of growth in each fish. The result is 150 observations; however, there are really only 10 observations for evaluating variation in effluent. The jar is the unit that receives the application and the fish in the jar represent subsamples or pseudoreplicates. The proper analysis is to average the measurements on fish in a given jar and then analyze the means or use a nested design (Sokal and Rohlf, 1981).

Local control or blocking may require collecting and analyzing additional samples. Because time and resources are limited and functional end points are often "accessory" rather than "core" information, it is rare to see gradients accommodated.

Clearly, the design and analysis of ecological monitoring studies are not simple tasks unless the effects of the suspected impact are so great that statistical analyses are superfluous. A number of papers have evaluated the design of ecological experiments in general and designs for impact assessment specifically (e.g., Eberhardt and Thomas, 1991; Green, 1979; Hurlbert, 1984; Peterman, 1990; Matson and Carpenter, 1990).

Green (1979) suggests that potential impacts could be assessed by collecting field data in a control zone and in the impact zone both before and after the potential impact begins. The impact could then be assessed using a two-factor analysis of variance (or the multivariate extension, with multivariate observations). Hurlbert (1984) has criticized ecological impact assessment methods, in particular the design of Green, as not being statistically sound, but Hurlbert does not suggest alternatives. The problem that Hurlbert identified as pseudoreplication is one of lack of replication and inability to randomize the impact to the sites. Ecological studies (even including some microcosm and mesocosm studies; Chapter 9) are typically not controlled experiments or may be only partially controlled and, hence, should not be analyzed as if they were controlled. The difficulty is due in part to the desire to assign causation (see Chapter 1, assumption of causality). In designed experiments in the laboratory (see chapters on toxicity testing), causation is inferred by controlling other potential factors or by randomization. With no randomization or control, there may be multiple causes, and additional evidence may be required to distinguish among these causes. The use of additional evidence to infer causation in impact analyses is best illustrated by the triad approach (Long and Chapman, 1985; see Chapter 8). Three types of information are needed for a convincing evaluation of environmental impact. First, the field survey provides the most realistic, relevant, and convincing evidence that environmental integrity has been harmed, but it does not establish a causal link to a chemical. Second, the chemical survey confirms the presence of the chemical in the impaired areas of the natural system. Third, toxicity tests with true replication establish a link between biological impact and the presence of the chemical, but sacrifice realism.

Stewart-Oaten et al. (1986), in a rejoinder to Hurlbert (1984), suggested a design using local control to assign cause. This design involves obtaining data on pairs of stations, one viewed as control and the other in the impact zone, collected both before and after the onset of the potential impact. This design differs from that of Green (1979) in that the below and above sites are paired. The design is quite similar to the one used by Thomas et al. (1978), and both research groups used the design to assess effects of nuclear power plants. Although this design can be useful for assessing change, not all the difficulties are resolved. Work by Stewart-Oaten et al. (1992) and Smith et al. (1993) suggests that potential confounding of effects is not completely removed.

How best to design ecosystem studies to assess functional (not structural) change is still a difficult question. However, some general guidelines are available. Eberhardt and Thomas (1991) point out that some of the difficulties in ecosystem field studies are due to inadequate sample size and lack of well-defined objectives. When objectives are well designed, it is possible to take advantage of the knowledge of uncertainty associated with the objectives to develop a design to minimize this uncertainty. Three recommendations arise, based on this discussion. First, it is better to view ecosystem studies as observational studies rather than controlled experiments—the researcher observes the system and does not control the system. Potential sources of confounding are possible and need to be addressed. Issues related to randomization, replication, and blocking become issues related to reliability, independence of observations, and control of variance. Reliability of measurements involves controlling as many of the components of the study as possible to maintain high quality control. Independence of the measurements is achieved by not sampling too closely in space or time. There are many approaches to the control of variance, such as matching control sites with treated sites, using confounding variates as model covariates, and using composite samples (take a number of samples, physically combine them, then sample from this combined sample). Thus, the ideas of statistical design are still useful. Second, when interest is in causation, it is best to have additional information to assess causation. Auxiliary information may be required to eliminate other sources of possible causation. Finally, it is better to have well-defined objectives for ecosystem studies. When one can form testable hypotheses, the design approach is most appropriate and can increase the sensitivity of the study. Without well-defined objectives or with a large number of objectives, many compromises are needed with regard to design and reducing all the components of uncertainty is difficult. Peterman (1990) points out that even with well-defined objectives, adherence to statistical principles is important. Peterman reviewed a number of impact assessments and noted that many of the impact assessments were lacking in power (ability to detect a difference with statistics). He recommended that a power analysis be an integral part of the planning of environmental studies.

Optimizing Functional Measures

Using statistical methods for reduction of confounding in ecosystem studies is sometimes difficult. The field study is likely to be affected by some form of bias. It is possible, for example, in a study on an estuary that the rainfall was great in the year sampling was undertaken. The large amount of rainfall may affect the survival of fish larvae and the impact of an effluent on the larvae. The bias imposed by sampling in that year is not removable without additional sampling. Another approach to assessing effects is based on statistical methodology for sampling. This approach can be quite useful for understanding and reducing variation in measurement. Some aspects of statistical sampling include the importance of a pilot study, stratification, and compositing (see Chapter 10).

Pilot studies are extremely important for addressing questions related to variation in measurement. These studies can be a great aid in resolving when to sample, how often to sample, and where to sample. Some examples of the usefulness of pilot studies are given in Green (1979).

Stratification in sampling is similar to blocking in design. In ecosystem studies, the researcher has some control over the variation by the selection of where, when, and how to sample and by measuring extraneous variables that are of indirect interest but may affect the functional measurement. For example, in some functional measurements, great variation may be associated with the depth of the sample. Thus, depth becomes a factor that is not of primary interest but has some control over the function measure. By stratifying on the depth variable, estimates of function with lower variance may be obtained. The set of possible sampling sites is divided according to depth. Then, within a given depth category, a number of samples are taken. By properly balancing the number of samples taken within each category, more efficient estimates of quantities such as the mean of the functional measure over a region may be obtained. Information about potential stratification factors comes from a knowledge of the system and the functional end point and frequently from a pilot study. Generally, the more information available about a particular functional measure and the ecosystem to be studied, the greater is the potential for reducing measurement uncertainty. Eberhardt and Thomas (1991) provide more information on types of sampling of use in ecological field studies.

Another useful tool is *compositing*. In ecosystem studies, there are frequently factors of interest that can affect the functional measurement. Although these factors are of importance, sampling budgets do not allow their inclusion in the sampling design. For example, if a functional measure is known to vary with water current velocity, interest is not in differences between velocities but in comparing sites or long-term change. Because of the change with velocity, water current becomes a confounding factor. One approach to dealing with this factor is to sample at similar current velocities at each site. Another approach is to sample at various current velocities but record the velocity, using it as a covariate (see Chapter 10) or averaging measurements over velocity. However, this may become prohibitively expensive because it requires additional determinations. A less expensive approach is compositing. Compositing is useful for obtaining an average value over levels of a factor when taking measurements over levels of the factor is not of interest. In compositing, the samples are taken, physically combined, and then the measurement is made. Compositing can greatly reduce expense as fewer measurements are made. The drawback to compositing is that information may be lost and sensitivity reduced. For example, if interest is in measuring photosynthesis, then five current velocities are selected in forming the composite sample. If interest is in average daily photosynthesis, compositing may be useful. However, if the most sensitive measure is the maximum photosynthesis rate, then actual measurements are needed. Compositing would result in an estimate with larger variance.

INCORPORATING FUNCTIONAL END POINTS INTO MANAGEMENT

Functional end points can be integrated into environmental impact assessment in the same way as structural end points are currently used. Predictions of the magnitude and extent of physical, chemical, and biological changes in the environment directly resulting from an action are made. The magnitude of the effects of these changes on environmental integrity is predicted by comparing the anticipated changes to concentration-response models for various important end points. When this comparison suggests that the end points of interest are impaired at anticipated levels of impact, an evaluation of system resilience will determine the reversibility of impairment and temporal extent of impairment. Finally, the acceptability of risk (see Chapter 28) is determined with information about (1) the magnitude, time span, and spatial extent of direct physical, chemical, and biological changes in the environment; (2) the environmental end points responding to these changes; (3) the magnitude of environmental response; (4) the reversibility of impairment; (5) the uniqueness of the resources at risk; and (6) who benefits and how much versus who loses and how much.

SUMMARY

Given the dependence of society on ecosystem services, some information about the effects of toxic substances on these functional attributes is essential. However, whether functional characteristics at the ecosystem level should be routinely screened depends on their uniqueness and cost-effectiveness. There has been great success in the use of ecosystem-level functions to diagnose and model several environmental problems, particularly cultural eutrophication and organic enrichment. These are both enrichment stresses. Will the same approach be as successful with toxics? Success will depend on several factors. First, until valid models are available, the correspondence between the end points measured and characteristics deemed worthy of protection must be maintained. So, if nutrient retention is a valued attribute in a particular system, direct measurement of nutrient removal can be made. Second, methods for sampling and measuring functional end points are not as well developed as those for structural end points. At present, methodological variations, inattention to scale, and suboptimal choice of expression of results undermine the usefulness of process measurements. Research to resolve these difficulties is needed. Third, after methods have been optimized, functional end points can be compared for generality, consistency, and sensitivity and their usefulness as indicators of ecosystem integrity evaluated.

LITERATURE CITED

Academy of Natural Sciences: Flint River-Lake Blackshear Ecosystem Studies. Philadelphia: Academy of Natural Sciences, 1984.

Aloi, J. E.: A critical review of freshwater periphyton field methods. *Can. J. Fish Aquat. Sci.* 47:656–670, 1990.

American Public Health Association, American Water Works Association, and Water Pollution Control Federation: Standard Methods for the Examination of Water and Wastewater, 17th ed. Washington, DC: American Public Health Association, American Water Works Association, and Water Pollution Control Federation, 1989.

American Society for Testing and Materials: Practice for algae growth potential testing with *Selenastrum capricornutum* D3978-80. Annual Book of ASTM Standards. Philadelphia: American Society for Testing and Materials, 1990.

Baker, J. P., and Christensen, S. W.: Effects of acidification on biological communities in aquatic ecosystems. Acidic Deposition and Aquatic Ecosystems: Regional Case Studies, edited by D. F. Charles, and S. Christie, pp. 83–106. New York: Springer-Verlag, 1991.

Bird, P. M., and Rapport, D. J.: State of the Environment Report for Canada. Quebec: Canadian Government Publishing Centre, 1986.

Bormann, F. H.: Factors confounding evaluation of air pollution stress on forests: pollution input and ecosystem complexity. Symposium on Acid Deposition, A Challenge for Europe. Karlsruhe, Germany, 1983.

Bott, T.; Brock, J. T.; Dunn, C. S.; Naiman, R. J.; Ovink, R. W.; and Petersen, R. C.: Benthic community metabolism in four temperate stream systems: An inter-biome comparison and evaluation of the river continuum concept. *Hydrobiologia* 123:3–45, 1985.

Boyle, T. P.; Finger, S. E.; Paulson, R. L.; and Rabeni, C. F.: Comparison of laboratory and field assessment of fluorene—Part II: Effects on the ecological structure and function of experimental pond ecosystems. Validation and Predictability of Laboratory Methods for Assessing the Fate and Effects of Contaminants in Aquatic Ecosystems, STP865, edited by T. P. Boyle, pp. 134–151. Philadelphia: American Society for Testing and Materials, 1985.

Brunskill, G. J.; Graham, B. W.; and Rudd, J. W. M.: Experimental studies on the effect of arsenic on microbial degradation of organic matter and algal growth. *Can. J. Fish Aquat. Sci.* 37:415–423, 1980.

Burton, G. A., Jr.; Stemmer, B. L.; Winks, K. L.; Ross, P. E.; and Burnett, L. C.: A multitrophic level evaluation of sediment toxicity in Waukegan and Indiana Harbors. *Environ. Toxicol. Chem.* 8:1057–1066, 1989.

Cairns, J., Jr.; Crossman, J. S.; Dickson, K. L.; and Herricks, E. E.: The recovery of damaged streams. ASB Bull 18:79–106, 1971.

Cairns, J., Jr.; Dickson, K. L.; and Herricks, E. E.: Recovery and Restoration of Damaged Ecosystems. Charlottesville, VA: University Press of Virginia, 1977.

Cairns, J., Jr.; Pratt, J. R.; Niederlehner, B. R.; and Bowers, N. J.: Structural and Functional Responses to Perturbation in Aquatic Ecosystems, AFOSR-TR-88, NTIS AD-192 071/9/GAR. Springfield, VA: National Technical Information Service, 1988.

Carpenter, S. R., and Kitchell, J. F.: The temporal scale of variance in limnetic primary production. *Am. Nat.* 129: 417–433, 1987.

Carpenter, S. R.; Kitchell, J. F.; Hodgson, J. R.; Cochran, P. A.; Elser, J. J.; Elser, M. M.; Lodge, D. M.; Kretchmen, D.; He, X.; and von Ende, C. N.: Regulation of lake primary productivity by food web structure. *Ecology*, 68:1863–1876, 1987.

Clark, J. R.; Dickson, K. E.; and Cairns, J., Jr.: Estimating aufwuchs biomass. Methods and Measurements of Periphyton Communities: A Review, STP 690, edited by R. L. Weitzel, pp. 116–141. Philadelphia: American Society for Testing and Materials, 1979.

Cooper, W. E., and Stout, R. J.: The Monticello experiment: a case study. Multispecies Toxicity Testing, edited by J. Cairns, Jr., pp. 96–116. New York: Pergamon, 1985.

Crossey, M. J., and LaPoint, T. W.: A comparison of periphyton community structural and functional responses to heavy metals. *Hydrobiologia*, 162:109–121, 1988.

DeAngelis, D. L.: Dynamics of Nutrient Cycling and Food Webs. New York: Chapman & Hall, 1992.

DeNoyelles, F., Jr.; Kettle, W. D.; Fromm, C. H.; Moffett, M. F.; and Dewey, S. L.: Use of experimental ponds to assess the effects of a pesticide on the aquatic environment. Using Mesocosms to Assess the Aquatic Ecological Risk of Pesticides: Theory and Practice, edited by J. R. Voshell, Jr., pp. 41–56. Lanham, MD: Miscellaneous Publications of the Entomological Society of America 75, 1989.

Downing, J. A., and Rigler, F. H.: A Manual on Methods for the Assessment of Secondary Productivity in Fresh Waters, International Biological Project 17. Oxford, UK: Blackwell, 1984.

Eaton, J.; Hermanutz, R.; Kiefer, R.; Mueller, L.; Anderson, R.; Erickson, R.; Nordling, B.; Rogers, J.; and Pritchard, H.: Biological effects of continuous and intermittent dosing of outdoor experimental streams with chlorpyrifos. Aquatic Toxicology and Hazard Assessment: Eighth

Symposium, STP891, edited by R. C. Bahner, and D. J. Hansen, pp. 85–118. Philadelphia: American Society for Testing and Materials, 1986.

Eberhardt, L. L., and Thomas, J. M.: Designing environmental field studies. *Ecol. Monogr.* 61:185–199, 1991.

Effler, S. W.; Litten, S.; Field, S. D., Tong-Ngork, T.; Hale, F.; Meyer, M.; and Quirk, M.: Whole lake responses to low level copper sulfate treatment. *Water Res.* 14:1489–1499, 1980.

Ehrlich, P. R., and Ehrlich, A. H.: Healing the Planet: Strategies for Solving the Environmental Crisis. New York: Addison-Wesley, 1991.

Fairchild, J. F.; Boyle, T.; English, W. R.; and Rabeni, C.: Effects of sediment and contaminated sediment on structural and functional components of experimental stream ecosystems. *Water Air Soil Pollut.* 36:271–293, 1987.

Fairchild, J. F.; LaPoint, T. W.; Zajicek, J. L.; Nelson, M. K.; Dwyer, F. J.; and Lovely, P. A.: Population-, community-, and ecosystem-level responses of aquatic mesocosms to pulsed doses of a pyrethroid insecticide. *Environ. Toxicol. Chem.* 11:115–130, 1992.

Forbes, A. M., and Magnuson, J. J.: Decomposition and microbial colonization of leaves in a stream modified by coal ash effluent. *Hydrobiologia* 76:263–267, 1980.

Freedman, B., and Hutchinson, T. C.: Smelter pollution near Sudbury, Ontario, Canada, and effects on forest litter decomposition. Effects of Acid Precipitation on Terrestrial Systems, edited by T. C. Hutchinson, and M. Havas, pp. 395–427. New York: Plenum, 1980.

Giddings, J. M., and Franco, P. J.: Calibration of laboratory bioassays with results from microcosms and ponds. Validation and Predictability of Laboratory Methods for Assessing the Fate and Effects of Contaminants in Aquatic Ecosystems, STP865, edited by T. P. Boyle, pp. 104–119. Philadelphia: American Society for Testing and Materials, 1985.

Giesy, J. P., Jr.; Kania, H. J.; Bowling, J. W.; Knight, R. L.; Mashburn, S.; and Clarkin, S.: Fate and Biological Effects of Cadmium Introduced into Channel Microcosms. EPA-600/3-79-039. Springfield, VA: National Technical Information Service, 1979.

Green, R. H.: Sampling Design and Statistical Methods for Environmental Biologists. New York: Wiley, 1979.

Griffiths, R. P.; Caldwell, B. A.; Broich, W. A.; and Morita, R. Y.: Long-term effects of crude oil on microbial processes in subarctic marine sediments: studies on sediments amended with organic nutrients. *Mar. Pollut. Bull.* 13:273–278, 1982.

Hall, J. C.; Raider, R. L.; and Grafton, J. A.: EPA's heavy metal criteria: strategies for obtaining reasonable limitations. *Water Environ. Technol.* 4:60–63, 1992.

Hall, R. J.; Likens, G. E.; Fiance, S. B.; and Hendrey, G. R.: Experimental acidification of a stream in the Hubbard Brook Experimental Forest, New Hampshire. *Ecology* 61:976–989, 1980.

Hammons, A. (ed): Methods for Ecological Toxicology. Ann Arbor, MI: Ann Arbor Science Publishers, 1981.

Harris, G. P.: Temporal and spatial scales in phytoplankton ecology. Mechanisms, methods, models, and management. *Can. J. Fish Aquat. Sci.* 37:877–900, 1980.

Hedtke, S. F.: Structure and function of copper-stressed aquatic microcosms. *Aquat. Toxicol.* 5:227–244, 1984.

Holling, C. S.: Resilience and stability of ecological systems. *Annu. Rev. Ecol. Syst.* 4:1–23, 1973.

Hunsaker, C. T., and Carpenter, D. E. (eds): Environmental Monitoring and Assessment Program: Ecological Indicators. Springfield, VA: National Technical Information Service, 1990.

Hurlbert, S. J.: Pseudoreplication and the design of ecological field experiments. *Ecol. Monogr.* 54:187–211, 1984.

Jones, R. D., and Hood, N. A.: The effects of organophosphorous pesticides on estuarine ammonia oxidizers. *Can. J. Microbiol.* 26:1296–1299, 1980.

Jones, R. A., and Lee, G. F.: Use of Vollenweider-OECD modeling to evaluate aquatic ecosystem functioning. Functional Testing of Aquatic Biota for Estimating Hazards of Chemicals, STP988, edited by J. Cairns, Jr., and J. R. Pratt, pp. 17–27. Philadelphia: American Society for Testing and Materials, 1988.

Karr, J. R.: Biological integrity: a long-neglected aspect of water resource management. *Ecol. Appl.* 1:66–84, 1991.

Kelly, J. R.: Ecotoxicology beyond sensitivity: a case study involving "unreasonableness" of environmental change. Ecotoxicology: Problems and Approaches, edited by S. A. Levin; M. A. Harwell; J. R. Kelly; and K. D. Kimball, pp. 473–496. New York: Springer-Verlag, 1989.

Kelly, J. R., and Harwell, M. A.: Indicators of ecosystem response and recovery. Ecotoxicology: Problems and Approaches, edited by S. A. Levin; M. A. Harwell; J. R. Kelly; and K. D. Kimball, pp. 9–40. New York: Springer-Verlag, 1989.

Kelly, J. R.; Duke, T. W.; Harwell, M. A.; and Harwell, C. C.: An ecosystem perspective on potential impacts of drilling fluid discharges on seagrasses. *Environ. Manage.* 11:537–562, 1987.

Kerr, A.: Canada's National Environmental Indicators Project: Background Report. Ottawa, Ontario: Environment Canada, Sustainable Development and State of the Environment Reporting Branch, 1990.

Kolasa, J., and Pickett, S. T. A.: Ecosystem stress and health: an expansion of the conceptual basis. *J. Aquat. Ecosys. Health* 1:7–13, 1992.

Levine, S. N.: Theoretical and methodological reasons for variability in the responses of aquatic ecosystem processes to chemical stress. Ecotoxicology: Problems and Approaches, edited by S. A. Levin; M. A. Harwell; J. R. Kelley; and K. D. Kimball, pp. 145–179. New York: Springer-Verlag, 1989.

Lewis, M. A.; Taylor, M. J.; and Larson, R. J.: Structural and functional response of natural phytoplankton and periphyton communities to a cationic surfactant with considerations on environmental fate. Community Toxicity Testing, STP908, edited by J. Cairns, Jr., pp. 241–268. Philadelphia: American Society for Testing and Materials, 1986.

Likens, G. E.; Bormann, F. H.; Johnson, N. M.; Fisher, D. W.; and Pierce, R. S.: Effects of forest cutting and herbicide treatment on nutrient budgets in the Hubbard Brook watershed-ecosystem. *Ecol. Monogr.* 40:23–47, 1970.

Long, E. R., and Chapman, P. M.: A sediment quality triad: measures of sediment contamination, toxicity, and infaunal community composition in Puget Sound. *Mar. Pollut. Bull.* 16:405–515, 1985.

Lozano, S. J.; O'Halloran, S. L.; Sargent, K. W.; and Brazner, J. C.: Effects of esfenvalerate on aquatic organisms in littoral enclosures. *Environ. Toxicol. Chem.* 11:35–48, 1992.

Macek, K.; Birge, W.; Mayer, F.; Buikema, A., Jr.; and Maki, A.: Discussion session synopsis. Estimating the Hazard of Chemical Substances to Aquatic Life, edited by J. Cairns, Jr.; K. L. Dickson; and A. W. Maki, pp. 27–32. Philadelphia: American Society for Testing and Materials, 1978.

Makarewicz, J. D., and Likens, G. E.: Structure and function of the zooplankton community of Mirror Lake, New Hampshire. *Ecol. Monogr.* 49:109–127, 1979.

Maki, A. W., and Johnson, H. E.: Evaluation of a toxicant on the metabolism of model stream communities. *J. Fish Res. Board Can.* 33:2740–2746, 1976.

Margalef, R.: Human impact on transportation and diversity in ecosystems; how far is extrapolation valid? Structure, Functioning, and Management of Ecosystems; Proceedings of the First International Congress of Ecology, pp. 237–243, Wageningen: Centre for Agricultural Publishing and Documentation, 1975.

Marker, A. F. H.: The benthic algae of some streams in southern England. II. The primary production of the epilithon in a small chalk-stream. *J. Ecol.* 64:359–373, 1976.

Marshall, J. S., and Mellinger, D. L.: Dynamics of cadmium-stressed plankton communities. *Can J. Fish Aquat. Sci.* 37:403–414, 1980.

Matson, P. A., and Carpenter, S. R. (eds): Statistical analysis of ecological response to large-scale perturbations. *Ecology* 71:2037–2060, 1990.

Minshall, G. W.; Petersen, R. C.; Cummins, K. W.; Bott, T. L.; Sedell, J. R.; Cushing, C. E.; and Vannote, R. L.: Interbiome comparison of stream ecosystem dynamics. *Ecol. Monogr.* 53:1–25, 1983.

Mount, D. I.: A changing profession. *Environ. Toxicol. Chem.* 6:83–84, 1987.

Naiman, R. J.: The annual pattern and spatial distribution of aquatic oxygen metabolism in boreal forest watersheds. *Ecol. Monogr.* 53:73–94, 1983.

National Eutrophication Research Program: Algal Assay Procedure: Bottle Test. Corvallis, OR: U.S. Environmental Protection Agency, Pacific Northwest Environmental Research Laboratory, 1971.

Niederlehner, B. R., and Cairns, J., Jr.: Community response to cumulative toxic impact: effects of acclimation on zinc tolerance of aufwuchs. *Can. J. Fish Aquat. Sci.* 49:2155–2163, 1992.

Niederlehner, B. R., and Cairns, J., Jr.: Effects of previous zinc exposure on pH tolerance of periphyton communities. *Environ. Toxicol. Chem.* 12:743–753, 1993.

Niemi, G. J.; DeVore, P.; Detenbeck, N.; Taylor, D.; Yount, J. D.; Lima, A.; Pastor, J.; and Naiman, R. J.: An overview of case studies on recovery of aquatic systems from disturbance. *Environ. Manage.* 14:571–585, 1990.

Odum, E. P.: The strategy of ecosystem development. *Science* 164:262–270, 1967.

Odum, E. P.: Trends expected in stressed ecosystems. *BioScience* 35:419–422, 1985.

Odum, H. T.: Primary production in flowing waters. *Limnol. Oceanog.* 1:102–117, 1956.

O'Neill, R. V.; Ausmus, B. S.; Jackson, D. R.; Hook, R. I.; Van Voris, P.; Washburn, C.; and Watson, A. P.: Monitoring terrestrial ecosystems by analysis of nutrient export. *Water Air Soil Pollut.* 8:271–277, 1977.

O'Neill, R. V.; DeAngelis, D. L.; Waide, J. B.; and Allen, T. F. H.: A Hierarchical Concept of Ecosystems. Monographs in Population Biology 22. Princeton, NJ: Princeton Univ. Press, 1986.

Organization for Economic Cooperation and Development: Eutrophication of Waters—Monitoring, Assessment, and Control. Paris, France, 1982.

Paul, B. J.; Corning, K. E.; and Duthie, H. C.: An evaluation of the metabolism of sestonic and epilithic communities in running waters using an improved chamber technique. *Freshwater Biol.* 21:207–215, 1989.

Peterman, R. M.: Statistical power analysis can improve fisheries research and management. *Can. J. Fish Aquat. Sci.* 47:2–15, 1990.

Pratt J. R.: Aquatic community response to stress: prediction and detection of adverse effects. Aquatic Toxicology and Risk Assessment: Thirteenth Volume, STP1096, edited by W. G. Landis, and W. H. van der Schalie, Jr., pp. 16–26. Philadelphia: American Society for Testing and Materials, 1990.

Pratt, J. R., and Bowers, N. J.: Effect of selenium on microbial communities in laboratory microcosms and outdoor streams. *Tox. Assess.* 5:293–307, 1990.

Rapport, D. J.; Regier, H. A.; and Hutchinson, T. C.: Ecosystem behavior under stress. *Am. Nat.* 125:617–640, 1985.

Regier, H. A., and Hartman, W. L.: Lake Erie's fish community: 150 years of cultural stresses. *Science* 180: 1248–1255, 1973.

Resh, V. H.; Brown, A. V.; Covich, A. P.; Gurtz, M. E.; Li, H. W.; Minshall, G. W.; Reice, S. R.; Sheldon, A. L.; Wallace, J. B.; and Wissmar, R. C.: The role of disturbance in stream ecology. *J. N. Am. Benthol. Soc.* 7: 433–455, 1988.

Ryder, R. A., and Edwards, C. J.: A conceptual approach for the application of biological indicators of ecosystem quality in the Great Lakes Basin. Report to the Great Lakes Science Advisory Board of the International Joint Commission, Windsor, ON, 1985.

Sayler, G. S.; Sherril, T. W.; Perkins, R. E.; Mallory, L. M.; Shiaris, M. P.; and Pedersen, D.: Impact of coal-coking effluent on sediment microbial communities: a multivariate approach. *Appl. Environ. Microbiol.* 44:1118–1129, 1982.

Scanferlato, V. S., and Cairns, J., Jr.: Effect of sediment-associated copper on ecological structure and function of aquatic microcosms. *Aquat. Toxicol.* 18:23–34, 1990.

Schaeffer, D. J.; Herricks, E. E.; and Kerster, H. W.: Ecosystem health: I. Measuring ecosystem health. *Environ. Manage.* 12:445–455, 1988.

Schindler, D. W.: Eutrophication and recovery in experimental lakes: implications for lake management. *Science* 184:897–899, 1974.

Schindler, D. W.: Detecting ecosystem responses to anthropogenic stress. *Can. J. Fish Aquat. Sci.* 44(Suppl 1):6–25, 1987.

Schindler, D. W.: Experimental perturbations of whole lakes as tests of hypotheses concerning ecosystem structure and function. *Oikos* 57:25–41, 1990.

Schindler, D. W.; Mills, K. H.; Malley, D. F., Findlay, D. L.; Shearer, J. A.; Davies, I. J.; Turner, M. A.; Linsey, G. A.; and Cruikshank, D. R.: Long-term ecosystem stress: the effects of years of experimental acidification on a small lake. *Science* 228:1395–1401, 1985.

Sedell, J. R., and Dahm, C. N.: Catastrophic disturbances to stream ecosystems: volcanism and clear-cut logging. Current Perspectives in Microbial Ecology, edited by M. J. Klug, and C. A. Reddy, pp. 531–539. Washington, D.C.: American Society for Microbiology, 1984.

Society for Environmental Toxicology and Chemistry (SETAC) Foundation for Environmental Education: Workshop on Aquatic Microcosms for Ecological Assessment of Pesticides. Workshop Report. Pensacola, FL, 1992.

Sokal, R. R., and Rohlf, F. J.: Biometry: The Principles and Practice of Statistics in the Biological Research, 2nd ed. New York: W. H. Freeman, 1981.

Smith, E. P.; Orvos, D. R.; and Cairns, J., Jr.: Impact assessment using the before-after-control-impact (BACI) model: concerns and comments. *Can. J. Fish Aquat. Sci.* 50:627–637, 1993.

Stay, F. S.; Larsen, D. P.; Katko, A.; and Rohm, C. M.: Effects of atrazine on community level responses in Taub microcosms. Validation and Predictability of Laboratory Methods for Assessing the Fate and Effects of Contaminants in Aquatic Ecosystems, STP865, edited by T. P. Boyle, pp. 75–90. Philadelphia: American Society for Testing and Materials, 1985.

Stay, F. S.; Katko, A.; Rohm, C. M.; Fix, M. A.; and Larsen, D. P.: The effects of atrazine on microcosms developed from four natural plankton communities. *Arch. Environ. Contam. Toxicol.* 18:866–875, 1989.

Stewart-Oaten, A.; Murdoch, W. W.; and Parker, K. R.: Environmental impact assessment: pseudoreplication in time? *Ecology* 67:929–940, 1986.

Stewart-Oaten, A.; Bence, J. R.; and Osenberg, C. W.: Assessing the effects of unreplicated perturbations: no simple solutions. *Ecology* 73:1396–1404, 1992.

Stream Solute Workshop: Concepts and methods for assessing solute dynamics in stream ecosystems. *J. N. Am. Benthol. Soc.* 9:95–119, 1990.

Suter, G. W., II: Endpoints for regional ecological risk assessments. *Environ. Manage.* 14:9–23, 1990.

Thomas, J. M.; Mahaffey, J. A.; Gore, K. L.; and Watson, D. G.: Statistical methods used to assess biological impact at nuclear power plants. *J. Environ. Manage.* 7: 269–290, 1978.

Torsvik, V.; Goksoyr, J.; and Daae, F. L.: High diversity in DNA of soil bacteria. *Appl. Environ. Microbiol.* 56:782–787, 1990.

Touart, L. W.: Aquatic Mesocosm Tests to Support Pesticide Registrations. EPA-540/9-88-035. Springfield, VA: National Technical Information Service, 1988.

U.S. Environmental Protection Agency: Toxic Substances Control Act test guidelines: proposed rule. *Fed. Regist.* 52:36333–36371, 1987.

U.S. Environmental Protection Agency: Technical Support Document for Water Quality-based Toxics Control. EPA/505/2-90-001. Springfield, VA: National Technical Information Service, 1991.

van Voris, P.; O'Neill, R. V.; Emanuel, W. R.; and Shugart, H. H.: Functional complexity and ecosystem stability. *Ecology* 6:1352–1360, 1980.

Vitousek, P. M.; Reiners, W. A.; Melillo, J. M.; Grier, C. C.; and Gosz, J. R.: Nitrogen cycling and loss following forest perturbation: The components of response. Stress Effects on Natural Ecosystems, edited by G. W. Barrett, and R. Rosenberg, pp. 115–127. New York: Wiley, 1981.

Vollenweider, R. A.: Scientific fundamentals of the eutrophication of lakes and flowing waters with particular reference to nitrogen and phosphorus as factors in eutrophication. Paris: Organization for Economic Cooperation and Development, 1968.

Vollenweider, R. A.: A Manual on Methods for Measuring Primary Production in Aquatic Environments, 2nd ed., IBP Handbook No. 12. London: Blackwell Scientific Publications, 1974.

Wallace, J. B.: Structure and function of freshwater ecosystems: Assessing the potential impact of pesticides. Using Mesocosm to Assess the Aquatic Ecological Risk of Pesticides: Theory and Practice, edited by J. R. Voshell, Jr., pp. 4–17. Lanham, MD: Miscellaneous Publications of the Entomological Society of America 75, 1989.

Webber, E. C.; Deutsch, W. G.; Bayne, D. R.; and Seesock, W. C.: Ecosystem-level testing of a synthetic pyrethroid insecticide in aquatic mesocosms. *Environ. Toxicol. Chem.* 11:87–106, 1992.

Webster, J. R.: Hierarchical organization of ecosystems. Theoretical Systems Ecology: Advances and Case Studies, edited by E. Halfon, pp. 119–129. New York: Academic Press, 1979.

Wilson, E. O. (ed.): Biodiversity. Washington, DC: National Academy Press, 1988.

Wissmar, R. C.; Devol, A. H.; Staley, J. T.; and Sedell, J. R.: Biological responses of lakes in the Mount St. Helens blast zone. *Science* 216:178–181, 1982.

Woodwell, G. M.: Effects of pollution on the structure and physiology of ecosystems. *Science* 168: 429–433, 1970.

Yount, J. D., and Niemi, G. J., eds.: Recovery of lotic communities and ecosystems from disturbance: theory and applications. *Environ. Manage.* 14:1990.

SUPPLEMENTAL READING

Kelly, J. R.: Ecotoxicology beyond sensitivity: A case study involving "unreasonableness" of environmental change. Ecotoxicology: Problems and Approaches, edited by S. A. Levin; M. A. Harwell; J. R. Kelly; and K. D. Kimball, pp. 473–496. New York: Springer-Verlag, 1989.

Levine, S. N.: Theoretical and methodological reasons for variability in the responses of aquatic ecosystem processes to chemical stress. Ecotoxicology: Problems and Approaches, edited by S. A. Levin; M. A. Harwell; J. R. Kelly; and K. D. Kimball, pp. 145–179. New York: Springer-Verlag, 1989.

O'Neill, R. V., and Waide, J. B.: Ecosystem theory and the unexpected: Implications for environmental toxicology. Management of Toxic Substances in our Environment, edited by B. W. Cornaby, pp. 43–73. Ann Arbor, MI: Ann Arbor Science Publishers, 1981.

Schindler, D. W.: Detecting ecosystem responses to anthropogenic stress. *Can. J. Fish Aquat. Sci.* 44(Suppl 1):6–25, 1987.

Chapter 20

STRUCTURE-ACTIVITY RELATIONSHIPS

R. L. Lipnick

INTRODUCTION

Definitions

The application of QSAR to environmental toxicology has been the subject of a number of symposium monographs and reviews (Kaiser, 1984, 1987a; Schüürmann, 1988; Turner et al., 1988; Hermens, 1989; Karcher and Devillers, 1990; Hermens and Opperhuizen, 1991; Nendza, 1991).

Changes in the structure of a chemical may influence the type and potency of its biological action. This principle is an extension of the concept that all chemical-biological effects are the result of either a reversible or a covalent interaction between the chemical and one or more components of the living system. Components of a living system that are capable of reversible or covalent binding with a foreign chemical (i.e., xenobiotic) are referred to as *receptors* or molecular sites of action. Such receptors are actually specific chemical moieties or ''active sites'' on molecules. The study of the type of chemical structure of a foreign substance which will interact or react with a receptor and produce a well-defined biological endpoint is commonly referred to as the study of *structure-activity relationships* (SARs). The objective of structure-activity relationship studies is to define as precisely as possible the limits of variation in the structure of a chemical moiety that are consistent with the production of a specific biological effect and the ways in which alterations in the environment of this moiety and overall properties of the molecule influence potency. If enough data related to this specific biological effect become available, a hypothesis can be developed regarding the molecular basis of interaction between the chemical or toxicant and active site (Albert, 1985; Horsfall, 1956; Fränkel, 1919; Sexton, 1963; Purcell et al., 1973). However, a narcosis or anesthetic effect represents a nonspecific response, in that no such moiety or molecular substructure requirement is involved and activity is totally dependent on the physicochemical properties of the whole molecule.

As indicated in Figure 1, structure is normally represented as a two-dimensional diagram by organic chemists, but implicit in this is the actual three-dimensional structure that can be obtained experimentally, for example by X-ray crystallography or by calculation by molecular mechanics. This three-dimensional structure provides information on additional electronic features (e.g., electron distribution), steric features (e.g., conformation, geometry, and shape), and molecular volume. In addition, biological activity can be related directly to measurable physicochemical and chemical properties, including physicochemical interactions, reactivity (e.g., electrophilicity), and reversible and irreversible binding to receptors or sites of interaction within the organism.

Quantitative structure-activity relationships (QSARs) are mathematical models that relate the biological activity (e.g., 96-h LC50) of molecules to their chemical structures and corresponding chemical and physicochemical properties. QSAR models were initially used in the design and development of new drugs and pesticides (Martin, 1978; Seydel and Schaper, 1979; Franke, 1984; Martin et al., 1990; Topliss, 1983; Tute, 1990; Stuper et al., 1979; Kubinyi, 1993). Although similar models relating physicochemical and chemical properties of molecules to chemical structure are known as quantitative structure-property relationships (QSPRs) (Varmuza, 1980; Exner, 1988; Shorter, 1982), these two categories of models are clearly interrelated. Van de Waterbeemd and Testa (1987) mention more than 200 physicochemical and structural descriptors used in QSAR studies; however only a small number appear in the aquatic toxicology literature.

For quantitative, continuously valued data for both the toxicological test and the physicochemical properties (e.g., water solubility or other *molecular descriptors*; see section later in chapter) a mathematical relationship is most commonly derived using regres-

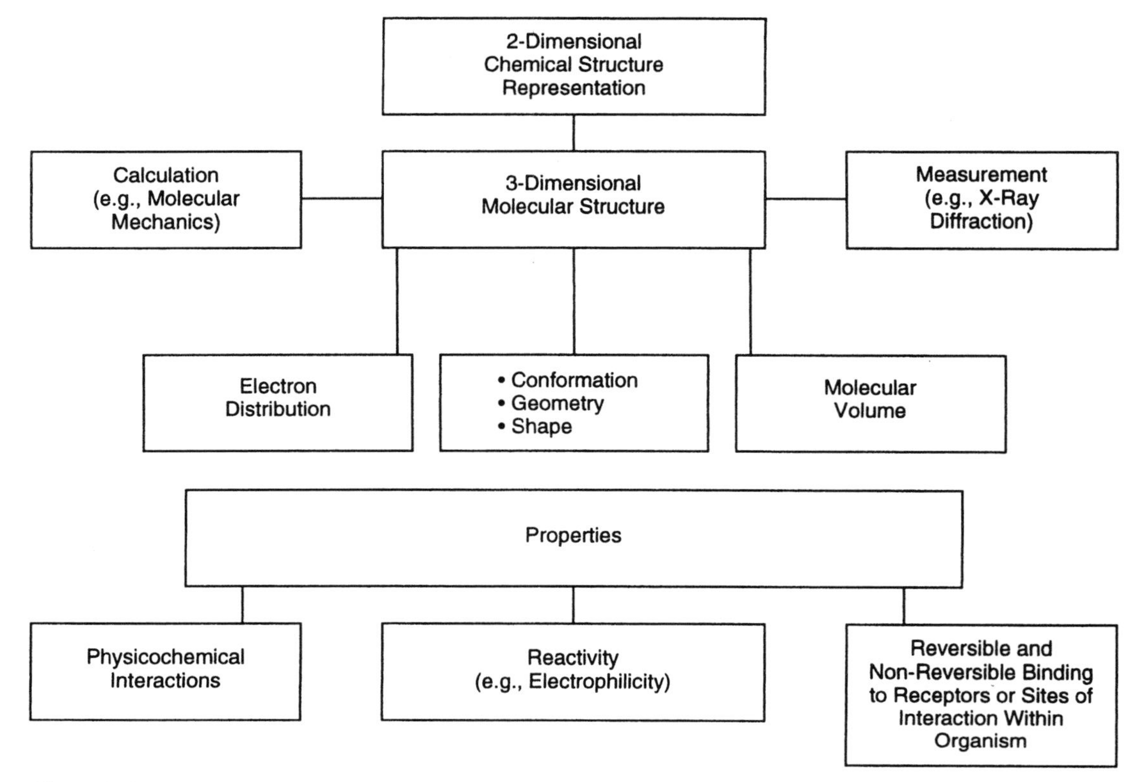

Figure 1. Structure and properties of molecules: their interrelationship and relationship to toxic response. A chemical substance is normally presented to the analyst as a two-dimensional structure representation of the molecule, while its toxicity is a manifestation of the actual three-dimensional structure, which can be measured (e.g., by X-ray crystallography) or calculated via molecular mechanics or quantum mechanics. Associated with the three-dimensional structure are an electron distribution, spatial disposition (conformation, geometry and shape), and molecular volume. In addition to these molecular properties, there are also associated physicochemical (e.g., physicochemical interactions such as log *P*) and chemical properties (e.g., electrophilicity) and molecular binding properties (e.g., to biological macromolecules).

sion analysis. Most regression models are based on the concept that changes in a property (e.g., partition coefficient) will be reflected in changes in free energy if such correlations are made on a double logarithmic scale. Such models are known as linear free energy relationships or LFERs (Leffler and Grunwald, 1969) or extrathermodynamic relationships (Fujita, 1990). The QSAR approach is based on the assumption, which should be intuitive, that the structure of a molecule (i.e., its geometric and electronic properties) must contain the features responsible for its physical, chemical, and biological properties, and on the ability to represent the chemical by one or more numerical molecular descriptors. The discovery of a new chemical moiety that induces certain biological effects will initiate the synthesis of related analogues of the substance in the hope of finding more useful or at least more effective agents capable of producing the same biological effects.

Biological actions of substances are also governed by their physicochemical properties (e.g., solubility, octanol-water partition coefficient), which in turn are determined by their chemical structure. There is thus an interrelationship of the chemical structure, physicochemical properties, and biological actions (Albert, 1985). This does not mean that knowledge of chemical structure alone is sufficient to predict action and potency of a substance. Examples are known of substances with similar biological toxic action but with dissimilar structures, and of different actions by substances with similar structures. Because a number of physicochemical properties are highly interrelated (Lazarev, 1944; Cramer, 1981), care must be taken in assigning significance to too many properties for a series of molecules. This is not unusual because the chemical structures of toxic substances are more varied than the responses they elicit in an organism. Also, the geometric and electronic properties of an active moiety can be produced by other molecular groupings. For a substance to trigger a biological response when administered to an aquatic organism, a number of processes must occur. The substance must first dissolve in water. Second, it must be transported across the external membranes (initial lipid compartments) and then internally dissolve in a body fluid. Last, it must be transported from the site of initial administration (or contact) to the site of action or receptor. It must then bind at this site at a certain concentration to produce a biological response. These factors are illustrated in Figure 2. This figure shows that certain substances must be converted (biotransformed; see Chapter 17) to become active metabolites that correspond to the actual toxicants. Aside from uptake and elimination and possible biotransformation (Sijm and Opperhuizen, 1989), binding to plasma proteins and further distribution are all important in determining the concentration of the toxicant at the site of action.

Absorption, transport, and distribution of the substance in a biological system are governed largely by its partitioning behavior between lipid and aqueous phases (hydrophobicity), and dissolution in body fluid is governed by its water solubility (which varies with the pH of the fluid for weak electrolytes). Although examples exist of active transport across membranes, transport of most xenobiotic substances within a biological system is a passive (i.e., passive diffusion) process of transfer from aqueous to lipid to aqueous phase, repeated until the receptor is reached. The rate of transport depends on molecular

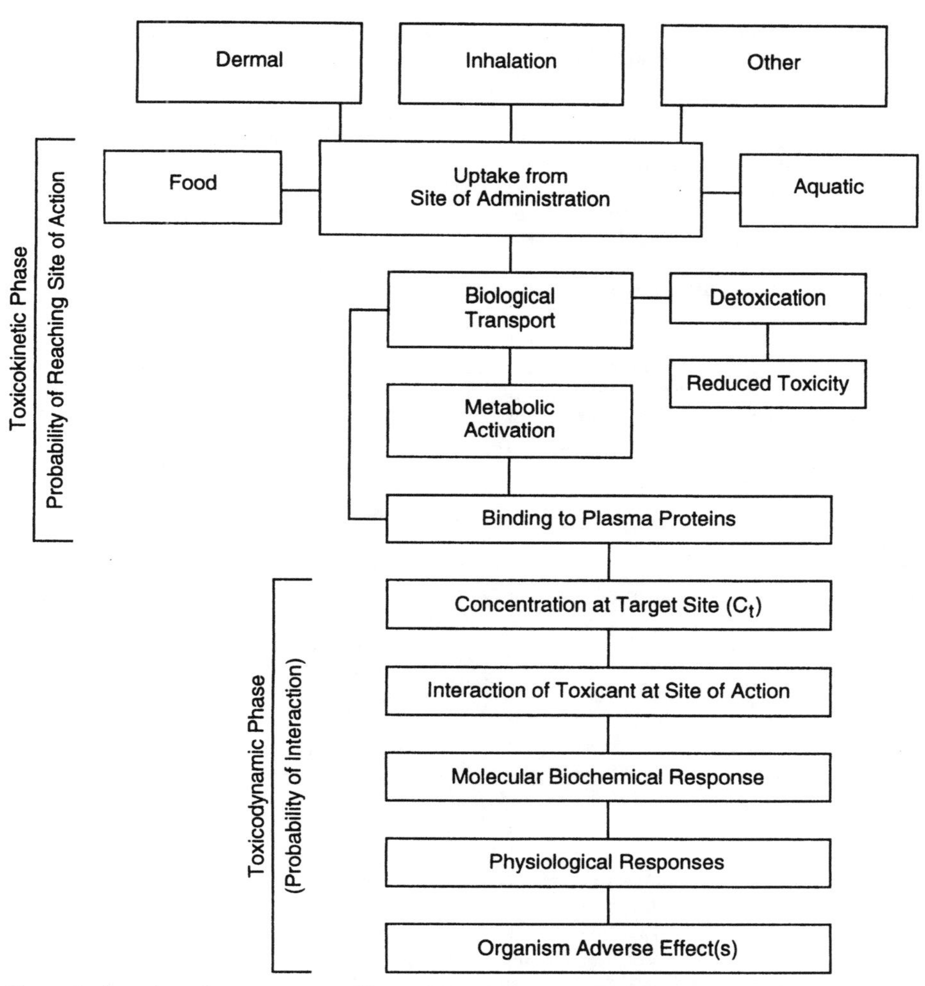

Figure 2. Overview of processes controlling toxicity within an organism. In the toxicokinetic phase, the organism is exposed to a substance by various routes: food, aquatic uptake, dermal, inhalation, etc. The substance is then transported within the organism and can be detoxified (leading to reduced toxicity), metabolically activated (leading to increased toxicity), or expressed directly as the original substance. In the process of reaching the target site (toxicodynamic phase, which involves probability of interaction), binding can occur to plasma proteins that reduces the blood concentration. Once reaching the target site (C_t), interaction can take place leading to a molecular biochemical response (receptor interaction), which produces a cascade of physiological responses leading ultimately to observed whole-organism adverse effect(s).

properties such as hydrophobicity, polarity, degree of ionization, molecular shape, and size. It is assumed that this partition of a chemical between a biophase and water is related to the octanol-water partition coefficient (K_{ow}), which is a measure of its hydrophobic or lipophilic properties. Other factors can control the rate at which the substance arrives at the receptor by controlling the rate at which it undergoes metabolism. Metabolism itself is controlled by electronic and steric factors governing interaction with the corresponding metabolizing enzymes and the energetics of covalent bond cleavage. Steric factors (size and shape) can also shield a bond from metabolic attack. These processes are illustrated in Figure 2.

Binding of the substance to the receptor depends on the forces of interaction between them (Andrews and Tintelnot, 1990; Bowden, 1990) and on the complementarity of their size and shape. Interaction forces can be ion-ion, ion-dipole, dipole-dipole, or hydrogen bonding. Dispersion forces are also important if the substance (foreign chemical or xenobiotic) and receptor possess nonpolar regions. Binding resulting from such forces is referred to as hydrophobic binding (Tanford, 1980). Size and shape of a molecule control how well it fits in an appropriate conformation and orientation to the receptor (Silipo and Vittoria, 1990; Blaney and Hansch, 1990).

The biological response to a xenobiotic can be controlled by three major physicochemical classes of properties—hydrophobic, electronic, and steric (Dearden, 1990a). This chapter reviews some of the most extensively used properties in each class to show how each is used in a quantitative description of the biological response.

Applications

The use of QSAR models (including QSPRs) in environmental chemistry and toxicology has been growing in the past 10 years. QSARs have been used for the following:

- Pharmaceutical/agricultural screening (deStevens, 1990; Martin, 1991); prediction of kinetics (McKim et al., 1985; Dearden, 1990); understanding the mechanism of biological activity of untested substances to reduce costs of product design synthesis and synthesis (Tute, 1990).
- Filling data gaps for large volumes of untested substances to meet regulatory requirements and to establish priorities for testing (Lipnick, 1985a; Hunter et al., 1986; ECETOC, 1986; Auer and Gould, 1987; Auer et al., 1990; Borman, 1990; Shimp et al., 1990; Tosato and Geladi, 1990; Hart, 1991; Clements et al., 1988; Lynch et al., 1991; Walker, 1990; Zeeman et al., 1993). The need to fill data gaps has been especially important under Section 5 (Premanufacturing Notification) of the U.S. Environmental Protection Agency's Toxic Substances Control Act (1976) (see Chapters 22 and 23).
- Estimating physicochemical properties relevant to exposure and serving as input parameters for transport models (see Chapter 18). These include octanol-water partition coefficient, water solubility, vapor pressure, hydrolysis, photolysis, soil sorption, and biodegradation. Models used for estimating some of these properties are discussed later.
- Estimating and predicting ecotoxicity (predictive toxicology) thresholds or end points [acute LC50s and chronic (MATC) toxicity] for fish, daphnia, and plant toxicity and effects on soil microorganisms (Zitko, 1975; Veith, 1981; Hermens, 1989; Lipnick, 1989a; Charton, 1985; Calamari and Vighi, 1988; Karcher, 1992). Models used for estimating these end points are discussed later.
- Estimating bioaccumulation potential of chemicals in aquatic organisms (Veith et al., 1979; Esser and Moser, 1982; Mackay, 1982; Connell, 1988; Schüürmann, 1988; Brooke et al., 1986; Nendza, 1991; Binstein et al., 1993). Models used for estimating this end point are discussed later.
- Use in risk assessment and ranking of chemicals according to their potential hazard (Veith et al., 1988; Calamari and Vighi, 1988; Barnthouse et al., 1988; Hart, 1988; Nendza et al., 1990; Eriksson et al., 1991; Nabholz, 1991; Russom et al., 1991).

Limitations

The following limitations need to be considered in the development and application of QSARs:

- Predictions apply only to the type of activity chosen as the basis for the correlation. A structure-activity relationship can predict potency with respect to the mechanism of action of a given group of substances, but it cannot predict an unanticipated type of activity (Hermens, 1990a; Lipnick, 1991a,c).
- Change in relationship or even position in series in which complete loss or diminution in toxicity (toxicity cutoff) (Lipnick, 1990a; Donkin et al., 1991) may occur outside the domain of physicochemical properties and reactivity of chemicals used to derive the model (i.e., beyond spanned substituent space or region of chemical space that has been explored with existing test data) (Hansch, 1977).
- Definition of the applicability of a model with respect to substitution by additional moieties (i.e., adding on a new functional group) cannot be explicitly described.
- Models derived from large databases of heterogeneous chemicals (little or no preselection of structure types) do not model a single mechanism, and chemicals acting by additional mechanisms are likely to fail.
- Models require biological data representing a standard response (e.g., 96-h LC50 for the fathead minnow of a specified age and size at a specific temperature and water parameters). Differences in

conditions such as temperature, pH, ionic strength, species, age, and test duration may be unacceptable for a relative comparison of biological response, although this is of less importance for the acute toxicity of chemicals to aquatic organisms acting by a narcosis mechanism (Sloof et al., 1983; Lipnick, 1985b; Thurston et al., 1985; Van Leeuwen et al., 1992).

- Experimental error in the dependent variable (e.g., LC50) is often neglected and models are overfitted (Martin, 1978).
- Too many independent variables (e.g., multiple quantum mechanical parameters) are used with small sets of substances with high probability of chance correlation (Topliss and Costello, 1972; Topliss and Edwards, 1979a,b; Hansch, 1977).
- Lack of experimental design in the selection of test chemicals for distribution within spanned substituent space (Hansch, 1977; Pleiss and Unger, 1990).
- Formal validation not normally employed (all available data usually used in model derivation).
- Systematic methods for choosing chemicals for validation are only poorly defined (Lipnick, 1985b).

HISTORICAL BACKGROUND

The qualitative relationship between chemical structure and biological activity has served as a stimulus for research in pharmacology and toxicology for some time. Advances in this area have commonly occurred as understanding has increased of the more fundamental relationships between molecular structure and chemical properties. Charles Frederic Gerhardt in France (1844–1845; Ihde, 1984) is considered the first to propose the concept of using a homologous series of substances to categorize groups of simple organic compounds and relating this to corresponding changes in properties.

This approach was extended by Cros in France, who reported (Cros, 1863; Lipnick, 1989a) that the toxicity of certain simple organic compounds increases with decreasing water solubility up to a limit governed by insufficient water solubility. Apparently unaware of the work of Cros, Richardson in England (Richardson, 1869; Lipnick, 1989a) published results 6 yr later concluding that the toxicity of alcohols increases with increasing molecular weight. Later authors cited Richardson's Rule (Munch and Schwartze, 1925) to account for such relationships within other series.

A similar finding was made independently in France by Rabuteau (1870), who concluded from studies with frogs immersed in aqueous solutions of toxicants that potency increases with an increase in the number of times the CH_2 group occurs in the molecule. Rabuteau's work may be considered the earliest aquatic toxicology SAR studies.

George Houdaille (1893), a medical student working under Charles Richet at the University of Paris, used tadpoles and small fish to test a number of hypnotic agents. Potency was found to vary inversely with water solubility (Table 1), and correlations of this type continue to be characterized as examples of Richet's Law (Richet, 1893; Lipnick, 1989a).

The next advances were directed toward a more specific understanding of what structural or physicochemical properties govern the potency of chemicals acting by an anesthetic or narcosis mechanism.

Beginning in the 1890s, Charles Ernest Overton at the University of Zürich (Lipnick, 1986) and Hans Horst Meyer at the University of Marburg (Lipnick, 1989d) independently began research that would refute the requirement for special structural moieties or water solubility dependence and provide a mechanistic proposal to account for the relationship between chemical structure and anesthetic potency. Overton, a botanist by training, was initially interested in defining which properties of a molecule govern its rate of permeation across plant cell membranes, using plasmolysis as an end point (Overton, 1895, 1896). He subsequently discovered that more reproducible results could be obtained using narcosis in tadpoles to measure transport across cell membranes, although he also tested a wide variety of other aquatic organisms. The results were presented in the classic work *Studien über die Narkose* (Overton, 1901), which is now available in English translation (*Studies of Narcosis*) (Lipnick, 1991b).

Daniel Diehl, working in Meyer's laboratory (1894), attempted to verify an earlier theory put forward by Baumann and Kast (1890) that the anesthetic ability of sulfones is due to the formation of active metabolites, and their potency is related to the number of ethyl groups present. Diehl was able to refute both of these hypotheses. Two (diisopropylsulfonedimethylmethane and diisobutylsulfonedimethylmethane) of the 14 sulfones he tested using

Table 1. Relationship between toxicity to small fish and water solubility

Chemical	Minimum lethal concentration (g/L)	Maximum lethal concentration (g/L)	Water solubility
Ethyl alcohol	40	20	∞
Ethyl ether	5.5	2	40 parts per 100
Amyl alcohol	1.0	0.5	Less soluble
Absinthe essence	0.005	0.0025	"Completely insoluble"

From Richet, 1893.

tadpoles contain no ethyl groups but were found to be quite potent anesthetic or hypnotic agents. Furthermore, of those containing ethyl groups, the potency was not found to be related to their number. Diehl concluded that the requirement for metabolic activation to the active anesthetic agent as proposed by Baumann and Kast was not supported by his own observation of anesthesia or narcosis in tadpoles taking place within only a few minutes of exposure to sulfones that were known to be stable chemically. Furthermore, tadpoles anesthetized for 24 h or longer returned to normal when transferred to fresh water. Finally, Diehl found the total quantity of sulfone in the test solution remained unchanged throughout the experiment, also inconsistent with Baumann and Kast's metabolism theory.

Frederick Adolph Bucholz (1895), also working in Meyer's laboratory, found that for amides, by contrast, when metabolism takes place to release the corresponding acid and ammonia, nonnarcotic effects alone were observed. Thus, formamide and acetamide both produced picrotoxin-like convulsions. Bucholz also showed that anesthetic agents produce a similar narcosis response in both warm-blooded and cold-blooded animals regardless of route of administration, demonstrating a common site of action, independent of the means of exposure.

Walter Dunzelt (1896), a third worker in Meyer's laboratory, used both tadpoles and whitefish to investigate a variety of anesthetic agents and to test Richet's Law regarding an inverse relationship between water solubility and potency. Dunzelt observed that bromal hydrate was more potent than expected from its water solubility but that the opposite was true for methyl urethane, thus providing the impetus to investigate other properties.

Meyer (1899) and Overton (1899) independently reported that narcosis potency was governed not by water solubility but by partition coefficient. Meyer's conclusions were based on careful measurements of partition coefficients in his laboratory by Fritz Baum (1899) for a series of 11 anesthetics of diverse chemical structure (Table 2), using purified olive oil. Olive oil was chosen as a readily available surrogate for the solubility properties of the postulated lipoid site of action; anesthetic potencies were compared on a molar basis. To further substantiate that partition coefficient and not the presence of specific substructures or degree of water solubility controlled anesthetic potency, Meyer (1901) showed that when tested with tadpoles at 3°C and 30°C, salicylamide, benzamide, and monoacetin were more potent at the lower temperature, whereas ethyl alcohol, chloral hydrate, and acetone responded oppositely. Olive oil–water partition coefficients measured at both temperatures for all six compounds showed a similar dependence, consistent with the lipoid theory and the mechanistic basis for employing partition coefficients for correlations between external dose and internal effect concentration at a lipoid or lipoid-like site of action.

Table 2. Relation between oil-water partition coefficient and threshold concentration producing narcosis in tadpoles

Chemical	$K_{oil/water}$	$C_{narcosis}$
Trional	4.46	0.0018
Tetronal	4.04	0.0013
Butyl chloral hydrate	1.59	0.0020
Sulfonal	1.11	0.0060
Bromal hydrate	0.66	0.0020
Triacetin	0.30	0.010
Diacetin	0.23	0.015
Chloral hydrate	0.22	0.020
Ethylurethane	0.14	0.040
Monoacetin	0.06	0.050
Methylurethane	0.4	0.40

Data from Baum (1899); $K_{oil/water}$, partition coefficient between olive oil and water; $C_{narcosis}$, minimum molar concentration in water producing narcosis in tadpoles.

Neither Meyer nor Overton made any attempt to graph these results or develop mathematical relationships between partition coefficient and potency. The first effort in this direction appears to have been made in Leningrad by Nikolai Vasilyevich Lazarev (1944), who developed a system for classifying toxicants by means of their partition coefficients expressed on a logarithmic scale from 1 to 9. From numerous graphs of partition coefficient versus molar toxicity on a double logarithmic scale, he derived equations of the form

$$\log C = a \log K_{\text{oil-water}} + b$$

where a and b are empirically derived constants and C is the molar concentration of a chemical. Lazarev derived such equations for both in vitro and in vivo biological effects (Lipnick and Filov, 1992). For example, Lazarev reported the following relationship:

$$\log C_{\text{tadpole narcosis}} = -0.69 \log K_{\text{oil-water}} + 1.06$$

where $C_{\text{tadpole narcosis}}$ (in mmol/L) data are from Overton (1901; Lipnick, 1991b) and $K_{\text{oil-water}}$ is the measured olive oil–water partition coefficient. Lazarev also showed that the partition coefficient increases with chain length within a variety of homologous series of chemicals and could be reliably estimated for compounds within the series for which measured data were not available. Moreover, he showed that substances for which toxicity was limited by insufficient water solubility could be identified from the relationships of partition coefficient to both water solubility and tadpole narcosis (and their simultaneous mathematical solution, either graphic or algebraic).

The development of modern QSAR methods was initiated in 1962 by Corwin Hansch and co-workers at Pomona College (Hansch et al., 1962; Hansch and Fujita, 1964; Martin, 1978). *n*-Octanol was selected as a standard lipophilic-hydrophobic phase for model partition coefficient measurements (Leo et al., 1971) and correlations with molar biological activity were performed using regression analysis. Many such regression equations were reported based solely on the logarithm of the octanol-water partition coefficient ($\log K_{ow}$ or $\log P$) using enzyme, cellular, and whole-animal test data from the literature (Leo et al., 1969; Hansch and Dunn, 1972).

The general form of the classical Hansch equation describes the influence of substituents (i.e., structure-related parameters for modeling hydrophobic, electronic, and steric properties) on the biological activity of a parent molecule,

$$\log\left(\frac{1}{C}\right) = a \sum \pi + b \sum \sigma + cE_s + d$$

where C = molar concentration or dose for a given chemical (with specified properties) that produces a defined biological response or effect (e.g., 96-h LC50)

$\Sigma \pi$ = sum of the substituent constant values for hydrophobic contributions for each substituent atom or functional group attached to the parent molecule

$\Sigma \sigma$ = sum of the Hammett substituent constant values for electronic contributions for each substituent atom or functional group attached to the parent molecule

E_s = Taft substituent constant for steric effects for ortho substituents

a,b,c,d = constants obtained by fitting experimental data to the above equations

This equation is solved by multiple linear regression for a set of chemicals for which both the biological data (C) and values of the corresponding substituent constants are available. Regression analysis is the most common statistical tool applied to QSAR studies.

Generally, hydrophobic substituent constants (i.e., the Hansch π values or $\log P$ values for entire molecules) are used for all such models and for most models represent the sole parameter used. Once such a QSAR equation has been derived, it can be used to make predictions for other related chemicals for which values of the required parameters are available.

Normally, biological activity increases with partition coefficient only up to some limiting value, beyond which potency is found to decrease with increasing partition coefficient. Although this nonlinear behavior has been attributed to the inability of high $\log P$ chemicals to leave membranes readily, it more likely reflects inability to achieve equilibrium between the site of administration and the site of action within the test duration (e.g., aquatic toxicology testing of high $\log P$ chemicals) or the limited transit time of the chemical within an organism (e.g., for oral dosing) prior to excretion. Limiting water solubility can also contribute to such behavior, but in this case, particularly for aquatic exposure, a cutoff will be seen at which no activity can be detected regardless of the amount of additional test substance present that is not in aqueous solution.

Some of the reported Hansch type QSAR equations were derived using data on fish and other aquatic organisms. Although this work seems to have been unnoticed by aquatic toxicologists (Lipnick, 1985b), it really presaged the current widespread interest in applying QSARs to predictive aquatic toxicology. Thus, using data of Cololian (1901) on the 24-h minimum lethal dose (MLD) of saturated monohydric alcohols,[1] QSAR models (Hansch and Dunn 1972) were derived from data for five species of fish. All of the QSAR relationships had the form of the general equation,

$$\log\left(\frac{1}{C}\right) = a \log P + b$$

where C is the concentration in mol/L producing a defined response (e.g., narcosis, lethality), P is the partition coefficient between n-octanol and water, and a and b are empirical constants derived from regression analysis. For example, for goby,

$$\log\left(\frac{1}{\text{MLD}}\right) = 0.943 \log P + 1.051$$
$$(n = 5,\ r = 0.993,\ s = 0.108)$$

where the MLD is in mol/L, n is the number of alcohols tested, r is the correlation coefficient, and s is the standard error.

Most QSAR studies today in environmental toxicology (including ecotoxicology) are based on the Hansch approach. It should be noted that the quadratic equation proposed by Hansch used to describe the biphasic variation of biological response with $\log P$ has been widely applied, although subsequently a bilinear model was proposed (Kubinyi, 1976, 1977, 1979; Kubinyi and Kehrhahn, 1978) to which such data from a variety of sources have been demonstrated to give a better fit, and this has become the model of choice in such circumstances. The fact that many QSARs show dependence of biological response on $\log P$ alone may indicate, but does not prove, nonspecific binding to a receptor. For example, the other properties that affect biological potency may be collinear with $\log P$ for all chemicals in the data set. In the case of narcosis or anesthetic response, there is strong evidence that the binding is truly nonspecific (i.e., no specific molecular features required to produce narcosis) given the range of structural types, including the rare gas xenon (for inhalation anesthesia, which represents the same mechanism but a different route of administration), that produce this response and fit such correlations. Although n-octanol may not be the best hydrophobic partitioning phase for all studies, its widespread use and large number of measured values make it the commonly accepted standard.

[1]Saturated monohydric alcohols are representative of nonreactive nonelectrolyte (nonionizable) chemicals that act by a narcosis mechanism.

BASIC CONCEPTS

Introduction

To understand QSAR and SAR models, it is important to understand the methods that are utilized in the development of a model. The general scheme for deriving a QSAR correlation is shown in Figure 3, and a hierarchical description of QSAR model types is shown in Figure 4. An essential requirement for the development and validation of meaningful QSAR applications is the availability of reliable experimental data. Such data can be obtained from numeric and factual databases and/or directly from the scientific literature. New toxicity testing can be performed where necessary. After acquisition of an initial data set from these sources, the first step involves the selection of a subset of the substances tested or relevant *training set* from these data. The appropriate structural parameters, properties, and other molecular descriptors (see later) need to be selected either on a mechanistic basis or based on the best QSAR correlation. Following a statistical analysis of these data sets, a QSAR model or regression equation can be derived between the chemical and biological data. The model can be validated using either additional dose- (or concentration-) response data similar to those used in the model derivation for additional chemicals, or using semiquantitative data, if available, for a more diverse group of chemicals. In the latter case, the process can be useful in establishing the predictive boundaries of the model through the identification of chemicals showing effects at concentrations below those predicted by the QSAR model (Lipnick et al., 1985c,d; Newsome et al., 1987).

Commonality of Molecular Mechanism

Commonality of molecular mechanism is perhaps the most important factor to consider in the development of QSAR models, yet it appears to be the one to which the least attention is given.

Receptor Binding

Although binding to a specific receptor represents an important mechanism for the action of drugs, it is of less importance with respect to the toxicology of industrial chemicals. Receptor binding can be studied using in vitro systems, and the structural requirements for such binding can be studied based on molecular modeling work using the three-dimensional structure of the receptor and its affinity for various substrates.

Symptomatology

A common set of symptoms or observed responses can be useful in suggesting commonality of mechanism. This approach has been used to develop a means of classification based on the transformation of multiple responses to a small set of orthogonal descriptors (Bradbury et al., 1990). However, caution needs to be used in that symptoms may arise from physiological events that occur simultaneously with those involving the toxicological effects of interest but are not relevant to that event itself.

QSAR Correlation

QSAR analysis of a data set can be useful in supporting a commonality of mechanism within a group of chemicals. The use of molecular descriptors (see later) and a mathematical formulation based on a putative mechanism of action from which a statistically satisfactory relationship can be derived can provide valuable support for the mechanism itself. In contrast, correlations based on ad hoc choice of molecular descriptors without regard to understanding some causal relationship do not provide this kind of support. The numerical values of the coefficients of these parameters can also provide information related to steric, electronic, and hydrophobic requirements for receptor interaction (Hansch, 1989).

Additivity of Biological Activity

It has been known for some time that the toxicities of chemicals acting by a common mechanism are additive (see Chapter 1). Although synergism can occur, normally the toxicities are additive for chemicals acting by a common mechanism, and toxicity values computed from QSAR estimates can be used to estimate the toxicity of the mixture (Broderius and Kahl, 1985; Hermens et al., 1984a,b, 1985a,d; Hermens and Leeuwangh, 1982; Könemann, 1980, 1981b; Deneer et al., 1988c).

Chemical Properties

The availability of a working hypothesis regarding molecular mechanism of toxicity provides an opportunity to compare substances that nominally (i.e., according to simple categorization) belong to different classes but may have the ability to exhibit similar reactivity. This also provides a possible means of extrapolating beyond an existing QSAR model if the relationship between a chemical reactivity descriptor and toxicity is well defined, and the same descriptor is suitable for additional classes of substances of similar reactivity. The ability to calculate reactivity (Karickhoff et al., 1991; Shorter, 1982) and other molecular descriptors directly from chemical structure provides a means of evaluating the relative toxicity of a series of substances even prior to their synthesis.

Ideally, statistical experimental design methods are used to select a group of chemicals for the analysis, such that the chemicals are well distributed within the domain of all of the molecular descriptors (spanned substituent space) used to derive the QSAR model. Figures 5–7 (Lipnick et al., 1985a) show the distribution of three data sets in log P–pK_a parameter substituent space. In the first set (Figure 5), chlorophenols were used in the derivation of the QSAR model, and there was a high degree of collinearity between log P and pK_a. This high degree of collinearity indicates that the extent to which each of these

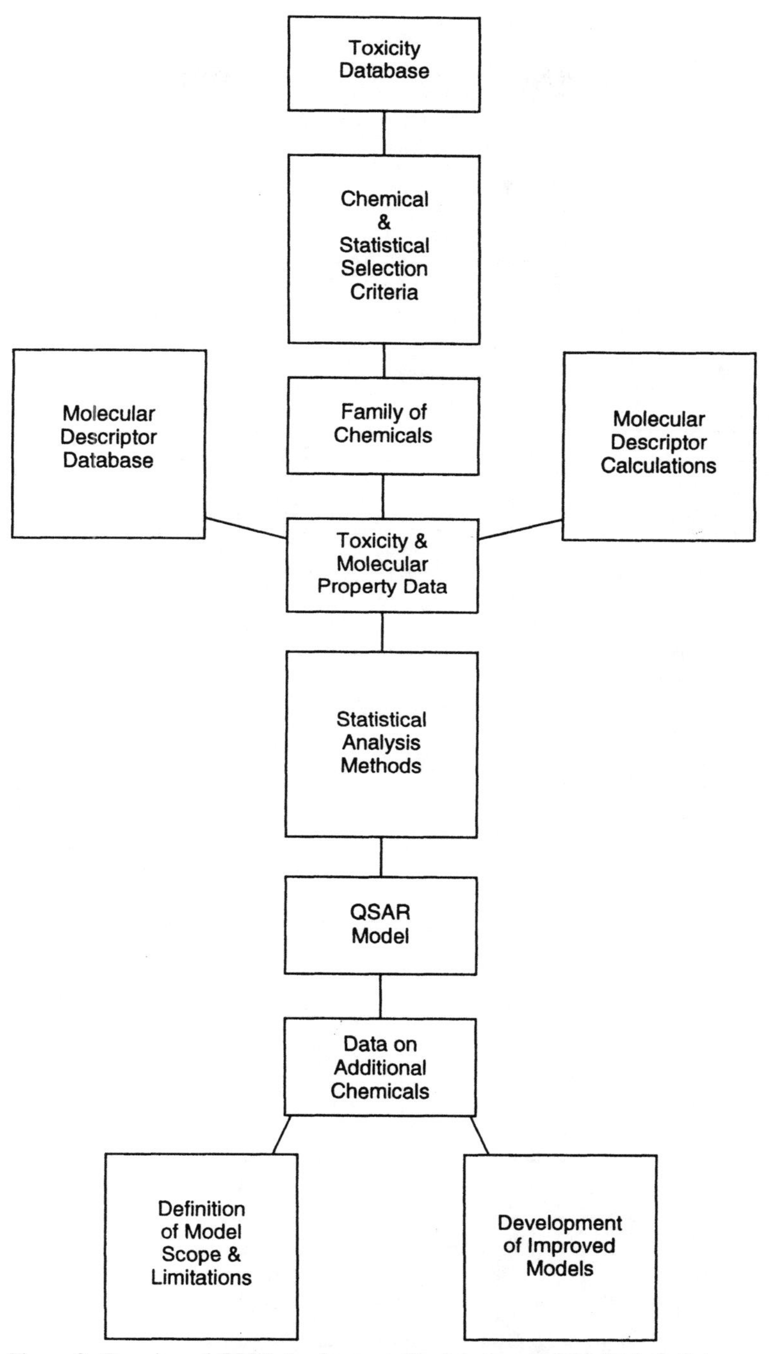

Figure 3. Overview of QSAR development. Needed are a toxicity database that provides a family of chemicals after a selection process. The family of chemicals requires in addition molecular property data (molecular descriptors) that are related to the toxicity data through statistical analysis methods to develop a QSAR model. The subsequent use of additional data provides a means to define the scope and limitations of the model and to develop improved, more robust models.

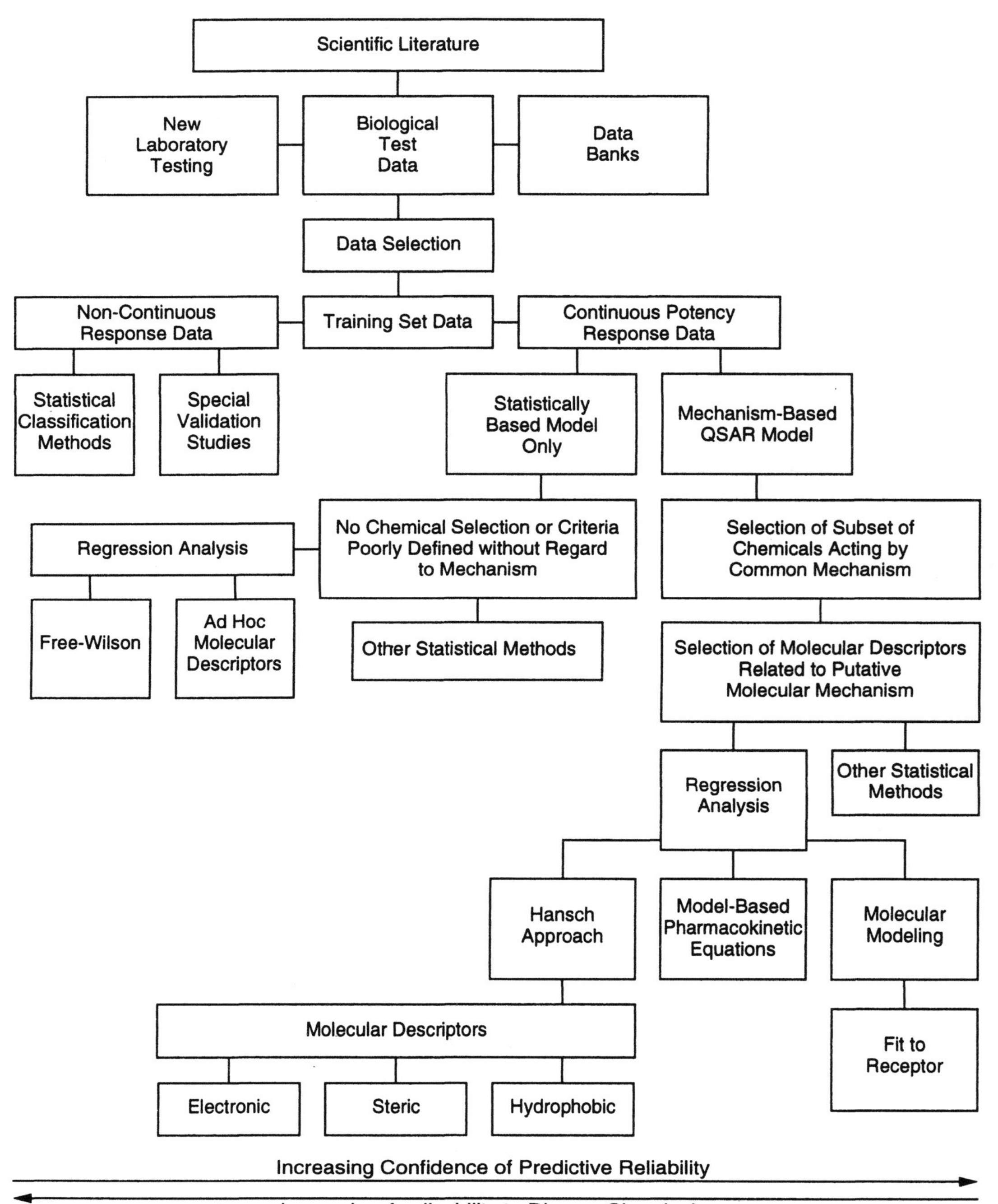
Scientific Literature
New Laboratory Testing
Biological Test Data
Data Banks
Data Selection
Non-Continuous Response Data
Training Set Data
Continuous Potency Response Data
Statistical Classification Methods
Special Validation Studies
Statistically Based Model Only
Mechanism-Based QSAR Model
Regression Analysis
No Chemical Selection or Criteria Poorly Defined without Regard to Mechanism
Selection of Subset of Chemicals Acting by Common Mechanism
Free-Wilson
Ad Hoc Molecular Descriptors
Other Statistical Methods
Selection of Molecular Descriptors Related to Putative Molecular Mechanism
Regression Analysis
Other Statistical Methods
Hansch Approach
Model-Based Pharmacokinetic Equations
Molecular Modeling
Molecular Descriptors
Fit to Receptor
Electronic
Steric
Hydrophobic
Increasing Confidence of Predictive Reliability
Increasing Applicability to Diverse Chemicals

Figure 4. Hierarchy of QSAR model types. Biological test data that can be used in the development of a QSAR can come directly from the scientific literature, from systematic data banks, or, if needed, from new laboratory testing. From these biological data, some selection is made to develop a "training set" that is composed of either continuous potency response data (e.g., fish LC50) or noncontinuous response data (e.g., lethal or nonlethal to fish at 5 mg/L). Continuous potency data can be used to derive either a mechanism-based QSAR model or one that is based on ad hoc descriptors and is only statistical. The noncontinuous data can be used with classification-type statistical methods (e.g., discriminant analysis) or in special validation studies. The mechanism-based model is based on selection of a subset of chemicals for the training set that are considered to act by a common molecular mechanism, along with a set of molecular descriptors that are considered appropriate to model this mechanism. The data are then analyzed using regression analysis or some other statistical method. Regression analysis can give rise to Hansch-type models in which the descriptors are assigned to account for electronic, steric, and hydrophobic contributions of the molecules to their activity; model-based pharmacokinetic equations (e.g., bilinear relationship between log *P* and distribution); and molecular modeling in which each toxicant is fit to a three-dimensional receptor. The mechanism-based QSAR model provides the greatest confidence in predictive reliability but the least applicability to diverse data sets (although three-dimensional molecular modeling is an exception in some ways). The models that are purely statistically based (which have potentially the greatest applicability to diverse data sets, e.g., by using molecular connectivity indices) involve either no chemical selection or use of selection criteria that are poorly defined without regard to mechanism. Regression analysis is normally done either using ad hoc molecular descriptors or a Free-Wilson approach in which the structure fragments themselves serve as descriptors (e.g., number of chlorine atoms).

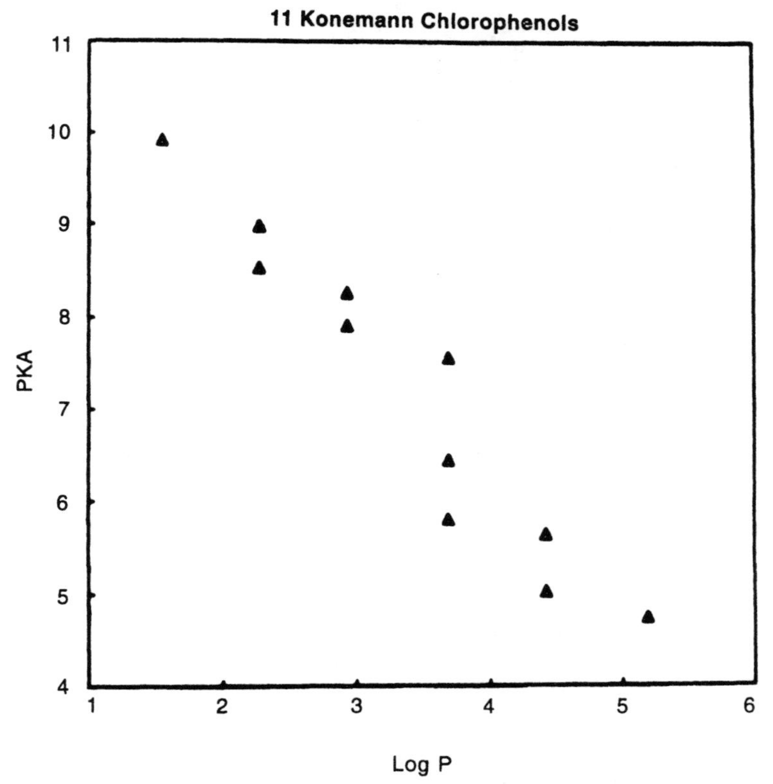

Figure 5. Distribution in log *P*, pK_a substituent space for nine chlorophenols. (From Lipnick et al., 1985c. Copyright ASTM. Reprinted with permission.) In this case, the chemicals chosen for the derivation of the model are those for which there is a high degree of collinearity between the two descriptors (log *P* and pK_a). This limitation makes it difficult to determine the independent effects of each descriptor and provides no information about other regions of descriptor space.

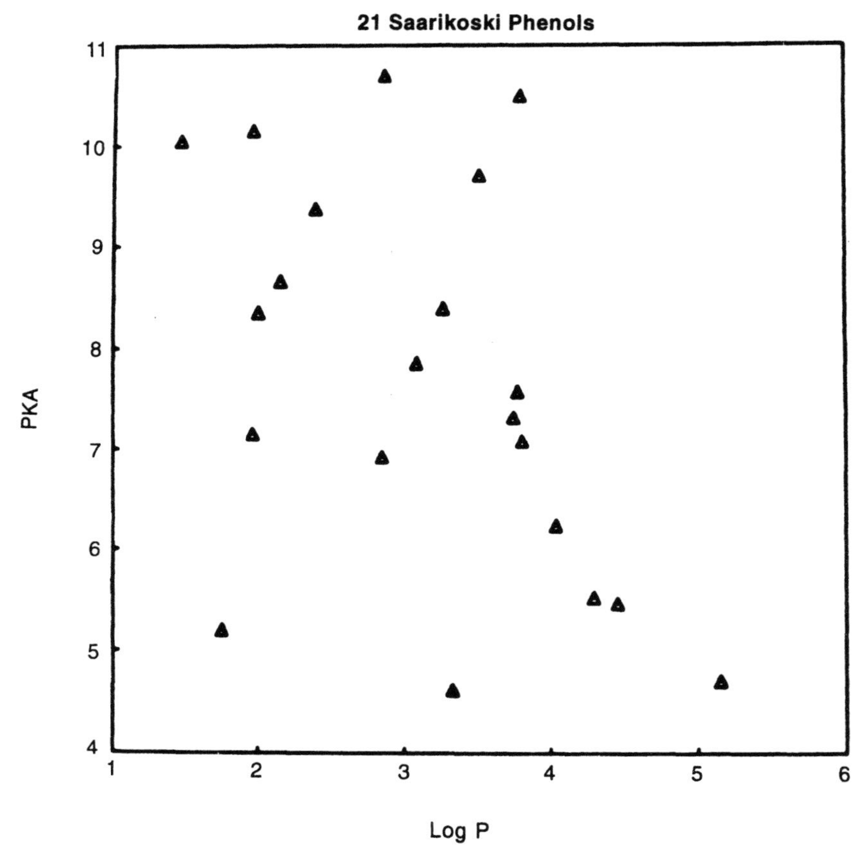

Figure 6. Distribution in log *P*, pK_a substituent space for 21 phenols. (From Lipnick et al., 1985c. Copyright ASTM. Reprinted with permission.) In this second example, the phenols chosen for QSAR model development are much better distributed in log *P*, pK_a space.

properties influences the toxicity is not well defined. In addition, the applicability of the resulting QSARs to substances lying outside the substituent space defined by these chlorophenols is in question. In the second and third sets (Figures 6 and 7), these two properties are increasingly well distributed and represent a larger range. If the derivation of such a QSAR model was conducted by experimental design, the substances representing the chosen training set would be selected based on their distribution within this domain. For example, if log *P* is the only such parameter used in a model, the chemicals should be chosen to be uniformly distributed within the desired log *P* region (e.g., 0–6). In practice, the chemicals for selection are usually restricted to a particular subgroup such as phenols. One approach to selection would be to satisfy the formal domain boundaries based on the molecular descriptors used, such as log *P* and pK_a, and discard additional chemicals from the training set only if they contain no phenolic hydroxyl group. This, of course, has the danger that the molecule may contain additional functional groups that themselves are active either by direct receptor interaction or through metabolic activation. For this reason, using mechanism as opposed to functionality alone in the selection process is the desired approach. In the most extreme case, no attempt is made regarding any preselection; for example, a model is derived using 96-h LC50 for rainbow trout based on all organic compounds for which test data are available. As noted in Figure 4, such models, which are of a nonmechanistic nature, have broader applicability (i.e., for more chemicals) but are also of lower predictive reliability.

Testing Set and Validation

It is desirable to divide an initial data set into two parts so that information is available for a testing set to assess the validity of the model. Because of the extreme difficulty of defining the boundaries of the model, which Hansch has referred to as spanned substituent space (SSS), the results of such validations and their applicability to other chemicals, particularly for large general QSAR models, should be viewed with caution, as Hansch (1977) has noted:

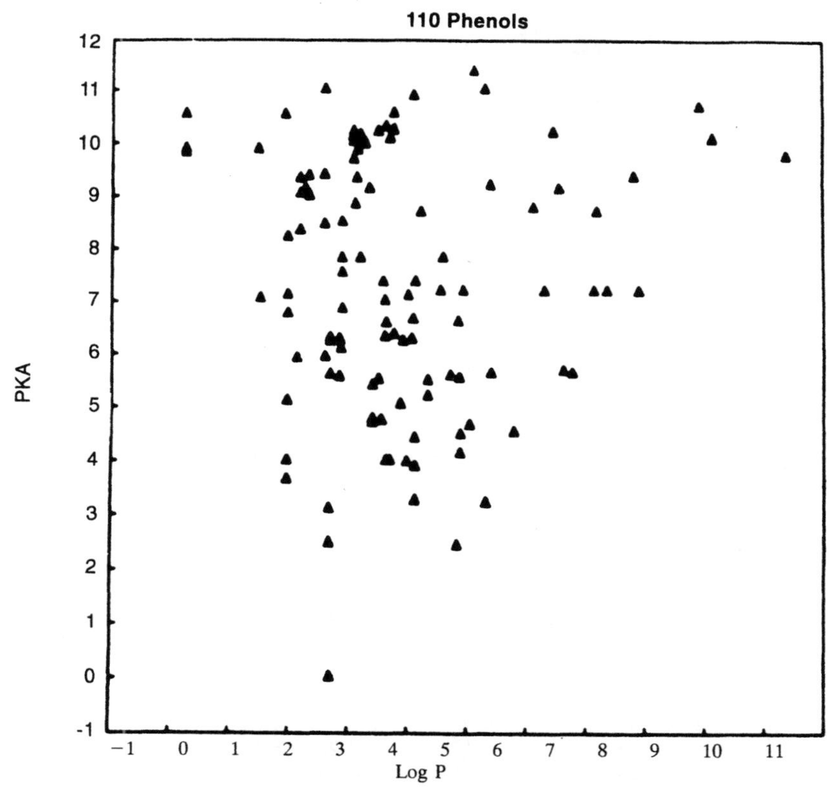

Figure 7. Distribution in log *P*, pK_a substituent space for 110 phenols. (From Lipnick et al., 1985c. Copyright ASTM. Reprinted with permission.) The phenols in this case are the best distributed in all data sets and provide the most information on the scope and limitation of the phenol model.

Put simply, the un-SSS is unknown and generally unknowable without some kind of exploration. One can consider a QSAR as a kind of map; for example, if one had contour maps of several sections of an unknown river flowing through a jungle, one might feel fairly secure in sketching in the regions between the known sections, especially if one knew they were not a great distance from one another. However, one would be chary of extending one's map far up- or downstream from the two end mapped regions. Heaven only knows what enormous waterfall might be around the next bend in the river valley or what large valleys might intersect our river valley. The problems of QSAR are much worse since we are dealing in *n*-dimensional instead of 3-dimensional space.

For QSAR models based on a more narrowly defined series of chemicals, limited data normally preclude its dichotomy into a training set and a testing set. Under these conditions, some validation can be conducted using semiquantitative data such as results of fish toxicity screening tests at fixed concentrations (Applegate et al., 1957; Hollis and Lennon, 1987; MacPhee and Cheng, 1989a,b; MacPhee and Ruelle, 1969; Wood, 1987). This approach has been used successfully for testing the limitations of fish toxicity QSARs for alcohols (Lipnick et al., 1985b), phenols (Lipnick et al., 1985a), and anilines (Newsome et al., 1987). Outliers could be detected when screening data showed toxicity at concentrations considerably lower than those predicted by the models.

Even in the absence of such testing, more confidence can probably be placed on the more narrowly defined models, especially if a mechanistic foundation exists for both chemical and molecular descriptor selection. This question is discussed in further detail in the next section.

Molecular Descriptors

Molecular descriptor is a generic term that applies to any parameter used in the formulation of either a QSAR or SAR to model a property (e.g., rate of hydrolysis) or biological response (e.g., 96-h LC50). Some examples of molecular descriptors that have been employed in QSAR are listed in Table 3. Ideally, such molecular descriptors are chosen to simulate the mechanisms of transport, metabolism, and interaction at the site of action.

Reviews of the literature (Dearden, 1990a,b; Van de Waterbeemd and Testa, 1987) show that a large number of molecular parameters have been employed or are available for QSAR studies. It is easy to be overwhelmed by the large number of available

Table 3. Molecular descriptors

Physicochemical
- Hydrophobicity (log P, water solubility)
- Henry coefficient (i.e., measure of partitioning between air and water)
- Electronic parameters (substituent parameters like Hammett constant σ and molar refractivity, whole-molecule parameters like pK_a and dipole moment, and quantum chemical parameters like atomic charge)
- Steric parameters (bulk parameters like molecular weight, volume, surface area; Taft constant E_S; shape parameters like kappa index, Austel's branching index)

Topological descriptors (or indices)
- Molecular connectivity
- Information content
- Autocorrelation vectors

parameters and their selection. Nevertheless, many of these parameters are collinear with one another and share the same "information content," and it is best to choose as small a number of such descriptors as possible, each of which can be assigned physical or chemical significance with respect to a common underlying toxic mechanism. Care in parameter selection also significantly aids in defining the boundaries of test substance selection and the corresponding prediction domain (Craig, 1990). Adrien Albert (1951), one of the foremost modern pioneers in relating chemical structure to biological activity, provided thoughtful advice that can be applied to this selection process:

> The most valuable approach to problems of how biologically active substances work is the method of *limiting factor*. The first step in this method is to investigate and record the various physical, chemical, and biochemical properties of active substances. Parenthetically, it may be pointed out that too little insight and imagination is usually brought to this elementary procedure so that the next step cannot be made effective for want of data that could be of *biological* relevance.

Examples of mechanism-based molecular descriptors include partition coefficient, pK_a (to model degree of dissociation), rate of reaction with a model nucleophile (to model propensity for covalent binding at site of action), and redox potential (to model rate of metabolic oxidation to active toxicant). Such physicochemical descriptors have a firm theoretical foundation, but measured or calculated values may not be available for all of the substances in a data set of interest or for new chemicals for which predictions are desired. Topological indices are numbers assigned to a structure that express in numerical form the topology (molecular connection graph) of a chemical species and usually reflect in varying degrees its size and shape. Topological indices (e.g., molecular connectivity parameters), while not having clear physical significance, can be of practical utility due to the ease with which they can be calculated directly from structure for diverse sets of chemicals (Basak et al., 1988; Sabljić, 1987; Protic and Sabljić, 1989; Dubois and Loukianoff, 1993). However, QSARs based on such indices do not provide direct information about relevant physicochemical properties and greater uncertainty exists regarding the extrapolation of models based on topological indices for predictive purposes.

Perhaps the simplest molecular descriptors are the structure fragments themselves, as employed in the Free-Wilson approach (Free and Wilson, 1964; Kubinyi, 1990). In the Free-Wilson approach, biological activity is described as a linear function of structure. For molecules that act by specific mechanisms in which precise interaction is required at the site of action, the most promising technique now under development is molecular modeling involving methods to relate the three-dimensional structural properties of the candidate drug or toxicant to the site of action (Martin, 1991).

Statistical Analysis

Derivation of a QSAR ideally requires the use of data representing a well-defined biological end point and values of a set of parameters (molecular descriptors) having a mechanistic relationship to the events of transport and molecular interaction leading to the observed biological response. The quality of such a QSAR is limited by the reproducibility of these biological data, the accuracy and precision in the measurement of the needed molecular descriptors, the mathematical form of the model, and the degree to which the chosen descriptors can satisfactorily model the molecular events. A somewhat different response is likely to yield a different value for the same chemical, and therefore the results for more than one chemical are not directly comparable. Statistical methods employed (Devillers, 1992) include regression analysis (Devillers and Lipnick, 1990; Draper and Smith, 1981; Newman, 1993), discriminant analysis (Borgstedt and Enslein, 1990; McFarland and Gans, 1990), cluster analysis (McFarland and Gans, 1990), principal components analysis (Niemi, 1990), and pattern recognition (Dunn and Wold, 1990). QSAR work based on fuzzy adaptive least-squares analysis has been reported (Liu et al., 1992).

Regression techniques (and principal component analysis) require precise and accurate input data, which tend to limit the number of samples in the testing set, but require only simple calculations and are easy to understand when a correlation is derived. The other statistical techniques do not demand high-precision input data and therefore can accommodate large data sets. However, these other methods demand substantial computing efforts and the results are more difficult to interpret. Multiple regression correlations whose quality is evaluated based on the correlation coefficient (r) and standard error of estimate (s) are most frequently used in environmental (or ecological) toxicology. The statistical analyses will lead to the desired model equation linking the parameter to be determined with the molecular descriptors.

Homogeneity of Variance

Theoretically, the variance of each of the data points used needs to be similar so that these can be directly compared. The quality of the resulting QSAR is limited by the reproducibility of the measurement. According to Martin (1978):

> If there is some prior estimate of the variance of the individual replicated measurements, one can compare this s^2_{repl} value with the s^2. If the two quantities are approximately equal, then it would be judged that the regression model estimated the data as precisely as could be expected.

The relationship of biological testing to the development of QSAR models has been reviewed (Martin et al., 1990).

Despite the strict requirements regarding the high quality of biological data needed for QSAR work, other data should be investigated, particularly when the predictive limits of a model are only poorly understood. For example, tests with small numbers of organisms have detected outliers, leading to additional understanding of underlying mechanism, and have been a source of guidance for future studies (Lipnick et al., 1985a,b; Newsome et al., 1987).

The Model: SAR and QSAR for Organic Substances

As an illustration of the development and application of a simple log P–dependent QSAR model, let us consider the data of Overton (Lipnick, 1989b; 1989d) in Table 4 on the lowest concentration of simple nonelectrolytes that induces anesthesia or narcosis in tadpoles of the species *Rana temporaria.*

A QSAR model has been derived (Lipnick, 1989b,d) from a subset of these data on saturated monohydric alcohols, saturated monoketones, and aromatic hydrocarbons. These classes were chosen because of their association with a narcosis-type toxicological response and are considered to act via a common molecular mechanism involving the attainment of a critical molar concentration (Connell and Markwell, 1992) or molar volume within a molecular site of action (McGowan, 1952; Mullins, 1954; Schultz, 1989; Franks and Lieb, 1987, 1990; Miller et al., 1983; McGowan and Mellors, 1986; Warne et al., 1991; Abernathy et al., 1988). This is the same as the mechanism of action of general gaseous anesthetic agents (Seeman, 1972), except for the dif-

Table 4. Overton's data on the lowest concentration producing narcosis in tadpoles (*Rana temporaria*)

Chemical	Calculated log P^a (measured in parentheses)	Lowest concentration producing narcosis in tadpoles (mg/L)	Lowest predicted concentration producing narcosis in tadpoles (mg/L)
	Saturated monohydric alcohols		
Methyl alcohol	−0.764 (−0.77)	18,200	29,700
Ethyl alcohol	−0.235 (−0.31)	13,300	14,100
n-Propyl alcohol	0.294 (0.25)	6,670	6,090
iso-Propyl alcohol	0.070	7,690	9,660
n-Butyl alcohol	0.823 (0.88)	2,850	2,480
iso-Butyl alcohol	0.693 (0.76)	3,330	3,260
tert-Butyl alcohol	0.473 (0.35)	10,000	5,170
iso-Amyl alcohol	1.222 (1.42)	2,000	1,280
tert-Amyl alcohol	1.002 (0.89)	5,000	2,030
Capryl alcohol (octyl alcohol)	2.939	50.0	52.1
	Saturated monoketones		
Acetone	−0.240 (−0.24)	16,700	18,000
Methyl ethyl ketone	0.261 (0.29)	6,670	7,830
Diethyl ketone	0.790	2,500	3,090
Methyl propyl ketone	0.790 (0.91)	1,670	3,090
	Aromatic hydrocarbons		
Benzene	2.142 (2.15)	166	166
Xylene	3.440 (3.15)	40.0	14.9
Naphthalene	3.316 (3.30)	8.33	23.3
Phenanthrene	4.490 (4.46)	2.00	2.78

[a]Calculated using CLOGP3.3; values in parentheses are measured.

Adapted from Table 2 in Lipnick, 1989e. Copyright ASTM. Reprinted with permission.

ference in route of administration. Such baseline narcosis models have been found to be of general importance for both aquatic and mammalian organisms (Hansch et al., 1989; Lipnick et al., 1985, 1987, 1993). A similar approach using internal volume fraction has been applied to chemicals acting by more specific mechanisms (Jaworska and Schultz, 1993).

First, the concentrations producing the standard biological response must be converted to molar concentrations. For example, in Table 4, methyl alcohol (CH_3OH) has a molecular weight of 32.04 ($1 \times 12.011 + 4 \times 1.008 + 1 \times 16.000$). To convert from mg/L to mol/L, 1 mol of methanol is equal to 32.04 g. Dividing the effect concentration (18,200 mg/L) by the molar conversion factor (32.04 g/mol):

$$\frac{(18{,}200\ \text{mg/L})(1\ \text{g}/1000\ \text{mg})}{32.043\ \text{g/mol}} = 0.570\ \text{g/mol}$$

To derive our QSAR model, we now convert this to a reciprocal logarithmic quantity[2]:

$$\log\left(\frac{1}{0.570}\right) = 0.246$$

Regression analysis of these 18 sets of data on toxicity and partition coefficient leads to the following QSAR model:

$$\log\left(\frac{1}{C}\right) = 0.909 \log P + 0.727$$

$$(n = 18,\ r = 0.986,\ s = 0.227)$$

Application of Model for Making Predictions

Let us examine how well this equation performs for predictive purposes. Overton tested over 100 simple organic nonelectrolytes belonging to various chemical classes and found that most produced a narcotic-type response in tadpoles and other aquatic organisms (Lipnick, 1986). The following examples (Lipnick, 1989b) have been chosen to illustrate the behavior of substances in several of these classes:

Tetrachloroethane

Tetrachloroethane (carbon tetrachloride) has molecular formula CCl_4, molecular weight 153.84, and log *P* 2.875.[3] Using the QSAR developed from Overton's data on saturated monohydric alcohols, saturated monoketones, and aromatic hydrocarbons:

$$\log\left(\frac{1}{C}\right) = 0.909 \times 2.875 + 0.727 = 3.340$$

$$\frac{1}{C} = 10^{(3.340)} = 2190$$

$$C = 4.567 \times 10^{-4}\ \text{mol/L}$$

To convert from mol/L to g/L, multiply by molecular weight:

$$C = 4.567 \times 10^{-4}\ \text{mol/L} \times 153.84\ \text{g/mol}$$
$$= 7.026 \times 10^{-2}\ \text{g/L}$$

To convert from g/L to mg/L, multiply by 1000:

$$C = 7.026 \times 10^{-2}\ \text{g/L} \times 1000\ \text{mg/g} = 70.3\ \text{mg/L}$$

This prediction is in good agreement with Overton's observation of narcosis with this substance at 111 mg/L.

Camphor

Similarly, for camphor ($C_{10}H_{16}O$), with a molecular weight of 152.23 and a calculated log *P* of 2.117, narcosis is predicted to take place at 340 mg/L. This compares favorably with the observed effect concentration of 200 mg/L.

Sulfonal

The next example from Overton's data illustrates the caution that must be exercised in basing predictions on calculated log *P* values. The calculated value for sulfonal ($C_7H_{16}O_4S_2$; molecular weight = 228.33) is −0.431, which yields a predicted narcotic toxicity for tadpoles of 106,000 mg/L. This prediction is in poor quantitative agreement with Overton's reported value of 2000 mg/L. In addition, it is inconsistent with Overton's observation that this substance produced a characteristic narcosis response. Unfortunately, in this case the calculated log *P* value could not be directly compared to an experimental value. As an alternative approach, however, log *P* could be independently estimated from water solubility (Banerjee et al., 1980):

$$\log P = 6.5 - 0.89 \log S - 0.015\text{MP}$$

$$(n = 27,\ r = 0.96)$$

where *P* is the *n*-octanol–water partition coefficient, *S* is the water solubility in micromoles per liter, and MP is the melting point in degrees centigrade (for substances that are liquid at room temperature a nominal melting point of 25°C is employed in the calculations). The two available water solubility values for sulfonal of 2000 and 2740 mg/L, yield estimated log *P* values from this equation of 1.10 and 0.980, respectively. These additional log *P* values yield estimates of narcosis concentration of 4290 and 5510 mg/L, in far better agreement with Overton's measured experimental toxicity data for sulfonal.

In general, it is advisable to estimate parameters used in QSAR work by more than one independent method as a cross-check. Even better, of course, is to obtain an experimental log *P* measurement.

USE OF QSAR MODELS FOR ESTIMATION OF PHYSICOCHEMICAL AND BIOLOGICAL PROPERTIES (EXPOSURE ASSESSMENT)

QSAR/QSPR models can be used for the estimation of rates of hydrolysis, photolysis, volatilization, soil

[2]The use of the reciprocal here follows the convention initiated by Corwin Hansch, providing a scale in which the dependent variable increases with increasing potency.

[3]Since log *P* values calculated using the CLOGP computer program (Leo and Weininger, 1985) were used in the QSAR derivation, calculated values were also used for consistency in making the predictions. The measured log *P* value of 2.83 for tetrachloroethane differs little from that calculated.

sorption, and biodegradation, and for determination of water solubility, boiling point, octanol-water partition coefficient, acid dissociation constant, and bioaccumulation and bioconcentration (Dearden, 1990a, 1991; Kuenemann et al., 1990; Degner et al., 1991; Desai et al., 1990; Howard et al., 1992; Klopman et al., 1993; Shimp et al., 1990; Tabak et al., 1992; Vaishnav and Babeu, 1987; Veith et al., 1979; Lyman et al., 1982; Mackay, 1982; Nendza et al., 1990; Mackay and Paterson, 1990; Sabljic, 1990; von Oepen et al., 1991; Chessels et al., 1992). The following section provides background on some of the molecular descriptors themselves, followed by information on the fate properties of environmental interest.

Octanol-Water Partition Coefficient

Partition coefficient is the most fundamental property governing the toxicity of simple nonelectrolytes; it is also the most important descriptor in modeling environmental fate and transport processes. It has been related to water solubility, soil-sediment sorption, and bioconcentration.

A liquid-liquid partition coefficient is defined as the ratio of the concentration (at equilibrium) of a substance (solute) when distributed between two immiscible liquid phases; the concentration in the more hydrophobic phase is in the numerator. For example, the octanol-water partition coefficient of azobenzene is determined by distributing this substance between *n*-octanol and water and calculating the ratio (unitless) of the measured concentrations in the two phases. For azobenzene, this ratio, $K_{\text{octanol-water}}$, is 6610. Partition coefficients are normally expressed as logarithmic quantities, and for azobenzene log $K_{\text{octanol-water}}$ = log(6610) = 3.82.[4] The quantity log $K_{\text{octanol-water}}$ is frequently referred to in the literature as log P or log K_{ow}.

A considerable literature exists related to the more fundamental molecular properties represented as a composite by partition coefficient (Leo et al., 1971; Tanford, 1980; Taylor, 1990). Log P values have traditionally been measured using the shake flask method (Leo et al., 1971; Dunn et al., 1986), but this method is of limited value for chemicals with partition coefficients greater than 4 or 5; alternatively, such chemicals can be measured by high-performance liquid chromatography (HPLC) (Könemann et al., 1979; Nowotnik et al., 1993) and relating the retention times to those of standard chemicals for which shake flask values are available. The slow-stirring method (De Bruijn et al., 1989; Brooke et al., 1990) provides an opportunity for equilibration between the two phases and limits the amount of solvent transferred to each phase via emulsion formation. This approach yields more reproducible values for very hydrophobic solutes [polychlorinated biphenyls (PCBs)] with very low water solubility. Leahy et al. (1989) have proposed use of propylene glycol dipelargonate (PGD)-water as another model partitioning system because it appears to better represent the hydrogen-bonding donor and acceptor properties of the target sites. It is known that the ability of a substance to penetrate the brain depends not only on partition coefficient but also specifically on hydrogen-bonding properties (Van de Waterbeemd and Kansy, 1992). Comparisons of different model partitioning systems have been reviewed (Leo, 1972; Seiler, 1974; Rekker, 1977). Compilations of measured log P values are available in the literature (Leo et al., 1971; Hansch and Leo, 1979).

Methods have been developed for the estimation of partition coefficients from chemical structure (Fujita et al., 1964; Nys and Rekker, 1973, 1974; Leo, 1975, 1983, 1987, 1990; Rekker, 1977; Hansch and Leo, 1979; Kamlet et al., 1988; Rekker and Mannhold, 1992). The estimation method now most widely employed is based on the ability to assign partial log P values to individual atoms or groups of atoms (fragments) within the molecule and sum these partial values (i.e., the fragment constant method). Multiple fragments containing heteroatoms (i.e., atoms other than carbon and hydrogen) or double or triple carbon-carbon bonds produce deviations from simple additivity rules and are accounted for by the use of special correction factors or interaction terms (Leo, 1975, 1983, 1990; Rekker, 1977; Hansch and Leo, 1979; Rekker and Mannhold, 1992). Thus, in general,

$$\log P = \sum a_i F_i + \sum b_j f_j$$

where F_i and f_j are the ith and jth terms (i.e., or values) for fragments and interaction terms, respectively, which occur a_i and b_j times in the molecule (i.e., a and b are numerical factors indicating the incidence of the fragment within the molecule). Lyman (1990) assessed the Fragment Constant method with a variety of chemicals and found the error was ± 0.12 log P unit.

These methods have been computerized (CLOGP) so that estimated log P values can be obtained in an automated form based on the Leo-Hansch (Chou and Jurs, 1979; Leo and Weininger, 1985) and Rekker (Van de Waterbeemd, 1986) approaches. The ability to perform such calculations has been assisted by the availability of various chemical structure representation algorithms, including the SMILES line notation (Weininger and Weininger, 1990; Weininger, 1988). If the log P value is calculated for the neutral (nonionized) form, the effects of ionization on transport and binding processes are not considered.

Partition coefficients can be estimated from water solubility (Hansch et al., 1968; Banerjee et al., 1980; Valvani and Yalkowsky, 1980), as in the foregoing example for sulfonal. In addition, they can be estimated from molecular surface area (Leahy, 1986; De Bruijn and Hermens, 1990), linear solvation energy parameters (Kamlet et al., 1988), and three-dimensional structure (Ghose and Crippen, 1986; Hopfinger and Battershell, 1976; Richards et al., 1991). Al-

[4] Log in this chapter refers to the logarithm to base 10.

though each of these approaches has limitations regarding the availability of data needed by the model, the fragment constant method for log K_{ow} estimation is currently the most practical because of the broad range of fragment constants available and its automation in the form of the CLOGP computer program.

Dissociation Constants

Compilations of dissociation constants are available in the literature (Perrin, 1972; Serjeant and Dempsey, 1979); and these parameters can also be estimated directly from chemical structure (Perrin et al., 1981), including via computer (Hunter, 1988).

Water Solubility

Water solubility (*S*), a measure of hydrophilicity, can be obtained from experimental measurement or literature data (Dannenfelser and Yalkowsky, 1991) or estimated from partition coefficient (as indicated above for the inverse) and melting point (as a second parameter to account for the enthalpy of fusion for compounds that are not liquids at room temperature)[5] (Banerjee et al., 1980), where *S* is in units of μmol/L and the melting point (MP) is in °C:

$$\log P = 6.5 - 0.89 \log S - 0.015\text{MP} \quad (n = 27,\ r = 0.96)$$

For example, in the calculations discussed earlier on estimating log *P* from the water solubility (*S*) of sulfonal:

$$\begin{aligned} \text{MW} &= 228.33 \\ S_{\text{water}} &= 2740 \text{ mg/L} \\ &= \frac{2740 \text{ mg/L} \times 1000\ \mu\text{g/mg}}{228.33\ \mu\text{g}/\mu\text{mol}} \\ &= 12{,}000\ \mu\text{g/L} \\ \text{Melting point} &= 124-126°\text{C} \\ \log P &= 6.5 - 0.89 \log(12{,}000\ \mu\text{g/L}) \\ &\quad - 0.015 \times 126° \\ &= 6.5 - 0.89 \times 4.079 - 1.89 = 0.980 \end{aligned}$$

Water solubility can also be estimated by a neural network method (Bodor et al., 1991) and a fragment method (Bodor and Huang, 1992). Because water solubility correlates directly with partitioning behavior only for liquid solutes, it has not been used widely as a QSAR parameter.

Bioaccumulation

Before a chemical can produce a response in an organism, it must accumulate to some extent in tissues. Understanding the process of bioaccumulation (see Chapter 16 and Appendix Chapter D) is thus fundamental in predictive aquatic toxicology. Application of QSARs for elucidating bioaccumulation is extensive for aquatic organisms (Nagel and Loskill, 1991; Donkin, 1994). The utility of such models and their limitations have been reviewed (Esser and Moser, 1982; Brooke et al., 1986; Connell, 1988; Donkin, 1994; Nendza, 1991).

Aquatic organisms can accumulate chemicals in their tissues from food, contaminated sediment, or by direct absorption from water (i.e., or from pore water by infaunal organisms) across body surfaces, particularly gills. The degree of accumulation varies with the relative affinity of the chemical for water and lipid or other hydrophobic constituents (i.e., largely lipids) within the organism. QSARs have been widely applied to laboratory data on biological uptake of chemicals from water, generally termed bioconcentration (see Chapter 16).

Although log *P* (or K_{ow}) has commonly been used for such models, it is well to keep in mind that the thermodynamic properties in the ordered lipid phase and in the nonordered *n*-octanol phase may be different for different compounds, limiting the reliability of this model substance (Opperhuizen et al., 1988). To illustrate qualitatively the utility of log *P* as a partitioning model, consider that carbon tetrachloride (log $P = 2.73$) is bioconcentrated in fish 30 times, whereas pentachlorobenzene (log $P = 4.94$) is bioconcentrated 3400 times (Veith and Kosian, 1982).

Most QSAR models for bioconcentration are in the form

$$\log \text{BCF} = a \log P + b$$

where BCF is the bioconcentration factor or ratio of the concentration in fish or other aquatic organism to that in the origin phase (e.g., water). References to some of these QSAR models can be found in Table 5. The preceding equation can be illustrated by a model developed by Veith and Kosian (1982) based on data for a broad group of industrial and agricultural chemicals,

$$\log \text{BCF} = 0.79 \log P - 0.40 \quad (n = 122,\ r^2 = 0.86)$$

These authors also examined the data by chemical class and fish species and the slope and intercepts of the resulting models were found to vary from 0.79 to 0.86 and from −0.40 to −0.80, respectively. The chemicals used to derive this model encompassed a log *P* range of 0.30 (dichloroethane) to 6.91 (Aroclor 1260). Most measurements were whole body in fish, with some in muscle, and test durations were from 4 to 300 d. Similar log-log relationships to that of Veith and Kosian have been reported for mussel (Geyer et al., 1982), oyster (Ogata et al., 1984), and mollusks (Hawker and Connell, 1986).

[5]For substances that are liquid solutes, the melting point term is not needed.

Table 5. QSAR studies of the bioaccumulation of chemicals by aquatic organisms

Organism/test conditions	Chemical class	Parameter(s)	Method	Reference
Salmo gairdneri	Organochlorine and aromatic chemicals	Log *P*	Regression analysis	Neely et al. (1974)
Daphnia pulex	7 polycyclic aromatic hydrocarbons	Log *P*	Regression analysis	Southworth et al. (1978)
Pimephales promelas (fathead minnow)	30 pesticides and industrial organic chemicals	Log *P*	Regression analysis	Veith et al. (1979)
Various species	358 chemicals (mostly pesticides)	Log *S*, log K_{oc} (K_{oc} = soil sorption coefficient)	QSAR prediction of BCF; comparison with measured BCF where data available	Kenaga and Goring (1980)
Various fish species	Various chemicals	Log *P*	Regression analysis	Kenaga and Goring (1980)
Poecilia reticulata	6 chlorobenzenes	Log *P*, $(\log P)^2$	Regression analysis	Könemann and van Leeuwen (1980)
Pimephales promelas (fathead minnow); *Lepomis cyanellus* (green sunfish); *Salmo gairdneri* (rainbow trout)	Aroclor 1254	Test temperature	Regression analysis (thermodynamic equation to derive enthalpy and entropy of process)	Matsuo (1980)
Lepomis macrochirus (bluegill sunfish)	Various organic compounds	Log *P*	Regression analysis	Veith et al. (1980)
Pseudorasbora parva (topmouth gudgeon)	15 pesticides	Log *P*, log *S*, acute toxicity	Regression analysis	Kanazawa (1981)
Cyprinus carpio (carp)	Chlorobenzenes and chloronaphthalenes	$\Sigma i/\Sigma o$ (inorganic/organic structure property)	Regression analysis	Matsuo (1981)
Carassius auratus (goldfish)	4-Aminoantipyrine and ethanol	Test temperature	Regression analysis (thermodynamic equation to derive enthalpy and entropy of process)	Matsuo (1981)
Chlorella fusca (algae)	34 industrial organic chemicals and pesticides	Log *S*	Regression analysis	Freitag et al. (1982)
Mytilus edulis (mussel)	16 agricultural and industrial organic compounds	Log *P*, log *S*	Regression analysis	Geyer et al. (1982)
Various fish species	Various organic compounds	Log *P*	Regression analysis	Mackay (1982)
Poecilia reticulata (guppy)	Substituted phenols	Log *P*	Regression analysis	Saarikoski and Viluksela (1982)
Pseudorasbora parva (topmouth gudgeon)	15 pesticides	Log *S*; log *P*; log MW (molecular weight)	Regression analysis	Kanazawa (1983)
Salmo gairdneri (rainbow trout)	10 chlorobenzenes, hexachlorobutadiene, hexachloroethane	Log *P*	Regression analysis	Oliver and Niimi (1983)
13 species of freshwater and marine fish	122 industrial and agricultural chemicals	Log *P*	Regression analysis	Veith and Kosian (1983)
Daphnia magna	5 polycyclic aromatic hydrocarbons and polycyclic sulfur heterocycles	Log *P*	Regression analysis	Eastmond et al. (1984)
Salmo gairdneri (rainbow trout)	22 chlorinated and brominated organic compounds	Log *P*, parachor[a]	Regression analysis	Oliver (1984)
Salmo gairdneri (rainbow trout)	10 chlorinated and brominated aromatic compounds (at environmental concentrations)	Log *P*; parachor[a]	Regression analysis	Oliver and Niimi (1984)
Poecilia reticulata, *Salmo gairdneri*	13 chlorobenzenes	Log *P*	Regression analysis	Chiou (1985)

Table 5. QSAR studies of the bioaccumulation of chemicals by aquatic organisms (*Continued*)

Organism/test conditions	Chemical class	Parameter(s)	Method	Reference
Guppies, goldfish, and trout	Chlorinated aromatic and aliphatic compounds	Log *P*, first-order clearance rate constants (also related to log *P*)	Regression analysis; pharmacokinetic model	Hawker and Connell (1985)
Salmo gairdneri (rainbow trout), *Pimephales promelas* (fathead minnow)	6 chlorinated dioxins	Log *P*	Nonlinear regression analysis	Muir et al. (1985)
Cyprinidon variegatus (sheepshead minnow); *Lagodon rhomboides* (pinfish); *Crassostrea virginica* (oyster); *Mytilus edulis* (mussel)	Various pesticides	Log *P*; log *S*	Regression analysis; comparison with predictions of other QSAR	Zaroogian et al. (1985b)
Poecilia reticulata, Salmo gairdneri, Carassius auratus	42 chlorinated hydrocarbons	Log *P*, $(\log P)^2$, $(\log P)^3$, $(\log P)^4$	Regression analysis	Connell and Hawker (1988)
Worst-case bioconcentration estimate	Various organic compounds	Linear and nonlinear terms in log *P*	Hypothetical relationship	Nendza (1991)
Fish (species not provided)	Various organic compounds	Linear and nonlinear terms in log *P*	Nonlinear regression analysis	Binstein et al. (1993)

[a]Parachor is a measure of molar volume and is related to K_{ow} (or *P*).

A second type of model for bioconcentration was developed by Mackay (1982), who derived a corresponding relationship (using a selected data set) in which he assumed that octanol was a good surrogate for the fish and a slope of unity could be assumed,

$$K_B = 0.048 \log P \qquad (r^2 = 0.95)$$

where K_B is the bioconcentration factor.

The literature (Connell, 1990) indicates that bioconcentration increases linearly for chemicals with log *P* values in the range of 2–6, but declines for chemicals with log *P* values above 6. The behavior of chemicals with log *P* values below 2 is not dominated by partitioning into lipid phases and therefore the relationship between bioconcentration and log *P* is less reliable in this range. Furthermore, models that take into account physiological characteristics of fish (Barron, 1990; Barber et al., 1991; Clark and MacKay, 1991; Clark et al., 1990; Erickson and McKim, 1990) indicate that uptake through the gills is an important route for low to high hydrophobic chemicals (log $P < 6$), but as the log *P* increases the dietary route increases in significance. The latter is especially relevant for exposure assessment of superhydrophobic chemicals (log $P > 6$) in the field.

Although the dominant factor influencing the degree of bioaccumulation of a chemical is its partitioning behavior, other factors need to be considered if such models are needed for a broad class of industrial organic chemicals. These factors include the following (some of which also need to be considered in toxicological testing):

- *Test duration.* Chemicals with 2 > log *P* (or K_{ow}) < 6 require a time to establish 99% equilibrium between substance in water and in fish that increases with increasing log *P* (Hawker and Connell, 1985). Under circumstances in which equilibrium is not achieved, relationships for higher log *P* (> 6) chemicals deviate from linearity and can be described by a nonlinear model (Connell and Hawker, 1988; Binstein et al., 1993).
- *Stability to biotransformation.* Metabolism or biotransformation normally leads to the production of less lipophilic, more water-soluble substances and lower bioaccumulation. Chemicals with a high proportion of halogenated aliphatic or aromatic substituents or highly branched aliphatic moieties have the highest persistence and are the most likely to bioaccumulate.
- *Lipid solubility and water solubility.* For nonelectrolytes (nonionizable organic compounds), both lipid solubility and water solubility decrease with increasing melting point and can lead to limiting body burdens; that is, the substance may be too insoluble in water to reach significant environmental concentrations or body burdens of concern may never be achievable due to limiting lipid solubility.

  Bioconcentration occurs in chemicals with water solubilities at approximately 18 − 0.02 mol/m^3 with a maximum at about 0.002 mol/m^3 with declining bioconcentration at lower values (Connell, 1988).
- *Molecular weight.* The rate of diffusion of a chemical at the aqueous interphase with the aquatic organism (e.g., the water layer in contact with the fish gills) is expected to decrease in relation to increasing molecular weight. Zitko (1980) and

Connell (1990) suggested that chemicals with molecular weights over 100 exhibit bioaccumulation with a maximum capacity at about 350, but this declines to a very low capacity at about 600 and is associated with decreased permeability of biological membranes. The importance of this factor has not been fully explored, however, due to the lack of test data on high-molecular-weight chemicals in which bioaccumulation is not already limited by low water solubility (i.e., which contain a sufficient number of hydrophilic functional groups).

- *Molecular cross section.* The inability of certain chlorinated naphthalenes to undergo significant bioconcentration has been interpreted in terms of a limiting molecular cross section of approximately 0.95 nm (Opperhuizen et al., 1985). This result can also be interpreted as arising from limiting water and/or lipid solubility (see above) arising from the high melting points of these highly symmetrical isomers (melting point within a series is highest for greatest symmetry).
- *Degree of ionization.* Chemicals with potential for bioaccumulation may be slightly ionized in water. However, the unionized form of a chemical is expected to have a lower polarity than the ionized form, and thus a higher K_{ow} value and lower water solubility (Connell, 1988).
- *Organism lipid content.* Because hydrophobic chemicals have their greatest affinity for lipid and fat, the total amount measured in an organism in a whole-body analysis strongly reflects this content. This is also dependent upon the organism life stage, season, and other factors affecting the organism.
- *Water temperature.* Generally, organisms become more active with increasing water temperature and therefore the rate of uptake of a chemical (with corresponding increased bioconcentration in a short-duration test) will be observed. This may be offset, however, by an increase in metabolic transformation rate. Water temperature also affects the partition coefficient, but not in a consistent fashion (increase or decrease in log P with increasing temperature). This will be a factor of only two or three times over a 30° temperature range.
- *Salinity.* All factors being equal, saltwater species are expected to undergo higher bioconcentration than the corresponding freshwater ones due to the increase in partition coefficient with salinity.

Biodegradation

QSAR has been applied at a rapidly growing rate to the study of biodegradation (Cambon and Devillers, 1993). Early studies included correlations between biodegradation and log P (Vaishnav et al., 1987), electronic properties (Pitter, 1985), steric effects (Paris et al., 1982), and connectivity indices (Boethling, 1986), but it was not possible to achieve validation of these QSARs for predicting biodegradation of additional classes of substances. Biodegradation is an important detoxification mechanism that occurs in aquatic ecosystems because of the natural occurrence of diverse populations of microorganisms capable of degrading organic chemicals (see Chapters 1 and 15). It is a complex multi-stage process involving uptake, intracellular transport, and enzymatic reactions. Biodegradation can lead to reduced accumulation of a chemical in the environment. Numerous factors affect biodegradation, such as water solubility, volatility, hydrolysis, sorption processes, partitioning, temperature, moisture concentrations, and microbial class and biomass. Biodegradation rates of chemicals (and polymers) are therefore dependent on chemical structure and environmental variables. The ability to optimize performance characteristics while maximizing the rate of biodegradation can be used to great advantage during the development of new industrial organic chemicals. The development of models relating chemical structure to rate of biodegradation can thus be valuable in limiting the release of persistent substances into the environment. A survey of such models for biodegradation is provided in Table 6.

Degner et al. (1991) modeled biodegradation data (measured as rates of oxygen consumption) for 112 disubstituted aromatic compounds using geometric and quantum mechanical parameters. Approximately 90% could be correctly classified as not readily degradable and readily biodegradable. Peijnenburg et al. (1991) developed QSARs for rates of abiotic and biotic sediment transformation. For aliphatic halogenated hydrocarbons they reported the following equation:

$$\log k = -0.142\text{BS} + 0.483\sigma^* + 0.039\text{BE} + 3.03 \qquad (n = 16,\ r^2 = 0.985)$$

where BS is the bond strength of the carbon-halogen bond being cleaved, σ^* is the Taft sigma star constant that reflects both field effects and effect on the reaction center of the number of halogens, and BE is a measure of the carbon-halogen bond energy. Roberts (1991) found that the initial rate of biodegradation (k) of linear alkylbenzene sulfonate (LAS) isomers and homologues can be modeled by a QSAR using the lengths (in number of carbon atoms) of the longest (L) and shortest (S) arms of the branched chain (L) as molecular descriptors, such that:

$$k = 0.909[(0.5 + L^{-1})^{-12} + (0.5 + S^{-1})^{-12}]^{1/2} - 8.376$$
$$(n = 19,\ r = 0.985,\ s = 0.59,\ F = 542)$$

Although there appears to be no general relationship between biodegradation and chemical structure, several studies indicate that the primary degradation step is often rate-limiting (Parsons and Govers, 1990). Little progress has been made in developing QSARs for biotransformation in higher organisms.

Table 6. QSAR studies of the biodegradation of organic chemicals

Chemical class	Parameter(s)	Method	Reference
Chlorophenols and related compounds	Log *P*	Regression analysis	Banerjee et al. (1984)
Alcohols, phenols, ketones, carboxylic acids, ethers, and sulfonates	Atomic charge difference across functional group bond; superdelocalizability	Regression analysis	Dearden and Nicholson (1986, 1987)
2,4-Dichlorophenoxyacetic acids	Log *P*	Regression analysis	Paris et al. (1984)
Anilines and phenols	Hammett substituent constant	Regression analysis	Pitter (1985)
Phenols	van der Waals radius of substituent	Regression analysis	Paris et al. (1982, 1983)
Alcohols, acids, esters, ketones, aromatics, and other compounds	Water solubility, log *P*, molar refractivity, molar volume, melting point, boiling point, number of carbon, hydrogen, and oxygen atoms, molecular weights, and theoretical BODs	Regression analysis of linear and second-order polynomials	Babeu and Vaishnav (1987)
2,4-D alkyl esters; *N*-3-chlorophenylcarbamates; dialkyl ethers; dialkyl phthalate esters, aliphatic acids	Molecular connectivity indices	Regression analysis	Boethling (1986)
112 disubstituted aromatic compounds	Geometric and quantum mechanical parameters	Regression analysis; discriminant analysis	Degner et al. (1991)
Alcohols, acids, ketones, amides, amines, hydrocarbons, and phenols	Group contributions	Regression analysis	Desai et al. (1990)
264 diverse organic chemicals	35 substructural fragments	Regression analysis	Howard et al. (1992)
283 compounds	Water solubility, structure fragments	Group contribution	Klopman et al. (1992)
283 diverse organic chemicals	Structure fragments	CASE method	Klopman et al. (1993)
PCBs	Hansch hydrophobic substituent and Hammett electronic substituents	Regression analysis	Parsons et al. (1991)
Linear alkylbenzene sulfonate isomers and analogues	Lengths of longest and shortest arms of branched LAS chain	Regression analysis	Roberts (1991)
Alcohols, acids, ketones, amides, hydrocarbons, and phenols	Group contribution	Neural network analysis	Tabak et al. (1992)
Alcohols, ketones, and alicyclic compounds	Log *P*	Regression analysis	Vaishnav et al. (1987)
Alcohols, ketones, and aromatics	van der Waals volume, log *P*, electronegativity, and hydrogen bonding (donor and acceptor)	Modified autocorrelation	Zakarya et al. (1993)
252–254 various organic compounds	Atom counts, molecular connectivity indices	Discriminant analysis	Enslein et al. (1984)
Halogenated aliphatic and aromatic compounds	Carbon-halogen bond strength; Hammett and Taft sigma constants; Taft steric parameters	Regression analysis	Peijnenburg et al. (1991)

USE OF QSAR MODELS FOR ECOTOXICITY (EFFECTS ASSESSMENT)

A large number of QSAR models have been reported for the effects of chemicals on aquatic organisms. An attempt has been made to summarize this literature in Tables 7–11. Table 7 contains references to QSARs derived for chemicals acting by a baseline or narcosis mechanism that is produced by nonelectrolytes of relatively low reactivity such as hydrocarbons, alcohols, ketones, and certain[6] aliphatic chlorinated hydrocarbons. In Table 8, references are provided on QSARs for various phenols; in Table 9, for aliphatic and aromatic amines and nitroaromatic

[6]In the absence of additional functional groups that cause the chlorine to be reactive, such as allyl chloride (aliphatic) or benzyl chloride (aromatic), or to be reactive following metabolic activation.

Table 7. QSAR studies for acute effects of non-reactive non-electrolyte organic chemicals to aquatic organisms

Organism	Effect (end point)[a]	Parameters[b]	Reference
Rana temporaria	Narcosis	Log $K_{\text{olive oil/water}}$	Lazarev (1944), Lipnick and Filov (1992)
Rana temporaria	Narcosis	Log *P*	Leo et al. (1969), Hansch and Dunn (1972)
Carp *Carassius auratus* Goby Roach Tench	24-h minimum lethal dose	Log *P*	Hansch and Dunn (1972)
Various aquatic organisms	Various end points	Log *P*	Hansch and Dunn (1972), Lipnick (1990)
Poecilia reticulata	14-d LC50	Log *P*	Könemann (1981a)
Daphnia magna	48-h EC50	Supercooled liquid solubility	Bobra et al. (1983)
Daphnia magna	24-h EC50	Log *P*	Calamari et al. (1983)
Selenastrum capricornutum	96-h EC50 (growth inhibition) 3-h EC50 (photosynthesis inhibition)	Log *P*	Calamari et al. (1983)
Daphnia magna	14-day EC50 (reproduction)	Log *P*	Calamari et al. (1983)
Salmo gairdneri *Brachydanio rerio*	48-h LC50	Log *P*	Calamari et al. (1983)
Leuciscus idus melanotus	48-h LC50	Log *P*	Lipnick and Dunn (1983), Lipnick (1985b)
Photobacterium phosphoreum	30 min EC50 (bioluminescence reduction)	Log *P*, no. symmetry planes	Ribo and Kaiser (1983)
Tetrahymena ellioti	18-h LC50	Log *P*	Rogerson et al. (1983)
22 freshwater species of bacteria, algae, protozoa, crustacea, insects, coelenterates, molluscs, fish, and amphibians	Various acute and sublethal tests	Log *P*	Slooff et al. (1983)
Pimephales promelas	96-h LC50	Log *P*	Veith et al. (1983)
Alburnus alburnus (bleak)	48-h LC50	Log *P*, number of carbon atoms	Bengtsson et al. (1984)
Nitocra spinipes Boeck	96-h LC50	Log *P*, number of carbon atoms	Bengtsson et al. (1984)
Daphnia magna *Tetrahymena ellioti* *Chlamydomonas* *Chlorella*	48-h LC50 48-h threshold 3-h EC50 3-h EC50	Water solubility	Bobra et al. (1984)
Artemia salina nauplii	24-h EC50 (immobilization)	Log *P*	Foster and Tullis (1984)
Daphnia pulex	96-h LC50	Log *P*, connectivity indices, HPLC, and TLC retention indices	Govers et al. (1984)
Daphnia magna	48-h EC50	Log *P*	Hermens et al. (1984b)
Salmo gairdneri	Intraperitoneal LD50	Log *P*	Kaiser et al. (1984)
Fish	LC50	SC	LeBlanc (1984)
Fish	MATC	SC	LeBlanc (1984)
Various fish species	LC50	Log *P*	Newsome et al. (1984)
Ankistrodesmus fulcatus	4-h EC50 (primary productivity)	Log *P*, log *S*	Wong et al. (1984)
Daphnia magna	48-h EC50	Log *P*, supercooled liquid solubility	Bobra et al. (1985)
Pimephales promelas	32-d MATC	Log *P*	Call et al. (1985)
Photobacterium phosphoreum	15 min EC50 (bioluminescence reduction)	Log *P*, molar refractivity, molar volume	Hermens et al. (1985b)
Daphnia magna	16-d NOEC (growth)	Log *P*	Hermens et al. (1985a)
Salmo gairdneri, *Poecilia reticulata*, *Pimephales promelas*, *Jordanella floridae* (flagfish)	96-h LC50	Log *P*	McCarty et al. (1985)
Salmo gairdneri, *Pimephales promelas*, *Jordanella floridae* (flagfish), *Salevlinus fontinalis*	Chronic toxicity	Log *P*	McCarty et al. (1985)
10 aquatic species	Toxicity	Log *P*	Thurston et al. (1985)

Table 7. QSAR studies for acute effects of non-reactive non-electrolyte organic chemicals to aquatic organisms (*Continued*)

Organism	Effect (end point)[a]	Parameters[b]	Reference
Daphnia magna	48-h EC50	Supercooled liquid solubility	Abernethy et al. (1986)
Brachydanio rerio	LC50	Molecular connectivity indices	Devillers and Chambon (1986)
Pimephales promelas	LC50	Atom frequencies, groups	Hall and Kier (1986)
Photobacterium phosphoreum	EC50 (bioluminescence)	V, π^*, β, α, log P	Kamlet et al. (1986)
Daphnia magna	EC50	Log P	Lipnick and Hood (1986)
Pseudomonas putida	EC3		
Scenedesmus quadricauda	EC3		
Microcystis aeruginosa	EC3		
Uronema parduczi	EC3		
Tetrahymena pyriformis	EC50	Molecular connectivity indices, physicochemical properties	Moulton and Schultz (1986)
Leuciscus idus melanotus	48-h LC50	V, π^*, β, α	Kamlet et al. (1987)
Carassius auratus	24-h LC50	Log P	Lipnick et al. (1987)
Various fish species	Acute and chronic end points	Log P (internal toxic concentration)	McCarty (1987a,b)
Pimephales promelas	96-h LC50	Molecular connectivity indices	Sabljic (1987)
Mytilus edulis	EC50	Log P, log S	Donkin et al. (1989)
Rana temporaria	Narcosis and toxicity	Log P	Lipnick (1989b,e)
Pimephales promelas	96-h LC50	Log P, log IGC50 (*Tetrahymena*)	Schultz et al. (1990a)
Tetrahymena pyriformis	IGC50	Log P, $\Sigma\sigma$	Schultz et al. (1990b)
Fish	NOEC	Log P	Van Leeuwen et al. (1990)
Cyclotella meneghiniana	48-h EC50 for DNA reduction	Log S	Del et al. (1991)
Oryzias latipes	LC50	Log P	Ikemoto et al. (1992)
Daphnia pulex			
Chlorella vulgaris			
Cyprinus carpio			
Pimephales promelas	Probability of lethal concentration	Log P, molar volume (V), pharmacokinetic parameters	Mackay et al. (1992)
Pimephales promelas	Elimination half-life	Log P	McCarty et al. (1992)
Pimephales promelas	Threshold 96-h LC concentration	Log P	McCarty et al. (1992)
25 species of bacteria, algae, fungi, protozoans, coelenterates, rotifers, mollusks, crustaceans, insects, fish, and amphibians	EC50, LC50, NOEC, and NOEC literature data; prediction of HC5 and NOEL	Log P	Van Leeuwen et al. (1992)
Poecilia reticulata	14-d LC50	Log P, molecular connectivity indices, log N	Koch (1984)
Daphnia magna	24-h EC50		
Cyprinodon variegatus	end point		

[a]LC50, lethal concentration with mortality in 50% of population; EC50, concentration with effect in 50% of population; NOEC, no observable effect concentration; MATC, maximum acceptable toxicant concentration from early life stage (ELS) test; IGC, inhibitory growth concentration.
[b]Parameters in QSAR equation: log P (or log K_{ow}), octanol-water partition coefficient; SC, structure coefficient; V, solute molar volume; π^*, β, α, solvatochromic parameters for dipolarity, polarizability, hydrogen-bond acceptor basicity, and hydrogen-bond donor acidity from Kamlet et al. (1986); log S, solubility in water; $\Sigma\sigma$, sum of the Hammett sigma substituent constants for substituted benzene derivatives to account for electronic effects; TI, topological indices such as molecular connectivity indices; log N, logarithm(10) of negentropy.

compounds; in Table 10, for electrophiles; and in Table 11 for various other classes including esters, organotin compounds, dithiocarbamates, pesticides, thioureas, and surfactants.

Baseline Narcosis

The class of relatively nonreactive nonelectrolyte organic chemicals (e.g., alcohols, ketones, ethers, chlorinated alkanes, and aromatics) is the largest group to which QSARs have been applied to different test species and effects (i.e., end points) (Table 7). These chemicals all act by a narcosis mechanism and most of the QSAR equations in Table 7 applied to toxicity data have the general form

$$\log 1/C = a \log K_{ow} + b$$

where K_{ow} is the sole chemical descriptor related to hydrophobicity that models the rate of uptake and ultimate internal concentration of the chemical at the site of action.

The following generalizations can be made from data in Table 7 and the general narcosis QSAR equation:

- The intercept *b* is a measure of test species or effect (LC50) sensitivity; it indicates that differences in species sensitivity are small for this series of chemicals.
- The values for *a* in most of these studies are around 1.0.
- *b* for this equation where acute and chronic data are available for the same species (fathead minnow, Veith et al., 1983; Call et al., 1985; *Daphnia*, Hermens et al., 1984a,b) shows that chronic values are higher than acute values. Intercept *b* values are similar for *Daphnia* and fathead minnow, indicating that differences in species sensitivity are small.
- The more hydrophobic chemicals are less toxic to fish in shorter (96-h) tests than in longer (14-d) tests. The time needed to establish equilibrium increases with increasing hydrophobicity (Hawker and Connell, 1985). The time to reach 99% of the equilibrium concentration in fish is related linearly to K_{ow} but holds only for chemicals with a log K_{ow} of 2 to 6. Hawker and Connell showed that for very hydrophobic chemicals 96 h through 14 d is not long enough to reach equilibrium and this will probably explain differences in toxicity (e.g., LC50s).
- At a certain K_{ow} it is not unusual to encounter a situation in which toxicity increases up to some point and then either decreases or has a sharp cutoff (cutoff effect) with no toxicity observed at saturation. The basis for this cutoff may be associated with limiting water or lipid solubility, receptor binding, and/or pharmacokinetics (Donkin et al., 1991).
- If the required external concentration to produce lethality within a test duration too short to reach equilibrium does not lead to a water solubility cutoff, the results can be modeled using a parabolic or bilinear model (Veith et al., 1983; Lipnick et al., 1987; Kubinyi, 1976, 1977, 1979, 1990; Kubinyi and Kehrhahn 1978).

Simple nonreactive nonelectrolytes act by a nonspecific mechanism and show dependence on their partition coefficient to produce the biological effect (e.g., LC50). Narcosis is a minimum effect or baseline effect because in principle every organic chemical can produce narcosis, except that it is masked in the presence of a more specific effect acting at lower concentrations. Equations from Table 7 then represent the minimal effect concentration of each organic chemical. Chemicals that act by specific mechanisms (e.g., electrophiles) will cause effects (e.g., LC50) at much lower concentrations. As previously mentioned, Overton (Lipnick, 1986) and Meyer (Lipnick, 1989d) at the turn of the century proposed that the anesthetic potency to tadpoles and aquatic organisms of diverse nonelectrolyte organics is governed by their partition coefficient (olive oil–water used then) between the phase used for administration (air or water) and a lipoid site of action. The exact mechanism of narcosis is unknown, but the two most important theories are the *membrane perturbation theory* and the *protein receptor theory*. The first states that changes in the lipid bilayer properties are responsible for the anesthetic effect because either the attainment of a critical molar concentration or molar volume of a chemical is present at the site of action. The protein receptor theory is based on the hypothesis that one or more proteins serve as anaesthetic receptor or binding sites (Franks and Lieb, 1987, 1990). Narcosis is often referred to as a nonspecific mode of action. Overton (Lipnick, 1991b, p. 85) described this process physiologically as follows:

> It should be kept in mind that, in order to reach the ganglia cells, the alcohol [in a discussion of ethyl alcohol] must first penetrate the outer surfaces of the epithelia of the gills into the epithelial cells of the gills, then must leave through their inner surfaces into the space between the epithelia of the gills and the capillaries of the gills. After that, it must pass through the outer surfaces of the capillaries into the body of the endothelial cells and through their inner surfaces into the bloodstream. After being transported by the bloodstream, the alcohol must then pass through the endothelia in the reverse direction in order to reach the intercellular lymph surrounding the ganglia cells and ultimately penetrate the ganglia cells themselves.

Internal Toxic Dose

Ferguson (1939) provided a thermodynamic explanation for correlations between depressant (or narcotic) activity and lipophilicity. Using data on the biological response of simple organic chemicals, he demonstrated that toxicity could be related quantitatively to chemical potential, which he defined as the ratio of concentration (C) producing the standard biological response to the chemical's solubility in water (C_S). The thermodynamic activity (A) needed to produce narcosis is $A = C/C_S$. He found that for chemicals exhibiting a narcotic or physical toxicity response, this ratio is a constant; it changes only in the case of chemicals acting by a more specific mechanism. The thermodynamic fugacity approach of MacKay and co-workers (Mackay and Paterson, 1990; see Chapter 18) is similar to Ferguson's use of chemical potentials. Although the Mackay thermodynamic approach uses water solubility, as was done in the early work (see Historical Background) by Houdaille and Richet, in this case corrections are made for solutes that are not liquids at room temperature to the supercooled liquid phase. This is equivalent to the melting point correction in correlating water solubility with log P (Banerjee et al., 1980).

The relationship between partition coefficient and toxic aqueous concentration is related to a concentration producing equivalent chemical potentials (Ferguson, 1939; Crisp et al., 1967) corresponding to critical internal molar concentration or molecular volume (Mullins, 1954; McGowan and Mellors, 1986). Narcosis in aquatic organisms is considered to reflect a common internal concentration or volume

(Mullins, 1954). In experiments with aquatic organisms and toxicants that act by a narcosis mechanism the slopes are all close to unity (as in Table 7), indicating that octanol is a reasonably representative model of the physicochemical properties of the biophase site of action, and that a pseudosteady equilibrium has been achieved with respect to partitioning between the aquatic donor phase concentration and the hydrophobic biophase site of action within the fish. This relationship between external and internal concentration is also being increasingly applied as a means of developing mechanistic pharmacokinetic models (McCarty et al., 1991; see Chapter 1).

McCarty et al. (1992) showed that the threshold critical body residue (CBR) for the acute toxicity (mortality) of hydrophobic neutral narcotic organic chemicals ($K_{ow} > 1.5$) to fathead minnows estimated from aquatic toxicity QSARS (LC50) and bioconcentration–log K_{ow} relationships is a constant of 4.4 (2.2–8.3) mmol kg^{-1} fish (assuming a fish density of 1.0 kg/L). However, for hydrophilic chemicals ($K_{ow} < 1.5$) the bulk of the chemical is in the water phase and not in the organic lipid phase of the fish. It is therefore likely that the whole-body residue is similar to the LC50 water concentration. The acutely toxic whole-body residues for narcotics (K_{ow} range 1.5 to 6) can be approximated by the QSAR-derived equation CBR (mM) = 2.5 mM + $50/K_{ow}$. Although the data set is small, both calculated and observed literature residue values for the acute toxicity of narcotic organic chemicals of various small fish and invertebrates appear to be similar to residues estimated by this method (see Chapter 1).

De Wolf et al. (1991) demonstrated that the lethal body burden of a chemical measured in fish may not be a constant value for chemicals acting by more specific mechanisms. The lethal body burden of 2,3,4,5-tetrachloroaniline to guppies was found to vary depending on the length of exposure and corresponding aqueous concentration of toxicant. This is consistent with the concept that the body burden concentration may have little relationship to the concentration of covalently bound toxicant or covalently bound toxicant metabolite at the actual molecular site of action.

Excess Toxicity

As previously mentioned, narcosis represents baseline (reversible) toxicity, which indicates that a nonelectrolyte chemical is at least as toxic as predicted by a narcosis QSAR model but can be more toxic if it is acting by a more specific mechanism. The ratio (T_e) between predicted narcosis toxicity and observed toxicity has been defined as the excess toxicity parameter:

$$T_e = \frac{LC50_{pred}}{LC50_{obs}}$$

Within a series of chemicals, T_e decreases with increasing partition coefficient. For this reason, particular caution should be exercised in using a narcosis QSAR for predictive purposes for chemicals having a partition coefficient less than 2, where the predicted hazard may be low but the degree of excess toxicity could be significant. The majority of nonelectrolyte industrial chemicals showing such behavior act by forming covalent bonds with sulfhydryl groups and other nucleophilic moieties present in cellular macromolecules (electrophile mechanism) or are metabolically activated to such reactive moieties (proelectrophile mechanism). These covalent binding mechanisms are discussed later.

Phenols

The toxicity data for phenols are also analyzed by QSAR studies (Table 8). Most QSAR equations for phenol include a descriptor for electronic effects that models both the influence of ionization of the chemical on its uptake (and consequent partition coefficient) and the significance of electronic effects with respect to interaction at the target site. The ionized form of a chemical is much poorer at traversing lipophilic membranes, consistent with the decline in uptake rate constants of several substituted phenols with guppies at different pH values (Saarikoski and Viluksela, 1982). LC50s for the guppies were dependent on the degree of ionization; therefore studies with ionizable chemicals must be conducted at constant and known pH.

Phenols are known to be uncouplers of oxidative phosphorylation (Corbett, 1974). Shultz et al. (1986) noted two separate groups of phenols with different QSAR equations for LC50 (96-h to fathead minnow) related to K_{ow}. It was suggested that one group of phenols acted by an uncoupler mechanism and the other group as narcotics. Veith and Broderius (1987) have proposed that weakly acidic phenols act by a polar narcosis mechanism, which they distinguish from narcosis produced by simple nonelectrolytes. These phenols and other compounds (e.g., some anilines), which have a strong hydrogen-bond donating group, are more toxic than predicted by a baseline narcotic QSAR and their toxicity is not concentration-additive with chemicals that fit such a QSAR (Veith and Broderius, 1990). These compounds have a log $K_{ow} < 2.9$ and are called polar or type II narcotics. See Chapter 1, Table 4.

Amines and Nitroaromatic Compounds

Several QSAR studies are summarized for the toxicity of aliphatic and aromatic amines and nitroaromatic chemicals to aquatic organisms in Table 9. Effect concentrations are correlated predominately with K_{ow} and electronic descriptors. A comparison of the intercepts of several of the equations for LC50 indicates that they cause mortality at concentrations 10-fold lower than those of unreactive nonelectrolytes. Several authors attribute the activity of this group of chemicals to the formation of reactive in-

Table 8. QSAR equations on the toxicity of phenols to aquatic organisms[a]

Organism	Effect[b]	Parameters[c]	Reference
Poecilia reticulata	96-h LC50	π, *F*, *R*	Kopperman et al. (1974)
Salmo salar (Atlantic salmon)	96-h LC50	Log *P*	Zitko et al. (1976)
Daphnia magna	24-h LC50	Hydrophobicity and others	Durkin (1978)
Crangon septemspinosa (shrimp)	96-h LC50	Log *P*, pK_a	McLeese et al. (1979)
Poecilia reticulata	14-d LC50	Log *P*, pK_a	Könemann and Musch (1981)
Poecilia reticulata	96-h LC50	Log *P*, pK_a	Saarikoski and Viluksela (1982)
Photobacterium phosphoreum	30-min EC50	Log *P*	Ribo and Kaiser (1983)
Lebistes reticulata (guppy)	24-h LC50	Steric parameter	Benoit-Guyod, et al. (1984)
Pimephales promelas	96-h LC50	Molecular connectivity	Hall and Kier (1984)
Salmo gairdneri	Intraperitoneal LD50	Log *P*; pK_a, molar refractivity (MR); resonance parameter (R)	Kaiser et al. (1984)
Lepomis machrochirus	96-h LC50	Structure coefficient	LeBlanc (1984)
Lepomis machrochirus; *Pimephales promelas*	MATC	Structure coefficient	LeBlanc (1984)
Various fish species	Screening data	Log *P*, pK_a	Lipnick et al. (1985a)
Salmo gairdneri (rainbow trout)	96-h LC50	Log *P*	McCarty et al. (1985)
Salmo gairdneri (rainbow trout)	Chronic toxicity	Log *P*	McCarty et al. (1985)
Daphnia magna	24-h LC50	Log *P*, pK_a, connectivity	Devillers and Chambon (1986)
Pimephales promelas	96-h LC50	Log *P*	Schultz et al. (1986)
Poecilia reticulata	LC50	Grouped accessible surface area	Gombar (1987)
Cyprinodon variegatus	96-h LC50	Molecular connectivity indices	Sabljic (1987)
Tetrahymena pyriformes	48/60-h EC50 for population growth	Log *P*, σ	Schultz and Cajina-Quezada (1987)
Tetrahymena pyriformes	60-h EC50 for population growth	Log *P*; electronic field effect (*F*); ability to serve as a hydrogen-bonding donor (H)	Schultz et al. (1987)
Pimephales promelas	96-h LC50	Log *P*	Veith and Broderius (1987)
Salmo gairdneri	LC50, LD50	Log *P*	Hodson et al. (1988)
Escherichia coli	IC50 (growth kinetics)	Log *P*	Nendza and Seydel (1988a,b)
Pimephales promelas	Narcosis	Log *P*, LUMO	Purdy (1988)
23 species of aquatic organisms	Narcosis	Log *P*	Lipnick (1989c)

[a]See Table 7 for most of the effect and parameter descriptions; only those not defined in Table 7 legend will be described here.
[b]IC50, inhibitory concentration for 50% effect on growth.
[c]DpKa, difference in pK_a between substituted phenol and phenol parent; LUMO, lowest unoccupied molecular orbital electron density, which is a quantum mechanical parameter for electronic effects; *F*, substituent constants for field effects; *R*, substituent constants for resonance effects.

termediates and consider that metabolism plays an important role in the correlation studies.

Electrophilic Toxicants

Reactive chemicals are more lethal than unreactive chemicals with similar K_{ow} values. Nucleophilic functional groups (—NH_2, —OH, —SH) in macromolecules such as proteins are likely sites for such reactive chemicals, analogous to the production of mutation effects by covalent binding of such molecules to DNA nucleic acid sites. This is illustrated for allyl bromide in Figure 8.

Chemicals whose toxicity results from covalent binding to one or more biological sites of action can be classified as acting by an electrophilic mechanism (Lipnick et al., 1987b; see Table 10). It has been known for some time that the action of lachrymators and vesicants is mediated by a covalent binding mechanism, the main target for acute toxic effects being the inhibition of enzymes containing sulfhy-

Table 9. QSAR equations on the toxicity of aliphatic and aromatic amines and nitroaromatic chemicals to aquatic organisms[a]

Organism	Effect	Chemical class	Parameters[b]	Reference
Pimephales promelas	96-h LC50	Benzene derivatives including nitro and amine substituents	Free-Wilson (de novo) substituent fragments	Hall et al. (1984)
Poecilia reticulata	14-d LC50	Alkyl- and chloroanilines	Log *P*	Hermens et al. (1984a)
4 species of yeast	EC50 for growth inhibition	Chloroanilines	Log *P*	Kwasniewska and Kaiser (1984)
Photobacterium phosphoreum	EC50	Chloroanilines	Log *P*	Ribo and Kaiser (1984)
Tetrahymena pyriformis	60-h IGC50	Nitroaromatics; aromatic amines; pyridines	Log *P*; molecular connectivity indices	Schultz and Moulton (1984)
Tetrahymena pyriformis	EC50 for population growth	Nitrogen heterocycles	Log *P*	Schultz and Moulton (1985)
Pimephales promelas	96-h LC50	Substituted benzenes	Free-Wilson substituent fragments	Hall and Kier (1986)
Tetrahymena pyriformis	EC50 for population growth	Pyridines	Log *P* and 8 substituent constants	Moulton and Schultz (1986)
Poecilia reticulata	14-d LC50	Nitrobenzene derivatives	Log *P*; redox potential	Deneer et al. (1987)
Poecilia reticulata; *Pimephales promelas*	14-d LC50 96-h LC50	Nitroaromatic compounds	$\Sigma\sigma$	Roberts (1987)
Various fish species	24-h screening data	Anilines	Log *P*, pK_a	Newsome et al. (1987)
Pimephales promelas	96-h LC50	Anilines and phenols	Log *P*	Veith and Broderius (1987)
Photobacterium phosphoreum	30-min EC50	Anilines	Log *P*, HOMO, LUMO	Kaiser and Gough (1988)
E. coli	I50 inhibition (multiplication growth kinetics	Anilines and phenols	pK_a, lipophilicity, Hammett σ constant	Nendza and Seydel (1988a,b)
Pimephales promelas	96-h LC50	Anilines	Log *P*	Purdy (1988)
Pimephales promelas	96-h LC50	Nitrobenzenes	Log *P*, HOMO, LUMO	Purdy (1988)
Pimephales promelas	96-h LC50	Aliphatic amines	Log *P*, pK_a, D $^1\chi_v$	Newsome et al. (1991)
Tetrahymena pyriformis	48-h static population growth impairment system	66 monosubstituted anilines	Log *P*, $\Sigma\sigma$	Schultz et al. (1991)
Tetrahymena pyriformis	48-h static population growth impairment system	24 aliphatic amines and alkylaniline derivatives	Log *P*	Schultz et al. (1991)

[a]See Tables 7 and 8 for most of the effect and parameter descriptions; only those not defined in Tables 7 and 8 will be described here.
[b]HOMO, highest occupied molecular orbital for electron density; *D*, steric parameter; χ, molecular connectivity indices.

$$CH_2{=}CH{-}CH_2Br \xrightarrow{\text{Enz SH}} CH_2{=}CH{-}CH_2{-}S{-}Enz$$

Allylic halide (allyl bromide) — Enzyme adduct with electrophile toxicant

Figure 8. Electrophile mechanism with covalent binding to enzyme sulfhydryl.

dryl groups associated with interaction at the site of action (Ross, 1962; Jocelyn, 1972). Some examples of chemicals acting by electrophilic mechanisms that have been documented in aquatic toxicology are the following (Hermens 1990b; Lipnick, 1991a; Verhaar et al., 1992):

- Nucleophilic substitution by allylic, propargylic, and benzylic halides; α-halo ketones; carbamoyl halides; diaryl disulfides; and α-halo nitriles (Figure 9).
- Ring opening of acid anhydrides, epoxides, and imines (Figure 10).
- Michael-type (1,3-conjugate) addition to α,β-unsaturated (double or triple bond) aldehydes, ketones, amides, esters, nitro compounds, sulfones, and nitriles (Figure 11).
- Schiff base formation with aldehydes (e.g., by covalent binding to an ε-amino group of lysine) (Figure 12).

Allyl Chloride and Allyl Bromide

As an example of an electrophilic mechanism, consider the toxicity of allyl chloride as expressed by the 24-h LC50 to goldfish (Lipnick et al., 1987). Baseline toxicity for this end point was found to be expressed by the following QSAR, where the LC50 is in units of mmol/L:

$$\log(1/\mathrm{LC50}) = 1.04 \log P - 2.17$$

Based on a log P of 1.450 and a molecular weight of 76.53, the predicted baseline toxicity is 345 mg/L, whereas the observed LC50 was reported to be 10 mg/L, yielding a calculated excess toxicity of 35 (345/10), consistent with a more specific mechanism. In the same study, the LC50 for allyl bromide (log P = 1.590; molecular weight = 120.98) was reported to be <0.8 mg/L, with a predicted baseline LC50 of 390 mg/L, yielding a calculated excess toxicity of greater than 490. The larger excess toxicity of the latter can be readily explained based on its greater electrophilicity, since bromide is a better leaving group than chloride in such a nucleophilic substitution reaction.

Reactive Halides

Hermens and co-workers studied three classes of electrophiles: reactive organic halides (Hermens et al., 1985c), epoxides (Deneer et al., 1988a), and aldehydes (Deneer et al., 1988b) (Table 10). Reactive organic halides, including allyl chloride and benzyl chloride, could be fit to a QSAR involving k, the pseudo-first-order reaction rate velocity for the reaction of the halide with the model nucleophile p-nitrobenzylpyridine (NBP), where LC50 is expressed in units of μmol/L:

$$\log(1/\mathrm{LC50}) = -1.30 \log(1604 + k_{\mathrm{nbp}}^{-1}) + 4.35 \text{ or } 10.4$$
$$(r = 0.938;\ s = 0.44)$$

With the addition of the log K_{ow} term, there was only a slight improvement in the result, suggesting that partition coefficient plays only a small role in governing the toxicity (LC50) of such electrophiles.[7] This reaction with NBP provides a model for the reactivity with the biological nucleophilic target site of action.

Epoxides

In the case of epoxides Deneer et al. (1988a) derived a QSAR equation for 14-d LC50s to guppies, also using log P and pseudo-first-order reaction rate (of the epoxides with NBP) as molecular descriptors. Although neither of the equations using either of these as a single descriptor led to satisfactory correlations, a model employing both log P and log K_{nbp} descriptors yielded a significant correlation:

$$\log(1/\mathrm{LC50}) = 0.39 \log P + 3.0 \log K_{\mathrm{nbp}} - 2.25 \text{ or } + 3.8$$
$$(n = 12;\ r = 0.945;\ s = 0.27)$$

Lipnick et al. (1987) compared LC50s of six epoxides with LC50s calculated with a QSAR for narcosis-type chemicals and observed similar effects.

Aldehydes

For aldehydes (Deneer et al., 1988b), data on 14 compounds for 14-d LC50s to guppies were found to give a good fit to an equation using solely log P:

$$\log(1/\mathrm{LC50}) = 0.36 \log P - 2.54 \text{ or } + 3.5$$
$$(n = 14;\ r = 0.923;\ s = 0.19)$$

The addition of a reactive parameter, which in this case is the second-order rate constant for reaction of the test chemicals with the model nucleophile L-cysteine (at pH 7), yields little improvement:[8]

$$\log(1/\mathrm{LC50}) = 0.36 \log P - 0.08 \log k_{\mathrm{cyst}} - 2.32$$
$$(n = 14;\ r = 0.938;\ s = 0.18)$$

This reaction was chosen because it quantifies differences in the rates of addition of nucleophiles at the C—O bond in aldehydes.

Proelectrophiles

Proelectrophile toxicants have been defined as those requiring metabolic activation to become active elec-

[7]The fact that most partition coefficients fell within a small range for the chemicals studied may have provided an inadequate opportunity to test the degree of influence of log P as a second parameter.

[8]The apparent small influence shown by the second reactivity parameter, k_{cyst}, may reflect the small variation in reactivities of the aldehydes studied.

Table 10. QSAR equations for the toxicity of electrophiles to aquatic organisms[a]

Organism	Effect	Chemical class	Parameters[b]	Reference
Pimephales promelas	96-h LC50	Esters	Log P	Veith et al. (1984–1985)
Pimephales promelas	96-h LC50	Esters	Physicochemical and topological parameters	Basak et al. (1984)
Poecilia reticulata	14-d LC50	Organic halides	Log P; log k_{NBP}	Hermens et al. (1985c)
Poecilia reticulata	14-d LC50	Phenyl substituted derivatives of *O,O*-dimethyl-*O*-phenyl phosphorothioate	π (hydrophobicity); log k_{NBP}; Hammett σ	Hermens et al. (1987)
Photobacterium phosphoreum	30 min EC50	Monosubstituted benzene derivatives	Log P; energy of UV absorption band; molar refractivity; indicator variable, B_3 Sterimol parameter.	Kaiser et al. (1987)
Photobacterium phosphoreum	30 min EC50	1,4-Disubstituted benzene derivatives	Log P; toxicity of monosubstituted compounds	Kaiser (1987b)
Leuciscus idus melanotus	48-h LC50	Esters	V, π^*, β, α, rate of hydrolysis	Kamlet et al. (1987)
Pimephales promelas	96-h LC50	Acrylates and methacrylates	Log P	Reinert (1987)
Carassius auratus	72-h LC50			
Poecilia reticulata	14-d LC50	Epoxides	Log P; log k_{NBP}	Deneer et al. (1988a)
Poecilia reticulata	14-d LC50	Aldehydes	Log P; log k_{cyst}	Deneer et al. (1988b)
Poecilia reticulata	14-d LC50	Organophosphorous	Quantum mechanical	Schüürmann (1990)
Pimephales promelas	96-h LC50	Acrylates	Log P	Reinert (1987)
Poecilia reticulata	14-d LC50	Epoxides	Log P; quantum mechanical	Purdy (1991)
Poecilia reticulata	14-d LC50	Organophosphorus	Log P; topological indices	Vighi et al. (1991)
Poecilia reticulata	14-d LC50	Organophosphorus	Log P; log k_{NBP}; log k_i; $\Sigma\sigma$	de Bruijn and Hermens (1993)

[a]See Tables 7–9 for most of the effect and parameter descriptions; only those not defined in Tables 7–9 will be described here.
[b]log k_{NBP}, reaction rate constant with model nucleophile 4-nitrobenzylpyridine (4NBP); log k_i, biomolecular reaction rate constant for oxon analogues for electric eel acetylcholinesterase inhibition; $\Sigma\sigma$, sum of the Hammett sigma constants for the corresponding phenyl substituents.

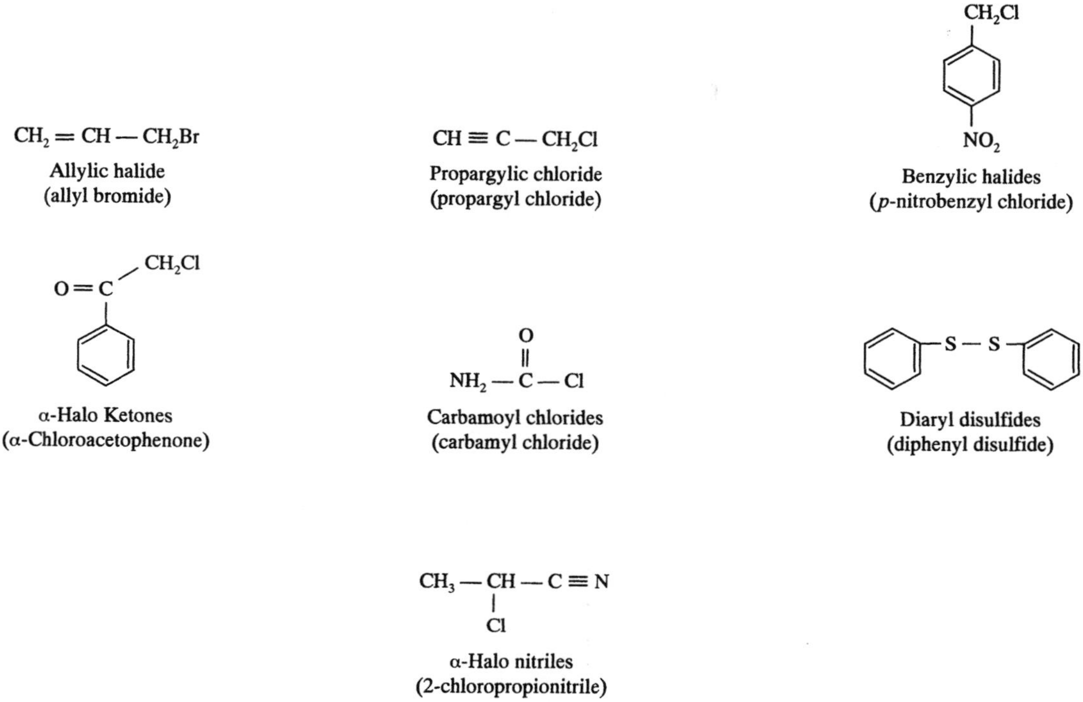

$CH_2=CH-CH_2Br$

Allylic halide
(allyl bromide)

$CH\equiv C-CH_2Cl$

Propargylic chloride
(propargyl chloride)

CH_2Cl

NO_2

Benzylic halides
(*p*-nitrobenzyl chloride)

$O=C$ CH_2Cl

α-Halo Ketones
(α-Chloroacetophenone)

O
||
NH_2-C-Cl

Carbamoyl chlorides
(carbamyl chloride)

S — S

Diaryl disulfides
(diphenyl disulfide)

$CH_3-CH-C\equiv N$
|
Cl

α-Halo nitriles
(2-chloropropionitrile)

Figure 9. Representative electrophile types: nucleophilic substitution. Examples of reactive electrophiles include allyl bromide, propargyl chloride, *p*-nitrobenzyl chloride, α-chloroacetophenone, carbamyl chloride, diphenyl disulfide, and 2-chloropropionitrile.

trophiles (Lipnick et al., 1985). Activation mechanisms of this type have been postulated to account for such excess toxicity in aquatic tests involving the enzymes alcohol dehydrogenase, monooxygenase, and glutathione transferase (Lipnick, 1991a) (Figure 13). For indications of additional possible mechanisms, the reader should consult a text on enzymatic reaction mechanisms (Walsh, 1979) and a review of what is known regarding the biotransformation of organic chemicals by fish (Sijm and Opperhuizen, 1989, see Chapter 17).

Acetylenic Alcohols

A very interesting example in the aquatic toxicology literature is related to the mechanism of toxicity of acetylenic alcohols (Lipnick et al., 1985d; Veith et al., 1989). Acetylenic alcohols whose triple bond is propargylic (attached to the carbon bearing the hydroxyl group) and for which the alcohol is primary or secondary show distinct excess toxicity, with the degree of excess toxicity (and specific physiological response) decreasing with increasing partition coefficient. In contrast, for LC50 data on the propargylic alcohols in which the alcohol is tertiary (and is unable to undergo such metabolic transformation), the observed values are in good agreement with the baseline narcosis predictions. The excess toxicity of the former compounds has been ascribed to their metabolic activation via the enzyme alcohol dehydrogenase to the corresponding α,β-unsaturated aldehydes and ketones, which can act as Michael-type electrophiles (see above under electrophile

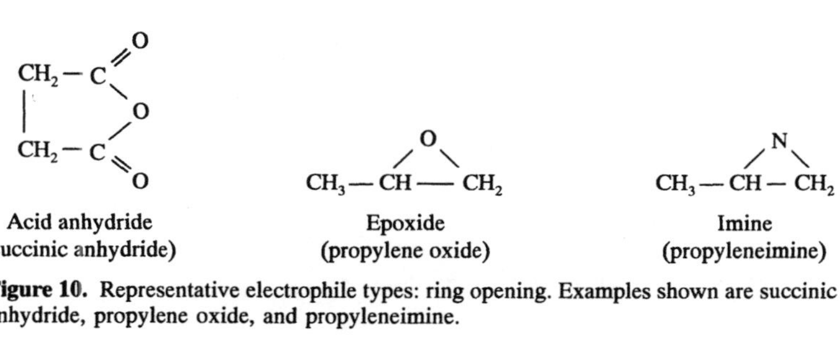

CH_2-C (=O)
| O
CH_2-C (=O)

Acid anhydride
(succinic anhydride)

O
$CH_3-CH-CH_2$

Epoxide
(propylene oxide)

N
$CH_3-CH-CH_2$

Imine
(propyleneimine)

Figure 10. Representative electrophile types: ring opening. Examples shown are succinic anhydride, propylene oxide, and propyleneimine.

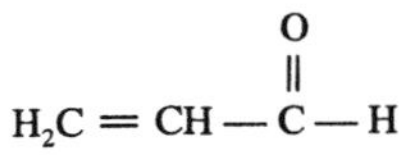

$H_2C=CH-C(=O)-H$

α,β-Unsaturated aldehyde
(acrolein)

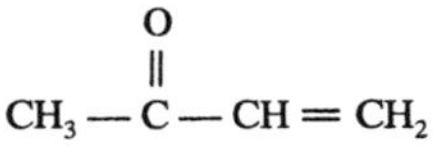

$CH_3-C(=O)-CH=CH_2$

α,β-Unsaturated ketone
(methyl vinyl ketone)

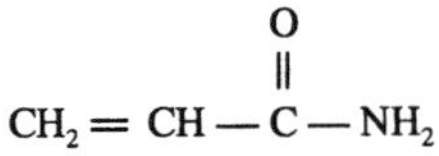

$CH_2=CH-C(=O)-NH_2$

α,β-Unsaturated amides
(acrylamide)

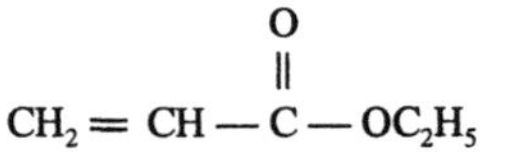

$CH_2=CH-C(=O)-OC_2H_5$

α,β-Unsaturated esters
(ethyl acrylate)

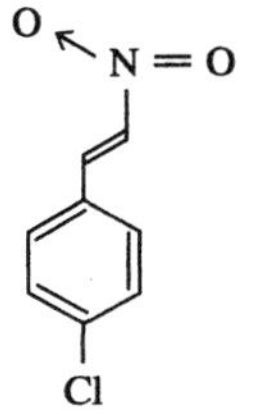

O ← N = O

Cl

α,β-Unsaturated nitro
(*p*-chloro-β-nitrostyrene)

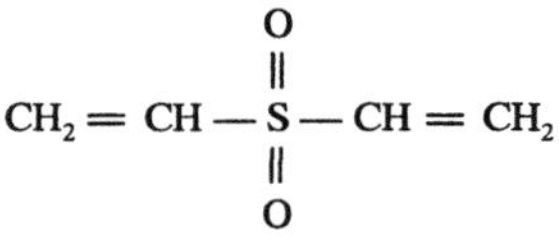

$CH_2=CH-S(=O)(=O)-CH=CH_2$

α,β-Unsaturated sulfone
(divinyl sulfone)

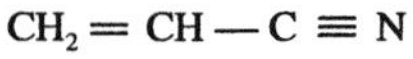

$CH_2=CH-C\equiv N$

α,β-Unsaturated nitrile
(acrylonitrile)

Figure 11. Representative electrophile types: Michael-type (1,3-conjugate addition). Examples shown are acrolein, methyl vinyl ketone, acrylamide, ethyl acrylate, *p*-chloro-β-nitrostyrene, divinyl sulfone, and acrylonitrile.

mechanism). This mechanism has been confirmed by study of the toxicity in which the action of the enzyme alcohol dehydrogenase was stopped by an inhibitor (Bradbury and Christensen, 1991). Mekenyan et al. (1993) have demonstrated that the toxicity of such proelectrophile acetylenic alcohols can be modeled using log *P* and a calculated quantum mechanical term (e.g., acceptor delocalizability) of the reactive metabolite. Dawson et al. (1990) have found that the acetylenic alcohols previously identified as proelectrophiles exhibited an embryo malformation (EmEC50) and lethality (EmLC50) end point ratio in *Xenopus laevis* larvae (see Chapter 7) consistent with a reactive mechanism, as found previously for acid hydrazides. In addition, whereas log *P* showed very poor correlation with these data, it showed excellent interspecies correlations with the 96-h fathead minnow (FM) data,

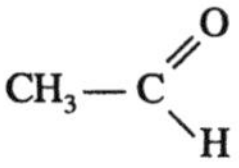

$$\log \text{EmLC50} = 1.045(\log \text{FMLC50}) + 0.374$$

where EmLC50 is the *Xenopus* embryo lethality and FMLC50 is the 96-h LC50 for the fathead minnow.

$CH_3-C(=O)H$

Aliphatic aldehyde
(acetaldehyde)

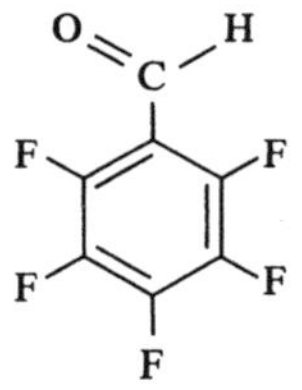

O=C–H, F, F, F, F, F

Aromatic aldehyde
(pentafluorobenzaldehyde)

Figure 12. Representative electrophile types: Schiff base formation. Examples shown are acetaldehyde and pentafluorobenzaldehyde.

Monooxygenase Activation

A second example of proelectrophiles is the oxidative cleavage of pentaerythritol triallyl ether to acro-

Table 11. QSAR studies of the effects of other classes of chemicals to aquatic organisms[a]

Organism	Effect	Chemical class	Parameters[b]	Reference
A. falcatus *S. quadricauda* *A. flos-aquae*	IC50 (reproduction and primary productivity)	Inorganic and organotin derivatives	Log *P*	Wong et al. (1984)
Pimephales promelas	96-h LC50	Esters	Log *P*	Veith et al. (1984–1985)
Rhithropoanoopeus harrissi (mud crab zoeae)	LC50	Organotin	Log *P*, surface area	Laughlin et al. (1985)
Daphnia magna; *Poecilia reticulata*; *Photobacterium phosphoreum*	48-h LC50 96-h LC50 15-min EC50	Dithiocarbamates	Log *P*	Van Leeuwen et al. (1985a)
Daphnia magna	21-d LRCT[c] for population growth	Dithiocarbamates	Log *P*	Van Leeuwen et al. (1985b)
Daphnia magna	LC50	Organotin	Log *P*, pK_a, connectivity	Vighi and Calamari (1985)
Cyprinidon variegatus; *Mysidopsis bahia* (mysid shrimp)	96-h LC50	Pesticides and industrial chemicals	Log *P*, log *S*	Zaroogian et al. (1985a)
Photobacterium phosphoreum	15 min EC50 (bioluminescence)	Thioureas	Log *P*, molecular connectivity indices	Govers, et al. (1986)
Orizias latipes (red killifish)	LC50	Insecticides, fungicides, herbicides, and industrial organic chemicals	Log *P*, connectivity	Yoshioka et al. (1986)
Daphnia pulex	LC50	Various classes	Log *P*, *S*, volume	Passino and Smith (1987)
Pimephales promelas	96-h LC50	Esters, nitriles, ketones, alcohols	Log *P*, σ_1	Purdy (1987)
Salmo gairdneri	96-h LC50	Various	Log *P*, i.p. LD50	Hodson et al. (1988)
Daphnia pulex	LC50	Various classes	Solvatochromic parameters	Passino et al. (1988)
Daphnia *Gammarus pulex*	Not provided	LAS surfactant	Specially calculated log *P*	Roberts (1988)
Pimephales promelas	96-h LC50	Substituted benzenes	Free-Wilson structure fragment substituents	Hall et al. (1989a)
Pimephales promelas	96-h LC50	Substituted benzenes	Connectivity indices	Hall et al. (1989b)
Xenopus laevis larvae (FETAX)	Embryotoxicity and lethality	Acetylenic alcohols	96-h LC50 fathead minnow	Dawson et al. (1990)
Mysidopsis bahia *Tetrahymena elliotti*	48-h LC50 motility MIC	Polyoxyethylene surfactants	Log *P* of parent compound and number of nonterminal EO bonds	Schüürmann (1990)
Pimephales promelas	96-h LC50	Various classes	Log *P*	Nendza and Russom (1991)
Carassius auratus	Not provided	Not provided	Specially calculated log *P*	Roberts (1991)
Japanese killifish	LC50	Nonylphenol ethoxylates; ethoxylated primary alcohols	Specially calculated log *P*; critical micelle concentration	Roberts (1991)
Photobacterium phosphoreum	30 min EC50	Benzene derivatives	Log *P*, *S*, connectivity	Romanowska (1992)
Pimephales promelas	96-h LC50	α,β-Unsaturated alcohols (proelectrophile)	Log *P*, S_1^N, f_3^{LUMO}	Mekenyan et al. (1993)

[a]See Tables 7–10 for most of the effect and parameter descriptions; only those not defined in Tables 7–10 will be described here.
[b]i.p. LD50, intraperitoneal LD50; S_1^N, acceptor superdelocalizability; f_3^{LUMO}, frontier charges for LUMO orbitals; EO, ethoxylates (e.g., polyoxyethylene surfactants).
[c]LRCT, lowest rejected concentration tested.

$HC \equiv C - CH_2OH$

Primary or Secondary
allyl or proparylic alcohols
(propargyl alcohol)
[Alcohol dehydrogenase activation]

$H - C \equiv C - CH_2CH_2 - OH$

Primary or Secondary
homopropargylic alcohols
(3-butyn-1-ol)
[Alcohol dehydrogenase activation]

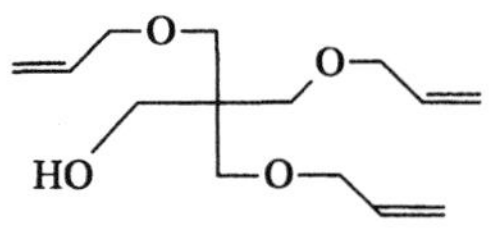

Monooxygenase activation
(Pentaerythritol triallyl ether)

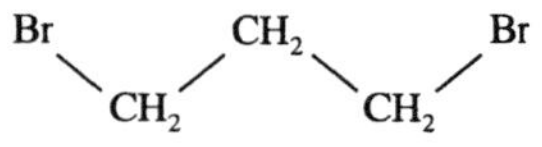

Glutathione transferase activation
(1,3-dibromopropane)

Figure 13. Representative proelectrophile types. Examples shown are propargyl alcohol (alcohol dehydrogenase activation), 3-butyn-1-ol (alcohol dehydrogenase activation), pentaerythritol triallyl ether (monooxygenase activation), and 1,3-dibromopropane (glutathione transferase activation).

lein by means of monooxygenase enzyme transformation. The acrolein product can also form covalent bonds as a Michael-type acceptor.

A final example in this category is the activation of 1,3-dibromopropane via the enzyme glutathione transferase (see Chapter 17) to a reactive four-member ring sulfonium electrophile intermediate. This substance was found to be 87 times more toxic to fish than predicted by baseline toxicity (Lipnick, 1988).

The LC50s of electrophile and proelectrophile chemicals such as reactive alkyl halides, epoxides, aldehydes, and unsaturated alcohols are lower than LC50s of unreactive chemicals. It is also evident that significant QSAR correlations for these reactive chemicals of a general nature require the use of rate constants as additional molecular descriptors. QSAR equations for aldehydes and epoxides show that ratios between observed LC50s and calculated LC50s (from a QSAR for narcosis-type chemicals) decrease with increasing values of K_{ow}. There thus appears to be a shift in mechanism of action toward narcosis for the more hydrophobic chemicals, consistent with greater likelihood of achieving a critical internal narcotic concentration.

SUMMARY OF QSAR MODELS FOR ECOTOXICITY

The summary of QSAR studies in ecotoxicology indicates that descriptors related to both measured and calculated physicochemical properties have been used extensively. A number of QSAR studies have employed topological or molecular connectivity indices as descriptors in models involving chemical, biological, and environmental properties of molecules (Sabljic, 1991). These parameters express quantitatively information about the interconnectedness of molecules and are considered to provide information on molecular shape and size. Although topological indices can easily be calculated from structure, they provide no information on mechanism or kinetics.

Descriptors used in correlations between biological activity and physicochemical properties will be important if they describe differences in rates of significant processes that are rate-limiting in the development of an effect, including uptake, elimination, biotransformation (i.e., metabolism—activation or deactivation), covalent binding, and receptor binding.

The dominant descriptors in the QSAR model provide information on these important rate-limiting steps. The QSAR equations for nonreactive nonelectrolytes employ K_{ow} to model the influence of hydrophobicity on rate and equilibrium of uptake and resulting molar concentration or molar volume at the site of action of this group of chemicals. This is also the case for phenols. However, both pK_a and log K_{ow} are needed in the QSAR equations to account for the effect of ionization on uptake, equilibrium and binding at the site of action. For epoxides and other electrophiles, potency is governed by the effect of log P on rate of uptake and concentration at the site of action, as well as ability to undergo covalent binding (as modeled by reaction rate with a model nucleophile or via theoretical molecular descriptors). The selection and use of molecular descriptors for modeling the aquatic toxicity of complex reactive chemicals, as well as chemicals such as certain pesticides that act by receptor binding, need to be made on a case-by-case basis through careful consideration of the mechanistic processes leading to the toxic event.

QSAR models with phenols have shown that certain descriptors can be removed and retain satisfactory correlations, particularly when weakly acidic phenols having similar pK_a values are considered. Two different QSAR equations have been used with phenols and represent two possibly different mechanisms for their activity and consequent effects (i.e., polar narcosis and uncoupling of oxidative phosphorylation). Because the acidity (and corresponding degree of ionization) of phenols can vary greatly, it is not surprising that this category needs further subdivision with respect to mechanism of action.

QSAR studies for aldehydes and epoxides showed that LC50s for the more hydrophobic chemicals within these classes are predicted by a QSAR derived from data for chemicals acting by a narcosis mechanism.

There are several problems with QSARs. First, the structural requirements in a QSAR are generally not well understood, making it difficult to define the limits of predictability of the QSAR for untested chemicals. The only case in which structural requirements are well defined (provided something is known about reactivity or the absence of more specific mechanisms) is that of nonreactive, nonelectrolyte organic substances. Second, the mechanism of action of chemicals in most aquatic organisms is not known. Normally, it is recommended that QSARs be derived only for groups of chemicals for which there is evidence of a common molecular mechanism.

QSARs have obvious utility in prediction, chemical priority setting, and analyzing data but it is also important to understand their limitations.

LITERATURE CITED

Abernathy, S.; Bobra, A. M.; Shiu, W. Y.; Wells, P. G.; and Mackay, D.: Acute lethal toxicity of hydrocarbons and chlorinated hydrocarbons to two planktonic crustaceans: the key role of organism-water partitioning. *Aquat. Toxicol.*, 8:163–174, 1986.

Abernathy, S. G.; Mackay, D.; and McCarty, L. S.: "Volume fraction" correlation for narcosis in aquatic organisms: the key role of partitioning. *Environ. Toxicol. Chem.*, 7:469–481, 1988.

Albert, A.: Selective Toxicity: With Special Reference to Chemotherapy. London: Methuen, 1951.

Albert A.: Selective Toxicity, 7th ed. Chapman & Hall, London, 1985.

Andrews, P. R., and Tintelnot, M.: Intermolecular forces and molecular binding. Hansch, C. (editorial board chairman): Comprehensive Medicinal Chemistry, Vol. 4. Pergamon Press, Oxford, 1990, pp. 321–347.

Applegate, V. C.; Howell, J. H.; Hall, Jr., A. E.; and Smith, M. A.: Toxicity of 4,346 chemicals to larval lampreys and fishes. Special Scientific Report, Fisheries No. 207. Washington, DC: U.S. Fish and Wildlife Service, Department of the Interior, 1957.

Auer, C. M., and Gould, D. H.: Carcinogenicity assessment and the role of structure-activity relationships (SAR) analysis under TSCA Section 5. *Environ. Carcinog. Rev.*, C5(1):29–71, 1987.

Auer, C. M.; Nabholz, J. V.; and Baetke, K. P.: Mode of action and assessment of chemical hazards in the presence of limited data: use of structure-activity relationships (SAR) under TSCA Section 5. *Environ. Health Perspect.*, 87:183–197, 1990.

Babeu, L., and Vaishnav, D. D.: Prediction of biodegradability of selected organic chemicals. *J. Ind. Microbiol.*, 2:107–115, 1987.

Banerjee, S.; Yalkowsky, S. H.; and Valvani, S. S.: Water solubility and octanol/water partition coefficients of organics. Limitations of the solubility-partition coefficient correlations. *Environ. Sci. Technol.*, 14:1227–1229, 1980.

Banerjee, S.; Howard, P. H.; Rosenberg, A. M.; Dombrowski, A. E.; Sikka, H.; and Tullis, D. L.: Development of a general kinetic model for biodegradation and its application to chlorophenols and related compounds. *Environ. Sci. Technol.*, 18:416–422, 1984.

Barber, M. C.; Suárez, L. A.; and Lassiter, R. R.: Modelling bioaccumulation of organic pollutants in fish with an application to PCBs in Lake Ontario salmonids. *Can. J. Fish Aquatic Sci.*, 48:318–337, 1991.

Barnthouse, L. W.; Suter, G. W. II; and Bartell, S. M.: Quantifying risks of toxic chemicals to aquatic populations and ecosystems. *Chemosphere*, 17:1487–1492, 1988.

Barron, M. G.: Bioconcentration. *Environ. Sci. Technol.*, 24:1612–1618, 1990.

Basak, S. C.; Gieschen, D. P.; and Magnuson, V. R.: A quantitative correlation of the LC50 values of esters in *Pimephales promelas* using physicochemical and topological parameters. *Environ. Toxicol. Chem.*, 3:191–199, 1984.

Basak, S. C.; Magnuson, V. R., Niemi, G. J.; and Regal, R. R.: Determining structural similarity of chemicals using graph-theoretic methods. *Discrete Appl. Math.*, 19: 17–44, 1988.

Baum, F.: Zur theorie der Alcoholnarkose. Zweite Mittheilung. Ein physikalisch-chemischer Beitrag zur Theorie der Narcotica. *Naunyn-Schmiedebergs Arch. Exp. Pathol. Pharmakol.*, 42:119–137, 1899.

Baumann, E., and Kast, A.: Ueber die Beziehungen zwischen chemischer Constitution und physiologischer Wirkung bei einigen Sulfonen. *Z. Physiol. Chem.*, 14: 52–74, 1890.

Bengtsson, B. E.; Renberg, L.; and Tarkpea, M.: Molecular structure and aquatic toxicity—an example with C1–C13 aliphatic alcohols. *Chemosphere*, 13:613–622, 1984.

Benoit-Guyod, J. L.; André, C.; Taillandier, G.; Rochat, J.; and Boucherie, A.: Toxicity and QSAR of chlorophenols on *Lebistes reticulatus*. *Ecotoxicol. Environ. Safety*, 8: 227–235, 1984.

Binstein, S.; Devillers, J.; and Karcher, W.: Nonlinear dependence of fish bioconcentration on *n*-octanol/water partition coefficient. *SAR QSAR Environ. Res.*, 1:29–39, 1993.

Blaney, J. M., and Hansch, C.: Application of molecular graphics to the analysis of macromolecular structures. Hansch, C. (editorial board chairman): Comprehensive Medicinal Chemistry, Vol. 4. Pergamon Press, Oxford, 1990, pp. 459–496.

Bobra, A. M.; Shiu, W. Y.; and Mackay, D.: A predictive correlation for the acute toxicity of hydrocarbons and chlorinated hydrocarbons to the water flea. *Chemosphere*, 12:1121–1129, 1983.

Bobra, A. M.; Shiu, W. Y.; and Mackay, D.: Structure-activity relationships for toxicity of hydrocarbons, chlorinated hydrocarbons and oils to *Daphnia magna*. Kaiser, K. L. E. (ed.): QSAR in Environmental Toxicology. Reidel, Dordrecht, 1984, pp. 3–16.

Bobra, A.; Shiu, W. Y.; and Mackay, D.: Quantitative structure-activity relationships for the acute toxicity of chlorobenzenes to *Daphnia magna*. *Environ. Toxicol. Chem.*, 4:297–305, 1985.

Bodor, N., and Huang, M.-J.: A new method for the estimation of the aqueous solubility of organic compounds. *J. Pharm. Sci.*, 81:954–960, 1992.

Bodor, N.; Harget, A.; and Huang, M.-J.: Neural network studies. 1. Estimation of the aqueous solubility of organic compounds. *J. Am. Chem. Soc.*, 113:9480–9483, 1991.

Boethling, R. S.: Application of molecular topology to quantitative structure-biodegradability relationships. *Environ. Toxicol. Chem.*, 5:797–806, 1986.

Borgstedt, H. H., and Enslein, K.: The development of practical discriminant models in toxicology. Karcher, W., and Devillers, J. (eds.): Practical Applications of Quantitative Structure-Activity Relationships (QSAR) in Environmental Chemistry and Toxicology. Kluwer Publishing, Dordrecht, 1990, pp. 145–152.

Borman, S.: New QSAR techniques eyed for environmental assessments. *Chem. Eng. News*, 68:20–23, 1990.

Bowden, K.: Electronic effects in drugs. Hansch, C. (editorial board chairman): Comprehensive Medicinal Chemistry, Vol. 4. Pergamon Press, Oxford, 1990, pp. 205–239.

Bradbury, S. P., and Christensen, G. M.: Inhibition of alcohol dehydrogenase activity by acetylenic and allylic alcohols: concordance with in vivo electrophile reactivity in fish. *Environ. Toxicol. Chem.*, 10:1155–1160, 1991.

Bradbury, S. P.; Henry, T. R.; Niemi, G. J.; Carlson, R. W.; and Snarski, V. M.: Use of respiratory-cardiovascular responses of rainbow trout (*Salmo gairdneri*) in identifying acute toxicity syndromes in fish: Part 3. Polar narcotics. *Environ. Toxicol. Chem.*, 8:247–261, 1989.

Bradbury, S. P.; Henry, T. R.; and Carlson, R. W.: Fish acute toxicity syndromes in the development of mechanism-specific QSARs. Karcher, W., and Devillers, J. (eds.): Practical Applications of Quantitative Structure-Activity Relationships (QSAR) in Environmental Chemistry and Toxicology. Kluwer Academic Publishing, Dordrecht, 1990, pp. 295–315.

Bradbury, S. P.; Carlson, R. W.; Niemi, G. J.; and Henry, T. R.: Use of respiratory-cardiovascular responses of rainbow trout (*Salmo gairdneri*) in identifying acute toxicity syndromes in fish: Part 4: Central nervous system seizure agents. *Environ. Toxicol. Chem.*, 10:115–131, 1991.

Broderius, S. J., and Kahl, M.: Acute toxicity of organic chemical mixtures to the fathead minnow. *Aquat. Toxicol.*, 6:307–322, 1985.

Brooke, D. N.; Dobbs, A. J.; and Williams, N.: Octanol: water partition coefficients (*P*): measurement, estimation, and interpretation, particularly for chemicals with $P > 10^5$. *Ecotoxicol. Environ. Safety*, 11:251–260, 1986.

Brooke, D.; Nielsen, I.; De Bruijn, J.; and Hermens, J.: An interlaboratory evaluation of the stir-flask method for the determination of octanol/water partition coefficients (log K_{ow}). *Chemosphere*, 21:119–133, 1990.

Bucholz, F. A.: Beiträge zur Theorie der Alkoholwirkung. Inaugural dissertation, University of Marburg, 1895.

Cajina-Quezada, M.: Structure-activity relationships for chemicals causing respiratory uncoupling. Turner, J. E.; England, M. W.; Schultz, T. W.; and Kvaak, N. J. (eds.): *QSAR88*. NTIS CONF-880520 (DE88013180), U.S. Department of Energy, Oak Ridge, TN, 1988, pp. 123–130.

Calamari, D., and Vighi, M.: Experiences on QSARs and evaluative models in ecotoxicology. *Chemosphere*, 17: 1539–1549, 1988.

Calamari, D.; Galassi, S.; Setti, F.; and Vighi, M.: Toxicity of selected chlorobenzenes to aquatic organisms. *Chemosphere*, 12:253–262, 1983.

Call, D. J.; Brooke, L. T.; Knuth, M. L.; Poirer, S. H.; and Hoglund, M. D.: Fish subchronic toxicity prediction model for industrial organic chemicals that produce narcosis. *Environ. Toxicol. Chem.*, 4:335–341, 1985.

Cambon, B., and Devillers, J.: New trends in structure-biodegradability relationships. *Quant. Struct.-Act. Relat.*, 12:49–56, 1993.

Charton, M.: Some methods of obtaining quantitative structure-activity relationships for quantities of environmental interest. *Environ. Health Perspect.*, 61:229–238, 1985.

Chessels, M.; Hawker, D. W.; and Connell, D. W.: Influence of solubility in lipid on bioconcentration of hydrophobic compounds. *Ecotoxicol. Environ. Safety*, 23: 260–273, 1992.

Chiou, C. T.: Partition coefficients of organic compounds in lipid-water systems and correlations with fish bioconcentration factors. *Environ. Sci. Technol.*, 19:57–62, 1985.

Chou, J. T., and Jurs, P. C.: Computer-assisted computation of partition coefficients from molecular structures using fragment constants. *J. Chem. Inf. Comput. Sci.*, 19:172–178, 1979.

Clark, K. E., and Mackay, D.: Dietary uptake and biomagnification of four chlorinated hydrocarbons by guppies. *Environ. Toxicol. Chem.*, 10:1205–1217, 1991.

Clark, K. E.; Gobas, F. A. P. C.; and Mackay, D.: Model of organic chemical uptake and clearance by fish from food and water. *Environ. Sci. Technol.*, 24:1203–1213, 1990.

Clements, R. G. (ed.); Johnson, D. W.; Lipnick, R. L.; Nabholz, J. V.; and Newsome, L. D.: Estimating toxicity of industrial chemicals to aquatic organisms using structure activity relationships. EPA-560-6-88-001. Washington, DC: Environmental Protection Agency, 1988.

Cololian, P.: La toxicité des alcools chez les poissons. *J. Physiol. Pathol. Gen. (Paris)*, 3:535–546, 1901.

Connell, D. W.: Bioaccumulation potential of persistent organic chemicals with aquatic organisms. *Rev. Environ. Contam. Toxicol.*, 101:117–154, 1988.

Connell, D. W.: Bioaccumulation of Xenobiotic Compounds. Boca Raton, Florida: CRC, 1990.

Connell, D. W., and Hawker, D. W.: Use of polynomial expressions to describe the bioconcentration of hydrophobic chemicals by fish. *Ecotoxicol. Environ. Safety*, 16:242–257, 1988.

Connell, D., and Markwell, R.: Mechanism and prediction of nonspecific toxicity to fish using bioconcentration characteristics. *Ecotoxicol. Environ. Safety*, 24:247–265, 1992.

Corbett, J. R.: The Biochemical Mode of Action of Pesticides. Academic Press, London, 1974, pp. 30–48.

Craig, P. N.: Substructural analysis and compound selection. Hansch, C. (editorial board chairman): Comprehensive Medicinal Chemistry, Vol. 4. Pergamon Press, Oxford, 1990, pp. 645–666.

Cramer, R. D. III: BC(DEF) parameters. 1. The intrinsic dimensionality of intermolecular interactions in the liquid state. *J. Am. Chem. Soc.*, 102:1837–1849; Errata *J. Am. Chem. Soc.*, 103:2143–2143, 1981.

Crisp, D. J.; Christie, A. O.; and Ghobashy, A. F. A.: Narcotic and toxic action of organic compounds on barnacle larvae. *Comp. Biochem. Physiol.*, 22:629–649, 1967.

Cros, A. F. A.: Action de l'alcool amylique sur l'organisme. Thesis, Faculté de Médecine de Strasbourg, 1863.

Dannenfelser, R. M., and Yalkowsky, S. H.: Database of aqueous solubility for organic nonelectrolytes. *Sci. Total Environ.*, 109/110:625–628, 1991.

Dawson, D. A.; Schultz, T. W.; Baker, L. L.; and Wilke, T. S.: Comparative developmental toxicity of acetylenic alcohols on embryos and larvae of *Xenopus laevis*. Landis, W. G., and van der Schalie, W. J. (eds.): Aquatic Toxicology and Risk Assessment: Thirteenth Volume, ASTM STP 1096, American Society for Testing and Materials, Philadelphia, 1990, pp. 267–277.

Dearden, J. C.: Molecular structure and drug transport. Hansch, C. (editorial board chairman): Comprehensive

Medicinal Chemistry, Vol. 4. Pergamon Press, Oxford, 1990b, pp. 375–411.

Dearden, J. C.: Physico-chemical descriptors. Karcher, W., and Devillers, J. (eds.): Practical Applications of Quantitative Structure-Activity Relationships (QSAR) in Environmental Chemistry and Toxicology. Kluwer Academic Publishing, Dordrecht, 1990a, pp. 25–59.

Dearden, J. C.: The QSAR prediction of melting point, a property of environmental relevance. *Sci. Total Environ.*, 109/110:59–68, 1991.

Dearden, J. C., and Nicholson, R. M.: Correlation of biodegradability with atomic charge difference and superdelocalizability. Kaiser, K. L. E. (ed.): QSAR in Environmental Toxicology-II. Reidel, Dordrecht, 1987, pp. 83–89.

Dearden, J. C., and Nicholson, R. M.: The prediction of biodegradability by the use of quantitative structure-activity relationships: correlation of biological oxygen demand with atomic charge difference. *Pestic. Sci.*, 17: 305–310, 1986.

De Bruijn, J., and Hermens, J.: Relationships between octanol/water partition coefficients and total molecular surface area and total molecular volume of hydrophobic organic chemicals. *Quant. Struct.-Act. Relat.*, 9:11–21, 1990.

De Bruijn, J., and Hermens, J.: Qualitative and quantitative modelling of toxic effects of organophosphorous compounds to fish. *Sci. Total Environ.*, 109/110:441–455, 1991.

De Bruijn, J., and Hermens, J.: Inhibition of acetylcholinesterase and acute toxicity of organophosphorous compounds to fish: a preliminary structure-activity analysis. *Aquat. Toxicol.*, 24:257–274, 1993.

De Bruijn, J.; Busser, F.; Seinen, W.; and Hermens, J.: Determination of octanol/water partition coefficients for hydrophobic organic chemicals with the "slow-stirring method." *Environ. Toxicol. Chem.*, 8:499–512, 1989.

De Bruijn, J.; Yedema, E.; Seinen, W.; and Hermens, J.: Lethal body burdens of four organophosphorous pesticides in the guppy (*Poecilia reticulata*). *Aquat. Toxicol.*, 20:111–122, 1991.

Degner, P.; Nendza, M.; and Klein, W.: Predictive QSAR models for estimating biodegradation of aromatic compounds. *Sci. Total Environ.*, 109/110:253–259, 1991.

Del, C.; Figueroa, I.; and Simmons, M. S.: Structure-activity relationships of chlorobenzenes using DNA measurement as a toxicity parameter in algae. *Environ. Toxicol. Chem.*, 10:323–329, 1991.

Deneer, J. W.; Sinnige, T. L.; Seinen, W.; and Hermens, J. L. M.: Quantitative structure-activity relationships for the toxicity and bioconcentration factor of nitrobenzene derivatives towards the guppy (*Poecilia reticulata*). *Aquat. Toxicol.*, 10:115–129, 1987.

Deneer, J. W.; Sinnige, T. L.; Seinen, W.; and Hermens, J. L. M.: A quantitative structure-activity relationship for the acute toxicity of some epoxy compounds to the guppy. *Aquat. Toxicol.*, 13:195–204, 1988a.

Deneer, J. W.; Seinen, W.; and Hermens, J. L. M.: The acute toxicity of aldehydes to the guppy. *Aquat. Toxicol.*, 12:185–192, 1988b.

Deneer, J. W.; Sinnige, T. L.; Seinen, W.; and Hermens, J. L. M.: The joint toxicity to *Daphnia magna* of industrial organic chemicals in low concentrations. *Aquat. Toxicol.*, 12:33–38, 1988c.

Desai, S. M.; Govind, R.; and Tabak, H. H.: Development of quantitative structure-activity relationships for predicting biodegradation kinetics. *Environ. Toxicol. Chem.*, 9:473–477, 1990.

deStevens, G.: Lead structure, discovery and development. Hansch, C. (editorial board chairman): Comprehensive Medicinal Chemistry, Vol. 1. Pergamon Press, Oxford, 1990, pp. 261–278.

Devillers, J.: Statistical analyses in drug design and environmental chemistry: basic concepts. Coccini, T.; Giannoni, L.; Karcher, W.; Manzo, L.; and Roi, R. (eds.): Quantitative Structure-Activity Relationships (QSAR) in Toxicology. Joint Research Center, Commission of the European Communities, Luxembourg, 1992, pp. 27–41.

Devillers, J., and Chambon, P.: Acute toxicity of chlorophenols to *Daphnia magna* and *Brachydanio rerio*. *J. Fr. Hydrol.*, 17:111–119, 1986.

Devillers, J., and Lipnick, R. L: Practical applications of regression analysis in environmental QSAR studies. Karcher, W., and Devillers, J. (eds.): Practical Applications of Quantitative Structure-Activity Relationships (QSAR) in Environmental Chemistry and Toxicology. Kluwer Publishing, Dordrecht, 1990, pp. 377–414.

De Wolf, W.; Opperhuizen, A.; Seinen, W.; and Hermens, J. L. M.: Influence of survival time on the lethal body burden of 2,3,4,5-tetrachloroaniline in the guppy, *Poecilia reticulata*. *Sci. Total Environ.*, 109/110:457–459, 1991.

Diehl, D.: Vergleichende Experimentaluntersuchungen über die Stärke der Narkotischen Wirkung eininger Sulfone, Säureamide und Glycerinderivate. Inaugural dissertation, University of Marburg, 1894.

Donkin, P.; Widdows, J.; Evans, S. V.; Worrall, C. M.; and Carr, M.: Quantitative structure-activity relationships for the effect of hydrophobic organic chemicals on rate of feeding by mussels (*Mytilus edulis*). *Aquat. Toxicol.*, 14: 277–294, 1989.

Donkin, P.; Widdows, J.; Evans, S. V.; and Brinsley, M. D.: QSARs for the sublethal responses of marine mussels (*Mytilus edulis*). *Sci. Total Environ.*, 109/110:461–476, 1991.

Donkin, P.: Quantitative structure-activity relationships. Calow, P. (ed.): Handbook of Ecotoxicology. Blackwell Scientific, London, 1994, pp. 321–347.

Draper, N., and Smith, H.: Applied Regression Analysis, 2nd ed., Wiley, New York, 1981.

Drummond, R. A., and Russom, C. L.: Behavioral toxicity syndromes: a promising tool for assessing toxicity mechanisms in juvenile fathead minnows. *Environ. Toxicol. Chem.*, 9:37–46, 1990.

Dubois, J. E., and Loukianoff, M.: DARC "logic method" for molar volume prediction. *SAR QSAR Environ. Res.*, 1:63–75, 1993.

Dunn, W. J. III, and Wold, S.: Pattern recognition techniques in drug design. Hansch, C. (editorial board chairman): Comprehensive Medicinal Chemistry, Vol. 4. Pergamon Press, Oxford, 1990, pp. 691–714.

Dunn, W. J. III; Block, J. H.; and Pearlman, R. S. (eds.): Partition Coefficient Determination and Estimation. Pergamon Press, New York, 1986.

Dunzelt, W.: Vergleichende Experimentaluntersuchungen über die Stärke der Wirkung einiger Narcotica. Inaugural dissertation, University of Marburg, 1896.

Durkin, P. R.: Biological impact of various chlorinated phenolics and related compounds on *Daphnia magna*. TAPPI Environ. Conf. Proc., pp. 165–169, 1978.

Eastmond, D. A.; Booth, G. M.; and Lee, M. L.: Toxicity, accumulation and elimination of polycyclic aromatic sulfur heterocycles in *Daphnia magna*. *Arch. Environ. Contam. Toxicol.*, 13:105–111, 1984.

ECETOC: Structure-activity relationships in toxicology and ecotoxicology: an assessment. European Chemical Industry Ecology and Toxicology Center, Brussels, Monograph No. 8, February 24, 1986.

Enslein, K.; Tomb, M. E.; and Lander, T. R.: Structure-activity models of biological oxygen demand. Kaiser, K. L. E. (ed.): QSAR in Environmental Toxicology. Reidel, Dordrecht, 1984, pp. 89–109.

Erickson, R. J., and Mckim, J. M.: A model for exchange of organic chemicals at fish gills: flow and diffusion limitations. *Aquat. Toxicol.*, 18:175–198, 1990.

Eriksson, L.; Jonsson, J.; Hellberg, S.; Lindgren, F.; Sjöström, M.; Wold, S.; Sandström, B. E.; and Svensson, I.: A strategy for ranking environmentally occurring chemicals. Part V: The development of two genotoxicity QSARs for halogenated organics. *Environ. Toxicol. Chem.*, 10:585–596, 1991.

Esser, H. O., and Moser, P.: An appraisal of problems related to the measurement and evaluation of bioaccumulation. *Ecotoxicol. Environ. Safety*, 6:131–148, 1982.

Exner, O.: Correlation Analysis of Chemical Data. Plenum Press, New York, 1988.

Ferguson, J.: The use of chemical potentials as indices of toxicity. *Proc. R. Soc. Lond. Ser. B*, 127:387–404, 1939.

Foster, G. D., and Tullis, R. E.: A quantitative structure-activity relationship between partition coefficients and the acute toxicity of napthalene derivatives in *Artemia salina nauplii. Aquat. Toxicol.*, 5:245–254, 1984.

Franke, R.: Theoretical Drug Design Methods. Elsevier: Amsterdam, 1984.

Fränkel, S.: Die Arzneimittel-Synthese: Auf Grundlage der Beziehungen zwischen Chemischen Aufbau und Wirkung, 4th ed. Springer, Berlin, 1919, pp. 26–135.

Franks, N. P., and Lieb, W. R.: What is the molecular nature of the general anesthetic target sites? *Trends Pharmacol. Sci.*, 8:169–174, 1987.

Franks, N. P., and Lieb, W. R.: Mechanisms of general anesthesia. *Environ. Health Perspect.*, 87:199–205, 1990.

Free, S. M., and Wilson, J. W.: A mathematical contribution to structure-activity studies. *J. Med. Chem.*, 1:395–399, 1964.

Freitag, D.; Geyer, H.; Kraus, A.; Viswanathan, R.; Kotzias, D.; Attar, A.; Klein, W.; and Korte, F.: Ecotoxicological profile analysis VII. Screening chemicals for their environmental behavior by comparative evaluation. *Ecotoxicol. Environ. Safety*, 6:60–81, 1982.

Fujita, T.: The extrathermodynamic approach to drug design. Hansch, C. (editorial board chairman): Comprehensive Medicinal Chemistry, Vol. 4. Pergamon Press, Oxford, 1990, pp. 497–560.

Fujita, T.; Iwasa, J.; and Hansch, C.: A new substituent constant, π, derived from partition coefficients. *J. Am. Chem. Soc.*, 86:5175–5180, 1964.

Gao, C.; Govind, R.; and Tabak, H. H.: Application of the group contribution method for predicting the toxicity of organic chemicals. *Environ. Toxicol. Chem.*, 11:631–636, 1992.

Gerhardt, C.: Précis de Chimie Organique, Chez Fortin, Masson, Paris, 1855–1845.

Geyer, H.; Sheehan, D.; Kotzias, D.; Freitag, D.; and Korte, F.: Prediction of ecotoxicological behaviour of chemicals; relationship between physicochemical properties and bioaccumulation of organic chemicals in the mussel. *Chemosphere*, 11:1121–1134, 1982.

Ghose, A. K., and Crippen, G. M.: Atomic physicochemical properties for three-dimensional structure–directed quantitative structure-activity relations. I. Partition coefficients as a measure of hydrophobicity. *J. Comput. Chem.*, 7:565–577, 1986.

Gombar, V. K.: Quantitative structure-activity relationship studies: acute toxicity of environmental contaminants. Kaiser, K. L. E. (ed.): QSAR in Environmental Toxicology—II. Reidel, Dordrecht, 1987, pp. 125–133.

Govers, H.; Ruepert, C.; and Aiking, H.: Quantitative structure-activity relationships for polycyclic aromatic hydrocarbons: correlation between molecular connectivity, physical-chemical properties, bioconcentration, and toxicity in *Daphnia pulex. Chemosphere*, 13:227–236, 1984.

Govers, H.; Ruepert, C.; Stevens, T.; and van Leeuwen, C. J.: Experimental determination and prediction of partition coefficients of thioureas and their toxicity to *Photobacterium phosphoreum. Chemosphere*, 15:383–393, 1986.

Hall, L. H., and Kier, L. B.: Molecular connectivity of phenols and their toxicity to fish. *Bull. Environ. Contam. Toxicol.*, 32:354–362, 1984.

Hall, L. H., and Kier, L. B.: Structure-activity relationship studies on the toxicities of benzene derivatives: II. An analysis of benzene substituent effects on toxicity. *Environ. Toxicol. Chem.*, 5:333–337, 1986.

Hall, L. H.; Kier, L. B.; and Phipps, G.: Structure-activity relationship studies on the toxicities of benzene derivatives: I. An additivity model. *Environ. Toxicol. Chem.*, 3: 355–365, 1984.

Hall, L. H.; Maynard, E. L.; and Kier, L. B.: Structure-activity relationship studies on the toxicity of benzene derivatives. III. Prediction and extension to new substituents. *Environ. Toxicol. Chem.*, 8:431–436, 1989a.

Hall, L. H.; Maynard, E. L.; and Kier, L. B.: QSAR investigation of benzene toxicity to fathead minnow using molecular connectivity. *Environ. Toxicol. Chem.*, 8:783–788, 1989b.

Hansch, C.: On the predictive value of QSAR. Keverling Buisman, J. A. (ed.): Biological Activity and Chemical Structure. Elsevier, Amsterdam, 1977, pp. 47–61.

Hansch, C.: Comparative structure-activity relationships. Fauchère, J. L. (ed.): Quantitative Structure-Activity Relationships in Drug Design. Alan R. Liss, New York, 1989, pp. 23–30.

Hansch, C., and Dunn, W. J. III: Linear relationships between lipophilic character and biological activity of drugs. *J. Pharm. Sci.*, 61:1–19, 1972.

Hansch, C., and Fujita, T.: RHO$\sigma\pi$ analysis. A method for the correlation of biological activity and chemical structure. *J. Am. Chem. Soc.*, 86:1616–1626, 1964.

Hansch, C., and Leo, A. J.: Substituent Constants for Correlation Analysis in Chemistry and Biology. Wiley, New York, 1979.

Hansch, C.; Maloney, P. P.; Fujita, T.; and Muir, R. M.: Correlation of biological activity of phenoxyacetic acids with Hammett substituent constants and partition coefficients. *Nature*, 194:178–180, 1962.

Hansch, C.; Quinlan, J. E.; and Lawrence, G. L.: The linear free-energy relationship between partition coefficients and the aqueous solubility of organic liquids. *J. Org. Chem.*, 33:347–350, 1968.

Hansch, C.; Kim, D.; Leo, A. J.; Novellino, E.; Silipo, C.; and Vittoria, A.: Toward a quantitative comparative toxicology of organic compounds. *CRC Crit. Rev. Toxicol.*, 19:185–226, 1989.

Hart, J. W.: Some Danish approaches to selection of existing chemicals. *Chemosphere*, 17:1411–1418, 1988.

Hart, J. W.: The use of data estimation methods by regulatory authorities. *Sci. Total Environ.*, 109/110:629–633, 1991.

Hawker, D. W., and Connell, D. W.: Prediction of bioconcentration factors under non-equilibrium conditions. *Chemosphere*, 14:1835–1843, 1985.

Hawker, D. W., and Connell, D. W.: Bioconcentration of lipophilic compounds by some aquatic organisms. *Ecotoxicol. Environ. Safety*, 11:184–197, 1986.

Hermens, J. L. M.: Quantitative structure-activity relationships of environmental pollutants. Hutzinger, O. (ed.): Handbook of Environmental Chemistry, Vol. 2E. Springer-Verlag, Berlin, 1989, pp. 111–162.

Hermens, J. L. M.: Quantitative structure-activity relationships for predicting fish toxicity. Karcher, W., and Devillers, J. (eds.): Practical Applications of Quantitative Structure-Activity Relationships (QSAR) in Environmental Chemistry and Toxicology. Kluwer Publishing, Dordrecht, 1990a, pp. 263–280.

Hermens, J. L. M.: Electrophiles and acute toxicity to fish. *Environ. Health Perspect.*, 87:219–225, 1990b.

Hermens, J. L. M., and Leeuwangh, P.: Joint toxicity of mixtures of 8 and 24 chemicals to the guppy (*Poecilia reticulata*). *Ecotoxicol. Environ. Safety*, 6:302–310, 1982.

Hermens, J. L. M., and Opperhuizen, A. (eds.): QSAR in Enviromental Toxicology. *Sci. Total Environ.*, 109/110: 706, 1991.

Hermens, J. L. M.; Leeuwangh, P.; and Musch, A.: Quantitative structure-activity relationships and mixture toxicity studies of chloro- and alkylanilines at an acute lethal toxicity level to the guppy (*Poecilia reticulata*). *Ecotoxicol. Environ. Safety*, 8:388–394, 1984a.

Hermens, J. L. M.; Canton, H.; Janssen, P.; and De Jong, R.: Quantitative structure-activity relationships and toxicity studies of mixtures of chemicals with anesthetic potency: acute lethal and sublethal toxicity to *Daphnia magna*. *Aquat. Toxicol.*, 5:143–154, 1984b.

Hermens, J. L. M.; Broekhuyzen, E.; Canton, H.; and Wegman, R.: QSARs and mixture toxicity studies of alcohols and chlorohydrocarbons. Effects on growth of *Daphnia magna*. *Aquat. Toxicol.*, 6:209–217, 1985a.

Hermens, J. L. M.; Busser, F.; Leeuwangh, P.; and Musch, A.: Quantitative structure-activity relationships and mixture toxicity of organic chemicals in *Photobacterium phosphoreum*. *Ecotoxicol. Environ. Safety*, 9:17–25, 1985b.

Hermens, J. L. M.; Busser, F.; Leeuwangh, P.; and Musch, A.: Quantitative correlation studies between the acute lethal toxicity of 15 organic halides to the guppy (*Poecilia reticulata*) and chemical reactivity towards 4-nitrobenzylpyridine. *Toxicol. Environ. Chem.*, 9:219–236, 1985c.

Hermens, J. L. M.; Leeuwangh, P.; and Musch, A.: Joint toxicity of groups of organic aquatic pollutants to the guppy (*Poecilia reticulata*). *Ecotoxicol. Environ. Safety*, 9:321–326, 1985d.

Hermens, J. L. M.; De Bruijn, J.; Pauly, J.; and Seinen, W.: QSAR studies for fish toxicity data of organophosphorus compounds and other classes of reactive organic compounds. Kaiser, K. L. E. (ed.): QSAR in Environmental Toxicology—II. Reidel, Dordrecht, 1987, pp. 135–152.

Hodson, P. V.; Dixon, D. G.; and Kaiser, K. L. E.: Estimating the acute toxicity of waterborne chemicals in trout from measurements of median lethal dose and octanol/water partition coefficient. *Environ. Toxicol. Chem.*, 7:443–454, 1988.

Hoeschle, J. D.; Turner, J. E.; and England, M. W.: Inorganic concepts relevant to metal binding, activity, and toxicity in a biological system. *Sci. Total Environ.*, 109/110:477–492, 1991.

Hoffman, A. D.; Bertelsen, S. L.; and Gargas, M. L.: An in vitro gas equilibration method for determination of chemical partitions in fish. *Comp. Biochem. Physiol.*, 101A:47–51, 1992.

Hollis, E. H., and Lennon, R. E.: The toxicity of 1085 chemicals to fish. U.S. Fish and Wildlife Service, Kearneysville, WV, 1953; in EPA Report No. 560/6-87-002, NTIS PB 87-200-275, 1987.

Hopfinger, A. J., and Battershell, R. D.: Application of SCAP to drug design. 1. Prediction of octanol/water partition coefficients using solvent-dependent conformational analysis. *J. Med. Chem.*, 19:569–573, 1976.

Horsfall, J. G.: Principles of Fungicidal Action. Chronica Botanica, Waltham, MA, 1956.

Houdaille, G.: Étude expérimentale et critque sur les nouveaux hypnotiques. Thesis, Faculté de Médecine, Paris, 1893.

Howard, P. H.; Boethling, R. S.; Stiteler, W. M.; Meylan, W. M.; Hueber, A. E.; Beauman, J. A.; and Larosche, M. E.: Predictive model for aerobic biodegradability developed from a file of evaluated biodegradation data. *Environ. Toxicol. Chem.*, 11:593–603, 1992.

Hunter, R. S.: Computer calculation of pKa. Turner, J. E.; England, M. W.; Schultz, T. W.; and Kvaak, N. J. (eds.): QSAR88. NTIS CONF-880520—(DE88013180), U.S. Department of Energy, Oak Ridge, TN, 1988, pp. 33–41.

Hunter, R. S.; Culver, F. D.; Hill, J. R. III; and FitzGerald, A.: QSAR System User Manual. EPA-ERL, Duluth, MN, and Montana State University, Bozeman, 1986.

Ihde, A. J.: The Development of Modern Chemistry, Dover Publications, New York, 1984, pp. 204–208.

Ikemoto, Y.; Motoba, K.; Suzuki, T.; and Uchida, M.: Quantitative structure-activity relationships of non-specific and specific toxicants in several organism species. *Environ. Toxicol. Chem.*, 11:931–939, 1992.

Jaworska, J. S., and Schultz, T. W.: Quantitative relationships of structure-activity and volume fraction for selected nonpolar and polar narcotic chemicals. *SAR QSAR Environ. Res.*, 1:3–19, 1993.

Jocelyn, P. C.: Biochemistry of the SH Group: The Occurrence, Chemical Properties, Metabolism and Biological Function of Thiols and Disulfides. Academic Press, London, 1972.

Kaiser, K. L. E. (ed.): QSAR in Environmental Toxicology. Reidel, Dordrecht, 1984.

Kaiser, K. L. E. (ed.): QSAR in Environmental Toxicology—II. Reidel, Dordrecht: 1987a.

Kaiser, K. L. E.: QSAR of acute toxicity of 1,4-disubstituted benzene derivatives and relationships with the acute toxicity of corresponding mono-substituted benzene derivatives. Kaiser, K. L. E. (ed.): QSAR in Environmental Toxicology—II. Reidel, Dordrecht, 1987b, pp. 169–188.

Kaiser, K. L. E., and Gough, K. M.: QSAR of the acute toxicity of *para*-substituted nitrobenzene and aniline derivatives to *Photobacterium phosphoreum*. Turner, J. E.; England, M. W.; Schultz, T. W.; and Kvaak, N. J. (eds.): QSAR88. NTIS CONF-880520—(DE88013180). U.S. Department of Energy, Oak Ridge, TN, 1988, pp. 111–121.

Kaiser, K. L. E.; Dixon, D. G.; and Hodson, P. V.: QSAR studies on chlorophenols, chlorobenzenes and para-substituted phenols. Kaiser, K. L. E. (ed.): QSAR in Environmental Toxicology. Reidel, Dordrecht, 1984, pp. 189–206.

Kaiser, K. L. E.; Palabrica, V. S.; and Ribo, J. M.: QSAR of acute toxicity of mono-substituted benzene derivatives to *Photobacterium phosphoreum*. Kaiser, K. L. E. (ed.): QSAR in Environmental Toxicology—II. Reidel, Dordrecht, 1987, pp. 153–168.

Kamlet, M. J.; Doherty, R. M.; Veith, G. D.; Taft, R. W.; and Abraham, M. H.: Solubility properties in polymers and biological media: 7. An analysis of toxicant properties that influence inhibition of bioluminescence in *Photobacterium phosphoreum* (the Microtox test). *Environ. Sci. Technol.*, 20:690–695, 1986.

Kamlet, M. J.; Doherty, R. M.; Taft, R. W.; Abraham, M. H.; Veith, G. D.; and Abraham, G. D.: Solubility properties in polymers and biological media: 8. An analysis of the factors that influence toxicities of organic nonelectrolytes to the golden orfe fish (*Leuciscus idus melanotus*). *Environ. Sci. Technol.*, 21:149–155, 1987.

Kamlet, M. J.; Doherty, R. M.; Abraham, M. H.; Marcus, Y.; and Taft, R. W.: Linear solvation energy equation for correlation and prediction of octanol/water partition coefficients of organic non-hydrogen bond donor solutes. *J. Phys. Chem.*, 92:5244–5255, 1988.

Kanazawa, J.: Measurement of the bioconcentration factors of pesticides by freshwater fish and their correlation with physicochemical properties and acute toxicities. *Pestic. Sci.*, 12:417–424, 1981.

Kanazawa, J.: A method of predicting the bioconcentration potential of pesticides by using fish. *Japan Agric. Res. Quarterly*, 17:173–179, 1983.

Karcher, W.: Basic concepts and uses of QSAR studies. Coccini, T.; Giannoni, L.; Karcher, W.; Manzo, L.; and Roi, R. (eds.): Quantitative Structure-Activity Relationships (QSAR) in Toxicology. Joint Research Center, Commission of the European Communities, Luxembourg, 1992, pp. 5–25.

Karcher, W., and Devillers, J. (eds.): Practical Applications of Quantitative Structure-Activity Relationships (QSAR) in Environmental Chemistry and Toxicology. Kluwer Publishing, Dordrecht, 1990.

Karickhoff, S. W.; McDaniel, V. K.; Melton, C.; Vellino, A. N.; Nute, D. E.; and Carreira, L. A.: Predicting chemical reactivity by computer. *Environ. Toxicol. Chem.*, 10: 1405–1416, 1991.

Kenaga, E. E., and Goring, C. A.: Relationship between water solubility, soil sorption, octanol-water partitioning and bioconcentration of chemicals in biota. Eaton, J. G.; Parrish, P. R.; and Hendricks, A. C. (eds.): Aquatic Toxicology. ASTM STP 707, American Society for Testing and Materials, Philadelphia, 1980, pp. 78–115.

Klopman, G.; Balthasar, D. M.; and Rosenkranz, H. S.: Application of the computer-automated structure evaluation (CASE) program to the study of structure-biodegradation relationships of miscellaneous chemicals. *Environ. Toxicol. Chem.*, 12:231–240, 1993.

Klopman, G.; Wang, S.; and Balthasar, D. M.: Estimation of aqueous solubility of organic molecules by the group contribution approach. Application to the study of biodegradation. *J. Chem. Inf. Comput. Sci.*, 32:474–482, 1992.

Koch, R.: Quantitative structure-activity relationships in ecotoxicology: possibilities and limits. Kaiser, K. L. E. (ed.): QSAR in Environmental Toxicology. Reidel, Dordrecht, 1984, pp. 207–222.

Könemann, H.: Structure-activity relationships and additivity in fish toxicity studies of environmental pollutants. *Ecotoxicol. Environ. Safety*, 4:415–421, 1980.

Könemann, H.: Quantitative structure-activity relationships in fish toxicity studies. Part 1: A relationship for 50 industrial pollutants. *Toxicology*, 19:209–221, 1981a.

Könemann, H.: Fish toxicity test with mixtures of more than two chemicals. A proposal for a quantitative approach and experimental result. *Toxicology*, 19:229–238, 1981b.

Könemann, H., and Musch, A.: Quantitative structure-activity relationships in fish toxicity studies. Part 2: The influence of pH on the QSAR of phenols. *Toxicology*, 19:223–228, 1981.

Könemann, H., and van Leeuwen, R.: Toxicokinetics in fish: accumulation and elimination of six chlorobenzenes in guppies. *Chemosphere*, 9:3–19, 1980.

Könemann, H.; Zelle, R.; Busser, F.; and Hammers, W. E.: Determination of log P_{oct} values of chloro-substituted benzenes, toluenes and anilines by high-performance liquid chromatography on ODS-silica. *J. Chromatogr.*, 178: 559–565, 1979.

Kopperman, H. L.; Carlson, R. M.; and Caple, R.: Aqueous chlorination and ionization studies. I. Structure-toxicity correlations of phenolic compounds to *Daphnia magna*. *Chem-Biol. Interact.*, 9:245–251, 1974.

Kubinyi, H.: Quantitative structure-activity relationships. IV. Nonlinear dependence of biological activity on hydrophobic character: a new model. *Arzneim. Forsch. Drug. Res.*, 26:1991–1997, 1976.

Kubinyi, H.: Quantitative structure-activity relationships. 7. The bilinear model, a new model for nonlinear dependence of biological activity on hydrophobic character. *J. Med. Chem.*, 20:625–629, 1977.

Kubinyi, H.: Nonlinear dependence of biological activity on hydrophobic character: the bilinear model. *Il Pharmaco*, 34:247–276, 1979.

Kubinyi, H.: The Free-Wilson method and its relationship to the extrathermodynamic approach. Hansch, C. (editorial board chairman): Comprehensive Medicinal Chemistry, Vol. 4. Pergamon Press, Oxford, 1990, pp. 589–643.

Kubinyi, H.: QSAR: Hansch Analysis and Related Approaches. VCH, Weinheim, 1993.

Kubinyi, H., and Kehrhahn, O. H.: Quantitative structure-activity relationships. IV. Nonlinear dependence of biological activity on hydrophobic character: calculation procedures for the bilinear model. *Arzneim. Forsch. Drug. Res.*, 28:598–601, 1978.

Kuenemann, P.; Vasseur, P.; and Devillers, J.: Structure-biodegradability relationships. Karcher, W., and Devillers, J. (eds.): Practical Applications of Quantitative Structure-Activity Relationships (QSAR) in Environmental Chemistry and Toxicology. Kluwer Publishing, Dordrecht, 1990, pp. 343–370.

Kwasniewska, K., and Kaiser, K. L. E.: Toxicities of selected chloroanilines to four strains of yeast. Kaiser, K. L. E. (ed.): QSAR in Environmental Toxicology. Reidel, Dordrecht, 1984, pp. 223–233.

Laughlin, R. B.; Johannesen, R. B.; French, W.; Guard, H.; and Brinckman, F. E.: Structure-activity relationships for organotin compounds. *Environ. Toxicol. Chem.*, 4:343–351, 1985.

Lazarev, N. V.: Neelektrolity. Voennomorskaya Medicinskaya Akademiya, Leningrad, 1944.

Leahy, D. E.: Intrinsic molecular volume as a measure of the cavity term in linear solvation energy relationships: octanol-water partition coefficients and aqueous solubilities. *J. Pharm. Sci.*, 75:629–636, 1986.

Leahy, D. E.; Taylor, P. J.; and Walt, A. R.: Model solvent systems for QSAR Part I. Propylene glycol dipelargonate (PGDP). A new standard solvent for use in partition coefficient determination. *Quant. Struct.-Act. Relat.*, 8:17–31, 1989.

LeBlanc, G. A.: Comparative structure-toxicity relationships between acute and chronic effects to aquatic organisms. Kaiser, K. L. E. (ed.): QSAR in Environmental Toxicology. Reidel, Dordrecht, 1984, pp. 235–260.

Leffler, J. E., and Grunwald, E.: Rates of Equilibria of Organic Reactions. Wiley, New York, 1969.

Leo, A. J.: Relationships between partitioning solvent systems. *Biological Correlations—The Hansch Approach.* Adv. Chem. Ser. 114, 1972, pp. 51–60.

Leo, A.: Calculation of partition coefficients useful in the evaluation of the relative hazards of various chemicals in the environment. Veith, G. D., and Konasewich, D. E. (eds.): Structure-Activity Correlations in Studies of Toxicity and Bioconcentration with Aquatic Organisms. Great Lakes Research Advisory Board, Windsor, Ontario, 1975, pp. 151–176.

Leo, A.: The octanol/water partition coefficient of aromatic solutes: the effect of electronic interactions, alkyl chains, hydrogen bonds and *ortho*-substitution. *J. Chem. Soc. Perkin Trans. II*, 825–838, 1983.

Leo, A. J.: Some advantages of calculating octanol/water partition coefficients. *J. Pharm. Sci.*, 76:166–168, 1987.

Leo, A.: Methods of calculating partition coefficients. Hansch, C. (editorial board chairman): Comprehensive Medicinal Chemistry, Vol. 4. Pergamon Press, Oxford, 1990, pp. 295–319.

Leo, A. J.: Calculating the hydrophobicity of chlorinated hydrocarbon solutes. *Sci. Total Environ.*, 109/110:121–130, 1991.

Leo, A., and Weininger, D.: Medchem Software Release 3.33. Medicinal Chemistry Project, Pomona College, Claremont, CA, 1985.

Leo, A.; Hansch, C.; and Church, C.: Comparison of parameters currently used in the study of structure-activity relationships. *J. Med. Chem.*, 12:766–771, 1969.

Leo, A. J.; Hansch, C.; and Elkins, D.: Partition coefficients and their uses. *Chem. Rev.*, 71:525–616, 1971.

Lipnick, R. L.: A perspective on quantitative structure-activity relationships in ecotoxicology. *Environ. Toxicol. Chem.*, 4:255–257, 1985a. Editorial.

Lipnick, R. L.: Validation and extension of fish toxicity QSARs and interspecies comparisons for certain classes of organic chemicals. Tichy, M. (ed.): QSAR in Toxicology and Xenobiochemistry. Elsevier, Amsterdam, 1985b, pp. 39–52.

Lipnick, R. L.: Charles Ernest Overton: narcosis studies and a contribution to general pharmacology. *Trends Pharmacol. Sci.*, 7:161–164, 1986.

Lipnick, R. L.: Quantitative structure-activity relationships and toxicity assessment in the aquatic environment. Richardson, M. L. (ed.): Risk Assessment of Chemicals in the Environment. Royal Society of Chemistry, London, 1988, pp. 379–397.

Lipnick, R. L.: Narcosis, electrophile and proelectrophile toxicity mechanisms: application of SAR and QSAR. *Environ. Toxicol. Chem.*, 8:1–12, 1989a.

Lipnick, R. L.: A QSAR study of Overton's tadpole data. Fauchère, J. L. (ed.): QSAR: Quantitative Structure-Activity Relationships in Drug Design. Alan R. Liss, New York, 1989b, pp. 421–424.

Lipnick, R. L.: Baseline toxicity predicted by quantitative structure-activity relationships as a probe for molecular mechanism of toxicity. Magee, P. S.; Henry, D. R.; and Block, J. H. (eds.): Probing Bioactive Mechanisms. ACS Symp. Ser. 413, 1989c, pp. 366–389.

Lipnick, R. L.: Hans Horst Meyer and the lipoid theory of narcosis. *Trends Pharmacol. Sci.*, 10:265–269, 1989d.

Lipnick, R. L.: A quantitative structure-activity relationship study of Overton's data on the narcosis and toxicity of organic compounds to the tadpole, *Rana temporaria*. Suter, G. W., and Lewis, M. A. (eds.): Aquatic Toxicology and Environmental Fate. ASTM STP 1007, ASTM, Philadelphia, 1989e, pp. 468–489.

Lipnick, R. L.: Selectivity. Hansch, C. (editorial board chairman): Comprehensive Medicinal Chemistry, Vol. 1. Pergamon Press, Oxford, 1990a, pp. 239–247.

Lipnick, R. L.: Narcosis: Fundamental and baseline toxicity mechanism for nonelectrolyte organic chemicals. Karcher, W., and Devillers, J. (eds.): Practical Applications of Quantitative Structure-Activity Relationships (QSAR) in Environmental Chemistry and Toxicology. Kluwer Publishing, Dordrecht, 1990b, pp. 281–293.

Lipnick, R. L.: Outliers: their origin and use in the classification of molecular mechanisms of toxicity. *Sci. Total Environ.*, 109/110:131–153, 1991a.

Lipnick, R. L. (ed.): Studies of Narcosis: Charles Ernest Overton. Chapman & Hall, London, 1991b.

Lipnick, R. L.: Molecular mechanism of action and quantitative structure-activity relationships in environmental toxicology. Silipo, C., and Vittoria, A. (eds.): QSAR: Rational Approaches to the Design of Bioactive Compounds. Elsevier, Amsterdam, 1991c, pp. 495–500.

Lipnick, R. L., and Dunn, W. J.: A MLAB study of aquatic structure-toxicity relationships. Dearden, J. (ed.): Quantitative Approaches to Drug Design. Elsevier, Amsterdam, 1983, pp. 265–266.

Lipnick, R. L., and Filov, V. A.: Nikolai Vasilyevich Lazarev, toxicologist and pharmacologist, comes in from the cold. *Trends Pharmacol. Sci.*, 13:56–60, 1992.

Lipnick, R. L., and Hood, M. T.: Correlation of chemical structure and toxicity of industrial organic compounds to daphnia, algae, bacteria, and protozoa. Abstracts of Papers, 7th Annual Meeting of the Society of Environmental Toxicology and Chemistry, Alexandria, VA, November 2–5, 1986.

Lipnick, R. L.; Bickings, C. K.; Johnson, D. E.; and Eastmond, D. A.: Comparison of QSAR predictions with fish toxicity screening data for 110 phenols. Bahner, R. C., and Hansen, D. J. (eds.): Aquatic Toxicology and Hazard Assessment: Eighth Symposium. ASTM STP 891, ASTM, Philadelphia, 1985a, pp. 153–176.

Lipnick, R. L.; Johnson, D. E.; Gilford, J. H.; Bickings, C. K.; and Newsome L. D.: Comparison of fish toxicity screening data for 55 alcohols with the quantitative structure-activity relationship predictions of minimum toxicity for nonreactive nonelectrolyte organic compounds. *Environ. Toxicol. Chem.*, 4:281–296, 1985b.

Lipnick, R. L.; Pritzker, C. S.; and Bentley, D. L.: A QSAR study of the rat LD_{50} for alcohols. Seydel, J. K. (ed.): QSAR and Strategies in the Design of Bioactive Compounds, VCH, Weinheim, 1985c, pp. 420–423.

Lipnick, R. L.; Pritzker, C. S.; and Bentley, D. L.: Application of QSAR to model the toxicology of industrial organic chemicals to aquatic organisms and mammals. Hadzi, D., and Jerman-Blazic, B. (eds.): Progress in QSAR. Elsevier, Amsterdam, 1987a, pp. 301–306.

Lipnick, R. L.; Watson, K. R.; and Strausz, A. K.: A QSAR study of the acute toxicity of some industrial organic chemicals to goldfish. Narcosis, electrophile and proelectrophile mechanisms. *Xenobiotica*, 17:1011–1025, 1987b.

Lipnick, R. L.; Bentley, J. L.; Bentley, D. L.; and Myers, R. C.: A QSAR study of the eye irritation of simple alcohols. *Toxicologist*, 13(1):444, 1993.

Liu, Q.; Hirono, S.; Matsushita, Y.; and Moriguchi, I.: QSARs based on fuzzy adaptive least-squares analysis for the aquatic toxicity of organic chemicals. *Environ. Toxicol. Chem.*, 11:953–959, 1992.

Lyman, W. J.; Reehl, W. F.; and Rosenblatt, D. H.: Handbook of Chemical Property Estimation Methods. McGraw-Hill, New York, 1982.

Lyman, W. J.: Octanol/water partition coefficient. Lyman, W. J.; Reehl, W. P.; and Rosenblatt, D. H. (eds.): Handbook of Chemical Property Estimation Methods, Environmental Behaviour of Organic Chemicals. American Chemical Society, Washington, DC, 1990, pp. 1.1.

Lynch, D. G.; Tirado, N. F.; Boethling, R. S.; Huse, G. R.; and Thom, G. C.: Performance of on-line chemical property estimation methods with TSCA premanufacture notice chemicals. *Ecotoxicol. Environ. Safety*, 22:240–240, 1991.

Mackay, D.: Correlation of bioconcentration factors. *Environ. Sci. Technol.*, 16:274–278, 1982.

Mackay, D., and Paterson, S.: Fugacity models. Karcher, W., and Devillers, J. (eds.): Practical Applications of Quantitative Structure-Activity Relationships (QSAR) in Environmental Chemistry and Toxicology. Kluwer Publishing, Dordrecht, 1990, pp. 433–460.

Mackay, D.; Puig, H.; and McCarty, L. S.: An equation describing the time course and variability in uptake and toxicity of narcotic chemicals to fish. *Environ. Toxicol. Chem.*, 11:941–951, 1992.

MacPhee, C., and Cheng, F. F.: Lethal effects of 964 chemicals upon steelhead trout and bridgelip sucker. Department of Wildlife and Fishery Resources, College of Forestry, Wildlife and Range Sciences, University of Idaho, Moscow, Idaho; in EPA Report 560/6-89-001, NTIS PB 89-156715, 1989a.

MacPhee, C., and Cheng, F. F.: Lethal effects of 2014 chemicals upon sockeye salmon, steelhead trout and threespine stickleback. Department of Wildlife and Fishery Resources, College of Forestry, Wildlife and Range Sciences, University of Idaho, Moscow, Idaho; in EPA Report 560/6-89-001, NTIS PB 89-156715, 1989b.

MacPhee, C., and Ruelle, R.: Lethal effects of 1,888 chemicals upon four species of fish from western North America, Bulletin No. 3, University of Idaho, Forest, Wildlife and Range Experiment Station, University of Idaho, Moscow, Idaho, 112 pp., 1969.

Martin, Y. C.: Quantitative Drug Design. Marcel Dekker, New York, 1978.

Martin, Y. C.: Overview of concepts and methods in computer-assisted rational drug design. *Methods Enzymol.*, 203:587–613, 1991.

Martin, Y. C.; Bush, E. N.; and Kyncl, J. J.: Quantitative description of biological activity. Hansch, C. (editorial board chairman): Comprehensive Medicinal Chemistry, Vol. 4. Pergamon Press, Oxford, 1990, pp. 349–373.

Matsuo, M.: A thermodynamic interpretation of bioaccumulation of aroclor 1254 in fish. *Chemosphere*, 9:671–675, 1980.

Matsuo, M.: i/o-Characters to describe bioconcentration factors of chloro-benzenes and -naphthalenes—meaning of the sign of the coefficients of $\Sigma i/\Sigma o$ in the correlating equations. *Chemosphere*, 10:1073–1078, 1981.

McCarty, L. S.; Hodson, P. V.; Craig, G. R.; and Kaiser, K. L. E.: The use of quantitative structure-activity relationships to predict the acute and chronic toxicity of organic chemicals to fish. *Environ. Toxicol. Chem.*, 4: 595–606, 1985.

McCarty, L. S.: Relationship between toxicity and bioconcentration of some organic chemicals I. Examination of the relationship. Kaiser, K. L. E. (ed.): QSAR in Environmental Toxicology—II. Reidel Publishing, Dordrecht, 1987a, pp. 207–220.

McCarty, L. S.: Relationship between toxicity and bioconcentration of some organic chemicals II. Application of the relationship. Kaiser, K. L. E. (ed.): QSAR in Environmental Toxicology—II. Reidel Publishing, Dordrecht, 1987b, pp. 221–229.

McCarty, L. S.; Mackay, D.; Smith, A. D.; Ozburn, G. W.; and Dixon, D. G.: Interpreting aquatic toxicity QSARs: the significance of toxicant body residues at the pharmacologic endpoint. *Sci. Total Environ.*, 109/110:515–525, 1991.

McCarty, L. S.; Mackay, D.; Smith, A. D.; Ozburn, G. W.; and Dixon, D. G.: Residue-based interpretation of toxicity and bioconcentration QSARs from aquatic bioassays. Neutral narcotic organics. *Environ. Toxicol. Chem.*, 11:917–930, 1992.

McGowan, J.: The physical toxicity of chemicals. II. Factors affecting physical toxicity in aqueous solutions. *J. Appl. Chem.*, 2:323–328, 1952.

McGowan, J. C., and Mellors, A.: Molecular volumes and the toxicities of chemicals to fish. *Bull. Environ. Contam. Toxicol.*, 36:881–887, 1986.

McFarland, J. W., and Gans, D. J.: Linear discriminant analysis and cluster significance analysis. Hansch, C. (editorial board chairman): Comprehensive Medicinal Chemistry, Vol. 4. Pergamon Press, Oxford, 1990, pp. 667–689.

McKim, J. M.; Schmieder, P.; and Veith, G.: Absorption kinetics of organic chemical transport across trout gills as related to octanol/water partition coefficient. *Toxicol. Appl. Pharmacol.*, 77:1–10, 1985.

McKim, J. M.; Bradbury, S. P.; and Niemi, G. J.: Fish acute toxicity syndromes and their use in the QSAR approach to hazard assessment. *Environ. Health Perspect.*, 71: 171–186, 1987a.

McKim, J. M.; Schmieder, P. K.; Carlson, R. W.; Hunt, E. P.; and Niemi, G. J.: Use of respiratory-cardiovascular responses of rainbow trout (*Salmo gairdneri*) in identifying acute toxicity syndromes in fish: Part 1: Pentachlorophenol, 2,4-dinitrophenol, tricaine methanesulfonate and 1-octanol. *Environ. Toxicol. Chem.*, 6:295–312, 1987b.

McKim, J. M.; Schmieder, P. K.; Niemi, G. J.; Carlson, R. W.; and Henry, T. R.: use of respiratory-cardiovascular responses of rainbow trout (*Salmo gairdneri*) in identifying acute toxicity syndromes in fish: Part 2. Malathion, carbaryl, acrolein and benzaldehyde. *Environ. Toxicol. Chem.*, 6:313–328, 1987c.

McLeese, D. W.; Zitko, V.; and Peterson, M. R.: Structure-lethality relationships for phenols, anilines and other aromatic compounds in shrimp and clams. *Chemosphere*, 2:53–57, 1979.

Mekenyan, O. G.; Veith, G. D.; Bradbury, S. P.; and Russom, C. L.: Structure-toxicity relationships for α,β-unsaturated alcohols in fish. *Quant. Struct.-Act. Relat.*, 12: 132–136, 1993.

Meyer, H.: Zur theorie der Alcoholnarkose. Erste Mittheilung. Welche Eigenschaft der Anasthetica bedingt ihre narkotische Wirkung? *Naunyn-Schmiedebergs Arch. Exp. Pathol. Pharmakol.*, 42:109–118, 1899.

Meyer, H.: Zur theorie der Alcoholnarkose. 3 Mittheilung. Der Einfluss wechselnder Temperatur auf Wirkungsstärke und Teilungscoefficient der Narcota. *Naunyn-Schmiedebergs Arch. Exp. Pathol. Pharmakol.*, 46:535–547, 1901.

Miller, K.; Paton, W. B. M.; Smith, R. A.; and Smith, E. B.: The pressure reversal of anesthesia and the clinical volume hypothesis. *Mol. Pharmacol.* 9:131–143, 1983.

Moulton, M. P., and Schultz, T. W.: Structure-activity relationships of selected pyridines. II. Principal components analysis. *Chemosphere*, 15:59–67, 1986.

Muir, D. C. G.; Marshall, W. K.; and Webster, G. R. B.: Bioconcentration of PCDDs by fish: effects of molecular structure and water chemistry. *Chemosphere*, 14:829–833, 1985.

Mullins, L. J.: Some physical mechanisms in narcosis. *Chem. Rev.*, 54:289–323, 1954.
Munch, J. C., and Schwartze, E. W.: Narcotic and toxic potency of aliphatic alcohols upon rabbits. *J. Lab. Clin. Med.*, 10:985–996, 1925.
Nabholz, V. N.: Environmental hazard and risk assessment under the United States Toxic Substances Control Act. *Sci. Total Environ.*, 109/110:649–665, 1991.
Nagel, R., and Loskill, R. (eds.): Bioaccumulation in Aquatic Systems: Contributions to the Assessment. VCH, Weinheim, 1991.
Neely, W. B.; Branson, D. R.; and Blau, G. E.: Partition coefficients to measure bioconcentration potential of organic chemicals in fish. *Environ. Sci. Technol.*, 8:1113–1115, 1974.
Nendza, M.: QSARs of bioconcentration: validity assessment of log P_{ow}/log BCF correlations. Nagel, R., and Loskill, R. (eds.): Bioaccumulation in Aquatic Systems: Contributions to the Assessment. VCH, Weinheim, 1991, pp. 43–66.
Nendza, M., and Russom, C. L.: QSAR modeling of the ERL-D fathead minnow acute toxicity database. *Xenobiotica*, 21:147–170, 1991.
Nendza, M., and Seydel, J. K.: Multivariate data analysis of various biological test systems used in the quantification of ecotox compounds. *Quant. Struct.-Act. Relat.*, 7:165–174, 1988a.
Nendza, M., and Seydel, J. K.: Quantitative structure-toxicity relationships for ecotoxicologically relevant biotestsystems and chemicals. *Chemosphere*, 17:1585–1602, 1988b.
Nendza, M.; Volmer, J.; and Klein, W.: Risk assessment based upon QSAR estimates. Karcher, W., and Devillers, J. (eds.): Practical Applications of Quantitative Structure-Activity Relationships (QSAR) in Environmental Chemistry and Toxicology. Kluwer Publishing, Dordrecht, 1990, pp. 213–240.
Newman, M. C.: Regression analysis of log-transformed data: statistical bias and its correction. *Environ. Toxicol. Chem.*, 12:1129–1133, 1993.
Newsome, L. D.; Lipnick, R. L.; and Johnson, D. E.: Validation of fish toxicity QSARs for certain non-reactive non-electrolyte organic compounds. QSAR in Environmental Toxicology, edited by K. L. E. Kaiser, pp. 279–299. Dordrecht: Reidel, 1984.
Newsome, L. D.; Johnson, D. E.; Cannon, D. J.; and Lipnick, R. L.: Comparison of fish toxicity screening data and QSAR predictions for 48 aniline derivatives. QSAR in Environmental Toxicology—II, edited by K. L. E. Kaiser, pp. 231–250. Dordrecht: Reidel, 1987.
Newsome, L. D.; Johnson, D. E.; Lipnick, R. L.; Broderius, S. J.; and Russom, C. L.: A QSAR study of the toxicity of amines to the fathead minnow. *Sci. Total Environ.*, 109/110:537–551, 1991.
Niemi, G. J.: Multivariate analysis and QSAR: applications of principal components analysis. Practical Applications of Quantitative Structure-Activity Relationships (QSAR) in Environmental Chemistry and Toxicology, pp. 153–170. Dordrecht: Kluwer, 1990.
Nowotnik, D. P.; Felt, T.; and Nunn, A. D.: Examination of some reversed-phase high-performance liquid chromatography systems for the determination of lipophilicity. *J. Chromatogr.*, 630:105–115, 1993.
Nys, G. G., and Rekker, R. F.: Statistical analysis of a series of partition coefficients with special reference to the predictability of folding of drug molecules. Introduction of hydrophobic fragmental constants (*f*-values). *Chim. Ther.*, 8:521–535, 1973.
Nys, G. G., and Rekker, R. F.: The concept of hydrophobic fragmental constants (*f*-values) II. Extension of its applicability to the calculation of lipophilicities of aromatic and heteroaromatic structures. *Eur. J. Med. Chem.*, 9: 336–375, 1974.
Ogata, M.; Fujisawa, K.; Ogino, Y.; and Mano, E.: Partition coefficients as a measure of bioconcentration potential of crude oil compounds in fish and shellfish. *Bull. Environ. Contam. Toxicol.*, 33:561–567, 1984.
Oliver, B. G.: The relationship between bioconcentration factor in rainbow trout and physical-chemical properties for some halogenated compounds. QSAR in Environmental Toxicology, edited by K. L. E. Kaiser, pp. 301–317. Dordrecht: Reidel, 1984.
Oliver, B. G., and Niimi, A. J.: Bioconcentration of chlorobenzenes from water by rainbow trout: correlations with partition coefficients and environmental residues. *Environ. Sci. Technol.*, 17:287–291, 1983.
Oliver, B. G., and Niimi, A. J.: Rainbow trout bioconcentration of some halogenated aromatics from water at environmental concentrations. *Environ. Toxicol. Chem.*, 3: 271–277, 1984.
Opperhuizen, A.; Velde, E. W.; Gobas, F. A.; Lem, D. A.; and Steen, J. M.: Relationship between bioconcentration in fish and steric factors of hydrophobic chemicals. *Chemosphere*, 14:1871–1896, 1985.
Opperhuizen, A.; Serné, P.; and Van der Steen, J. M. D.: Thermodynamics of fish/water and octan-1-ol/water partitioning of some chlorinated benzenes. *Environ. Sci. Technol.*, 22:286–292, 1988.
Overton, E.: Über die osmotischen Eigenschaften der lebenden Pflanzen- und Tierzelle. *Vierteljahrsschr. Naturforsch. Ges. Zuerich*, 40:159–201, 1895.
Overton, E.: Über die osmotischen Eigenschaften der Zelle in ihrer Bedeutung für die Toxikologie und Pharmakologie. *Vierteljahrsschr. Naturforsch. Ges. Zuerich*, 41: 383–406, 1896.
Overton, E.: Ueber die algemeinen osmotischen Eigenschaften der Zelle, ihre vermutlichen Ursachen und ihre Bedeutung für die Physiologie. *Vierteljahrsschr. Naturforsch. Ges. Zuerich*, 44:88–135, 1899.
Overton, E.: Studien über die Narkose zugleich ein Beitrag zur allgemeinen Pharmakologie. Jena: Gustav Fischer, 1901.
Paris, D. F.; Wolfe, N. L.; and Steen, W. C.: Structure-activity relationships in microbial transformation of phenols. *Appl. Environ. Microbiol.*, 44:153–158, 1982.
Paris, D. F.; Wolfe, N. L.; Steen, W. C.; and Baughman, G. L.: Effect of phenol molecular structure on bacterial transformation rate constants in pond and river samples. *Appl. Environ. Microbiol.*, 45:1153–1155, 1983.
Paris, D. F.; Wolfe, N. L.; and Steen, W. C.: Microbial transformation of esters of chlorinated carboxylic acids. *Appl. Environ. Microbiol.*, 47:7–11, 1984.
Parsons, J. R., and Govers, H. A. J.: Quantitative structure-activity relationships for biodegradation. *Ecotoxicol. Environ. Safety*, 19:212–227, 1990.
Parsons, J. R.; Commandeur, L. C. M.; Van Eyseren, H. E.; and Govers, H. A. J.: QSARs and PARs for biodegradation of PCBs. *Sci. Total Environ.*, 109/110:275–281, 1991.
Passino, D. M. and Smith, S. B.: Quantitative structure-activity relationships (QSAR) and toxicity data in hazard assessment. QSAR in Environmental Toxicology—II, edited by K. L. E. Kaiser, pp. 261–270, Dordrecht: Reidel, 1987.
Passino, D. R. M.; Hickey, J. P.; and Frank, A. M.: Linear solvation energy relationships for toxicity of selected or-

ganic chemicals to *Daphnia pulex* and *Daphnia magna*. QSAR88, edited by J. E. Turner; M. W. England; T. W. Schultz; and N. J. Kvaak, pp. 131–146. Oak Ridge, TN: U.S. Dept. of Energy, NTIS CONF-880520—(DE88013180), 1988.

Peijenburg, W. J. G. M.; Hart, M. J. T.; den Hollander, H. A.; van de Meent, D.; Verboom, H. H.; and Wolfe, N. L.: QSARs for predicting biotic and abiotic reductive transformation rate constants of halogenated hydrocarbons in anoxic sediment systems. *Sci. Total Environ.*, 109/110:283–300, 1991.

Perrin, D. D.: Dissociation Constants of Organic Bases in Aqueous Solution. London: Butterworth, 1965; Perrin, D. D.: Supplement, 1972.

Perrin, D. D.; Dempsey, B.; and Serjeant, E. P.: pKa Prediction for Organic Acids and Bases. London: Chapman & Hall, 1981.

Pitter, P.: Correlation of microbial degradation rates with chemical structure. *Acta Hydrochim. Hydrobiol.*, 13: 453–460, 1985.

Pleiss, M. A., and Unger, S. H.: The design of test series and the significance of QSAR relationships. Comprehensive Medicinal Chemistry, editorial board chairman C. Hansch, Vol. 4, pp. 561–587. Oxford: Pergamon, 1990.

Protic, M., and Sabljic, A.: Quantitative structure-activity relationships of acute toxicity of commercial chemicals on fathead minnows: effect of molecular size. *Aquat. Toxicol.*, 14:47–64, 1989.

Purcell, W. P.; Bass, G. E.; and Clayton, J. M.: Strategy of Drug Design: A Guide to Biological Activity. New York: Wiley-Interscience, 1973.

Purdy, R.: Taft sigma and sigma$_I$ constants improve log octanol/water coefficient based QSAR for fathead minnow toxicity. QSAR in Environmental Toxicology—II, edited by K. L. E. Kaiser, pp. 271–280. Dordrecht: Reidel, 1987.

Purdy, R.: Quantitative structure-activity relationships for predicting toxicity of nitrobenzenes, phenols, anilines, and alkylamines to fathead minnows. QSAR88, edited by J. E.Turner; M. W. England; T. W. Schultz; and N. J. Kvaak, pp. 99–110. Oak Ridge, TN: U.S. Dept. of Energy, NTIS CONF-880520—(DE88013180), 1988.

Purdy, R: The utility of computed superlocalizability for predicting the LC50 values of epoxides to guppies. *Sci. Total. Environ.*, 109/110:553–556, 1991.

Rabuteau: De quelques propriétés nouvelles ou peu connues due l'alcool du vin ou alcool ethylique; deductions thérapeutiques de ces propriétés—des effets toxiques des alcools butylique et amylique—application à l'alcoolisation du vin improprement appelée vinage. *Union Med. (3rd Ser)* 10:165–173, 1870.

Reinert, K. H.: Aquatic toxicity of acrylates and methacrylates: quantitative structure-activity relationships based on Kow and LC50, *Regul. Toxicol. Pharmacol.*, 7:384–389, 1987.

Rekker, R. F.: The Hydrophobic Fragmental Constant. Pharmacochemistry Library, Vol. 1. Amsterdam: Elsevier, 1977.

Rekker, R. F., and Mannhold, R.: Calculation of Drug Lipophilicity. Weinheim: VCH, 1992.

Ribo, J. M., and Kaiser, K. L. E.: Effects of selected chemicals to photoluminiscent bacteria and their correlation with acute and sublethal effects on other organisms. *Chemisphere*, 12:1421–1442, 1983.

Ribo, J. M., and Kaiser, K. L. E.: Toxicities of chloroanilines to *Photobacterium phosphoreum* and their correlations with effects on other organisms and structural parameters. Kaiser, K. L. E. (ed.) QSAR in Environmental Toxicology, edited by K. L. E. Kaiser, pp. 319–336. Dordrecht: Reidel, 1984.

Richards, N .G. J.; Williams, P. B.; and Tute, M .S.: Empirical methods for computing molecular partition coefficients. I. Upon the need to model the specific hydration of polar groups in fragment-based approaches. *Int. J. Quant. Chem.*, Quantum Biol. Symp. 18:299–316, 1991.

Richardson, B. W.: Physiological research on alcohols. *Med. Times Gazette (Lond.)* 2 (December 18);703–706, 1869.

Richet, C.: Sur le rapport entre la toxicité et les propriétés physiques des corps. *C. R. Soc. Biol. Paris*, 54:775–776, 1893.

Roberts, D. W.: An analysis of published data on fish toxicity of nitrobenzene and aniline derivatives. QSAR in Environmental Toxicology—II, edited by K. L. E. Kaiser, pp. 295–308. Dordrecht: Reidel, 1987.

Roberts, D. W.: Aquatic toxicity of linear alkyl benzene sulphonates (LAS)—a QSAR analysis. QSAR 88, Proceedings Third International Workshop, Quantitative Structure-Activity Relationships in Environmental Toxicology, May 22–26, 1988, Knoxville, TN, edited by J. E. Turner; M. W. England; T. W. Schultz; and N. J. Kwaak, pp. 91–98. NTIS CONF-880520—(DE88013180), 1988.

Roberts, D. W.: Application of QSAR to biodegradation of linear alkylbenzene sulphonate (LAS) isomers and homologues. *Sci. Total. Environ.* 109/110:301–306, 1991.

Rogerson, A.; Shiu, W. Y.; Huang, G. L.; Mackay, D.; and Berger, J.: Determination and interpretation of hydrocarbon toxicity to ciliated protozoa. *Aquat. Toxicol.*, 3:215–228, 1983.

Romanowska, K.: The application of the graph-theoretical method in the QSAR scheme. Possibilities and limits. *Int. J. Quant. Chem.*, 43:175–195, 1992.

Ross, W. C. J.: Biological Alkylating Agents: Fundamental Chemistry and the Design of Compounds for Selective Toxicity. London: Butterworths, 1962.

Russom, C. L.; Anderson, E. B.; Greenwood, B. E.; and Pilli, A.: ASTER: an integration of the AQUIRE database and the QSAR system for use in ecological risk assessments. *Sci. Total Environ.*, 109/110:667–670, 1991.

Saarikoski, J., and Viluksela, M.: Relation between physicochemical properties of phenolics and their toxicites and accumulation in fish. *Ecotoxicol. Environ. Safety*, 6: 501–512, 1982.

Saarikoski, J.; Lindström, R.; Tyynelä, M.; and Viluksela, M.: Factors affecting adsorption of phenolics and carboxylic acids in the guppy (*Poecilia reticulata*). *Ecotoxicol. Environ. Safety*, 11:158–173, 1986.

Sabljic, A: Nonempirical modeling of environmental distribution and toxicity of major organic pollutants. QSAR in Environmental Toxicology—II, edited by K. L. E. Kaiser, pp. 309–332. Dordrecht: Reidel, 1987.

Sabljic, A.: Topological indices and environmental chemistry. Practical Applications of Quantitative Structure-Activity Relationships (QSAR) in Environmental Chemistry and Toxicology, edited by W. Karcher and J. Devillers, pp. 61–82. Dordrecht: Kluwer, 1990.

Sabljic, A.: Chemical topology and ecotoxicology. *Sci. Total Environ.*, 109/110:197–220, 1991.

Schultz, T. W.: Nonpolar narcosis: A review of the mechanism of action for baseline toxicity. Aquatic Toxicology and Environmental Fate: 12th Vol., ASTM STP 1027, edited by U. M. Cowgill and L. R. Williams, pp. 104–109. Philadelphia: American Society for Testing and Materials, 1989.

Schultz, T. W., and Cajina-Quezada, M.: Structure-activity relationships for monoalkylated or halogenated phenols. *Toxicol. Lett.*, 37:121–130, 1987.

Schultz, T. W., and Moulton, B. A.: Structure-activity correlations of selected azaarenes, aromatic amines, and heteroaromatics. QSAR in Environmental Toxicology, edited by K. L. E. Kaiser, pp. 337–357. Dordrecht: Reidel, 1984.

Schultz, T. W., and Moulton, B. A.: Structure-activity relationships for nitrogen-containing aromatic molecules. *Environ. Toxicol. Chem.*, 4:353–359, 1985.

Schultz, T. W.; Halcombe, C. S.; and Phipps, G. L.: Relationships of quantitative structure-activity to comparative toxicity of selected phenols in the *Pimephales promelas* and *Tetrahymena pyrformis* test systems. *Ecotoxicol. Environ. Safety*, 12:146–153, 1986.

Schultz, T. W.; Riggin, G. W.; and Wesley, S. K.: Structure-activity relationships for para-substituted phenols. QSAR in Environmental Toxicology—II, edited by K. L. E. Kaiser, pp. 333–345. Dordrecht: Reidel, 1987

Schultz, T. W.; Arnold, L. M.; Wilke, T. S.; and Moulton, M. P.: Relationships of quantitative structure-activity for normal aliphatic alcohols. *Ecotoxicol. Environ. Safety*, 19:243–253, 1990a.

Schultz, T. W.; Lin, D. T.; Wilke, T. S. and Arnold, L. M.: Quantitative structure-activity relationships for the *Tetrahymena pyriformis* population growth endpoint: a mechanism of action approach. Practical Applications of Quantitative Structure-Activity Relationships (QSAR) in Environmental Chemistry and Toxicology, edited by W. Karcher and J. Devillers, pp. 241–262. Dordrecht: Kluwer, 1990b.

Schultz, T. W.; Lin, D. T.; and Arnold, L. M.: QSARs for monosubstituted anilines eliciting the polar narcosis mechanism of action. *Sci. Total Environ.*, 109/110:569–580, 1991a.

Schultz, T. W.; Wilke, T. S.; Bryant, S. E.; and Hosein, L. M.: QSARs for selected aliphatic and aromatic amines. *Sci. Total Environ.*, 109/110:581–587, 1991b.

Schüürmann, G.: Advances in bioconcentration prediction. *Chemosphere*, 17:1551–1574, 1988.

Schüürmann, G.: Quantitative structure-property relationships for the polarizability, solvatochromic parameters and lipophilicity. *Quant. Struct.-Act. Relat. Chem.*, 9: 326–333, 1990a.

Schüürmann, G.: QSAR analysis of the acute fish toxicity of organic phosphorothionates using theoretically derived molecular descriptors. *Environ. Toxicol. Chem.*, 9: 417–428, 1990b.

Schüürmann, G.: QSAR analysis of the acute toxicity of oxyethylated surfactants. *Chemosphere*, 21:467–478, 1990c.

Schüürmann, G.: Acute aquatic toxicity of alkyl phenol ethoxylates. *Ecotoxicol. Environ. Safety*, 21:227–233, 1991.

Seeman, P.: The membrane actions of anesthetics and tranquilizers. *Pharmacol. Rev.*, 24:583–655, 1972.

Seiler, P.: Interconversion of lipophilicities from hydrocarbon/water systems into the octanol/water system. *Eur. J. Med. Chem. Chim. Ther.*, 9:473–479, 1974.

Serjeant, E. P. and Dempsey, B.: Ionization Constants of Organic Acids in Aqueous Solution. New York: Pergamon, 1979.

Sexton, W. A.: Chemical Constitution and Biological Activity, 3rd ed. Princeton, NJ: Van Nostrand, 1963.

Seydel, J. K., and Schaper, K. J.: Chemische Struktur und Biologische Activität von Wirkstoffen: Methoden der Quantitaven Struktur-Wirkung Analyse. Weinheim: Verlag Chemie, 1979.

Shimp, R. J.; Larson, R. J.; and Boethling, R. S.: Use of biodegradation data in chemical assessment. *Environ. Toxicol. Chem.*, 9:1369–1377, 1990.

Shorter, J.: Correlation Analysis of Organic Reactivity: With Particular Reference to Multiple Regression. Chichester: Research Studies, Press, 1982.

Sijm, T. H. M. and Opperhuizen, A.: Biotransformation of organic chemicals by fish: a review of enzyme activities and reactions. Handbook of Environmental Chemistry, Vol. 2E, Reactions and Processes, edited by O. Hutzinger, pp. 163–235. Heidelberg: Springer Verlag, 1989.

Sijm, T. H. M.; Schipper, M.; and Opperhuizen, A.: Toxicokinetics of halogenated benzenes in fish: Lethal body burden as a toxicological end point. *Environ. Toxicol. Chem.*, 12:1117–1127, 1993.

Silipo, C., and Vittoria, A.: Three-dimensional structure of drugs. Comprehensive Medicinal Chemistry, editorial board chairman C. Hansch, Vol. 4, pp. 153–204. Oxford: Pergamon, 1990.

Slooff, W.; Canton, J. H.; and Hermens, J. L. M.: Comparison of the susceptibility of 22 freshwater species to 15 chemical compounds. I. (Sub)acute toxicity tests. *Aquat. Toxicol.* 4:113–128, 1983.

Southworth, G. R.; Beauchamp, J. J.; and Schmieder, P. K.: Bioaccumulation potential of polycyclic aromatic hydrocarbons in *Daphnia pulex. Water Res.*, 12:973–977, 1978.

Stuper, A. J.; Brügger, W. E.; and Jurs, P. C.: Computer Assisted Studies of Chemical Structure and Biological Function. New York: Wiley, 1979.

Tabak, H. H.; Govind, R.; Gao, C.; and Yu, X.: Prediction of biodegradation kinetics using a nonlinear group contribution method. *Environ. Toxicol. Chem.*, 12:251–260, 1992.

Tanford, C.: The Hydrophobic Effect: Formation of Micelles and Biological Membranes, 2nd ed. New York: Wiley-Interscience, 1980.

Taylor, P. J.: Hydrophobic properties of drugs. Comprehensive Medicinal Chemistry, editorial board chairman C. Hansch, Vol. 4, pp. 241–294. Oxford: Pergamon, 1990.

Thurston, R. F.; Gilfoil, T. A.; Meyn, E. L.; Zajdel, R. K.; Aoki, T. I.; and Veith, G. D.: Comparative toxicity of ten organic chemicals to ten common aquatic species. *Water Res.*, 19:1145–1155, 1985.

Topliss, J. G.: Quantitative Structure-Activity Relationships of Drugs. New York: Academic Press, 1983.

Topliss, J. G., and Costello, R. J.: Chance correlations in structure-activity studies using multiple regression analysis. *J. Med. Chem.*, 15:1066–1068, 1972.

Topliss, J. G., and Edwards, P.: Chance factors in studies of quantitative structure-activity relationships. *J. Med. Chem.*, 22:1238–1244, 1979a.

Topliss, J. G., and Edwards, R. P.: Chance factors in QSAR studies. Computer-Assisted Drug Design, edited by E. D. Olson and R. E. Christofferson, ACS Symp. Ser. 112, pp. 131–145. Washington, American Chemical Society, 1979b.

Tosato, M. L. and Geladi, P.: Design: A way to optimize testing programmes for QSAR screening of toxic substances. Practical Applications of Quantitative Structure-Activity Relationships (QSAR) in Environmental Chemistry and Toxicology, edited by W. Karcher and J. Devillers, pp. 317–341. Dordrecht: Kluwer, 1990.

Toxic Substances Control Act, Public Law 94-469, October 11, 1976.

Turner, J. E.; England, M. W., Schultz, T. W.; and Kvaak, N. J. (eds.): QSAR88, Oak Ridge, TN: U.S. Dept. of Energy, NTIS CONF-880520—(DE88013180), 1988.

Tute, M. S.: History and objectives of quantitative drug design. Comprehensive Medicinal Chemistry, editorial board chairman C. Hansch, Vol. 4, pp. 1–31. Oxford: Pergamon, 1990.

Vaishnav, D. D., and Babeu, L.: Comparison of occurrence and rates of chemical biodegradation in natural waters. *Bull. Environ. Contam. Toxicol.*, 39:237–244, 1987.

Vaishnav, D. D.; Boethling, R. S.; and Babeu, L.: Quantitative structure-biodegradability relationships for alcohols, ketones and alicyclic compounds. *Chemosphere*, 16:695–703, 1987.

Valvani, S. C., and Yalkowsky, S. H.: Physical Chemical Properties of Drugs, edited by S. H. Yalkowsky; A. A. Sinkula; and S. C. Valvani. New York: Marcel Dekker, 1980.

Van de Waterbeemd, H.: Hydrophobicity of Organic Compounds. How to Calculate It by Personal Computers. Vienna: Compudrug International, 1986.

Van de Waterbeemd, H., and Kansy, M.: Hydrogen-bonding capacity and brain penetration. *Chimia*, 46:299–303, 1992.

Van de Waterbeemd, H., and Testa, B.: The parameterization of lipophilicity and other structural properties in drug design. *Adv. Drug Res.*, 16:85–225, 1987.

Van Leeuwen, C. J.; Maas-Diepeveen, J. L.; Niebeek, G.; Vergouw, W. H. A.; Griffioen, P. S.; and Luijken, M. W.: Aquatic toxicological aspects of dithiocarbamates and related compounds. I. Short-term toxicity tests. *Aquat. Toxicol.*, 7:145–164, 1985a.

Van Leeuwen, C. J.; Moberts, F.; and Niebeek, G.: Aquatic toxicological aspects of dithiocarbamates and related compounds. II. Effects on survival, reproduction, and growth of *Daphnia magna. Aquat. Toxicol.*, 7:165–175, 1985b.

Van Leeuwen, C. J.; Adema, D. M. M.; and Hermens, J. L. M.: Quantitative structure-activity relationships for fish early life stage toxicity. *Aquat. Toxicol.*, 16:321–334, 1990.

Van Leeuwen, C. J.; Van der Zandt, P. T. J.; Aldenberg, T.; Verhaar, H. J. M.; and Hermens, L. M.: Application of QSARs, extrapolation and equilibrium partitioning in aquatic effects assessment. I. Narcotic industrial pollutants. *Environ. Toxicol. Chem.*, 11:267–282, 1992.

Varmuza, K.: Pattern Recognition in Chemistry. Berlin: Springer-Verlag, 1980.

Veith, G. D.: State-of-the-art of structure-activity methods development. U.S. Environmental Protection Agency, Environmental Research Laboratory, Duluth, MN. EPA-560/81-029; NTIS PB 81-187-239, 1981.

Veith, G. D., and Broderius, S. J.: Structure-toxicity relationships for industrial chemicals causing type (II) narcosis syndrome. QSAR in Environmental Toxicology—II, edited by K. L. E. Kaiser, pp. 385–391. Dordrecht: Reidel, 1987.

Veith, G. D., and Broderius, S. J.: Rules for distinguishing toxicants that cause type I and type II narcosis syndromes. *Environ. Health Perspect.*, 87:207–211, 1990.

Veith, G. D., and Kosian, P.: Estimating bioconcentration potential from octanol/water partition coefficients. Physical Behavior of PCBs in the Great Lakes, edited by D. Mackay; S. Paterson; and S. J. Eisenreich, pp. 269–282. Ann Arbor, MI: Ann Arbor Science, 1982.

Veith, G. D.; Defoe, D. L.; and Bergstedt, B. V.: Measuring and estimating the bioconcentration factor of chemicals in fish. *J. Fish. Res. Board Can.* 36:1040–1048, 1979.

Veith, G. D.; Macek, K. J.; Petrocelli, S. R.; and Carroll, J.: An evaluation of using partition coefficients and water solubility to estimate bioconcentration factors for organic chemicals in fish. Aquatic Toxicology, STP 707, edited by J. G. Eaton; Parrish, P. R.; and Hendricks, A. C., pp. 116–129. Philadelphia: American Society for Testing and Materials, 1980.

Veith, G. D.; Call, D. J.; and Brooke, L. T.: Structure-toxicity relationships for the fathead minnow, *Pimephales promelas*: narcotic industrial chemicals. *Can. J. Fish. Aquat. Sci.*, 40:743–748, 1983.

Veith, G. D.; Defoe, D. L.; and Knuth, M.: Structure-activity relationships for screening organic chemicals for potential ecotoxicity effects. *Drug. Metab. Rev.*, 15: 1295–1303, 1984–1985.

Veith, G. D.; Greenwood, B.; Hunter, R. S.; Niemi, G. J.; and Regal, R. R.: On the intrinsic dimensionality of chemical structure space. *Chemosphere*, 17:1617–1630, 1988.

Veith, G. D.; Lipnick, R. L.; and Russom, C. L.: The toxicity of acetylenic alcohols to the fathead minnow, *Pimephales promelas*: narcosis and proelectrophile mechanisms. *Xenobiotica*, 19:555–565, 1989.

Verhaar, H. J. M.; Van Leeuwen, C. J.; and Hermens, J. L. M.: Classifying environmental pollutants. 1. Structure-activity relationships for prediction of aquatic toxicity. *Chemosphere*, 25:471–491, 1992.

Vighi, M., and Calamari, D.: QSAR for organotin compounds on *Daphnia magna*. *Chemosphere*, 14:1925–1932, 1985.

Vighi, M.; Garlanda, M. M.; and Calamari, D.: QSARs for toxicity of organophosphorous pesticides to Daphnia and honeybees. *Sci. Total. Environ.*, 109/110:605–622, 1991.

von Oepen B.; Kördel, W.; Klein, W.; and Schüürmann, G.: Predictive QSPR models for estimating soil sorption coefficients: potential and limitations based on dominating processes. *Sci. Total Environ.*, 109/110:343–354, 1991.

Walker, J. D.: Chemical selection by the Interagency Testing Committee: use of computerized substructure searching to identify chemical groups for health effects, chemical fate and ecological effects testing. Practical Applications of Quantitative Structure-Activity Relationships (QSAR) in Environmental Chemistry and Toxicology, edited by W. Karcher, and J. Devillers, pp. 691–700. Dordrecht: Kluwer, 1990.

Wallace, K.B.: Glutathione-dependent metabolism in fish and rodents. *Environ. Toxicol. Chem.*, 8:1049–1055, 1989.

Walsh, C.: Enzymatic Reaction Mechanisms. New York: Freeman, 1979.

Warne, M. S. J.; Connell, D. W.; and Hawker, D. W.: Comparison of the critical concentration and critical volume hypotheses to model non-specific toxicity of individual compounds. *Toxicology*, 66:187–196, 1991.

Weininger, D.: SMILES, a chemical language and information system. 1. Introduction to methodology and encoding rules. *J. Chem. Inf. Comput. Sci.*, 28:31–36, 1988.

Weininger, D., and Weininger, J. L.: Chemical structures and computers. Comprehensive Medicinal Chemistry, editorial board chairman C. Hansch, Vol. 4, pp. 59–82. Oxford: Pergamon, 1990.

Wong, P. T. S.; Chau, Y. K.; Rhamey, J. S.; and Docker, M.: Relationship between water solubility of chlorobenzenes and their effects on a freshwater green algae. *Chemosphere*, 13:991–996, 1984.

Wood, E. M.: The toxicity of 3400 chemicals to fish. U.S. Fish and Wildlife Service, Kearneysville, WV, 1953; in EPA Report 560/6-87-002, NTIS PB 87-200-275, 1987.

Yoshioka, Y.; Mizuno, T.; Ose, Y.; and Sato, T.: The estimation of toxicity of chemicals in fish by physicochemical properties. *Chemosphere*, 15:195–203, 1986.

Zakarya, D.; Belkhadir, M.; and Fkih-Tetouani, S.: Quantitative structure-biodegradability relationships (QSBRs) using modified autocorrelation method (MAM). *SAR QSAR Environ. Res.*, 1:21–27, 1993.

Zaroogian, G.; Heltshe, J. F.; and Johnson, M.: Estimation of toxicity to marine species with structure-activity models developed to estimate toxicity to freshwater fish. *Aquat. Toxicol.*, 6:251–270, 1985a.

Zaroogian, G.; Heltshe, J. F.; and Johnson, M.: Estimation of bioconcentration in marine species using structure-activity models. *Environ. Toxicol. Chem.*, 4:3–12, 1985b.

Zeeman, M. G.; Nabholz, J. V.; and Clements, R. G.: The development of SAR/QSAR for use under EPA's Toxic Substances Control Act (TSCA): an introduction. Environmental Toxicology and Risk Assessment—2nd volume, ASTM STP 1173, edited by J. S. Gorsuch; F. J. Dwyer; C. G. Ingersall; and T. W. LaPoint, pp. 523–539. Philadelphia: American Society for Testing and Materials, 1993.

Zitko, V.: Structure-activity relationships in fish toxicology. Structure-Activity Correlations in Studies of Toxicity and Bioconcentration with Aquatic Organisms, edited by G. D. Veith, and D. E. Konasewich, pp. 7–24. Windsor, Ontario: Great Lakes Research Advisory Board, 1975.

Zitko, V.: Metabolism and distribution by aquatic animals. Handbook of Environmental Chemistry, edited by O. Hutzinger, pp. 221–229. Berlin: Springer-Verlag, 1980.

Zitko, V.; McLeese, D. W.; Carson, W. G.; and Welch, H. E.: Toxicity of alkyldinitrophenols to some aquatic organisms. *Bull. Environ. Contam. Toxicol.*, 16:508–515, 1976.

Chapter 21

THE FEDERAL INSECTICIDE, FUNGICIDE, AND RODENTICIDE ACT

L. W. Touart

INTRODUCTION

The Federal Insecticide, Fungicide, and Rodenticide Act (FIFRA) is the predominant statute for regulating economic poisons commonly referred to as pesticides. Pesticides, as defined by FIFRA, include any substance or mixture of substances intended for preventing, destroying, repelling, or mitigating insects, rodents, fungi, weeds, and other forms of plant or animal life including bacteria and viruses (except in humans or animals). The legislation is curious among environmental protection statutes in that it licenses materials known to be toxic for *intentional* release into the environment. In contrast, other statutes (e.g., Clean Air Act, Clean Water Act, Toxic Substances Control Act) are designed to prevent or reduce the *unintentioned* environmental release of toxic substances. FIFRA is arguably the most forceful environmental statute, and deservedly so, for regulating toxic releases expected to kill or deter pest organisms and potentially other nontarget organisms.

Background

National pesticide control was first attempted by the Insecticide Act of 1910, a limited law whose essence was to prevent the sale of any adulterated or misbranded fungicide or insecticide. The Act was concerned primarily with product effectiveness and protected against misbranded or otherwise fraudulent products in commerce rather than misuse of poisons. FIFRA, as first enacted in 1947 and amended in 1959 and 1961, emphasized consumer protection and product efficacy. This law provides that any "economic poison" or chemical pesticide must be registered before being sold or distributed in interstate commerce. Partially in response to public pressures arising from Rachel Carson's *Silent Spring* (Carson, 1962), Congress amended FIFRA in 1964 to recognize the value of pesticides to the economy and human health but also the potentially harmful environmental effects that could compromise these interests. The amendments were directed at improving labeling statements with precautionary information.

Prior to 1970, the Department of Agriculture controlled approval to market pesticides under FIFRA. This function was passed by Executive Order to the Environmental Protection Agency (EPA), which was established in 1970 to regulate matters concerning air and water pollution, the use of pesticides, and other matters regarding the environment. Following the upsurge in environmental awareness typified by Earth Day and the establishment of the EPA, in 1972 FIFRA was thrust into the forefront of environmental legislation by a major overhaul with far-reaching amendments mandating the protection of both public health and the environment. Specifically, pesticide registration (approval to market) was restricted to pesticides that "will perform their intended purpose without unreasonable adverse effects on the environment" (FIFRA). These amendments placed the burden of generating information to support safety claims directly on the manufacturer seeking registration (registrant). In 1978 Congress passed further amendments to streamline the registration process with generic reregistration reviews, add the authority to grant conditional registrations (interim registration to permit time to fulfill data requirements under FIFRA), and resolve data issues of propriety and compensation. Finally, the current FIFRA was amended in 1988 to require the expedited reregistration of existing pesticides to ensure that registrations are supported by adequate data.

FIFRA as a Balancing Statute

The strength of FIFRA, therefore, is the presumption that pesticides are hazardous and that sufficient human health, environmental fate, and ecological effects data must be provided to assess potential adverse effects to humans and the environment. The inherent risks associated with pesticides and their deliberate release into the environment lead to wide public acceptance with little controversy that registrants should shoulder the burden of these data requirements. Before registration, the registrant must show that a pesticide "when used in accordance with widespread and commonly recognized practice will not generally cause unreasonable adverse effects on the environment" [FIFRA, Section 3(c)(5)(D)]. The criterion for protection was established as "unreasonable adverse effects," which were defined as "any unreasonable risk to man or the environment, taking into account the economic, social, and environmental costs and benefits of the use of any pesticide" (PL 95396, §2 (bb)]. The data provided by the registrant are used by EPA to evaluate a pesticide's risks and benefits. The fact that risks are compared with benefits recognizes that absolute safety is not a viable concept. Some risks are acceptable in order to obtain substantial benefits.

Special Review

The term "special review" is now used by the EPA for a process previously called rebuttable presumption against registration (RPAR). Among the modifications to the process are new risk assessment criteria. Before a product can be registered, the registrant must prove that the material can be used without "unreasonable adverse effects to humans or the environment." If at any time the EPA determines that a pesticide no longer meets the standard for registration, the administrator of the EPA may cancel the registration under Section 6 of FIFRA.

The EPA has developed an administrative process for evaluating whether a pesticide satisfies or continues to satisfy the statutory standard for registration. Special review provides a formal process through which the EPA may gather and evaluate information about the risks and benefits of a pesticide's use. It also provides a means by which the public may comment on and participate in the EPA's decision-making process. The regulations governing this process are set forth in 40 CFR 154.

A special review is triggered when the EPA determines that a pesticide meets or exceeds one or more of the risk criteria listed in the regulations [40 CFR 154.7(a)(3)]. The criteria are designed to assure that a determination of unreasonable risk will be based on both the toxic effects associated with the pesticide and the actual or projected exposure of humans and other nontarget organisms to the pesticide.

EPA can initiate a special review when, based on a validated test or other significant evidence, use of the subject pesticide poses a risk of serious injury to humans or domestic animals. Risks that can trigger a special reivew include oncogenic, heritable genetic, teratogenic, fetotoxic, reproductive, chronic, or delayed toxic effects on humans. Environmental risk that can cause a special review involves pesticide use that may result in residues at or above toxic levels to nontarget organisms or that produce adverse reproductive effects. Included in protection are endangered or threatened species. In addition, pesticide use that threatens habitat destruction or adverse modification can be considered in a special review, as can any other use that poses a risk to humans or the environment to determine whether the risk is offset by social, economic, or environmental benefits.

The EPA generally announces a special review in the Federal Register. Registrants and other interested parties are invited to review the basis for the EPA's decision to initiate the special review and to submit data and information that rebut or support the EPA's determination of risk. Commenters may also suggest methods to reduce risk of the pesticide.

If risk issues are not satisfactorily resolved, the EPA will proceed to evaluate the risks and benefits of the pesticide in order to determine whether to propose regulatory actions to reduce the risks. If the EPA determines that the risk of use exceeds the benefits, it will issue a Notice of Intent to Cancel the registration of products intended for such use. The notice may state the intention to cancel registrations outright or may require certain changes in the composition, packaging, application methods, and/or labeling of the product. The changes would be intended to reduce risks to levels that, when considered against the benefits, will not pose "unreasonable adverse effects to man or the environment."

GENERAL ECOTOXICOLOGY AND ENVIRONMENTAL TESTING REQUIREMENTS UNDER FIFRA

In July 1975, the EPA promulgated final regulations (40 CFR 162, Subpart A) that established the basic testing requirements for registration of pesticide products under FIFRA. From 1975 to 1981, EPA issued or made available several subparts of the Guidelines for Registering Pesticides in the United States which described the kinds of data that must be submitted to satisfy the requirements of the registration regulations. These guidelines detail what data are required and when, the standards for conducting acceptable tests, guidance on the evaluation and reporting of data, and examples of acceptable protocols.

In 1981 the EPA decided it was impractical to include all technical data contained in the guidelines, because there were several acceptable procedures and methods that could be used for each test. Furthermore, ecotoxicology was rapidly evolving as a scientific discipline with continuous changes in state-of-the-art techniques. EPA consequently reorganized the guidelines to limit regulation to a concise presentation of the data requirements and when they are required. Therefore, the full range of data requirements for registration/reregistration or experimental use of each pesticide pertaining to all former subparts of the guidelines are now specified in Part 158. Part 158 specifies the types of data and information the EPA requires to make regulatory judgments regarding the safety of each pesticide proposed for registration or experimental use. Part 158 also specifies the test substance to be used in tests conducted to fulfill the data requirements. The standards for conducting acceptable tests, guidance on evaluation and reporting of data, further guidance on when data are required, and examples of protocols are not specified in Part 158 but are available as advisory documents through the National Technical Information Service (NTIS).

The data requirements for registration presented in Part 158 are intended to generate information necessary to address concerns pertaining to the identity, composition, potential adverse effects, and environmental fate of each pesticide. Subdivisions E (Hazard Evaluation: Wildlife and Aquatic Organisms) and N (Chemistry: Environmental Fate) of the pesticide assessment guidelines are briefly discussed below because of their relevance in ecological risk assessment for aquatic organisms in the registration process.

Tiered Ecological Testing

Subdivision E provides guidelines for testing the effects of pesticides on wildlife and aquatic organisms. The data requirements in 40 CFR Section 158.145 pertain to Subdivision E (see Table 1). These data will be used by the EPA for determining potential hazards to nontarget birds, wild mammals, fish, and aquatic invertebrates. Potential adverse ecological effects of a pesticide are assessed from data developed in a tiered fashion beginning with less expensive acute toxicity tests of a few representative organisms and progressing through (if necessary) more expensive chronic toxicity tests to a final tier of simulated and/or actual field tests, which are the most costly (Figures 1 and 2). The minimum ecological effects data required to support the registration of an outdoor use pesticide include six basic studies: an avian single-dose oral toxicity test, an avian dietary toxicity test with an upland gamebird, an avian dietary test with a waterfowl species, a freshwater fish acute toxicity test with a warm-water species, a freshwater fish acute toxicity test with a cold-water species, and an aquatic invertebrate acute toxicity test with an immature life stage. These tests are used to screen pesticides for potential adverse effects to nontarget organisms. Related information such as use pattern, environmental chemistry, and mammalian toxicology is also considered to determine whether additional information is required to evaluate possible problems. Other short-term nontarget organism testing may be required in the initial tier, which includes tests on estuarine species (finfish, crustaceans, and mollusks), beneficial insects (e.g., honeybees), and nontarget plants.

The initial toxicity data are, therefore, used for the following:

- To define acute toxicity (i.e., LC50, EC50, LD50) of the active ingredient of each chemical to different aquatic and terrestrial oganisms.
- To compare initial acute toxicity data with actual or estimated environmental concentrations (EECs) to assess potential biological impact.
- To provide data that determine the need for precautionary label statements to minimize the potential adverse effects to wildlife and aquatic organisms.
- To indicate the need for further laboratory and/or field studies.

Understanding that pesticides can be beneficial to society, concerns identified in this first tier do not halt the registration process. Registrants can obviate concens such as potential long-term problems raised by the initial tier by providing information at higher tiers that contribute to a more comprehensive assessment. Concerns are typically based on ecotoxicity and fate data (see below). Additional higher-tier testing includes chronic and reproduction toxicity tests with aquatic organisms and/or birds that allow the estimation of a no-effect or lowest effect concentration for the toxicant that can be compared with expected (or estimated) environmental concentrations or residues for determining whether adverse effects are likely to occur. At this stage in the registration and assessment process, a pesticide is considered, for regulatory purposes, as either acceptably safe or potentially harmful. Where harm is anticipated and the proposed use warrants, registrants may be given the opportunity to obviate concerns by conducting simulated (e.g., replicated aquatic mesocosm studies) and/or actual (e.g., pond studies) field testing to negate the impacts predicted by interpretation of the laboratory-generated information. The Agency intends to regulate on the basis of laboratory and exposure information without requiring or waiting for field studies except in rare circumstances.

When available, field studies are used to confirm experimentally the safety of a chemical under anticipated conditions of use (i.e., at the label use rate, frequency, and typical method of application). The necessity for conducting field studies is greatest if the margin between laboratory-demonstrated toxicological effects and environmental exposure concentrations is narrow (i.e., margin of safety is low). In addition, if a chemical is directly applied to water, field data may be necessary to confirm that the chemical has no significant adverse effect on the ecosystem structure or function, which cannot be ascertained by single-species laboratory testing alone.

Environmental Fate Testing

The data requirements for environmental fate (Subdivision N) are presented in 40 CFR §158.130 (see Table 2). Subdivision N provides guidance concerning data development for environmental fate of pesticides through degradation, metabolism, dissipation, accumulation, and transport. Such data provide information on:

- The most significant degradation routes
- The primary degradates

Table 1. 158.145 Wildlife and Aquatic Organism Data Requirements[a]

Kind of data required	Wildlife and aquatic organisms data requirements: general use patterns									Test substance		
	Terrestrial		Aquatic		Greenhouse							
	Food crop	Nonfood	Food crop	Nonfood	Food crop	Nonfood	Forestry	Domestic outdoor	Indoor	Data to support MP	Data to support EP	Guidelines reference number
Avian and mammalian testing												
Avian oral LD_{50}[b]	(R)	(R)	(R)	(R)	(CR)	(CR)	(R)	(R)	(CR)	TGAI	TGAI	71-1
Avian dietary LC_{50}[b]	(R)	(R)	(R)	(R)	(CR)	(CR)	(R)	(R)	(CR)	TGAI	TGAI	71-2
Wild mammal toxicity[c]	(CR)	(CR)	(CR)	(CR)			(CR)	(CR)		TGAI	TGAI	71-3
Avian reproduction[d]	(CR)	(CR)	(CR)	(CR)			(CR)	(CR)		TGAI	TGAI	71-4
Simulated and actual field testing[b]—mammals and birds	(CR)	(CR)	(CR)	(CR)						TEP	TEP	71-5
Aquatic organism testing												
Freshwater fish LC_{50}[b,e]	(R)	(R)	(R)	(R)	(CR)	(CR)	(R)	(R)	(CR)	TGAI	TGAI	72-1
Acute LC_{50} freshwater invertebrates[b,e]	(R)	(R)	(R)	(R)	(CR)	(CR)	(R)	(R)	(CR)	TGAI	TGAI	72-2
Acute LC_{50} estuarine and marine organisms[e,f]	(CR)	(CR)	(CR)	(CR)			(CR)	(CR)		TGAI	TGAI	72-3
Fish early life stage and aquatic invertebrate life cycle[g]	(CR)	(CR)	(CR)	(CR)			(CR)	(CR)		TGAI	TGAI	72-4
Fish, life cycle[h]	(CR)	(CR)	(CR)	(CR)			(CR)	(CR)		TGAI	TGAI	72-5
Aquatic organism accumulation[c]	(CR)	(CR)	(CR)	(CR)			(CR)	(CR)		TGAI, PAIRA, or degradation product	TGAI, PAIRA, or degradation product	72-6
Simulated or actual field testing aquatic organisms[i]	(CR)	(CR)	(CR)	(CR)			(CR)	(CR)		TEP	TEP	72-7

[a]Sections 158.50 and 158.100 describe how to use this table to determine the environmental fate data requirements and the substances to be tested. *Key*: R = required; CR = conditionally required; parentheses (i.e., (R), (CR)) indicate requirements that apply when an experimental use permit is being sought; TGAI = technical grade of the active ingredient; PAIRA = "pure" active ingredient—radiolabeled; TEP = typical end-use product; EP = end-use product; MP = manufacturing use product.

[b]Tests for pesticides intended solely for indoor application will be required on a case-by-case basis, depending on use pattern, production volume, and other pertinent factors.

[c]Tests required on a case-by-case basis depending on the results of lower tier studies such as acute and subacute testing, intended use pattern, and pertinent environmental fate characteristics.

[d]Data required if one or more of the following criteria are met: (1) Birds may be subjected to repeated or continued exposure to the pesticide or any of its major metabolites or degradation products, especially preceding or during the breeding season. (2) The pesticide or any of its major metabolites or degradation products are stable in the environment to the extent that potentially toxic amounts may persist in avian feed. (3) The pesticide or any of its major metabolites or degradation products is stored or accumulated in plant or animal tissues, as indicated by its octanol–water partition coefficient, accumulation studies, metabolic release and retention studies, or as indicated by structural similarity to known bioaccumulative chemicals. (4) Any other information, such as that derived from mammalian reproduction studies, that indicates that reproduction in terrestrial vertebrates may be adversely affected by the anticipated use of the pesticide product. *Note*: Prior to conducting this test to support the registration of an avicide, the applicant should consult the U.S. EPA.

[e]Data from testing with the applicant's end-use product or a typical end-use product is required to support the registration of each end-use product which meets any one of the following conditions: (1) the end-use pesticide will be introduced directly into an aquatic environment when used as directed; (2) the LC_{50} or EC_{50} of the technical grade of active ingredient is equal to or less than the maximum expected environmental concentration (MEEC) or the estimated environmental concentration (EEC) in the aquatic environment when the end-use pesticide is used as directed; or (3) an ingredient in the end-use formulation other than the active ingredient is expected to enhance the toxicity of the active ingredient or to cause toxicity to aquatic organisms.

[f]Data required if the product is intended for direct application to the estuarine or marine environment, or the product is expected to enter this environment in significant concentrations because of its expected use or mobility pattern.

[g]Data from fish early life-stage tests or life-cycle tests with aquatic invertebrates (on whichever species is most sensitive to the pesticide as determined from the results of the acute toxicity tests) are required if the product is applied directly to water or expected to be transported to water from the intended use site, and when any one or more of the following conditions apply: (1) if the pesticide is intended for use such that its presence in water is likely to be continuous or recurrent regardless of toxicity; or (2) if any LC50 or EC50 value determined in acute toxicity testing is less than 1 mg/L; or (3) if the estimated environmental concentration in water is equal to or greater than 0.01 of any EC50 or LC50 determined in acute toxicity testing; or (4) if the actual or estimated environmental concentration in water resulting from use is less than 0.01 of any EC50 or LC50 determined in acute toxicity testing and if the following conditions exist: (a) studies of other organisms indicate the reproductive physiology of fish and/or invertebrates may be affected; or (b) physicochemical properties indicate cumulative effects; or (c) the pesticide is persistent in water (e.g. half-life in water greater than 4 d).

[h]Data are required if end-use product is intended to be applied directly to water or expected to transport to water from the intended use site, and when any of the following conditions apply: (i) if the estimated environmental concentration is equal to or greater than one-tenth of the no-effect level in the fish early life-stage or invertebrate life-stage or invertebrate life-cycle test; or (ii) if studies of other organisms indicate the reproductive physiology of fish may be affected.

[i]Required if significant concentrations of the active ingredient and/or its principal degradation products are likely to occur in aquatic environments and may accumulate in aquatic organisms.

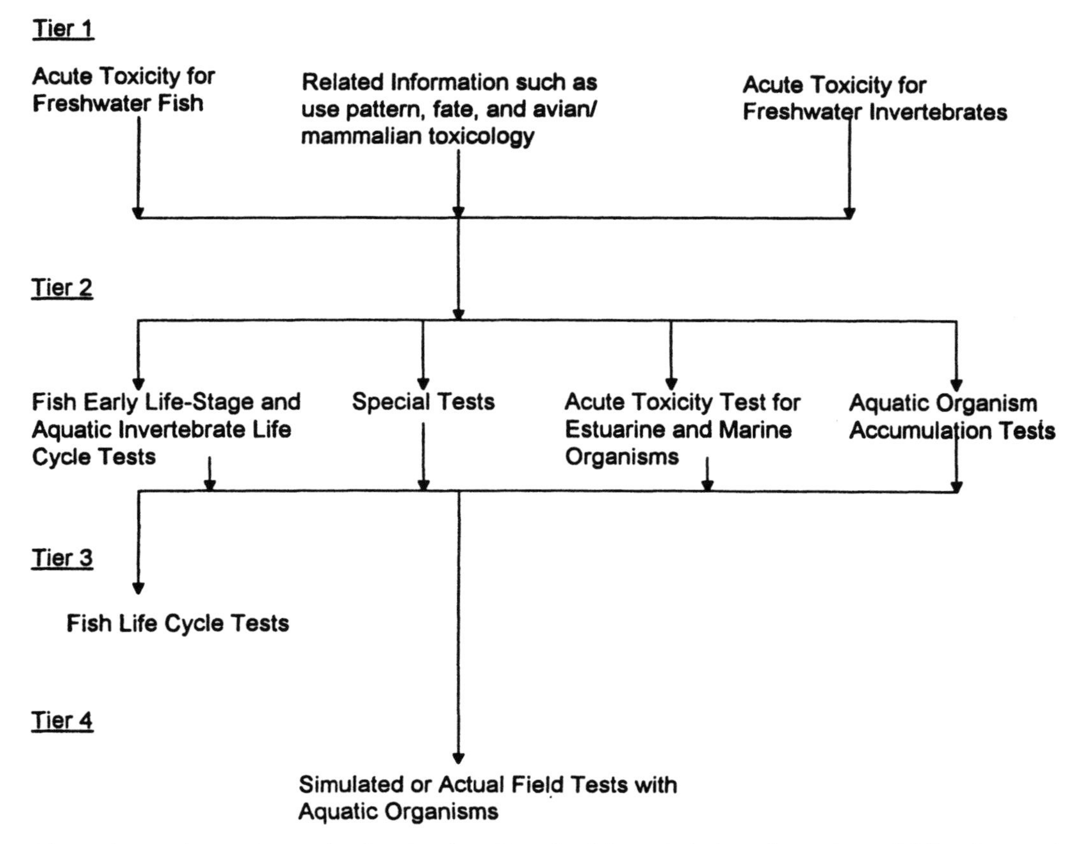

Figure 1. Aquatic organism testing flowchart in FIFRA. Special tests include studies on bioavailability (i.e., pond water studies) and/or small microcosm tests with a multispecies approach.

- The degradation/metabolic pathway
- the potential half-life and persistence in different environmental compartments (soil, sediment, water, air)
- The dissipation and mobility of the chemical, which environmental compartments will most likely be exposed, and potential environmental concentrations

For outdoor use products, the environmental fate information must be adequate to allow determination of an EEC appropriate for use in an ecological risk assessment (see Chapter 28). Environmental fate data are used as part of the criteria along with the effects data to trigger or move from one tier to the next higher tier.

Degradation tests performed in the laboratory include hydrolysis and photodegradation in both soil and water. Data on photodegradation in air are required on a case-by-case basis depending on the product use pattern and other factors. Laboratory aerobic and anaerobic metabolism tests are performed in soil and water. Anaerobic soil metabolism data are not required if an anaerobic aquatic metabolism study has been conducted. Anaerobic aquatic data are required for aquatic food crop and nonfood uses.

Environmental transport is determined through leaching, adsorption/desorption, volatility, and spray drift studies. Leaching and adsorption/desorption data are required for all outdoor use registration actions. Laboratory and field data on volatility are required on a case-by-case basis depending on product chemistry, use patterns, and other pertinent factors. These tests are considered basic to an understanding of likely field dissipation routes.

Field dissipation studies are done under actual use conditions at maximum label rates and include studies in soil (field and vegetable crops), water, and forests. Field soil dissipation data are required for permits for terrestrial food crop, nonfood, and domestic outdoor uses of pesticides. Aquatic sediment dissipation data are required for aquatic food crop and nonfood uses. Combination and tank mix dissipation data are required on a case-by-case basis depending on product use patterns and other pertinent factors. Long-term soil dissipation data are required for pesticide residues that persist in soil for more than one growing season.

Accumulation tests can be done for rotational crops, irrigated crops, fish, and aquatic invertebrates. A confined accumulation study on rotational crops is

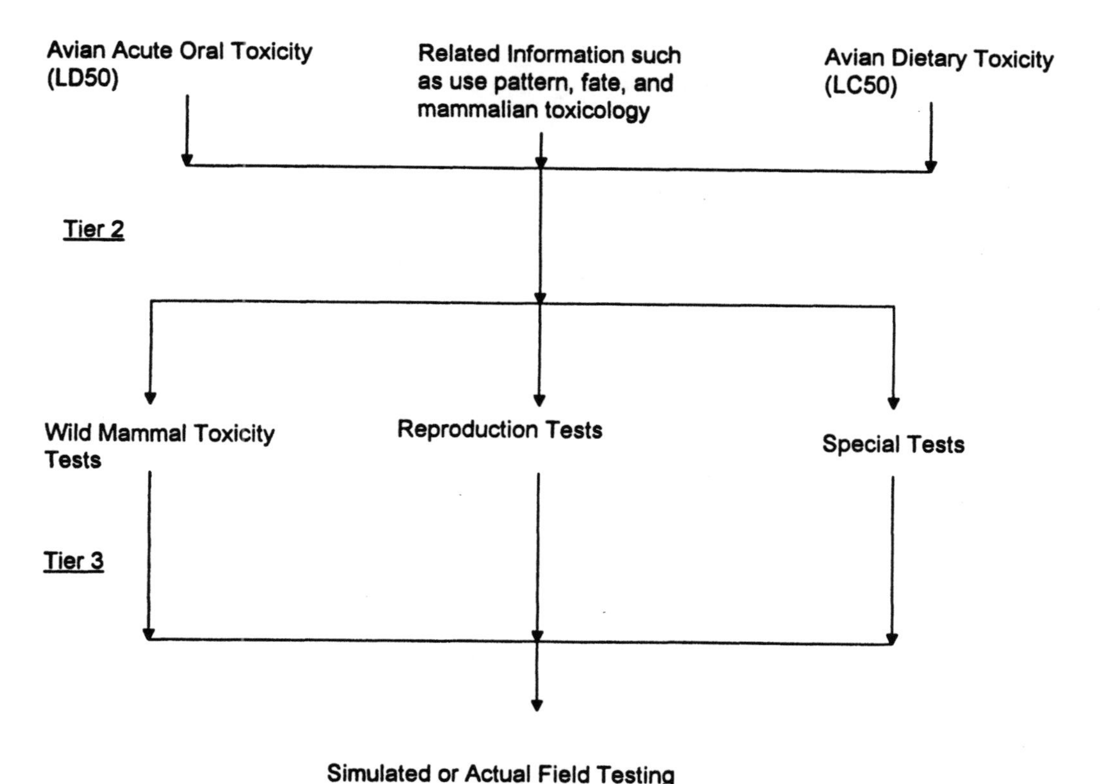

Figure 2. Terrestrial wildlife testing flowchart. Special tests include studies on bioconcentration, behavior, etc.

required when any food or feed crop may be planted after the treated crop. A field accumulation study on rotational crops may be required if significant pesticide accumulation in rotational crops is evidenced by residue data obtained from a confined accumulation study. Data are required on accumulation in irrigated crops if water at a treated site may be used for irrigation. Data on fish accumulation are required if aquatic biota are likely to be exposed to significant concentrations of the active ingredient and/or its principal degradation products.

RISK ASSESSMENT UNDER FIFRA

The Office of Pesticide Programs (OPP) within EPA administers the statutory requirements under FIFRA. Ecological risk assessments are required to determine whether use of a pesticide will present an unreasonable risk to the environment. Ecological risk assessments are used in several ways within OPP. EPA evaluates ecological risk to (1) determine whether to register or reregister a pesticide and whether its use should be restricted to certified applicators; (2) trigger a special review of a pesticide, which may lead to suspension or cancellation of the pesticide's registration; and (3) determine whether use limitations are necessary in certain areas to protect endangered species.

The Risk Assessment Process

The potential aquatic risks of a proposed pesticide to nontarget finfish and invertebrates in freshwater and estuarine environments are examined by comparing toxicological and exposure data using regulatory risk criteria (Table 3). The toxicological data may include acute LC50 values or chronic no-effect levels. Exposure data may consist of aquatic field pesticide residue data or an estimate of pesticide residues to which aquatic organisms may be exposed. The approach taken by the EPA to estimate exposure (i.e., how much pesticide will be in water, the estimated environmental concentration) is similar to the tier testing scheme employed by EPA (Table 4). Lentic systems (e.g., ponds) are more easily modeled, so this type of system is evaluated first.

At levels 1 and 2 mathematical calculations are used to estimate an EEC. The objective is to estimate a concentration that the Agency is confident will not be exceeded by the labeled use and is generally referred to as a maximum worst-case EEC. In level 3, aquatic EECs are estimated using computer runoff

Table 2. 158.130 Environmental Fate Data Requirement[a]

	Environmental fate data requirements: general use patterns									Test substance		
	Terrestrial		Aquatic		Greenhouse							
Kind of data required	Food crop	Nonfood	Food crop	Nonfood	Food crop	Nonfood	Forestry	Domestic outdoor	Indoor	Data to support MP	Data to support EP	Guidelines reference number
					Degradation studies—laboratory							
Hydrolysis	(R)	(R)	(R)	(R)	(R)	(R)	(R)	(R)		TGAI or PAIRA	TGAI or PAIRA	161-1
Photodegradation												
In water	R	R	R	R			R			TGAI or PAIRA	TGAI or PAIRA	161-2
On soil[b]	CR CR						CR			TGAI or PAIRA	TGAI or PAIRA	161-3
In air[c]										TGAI or PAIRA	TGAI or PAIRA	161-4
					Metabolism studies—laboratory							
Aerobic soil	(R) R	(R)			R	R	(R)	R		TGAI or PAIRA	TGAI or PAIRA	162-1
Anaerobic soil[d]										TGAI or PAIRA	TGAI or PAIRA	162-2
Anaerobic aquatic			R	R						TGAI or PAIRA	TGAI or PAIRA	162-3
Aerobic aquatic			(R)	(R)						TGAI or PAIRA	TGAI or PAIRA	162-4
					Mobility studies							
Leaching (adsorption/desorption)	(R)	(R)	(R)	(R)	(R)	(R)	(R)	(R)		TGAI or PAIRA	TGAI or PAIRA	163-1
Volatility												
(Lab)[c]	(CR)				(CR)	(CR)				TEP	TEP	163-2
(Field)[c]	(CR)				(CR)	(CR)				TEP	TEP	163-3
					Dissipation studies—field							
Soil	(R)	(R)						(R)		TEP	TEP	163-1
Aquatic (sediment)			(R)	(R)						TEP	TEP	164-2
Forestry							R			TEP	TEP	164-3
Combination and tank mixes[c]												164-4
Soil, long-term[e]	CR		(CR)							TEP	TEP	164-5

Accumulation studies

Rotational crops												
(Confined)[f]	(CR)		(CR)							PAIRA	PAIRA	165-1
(Field)[g]	(CR)		(CR)							TEP	TEP	165-2
Irrigated crops[h]			(CR)	(CR)						TEP	TEP	165-3
In fish[i]	(CR)	(CR)	(CR)	(CR)				(CR)		TGAI or PAIRA	TGAI or PAIRA	165-5
In aquatic nontarget organisms[i,j]				(CR)				(CR)		TEP	TEP	165-5

[a]Sections 158.50 and 158.100 describe how to use this table to determine the environmental fate data requirements and the substances to be tested. *Key*: R = required; CR = conditionally required: parentheses (i.e., (R), (CR)) indicate requirements that apply when an experimental use permit is being sought; TGAI = technical grade of the active ingredient; PAIRA = "pure" active ingredient—radiolabeled; TEP = typical end-use product; EP = end-use product; MP = manufacturing use product.

[b]Not required if use involves application to soils solely by injection of the product into the soil or by incorporation of the product into the soil upon application.

[c]Required on case-by-case basis depending on product use pattern and other pertinent factors.

[d]Not required if anaerobic aquatic metabolism study has been conducted.

[e]Required if pesticide residues do not readily dissipate in soil.

[f]Confined accumulation study is required when it is reasonably foreseeable that any food or feed crop may be subsequently planted on the site of pesticide application.

[g]Field accumulated study is required if significant pesticide residue is likely to be present in soil at time of plant crop, as evidence by residue data obtained from confined accumulation study.

[h]Required if it is reasonably foreseeable that water at treated site may be used for irrigation purposes.

[i]Required if significant concentrations of the active ingredient and/or its principal degradation products are likely to occur in aquatic environments and may accumulate in aquatic organisms.

[j]Required unless tolerance or action level for fish has been granted.

Table 3. Regulatory Risk Criteria—Aquatic Organisms

Presumption of no risk	Presumption of risk that may be mitigated by restricted use	Presumption of unacceptable risk	
		Nonendangered	Endangered
Acute toxicity EEC < 1/10 LC50	1/10 LC50 ≤ EEC ≤ 1/2 LC50	EEC ≥ 1/2 LC50	EEC ≥ 1/20 LC50 or EEC ≥ 1/10 LC10
Chronic toxicity EEC ≤ Chronic NOEC	N/A	EEC > NOEC	EEC > NOEC

Adapted from Urban and Cook, 1986.

and exposure models with a reference scenario and/or simulation modeling is undertaken using an environmental scenario based on information from a typical high-production growing area for the crop or use situation. For level 4, which is undergoing development, simulation modeling is done as in level 3 but with the addition of stochastic processing for significant parameters that are known to vary appreciably. Estimates of environmental concentrations are combined with probability functions to allow statements with expressed degrees of confidence (e.g., an expected concentration greater than a specific value may occur in 5% of 1000 applications made during a 10-yr period).

Table 4. Aquatic EEC Tier System

Level 1: Direct application model
Nondynamic shallow-water system (< 15 cm) receiving direct application which is evenly mixed and available
If EEC < risk criteria = minimal or no risk
If EEC > risk criteria = proceed to level 2 if direct application scenario unlikely
Level 2: Simple drift or runoff model
Exposure variables such as drainage basin size, surface area or receiving water, average depth, pesticide solubility, surface runoff, or spray drift loss rate are considered which attenuate the direct application model estimate.
If EEC < risk criteria = minimal or no risk
If EEC > risk criteria = risk perceived, proceed to level 3
Level 3: Computer runoff and aquatic exposure simulation model
A loading model (SWRRB, PRZM, etc.) is used to estimate field losses of pesticide associated with surface runoff and erosion, which then serves as input into a partitioning model (EXAMS II) to estimate sorbed and dissolved residue concentrations. Simulations are based on either reference environment scenarios or environment scenarios derived from typical use circumstances.
If EEC < risk criteria = minimal risk
If EEC > risk criteria = risk perceived, proceed to level 4
Level 4: Stochastic modeling
EECs expressed as exceedance probabilities for the environment, field, and/or cropping conditions

Of the commonly used models, the Simulator for Water Resources in Rural Basins–Water Quality (SWRRB-WQ), Pesticide Root Zone Model (PRZM), Groundwater Loading Effects of Agricultural Management Systems (GLEAMS), and Erosion/Productivity Impact Calculator (EPIC) estimate pesticide concentration (by mass) in runoff, and the Exposure Analysis Modeling System (EXAMS II) estimates pesticide concentrations in aquatic systems receiving runoff. The models use the physical/chemical characteristics and environmental fate data (Subdivision N) for the pesticide. The models yield estimated concentrations in the sediment and water compartments and a series of tables and graphs estimating residue concentrations over time.

An actual aquatic field residue study may be performed to verify or refute the model estimates. Typically, if EECs exceed the aquatic toxicological levels of concern based on the regulatory risk criteria, hazard to aquatic organisms is presumed. In some cases risk may be mitigated by restricted use or changes to the use label (e.g., application rate reduction, reduced freqency of application). In a few cases, the risk may be too great to mitigate and would require scrutiny under special review risk-benefit analysis to determine if it is indeed unacceptable.

Risk Assessment End Points

Ecological end points are those that are directly related to observable changes in the abiotic and biotic components of an aquatic ecosystem. Kelly and Harwell (1989) suggest that ecological end points should be limited to those with "relevance to issues of concern to humans," so that the myriad potential measures of ecosystem health may be narrowed to a manageable level. This suggestion implies that ecosystem changes are considered relevant only if they are directly or indirectly related to something of concern to humans. Without apparent human relevance the change would not constitute an end point. Another perspective could be that if an impact is relevant to the health of an ecosystem, then it is necessarily significant and relevant to humankind.

Harte et al. (1980, 1981) identified direct chemical threats to the quality of drinking water, impairment of sport fish populations, aesthetic loss from increasing turbidity or eutrophication, enhanced odor-producing biological activity, and increased likelihood of disease-bearing vectors and pathogens as areas of human concern. Any observed effects that degrade or threaten to alter an ecosystem such as might be observed in an aquatic field study in tier 4 for aquatic testing should be considered a significant and relevant observational end point. Ecosystem stress is manifested through changes in nutrient cycling, productivity, the size of dominant species, species diversity, and/or a shift in species dominance to opportunistic shorter-lived forms (Rapport et al., 1985; see Chapter 19).

Levels of biological organization range from that of biochemical structure and activity to that of the ecosphere, which embodies all living systems (see Chapter 19). The organizational levels most relevant for consideration within an aquatic mesocosm or other aquatic field study are those of individual species, populations, communities, and the ecosystem. Emergent properties of individuals, populations, and communities presented by Meyerhoff (1990) are relevant to this discussion and are discussed below.

Survival is the fundamental emergent property of an individual; an organism is either alive or dead. Without life, one cannot possess any other property at the individual level or contribute to a property at another organizational level. Other major properties at the individual level include behavior, growth, and reproduction. Behavior measures are somewhat species specific in that one may measure swimming speed for a finfish or siphon rate for a sedentary bivalve, and the relevance of these measures also varies by their importance to the survival, growth, or reproduction of the given individual. Growth is usually determined as a change in size, length, and/or weight over a set period of time or as a change in morphological development. Measures of reproduction (e.g., fecundity, number of offspring) determine the success or adaptive fitness of the individual.

Populations are defined by their persistence, production, and structure. Measures of abundance and distribution over time are used in determining the persistence of a population. Changes in biomass and density (population number per unit area) determine population production. The structure of a population is characterized by size/age class distribution, recruitment, and spatial distribution.

Emergent properties of communities are expressed in their composition, organization, and productivity. Community composition is expressed in species diversity, species dominance, and succession (change in species dominance). Organization is measured by functional groups or guild structure and complexity of the trophic structure (linkages). The productivity of a community is expressed by measures of energy or material flow (trophic dynamics) or production capacity of the functional groups.

Jenkins et al. (1989) identified several methods sufficiently developed for measuring aquatic ecosystem functioning (see Chapter 19). These include primary production, secondary production of zooplankton and benthic macroinvertebrates, respiration rates, and leaf degradation rates.

Socioeconomic end points relate directly to an aesthetic, human health, or economic condition resulting from a change in an aquatic ecosystem. Although seemingly defined in much the same way as an ecological end point of human concern, the distinction is in the direct nature or influence of an ecosystem change on the human condition. Examples include reductions in abundance and production of commercial or sport (game) fish populations or development of algal populations that detract from water use (Barnthouse et al., 1986).

Regulatory Actions Under FIFRA

Impacts of regulatory concern include impacts of both ecological and socioeconomical relevance. FIFRA guides three distinct actions in its licensing of pesticides: (1) labeling, (2) classification, and (3) registration.

Assessment end points associated with labeling are those used in establishing appropriate label precautions. For example, the precautionary statement "This pesticide is toxic to fish" is based on acute lethality and is required when a laboratory-derived LC50 is less than 1 ppm for the most sensitive fish surrogate tested. Stronger language, such as the statement "This pesticide is extremely toxic to fish," is based on laboratory data as above and also on lethality confirmed in a simulated or actual field study or under normal use. That is, if direct mortality to finfish is confirmed (as opposed to expected) under conditions akin to normal use of a pesticide, the adjective "extremely" is used to accentuate the concern. Therefore, the impacts of regulatory concern that have been identified are incorporated in label precautions to reduce risks to the environment and/or warn the user of expected or possible adverse effects to the environment.

Under FIFRA, pesticides may be classified for *general use* or *restricted use* or simply left unclassified. A general-use or unclassified product is one available for use to the general public, whereas a restricted-use product is available for use only by or under the supervision of an applicator certified to have completed pesticide safety training.

Criteria have been established to identify pesticides that pose sufficient risk to warrant a restricted-use classification [40 CFR 152.170(c)]. A pesticide product intended for outdoor use will be considered for restricted-use classification, on ecological grounds, if it may cause discernible adverse effects in nontarget organisms, such as significant mortality or effects on the physiology, growth, population levels, or reproduction rates of such organisms. These criteria apply whether from direct or indirect exposures to the pesticide, its metabolites, or its degradation products. The classification for restricted use is also contingent on the condition that restricting the use will result in the associated risks being acceptable after risk-benefit analysis. Classification of a pesticide is generally done (1) for new active ingredients, (2) for new uses of a previously restricted pesticide, (3) as part of reregistration, and (4) as part of special review.

For ultimate regulatory action on registration, there are three possible decisions—register, deny registration, or suspend or cancel an existing registration. The Agency under FIFRA [Section 3(c)(5)] may register a pesticide when it determines, among other findings, that the pesticide will perform its intended function when used in accordance with widespread and commonly recognized practice and will not generally cause unreasonable adverse effects on the environment.

Criteria have been established for identifying potentially unreasonable adverse effects (40 CFR Part 154.7), albeit a determination that adverse effects are indeed *unreasonable* is made only after risk-benefit analysis in a process termed special review, as previously described. Briefly, a pesticide use exceeds the criteria for unreasonable adverse effects and could be considered for special review if environmental concentrations equal or exceed concentrations that are toxic to nontarget organisms or impair reproduction, if it poses a risk to the continued existence of any endangered or threatened species, or if it otherwise poses a risk to the environment of sufficient magnitude to warrant risk-benefit analysis.

Implicit within FIFRA and explicit in the regulations (40 CFR Part 154.5), "the burden of persuasion that a pesticide product is entitled to registration or continued registration for any particular use or under any particular set of terms and conditions of registration is always on the proponent(s) of registration." Therefore, a registration decision hinges on the ability of the registrant (proponent of registration) to demonstrate that a given product will not result in *unreasonable* adverse effects.

LITERATURE CITED

Barnthouse, L. W.; and Suter, G. W., II; Bartell, S. M.; Beauchamp, J. J.; Gardner, R. H.; O'Neill, R. V.; and Rosen, A. E.: User's Manual for Ecological Risk Assessment. ORNL-6251. Oak Ridge, TN: Environmental Sciences Division Publication No. 2679, Oak Ridge National Laboratory, 1986.

Carson, R.: *Silent Spring*. New York: Houghton Mifflin, 1962.

Harte, J.; Levy, D.; Rees, J.; and Sagebarth, E.: Making microcosms an effective assessment tool. Microcosms in Ecological Research, edited by Giesy, J. P. Jr., pp. 105–137. Symposium Series 52, Conference 781101. Washington, DC: U.S. Department of Energy, 1980.

Harte, J.; Levy, D.; Rees, J.; and Sagebarth, E.: Assessment of optimum aquatic microcosm design for pollution impact studies. Report EA-1989, Electric Power Research Institute, Palo Alto, CA, 1981.

Harwell, M. A., and Harwell, C. C.: Environmental decision making in the presence of uncertainty. Ecotoxicology: Problems and Approaches, edited by S. A. Levin; M. A. Harwell; J. R. Kelly; and K. D. Kimball; pp. 515–540. New York: Springer-Verlag, 1989.

Jenkins, D. G.; Layton, R. J.; and Buikema, A. L., Jr: State of the art in aquatic ecological risk assessment, MPPEAL 75:18–32. Using Mesocosms to Assess the Aquatic Ecological Risk of Pesticides: Theory and Practice, edited by J. R. Voshell, Lanham, MD: Entomological Society of America, 1989.

Kelly, J. R., and Harwell, M. A.: Indicators of ecosystem response and recovery. Ecotoxicology: Problems and Approaches, edited by S. A. Levin; M. A. Harwell; J. R. Kelly; and K. D. Kimball; pp. 9–35. New York: Springer-Verlag, 1989.

Meyerhoff, R.: Biological systems of convern. Presentation to the Aquatic Effects Dialogue Group, May 15, 1990, The Conservation Foundation, Washington, DC, 1990.

Rapport, D. J.; Regier, H. A.; and Hutchinson, T. C.: Ecosystem behavior under stress. *Am. Nat.*, 125:617–640, 1985.

Urban, D. J., and Cook, N. J.: Hazard Evaluation Division Standard Evaluation Procedure: Ecological Risk Assessment. EPA-540/9-86-167. NTIS PB86-247657. Springfield, VA: National Technical Information Service, 1986.

Chapter 22

TOXIC SUBSTANCES CONTROL ACT (TSCA) INTERAGENCY TESTING COMMITTEE: DATA DEVELOPED UNDER SECTION 4 OF TSCA

J. D. Walker

INTRODUCTION

The Toxic Substances Control Act (TSCA; Public Law 94-469, 90 Stat 2003 et seq., 15 USC 2601 et seq.) was promulgated in 1976 to regulate the manufacture, processing, use, transportation, or disposal of certain industrial chemical substances and to protect human health and the environment by requiring testing and necessary use restrictions on these chemical substances. In general, these chemicals do not include substances that are used only as pesticides, tobacco products, nuclear materials, foods, food additives, drugs, cosmetics, and medical devices.

This chapter describes TSCA and the roles of the TSCA Interagency Testing Committee (ITC), the U.S. Environmental Protection Agency (EPA), and chemical manufacturers in recommending, implementing, and conducting aquatic toxicity, bioconcentration, and chemical fate testing under section 4 of TSCA. It provides readers with an understanding of how TSCA is used to facilitate aquatic toxicity, bioconcentration, and chemical fate testing of commercially available industrial chemicals (*existing chemicals*) and what aquatic toxicity, bioconcentration, and chemical fate data have been developed as a result of this testing.

CHEMICAL TESTING UNDER THE TOXIC SUBSTANCES CONTROL ACT

There are 31 sections in TSCA. Sections 2 and 4 of TSCA discuss the development of aquatic toxicity, bioconcentration, or chemical fate data for existing chemicals.

Existing Chemicals

Existing chemicals are chemicals that were produced in or imported into the United States before TSCA became effective in 1977 or chemicals for which notice to commence production were issued after 1977 by EPA under TSCA section 5.

Section 2 of TSCA states that the U.S. Congress finds that humans and the environment are exposed to large numbers of chemicals and that some of these chemicals may present an unreasonable risk of injury to health or the environment. Section 2(b) states that it is the policy of the United States that "adequate data should be developed with respect to the effect of chemical substances and mixtures on health and the environment and that the development of such data should be the responsibility of those who manufacture and those who process such chemical substances and mixtures."

Section 4 requires the testing of existing chemical substances and mixtures ("chemicals"). Section 4 of TSCA was enacted by Congress in response to concerns that the effects of existing chemical substances and mixtures on human health and the environment were inadequately documented and understood. To alleviate these concerns, Congress created the ITC to screen, prioritize, and recommend existing chemicals for testing to the EPA Administrator and empowered the EPA Administrator with authority to require that manufacturers or processors test their chemicals to develop adequate data. These data are used by EPA and other U.S. government, foreign government, and international organizations, as well as state and local governments, to develop hazard and exposure assessments (see Chapters 28 and 29) that are necessary to promote pollution prevention or chemical regulation (Walker, 1993a).

Section 4(a) ensures that existing chemicals that may present an "unreasonable risk" to human health or the environment, or may involve substantial production or exposure, receive priority testing consideration and that manufacturers or processors of these chemicals test them to assure that adequate data are developed to assess their potential risk to humans or the environment. Section 4(a) requires that the EPA Administrator make three findings before requiring the manufacturers or processors of a chemical to conduct testing. These findings (especially as they relate to aquatic toxicity, bioconcentration, and chemical fate testing) are described in the third part of this chapter, on the EPA's implementation of ITC's testing recommendations. Historical details of these findings are discussed in EPA's first two TSCA section 4(a) test rules (U.S. EPA, 1980, 1981a).

Section 4(b) requires that EPA publish standards for development of test data and review the standards at least every 12 mo. The test standards that were developed for aquatic toxicity, bioconcentration, and chemical fate testing are described in the fourth part of this chapter, on TSCA test guidelines. Section 4(c) allows manufacturers and processors of chemicals to apply for exemptions for testing under section 4(a). Section 4(d) requires EPA to publish in the Federal Register the receipt of any data developed under section 4(a) within 15 days of the receipt.

Section 4(e) describes the statutory responsibilities of the ITC. The ITC is described in the next part of this chapter. Section 4(f) requires EPA to prevent or reduce risks of substances that present a significant risk of serious or widespread harm from cancer, gene mutations, or birth defects. Section 4(g) allows manufacturers to petition the EPA to prescribe standards for the development of data.

New Chemicals

Chapter 23 describes EPA's ability to regulate the manufacturer of new chemicals under TSCA Section 5. Section 5 requires manufacturers that want to produce new chemicals to submit premanufacturing notices to the EPA before initiating commercial production. New chemicals are those that were not produced in or imported into the United States before 1977. EPA has 90 d to approve a premanufacturing notice. EPA can approve these notices with no contingencies, approve them with contingencies for chemical testing or pollution prevention, or not approve them, thereby banning the chemical from production. EPA approves a premanufacturing notice by issuing a commencement notice. Chemicals for which commencement notices are issued and for which commercial production is initiated become existing chemicals and are subject to the requirements of TSCA section 4 including ITC review.

THE TSCA INTERAGENCY TESTING COMMITTEE

The TSCA Interagency Testing Committee is an independent advisory committee to the EPA Administrator that is responsible for establishing testing priorities for existing chemicals under section 4(e) of TSCA (Walker and Brink, 1989; Walker, 1991a, 1992, 1993a,b). The ITC includes representatives from most U.S. government organizations that regulate or establish health and environmental policies for existing chemicals (Table 1).

The ITC recommends testing priorities for EPA by submitting semiannual reports to the EPA Administrator. These reports are required by TSCA section 4(e). They contain the TSCA section 4(e) ***Priority Testing List*** of chemicals and chemical groups for which the EPA Administrator must consider implementing the testing recommended by the ITC. TSCA requires that the EPA Administrator publish the ITC's reports and EPA's decisions to implement or not implement the ITC's testing recommendations in the Federal Register.

If the EPA Administrator determines that the chemicals or chemical groups recommended for testing by the ITC satisfy the criteria of TSCA section 4(a), as described in the latter part of this chapter. then the testing recommendations of the ITC are implemented. As a result of the EPA Administrator's decisions to implement the ITC's testing recommendations, about 400 health effects and about 500 environmental tests were conducted by the manufacturers of the recommended chemicals, under section 4(a) of TSCA. This chapter discusses only the aquatic toxicity, bioconcentration and chemical fate data developed under TSCA section 4. The types of environmental tests that have been conducted under section 4(a) of TSCA and the methods and decision criteria for conducting these tests were previously described (Walker, 1988, 1990a,b,c, 1991b, 1993c,d).

The ITC's processes for selecting chemicals of concern and establishing testing priorities for the

Table 1. U.S. government organizations represented on the TSCA Interagency Testing Committee

Statutory organizations	Liaison organizations
Council of Environmental Quality	Agency for Toxic Substances and Disease Registry
Department of Commerce	Consumer Product Safety Commission
Environmental Protection Agency	Department of Agriculture
National Cancer Institute	Department of Defense
National Institute of Environmental Health Sciences	Department of the Interior
National Institute for Occupational Safety and Health	Food and Drug Administration
National Science Foundation	National Library of Medicine
Occupational Safety and Health Administration	National Toxicology Program

EPA Administrator have been described in detail (Walker, 1993a). The process as it is used to meet the data needs of the U.S. government is illustrated in Figure 1.

The ITC has made testing decisions on about 40,000 of 70,000 existing chemicals in the TSCA Inventory (Figure 2). The ITC has recommended testing for about 3000 chemicals. The ITC has deferred testing for about 37,000 chemicals because the chemicals were not TSCA regulable (see the chapter introduction for a list of non-TSCA-regulable chemicals), because they were not produced in or imported into the United States or were produced in or imported into the United States in <10,000 pounds per year, because their health effects were well characterized or well regulated at the time ITC reviewed them, because they were used as captured on-site intermediates and did not have physical-chemical properties that might result in accidental release, or because they had a low potential for causing adverse health or ecological effects or low potential for occupational, consumer, or environmental exposure. The ITC has not considered testing for about 30,000 polymers and complex mixtures because of uncertainties related to the ability of these chemicals to penetrate biological membranes and ability to interpret the data from testing such complex substances and mixtures.

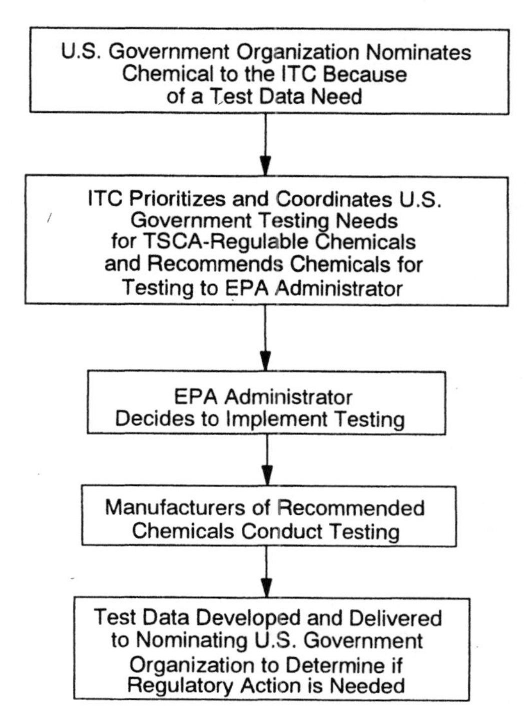

Figure 1. Meeting the U.S. government's data needs for TSCA-regulable chemicals through the TSCA Interagency Testing Committee.

To determine the most important aquatic toxicity, bioconcentration, and chemical fate tests that need to be conducted, the process developed by Walker (1993e) is used (Figure 3). The process is designed to save money in testing chemicals by obligating resources toward conducting only the most important tests based on the tested chemicals' physical-chemical properties or structural characteristics (Walker, 1993e).

Every spring and fall, ITC reports to the EPA Administrator are published in the Federal Register. A table listing the chemicals that were included in ITC reports from 1977 to 1992 has been published (Walker, 1993a).

THE EPA'S IMPLEMENTATION OF ITC'S TESTING RECOMMENDATIONS

After the ITC recommends a chemical or chemical group for testing, the EPA must decide whether the chemical meets certain criteria under TSCA section 4(a) before the testing can be implemented. The general process for implementing ITC's testing recommendations is described in Figure 4. About 80% of ITC's testing recommendations have been implemented and the manufacturers have developed the data (Walker, 1993a).

To implement testing under section 4(a) of TSCA, EPA must find:

1. That a chemical may present an "unreasonable risk" to human health or the environment (the "A" finding);

   or

   that a chemical is produced in "substantial" quantities and it may result in "substantial" or "significant" human exposure or "substantial" environmental release (the "B" finding);

   and
2. That insufficient data or knowledge exist to reasonably determine or predict the impacts of its manufacture, processing, distribution, use and/or disposal (the "A" and "B" finding);

   and
3. that testing is necessary to develop the insufficient data (the "A" and "B" finding).

The "A" Finding

Congress intended that the "A" finding be applied to existing chemicals about which there is a basis for concern but for which there is inadequate information to reasonably determine or predict adverse effects (Library of Congress, 1976). The concern for chemicals that may be released or transported to the environment and have potential to cause adverse effects may be based on several factors identified by Walker (1990b).

1. Available, but insufficient toxicity data
2. Data which suggest that toxicity increases with exposure (potential cumulative toxicity)

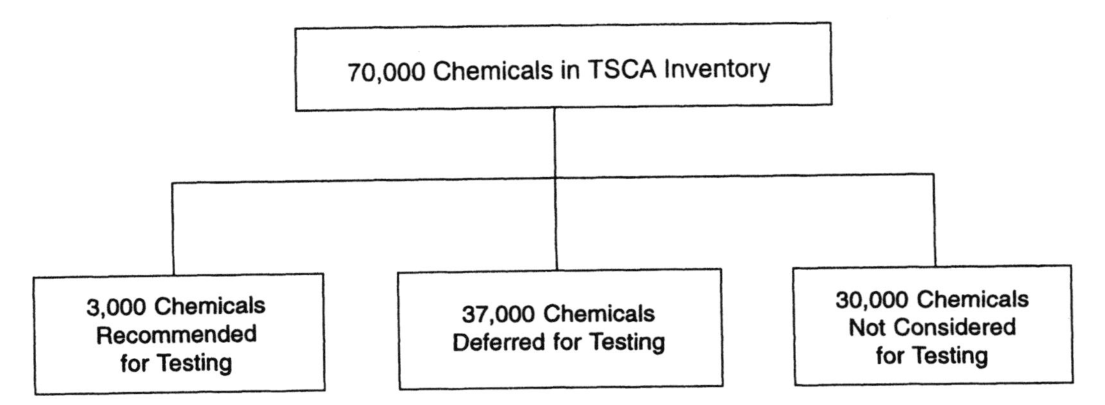

Figure 2. ITC's testing decisions on existing chemicals.

3. Data on homologous chemicals that are known to cause adverse ecological effects
4. Quantitative structure-activity relationships (QSARs) that use data on physical or chemical properties to predict toxicity
5. Knowledge that the chemical causes adverse health effects
6. Information indicating that the chemical has the potential to bioconcentrate, persist in tissues, and subsequently affect consumers (secondary toxicity)
7. Data suggesting that the chemical may sequester in reproductive tissues and produce adverse effects to subsequent generations
8. Predictions that the chemical may cause indirect ecological effects
9. Data which suggest that mortalities continue to occur or symptoms continue to persist after exposure to the test substance is discontinued.

If none of these concerns are alleviated by available data, EPA may require aquatic toxicity, bioconcentration, or chemical fate testing under an "A" finding for chemicals that may be released or transported to aquatic environments.

The "B" Finding

The "B" finding reflects Congress' belief that testing of certain existing chemicals should be conducted on the basis of potential exposure (Library of Congress, 1976). Generally, when production is substantial and release may be anticipated, the "B" finding is applied to chemicals that may be in the environment (Walker, 1990b). Release may generally be anticipated unless the chemical is used as a captured, on-site intermediate and is quantitatively consumed in a closed process (Walker, 1990b). EPA has quantified "substantial" production, "substantial" and "significant" human exposure, and "substantial" environmental release for testing under the "B" finding (U.S. EPA, 1991a, 1993).

Chemicals that are considered for priority aquatic toxicity, bioconcentration, and chemical fate testing are those that have the potential to be released or transported to the aquatic environment (Figure 5; see Chapter 15). Chemicals may be released directly to the aquatic environment from manufacturing, processing, use, distribution, disposal, or treatment or be transported to the surface waters from air by wet or dry deposition, from soil by runoff, or from sediment by desorption. Chemicals may be transported to sediments from water by adsorption or be transported to water and sediment by depuration from aquatic and benthic organisms, respectively. Transport and transformation are discussed in detail in Chapter 15.

Because many existing chemicals may have the potential to be released or transported to aquatic (including sediment) environments, Walker (1990b) suggested that less consideration should be given to existing chemicals that are:

1. Produced in only small quantities (less than 10,000 pounds per annum)
2. Biologically inactive under environmental conditions
3. Likely to decompose in a few hours to an innocuous substance(s)

Walker (1990b) also suggested that less consideration should be given to existing chemicals with reliable acute EC50 or LC50 values (for aquatic or benthic organisms) equal to or greater than 1000 mg/L or reliable EC50 or LC50 values equal to or greater than 1000 times reliable measured or predicted environmental (i.e., surface waters or sediments) concentrations (PECs). These criteria were recommended by Kimerle et al. (1978) and by participants in an EPA triggers testing workshop (Life Systems, 1983) for not conducting additional acute toxicity tests. Walker (1990b) suggested that reliable EC50 or LC50 values are generally produced from tests (conducted with at least three phylogenetically

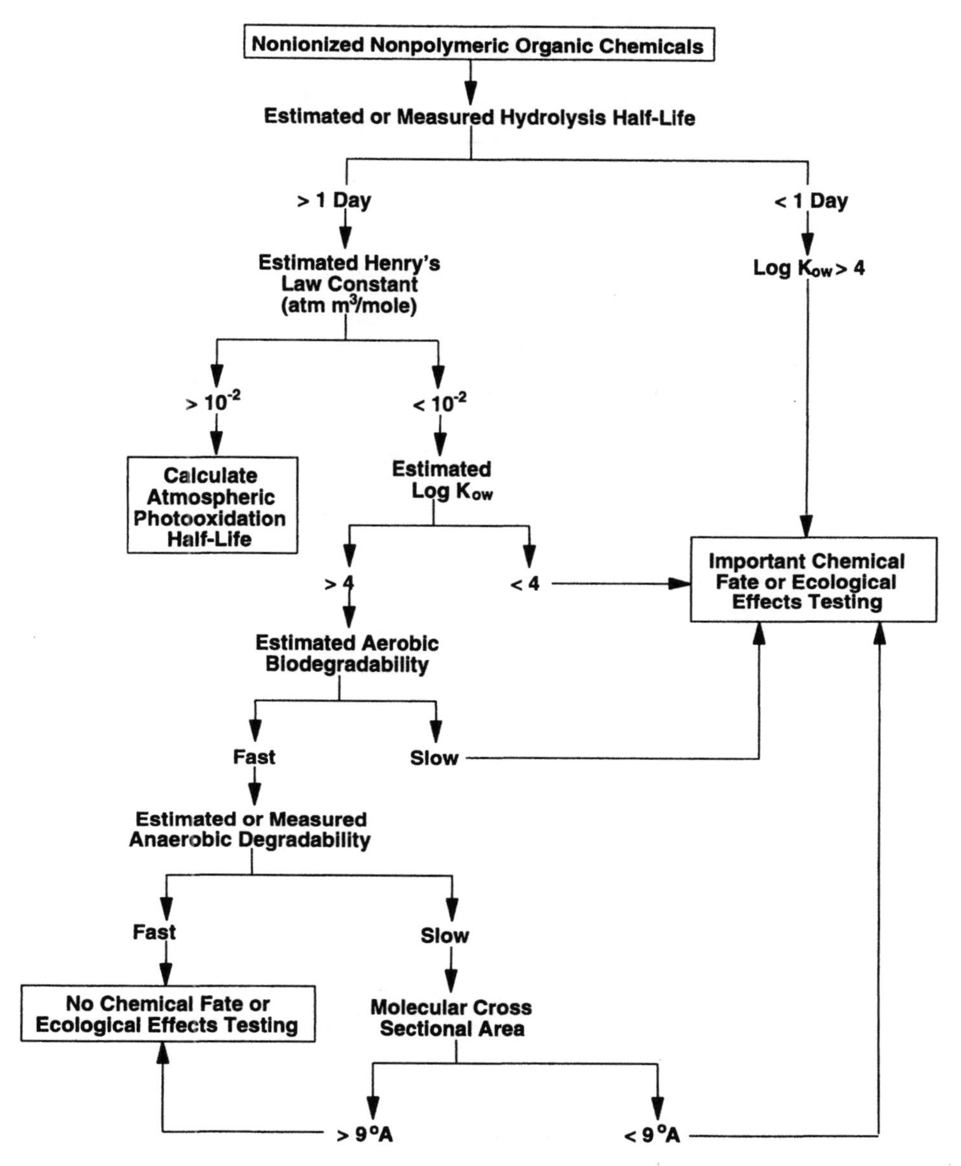

Figure 3. Process for determining important chemical fate or ecological effects tests that should be conducted for nonionized, nonpolymeric organic chemicals (From Walker, 1993e).

distinct organisms, e.g., algae, fish, invertebrates; see Chapters 2 to 8):

1. That are conducted using accepted protocols developed by the EPA's Office of Pollution Prevention and Toxics (OPPT), the American Society for Testing and Materials (ASTM), the American Public Health Association, the International Standards Organization, or the Organization for Economic Cooperation and Development (OECD)
2. For which the chemical concentrations have been measured before, during and after the test period
3. That are conducted using static-renewal or flow-through conditions for chemicals that are likely to sorb, hydrolyze, volatilize, or biodegrade
4. That satisfy the EPA's good laboratory practice (GLP) standards.

Reliable PECs are most likely to be produced by using statistically sound environmental surveys or using measured physical or chemical properties (or measured properties from which a few other prop-

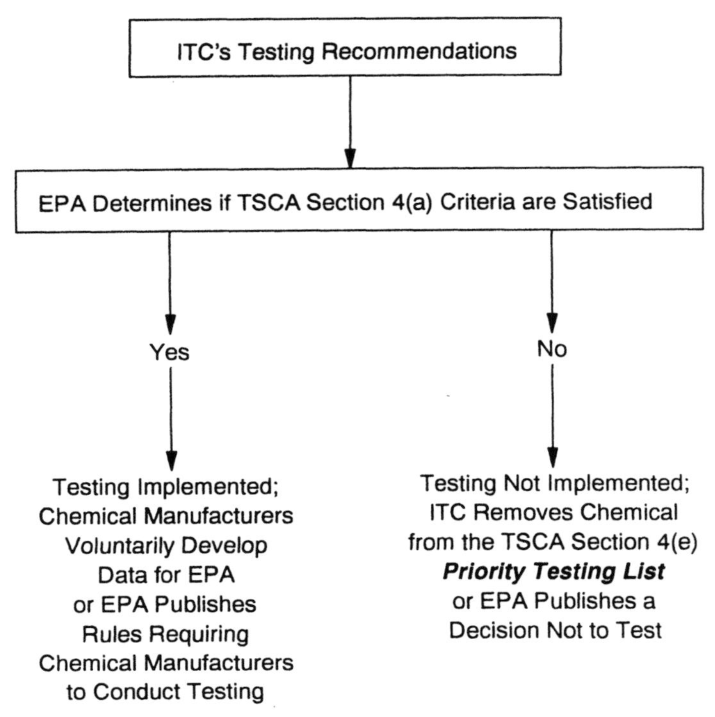

Figure 4. EPA's process for implementing ITC's testing recommendations.

erties can be estimated) and mathematical models (see Chapter 18) that account for fate of chemicals during wastewater treatment and after dilution by receiving streams (Walker, 1990b). For some chemicals, data on measured concentrations in aquatic systems are available. Walker (1990b) suggested that these data should be analyzed to estimate:

1. The reliability of the analytical methods used to generate the data

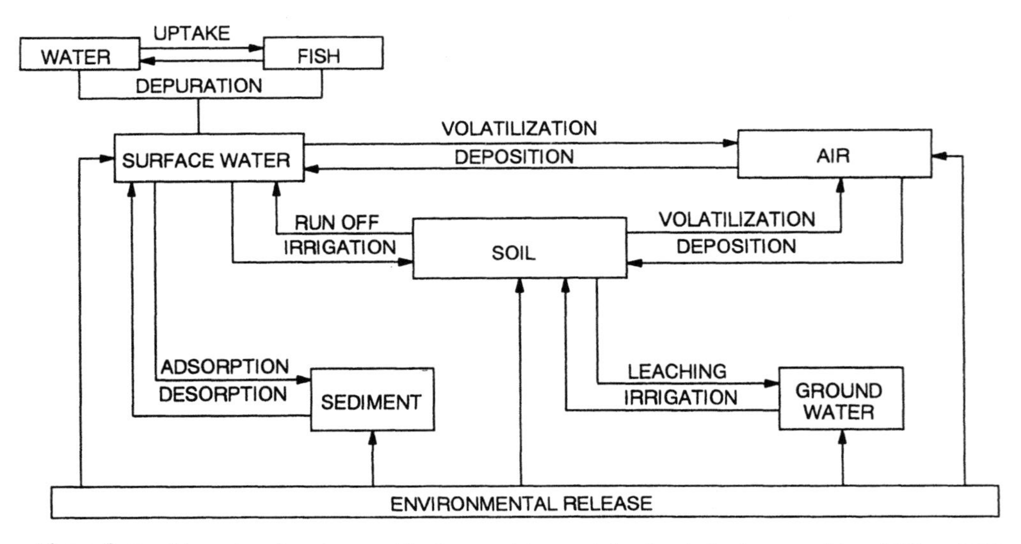

Figure 5. Possible routes of environmental release and transport for chemical substances (From Walker, 1986).

2. The probability that the chemical's detection in an aquatic system resulted from and continues to result from nearby manufacturing, use, processing, disposal, or treatment
3. The possibility that the chemical was detected in an aquatic system because it was an industrial by-product or an in situ transformation product of a chemical that was and continues to be manufactured, used, and processed

TSCA Section 4 Federal Register Notices

EPA publishes several different TSCA Section 4(a) Federal Register notices, including decisions not to test, negotiated testing agreements, proposed rules, final rules, and consent orders. The numbers and types of chemicals included in these notices have been published (Walker, 1990a). The EPA issues decisions not to test if:

1. There are insufficient exposures to justify testing under TSCA section 4(a).
2. There are adequate data to reasonably determine or predict adverse effects.
3. There is ongoing testing by industry or government.
4. Controls to limit exposures are being installed.

In the early 1980s, the EPA developed negotiated testing agreements with industry in lieu of publishing proposed or final rules. In August 1984, as a result of a lawsuit initiated by the Natural Resources Defense Council (NRDC), the U.S. District Court, Southern District of New York, ruled that negotiated testing agreements were not a legal substitute for rulemaking under TSCA section 4 [*NRDC v EPA*, 595 F Supp 1255 (SDNY) 1984]. In 1986, the EPA, NRDC, and the Chemical Manufacturers Association (CMA) recommended developing consent orders as a legally enforceable alternative to developing negotiated testing agreements (U.S. EPA, 1986a, 1987a).

TSCA Test Guidelines

Under TSCA section 4(b) EPA is required to prepare test guidelines to develop test data under TSCA section 4(a). Many test guidelines were developed to conduct the testing recommended by the ITC (Walker, 1993d). TSCA test guidelines contain "shall" and "should" statements, e.g., "water temperatures shall be maintained at 12°C" or "fish food should be added daily." When the TSCA test guidelines are incorporated into TSCA section 4(a) test rules or consent orders, they become enforceable test standards and laboratories conducting tests must comply with all "shall" statements. Studies are reviewed and audited to ensure compliance with the standards and with GLPs (see Part 792 of Volume 40 of the Code of Federal Regulations for additional information on EPA's GLPs; also see Chapter 11).

Aquatic Toxicity and Bioconcentration

The 20 TSCA aquatic toxicity and bioconcentration test guidelines are listed in Table 2. Five of twenty of these test guidelines were modified from OECD guidelines (see Chapter 26). Seventeen were published in the Federal Register in 1985 after being made publicly available in 1982 through the National Technical Information Service (NTIS). TSCA test guidelines were prepared for incorporation into test rules as test standards to facilitate the development of reliable data as monitored by compliance with "shall" statements. The aquatic toxicity and bioconcentration test guidelines were developed to measure the acute and chronic effects of chemicals on and partitioning of chemicals into aquatic organisms under controlled laboratory conditions. Two microcosm test guidelines were developed to measure the effects of chemicals on aquatic communities under simulated environmental conditions. Using the TSCA aquatic toxicity and bioconcentration test guidelines, toxicity tests can be conducted for six algae, one aquatic plant, nine aquatic invertebrates, and eight fish and bioconcentration tests can be conducted for one aquatic invertebrate and 10 fish (Table 3). In general, the laboratory methods will provide reliable data for chemicals with water solubilities >1 mg/L and Henry's law constants $>10^{-3}$ atm m^3/mol (Walker, 1993c). A list of chemicals for which these test guidelines were used to develop test data recommended by the ITC has been published (Walker, 1993c).

Chemical Fate

The 31 TSCA chemical fate test guidelines are listed in Table 4. For the most recent versions of these guidelines, the Code of Federal Register (CFR) should be consulted. Fourteen were from the OECD (see Chapter 26). Most were made publicly available through the NTIS in 1982. These test guidelines were prepared for incorporation into test rules as test standards to facilitate the development of reliable data as monitored by compliance with "shall" statements. The chemical fate test guidelines were developed to measure the physical and chemical properties, transport, and transformation of chemicals under controlled laboratory conditions. Chemicals for which these test guidelines were used to develop test data recommended by the ITC have been published (Walker, 1990c).

TSCA Section 4 Aquatic Toxicity Testing Scheme

The TSCA section 4 aquatic toxicity testing scheme is illustrated in Figure 6. It has been described in detail (Walker, 1990b).

The TSCA section 4 aquatic toxicity testing scheme developed by Walker (1990b) provides a cost-effective process for consistently developing reliable chemical fate and aquatic toxicity data for existing chemicals under TSCA section 4 by:

1. Recommending similar chemical fate tests for most existing chemicals that will consistently estimate persistence and modify PECs

Table 2. TSCA aquatic toxicity and bioconcentration test guidelines

40 CFR No.	Guideline title	Source	Federal Register reference	Date
797.1050	Algal acute toxicity test	OPPT	50 FR 39321	09/27/85
797.1060	Freshwater algae acute toxicity test	OECD	50 FR 39323	09/27/85
797.1075	Freshwater and marine algae acute toxicity test	OECD	50 FR 39327	09/27/85
797.1160	*Lemna* acute toxicity test	OPPT	50 FR 39331	09/27/85
797.1300	Daphnid acute toxicity test	OPPT	50 FR 39333	09/27/85
797.1310	Gammarid acute toxicity test	OPPT	51 FR 472	01/06/86
797.1330	Daphnid chronic toxicity test	OPPT	50 FR 39336	09/27/85
797.1350	Daphnid chronic toxicity test	OECD	50 FR 39339	09/27/85
797.1400	Fish acute toxicity test	OPPT	50 FR 39342	09/27/85
797.1440	Fish acute toxicity test	OECD	50 FR 39345	09/27/85
797.1520	Fish bioconcentration test	OPPT	50 FR 39347	09/27/85
797.1560	Fish bioconcentration test	OECD	50 FR 39351	09/27/85
797.1600	Fish early life stage toxicity test	OPPT	50 FR 39355	09/27/85
797.1800	Oyster acute toxicity test	OPPT	50 FR 39361	09/27/85
797.1830	Oyster bioconcentration test	OPPT	50 FR 39364	09/27/85
797.1930	Mysid shrimp acute toxicity test	OPPT	50 FR 39367	09/27/85
797.1950	Mysid shrimp chronic toxicity test	OPPT	50 FR 39369	09/27/85
797.1970	Penaeid shrimp acute toxicity test	OPPT	50 FR 39372	09/27/85
797.3050	Generic freshwater microcosm test	OPPT	52 FR 36344	09/28/87
797.3100	Site-specific aquatic microcosm test	OPPT	52 FR 36352	09/28/87

Table 3. Species for aquatic toxicity and bioconcentration testing listed in TSCA test guidelines

TSCA test guideline		
Title	40 CFR No.	Species
797.1050	Algal acute toxicity test	*Selenastrum capricornutum* *Skeletonema costatum*
797.1060	Freshwater algae acute toxicity test	*Selenastrum capricornutum* *Scenedeamus quadricauda* *Chlorella vulgaris*
797.1075	Freshwater and marine algae acute toxicity test	*Selenastrum capricornutum* *Scenedeamus quadricauda* *Chlorella vulgaris* *Skeletonema costatum* *Thallassiosira pseudonana* *Isochrysis galbana*
797.1160	*Lemna* acute toxicity test	*Lemna gibba* G3
797.1300	Daphnid acute toxicity test	*Daphnia magna* *Daphnia pulex*
797.1310	Gammarid acute toxicity test	*Gammarus fasciatus* *Gammarus pseudolimnaeus* *Gammarus lacustris*
797.1330	Daphnid chronic toxicity test	*Daphnia magna* *Daphnia pulex*
797.1350	Daphnid chronic toxicity test	*Daphnia magna*
797.1400	Fish acute toxicity test	Rainbow trout Bluegill Fathead minnow
797.1440	Fish acute toxicity test	Zebra fish Fathead minnow Common carp Red killifish Guppy Bluegill Rainbow trout
797.1520	Fish bioconcentration test	Fathead minnow
797.1560	Fish bioconcentration test	Rainbow trout Bluegill Fathead minnow Spot Sheepshead minnow Silversides Shiner perch English sole Staghorn sculpin Three-spine stickleback
797.1600	Fish early life stage toxicity test	Fathead minnow Sheepshead minnow Brook trout Rainbow trout Atlantic silverside Tidewater silverside
797.1800	Oyster acute toxicity test	*Crassostrea virginica*
797.1830	Oyster bioconcentration test	*Crassostrea virginica*
797.1930	Mysid shrimp acute toxicity test	*Mysidopsis bahia*
797.1950	Mysid shrimp chronic toxicity test	*Mysidopsis bahia*
797.1970	Penaeid shrimp acute toxicity test	*Penaeus aztecus* *Penaeus duorarum* *Penaeus setiferus*

Table 4. TSCA chemical fate test guidelines

40 CFR No.	Guideline title	Source	Federal Register reference	Date
796.1050	Absorption in aqueous solution: ultraviolet/viable spectra	OECD	50 FR 39472	09/27/85
796.1220	Boiling point/boiling range	OECD	50 FR 39473	09/27/85
796.1370	Dissociation constants in water	OECD	50 FR 39477	09/27/85
796.1520	Particle size distribution/fiber length and diameter distributions	OECD	50 FR 39479	09/27/85
796.1550	Partition coefficient (*n*-octanol/water)	OPPT	50 FR 39252	09/27/85
796.1570	Partition coefficient (*n*-octanol/waters) estimation by liquid chromatography	OPPT	50 FR 39255	09/27/85
796.1720	Octanol/water partition coefficient generator column method	OPPT	50 FR 39257	09/27/85
796.1840	Water solubility	OPPT	50 FR 39263	09/27/85
796.1860	Water solubility generator column method	OPPT	50 FR 39265	09/27/85
796.1950	Vapor pressure	OPPT	50 FR 39271	09/27/85
796.2700	Soil thin-layer chromatography	OPPT	50 FR 39273	09/27/85
796.2750	Sediment and soil adsorption isotherm	OPPT	50 FR 39275	09/27/85
796.3100	Aerobic aquatic biodegradation	OPPT	50 FR 39277	09/27/85
796.3140	Anaerobic biodegradability of organic chemicals	OPPT	50 FR 39280	09/27/85
796.3180	Ready biodegradability: modified AFNOR test	OECD	50 FR 39483	09/27/85
796.3200	Ready biodegradability: closed bottle test	OECD	50 FR 39485	09/27/85
796.3220	Ready biodegradability modified MITI test (1)	OECD	50 FR 39488	09/27/85
796.3240	Ready biodegradability: modified OECD screening test	OECD	50 FR 39495	09/27/85
796.3260	Ready biodegradability: modified Sturm test	OECD	50 FR 39499	09/27/85
796.3300	Simulation test-aerobic sewage treatment: coupled units test	OECD	50 FR 39502	09/27/85
796.3340	Inherent biodegradability: modified SCAS test	OECD	50 FR 39506	09/27/85
796.3341	Inherent biodegradability: modified SCAS test for chemical substances that are water insoluble or water insoluble and volatile	OPPT	50 FR 46793	11/13/85
796.3360	Inherent biodegradability: modified Zahn-Wellens tests	OECD	50 FR 39508	09/27/85
796.3400	Inherent biodegradability in soil	OECD	50 FR 39512	09/27/85
796.3480	Complex formation ability in water	OECD	50 FR 39514	09/27/85
796.3500	Hydrolysis as a function of pH at 25°C	OPPT	50 FR 39283	09/27/85
796.3510	Hydrolysis as a function of pH and temperature	OPPT	52 FR 36334	09/28/87
796.3700	Photolysis in aqueous solution in sunlight	OPPT	50 FR 39285	09/27/85
796.3765	Indirect photolysis screening test: sunlight photolysis in water containing dissolved humic substances	OPPT	51 FR 483	01/06/86
796.3780	Laboratory determination of the direct photolysis reaction quantum yield in aqueous solution and sunlight photolysis	OPPT	50 FR 39296	09/27/85
796.3800	Gas-phase absorption spectra and photolysis	OPPT	50 FR 39311	09/27/85

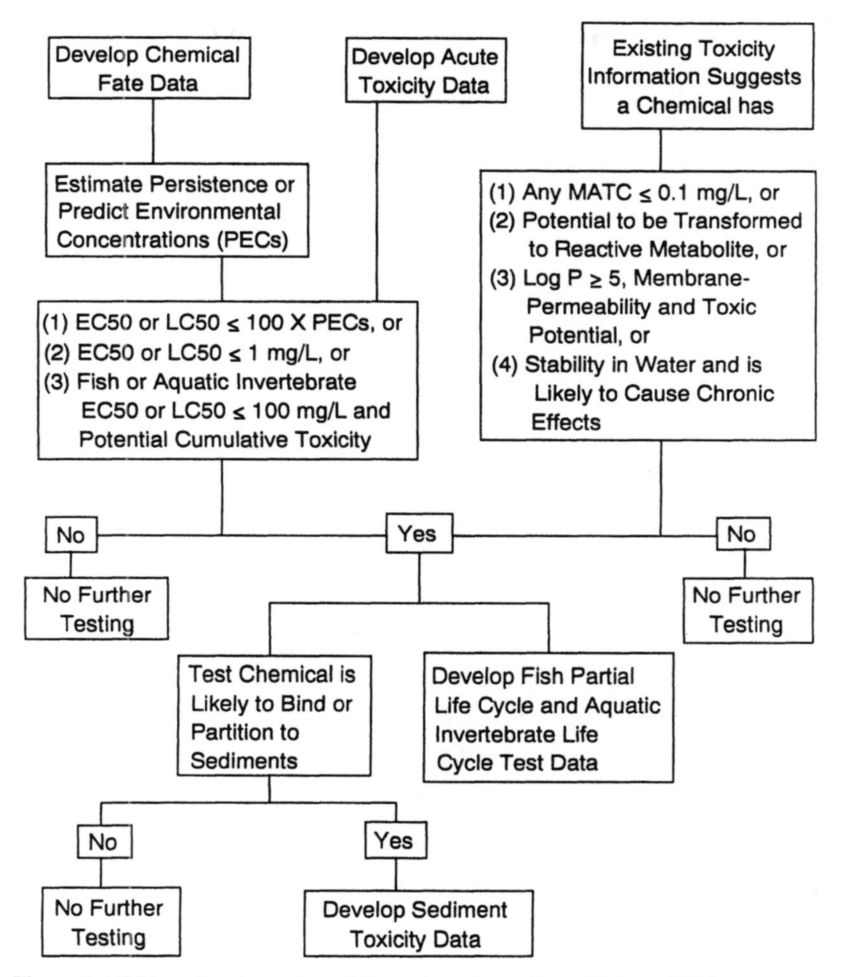

Figure 6. TSCA section 4 aquatic toxicity testing scheme (From Walker, 1990b).

2. Using test species for acute toxicity tests that will consistently be among the most sensitive to a large number of existing chemicals and will identify the test chemical's potential to cause acute effects
3. Applying decision criteria that will consistently and appropriately promote the development of fish partial life cycle toxicity tests, aquatic invertebrate life cycle tests, and sediment toxicity (i.e., bioassay) data.

The scheme does not include decision criteria for development of bioconcentration data. Walker (1990b) suggested that bioconcentration data should be developed under TSCA section 4 when there are concerns for:

1. Secondary toxicity or
2. Potential for chemicals to concentrate in fish ova and adversely effect the survival of offspring

Secondary toxicity occurs when higher-trophic-level organisms are indirectly and adversely affected as a result of consuming sufficient quantities of lower-trophic-level organisms that have bioconcentrated but not been adversely affected by a potentially toxic chemical (Walker, 1990b).

The TSCA section 4 testing scheme promotes the rapid development of screening-level chemical fate data and aquatic toxicity data that are limited to evaluating direct acute effects of chemicals, potential chronic toxicity, sensitivity of fish and aquatic invertebrate life stages, and impairment of reproduction potential in aquatic invertebrates as a result of conducting single-species tests that last days to weeks. Chemical fate and short-term aquatic toxicity data are developed simultaneously; longer-term aquatic toxicity tests are conducted subsequently, allowing the chemical fate data to be used during the selection/design of these longer-term toxicity tests (Walker, 1990b).

The decision criteria for developing aquatic toxicity data under TSCA section 4 should be used as guidelines. Professional judgment should always be used to interpret unusual and complex data and to determine the need for subsequent testing based on the proposed decision criteria (Walker, 1990b).

The decision criterion that compares the lowest EC50 or LC50 value with PECs (EC50 or LC50 $\leq 100 \times$ PECs) was used for *m*-, *o*-, and *p*-phenylenediamines, 2,6-di-*tert*-butylphenol, and tributyl phosphate (U.S. EPA, 1986b, 1987b,c, 1988a). It was recommended for conducting fish partial life cycle toxicity tests by Akerman and Coppage (1979) in their review of criteria that were used by EPA's Office of Pesticide Programs. It was also recommended by American Institute of Biological Sciences (AIBS, 1978). It was recommended by participants of an EPA-sponsored Triggers Testing Workshop who suggested that integrated decision criteria containing reliable PECs should be used as testing decision criteria (Life Systems Inc., 1983). The factor of "100 ×" accounts for two orders of magnitude difference that could easily be observed between acute effects of a chemical on a single species tested in the laboratory and long-term effects that the same chemical might produce in situ (Walker, 1990b). If no differences between laboratory acute toxicity data and long-term in situ effects are anticipated, it also could account for up to two orders of magnitude differences that can be observed between acute toxicity values for different species (Sloof et al., 1986). Finally, if no differences between laboratory long-term effects and long-term in situ effects are anticipated, it can account for the two orders of magnitude differences that have been observed between laboratory-generated short-term and long-term effects levels (Walker, 1990b).

The decision criterion of an EC50 or LC50 ≤1 mg/L was used previously for bisphenol A, 1,2,3-trichlorobenzene, 2,6-di-*tert*-butylphenol, tributyl phosphate, and *p*-phenylenediamines (U.S. EPA, 1985a, 1986c, 1987b,c, 1988a). It was recommended by Kimerle et al. (1978) for conducting additional acute toxicity tests and by Akerman and Coppage (1979) for conducting fish partial life cycle toxicity and aquatic invertebrate life cycle tests. It is based on a system to classify the toxicity of chemicals based on their acute lethal threshold (Dryssen et al., 1969). The system was modified to characterize toxicity of chemical substances to aquatic organisms based on effective concentrations (ECs) or lethal concentrations (LCs), where EC50 or LC50 <1 mg/L would be considered "very toxic," EC50 or LC50 = 1–100 mg/L would be considered "toxic," EC50 or LC50 = 100–1000 mg/L would be considered "moderately toxic," EC50 or LC50 = 1000–10,000 mg/L would be considered "slightly toxic," and EC50 or LC50 = 10,000 mg/L would be considered "practically nontoxic." These terms are used in the next section of this chapter to describe the relative toxicity of the chemicals tested by industry under TSCA section 4(a).

The decision criterion that uses fish and aquatic invertebrate EC50 or LC50 values ≤ 100 mg/L and information on potential cumulative toxicity was used previously for 2,6-di-*tert*-butylphenol, tributyl phosphate, and *p*-phenylenediamines (U.S. EPA, 1987b,c, 1988a). It reflects a concern that certain chemicals that are toxic should be examined for their potential to produce long-term effects based on the shape and slope of the concentration-response curve (Lloyd, 1979). The decision criteria for conducting fish partial life cycle toxicity or aquatic invertebrate life cycle tests based on a maximum acceptable toxicant concentration (MATC) ≤0.1 mg/L was used previously for 2,6-di-*tert*-butylphenol, tributyl phosphate, and *p*-phenylenediamines (U.S. EPA, 1987b,c, 1988a). Conducting fish partial life cycle toxicity or aquatic invertebrate life cycle testing based on a MATC ≤0.1 mg/L or on a water-stable chemical's potential to cause cancer or other chronic effects reflects a concern that long-term toxicity data are needed to determine the potential of these types of chemicals to cause chronic effects in aquatic organisms (Walker, 1990b).

The decision criterion to develop sediment bioassay data (see Chapter 8) for chemicals that are being evaluated in long-term toxicity tests and are likely to be associated with sediment was used previously for 2,6-di-*tert*-butylphenol and tributyl phosphate (U.S. EPA, 1987b,c). It reflects a concern that certain chemicals will accumulate in sediments and be potentially bioavailable for sediment-feeding/ingesting organisms (Walker, 1990b).

THE CHEMICAL INDUSTRY'S DEVELOPMENT OF AQUATIC TOXICITY, BIOCONCENTRATION, AND CHEMICAL FATE TEST DATA UNDER TSCA SECTION 4

The manufacturers of chemicals recommended for testing by the ITC have conducted about 500 aquatic toxicity, bioconcentration, and chemical fate tests. Most of these chemicals and the methods used to conduct the tests have been described (Walker, 1990c, 1993c). The data developed by conducting these tests are described below. To facilitate comparisons of aquatic toxicity, bioconcentration, and chemical fate data, the chemicals that were tested by the manufacturers are organized into structural groups.

Alkyl and Aryl Benzenes

Structures of some of the alkyl and aryl benzenes for which the chemical industry has developed data under TSCA section 4 are presented in Figure 7.

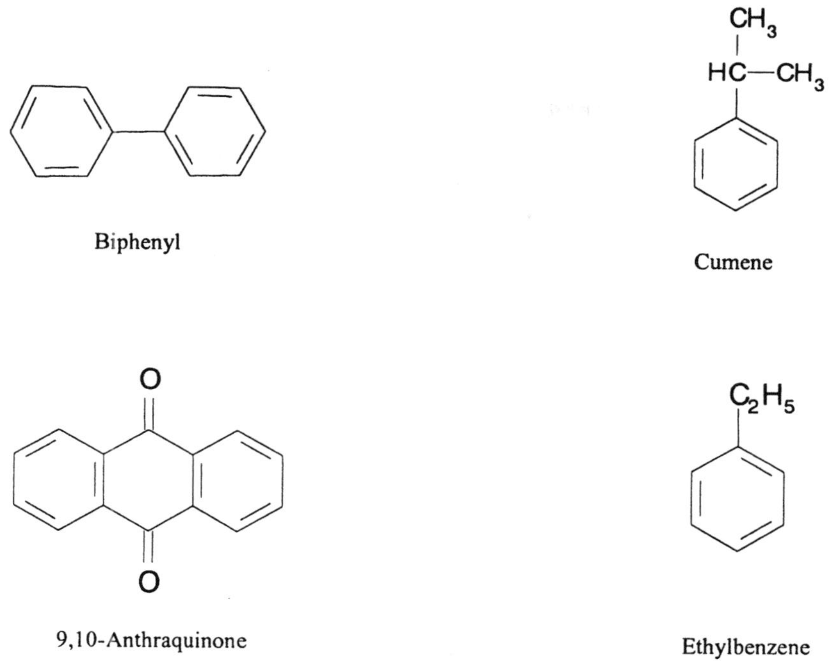

Biphenyl

Cumene

9,10-Anthraquinone

Ethylbenzene

Figure 7. Structures of some of ITC's alkyl and aryl benzenes tested by the chemical industry under section 4 of TSCA.

Biphenyl

The ITC designated biphenyl (CAS Registry number 92-52-4) for testing in its Tenth Report (ITC, 1982a). The ITC designated biphenyl for testing because 700 million pounds were produced per year and its use as a dye carrier and heat transfer fluid suggested that it would be released to the environment through wastewater effluents as a result of use or disposal. The ITC was concerned about the persistence and toxicity of biphenyl in aquatic systems and the potential for production of chlorinated biphenyls as a result of chlorination at wastewater treatment plants. The structure of biphenyl is illustrated in Figure 7.

In response to ITC's designation, the EPA issued proposed and final rules in 1983 and 1985, respectively, to test biphenyl under conditions that would maintain stable measured concentrations (U.S. EPA, 1983a, 1985b). The Synthetic Organic Chemical Manufacturers Association (SOCMA) organized a task force to coordinate and sponsor this testing.

Biphenyl has a water solubility of 7.1 mg/L at 25°C, a vapor pressure of 0.0089 mm Hg at 25°C, a calculated Henry's law constant of 3×10^{-4} atm m^3/mol, and a log octanol-water partition coefficient (log K_{ow}) of 4.09 (Walker, 1993c). The moderate Henry's law constant suggested the need to measure concentrations of biphenyl during the environmental tests to account for biphenyl that could evaporate from test systems.

In acute and chronic toxicity tests, biphenyl's LC50 and MATC values were less than 1 mg/L (Table 5). Results of the *Daphnia magna* chronic tests were published by Gersich et al. (1989). Attempts to measure an exact EC50 value with the Eastern oyster in an acute toxicity were unsuccessful even though the test was repeated three times in two different laboratories. The tests were unsuccessful because there was insufficient growth in control oysters (Walker, 1993c). Biphenyl's bioconcentration factor in the Eastern oyster was 110 and the depuration half-life was 14 d.

To simulate environmental biodegradation of biphenyl, an ecocore microcosm was selected as the test system. The system was based on the ecocore described by Bourguin et al. (1977). An ecocore microcosm chamber with threaded glass joints had to be designed to measure biphenyl biodegradation, because biphenyl diffused into rubber components and made it difficult to determine a mass balance.

In aerobic river microcosms that contained only water and 0.08 or 1.0 mg/L biphenyl, 1% or less biphenyl remained in test systems under static or aerated conditions after 10 or 11 d, compared to 32.5 to 56.8% that remained in sterile controls. In river microcosms that contained water and sediment and 0.08 or 1.0 mg/L biphenyl, 6 to 10% biphenyl remained in test systems under static or aerated conditions after 10 or 11 d, compared to 13.4 to 19% that remained in sterile controls. Similar results were observed with aerobic lake microcosms. The differences between the water and sediment systems can probably be explained by partitioning of biphenyl to sediment. In aerated aquatic and sediment microcosms most biphenyl was removed by d 2.

In anaerobic lagoon microcosms that contained only water and 1.0 mg/L of biphenyl, about 2% biphenyl remained in test systems after 84 d, compared to about 3% that remained in sterile controls. In anaerobic lagoon microcosms that contained water and sediment and 1.0 mg/L of biphenyl, 30 to 50% biphenyl remained in test systems after 84 d, compared to 46% that remained in sterile controls. The differences between the water and sediment systems can probably be explained by partitioning of biphenyl to sediment.

Cumene

The ITC designated cumene (CAS Registry number 98-82-8) for testing in its Fifteenth Report (ITC, 1984b). The ITC designated cumene for testing because of potential occupational and environmental exposures resulting from very high production volumes (3.3 billion pounds of cumene were produced in 1983). The structure of cumene is illustrated in Figure 7.

In response to ITC's designation, the EPA issued proposed and final rules in 1985 and 1988, respectively, to test cumene under conditions that would maintain stable measured concentrations (U.S. EPA, 1985c, 1988b). The Chemical Manufacturers Association organized a panel to coordinate and sponsor this testing.

Cumene has a water solubility of 50 mg/L at 20°C, a vapor pressure of 3.2 mm Hg at 20°C, a measured Henry's law constant of 1.46×10^{-2} atm m^3/mol, and a log K_{ow} of 3.6. The high Henry's law constant suggested the need to measure concentrations of cumene during the environmental tests to account for cumene that could evaporate from test systems.

In acute toxicity tests, cumene's lowest LC50 value was about 1 mg/L (Table 5).

In aquatic microcosm tests (that were conducted using ecocores similar to those designed for biphenyl), less cumene but more carbon dioxide evolved with decreasing cumene concentrations. Most of the cumene evaporated from the systems within 10 d. Details of the ecocore tests were published by Williams et al. (1993).

Compared to biphenyl, slightly less cumene was removed from the test systems, but comparable quantities of carbon dioxide were evolved from both systems, suggesting that both chemicals would volatilize from or biodegrade in aquatic systems.

Anthraquinone

The ITC designated 9,10-anthraquinone (CAS Registry number 84-65-1) for testing in its Fifteenth Report (ITC, 1984b). The ITC designated anthraquinone for testing because it was measured in several aquatic environments and because it was difficult to determine if anthraquinone was persistent or toxic at its low water solubility level, because all testing had been conducted at concentrations exceeding the water solubility level. The structure of anthraquinone is illustrated in Figure 7.

In response to the ITC's designation, the EPA issued proposed and final rules in 1985 and 1987, respectively, to test anthraquinone under conditions that would maintain stable measured concentrations (U.S. EPA, 1985d, 1987d). Mobay Corporation and C-I-L Corporation of American formed a consortium to coordinate and sponsor this testing.

Anthraquinone has a water solubility of 0.125 mg/L at 22°C, a estimated vapor pressure of 1×10^{-7} mm Hg at 25°C, a calculated Henry's law constant of 2.19×10^{-7} atm m^3/mol, and a log K_{ow} of 3.4 (Walker, 1993c). The low water solubility and moderate log K_{ow} suggested the need to measure concentrations of anthraquinone during the environmental tests to account for anthraquinone that could be adsorbed during the tests.

In acute toxicity tests, the toxicity of anthraquinone to fish and invertebrates could not be determined because exact EC50 and LC50 values could not be calculated (Walker, 1993d). EC50 and LC50 values for daphnids, oysters, salmon, trout, and bluegills were all >0.05 mg/L. These values could not be calculated because anthraquinone was not toxic at the solubility of anthraquinone in these test systems; e.g., in fish toxicity test systems, anthraquinone's solubility ranged from 0.050 to 0.250 mg/L (Walker, 1993c). Using acetone to increase water solubility and extending the test to 14 d to increase exposure did not produce exact EC50 or LC50 values.

Ethylbenzene

The ITC recommended ethylbenzene (CAS Registry number 100-41-4) for testing in its Twentieth Report (ITC, 1987). The ITC recommended ethylbenzene for testing at the request of EPA because of its very high production volume (7.6 billion pounds in 1986), widespread environmental release, and a need for more reliable data to revise a water quality advisory. There was a need for aquatic toxicity data that were developed using flow-through systems and measured ethylbenzene concentrations, because most of the available data that were developed using static systems and nominal ethylbenzene concentrations underestimated ethylbenzene's toxicity. The structure of ethylbenzene is illustrated in Figure 7.

In response to ITC's testing recommendation, SOCMA's Styrene and Ethylbenzene Association sponsored the studies and submitted them directly to the ITC.

Ethylbenzene has a water solubility of 161 mg/L at 25°C, a vapor pressure of 9.57 mm Hg at 25°C, a measured Henry's law constant of 8.43×10^{-3} atm m^3/mol, and a log K_{ow} of 3.2. The high Henry's law constant suggested the need to measure concentrations of ethylbenzene during the environmental tests to account for ethylbenzene that could evaporate from test systems.

In acute toxicity tests, ethylbenzene's lowest LC50 value was about 2 mg/L for *Mysidopsis bahia* (Table 5). Aquatic toxicity data have been published (Walker, 1995).

Table 5. Summary of aquatic toxicity, bioconcentration, and chemical fate data developed for ITC chemicals by the U.S. chemical industry under TSCA section 4

ITC Chemicals	Lowest EC50 or LC50 value for aquatic organisms (mg/L)	Lowest MATC value for aquatic organisms (mg/L)	BCF	Henry's law constant (atm m^3/mol)	Log K_{ow}	Degradation
Alkyl and aryl benzenes						
Biphenyl	0.36	0.23	110	3×10^{-4}	4.1	Microbial > 90% in 10 d
Cumene	1.2			1.5×10^{-2}	3.6	Microbial 13 to 95% in 45 d
Anthraquinone	ND[a]	ND		2.2×10^{-7}	3.4	
Ethylbenzene	2.6			8.4×10^{-3}	3.2	
Alkyl phenols						
Tetramethylbutylphenol	0.12	0.008		1×10^{-11}	3.7	
Di-*tert*-butylphenol	0.45			7.6×10^{-3}	5.4	
Branched 4-nonylphenol	0.04	0.005	344	1.5×10^{-6}	>4.7	Microbial—negligible
Aromatic amines						
Anilines						
Aniline	2.3	0.02		1.9×10^{-6}	0.9	Stability—20 to 100% lost in 4 d
2-Chloroaniline	1.0	0.007		5.4×10^{-6}	1.9	Stability—22 to 78% lost in 4 d
2,6-Dichloro-4-nitroaniline	2.6	0.02		4.7×10^{-8}	2.8	Stability—9 to 45% lost in 4 d
Phenylenediamines						
ortho	0.16	0.004		6.7×10^{-10}	0.2	Photolysis half-life 0.26 d
meta	2.4	0.07		9.5×10^{-11}	−0.4	Photolysis half-life at 0.51 d
para	0.28	0.11		6.7×10^{-10}	−0.7	Photolysis half life 0.12 d
Benzene carboxylic acid esters						
Alkyl phthalates						
Dimethyl	29	15			1.9	Primary biodegradation half-life 1.9 d
Diethyl	15	38			2.3	Primary biodegradation half-life 2.2 d
Butyl-2-ethylhexyl	ND	0.09			3.7–7.9	Primary biodegradation half-life 19.4 d
Di-*n*-butyl	0.75	0.14			3.7	Primary biodegradation half-life 15.4 d
Dihexyl	ND	0.11			5.7–5.9	Primary biodegradation half-life 2.9 d
Di-2-ethylhexyl	ND	0.11			7.9	Primary biodegradation half-life 5.3 d
Di-(*n*-hexyl,*n*-octyl,*n*-decyl)	ND	0.15			5.9–8.6	Primary biodegradation half-life 5.3 d
Diisooctyl	ND	0.09				Primary biodegradation half-life 8.8 d
Diisononyl	ND	0.06				Primary biodegradation half-life 5.3 d
Di-(heptyl,nonyl,undecyl)	ND	0.13				Primary biodegradation half-life 5.0 d
Diisodecyl	ND	0.04				Primary biodegradation half-life 9.6 d
Diundecyl	ND	0.06				Primary biodegradation half-life 6.2 d
Ditridecyl	ND	0.07				Primary biodegradation half-life 27.7 d
Butybenzyl phthalate	0.2	0.11	380	1.3×10^{-6}	4.9	Primary biodegradation half-life 19.4 d
Bis(2-ethylhexyl) terephthalate	ND	ND	314	1×10^{-5}	5.7	Microbial—95% in 28 d
Tris(2-ethylhexyl)trimellitate	ND	ND			4.4	Microbial—68% in 28 d

Bisphenols						
Bisphenol A	1.0			1×10^{-11}	3.3	
Tetrabromobisphenol A	0.13	0.07		4.5×10^{-5}	4.5	Primary biodegradation half-life 48 to 84 d
Chloroalkanes						
59% Chlorinated *n*-undecane	0.009	0.006	40,900			Activated sludge removal—93%
1,2-Dichloropropane	55.9	11.4		2.8×10^{-3}	2.0	
Halogenated benzenes						
Trichlorobenzenes						
1,2,3-Trichlorobenzene	0.35	0.022		1.25×10^{-3}	4.0	
1,2,4-Trichlorobenzene	0.49	0.046		1.41×10^{-3}	4.0	
2-Chlorotoluene	1.1	0.11	890	1.53×10^{-3}	3.4	
4-Chlorobenzotrifluoride			169	6×10^{-2}	3.8	Aquatic volatilization half-life 0.3 to 3 h
3,4-Dichlorobenzotrifluoride	1.7	0.18		2.6×10^{-2}	4.4	
Acrylamide	78	0.08		1×10^{-9}	−0.7	
Diethylenetriamine				3.1×10^{-13}	−2.1	Primary biodegradation 6 to 90% in 14 d
2-Mercaptobenzothiazole		0.6		3.4×10^{-8}	2.42	Aquatic photolysis half-life 0.5 h
Octamethylcyclotetrasiloxane	0.01	0.005		0.4	4.45	Aquatic volatilization half-life 1 to 6 d
Tri-*n*-butylphosphate	2.4	1.2		3.2×10^{-6}	4.0	
Crotonaldehyde	0.6			1.7×10^{-5}	0.6	
4-Vinylcyclohexene	0.6			2.9×10^{-2}	2.7	Aquatic volatilization half-life 1.4 to 6 d

[a]ND: An exact EC50, LC50, or MATC could not be determined because the test chemical was not inherently toxic at concentrations that equalled water solubility values. Walker, 1993c.

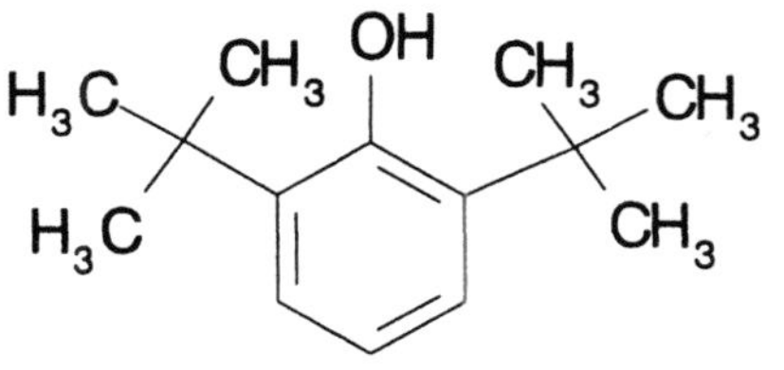

2,6-Di-*tert*-butylphenol

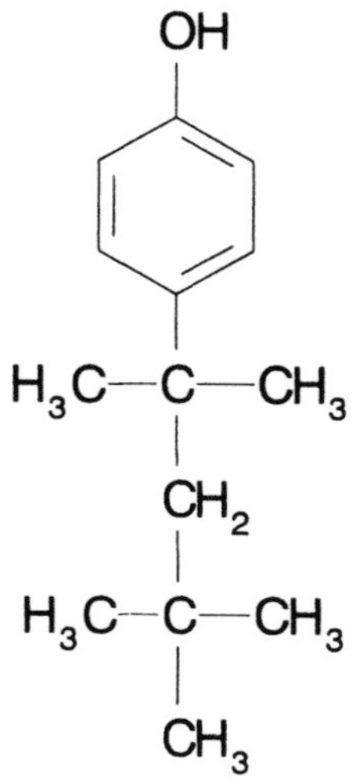

4-(1,1,3,3-Tetramethylbutyl) phenol

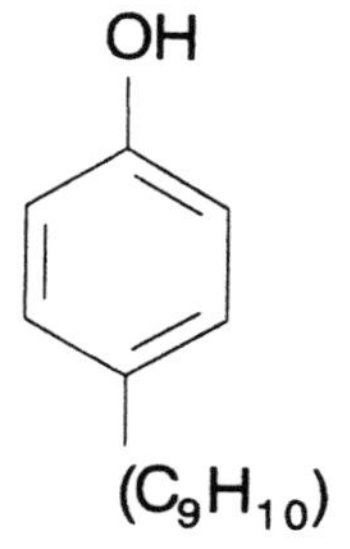

Branched 4-nonylphenol

Figure 8. Structures of some of ITC's alkylphenols tested by the chemical industry under section 4 of TSCA.

Alkylphenols

Structures of some of the alkylphenols for which the chemical industry has developed data under TSCA section 4 are presented in Figure 8.

4-(1,1,3,3-Tetramethylbutyl)phenol

The ITC designated 4-(1,1,3,3-tetramethylbutyl)phenol (CAS number 140-66-9) for testing in its Eleventh Report (ITC, 1982b). The ITC designated tetramethylbutylphenol for testing because 50 million pounds a year were produced, there was potential for occupational exposure and environmental release and persistence, there were measured environmental concentrations that exceeded acute toxicity levels, and there were insufficient data to assess its persistence and toxicity. The structure of 4-(1,1,3,3-tetramethylbutyl)phenol is illustrated in Figure 8.

In response to ITC's designation, the EPA published proposed and final negotiated testing agreements in 1983 and 1984, respectively (U.S. EPA, 1983b, 1984a). CMA's Alkylphenols and Ethoxylates panel coordinated and sponsored this testing.

Tetramethylbutylphenol has a water solubility of 18 mg/L at 25°C, an estimated vapor pressure of 4.7 $\times 10^{-4}$ mm Hg at 25°C, a calculated Henry's law constant of 1×10^{-11} atm m^3/mol, and a log K_{ow} of 3.7 (Walker, 1993c).

Tetramethylbutylphenol's toxicity values were substantially lower than water solubility (Table 5).

2,6-Di-tert-butylphenol

The ITC designated 2,6-di-*tert*-butylphenol (CAS number 128-39-2) for testing in its Seventeenth Report (ITC, 1985b). The ITC designated 2,6-di-*tert*-butylphenol for testing because of its multimillion pound production volume, potential occupational exposure, release from wastewater and because it was predicted to be highly toxic to aquatic organisms. The structure of 2,6-di-*tert*-butylphenol is illustrated in Figure 8.

In response to ITC's designation, the EPA published a proposed rule in 1987 (U.S. EPA, 1987b). After this rule was published, Ethyl Corporation initiated a voluntary program to conduct the tests designated by the ITC.

2,6-Di-tert-butylphenol has a water solubility of 4.1 mg/L at 25°C, a vapor pressure of 7.6×10^{-3} mmHg at 25°C, a calculated Henry's Law Constant of 5×10^{-4} atm m^3/mole and an estimated log K_{ow} of 5.4. The water solubility and vapor pressure data were developed by Ethyl Corporation. Ethyl also measured the water solubility of 2,6-di-tert-butylphenol at pH 5 (3.99 mg/L) and pH 9 (4.69 mg/L) and determined that 2,6-di-tert-butylphenol would be immobile in soil because it had a soil organic carbon partition coefficient of 2570 to 6960.

2,6-Di-tert-butylphenol's toxicity values were lower than water solubility (Table 5). Aquatic toxicity and chemical fate data have been published (Walker, 1995). The LC50 value of 0.45 mg/L, listed in Table 5 is for *Daphnia magna*; it is similar to the LC50 values that were generated for *Gammarus fasciatus* (0.60 mg/L) and rainbow trout (0.74 mg/L) (Walker, 1995).

Branched 4-nonylphenol

EPA published a consent order for branched 4-nonylphenol (CAS number 84852-15-3) in 1990 (U.S. EPA, 1990). CMA's Alkylphenols and Ethoxylates panel coordinated and sponsored this testing. The structure of branched 4-nonylphenol is illustrated in Figure 8.

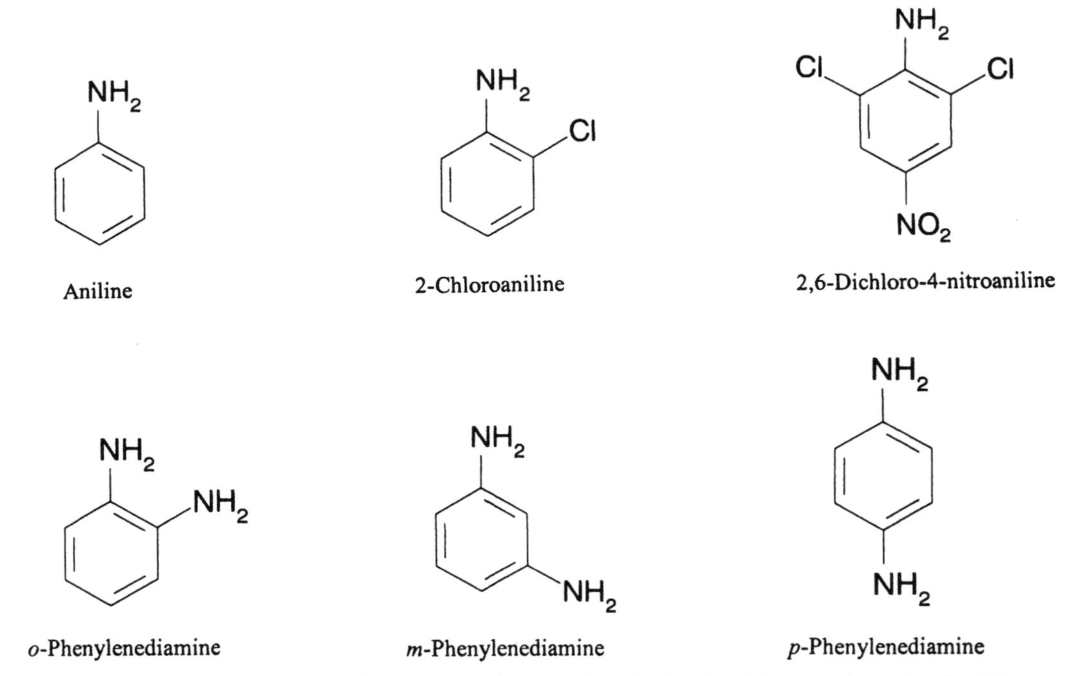

Figure 9. Structures of some of ITC's aromatic amines tested by the chemical industry under section 4 of TSCA.

Branched 4-nonylphenol has a water solubility of 6.2 mg/L at 25°C, a vapor pressure of 3.4×10^{-5} mm Hg at 25°C, a calculated Henry's law constant of 1.5×10^{-6} atm m^3/mol, and a log K_{ow} >4.7. These physical and chemical property data as well as other chemical fate data for branched 4-nonylphenol were developed by CMA's Alkylphenols and Ethoxylates Panel.

Toxicity values for branched 4-nonylphenol were substantially lower than water solubility (Table 5). The bioconcentration factors of branched 4-nonylphenol for fathead minnows were 271 and 344 when they were exposed to 0.005 and 0.023 mg/L branched 4-nonylphenol, respectively.

Aromatic Amines

Structures of some of the aromatic amines for which the chemical industry has developed data under TSCA section 4 are presented in Figure 9.

Anilines

The ITC designated aniline and chloro-, bromo-, and/or nitroanilines for testing in its Fourth Report (ITC, 1979). The ITC designated anilines because none of the chemicals has been adequately tested for human health or environmental effects, because results from limited studies raised concern about the potential to induce methemoglobinemia or mutations or cause cancer, and because of high production volumes and potential for environmental release and human exposure. The structures of aniline, 2-chloroaniline, and 2,6-dichloro-4-nitroaniline are illustrated in Figure 9.

In response to that designation, the EPA published an advanced notice of proposed rulemaking in 1984 and a consent order in 1988 to conduct aquatic toxicity tests for aniline (CAS number 142-04-1), 2-chloroaniline (CAS number 95-51-2), and 2,6-dichloro-4-nitroaniline (CAS number 99-30-9) (U.S. EPA, 1984b, 1988c). SOCMA's Anilines Association and Substituted Anilines Task Force coordinated and sponsored this testing.

Aniline has a water solubility of 36,000 mg/L at 25°C, a vapor pressure of 4.9×10^{-1} mm Hg at 25°C, a calculated Henry's law constant of 1.9×10^{-6} atm m^3/mol, and a log K_{ow} of 0.9. 2-Chloroaniline has a water solubility of 8160 mg/L at 25°C, a vapor pressure of 2.04×10^{-1} mm Hg at 25°C, a calculated Henry's law constant of 5.4×10^{-6} atm m^3/mol, and a log K_{ow} of 1.9. 2,6-Dichloro-4-nitroaniline has a water solubility of 7 mg/L at 20°C, a vapor pressure of 1.2×10^{-6} mm Hg at 25°C, a calculated Henry's Law Constant of 4.7×10^{-8} atm m^3/mol, and an estimated log K_{ow} of 2.76.

In addition to the aquatic toxicity tests listed in the Consent Order, SOCMA's Anilines Association and Substituted Anilines Task Force sponsored testing to determine the stability of the anilines in well water after 4 d in the presence and absence of food used during aquatic toxicity tests. In the absence of this food 81% aniline remained in the well water; in its presence none remained. In the absence of this food 78% 2-chloroaniline remained in the well water; in its presence 22% remained. In the absence of this food 91% 2,6-dichloro-4-nitroaniline remained in the well water; in its presence 55% remained. As a result of these tests, all toxicity tests were conducted using flow-through systems and measured concentrations of anilines.

MATC values for aniline, 2-chloroaniline, and 2,6-dichloro-4-nitroaniline were 100 times less than LC50 values (Table 5). For industrial chemicals with a nonspecific mode of toxic action, MATC values are generally less than 10 times LC50 values. The LC50/MATC ratio for these anilines suggest that they are toxic by a specific mode of action, probably polar narcosis (see Chapters 1 and 20). Aquatic toxicity data have been published (Walker, 1995). For aniline, the LC50 is for *G. fasciatus* and the MATC is for *D. magna*; for 2-chloroaniline the LC50 and MATC are for rainbow trout; for 2,6-dichloro-4-nitroaniline, the LC50 is for *Selenastrum capricornutum* and the MATC is for rainbow trout (Table 5; Walker, 1995).

Phenylenediamines

The ITC designated phenylenediamines for testing in its Sixth Report (ITC, 1980a). These included nitrogen-unsubstituted phenylenediamines with zero to two halo, nitro, hydroxy, and lower (less than four carbons) alkoxy or lower alkyl groups. The ITC designated phenylenediamines because many of the phenylenediamines have demonstrated specific biological activities associated with adverse health effects including methemoglobinemia and cancer and because few of the chemicals had been adequately tested for human health or environmental effects. The structures of *o*-phenylenediamine, *m*-phenylenediamine, and *p*-phenylenediamine are illustrated in Figure 9.

In response to that designation, the EPA published an advanced notice of proposed rulemaking in 1982 and issued proposed and final rules in 1986 and 1989, respectively, to test *o*-phenylenediamine (CAS Registry number 95-54-5), *m*-phenylenediamine (CAS Registry number 108-45-2), and *p*-phenylenediamine (CAS Registry number 106-50-3) under conditions that would maintain stable measured concentrations (U.S. EPA, 1982a, 1986b, 1989a). Du Pont coordinated and sponsored this testing.

o-Phenylenediamine has an estimated water solubility of 36,000 mg/L, an estimated vapor pressure of 1.5×10^{-3} mm Hg at 25°C, a calculated Henry's law constant of 6.7×10^{-10} atm m^3/mol, and a log K_{ow} of 0.15. *m*-Phenylenediamine has a water solubility of 240,000 mg/L at 25°C, a vapor pressure of 1.4×10^{-3} mm Hg at 25°C, a calculated Henry's law constant of 9.5×10^{-11} atm m^3/mol, and an estimated log K_{ow} of -0.36. *p*-Phenylenediamine has a water solubility of 37,000 mg/L at 25°C, a vapor pressure of 1.8×10^{-4} mm Hg at 25°C, a calculated Henry's law constant of 6.7×10^{-10} atm m^3/mol, and an estimated log K_{ow} of -0.66.

Maintaining stable measured concentrations of phenylenediamines for aquatic toxicity testing was a concern of the ITC. Ten years after the ITC designated phenylenediamines for testing, Stahl et al. (1990) reported that the oxidative half-lives for *para*, *ortho*, and *meta* phenylenediamine were 0.25 to 0.35 d, 25 to 40 d, and 133 to 330 d, respectively. Despite its short half-life, the toxic action of *p*-phenylenediamine or its oxidation products was rapid enough to cause adverse effects at low concentrations (Table 5). During acute toxicity tests, *p*-phenylenediamine was not even detected at the beginning of the test (Walker, 1993c).

When the ITC designated phenylenediamines for environmental testing, there was a concern that phenylenediamines might be rapidly transformed by direct and indirect photolysis. Testing sponsored by Du Pont revealed that *p*-phenylenediamine had a direct photolysis rate constant of 4.2 d^{-1}, an indirect photolysis rate constant of 1.7 d^{-1}, and an overall photolysis half life of 0.12 d; *o*-phenylenediamine had a direct photolysis rate constant of 1.0 d^{-1}, an indirect photolysis rate constant of 1.7 d^{-1}, and an overall photolysis half-life of 0.26 d; *m*-phenylenediamine had a direct photolysis rate constant of 0.5 d^{-1}, an indirect photolysis rate constant of 0.86 d^{-1}, and an overall photolysis half-life of 0.51 d.

Benzene Carboxylic Acid Esters

Structures of some of the benzene carboxylic acid esters for which the chemical industry has developed data under TSCA section 4 are presented in Figure 10.

Alkyl Phthalates

The ITC designated alkyl phthalates for testing in its First Report (ITC, 1977). The ITC designated alkyl phthalates for testing because many were produced in over 100 million pounds per year and their use as plasticizers in a wide variety of products was likely to result in release to aquatic environments as waste from processing plants or from use and disposal of end products. The ITC was concerned about their persistence and toxicity in aquatic systems. The structures of three simple alkyl phthalates (dimethyl, di-*n*-butyl, and di-2-ethylhexyl) are illustrated in Figure 10.

In response to the ITC's designation, the EPA published proposed and final negotiated testing agreements in 1981 and 1982 to implement the ITC's testing recommendations (U.S. EPA, 1981b, 1982b). CMA's Alkyl Phthalates Panel coordinated and sponsored the testing designated by the ITC.

The alkyl phthalates' physical-chemical properties and other chemical fate data developed under the final negotiated testing agreement are listed in Table 6. Details of the results of water solubility, octanol-water partition coefficient (K_{ow}), and vapor pressure tests and problems associated with conducting octanol-water partition coefficient and vapor pressure tests for moderate to long-chain alkyl phthalates were discussed by Howard et al. (1985) and Walker (1990c). Details of the results of biodegradation tests

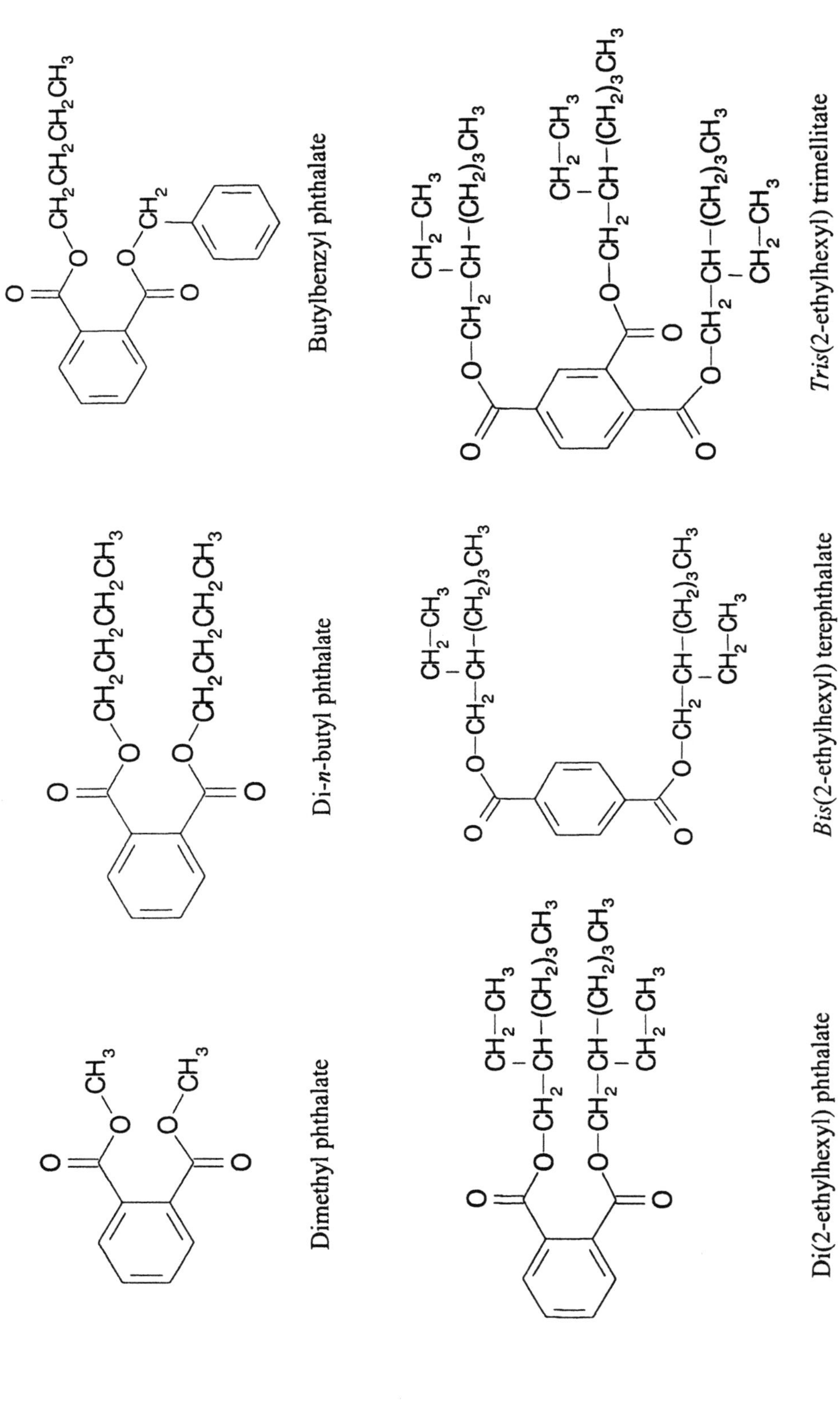

Dimethyl phthalate

Di-*n*-butyl phthalate

Butylbenzyl phthalate

Di(2-ethylhexyl) phthalate

Bis(2-ethylhexyl) terephthalate

Tris(2-ethylhexyl) trimellitate

Figure 10. Structures of some of ITC's benzene carboxylic acid esters tested by the chemical industry under section 4 of TSCA.

Table 6. Alkyl phthalates chemical fate data

CAS number	Alkyl phthalate	Water solubility (mg/L)	Log K_{ow}	Log K_{oc}	Vapor pressure (mm Hg)	Ultimate biodegradation (%TCO_2)	Primary biodegradation (% LPC)	Primary biodegradation ($T_{1/2}$, d)
131-11-3	Dimethyl (1)[a]	4000	1.90		1.65×10^{-3}	85.9	>99	1.9
84-66-2	Diethyl (2)	1080	2.29		1.65×10^{-3}	94.6	>99	2.21
84-74-2	Di-*n*-butyl (3)	11.2	3.74		7.28×10^{-5}	57.4	89.8	15.4
146-50-9	Dihexyl (4)	0.24	5.65–5.93	4.7	1.43×10^{-5}	77.1	>99	2.93
85-69-8	Butyl-2-ethylhexyl (5)	0.69	3.70–7.88		8.25×10^{-6}	42.5	77.7	19.4
117-81-7	Di-2-ethylhexyl (6)	0.34	7.94	5.3–5.9	6.45×10^{-6}	85.5	>99	5.25
25724-58-7	Di-(*n*-hexyl, *n*-octyl,*n*-decyl) (7)	0.9	5.90–8.61	5.0–5.8	5.025×10^{-6}	90.3	>99	5.30
27554-26-3	Diisooctyl (8)	0.09			5.55×10^{-6}	56.7	>99	8.82
28553-12-0	Diisononyl (9)	0.2			5.40×10^{-7}	61.5	>99	5.31
39393-37-8	Di(heptyl,nonyl,undecyl) (10)	<1				98.4	>99	5.03
26761-40-0	Diisodecyl (11)	1.19				56.2	>99	9.6
3648-20-0	Diundecyl (12)	1.11				76.0	>99	6.17
119-06-2	Ditridecyl (13)	<0.3		4.0–6.5		37.0	>50	27.7

[a]Numbers correspond to those in Figure. 11.

were discussed by Sugatt et al. (1984) and O'Grady et al. (1985).

Under the negotiated testing agreement, aquatic toxicity testing was also conducted. All tests were conducted using measured concentrations of alkyl phthalates. Most tests were conducted under static conditions. For alkyl phthalates with carbon chain lengths greater than four (di-*n*-butyl), the EC50 and LC50 values exceeded water solubility values (Table 5). The relationship between toxicity and water solubility of alkyl phthalates is illustrated in Figure 11. QSAR was developed to predict the toxicity of other alkyl phthalates: log MATC = 0.65 log water solubility − 0.58. For alkyl phthalates, listed in Table 5 the LC50 values were for sheepshead minnow (dimethyl), fathead minnow (diethyl) and *S. capricornutum* (di-*n*-butyl) and MATC values were for *D. magna* (Walker, 1995).

Butylbenzyl Phthalate

The ITC designated butylbenzyl phthalate (CAS number 85-68-7) for testing in its Seventh Report (ITC, 1980b). The ITC designated butylbenzyl phthalate because of its structural similarity to alkyl phthalates, because its production volume ranged from 100 to 500 million pounds in 1977, and because its use as a plasticizer could result in environmental release. The ITC was concerned about the potential for butylbenzyl phthalate to cause cancer and birth defects. Butylbenzyl phthalate is an alkyl aryl phthalate that is similar to di-*n*-butyl phthalate except a benzene ring is substituted for one of the *n*-butyl groups (Figure 10).

In response to the ITC's designation, the EPA included butylbenzyl phthalate in the proposed and final negotiated testing agreements for alkyl phthalates that were published in 1981 and 1982 (U.S. EPA, 1981b, 1982b). CMA's Alkyl Phthalates Panel coordinated and sponsored the testing recommended by the ITC.

The physical properties of butylbenzyl phthalate are similar to those of di-*n*-butyl phthalate (Walker, 1993c). Testing to measure the water solubility (2.7 mg/L) and log octanol-water partition coefficient (4.9) were conducted under the negotiated testing agreement. Butylbenzyl phthalate has a vapor pressure of 8.2×10^{-6} mm Hg at 25°C and a calculated Henry's law constant of 1.3×10^{-6} atm m^3/mol (Walker, 1993c).

Under the negotiated testing agreement, 28-d shake-flask biodegradation tests were conducted. Greater than 99% of the butylbenzyl phthalate was removed through primary biodegradation (see Chapter 15) with a half-life of 4.6 d and 86.5% was mineralized to CO_2 through ultimate biodegradation. The results of butylbenzyl phthalate biodegradability testing were discussed by Adams et al. (1989), Carson et al. (1990), and Walker (1995).

Under the negotiated testing agreement, tests were also conducted to measure the toxicity of butylbenzyl phthalate to aquatic organisms (Table 5). Aquatic toxicity data have been published (Walker, 1995). Butylbenzene phthalate's LC50 value was for *S. capricornutum*; the MATC was for *M. bahia* (Table 5; Walker, 1995).

Bis(2-ethylhexyl) terephthalate

The ITC designated bis(2-ethylhexyl) terephthalate (CAS number 6422-86-2) for testing in its Eleventh Report (ITC, 1982b). The ITC designated bis(2-ethylhexyl) terephthalate because 1 to 10 million pounds were produced in 1977 and because it was used as a plasticizer when alkyl phthalates were in short supply. The ITC was concerned that bis(2-ethylhexyl) terephthalate would leach out of polyvinyl chloride resins into the environment. The ITC was also concerned about the metabolism of bis(2-ethylhexyl) terephthalate to 2-ethylhexyl alcohol and any resulting hepatic peroxisome proliferation that might be caused. Bis(2-ethylhexyl) terephthalate is similar in structure to di(2-ethylhexyl) phthalate except that the 2-ethylhexyl esters are in the *para*, not *ortho*, position on the benzene ring (Figure 10).

In response to the ITC's designation, the EPA published proposed and final negotiated testing agreements in 1983 and 1984 (U.S. EPA, 1983c, 1984c). Eastman Kodak Company coordinated and sponsored the testing designated by the ITC.

The water solubility and octanol-water partition coefficient data for bis(2-ethylhexyl) terephthalate ranged from 0.3 to 1.5 mg/L and 5.3 to 5.7, respectively. Bis(2-ethylhexyl) terephthalate has a vapor pressure of 4×10^{-7} mm Hg at 25°C and a measured Henry's law constant of 1×10^{-5} atm m^3/mol (Walker, 1993c).

The relatively low water solubility of bis-(2-ethylhexyl) terephthalate made it difficult to establish exact acute and chronic toxicity values. The rainbow 7-d LC50 value and 71-d MATC values were greater than 0.25 mg/L. The Eastern oyster EC50 value was greater than 0.62 mg/L. However, the low water solubility of bis(2-ethylhexyl) terephthalate did not preclude development of bioconcentration and biodegradation data. In Eastern oysters, bis(2-ethylhexyl) terephthalate had a bioconcentration factor of 314 and a depuration half-life of 5 d. During 28-d shake-flask tests, most of the bis(2-ethylhexyl) terephthalate was biodegraded and 40% of it was mineralized to CO_2.

Compared to di(2-ethylhexyl) phthalate, bis(2-ethylhexyl) terephthalate was equally insoluble in water but had a lower octanol-water partition coefficient and ultimate biodegradability to CO_2. It was difficult to compare the toxicity of the two chemicals because when similar species were tested the LC50 values exceeded water solubility values.

Tris(2-ethylhexyl) trimellitate

The ITC designated tris(2-ethylhexyl) trimellitate (CAS number 3319-31-1) for testing in its Eleventh Report (ITC, 1982b). The ITC recommended tris(2-ethylhexyl) trimellitate because more than 2 million pounds was produced in 1977, more than 26 million

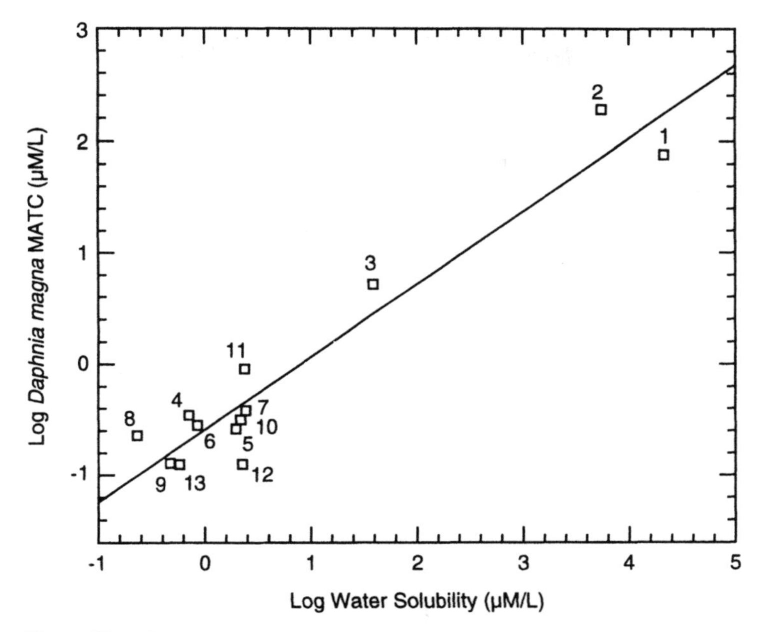

Figure 11. Influence of water solubility on chronic toxicity of alkyl phthalates to *Daphnia magna*. (Chemicals corresponding to the numbers are listed in Table 6.)

pounds of trimellitate ester plasticizers were produced in 1977, and production volumes were expected to increase. The trimellitate was used as a plasticizer for polyvinyl chloride resins and could be released into the environment. The ITC was also concerned about the metabolism of tris-(2-ethylhexyl) trimellitate to 2-ethylhexyl alcohol and any resulting hepatic peroxisome proliferation that might be caused. Tris(2-ethylhexyl) trimellitate is similar in structure to bis(2-ethylhexyl) terephthalate and di(2-ethylhexyl) phthalate, except that the trimellitate has three 2-ethylhexyl esters (Figure 10).

In response to the ITC's designation, the EPA published proposed and final negotiated testing agreements in 1983 and 1984 (U.S. EPA, 1983d, 1984d). CMA's Trimellitates Panel coordinated and sponsored the testing designated by the ITC.

The water solubility of tris(2-ethylhexyl) trimellitate was as low (0.385 µg/L) as that of di(2-ethylhexyl) phthalate and bis(2-ethylhexyl) terephthalate. The measured log octanol-water partition coefficient (4.35) was not as high as that of di(2-ethylhexyl) phthalate or bis(2-ethylhexyl) terephthalate. Based on the studies of De Bruijn et al. (1989) on di(2-ethylhexyl) phthalate, the log octanol-water partition coefficient for tris(2-ethylhexyl) trimellitate should be about three orders of magnitude higher.

The water solubility of tris(2-ethylhexyl) trimellitate was low enough to prevent exact acute and chronic toxicity values from being developed. Primary biodegradation studies demonstrated that after 28 d at a concentration of 0.28mg/L about 68% of the parent compound was removed. Ultimate biodegradation was affected by the concentration of tris(2-ethylhexyl)trimellitate; at 0.28 mg/L the half-life was less than 3 d, and at 0.028 mg/L the half-life was 49 d. Using primary biodegradation data, *tris*(2-ethylhexyl) trimellitate was less biodegradable than bis(2-ethylhexyl) terephthalate and di(2-ethylhexyl) phthalate.

Bisphenols

Structures of the two bisphenols for which the chemical industry developed data under TSCA section 4 are presented in Figure 12.

Bisphenol A

The ITC designated bisphenol A for testing in its Fourteenth Report (ITC, 1984a). The ITC designated bisphenol A for testing because it was manufactured in very high production volumes (500 million pounds per year) and used at many sites where there were opportunities for potential environmental release. The structure of bisphenol A is illustrated in Figure 12.

In response to the ITC's designation, the EPA published proposed and final test rules in 1985 and 1986 (U.S. EPA, 1985a, 1986d). The Society of Plastics Industry Inc., ad hoc Bisphenol A Task Group, sponsored the testing designated by the ITC.

Bisphenol A has a water solubility of 120 mg/L at 25°C, a vapor pressure of 18.4×10^{-3} mm Hg at 170°C, a calculated Henry's law constant of 1×10^{-11} atm m^3/mol, and a log K_{ow} of 3.3 (Walker, 1993c).

Biphenol A's lowest LC50 value was 1 mg/L (Table 5). Details of the bisphenol A toxicity tests were published by Alexander et al. (1988). There was some loss of bisphenol A during some of the toxicity tests. Some of this loss may have resulted from biodegradation, as biodegradation accounted for about 90% of bisphenol A removal during an OECD ready biodegradability test (Dorn et al., 1987).

Tetrabromobisphenol A

The ITC designated tetrabromobisphenol A for testing in its Sixteenth Report (ITC, 1985a). The ITC designated tetrabromobisphenol A for testing because it was manufactured in over a million pounds per year and detected at production sites where there were opportunities for environmental release. The structure of tetrabromobisphenol A is illustrated in Figure 12.

In response to the ITC's designation, the EPA published proposed and final test rules in 1986 and 1987 (U.S. EPA, 1986e, 1987f). CMA's Brominated Flame Retardants Industry Panel sponsored the testing designated by the ITC.

Tetrabromobisphenol A has a water solubility of 4.2 mg/L at 25°C, an estimated vapor pressure of 4.15×10^{-5} mm Hg, a calculated Henry's law constant of 4.5×10^{-5} atm m^3/mol, and a log K_{ow} of 4.5 (Walker, 1993c).

In microcosm biodegradation tests, tetrabromobisphenol A had a 48- to 84-d half-life and less than 8% of total tetrabromobisphenol A was mineralized to carbon dioxide. In chronic aquatic toxicity tests tetrabromobisphenol A had MATC values of 0.54 mg/L (fathead minnows) and 0.22 mg/L (*Daphnia magna*). In tests designed to measure the toxicity and bioconcentration of chemicals added to sediments, tetrabromobisphenol A had a bioconcentration factor of 3190 in *Chironomus tentans*, an LC50 of 0.13 mg/L, and an MATC of 0.07 mg/L in aqueous-phase studies (Walker, 1993d). In sediment-phase studies, it was difficult to determine whether tetrabromobisphenol A was toxic to *C. tentans*, because control survival was lower than survival of the midge in test systems (Walker, 1993d).

Chloroalkanes

The ITC recommended testing for a group of long-chain chloroalkanes (chlorinated paraffins) and one short-chain chloroalkane (1,2-dichloropropane).

Chlorinated Paraffins

The ITC designated chlorinated paraffins for testing in its First Report (ITC, 1977). The ITC designated chlorinated paraffins because in 1972 their annual production volume was about 80 million pounds and their use in a wide variety of household and paint products, adhesives, and flame retardants resulted in an estimated release of about 50 million pounds per year. The ITC was concerned because residues of chlorinated paraffins were in fish and because these mixtures caused degenerative liver and spleen changes in mice.

Mixtures of chlorinated paraffins were designated for testing by the ITC. These included two chlorinated paraffins of $C_{22}-C_{26}$ that contained 40–50% chlorine (43% chlorinated *n*-pentadecane-8-^{14}C, a synthetic Chlorowax 40) or 60–70% chlorine (70% chlorinated *n*-pentacosane-13-^{14}C, a synthetic Electrofine S70), one chlorinated paraffin of $C_{14}-C_{19}$ that contained 50–60% chlorine (51% chlorinated *n*-pentacosane-13-^{14}C, a synthetic Cereclor S52) and one chlorinated paraffin of $C_{10}-C_{12}$ that contained 60–70% chlorine (59% chlorinated *n*-undecane-6-^{14}C, a synthetic Chlorowax 500C) (Table 7).

In response to the ITC's designation, a consortium of chlorinated paraffin manufacturers presented EPA with a list of completed, ongoing, and planned studies for chlorinated paraffins. As a result of this presentation, EPA issued a decision not-to-test in 1982 that did not require this consortium to conduct additional tests (U.S. EPA, 1982c).

The testing conducted by the consortium used the ^{14}C-labeled chlorinated paraffins listed above. The consortium proposed to conduct water solubility, bioconcentration, and aquatic toxicity tests on the four chlorinated paraffins listed above and then conduct more extensive testing on the most toxic chlorinated paraffin. The chlorinated paraffin of $C_{10}-C_{12}$ that contained 60–70% chlorine (59% chlorinated *n*-undecane-6-^{14}C, a synthetic Chlorowax 500C) was the most toxic and had the highest bioconcentration factors (BCFs), especially in the mussel, *Mytilis edulis*; the 60-day LC50 was 0.074 mg/L and the 60-day BCF was 7,923 (Table 7).

Chemical fate testing demonstrated that 59% chlorinated *n*-undecane-6-^{14}C was not anaerobically biodegraded. However, a vacuum was created in the

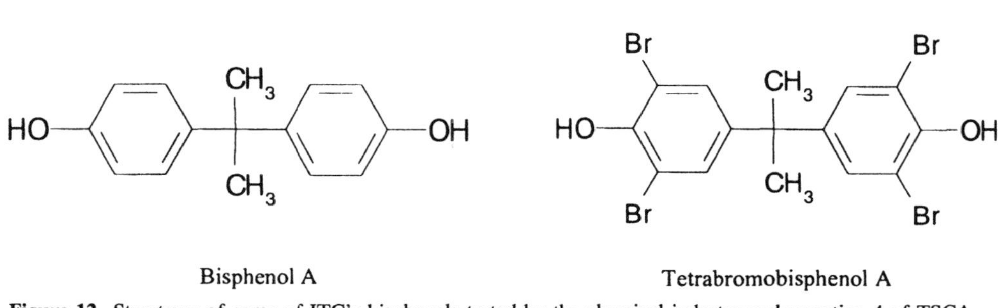

HO–C6H4–C(CH3)2–C6H4–OH

Bisphenol A

HO–C6H2Br2–C(CH3)2–C6H2Br2–OH

Tetrabromobisphenol A

Figure 12. Structures of some of ITC's bisphenols tested by the chemical industry under section 4 of TSCA.

test system that made it difficult to assess anaerobic biodegradation (Walker, 1990c).

Long-term bioconcentration testing of 59% chlorinated n-undecane-6-^{14}C with rainbow trout and mussels confirmed the high BCFs that were observed in 60-d tests (Table 5). Long-term toxicity testing of 59% chlorinated n-undecane-6-^{14}C also confirmed the toxicity that was observed in 60-d tests (Table 5). Aquatic toxicity and chemical fate data have been published (Walker, 1995). The LC50 value was for *Mytilis edulis* and the MATC was for *D. magna* (Table 5; Walker, 1994).

1,2-Dichloropropane

The ITC designated 1,2-dichloropropane (CAS Registry number 78-87-5) for testing in its Third Report (ITC, 1978b). The ITC designated 1,2-dichloropropane because of its large production volume (71 million pounds in 1976), widespread use as a solvent, and high potential for occupational and consumer exposure. 1,2-Dichloropropane is a short-chain chloroalkane that is considerably more volatile than any of the chlorinated paraffins.

In response to the ITC's designation, the EPA issued proposed and final rules in 1984 and 1986, respectively, to test 1,2-dichloropropane under conditions that would maintain stable measured concentrations (U.S. EPA, 1984e, 1986f). Dow Chemical Company sponsored the testing designated by the ITC.

1,2-Dichloropropane has a water solubility of 2700 mg/L at 25°C, a vapor pressure of 53 mm Hg, a calculated Henry's law constant of 2.8×10^{-3} atm m^3/mol, and a log K_{ow} of 2 (Walker, 1993c).

The high Henry's law constant of 1,2-dichloropropane made it difficult to keep the chemical in aqueous test systems (Walker, 1993c). However, in flow-through systems using measured concentrations of 1,2-dichloropropane, an LC50 and an MATC were established for *D. magna* (Table 5). In addition, there was an increase in mortality of *M. bahia* from 10% (controls) to 28% (0.4 mg/L) and a decrease in the number of young per female *M. bahia* from 11.4 (controls) to 6.8 (0.4 mg/L). In 2-day flow-through toxicity tests with the freshwater alga, *S. capricornutum*, increased concentrations of 1,2-dichloropropane from 86 to 676 mg/L decreased the number of cells from 61,000 to 15,000. In 2-day flow-through toxicity tests with the more sensitive marine alga, *Skeletonema costatum*, increased concentrations of 1,2-dichloropropane from 6 to 38 mg/L decreased the number of cells from 16,600 to 9,200.

Halogenated Benzenes

The ITC designated testing for several halogenated benzenes (Figure 13).

Trichlorobenzenes

The ITC designated trichlorobenzenes for testing in its Third Report (ITC, 1978b). The ITC designated trichlorobenzenes because of their potential for widespread occupational exposure and environmental release from use as solvents and transformer oils and because of their environmental occurrence as pesticide contaminants and degradation products and products of effluent chlorination. The structures of two simple trichlorobenzenes designated by the ITC are illustrated in Figure 13.

In response to the ITC's designation, the EPA issued proposed and final rules in 1984 and 1986, respectively, to test 1,2,3-trichlorobenzene (CAS Registry number 87-61-6) and 1,2,4-trichlorobenzene (CAS Registry number 120-82-1) under conditions that would maintain stable measured concentrations (U.S. EPA, 1984f, 1986c). CMA's Chlorobenzenes Panel sponsored the testing designated by the ITC.

1,2,3-Trichlorobenzene has a water solubility of 18 mg/L at 25°C, an estimated vapor pressure of 0.21 mm Hg, a calculated Henry's law constant of 1.25

Table 7. Chlorinated paraffins nomenclature, water solubility, bioconcentration, and aquatic toxicity data

	Chlorinated paraffin trade name			
	Chlorowax 500C	Cereclor S52	Chlorowax 40	Electrofine S70
Carbon chain length	$C_{10}-C_{12}$	$C_{14}-C_{19}$	$C_{22}-C_{26}$	$C_{22}-C_{26}$
Percent chlorine	60–70	50–60	40–50	60–70
CAS number	63449-39-8	63449-39-8	None	61788-76-9
Chemical name associated with CAS number	Chlorinated paraffin waxes	Chlorinated paraffin waxes	None	Chlorinated alkanes
Test chemical	59% chlorinated n-undecane-6-^{14}C	51% chlorinated n-pentacosane-13-^{14}C	43% chlorinated n-pentacosane-8-^{14}C	70% chlorinated n-pentacosane-13-^{14}C
91-d water solubility (mg/L)	0.13	0.0269	0.0064	0.0059
Rainbow trout 60-d BCF[a]	574	32.3	9	42.8
Mytilis edulis 60-d BCF	7923	339	87.2	105
Rainbow trout 60-d LC50 (mg/L)	0.34	>4.5	>4.0	>3.8
Mytilis edulis 60-d LC50 (mg/L)	0.074	>3.8	>2.18	>1.33

[a]BCF, bioconcentration factor.

$\times\ 10^{-3}$ atm m^3/mol, and a log K_{ow} of 4 (Walker, 1993c).

1,2,4-Trichlorobenzene has a water solubility of 49 mg/L at 25°C, a vapor pressure of 0.29 mm Hg at 25°C, a calculated Henry's law constant of 1.41 $\times\ 10^{-3}$ atm m^3/mol, and a log K_{ow} of 4 (Walker, 1993c).

In flow-through acute toxicity tests, LC50 and MATC values for 1,2,3- and 1,2,4-trichlorobenzenes were established for *M. bahia* (Table 5). Aquatic toxicity data have been published (Walker, 1995).

2-Chlorotoluene

The ITC designated 2-chlorotoluene (CAS number 95-49-8) for testing in its Eighth Report (ITC, 1981a). The ITC designated 2-chlorotoluene because of its high production volume (11–60 million pounds in 1977) and because mixtures containing high concentrations of 2-chlorotoluene are used in solvent and other applications that could result in occupational or consumer exposures. The structure of 2-chlorotoluene is illustrated in Figure 13.

In response to the ITC's designation, the EPA published proposed and final negotiated testing agreements in 1982 (U.S. EPA, 1982d,e). Hooker Chemical Company sponsored the testing designated by the ITC.

2-Chlorotoluene has a water solubility of 374 mg/L at 25°C, a vapor pressure of 3.4 mm Hg at 25°C, a calculated Henry's law constant of 1.53 $\times$ 10^{-3} atm m^3/mol, and a log K_{ow} of 3.4 (Walker, 1993c).

The MATC for 2-chlorotoluene was about 10 times less than the LC50 value (Table 5). In addition, the bioconcentration factor was 890 $\pm$ 340, but the depuration half-life was less than a day.

4-Chlorobenzotrifluoride

The ITC designated 4-chlorobenzotrifluoride (CAS number 98-56-6) for testing in its Ninth Report (ITC, 1981b). The ITC designated 4-chlorobenzotrifluoride because it had a production volume of 10–50 million pounds per year and because it appeared to be released to wastewater from production and use sites. The structure of 4-chlorobenzotrifluoride is illustrated in Figure 13.

In response to that designation, the EPA published proposed and final negotiated testing agreements in 1982 and 1983 (U.S. EPA, 1982f, 1983e). Hooker Chemical Company sponsored the testing designated by the ITC.

4-Chlorobenzotrifluoride has a water solubility of 29.1 mg/L at 25°C, a vapor pressure of 5.6 mm Hg at 25°C, a calculated Henry's law constant of 6 $\times$ 10^{-2} atm m^3/mol, and a log K_{ow} of 3.8 (Walker, 1993c).

4-Chlorobenzotrifluoride has a bioconcentration factor of 169 and a depuration half-life of 8 h. In the atmosphere the photooxidation half-life of 4-chlorobenzotrifluoride was determined by the type of oxidizing agent: nitrate (6 d), hydroxyl radicals (50 d), or ozone (9 yr). 4-Chlorobenzotrifluoride has an aquatic volatilization half-life of 0.3 to 3 h. Volatilization of 4-chlorobenzotrifluoride during the bio-

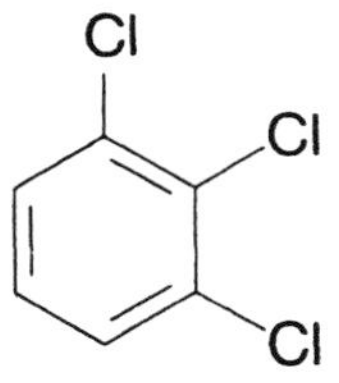

1,2,3-Trichlorobenzene 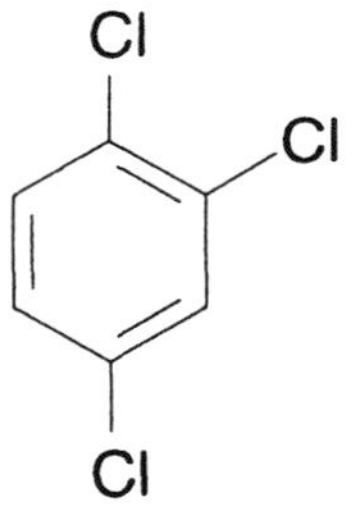 1,2,4-Trichlorobenzene 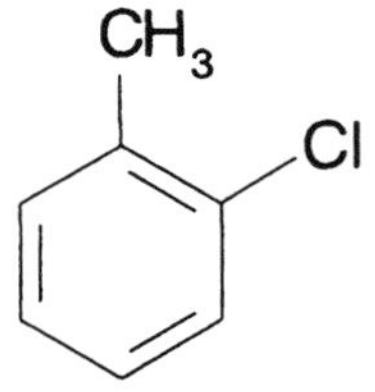 2-Chlorotoluene

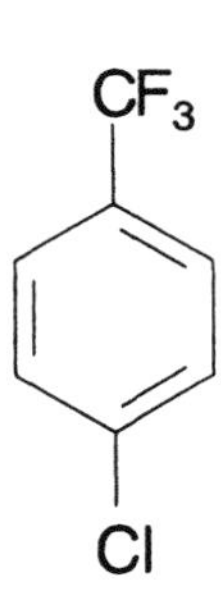

4-Chlorobenzotrifluoride 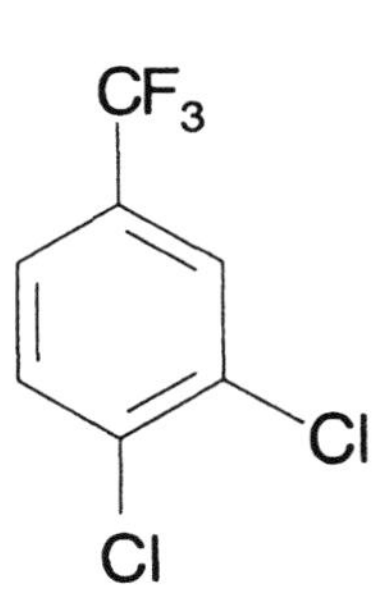 3,4-Dichlorobenzotrifluoride

Figure 13. Structures of some of ITC's halogenated benzenes tested by the chemical industry under section 4 of TSCA.

degradation tests provided some technical problems that made it difficult to interpret the results of biodegradation testing (Walker, 1990c).

3,4-Dichlorobenzotrifluoride

The ITC recommended 3,4-dichlorobenzotrifluoride (CAS number 328-84-7) for testing in its Fourteenth Report (ITC, 1984a). The ITC recommended 3,4-dichlorobenzotrifluoride because it was imported into the United States in over 1 million pounds per year and because it is likely to be released from landfills. The structure of 3,4-dichlorobenzotrifluoride is illustrated in Figure 13.

In response to that recommendation, the EPA negotiated a consent order in 1987 (U.S. EPA, 1987e). Hooker Chemical Company sponsored the testing recommended by the ITC.

3,4-Dichlorobenzotrifluoride has a water solubility of 11.6 mg/L at 25°C, a vapor pressure of 1.6 mm Hg at 25°C, a calculated Henry's law constant of 2.6×10^{-2} atm m^3/mol, and a log K_{ow} of 4.4 (Walker, 1993c).

As with 4-chlorobenzotrifluoride, volatilization of 3,4-dichlorobenzotrifluoride during the biodegradation tests caused some technical problems that made it difficult to interpret the results of biodegradation testing (Walker, 1990a). Despite these volatilization problems, LC50 and MATC values were established for 3,4-dichlorobenzotrifluoride and shown to be toxic to aquatic organisms (Table 5).

Other Chemicals

Structures of other chemicals designated or recommended for testing by the ITC are illustrated in Figure 14.

Acrylamide

The ITC designated acrylamide (CAS Registry number 79-06-1) for testing in its Second Report (ITC, 1978a). The ITC designated acrylamide because of its very high annual production volume (65 million pounds), potential for increased annual production (12%) and occupational exposure, and environmental release at manufacturing and use sites. The structure of acrylamide is illustrated in Figure 14.

In response to ITC's designation, a group of American acrylamide manufacturers agreed to conduct the testing and EPA issued a decision not-to-test in 1983 (U.S. EPA, 1983f).

Acrylamide has a water solubility of 640,000 mg/L at 25°C, a vapor pressure of 7×10^{-3} mm Hg at 25°C, a calculated Henry's law constant of 1×10^{-9} atm m^3/mol, and a log K_{ow} of −0.67 (Walker, 1993c).

The lowest LC50 and MATC values for acrylamide listed in Table 5 are for *M. bahia* (Walker, 1991b). Details of aquatic toxicity testing were published by Krautter et al. (1986) and Walker (1991b).

Diethylenetriamine

The ITC designated diethylenetriamine (CAS Registry number 111-40-0) for testing in its Eighth Report (ITC, 1981a). The ITC designated diethylenetriamine because it was produced in excess of 10 million pounds per year; because of potential for human exposures resulting from its use in a wide variety of products, such as resins for adhesives and coatings, surfactants in shampoos, fabric softeners, and corrosion inhibitors; and because of inadequate data on potential chronic effects, reproductive effects, and developmental toxicity. The structure of diethylenetriamine is illustrated in Figure 14.

In response to that designation, the EPA published proposed and final rules in 1982 and 1985 to determine the environmental fate of diethylenetriamine (U.S. EPA, 1982g, 1985h). Dow Chemical Company sponsored the testing designated by the ITC.

Diethylenetriamine is miscible in water. It has a vapor pressure of 2.35×10^{-1} mm Hg at 25°C, a calculated Henry's law constant of 3.1×10^{-13} atm m^3/mol, and a log K_{ow} of −2.13.

The EPA implemented testing to determine the environmental fate of diethylenetriamine, because of a concern that diethylenetriamine, as a secondary amine, could form a persistent nitrosamine. In flasks with sewage like water or soil inocula, the extent of primary biodegradation ranged from 6 to 90% after 14 d (Table 5). There was no evidence of nitrosamine formation in any of the studies.

2-Mercaptobenzothiazole

The ITC designated 2-mercaptobenzothiazole (CAS Registry number 149-30-4) for testing in its Fifteenth Report (ITC, 1984b). The ITC designated 2-mercaptobenzothiazole because of its high annual production volume (2 million pounds), the high annual production volumes of the zinc salt (4 million pounds) and sodium salt (40 million pounds), and potential environmental release from use as a rubber vulcanization accelerator, antioxidant, and corrosion inhibitor. The structure of the reduced form of 2-mercaptobenzothiazole is illustrated in Figure 14.

In response to that designation, the EPA published proposed and final rules in 1985 and 1988 (U.S. EPA, 1985i, 1988d). CMA's Rubber Additives Panel sponsored the testing designated by the ITC.

2-Mercaptobenzothiazole has a water solubility at 25°C of 51 mg/L at pH 5, 118 mg/L at pH 7, and 900 mg/L at pH 9, an estimated vapor pressure of 1.8×10^{-5} mm Hg at 25°C, a calculated Henry's law constant of 3.4×10^{-8} atm m^3/mol at pH 7, and a log K_{ow} of 2.42.

In chemical fate tests, 2-mercaptobenzothiazole was not biodegradable but had an aquatic photolysis half-life of 30 min. Soil and sediment organic carbon partition coefficients were 700–1400 and 2100–3000, respectively. In chronic toxicity tests, 2-mercaptobenzothiazole's lowest MATC value of 0.06 mg/L was for rainbow trout (Table 5). The MATC for *D. magna* was 0.34 mg/L.

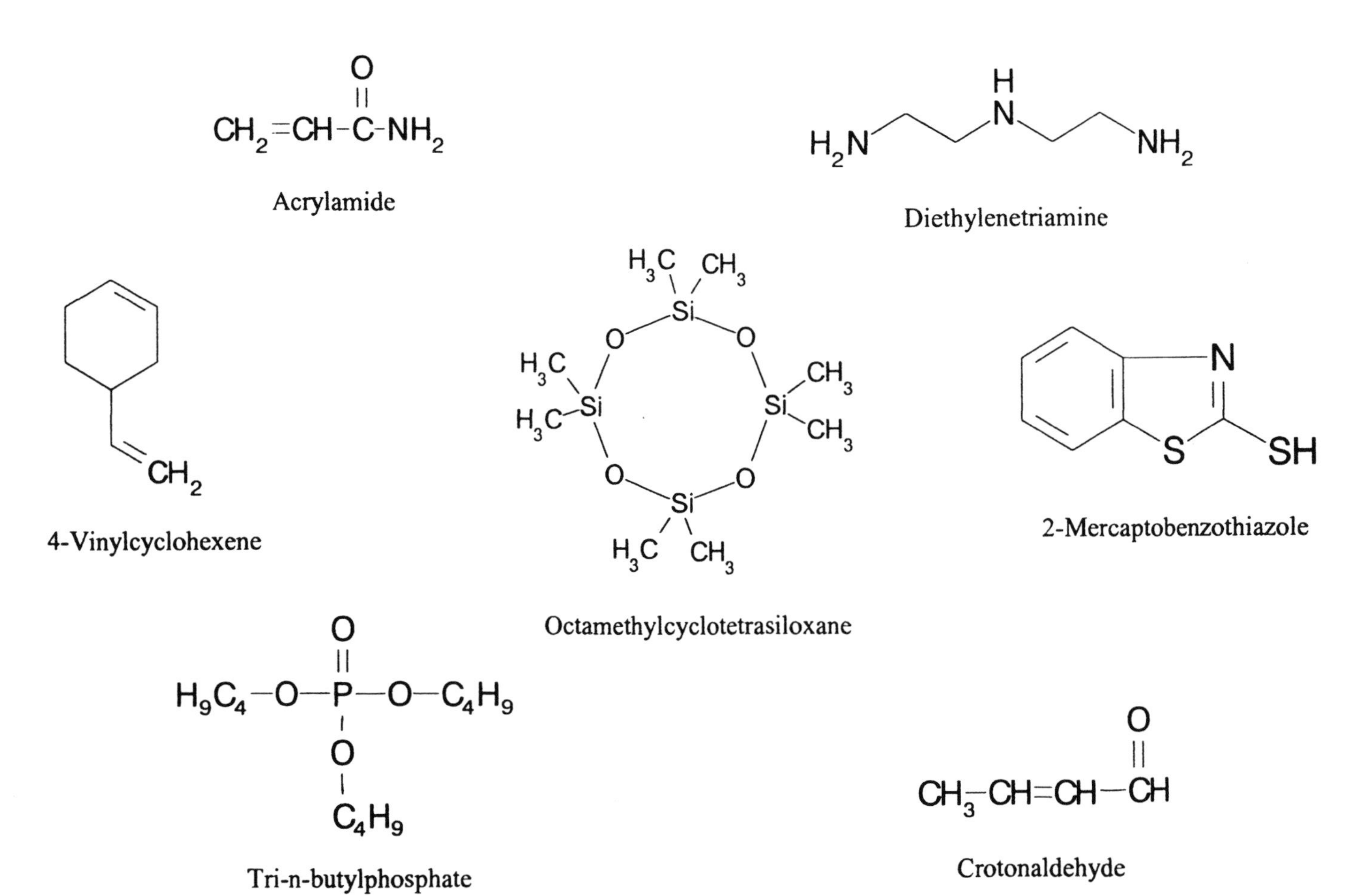

Figure 14. Structures of some of ITC's other chemicals tested by the chemical industry under section 4 of TSCA.

Octamethylcyclotetrasiloxane

The ITC designated octamethylcyclotetrasiloxane (CAS Registry number 556-67-2) for testing in its Fifteenth Report (ITC, 1984b). The ITC designated octamethylcyclotetrasiloxane because of its high annual production volume (25 million pounds) and potential for widespread environmental release. The structure of octamethylcyclotetrasiloxane is illustrated in Figure 14.

In response to that designation, the EPA published a proposed rule in 1985 and a consent order in 1989 (U.S. EPA, 1985j, 1989b). SOCMA's Silicones Environmental Health and Safety Council coordinated and sponsored the testing designated by the ITC.

Octamethylcyclotetrasiloxane has a solubility of 0.033 to 0.074 mg/L at 25°C and a measured Henry's law constant of 0.4 atm m^3/mol. It has a vapor pressure of 1 mm Hg at 25°C and a log K_{ow} of 4.45. It has relatively short aquatic and atmospheric half-lives and was not biodegradable (Table 5).

Octamethylcyclotetrasiloxane's lowest LC50 and MATC values of 10 and 5 μg/L, respectively, were for rainbow trout (Table 5). Octamethylcyclotetrasiloxane was also tested for toxicity to *Chironomus tentans* in sediment tests; details of these tests have been previously described (Walker, 1993d).

Tri-n-butyl phosphate

The ITC recommended tri-*n*-butyl phosphate (CAS Registry number 126-73-8) for testing in its Eighteenth Report and designated it in its Nineteenth Report after determining that unpublished studies submitted under section 8(d) of TSCA did not satisfy their testing recommendations (ITC, 1986a,b). The ITC designated tri-*n*-butyl phosphate because of its high annual production volume (3 million pounds); potential widespread occupational and environmental exposure to low concentrations (parts per billion); potential to cause adverse health effects to neurological, circulatory, and excretory systems; and potential to cause chronic ecological effects. The structure of tri-*n*-butyl phosphate is illustrated in Figure 14.

In response to the ITC's designation, the EPA published proposed and final rules in 1987 and 1989 (U.S. EPA, 1987c, 1989c). SOCMA's Tributyl Phosphate Task Force coordinated and sponsored the testing designated by the ITC.

Tri-*n*-butyl phosphate has a water solubility of 420 mg/L at 25°C, a vapor pressure of 2×10^{-6} mm Hg at 25°C, a calculated Henry's law constant of 3.19 $\times 10^{-6}$ atm m^3/mol, and a log K_{ow} of 4. Tri-*n*-butyl phosphate also has a hydrolysis half-life of 30 d at pH 3.7 and 11 and soil organic carbon partition coefficients ranging from 380 to 1460.

In acute and chronic toxicity tests, LC50 and MATC values were established for tri-*n*-butyl phosphate (Table 5).

Crotonaldehyde

The ITC recommended crotonaldehyde (CAS Registry number 4170-30-3) for testing in its Twenty-second Report and designated it in its Twenty-third Report after determining that unpublished studies submitted under section 8(d) of TSCA did not satisfy their testing recommendations (ITC, 1988a,b). The ITC designated crotonaldehyde because of anticipated environmental releases through wastewaters and because there were inadequate data to characterize persistence and toxicity in aquatic environments. The structure of crotonaldehyde is illustrated in Figure 14.

In response to that designation, the EPA published a consent order in 1989 (U.S. EPA, 1989e). The Eastman Kodak Company sponsored the testing designated by the ITC.

Crotonaldehyde has a water solubility of 155,000 mg/L at 20°C, a vapor pressure of 19 mm Hg at 20°C, a calculated Henry's law constant of 1.7 $\times$ 10^{-5} atm m^3/mol, and a log K_{ow} of 0.6.

In acute toxicity tests, LC50 values were established for crotonaldehyde (Table 5).

4-Vinylcyclohexene

The ITC recommended 4-vinylcyclohexene (CAS Registry number 100-40-3) for testing in its Twenty-fifth Report and designated it in its Twenty-seventh Report after determining that studies submitted under sections 8(a) and 8(d) of TSCA did not satisfy their testing recommendations (ITC, 1989, 1990). The ITC designated 4-vinylcyclohexene because it was produced in substantial quantities, inadvertently produced by the spontaneous dimerization of butadiene during manufacture of butadiene polymers, and detected at wastewater facilities; because of the potential for inhalation and dermal exposure; and because there were inadequate data on aqueous volatilization rates, pharmacokinetics, and oncogenicity testing. The structure of 4-vinylcyclohexene is illustrated in Figure 14.

In response to that designation, the EPA published a consent order in 1991 (U.S. EPA, 1991b). CMA's Butadiene Panel sponsored the testing designated by the ITC.

4-Vinylcyclohexene has a water solubility of 50 mg/L at 20°C, a vapor pressure of 25.8 mm Hg at 38°C, a calculated Henry's law constant of 2.9 $\times$ 10^{-2} atm m^3/mol, and a calculated log K_{ow} of 2.7.

The volatilization rate of 4-vinylcyclohexene depends on the turbulence of the water body. In lakes and ponds, 4-vinylcyclohexene should have a half-life of 6 d; in rivers, 4-vinylcyclohexene should have a half-life of 1.4 d (Table 5).

Comments on TSCA Section 4 Acute Toxicity, Bioconcentration, and Chemical Fate Data

The aquatic toxicity, bioconcentration, and chemical fate data developed by the U.S. chemical industry under TSCA section 4 in response to the ITC's testing recommendations are summarized in Table 5. The lowest EC50, LC50, and MATC for any aquatic organism that was tested under TSCA section 4 indicate the chemical concentrations necessary to produce these effects under conditions that maintained constant exposure of aquatic organisms to the test chemical. The measured BCF indicates the potential for a chemical to bioconcentrate in fish. The Henry's law constant indicates the potential for a chemical to evaporate from an aqueous environment, and log K_{ow} indicates the potential for a chemical to partition to organic matter (e.g., lipid) in an aqueous environment. Degradation indicates the potential for a chemical to be transformed in an aqueous environment.

Using criteria for toxicity (EC50 or LC50 values of 1 mg/L and MATC values 0.01 mg/L), bioconcentration (BCF of 1,000) and persistence (negligible biodegradation, volatilization, hydrolysis or photolysis under propitious conditions), it was possible to estimate the hazard of some of the chemicals tested under TSCA section 4. Cumeme, ethylbenzene, dimethyl phthalate, diethyl phthalate, 1,2-dichloropropane, 2-chlorotoluene, 3,4-dichlorobenzotrifluoride and tri-*n*-butyl phosphate are likely to be considered less hazardous because the EC50 or LC50 values were greater than 1 mg/L, MATC values (if available) were greater than 0.1 mg/L and partitioning or degradation data that suggest that these chemicals will be rapidly transported from water or biodegraded. Tetramethylbutyl phenol, branched 4-nonylphenol and 59% chlorinated-*n*-decane are likely to be considered more hazardous because the EC50 or LC50 values were less than 1 mg/L, the MATCs were less than 0.01 mg/L, the Henry's Law Constants were greater than 10^{-6} atm m^3/mole, biodegradation was negligible and the BCF for one of the chemicals was greater than 10,000.

Public Availability of Chemical Industry TSCA Studies

All TSCA studies are available from the Public Docket, (TS-790), Office of Pollution Prevention and Toxics, Environmental Protection Agency, Room ET G-102, 401 M St., SW, Washington, DC 20460; National Technical Information Service (1-800-336-4700); and Chemical Information Systems, Inc. (1-800-CIS-USER).

LITERATURE CITED

Adams, W. J.; Renaudette, W. J.; Doi, J. D.; Stepro, M. G.; Tucker, M. W.; Kimerle, R. A.; Franklin, B. R.; and Nabholtz, J. V.: Experimental freshwater microcosm biodegradability of butylbenzyl phthalate. Aquatic Toxicology and Hazard Assessment: Eleventh Symposium, edited by Suter, G. W., and Lewis, M. A. pp. 19–40. ASTM STP 1007. Philadelphia: American Society for Testing and Materials, 1989.

AIBS: American Institute of Biological Sciences, Aquatic Hazards of Pesticides Task Group: Criteria and rationale for decision making in aquatic hazard evaluation (third draft). Estimating the Hazard of Chemical Substances to Aquatic Life, edited by Cairns, J.; Dickson, K. L.; and Maki, A. W. pp. 241–273. ASTM STP 657. Philadelphia: American Society for Testing and Materials, 1978.

Akerman, J. W., and Coppage, D. L.: Hazard assessment philosophy: a regulatory viewpoint. Analyzing the Hazard Evaluation Process, edited by Dickson, D. L.; Maki, A. W.; and Cairns, J. pp. 68–73. Washington, DC: American Fisheries Society, 1979.

Alexander, H. C.; Dill, D. C.; Smith, L. W.; Guiney, P. D.; and Dorn, P.: Bisphenol A: Acute aquatic toxicity. *Environ. Toxicol. Chem.* 7:19–26, 1988.

Bourquin, A. W.; Hood, M. A.; and Garnas, R. L.: An artificial microbial ecosystem for determining effects and fate of toxicants in a salt-marsh environment. *Dev. Ind. Microbiol.*, 18:185–191, 1977.

Carson, D. B.; Saeger, V. W.; and Gledhill, W. E.: Use of microcosms versus conventional biodegradation testing for estimating chemical persistence. Aquatic Toxicology and Risk Assessment: Thirteenth Volume, ASTM STP 1096, edited by Landis, W. G., and van der Schalie, W. H. pp. 48–59. Philadelphia: American Society for Testing Materials, 1990.

De Bruijn, J.; Busser, F.; Seinen, W.; and Hermens, J.: Determination of octanol/water partition coefficients for hydrophobic organic chemicals with the "slow-stirring" method. *Environ. Toxicol. Chem.*, 7:499–512, 1989.

Dorn, P. B.; Chou, C. S.; and Gentempo, J. J.: Degradation of bisphenol A in natural waters. *Chemosphere*, 16: 1501–1507, 1987.

Dryssen, D.; Foyn, E.; Halstead, B.; Haug, O.; Korringa, P.; Lopuski, J.; Portmann, J. E.; Simonov, A. I.; Tendron, G.; and Freitag, D.: Joint IMCO/FAO/UNESCO/WMO group of experts on the scientific aspects of marine pollution. *Water Res.*, 3:995–1005, 1969.

Gersich, F. M.; Bartlett, E. A.; Murphy, P. G.; and Milazzo, D. P.: Chronic toxicity of biphenyl to *Daphnia magna Straus*. *Bull. Environ. Contam. Toxicol.*, 43:355–362, 1989.

Howard, P. H.; Banerjee, S.; and Robillard, K. H.: Measurement of water solubilities, octanol/water partition coefficients and vapor pressures of commercial phthalate esters. *Environ. Toxicol. Chem.*, 4:653–661, 1985.

ITC: Initial report of the TSCA Interagency Testing Committee (October 1, 1977) to the Administrator; receipt of report and request for comments regarding priority testing list of chemicals. *Fed. Regist.*, 42:55026–55080, 1977.

ITC: Second report of the TSCA Interagency Testing Committee (April 10, 1978) to the Administrator; receipt of report and request for comments regarding priority testing list of chemicals. *Fed. Regist.*, 43:16684–16688, 1978a.

ITC: Third report of the TSCA Interagency Testing Committee (October 2, 1978) to the Administrator; receipt of report and request for comments regarding priority testing list of chemicals. *Fed. Regist.*, 43:50630–50635, 1978b.

ITC: Fourth report of the TSCA Interagency Testing Committee (April 30, 1979) to the Administrator; receipt of report and request for comments regarding priority testing list of chemicals. *Fed. Regist.*, 44:31866–31899, 1979.

ITC: Sixth report of the TSCA Interagency Testing Committee (April 9, 1980) to the Administrator; receipt of report and request for comments regarding priority testing list of chemicals. *Fed. Regist.*, 45:35897–35910, 1980a.

ITC: Seventh report of the TSCA Interagency Testing Committee (October 24, 1980) to the Administrator; receipt of report and request for comments regarding priority testing list of chemicals. *Fed. Regist.*, 45:78432–78446, 1980b.

ITC: Eighth report of the TSCA Interagency Testing Committee (April 24, 1981) to the Administrator; receipt of report and request for comments regarding priority testing list of chemicals. *Fed. Regist.*, 46:28138–28144, 1981a.

ITC: Ninth report of the TSCA Interagency Testing Committee (October 31, 1981) to the Administrator; receipt of report and request for comments regarding priority testing list of chemicals. *Fed. Regist.*, 47:5456–5463, 1981b.

ITC: Tenth report of the TSCA Interagency Testing Committee (May 10, 1982) to the Administrator; receipt of report and request for comments regarding priority testing list of chemicals. *Fed. Regist.*, 47:22585–22596, 1982a.

ITC: Eleventh report of the TSCA Interagency Testing Committee (November 3, 1982) to the Administrator; receipt of report and request for comments regarding priority testing list of chemicals. *Fed. Regist.*, 47:54625–54644, 1982b.

ITC: Fourteenth report of the TSCA Interagency Testing Committee (May 29, 1984) to the Administrator; receipt of report and request for comments regarding priority testing list of chemicals. *Fed. Regist.*, 49:22389–22407, 1984a.

ITC: Fifteenth report of the TSCA Interagency Testing Committee (November 6, 1984) to the Administrator; receipt of report and request for comments regarding priority testing list of chemicals. *Fed. Regist.*, 49:46931–46949, 1984b.

ITC: Sixteenth report of the TSCA Interagency Testing Committee (May 2, 1985) to the Administrator; receipt of report and request for comments regarding priority testing list of chemicals. *Fed. Regist.*, 50:20930–20939, 1985a.

ITC: Seventeenth report of the TSCA Interagency Testing Committee (November 1, 1985) to the Administrator; receipt of report and request for comments regarding priority testing list of chemicals. *Fed. Regist.*, 50:47603–47612, 1985b.

ITC: Eighteenth report of the TSCA Interagency Testing Committee (May 1, 1986) to the Administrator; receipt of report and request for comments regarding priority testing list of chemicals. *Fed. Regist.*, 51:18368–18375, 1986a.

ITC: Nineteenth report of the TSCA Interagency Testing Committee (October 31, 1986) to the Administrator; receipt of report and request for comments regarding priority testing list of chemicals. *Fed. Regist.*, 51:41417–41432, 1986b.

ITC: Twentieth report of the TSCA Interagency Testing Committee (May 1, 1987) to the Administrator; receipt of report and request for comments regarding priority testing list of chemicals. *Fed. Regist.*, 52:19020–19026, 1987.

ITC: Twenty-second report of the TSCA Interagency Testing Committee (May 2, 1988) to the Administrator; receipt of report and request for comments regarding priority testing list of chemicals. *Fed. Regist.*, 53:18196–18210, 1988a.

ITC: Twenty-third report of the TSCA Interagency Testing Committee (November 1, 1988) to the Administrator; receipt of report and request for comments regarding priority testing list of chemicals. *Fed. Regist.*, 53:46262–46278, 1988b.

ITC: Twenty-fifth report of the TSCA Interagency Testing Committee (November 1, 1989) to the Administrator; receipt of report and request for comments regarding priority testing list of chemicals. *Fed. Regist.*, 54:51114–51130, 1989.

ITC: Twenty-seventh report of the TSCA Interagency Testing Committee (November 19, 1990) to the Administrator; receipt of report and request for comments regarding priority testing list of chemicals. *Fed. Regist.*, 56:9534–95729, 1991.

Kimerle, R. A.; Gledhill, W. E.; Levinskas, G. J.: Environmental safety assessment of new materials. Estimating the Hazard of Chemical Substances to Aquatic Life, edited by Cairns, J.; Dickson, K. L.; and Maki, A. W., pp. 132–146. ASTM STP 657. Philadelphia: American Society for Testing and Materials, 1978.

Krautter, G. R.; Mast, R. W.; Alexander, H. C.; Wolf, C. H.; Friedman, M. A.; Koshier, F. J.; and Thompson, C. M.: Acute aquatic toxicity tests with acrylamide monomers and macroinvertebrates and fish. *Environ. Toxicol. Chem.*, 5:373–377, 1986.

Library of Congress: Legislative History of the Toxic Substances Control Act. Washington, DC: U.S. Government Printing Office, 1976.

Life Systems, Inc.: Testing triggers workshop: recommendations report. Life Systems, Inc., Contract 68-01-6554. Washington, DC: Office of Toxic Substances, U.S. Environmental Protection Agency, 1983.

Lloyd, R.: The use of the concentration-response relationship in assessing acute fish toxicity data. Analyzing the Hazard Evaluation Process, edited by Dickson, K. L.; Maki, A. W.; and Cairns, J., pp. 58–61. Washington, DC: American Fisheries Society, 1979.

O'Grady, D. P.; Howard, P. H.; and Werner, A. F.: Activated sludge biodegradation of 12 commercial phthalate esters. *Appl. Environ. Microbiol.*, 49:443–445, 1985.

Sloof, W.; Van Oerr, J. A. M.; and De Zwart, D.: Margins of uncertainty in ecotoxicological hazard assessment. *Environ. Toxicol. Chem.*, 5:841–852, 1986.

Stahl, R. G. Jr.; Lieder, P. H.; and Hutton, D. G.: Relationship between aquatic toxicity and oxidative degradation of unsubstituted phenylenediamines. *Environ. Toxicol. Chem.*, 9:485–488, 1990.

Sugatt, R. H.; O'Grady, D. P.; Banerjee, S.; Howard, P. H.; and Gledhill, W. E.: Shake flask biodegradation of 14 commercial phthalate esters. *Appl. Environ. Microbiol.*, 47:601–606, 1984.

U.S. Environmental Protection Agency: Chloromethane and chlorinated benzenes proposed rule; amendment to the proposed health effects standards. *Fed. Regist.*, 45: 48524–48566, 1980.

U.S. Environmental Protection Agency: Dichloromethane, nitrobenzene and 1,1,1-trichloroethane; proposed test rule. *Fed. Regist.*, 46:30300–30320, 1981a.

U.S. Environmental Protection Agency: Alkyl phthalates and benzyl butyl phthalate; response to the Interagency Testing Committee. *Fed. Regist.*, 46:53775–53777, 1981b.

U.S. Environmental Protection Agency: Phenylenediamines; response to the Interagency Testing Committee. *Fed. Regist.*, 47:973–983, 1982a.

U.S. Environmental Protection Agency: Alkyl phthalates and benzyl butyl phthalate; follow-up response to the Interagency Testing Committee. *Fed. Regist.*, 47:335–336, 1982b.

U.S. Environmental Protection Agency: Chlorinated paraffins; response to the Interagency Testing Committee. *Fed. Regist.*, 47:1017–1019, 1982c.

U.S. Environmental Protection Agency: 2-Chlorotoluene; negotiated testing plan and request for public comment. *Fed. Regist.*, 47:3596, 1982d.

U.S. Environmental Protection Agency: 2-Chlorotoluene; response to the Interagency Testing Committee. *Fed. Regist.*, 47:18172–18175, 1982e.

U.S. Environmental Protection Agency: Toxic substances; 4-chlorobenzotrifluoride; response to the Interagency Testing Committee. *Fed. Regist.*, 47:50555–50558, 1982f.

U.S. Environmental Protection Agency: Diethylenetriamine; proposed test rule. *Fed. Regist.*, 47:18386–18391, 1982g.

U.S. Environmental Protection Agency: Biphenyl; proposed test rule. *Fed. Regist.*, 48:23080–23086, 1983a.

U.S. Environmental Protection Agency: 4-(1,1,3,3-Tetramethylbutyl)phenol; response to the Interagency Testing Committee. *Fed. Regist.*, 48:51971–51977, 1983b.

U.S. Environmental Protection Agency: Bis(2-ethylhexyl) terephthalate; response to the Interagency Testing Committee. *Fed. Regist.*, 48:51845–51848, 1983c.

U.S. Environmental Protection Agency: Tris(2-ethylhexyl) trimellitate; response to the Interagency Testing Committee. *Fed. Regist.*, 48:51842–51845, 1983d.

U.S. Environmental Protection Agency: 4-Chlorobenzotrifluoride; decision to adopt negotiated testing program. *Fed. Regist.*, 48:32730–32732, 1983e.

U.S. Environmental Protection Agency: Acrylamide; response to the Interagency Testing Committee. *Fed. Regist.*, 48:725–727, 1983f.

U.S. Environmental Protection Agency: 4-(1,1,3,3-Tetramethylbutyl)phenol; decision to adopt negotiated testing program. *Fed. Regist.*, 49:29449–29450, 1984a.

U.S. Environmental Protection Agency: Anilines; response to the Interagency Testing Committee. *Fed. Regist.*, 49: 108–126, 1984b.

U.S. Environmental Protection Agency: Bis(2-ethylhexyl) terephthalate; decision to adopt negotiated testing program. *Fed. Regist.*, 49:23110–23112, 1984c.

U.S. Environmental Protection Agency: Tris(2-ethylhexyl) trimellitate; decision to adopt negotiated testing program. *Fed. Regist.*, 49:23116–23117, 1984d.

U.S. Environmental Protection Agency: 1,2-Dichloropropane; proposed test rule. *Fed. Regist.*, 49:899–908, 1984e.

U.S. Environmental Protection Agency: Mono-, di-, and trichlorinated benzenes; proposed environmental effects test rule. *Fed. Regist.*, 49:1760–1770, 1984f.

U.S. Environmental Protection Agency: Bisphenol A; proposed test rule. *Fed. Regist.*, 50:20691–20703, 1985a.

U.S. Environmental Protection Agency: Biphenyl; final test rule. *Fed. Regist.*, 50:37182–37189, 1985b.

U.S. Environmental Protection Agency: Cumene; proposed test rule. *Fed. Regist.*, 50:46104–46121, 1985c.

U.S. Environmental Protection Agency: Anthraquinone; proposed reporting and recordkeeping requirements and test rule. *Fed. Regist.*, 50:46090–46103, 1985d.

U.S. Environmental Protection Agency: 2-Chlorotoluene; decision not to test. *Fed. Regist.*, 50:40445–40449, 1985e.

U.S. Environmental Protection Agency: 4-Chlorobenzotrifluoride; decision not to test. *Fed. Regist.*, 50:42216–42221, 1985f.

U.S. Environmental Protection Agency: Diethylenetriamine; identification of specific chemical substance and mixture testing requirements; final test rule. *Fed. Regist.*, 50:21398–21412, 1985h.

U.S. Environmental Protection Agency: 2-Mercaptobenzothiazole; proposed test rule. *Fed. Regist.*, 50:46121–46133, 1985i.

U.S. Environmental Protection Agency: Octamethylcyclotetrasiloxane; proposed test rule. *Fed. Regist.*, 50: 45123–45133, 1985j.

U.S. Environmental Protection Agency: Procedures governing testing consent agreements and test rules under the Toxic Substances Control Act; interim final rule. *Fed. Regist.*, 51:23706–23718, 1986a.

U.S. Environmental Protection Agency: Unsubstituted phenylenediamines; proposed test rule. *Fed. Regist.*, 51: 472–494, 1986b.

U.S. Environmental Protection Agency: Certain chlorinated benzenes; final test rule. *Fed. Regist.*, 51:11728–11759, 11843, 1986c.

U.S. Environmental Protection Agency: Bisphenol A; final test rule. *Fed. Regist.*, 51:33047–33054, 1986d.

U.S. Environmental Protection Agency: Tetrabromobisphenol A; proposed test rule. *Fed. Regist.*, 51:17872–17883, 1986e.

U.S. Environmental Protection Agency: Toxic substances; 1,2-dichloropropane; testing requirements. *Fed. Regist.*, 51:32079–32087, 1986f.

U.S. Environmental Protection Agency: Procedures governing modification of test standards and schedules for tests required under test rules and testing consent agreements; interim final rule. *Fed. Regist.*, 52:36569–36572, 1987a.

U.S. Environmental Protection Agency: 2,6-Di-*tert*-butylphenol; proposed test rule. *Fed. Regist.*, 52:23862–23873, 1987b.

U.S. Environmental Protection Agency: Tributylphosphate; proposed test rule. *Fed. Regist.*, 52:43346–43366, 1987c.

U.S. Environmental Protection Agency: Anthraquinone; final reporting, recordkeeping requirements and test rule. *Fed. Regist.*, 52:21018–21031, 1987d.

U.S. Environmental Protection Agency: Testing consent order on 3,4-dichlorobenzotrifluoride and response to the Interagency Testing Committee. *Fed. Regist.*, 52:23547–23548, 1987e.

U.S. Environmental Protection Agency: Tetrabromobisphenol A; final test rule. *Fed. Regist.*, 52:25219–25226, 1987f.

U.S. Environmental Protection Agency: Unsubstituted phenylenediamines; reopening of the comment period. *Fed. Regist.*, 53:913–922, 1988a.

U.S. Environmental Protection Agency: Cumene; final test rule. *Fed. Regist.*, 53:28195–28206, 1988b.

U.S. Environmental Protection Agency: Testing consent orders on aniline and seven substituted anilines. *Fed. Regist.*, 53:31804–31813, 1988c.

U.S. Environmental Protection Agency: 2-Mercaptobenzothiazole; final test rule. *Fed. Regist.*, 53:39514–34532, 1988d.

U.S. Environmental Protection Agency: Unsubstituted phenylenediamines; final test rule. *Fed. Regist.*, 54: 49285–49296, 1989a.

U.S. Environmental Protection Agency: Testing consent order for octamethylcyclotetrasiloxane. *Fed. Regist.*, 54: 818–821, 1989b.

U.S. Environmental Protection Agency: Tri-*n*-butylphosphate: final test rule. *Fed. Regist.*, 54:33400–33415, 1989c.

U.S. Environmental Protection Agency: Testing consent order for crotonaldehyde. *Fed. Regist.*, 54:47062–47066, 1989d.

U.S. Environmental Protection Agency: Testing consent order on branched 4-nonylphenol. *Fed. Regist.*, 55:5991–5994, 1990.

U.S. Environmental Protection Agency: TSCA section 4(a)(1)(B) proposed statement of policy. *Fed. Regist.*, 56: 32299–32300, 1991a.

U.S. Environmental Protection Agency: Testing consent order for 4-vinylcyclohexene. *Fed. Regist.*, 56:17912–17915, 1991b.

U.S. Environmental Protection Agency: TSCA section 4(a)(1)(B) final statement of policy. Criteria for evaluating substantial production, substantial release and substantial or significant human exposure. *Fed. Regist.*, 58: 28736–28749, 1993.

Walker, J. D.: A U.S. EPA perspective on ecotoxicity testing using microorganisms. Toxicity Testing Using Microorganisms, Vol. II, edited by Dutka, B. J., and Bitton, G., pp. 175–186. Boca Raton, FL: CRC Press, 1986.

Walker, J. D.: Relative sensitivity of algae, bacteria, invertebrates, and fish to phenol: analysis of 234 tests conducted for more than 149 species. *Toxicity Assess.*, 3: 415–447, 1988.

Walker, J. D., and Brink, R. H.: New cost-effective, computerized approaches to selecting chemicals for priority testing consideration. Aquatic Toxicology and Hazard Assessment: Eleventh Symposium, ASTM STP 1007, edited by Suter, G. W., and Lewis, M. A., pp. 507–536. Philadelphia: American Society for Testing and Materials, 1989.

Walker, J. D.: Bioconcentration, chemical fate and environmental effects testing under Section 4 of the Toxic Substances Control Act. *Toxicity Assess.*, 5:61–75, 1990a.

Walker, J. D.: Chemical fate, bioconcentration and environmental effects testing: proposed testing and decision criteria. *Toxicity Assess.*, 5:103–134, 1990b.

Walker, J. D.: Review of chemical fate testing conducted under Section 4 of the Toxic Substances Control Act: chemicals, tests, and methods. Aquatic Toxicology and Risk Assessment: Thirteenth Volume, ASTM STP 1096, edited by Landis, W. G., and van der Schalie, W. H., pp. 77–90. Philadelphia: American Society for Testing and Materials, 1990c.

Walker, J. D.: Chemical selection by the TSCA Interagency Testing Committee: use of computerized substructure searching to identify chemical groups for health effects, chemical fate and ecological effects testing. *Sci. Total Environ.*, 109/110:691–700, 1991a.

Walker, J. D.: Ecological effects testing under the Toxic Substances Control Act: acrylamide. *Environ. Toxicol. Water Qual.*, 6:363–369, 1991b.

Walker, J. D.: Prioritizing and coordinating chemical testing. Proceedings of the Summer Toxicology Forum, Aspen, CO (July 13–17, 1992). Washington, DC: Toxicology Forum, 1992.

Walker, J. D.: The TSCA Interagency Testing Committee, 1977–1992: creation, structure, functions and contributions. Environmental Toxicology and Risk Assessment: Second Volume, ASTM STP 1216, edited by Gorsuch, J.; Dwyer, J.; La Point, T.; and Ingersoll, C., pp. 451–509. Philadelphia: American Society for Testing and Materials, 1993a.

Walker, J. D.: The TSCA Interagency Testing Committee's approaches to screening and scoring chemicals and chemical groups: 1977–1983. Symposium on Access and Use of Information Resources in Assessing Health Risks from Chemical Exposure, edited by Po-Yung Lu, pp. 71–93. Oak Ridge, TN: Oak Ridge National Laboratories, (June 27–29, 1990), 1993b.

Walker, J. D.: Review of ecological effects and bioconcentration testing recommended by the TSCA Interagency Testing Committee and implemented by EPA under the Toxic Substances Control Act: chemicals, tests, and methods. Environmental Toxicology and Risk Assessment: First Volume, ASTM STP 1179, edited by Landis, W. G., and Hughes, J., pp. 92–115. Philadelphia: American Society for Testing and Materials, 1993c.

Walker, J. D.: The TSCA Interagency Testing Committee's role in facilitating development of test methods: benthic invertebrate sediment toxicity testing. Environmental Toxicology and Risk Assessment: Second Volume, ASTM STP 1216, edited by Gorsuch, J.; Dwyer, J.; La Point, T.; and Ingersoll, C.; pp. 688–722. Philadelphia: American Society for Testing and Materials, 1993d.

Walker, J. D.: Can chemical structures and physical properties be used to define required chemical fate and ecological tests? American Society for Testing and Materials 3rd Symposium on Environmental Toxicology and Risk Assessment: Aquatic, Plant and Terrestrial, April 26–29, Atlanta, GA, 1993e.

Walker, J. D.: Testing decisions of the TSCA Interagency Testing Committee for chemicals on the Canadian Environmental Protection Act Domestic Substances List and Priority Substances List: di-*tert*-butylphenol, ethyl benzene, brominated flame retardants, phthalate esters, choloroparaffins, chlorinated benzenes and anilines. Environmental Toxicology and Risk Assessment: Fourth Volume, ASTM STP 1241, edited by LaPoint, T. W., Price, F. T., and Little, E. E. Philadelphia, PA: American Society for Testing and Materials, in press.

Williams, R. T.; Ziegenfuss, P. S.; and Lee, C. M.: Aerobic biodegradation of cumene: an ecocore study. *Environ. Toxicol. Chem.*, 12:485–492, 1993.

Chapter 23

ECOTOXICITY TESTING AND ESTIMATION METHODS DEVELOPED UNDER SECTION 5 OF THE TOXIC SUBSTANCES CONTROL ACT (TSCA)

M. G. Zeeman

INTRODUCTION

In 1976 the Toxic Substances Control Act (TSCA) was enacted into law and it established the Office of Toxic Substances (now the Office of Pollution Prevention and Toxics, OPPT). This law empowered OPPT to take actions to prevent unreasonable risks to the organisms in the environment before they occur due to the manufacture, import, processing, use, or disposal of industrial chemicals. An initial step that needed to be taken by OPPT was to identify the industrial chemicals that were already in commerce in the United States.

This led to the publication in mid-1979 of the TSCA Inventory of existing chemicals, which was a listing of the approximately 60,000 chemicals in commercial production in the United States as of the previous year. The TSCA Inventory has been periodically updated to include the new industrial chemicals produced in the United States. In 1991 the TSCA Inventory contained over 70,000 chemicals (EPA, OPPT, 1991a) and it currently contains about 72,000 chemicals (GAO, 1993).

Under section 5 of TSCA, anyone intending to manufacture in or import into the United States a chemical not on the TSCA inventory and not exempted from TSCA is required to submit a formal notice to EPA/OPPT on this new industrial chemical (as defined under TSCA section 5). The notice is called a new chemical premanufacture notice (PMN) and must be submitted to OPPT at least 90 d before the commercial manufacture or import of that chemical. OPPT uses the data contained in such a notice to make rapid decisions about the hazards, exposures, and risks to organisms in the environment of allowing the manufacture or import of that chemical. After suitable deliberations in the new chemical review process, OPPT may determine to request ecotoxicity test data that are needed to help make this decision.

This chapter initially describes the parts of the OPPT new chemical review process in which ecotoxicity data (or estimations) are used to provide evaluations of environmental hazard and risk. Only a portion of the new chemical process is described here. Those interested in a more detailed description, including the environmental engineering and fate evaluations that result in predicted environmental concentrations (PECs), are encouraged to review Nabholz (1991), Nabholz et al. (1993a), and Zeeman et al. (1993a).

THE NEW CHEMICAL REVIEW PROCESS

The number of TSCA section 5 new chemical notices received by OPPT since 1979 has increased rapidly (Figure 1). The average number of notices to be examined over the 9 yr from 1986 through 1994 was over 2200 per year. This large number of chemicals to review and evaluate in the required 90-d time frame put considerable pressure on the Environmental Protection Agency to establish a process that allowed rapid test data evaluations and quick decisions. Such a new chemical process was established (EPA, 1986) and is routinely used by OPPT (Figure 2).

Since its inception, this process has been used to review more than 26,000 new chemical substances. More than 10,000 of these chemicals made it through the new chemical process and then went into commercial production and now reside on the TSCA Inventory of existing chemicals. Another 10,000 or so of these chemicals also made it through the new chemical process but for various reasons appear not to have been manufactured in or imported into the

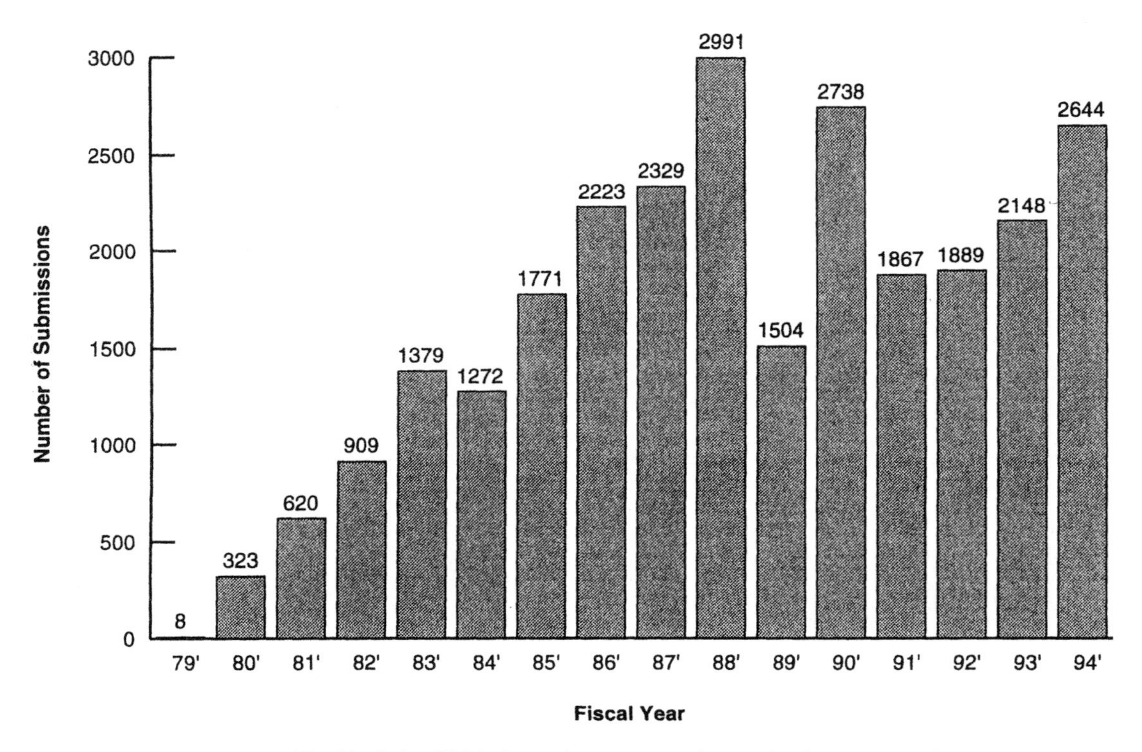

Figure 1. Number of new chemical notices received by EPA/OPPT (1994).

United States by the submitters (e.g., industrial chemical companies) of those PMNs.

Each notice is to be reviewed by the Agency within a 90-d time frame, starting the day the PMN is received. In order to propose a denial of the notice or any form of regulation, OPPT's evaluation must be able to show that unreasonable risks to the environment may occur as a result of the manufacture, import, processing, use, or disposal of that chemical. If no action is taken by the Agency in this 90-d time frame, the chemical may start to be manufactured or imported after a commencement notice has been filed with OPPT by the sponsor. Those 10,000+ new chemicals for which commencement notices were issued and commercial production was initiated then became existing chemicals (see Chapter 22).

It is important to understand what types of information are useful in PMN notices for the evaluation of potential environmental effects posed by new chemicals under section 5 of TSCA and the process by which these data are considered. After that discussion, it will be useful to examine the kinds and amounts of ecotoxicity test data that have actually been submitted by industry for evaluation by OPPT in such new chemical notices.

PMN Requirements and New Chemical Evaluations

The PMN form requires submission of all available data on the structure and identity of the chemical, intended production volume and use(s), by-products, environmental release, disposal practices, and potential human exposure (EPA, 1991a,b). It also requires the submission of all existing environmental test data in the possession of or that are reasonably ascertainable by the submitter, its parent company, or affiliates. Under section 5 of TSCA, up-front testing is not required for new industrial chemicals.

The data submitted by the sponsor (and any other pertinent data available to OPPT) are used in the new chemical review process (Figure 2) to assess the potential hazards, exposures, and risks to the environment of allowing the manufacture, import, processing, use, and disposal of that chemical. Within 2 wk of receipt of each PMN, several reviews are made

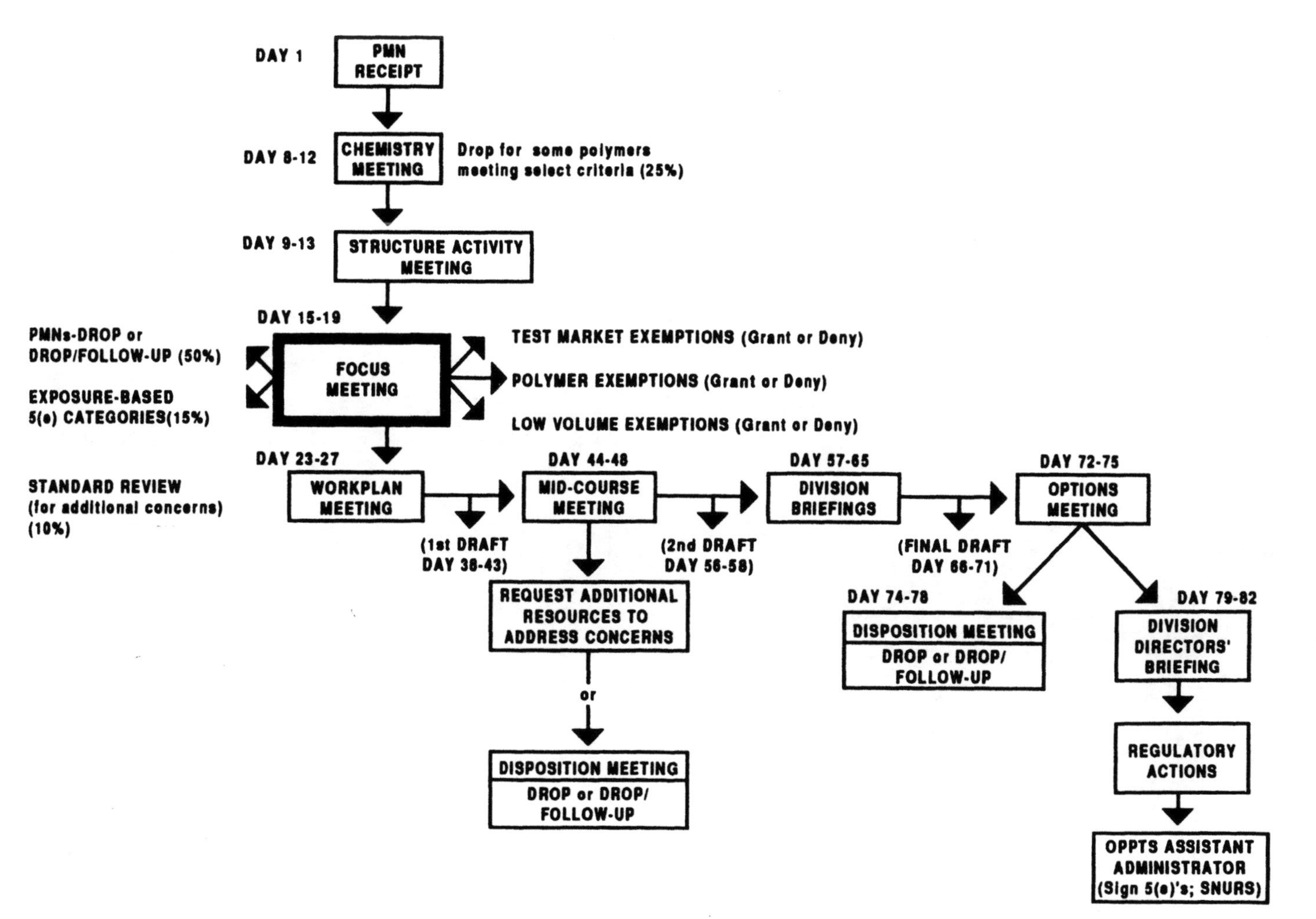

Figure 2. The new chemical review process at EPA/OPPT.

and meetings held to assess the chemistry, hazards, and exposures of these new chemicals (Auer et al., 1990; Nabholz, 1991; Nabholz et al., 1993a).

The chemistry review ensures that the information submitted is consistent with itself and the principles of chemistry. The initial hazard assessment performed by experts in structure-activity relationships identifies the degree of potential effects to organisms in the environment. If a chemical is determined to be of low hazard (e.g., acute aquatic LC50 > 100 mg/L) and the production volume is to be less than 100,000 kg/yr, it can be eliminated from further review. If concerns remain about a new chemical's potential toxicity or there is a significant production volume, the potential exposures of the chemical to organisms in the environment is then assessed.

This exposure assessment identifies the potential releases of a chemical throughout its entire life cycle (manufacturing, processing, use, and disposal) and quantifies each release. The engineering report estimates chemical releases from each site, and the environmental exposure report estimates the expected (or predicted) environmental concentrations (Nabholz, 1991).

Within 3 wk of receipt, at the first risk assessment meeting (the focus meeting), the initial assessment of the new chemical's potential hazard to organisms in the environment [as modified by assessment (uncertainty) factors—see section on assessment factors below], and their predicted environmental concentrations are contrasted with each other to characterize the potential ecological risks (EPA, 1986; Auer et al., 1990; Nabholz, 1991; Nabholz et al., 1993a,b; Zeeman and Gilford, 1993; Zeeman et al., 1993a).

New chemicals with no or reasonable risks predicted at the focus meeting are eliminated from further review. If sufficient risks are identified at the focus meeting, the chemical could be regulated directly, often with a request for testing [e.g., see 5(e) Chemical Categories below], or it could be subjected to a more in-depth analysis (e.g., using a more refined exposure analysis) called a standard review (Figure 2).

When this review supports the risk assessment findings from the focus meeting, the case is taken to senior decision makers in OPPT. They also take into consideration economic and relative risk factors, such as the costs and benefits of the chemical, the cost of any additional testing, and the relative hazards and exposures of substitute chemicals on the TSCA Inventory, to decide whether or not the new chemical presents an "unreasonable" risk to the environment. Depending on the strength of the case, the decisions made can vary from eliminating from further review for weak cases to direct controls for cases in which the chemical "will present" an unreasonable risk to the environment (Nabholz et al., 1993a).

Ecotoxicity Data Submitted for New Chemicals

When environmental data are not available for a new chemical, there is no requirement for any testing as a precondition for approval. As a result of this, the consistent experience of OPPT has been that the great majority of new chemical notices contain no data at all on the properties of the chemical, its environmental fate, or on its toxicity to organisms in the environment.

Nevertheless, decisions on the potential risks of each chemical must be made quickly or the manufacture (or import) of these PMN chemicals could ensue after the 90-d period is passed. This has forced OPPT technical staff to become resourceful, both in making a case for obtaining environmental test data from industry (see chemical categories below) and in developing structure-activity relationships (SARs) and quantitative structure-activity relationships (QSARs) (see Chapter 20) for rapidly estimating the ecotoxicity potential of chemicals for which little or no test data was submitted in the notice.

TSCA 5(e) Chemical Categories

Section 5(e) of TSCA allows the Agency to propose regulation of new chemical substances pending the development of test data when the manufacture, processing, distribution, use, or disposal of these chemicals may present an unreasonable risk to the environment. For example, between 1988 and 1993, section 5(e) of TSCA was used to request test data on new chemicals without the 45-d standard review (Figure 2) because they were in what OPPT has termed 5(e) New Chemical Categories and unreasonable risk was predicted. At present, there are chemical categories for 40 different groups of industrial chemicals. Each separate category is composed of chemical groups having related chemical structures (Table 1).

Each of these 5(e) categories has resulted from OPPT's New Chemicals Program repeatedly doing standard reviews examining the ecotoxicity, environmental fate, and/or human health concerns for the PMNs submitted in these chemical groups. Note that 32 of these chemical categories resulted solely from environmental concerns, 15 resulted solely from human health concerns, and 7 were developed from concerns for both the environment and human health.

OPPT used both the test data available from the scientific literature and the test data submitted to the Agency to define the relationship between the toxicity and chemical structure in these chemical categories and to establish the boundaries of concern in each group (EPA, 1993a). Each of these chemical categories provides (1) a description of the chemical class comprising the category, (2) the basic chemical group that causes toxicity, (3) typical toxicity testing requirements (e.g., the OPPT aquatic "base set" of acute toxicity tests requires the testing of fish, invertebrates, and algae), and (4) the boundaries of environmental and health concerns [e.g., certain molecular weights, carbon chain lengths, or *n*-octanol/water partition coefficients that have resulted in concerns (or nonconcerns) for ecotoxicity].

As a result of this approach, OPPT has been able to routinely support the predictions of toxicity for

Table 1. TSCA 5 (e) new chemical categories

Category	Environment	Health
Acid chlorides	X	
Acid dyes	X	
Acrylamides	X	X
Acrylates/methacrylates	X	X
Aliphatic amines	X	
Alkoxysilanes	X	X
Aminobenzothiazole azo dyes	X	X
Amphoteric dyes	X	
Anhydrides, carboxylic acid	X	
Anilines	X	
Anionic surfactants	X	
Benzotriazoles	X	
Borates	X	
Cationic dyes	X	
Cationic (quaternary ammonium) surfactants	X	
Dianilines	X	
Diazoniums	X	
Dithiocarbamates	X	
Epoxides	X	X
Esters	X	
Ethylene glycol ethers		X
Hindered amines		X
Hydrazines and related compounds	X	X
Imides	X	
Isocyanates		X
Beta-naphthylamines, monosulfonated		X
Neutral organics	X	
Nickel compounds	X	X
Nonionic surfactants	X	
Peroxides		X
Phenols	X	
Polyanionic polymers and monomers	X	
Polycationic polymers	X	
Polynitroaromatics	X	
Stilbene, 4,4-bis(triazin-2-ylamino), derivatives of		X
Substituted triazines	X	
Vinyl esters		X
Vinyl sulfones		X
Soluble complexes of zinc	X	
Zirconium compounds	X	

these 32 chemical groups to organisms in the environment. When the initial chemical release and exposure predictions and SAR/QSAR hazard estimates (as modified by assessment factors) result in predictions of unreasonable risks to organisms in the environment, these chemical categories then provide the technical support for being able to request routinely the acute base set (or other) ecotoxicity tests for such chemical groups at the focus meeting rather than performing a 45-d standard review.

Specific information on each of these OPPT chemical categories is available from the EPA OPPT New Chemicals Program (EPA, 1993a). Publicly available information on each of these chemical categories has also been provided to the Chemical Manufacturers Association (CMA) and the Synthetic Organic Chemicals Manufacturers Association (SOCMA), both located in Washington, DC.

HAZARD ASSESSMENT AND ECOTOXICOLOGY

This chapter focuses both on the ecotoxicity data actually submitted by industry for evaluation in chemical PMNs and on the ecotoxicity data that must be estimated by OPPT (by using SAR/QSAR) in the evaluations of the large number of new chemical notices submitted with little or no ecotoxicity test data.

Anyone interested in how various genetically engineered microorganisms (GEMs) and their products are subject to regulation and hazard evaluation as new chemicals under section 5 of TSCA should see Sayre and Kough (1993). Similarly, anyone interested in summaries of the number and types of physical-chemical and environmental fate data submitted in TSCA section 5 PMNs and in OPPT's use of SAR/QSAR for estimation of these properties should see Boethling (1993) and Zeeman et al. (1993a).

Table 2. OPPT criteria for high, medium, and low concerns for aquatic toxicity and fish bioconcentration

Hazard parameters	Criteria for concern		
	High	Medium	Low
Aquatic acute toxicity (mg/L)	≤1.0	>1.0≤100	>100
Aquatic chronic toxicity (mg/L)	≤0.1	>0.1≤10	>10
Fish bioconcentration factor (BCF)	≥1000	≥100<1000	<100

Historical Background

In brief, over the past decade, the Environmental Effects Branch (EEB) in the Health and Environmental Review Division (HERD) of OPPT has provided significant direction to and rationale for how ecological testing and hazard assessment activities for new chemicals can be accomplished within the confines of section 5 of TSCA. To assure that adequate hazard data were being developed, this group established the procedures and guidelines for developing ecotoxicological test data that are appropriate and adequate for assessing the ecological hazards and risks of new chemical substances.

For new industrial chemicals this required the development of (1) appropriate ecological end points (e.g., fish mortality, inhibition of growth, development and/or reproduction of fish), (2) a tier-testing scheme for sequencing the acute and chronic hazard testing for such ecological end points, (3) ecotoxicological testing guidelines, (4) hazard criteria for determining the thresholds for high, medium, or low aquatic toxicity (Table 2), (5) techniques for estimating ecotoxicity from chemical structure (SAR/QSARs), (6) assessment factors for establishing chemical concentrations of concern (environmental risk), and (7) risk assessment methods that characterize the ecological risks of new chemicals by incorporating the hazard (ecotoxicity, as modified by assessment factors) and exposure data (Clements, 1983; EPA, 1983a, 1984, 1985; Rodier, 1987; Clements et al., 1988, 1993a,b; Auer et al., 1990; Nabholz, 1991; Nabholz et al., 1993a,b; Rodier and Mauriello, 1993; Smrchek et al., 1993; Zeeman and Gilford, 1993, Zeeman et al.; 1993a; Rodier and Zeeman, 1994; Auer et al., 1994).

In summary, the techniques, criteria, methods, and guidelines developed over the past decade by this group in OPPT allow estimation of the potential adverse effects of new industrial chemicals on ecological systems, including animal and plant populations. Those interested in a more comprehensive look at the development of this approach in assessing the ecotoxicological hazards and risks of new industrial chemicals are especially encouraged to see the OPPT publications by Zeeman and Gilford (1993), Zeeman et al. (1993a), and Smrchek et al. (1993).

Approach for Hazard Assessment (Ecotoxicity)

The primary concern about the toxicity of PMN chemicals to organisms in the environment has historically focused on an industrial chemical's toxicity to aquatic organisms due to potential releases to water from chemical manufacture (Auer et al., 1990; Nabholz, 1991; Zeeman and Gilford, 1993). However, not much information has been published with regard to the chemical industries' submission of data on the toxicity of PMN chemicals to organisms in the aquatic environment.

For example, Auer et al. (1990) reported that, up to September 1985, acute toxicity test data for aquatic vertebrates or invertebrates were submitted for only 6 or 3% of all PMNs, respectively. A survey of the ecotoxicity data reports found in the files from over 13,000 notices received from 1986 through 1991 indicated that, on average, only 4.8% of these notices contained any submitted data on toxicity to organisms in the environment (Zeeman et al., 1993a). Inclusion of the ecotoxicity data in the files from the approximately 1900 new chemical notices submitted in fiscal year 1992 further support that only about 5% of the time will a notice filed with OPPT contain any ecotoxicity data (Table 3).

Nabholz et al. (1993b) reviewed a large portion of the validated ecotoxicity data that have been submitted to the OPPT PMN program. Several hundred acceptable aquatic toxicity test results from hundreds of different chemical PMNs have been received from industry. Most of the data received by OPPT have been acute toxicity data, primarily for fish, aquatic invertebrates, and/or algae (Auer et al., 1990; Nabholz, 1991; Nabholz et al., 1993a,b).

In a study designed to determine the adequacy of QSARs, Nabholz et al. (1993b) selected validated aquatic toxicity test data submitted to OPPT from 1981 to 1992 (mostly under section 5 of TSCA) for

Table 3. New chemical notices with ecotoxicity data

Year	Number of notices	Number of ecotoxocity data reports	Percent
1986	2223	84	3.8
1987	2329	116	4.9
1988	2991	163	5.4
1989	1504	93	6.2
1990	2738	88	3.2
1991	1867	115	6.2
1992	1889	140	7.4
Total	15,541	799	5.1[a]

[a]Average.

462 different chemicals in several different chemical classes (e.g., neutral organics, phenols, surfactants). The submissions for several of these chemicals included the results of tests for more than one species, with almost 900 separate ecotoxicity tests submitted for these 462 chemicals. From these data the following preliminary conclusions are drawn (Table 4).

From this small subset of submissions to OPPT that included test data, fish and daphnid acute test results were, by far, the most frequently received ecotoxicity tests for industrial chemicals. Acute fish, daphnid, and green algal tests were submitted, respectively, for about 90, 66, and 31% of the 462 chemicals examined with ecotoxicity data. Results of chronic/subchronic tests are very infrequently received by OPPT, with test results for fish and daphnids submitted for only about 2 and 3% of these 462 industrial chemicals, respectively.

Hazard Profile

OPPT prefers accurate experimentally determined ecotoxicity data when assessing the hazards of new industrial chemicals. However, ecotoxicity data are typically not provided by the PMN sponsors. Therefore, estimation of the aquatic toxicity of a chemical is essential if OPPT is to be able to rapidly predict the risks of these chemicals to organisms in the aquatic environment.

Where ecotoxicity data are not provided, the toxicity of all new chemicals to surrogate aquatic species is now routinely predicted by OPPT, with QSAR methods being used most often to estimate ecotoxicity. The aquatic toxicity of many structurally discrete chemicals can be estimated with a reasonable degree of confidence from the numerous QSARs developed by OPPT (Clements et al., 1988, 1993a,b; Nabholz et al., 1993a,b; Zeeman et al., 1993a,b). However, some chemical classes lack a method for a quantitative (QSAR) estimation of toxicity via regression equations (e.g., cationic dyes, anionic polymers). These chemicals require searching for toxicity data on chemical analogues or the chemical classes that are closest to the PMN substance (i.e., qualitative SAR estimates of toxicity).

The data submitted and the OPPT predictions of aquatic toxicity and bioconcentration for new industrial chemicals are routinely compiled into a hazard profile for ecotoxicity (Table 5) listing these test data along with the most appropriate toxicity values from SAR and from QSARs. The most useful aquatic hazard profile contains data from the fish 96-h LC50, daphnid 48-h LC(EC)50, green algae 96-h EC50 values, and fish, daphnid, and algal chronic/subchronic values. The fish and daphnid chronic values are most often the maximum acceptable toxicant concentrations (MATCs) (geometric mean of) from fish 28–90-d early life stage tests and daphnid 14–21-d three-brood partial life cycle tests, respectively. As the green algal test allows for the reproduction of several generations in 96 h, the algal chronic value is typically the 96-h no-effect or no-observed-effect concentration (Nabholz et al., 1993b).

Table 4. Types of ecotoxicity tests submitted to OPPT for 462 industrial chemicals with test data

Test description	Number	Percent
Fish 96-h LC50	414	89.6
Daphnid 48-h LC50	307	66.4
Green algal 96-h EC50	145	31.4
Fish early life stage	8	1.7
Daphnid partial life cycle	14	3.0

Table 5. OPPT standard hazard profile for aquatic ecotoxicity

Freshwater test descriptions
Fish acute toxicity (96-h LC50)
Daphnid acute toxicity (48-h LC50)
Algal toxicity (96-h EC50)
Fish early life stage (28–90-d MATC)
Daphnid partial life cycle (14–21-d MATC)
Algal toxicity (96-h NOEC)

This PMN hazard profile of fish, daphnid, and algal acute values and (sub)chronic values is completed (as much as it can be) by all means possible. The preferred methods, in descending order, for completing the hazard profile are: (1) measured test data, (2) QSAR predictions, (3) SAR nearest chemical analogue data, and (4) SAR chemical class data.

QSAR DEVELOPMENT

The dilemma of routinely not being provided with ecotoxicity test data and still having to quickly assess the ecological risks of numerous industrial chemicals required that OPPT scientists become active in the practical development and use of SAR and QSAR analyses. The OPPT SAR/QSARs that have been developed over the past decade frequently allow rapid estimations of the ecotoxicity of many new industrial chemicals submitted for evaluation by the OPPT New Chemicals Program (Clements et al., 1988, 1993a,b; Auer et al., 1990, 1994; Nabholz, 1991; Nabholz et al., 1993a,b; Zeeman and Gilford, 1993; Zeeman et al., 1993a).

OPPT QSAR Manual

Ecotoxicity data from a variety of sources (the industrial chemical PMN data, the scientific literature, and the acute toxicity database for the fathead minnow from the EPA Environmental Research Laboratory in Duluth, MN) were compiled by OPPT staff into the OPPT SAR manual (Clements et al., 1988). These data were used to develop 49 different QSARs for aquatic toxicity using physical-chemical properties like molecular weight and log P (K_{ow}) for estimating the aquatic toxicity (or bioconcentration potential) for about 30 classes or subclasses of industrial chemicals (e.g., various neutral organics, acrylates, anilines, esters, and phenols).

The first OPPT SAR/QSAR manual was internally available in 1983. It was revised and made available to the public in 1988 and has been distributed to more than 800 scientists in industry, academia, and regulatory agencies in the United States and overseas. It has also been the subject of substantial comment and evaluation by Suter (1993) and the Organization for Economic Cooperation and Development (OECD, 1992a,b; Balk et al., 1993).

Suter (1993) published one of the first comprehensive texts on ecological risk assessment (see Chapter 28). In the section on organism-level effects in the book, he generally discussed the large number of QSARs in the literature and stated that the list of QSARs used by OPPT "constitutes the nearest thing to a standard QSAR set for ecological toxicology." Although Suter cautioned that not every one of the models may be of the highest quality, the OPPT QSAR manual was included as a (three-page) table that abstracted each of the QSAR equations for every one of the chemical classes in the manual (Table 7.1, pp. 189–191).

Internationally, large numbers of new and existing industrial chemicals are being produced. Because of testing limitations, many of these chemicals must probably be screened by QSAR for ecotoxicity. Therefore the international OECD has also found the OPPT QSAR manual by Clements et al. (1988) and other more recent QSARs of OPPT (see below) to be of interest.

The OECD effort on QSARs is being led by experts from The Netherlands. Their OECD Workshop on QSARs in Aquatic Effects Assessment (OECD, 1992a) considered the use of certain QSARs by experts in the field acceptable. The chemical class acceptance by these experts focused principally on the ecotoxicity QSARs that can be used for neutral organics, anilines, and phenols.

The OECD acknowledged the ongoing use of the OPPT QSAR manual by the U.S. EPA for the regulatory purpose of assessing the ecotoxicity of new industrial chemicals under section 5 of TSCA. However, some of these experts recommended that the QSARs currently used by the EPA needed to be further validated by actual data supplied by industry.

This validation has been undertaken and it has confirmed that most of the predictions were not only reasonable but also often accurate (Nabholz et al., 1993b). The OPPT SAR/QSAR predictions used to estimate the aquatic toxicity of hundreds of new chemicals were contrasted with validated data received from submitters (i.e., industry). Comparisons were made by calculating the ratio of predictions to the actual data (prediction/data). Ratios near 1.0 represented high accuracy. Acceptable ratios were determined to be within an order of magnitude of the ideal ratio (i.e., from 0.1 to 1.0 or from 10 to 1.0). Test data for 920 individual estimates (e.g., acute daphnid or fish toxicity) for 462 chemicals were compared with the 920 SAR/QSAR predictions for these chemicals. Eighty-five percent of these predictions (782/920) were within the acceptable range, only 9% (83/920) had ratios <0.1 (overpredictions of toxicity), and only 6% (55/920) were >10 (underpredictions of toxicity).

The validation of the use of a variety of ecotoxicity QSARs by OPPT appears to have been further substantiated by the results of a collaborative "SAR/MPD" study undertaken jointly by the EPA and the European Community (now the European Union) to compare the OPPT SAR/QSAR predictions with the minimum premarketing dataset (MPD) received for 175 new chemicals by the European Community (EPA, 1993b; Auer et al., 1994; OECD, 1994). The same general proportions of acceptable SAR/QSAR predictions, underpredictions, and overpredictions as in the prior PMN validation study (Nabholz et al., 1993b) also hold for the SAR/MPD study.

A subsequent report on the OECD Workshop on the Extrapolation of Laboratory Aquatic Toxicity Data to the Real Environment (OECD, 1992b) also mentions routine use of the QSAR manual by EPA (OPPT) as an ecotoxicity prioritization tool that has a wide range of QSARs. This OECD report also mentioned that EPA (OPPT) is continuing to develop new QSARs. The OECD assessment of the original EPA manual and the subsequent addition of over 50 new QSARs is also discussed in the recent Aquatic Effects Assessment Guidance Document prepared by the Dutch for the OECD (Balk et al., 1993).

Revised OPPT QSAR Manual and Personal Computer Version ("ECOSAR")

The QSARs originally published by OPPT in the 1988 volume are routinely used in OPPT's new chemical evaluation process to predict quantitatively the aquatic toxicity of PMN chemicals in specific chemical classes or subclasses. In addition, many new QSARs developed by OPPT are now being used and will be included in the revision of the OPPT QSAR manual that is currently under way and should be completed in 1995 (EPA, 1994a).

The revised OPPT QSAR manual will contain approximately 120 QSARs suitable for estimating aquatic toxicity to at least one surrogate aquatic species (e.g., fish, aquatic invertebrates, and/or algae) for 40 or more classes or subclasses of industrial chemicals (Table 6). A PC version of this QSAR manual ("ECOSAR") is publicly available (EPA, 1993c, 1994b; Wedge et al., 1993).

Use of QSARs to Screen the TSCA Inventory of Existing Chemicals

The QSARs developed for use in the OPPT New Chemicals Program have also been finding applications in the OPPT Existing Chemicals Program. This is because there are about 72,000 chemicals in the TSCA Inventory and most of these also appear not

Table 6. Chemical classes or subclasses with aquatic toxicity QSARs

	Number of QSARs[a]
Neutral organics, nonreactive	9
Acid chlorides	1
Acrylates	4
Aldehydes	5
Amines, aliphatic	4
Anilines	6
Aromatic diamines, *meta*	4
Aromatic diamines, *ortho*	3
Aromatic diamines, *para*	3
Aziridines	3
Benzotriazoles	3
Carbamates	1
Diazoniums, aromatic	1
Diepoxides	3
Diesters, aliphatic	1
Diketones, aliphatic	4
Dinitroanilines	3
Dinitrobenzenes	6
Dinitrophenols	4
Esters	4
Esters, phosphate	1
Esters, phthalate	3
Hydrazines	3
Imides	1
Malononitriles	1
Methacrylates	1
Monoepoxides	3
Monoesters, aliphatic	1
Peroxy acids	2
Phenols	7
Polycationic polymers	3
Propargyl alcohols	1
Semicarbazides, alkyl substituted	1
Semicarbazides, *meta/para* substituted	1
Semicarbazides, *ortho* substituted	1
Surfactants, ethomeen	1
Surfactants, anionic	6
Surfactants, nonionic	2
Surfactants, cationic	3
Thiols (mercaptans)	2
Ureas, substituted	1
Total	117

[a]Fish, aquatic invertebrate, and/or algae (acute and/or subchronic).

to have been previously tested for aquatic and/or terrestrial toxicity. A similar problem is being founded by the Europeans for their inventory of existing high-production-volume chemicals (HPVC) and is being addressed in a similar fashion (Bol et al., 1993a,b).

The use of QSARs by OPPT to screen large sets of existing chemicals and then rank them into categories of high, medium, or low acute and chronic toxicity to fish, daphnids, and algae has been initiated (Clements et al., 1993b; Zeeman et al., 1993b). This effort will subsequently allow OPPT to prioritize concerns on the basis of estimated ecological toxicity. As a result of ecotoxicity screening of over 8000 discrete organic chemicals, OPPT will be able to focus its initial efforts on the chemicals that appear to have the most potential for resulting in risks to organisms in the aquatic environment.

One conclusion from this study by OPPT (Zeeman et al., 1993b) was that only seven chemical classes, cumulatively, represented almost 95% of the discrete organic chemicals in this subsample of the TSCA Inventory (Table 7). If this screening of 8234 discrete organic chemicals is at all representative of the discrete (nonpolymeric) chemicals in the TSCA Inventory, focusing QSAR developmental and validation efforts on these seven classes of chemicals makes a great deal of sense.

In addition, an initial examination of the cumulative acute aquatic toxicity distributions for these 8234 industrial chemicals allowed OPPT to conclude that about 10–20% of TSCA discrete organic chemicals could have high acute toxicity to aquatic organisms (LC50 or EC50 $\leq$ 1 mg/L). Similarly, the cumulative chronic aquatic toxicity distributions showed that about 15–18% of such industrial chemicals could have high chronic toxicity to aquatic organisms (chronic values $\leq$ 0.1 mg/L).

ASSESSMENT FACTORS FOR ECOTOXICITY

Assessment factors were developed about a decade ago by the ecological scientists in OPPT as a way of accounting for the several levels of uncertainty found when evaluating the limited ecotoxicity data

Table 7. The seven most frequent discrete organic chemical classes in a subset of the TSCA inventory

QSAR class	Number of chemicals	Percent	Cumulative percent
Neutral organics	3679	44.7	44.7
Esters	1433	17.4	62.1
Acids	1007	12.2	74.3
Amines	704	8.6	82.9
Phenols	465	5.6	88.5
Aldehydes	254	3.1	91.6
Anilines	247	3.0	94.6
Other classes	445	5.4	100.0
Total	8234	100.0	

submitted for new industrial chemicals (EPA, 1984). The OPPT assessment factors were developed to account for uncertainties such as estimating chronic toxicity from acute toxicity, accounting for species-to-species differences, and extrapolating from laboratory toxicity tests to field toxic effect levels.

The primary purposes of the OPPT ecotoxicity assessment factors have always been to provide a consistent regulatory basis for assessing the potential for ecological risks and to provide a rationale for possible subsequent testing. When the PMN ecotoxicity concern level (an acute or chronic toxicity value adjusted by an assessment factor) results in an inference of ecological risks (by contrasting this level with the PEC) in the OPPT New Chemical Program, the need for acute or chronic ecotoxicity testing is demonstrated. The use of assessment factors to account for uncertainty has resulted in the development of acute and chronic ecotoxicity test data for some new industrial chemicals.

These assessment factors are used to help OPPT in setting what are termed chemical concentrations of concern or concern levels. In practice, the hazard estimates available for new chemicals (e.g., fish LC50) are divided by an appropriate assessment factor (Table 8) to set a level of concern for exposures in the environment. In essence, this concern level is the level of chemical exposure in the environment at which, if it is reached or exceeded, significant risks to aquatic organisms have a higher probability of occurring (EPA, 1984; Auer et al., 1990; Nabholz, 1991; Smrchek et al., 1993; Zeeman and Gilford, 1993).

Environmental exposure PECs below this concern level or concern concentration are *not* presumed by OPPT to be "safe" or without risk (EPA, 1984). For practical purposes, however, if the concern concentration is not exceeded, it has usually been assumed by OPPT that the likelihood of a significant environmental risk is probably too low to warrant taking any regulatory actions (Nabholz, 1991; Nabholz et al., 1993a; Zeeman and Gilford, 1993).

The simplicity of the OPPT ecotoxicity assessment factors and their ready utility in predicting chemical thresholds of ecological concern appear to be important factors considered in a current debate

Table 8. OPPT assessment factors used in setting "concern levels" for new chemicals

Available data on chemical or analogue	Assessment factor
Limited (e.g., only one acute LC50 via SAR/QSAR)	1000
Base set acute toxicity (e.g., fish and daphnid LC50s and algal EC50)	100
Chronic toxicity MATCs	10
Field test data for chemical	1

in the ecotoxicology scientific and regulatory literature (Okkerman et al., 1991, 1993; OECD, 1992b; Balk et al., 1993; Calabrese and Baldwin, 1993; Forbes and Forbes, 1994).

Utility of Ecotoxicity Assessment Factors Beyond EPA

The OECD Workshop on the Extrapolation of Laboratory Aquatic Toxicity Data to the Real Environment (OECD, 1992b) also recommended a set of assessment factors for small data sets that came from the Dutch National Institute of Public Health and Environmental Protection (RIVM). These factors are virtually identical to the OPPT assessment factors (EPA, 1984). The Dutch subsequently acknowledged that the RIVM assessment factor scheme was simply a modification of the method originally developed by the EPA (1984), which they term the "modified EPA method" in their later publications (Balk et al., 1993; Emans et al., 1993; Okkerman et al., 1993; Romijn et al., 1993).

The Dutch proposed that the use of assessment factors was adequate only for a preliminary effects assessment and that a more complex extrapolation method using statistical models that require acute and chronic data for several species was necessary for performing a refined effects assessment that could be suitable for the protection of 95% of the species in ecosystems (Okkerman et al., 1991; OECD, 1992b; van den Berg, 1992/93; Balk et al., 1993; Emans et al., 1993; Okkerman et al., 1993).

The scientific concern about the use of assessment factors appeared to focus on their simple methodology, which seems, to some, to have little or no theoretical or scientific foundation (Okkerman et al., 1991,1993; van den Berg, 1992/93; Emans et al., 1993). However the support document included as part of the OPPT assessment factors paper contains over 20 pages of detailed technical analyses of the rationale for and development and suitability of these assessment (uncertainty) factors (EPA, 1984).

It is interesting that van den Berg (1992/93) stated that the Dutch "model provides a more scientific based extrapolation factor" and that the EPA method has fewer input data and greater uncertainty that "finally results in a lower 'safe' concentration for the ecosystem." Romijn et al. (1993) also reported that a modified EPA method led to a more conservative value (concern level) than did a Dutch statistical method.

Okkerman et al. (1993) attempted to compare and validate these various methods by contrasting the no-observed-effect concentrations (NOECs) from multiple-species ecotoxicity test results for 13 different pesticides and industrial chemicals with the preliminary assessment factor values and the refined statistical extrapolation values also calculated for these 13 chemicals. Even though they preferred the Dutch methods on theoretical grounds, they determined that they could not give preference to the concern levels predicted by the Dutch statistical methods over those predicted by the modified EPA method and concluded that "there is no reason to choose for one or the other method."

Two books on ecotoxicity and ecological risk assessment (Calabrese and Baldwin, 1993; Forbes and Forbes, 1994) have also commented extensively on the use of the EPA assessment factors (EPA, 1984) and contrasted and compared the results of various analyses of the Dutch statistical methods with the EPA method. Calabrese and Baldwin (1993) stated that the EPA assessment factor method and the Dutch statistical methods were both "methodologies [that] represent major contributions to various aspects of the ecological risk assessment process."

Although Calabrese and Baldwin (1993) also recommended the more complex Dutch system for its "theoretically sound approach for the derivation of chemical-specific ecosystem MATCs," they noted that their brief analysis of the experimental data in Okkerman et al. (1991) comparing these two methods by using the multispecies ecotoxicity results for eight chemicals showed that "the two approaches were essentially equivalent." Calabrese and Baldwin concluded that they could not determine whether "this striking similarity [in the results of the two methods] is a coincidence or a generalizable pattern."

In a more detailed assessment, Forbes and Forbes (1994) also focused on the EPA assessment factor approach (EPA, 1984) and the Dutch positions in (1) the Report of the OECD Workshop on Extrapolation of Laboratory Aquatic Toxicity Data to the Real Environment (OECD, 1992b) and (2) the published literature.

Forbes and Forbes performed a statistical analysis of these two extrapolation methods using the multispecies NOEC data for eight chemicals that were presented in the OECD Workshop and that had come from the published literature. They asked, "Are the newer and much more complex extrapolation methods an improvement over the previous simple and arbitrary methods?" They likewise concluded that "these [statistical] methods have not been shown to be either more accurate or more conservative than the simpler assessment factor approaches."

In addition, Forbes and Forbes (1994) state that "If the predictions of two or more methodologies are indistinguishable, parsimony demands that we not use increasingly complex methods unless they have clearly demonstrated an increased usefulness. . . . The more complex extrapolation methods have not provided improved predictive power over the cruder methods." They conclude that "Most ecotoxicologists justifiably abhor the use of arbitrary assessment factors, but the wish for improved chemical evaluation methods must not obscure critical scientific evaluation of their use. Thus we recommend that the older methods remain in use until the assumptions of these new models have been tested rigorously."

Clearly, the simple assessment factor method developed by OPPT and routinely used in the TSCA section 5 new chemical program since the early 1980s still remains a very effective tool for estimating the levels of concern (risk) for chemicals released into the aquatic environment.

CONCLUSIONS

Section 5 of TSCA provides a stimulus for the development of ecological hazard and risk assessment methods and tools that end up focusing on evaluating the potential ecotoxicity of new industrial chemicals. Up-front ecotoxicity testing is not required for submitting a new chemical PMN, and therefore OPPT ecological experts have been forced to be very innovative in using the limited ecotoxicity test data that are available for most industrial chemicals. The defining of 32 chemical categories that set a basis for routinely considering ecotoxicity testing, the development and routine use of validated SAR/QSAR techniques to estimate ecotoxicity, and the pragmatic development and use of assessment factors to set reasonable environmental "concern levels" all demonstrate the creativity of EPA professionals dedicated to protecting organisms living in the environment from the adverse effects of industrial chemicals.

ACKNOWLEDGMENTS

This paper is dedicated to the past and present outstanding professionals in the Environmental Effects Branch who have for over a decade actively participated in the practical development of all of the ecological hazard assessment and risk assessment methodologies used in OPPT. In particular, the former Branch Chief, Dr. James A. Gilford, deserves mention for providing the leadership to accomplish the tasks involved. This paper is also dedicated to the memory of David W. Johnson, an outstanding EEB chemist, without whose expert work (and generous spirit) much of our SAR/QSAR activities would have faltered.

LITERATURE CITED

Auer, C. M.; Nabholz, J. V.; and Baetke, K. P.: Mode of action and the assessment of chemical hazards in the presence of limited data: use of structure-activity relationships (SAR) under TSCA, Section 5. *Environ. Health Perspect.*, 87:183–197, 1990.

Auer C. M.; Zeeman, M.; Nabholz, J. V.; and Clements, R. G.: SAR—the U.S. regulatory perspective. *SAR & QSAR in Environ. Res.*, 2(1–2):29–38, 1994.

Balk, F.; de Bruijn, J. H. M.; and van Leeuwen, C. J.: Guidance Document for Aquatic Effects Assessment. Report prepared for the Organization for Economic Co-operation and Development (OECD). Paris: OECD, 1993.

Boethling, R. S.: Structure/activity relationships for evaluation of biodegradability in the EPA's Office of Pollution Prevention and Toxics. Environmental Toxicology and Risk Assessment, 2nd Vol., edited by J. W. Gorsuch, F. J. Dwyer, C. G. Ingersoll, T. W. LaPoint, pp. 540–554. ASTM STP 1216. Philadelphia: American Society for Testing and Materials, 1993.

Bol, J.; Verhaar, H. J. M.; van Leeuwen, C. J.; and Hermens, J. L. M.: Predictions of the Aquatic Toxicity of High-Production-Volume Chemicals. Part A: Introduction and Methodology. Dutch Ministry of Housing, Physical Planning and Environment, Directorate-General for Environmental Protection, PO Box 30945, 2500 GX The Hague, The Netherlands. The Hague: Dutch Ministry, 1993a.

Bol, J.; Verhaar, H. J. M.; van Leeuwen, C. J.; and Hermens, J. L. M.: Predictions of the Aquatic Toxicity of High-Production-Volume Chemicals. Part B: Predictions. Dutch Ministry of Housing, Physical Planning and Environment, Directorate-General for Environmental Protection, PO Box 30945, 2500 GX The Hague, The Netherlands. The Hague: Dutch Ministry, 1993b.

Calabrese, E. J., and Baldwin, L. A.: Performing Ecological Risk Assessments. Boca Raton, FL: Lewis, 1993.

Clements, R.G.: Environmental effects of regulatory concern under TSCA—a position paper. U.S. Environmental Protection Agency, Office of Toxic Substances, Environmental Effects Branch. Washington, DC: U.S. EPA, 1983.

Clements, R. G. (ed.); Johnson, D. W.; Lipnick, R. L.; Nabholz, J. V.; Newsome, L. D.: Estimating Toxicity of Industrial Chemicals to Aquatic Organisms Using Structure Activity Relationships. U.S. Environmental Protection Agency, Office of Toxic Substances, Health & Environmental Review Division, Environmental Effects Branch. EPA-560-6-88-001. Washington, DC: U.S. EPA, 1988.

Clements, R. G.; Nabholz, J. V.; Johnson, D. W.; and Zeeman, M.: The use and application of QSARs in the Office of Toxic Substances for ecological hazard assessment of new chemicals. Environmental Toxicology and Risk Assessment, edited by W. G. Landis, J. S. Hughes, M. A. Lewis, pp. 56–64. ASTM STP 1179. Philadelphia: American Society for Testing and Materials, 1993a.

Clements, R. G.; Nabholz, J. V.; Johnson, D. W.; and Zeeman, M.: The use of quantitative structure-activity relationships (QSARs) as screening tools in environmental assessment. Environmental Toxicology and Risk Assessment, 2nd Vol., edited by J. W. Gorsuch, F. J. Dwyer, C. G. Ingersoll, T. W. LaPoint, pp. 555–570. ASTM STP 1216. Philadelphia: American Society for Testing and Materials, 1993b.

Emans, H. J. B.; van der Plassche, E. J.; Canton, J. H.; Okkerman, P. C.; and Sparenburg, P. M.: Validation of some extrapolation methods used for effect assessment. *Environ. Toxicol. Chem.*, 12:2139–2154, 1993.

EPA: Testing for environmental effects under the Toxic Substances Control Act. U.S. Environmental Protection Agency, Office of Toxic Substances, Health & Environmental Review Division, Environmental Effects Branch. Washington, DC: U.S. EPA, 1983a.

EPA: Technical support document for the environmental effects testing scheme. U.S. Environmental Protection Agency, Office of Toxic Substances, Health & Environmental Review Division, Environmental Effects Branch. Washington, DC: U.S. EPA, 1983b.

EPA: Estimating 'Concern Levels' for concentrations of chemical substances in the environment. U.S. Environmental Protection Agency, Office of Toxic Substances, Health & Environmental Review Division, Environmental Effects Branch. Washington, DC: U.S. EPA, 1984.

EPA: Toxic Substances Control Act Test Guidelines; final rules. *Fed. Regist.*, 50:39252–39516, 1985. (Part 797—Environmental Effects Guidelines, pp. 39321–39397).

EPA: New Chemical Review: Process Manual. U.S. Environmental Protection Agency, Office of Toxic Substances, Chemical Control Division. EPA-560/3-86-002. Washington, DC: U.S. EPA, 1986.

EPA, OPPT: New chemicals program. *Chem. Prog. Bull.*, 12(4):6–7, 30, 1991a.

EPA: New Chemicals Program. U.S. Environmental Protection Agency, Office of Toxic Substances, Environmental Assistance Division. EPA-560/1-91-005. Washington, DC: U.S. EPA, 1991b.

EPA, OPPT: Categories provide guidelines for premanufacture notice submitters. *Chem. Prog. Bull.*, 14(2):35–36, 1993a.

EPA: U.S. EPA/EC joint project on the evaluation of (quantitative) structure activity relationships. Final report (July 1993). U.S. Environmental Protection Agency, Office of Pollution Prevention & Toxics, Chemical Control Division. Washington, DC: U.S. EPA, 1993b.

EPA: ECOSAR, a computer program for estimating the ecotoxicity of industrial chemicals based on structure activity relationships. U.S. Environmental Protection Agency, Office of Pollution Prevention & Toxics, Health & Environmental Review Division, Environmental Effects Branch. EPA-748-F-93-001. Washington, DC: U.S. EPA, 1993c.

EPA: Estimating Toxicity of Industrial Chemicals to Aquatic Organisms Using Structure Activity Relationships. U.S. Environmental Protection Agency, Office of Pollution Prevention & Toxics, Health & Environmental Review Division, Environmental Effects Branch. EPA-748-R-93-001. Washington, DC: U.S. EPA, 1994a.

EPA: ECOSAR: Computer Program and User's Guide for Estimating the Ecotoxicity of Industrial Chemicals Based on Structure Activity Relationships. U.S. Environmental Protection Agency, Office of Pollution Prevention & Toxics, Health & Environmental Review Division, Environmental Effects Branch. EPA-748-R-93-002. Washington, DC: U.S. EPA, 1994b.

Forbes, V. E., and Forbes, T. L.: Ecotoxicology in Theory and Practice. New York: Chapman & Hall, 1994.

GAO: EPA Toxic Substances Program: Long-standing information planning problems must be addressed. U.S. General Accounting Office, Accounting and Information Management Division. GAO/AIMD-94-25. Washington, DC: U.S. GAO, 1993.

Nabholz, J. V.: Environmental hazard and risk assessment under the United States Toxic Substances Control Act. *Sci. Total Environ.*, 109/110:649–665, 1991.

Nabholz, J. V.; Miller, P.; and Zeeman, M.: Environmental risk assessment of new chemicals under the Toxic Substances Control Act TSCA section five. Environmental Toxicology and Risk Assessment, edited by W. G. Landis, J. S. Hughes, M. A. Lewis, pp. 40–55. ASTM STP 1179. Philadelphia: American Society for Testing and Materials, 1993a.

Nabholz, J. V.; Clements, R. G.; Zeeman, M.; Osborn, K.C.; and Wedge, R.: Validation of structure activity relationships used by the USEPA's Office of Pollution Prevention and Toxics for the environmental hazard assessment of industrial chemicals. Environmental Toxicology and Risk Assessment, 2nd Vol., edited by J. W. Gorsuch, F. J. Dwyer, C. G. Ingersoll, and T. W. LaPoint, pp. 571–590. ASTM STP 1216. Philadelphia: American Society for Testing and Materials, 1993b.

OECD (Organization for Economic Cooperation and Development): Report of the OECD Workshop on Quantitative Structure Activity Relationships (QSARS) in Aquatic Effects Assessment. *OECD Environ. Monogr. No. 58*. Paris: OECD, 1992a.

OECD (Organization for Economic Cooperation and Development): Report of the OECD Workshop on the Extrapolation of Laboratory Aquatic Toxicity Data to the Real Environment. *OECD Environ. Monogr. No. 59*. Paris: OECD, 1992b.

OECD (Organization for Economic Cooperation and Development): U.S. EPA/EC Joint Project on the Evaluation of (Quantitative) Structure Activity Relationships. *OECD Environ. Monogr. No. 88*. Paris: OECD, 1994.

Okkerman, P. C.; van der Plassche, E. J.; Sloof, W.; van Leeuwen, C. J.; and Canton, J. H.: Ecotoxicological effects assessment: A comparison of several extrapolation procedures. *Ecotoxicol. Environ. Safety*, 21:182–193, 1991.

Okkerman, P. C.; van der Plassche, E. J.; Emans, H. J. B.; and Canton, J. H.: Validation of some extrapolation methods with toxicity data derived from multiple species experiments. *Ecotoxicol. Environ. Safety*, 25:341–359, 1993.

Rodier, D. J.: Ecological risk assessment in the Office of Toxic Substances: problems and progress, 1984–1987. U.S. Environmental Protection Agency, Office of Toxic Substances, Health & Environmental Review Division, Environmental Effects Branch. Washington, DC: U.S. EPA, 1987.

Rodier D. J., and Mauriello, D. A.: The quotient method of ecological risk assessment and modeling under TSCA: a review: Environmental Toxicology and Risk Assessment, edited by W. G. Landis, J. S. Hughes, and M. A. Lewis, pp. 80–91. ASTM STP 1179. Philadelphia: American Society for Testing and Materials, 1993.

Rodier, D. J., and Zeeman, M.: Ecological risk assessment. Basic Environmental Toxicology, edited by L. G. Cockerham, and B. S. Shane, pp. 581–604. Boca Raton, FL: CRC Press, 1994.

Romijn, C. A. F. M.; Luttik, R.; van der Meent, D.; Sloof, W.; and Canton, J. H.: Presentation of a general algorithm to include effect assessment on secondary poisoning in the derivation of environmental quality criteria: Part 1. Aquatic food chains. *Ecotoxicol. Environ. Safety*, 26:61–85, 1993.

Sayre, P. G., and Kough, J. L.: Assessment of genetically engineered microorganisms under TSCA: considerations prior to use in fermentors or small-scale field release. Environmental Toxicology and Risk Assessment, edited by W. G. Landis, J. S. Hughes, M. A. Lewis, pp. 65–79. ASTM STP 1179. Philadelphia: American Society for Testing and Materials, 1993.

Smrchek, J.; Clements, R.; Morcock, R.; and Rabert, W.: Assessing ecological hazard under TSCA: methods and evaluation of data. Environmental Toxicology and Risk Assessment, edited by W. G. Landis, J. S. Hughes, M. A. Lewis, pp. 22–39. ASTM STP 1179. Philadelphia: American Society for Testing and Materials, 1993.

Suter, G. W., II (ed.): Ecological Risk Assessment. Boca Raton, FL: Lewis, 1993.

van den Berg, M.: Ecological risk assessment and policy-making in the Netherlands: dealing with uncertainties. *Network* 6(3):8–11, 1992/1993.

Wedge, R. M.; Osborn, K.; Sak, S.; Clements, R. G.; Nabholz, J. V.; and Zeeman, M.: Structure-activity relationships (SAR) program for ecotoxicity: a PC version. SETAC Abstract Book for the 14th Annual Meeting, Houston, TX, p. 218, P304. Pensacola, FL: Society of Environmental Toxicology and Chemistry, 1993.

Zeeman, M., and Gilford, J.: Ecological hazard evaluation and risk assessment under EPA's Toxic Substances Control Act (TSCA): an introduction. Environmental Toxicology and Risk Assessment, edited by W. G. Landis, J. S. Hughes, M. A. Lewis, pp. 7–21. ASTM STP 1179. Philadelphia: American Society for Testing and Materials, 1993.

Zeeman, M.; Nabholz, J. V.; and Clements, R. G.: The development of SAR/QSAR for use under EPA's Toxic Substances Control Act (TSCA): an introduction. Environmental Toxicology and Risk Assessment, 2nd Vol., edited by J. W. Gorsuch, F. J. Dwyer, C. G. Ingersoll, T. W. La Point, pp. 523–539. ASTM STP 1216. Philadelphia: American Society for Testing and Materials, 1993a.

Zeeman, M.; Clements, R. G.; Nabholz, J. V.; Johnson, D.; and Kim, A.: SAR/QSAR ecological assessment at EPA/OPPT: ecotoxicity screening of the TSCA Inventory. SETAC Abstract Book for the 14th Annual Meeting, Houston, TX, P312A. Pensacola, FL: Society of Environmental Toxicology and Chemistry, 1993b.

Chapter 24

THE CLEAN WATER ACT

G. W. Hudiburgh, Jr.

INTRODUCTION

The Clean Water Act, 33 USC §§1251 et seq. (CWA or Act), as it is currently structured, was enacted in 1972 as the Federal Water Pollution Control Act Amendments of 1972 (FWPCA). Major revisions of the CWA were enacted in 1977 and in 1987; other revisions to the Act were enacted at various times.[1] The CWA has had a major impact on the implementation of aquatic toxicity.

It is the purpose of this chapter to discuss in summary the various provisions of the statute that are related to aquatic toxicity issues; the implementation of the statute through regulations, policy, and guidance issued by the U.S. Environmental Protection Agency (EPA or Agency); the Agency's implementation of aquatic toxicity activities under the CWA; the states' implementation of aquatic toxicity requirements; and EPA's oversight of the states. It should be noted that section 510 of the CWA allows states to be more stringent than the federal regulations require (except in certain limited circumstances). Thus, each state has its own statutes and regulations for the control of water pollution, and those statutes and regulations almost always contain some different or additional requirements. Certain federal regulations are binding on the states and represent the minimum requirements in any state.

The major regulatory mechanism available to establish controls on aquatic toxicity is the National Pollutant Discharge Elimination System (NPDES) permit program, which is established under section 402 of the Act. It is through an NPDES permit that EPA or an authorized state[2] establishes controls on the "discharge of pollutants" from "point sources" into "waters of the United States." In general, any material that is added to water is a "pollutant" and any discrete point such as a pipe, ditch, container, or vehicle is considered a "point source." "Waters of the United States" include most surface waters or wetlands.[3] NPDES permits are not required for the introduction of pollutants into a publicly owned treatment works [POTWs, which may nevertheless be regulated by the pretreatment program, 40 CFR Part 403 (1994)] or for certain agricultural and silvicultural discharges; however, they are required for point source discharges of pollutants including noncontact cooling water and many types of storm water runoff. For this reason, a major part of the discussion in this chapter relating to aquatic toxicity will focus on the NPDES program. However, water quality standards (WQS), total maximum daily load and waste load allocations (TMDL/WLA), water quality data management, and assessment activities will also be mentioned. The actions taken by Congress to establish specific actions related to "toxic hot spots" in its establishment of section 304(1) of the CWA in its 1987 amendments, as well as its amendment to section 118 to the CWA by the Great Lakes Critical Programs Act of 1990, will also be discussed. Finally, the discussion of the implementation of the CWA related to aquatic toxicity will close with a review of some of the emerging activities, including the development of sediment criteria and biocriteria.

TERMINOLOGY[4]

Administrator. The administrator of the United States Environmental Protection Agency, or designee.

[1]At the time this chapter was being prepared, potential revisions to the CWA were under consideration by the 103rd and 104th Congress.

[2]As of March 1995, there are 40 states (39 states and the territory of the U.S. Virgin Islands) that are authorized to implement the NPDES program within their State. EPA issues NPDES permits within the other 11 states, five territories, and on federal Indian reservations. Federal Indian tribes can also be authorized to operate various CWA programs, including the NPDES program.

[3]The terminology section contains a more detailed explanation of the definition of these words based upon the CWA and NPDES regulations.

[4]In certain cases, a citation to the CWA or EPA regulations is provided.

Anti-backsliding. Prohibition on less stringent effluent limitations in modified or reissued permits, except under limited circumstances as specified in CWA §402(o) and/or 40 CFR §122.44(1) (1994).

Antidegradation. Provision required in state water quality standards by 40 CFR §131.12 (1994) to deal with the question of if and when water quality may be reduced.

Best available technology economically achievable (BAT). The level of technology-based control for toxic and nonconventional pollutants for industrial dischargers to water of the United States to be achieved within 3 yr of promulgation but not later than March 31, 1989, as defined by CWA §§301(b)(2)(A) and 304(b)(2).

Best conventional pollutant control technology (BCT). The level of technology-based control for conventional pollutants for industrial dischargers to waters of the United States to be achieved within 3 yr of promulgation but not later than March 31, 1989, as defined by CWA §§301(b)(2)(E) and 304(b)(4). The methodology for establishing BCT guidelines was published at 51 Fed. Reg. 24974 (July 9, 1986).

Best management practices (BMPs). Practices to prevent or reduce pollution of waters of the United States, including treatment requirements, operating procedures and practices to control plant site runoff, spillage, or leaks, sludge or waste disposal, or drainage from raw material storage established under CWA §304(e) and/or 40 CFR §122.44(k) (1994).

Best practicable control technology currently available (BPT). The level of technology-based control for toxic, conventional, and nonconventional pollutants for industrial discharges to waters of the United States to be achieved by July 1, 1977, as defined by CWA §§301(b)(1)(A) and 304(b)(1). If guidelines established after January 1, 1982 require substantially greater treatment than was generally in place in the industrial category or based on fundamentally different control technology than in permits, then the deadline is 3 yr after promulgation, but no later than March 30, 1989.

Best professional judgment (BPJ). The basis for establishing permit limitations [under CWA §402(a)(1)(B)] for industrial dischargers for whom, or pollutants for which, final effluent limitations guidelines have not yet been issued (see Chapter 33).

Bypass. Intentional diversion of waste streams from any portion of a treatment facility. 40 CFR §122.41(m) (1994).

Category. Also, point-source category. A segment of industry for which a set of effluent limitations guidelines or pretreatment standards has been established.

Clean Water Act. Redesignated name for the Federal Water Pollution Control Act Amendments of 1972, following the 1977 amendments thereto; 33 USC §§1251 et seq.

Conventional pollutant. Biochemical Oxygen Demand (BOD_5), Total Suspended Solids (TSS), pH, oil and grease, or fecal coliform. CWA §304(a)(4), 40 CFR §401.16 (1994).

Cooling water. Water that is used to absorb and transport heat generated in a process or machinery.

Direct discharger. A facility that discharges all or part of its wastewater directly to waters of the United States through a point source. (Compare with Indirect discharger.)

Discharge monitoring reports. EPA uniform national forms used for the reporting of self-monitoring results by a permittee. 40 CFR §122.2 (1994).

Effluent. The water and the quantities, rates, and concentrations of chemical, physical, biological, and other constituents that are discharged from a point source (see Chapters 1 and 33).

Effluent limitation. Any restriction (including schedules of compliance) established by a state or the EPA on quantities, discharge rates, and concentrations of pollutants (chemical, physical, biological, and other constituents) that are discharged from a point source into navigable waters, waters of the contiguous zone, or the ocean. CWA §502(11), 40 CFR §122.2 (1994).

Effluent limitations guidelines. National guidelines on the technology-based limitations achievable by industrial categories and subcategories, representing the minimum level of treatment required for an affected industry unless a variance is obtained; published by the Administrator under CWA §304(b).

Environmental Protection Agency (EPA or Agency). The United States Environmental Protection Agency.

Fact sheet. Explanation of the basis for proposed NPDES permit conditions; required for all major dischargers.

FWPCA. Federal Water Pollution Control Act Amendments of 1972, redesignated the Clean Water Act in 1977 amendments.

Grab sample. A single sample of wastewater taken without regard to time or flow.

Hazardous substance. Any substance designated under 40 CFR Part 116 (1994) as having the potential to present an imminent and substantial danger to the public health or welfare, pursuant to CWA §311.

In-plant control technology. Includes the regulation and the conservation of chemicals, the reduction of water usage throughout the operations, and improved management practices and plant control, as opposed to end-of-pipe treatment.

Indicator limit. Limitation on a more easily monitored conventional or nonconventional pollutant as an indicator of proper treatment of one or more toxic pollutants. 40 CFR §125.3(h) (1994).

Indirect discharger. A facility (nondomestic discharger) that introduces pollutants into a publicly owned treatment works, either directly or through a publicly owned sewer system. These dischargers are or may be subject to regulation under the pretreatment program under CWA §307. [See 40 CFR Part 403 (1994).]

Individual control strategy (ICS). Program of effluent limitations (and water quality standards) designed by a state to control toxic effects from a discharger pursuant to CWA §304(1). [See 40 CFR §123.46 (1994).]

Mixing zone. The portion of the receiving water that may be set aside for mixing of effluent with the receiving waters. An optional part of state water quality standards. 40 CFR §131.13 (1994).

Monitoring. The measurement, sometimes continuous, of some aspect of a facility's discharge or operations, including water quality. See biomonitoring in Chapter 1.

Navigable waters. The waters of the United States, including the territorial seas. (In the context of the Corps of Engineers authority under the Rivers and Harbors Act, this term has a more limited meaning related to use or potential use for navigation.)

NEPA. National Environmental Policy Act of 1969, 42 USC §§4321 et seq. In the NPDES program, applicable only to "new source" permits issued by EPA.

Noncontact cooling water. Water used for cooling that does not come into direct contact with any raw material, intermediate product, waste product, or finished product.

Nonconventional pollutant. A parameter or chemical that has not been listed either as a conventional pollutant or as a toxic pollutant.

Nonpoint source. Any source of water pollution or pollutants not associated with a discrete conveyance or point source, including, in certain cases, runoff from fields, forest lands, mining, construction activity, and saltwater intrusion (see Chapter 1).

NPDES. National Pollutant Discharge Elimination System. The system of permits for point source discharges to surface waters issued by EPA or by states/tribes with EPA approved programs under CWA §402.

Operation and maintenance costs. The costs of running the wastewater treatment equipment. This includes labor costs, material and supply costs, and energy and power costs. (It generally does not include amortization of capital costs.)

Outfall. The point or location where sewage, effluent, or drainage discharges from a sewer, drain, or conduit to the receiving water.

Point source. Any discernible, confined, and discrete conveyance including, but not limited to, any pipe, ditch, channel, tunnel, conduit, well, discrete fissure, container, rolling stock, concentrated animal feeding operation, landfill leachate collection system, or vessel or other floating craft from which pollutants are or may be discharged. The term does not include return flows from irrigated agriculture or agricultural storm water runoff (see Chapter 1). CWA §502(14), 40 CFR §122.2 (1994).

Pollutant. Dredged spoil, solid waste, incinerator residue, filter backwash, sewage, garbage, sewage sludge, munitions, chemical wastes, biological materials, radioactive materials [except those regulated under the Atomic Energy Act of 1954, as amended (42 USC §§2011 et seq.)], heat, wrecked or discarded equipment, rock, sand, cellar dirt, and industrial, municipal, and agricultural waste discharged from water (see Chapter 1). (Note there are two exclusions for sewage from vessels and certain well injection activities.) CWA §502(6), 40 CFR §122.2 (1994).

Primary treatment. The first stage in wastewater treatment in which floating or settleable solids are mechanically removed by screening and sedimentation (see Chapter 33).

Priority pollutant. One of 65 pollutants or groups of pollutants that are, or may be, toxic; EPA has interpreted this list to include, for the purpose of analysis, 126 chemical compounds, which are also referred to as "priority pollutants." CWA §§301(b)(2)(C) and 307(a)(1), 40 CFR §401.15 (1994).

Privately owned treatment works. Any device or system used to treat wastes from any facility whose operator is not the operator of the treatment works (but not a POTW). 40 CFR §122.2 (1994).

Process wastewater. Any water that, during manufacturing or processing, comes into direct contact with, or results from the production or use of, any raw material, intermediate product, finished product, by-product, or waste product. 40 CFR §122.2 (1994).

Publicly owned treatment works (POTW). System used in the treatment of municipal sewage or industrial liquid wastes which is owned by a state or municipality. 40 CFR §122.2 (1994).

Raw water. Plant intake water prior to any treatment or use.

Receiving waters. Rivers, lakes, oceans, or other water courses that receive treated or untreated wastewaters.

Reopener. A provision allowing a permit to be reopened and modified in certain circumstances (e.g., promulgation of BAT effluent limitations guideline).

Secondary treatment. Limitations, based on biological treatment, required as a minimum technology-based level of treatment for all POTWs (see Chapter 33). Defined in 40 CFR Part 133 (1994).

Sewage sludge. Any solid, semisolid, or liquid residue generated by a wastewater treatment plant (public or private) that treats any domestic wastewater. May be regulated under the sludge management program under CWA §405(f) and 40 CFR Parts 122, 123, 501, and 503 (1994).

Silvicultural point sources. Any discernible, confined, and discrete conveyance related to rock crushing, gravel washing, log sorting, or log storage facilities that are operated in connection with silvicultural activities and from which pollutants are discharged into water of the United States. 40 CFR §122.27(b)(1) (1994).

Sludge. Residue produced in a waste treatment process.

Storm water runoff. Point source discharge of storm water runoff that has been designated as part of the NPDES program (includes runoff from urban areas or areas associated with industrial activity). CWA §402(p), 40 CFR §122.26 (1994).

Subcategory or subpart. A segment of an industrial point source category for which specific effluent limitations have been established.

Total maximum daily load (TMDL). The total allowable pollutant load to a receiving water (individual waste load allocations for point sources and load allocations for nonpoint sources) such that any additional loading will produce a violation of water quality standards. CWA §303(d), 40 CFR §§130.2(i) and 130.7 (1994).

Toxic hot spots. Stream segments adversely affected by point source discharges of toxic pollutants, required to be listed by states by CWA §304(1).

Toxic pollutant. Any pollutant listed as toxic under CWA §307(a)(1), 40 CFR §122.2 (1994).

Toxic pollutant standard. A limitation or prohibition, applicable to some or all categories of sources, on the discharge of a toxic pollutant. Such across-the-board standards, issued pursuant to CWA §307, have been promulgated for only six pollutants. See 40 CFR Part 129 (1994).

Toxicity reduction evaluation (TRE). A site-specific investigation using toxicity testing and physical and chemical analysis of an effluent to identify causative toxicants or treatment methods to reduce effluent toxicity (see Chapters 2 and 33).

Waste load allocation (WLA). The portion of a receiving water's total maximum daily pollutant load that is allocated to one of its existing or future point sources of pollution. 40 CFR §130.2(h) (1994).

Waters of the United States:

1. All waters that are currently used, were used in the past, or may be susceptible to use in interstate or foreign commerce, including all waters that are subject to the ebb and flow of the tide;
2. All interstate waters, including interstate ''wetlands'';
3. All other waters such as intrastate lakes, rivers, streams (including intermittent streams), mudflats, sandflats, ''wetlands,'' sloughs, prairie potholes, wet meadows, playa lakes, or natural ponds the use, degradation, or destruction of which would affect or could affect interstate or foreign commerce, including any such waters:

    (a) Which are or could be used by interstate or foreign travelers for recreational or other purposes;
    (b) From which fish or shellfish are or could be taken and sold in interstate or foreign commerce; or
    (c) Which are used or could be used for industrial purposes by industries in interstate commerce;

4. All impoundments of waters otherwise defined as waters of the United States under this definition;
5. Tributaries of waters identified in paragraphs 1–4 of this definition;
6. The territorial sea; and
7. ''Wetlands'' adjacent to waters (other than waters that are themselves wetlands) identified in paragraphs 1–6 of this definition.

Note: Waste treatment systems, including treatment ponds or lagoons designed to meet the requirements of CWA [other than cooling ponds defined in 40 CFR §423.11(m), which also meet the criteria of this definition] are not waters of the United States. Waters of the United States do not include prior converted cropland. Notwithstanding the determination of an area's status as prior converted cropland by another federal agency for the purposes of the Clean Water Act, the final authority regarding the Clean Water Act jurisdiction remains with EPA. 40 CFR §122.2 (1994).

Water Quality Act of 1987 (WQA). The amendments to the federal Clean Water Act, PL 100-4, enacted on February 4, 1987.

Water quality criteria. The portion of water quality standards specifying the parameters that must be met to achieve and protect the designated use. They may specify concentration of specific chemicals, provide narrative proscriptions (e.g., ''free from toxic pollutants in concentrations acutely toxic to aquatic life''), or be expressed in terms of biological monitoring methods. CWA §303(c), 40 CFR Part 131 (1993).

Water quality standards. The regulations specifying, through a combination of use designations and water quality criteria to protect those uses and antidegradation, the quality to be achieved and maintained for each surface water in the state, required by CWA §303.

Whole-effluent toxicity (WET). The aggregate toxic effect of an effluent measured directly by an aquatic toxicity test. 40 CFR §122.2 (1994).

THE CLEAN WATER ACT AND ITS IMPLEMENTATION

The CWA is structured into six major parts:

Title I—Research and Related Programs
Title II—Grants for Construction of Treatment Works
Title III—Standards and Enforcement
Title IV—Permits and Licenses
Title V—General Provisions
Title VI—State Water Pollution Control Revolving Funds

A review of the various provisions of the Act and their implementation related to aquatic toxicity follows.

Objective and Goals of the Clean Water Act

Section 101(a) of the Clean Water Act, contained in Title I, indicates that ''[t]he objective of this Act is to restore and maintain the chemical, physical and biological integrity of the Nation's waters.'' The implementation of controls on and monitoring of aquatic toxicity relate specifically to the stated objective of the Act. Section 101(a) also establishes seven goals or policies to achieve this objective, consistent with the provisions of the Act[5]; again, either the goals or policies specifically relate to aquatic toxicity or implementation of the goal or policy would relate to the objective of restoring and maintaining the integrity of the nation's waters. The seven goals or policies contained in section 101(a) are

1. It is the national goal that the discharge of pollutants into the navigable waters be eliminated by 1985;
2. It is the national goal that wherever attainable, an interim goal of water quality which provides for the protection and propagation of fish, shellfish, and wildlife and provides for recreation in and on the water be achieved by July 1, 1983.
3. It is the national policy that the discharge of toxic pollutants in toxic amounts be prohibited.
4. It is the national policy that federal financial assistance be provided to construct publicly owned waste treatment works.
5. It is the national policy that areawide treatment management planning processes be developed and implemented to assure adequate control of sources of pollutants in each state.
6. It is the national policy that a major research and demonstration effort be made to develop technology necessary to eliminate the discharge of pollutants into the navigable waters, waters of the contiguous zone, and the oceans.
7. It is the national policy that programs for the control of nonpoint sources of pollution be developed and implemented in an expeditious manner so as to enable the goals of this Act to be met through the control of both point and nonpoint sources of pollution.

Other Provisions of Title I

Section 104 of the Act contains various provisions related to ''research, investigations, training, and information.'' Included in section 104(e) are specific provisions related to the establishment of various field laboratories and research facilities. Many of these facilities focus their activities on water research activities, including the Environmental Research Laboratories in Narragansett, Rhode Island, Duluth, Minnesota, Gulf Breeze, Florida and Corvallis, Oregon, as do laboratories operated by states, universities, industry, and environmental groups. Section 105 specifically provides statutory authority for ''grants for research and development;'' in many cases these grants (if funds are provided within the appropriation) can include work on aquatic toxicity.

These various provisions are some of the specific provisions contained in the Act to implement the national policy contained in section 101(a)(6) of ''a major research and demonstration effort . . . to develop technology necessary to eliminate the discharge of pollutants into the navigable waters, waters of the contiguous zone and the oceans.''

In 1990, the Great Lakes Critical Programs Act of 1990 was enacted. Among other provisions, this law amended section 118 of the CWA to provide [in section 118(c)(2)] for issuance by the Agency of Great Lakes Water Quality Guidance which

> shall specify numerical limits on pollutants in ambient Great Lakes waters to protect human health, aquatic life, and wildlife, and shall provide guidance to the Great Lakes States on minimum water quality standards, antidegradation policies, and implementation procedures for the Great Lakes System.

The Great Lakes states are to adopt and implement such guidance.[6]

Waste Treatment Management Plans

Title II of the CWA is entitled ''Grants for Construction of Treatment Works'' and its primary focus is on various provisions related to the construction grants program in which the federal government provided grants to states and municipalities for the construction of publicly owned treatment works (refers to municipal sewer systems that convey wastewater to treatment or to sewage treatment itself). The various provisions of Title II related to the construction grants program, along with the current emphasis contained in the CWA under Title VI for the state revolving fund program, are specific actions related to the national policy contained in section 101(a)(4) for federal financial assistance to construct POTWs.

However, various provisions of Title II are more generally applicable. The purpose section of Title II, contained in section 201, refers to various activities regarding waste treatment management plans and practices. Section 208 is entitled ''Areawide Waste Treatment Management.'' Section 208(a) itemizes various activities to be undertaken by either the Agency or the states ''[f]or the purpose of encouraging and facilitating the development and implementation of areawide waste treatment management plans.'' Specific statutory provisions related to a continuous areawide waste treatment management planning process are contained in section 208(b). EPA has issued regulations that are contained in 40 CFR Part 130 (1994) related to ''Water Quality Planning and Management''; as provided in 40 CFR §130.0(a) (1994) of the regulation, ''[t]he Water Quality Management (WQM) process described in the Act [including under 208 and 303] and in this regulation provides the authority for a consistent national approach for maintaining, improving and protecting

[5]It is generally agreed that the goals and policies established in section 101(a) are not self-implementing and are implemented consistent with the other provisions of the Clean Water Act.

[6]EPA proposed the Guidance on April 16, 1993 (58 Fed. Regist. 20802). The final guidance was signed by the Administration on March 13, 1995, and published shortly thereafter.

water quality while allowing States to implement the most effective individual programs." The process is implemented by EPA, the states, interstate agencies, and various planning organizations; the WQM process can include controls both on point sources through the NPDES program, total maximum daily load/waste load allocation development, monitoring, and reporting, and on nonpoint source activities. Specifics of these various actions are discussed later in this chapter. 40 CFR §130.6 (1994) specifically relates to water quality management plans produced in accordance with sections 208 and 303(e) of the Act. 40 CFR §130.12(a) (1994) provides that under section 208(e) of the Act, "no NPDES permit may be issued which is in conflict with an approved Water Quality Management (WQM) plan." These various provisions contained in Title II (and subsequent provisions contained in Title III) related to the areawide treatment management planning processes are specific actions related to the national policy contained in section 101(a)(5) that these processes be developed and implemented to assure adequate control of sources of pollutants in each State.

Title III—Standards and Limitations

Title III of the Clean Water Act contains several sections specifically related to the issue of aquatic toxicity. The sections, which will be discussed in more detail, follow. (Not all sections of Title III are specifically related to aquatic toxicity.)

Section 301—Effluent Limitations
Section 302—Water Quality Related Effluent Limitations
Section 303—Water Quality Standards and Implementation Plans
Section 304—Information and Guidelines
Section 305—Water Quality Inventory
Section 307—Toxic and Pretreatment Effluent Standards
Section 308—Inspections, Monitoring, and Entry
Section 311—Oil and Hazardous Substances Liability
Section 314—Clean Lakes
Section 316—Thermal Discharges
Section 319—Nonpoint Source Management Programs
Section 320—National Estuary Program

These sections contained in Title III, along with the section 402 NPDES program and other sections of Title IV discussed below, are some of the specific provisions contained in the Act to implement the national goal contained in section 101(a)(1) that "the discharge of pollutants into the navigable waters be eliminated by 1985"; the national goal contained in section 101(a)(2) that "wherever attainable, an interim goal of water quality which provides for the protection and propagation of fish, shellfish, and wildlife and provides for recreation in and on the water be achieved by July 1, 1983"; and the national policy contained in section 101(a)(3) "that the discharge of toxic pollutants in toxic amounts be prohibited."

A summary of the various technology-based provisions of the CWA, which provide a minimum basis of control, follows.

Under provision of Title III of the CWA, EPA is required to promulgate nationwide effluent guidelines or standards for certain types of discharges. With the exception of toxic pollutant standards for certain pollutants that are contained in 40 CFR Part 129 (1994), these federal guidelines and standards are not self-implementing.[7] In other words, they do not apply to a discharge unless and until they have been incorporated into (used as a basis for) effluent limitations contained in the NPDES permit. The deadlines for achieving permit limitations based on national guidelines are established by statute, however, and generally do not depend on when the limitations are incorporated in the permit. The first level of effluent limitations in NPDES permits is based on a "technology basis." The limitations do not specify use of a specific technology at a facility but are based on what a technology can achieve.

POTWs were supposed to achieve limitations based on secondary treatment, as currently defined in 40 CFR Part 133 (1994), by July 1, 1977. In some cases, extensions of this deadline up to July 1, 1988 were available, and under EPA's national municipal policy (January 30, 1984, 49 Fed. Regist. 3832), 1988 was the effective compliance date. For most types of sewage treatment plants, secondary treatment limitations are 30 mg/L of BOD_5 (biochemical oxygen demand) and 30 mg/L of TSS (total suspended solids) on a monthly average, along with an 85% removal requirement, 7-d limitations for BOD_5 and TSS, and pH limitations of 6.0 to 9.0.

Unlike secondary treatment limitations for POTWs, most effluent limitations guidelines for industrial categories are expressed in terms of the allowable mass of pollutants that can be discharged per unit of production. EPA promulgates effluent limitations guidelines for different industrial categories. When a permit is written for a facility that falls within one of these categories, the permit limitation is based on the applicable effluent limitations guideline, multiplied by the amount of production that has been achieved (or, in some cases, is anticipated for new sources and new discharges) by that facility. The permit limitations do not rise and fall with production. Rather, if there has been a significant increase or decrease in production this may be grounds for modifying the permit. (EPA regulations do allow the permit writer to include alternative effluent limitations in the permit that will apply if production increases or decreases to a certain level on a tiered basis. Such alternative limitations are still the exception rather than the rule, however.)

[7]The standards for the use or disposal of sewage sludge (40 CFR Part 503) which EPA issued on February 19, 1993 (58 Fed. Regist. 9248), as revised 59 Fed. Regist. 9095 (February 25, 1994) under sections 405(d) and (e) of the CWA, are also self-implementing.

Under the CWA, EPA or an authorized state can impose case-by-case technology-based effluent limitations based on best professional judgment (BPJ) under section 402(a)(1)(B) of the CWA, in the absence of national effluent limitations guidelines. BPJ limits are used where national effluent limitations guidelines have not yet been issued for the appropriate industry category, do not cover the particular processes involved, or have been withdrawn or remanded. BPJ limitations are used for the vast number of industrial facilities that do not fall within any industrial category for which effluent limitations guidelines have been promulgated. In establishing BPJ limitations, the permit writer must apply the same factors that the statute requires to be considered in establishing national guidelines. The "building block approach" is used to combine the allowances under effluent limitation guidelines for different industrial categories or subcategories.

Industrial dischargers were to have met effluent limitations to achieve the first level of technology-based control for all pollutants based on the best practicable control technology currently available (BPT) no later than July 1, 1977. If, however, the BPT effluent limitations guidelines were established after January 1, 1982, and they are substantially more stringent or require a different type of technology than was generally required of the industry prior to that time, the Water Quality Act of 1987 allows compliance within 3 yr after the guidelines are established, but no later than March 31, 1989.

The second level of technology-based control for conventional pollutants (BOD_5, TSS, pH, fecal coliform, and oil and grease) is called best conventional pollutant control technology (BCT). Dischargers must meet effluent limitations based on BCT within 3 yr of promulgation of the guidelines, but not later than March 31, 1989. Treatment beyond BPT is required only if "reasonable," in comparison to the cost per pound of pollutant removed for secondary treatment by POTWs and for BPT treatment.

For designated toxic pollutants and all other pollutants that are neither toxics nor conventionals (the "nonconventionals"), the second level of technology-based control is based on the best available technology economically achievable (BAT) and must be met within 3 yr of promulgation of the guidelines, but not later than March 31, 1989. Several variances are available from BAT limits for nonconventionals, but only the fundamentally different factors (FDF) variance (discussed below) is available for toxics. (There may be some confusion about references to 65 "toxic pollutants" or 126 "priority pollutants." Congress in 1977 designated 65 "toxic pollutants" that included several groups of chemicals. EPA has defined the list of 65 pollutants and groups of pollutants to include 126 chemical substances, frequently referred to as the " priority pollutants.")

The level of technology-based controls applicable to new sources is new source performance standards (NSPS), which are based on best available demonstrated technology. Under the CWA, dischargers subject to NSPS are provided a 10-yr "protection period" from new technology-based limitations but not from new water quality–based limitations or case-by-case limits on pollutants not regulated by NSPS.

Although the CWA provides for a number of variances from technology-based limitations (other than NSPS), variances are really very limited. The variances include §301(c) variances (variance from BAT for nonconventionals based on the economic capability of the owner/operator), §301(g) variances [variance from BAT for ammonia, chlorine, color, iron, and total phenols (4AAP)[8] (when not a toxic pollutant) based on lack of impact of water quality], §301(h) variances (variance from secondary treatment requirements for POTWs discharging to certain marine waters), §301(k) time extensions (time extension from otherwise applicable BAT/BCT compliance deadlines based on innovative technology), and §301(n) (FDF) variances (fundamentally different factors variances from otherwise applicable technology-based effluent limitations guidelines). In general, variances must be approved by EPA and have been issued very infrequently. Deadlines for requesting variances vary; they may have to be requested within 180 d after promulgation of the effluent limitations guidelines.

The specific provisions of Title III are discussed below.

Section 301—Effluent Limitations

Section 301 of the Act contains numerous provisions regarding "effluent limitations" that are implemented through NPDES permits issued under section 402 of the Act. NPDES permits can contain both technology-based effluent limitations (established under various provisions of the CWA including sections 301, 304, 306, and 307 and discussed earlier in this chapter) as well as water quality–based effluent limitations (established under various provisions of the CWA including sections 301, 302, 303, and 307).

Where the technology-based permit effluent limitations are insufficient to assure that state water quality standards will be met, additional water quality–based limitations are included in the NPDES permit in order to meet the water quality standards, including the narrative criterion of "no toxic in toxic amounts." As will be explained in more detail later, water quality–based limitations may include limits on whole effluent toxicity (acute and/or chronic) and/or limits on individual chemical constituents. These limits are intended to protect beneficial uses of water bodies and consider factors such as dilution, environmental fate, and the sensitivity of the resident aquatic community.

EPA interprets the requirement that NPDES permits contain limitations to assure compliance with water quality standards as an absolute obligation. No variance is available from water quality–based lim-

[8] Method for total phenol analysis.

itations necessary to meet WQS (other than a change in the WQS themselves, including "downgrades of use"). Unlike technology-based limitations, water quality–based limitations vary with the individual stream and the individual discharger. They can be based on some type of model (frequently just a mass balance calculation) to determine the discharge loading that can be allowed without violating water quality standards. The assumptions used in the modeling are critical. In many cases, limitations are calculated to assure that WQS will be met at the maximum discharge rate, the critical low stream flow (often the 7-d low flow that occurs once every 10 yr), and some sort of reasonable worst-case assumption as to the nature of the discharge and the characteristics of the receiving stream. Since water quality–based effluent limitations are derived from the state WQS, and frequently from a waste load allocation for the receiving stream prepared by the state, it is advantageous that the discharger or concerned party get involved in the state's water quality planning and management efforts long before negotiations on the individual discharge permit.

In certain cases, there may be some flexibility in how water quality–based effluent limitations are met. For example, installation of a high-rate diffuser or moving the outfall pipe may increase the discharge allowance by increasing the effective mixing of the discharge with the receiving stream. EPA and the NPDES states may make use of variable effluent limitations, which may vary according to the time of year, the receiving stream flow or temperature, or other factors that influence the effect of the discharge on the receiving stream.

The regulations implementing many of the provisions contained in section 301 are contained in 40 CFR Parts 122–125 (1994). Various aspects of implementation of aquatic toxicity through NPDES permits will be discussed later in the chapter in connection with section 402.

Section 301(a) contains the prohibition that the discharge of any pollutant by any person shall be unlawful, except in compliance with sections 301, 302, 306, 307, 318, 402, and 404.[9] The NPDES program can be used to authorize such discharges.

Section 301(b) of the Act contains compliance dates for various technology-based and water quality–based levels of control. The compliance dates for various technology-based limitations have been discussed earlier in this chapter. Section 301(b)(1)(C) of the CWA provides the statutory basis for inclusion in NPDES permits of the provisions of state law or regulation, including state water quality standards. Section 301(b)(1)(C) provides for a July 1, 1977 compliance date for

> any more stringent limitations, including those necessary to meet water quality standards . . . established pursuant to any State law or regulations . . . or required to implement any applicable water quality standard established pursuant to the Act.

The NPDES regulations contain specific provisions related to establishment of effluent limitations, methods of establishing these technology-based and water quality–based effluent limitations, and compliance schedules and are contained in 40 CFR §§122.44, 122.47, and 125.3 (1994).

The question of compliance dates for water quality–based limitations has been resolved in an NPDES permit appeal. In an April 16, 1990 decision on an NPDES permit appeal, the EPA Administrator issued a decision (*In the Matter of Star-Kist Caribe, Inc.*)[10] on the question of the ability to establish compliance schedules in NPDES permits for effluent limitations based on post–July 1, 1977 WQS. In the 1990 decision the Administrator held that

> the Clean Water Act does not authorize EPA to establish schedules of compliance in the permit that would sanction pollutant discharges that do not meet applicable state water quality standards. In my opinion, the only instance in which the permit may lawfully authorize a permittee to delay compliance after July 1, 1977, pursuant to a schedule of compliance, is when the water quality standard itself (or the State's implementing regulations) can be fairly construed as authorizing a schedule of compliance.

The *Star-Kist* decision means that NPDES permits must require immediate compliance with post-1977 WQS and cannot establish a compliance schedule for the water quality–based effluent limitations in the NPDES permit [except in certain cases related to establishment of water quality–based effluent limitations for toxic pollutants under section 304(1) of the Act], unless there is some provision of a state's WQS or other implementing regulations (including NPDES regulations) that allows for establishment of a schedule of compliance.

Section 301(f) contains a prohibition that it is unlawful to discharge any radiological, chemical, or biological warfare agent, high-level radioactive waste, or medical waste into navigable waters. Section 301(g) provides for a variance from technology-based BAT effluent limitations for five specifically listed nonconventional pollutants [ammonia, chlorine, color, iron, and total phenols (4AAP, when not a toxic pollutant)] or other pollutants listed by the Administrator based on a lack of impact on water quality; this variance has been of limited use, with relief, if granted, being provided for iron and steel and steam electric facilities. Section 301(h) provides for a variance from otherwise applicable secondary treatment requirements for POTWs to marine waters based on rigorous demonstrations that are required

---

[9]The §404 program under the CWA regulates the discharges of dredged or fill material. The U.S. Army Corps of Engineers issues §404 permits, unless a state has been authorized to operate the program. EPA also has a significant role to play in the dredge and fill program. Applicable EPA regulations are contained in 40 CFR Parts 230–233 (1994); the Corps regulations are contained in 33 CFR Parts 320–330 (1994).

[10]The Administrator's April 1990 decision had been stayed on September 4, 1990 to respond to a request to modify the April 1990 decision; the modification request was denied on May 26, 1992.

by the terms of the Act and EPA regulations contained in 40 CFR Part 125, Subpart G (1994).

Section 302—Water Quality-Related Effluent Limitations

The primary focus of the authority for implementation of water quality-based effluent limitations for a very large majority of NPDES permits is sections 301(b)(1)(C), 303 and 304 of the CWA. Section 302(a), entitled "water quality-related effluent limitations," provides a mechanism for the Administrator to establish more stringent effluent limitations for point sources to facilitate attainment of the fishable, swimmable water quality goal, where those sources still interfere with the attainment of such water quality after application of technology-based limitations required by section 301(b)(2); section 302(b) provides for public hearing before such modification.[11]

Section 303—Water Quality Standards and Implementation Plans

Section 303 is the section of the CWA that contains the statutory basis for water quality standards and implementation plans, including provisions relating to waste load allocations and total maximum daily loads. A more detailed explanation of these provisions follows. Specific provisions related to water quality criteria development are contained in section 304.

Sections 303(a), 303(b), and 303(c) contain provisions related to development of WQS by the states, review and approval or disapproval of those standards by the Administrator, and the triennial review of WQS by the states. Section 303(c)(2)(A) contains direction to the states when the states revise or adopt new standards and provides that

> [s]uch revised or new water quality standard shall consist of the designated uses of the navigable waters involved and the water quality criteria for such waters based upon such uses. Such standards shall be such as to protect the public health or welfare, enhance the quality of the water and serve the purposes of the Act. Such standards shall be established taking into consideration their use and value for public water supplies, propagation of fish and wildlife, recreational purposes, and agricultural, industrial, and other purposes, and also taking into consideration their use and value for navigation.

EPA's regulations dealing with WQS and the WQS process are contained in 40 CFR Parts 130 and 131 (1994).[12] The provisions contained in 40 CFR Part 131 (1994) specifically lay out the WQS process, including the purpose of WQS [40 CFR §131.2 (1994)], procedures for the states' establishment of WQS [40 CFR Part 131, Subpart B (1994)], EPA's procedures for review and revision of WQS [40 CFR Part 131, Subpart C (1994)], and the limited situations in which the agency has issued federally promulgated WQS [40 CFR Part 131, Subpart D (1994)].[13] Section 303(c)(2)(B) was amended in 1987 to provide specifically for the states to adopt criteria for toxic pollutants. Most states have taken action to comply with this provision. On December 22, 1992 EPA took action and promulgated in 40 CFR §131.36 (1994) toxic pollutant criteria (or other WQS provisions) for the 14 states and territories that had not acted or if EPA had disapproved the states' action.[14]

Section 303(d) of the Act provides for the states to identify waters for which technology-based limitations are not stringent enough to achieve WQS, develop a priority ranking of waters, and develop total maximum daily loads (TMDL) for the pollutants identified by the Agency under section 304(a). Section 303(d)(1)(C) provides that

> [s]uch load shall be established at a level necessary to implement the applicable water quality standards with seasonal variations and a margin of safely which takes into account any lack of knowledge concerning the relationship between effluent limitations and water quality.

These TMDLs are to be submitted by the states to EPA for approval or disapproval; if EPA disapproves, it has the ability to establish TMDLs for the state. EPA's regulations at 40 CFR §130.7 (1994) provide more specificity for establishment of TMDLs [a TMDL is the sum of waste load allocations (WLA), load allocations (LA), and natural background], WLA (the portion of a receiving water's loading capacity allocated to an existing or future point source), and LA (the portion of a receiving water's loading capacity allocated to an existing or future nonpoint source or natural background source), and the eventual implementation of the TMDL/WLA/LA through a water quality management plan into effluent limitations.[15]

Section 303(e) of the CWA provides for the states to establish and submit to EPA for approval a continuing planning process for management of their overall water quality program; the process must meet numerous requirements contained in section 303(e)(3) including various effluent limitations and schedules of compliance, incorporation of elements of areawide waste management plans, TMDLs, and WQS. EPA regulations in 40 CFR §130.5 (1994) deal with the continuing planning process. As noted earlier, water quality management plans produced in

---

[11]For additional discussion of section 302, see December 18, 1974 letter from Alan G. Kirk, II, Assistant Administrator for Enforcement and General Counsel, EPA, to Edward Dunkalberger, and September 21, 1984 letter from Colburn T. Cherney, Associate General Counsel, EPA, to Jerry W. Raisch.

[12]EPA has considered proposing changes to the Part 131 regulations. EPA has also issued guidance, including Water Quality Standards Handbook, October 1983, dealing with WQS issues. An updated second edition was issued in August 1994.

[13]As of July 1, 1993, the only currently effective federally promulgated WQS, except for those based on section 303(c)(2)(B), are for Arizona [40 CFR §131.31 (1994)] and the Colville Confederated Tribes Indian Reservation [40 CFR §131.35 (1994)].

[14]December 22, 1992 (57 Fed Regist 60848), revised.

[15]EPA has issued guidance on implementation of the section 303(d) process: Guidance for Water Quality-based Decisions: The TMDL Process (EPA 440/4-91-001).

accordance with sections 208 and 303(e) (including the continuing planning process) can include controls both on point sources through the NPDES program, TMDL/WLA development, monitoring, and reporting and on nonpoint source activities.

Section 304—Information and Guidelines

Section 304 of the Act is another section that contains the statutory authority for the Agency to, among other requirements, develop, publish, and periodically update ambient water quality criteria [section 304(a)], test procedures for analysis of pollutants [section 304(h)], and various requirements concerning individual control strategies (ICSs) for toxic pollutants [section 304(1)].

The authority for the Agency to develop and publish ambient water quality criteria is contained in section 304(a). As provided in section 304(a)(1), these criteria are to reflect accurately the latest scientific knowledge on the kind and extent of all identifiable effects on public health and welfare including, but not limited to, plankton, fish, shellfish, wildlife, plant life, shorelines, beaches, aesthetics, and recreation that may be expected from the presence of pollutants in any body of water; on the concentrations and dispersal of pollutants, or their byproducts, through biological, physical, and chemical processes; and on the effects of pollutants on biological community diversity, productivity, and stability including information on the factors affecting rates for eutrophication and rates for organic and inorganic sedimentation for various types of receiving waters. EPA, or its predecessor agencies, has developed numerous water quality criteria and published this information. Such criteria have been issued periodically in the Green Book (1968), the Blue Book (1972), the Red Book (1976), and the Gold Book (1986). The criteria themselves are also published in the Federal Register. In 1980 (45 Fed Regist 79318; November 28, 1980) and 1984 (49 Fed Regist 5831; February 15, 1984), EPA published 65 individual ambient water quality criteria for toxic pollutants (termed priority pollutants) listed under section 307(a)(1). Criterion for additional priority pollutants have been added since that time. The development and publication of ambient water quality criteria has been pursued over the past 10 yr and is an ongoing process. EPA expects to publish additional final criteria each year. Some of these will update and revise existing criteria and others will be issued for the first time. As of July 1993, EPA has published water quality criteria for numerous pollutants, including 36 freshwater acute criteria, 47 freshwater chronic criteria, 32 marine acute criteria, 41 marine chronic criteria, and 119 human health criteria. These criteria themselves are not self-implementing under federal law. They can be used by the states in establishing criteria as a part of their WQS and subsequently in the TMDL process. Finally, the criteria can be applied to a discharge by incorporating effluent limitations in a permit.

Section 304(h) provides for the Agency to promulgate guidelines for the establishment of test procedures for the analysis of pollutants to be used under section 401 certification and section 402 permit applications. EPA has established numerous test procedures that are contained in 40 CFR Part 136 (1994).

Section 304(1) of the CWA was added by Congress in 1987 to require the states and EPA to take various actions regarding listings of waters and issuance of individual control strategies[16] for toxic hot spots. These requirements, among other things, require states to identify stream segments primarily affected by point source discharges and prepare ICSs for those sources. Those ICSs must be sufficient, in combination with existing controls on point and nonpoint sources, to achieve compliance with water quality standards as soon as possible, but not later than 3 yr after the strategies are approved by EPA. These new requirements placed even greater emphasis on water quality–based limitations for toxics. EPA issued regulations implementing section 304(1) on June 2, 1989 (54 Fed Regist 23686); these regulations amended various EPA regulations dealing with surface water toxics issues, including NPDES regulations contained in 40 CFR Parts 122 and 123 and water quality planning and management regulations contained in 40 CFR Part 130. There has been litigation on the regulations themselves and on EPA's listing and ICS decisions. In *NRDC v. U.S. EPA*, 915 F2d 1314 (9th Cir 1990), the court remanded various portions of the regulation to EPA for further consideration or action.[17] In *API v. U.S. EPA*, 996 F2d 346 (D.C. Cir 1993), the court upheld all regulations that were challenged by the petitioners, including EPA's regulation in 40 CFR §122.44(d)(1)(vi) (1994) that requires permit writers to interpret state WQS containing narrative criteria to establish, in certain situations, chemical-specific limitations in the NPDES permits.

Section 305—Water Quality Inventory

Section 305(b) of the Act and EPA's regulations in 40 CFR §130.8 (1994) provide for biennial reporting by each state to EPA of its water quality using the criteria contained in the statute and regulations. The report is the primary assessment of state water quality; problems identified in the 305(b) report are to be analyzed through the water quality management planning process.

Section 307—Toxic and Pretreatment Effluent Standards

Section 307(a)(1) of the CWA provides for establishment of a list of toxic pollutants for various purposes

[16]In its regulations [40 C.F.R. §123.46(c) (1994)], EPA has defined an ICS as a draft or final NPDES permit or CERCLA (Comprehensive Environmental Response, Compensation and Liability Act) decision document.

[17]EPA issued final regulations on some aspects of the *NRDC* remand on July 24, 1992 (57 Fed Regist 33040).

under the Act. The Agency has codified the list in 40 CFR §401.15 (1994).

Section 307(a)(2) provides that the Agency can establish requirements for toxic pollutants that can be composed of effluent standards or prohibitions and that EPA shall consider various matters, including the characteristics of the pollutant, the impact of the pollutant, and other potential regulatory authorities. The Agency has established such toxic pollutant effluent standards in 40 CFR Part 129 (1994) for a limited number of toxic pollutants.[18] The scope of the regulation in 40 CFR §129.1(b) (1994) provides that toxic pollutant standards established under the provision are directly applicable to the sources and pollutants within the time period established in the regulation, even if not incorporated into an NPDES permit, but are also to be incorporated into NPDES permits. EPA has not established any new toxic pollutant effluent standards under 40 CFR Part 129 (1994) since 1977, but instead established numerous effluent limitations guidelines and standards regulations for various industrial categories and subcategories, which can include effluent limitations on toxic pollutants, in 40 CFR Subchapter N—Effluent Guide lines and Standards (1994).

Section 308—Inspections, Monitoring, and Entry

Section 308 contains numerous provisions related to the authority of the Administrator to collect information, have right of entry to collect information, and require recording and reporting of various data, including effluent data, and requirements for states to develop similar provisions. These statutory provisions are implemented through various EPA regulations, including specific NPDES regulations that provide for section 308 provisions. These section 308 requirements (and those contained in section 402, discussed below) include the ability for EPA or a state (through similar state law) to collect or require NPDES permit applicants and permittees to collect aquatic toxicity, chemical, and in-stream biological data necessary to assure compliance with standards.

Other Provisions of Title III

Certain other provisions of Title III also have a relationship to aquatic toxicity and are noted in passing. These provisions include section 311, oil and hazardous substances liability; section 314, clean lakes; section 316, thermal discharges; section 319, nonpoint source management programs; and section 320, national estuary program. Appropriate and timely implementation of these sections is focused, in part, on improvement of water quality.

Title IV—Permits and Licenses

Title IV of the CWA contains three significant sections specifically related to the issue of aquatic toxicity. These various provisions, that will be discussed in more detail, follow.

Section 401—Certification
Section 402—National Pollutant Discharge Elimination System
Section 403—Ocean Discharge Criteria

These provisions contained in Title IV, along with the provisions of Title III discussed above, are some of the specific provisions contained in the Act to implement the national goals contained in sections 101(a)(1) and 101(a)(2) and the national policy contained in section 101(a)(3).

Section 401—Certification

Section 401 of the CWA provides that state certification is required before issuance of federal license or permit (including NPDES) that the discharge will comply with sections 301 (including any more stringent state requirements), 302, 303 (including state WQS), 306, and 307.[19] Under section 401(a)(1) and EPA regulations, EPA may not issue a permit until a certification is granted, the permit contains the more stringent provision of certification, or certification is waived. The statute also provides for procedures regarding waiver and if a discharge affects waters of other states. EPA has established general regulations in 40 CFR Part 121 (1994) dealing with state certification of activities requiring a federal license or permit and more specific regulations regarding state certification of NPDES permits issued by EPA in 40 CFR §§124.53–124.55 (1994).[20] Review and appeals of conditions of state certification are through the state's process.

Section 402—National Pollutant Discharge Elimination System

As noted earlier in this chapter, the major regulatory mechanism available to establish controls on aquatic toxicity is the NPDES permit program, established under section 402 of the Act. It is through an NPDES permit that EPA or an authorized state/tribe establishes controls on the discharge of pollutants from point sources to waters of the United States.[21]

Section 402(a)(1) provides that, except as provided in sections 318 and 404, EPA may, after opportunity for public hearing, issue a permit for discharge of any pollutant. By statute and regulation, the NPDES program requires permits for the dis-

[18] Standards are established in 40 CFR Part 129 (1994) for aldrin/dieldrin; DDT, DDE, and DDE; endrin; toxaphene; benzidine; and polychlorinated biphenyls (PCBs).

[19] In *Arkansas v. Oklahoma*, 112 SCt 1046 (1992), the Supreme Court upheld EPA's interpretation of the Clean Water Act that point sources in an upstream state cannot violate water quality standards in the downstream state at the state line.

[20] A letter from LaJuana S. Wilcher, Assistant Administrator for Water, EPA, to Lois D. Cashell, FERC, provides other recent EPA interpretations of certain section 401 certification issues, January 18, 1991.

[21] In 40 CFR Part 123 (1994), EPA has established requirements for authorized state programs and provisions establishing requirements for the states to maintain their programs, which are based upon state law, consistent with the minimal federal requirements. EPA has oversight responsibilities over authorized programs, including permit review.

charge of "pollutants" from any "point source" into "waters of the United States," 40 CFR §122.1(b)(1) (1994). The NPDES program includes, along with discharges from conventional types of point sources, discharges from concentrated animal feeding operations, concentrated aquatic animal production facilities, discharges into aquaculture projects, discharges of storm water as set forth in 40 CFR §122.26 (1994), and silvicultural point sources, 40 CFR §122.1(b)(2) (1994).[22] Section 402(k) of the CWA and 40 CFR §122.5(a)(1) (1994) are generally interpreted to provide that compliance with an NPDES permit is compliance with various specifically enumerated provisions of the Act (and a "shield from any other enforcement action"), except for any toxic pollutant effluent standards and prohibitions imposed under section 307 and standards for sewage sludge use or disposal under section 405(d).[23]

Section 402(b) of the CWA contains nine specific, detailed requirements of state programs; these provisions form the basis of NPDES program requirements because section 402(a)(3) provides that the EPA permit program be subject to the same minimum requirements as state programs. Even if EPA has authorized a state to operate the NPDES program within its jurisdiction, EPA has oversight responsibility under sections 402(c)–402(e) of the Act and 40 CFR Part 123 (1994). Section 402(j) specifically provides for public availability of permit applications and permits.

Throughout the 20 plus years of operation of the NPDES program by EPA and the 40 states that are authorized to implement the NPDES program within the respective states, the program has issued numerous regulations, guidance, and policy and been subject to various court challenges. Important provisions related to aquatic toxicity are discussed below.[24]

On February 3, 1984, Jack E. Ravan, Assistant Administrator for Water, EPA, issued a "Policy for Development of Water Quality-based Permit Limitations for Toxic Pollutants."[25] To control pollutants beyond BAT, secondary treatment, and other CWA technology-based requirements in order to meet water quality standards, EPA developed a policy that involves use of an integrated approach of both biological and chemical methods to address toxic and nonconventional discharges from industrial and municipal sources. Where state WQSs contain numerical criteria for pollutants, NPDES permits will contain limitations to assure compliance with the standards. The policy contains specific provisions regarding testing requirements, use of data, setting permit limitations, and monitoring by both EPA and the states.

The policy indicates that, in addition to enforcing specific numerical criteria for toxic pollutants, EPA will use biological techniques or whole-effluent toxicity (WET) testing as a complement to chemical-specific analysis to assess effluent discharges and express permit limitations. Biological techniques have several advantages over strictly chemical methods for assessing water quality impact: the toxicological effects of complex discharges of many known and unknown constituents in an effluent can be predicted only with a biological method, the bioavailability of pollutants after discharge may be measured more accurately by toxicity testing, and pollutants for which there are inadequate chemical methods, criteria, and aquatic toxicity data can be addressed.

The policy states that

> [w]here violations of water quality standards are identified or projected, the State will be expected to develop water quality–based effluent limits for inclusion in any issued NPDES permit. Where necessary, EPA will develop these limits in consultation with the State. Where there is a significant likelihood of toxic effects to biota in the receiving water, EPA and the States may impose permit limits on effluent toxicity and may require an NPDES permittee to conduct a *toxicity reduction evaluation* [emphasis added, discussed below]. Where toxic effects are present but there is a significant likelihood that compliance with technology-based requirements will sufficiently mitigate the effects, EPA and the States may require chemical and toxicity testing after installation of treatment and may reopen the permit to incorporate additional limitations if needed to meet water quality standards. March 9, 1984, 49 Fed Regist 9017.

Standardized methods have been developed by EPA to determine the toxicity of effluents to aquatic life (U.S. EPA, 1991a; 1991b). These methods facilitate the inclusion of whole-effluent toxicity limits and biomonitoring conditions (including in-stream monitoring) in NPDES permits for facilities suspected of violating state WQSs and to ensure continuing compliance. At times, inability to meet toxicity-based effluent limitations in two or more tests will trigger a requirement for the discharger to conduct a toxicity reduction evaluation (TRE, discussed below), along with possible enforcement even for a single test violation.

Effluent toxicity testing is an important aspect of the water quality–based approach for controlling toxics because it can be used to establish different control priorities, assess compliance with state water quality standards, and set permit limitations to

[22]There are also limited specific "discharges" that are excluded from the NPDES program which are contained in 40 CFR §122.3 (1994).

[23]Indirect dischargers, or facilities that introduce pollutants into a POTW, may be subject to regulation under the national pretreatment program established under section 307 of the Act. See 40 CFR Part 403 (1994). Treatment works treating domestic sewage (TWTDS) may be subject to regulation under the sewage sludge management program established under section 405 of the Act. See 40 CFR Parts 122, 123, 501, and 503 (1994).

[24]In a memorandum dated January 11, 1979, James A. Rogers, Associate General Counsel, EPA, to Deputy Assistant Administrator for Water Enforcement, EPA, "Use of Biomonitoring in the NPDES Permit Program," various issues regarding the legal authority of EPA to impose toxicity test requirements in second-round permits were discussed. EPA has subsequently issued regulations, policy, and guidance in this area.

[25]The policy was published in the Federal Register on March 9, 1984 (49 Fed Regist 9016).

achieve those standards. States have narrative statements for WQSs (including a narrative "no toxics" statement) and some states have specific numerical criteria for specific toxic pollutants or toxicity criteria to achieve certain designated uses. When numerical criteria are not specified, the regulatory authority can develop quantitative water quality–based limits on chemical and effluent toxicity to assure compliance with WQSs.

On January 25, 1989 Rebecca W. Hanmer, Acting Assistant Administrator for Water, EPA, issued a memorandum to EPA's Regional Administrator entitled "Whole Effluent Toxicity Basic Permitting Principles and Enforcement Strategy," which established basic permitting principles for establishing whole-effluent toxicity limitations in NPDES permits, as well as an enforcement strategy for toxics control.[26]

EPA regulations establish permit application requirements in 40 CFR Part 122, Subpart B (1994).[27] Depending on the individual discharger, these requirements contained in 40 CFR §122.21 (1994) can include information characterizing the effluent, including chemical-specific information. Included in these requirements is the requirement in 40 CFR §122.21(g)(11) (1994), for existing industrial discharges, that the applicant indicate whether any biological toxicity tests have been conducted on the discharge or receiving water in relation to the discharge within the past 3 yr; in addition, using the provision contained in 40 CFR §122.21(g)(13) (1994), the Agency or state can request additional information, including quantitative data and bioassays to assess the relative toxicity of discharges to aquatic life and requirements to determine the cause of toxicity.[28] In addition, application requirements for new and existing POTWs contained in 40 CFR §122.21(j) (1994) require certain POTWs to submit with their permit application valid whole-effluent biological toxicity testing.

NPDES permits issued by either EPA or an authorized state must contain certain standard conditions specified in 40 CFR §122.41 (1994), including a duty to comply, proper operation and maintenance, duty to provide information, inspection, and entry provisions, monitoring and records provisions, and reporting requirements. Violations of the permit can be subject to enforcement action.

EPA's NPDES regulations contain provisions for establishing limitations, standards, and other permit conditions in 40 CFR §§122.44 and 125.3 (1994). Contained in 40 CFR §122.44(d) (1994) are provisions concerning establishment of permit limitations dealing with water quality standards and state requirements. 40 CFR §122.44(d)(1) (1994) contains detailed provisions concerning establishment of limitations, either chemical specific or narrative, to achieve compliance with WQSs established under section 303, including a narrative criterion.[29] Other provisions of 40 CFR §122.44(d) (1994) concern establishing limitations due to section 302, 401 certification, section 401(a)(2) requirements, other state requirements, section 208(b), or ocean discharge criteria under section 403(c). In addition, 40 CFR §125.3(c)(4) (1994) provides specific regulatory authority to include in permits limitations expressed in terms of toxicity, where it is demonstrated that the limitations reflect the appropriate technology-based requirements or WQSs under the Act.[30]

In March 1991, EPA's Office of Water issued a revised "Technical Support Documents for Water Quality-Based Toxics Control" (EPA/505/2-90-001) (TSD). The document contains technical guidance for assessing and regulating the discharge of toxic substances to waters of the United States and contains information on approaches to water quality–based toxics control, water quality criteria and standards, effluent characterization, exposure and waste load allocation, permit requirements (including toxicity reduction evaluations), compliance monitoring and enforcement, and case examples. The process of water quality–based toxics control and effluent toxicity assessment is illustrated in Figure 1, which is from the TSD. The illustration outlines the steps involved, from definition of the water quality objectives, criteria, and standards to the setting of the final permit conditions with monitoring requirements. As mentioned above, when an NPDES permittee cannot achieve effluent limitations for toxicity, the EPA and/or a state regulatory authority responsible for the per-

[26]The permitting principles were essentially codified in revisions to EPA's NPDES regulations at 40 CFR §122.44(d) issued on June 2, 1989, on establishing permit limitations to assure compliance with WQSs.

[27]Authorized states may have established additional, more stringent, requirements for their implementation of the NPDES program within their state under state law. Section 510 of the Act and 40 CFR §123.1(i) (1994).

[28]This information can include toxicity identification evaluations (TIEs) or toxicity reduction evaluations (TREs); in the alternative, TIE/TRE requirements can be established by the Agency or state in permits with a reopener clause to establish limitations based on the results of the evaluations.

[29]The regulations contained in 40 CFR §122.44(d) (1994), which establish the specific provisions on establishing limitations to assure compliance with state WQSs, were made more specific when the regulations were revised on June 2, 1989 (54 *Fed. Reg.* 23868); however, the substantive authority was not changed. An August 21, 1989 memorandum from James R. Elder, Director, Office of Water Enforcement and Permits, EPA, to the EPA Regional Water Management Division Directors, entitled "New Regulations Governing Water Quality-Based Permitting in the NPDES Permitting Program," discusses in detail the various provisions of the regulations. In addition, an August 14, 1992 memorandum from Michael B. Cook, Director, Office of Wastewater Enforcement and Compliance, and Robert H. Wayland, III, Director, Office of Wetlands, Oceans, and Watersheds, entitled, "Clarification Regarding Certain Aspects of EPA's Surface Water Toxics Control Regulations," discusses some clarification of the regulations. Finally, as previously indicated, in *API v. U.S. EPA*, 996 F2d 346 (DC Cir, slip opinion June 22, 1994), the court upheld all regulations that were challenged by the petitioners, including EPA's regulations in 40 CFR §122.44(d)(1)(vi) (1994) which requires permit writers to interpret state WQSs containing narrative criteria to establish, in certain situations, chemical-specific limitations in the NPDES permits.

[30]This provision was upheld by the U.S. Court of Appeals for the District of Columbia Circuit in *NRDC v. U.S. EPA*, 859 F2d 156 (DC Cir 1988).

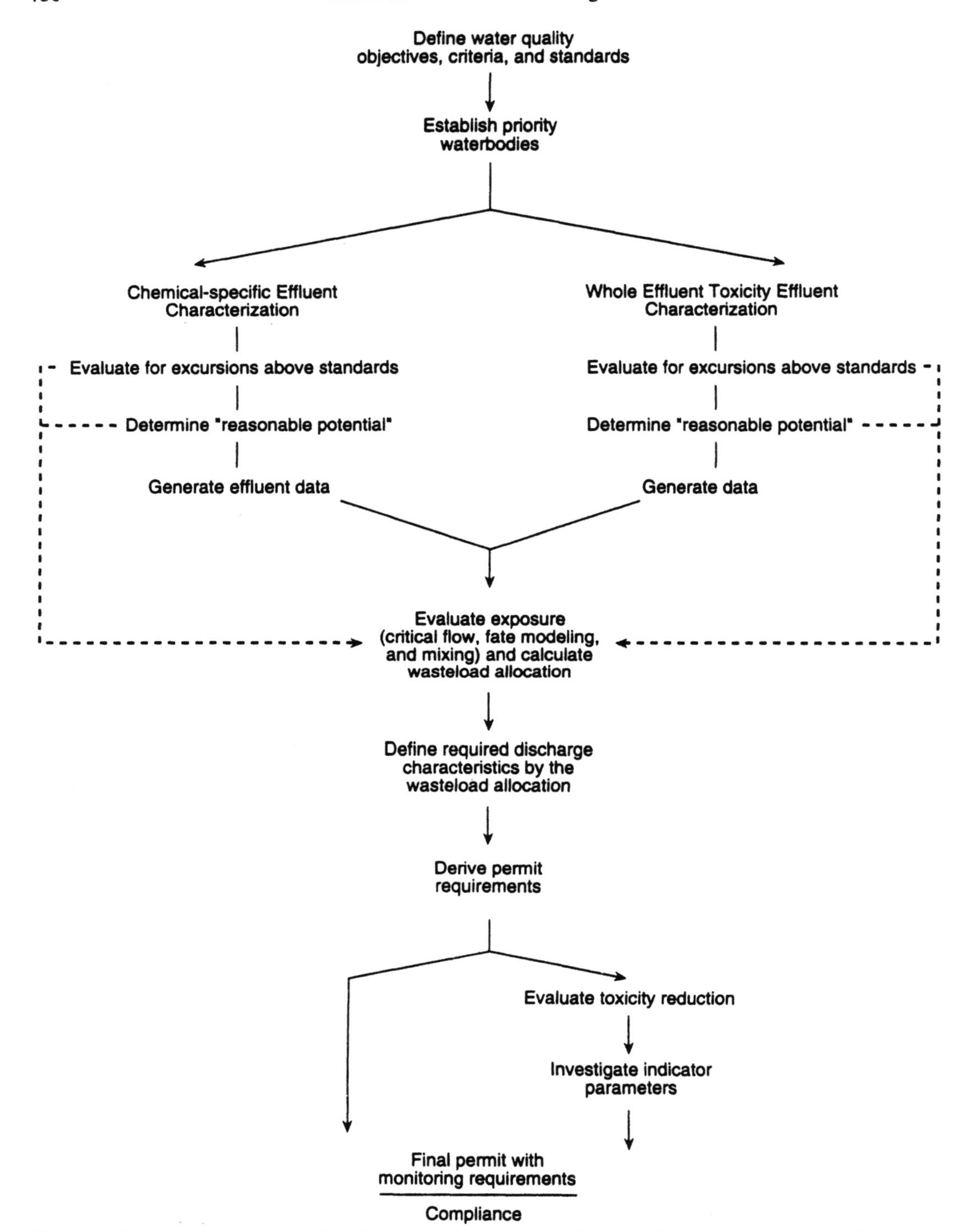

Figure 1. Overview of the water quality–based "standards to permits" process for toxics control. (From U.S. EPA, 1991c.)

mit may require the permittee to reduce toxicity so that no harmful effects occur, along with possible enforcement for the violation itself. A toxicity reduction evaluation is a means of ensuring that WQSs are met and can be used either before permit issuance, as a term of the permit, or as an enforcement tool upon violation of a permit. A TRE focuses attention on toxicity reduction procedures that will enable the permittee to come into compliance with WQS.

EPA has several documents regarding TRE. A TRE is a stepwise process that combines toxicity testing and analysis of the physical and chemical characteristics of causative toxicants to focus on the toxicants causing effluent toxicity and/or on treatment methods that will reduce the effluent toxicity. A summary of how a TRE should be performed is provided below (from TSD, pp. 114–120) and is illustrated in Figure 2.

Step 1—TRE regulatory requirements. Establishment of a TRE, its objectives, and development of a TRE plan.

Step 2—Information and data acquisition. Evaluation of existing site-specific information relevant to effluent toxicity.

Step 3—Facility operation and maintenance evaluation. Evaluation of whether the facility is well operated and whether the effluent toxicity is a result of upsets, bypasses, or other operational difficulties.

Step 4—Toxicity identification evaluation. A TIE is conducted in three phases. Phase one is characterization: a determination of the general nature of the causative agents. Phase two is identification of the specific chemical causing the toxicity. Phase three is confirmation of the identified toxics.

Step 5—Toxicity reduction evaluation. Based on the results of the TIE, a decision is made whether to conduct the TRE using source identification (which will lead to control methods) and/or treatability studies on the final effluent itself (which will lead to identification of possible treatment methods for the effluent).

Step 6—Control method selection and implementation. After completion of the TRE, a control method is selected and implemented.

Step 7—Follow-up and confirmation. After implementation of the control method, there is follow-up and confirmation of its effectiveness, or other steps, as necessary.

In conclusion, in dealing with issuance of NPDES permits and aquatic toxicity, EPA is continuing to move forward in implementation of these various provisions in its issuance of NPDES permits and in its oversight of issuance of permits by the authorized states.

Section 403—Ocean Discharge Criteria

Section 403 of the ACT and EPA regulations at 40 CFR Part 125, Subpart M (1994) contain provisions related to aquatic toxicity that are applicable for NPDES permits issued for discharges from a point source into the territorial seas, contiguous zone, or the oceans.

Title V

The definition of various terms comes from section 502 of the CWA, which contains definitions for, among other terms, effluent limitation [section 502(11)] and biological monitoring [section 502(15)].

Section 504, contained in Title V, contains a powerful (but rarely used) emergency powers provision that provides for the Administrator to sue by going into a U.S. District Court to obtain an order to immediately restrain any person from causing or contributing to alleged pollution to stop discharge if there is evidence of imminent and substantial endangerment to the health of persons or to the welfare of persons where the endangerment is to the livelihood of such person, such as the ability to market shellfish.[31]

EMERGING ACTIVITIES

EPA'S Office of Water has established various goals for moving forward in the aquatic toxicity area, along with actions to continue implementation of the ongoing base program activities to further improve water quality discussed above.[32] Included are provisions for the states to adopt by September 1993 (1) narrative biological criteria and (2) WQSs for wetlands in accordance with Agency guidance or other scientifically valid methods. In addition, the Agency is increasing the emphasis on water quality impaired by contaminated sediments (see Chapter 8) and is initiating action toward controlling sources of sediment contamination. The first step in dealing with this issue is the assessment of sediment quality and the development of sediment quality standards with an eventual goal of NPDES sediment quality–based permit limitations. Finally, EPA is initiating action to view the integrity of the water environment holistically and is initiating action to develop ecological criteria guidance.

SUMMARY AND CONCLUSIONS

The Clean Water Act provides statutory authority for EPA to establish provisions regarding aquatic toxicity. EPA has taken various actions to deal with aquatic toxicity through (1) issuance of regulation, guidance, and policy; (2) implementation of its various responsibilities through issuance of NPDES per-

[31]This authority is in addition to the otherwise applicable civil, criminal, and administrative enforcement authority available to EPA and civil and criminal enforcement authority available to a state.

[32]These and other activities are discussed in "A Guide to the Office of Water Accountability System and Regional Evaluations, Fiscal Year 1992." These activities are accountability measures used in conjunction with the Agency's operating guidance.

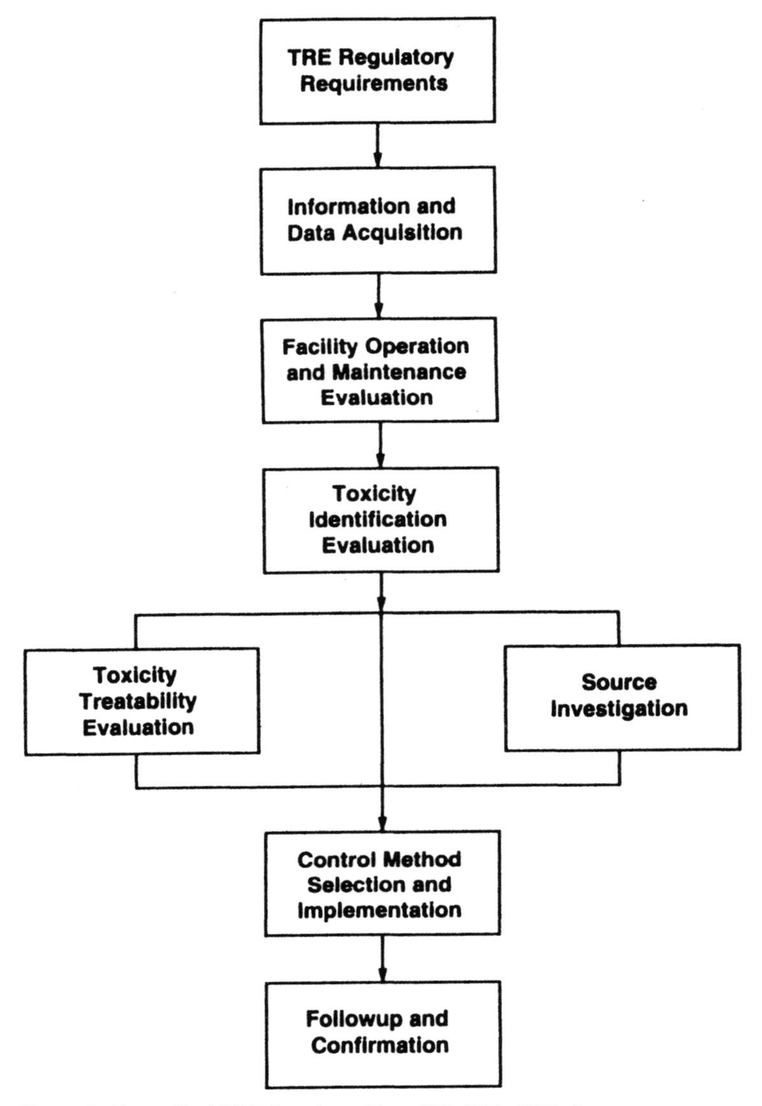

Figure 2. Generalized TRE flowchart. (From U.S. EPA, 1991c.)

mits; and (3) oversight of the states, where the state must take action. These various provisions that EPA and the states deal with that affect aquatic toxicity include water quality management plans, water quality standard development, water quality criteria, waste load allocations and total maximum daily loads, water quality assessment, 401 certification, and National Pollutant Discharge Elimination System permit issuance. Continued implementation of these base CWA activities, as well as the various actions that EPA is undertaking to move the program forward, will continue progress toward attainment of the various goals and compliance with policies established in section 101(a) of the CWA.

LITERATURE CITED

API v. U.S. EPA, 996 F2d 346 (DC Cir, slip opinion, June 22, 1993).

Arkansas v. Oklahoma, 112 S.Ct. 1046 (1992).

Cherney, C. T. (U.S. EPA): Letter to Jerry W. Raisch regarding section 302. 1984.

Cook, M. B., and Wayland, R. H., III (U.S. EPA): Memorandum to EPA Regional Water Management Division directors on clarification regarding certain aspects of EPA's surface water toxics control regulations, 1992.

Elder, J. R. (U.S. EPA): Memorandum to EPA Regional Water Management Division directors on new regulations governing water quality–based permitting in the NPDES permitting program, 1989.

Hanmer, R. W. (U.S. EPA): Memorandum to regional administrators on whole effluent toxicity basic permitting principles and enforcement strategy, 1989.
Kirk, A. G., II (U.S. EPA): Letter to Edward Dunkalberger regarding section 302, 1974.
NRDC v. U.S. EPA, 859 F2d 156 (DC Cir 1988).
NRDC v. U.S. EPA, 915 F2d 1314 (9th Cir 1990).
Ravan, J. E. (U.S. EPA): Policy for development of water quality-based permit limits for toxic pollutants. *Fed. Regist.*, 49:9016 et seq., 1984.
Rogers, J. A. (U.S. EPA): Memorandum to Deputy Assistant Administrator for Water Enforcement on the use of biomonitoring in the NPDES program, 1979.
Secretary of the Interior: Water Quality Criteria [Green Book], 1968.
U.S. EPA: Water Quality Criteria 1972 [Blue Book], 1972.
U.S. EPA: Quality Criteria for Water [Red Book], 1976.
U.S. EPA: Water quality criteria. *Fed. Regist.*, 45:9318 et seq., 1980.
U.S. EPA: Water Quality Standards Handbook, 1983.
U.S. EPA: National municipal policy. *Fed. Regist.*, 49:3832 et seq., 1984.
U.S. EPA: Water quality criteria. *Fed. Regist.*, 49:5831 et seq., 1984.
U.S. EPA: BCT guidelines and methodology. *Fed. Regist.*, 51:24974 et seq., 1986.
U.S. EPA: Quality Criteria for Water 1986 [Gold Book], 1986.
U.S. EPA: Surface water toxics regulation. *Fed. Regist.*, 54: 23686 et seq., 1989.
U.S. EPA: Decision of the Administrator, In the Matter of Star-Kist Caribe, Inc., 1990, as modified 1990 and 1992.
U.S. EPA: Methods for Measuring the Acute Toxicity of Effluents to Aquatic Organisms, 4th ed. Office of Research and Development, Cincinnati, OH. EPA-600/4-90-027, 1991a.
U.S. EPA: Short-term methods for Estimating the Chronic Toxicity of Effluents and Receiving Waters to Freshwater Organisms. 3rd ed. Office of Research and Development, Cincinnati, OH. EPA-600/4-91/002, 1991b.
U.S. EPA: A Guide to the Office of Water Accountability System and Regional Evaluations, Fiscal Year 1992. 1991.
U.S. EPA: Technical Support Document for Water Quality-based Toxics Control. EPA 505-2-90-001, 1991c.
U.S. EPA: Revised surface water toxics regulation. *Fed. Regist.*, 57:33040 et seq., 1992.
U.S. EPA: Water quality standards; establishment of numeric criteria for priority toxic pollutants; states' compliance. *Fed. Regist.*, 57:60848 et seq., 1992.
U.S. EPA: Proposed water quality guidance for the Great Lakes system. *Fed. Regist.*, 58:20802 et seq., 1993.
U.S. EPA: Standards for the use or disposal of sewage sludge. *Fed. Regist.*, 58:9248 et seq., 1993, as revised *Fed. Regist.*, 59:9095 et seq., 1994.
U.S. EPA: Water Quality Standards Handbook: Second Edition. Office of Water, Washington, D.C. EPA-823-B-94-005a/b. 1994.
U.S. EPA: Guidance for Water Quality-based Decisions: The TMDL Process. EPA 440/4-91-001. Date unknown.
Wilcher, L. S. (U.S. EPA): Letter to Lois D. Cashell regarding section 401. 1991.

Chapter 25

FDA's IMPLEMENTATION OF THE NATIONAL ENVIRONMENTAL POLICY ACT

P. G. Vincent

INTRODUCTION

Background

The Food and Drug Administration (FDA), under the Federal Food, Drug and Cosmetic (FD&C, 1938) Act and other public health statutes, is responsible for ensuring that food is safe and wholesome; human and animal drugs, biological products, and medical devices are safe and effective; cosmetics are safe; radiological products do not result in unnecessary radiation exposure; and products are honestly and informatively labeled. FDA is able to carry out its responsibilities by ensuring compliance by approving products before they are marketed, promulgating regulations [21 CFR (Code of Federal Regulations) Part 25], and taking necessary compliance actions when problems are identified.

The National Environmental Policy Act [NEPA; 42 USC § (i.e., section) 4321 et seq., 1970] is the Magna Carta of U.S. environmental laws. NEPA is intended to ensure that all federal agencies consider every significant aspect of the environmental impact of a proposed action, and the public has an opportunity to review the agency's evaluation. NEPA therefore requires all federal agencies to prepare a detailed environmental analysis, known as an environmental impact statement (EIS), for all "major Federal actions significantly affecting the quality of the human environment" [42 USC §4332(c)]. An EIS need not be prepared if the agency concludes that the major action will have "no significant impact." FDA's conclusions are based on a thorough evaluation of a claim for categorical exclusion or an environmental assessment (EA). Regulations promulgated under NEPA by the Council on Environmental Quality (CEQ) and binding on all agencies (*Andrus v. Sierra Club*, 442 US 347,351, 1979) require, however, that unless the action falls within one of the "categorical exclusions" established by each agency [see 21 CFR §§ (i.e. subsections) 25.23 and 25.24 for FDA's excluded actions], the agency must support each finding of no significant impact with a publicly available document called an environmental assessment (EA) [40 CFR §§1501.4 (a)–(b) and 1508.9]. CEQ regulations also specify the factors agencies should take into account in determining whether particular actions will have a "significant" impact on the environment (40 CFR §§1508.27). Among other things, agencies are to consider "[t]he degree to which the effects on the quality of the human environment are highly uncertain or involve unique or unknown risks" (40 CFR §§1508.27(b)(4) and (5)).

The purposes of NEPA are stated in NEPA §§2 and 101, and its procedures set NEPA apart from all other environmental statutes, which regulate specific aspects of the human environment. NEPA represents an "umbrella" law encompassing all existing environmental law. The congressional declaration of NEPA's purpose was to declare a national policy that will encourage productive and enjoyable harmony between humans and their environment; to promote efforts that will prevent damage to the environment and biosphere and stimulate human health and welfare; to enrich the understanding of the ecological systems and natural resources important to the nation; and to establish a Council on Environmental Quality. The congressional declaration of national environmental policy states

> The Congress, recognizing the profound impact of man's activity on the interactions of all components of the natural environment, particularly the profound influences of population growth, high-density urbanization, industrial expansion, resource exploitation, and new and expanding technological advances and recognizing further the critical importance of restoring and maintaining environmental quality to the overall welfare and development of man, declares that it is the continuing policy of the Federal Government, in

cooperation with State and local governments, and other concerned public and private organizations, to use all practicable means and measures, including financial and technical assistance in a manner calculated to foster and promote the general welfare, to create and maintain conditions under which man and nature can exist in productive harmony, and fulfill the social, economic, and other requirements of present and future generations of Americans.

TERMINOLOGY

The following definitions were abstracted from 40 CFR Part 1508. [*Authority*: NEPA, the Environmental Quality Improvement Act of 1970, as amended (42 USC 4371 et seq.), section 309 of the Clean Air Act, as amended (42 USC 7609), and EO 11514 (March 5, 1970, as amended by EO 11991, May 24, 1977). *Source*: 43 FR 56003, November 29, 1978, unless otherwise noted.]

§1508.1 Terminology. The terminology of this part is uniform throughout the federal government.

§1508.2 Act. Act means the National Environmental Policy Act, as amended (42 USC 4321, et seq.) which is also referred to as "NEPA."

§1508.3 Affecting. "Affecting" means will or may have an effect on.

§1508.4 Categorical exclusion. "Categorical exclusion" means a category of actions that do not individually or cumulatively have a significant effect on the human environment and have been found to have no such effect in procedures adopted by a federal agency in implementation of these regulations (§1507.3) and for which, therefore, neither an environmental assessment nor an environmental impact statement is required. An agency may decide in its procedures or otherwise to prepare environmental assessments for the reasons stated in §1508.9 even though it is not required to do so. Any procedures under this section shall provide for extraordinary circumstances in which a normally excluded action may have a significant environmental effect.

§1508.5 Cooperating agency. "Cooperating agency" means any federal agency other than a lead agency that has jurisdiction by law or special expertise with respect to any environmental impact involved in a proposal (or a reasonable alternative) for legislation or other major federal action significantly affecting the quality of the human environment. The selection and responsibilities of a cooperating agency are described in §1501.6. A state or local agency of similar qualifications or, when the effects are on a reservation, an Indian tribe, may by agreement with the lead agency become a cooperating agency.

§1508.6 Council. "Council" means the Council on Environmental Quality established by Title II of the Act.

§1508.7 Cumulative impact. "Cumulative impact" is the impact on the environment that results from the incremental impact of the action when added to other past, present, and reasonably foreseeable future actions regardless of what agency (federal or nonfederal) or person undertakes such other actions. Cumulative impacts can result from individually minor but collectively significant actions taking place over a period of time.

§1508.8 Effects. "Effects" include:

1. Direct effects, which are caused by the action and occur at the same time and place.
2. Indirect effects, which are caused by the action and are later in time or farther removed in distance, but are still reasonably foreseeable. Indirect effects may include growth-inducing effects and other effects related to induced changes in the pattern of land use, population density, or growth rate and related effects on air and water and other natural systems, including ecosystems.

Effects and impacts as used in these regulations are synonymous. Effects include ecological (such as the effects on natural resources and on the components, structures, and functioning of affected ecosystems), aesthetic, historic, cultural, economic, social, or health, whether direct, indirect, or cumulative. Effects may also include those resulting from actions which may have both beneficial and detrimental effects, even if on balance the agency believes that the effect will be beneficial.

§1508.9 Environmental assessment. "Environmental assessment":

1. Means a concise public document for which a federal agency is responsible that serves to:
   (a) Briefly provide sufficient evidence and analysis for determining whether to prepare an environmental impact statement or a finding of no significant impact.
   (b) Aid an agency's compliance with the Act when no environmental impact statement is necessary.
   (c) Facilitate preparation of a statement when one is necessary.
2. Shall include brief discussions of the need for the proposal, of alternatives as required by section 102(2)(E), of the environmental impacts of the proposed action and alternatives, and a listing of agencies and persons consulted.

§1508.10 Environmental document. "Environmental document" includes the documents specified in §1508.9 (environmental assessment), §1508.11 (environmental impact statement), §1508.13 (finding of no significant impact), and §1508.22 (notice of intent).

§1508.11 Environmental impact statement. "Environmental impact statement" means a detailed written statement as required by section 102(2)(C) of the Act.

§1508.12 Federal agency. "Federal agency" means all agencies of the federal government. It does not mean the Congress, the Judiciary, or the President, including the performance of staff functions for

the President in his Executive Office. It also includes, for purposes of these regulations, states and units of general local government and Indian tribes assuming NEPA responsibilities under section 104(h) of the Housing and Community Development Act of 1974.

§1508.13 Finding of no significant impact. "Finding of no significant impact" (FONSI) means a document by a federal agency briefly presenting the reasons why an action, not otherwise excluded (§1508.4), will not have a significant effect on the human environment and for which an environmental impact statement therefore will not be prepared. It shall include the environmental assessment or a summary of it and shall note any other environmental documents related to it [§1501.7(a)(5)]. If the assessment is included, the finding need not repeat any of the discussion in the assessment but may incorporate it by reference.

§1508.14 Human environment. "Human environment" shall be interpreted comprehensively to include the natural and physical environment and the relationship of people with that environment. [See the definition of "effects" (§1508.8).] This means that economic or social effects are not intended by themselves to require preparation of an environmental impact statement. When an environmental impact statement is prepared, and economic or social and natural or physical environmental effects are interrelated, then the environmental impact statement will discuss all of these effects on the human environment.

§1508.15 Jurisdiction by law. "Jurisdiction by law" means agency authority to approve, veto, or finance all or part of the proposal.

§1508.16 Lead agency. "Lead agency" means the agency or agencies preparing or having taken primary responsibility for preparing the environmental impact statement.

§1508.17 Legislation. "Legislation" includes a bill or legislative proposal to Congress developed by or with the significant cooperation and support of a federal agency but does not include requests for appropriations. The test for significant cooperation is whether the proposal is in fact predominantly that of the agency rather than another source. Drafting does not by itself constitute significant cooperation. Proposals for legislation include requests for ratification of treaties. Only the agency that has primary responsibility for the subject matter involved will prepare a legislative environmental impact statement.

§1508.18 Major federal action. "Major federal action" includes actions with effects that may be major and that are potentially subject to federal control and responsibility. Major reinforces but does not have a meaning independent of significantly (§1508.27). Actions include the circumstance in which the responsible officials fail to act and that failure to act is reviewable by courts or administrative tribunals under the Administrative Procedure Act or other applicable law as agency action.

1. Actions include new and continuing activities, including projects and programs entirely or partly financed, assisted, conducted, regulated, or approved by federal agencies; new or revised agency rules, regulations, plans, policies, or procedures; and legislative proposals (§§1506.8, 1508.17). Actions do not include funding assistance solely in the form of general revenue sharing funds, distributed under the State and Local Fiscal Assistance Act of 1972, 31 USC 1221 et seq., with no federal agency control over the subsequent use of such funds. Actions do not include bringing judicial or administrative civil or criminal enforcement actions.
2. Federal actions tend to fall within one of the following categories:
   - (a) Adoption of official policy, such as rules, regulations, and interpretations adopted pursuant to the Administrative Procedure Act, 5 USC 551 et seq.; treaties and international conventions or agreements; formal documents establishing an agency's policies that will result in or substantially alter agency programs.
   - (b) Adoption of formal plans, such as official documents prepared or approved by federal agencies that guide or prescribe alternative uses of federal resources, upon which future agency actions will be based.
   - (c) Adoption of programs, such as a group of concerted actions to implement a specific policy or plan; systematic and connected agency decisions allocating agency resources to implement a specific statutory program or executive directive.
   - (d) Approval of specific projects, such as construction or management activities located in a defined geographic area. Projects include actions approved by permit or other regulatory decision as well as federal and federally assisted activities.

§1508.19 Matter. "Matter" includes for purposes of part 1504:

1. With respect to the Environmental Protection Agency, any proposed legislation, project, action, or regulation as those terms are used in section 309(a) of the Clean Air Act (42 USC 7609).
2. With respect to all other agencies, any proposed major federal action to which section 102(2)(C) of NEPA applies.

§1508.20 Mitigation. "Mitigation" includes:

1. Avoiding the impact altogether by not taking a certain action or parts of an action.
2. Minimizing impacts by limiting the degree or magnitude of the action and its implementation.
3. Rectifying the impact by repairing, rehabilitating, or restoring the affected environment.

4. Reducing or eliminating the impact over time by preservation and maintenance operations during the life of the action.
5. Compensating for the impact by replacing or providing substitute resources or environments.

§1508.21 NEPA process. "NEPA process" means all measures necessary for compliance with the requirements of section 2 and title I of NEPA.

§1508.22 Notice of intent. "Notice of intent" means a notice that an environmental impact statement will be prepared and considered. The notice shall briefly:

1. Describe the proposed action and possible alternatives.
2. Describe the agency's proposed scoping process including whether, when, and where any scoping meeting will be held.
3. State the name and address of a person within the agency who can answer questions about the proposed action and the environmental impact statement.

§1508.23 Proposal. "Proposal" exists at that stage in the development of an action when an agency subject to the Act has a goal and is actively preparing to make a decision on one or more alternative means of accomplishing that goal and the effects can be meaningfully evaluated. Preparation of an environmental impact statement on a proposal should be timed (§1502.5) so that the final statement may be completed in time to be included in any recommendation or report on the proposal. A proposal may exist in fact as well as by agency declaration that one exists.

§1508.24 Referring agency. "Referring agency" means the federal agency that has referred any matter to the Council after a determination that the matter is unsatisfactory from the standpoint of public health or welfare or environmental quality.

§1508.25 Scope. Scope consists of the range of actions, alternatives, and impacts to be considered in an environmental impact statement. The scope of an individual statement may depend on its relationships to other statements (§§1502.20 and 1508.28). To determine the scope of environmental impact statements, agencies shall consider three types of actions, three types of alternatives, and three types of impacts. They include:

1. Actions (other than unconnected single actions) which may be:
   - (a) Connected actions, which means that they are closely related and therefore should be discussed in the same impact statement. Actions are connected if they:
     - (i) Automatically trigger other actions which may require environmental impact statements.
     - (ii) Cannot or will not proceed unless other actions are taken previously or simultaneously.
     - (iii) Are interdependent parts of a larger action and depend on the larger action for their justification.
   - (b) Cumulative actions, which when viewed with other proposed actions have cumulatively significant impacts and should therefore be discussed in the same impact statement.
   - (c) Similar actions, which when viewed with other reasonably foreseeable or proposed agency actions have similarities that provide a basis for evaluating their environmental consequences together, such as common timing or geography. An agency may wish to analyze these actions in the same impact statement. It should do so when the best way to assess adequately the combined impacts of similar actions or reasonable alternatives to such actions is to treat them in a single impact statement.
2. Alternatives, which include:
   - (a) No action alternative.
   - (b) Other reasonable courses of actions.
   - (c) Mitigation measures (not in the proposed action).
3. Impacts, which may be direct, indirect, cumulative.

§1508.26 Special expertise. "Special expertise" means statutory responsibility, agency mission, or related program experience.

§1508.27 Significantly. "Significantly" as used in NEPA requires considerations of both context and intensity:

1. Context. This means that the significance of an action must be analyzed in several contexts, such as society as a whole (human, national), the affected region, the affected interests, and the locality. Significance varies with the setting of the proposed action. For instance, in the case of a site-specific action, significance would usually depend on the effects in the locale rather than in the world as a whole. Both short- and long-term effects are relevant.
2. Intensity. This refers to the severity of impact. Responsible officials must bear in mind that more than one agency may make decisions about partial aspects of a major action. The following should be considered in evaluating intensity:
   - (a) Impacts that may be both beneficial and adverse. A significant effect may exist even if the federal agency believes that on balance the effect will be beneficial.
   - (b) The degree to which the proposed action affects public health or safety.
   - (c) Unique characteristics of the geographic area such as proximity to historic or cultural resources, parklands, prime farmlands, wetlands, wild and scenic rivers, or ecologically critical areas.

(d) The degree to which the effects on the quality of the human environment are likely to be highly controversial.
(e) The degree to which the possible effects on the human environment are highly uncertain or involve unique or unknown risks.
(f) The degree to which the action may establish a precedent for future actions with significant effects or represents a decision in principle about a future consideration.
(g) Whether the action is related to other actions with individually insignificant but cumulatively significant impacts. Significance exists if it is reasonable to anticipate a cumulatively significant impact on the environment. Significance cannot be avoided by terming an action temporary or by breaking it down into small component parts.
(h) The degree to which the action may adversely affect districts, sites, highways, structures, or objects listed in or eligible for listing in the National Register of Historic Places or may cause loss or destruction of significant scientific, cultural, or historical resources.
(i) The degree to which the action may adversely affect an endangered or threatened species or its habitat that has been determined to be critical under the Endangered Species Act of 1973.
(j) Whether the action threatens a violation of federal, state, or local law or requirements imposed for the protection of the environment.

(43 FR 56003, November 29, 1978; 44 FR 874, January 3, 1979)

§1508.28 Tiering. "Tiering" refers to the coverage of general matters in broader environmental impact statements (such as national program or policy statements) with subsequent narrower statements or environmental analyses (such as regional or basinwide program statements or ultimately site-specific statements) incorporating by reference the general discussions and concentrating solely on the issues specific to the statement subsequently prepared. Tiering is appropriate when the sequence of statements or analyses is:

1. From a program, plan, or policy environmental impact statement to a program, plan, or policy statement or analysis of lesser scope or to a site-specific statement or analysis.
2. From an environmental impact statement on a specific action at an early stage (such as need and site selection) to a supplement (which is preferred) or a subsequent statement or analysis at a later stage (such as environmental mitigation). Tiering in such cases is appropriate when it helps the lead agency to focus on the issues that are ripe for decision and exclude from consideration issues already decided or not yet ripe.

The following additional terms are defined solely for the purpose of implementing the supplemental procedures provided by 21 CFR Part 25 and are not necessarily applicable to any other statutory or regulatory requirements:

Agency means the Food and Drug Administration (FDA).

Emissions requirements specify the limits on the quantities of pollutants allowed to be released into the workplace and the area outside a production site or facility. These requirements or standards are set and enforced by local, state, and federal government components, e.g., Environmental Protection Agency, Occupational Safety and Health Administration.

Environmental assessment technical guide means that technical guidance prepared by FDA and intended to assist applicants and petitioners in preparing their environmental assessments.

Production includes manufacture, processing, and packaging operations for FDA-regulated articles that are the subject of proposed actions.

Responsible agency official means the agency decision maker designated in 21 CFR Part 5, the Commissioner of Food and Drugs, or the Commissioner's designated representative.

Toxic substance means any substance that is harmful to some biological mechanism or system. Although it is recognized that any substance may produce damage to biological mechanisms or systems under specific conditions, for the purpose of these regulations, a substance is considered to be a toxic substance if it is harmful to appropriate test organisms at expected environmental concentrations even though it may be without effect to humans or other organisms at these concentrations and may even be used by humans because of its toxic properties. A substance is considered toxic in the environment if the maximum concentration of the substance at any point in the environment, i.e., either at any point of entry or any point where higher concentrations are expected as a result of bioaccumulation or other types of concentration processes, exceeds the concentration of the substance that causes any adverse effect in a test organism species (minimum effect level) or exceeds 1/100 or the concentration that causes mortality in 50% of the test organism species, whichever concentration is less.

The following acronyms are used in this part:

ACO, air contaminant objective
ANDA, abbreviated new drug application
CAS, Chemical Abstracts Service
CEQ, Council on Environmental Quality
CFR, Code of Federal Regulations
CGMP, current good manufacturing practice
EA, environmental assessment
EIS, environmental impact statement
EPA, Environmental Protection Agency
FDA, Food and Drug Administration
FD&C Act, Federal Food, Drug, and Cosmetic Act
FIFRA, Federal Insecticide, Fungicide, and Rodenticide Act

FONSI, finding of no significant impact
GLP, good laboratory practice
GRAS, generally recognized as safe
HHS, Department of Health and Human Services
IDE, investigational device exemption
INAD, notice of claimed investigational exemption for new animal drug
IND, investigational new drug application
NADA, new animal drug application
NDA, new drug application
NEPA, National Environmental Policy Act of 1969
OTC, over-the-counter
PMA, premarket approval application
PDP, product development protocol
TSCA, Toxic Substances Control Act
USC, United States Code.

(50 FR 16656, April 26, 1985, as amended at 54 FR 9038, March 3, 1989)

Additional acronyms used in this chapter include the following:

ASTM, American Society for Testing and Materials
BCP, bulk chemical plant
BOD, biological oxygen demand
COD, chemical oxygen demand
EEC, expected environmental concentration
EPA, Environmental Protection Agency
MAC, maximum allowable concentration
MEEC, maximum expected emitted concentration
MSDS, material safety data sheet
NCE, new chemical entity
OECD, Organization for Economic Cooperation and Development
PEL, permissible exposure limit
POTW, publicly owned treatment works
TAD, technical assistance document
ThOD, theoretical oxygen demand
TLV, threshold limit value

FDA's Responsibility under NEPA

NEPA provides FDA with supplementary authority to consider environmental factors in its decision making, and it establishes a broad mandate for federal agencies to protect and enhance all aspects of the human environment. The Act declares a national environmental policy and states goals and a method for accomplishing this policy and these goals through action-forcing provisions. The heart of the action-forcing provisions is in NEPA §102(2)(C), which requires that, before taking an action, federal agencies must prepare a detailed environmental impact statement for every major action significantly affecting the quality of the human environment.

FDA began implementing NEPA in 1971 with the preparation of an EIS on modifications for its toxicological research facility in Arkansas. An environmental policy staff was established in the FDA Commissioner's office in 1972 to provide direction and leadership for FDA's new environmental responsibilities.

FDA's first NEPA regulations were published in the Federal Register on March 15, 1973 [38 FR (Federal Register) 7001, 1973], which were eventually promulgated as 21 CFR Part 25. The rule listed agency actions requiring environmental review and actions that were exempt from the environmental review process. In 1974, the NEPA operating function was transferred to a focal point in each operating component of FDA to effectively implement the new regulations. The focal point would be staffed with individuals who are environmental generalists having interest and motivation necessary to conduct environmental assessments and to raise appropriate questions regarding the adequacy of environmental documents; have a thorough understanding of the provisions of NEPA, CEQ guidelines, DHHS compliance procedures, and FDA regulations; and are not overburdened with other responsibilities to the extent that environmental matters are not given the necessary consideration.

FDA issued revised NEPA-implementing regulations on April 15, 1977 to exempt conditionally certain classes of actions and explain in greater detail the types of information required for environmental documents (42 FR 19985). The revised documents also responded to 1973 guidelines issued by the CEQ. CEQ issued revised guidelines in 1978 as regulations (40 CFR 1500–1508), which required FDA to publish proposed revisions of 21 CFR Part 25 (50 FR 15536, April 26, 1985).

FDA's Implementation of the CEQ Regulations

FDA's program for implementing NEPA includes the following:

- Adoption of necessary procedures to supplement the CEQ regulations [40 CFR 1507.3(a)]. FDA's final NEPA-implementing procedures (21 CFR Part 25) were published in the Federal Register in 1985.
- Specification in agency procedures (21 CFR Part 25) of the criteria for and identification of the typical classes of actions that normally require an EIS, are categorically excluded, or normally require an environmental assessment but not necessarily an EIS [40 CFR 1507.3(b)(2)]. There are no categories of decisions made by the FDA that routinely require the preparation of an EIS. An EIS is prepared if the evaluation of the data in an EA shows that a proposed action may significantly affect the quality of the human environment (21 CFR 25.21). Before the FDA makes a decision on a requested action, the agency requires an EA to be prepared and submitted for review unless the action qualifies for a categorical exclusion (21 CFR 25.23 and 25.24). An action that qualifies for a categorical exclusion does not require the preparation of either an EA or an EIS because the action will not result in the introduction of any toxic substance into the environment, meets specific criteria that are intended to ensure that it will not cause a significant environmental effect, or can be

classified as routine maintenance or minor leasing or construction activities conducted for or contracted for by FDA. However, claims for categorical exclusion must be accompanied by documentation supporting the claim.

• Preparation or oversight of the preparation of the required environmental documents—FONSI, EA, and EIS (draft, final, and supplemental) (40 CFR §§1501.3, 1501.4, and 1502). FDA's procedures and formats for preparing and submitting environmental documents are provided in 21 CFR Part 25 (§§25.30, 25.31, 25.32, 25.33, 25.34, 25.40, and 25.42). Most FDA actions, including those initiated by industry sponsors (i.e., pharmaceutical companies), do not require a detailed EA; they belong to one or more classes of actions that are categorically excluded from environmental review under appropriate regulatory provisions (21 CFR §25.24) or qualify for an abbreviated EA [21 CFR §25.31a(b)].

• Provision for public involvement in the environmental review and decision-making process (40 CFR §1506.6). Procedures for involving the public are specified in 21 CFR §§25.41(b) and 25.42(b). The Food and Drug Administration, Center for Drug Evaluation and Research implements its procedures for making EAs and FONSIs available to the public, e.g., by periodically (two to four times per year) publishing a list of available EAs and FONSIs in the Federal Register. If, after publishing the list of EAs and FONSIs, new information becomes available to FDA suggesting that a human drug approval may significantly affect the environment, the agency may consider the need for an EIS under the retroactive provision (21 CFR §25.25).

• Adoption of procedures for incorporating environmental considerations as an integral part of the decision-making process (40 CFR §1505.1). FDA's procedures for incorporating environmental factors in the decision-making process are specified (21 CFR §25.40 and §25.42(b)(5)(i)). It is the policy of the agency that environmental considerations are evaluated early in the review process and concurrently with other considerations.

• Assistance for applicants required to submit environmental information by outlining the types of information required [40 CFR §1506.5(a)]. The information required to be submitted by a sponsor to claim a categorical exclusion is specified in the regulations (21 CFR §25.23). 21 CFR Section 25.31a provides formats for full and abbreviated EAs for environmental actions. FDA prepared an assistance document in 1987 entitled Environmental Assessment Technical Handbook (FDA, 1987), which describes approaches for the efficient preparation of environmental assessments, protocols for environmental tests, and the interpretation of results of such tests. FDA also encourages meetings with applicants (e.g., pharmaceutical companies) to discuss preparation of NEPA documents.

• Ensurance of the completeness and accuracy of information provided by applicants in EAs and claims for categorical exclusion [40 CFR §1506.5(a) and (b)]. FDA procedures for meeting these requirements are specified in 21 CFR §§25.22(d), 25.32(c), and 25.41(c). For actions requiring a FONSI, the agency must ensure that information in an applicant's EA is complete and accurate, and the agency must prepare the FONSI. The agency officials responsible for preparing and approving the FONSI must sign the document to show that they approve the conclusions not to prepare an EIS. The general approach for ensuring data validity is to recommend compliance with good laboratory practice procedures in 21 CFR Part 58, conduct periodic audits of laboratories, and require reports to contain certification. Manufacturing site compliance may be monitored during FDA's site inspections.

• Provision in agency procedures for extraordinary circumstances in which a normally excluded action may have a significant environmental impact (40 CFR §1508.4). FDA [21 CFR §25.25(b)] requires that the applicant submit an EA as described in 21 CFR 25.31 for the approval, with notification of this requirement provided to the applicant in writing.

• Utilization of a systematic, interdisciplinary approach that will ensure the integrated use of natural and social sciences in planning and decision making (40 CFR §1507.2). Most FDA actions with potential for adverse environmental effects involve the introduction of chemicals (drugs) into the environment. Utilization of FDA personnel trained in environmental chemistry, environmental engineering, ecotoxicology, and ecology in the review process is paramount.

• Identification of methods and procedures required by EPA [40 CFR §1507.2(b), Section 102(2)(B)] to ensure that presently unquantified environmental accommodations and values may be given appropriate consideration. Environmental considerations (such as potential toxic impact on ecosystems, use of energy and resources, effects on national historic sites) are an integral part of FDA's decision-making process.

• Provision of comments on EISs and environmental documents prepared by other federal agencies in cases in which FDA has jurisdiction by law or special expertise or is authorized to develop and enforce environmental standards [40 CFR §1507.2(c)]. FDA is continuously involved in this activity. An example is the review of the draft EIS for taxol prepared by the U.S.D.A. Forest Service, which FDA used to discuss natural resource issues in the preparation of the FONSI for drug approval (1992).

• The study, development, and description of alternatives so that courses of action can be recommended by the agency regarding any proposal that involves unresolved conflicts concerning alternative uses of available resources [40 CFR §1507.2(c)]. FDA routinely considers the impact of its actions on natural resources, including actions that require an EA or an EIS.

• Recognition of the global and long-range character of environmental problems and, where consistent with the foreign policy of the United States, the provision of appropriate support to initiatives, resolutions, and programs designed to maximize international cooperation in anticipating and preventing a decline in the quality of the world environment [40 CFR §1507.2(f)]. FDA's procedures for evaluating the environmental effects of its actions abroad are provided in 21 CFR §25.50. FDA routinely assesses the impact of manufacturing FDA-regulated products in other countries and the impact of actions that could affect the global commons.

- Initiation and utilization of ecological information in the planning and development of resource-oriented projects [40 CFR §1507.2(e)]. Although most FDA actions do not involve resource-oriented projects, FDA does use ecological information in its planning and decision making.
- Provision, to states, counties, municipalities, institutions, and individuals of advice and information useful in restoring, maintaining, and enhancing the quality of the environment [40 CFR §1507.2(f)]. FDA meets this requirement by allowing individuals and nonfederal public entities to participate in the environmental review process, review and comment on FDA's NEPA documents, and obtain copies of FDA's environmental procedures and assistance documents.
- Provision of assistance to the Council on Environmental Quality [40 CFR §1507.2(f)].

Organization of NEPA Function in the Center for Drug Evaluation and Research (CDER)

The center's environmental assessment officer (EAO) is responsible for coordinating NEPA functions. This includes the Offices of Drug Evaluation I and II, Pilot Drug, Office of Generic Drugs, and Office of Compliance. The individual review chemist or microbiologist and his or her immediate supervisor are responsible for evaluating information contained in environmental assessments for determining whether exclusions are warranted and for completeness of the document. The environmental assessment is forwarded to the EAO for review. Based on the EAO's conclusion as to whether the proposed action may significantly affect the quality of the human environment, the EAO either prepares a FONSI or initiates a notice of intent to prepare an EIS.

Types of Agency (FDA) Action Requiring Environmental Consideration and When an EA or an EIS Is Required

All agency actions are subject to environmental consideration. Actions are individually examined for potential environmental impact unless excluded as a class by categorical exclusion. An EIS is prepared for agency actions when the evaluation of the data in an EA leads to a finding that a proposed action may significantly affect the quality of the human environment. FDA may consider the need to prepare an EIS for an existing regulation, approval, or other action, whether or not previously subject to environmental analysis, when there is new information before the agency that suggests that the action may significantly affect the quality of the human environment. Actions requiring the preparation of an environmental assessment include major recommendations or reports made to Congress on proposals for legislation in instances where the agency has primary responsibility for the subject matter involved; destruction or other disposition of articles condemned after seizure or whose distribution or use has been enjoined; destruction or other disposition of articles following detention or recall at agency request; disposition of FDA laboratory waste materials; intramural and extramural research support in whole or in part through contracts, other agreements, or grants; establishment by regulation of labeling requirements, a standard, or a monograph, amendments to, or an exemption or variance from, requirements of existing FDA regulations; approval of supplements to existing approvals of FDA-approved articles; withdrawal of existing approvals of FDA-approved articles; approval of new drug applications (NDAs) and abbreviated new drug applications (ANDAs) and actions on investigational new drug applications (INDs); approval of antibiotic applications; and action other than that listed above.

Adequacy of EA for the NDA

An EA adequate for approval is one that contains sufficient information to enable the agency to determine whether the proposed action may significantly affect the quality of the human environment. Failure to submit an adequate EA, if one is required for such an action, is sufficient grounds for FDA to refuse to file or approve the drug application.

There are no categories of FDA actions that routinely require the preparation of an EIS. An EIS is prepared if an evaluation of the data in an EA shows that a proposed action may significantly affect the quality of the human environment (21 CFR §25.21). Therefore, FDA actions ordinarily require an EA, unless they qualify for a categorical exclusion (as specified in 21 CFR §§25.23 and 25.24).

An action that qualifies for categorical exclusion does not require either an EA or an EIS because the action will not result in introduction of any toxic substance into the environment, meets specific criteria that are intended to ensure that it will cause no significant environmental effects, or is a routine maintenance or minor leasing or construction activities conducted for or contacted for by FDA. 21 CFR §25.24 lists the excluded actions along with any applicable criteria for the exclusion.

PERTINENT NEPA LEGAL ISSUES

The following litigation case emphasizes the importance of NEPA in FDA decision making: *Environmental Defense Fund v. Mathews*, 410F Supp. 336 (DDC 1976).

The Commissioner of the Food and Drug Administration issued a Federal Register notice (38 FR

7001, 1973) to promulgate its regulations implementing its obligations under NEPA. In recognition of the breadth of the NEPA mandate, the Commissioner declared that

> [T]he National Environmental Policy Act, as interpreted by the courts, amends the Federal Food, Drug, and Cosmetic Act (1938) to the extent that it requires the FDA to give full consideration without restrictions of time to all environmental issues relevant to FDA approval of food additive petitions, new drugs and new animal drugs.

In April 1975, FDA promulgated an amendment to this regulation that reads:

> A determination of adverse environmental impact has no legal or other regulatory effect and does not authorize the Commissioner to take or refrain from taking any action under the laws he administers. The Commissioner may take or refrain from taking action on the basis of a determination of an adverse environmental impact only to the extent that such action is independently authorized by the laws he administers [21 CFR 6.1(a)(3)]

This amendment was challenged by the Environmental Defense Fund, charging that the regulation promulgated in April 1975, as an amendment to FDA's existing regulations governing its obligations under NEPA, unlawfully limits the scope of those obligations in violation of NEPA.

On March 26, 1976, the District Court for the District of Columbia declared that the amending regulation was contrary to and in violation of NEPA and ordered FDA to withdraw the amending regulation. The court said, in part, that

> It appears clear to us that, contrary to defendants' contention, FDA's existing statutory duties under the FDCA [FD&C] and its other statutes are not in direct conflict with its duties under NEPA. The FDCA does not state that the listed considerations are the only ones which the Commissioner may take into account in reaching a decision. Nor does it explicitly require that product applications be granted if the specified grounds are met. It merely lists criteria which the Commissioner must consider in reaching his decision. In the absence of a clear statutory provision excluding consideration of environmental factors, and in light of NEPA's broad mandate that all environmental considerations be taken into account, we find that NEPA provides FDA with supplementary authority to base its substantive decisions on all environmental considerations including those not expressly identified in the FDCA and FDA's other statutes. This conclusion finds support in the legislative history, the precise statutory language, the holdings of the courts, and the construction adopted by other Federal agencies.
>
> This is not to say that NEPA requires FDA's substantive decisions to favor environmental protection over other relevant factors. Rather, it means that NEPA requires FDA to consider environmental factors in its decision-making process and supplement its existing authority to permit it to act on those considerations. It permits FDA to base a decision upon environmental factors, when balanced with other relevant considerations. Since the contested regulation prohibits FDA from acting on the basis of such environmental considerations, it is directly contrary to the letter and spirit of NEPA.

Subsequently, the agency revoked the regulation on May 28, 1976 (41 FR 21768) acknowledging that the court's decision represents a specific judicial ruling regarding the broad authority of the FDA under NEPA to act or to withhold action on the basis of a determination of an adverse environmental impact.

THE ENVIRONMENTAL ASSESSMENT

FDA requires applicants or petitioners to submit an environmental assessment with their application or petition (unless the claim of a categorical exclusion is made). FDA then reviews the NEPA document to determine whether FDA's action is categorically excluded, has no significant impact and that a FONSI can be prepared, or has identified significant environmental impacts and that an environmental impact statement needs to be prepared.

The environmental assessment is usually in the following format.

Format

21 CFR §25.31a Environmental assessment for proposed approvals of FDA-regulated products—Format 1.

(a) For proposed actions to approve food or color additives, drugs, biological products, animal drugs, and class III medical devices, and to affirm food substances as generally recognized as safe (GRAS), the applicant or petitioner shall prepare an environmental assessment in the following format:

Environmental assessment

1. Date:
2. Name of applicant/petitioner:
3. Address:
4. Description of the proposed action: Briefly describe the requested approval; need for the action; the locations where the products will be produced; to the extent possible, the locations where the products will be used and disposed of; and the types of environments present at and adjacent to those locations.
5. Identification of chemical substances that are the subject of the proposed action: Provide complete nomenclature, CAS Reg. No. (if available), molecular weight, structural formulae, physical description, additives, and impurities. This information is required to be adequate to allow accurate location of data about chemicals in the scientific literature and to allow identification of closely related chemicals.
6. Introduction of substances into the environment: For the site(s) of production: list the substances expected to be emitted; state the controls exercised; include a citation of, and statement of compliance with, applicable emissions requirements (including occupational) at the federal,

state, and local level; and discuss the effect the approval of the proposed action will have on compliance with current emissions requirements at the production site(s). Through use of calculations and/or direct measures, estimate to the extent possible the quantities and concentrations of substances expected to enter the environment as a result of use and/or disposal of products affected by the action.

7. Fate of emitted substances in the environment: Predict environmental concentrations of and exposures to substances entering the environment as a consequence (direct or indirect) of the use and/or disposal of the products affected by the action for the following environmental compartments, including consideration of the major environmental transport and transformation processes involved:

   (a) Air—taking into account, to the extent possible, factors such as volatilization, photochemical and chemical degradation, rainout, and dispersion;
   (b) Freshwater, estuarine, and marine ecosystems—taking into account, to the extent possible, factors such as chemical and biological degradation, exchange between the water column and sediments via sorption/desorption and biological processes, accumulation in animals, plants, and other organisms, introductions due to rainfall, and losses due to volatilization;
   (c) Terrestrial ecosystems—taking into account, to the extent possible, factors such as chemical and biological degradation, sorption/desorption and leaching in soils, accumulation in animals and plants, introductions due to rainfall, losses due to volatilization, and entry into groundwater.

8. Environmental effects of released substances: Given the information developed on the introduction (item 6) and fate (item 7) of substances that would be released as a consequence of the use and/or disposal of the products affected by the action, use any relevant toxicological data or other appropriate measures to predict, to the extent applicable, effects on animals, plants, humans, other organisms, and effects at the ecosystem level in each of the environmental compartments listed in item 7.
9. Use of resources and energy: Specify the natural resources, including land use, minerals, and energy, required to produce, transport, use, and/or dispose of a given amount of any product that is the subject of the action, including the resources and energy required to dispose of wastes generated from production, use, and/or disposal. Effects, if any, on endangered or threatened species and on property listed in or eligible for listing in the National Register of Historic Places must be discussed.
10. Mitigation measures: Describe measures taken to avoid or mitigate potential adverse environmental impacts associated with the proposed action.
11. Alternatives to the proposed action: If potential adverse environmental impacts have been identified for the proposed action, describe in detail the environmental impact of all reasonable alternatives to the proposed action (including no action, and including measures that FDA or another government agency could undertake as well as those the applicant/petitioner would undertake). Describe particularly those alternatives that will enhance the quality of the environment and avoid some or all of the adverse environmental impacts of the proposed action. Discuss the environmental benefits and risks of the proposed action. Discuss the environmental benefits and risks of each alternative.
12. List of preparers: The persons preparing the assessment together with their qualifications (expertise, experience, professional disciplines) shall be listed. Persons and agencies consulted shall also be listed.
13. Certification: The undersigned official certifies that the information presented is true, accurate, and complete to the best of the knowledge of the firm or agency responsible for preparation of the environmental assessment.
    (Date)
    (Signature of responsible official)
    (Title)
14. References: List complete citations for all referenced material. Copies of referenced articles not generally available should be attached.
15. Appendices:

    (a) Data summary charts (e.g., structural formula, vapor pressure, water solubility, *n*-octanol/water partition coefficient, biodegradation half-life, LC50 for each species tested).
    (b) Test reports [for each experiment: research objective, experimental design and procedure, all data relevant to interpretation of the test result given in item [15(a), sample calculations, and statistical analyses].

Detailed Guidance

The following discussions provide more detailed descriptions of examples of acceptable methods for meeting the FDA's requirements for pivotal sections of the EA. Italicized text refers to the regulation in question. These descriptions are elaborations of guidance that was developed by the Pharmaceutical Manufacturers Association in collaboration with FDA and presented to the industry in July 1991 (PMA, 1991). Note that the EA reflects the state of knowledge at the time the NDA is submitted.

Item 6: Introduction of Substances into the Environment

Introduction from Production: Background

The point in the manufacturing process at which the environmental evaluation begins (item 6) depends on what is being asked for in the NDA. The FDA does not require that the environmental impact of the production of raw materials used in the drug manufacturing process be addressed. However, for a material to be defined as a raw material, it must be commercially available to all parties in the open marketplace. For example, if the NDA includes a proposal to purchase a commercially available intermediate product as a raw material from an outside vendor that will serve as the starting point in the manufacturing process of the NDA applicant, only the manufacturing operations of the NDA applicant need to be addressed in the EA. If, on the other hand, the purchased intermediate is a proprietary product, it loses its classification as a commercially available raw material, and the operations of both the applicant and the intermediate product manufacturer must be evaluated in the EA.

There are some cases in which the environmental impact of providing raw materials must be included in the environmental assessment. Examples include assessing the impact that harvesting raw materials derived from natural resources (e.g., bark of a tree) may have on threatened or endangered species. This requirement extends to both domestic and international resources including the African and Amazon rain forests.

Many NDA applications will propose to synthesize the drug substance at a bulk chemical plant (BCP) manufacturing site and to formulate and package the drug product at a pharmaceutical finishing and packaging manufacturing plant. The environmental evaluation must be done for both locations. Other manufacturing scenarios may propose to use a third party contractor as an alternative. An environmental evaluation will still need to be performed for the third party location, even if the third party contractor is never utilized.

Each of the topics below must be covered in detail for the environmental assessment to pass the adequacy review. Failure to provide an adequate environmental assessment is sufficient grounds for the FDA to refuse to file or to approve the application or petition [21 CFR 25.22(b)]. Since the process information provided in this section is of a confidential nature, the submission should be so constructed that all confidential information is presented as separate attachments, with only summary information provided in the actual EA itself. This facilitates the generation of the nonconfidential (freedom of information or FOI) copy that is required by law to be available to the public after drug approval.

For the site(s) of production: list the substances expected to be emitted;

The example provided here is information related to the manufacture of the drug substance at a bulk chemical plant. The evaluation for the drug product formulation and packaging facility is handled in a similar manner.

The first task for generating the incremental waste load is to obtain the 5-yr marketing plan and identify which of these years has the largest production projection. An example of how to report the 5-yr marketing data is shown in Table 1.

Raw materials, process descriptions, batch sizes, and production times then need to be described and summarized. Engineering calculations (or actual measurements when available) are used to estimate the quantity of each chemical and intermediate product used or generated in the various production steps of manufacturing the drug substance. Appropriate allowances should be taken for on-site recycling and reuse activities. The engineering evaluation needs to assign each waste stream to an appropriate disposition route, such as air emissions from process point sources and/or from on-site incinerators or other thermal destruction equipment; wastewater discharges leaving the production building directly or leaving a site biotreatment facility; hazardous wastes requiring destruction or disposal either on site or by outside contractors; and nonhazardous solid wastes transported to off-site recycling or landfill facilities. These estimations for waste leaving the site are to be added to the current waste levels and compared to the current permit limits at the manufacturing facility to demonstrate that a permit violation will not result if the FDA approves the NDA. Appropriate adjustments should be made for current operations that are to be discontinued (e.g., to make room for manufacturing the NDA approved drug substance). As a starting point to environmental evaluation, the address of the specific manufacturing site should be provided along with a short general description of the manufacturing process.

The following tables constitute one example of how the required information can be summarized for regulatory review. Alternative approaches may be considered. These tables should be assembled in a separate confidential attachment in the EA so that they may more easily be excluded from the nonconfidential (FOI) copy of the EA. The process flowchart, the production batch size of the drug substance and its intermediates, and the process time required to manufacture a batch should also be included in

Table 1. Five-year marketing plan for the (insert name here) drug substance

Year	Projected volume
1st year	... kg
2nd year	.
3rd year	.
4th year	.
5th year	... kg

Table 2. Input chemicals for production of (insert name of drug substance)

Chemical names	CAS number
Chemical 1	...
Chemical 2	Mixture
Chemical 3	N/A[a]
⋮	⋮
Chemical *n*	...

[a]N/A = Not Available

the confidential attachment. The discussion sections around the tables should be included in the body of the environmental assessment itself.

Table 2 is a list of all the input chemicals used in the manufacturing process. Chemical Abstract Services Registry Numbers (CAS numbers) are listed when known and mixtures are noted as such. Input chemicals include those used in both the chemical reaction and in the product purification steps, and should include any water used in the process.

Table 3A. Process feed streams for production of (insert name of drug substance)

Chemical	Weight (kg)
Stage 1	
Chemical 1	...
Chemical 2	.
Chemical 3	.
⋮	...
Chemical *n*	...

Table 3B. Process feed streams for production of (insert name of drug substance)

Chemical	Weight (kg)
Stage 2	
Chemical 1	...
Chemical 2	.
Chemical 3	.
⋮	...
Chemical *n*	...

Table 3*n*. Process feed streams for production of (insert name of drug substance)

Chemical	Weight (kg)
Stage *n*	
Chemical 1	...
Chemical 2	.
Chemical 3	.
⋮	...
Chemical *n*	...

Table 3 (designated as 3A, 3B, etc.) should include the process feed streams for each stage of production.

Table 4 should include a list of all permitted effluent chemicals from production, again with their CAS numbers. The CAS numbers allow FDA reviewers to more easily locate information on the chemical in the literature.

A summary of the permitted waste streams generated by each stage of the production process should be provided. Table 5 (designated as 5A, 5B, etc.) provides an example of an acceptable format for this information.

Waste stream composition should be provided, preferably on the basis of one lot of drug substance. Table 6 presents a typical example.

Estimated material balance information for the production of the drug substance on both a kg/kg of final product basis and on a per year basis should be provided. Table 7 provides one type of format for presenting this type of information.

state the controls exercised;

Once the material balance for the production process has been presented, the material balances around each effluent disposition method must be provided. The issue here is to provide calculations showing that introduction of the process into the production facility will not cause the facility to go out of compliance with its discharge permits. Disposition of wastes through a site biotreatment facility, a site incinerator, gas scrubbing systems, sanitary and hazardous landfills, etc. should be presented as appropriate. Also included should be any direct discharge of materials such as innocuous gases. Note that many times permit levels are in terms of composite parameters such as chemical oxygen demand (COD) or biochemical oxygen demand (BOD) rather than as specific chemical entities. Information on the specific chemicals will be needed in order to calculate their contribution to the composite parameter. It should also be noted that the calculated concentration of specific permitted chemicals in the aqueous effluent after treatment provides an estimate of the maximum

Table 4. Permitted effluent chemicals from production of (insert name of drug substance)

Chemical	CAS number
Chemical 1	.
Chemical 2	.
Chemical 3	.
⋮	⋮
Chemical *n*	.

Table 5. Waste stream summary for production of (insert name of drug substance here)

Stage	Product code	Waste number	kg/kg final product	Destination[a]	Description
1	AAA	1	...	I	Spent solvent waste
		2	...	R	Recoverable solvent waste
2	XXX	3	...	B	Dilute aqueous salt
		4	...	D	Spent filter aid
		5	...	V	Innocuous gas
.	.	.	...	.	.
.	.	.	...	.	.
n	*NNN*	*n*	...	.	.

[a]Destination code: I, site incinerator; R, on-site recovery; B, site bioplant; D, disposal by off-site contractor; V, vent to atmosphere.

Table 6. Waste stream composition for production of one lot of (insert name of drug substance here)

Stream number	Description	Weight (kg)	Stream number	Description	Weight (kg)
	Stage 1			Stage 3	
1	Spent solvent A	...	6	Aqueous mother liquors	.
	Solvent	...		Water	
	By-products	...		Inorganic salts	
2	Spent aqueous caustic	...	.	...	.
	Water	...		...	.
	Sodium hydroxide	...		...	.
	Tar residues	...	*n*	*nnn*	
3	Spent filter aid	...		.	
			.	...	.
	Stage 2			...	.
4	Solvent distillate	...		Stage *N*	
	Solvent	...			
	Water	...	.	...	.
	Other	...		...	.
5	Carbon dioxide off-gas	...		...	.

Table 7. Material balance for (insert name of drug substance here)[a]

Chemical	Input		Output		Waste stream number
	kg/kg intermediate	kg/yr	kg/kg final product	kg/yr	
Chemical 1	...	...	...	...	3
Chemical 2	...	...	...	...	1,4,6,...,*n*
.	.	.	.	.	.
.	.	.	.	.	.
.	.	.	.	.	.
Chemical *n*	...	...	...	...	1,2,*l*,*m*,*n*

[a]Footnotes should be provided to give any additional relevant information. For example, yearly production basis for the mass balance should be provided.

expected emitted concentration (MEEC) of that substance via that route into the environment. The basis and references for BOD, COD, ThOD (theoretical oxygen demand) calculations should be provided in the footnotes to the table, along with other relevant information such as treatment plant flow. For example, "Theoretical oxygen demand (ThOD) was calculated from the molecular structure, BOD is assumed equal to ThOD, or based on actual experimental or literature data, COD = xx% of ThOD." Where substances have reported BODs, some biodegradation of the substance can be assumed, and the

Table 8. Disposition of major effluents in aqueous effluent from site biotreatment at (insert site name here) for (insert name of drug substance here) production process

Component	kg/d to site bioplant	mg/L due to process	BOD (mg/L)	COD (mg/L)	ThOD (mg/L)	Regulated component (mg/L)	Footnotes
Chemical 1	...	...	...	...	...	.	*a*
Chemical 2	...	...	...	...	...	...	*b,d*
.	.	.	.	.	.	.	.
.	.	.	.	.	.	.	.
Chemical *n*	...	...	...	...	...	.	*c*
Total			AA.A	BB.B		QQ.Q	

concentration of the compound in the final effluent can be calculated as some fraction of the original concentration. The contributions of other compounds, such as chloride from chlorine-containing compounds, to specific discharge parameters should be calculated from their molecular structures. Tables 8 through 10 provide typical formats for presenting these data.

Air emissions are normally calculated for emissions entering the atmosphere at the end of pipe following pollution control devices. For the pharmaceutical industry, these may be as simple as condensers, scrubbers, carbon, and HEPA filters or they may be as complex as catalytic thermal oxidation units. Table 9 is a list of chemicals emitted into the atmosphere from process point sources.

The individual point sources summarized in Table 9 should be clearly stated along with the pollution controls used on each of these point sources. The performance of air pollution control equipment is both site and process specific. Table 9 should reflect the performance efficiencies for each control device.

Table 9. Estimated emissions from the manufacturing of (insert name of drug substance here)

Process air emissions	Estimated air emissions
Chemical 1	... kg/batch
Chemical 2	... kg/batch
.	...
.	...
.	...
Chemical *n*	... kg/batch

Table 10. Air pollution control equipment in use at the (insert name here) facility

Control equipment	Efficiencies (%)
Process condensers (at 8°C)	85
Vent condensers (at −20°C)	85
Baghouse filters	95
HEPA filters	99.9
Mist eliminators	99

The control equipment and tabulated efficiencies are provided in Table 10 only as a guide.

Additional control measures used to reduce air emissions should also be described. These may include the use of closed containers to transport dry material to various vessels and vacuum loading of inert ingredients into the process equipment.

Contributions to the site's air emissions will also come about from emissions from site incinerators, if any. The inputs and outputs around such incinerators for the drug process wastes should be presented, as illustrated in Table 11. Footnotes should provide such information as destruction efficiency of incinerator, removal efficient of scrubbers, and relevant basis for calculations. The flue gas discharge rate should also be included.

Liquid hazardous wastes may be containerized and transported off site to a permitted and government-approved liquid hazardous waste incinerator for thermal destruction. Solid hazardous materials may be containerized and sent off site to a cement kiln for thermal destruction. This information should be summarized as shown in Table 12.

Some solid wastes may be handled on site. For example, sludge from a site biotreatment plant may be dewatered and disposed of by on-site incineration. Details of these types of operations should be presented in discussion format.

include a citation of, and statement of compliance with applicable emissions requirements (include occupational) at the federal, state, and local level.

The environmental permits for each site need to be identified for each applicable environmental matrix. Copies of the permits do not need to be included in the appendices but they should be available for agency review on request. As legislation and regulatory requirements require new permits to be issued, it is not necessary to advise the FDA of these changes until either requested or audited. Table 13 is constructed to provide the FDA with administrative information related to site environmental permits.

Table 14 complements Table 13 by listing the compliance criteria for each permit and the reporting mechanism used to report data back to the issuing agency. Copies of these reports should be maintained in company files until either requested or audited by

Table 11. Disposition of major effluents in site incinerator at (insert site name here) for (insert name of drug substance here) production process

Component	kg/d to treatment	Process contribution to effluent (μg/L)	Particulate contribution from process (kg/d)	Footnotes
Chemical 1	...	...	...	*a*
Chemical 2	...	...	...	*b*
.	.	.	.	.
.	.	.	.	.
Chemical *n*	...	...	...	*n*

Table 12. Disposition of major components in solid wastes disposed of off site for (insert name of drug substance here) production process

Component	kg/kg final product	kg/d
Component 1	...	...
.	.	.
Component *n*	...	...

Table 13. List of environmental permits for the XYZ facility

Matrix	Agency	Permit number
Wastewater	Name of issuing agency	Permit number
Hazardous waste	Address	Date of issue
Air	Telephone number	Expiration date

the FDA. This may require revising company retention policies for plant records. As long as the NDA is in effect, both the data used to determine initial compliance and the data generated during the life of the NDA must remain available for agency review. Company retention policies should reflect this consideration.

A more detailed example of using the material balance information derived above to show that the implementation of the production process will not cause the facility to go out of compliance with its discharge permits is shown in Table 15.

A statement regarding compliance of the plant with environmental permits should be provided at the time that an NDA submission is made to the FDA. A typical compliance statement for a manufacturing location in the U.S. might read as follows:

> The XYZ facility currently operates in compliance with all state and federal environmental regulations.

Table 14. Permit compliance criteria for the XYZ facility

Permit	Compliance criteria	Report requirements
Wastewater	Chemical parameters	Agency
Hazardous waste	(list here)	Address
Air	Operational requirements	Reporting period
	(list here)	(e.g., semiannual)

Table 15. Regulated components, permit levels, and projected impact on aqueous site effluent from production of (insert drug substance name here)

Regulated component	Contribution from process (mg/L)	Current level in effluent (mg/L)	Total projected level in effluent (mg/L)	Permit level (mg/L)
Total suspended solids	...	...	...	...
Total dissolved solids	...	...	...	...
COD	...	...	...	...
BOD_5	...	...	...	...
Cl	...	...	...	...
.	...	...	...	...
.	...	...	...	...

The environmental permits governing the operations at the XYZ Plant are listed in Table 13. The permit specifics detailing compliance requirements are listed in Table 14. Appendix 1 contains a letter from the plant manager on company letterhead certifying compliance with all environmental permits and regulations.

Occupational controls need to be addressed to assure the FDA that factory workers and support staff are not being exposed to dangerous levels of chemicals. As new chemical entities (NCEs) are developed and they appear to have efficacy for treating specific diseases, biological toxicity testing is normally performed on the drug substance and all precursor intermediate species. These results are used in developing material safety data sheets (MSDSs), which contain relevant health-based limits (e.g., TLVs, PELs, MACs, ACOs). Copies of these MSDSs should be provided. Based on the limit value, plant controls and personnel protection levels are defined. An example occupational control statement is as follows:

The following controls will be used during the manufacture of (insert name here) to minimize the exposure of plant operators and support personnel to the drug substance and intermediate species. Local exhaust ventilation will be used to capture both dust and vapor emissions in the chemical process building during chemical transfer operations. Where possible, closed systems and vacuum loading of vessels will be utilized. Plant operators will be issued personal protective equipment to prevent both bodily contact and inhalation of chemical dusts and vapors.

An industrial hygiene monitoring program is operational at the facility to measure and control employee chemical exposure. An ACO has been determined for each intermediate species and for the final drug substance and is included in the material safety data sheet (MSDS) for each chemical. Hazard communication training has been given to employees to alert and educate them on specific chemical hazards. Appendix 2 contains a set of MSDSs for the (insert name here) drug substance and intermediate precursor drug products to be used or manufactured at this site.

Introduction from Use and Disposal

Through use of calculations and/or direct measures, estimate to the extent possible the quantities and concentrations of substances expected to enter the environment as a result of use and/or disposal of products affected by the action.

Introduction from use. Usage of the drug product by humans moves the concern of environmental impact away from the manufacturing site to the locations where people reside. The data of interest are related to the quantity of the drug substance that will be marketed rather than to the drug product, which incorporates additional inert excipients in the formulation.

This requires identifying the locations where the drug product will be marketed and sold. If it is across the entire United States, one can simply assume, for the purposes of the environmental evaluation, that the product is evenly distributed across the population. If regional sales are projected, it is important to identify the percentage of the U.S. population that will have access to the drug product.

People who use drug products will normally expel a portion of the drug substance through bodily excretions. This begins a cycle of introducing the drug substance into the environment. Seventy-four percent of the U.S. population is serviced by sewer systems and publicly owned treatment works (POTW; EPA, 1984). The remaining 26% are considered to use septic tank systems whose sludge contents are periodically pumped, trucked, and discharged to the front end of the sewage treatment plants.

The maximum expected environmental concentration is the sewer concentration reaching the front end of a wastewater treatment plant. The MEEC is based on patients' use of the drug product and does not account for any depletion mechanisms (e.g. hydrolysis, biodegradation, photolysis). The relationship for determining the MEEC is as follows:

$$\text{MEEC (ppm)} = (A \times B \times C \times D \times E) = (1.96 \times 10^{-8}) \times (\text{maximum annual production, kg})$$

where A = kg/yr production
B = yr/365 d
C = day person/567.75 L (daily sewer usage)
$D = 1/246 \times 10^6$ people (U.S. population)
$E = 10^6$ mg/kg (conversion factor)

For example, a drug substance with an anticipated maximum yearly production of 20,000 kg would have an MEEC calculated as follows:

$$\begin{aligned}\text{MEEC} &= 1.96 \times 10^{-8} \times 20{,}000\\ &= 4.0 \times 10^{-4}\ \text{mg/L (ppm)}\\ &= 0.4\ \mu\text{g/L (ppb)}\end{aligned}$$

Depending on how much is known about human metabolism of the drug substance, one can modify this relationship. If the drug substance does not sorb to sludge, the MEEC can be taken as 74% of the calculated value, since the pumped-out sludges from septic tanks that are trucked and discharged into POTWs will not have the drug substance sorbed to the sludge.

Introductions from disposal. Some consideration needs to be given to the potential introduction of drug substance into the environment from the disposal of unused, expired, or returned goods. Although information on disposal practices by patients themselves would be difficult to quantify, for drugs used in controlled settings such as hospitals, information on standard disposal practices in such settings should be provided as they relate to the particular dosage form of the drug substance in the application. Information on company practices regarding the disposal of returned or expired goods should also be provided.

Item 7: Fate of Emitted Substances in the Environment

Predict environmental concentrations of and exposures to substances entering the environment as a consequence, direct and indirect, of the use and/or disposal of the products affected by the action for the following environmental compartments, including consideration of the major environmental transport and transformation processes involved: a) air . . . , b) fresh water, estuarine and marine ecosystems. . . , c) terrestrial ecosystems. . .

This section deals with predicting the probable environmental fate (see Chapter 15) of any substances emitted into the environment from the use and/or disposal of the drug substance.

Emission from Use

A primary focus here is the fate of substances discharged into the environment from patients, and several important issues need to be addressed before any testing is actually carried out.

Identification of the actual chemical species discharged from patients' use. Unlike many other chemicals that enter the environment from use in an unchanged state, human drug substances may undergo transformations in the patient prior to excretion; hence the actual chemical species discharged into the environment may differ from the parent drug substance itself, and form the following:

- Charged species resulting from ionization of salts
- Metabolic conjugates, such as glucuronides or sulfates
- New chemical species resulting from specific metabolic action

In addition, many drug substances are prepared in specific crystalline forms, as specific polymorphs, or as specific hydrates. Physical properties of such compounds in the solid state may bear little relation to the behavior of the species that results from the solubilization of the chemical in the essentially aqueous environment of conventional domestic sewer systems.

In general, studies carried out on parent drug substance can be extrapolated to the human metabolites when such metabolites contain the same backbone chemical structure as the parent compound and are more polar analogues. More polar analogues, in general, tend to be more readily biodegradable and less toxic than the parent compound. However, studies on a major metabolite might be carried out if the nonpersistence of the parent cannot be established and the major metabolite is known to represent the majority of the dose. In addition, FDA may require additional testing on metabolites that are excreted at greater than 10% of the administered dose or that could be discharged or be present in the environment at concentrations greater than 1 mg/L. Discussions with FDA are recommended before extensive testing on human metabolites is initiated.

Physical property testing. The physical properties of a chemical are useful in predicting its likely distribution in the environment and its potential mobility and accumulation characteristics. While FDA recommends that all testing be conducted in accordance with good laboratory practices (see Chapter 11) or its equivalent, physical property data on the drug substance that are provided in other portions of the NDA may be acceptable in the environmental assessment. However, as noted above, care should be taken that the data reported are meaningful in terms of the chemical species actually emitted into the environment, particularly if any of the data are used for estimation or modeling purposes (see Chapter 18). For example, the water solubility of a sparingly soluble calcium salt may bear little relation to that of the chemical when it is in a dissociated state.

Transformations and depletions. Because the majority of patients will excrete any unmetabolized or incompletely metabolized drug substance into municipal sewer systems, initial emphasis in any testing strategy should be on the chemical's fate in POTWs. A base set of data should allow determination of the likelihood that the chemical will be transformed (e.g., through hydrolysis, biodegradation, biotransformation) or transported (e.g., volatilization through stripping, sorption to sludge) in a sanitary sewer/POTW system and that no discharge into the environment is likely. Additional tests may be required if treatability and environmental fate cannot be adequately inferred from the base set data.

Environmental assessment technical test matrix. To assist in determining the most efficient test strategy, an environmental assessment technical test matrix was devised with decision criteria established from the scientific literature. The matrix comprises four tiers and is intended to represent the logic of the testing, not necessarily the chronology. The test methods used should be those provided in the FDA Environmental Assessment Technical Assistance Handbook (FDA, 1987), equivalent tests such as those provided by EPA, ASTM, or OECD, or other validated methods (see Chapters 2 to 8) as required. All test methods must fulfill FDA reporting requirements as detailed in the appropriate FDA technical assistance documents (TADs) in the handbook (FDA, 1987). This testing scheme is intended to serve only as a guide. Other approaches may be used; however, FDA's concurrence with alternative strategies should be sought before initiating these alternatives. An outline of the tiers follows.

Tier 0. Tier 0 represents the minimum base set of physical property tests needed to determine the environmental compartment(s) most likely to be affected by the target compound and the criterion values for determining which additional tier or tiers to include. Tier 0 is shown in Figure 1 and described below. The importance of physical properties is also discussed in Chapter 15.

Water solubility/hydrolysis. The water solubility of a chemical is an important parameter in predicting environmental fate. However, it is relevant only if the compound itself is hydrolytically stable. If the

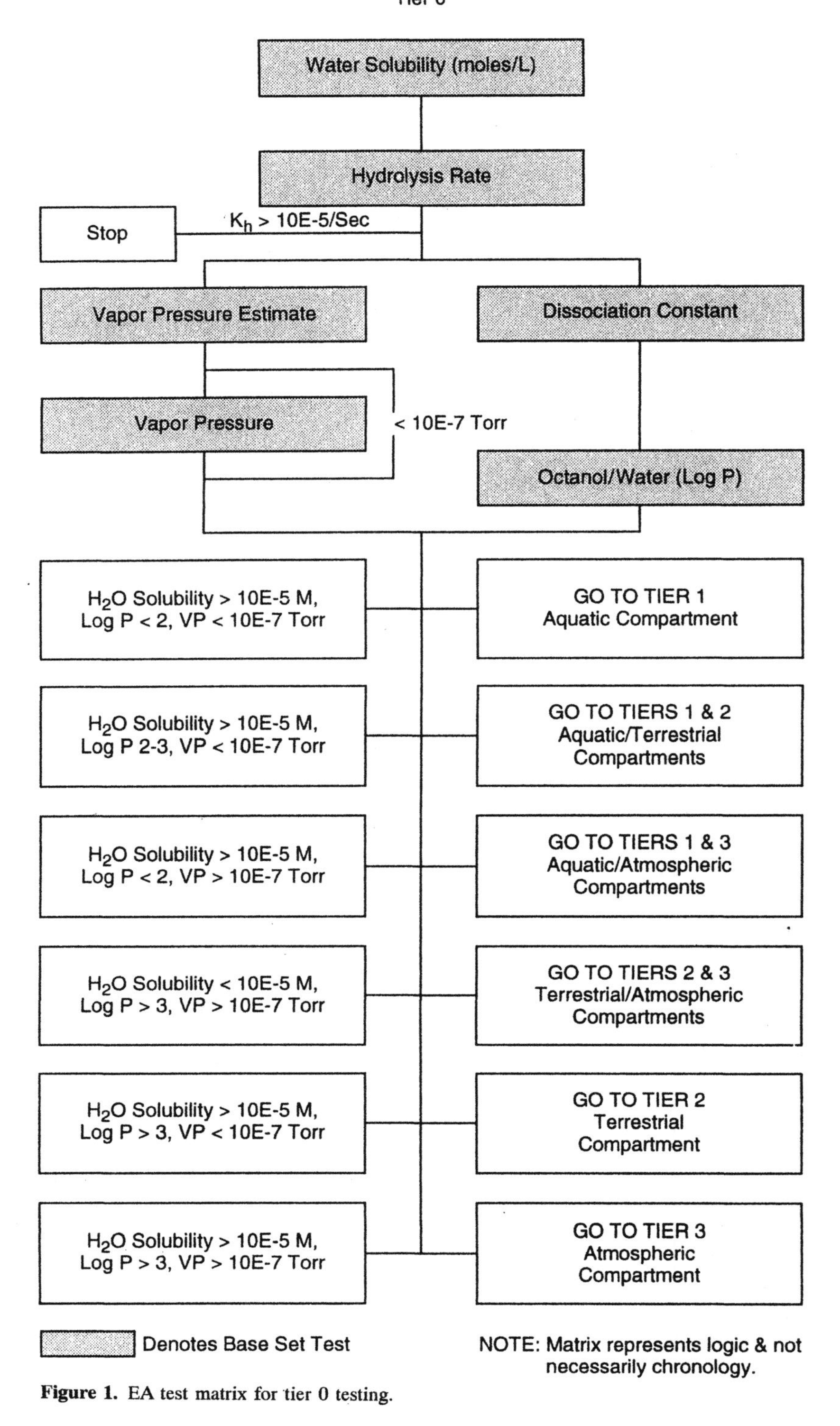

Figure 1. EA test matrix for tier 0 testing.

compound hydrolyzes at such a rate that the concentration after 5 d is less than 90% of the initial concentration (a rate constant greater than 10E-5/s or $10^{-5}\ s^{-1}$), then hydrolysis may be considered a primary removal mechanism. However, consideration must be given to the nature and extent of the hydrolysis. Where the hydrolysis products are likely to be relatively simple polar by-products, no additional work may be needed. Conversely, where hydrolysis transforms only a small portion of the parent structure, additional work may be required to determine potential biological effects from these hydrolysis products. Where these hydrolysis products are themselves hydrolytically stable, it may be necessary to carry out fate testing on these compounds, because they represent the species actually being emitted into the environment from use. Consultation with FDA prior to test initiation is recommended in these cases.

In carrying out tests on the water solubility of a compound, particular consideration must be given to the specific form of the solid test material. As noted above, many drug substances may be manufactured in salt or crystal forms that would not exist once the chemical has been ingested and has passed through the patient. Therefore, relying only on data on the parent drug substance may lead to erroneous conclusions. In addition, some testing protocols may lead to difficulty if unachievable equilibria are assumed. For example, it is highly unlikely that the dissolved form of a chemical in water will ever come to equilibrium with its hemihydrate crystalline form.

Dissociation constant(s). The presence of ionizable groups on a molecule can make a profound difference in its environmental behavior compared with a neutral analogue (see Chapter 20). Solubility and partitioning characteristics may become highly dependent on the pH and/or ionic strength of the medium, and studies of these properties must be done at appropriate pH (pH 5, 7, and 9) and ionic strength levels (<0.1 M) for the data to be meaningful. Conflicts can arise with compounds that are relatively strong acids or bases, where the addition of sufficient buffer to maintain pH increases the ionic strength to unrealistically high levels. Preliminary screening data will need to be carefully reviewed so that an appropriate testing regimen can be devised.

Octanol-water partition coefficient. The octanol-water partition coefficient, K_{ow}, is a measure of a compound's lipophilicity, or tendency to distribute into organic or lipid materials (see Chapters 1, 16 and 20). The base 10 logarithm of this value, called log P, is a commonly used parameter for predicting environmental fate. It can be a useful tool for estimating sorption to biomass and bioconcentration in animals (see Chapters 15 and 16). However, much of the work on correlating log P to environmental fate was carried out using neutral molecules. As noted above, compounds that will exist as ionized species at environmental pH levels may behave quite unexpectedly, and care must be used in extrapolating results based on the neutral species. An alternative to K_{ow} is K_p, the distribution coefficient between water and a biosolid phase such as sewage sludge. K_p can be estimated from K_{ow}; however, for ionic species it is best determined experimentally. For compounds whose K_{ow} and K_p differ significantly, K_p may be a better descriptor of a compound's distribution in a wastewater treatment.

Vapor pressure. The vapor pressure of a compound is used to estimate its potential to be air-stripped from a wastewater treatment plant or to volatilize from other environmental compartments. Drug substances of relatively high molecular weight would be expected to have very low vapor pressures, for example, $< 1 \times 10^{-7}$ torr (mm Hg). In these cases, it may be very difficult to determine the vapor pressure experimentally, and an estimate of the vapor pressure from volatility determinations may be acceptable. For compounds that are acids or bases, the vapor pressure should be determined for the molecular species. In cases in which this is not possible or desirable, the air-water partition coefficient, or Henry's law constant (see Chapter 15), may be determined experimentally at appropriate environmental pH levels.

Based on the values obtained from this preliminary set of tests, decide which environmental compartment(s) the compound is likely to distribute into. Tier 0 presents a number of scenarios based on the values for water solubility, log P (or log K_p or log K_{oc}) and vapor pressure (or Henry's law constant). Depending on the particular combination of values, a compound may be predicted to enter the aquatic, terrestrial, or atmospheric environmental compartments, or some combination. Tiers 1, 2, and 3 represent testing strategies for these compartments.

Tier 1. Tier 1 provides testing guidance logic for the *aquatic compartment.* Given the nature of most human drug substances and their disposition from patients through POTWs, the aquatic matrix is the environmental compartment in which the compound will most likely be found. Tier 1 testing for the aquatic compartment is shown in Figure 2 and discussed below.

Biodegradability. This is a key test for estimating environmental persistence (see also Chapters 15 and 31). If the experimentally determined biodegradation half-life ($t_{1/2}$) is less than 8 h for aerobic biodegradation or less than 24 h for anaerobic biodegradation, no significant amounts of the compound should be emitted from the POTW and nonpersistence of the compound in the environment may be inferred. The time criteria are based on data for typical retention times in the aerobic and anaerobic portions of POTWs, considering sewer collection systems, the primary clarifier, the activated sludge basin, and the secondary clarifier. Anaerobic digestion systems are not included.

The biodegradability in water may be determined by a variety of test methods, including simple BOD or CO_2 evolution, respirometric, acclimation, enrich-

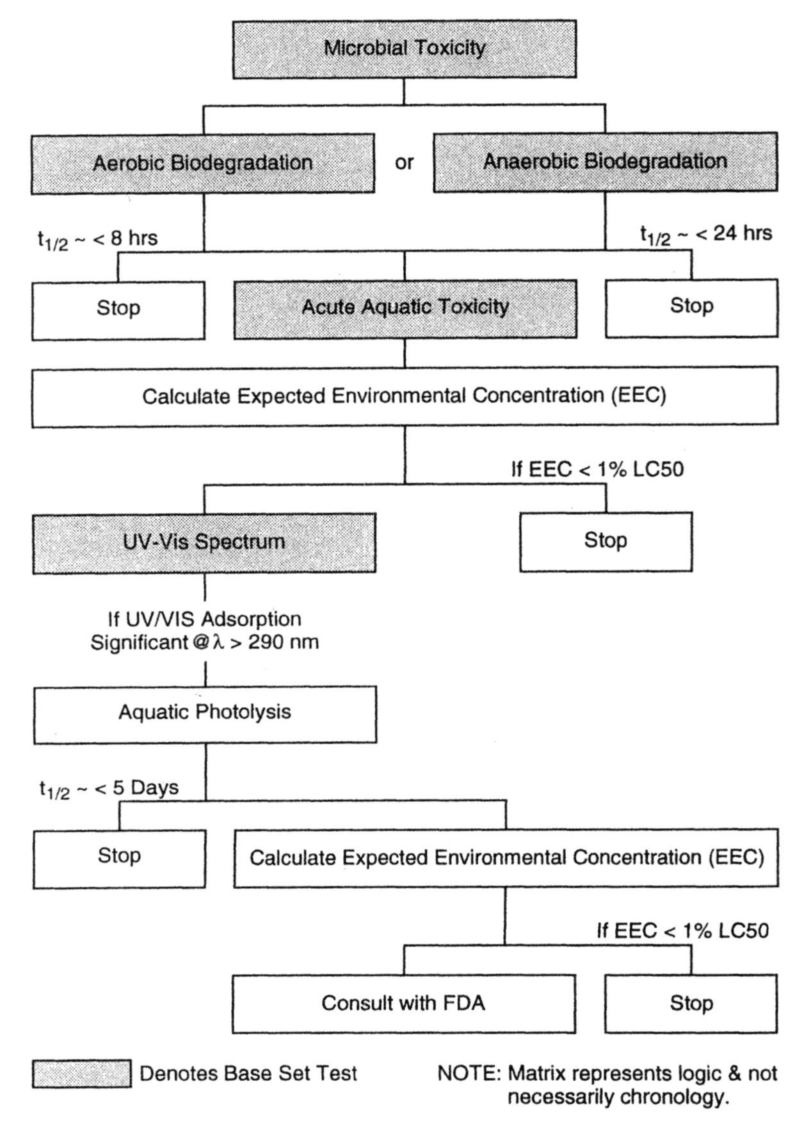

Figure 2. EA testing matrix for tier 1 testing (aquatic compartment).

ment, anaerobic tests, or other validated methods (see Chapters 25 and 31). (In general, it is advisable to carry out a microbial toxicity test prior to initiating biodegradability studies to assure that the compound is tested at concentrations that are not inhibitory or toxic to the inocula used.) If the compound is not biodegraded in tests used to determine ready biodegradability, then other biodegradation methods may be required to allow longer acclimation periods or expression of specific biotransformation pathways. Methods using concentrations of biomass typical of those found in POTWs (~2500 ppm) may be employed if coupled with compound monitoring by a specific method such as high-pressure liquid chromatography (HPLC). While many drug substances do not mineralize (see Chapters 15 and 31), many are biotransformed (see Chapter 17) into more polar, less active, or less toxic species.

If the compound is not readily biodegraded, tier 1 calls for calculation of the expected environmental concentration (EEC) and comparison of this with some aquatic toxicity measure (e.g., LC50). Should the EEC be shown to be less than 1% of the lowest acute aquatic toxicity measure, then, other tier 1 criteria being met, additional testing may not be required. For the aquatic compartment, the EEC may be calculated as follows:

$$\text{EEC} = \text{MEEC} - \text{depletion due to aquatic biodegradation processes}$$

Given a calculated MEEC of 10 mg/L, a biodegradation rate of 0.1 h^{-1}, and a hydraulic retention (i.e., residence) time in the POTW of 8 h, the EEC calculation can be carried out using standard first-order kinetic models as shown below.

$$\text{EEC} = C_t = (C_0)(e^{-kt})$$

where C_t = EEC at time t
C_0 = MEEC (initial concentration of drug discharged into the environment)
e^{-kt} = first-order rate expression, where k = first-order rate constant, t = time

Then

$$\begin{aligned}\text{EEC} &= (10\ \text{mg/L})\ e^{-(0.1\ \text{h-1}\times 8\ \text{h})}\\ &= (10\ \text{mg/L})e^{-0.8}\\ &= (10\ \text{mg/L})(0.45)\\ &= 4.5\ \text{mg/L}\end{aligned}$$

If the EEC criterion cannot be met, investigation of other possible abiotic depletion processes should be considered.

Aqueous photolysis. The propensity of a compound to undergo photolytic decomposition should be determined from its ultraviolet/visible spectrum (see Chapter 15). If there is significant absorption at wavelengths greater than 290 nm, determination of the rate of aqueous photolysis should be considered. If the photolytic half-life is less than 5 d, additional testing may not be needed, as direct photolysis may be considered as a primary depletion mechanism, provided that the compound will distribute into an environmental compartment where exposure to direct sunlight is probable (e.g., aquatic compartment, as compared to terrestrial compartment). Another EEC calculation may be made taking into account depletion by photolysis as well as any biodegradation.

Tier 2. Tier 2 presents the logic matrix for the *terrestrial compartment.* Tier 2 testing for the terrestrial compartment is shown in Figure 3 and discussed below.

Sorption coefficient (K_{oc}). For compounds that are not readily biodegradable, the K_{oc} may be used to predict whether the compound will sorb to sewage sludge or remain in the aqueous effluent from the POTW. K_{oc} is defined as

$$K_{oc} = \frac{\mu\text{g chemical sorbed/g soil organic carbon}}{\mu\text{g chemical dissolved at equilibrium/g solution}}$$

If K_{oc} is greater than about 100, sorption to sludge in the POTW is likely, and fate in the terrestrial environmental compartment should be considered. As discussed above, K_p, the water-biomass distribution, is also a measure of sludge sorption, and either K_{oc} or K_p can be used as a criterion for fate prediction. In either case, for the terrestrial compartment, aerobic and/or anaerobic soil biodegradation may become an important depletion mechanism. If the soil biodegradation half-life is less than about 5 d, additional testing may not be required. If the compound is not readily biodegradable, the EEC-toxicity comparison should be carried out in tier 2 as in tier 1.

Tier 3. Tier 3 presents the logic matrix for the *atmospheric compartment.* In this compartment, the only requisite data are the ultraviolet/visible spectrum and, if there is significant absorbance above 290 nm, the rate of vapor-phase photolysis. The EEC calculation and toxicity comparison can then be carried out as in the prior tiers. Tier 3 testing for the atmospheric compartment is shown in Figure 4 and discussed below.

In general, if the following criteria are met, further testing may not be required:

- Log P, log K_p, or log K_{oc} is less than 2 (i.e., no significant accumulation should occur).
- Vapor pressure is less than 1×10^{-7} torr and water solubility is greater than about 4×10^{-1} M or Henry's law constant is less than 3.7×10^{-7} atm m^3/mol (i.e., no significant air stripping or volatilization should occur).
- Biodegradation half-life is less than about 8 h (i.e., substance will not be discharged into the environment in significant amounts)

  or
- Biodegradation half-life is less than about 28 d and K_{oc} or K_p is less than about 100 (i.e., compound will not sorb to sludge and will not persist in the environment)

  and
- MEEC or EEC is less than about 1% of an acute LC50 (i.e., substance is not toxic).

Proceeding according to the logic of the matrix should allow the most economical testing strategy while still providing sufficient relevant information on the environmental fate of the drug substance for FDA evaluation.

Beyond the tiers. In cases in which environmental fate data suggest that more complex partitioning phenomena may occur, more detailed evaluations of the results may be required. For example, where K_{ow}, K_p, or K_{oc} values suggest that the compound will distribute itself between the aqueous and biosolids phases of a POTW, it may be useful to carry out a sample calculation of the distribution to allow more realistic comparisons with toxicity values in various compartments.

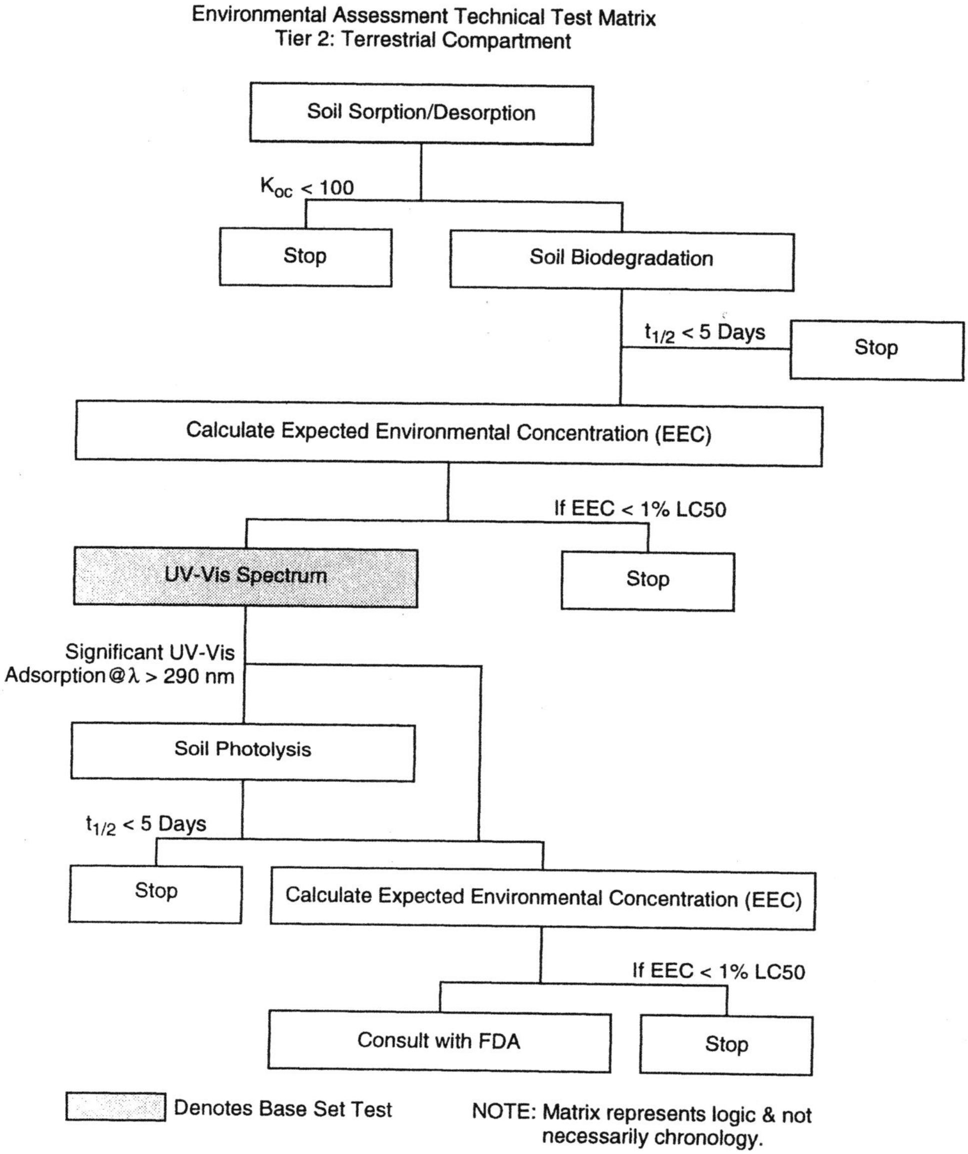

Environmental Assessment Technical Test Matrix
Tier 2: Terrestrial Compartment

Soil Sorption/Desorption
$K_{oc} < 100$
Stop
Soil Biodegradation
$t_{1/2} < 5$ Days
Stop
Calculate Expected Environmental Concentration (EEC)
If EEC < 1% LC50
UV-Vis Spectrum
Stop
Significant UV-Vis Adsorption @ $\lambda > 290$ nm
Soil Photolysis
$t_{1/2} < 5$ Days
Stop
Calculate Expected Environmental Concentration (EEC)
If EEC < 1% LC50
Consult with FDA
Stop

Denotes Base Set Test

NOTE: Matrix represents logic & not necessarily chronology.

Figure 3. EA test matrix for tier 2 testing (terrestrial compartment).

Concentration distribution approach. Given the following assumptions:

Total annual market volume, 5th yr (MV_{total})	= 10,000 kg
Fraction of U.S. population using POTW ($F_{sewered}$)	= 0.74
Fraction of drug substance not excreted (F_m)	= 0
Annual U.S. sewage flow (SF_{annual})	= 3.7×10^{13} L/yr
Annual yield of dry sludge from POTW (DS_{annual})	= 5.9×10^9 kg/yr
Dilution rate from land application of sludge (DR_{annual})	= 0.025
K_p of drug substance (or K_{oc} from soil adsorption study)	= 8×10^2

Then the maximum expected emitted concentration of drug substance in the environment from use is

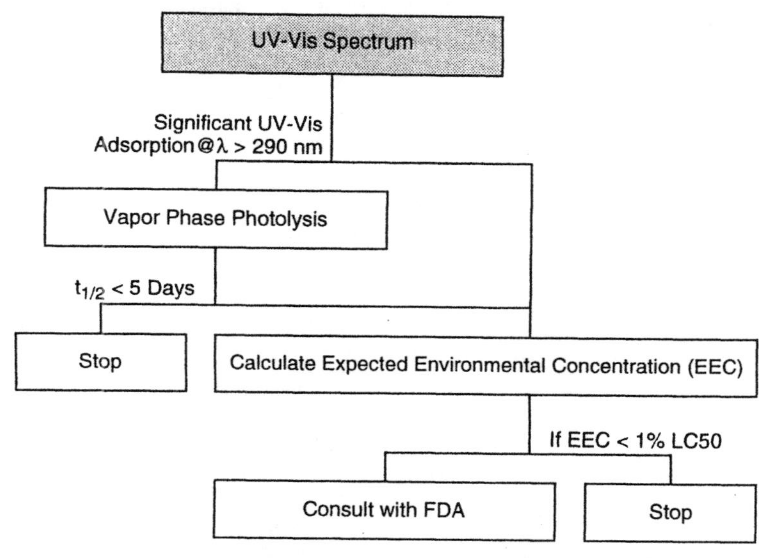

Figure 4. EA test matrix for tier 3 testing (atmospheric compartment).

calculated as follows, assuming 100% excretion in 1 d:

$$\text{Quantity emitted into POTW} = (MV_{total})(F_{sewered}) = 10{,}000 \times 0.74 = 7400 \text{ kg}$$

Based on the distribution of the drug substance between sludge and water,

$$K_p = \frac{\text{concentration in sludge } (C_2)}{\text{concentration in effluent } (C_1)}$$

$$C_2 = (8 \times 10^2)(C_1) \quad (1)$$

As the fraction emitted (0.74) will be distributed between the sludge and the sewer effluent, the balance around the POTW is

$$(C_1)(SF_{annual}) + (C_2)(DS_{annual}) = 0.74 \quad (2)$$

Substitution of the equation for C_2 [Eq. (1)] into Eq. (2) gives

$$(3.7 \times 10^{13}C_1) + (5.9 \times 10^9)(8 \times 10^2C_1) = 0.74$$

$$(3.7 \times 10^{13}C_1) + (4.7 \times 10^{12}C_1) = 0.74$$

$$4.17 \times 10^{13}C_1 = 0.74$$

$$C_1 = 1.77 \times 10^{-14}$$

$$C_2 = 8 \times 10^2C_1$$

$$C_2 = (8 \times 10^2)(1.77 \times 10^{-14})$$

$$C_2 = 1.41 \times 10^{-11}$$

Using the relationships of Eq. (2) gives

$$3.7 \times 10^{13}(C_1) = (3.7 \times 10^{13})(1.77 \times 10^{-14})$$

$C_1 = 0.655$ (88.5% of POTW total) [concentration in effluent]

$$5.9 \times 10^9(C_2) = (5.9 \times 10^9)(1.41 \times 10^{-11})$$

$C_2 = 0.0832$ (11.5% of POTW total) [concentration in sludge]

For estimated environmental concentration calculations,

Aquatic compartment:

$$(MV_{total})(F_{sewered})(1 \times 10^6 \text{ mg/kg}) \times (\text{aqueous fraction})/(SF_{annual}) = \text{MEEC (aquatic)}$$

$$(10{,}000 \text{ kg/yr})(0.74)(1 \times 10^6 \text{ mg/kg}) \times (0.885)/(3.7 \times 10^{13} \text{ L}) = 0.0002 \text{ mg/L (ppm)}$$

Terrestrial compartment:

$$(MV_{total})(F_{sewered})(1 \times 10^6 \text{ mg/kg})(\text{sludge fraction}) \times (\text{dilution})/(SF_{annual}) = \text{MEEC (terrestrial)}$$

$$(10{,}000 \text{ kg/yr})(0.74)(1 \times 10^6 \text{ mg/kg})(0.115) \times (0.025)/(5.9 \times 10^9 \text{ kg}) = 0.0036 \text{ mg/kg (ppm)}$$

As with the EEC calculations given above, these steady-state concentrations can be modified using the rate constants determined for various depletion mechanisms. Then comparison of the compartment concentrations with toxicity values in appropriate species should allow prediction of potential adverse environmental effects.

Emissions from Disposal

In addition to predicting the fate of the drug substance entering the environment from use, the EA regulations require that consideration be given to the fate arising from disposal of unused drug substance, for example, returned goods, material not meeting product specifications (off-spec material), or residual material from administration or testing. In many cases disposition of these materials will be by high-temperature incineration; however, in some instances, disposal in landfills may be an option. Some discussion of the likely routes of disposal and the potential impacts on the environment should be included in the EA.

Item 8: Effect of Released Substances

Given the information developed on the introduction (item 6) and fate (item 7) of the substances that would be released as a consequence of the use and/or disposal of the products affected by the action, use any relevant toxicological data or other appropriate measures to predict to the extent applicable effects on animals, plants, humans, or other organisms and the effects at the ecosystem level in each of the environmental compartments listed in item 7.

This section should include a summary of available toxicological information from human, mammalian, aquatic, and terrestrial organism studies, as appropriate. From the information generated in item 7, the environmental compartments likely to be affected by the compound entering the environment from use can be identified. For each of these compartments, information on toxicological or other effects on appropriate species should be provided. The compartments to be considered include occupational, aquatic, terrestrial, and atmospheric.

Environmental Assessment Effects Test Matrix

An environmental assessment effects test matrix has been devised to assist in identifying appropriate biological effects tests for each compartment, depending on the predicted partitioning pattern of the compound. The matrix is shown in Figure 5 and discussed below.

Discussion

Human and Mammalian. Acute and chronic toxicity measures on the parent drug substance from studies in humans and mammals are a measure of the potential acute effects of the substance in the occupational compartment. This information will generally be available from the substance's Material Safety Data Sheet, and it is important to include it as part of predictions of the effects of the substance from use in terms of administration, from disposal of returned goods, and in the production of the bulk pharmaceutical. Summaries of the results of studies such as the following should be included:

Oral and/or intravenous LD50 values
Skin irritation
Eye irritation
Sensitization
Mutagenicity and/or carcinogenicity
Reproductive toxicology

Aquatic Organisms. Because human drugs or their metabolites will be excreted into POTWs from use, acute aquatic toxicity studies are required for all compounds except for those that have been shown to be at least 65% mineralized with a half-life <2 h (i.e., so that two half-lives = 8 h retention time in POTW). In general, studies of microorganisms (i.e., mixed or single culture), an invertebrate (e.g., various *Daphnia* species), and a vertebrate (e.g., various fish species) should be sufficient. Summaries of the results of studies such as the following should be included:

Microbial EC50
Microbial respiration inhibition
48-h LC50 in *Daphnia* species (*magna, pulex*, etc.)
96-h LC50 in fish (fathead minnow, trout, bluegill, etc.)

Where data from item 7 indicate that the substance will be persistent in the environment, and will tend to remain in the aquatic compartment, subchronic (e.g., fish early life cycle test) or chronic aquatic toxicity tests may be required. It is recommended that FDA be consulted before any of these types of tests are initiated.

Terrestrial Organisms. Where data from item 7 indicate that the substance is likely to persist and be introduced into the terrestrial compartment, acute and subchronic tests may be required. Acute tests include those using amphipod crustaceans such as *Hyalella azteca*, an aquatic benthic organism. Subchronic tests include those using various species of earthworm. In general, consultation with FDA is recommended before these types of tests are initiated.

Item 9: Use of Resources and Energy

Specify the natural resources, including land use, minerals, and energy, required to produce, transport, use, and/or dispose of a given amount of any product that is the subject of the action, including the resources and energy required to dispose of wastes generated from production, use, and/or disposal. Effects, if any, on endangered or threatened species and on property listed in or eligible for listing in the National Register of Historic Places must be discussed.

NEPA requires that energy and natural resource requirements for manufacturing, recycling, transporting, and disposing of pharmaceutical products be predetermined before the FDA issues either a FONSI or an EIS mandate. Natural resources include both the fuels and raw materials used in producing and disposing of pharmaceutical drug products. Both drug substance and drug dosage manufacturing fa-

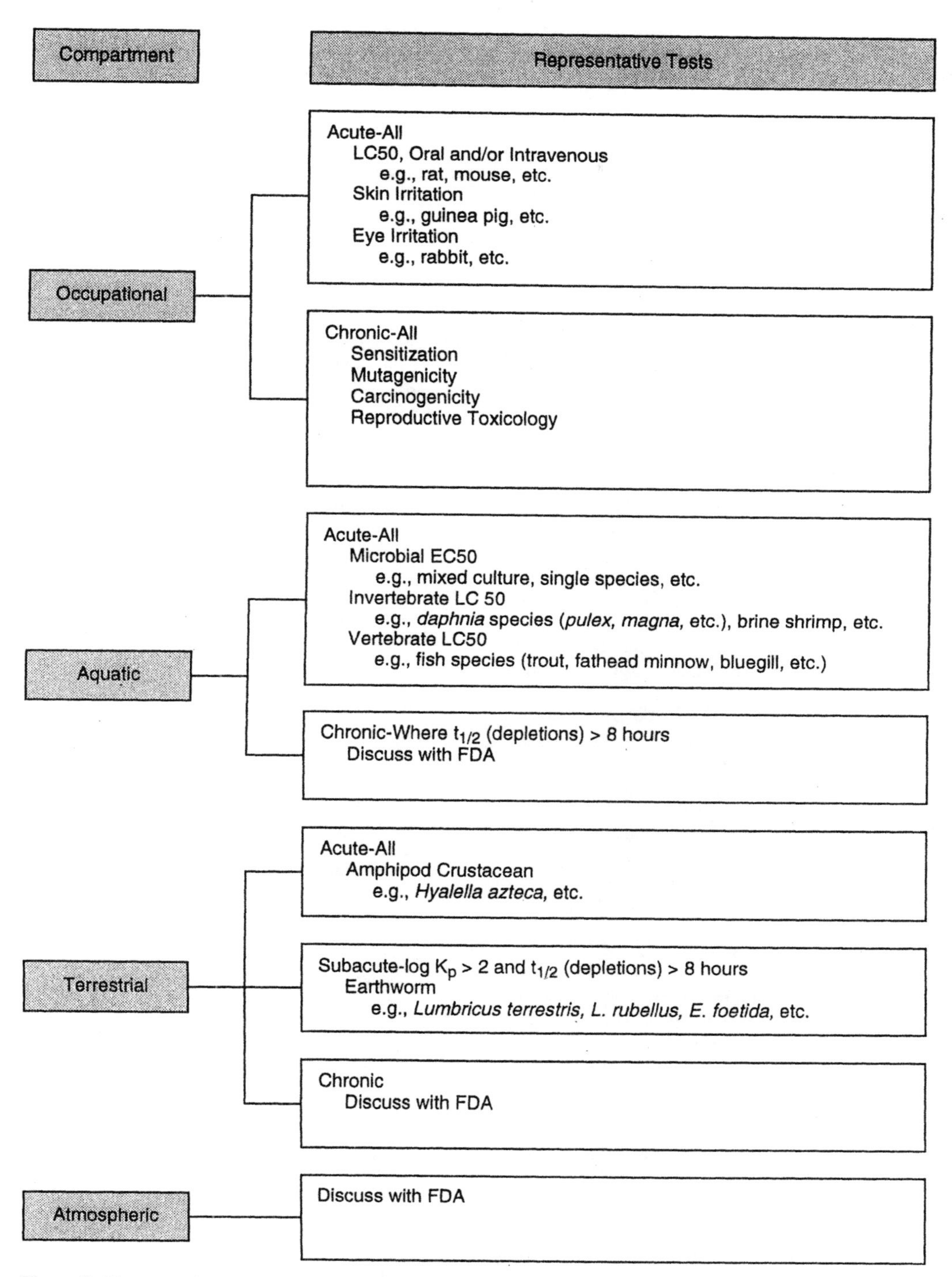

Figure 5. The general EA biological effects test matrix.

Table 16. Projected natural resource usage for the manufacture of the drug product

Item	Elec. (kW/d)	Oil (gal/d)	Propane (m^3/d)	Other (/d)
Daily energy usage	...	...	...	N/A[a]
Drug product as a percent of total manufacturing	...	...	...	
Drug product daily energy usage	...	...	...	N/A

[a]N/A, not applicable.

Table 17. Natural resource use as raw material

Raw material	Location(s) in nature	Quantity required
AAA...	City, state	... Trees
BBB...	Region, country	... Plants
CCC...	Ocean	... Shellfish

cilities need to be separately assessed, as well as all third party sites. The focus of the assessment is generally the manufacturing site, as pharmaceutical raw materials are typically commodity chemicals readily available in the marketplace. However, where unique raw materials are required for drug production, the scope of the evaluation is enlarged to include the impact on the location where unique raw materials are obtained and processed.

NEPA also requires the FDA to determine what the impact will be on endangered and threatened species from the proposed manufacturing operations. Federal, state, and local agencies can generally provide lists of endangered and threatened species within a 15-km radius of the manufacturing site. For existing sites that are already producing drug products, it is generally a straightforward procedure to ensure that sufficient environmental controls and material-handling procedures are in place to prevent the occurrence of an adverse impact on local wildlife. However, when a company plans to construct a new facility or to expand manufacturing capacity at an existing site, a more detailed evaluation of proposed environmental controls needs to be undertaken to demonstrate no adverse impact on local wildlife.

Resources and Energy

Energy usage for production of a drug product can be summarized in relation to the total energy usage at the manufacturing facility. The requirements to run administrative buildings, analytical laboratories, warehouses, etc. are also included in the total energy number. At existing facilities making similar drug products, energy requirements for production may be estimated by the percentage of the manufacturing capacity a drug product occupies. When this is not true, more detailed calculations will be necessary to determine the appropriate levels of energy usage. Table 16 presents the per diem energy (kW/d) requirements as a percentage of total daily energy usage at a drug manufacturing facility.

If raw materials are unique substances not widely available in the marketplace, they must be summarized along with the energy requirements used to transform them from their natural state to the state and purity required for use in manufacturing operations. Table 17 is an example of a summary table with numerical values based on the requirements to produce 1 kg of drug substance.

Table 18 lists the energy requirements for obtaining, processing, and transporting unique raw materials to the manufacturing site. Numerical values are based on the requirements to produce 1 kg of the drug substance.

Natural Resources Required to Dispose of Wastes Generated from Production

Manufacturing operations producing high-purity pharmaceutical drug products generate solvent, aqueous, and solid wastes requiring either recycling or

Table 18. Raw material processing and transporting energy requirements

Raw material	Energy to obtain raw material	Energy to process raw material	Energy to transport raw material
Comp 1	Elec: Oil: Propane: Gas: Other:	Elec: Oil: Propane: Gas: Other:	Oil: Propane: Gas: Other:

Table 19. Energy requirements for recycling and waste disposal

Item	Elec. (kW/d)	Oil (gal/d)	Propane (m^3/d)	Other (/d)
Recycling	N/A	...	...	N/A
Incineration	N/A	N/A	...	N/A
Wastewater treatment plant	...	...	...	N/A
Transportation	N/A	...	N/A	Gas: ... gal

disposal. The energy requirements for air pollution control devices such as condensers and scrubbers are typically evaluated as part of the manufacturing process. Table 19 shows the energy requirements associated with recycling and waste disposal operations. Numerical values are based on the requirements to produce 1 kg of the drug substance.

Endangered or Threatened Species

Whether a manufacturing facility is existing, undergoing expansion, or being newly constructed from the ground up, the impact of drug manufacturing on endangered and threatened species must be evaluated.

The first statement should reflect the status of the manufacturing facility that will manufacture the drug product. If an expansion project is under way or if a new facility is being constructed, additional details of proposed pollution abatement and material-handling procedures will need to be presented and discussed in Section 10 (Mitigation Measures). For existing facilities, discussions presented in Section 6 (Introduction of Substances into the Environment) should suffice.

Table 20 lists the endangered and threatened species in the vicinity of the manufacturing facility. A 15-km radius might be taken as a reasonable area to consider based on air emission dispersion models. When written correspondence is available to document the contents of Table 20, it should be included in the appendix. Foreign sites are handled similarly to U.S. sites but a letter from the appropriate governmental agency is desired, even if it provides a negative declaration on the existence of endangered and threatened species. Both the original and a copy of the letter translated into English should be included in the appendix.

A similar evaluation needs to be made for areas where unique raw materials are obtained for use in the drug manufacturing process. The area to be covered is site specific and may extend beyond a 15-km radius.

Table 20. List of threatened and endangered species within a 15-km radius of the manufacturing facility

Latin name	Common name
xxxxxxxxxx	...
yyyyyyyyyy	...
zzzzzzzzzzz	...

Table 21. List of historic sites near the manufacturing facility

Location	Historic site
xxxxxxxxxx	...
yyyyyyyyyy	...
zzzzzzzzzzz	...

Property Listed in or Eligible for Listing in the National Register of Historic Places

Table 21 should contain any properties that are eligible for listing in the National Register of Historic Places and that might be affected by the manufacturing facility. When written correspondence is available to document the contents of Table 21, it should be included in the appendix. Foreign sites are not covered by this requirement, as the National Register lists only historic places located in the United States.

Item 10: Mitigation Measures

Describe measures taken to avoid or mitigate potential adverse environmental impacts associated with the proposed action.

The focus of this section is to demonstrate to FDA how the environmental controls and material-handling procedures will either avoid or mitigate adverse environmental impacts from the manufacturing and disposal of drug products.

The discussion should demonstrate how environmental controls and material-handling procedures will be integrated to ensure no adverse environmental impacts from manufacturing processes, recycling, wastewater treatment operations, air emission abatement, destruction of hazardous wastes, and disposal of solid waste materials. When unique raw materials are used, the discussion should be enlarged to include the measures in place around the operations to control the releases of hazardous materials while obtaining, processing, and transporting them to the drug product manufacturing facility.

The mitigation measures previously described in the section entitled "Item 6: Introduction of Substances into the Environment" can be cited and cross-referenced in the assessment. To complement the discussion, waste minimization and pollution prevention programs specific to the production of the drug product should also be highlighted. Particular attention should be given to process modifications made during synthesis development to accommodate the use of less toxic materials in the manufacturing process, especially where they support the voluntary toxic reduction program (e.g., EPA 35/50 program). Accidental release contingency plans should also be summarized to demonstrate that the applicant has an effective spill prevention program in place to prevent the release of hazardous materials into the environment. This need not be included in the EA but it should be available for review by FDA inspectors who visit the plant.

The manufacture of cytotoxic agents and radiopharmaceuticals generates wastes that require special handling and disposal. Specific details need to be provided on the instructions and in-place procedures for handling and disposing of these materials and their waste products.

LITERATURE CITED

EPA: 1984 need survey: report to Congress: assessment of needed publicly owned wastewater treatment facilities in the U.S. EPA 430-9-84-011. U.S. EPA, 1984.

FDA: Environmental Assessment Technical Assistance Handbook. NTIS PB87-175345. Washington, DC: Food and Drug Administration, 1987.

NEPA: The National Environmental Policy Act of 1969 [sec. 102(2), 83 Stat 853 (42 USC 4332)], 1969.

PMA: Interim Guidance to the Pharmaceutical Industry for Environmental Assessment Compliance Requirements for the FDA, Vol. 7. Pharmaceutical Manufacturers Association, 1991.

Chapter 26

ROLE OF THE OECD IN CHEMICALS CONTROL AND INTERNATIONAL HARMONIZATION OF TESTING METHODS

N. J. Grandy

INTRODUCTION—THE OECD AND ITS ENVIRONMENT PROGRAMME

The Organization for Economic Cooperation and Development (OECD) is an intergovernmental organization of 25 democratic nations (member countries) with advanced market economies.[1]

The OECD was established in 1960. Its basic aims, enshrined in its founding Convention, are to:

- Achieve the highest sustainable economic growth and employment
- Promote economic and social welfare throughout the OECD region by coordinating the policies of its member countries
- Stimulate and harmonize its members' efforts in favor of developing countries

Most of the Organization's work is carried out by over 200 specialized committees and subsidiary groups, composed of representatives designated by the governments of member countries. These bodies are serviced by the OECD Secretariat (some 1800 individuals), which is located in Paris and headed by the Secretary-General.

The OECD is not a regulatory body. It can, however, take formal actions in the form of decisions and recommendations of the Council (the highest authority of OECD). Council decisions are legally binding on all the member countries. Council recommendations, whilst not legally binding, carry a strong moral obligation.

The OECD is perhaps best known for its work in monitoring and forecasting trends in the world's economies. However, OECD member countries account for 70% of the world gross domestic product, 70% of world trade, and most of the world production of chemicals. They therefore have a special responsibility with regard to the state of the environment and there has been a strong OECD Environment Programme, overseen by the Environment Committee, since 1970.

The Environment Committee, now called the Environment Policy Committee (EPOC), was created in order to bring governments together to work cooperatively toward shared environmental objectives. The role of the committee is to promote sustainable development (the interdependence between economic growth and environmental sustainability of resources), as well as a high standard of living throughout OECD countries.

Through EPOC, experts on health and the environment are brought together and provide OECD with a strong capability for examining environmental issues and for developing harmonized policies and practical tools to address them. EPOC is supported by several subsidiary groups, although its structure is constantly evolving in response to member countries' major environmental concerns. Some groups are long-standing and address general areas of environmental policy, such as waste management and chemicals control. Others are established for short periods to deal with specific problems.

The Environment Directorate of the OECD Secretariat serves EPOC and carries out activities in the following areas:

- Pollution prevention and control

The opinions presented in this paper do not necessarily represent the opinions of the OECD or its Member countries and should therefore be viewed solely as those of the author.

[1]The OECD member countries are Australia, Austria, Belgium, Canada, Denmark, Finland, France, Germany, Greece, Iceland, Ireland, Italy, Japan, Luxembourg, Mexico, the Netherlands, New Zealand, Norway, Portugal, Spain, Sweden, Switzerland, Turkey, United Kingdom, United States of America. The Commission of the European Community also participates in the work of the OECD.

- Environmental health and safety
- Environment and economics
- Climate change
- Waste management
- Natural resource management
- Assessment of environmental performance

It is OECD's work in the area of environmental health and safety that has particular relevance in this book.

CHEMICAL SAFETY AND THE OECD

OECD work on chemical safety began in the early 1970s as a result of concern over widespread contamination of the environment and accompanying adverse environmental effects. Chemicals causing particular concern included persistent organochlorine pesticides, polychlorinated biphenyls (PCBs), and heavy metals like mercury, cadmium, and lead. The initial focus was on specific chemicals, such as those mentioned, but was later broadened to an active role in initiating the development, within OECD member countries, of harmonized policies and practical tools for protecting human health and the environment, that is, chemicals management. This work is organized under the umbrella of the OECD Chemicals Programme, the main activities of which are test guidelines, good laboratory practice, hazard assessment, harmonization of classification and labeling systems, and cooperative activities on existing chemicals, risk reduction and pesticides. The work of the Chemicals Programme is overseen and managed by the Joint Meeting of the Chemicals Group and Management Committee. The Joint Meeting reports to EPOC.

The United Nations Conference on Environment and Development (UNCED) held in Rio de Janeiro, Brazil, in June 1992 has had, not surprisingly, an impact on the OECD Chemicals Programme, particularly in relation to the implementation of UNCED's Agenda 21 (Chapter 19: Environmentally Sound Management of Toxic Chemicals, including Prevention of Illegal International Traffic In Toxic and Dangerous Products). The OECD Chemicals Programme has always worked closely with other international organizations having complementary programs, especially UN agencies such as the International Programme on Chemical Safety (IPCS), as well as various nongovernmental organisations. However, in response to the implementation of Agenda 21, increased efforts are being made to improve the integration of policies for control for all types of chemicals while emphasizing the links between economic and environmental considerations.

MANAGEMENT OF CHEMICALS

There are tens of thousands of chemicals in commerce today (existing chemicals), many of which are produced in very large volumes. More than 2000 new chemical substances (including industrial chemicals, pesticides, food additives, and pharmaceuticals) reach the world market each year. If human health and the environment are to be adequately protected, there is a clear need to understand the potential effects of these substances in order that they can be managed safely. Many countries now have laws requiring that chemicals and formulated products be tested and assessed before they are put on the market (see Chapters 21, 22, 23 and 27). In addition, regulations exist or are under development at both national and international levels that call for (additional) testing and assessment of chemicals already on the market (e.g., reregistration of pesticides under FIFRA in the United States and the evaluation and control of the risks of existing substances within the European Community under EC Council Regulation No. 793/93, March 1993).

The management of chemicals is a process involving several steps (Figure 1).[2] The initial step is the identification of a chemical's intrinsic properties (i.e., hazard identification), which include its physical-chemical characteristics, potential to degrade or accumulate, and toxicity to humans and the environment. The second step, hazard assessment, integrates information on toxicity and possible exposure to determine the nature and magnitude of the adverse effects resulting from release of the chemical into the environment. This can be followed by risk assessment, a quantitative or semiquantitative step in which the probability of clearly defined environmental effects occurring as a result of the exposure to a chemical is estimated. The final step is risk management (see Chapter 28 for an overview of ecological risk assessment).

The following sections provide an overview of the OECD Chemicals Programme and its role in chemicals management (some useful OECD references from the Chemicals Programme are listed at the end of this chapter). The focus is principally on the development of internationally harmonized testing methods, which complements other chapters in this book. This is a very important area of work because, as can be seen from Figure 1, the basis for hazard assessment, and ultimately risk assessment, is the methods applied in hazard identification.

OECD GUIDELINES FOR TESTING OF CHEMICALS

A large amount of work is currently being done by governments (including intergovernmental organizations like OECD and the Commission of the European Communities), industry, and others to develop internationally harmonized hazard and risk assess-

[2]Please note that the terminology used here is slightly different from that used in the United States. The U.S. equivalent for the terms used in this chapter are as follows:

This chapter		U.S. terminology
hazard identification	=	hazard assessment
hazard assessment	=	risk assessment
risk assessment	=	quantitative risk assessment

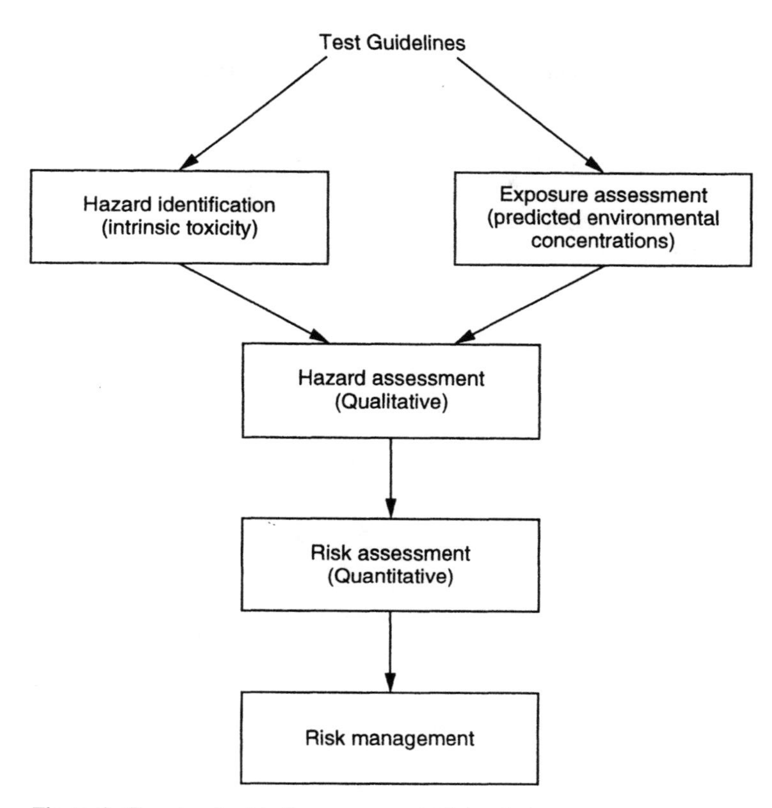

Figure 1. Steps involved in the management of chemicals.

ment procedures (see Chapter 27 on the European Community). In parallel, there is a need for further international harmonization of the test methods used to generate the data on which assessments are made, in order that countries start from a common basis. Increased availability of internationally agreed test methods also helps to:

- Increase the validity and international acceptance of data, thus avoiding duplication of testing and hence the unnecessary use of laboratory animals
- Facilitate international trade by avoiding or minimizing barriers caused by differences in national legislation and thereby creating a level playing field for chemical industries in member countries

The OECD Guidelines for Testing of Chemicals are internationally agreed methods. They were first published in 1981 and cover testing in the areas of physical-chemical properties, effects on biotic systems (ecotoxicity), environmental fate (degradation or accumulation), and human health effects (toxicity).

In order to formalize the international acceptance of data, the OECD "Decision Concerning the Mutual Acceptance of Data in the Assessment of Chemicals" (i.e., the MAD decision) was adopted in 1981 and is legally binding on all member countries. The MAD decision states that "data generated in the testing of chemicals in an OECD Member country in accordance with OECD Test Guidelines and OECD Principles of Good Laboratory Practice [see Chapter 11 on good laboratory practices] shall be accepted in other Member countries for purposes of assessment and other uses relating to the protection of man and the environment." This means that, for example, data generated on a chemical in a European laboratory should be accepted under the Toxic Substances Control Act (TSCA) or under the Federal Insecticide, Fungicide, and Rodenticide Act (FIFRA) by the U.S. Environmental Protection Agency (see Chapters 21, 22 and 23), provided OECD test guidelines and OECD GLPs have been followed. Likewise, data generated under the same conditions in the United States should be accepted in Europe.

There are now over 70 OECD test guidelines (see chapter appendix), which are available in English and French. However, there is a continuous process of revision of existing guidelines and the development of new guidelines when new scientific concerns indicate the need.

With respect to ecotoxicity, degradation, and fate in the aquatic environment, the following OECD methods, as of December 1994, exist:

Ecotoxicity
201 Alga, Growth Inhibition Test
202 *Daphnia* sp. Acute Immobilisation Test and Reproduction Test
203 Fish, Acute Toxicity Test
204 Fish, Prolonged Toxicity Test: 14-Day Study
209 Activated Sludge Respiration Inhibition Test
210 Fish Early-Life Stage Toxicity Test

Degradation/accumulation
301 Ready Biodegradability
302 Inherent Biodegradability
303 Simulation Test—Aerobic Sewage Treatment: Coupled Units Test
305 Bioaccumulation in Fish
306 Biodegradability in Seawater

Methods in draft form
- Fish, Egg and Sac-Fry Toxicity Test
- Fish, Juvenile Growth Test
- Aerobic/Anaerobic Transformation in Aquatic Water/Sediment Systems

Structure and Content of the Guidelines

The OECD test guidelines contain general procedures for the testing of a chemical property or effect considered important for the evaluation of the health and environmental hazards of a chemical. The guidelines vary somewhat in detail but include all the essential elements which, assuming good laboratory practice, should enable an operator to perform the required test. It is important to note that OECD test guidelines are not designed to serve as rigid protocols. They are instead designed to allow some flexibility for expert judgment and adjustment to new developments.

All OECD test guidelines have a similar structure. For guidelines published after 1992, the principal sections are indicated below.

The *Introduction* provides a background to the particular test, providing information on the origin of the proposal for guideline development or revision (e.g., from a particular member country). When a guideline is revised, the introduction outlines the main differences from the previous version.

The *Principle of the Test* gives a brief outline of the test, indicating the end points to be measured. For example, for Guideline 210, Fish, Early-Life Stage Toxicity Test, the principle of the test is as follows:

> The early-life stages of fish are exposed to a range of concentrations of the test substance dissolved in water, preferably under flow-through conditions. The test is begun by placing fertilised eggs in the test chambers and is continued at least until all the control fish are free-feeding. Lethal and sub-lethal effects are assessed and compared with control values to determine the lowest observed effect concentration and hence the no observed effect concentration.

The information that should be available and the information that would be of use prior to starting the test are indicated in *Information on the Test Substance.* This could include the results from other OECD tests. For example, Guideline 210 states:

> Results of an acute toxicity test (see Guideline 203), preferably performed with the species chosen for this test, should be available. This implies that the water solubility and the vapour pressure of the test substance are known and a reliable analytical method for the quantification of the substance in the test solutions with known accuracy and limit of detection is available.
>
> Useful information includes the structural formula, purity of the substance, stability in water and light, pK_a, P_{ow} and results of a test for ready biodegradability (see Guideline 301).

Information on how to prepare for the test is provided in the *Description of the Method.* For example, for toxicity tests with fish, it will indicate the type of apparatus required, the species that can be used and how they should be held before the test, suitable dilution water in which to perform the test, and how solutions of the test substance should be prepared.

The *Procedure* actually provides instructions and guidance on how the test should be performed. Again using fish toxicity tests as an example, this section will indicate the test duration, loading, light and temperature conditions, feeding (where appropriate), number of fish to be placed in each tank, number of different test substance concentrations and controls, frequency of analytical determinations and water quality measurements, and type of biological observations to be made.

The *Data and Reporting* section indicates how the test data should be analyzed and the information that must be included in the test report.

The references cited in the test guideline are listed in the *Literature* section. Generally the references are limited in number but they will represent those considered by the experts consulted to be the most important at the time the guideline was published.

In addition to the sections described above, the guidelines in the area of aquatic toxicology and some of those for degradation and accumulation also include a section on validity of the test. This section indicates certain conditions that the test must fulfill for it to be valid. For toxicity tests with fish, these conditions may include acceptable control mortalities, limits for dissolved oxygen concentration, and acceptable fluctuations in test substance concentrations.

OECD Test Guideline Development

The OECD Secretariat is responsible for the daily management and administration of the test guidelines work. It directs routine activities, including the drafting of proposals and other documents and the organization of expert meetings and workshops.

The development and/or revision of an OECD test guideline takes place in two major phases: defining the need for development (Figure 2), followed by producing the product—the test guideline itself (Figure 3). Each phase comprises of several steps, as described below.

Defining the Need (Figure 2)

Initiatives

Proposals for the development and/or revision of a test guideline can be made by OECD member countries, the Secretariat, and the international scientific community. Proposals from member countries and the scientific community are made via a national coordinator.

Group of National Coordinators

A key role in the test guideline work is played by the national coordinators. Each member country has a national coordinator, designated by its government, who acts as a focal point for test guidelines work in that country. There is also a national coordinator representing the Commission of the European Communities. Each national coordinator maintains an efficient network of technical experts (usually drawn from government, academia, and industry) for consultation purposes. National coordinators meet at least once a year to propose the work program and the priorities and to decide on the approach to be taken with work in specific areas.

The Approach

Any proposal to develop or revise test guidelines has to be justified by valid arguments [e.g., need for such a test (or update), ethical arguments indicating the advantages of the proposed test/procedure with respect to animal use/discomfort without loss of essential information].

Once the proposal for work in a particular area is approved, it can be developed either by starting with a **detailed review paper** or, more directly, by producing a **draft test guideline proposal**.

Detailed Review Papers

In order to identify the specific needs for data in a particular area of hazard identification or to confirm the need to update a specific cluster of existing test guidelines or even a single guideline, it is essential

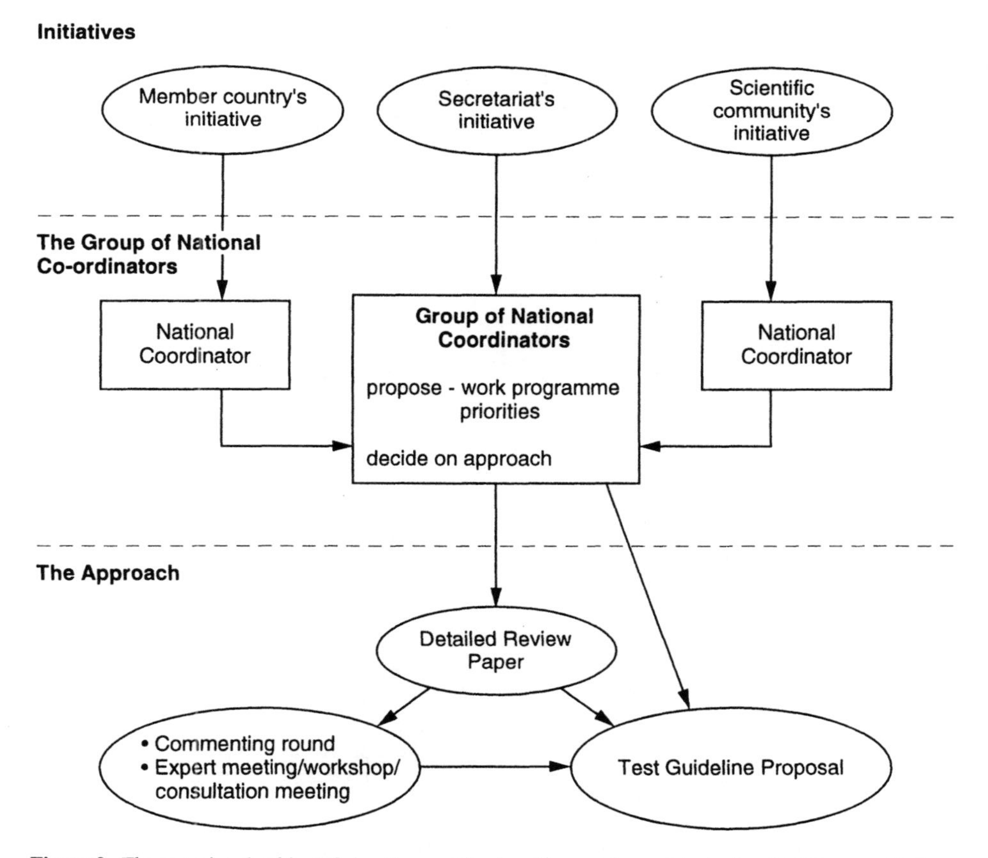

Figure 2. The steps involved in defining the need for development (or revision) of an OECD test guideline.

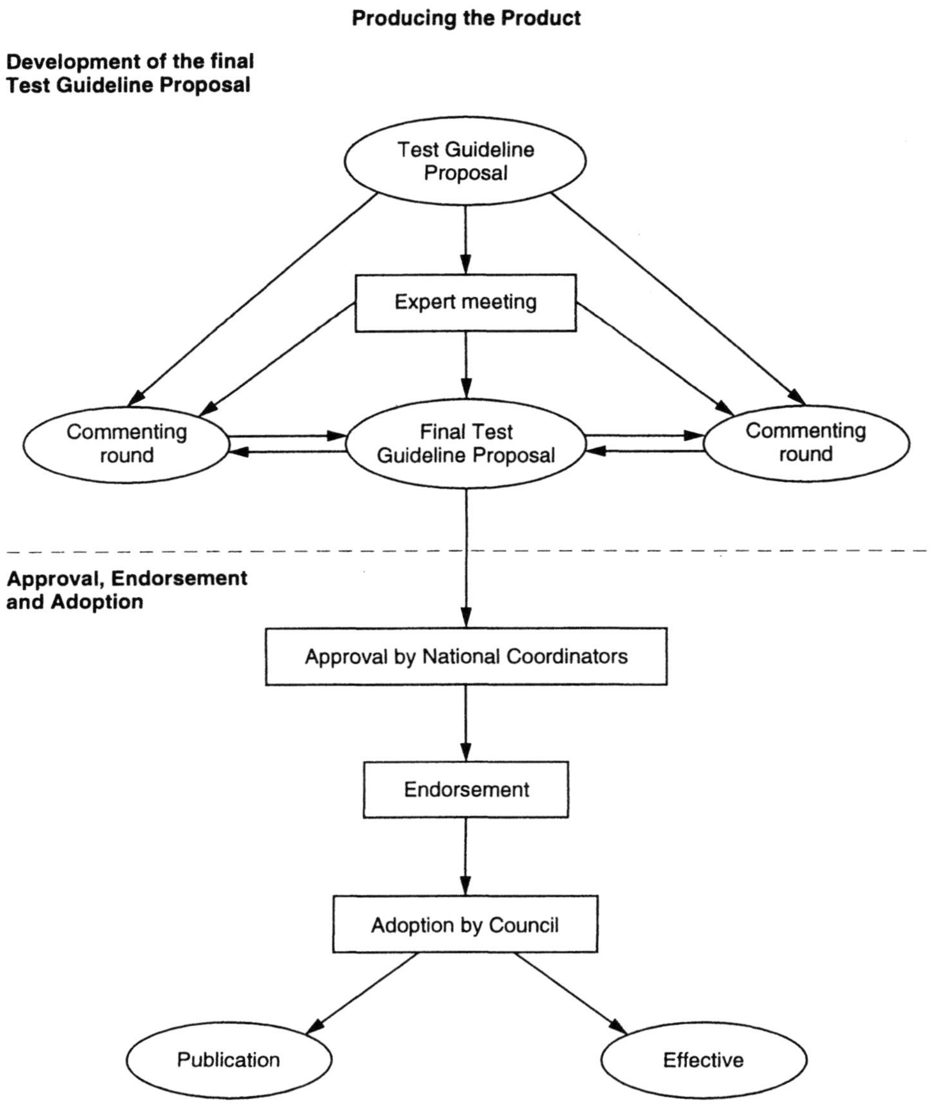

Producing the Product

Development of the final Test Guideline Proposal

Test Guideline Proposal

Expert meeting

Commenting round

Final Test Guideline Proposal

Commenting round

Approval, Endorsement and Adoption

Approval by National Coordinators

Endorsement

Adoption by Council

Publication

Effective

Figure 3. The steps involved in producing the final OECD test guidelines.

that the state of the art in that area is first assessed. This is done using a detailed review paper (DRP). Reviews for DRPs should be extensive and should include:

- A description of the scientific progress and new techniques available in the area under review
- An inventory of existing test methods in that area, together with an appreciation of, among other things, the scientific validity, sensitivity, specificity, and reproducibility of these methods
- An inventory of (inter)national data requirements with respect to the area under review, including the data used as part of existing hazard assessment procedures
- Identification of gaps with respect to significant end points not yet sufficiently covered by OECD test guidelines
- Identification of methods that are currently covered by OECD test guidelines but are to be replaced or updated in order to comply with current scientific views
- Proposals with respect to the development of new test guidelines and/or the revision of existing ones
- Indication of the relationship between the proposed and existing tests and of their limitations of use

National coordinators may take the responsibility for the preparation of a DRP in their member country. Alternatively, the Secretariat may request the assistance of recognized (international) scientific societies, accept offers from industry, including their (international) associations, or hire a consultant, in order to prepare the DRP.

DRPs are reviewed in member countries by nominated technical experts and their opinions are fed back to the Secretariat via the national coordinators. It may be necessary to hold an expert meeting or a workshop to discuss the DRP and to decide on further action. The outcome of a DRP will usually be to propose that (a) draft test guideline(s) is/are prepared.

Draft Test Guideline Proposals

If the need to develop or revise a test guideline is clear-cut, the Secretariat may arrange for the preparation of a draft test guideline proposal without a DRP. A critical appraisal of the test concerning its scientific justification, its sensitivity, and its reproducibility must be made prior to submitting the proposal for consideration by the Group of National Coordinators. In addition, and where feasible and relevant, a comparative study (e.g., a ring test or round-robin test; see Chapter 1) supporting the validity of the test should be included. A draft test guideline can be prepared by a member country, a scientific organization, industry associations, or a consultant.

Producing the Product (Figure 3)

Development of the Final Test Guideline Proposal

Once it is agreed that work on a draft test guideline proposal should proceed, the proposal is circulated to member countries for comment in the same way as is done with DRPs. Depending on the nature and extent of the comments received, the Secretariat will either circulate an updated draft or propose that a meeting be held. Commenting rounds continue until consensus by the member countries is reached and a final test guideline proposal is developed.

Approval, Endorsement and Adoption

The final test guideline proposal is first submitted to the Group of National Coordinators for approval. Once approved, it must then be endorsed by the Joint Meeting and the Environment Policy Committee before final adoption by the OECD Council. Publication of the test guideline then follows, although it is effective once it is adopted by the Council.

The OECD Pesticide Programme and Test Guidelines

As part of its Pesticide Programme, OECD is expanding its test guideline activities to increase the availability of methods suitable for the testing of pesticides for registration purposes. The priority areas for method revision and development are ecotoxicology (for both the aquatic and terrestrial environments), environmental fate, and human health and exposure. Projects also exist in the areas of registration data requirements, hazard/risk assessment, reregistration of existing pesticides, and risk reduction.

The Pesticide Programme itself has three major goals:

1. Promoting sound, consistent national pesticide registration procedures
2. Burden sharing in the reregistration of old pesticides
3. Reducing risks to human health and the environment associated with pesticide use

It was initiated at the request of member countries who perceive significant benefits from increased international coordination and cooperation in pesticide assessment and control. Harmonization of the data required for pesticide registration and of pesticide review procedures saves resources for both government and industry by minimizing duplicative testing, by enabling governments to share the burden of pesticide review, and by helping to reduce trade barriers caused by existing differences. International sharing of information and advice on policy and technical matters contributes to a wider adoption of risk reduction practices, including phaseouts of the most dangerous pesticides, reduction in pesticide use overall, and increased use of alternative methods of pest control such as integrated pest management.

OTHER OECD CHEMICALS PROGRAMME ACTIVITIES

Good Laboratory Practice

As the work of the Chemicals Programme was extended to cover the harmonization of testing procedures (i.e., OECD test guidelines), concerns were expressed by member countries regarding the reliability of test data. In response, the OECD principles of good laboratory practice (GLP) were developed and adopted by the Council in 1981.

Good laboratory practice is concerned with the organization of test laboratories and the conditions under which laboratory studies are planned, performed, monitored, recorded, and reported (see Chapter 11 for an overview of GLPs). The purpose of the OECD principles of GLP is to promote the development of good-quality test data. The GLP principles address general aspects of the organization and management of all types of laboratory and field studies and other issues not specific to particular tests.

A system of compliance-monitoring procedures has also been established to ensure that laboratories are complying with the GLP principles. The harmonization and mutual recognition of compliance-monitoring methods by member countries has been a crucial step in the international acceptance of data.

Hazard Assessment

In the Chemicals Programme, special attention is given to exchanging information on, reviewing, updating, expanding, and promoting the harmonization of methods used for hazard assessment. Specific work has been directed to the assessment of effects of chemicals in the aquatic environment and to exposure of consumers, workers, and the environment. A considerable amount of work is also being done on the development and use of data estimation methods [i.e., (quantitative) structure-activity relationships, (Q)SARs; see Chapter 20]. The assessment of effects in the terrestrial environment is an additional area of new work.

In estimating concentrations of chemicals that would not cause adverse effects in the environment, the use of data derived from field or semifield tests (e.g., mesocosms; see Chapter 9) or from field observations would be preferable. However, these data are seldom available, and even if they are available, they are often difficult to interpret. In most cases, therefore, extrapolation procedures are applied to data from acute and/or chronic laboratory tests to predict environmentally "safe" concentrations. When data on certain physicochemical properties and effects on certain species are absent, data derived using (Q)SARs may be used.

With respect to hazard assessment for the aquatic environment, OECD activities have included workshops on the application of QSARs to estimate aquatic toxicity data, the extrapolation of laboratory data to the real environment, and the effects of chemicals in sediments. The outcome of these workshops has been integrated into a guidance document for aquatic effects assessment.

The work on hazard assessment is being extended to include the new evolving life cycle risk assessment methods in close cooperation with OECD's Pollution Prevention and Control Group. Work on hazard assessment is closely linked to work on test guidelines, existing chemicals and pesticides.

Harmonization of Classification and Labeling Systems

Classification and labeling systems provide information on the properties (e.g., physical-chemical properties such as flammability and toxicity) and the safe use of chemicals. They are used principally for transport purposes and to provide a warning for users of all kinds (e.g., industry and the general public).

At present, various classification systems exist (e.g., within the UN, the European Community, and the United States). In cooperation with other organizations and particularly with IPCS, OECD is working toward the harmonization of classification systems. Emphasis is being placed on the criteria underpinning the classification of hazards to human health and the environment.

Existing Chemical Activities

As mentioned earlier, many countries have laws requiring that chemicals be tested and assessed before they are put on the market. However, this has not always been the case and there are many existing chemicals for which adequate safety data or hazard assessments are not available. Since the early 1980s the OECD has been working to strengthen and harmonize policies to ensure that the safety of these existing chemicals is systematically assessed. This was formalized in a council decision, adopted in 1987. OECD is an appropriate forum for taking on this work on existing chemicals—the task is too large for any one country to tackle on its own.

Priority for testing and assessment is being given to chemicals that are produced in high volumes (i.e., 10,000 tonnes per year in one country or 1000 tonnes per year in more than one country) and for which adequate data are not available (i.e., over 600 chemicals). OECD member countries are sharing the burden of testing by portioning out the workload among themselves, with industry playing a significant role. Cooperative assessments are made on the data generated in this way, as are decisions on whether follow-up activities are necessary (e.g., further testing or development of more exposure data). This work is being done in accordance with the 1991 Council Decision-Recommendation on the Cooperative Investigation and Risk Reduction of Existing Chemicals.

The benefits derived from this work are many. The protection of human health and the environment is improved, the resource burdens are decreased by sharing the workload among governments and industry, and the amount of animal testing is reduced by avoiding duplication.

Risk Reduction

The ultimate aim of the testing and assessment of chemicals is to prevent or reduce undue risks to human health and the environment. As part of the Chemicals Programme, and strengthened by the 1991 Council Decision-Recommendation on the Cooperative Investigation and Risk Reduction of Existing Chemicals, OECD is supporting member countries' efforts to develop national policies and actions for risk reduction and the development and implementation of international measures, where appropriate.

Several projects on specific chemicals or groups of major concern have been initiated (i.e., lead, cadmium, mercury, methylene chloride, and brominated flame retardants). The work on lead is, at present, in a more advanced stage than that for the other substances and member countries are considering the possible development of a council decision aimed at reducing the release of lead into the environment and exposure to humans.

SUMMARY

OECD provides a forum for the governments from its 25 member countries to meet and express their points of view, share their experiences, and search for common ground. Although the OECD is not a regulatory body, the work of the OECD Chemicals Programme helps member countries to implement their own national chemical control procedures. It does this by working with countries to develop harmonized policies and practical tools (such as the OECD test guidelines) for protecting human health and the environment. This cooperation enables countries to learn from each others' experiences, to share the burden with respect to the testing and assessment of both new and existing chemicals, and to work toward identifying, preventing, and reducing risks associated with use of chemicals.

The OECD Chemicals Programme has evolved over the years to reflect the change in member countries from the "react and cure" policies of the 1970s to the "anticipate and prevent" policies of the 1990s. Important goals of future work are to address the needs of sustainable environmental management within member countries and to increase the usefulness of the Chemicals Programme work for countries outside OECD.

The activities on test guidelines, GLPs, hazard assessment, classification and labeling, cooperative investigation of existing chemicals and risk reduction and pesticides described above are continuing. With respect to test guidelines, the work is broadening to address, in more detail, issues such as test design and data analysis, problems with testing difficult substances (e.g., particularly those of low solubility), and exposure assessment in relation to human health.

Appendix: OECD GUIDELINES FOR TESTING OF CHEMICALS: LIST OF GUIDELINES CURRENTLY AVAILABLE (December 1994)

Section 1: Physical-Chemical Properties

Summary of considerations in the report from the OECD Expert Group on Physical Chemistry:

101 UV-VIS Absorption Spectra[1]
102 Melting Point/Melting Range[1]*
103 Boiling Point/Boiling Range[1]*
104 Vapour Pressure Curve[1]*
105 Water Solubility[1]*
106 Adsorption/Desorption[1]
107 Partition Coefficient (*n*-octanol/water)[1]*
108 Complex Formation Ability in Water[1]*
109 Density of Liquids and Solids[1]*
110 Particle Size Distribution/Fibre Length and Diameter Distributions[1]
111 Hydrolysis as a Function of pH[1]
112 Dissociation Constants in Water[1]
113 Screening Test for Thermal Stability and Stability in Air[1]
114 Viscosity of Liquids[1]
115 Surface Tension of Aqueous Solutions[1]*
116 Fat Solubility of Solid and Liquid Substances[1]
117 Partition Coefficient (*n*-octanol/water), HPLC Method[8]

Section 2: Effects on Biotic Systems

Summary of considerations in the report from the OECD Expert Group on Ecotoxicology:

201 Alga, Growth Inhibition Test[5]
202 *Daphnia* sp. Acute Immobilisation Test and Reproduction Test[3]
203 Fish, Acute Toxicity Test[9]
204 Fish, Prolonged Toxicity Test: 14-Day Study[4]
205 Avian Dietary Toxicity Test[4]
206 Avian Reproduction Test[4]
207 Earthworm, Acute Toxicity Tests[4]
208 Terrestrial Plants, Growth Test[4]
209 Activated Sludge, Respiration Inhibition Test[4]
210 Fish, Early-Life Stage Toxicity Test[10]

Section 3: Degradation and Accumulation

Summary of considerations in the report from the OECD Expert Group on Degradation/Accumulation:

301 Ready Biodegradability[9]
- 301 A: DOC Die-Away Test
- 301 B: CO_2 Evolution Test
- 301 C: Modified MITI Test (I)
- 301 D: Closed Bottle Test
- 301 E: Modified OECD Screening Test
- 301 F: Manometric Respirometry Test

302 A Inherent Biodegradability: Modified SCAS Test[1]
302 B Inherent Biodegradability: Zahn-Wellens/EMPA Test[9]
302 C Inherent Biodegradability: Modified MITI Test (II)[1]
303 A Simulation Test—Aerobic Sewage Treatment: Coupled Units Test[1]
304 A Inherent Biodegradability in Soil[1]
305 A Bioaccumulation: Sequential Static Fish Test[1]
305 B Bioaccumulation: Semi-Static Fish Test[1]
305 C Bioaccumulation: Degree of Bioconcentration in Fish[1]
305 D Static Fish Test[1]

[1]Original guideline, adopted May 12, 1981.
[2]Original guideline, adopted May 26, 1983.
[3]Updated guideline, adopted April 4, 1984.
[4]Original guideline, adopted April 4, 1984.
[5]Updated guideline, adopted June 7, 1984.
[6]Original guideline, adopted October 6, 1986.
[7]Updated guideline, adopted February 24, 1987.
[8]Original guideline, adopted March 30, 1989.

305 E Flow-Through Fish Test[1]
306 Biodegradability in Seawater[10]

Section 4: Health Effects

Summary of considerations in the report from the OECD Expert Groups on Short and Long Term Toxicology:

401 Acute Oral Toxicity[7]
402 Acute Dermal Toxicity[7]
403 Acute Inhalation Toxicity[1]
404 Acute Dermal Irritation/Corrosion[9]
405 Acute Eye Irritation/Corrosion[7]
406 Skin Sensitisation[9]
407 Repeated Dose Oral Toxicity—Rodent: 28/14-Day[1]*
408 Subchronic Oral Toxicity—Rodent: 90-Day[1]
409 Subchronic Oral Toxicity—Non-Rodent: 90-Day[1]
410 Repeated Dose Dermal Toxicity: 21/28-Day[1]
411 Subchronic Dermal Toxicity: 90-Day[1]
412 Repeated Dose Inhalation Toxicity: 28/14-Day[1]
413 Subchronic Inhalation Toxicity: 90-Day[1]
414 Teratogenicity[1]
415 One-Generation Reproduction Toxicity[2]
416 Two-Generation Reproduction Toxicity[2]
417 Toxicokinetics[4]
418 Acute Delayed Neurotoxicity of Organophosphorus Substances[4]*
419 Subchronic Delayed Neurotoxicity of Organophosphorus Substances: 90-Day[4]*
420 Acute Oral Toxicity—Fixed Dose Method[10]
421 Reproduction/Developmental Toxicity Screening Test▲
451 Carcinogenicity Studies[1]
452 Chronic Toxicity Studies[1]
453 Combined Chronic Toxicity/Carcinogenicity Studies[1]

Introduction to the OECD guidelines on genetic toxicology testing and guidance on the selection and application of assays:

471 Genetic Toxicology: *Salmonella typhimurium*, Reverse Mutation Assay[2]
472 Genetic Toxicology: *Escherichia coli*, Reverse Mutation Assay[2]
473 Genetic Toxicology: *In vitro* Mammalian Cytogenetic Test[2]
474 Genetic Toxicology: Micronucleus Test[2]
475 Genetic Toxicology: *In vivo* Mammalian Bone Marrow Cytogenetic Test—Chromosomal Analysis[4]
476 Genetic Toxicology: *In vitro* Mammalian Cell Gene Mutation Tests[4]
477 Genetic Toxicology: Sex-Linked Recessive Lethal Test in *Drosophila melanogaster*[4]
478 Genetic Toxicology: Rodent Dominant Lethal Test[4]
479 Genetic Toxicology: *In vitro* Sister Chromatid Exchange Assay in Mammalian Cells[6]
480 Genetic Toxicology: *Saccharomyces cerevisiae*, Gene Mutation Assay[6]
481 Genetic Toxicology: *Saccharomyces cerevisiae*, Mitotic Recombination Assay[6]
482 Genetic Toxicology: DNA Damage and Repair, Unscheduled DNA Synthesis in Mammalian Cells *in vitro*[6]
483 Genetic Toxicology: Mammalian Germ Cell Cytogenetic Assay[6]
484 Genetic Toxicology: Mouse Spot Test[6]
485 Genetic Toxicology: Mouse Heritable Translocation Assay[6]

[9]Updated guideline, adopted July 17, 1992.
[10]Original guideline, adopted July 17, 1992.
*Updated guideline, expected to be adopted in 1995.
▲New guideline, expected to be adopted in 1995.

SUPPLEMENTAL READING

Publications

OECD Guidelines for Testing of Chemicals, continuing series, first published 1981.
Existing Chemicals: Systematic Investigation—Priority Setting and Chemicals Reviews, 1986.

OECD Environment Monographs

No. 26: Report of the OECD Workshop on Ecological Effects Assessment, 1989.
No. 27: Compendium of Environmental Assessment Methods for Chemicals, 1989.
No. 35: A Survey of New Chemicals Notification Procedures in OECD Member Countries, 1990.
No. 37: Integrated Pollution Prevention and Control, 1991.
No. 45: The OECD Principles of Good Laboratory Practice, 1992.
No. 46: Guides for Compliance Monitoring Procedures for Good Laboratory Practice, 1992.
No. 47: Guidance for the Conduct of Laboratory Inspection and Study Audits, 1992.
No. 48: Quality Assurance and GLP, 1992.
No. 49: Compliance of Laboratory Suppliers with GLP Principles, 1992.
No. 50: The Application of the GLP Principles to Field Studies, 1992.
No. 58: Report of the OECD Workshop on Quantitative Structure Activity Relationships (QSARS) in Aquatic Effects Assessment, 1992.
No. 59: Report of the OECD Workshop on the Extrapolation of Laboratory Aquatic Toxicity Data to the Real Environment, 1992.
No. 60: Report of the OECD Workshop on Effects Assessment of Chemicals in Sediment, 1992.
No. 65: Risk Reduction Monograph 1: Lead—Background and National Experience with Reducing Risk, 1993.
No. 67: Application of Structure Activity Relationships to the Estimation of Properties Important in Exposure Assessment, 1993.
No. 68: Structure Activity Relationships for Biodegradation, 1993.

No. 69: Report of the OECD Workshop on Application of Simple Models for Exposure Assessment, 1993.
No. 70: Report of the OECD Workshop on Occupational and Consumer Exposure Assessments, 1993.
No. 76: Guidance Document for the Development of OECD Guidelines for Testing of Chemicals, 1993.
Guidance on Aquatic Effects Assessment, in preparation.

Relevant OECD Council Acts

C(81)30: Decision on the Mutual Acceptance of Data in the Assessment of Chemicals.
C(83)95: Recommendation concerning the Mutual Recognition of Compliance with Good Laboratory Practice.
C(87)90: Decision/Recommendation on the Systematic Investigation of Existing Chemicals.
C(90)163: Decision/Recommendation on the Co-operative Investigation and Risk Reduction of Existing Chemicals.

How to Obtain Copies of OECD Documents

Publications should be ordered from:

OECD Publications
Environmental Health and Safety Division
2, rue André Pascal
75775 Paris Cedex 16, France

The **OECD Environmental Monographs and Council Acts** are available at no charge on a limited basis from the OECD Environment Directorate, Environmental Health and Safety Division at the address above.

Chapter 27

ENVIRONMENTAL REGULATION IN THE EUROPEAN COMMUNITY

J. Blok and *F. Balk*

INTRODUCTION—ENVIRONMENTAL AWARENESS FROM A EUROPEAN HISTORICAL PERSPECTIVE

An outbreak of waterborne epidemic diseases, cholera, and typhus in several European countries occurred about 100 to 150 years ago (e.g., in 1832 and 1877 along the Thames). As a result, the first environmental law in Europe, the Rivers Pollution Prevention Act, was enacted in the United Kingdom in 1876. Since then, slow sand filtration and river bank filtration have been effectively applied for the preparation of drinking water. A United Kingdom Royal Commission studied sewage disposal from 1896 to 1915 and promulgated standards for the discharge of sewage to rivers. This Commission used biochemical oxygen demand (BOD) as the main criterion for quantification of water pollution. The Commission believed that the longest retention time for any particle in a British river would be about 5 d; the famous BOD_5 was therefore introduced (Haigh, 1984). At this time no one realized that a 5-d period for BOD would cause practical inconvenience in laboratories. The United Kingdom Royal Commission recommended that effluents with a BOD_5 of less than 20 mg/L were sufficiently purified. The recommendation had no legal status, and it was not until 1951 that the Rivers Pollution Prevention Act included river water quality objectives. However, the resistance against clear and precise water quality objectives was so strong that the law was ineffective. Figure 1 shows that in the period 1950 to 1955 the concentration of dissolved oxygen in the Thames River reached almost zero. It was not until 1973 and 1974 that the Water Act and the Control of Pollution Act, respectively, became effective.

In The Netherlands the situation was not much better. A special governmental commission proposed prohibiting the discharge of industrial wastewater to rivers in 1901. However, it took the government 69 yr to prepare a law on surface water pollution. In 1970 only 20% of the Dutch urban and industrial wastewater was purified.

In France the Water Law of 1964 was not very effective until the Ministerial Circular of 1971 set precise water quality objectives. Other countries showed similar developments in environmental awareness. In 25 Organization for Economic Cooperation and Development (OECD) countries, 15 environmental laws were enacted before 1967, whereas in the period from 1967 to 1979, 73 laws on water, waste, air, and chemicals were prepared (OECD, 1979).

Figure 2 shows the historical picture of environmental concerns during a period of about 150 yr in western Europe. It is worth noting that this was during a period of strong economic growth. The economic growth was considerably lower in the countries of southern Europe and in the eastern European countries. Thus at the time (1981–1983) when these countries of southern Europe became members of the European Community (EC), they were at a different stage of environmental awareness. The unification of Germany and the opening of the European market to and from Hungarian, Polish, Czech, and Slovakian markets in 1990 revealed significant differences in the level of environmental awareness.

Table 1 shows a list of major events that have influenced environmental policy in Europe since 1960. In the 1960s widespread public concern with environmental issues was roused by several books with imaginative titles such as *Silent Spring, Spaceship Earth,* and *The Population Bomb.* Many dramatic incidents underpinned the message of the environmental movement.

The environment became more seriously accepted as a political issue in the 1970s. The Council of Europe designated the year 1970 as European Conservation Year, and in 1972 the EC declared the first environmental policy. Also in 1972, at the London

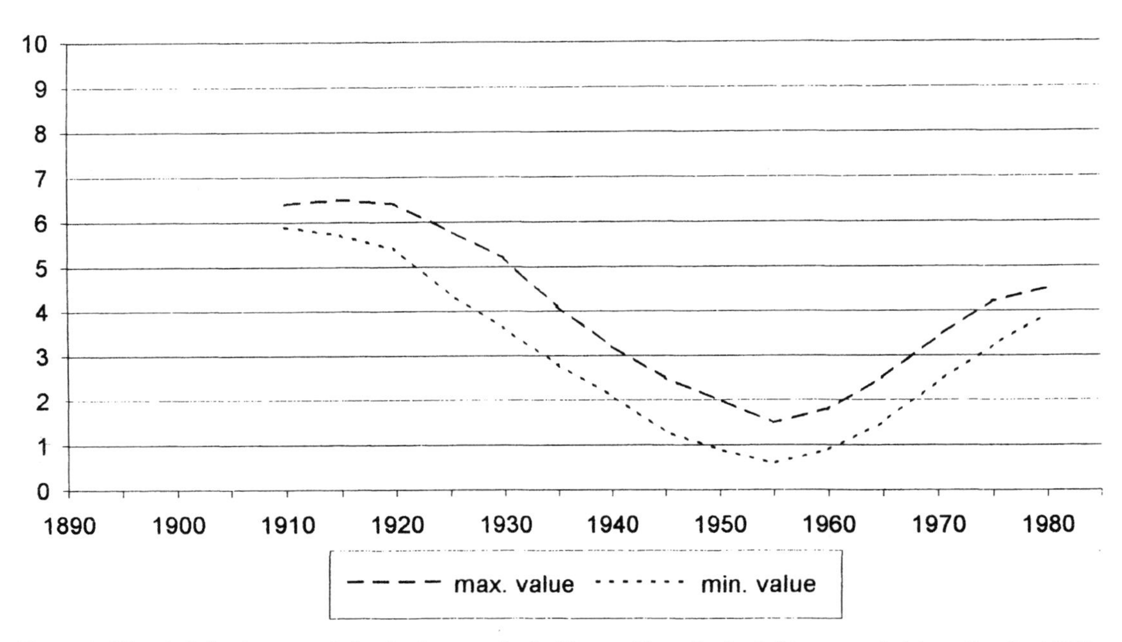

Figure 1. Historical development of dissolved oxygen in the Thames River, England. (Data compiled from Goudie, 1989.)

Convention Against Marine Pollution, it was agreed to stop the dumping of waste in the seas.

Despite many agreements, conventions, and regulations, the actions against pollution took time. Technology had to be adopted and finances had to be generated. This can be illustrated by the percentage of the European population served by complete biological treatment of domestic wastewater. Table 2 gives these figures for the EC member states. At the end of the 1980s Luxembourg, Denmark, West Ger-

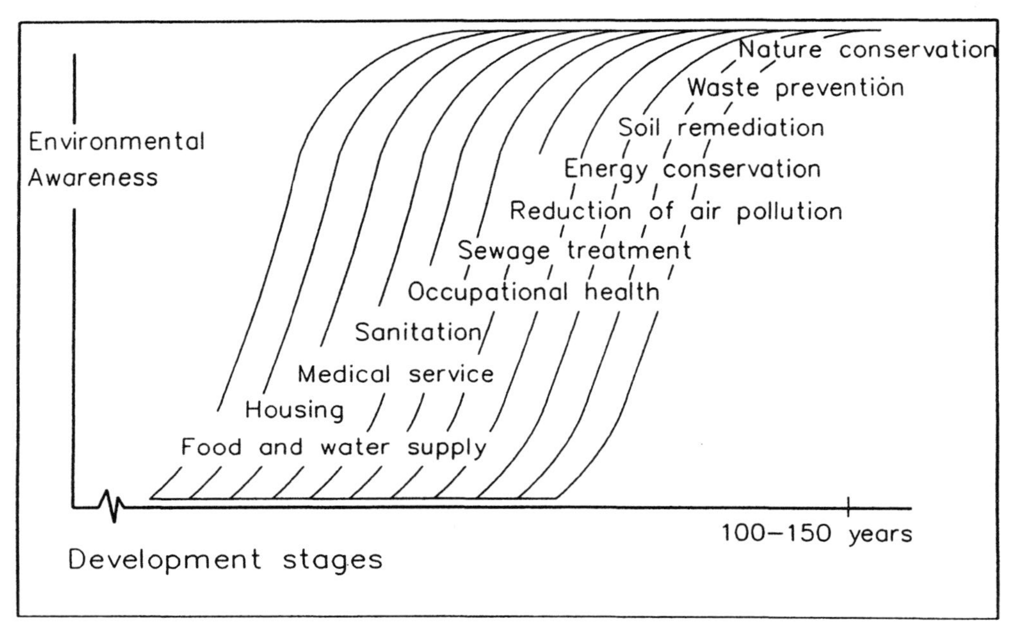

Figure 2. Sequence of environmental items in a period of 150 yr.

Table 1. Important events that influenced environmental awareness in Europe

	The 1960s, decade of prophets
1960	Invention of the electron capture detector by Lovelock.
	An extremely sensitive device to identify chlorinated hydrocarbons in environmental samples.
1962	*Silent Spring* by Rachel Carson warns against the worldwide pollution with chlorinated pesticides and becomes a bestseller.
1963	The yearly production of DDT in the United States reaches its peak at 80,000 tons. Explosion of Philips Duphar in Amsterdam with massive release of TCDD (tetrachlorodibenzodioxin), the most toxic substance ever made by man.
	The International Rhine Commission starts its struggle for a healthy river by a treaty between the five bordering countries.
1964	The French issue their Water Law, one of the first environmental regulations in Europe.
1965	In Japan the Minamata disease appears to be caused by mercury-poisoned fish.
1966	*Spaceship Earth* is written by Kenneth Boulding, focusing on the limitations on the economy of the planet.
1967	The wreck of the giant oil tanker *Torrey Canyon* shocks the world by tremendous oil pollution.
1968	The painful Itai-Itai disease in Japan is caused by cadmium poisoning.
	The oil disease or Yusho in Japan is caused by PCB-contaminated rice oil.
	The Population Bomb by Paul Ehrlich becomes a bestseller.
1969	In Santa Barbara the first blowout of an oil well causes dramatic pollution.
	Massive fish kill in the Rhine River over a distance of 800 km is caused by accidental release of endosulfan by Hoechst am Main in Germany.
	Massive kill of seabirds in the Irish Sea is caused by PCB and other poisons accumulated in fat tissue.
	The 1970s, decade of political awareness
1970	The Council of Europe proclaims the European Conservation Year.
	After 69 years of preparation the Dutch government presents the law on surface water protection.
1971	The French government produces the first precise quality objectives for surface water.
	In Iraq the consumption of grain disinfected with mercury compounds causes several thousands of victims.
1972	The EC declares its environmental policy.
	The London Marpol Convention declares an end to dumping waste in the sea.
	The "Club of Rome" prepares *The Limits of Growth* by Meadows et al.
1973	The British government produces its Water Act after 80 years of preparation by several commissions.
	The EC notifies the first directive on biodegradability of detergents.
1974	The UK follows with the Control of Pollution Act.
	The carcinogenity of vinyl chloride is discovered.
1976	At Seveso, Italy, a herbicide (2,4,5-T) factory explodes and causes serious dioxin (TCDD) pollution.
	The EC notifies the "black" and "gray" Lists, regulating elimination and reduction, respectively, of dangerous substances in water.
1978	The EC publishes water quality standards for freshwater fish.
1979	The EC publishes water quality standards for shellfish.
	The EC amends for the sixth time a directive of 1967 by adding environmental criteria for regulating the classification, packaging, and labeling of dangerous substances.
	The relation between minimal brain disease and lead poisoning becomes evident.
	The EC prohibits the use of PCB in open systems.
	The 1980s, decade of action
1980	Discovery of hormonal disorder in oysters caused by accumulation of the antifoulant tributyl/tin oxide (TBTO).
1982	Major accidental release of chloronitrobenzene in Rhine.
	The International Rhine Tribunal in Rotterdam takes to court industrial polluters and the International Rhine Commission.
	The EC directive on detergents is completed with methods on biodegradability measurement.
	Mercury use and release from the chloralkali industry is regulated by EEC Directive 82/176.
1983	Waste dumps from the production of titanium dioxide are regulated by EC directives.
1984	In Bhopal, India, an accident with methylisocyanate causes a high number of human casualties.
1985	The first unleaded petrol appears on the European market.
1986	A fire in a pesticide storage plant in Sandoz (Basel) causes massive release of pesticides into the Rhine.
	After the Chernobyl accident a considerable part of Europe is covered with radioactive fallout from the nuclear power plant.
1990	The borders between western and eastern European countries are opened. After 40 years of communism the facts on heavily polluted areas are revealed.

Table 2. Percentage of the population served by biological wastewater treatment

Country	1970	1975	1980	1985	1987	1989
Belgium	4	6	23			
Denmark	54	70	85	90		
West Germany	61	75	82	86	90	
Spain		14	18	29		48
France	19	31	43	49	50	52
Ireland			11			
Italy	14	22	30		60	
Luxembourg	28		81	83		91
Netherlands		45	72	85	89	92
Portugal	3	6	9	9	11	
United Kingdom			82	83	84	
United States	42	67	70	74		
Japan	16	23	30	36	39	

From Eurostat, OECD, 1992.

many, and The Netherlands were ahead, with more than 90% of the wastewater being completely treated. Results of the efforts are also illustrated in Figure 3*a* and *b*, which show decreasing pollution levels for heavy metals in the Rhine River.

THE STORY OF THE SOAPERS

Because of the high production volumes and general widespread use of surfactants, the soap and detergent industry has been facing health and environmental concerns about its products since the early 1960s. For this reason the development of testing methods and regulations was triggered by this industry, and therefore the story of these front-runners is described here.

Although natural ''soaps'' have been extracted from soapwort and other plants, commercial soap is derived by ''soaping'' of natural fats and oils with sodium or potassium. Boiling under alkalic conditions yields the hard sodium salts or the soft potassium salts of fatty acids. The original small-scale manufacturers were called soapers. Soaping was practiced 4500 yr ago (Richtler and Knaut, 1991). Although the raw materials (tallow, line oil, sunflower oil, olive oil, cocos oil, ricinus oil, soy oil, and palm oil) are natural, soaps do not occur in nature and thus are synthetic products in the strict sense of the word. In 1987 the world production of soap for all applications was about 8.5 million metric tonnes a year (Richtler and Knaut, 1991), and soap is the world's most important oleochemical.

Cleaning is effectuated by the surface activity of molecules with a hydrophobic part connected to a hydrophilic counterpart. These surfactants are able to form monomolecular layers at interfaces and emulsions on themselves or around particles. For the process of cleaning, protein, fat droplets, and soil particles are to be dispersed and emulsified. The creation of foam in solutions containing as little soap as 0.5 mg/L is another consequence of surface activity that can be useful in certain flotation processes but is not necessarily valuable for washing.

Soaps or other surface-active substances can be mixed with abrasives, disinfectants, chelating agents, bleaching agents, perfumes, and fillers to tailor detergent formulations for a variety of applications. Of these the laundry detergents are the most well known.

Since about 1930 fatty alcohols from natural oils and fats have been sulfonated and alcohol ethoxylates produced as half-synthetic anionic and nonionic surfactants (the fatty alcohols are of natural origin, whereas the polyethylene-oxide chain is synthetic). Since about 1950 so-called synthetic alcohols have been derived from mineral oil. One of the synthetic products, tetrapropylenebenzenesulfonate or TPBS, was used as a versatile replacement for soap. During the 1950s the production volume of TPBS increased until the early 1960s, when rivers and creeks were periodically covered with huge layers of foam that accumulated at places of turbulence. In some places in Germany the foam on waterways became a hazard to navigation. The foam problem became more dramatic as a result of reduced oxygenation of the surface water. The cause was easily identified as poor biodegradability of the branched alkyl chain with four isopropylene groups in TPBS. In 1958 full-scale trials in a treatment plant with a linear product gave encouraging results. In 1964 the chemical industry was ready to replace TPBS with linear alkylbenzenesulfonate, or LAS (see Chapter 31). LAS is produced on the basis of linear alpha-olefins reacting with benzyl chloride, or chloroparaffins reacting with benzene, followed by chemical sulfonation.

In 1961, almost parallel to the production of LAS, the German government passed an act specifying that any surfactant in household laundry detergents should be degradable. A committee chaired by Professor Hussmann was set up to define a standard and suitable test procedure for measuring biodegradability. The test procedure that was developed was a laboratory-scale simulation of the activated sludge process for wastewater treatment. The equipment became well known as the Hussmann apparatus. It was found that the cationic dye methylene blue differ-

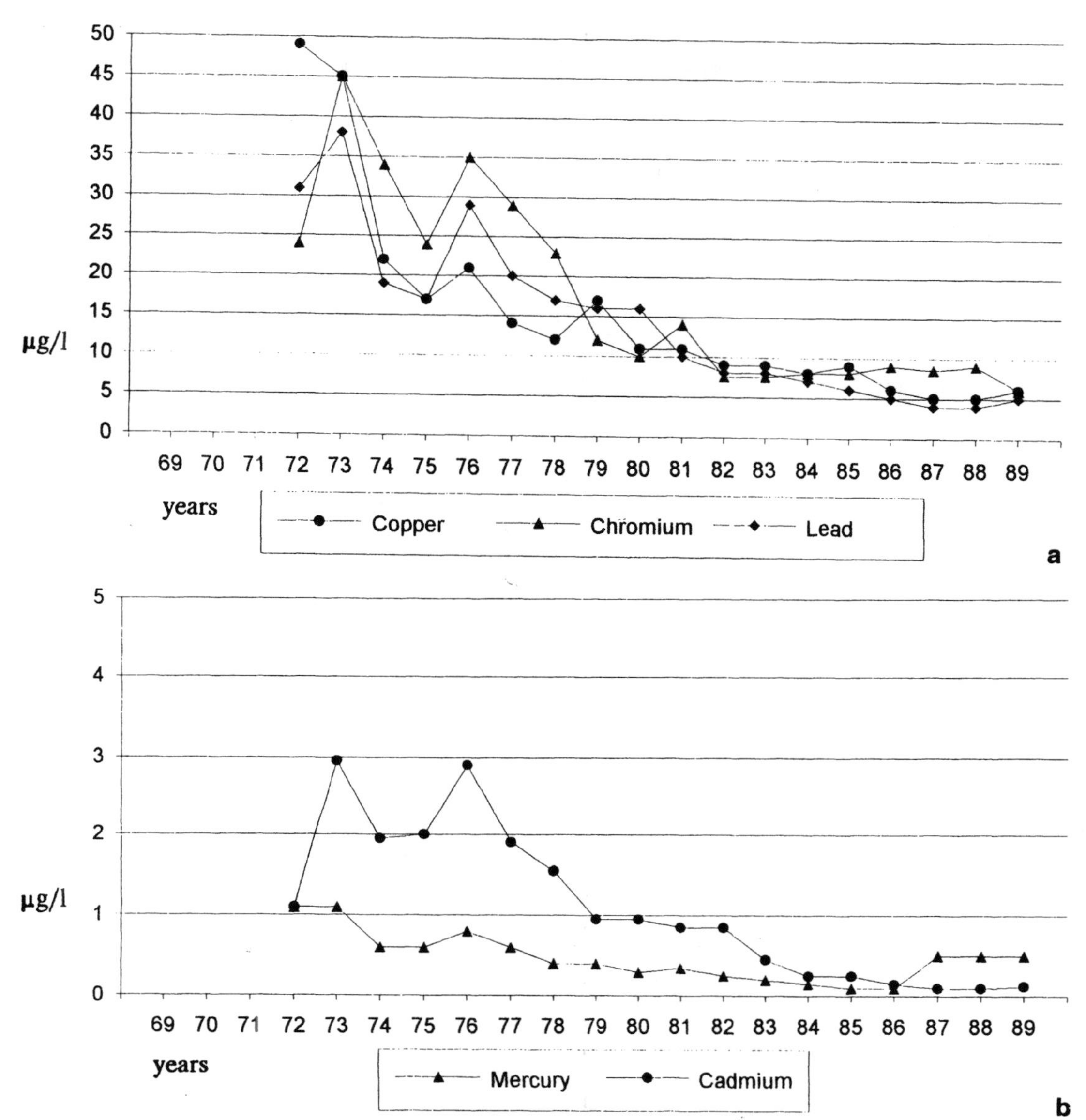

Figure 3. (*a*) Decrease of heavy metal pollution in the Rhine River from about 40 to 5 µg/L for copper, chromium, and lead. (*b*) Decrease of heavy metal pollution in the Rhine River from about 3 to 0.1 µg/L for cadmium and from about 1 to 0.05 µg/L for mercury. (Data compiled from RIWA, 1992.)

entiates between anionics and soap by forming chloroform-extractable complexes. On this basis a method for testing and measuring the removal could be described with a standard of 80% disappearance of methylene blue–active substance (MBAS). In 1964 the German act could be enforced and almost at the same time all soap and detergent industries in Europe agreed voluntarily on the replacement of TPBS with LAS; therefore no other countries felt it necessary to enact special detergent laws.

The Council of Europe achieved partial agreement in 1968 on restricted use of certain surfactants. In 1971 the OECD came up with a specific guideline on the biodegradation of surfactants based on the Hussmann test and MBAS removal. Certain non-degradable anionic surfactants were then prohibited in France, and in the same year an EC directive was proposed. This directive became enforceable in 1973, 9 yr after the TPBS problem had been solved.

The disappearance of the foam problem was as spectacular as its appearance. From measurements of the total amount of MBAS in the Rhine River, the increase starting in 1958 and the decrease starting in 1964 are clearly visible in Figure 4. Further decrease

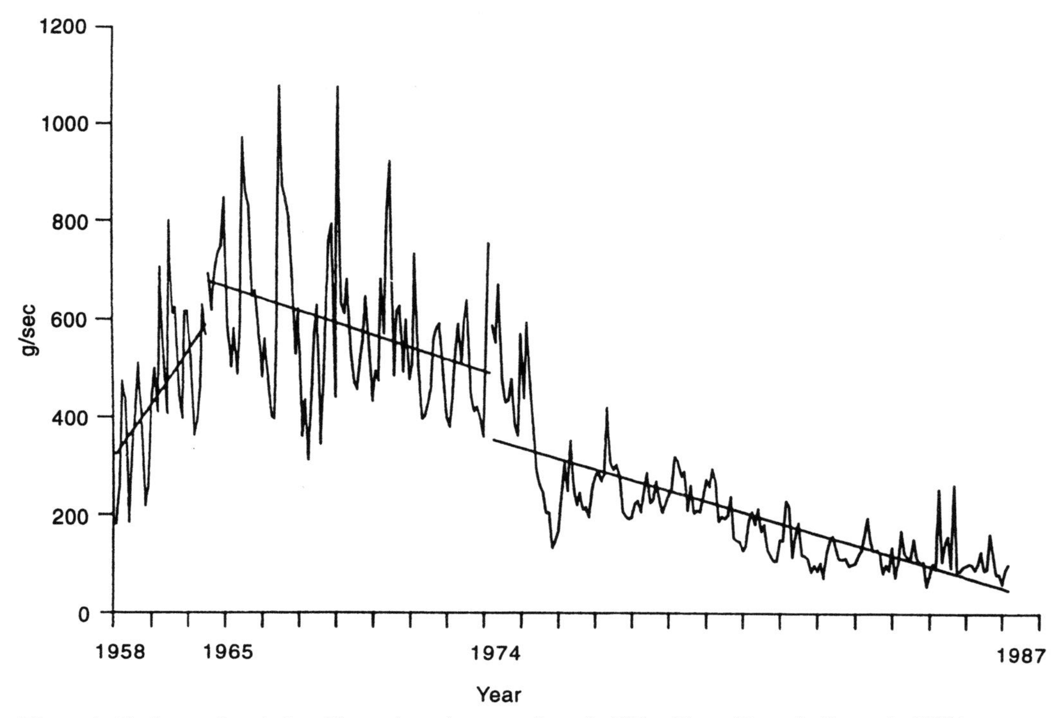

Figure 4. Discharge of methylene blue active substances along the Rhine River. (From Gerike et al., 1989.)

since 1972 is due to the increased number of sewage purification plants built in that period. In 1973 a plant with a capacity of 1.5 million inhabitant equivalents came into operation (Gerike et al., 1989).

In the meantime, about three dozen different synthetic surfactants have been introduced on the market without creating a foam problem. In 1982 the total world output of synthetic surfactants was 5 million metric tonnes a year. In 1988 the production of LAS in the United States, Europe, and Japan was estimated at about 1.4 million tonnes a year.

EC Directive 73/404 covers many types of surfactants: anionic, cationic, nonionic, and ampholytic, specifying an average of 90% biodegradability without description of a test method. The directive is not concerned with the biodegradability of any constituent other than surfactants. In the "daughter" Directive 73/405 specific methods for testing biodegradability of anionic surfactants were promulgated. The methods were taken from the German and the French legislation and from the OECD method (see Chapter 26). In Directives 82/242 and 82/243 the former directives are amended with an analytical method for nonionic surfactants and a British test method called the "porous pot." It should be mentioned here that at that time no regulation on the aquatic toxicity of these "bulk chemicals" was proposed in any country.

The regulation of detergents in Europe had a few interesting consequences. It did not influence the use of TPBS, because this product had been replaced voluntarily by LAS before the regulation came into force. However, such regulation appeared rather successful in preventing any of the three dozen newly developed synthetic surfactants from causing comparable problems. The most significant effect of the regulation was on the early development of testing for biodegradation and aquatic toxicity. Since the TPBS situation, researchers in the laboratories of the soap and detergent industry have been the frontrunners in routine testing as well as in methodological development. The soap and detergent producers of competing companies worked closely together in environmental task forces of the Soap and Detergent Association (SDA; United States), Association Internationale de la Savonnerie (AIS; Europe), and Comité Européen des Agents de Surface et leurs Intermédiaires Organiques (CESIO) in a remarkably cooperative way. This is even more remarkable because the same companies competed and still compete with one another in commercials on television with such intensity that the easily digestible family television programs designed to attract millions of people, the "soap operas," were exclusively financed by these detergent commercials. The reason for this remarkable cooperation of these otherwise fiercely competing companies is to sustain maximum freedom of detergent formulation on the basis of the emerging environmental science.

A second interesting feature was the invention of the tiered approach of testing (see Chapter 1). In this approach, testing is done in a series of steps, or tiers, of increasing complexity. Testing for biodegradation starts with a screening method and is followed by a confirmatory test if the results of the screening were not sufficiently clear. The screening methods are simple and easy to perform, using static systems with open flasks. The confirmatory tests use systems designed to simulate realistic conditions in biological sewage treatment works. After 20 yr this tiered approach is still a guiding principle for environmental risk assessment.

Finally, it is interesting that for the preparation of the OECD guideline on detergents of 1971, experts from 29 different laboratories representing 13 countries worked together to develop the methods in parallel trials in interlaboratory ''round-robin'' or ring tests (see Chapter 1). Later, interlaboratory testing became almost ''epidemic'' among environmental scientists and greatly influenced the further environmental regulation of chemicals.

Only a few other surfactants have been withdrawn from the market for environmental reasons. Among these are some nonionic surfactants. An early difficulty was the lack of a suitable analytical method for measuring traces of a variety of nonionic surfactants analogous to the MBAS procedure for anionics. After the analytical problems were solved with the bismuth iodide method, a group of branched alkylphenol ethoxylates (APEs) was found to be poorly degradable. Quite remarkably, tests in the United States showed higher degrees of biodegradability for APEs, than had been found in Europe. Then, from about 1970 onward, laboratory tests in Europe also began to give much higher results, presumably because of the development of adapted microorganisms in water purification systems. At the time EC Directive 82/242 became effective, the tests in Europe routinely showed results higher than 80% removal.

In 1984 scientific publications on nonylphenol-ethoxylates (NPOs) demonstrated some transient biodegradation intermediates that were relatively more recalcitrant to further mineralization at lower temperatures (Giger et al., 1987). These NPO intermediates with about two ethoxylate groups were detectable in sediment of rivers and estuaries and appeared to be more toxic than the original NPO. The intermediates were strongly adsorbed on sludge of treatment plants and converted to nonylphenol under anaerobic conditions in the digester. The NPO intermediates slipped through the screens of the testing systems because these methods measured only so-called primary degradation or removal and not complete mineralization (see Chapter 15). A few years later the soap and detergent industries agreed to phase out NPO from all household products and, as far as technically possible, from industrial products as well.

In the period 1980 to 1990 the environmental acceptability of the quaternary ammonium salts of dimethyl distearic acid or ditallow acid (DSDMAC or DTDMAC) was discussed. These cationic surfactants used for fabric softening were almost completely removed in the confirmatory test systems but biodegraded slowly. Because of their strong adsorption and lower mineralization rate, they were found at relatively high levels on the activated sludge of treatment plants. The analytical determination of DSDMAC in environmental samples is rather complicated. Moreover, DSDMAC as such is insoluble in water and sorbs to anionic surfactants and clay to become poorly bioavailable. The toxicity to aquatic organisms is difficult to measure and results from tests with artificial ''solutions'' are even more difficult to extrapolate to river conditions. The soap and detergent industry spent much effort to defend the use of this type of chemical in fabric softener products in The Netherlands. Despite the technical and methodological problems, calculations could predict environmental concentrations (PECs) safely below the lowest no-effect concentration (NEC) for aquatic organisms. A 90th percentile value for the concentration in surface water in The Netherlands is estimated as 0.045 mg/L and the lowest no observed effect concentration (NOEC) is 0.2–0.3 mg/L (Versteeg et al., 1992). The discussion focused, however, on the application of a safety factor based on a newly proposed method for statistical extrapolation of laboratory-derived toxicity data to the field. As a result of the continuing discussions, a market was created for a less suspicious alternative. In 1990 a less disputable alternative came on the market, and since then DSDMAC has been gradually disappearing.

Another group of surfactants escaped the regulations on household detergents because their application was mainly in industrial dishwashing machines. These products are tailored to have a low degree of foaming and contain a copolymer chain of ethylene oxide and propylene oxide units. The copolymer structure of this group of products, however, is poorly degradable. Through an agreement between the chemical industry and the German government it was decided that these products should be replaced by degradable alternatives within a specific time schedule.

Although there was no doubt about the better environmental performance of LAS as compared to TPBS, the environmental fate and behavior of LAS have been researched extensively since 1964. In 1988 the European soap and detergent industries sponsored a comprehensive literature review on LAS. The review refers to more than 500 scientific papers. A few months later a special seminar on LAS was held in Germany (Aachen, 1989) showing many new results. In spite of all evidence on the environmental safety of LAS, some results triggered new questions and research proceeded. The development of improved analytical procedures based on high-performance liquid chromatography (HPLC) and fluorometric detection in 1987 and the method based on derivation and gas chromatography–mass spectrometry (GC-MS) enabled the measurement of specific homologues and isomers at a level of 1 μg/L in environmental samples. The latter motivated several laboratories to conduct monitoring studies on sewage treatment plants. The availability of radiolabeled material, on the other hand, triggered a series of detailed studies on mass balance and on the kinetics of biodegradation at low concentration levels. As a result, LAS has become the most extensively environmentally researched chemical (De Oude, 1992).

In a survey including the "gray literature" (or nonpublished data) from the industrial laboratories, more than 650 records on aquatic toxicity were filed and statistically evaluated. The review of toxicity data and the most recent studies on the environmental fate of LAS using sensitive specific analytical methods yielded some interesting results (Blok et al., 1993):

- Commercial LAS consists of about 26 homologues and isomers with varying alkyl chain lengths and variable phenyl positions. Commercial mixtures also contain a small percentage of Tetralin derivatives in about 70 isomer structures and some unsulfonated alkylbenzenes. Adsorption, degradation, and toxicity of these molecules differ significantly and, in sewage treatment, on the way through the primary settler and the biological oxidation tank and farther down to the river, the composition of the mixture changes. This means that testing based on the commercial parent material is not fully adequate for environmental hazard assessment (see Chapters 28 and 29). This phenomenon has been largely underestimated in the past and possibly also occurs with many other industrial products consisting of a diverse mixture of homologues and isomers of a certain chemical. For this reason aquatic toxicity testing should predict environmental effects for such mixtures either on the basis of knowledge of all individual homologues or on the basis of structure-activity relationships (see Chapter 20) and the calculation of the combined toxic effects for the mixtures.
- LAS has been tested with about 75 aquatic species. The distribution pattern of the mean toxicity for the 75 species is almost equal to the distribution pattern of replicate toxicity data for the most frequently tested organism, *Daphnia magna* (Figure 5*a* and *b*). This indicates that influences other than species diversity determine the variability of data. It also implies that statistical extrapolation of literature data to calculate the concentration at which a high percentage of the natural species can be assumed to be protected is not justified.
- Not surprisingly, the distribution of tested species over the main taxonomic groups does not at all correspond to their occurrence in natural ecosystems of interest (Figure 6*a* and *b*). The reason is that some species are extremely suitable for testing and therefore became "popular" test species for which international standardized methods have been made. The question of representative test species and the need to cover a variety of taxonomic groups create a conflict of interest with the need for standardization for intercomparability.

In spite of all research thus far and in spite of repeated validation of predictive models, new questions are being raised continually. Detailed monitoring data and field toxicity tests are expected to present a more refined picture. The attention to the environmental impact of LAS contrasts strongly with the ease with which soaps are being accepted. Although world production volumes for soaps are about two times higher than for all synthetic surfactants and their use is much more pronounced in countries without proper sewage treatment, the toxicity or the biodegradability of soap is not questioned, nor is there adequate scientific literature to answer such questions satisfactorily if they were posed. Presumably the incorrect use of the words "synthetic" for LAS versus "natural" for soap is to be blamed for this remarkable contrast.

Between 1970 and 1990 the detergent industry was involved in a debate on the environmental acceptability of a major detergent ingredient, phosphate, because of its nutrient value. The introduction of alternatives such as nitrilotriacetic acid (NTA) was judged on the possibility of heavy metal mobilization by forming complexes.

The story of the soap and detergent industry illustrates that aquatic toxicity is only a part of environmental acceptability. Biodegradability, nutrient value, and complexing of heavy metals are also important aspects to consider.

STRUCTURE OF THE EUROPEAN COMMUNITY WITH REGARD TO ENVIRONMENTAL LEGISLATION

The European Community (EC) was formed in 1957 by the Treaty of Rome. It unified the European Economic Community (EEC), the European Atomic Energy Community (EURATOM), and the European Coal and Steel Community (ECSC). At that time there were six member states. At present the European Community has 15 member states: Belgium, Luxembourg, The Netherlands, France, Germany, Italy, United Kingdom (1972), Ireland (1973), Denmark (1973), Greece (1981), Spain, and Portugal (1983). As of 1995, Finland, Sweden and Austria joined.

The European Community is a unique political entity. It is neither a national government nor just an international organization. The EC member states are not subnational entities but completely sovereign states. Still, these states have given the Community law-making and law-enforcing powers that go far beyond those of international organizations such as the United Nations (UN) or the OECD. The European Community can monitor, coordinate, and enforce the implementation of its laws by member states. Detailed outlines of the EC and its environmental legislation are given by Haigh (1984) and Whitehead (1988).

Laws and regulations are exclusively proposed and administered by the Commission and legislated by a council of representative ministers. The consti-

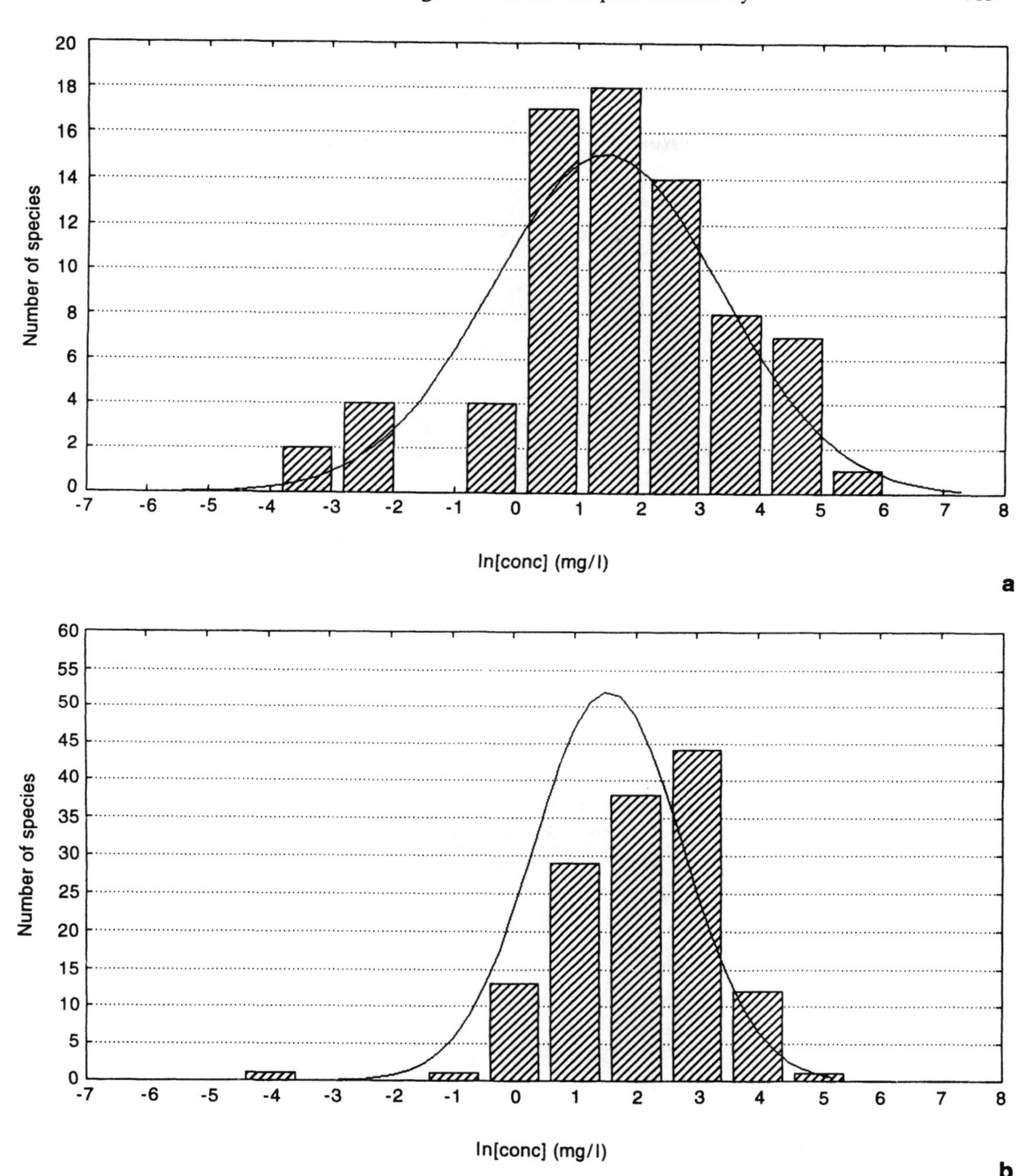

Figure 5. (*a*) Frequency distribution of average LC50 values for LAS toward 75 aquatic species. (*b*) Frequency distribution of replicate LC50 values for LAS to *Daphnia magna* (n = 139). (From Blok et al., 1993.)

tution of the councils differs with the subject area. Environmental legislation typically falls under the competency of the Environmental Council, but some issues fall under the competency of the Council for Internal Market and Industrial Affairs or the Council for Agriculture.

The Commission is composed of 17 commissioners and is supported by a number of Directorates General (DG) and services as the main administrative units. DG XI is responsible for environment, consumer protection, and nuclear safety. Product safety legislation, however, is covered by DG III, Internal Market and Industrial Affairs, and pesticide regulation is covered by DG VI, Agriculture. The Commission proposes new legislation, which is to be adopted by the Council. However, adaptations of

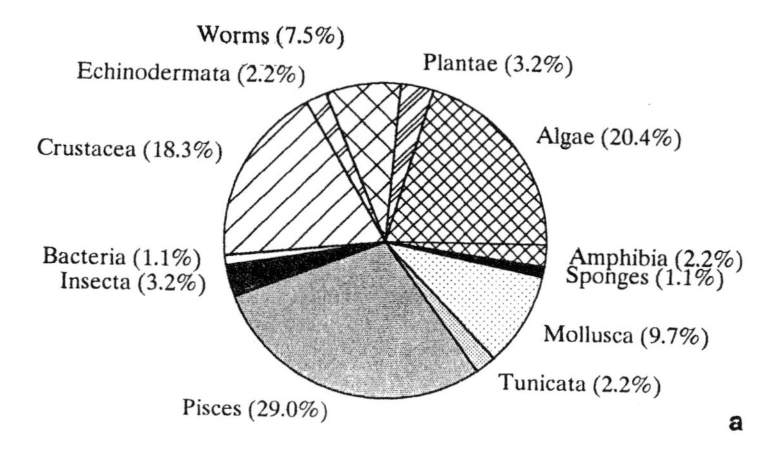

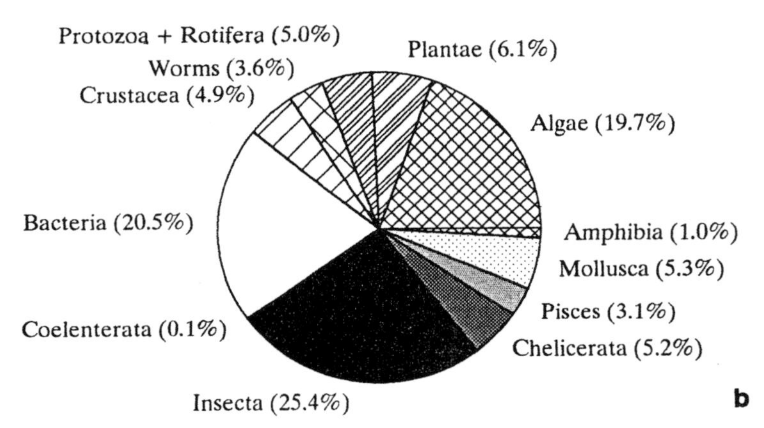

Figure 6. (*a*) Distribution of available toxicity data on LAS for 93 species over taxonomic groups compared to (*b*) the distribution of species over taxonomic groups in the Rhine River.

environmental laws to scientific or technical progress set out in annexes to the legislation may be adopted by the Commission. In most cases the annexes contain descriptions of methods, lists of substances, and several types of limiting values, critical quantities, etc.

An elected European Parliament has a largely advisory role but no legislative power. Other advisory bodies are the Economic Committee and the Social Committee. Cases of conflict between EC institutions or between the Commission and the member states are judged by the Court of Justice.

The Council can adopt:

- Nonbinding recommendations and resolutions
- Regulations directly applicable and binding in member states
- Decisions that are directly binding on member states, legal persons, and individuals
- Directives to be implemented by national laws of the member states

A directive is binding as to the result to be achieved but leaves to the member states the choice of form and methods. A directive, also frequently called a guideline, is the main tool for environmental policy. It empowers the Community to define objectives, standards, and procedures that allow the member states some flexibility in implementation, so they can use their accustomed national regulative processes. Figure 7 gives a schematic presentation of the European Community structure.

In 1973, 1977, 1982, and 1987 the European Community adopted rather comprehensive Programmes of Action, but the instruments for putting them into effect were few. The Community can enunciate principles; it can commission studies and research; it can issue grants from the small experimental fund for

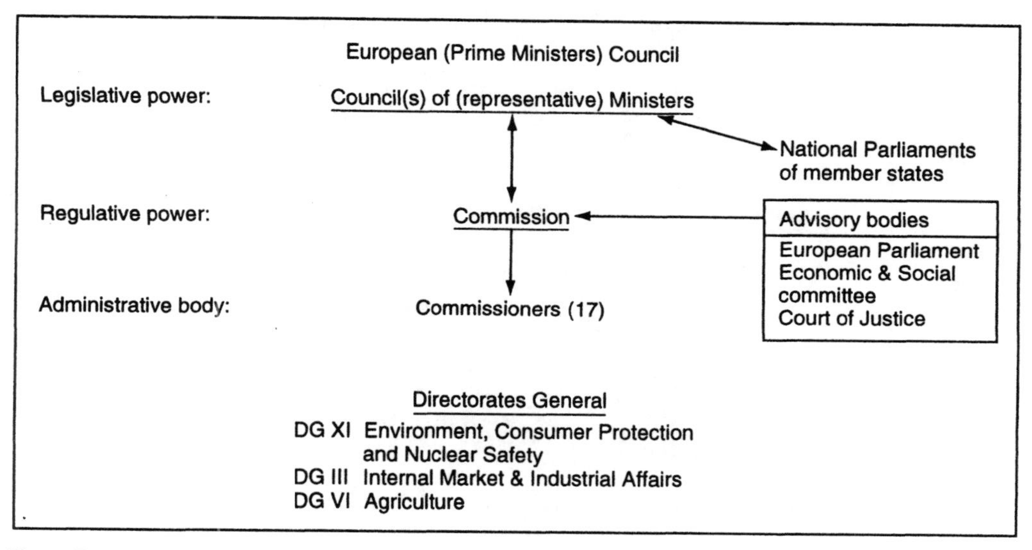

Figure 7. Structure of the European Community.

environmental purposes; it can set up advisory committees and otherwise exhort member states and people within them by promoting meetings, conferences, and publications; and above all, it can legislate. It cannot by itself, or through an agency, directly administer its environmental policies. That must be done by the member states.

The European Community's legislative process involves numerous close consultations with the member states, the European Parliament, and the Economic and Social Committees, as well as with private organizations at the national and EC levels. Member states must often formally consult their national parliaments and other interest groups.

In each member state each environmental issue normally follows a cycle of development that is mostly triggered by some disaster or public scandal. Full integration of measures at legal, technical, institutional, and commercial levels is reached after a period of about 25 yr. Figure 8 shows the stages in the cycle leading to protection of the aquatic environment. As a result of different stages in development, national environmental laws could deviate from each other and form barriers to trade or create "false" competition. Most of the EC directives are, in fact, intended to prevent unequal national requirements. For registration and notification of new substances, for example, the producer has to address one of the member states. Once a new substance is accepted, other member states will be informed by the commission and will have to accept that notification.

As can be concluded from Figures 2 and 8, aquatic toxicology is only a minor item in the broad field of environmental protection. Most of the items mentioned in Figure 2 are the responsibility of national or local authorities and the rules set by these bodies may vary significantly from place to place. Nevertheless, the EC has a major influence on environmental regulation and on methodology development within aquatic toxicology.

The next section will focus on several topics in the development of EC regulations that influenced aquatic toxicology as a science.

A SHORT REVIEW OF ENVIRONMENTAL DIRECTIVES IN THE EUROPEAN COMMUNITY

Protection of the environment can be achieved in different ways. First, the different environmental compartments (i.e., air, soil, and water) should have a minimum quality to serve their various functions. Thus quality objectives can be set for water to be used for bathing, fishing, or drinking. Second, the sources of pollution have to be kept under control by regulation of substances. Finally, accidental spills have to be prevented and disposal of wastes has to be controlled by proper regulation.

Environmental Protection by Water Quality Objectives

Directive 75/440 for Surface Water Quality

Within the First EC Action Programme on the environment the first directive was adopted in 1975 and concerned "surface water for drinking" (75/440). The directive has two purposes:

- To ensure that surface water is treated adequately before being put into public supply as drinking water
- To improve quality of rivers and other surface waters used as sources of drinking water

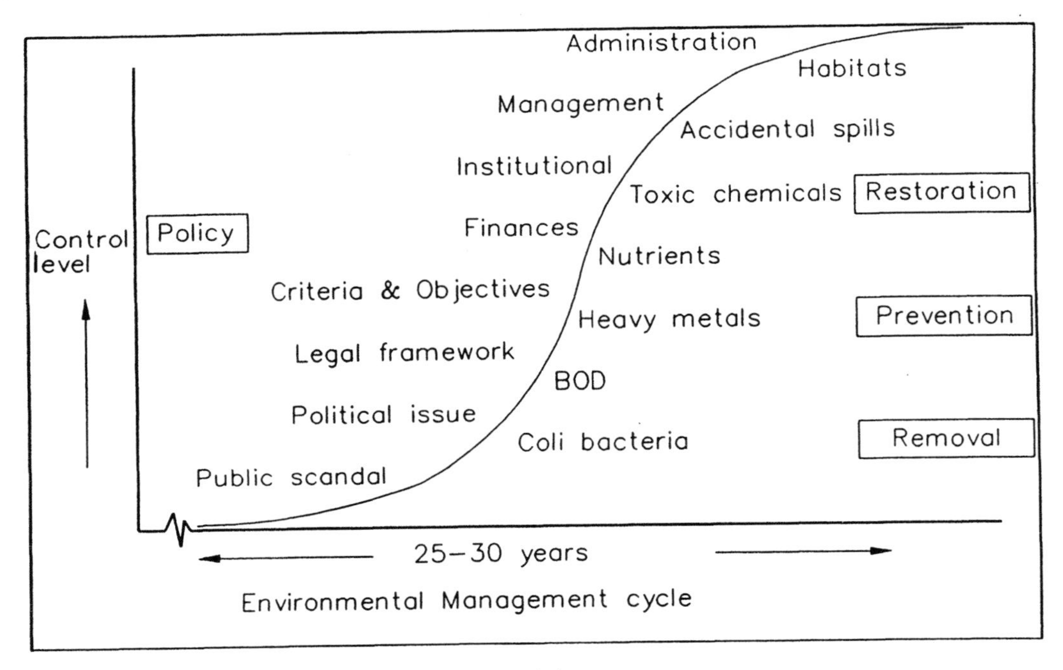

Figure 8. Environmental management of surface water pollution.

Surface water can be classified according to the need for increasingly complex standard methods of treatment required to transform it into drinking water. The best quality (A 1) requires only filtration, whereas the worst quality (A 3) requires intensive physical and chemical treatment and disinfection. The use of surface water worse than A 3 quality is prohibited. The directive requires a plan of action, including a timetable, for the improvement of low-quality surface water. Imperative and guide numerical values for 46 water quality parameters, including physical, chemical, and microbiological characteristics, are listed in the annex.

Directive 80/778—Drinking Water

Although not strictly environmental, Directive 80/778—Drinking water can be mentioned in this context. Annex I in this directive gives a list of more than 60 parameters with a guide level and maximum admissible concentration for drinking water. It covers organoleptic, physiochemical, and microbiological parameters, undesirable and toxic substances, and minimum required hardness.

Other Directives

Further directives on water quality standards came in 1976 (76/160—Quality of bathing water), in 1978 (78/659—Freshwater fish), and in 1979 (79/923—Shellfish waters). Standards for bathing water, the most important of which are the coliform counts, should be met in waters where bathing is explicitly authorized or traditionally practiced and not prohibited. For freshwater fish, member states are to designate stretches of river or other surface waters that should meet the quality standards, and they should establish pollution reduction programs to ensure that within 5 yr the designated waters conform to the values set. Standards are given for salmonid and cyprinid fish species. Likewise, in the shellfish directive member states are to designate coastal or brackish waters that need improvement to support shellfish.

Council Directive 76/464—Dangerous Substances

Although the title Directive 76/464—Dangerous substances suggests that it is regulating substances, it provides a framework for the elimination or reduction of water pollution by particularly dangerous substances. Member states are to take appropriate steps to eliminate pollution by substances of list I ("black list") and to reduce pollution by substances of list II ("gray list"). List I contains organohalogen and organophosphorus compounds, organotin compounds, carcinogenic substances, mercury, and cadmium compounds. List II includes biocides not included in list I; metalloids/metals and their compounds; toxic or organic compounds of silicon; inorganic compounds of phosphorus, ammonia, and nitrites; cyanides and fluorides; and nonpersistent mineral oils and hydrocarbons of petroleum origin. Member states must establish a prior authorization system containing emission standards for list I substances. For list II substances pollution reduction programs must be established in combination with quality objectives and a prior authorization requirement.

A couple of "daughter" directives concern mercury discharged from the chloralkali industry and from other industrial processes (82/176 and 84/156, respectively). Other daughter directives concern discharges of cadmium (83/513) and hexachlorocyclohexane (84/491). According to these daughter directives, member states must prepare limit values or quality objectives and require regularly reviewed authorizations for existing plants. For new plants the best technical means available for preventing discharge should be applied.

Directive 86/280 is another daughter directive that intends to accelerate implementation of Directive 74/464. It requires that a comparative assessment of the implementation be reported by the Commission and submitted to the Council every 5 yr. This directive was also extended for non-point-source pollution and certain chemicals were added to the annexes.

Environmental Protection by Regulating Substances

Directive 76/769—Marketing and Use of Dangerous Substances

Substance regulation started with 76/769 concerning the marketing and use of dangerous substances. This directive created a framework to ban or restrict the use of specific dangerous chemicals or preparations. The specific chemicals are listed in an annex, and this annex has been amended several times by new daughter directives. Polychlorinated biphenyls, polychlorinated triphenyls, and vinyl chloride are regulated in 76/769, 82/828, and 85/467, and the tri- and tetrachlorinated carbon solvents are regulated in 79/663. Benzene is restricted in 82/806, several flame retardants and other substances in 83/264, and asbestos in 83/478, 85/610, and 87/217.

Council Directive 67/548—Classification, Packaging, and Labeling of Dangerous Substances

The great breakthrough of environmental regulation came in 1979 with Council Directive 79/831, which was the "Sixth Amendment" of Council Directive 67/548. The purpose of the original directive was to harmonize the laws of the member states on the testing, classification, packaging, and labeling of chemicals that are dangerous to people or the environment. The purpose was more commercially oriented than ecologically oriented. The Sixth Amendment introduced a premarket testing and notification system for new chemicals placed on the Community market. The amount and type of information to be provided depend on the level of production of the chemical in question and increase as production increases. Testing in this scheme comprises three levels (see Table 3). Level 0 (or base level) is for production levels up to 10 tonnes/yr. In Annex VII this base set of required information is specified, and for the first time in history it includes toxicological and ecotoxicological data. Level 1 is for production levels up to 1000 tonnes/yr, and level 2 is for higher production levels. More thorough toxicological and ecotoxicological testing is required at levels 1 and 2; although the testing requirements are more flexible, they take into account the nature of the substance, its use pattern, likely exposure scenarios, and results from tests at the preceding level(s). These testing requirements are set out in Annex VIII. The test methods themselves are described in Annex V and are based on the test methods adopted by the OECD Council in 1981.

The directive distinguishes between "new" and "existing" chemicals. Existing chemicals are those placed on the Community market before September 18, 1981 and are listed in the European Inventory of Existing Commercial Chemical Substances (EINECS). Annex I of the directive gives a list of about 1500 dangerous substances that have been classified under the directive and is updated regularly. Annex II gives 14 different symbols to be used for the labeling of the substances according to dangerous properties, such as explosive, flammable, corrosive, teratogenic, and toxic, whereas Annex III gives standard risk (R) phrases to be used for describing the substance's dangerous properties on the labels. Annex IV gives standard safety advice (S) phrases describing the necessary precautions in handling the substances. The classification criteria are given under Annex VI. Adaptations to technical progress and updates of Annex I are made by amendments or commission directives.

Council Regulation 793/93—Existing Chemicals

Since 1990 a council regulation was in preparation concerning environmental risk assessment and restrictive measures for existing chemicals. The purpose of Regulation 793/93 is to harmonize the different national systems for risk assessment within the member states to remove barriers to trade in the Community. The regulation became effective in 1993 at the same time and in the same way in all member states. The regulation covers about 100,000 chemical substances listed in the EINECS.

Under the regulation, information has to be provided by industry for substances produced or imported at quantities more than 1000 tonnes/yr, within 6 mo of enforcement. These "high-volume" substances are listed in Annex I of the regulation and cover about 1500 substances. For other high-volume substances not on the list a period of 18 mo is accepted for submission of the dossiers. For other substances produced in quantities between 10 and 1000 tonnes/yr a limited dossier has to be submitted within 4.5 yr.

The information collected by the Commission will be used for setting priorities on the basis of a preliminary risk assessment. The priority substances will be assessed by distributing them among "reporting" member states, which may ask for additional information if necessary to assess the risk and to rec-

Table 3. Information required by *Council Directive 67/548—Classification, packaging and labeling of dangerous substances* as described in the Seventh Amendment (*92/32*). The ecotoxicological studies are specified

Annex VII Information required for the technical dossier of the "base set"

1. Identity of the substance (names, numbers, formula, composition, analytical methods)
2. Information on production, uses, precautionary measures, emergency measures, and packaging
3. Physicochemical properties (e.g., vapor pressure, surface tension, water solubility, *n*-octanol/water partition coefficient)
4. Toxicological studies (e.g., acute toxicity, repeated dose, mutagenicity)
5. Ecotoxicological studies
   - 5.1 Effects on organisms
     - LC50 (96 h) for fish (mortality)
     - EC50 (48 h) for *Daphnia* (immobilization)
     - IC50 (72 h) for algae (inhibition of growth)
     - Bacterial inhibition (in relation to biodegradation)
   - 5.2 Degradation
     - Biotic (screening test for "ready biodegradability")
     - Abiotic (e.g., hydrolysis)
   - 5.3 Absorption/desorption screening test
6. Possibility of rendering the substance harmless (e.g., recycling, destruction)

Annex VIII Additional information and tests required for level 1 and level 2

Level 1 (≥10 tonnes/yr per manufacturer or total quantity on the market >50 tonnes per manufacturer)

Taking into account the base set data and other knowledge of the substance, some or all of the following studies may be required (for ≥10 tonnes) or the following studies shall be required (for ≥100 tonnes):

Further physicochemical studies (e.g., including development of analytical detection methods for transformation products)

Toxicological studies [e.g., (sub-)chronic, fertility, carcinogenesis screening, toxicokinetics]

Ecotoxicological studies

- Prolonged toxicity study with *Daphnia magna*, 21 d
- Test on higher plants
- Test on earthworms
- Further toxicity studies with fish
- Species accumulation, preferably fish
- Supplementary degradation study if sufficient degradation has not been proved by the base set studies
- Further studies on absorption/desorption

Level 2 (≥1000 tonnes/yr or a total of 5000 tonnes)

The test program shall cover the following aspects:

Toxicological studies (e.g., chronic toxicity, carcinogenicity, development toxicity, teratology)

Ecotoxicology

- Additional tests for accumulation, degradation, mobility and absorption/desorption
- Further toxicity studies with fish
- Toxicity studies with birds
- Additional toxicity studies with other organisms

ommend adequate measures. Manufacturing or importing industries must provide this additional information if a substance is suspected to be dangerous. On the basis of these data and more refined risk assessments, substances may be designated dangerous by the Commission and the Community may ban or restrict substances on the basis of the Directive on Dangerous Substances, 76/769.

The data required are largely comparable with the data required for new substances under the Sixth Amendment (79/831; see Table 3). In contrast to the test requirements for new substances, however, test results derived with methods other than those specified in Annex V of the Sixth Amendment may be submitted. In this way, industries are not obliged to repeat the testing of a large number of chemicals according to the new protocols and the need to minimize testing on vertebrate animals is taken into account. This acceptance of a variety of test methods may cause problems for the comparability of test results.

The data are collected on the so-called Harmonized Electronic Data Set (HEDSET) and industry is allowed to cooperate or to be represented by their trade associations or ad hoc combinations.

Council Directives on Pesticides

Pesticides have been regulated by the individual member states in various ways. In 1976 Council Directive 76/895 on the maximum pesticide residue levels was implemented. The directive was amended by 80/428, 81/36, and 82/528. In 1978 the classification system of the Sixth Amendment (79/831) was made applicable for pesticides by Council Directive 78/631 and amended by 81/187.

In 1979 several pesticides with unacceptable dangerous properties were prohibited by Council Directive 79/117. The prohibited substances are mercury compounds, DDT, aldrin, dieldrin, endrin, chlordane, hexachlorocyclohexane (HCH), heptachlor, and hexachlorocyclobenzene (HCB). This directive was amended by 90/533.

The most important directive on pesticides is Council Directive 91/414 on the use, marketing, and registration of pesticides. In Annex I this directive gives a list of active ingredients that are registered and accepted for use. New active ingredients may be added to Annex I on the basis of a data set to be submitted (described in Annex II of the directive). Also for the registration of a new pesticide a technical dossier (Annex III) is to be submitted before marketing. The data sets are evaluated in order to assure that with normal use, no unacceptable effects on the environment and ground water are expected. The data sets (Annex II and III; see Table 4) are more extensive than the base set of the Sixth Amendment (see Table 3). Information on fate and behavior in air, water, and soil is required, including the degradation rate and pathway and the identification of metabolites, if present. Data on adsorption and mobility in a variety of soil types and in sediment are to be presented. A complete data set for human toxicity and a rather extensive data set for ecotoxicity have to be submitted. For example, chronic toxicity and effects on reproduction and growth are needed for fish and *Daphnia*. Effects on honeybees, earthworms, and other relevant nontarget organisms are included as well.

Environmental Protection by Prevention of Accidents and Regulation of Waste Handling

Post-Seveso Guideline

In 1982, 6 yr after the dioxin accident in the Italian town of Seveso near Milan, the so-called "post-Seveso guideline" or "Seveso directive" was adopted. The purpose of this directive (82/501—Major accident hazards) is to prevent major accidents in industrial installations. A major accident is defined as an occurrence such as a major emission, fire, or explosion resulting from uncontrolled developments in the course of an industrial activity, leading to a serious danger to humans, immediate or delayed, inside or outside the establishment, and/or to the environment, involving one or more dangerous substances. The plant operator must take "all measures necessary" to prevent major accidents and to limit their consequences for people and the environment. Manufacturers must be able to prove to the national authorities at any time that they have identified existing major accident hazards, adopted the necessary safety measures, and provided the persons working on the site with information, training, and equipment in order to ensure their safety.

If one of the 190 dangerous chemicals (flammable, explosive, irritating, or toxic) on the lists in Annex II and Annex III of the directive is or may be present in the designated quantity, the industrial activity is subject to a notification procedure. The designated quantities depend on the character of the substance in question and vary from 1 kg to 50,000 tonnes. The notification must contain detailed information on:

- Substances and processes and corresponding hazards and risks, safety precautions, and emergency procedures
- Maps of the industrial plant, sources of danger, and siting of exposed groups or environments
- Possible major-accident situations, including emergency plans, safety equipment, safety precautions, emergency procedures, alarms, and resources

Most member states have implemented the post-Seveso guideline in their own specifications for safety plans and precaution standards. In comparison to the quantification procedures used to assess possible human health effects and hazards, the environmental impact is usually treated in a rather superficial way.

In a study by Blok et al. (1992) the quantities designated by the directive were assessed with respect to their environmental safety, for example, the possibility of reaching a concentration as high as the LC50 for aquatic organisms over a large distance downstream. Based on a dilution model of the Rhine River and the ecotoxicity data found, critical quantities M were calculated for 87 priority chemicals. These critical quantities were compared to the quantities designated in the EC directive (the EEC, or estimated environmental concentration). It was concluded that for some substances, especially pesticides, the risks to the aquatic ecosystem were not sufficiently reduced by the quantities designated by the EC directive.

In Figure 9 the ratios between M and EEC for 87 substances are presented in a frequency distribution. If the ratio M/EEC = 1, discharge of the EEC quantity may cause a concentration as high as the LC50 for fish more than 300 km downstream in the river. If the ratio M/EEC > 1, the EEC quantity is relatively safe (i.e., the concentration will be below the LC50 over that distance). If the ratio M/EEC < 1, the EEC quantity is not safe.

It should be realized that concentrations at the LC50 level cause serious fish mortality and that safety precautions should be designed to prevent any unacceptable effect. Therefore a criterion safely below the LC50 should be used. In Figure 9 the consequences of a criterion lower than the LC50 can be seen. For example, if the criterion for environmental safety is chosen at 1/1000 * LC50, the EEC quantities set by the directive are not strict enough for the majority of these substances.

Council Directive 75/442—Framework for Waste

Environmental pollution by wastes cannot be prevented by water quality objectives, substance regulation, or process regulation. Therefore member states must take necessary measures to ensure that

Table 4. Information required by *Council Directive 91/414—Use, marketing and registration of pesticides.* The ecotoxicological and environmental fate studies are specified

Annex II Information required for the technical dossier for inclusion of an active ingredient (a.i.) in Annex I (list of *accepted* ingredients)

1. Identity of the a.i. (e.g., names, numbers, formula, composition, synthesis, purity, and impurities)
2. Physicochemical properties (e.g., vapor pressure, water solubility, including effects of pH, *n*-octanol/water partitioning coefficient, stability in water and in air: photochemical degradation, transformation products; thermal stability, surface tension)
3. Other data (e.g., fields of application, purpose, target organisms, mode of action, resistance, emergency measures)
4. Analytical methods (e.g., residues in food, soil, water, air and body tissue, detection limits)
5. Toxicological studies on the a.i. (e.g., acute, subacute, chronic toxicity, reproduction, metabolism, medical data)
6. Residues in products (e.g., characteristics of the metabolites and decomposition products, effects on public health, environment, and livestock)
7. Distribution and fate in the environment
   - 7.1 Distribution and fate in soil
     - Degradation rate and pathways (to 90% degradation) in at least three soil types
     - Adsorption/desorption in at least three soil types of the a.i. and of its metabolites and decomposition and decomposition products, where relevant
     - Mobility in at least three soil types of the a.i. and of its metabolites and decomposition products, where relevant
     - Characterization of bound residues
   - 7.2 Distribution and fate in water and air
     - Degradation rate and pathways in the aquatic environment—biodegradation, hydrolysis, photolysis
     - Adsorption/desorption in water (sediment) of the a.i. and its metabolites and decomposition products, where relevant
     - Degradation rate and pathways in air (for desinfectantia and other volatile active ingredients)
8. Ecotoxicological studies of the a.i.
   - 8.1 Effects on birds
     - Acute oral toxicity
     - Subacute toxicity; 8 d diet study on at least one bird species (other than chicken)
     - Effects on reproduction
   - 8.2 Effects on aquatic organisms
     - Acute toxicity for fish
     - Chronic toxicity for fish
     - Effects on rates of reproduction and growth for fish
     - Bioaccumulation in fish
     - Acute toxicity for *Daphnia magna*
     - Growth inhibition test with an alga
   - 8.3 Effects on other nontarget organisms
     - Acute toxicity for honeybees and other beneficial arthropods
     - Toxicity for earthworms and other soil organisms
     - Effects on soil microorganisms
     - Effects on other nontarget organisms (flora and fauna) that may be exposed
     - Effects on biological water treatment processes
9. Summary and evaluation of parts 7 and 8
10. Proposals for the classification and labeling of the a.i. according to Directive 67/548/EEC
11. Dossier as meant in Annex III of the directive for a representative pesticide

Annex III Information required for the technical dossier for registration of a pesticide

1. Identity of the pesticide (e.g., names, composition, physical form, fields of application)
2. Physicochemical and technical properties (e.g., appearance, surface tension, stability, and effects of environmental conditions)
3. Data on use (e.g., fields of application, effects on harmful organisms, application methods, dose rates, instruction manual)
4. Other data (e.g., packaging, cleaning, preventive and emergency measures, destruction, and decontamination)
5. Analytical methods
6. Data on the activity (e.g., field tests, resistance, quality and yield of treated plants or products, phytotoxicity, unwanted side effects)
7. Toxicological studies (acute toxicity, exposure of the user, toxicity of the nonactive ingredients)
8. Residues in products (e.g., maximum residue levels, preharvest waiting times)
9. Distribution and fate in the environment (in addition to data in Annex II-7)
   - Experiments for distribution and dissipation in soil, water and air
10. Ecotoxicological studies
    - 10.1 Effects on birds
      - Acute oral toxicity
      - Studies to assess the risks for birds under field conditions
      - Studies on the acceptance of bait, granules, or treated seeds, where appropriate

Table 4. Information required by *Council Directive 91/414—Use, marketing and registration of pesticides.* The ecotoxicological and environmental fate studies are specified (*Continued*)

10.2 Effects on aquatic organisms
Acute toxicity for fish
Acute toxicity for *Daphnia magna*
Overspray study (if toxic for aquatic organisms and persistent in water)
For application in/near surface water: special studies on aquatic organisms, residues in fish and studies listed in Annex II-8.2
10.3 Effects on other nontarget organisms
Effects on terrestrial invertebrates (other than birds)
Toxicity for bees and other beneficial arthropods
Toxicity for earthworms and other soil organisms
Effects on soil microorganisms
Other available biological data
11. Summary and evaluation of parts 9 and 10
12. Further information (e.g., registration in other countries, proposals for the classification and labeling, risk, and safety advice phrases according to Directive 67/548/EEC)

waste is disposed of without endangering human health or harming the environment. Member states must:

- Designate national competent authorities to be responsible for waste management
- Draw up waste disposal plans
- Subject installations that treat, store, or dispose of wastes for third parties to a prior permit
- Apply the "polluter pays" principle

In addition, the member states must encourage recycling and submit status reports showing progress to the commission every 3 yr. The plans to be drawn up must cover the type and quantity of wastes, technical requirements and precautions, and information concerning origin, destination, and treatment.

Within the framework of this directive, "toxic and dangerous waste" is specifically regulated by Council Directive 78/319. Toxic and dangerous waste is defined as any waste containing or contaminated by the substances or materials, listed in the annex to the directive, of such a nature, in such quantities, or in such concentrations as to constitute a risk to human health or the environment. These wastes may be

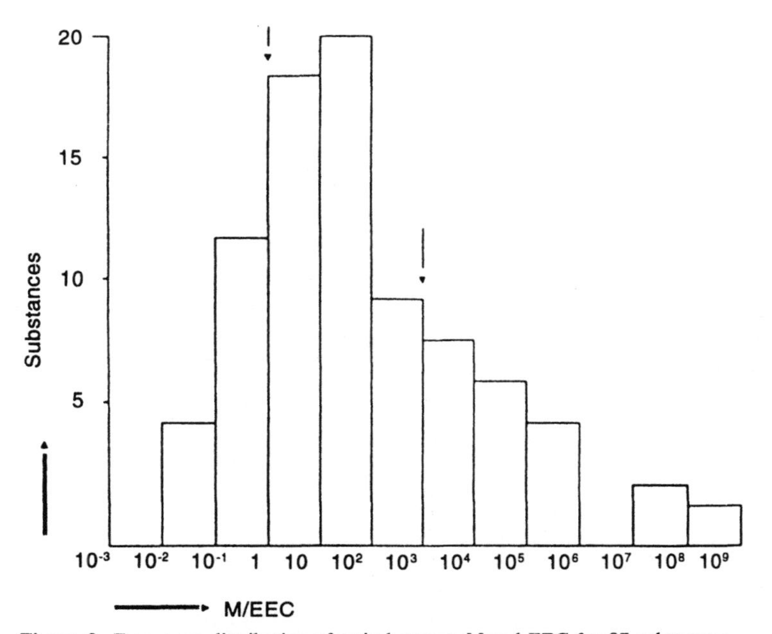

Figure 9. Frequency distribution of ratio between *M* and EEC for 87 substances, where *M* is the critical mass discharge quantity based on reaching median lethal concentrations (LC50s) downstream of the Rhine River and EEC is the quantity given in Directive 82/501 of the European Community.

stored, treated, and/or disposed of only by authorized companies. Companies that produce, hold, and/or dispose of these wastes must keep records and make them available to the competent authority on request. A form must accompany any shipment of these wastes. The form must give the name and address of the producer, previous holder, and final disposer of the wastes. These wastes must be kept separate from other matter and residues when being collected, transported, stored, or deposited. In a special supplementary daughter directive (84/631—Transfrontier shipment) shipment of toxic and dangerous wastes across national frontiers, both within and outside the Community, is regulated.

In 1991 the directive on toxic and dangerous waste was technically updated with an amendment, 91/689—Dangerous waste.

Council Directive 75/439—Waste Oils

The "waste oils" directive is designed to prevent damage to the environment from the uncontrolled disposal of waste oils and seeks to ensure that financial arrangements adopted to promote safe disposal and recycling do not create barriers to the common market. Member states must ensure the safe collection and disposal of waste oils and ensure that they are recycled "as far as possible."

The directive prohibits:

- Discharge of waste oils to water and drainage systems
- Any deposit and/or discharge harmful to the soil
- Any uncontrolled discharge of residues from processing
- Any processing of waste oils causing air pollution exceeding the level prescribed by existing provisions

Companies disposing of waste oil must obtain a permit and companies collecting waste oil must be registered and adequately supervised. Companies that regenerate or burn waste oils must take appropriate preventive measures in order to obtain permits from the competent authority.

ENVIRONMENTAL LEGISLATION IN VIEW OF THE BASIC PRINCIPLES OF ENVIRONMENTAL IMPACT ASSESSMENT

Environmental legislation in the EC can be seen in the historical perspective of growing environmental awareness, because the different items under regulation followed a sequence illustrated in Figures 2 and 8. Environmental legislation can also be related to the basic principles of impact assessment. Environmental impact assessment for chemical substances is a process involving four levels. The first three levels can be described on the basis of the "source-path-object" concept; the last level concerns a value judgment. Table 5 gives a schematic presentation of the levels of environmental impact assessment.

Table 5. Levels of environmental impact assessment

Source	Path probability	Object	Value judgment
Hazard			
	Exposure risk		
		Impact (effect)	
			Gravity

In this concept the substances are regarded as the *source*, their distribution and environmental fate as the *path*, and the effects they may cause to *objects* cover the environmental impact. At the source level the hazard is identified on the basis of the inherent (intrinsic) properties of the substance. Substances are measured or tested with standard test methods that give results with a comparative value. Substances are compared to each other in order to rank them according to their hazardous properties, such as persistence, accumulation potential, genotoxicity, and aquatic toxicity. The standard tests have no, and do not need, great predictive power. The number of different test species can be limited. In most cases the hazard is assessed on the basis of acute toxicity data. Tests on chronic toxicity are to be preferred if available, if there is any indication of poor degradability, if high production volumes are involved, or if the differences between acute and chronic toxicities indicate a different mechanism (or mode) of action.

At the path level the potential exposure is identified on the basis of the interaction of the properties of the substance and the environment. Starting with the release pattern and the discharge volume, the probability of dilution, dispersion, and concentration is assessed. The environmental compartments are modeled in a standard way and have a generic character. Based on water solubility, adsorption, volatility, and photochemical and biological degradability the distribution over the main environmental compartments is predicted, with predicted environmental concentrations (PECs) as the result.

So far, hazard and exposure assessments have a potential character. At the path level statistical probability can be introduced to quantify the risk. Hazard/exposure assessment and risk are discussed in Chapters 28 and 29. After exposure assessment more data may be needed on the properties of the substance (OECD, 1989a).

At the level of the object, consisting of the biota in a specific location, the impact or effect is assessed. At this level the assessment is no longer based on potential effects and generic circumstances but on actual effects and therefore test results should have a high predictive value. This implies the use of test species representative of the variety of local species. The possible selective effect of the substance on species diversity and the possible specific sensitivity of the particular ecosystems should be identified. At this level also the exposure should be specified in more detail according to the extent (local or regional), the distribution pattern (point sources or diffuse), and the periodicity (incidental or at steady state). Instead of standard laboratory testing, ecosystem modeling, simulation studies (e.g., microcosms), field tests, and biomonitoring studies are used to cover the gap between laboratory and field (see Chapters 1 and 9). The remaining uncertainty of the predicted effect depends on the inclusion of five principal extrapolations in ecological impact assessment:

1. From acute lethal effects to chronic sublethal effects
2. From effects on individuals to effects on populations
3. From single species to multiple species
4. From standard laboratory test conditions to variable field conditions, including the interaction of other stress factors and other pollutants
5. From multiple species to complex ecosystems including their unique structures, interactions, and self-regulative potential

Without detailed experimental data for these extrapolations, *application factors* may be used to cover the gaps of uncertainty. The final result of the assessment at the level of the object is a best achievable no-effect concentration (NOEC or NEC) and knowledge of the ratio between the predicted environmental concentration and the no-effect concentration (PEC/NOEC or NEC) in a particular environment.

At each of these levels the assessment can be carried out in phases. In a preliminary assessment only base set data and generic assumptions or worst-case scenarios are applied, whereas with more data available the assessment progresses to an investigative phase and, if regarded as necessary, ends in a definitive or confirmatory phase (ECETOC, 1993; OECD, 1989b). In most schemes the process follows a cycle and is repeated when new data are collected. The start of the investigative phase is triggered by the risk predicted in the preliminary phase. A complete assessment at three principal levels up to the confirmatory phase is, however, rarely performed. The high costs to generate the required data may be prohibitive. Only high-volume products of great economic value can support the financial resources to conduct all the testing. The uncertainty in PEC/NOEC or NEC ratios may be reduced by additional studies, but more detailed work will not reduce the natural variability. A safety margin between PEC and NOEC or NEC can be used to trigger the decision on the need for more data.

At the level of value judgment the gravity of the impact is assessed. Environmental damage can be more or less serious, varying between catastrophic and acceptable hindrance. The value judgment can be rather subjective and dependent on personal opinion unless objective ranking criteria are set. For this purpose various sets of criteria can be used in a hierarchic order for gravity ranking. Three different starting points may be possible:

1. Ecological ranking criteria
    - Effects on the global biosphere
    - Ecosystem effects
    - Effects at species levels
    - Effects on population level
    - Effects on individual level
2. Anthropocentric or emotional ranking criteria
    - Effects on the human genome
    - Effects on human reproduction
    - Effects on health of children
    - Acute lethal effects
    - Chronic sublethal effects
    - Allergies, irritation, taste, smell, and hindrance effects
    - Health effects on mammals and birds
    - Loss of agricultural productivity
    - Population effects on fish
    - Species diversity effects on invertebrates
    - Functional system effects on algae
    - Ecosystem or biosphere effects on bacteria
3. Sustainability ranking criteria
    - Irreversible effects
    - Duration of effects beyond one generation
    - Extent of effects beyond national borders
    - Extent of effects beyond regional borders
    - Local effects and temporary effects
    - Effects that may be prevented by risk management
    - Individually and voluntarily taken risks
    - Effects that are regarded as beneficial for survival of humans

The value judgment or acceptability is subject to national political influences. National quality standards and quality objectives can be formulated and a variety of threshold values can be used. If EC member states develop different threshold values, these values and criteria may create barriers to international trade and interfere with sound economic competition. In the policy of the community this should be prevented by regulation and therefore a harmonized "gravity scaling" will have to be prepared.

The course of development in ecotoxicology and the impact on regulation can be elucidated with the scheme in Table 5. Council Directive 76/464 for eliminating dangerous substances from the market and Council Directive 79/117 on prohibition of certain persistent pesticides are a typical result of early value judgment. The environmental effects of the black and gray list chemicals were so unacceptable that decisions were taken immediately. In fact the black and gray list chemicals had already shown their lethal effects in the environment, so the regulation was not subject to questions regarding the predictive value of laboratory tests and assessment procedures. Although very conclusive from a scientific point of view, such "real-life" experiments are unacceptable and should be prevented.

The Sixth Amendment or Council Directive 79/831 is a typical source-oriented regulation for hazard assessment. The comparative labeling of substances is based on the minimal premarketing data set of Annex VII, which contains mainly intrinsic properties of the substance. As far as toxicity testing is involved, the methods are standardized for intercomparability. Shortly after the directive and throughout the 1980s, standardization and certification for good laboratory practices (see Chapter 11) became key items to ensure reproducibility of test data.

For high-volume chemicals or if the base set data indicates a high risk, a second cycle of assessment at the investigative phase is foreseen. Additional data such as those specified in Annex VIII (see Table 3) are required in the second phase. The priority listing envisaged for high-volume chemicals in the Council Regulation of 1993 (see previous section in this chapter) is a typical example of assessment at the risk level. Priority setting may be achieved by scoring systems or generic modeling. The Mackay box models (Mackay, 1991; see Chapter 18) for compartment distribution seem a valuable tool at this level. In the 1990s a broader experience with this type of path-oriented assessment focusing on the prediction of environmental fate and behavior of chemicals will be realized. Scoring systems based on expert opinions are less desirable at this level because they are vulnerable to subjective value judgment, thus mixing two principal levels of assessment.

For some of the priority chemicals to be selected on the basis of the HEDSET data (containing results of the required tests; see previous section), the level of impact assessment can be envisaged within a few years. The methodology to be applied is described in technical guidance documents for risk assessment of new and existing substances (EC, 1995). Several high-volume chemicals such as surfactants (linear alkylbenzene sulfonates or LAS, alkyl or alcohol ethoxylates or AE) and other detergent components (the calcium builders NTA and sodium aluminum silicate or NAS) have already arrived at this level. For pesticides, impact assessment is routinely requested by national laws and regulations of the member states. Pesticides are usually not marketed before extensive field testing has shown the performance for several types of application and elucidated possible side effects. Council Directive 91/414 on the registration of pesticides states that it includes impact assessment but specifies only the extensive data set and not the methodology.

Because of the expense of field tests, extrapolation methods are proposed and under discussion. For example, OECD prepared guidelines for aquatic effects assessments (Balk et al., 1993; OECD, 1995; also see Chapter 26). It can be concluded that the number of difficulties is still enormous but progress can be expected from the use of quantitative structure-activity relationships (QSARs; see Chapter 20), statistical evaluation of variability between species, and soundly based application factors for the above-mentioned steps of uncertainty.

THE "KEY" TO CLASSIFICATION

According to EC Directive 67/548, substances will be labeled because of their explosive, oxidizing, flammable, corrosive, toxic, harmful, or irritant character. The classification and labeling as dangerous for the environment are described in the Sixth Amendment of the Council Directive (Directive 79/831, amended by the Seventh Amendment, 92/32). The primary objective of classifying substances as dangerous for the environment is to alert the user to the hazards these substances present to ecosystems and to give safety instructions. The present criteria are limited to aquatic ecosystems but criteria for the terrestrial environment will include definitions for toxicity to flora, fauna, soil organisms, and bees and for long-term adverse effects in the environment. Furthermore, substances may be classified because they may present a danger to the structure and/or functioning of the stratospheric ozone layer.

Substances are classified and labeled as dangerous to the aquatic environment and risk phrases are assigned in accordance with the criteria illustrated in Figure 10. The classification for adverse effects in the aquatic ecosystem is directed mainly by the toxicity of the substance. Four toxicity classes may lead to classification. The toxicity criterion is based primarily on results from the base data set tests in Annex VII (EC 79/831), including 96-h LC50 for fish, 48-h EC50 for *Daphnia*, and 72-h IC50 for algae (see Table 3). A substance is classified as "very toxic to aquatic organisms" if the acute toxicity level in one or more of these tests is below 1 mg/L. If the acute toxicity level is between 1 and 10 mg/L, the substance is classified as "toxic to aquatic organisms," whereas an acute toxicity level between 10 and 100 mg/L is classified as "harmful to aquatic organisms." A substance may also be classified "harmful to aquatic organisms" if evidence is available from other tests than the base data set methods that its toxicity may present a danger to the structure and/or functioning of aquatic ecosystems.

Another criterion is based on the potential of the substance to cause long-term adverse effects in the aquatic environment. The risk of long-term adverse effects from a substance may be due to its bioaccumulation potential (see Chapter 16) or to its persistence (see Chapter 15). The bioaccumulation potential of a substance is derived from its octanol-water partition coefficient: if $\log P_{ow} \geq 3$, long-term adverse effects due to bioaccumulation may be expected. However, if the experimentally determined bioconcentration factor (BCF) in fish is below 100 (BCF ≤ 100), it is accepted that the substance does not bioaccumulate.

A substance can be classified as readily degradable if (1) it passes one of a series of defined screening tests for ready biodegradability (OECD Test Guideline 301 A–F; OECD, 1992), (2) the ratio $BOD_5/COD \geq 0.5$, or (3) convincing evidence is available that the substance can be degraded in the aquatic environment, biotically or abiotically, to a level of more than 70% degraded within a 28-d period. The risk phrase on long-term adverse effects may be deleted if there exists sufficient evidence to provide

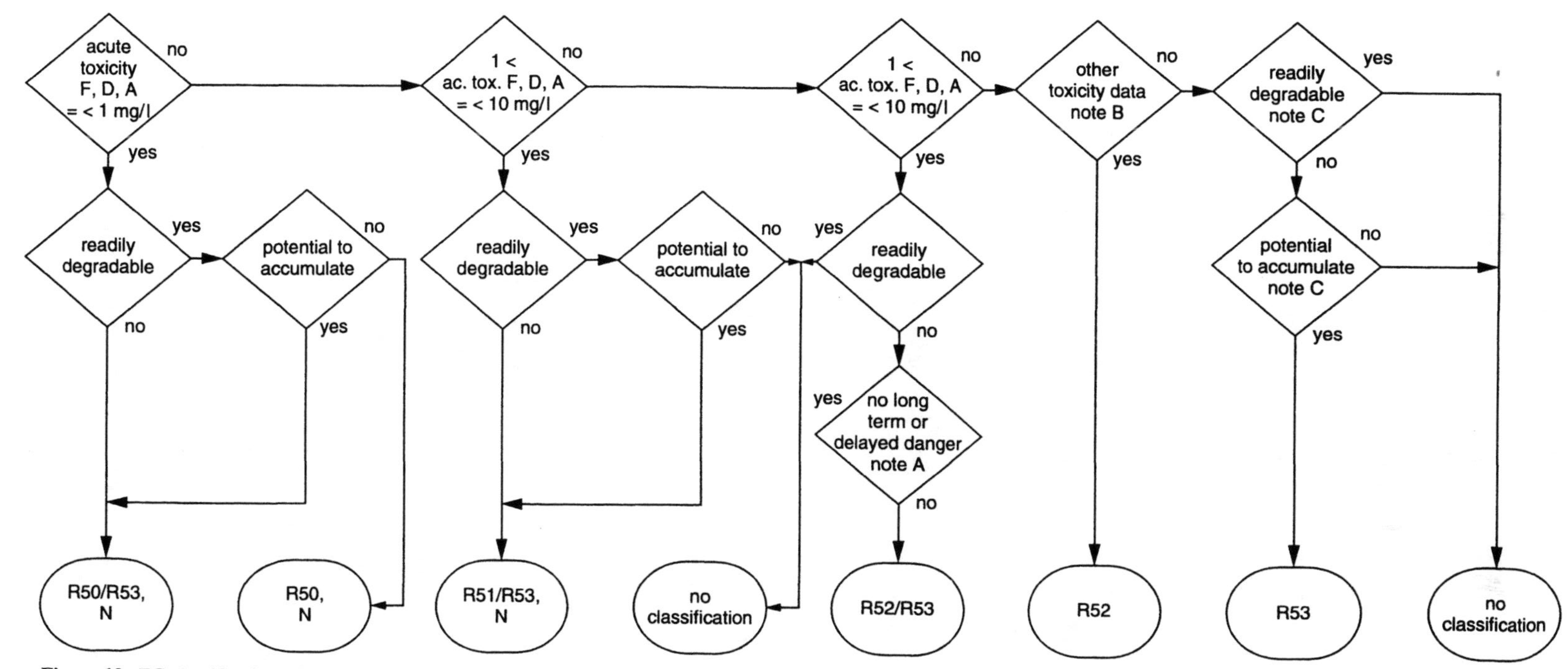

Figure 10. EC classification scheme—dangerous for the aquatic environment. Note A: rapid degradation of substance and its degradation products in the aquatic environment, or absence of toxicity at 1 mg/L (NOEC fish or *Daphnia* in prolonged study > 1 mg/L). Note B: Other toxicity data, e.g., other test species, other test methods. Note C: see note A; e.g., for poorly water soluble substances, absence of chronic toxicity effects at solubility limit (NOEC fish or *Daphnia* in prolonged study > solubility limit). N: labeled—dangerous for the environment. Risk phases: R 50, very toxic to aquatic organisms; R51, toxic to aquatic organisms; R52, harmful to aquatic organisms; R53, may cause long-term adverse effects in the aquatic environment. A,D,F: algae, *Daphnia*, fish.

adequate assurance that neither the substance nor its degradation products will constitute a potential long-term and/or delayed danger to the aquatic environment. Such evidence could include (1) a proven potential to degrade rapidly in the aquatic environment or (2) the absence of chronic toxicity effects at a concentration of 1 mg/L, for example, NOEC > mg/L in a prolonged toxicity study with fish or *Daphnia*.

For substances labeled as dangerous for the environment safety advice phrases are selected in order to supply the user (either the general public or specialized manufacturers) with advice on the safe use, storage, and disposal of the material.

Practical Experience

Classification of so-called new substances is rather easy because the proper base data set is generated and the classification scheme can be followed straightforwardly. For "existing" chemicals, however, experimental data may be available but the experiments hardly ever match the standard test protocols. This causes difficulties in interpreting the results. In particular the interpretation of biodegradation data from nonstandard test methods is difficult due to the wide variety of test methods and the criteria for ready biodegradability (70% degradation in the aquatic environment within 28 d).

When more toxicity data are available it is not clear which data should be taken: the lowest, with the risk of using outliers; a mean value; data for the standard species; or data for the most sensitive species. Moreover, the distance between the various toxicity classes (one order of magnitude) is rather small compared with the variability of test results experienced when tests are repeated (see the next section on ring tests). As a result, a substance may be classified as "very toxic" by one laboratory, whereas it would be considered only "harmful" by another laboratory.

About 1500 substances in Annex I of Directive 67/548 that have been labeled for other characteristics or dangerous properties have to be classified for their effects on the environment within the next few years. Due to a lack of suitable experimental data, many substances will be considered as persistent and accumulative until the standard test data have been made available. As a result, a large fraction of these substances are expected to be labeled as dangerous for the environment or to carry risk and safety advice phrases. If products are overlabeled because of a lack of data this may weaken the effect of labeling. Another dilemma that arises is the limitation in the number of labels on the package. If a substance already carries a number of risk phrases due to its explosive, oxidizing, or other characteristics, it is subject to debate which of the labels, risk, and safety phrases are finally selected for the package. Thus fear of "overlabeling" may restrict the labels and risk phrases for the environment.

LESSONS FROM INTERLABORATORY RING TESTS

The ecotoxicological test methods laid down in Annex V of the Sixth Amendment in 1979 and described by the OECD in 1981 were developed with the ecotoxicology experience and knowledge that existed within the member states and within the OECD countries. For most of the items some preliminary national or institutional test protocols were already available. However, validated international standards were not ready at that time. The state of the art at the time on methods for testing degradation, ecotoxicology, and bioaccumulation was recorded by The Netherlands Research Institute TNO (TNO, 1977).

Although all regular users of test methods and the original test designers wanted their methods officially accepted by the EC and OECD, the only suitable procedure for getting a method on the official list was via interlaboratory ring testing (also called round-robin testing; see Chapter 1). Over a few years several contract research institutes organized these ring tests for the OECD and EC. Any laboratory that wanted to be involved participated on a voluntary basis. The purpose of ring testing is to validate existing protocols and to achieve international agreement on the methods. Because several protocols were not yet sufficiently evolved, some methodological development and repeated ring testing was inevitable. However, some harmonization of similar existing national test methods was necessary. Unlike the normal procedure in the International Standard Organization (ISO), there was no possibility to select the best of the available protocols and to modify this further until a really sophisticated method was developed. The time pressure and national interests prompted compromises and protocols based on the state of the art.

Test Methods for Ready Biodegradability

For biodegradation there were several methods available, including American Society for Testing and Materials (ASTM) test methods (e.g., Sturm), Japanese MITI methods, a French AFNOR method, a German closed-bottle method, and an OECD screening test used for testing surfactants. These methods differed with respect to the analytical technique applied to measure biodegradation, the constitution of test medium, the temperature, the shaking or stirring, the way of "seeding" with bacteria, the test duration, the use of controls and blanks, and the type of equipment needed.

The first agreement on harmonization consisted of a uniform organization of the test protocol. The second agreement consisted of some limitations regarding the seed density and its origin. It was generally felt that a relatively simple screening system was needed to select only substances that were rapidly and easily degradable by naturally occurring bacteria. Therefore tests with a high density of seed bacteria or with already adapted seed bacteria or bacteria from industrial wastewater treatment plants were not allowed.

After a first series of ring tests, the results were rather disappointing because of poor interlaboratory reproducibility. Further steps toward harmonization were proposed and tested and this process was repeated several times. As a compromise, the test duration was prolonged for some methods and shortened for others and the seed density was increased to promote the chances for adaptation. The latest set of harmonized OECD methods, adopted by the EC in 1992, was prepared by a task force of ECETOC, the European Center for Ecotoxicology and Toxicology of Chemicals (Blok et al., 1985). The methods prescribe a comparable seed density, a similar test medium, a fixed test duration of 28 d, and similar pass/fail levels for mineralization. Several superfluous items such as sterile controls, viable cell counts, and complicated seed preparations have been deleted. The main problem, poor interlaboratory reproducibility, however, remained unchanged. This problem is illustrated in Table 6.

Presumably the chance in each of about 25 different laboratories to get one of the chosen compounds through the screen as a positively degradable substance is about 50%, whereas in most cases repetitive testing of a substance by one laboratory gave reproducible results. The origin of the seed samples was responsible for this. Thus testing xenobiotic chemicals for ready biodegradation seems to yield more data about adaptation of the environment where the inoculum was collected than about inherent substance properties.

Although scientifically disputable, the practical solution to this problem was the strict use of the definition for "ready biodegradability" in such a way that only positive results were taken seriously. Thus negative results are normally neglected and ascribed to bad luck, and tests are repeated with seed bacteria of a different origin until a positive test result is achieved. In other words the expression "readily biodegradable" can be used only in a positive way (i.e., if a positive result is available) and, unless repeatedly negative, the tests are unable to show that a substance is "not readily degradable."

Unfortunately, test results for existing substances conducted by one of the modified test methods for "ready biodegradability" are extremely scarce. If no result with one of these specific methods is available,

Table 6. Percentage of positives (readily degradable) in three interlaboratory ring tests

Method	6 substances	29 substances	4 substances
Sturm	67	52	—
AFNOR	58	27	—
MOECD[a]	39	38	36
MITI	38	25	40
Closed bottle	10	14	11

From Blok and Booy, 1984.
[a]MOECD; modified OECD screening test. (OECD test guideline 301E; see Chapter 26.)

it is easily stated that the substance is "not readily degradable." This expression, however, is exactly the type of misuse that the scientists wanted to prevent. For this reason the expression "inherently biodegradable" is used for any test result from any method showing some loss of parent compound.

Positive test results in a test for ready biodegradability should correlate with nearly complete removal in real sewage treatment plants and with short biological half-lives in surface water. If not, the result is called a false positive.

Up to 1995, examples of false positives are unknown, whereas there are plenty of examples of false negatives. The screening system may be too stringent or "on the safe side" for some chemicals, but unless a good understanding of the processes involved in adaptation and predictive models for the biodegradation rate are available, it is the best alternative.

Toxicity Test for Reproduction of *Daphnia magna*

Daphnia magna, the water flea, can be regarded as the choice out of millions of aquatic species. Comparable to the fruit fly *Drosophila melanogaster* for the science of genetics, *Daphnia magna* is the "parade horse" for ecotoxicology. The reasons are evident. The animal is easily grown in captivity, and because of asexual reproduction genetically identical clones are available all year round. It is extremely suitable to use for 24-h acute toxicity tests as well as for 21-d chronic life cycle tests. The species occurs all over the world, especially in temperate zones, and represents an important link in the food web, between phytoplankton and fish. Its sensitivity to toxic substances appears to be rather representative for a wide spectrum of aquatic organisms. The main advantage, however, is that is has been used in laboratories all over the world for many years. Unlike fish and algae, for which it was difficult to achieve international agreement on a particular test species, international agreement on *D. magna* was not a problem at all.

A synthetic medium for 24-h acute toxicity testing of *D. magna* was described by ISO 7346 in 1979. On the basis of a similar simple synthetic medium, Dutch test laboratories standardized a medium and a full protocol for a chronic test in 1980. The OECD described a principal guideline in 1981, the EPA designed a guideline in 1982, and in Germany a protocol was in preparation in 1982. It seemed logical to use all existing experience and prepare an EC guideline for a chronic test on *Daphnia*. Two EC ring tests were organized in 1983 and 1984 (Cabridenc, 1986). It appeared that the media used for acute tests were not suitable for chronic tests and the reproducibility was lower than expected. A third ring test was organized by the University of Sheffield (UK) with participants from the EC and OECD in 1992.

Most experienced laboratories used to culture *Daphnia* on systems with local surface water or ground water, in large aquaria with biological filtration units, or at least in big storage tanks or ponds with some detritus on the bottom and attached growth of algae and bacteria on the walls. There was a "standardization fever" at that time that required clean containers and purely synthetic media with food consisting of axenic cultured algal species under the assumption that interlaboratory differences would be reduced in this way. This optimism was immediately changed by the first preliminary ring test results.

Under normal conditions a daphnid matures in 9 d and in the next 12 d it normally can produce between 40 and 80 juveniles. The animals in the synthetic ISO medium of the first ring tests failed to reproduce or reproduced at highly variable rates of 0 to 30 juveniles. The observed phenomenon has been described by Blok (1981) as the "brood pouch disease." The eggs that normally develop within 3 to 4 d in free-swimming juveniles suddenly become white because the embryos die in the brood pouch and are overgrown by bacteria. The cause is not an infectious bacterial disease but could be identified as deficiency in the food and in the medium. Addition of a small amount of a suitable mixture of trace elements, preferably of natural origin, solved the problem. Effective additives could be yeast extract or yeast cells, extract of activated sludge, peat extract, or dialysate of compost leachate, matured water from old aquaria, etc. Blok et al. (1989) found that a combination of two trace elements, selenium and bromide, alone or in combination with iodide, barium, lithium, rubidium, strontium, molybdenum, cobalt, and vanadium at a microgram per liter level, appeared responsible for the effect. Other successful media such as the trace element mixture of Keating (1985) and minor modifications of that medium proposed by Elendt (1990) contain about 100 times the needed concentrations in a combination with EDTA to reduce the toxicity of some elements at that level. Several variants of these media were ring tested between 1989 and 1993. Up to 1994, however, a purely synthetic additive trace element mixture with good long-term performance has not yet been found, although several proposals look promising (Keating, 1985; Blok et al., 1989).

Although there is no strict need for a large number of juveniles to test chronic toxicity, the main problem with deficient growth conditions is a statistical problem. With deficient ISO medium and aseptically cultured *Chlorella* as food the coefficient of variation (CV; standard deviation/mean) in the mean offspring of individual daphnids is typically 1.25, whereas in a good medium the CV is around 0.2. As a consequence, it is hardly possible to detect small differences between reproduction in the control and the test concentrations. For example, with a CV of 1.25 one would need a series of 100 individually observed daphnids per test concentration to detect reproduction 25% below normal. With the usual number of 10 to 20 animals per concentration the detection level would be not lower than a reduction of 50%. As a consequence, the determination of a NOEC value will be much more subject to interlaboratory variation than the determination of an EC50 level.

In the second EC ring test about 40 laboratories participated, with NaBr as a conservative, nonproblematic test chemical, to find the NOEC. The ISO medium was supplemented by one of the "magic factors" of natural origin mentioned above to prevent deficiency. The results of the control test runs showed a CV in the offspring over 21 d of typically between 0.1 and 0.2, whereas in the first effective concentration the CV was usually higher (up to 0.6).

Under these conditions the use of 20 animals per concentration observed in at least four separate vessels would enable the detection of a 25% lower reproduction. Based on this criterion, the NOEC values in 44 test runs were compared. The frequency distribution of NOEC values over the laboratories is given in Figure 11. The geometric average was 9.7 mg/L, whereas 10% of the resulting NOEC values exceeded the preset range of test concentrations from 3 to 117 mg/L. The difference between these extreme NOECs was a factor of about 240.

From 1989 to 1993 several workshops and ring tests were organized by the University of Sheffield (UK). In these tests the reproduction of daphnids in various synthetic media was compared. The media were variants of the trace element mixture of Keating (1985) containing 22 mineral salts, a substantial amount of EDTA, and three vitamins. In the same ring tests the influence of different genetic clones of *Daphnia* was tested. The results showed that different genetic clones reproduce similarly and have similar sensitivity toward dichloroaniline (DCA). Both variants of the mineral medium showed similar performance. Most of the 59 test results from 18 different laboratories showed more than 60 juveniles per parent and only 12 tests showed mortality in the control below 20%. In the chronic toxicity tests with DCA, EC50 values were within the range between 5 and 25 μg/L for 95% of the cases.

Although these ring test results cannot be compared to the former ring test results with sodium bromide, it seems that the use of a nondeficient medium can reduce the interlaboratory variability. Another reason for the improvement is the use of the end point EC50 versus the use of NOEC (or NEC).

The conclusion might be surprising in view of the experience with *Daphnia*, but it has to be accepted that even with state-of-the-art methodology, ecotoxicologists have to focus on significant interlaboratory variability between test results. The easiest to perform chronic test with a single species at the best achievable level of standardization gives a detection limit of 25% effect on *Daphnia* reproduction with an interlaboratory reproducibility of a factor 5 at best between the highest and lowest 5th percentile values.

Tests to Determine a Chronic NOEC for Fish

Most scientists agree that decisions on high-production-volume chemicals should not be based on acute toxicity data. Therefore in the Sixth Amendment so-

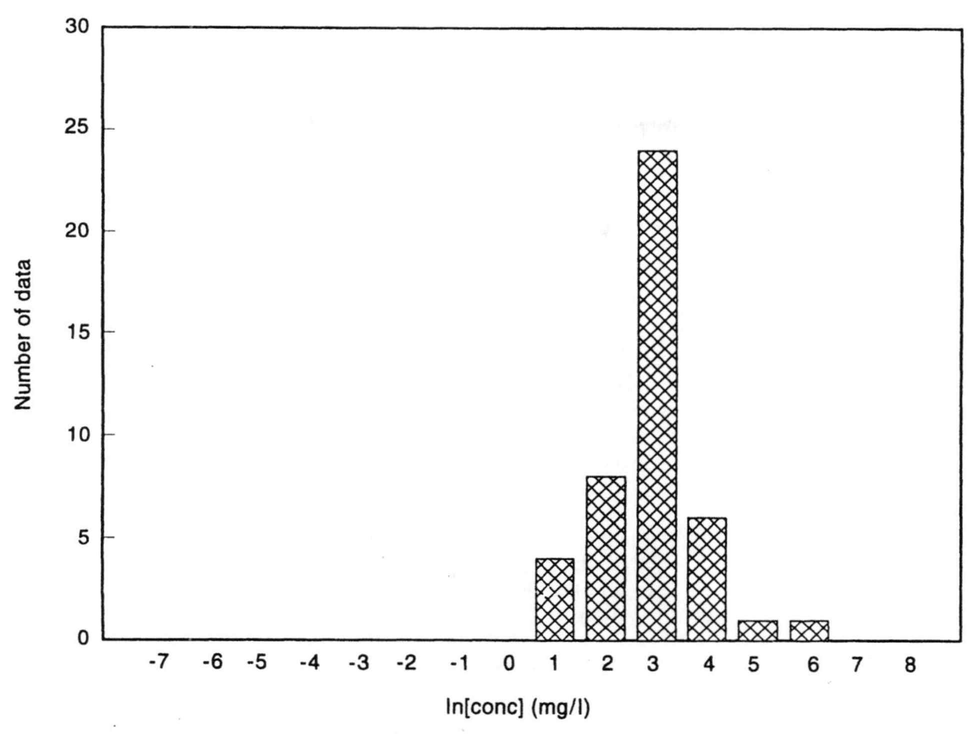

Figure 11. Frequency distribution of NOECs in 44 tests with NaBr and *D. magna* in EC ring test (of 1984).

called level 1 and level 2 tests are required at higher production volumes (see Table 3). Ideally an NOEC based on full life cycle tests should be determined but for practical reasons several substitutes with shorter exposure time have been proposed. Among these, a growth test on juvenile fish lasting 28 d has been ring tested by 14 laboratories from 1989 to 1990. Alternative tests focusing on the sensitive egg and fry stages in the life cycle of fish have also been proposed. These tests can be restricted to the embryo development (short-term early life stage, stELS; see Chapters 1, 2, and 3) or be extended to include the external feeding stage and early growth. The first type can easily be performed within 14 d (for embryo development), whereas complete ELS tests need 32 to 60 d and suitable cultures of *Brachionus* or other small live food (Kristensen, 1990).

The stELS has been a subject of intensive discussions among scientists and national coordinators. The test was believed to be too complicated for level 0 testing and not long enough for level 1. Although the early life stage is sensitive to chemicals, it has been controversial whether to include it as part of the testing scheme. One of the strong arguments for an early life stage test is that demonstration of teratogenic effects has a relatively strong influence on decision makers. The political character of the discussions between scientists representing different political and national views was a typical example for EC rule making. On the other hand, the growth test for fish is accepted without much discussion, although it is not proved by the ring tests or any other scientific report that the growth test is more sensitive than the acute toxicity test and the test does not show teratogenic effects (Kristinsen, 1990).

The EEC ring test on fish growth (Ashley and Mallet, 1990) gives a good illustration of the type of problems met in determining a chronic NOEC. In this test method the growth of rainbow trout (*Oncorhynchus mykiss*, previously *Salmo gairdneri*) was followed for 28 d from a typical weight of about 4 g to about 8–10 g. Before the ring test a few basic water quality criteria such as pH, temperature, and dissolved oxygen and baseline mortality in the control, variation of individual weights, and maintenance of test concentrations were set, but no single laboratory of the 14 participants complied with all requirements. Thus it can be concluded that the ranges set were too optimistic. Measured concentrations of dichloroaniline varied from 12 to 172% of the nominal value and measured concentrations of LAS from 15 to 112% of nominal. In most cases, however, measured values were between 75 and 115% of nominal. The NOECs for DCA varied between 7 and 600 μg/L. The majority of the results (95%) were between 60 and 600 μg/L. The reason for this wide range is clearly not hidden in different sensitivity of fish but in the definition of the NOEC in combination with the test design, i.e., the steps between the test concentrations.

Although at certain times during the test statistically significant differences between test concentrations and controls could be observed, this does not necessarily mean that a lowest observed effect concentration (LOEC) is found. In some cases a concentration-effect relationship did not exist: growth in a next higher concentration appeared not to be significantly lower, whereas in other cases the "effect" disappeared in a later period. In some tests a difference of 5–10% could be shown to be statistically significant, whereas in other tests a difference of 25% fell within the range of control variability. In all of the 14 tests a serious effect of about 50% was not observed below 500 μg/L. Thus the EC50 was not found and it cannot be concluded that this test is more sensitive than the acute LC50 for fish (between 2 and 10 mg/L). With the nominal test concentrations 6, 20, 60, 200, and 600 μg/L and variable LOEC levels, the poorly defined NOEC can easily vary between 6 and 600 μg/L.

Compared to the results of a ring test that includes all life stages of *D. magna* with DCA, the fish growth test is relatively insensitive. The reproduction of *Daphnia* is already reduced by 50% at about 10 μg/L.

For LAS the fish growth ring test with seven participating laboratories showed none or only marginal effects up to the level of 2 mg/L. In no case was a serious effect on the order of 50% inhibition found. Statistically significant differences on the order of 15% reduction of growth were found at 1 mg/L. The result of this ring test for LAS can be compared to acute and chronic toxicity data for LAS from the literature. In a data review by Blok et al. (1993) the average of 88 LC50 values for *Lepomis* was 3.5 mg/L and 90% of the values were between 1.6 and 12. On the other hand, 25 reported NOEC data for *Pimephalus* sp. varied between 0.08 and 55 mg/L. Thus in the case of LAS an NOEC based on growth of fish gives no information in addition to acute toxicity data other than just more variability around the LC50. Although the test protocol appeared suitable, these ring tests did not provide evidence that fish growth is a sensitive parameter that can replace full life cycle tests. Nevertheless, the report on this ring test states: "The test is found suitable for inclusion at Level 1 of the Sixth Amendment."

Tests for Measuring 50% Growth Inhibition of Algae

In a series of ring tests (Hanstveit, 1979, 1980, 1981, 1982, 1991; Adema et al., 1981; WQI, 1992) a growth inhibition test for algae has been developed (see Chapter 4 for discussion of toxicity testing with algae). The results are summarized in Table 7. Compared with the observed variability in chronic tests for *Daphnia* and fish, variability for the algal EC50 was low. The results show a very narrow range for 90% of the EC50 values. This is even more remarkable when it is realized that inoculum density, lighting conditions, test vessels, test duration, calculation method, and even test species were not strictly prescribed. Except for the toxicity of potassium dichromate to *Phaeodactylum tricornutum*, all other EC50 values with two test substances on eight species including some marine species, tested in six ring tests, were between 0.13 and 6.12 mg/L for potassium dichromate and between 0.65 and 5.5 mg/L for 3,5-dichlorophenol.

Several hypothetical explanations for the relatively low variability can be postulated. First, the 50% effective concentration is much better defined than the NOEC. The EC50 value is found by interpolation on a plot with the results for all test concentrations, whereas the NOEC is a point estimate influenced by the test concentrations and the level of the lowest observed effect. The detectability of this LOEC in turn depends on the variability of the test and is influenced by test conditions and test design.

A second explanation is that in algal tests a mean value for a total population is measured, whereas in fish and *Daphnia* only 10 to 20 individuals are ob-

Table 7. Summary of ring tests with algae

Species	Range of EC50 in ring test (90th percentile)		Number of tests (*n*) or laboratories	
	$K_2Cr_2O_7$	3,5-DCP	*n*	Laboratories
Selenastrum capricornutum	0.29–1.70	0.94–4.40	13	11
Scenedesmus subspicatus	0.46–1.07	1.30–5.20	13	11
Scenedesmus subspicatus	0.13–1.00	1.00–5.50	19	20
Scenedesmus pannonicus	1.03–2.66		14	3
Chlorella pyrenoidosa	1.10–1.95		4	3
Chlorella vulgaris	1.10–1.95		10	3
Selenastrum capricornutum	2.76–6.12		4	3
Skeletonema costatum	1.4–4.1	1.3–2.4	11	10
		0.65–1.95	9	11
Phaeodactylum tricornutum	8.4–29	2.0–2.8	11	10
		1.38–3.14	9	11

served. Individual variability in algal tests is masked by the high number of cells.

The third explanation is the use of unbalanced synthetic test media for *Daphnia* and fish. This effect can be seen from the considerable variability and high mortality of control tests if such media are used, whereas in natural water the mortality is relatively low. Similar problems occur with marine algae and synthetic seawater but not in natural seawater.

Conclusions

Unexpected and sometimes unacceptable interlaboratory variability became visible as a result of the ring testing. Many reasons for this variability were found and eliminated by modification of test protocols. In some cases (i.e., biodegradability) inevitable variability caused by the nature and origin of the bacteria seed has to be accepted. In other cases (i.e., *Daphnia* reproduction) it appeared that the trend to use fully specified synthetic media was initially counterproductive in the attempts to reduce variability. Optimization of synthetic media by including just the correct amount of trace elements may need more research. Finally, in the case of prolonged fish tests the main controversial question of either an ELS or a growth test remains unanswered. Ring tests are not the proper instrument to answer this question. In all cases determination of the NOEC or LOEC introduces unnecessary problems and variability as compared to EC50 levels. The main lesson to be learned from the many interlaboratory ring tests is that toxicity is influenced by many factors not controlled by any GLP system and therefore toxicity data show an intrinsic variability that has to be taken into account by regulators.

OUTLOOK

Aquatic toxicology as a science is not much older than 20 yr. It evolved from the need to test the performance of new chemicals including pesticides and developed fast as a direct consequence of environmental regulations. This development in turn is highly influenced by the urgent need for standard test protocols.

Although some substances have been tested hundreds of times, more than 99% of the chemicals have never been tested. Several test species have evolved as "popular" test species, whereas many thousands of species have never been tested.

In view of this tremendous gap of knowledge and the urgent need for environmental regulation of substances, the temptation is great to cover the gap by extrapolation methods. The unknown toxicity of many related chemicals can be approximately predicted by extrapolation from the measured toxicity of a few chemicals. For the large class of chemicals that act via the mechanism of narcosis (see Chapter 20) such predictions can be based on the lipophilicity or the octanol-water partition coefficient. For other modes of action, however, such quantitative structure-activity relationships have not yet been sufficiently validated.

Prediction of the unknown toxicity toward many species on the basis of the measured toxicity for a few species is another type of extrapolation. Proposals on the basis of logistic distribution of the sensitivity for many species have been tested but appear of limited practical use.

One of the main problems in ecotoxicology is the observed high variability of results even with standard species in standard test protocols. The other nearly inevitable problem is the uncertainty of extrapolation of data from laboratory tests to the field. In view of the established standards and the need for comparable data for more chemicals, new methods or even new concepts will not be welcomed by regulatory bodies. On the contrary, it can be expected that further development of new methodology may be hindered by the pressure for strict application to conservative protocols according to good laboratory practice. This does not mean that GLP is unnecessary in general. Attempts should be made to register the relevant parameters that ensure the reliability and integrity of data, and research is necessary to identify those parameters.

For the future, academic scientists will have to take initiatives for more in-depth research. This research will have to focus not only on the understanding of mechanisms of action such as teratogenic effects, hormonal effects, and metabolic activation but also on ecological adaptive mechanisms such as development of resistance by selection at the population level and by metabolic inactivation. Instead of the worldwide hunt to find the most toxic chemical and the most sensitive organism, scientists should try to understand why some species are tolerant to certain toxicants and others not. Until now, scientists have tried to reduce variability by standardization. In the future, variability has to be accepted as an inevitable natural phenomenon. The reasons for variability should be understood in view of the variety of environmental conditions. Normal and abnormal variability should be recognized, and the combined effect of many variable parameters has to be statistically managed.

Toxicological testing is expensive and any predictive methods based on better theoretical knowledge will be welcomed. If regulatory bodies are able to accept some flexibility for new concepts, this will stimulate the science of aquatic toxicology in the right direction.

LITERATURE CITED

Aachen Symposium on LAS: Tenside Surfactants Detergents 26(2), 1989.

Adema, D. M. M.; Canton, J. H.; Slooff, W.; and Hanstveit, A. O.: Research on the most suitable combination of test methods for determining the aquatic toxicity of environmental chemicals. TNO Report CL 81/100. Delft, Netherlands: TNO, 1981.

Ashley, S., and Mallett, M. J.: EEC ring test of a method for determining the effects of chemicals on the growth rate of fish. Report EEC 2600 M, CEC ref B 86614/88/37. Brussels, Belgium: European Community, 1990.

Balk, F.; Okkerman, P. C.; and Dogger, J. W.: OECD guidelines for aquatic effects assessment. Report to the Netherlands Ministry of Housing, Physical Planning, and Environment, 1993.

Blok, J.: Measuring the reproduction rate of *Daphnia magna*. Lecture at Toxicity and Standardization Symposium, Lille, France, 1981. *Inserm*, 106:337–346. ISBN 2-85598-2227.

Blok, J., and Booy, M.: Biodegradability test results related to quality and quantity of the inoculum. *Ecotoxicol. Safety*, 8:410–422, 1984.

Blok, J.; Gerike, P.; de Morsier, A.; Wellens, H.; and Reynolds, L.: Harmonization of ready biodegradability tests. *Chemosphere*, 14(11/12):1805–1985, 1985.

Blok, J.; Balk, F.; Klijnstra, J. W.; and Span, A. S. W.: The contribution of humic compounds, EDTA, and trace elements to the reproduction of *Daphnia magna*. Akzo report CRL E 89004. Arnhem, Netherlands: Akzo, 1989.

Blok, J.; Oostergo, H. D.; Wondergem, A. C.; and van Leeuwen, C. J.: Data search and environmental hazard assessment for post-Seveso chemicals. *Environ. Manage.*, 16(3):317–321, 1992.

Blok, J.; Balk, F.; Edelman, M. H.; Noppert, F.; and Okkermand, P. C.: The use of existing toxicity data for estimation of the maximum tolerable environmental concentration of linear alkyl benzene sulfonate. Part 1 of BKH report for ECOSOL, a product group of CEFIC. Brussels. Delft, Netherlands: BKH Consulting Engineers, 1993.

Boulding, K. E.: The economics of the coming spaceship earth. Environmental Quality in a Growing Economy. Resources for the Future edited by H. Jarett, pp. 3–14. Baltimore: Johns Hopkins Press, 1966.

Cabridenc, R.: Intercalibration exercise relating to a method of determining medium term ecotoxicity of chemical substances in *Daphnia*. EEC Contract W/83/476 5214, Institut National de Recherche Chimique Appliquée. IRCHA Report D 8523. France: IRCHA, 1986.

Carson, R.: Silent Spring. New York: Houghton Mifflin, 1962.

EC: Technical Guidance Document in Support of the Commission Regulation (EC/No. 1488/94 on risk assessment) for Existing Substances in accordance with Council Regulation (EEC) No. 793/93. 1995.

ECETOC: The 6th amendment: prolonged fish toxicity test. Technical Report 24. Brussels, Belgium: European Centre for Ecotoxicology and Toxicology of Chemicals, 1986.

ECETOC: Environmental hazard assessment of substances. Technical report 51. Brussels, Belgium: European Centre for Ecotoxicology and Toxicology of Chemicals, 1993.

Ehrlich, P.: The Population Bomb. New York: Ballantine, 1968.

Elendt, B. P., and Bias, W. R.: Trace nutrient deficiency in *Daphnia magna* cultured in standard medium for toxicity testing. Effects on the optimization of culture conditions on the life history parameters of *D. magna*. *Water Res.*, 24:1157–1168, 1990.

EUROSTAT: Paris: OECD, 1992.

Gerike, P.; Winkler, K.; Schneider, W.; and Jacob, W.: Residual LAS in German rivers. Tenside Surfactants Detergents, 26(2):136–140, 1989.

Giger, W.; Brunner, P. H.; Ahel, M.; McEvoy, J.; Marcomini, A.; and Schaffner, C.: Organische Waschmittelinhaltsstoffe und deren Abbauprodukte in Abwasser und Klaerschlamm Gas. *Wasser Abwasser*, 67(3):111–122, 1987.

Goudie, A.: The Nature of the Environment. Oxford: Basil Blackwell, 1989.

Haigh, H.: EEC Environmental Policy and Britain. London: Institute of European Environmental Policy and International Institute for Environment and Development, 1984.

Hanstveit, A. O.: Evaluation of the results of the ISO interlaboratory studies with an algal toxicity test. TNO Reports CL 79/46, 80/44, 81/111, and 82/128. Delft, Netherlands: TNO, 1979, 1980, 1981, 1982.

Hanstveit, A. O.: The results of an international ring test on the marine algal growth inhibition test according to ISO/DP 10253. TNO Report R 91/236. Delft, Netherlands: TNO, 1991.

Keating, K. I.: A system of defined media for daphnial culture. *Water Res.*, 19(1):3–78, 1985.

Kristensen, P.: Evaluation of the sensitivity of short term fish early life stage tests in relation to other FELS test methods. Draft Final Report, EEC Contract 136614-43-89, WQI Report 305019. Hørsholm, Denmark: Water Quality Institute, 1990.

Mackay, D.: Multimedia Environmental Models, the Fugacity Approach. Chelsea, MI: Lewis, 1991.

Meadows, D.: The Limits to Growth. Guilford, United Kingdom: PAN Books, IPC Science and Technology Press, 1972.

OECD: The State of the Environment. Paris: OECD, 1979.

OECD: Compendium of environmental exposure assessment methods for chemicals. Environment Monographs No. 27. Paris: OECD, 1989a.

OECD: Report of the OECD workshop on ecological effects assessment. Environment Monographs No. 27. Paris: OECD, 1989b.

OECD: Guidelines for testing of chemicals. TG 301 A–F: Ready Biodegradability. Updated July 1992. Paris: OECD, 1992.

OECD: Draft report of the workshop on *Daphnia magna* pilot ring test, Sheffield University, UK, March 20–21, 1993. OECD Test Guideline Programme.

OECD: Guidance Document for Aquatic Effects Assessment. OECD Environment Monographs, No. 92. Paris: OECD, 1995.

De Oude, N. T.: Anthropogenic compounds: Detergents. Handbook of Environmental Chemistry, Vol. 3. Part F, edited by O. Hutzinger. Berlin: Springer Verlag, 1992.

Richtler, H. J., and Knaut, J.: World prospects for surfactants. Henkel Oleochemical Division. Düsseldorf, Germany: Henkel, 1991.

RIWA: Samenstelling van Rijnwater. Yearly reports 1988–1989. Amsterdam: RIWA, 1992.

TNO Delft: Degradability, Ecotoxicology and Bioaccumulation. The Hague, Netherlands: Governmental Publishing Office, 1977.

Versteeg, D. J.; Feijtel, T. C. J.; Cowan, C. E.; Ward, T. E.; and Rapaport, R. A.: An environmental risk assessment for DTDMAC in The Netherlands. *Chemosphere*, 24(5):614–662, 1992.

Water Quality Institute (WQI) Denmark: Intercalibration and comparison of test laboratories for evaluation and approval of offshore chemicals and drilling muds. Paris Commission Ring Test. Hørsholm, Denmark: WQI, 1992.

Whitehead, C. (ed.): European Community Environmental Legislation Doc. No. XI/989/87. Brussels, Belgium: Commission of the European Community, 1988.

Chapter 28

INTRODUCTION TO ECOLOGICAL RISK ASSESSMENT FOR AQUATIC TOXIC EFFECTS

G. W. Suter II

INTRODUCTION

Ecological risk assessment is the process of estimating and characterizing the likelihood that adverse effects of human actions on the nonhuman environment will occur, are occurring, or have occurred. It is a rapidly evolving discipline that attempts to perform assessment of effects on nonhuman organisms that are equivalent to the health risk assessments that came to dominate regulatory practice in the 1980s. Risk assessment originated in the insurance industry, which, since the late seventeenth century, has attempted to estimate the probabilities of death, fire, or other undesired events. Since then risk assessment has been adopted by business as a means of estimating probabilities of losses, by engineers as a means of estimating probabilities of failures of complex systems, and by health professionals as a means of estimating the probability of death, disease, or injury. Ecological risk assessment resembles these activities, and differs from prior environmental assessment approaches, in having clear end points, probabilistic methods, clear consistent procedures and assumptions, and an emphasis on quantitative methods and results (Suter, 1990a).

Although some recent regulations call for risk assessments (EPA, 1990), ecological risk assessment, like health risk assessment, has no specific legal mandate. Rather, it is a general methodology for basing regulatory decisions on the expected magnitude of effects and the uncertainty concerning those effects rather than on technology (e.g., best available control technology), on analytical chemistry (e.g., no detectable amounts), or on categorical exclusions (e.g., no dumping of medical wastes).

Although ecological risk assessment can be structured like human health risk assessments, it can be considerably more complex. This is because there are considerably more potential end points as a result of there being more species and more levels of biological organization, more potential pathways of exposure and toxicological modes of action (e.g., inhibition of photosynthesis), and indirect effects (e.g., loss of habitat or prey). This complexity means that there are more measurements, testing, and modeling methods that may be used and more decisions that must be made by the assessor. However, there is considerably less guidance from regulatory agencies or consensus from the scientific community.

Some authors have argued that ecological risk assessment is not worthy of significant effort because human health risks and regulatory standards inevitably dominate decision making (Fink, 1992). However, this attitude is incorrect for two reasons. First, regulatory agencies are placing greater emphasis on protection of the nonhuman environment (Reilly, 1990). Second, nonhuman organisms, populations, and ecosystems can be much more sensitive to environmental contamination than humans. This is because nonhuman organisms have higher exposures because they are more immersed in the environment, they have modes of exposure such as respiring water and consuming sediments that humans do not, and it is highly likely that some of the thousands of nonhuman organisms that are exposed to a contaminant will be inherently more sensitive than humans (Suter, 1992).

Ecological risk assessments can be categorized according to the problems addressed.

- *Predictive risk assessments* estimate risks from proposed future actions such as marketing a new pesticide or commercial chemical, operating a new aqueous or atmospheric emission source, or permitting a wetland to be dredged.

- *Retrospective risk assessments* estimate the risks posed by actions such as disposal of hazardous wastes, spills, existing effluents, and landscape modifications that occurred in the past and may have ongoing consequences. Retrospective risk assessments can be more complex than predictive assessments because the existence of a contaminated site permits more types of studies to be performed. These include biological surveys, in situ toxicity tests, and laboratory tests of ambient media. However, the availability of these data presents additional problems in inference, particularly concerning the cause of observed degradation. In particular, retrospective assessments must be concerned with determining whether the apparent effects observed in the field are due to the contaminants of concern, to some other anthropogenic cause, or to natural variability.

Ecological risk assessment requires the participation of individuals with diverse expertise including environmental chemistry, toxicology, and ecology (see Chapter 29). In addition, there is an increasing need for people trained in risk assessment. Although hazard assessment (defined below), the prevailing paradigm for assessing ecotoxicological effects in the 1980s, was developed by aquatic toxicologists (Cairns et al., 1978), the day is passing when an individual trained only in toxicology, chemistry, or ecology can perform state-of-the-art assessments. Increasingly, ecological assessments must involve multidisciplinary teams that are coordinated by an individual with broad experience in applied environmental science who is familiar with available assessment tools and concepts. Inevitably, academic programs will begin to train students in ecological risk assessment just as students are now trained in health risk assessment.

The following discussion emphasizes concepts rather than procedures for two reasons. First, the scope of the topic is too broad relative to the length of a chapter to allow discussion of specific methods. Second, ecological risk assessment is a new field that is only beginning to develop a distinct set of tools and methods. However, the newness and unfamiliarity of ecological risk assessment make it important to explore new concepts and agree upon common concepts and terminology.

TERMINOLOGY

Ecological risk assessment. The process of identifying and quantifying risks (see the definition of risk) to nonhuman biota and determining the acceptability of those risks.

Effects assessment. The component of an ecological risk analysis that is concerned with quantifying the manner in which the frequency and intensity of effects increase with increasing exposure to a contaminant or other hazardous agent. Equivalent but less general terms include dose-response assessment and toxicity assessment.

End point, assessment. A quantitative or quantifiable expression of the valued environmental component that is considered to be at risk in a risk analysis. Examples include a 25% reduction in game fish biomass or local extinction of an avian species.

End point, measurement. A quantitative summary of the results of a biological monitoring study, a toxicity test, or other activity intended to reveal the effects of a hazard. Examples include the mean species per sample and the LD50.

End point, test. A type of measurement end point. The numeric summary of the results of a toxicity test. Examples include the LC50 and the NOEC.

Epidemiology, ecological. The study of the incidence, distribution, and causes of harmful effects of chemical, physical, or biological agents on natural populations and communities.

Exposure. (a) The process by which the temporally and spatially distributed concentrations of a chemical or physical agent in the environment are converted to a dose. (b) The degree of contact of organisms to a chemical or physical agent (see Chapter 1).

Exposure assessment. The component of an environment risk analysis that estimates the exposure resulting from a release or occurrence of a chemical, physical, or biological agent in a medium. The exposure assessment strives to quantify, using measurement or predictive tools, the magnitude, duration, and extent of exposure. For chemicals, it includes estimation of transport, fate, and uptake.

Extrapolation. Use of data derived from observations to estimate values for unobserved entities or conditions.

Hazard. A state that may result in an undesired event, the cause of risk. In ecotoxicology, the potential for exposure of organisms to chemicals at potentially toxic concentrations constitutes the hazard.

Hazard assessment (analysis). Determination of the existence of a hazard. (a) In risk assessments, a preliminary activity that helps define the assessment problem by determining which environmental components are potentially exposed to toxic concentrations and how those components might be affected by the exposure. (b) An alternative assessment method that determines whether a hazard exists by comparing the magnitudes of estimated environmental concentrations to toxicological test end points for a contaminant.

Health risk analysis. Determination of the probability of human morbidity and mortality.

Risk. The probability of a prescribed undesired effect. If the level of effect is treated as an integer variable, risk is the product of the probability and frequency of effect [e.g., (probability of an accident) × (the number of expected mortalities)].

Risk characterization. The process of (a) integrating the exposure and effects assessments to estimate risks and (b) summarizing and describing the results of a risk analysis for a risk manager or for the public and other stakeholders.

Risk management. The process of deciding what actions to take in response to a risk.

Source term. An estimate of the total amount released, or the temporal pattern of the rate of release of a chemical from a source.

RISK ASSESSMENT PARADIGMS

The Traditional Predictive Paradigm

In the United States, the standard paradigm for risk assessment is the one published by the National Academy of Sciences (NRC, 1983). It has been implemented in the U.S. Environmental Protection Agency's guidance for health risk assessment (EPA, 1987). It has also been adapted for ecological risk assessment by Oak Ridge National Laboratory (Barnthouse and Suter, 1986), the EPA (Risk Assessment Forum, 1992), and the National Academy of Sciences (NRC 1993). A version of the NAS's diagram, modified for clarity and applicability to ecological assessments, is presented in Figure 1. It defines the hazard to be assessed, proceeds to estimate exposure and the relationship of effects to exposure, characterizes the risk resulting from the conjunction of a particular estimated exposure and an estimated exposure-response relationship, and conveys the results to a risk manager.

The sources of risk described here are chemical contaminants, which is in keeping with the topic of this text. However, other hazardous agents such as dredges, introduced organisms, or heat can also be assessed by adapting the same risk paradigm. Chemicals and these other agents are often referred to as stressors (Risk Assessment Forum, 1992). However, it must be remembered that hazardous agents are not necessarily stressors. They may act as a subsidy, as a stressor, or have no effect, depending on the level of exposure.

Hazard Definition

For human health risk assessments, it is sufficient to identify the hazard by determining that humans will be exposed to a source of a chemical with potentially adverse properties (i.e., a teratogen in drinking water). However, hazard definition is more complex for ecological risk assessments because of the multiplicity of possible end points and modes of exposure and because of the importance of characteristics of the environment in which the exposure occurs. Therefore, it is necessary to define the sources, describe the exposed environments, and choose the end points. This phase of ecological risk assessment is termed problem definition by the EPA (Risk Assessment Forum, 1992).

Source definitions include a description of the material released, its matrix, the rate of release, and any temporal variability in the composition and rate of the release including spills or failures of treatment facilities. Source terms (the quantification of all of those characteristics of the releases) are usually provided to assessors by engineers or applicants for a permit or license. In such cases, it is important but often difficult to obtain estimates of variance in the source term so that variance in exposure can be estimated.

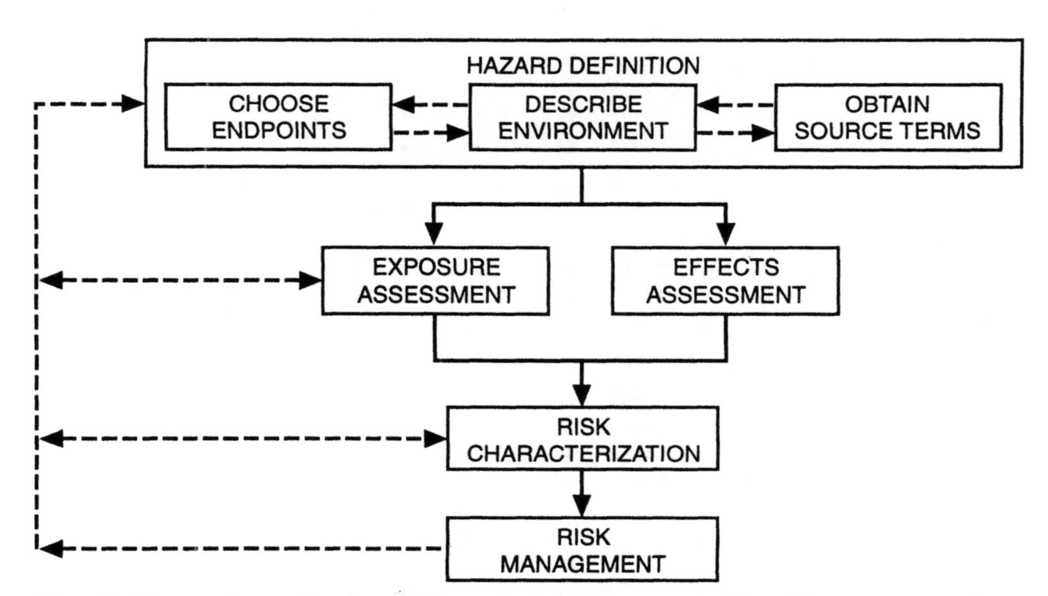

Figure 1. Diagram of a paradigm for predictive ecological risk assessments. The solid arrows represent the sequential flow of the procedure. The dashed arrows represent feedback and other constraints on one assessment process by other processes. (From Suter, 1992.)

The environmental description determines the transport and fate processes that lead to exposure and the populations and communities that are potentially exposed. The receiving environment may be an actual site such as a proposed landfill site. However, for assessment of new chemicals or technologies a generic reference environment is used. For example, a generic description of the soils, hydrology, meteorology, and biota of the corn belt can be used to assess the risks of a pesticide proposed for use on corn (Beck et al., 1981). The description of the environment must include all of the descriptive parameters such as flow rates and temperatures that are needed for the exposure and effects models.

The *assessment end points* are formal descriptions of the specific environmental components that are at risk and deemed to be worthy of protection. An assessment end point includes an entity such as a population of fish or an entire aquatic community and some property of that entity such as abundance or productivity. Assessment end points should have the following properties (Warren-Hicks et al., 1989; Suter, 1990b):

- Social relevance
- Biological relevance
- An unambiguous operational definition
- Accessibility to prediction and measurement
- Susceptibility to the hazardous agent

End point choice is the most difficult part of hazard definition. Clearly, one cannot measure or estimate effects on every property of all potentially exposed organisms, populations, and communities. Just as health assessments devote considerable effort to estimating cancer risks to humans but neglect less dreaded health effects such as rashes, ecological risk assessors must choose a limited number of socially and biologically significant end points that can be operationally defined (i.e., defined in a way that is reasonably unambiguous and that indicates what specific properties must be measured or predicted) and that can be measured or predicted using available methods. The risk manager's input is important to the process of selecting assessment end points. The utility and importance of the assessment to decision making depend on a mutual education of the assessor and the risk manager concerning societal and biological significance. That is, the risk manager must convey to the assessor his ideas about what end points, because of their societal relevance, are important to his decision. In turn, the assessor must explain to the risk manager what end points should be considered even though they do not have obvious societal relevance, or how the risk manager's end points might be modified to be more biologically relevant or more amenable to measurement or prediction.

In addition, assessment end points are relevant only if they are susceptible to the hazardous agent. Susceptibility is a function of exposure, sensitivity to the agent, and the potential for indirect effects. Consideration of susceptibility requires a qualitative or simple quantitative hazard assessment. This assessment should include a description of transport and fate processes sufficient to identify significantly contaminated areas and media, a summation of knowledge concerning the relative sensitivity of taxa and ecological processes to the chemical or other agent, and a description of the trophic interactions and other ecological processes that can cause indirect effects.

The result of this hazard definition phase is a conceptual model of the situation to be assessed. That is, a specific release of a contaminant into a specific environment will, through certain processes of transport and uptake, result in a hazardous exposure to the contaminant of certain valued components of the environment (Risk Assessment Forum, 1992).

Exposure Assessment

Exposure assessment estimates the temporal dynamics of the concentration of a contaminant at the point of uptake by organisms. It consists of (1) estimation of the properties of the contaminant that determine its transport and fate, (2) modeling of the processes of transport and fate, and (3) estimation of the degree of contact with and uptake of the contaminant. The first two steps have relatively well-developed methods. However, the role of the biotic responses in the process of exposure is poorly understood and therefore is not well incorporated in assessment models.

Given the estimated contaminant levels in various media, the assessor must determine to which media the end point species are exposed, by what route and at what rate they are exposed, and how initial exposures modify subsequent exposures (i.e., avoidance, attraction, or modification of consumption rates). Issues to be considered include (1) the bioavailability of various chemical fractions and species, (2) the need to combine multiple routes of exposure such as diet, respiration, and direct contact, and (3) the degree to which avoidance behaviors in the laboratory actually protect organisms in the field.

Effects Assessment

Effects assessment establishes the relationship between the degree of exposure and the nature, severity, and duration of effects. The results of toxicity tests, expressed as test end points such as LC50s, are the principal input to effects assessments for chemicals. The chief problem in ecological effects assessment is extrapolation from test end points to assessment end points. In general, test species are not the same as end point species, the tested life stages do not include all of the exposed life stages, the test durations are not the predicted exposure durations, the test conditions are not the expected field conditions, the reported responses do not include all of the responses of concern, and the test end points are at a different level of biological organization (usually the organism level) than the assessment end points (usually population or ecosystem levels).

There are three approaches to these extrapolations. The first and most common approach is to multiply the test end points by factors that are variously termed safety factors, uncertainty factors, correction factors, or assessment factors (Mount, 1977). This approach has the virtue of simplicity, but the factors have little technical support and obscure the expected level of effect, the magnitude of uncertainty, and the nature of the assessor's assumptions. The second approach is statistical models. These models either regress one class of responses against anotther (Figure 2), or fit distributions to sets of response data (Stephan et al., 1985; Kooijman, 1987; Van Straalen and Denneman, 1989). The third approach is to develop mathematical models that simulate the assessment end point response and use the test end points as input parameters. These include population models (Barnthouse et al., 1987, 1990; Hallam et al., 1990) and ecosystem models (O'Neill et al., 1982; Bartell et al., 1988).

Risk Characterization

Risk characterization is the process of (1) estimating the nature and likelihood of effects by combining the exposure estimates from the exposure assessment with models of the relationship between exposure and effects from the effects assessment and (2) providing an explanation and conceptual context for the analytical results. The integration of the exposure and effects assessments to generate estimates of risk requires expressing exposure and effects in common dimensions. This can be thought of as locating the expected exposure and response in a four-dimensional state space with dimensions of concentration, duration, proportion responding, and severity of the response. That is, a particular duration of exposure to a particular concentration of a substance will cause a particular proportion of organisms, populations, or communities to experience effects of a particular severity (Suter, 1992). These four are the dimensions that are most commonly treated in toxicological assessments, but other dimensions such as space should also be considered when relevant.

Commonly only one or a few of these dimensions are considered in ecological assessments. For example, when an estimated environmental concentration (EEC) is compared to an LC50, only the concentration dimension is considered (Figure 3). The duration dimension is lost by considering only the duration at the end of the test, the severity dimension is lost by considering only lethality, and the proportion responding dimension is lost by considering only the 50th percentile. Assessors should consider which dimensions need to be modeled explicitly in order to characterize the assessment in terms appropriate to the decision.

For example, if a power plant periodically releases chlorinated water, it is necessary to identify what combination of chlorine concentration and release

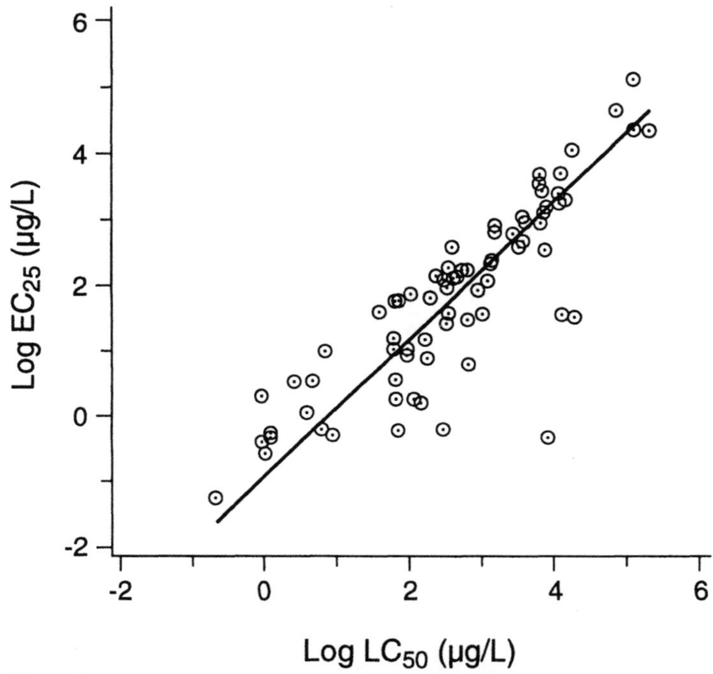

Figure 2. A regression model to estimate a concentration that would reduce larval survival by 25% in exposures extending over the entire larval stage from 96-h LC50 values for juveniles or adults of the same species tested with the same chemical. (See Suter et al., 1987, for derivation of the model.)

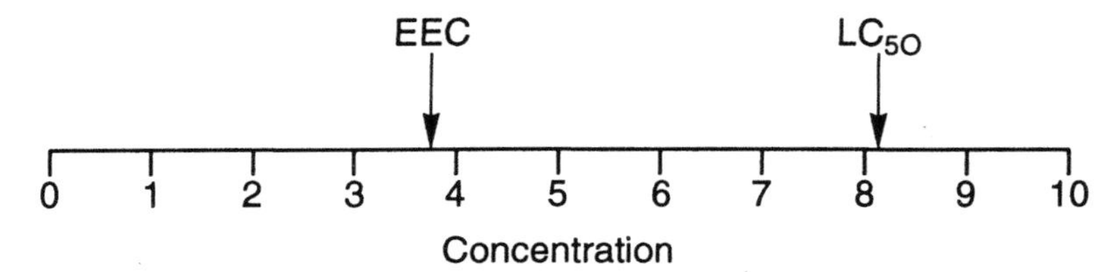

Figure 3. Exposures (EEC) and effects (LC50) are commonly arrayed on a single dimension, concentration.

duration will result in less than some prescribed threshold level of effect. The assessor may decide that only lethality is important because the releases are infrequent relative to the rate of recovery of organisms from sublethal effects. This decision eliminates severity as a variable by reducing it to one level (death). The dimension of proportion responding may be eliminated similarly by picking a single value such as 1% as a threshold of acceptability. Therefore, those two dimensions (severity and proportion responding) can be held constant, and the results of the assessment can be expressed as a concentration-duration curve (Figure 4). Combinations of chlorine concentration, release duration, and dilution in the receiving stream that result in exposures above the central curve are expected (i.e., >50% probability) to cause greater than 1% mortality in the end point species. In other cases the exposure may be essentially constant and continuous, and the decision would hinge on the severity of effects (e.g., the estimated reduction in fish abundance) or frequency (the proportion of sites at which fish abundance would be reduced by x%). The point is that the risk characterization inevitably involves some simplification of the processes of exposure and effects induction, and a major aspect of that simplification is elimination of some variables.

The assessor must always acknowledge the uncertainties in the assessment and present the results as either qualitative probabilities (e.g., the expected severity of effects is x but effects as severe as $x + u_x$ are credible) or as quantitative probabilities (e.g., the median predicted severity of effects is x but the risk of effects as severe as $x + u_x$ is 10%). Quantification of uncertainty increases the credibility of results by forcing the assessor to be rigorous in accounting for uncertainty and by opening to review the assumptions made in estimating and combining the uncertainty parameters.

A simple example of use of quantitative uncertainty to estimate risks is presented in Figure 5. Only the concentration dimension is expressed. The distribution of the estimated environmental concentration is generated by Monte Carlo analysis of a chem-

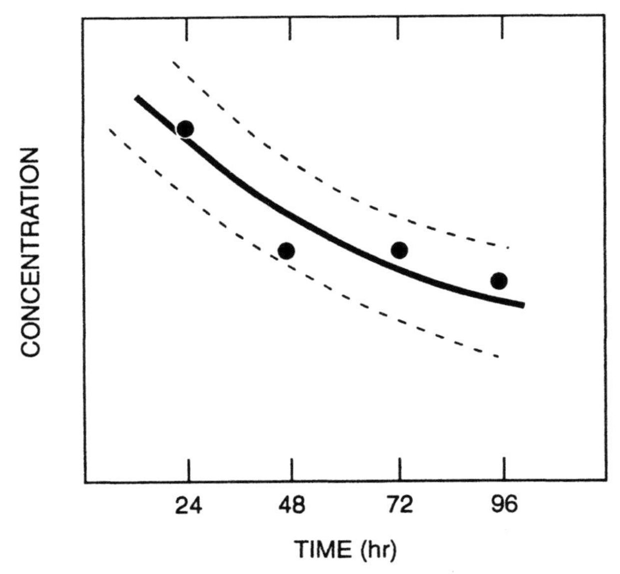

Figure 4. A function expressing the relationship between the concentration and duration of exposure necessary to cause a specific effect.

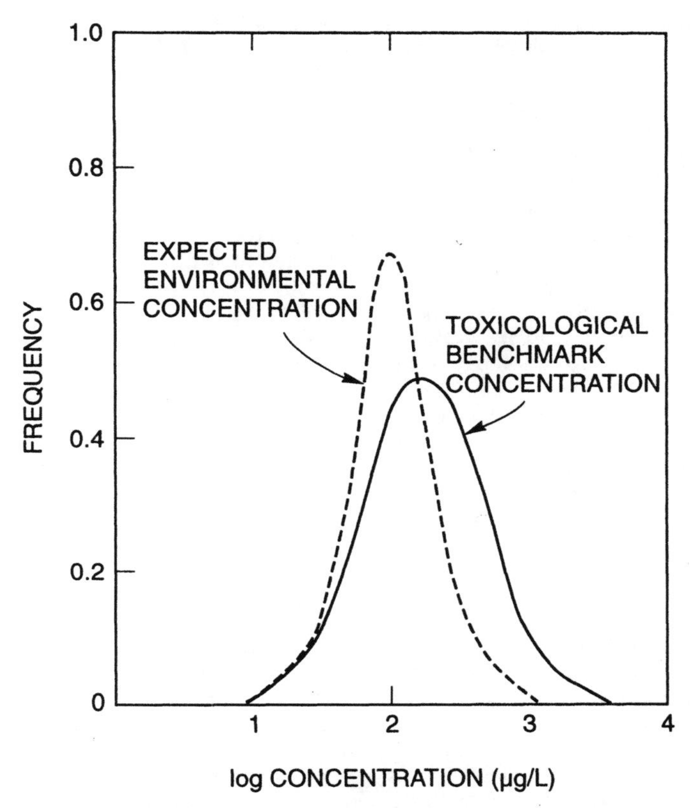

Figure 5. If exposure and response are represented as distributions on the concentration scale rather than point estimates, risk can be expressed as the probability that the exposure concentration is greater than the effective concentration.

ical fate model. This technique involves (1) assigning probability density functions to each of the variables of the environmental transport and fate model, (2) parameterizing the model by selecting values from each of the probability density functions, (3) running the model to generate an output value (i.e., an estimate of the EEC), (4) repeating steps 2 and 3 to generate a large set of alternative output values, and (5) determining percentiles of that set of output values. Software for Monte Carlo analysis is commercially available (e.g., @RISK and Crystal Ball).

As an example of uncertainty analysis for effects assessment, consider as an end point the concentration that reduces the weight of juvenile fish per egg (i.e., integrated effects on early life stages) by 25%. The distribution of the EC25 for weight/egg could be simply calculated from the variance in early life stage test results, could be a subjective estimate of the uncertainty concerning loss of early life stage production in the field, or could be estimated from statistical models (Suter et al., 1987), depending on the data available and the assessor's judgment.

The source of the uncertainty quantified in these sorts of uncertainty analyses is the uncertainty concerning the individual parameters, as they are propagated through the assessment models. The parameter uncertainty results from environmental stochasticity (e.g., the distribution of dilution volume resulting from variation in the flow rate of a receiving river or variation in the size of rivers into which a chemical may be released) and from ignorance concerning measurable values that were not measured [e.g., the biodegradation rate of the chemical, which may be assumed to be the same as that of a similar chemical or may be derived from a quantitative structure-activity relationship (QSAR)].

This sort of uncertainty analysis is important and useful, but it captures only a portion of the uncertainty concerning the effects that will occur. Another source is uncertainty associated with the fundamental assumptions of the models or tests used in the assessment. Use of organism-level test data alone to estimate population-level responses in the field implies an assumption that population and ecosystem processes do not influence the response (i.e., that the population response is simply the summation of the responses of the individual member organisms). Use of a demographic population model implies an assumption that ecosystem processes are not influential and organismal processes other than birth and death are not influential. Use of pond mesocosm data implies that the essential features of all exposed aquatic populations and ecosystems are contained in a small static pond. This uncertainty due to fundamental assumptions can be addressed only by using multiple lines of evidence and careful exploration of the reasons for convergence or divergence of results.

The use of uncertainty analyses can be illustrated by returning to the example of intermittent chlorine emissions. If the dilution volume is expressed as a frequency distribution derived from historic frequencies of flow rates in the receiving river, the risk could be expressed as an annual risk of exceeding the threshold exposure level due to inadequate dilution. If the uncertainty concerning the concentration-duration relationship is included in the risk characterization (e.g., the confidence intervals in Figure 4), the risk could be expressed as the credibility of effects. Either of these risk might be deemed important, or, if they are both important, a joint probability might be calculated (the annual risk that effects will actually occur).

The second stage of risk characterization is the preparation of a summary of results for the risk manager. Elements that should appear in this risk characterization report are listed in Table 1. The critical component of this report is the assessors' analysis of the weight of evidence concerning the risks. This analysis depends on a thorough understanding of the data and models used in the assessment. It is tempting but inappropriate to include only data and results that support the most liberal or conservative conclusion (depending on one's predilections) or only the results from one's favorite method.

Risk Management

Risk management is the process of reaching a decision concerning the appropriate course of action in a situation involving risks. The risk manager must consider not only the risks reported in the health and ecological risk characterizations but also the costs and benefits of avoiding the risks, the legal and regulatory constraints, the social and political climate, the technical feasibility of treatment or remediation, and the risks associated with alternative products, remedial actions, or waste streams. Risk management is performed independently of risk assessment (NRC, 1983), but the risk manager must communicate his managerial goals and information needs to the assessors, and the assessors must present their results in a form that is comprehensible to the risk manager. For example, if the risk manager requires a formal cost-benefit analysis, the assessor must be able to assign monetary values to the resources at risk. In particular, the assessor may be required to demonstrate that the resources at risk are worth more than the cost of requiring additional waste treatment.

Retrospective Risk Assessment and the Epidemiological Paradigm

Retrospective risk assessments differ from predictive assessments in that the environment has already been contaminated, so that, potentially, the source, exposure, and effects can all be observed and measured. Therefore, in addition to the toxicity testing and measurement of chemical properties that support predictive assessments, retrospective assessments can use data from measurements of effluent concentrations and other measurements of sources; of concentrations in media, body burdens, and other measures of exposure; and of population abundance, growth rates, and other measures of effects in the field. In addition, unlike predictive risk assessments that always begin with a proposed source and estimate exposure and effects, retrospective assessments may begin with a source (e.g., a hazardous waste site), evidence of exposure (e.g., elevated body burdens), or evidence of effects (e.g., declines in fish abundance). Therefore, a slightly different paradigm is required for retrospective assessments that begins by identifying the impetus for the assessment and that may estimate the nature and magnitude of sources as well as exposures and effects (Figure 6).

If ecological effects can be measured in a contaminated environment, then it is possible to perform risk assessments using a methodology that is equivalent to epidemiology. Therefore it is termed ecological epidemiology or ecoepidemiology (Bro-Rasmussen and Lokke, 1984; Suter, 1990c, 1992;

Table 1. Information that should be conveyed to the risk manager along with the numeric results of the risk analysis as part of the risk characterization

1. A description of the models used including their major assumptions, the evidence for their validity, and their degree of acceptance by the scientific community.
2. A description of the sources of the test data and other input parameters and of the quality of the data.
3. A description of the sources of uncertainty in the analysis and of the extent to which they have been quantified.
4. Any conflicting evidence and explanations of its existence.
5. Alternative credible assumptions and interpretations of conflicting evidence.
6. Alternative results that are obtained if alternative credible assumptions are used.
7. A description of any research that would resolve major uncertainties.
8. A description of precedent assessments and analogous situations (e.g., similar chemicals with well-characterized risks) that would help to clarify the decision.

From Suter, 1992.

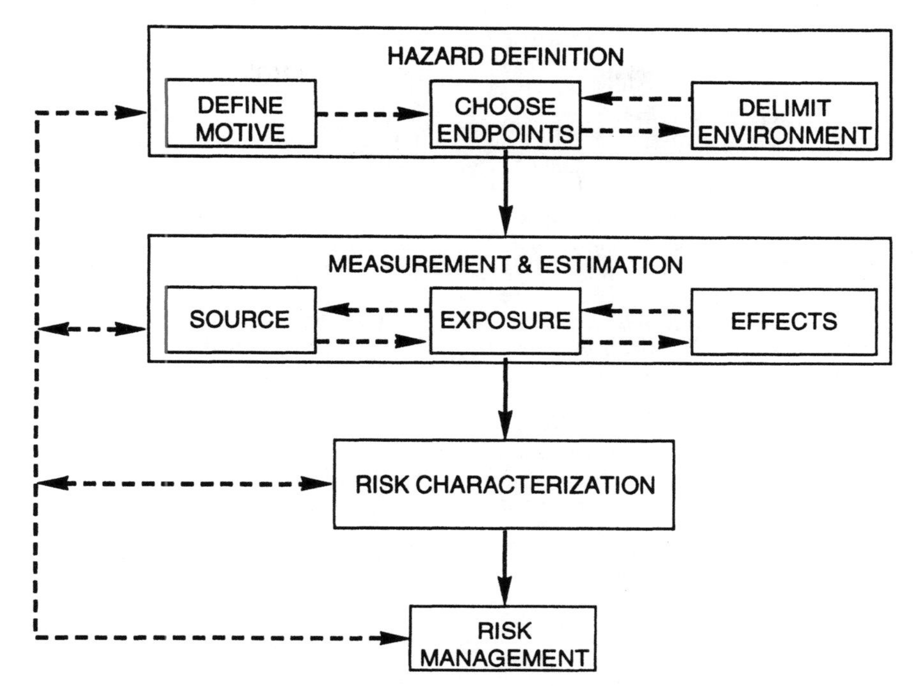

Figure 6. Diagram of a paradigm for retrospective ecological risk assessments. The solid arrows represent the sequential flow of the procedure. The dashed arrows represent feedback and other constraints on one assessment process by other processes. (From Suter, 1992.)

Fox, 1991). It uses field and laboratory data with statistical and logical inference to answer the questions:

1. What is the likelihood that the apparent effect is real and not simply a result of natural variation?
2. What is its likely cause?
3. What is its magnitude?
4. What are its ultimate consequences?

The question of reality of effects arises because an exposed population or ecosystem may appear to be degraded relative to a preexposure state due to natural temporal variability or degraded relative to an unexposed population or ecosystem due to natural spatial variability. Unless effects are catastrophic (e.g., a large fish kill), it is not obvious whether, when an exposed site differs from an unexposed site, the difference is real or random. Either site, neither site, or both sites may be degraded relative to some ideal undisturbed state. In epidemiology, public health records are used to determine normal rates of diseases and injuries. In ecological epidemiology, it is necessary to (1) establish a local or regional baseline for particular classes of ecosystems (Ohio EPA, 1988), (2) apply intervention analysis (which determines whether there is a significant change in the temporal variance of a system following imposition of a contaminant or some other intervention) or some other form of time series analysis if a time series of data is available for the site (Carpenter et al., 1989), or (3) limit the analysis to manifestly aberrant states such as fishless lakes or oiled birds. The first two approaches allow the assessor to calculate the probability that contaminated sites do not belong to the population of uncontaminated sites or do not belong to the precontamination time series. The third approach is an application of Descarte's method of doubt. That is, if you can doubt that the contaminated site is degraded, then you cannot accept that a real effect has occurred.

The question of causation is even more complex and subject to error than the question of reality of effects. A common practice for determining causation is to use statistical hypothesis testing to compare a contaminated site with a "control" site and to then infer that, if the sites are statistically significantly different, the difference is caused by the contamination. This procedure is statistically and logically illegitimate (Warren-Hicks et al., 1989; Suter, 1990c). Samples from a site are pseudoreplicates (Hurlbert, 1984), and treatments are not randomly assigned to the sites. It is an unintended but nevertheless ironic commentary on environmental science that epidemiologists refer to the assumption that a measured difference between the environments of subject populations is the cause of differences in the frequency of disease as the "ecological fallacy" (Morgenstern, 1982).

As in epidemiology, the solution to the problem of inferring the cause of ecological damage is to demonstrate that not only is the putative cause consistently associated with the effect in the field but also the same relationship between cause and effect is found in toxicity tests or other controlled exposures. For example, if a chlorine release is said to be responsible for fish kills, then chlorine should kill fish in toxicity tests at the concentrations observed in the field. This requirement is an adaptation of Koch's postulates (a criterion for proving causation in pathology) to environmental toxicology (Woodman and Cowling, 1987; Suter, 1990c). Experienced readers will realize that fulfilling this requirement can be difficult in practice. The effects of varying environmental and biological conditions and the problem of estimating an exposure concentration and temporal dynamics that correspond to the standard laboratory test conditions may obscure the concordance between laboratory and field results. Therefore, the success of this approach may depend on the use of extrapolation models to relate test conditions to field conditions. In any case, neither field surveys nor laboratory toxicity tests alone can be used to infer confidently that a contaminant has caused an effect.

The process of establishing a causal relationship is portrayed by the diagram in Figure 7. Field studies, conventional laboratory toxicity tests, and tests of contaminated water and sediment all have a contaminant source, an exposure process, and an array of potential effects. The three components of the causal evidence (i.e., sources, indicators of exposure, and indicators of effects) are measured in different ways in different systems. For example, effects on fish may be measured as mortality in a laboratory test with the contaminant, a no-observed-effect concentration in a toxicity test of ambient water, and population abundance in the field. Also, the processes by which sources induce exposure and exposure causes effects are modified by differences in water chemistry (e.g., toxicity of many metals is a function of hardness), the condition of the exposed organisms (e.g., nutritional state), and other biological and nonbiological factors. Concordance between the results of toxicity tests and field biological surveys can be demonstrated by extrapolating between the different expressions of source, exposure, and effects to achieve equivalent source-exposure-effect relationships. For example, pH, hardness, or suspended-solids data might be used to convert total contaminant concentrations measured in samples from the field into an estimate of the exposure concentration that is equivalent to a concentration experience in the laboratory toxicity test. If the converted field concentration is approximately equal to a laboratory concentration that causes effects equivalent to those observed in the field, then concordance between the studies has been demonstrated. Similarly, standard test end points and species could be extrapolated to population abundance estimates using statistical and mathematical models. If the effects predicted to occur at field concentrations by the extrapolation model are the same as those observed in the field (allowing for uncertainties in the data and models), then the laboratory and field studies are concordant.

Establishing the magnitude of effects is the point in an epidmiologically based risk assessment at which risk, in a conventional sense, begins to be estimated. That is, given that a real effect has been detected (e.g., fish kills) and this effect has been attributed to a cause (e.g., release of chlorinated cooling water during periods of low flow), what is the risk that these kills are sufficiently large and frequent to exceed some end point (e.g., greater than *x*% re-

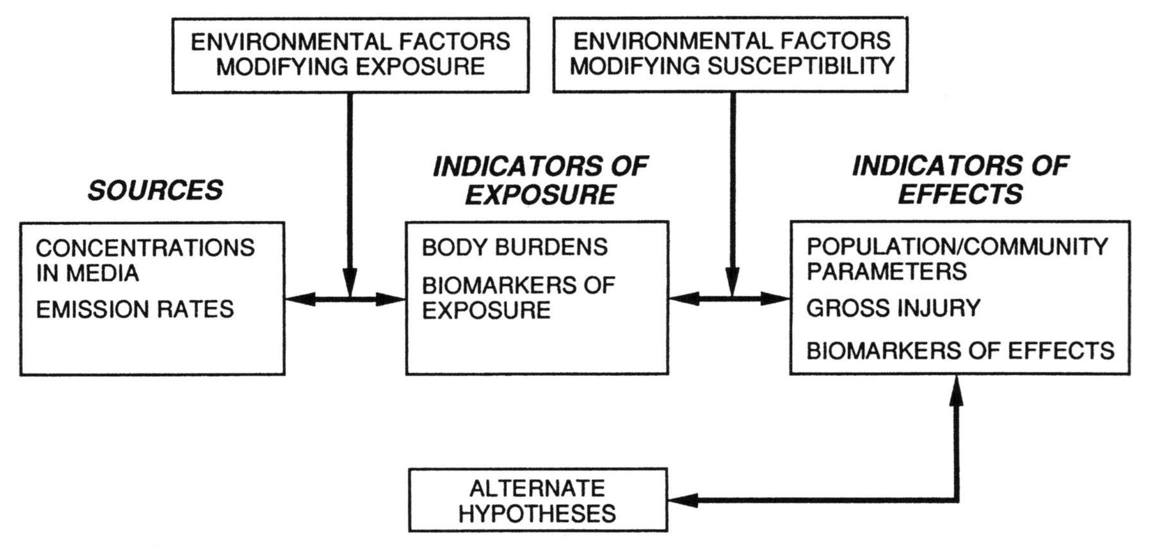

Figure 7. A diagram of the relationship among the components of the risk characterization stage of retrospective assessments based on ecological epidemiology. (From Suter 1990c.)

duction in local populations of any native fish species)? Estimating the magnitude of effects requires that the relationship between the source and the effects be quantified and used to establish the distribution of effects in space and time. For a single point source this may simply be a curve fit to the relationship between the measurement end point (e.g., density of mussels) and distance from the source. If biological effects are measured at sampling points in an area, statistical mapping techniques may be used to establish areas experiencing particular classes or level of effects (Thomas et al., 1986). Alternatively, patterns of contamination may be monitored and used to model effects.

Estimating the ultimate consequences of contamination requries some sort of model to project the consequences of either continued operation of a source or discontinuation of the source and subsequent recovery. These models are analogous to the models used in predictive risk assessments except that they may be empirical models derived by quantitatively describing the current situation or may be mechanistic models that have been validated against the existing situation.

QUANTITATIVE METHODS OF RISK CHARACTERIZATION

The simplest quantitative technique for estimating ecological risks is the quotient method (Barnthouse et al., 1982). The quotient method is simply the division of the estimated environmental concentration by some toxicological benchmark concentration (TBC). If the quotient is greater than one,

$$\mathrm{EEC/TBC} > 1$$

then a toxic effect is expected to occur. In most cases, the TBC is simply some test end point such as a no-observed-effect concentration (NOEC). The EEC may be equally simple (e.g., a release rate divided by the flow rate of a receiving river) or it may be estimated by a mathematical model that simulates the transport and fate of a chemical, such as EXAMS (Burns et al., 1981; see Chapter 18).

An elaboration of the quotient method applies factors to the EEC and TBC to account for extrapolations and uncertainties. In the most general form of the quotient method, effects are expected if

$$\frac{F_{e1} \times F_{e2} \cdots \times F_{en} \times \mathrm{EEC}}{F_{b1} \times F_{b2} \cdots \times F_{bn} \times \mathrm{TBC}} > 1$$

where the F_e factors are factors affecting the exposure estimate and F_b factors are factors affecting the toxicological benchmark. For example, the EPA Office of Water Regulation and Standards (1985) uses factors for variation in species sensitivity, effluent variability, and the acute/chronic extrapolation. The factors may be derived by data analysis but more commonly they are based on the judgment of the assessor. Commonly, a single factor is used to express all uncertainties and extrapolations,

$$F(\mathrm{EEC/TBC}) > 1$$

For example, when comparing an LC50 to an EEC, the EPA Office of Toxic Substances would multiply the quotient by a factor of 0.001 (or equivalently divide the LC50 by a factor of 1000) to estimate a concern level (Environmental Effects Branch, 1984).

Currently, the quotient method is the only assessment method that is used with any frequency. It has the considerable advantage it does not require any skill in mathematics or statistics. However, it does not estimate the type, level, or probability of effects.

One alternative is to use statistical models to estimate exposure and effects and then to estimate risks as the probability that EEC exceeds TBC. As an example of a possible statistical exposure model, consider the distributions of linear alkylbenzene sulfonate (LAS) concentrations depicted in Figure 7 of Chapter 31. If one assumes that exceeding a chronic toxic threshold for 7 d with an overage frequency of once in 10 y is acceptable, then the 7Q10 concentration could be used as the measure of exposure. Given that assumption, a cumulative probability function fit to the 7Q10 data for the various locations could be used to estimate either (1) the annual probability of occurrence of concentrations greater than some specified value or (2) the number of sites per annum out of the expected number of release points having concentrations greater than some specified value. If the assessment is being performed for a particular site rather than an entire nation, then the distribution of flows used to estimate the 7Q10 could be used to estimate the probability that some specified concentration will be exceeded for some specified duration during some specified period of release.

If a toxicity test has been conducted for the expected exposure conditions and durations, and if the test species or communities and measured responses include the assessment end point, then the exposure-response model fit to the test data constitutes a statistical model of effects. Similarly, if a contaminated site already exists, then a model relating concentrations measured in the field to effects levels measured in the field constitutes a statistical model of effects. Otherwise, one of the various statistical extrapolation models discussed above can be used to estimate effects from standard toxicity test end points. Depending on the type of model, statistical extrapolation models may be used to estimate concentrations that would be protective of some specified percentage of species (Stephan et al., 1985), all species with some probability (Kooijman, 1987), some specified percentage of species with some probability (van Straalen and Denneman, 1989), some particular species with some probability (Suter et al., 1982), or some ecosystem property (Sloof et al., 1986).

These statistical models could be used simply to provide mean or median estimates of exposure and effects levels that could then be used in the quotient method. However, because they can generate estimates of variance as well as means or medians, they can be used to generate probabilistic estimates of risk.

In one formulation of this method, risk is defined as

$$\text{Risk} = \text{Prob}(\text{EEC} > \text{TBC})$$

If we assume that the EEC and TBC are independent and lognormally distributed, then

$$\begin{aligned}\text{Risk} &= \text{Prob}(\log \text{TBC} - \log \text{EEC} < 0)\\ &= \text{Prob}[Z < [0 - (m_t - m_e)/(s_t^2 + s_e^2)^{1/2}]\\ &= d_z[(m_e - m_t)/(s_t^2 + s_e^2)^{1/2}]\end{aligned}$$

where (m_t, s_t^2) and (m_e, s_e^2) are the mean and variance of the log TBC and log EEC, respectively, and

$$Z = \frac{(\log \text{TBC} - \log \text{EEC}) - (m_t - m_e)]}{(s_t^2 + s_e^2)^{1/2}}$$

Z is a standard normal random variable with d_z as its cumulative distribution function. This method of estimating risk is described and demonstrated in Suter et al. (1983), Suter (1992), Volmer et al. (1988), and Nendza et al. (1990) and illustrated in Figure 7.

Like the statistical models, mechanistic mathematical models can be used to generate point estimates of exposure and effects to be used in the quotient method. A more sophisticated approach is to use the estimated exposure dynamics from a chemical fate model as input to an effects model, which then estimates the nature and magnitude of effects on the assessment end point. If the exposure and effects models are probabilistic, then the probability (risk) that a particular effect will occur can be estimated (O'Neill et al., 1982). An example of results from such an analysis is shown in Figure 8.

CONCLUSIONS

Currently, the U.S. Environmental Protection Agency and other regulatory agencies are in a transitional period between use of diverse ad hoc ecological assessment approaches and adoption of an ecological risk assessment approach. The EPA's framework for ecological risk assessment will lead to guidance for ecological risk assessment [guidance for health risk assessment has been available for several years (EPA, 1987)]. The administrator of the EPA has given ecological risks the same priority as health risks (Reilly, 1990). The U.S. National Academy of Sciences Board on Environmental Studies and Toxicology is preparing a report concerning development of ecological risk assessment. As a result, the next few years will be crucial to determining whether ecological risk assessment will constitute a rigorous approach to regulation based on use of the best available science to estimate the effects of alternative actions or simply a new set of jargon applied to the existing practices.

A critical step in the development of ecological risk assessment will be selection of a limited number of end points that are deemed to represent important environmental components that are valued commonly at risk. These end points would serve to focus the efforts of environmental scientists on developing, applying, and validating a basic ecorisk tool kit. The process would be analogous to the emphasis that health risk assessors placed on cancer assessment

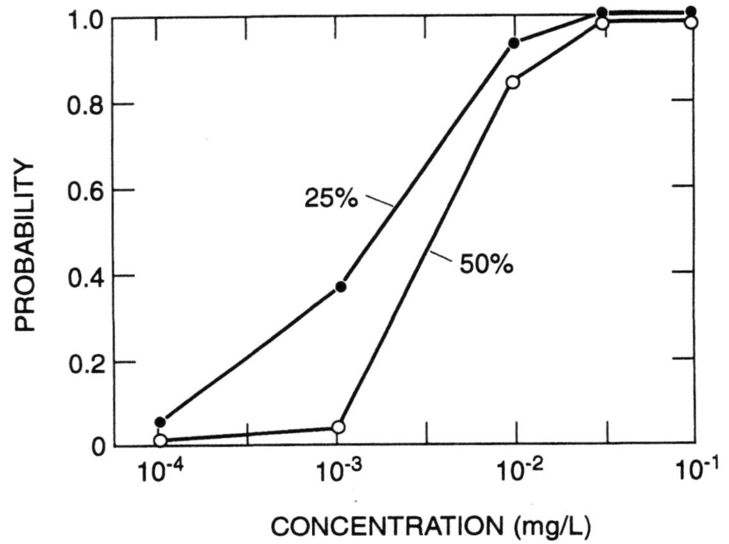

Figure 8. Estimated relationships between the risk of a 25% or 50% reduction in annual production of a piscivorous fish and exposure concentration (from Bartell, 1990). The probabilities were calculated by Monte Carlo simulations of effects of chloroparaffins in a bioenergetic model of a temperate lake.

methods in the 1980s. The development and acceptance of simple standard cancer risk models have led to both the development of risk models for other health effects and the development of more sophisticated models of cancer risk. Development of better toxicity tests, field measurement methods, and mathematical or statistical models cannot occur without some clear indication of what environmental properties must be tested, measured, and modeled.

Individuals involved in ecological risk assessment can, through their individual practices, improve the general practice. They should clearly state their assessment end points and the criteria for their selection, define the relationships between test end points and assessment end points and the model used to extrapolate between them, estimate uncertainties in all components of the assessments, and otherwise attempt to estimate ecological risks clearly and logically and communicate them to decision makers and the public. Also, they should insist on the importance of assessing ecological risks despite the overwhelming emphasis on health risks. The legal protections for the nonhuman environment are every bit as binding as those for human health.

As health risk assessment practices advance, the extremely conservative assumptions of the past are being replaced by more realistic assumptions about exposure and effects. As a result, it is becoming increasingly clear that risks to nonhuman organisms are often greater than those to humans. For example, the proposed revisions in the U.S. EPA's model of cancer risk from 2,3,7,8-TCDD suggest that aqueous concentrations of this compound that are acceptable for human exposure are higher than those that are highly toxic to fish (Mehrle et al., 1988; Walker et al., 1991). This realization and the intensifying regulatory emphasis on ecological risks (Reilly, 1990; EPA, 1990) are increasing the demand for ecological risk assessors and are increasing the demands placed upon ecological risk assessors. The challenges and opportunities are enormous.

LITERATURE CITED

Barnthouse, L. W., and Suter, G. W., II (eds.): User's Manual for Ecological Risk Assessment. ORNL-6251. Oak Ridge, TN: Oak Ridge National Laboratory, 1986.

Barnthouse, L. W.; DeAngelis, D. L.; Gardner, R. H.; O'Neill, R. V.; Suter, G. W., II; and Vaughan, D. S.: Methodology for Environmental Risk Analysis. ORNL/TM-8167. Oak Ridge, TN: Oak Ridge National Laboratory, 1982.

Barnthouse, L. W.; Suter, G.W., II; Rosen, A. E.; and Beauchamp, J. J.: Estimating responses of fish populations to toxic contaminants. *Environ. Toxicol. Chem.*, 6:811–824, 1987.

Barnthouse, L. W.; Suter, G. W., II; and Rosen, A. E.: Risks of toxic contaminants to exploited fish populations: influence of life history, data uncertainty, and exploitation intensity. *Environ. Toxicol. Chem.*, 9:297–311, 1990.

Bartell, S. M.: Ecosystem context for estimating stress-induced reductions in fish populations. *Am. Fish Soc. Symp.*, 8:167–182, 1990.

Bartell, S. M.; Gardner, R. H.; and O'Neill R. V.: An integrated fate and effects model for estimation of risk in aquatic systems. Aquatic Toxicology and Hazard Assessment: 10th Volume, edited by W. J. Adams, G. A. Chapman, W. G. Landis, pp. 261–274. Philadelphia: American Society for Testing and Materials, 1988.

Beck, L. W.; Maki, A. W.; Artman, N. R.; and Wilson, E. R.: Outline and criteria for evaluating the safety of new chemicals. *Regul. Toxicol. Pharmacol.*, 1:19–58, 1981.

Bro-Rasmussen, F., and Lokke, H.: Ecoepidemiology—a caustic discipline describing ecological disturbances and damages in relation to their specific causes: exemplified by chlorinated phenols and chlorophenoxy acids. *Reg. Toxicol. Pharmacol.*, 4:391–399, 1984.

Burns, L. A.; Cline, D. M.; and Lassiter, R. R.: Exposure Analysis Modeling Systems (EXAMS): User Manual and System Documentation. Athens, GA: U.S. EPA, Environmental Research Laboratory, 1981.

Cairns, J. J.; Dickson, K. L.; and Maki A. W. (eds.): Estimating the Hazard of Chemical Substances to Aquatic Life. STP 657. Philadelphia: American Society for Testing and Materials, 1978.

Carpenter, S. R.; Frost, T. M.; Hiesey, D.; and Kratz, T. K.: Randomized intervention analysis and the interpretation of whole-ecosystem experiments. *Ecology*, 70:1142–1152, 1989.

Environmental Effects Branch: Estimating "concern levels" for concentrations of chemical substances in the environment. Washington, DC: U.S. Environmental Protection Agency, 1984.

EPA (U.S. Environmental Protection Agency): The risk assessment guidelines of 1986. EPA/600/8-87/045. Washington, DC: Office of Health and Environmental Assessment, 1987. (Also 51 FR 33992-34045.)

EPA (U.S. Environmental Protection Agency): National oil and hazardous substances pollution contingency plan; final rule. *Fed. Regist.*, 55:8666–8873, 1990.

Fink, L. E.: 1992. Hazardous waste sites. *Environ. Sci. Technol.*, 26:1470–1471, 1992.

Fox G. A.: Practical causal inference for ecoepidemiologists. *J. Toxicol. Environ. Health*, 33:359–373, 1991.

Hallam, T. G.; Lassiter, R. R.; Li, J.; and McKinney, W.: Toxicant-induced mortality in models of *Daphnia* populations. *Environ. Toxicol. Chem.*, 9:597–621, 1990.

Hurlbert, S. H.: Pseudoreplication and the design of ecological field experiments. *Ecol. Monogr.*, 54:187–211, 1984.

Kooijman, S. A. L.: A safety factor for LC_{50} values allowing for differences in sensitivity among species. *Water Res.*, 21:269–276, 1987.

Mehrle, P. M.; Buckler, D. R.; Little, E. E.; Smith, L. M.; Petty, J. D.; Peterman, P. H.; Stalling, D. L.; DeGraeve, G. M.; Coyle, J. J.; and Adams, W. J.: Toxicity and bioconcentration of 2,3,7,8-tetrachlorodibenzodioxin and 2,3,7,8-tetrachlorodibenzofuran in rainbow trout. *Environ. Toxicol. Chem.*, 7:47–62, 1988.

Morgenstern, H.: Uses of ecological analysis in epidemiologic research. *Am. J. Public. Health*, 72:1336–1344, 1982.

Mount, D. I.: An assessment of application factors in aquatic toxicology. Recent Advances in Fish Toxicology, edited by R. A. Tubb, pp. 183–190. EPA-600/3-77-085. Corvallis, OR: U.S. Environmental Protection Agency, 1977.

Nendza, M.; Volmer, J.; and Klein, W.: Risk assessment based on QSAR estimates. W. Karcher and J. Devillers (eds.), Practical Applications of Quantitative Structure-

Activity Relationships, edited by W. Karcher, J. Devillers, pp. 213–240. Brussels: ECSC, EEC, EAEC, 1990.

NRC (National Research Council): Risk Assessment in the Federal Government: Managing the Process. Washington, DC: National Academy Press, 1983.

NRC (National Research Council): Issues in Risk Assessment: Vol 3, Ecological Risk Assessment. Washington, DC: National Academy Press, 1993.

Office of Water Regulations and Standards: Technical Support Document for Water Quality-Based Toxics Control. EN-336. Washington, DC: U.S. Environmental Protection Agency, 1985.

Ohio EPA: Biological Criteria for the Protection of Aquatic Life. Columbus, OH: Division of Water Quality Monitoring and Assessment, Surface Water Section, 1988.

O'Neill, R. V.; Gardner, R. H.; Barnthouse, L. W.; Suter, G. W.; Hildebrand, S.G.; and Gehrs, C. W.: Ecosystem risk analysis: a new methodology. *Environ. Toxicol. Chem.*, 1:167–177, 1982.

Reilly, W. K.: Aiming Before We Shoot: The Quiet Revolution in Environmental Policy. 20Z-1011. Washington, DC: U.S. Environmental Protection Agency, 1990.

Risk Assessment Forum: Framework for Ecological Risk Assessment. EPA/630/R-92/001. U.S. Washington, DC: Environmental Protection Agency, 1992.

Sloof, W.; van Oers, J. A. M.; and de Zwart, D.: Margins of uncertainty in ecotoxicological hazard assessment. *Environ. Toxicol. Chem.*, 5:841–852, 1986.

Sprague, J. B.: Measurement of pollutant toxicity to fish. I. Bioassay methods for acute toxicity. *Water Res.*, 3:793–821, 1969.

Stephan, C. E.; Mount, D. I.; Hanson, D. J.; Gentile, J. H.; Chapman, G. A.; and Brungs, W. A.: Guidelines for Deriving Numeric National Water Quality Criteria for the Protection of Aquatic Organisms and Their Uses, PB85-227049. Duluth, MN: U.S. Environmental Protection Agency, 1985.

Suter, G. W., II: Environmental risk assessment/environmental hazard assessment: similarities and differences. Aquatic Toxicology and Risk Assessment: Thirteenth Volume, edited by W. G. Landis, and W. H. van der Schalie, pp. 5–15. Philadelphia: ASTM, 1990a.

Suter, G. W., II: Endpoints for regional ecological risk assessments. *Environ. Manage.*, 14:9–23, 1990b.

Suter, G. W., II: Use of biomarkers in ecological risk assessment. Biomarkers of Environmental Contamination, edited by J. F. McCarthy, and L. R. Shugart, pp. 419–426. Boca Raton, FL: Lewis, 1990c.

Suter, G. W. (ed.): Ecological Risk Assessment. Boca Raton, FL: Lewis, 1992.

Suter, G. W., II; Vaughan, D. S.; and Gardner, R. H.: Risk assessment by analysis of extrapolation error, a demonstration for effects of pollutants on fish. *Environ. Toxicol. Chem.*, 2:369–378, 1983.

Suter, G. W., II; Rosen, A. E.; Linder, E.; and Parkhurst, D. F.: Endpoints for responses of fish to chronic toxic exposures. *Environ. Toxicol. Chem.*, 6:793–809, 1987.

Thomas, J. M.; Skalski, J. R.; Cline, J. F.; McShane, M. C.; Simpson, J. C.; Miller, W. E.; Peterson, S. A.; Callahan, C. A.; and Greene, J. C.: Characterization of chemical waste site contamination and determination of its extent using bioassays. *Environ. Toxicol. Chem.*, 5: 487–501, 1986.

van Straalen, N. M., and Denneman, G. A. J.: Ecological evaluation of soil quality criteria. *Ecotoxicol. Environ. Safety*, 18:241–245, 1989.

Volmer, J.; Kordel, W.; and Klein, V.: A proposed method for calculating taxonomic-group-specific variances for use in ecological risk assessment. *Chemosphere*, 17: 1493–1500, 1988.

Walker, M. K.; Spitsbergen, J. M.; Olson, J. R.; and Peterson, R. E.: 2,3,7,8-Tetrachlorodibenzo-*p*-dioxin (TCDD) toxicity during early life stage development of lake trout (*Salvelinus namaycush*). *Can. J. Fish. Aquat. Sci.*, 48: 875–883, 1991.

Warren-Hicks, W.; Parkhurst, B. R.; and Baker, S. S., Jr. (eds): Ecological Assessment of Hazardous Waste Sites: A Field and Laboratory Reference Document. EPA 600/3-89/013. Corvalis, OR: Environmental Research Laboratory, 1989.

Woodman, J. N., and Cowling, E. B.: Airborne chemicals and forest health. *Environ. Sci. Technol.*, 21:120–126, 1987.

SUPPLEMENTAL READING

Cohrssen, J. J., and Covello, V. T.: Risk Analysis: A Guide to Principles and Methods for Analyzing Health and Environmental Risks. Washington, DC: Council on Environmental Quality, 1989.

Fava, J. A.; Adams, W. J.; Larson, R. J.; Dickson, G. W.; Dickson, K. L.; and Bishop, W. E. (eds.): Research Priorities in Environmental Risk Assessment. Pensacola, FL: Society for Environmental Toxicology and Chemistry, 1987.

Paustenbach, D. J. (ed.): The Risk Assessment of Environmental and Human Health Hazards: A Textbook of Case Studies. New York: Wiley, 1990.

Risk Assessment Forum: Framework for Ecological Risk Assessment. EPA/630/R-92/001. Washington, DC: U.S. Environmental Protection Agency, 1992.

Suter, G. W., II (ed): Ecological Risk Assessment. Boca Raton, FL: Lewis, 1992.

Chapter 29

GENERAL TYPES OF AQUATIC ASSESSMENTS

R. K. Markarian

INTRODUCTION—WHAT IS AQUATIC ASSESSMENT?

In today's society, rich in regulatory initiatives and environmental awareness, assessment of the ecological health of aquatic systems can serve many purposes. Based on the goals and nature of the effort, assessments can take the form of a chemical-specific hazard assessment or can have a more ecological emphasis and be termed an ecological risk assessment (see Chapter 28). Regardless of the specific terms used, efforts in this area often center around the prediction and understanding of the short- and long-term impacts of man-made perturbations to aquatic systems.

Activities that trigger the need for aquatic assessments are diverse. Often assessments are associated with effluent discharge effects, cleanup and remediation requirements of contaminated sites, or proposals for developmental activities in aquatic systems. Competing interests in the same aquatic resource (e.g., wetland development) call for comparative assessments and study of various proposals in order to compare risks and benefits associated with each alternative. The development of new chemicals (e.g., pesticides and drugs) also calls for studies associated with aquatic habitats because new compounds are often assessed in terms of their aquatic toxicity, fate, and biodegradability. These are the major categories of activities that often initiate an aquatic assessment.

Aquatic assessments have been required or are an implicit part of various laws enacted during the past 20 yr and a variety of types can be found throughout the literature. Regulatory impetus for aquatic assessments include the Clean Water Act (see Chapter 24), the Toxic Substances Control Act (see Chapter 22 and 23), Federal Insecticide and Fungicide and Rodenticide Act (see Chapter 21), and Superfund legislation under the Comprehensive Environmental Response, Compensation and Liability Act (CERCLA, 1980). These major legislative requirements have driven the need for assessments in aquatic systems.

Key Issues and Questions Associated with Assessments

The regulatory acts noted above drive certain issues and questions that become basic to aquatic assessments. Protection of aquatic systems is the central theme for many of the studies that are undertaken in the laboratory and the field. The understanding implicit in the efforts, however, is that complete protection is not always feasible and that man-made activities will have impacts on aquatic habitats. Furthermore, not all aquatic species have the same value from a societal perspective. Protection, or loss, of *Daphnia* spp. from an aquatic system will probably not stir much concern. However, linking the loss of this small and relatively obscure crustacean to the well-being and productivity of higher, more valued organisms (e.g., trout) can spur more interest.

Finally, as you read the text of this chapter, bear in mind that the true goal from an ecological perspective is not protection of the individual organism but protection of the ecosystem. Because ecological science has not developed well-accepted parameters to predict impact at the community or ecosystem level, we utilize the population level as the most practical level of organization in our ecological assessments. As a group of interacting populations is essential to the community, protection of the populations' reproductive and survival potential is seen as essential in protecting community and ecosystem integrity.

Central questions associated with aquatic assessments often include the following:

- How large an impact will perturbation cause? Is it localized to a small area? Will it affect many populations and eventually the community/ecosystem? Can we actually predict the impacts of a proposed activity or perturbation on the aquatic system?
- Is the perturbation likely to be short or long term in its impact? Will chemicals biodegrade rapidly or will they persist and reach a concentration that can pose a chronic threat to a population? Is it a single, one-time perturbation (e.g., dredging a har-

bor or a single release or a specific contaminant) or will it recur over time?

- In already perturbed areas, is remediation warranted? This becomes a central issue in decision making for those who must concern themselves with cleanup of contaminated sites. Is the potential cost of the cleanup or remediation alternative justified; that is, does it yield real ecological benefit? Will there be other recreational or aesthetic benefits? What are the perceived or actual environmental benefits? Is there a future threat if the perturbed area was not remediated? Would the impacted area reclaim itself through natural processes? If so, how long will the recovery take?
- Is the perturbation a threat to humans or is the impact primarily a threat to nonhuman target organisms? Concerns and emotions regarding contaminant impacts on human health can become driving forces to an assessment when food chains are threatened.

Questions such as these can be difficult to answer in aquatic assessments. The tools for the assessments are not always available and some of the available tools can be expensive and time-consuming to use. When an assessment is planned, it becomes immediately obvious that a number of professional skill areas are required in the areas of ecology, toxicology, chemistry, and biology (see Chapter 28). Other skills often required are associated with predicting movements of contaminants in aquatic systems through mathematical models and predicting risks to humans from the consumption of contaminated fish and invertebrates.

A team approach, with careful planning from the outset, is required in an aquatic assessment. Anything short of this approach can easily lead to an incomplete effort yielding a poor basis for both predictive and remedial assessments. A purely toxicological rather than an ecological perspective in aquatic assessments can also result in unnecessary and wasteful utilization of resources to protect the individual rather than the ecosystem. These complexities require the utilization of a scientific team in order to provide the most rational and realistic assessment.

TYPES OF ASSESSMENTS

General

Hazard or risk assessments are performed for many reasons. The purpose of the assessment drives the approach, the data needs, and, of course, the cost. Current regulatory and business needs often require aquatic assessments in the following areas:

- Baseline assessments
- Spill or injury assessments
- Chemical assessments
- Effluent assessments
- Remediation assessments

Some of these assessments have more of a "hazard" than an "ecological risk" orientation. A general description of the purpose, approach, and typical methodologies utilized in each assessment is provided below. Effluent assessments are reviewed in Chapter 33 and will not be discussed here.

Baseline Assessment

This type of study has been performed in many cases to satisfy environmental impact assessments required by the National Environmental Policy Act (1972). This legislation requires not only the documentation of existing conditions but also a prediction of what future effects a particular activity may have on local aquatic and/or terrestrial systems. Outside of this legislation, baseline studies are required in order to document existing ecological conditions only. These studies serve to establish a baseline database that is necessary in the event that owners or operators of a facility are accused of causing harmful impacts in the future.

The use of baseline data varies but the basic question is the same. What are the existing ecological conditions in the area of concern? Or, stated another way, how healthy is the system and where are the problem areas? The assessment normally requires a certain amount of biological sampling to identify and quantify the populations present in the area of concern. The effort expended in a baseline assessment varies with the ultimate use and purpose of the effort. Some forms of legislation may require only a cursory examination of the existing situation, whereas major impact assessments often require a detailed examination with cataloguing and evaluation of the vulnerability of potentially affected systems.

Aquatic assessments often use population densities and species composition of benthic invertebrates, fish, and plankton as a tool to describe and document important communities for baseline reference in the future. Other background water column parameters such as temperature, dissolved oxygen, conductivity, salinity, and pH are also part of the effort. The need for more extensive documentation of the physical habitat, such as currents, water depths, and tidal influences, would be determined based on the purpose of the baseline study. If, for example, the purpose of the baseline study involved evaluating the impact of a new bridge, such physical data could become important in future studies.

As it is likely that some man-made activity is already influencing the area of study, chemical analysis of the water column is often performed during a baseline study. Even in relatively pristine areas where no contamination is suspected, the study needs to consider that some areas have higher than expected, naturally occurring contamination (e.g., natural oil seeps) that needs to be documented. The presence of other facilities in an area would also prompt the investigator to sample aquatic environments for suspected contaminants.

In the past, sediment quality was not a major area of concern and therefore chemical analyses of this medium were not routinely performed. With the development of sediment criteria (see Chapter 8) in some states in the United States (e.g., Washington) and the U.S. EPA's (i.e., the Agency's) current approach to sediment quality criteria development, concerns regarding sediment quality have increased in the past 5 to 10 yr (Science Advisory Board, 1989). Sediment sampling and analysis should now receive increased effort during baseline studies in order that it too is documented for future reference.

Species of special concern (e.g., endangered or commercially valuable species) may require special efforts during the assessment in terms of quantification or understanding their habitat needs. Habitats that are especially important are those that serve as either breeding or food production areas. These areas must be carefully assessed in aquatic habitats because dissolved oxygen, sediment loading, bottom characteristics, water depth, and flow can alter the reproductive suitability of a habitat. Other special habitats associated with ground water recharge or filtration of surface water also need to be recognized for their role during baseline studies. These "cleansing systems," often serving a less obvious role in wetland areas, must also be recognized and noted while establishing the baseline character of the habitat.

Spill or Injury Assessments

The purpose of this type of assessment is at least twofold. The first purpose is to ascertain and document the degree of the biological and habitat damage that was caused by a spill incident. The second purpose is to predict the short- and long-term potential for recovery of the injured resource. These data will basically answer two simple questions, "How much injury did the spill cause?" and "How long will the damage last?" Another important use of the data from a "preliminary injury assessment" is to help cleanup response teams decide on the best way to conduct cleanup operations (i.e., to avoid additional injury).

Data from this type of assessment, depending on the size/scope of the incident, are often destined for use under CERCLA (1980) or the Clean Water Act (CWA; see Chapter 24). Under Department of Interior provisions of Natural Resource Damage Assessment (NRDA, 1986), some financial estimate must be attached to the natural resource impacts caused by an incident and these damages can be based on rather extensive economic variables in major accidents (e.g., Exxon *Valdez*). The "damages" in NRDA are economic, not biological, in nature. Under NRDA, biologically harmed resources are referred to as injured and the "injury assessment phase" examines the extent of such biological harm or injury. In the "damage assessment phase," economic studies yield not only a dollar value estimate of the biological services lost or injured but also dollar value estimates for reductions in recreational services, aesthetic values associated with an area, and other factors that are equally difficult to assess (Katz and Ogden, 1991; Ward and Duffield, 1992). NRDA studies require interaction between social and biological scientists and are good examples of how the complex regulatory environment has increased the importance of clear communications among diverse technical disciplines.

There are many instances in which small spills do not require major federal intervention or resources but do require responsible and responsive actions in order to efficiently assess and quantitate damage. Once human safety concerns are under control, attention often focuses on environmental damages associated with the spill. Cleanup teams will quickly attempt to either contain or remove the spilled material. In association with the cleanup crew, natural resource advisors are often present to ensure that the cleanup methods are not excessively harsh or more harmful to the environment than the spilled product itself. This requires an immediate "assessment" of the injured or potentially injured populations. These data can then be used to provide guidance as to where to focus the cleanup crew efforts first and what methods of cleanup are least damaging to affected areas. This phase of the assessment should involve individuals familiar with both local biological habitats and the ecotoxicity of the spilled material.

The period immediately after the spill is important for several reasons. Physical-chemical properties of the spilled material such as solubility, vapor pressure, and density will immediately influence the environmental fate of the materials as well as potential cleanup and containment methods employed on site. These factors will also determine which aquatic habitats are under immediate threat from the spilled material. For example, materials with low solubilities and low densities will float in aquatic systems and, depending on their vapor pressure, may volatilize quickly. Other materials with higher solubilities or with components that are insoluble may create both a flotation threat to wildlife and a dissolved threat to fish and other aquatic life.

The spill assessment must consider the environmental fate of the material if a rapid effects assessment is to be accomplished. For instance, water column samples taken immediately after a spill event help document where the material went and the environmental concentrations that actually developed during the event. These data help document the extent and severity with which the spilled substance may have created acutely toxic conditions during its dissipation and movement within the spill zone. Other obvious and immediate damage caused by the spill would be enumerated in terms of mortality counts of birds, aquatic mammals, and fish/invertebrates. Logistically, however, certain spill events do not allow rapid mobilization of sampling teams and equipment (e.g., Exxon *Valdez*).

Aside from the obvious damage caused by a spill, more detailed investigations are needed to assess comprehensively short-term and long-term damages to aquatic communities. Assessing damage requires a baseline or point of comparison to be able to make some quantitative estimate of the percentage of a particular resource that was affected. Typically, few data are available regarding the species composition and population densities of the affected organisms prior to the spill. Without the background or "baseline" data, the assessor must rely on samples taken from presumably unaffected and adjacent areas, similar to those at the spill site. This method allows some degree of collection of both qualitative and quantitative data for comparisons during the damage assessment process.

Depending on the situation and the nature of the spilled material, samples of sediment, water, fish, and other dead or moribund organisms are enumerated and taken for future analyses. These samples help define actual exposure levels in a variety of media. Estuarine habitats may require extensive sampling in marsh, intertidal, and subtidal areas. Quantification is useful from the standpoint of monitoring future recovery, relating biological impact to exposure levels, and providing data to various parties as to actual versus perceived quantities of spilled product.

Estimation of future levels of product and the natural removal rates is very important during the injury quantification/assessment phase of a natural resource damage assessment. This information becomes important because, under Department of Interior rules, monetary damages are computed based on both the cost of restoring the services provided by the injured resources and the time it takes for recovery to occur (related to the period of time recreational services are impaired).

Without some effort to develop a mass balance and an estimate of the fate of the spilled product, it is difficult to know where to focus the evaluation of long-term impacts. If, for example, the spilled material was relatively dense and sank to the sediments, questions should arise about the material's bioavailability, biodegradability, solubility, invertebrate toxicity, ingestibility, bioaccumulation potential, mobility, and, finally, habitat fouling characteristics. All of these aspects will influence the future harm the material represents to aquatic organisms. This information may not be relevant from a cleanup standpoint, because it may be impractical or impossible to remove the material from the sediments. However, in the computation of monetary damages under current NRDA rules, understanding the role of various biological and physical processes will help determine the long-term impacts of any spilled product.

Understanding the fate and effects of a released material will also allow better estimation of the extent to which biological and recreational services have been affected by the spill. Estimates regarding recovery rates and chronic or long-term impacts will also be influenced by the studies undertaken at the time of the spill. The quality and technical merit of spill-related studies are crucial because they may come under careful scrutiny by opposing parties in either settlement reviews or litigation. Aside from the importance of carefully measuring and documenting resource injuries, the quality and credibility of technical studies can affect the total monetary damages assessed against the responsible party in a spill situation.

The technical (and legal) validity of data for water, sediment, and tissue samples is directly related to sampling, storage, and analytical methodologies used subsequent to the spill itself. The fact that a significant level of confusion exists after a spill will have little mitigating effect on challenges regarding the quality and control associated with these procedures. The program leaders responsible for planning and managing the assessment must be familiar with appropriate procedures and have at their disposal qualified laboratories and trained sample collection personnel. Quality data collection requires planning. If spill-related data are to survive the scrutiny that both the natural resource trustees and the parties responsible for the spill will subject the data to, preplanning regarding natural resource response is critical. This brings home the importance of having spill response and field assessment plans properly prepared before an incident. The identification of quality analytical and biological services prior to the spill is crucial in developing a quality data set on short notice.

Chemical Assessments

When a new substance is synthesized by a manufacturer, it typically needs to go through some form of safety assessment (similar to that described in Chapter 31). The exact nature of the assessment is normally a function of the regulating entity, the intended use of the material, where it will be made and sold, and the volume expected to reach the marketplace (and, by extension, the environment). If it is a pesticide, it would need to satisfy the Federal Insecticide, Fungicide and Rodenticide Act (FIFRA) statutes (see Chapter 21). An industrial chemical would need to satisfy the Agency's premanufacture notification requirements under the Toxic Substances Control Act (TSCA; see Chapters 22 and 23). If the material was bound for member countries of the European Community, tests would have to conform to specific guidelines developed under these auspices (OECD, 1981).

Assessments performed for a substance often have a fairly rigorous testing component. In the United States, the Environmental Protection Agency is responsible for publishing testing methods and assessment considerations (Environmental Protection Agency, 1982; see Chapters 2 and 3). European countries often rely on testing guidelines developed under the Organization for Economic Cooperation and Development (OECD, 1981; see Chapter 26). Authorities for the approval of new chemicals typically require strict adherence to protocols and methods that they have defined. When a product is to be sold or manufactured in different countries it is sometimes necessary to satisfy multiple, country-specific testing requirements even if they have a high degree of redundancy.

Regardless of the specific testing method, country in which approval is required, or where or for whom the work is intended, the basic principles associated with hazard assessment in aquatic systems are the same. Chemical assessments for aquatic systems are more properly considered a hazard assessment than a risk assessment because they typically attempt to determine whether harm could occur from the intended use of the substance rather than the extent of the damage the substance would inflict on various aquatic systems. A number of texts and references are available that describe testing and decision trees available for chemical hazard assessment (Cairns et al., 1978).

Data on intended production, use, and disposal of the new substance become important if potential exposures associated with it are to be understood. Background and existing data on substances that are similar in structure are collected and evaluated. The chemical-physical properties of the substance are established in order to help determine the fate of the substance during production, use, and disposal. Necessary toxicity data are also collected. Most aquatic assessments for a chemical require a baseline effort to determine toxicity. This typically includes data on acute toxicity to a fish, an invertebrate (e.g., *Daphnia* sp.), and an algal species. Depending on the results of the initial acute tests, the need for further testing is determined.

Part of the species selection process for toxicity testing may be driven by environmental fate considerations. A careful exposure assessment studies the movement of the substance through the environment during the active product life of the substance. This assessment identifies the environmental media with which the product will come in contact. The fate assessment will attempt to determine the form, duration, and concentration of the product in a specific medium during the ''life'' of the product. This ''life cycle'' approach includes the final disposal or recycling consideration as well, because the potential harm or liability a product represents does not end when the intended use has been accomplished. Questions such as the following become important to a comprehensive chemical assessment.

- Will the substance be present in effluents during the production process? If so, what concentrations and what total amount will be discharged?
- During a spill incident, how will the substance behave? Is it soluble enough to reach toxic concentrations in the water column? Will it volatilize quickly? Does it have characteristics that will allow the material to bind to sediments?
- When used as intended, how could the substance come in contact with the environment and aquatic sytems?
- Does the substance represent a concern regarding improper disposal techniques and subsequent contamination of aquatic systems through leaching?

Certain evaluation programs dictate additional subchronic testing if acute toxicity LC50 values are below a certain threshold (e.g., <1 mg/L). These thresholds and decision points vary depending on the regulatory source. If a substance is to be present in an effluent, then in-stream waste, estimates need to be made and, depending on how these compare to acute values, additional chronic toxicity concerns may need to be addressed. Results of the assessment may indicate either that the substance is safe for the intended uses or that some restriction must be made on a particular aspect in the life of the substance. Restrictions can include additional control during production to reduce effluent loads and restrictions on the way the substance is shipped and disposed. Sometimes regulatory agencies find that a new substance is too hazardous based on a particular use. The U.S. EPA can rule to place a restriction on the sale of the substance so that it can be sold only for uses deemed safe by the Agency. After the restrictive decision has been made for a new product, any new uses of the same substance must be readdressed and go through the Agency for approval.

Remediation Assessments

Perhaps the most complex and difficult assessments are those intended to help decide the actions warranted at a contaminated site (see also Chapter 32). There are a number of reasons for this. First, decisions regarding the need to clean up a contaminated site involve various technical, sociological, economic, and political considerations. Regarding the technical aspect of the assessment, contaminated sites are seldom contaminated with only a single substance. This makes the exposure and toxicity assessments more complex and increases the uncertainty levels associated with the risk assessments.

Remediation assessments, whether for sites covered by CERCLA (1980) statutes in the United States or for less hazardous sites anywhere in the world, have similar requirements. Initially, a review of the historical background of the site as it might have affected aquatic systems is required. This gives the assessor key information regarding the type, complexity, chemistry, and volume of contaminants. These data allow an initial estimate of the expected fate of the contaminants. Depending on the quality of the available information, an estimate of the area potentially affected by the discharges may also be possible.

Once a historical background study has been completed, a ''field scoping'' effort is often required. The field scoping study will provide the basic data to decide on the seriousness of the contamination. Is an immediate human health risk associated with the site? Are health hazards serious enough to require evacuation? Is the site contaminating drinking water or food sources of local human populations? Were any unexpected contaminants found and, if so, what does this imply?

The field scoping study is intended to validate expectations regarding contaminant types and presence, uncover "hot spots," yield an estimate of contaminant concentrations in the area, and estimate the horizontal extent of the contaminated areas. This scoping effort can go by a variety of names, such as phase 1 or preliminary investigation, but the intent is the same. Sometimes samples are taken as a screening tool; in other instances samples are taken at a later stage in the investigation. The worker protection required during this preliminary assessment must be based on the nature of the health hazards associated with the expected contaminants on site.

Depending on the nature of the site (i.e., estuary, soil plot, river, or stream), additional, more detailed sampling may be required. The goal of the additional sampling will be to answer the key questions posed by the scoping study. Often questions are related to the extent (horizontal and/or vertical) and the form of the chemical contamination. The human health aspects of the site may require more specific analytical methods focused on selected carcinogens of concern. Tissue samples from resident fish or invertebrate populations may be required if substances known to bioaccumulate are found in soil, water, or sediments on the site.

Once the first "detailed" sampling effort has been accomplished, some form of hazard and/or risk assessment effort will be required. The hazard assessment will involve comparing exposure concentrations found in the environment to known toxic thresholds for the contaminants. It provides an estimate of whether sufficient toxicity and exposure exist for the contaminant to have a negative impact on a given type of organism. Typically, toxicity data on the species and contaminants found on the site will be readily available in the literature. Initial assessment will be based on extrapolations with closely related organisms where data do not exist. These types of contaminant-specific assessments will give the investigator some indication of whether site contaminants exist at levels hazardous to aquatic populations.

While toxicologists are examining the toxicity portion of the hazard assessment, fate and exposure analysts must determine the expected stability, mobility, persistence, and ultimate fate of the contaminants. How long will they reside where they are? Will they be in the same form or will they become more or less bioavailable due to oxidation or biodegradation? Are the contaminants confined to a single medium (e.g., water column) or are they found in multiple media? Are the contaminants in a bioavailable form or they bound to highly organic sediments? These types of questions will influence the outcome of the hazard assessment because they concern the types of organisms that may ultimately be exposed as well as the duration and nature of the exposure.

Once these hazards are assessed, the investigator must attempt to quantitate the impact and risk that they pose to aquatic populations at or adjacent to the site (Bartell, 1992). The level of uncertainty associated with this portion of the assessment is often very high. This is due to a number of factors. Actual impacts of contaminants on field populations are known only in a few well-studied areas (Geckler et al., 1976), so they are site-specific in nature and difficult to relate to different locales. We have a limited capability to predict accurately potential impacts where numerous contaminants are present, which adds to the uncertainty. Also, toxicology is a laboratory-based science. Comparisons between laboratory toxicity results and field impacts are often difficult because few contaminants have been field tested in a controlled manner. Thus, although toxicologists can help explain sources, certain impacts, and potency of contaminants, they are currently limited in their ability to predict ecosystem effects with laboratory data.

Because of these difficulties associated with predicting the impacts of certain levels of contaminants on an ecosystem, assessors often study the affected system to see if the expected effects actually exist. Ecological impacts of contaminants are often assessed by comparing population densities and structural end points in presumably unaffected areas (based on where contamination was not found in the detailed field study) to those in contaminated areas. This approach can provide the assessor with information on the relative bioavailability of the contaminants. If, for example, high contaminant levels were found in the field study area but no apparent effects on the populations were found, the investigator would probably conclude that the contaminants, at quantities known to be toxic in laboratory studies, were not bioavailable at the site. This can be an extremely important finding, as costs associated with site remediation can be extremely high and removal of contaminants that are not truly bioavailable can be a waste of effort. Unfortunately, control areas suitable for comparison with affected areas are not always easy to find and the costs associated with extensive field studies often discourage companies from utilizing site comparison in hazard assessment determination.

In some aquatic assessments, ecological impacts can be shown to be minimal but human health concerns become the driving force for further study of risk. These assessments often involve food chain contamination concerns. Persistent contaminants such as polychlorinated biphenyls (PCBs) are found in estuaries where thriving invertebrate and fish populations exist (Bartell, 1992). These contaminants may accumulate in food chains and persist within resident organisms but have little or unknown impact on the populations themselves. Populations in contaminated waters often thrive because they are untouched by recreational or commercial fishing due to advisories banning collection issued by local authorities. Often the data available on human health impacts of the contaminant are limited and no federal standards on "safe" concentrations exist. In these cases, the best available data are used by toxicologists to estimate health impacts and develop a safe food concentration for the contaminant of concern.

PLANNING THE ASSESSMENT

General

From the preceding discussion we can see that terms, goals, and methods are important when planning the conduct of an aquatic assessment. Terminology becomes extremely important because the individuals involved in discussing, planning, or paying for the assessment may not be familiar with the state of the science of ecological risk assessment or the real output possible from an assessment. This requires that the assessor be able to discuss the concepts and realities associated with an assessment so that false expectations are not developed by any of the participating parties. Good communication early in the planning process is important so that all parties understand the true goals and objectives of the work.

Team Development

As noted earlier, a multidisciplinary team approach is normally required in an assessment (see Chapter 28). Figure 1 shows an example of how such a team can be organized as well as the team members often required in an aquatic assessment.

Teams developed for aquatic assessments will vary in structure based on the specific task at hand. Figure 1 notes some of the key services and professionals that would typically be required. Most of the services are self-explanatory; however, special note should be made in two areas. The quality assurance (see Chapter 11) task or officer has the unenviable task of ensuring that all of the data collection, sample handling, and analyses are within the specifications noted in the field quality assurance plan. This plan can be a separate document or can be incorporated in a field study plan and used to ensure that data are collected and documented according to agreed-upon protocols. This plan, often referred to as a quality assurance plan or QAP, is a common requirement under EPA, CERCLA, and RCRA (Resource Conservation and Recovery Act) type assessments (Environmental Protection Agency, 1989).

Other Available Tools for the Investigator

Many questions arise as an aquatic habitat is investigated. An appreciation of the basic concepts in ecology is fundamental to the investigator attempting to understand the potential significance of toxic materials in our environment. The movement of nutrients, the impacts of buffered and nonbuffered aquatic systems on toxic compounds, the dependence of populations on each other, and a multitude of other considerations make planning the assessment and the interpretation of the data a challenge. Tools are available to answer some of our questions, but in many instances investigators can provide only "best professional judgment." The ability to answer questions regarding potential impacts and fate of contaminants has grown significantly over the past 20 yr. A decade or two ago the science of aquatic toxicology received little attention and had an extremely limited database. The Clean Water Act in the United States helped trigger the need for investigations into elemental and compound toxicity and the science progressed beyond the realm of pesticide toxicology, which had been the dominant area of study for many years. As more and more toxicity data became available, investigators soon realized that their understanding of fate and exposure needed additional emphasis if a true understanding of hazard and ecological risk was to be developed.

Fortunately, as environmental information and publications began to increase in volume, so did the use of the personal computer. Information that was once locked in the archives of the reference library

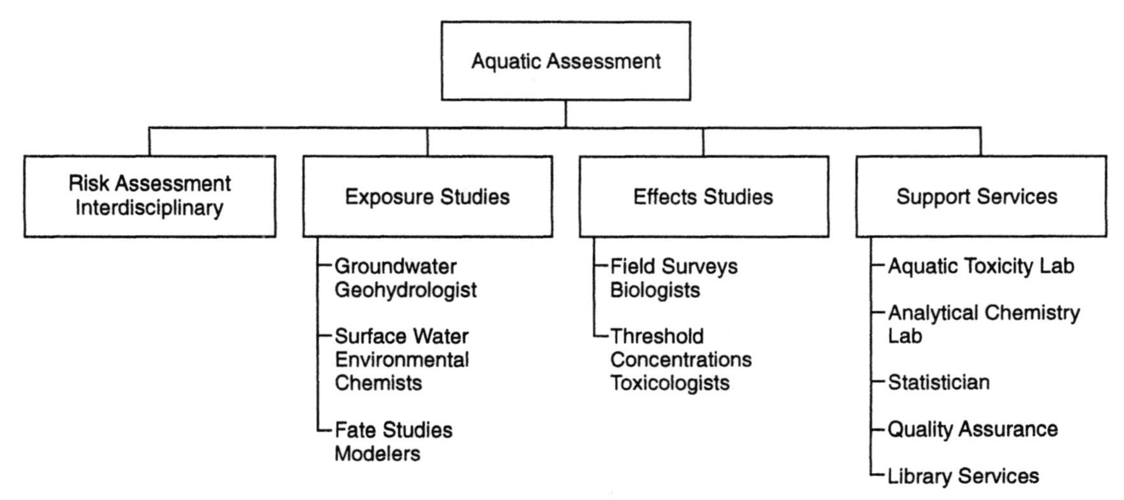

Figure 1. Interdisciplinary team structure for typical aquatic assessment.

soon found its way to on-line databases that allowed ready access to abstracts and bibliographies. This allowed better, more comprehensive searches to be performed in a more efficient manner. As the power of the personal computer grew, so did the sophistication and utility of the programs available to the aquatic assessor. Programs are now available that help predict toxicity, fate, and biodegradability of known and new compounds (Barnthouse, 1986). Computer programs and their predictions in the area of ecotoxicology are useful tools in the right hands but can be misused by those unfamiliar with the limitations associated with the "input data" on which they are based. However, when used with proper judgment, these tools can provide insights and aid in prioritization when many chemicals, some with known and other unknown toxicities, are involved.

When predictive tools are seen as inadequate or when literature data are unavailable on the contaminant or circumstance in question, the laboratory provides the final source of predictive data in the assessment process. The "laboratory" can be the study site itself or, in fact, a true controlled laboratory indoors. Toxicologists in the ecological field have the advantage over their mammalian/human health counterparts that they can use the actual organism of interest in laboratory studies. If fish populations are under threat due to the complex combination of effluents, a variety of fish can be tested and the no-effect concentration can be determined. This important advantage that ecotoxicologists have over human health toxicologists is offset by the number of species ecologists need to consider and the complex movement of chemicals among a variety of media (i.e., exposure) within any given ecosystem.

Laboratory toxicity studies have become a more common tool in aquatic assessments involving complex, multicontaminant sites. The ability to predict the bioavailability of multiple constituents in sediments, for example, is now improving with the EPA's emphasis on sediment quality criteria, but toxicity investigations are still needed to understand the effects of contaminated sediments (Science Advisory Board, 1989). The toxicity test in a sediment-water dual-media exposure system provides an integrative method for measuring the potential impacts associated with a sediment, even when not all the factors involved are understood.

CONCLUSIONS

Aquatic assessment methodologies have progressed significantly over the past 15 to 20 yr. Much of the progress can be credited to remedial investigations spurred by CERCLA. Complex contamination problems often require fairly advanced techniques in modeling, exposure, and toxicity assessment before a decision as to remediation can be agreed upon by all parties. Ecological assessments associated with CERCLA and remedial studies have, in the recent past, received less attention than health considerations. Reasons for this are many, but one major factor is that the health risk assessment has developed into a more exact science than the ecological risk assessment. Investigators and government program managers are more comfortable assessing the quality and validity of assessments that have specific guidelines upon which they can be evaluated. Ecological assessment methodologies are not readily available and the "risk to an ecosystem" cannot be easily calculated.

The complexities of organism, population, and community interactions are so great that our understanding is only at the most basic levels when we try to describe undisturbed systems. The ability to predict risk based on chemical impacts on an ecosystem will have to await better levels of understanding of undisturbed systems and the mechanisms that are most important to their stability.

The continued impact and advances associated with personal computers have allowed scientists with modest budgets access to large information networks where predictive assessment programs can be downloaded to personal systems without charge. Programs that assist in predicting impacts of chemicals and movements of compounds in complex soil-water systems have made the science of aquatic risk assessment more complex and, with continued refinement, more accurate.

A number of challenges still lie ahead. Understanding the impacts of chemicals on ecosystems is an area where the science is only beginning to concentrate effort. Making the leap from the aquatic toxicology laboratory to the ecosystem in predicting impacts is probably the challenge of the next two decades. The EPA has recognized the need for specific guidance in ecological risk assessment in terms of providing specific procedures for remediation work in both RCRA and CERCLA investigations. Significant efforts between trustees of our natural resources (e.g., U.S. Fish and Wildlife Service, National Oceanic and Atmospheric Administration) and the U.S. EPA have resulted in renewed cooperation in focusing on ecosystem-level concerns when measuring injuries associated with CERCLA sites. These efforts and regulations will continue to drive the science of aquatic and ecological assessments toward more accurate and predictive results to the benefit of all parties.

LITERATURE CITED

Barnthouse, L. W.; Suter, G. W., II; Bartell, S. M.; Beauchamp, J. J.; Gardner, R. H.; Lindner, E.; O'Neill, R. V.; and Rosen, A.E.: User's Manual for Ecological Risk Assessment. ORNL-6257, Oak Ridge, TN: Oak Ridge National Laboratory, 1986.

Bartell, S. M.; Gardner, R. H.; and O'Neill, R. V.: Ecological Risk Estimation. Boca Raton, FL: Lewis, 1992.

Cairns, J., Jr.; Dickson, K. L.; and Maki, A. W. (eds.): Estimating the Hazard of Chemical Substances to Aquatic Life. STP657. Philadelphia: American Society for Testing and Materials, 1978.

CERCLA: Comprehensive Environmental Response Compensation and Liability Act, CERCLA includes the Su-

perfund Amendments and Reauthorization Act of 1986 (SARA). 42 USC §9601 et seq., 1980.
CWA: Federal Water Pollution Control Act and Clean Water Act of 1977. 33 USC §1251 et seq., 1972.
Environmental Protection Agency: Environmental Effects Test Guidelines. EPA 560/6-82-01. Office of Pesticides and Toxic Substances, 1982.
Environmental Protection Agency: Region II CERCLA Quality Assurance Manual. EPA, 1989.
Federal Register: Natural resource damage assessments (type B); final rule. *Fed. Regist.*, 27674–27753, August 1, 1986.
FIFRA: Federal Insecticide Fungicide and Rodenticide Act. 7 USC §136 et seq., 1972.
Geckler, J. A.; Horning, W. B.; and Neiheisel, T. M.: Validity of laboratory tests for predicting copper toxicity in streams. EPA 600/3-76-116: 208, 1976.
Katz, S., and Ogden, E. M.: DOI's natural resource damage assessments: new rules and new challenges. *Federal Facilities Environmental Journal*, Autumn 1991.
NEPA: National Environmental Policy Act, 1972.
OECD: Guidelines for Testing of Chemicals. Organization for Economic Cooperation and Development, 1981.
Science Advisory Board: Report of the Sediment Subcommittee of the Ecological Processes and Effects Committee: Evaluation of the Apparent Effects Threshold Approach for Assessing Sediment Quality. SAB-EETFC-89-027. Washington, DC: EPA, 1989.
TSCA: Toxic Substances Control Act. 15 USC §§2601–2629, 1976.
Ward, K. M., and Duffield, J. W.: Natural Resource Damages: Law and Economics. New York: Wiley, 1992.

SUPPLEMENTAL READING

Bartell, S. M.; Gardner, R. H.; and O'Neill, R. V.: Ecological Risk Estimation. Chelsea, MI: Lewis, 1992.
Cairns, J.; Dickson, K. L.; and Maki, A. W. (eds.): Estimating the Hazard of Chemical Substances to Aquatic Life. STP657. Philadelphia: American Society for Testing and Materials, 1978.
Carpenter, S. R. (ed.): Complex Interactions in Lake Communities. New York: Springer-Verlag, 1987.
Dickson, K. L.; Cairns, J.; and Maki, A. W. (eds): Analyzing the Hazard Evaluation Process. Washington, DC: American Fisheries Society, 1979.
Duthie, J. R.: The importance of sequential assessment in test programs for estimating hazard to aquatic life. Mayer, F. M., and Hamelink, J. L. (eds.): Aquatic Toxicology and Hazard Evaluation, STP634, pp. 17–35. Philadelphia: ASTM, 1977.
Environmental Protection Agency: Technical Support Document for Water Quality Based Toxics Control. EPA-440/4-85-032. Office of Water, 1985.
Environmental Protection Agency: Risk Assessment Guidance for Superfund: Environmental Evaluation Manual (Interim Final). EPA-540/1-89/001A. Washington, DC: Office of Solid Waste Emergency Responses, 1989.
Katz, S., and Ogden, E. M.: DOI's natural resource damage assessments: new rules and new challenges. *Federal Facilities Environmental Journal*, 2(3):253–266, 1991.
Sloof, W.; van Oers, J. A. M.; and De Zwart, D.: Margins of uncertainty in ecotoxicological hazard assessment. *Environ Toxicol. Chem.*, 5:841–852, 1986.
Suter, G. W., II, and Loar, J. M.: Weighing the ecological risk of hazardous waste sites: the Oak Ridge case. *Environ. Sci. Technol.*, 26(3):432–439, 1992.
Suter, G. W., II: Environmental risk assessment/environmental hazard assessment: similarities and differences. Edited by Landis, W. G., and van der Schalie, W. S., Aquatic Toxicology and Risk Assessment, STP1096, pp. 5–15. Philadelphia: American Society for Testing and Materials, 1990.

Chapter 30

PREDICTING THE EFFECTS OF INSECTICIDES ON AQUATIC SYSTEMS AND THE WATERFOWL THAT USE THEM

P. J. Sheehan, A. Baril, P. Mineau, and *D. J. Paustenbach*

INTRODUCTION

Although the importance of protecting ecological systems is not a new concept to ecologists, the idea of using ecological risk assessment in environmental decision making is relatively new. The development of quantitative ecological assessment methods started with the early studies of the effects of pesticides on wildlife (e.g., Moore, 1966) and of chemicals in surface water on aquatic life (e.g., Cairns and Dickson, 1978). In the 1980s, a synthesis of published assessment methods was presented in the first textbooks on ecotoxicology (Hammons, 1981; Moriarty, 1983, 1988; Connell and Miller, 1984; Sheehan et al., 1984; Levin et al., 1989). Ecotoxicity and ecotoxicological risk assessment were reviewed more recently in a four-part series of articles in Environmental Science and Technology (Bascietto et al., 1990; Cairns and Mount, 1990; Hoffman et al., 1990; Harris et al., 1990) and in two books (Bartell et al., 1992; Suter, 1993). U.S. Environmental Protection Agency (EPA) guidance documents on ecological assessments have followed (EPA, 1988, 1989a,b,c,d, 1992; TRI, 1988). There are however, few examples of applied quantitative ecotoxicological risk assessments describing the hazards associated with chemical exposure. Most of the well-documented case studies involve pesticide exposures. This case study describes an approach for assessing ecological risks of aerial insecticide spraying in areas where agriculture and waterfowl habitat overlap. It is based on an assessment conducted to evaluate the potential direct toxic effects of common agricultural insecticides on aquatic macroinvertebrates in sloughs in the pothole region of Canada and subsequent indirect impact of macroinvertebrate mortality (loss of food resource) on duckling survival and recruitment. This case study is largely taken from a comprehensive evaluation of the impacts of pesticides (both herbicides and insecticides) on the ecology of duck populations in the Canadian prairies (Sheehan et al., 1987).

PROBLEM DEFINITION

The pothole region of the prairies of Canada and the United States is the principal grain-growing region of North America. This unique region is dotted with a vast number of freshwater ponds and lakes that are primarily small, shallow, and biologically productive sloughs ringed with higher aquatic plants. These sloughs are critical to a wide variety of aquatic organisms and wildlife, the most notable of which are an estimated 16 million ducks that breed there annually. There is an intimate relationship between the ecology of the sloughs and the essential resources they provide to the waterfowl species that inhabit them for part of each year. The overlap between agriculture and waterfowl over much of this habitat has been of great concern to wildlife regulatory agencies for some years. Drainage programs to increase the quantity of arable land have resulted in the loss of valuable duck breeding habitat, and agricultural practices such as the cultivation and burning of fields, as well as their margins, have contributed to the degradation of much of the remaining waterfowl habitat in agricultural regions. The potential impact of the agricultural use of pesticides on the reproductive success of duck populations had received only scant attention prior to studies by the Canadian Wildlife Service (Shaw et al., 1984; Mineau et al., 1987; Neal, 1987; Sheehan et al., 1987; Forsyth, 1989) and U.S. Fish and Wildlife Service (Grue et al., 1986, 1989).

Direct Versus Indirect Effects

As pesticide manufacturers endeavor to expand the margin of safety between the median lethal response to target invertebrates and levels toxic to humans and other vertebrates, concern over direct effects (see Chapter 1) on waterfowl in agricultural areas should diminish. However, indirect effects (see Chapter 1) such as reduction or elimination of food resources (e.g., nontarget aquatic invertebrates) for these species may become a more important concern. The extreme toxicity to aquatic macroinvertebrates of the newer insecticides such as the synthetic pyrethroids and the high toxicity of many other insecticides used currently in Canada and the United States have raised the specter of macroinvertebrate kills in contaminated sloughs in agricultural areas. The likely aerial application of the synthetic pyrethroids and other potent insecticides for control of grasshoppers and other prairie insect pests provided the impetus for a comprehensive evaluation of the threat of broad-scale insecticide-induced mortality within the sloughs' macroinvertebrate community and the subsequent impact on duck reproductive success. This assessment was one of the first to evaluate both the direct toxic effects of insecticides on nontarget aquatic organisms and the indirect (ecological) impacts on duck populations as a result of reductions in macroinvertebrate food sources.

Waterfowl Habitat and Agriculture Overlap

Assessing the hazards posed by insecticide use to waterfowl requires a full understanding of the habitat and food requirements of ducks and the structure and function of the slough ecosystem. Key aspects of this study include the identification of the vulnerable periods in the life cycle of ducks during their stay in the prairies and the spatial and temporal overlap between the duck's requirements and insecticide use.

The prairie pothole region of North America covers about 300,000 square miles (770,000 km^2) in south-central Canada and north-central United States (Figure 1). This swath, 300 miles (483 km) wide and 1000 miles (1609 km) long, crosses the provinces of Alberta, Saskatchewan, and Manitoba and parts of the states of Minnesota, North Dakota, South Dakota, Iowa, and Montana. Approximately two-thirds of this region is located in Canada with 64,000 square miles in Alberta, 114,000 in Saskatchewan, and 39,000 in Manitoba (165,750, 295,250, and 101,000 km^2, respectively) (Gallop, 1965).

The average slough density in the prairies is 12/km^2, with values ranging from 4 to 40/km^2 (Smith et al., 1964). In a wetland habitat study in Saskatchewan, Millar (1969) found basin densities to range between 31 and 36/km^2, occupying 8 to 12% of the total land surface; approximately 80% of the basins were smaller than 1 acre (0.4 ha) in size. The majority of sloughs at Lousana (Alberta) and at Redvers (Saskatchewan) were smaller than 0.5 acres (0.2 ha) in area (Smith and Stoudt, 1968).

Seasonal and annual variability in slough density is high. The average number of sloughs at Lousana and Redvers decreased by 18 and 31%, respectively, between May 1 and July 1; over 80% of the sloughs lost were less than 0.5 acre (0.2 ha) in size. These seasonal reductions vary over the pothole region: although densities in the parklands are twice those in grasslands in May, in July the parklands have three times as many sloughs due to much greater rates of evaporation and the subsequent drying of sloughs in the grasslands (Smith et al., 1964).

A substantial portion of North American waterfowl populations use the wetland habitat of the pothole region of the prairies in Canada and the northern United States during the critical reproductive phase of their life cycle (Bellrose, 1979; Batt et al., 1989). Most ducks using the slough habitat are dabblers (puddle ducks). They customarily prefer semipermanent shallow ponds and feed in the pond either at the surface or by "tipping up" so that their tails show above the water. The divers tend to occupy deeper, more stable sloughs and feed in water by diving below the water surface. The duck species with large breeding populations in the prairie pothole region are shown in Table 1. The mallard is a dabbler and the most abundant species. The lesser scaup is the most abundant of the divers.

The work of wildlife ecologists conducted over the past four decades on duck populations in the pothole region has suggested that both the abundance and distribution of ducks are dependent on the quantity and quality of both wetland and upland habitat (Swanson and Duebbert, 1989). To assess the overlap of waterfowl habitat and agriculture in Canada, the Canada Land Inventory (CLI) was used. The CLI data represent the potential for a given resource (such as waterfowl or agriculture) and not actual use (Lands Directorate, 1978). This potential was assessed in the early 1960s. However, with the aid of additional information such as agricultural or waterfowl production surveys and the "current" land use data (assessed in 1967), it is possible to make some judgments as to the actual use of the land by these interests. The computerized CLI is the only database that includes the entire parkland and grassland region for both agriculture and waterfowl production potential, the primary foci of this report. The CLI generally recognizes eight primary classes of resource potential. Class 1 represents areas that have no significant limitations to a particular resource, whereas class 8 areas have such severe limitations that little or no potential exists for that resource. Within the primary waterfowl classes 1 through 3, there are special designations for habitat that also serves as important migration stops or wintering areas. With the exception of class 1, subclasses can also be used to identify significant limiting factors for a particular resource. Because this level of detail is beyond the needs of our analysis, only the primary classification for both the waterfowl and agriculture potential is utilized.

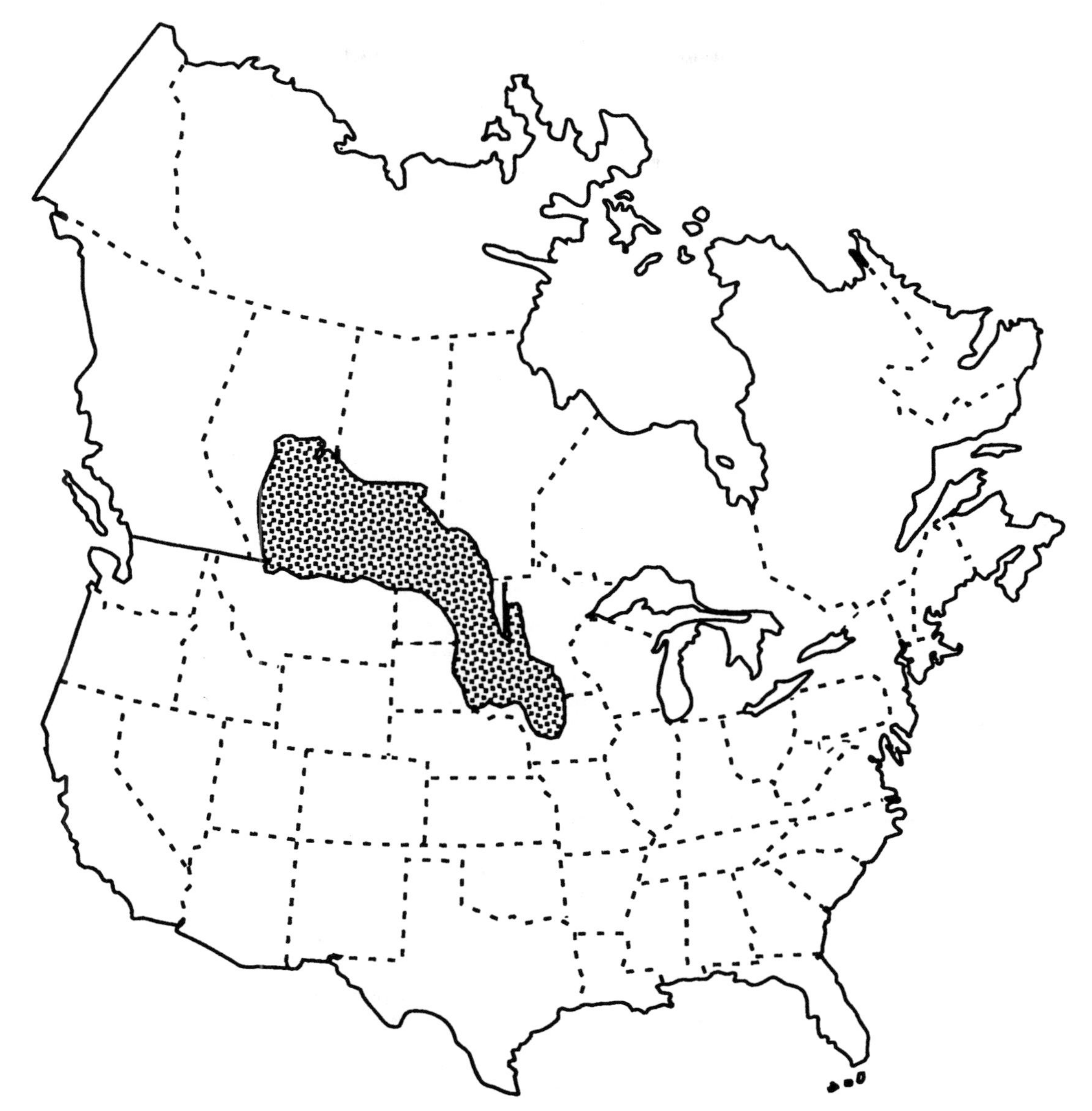

Figure 1. Area of overlap between duck brood-rearing habitat and agriculture in the prairie pothole region of North America.

Approximately 93% of breeding ducks in the prairies were found to nest within CLI waterfowl classes 1–5. Most of the area within CLI agriculture classes 1–3 was already in production by 1967, and most of the perceived expansion in agricultural land use, if it hasn't taken place already, will take place in agricultural class 4. In fact, the amount of land in agricultural classes 1–3, which was not yet in crop but was in pasture or some other agricultural or non-agricultural use, approximated the area of land in class 4 and up which was in crop. It shall be assumed that pesticide use on pasturelands and other non-croplands is minimal, and therefore the analysis will be restricted to cropland. The main exception is grasshopper spraying, which also takes place on non-cropland. An analysis of the overlap of waterfowl classes 1–5 with agricultural classes 1–3 provides a reasonable index of the extent of impingement of cropping practices on duck habitat in the late 1960s. The overlap between the same waterfowl classes with agricultural classes 1–4 should reflect the expected impact of agricultural practices on waterfowl habitat in the late 1980s and early 1990s, given recent trends in land use as well as prevailing agricultural policy (Statistics Canada, 1951–1981; Weaver et al., 1982).

Table 1. Duck species with breeding populations of greater than 100,000 in the prairie pothole region[a]

Common name	Scientific nomenclature
Dabblers	
Mallard	*Anas plathyrhynchos*
Pintail	*Anas acuta*
Blue-winged teal	*Anas discors*
Northern shoveler	*Spatula clypeata*
American widgeon	*Mareca americana*
Gadwall	*Anas strepera*
Green-winged teal	*Anas carolinensis*
Divers	
Lesser scaup	*Aythya affinis*
Redhead	*Aythya americana*
Canvasback	*Aythya valisneria*
Ruddy duck	*Oxyura jamaicensis rubida*

[a]Species within the dabbler and diver groups are listed in order of declining abundance. Cooch (1981).

Based on potential land use on an area basis, from 60 to 80% of the best waterfowl breeding and rearing areas overlap with the best agricultural land in the pothole region (Sheehan et al., 1987). Based on actual duck density figures, the overall prairie overlap between duck breeding populations and agriculture is similar to that predicted potential land use on an area basis. In the late 1960s, an estimated 56% of all prairie (grassland-parkland) breeding ducks (approximately 9.3 million) coexisted with agricultural development. This figure is now expected to have risen to 75% with most of the marginal agriculture land (class 4) having been placed in production.

Insecticide Use in the Canadian Prairies

Of the economically important insect pests of prairie crops, grasshoppers are potentially the most important because they occasionally give rise to outbreaks that cover areas greater than those covered by any other insect pest species. Because of the wide fluctuation in their year-to-year importance, they are the only pests for which investigations have been conducted to identify regions where large infestations might appear (Gage and Mukerji, 1977). Areas are rated as light, moderate, or severe on the basis of predicted infestation.

The total land area (cropland as well as pasture rangeland) placed in the moderate-severe category (and hence potentially requiring control) for the years 1973–1983 in all prairie provinces is given in Table 2. Of note is the large relative importance of Saskatchewan as a potential center of infestation. An area of southern Saskatchewan (part of the "Palliser Triangle") has traditionally had the most severe outbreaks. The actual areas sprayed are more difficult to determine. For years for which spray information was available, the proportion of "menaced" land that was actually sprayed varied by three orders of magnitude (from 0.01 to 18%). In both 1974 and 1982, the total area sprayed in Saskatchewan alone was in the vicinity of 1 million ha. For comparison, the area sprayed for grasshoppers in Saskatchewan in the outbreak of 1962 was 2 million ha and in 1985 an estimated 2.0–2.8 million ha was sprayed at least once; 2.8–3.6 million ha is the "extended" estimate (including reapplication). Typically, whereas crop areas or pastures may be sprayed once, noncrop areas such as fence lines, roadsides, and slough margins may be sprayed from two to seven times. It is interesting to note that, although grasshopper infestations often coincide with drought conditions, some of the heaviest spray-years on record (e.g., 1974 and 1982) were years with record-high numbers of ponds in southern Saskatchewan, indicating very wet conditions. This has a large bearing on this assessment.

Information on the other spray programs, notably outbreaks in oilseed crops (canola or double zero rape), has to be gleaned from numerous sources (Sheehan et al., 1987). On the basis of multiyear records, it is possible to piece together "typical" insecticide use patterns for the prairies, keeping in mind the highly variable nature of both pest infes-

Table 2. Grasshopper forecasts for the prairie province 1973–1983 (excluding light infestation category)

Year	Total area[a] (ha)			
	Alberta	Saskatchewan	Manitoba	Prairies
1973	2,706,540	4,829,810	718,200	8,254,550
1974	2,330,990	9,146,810	524,470	12,002,270
1975	3,202,170	10,498,780	697,220	14,398,170
1976	2,427,870	11,178,440	228,470	13,887,780
1977	683,650	7,440,290	0	8,120,950
1978	123,800	2,954,150	0	3,077,960
1979	216,520	4,352,750	68,480	4,637,760
1980	45,500	12,706,540	159,540	12,911,550
1981	999,900	9,541,200	89,300	10,630,400
1982	1,480,900	12,354,300	47,900	13,883,100
1983	3,086,700	17,808,000	50,500	20,961,100

[a]Land area that would potentially require chemical control and includes pasture rangeland, rough land and seeded acreage.
Source: Modified from the Canadian Agricultural Insect Pest Review (Agriculture Canada) unless otherwise specified.

tations and control programs. At the present time, the estimate is that normal (average) areas sprayed annually with insecticides are in the range of 1–1.5 million ha for the prairies. Another 2 million ha may be added to this estimate for a total of 3–3.5 million ha during years of serious grasshopper outbreaks. Should there be a concidental infestation of bertha armyworm or diamondback moth, two pests of canola, a further 0.4–0.5 million ha could be added to the total area sprayed. If it is assumed that most of the insecticide use takes place on agriculturally cultivated lands, estimated at 37.7 million ha in 1981 (Statistics Canada—1981 Census of Canada), insect control programs would cover 3–4% of improved lands during normal years, 8–9% during serious grasshopper outbreaks, and 10–11% if a grasshopper outbreak was to coincide with another serious outbreak such as by bertha armyworm or diamondback moth. In 1985, when there was one of the worst grasshopper outbreaks on record, it is estimated that 6.5–7 million ha was sprayed with insecticides—an area that corresponds to 17–19% of all improved lands or roughly 21–23% of the best cropland (land classified in CLI agriculture classes 1–3) in the Canadian prairies.

The insecticide products used in Canada to control pests on cereal and oil crops are the synthetic pyrethroids cypermethrin, deltamethrin, fenvalerate, and permethrin; the organophosphorus compounds azinphos methyl, chlorpyrifos, phosmet, dimethoate, and diazinon; the organochlorine compound methoxychlor; and the carbamate insecticides carbaryl, methomyl, and carbofuran. The primary insecticides used during the severe grasshopper infestation of 1985 were carbofuran, carbaryl, and deltamethrin, followed by chlorpyrifos and dimethoate. From 10 to 25% of the grasshopper spraying in 1985 was done aerially. Typically, most of the spraying for pest outbreaks in oilseed crops is carried out by air.

Temporal Overlap

Migratory waterfowl fly north to the pothole region each spring via flyways that span the North American continent. Their arrival in the Canadian prairies is generally associated with the early melt of ice on the potholes, but the exact timing of arrival is species specific and highly dependent on spring weather conditions, as well as geographic region within the Canadian prairies (Smith and Stoudt, 1968). Mallards and Pintails are the first dabblers to arrive each spring, and canvasbacks are the first divers to mass at the potholes. These species reach the pothole region before the end of March and the majority of their breeding populations have completed the spring migrations by the end of April. Large numbers of green-winged teal, gadwall, American widgeon, northern shoveler, lesser scaup, and redhead begin arriving at Canadian potholes after the second week of April, and most of their in-migration is finished by mid-May. Blue-winged teals do not reach the prairie habitat until late April. The ruddy duck is the last to arrive; the majority of individuals do not reach nesting areas until June.

Mallards are the first to nest; the early-arriving flock initiates nest preparation by the second week of April, prior to the complete in-migration of the bulk of the population. Similarly, the majority of pintail initiate nest attempts between mid-April and the first week of May. Canvasbacks, the first divers to nest, follow the pintails by approximately 1–2 wk. The lesser scaup and ruddy duck are late nesters; nest initiation does not begin in earnest until the latter half of June and may continue into the first week of July. These late nesters continue to select mates on the nesting grounds and may delay initiation of their nesting activity for 30–40 d after arrival. Most species renest if early attempts fail (Stoudt, 1971; Smith, 1971).

Incubation periods for the major divers and dabblers range from 20 to 30 d (Hochbaum, 1944). The nesting period places female ducks and embryos at great risk. The female and incubating eggs are away from the comparative security of the water and therefore more subject to predation and damage from cultivation.

The chronology of brood rearing for ducks in the Canadian prairies is shown in Figure 2. Mallard and pintail eggs that survive predation and destruction by agricultural machinery hatch between the last week in May and middle to late June. Blue-winged teal, northern shoveler, and canvasback broods hatch during the latter half of June to early July. Gadwall hatch is from the last week in June through mid-July. Lesser scaup and ruddy duck broods hatch between the first and last weeks in July. The redhead hatch is intermediate between the early- and late-nesting divers. Hatch of successful renesting efforts occurs primarily for the early nesters.

Upon hatch, broods are flightless and will remain that way for the next 40–75 d (see Hochbaum, 1944, p. 108, for species-specific data on age to attain flight). During the flightless period of rearing, broods are highly dependent on wetland habitat for protective cover and food resources. The majority of pintail, mallard, and teal broods fledge from the latter half of July into the initial week of August. Shoveler broods fledged in mid-August are followed in late August through mid-September by the peak fledging of gadwall, lesser scaup, canvasback, and redhead. All dabbler broods typically attain flight by the end of September. Broods of diving species may still be flightless in October; broods that are late to fledge are at risk of being caught by an early freeze.

Figure 3 summarizes the timing of insecticide applications for some of the major prairie crop pest infestations. The temporal overlap between the likely period of insecticide application and the period during which ducks are nesting and rearing broods is extensive. If it is assumed that the time between hatch and fledgling is the exposure period most sensitive to insecticide sprays, then late May to mid-July is the critical period for most duck species. On this basis, some pest control programs are more likely to cause problems than others. For example, an insecticide foliar spray for flea beetle control (late May–June) has the potential to have more impact on the bulk of nesting duck species than a similar spray for bertha armyworm (August). On the other hand, it could be argued that late sprays have the potential to have more impact on the less numerically important species such as diving ducks and that these species require greater protection as a result of their small population sizes and slower "rebound" capability (Cooch, 1981). Certainly, spraying for grasshoppers may overlap with critical brood rearing periods for all species. There does not, therefore, appear to be any way to reduce the temporal overlap between waterfowl nesting and insecticide use, particularly in years of large grasshopper infestations.

Food Resources for Ducks

The importance of the quantity and quality of food resources in the pothole habitat during the nesting and brood-rearing period to the reproductive success of ducks is well documented. During the reproductive period, adults and ducklings are highly dependent on food, particularly aquatic macroinvertebrates, from the aquatic habitat. The food habits of a reproductively active female and newly hatched duckling are, in many cases, substantially different from those of other adults and immatures. For example, dabblers such as pintails feed primarily on plant food for most of the year but select animal food (i.e., aquatic macroinvertebrates) during the period of egg formation (Krapu, 1974a,b) and early duckling growth (Sugden, 1973). Feeding ecology studies on a number of dabbling and diving ducks indicate that these periods of high dependence on aquatic macroinvertebrate prey are a normal part of the reproductive cycle of all duck species and are largely governed by the nutritional requirements of each stage.

Laying Female

During the period of egg formation, food intake by female ducks must be sufficient to supply normal metabolic needs and to provide the nutrients necessary to produce viable clutches of eggs. Laying mallards, pintails, and gadwalls under wild breeding conditions generally select a diet providing about 28% protein (Krapu, 1979). Shoveler, blue-winged teal, lesser scaup, canvasback, and redhead hens consume a diet exceeding 30% protein (Krapu, 1979; Noyes and Jarvis, 1985).

Inadequate nutrition adversely affects reproduction. Krapu (1979) pointed out the importance of a stable source of high-quality foods during the nesting season. Under controlled conditions, mallard hens fed a diet 45% lower in protein laid 46% fewer eggs during the nesting season than a control group of birds fed a normal animal diet. Calcium-deficient diets may also cause reduced egg production (Grau, 1968). Calcium must come primarily from foods such as snails.

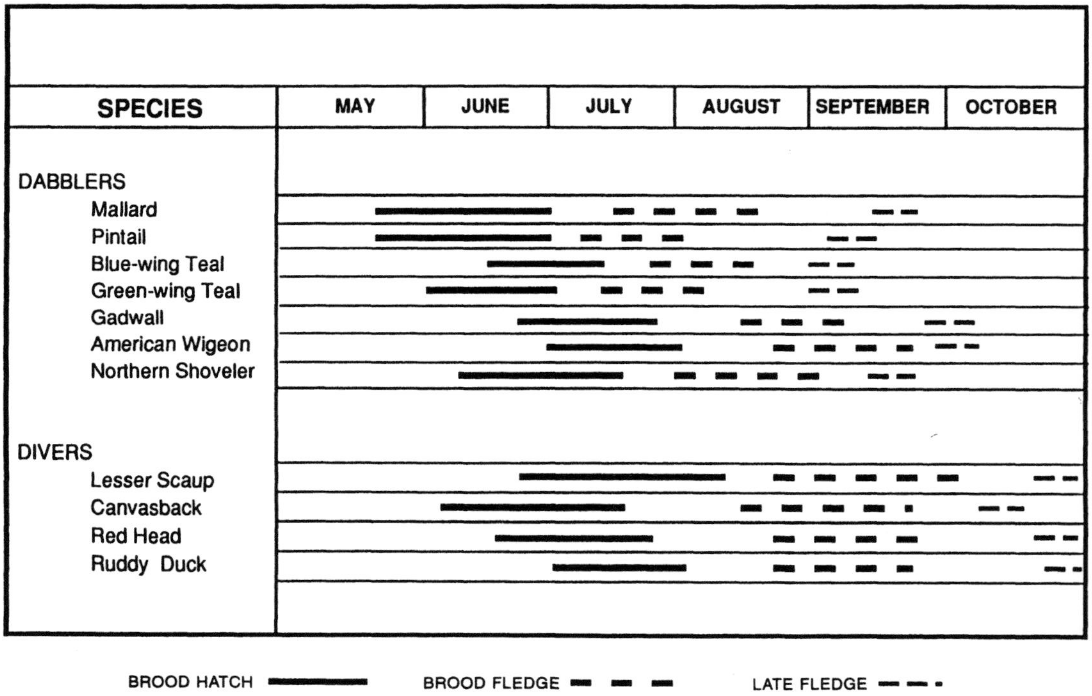

SPECIES | MAY | JUNE | JULY | AUGUST | SEPTEMBER | OCTOBER

DABBLERS
Mallard
Pintail
Blue-wing Teal
Green-wing Teal
Gadwall
American Wigeon
Northern Shoveler

DIVERS
Lesser Scaup
Canvasback
Red Head
Ruddy Duck

BROOD HATCH — BROOD FLEDGE — LATE FLEDGE

Figure 2. Chronology of brood rearing for ducks in the Canadian prairies.

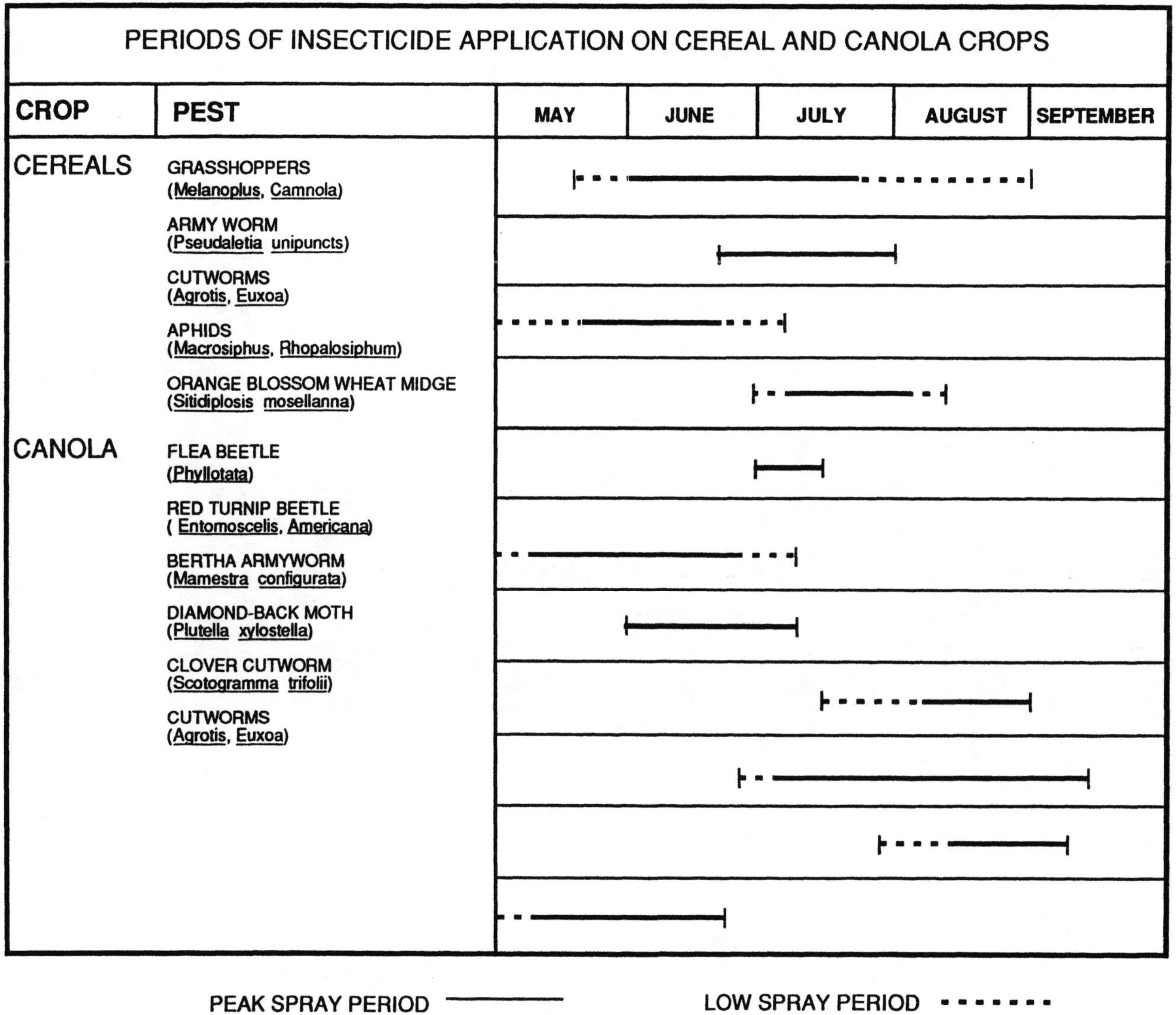

PERIODS OF INSECTICIDE APPLICATION ON CEREAL AND CANOLA CROPS

CROP	PEST	MAY	JUNE	JULY	AUGUST	SEPTEMBER
CEREALS	GRASSHOPPERS (Melanoplus, Camnola)					
	ARMY WORM (Pseudaletia unipuncts)					
	CUTWORMS (Agrotis, Euxoa)					
	APHIDS (Macrosiphus, Rhopalosiphum)					
	ORANGE BLOSSOM WHEAT MIDGE (Sitidiplosis mosellanna)					
CANOLA	FLEA BEETLE (Phyllotata)					
	RED TURNIP BEETLE (Entomoscelis, Americana)					
	BERTHA ARMYWORM (Mamestra configurata)					
	DIAMOND-BACK MOTH (Plutella xylostella)					
	CLOVER CUTWORM (Scotogramma trifolii)					
	CUTWORMS (Agrotis, Euxoa)					

PEAK SPRAY PERIOD ———— LOW SPRAY PERIOD - - - - - - -

Figure 3. Periods of insecticide application for several common insect pests in the Canadian prairies.

Diets with inadequate protein also cause hens to ovulate at irregular intervals (Krapu, 1974a). Fed a plant diet low in protein, hens skipped 1 d every two eggs laid whereas on an animal diet they failed to ovulate on average only after every 4.6 eggs laid (Krapu, 1979). An extended laying period increases the risk of nest destruction. In addition, Krapu (1979) reported that diet quality influenced egg size and lipid reserves. These findings suggest that the quantity and quality of food reserves available during nesting can significantly affect the quantity of nutrient reserves available to ducklings at hatch. Nutrient reserves provide energy to the newly hatched ducklings and are crucial to their survival if adverse environmental conditions prevent them from obtaining adequate food in the initial few days after hatch. Animal matter in the diet of laying female dabblers is also necessary to assure adequate hatchability of eggs (Krapu and Swanson, 1975; Krapu, 1979).

The nutritional demands of egg formation are met by the inclusion of high quantities of aquatic macroinvertebrates in the diet of laying hens (Krapu 1974a,b, 1979; Noyes and Jarvis, 1985). Aquatic insects, crustaceans, and snails are highly selected, presumably because of their nutritional value, digestibility, and availability. Although the composition of the animal portions of the diet of laying females varies among the duck species listed, all hens consume more than 70% animal food (Noyes and Jarvis, 1985). Diving species and the northern shoveler and blue-winged teal are almost totally dependent on macroinvertebrates.

Although laying hens are opportunistic feeders and will switch prey with seasonal changes in macroinvertebrate species abundance, laying efforts generally terminate when selected aquatic food resources start to decline (Pehrsson, 1984). Reduced food availability also causes females to terminate renesting attempts (Swanson and Meyer, 1977).

Ducklings

Although there are less data available on the diet of ducklings than on the diet of laying hens, evidence indicates a high dependence on animal food, primarily aquatic macroinvertebrates, for periods of 1 to 7 weeks posthatch (Chura, 1961; Bartonek and Hickey, 1969b; Sugden, 1973). Street (1978) showed that the protein content of food appeared to be the major factor influencing the growth rate of mallard ducklings, particularly over the initial 4 d of growth. Diets of barley meal or seeds from aquatic plants were insufficient to maintain the ducklings; after 4–5 d of feeding solely on a plant food diet, when lipid reserves were depleted, birds lost weight and became weak. Street concluded that a diet of greater than 50% animal protein was required to sustain a satisfactory growth rate in wild mallard ducklings. He pointed out that because growth rate is correlated with fat deposit (Marcström, 1966) and plumage development (both important in temperature control) and with strength and feeding activity, growth is roughly equivalent to survival. Therefore, fast growth of ducklings contributes to their success.

The contribution of major food groups to the natural diet of selected species of ducklings foraging in the prairie pothole region is summarized in Table 3. Animal matter, predominantly aquatic macroinvertebrates, composed an average of 91% of the diet of mallard ducklings and 67% of the food of pintail ducklings in the prefledgling stage. Although gadwall and American widgeon ducklings consumed an average of nearly 90% plant foods, animal food was eaten in the critical first 20-d growth period (Sugden, 1982). There was generally a greater dependence on animal prey in the diet of diving ducks than in dabbling species. Of the diving species listed (Table 1), only the redhead ducklings had an average intake of animal food substantially less than 90% (Bartonek and Hickey, 1969a,b). Where data were available on dietary consumption with age, class I ducklings (0–18 d) showed a much higher than average dependence on animal protein (Table 3).

The period of dependence on macroinvertebrates varies for duckling species (Table 4). Gadwall ducklings have a diet composed of more than 70% macroinvertebrate foods for only the initial 5 d of growth and, by 3 wk of age, take 5% or less macroinvertebrate foods. In contrast, macroinvertebrates dominate the diets of widgeons, pintails, and mallards for approximately the first 20 d of growth. Perret (1962) reported that juvenile mallards in the Minnedosa, Manitoba, area were dependent on macroinvertebrate food resources for a much longer period. Canvasbacks (Bartonek and Hickey, 1969b) and ruddy ducks (Siegfried, 1973) are highly dependent on macroinvertebrate food resources for the entire prefledgling period.

Class I mallard ducklings fed heavily on flying insects, particularly chironomids, as well as the larval and pupal stages of midges (Perret, 1962). Larvae of cased caddisflies (particularly Limnephilidae), aquatic snails (particularly Lymnaeidae), dragon- and damselfly naiads, water boatmen (Corixidae), and aquatic beetles (Hydrophilidae and Dytiscidae) were taken in moderate amounts. Selected macroinvertebrate prey for mallard ducklings in approximate order of importance are shown in Table 5. Older mallard ducklings selected more plant foods including grass, seeds, nutlets of *Scirups* and seeds and tubers of *Potamogeton* (Chura, 1961).

Ducklings of diving species are predominantly macroinvertebrate feeders throughout the flightless growth period. The macroinvertebrate prey selected by lesser scaup ducklings are shown in Table 5. Amphipods were the most important food for scaup ducklings (52% of the diet) (Sugden, 1973). Chironomid larvae and pupae were also consumed in substantial quantities (15% of diet), particularly by ducklings in the first age class. Snails were preyed upon in appreciable amounts (16% of diet). Hemipterans, zygoptera naiads, leptocerid caddisfly larvae, and haliphid and dytiscid beetles all represented

Table 3. Percentage of major foods consumed by ducklings prior to attaining flight in brood foraging areas in the prairie pothole region. Data are based on volumetric measurements of esophageal contents

Species (duckling class)[a]	Percent food type: Gastropoda	Oligochaeta	Crustacea	Insecta	Seeds	Plant parts	Percent total: Animal	Plant	References
Dabblers									
Mallard (I–III) (I)	11	1	3	72[b]	9	—	91	9	Perret (1962)
Pintail (I–III) (I)	36	—	4	26[b]	32	1	67	33	Sugden (1973)[c]
							>91	<9	Sugden (1973)[c]
Gadwall (I–III) (I)	—	—	2	8	18	72	10	90	Sugden (1973)[c]
							>71	<29	Sugden (1973)[c]
American (I–III) widgeon (I)	2	—	—	9	13	76	11	89	Sugden (1973)[c]
							>89	<11	Sugden (1973)[c]
Divers									
Lesser (I–III) scaup (I)	16	1	53	26	2	2	96	4	Sugden (1973)[c]
							>98	<2	Sugden (1973)[c]
Canvasback (I–III)	15	—	—	79	1	5	94	6	Bartonek and Hickey (1969b)[c]
Redhead (I–III)	—	—	—	43	45	12	43	57	Bartonek and Hickey (1969b)[d]
Ruddy (I–III) duck	—	—	10	78	10	2	88	12	Siegfried (1973)

[a]I, 1–18 d of age; young down-colored; II, 19–42 d of age; feathers first appear and cover most of the body; III, 43–55 d of age; young fully feathered by juvenile plumage, prior to attaining the ability to fly.
[b]Remaining proportion of animal matter not categorized.
[c]Data based on percent dry weight.
[d]Only 5 of 79 ducklings were 2 weeks of age or younger; thus, proportion may not adequately represent the diet of young ducklings. Age class description from Bellrose (1976).

greater than 1% of the food of scaup ducklings (Sugden, 1973). Ducklings of all ages ate only 4% plant food of which most was *Chara, Oögonia*, and *Myriophyllum* nutlets (Sugden, 1973).

Figure 4 shows the timing of reproductive phases in relation to available food resources for mallards and stresses the importance of macroinvertebrate food availability during the critical periods of egg laying and posthatch duckling feeding. Seed and vegetative foods are available later for older ducklings (Pehrsson, 1984). The general relationship between nutritional demands of laying hens and duck-

Table 4. Period of duckling growth during which macroinvertebrates account for more than 70% of food consumed

Species	Age group (d)	Reference
Mallard	0–18	Chura (1961)
Pintail	0–20	Sugden (1973)
Gadwall	0–5	Sugden (1973)
American widgeon	0–15	Sugden (1973)
Lesser scaup	0–50+	Sugden (1973)
Canvasback	0–50+	Bartonek and Hickey (1969b)
Redhead	*a*	Bartonek and Hickey (1969b)
Ruddy duck	0–50+	Siegfried (1973)

[a]Insufficient age species dietary data, particularly for birds less than 2 weeks of age.

Table 5. Macroinvertebrate prey selected by flightless ducklings foraging in the prairie pothole region

Species	Selected macroinvertebrate prey in approximate order of importance	Reference
Dabblers		
Mallard	True fly larvae, pupae, adults (primarily Chironomidae)	Chura (1961)
	Caddisfly larvae (primarily Limnophilidae)	Perret (1962)
	Aquatic snails (particularly Lymnaeidae)	
	Dragon- and damselfly nymphs (Odonata)	
	Water boatman adults, nymphs (Orixidae)	
	Beetle larvae, adult (Hydrophiliadae, Dytiscidae)	
Divers		
Lesser scaup	Scuds (Amphipoda)	Bartonek and Hickey (1969a)
	Aquatic snails (Lymnaeidae)	Sugden (1973)
	Midge larvae, pupae (Chironomidae)	
	Dragonfly naiads (Zygoptera)	
	Caddisfly larvae (Leptoceridae)	
	Beetle larvae (Haliplidae, Dytiscidae)	

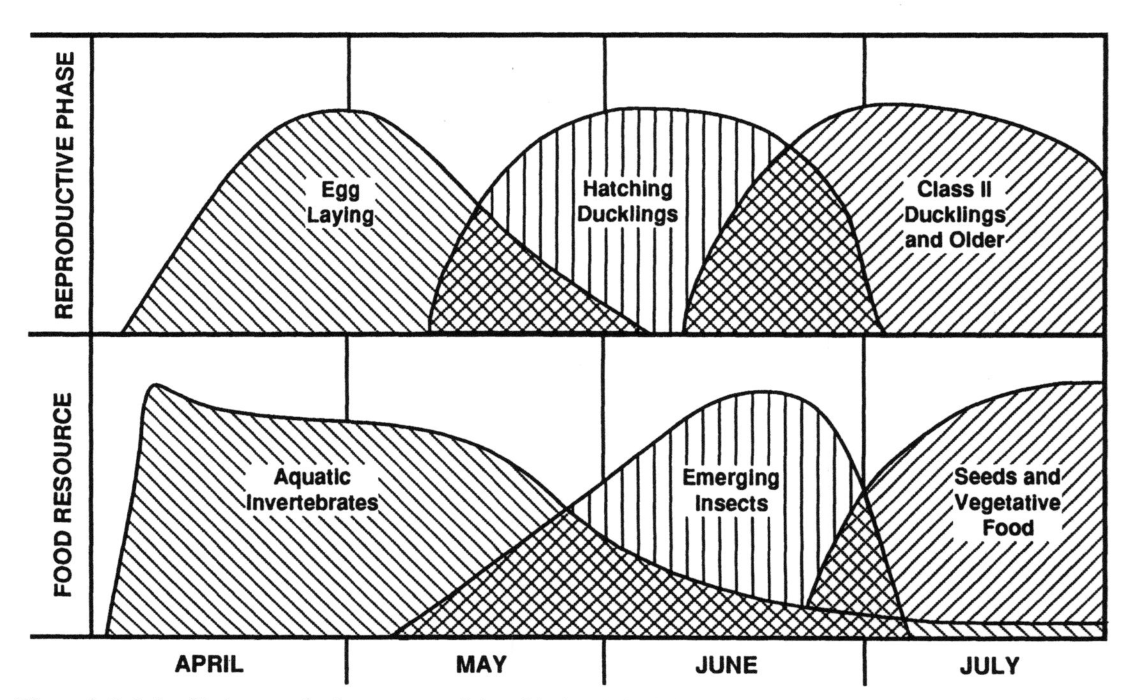

Figure 4. Relationship between food resources and the critical periods of demand by laying hens and duckling broods.

lings may be applied to early nesting dabblers and divers in the pothole region of the Canadian prairies. The increased nutritional demand during the reproductive phase of the duck life cycle has critically linked breeding pairs and small ducklings to the protein-rich macroinvertebrate food resources of their aquatic habitat.

RISK ASSESSMENT APPROACH

It is clear from the preliminary evaluation that there is a temporal and spatial overlap between insecticide spraying and duckling rearing. Quantifying possible adverse impacts is, however, a complex process.

To assess the potential impact of spraying of insecticides on the slough ecosystem and duck recruitment success, a number of risk assessments tasks were completed. These include:

- Prediction of the concentration of aerially sprayed insecticides in slough water and sediment based on application rates and slough conditions.
- Analysis of insecticide toxicity to key aquatic macroinvertebrate species and the ranking of insecticides according to their toxicity to macroinvertebrates.
- Quantitative analysis of the potential reduction in macroinvertebrate numbers and biomass in sloughs due to exposure to expected levels of the most potent insecticides.
- Evaluation of the dependence of duckling growth and survival on the availability of macroinvertebrates.
- Calibration of growth model predictions using data from controlled field studies of duckling growth and survival on insectide-treated ponds.
- Estimation of the size of the duckling population at risk from aerial insecticide spraying.
- Analysis of duckling recruitment in insecticide-sprayed and unsprayed areas based on field census data.

The methods used to evaluate each of these risk assessment components and the results of these evaluations are summarized in the following discussions.

Prediction of Exposure Concentration

The entry of insecticides into wetlands is expected to occur through either direct application, droplet drift, runoff, or seepage (Sheehan et al., 1987). The likelihood of direct contamination is highly dependent on the selected mode of application. The method of application with the greatest probability of resulting in direct contamination is aerial application.

Major sources of contamination are expected to be direct overspray and drift droplet deposition from target swaths close to the slough. Analysis of the likelihood of direct overspray in such pothole regions depends on assumptions about applicator ability, specific limitations in aircraft control, the effectiveness of buffer zones, and the size and dispersal pattern of wetlands. An applicator's desire for complete coverage and perception of sloughs as something requiring protection from pesticide contamination or as pest refuges determine, in part, whether sloughs are purposely oversprayed. In cases such as grasshopper infestations, effective control dictates the spraying of all possible refuges. For small sloughs (<0.4 ha) with narrow margins, common in cultivated lands, partial overspray is highly probable. These ponds may account for 60 to 80% of the wetland habitat available to ducks in May–June when insecticide spraying is likely.

In the worst case of direct aerial overspray, 100% of the applied amount is assumed to be deposited on ponds and margins. This "worst-case" assumption provides a basis for comparisons between chemicals but is not valid for actual application conditions. Under a variety of meteorological and application conditions, on-target deposits have been found to vary substantially and rarely approach 100%. In fact, an average on-target deposit of approximately 50% of applied quantities from a single swath may be realistic (Ware et al., 1970) can be assumed to fall on the surface of an oversprayed slough. Each additional complete overflight of a slough can potentially increase contamination. Overspray deposits of approximately 80% of applied quantities are expected from multiswath coverage, which is common in insect control (N. B. Akesson, personal communication). Therefore, in areas being sprayed aerially, the concentration of the insecticides reaching pond water is largely a function of the application rate and the number of swaths sprayed and the physical-chemical characteristics of the insecticide.

The second substantial source of insecticide contamination is drift droplet deposition from upwind applications. The concentration of pesticide deposited off-target from aerial spray at distances greater than 5 m decreased rapidly with distance. Studies show that deposits at 50 m downwind of the direct area of application ranged between 1 and 10% of applied quantities (Argauer et al., 1968; Ware et al., 1969a,b; Currier et al., 1982; Rennee and Wolf, 1979). Indirect sources of insecticide contamination of prairie sloughs such as runoff are expected to be relatively small contributors in areas of aerial application, perhaps <1% of applied amounts (Wauchope, 1978).

The recommended application rates for insecticides are crop and pest dependent and are based on information provided in insect control pamphlets for Alberta, Manitoba, and Saskatchewan. The recommended application rates for the insecticides of concern in the Canadian prairies and the maximum expected initial concentration, based on a worst-case thorough mixing of 100% of applied material into a hypothetical shallow pond (surface area 4550 m^2 and mean depth 1 m), are shown in Table 6. Where possible, the rates shown in the table were selected to reflect the control of grasshoppers on cereal crops. If a product was not recommended for the control of grasshoppers, rates for other pests on grain or oil crops were selected. By using the application rates for these insecticides, comparable treatment levels under field conditions were approximated to assess the relative toxicity of these products to aquatic organisms in ponds hypothetically contaminated due to aerial spraying. However, because of expected on-target deposits of less than 100%, the surface film effect, and incomplete mixing, the highest expected insecticide concentrations in pond water due to aerial sprays are substantially less, perhaps 50% of the theoretical maximum reported.

The exposure scenario used in this analysis is based on a single aerial application of an insecticide to the pond (Figure 5). Two distinctly different phases of bioavailability of the applied compound are expected in water under such conditions. During the initial phase of dissipation, concentrations are high and losses rapid. This is followed by a phase where distribution between environmental compartments has equilibrated. During this second equilibrated phase, concentrations are low and loss rates are low. The initial phase is descriptive of an acute exposure scenario and the equilibrated distribution phase of a chronic exposure scenario.

The concentrations of insecticides in the two exposure scenarios were predicted using the Persistence Model (NRCC, 1981). The Persistence Model involves a generalized view of the aquatic environment with four compartments (water, sediment, fish, catchall); it is similar to the Fugacity Model (Mackay and Paterson, 1982) and EXAMS (Burns et al., 1982). For an in-depth discussion of exposure fate models, see Chapter 18. First-order transfer and removal coefficients and partition coefficients between each compartment and water are indicated. The catchall compartment is included to consider sorptive components such as algae and suspended solids.

The following constraints and assumptions were used:

- Only well-characterized chemical and physical processes are included in the analysis of chemical fate. The intent of the model is to allow the user to determine what is predictable using valid data and then to predict the rate at which other processes must occur to influence significantly the behavior patterns of the chemicals.
- Hydrodynamic properties and settling effects, which cause dilution and redeposition of the compound but do not affect degradation patterns, were not included. This constraint may lead to an overestimate of soluble concentrations but the model is conservative.
- The sizes of the compartments were assumed to remain constant and the rate of mixing of the chemical was assumed to be rapid relative to the rates of all other processes. This should provide a basis for the determination of relative behavior patterns.
- For comparative purposes, all processes were assumed to follow simple kinetics and were approximated by first-order or pseudo-first-order rate constants.
- It was assumed that equilibrium will eventually describe the distribution of the chemical.

The Persistence Model was used to predict the distribution of each insecticide in a shallow pond and the relative importance of the various removal processes. Outputs of the algebraic solution included partition coefficients between each compartment and water, first-order transfer and removal rate constants

Table 6. Insecticide application rates[a] and expected initial concentration in pond water used to assess exposure of aquatic invertebrates

Insecticide	Recommended application rate[b] (g AI ha^{-1})	Expected initial concentration in slough water[c] ($\mu g\ L^{-1}$)
Synthetic pyrethroids		
Cypermethrin	28	3
Deltamethrin	7.5	1
Fenvalerate	97.5	12
Permethrin	140	17
Other insecticides		
Azinphos methyl	420	52
Carbaryl	1100	135
Carbofuran	140	17
Chlorpyrifos	560	69
Diazinon	550	68
Dimethoate	490	60
Malathion	840	103
Methoxychlor	1500	184
Phosmet	1125	138

[a]Applicable rates are typically based on control of grasshoppers on cereal crops.
[b]Recommended application rates based on those provided in insect control pamphlets for Alberta, Manitoba, and Saskatchewan.
[c]Concentration calculated by multiplying recommended application rate times surface area/volume ratio of the model pond.

for each compartment, retentive capacity of the system, half-life of the chemical in the system, fractional retention in each compartment, and fractional removal by each process. The outputs of the graphic or tabular solution described the dynamics of distribution and decay of a single pulse exposure of each insecticide. The one output requiring further definition is retentive capacity (RC). In mathematical terms, retentive capacity is

$$RC = \frac{[dQ_t]^{-1}}{[Q_t\, dt]}$$

or simply the inverse of the specific removal rate of the chemical from the system at any time, t, where Q_t is the total amount of the chemical in the system at any time. Retentive capacity, therefore, describes the residence time (see Chapter 1) of a chemical in an aquatic ecosystem as a function of the characteristics of the system and characteristics of the chemical (Roberts and Marshall, 1980).

The computer model–predicted distribution and persistence of selected insecticides are shown in Table 7. The synthetic pyrethroid compounds are predicted to be highly persistent and sequestered primarily in benthic sediments. In contrast, carbaryl, carbofuran, dimethoate, and malathion are expected to be found primarily in pond water and are rapidly degraded if the water is slightly alkaline. It is clearly important to know the pH of the slough to characterize properly the persistence of these compounds. The organochlorine compound methoxychlor is expected to sequester in benthic sediments and is relatively persistent.

Ranking of Insecticide Toxicity

Two indices of relative hazard were derived to assess potential impacts of insecticides on aquatic fauna under acute and chronic exposure conditions in the simulated pond ecosystem. The relative hazard index (RHI) is a dimensionless index of direct hazard; high RHI values indicate high risk of substantial mortality in exposed populations.

Acute Exposure Scenario

$$RHI = \frac{AR \times [1/K_{\text{phytolysis}} + 1/K_{\text{volatilization}} + 1/K_{\text{hydrolysis}} + 1/K_{\text{suspended sediments}} + 1/K_{\text{sediment}}]}{V_{\text{mixing}} \times \text{LC50t}} \qquad (1)$$

Where AR is the recommended application rate for the insecticide in g ha^{-1} [the model accounts for pond surface area and computes loading based on 100% of applied quantities (mg)]; $k_{\text{photolysis}}$, $k_{\text{volatilization}}$, and $k_{\text{hydrolysis}}$ are removal constants and $k_{\text{suspended particulates}}$ and $k_{\text{sediments}}$ are transfer rate constants (d^{-1}); V_{mixing} is equivalent to the volume at a specific depth below the surface or the entire volume of a shallow pond; and LC50t is the estimated integrated mean lethal exposure in the exposure units (mg L^{-1} d).

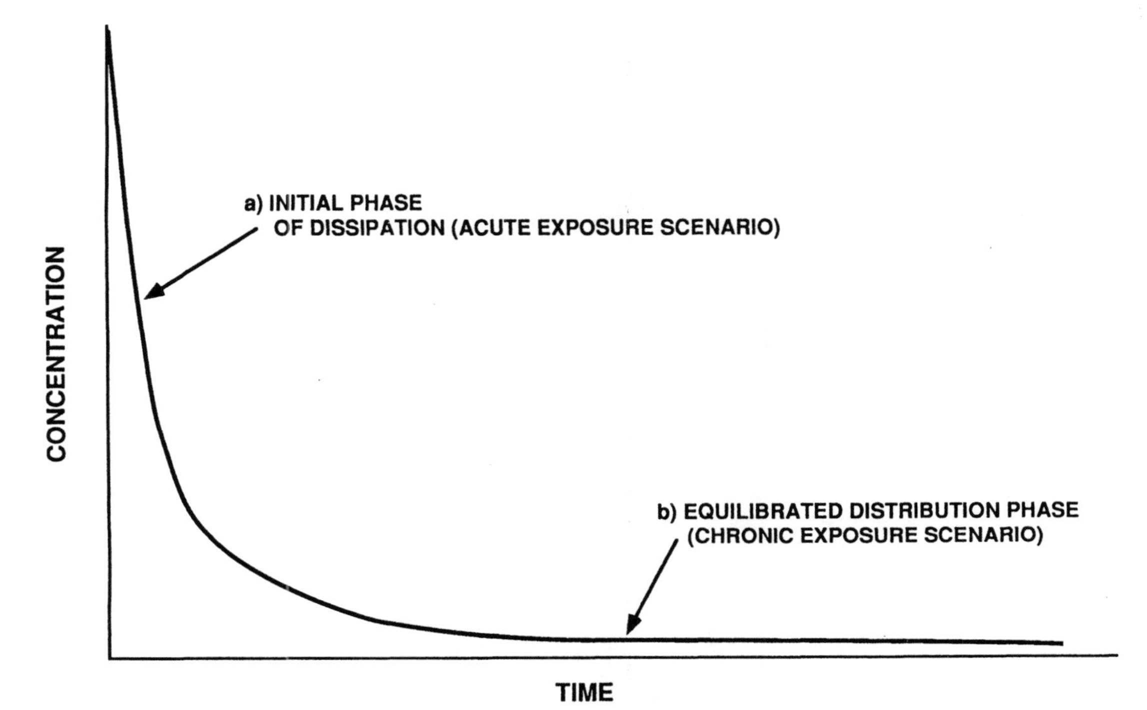

Figure 5. Hypothetical dissipation curve for a single application of a slightly soluble organic insecticide in water of a shallow pond.

Table 7. Computer model[a] predicted distribution and degradation of selected insecticides in a simulated shallow pond

	Deltamethrin		Permethrin		Carbaryl		Carbofuran		Dimethoate		Malathion		Methoxychlor	
	pH 6	pH 8	pH 5	pH 9	pH 5	pH 9	pH 5	pH 9	pH 5	pH 9	pH 5	pH 9	pH 5	pH 9
Retentive capacity (RC) (d)	17,058	9,913	3,520	3,175	8	0.2	2,444	0.7	222	4	842	0.2	298	297
First-order *t*1/2 (d)	11,819	6,871	2,440	2,200	6	0.1	1,694	0.5	153	3	584	0.1	206	297
Percent of total load sequested in:														
Water	0.05	0.5	0.8	0.8	88.0	88.0	97.3	97.3	73.8	73.8	64.9	64.9	2.9	2.9
Suspended solids	2.5	2.5	2.5	2.5	0.3	0.3	0.1	0.1	0.6	0.6	0.9	0.9	2.4	2.4
Benthic sediments	97.0	97.0	96.7	96.7	11.7	11.7	2.6	2.6	25.6	25.6	34.2	34.2	94.6	94.6
Percent removal due to:														
Volatilization	50.9	29.6	81.8	73.8	77.6	1.9	1.0	0	0	0	45.8	0	80.1	79.8
Photolysis	0.1	0.1	0	0	3.9	0.1	0	0	0	0	0	0	0.1	0.1
Hydrolysis	52.2	52.2	0	8.0	0.3	79.8	80.8	81.8	81.8	81.8	36.0	81.8	1.7	1.9
Biolysis[b]	18.2	18.2	18.2	18.2	18.2	18.2	18.2	18.2	18.2	18.2	18.2	18.2	18.2	18.2

[a]Based on the NRCC Persistence Model.
[b]The percent of total degradation that is microbial was arbitrarily chosen and input into the model.

Chronic Exposure Scenario

$$\mathrm{RHI} = \frac{F_1 \times \mathrm{AR} \times \mathrm{RC}}{V_{\mathrm{mixing}} \times \mathrm{LC50t}} \qquad (2)$$

where AR, V_{mixing}, and LC50t are as defined above; F_1 is the fraction of the insecticide in the water at equilibrium; and RC is the residence time of the compound in the aquatic ecosystem.

The definition of a toxicity model requires the selection of appropriate comparable toxicological measures and the analysis of available test data to determine whether the effect appears to be a function of estimated integrated exposure or estimated maximum exposure.

Several possible toxicological measures that could be incorporated into relative hazard calculations are available from laboratory toxicity tests (Rand and Petrocelli, 1985). Among available water quality criteria are the EC50 (concentration causing a designated sublethal effect on 50% of the test organisms within a specific exposure period), the LC50 (concentration at which 50% of the test organisms survive within a specific exposure period), the NOEL (no-observed-effect level defined as the highest test concentration at which no effects are observed), the LOEL (the lowest effect level, defined as the lowest test concentration at which an effect is observed), and the MATC (maximum acceptable toxicant concentration range defined by the highest tested value leading to no observed harmful effect). EC50 and LC50 define an acute toxic response, and the MATC defines the boundary for chronic toxicity. The most widely reported toxicological benchmark is the LC50, and as LC50 data sets collected under standardized conditions are generally available for various species and pesticides, this measure offers the best opportunity for broad-scale data comparisons. There are few reported MATC, NOEL, and LOEL values for aquatic species exposed to pesticides, which reduces the utility of these measures in hazard models at the present time.

The theory of the concentration-response relationship producing LC50 values in a toxicity test assumes that death is a positive function of exposure, response magnitudes are proportional to the logarithm of concentration, and randomly selected test animals are normally distributed in their sensitivity to a toxicant (see Chapter 1). These assumptions are largely met in most toxicity tests if close attention is paid to controlling variables that influence contact and absorption of the toxic compound. For example, carriers, surfactants, and formulation can greatly alter the adsorption and apparent toxicity of pesticides (e.g., Stratton et al., 1980). Alteration of the temperature or pH of the test water can also alter adsorption and apparent toxicity (e.g., Eisler, 1970). In addition, loss of pesticide from the test container through volatilization, sorption or bioaccumulation by overdense test populations, or rapid transformation of the compound through biological or chemical processes can result in a reduction in real toxicant concentrations below nominal dosage levels either invalidating the test or grossly underestimating the toxicity. The numerous examples of experimental and environmental variables influencing toxicity test results (see Tucker and Lietzke, 1979; Sprague, 1985) suggest that the comparative toxicology of pesticides to aquatic macroinvertebrates must be obtained from experimental data sets obtained with methods that are well defined and comparable.

Theoretically, for organisms exposed to a single concentration of a toxicant in a toxicity test, lethal toxicity is usually viewed as a function of percent mortality and survival time. At each higher concentration more organisms will die. Results can be plotted on a three-dimensional graph of concentration, percent mortality, and survival time (see Skidmore, 1974, for discussion). Discounting any lag period or delayed effect, which would occur if adsorption and transport to the site of action were other than first-order processes, the relationship between median lethal concentration of the test compound and duration of exposure can be expressed as $\mathrm{LC50}(t) = k$, where t is the duration of exposure and k is a toxicity constant that approximates the integrated median lethal exposure. This expression implies that a low concentration over a long treatment period can produce the same effect as a high concentration over a short period. Although the fixed relationship between lethality and time does not hold for all pesticides and species, it can serve as an appropriate model for compound comparisons. Correlation coefficients for the regression of LC50 values against test duration for toxicity tests conducted on the amphipod *Gammarus lacustris* and the stone fly *Pteronarcys californica* for several insecticides were greater than $r = 0.87$ (Sanders and Cope, 1968; Sanders, 1969). If doubling the exposure time cuts the median lethal concentration in half, the model is one of perfect chronicity. Although this time-integrated exposure model may not accurately fit all toxicity test values, it is not unusual to see a 24-h LC50 that is twice the value of the 48-h LC50, which is, in turn, twice the toxicity reported for 96-h exposure if there is no substantial loss or transformation of the toxicants throughout the exposure period.

A number of studies have found correlations between reported LC50s and the estimated time-integrated exposure. Allison (1977) suggested that an exposure unit (mg L^{-1} d) may have value in estimating the environmental effects of pesticides on fish over a range of exposure conditions. LC50 data for fish exposed to chlorpyrifos also suggest the applicability of the median lethal exposure model to explain toxicity, whereas the reported maximum concentrations did not demonstrate any apparent relationship with toxicity to fish (NRCC, 1981). The time-integrated exposure model is particularly appealing because the value of LC50t for short exposure time periods approaches that of the alternative toxicity model based on maximum concentration. This time-integrated exposure model is used in this evaluation.

Choice of Test Species

The availability of appropriate data sets for application of the integrated median lethal exposure model is limited, particularly for macroinvertebrate species important to the feeding of ducklings and laying hens. There are comparative LC50 data for the amphipod genus *Gammarus* (e.g., Sanders 1969, 1972), which is a choice prey in the early diet of diving duck species and is highly susceptible to chemicals. Sloof (1983) reported that *Gammarus pulex* was the most sensitive of 11 macroinvertebrate species exposed to 15 chemicals in acute laboratory toxicity tests. Toxicity test protocols for gammarids are well described (reviewed by Arthur, 1980), and the reporting of LC50 data for amphipods in pesticide registration submissions is common but not universal.

A broad base of toxicity data is available for the midge family, Chironomidae, a second group important as prey for both diving and dabbling ducks. Acute LC50 values for chironomids come largely from efficacy studies with potential pest species (e.g., Ali and Mulla, 1978) and chironomids appear to be relatively less sensitive to toxicants than amphipod species (Sloof, 1983). Standard toxicity test procedures for midge species have been described (Anderson, 1980). Efficacy studies also provide a broad comparative database for mosquito larvae. As noted earlier, mosquito larvae make up a small proportion of duck diets, although they can be important prey in isolated cases when larval density is high.

The only internationally recognized invertebrate taxon for aquatic toxicity testing is the cladoceran *Daphnia* (OECD, 1981). However, caldocerans appear to be largely incidental selections as prey by ducks. *Daphnia* is perhaps the most widely chosen aquatic invertebrate for toxicity testing because test protocols are highly developed (Buikema et al., 1980; OECD, 1981). *Daphnia* is often the only zooplankton genus for which comparative pesticide test data are available (e.g., Sanders and Cope, 1966).

Comparative LC50 data for insecticides gathered under the same test conditions are also available for the stone fly *Pteronarcys californica* (Sanders and Cope, 1968). This aquatic insect is relatively tolerant of insecticides and does not contribute to the diet of ducks, but it may be representative of other large-sized insects (mayflies and caddisflies) that are important prey.

It is important to note that no one taxon met all the criteria of an ideal macroinvertebrate species for our relative hazard model, including availability of comparative data, exposure duration–effect relationship, dietary importance, and sensitivity to insecticides. The work of Sloof (1983), however, indicates that the maximum difference in relative tolerance among species is likely to be relatively low: less than a factor of 5 for 15 toxicants tested on 11 invertebrates. Hazard analysis can, therefore, be based on toxicity data for the most sensitive species or the most ecologically relevant species, or on the range of values across species for which data are available.

The data summarized in Table 8 show the range of time-integrated median lethal exposure values for *Daphnia magna*, *Gammarus lacustris*, *Pteronarcys californica*, *Chironomus* spp., and *Culex* spp. calculated for 1- to 4-d exposure periods. For *Gammarus* and *Pteronarcys* the LC50t range is quite narrow. Insufficient data are available to determine the goodness of fit of data summaries for other invertebrate taxa in the integrated exposure model. Within the data sets for individual species, the range of integrated median lethal exposure values for specific insecticides does not exceed a factor of 10. The range among species rarely exceeds 100. The ranges are relatively narrow considering that differences in both exposure period and laboratory procedure are included in the data. Kenaga (1978) stated that the limits of reliability of acute toxicity values are ±1 order of magnitude.

Although the time-integrated median lethal exposure model appears to explain adequately toxicity data (in cases in which evaluation can be made) for acute exposure conditions as would occur immediately after insecticide release to a pond, the utility of this toxicity model in defining response under chronic exposure conditions as would occur when the insecticide equilibrated with the aquatic environment is unclear. Kenaga (1982a,b) concluded that there appeared to be no consistent pattern of ratios between acute and chronic toxicity and this was particularly evident for pesticides tested on aquatic organisms. Therefore, this case study focuses on the acute exposure scenario.

Relative Hazard Ranking

A listing of the weighted average ranking of potential short-term hazard of chemicals to aquatic macroinvertebrates based on calculated RHI values from the acute exposure scenario is presented in Table 9. The values in parentheses (e.g., <1, <10, 10–100, >100) identify the calculated range of the relative hazard index. The larger the index, the greater the potential hazard. The acidic and alkaline ponds show similar rankings of compounds. With the exception of the pyrethroid fenvalerate, the pyrethroids are in the top half of the risk lists along with azinphos methyl, chlorpyrifos, and methoxychlor. They are ranked as potentially high-hazard compounds with RHI value > 100. These are followed by phosmet, malathion and fenvalerate, carbaryl, and diazinon with RHI between 10 and 100 in the acid pond simulation. Phosmet, malathion, and carbaryl are more rapidly degraded and therefore potentially less hazardous under alkaline pond conditions. Carbofuran and dimethoate are also affected by pH but based on toxicity are not expected to affect aquatic macroinvertebrates substantially under worst-case mixing at normal application rates. The value of the model in ranking relative hazard is most apparent for compounds that are substantially affected by environmental conditions (e.g., pH).

Table 8. Summary of acute toxicity values for the time-integrated median lethal exposure model reported as range from available data sets[a]

Insecticide	LC50' (mg L^{-1} d) (number of data sets)				
	Daphnia magna	*Gammarus lacustris*[b]	Aquatic insects[c]	*Chironomus* spp.	*Culex* spp.
Synthetic pyrethroids					
Cypermethrin	0.002 (1)	0.00006–0.00013 (2)	0.0006Cd (1)	0.0002 (1)	0.00007 (1)
Deltamethrin	0.008–0.01 (2)	0.00002p (1)	0.000028Bp (1)	0.00023–0.00029 (2)	0.00002–0.00019 (2)
Fenvalerate	0.0066 (1)	0.00012–0.00014ps (3)	0.00039–0.0037E, Pd (2)	0.0042–0.015 (2)	0.0040–0.0047 (2)
Permethrin	0.0004 (1)	0.0005p (1)	0.00011Br (1)	NA	0.0014–0.0030 (2)
Other insecticides					
Azinphos methyl	0.0023–0.0040 (2)	0.00056–0.0006 (3)	0.006–0.016, Pc (3)	NA	NA
Carbaryl	0.01–0.014 (2)	0.040–0.044 (3)	0.019–0.030 Pc (3)	0.020 (1)	0.480 (1)
Carbofuran	0.020 (1)	NA	NA	NA	NA
Chlorpyrifos	0.016 (1)	0.00044–0.0008 (3)	0.036–0.050, Pc (3)	0.0005–0.0012 (2)	0.0012–0.0016 (3)
Diazinon	0.0025–0.003 (2)	0.800–1.0	0.100–0.155, Pc (3)	NA	0.024–0.031 (2)
Dimethoate	5.000–12.800 (3)	0.800–0.900 (3)	0.172–0.510, Pc (3)	NA	NA
Malathion	0.0009–0.002 (2)	0.0036–0.004 (3)	0.035–0.040, Pc (3)	0.0030–0.0074 (2)	0.032–0.034 (5)
Methoxychlor	0.0037 (1)	0.0026–0.0047 (3)	0.0056–0.030, Pc (3)	0.0065 (1)	0.0089–0.0189 (2)
Phosmet	0.0112 (1)	0.003–0.008f (2)	NA	NA	NA

[a]NA, no appropriate data set available.
[b]Index calculated from toxicity data for *Gammarus fascitis* (f), *Gammarus pulex* (p), *Gammarus pseudolimnaeus* (ps).
[c]Index calculated from toxicity data for *Cloeon dipterum* (Cd). *Baetis parvus* (Bp), *Baetis rhodani* (Br), *Ephemerella* sp. (E), *Pteronarcys dorsata* (Pd), *Pteronarcys californica* (Pc).

The relative hazard ranking approach provides a mechanism for quantitatively ranking toxicity under simulated conditions. In this case, the insecticides can be divided into three groups based on RHI values. Compounds with RHI values < 10 probably pose a low hazard to aquatic macroinvertebrates at recommended application rates. Insecticides with RHI values from 10 to 100 probably pose a moderate hazard to exposed macroinvertebrates. Compounds with calculated RHI values > 100 probably pose a high hazard under the acute exposure model.

Estimating the Magnitude of Impact Using Benchmarks

Although the hazard ranking index is effective in grouping chemicals by the potential hazard they pose, the index itself does not provide a measure of the possible severity of toxic impact under the described exposure conditions. Microcosm, mesocosm, and field studies (see Chapter 9) are required to provide quantitative estimates of impact. Carbaryl and permethrin, which have moderate and high relative hazard rankings, respectively, are possible benchmarks to be used to calibrate the expected impact under field conditions. Although these compounds have not previously been considered in a benchmark capacity, they have been subjected to long-term controlled field studies that provided data relating exposure to the magnitude and duration of effects. A summary of field test data is presented in Table 10. A moderately toxic insecticide such as carbaryl is expected to deplete macroinvertebrate numbers to approximately 30–80% of normal levels and biomass to 60–75% of pretreatment standing crop when applied to ponds at levels similar to those expected from aerial application. Certain species that are important to duckling feeding (e.g., chironomids) will remain relatively unaffected. The lag period (recovery) before the reestablishment of normal numbers and biomass is expected to be 3–6 mo. A high-impact compound such as permethrin is expected to reduce pelagic invertebrate population abundance by >80%. Benthic macroinvertebrate biomass in treated systems may be reduced by >70%. Similar dramatic declines in macroinvertebrate populations were also found in controlled field studies with deltamethrin. Morril and Neal (1990) reported a 99% decrease in chironomid larvae following application of deltamethrin at the recommended agricultural application rate. It appears that only aquatic worms, mollusks, and rotifers will be relatively unaffected by the highly toxic pyrethroid compounds at their recommended application levels. Full recovery of community numbers and biomass is expected to be delayed for more than 6 mo and may take longer than 1 y depending on the availability of mobile colonizers.

Insecticides with RHI values for acute exposure conditions consistently greater than 100 are expected to produce effects equivalent in severity to those summarized for permethrin. Cypermethrin, deltamethrin, permethrin, azinphos methyl, and chlorpyrifos are associated with the high-impact benchmark and malathion is predicted to induce similar effects under acidic conditions. Compounds with values between 10 and 100, such as methoxychlor, should produce moderate impacts similar to those described for carbaryl. Carbofuran and dimethoate (RHI < 10) present little or no risk. It should be cautioned, how-

Table 9. Lists of insecticides reflecting relative hazard rankings presented as a range in parentheses. The RHI is based on the acute exposure scenario predictions and short-term laboratory toxicity test data[a] for acidic and alkaline ponds

Acute exposure risk scenario	
Acidic pond (RHI)	Alkaline pond (RHI)
Permethrin (> 100)	Permethrin (> 100)
Azinphos methyl (> 100)	Chlorpyrifos (> 100)
Chlorpyrifos (> 100)	Deltamethrin (> 100)
Deltamethrin (> 100)	Methoxychlor (> 100)
Methoxychlor (> 100)	Cypermethrin (> 100)
Cypermethrin (> 100)	Azinphos methyl (> 100)
Phosmet (10 to 100)	Fenvalerate (10 to 100)
Malathion (10 to 100)	Diazinon (10 to 100)
Fenvalerate (10 to 100)	Phosmet (< 10)
Carbaryl (10 to 100)	Malathion (< 10)
Diazinon (10 to 100)	Carbaryl (< 1)
Carbofuran[b] (< 10)	Dimethoate (< 1)
Dimethoate (< 1)	Carbofuran[b] (< 1)

[a]Toxicity ranking based on average 24- and 48-h LC50 values for five taxa; model rankings are a weighted average or ranking for individual taxa accounting for the number of compounds for which toxicity data was available (the weighing factor is the ratio of the total number of compounds over the number of compounds for which we have RHI value).

[b]Ranking based solely on toxicity data for *Daphnia*.

Table 10. Summary of expected adverse response of aquatic invertebrate populations in treated ponds for suggested benchmark insecticides. Carbaryl impacts are classified as moderate and permethrin impacts as severe

Insecticide	Reductions in aquatic invertebrates[a]: Pretreatment (control)	Posttreatment	Percent reduction	Time required to recover	General trends	Comments	References
Carbaryl (840 g ha^{-1})[b]	Benthic						
	~300–1000 ind. m^{-2}	~200 ind. m^{-2}	~30–80	Little effect on chironomids; 3 wk for ephemeropterans; up to 1 yr amphipods; 20 wk for total numbers of benthic organisms; no data on recovery of open-water community	Elimination of sensitive populations (amphipods) for an extended period; chironomids become dominant; seasonal reductions in available macroinvertebrate biomass	Ducklings on treated ponds weighted 30% less than those on untreated ponds and had only 40% the growth rate of non-food-stressed ducklings over the initial 10 d posttreatment associated with treatment	Gibbs et al. (184); Hunter et al. (1984)
	~1.5 g dry wt. m^{-2}	~0.4 g dry wt. m^{-2}	~75				
	Surface water						
	~60–180 ind. m^{-3}	~20–30 ind. m^{-3}	~70				
	~0.008 g m^{-3}	~0.003 g m^{-3}	~60				
Permethrin (35–140 g h^{-1})[b]	Benthic						
	~2,800 ind. m^{-2}	~200 ind. m^{-2}	>90	Oligochaetes and mollusks least affected; up to a year for many crustaceans and insects; 16 wk for macrozooplankton; rotifers actually increased in numbers	Elimination of most free-swimming and benthic arthropods; invertebrate population may remain depleted for a year or more	Alterations of reproduction potential and transient effects on growth rates of fish in treated systems	Kingsbury (1976); Kingsbury and Kreutzeizer (1979); Kaushik et al. (1985)
	1.4 g dry wt. m^{-2}	~0.1 g dry wt. m^{-2}	>90				
	Open water						
	~300–1500 ind. m^{-3}	~100–150 ind. m^{-3}	80				
	Near macrophytes						
	~0.5–2.8 g dry wt. m^{-3}	>0.3 g dry wt. m^{-3}	70				
	Zooplankton						
	~100–200 ind. L^{-1}	>1 ind. L^{-1} (cladocerans, copepods)	99				

[a] Estimated average reduction for the initial 30 d posttreatment; biomass if unreported was estimated from taxonomic composition and abundance data from authors and mean dry wt. data from Driver et al. (1974).
[b] Treated at application rates approximately equivalent to those presently recommended for crop protection.

ever, that carbofuran was ranked with a very inadequate database.

Hazard to Ducklings

The data discussed earlier established that duckling growth in natural habitats during the initial 2 wk for most species is highly dependent on protein provided by the consumption of aquatic macroinvertebrates. Dependence is much longer for species such as the lesser scaup and pintail (Sugden, 1973, 1982). Therefore, a direct relationship between aquatic macroinvertebrate abundance (biomass) and duckling growth is expected during the initial 2 to 4 wk. In addition, the chances of survival in slow-growing individuals are low because resistance to chilling increases in proportion to bird weight (Cross, 1966) and plumage development (Sturkie, 1954). Also, increased searching activity to compensate for low prey availability is energetically inefficient and may place the ducklings at an increased risk of predator attack while feeding (J.-L. Desgranges, CWS Quebec Region, personal communication). Overland migrations by hungry ducklings in search of sloughs with adequate food resources also make broods vulnerable to predation losses. Hence, fast growth in ducklings is ecologically beneficial, and conversely, slow or no growth places broods at high risk of mortality.

Experimental Evidence of Indirect Impacts of Insecticides on Duckling Growth

Carbaryl has been implicated as an indirect cause of reduced duckling growth because of its effect on the quality and quantity of duckling food resources (Hunter et al., 1984). Therefore, insecticides of equal or greater relative hazard are expected to affect ducklings adversely. The work by Hunter et al. (1984) is significant in that it was the first published study that looked at the link between aquatic system response to insecticide stress and the impact on ducklings feeding in the stressed ecosystem. Hunter and colleagues aerially sprayed a pond with 840 g carbaryl/ha and placed a brood of 7-wk-old ducklings on this treated pond and a similar unsprayed reference pond. The ducklings received no food other than that obtained from natural foraging on the ponds. Food-stressed ducklings on a carbaryl-treated pond grew at only half the rate of the matched brood on a reference pond and at 30% of their optimal rate.

Recent evidence (M. Haramis, U.S. Fish and Wildlife Service, Patuxent, Maryland, personal communication) suggests that, under conditions of more massive resource depletions, ducklings may die. On experimentally acidified artificial ponds, 50% of black ducklings lost weight and 25% died; only 25% were able to maintain their weight over a 10-d experimental period. In contrast, 85% of the control brood gained weight, and the remainder maintained their weight. Black ducklings on a natural lake with a large fish population and low abundance of open-water macroinvertebrates also grew less rapidly than experimental broods on more productive lakes without fish (J.-L. Desgranges, CWS Quebec Region, personal communication). In all of these experimental studies, there was documentation of substantially increased searching-moving activity in stressed ducklings and of loss of cohesiveness within the broods.

In another study, the growth and survival of mallard ducklings were monitored in enclosures on deltamethrin-treated and control ponds (Neal, 1987). Neal aerially applied 5.2 and 5.7 g deltramethrin/ha, respectively, to the two treated ponds. Broods of seven 2-wk-old ducklings were placed on the treated ponds and on unsprayed reference ponds. For the first 5 d after spraying, the ducklings on the treatment pond gained weight approximately twice as fast as those on the control pond. This artifact was possibly because of the abundance of dead and dying macroinvertebrates in the treatment pond. For the next 2 wks, the ducklings on the treatment pond either lost or just maintained their weight, and most began to die toward the end of this period. In contrast, the ducklings on the control pond continued to increase in weight throughout the study period, and all survived. A second growth trial that started at 12 d after spraying resulted in ducklings from the treated pond losing weight and dying within 6 d. In contrast, all control birds survived and four of the five steadily increased in weight.

Although the direct experimental evidence is limited, the ecological concept of carrying capacity (that there is an upper limit of population size that the environment can sustain) strongly suggests a relationship between low macroinvertebrate standing crop (resulting from chemical stress) and reduced duckling growth and survival that may allow crude predictions of secondary trophic impacts based on RHI values. As a first approximation, the sample estimates of macroinvertebrate biomass from Hunter et al. (1984) and Neal (1987) were regressed against duckling growth rate. This approach is, at present, unsatisfactory for several reasons. The most obvious problem is lack of sufficient data to compare growth with food supply at or below maintenance levels. Also problematic are the very low macroinvertebrate biomass estimates that Hunter and colleagues associated with their uncontaminated test ponds. Their techniques of sweeping the surface and to a depth of only 15 cm provided biomass estimates a factor of 10 lower than others reported for surface and water column samples. Differentiation between treatment and control biomass would be exceedingly difficult if the low estimates were correct considering inherent natural variability in macroinvertebrate community standing crop. Under such conditions, relating estimates of prey availability to growth is difficult. However, the value of developing such an empirical relationship is obvious and should be further explored experimentally with the appropriate considerations of sampling design.

Duckling Growth Models

Alternatively, the problem of estimating hazards to ducklings may be approached through energetic considerations. Bioenergetic requirements have been determined for some of the major species, like mallards (Sugden et al., 1981), lesser scaups (Sugden and Harris, 1972; Lightbody and Ankney, 1984), and black ducks (Penney and Bailey, 1970; Reinecke, 1979). These studies have established relationships between dietary intake, growth rate, live weight, and age of ducklings for laboratory-reared birds. There are some striking similarities in shape among the growth curves for various species, although the slope of the growth plots can differ slightly among species. Ducklings gain weight rapidly until about 7 wks of age. Increases in live weight are generally linear during the maximum growth period age of 1 to 4 weeks. At the age of 2 to 3 weeks the maximum rate of weight change is achieved. This is also the age at which the greatest quantities of invertebrates are eaten by many species (Sugden, 1973). Food intake increases to a maximum in week 4 or 5 and declines thereafter. The efficiency of converting food to growth, expressed as the ratio of live weight change to dry matter intake, decreases linearly with increasing age and ranges from a low of 0.35 for week 4 to a high of 0.60 for week 2 for the mallard ducklings' critical first 4 wk of growth. Bioenergetic efficiencies do differ between species but differences are not large and appear to be related to live weight as well as to the age of the ducklings.

If the relationships between age, live weight, dry matter intake, and optimal growth rate are known and relatively comparable for a number of species, a generalized growth rate can be calculated as a function of dietary intake as long as certain assumptions on the efficiency of converting dry weight of food into live weight of duckling are made. Estimates of necessary dietary intake to meet specified growth requirements can then be related to available macroinvertebrate biomass using relationships involving average time spent feeding, feeding success per unit time, consumption rate per prey density, and energy-biomass conversions for macroinvertebrates.

The suggested duckling growth models are based on growth, consumption, and feeding success data for the mallard as a representative dabbler and the lesser scaup as a representative diver with a long period of macroinvertebrate dependence (Sugden and Harris, 1972; Pehrsson, 1979).

Mallard Growth

Weight is related to age for weeks 1 to 4 through the following regression equation:

$$\text{Weight (g)} = 187.55 \text{ age (wk)} - 86.18, \quad r = 0.99 \tag{3}$$

Dry matter intake is related to age for weeks 1 to 4 through the following regression equation:

$$\text{Dry matter intake (g dry wt. wk}^{-1}\text{)} = 109.40 \text{ age (wk)} + 60.43, \quad r = 0.98 \tag{4}$$

Food conversion efficiency is calculated from the ratio of live weight change (LWC) to dry matter intake (DMI) as a function of age:

$$\frac{\text{LWC (g live wt. wk}^{-1}\text{)}}{\text{DMI (g dry wt. wk}^{-1}\text{)}} = -0.10 \text{ age (wk)} + 0.80, \quad r = -0.99 \tag{5}$$

Energetic costs in terms of dry matter intake at a specific age are estimated as

$$\text{Energetic costs} = \left(1 - \frac{\text{LWC}}{\text{DMI}}\right) \times \text{DMI} \tag{6}$$

Mallard ducklings are assumed to feed exclusively on aquatic macroinvertebrates until at least 2 wk of age.

Lesser Scaup Growth

Weight is related to age for weeks 1 to 4 through the following regression equation:

$$\text{Weight (g)} = 101.50 \text{ age (wk)} - 58.08, \quad r = 0.99 \tag{7}$$

Dry matter intake is related to age for weeks 1 to 4 through the following regression equation:

$$\text{Dry matter intake (g dry wt. wk}^{-1}\text{)} = 104.00 \text{ age (wk)} - 23.67, \quad r = 0.99 \tag{8}$$

Food conversion efficiency is calculated from the ratio of live weight change to dry matter intake as a function of age:

$$\frac{\text{LWC (g live wt. wk}^{-1}\text{)}}{\text{DMI (g dry wt. wk}^{-1}\text{)}} = -0.095 \text{ age (wk)} + 0.675, \quad r = 0.99 \tag{9}$$

Energetic costs in terms of dry matter intake are estimated as with the mallard. Lesser scaup ducklings are assumed to feed exclusively on aquatic macroinvertebrates until at least 4 weeks of age.

The relationship between consumption and prey density was developed from work by Sugden (1971), who measured feeding rates for 3-week-old lesser scaup ducklings in laboratory tanks under conditions of constant prey (*Gammarus*) density. He reported that the consumption rate increased linearly with increasing *Gammarus* densities from 25 to 200 individuals m^{-3} (0.1 to 0.8 g dry weight m^{-3}). The number of prey taken per dive increased linearly with increased prey density; however, the number of dives increased with density only between 25 and 50 individuals m^{-3} and then plateaued at >4 individuals min^{-1}. Below prey densities of 25 individuals/m^3 the behavioral reinforcement for feeding was weak and ducklings tended not to seek prey and feed.

The consumption rate for feeding ducklings can be expressed in terms of prey density as follows:

$$\text{Consumption rate (prey min}^{-1}\text{)} = 0.165 \text{ prey density (prey m}^{-3}\text{)} + 3.465, \quad r = 0.95 \tag{10}$$

or

$$\text{Consumption rate (mg dry wt. min}^{-1}) = 0.165 \text{ standing biomass (mg dry wt. m}^{-3}) + 13.86, \quad r = 0.96 \qquad (11)$$

Based on estimated energy requirements, Sugden (1971) concluded that 25 individuals m^{-3} averaging 4 mg dry wt. per individual (~12 calories per individual) was insufficient to support a 3-week-old duckling even if it fed continuously for 24 h. He estimated that a duckling could conceivably obtain sufficient food if it consumed prey approximately 40% of the time at a prey density of 50 individuals m^{-3} (0.2 g dry wt. m^{-3}). At 50 individuals m^{-3}, approximately 12 prey are consumed per minute according to the regression (8) from Sugden's data. Pehrsson (1979) estimated that wild mallard ducklings less than 20 d old captured between 11 and 13 prey items per minute in productive lakes without fish.

The estimate of 4 mg average dry weight for prey used by Sugden (1971) may be appropriate for aquatic macroinvertebrate prey in prairie potholes according to mean dry weight data for taxa in prairie sloughs (Driver et al., 1974). Armstrong and Nudds (1985) reported a high frequency of large macroinvertebrates 7 to 12 mm in length associated with aquatic vegetation in duck feeding areas.

Ducklings between 1 and 4 weeks of age under natural and captive conditions spend between 40% and 60% of their time on the water feeding (E. Driver, CWS Prairie Migratory Bird Research Center, personal communication). As they are visual predators, most feeding is assumed to occur during daylight hours, although Swanson and Sargeant (1972) reported that mature birds and ducklings may feed actively at night during periods of insect emergence. For the model, time spent feeding is assumed to be constant and equal to 10 h daily, a value representing 60% of the nearly 16 h of summer sunlight. There are some data to suggest that ducklings may feed longer on food-depleted aquatic ecosystems (J. L. Desgranges, CWS Quebec Region, personal communication). Hunter et al. (1984) reported that broods on carbaryl-treated ponds spent approximately 10% more time searching for food than did a brood on an uncontaminated pond. More time spent searching for prey means greater energetic costs expended by ducklings. However, the assumption of a constant daily feeding period may help balance the assumption of constant maintenance costs.

Assumptions used in the duckling growth models are as follows:

- There is a direct relationship between aquatic-macroinvertebrate density/standing biomass and duckling growth and survival.
- Mallard ducklings consume only aquatic macroinvertebrates for the initial 2 weeks of growth and lesser scaup ducklings are entirely dependent on aquatic macroinvertebrates during the first 4 weeks of growth.
- Growth rate in captive ducklings fed ad libitum under laboratory conditions is approximately equal to the optimal growth rate of ducklings in the wild.
- Consumption rate is related to prey density/biomass according to the linear regression (10) or (11).
- Time spent feeding by ducklings is constant and equal to 10 h daily.
- Potential dietary intake is directly related to age as in Eqs. (4) and (8).
- Energetic costs (g wk^{-1}) are constant regardless of available food supply and decrease with the age and weight of the duckling between the second and fourth weeks.

The standing crop of macroinvertebrates in a pond is biomass or density per unit volume of pond water at any instant in time. If the additions to the community through growth, reproductive recruitment, and immigration are equal to the losses through death, predation, and emigration, the standing crop will remain stable over time. If consumption rate is constant over the feeding period and standing macroinvertebrate biomass is not readily depleted by feeding activity, then

$$\text{Dry matter intake (g dry wt. per wk)} = 4200 \text{ (min wk}^{-1} \text{ feeding time)} \times \text{consumption rate (mg dry wt. min}^{-1}) \qquad (12)$$

In order to predict the macroinvertebrate standing crop necessary to support a specific growth rate, consumption rate (dry matter intake) is used to estimate macroinvertebrate standing crop based on the previously stated assumptions:

$$\text{Macroinvertebrate standing biomass (g dry wt. m}^{-3}) = 0.0013\text{DWI(g dry wt. m}^{-3}) - 0.044 \qquad (13)$$

and

$$\text{Macroinvertebrate density (numbers m}^{-3}) = 0.329\text{DWI(g dry wt. m}^{-3}) - 10.99 \qquad (14)$$

Two examples of predictions from the duckling growth models are presented in Table 11 and plotted for mallards in Figure 6. The plot shows optimal growth, marginal growth, and no-growth lines and data for broods reared on insecticide-treated and untreated (natural) ponds previously discussed. The optimal growth rate for a 2-wk-old mallard, 167 g live wt. wk^{-1}, can be supported by a continuous supply of macroinvertebrates at approximately 80 individuals m^{-3} or 0.32 g dry wt. m^{-3}. Approximately 39 individuals m^{-3} or 0.16 g dry wt. m^{-3} are required to support marginal growth (25% of potential) of 42 g wk^{-1}, which is similar to the growth estimated by Hunter et al. (1984) for broods on a carbaryl-treated pond. After an initial week of rapid growth, ducklings on the deltamethrin-treated pond grew at less than the marginal rate and began to die (Neal, 1987). At macroinvertebrate densities below 26 individuals

m^{-3} (0.10 g dry wt. m^{-3}), no growth or weight loss is predicted to occur, and ducklings are expected to lose weight and die rapidly if maintained under the depleted resource conditions.

The optimal growth rate for a 4-wk-old lesser scaup is less than that of the 2-week-old mallard due to additional maintenance costs. The estimated standing crop of macroinvertebrate prey necessary to support the scaup (118 individuals m^{-3} or 0.47 g dry wt. m^{-3}) is approximately 1.5 times the food resource base of the 2-week-old mallard. Marginal growth requires 89 individuals m^{-3} or 0.36 g dry wt. m^{-3} and maintenance requires 80 individuals m^{-3} or 0.32 g dry wt. m^{-3}. The model suggests that a continuous supply of prey at a standing biomass of approximately 4.2 g dry wt. m^{-3} would be necessary to support optimal growth for an average brood of nine lesser scaup ducklings 4 weeks of age. Based on available data, 4.2 g dry wt. m^{-3} is near the middle of the range of reported average biomass values for prairie sloughs. Gerking (1962) estimated that the production of benthic fauna in a shallow slough during summer months is twice the standing biomass. Based on a turnover ratio (production of biomass/standing biomass) of two per month or approximately 0.5 per week, approximately 2.1 g dry wt m^{-3} macroinvertebrate biomass can be generated within a week. If it is assumed that duckling feeding equals production and therefore does not deplete the resource supply, the brood would have to search approximately 1680 m^3 (392 g dry wt. wk^{-1} individual^{-1} $\times$ 9 individuals $\div$ 2.1 g dry wt. wk^{-1} m^{-3} = 1680 m^3) of pond per week to obtain the necessary dry matter intake for optimal growth (see Table 11 for DMI requirements). This estimate of habitat food capacity appears reasonable considering the dimensions of semipermanent potholes (1000 m^2 average surface area $\times$ 2 m average depth = 2000 m^3 water volume) preferred by Scaups. Approximately 1195 m^3 (279 g dry wt. wk^{-1} individual^{-1} $\times$ 9 individuals $\div$ 2.1 g dry wt. wk^{-1} m^{-3} $\approx$ 1195 m^3) of pond habitat would be required for optimal growth of a mallard brood of nine ducklings 2 weeks of age under the same conditions as described above for the lesser scaup.

A crude estimate of the available standing crop biomass in the carbaryl-treated ponds in this assessment is approximately 0.4 g dry wt. m^{-3} based on extrapolation of data from Gibbs et al. (1984). Macroinvertebrates of less than 0.1 g dry wt. m^{-3} were found in ponds treated with permethrin (Kingsbury, 1976; Kingsbury nad Kreutzweiser, 1979) and deltamethrin (Neal 1987). Based on the estimated 0.32 g dry wt. m^{-3} (see Table 11) necessary to support a single duckling at the rate of optimal growth, available resources after the macroinvertebrate kill associated with insecticide exposure were far short of the requirements for five to seven member test broods. The brood on the treated carbaryl pond would be required to search nearly 5320 m^3 1 wk (152 g dry wt. wk^{-1} individual^{-1} $\times$ 7 individuals $\div$ 0.2 g dry wt. wk^{-1} m^{-3} = 5320 m^3) to obtain sufficient food to grow at 25% of their potential and not to further deplete the biomass of macroinvertebrates.

Relating predictions of necessary macroinvertebrate biomass from the duckling growth models to insecticide stress may be crudely done with the benchmarks in Table 10 assuming an average macroinvertebrate standing biomass of approximately 4.0 g dry wt. m^{-3} and a turnover rate of 0.5 per week for an undisturbed pond (Gerking, 1962). Compounds with an RHI equivalent to that of carbaryl should reduce standing biomass levels of macroinvertebrates in productive potholes to less than 1.0 g dry wt. m^{-3} (0.5 g dry wt. m^{-3} new biomass production per week). This prey resource level is predicted to be insufficient to support marginal growth in an average brood of nine lesser scaups (or mallards) in a pothole of less than 5490 m^3 (305 g dry wt. wk^{-1} individual^{-1} $\times$ 9 individuals $\div$ 0.5 g dry wt. wk^{-1} m^{-3} = 5490 m^3). Compounds calculated to have a relative hazard equivalent to that of permethrin are expected to deplete macroinvertebrate standing stocks by >90%, reducing expected biomass to <0.4 g dry wt. m^{-3}. A standing biomass of 0.4 g dry wt. m^3 (or biomass production of 0.2 g dry wt. m^{-3} wk^{-1}) is predicted to be just sufficient to support marginal growth in a single lesser scaup (or mallard) duckling and totally inadequate for a normal-size brood (1 wk old) in a pond with anything less than

Table 11. Predicted macroinvertebrate standing crop necessary to support various levels of growth rate for mallard and lesser scaup ducklings

Duck species	Dry matter intake (g dry wt. wk^{-1})	Growth rate (g live wt. wk^{-1})	Macroinvertebrate standing crop necessary to support growth rate (number m^{-3})	(g dry wt. m^{-3})
Mallard, 2 wk old (< 285 g)	279	167 (optimal)	80	0.32
	152	42 (marginal)[a]	39	0.16
	112	0 (no growth)	26	0.10
Lesser scaup, 4 wk old (<348 g)	392	116 (optimal)	118	0.47
	305	29 (marginal)[a]	89	0.36
	276	0 (no growth)	80	0.32

[a] Marginal growth is defined as approximately 25% of potential growth.

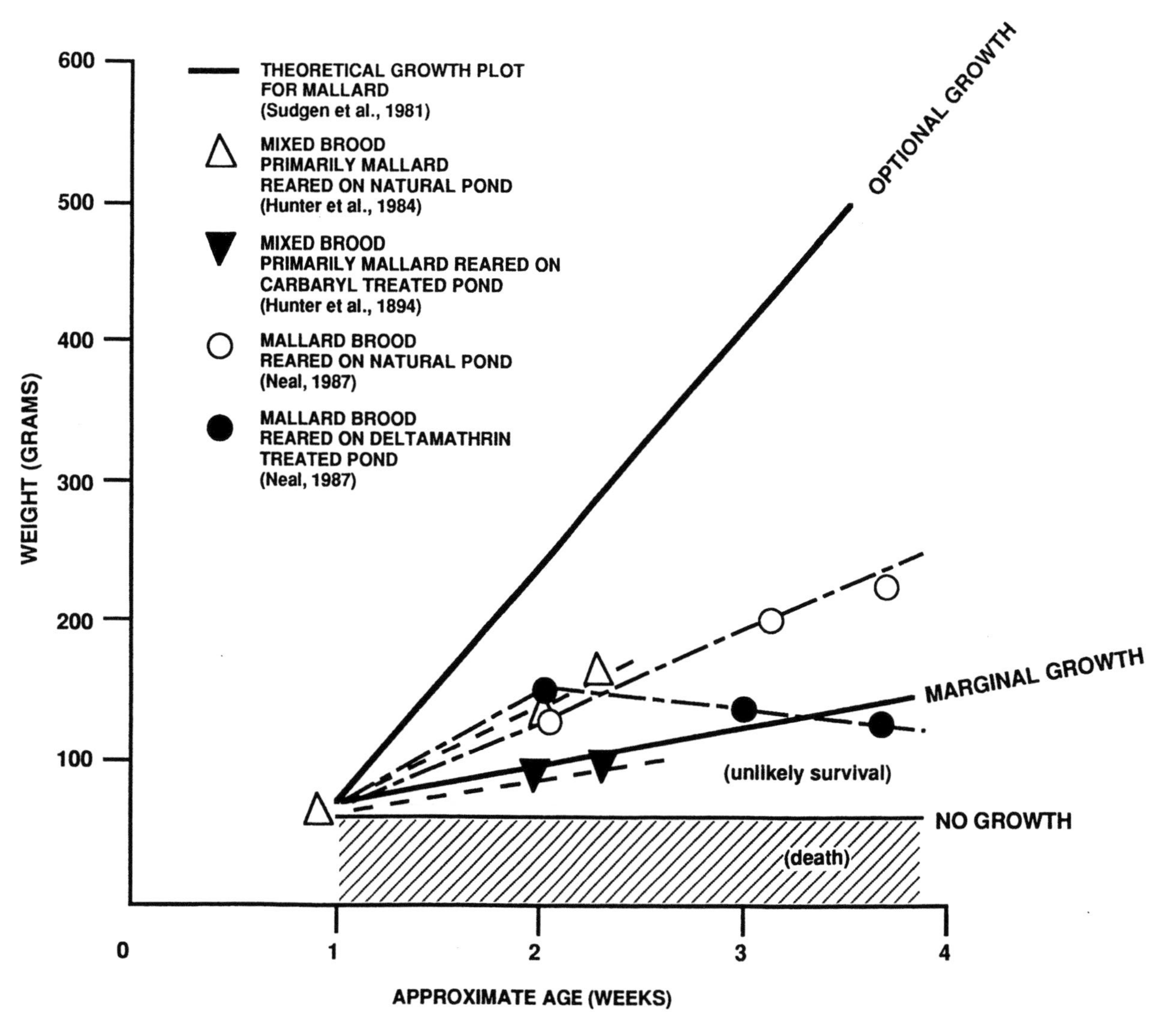

Figure 6. Plot of optimal and marginal growth for mallard duckling and growth data for broods reared on natural and insecticide-treated ponds.

1525 m^3 (305 g dry wt. wk^{-1} $individual^{-1}$ × 1 individual ÷ 0.2 dry wt. m^{-3} wk^{-1} = 1525 m^3) of productive littoral habitat.

Duck Population at Risk

To estimate the size of the duck population at risk from insecticide exposure, it is necessary to refine past estimates of the overlap of ducks with crop type and insecticide use. Since basic information on crop type as well as information on pesticide use is available by census units from agriculture surveys, this level of resolution for the waterfowl data is required. This task is facilitated by the published analysis of CLI capability classes by census divisions (Lands Directorate, 1980).

The ultimate goal is to obtain the number of ducks "associated" (or overlapping) with any one crop type within any one census division. As a first step, the amount of each waterfowl land class (from class 1–5) in each census district was noted. These areas were then corrected to obtain a measure of overlap with agricultural land classes 1–3. These corrected areas provide a very conservative estimate of the overlap between the waterfowl habitat and the best agricultural lands already under cultivation by the late 1960s (the lands most likely to be cropped rather than used as pasture). To each corrected area, the class-specific duck breeding density estimate obtained from Patterson (1978) was then applied. Because of the high degree of data manipulation and high uncertainty involved in setting these density estimates, it was possible to perform this exercise only for the most abundant species, mallards, or for all ducks. The results of this exercise are given in Table 12.

Assigning breeding ducks to a given crop type within any one census division is very difficult. In the absence of the data just mentioned, further assumptions must be made. Within the agricultural landscape, nesting preferentially takes place within the vegetative margin left intact around sloughs or in suitable "upland sites" such as idle grassland, field edges, and roadsides. A limited amount of nesting by dabblers takes place in the cropland itself, depending on crop phenology and timing of cultivation practices. Based on literature values, two crop types appear to differ from the others in their attractiveness to nesting ducks: tame hay (seeded forage) has been reported to support nest densities 10-fold those found in other field crops, whereas fallow fields are typically only a third as attractive as cropped fields. It is useful here to remember that pastureland and its attending nesting duck populations have been excluded from this analysis. The nesting duck populations given in Table 12 are strictly those that are estimated to be nesting in and around cropland and are, therefore, more likely to be affected by agricultural spraying of insecticides.

To calculate the number of ducks at risk with respect to any given spray operation, it is important to distinguish between the birds actually nesting within the crop proper and those nesting in noncrop habitat. Diving ducks do not usually nest in upland habitat and must be treated separately from dabblers. For this analysis it must be assumed that their overlap with agriculture is similar to that of the dabblers but that they nest only in close association with the sloughs and that crop type is independent of slough density. Therefore, divers can be "assigned" to the various crop types on the basis of crop area only.

The assignment of dabblers to various crops is more complex. Whereas cropland and summer fallow in 1981 made up 65% of the total farm area, a more conservative estimate of the noncrop area available to nesting ducks within an intensively cropped landscape is 10% (Higgins, 1977; Sugden and Beyersbergen, 1984). This area comprises lands of poorer agricultural capability hidden within complexes of good agricultural land (CLI classes 1–3) under crop or fallow. Obviously, this 90:10 ratio will vary across the prairies, but in the absence of any hard data, it will be assumed that it applies equally to all census divisions.

After dividing each census division into the seven major crop types (wheat, oats, barley, rye, flax, canola, tame hay) and summer fallow, these areas were in turn divided into crop and "margin." For most census division, these seven crops and summer fallow account for over 95% of the duck populations that are estimated to breed in association with cropland. The only exceptions are a few districts in Man-

Table 12. Estimated number of ducks (thousands) of all species breeding on land currently under crop[a] in the grasslands-parklands region of the Canadian prairie provinces

Census division	Number of ducks (thousands)		
	Alberta	Saskatchewan	Manitoba
1	159	475	NA[b]
2	140	338	NA
3	90	280	31
4	295	250	56
5	250	480	173
6	128	509	113
7	622	348	44
8	189	301	30
9	NA	358	21
10	671	425	16
11	228	388	3
12	NA	293	NA
13	NA	359	29
14	NA	511	27
15	NA	406	51
16	NA	406	51
17	NA	423	52
18	NA	NA	71
19	NA	NA	NA
20	NA	NA	36
Total	2773	6607	989

[a] This includes area in summer fallow.
[b] NA, not applicable.

itoba where other crop types (mainly row crops) account for 20–30% of the cropland and hence the estimated nesting duck population. Realistically, duck nesting in row crops is probably minimal. Nest density factors were used to apportion the nesting birds to different habitat types. The actual values used in the computations are given in Table 13. All margin types were assumed to have the same attractiveness to nesting ducks. The calculation steps described above that are used to estimate the number of dabblers and divers in the various agricultural crops are summarized in Figure 7. The results of these computations are presented in Table 14.

Not surprisingly, the largest proportion of breeding ducks are associated with wheat because it is the dominant prairie crop and it is broadly distributed throughout the region. Provided that the data of Duebbert and Kantrud (1974) on the attractiveness of forage crops are valid, tame hay (e.g., alfalfa) is used by dabbling ducks to a much greater extent then would be expected based on the relatively small area in forage crops in the prairie region. Tame hay and canola hold the largest numbers of nests of dabblers and divers, respectively, when expressed on a "per area" basis. The association of canola and diving ducks is explained by the cocorrelation of the two within the parklands region where canola is a primary crop and there are many permanent and semipermanent ponds, which are the preferred habitat of diving ducks.

This approach, coupled with information on pests and insecticides used on each crop type, was used to estimate the duck populations at risk under three specific insecticide exposure scenarios: a nonoutbreak situation, a major grasshopper outbreak, and an outbreak of a major canola pest. The results of this analysis are presented in Table 15. In an average "nonoutbreak" year, from 2 to 3% of the total Canadian breeding duck population nests in the prairie cropland sprayed with insecticides. Approximately 6% of the total breeding population or over 1,000,000 ducklings (given realistic rates of nest success) are expected to be within the cropland sprayed with insecticide during a major grasshopper outbreak. The consequences of this exposure depend on the toxicity of the insecticide, both to the ducks directly and to their macroinvertebrate food resource, as well as on the extent of contamination of their aquatic habitat. Direct overspray of the sloughs is very likely in the course of aerial application. Sheehan et al. (1987) estimated that 10–25% of the total insecticide use was applied aerially during the last major grasshopper outbreak.

One can get a better sense of the potential effects of insecticide use on prairie waterfowl if these numbers are compared to harvest levels (i.e., mortalities from hunting), a much better studied source of mortality. It is estimated, based on the National Harvest Survey carried out by the Canadian Wildlife Service (Metras, 1986), that the retrieved kill of 11 species of ducks studied in this report averaged just under 1 million birds per year between 1982 and 1985 in the three prairie provinces. Therefore the number of these ducks harvested in the prairies by hunters represents approximately 6% of the Canadian prairie population, which is well within the range of the

Table 13. Nest densities of ducks for various habitats. The value in brackets is the correction factor used for assigning duck nests to the available upland nesting habitat

Habitat type	Nest densities (per ha)	Source
Grain crops and stubble		
Grain stubble	0.05	Duebbert and Kantrud (1974); Higgins (1977)
Mulched stubble	0.02	Duebbert and Kantrud (1974); Higgins (1977)
Standing stubble	0.04	Duebbert and Kantrud (1974); Higgins (1977)
Growing grain	0.01	Duebbert and Kantrud (1974); Higgins (1977)
	Mean = 0.03 (0.03)	
Forage		
Hay	0.37	Duebbert and Kantrud (1974)
Tame hay	0.25	Duebbert and Kantrud (1974)
	Mean = 0.31 (0.3)	
Fallow		
Summer fallow	0.01 (0.1)	Higgins (1977)
Margins (undisturbed land)		
Native grassland	0.11	Kirsh (1969)
Undisturbed grass legume cover	0.84	Duebbert and Kantrud (1974)
Idle grassland	0.16	Higgins (1977)
Idle pasture	0.15	Higgins (1977)
Roadsides	0.36	Higgins (1977)
Idle alfalfa	0.17	Higgins (1977)
Dry wetland basins	0.14	Higgins (1977)
	Mean = 0.28 (0.3)	

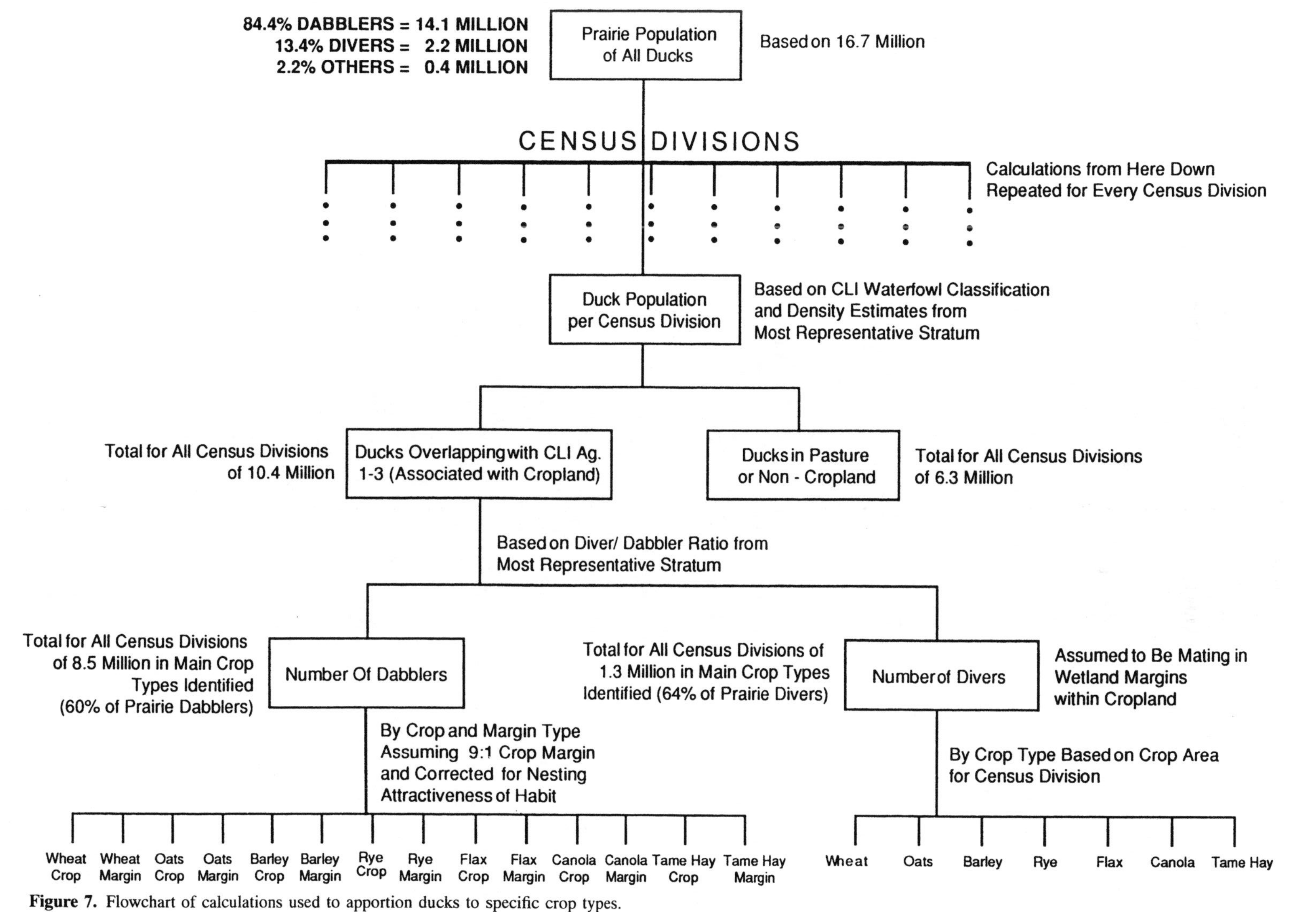

Figure 7. Flowchart of calculations used to apportion ducks to specific crop types.

Table 14. Major crop areas in the grassland-parkland region of the prairie provinces and estimated numbers of ducks nesting within or in association with these crops (expressed as absolute numbers, densities and proportions of the total nesting within these particular crop types)

Category	Wheat	Oats	Barley	Rye	Flax	Canola	Tame hay (alfalfa)	Summer fallow	Total[b]
Area in grassland-parkland region (km^2)[a]	116,194	10,223	42,090	3,898	4,486	11,410	19,718	90,445	298,464
Dabblers, number	3,017,01	259,240	916,748	93,369	86,137	301,916	2,153,502	1,652,79	8,486,713
Dabblers, proportion by crop type	35.5%	3.05%	10.8%	1.10%	1.01%	3.56%	25.4%	19.5%	100%
Dabblers, density (per km^2 of crop)	26.0	25.4	21.8	24.0	19.2	26.5	110	18.3	28.4
Divers, number	485,432	57,337	215,935	15,274	16,306	76,026	97,190	270,356	1,33,856
Divers, proportion by crop type	36.4%	4.30%	16.2%	1.15%	1.22%	5.70%	7.29%	27.8%	100%
Divers, density (per km^2 of crop)	4.18	5.61	5.13	3.92	3.63	6.66	4.93	4.09	4.47

[a] From 1981 Statistics Canada Census of Agriculture. Equals the sum of all census divisions falling within the grassland-parkland area as defined from waterfowl survey strata, Bellrose (1979).

[b] Excludes other minor crops (2.8% of total area in crop or summer fallow).

proportion of prairie birds in insecticide-sprayed areas (Table 15). It is thus clear that the potential effect of pesticide use on prairie duck populations may be on the same scale as the effect of Canadian prairie hunters on these same populations.

Measuring Insecticide Effects on Recruitment

The present assessment establishes a link between aerial spray of insecticides for pest control in the Canadian prairie region, contamination of sloughs, macroinvertebrate mortality, and indirect impacts on duckling growth and survival. This linkage was established with laboratory and field toxicity studies and duckling growth models. Although these studies predict an insecticide hazard under certain conditions, they do not provide a measure of actual impact of insecticide use on duckling recruitment.

Current efforts are under way in the United States and Canada to assess the magnitude of the impact that insecticides have on duckling recruitment when viewed in the context of other factors limiting recruitment, such as drought, loss of acceptable nesting habitat, and predation on eggs and ducklings. Possible monitoring efforts include comparison of duckling recruitment (clutch size and survival) during years of average and heavy insecticide use; modification of the collection of the annual waterfowl transect data to include surveys of farmers for information on pesticide use within the transect (Forsyth, 1989; Grue et al., 1989); and analysis of the geographic overlap between insecticide sales and waterfowl transect data in some parts of the prairie. A comprehensive monitoring program, which tracks other factors limiting recruitment as well as pesticides, will be necessary to determine the magnitude of the reduction in recruitment specifically associated with aerial application of highly toxic insecticides during the critical brood-rearing period for ducklings.

Table 15. Summary of ducks affected under three scenarios of insecticide use

	Scenario of insecticide use		
	Annual use/ no pest outbreak	Grasshopper outbreak	Canola pest outbreak
Dabblers			
Number of birds sprayed directly	320,000	930,000	174,000
Percent of Canadian prairie population	2.3%	6.6%	1.2%
Number of broods sprayed	50,000	142,235	22,500
Divers			
Number of birds sprayed directly	45,000	130,000	43,000
Percent of Canadian prairie population	2.0%	6.2%	2.0%
Total			
Number of birds sprayed directly	495,000	1,060,000	217,000
Percent of Canadian prairie population	3.0%	6.3%	1.3%

LITERATURE CITED

Ali, A., and Mulla, M. S.: Declining field efficacy of chlorpyrifos against chironomid midges and laboratory evaluation of substitute larvicides. *J. Econ. Entomol.*, 71: 778–782, 1978.

Allison, D. T.: Use of Exposure Units for Estimating Aquatic Toxicity of Organophosphate Pesticides. EPA600/3-77/077. Washington, DC: U.S. EPA, 1977.

Anderson, R. L.: Chironomidae toxicity tests, biological background and procedures. In: Aquatic Invertebrate Bioassays, edited by A. L. Buikema Jr., J. Cairns Jr., pp. 70–80. ASTM STP 715. Philadelphia: American Society for Testing and Materials, 1980.

Argauer, R. J.; Mason, H. C.; Corley, C.; Higgins, A. H.; Sauls, J. N.; and Liljedahl, L. A.: Drift of water-diluted and undiluted formulations of malathion and azinphos-methyl applied by airplane. *J. Agric. Entomol.*, 61(4): 1015–1020, 1968.

Armstrong, D. P., and Nudds, T. D.: Factors influencing invertebrate size distributions in prairie potholes and implications for coexisting duck species. Unpublished manuscript, 1985.

Arthur, J. W.: Review of freshwater bioassay procedures for selected amphipods. In: Aquatic Invertebrate Bioassays, edited by A. L. Buikema Jr., J. Cairns Jr., pp. 98–108. ASTM STP 715. Philadelphia: American Society for Testing and Materials, 1980.

Bartell, S. M.; Gardner, R. H.; and O'Neill, R. V.: Ecological Risk Estimation. Boca Raton, FL: Lewis, 1992.

Bartonek, J. C., and Hickey, J. J.: Food habits of canvasbacks, redheads and lesser scaup in Manitoba. *Condor*, 71:280–290, 1969a.

Bartonek, J. C., and Hickey, J. J.: Selective feeding by juvenile diving ducks in summer. *Auk*, 86:443–457, 1969b.

Bascietto, J.; Hinckley, D.; Plafkin, J.; Slimak, M.: Ecotoxicity and ecological risk assessment. Regulatory application at EPA. *Environ. Sci. Technol.*, 24(1):10–15, 1990.

Batt, B. D. J.; Anderson, M. G.; Anderson, C. D.; and Caswell, F. D.: The use of prairie potholes by North American ducks. In: Northern Prairie Wetlands, edited by A. van der Valk, pp. 204–227. Ames: Iowa State University Press, 1989.

Bellrose, F. C.: Ducks, Geese and Swans of North America. Harrisburg, PA: Stackpole Books, 1976.

Bellrose, F. C.: Species distribution, habitats and characteristics of breeding dabbling ducks in North America. Waterfowl and Wetlands an Integrated Review. Madison, WI: Proceedings of a symposium held at the 39th Midwest Fish and Wildlife Conference, 1979.

Buikema, A. L. Jr.; Geiger, J. G.; and Lee, D. R.: *Daphnia* toxicity tests. Aquatic Invertebrate Bioassay, edited by A. L. Buikema Jr., and J. Cairns, Jr., pp. 48–69. ASTM STP 715. Philadelphia: American Society for Testing and Materials, 1980.

Burns, L. A.; Cline, D. M.; and Lassiter, R. R.: Exposure Analysis Modeling System (EXAMS). User Manual and System Documentation. EPA 600/3-82-023. Environmental Research Laboratory. Athens, GA: U.S. EPA, 1982.

Cairns, J. J., and Dickson, K. L.: Field and laboratory protocols for evaluating the effects of potentially toxic wastes on aquatic life. *J. Test. Eval.*, 6:85–94, 1978.

Cairns, J. Jr., and Mount, D. I.: Aquatic toxicology. *Environ. Sci. Technol.*, 24(2):154–161, 1990.

Canadian Agricultural Insect Pest Review. Research program. Research Branch, Agriculture Canada, annual.

Chura, N. J.: Food availability and preferences of juvenile mallards. *Trans. North Am. Wildl. Conf.*, 26:121–134, 1961.

Connell, D. W., and Miller, G. J.: Chemistry and Ecotoxicology of Pollution. New York: Wiley, 1984.

Cooch, F. G.: A continental duck population balance for North America. Unpublished report. Canadian Wildlife Service, Environment Canada, 1981.

Cross, D. J.: Approaches toward an assessment of the role of insect food in the ecology of game birds, especially the partridge (*Perdix perdix*). Ph.D. thesis, Imperial College, University of London, London, England, 1966.

Currier, W. W.; Maccollom, G. B.; and Baumann, G. L.: Drift residues of air-applied carbaryl in an orchard environment. *J. Econ. Entomol.*, 75:1065–1068, 1982.

Driver, E. A.; Sugden, L. G.; and Kovach, R. J.: Calorific, chemical physical values of potential duck foods. *Freshwater Biol.*, 4:281–292, 1974.

Duebbert, H. F., and Kantrud, H. A.: Upland duck nesting related to land use and predator reduction. *J. Wildl. Manage.*, 38:257–265, 1974.

Eisler, R.: Factors affecting pesticide induced toxicity to estuarine fish. U.S. Dep. Int. Fish. Wildl. Ser. Tech. Paper. No. 15, 1970.

EPA: Review of Ecological Risk Assessment Methods. EPA 230/10-88-041. Office of Policy, Planning and Evaluation. Washington, DC: U.S. EPA, 1988.

EPA: Risk Assessment Guidance for Superfund Volume II Environmental Evaluation Manual. EPA 540/1-89-001. Office of Emergency and Remedial Response. Washington, DC: U.S. EPA, 1989a.

EPA: Ecological Assessment of Hazardous Waste Sites: A Field and Laboratory Reference. EPA 600/3-89-03. Environmental Research Laboratory. Corvallis, OR: U.S. EPA, 1989b.

EPA: Summary of Ecological Risks, Assessment Methods and Risk Management Decisions in Superfund and RCRA. EPA 230/3-89-004. Office of Policy, Planning and Evaluation. Washington, DC: U.S. EPA, 1989c.

EPA: Ecological Risk Management in the Superfund and RCRA Programs. EPA 230/3-89-045. Office of Policy, Planning and Evaluation. Washington, DC: U.S. EPA, 1989d.

EPA: Framework for Ecological Risk Assessment. EPA 630/12-92-001. Risk Assessment Forum. Washington, DC: U.S. EPA, 1992.

Forsyth, D. J.: Agricultural chemicals and prairie pothole wetlands: measuring the needs of the resource and the farmer—Canadian perspective. Trans. 54th North Am. Wildl. Nat. Res. Conf., pp. 5–66, 1989.

Gage, S. H., and Mukerji, M. K.: A perspective of grasshopper population distribution in Saskatechewan and interrelationship with weather. *Environ. Entomol.*, 6(3): 469–479, 1977.

Gerking, S. D.: Production and food utilization in a population of bluegill sunfish. *Ecol. Monogr.*, 32:31–87, 1962.

Gibbs, K. E.; Mingo, T. M.; and Courtemanch, D. L.: Persistence of carbaryl (Sevin-4-oil) in woodland ponds and its effects on pond macroinvertebrates following forest spraying. *Can. Entomol.*, 116:203–213, 1984.

Gollop, J. B.: Wetland inventories in western Canada. In International Union of Game Biologists. Transactions of the 6th Congress, October 7–12, 1963, pp. 249–264. London: The Nature Conservancy, 1965.

Grau, C. R.: Avian embryo nutrition. *Fed. Proc.*, 27(1): 185–192, 1968.

Grue, C. E.; DeWeese, L. R.; Mineau, P.; Swanson, G. A.; Foster, J. R.; Arnold, P. M.; Huckins, J. N.; Sheehan,

P. J.; Marshall, W. K.; and Ludden, A. P.: Potential impacts of agricultural chemicals on waterfowl and other wildlife inhabiting prairie wetlands: and evaluation of research needs and approaches. Trans. 51st North Am. Wildl. Nat. Res. Conf., pp. 357–383, 1986.

Grue, C. E.; Tome, M. W.; Messmer, T. A.; Henry, D. B.; Swansen, G. A.; and DeWeese, L. R.: Agricultural chemicals and prairie pothole wetlands: meeting the needs of the resource and the former-U.S. Perspective. Trans. 54th North Am. Wildl. Nat. Rec. Conf., pp. 43–58, 1989.

Hammons, A. S.: Methods for Ecological Toxicology. A Critical Review of Laboratory Multispecies Tests. Ann Arbor, MI: Ann Arbor Science Publishers, 1981.

Harris, H. J.; Sager, P. E.; Regier, H. A.; and Francis, G. R.: Ecotoxicology and ecosystem integrity: the Great Lakes examined. *Environ. Sci. Technol.*, 24(5):598–603, 1990.

Higgins, K. F.: Duck nesting in intensively farmed areas of North Dakota. *J. Wildl. Manage.* 41(2):232–242, 1977.

Hochbaum, H. A.: The Canvasback on a Prairie Marsh. Washington, D.C.: The American Wildlife Institute, 1944.

Hoffman, D. J.; Ratter, B. A.; and Hall, R. J.: Wildlife toxicology. *Environ. Sci. Technol.*, 24(3):276–283, 1990.

Hunter, M. L., Jr.; Witham, J. W.; and Dow, H.: Effects of carbaryl-induced depression in invertebrate abundance on the growth and behavior of American black duck and mallard ducklings. *Can. J. Zool.*, 62:452–456, 1984.

Kaushik, N. K.; Stephenson, G. L.; Solomon, K. R.; and Day, K. E.: Evaluation of the impacts of permethrin on zooplankton communities using limnocorrals. II. Impact on zooplankton. *Can. J. Fish. Aquat. Sci.*, 42:77–85, 1985.

Kenaga, E. E.: Test organisms and methods useful for early assessment of acute toxicity of chemicals. *Environ. Sci. Technol.*, 12(12):1322–1329, 1978.

Kenaga, E. E.: The use of the environmental toxicology and chemistry data in hazard assessment: progress, needs, challenges. *Environ. Toxicol. Chem.*, 1:69–79, 1982a.

Kenaga, E. E.: Predictability of chronic toxicity of chemicals in fish and aquatic invertebrates. *Environ. Toxicol. Chem.*, 1:347–358, 1982b.

Kingsbury, P. D.: Studies of the impact of aerial applications of the synthetic pyrethroid NRDC-143 on aquatic ecosystems. Chemical Control Research Institute Information Report CC-X-127, Canadian Forestry Service, 1976.

Kingsbury, P. D., and Kreutzweiser, D. P.: Impact of double applications of permethrin on forest streams and ponds. Forest Pest Management Institute. Report FPM-X-27, Canadian Forestry Service, 1979.

Kirsh, L. M.: Waterfowl production in relation to grazing. *J. Wildl. Manage.*, 33(4):821–828, 1969.

Krapu, G. L.: Feeding ecology of pintail hens during reproduction. *Auk*, 91:278–290, 1974a.

Krapu, G. L.: Foods of breeding pintails in North Dakota. *J. Wildl. Manage.*, 38(3):408–417, 1974b.

Krapu, G. L.: Nutrition of female dabbling ducks during reproduction. In: Waterfowl and Wetlands. An Integrated Review, edited by T. A. Bookout. Madison, WI. Proc. Symp. 39th Midwest Fish and Wildlife Conf., pp. 59–69, 1979.

Krapu, G. L., and Swanson, G. A.: Some nutritional aspects of reproduction in prairie nesting pintails. *J. Wildl. Manage.*, 39(1):156–162, 1975.

Lands Directorate: The Canada Land Inventory. Objectives, Scope, Organization. Environment Canada, 1978.

Lands Directorate: Waterfowl capability by province, census division breakdown. Land and Water. Man./Sask./Alta./B.C. Publication No. R002250, Environment Canada, 1980.

Levin, S. A.; Harwell, M. A.; Kelly, J. R.; and Kimball, R. D. (eds): Ecotoxicology Problems and Approaches. New York: Springer-Verlag, 1989.

Lightbody, J. P., and Ankney, C. D.: Seasonal influences on the strategies of growth and development of canvasback and lesser scaup ducklings. *Auk*, 101:121–133, 1984.

Mackay, D., and Paterson, S.: Fugacity models for predicting environmental behavior of chemicals. Report to Environment Canada under contract OSU 81-00163, 1982.

Marcström, V.: Mallard ducklings (*Anas platyrhynchos* L.) during the first days after hatching. A physiological study with ecological considerations and comparison with capercaillie chicks (*Tetrao urogallus* L.). *Vittrevy*, 4(5):342–370, 1966.

Metras, L.: Migratory birds killed in Canada during the 1985 season. Canadian Wildlife Service Progress Note No. 166, Environment Canada, 1986.

Millar, J. B.: Some characteristics of wetland basins in central and southwestern Saskatchewan. Saskatoon Wetland Seminar. Canadian Wildlife Service Report Sales No. 6, pp. 73–101, 1969.

Mineau, P.; Sheehan, P. J.; and Baril, A.: Pesticides and water fowl on the Canadian Prairies: A pressing need for research and monitoring. ICBP Tech Publ No. 6:133–147, 1987.

Moore, N. W.: Pesticides in the environment and their effects on wildlife. J. Appl. Ecol. 3(suppl), 1966.

Moriarty, F.: Ecotoxicology. The Study of Pollutants in Ecosystems. London: Academic Press, 1983.

Moriarty, F.: Ecotoxicology. The Study of Pollutants in Ecosystems, 2nd ed. London: Academic Press, 1988.

Morril, P. K., and Neal, B. R.: Impacts of deltamethrin insecticide on choronomidae (Diptera) of prairie ponds. *Can. J. Zool.*, 68:289–296, 1990.

Neal, D. R.:, The effects of deltamethrin on prairie ponds and duck populations. Proceedings Recent Research on the Environmental Chemistry and Environmental Toxicology of Synthetic Pyrethroid Insecticides. Ottawa, Ontario: Environment Canada, 1987.

Noyes, J. H., and Jarvis, R. L.: Diet and nutrition of breeding female redhead and canvasback ducks in Nevada. *J. Wildl. Manage.*, 49(1):203–211, 1985.

NRCC: A screen for the relative persistence of lipophilic organic chemicals in aquatic ecosystems. An analysis of the role of a simple computer model in screening. NRC Associate Committee on Scientific Criteria for Environmental Quality, NRCC 18570, National Research Council of Canada, 1981.

OECD: OECD guidelines for testing of chemicals. The Organization for Economic Cooperation and Development, 1981.

Patterson, J. H.: Canadian Waterfowl Management Plan. Unpublished Report. Canadian Wildlife Service. Environment Canada, 1978.

Pehrsson, O.: Feeding behavior, feeding habitat utilization, and feeding efficiency of mallard ducklings (*Anas platyrhynchos* L.) as guided by a domestic duck. Viltrevy (Swedish Wildlife) 10(8):193–215, 1979.

Pehrsson, O.: Relationships of food to spatial and temporal breeding strategies of mallard in Sweden. *J. Wildl. Manage.*, 48(2):322–339, 1984.

Penney, J. G., and Bailey, E. D.: Comparison of energy requirements of fledgling black ducks and coots. *J. Wildl. Manage.*, 34(1):105–114, 1970.

Perret, N. G.: The spring and summer foods for the common mallard (*Anas platyrhynchos* L.) in south-central Manitoba. M.Sc. thesis, University of British Columbia, Vancouver, British Columbia, 1962.

Rand, G. M., and Petrocelli, S. R. (eds.): Fundamentals of Aquatic Toxicology. Washington, DC: Hemisphere, 1985.

Reinecke, N. J.: Feeding ecology and development of juvenile black ducks in Maine. Auk 96:737–745, 1979.

Renne, D. S., and Wolf, M. A.: Experimental studies of 2,4-D herbicide drift characteristics. *Agric. Meteorol.*, 20:7–24, 1979.

Roberts, J. R., and Marshall, W. K.: Retentive capacity: an index of chemical persistence expressed in terms of chemical-specific and ecosystem-specific parameters. *Environ. Toxicol. Safety*, 4:158–171, 1980.

Sanders, H. O.: Toxicity of pesticides to the crustacean *Gammarus lacustris.* U.S. Fish and Wildlife Service. Bureau of Sport Fisheries and Wildlife Technical Paper No. 25, 1969.

Sanders, H. O.: Toxicity of some insecticides to four species of malacostracan crustaceans. U.S. Fish and Wildlife Service Technical Paper No. 66, 1972.

Sanders, H. O., and Cope, O. B.: Toxicities of several pesticides to two species of cladscerons. *Trans. Am. Fish. Soc.*, 95:165–160, 1966.

Sanders, H. O., and Cope, O. B.: The relative toxicities of several pesticides to naiads of three species of stoneflies. *Limnol. Oceanogr.*, 13:112–117, 1968.

Shaw, G. C.; Smith, D. K.; Sheehan, P. J.; and Mineau, P.: Environmental concerns. In: Proceedings of the Symposium on the Future Role of Aviation in Agriculture, edited by PW Voisey, pp. 47–63. AFA-TN-17, NRC No. 23501. Associate Committee on Agriculture and Forestry Aviation, National Research Council of Canada, 1984.

Sheehan, P. J.; Miller, D. R.; Butler, G. C.; and Bourdeau, P. H.: Effects of Pollutants at the Ecosystem Level. Scope 22. Chichester: Wiley, 1984.

Sheehan, P. J.; Baril, A.; Mineau, P.; Smith, D. K.; Hartenist, A.; and Marshall, W. K.: The Impact of Pesticides on the Ecology of Prairie Nesting Ducks. Technical Reports Series No. 19. Canadian Wildlife Service, 1987.

Siegfried, W. R.: Summer food and feeding of the ruddy duck in Manitoba. *Can. J. Zool.*, 51:1293–1297, 1973.

Skidmore, J. F.: Factors affecting the toxicity of pollutants to fish. *Vet. Rec.*, 94:456–458, 1974.

Sloof, W.: Benthic macroinvertebrates and water quality assessment: some toxicological considerations. *Aquat. Toxicol.*, 4:73–82, 1983.

Smith, A. G.: Ecological factors affecting waterfowl production in the Alberta parklands. U.S. Fish and Wildlife Service. Bureau of Sport Fisheries and Wildlife Resource Publication No. 98, 1971.

Smith, A. G., and Stoudt, J. H.: Ecological factors affecting waterfowl production in the Canadian parklands. U.S. Fish and Wildlife Service. Bureau of Sport Fisheries and Wildlife. Unpublished report, 1968.

Smith, A. G.; Stoudt, J. H.; and Gollop, J. B.: Prairie potholes and marshes. Waterfowl Tomorrow, edited by J. P. Linduska, and A. L. Nelson, pp. 39–50. U.S. Fish and Wildlife Service, 1964.

Sprague, J. B.: Factors that modify toxicity. In: Fundamentals of Aquatic Toxicology, edited by G. M. Rand, S. R. Petrocelli, pp. 124–163. Washington, DC: Hemisphere, 1985.

Statistics Canada: Census of Canada, Agriculture Ottawa: 1951, 1981.

Stoudt, J. H.: Ecological factors affecting waterfowl production in the Saskatchewan parklands. U.S. Fish and Wildlife Service. Bureau of Sport Fisheries and Wildlife Resource Publication No. 99, 1971.

Stratton, G. W.; Burrell, R. E.; Kurp, M. L.; and Corke, C. T.: Interactions between the solvent acetone and the pyrethroid insecticide permethrin on activities of the blue-green alga *Anabaena. Bull. Environ. Contam. Toxicol.*, 24:562–569, 1980.

Street, M.: The role of insects in the diet of mallard ducklings, an experimental approach. *Wildfowl*, 29:93–100, 1978.

Sturkie, P. D.: Avian Physiology. Ichaca: Comstock Publishing Association, 1954.

Sugden, L. G.: Feeding activity of captive lesser scaup. Canadian Wildlife Service. Progress Note No. 24, 1971.

Sugden, L. G.: Feeding ecology of pintail, gadwall, American Wigeon and lesser scaup ducklings. Canadian Wildlife Service Report Series No. 24, 1973.

Sugden, L. G.: Feeding ecology of pintail, gadwall, American wigeon and lesser scaup ducklings in southern Alberta. Canadian Wildlife Service Report Series No. 24, 1982.

Sugden, L. G., and Beyersbergen, G. W.: Farming intensity on waterfowl breeding grounds in Saskatchewan parklands. *Wildl. Soc. Bull.*, 12(1):22–26, 1984.

Sugden, L. G., and Harris, L. E.: Energy requirements and growth of captive lesser scaup. *Poultry Sci.*, 51:625–633, 1972.

Sugden, L. G.; Driver, E. A.; and Kingsley, M. C. S.: Growth and energy consumption by captive mallards. *Can. J. Zool.*, 59(8):1567–1570, 1981.

Suter, G. W., II: Ecological Risk Assessment. Boca Raton, FL: Lewis, 1993.

Swanson, G. A., and Duebbert, H. F.: Wetland habitats of waterfowl in the prairie pothole region. In Northern Prairie Wetlands, edited by A. Vander Valk, pp. 228–267. Ames: Iowa State Univ. Press, 1989.

Swanson, G. A., and Meyer, M. I.: Impact of fluctuating water levels on feeding ecology of breeding blue-winged teal. *J. Wildl. Manage.*, 41:426–433, 1977.

Swanson, G. A., and Sargeant, A. B.: Observation of nighttime feeding behavior of ducks. *J. Wildl. Manage.*, 36(3): 956–961, 1972.

TRI: Endpoints for Ecological Toxicity. Contract No. 68-024199. Technical Resources Incorporated, Rockville, MD, 1988.

Tucker, R. K., and Lietzke, J. S.: Comparative toxicology of insecticides for vertebrate wildlife and fish. *Pharmac. Theor.*, 6:167–220, 1979.

Ware, G. W.; Apple, E. P.; Cahill, W. P.; Gerhardt, P. D.; and Frost, K. R.: Pesticide drift, II. Mist-blower vs. aerial application of sprays. *J. Econ. Entomol.*, 62(4):844–846, 1969a.

Ware, G. W.; Estesen, B. J.; Cahill, W. P.; Gerhardt, P. D.; and Frost, K. P.: Pesticide drift I. High-clearance vs. aerial application of sprays. *J. Econ. Entomol.*, 62(4):840–843, 1969b.

Ware, G. W.; Cahill, W. P.; Gerhardt, P. G.; and Witt, J. M.: Pesticide drift IV. On-target deposits from aerial application of insecticides. *J. Econ. Entomol.*, 63(6): 1982–1983, 1970.

Wauchope, R. D.: The pesticide content of surface water draining from agricultural fields, a review. *J. Environ. Qual.*, 7(4):459–472, 1978.

Weaver, G. D.; Nilsson, M. J.; and Turney, R. E.: Prospects for the prairie grain industry 1990. Winnipeg: Canada Grains Council, 1982.

Chapter 31

LINEAR ALKYLBENZENE SULFONATE (LAS)

R. J. Larson and *D. M. Woltering*

INTRODUCTION

Background

Because of their high-volume use in consumer products, laundry and cleaning (L&C) chemicals have the potential for broad-scale release into aquatic and terrestrial environments. This chapter presents an environmental risk assessment case study for one of the most widely used (and studied) L&C chemicals, the anionic surfactant linear alkylbenzene sulfonate (LAS). For the past 25 yr, LAS has been the major anionic surfactant used in commercial laundry and cleaning products. Approximately 800 million pounds of LAS are now used annually in the United States. It is estimated that almost 18 billion pounds of LAS have been used in the United States since the introduction of LAS in the mid-1960s (Chemical Economics Handbook, 1992). Currently, LAS accounts for approximately 28% of all synthetic surfactants used worldwide, with annual production volumes in the United States, Japan, and Western Europe totaling about 2.8 billion pounds (Painter and Zabel, 1988). The large volumes of LAS used worldwide, coupled with its potential for broad-scale distribution in the environment, make accurate determinations of the fate and effects of LAS a key requirement for assessing its overall environmental safety.

In recent years, considerable attention has been focused on L&C chemicals and the environmental risks associated with the manufacture, use, and disposal of these chemicals. Clearly, large-volume L&C chemicals can reach the environment by many different routes, including those associated with raw material sourcing and manufacturing operations, processing and formulation activities, and transportation/distribution to the consumer. The vast majority of L&C chemicals, however, ultimately reach the environment as components of domestic and municipal wastewater after consumer use of laundry and cleaning products. These wastewaters typically receive treatment in on-site or municipal wastewater treatment systems and are then discharged as treated effluents to either terrestrial or aquatic receiving environments (Rapaport and Eckhoff, 1990).

The detergent industry has been proactive in the development and application of environmental risk assessment strategies. The exact approaches may vary somewhat by company, but the general components are similar. Such a generalized environmental risk assessment program is depicted in Figure 1. The process is shown as a flowchart consisting of several phases. The phases, which begin with a review of usage patterns including how much and where a product or ingredient is to be used and end with a decision to use, not use, or restrict use to limit exposures, are interrelated to yield a scientifically defensible but practical assessment of risks. Initial consideration is given to the physical and chemical properties of the material and to the environmental compartments (e.g., surface water, sediments, soil) where exposure is likely to occur. Then environmental fate and adverse ecological effects are evaluated using predictive models, laboratory studies, and/or field investigations. As the information is developed, a point is reached at which a prediction can be made regarding the safety of a material. Scientific judgment is used in deciding which tests and analyses to conduct and when sufficient data are available to make an overall decision regarding safety. Assumptions and uncertainties in the analysis, which are always present in risk assessment, can be strengthened by soliciting review and evaluation from additional ("outside") experts and by monitoring studies to confirm predictions of fate, exposure, and effects under actual use conditions. The steps in the ecological risk assessment process are presented in more detail as part of the case study of linear alkylbenzene sulfonate.

Figure 2 presents a flow diagram of the major routes of entry of LAS and other detergent chemicals into the environment, showing their possible distribution in various environmental compartments. The

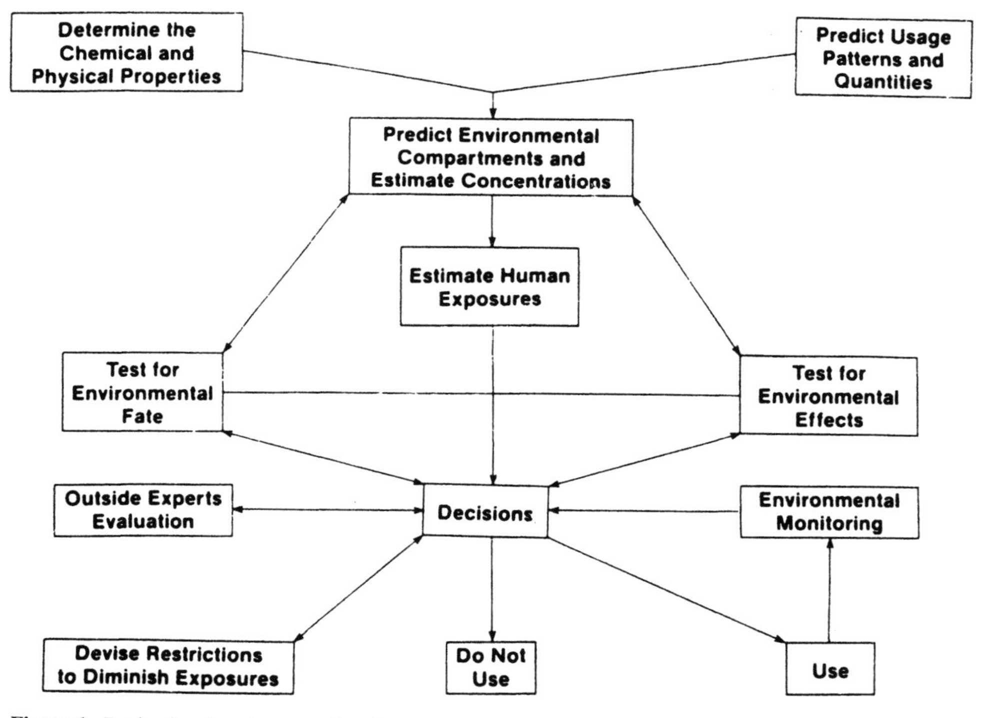

Figure 1. Predominant environmental pathways and compartments for L&C chemicals. (From Woltering DM, Bishop WE, Evaluating the environmental safety of detergent chemicals: A case study of cationic surfactants, in *The Risk Assessment of Environmental Hazards*, ed. DJ Pastenbach, pp. 345–389, copyright © 1989 John Wiley & Sons, Inc. Reprinted by permission of John Wiley & Sons, Inc.)

environmental concentration of LAS in these compartments will be dependent on a number of factors, including the percent of LAS used in the product and the quantity of product marketed, household and industrial water use practices, type and removal efficiency of wastewater treatment practices, dilution and dispersion into receiving environments, and removal and degradation in posttreatment environmental compartments. Using assumptions specific for the United States and information presented by McAvoy et al. (1993) on LAS concentrations in sewage effluents and sludges, surface waters, sediments, and soils, it is possible to construct a generalized accounting or mass balance (see Chapter 18) for LAS removal and biodegradation (see Chapters 1, 15, 20, and 27) in various environmental compartments. This mass balance is shown in Figure 3.

In the United States, the majority of wastewater (>98%) undergoes some form of secondary treatment, with the percentage of treatment by municipal wastewater treatment plants (MWTPs) and on-site disposal systems (OSDSs) being approximately 75% and 25%, respectively. In general, the percentage of LAS removal during municipal wastewater treatment is high (>98%), and the majority of this removal is due to ultimate biodegradation or mineralization (77%). Approximately 60% of the total volume of LAS used in the United States ($75 \times 0.77 = 58\%$) is therefore biodegraded during municipal wastewater treatment and never reaches natural receiving environments. However, significant amounts of LAS are also released to terrestrial environments as excess sludge from municipal treatment plants and as effluents and sludge from OSDSs. Biodegradation of LAS released in this manner continues in surface soils (Ward and Larson, 1989; Knaebel et al., 1990) and subsurface soils and ground water (Larson et al., 1989), resulting in overall mass reductions of 16% and 23%, respectively. The total amount of LAS degradation occurring in terrestrial environments (39%) is approximately two-thirds of the amount occurring during activated sludge wastewater treatment (58%). Terrestrial environments, therefore, are important reservoirs of LAS degradation capability in the environment.

Although the majority of LAS biodegradation occurs during municipal wastewater treatment and in surface and subsurface soils (97% across all three compartments), small amounts of LAS are also released to freshwater and marine systems as treatment plant effluents and untreated waste streams. Biodegradation of LAS continues in these systems, however, resulting in further reductions in exposure concentrations (Larson, 1990; Shimp, 1989). In general, the majority of LAS biodegradation occurs in aerobic environments. Anaerobic environments like digester

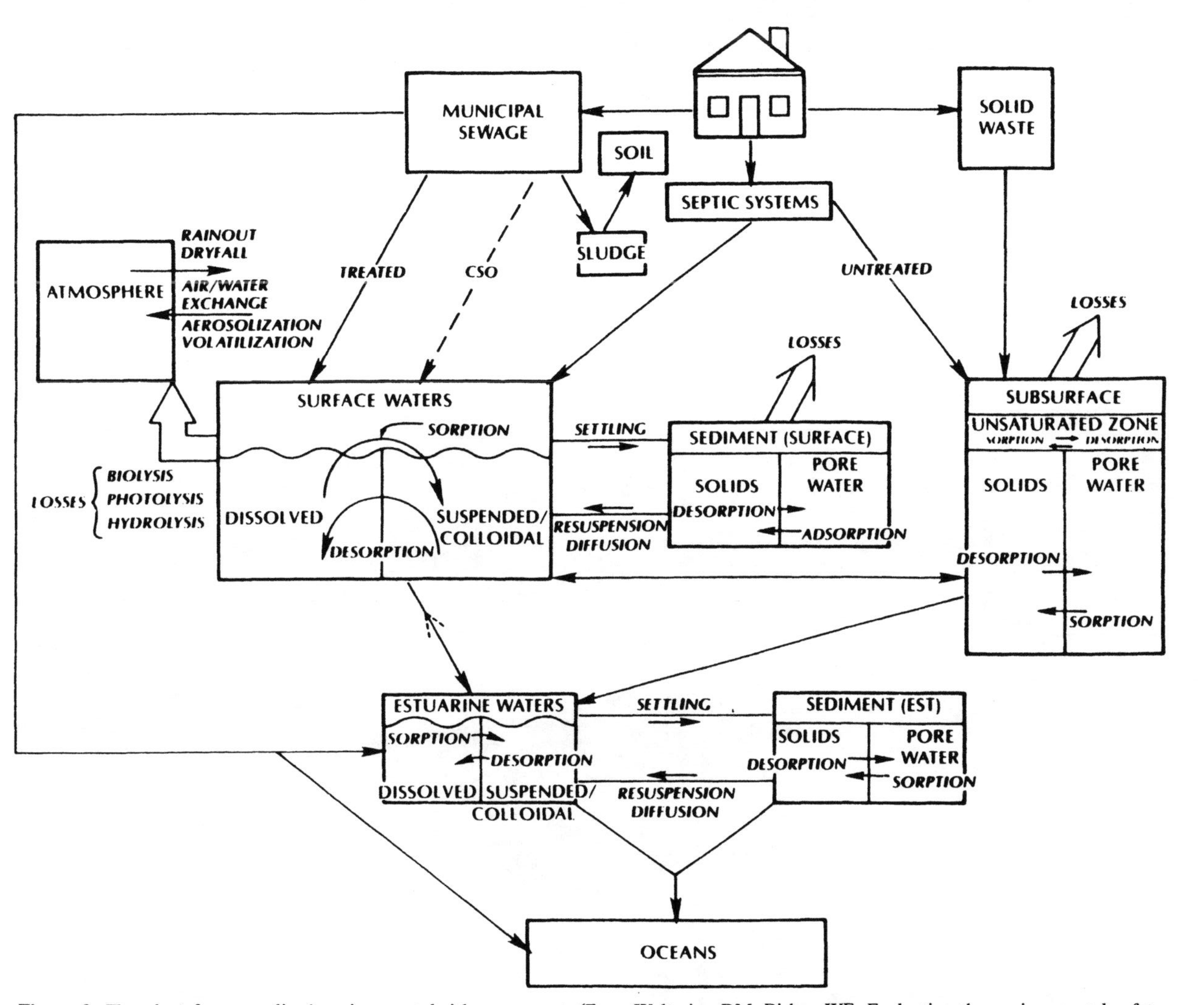

Figure 2. Flowchart for generalized environmental risk assessment. (From Woltering DM, Bishop WE, Evaluating the environmental safety of detergent chemicals: A case study of cationic surfactants, in *The risk Assessment of Environmental Hazards*, ed. DJ Pastenbach, pp. 345–389, copyright © 1989 John Wiley & Sons, Inc. Reprinted by permission of John Wiley & Sons, Inc.)

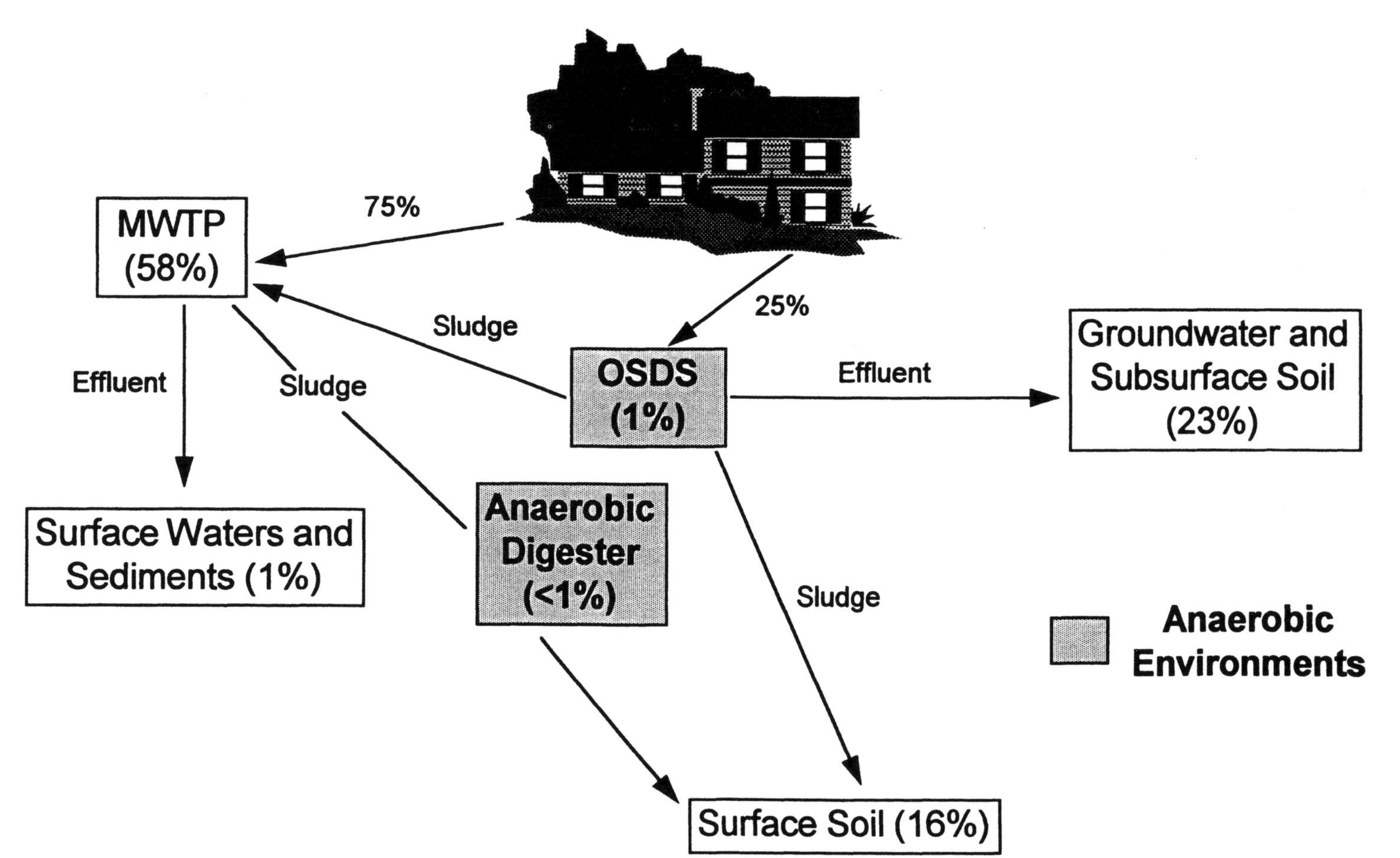

Figure 3. Schematic diagram showing the distribution of LAS in various environmental compartments and the amount of biodegradation occurring in each of these compartments.

sludge and OSDSs do not readily support the biodegradation of LAS and related L&C chemicals (Larson et al., 1993), and, as Figure 3 indicates, these environments play a relatively minor role in the removal of L&C chemicals. Overall, the major conclusion from LAS mass balance calculations is that while treatment processes play a major role in reducing LAS exposure concentrations in aquatic environments, posttreatment removal and degradation processes also lead to significant reductions in LAS levels in terrestrial and subsurface environments. Both types of processes are clearly necessary to maintain LAS exposure concentrations at acceptable environmental levels.

L&C Chemical Risk Assessment

Procedures to evaluate the environmental safety of L&C chemicals in aquatic environments generally rely on a sequential testing approach. A key assumption of this approach is that as testing proceeds through successive stages, more informed decisions can be made about expected environmental concentrations and the threshold concentrations producing adverse biological effects. Figure 4 provides a conceptual diagram of the sequential testing approach, illustrating the decrease in uncertainty that evolves over time regarding the magnitude of the difference between environmental concentrations and adverse biological effect concentrations. At each stage of the testing program, specific protection margins or safety factors are calculated that relate the exposure concentration of a chemical to its measured or predicted toxicity concentration. These factors allow decisions to be made regarding the need for additional testing and also dimension the potential environmental risk associated with the intended use of the chemical.

Accurate predictions of environmental concentrations for L&C chemicals require knowledge of a number of factors, including usage concentration and disposal routes, removal efficiency during wastewater treatment, dilution into receiving environments, and subsequent degradation, partitioning, and transport in key environmental compartments. Testing in laboratory systems using sensitive species from appropriate compartments provides information on the inherent toxicity of a chemical and on the most sensitive species and biological end points (see Chapters 2–8). A comparison between the predicted environmental concentration (PEC) in relevant environmental compartments and the concentration exerting negligible or no effects on organisms (NEC) can then be used to dimension the potential environmental risks associated with the intended use of the chemical. Figure 4 represents the two concentrations by parallel lines and demonstrates that increasingly accurate estimates of these concentrations will result from a sequential series of tests.

In the early phases of the environmental risk assessment process, biological effects levels and environmental concentrations are estimated from simple exposure models (see Chapter 18), quantitative structure-activity relationships (QSARs; see Chapter 20), and standardized laboratory toxicity tests (see

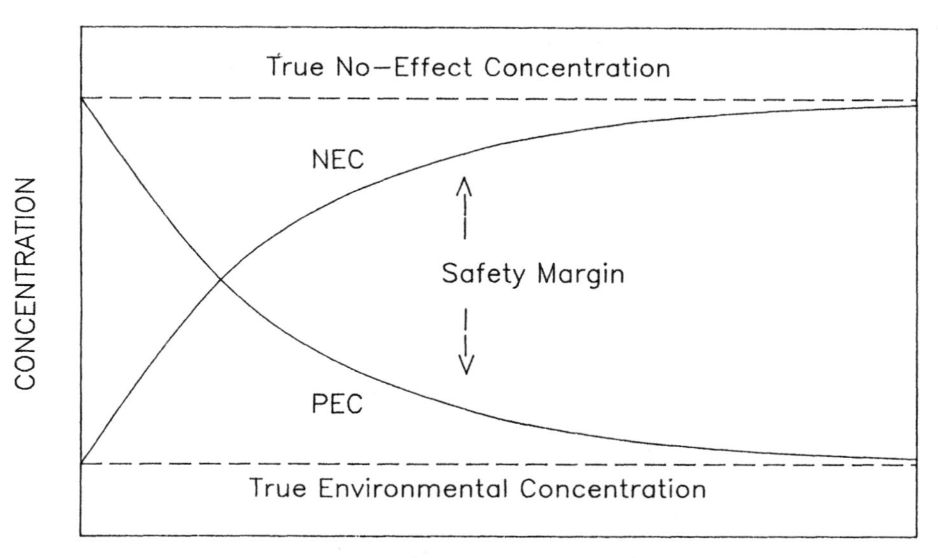

Figure 4. Diagrammatic representation of the environmental risk assessment process, showing the increased level of certainty in NEC and PEC values over time as environmental testing progresses. (See also Figure 1, Chapter 33 and accompanying explanation on pages 905–906.)

Chapters 2–8). At this point, there is a high degree of uncertainty about the accuracy of PEC and NEC values and a relatively wide margin is needed to conclude that minimal risk exists. As testing proceeds toward in situ measurements of chemical concentrations and ecological effects in the field (see Chapter 9), this uncertainty in PEC and NEC values decreases, and a narrower margin is acceptable. The U.S. Environmental Protection Agency (EPA) has developed guidelines for acceptable uncertainty factors in aquatic environments, based on the amount of toxicity data available and the type of biological end point (acute or chronic) measured (U.S. EPA, 1984). These uncertainty factors (often called application factors) essentially adjust PEC values for the uncertainty in NEC values in a standard (although somewhat arbitrary) fashion, allowing the calculation of specific safety factors from a wide range of databases.

Application factors vary significantly, from 100 to 1000, when only limited acute toxicity data are available, to as low as 10, when extensive chronic toxicity information on a variety of species exists. The uncertainty of extrapolation can be essentially eliminated (i.e., a factor of 1) when data from a comprehensive field-scale examination are available. To date, few comprehensive laboratory and field studies have been conducted on any given chemical that confirm the need for uncertainty factors or verify the assumptions used to justify the specific numerical values (i.e., 1000, 100, 10). Such information does exist for certain L&C chemicals, however, including LAS (Woltering and Bishop, 1989; Belanger, 1994). This information indicates that a significant amount of conservatism is already built into safety factors developed from chronic toxicity testing of single species in laboratory tests. For L&C chemicals at least, single-species tests are as sensitive as community or higher-level tests in predicting adverse biological effect concentrations. Although the database to support this conclusion is continually being developed, it is apparent from our current understanding that results of single-species tests are sufficiently protective for safety factor calculations and additional application factors are unnecessary.

TERMINOLOGY

Predicted environmental concentration (PEC). Concentration of chemical in a particular environmental compartment that is based on estimates or modeling simulations of the use, disposal, transport, and transformation (or fate) of the chemical.

Measured environmental concentration (MEC). Concentration of a chemical in a particular environmental compartment that has been confirmed by a specific analytical method at an appropriate level of sensitivity.

Negligible effect concentration (NEC). Concentration producing negligible or no effects; that is, a toxicological threshold.

Safety (protection factor) margin. Ratio of the PEC or MEC to the NEC.

Mixing zone. Allocated impact zone where numeric water quality can be exceeded as long as acutely toxic conditions are prevented (see Chapter 24).

Ultimate biodegradation (mineralization). Metabolism and complete breakdown of a chemical by microorganisms to CO_2, biomass, and simple organic (e.g., CH_4) and inorganic end products (see Chapters 1, 15 and 27).

Practical biodegradation. Ultimate biodegradation or mineralization of a chemical at a rate that is sufficient to prevent accumulation and minimize concentrations in the key environmental compartments where exposure occurs. Practical biodegradation allows the change in environmental exposure concentrations to be quantified by relating the biodegradation half-life of a chemical to its residence time in a particular environmental compartment.

Application (uncertainty) factor. Numerical factor (typically ranging from 1 to 1000) used in safety factor calculations to adjust PEC values for uncertainty associated with NEC values.

INTRODUCTION TO LAS

Physicochemical Properties

As used in commercial L&C formulations, LAS is a complex mixture of homologues with different alkyl chain lengths and phenyl isomers with different points of attachment of the benzene ring to the alkyl chain (Figure 5). Commercial LAS is produced by the sulfonation of linear alkylbenzene (LAB) and neutralization of the resulting sulfonic acids as the sodium salts. The starting LAB, which is derived by reacting straight-chain olefins with benzene, largely dictates the composition of the sulfonated product. The proportions of the homologues and phenyl isomers produced are dependent on the sulfonation process, as are the concentrations of non-LAS by-products. The alkyl chains of commercial mixtures generally range from C_{10} to C_{14}, with an average chain length of about C_{12} and an approximately equal distribution of phenyl groups along with alkyl chain. Major impurities consist of dialkyl indanes, tetralins, naphthalenes, and branched materials, which are produced at very low concentrations by modern sulfonation processes (Rapaport et al., 1992).

Given the hydrophobic and hydrophilic character of the alkyl and benzenesulfonic moieties of LAS, respectively, all LAS homologues are moderately soluble in both water and organic solvents. They also have comparable chemical reactivities, based on the similarities of the functional groups present. LAS is not volatile under normal conditions of temperature and pressure and is not readily hydrolyzed by either acids or bases. Because of the presence of the aromatic ring, LAS does absorb ultraviolet light and is considered moderately UV active. This UV activity,

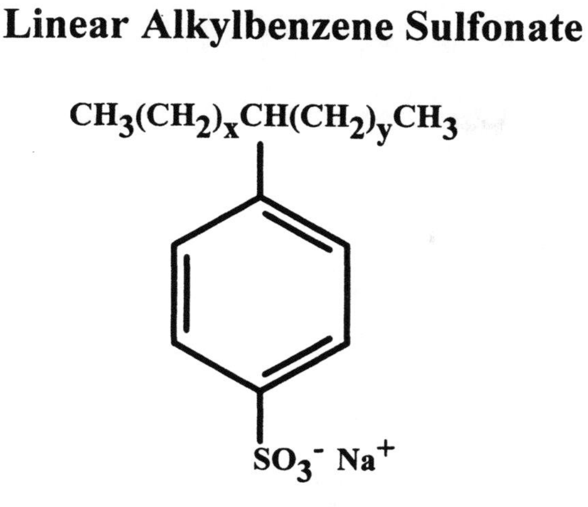

Linear Alkylbenzene Sulfonate

$CH_3(CH_2)_xCH(CH_2)_yCH_3$

$SO_3^- \ Na^+$

where: x + y = n, and n = 7-11 carbon units

Figure 5. Structure of linear alkylbenzene sulfonate (LAS). For C_{12} LAS, $x + y = 9$.

however, does not result in significant photoinstability, and LAS photodegradation is quite slow under typical environmental conditions of light, temperature, and pH. More detailed information on the physiochemical properties of LAS is given later in the section on environmental fate.

In contrast to the other chemical properties, the partitioning or sorption behavior of different LAS homologues is highly variable and dependent on the length of the alkyl chain and, to a lesser extent, on the point of attachment of the benzene ring to the alkyl chain (Hand and Williams, 1987). Equilibrium sorption coefficients (K_d values; see Chapter 15), defined as the mass ratio of LAS solid-phase concentrations to liquid-phase concentrations (L/kg), increase approximately 10-fold for every two-carbon increase in the length of the alkyl chain and approximately 2-fold as the phenylsulfonate group moves from a middle to a more terminal location. Sorption and desorption are quite rapid (<3 and <8 h, respectively) and almost completely reversible in single-step desorption studies conducted in river sediments. Equilibrium K_d values for LAS homologues in river sediments vary widely depending on sediment type and alkyl chain length, ranging from 3 to 26,000 L/kg. The mechanism of LAS sorption appears to involve both hydrophobic and ionic components, with the hydrophobic mechanism dominating in river water and sediment systems (Hand and Williams, 1987).

Biodegradation

Given the overall chemical stability for LAS, microbial degradation (biodegradation; see Chapter 15 for a discussion of microbial degradation) is the major removal mechanism for LAS in the environment. Studies with pure microbial cultures have characterized the major pathway for LAS degradation (Figure 6), which involves omega/beta oxidation of the alkyl chain followed by desulfonation and cleavage of the benzene ring (Swisher, 1987). Ring degradation is the final step in the LAS degradation pathway and is therefore rate-limiting for complete mineralization. Early laboratory studies, which utilized specific analytical methods to follow the disappearance of individual LAS homologues, suggested that the rates of primary degradation of LAS homologues and phenyl isomers were variable. Rates of primary degradation of individual homologues appeared to increase as the length of the alkyl chain increased or as the phenylsulfonate moiety assumed a more terminal location. The enhanced biodegradation of longer homologues and external phenyl isomers led to the development of the "distance principle" for LAS degradation, a principle that has been generally accepted since the early 1960s (Swisher, 1987).

More recent studies, however, focusing on the rate of mineralization of the LAS benzene ring, have shown that the distance principle is not applicable to the final, rate-limiting step in the LAS biodegradation pathway (Larson, 1990; Larson et al., 1993). These studies, which were conducted with a variety of environmental samples at LAS concentrations approximating realistic environmental concentrations, indicate that degradation rates for individual LAS homologues and isomers remained constant and independent of alkyl chain length and phenyl position over the range of C_{10} to C_{14} and C_2 to C_7, respectively. Because ring cleavage is ultimately the rate-limiting step for LAS degradation in the environment, these findings indicate that complete mineralization proceeds at equivalent rates for all homologues and isomers. Recent monitoring studies

LAS AEROBIC BIODEGRADATION PATHWAY

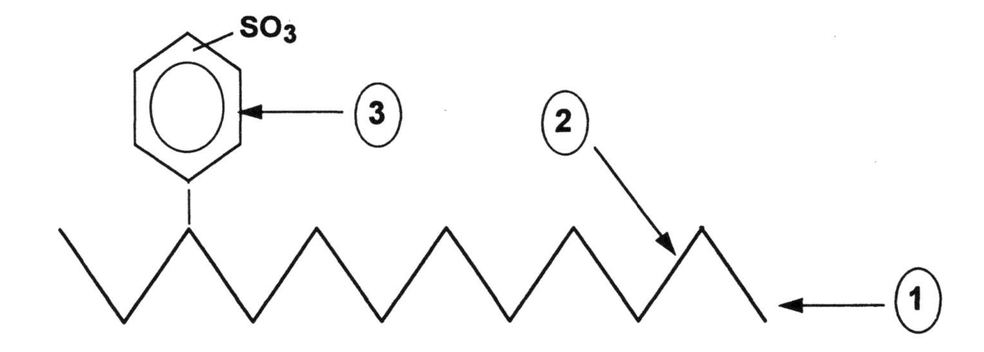

(1) **Omega-oxidation** ⟶ $\phi(CH_2)_nCOOH$

(2) **Beta-oxidation** ⟶ **acetate units** ⟶ $CO_2 + H_2O$

(3) **Ring desulfonation / cleavage** ⟶ $CO_2 + H_2O + SO_4^{=}$

Figure 6. Aerobic biodegradation pathway for LAS showing the major steps leading to mineralization to CO_2.

lend further support to the equivalence of primary degradation rates for LAS homologues and phenyl isomers. These studies indicate that the mean LAS chain length present in wastewater effluents and river water ranges from $C_{11.8}$ to $C_{11.9}$, which is essentially identical to the mean chain length used in commercial LAS formulations (McAvoy et al., 1993). As measured under actual use conditions, therefore, removal and biodegradation of parent LAS in MWTPs and receiving waters is not related to alkyl chain length and is essentially constant over the range of homologues used in commercial products.

Analytical Methods

A number of analytical techniques have been developed to measure the concentration of LAS in environmental samples. Many of these techniques, including those based on relatively simple and nonspecific colorimetric techniques as well as sensitive and highly specific techniques utilizing both gas chromatography and high-performance liquid chromatography, have been reviewed by Painter and Zabel (1988). Each technique has unique applications and limitations.

The most commonly accepted nonspecific method for LAS is the methylene blue–active substances (MBAS) method (see Chapter 27), originally reported by Longwell and Maniece (1955) and modified by Abbott (1962). The MBAS method is a colorimetric method with a detection limit of about 0.05 mg/L. It is based on the formation of a chloroform-extractable complex between LAS and the dye methylene blue, followed by spectrophotometric detection of the LAS-dye complex at 652 nm. The complex is soluble in chloroform whereas the dye is not. Because methylene blue reacts with a variety of anionic materials, the MBAS method is not specific for LAS and is subject to numerous interferences. The contribution of LAS to the MBAS response is highly variable, ranging from 4 to 93%, depending on the environmental sample analyzed (Hennes and Rapaport, 1989). Additional solvent extraction techniques have been developed to separate LAS from major interferences prior to complex formation (Osburn, 1986). These extractions significantly improve the specificity of the MBAS method for LAS, but they do not completely eliminate interferences from unknown anionic compounds.

The most accurate and sensitive techniques for analyzing LAS in environmental samples are based on gas and high-performance liquid chromatography. Gas chromatographic (GC) techniques, which involve the derivatization of LAS followed by ion-selective or mass-selective detection, have a high degree of selectivity and sensitivity. GC techniques have been developed to analyze LAS in a variety of environmental samples, including influent and effluent wastewaters, river water and sediments, sewage sludges, and sludge-amended soils (Waters and Garrigan, 1983; McEvoy and Giger, 1986; Osburn, 1986; Trehy et al., 1990). These techniques are particularly useful in providing quantitative information on individual LAS homologues and phenyl isomers in complicated environmental matrices such as sewage sludges and sediments.

A major limitation of GC techniques is the amount of sample preparation time required for analysis, which makes them prohibitive for routine analytical work or large numbers of samples. For the latter types of work, high-performance liquid chromatography (HPLC) is a simpler, more efficient technique for measuring LAS in environmental samples. HPLC methods have been developed to resolve LAS homologues in a variety of different environmental matrices (Marcomini and Giger, 1987; Matthijs and DeHenau, 1987; Castles et al., 1989). They offer a combination of speed, sensitivity, and selectivity that makes them ideal for large numbers of samples and routine analytical work. The use of HPLC also avoids the requirement for preliminary derivitization and offers a practical tool for obtaining LAS monitoring data when detailed information on phenyl isomer distribution is not required.

LEVEL I RISK ASSESSMENT

The traditional ''level I'' approach to evaluating the potential environmental risks of consumer product chemicals like L&C surfactants involves initial estimates of exposure in relevant environmental compartments and of toxicity to representative, sensitive biota. Because of their discharge in wastewater, the primary focus for assessing environmental risks for L&C chemicals has historically been the aquatic environment (i.e., surface waters that receive treated municipal wastewater). The case study assessment for LAS begins with predicted use or loading concentrations, chemical and physical property data, and results from laboratory treatability and aquatic toxicity studies.

Environmental Loading

Approximately 800 million pounds of LAS are used annually in the United States in laundry and cleaning products. This maximum loading value is based on the amount of products sold and the concentrations of LAS in the product formulations. Other uses of LAS that could lead to releases to the environment (and therefore could be included in an overall environmental assessment) are very small (<5%) in comparison to their use in laundry and cleaning products.

Environmental Fate

Initial estimates of the environmental fate of LAS can be determined from the use and disposal patterns of laundry and cleaning products and the physical and chemical properties of LAS in particular. Of the approximately 98% of residential and institutional wastewater that receives treatment, about three-quarters (75%) is treated via municipal wastewater treatment systems (often referred to as publicly owned treatment works, POTWs; see Chapter 24) and one-quarter (25%) is treated in on-site disposal systems (OSDS) like septic tanks. This means that about 588 million pounds per year of LAS are treated via POTWs.

$$(800 \times 10^6 \text{ lb/yr}) \times (98\% \text{ treated}) \times (75\% \text{ via POTWs}) = 588 \times 10^6 \text{ lb/yr}$$

An additional 196 million pounds per year of LAS are treated via OSDS.

$$(800 \times 10^6 \text{ lb/yr}) \times (98\% \text{ treated}) \times (25\% \text{ via OSDSs}) = 196 \times 10^6 \text{ lb/yr}$$

Knowledge of LAS's physical and chemical properties, such as water solubility, octanol-water partition coefficient (K_{ow}), and sorption characteristics, can be used early in the assessment to predict its fate and the environmental compartments into which it is most likely to partition.

As detailed in the section on physiochemical properties, commercial LAS is not a pure compound but a complex mixture of homologues with alkyl chain lengths ranging from C_{10} to C_{14} for most commercial products. The physicochemical properties of LAS vary somewhat by chain length, but the average C_{12} LAS provides a reasonable point of reference for predicting environmental fate. LAS is soluble in water (g/L), with a critical micelle concentration (CMC) in the 1–10 mM range (Di Toro et al., 1990). Estimated octanol-water partition coefficients (log K_{ow}) for LAS are relatively low, in the range of 2 to 3 (Holysh et al., 1986). As with all surface-active chemicals, both solubility and log K_{ow} values for LAS are highly dependent on experimental conditions, due to the formation of micelles in both polar and nonpolar solvents at the CMC. In general, micelle formation precludes the accurate determination of solubility and K_{ow} values for LAS and other L&C chemicals and limits the environmental relevance of these measurements as predictive tools.

Given its high aqueous solubility and ionic character, LAS is not volatile under normal environmental conditions (vapor pressure, ~0.01 Pa). It is not readily hydrolyzed by either acids or bases and is not readily photodegraded. As mentioned earlier in this section, adsorption and desorption of C_{12} LAS are quite rapid, with sorption coefficients (K_d) in the range 1000–5000 for sewage sludge, 50–500 for river sediments, and 2–20 for soils. The rapid sorption and desorption of LAS tend to minimize the effect of environmental solids on the kinetics of other reactions like biodegradation.

Based on the foregoing properties, a fraction of LAS discharged from the household to wastewater treatment would be expected to be sorbed to the sludge, with the remainder staying in the wastewater. Because of its low vapor pressure, there would be essentially no loss to the atmosphere via volatilization. Some partitioning to suspended solids would occur during primary (~26%) and secondary (~7%) wastewater treatment (Cowan et al., 1993). However, most of the LAS that reached a surface water via effluent would be in solution (>90%), with a small percentage (<10%) eventually coming into equilibrium with the suspended and settled solids (sediments) in the stream (Hennes and Rapaport, 1989). LAS would not be likely to accumulate in biota. Using one of the available correlation equations relating K_{ow} to bioconcentration potential (e.g., log BCF = 0.76 log K_{ow} − 0.23 from Lyman et al., 1982; see Chapter 20) yields a relatively low bioconcentration factor (BCF) of 112. Thus the compartments of most interest for the level I risk assessment include sludges and soils that receive sludges (sludge-amended soils) and surface waters. Sediments would be a lesser concern, due to the relatively low concentrations of LAS present in wastewater effluents and its moderate sorption properties (this is discussed later in the section on level II analysis of environmental risk).

Biodegradation is the major fate process operative on LAS once it reaches wastewater, sludge, surface water, sediment, or soil compartments. LAS is readily and completely biodegraded over relatively short periods of time (days to weeks) in all of these environmental compartments (Larson et al., 1993). The rapid mineralization of LAS to carbon dioxide and microbial biomass means that LAS will not accumulate or persist in the environment and that relative safety margins will increase over time in direct proportion to rates of biodegradation. The quantitative effects of LAS biodegradation rates on safety factors will be discussed later in more detail in the section on level II analysis of environmental risk.

Predicted Environmental Exposure Concentrations

Initial estimates of environmental concentrations of consumer product chemicals like LAS can be made following the published methods of the L&C industry (e.g., Holman, 1981; Rapaport, 1988). The basic inputs for a national average scenario include:

- The amount of the chemical being used that is likely to reach the environment (approximately 800 million lb/yr of LAS in the United States).
- The average per person sewage flow going to POTWs (about 500 L/capita per day).
- The U.S. population size (~230 million).
- The proportion of POTW sewage receiving various levels of treatment; in the United States about 75% goes to POTWs, the remainder to OSDSs.
- About 95% of the POTW sewage receives biological secondary treatment (~75% as activated sludge) and the remaining 5% receives only primary treatment (gravitational settling) prior to discharge to receiving waters.
- The treatment efficiency for the chemical of interest. [Rapaport and Eckhoff (1990) report LAS removals of about 30% for primary treatment and 98+% for secondary treatment.]
- A reasonable minimum dilution factor of the receiving streams and/or the incorporation rates for sludge-amended soils. [Rapaport (1988) determined that approximately 90% of the POTW effluent discharged to surface waters is diluted more than threefold; Holman (1981) reported a typical sludge incorporation rate of 1.22×10^{-2} kg sludge solids per kilogram soil per year, which results in a "dilution" of digested sludge in soil of about 80-fold at application.]
- A sediment (or other solids) sorption coefficient (K_d).

A complete set of equations and example calculations for each of these input parameters has been provided by Woltering and Bishop (1989). Using their approach, the following environmental concentration estimates can be made for LAS in influent and effluent wastewaters, surface waters and sediments, sewage sludge, and sludge-amended soils. Most of these estimates have been confirmed by monitoring studies conducted by the L&C industry (Painter and Zabel, 1989).

Influent sewage:	3–4 mg/L
Effluent sewage:	0.5–2 mg/L (activated sludge —primary only)
Surface water:	0.03–0.1 mg/L (below outfall)
Sediments:	1–300 mg/kg dry (below outfall)
POTW sludge:	<1–10 g/kg dry
Sludge-amended soil:	<1–40 mg/kg dry

Toxicity

An environmental assessment for most consumer product chemicals will typically require some environmental toxicity data. Exceptions might be chemicals for which the exposure estimates are very low (e.g., <10 ppb). Very few chemicals exhibit toxicity at exposure concentrations in this range, and those that do (like pesticides) are typically selected for their biological activity. Other exceptions include chemicals for which bioconcentration is unlikely given the structure of the chemical, significant degradation is expected, and usage is confined so that exposure will not occur in terrestrial or aquatic environments. Because of their relatively high volume and potential discharge into surface waters as a part of wastewater effluent (even though 90+% removal occurs during treatment), a toxicity evaluation is warranted for most L&C chemicals. The initial focus of this evaluation will be on aquatic organisms.

However, a chemical that is readily removed during wastewater treatment by sorption to sludge (as opposed to biodegradation) may also require toxicity evaluation for terrestrial plants and animals that could be exposed via sludge-amended soils.

Screening-level aquatic toxicity tests utilize sensitive, or otherwise representative, species of fish, invertebrates, and algae (freshwater and/or saltwater; see Chapters 2–8). Soil invertebrates (e.g., earthworms) and several crop plants are tested if, in fact, the soil compartment needs to be evaluated. Microbial toxicity can also be assessed for both aquatic and terrestrial systems. In general, small mammals and birds are unlikely to be exposed to significant concentrations of L&C chemicals (unless they are expected to be bioaccumulated). Toxicity protocols, however, are available for birds, and data from small mammal tests (i.e., rats and mice) used to assess human health can be used to evaluate potential adverse effects on mammals in terrestrial ecosystems.

Toxicity is assessed in both acute and chronic toxicity tests. Acute toxicity is assessed over a relatively short exposure period, using survival and serious incapacitation as the major end points. Chronic tests have a longer exposure period (i.e., either an early life stage or the full life cycle of an organism) and use sublethal as well as lethal end points. Both acute and chronic toxicities of LAS are related to the length of the alkyl chain, with higher toxicity at higher chain lengths. Table 1 summarizes representative, laboratory-derived acute and chronic toxicity data for the average chain length C_{12} LAS as either a pure homologue or a typical commercial blend (Painter and Zabel, 1988; Kimerle, 1989).

Based on the toxicity data for commercial LAS shown in Table 1 (which are considerable compared with those available for most consumer product or L&C chemicals at the initial stages of environmental assessment), it can be concluded that sewage treatment microbes and operations should not be adversely affected below about 20 ppm in sewage influent, soil microbes below about 250 ppm, and terrestrial plants below 100 ppm. Fish and invertebrates are about equally sensitive to the acute effects of LAS (lowest reported LC50s of about 1 ppm). Fish, however, are more sensitive than freshwater invertebrates to chronic effects (lowest NOECs of about 0.2 ppm) and marine invertebrates are probably as sensitive as fish (mysid shrimp NOEC of 0.2 ppm).

Initial Estimate of Environmental Risk

The information regarding environmental loading and fate, estimated maximum exposures in key compartments, and toxicity to sensitive species is collectively evaluated to form an initial assessment of the potential for significant environmental risk. The data for this evaluation, including a brief description of the types of tests conducted and end points used, are given in Table 2.

The maximum predicted influent sewage concentrations of LAS are about an order of magnitude less than the adverse effect threshold concentrations for microbial inhibition and impacts on aerobic and anaerobic wastewater treatment plant operations. The highest treated wastewater effluent concentrations of LAS approach the acute toxicity thresholds for sensitive aquatic species. Maximum surface water concentrations, however, are about one-tenth of the acute toxicity (LC50) concentrations for fish, algae, and invertebrates and about the same as the chronic toxicity no-effect thresholds for the most sensitive aquatic species (i.e., rainbow trout and mysid shrimp). There are no data available on the toxicity

Table 1. Acute and Chronic Laboratory Toxicity Data for C_{12} LAS

Organism	Toxicity concentration (mg/L)
Sewage treatment microbes and operations (e.g., activated sludge and anaerobic digester)	No effects at up to 20–30
Aquatic microbes (acclimated to LAS)	No effects at up to 5
Soil microbes	No effects at up to 250[a]
Algae	
Green, bluegreen, and diatom	4-d EC50 0.9–29
Marine diatom	4-d NOEC 2.0
Aquatic macrophyte (duckweed)	Chronic EC50 of 2.7
Aquatic invertebrates	
Daphnia magna	48-h EC50 ranges from about 1 to 10
	21-d chronic NOEC of 1.2
Marine mysid shrimp	96-h LC50 of 1.4
	28-d NOEC of 0.2
Freshwater fish (e.g., fathead minnow and rainbow trout)	96-h LC50 ranges from about 1 to 5
	30-d early life stage NOEC of about 0.2–1.0
Marine fish (e.g., sheepshead minnow and mummichog)	96-h LC50 of about 1–5
Crop plants (e.g., sorghum, sunflower, and mung bean)	21-d EC50 of 167–316 in soil[a]
	21-d NOEC of 100 in soil[a]
Earthworms	14-d NOEC of 500 in soil[a]

[a] Concentrations expressed as mg/kg.

Table 2. Fate, Effects, and Exposure Data for Level I Risk Assessment

Loading: Assume 800 million pounds per year in the United States	
Fate	
Water solubility	Relatively high (hundreds of mg/L)
Bioconcentration potential	Low (BCF of about 112)
Biodegradability test	>70% CO_2 production in screening
Treatability	>90% organic carbon removal in sludge
Adsorption	K_d in thousands for sludge, hundreds for sediments, and tens for soil
Volatilization	<0.1 Pa
Exposure (maxima)	
Influent sewage	3–4 mg/L
Effluent sewage	0.5–2.0 mg/L
Surface water	0.03–0.10 mg/L
River sediments	100–300 mg/kg
POTW digester sludge	10,000 mg/kg
Sludge-amended soils	1–40 mg/kg
No-adverse-effect concentrations	
Wastewater treatment	20–30 mg/L
Aquatic biota	Acute 1.0 mg/L
	Chronic 0.2 mg/L
Terrestrial biota	Microbes 250 mg/kg
	Plants 100 mg/kg
	Worms 500 mg/kg

of LAS sorbed to sediments to biota that live in and/or ingest sediments.

The maximum estimated concentrations of LAS in soil (immediately following amendment of LAS-containing sewage sludge) are below the no-effect threshold for soil microbes and sensitive crop plants and considerably below the no-effect threshold for earthworms. Thus it appears, based on this initial assessment, that LAS has an acceptable environmental profile at current usage concentrations. It is not persistent, it has very minimal potential to bioaccumulate, and LAS exposures are not likely to exceed adverse effect thresholds for sensitive species or systems. However, the ''margins of safety'' between exposure concentrations and adverse effect thresholds are not large (they are typically less than 10), and because of the broad and continuous use of LAS, additional environmental studies seem warranted to reduce the uncertainties of the assessment and to assure the safe use of this material.

A number of uncertainties are associated with the environmental risk assessment methodology used to evaluate LAS; however, it does follow the generally accepted approach (i.e., EPA and the general scientific community), which is acknowledged to have a relatively high degree of conservatism built in. Key sources of uncertainty in this LAS assessment that are amenable to a second level of assessment include the following:

- The accuracy of the exposure estimates beginning with wastewater influent, effluent, and sludge concentrations and going through surface water, sediment, and soil concentrations. (More sophisticated procedures can be used to predict exposure, and because a sensitive and specific analytical method is available for LAS it is possible to monitor real-world concentrations in key environmental compartments.)
- The influence of degradation in various environmental compartments and the resultant exposures of LAS as a function of time (e.g., in soil) and distance (e.g., downstream). (Potential risks should include considerations of extent and duration of exposure and effects.)
- The bioavailability and toxicity of sediment-sorbed LAS to benthic organisms. (This compartment needs to be evaluated because of the amount of LAS released in effluent and its moderate to high sorption properties.)
- Likewise, the accuracy of the toxicity values, which include the rather large extrapolation from laboratory studies to the bioavailability and population-level effects in the field. (Microcosm and/or field-type studies that combine more realistic fate and effect scenarios can be used to confirm or refute the laboratory-based assessment.)

LEVEL II RISK ASSESSMENT

Introduction

As introduced at the end of the level I assessment, there are acknowledged limitations and uncertainties associated with the first-level risk assessment. Exposure estimates can be rough approximations and toxicity thresholds are based on relatively simple laboratory-scale studies designed to determine the inherent toxicity of a material and the relative sensitivity of a wide range of biota. Often, the form and fate of the chemical under study are not necessarily the same as would occur in the real world, and therefore bioavailability and toxicity are uncertainties for field assessment. Finally, the biological end points in standard toxicity tests, although related to a population's health and survival, are measures of the effects on individuals and do not necessarily correlate directly to how a population of individuals or a community would respond in the field. The initial assessment of LAS has many of these uncertainties associated with it, and because of the relatively narrow safety margin between the exposure and toxicity values, LAS is a good candidate chemical for a more thorough and rigorous environmental assessment.

Level II risk assessments involve the use of more sophisticated models, novel test designs, in situ or field studies, and monitoring of chemical behavior in key environmental compartments using a specific and sensitive analytical method. The next section details the modeling and monitoring efforts to strengthen the exposure values for LAS. Following this section, biodegradation is discussed, along with its role in regulating the temporal and spatial exposure concentrations of LAS following discharge to surface waters and disposal of sludge to soils. A more thorough evaluation of the effects of LAS in surface water, sediments, and soils is then covered, followed by a final analysis of the environmental risks of LAS. In general, the goal of level II risk assessments is to utilize comprehensive determinations of the fate, effects, and exposure potential of L&C chemicals to generate more reliable and accurate estimates of environmental safety margins. In the specific case of LAS, these safety margins have been refined by a comprehensive environmental database (one of the largest assembled for any synthetic chemical) and verified by a 30-yr history of safe LAS use.

Exposure Analysis

Exposure Concentrations

Two approaches can be used in level II assessments to determine the concentration of LAS in receiving environments. The first approach, initially described by Holman (1981) and later refined and expanded by Rapaport (1988), involves the use of global exposure models and computerized databases to estimate predicted environmental concentrations (PECs) in appropriate receiving environments. Several models (see Chapter 18) are currently available to predict concentrations of LAS and other L&C chemicals in aquatic, benthic, and terrestrial environments. These models allow the calculation of LAS concentrations after wastewater treatment, utilizing LAS usage data, physiochemical and biodegradation data, and measured or predicted removal values under specific treatment conditions (Rapaport, 1988; Cowan et al., 1993; McAvoy et al., 1993).

The second approach for estimating exposure concentrations in level II assessments involves the use of monitoring studies to determine analytically the concentration of chemicals in specific environmental compartments. These determinations typically utilize analytical methods that are highly specific and selective for the chemical. In some cases, however, nonspecific methods are used to expedite the monitoring studies. This is the case for LAS, for which much of the monitoring data prior to 1980 is based on the MBAS method. As mentioned earlier, MBAS is not specific for LAS and subject to numerous interferences. More recent monitoring studies have utilized GC or HPLC methods to confirm LAS concentrations analytically. These methods provide detailed information on both LAS chain length and phenyl isomer distribution and are highly specific and sensitive for parent LAS.

Exposure Modeling

Exposure models to predict environmental concentrations of LAS and other L&C chemicals have improved significantly in recent years. A computer model, PG-GRiDS (formally called USTEST), developed by Rapaport (1988) to predict the concentration of consumer product chemicals in nonestuarine rivers and streams, illustrates the value of current exposure models in determining PEC values. Unlike most exposure models, which are useful in predicting chemical concentrations in specific water bodies or localized geographic areas, PG-GRiDS predicts concentrations in rivers and streams below treatment plants across the entire United States. The model utilizes national U.S. EPA databases covering 11,500 publicly owned treatment plants and approximately 600,000 river miles. It considers removal by different types of wastewater treatment (e.g., activated sludge, trickling filter, primary settling) and surface water dilution under both average flow and critical low-flow conditions to estimate final stream concentrations (McAvoy et al., 1993).

The output of PG-GRiDS is a frequency distribution of concentrations in U.S. rivers and streams below the mixing zone of POTW effluent discharges. This frequency distribution, which can be determined as a function of either total POTW discharge or the total number of treatment plants, is used to predict the probability that a stream concentration will equal or exceed a specific NEC value. Output predictions expressed on the basis of total treatment plant discharge (flow) are generally preferable to those based on treatment plant number. This is because flow-based estimates reflect the total mass of LAS treated and are more conservative, especially under average river flow conditions.

Figure 7 shows predicted LAS concentrations under both average or mean flow conditions and critical low-flow conditions, represented by 7Q10. Removal rates for LAS during wastewater treatment are based on measured average values in activated sludge, trickling filters, and primary settling across the United States. The 7Q10 flow conditions (7 consecutive days/10-yr low flow) are used to estimate maximum LAS concentrations under conditions of extremely low flow and represent somewhat of a worst-case situation. They have, however, been validated by national monitoring studies on LAS (monitoring results in Figure 7), which indicate that the

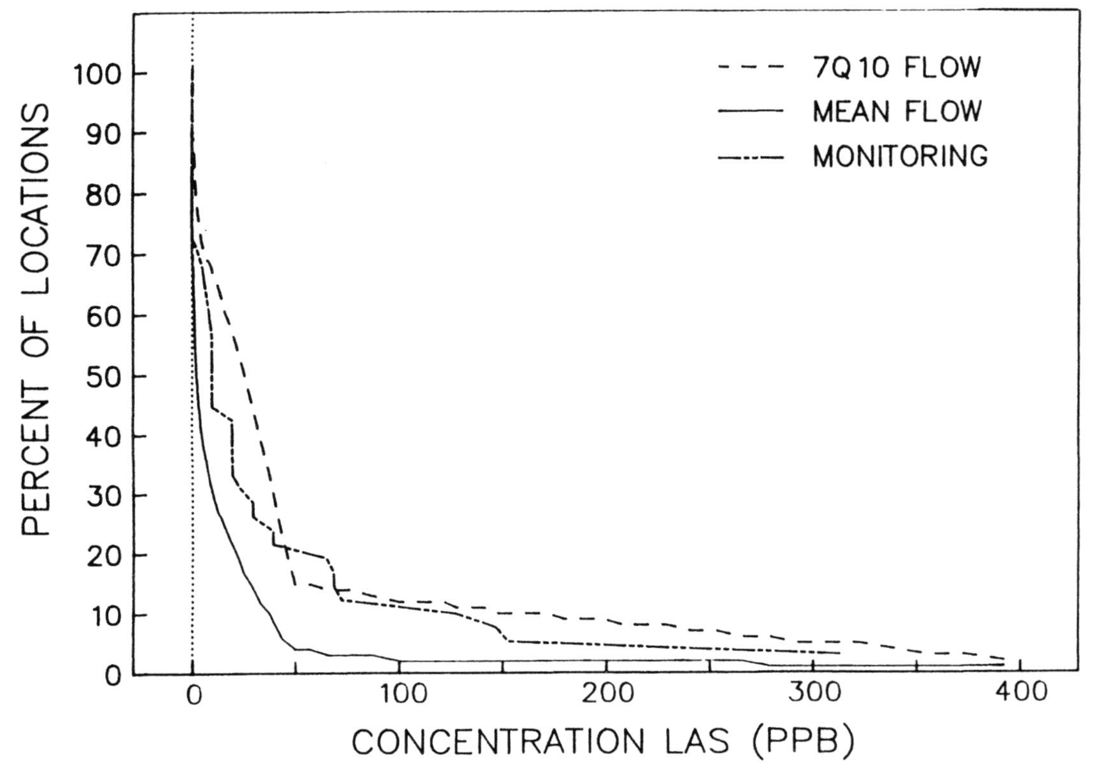

Figure 7. Comparison of mean and critical low-flow concentrations of LAS predicted by PG-GRiDS with measured LAS concentrations in the mixing zone of low-dilution rivers in a national monitoring study. (Reprinted with permission from *Environmental Toxicology and Chemistry*, 12(6), McAvoy DC, Eckhoff WS, Rapaport RA, Fate of linear alkylbenzene sulfate in the environment. Copyright 1993, SETAC.)

distribution of LAS in low-dilution receiving waters closely approximates the 7Q10 predictions (McAvoy et al., 1993).

The frequency predictions obtained from PG-GRiDS using flow-based analysis indicate that >95% of the LAS concentrations in rivers and streams are less than 50 ppb. Under critical low-flow conditions (7Q10), this figure drops to about 85%. Based on NEC values for a range of aquatic species (see level I risk assessment), predicted safety margins at these exposure concentrations are 10-fold or greater. These safety factors assume broad usage, distribution, and national exposure conditions and therefore cover a range of possible exposure scenarios. They do not, however, consider the mitigating effects of in-stream removal processes like sorption or biodegradation. As discussed later, these processes have the potential to reduce LAS exposure concentrations significantly, thereby significantly increasing safety factors in environmental compartments located outside the zone of immediate impact.

Chemical Monitoring

When suitable analytical methods are available, field monitoring studies clearly provide the most accurate exposure information on the actual concentration of chemicals in a specific environment. As discussed earlier, specific and sensitive analytical techniques are available to measure LAS. These methods have been used with considerable success to measure LAS concentrations in a variety of environmental compartments, including wastewaters, river and stream water, sediments, and sludge-amended soils (De-Henau et al., 1986; Giger et al., 1989). Figure 8 summarizes the results of one of the most recent studies, an extensive national monitoring study conducted by McAvoy et al. (1993). This study measured the concentration of LAS in surface waters below a number of wastewater treatment plants having low effluent dilution factors. The low dilution factor plants were specifically chosen to maximize the accuracy of LAS analytical measurements, which were performed using a sensitive HPLC technique specific for individual LAS homologues (Castles et al., 1989). The study also measured LAS concentrations in influent and effluent wastewaters, sewage sludges, river sediments, and sludge-amended soils.

In general, the surface water concentrations of LAS in the mixing zone below wastewater treatment plants were generally less than 0.05 mg/L, even

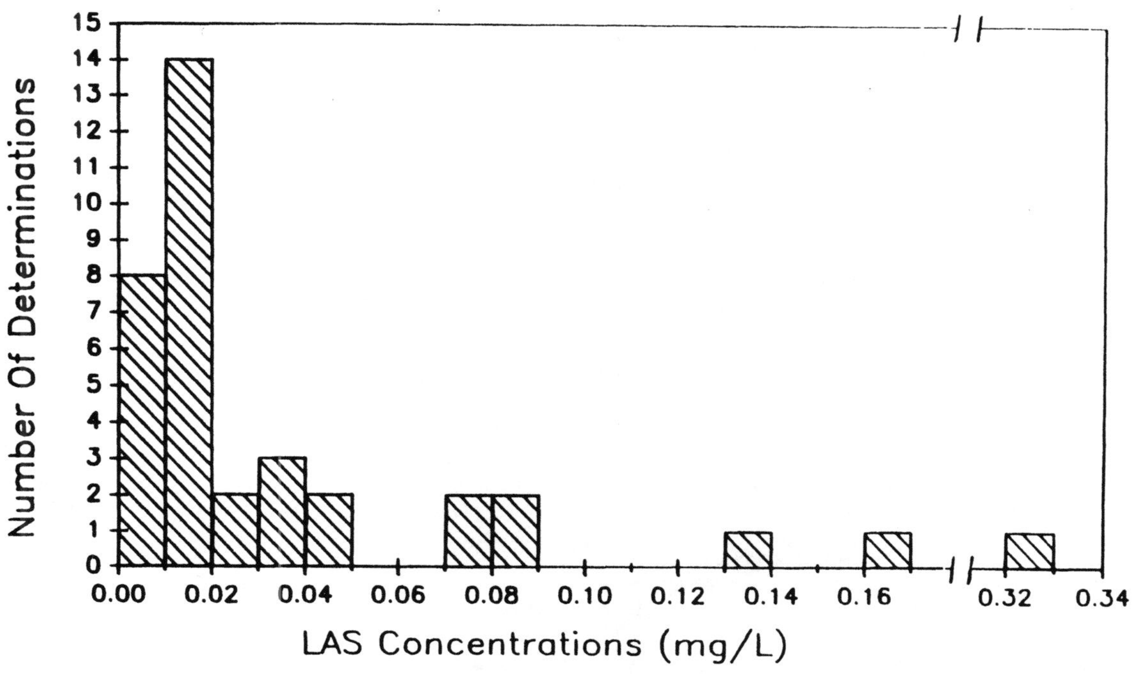

Figure 8. Frequency distribution of LAS concentrations in the mixing zone of low-dilution rivers below MWTP outfalls. (Reprinted with permission from *Environmental Toxicology and Chemistry*, 12(6), McAvoy DC, Eckhoff WA, Rapaport RA, Fate of linear alkylbenzene sulfate in the environment. Copyright 1993, SETAC.)

though the water samples were collected under low-flow conditions at locations having minimal effluent dilutions. The measured concentrations of LAS in this national monitoring study also compared quite well with PG-GRiDS model predictions under 7Q10 flow conditions (Figure 7), indicating that the model was a good predictor of actual concentrations of LAS in the environment. Good agreement between measured and predicted concentrations of LAS has also been obtained in influent and effluent wastewaters, river sediments, and sludge-amended soils (Hennes and Rapaport, 1989). In the case of sediment and soil systems, the exposure models used are much simpler than those represented by PG-GRiDS, and the monitoring database is less extensive. However, the good correlations obtained between measured and predicted concentrations provide reasonable assurance that LAS exposure concentrations can be accurately established in nonaqueous compartments. Good correlations between model predictions and monitoring results have also been obtained in subsurface environments exposed to LAS in OSDS effluents (Robertson et al., 1989; McAvoy et al., 1993).

Biodegradation

Role of Biodegradation

As shown in Figure 3, biodegradation (see Chapters 1, 15 and 27) plays a major role in the removal of LAS during wastewater treatment. This removal results in significant reductions in LAS influent and effluent wastewater concentrations and plays a key role in maintaining environmentally acceptable safety factors. Biodegradation also plays a major role in the removal of LAS in posttreatment environmental compartments, including fresh and marine surface waters, freshwater and marine sediments, and surface and subsurface soils. Unlike removal in POTWs, however, removal in posttreatment environments has not been traditionally considered in level I risk assessments, even though this removal also has the potential to reduce LAS exposure concentrations significantly. The lack of consideration of biodegradation and other posttreatment removal processes in risk assessment is not due to the absence of biodegradation data, because these data are often generated at the same time as other effects information. Rather, it is due to the absence of a conceptual framework for how to incorporate biodegradation information more quantitatively into posttreatment environmental risk assessments.

The following sections outline an approach and provide specific criteria for incorporating biodegradation information, specifically biodegradation rate information, into environmental exposure calculations used for risk assessments. The approach is a generic one that relates the biodegradation half-lives of chemicals to their residence times in specific environmental compartments. This generic approach is broadly applicable to any chemical in any environment, and it provides the basic contextual framework for extending environmental risk assessments beyond the mixing zone (i.e., end of the sewage pipe).

Types of Biodegradation

In broad terms, biodegradation can be defined as any process mediated by living organisms that results in the conversion of an organic chemical into organic and/or inorganic end products that are chemically distinct from the parent material. A more precise technical definition of biodegradation would be the metabolism of organic chemicals as sources of carbon and energy by heterotrophic microorganisms (primarily bacteria and fungi) to form microbial biomass, inorganic end products like carbon dioxide and water, and simple organic end products like methane (i.e., from anaerobic microorganisms). This latter type of biodegradation, termed ultimate biodegradation or mineralization, is highly significant from an environmental standpoint. It leads to a total loss of molecular identity and the complete reassimilation of a synthetic organic chemical into natural elemental cycles. Chemicals that undergo only primary biodegradation (i.e., loss of parent compound), or are partially degraded or biotransformed to persistent organic intermediates, do not experience significant mass reductions and are not converted into simple inorganic building blocks. For these materials, biodegradation is not a practically significant environmental removal mechanism.

Biodegradation Rate and Chemical Residence Time

Although the extent or completeness of mineralization is a key factor in assessing the true biodegradation potential of a chemical, it is not the most important factor controlling the environmental significance of biodegradation as a removal mechanism. Two factors determine whether biodegradation will be a meaningful removal mechanism in the environment: (1) the kinetics or rate of biodegradation in a particular environmental compartment and (2) the time available for biodegradation to occur, based on the residence time of a chemical in that compartment. Residence times are important because they can vary significantly from hours to years depending on the compartment and the distribution of the chemical in different compartments.

For LAS and a variety of other L&C chemicals, an extensive database has been developed over the years showing that biodegradation rate processes can be adequately described by a simple first-order reaction, $C/C_0 = \exp^{-kt}$, where C and C_0 are the concentration at time t and the initial concentration, respectively, and k is the first-order rate constant with units of reciprocal time (Shimp et al., 1990; Larson, 1991; Cowan et al., 1993). During first-order degradation, the rate of biodegradation is directly proportional to chemical concentration. The time required to reduce the concentration of parent material to 50% of its original value, or to yield half the maximal amount of an end product like CO_2, is termed the half-life value. Mathematically, the biodegradation half-life (BHL) value is derived from the first-order rate constant by the equation $BHL = \ln(2)/k = 0.693/k$. Although the actual rate of degradation varies directly with concentration, the biodegradation half-life remains constant. This is shown in Figure 9, which shows the change in biodegradation rate as a function of chemical concentration over several consecutive half-lives. At each time point on the x axis, the concentration changes by exactly 50% from the previous value. From an environmental exposure standpoint, the time frame represented by the biodegradation half-life is an extremely useful parameter. It can be directly related to the residence time of chemicals in specific environmental compartments and used to calculate exposure concentrations as a function of time.

As alluded to above, the second major factor determining whether biodegradation will be practical removal mechanism in the environment is the relationship between the biodegradation rate or half-life of a chemical and its residence time in specific environmental compartments. Residence time (see Chapter 1) is defined quite broadly as the time available for biodegradation to occur, relative to such factors as chemical application or loading rates, transport velocities within an environmental compartment, or travel times between multiple inputs. The relationship between the biodegradation half-life and residence time is shown in Figure 10, which indicates the amount of biodegradation that can occur in a given environmental compartment as a function of the biodegradation half-life/residence time (BHL/RT) ratio.

As indicated in Figure 10, significant biodegradation (50%) occurs when the biodegradation half-life for a chemical equals its residence time in a particular environment (BHL/RT = 1). The total amount of biodegradation occurring increases still further as the biodegradation half-life becomes a smaller fraction of the chemical residence time (BHL/RT <1). For example, at BHL/RT ratios of 0.1 or less (which represents 10 biodegradation half-lives within a given residence time), biodegradation is essentially complete (i.e., >99.9%). By contrast, biodegradation is much less effective as a removal mechanism when the biodegradation half-life exceeds the residence time of a chemical in a given environmental compartment (i.e., BHL/RT > 1). At BHL/RT ratios of about 4 or greater, less than 15% removal occurs due to biodegradation. Materials that exhibit this range of BHL/RT ratios are nondegradable in a practical sense and will accumulate in natural environments at rates that are directly proportional to their annual usage rates and the total length of use.

Accumulation and Persistence

Accumulation and persistence are two key issues that often arise when conducting risk assessments for nonbiodegradable organic chemicals. Underlying these issues is the very real concern that nondegrad-

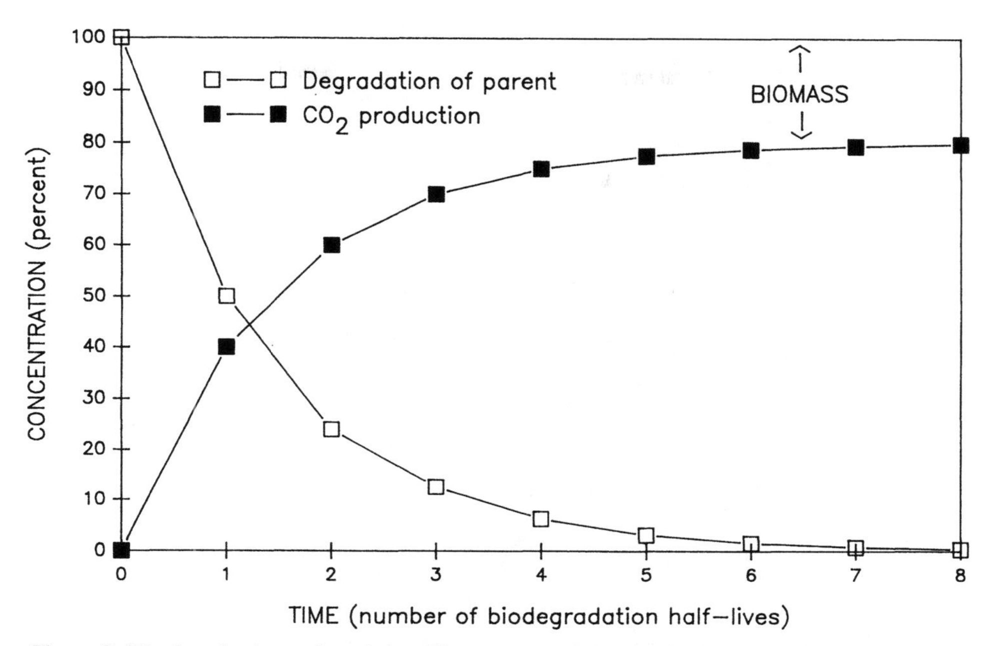

Figure 9. Kinetics of primary degradation (disappearance of parent compound) and ultimate degradation (mineralization to CO_2) during first-order degradation. The x axis is plotted as a function of biodegradation half-life to illustrate the time independence of half-life values.

able chemicals constitute an unmanageable load or burden to the environment because they have the potential to accumulate in specific environmental compartments. Once present, these accumulated materials can persist for extended periods of time with unknown or uncertain ecological consequences. Although nonbiodegradability is not, in and of itself, evidence of an environmental problem, it is indicative of a chronic environmental exposure that cannot be quickly ameliorated by discontinued use. Such chronic exposure is intuitively difficult to accept because it isn't possible to predict with absolute certainty the long-term consequences.

Although accumulation and persistence are often defined in qualitative terms, it is relatively easy to dimension quantitatively the potential for materials to accumulate and persist in the environment using the BHL/RT approach. To prevent the buildup or accumulation of a chemical in a specific environmental compartment, the biodegradation half-life must be equal to (or less than) the residence time for the chemical in that compartment. When the biodegradation half-life is equal to the chemical's residence time, the maximum mass loading of the chemical in the environment at steady state will *never* exceed the input rate, and hence no net buildup above this rate will occur. When the half-life is less than the residence time, the maximum mass loading will obviously be some fraction of the annual usage rate. By contrast, when the BHL/RT ratio is greater than one, buildup or accumulation of that material will necessarily occur as a function of the usage rate. At a BHL/RT ratio of 10, for example, accumulation is essentially linear and directly proportional to usage rate. At high BHL/RT ratios, accumulation of a chemical can result in environmental concentrations orders of magnitude higher than the annual usage rate. These high concentrations also persist for extended periods after the input has ceased and have an increased potential to exert chronic toxicological effects.

Based on their potential for accumulation and persistence in specific environmental compartments, four different categories of biodegradable chemicals can be described. These categories are shown in Figure 11.

Rapidly biodegradable ("transient") chemicals have biodegradation half-life/residence time ratios significantly less than one (i.e., 0.1 to 0.3). They reach equilibrium or steady-state concentrations essentially instantaneously in the environment, and steady-state concentrations represent only a small fraction (<10%) of the input concentrations. Once the input of these types of chemicals is stopped, steady-state concentrations decrease to zero almost immediately. Rapidly biodegradable materials, which include LAS and a number of other L&C chemicals, do not build up or persist in the environment and have a very limited potential to exert chronic ecotoxicological effects.

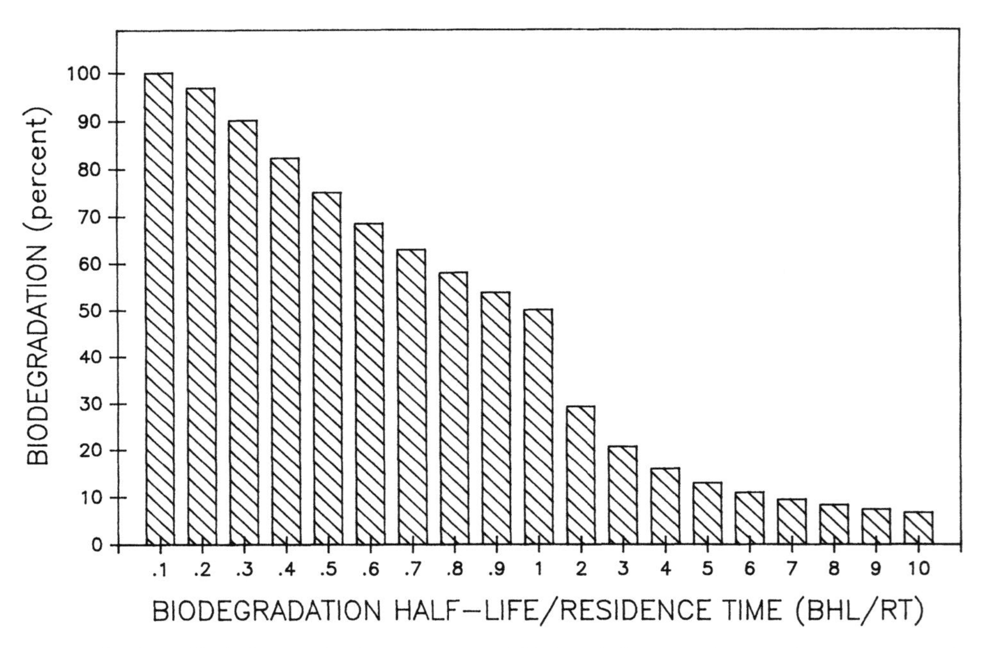

Figure 10. Generalized diagram showing the amount of biodegradation occurring in a given environmental compartment as a function of biodegradation half-life/residence time (BHL/RT) ratio. At BHL/RT = 1, 50% biodegradation occurs and steady-state concentrations will never exceed chemical input rates (i.e., no accumulation occurs).

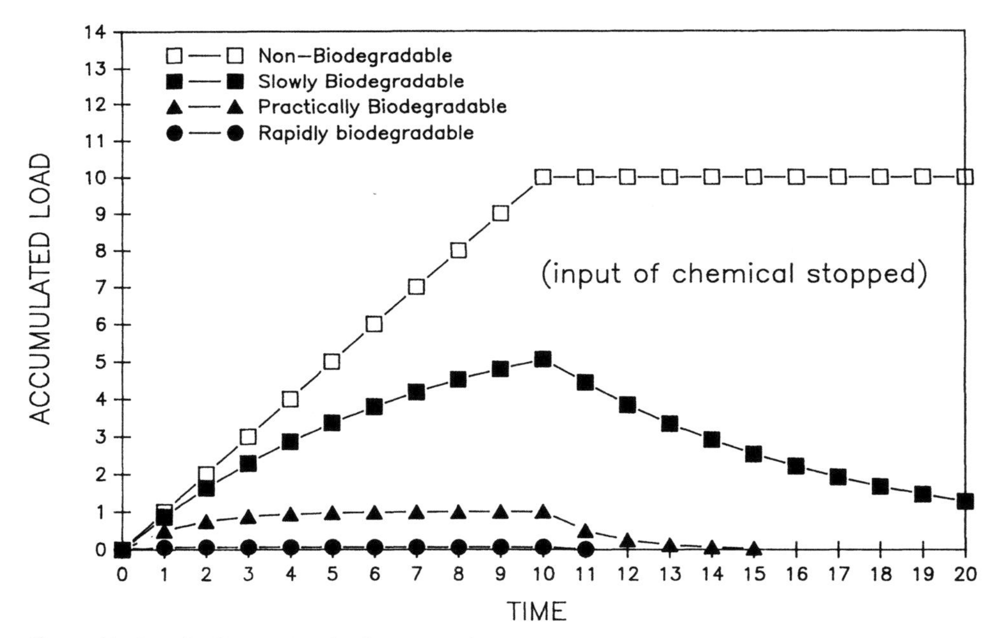

Figure 11. Generic diagram for classifying the biodegradability of chemicals based on their potential for accumulation and persistence in the environment. The diagram is applicable to any environmental compartment in which first-order biodegradation kinetic are occurring.

Practically biodegradable ("biodegradable") chemicals have biodegradation half-lives that do not exceed their residence time in specific environmental compartments (BHL/RT ratio less than or equal to one). They also reach steady-state concentrations quickly in the environment and do not accumulate to concentrations above the input rate once steady state has been established. If the input of these chemicals is stopped, steady-state concentrations decrease relatively quickly to baseline values. Practically biodegradable chemicals, therefore, do not accumulate above their input rate and would not be considered persistent in the environment.

Slowly biodegradable ("persistent") chemicals have biodegradation half-lives that exceed their residence times in specific environmental compartments (a BHL/RT ratio of five is used in Figure 11). These chemicals accumulate as a function of their usage and removal rates and persist for extended periods of time after their use has been discontinued. Slowly biodegradable chemicals will not persist indefinitely, because finite rates of degradation do occur. They can, however, reside in environmental compartments for extended periods of time after their input has been stopped and have an increased potential to exert chronic toxicological effects.

Nonbiodegradable ("recalcitrant") chemicals have a biodegradation rate approaching zero and show no evidence of significant mineralization by environmentally realistic assemblages of microorganisms in specific compartments. They accumulate as a direct function of their usage rate and do not reach equilibrium or steady-state concentrations. Recalcitrant chemicals persist indefinitely in the environment after their input has been stopped at concentrations determined by the rate and duration of their use.

In ranking the relative exposure potential of the various biodegradation categories, rapidly biodegradable (transient) and practically biodegradable (biodegradable) materials are ideal, because they are rapidly removed from the environment and allow upper bounds to be defined for safety factor calculations. Safety factors cannot be conclusively established for slowly biodegraded (persistent) and nonbiodegradable (recalcitrant) chemicals, because these chemicals accumulate and persist in the environment. With long-term use, safety factors for persistent and recalcitrant chemicals actually decrease in direct proportion to the rate and duration of their use. This decrease in safety factors is due primarily to a continuous increase in steady-state concentrations as a result of slow or ineffective biodegradation.

To reduce the number of biodegradation categories described above and simplify the process for incorporating biodegradation data into environmental exposure analysis, Rapaport (1993) has proposed that three criteria be used for separating L&C chemicals into practically biodegradable and nondegradable categories. The three criteria for practically biodegradable materials focus on removal and biodegradation in activated sludge wastewater treatment and are (1) the potential for complete mineralization and/or conversion to microbial biomass, (2) >90% removal in activated sludge, with (3) a significant fraction of this removal (40–50%) due to mineralization. These criteria are intended to minimize the exposure and dispersion of L&C chemicals in the environment and to be consistent with regulatory guidelines concerning biodegradability claim support. As such, they provide useful and transparent "ground rules" for classifying specific materials that are easy to communicate and consistent with the kinetic principles outlined above.

Effects

A number of additional studies and analyses have been carried out in order to reduce the uncertainties in the initial (level I) risk assessment regarding ecotoxicological effects (see list in Table 3). The sensitivity of a wider range of aquatic and terrestrial species was assessed, bioavailability and toxicity were addressed in the sediment compartment, the estimates of bioconcentration based on octanol-water partitioning were confirmed in laboratory studies with fish, and the toxicity of LAS degradation intermediates was evaluated. These new data often involved novel test designs including laboratory (e.g., sediment toxicity assays; Pittinger et al., 1989), in situ, and field evaluations (e.g., phytoplankton and periphyton assays; Lewis, 1986; Lewis and Hamm, 1986). The following sections provide the basic new information and also references for studies that have been published. Painter and Zabel (1988) and Kimerle (1989) have provided a thorough review of all the available freshwater, marine, and terrestrial toxicity data for LAS.

To assess whether *Daphnia* is a good surrogate (i.e., protective) to assess the sensitivity of other aquatic invertebrates, Lewis and Suprenant (1983) compared the acute toxicity of LAS for six freshwater species (48-h LC50 for isopod = 270 mg/L, for midge = 23, for nematode = 16, for amphipod =

Table 3. Effects Studies Conducted as Part of Level II Risk Assessment for LAS

- Comparative acute toxicity for seven aquatic invertebrate species
- Terrestrial plant toxicity evaluation
- Comparison of laboratory algal assays to field exposures of phytoplankton and periphyton
- Toxicity of sediment-sorbed material to benthic macroinvertebrates
- Fish bioconcentration evaluation
- Loss of aquatic toxicity as a function of biodegradation
- Estimate of a "water quality criterion" value using U.S. EPA approach

3.3, for flatworm = 1.8, and for oligochaete = 1.7) to that for *Daphnia magna* (3.7 mg/L). *D. magna* was as sensitive as the more sensitive species tested and about 70 times more sensitive than the least sensitive invertebrate. Thus *Daphnia* appears to be a conservative predictor of the sensitivity of a wider range of aquatic invertebrates, at least for LAS.

Kimerle (1989) presented a summary of previously unpublished terrestrial toxicity data for LAS. Over three dozen species of plants have been tested using hydroponic, foliar, or soil exposures. No-effect concentrations ranged from about 10 to 1000 ppm with approximately 90% of the values at or above 100 ppm.

Lewis (1986) and Lewis and Hamm (1986) report that the toxicity threshold levels (i.e., the lowest concentration at which there was a statistically reduced number of algal cells) for LAS in standard 96-h laboratory algal assays (''first-effect concentration'' of about 1.0 mg/L) are 27 to 108 times lower than the toxicity threshold levels of the same LAS in 21-d field studies of natural phytoplankton communities. The toxicity end points in the field studies were community structure (i.e., species diversity and density; see Chapters 1 and 19) and the first-effect concentration was between 27 and 108 mg/L. Lewis et al. (1986) also conducted 21-d field studies of periphyton communities exposed to LAS. The first-effect concentrations were 3 to 10 times lower in the laboratory than in the field. The toxicity end points in the field were again community structure and the first-effect concentration was between 3 and 10 mg/L LAS. Therefore, algae, including both free-floating and attached varieties, appear not to be particularly sensitive to LAS and would be protected based on a LAS aquatic toxicity threshold developed for a more sensitive trophic level, for example, freshwater fish (chronic toxicity threshold of about 0.2 mg/L).

The bioavailability and chronic toxicity of LAS in sediments were studied using a relatively new benthic toxicity test procedure. Pittinger et al. (1989) reported the chronic no-effect and first-effect concentrations for the midge, *Chironomus riparius,* to be 319 and 993 ppm, respectively. These values are much higher than the 2.4 ppm chronic no-effect concentration for the same species in a water-only exposure. Thus the sorbed LAS is apparently less bioavailable to benthic organisms. Other researchers have confirmed the reduced bioavailability of LAS in sediments. Stream field data reported by Ladle et al. (1989) showed no adverse effects of LAS on benthic invertebrates (segmented aquatic worms, midge, mayfly, stone fly, caddis fly) dwelling in river sediments containing LAS concentrations as high as 40 mg/kg (ppm). Additional studies from Bressan et al. (1989) on marine benthic invertebrates (mussels, benthic copepods) showed similar no-effect concentrations.

The estimate of the low bioconcentration potential of LAS (a BCF of 112 based on its octanol-water partition coefficient) has been confirmed in laboratory fish bioconcentration studies with bluegill. Intact LAS accumulated from approximately 35- to 220-fold in whole fish and in muscle (Painter and Zabel, 1988). There was evidence of significant metabolism and rapid elimination (within 5 to 10 d) in the same fish.

Finally, it has been demonstrated that partially degraded LAS is less toxic than intact LAS. Kimerle and Swisher (1977) used a combined river water die-away test and *Daphnia* acute toxicity test to show a 10-fold reduction in toxicity of the carboxylated degradation intermediates of LAS (sulfophenylcarboxylates).

The relatively large ecotoxicological database for LAS does provide the basis for one additional approach to assessing potential risk to freshwater aquatic systems. That is to approximate a national water quality criterion (WQC) value using the U.S. EPA procedures applied to many of the ''priority pollutants'' (U.S. EPA, 1986). The goal of the criterion is to provide protection to 95% of the species and it is assumed that the available toxicity data cover many of the most sensitive species. Criterion values were estimated for C_{12} LAS following the guidelines established by the U.S. EPA for calculating national water quality for protecting aquatic organisms (Stephan et al., 1985). Data for seven species of fish, ten species of invertebrates, and three algae were available to develop the criterion estimates. The criterion maximum concentration (CMC), the hourly average exposure concentration which if not exceeded more than once every 3 yr should be protective, was estimated to be 0.55 ppm; the criterion continuous concentration (CCC), the 4-d average exposure concentration which if not exceeded more than once every 3 yr should be protective, was 0.21 ppm. These estimated criterion values are in fact very close to the lowest acute LC50 (~1 ppm) and lowest chronic NOEC (~0.2 ppm) used above to estimate ''safe concentrations.'' The WQC approach is considered to be a conservative and protective approach to evaluating aquatic risk on a nationwide average basis. It also provides an added measure of comfort in the risk assessment for LAS, although its use is restricted to a few chemicals due to the large database required.

A field study was conducted to confirm that the estimated CCC value (0.21 ppm) would be protective of aquatic life. An outdoor stream community, consisting of a diverse invertebrate benthic population and *Hyallela azteca* and *Pimpephales promelas,* was exposed to 0.35 ppm of C_{12} LAS for a 45-d period (Fendinger et al., 1994). No effects were noted on the biota in this study, supporting the CCC estimated values as a safe level.

Level II Analysis of Environmental Risk

Introduction

This section continues the level I risk assessment process by providing a more thorough and rigorous environmental analysis. The impetus for this effort is the relatively small margin between the environmental exposure estimates and the predictions of toxicity thresholds for sensitive species. The standard (level I) ecological risk assessment methodology is generally considered to be adequately conservative (i.e., protective of most species/situations). However, the broad-scale use of LAS, the uncertainties in the model- and/or laboratory-to-field extrapolations, the lack of data for some compartments (e.g., sediments), and the opportunity to develop and use novel approaches (including exposure modeling and field monitoring, biodegradation, and toxicity) were all incentives for the level II assessment. Three key environmental compartments for LAS—surface water, sediments, and agricultural soils—are analyzed below.

Surface Water

The majority (90+%) of the volume of LAS used in laundry and cleaning products (mostly L&Cs) is removed and biodegraded during the wastewater treatment process. Some residual LAS (approximately 0.5–2.0 mg/L) is discharged in treatment effluent. Kimerle (1989) and McAvoy et al. (1993) reported the results of a national monitoring study indicating that measured LAS concentrations in rivers and streams below wastewater outfalls range from less than 0.01 ppm under mean river flow conditions to about 0.3 ppm LAS at critical low flow (7Q10). The monitoring sites were specifically chosen to represent worst-case situations. As shown in Figure 8, approximately 80% (29 of 36 locations) of the surface water LAS concentrations fell between 0.01 and 0.05 ppm and approximately 90% (33 of 36 locations) between 0.01 and 0.1 ppm. The highest few concentrations were for sites where there was inadequate treatment or for sites very near an outfall. As discussed later, the extent and duration of downstream exposure are limited due to the rapid and complete biodegradation of LAS.

The most encompassing estimate of LAS toxicity thresholds for aquatic organisms exposed in the water column is based on the national water quality criterion methodology, which integrates the data for a wide range of fish, invertebrates, and algal species and is intended to protect the most sensitive species. The acute toxicity (maximum 1-h; CMC) threshold is approximately 0.5 mg/L LAS. The chronic toxicity (maximum 4-d average; CCC) threshold is approximately 0.2 ppm. Comparing the maximum exposures of ~0.01–0.3 mg/L below wastewater outfalls to the toxicity thresholds of ~0.2–0.5 mg/L and considering the extensive database from which these values are derived (Painter and Zabel, 1988), plus the results of the outdoor stream study reported by Fendinger et al. (1994), it is concluded that the risks to aquatic species are minimal. The overlap between the chronic toxicity threshold (for the most sensitive species) and the maximum measured LAS exposure concentration (0.3 mg/L) is not considered to represent a serious risk. This is because of the extreme unlikelihood that areas immediately below wastewater outfalls would support trout (or another equally chemical-sensitive species) considering that these are often areas of high biological oxygen demand (BOD) and ammonia, low oxygen, high solids, anaerobic sediments, pathogenic microbes, etc. Trout (chronic NOEC of 0.2 to 1.0) were about five times more sensitive than the next aquatic species. In addition, Rapaport and Eckhoff (1990) have shown that the sorptive nature and rapid biodegradation of LAS result in rapid decreases in LAS exposure concentrations with distance downstream. This would result in increased safety factors for trout and other sensitive organisms.

Sediments

A small portion (<10%) of the LAS discharged as part of treated wastewater will be associated with the solids. These solids will eventually settle out in the receiving water. LAS in solution will also adsorb to suspended and settled solids in the surface water. In recent monitoring studies, measured sediment concentrations were <4 μg/g (ppm) directly below activated sludge plants and <200 μg/g below two of three trickling filter plants (McAvoy et al., 1993). Furthermore, LAS concentrations in sediments have been observed to decrease by more than an order of magnitude within a 7 km of a sewage outfall (Rapaport and Eckhoff, 1990; Hennes and Rapaport, 1989; Kimerle, 1989). A substantial margin of protection is therefore expected for sediment benthos. This results in increased safety factors for downstream benthic communities. The laboratory sediment toxicity study (Pittinger et al., 1989) indicated that the chronic toxicity threshold is between 319 (NOEC) and 993 μg/g LAS. Comparing the sediment monitoring results to the chronic toxicity threshold, it can be concluded that LAS does not pose a significant risk in the sediments.

Agricultural Soils

Sludges from both primary and secondary wastewater treatment facilities will contain some LAS. Approximately 25–30% of the LAS that is removed in treatment is associated with the sludge (Painter and Zabel, 1988). It is estimated that the LAS concentration in soil immediately following sludge application (usually limited to once per year) is typically between 10 and 50 mg/kg (ppm). These concentrations have been confirmed in published monitoring studies (DeHenau et al., 1986; Marcomini et al., 1989; Holt et al., 1989). Tests with plants grown on sludge-amended soils indicate that toxicity thresholds are above 100 mg/kg LAS (Kimerle, 1989). Tests with earthworms show that their toxicity threshold is above 500 mg/kg (Kimerle, 1989). LAS present in sludge-amended soils degrades at relatively high rates with half-lives ranging from 3 to 35 d (Painter and Zabel, 1988). These data suggest that there is no significant risk to terrestrial biota from LAS associated with sludge in soils.

Effect of Biodegradation on LAS Safety Factors

Based on the half-life values determined for LAS biodegradation in both engineered systems (wastewater treatment) and a variety of natural environmental compartments, biodegradation is a significant removal mechanism for LAS (Larson et al., 1993). Half-lives for LAS in activated sludge treatment systems, where the sludge residence time is 1 to 2 wk, are 1 to 2 d. Half-lives in aquatic and benthic compartments, where the residence time can vary from days to weeks, are 1 d or less. Half-lives in terrestrial and subsurface compartments, where residence times can vary from months to years, range from less than 1 d to a few weeks. Conservative estimates of the BHL/RT ratio for LAS range from 0.5 to less than 0.1. These ratios translate to a reduction in LAS steady-state exposure concentrations by 75% to >>99% (several orders of magnitude; see Figure 10). These steady-state concentrations are well below those having adverse ecological impacts.

Figure 12 more directly illustrates the effect of biodegradation on LAS safety factors. The plot is a generic one that relates the relative increase in LAS safety factor to the ratio of its biodegradation half-life to residence time. It shows that biodegradation can significantly increase relative safety factors in any environmental compartment by decreasing LAS exposure concentrations. Clearly, relative safety factors for LAS multiply sharply as the BHL/RT ratio becomes less than 1. At a BHL/RT ratio of 0.3, for example, safety factors are multiplied by a factor of 10. At a ratio of 0.1, they are multiplied a thousandfold, and at 0.05, they are more than six orders of magnitude higher than at BHL/RT ratios where biodegradation is not a practical removal mechanism (i.e., BHL/RT > 5). These increased safety factors clearly indicate that biodegradation plays a key role in maintaining LAS exposure concentrations well below those predicted to have adverse environmental impacts. They also help to explain the long history of safe LAS use in laundry and cleaning products and the lack of accumulation of this material in key environmental compartments, even given usage levels in the hundreds of millions of pounds per year.

SUMMARY

Traditional (level I) approaches for evaluating the environmental safety of L&C chemicals rely on the development of specific safety factors, which compare the expected environmental concentration to the measured or estimated chronic no-effect concentration for the most sensitive test species. Safety factors derived from level I approaches typically reflect mixing zone or "end of the sewage pipe" situations and the numerous limitations and uncertainties associated with level I risk assessments. Level II risk assessments, on the other hand, utilize comprehensive determinations of the fate, effects, and exposure potential of specific L&C chemicals to generate more reliable and accurate environmental safety factors.

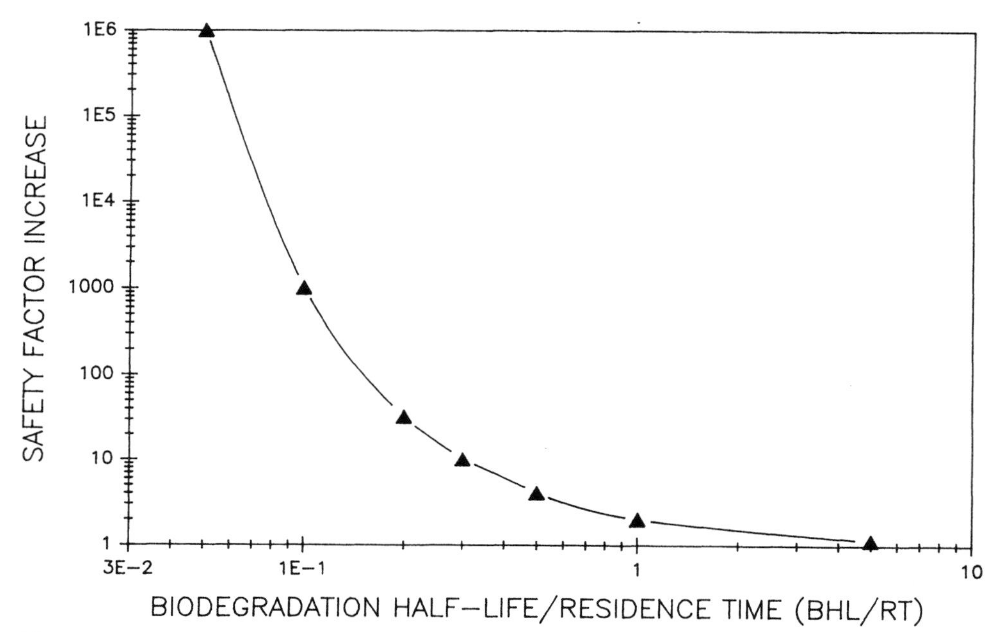

Figure 12. Change in relative LAS safety factors as a function of biodegradation half-life/residence time (BHL/RT) ratio. Note that at low BHL/RT ratios, relative safety factors are increased by several orders of magnitude.

They also consider chemical or biological removal mechanisms that may be operative in the environment and can significantly decrease actual environmental exposure concentrations.

In this chapter, a risk assessment case study was presented for a specific, large-volume L&C chemical, LAS, with an illustration of level I and II approaches to estimate the environmental safety factors for this material. In the case of LAS, these safety factors have been developed and refined by an environmental database that is one of the most extensive developed for any synthetic chemical. They have also been verified by history of safe use spanning almost 30 yr, during which time no adverse environmental impacts have been identified that are attributable to the use of LAS in consumer products.

LITERATURE CITED

Abbott, D. C.: Colorimetric determination of anionic surfactants in water. *Analyst*, 82:266–293, 1962.

Belanger, S. E.: Use of experimental stream mesocosms to assess ecosystem risk: a case study with a cationic surfactant. In: Predicting Ecosystem Risk, edited by J. Cairns, Jr., B. R. Niederlehner, D. R. Orvos, pp. 263-287. Princeton, NJ: Princeton Univ. Press, 1992.

Belanger, S. E.: Review of experimental microcosm, mesocosm and field tests used to evaluate the potential hazard of surfactants to aquatic life and the relation to single species data. Freshwater Field Tests for Hazard Assessment of Chemicals, edited by I. R. Hill, F. Heimbach, P. Leenwangh, P. Matthiessen, pp. 287-314. Boca Raton, FL: CRC, 1994.

Bressan, M.; Brunetti, R.; Casellato, S.; Fava, G. C.; Giro, P.; Marin, M.; Negrisolo, P.; Tallandini, L.; Thomann, S.; Tosoni, L.; and Turchetto, M.: Effects of linear alkylbenzene sulfonate (LAS) on benthic organisms. *Tenside Surf. Det.*, 26:148–158, 1989.

Castles, M. A.; Moore B.L.; and Ward S.R.: Measurement of linear alkylbenzene sulfonates in aqueous environmental matrices by liquid chromatography with fluorescence detection. *Anal. Chem.*, 61:2534–2540, 1989.

Chemical Economics Handbook. Menlo Park, CA: Stanford Research Institute, 1992.

Cowan, C. E.; Larson, R. J.; Feijtel, T. C. J.; and Rapaport R.A.: PG-TREAT: an improved model for predicting the fate of consumer product chemicals in wastewater treatment plants. *Water Res.*, 27:561–573, 1993.

DeHenau, H.; Hopping, W. D.; and Matthijs, E.: Linear alkylbenzene sulfonates (LAS) in sewage sludges, soils and sediments: analytical determinations and environmental safety considerations. *Int. J. Environ. Anal. Chem.*, 26:279–293, 1986.

Di Toro, D. M.; Dodge, L. J.; and Hand, V. C.: A model for anionic surfactant sorption. *Environ. Sci. Technol.*, 24 (7): 1013–1020, 1990.

Fendinger, N. J.; Versteeg, D. J.; Weeg, E.; Dyer, S.; and Rapaport, R. A.: Environmental behavior and fate of anionic surfactants. Environmental Chemistry of Lakes and Reservoirs, edited by L. A. Baker, pp. 527-557. Advances in Chemistry Series (Vol. 237). Washington, D.C.: *Am. Chem. Soc.*, 1994.

Giger, W.; Alder, A. C.; Brunner, P. H.; Marcomini, A.; and Siegrist, H.: Behavior of LAS in sewage and sludge treatment and in sludge-treated soil. *Tenside Surf. Det.*, 26:95–99, 1989.

Hand, V. C., and Williams, G. K.: Structure-activity relationship for sorption of linear alkylbenzene sulfonates. *Environ. Sci. Technol.*, 21:370–373, 1987.

Hennes, E. C., and Rapaport, R. A.: Calculation and analytical verification of LAS concentrations in surface water, sediment and soil. *Tenside Surf. Det.*, 26:141–147, 1989.

Holman, W. F.: Estimating the environmental concentrations of consumer product components. Aquatic Toxicology and Hazard Assessment: Fourth Conference, edited by D. R. Branson and D. L. Dickson, pp. 159–182. ASTM STP 737. Philadelphia: ASTM, 1981.

Holt, M. S.; Matthijs, E.; and Waters, J.: The concentrations and fate of linear alkylbenzene sulphonate in sludge amended soils. *Water Res.*, 23(6):749–759, 1989.

Holysh, M.; Paterson, S.; and Mackay, D.: Assessment of the environmental fate of linear alkylbenzenesulphonates. *Chemosphere*, 15(1):3–20, 1986.

Kimerle, R. A.: Aquatic and terrestrial ecotoxicology of linear alkylbenzene sulfonate. *Tenside Surf. Det.*, 26: 169–176, 1989.

Kimerle, R., and Swisher, R.: Reduction of aquatic toxicity of linear alkylbenzene sulfonate (LAS) by biodegradation. *Water Res.*, 11:31–37, 1977.

Knaebel, D. B.; Federle, T. W.; and Vestal, J. R.: Mineralization of linear alkylbenzene sulfonate (LAS) and linear alcohol ethoxylate (LAE) in 11 contrasting soils. *Environ. Toxicol. Chem.*, 9:981–988, 1990.

Ladle, M.; House, W. A.; Armitage, P. D.; and Farr, I. S.: Faunal characteristics of a site subject to sewage plant discharge. *Tenside Surf. Det.*, 26:159–168, 1989.

Larson, R. J.; Federle, T. W.; Shimp, R. J.; and Ventullo, R. M.: Behaviour of linear alkylbenzene sulfonate (LAS) in soil infiltration and groundwater. *Tenside*, 26:116–121, 1989.

Larson, R. J.: Structure-activity relationships for biodegradation of linear alkylbenzene sulfonates. *Environ. Sci. Technol.*, 24:1241–1246, 1990.

Larson, R. J.: Biodegradation in regulating the environmental exposure of detergent chemicals. *Chem. Times Trends*, 14:47–55, 1991.

Larson, R. J.; Rothgeb, T. M.; Shimp, R. J.; Ward, T. E.; and Ventullo, R. M.: Kinetics and practical significance of biodegradation of linear alkylbenzene sulfonate (LAS) in the environment. *J. Am. Oil Chem. Soc*, 70(7):645–657, 1993.

Lewis, M. A.: Comparison of the effects of surfactants on freshwater phytoplankton communities in experimental enclosures and on algal population growth in the laboratory. *Environ. Toxicol Chem.*, 5:319–332, 1986.

Lewis, M. A., and Hamm, B. G.: Environmental modification of the photosynthetic response of lake plankton to surfactants and significance to a laboratory-field comparison. *Water Res.*, 20:1575–1583, 1986.

Lewis, M. A., and Suprenant, D.: Comparative acute toxicities of surfactants to aquatic invertebrates. *Ecotoxicol. Environ. Safety*, 7:313–322, 1983.

Lewis, M. A.; Taylor, M. J.; and Larson, R. J.: Structural and functional response of natural phytoplankton and periphyton communities to a cationic surfactant with considerations of environmental fate. Community Toxicity Testing, edited by J. Cairns Jr., pp. 241–268. ASTM STP920. Philadelphia: ASTM, 1986.

Longwell, J., and Maniece, W. D.: Determination of anionic detergents in sewage effluents and river waters. *Analyst*, 80:167–171, 1955.

Lyman, W. J.; Ruhl, W. F.; and Rosenblatt, D. H.: Handbook of Chemical Property Estimation Methods. New York: McGraw-Hill, 1982.

Maki, A. W., and Duthie, J. R.: Summary of proposed procedures for the evaluation of aquatic hazard. *ASTM Spec. Tech. Publ.*, STP 657, 153–163, 1978.

Marcomini, A., and Giger, W.: Simultaneous determination of linear alkylbenzene sulphonates, alkylphenol polyethoxylates and nonylphenol by high-performance liquid chromatography. *Anal. Chem.*, 59:1709–1715, 1987.

Marcomini, A.; Capel, P. D.; Lichtensteiger, T. H.; Brunner P. H.; and Giger W.: Behavior of aromatic surfactants and PCBs in sludge-treated soil and landfills. *J. Environ. Qual.*, 18:523–528, 1989.

Matthijs, E., and DeHenau, H.: Determination of LAS. *Tenside Surf. Det.*, 24:193–199, 1987.

McAvoy, D. C.; Eckhoff, W. E.; and Rapaport, R. A.: Fate of linear alkylbenzene sulfonate in the environment. *Environ. Toxicol. Chem.*, 12:977–987, 1993.

McEvoy, J., and Giger, W.: Determination of linear alkylbenzene sulfonates in sewage sludge by high-resolution gas chromatography/mass spectrometry. *Environ. Sci. Technol.*, 20:376–383, 1986.

Osburn, Q. W.: Analytical methodology for linear alkylbenzene sulfonate (LAS) in waters and wastes. *J. Am. Oil Chem. Soc.*, 63:257–263, 1986.

Painter, H. A., and Zabel, T. F.: Review of the environmental safety of LAS. Co 1659-m/1/EV 8658. Medmenham, UK: Water Research Center, 1988.

Painter, H. A., and Zabel, T. F.: The behavior of LAS in sewage treatment. *Tenside Surf. Det.*, 26:108–115, 1989.

Pittinger, C. A.; Woltering, D. M.; and Masters, J. A.: Bioavailability of sediment-sorbed and aqueous surfactants to *Chironomus riparius* (midge). *Environ. Toxicol. Chem.*, 8:1023–1033, 1989.

Rapaport, R. A.: Prediction of consumer product chemical concentrations as a function of publicly owned treatment works, treatment type and riverine dilution. *Environ. Toxicol Chem.*, 7:107–115, 1988.

Rapaport, R. A.: Biodegradability case Study. *Happi*, 30(5): 120–126, 1993.

Rapaport, R. A., and Eckhoff, W. S.: Monitoring linear alkylbenzene sulfonate in the environment. *Environ. Toxicol. Chem.*, 9:1245–1257, 1990.

Rapaport, R. A.; Larson, R. J.; McAvoy, D. C.; Mielsen, A. M.; and Trehy, M.: The fate of commercial LAS in the environment. Proceedings of the Third CESIO International Surfactants Congress and Exhibition—A World Market. Section E—Environment, pp. 78–87, London, 1992.

Robertson, W. D.; Sudicky, E. A.; Cherry, J. A.; Rapaport, R. A.; and Shimp, R. J.: Impact of a domestic septic system on an unconfined sand aquifer. Proceedings of the International Symposium on Contaminant Transport in Groundwater, pp. 105–112. Stuttgart, Germany, April 4–6, 1989.

Shimp, R. J.: LAS biodegradation in estuaries. *Tenside Surf. Det.*, 26:390–393, 1989.

Shimp, R. J.; Larson, R. J.; and Boethling, R. S.: Use of biodegradation data in chemical assessment. *Environ. Toxicol. Chem.*, 9:1369–1377, 1990.

S.R.I. International: Linear alkylbenzene sulfonates. Chemical Economics Handbook. Report No. 583.8000A. Menlo Park, CA: Chemical Information Services, 1992.

Stephan, C. E.; Mount, D. I.; Hanson, D. J.; Gentile, J. H.; Chapman, G. A.; and Brungs, W. A.: Guidelines for Deriving Numeric National Water Quality Criteria for the Protection of Aquatic Organisms and Their Uses. PB85-227049. Duluth, MN: U.S. Environmental Protection Agency, 1985.

Swisher, R. D.: Surfactant Biodegradation. Second edition. New York: Marcel Dekker, 1987.

Trehy, M. L.; Gledhill W. E.; and Orth R. G.: Determination of linear alkylbenzene sulfonates and dialkyltetralinsulfonates in water and sediment by gas chromatography/mass spectrometry. *Anal. Chem.*, 62:2581–2586, 1990.

U.S. Environmental Protection Agency: Estimating Concern Levels for Concentrations of Chemical Substances in the Environment. Environmental Effects Branch. Washington, D.C.: U.S. EPA, 1984.

U.S. Environmental Protection Agency: Quality Criteria for Water, 1986. EPA 440/5-86-001. Office of Water Regulations and Standards, May 1, 1986.

Ward, T. E., and Larson, R. J.: Biodegradation kinetics of linear alkylbenzene sulfonate in sludge-amended agricultural soils. *Ecotoxicol. Environ. Safety*, 17:119–130, 1989.

Waters, J., and Garrigan, J. T.: An improved microdesulfonation/gas liquid chromatography procedure for the determination of linear alkylbenzene sulfonates in U.K. rivers. *Water Res.*, 17:1549–1562, 1983.

Woltering, D. M., and Bishop, W. E.: Evaluating the environmental safety of detergent chemicals: a case study of cationic surfactants. The Risk Assessment of Environmental Hazards, edited by D. J. Pastenbach, pp. 345–389. New York: Wiley, 1989.

Chapter 32

A HAZARDOUS WASTE SITE AT THE NAVAL WEAPONS STATION, CONCORD, CA.

K. D. Jenkins, C. R. Lee, and *J. F. Hobson*

INTRODUCTION

As indicated in the introductory chapter of this section (Chapter 28), ecological risk assessments (ERAs) can be used to address a wide range of issues and are generally classified as predictive or retrospective. This chapter presents a case study of a retrospective ERA of a hazardous waste site located at the Concord Naval Weapons Station (NWS), Concord, California. The goal of a retrospective ERA is to establish and define the relationship between the sources(s) of contaminants, the distribution of contaminants, exposure of biological end points, and level of effect of this exposure on the ecosystem. Retrospective ERAs take advantage of field data, when available, to define contaminant sources and measure adverse biological effects. This type of evaluation, however, can be complicated by a number of factors, including:

- *Multiple biological end points:* These could include multiple species and various levels of biological organization (e.g., individual, population, community, and ecosystem).
- *Complex exposure pathways:* These are determined by the biological receptors that are of concern.
- *Indirect effects:* Indirect effects such as habitat impairment may be more important than direct exposure to chemicals.
- *Evaluating impacts on ecosystems:* Ecosystems are complex. Their function is often not tightly coupled to stressor inputs, and they show resilience and recovery to varying degrees of stress.

This added complexity associated with ERAs results in a higher degree of uncertainty than is normally associated with human health–based risk assessments, requiring more effort in the initial planning stages so that the final assessment is well focused.

Retrospective Assessments of Hazardous Waste Sites

The cleanup of sites containing historical hazardous waste (e.g., hazardous waste sites) is controlled by *regulatory statute.* The main federal statute is the Comprehensive Environmental Response, Compensation, and Liability Act (CERCLA) as amended by the Superfund Amendments and Reauthorization Act of 1986 (SARA). This act calls upon the Environmental Protection Agency (EPA) to protect human health and the environment with respect to the release or potential release of contaminants from abandoned hazardous waste sites. It makes spills and dumping of hazardous substances less likely through liability. The main provision of CERCLA is the National Contingency Plan (NCP), which provides for responses to the discharges of oil and releases of hazardous substances. The processes of evaluating and remediating hazardous waste sites are broadly defined in this statute and EPA has developed policy and guidance documents to formalize the requirements set forth in this statute (e.g., U.S. EPA, 1989a,b,c).

In evaluating a hazardous waste site the initial step usually consists of a preliminary site investigation (SI) in which the history and information regarding the current state of the site is investigated. The second step is referred to as the remedial investigation (RI). In the remedial investigation, new data are gathered as required to define the nature and extent of the contamination and its potential effects on human health and the ecosystem. Ecological assessments and risk assessments are usually integrated into this phase of the program. The *ecological assessment* (EA) serves to define the current state of the ecosystem and its relationship to contamination on the site. The *ecological risk assessment* (ERA) represents a formal evaluation of the risk associated with the contaminants at the site. The ERA is based on data derived from the RI, EA, and relevant data from the literature. The final step is the development of a feasibility study (FS). The feasibility study is the phase in which various remediation scenarios are defined and evaluated and a final remedial action is chosen.

In this chapter the EA and ERA carried out on the NWS, Concord will be discussed. The EA and ERA were designed to evaluate the impact of historical contamination and to provide information for the development of a plan to remediate the contaminated areas of the site. The initial question here is, "Does the contamination present at the site, if left in place, pose a current or potential threat to the environment?"

Background Information on the Site

The site consists of approximately 124 ha including a freshwater creek that flows from rolling hills down through pastures, under railroad tracks, into a freshwater wetland, then under another railroad trestle and into a brackish water wetland tidal creek that empties into Suisun Bay (Figure 1). Preliminary evaluations of the soil from these areas indicated that the stressors of concern (SOC) were limited to specific heavy metals (i.e., cadmium, copper, lead, and zinc) and metalloids (i.e., arsenic and selenium). In localized areas of the site the concentrations of these SOC were several orders of magnitude higher than those of uncontaminated sites (Lee et al., 1986).

Several potential sources of this contamination were identified in the initial studies of the site. The first was a chemical manufacturing facility located to the east of the brackish tidal wetland (Figure 1). This plant maintained lagoons containing chemical wastes that were normally separated from the tidal wetland by an earthen levee. However, the levee that separated the lagoons from the tidal wetland was breached in the 1960s, resulting in spillage of materials from the waste lagoon into the tidal wetland (Figure 1). Preliminary chemical analysis of materials released from the lagoon indicated that it represented a major source of arsenic and several metals for the tidal wetland. Shortly after the spill, affected soils in the tidal wetland were treated with lime to increase the pH and reduce the mobility of the metals. In spite of this treatment, areas of the tidal wetland affected by the spill were still devoid of plant life or showed reduced plant growth some 20 yr after the spill. Arsenic and trace metal contamination was also identified in several sloughs and mosquito ditches draining the area of the tidal marsh where the breech of the levee occurred. These channels feed into the tidal creeks that ultimately empty into Suisun Bay.

A chemical processing facility located upland of the freshwater wetland was also identified as a potential source of contamination (Figure 1). This facility had used several heavy metals including cadmium, lead, and zinc in the development of chemical pigments for paints. The concentrations of these metals were substantially elevated in the bed of the creek that runs adjacent to this chemical facility and into the freshwater wetland that is fed by that creek (Figure 1). Concentrations of cadmium, lead, and zinc were also significantly elevated in the areas of the tidal wetland that receive drainage from the freshwater wetland. During storm events this creek also flowed over its banks upstream of the freshwater wetland and discharged into the tidal wetland via a culvert beneath the railroad tracks.

These preliminary studies indicated that a variety of ecosystems were at risk at this site, including the creek, a freshwater wetland, a brackish wetland, and Suisun Bay. Because of the complexity of the site and variety of ecosystems, the discussion will focus on only one of these ecosystems, the tidal wetland. In order to evaluate the potential ecological risk of SOC to the tidal wetlands, it was first necessary to establish a conceptual framework for this site-specific assessment.

THE CONCEPTUAL FRAMEWORK

The first step in this process involves defining the specific purpose of the ERA. Although this may seem trivial, ERAs are often very complex and a lack of clear focus can result in ambiguous and misleading conclusions. In the case of the NWS Concord, the purpose of the ERA was twofold:

- To assess the degree of risk that the SOC posed for the tidal wetland ecosystem
- To provide appropriate data for the development of an ecologically sound remedial action plan for the site

Several site-specific issues had to be addressed before the formal ERA could be implemented. These included:

- Defining the hazard
- Characterizing the ecosystem of concern and identifying the presence and location of sensitive biological resources including sensitive species, habitats, and communities
- Delineating major exposure pathways for sensitive biological resources
- Choosing appropriate measurement end points for the assessment study
- Establishing a sampling design for gathering the required data

Hazard Definition

As indicated in the initial chapter of this section, the ecological risk assessment paradigm includes a component that focuses on defining the nature of the hazards. The first step in this process is the identification of the potential stressors. In the case of the NWS Concord wetlands the SOC were defined through a

Figure 1. Aerial photograph of tidal and freshwater wetland sites at NWS Concord. Three dark areas represent the area where the breach in the levee occurred resulting in release of contaminants from the waste lagoon into the tidal wetland. The creek passes by the upland chemical facility and enters into the tidal wetland that runs along the railroad tracks. The tidal creek drains the tidal wetland and releases water into Suisun Bay via the channel at the left in the photograph.

series of field studies that were previously summarized in the section on the background information on this site. Once the stessors are defined, the next step is defining the types of hazards (e.g., toxic effects) associated with exposure to the given stressor(s). The toxicities of the metal (i.e., cadmium, copper, lead, nickel, and zinc) and metalloid (i.e., arsenic and selenium) SOC at the NWS Concord sites are well documented in mammalian, avian, and aquatic species and exposure to elevated concentrations of these SOC constitutes a potential hazard (Eisler, 1985a,b, 1987, 1988a,b). The modes of action of these SOC are also well characterized in mammalian species and more limited data are available for avian and aquatic species (Eisler, 1985a,b, 1987, 1988a,b). The relative toxicity and modes of action of these SOCs are discussed in more depth in subsequent sections of this chapter.

With the potential hazards identified, the relationship between the level of exposure to the stressor or stressors and the deleterious biological response must be established. The defining of concentration-response relationship constitutes the second part of the hazard assessment. The concentration-response relationships are often difficult to define in an ecological risk assessment because of the variety of biological end points and overall complexity of the system. The first step in this process is characterizing the ecology of the site and identifying potential receptor species for which concentration-response relationships must be established.

Biology of the Wetlands

The wetlands at NWS Concord consisted of a mosaic of community types. Two major factors that influenced the distribution of vegetation across the wetlands were the establishment of a network of mosquito ditches across the wetlands and the discharge and overflow of waste materials from adjacent waste lagoons of the chemical manufacturing and processing plants. Pickleweed (*Salicornia virginica*) and salt grass (*Distichlis spicata*) communities were most common and usually occurred immediately adjacent to the upland areas. Pickleweed occurred at higher elevations in the wetlands where soil salinities are high due to the concentration and accumulation of salts at the soil surface by evaporation. Salt grass usually occurred in slightly more moist areas where soil salinities were slightly lower than those in the pickleweed communities, but it was also scattered throughout the wetlands in colonies of varying sizes. Rush (*Juncus* spp.), bull rush (*Scripus* spp.), and cattail (*Typha angustifolia*) communities were also present but were usually confined to areas of deeper water nearer to Suisun Bay. Cattail communities were also scattered as small clumps throughout the wetland area in deepwater and along stream channels.

The wetlands provide cover and nesting materials for small mammals and birds. They also serve as a food source for a number of mammals and birds. More than 100 species of birds, mammals, reptiles, and amphibians are known to occupy the salt marsh, grassland, and freshwater marsh habitats on site. In addition, the sediments of the slough channels contain a number of benthic species that serve as a food source for waterfowl and shorebirds. These sloughs also provide food and protection for juvenile fish and small nongame fish that provide a food source for shorebirds and wading birds inhabiting the tidal marsh.

At the initiation of this study, a review of available literature indicated that several rare or endangered species may be of concern at this site. These included the salt marsh harvest mouse (*Reithrodontomys raviventris halicoetes*), the California least tern (*Sterna antillarum browni*), the California clapper rail (*Rallus longirostris obsoletus*), and the bald eagle (*Haliaeetus leucocephalus*). Candidates for listing as endangered species included the Suisun shrew (*Sorex sinuosus*), the California black rail (*Laterallus jamaicensis conturniculus*), and salt marsh yellowthroat (*Geothlypis trichas sinuosa*). A species listed by the California Department of Fish and Game as a rare species, the giant garter snake (*Thamnophis couchii gigas*), was also found at the site.

Subsequent studies confirmed that the salt marsh harvest mouse was present at the contaminated site. The California black rail and California clapper rails were also observed in the vicinity of the site. The salt marsh harvest mouse is herbivorous and is usually found in close association with pickleweed vegetation. The pickleweed appears to provide both habitat and a food source for this mammal. The black rail nests along the sloughs of the marshes. Its favored habitat consists of cordgrass, pickleweed, and gum plant and it feeds on benthic invertebrates found in the slough sediments. The California least tern was also observed in the vicinity of the site but was considered of less concern because it nests in open sandy areas and does not make extensive use of marsh habitats.

Exposure Pathways and Receptor Species

Based on the available information, three primary exposure pathways were identified for the species inhabiting the tidal wetland. These included:

- Direct exposure to SOC in the soils and sediments of the wetland
- Direct exposure to SOC in the surface waters of the wetland
- Indirect exposure to SOC via the food web

A simplified food web for the wetland and adjacent upland sites is presented in Figure 2. This food web emphasizes endangered species, candidates for listing, and other prominent species that could serve as receptors in the ecological assessment.

The soils and sediments serve as the primary sink for all of the SOC at the NWS Concord site. Thus, contact with these media was considered a primary exposure pathway. This pathway is of particular concern for organisms such as benthic or soil inverte-

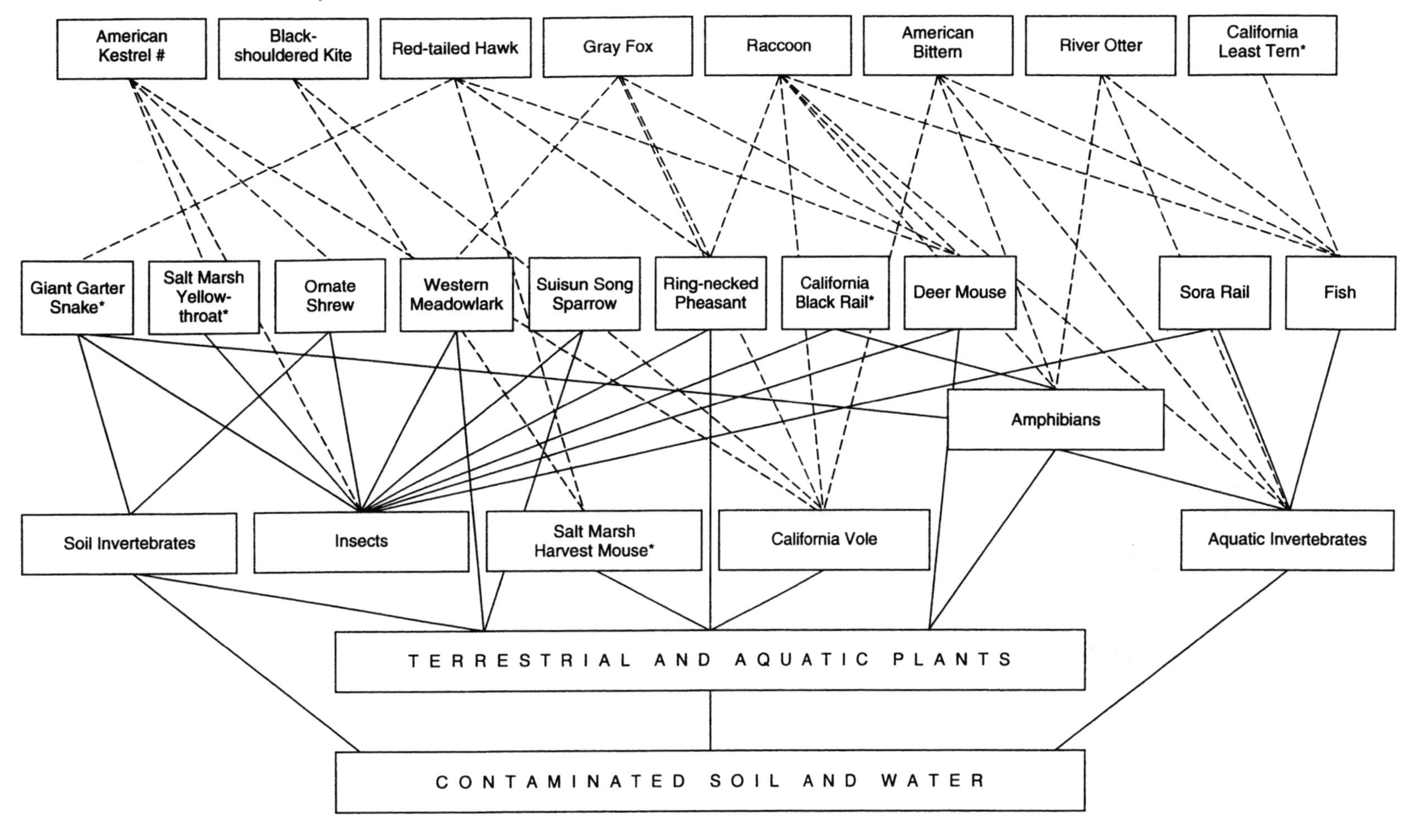

Figure 2. Simplified food web showing the potential migration pathways for the tidal and freshwater wetlands and adjacent upland areas.

brates and plants that are in continuous contact with these media. Initial studies indicated that the concentrations of several of the SOC in waters from the sloughs and ditches were above EPA chronic water quality criteria for the protection of aquatic life. Therefore, exposure of aquatic organisms to SOC in surface waters was considered an important exposure pathway. Indirect exposure to stressors via the food chain was also considered. The major SOC at this site (e.g., arsenic, lead, cadmium, copper, and zinc) do not tend to biomagnify with increasing trophic levels. Thus, the primary focus was short food chains with sensitive receptor species (Figure 2).

Based on a review of the available data, two food chain exposure pathways were considered to be of particular concern. The first involved exposure via the ingestion of benthic or soil invertebrates. This food chain was important because these organisms are not only in intimate contact with the contaminated soils and sediments but often contain substantial quantities of these materials in their digestive tracts. They represent a primary exposure pathway for receptor species at the next trophic level. For this site the most sensitive receptors for this food chain were avian species feeding on benthic invertebrates (Figure 2). Of these, the California black rail and California clapper rail were considered to be of greatest concern because of their threatened or endangered status. The second food chain pathway involved the ingestion of plants. For this pathway, herbivorous small mammals were considered to represent the most sensitive receptors. At this site the primary receptor species for this exposure pathway were the California vole (*Microtus californicus*) and the endangered salt marsh harvest mouse (Figure 2).

Design of the Sampling Program

Once the exposure pathways and receptors species were defined, a field sampling program was developed to acquire the necessary data for the ecological risk assessment. In designing a field sampling program primary issues to be addressed were:

- What kinds of parameters should be measured?
- Where should the measurements be taken?
- When should the measurements be taken?

Clearly, the data gathered in this phase should provide a basis for a rigorous evaluation of the ecological risk associated with the contaminants present at the NWS Concord site. As indicated in the introductory chapter of this section (Chapter 28), the final characterization of risk is based on both an exposure assessment and an effects assessment. Therefore, the parameters to be measured should provide a basis for a site-specific evaluation of both exposure and effects.

In ecological assessments the distinction is often made between assessment end points and measurement end points (see Chapter 28 and Warren-Hicks et al., 1989). Assessment end points represent the ultimate resource(s) or "final environmental values" to be protected. They should have social and/or biological relevance, be quantifiable, and provide useful information for resource management or regulatory decisions (Table 1). In contrast, measurement end points represent the specific parameters measured in a given assessment. They are often chosen based on practical considerations such as availability and ease of measurement. They should be well characterized and take into account exposure pathways and temporal factors. Ideally, measurement end points should be chosen so that the data for these end points can be linked directly or indirectly to appropriate assessment end points. This latter issue is often the most difficult to address in ERAs. In the following section the types of measurements and assessment end points selected will be reviewed, along with their role in assessing exposure and biological effects.

Table 1. Potential measurement and assessment end points

Level of biological organization	Measurement end points	Assessment end points
Individual	Mortality	
	Growth	
	Fecundity	
	Biomarkers	
	Tissue concentrations	
Population	Occurrence	Extinction
	Abundance	Abundance
	Age/size class	Age/size class
	Yield/productivity	Yield/productivity
	Reproductive performance	
Community	Number of species	Market/sport value
	Species evenness	Recreational quality
	Species diversity	
	Community quality	
Ecosystem	Biomass	Productive capability
	Productivity	
	Nutrient dynamics	

Adapted from U.S. EPA (1989c).

Measurement and Assessment End Points

The measurement end points that were utilized for the NWS Concord ERA are presented in Table 2. These measurement end points fall into several categories. The first focused on defining the existing distributions of SOC in soils, sediments, and surface waters. This information is essential for developing estimates of the potential for exposure to SOC throughout the wetland. However, establishing precise estimates of exposure based on the concentration of an SOC in soils/sediments or surface waters is not always straightforward. A number of factors can effect the bioavailability of stressors and thus modify the actual exposure to the receptor species. For the SOC of concern at this site (e.g., metals and metalloids), salinity, pH, sulfides, acid volatile clay and organic content of the sediments can all affect exposure (Eisler, 1985a,b, 1987, 1988a,b; see Chapter 1).

The issue of bioavailability was addressed directly by evaluating contaminant bioaccumulation in selected species (Table 2). The bioavailability of SOC to soil invertebrates was evaluated using the earthworm *Eisenia foetida*, which was exposed to soil samples from the NWS Concord for 30 d under controlled laboratory conditions (Simmers et al., 1984). A similar strategy was utilized to evaluate the bioavailability of SOC to plants. In this case, the yellow nutsedge (*Cyperus esculentus*) was exposed to field-collected soil samples for 45 d under controlled greenhouse conditions as described by Folsom and Lee (1981). Bioaccumulation of SOC was also evaluated in native cattails, but these data were limited due to the spotty distribution of this species. The limitations inherent in collecting field data emphasize an important advantage of laboratory-based evaluations of bioaccumulation: they allowed SOC bioavailability to be evaluated and compared in soils and sediments throughout the site regardless of the native species present at each location.

Bioavailability of SOC in surface waters was evaluated in the field using caged Asiatic clams (*Corbicula fluminea*) exposed for 30 d throughout the wetlands using the methods of Marquenie (1981). Field exposure were considered to be more appropriate in this instance because the surface waters of the wetland have a more dynamic state of flux than the sediments or soil. The field exposure integrates variations in SOC concentrations in surface waters that occur due to the tidal cycle and surface runoff from storm events. This complex exposure pattern would be exceedingly difficult to duplicate in the laboratory. Uneven distributions of native clams required the use of caged-clam biomonitoring methods to assure adequate biomass for bioaccumulaion studies throughout the site.

The data from the bioaccumulation studies also provided a basis for evaluation of exposure to SOC via the food chain. The earthworm and clam data were used to estimate exposures to organisms ingesting soil and benthic invertebrates, respectively. These data were particularly important in evaluating the potential for exposure of black rails and clapper rails to SOC. Similarly, the plant data provided a basis for estimating the exposure of herbivorous small mammals, such as the salt marsh harvest mouse, to SOC. Because of concerns for the salt marsh harvest mouse, bioaccumulation of SOC was also evaluated in the tissues of native small mammals trapped at various locations in the tidal wetland. As small mammals have fairly extensive ranges and may avoid the most contaminated areas, it was felt that this periodic trapping of native small mammals provided a more accurate measure of exposure potential than would laboratory or caged field studies. Bioaccumulation of SOC could not be evaluated directly in the salt marsh harvest mouse because its endangered status precluded sacrificing individuals for chemical analysis. As a consequence, these studies focused on the house mouse (*Mus musculus*) and California vole, which were the most abundant small mammalian species captured at the site. The California vole was of particular interest because it is a

Table 2. Measurement end points employed for the NWS Concord wetlands ERA

Measurement end points
SOC distributions in physical media
Concentrations of contaminants in sediments/soil
Concentrations of contaminants in surface waters
Bioaccumulation of SOC in representative species
Bioaccumulation of contaminants in indicator soil invertebrates
Bioaccumulation of contaminants in resident and indicator plant species
Bioaccumulation of contaminants in caged clams exposed to surface waters
Bioaccumulation of contaminants in native small mammals
Evaluation of toxicity
Plant tests to evaluate soil toxicity
Earthworm tests to evaluate soil toxicity
Clam tests (biomonitoring) to evaluate surface water toxicity
Histopathological evaluation of native small mammals
Population and community studies
Soil invertebrate population and community structure
Plant population and community structure within sample locations
Small mammal population survey for sample locations

herbivore that competes with the salt marsh harvest mouse for both habitat and food, making it an excellent surrogate for the endangered species. The house mouse is an omnivore and although it is not an ideal surrogate for the salt marsh harvest mouse, it still provides an indication of the potential for exposure of omnivorous small mammals to SOC via the food chain.

Measurement end points that evaluated the effects of exposure to SOC were also incorporated in the study design. As an example, growth and survival were monitored in the laboratory exposures of earthworms and plants described above. This provided a basis for evaluating the toxicity of these media and relating it to SOC exposure and bioaccumulation. Similarly, condition indices (i.e., soft tissue weight relative to shell diameter) and survival were monitored in the caged clam, allowing surface water toxicity to be evaluated in parallel with SOC bioaccumulation. Finally, the small mammals used to evaluate SOC bioaccumulation were subjected to detailed histopathological evaluations to determine if they showed any contaminant-related pathology.

Sublethal impacts on the organism can also affect higher levels of biological organization through inhibition of such physiological processes as growth and reproduction. Changes in these parameters can impinge upon populations and communities. In addition, if the marsh plant populations are adversely affected, the ecological impact could result in habitat disturbance for birds and small mammals, including the endangered species. Impact at these higher levels of biological organization was evaluated directly. For this purpose, the population and community structure of soil invertebrates and plants were evaluated in areas with elevated concentrations of soil SOC and compared with reference areas. Small mammal populations surveys were also carried out throughout the wetland.

Because of the complexity of the linkages between SOC distribution and ecosystem effects, this study was designed to facilitate correlations between the various measurement end points. The measurement end points were chosen to provide integrated data at a number of levels of physical and biological organization. This integration facilitates the establishment of relationships between the distributions of SOC and observed biological effects. For example, the effects of soil SOC on plants were evaluated by comparing the SOC concentrations in soils with:

- The bioaccumulation of SOC in plants exposed to those soils
- Toxicity of the soils to plants
- The relationship between the distributions of elevated concentrations of SOC in the soils and plant community structure

In terms of assessment end points, the maintenance of the endangered species populations was of particular concern. Here, the plant survey data provided information on habitat suitability for both the salt marsh harvest mouse and the California clapper rail. The plant bioaccumulation data provide information on the exposure of the herbivorous salt marsh harvest mouse to contaminants via the food chain. The earthworm toxicity tests and the clam biomonitoring data provided information on contaminant exposure via the food chain for other species feeding on soil invertebrates and for filter feeders in the streams, ditches, tidal creek of the wetland areas. The small mammal survey provided direct data on the distribution, population dynamics, and sex ratios of the endangered salt marsh harvest mouse. The other small mammal species served as surrogates for the salt marsh harvest mouse for evaluating contaminant bioavailability and toxicity via necropsy and histopathology studies.

Spatial and Temporal Sampling

A careful consideration of the spatial and temporal aspects of the study design was also essential to the ERA. Spatial factors include potential routes of exposure, other sources of stressors, location of sensitive biological resources, and factors that may modify contaminant mobility or availability (e.g., changing composition of sediments). Temporal factors include seasonal changes in physical, chemical, or biological aspects of ecosystems that may affect the potential for exposure or biological sensitivity to that exposure. For example, increased surface water movement during the wet season can dramatically affect contaminant migration and the potential for exposure. Also, seasonal variation in physical or chemical parameters (e.g., temperature or pH) can modify the bioavailability of contaminants and thus change the nature of the exposure. The biological characteristics of the ecosystem may also change over time and with seasons. Species may be present only seasonally (e.g., migratory waterfowl, anadromous fish, or species that migrate seasonally within a locale), or exposure to a given species, population, or community may vary with seasonal changes in life history habits (e.g., seasonal feeding patterns). Thus, temporal changes may influence the distribution of the stressor, or the biological end point, within the ecosystem and the ultimate risk assessment. An overview of the spatial and temporal sampling of the various measurement end points is presented below.

Distributions of SOC

Soils and sediment. A total of 637 soil/sediment samples were collected from 414 locations across the site (Figure 3) (Lee et al., 1986). Locations were selected based on previous soil data and potential contaminant pathways for migration within the site, and to represent a range of contamination from the lowest to the highest levels present at the site. Soils/sediments from all of these locations were analyzed for arsenic, cadmium, copper, lead, zinc, and selenium concentrations. Sampling of sediments took place over a several-year period and sampling activities were arranged to encompass both the wet season (winter months) and the dry season (summer

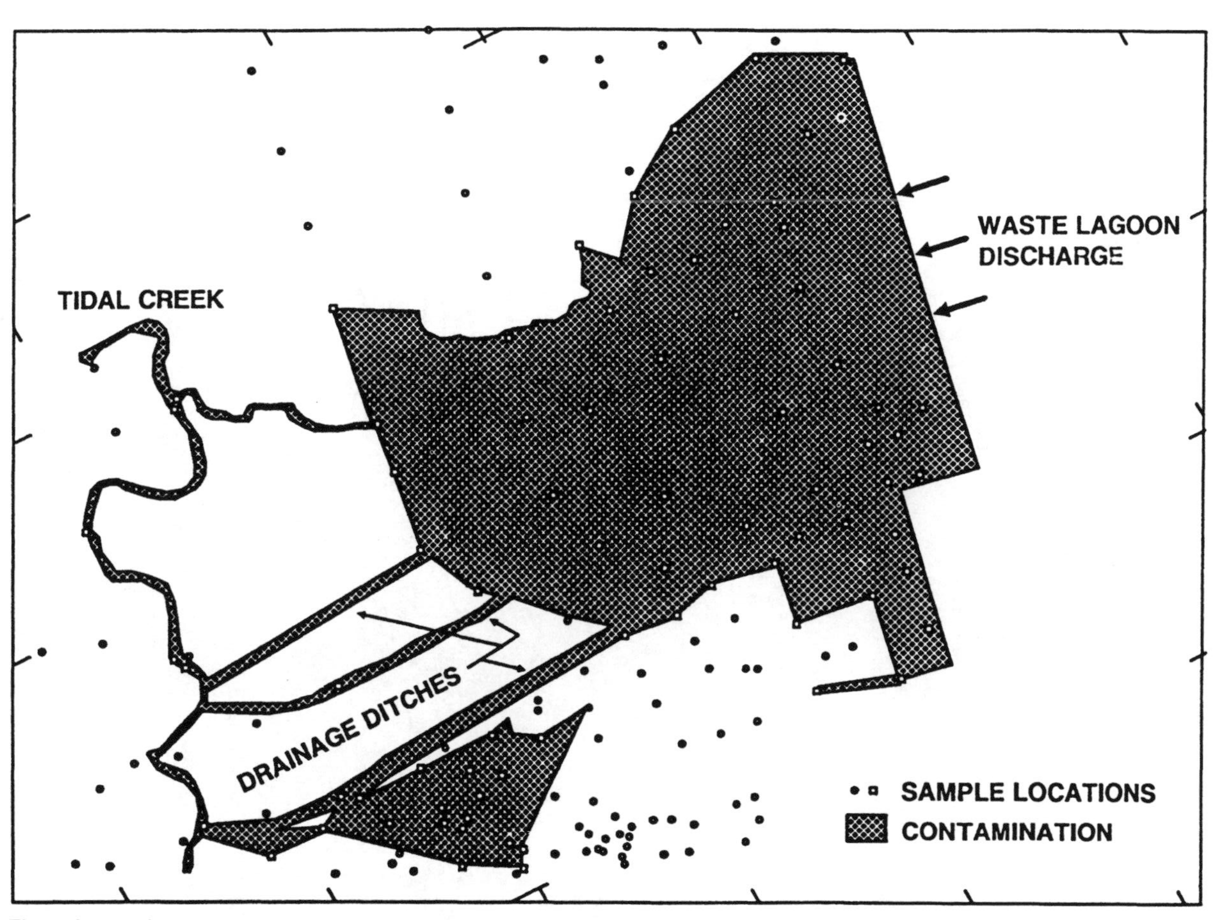

Figure 3. Locations where sediment samples were taken in the wetlands of the NWS, Concord. Sediment sampling locations at which there was significant plant and earthworm toxicity are also shown.

months). Temporal sampling was particularly important at the NWS Concord site because the wet season is discrete and the vast majority of the yearly precipitation occurs during this period. It is also a period of exceptionally high tides. Together, these factors may result in a substantial increase in the mobility of SOC within the site.

Soils from 10 locations in wetland and upland reference areas were also sampled in this phase of the ERA to define the ambient concentrations of the SOC along the southern shore of Suisun Bay. The upland reference sites were located along the creek that feeds the freshwater wetland but upstream of any known source of contamination. The wetland reference site was located in the tidal wetland well to the west of the contaminated area. These sites were initially chosen because of their similarity to the contaminated wetland and upland areas. Analysis of soils indicated that the concentrations of SOC were consistent with other uncontaminated sites in the San Francisco Bay area.

Surface waters. Surface water was sampled at 21 locations in the tidal wetlands, freshwater wetlands, and the creek that feeds these wetlands (Figure 4). Locations were selected based on previous soil data and to evaluate potential areas of SOC accumulation and pathways of migration within the site. Surface waters from all of these locations were analyzed for total concentrations of arsenic, cadmium, copper, lead, zinc, and selenium.

SOC Bioaccumulation and Toxicity Tests

Plants. Sediments and soils for laboratory evaluations of plant bioaccumulation and toxicity were collected from a subset of the locations used to characterize the distributions of SOC in sediments (Figure 5). Specific locations were chosen to reflect the full range of potential exposures based on previous

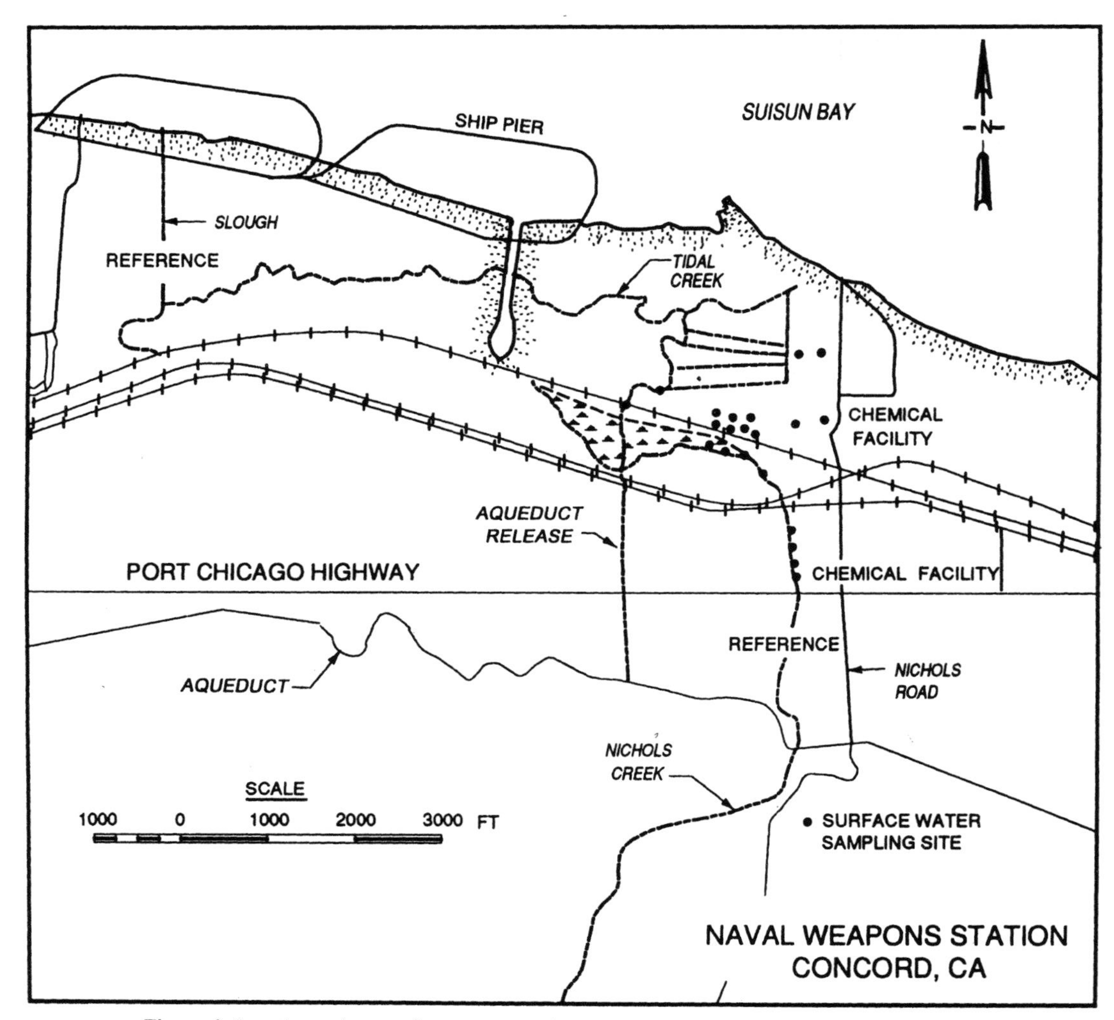

Figure 4. Locations where surface water samples were taken in the wetlands/creeks of the NWS.

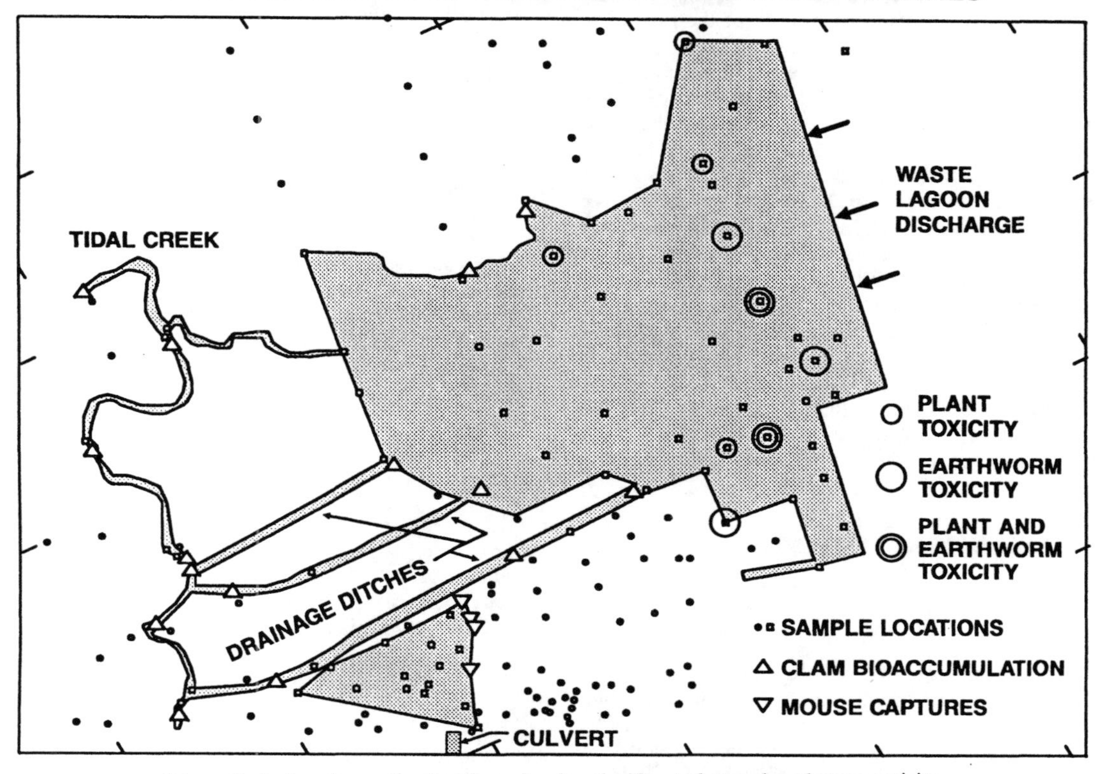

Figure 5. Sediment sampling locations showing significant plant and earthworm toxicity.

soil data. Samples from the wetland and upland reference sites were also included in these tests to establish baseline levels of bioaccumulation and toxicity. A 45-d plant toxicity test using the yellow nutsedge (*Cyperus esculentus*) was conducted with each sample. At the end of the testing period, plant survival, plant biomass (tissue weight in grams), and the concentrations of SOC in plant tissues were measured.

Earthworms. Sediments and soils used in laboratory evaluations of SOC bioaccumulation and toxicity to soil invertebrates were collected from a subset of the locations used to characterize the distributions of SOC in sediments (Figure 5). These locations were chosen to reflect the full range of potential exposures based on previous soil data and to correspond to the plant toxicity tests. As with the plant toxicity tests, the wetland and upland reference site soils were also included to establish baseline levels of bioaccumulation and toxicity in these tests. The bioavailability and toxicity of SOC to soil invertebrates were evaluated using a 30-d earthworm tests. Following the 30-d exposure, earthworm survival and biomass and the concentrations of SOC in earthworm tissues were measured.

Clams. To evaluate the bioavailability and toxicity of SOC in surface waters, caged Asiatic clams (*Corbicula fluminea*) were placed at a number of sites along the ditches and tidal channels of the tidal wetland, in the freshwater wetland, and along the creek that feeds these wetlands. These locations were selected based on previous soil and surface water data and to evaluate potential areas of SOC bioaccumulation and pathways of migration within the site. Three locations were also selected along the shores of Suisun Bay to evaluate the potential impact of SOC on biota in the bay. Clams were also placed in the wetland and upland reference areas to define the ambient potential for SOC bioaccumulation. Clams in the upland reference sites were located in the creek that feeds the freshwater wetland but upstream of any known source of contamination. Clams in the wetland reference site were placed in tidal channels well to the west of the contaminated area. Clams from all of these locations were analyzed for total concentrations of arsenic, cadmium, copper, lead, zinc, and selenium. Separate studies were carried out during the dry season and wet season to allow comparisons between results at different levels of surface runoff and tidal inundation. Both runoff and tidal inundation could affect the mobility and distributions of SOC within the site.

Small mammals. Small mammal trapping was carried out in the contaminated areas of the tidal wetland and in the reference wetlands. These trapping activities resulted in the capture of a number of small mammal species including the house mouse (*Mus musculus*) the California vole, and the endangered salt marsh harvest mouse. Salt marsh harvest mice were sexed, weighed, and their overall condition evaluated prior to release. These data and the relative trapping success (e.g., number of individuals per trap-night), provided information on the current status of the salt marsh harvest mouse populations in contaminated and reference marshes. The house mice and voles from contaminated and reference wetlands were sacrificed and evaluated for gross pathology (necropsy) and tissue and cellular pathology (histopathology). Following pathological evaluations, the liver, kidney, and femur tissues were digested and concentrations of SOC determined.

EXPOSURE ASSESSMENT

Sediment Exposure

The concentrations of SOC in soils and sediments from the contaminated sites were highly variable. The ranges of SOC concentrations in soils and sediments collected for the 1986 remedial investigation (RI) (Lee et al., 1986) are summarized in Table 3. Data from the various sampling stations were compared statistically using analysis of variance (ANOVA) and Duncan's new multiple range test (see Chapter 10). All significant differences were obtained at $P<.05$. Concentrations of arsenic, cadmium, copper, and zinc were statistically elevated in soils/sediments from a number of sites in the tidal wetland relative to those from the reference sites. Cadmium and zinc were also significantly elevated at a number of locations in the freshwater wetland.

For the multiyear sampling period, concentrations of SOC in individual soil/sediment samples ranged as high as 2500 mg/kg for arsenic, 89 mg/kg for cadmium, 3050 mg/kg for copper, 7600 mg/kg for lead, 85,490 mg/kg for zinc, and 138 mg/kg for selenium. The higher concentrations were observed near the two chemical facilities and in depositional areas downstream from these facilities. The ranges of SOC concentrations for soils/sediments from the tidal and freshwater wetlands, tidal wetland, and tidal creeks are presented in Table 3. The ranges of SOC concentrations in soils from the reference sites are also summarized. The concentrations of SOC in contaminated soils/sediments are several orders of magnitude higher than those found at the reference sites, indicating that contact with the soils in localized areas would result in a greatly increased potential for exposure to SOC.

Water Column Exposures

The ranges of SOC concentrations in surface waters from the wetlands and creek are presented in Table 4. Concentrations of zinc were as high as 20 mg/L, and those for the other SOC were generally one to two orders of magnitude lower. The concentrations of zinc at 80% of the locations in the creek and freshwater wetland exceeded at the acute water quality criteria (WQC). Zinc concentrations exceeded the chronic WQC at a number of additional locations in the tidal wetland. Copper, cadmium, nickel, and lead concentrations also exceeded the acute or chronic WQC at several locations in each of the sampling areas. Arsenic concentrations exceed the acute WQC at over 40% of the stations sampled in the tidal wetland. These data suggested a substantial potential for exposure of aquatic organisms to elevated concentrations of SOC in the surface waters of the NWS Concord site.

Food Chain Exposure

The concentrations of SOC in the tissues of soil invertebrates, aquatic invertebrates, and plants provided a basis for estimating exposure to higher-level consumers in these food chain pathways. The specific exposure scenarios for small mammals and avian species ingesting these organisms will be discussed in subsequent sections.

Soil Invertebrates

The ranges of SOC concentrations in earthworms exposed to soils or sediments in the laboratory are presented in Table 5. Concentrations of SOC in earthworms exposed to contaminated soils/sediments were statistically elevated at a number of locations relative to earthworms exposed to reference sediments. In the tidal wetland, arsenic bioaccumulation was significantly elevated at many of the sampling points where the levee had been breached (Figure 3). Copper bioaccumulation was also significantly increased at one location in the tidal wetland that showed elevated copper in the sediments. Statistically significant increases in the bioaccumulation of cadmium, copper, and lead were also observed in earthworms from one sampling location in the tidal wetland and several locations in the freshwater wetland. These data indicated that the SOC were biologically available to soil-dwelling invertebrates and provided an indication of the accumulated dose of SOC relative to concentrations in the soils/sediments. They also provided a basis for estimating the exposure to consumers ingesting soil invertebrates.

Aquatic Invertebrates

The ranges of SOC concentrations in tissues of clams exposed to surface waters in the tidal and freshwater wetlands are presented in Table 6. For the tidal wetland, cadmium was significantly increased in clams at several stations relative to those from the reference sites. Although statistically significant, the maximum

Table 3. Representative concentrations of stressors of concern in soil/sediments

Stressor	Range of concentrations (mg/kg, dry weight)[a]							
	Tidal wetland		Tidal reference		Freshwater wetland/ creek		Freshwater reference	
	High	Low	High	Low	High	Low	High	Low
Arsenic	**2024.9**[b]	6.4	5.6	5.5	5.3	1.4	2.4	0.2
Cadmium	**23.92**	0.4	1.5	1.1	**20.6**	0.6	0.6	0.1
Copper	**780.1**	40.1	54.5	39.6	161.8	17.8	18.7	8.3
Nickel	82.7	16.2	82.3	72.8	32.7	3.2	22.7	11.5
Lead	226.8	1.7	33.4	17.2	221.9	3.5	33.8	2.2
Selenium	4.75	ND[c]	ND	ND	0.2	ND	11.2	ND
Zinc	**2552.6**	86.8	286.6	127.6	**6514.1**	62.8	141.6	29.3

[a] Mean values (n = 3) adapted from table 2-A1, Lee et al. (1986).
[b] Bold values are statistically higher than reference values.
[c] ND, not detected.

Table 4. Representative concentrations of stressors of concern in surface waters

Stressor	Range of concentrations in wetlands/creek (mg/L)[a]		National water quality criteria (mg/L)[b]			
	High	Low	Acute-FW	Acute-SW	Chronic-FW	Chronic-SW
Arsenic	**3.6**[c]	<0.001	0.360	0.069	0.190	0.036
Cadmium	**0.1**	<0.01	0.004	0.043	0.001	0.009
Copper	**2.1**	<0.01	0.018	0.003	0.012	0.003
Nickel	**0.3**	0.02	1.400	0.075	0.160	0.008
Lead	**0.4**	<0.05	0.082	0.140	0.003	0.006
Zinc	**20.0**	<0.01	0.120	0.095	0.110	0.095

[a] Metal concentrations are based on analysis of unfiltered (total) water samples.
[b] Freshwater (FW) and saltwater (SW) criteria are presented from U.S. EPA (1986). FW criteria are based on a hardness of 100 mg/L; arsenic criteria are based on trivalent arsenic.
[c] Bold values exceed national water quality criteria.

Table 5. Representative concentrations of stressors of concern in tissues of earthworms exposed to soils in the laboratory

Stressor	Range of concentrations (mg/kg, dry weight)[a]							
	Tidal wetland		Tidal reference		Freshwater wetland/ creek		Freshwater reference	
	High	Low	High	Low	High	Low	High	Low
Arsenic	**120.84**[b]	2.64	2.95	2.78	14.8	3.56	6.65	1.42
Cadmium	5.94	1.6	4.42	4.05	**19.77**	3.2	9.7	2.87
Copper	**53.53**	9.61	11.84	8.96	**65.17**	9.82	13.2	9.51
Nickel	7.87	0.55	5.41	3.27	4.56	1.23	6.12	1.51
Lead	9.37	0.08	0.64	0.53	**33.36**	1.32	4.06	0.03
Selenium	4.67	2.78	3.77	3.4	8.09	4.07	7.78	2.71
Zinc	273.81	111.16	124.36	118.13	208.42	119.19	146.78	113.03

[a] Mean values (n = 3) adapted from table 2-A9, Lee et al. (1986).
[b] Bold values are statistically higher than reference values.

Table 6. Representative concentrations of stressors of concern in clam tissues

Stressor	Tidal wetland High	Tidal wetland Low	Tidal reference High	Tidal reference Low	Freshwater wetland/creek High	Freshwater wetland/creek Low	Freshwater reference High	Freshwater reference Low
	Range of concentrations (mg/kg, dry weight)[a]							
Arsenic	2.0	1.3	2.0	1.9	2.0	1.2	2.3	1.7
Cadmium	**2.0**[b]	0.7	0.9	0.7	1.2	0.6	0.8	0.6
Copper	78.9	51.6	80.1	64.3	76.3	51.0	71.4	48.0
Nickel	4.2	0.2	1.4	0.5	3.4	0.1	1.5	0.8
Lead	0.5	0.1	0.3	0.3	**6.1**	0.1	1.0	0.8
Selenium	1.4	1.0	1.1	1.0	1.9	0.8	2.0	1.5
Zinc	152.6	107.1	129.4	128.9	**255.8**	125.2	125.5	106.1

[a] Mean values (n = 3) adapted from table 2-A2, Lee et al. (1986).
[b] Bold values are statistically higher than reference values.

concentrations of cadmium were only twice those at the reference site, resulting in a modest increase in exposure potential via this route. In the freshwater wetland, statistically significant increases in lead and zinc were observed at several stations. As with cadmium, the maximum concentrations of zinc were only twice those of the reference site; however, concentrations of lead were as much as eight times higher than at the reference site, indicating a potential for increased exposure to lead via this pathway. These data provided an indication of the accumulated dose of SOC relative to concentrations in the surface waters and provided a basis for estimating the exposure to consumers ingesting benthic invertebrates. Increased potential for exposure to lead via this pathway was of particular concern for the California clapper rail and black rail, which feed on benthic invertebrates from these wetlands.

Wetland and Upland Plants

The ranges of SOC concentrations in plants exposed to soils or sediments in the laboratory are presented in Table 7. These data are based on the flooded plant bioaccumulation test data. Statistically significant increases in arsenic bioaccumulation were observed for soils from several sites in the tidal wetland, with the highest over 40 times higher than those of the reference site. Concentrations of cadmium, copper, and zinc in plants exposed to soils from several sites in the freshwater wetland were also significantly elevated relative to the reference sites (Table 7). Based on these data, the ingestion of plants grown on soils from the tidal wetland could result in 17-, 10-, and 4-fold increases in exposures to zinc, cadmium, and copper, respectively.

The worst case for exposures was estimated from the highest SOC concentrations in individual samples of plant tissues (rather than a mean of three replicates). In the tidal wetland the highest concentrations of arsenic, cadmium, copper, lead, and zinc were 26.74, 1.58, 14.59, 4.46, and 97.55 mg/kg, respectively (Lee et al., 1986). The maximum potential for exposure via this route was higher in the freshwater wetland for lead and zinc with concentrations of 10.5 and 393.5 mg/kg, respectively. Based on these data, it is clear that there is potential for substantial increases in food chain exposure to SOC for organisms ingesting plants from the contaminated areas of the NWS Concord. This increased potential for exposure is of particular concern for the herbivorous salt marsh harvest mouse, which inhabits areas in the tidal wetland.

EFFECTS ASSESSMENT

Effects Due to Direct Exposure to Surface Waters

In evaluating the potential effects of SOC in surface waters, the concentrations of arsenic, cadmium, copper, nickel, lead, selenium, and zinc were compared with both the acute and chronic national water quality criteria for those metals. The national water quality criteria were developed by the U.S. Environmental Protection Agency to protect aquatic life and human health from exposure to excess concentrations of pollutants (Chapters 1 and 24). The acute criteria are designed to protect against toxicity resulting from short-term exposure to high levels of pollutants. The chronic criteria are designed to protect against sublethal toxicity due to long-term, low-level exposure to pollutants.

In reviewing the surface water data, a number of the metals of concern, including zinc, copper, and cadmium, exceeded both acute and chronic water quality criteria in surface waters from a number of sites in the tidal and freshwater wetlands/creek. Arsenic, nickel, and lead also exceeded the acute and chronic water quality criteria in the tidal wetland. Based on the numerous instances of exceeding the water quality criteria by arsenic, cadmium, copper, nickel, lead, selenium, and zinc, it was concluded that there was a substantial potential for both acute and chronic toxicity to aquatic organisms in contact with the surface waters in localized areas of the wetlands of the NWS Concord site.

Table 7. Representative concentrations of stressors of concern in tissues of plants exposed in the laboratory

	Range of concentrations (mg/kg, dry weight)[a]										
	Tidal wetland		Tidal reference		Freshwater wetland		Freshwater reference		Toxic thresholds		
Stressor	High	Low	High	Low	High	Low	High	Low	CC[b]	10% yield reduction[c]	Toxic[d]
Arsenic	**10.90**	ND[e]	0.26	0.26	0.12	ND	ND	ND	—	—	3
Cadmium	0.71	0.06	0.04	0.04	2.27	0.25	0.21	0.08	8	15	5
Copper	7.11	1.99	3.73	3.73	**14.71**	2.16	3.95	1.65	20	20	25
Nickel	1.14	0.35	0.45	0.45	3.14	0.78	0.76	0.05	11	26	500
Lead	1.94	0.07	0.16	0.16	5.34	0.32	0.58	0.21	—	—	
Selenium	ND	ND	ND	ND	ND	ND	ND	ND	—	—	100
Zinc	36.02	16.93	25.26	25.26	**659.97**	23.95	39.01	22.31	200	260	500

[a] Mean values (n = 3) adapated from table 2-A5, Lee et al. (1986).
[b] Critical concentration. Content of metals in leaves above which toxic effects have been observed Lee et al. (1986).
[c] Concentrations in leaves correlating with a 10% reduction in crop yield Lee et al. (1986).
[d] Phytotoxic concentration: lowest content in leaves of plants dying due to exposures to specific SOCs Lee et al. (1986).
[e] ND, not detected.

Effects Due to Direct Exposure to Soil/Sediments

Evaluating the exposure-response relationships due to direct exposure to soil/sediment SOC was complicated by the lack of relevant soil or sediment criteria for these ecosystems. In the absence of soil/sediment criteria, the potential toxicities of arsenic, cadmium, copper, lead, selenium, and zinc were evaluated based on the site-specific toxicity tests and field data gathered during this RI and a review of relevant toxicological data in the literature.

Soil and Sediment Invertebrates

Several site-specific data sets from the RI were used in evaluating the toxicity of SOC to soil invertebrates, that is, the accumulation of SOC by earthworms during the laboratory exposures and earthworm toxicity data from the same tests. In evaluating these data, a clear relationship was observed between exposure to SOC in the sediments, accumulation of SOC by earthworms, and reduced growth and survival of earthworms. Significant reductions in growth and increased mortality were commonly observed in soils/sediments where concentrations of SOC were significantly elevated relative to the remote reference sites. Based on these correlations between metal concentrations in sediments, metal bioaccumulation, and sediment toxicity, it was concluded that the observed toxicity of specific soils and sediments was likely due to the elevated concentrations of SOC in those samples.

The data from the laboratory exposures to earthworms were consistent with studies of the native soil invertebrate populations from the tidal wetland. In this study, the soil invertebrate community from a site within the contaminated area of the tidal wetland was compared with that at a remote reference site to the west of the contaminated area. The total numbers of macroinvertebrates were statistically reduced in the contaminated site. Moreover, the numbers of three species of oligochaete worms were significantly reduced at this site relative to the remote reference site (Table 8). The soils within the invertebrate community sampling site had statistically elevated concentrations of SOC, and earthworms exposed to these soils showed increased SOC bioaccumulation, reduced growth, and significant mortality.

The site-specific data demonstrated a clear exposure-response relationship with higher concentrations of SOC in soils/sediments correlating with increased toxicity of earthworms in laboratory exposures and reduced numbers of native soil invertebrates. Moreover, the concentrations of SOC in sediments showing significant toxicity are consistent

Table 8. Summary of macroinvertebrate community data

	Mean number of individuals/square meter[a]	
Dominant species	Tidal wetland	Reference wetland
Quistradrilus multisetosus[b]	**0**[e]	388.1
Telematodrillus vejdovskyi[b]	**35.3**	564.5
Enchytraeus albidus[b]	**0**	35.3
Paranais litoralis[c]	**35.3**	1323
Orchestia traskiana[d]	**35.3**	105.8
Assiminea californica[c]	**35.3**	17.6
All species	**176.4**	2610.7

[a] Mean values (n = 3) adapted from table 3-16, Lee et al. (1986).
[b] Oligochaeta.
[c] Amphipoda.
[d] Gastropoda.
[e] Bold values are statistically different from reference values.

with data published in the literature (Lee et al., 1986). Based on these observations, it was concluded that the concentrations of SOC in soils at the higher range of exposures were toxic to soil- and sediment-dwelling organisms. However, because the concentrations of the various SOC tended to covary (e.g., increase or decrease together), it was difficult to establish precise exposure-response relationships for individual SOC from these data alone.

Plants

Initial evaluation of the effects of soil/sediment SOC on plants focused on SOC accumulation and plant growth and mortality data derived from the laboratory toxicity tests. As in the earthworm laboratory exposures, a clear relationship was observed between exposure to SOC in the sediments, accumulation of SOC by plants, and reduced growth and survival of the plants. Significant reductions in growth and increased mortality were commonly observed where the soil concentrations of SOC were significantly elevated relative to the remote reference sites. Several areas of the tidal wetland with highest concentrations of SOC in the soils were devoid of any plant life, implying plant toxicity at the highest exposure levels.

The correlations between SOC concentrations in sediments and SOC bioaccumulation and toxicity in plants suggested that the observed toxicity of specific soils and sediments was most likely due to the elevated concentrations of SOC in those samples. Data from other studies are consistent with these observations. Table 7 includes data from several studies that show the toxic thresholds for concentrations of SOC in plant tissues and the onset of plant toxicity. Based on these data, it was predicted that the levels of arsenic bioaccumulation in samples from the tidal wetland and zinc bioaccumulation in the freshwater wetland would result in plant toxicity.

Effects Due to Exposure via the Food Chain

Small Mammals

Toxicological effects of the increased SOC exposure of small mammals were evaluated using several different strategies. The first compared the predicted exposure scenarios against published dose-response data for SOC in small mammals. In this evaluation, the primary exposure scenario focused on ingestion of plants by herbivorous species such as the salt marsh harvest mouse and California vole. Concentrations of SOC in plant tissues thus provide an indication of the potential for exposure in these species. Based on the plant tissue data, arsenic, copper, and zinc were statistically elevated relative to remote references sites (Table 7). When bioaccumulation of SOC was evaluated directly in native voles and house mice, significant increases in bioaccumulation were observed only for lead and cadmium (Table 9). These data indicate that concentrations of SOC in plant foods may not always provide an accurate estimate of their potential for accumulation in small mammals. The lack of significant accumulation of arsenic appears to reflect the ability of mammalian species to metabolize and excrete this SOC rapidly (NAS, 1977). As for lead and cadmium, the concentrations in plants from the contaminated wetlands were as much as an order of magnitude higher than those from the reference wetlands. Although these differences in plant tissues were not statistically significant, they clearly contributed to the increased accumulation of these SOC in the tissues of mammalian species that ingest plants.

Table 9. Concentrations of stressors of concern in tissues of small mammals

			Range of concentrations (mg/kg, dry weight)[a]			
			Tidal wetland		Tidal reference	
Species	Tissue	Stressor	High	Low	High	Low
Mus musculus	Liver	Arsenic	ND[b]	ND	ND	ND
		Cadmium	2.5	0.016	0.18	0.013
		Lead	16	0.4	2.9	0.12
	Kidney	Arsenic	ND	ND	ND	ND
		Cadmium	2.6	0.041	0.11	0.026
		Lead	180	5.1	1.5	0.18
	Femur	**Lead**	800	14	110	2.1
Microtus californicus	Liver	Arsenic	0.23	ND	ND	ND
		Cadmium	0.41	0.023	0.035	0.001
		Lead	29	0.05	3.3	19
	Kidney	Arsenic	11	ND	ND	ND
		Cadmium	0.55	0.079	0.46	0.017
		Lead	3.7	0.14	3.5	0.18
	Femur	Lead	35	2.2	12	1.6

[a] Mean values (*n* varies with site).
[b] ND, not detected.
[c] Metals shown in bold are statistically higher at the contaminated sites than the reference sites.

The potential effects of exposure of small mammals to SOC were initially evaluated by comparing data on SOC concentrations in plants with dose-response data from the literature. Because arsenic is rapidly metabolized and excreted, it does not accumulate to high levels and toxicity is usually limited to acute or subacute levels of exposure (NSA, 1977). Arsenic, however, is a known mammalian teratogen (Eisler, 1988a) and reproductive effects were considered to be the most sensitive and relevant end point from an ecological standpoint. Mice fed a diet of 5 mg/kg sodium arsenite in the diet for three generations have shown significant reductions in litter size (Pershagen and Vahter, 1979). The highest concentration of arsenic observed in plants from the laboratory was 10.9 mg/kg (Table 7). These data suggested that chronic exposure to arsenic at the levels found in plants from the tidal wetland may be sufficient to cause reproductive impairment in small mammals ingesting those plants.

Lead is a potent inhibitor of reproduction and sublethal exposures affect a number of tissues, including red blood cells, kidneys, and the central nervous system (Eisler, 1988b). For small mammals, the diet provides the major pathway of exposure for lead (Eisler, 1988b). The lowest observable adverse effect level (LOAEL) was based on reproductive failure in mice following a chronic exposure to a diet containing 25 mg/kg lead (Schroedner and Mitchener, 1971). Although the highest average concentration observed in plant tissues exposed to sediments from the freshwater wetland was well below the LOAEL (5.34 mg/kg, Table 7), individual plant samples contained concentrations of lead in excess of 200 mg/kg (Lee et al., 1986). These data suggest a localized potential for chronic toxicity to small mammals due to the ingestion of lead that had been taken up by plants.

Elevated concentrations of lead in tissues of mammals from the tidal wetland confirmed the bioavailability of lead at this site and provided an independent measure of the effects of lead on small mammals. As an example, Mouw et al. (1975) demonstrated a relationship between concentrations of lead in the kidneys of urban rats and kidney function. Kidneys with concentrations of lead in excess of 15 μg/g were oversized, accumulated lead in intranuclear inclusion bodies, and showed reduced function. Intranuclear inclusion bodies are dense complexes of lead and proteins that accumulate in the nuclei of cells experiencing sublethal effects due to lead exposure. By comparison, the highest concentration of lead in house mice captured from the tidal wetland was more than 10 times higher than lead levels in the rat study (e.g., 180 mg/kg; Table 9). Although these data are based on another species, they suggested that exposure to excess concentrations of lead in the contaminated areas of the tidal wetland could result in substantial impairment of kidney function in native small mammals.

Cadmium is also an inhibitor of reproduction, a strong teratogen, and a renal toxin (Eisler, 1985a). Evaluations of the exposure potential for cadmium indicated that it was consistently below levels that would be expected to result in chronic effects on small mammals. There was concern, however, about possible synergistic effects of cadmium and lead on reproduction, embryonic development, and renal function.

This evaluation of various exposure scenarios suggested that there was a substantial potential for sublethal effects of lead on small mammals in the contaminated areas of the wetlands of the NWS Concord. Moreover, the additive effects of lead, cadmium, and arsenic on reproduction and embryonic development were of particular concern. Because of these concerns, a parallel study was carried out to assess directly gross pathology and histopathology in native house mice and California voles. In contrast to the above exposure scenarios, these direct evaluations showed no evidence of contaminant-related pathology in any of the organisms from the contaminated sites or reference sites. Thus, even the organisms with order of magnitude increases in cadmium and lead accumulation (Table 9) showed no observable pathology. The discrepancies between these two approaches could be reconciled in one of two ways: the exposure scenarios and toxicity evaluations overestimated the actual exposure level or toxicity of SOC or the histopathological evaluations were not capable of resolving critical sublethal effects in the native organisms (e.g., reproductive impairment). After carefully evaluating the data, it was concluded that the exposure scenarios presented an appropriate and conservative basis for developing a restoration plan for protecting the endangered salt marsh harvest mouse.

Avian Species

For avian species the primary focus was on the impact of SOC on the black rail and the California clapper rail. However, no data were available on the toxic effects of food chain exposures to the SOCs for these rail species. Therefore, it was necessary to extrapolate from studies on other avian species. Existing data indicate that birds are quite capable of metabolizing and excreting arsenic and are generally less sensitive than small mammals (Eisler, 1988a). For sodium arsenite, the reported acute LD50 values for birds ranged from 323 to 500 mg/kg in the diet for the mallard (NAS, 1977; Eisler, 1988a). The LD50 for a single oral dose was 47.5 μg/g body weight (BW) for the California quail (NAS, 1977; Eisler, 1988a). These concentrations are well above the dose of arsenic delivered via the ingestion of plants, earthworms, or clams from the tidal wetland, suggesting the potential for toxic effects in both rail species (Tables 5, 6, and 7).

For lead, reproductive impairment was demonstrated in the Japanese quail (*Coturnix coturnix japonica*) exposed to 10 μg lead/g per day (Eisler, 1988b). When the average size (240 g) and feeding rates of quail (18 g/d) are taken into account, this

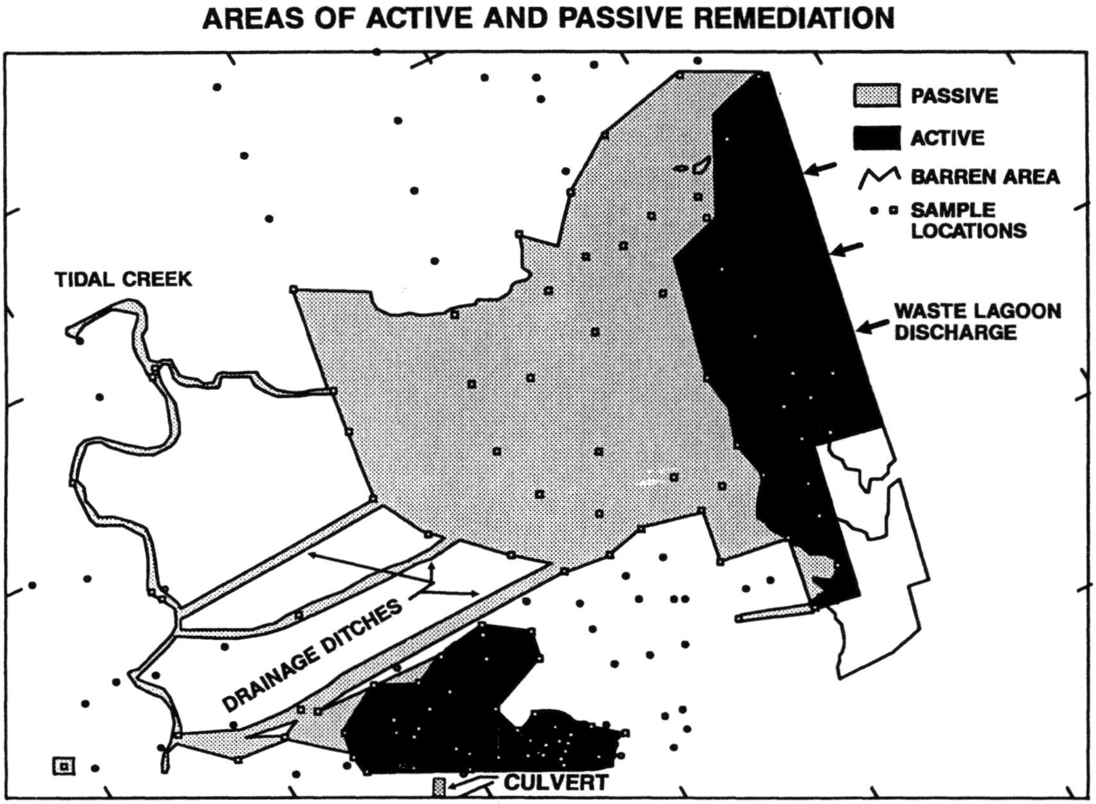

AREAS OF ACTIVE AND PASSIVE REMEDIATION

Figure 6. Areas of active and passive remediation. Active areas are to be remediated by removing the contaminated soils; passive areas are to be subjected to extensive long-term monitoring.

represents an LOAEL of 0.75 mg lead/kg BW/d (Lee et al., 1986). For comparison, the maximum concentration of lead in clams from the freshwater wetland was in the 6 mg/kg range. A black rail with an average body weight of 100 g (0.1 kg) and a feeding rate of 8 g/d (0.008 kg/d) would experience an exposure rate of approximately 0.48 mg/kg BW/d when ingesting clams containing 6 mg/kg lead (e.g., 6 mg/kg $\times$ 0.008 kg/d = 0.048 mg/d; 0.048 mg/d/0.1 kg = 0.48 mg/kg BW/d). This maximum estimated level of exposure via the benthic invertebrate pathway is below that causing reproductive impairment in quails. It is important to note that the clams were maintained in cages and thus were exposed to SOC primarily via the surface water pathway. The earthworm data may provide a better estimate of the potential exposure to lead via the ingestion of sediment-dwelling organisms. Using the highest average concentrations for bioaccumulation of lead by earthworms (33.36 mg/kg, from Table 5) an exposure rate of 2.67 mg/kg BW/d is obtained, which is substantially higher than the 0.75 mg lead/kg BW/d LOAEL obtained for lead in quail.

These data imply that the level of exposure of avian species to lead via the food chain may result in chronic impairment of reproduction. However, no toxicity data are available for the two rail species and there is substantial uncertainty associated with extrapolating from data on other species. This uncertainty can be accounted for by incorporating an uncertainty factor for comparisons between species. By convention, an uncertainty factor of 10 is usually applied in comparisons between species. This would result in a predicted LOAEL of 0.075 mg/kg BW/d (e.g., LOAEL/10). This predicted exposure via both the sediment and soil invertebrate pathways is well above this calculated LOAEL, suggesting some potential for chronic impairment of reproduction in avian species exposed to SOC via the food chain.

RISK EVALUATION

The evaluation of the RI data in the previous two sections indicates that the concentrations of SOC in the soils/sediments and surface waters were of sufficient magnitude to affect the tidal ecosystems at a number of levels. The observed and predicted impacts included:

- Chronic toxicity for plants and soil invertebrates exposed to soils/sediments with elevated concentrations of arsenic and zinc.

- Acute and chronic toxicity for aquatic species exposed to surface waters of the tidal and freshwater wetland with elevated concentrations of arsenic, cadmium, copper, nickel, lead, selenium, and zinc.
- Chronic effects on small mammals (e.g., birth defects, reproductive impairment, and reduced kidney function) exposed to elevated concentrations of arsenic, lead, and cadmium via the food chain.
- Chronic effects on birds (e.g., birth defects, reproductive impairment) in birds exposed to elevated concentrations of lead and cadmium via the food chain.

Given the broad range of ecological effects and the fact that several of the important receptor species (i.e., the salt marsh harvest mouse, black rail, and California clapper rail) are endangered, the risks associated with the historic contaminants in the tidal and freshwater tidal wetlands of the NWS Concord were considered to be ecologically unacceptable.

RISK MANAGEMENT

As this contamination was due to historic activities, management of the risk associated with this contamination focused on strategies that would reduce the potential for exposure to SOC at the site. As with other hazardous waste sites, the various alternatives to be used in reducing risk were evaluated formally as part of the FS (Cullinane et al., 1988). Strategies evaluated during the FS included no action, monitoring of the site, in situ treatment of the contaminated soils to reduce SOC mobility and bioavailability, and removal of contaminated soil and sediments (Cullinane et al., 1988). As a result of the deliberations carried out during the FS, it was ultimately decided that soils and sediments should be removed from the contaminated areas of the site to reduce the risk associated with exposure to SOC. Following remediation, clean soil would be used to achieve the appropriate elevations required for reestablishment of the wetland. Based on the sediment chemistry and toxicological evaluations, a total of 18 ha was designated as contaminated and potentially requiring some form of remediation (Figure 6). However, the total acreage to be remediated was ultimately reduced because of concern about the impact of the excavation and remediation activities on the endangered species utilizing the site. Removal of sediments was focused on the areas with the highest concentrations of SOC and the highest demonstrated toxicity based on the plant and earthworm bioassays described previously in this chapter. As this strategy would leave elevated concentrations of SOC in the wetland ecosystems, the remedial strategy also included extensive long-term monitoring to evaluate the success of the remediation program. The contaminated areas that were to be monitored are referred to as the passive remediation areas to distinguish them from the areas where active remediation (e.g., soil/sediment removal) would take place (Figure 6).

LITERATURE CITED

Cullinane, M. J.; Lee, C. R.; and O'Neil L. J.: Feasibility study of contaminant remediation at Naval Weapons Station, Concord, California. Vol. I. Miscellaneous Paper EL-86-3. U.S. Army Engineers, Waterways Experiment Station, Vicksburg, MS, 1988.

Edens, F. W.; Banton, E.; Bursian, J. J.; and Morgan, C. W.: Effect of dietary lead on reproductive performance in Japanese quail *Coturnix coturnix japonica*. *Toxicol. Appl. Pharmacol.*, 38:307–314, 1976.

Eisler, R.: Cadmium hazards to fish, wildlife and invertebrates: A synoptic review. U.S. Fish and Wildlife Service, Laurel, MD. Contaminant Hazard Review No. 2, 1985a.

Eisler, R.: Selenium hazards to fish, wildlife and invertebrates: A synoptic review. U.S. Fish and Wildlife Service, Laurel, MD. Contaminant Hazard Review No. 5, 1985b.

Eisler, R.: Mercury hazards to fish, wildlife and invertebrates: A synoptic review. U.S. Fish and Wildlife Service, Laurel, MD. Contaminant Hazard Review No. 10, 1987.

Eisler, R.: Arsenic hazards to fish, wildlife and invertebrates: A synoptic review. U.S. Fish and Wildlife Service, Laurel, MD. Contaminant Hazard Review No. 12, 1988a.

Eisler, R.: Lead hazards to fish, wildlife and invertebrates: A synoptic review. U.S. Fish and Wildlife Service, Laurel, MD. Contaminant Hazard Review No. 14, 1988b.

Folsom, B. L., and Lee, C. R.: "Development of a Plant Bioassay for Determining Toxic Metal Uptake from Dredged and/or Fill Material," Proceedings of Heavy Metals in the Environment, International Conference, Amsterdam, NL, 1981.

Lee, C. R. et al.: Remedial Investigation of contaminant mobility at Naval Weapons Station, Concord, California. Miscellaneous Paper EL-86-2. U.S. Army of Engineers, Waterways Experiment Station, Vicksburg, MS, 1986.

Marquenie, J. M.: The freshwater mollusc *Dreissena polymorpha* as a potential tool for assessing bio-availability of heavy metals in aquatic systems. Proceedings of Heavy Metals in the Environment, International Conference, Amsterdam, 1981.

Mouw, D.; Kalitis, K.; Anver, M.; Schwartz, J.; Constan, A.; Hartung, R.; Cohen, B.; and Ringler, D.: Lead: possible toxicity in urban vs. rural rats. *Arch. Environ. Health*, 30:276–280, 1975.

NAS: Arsenic. National Academy of Sciences, Washington, DC, 1977.

Pershagen, G., and Vahter, M.: Arsenic: a toxicological and epidemiological appraisal. Naturvardsverket Rapp. SNV PM 1128, Liber Tryck, Stockholm, 1979.

Schroedner, H. A., and Mitchener, M.: Toxic effects of trace elements on the reproduction of mice and rats. *Arch. Environ. Health*, 23:102–106, 1971.

Simmers, J. W.; Lee, C. R.; and Marquenie, J. M.: Use of bioassays to predict the potential bioaccumulation of contaminants by animals colonizing upland and wetland habitats created with contaminated dredged material. Proceedings of the Third International Symposium on Interactions Between Sediments and Water, Geneva, 1984.

U.S. EPA: Quality Criteria for Water. EPA 440/5-86-001, 1986.

U.S. EPA: Risk Assessment Guidance for Superfund: Human Health Evaluation Manual, Part A. No. 9285.702A, 1989a (Interim Final).

U.S. EPA: Risk Assessment Guidance for Superfund, Vol. II: Environmental Evaluation Manual. EPA/540/1-89/001, March 1989b (Interim Final).

U.S. EPA: Ecological Assessment of Hazardous Waste Sites: A Field And Laboratory Reference Manual. EPA/600/3/89/013, 1989c.

Chapter 33

EFFLUENTS

P. B. Dorn and *R. van Compernolle*

INTRODUCTION

The federal regulation of effluents during the past 25 yr, by legislation such as the Federal Water Pollution Control Act amendments of 1972 and the Clean Water Act of 1987 (see Chapter 24), has resulted in the reduction of pollutants discharged to receiving waters. Resulting progress has been observed in the reclamation of many streams, rivers, and lakes to higher orders of use. The continuous improvement in the receiving waters of the United States has been due in part to the control of point source discharges from industrial manufacturing and municipal wastewater treatment facilities. Much remains to be done for non-point-source pollution (e.g., agriculture runoff, urban runoff, dredged sediment disposal). The aqueous discharge wastes produced in such point source facilities are termed *effluents.* Webster's defines effluents as something that flows out; liquid discharged as waste (as water used in an industrial process or sewage). Effluents, whether from industrial or municipal sources, may contain hundreds to thousands of chemicals with only a few being responsible for aquatic toxicity. Furthermore, the matrix of effluent may change significantly over time with cyclic changes occurring daily, weekly, monthly, and seasonally and may be affected by weather conditions or other unpredictable events. Many parameters of the matrix including total organic carbon (TOC), total suspended solids (TSS), pH, hardness, and ionic strength can strongly affect toxicity. In addition, when multiple chemical toxicants are present, the proportion of the overall toxicity due to each chemical may vary over time because of interactions.

Effluents in the United States are permitted under the National Pollutant Discharge Elimination System (NPDES) authorized under the Clean Water Act (Title IV), which set performance criteria for effluent quality (see Chapter 24). NPDES enables chemical and toxicity limits to be set for point source dischargers through permit programs administered by the U.S. Environmental Protection Agency (EPA) or authorized by state agencies. Effluents may be from any industrial or municipal source and can include those from petroleum refining, marketing and production discharges, metal plating, organic chemical synthesis, sewage treatment plants, power plants, pulp and paper mills, textile plants, and so forth. The NPDES process requires that effluent dischargers apply for permits to discharge, a process in which there is participation from the regulatory authority, discharger, and the public. NPDES monitoring is a self-monitoring program, and the discharger is responsible for performing the necessary testing according to specified procedures and schedule. Permit violations are voluntarily reported to regulatory authorities for remedial action, and periodic regulatory audits of data records are conducted.

The NPDES permit sets both technology-based performance criteria for pollutant chemical discharge and water quality–based biological parameters (see Chapter 24). Technology-based limits are developed from the ability of well-designed and operated facilities to remove specific chemicals from wastewater and, hence, are derived from the results of chemical analysis. Water quality–based limits, on the other hand, are based on the ability of living organisms to survive and/or reproduce when exposed to the whole effluent and, hence, are derived from the results of biological or whole-effluent toxicity (WET) testing. The purpose of this chapter is to discuss testing procedures for evaluating effluent quality with a focus on effluent toxicity. Current knowledge used in testing of effluent will be discussed from the hazard evaluation perspective with examples of toxicity testing of effluents and identification and control of toxic discharges. Although effluent toxicity is the primary focus, consideration is also given to deposition of effluent chemicals in sediments and bioconcentration of effluent chemicals in aquatic life. The discussion of case studies will draw heavily on experiences with effluent from petroleum refineries, chemical plants, and production platforms.

TERMINOLOGY

Bench-scale studies. Studies involving process simulations in small reactors (1 to 10 L in volume) on a laboratory bench.

Best professional judgment. A decision-making process that accommodates uncertainty and incomplete facts by synthesizing available facts with knowledge based on professional experience to provide a solution (see Chapter 24).

Chloroethers. Any of a class of chemical compounds containing an ether linkage (i.e., carbon-oxygen-carbon) and one or more chlorine atoms bonded to one or more of the carbon atoms.

Dye-dilution tracer studies. A procedure in which a nondegradable, nonadsorbable dye is added to the effluent just prior to discharge and the dye concentration in the mixing zone is monitored over time. The resultant dye concentration profile is used to determine the dilution patterns within an effluent and receiving water mixing zone and thus to define the mixing zone.

Effluents. Refers to aqueous wastewaters that have been treated in the wastewater treatment facility and are thus ready to discharge to the receiving water (see Chapters 1 and 24).

Feeds. Refers to aqueous wastewaters that have not yet been treated to remove pollutants. Feed refers to the fact that it is the feed to the wastewater treatment facility.

FM. Food chain multiplier. A multiplier that is intended to be applied to the bioconcentration factor for a chemical to estimate the bioaccumulation factor, thereby taking into account accumulation of the chemical up the food chain due to predation.

Full-scale studies. Studies performed using the full-scale commercial unit.

GC/MS. Gas chromatography/mass spectrometry.

Granular activated carbon (GAC) treatment. The use of granular activated carbon to adsorb organic chemicals from water. The carbon must be reprocessed periodically to remove adsorbed organics and then returned to service. This is usually treatment that is applied to effluents after secondary treatment when secondary treatment has failed to remove chemicals of concern. GAC treatment is often referred to as tertiary treatment.

HPLC. High-performance liquid chromatography.

Interlaboratory experiments. An experimental program in which several laboratories perform the same set of tests (i.e., bioassays) in order to determine the variability associated with the test (see Chapter 1). Results are used to improve testing and to guide regulatory officials who must account for such variability in setting criteria and limits.

Naphthenic acids. A large group of carboxylic acids present in crude oil and characterized by the carboxylate group being attached to aliphatic saturated and/or unsaturated cyclic carbon chains. Examples include cyclohexane and carboxylic acid.

Noncarcinogen. A chemical that has been tested and shows no potential to cause cancer in the test organism.

Pollution abatement program. See Source control.

Powdered carbon treatment. An alternative carbon adsorption process in which finely ground powdered carbon is added to the biological wastewater treatment tank, thus achieving biodegradation and adsorption in the same vessel.

Primary treatment. The first steps in wastewater treatment, in which insoluble oil and grease and suspended solids are removed. It may also include pH adjustment (see Chapter 24).

Process modification. A mechanical and/or procedural change in a manufacturing process that reduces the discharge of pollutants without affecting product yield or quality.

Process streams. The various aqueous streams that are produced in the manufacturing process and must subsequently be treated to remove pollutants. These streams are formed by water from chemical reactions, water used as the reaction solvent, condensed steam, etc. (see process wastewater in Chapter 24).

Reference ambient concentration. The concentration in an effluent below which a chemical would not be expected to have an effect on human health resulting from consumption of fish or shellfish living in a receiving water into which that effluent discharges.

RfD. Reference dose. The estimated dose at which a chemical would be expected to show noncancer effects in a test organism.

Secondary treatment. The steps in wastewater treatment in which soluble organics and ammonia are removed from wastewater, usually using a biological treatment process. This step includes subsequent removal of bacteria from the treated effluent (see Chapter 24).

Sewer lateral. A tributary sewer line that conducts aqueous effluents from individual manufacturing processes to the main plant sewer line.

Source control. The process of reducing the discharge of pollutants in aqueous effluents at the point of generation. This is usually accomplished by process modifications as defined earlier, but it could also involve some treatment process for a specific process stream or sewer lateral.

Substituted PAH. A polynuclear aromatic hydrocarbon that contains an aliphatic carbon side chain.

Synthetic effluent. An effluent made by mixing chemicals in dilution water in proportions representative of the actual effluent. Synthetic effluents are usually used when actual effluent is not available, as would be the case with a new process under development or when the effect of only one fraction of the effluent composition is of interest.

Regulation of Effluents

Technology Based

As stated above and in Chapter 24, all effluents in the United States are regulated through the NPDES permit process, which is authorized by the Federal

Clean Water Act (CWA). The specified effluent limits basically require the application of best available technology (BAT) for controlling effluent pollution discharge. Based on the performance of this technology, EPA and state regulatory authorities stipulate compliance with many effluent chemical parameters, which include conventional pollutants (e.g., BOD, COD, total suspended solids, oil and grease, pH, ammonia), and toxic priority pollutants. In addition, the language of the Clean Water Act requires the removal of "toxic substances in toxic amounts," which often goes beyond technology-based performance criteria. In this case, the best integrator of measuring toxic amounts of substance is a biological test response. Although considerable activity has been expended in performing ecological field studies (see Chapter 9) to assess stream, pond, estuary, and river quality, it is not practical to monitor a point source's discharge compliance by performing routine receiving water body surveys (or field studies). In the 1970s and early 1980s extensive receiving water body surveys were conducted to assess the effect of power plant thermal discharge plumes and the entrainment of aquatic life in power plant cooling water intakes required under section 316 (a) and (b) of the CWA. After many millions of dollars worth of effort in data collection and analysis, it was difficult to conclude that there were significant effects from all but a handful of power plant sites. Using surrogates such as laboratory toxicity tests has been preferable to expensive receiving water surveys. Within the past 10 yr, many permit requirements have included biomonitoring using a variety of species for acute and chronic toxicity effects measurements (see Chapters 1, 2, and 3). The assumption are that laboratory toxicity tests can be predictive of receiving water toxicity if the effluent exposures are correctly determined.

Technology-based toxicity controls remain in effect for some applications such as oil and gas production. Drilling fluid used in a well, however, cannot be discharged offshore without meeting a toxicity limit (U.S. Environmental Protection Agency, 1985, 1986a).

Water Quality–Based Approach

A traditional hazard assessment approach that couples in-stream receiving water dilution and the acute and chronic effluent effect can be applied to determine acceptable effluent discharge concentrations (Figure 1) (Bergman et al., 1986; Cairns et al., 1979). The theory states that as long as the receiving water concentration is less than the effect level of interest, no adverse effect would be observed. The approach is conceptually attractive and has been applied to

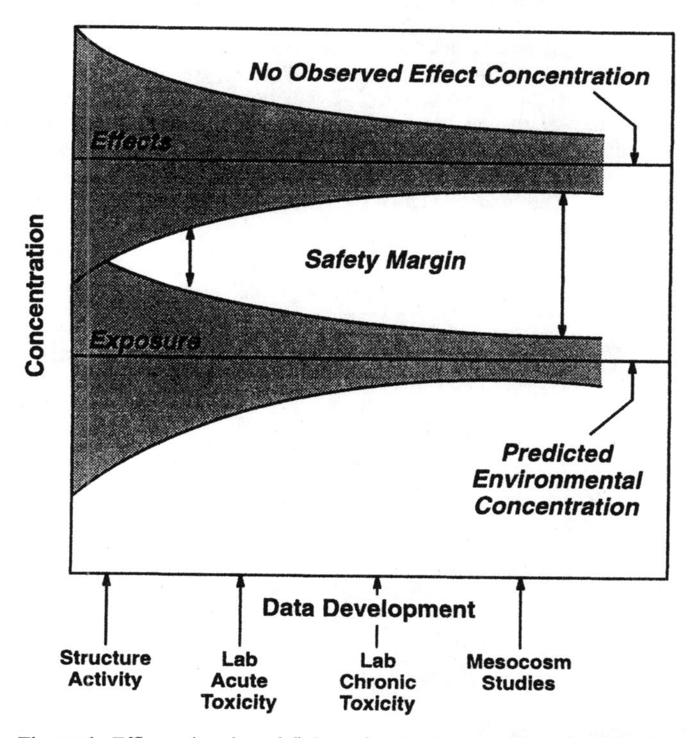

Figure 1. Effluent (or chemcial) hazard assessment conceptual approach. The safety margin is reduced as more data are developed on the chemical or effluent moving from structure-activity predictions toward scaled mesocosm studies. Most effluent studies are laboratory based to include acute and chronic toxicity and chemistry measurements. (See also Figure 4, Chapter 31 and accompanying explanation on pages 863–864.)

many sites for validation with successful results. The effluent hazard assessment is predicated on predicting environmental concentrations of effluent that will not result in unacceptable biological effects. As more data are generated, the ability to assess the "safety" of the effluent in receiving waters is increased. Thus, prediction of the margin of safety in the hazard evaluation is enhanced with more data. As we move from predictions using only structure-activity relationships (see Chapter 20) for specific chemicals to laboratory testing for acute and chronic effects, predictive abilities increase. Scale experiments using effluent or effluent components in stream mesocosm experiments would give considerable predictive capabilities. Receiving water monitoring would be assessing the actual environment; however, experiment control would be minimized and detection and prediction of effects would be least powerful unless a complex monitoring program was used.

The water quality–based permitting approach incorporates exposure and effects in the permit to provide a mixing zone whereby effluent is mixed with receiving water (U.S. Environmental Protection Agency, 1984, 1985, 1991e). This approach assumes that at the end of the discharge pipe, there will be no acute toxicity, and at the edge of the defined or allowed "mixing zone," there will be no chronic toxicity. A *mixing zone* is an area surrounding an effluent discharge that is allowed to combine the total effluent (100%) with the receiving waters so that a mixed condition results. This area is not typically considered as high quality as the receiving water and local regulatory authorities may grant this area for specific discharges. The mixing zone would be allowed if the effluent does not possess highly toxic and nondegradable chemicals. The water quality standards of the specific state would be met at the edge of the mixing zone. Mixing zones are hypothetical areas that can be defined by dye studies and mathematically calculated. A definition used may be such that the mixing zone is defined as an area where the discharge plume ceases to have velocity or ceases to be buoyant or reaches within 1 ft of the surface. Different definitions may be used by different states to meet site specific conditions. The approach assumes an "integrated strategy" to achieve and maintain water quality standards for the protection of aquatic life. This involves a combination of three controls: chemical-specific measurements, whole-effluent toxicity, and receiving water biological criteria/bioassessments (or biosurveys) for water quality monitoring and control. Each of the three components has capabilities and limitations (Table 1).

States have water quality standards that comprise *chemical-specific numeric criteria* (i.e., scientifically derived to protect aquatic life) for individual toxicants and narrative criteria (i.e., statements that describe the desired goal) such as "free from toxics in toxic amounts." Some states have also implemented biological criteria in water quality standards. In many instances states have adopted EPA-recommended chemical-specific numeric water quality criteria on the basis of which specific chemicals are limited in permits. States that do not develop chemical-specific numeric criteria may use their narrative standards along with quantitative risk values.

The integrated strategy relies on the use of a combination of these three controls. Each control approach complements the other two and provides greater protection of water quality.

The *chemical-specific* approach to toxics control for protection of aquatic life uses specific chemical

Table 1. Components of an integrated approach to water quality–based toxics control

Control approach	Capabilities	Limitations
Chemical-specific	Human health protection	Does not consider all toxics present
	Complete toxicology	Bioavailability not measured
	Straightforward treatability	Interactions of mixtures (e.g., additivity) unaccounted for
	Fate understood	Complete testing can be expensive
	Less expensive testing if only a few toxicants are present	Direct biological impairment not measured
	Prevent impacts	
Whole-effluent toxicity	Aggregate toxicity measured	No direct human health protection
	Unknown toxicants addressed	Incomplete toxicity (few species may be tested)
	Bioavailability measured	No direct treatment
	Accurate toxicology	No persistence or sediment coverage
	Prevents impacts	Conditions in ambient may be different
		Incomplete knowledge of causative toxicant
Bioassessments	Measures actual receiving water effects	Critical flow effects not always assessed
	Historical trend analysis	Difficult to interpret impacts
	Assesses quality above standards	Cause of impact not identified
	Total effect or all sources, including unknown sources	No differentiation of sources
		Impact has already occurred
		No direct human health protection

effluent limits in the NPDES permit to control toxic discharge. These may be developed via laboratory-derived biologically based water quality criteria. The limits are based on a site-specific evaluation of the discharge and its effects on the receiving water. This may include collection of effluent and receiving water data to generate the water load allocation (WLA) and total maximum daily load (TMDL) through modeling, a mixing zone analysis, and calculation of permit limits. After the numeric water quality criterion is adopted, chemical-specific limits are developed in the NPDES permits to ensure that a permittee's discharge does not exceed acute or chronic water quality criteria for the pollutant in a receiving water. Water quality criteria for the protection of aquatic life are developed by the EPA [under the Clean Water Act 304(a)(1)] and are published in individual criterion documents and summarized in the Quality Criteria for Water (U.S. EPA, 1986b). Water quality criteria are derived from scientific studies that evaluate a wide range of acute and chronic toxicity end points (see Chapters 2 to 8) including bioaccumulation (see Chapter 11). Each criterion consists of an acute and a chronic value. Criteria are developed using a variety of aquatic organisms including plankton, fish, shellfish, and invertebrates.

The *whole-effluent toxicity* (WET) approach to control toxics for the protection of aquatic life involves the use of acute and chronic toxicity tests to measure wastewater toxicity. Whole-effluent toxicity tests, when related to ambient conditions, yield a valid assessment of receiving water impact.

The EPA recommends that states also implement *biological criteria* (or *biocriteria*) in their water quality standards. These are numerical values or narrative statements that describe the biological or ecological integrity of communities that inhabit aquatic systems of a given designated aquatic life use and that can be used as legal end points for determining nonattainment. Biocriteria can supplement existing chemical-specific criteria or provide an alternative when chemical-specific criteria are not available. Quantitative biocriteria can be developed from unimpaired reference waters that represent best possible conditions. *Biological assessment* (or *bioassessment*) is an evaluation of the biological condition of an aquatic system using biological surveys. Integrated biological surveys (or biosurveys) are typically used to characterize the structure and function of resident aquatic biotic communities after reference sites are selected. Deviations from biological integrity are measured with biosurveys when the affected community is compared with the reference site. If there are improper quality controls, biosurveys may tend to underestimate impairment, especially if reference conditions are not adequate. Criteria developed from these surveys should consider both the structure and function (see Chapter 1 and 19) of aquatic systems.

Biological criteria also require an evaluation of physical integrity (i.e., by protecting conventional pollutants like pH, temperature, and dissolved oxygen; see Chapter 24) and provide a monitoring tool to assess the effectiveness of current chemically based criteria. Biosurveys tend to measure the aggregate effect of a multiplicity of factors on a community and therefore do not measure the impact of one effluent being discharged to the receiving water. Chemical-specific analyses of pollutants and whole-effluent toxicity tests are better predictor tools of biological integrity and water quality assessment. At present, it is difficult to use biological assessments as predictive water quality tools.

Effluent Assessment

For effluent toxicity determinations, the in-stream dilution of the effluent is critical to estimating receiving water effects. Analysis of the national distribution of effluent dilution conditions at the 7-d low flow during a 10-yr period (7Q10) and at annual mean flow revealed that at low flow dilution is largely in the <10:1 range, and at annual mean flow there is a higher distribution toward higher dilution in the range of 100–1000:1 (Figure 2). From this distribution it is apparent that in low flow conditions, many streams may become effluent dominated or impacted if toxicity is high.

Water quality–based permitting is based on the assumption that the laboratory toxicity test results for acute or chronic toxicity relate to receiving water effects when the exposure is considered. The estimation of exposure is complicated and may require at least a detailed water quality modeling study to include plume dilution and spatial dimensional description, a receiving water flow weighted analysis of the dilution of the effluent in different receiving water conditions that would include frequency, and duration of different effluent receiving water concentrations. This then helps define low flow and one-half critical low flow, etc. For example, the 7-d low-flow rate for 10 yr may be used (7Q10). The probabilistic dilution modeling approach is described in the Technical Support Document for Water Quality–based Toxics Control (TSD) (EPA, 1985b, 1991e) but has seen few actual applications in state or regional NPDES permits. The ability of this approach to predict effects in saltwater environments is illustrated in studies incorporating four discharges and 79 ambient field stations, where 89% of the prediction of field effects using effluent toxicity were correct (U.S. Environmental Protection Agency, 1991a).

Most applications of water quality–based permitting utilize the simplistic approach of estimating receiving water flow from gauging data to determine the critical flow rate and the one-half critical flow rate for compliance and calculating dilution of the effluent into the water body. River and stream flow rates are calculated by gauging data at specific monitoring stations along a water body that are recorded over long periods of time. The U.S. Geological Survey has river stations used to measure flow. Mea-

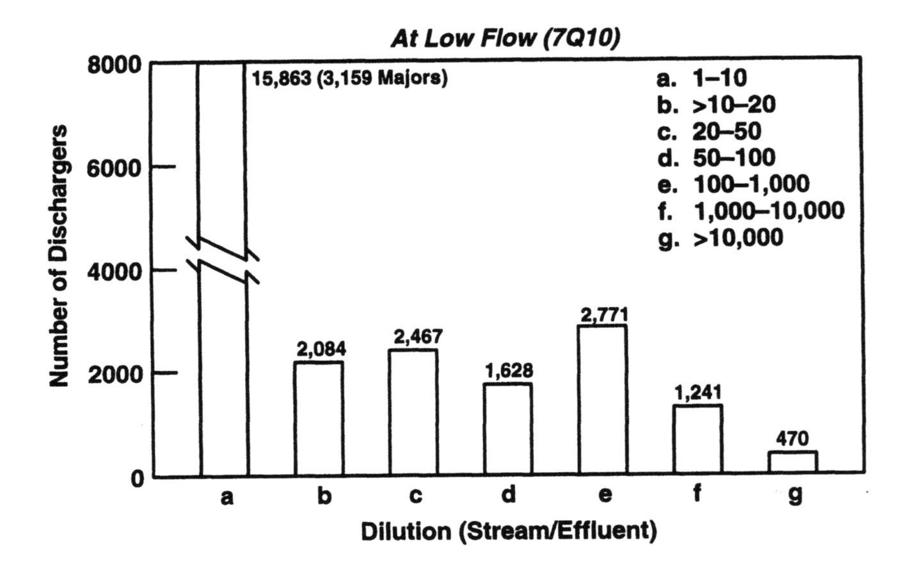

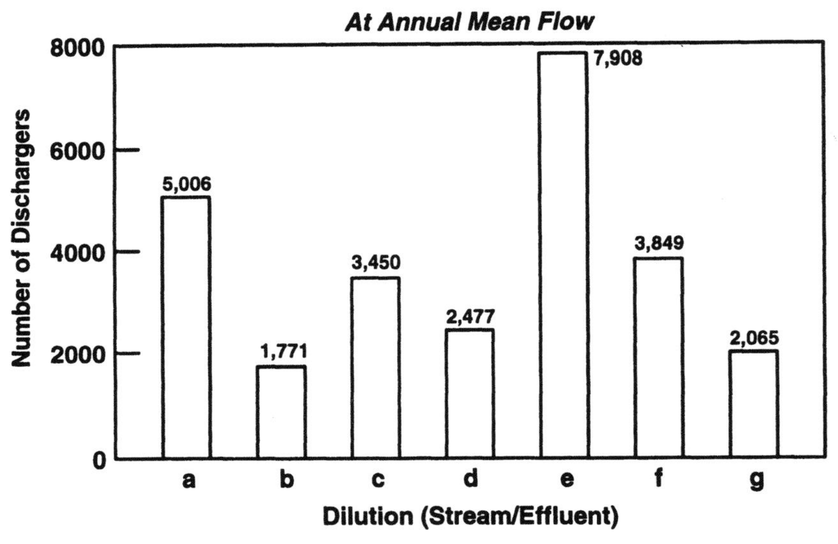

Figure 2. The distribution of NPDES dilution conditions at low flow and mean flow conditions.

surements are used to calculate river flow rates over specific periods such as the 7-d average low flow during a 10-yr period (7Q10). Effluent discharge permit limits are often based on a conservative estimate of the one-half (critical) flow rate during the 7Q10 period. For example, the discharge permit may state that the effluent limit for chemical "A" based on the one-half critical dilution at 7Q10 conditions should be 0.005 μ/L. Effluent limits are then determined based on acute or chronic toxicity.

If the toxicity compliance limit is exceeded for a period of monitoring such as three test periods, the NPDES permit holder is usually required to take corrective action. In previous permits, in some regions, noncompliance with an acute toxicity limit that could have been based on best available technology (BAT) for that category of discharger could have resulted in a fine of as much as $25,000 for each day of the violation. Current practice, however, is directed toward water quality–based permitting and if permits are exceeded, the discharger is required to conduct a study to determine the cause of the toxicity and develop a remediation strategy. These studies are termed toxicity identification evaluations (TIEs) and toxicity reduction evaluations (TREs), respectively. See Chapter 24 for a general description of these terms.

Test Methods In Effluent Assessments

Effluent assessments are conducted utilizing a suite of species and analyzing for acute and chronic (sublethal) effects. The toxicity and effluent plume spatial and temporal distribution are then combined to determine if there is reason to believe that there are ecological effects on the receiving waters. This is in effect a hazard evaluation (Bergman et al., 1986) and the need for effluent controls would be assessed based on the analysis.

Biological assessments may include both whole-effluent toxicity testing and receiving water biological surveys and water chemistry. However, this discussion will be limited to a description of the toxicity testing strategy. Toxicity testing methods applied to effluents are described in EPA guidance documents for acute (Weber, 1991) and sublethal (U.S. Environmental Protection Agency, 1985, 1988, 1991b) measurements. General aquatic toxicity testing methods are also discussed in Chapters 2 to 8. An effluent assessment should include estimation of freshwater and marine/estuarine toxicity to plants, invertebrates, and fish, at a minimum during some specified planned duration (U.S. Environmental Protection Agency, 1991a). EPA recommends for acute and chronic tests a dilution series (six) of five effluent concentrations (i.e., 100, 50, 25, 12.5, 6.25) and a control to determine the magnitude of toxicity. When conducting compliance monitoring, an option is to choose five concentrations that bracket the receiving water concentration (RWC; two above and two below it) under investigation. This monitoring would determine compliance status as well as an estimate (i.e., statistical) of the no-observed-effect concentration (NOEC). A year of effluent sampling from the actual site on a monthly basis followed by toxicity testing using several species may be recommended for water bodies that are affected and in need of remediation to achieve the specified water quality standard. However, a lesser frequency would be used if the water body is achieving designated standards and no impact from the dischargers is expected. Historical records outlining industrial and urban activities and biological quality of the water body are first indications of water quality changes and potential impacts.

If effluent toxicity methods are to be used for compliance, it is necessary to have methods carefully developed and laboratory data on test reproducibility should be available. Selection of test species for effluent toxicity assessment has been a subject of considerable debate. The list of acceptable effluent toxicity test species may be different depending on regional or site-specific concerns. Several recommended species are listed in Table 2 and described in detail in toxicity testing chapters in this book. Some species are only seasonally available and may be seasonally sensitive to stress. If the receiving water is to be protected, species selection should be consistent with that environment. However, if the effluent possesses higher total dissolved solid or different ionic composition or salinity, test methods may have to be modified to incorporate potential toxic effects from effluent quality not associated with "toxics" (Dorn and Rogers, 1989). Measurement of "salinity" or total dissolved solids alone may not determine whether one species or another is a better choice for testing. If the ionic balance is markedly different from the environmental tolerance of the organism, toxicity may result (American Petroleum Institute, 1989a,b). Test conditions should provide a tolerable environment for the test organisms, so the "toxics" in the effluent may be expressed independent of other sources of effect. EPA guidance rec-

Table 2. "Short-term" chronic effluent toxicity methods

Species/common name	Test duration	Test end points
Freshwaer species		
Ceriodaphnia dubia, cladoceran	Approximately 7 d (until 60% of control have three broods)	Survival, reproduction
Pimephales promelas, fathead minnow	7 d	Larval growth, survival
Pimephales promelas, fathead minnow	7–9 d	Embryo-larval survival, percent hatch, percent abnormality
Selenastrum capricornutum, freshwater algae	96 h	Growth
Marine/estuarine species		
Arbacia punctulata, sea urchin	1.5 h	Fertilization
Champia parvula, red macroalgae	7–9 d	Cystocarp production (fertilization)
Mysidopsis bahia, mysid	7 d	Growth, survival, fecundity
Cyprinodon variegatus, sheepshead minnow	7 d	Larval growth, survival
Cyprinodon variegatus, sheepshead minnow	7–9 d	Embryo-larval survival, percent hatch, percent abnormality
Menidia beryllina, inland silverside	7 d	Larval growth, survival

ommends that control data be carefully evaluated using both positive (i.e., from reference toxicants) and negative (i.e., from untreated dilution water) control data (see Chapter 1). The dilution water, as well as laboratory reference toxicity tests, should be used to determine organism health before decisions on effluent toxicity are made.

Presently, standard species for freshwater toxicity testing for sublethal effects are focused on the daphnid, *Ceriodaphnia dubia*, and the fathead minnow, *Pimephales promelas* (U.S. Environmental Protection Agency, 1991b; see Chapter 2). Measured end points include survival and reproduction for the daphnid and survival and growth (weight) for the fish. Common marine species currently in place for sublethal effects testing are mysid (*Mysidopsis bahia*), sheepshead minnow (*Cyprinodon variegatus*), and silverside (*Menidia beryllina*) (U.S. Environmental Protection Agency, 1988) (see Chapter 3). Other species being used, especially on the West Coast of the United States, include sea urchins (see Chapter 6), abalone, and giant kelp (see Chapter 5).

Examples of Effluent Toxicity Testing Results

Before industrial or municipal category controls for treatment were formalized, effluent toxicity could be estimated in some cases based on the discharge. For most categories of industrial manufacturing, there are EPA guidelines for effluent limits and the specific technologies that are demonstrated to meet those limits. Examples of some industries that have specific effluent treatment technologies that are demonstrated as best available technology are synthetic organic chemicals, petroleum refining, electric generating, poultry processing, and metal finishing. For example, organic chemical manufacturers would typically produce a degrading effluent that showed acute toxicity and a small ratio of acute-to-chronic (ACR) toxicity. Pesticide manufacturers would potentially discharge small quantities of the product after cursory treatment that would be target specific, affecting a specific phylum, such as invertebrates, and the ratio of acute to chronic toxicity would be large, denoting the specific mode of action of the chemical. However, treatment technology has progressed to a point where effluent quality is high for most industrial discharges, and these discharges are not distinguished by toxicity test response. For that reason it is not appropriate to categorize discharge toxicity by industry.

A study to assess the toxicity of production waters resulting from oil and gas production activities in the Gulf of Mexico illustrates the selection concerns for a monitoring program (Moffitt et al., 1992). The pilot study evaluated eight production effluents, which received moderate treatment and were in compliance with all permit parameters. During three time periods each of the eight production effluents selected for geographic difference and flow difference were tested for toxicity to sheepshead minnows, mysids, and Microtox. The Microtox toxicity test measures the effects of chemicals or effluents on the luminescent bacteria, *Photobacterium phosphoreum.* The more toxic the substances tested, the lower the light produced by the bacteria relative to the control. In the production platform effluent study, results showed no significant effects variation due to time, flow, or location. Species sensitivity showed that mysid growth was the most sensitive end point (Table 3). The data show that the most sensitive end point measured was mysid growth, followed by mysid fecundity. The mysids were the most sensitive of the two species in comparing data from two studies. The Louisiana database contains more than 400 test results compiled after an emergency rule was developed to require rapid collection of produced water toxicity information by produced water dischargers of the state. The similarity of the two sets of data is remarkable, and statistical analysis of "important factors" found no apparent relative factors affecting toxicity (i.e., location, depth, time period).

Variation In Effluent Toxicity Tests

We are often asked to make an assessment of the quality of a measurement, regarding accuracy and/or precision. The accuracy of biological testing cannot be assessed because there is no exact baseline result that it can be calibrated against. However, the precision of that data may be assessed in several ways, such as averages, standard deviations, and coeffi-

Table 3. Summary data for toxicity of produced waters to mysids and sheepshead minnows

	Produced water study		Existing Louisiana database	
Overall mean[a]	Mysid	Sheepshead	Mysid	Sheepshead
96-h LC50	7.08	21.55	10.36	14.82
7-d LC50	5.77	19.72	[b]	[b]
NOEC survival	3.14	11.70	3.39	6.28
NOEC growth	1.60	2.75	2.44	5.23
NOEC fecundity	2.20	[b]	2.67	[b]
Number of tests	24	24	>400	>400

[a]In % produced water.
[b]No data.

Table 4. Interlaboratory variability of 7-d *C. dubia* and fathead minnows exposed to reference toxicants (7-d LC50) and effluents in 10 laboratory evaluations

	Ceriodaphnia				Fathead minnow			
	Set 1		Set 2		Set 1		Set 2	
	Mean	CV	Mean	CV	Mean	CV	Mean	CV
$K_2Cr_2O_7$ (μg/l)	29.6	70.8	49.6	37.7	2000	22.8	2000	23.6
NaCl (g/L)	1.77	6.7	1.33	49.9	—	—	—	—
Utility effluent (%)	3.7	80.6	100	0	3.7	36.9	—	—
Pulp effluent (%)	—	—	70.0	12.0	—	—	—	—
Refinery 301 (%)	—	—	—	—	1.0	30.9	—	—
					0.9	33.1	—	—
Refinery 401 (%)	—	—	—	—	10.0	27.9	—	—
					9.0	26.0	—	—
NaPCP (μg/L)	—	—	—	—	228	54	293	40.3
					288	43.5	285	44.7
								—

cients of variation (see Chapter 10). Because effluent toxicity testing is often conducted in direct response to a regulatory compliance order, the permittee is concerned about the precision of the test estimate obtained. Effluent toxicity testing has been conducted for compliance for more than 20 yr, but it "how good the numbers are" has only recently become a concern. A common practice in the United States in assessing method precision and reproducibility is the interlaboratory or "round-robin" test (in Europe referred to as the "ring test"; see Chapters 1 and 27) that incorporates a number of laboratories in a program to test a specified sample (e.g., effluent) with the same species and method and reports results. The test program assesses the ability of laboratories to conduct testing correctly and to produce the correct results (such as in chemical testing) or allow toxicity to be determined. Results are analyzed and precision estimates are derived.

Few effluent toxicity test round-robin exercises have been conducted, but the reported results demonstrate that there is a variation between laboratories, in some cases so much that the ability to determine compliance with a toxicity requirement would be questionable. In the last few years, interlaboratory experiments have been conducted using NPDES compliance test methods for *Daphnia*, *Ceriodaphnia dubia*, *Pimephales promelas*, and *Mysidopsis bahia*. These programs have investigated interlaboratory response to reference toxicants (Dorn et al., 1987; Jop et al., 1986) in conjunction with effluent testing. Reference toxicants used include potassium dichromate ($K_2Cr_2O_7$), sodium chloride (NaCl), and sodium pentachlorophenol (NaPCP). Effluents tested have been from electric generating utilities (Degraeve et al., 1988, 1992), pulp and paper mills (DeGraeve et al., 1992), refineries (Degraeve et al., 1988), chemical plants (Grothe and Kimerle, 1985), and drilling muds (Ray et al., 1989). In studies to determine interlaboratory variation of *Ceriodaphnia dubia* test methods, 10 laboratories participated and tested utility and paper mill effluents. Test variation (coefficient of variation) ranged from 12 to 80%, and variation using reference toxicants ranged from 6 to 70%. Similar results were obtained in a second study using the fathead minnow (Table 4). In a summary of acute and chronic round-robin data (Parkhurst et al., 1992), the mean coefficient of variation was between 7 and 34% (Table 5).

Toxicity Testing for Permit Compliance

For screening purposes, effluent toxicity testing is a valuable tool and can enable a discharger to determine the need for additional treatment, process changes, point source control, etc. Effluent toxicity testing is routinely being used for compliance with toxicity requirements (i.e., NOEC >90% effluent). Although the interlaboratory precision of effluent toxicity testing has been shown as demonstrated above to be approximately 30% variation, the variation has been calculated on LC50 or IC50 point estimates and not on specific variability in complying with specific concentration. Effluent performance characteristics and implications for regulatory testing were evaluated (Parkhurst et al., 1992; Warren-Hicks and Parkhurst, 1992), and were found to show higher variation in one-concentration comparisons than for

Table 5. Intra- and interlaboratory variation in effluent toxicity as shown in coefficient of variation (CV)

	Species	Effluents	Mean CV	CV range
Intralaboratory				
Acute	5	13	17	0–135
Chronic	2	7	7	0–20
Interlaboratory				
Acute	4	13	34	0–166
Chronic	2	7	34	0–83

a point estimate such as LC50. Data analyzed by Warren-Hicks and Parkhurst (1992) from the fathead minnow round-robin (7-d) test on two oil refineries conducted by Degraeve et al. (1988) illustrate the variation in response. If the same concentration/mortality data from "refinery 301" and "refinery 401" are plotted for each laboratory used in the evaluation and that plot is compared with a plot of predicted survival (Figure 3), the difference in variation is dramatic. The data in the figure show that individual laboratories show a high variation in survival response to the same effluent and same effluent concentration. Comparison of the individual data (points) with the predictions (lines) from all of the data shows high variation.

Although interlaboratory precision is acceptable for general research testing, permit compliance with specific limits may have greater variation than the discharger can accept. Since the NPDES program is premised on self-monitoring and reporting to regulatory authorities, it is intuitive that dischargers want to comply with requirements. Thus, additional effort ought to be applied by dischargers and regulators to move toward more reliable, yet protective, approaches to effluent toxicity control.

Technology-Based Toxicity Limits

Before the water quality–based permitting approach was used, acute toxicity limits were set for individual dischargers based on the category of manufacturing (e.g., metal finishing, poultry processing) and the best available technology for effluent control. An example of a BAT toxicity control is presently in use in the oil and gas extraction industry. In the process of oil field development, wells are drilled to great depths in the earth using a rotary drilling apparatus. The drill bit is cooled, and the drilling cuttings are removed to the surface by a circulating drilling mud system (Engelhard et al., 1989). These systems comprise generally nontoxic materials including barite, clay, xanthum gums, water, and sometimes alcohols. Various packages are also added that may include mineral oils and biocides to prevent growth of bacteria. The present U.S. requirement for drilling mud toxicity is that a drilling mud is in compliance with offshore disposal if it can pass an acute test with the mysid, *Mysidopsis bahia*. The specific methodology for that test requires that an aliquot of the drilling mud be diluted in seawater to a 1:10 (v/v) mud-seawater mixture, mixed for 5 min. and allowed to settle for 60 min, after which the supernatant is decanted. This fraction is called the suspended particulate phase (SPP) and is used to test toxicity to mysids in 96-h static nonreplacement conditions (Figure 4). The 96-h LC50 corrected for control mortality must be >30,000 ppm (3%) of the SPP for the discharger to be in compliance. Alternative methods using Microtox were also shown to be predictive of mysid effects in decision-criteria analysis (Dorn et al.,

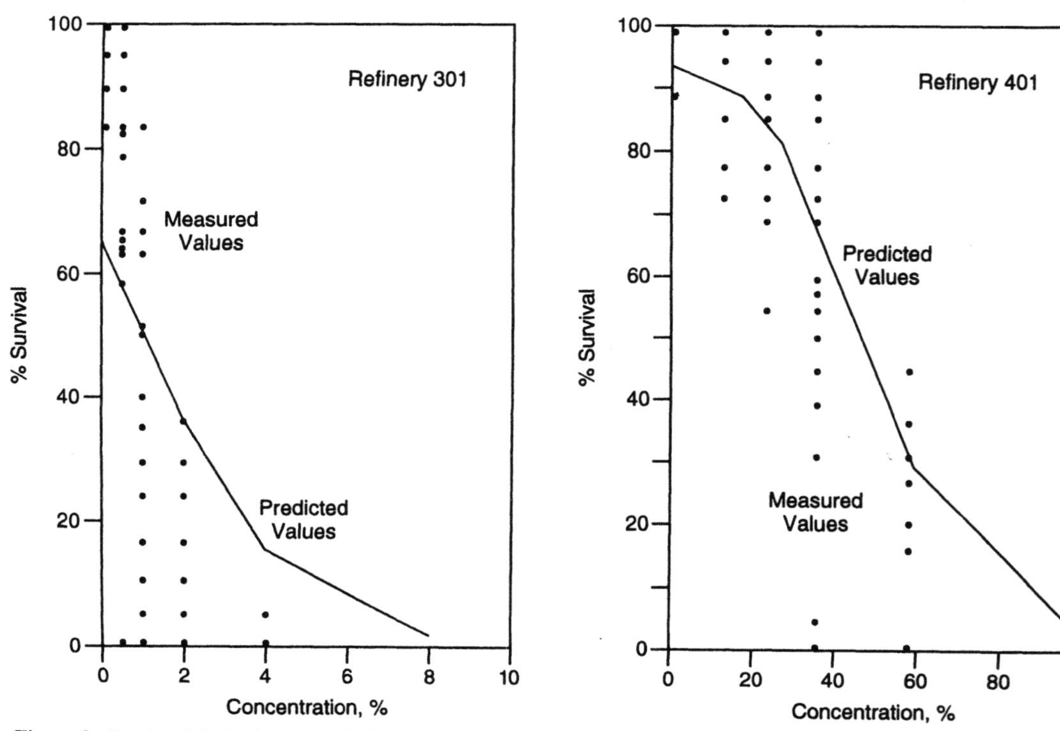

Figure 3. Results of fathead minnow 7-d chronic effluent test results for two refinery effluents.

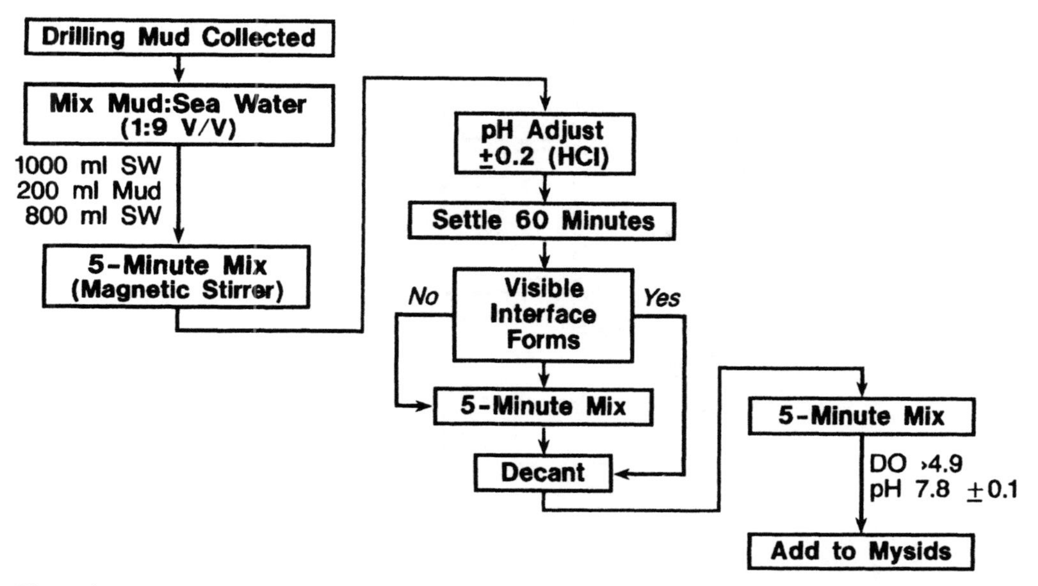

Figure 4. Preparation of the suspended particulate phase (SPP) of drilling mud for mysid toxicity testing.

1990). In a study to assess the interlaboratory variation of drilling mud toxicity to mysids, 14 laboratories participated with two generic drilling muds, of which 12 completed testing, one with no additives and low toxicity and one with diesel added for higher toxicity (Ray et al., 1989). The results of the study showed that variation between laboratories was within a factor of 12 (Figure 5).

Effluent Treatment

The typical wastewater treatment plant today contains equipment for primary and secondary treatment. In primary treatment, large solid materials, free oils, and suspended solids are removed. Secondary treatment utilizes biological treatment to remove soluble oils and organics. This process involves nitrification, whereby ammonia is removed. After clarification for biological solids separation, the effluent may be chlorinated and dechlorinated and then discharged to the receiving water (Figure 6).

Removal and fate of organics (see Chapter 15) found in refinery waste streams may be categorized based on partitioning into air, water, and solids phases. Those with highest partitioning components into air are contained in covered tanked areas and fed into a biotreater whereby biodegradation is complete to mineralization (see Chapters 1, 15, and 31). Those with high partitioning coefficients with affinity to solids sorption such as polycyclic aromatic hydrocarbons (PAHs) are biodegraded to a high degree and sorbed onto biosolids that are removed and either incinerated or disposed of according to regulations. The general characteristics of these chemicals that make them amenable to different removal processes are shown in Figure 7. The characteristics of physical partitioning into lipids/solids, water, and air are indicative of the mode of treatment most appropriate. Those toxics, such as benzene, naphthalene, phenol, and cresol, are water soluble and readily biodegradable and are effectively removed by biological treatment. Chemicals that partition into solids and are absorbable, such as PAHs, are partially degradable and will partition onto biosolids in the treatment system. The removed solids are typically incinerated or disposed of in approved landfills.

Removal and fate of organics in wastewater treatment systems may be categorized based on physical-chemical properties, such as water solubility, log octanol-water partition coefficient (log K_{ow}), vapor pressure (P_v) and Henry's law constant (H_0). Figure 7 relates these properties to fate within a wastewater treatment present in the system. This partitioning will make the organics unavailable for biological degradation and their removal is dependent only on physical separation due to absorption or adsorption. If these chemicals have sufficiently high vapor pressures, they will escape from the free organic layer into the atomosphere unless contained in covered tanks with vapor control.

Chemicals with log K_{ow} less than about 4 will tend to be in aqueous solution and available for biodegradation. Biodegradation of these chemicals will also depend on their Henry's law constant, which is the air water partitioning coefficient. Chemicals such as benzene are biodegradable but also have high H_0 values. This results in a competition between biodegradation and stripping (volatilizing) for removal from the wastewater. If the biotreatment system is too severely mixed and/or aerated, the chemical will be removed mostly by stripping to the atmosphere. Chemicals such as phenol, alcohols, and amines have

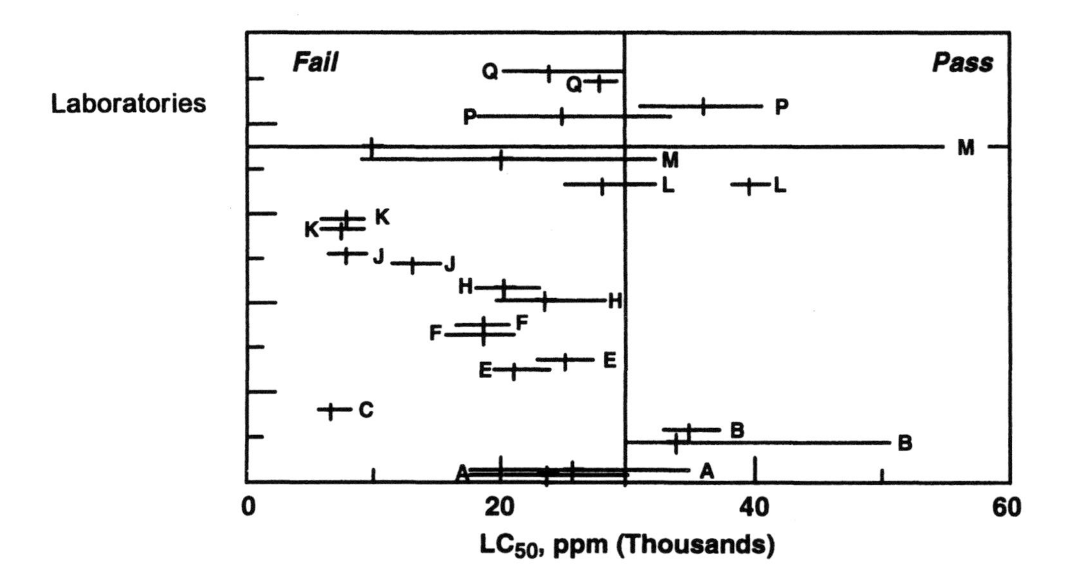

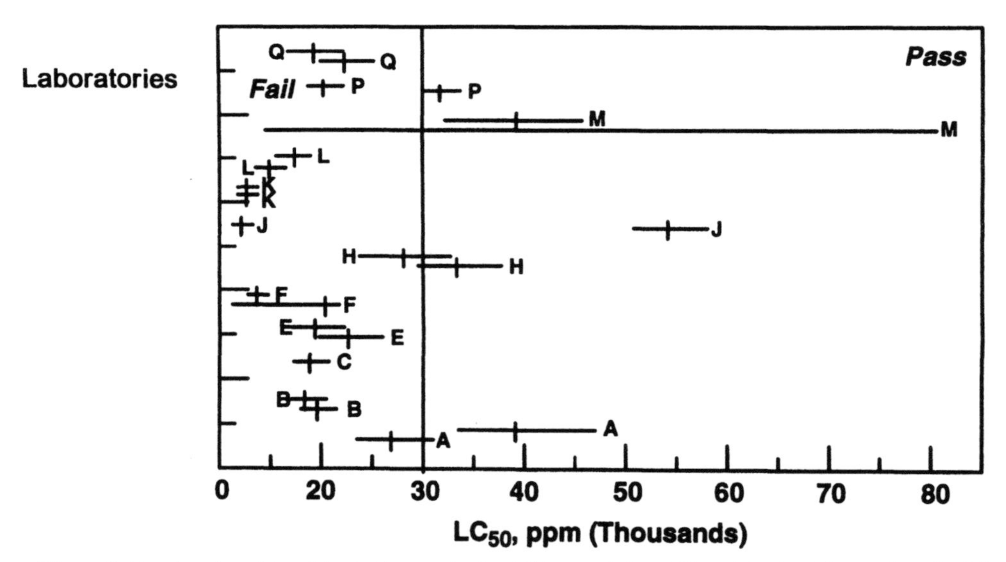

Figure 5. Results of mysid toxicity testing results for drilling muds from 12 participating laboratories submitting data. Each part of the figure represents a separate drilling mud tested.

low H_0 values and will remain in aqueous solution and be removed by biodegradation.

The effectiveness of biological secondary treatment of industrial waste after primary solids removal can be illustrated in a comparison of acute toxicity measured before and after activated sludge biological treatment for two chemical plants and a refinery (Table 6). The pretreatment acute toxicity in chemical plant 1 (site A) ranged between 6 and 10% during 3 wk of sampling and was not appreciably decreased after treatment. However, in chemical plant 2 (site B) and refinery 1 (site C) samples, acute LC50 values < 6.25% feed effluent were reduced to nonacutely toxic levels (LC50 > 100%) after secondary treatment. The cause of toxicity in chemical plant 1 will be discussed in this chapter in the development of toxicity identification evaluation (TIE) studies.

Another example of treatment effectiveness may be illustrated in laboratory tests to identify the removal of toxicity following bench-scale activated sludge treatment (Figure 8). In the study, five nonionic surfactants with different degrees of branching on the alkyl chain backbone were assessed for acute and chronic toxicity of neat chemicals and their bio-

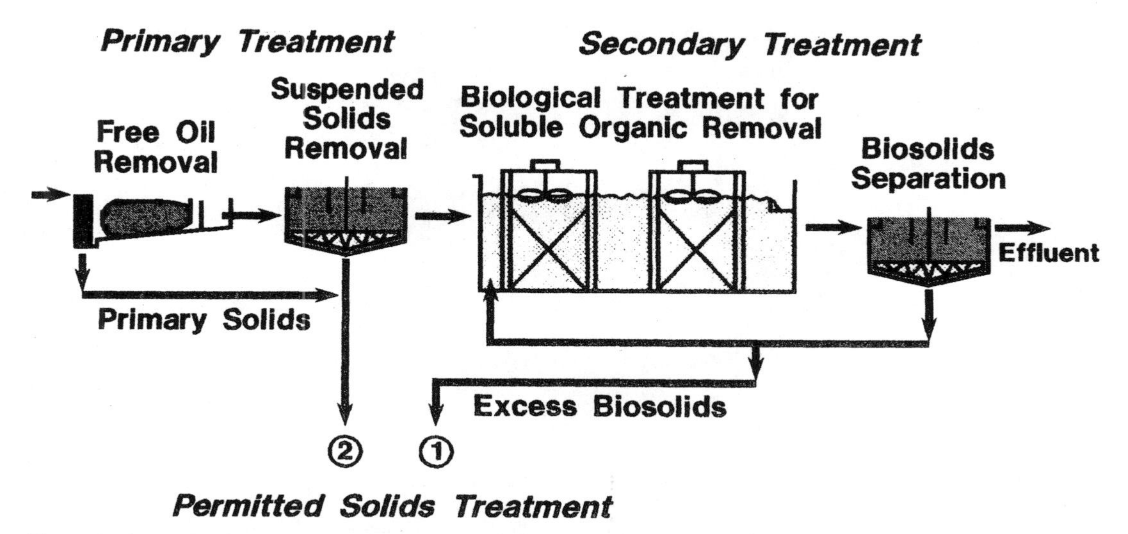

Figure 6. Conventional wastewater treatment system found at a major industrial site.

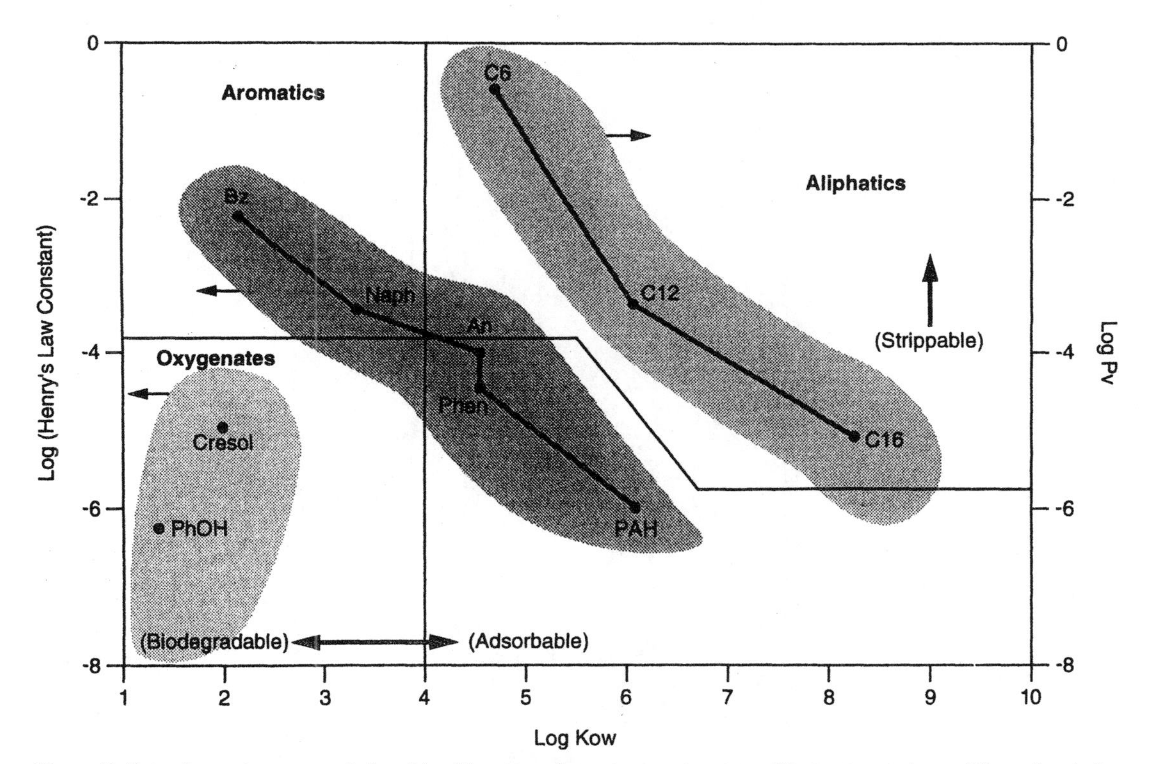

Figure 7. Fate of organic compounds found in effluent to refinery treatment systems. The treatment phase of these chemicals is determined by the fate compartment taken. For example, PAH (polycyclic aromatic hydrocarbons) will adsorb to solids and are biodegraded and not present in effluent discharge or are adsorbed on solids and removed by separation and disposed of in permitted facilities or incinerated.

Table 6. Secondary treatment effectiveness in toxicity reduction (48-h EC50 as percent effluent)

	Week	EC50 before treatment	EC50 after treatment
Site A[a] (chemical plant 1)	1	6.89 (6.2–12.5)[c]	8.56 (7.3–9.9)
	2	9.57 (8.6–10.8)	9.29 (8.4–10.2)
	3	7.89 (6.5–9.1)	8.21 (6.6–9.7)
Site B[b] (chemical plant 2)	1	37.7 (33.6–42.6)	>100
	2	<6.25	>100
	3	9.5 (7.3–13.6)	>100
Site C[b] (refinery 1)	1	<6.25	>100

[a]*Mysidopsis bahia.*
[b]*Daphnia pulex.*
[c]95% confidence limits in parentheses.

treated effluents. The results shown in Table 7 illustrate that detergent effluent acute toxicity is removed in all but one effluent after activated sludge treatment and chronic toxicity is reduced considerably in effluents. However, the more highly branched materials had more chronic toxicity (Table 7). The data compared the toxicity of the effluents produced by the laboratory biotreater units to measured concentrations of intact (parent) surfactants measured by CTAS, a colorometric analytical tool. Benzoate is used as a control because it is highly degraded in sewage treatment systems. Benzoate is also acting as a carbon source for bacteria in the treater to maintain healthy populations of degraders and a good mixed suspension. Essentially all acute toxicity was removed after treatment with the exception of the nonylphenol ethoxylate. Chronic toxicity was also removed after biotreatment of the essentially linear alcohol ethoxylate, but not well removed from biotreater units dosed with the other surfactants. This laboratory treatment unit is used to model the activity of municipal sewage treatment plants that receive household "down the drain" discharges of surfactants from everyday use. The bench-scale treatment unit is a very useful tool in determining the treatability of pure chemicals (Dorn et al., 1993).

Toxicity Identification in Effluents

If the effluent is found to be toxic at concentrations exceeding the allowable instream dilution or other permit stipulation, the discharger is required to conduct a toxicity identification evaluation and toxicity reduction evaluation (see Chapter 24). The general

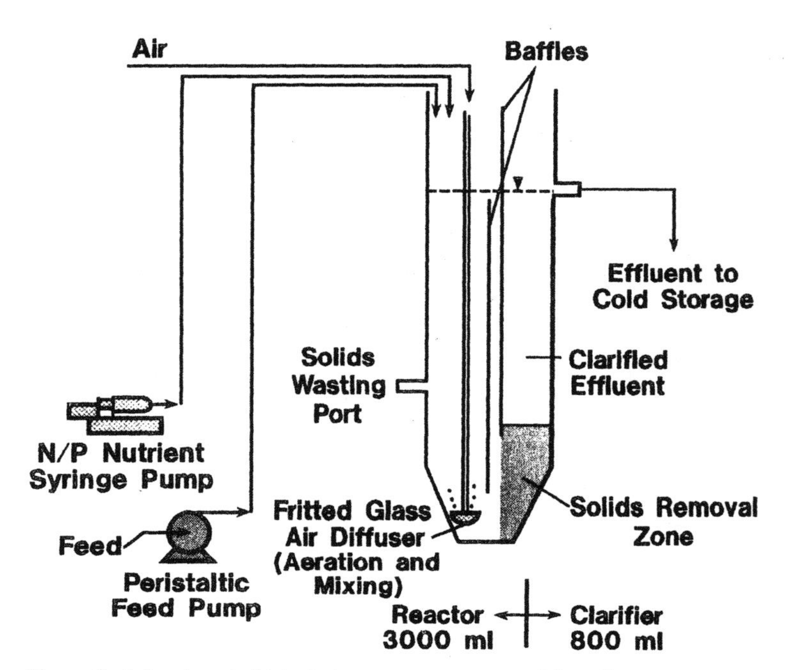

Figure 8. A bench-scale biological treatment system used for effluent treatability studies.

procedures for conducting a TIE have been outlined in EPA guidance (U.S. Environmental Protection Agency, 1988, 1989a,b, 1991c,d,e) and can be characterized as procedures whereby the effluent is manipulated to measure changes in toxicity. The manipulations performed are chemical and physical separation techniques that basically utilize processes of stripping, sorbing, filtering, and aging under different pH conditions to effect toxicity changes. Chemical manipulation techniques can be used for identification and tracking of oxidants and metals. Several of the techniques can be used in sequence to develop fractions for toxicity studies (Figure 9). The TIE approach is divided into three phases (see Chapters 2 and 24). Phase I is illustrated in Figure 9 and consists of methods to identify the physical and chemical nature of constituents causing active toxicity. Phase I can also be used to develop treatment methods to remove toxicity. The TIE, in general, is performed to try to identify the fraction or procedure that has been effective in removing toxicity or affecting toxicity changes. Once the fraction(s) has been identified, chemical identification is initiated, and subsequent spiking or back addition of the identified chemical is performed for confirmation. The TIE by its nature is a single-sample analysis and should be repeated several times and confirmed for several test species, including those of regulatory concern. At that point, toxicity reduction schemes can be initiated.

The TRE phase of the process may take different courses. Experience has demonstrated that after changing housekeeping practices and process controls, the toxicity may be eliminated.

For example, one can recall the story of the environmental scientist performing a discharge study to identify chemicals in a discharge plume on a diurnal basis, who kept identifying industrial chemicals not used in any process. Upon investigation and discussion with the process unit operators, it was discovered that on nightly rounds one of the plant staff, on his way to a coffee break, found it convenient to pour his spent metal cleaning fluid into the effluent well rather than carry it back to the control room. His uninformed response was that the dilution rendered the solvents harmless to the environment. A lesson learned in TRE: education of plant personnel may result in increased awareness and understanding of potential problems.

Another lesson learned pertains to the industrial plant that conducted a multimillion dollar TIE/TRE on a marine species and invoked a treatment scheme to remove toxicity. Several years later, the regulatory agency changed the compliance test species that was sensitive to the new effluent, requiring additional treatment. In conducting these studies, many samples should be tested on a suite of organisms. The technical support document (EPA, 1991e) recommends at a minimum a plant, an invertebrate, and a fish species. This recommendation would have protected the discharger and identified specific toxicity to one phylum and would perhaps have resulted in a solution protective of all test species.

Other possible noncapital solutions may include product substitution, such as water treatment chemicals, or using nontoxic additives. On several occasions, chemical plant process engineers found a process modification beneficial to product quality and also produced a low-toxicity effluent, which benefited both the environment and the business. Once

Table 7. Acute and chronic toxicity of biologically treated surfactant effluent (% effluent)

	Acute toxicity		
Surfactant unit	Fathead minnow 96-h LC50	*Daphnia magna* 48-h EC50	CTAS (mg/L)[b,c]
Benzoate control	>100	>100	.24
80% linear alcohol ethoxylate	>100	>100	.25
Nonylphenol ethoxylate	21 (12.5–25)[a]	40 (25–40)	.62
Branched alcohol ethoxylate	>100	>100	.39
Branched alcohol ethoxylate	>100	>100	.28

	Chronic toxicity								
	Fathead minnow				*Daphnia magna*				
	Survival		Growth		Survival		Growth		
Surfactant unit	NOEC	LOEC	NOEC	LOEC	NOEC	LOEC	NOEC	LOEC	CTAS[c] (mg/L)
Benzoate control	100	>100	100	>100	100	>100	50	100	.24
80% linear alcohol ethoxylate	100	>100	50	100	100	>100	100	100	.25
Nonylphenol ethoxylate	12.5	25	6.25	12.5	24	50	25	50	.62
Branched alcohol ethoxylate	12.5	25	6.25	12.5	50	100	50	100	.39
Branched alcohol ethoxylate	25	50	25	50	100	>100	50	100	.28

[a]95% confidence interval.
[b]Intact surfactant measured in 100% effluent. Data apply to both test species.
[c]Cobalt thiocyanate active substance.

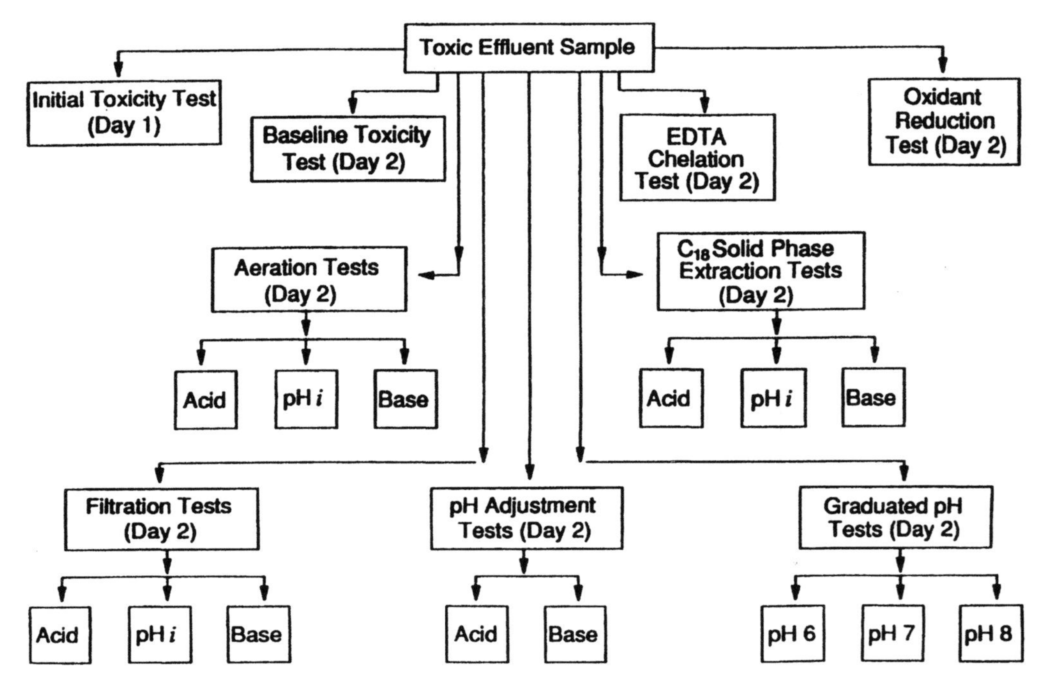

Figure 9. A schematic diagram of the EPA phase I procedures for effluent toxicity identification evaluations (pHi = initial pH)

process streams are diluted, the toxicants are more difficult to treat. Figure 10 illustrates the ability of several treatments to eliminate pollutants. Generally, the more concentrated the waste stream, the easier it is to remove the toxicant. The ability of a specific treatment technology to remove chemical oxygen demand (COD) from a waste stream is illustrated in Figure 10. Aerobic treatment such as a sewage treatment system is most effective at removing COD in influent from trace concentrations to up to about 6000 mg/L, compared to a range of 1000–100,000 mg/L for anaerobic treatment. The concentration of the COD stream is very important and dictates the cost-effectiveness of the selected treatment.

THE TOXICITY IDENTIFICATION EVALUATION PROCESS

Case Examples

The following examples illustrate successful toxicant identification studies from chemical plants and refineries. The examples also show that scientific understanding must be applied to each individual evaluation and biological knowledge alone cannot solve complex toxicant identification problems. A team approach utilizing biologists, chemists, and engineers is the appropriate method of attack. The examples will identify a variety of toxicant specific to the particular effluent and will include organic halides, organic acids, polymeric water treatment chemicals, ammonia, calcium, ''ionic strength,'' and vanadium. A summary study (Steen, 1988, 1990) will be used to illustrate the diversity of toxicants identified in effluents from many types of dischargers. In a majority of TIE studies conducted during the past 5 yr, several toxicants have frequently been identified: ammonia, chlorine, pesticides, surfactants, polymeric treatment chemicals, and ''salt.''

The studies were performed using a variety of methods most attributable to best professional judgment (BPJ) (see Chapter 24)—that is, in essence, the most successful approach to effluent toxicity identification evaluations. Beginning with some of the EPA guidance (U.S. Environmental Protection Agency, 1989a,b, 1991c,d), a creative approach to the TIE will usually yield successful results.

Petroleum Refinery—Multiple Effects

This example is shown to illustrate that the toxicity identification process is not necessarily direct and interacting factors in addition to the toxicant may play a significant role in the story (Dorn et al., 1992).

This refinery example is from the Shell Martinez Refinery located on the Carquinez Strait between San Pablo and Suisin Bay on the eastern reaches of San Francisco Bay. Effluent toxicity testing had begun in the late 1960s with acute toxicity testing utilizing the three-spine stickleback, *Gasterosteus aculeatus*, in 96-h LC50 determinations. Over time, toxicity tests requirements were changed to lower toxicity for

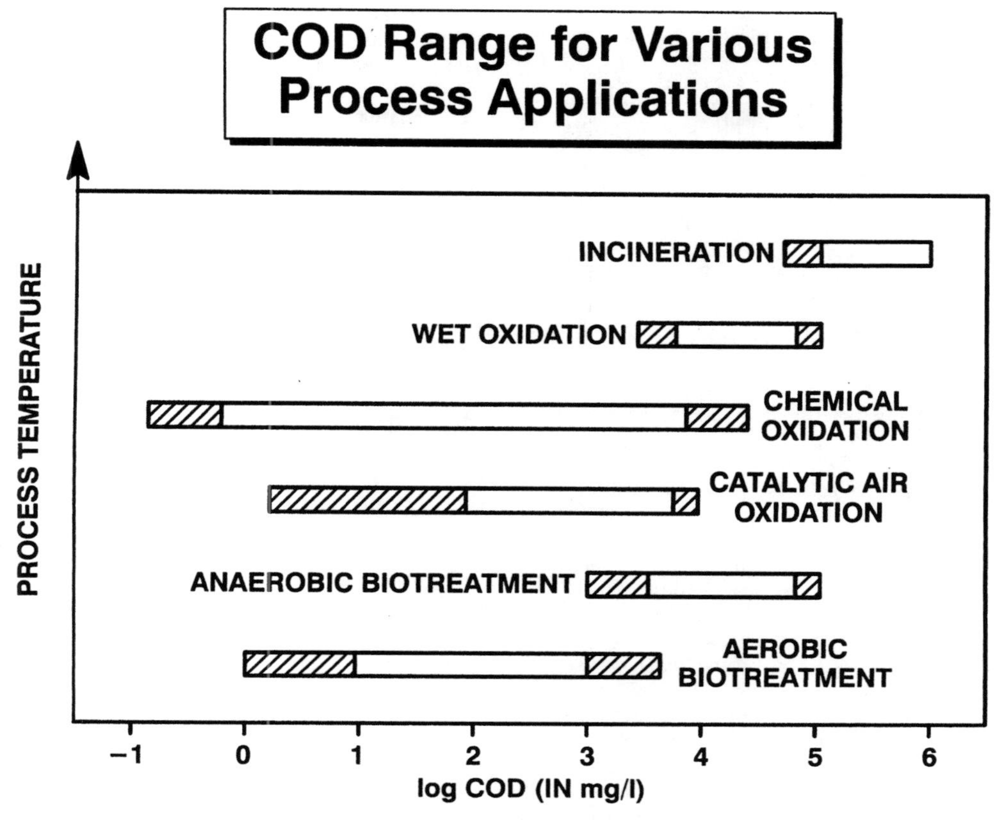

Figure 10. Organic loading removal with different treatment technologies.

compliance and although the refinery biotreater performance was good, performance in effluent toxicity varied. Toxicity testing requirements developed in the late 1960s set an effluent compliance limit such that the LC50 = 50% effluent to stickleback fish. The requirement was changed to require fathead minnows and an LC50 of 50% and later 100% effluent. Recently the requirement was changed to rainbow trout survival at >90% in 100% effluent in a median of 11 tests. Historical studies at the location in the 1970s using process sewer lateral isolation and treatment identified several sources of toxicity. Sewers receiving process unit wastes were isolated and sampled at each intersecting trunk line. Each of the samples was biotreated in a laboratory unit (see Figure 6) and tested for toxicity. Sewer samples not amenable to biotreatment and resulting in toxicity were isolated for further study. The identified sources of toxicity were naphthenic acids, which were quantified as ''soluble oil and grease,'' ammonia, total dissolved solids, organic amines termed ''process chemicals,'' and a water treatment flocculent. A review of the site operations revealed that, in addition to earlier work, vanadium was a potential contributor to the effluent toxicity.

Laboratory bench-scale treatment units and a site study with pilot units were used to identify the causes of toxicity and implement a control strategy (see Figure 8). Laboratory bench-scale units (similar to those in Figure 8) were used to determine the treatability and removal of toxicity. Larger ''pilot'' treatment units up to 40 gallons in volume, closely simulating the actual treatment plant, were conducted on the refinery site. Following the site toxicity identification study, a water quality–based effluent study was performed using dye-dilution tracer studies to identify the effluent plume and detailed acute and chronic toxicity testing.

The toxicity identification strategy was to perform experiments in which the suspect toxicant was spiked into clean laboratory water or receiving water and then results were compared with effluent effects. Manipulation of the existing effluent by pH adjustment and chelation experiments were also used to determine whether toxicity was affected.

Experimentation showed that pH had a profound effect on effluent toxicity, and a pH change from 6.5 to 8.0 in effluent resulted in a loss of lethality (Table 8). This was attributed to protonation of naphthenic acids at low pH and conversion into a toxic bioavailable form. The table shows the concentration of effluent compared to pH and the survival of fish. Highest survival is observed with a more basic pH, with lowest survival at acid conditions. Survival is

Table 8. Effect of pH on the toxicity of refinery biotreater effluent to stickleback

	Percent survival (96-h)		
Dilution (%)	pH 6.5	pH 7.5	pH 8.0
100	0 (3 h)	0 (48 h)	60
75	0 (24 h)	20	NA[a]
50	0 (72 h)	100	NA
30	80	100	NA
Control	100	100	100
Estimated LC50	35	66	>100

[a] NA, not applicable or tested.

highest at lowest effluent concentrations regardless of pH.

Vanadium toxicity experiments indicated that at effluent discharge concentrations toxicity would be partially attributable to this metal; however, in effluent spiking there was no contribution of vanadium (Table 9). The effect of different species of vanadium, on the effluent toxicity was evaluated using two forms of vanadium, and effluent. The sodium vanadate was most acutely toxic to stickleback with the LC50 as low as 6.0 mg/L, and when it was added to effluent at concentrations approximating that typically identified, no additive effect was observed. The LC50 values of vanadium solutions alone ranged from 6.0 to 15.8 mg/L vanadium. The effluent LC50 was 45% without added vanadium and 59% and 54% when 0.1–4.3 mg/L vanadium was added, respectively.

Effects of the water treatment chemical were profound and comparative tests showed that toxicity of the available polyethyleneimine (PEI) and diethylaminoethylmethylacrylate/acrylamide (DMAEM/AM) flocculents showed LC50 values of 5–15 and 30–50 mg/L, respectively. In experiments using filtered effluent, unfiltered effluent, and dilution water, the effects of polymer addition are seen to affect fish survival (Figure 11). In addition, the water-treating polymer was found to inhibit the ability of the biotreater to nitrify ammonia.

Treatability studies with powdered activated carbon (PAC) showed that 50 mg/L (basis feed flow) completely removed acute toxicity to stickleback after addition of 20 mg/L naphthenic acid to biotreater feed. In contrast, conventional biotreater fed the same naphthenic acid spike yielded an effluent with an LC50 of about 60%.

The summary data identified the specific toxicants as water treatment polymer, ammonia, and naphthenic acids. The pH also had a profound effect on the toxicity of naphthenic acid, and a pH <7.5 would show toxicity attributable to these toxicants. The effect of these factors can be illustrated as a flow diagram of causative agents (Figure 12). Water treatment polymer resulted in biotreater bacteria toxicity. This resulted in decreased ability to nitrify the feed nitrogen load, resulting in effluent ammonia. Shifts in pH resulted in naphthenic acid toxicity at lower pH.

The TIE/TRE study was conducted for the most part on the refinery site using a mobile laboratory equipped with bench-scale biotreater units that could be manipulated to change conditions of pH, ammonia, and water treatment chemicals, as well as sludge age. Studies were also conducted for carbon treatment to evaluate the effectiveness of tertiary treatment. The site study utilizing specific experiments to identify key factors affecting toxicity coupled with biotreater units enabled the identification and control of the effluent toxicants. Possibly the most important control method was an awareness program for unit operators in the refinery. Source control and pollutant abatement programs resulted in a more consistent effluent with less toxicity.

Derivation of Water Quality Based Limit

As a result of dye-dilution studies, an effluent dilution was calculated to be 33:1 on the average and a target zone of initial concentration (ZID) similar to a mixing zone (see Chapter 24) was determined for toxicity (U.S. Environmental Protection Agency, 1979). A toxicity testing program was conducted for 12 mo using six species for acute and three species for chronic toxicity.

The results of the acute toxicity testing program indicate that the mysid shrimp is, on the average, the most sensitive species tested (Figure 13). According to U.S. EPA guidance, *M. bahia* should be approximately as sensitive as the most sensitive species in

Table 9. Acute toxicity of vanadium and biotreated effluent to stickleback, *Gasterosteus aculeatus*

Test		Concentration range[a]	96-h LC50 (mg/L)
$VOSO_4$ (vanadyl sulfate)		13–79	15.8[a]
$NaVO_3$ (sodium vanadate)	(1)	2–19	6[a]
	(2)	2–10	>9.7[a]
	(3)	2–28	14[a]
Effluent		—	45%[b]
Effluent + $NaVO_3$	(1)	0.1–1.8	59%[b]
	(2)	0.7–4.3	54%[b]

[a] mg/L.
[b] % effluent.

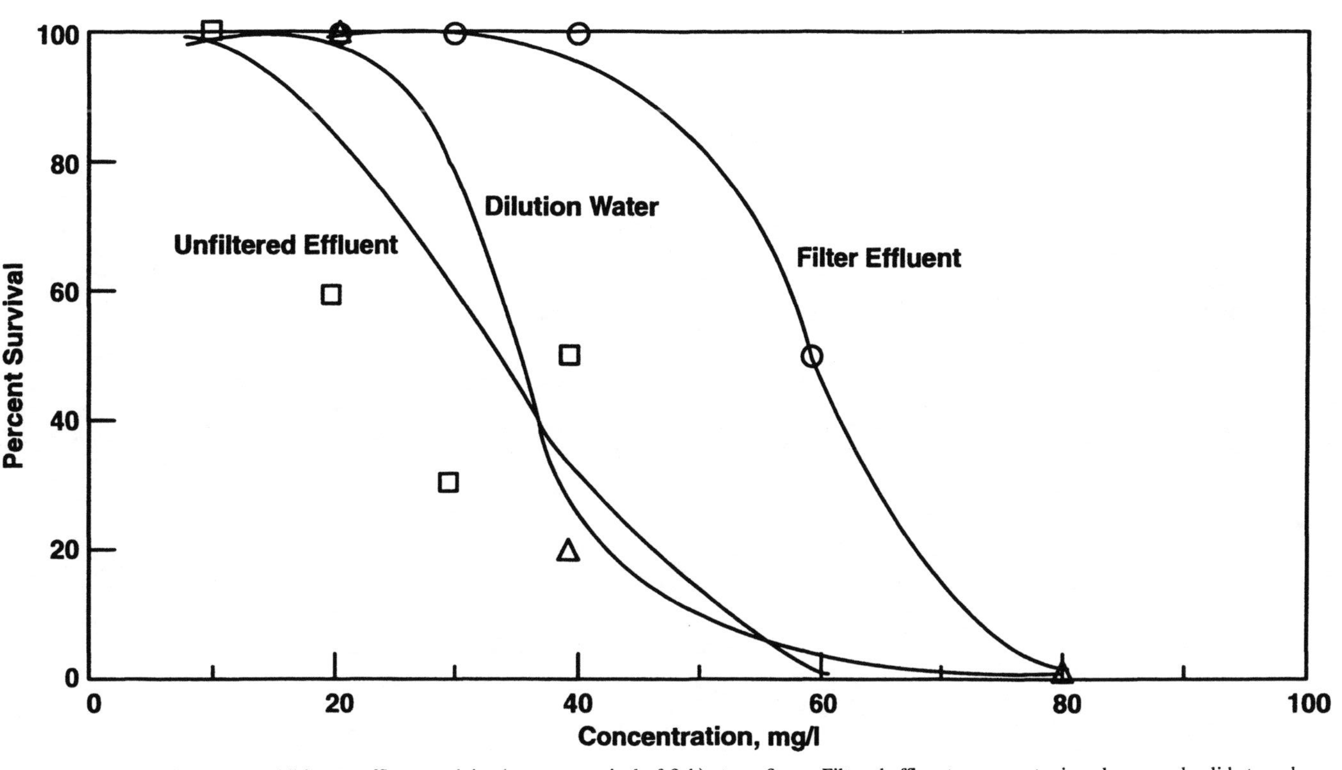

Figure 11. Effect of polymer addition on effluent toxicity (percent survival of fish) at a refinery. Filtered effluent removes toxic polymer and solids to reduce effluent toxicity. Product substitution is a way to remove effluent toxicity.

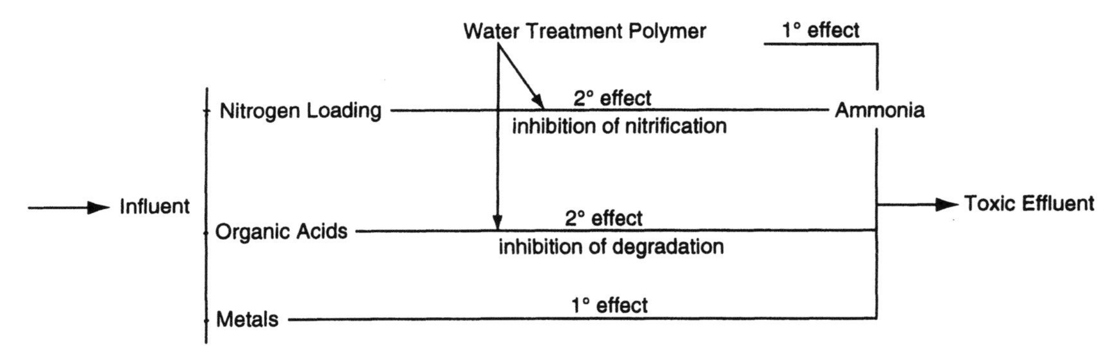

Figure 12. The secondary toxic effects in a refinery biotreatment system. The primary toxicity is caused by organic and polymer toxicity. As a result of polymer toxicity to the biotreater bacteria, some toxicity is observed that results in poor nitrification leading to un-ionized ammonia production, which in turn has a toxic effect, reducing degradation of organic loading. The resulting effects are demonstrated in a toxic effluent.

the Carquinez Strait receiving water community according to local water pollution control authorities. Consequently, by protecting *M. bahia*, the entire community should be protected from acute toxicity. The most direct way to provide such protection would be to use *M. bahia* as the biomonitoring species. The refinery effluent was quite consistent. This is illustrated by the relatively small standard deviations, with a maximum of 6% of the mean. In addition, the ratio of the maximum to minimum values observed for each species over the entire testing period ranged from 1.68 to 5.97, with a geometric mean of 3.61. This mean ratio can be considered a good estimate of effluent variability as it pertains to acute toxicity (interpreted from Figure 13).

Chronic toxicity NOECs ranged between 5.6 and 31.7% effluent (Table 10). The acute-to-chronic ratio (ACR), however, was small and ranged as a geometric mean for each of the three species from 5.19 to 1.63 (Table 11). This low ACR shows that the EPA TSD (U.S. Environmental Protective Agency, 1991e) gives considerable conservatism in estimat-

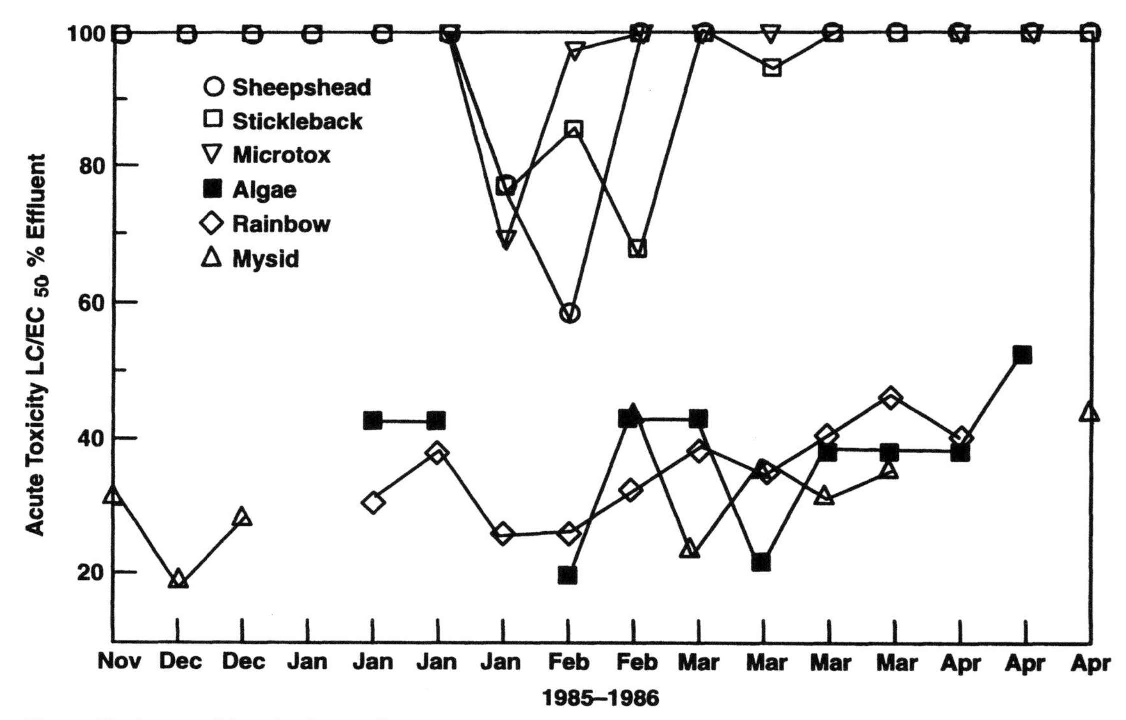

Figure 13. Acute toxicity of refinery effluent to six species during a 1-yr temporal monitoring period. Mysids, algae, and rainbow trout were observed to be most sensitive with less than threefold variation in toxicity during the monitoring period.

ing the ACR at 20 for whole effluents for different categories of discharges (i.e., POTWs, oil refineries, and chemical manufacturers).

A toxicity effluent limit of 3% should ensure that following initial dilution, the biological community in the receiving water will not suffer any acute or chronic effects. This value is conservative and takes into account effluent variability.

The refinery conducted effluent variability studies from 1988 to 1990 using fathead minnows, *Ceriodaphnia*, and mysids. The new data compared with those in Tables 10 and 11, showed that the NOEC range for the most sensitive end points for each of the three species was 3.1–25, 12.5–100, and 3.1–50% effluent for *Ceriodaphnia*, fathead minnows, and mysids, respectively. Based on the calculated dilution data in the ZID, all test results would meet compliance on an average 33:1 estimate receiving water to effluent dilution. Current plans for the refinery treatment system are for a continuously operated granular activated carbon (GAC) treatment system following activated sludge treatment.

Chemical Plant—Salt and Organics

This example evaluated toxicity tests using more than one species in revealing the interference of a nontoxic entity in masking the response of a toxic fraction.

Regulatory authorities have designed many of the effluent toxicity testing methods to be cost-effective. In some instances, tests have been streamlined to save time and money for the permittee. Simplification may lead to problems. Acute tests were run on either a freshwater or saltwater species, the selection of which was based on effluent "salinity." For example, if salinity was below 5‰, *Daphnia* spp. were to be used; if salinity was above 5‰, *Mysidopsis* sp. was required.

The subject chemical plant had a very toxic influent and equally toxic effluent on three successive weekly samplings. The treatment system appeared to be ineffective and resulted in an acutely toxic effluent (Table 6). There was little difference between untreated feed and effluent (after treatment) EC50 numbers for *Mysidopsis* as observed in chemical plant 1 in this 3-wk study. Toxicity ranged between 8 and 9.6% concentration in the feeds and between 8.2 and 9.2% concentration in the effluents.

Cyprinodon showed the effluents to be less toxic. The feed EC50 values were lower than the effluent values, indicating treatment effectiveness. Feed toxicity for *Cyprinodon* ranged between 11.7 and 37.9% concentration and between 76.1 and >100% concentration of effluents. Microtox toxicity was lower in effluents than feeds. Toxicity to feeds ranged from 15.5 to 62.9% concentration and between 52.9 and >100% concentration in effluents. In addition to total organic halide in the waste stream, high $CaCl_2$ con-

Table 10. Maximum concentrations of effluent that do not cause chronic effects: results reported a no observable effect concentration (NOEC) in % effluent compared to control.

	Species						
	Ceriodaphnia		Fathead minnow		Opossum shrimp		
Test initiation date	Survival	Reproduction[a]	Survival (%)	Growth[b]	Survival	Reproduction[c]	Growth[b]
1/24	—[e]	—	14.1[d]	—	—	—	—
2/26	10.1[d]	—	7.07[d]	7.07	—	—	—
3/5	5.6[d]	—	—	—	—	—	—
3/12	10.1[d]	—	14.1[d]	14.1	—	—	—
3/19	10.1[d]	10.1	24.5	14.1[d]	—	—	—
4/2	18.0[d]	—	14.1[d]	24.5	—	—	—
4/9	10.1[d]	10.1	14.1[d]	14.1	—	—	—
4/17	10.1[d]	10.1	7.07[d]	38.7	—	—	—
4/23	10.1[d]	10.1	7.07[d]	24.5	24.0[d]	—	24
5/14	—	31.7[d]	24.5	14.1[d]	—	—	24[d]
5/28	—	18.0[d]	7.07[d]	24.5	—	—	—
6/11	—	31.7[d]	—	—	13.4[d]	—	—
Geometric mean NOEC	12.2	—	19.8	—	—	—	—
Standard deviation	1.7	1.4	1.4	—	—	—	—
Maximum value	31.7	14.1	24.0	—	—	—	—
Minimum value	5.6	7.1	13.4	—	—	—	—
Ratio max/min	5.7	—	1.8	—	—	—	—

[a] Number of neonates.
[b] Pooled dry weight gain.
[c] Egg production.
[d] Indicates the value used for determined NOEC.
[e] —, no test data calculated.

Table 11. Calculation of acute-to-chronic ratios for Shell Martinez refinery effluent

Species	Date	48-h LC50 (% effluent)	NOEC (% effluent)	ACR
Fathead minnow	1/24	26.6[a]	14.1	1.89
	2/26	24.5[a]	7.07	3.47
	3/12	36.6[a]	14.1	2.60
	3/19	40.5	14.1	2.87
	4/2	32.8[a]	14.1	2.33
	4/9	34.2[a]	14.1	2.43
	4/17	37.2	7.07	5.26
Geometric mean ACR = 3.02	4/23	35.9	7.07	5.03
Standard deviation = 1.39	5/14	38.7	14.1	2.74
Upper limit 95% CI[b] = 3.71	5/28	22.0[a]	7.07	3.11
Species	**Date**	**48-h LC50 (% effluent)**	**NOEC (% effluent)**	**ACR**
Ceriodaphnia	3/5	50.9[a]	5.6	9.09
	3/12	39.3[a]	10.1	3.89
	3/19	66.5	10.1	6.58
	4/2	65.4	18.0	3.63
	4/9	69.8	10.1	6.91
	4/17	71.2[a]	10.1	7.05
Geometric mean ACR = 5.19	4/23	71.8[a]	10.1	7.11
Standard deviation = 1.53	5/28	65.4	18.0	3.63
Upper limit 95% CI[b] = 6.73	6/11	82.0	31.7	2.59
Species	**Date**	**48-h LC50 (% effluent)**	**NOEC (% effluent)**	**ACR**
Mysid	4/23	38.0[a]	24.0	1.58
	5/14	35.8[a]	24.0	1.49
Geometric mean ACR = 1.63	6/11	24.7	13.4	1.84
Standard deviation = 1.12				
Upper limit 95% CI[b] = 1.90				

[a] Corrected for control mortality via Abbott's formula.
[b] The 95% CI assumes a single-tailed test with all 5% distributed in the upper tail of the lognormal distribution.

centrations were identified. Calcium concentrations ranged between 5000 and 6000 mg/L.

Part of the study involved identifying the specific role of calcium in affecting the effluent toxicity. Feeds and effluents were fractionated into component mixtures of total organic halide and calcium to create a two-factor experiment with four levels of each factor. A matrix of 0, 100, 1000, and ~6000 mg/L Ca and 0, 10, 20, and ~50 mg/L total organic halide was created by carbon sorption fractionating and Na_2CO_3 precipitation of Ca. A toxicity identification study was initiated using 13 *Mysidopsis bahia*, 11 *Cyprinodon variegatus*, and 12 Microtox tests. Feeds and effluents from the plant and an experimental pilot plant were tested for aquatic toxicity and compared to the total organic halide concentration in the solutions (Figure 14). The concentration of the total organic chloride represents concentrations of the influent into the treatment system and of the effluent. The higher concentrations are influent ranging from approximately 50 to 100 mg/L. The concentration-response curves show that a decrease in organic chlorides results in a decrease in toxicity (LC50 goes up) for sheepshead minnows and Microtox, but not the mysids. This observation led to investigations to understand what factor other than organic chloride concentration was responsible for mysid toxicity in the effluent.

In separate experiments, the toxicity of calcium chloride to five aquatic species was tested and the acute toxicity to *Mysidopsis* was found to be 1140 and 1178 mg/L in two tests, which is approximately one-fifth of the effluent concentration. Calcium toxicity to *Cyprinodon* was 2766 and 5640 mg/L in two tests using diluent water and 5300 mg/L in effluent. The apparent toxicity to mysid was due to the effect of calcium, with sheepshead toxicity attributed mainly to total organic halide, with some calcium toxicity. Figure 15 shows the specific effect of calcium concentration and organic halides on toxicity to mysids.

The laboratory fractionation experiments measured toxicity of 28 combinations of calcium and total organic halide (i.e., as total organic chlorides) to *Mysidopsis* and *Cyprinodon*. At 200 mg/L Ca^{2+}, mysid response to different concentrations of total organic halide was observed with mortality ranging from 5% at 2 mg/L TOCL (total organic chloride) to

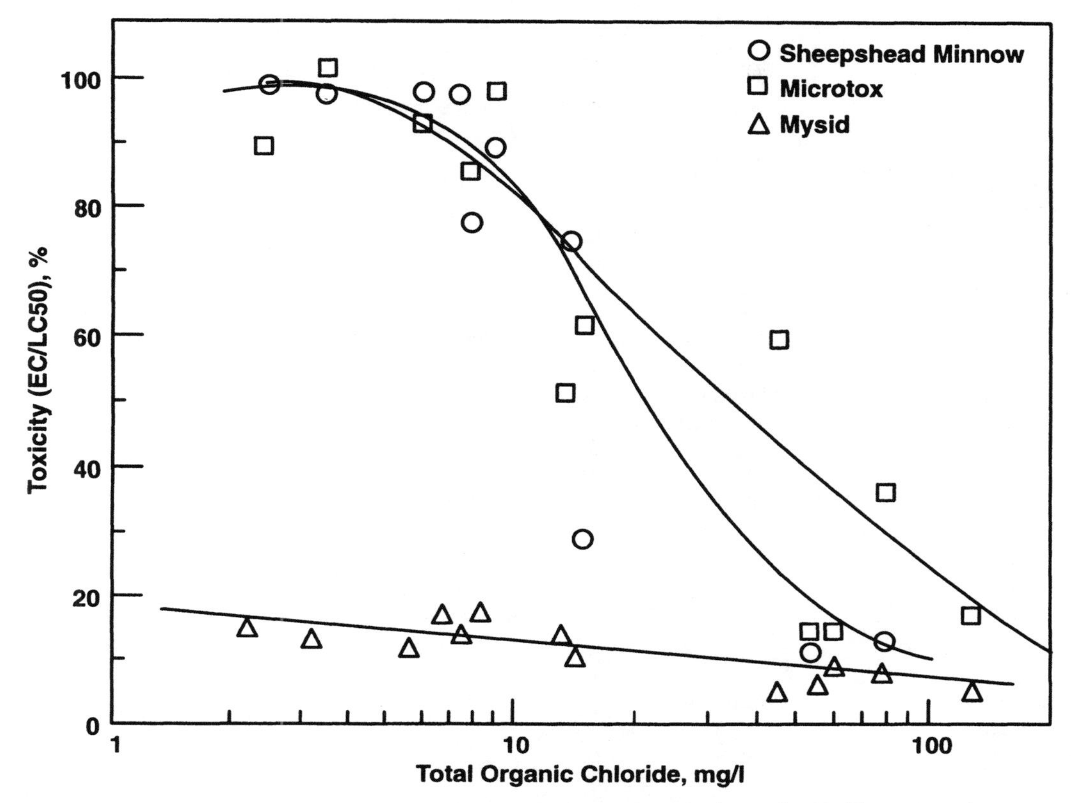

Figure 14. Interspecies comparison of effluent toxicity compared to effluent total organic chloride concentration.

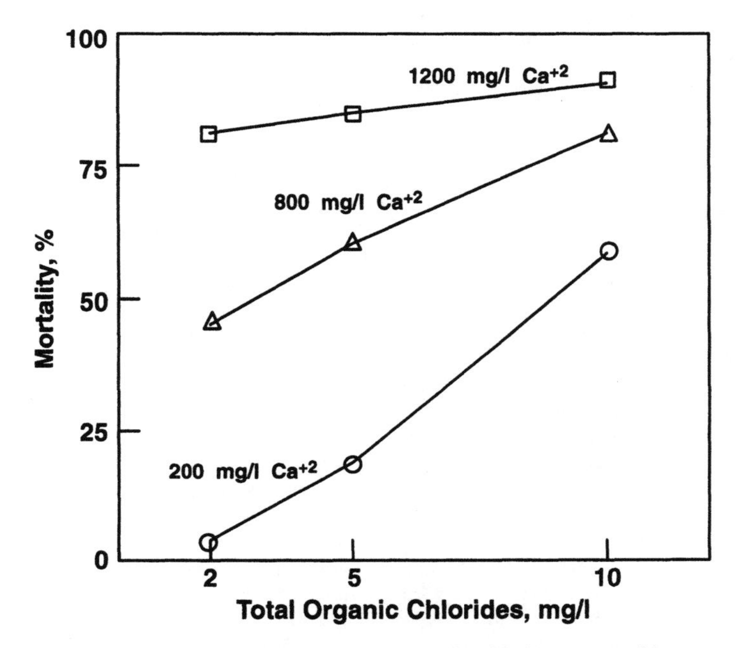

Figure 15. Effect of calcium and total organic chlorides on mysid acute toxicity.

60% at 10 mg/L TOCL. As calcium concentrations increased, the differential response to TOCL was lessened. For example, at 1200 mg/L Ca^{2+} there was 77% mortality at 2 mg/L TOCL and 85% mortality at 10 mg/L TOCL (Figure 16).

A comparison of the responses of *Mysidopsis* and *Cyprinodon* to combinations of Ca^{2+} and TOCL reveals that both respond similarly to TOCL. The 48-h EC50 for *Mysidopsis bahia* is 13.7 mg/L TOCL and the 96-h LC50 for *Cyprinodon* is 9.7 mg/L TOCL. The minimal difference in the response is due to response to calcium toxicity. *Cyprinodon* is affected to a lesser degree than *Mysidopsis.*

The calcium masked the TOCL effect and the effectiveness of the plant treatment in reducing process stream toxicity. In this instance, had detailed investigations been made, incorrect conclusions would have been drawn concerning "toxics" in the effluent and the plant treatment system performance. The simple measurement of "salinity" is not representative of the true balance of cationic and anionic species in the effluent.

Mysidopsis is unable to ion regulate to calcium changes, whereas *Cyprinodon* has good capabilities. Salinity cannot be considered a broad term when measuring effluents. Complex effluents by their nature contain different balances of cationic and anionic species that are measured as total dissolved solids (TDS), conductivity, and sometimes salinity. If salinity is to be used as a determinant for species selection, specific ionic composition and ratios consistent with standard seawater should be used for decision making.

In additional follow-up investigations, the toxic fraction observed in the total organic halide after the overlying calcium toxicity was removed was separated and evaluated for acute toxicity, chronic toxicity, toxicity in sediments and sediment sorption (see Chapter 15), and analysis of the mixture on stream mesocosms. The mixture was found to consist of three chloroethers (Figure 17) present in the process used to manufacture epichlorohydrin (Dorn et al., 1991; Crossland et al., 1992). The acute toxicity to nine species was observed to range from 1.5 to 20 mg/L of the mixture and from 4.0 to 9.3 mg/L for TCPE to five species (Table 12). Chronic effects were between 1.3 and 6.3 mg/L depending on the end point used (Table 13). Stream mesocosm studies with chloroether mixture showed that the laboratory tests were good predictors of stream effects with chronic NOEC concentrations between 0.26 and 1.6 mg/L, with some parameters >3.0 mg/L. The ACR for chloroether mixture ranged from 3 to 14 and was typical of chemicals exhibiting a narcosis effect, with small ACR (Table 14). These data, when compared to receiving water concentrations around the chemical plant, indicate that concentrations several orders of magnitude from the NOEC were present, affording considerable in-stream protection. This study also addressed possible sediment effects from a chemical mixture discharge and found that the sediment sorption of these chloroethers (K_p) ranged between 1.4 and 43 for three sediments, and when normalized for organic carbon content the K_{oc} was found to be 800 ($\log K_{oc} = 2.9$) (Figure 18). Toxicity of spiked sediments ranged from 1.3 to 10.7 mg/L chloroether as interstitial water to *Daphnia, Chironomus,* and *Hyalella* (Meyer et al., 1993).

Based on the findings of the TIE study, intensive laboratory and plant-scale process research resulted in a chemical plant process modification that reduced the process waste concentration discharged to several orders of magnitude lower than effect concentrations. Simultaneously, additional treatment enhancements including hydrolysis, powdered carbon treatment, and activated sludge treater optimization were identified and proved in bench and full-scale studies. Because of the effectiveness of the process change—which is the preferred option—the existing treatment option did not require modification.

Refinery "Salt Toxicity"—Salts

This plant study shows how the use of inappropriate test species can affect aquatic toxicity results and illustrates the problem of wide differences between receiving water and effluent salinity. A Texas refinery currently required to conduct 48-h acute effluent biomonitoring using *Pimephales promelas* and *Daphnia pulex* observed no mortality in fish; however, mortality to *D. pulex* was observed in all effluent tests at the critical concentration of 100% effluent. The permit requires that the source of toxicity to be identified if either species exhibits mortality in 100% effluent. The salinity of the refinery effluent exceeds the tolerance range for *D. pulex* and was thought to cause the observed toxicity, rather than other toxics.

Sodium chloride toxicity as the 48-h LC50 was found to be 2019 mg/L for *C. dubia* and 6034 mg/L for *D. magna* (Cowgill and Milazzo, 1990). The NOEC for these tests was found to be approximately 1200 mg/L for both species, and it was recommended that toxicity test salinity be kept below this value to avoid interferences due to osmotic toxicity. The U.S. Environmental Protection Agency (1991c,d) reports the acute 48-h LC50 for sodium chloride to be 2100 to 2800 mg/L for *C. dubia* and 3100 to 5900 mg/L for *D. magna.* Additional data using sodium chloride as reference toxicant for *C. dubia* 7-d tests give an average lowest observed effect concentration (LOEC) for mortality of 720 mg/L over a 10-mo monitoring period. The two organisms in the above discussion bracket the response of *D. pulex* to salinity.

A refinery site study was designed to characterize and confirm the source of effluent toxicity as salinity. A 5-wk sampling, analytical, and toxicity testing program was conducted. Refinery intake water and refinery biotreater effluent were collected and analyzed for commonly occurring inorganic ions. During weeks 1, 3, and 5, *D. magna* 48-h effluent toxicity tests (with 24-h renewal) and synthetic salt

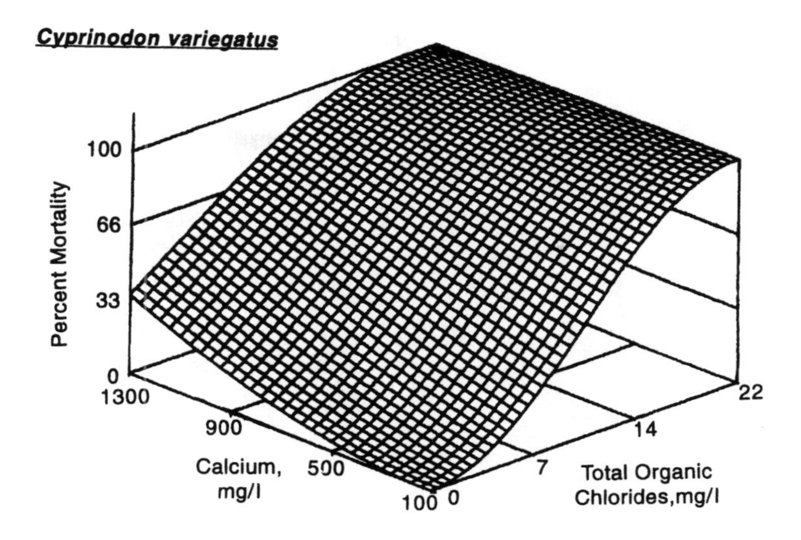

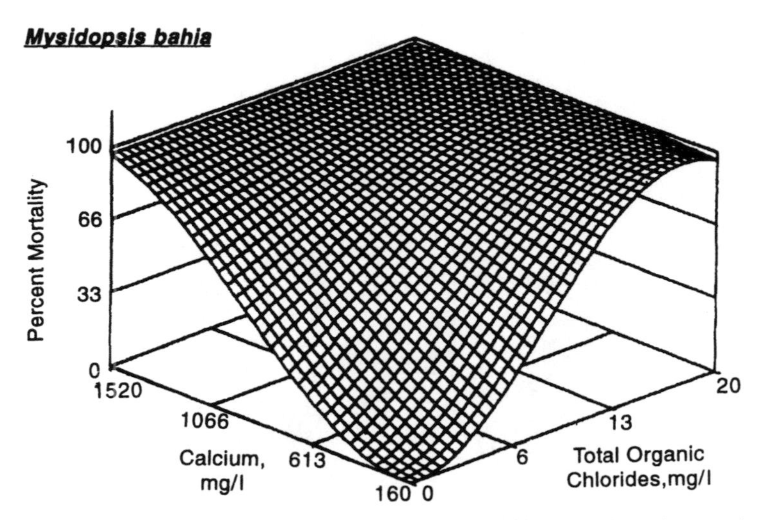

Figure 16. Response of sheepshead minnows and mysids to varying mixtures of calcium and organic halide (total organic chloride) concentrations.

effluent toxicity tests were conducted. During weeks 1 and 5, *D. pulex* toxicity tests were also performed on the same samples for species sensitivity comparison. Synthetic effluent samples were also analyzed for the same ions as the intake and effluent water samples. The major ion (sodium, calcium, magnesium, and sulfate) concentrations in the synthetic solution matched within 20% those in the refinery effluent. Notable exceptions were higher potassium and nitrate in the effluent and higher alkalinity (due to the use of sodium bicarbonate) in the synthetic effluent.

The approximate material balance of the ions analyzed in this study is illustrated in Table 15. Intake and effluent average flow rates of 400 and 185 gpm, respectively, were used to calculate the mass flows. In both intake and effluent, the major salts are sodium chloride and sodium sulfate. Calcium and magnesium (hardness), barium, fluoride, nitrate, phosphate, thiocyanate, and total cyanide appear to increase due to various refinery operations. None of these components appear to be present in the effluent in concentrations high enough to cause toxicity to *D. pulex* (Table 16). *D. pulex* and *D. magna*

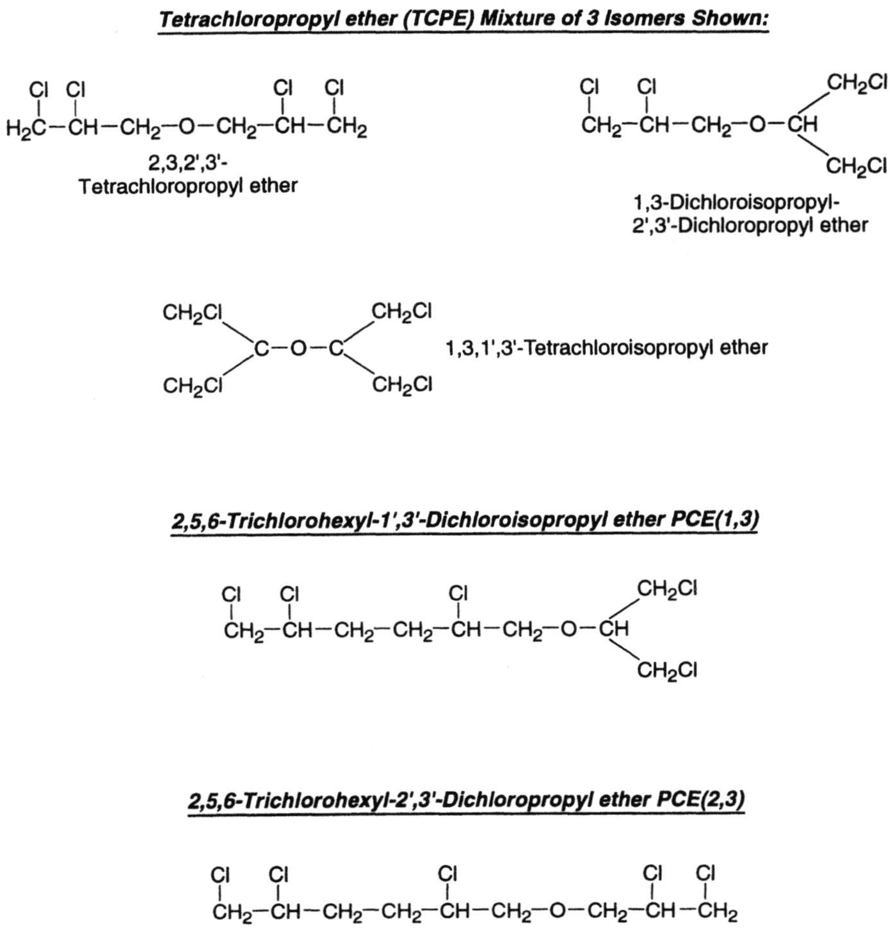

Tetrachloropropyl ether (TCPE) Mixture of 3 Isomers Shown:

Cl Cl Cl Cl
$H_2C-CH-CH_2-O-CH_2-CH-CH_2$
2,3,2',3'-
Tetrachloropropyl ether

Cl Cl CH_2Cl
$CH_2-CH-CH_2-O-CH$
CH_2Cl
1,3-Dichloroisopropyl-
2',3'-Dichloropropyl ether

CH_2Cl CH_2Cl
$C-O-C$
CH_2Cl CH_2Cl
1,3,1',3'-Tetrachloroisopropyl ether

2,5,6-Trichlorohexyl-1',3'-Dichloroisopropyl ether PCE(1,3)

Cl Cl Cl CH_2Cl
$CH_2-CH-CH_2-CH_2-CH-CH_2-O-CH$
CH_2Cl

2,5,6-Trichlorohexyl-2',3'-Dichloropropyl ether PCE(2,3)

Cl Cl Cl Cl Cl
$CH_2-CH-CH_2-CH_2-CH-CH_2-O-CH_2-CH-CH_2$

Figure 17. Structural configuration of a mixture of organic halides present in an effluent.

Table 12. Acute toxicity of chloroethers and chloroether mixtures

Species	Component (mg/L)	
	TCPE	Chloroether mixture
C variegatus	9.3 (9–15)	20 (14–20)
Microtox[a]	4.0 (3.8–4.3)	3.0 (2.4–3.6)
D. pulex	7.1 (5.8–7.3)	3.6 (3.1–4.3)
M. bahia	5.8 (5.2–6.8)	1.5 (1.3–1.7)
P. promelas	6.7 (4.5–9.0)	8.2 (7.6–8.8)
Microtox		8.6 (7.2–10.4)
S. capricornutum		6.6 (6.1–7.1)
D. magna		3.6 (2.8–4.7)
		3.3 (2.2–5.1)
G. pulex		6.2 (4.3–9.0)
S. gairdneri		3.1 (2.5–3.9)

[a] Microtox is a trademark for a specific bacterial effluent toxicity test.

concentration-response data on a TDS basis are similar for both effluent and synthetic salt water, indicating that salinity is the major source of toxicity (Table 17). As anticipated, *D. magna* is more tolerant to salt in both effluent and synthetic water but still exhibits significant mortality (i.e., greater than control) at the critical effluent concentration of 100%.

Determinations of fluoride toxicity at the maximum observed effluent concentration (110 mg/L) in intake water were augmented by two determinations of fluoride LC50 in EPA dilution water. The LC50 for fluoride was found to be 135 and 174 mg/L as fluoride in each of the two tests. The results of the LC50 determinations in EPA water agree with and confirm the results of the fluoride spiking tests done using refinery intake water, which found no toxicity due to fluoride. In both tests, no significant difference in survival versus control was observed at a fluoride concentration of 100 mg/L.

The effects of salinity toxicity were further differentiated from other potential sources in the plant effluent by determining the LC50 of a series of solu-

Table 13. Seven-day exposure toxicity of chloroether mixture to fathead minnows and *Daphnia magna*

	P. promelas			*D. magna*		
	Exposure (mg/L)	Percent mortality	Growth weight (mg/L)	Exposure (mg/L)	Percent mortality	Growth length (mm)
	0	2	0.287	0	12	54.1
	0.3	3.3	0.274	0.2	15	53.2
	0.6	0	0.247	0.5	20	54.6
	1.3	0	0.274	1.0	15	53.6
	3.2	10	0.184	2.5	30	—
	6.3	37	—	5.0	45	—
	12.7	100	—	7.5	75	—
7-d LC50 (mg/L)		6.3			4.4	
95% confidence intervals		(5.5–7.2)	—		(3.1–7.3)	—
NOEC[a] (mg/L)		3.2	1.3		2.5	1.0
LOEC[b] (mg/L)		6.3	3.2		5.0	1 < LOEC < 2.5
CHV[c] (mg/L)		4.5	2.0		3.5	1 < CHV < 2.5

[a] No-observable-effect concentration.
[b] Lowest observable effect concentration.
[c] Chronic value = $\sqrt{NOEC * LOEC}$.

tions of plant effluent and a synthetic salty effluent in percentage ratios 100:0, 70:30, 50:50, 30:70, and 0:100. The synthetic effluent was prepared to match as closely as possible the major ion, hardness, and salinity composition of the plant effluent in the absence of other potential toxicants. As it is not possible to match totally the ionic composition of the plant effluent, determination of LC50 values for the mixtures provides a continuum of effects of both synthetic effluent and plant effluent in the same test. If salinity is the sole source of toxicity in the plant effluent, then the LC50 values should remain constant as the percentage of plant effluent increases. Only if the LC50 values decrease as the percentage of plant effluent increases would one suspect that the plant effluent contains toxicity other than that due to salinity (Figure 19). The plots for January and December effluent toxicity data compare to the "predicted (theoretical)" data line. The predicted line toxicity is based on effects of salt without any other pollutants present. The December data show less toxicity than the predicted data, and the January sample is more toxic at lower effluent concentrations. However, at 100% effluent concentration, the toxicity is equal for both samples and the predicted value based on the effects of salinity alone.

The LC50 results for two sets of effluent synthetic effluent mixtures, in units of percent effluent, are summarized in Figure 19. The LC50 of the mixtures increases with the proportion of plant effluent. This is due to a greater salt concentration in the synthetic effluent sample. Both sets of data conform to the response expected from salinity as the sole source of toxicity, as illustrated in Figure 19.

The investigation showed that (1) osmotic stress due to the high salinity of the effluent is the major cause of *D. pulex* toxicity, (2) ion analyses and material balance indicate that some ions are added to the wastewater due to refinery processes but none are present at toxic concentrations or in excess of any applicable state limit, and (3) *D. magna* is less sensitive to salinity than *D. pulex* but does not offer an alternative test species because it still exhibits significant toxicity at the critical effluent concentration of 100%.

Table 14. Comparison of effects data for chloroether mixture

	LC50 or EC50[a]		NOEC[b]		
Laboratory	2–4 d	7 d	Mortality	Growth	ACR[c]
Daphnia	3.0	4.4	2.5	1.0	3
Fathead minnow	8.2	6.3	3.2	1.3	6.3
Gammarus	6.2	—	—	0.44	14
Rainbow trout	3.1	—	—	0.26	12

[a] Calculated in mg/L.
[b] No-observable-effect concentration in mg/L significant at α = 0.05.
[c] (2- or 4-d LC50 or EC50)/lowest NOEC.

Summary of TIE Studies

In a summary evaluation of toxicity identification and reduction investigations (TIE/TRE) conducted until 1990 in the United States, 52 facilities had conducted evaluations. The group represented 5 oil refineries, 5 metal industries, 4 textile plants, 16 chemical plants, 16 municipal sewage treatment plants, and 6 "other" industries (Steen and Wood, 1988; Steen, 1990). The length of the studies ranged from 3 mo to 2 yr, depending on the complexity of the waste streams and effluent variability, and included the use of 14 species, with *Daphnia pulex* most fre-

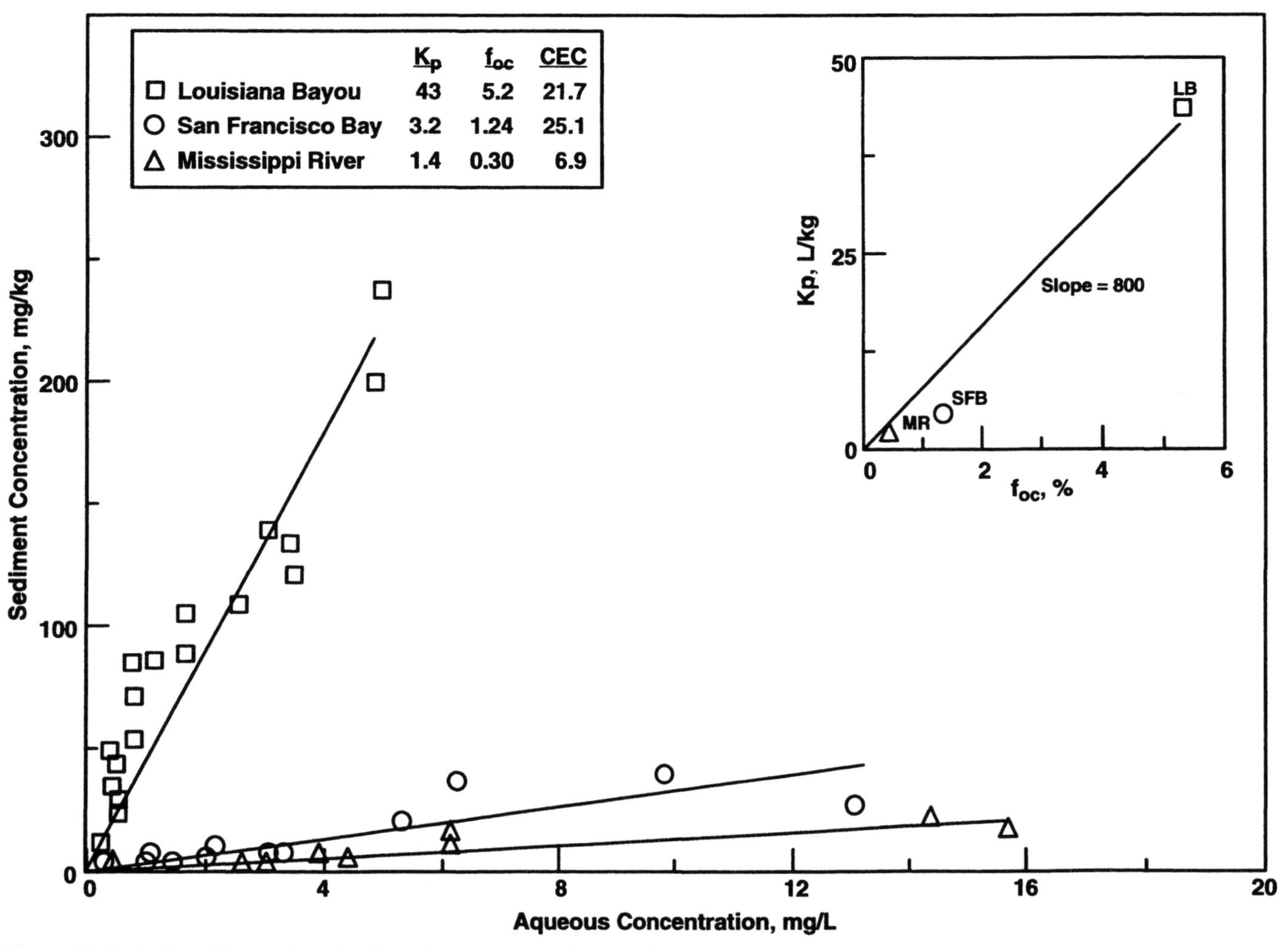

Figure 18. Evaluation of the sorption of a chloroether mixture on different sediments.

Table 15. Material balance of common inorganic ions in refinery intake and effluent waters

Stream parameter	Mass flows (lb/d)									
	Week 1		Week 2		Week 3		Week 4		Week 5	
	Int.[a]	Eff.[b]	Int.	Eff.	Int.	Eff.	Int.	Eff.	Int.	Eff.
Alkalinity	NA[c]	NA	355	84	653	73	542	73	706	47
Hardness	144	710	53	744	341	790	120	766	96	895
TDS	NA	15296	12720	13964	12720	14719	12864	14608	12912	15917
Calcium	31	152	9	173	28	180	13	167	9	202
Magnesium	16	80	7	76	66	83	22	85	18	95
Sodium	4656	4862	4320	5150	4013	4573	4368	4751	4464	5395
Barium	ND[d]	0.2	ND	0.2	ND	0.2	ND	0.2	ND	0.2
Iron	5	1	7	1	1	1	4	2	2	2
Potassium	120	362	101	135	120	113	120	120	115	135
Chloride	NA	5062	4944	5328	4723	5506	4848	5150	4618	6038
Sulfate	3091	3397	2923	3219	3240	3441	3024	3707	2803	3818
Bromide	NA	41	85	8	91	6	17	56	20	55
Fluoride	1.4	160	1.9	29	1.4	32	1.4	69	2.4	38
Iodide	NA	11	29	7	19	9	38	7	53	9
Nitrate-N	NA	75	7	55	7	77	2	101	5	109
Phosphate-P	0.3	3	1	7	1	14	1	10	1	6
Thiocyanate	ND	.2	ND	ND	ND	ND	2	2	1	1
Total cyanide	ND	.1	ND	.1	ND	.1	ND	.2	ND	.2

[a] Int., intake water (400 gpm).
[b] Eff., effluent water (185 gpm).
[c] NA, not analyzed.
[d] ND, not detected.

quently tested. At 27 of the facilities, toxicants were identified in the effluents. Toxicants identified were organics at 22 facilities, metals at 12 facilities, ammonia at 12 facilities, biocides or pesticides at 10 locations, inorganics at 2 sites, and chlorine at 2 sites (Table 18). Some of the facilities identified multiple toxicants, which are often complicating factors in unraveling problems with toxicity control in the treatment system. One toxicant may inhibit a treatment process such as nitrification, and by eliminating ammonia, the pH of the system changes (downward), resulting in activating organic acids and consequent toxicity. Twelve of the facilities did not conduct a TIE but only the reduction phase of a changing treatment without confirming the toxicants responsible for effluent effects, and as a result only 4 of the 12 are currently meeting permit limits. It has been apparent that in the last few years, with increasing site experience, toxicant identification is critical before deciding on the appropriate control strategy. In the summary evaluation, the most effective remedial actions to control toxicity were product substitution, facility upgrades, and the use of powdered activated carbon treatment (PACT) (Table 19).

SEDIMENTS

Chemicals with moderate to high partitioning coefficients (log K_{ow}3.5) may sorb to sediments after entering receiving waters. Depending on the degree of effluent controls, it is possible that some of these chemicals may escape treatment, become captured in

Table 16. Average effluent ion concentrations vs. LC50 for ions found to increase across the refinery

Ion	Average effluent conc. (mg/L)	LC50, mg/L (species)
Calcium	79	1000 (*D. magna* and *C. dubia*)
Magnesium	38	340 (*D. magna*)
Barium	0.1	Not available
Fluoride	30	>100 (*D. magna*)
Nitrate-N	34	Not available
Phosphate-P	3.6	Not available
Thiocyanate	<0.1 to 0.8	Not available
Total cyanide	0.070	Depends on free cyanide and pH

Table 17. Summary of salinity-survival data for refinery effluent toxicity study

		Actual effluent			Synthetic effluent		
			Percent survival			Percent survival	
Week	% Effluent	TDS/salinity (mg/L)	*D. magna*	*D. pulex*	TDS/salinity (mg/L)	*D. magna*	*D. pulex*
1	Control	448.2	100	100	442.2	100	100
	6	763.8	100	95	882	100	90
	12	1110	100	100	1416	100	100
	25	1809	100	90	2434.8	100	0
	50	3117	100	25	4347	35	0
	100	5679	15	0	7923	0	0
3	Control	336.3	96		356.4	84	
	6	632.7	100		736.2	100	
	12	961.2	70	Not	1097.4	90	Not
	25	1614	100	tested	1887	95	tested
	50	1914	100		3399	85	
	100	5196	65		5868	0	
5	Control	301.2	100	96	315.6	100	100
	6	619.8	90	100	636	100	100
	12	960	95	100	978	100	85
	25	1656	100	100	1698	95	90
	50	2982	100	40	3051	65	0
	100	5532	35	0	5553	0	0

the system, and enter receiving water sediments. However, for the most part, modern treatment systems utilizing biological treatment and activated carbon systems produce effluent with trace quantities of pollutants. There has been concern that sediments receiving effluent discharge chemicals may become contaminated to levels resulting in toxicity to aquatic life or levels that exceed acceptable tissue residue levels in aquatic life. If sediment sorbing chemicals are discharged, they may be identified through sediment quality criteria assessments using sediment toxicity (see Chapter 8), sediment chemical identification, and benthic infaunal analysis.

BIOCONCENTRATION

Besides the concern about the effect chemicals discharged in effluents may have on aquatic organisms, there is the concern that these chemicals may accumulate in these organisms and have effects up through the food chain to humans. The EPA has issued draft guidance (U.S. Environmental Protection Agency, 1991f) for identifying unknown chemicals in effluents, sediments, and fish tissue that have the potential to bioaccumulate and have some effect on human consumers. For conservatism, the EPA states that a chemical having a log *n*-octanol–water partition coefficient (log K_{ow}) > 3.5 which is detected at > 0.1 μg/L (U.S. Environmental Protection Agency, 1991g) is a concern in terms of bioconcentration (see Chapter 16). For discharges with effluents containing bioaccumulative chemicals above 0.1 μ/L as determined by HPLC fractionation and GC/MS analysis of the fractions, the potential for these chemicals to have an effect on human consumers would have to be determined by comparing the in-stream chemical concentration with a calculated reference ambient concentration (RAC). For noncarcinogens (see Chapter 14 for discussion of carcinogens), the calculation of RAC is shown in Table 20.

Table 21 gives a representative list of the kinds of chemicals having log K_{ow} greater than 3.5 that are found in treated refinery effluents, along with their reported effluent concentrations. Substituted PAHs (e.g., benzo[*a*]anthracene) have also been found, but at concentrations at or below those of the parent chemicals and, for this initial screen, the parent chemicals can be used. Normal alkanes were ignored because they are not of concern for bioconcentration effects (Table 21). RFD (reference dose for humans) values, needed for RAC calculations, are available only for acenaphthene, fluorene, anthracene, pyrene, and chrysene. For these chemicals, the effluent concentrations observed are one to two orders of magnitude lower than the calculated RAC values. Therefore, these effluents would be of low concern for further action to control bioconcentration.

These methods are most effective in identifying hydrophobic organic chemicals that may accumulate in aquatic life and are recalcitrant in the environment. The prevention of chemical manufacture and discharge of environmentally incompatible chemicals is the goal of several programs (i.e., Toxic Substances Control Act, Federal Insecticide, Fungicide and Rodenticide Act, Food and Drug Administration implementation of NEPA; see Chapters 21, 22, 23, and 25 for further discussion), and the ultimate exposure through effluent discharges is assessed in effluent control programs. Industries with few organic pollutants in effluents, such as utility industry power plants, would be of low concern,

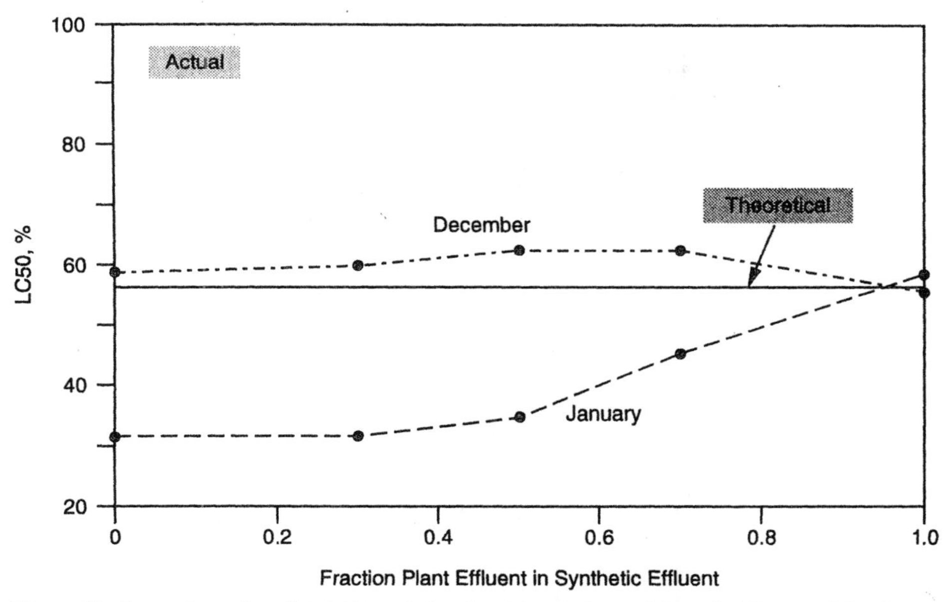

Figure 19. Comparison of predicted (theoretical; salt only) and observed (actual) effluent toxicity due to salt concentrations.

whereas organic chemical manufacturing operations without advanced treatment may be of higher concern.

This also indicates that EPA should adopt a tiered approach in implementing the regulation of bioconcentratable chemicals. This approach should be based on both type of industry (i.e., chemicals produced) and the observation of effects in the receiving streams. This would allow optimum use of limited laboratory resources to address this issue.

SUMMARY

The effluent assessment discussion has highlighted the general procedures for assesssing effluent hazard for acute and chronic toxicity and for toxicity identification. Effluent chemicals being bioconcentrated in edible tissue or in sediments was discussed as alternative routes of exposure. In toxicity assessment and compliance determinations, caution is raised in the application of concrete pass/fail criteria for action before best professional judgment is used. There are no "magic bullets" in the TIE process, and an interdisciplinary team is certainly required for any toxicity reduction or treatability study. However, current technology for effluent controls would be protective of large-scale pollution, and advanced measures have the ability to control toxicity, sediment sorption, and bioconcentration potential.

Table 18. Toxicants contributing to whole-effluent toxicity (WET)

Facility[a]	Organic[b]	Surfactant	Metal	NH_3	Biocide/ pesticide	SS[c]	Inorganic	Cl[d]
POTW (n = 16)	9	2	1	3	4	1	1	1
Chemical manufacture (n = 16)	6	—	3	5	2	—	1	1
Oil refinery (n = 5)	2	1	—	2	1	1	—	—
Metal industry (n = 4)	2	—	4	—	—	—	—	—
Textile industry (n = 4)	—	2	4	—	1	—	—	—
Miscellaneous (n = 6)	3	1	—	2	2	—	—	—
Total (n = 52)	22	6	12	12	10	2	2	2

[a] Facility totals may not match toxicant totals due to multiple toxicants.
[b] Not including surfactants, biocides, or pesticides.
[c] Suspended solids.
[d] Chlorine.

Table 19. Control options selected

Facility[a]	Source control	Product sub.	Process change	System upgrade	New (non-PACT)[b] treatment	Ensure BMPs[c]	PACT[d]	Local limit[e]	Banned compounds[f]	POTW hookup	Closed outfall	No controls/controls
POTW (n = 16)	6	—	1	2	1	3	—	2	2	—	—	11/5
Chemical manufacture (n = 16)	3	1	2	2	3	2	2	—	—	—	3	15/1
Oil refinery (n = 5)	—	1	—	1	1	—	3	—	—	—	—	4/1
Metal industry (n = 5)	1	1	—	1	2	1	1	—	—	—	—	3/2
Textile industry (n = 4)	—	2	—	2	—	1	—	—	—	—	—	3/1
Miscellaneous (n = 6)	—	1	—	—	1	2	1	—	—	1	—	4/2
Total (n = 52)	10	6	3	8	8	9	7	2	2	1	3	40/12

[a] Facility totals may not match option totals.
[b] Powdered activated carbon treatment.
[c] Best management practices.
[d] Includes other activated carbon treatments.
[e] Local toxicity limit in place.
[f] Identified discharge of restricted or banned chemicals.

Table 20. Equation for estimating reference ambient concentrations (RACs) of noncarcinogenic bioconcentration chemicals[a]

$$\text{RAC (mg/L)} = \frac{(\text{RfD} \times \text{WT}) - (\text{DT} + \text{IN}) \times \text{WT}}{\text{WI} + [\text{FC} \times \text{L} \times (\text{FM} \times \text{BCF})]}$$

Definition	Value applied/source
WI = water intake, L/d	2 L/d
WT = weight of average human	70 kg
FC = daily fish consumption, kg fish/d	0.020 kg/d
L = ratio of lipid fraction of tissue to 3%	1
DT = dietary intake of chemical, mg toxicant/kg body weight of human/d	0
IN = inhalation exposure, mg toxicant/kg body weight of human/d	0
RfD = estimate of daily exposure to human without causing adverse effects, mg toxicant/kg body weight of human/d	IRIS database
FM = food chain multiplier	1 (Table 4.1 of guidance document)[a]
BCF = bioconcentration factor (mg toxicant/kg fish divided by mg toxicant/L water)	From IRIS or estimated from log K_{ow} value using the equation by Veith et al. (1979)
IRIS = integrated risk information system	Appendix D of guidance document

[a] U.S. Environmental Protection Agency, 1991g.

Table 21. Characterization of refinery-treated effluents for chemicals with log K_{ow} > 3.5

Name	log K_{ow}	BCF	Reported effluent concentration (μg/L)	RAC RfD basis (μg/L)	RfD value from IRIS
Acenaphthylene	4.07	260[a]	<0.1–<1		
Acenaphthene	3.92	198[a]	<0.1–6	700	0.06
Fluorene	4.18	317[a]	<0.1–0.6	300	0.04
Anthracene	4.45	550	<0.01–<2	1600	0.3
Phenanthrene	4.46	528[a]	<0.2–<0.5		
Pyrene	5.18	1280[b]	<0.5–7	80	0.03
Fluoranthene	5.33	1280[b]	<0.1–<10	100	0.04
Chrysene	5.61	4285[a]	<0.02–1.4		
Benzo[*a*]anthracene	5.61	4285[a]	<0.02–0.08		
Benzo[*a*]pyrene	5.98	10960[b]	0.1–2.9		
Benzo[*b*]fluoranthene	6.57	24564[a]	<0.06–0.3		
Benzo[*k*]fluoranthene	6.84	40142[a]	<0.3–0.3		
Benzo[*ghi*]perylene	7.23	81601[a]	<0.2–0.2		
Indeno(1,2,3-*cd*)pyrene	7.66	178402[a]	<1		
Dibenzo[*a,h*]anthracene	5.97	8247[a]	<0.3		
Dibenzothiophene	4.42	491[a]	<1		
Dimethyl indan	3.8	159[a]	<1		
Methyl/ethyl tetralins	4	229[a]	<1		
Benzenes	3.8	159[a]	<1		

[a] Calculated as log BCF = 0.79 log P − 0.40.
[b] From IRIS database.

LITERATURE CITED

American Petroleum Institute: Species Tolerances for NPDES Bioassays. Vol. I, Freshwater Species. API Publication N. 4483, Washington, DC, 1989a.

American Petroleum Institute: Species Tolerances for NPDES Bioassays. Vol. II, Marine Species. API Publication No. 4482, Washington, DC, 1989b.

Beckman. 1982. Microtox. System Operating Manual. Carlsbad, CA: Beckman Instruments.

Bergman, H. L.; Kimerle, R. A.; and Maki, A. W. (eds.): Environmental Hazard Assessment of Effluents. Elmsford, NY: Pergamon, 1986.

Buikema, A. L.: A variation in static acute toxicity test results with *Daphnia magna* exposed to refinery effluents and reference toxicant. *Oil Petrochem. Pollut.* 1(3):189–198, 1983.

Cairns, J. Jr.; Dickson, K. L.; and Maki, A. W. (eds.): Estimating The Hazard of Chemical Substances to Aquatic Life. Philadelphia: American Society for Testing and Materials, 1979.

Cowgill, U. M. and Milazzo, D. P.: The sensitivity to two cladocerans to water quality variables: salinity and hardness. *Arch. Hydrobiol.* 120(2):185–196, 1990.

Crossland, N. O.; Mitchell, G. C.; and Dorn, P. B.: Use of outdoor artificial streams to determine threshold toxicity concentrations for a petrochemical effluent. *Environ. Toxicol. Chem.* 11:49–59, 1992.

DeGraeve, G. M.; Cooney, J. D.; Marsh, B. H.; Pollock, T. L.; and Reichenbach, N. G.: Variability in the performance of the 7-d *Ceriodaphnia dubia* survival and reproduction test: an intra- and interlaboratory study. *Environ. Toxicol. Chem.* 11:851–866, 1992.

DeGraeve, G. M.; Cooney, J. D.; Pollock, T. L.; Reichenbach, N. G.; Dean, J. H.; Marcus, M. D.; and McIntyre, D. O.: 1988. Fathead minnow 7-day test: intra- and interlaboratory study to determine the reproducibility of the seven-day fathead minnow larval survival and growth test. Report to the American Petroleum Institute, Publication 4468, Washington, DC.

Dickson, K. L.; Maki, A. W.; and Cairns, J. Jr.: Analyzing the Hazard Evaluation Process. Bethesda, MD: American Fisheries Society, 1979.

Dorn, P. B.; Raia, J. C.; Evans, S. H.; Balderas, R. A.; and Chai, E. Y.: Rapid methods for predicting compliance to drilling mud toxicity tests with mysids for offshore NPDES general permits. *Environ. Auditor* 1:221–227, 1990.

Dorn, P. B., and Rodgers, J. H., Jr.: Variability associated with identification of toxics in National Pollutant Discharge Elimination System (NPDES) effluent toxicity tests. *Environ. Toxicol. Chem.* 8:893–902, 1989.

Dorn, P. B.; Salanitro, J. P.; Evans, S. H.; and Kravetz, L.: Assessing the aquatic hazard of some branched and linear nonionic surfactants by biodegradation and toxicity. *Environ. Toxicol. Chem.* 12:1751–1762, 1993 in press.

Dorn, P. B.; Van Compernolle, R.; Meyer, C. L.; and Crossland, N. O.: Aquatic hazard assessment of the toxic fraction from the effluent of a petrochemical plant, *Environ. Toxicol. Chem.* 10:691–703, 1991.

Dorn, P. B.; Raia, J. C.; Evans, S. H.; Balderas, R. A. and Chai, E. Y.: Rapid methods for predicting compliance to drilling mud toxicity tests with mysids for offshore NPDES general permits. *Environ. Auditor* 1(4):221–227, 1990.

Dorn, P. B.; Rodgers, J. H. Jr.; Jop, K. M.; Raia, J. C.; and Dickson, K. L.: Hexavalent chromium as a reference toxicant in effluent toxicity tests. *Environ. Toxicol. Chem.* 6:435–444, 1987.

Dorn, P. B.; Van Compernolle, R.; Mueller, G. R.; Sun, P. T.; Glaze, D. E.; Hwang, J. C.; and Hansen, S. R.: Toxicity identification and derivation of a water quality based effluent limit for a West Coast refinery. Ford, D. L. (ed.). Toxicity Reduction: Evaluation and Control, edited by D. L. Ford, pp. 183–223. Lancaster, PA: Technomic Publishing Co., 1992.

Engelhardt, F. R.; Ray, J. P.; and Gillam, A. H. (eds.): Drilling Wastes. New York: Elsevier Applied Science, 1989.

Ford, D. L. (ed.): Toxicity Reduction: Evaluation and Control, Vol. 3. Lancaster, PA: Technomic Publishing Co., 1992.

Girling, A. E.: Acute and chronic toxicity of produced water from a North Sea oil production platform to the calanoid copepoda *Acartia tonsa*. *Bull. Environ. Contam. Toxicol.* 8:231–245, 1989.

Grothe, D. R., and Kimberle, R. A.: Inter- and intra-laboratory variability in *Daphnia magna* effluent toxicity test results. *Environ. Toxicol. Chem.* 4(2):189–192, 1985.

Jop, K. M.; Parkerton, R. F.; Rodgers, J. H.; Dickson, K. L.; and Dorn, P. B.: Comparative toxicity and speciation of two hexavalent chromium salts in acute toxicity tests. *Environ. Toxicol. Chem.* 6:697–703, 1986.

Jop, K. M.; Rodgers, J. H. Jr.; Dorn, P. B.; and Dickson, K. L.: Use of hexavalent chromium as a reference toxicant in aquatic toxicity tests. Aquatic Toxiciology and Environmental Fate, edited by T. M. Poston, R. Purdy. ASTM STP 921. Philadelphia: American Society for Testing and Materials, 1986.

Khangarot, B. A., and Ray, P. K.: Investigation of correlation between physicochemical properties of metals and their toxicity to the water flea *Daphnia magna Straus*. *Ecotoxicol. Environ. Safety* 18(2):109, 1989.

Meyer, C. L.; Suedel, B. C.; Rodgers, J. H. Jr.; and Dorn P. B.: Bioavailability of sediment-sorbed chlorinated ethers. *Environ. Toxicol. Chem.* 12(3):493–506, 1993.

Moffitt, C. M.; Rhea, M. R.; Dorn, P. B.; Hall, J. F.; Bruney, J. M.; and Evans, S. H.: Short-term chronic toxicity of produced water and its variability as a function of sample time and discharge rate. Produced Water. Technological/Environmental Issues and Solutions, Environmental Science Research Series, Vol. 46, edited by J. P. Ray, F. R. Engelhardt, pp. 235–244. New York: Plenum, 1992.

Neff, J. M.: Biological effects of drilling fluids, drill cuttings, and produced waters. Long Tern Environmental Effects of Offshore Oil and Gas Development, D. F. Boesch, N. N. Rabalais, pp. 469–538. New York: Elsevier Applied Science, 1987.

Organization for Economic Cooperation and Development: The use of biological tests for water pollution assessment and control. Environmental Monograph N. 11. OECD Water Management Policy Group, Paris, 1987.

Parkhurst, B. R.; Warren-Hicks, W.; and Noel, L. E.: Performance characteristics of effluent toxicity tests: summarization and evaluation of data. *Environ. Toxicol. Chem.* 11:771–791, 1992.

Peltier, W: Methods for Measuring the Acute Toxicity of Effluents to Aquatic Organisms. EPA 600/4-78-012. Springfield, VA: National Technical Information Service, 1978.

Peltier, W. H., and Weber, C. I. (eds.): Methods for Measuring Acute Toxicity of Effluents to Freshwater and Marine Organisms, 3rd ed. Environmental Monitoring and Support Laboratory, EPA 600-4-85-014. Cincinnati, OH: U.S. Environmental Protection Agency, 1985.

Ray, J. P.; Fucik, K. W.; O'Reilly, J. E.; Chai, E. Y.; and LaMotte L. R.: Drilling fluid toxicity test variability in US commerical laboratories. Drilling Wastes, edited by

F. R. Engelhardt, J. P. Ray, A. H. Gillam, pp. 731–755, New York: Elsevier Applied Science, 1989.

Rue, W. J.; Fava, J. A.; and Grothe, D. R.: A review of inter- and intra-laboratory effluent toxicity test method variability. American Society for Testing and Materials Tenth Aquatic Toxicology Symposium Abstracts, May 4–6, 1986, New Orleans, LA, 1986.

Steen, A. E.: A summary review of TRE's findings and effectiveness. Abstracts of the AICHE Spring National Meeting, Orlando, FL, March 19–21, 1990.

Steen, A. E., and Wood, S. G.: Toxicity identification and reduction evaluations. Abstracts of the 9th Annual Meeting of the Society of Environmental Toxicology and Chemistry, Arlington, VA, 1988.

U.S. Environmental Protection Agency: Prediction of Initial Mixing for Municipal Ocean Discharges. CERL Publication 043. Corvallis, OR: Corvallis Environmental Research Laboratory, Office of Research and Development, 1979.

U.S. Environmental Protection Agency: Final water quality standards. *Fed. Regist.* 48:51400–51413, 1983.

U.S. Environmental Protection Agency: Development of water quality based permit limitations for toxic pollutants: national policy, *Fed. Regist.* 49(48):9016–9019, 1984.

U.S. Environmental Protection Agency: Drilling Fluids Toxicity Test Method, Appendix 3—Drilling Fluids Toxicity Test. Proposed regulation for the offshore subcategory of the oil and gas extraction point source category. Washington DC: EPA Industrial Technology Division, 1985a.

U.S. Environmental Protection Agency: Technical Support Document for Water Quality Based Toxics Control. EPA-440-4-85-032. Washington, DC: Office of Water, 1985b.

U.S. Environmental Protection Agency: Short Term Methods for Estimating the Chronic Toxicity of Effluents and Receiving Waters to Freshwater Organisms. EPA-600-4-85-014. Cincinnati, OH: U.S. EPA, 1985.

U.S. Environmental Protection Agency: Final NPDES general permit for the outer continental shelf (OCS) of the Gulf of Mexico. *Fed. Regist.* 51(131):24897–24927, 1986a.

U.S. Environmental Protection Agency: Quality Criteria for Water 1986. EPA 440/5-86-001. Office of Water Regulations and Standards, 1986b.

U.S. Environmental Protection Agency: Short-Term Methods for Estimating the Chronic Toxicity of Effluent and Receiving Waters to Marine and Estuarine Organisms. EPA-600-4-87-028. Springfield, VA: National Technical Information Service, 1988.

U.S. Environmental Protection Agency: Methods for Aquatic Toxicity Identification Evaluations. Phase II Toxicity Identification Procedures. EPA-600-3-88-035. Duluth, MN: Environmental Research Laboratory, 1989a.

U.S. Environmental Protection Agency: Methods for Aquatic Toxicity Identification Evaluations. Phase III Toxicity Confirmation Procedures. EPA-600-3-88-036. Duluth, MN: Environmental Research Laboratory, 1989b.

U.S. Environmental Protection Agency: Proposed reissuance of the NPDES general permit for the western portion of the outer continental shelf (OCS) of the Gulf of Mexico (GMG 290000). *Fed. Regist.* 56(73):15349–15372. 1991a.

U.S. Environmental Protection Agency: Short-Term Methods for Estimating the Chronic Toxicity of Effluents and Receiving Waters to Freshwater Organisms, 3rd ed. EPA-600-4-91-002. Springfield, VA: National Technical Information Service, 1991b.

U.S. Environmental Protection Agency: Methods for Aquatic Toxicity Identificaton Evaluations. Phase I Toxicity characterization Procedures, 2nd ed. EPA-600-6-91-003. Springfield, VA: National Technical Information Service, 1991c.

U.S. Environmental Protection Agency: Toxicity Identification Evaluation: Characterization of Chronically Toxic Effluents, Phase I. EPA-600-6-91-005. Duluth, MN: Environmental Research Laboratory, 1991d.

U.S. Environmental Protection Agency: Technical Support Document for Water Quality Based Toxics Control. EPA-505-2-90-001. Washington, DC: Office of Water, 1991e.

U.S. Environmental Protection Agency: Assessment and control of bioconcentratable contaminants in surface waters: draft guidance availability. *Fed. Regist.* 56(61): 13150-13151, 1991f.

U.S. Environmental Protection Agency: Assessment and Control of Bioconcentratable Contaminants in Surface Waters. Draft. Washington, DC: Office of Water, 1991g.

Veith, G. D.; Defoe, D. L.; and Bergstedt, B. V.: Measuring and estimating the bioconcentration factors of chemicals in fish. *J. Fish Res. Bd. Can.* 36:1040-1048, 1979.

Warren-Hicks, W.; and Parkhurst, B. R.: Performance characteristics of effluent toxicity tests: variability and its implications for regulatory policy. *Environ. Toxicol. Chem.* 11:793–804, 1992.

Weber, C. I. (ed.): Methods for Measuring the Acute Toxicity of Effluents and Receiving Waters to Fresh Water and Marine Organisms, 4th ed. EPA-600-4-85-013. Springfield, VA: National Technical Information Service, 1991.

GLOSSARY

Acclimation (1) Steady-state compensatory adjustments by an organism to the alteration of environmental conditions. Adjustments can be behavioral or physiological/biochemical. (2) An adaptation of aquatic organisms to some selected experimental conditions, including adverse stimuli. (3) Acclimation also refers to the time period prior to the initiation of a toxicity test in which aquatic organisms are maintained in untreated, toxicant-free dilution water with physical and chemical characteristics (e.g., temperature, pH, hardness) similar to those to be used during the toxicity test.

Acid volatile sulfide (AVS) An extractable reactive pool of solid-phase sulfide that is associated with and available from the mineral surfaces of sediment to bind metals and may render that portion unavailable and non-toxic to biota. Metals associated with the sulfide fraction of suspended matter and sediments in anaerobic environments include zinc, lead, copper, cobalt, nickel, cadmium, arsenic, antimony, mercury, manganese, and molybdenum.

Accumulation factor (AF) See Biota-sediment accumulation factor (BSAF).

Acute Having a sudden onset, lasting a short time. Of a stimulus, severe enough to induce a response rapidly. Can be used to define either the exposure or the response to an exposure (effect). For clarity, the length of the exposure (short, medium, or long) and the nature of the effect end point (lethal or nonlethal) should be specified. The duration of an acute aquatic toxicity test is generally 4 d or less and mortality is the response measured.

Acute-to-chronic toxicity ratio (ACR) A numerical, unitless value, that is the ratio of an acute toxicity test result (i.e., LC50) to a chronic toxicity test result (i.e., MATC) where both are expressed in the same units (e.g., mg/L). Ideally, the data are for the same species and chemical. It is the inverse of the application factor and is used in a similar manner. The ACR is commonly used for estimating chronic toxicity of a chemical on the basis of its acute toxicity. The ACR should be greater than one because the ratio compares an acute to a chronic value. Also called acute-to-chronic toxicity ratio.

Additive toxicity The toxicity of a mixture of chemicals that is approximately equivalent to that expected from a simple summation of the known toxicities of the individual chemicals present in the mixture (i.e., algebraic summation of effects).

Advection The transport of dissolved and suspended solutes (including fine particulates) in natural waters by the bulk movement of water (i.e., at water current velocity). In open-ocean marine systems advective transport of chemicals into the water column from sediments is modest compared with diffusive processes, whereas in estuarine systems, freshwater rivers, and lakes, advective processes can contribute substantially to system transport.

Alkalinity The acid-neutralizing (i.e., proton-accepting) capacity of water; the quality and quantity of constituents in water that result in a shift in the pH toward the alkaline side of neutrality.

Antagonism A phenomenon in which the toxicity of a mixture of chemicals is less than that which would be expected from a simple summation of the toxicities of the individual chemicals present in the mixture (i.e., algebraic subtraction of effects).

Application factor (AF) A numerical, unitless value, calculated as the threshold chronically toxic concentration of a chemical divided by its acutely toxic concentration. An AF is generally calculated by dividing the limits [no-observed-effect concentration (NOEC) and lowest observed effect concentration (LOEC)] on the maximum acceptable toxicant concentration (MATC) by the time-independent LC50, if available, or the 96-h LC50 (or 48-h EC50 or 48-h LC50 for daphnids) from a flow-through acute toxicity test. The AF is usually reported as a range and is multiplied by the median lethal concentration of a chemical, as determined in a short-term (acute) toxicity test, to estimate an expected no-effect concentration under chronic exposure.

Asymptotic threshold concentration The concentration of a chemical at which some percentage of a population of test organisms is in a state of approximate homeostasis for some prolonged period of time (not necessarily absolute). It may be demonstrated as the concentration at which the toxicity curve is approximately asymptotic (parallel) to the time axis. The asymptotic LC50 is the concentration at which acute mortality has essentially ceased and will not change substantially with exposure time. That is, there is no evidence of significantly increasing effects resulting from increasing duration of exposure.

Bioaccumulation General term describing a process by which chemicals are taken up by aquatic organisms directly from water as well as through exposure through other routes, such as consumption of food and sediment containing the chemicals.

Bioaccumulation factor (BAF) A value that is the ratio of tissue chemical residue to chemical concentration in an external environmental phase (i.e., sediment or food). BAF is measured at steady state in situations where organisms are exposed from multiple sources (i.e., water, sediment, food), unless noted otherwise.

Bioassay "A biological assay (or bioassay) is an experiment for estimating the nature, constitution, or potency of a material (or of a process), by means of the reaction that follows its application to living matter" (Finney, 1978[a]). This general definition includes both the aquatic toxicology and the pharmaceutical usage. The pharmaceutical definition of bioassaay is a test used to evaluate the relative potency of a chemical or mixture of chemicals by comparing its effect on a living organism with the effect of a standard preparation on the same type of organism. Bioassays are frequently used in the pharmaceutical industry to evaluate the potency of vitamins and drugs.

Biochemical (or biological) oxygen demand (BOD) A measure of the rate at which molecular oxygen is consumed by microorganisms during oxidation of organic matter. The standard test is the 5-day BOD test, in which the amount of dissolved oxygen required for oxidation over a 5-day period is measured. The results are measured in mg of oxygen/L (mg/L).

Bioconcentration A process by which there is a net accumulation of a chemical directly from water into aquatic organisms resulting from simultaneous uptake (e.g., by gill or epithelial tissue) and elimination.

Bioconcentration factor (BCF) A term describing the degree to which a chemical can be concentrated in the tissues of an organism in the aquatic environment as a result of exposure to water-borne chemical. At steady state during the uptake phase of a bioconcentration test, the BCF is a value which is equal to the concentration of a chemical in one or more tissues of the exposed aquatic organisms divided by the average exposure water concentration of the chemical in the test.

Biodegradation The transformation of a material resulting from the complex enzymatic action of microorganisms (e.g., bacteria, fungi). It usually leads to disappearance of the parent structure and to the formation of smaller chemical species, some of which are used for cell anabolism. Although typically used with reference to microbial activity, it may also refer to general metabolic breakdown of a substance by any living organism.

Biomagnification Result of the processes of bioconcentration and bioaccumulation by which tissue concentrations of bioaccumulated chemicals increase as the chemical passes up through two or more trophic levels. The term implies an efficient transfer of chemical from food to consumer, so that residue concentrations increase systematically from one trophic level to the next.

Biomarker The use of physiological, biochemical, and histological changes as indicators of exposure and/or effects of xenobiotics at the suborganismal and organismal level.

Biomonitoring (or biological monitoring) Use of living organisms as "sensors" in water/sediment quality surveillance and compliance to detect changes in an effluent or water body and to indicate whether aquatic life may be endangered.

Biota-sediment accumulation factor (BSAF) A specific type and form of bioaccumulation factor that is the ratio of lipid-normalized tissue chemical residue to organic carbon-normalized sediment chemical concentration. Note that various other terms for this are used in the literature, including accumulation factor (AF).

Biotransformation Changes in the structure/nature of a material mediated by enzymatic action of microorganisms, fish, invertebrates, etc.

Bottom-up The approach to investigating ecotoxicological effects that starts with a determination of the presence and nature of any adverse effects via responses at the suborganismal (cellular and biochemical) levels of organization rather than the community and/or ecosystem levels of organization.

Chemical oxygen demand (COD) When organic materials are not easily degraded by microorganisms, strong oxidizing agents (e.g., potassium permanganate) are used to enhance oxidation. COD is thus measured instead of BOD (see BOD). COD values will be larger than BOD values.

Chronic Involving a stimulus that is lingering or continues for a long time; often signifies periods from several weeks to years, depending on the reproductive life cycle of the aquatic species. Can be used to define either the exposure or the response to an exposure (effect). For clarity the length of the exposure and the nature of the effect end point should be specified. Chronic exposure typically induces a biological response of relatively slow progress and long continuance. The chronic aquatic toxicity test is used to study the effects of continuous, long-term exposure to a chemical or other potentially toxic material on aquatic organisms.

Concentration The quantifiable amount of chemical in a specific volume or mass of water, food, tissue, or sediment. The choice of units for concentration depends on the medium that is being measured or de-

[a]Finney DJ: Statistical Method in Biological Assay, 3rd ed. London: Griffin, 1978.

scribed. In water, a common expression of concentration is mass of chemical (μg or mg) per unit volume (L) of water or μg/L or mg/L. In sediment, a common expression is mass of chemical per unit mass (mg/kg) of sediment.

Concentration-response curve A curve describing the relationship between different exposure concentrations of an agent or material and percentage response of the exposed test population.

Contaminant A foreign agent that is present (e.g., in water, sediment) which may produce a physical or chemical change but may not cause adverse biological effects.

Control A treatment in a toxicity test that duplicates all the conditions of the exposure treatments but contains no test material. The control is used to determine the absence of toxicity of basic test conditions (e.g., health of test organisms, quality of dilution water).

Conventional pollutant See Pollutant.

Convection A type of fluid flow like advection but likely to occur in high-energy environments (e.g., streams) where sand ripples may be found and contaminated sediments are not likely to be deposited.

Critical body residue (CBR) The whole-body concentration of a chemical that is associated with a given adverse biological response. This assumes organisms are a single compartment, rather than the multiple compartments that they actually are, but it has considerable utility as a first approximation of dose.

Criteria for effects See End point.

Cumulative Brought about, or increased in strength, by successive additions at different times or in different ways.

Delayed effects Effects or responses that occur some time after exposure. Carcinogenic effects of chemicals typically have a long latency period; the occurrence of a tumor may take years after the initial exposure.

Depuration A process that results in elimination of a material from an aquatic organism.

Diffusion (nonbiological) Nonadvective transport due to migration and mixing of dissolved suspended solutes (including particulates) in natural waters in response to concentration gradients. Diffusion can be at the molecular level, due to Brownian motion producing random movements of the solute molecules (molecular diffusion), or it can be movements of solutes (including particles) due to turbulent eddies, velocity shear, or bioturbation (turbulent diffusion). The two types of diffusion result in mixing and dispersal of dissolved and bound chemicals.

Dilution water (diluent) Water used to dilute the test material in an aquatic toxicity test in order to prepare either different concentrations of a test chemical or different percentages of an effluent for the various test treatments. The water (negative) control in a test is prepared with dilution water only.

Direct toxicity Toxicity that results from and is readily attributable to the toxic agents(s) acting more or less directly at the sites of toxic action in and/or on the exposed organisms that are exhibiting the adverse biological response in question.

Dissolved organic carbon (DOC) The fraction of the organic carbon pool that is dissolved in water and that passes through a 0.45 μm glass fiber filter. DOC quantifies the chemically reactive organic fraction and is an accurate measure of the simple and complex organic molecules making up the dissolved organic load. The majority of the DOC is humic substances.

Dissolved organic matter (DOM) It is analogous to DOC (see DOC), but it refers to the entire organic pool that is dissolved in water.

Dose The quantifiable amount of a material introduced into the animal by injection or ingestion is considered to be the exposure dose. The amount of a material at receptor or target sites in the animal that elicits a response is the target dose.

Dose-response curve Similar to concentration-response curve except that the exposure dose (i.e., the quantity) of the chemical administered (e.g., by injection) to the organisms is known. The curve is plotted as dose versus response.

Effluent A complex waste material (e.g., liquid industrial discharge or sewage) that may be discharged into the environment.

End point In toxicity testing and evaluation it is the adverse biological response in question that is measured. End points vary with the level of biological organization being examined but include changes in biochemical markers or enzyme activities, mortality or survival, growth, reproduction, primary production, and changes in structure (and abundance) and function in a community. End points are used in toxicity tests as criteria for effects.

Environmental availability The portion of the total chemical material present in all or part of the environment that is actually involved in a particular process or processes and is subject to all physical, chemical, and biological modifying influences. It defines the total amount of material potentially available to organisms.

Environmental bioavailability The ratio of the uptake clearance divided by the rate at which an organism encounters a given contaminant in a medium (e.g., water, food) being processed by the organism. This is a measure of an organism's extraction efficiency, via respiratory, dietary, and surface absorption processes, from the environmentally available portion of a material.

Equilibrium In a thermodynamic sense this indicates that both a steady state of flux and an equivalence in chemical activity have been reached in compartments or phases separated by a membrane or boundary across which the chemical fluxes occur.

Equilibrium partitioning (EqP) An approach for estimating the fate of chemicals (primarily organics) in the aquatic environment that is based on the assumption that a steady-state can be achieved, and usually is achieved, between the activity of chemicals (usually approximated as concentration) in the various component phases—water, sediment, organisms. The EqP approach is often exploited, for interpretation and extrapolation purposes, by normalizing chemical concentrations based on the lipid content of the aquatic organisms and the organic carbon content of the sediments. These normalized BSAF values are considered to be independent of particular sediments and species.

Estimated (or expected) environmental concentration (EEC) The concentration of a material estimated as being likely to occur in environmental waters to which aquatic organisms are exposed as a result of planned manufacture, use, and disposal. Also called predicted environmental concentration (PEC).

Exposure The contact reaction between a chemical or physical agent and a biological system.

Fate Disposition of a material in various environmental compartments (e.g., soil or sediment, water, air, biota) as a result of transport, transformation, and degradation.

Federal register (Fed Reg or FR) The daily publication of federal regulations, proposed regulations, and public notices in the United States.

Final acute value (FAV) An estimate of the concentration of the material corresponding to a cumulative probability of 0.05 in the acute toxicity values for all genera for which acceptable acute tests have been conducted on the material.

Final chronic value (FCV) An estimate of the concentration of the material corresponding to a cumulative probability of 0.05 in the chronic toxicity values for all genera for which acceptable chronic tests have been conducted on the material. The FCV can also be calculated by dividing the FAV by the Final ACR.

Fish acute toxicity syndromes (FATS) Behavioral, physiological, and biochemical responses of fish used to classify chemicals by mode of action.

Flow-through system An exposure system for aquatic toxicity tests in which the test material solutions and control water flow into and out of test chambers on a once-through basis either intermittently or continuously.

Fulvic acid Humic substance that is soluble at all pHs. Most of the humic substance in natural water is fulvic acid.

Good laboratory practices (GLPs) A set of standards governing the experimental design and conduct of toxicological and analytical studies.

Half-life Time required to reduce by one-half the concentration of a material in a medium (e.g., soil or water) or organism (e.g., fish tissue) by transport, degradation, transformation, or depuration.

Hardness The concentration of all metallic cations, except those of the alkali metals, present in water. In general, hardness is a measure of the concentration of calcium and magnesium ions in water and is frequently expressed as mg/L calcium carbonate equivalent.

Hazard Likelihood that exposure to a chemical will cause an injury or adverse effect under the conditions of its production, use, or disposal.

Henry's law constant (H) A partition coefficient defined as the ratio of a chemical's concentration in air to its concentration in water at steady state. The constant can be dimensionless or with units. Occasionally, the constant is interpreted in an inverse fashion.

Hazard evaluation Identification and assessment of the potential adverse effects that could result from manufacture, use, and disposal of a material in a specified quantity and manner.

Humic acid Humic substance that is insoluble at acidic pHs.

Humic substances High-molecular-weight (500–5000), non-volatile organic, polyelectrolytic compounds which have complex structures and are variable in composition. They occur naturally as deposits on sediment and soil particles and constitute 30–50% of the DOC in natural waters. Humic substances are classified by solubility and may be composed of humic acid, fulvic acid and humin.

Incipient LC50 The concentration of a chemical that is lethal to 50% of the test organisms as a result of exposure for periods sufficiently long (time-independent) that acute lethal action has essentially ceased. The asymptote (part of the toxicity curve parallel to the time axis) of the toxicity curve indicates the value of the incipient LC50, approximately.

Indirect toxicity Adverse effects or toxicity that results from the agent(s) acting on and producing changes in the chemical, physical, and/or biological environment external to the organisms under study (e.g., decrease in food for predatory species due to direct toxicity from a chemical to prey may produce adverse effects in the predator species due to starvation rather than inducing any direct chemical toxicity in predator organisms).

Inhibition concentration (IC) A point estimate of the chemical concentration that would cause a given percent reduction (e.g., IC25) in a nonlethal biological measurement of the test organisms, such as reproduction or growth.

Joint action Two or more chemicals exerting their effects simultaneously.

Lethal Causing death by direct action. Death of aquatic organisms is the cessation of all visible signs of biological activity.

Lethal body burden (LBB) The body residue of chemical that is associated with mortality in short-term exposures.

Lethal time See Survival time.

Life cycle study A chronic (or full chronic) study in which all the significant life stages of an organism are exposed to a test material. Generally, a life cycle test involves an entire reproductive cycle of the organism.

Loading Ratio of the animal biomass to the volume of test solution in an exposure chamber.

Lordosis An anteroposterior curvature of the spine, generally in the lumbar region. Also called hollow back or saddle back.

Lowest observed adverse effect level (LOAEL) See Lowest observed effect concentration.

Lowest observed effect concentration (LOEC) The lowest concentration of a material used in a toxicity test that has a statistically significant adverse effect on the exposed population of test organisms compared with the controls. When derived from a life cycle or partial life cycle test, it is numerically the same as the upper limit of the MATC. Also called lowest observed adverse effect level (LOAEL).

Maximum acceptable toxicant concentration (MATC) The hypothetical toxic threshold concentration lying in a range bounded at the lower end by the highest tested concentration having no observed effect (NOEC) and at the higher end by the lowest tested concentration having a significant toxic effect (LOEC) in a life cycle (full chronic) or partial life cycle (partial chronic) test. This may be represented as NOEC < MATC < LOEC. The MATC may be calculated as the geometric mean of the LOEC and NOEC. Calculation of an MATC requires quantitative life cycle toxicity data on the effects of a material on survival, growth, and reproduction.

Median effective concentration (EC50) The concentration of material in water to which test organisms are exposed that is estimated to be effective in producing some sublethal response in 50% of the test organisms. The EC50 is usually expressed as a time-dependent value (e.g., 24-h or 96-h EC50). The sublethal response elicited from the test organisms as a result of exposure to the test material must be clearly defined. For example, test organisms may be immobilized, lose equilibrium, or undergo physiological or behavioral changes.

Median effective dose (ED50) The exposure dose of material estimated to be effective in producing some sublethal response in 50% of the test organisms. It is appropriately used with test animals such as rats, mice, and dogs, but it is rarely applicable to aquatic organisms because it indicates the quantity of a material introduced directly into the body by injection or ingestion rather than the concentration of the material in water in which aquatic organisms are exposed during toxicity tests.

Median effective time (ET50) The time required for half of the organisms in a toxicity test to exhibit a given nonlethal response end point at a given exposure concentration.

Median lethal concentration (LC50) The concentration of material in water to which test organisms are exposed that is estimated to be lethal to 50% of the test organisms. The LC50 is often expressed as a time-dependent value (e.g., 24-h or 96-h LC50; the concentration estimated to be lethal to 50% of the test organisms after 24 or 96 h of exposure). The LC50 may be derived by *observation* (i.e., 50% of the test organisms may be observed to be dead in one test material concentration), by *interpolation* (i.e., mortality of more than 50% of the test organisms occurred at one test concentration and mortality of fewer than 50% of the test organisms died at a lower test concentration, and the LC50 is estimated by interpolation between these two data points), or by *calculation* (i.e., the LC50 is statistically derived by analysis of mortality data from all test concentrations).

Median lethal dose (LD50) The exposure dose of material that is estimated to be lethal to 50% of the test organisms. It is appropriately used with test animals such as rats, mice, and dogs, but it is rarely applicable to aquatic organisms because it indicates the quantity of a material introduced directly into the body by injection or ingestion rather than the concentration of the material in water in which aquatic organisms are exposed during toxicity tests.

Median lethal time (LT50) The time required for half of the organisms in a toxicity test to die at a given exposure concentration.

Median tolerance limit (TLm or TL50) The concentration of material in water at which 50% of the test organisms survive after a specified time of exposure. The TL50 (equivalent to the TLm) is usually expressed as a time-dependent value (e.g., 24-h or 96-h TL50; the estimated concentration at which 50% of the test organisms survive after 24 or 96 h of exposure). Unlike lethal concentration and lethal dose, the term tolerance limit is applicable in designating a level of any measurable lethal condition (e.g., extremes in pH, temperature, dissolved oxygen). TLm and TL50 have been replaced by median lethal concentration (LC50) and median effective concentration (EC50).

Mixing zone An area where an effluent discharge undergoes initial dilution and is extended to cover the secondary mixing in the ambient water body. A mixing zone is an allocated impact zone where water quality criteria can be exceeded as long as acutely toxic conditions are prevented.

Mole The SI (Système International) metric unit for reporting amounts of chemicals whose relative molecular mass is known. One mole of a chemical is Avogadro's number (6.023×10^{23}) of molecules of that chemical. Metric mass–based units can be converted to molar units by dividing the former by the gram-molecular weight of the chemical in question. Molarity refers to the number of moles of chemical per liter of solution and is denoted as M.

Monitoring test A test designed to be applied on a routine basis, with some degree of control, to ensure that the quality of water or effluent has not exceeded some prescribed criterion range. In a biomonitoring test, aquatic organisms are used as "sensors" to detect changes in the quality of water or effluent. A monitoring test implies generation of information, on a continuous or other regular basis.

Narcosis A general, nonspecific, reversible mode of toxic action that can be produced in most living organisms by the presence of sufficient amounts of many organic chemicals. Effects result from the general disruption of cellular activity. The mechanism producing this effect is unknown, with the main theories being binding to proteins in cell membranes and "swelling" of the lipid portion of cell membranes resulting from the presence of organic chemicals. Hydrophobicity dominates the expression of toxicity in narcotic chemicals.

No observed adverse effect level (NOAEL) See No observed effect concentration.

No observed effect concentration (NOEC) The highest concentration of a material in a toxicity test that has no statistically significant adverse effect on the exposed population of test organisms compared with the controls. When derived from a life cycle or partial life cycle test, it is numerically the same as the lower limit of the MATC. Also called no observed adverse effect level (NOAEL) or no observed effect level (NOEL).

No observed effect level (NOEL) See No observed effect concentration.

Nonconventional pollutant See Pollutant.

Octanol-water partition coefficient (K_{ow}) The ratio of a chemical's solubility in *n*-octanol and water at steady state; also expressed as P. The logarithm of P or K_{ow} (i.e., log P or log K_{ow}) is used as an indication of a chemical's propensity for bioconcentration by aquatic organisms.

Particulate organic carbon (POC) The fraction of the organic carbon pool that is not dissolved in water, but is retained on a 0.45 μm glass fiber filter. POC is identical to suspended organic carbon (SOC) and is composed of plant and animal organic carbon and organic coatings on silt and clay.

Particulate organic matter (POM) It is analogous to POC (see POC) but it refers to the entire organic pool that is not dissolved in water.

Partition coefficient The ratio of chemical concentrations in two different compartments or phases under steady state conditions.

Parts per billion (ppb) One unit of chemical (usually expressed as mass) per 1,000,000,000 (10^9) units of the medium (e.g., water) or organism (e.g., tissues) in which it is contained. For water, the ratio commonly used is micrograms of chemical per liter of water, 1 μg/L = 1 ppb; for tissues, 1 μg/kg = 1 ng/g = 1 ppb.

Parts per million (ppm) One unit of chemical (usually expressed as mass) per 1,000,000 (10^6) units of the medium (e.g., water) or organism (e.g., tissues) in which it is contained. For water, the ratio commonly used is milligrams of chemical per liter of water, 1 mg/L = 1 ppm; for tissues, 1 mg/kg = 1 μg/g = 1 ppm.

Parts per thousand (ppt) One unit of chemical (usually expressed as mass) per 1000 (10^3) units of the medium (e.g., water) or organism (e.g., tissues) in which it is contained. For water, the ratio commonly used is grams of chemical per liter of water, 1 g/L = 1 ppt; for tissues, 1 g/kg = 1 ppt. This ratio is also used to express the salinity of seawater, where the grams of chloride per liter of water is denoted by the symbol ppt. Full-strength seawater is approximately 35 ppt.

Parts per trillion (pptr) One unit of chemical (usually expressed as mass) per 1,000,000,000,000 (10^{12}) units of the medium (e.g., water) or organism (e.g., tissues) in which it is contained. The ratio commonly used is nanograms of chemical per liter of water, 1 ng/L = 1 pptr; for tissues, 1 ng/kg = 1 pptr.

Percentage (%) One unit of material (usually a liquid) per 100 units of dilution water. In tests with industrial wastes (effluents), test concentrations are normally prepared on a volume-to-volume basis and expressed as percent of material.

Pesticide A substance used to kill undesirable and unwanted fungi, plants, insects, or other organisms. This generic term is used to describe fungicides, algicides, herbicides, insecticides, rodenticides, and other substances.

Pollutant It is a general term for a chemical or nonchemical (e.g., increased suspended solids) agent that is present in the environment and causes adverse effects. Pollutants are categorized in the Clean Water Act as conventional, nonconventional, and toxic. Conventional pollutants are biochemical oxygen demand, total suspended solids, pH, oil and grease, or fecal coliform. A nonconventional pollutant is a chemical or parameter not listed as either conventional or as a toxic pollutant. A toxic pollutant is anything listed under the Clean Water Act 307 (a) (1) of the United States.

Pore water (or insterstitial water) Water found in spaces between particles of soil or sediment.

Predicted environmental concentration (PEC) See Estimated or expected environmental concentration (EEC).

Preliminary test See Screening test.

Quality assurance (QA) A program organized and designed to provide accurate and precise results. Included are selection of proper technical methods, tests, or laboratory procedures; sample collection and preservation; selection of limits; evaluation of data; quality control; and qualifications and training of personnel.

Quality control (QC) Specific actions required to provide information for the quality assurance program. Included are standardization, calibration, replicates, and control and check samples suitable for statistical estimates of confidence of the data.

Quantitative structure-activity relationships (QSARs) See Structure-activity relationships (SARs).

Range-finding test See Screening test.

Reference toxicant A chemical used in an aquatic toxicity test as a positive control (i.e., deliberately causes known adverse effects) in contrast to the negative control provided by exposure water without the test chemical. Information collected is used to determine the general health and viability of the test organisms and assess consistency in testing protocol implementation.

Resistance time The period of time for which an aquatic organism can live beyond the incipient lethal level.

Ring test (or round-robin test) (1) A conjoint test conducted under strictly standardized and uniformly applied conditions to assess the precision and accuracy with which different laboratories can determine the toxicity of a chemical or effluent. (2) A test designed to measure statistically the reproducibility of a test method, or to compare the results obtained from the use of different test methods.

Risk A statistical concept defined as the expected frequency or probability of undesirable effects resulting from a specified exposure to known or potential environmental concentrations of a material. A material is considered safe if the risks associated with its exposure are judged to be acceptable. Estimates of risk may be expressed in absolute or relative terms. Absolute risk is the excess risk due to actual exposure. Relative risk is the ratio of the risk in the exposed population to the risk in an unexposed population.

Round-robin test See Ring test.

Safe concentration (SC) Concentration of material to which prolonged exposure will cause no appreciable adverse effect.

Safety The practical certainty that adverse effects or injury will not result from exposure to a material when used in the quantity and the manner proposed for its use. "Practical certainty" must be defined—for example, in terms of a numerically specified low risk. Another view is that "safety" is a value judgment of the acceptability of risk.

Scoliosis Lateral curvature of the spine.

Screening test (Preliminary test or Range-finding test) (1) A test conducted to estimate the concentrations to be used for a definitive test. (2) A short-term test used early in a testing program to evaluate the potential of a chemical (or other material) to produce some selected adverse effect (e.g., mortality).

Solvent or carrier An agent (other than water) in which the test chemical is mixed to make it miscible with dilution water before distribution to test chambers. Solvents or carriers are used in toxicity test where the concentrations of the test chemical are extremely low and a very small amount of test material must be added to the test chambers.

Static system An exposure system for aquatic toxicity tests in which the test chambers contain still solutions of the test material or control water.

Statistically significant effects Effects (responses) in the exposed population that are different from those in the controls at a given statistical probability level, typically $P \leq .05$. Biological end points that are important for the survival, growth, behavior, and perpetuation of a species are selected as criteria for effects. The end points differ depending on the type of toxicity test conducted and the species used. The statistical approach also changes with the type of toxicity test conducted.

Steady state The state in which fluxes of material moving bidirectionally across a membrane or boundary between compartments or phases have reached a balance. An equilibrium between the phases is not necessarily achieved.

Structure-activity relationships (SARs) Studies relating the structure (and related properties) of chemicals to their activities in biological systems. SARs are used to assist in explaining (and predicting) the occurrence and mechanism of biological responses to chemicals and to aid in prediction of incidence and magnitude. Although lethal and sublethal toxicity test results are often employed, other biological activities, such as bioaccumulation, biodegradation, and biotransformation, may also be used. SARs can also be used to explain or predict physicochemical characteristics of chemicals. Also called quantitative structure-activity relationships (QSARs).

Sublethal Below the concentration that directly causes death. Exposure to sublethal concentrations of a material may produce less obvious effects on behavior, biochemical and/or physiological function, and histology of organisms.

Survival time The time interval between initial exposure of an aquatic organism to a harmful chemical and death.

Synergism A phenomenon in which the toxicity of a mixture of chemicals is greater than that which would be expected from a simple summation of the toxicities of the individual chemicals present in the mixture.

Test material (or substance) A chemical, formulation, effluent, sludge, or other agent or substance that is under investigation in an aquatic toxicity test.

Test solution (or test treatment) Medium containing the material to which organisms are exposed in a toxicity test. Different test solutions contain different concentrations of the test material.

Threshold concentration A concentration above which some effect (or response) will be produced and below which it will not.

Time-independent (TI) test An acute toxicity test with no predetermined temporal end point. This type of test, sometimes referred to as a "threshold" or "incipient" lethality test, is allowed to continue until acute toxicity (mortality or a defined sublethal effect) has ceased or nearly ceased and the toxicity curve (median survival time versus test material concentration) indicates a threshold or incipient concentration. With most test materials, this point is reached within 7–10 d, but it may not be reached within 21 d. Practical or economic reasons may dictate that the test be stopped at this point and a test be designed for longer duration.

Top-down The approach to investigating ecotoxicological effects that starts with a determination of the presence and nature of any adverse effects via responses at community and ecosystem levels of organization rather than the suborganismal levels of organization.

Total organic carbon (TOC) The sum of dissolved organic carbon (DOC) and particulate organic carbon (POC) or suspended organic carbon (SOC).

Total organic matter (TOM) The sum of dissolved organic matter (DOM) and particulate organic matter (POM) or suspended organic matter (SOM).

Toxic pollutant See Pollutant.

Toxicant An agent or material capable of producing an adverse response (effect) in a biological system, seriously injuring structure and/or function or producing death.

Toxicity The inherent potential or capacity of an agent or material to cause adverse effects in a living organism when the organism is exposed to it. See Direct toxicity and Indirect toxicity.

Toxicity curve The curve obtained by plotting either the median survival times of a population of test organisms against test material concentrations (preferred) or median effective concentrations against exposure times. The most important feature of a toxicity curve is that it provides an indication of the threshold median lethal concentration. Unless toxicity curves are obtained and lethal thresholds established, the results of any toxicity test must be interpreted with caution.

Toxicity identification evaluation (TIE) A set of procedures to identify the specific chemicals responsible for the toxicity of effluents.

Toxicity reduction evaluation (TRE) A site-specific study conducted in a stepwise process designed to identify the causative agents in a toxic effluent, isolate the sources of toxicity, evaluate the effectiveness of toxicity control options, and then confirm the reduction in effluent toxicity.

Toxicity test The means by which the toxicity of a chemical or other test material is determined. A toxicity test is used to measure the degree of response produced by exposure to a specific level of stimulus (or concentration of chemical).

Toxicodynamics The phase of toxic action that consists of the biological response resulting from the interaction of the chemical at the site of toxic action in the exposed organism and the resultant sequelae.

Toxicokinetics The phase of toxic action that comprises the uptake (absorption), distribution, metabolism (and/or biotransformation), and elimination (excretion) of bioavailable chemical by the exposed organism.

Toxicological bioavailability The classical pharmacological/toxicological definition of bioavailability. It is the fraction of the total exposure dose extracted from the medium that is actually adsorbed/absorbed by the organism, is distributed by the system circulation, and is ultimately presented to the receptor sites or sites of toxic action.

Toxic unit (TU) The strength of a chemical (measured in some unit) expressed as a fraction or proportion of its threshold effect concentration (measured in the same unit). The strength may be calculated as follows: toxic unit = actual concentration of chemical in solution/threshold effect concentration. For the case where 1.0 toxic unit equals the incipient LC50, a TU greater than 1.0 should cause the death, during long exposures, of more than half of a group of aquatic organisms. If the TU is less than 1.0, less than half of the organisms should be killed.

Ultimate median tolerance limit The concentration of a chemical at which acute toxicity ceases. Also called incipient lethal level, lethal threshold concentration, and asymptotic LC50.

Uptake A process by which materials are transferred into and onto an aquatic organism.

Van der Waals forces Weak mutual attractions between molecules which can contribute to bonding between atoms.

Water quality criterion An estimate, based on scientific judgments, of the concentration of a chemical or other constituent in water which, if not exceeded, will protect an organism, an organism community, or a prescribed water use or quality with an adequate degree of safety. In some jurisdictions terminology may be different and the term guideline or objective may be substituted for criterion. Criteria may also refer to the scientific principles and data that are used to formulate recommended guidelines or limits.

Water quality guideline or limit A scientifically based numerical concentration limit or narrative statement recommended to support and maintain a designated water use.

Water quality objective A scientifically based numerical concentration limit or narrative statement (e.g., receiving water should sustain a healthy population of salmonids) that has been established to support and protect designated uses of water, often at a specified site or sites. Objectives reflect desired conditions in receiving waters, usually do not carry the force of law (i.e., are not water quality standards), and often consider various receiving waters separately.

Water quality standard Standards are established for regulatory purposes and have a legal meaning that may vary between jurisdictions. A standard, numerical criterion or narrative that is based on the limiting concentration of a chemical (or degree of intensity of some adverse condition, e.g., pH) which is permitted in an effluent or waterway. It is usually formulated with reference to specific enabling legislation and is the product of both legal and scientific judgments. A water quality standard is typically dependent on the use (e.g., potable, agricultural) of the water to be protected.

Whole-effluent toxicity (WET) The total toxic effect of an effluent measured directly with aquatic organisms in a toxicity test.

Xenobiotic A foreign chemical or material usually not produced in nature and not normally considered a constitutive component of a specified biological system. This term is usually applied to manufactured chemicals.

The four chapters of these appendices are reprinted, unchanged from the first edition of *Fundamentals of Aquatic Toxicology* © 1985. The editor and publisher include them here in the second edition as classic chapters in aquatic toxicology that should be kept in print and available for reference by more recent works.

Appendix A

ACUTE TOXICITY TESTS

P. R. Parrish

INTRODUCTION

In the period since World War II, acute toxicity tests have been used extensively to determine the effects of potentially toxic materials (e.g., pesticides, metals, and industrial effluents) on aquatic organisms during short-term (usually 4 d or less) exposure. Most of the earlier acute aquatic toxicity tests were conducted with freshwater organisms, especially fishes, but the requirement for assessing the impact of test materials in the estuarine and marine environment has provided the impetus for the development of test methods for saltwater organisms, both invertebrates and fishes. This development has been most pronounced during the past 10–15 yr.

Toxicity tests with unicellular algae (phytotoxicity tests) have also been widely employed to screen chemicals, especially chemicals known to be herbicides or suspected of having phytotoxic activity. Such tests have been conducted according to the same principles as invertebrate and fish tests. However, even a 4- or 5-d exposure represents a chronic exposure of the algal population because growth rate and cell density have probably reached their maximum during this time. Thus, although they are short-term tests, phytotoxicity tests are based on the life cycle of the test organism and will be discussed only briefly in this chapter. Appropriate references for algal test methods are cited in the supplemental reading section.

The trend toward conducting acute toxicity tests according to uniform, detailed methods is relatively new, having been promoted most vigorously by the Committee on Methods for Toxicity Tests with Aquatic Organisms. The committee was formed in 1971, and the result of its efforts was a

Originally appeared as Chapter 2 in Rand and Petrocelli: *Fundamentals of Aquatic Toxicology*, © 1985 Taylor & Franicis.

U.S. Environmental Protection Agency (U.S. EPA) publication on acute toxicity tests (Committee on Methods for Toxicity Tests with Aquatic Organisms, 1975). The committee was dissolved because it was logical that the movement toward standardization of toxicity test methods be brought under the auspices of a consensus-type organization. Thus, the American Society for Testing and Materials (ASTM) has been the most recent focal point for test method standardization. At the time of this writing, two "standard practices" on acute test methods have been published (ASTM, 1980a,b), and several other standard practices on long-term test methods are near completion. A second contribution to the standardization of test methods has been the publication of Standard Methods for the Examination of Water and Wastewater by the American Public Health Association, American Water Works Association, and Water Pollution Control Federation (APHA, AWWA, and WPCF, 1981).

The application of routine acute toxicity test methods to evaluate the effects of test materials other than individual chemicals is valid. Tests have been conducted with industrial effluents for more than 35 yr, and more recently tests with dredged materials, drilling muds, and mine tailings have been conducted.

Because the principles for conducting acute toxicity tests with freshwater and saltwater organisms are essentially the same, the basic methods for testing will be discussed in this chapter with appropriate consideration given to special conditions for each aquatic medium.

GENERAL PRINCIPLES OF ACUTE TOXICITY TESTS

The objective of an acute aquatic toxicity test is to determine the concentration of a test material (e.g., a chemical or effluent) or the level of an agent (e.g., temperature or pH) that produces a deleterious effect on a group of test organisms during a short-term exposure under controlled conditions. Although toxicity tests with aquatic organisms can be conducted by administering the material directly by injection or incorporating it into food, most tests are conducted by exposing groups of organisms to several treatments in which different concentrations of the material are mixed in water. Because death is an easily detected deleterious response, the most common acute toxicity test is the acute lethality test. The criteria for death are usually lack of movement (especially gill movement in fishes) and lack of reaction to gentle prodding. Experimentally, a 50% response is the most reproducible measure of the toxicity of a test material, and 96 h (or less) is the standard exposure time because it usually covers the period of acute lethal action. Therefore, the measure of acute toxicity most frequently used with fish and macroinvertebrates is the 96-h median lethal concentration (96-h LC50). However, because death is not easily determined for some invertebrates, an EC50 (median effective concentration) is estimated rather than an LC50. The effect used for estimating the EC50 with some invertebrates (e.g., daphnids and midge larvae) is immobilization, which is defined as lack of movement. The effects generally used for estimating the EC50 with crabs, crayfish, and shrimp are immobilization and loss of equilibrium, defined as inability to maintain normal posture.

A common acute testing approach is to structure a test so that a quantal response (i.e., an all-or-none response: dead or alive) is elicited. The relation between test material concentration and the percentage of exposed organisms affected can then be determined and a concentration-mortality curve plotted. Results of short-term tests have generally been expressed as (1) the percentage of organisms killed or immobilized in each test concentration, and (2) the LC50 or EC50 derived by observation (i.e., where 50% of the test organisms were killed in a concentration), interpolation, or calculation. When the median lethal concentration is calculated, the 95% confidence limits associated with that value are usually also reported.

Test organisms in acute toxicity tests may be exposed to the test materials by one of four techniques—static, recirculation, renewal, or flow-through—as described in Chapter 1.

Acute lethality tests have been useful in providing rapid estimates of the concentrations of test materials that cause direct, irreversible harm to the test organism. The static and flow-through techniques are the most widely used in acute toxicity tests. As stated by Macek et al. (1978), static acute tests provide practical means for (1) deriving estimates of the upper limit of concentrations that produce toxic effects, (2) evaluating the relative toxicity of large numbers of test materials, (3) evaluating the relative sensitivity of different aquatic organisms to test materials, (4) evaluating the effects of water quality (e.g., dissolved oxygen concentration, pH, salinity, hardness, suspended particulates) on the toxicity of the test materials, and (5) developing an understanding of the concentration-response relationship and of the significance of duration of exposure to the test material. Flow-through acute tests enhance the ability to maintain satisfactory test conditions and are not limited in duration; thus they allow a more complete assessment of the relationship between exposure duration and effect.

METHODOLOGY FOR FISH AND INVERTEBRATES

Test Material

The test material may be one or more pure chemical compounds or a complex mixture such as an effluent or formulation. It should be representative of the material as it is used or as it enters the aquatic environment. The sample of test material may be collected by a single "grab" if the composition of the material remains relatively constant. Composite grab sampling is recommended for static tests with effluents whose composition is relatively constant. A composite effluent sample consisting of grab samples collected at each of four consecutive 6-h intervals is recommended. For effluents that are not constant in composition, samples should be collected at different times. They should not be combined because knowledge of the maximum toxicity level and its potential frequency of occurrence is more important than knowledge of the average toxicity of a variable effluent. With an effluent of variable quality it is better to conduct a flow-through toxicity test, if possible. If a flow-through test is conducted and continuous sampling is impractical, a composite sample should be collected and transported to the effluent delivery system at least every 6 h. Each new sample should be used only during the time until the next sample delivery. An effluent sample should completely fill the sample container to exclude as much air space as possible, and the container should be sealed.

Test materials that are stable can be transported and stored in appropriate containers for reasonable periods. Test materials other than effluents should be stored in a place that is cool (to minimize chemical or biochemical degradation) and dark (to minimize photodegradation). Volatile test materials should be kept in tightly sealed, refrigerated containers (to minimize loss to atmosphere), but allowed to come to room temperature before testing. Effluents should be transported at a low (but above freezing) temperature and delivered to the testing facility within 24 h after collection. Effluents, especially those containing organic matter subject to bacterial decomposition, should be refrigerated at approximately 4°C but should never be frozen.

If required, stock solutions of the test material should be prepared in advance so that the appropriate volume can be easily added to the dilution water (directly in static tests or via some delivery system in flow-through tests) to obtain the different test concentrations. All solutions needed for a toxicity test should be prepared from the same sample of test material. Any undissolved material should be uniformly dispersed by shaking or mixing the stock solution before a sample is withdrawn and added to dilution water in the test containers. If possible, the test material should be added directly to the dilution water, without the use of any solvent or carrier other than water. If a solvent/carrier is necessary, the organic chemicals triethylene glycol (TEG) and dimethyl formamide (DMF) are preferred because of their low toxicity to aquatic organisms, low volatility, and ability to dissolve many organic materials. However, other solvent/carriers such as acetone, methanol, and ethanol may be used. When a solvent/carrier other than water is used, it is recommended that its concentration in the test solution be kept at a minimum and not exceed 0.5 ml/l. There may be tests in which this limit cannot be met, and in those tests higher concentrations are permitted as long as an appropriate solvent/carrier control is used and the test organisms show no sign of stress or toxicity in the solvent/carrier control.

Effluent tests should be conducted with a sample that has not been aerated or altered in any way. Undissolved materials should be dispersed by gentle agitation immediately before removal of a subsample and addition to dilution water. In flow-through tests, undissolved material may be kept in suspension by using a stirring device. The effluent sample should be used directly as a stock solution, but it may be necessary to prepare a stock solution by diluting the effluent with filtered freshwater or saltwater to obtain the desired concentration and volume for mixing with the dilution water. Stock solutions or test solutions should be prepared as a volume-to-volume ratio of dilution water and liquid effluent so that the percentage of effluent in each of the test solutions can be calculated. If the effluent is a solid waste, dilution should be on a weight-to-volume basis. If a volume-to-volume basis is used and the concentration is to be expressed as weight per unit volume, a correction should be made for specific gravity.

Dilution Water

Adequate dilution water for the test organisms and for the purpose of the test must be available. For an acute toxicity test, the minimal criterion for acceptable dilution water is that test organisms will live in it during acclimation to test conditions and testing without being stressed. Dilution water should be well aerated before organisms or test material are added in order to bring the pH and concentrations of dissolved oxygen and other gases into equilibrium with the ambient air and thus minimize oxygen demand and concentration of volatiles. The initial concentration of dissolved oxygen in the dilution water should be greater than 90% of the saturation value. Dilution water that may contain undesirable microorganisms should be passed through a properly maintained ultraviolet sterilizer.

For effluent tests, the dilution water should be representative of the receiving water and should be obtained as close to the point of effluent discharge as possible, but upstream from the point of discharge or outside the zone of influence of the effluent. If acceptable dilution water cannot be obtained from the receiving water, some other uncontaminated surface, ground, or reconstituted water may be used; however, the water quality characteristics (e.g., hardness, alkalinity, salinity, and specific conductance) should be within approximately 25% and 0.2 pH unit of those of the receiving water at the time of the test. The dilution water should be uncontaminated, of constant quality, and should meet the specifications outlined by the ASTM (1980a).

Dilution water for tests with freshwater organisms should be obtained from an uncontaminated well or spring, if possible. Dechlorinated tap water should be used only if no other water supply is available. Whenever possible, dilution water with a hardness of 40–48 mg/l as $CaCO_3$ should be used for acute tests (ASTM, 1980a).

Dilution water for tests with saltwater organisms should be obtained from an uncontaminated natural body of water and filtered through a non-contaminating filter of 5-μm porosity before use. Reconstituted saltwater (ASTM, 1980a) or commercial sea salts may also be used for acute toxicity tests. Whenever possible, water of higher salinity—30–34 parts per thousand (ppt)—should be used when testing stenohaline species, and water of lower salinity (20–25 ppt) when testing euryhaline species.

Test Animals

Test animals should be sensitive, important species indigenous to the area of impact of a test material. They can be collected from wild populations in relatively unpolluted areas, purchased from commercial suppliers, or amenable to culture in the laboratory. Animals should not be collected by electroshocking or use of chemicals because of the physiological stress induced by these methods. All animals in a test should be from the same source. Test species may also be selected on the basis of availability and commercial, recreational, and ecological importance. The most commonly used species in acute aquatic toxicity tests are listed in Table 1.

Collecting, transporting, and handling animals should be done in a manner that minimizes injury and physiological stress. For example, many invertebrates and fishes can be collected by seining. It is far less harmful to the animals to scoop them from the water with a soft net and place them in a water-filled container while they are still within the enclosed seine in shallow water than to beach the seine. Also, slow tows of very short duration are recommended when trawls or other towed devices must be used to collect animals.

Containers for transporting animals should be circular or elliptical to prevent the animals from crowding in corners or damaging themselves by striking the walls (Cox, 1974), but conventional styrofoam coolers have proved acceptable. Other arrangements, such as the use of water-filled plastic bags inside the container, have also been successful. The density of animals in a container will vary according to the kind of animal, but it should never be great enough to stress the animals. When temporary containment and transportation time will be more than 30 min, additional measures should be taken to promote optimal conditions. For example, it may be necessary to shield the animals from sunlight, to protect them from extreme heat or cold, or to aerate water in order to maintain dissolved oxygen concentrations. For some sessile or slow-moving animals, large volumes of water may not be needed and materials such as wet seaweed or cloth may be used to provide moisture during transportation.

Table 1 Fish and Invertebrate Species Commonly Used for Acute Toxicity Tests

Freshwater
Vertebrates
Rainbow trout, *Salmo gairdneri*
Brook trout, *Salvelinus fontinalis*
Fathead minnow, *Pimephales promelas*
Channel catfish, *Ictalurus punctatus*
Bluegill, *Lepomis macrochirus*
Invertebrates
Daphnids, *Daphnia magna, D. pulex, D. pulicaria*
Amphipods, *Gammarus lacustris, G. fasciatus, G. pseudolimnaeus*
Crayfish, *Orconectes* sp., *Cambarus* sp., *Procambarus* sp., or *Pacifastacus leniusculus*
Midges, *Chironomus* sp.
Snails, *Physa integra*
Saltwater
Vertebrates
Sheepshead minnow, *Cyprinodon variegatus*
Mummichog, *Fundulus heteroclitus*
Longnose killifish, *Fundulus similis*
Silverside, *Menidia* sp.
Threespine stickleback, *Gasterosteus aculeatus*
Pinfish, *Lagodon rhomboides*
Spot, *Leiostomus xanthurus*
Sand dab, *Citharichthys stigmaeus*
Invertebrates
Copepods, *Acartia tonsa, Acartia clausi*
Shrimp, *Penaeus setiferus, P. duorarum, P. aztecus*
Grass shrimp, *Palaemonetes pugio, P. vulgaris*
Sand shrimp, *Crangon septemspinosa*
Mysid shrimp, *Mysidopsis bahia*
Blue crab, *Callinectes sapidus*
Fiddler crab, *Uca* sp.
Oyster, *Crassostrea virginica, C. gigas*
Polychaetes, *Capitella capitata, Neanthes* sp.

A separate, uncontaminated area at the testing facility should be set aside for maintaining test animals. When organisms are brought there, they should be quarantined at least until it is apparent that they are disease-free. Animals should be maintained in good condition, that is, without crowding and with adequate dissolved oxygen concentrations. To minimize stress, organisms should not be subjected to rapid changes in temperature or water quality. As a rule, organisms should not be subjected to more than 5°C gradual change in water temperature in any 24-h period, and saltwater organisms should not be subjected to more than a 5 ppt gradual salinity change in 24 h. Organisms should be fed at least once a day and tanks cleaned as needed. Recently collected animals may not eat, but this initial nonfeeding period may be shortened by "social facilitation," that is, by adding animals of the same species which are already feeding to the holding tank. One must not indiscriminately mix animals from different lots because of possible differences in age, condition, and stage of acclimation. When animals do not begin to eat, or stop eating after having begun, unhealthy conditions may be indicated. Animals should be carefully observed for signs of disease, stress, physical damage, and mortality. Dead, diseased, and abnormal individuals must be discarded.

During collection, transportation, and maintenance, organisms must be handled as little as possible. When necessary, transferring animals from one container to another should be accomplished gently and quickly. Small, soft nylon dipnets are generally best for moving larger organisms. Such nets are commercially available, or can be made from small-mesh nylon netting, nylon or silk bolting cloth, plankton netting, or similar material. Smooth-bore glass tubes with fire-polished ends and rubber bulbs should be used for transferring smaller organisms such as daphnids, midge larvae, copepods, and mysid shrimp.

It is generally recommended that fish be maintained for 7-14 d and invertebrates for 2-4 d after collection or transportation. All animals should be held in uncontaminated dilution water under stable conditions of temperature and water quality in a flow-through system with a flow rate of at least two water volume additions per day, or in a recirculating system in which the water flows through an appropriate filter.

Test organisms amenable to laboratory culture should be reared in dilution water at test temperature. Other animals should be acclimated simultaneously to the dilution water and test temperature in a flow-through system after transferring an appropriate number of similar sized animals to an acclimation tank by (1) gradually changing the water in the acclimation tank from 100% holding water to 100% dilution water over a period of 2 or more days, and (2) changing the water temperature at a gradual rate (not more than 5°C in 24 h) until the test temperature range is reached. Animals should be maintained for at least 2 d in the dilution water at the test temperature before a test is begun. Longer acclimation is generally desirable for fishes but may be unnecessary for invertebrates (Tatem et al., 1976).

In view of the importance of beginning a toxicity test with healthy organisms, a group of test animals must not be used for a test if they appear to be diseased or otherwise stressed or if more than 10% died during the 48 h immediately before the test. If a group fails to meet these criteria, it must be either discarded or treated for disease and reacclimated as described above.

Organisms should be uniform in size, age, and physiological condition to minimize variability of response to the test material. The small, early life stages of invertebrates and fishes (approximately 0.5 g each) and invertebrates with relatively short life cycles (approximately 60 d or less) must be fed up to the beginning of the test. Other organisms should not be fed for 24 h before the beginning of the test, nor should they be fed during the test to minimize variability due to nutritional and metabolic condition.

Most freshwater and some saltwater fishes may be chemically treated to prevent or cure diseases by using recommended treatments (ASTM, 1980a). If the fishes are severely diseased, however, it is often better to destroy the entire group. All diseased animals other than fishes should be discarded unless effective treatment is available. Since fishes have probably been stressed during collection or transportation (and because some are treated during transit), they should not be treated during the first 12-24 h after they arrive at the facility. Further, treated animals should be maintained for at least 4 d after treatment before they are used in a test.

If diseases occur among animal populations in a facility, care should be taken to isolate the diseased populations and to adequately clean the containers in which the diseased animals were maintained.

Exposure Apparatus

Static Acute Tests

The equipment necessary to conduct static acute toxicity tests includes a constant-temperature room or temperature-controlled water bath in which to place the test chambers, as well as an appropriate number of test chambers. The size and configuration of test chambers will vary according to the size of the test animals; the chambers can be glass finger bowls, jars, tanks, aquaria, beakers, crystallizing dishes, and so on. If the test containers are constructed in the testing facility, they should be made from materials that do not contain any substances that can leach or dissolve in the water and adversely affect the test animals. Likewise, materials and equipment that come in contact with stock solutions of test material should be chosen to minimize sorption of the test material. To minimize leaching, dissolution, and sorption, containers made of glass, no. 316 stainless steel, or perfluorocarbon plastics (e.g., Teflon) should be used whenever possible. Plasticized materials (nonrigid, soft plastics), rubber, copper, brass, or lead should not be used in any water delivery system as a test chamber, or in any exposure apparatus.

Flow-through Acute Tests

As with static tests, a constant-temperature room or temperature-controlled water bath is necessary to hold test chambers. There should also be a dilution water tank placed so that dilution water can flow into the test material delivery system. Temperature control and aeration for the dilution water tank are desirable. The heating/chilling device should be impervious to corrosion by the dilution water and the air used for aeration should be free of oil and fumes. Filters to remove oil and water are also desirable. Materials for construction of other flow-through test system parts should be the same as described above.

Although many toxicant delivery systems have been used successfully, the proportional diluter (Mount and Brungs, 1967; Lemke et al., 1978) is probably best for routine use. Diluters can be constructed in the testing facility or can be purchased from a commercial source. A diluter consists of water cells of known volume which receive various volumes of uncontaminated dilution water and chemical cells of known volume which receive various volumes of test solution. Water from the various dilution water cells siphons down and mixes with the test solution from the various chemical cells to produce the test material concentrations. Chambers to promote mixing of the test solution and the dilution water are located between the diluter and the test chambers (Fig. 1).

The proportional diluter can generally deliver five to seven different test concentrations and uncontaminated water for a control at any desired flow rate up to 400 ml/min and with dilution factors from 0.75 to 0.50. If necessary, metering cells can be exchanged to provide dilution factors outside this range. The flow rates through the test chambers should be at least five tank volume additions per 24 h, but the size of the test animals, the size of the test chambers, and the loading (ratio of test animal biomass to total volume of test water in the chamber) should also be considered. The rates should be able to maintain desirable temperatures, oxygen concentrations, and safe concentrations of metabolites.

For effluent tests in which the highest test concentration might be nearly 100% effluent, the proportional diluter can be modified by introducing 100% effluent directly into the chemical cells and eliminating the first water and mixing cells (Fig. 2). A gradient of effluent concentrations, based on the dilution factor of the system, results.

The calibration of the toxicant delivery system should be checked before each test. This includes

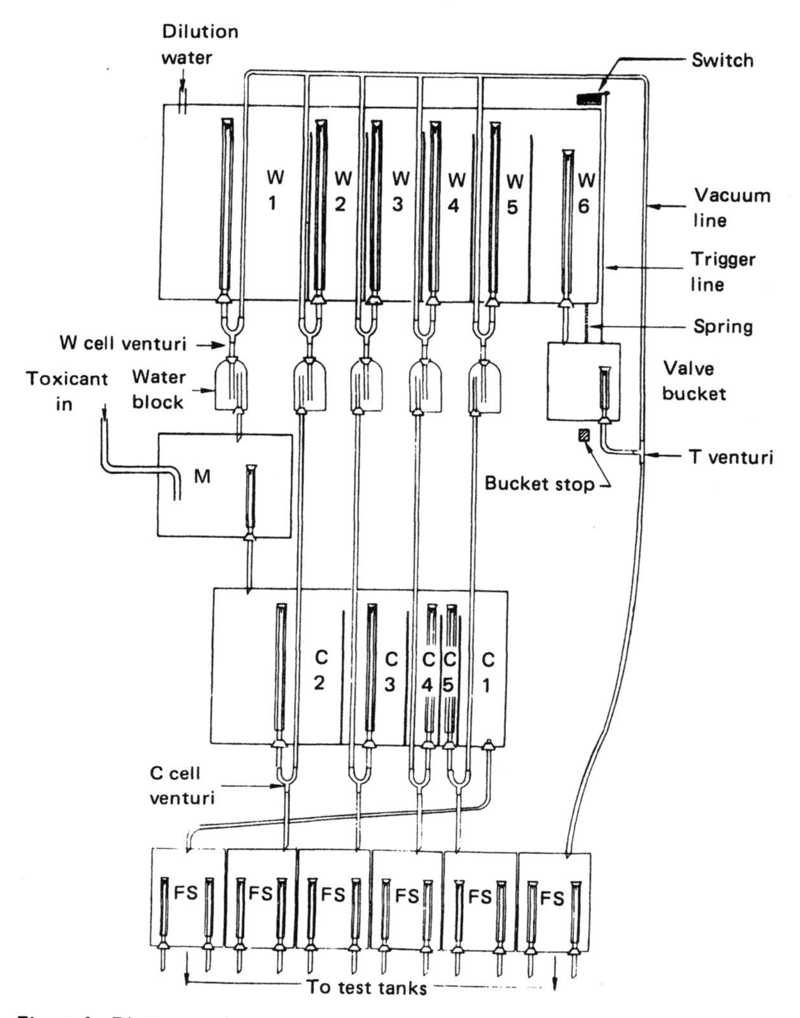

Figure 1 Diagrammatic representation of a proportional diluter designed to deliver serial concentrations of a chemical in clean dilution water to replicate test aquaria. Abbreviations: W1-W6, dilution water cells; M, mixing chamber in which the chemical is mixed with dilution water to produce a concentration equivalent to the highest to be tested; C1-C5, chemical cells, each of which contains the same concentration of the chemical as the other cells and the chemical mixing chamber; FS, flow-splitting cells, which divide the combined volume from the complementary water and chemical cells (e.g., W2 + C2) into two or more equal flows to the replicate test aquaria. *[Modified from Lemke et al. (1978).]*

the determination of the flow rate through each test chamber and measurement of either the concentration of toxicant in each test chamber or the volume of solution used in each portion of the delivery system. The general operation of the system should be checked at least daily during a test.

Experimental Design

The minimum number of test animals to be exposed to each treatment is 10 for a static test and 20 for a flow-through test. Depending on size and requirements of the test species, animals may be divided between two or more test chambers. Use of more organisms and replicate test chambers for each treatment is generally recommended, but the effort (and cost) necessary to test extremely large numbers of animals may not be worth the diminishing confidence in the data obtained (Jensen, 1972).

Distribution of test animals in the test chambers is important. A representative subsample of the test organism population should be impartially distributed, by adding either 2 animals (if there are to be 10 or fewer animals per chamber) or 4 animals (if there are to be more than 10 animals per chamber) to each chamber, and then adding 2 or 4 more, and repeating the process until each chamber has the correct number of test organisms in it. For very small organisms (e.g., daphnids or copepods) it is often convenient to place animals

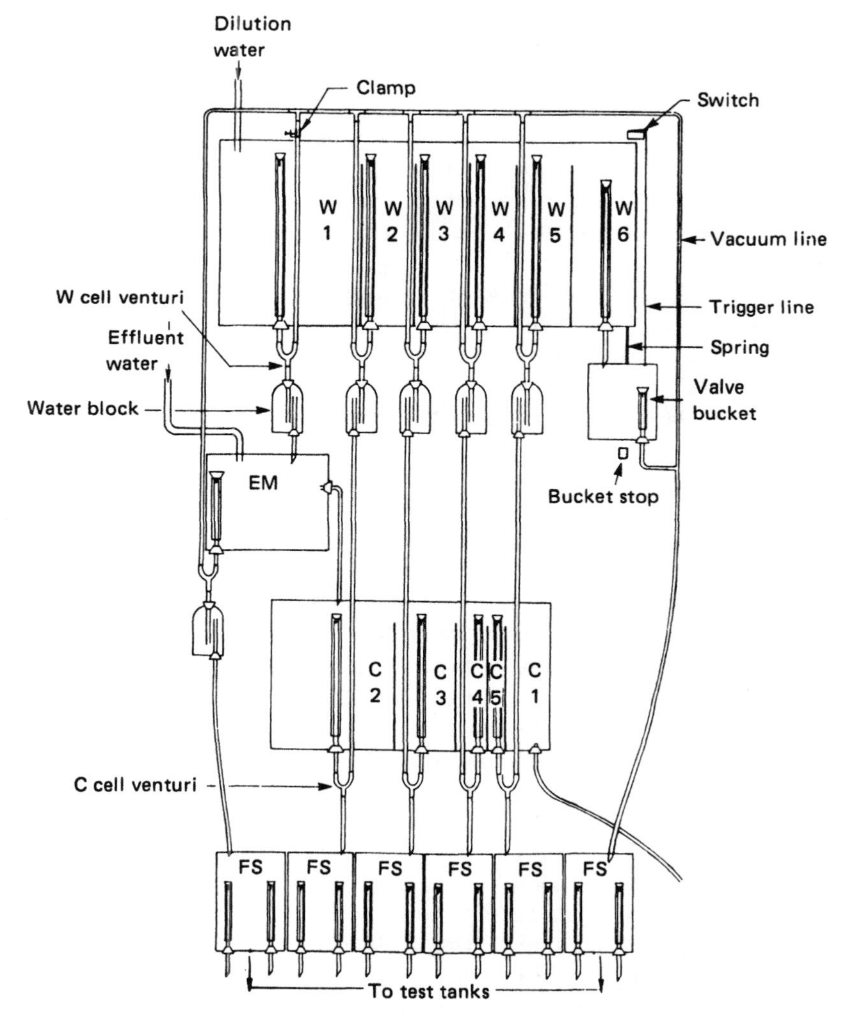

Figure 2 Diagrammatic representation of a proportional diluter designed to deliver serial concentrations of an effluent in clean dilution water to replicate test aquaria. Abbreviations: W1–W6, dilution water cells; EM, effluent mixing chamber; C1–C5, effluent cells; FS, flow-splitting cells. *[Modified from Lemke et al. (1978)]*

in interim containers during the distribution process and then gently add them to the test chambers as a group.

Weights and lengths or widths of test animals should be determined by measuring a group of representative animals from the test population immediately before the test or by measuring the control animals immediately after the test. If the former practice is followed, the animals that were weighed and measured should not be used in the test.

Every test requires a control. The control is a treatment consisting of the same dilution water, conditions, procedures, and organisms as used in the remainder of the test, except that no test material is added. If a solvent/carrier or other additive is used, a solvent/carrier or additive control should be maintained. This additional control is the same as the dilution water control except that the greatest volume of solvent/carrier or additive present in any other test chamber is added to this treatment. A test is generally not considered acceptable if more than 10% of the animals die in any control in a test to determine an LC50 or show the effect in a test to determine an EC50.

Test Conditions

Dissolved Oxygen Concentration

Test solutions in the test chambers or in the test material delivery system should not be aerated. If the dilution water is aerated before it enters the test chamber or delivery system (static or flow-through) and proper loading is ensured (discussed below), the dissolved oxygen concentration should remain ⩾60% of saturation during an entire flow-through test and the first 48 h of a static test and ⩾40% of saturation after 48 h in a static test. In situations where the dissolved oxygen concentration decreases substantially, test animals should be carefully observed for signs of stress. In tests with materials of known high biochemical oxygen demand (e.g., effluents), it is often useful to set up an additional test chamber containing the highest concentration of test material but no test animals. The effect of the test material alone on the dissolved oxygen concentration can be monitored in this treatment and a judgment can be made at the conclusion of the test about the relative contribution of low dissolved oxygen concentration to toxicity. Alternatively, one can test materials known to deplete the dissolved oxygen concentration under both aerated and unaerated conditions in order to determine the toxicity of the test material in adequate and low dissolved oxygen concentrations.

Test Temperature

Test temperatures generally recommended are from the series 7, 12, 17, 22, and 27°C. The derivation of these increments is interesting—they come from the temperatures of well water at major industry and government freshwater laboratories. Whatever temperature is chosen, the actual test temperature should not instantaneously deviate from the selected test temperature by more than 2° during the test. The mean temperature during the test should be within ±1° of the selected temperature.

For flow-through tests with effluents, the test temperature should be between the daily low and the daily high temperature of the receiving water measured just outside the zone of influence of the effluent at the time of the test. Depending on the purpose of the tests, the temperature may remain constant or may fluctuate within the same temperature range as the receiving water.

Loading

The grams of test animals (biomass) per liter of test solution in the test chambers should not be great enough to unduly influence the results of the test. The loading must be low enough to ensure that the concentrations of dissolved oxygen and toxicant are not decreased below acceptable limits, that the concentrations of metabolic products are not increased above acceptable limits, and that crowding does not stress the test animals. Maximum recommended loading is generally 0.5–0.8 g/l for static tests and 1–10 g/l of test solution passing through the test chamber in 24 h for flow-through tests, based on the test temperature (ASTM, 1980a).

Test Procedures

Range-finding Test

When test materials of unknown toxicity are tested, it is almost always necessary to conduct a range-finding test to determine the concentrations of test material that should be used in the full-scale definitive test. A range-finding test requires less effort than a definitive test, but the results of a good range-finding test go far toward ensuring success in the definitive test. Generally, groups of 3–5 animals are exposed to at least three test material concentrations spaced at order-of-magnitude intervals based on a logarithmic ratio (e.g., 0.01, 0.1, 1, 10, and 100 mg/l or ppm); a control is maintained concurrently. If the material is an effluent, the test concentrations are set up as a percentage of the effluent in the test concentration by volume (e.g., 0.01, 0.1, 1, 10, and 100%). Ideally, the duration of the range-finding test is the same as that planned for the definitive test, but in certain instances (if all acute toxic effect has been exerted) the test may be terminated. It is emphasized, however, that the range-finding test serves its function best when it most closely follows the conditions planned for the definitive test.

Definitive Test

For the determination of an LC50 or an EC50, the appropriate number of test animals should be exposed to at least five concentrations of test material or effluent in a geometric progression (that is, the sequence of concentrations should be such that the ratio of a concentration to its predecessor is always the same, as in 10, 5, 2.5, 1.25, and 0.62 ppm). Use of more than five concentrations may provide additional data and allow calculation of narrower confidence limits around the median effect concentration, but the effort (cost) to set up and monitor the extra concentrations should be weighed against the potential gain.

Ideally, a definitive test should meet the following criteria so that an LC50 or EC50 and reasonable confidence limits can be calculated:

- Each test material concentration should be at least 50% of the next higher test material concentration (for example, 60, 30, 15, 7.5, and 3.75 ppb).
- Fewer than 35% of the test animals in one treatment (other than the control) should have been killed or affected, and more than 65% of the test animals in one treatment should have been killed or affected. It is best to have several test concentrations in which some animals are killed or affected (ASTM, 1980a). Obviously, the effect should be related to the concentration of test material.

Beginning a Test

Static tests are begun by (1) adding the appropriate amount or volume of test material to the dilution water in the test chambers, (2) stirring the mixture, (3) making necessary chemical and physical measurements (discussed in detail below), and (4) adding the test animals. The animals should be placed in the test chambers within 1 h after addition of the test material.

Flow-through tests are begun by following essentially the same procedure, except that the test animals are placed in the chambers after the test solutions have been flowing through the exposure apparatus long enough to ensure equilibrium within the system and relatively constant concentrations of test material. Equilibrium can be verified by chemical analyses of test material in water samples from test chambers, if methods are available.

Feeding

Test organisms should not usually be fed while in the test chambers. Minimal feeding may be necessary, however, if (1) cannibalistic animals cannot be separated or restrained, or (2) the test duration constitutes a relatively large portion of the organism's life span. If feeding is necessary, all uneaten food should be removed so that dissolved oxygen and test material concentrations will not be reduced.

Duration

A test begins at the time the test animals are first placed in the test material or control. The normal duration of the acute toxicity test is 96 h, but some animals (daphnids, oyster larvae, and others) are exposed for only 48 h because of problems associated with longer exposure times or because the intent is to test up to a specific point in the developmental process of the animal.

Biological Data

The number of dead or affected animals in each test chamber should be counted every 24 ± 1 h from the beginning of the test until the termination. Death is the criterion for effect most often used in acute toxicity tests to estimate the LC50. The criteria for death are usually lack of movement and lack of reaction to gentle prodding. Since death is not easily determined for some invertebrates, effects such as immobilization and loss of equilibrium are used to estimate the EC50. Other effects can be used to determine an EC50, but the effect and its definition must always be reported. During all tests, general observations of erratic swimming, loss of reflex, discoloration, changes in behavior, excessive mucus production, hyperventilation, opaque eyes, curved spine, hemorrhaging, molting, and cannibalism should be quantified and reported.

Chemical and Physical Data

If a freshwater dilution water is used, the hardness, alkalinity, pH, suspended particulate matter, and specific conductance should be measured. If a saltwater dilution water is used, its salinity, suspended particulate matter, and pH should be measured. The dissolved oxygen concentration should be measured at the beginning of the test and every 24 ± 1 h thereafter until the end of the test in all treatments in which there are living test animals. The pH should be measured at least at the beginning and end of the test in all treatments. Temperature, salinity or hardness, alkalinity, and specific conductance in the control should be measured and recorded daily.

If possible (if methods are available and costs are not prohibitive), the concentration of test material should be measured at least at the beginning and end of the test in all test chambers, especially in flow-through toxicity tests. Measurements at 24-h intervals are recommended. When test material concentrations are not measured in flow-through tests, the usefulness of the technique may be greatly diminished because the results of this test may lead to a need for further testing. The analytical method for measuring test material concentrations should be validated before the test. The accuracy of the method can be determined by the method of known additions, using dilution water from the tanks that contain test organisms or from the diluter. An analytical chemist should provide detailed information about the efficiency of the method. It is recommended, however, that standard solutions corresponding to the low, middle, and high concentrations plus a control (blank) be analyzed in determining percentage recovery of a method. This procedure allows quantification of the efficiency of the analytical method and the results assist in interpreting biological results. Generally, atomic absorption spectrophotometric methods are used for metals and gas chromatographic methods for organic test materials; colorimetric methods are not usually recommended (ASTM, 1980a). Techniques for chemical analyses are discussed in Chapter 16.

When water samples are taken from test chambers for chemical analyses, it is important that they be representative of the concentration to which the test animals are being exposed. Samples should be taken midway between the top, bottom, and sides of the test chambers and should not include surface scum or material stirred up from the bottom or sides. Samples should be treated immediately and appropriately to preserve the material being tested and the analytical method being used.

Data Analysis and Interpretation

Calculations

Based on the data from a test, time-dependent LC50s or EC50s and their 95% confidence limits should be calculated for each 24-h period by using the average measured concentrations of test material for flow-through tests, if concentrations are measured. If test material concentrations are not measured, the nominal (intended) initial concentrations should be used. For effluents, the calculated initial volume percentage of the effluent in the dilution water should be used. A variety of statistical methods can be used to calculate an LC50 or EC50, but the most widely used are the probit (Finney, 1971), logit, moving average, and Litchfield-Wilcoxon (Litchfield and Wilcoxon, 1949) methods. Some of these methods are discussed in Chapter 5. The percentage of test organisms that die or display the effect in the control treatment must not be used in calculating the results. It is not recommended that corrections be made for control mortality or control effect, but Abbott's formula (Abbott, 1925) may be used to do so, if necessary.

According to Stephan (1977), almost all concentration-mortality data generated during acute tests can be analyzed to produce a statistical "best estimate" of the LC50 and its 95% confidence limits by using the moving average method and logarithmic transformation of the concentration data. The exception is when there are no so-called partial kills—that is, when none of the test concentrations have test animal mortality $> 0 < 100\%$. The binomial method can be used in such cases to estimate the LC50 as follows: $LC50 = (AB)^{1/2}$, where A is the highest test material concentration in which none of the test animals were killed and B is the lowest concentration in which all the animals were killed. If a logarithmic transformation is not used, the formula $LC50 = (A + B)/2$ will give the same result as the moving average. Although the 95% confidence limits cannot be calculated in the absence of test material concentrations in which test animal mortality was $> 0 < 100\%$, the confidence limits are generally between A and B. The level of confidence associated with these limits depends on the number of animals in each treatment and can be calculated from the formula: confidence level $= 100[1 - 2(\frac{1}{2})N]$, where N is the number of animals. Thus, if five or fewer animals are used in each treatment the confidence level is less than 95%, but if six or more animals are used per treatment the confidence level represented by A and B is always greater than 95%.

Ecological Significance of Test Results

Although laboratory toxicity tests are conducted under "unnatural" conditions, the results are generally accepted as a conservative estimate of the potential effects of test materials in the field. Laboratory tests may give conservative estimates of field effects because of the absence of mitigating factors (sunlight, sediment, microbes) that would decrease the toxicity of the test material under "natural" conditions. There is great interest in conducting more environmentally realistic tests in the laboratory, and the ultimate goal is to conduct tests in the field to validate the results of laboratory tests.

As stated by Butler et al. (1977), any detectable and measurable response of organisms in a toxicity test should not be interpreted as an ecologically significant event. The effect must be related to the expected environmental concentration (Cairns et al., 1978) if one is to make a judgment about the risk posed by a material in the environment and thus estimate its ecological significance.

EXAMPLES OF TEST DATA

It is essential in any acute toxicity test that orderly, legible, and complete records be maintained. Information about the test material, test conditions, test animals, test procedures, physical and chemical measurements, biological results, date(s) of testing, and person(s) who performed the test should be permanently recorded, either in a laboratory notebook or on looseleaf sheets that are assembled into a project or test file. Toxicity test data sheets such as the ones shown in Figs. 3 and 4 have proved effective in data recording for static acute tests. They "force" the experimenter to provide all the information about the test because all blanks must be completed. Such a format is highly recommended. These forms are only one example of the types of forms that should comprise a complete quality assurance (QA) package for any aquatic toxicity test. Data sheets for flow-through acute tests are similar, except that additional information is included on the functioning and operation of the flow-through system.

After a test has been completed, it is important to present the data in tabular form. The data in Table 2 are the results of chemical analyses performed during a 96-h flow-through acute test in which sheepshead minnows (*Cyprinodon variegatus*) were exposed to sodium selenite. Similarly, the mortality (or other effect criterion) data can be grouped as shown in Table 3. Based on the mortality data, time-dependent LC50s can be calculated by one of the statistical methods previously described (see Table 4). Finally, physical or chemical data can be assembled as shown in Tables 5 and 6.

In the absence of a calculator or computer program to statistically estimate the LC50 (or EC50) value and confidence limits, the LC50 can be determined by graphical interpolation (APHA, AWWA, and WPCF, 1981). The mortality data from Table 3 have been used to plot the graph in Fig. 5. The LC50 value can be interpolated from the percentages of test organisms dying in two or more concentrations–in one less than 50% mortality and in another more than 50% mortality. Estimation of the LC50 by interpolation involves plotting the mortality data on semilogarithmic coordinate paper with concentrations (average measured) of sodium selenite on the log scale and percent mortality on the arithmetic scale. A straight line may be drawn between the two points that represent the percent mortality at the two successive concentrations that were lethal to more than 50% and less than 50% of the test organisms. The point at which the line crosses the 50% mortality line is the LC50 value for the test. Figure 5 illustrates the use of this procedure (straight-line graphical interpolation) to determine the LC50s for the 48-, 72-, and 96-h exposures of sheepshead minnows to sodium selenite. The LC50 values for these exposures are 18, 11, and 8 ppm, respectively. The 72- and 96-h LC50s are different from the calculated values but within the 95% confidence limits. The graphically interpolated 48-h LC50 does not fall within the calculated 95% confidence limits. This can be explained by the fact that only two data points were used in the graphical interpolation but all five data points were used to statistically calculate the LC50s. An alternative to graphical interpolation is to plot the data on logarithmic-probability paper with toxicant concentrations on the logarithmic scale and percent mortality on the probability (probit) scale.

METHODOLOGY FOR PHYTOPLANKTON

Phytoplankton (microalgae) form the basis of food chains in the aquatic environment. These organisms store energy through photosynthesis and thus are the primary producers for the food webs in the aquatic environment. It is important that optimum conditions exist for their growth and reproduction because changes in abundance or diversity may affect the animals that use algae as a primary source of food. Furthermore, algal density and condition can affect the dissolved oxygen concentration, pH, color, alkalinity, clarity, and taste of surface waters. Therefore, the effects of potentially toxic test materials on algal cells should be of interest.

Since algal species and communities are sensitive to environmental changes, growth may be either inhibited or stimulated by the presence of various chemicals. Therefore, the response of algae must be considered when assessing the potential ecological effects of chemicals and other toxic agents on the aquatic environment.

The purpose of an algal toxicity test is to

EG&G BIONOMICS

QUALITY ASSURANCE FORM

NO. BW-QAF-001
PAGE SIDE 1
REVISION #3
EFFECTIVE 10 June 1981

SUBJECT: STATIC ACUTE TOXICITY TEST DATA SHEET

FISH SPECIES	FISH SOURCE	LENGTH AND WEIGHT	IN THE 48 HOURS PRIOR TO TESTING
____________	TANK ____________	MEAN RANGE L. ____________ (mm)	1. % MORTALITY WAS ____
____________	LOT NO. ____________	W. ____________ (g)	2. FISH WERE ☐ NOT FED ☐ FED
____________	STATE ____________	DATA FOUND ON PAGE ______ VOL: ______ / DATE MEAN DETERMINED	3. TEMPERATURE RANGE IN TANK WAS: ____________
DATA TAKEN FROM MONTHLY FISH CULTURE SHEET			

WATER BATH ID	REFERENCE TEST DATA	DATA IN REFERENCE TEST LOG	SIGNATURE	INITIALS
CLIENT STUDY NO.	LC50 ________ mg/l	DATE ________		

COMMENTS:

pH METER USED: D.O. METER USED:

OBSERVATION KEY

NONE :- OBSERVATION WAS MADE AND NOTHING OUT OF THE ORDINARY WAS OBSERVED

GENERAL BEHAVIOR		SWIMMING		RESPIRATION	
AS	AT THE SURFACE	ERR	ERRATIC	RA	RAPID
MSP	MUSCLE SPASM	PIGMENTATION		RE	REDUCED
CLE	COMPLETE LOSS OF EQUILIBRIUM	LT	LIGHT	SOLUTION	
PLE	PARTIAL LOSS OF EQUILIBRIUM	DRK	DARK	CLDY	CLOUDY
LETH	LETHARGIC	INTEGUMENT		PRE	PRECIPITATE
PFAE	PECTORAL FINS ANTERIORLY EXTENDED	EMP	EXCESSIVE MUCUS	FOS	FILM ON SURFACE
EXO	EXOPHTHALMUS	HEM	HEMORRHAGIC	UN	UNDISSOLVED CHEMICAL
EA	EXTENDED ABDOMEN			PM	PARTICULATE MATTER
HYP	HYPERACTIVE				

(*a*)

Figure 3 Quality assurance form for static acute toxicity tests with fish. This is one of the series of 10–15 forms that comprise a typical QA package for any test. (*a*) Species identification, source and measurements of test fish and general observations. *[Courtesy of EG&G Bionomics Aquatic Toxicology Laboratory (1981)]*

SAMPLE IDENTIFICATION	SPONSOR	SAMPLE LOT. NO.	% ACTIVITY	TIME: CHEMICAL/FISH ADDED	PRINCIPAL INVESTIGATOR

PROJECT NO.	DILUTION WATER: INFO BELOW NOT RAW DATA			
FOUND IN: RECON WATER QUALITY NOTEBOOK VOL:__ PG:____	SOURCE	pH	TOTAL ALKALINITY (mg/l $CaCO_3$)	
			CONDUCTIVITY (μmhos/cm)	
	BATCH NO.	SALINITY ‰	TOTAL HARDNESS (mg/l $CaCO_3$)	

TYPE TEST PERFORMED	TEST SYSTEM
☐ PRELIMINARY ☐ DEFINITIVE	☐ OPEN ☐ CLOSED
NUMBER OF ANIMALS PER JAR	NUMBER OF REPLICATES PER CONC.
3 5 10 20	0 2 4 THIS IS REP.______
TEST CHAMBER VOLUME (LITERS)	DILUENT VOLUME (LITERS)
3.8 19.6	3 15

STOCK CONC	SOLVENT USED		TEST CONCENTRATIONS											
PREPARED BY		DATE PREP.	% mg/l μg/l HIGH → LOW	CONT	SOLV. CONT									
CALCULATION APPROVED BY		DATE	AMOUNT OF CHEMICAL/STOCK ADDED ( )	N.A.	N.A.									
			AMOUNT OF SOLVENT ADDED (ml)	N.A.			N.A.	N.A.	N.A.	N.A.	N.A.	N.A.	N.A.	N.A.
CIRCLE ONE: BG, CCF, Fhm, RBT							96 HOUR NO DISCERNIBLE EFFECT CONC.							

	0-HOUR				24-HOUR				48-HOUR				72-HOUR				96-HOUR			
DATE			CONT. TEMP. °C				CONT. TEMP. °C				CONT. TEMP. °C				CONT. TEMP. °C				CONT. TEMP. °C	
TIME																				
DATA BY																				
TEST CONC. % mg/l μg/l	OBSERVATION	NO. DEAD	pH	DO	OBSERVATION	NO. DEAD / CUM	pH	DO	OBSERVATION	NO. DEAD / CUM	pH	DO	OBSERVATION	NO. DEAD / CUM	pH	DO	OBSERVATION	NO. DEAD / CUM	pH	DO
CONTROL																				
SOLV. CONT.																				

(*b*)

Figure 3 Quality assurance form for static acute toxicity tests with fish. This is one of the series of 10–15 forms that comprise a typical QA package for any test. (*Continued*) (*b*) Characteristics of dilution water quality, verification of test chemical solution preparation, observations of fish mortality, and water quality measurements (temperature, pH, and dissolved oxygen) made during the toxicity test. *[Courtesy of EG&G Bionomics Aquatic Toxicology Laboratory (1981)]*

EG&G BIONOMICS

QUALITY ASSURANCE FORM

NO:	BW-QAF-I001
PAGE	Side 1
REVISION	Original
EFFECTIVE	October 1, 1978

SUBJECT: STATIC ACUTE TOXICITY TEST DATA SHEET — INVERTEBRATE LABORATORY

SAMPLE IDENTIFICATION: ______ SPONSOR: ______

SAMPLE LOT NO. ______ % ACTIVITY ______ JOB NO. ______

DENSITY gram/ml ______

PRINCIPAL INVESTIGATORS: ______

TIME ADDED CHEMICAL/ANIMALS	NO. OF ANIMALS PER JAR	NO. OF REPLICATES PER CONCENTRATION	TYPE TEST VESSEL	TEST SYSTEM USED
	5, 15, ______	0, 3, ______		open closed

TEST CHAMBER VOLUME (Liters)	TOTAL SOLUTION VOLUME (ml)	SPECIES	SOLUTION VOLUME PER CHAMBER (Liters)	AGE/SIZE
0.25, 1, 2, ______	500, ______		0.15,0.5,______	

CHEMICAL QUALITY OF DILUTION WATER: Information below is not raw data

DATA TRANSCRIBED NOTEBOOK ______	SOURCE BATCH NO. ______	TOTAL ALKALINITY (mg/l $CaCO_3$)
PAGE NO. ______	pH ______	TOTAL HARDNESS (mg/l $CaCO_3$)
LOCATION ______	SALINITY ______	CONDUCTIVITY (μmhos/cm)

COMMENTS ______

NO DISCERNIBLE EFFECT LEVEL AT 96 HOURS IS ______. DETERMINED BY:

OBSERVATIONS KEY

D - Daphnia magna
OS - ON SURFACE
OB - ON BOTTOM
LETH - LETHARGIC
ERR - ERRATIC SWIMMING
FC - FLARED CARAPACE
SC - SWIMMING, CARRYING
CO - CAUGHT ON
CLDY - CLOUDY SOLUTION
PRE - PRECIPITATE
UM - UNDISSOLVED MATERIAL
PM - PARTICIPATE MATTER (EFFLUENTS)
F - FILM

SIGNATURE	INITIAL

(a)

Figure 4 Quality assurance form for static acute toxicity tests with aquatic macroinvertebrates. This is one of the series of 10–15 forms that comprise a typical QA package for any test. (*a*) Test chemical characteristics, test design, dilution water quality characteristics, and general observations. *[Courtesy of EG&G Bionomics Aquatic Toxicology Laboratory, 1978.]*

		SAMPLE ID
PRELIMINARY/DEFINITIVE TEST (circle one)		

		0 HOUR				48 HOUR		
	DATE							
	TIME							
	DATA BY							
	REP	CONC	DO	pH	TEMP	DO	pH	TEMP
CONTROL	A							
SOL. CONT.	A				NA			NA
HIGH	A				NA			NA
MIDDLE	A				NA			NA
LOW	A				NA			NA

STOCK CONCENTRATION: ______ SOLVENT: ______

PREPARED BY: ______ DATE PREPARED: ______

TEST CONCENTRATIONS			HIGH → LOW						
mg/l µg/l %	CONT	S. CONT							
AMOUNT OF ( ) STOCK/CHEM ADDED	NA	NA							
AMOUNT OF (ml) SOLVENT ADDED	NA		NA	NA	NA	NA	NA	NA	NA

	0 HOUR			24 HOUR							48 HOUR						
DATE																	
TIME																	
DATA BY																	
REPLICATE	A	B	C	A		B		C		CUM NO. DEAD	A		B		C		CUM NO. DEAD
CONC. % mg/l µg/l	OBSER.	OBSER.	OBSER.	OBSER.	NO. DEAD	OBSER.	NO. DEAD	OBSER.	NO. DEAD		OBSER.	NO. DEAD	OBSER.	NO. DEAD	OBSER.	NO. DEAD	
CONTROL																	
SOL. CONT.																	

(*b*)

Figure 4 Quality assurance form for static acute toxicity tests with aquatic macroinvertebrates. This is one of the series of 10–15 forms that comprise a typical QA package for any test. (*b*) Observations of invertebrate mortality and water quality measurements (temperature, pH, and dissolved oxygen) made during the toxicity test. *[Courtesy of EG&G Bionomics Aquatic Toxicology Laboratory, 1978.]*

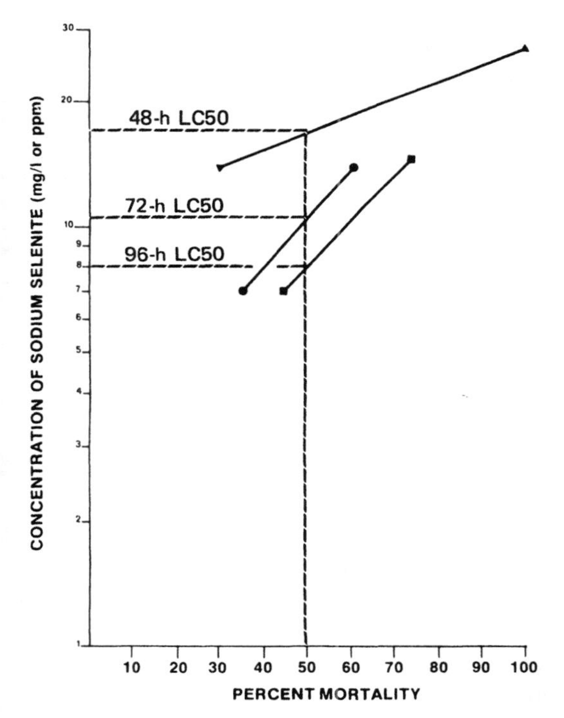

Figure 5 Estimation of time-dependent LC50s for sodium selenite by the graphical interpolation method.

Table 2 Nominal and Measured Selenium Ion (Se^{2+}) Concentrations during a 96-h Exposure of Sheepshead Minnows to Sodium Selenite ($Na_2Se_3 \cdot 5H_2O$) in Flowing, Natural Seawater[a]

Nominal concentration (mg/l)	Measured concentration (mg/l)			Percent of nominal
	0-h	96-h	Average	
Control	ND[b]	0.17	ND	–
1.9	1.9	2.1	2.0	105
3.8	4.0	3.3	3.6	95
7.5	6.6	7.6	7.1	95
15	15	13	14	93
30	26	28	27	90

[a]Salinity was 25 ppt and temperature was 20°C.
[b]None detected.

Table 3 Test Concentrations and Mortality of Sheepshead Minnows ($n = 20$) Exposed to Sodium Selenite in Flowing, Natural Seawater[a]

Average measured concentration (mg/l)	Mortality (%)			
	24-h	48-h	72-h	96-h
Control	0	0	0	0
2.0	0	0	0	0
3.6	0	5	10	20
7.1	0	25	35	45
14	10	30	60	75
27	40	100	100	100

[a]Salinity was 25 ppt and temperature was 20°C. Concentrations are given as selenium ion (Se^{2+}) in seawater.

determine the relative toxicity of test materials to representative algal species. Algal toxicity may be expressed in terms of a broad range of responses, encompassing those that are inhibitory (50% reduction in cell numbers at specified time intervals), algistatic (halting of cell division), and algicidal (cell death). All three degrees of algal toxicity can be determined by the algal toxicity test.

General Methods

Algal toxicity tests are conducted according to the same general principles discussed earlier for fish and invertebrate testing. Cultures of a unicellular alga of known age and density are exposed to a range of test material concentrations in flasks

Table 4 Calculated LC50s of Sodium Selenite to Sheepshead Minnows Exposed in Flowing, Natural Seawater[a]

Time (h)	LC50 (mg/l)	95% confidence limits (mg/l)
24	>27	–
48	8.4	5.2–14
72	7.7	5.4–11
96	7.0	5.2–9.5

[a]Salinity was 25 ppt and temperature was 20°C. Calculations were based on measured concentrations of selenium ion (Se^{2+}) in seawater.

Table 5 Measured Concentrations of Dissolved Oxygen During a 96-h Exposure of Sheepshead Minnows to Sodium Selenite in Flowing, Natural Seawater[a]

Mean measured concentration (mg/l; ppm)	Dissolved oxygen concentration, mg/l (% of saturation)			
	24-h	48-h	72-h	96-h
Control	5.4 (83)	5.4 (83)	5.2 (80)	6.4 (98)
2.0	5.6 (86)	5.4 (83)	5.5 (85)	6.2 (95)
3.6	5.7 (88)	5.4 (83)	5.4 (83)	6.3 (97)
7.1	5.7 (88)	5.5 (85)	5.3 (82)	6.2 (95)
14	5.8 (89)	5.6 (86)	5.6 (86)	6.4 (98)
27	5.9 (91)	5.5 (86)	—[b]	—

[a]Salinity was 25 ppt and temperature was 20°C. Dissolved oxygen concentrations are averages of measurements in replicate test chambers A and B. Concentrations are given as selenium ion (Se^{2+}) in seawater.
[b]No measurement because all fish had died.

under defined static conditions in a temperature- and photoperiod-controlled environment. An untreated control flask (and/or solvent control) is maintained to measure the normal growth of the alga. The effects of the test materials are evaluated by measuring the growth of treated and untreated cultures and comparing the results. Because of the inherent variability of the test, the experimental design should include sufficient replication to permit statistical evaluation of the results.

General algal culture and exposure conditions including light intensity, photoperiod, temperature, pH, growth medium composition, and incubation are carefully controlled and are detailed in standard methods (APHA, AWWA, and WPCF, 1981). Test algae may be selected from those in Table 7. Several algal species should be tested to determine the relative toxicity of the test material. Initiation of an algal toxicity test consists of two major phases, the preparation of the inoculum in the log growth phase and the preparation of test solutions.

Inoculum

Stock cultures of algal test species should be grown on suitable culture medium through at least

Table 6 Measured pH of Test Solutions during a 96-h Exposure of Sheepshead Minnows to Sodium Selenite in Flowing, Natural Seawater[a]

Mean measured concentration (mg/l)	pH							
	24-h		48-h		72-h		96-h	
	A	B	A	B	A	B	A	B
Control	8.0	8.1	8.2	8.1	8.1	8.1	8.2	8.2
2.0	8.1	8.2	8.1	8.2	8.1	8.2	8.2	8.2
3.6	8.1	8.2	8.2	8.2	8.1	8.2	8.2	8.2
7.1	8.2	8.2	8.3	8.2	8.2	8.3	8.2	8.2
14	8.2	8.2	8.3	8.3	8.3	8.3	8.3	8.3
27	8.3	8.3	8.2	8.3	—[b]	—	—	—

[a]Salinity was 25 ppt and temperature was 20°C. Concentrations are given as selenium ion (Se^{2+}) in seawater.
[b]No measurement because all fish had died.

Table 7 Recommended Algae Species for Toxicity Tests

Freshwater
Chlorophyta (green algae)
Selenastrum capricornutum
Cyanophyta (blue-green algae)
Anabaena flos-aquae
Microcystis aeruginosa
Chrysophyta (brown algae and diatoms)
Navicula pelliculosa
Cyclotella sp.
Synura petersenii
Saltwater
Chlorophyta (green algae)
Chlorella sp.
Chlorococcum sp.
Dunaliella tertiolecta
Chrysophyta (brown algae and diatoms)
Isochrysis galbana
Nitzschia closterium
Pyrmnesium parvum
Skeletonema costatum
Thalassiosira pseudonana
Rhodophyta (red algae)
Porphyridium cruentum

two complete growth cycles before use as a source of inoculum for the toxicity test. This is necessary since the nutritional history of the culture can have marked effects on growth responses.

The inoculum is usually prepared by transferring between 0.1 and 1.0 ml of the stock algal culture into triplicate flasks containing nutrient-enriched saltwater or freshwater, depending on the species being tested. Three new flasks should be inoculated at the point of inflection of the growth curve, that is, at the start of the log growth phase. The second growth curve should be followed and cells from the second or later transfers can be used for the toxicity test because they have become adapted to the ambient nutrient levels. Algae acclimated in this manner will tend to give more reproducible results in toxicity tests. The algal cultures should also be checked to make sure the stock cultures are axenic (unispecies).

Test Solutions

Filter-sterilized, nutrient-enriched water is dispensed into presterilized flasks. An inoculum of alga is added to yield a specific cell density or concentration (number of cells per milliliter). Initial cell density or biomass (dry weight) is measured in an aliquot.

Test material stock solutions are prepared in water, or in solvents if the material is insoluble in water. Stock solutions should be prepared to ensure that the same volume is added to all treatments. Test material is added to the flasks containing the inoculated enriched water and the flasks are placed in an incubator.

Test Design

The general test design is similar to that for fish and invertebrates in that it includes a preliminary range-finding test to select concentrations for the final definitive test. The preliminary test should include concentrations of several orders of magnitude (e.g., 0.1, 1.0, 10, 100, and 1000 mg/l) and duplicate culture flasks should be used at each concentration. The flasks are first inoculated with a unialgal culture and then incubated under standard conditions. Typically, algal toxicity tests are conducted for 96 h. However, protocols are available for evaluations of longer duration. Biomass is determined in each flask and compared with that of the untreated controls. These results are used to determine the test concentrations for the definitive test. The criteria for selecting these concentrations are described in standard methods (APHA, AWWA, and WPCF, 1981) and are similar to those used for fish and invertebrates. The definitive toxicity test should include five or six test material concentrations and an untreated control, with three or four replicates of each. Three replicates may be inoculated with algae while the fourth serves as a blank. Three replicates are needed for statistical analysis, and the blank may be used, if necessary, to correct biomass measurements for particulates present in the treatments. If possible, the five or six test concentrations should include an algicidal concentration, an algistatic concentration, a slightly inhibitory concentration, and a concentration in which growth is similar to that of the control (no observed effect concentration). The flasks should be incubated for 4 d.

Growth should be monitored and measured once every 24 h for the duration of the study to determine the effects of the test material. Ideally, the experiment should continue until the control population of algae can complete its logarithmic growth phase (maximum standing crop) and reach a stationary growth rate (Fig. 6); however, the duration of these tests is usually predetermined to be 96 h.

Some test methods (ASTM, 1981) include a postexposure observation period in which living cells from each test concentration are transferred to growth medium without test material and the rate of their growth is measured. The purpose of the recovery period is to determine the ability of the previously exposed cultures to regain their normal rate of growth.

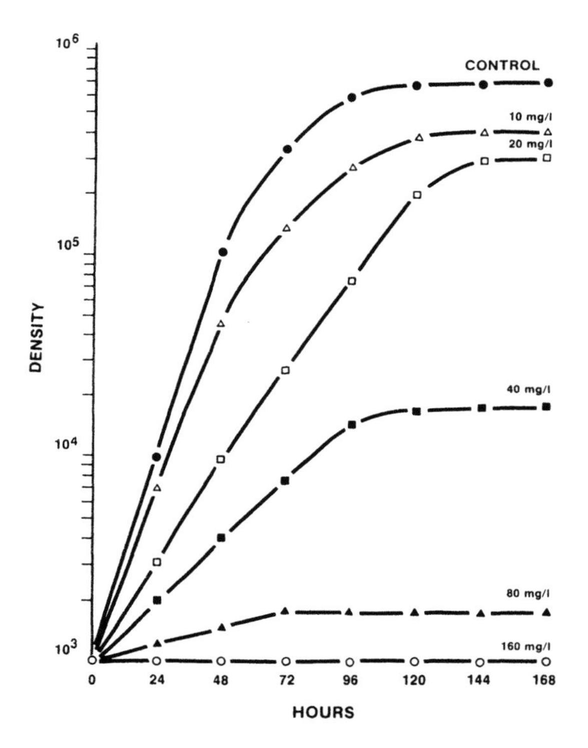

Figure 6 Hypothetical relationship between time-dependent algal growth measured as cell density and concentrations of test chemical to which algal cultures are exposed.

Criteria for Effect

There are a variety of parameters that characterize the growth response of algal cultures. These parameters are measures (or estimates) of biomass of treated and untreated cultures which, when plotted against time, produce a growth response curve (Fig. 6). This curve can be used to determine the rate of log growth and a maximum population density for control and exposed cultures.

Cell Counts

Microscopic measurements of cell numbers can be made with a hemacytometer, Palmer-Maloney chamber, Sedgewick-Rafter plankton counting chamber, or inverted microscope. Some of these methods are discussed by Lund et al. (1958) and Palmer and Maloney (1954).

Direct counting methods are time-consuming and their statistical significance decreases at low cell densities ($<1 \times 10^4$). If large numbers of tests and replicates are needed, it becomes highly impractical to use cell counts for growth assessment of algae. In this case, it is rapid, practical, and reasonably accurate to use an electronic particle counter.

Chlorophyll

All algae contain chlorophyll *a,* and measuring the amount of this pigment can give some insight into the relative amount of algal biomass present. Chlorophyll may be measured *in vivo* or *in vitro* by fluorescence or spectrophotometric techniques. Fluorescent systems are more sensitive and can be used at cell densities $<1 \times 10^4$ cells/ml. Furthermore, the *in vivo* technique is sensitive, accurate and does not require pre-analysis extraction. A limitation of the chlorophyll *a* measurement is that it varies as a function of nutrition and environmental variables. For additional information on this method, see Forenzen (1966, 1967), Moss (1967), Odum et al. (1959), Yentsch and Ryther (1957), and Yentsch and Menzel (1963).

Carbon-14 Assimilation

The method based on carbon-14 assimilation is used for productivity measurements for freshwater and marine algae. It is used as a measure of photosynthetic activity, and it yields a biomass index that correlates well with growth rates. For more information on this technique, see Jenkins (1965), Jitts (1963), and McAllister (1961).

Adenosine Triphosphate Concentration

Adenosine triphosphate (ATP) measurements are a sensitive indicator of living biomass because of the constancy of the cellular ATP/carbon ratio. There is a good correlation between ATP and direct measures of biomass (counting) and labeling with carbon-14. Since ATP is a measure of living biomass, it may not be appropriate for contaminated wastes (e.g., sludge), which may provide interference. For more information on this method, see Hamilton and Holm-Hansen (1967), Holm-Hansen and Booth (1966), and Holm-Hansen (1969).

All the above techniques have certain advantages and disadvantages that depend on the test design, type of test material, and facilities.

Biological Data

Growth effects are most often used to define the toxicity of test materials to algae. Since growth is being measured rather than death, an EC50 is calculated. The EC50 is the test material concentration estimated to cause a 50% reduction in number of cells or chlorophyll *a* concentration relative to the control. Biomass measurements may be estimated after 24, 48, 72, 96, and 120 h of exposure by one or two of the techniques previously discussed (e.g., cell count, chlorophyll *a*). If the range of concentrations of the chemical tested results in average algal cell counts that bracket the 50% response level as compared with the controls, an EC50 value can be estimated. Semilogarithmic coordinate paper can be used to plot the average cell count for a concentration that yielded more than 50% and one that yielded less than 50% of the average cell count of control flasks (Fig. 7).

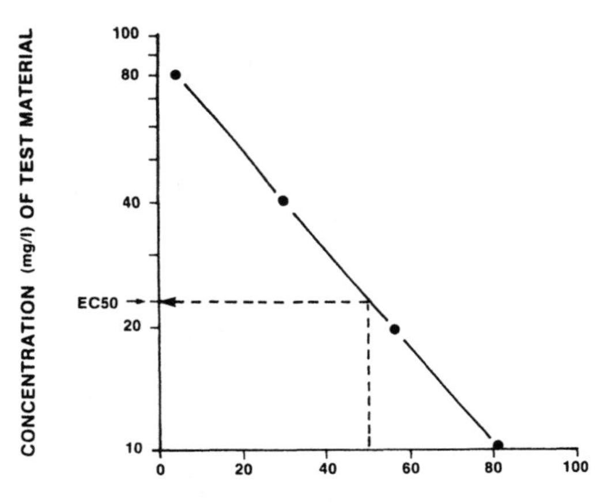

Figure 7 Estimation of EC50 for a chemical based on the degree of growth inhibition of exposed cultures relative to the growth of control culture.

The concentration should be plotted on the logarithmic axis and the percentage of growth in relation to the control on the arithmetic axis. A straight line should be drawn between these two points. The EC50 can be interpolated from the graph by determining the concentration at which this line crosses the 50% point on the percent growth response axis.

Data Analysis and Interpretation

Based on the data from a test, time-dependent EC50s and their 95% confidence limits should be calculated for each 24-h period through test termination. The statistical methods used are similar to those for fish and invertebrates. In some tests, stimulation of algal biomass to levels greater than that of controls may result from exposure to one or more concentrations of a chemical that can serve as a carbon source for the algae. Generally, these stimulatory effects are not used in the calculation of an EC50 since the purpose of the test is to estimate the median adverse effect (i.e., producing a decrease in biomass relative to controls) rather than a median change relative to controls. The results of each test can also include the algistatic concentration with its confidence intervals and the no-observed-effect-concentration. The algistatic concentration is the test material concentration that produces no change in number of algal cells during exposure but allows recovery to logarithmic growth when the cells are transferred to medium without the test material. The no-observed-effect-concentration (or range) is the highest test material concentration that produces no statistically different growth response compared to controls throughout the exposure period. A description should also be included of the test conditions, including culture medium, species, temperature, light intensity, photoperiod, and any analytic procedures used.

SPECIAL CONSIDERATIONS

Safety

Many test materials used in aquatic toxicity tests are biocides. They are intended to kill microbes, plants, insects, rodents, birds, and other organisms. It is extremely important, therefore, to treat all test materials with care. Before any testing is begun, the testing facility should have information on safe handling of the test material, on its known toxicity, and on proper cleanup and disposal methods. This information should be used to decide whether the testing facility is able to test the material safely. It may also provide for the assignment of a hazard rating and for the stipulation of special precautions necessary to handle or test the material (e.g., wearing a respirator while working with the technical chemical). Finally, and most important, these safety data must be conveyed to the scientists or technicians who will perform the test.

Disposal

Closely related to safe handling of test material in the laboratory is proper disposal of the unused material and stock solutions and the contaminated animals and water. A testing facility must meet legal requirements for discharge of test materials into sewers, on-site waste treatment systems, deep wells, or surface waters. Fortunately, because of the recent emphasis on "cradle-to-grave" custody of chemicals, commercial waste handling and storage contractors are becoming more plentiful and, we hope, reliable. Such services may be considered for use by the testing facility.

Exotic Organisms

Control of test animals not indigenous to the geographic area of the testing facility is essential. Exotic or nonstandard test animals may be shipped to a facility from another area or even another country. The testing facility must prevent the escape of any living exotic animals. Instances of harm from the importation of plants and animals are numerous.

Good Laboratory Practices and Quality Assurance

No matter how well qualified the scientists and technicians who perform a test, the work will be meaningless if the conditions and results of testing cannot be reconstructed and verified in the future. Thus, the testing facility should be aware of requirements for standard operating procedures, test protocols, record retention, sample custody, and other basics of good laboratory practices and quality assurance. Good laboratory practice regulations for nonclinical laboratory studies have been published by the U.S. Department of Health, Education and Welfare (1978), and requirements for aquatic and environmental toxicity testing have been published (as proposed standards) by the U.S. Environmental Protection Agency (1980).

SUMMARY

Acute toxicity tests constitute only one of the many tools available to the aquatic toxicologist, but they are the basic means of providing a quick, relatively inexpensive, and reproducible estimate of the toxic effects of a test material. They are an indispensable "first-look" method, at least at this point in the development of the science of aquatic toxicology.

Acute tests are also useful in screening large

numbers of chemicals and in evaluating the relative sensitivity of different organisms to the same chemical. Thus they have been highly rated in "their present utility for use in assessing the hazard to aquatic environments" (Macek et al., 1978). They are the first step toward understanding the toxic effects of materials in aquatic systems.

There are limitations of acute tests that should be recognized. The results of acute toxicity tests usually do not provide substantive information about the sublethal or cumulative effects of a test material. Furthermore, these tests are not predictive of potential chronic toxicity (Macek et al., 1978). These limitations should be considered when planning testing programs for specific test materials. The chapters of this book are intended to inform and assist the reader in understanding such testing programs.

LITERATURE CITED

Abbott WS: A method of computing the effectiveness of an insecticide. J Econ Entomol 18:265–267, 1925.

APHA, AWWA, and WPCF: Standard Methods for the Examination of Water and Wastewater, 15th ed. Washington, D.C., 1981.

American Society for Testing and Materials: Standard practice for conducting toxicity tests with fishes, macroinvertebrates, and amphibians. ASTM E 729-80. Philadelphia: ASTM, 1980a.

American Society for Testing and Materials: Standard practice for conducting toxicity tests with larvae of four species of bivalve molluscs. ASTM E 724-80. Philadelphia: ASTM, 1980b.

American Society for Testing and Materials: Proposed standard practice for conducting toxicity tests with freshwater and saltwater algae. ASTM E47.01, Draft. Philadelphia: ASTM, 1981.

Brown VW: Concepts and outlooks in testing the toxicity of substances to fish. In: Bioassay Techniques and Environmental Chemistry, edited by GE Glass, pp. 73–96. Ann Arbor, Mich.: Ann Arbor Science, 1973.

Butler PA, Doudoroff P, Fontaine MA, Fujiya M, Lange R, Lloyd R, Reish DJ, Sprague JB, Swedmark M: Manual of Methods in Aquatic Environment Research. Part 4: Bases for Selecting Biological Tests to Evaluate Marine Pollution. Rome, Italy: U.N. Food and Agricultural Organization, 1977.

Cairns J, Jr, Dickson KL, Maki AW (eds): Estimating the hazard of chemical substances to aquatic life. ASTM STP 657. Philadelphia: American Society for Testing and Materials, 1978.

Committee on Methods for Acute Toxicity Tests with Aquatic Organisms: Methods for Acute Toxicity Tests with Fish, Macroinvertebrates, and Amphibians. EPA-660/3-75-009, 1975.

Cox, GV (ed): Marine Bioassays Workshop Proceedings 1974. American Petroleum Institute, Environmental Protection Agency, Marine Technology Society. Washington, D.C.: Marine Technology Society, 1974.

Finney DJ: Probit Analyses. London: Cambridge Univ. Press, 1971.

Hamilton RD, Holm-Hansen O: Adenosine triphosphate content of marine bacteria. Limnol Oceanogr 12:319–324, 1967.

Holm-Hansen O, Booth CR: The measurement of adenosine triphosphate in the ocean and its ecological significance. Limnol Oceanogr 11:510–519, 1966.

Holm-Hansen O: Determination of microbial biomass in ocean profiles. Limnol Oceanogr 14: 740–747, 1969.

Jenkins D: Determination of primary productivity of turbid waters with carbon-14. J Water Pollut Control Fed 37:1281–1288, 1965.

Jensen AL: Standard error of LC50 and sample size in fish bioassays. Water Res 6:85–89, 1972.

Jitts HR: The standardization and comparison of measurements of primary production by the carbon-14 technique. In: Proceedings of a Conference on Primary Productivity Measurement, Marine and Fresh Water, Univ. of Hawaii, Aug–Sept 1961, edited by MS Doty, pp. 103–113. U.S. AEC Div Tech Inf TID 7633, 1963.

Lemke AE, Brungs WA, Halligan BJ: Manual for Construction and Operation of Toxicity-Testing Proportional Diluters. EPA-600/3-78-072, 1978.

Litchfield JT, Jr, Wilcoxon F: A simplified method of evaluating dose-effect experiments. J Pharmacol Exp Ther 96:99–113, 1949.

Lorenzen CJ: A method for the continuous measurement of *in vivo* chlorophyll concentration. Deep Sea Res 13:223–227, 1966.

Lorenzen CJ: Determination of chlorophyll and pheopigments: Spectrophotometric equations. Limnol Oceanogr 12:343–346, 1967.

Lund JW, Kipling C, Lecren ED: The inverted microscope method of estimating algae numbers and the statistical basis of estimations by counting. Hydrobiologia 11:143–170, 1958.

Macek K, Birge W, Mayer FL, Buikema AL, Jr, Maki AW: Toxicological effects. In: Estimating the Hazard of Chemical Substances to Aquatic Life, edited by J Cairns, Jr, KL Dickson, AW Maki, pp. 27–32. ASTM STP 657. Philadelphia: American Society for Testing and Materials, 1978.

McAllister CD: Decontamination of filters in the C-14 method of measuring marine photosynthesis. Limnol Oceanogr 6:447–450, 1961.

Moss B: A spectrophotometric method for the estimation of percentage degradation of chlorophylls to pheopigments in extracts of algae. Limnol Oceanogr 12:335–340, 1967.

Mount DI, Brungs WA: A simplified dosing apparatus for fish toxicological studies. Water Res 1: 21–29, 1967.

Odum HT, McConnel W, Abbot W: The chlorophyll "a" of communities. Publ Texas Inst Mar Sci 5:65–95, 1959.

Palmer CM, Maloney TE: A new counting slide for nannoplankton. Am Soc Limnol Oceanogr Spec Publ No 21, 1954.

Stephan CE: Methods for Calculating an LC50. In: Aquatic Toxicology and Hazard Evaluation, Proceedings of the First Annual Symposium on Aquatic Toxicology, edited by FL Mayer and JL Hamelink, pp. 65–84. ASTM STP 634. Philadelphia: American Society for Testing and Materials, 1977.

Tatem HE, Anderson, JW, Neff JM: Seasonal and laboratory variations in the health of grass shrimp *Palaemonetes pugio:* Dodecyl sodium sulfate bioassay. Bull Environ Contam Toxicol 16:368–375, 1976.

U.S. Environmental Protection Agency: Proposed environmental standards; and proposed good laboratory practice standards for physical, chemicals, persistence, and ecological effects testing. Fed Regist 45:77332–77365, 1980.

U.S. Department of Health, Education and Welfare, Food and Drug Administration: Nonclinical laboratory studies. Good laboratory practice regulations. Fed Regist 43:59986–60025, 1978.

Yentsch CS and Ryther JH: Short-term variations in phytoplankton chlorophyll and their significance. Limnol Oceanogr 2:140–142, 1957.

Yentsch CS and Menzel DW: A method for the determination of phytoplankton chlorophyll and phaeophytin by fluorescence. Deep Sea Res 10: 221–231, 1963.

SUPPLEMENTAL READING

Brungs WA: Continuous-Flow Bioassays with Aquatic Organisms: Procedures and Applications. In: Biological Methods for the Assessment of Water Quality, edited by J Cairns, Jr, KL Dickson, pp. 117–126. ASTM STP 528. Philadelphia: American Society for Testing and Materials, 1973.

Brusick DJ, Young RR, Hutchinson C, Vilkas, AG, Gezo TA: Aquatic ecological assays. In: IERL–RTP Procedures Manual: Level 1 Environmental Assessment Biological Tests, pp. 69–112. EPA 68-02-2681. U.S. EPA Office of Research and Development, 1980.

David HS: Culture and Diseases of Game Fishes. Berkeley: Univ. of California Press, 1953.

Duke KM, Davis ME, Dennis AJ: Ecological effects tests. In: IERL–RTP Procedures Manual: Level 1 Environmental Assessment Biological Tests for Pilot Studies. pp. 40–76. EPA-600/7-77-043. Washington, D.C.: U.S. EPA, Office of Research and Development, 1977.

Hoffman GL, Meyer FP: Parasites of Freshwater Fishes. Neptune City, N.J.: TFH Publications, 1974.

Kester E, Dredall I, Conners D, Pytowicz R: Preparation of artificial seawater. Limnol Oceanogr 12:176–178, 1967.

Miller WE, Greene JC, Shiroyama T: *Selenastrum capricornutum* Printz Algal Assay Bottle Test: Experimental Design, Application, and Data Interpretation Protocol. EPA-600/9-78-018. Corvallis, Ore.: U.S. EPA, 1978.

Organization for Economic Cooperation and Development: OECD Guidelines for Testing of Chemicals. Paris: OECD, 1981.

Payne AG, Hall RH: A method for measuring algal toxicity and its application to the safety assessment of new chemicals. In: Aquatic Toxicology, Proceedings of the Second Annual Symposium on Aquatic Toxicity, edited by LL Marking, RA Kimerle, pp. 171–180. ASTM STP 667. Philadelphia: American Society for Testing and Materials, 1979.

Peltier W: Methods for Measuring the Acute Toxicity of Effluents to Aquatic Organisms. EPA-600/4-78-012, 1978.

Reichenbach-Klinke H, Elkan E: The Principal Diseases of Lower Vertebrates. New York: Academic, 1965.

Snieszko SF (ed): A Symposium on Diseases of Fish and Shellfishes. Washington, D.C.: American Fisheries Society, 1970.

Sprague JB: Measurement of pollutant toxicity to fish. I. Bioassay methods for acute toxicity. Water Res 3:793–821, 1969.

Sprague JB: Measurement of pollutant toxicity to fish. II. Utilizing and applying bioassay results. Water Res 4:3–32, 1970.

Sprague JB: Measurement of pollutant toxicity to fish. III. Sublethal effects and "safe" concentrations. Water Res 5:245–266, 1971.

Sprague JB: The ABC's of pollutant bioassay using fish. In: Biological Methods for the Assessment of Water Quality, edited by J Cairns, Jr, KL Dickson, pp. 6–30. ASTM STP 528. Philadelphia: American Society for Testing and Materials, 1973.

Steel RGD, Torrie JH: Principles and Procedures of Statistics. New York: McGraw-Hill, 1960.

U.S. Environmental Protection Agency, Corps of Engineers, Technical Committee on Criteria for Dredged and Fill Material: Ecological Evaluation of Proposed Discharge of Dredged Material into Ocean Waters. Vicksburg, Miss.: Environmental Effects Laboratory, U.S. Army Engineer Waterways Experiment Station, 1977.

U.S. Environmental Protection Agency, Environmental Research Laboratory, Office of Research and Development: Bioassay Procedures for the Ocean Disposal Permit Program. EPA-600/9-78-010, 1978.

Van Duijn C, Jr: Diseases of Fishes, 3rd ed. Springfield, Ill.: Thomas, 1973.

Appendix B

EARLY LIFE STAGE TOXICITY TESTS

J. M. McKim

INTRODUCTION

The life cycle toxicity test is considered by most aquatic toxicologists to be the ultimate test in establishing long-term "safe" environmental concentrations of toxic chemicals for both vertebrate and invertebrate aquatic populations.

The first aquatic vertebrate life cycle toxicity study with fish was performed with the fathead minnow, *Pimephales promelas,* by Mount and Stephan (1967); this was followed by studies with the bluegill, *Lepomis macrochirus* (Eaton, 1970); the brook trout, *Salvelinus fontinalis* (McKim and Benoit, 1971, 1974); the flagfish, *Jordanella floridae* (Smith, 1973); and the only saltwater species to date, the sheepshead minnow, *Cyprinodon variegatus* (Hansen et al., 1977; Hansen and Parrish, 1977). The development of life cycle toxicity tests for invertebrates has fallen behind that for fish, mainly because of lack of emphasis and low research funding. However, a number of important invertebrates can be used in life cycle toxicity tests (Buikema and Cairns, 1980). These include the water flea, *Daphnia magna* (Biesinger and Christensen, 1972); the amphipod *Gammarus pseudolimnaeus* (Arthur, 1970); the gastropods *Physa integra* and *Campeloma* sp. (Arthur, 1970); and the midges *Chironomus tentans* (Derr and Zabik, 1972a,b) and *Tanytarsus dissimilis* (Nebeker, 1973). Saltwater invertebrates are represented by the opossum shrimp, *Mysidopsis bahia* (Nimmo et al., 1977); the grass shrimp, *Palaemonetes pugio* (Tyler-Schroeder, 1979); and the polychaetous annelids *Neanthes arenaceodentata, Capitella capitata,* and *Ctenodrilus serratus* (Reish, 1980).

A life cycle test demands a minimum laboratory exposure of the animal from "embryo to embryo," which for many animals, especially vertebrates (fish), requires a minimum of 6-12 mo of concentrated effort. The Toxic Substances Control Act of 1976, which required the Environmental Protection Agency (EPA) and industry to evaluate the environmental impact of new chemicals before commercial production, and the manufacture or marketing of an estimated 1000 new chemicals each year created the need for a more rapid, less costly, and less risky vertebrate test than the fish life cycle test for determining safe environmental concentrations of toxic chemicals.

During life cycle tests with several species of fish and a variety of toxicants, certain developmental stages have consistently been more sensitive than others. The possibility of focusing research efforts on these more sensitive stages promises success in searching for quicker and less costly ways of predicting chronic toxicity of chemicals to fish. Several investigators proposed that chronic toxicity to fish might be predicted by use of shorter tests with early developmental stages. In studies with selected toxicants, these early stages were shown to be among the most sensitive in the life cycle (Pickering and Thatcher, 1970; Pickering and Gast, 1972; McKim et al., 1975, 1978; Eaton et al., 1978; Sauter et al., 1976). It was emphasized that, with a relatively short exposure (several months) of the embryo-larval and early juvenile stages of fish to a toxicant, an estimate of the maximum acceptable toxicant concentration (MATC) could be obtained without a complete life cycle test.

The need for toxicity tests of short duration was not as great for the more routinely tested invertebrates (daphnids and mysids), since most invertebrate life cycle tests require only 1-2 mo. Shorter, partial life cycle tests with certain invertebrates have, in some cases, resulted in less sensitive responses than the complete life cycle or lifetime test (Buikema and Cairns, 1980).

Reviews of life cycle toxicity test data on freshwater fish, from more than 60 chronic tests with more than 40 organic and inorganic chemicals, show that tests with early life stages of four species of fish can be used to estimate the MATC within a factor of 2 in most cases (McKim, 1977; Macek and Sleight, 1977). A comparison of chronic toxicity data for saltwater fish also indicates that early life stage exposures provide a close estimate of the MATC (Ward and Parrish, 1980; D. I. Mount, personal communication). The greater sensitivity of the early life stages compared to the later stages provides aquatic toxicologists with an accurate, efficient tool for predicting chronic effects of environmental pollutants in 1-2 mo of testing. It also makes it feasible to test other important fish species that cannot be studied in life cycle tests because of size or age at maturity, spawning requirements, or other factors inconducive to laboratory culture.

This chapter will provide more details on terminology, fish developmental events, end points used in determining effects, general methodologies employed, and the interpretation and utility of early life stage (ELS) tests with aquatic animals.

TERMINOLOGY

The terminology used in life cycle, partial life cycle, and early life stage aquatic toxicology and in developmental fish biology is reviewed below. For more general aquatic toxicology terminology, the reader is referred to the glossary in this book and to ASTM (1980a).

Aquatic Toxicology

Fish Life Cycle Toxicity Test

Each of several groups of individuals of one species is exposed to a different concentration of a toxicant throughout a life cycle in order to study the effect of the toxicant on the survival, growth, and reproduction of the species. To ensure that all life stages and life processes are exposed, the test begins with embryos or newly hatched young fish less than 8 h old, continues through maturation and reproduction, and ends not less than 28 d (60 d for salmonids) after the hatching of the next generation. Figure 1 shows the chronology of a life cycle test with brook trout which began with

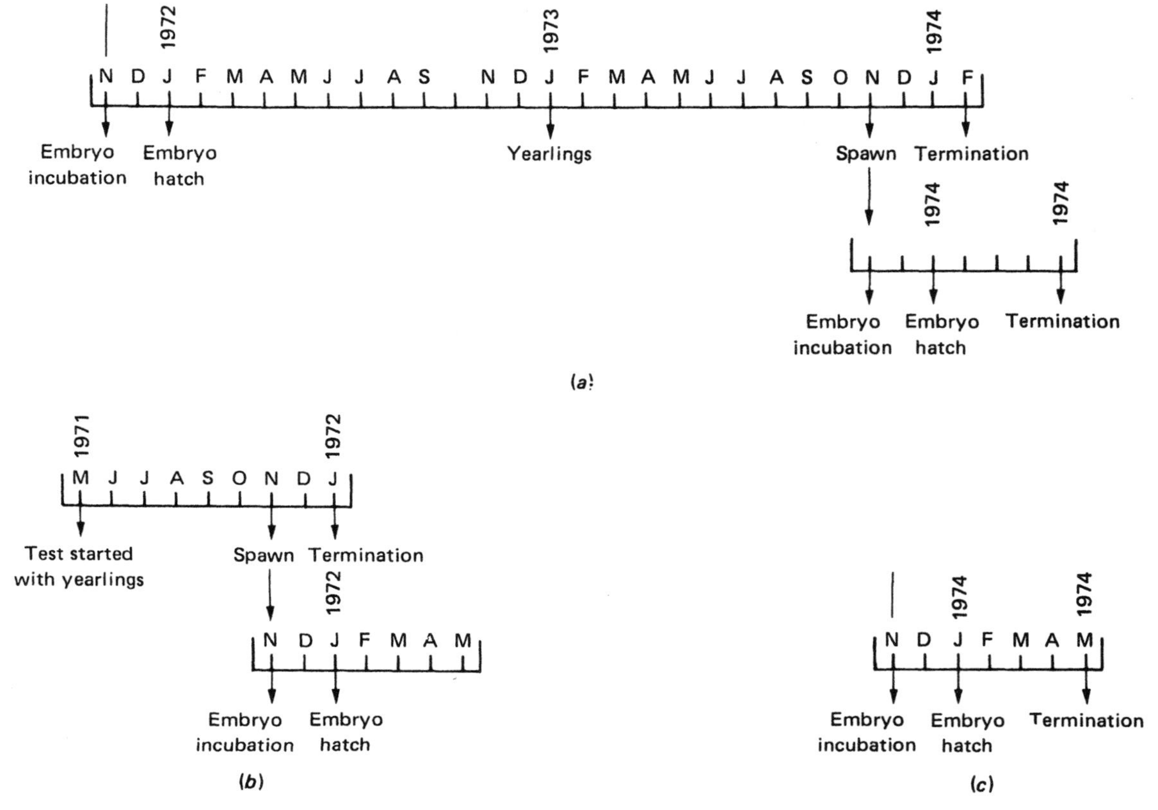

Figure 1 Chronology for (*a*) brook trout life cycle toxicity test; (*b*) partial life cycle toxicity test; and (*c*) early life stage (ELS) toxicity test. *[Modified from Holcombe et al. (1979)]*

embryo incubation in November 1971 and terminated 60 d after embryo hatch in May 1974. The trout requires 30 mo for a life cycle test, while the fathead minnow requires 12 mo.

Fish Partial Life Cycle Toxicity Test

Each of several groups of individuals of one species is exposed to a different concentration of a toxicant through part of a life cycle, which includes life stages observed to be especially sensitive to chemical exposure. Partial life cycle tests are conducted with fish species that require more than 1 yr to reach sexual maturity (e.g., brook trout and bluegills). The test can be completed in less than 15 mo and all major life stages are still exposed to the toxicant. With fish, exposure to the toxic agent begins with immature juveniles at least 2 mo before active gonad development, continues through maturation and reproduction, and ends not less than 30 d (60 d for salmonids) after the hatching of the next generation. The chronology of a partial life cycle test with brook trout is also illustrated in Fig. 1. The exposure began in May 1971 with first-generation yearlings and ended in May 1972 with second-generation offspring.

Fish Early Life Stage (ELS) Toxicity Test

The early life stages of a species of fish are exposed for 28–32 d (60 d posthatch for salmonids), from ova fertilization through embryonic, larval, and early juvenile development. The major toxic effects measured are on survival and growth. The time intervals involved in a brook trout ELS toxicity test are also illustrated in Fig. 1.

Teleost Early Life Stage Development

The following terms describe the anatomy, developmental periods, and developmental phases of a typical teleost which are essential in ELS toxicity tests. The major anatomic features clearly recognizable at various stages of fish embryonic and larval development are illustrated in Fig. 2. Those not familiar with these fish embryological terms may consult the glossary in Jones et al. (1978). The description of periods and phases presented here is drawn in part from Balon (1975) and Snyder et al. (1977).

Embryonic Period

Begins with fertilization or union of the gametes and is characterized by an endogenous food source or yolk. The period ends at hatching. The embryonic period is divided into two phases:

1 The cleavage phase involves the first interval of development within the egg membranes, from the beginning of development to organogenesis. Most ELS toxicity tests begin early in this phase, although some begin at fertilization. The start of this phase corresponds to the early cleavage stage in Fig. 2.

2 The embryonic phase involves the interval of intense organogenesis within the egg membranes and continues until hatching is completed. The start of this phase corresponds to the early embryo stage illustrated in Fig. 2.

Larval Period

Begins with hatching of the egg and lasts until the disappearance of the last vestige of the embryonic median fin fold and the appearance of a full complement of fin rays and spines. This period is divided into three phases (Fig. 3):

1 Protolarva is the larval phase with no dorsal, anal, or caudal spines or rays apparent. The only median fin elements present are the dorsal and ventral fin folds.

2 Mesolarva is the larval phase where at least one of the principal rays is apparent in the median fin (see incipient rays of dorsal and anal fins in Fig. 3, mesolarva) or, if all fin rays are present in the median fin and the adult has pelvic fins, the pelvic buds or fins are not yet apparent.

3 Metalarva is the larval phase where all principal median fin rays are apparent. If the adult has pelvic fins, the pelvic buds or fins are apparent.

The salmonids do not go through the three phases described above. Their larval period is represented by the yolk-sac or alevin period, which begins at hatching and ends, after complete absorption of the yolk, with the juvenile.

Juvenile Period

Begins when all fins are fully differentiated and no median fin fold is apparent. The body is entirely scaled. The juvenile at this point is a miniature adult from all outward appearances (Fig. 3, juvenile). Most ELS toxicity tests with fish end shortly after metamorphosis from the larval to the juvenile period.

Adult Period

Commences with the first maturation of the gamete and is accompanied by secondary sexual characteristics and spawning behavior.

Senescent Period

Old age, accompanied by slow to no growth and few to no gametes.

END POINTS FOR FISH EARLY LIFE STAGE TOXICITY

Early life stages are sensitive parts of the life cycle of a fish because of the many critical events that take place in a very short span of time. For example, in a period of 7–8 d from fertilization through hatching to the first exogenous food consumption, the fathead minnow goes from an initial two cells to a swimming, feeding, functional animal with well-developed organ systems. If at

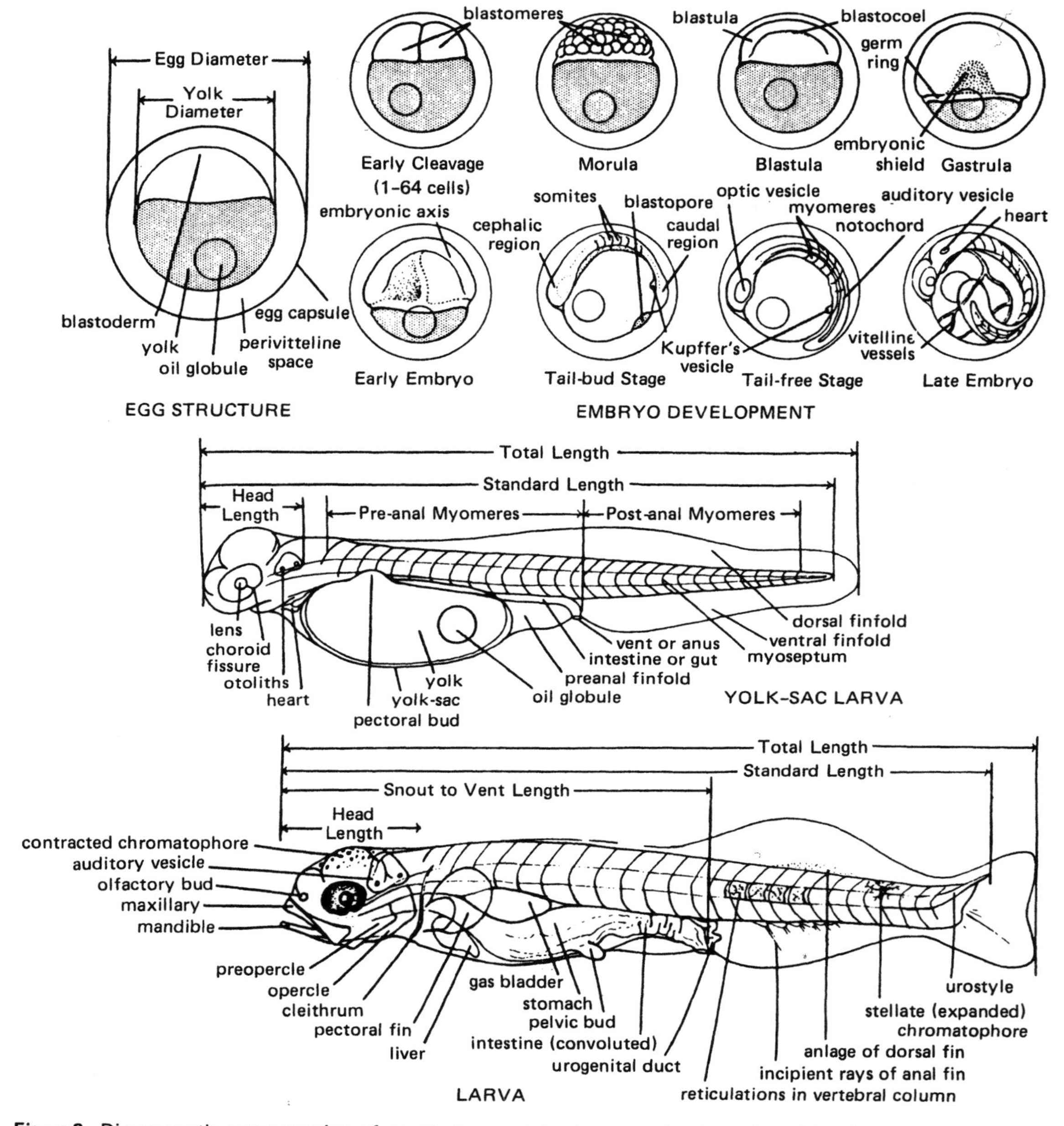

Figure 2 Diagrammatic representation of morphology and development of embryonic and larval periods of a typical teleost. *[Modified from Jones et al. (1978).]*

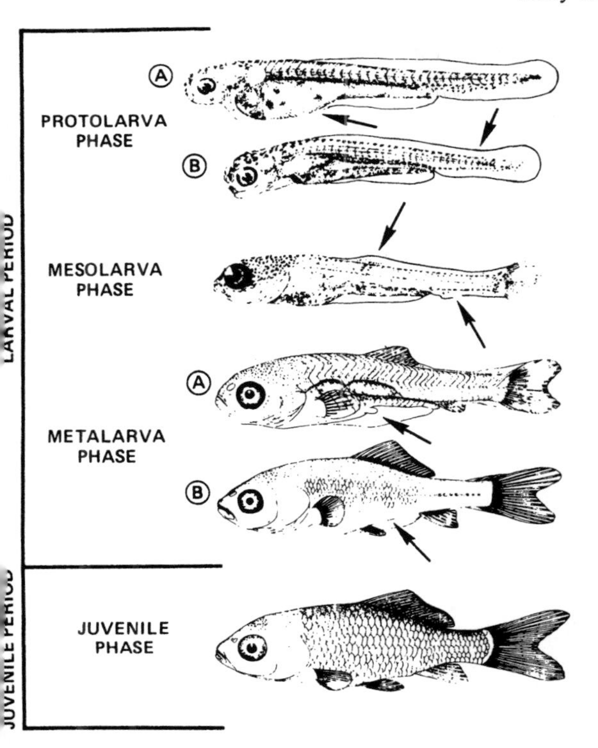

Figure 3 Sequence showing morphological changes separating larval fish into phases. *[Drawings from Jones et al. (1978).]* Protolarva: (*a*) Just hatched with large yolk sac and complete median fin fold; (*b*) yolk almost gone approaching shift to exogenous food, median fin fold still complete. Mesolarva: Feeding on exogenous food and distinct fin rays apparent in median fin folds. Metalarva: (*a*) Rays in median fins well developed and pelvic fin bud just apparent; (*b*) fins well developed, but preanal median fin fold still present. Juvenile: Full complement of fins and no median fin folds remaining.

any point during this period an environmental stress induces a change in the timing of these developmental events, the animal's chances of survival are reduced. For example, if the biochemistry for digesting exogenous food is not in place when the yolk is gone, the animal will starve. Or if the development of the eye is abnormal and sight is imperative for feeding at swim-up, again the animal is doomed.

The rapidity of growth and morphological changes during early fish development is a prime factor in the usefulness of these early life stage tests to the aquatic toxicologist seeking a short, sensitive, predictive toxicity test. At present, standardized ELS toxicity tests deal primarily with survival and growth. However, there are many possible toxic end points available for estimating the chronic toxicity of chemicals in the aquatic environment (see Table 1).

Rosenthal and Alderdice (1976) used an ontogenetic sequence to group and discuss the sublethal effects of toxic chemicals on the early life stages of marine fish. The sequence included fertilization, embryonic development, hatching, larval development, and morphogenesis. Effects on gonadal tissue were also considered. Since sublethal effects by definition do not cause mortality, the sublethal effect of a stress on a particular early life stage will not be seen in that life stage. Instead, the effect will show up at some later stage of development, either in behavioral impairment, poor growth, or death. Rosenthal and Alderdice related this altered structure-function to stage of development and to the observed or suspected detrimental effects on the individual and/or population. Table 1 shows sublethal effects (toxic end points) according to stage of development; the last column gives the observed or suspected consequences of the effects or responses to stress. This list of observed effects at different stages of development was included to show that there are many possible adverse effects other than decreases in survival and growth. Many of the effects in Table 1 were observed at much higher water concentrations than would be tolerated in an exposure of 30-90 d, such as required in an ELS toxicity test with fish. However, many of the responses are commonly seen in ELS toxicity tests, as indicated in the third column of Table 1. The toxic effects most readily observed in ELS toxicity tests occur at hatching and/or during larval and early juvenile development. There are several reasons for this. First, most investigators do not make detailed examinations of embryonic development, and record gross abnormalities only after hatching. Second, the chorion (eggshell) in most cases provides some protection to the developing embryo;

Table 1 Summary of Some Observed Responses to Environmental Alteration Considered as Sublethal Effects[a]

Stage of organization where stress is imposed or recognized	Observed or deduced response to stress (altered structure or function)	Toxic effects/observed in 30-90 d ELS toxicity tests[b]	Observed or suspected consequences of response to stress
Sex products:			
Sperm (prior to fertilization)	Reduction in motility		Reduced fertility
	Reduced fertility		Reduced rate of fertilization
	Gene damage		Embryonic malformations
Eggs (unfertilized and during fertilization)	Changes in properties of egg:		
	Membranes, related to surface structure capsule	x	Reduced strength of chorion Reduced rates of gas diffusion, respiration
	Water uptake during and after fertilization	x	Changes in osmoregulatory capacity Changes in buoyancy of pelagic eggs (changes in transport, distribution and location in water column)
	Rate of fertilization		Reproductive success
Embryonic development (early and advanced)	Biochemical effects:		
	Changes in ATP levels		Energy deficit Retarded development Necrotic tissue Dedifferentiation Organ malformation
	Changes in enzyme activity		Interference with general metabolism and biosynthetic processes, retarded development, reduced yolk energy conversion, smaller larval size at hatching, reduced hatching success
	Physiological effects:		
	Respiration changes		Changes in embryonic growth rates Changes in incubation rate
	Embryonic heart rate		Retarded development
	Morphological effects:		
	Unusual shape of blastodisc		Embryonic malformations(?)
	Deformation of blastomeres		Embryonic malformations(?)
	Irregular cleavage of blastomeres		Embryonic malformations(?)
	Amorphous embryonic tissue (no definite embryo formed)	x	No visible hatch
	Yolk deformation		Impaired yolk utilization, respiration(?)

Note: See footnotes on page 66.

Table 1 Summary of Some Observed Responses to Environmental Alteration Considered as Sublethal Effects[a] (*Continued*)

Stage of organization where stress is imposed or recognized	Observed or deduced response to stress (altered structure or function)	Toxic effects/observed in 30-90 d ELS toxicity tests[b]	Observed or suspected consequences of response to stress
	Yolk-sac blood circulation not well developed		Impaired yolk utilization(?) respiration(?)
	Organ malformations		
	Bent body axis		Impaired swimming, feeding, escape reactions
	Elongated heart tube	x	Impaired blood circulation(?)
	Eye malformations (see also yolk sac larvae)	x	Impaired vision, prey hunting, photoaxis
	Malformed otoliths and/or otic capsules		Impaired equilibrium, swimming, prey hunting
	Behavioral effects:		
	Embryonic activity reduced		Reduced mixing of perivitelline fluid affecting respiration, distribution of hatching enzyme; retarded development, abnormal hatching process
	Pectoral fin movements reduced		As above (embryonic activity)
Larvae at hatching	Altered hatching parameters:		
	Change in duration of hatching period	x	Altered distribution, density of larvae in time and space
	Increased or decreased incubation time	x	Desynchronizing of food availability at time of first feeding
	Reduced viable hatch	x	Reduced survival potential at population level
	Smaller larval size at hatching	x	Reduced biomass, increased susceptibility to predation
	Head first hatching	x	
Yolk-sac larvae (protolarvae)	Buoyancy anomalies		Difficulty in maintaining position in water column
	Swimming behavior:		
	Inability to swim		Impaired ability to maintain position in water column
	Equilibrium anomalies	x	Inability to capture food
	Loss of avoidance reaction		Reduced escape reaction, increased susceptibility to predation
	Altered yolk utilization:		
	Reduced conversion efficiency	x	Reduced biomass production, reduced larval size at time of first food intake, increased susceptibility to predation

Note: See footnotes on page 66.

Table 1 Summary of Some Observed Responses to Environmental Alteration Considered as Sublethal Effects[a] (*Continued*)

Stage of organization where stress is imposed or recognized	Observed or deduced response to stress (altered structure or function)	Toxic effects/observed in 30-90 d ELS toxicity tests[b]	Observed or suspected consequences of response to stress
	Changed rate of yolk utilization	x	Shift in timing to first food intake
	Malformations:		
	Serrated fins	x	Altered dermal respiration, swimming ability, susceptibility to disease
	Eye defects (anophthalmia, microphthalmia, monophthalmia, cyclopia)	x	Reduced visual perception, phototaxis, prey hunting ability
	Mouth, lower jaw, branchial apparatus	x	Impaired respiration, feeding success
	Vertebral column	x	Impaired swimming ability, escape reaction, prey hunting
Larvae (post yolk sac) (meso-meta larvae and early juvenile)	Altered food relations:		
	Reduced food availability, conversion efficiency, growth		Starvation; increased susceptibility to disease, predation; reduced survival potential at population level
	Reduced swimming capacity	x	Increased susceptibility to predation
	Behavioral effects:		
	Changes in opercular rates coughing response		Impaired gill function, respiration
Gonadal tissue (adult)	Success and timing of gamete production:		Desynchronization of spawning, larval production, and food supply
	Reduced fecundity		Impaired reproductive success
	Inhibition of maturation of ova		Impaired reproductive success
	Destruction of male germinal tissue		Impaired reproductive success
	Passage of contaminants through eggs to next generation		Deleterious effects on offspring

[a]These are shown in relation to ontogenetic stages of development in which they are observed and to known or possible consequences at later stages of development. For example, response to stress imposed at fertilization may be observed first as tissue anomalies in advanced embryos, whose significance to survival is expressed later as altered behavior in hatched larvae *[modified from Rosenthal and Alderdice (1976), reproduced by permission of the Ministry of Supply and Services Canada]*.

[b]The effects listed in this column were taken from the references listed in Table 6 of this chapter.

this is lost at hatching, exposing the larval fish to the full impact of the toxic chemical. Third, subtle alterations of the embryo are often missed by the untrained eye and show up later as gross abnormalities. Several examples of the toxic effects listed in Table 1 are given in Tables 2 and 3 and Figs. 4–14.

Tables 2 and 3 represent effects caused by zinc at fertilization and during early developmental changes in the membranes involved in the surface structure of the chorion (Benoit and Holcombe, 1978; Holcombe et al., 1979). Similar changes in chorion structure occurred in marine fish exposed to cadmium (Rosenthal and Alderdice, 1976) and freshwater fish exposed to copper and nickel (J. M. McKim, unpublished data). This is a very sensitive end point, which develops quickly and can in most cases be related to sublethal or chronic effects seen later in the life cycle. The effects on chorion structure and the force required to rupture the chorion seem to be more severe if embryos are exposed before water hardening (Table 2). The force required to rupture the chorion also seems to be inversely related to the toxicant exposure concentration (Table 3).

Both gross and microscopic abnormalities are useful in determining the sublethal effects of toxicants on embryonic development. Figure 4 shows the teratogenic effect of low concentrations of methyl mercury on eye development in *Fundulus* sp. shortly before hatching. Further teratogenic effects of methyl mercury are seen after hatching (Fig. 5) and consist of severe scoliosis and jaw abnormalities, which preclude feeding or swimming. Newly hatched pike larvae exposed to extremely low concentrations of 2,3,7,8-tetrachlorodibenzo-*p*-dioxin (TCDD) show reduced size at hatching and edema in the heart area (Fig. 6). The intensity of these effects is directly related to the toxicant concentration. Photomicrographs of 6-d-old bluegill larvae exposed to cadmium (Fig. 7) showed pericardial and abdominal edema. In addition, they showed lordosis dorsally of more than 90° from normal and delayed yolk sorption. Other sublethal effects of cadmium on embryonic development are seen in Fig. 8, which shows the serrated caudal fins of newly hatched garpike larvae.

Other common effects seen after hatching, during the larval phases of development, are reduced growth and poor yolk utilization (Fig. 9). The copper-exposed brook trout alevins are twice as old as the controls, yet have used almost no yolk and are half as large. These gross effects are indicative of subtle effects on metabolism, possibly on

Table 2 Fathead Minnow Embryos Spawned in Control Water and Transferred Intact (Attached to the Substrate) before and after Water Hardening to Water Containing Zn at 295 μg/l[a]

Item	Before water hardening				After water hardening[b]	
Eggs spawned on substrate and transferred intact	14	20	39	68	54	17
Hours incubated after transfer	4	4	4	24	4	24
Chorions ruptured during removal from substrate (%)	21[c]	30[c]	15[c]	13[c]	0[d]	6[d]

[a]With permission from Benoit and Holcombe (1978); copyright the Fisheries Society of the British Isles.

[b]Embryos transferred were approximately 25–35 min old. At 25°C embryos became water-hardened approximately 20 min after they were spawned.

[c]Embryos adhered poorly to spawning substrate and fell off at the slightest touch.

[d]Embryos adhered normally to spawning substrate.

Table 3 Force Required to Burst Brook Trout Chorions Exposed to Various Concentrations of Zinc Prior to Water Hardening[a]

Mean total zinc concentration (μg/l)	Duplicates	Mean weight required to rupture the chorion after 11 d exposure to zinc (g)[b]
1360	A	182 ± 121 (6)
	B	
534	A	971 ± 111 (15)
	B	
266	A	993 ± 179 (7)
	B	
144	A	1155 ± 64 (2)
	B	
69	A	
	B	
39	A	
	B	
Control 2.6	A	1223 ± 107 (11)
	B	

[a]Modified from Holcombe et al. (1979).
[b]Mean in grams ±95% confidence interval. Numbers of groups of five embryos examined are in parentheses. Missing entries indicate that fragility was not measured.

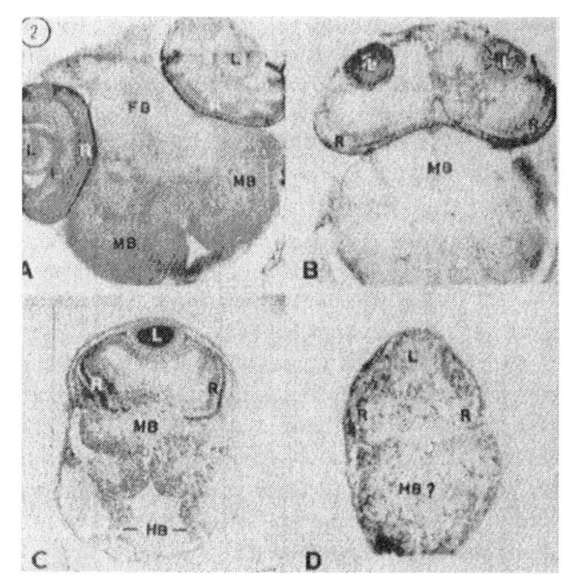

Figure 4 Photomicrographs of *Fundulus* embryos fixed near time of hatching (stage 34.14 d), sectioned coronally and stained with toluidine blue. (*a*) Control embryo; (*b*) third-degree (partially fused) synophthalmic embryo; (*c*) cyclopic embryo; (*d*) severely retarded and cyclopic embryo. Key: L, lens; R, retina; FB, MB, and HB, forebrain, midbrain, and hindbrain, respectively. Magnifications: (*a–c*) ×60; (*d*) ×100. *[From Weis and Weis (1977).]*

specific enzyme or hormonal systems. They are commonly seen in ELS toxicity tests. Another common anomaly is scoliosis or the broken back syndrome (Figs. 10 and 11), which occurs at different stages of development, depending on the chemical used and the exposure regime. Again, these abnormalities are probably tied to metabolic disruption.

Enzymatic and hormonal monitoring, which has received little attention in ELS testing, would provide a measure of the disruption of normal enzyme and/or hormone development during organogenesis in the growing embryo and larval fish. Figures 12, 13, and 14 show the normal development of three essential biochemicals in fish embryos and larvae. Alteration of these development patterns and timing sequences by low concentrations of toxic chemicals could greatly reduce the survival potential of rapidly growing fish. Future research should be aimed at studying changes in

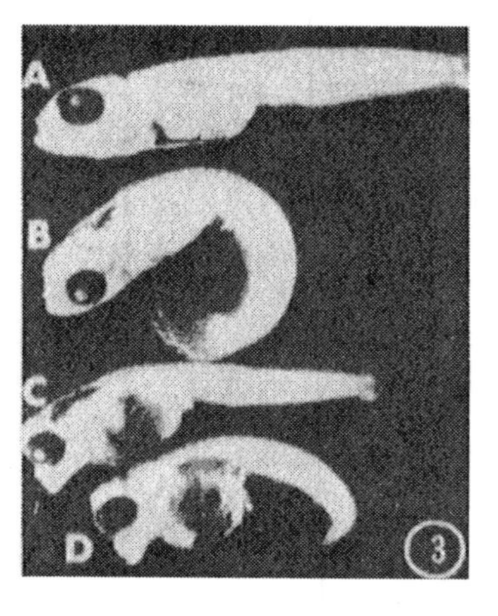

Figure 5 One-day posthatch larval *Fundulus,* approximately ×13. (*a*) Control; (*b–d*) treated with 0.02 mg/l methyl mercury *in ovo,* showing that inability to uncurl after hatching (*b* and *d*) can occur independently of craniofacial anomalies (*c* and *d*). *[Modified from Weis and Weis (1977).]*

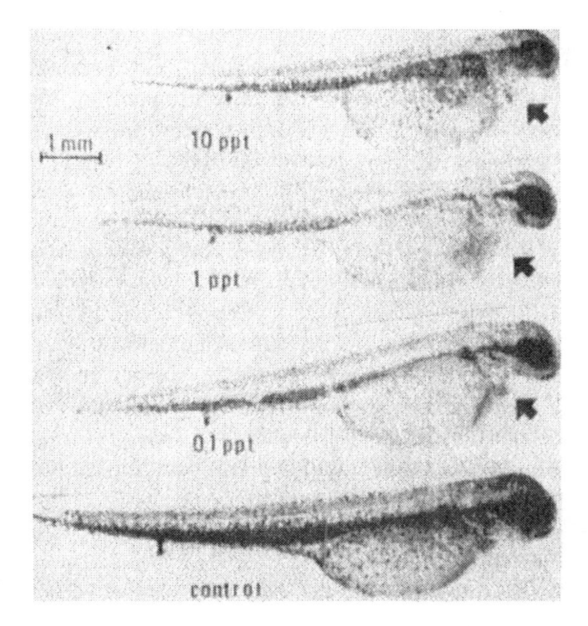

Figure 6 Appearance of pike yolk-sac fry exposed to different concentrations of TCDD, 11 d after fertilization. Arrows indicate the onset of edema. *[From Helder (1980).]*

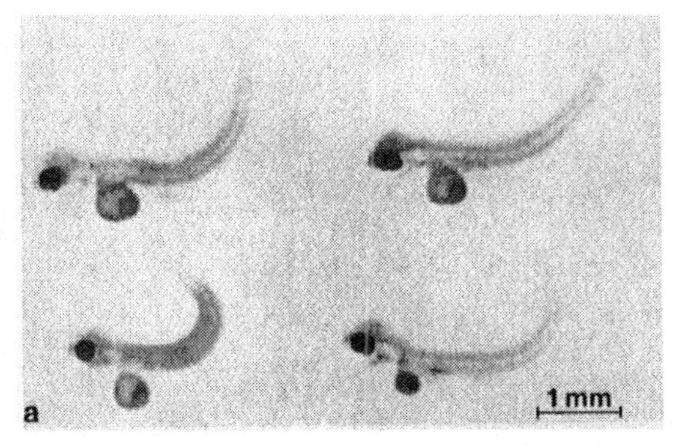

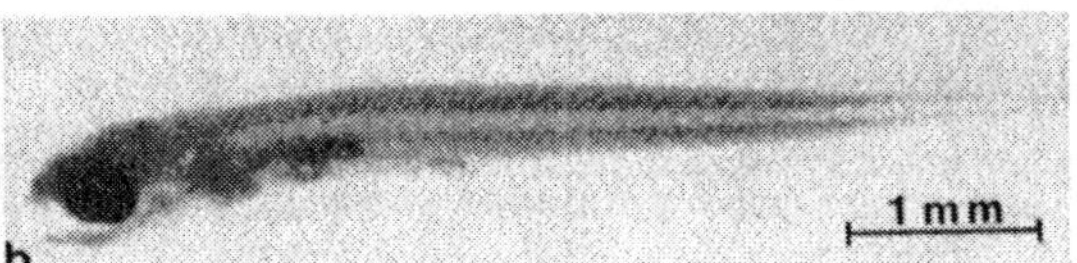

Figure 7 Photomicrographs of (*a*) crippled and (*b*) normal 6-d-old bluegill larvae from the 239 μg/l cadmium and control tanks, respectively. *[From Eaton (1974).]*

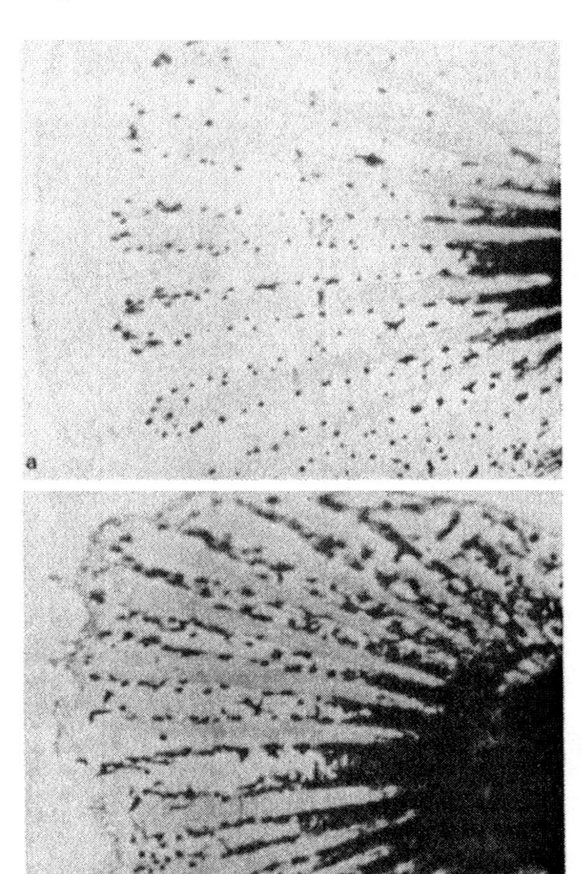

Figure 8 Serrated caudal fin of a newly hatched larva incubated in cadmium-contaminated seawater (*b*) compared to the undamaged primordial fin of a control specimen (*a*). *[From Von Westernhagen et al. (1975).]*

hormonal and enzymatic development at low levels of toxic chemicals.

Effects used as critical end points in life cycle and ELS toxicity tests to establish MATCs are discussed further in a later section.

METHODOLOGY FOR EARLY LIFE STAGE TOXICITY TESTS IN FISH

The methodology for acute and chronic toxicity tests in fish has developed rapidly over the past decade. Protocols for both types of tests are presented in Chapters 2 and 4. A few points related to

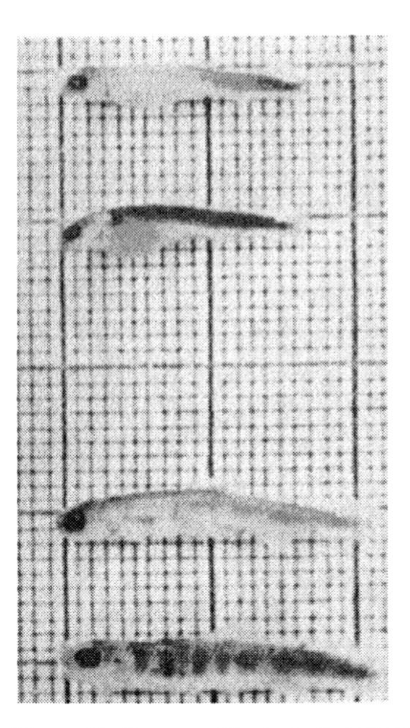

Figure 9 Copper effects on brook trout development and yolk sac absorption: (top) 32.5 μg/l copper, 553 degree-days after hatch; (bottom) control (1.9 μg/l copper), 276 degree-days after hatch (each division equals 1 mm) (degree-days = mean daily temperature × number of days). *[From McKim and Benoit (1971); reproduced by permission of the Ministry of Supply and Services Canada.]*

chronic testing are mentioned here as a lead-in to ELS tests. Fish life cycle toxicity tests were initially performed only with fathead minnows. The need for partial life cycle tests arose in order to test more species, so that sensitivities could be compared with those found in the earlier life cycle tests with fathead minnows.

The fish selected for these comparisons were the brook trout, a salmonid, the bluegill, a centrarchid, and the flagfish, a cyprinodontid. The first two species require 2 yr to reach sexual maturity, but this period was shortened by starting the fish as yearlings just prior to gonad development. Thus the partial life cycle toxicity test was created. It required 12 mo to complete with the bluegill and the brook trout. All important aspects of the life cycle, including reproduction, were exposed to the toxicant.

The similarity in sensitivity between partial life cycle tests and life cycle tests was established for four toxicants (copper, methyl mercury, cadmium, and lead) by conducting both types of tests with brook trout (McKim and Benoit, 1971, 1974; McKim et al., 1976; Benoit et al., 1976; Holcombe et al., 1976) and fathead minnows (Pickering et al., 1977). Partial life cycle tests were considered

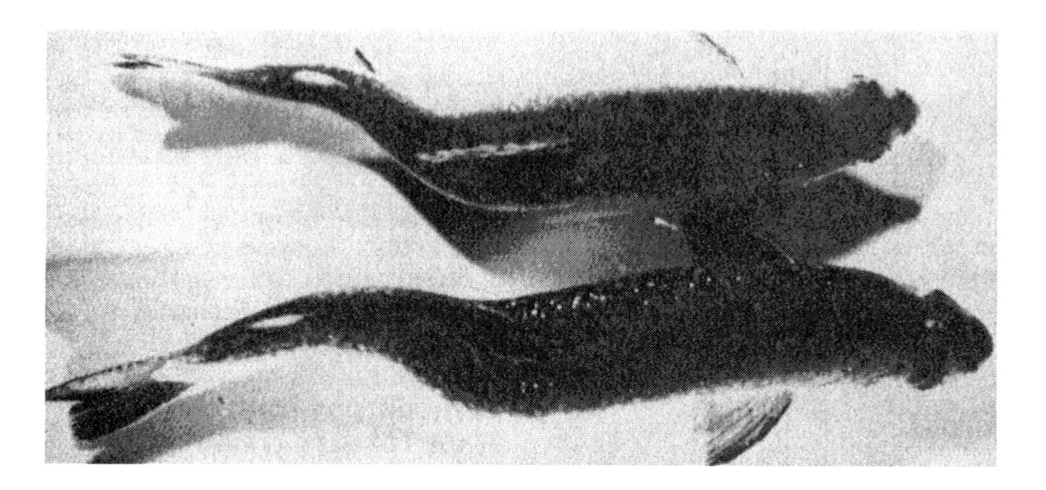

Figure 10 Deformed 65-wk-old second-generation brook trout exposed to 125 μg/l lead. *[From Holcombe et al. (1976); reproduced by permission of the Ministry of Supply and Services Canada.]*

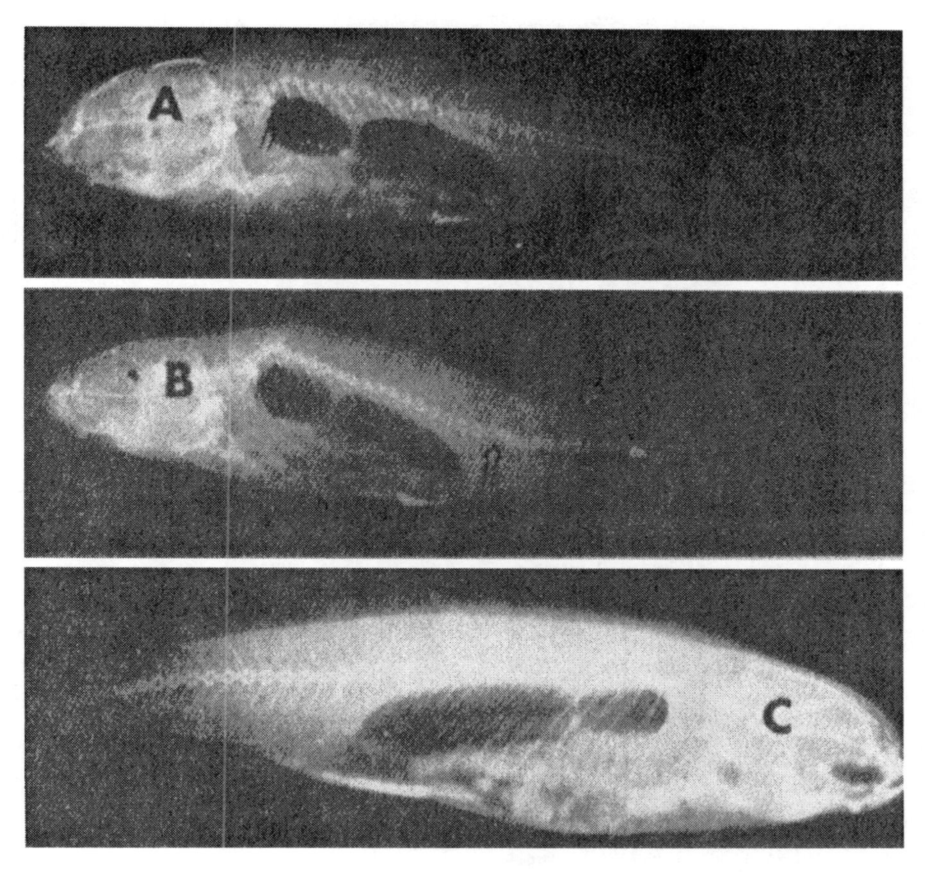

Figure 11 Effects of toxaphene on structure of fathead minnow backbones. (*a* and *b*) Radiographs representative of fish exposed to 55 ng/l toxaphene; (*c*) control fish. Arrows point to areas of backbone affected. *[From Mehrle and Mayer (1975); reproduced by permission of the Ministry of Supply and Services Canada.]*

equivalent to life cycle tests in the development of water quality criteria for aquatic animals. The desire to have even shorter tests for predicting the chronic toxicity of chemicals to fish led to the development of ELS toxicity tests over the past decade. This evolution was a logical outgrowth of life cycle and partial life cycle testing, which showed the great sensitivity of these stages in the life cycle.

Specific methodologies for conducting ELS tests with fish are documented in the references cited at the end of this chapter, which should be consulted for detailed information on culture techniques for selected species. References of particular interest are ASTM (1981) and USEPA (1981b). The rest of this section describes the general methods used at present and some innovations available for conducting ELS toxicity tests with fish.

Physical-Chemical System

Exposure System

Various designs have been used for toxicant delivery systems, but all of them are offshoots of the serial diluter (Mount and Brungs, 1967) and the

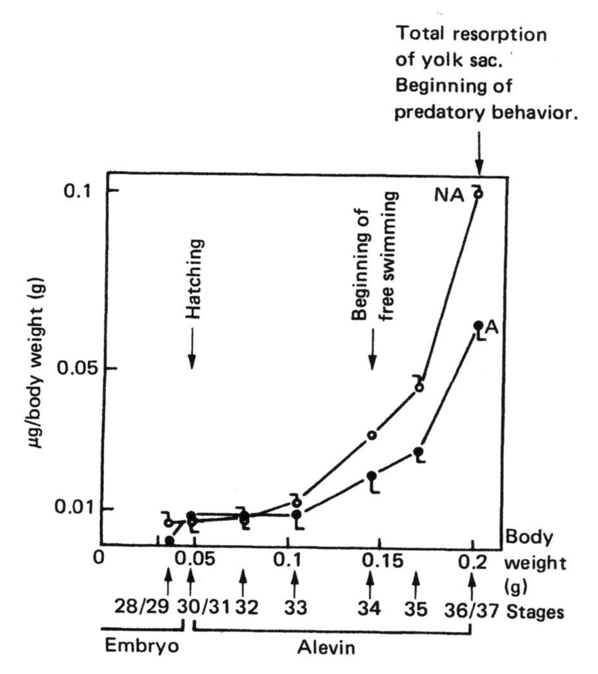

Figure 12 Adrenalin (A) and noradrenalin (NA) contents in embryos and larvae of rainbow trout in relation to body weight (wet). Each point represents the average for 6–12 animals. *[From Meyer and Sauerbier (1977) with permission; copyright 1977 Fisheries Society of the British Isles.]*

continuous flow diluter (Warner, 1964). Complete design and construction details are available for diluters (Lemke et al., 1978), and these diluter systems can be adapted for ELS toxicity studies with fish. They are discussed in detail in Chapters 2 and 4.

Recently, an integrated system designed especially for use in ELS tests with fish and invertebrates was built and evaluated for use both in the laboratory and in the field (Benoit et al., 1981a,b). This is a relatively small, space-saving, portable, flow-through diluter. It is adaptable for use with either single toxicants or complex effluents and requires only 24 l of test water per hour (Figs. 15 and 16). The system handles water containing suspended solids. It continues normal operation for 10 h if the effluent supply is cut off and up to 4 h if the diluent water is cut off. The smaller volumes required with this system allow easier shipment of effluent samples for testing. They also facilitate the removal of hazardous chemicals from test water by laboratory filtration prior to discharge into the sewer system. Although this minidiluter system was designed primarily for ELS studies with fathead minnows, it would perform equally well with other fish species routinely used in ELS studies except salmonids. Salmonids are too large and active 60 d after hatching for the small exposure chambers of the minidiluter (Fig. 17). The system has also been used successfully with invertebrates (insects and amphipods).

In addition, Benoit et al. (1981a) designed a stationary vented exposure system for containing and exhausting the fumes of hazardous volatile chemicals used in ELS tests (Figs. 18 and 19).

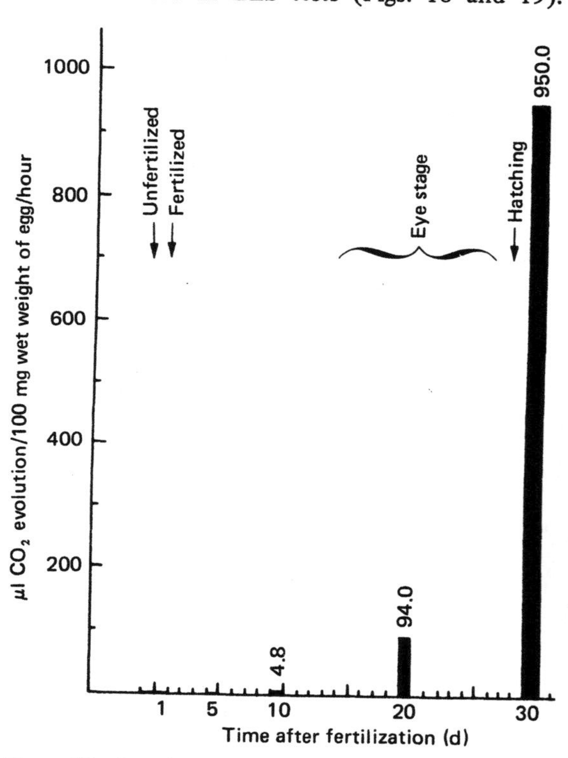

Figure 13 Acetylcholinesterase activity of rainbow trout embryo homogenate during development. Activity is expressed as microliters of CO_2 evolved per 100 mg of egg (wet weight) per hour. *[From Uesugi and Yamazoe (1964).]*

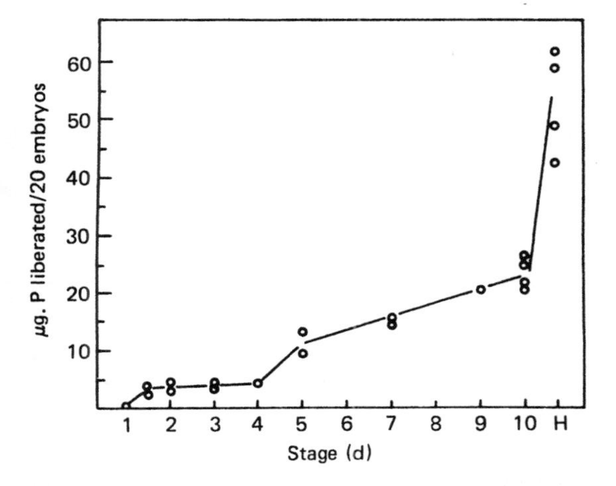

Figure 14 Change in adenosinetriphosphatase activity of fish embryos during embryonic development. Conditions: 0.05 *M* histidine buffer, 1 m*M* ATP, 0.1 *M* KCl. Activity is expressed as micrograms of P split per 20 embryos in 15 min at pH 9.1 and 37°C. *[From Ishida et al. (1959).]*

The diluter system is the same minidiluter just described hung inside an enclosure that has ample viewing and sampling ports, a lower watertight shelf for saturators (Veith and Comstock, 1975; Gingerich et al., 1979), metering pumps, and a carbon filtration device for waste water before it enters the sanitary sewer. The lower portion of the unit is watertight and can hold up to 100 l of water—enough to protect the room from contamination in case a leak develops in the system. To date, 32 ELS studies with fathead minnows and several invertebrates have been performed with this system.

The toxicant delivery system selected should be designed to accommodate a minimum of five exposure concentrations plus controls. A dilution factor not greater than 50% between concentrations should be used. There must be at least two separate exposure chambers for each toxicant concentration, and the test water delivery tubes from the toxicant delivery system to these individual exposure chambers should be positioned by stratified random assignment (Fig. 20). A minimum of 5-10 water volume turnovers per 24 h in

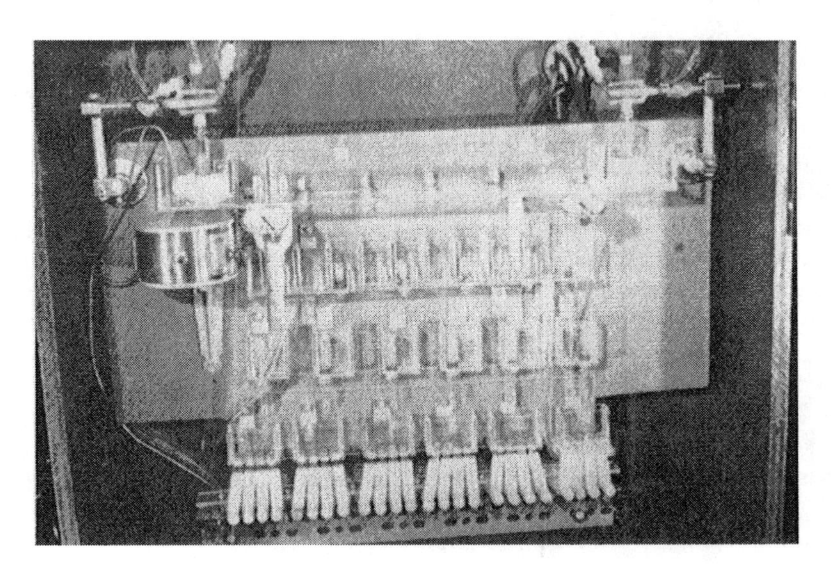

Figure 15 Continuous flow minidiluter for use with either single chemicals or treated complex effluents. *[Reprinted with permission from Benoit et al. (1981a); copyright 1981, Pergamon Press, Ltd.]*

Figure 16 Portable early life stage exposure system. *[Reprinted with permission from Benoit et al. (1981a); copyright 1981, Pergamon Press, Ltd.]*

each exposure chamber is usually required to maintain proper toxicant concentrations, dissolved oxygen concentrations above 60% of saturation, and test temperatures within ±1°C, and to prevent the buildup of dissolved waste materials, such as ammonia, in the experimental chambers. Once daily feeding of larval fish begins, all exposure chambers should be siphoned daily to remove solid waste materials and uneaten food. The photoperiod should be held constant throughout the test at 16 h light and 8 h dark, while light intensity at the water surface should be 30–60 lumens.

Before starting an ELS study, the system should be calibrated and run without animals for several days to check toxicant delivery. It should be checked daily during a test and recalibrated if necessary.

Water Characteristics

Dilution water used in toxicity tests must be uncontaminated and of constant quality. The best source of dilution water for freshwater studies is spring or well water. The second choice is uncontaminated surface water; and the least desirable source is dechlorinated city water, which contains small amounts of residual chlorine that are very difficult to remove. For saltwater tests, an uncontaminated surface source of constant salinity ranging between 15 and 35 g/kg (parts per thousand) is recommended. Saltwater from a natural source must be filtered through a filter with a pore size <10 μm to remove the larval stages of predators. Dilution water for both freshwater and saltwater studies should be filtered, ultraviolet-sterilized, brought to the desired test temperature, and vigorously aerated to stabilize oxygen and pH before use.

Certain water chemistry measurements should be made at the beginning and end of all freshwater and saltwater tests to make sure the quality of the dilution water remains constant. Total hardness, alkalinity, pH, conductivity, total organic carbon, and particulate matter should be measured for freshwater. If possible, calcium, magnesium, sodium, potassium, chloride, and sulfate should be determined. Salinity, pH, and particulate matter (total suspended solids) should be measured for saltwater. For both saltwater and freshwater tests, dissolved oxygen should be measured at least twice a week in all test concentrations throughout the test period. Temperature and salinity should be recorded continuously for the dilution water, and temperatures at each exposure concentration should be taken at least twice a week. All chemical measurements should be performed as described by the ASTM (1980b), American Public Health Association (APHA) et al. (1975), and Strickland and Parsons (1968).

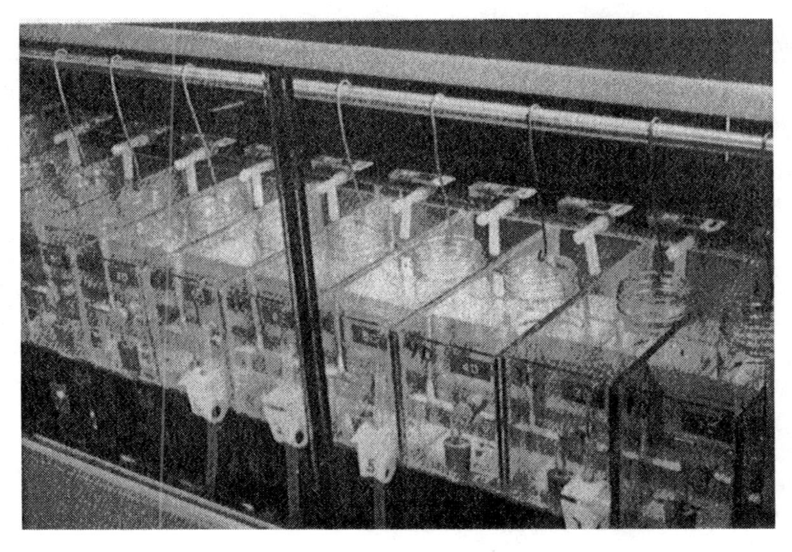

Figure 17 Exposure chambers with egg cups on rocker arm. *[From Benoit et al. (1981c).]*

Figure 18 Stationary vented early life stage exposure system. *[Reprinted with permission from Benoit et al. (1981a); copyright 1981, Pergamon Press, Ltd.]*

Toxicant Solutions

Stock solutions of test chemicals to be metered into the dilution water in a toxicity test are at best saturated water solutions of the test chemical. Use of water as the solvent avoids the possible toxic effects of an organic carrier and is generally more realistic. Several excellent devices for saturating water with chemicals have been designed by Veith and Comstock (1975) and Gingerich et al. (1979). If solvents (carriers) other than water must be used, dimethyl formamide and triethylene glycol are preferred. No organic carrier should exceed 0.5 ml/l in the test water. If an organic carrier is used, it is usually necessary to run a separate carrier control along with the regular (negative) control to make sure there are no effects of the carrier on the test animals. The amount of carrier added should be the same at all toxicant concentrations tested. Metering pumps, mechanical injectors, and Mariotte bottles are commonly used to deliver the toxicant stock solution to the diluter system (Mount and Brungs, 1967; Warner, 1964; Benoit et al., 1981a). Exposure concentrations are assigned in a stratified random manner to the exposure chambers in order to eliminate position effects. In starting a new toxicity test, the exposure concentrations should be checked immediately before placing animals in the system. Exposure concentrations are routinely monitored twice a week in alternate replicate exposure chambers.

Analytical methods should be standardized accepted procedures such as those found in the Annual Book of ASTM Standards (ASTM, 1980b),

Figure 19 Vented enclosure for testing hazardous volatile chemicals. *[Reprinted with permission from Benoit et al. (1981a); copyright 1981, Pergamon Press, Ltd.]*

Figure 20 Delivery tubes showing stratified random assignment. *[From Benoit et al. (1981c).]*

Methods for Chemical Analysis of Water and Wastes (USEPA, 1979), and Manual of Analytical Methods for the Analysis of Pesticides in Human and Environmental Samples (USEPA, 1980). All analytical methods must be validated before animals are placed in the system. As a minimum, accuracy should be checked by the method of known additions, using dilution water from the diluter apparatus. Precision should be evaluated by analyzing three samples collected on two separate days from one of the lower exposure concentrations. The analytical method should be further checked against reference or split samples and, if possible, against other methods.

Biological System

A number of freshwater fish species and several saltwater fish species (Table 4) have been used in life cycle or ELS toxicity tests (Eaton et al., 1978; Nebeker et al., 1977; Chapman and Shumway, 1977; McKim et al., 1975, 1978; Hokanson et al., 1973a,b; Benoit et al., 1981b; McCormick et al., 1977; Eaton, 1970; Mayer et al., 1977; Schimmel and Hansen, 1974; Hansen and Parrish, 1977; Goodman et al., 1976; Parrish et al., 1978; Middaugh and Lempesis, 1976; Middaugh and Dean, 1977). The species listed in Table 4 were selected on the basis of availability; commercial, recreational, and ecological importance; and past successful use or ease of culture in the laboratory. Test temperatures were taken from the references cited above or from Brungs and Jones (1977). The duration of ELS tests is 28 d posthatch for all the species in Table 4 except the salmonids, which require 60 d, and the sheepshead minnow, which require 24 d. These periods allow exposure from fertilization through embryonic, larval, and early juvenile stages of development.

Before performing an ELS toxicity test with a particular species, a 6–10 d range-finding test should be conducted with 24-h-old (newly hatched) larvae. The lowest concentration that has obvious effects on the larvae should be used as the high exposure concentration in the final definitive test. All other exposure concentrations should be diluted downward from this concentration by using a set dilution factor of not less than 0.5.

Embryonic Period

The initial step in setting up an ELS toxicity test is obtaining high-quality test animals. Fish embryos can generally be obtained by (1) collecting mature, ripe adults in the field and stripping gametes, (2) buying mature, ripe adults from a commercial hatchery and stripping gametes, (3) obtaining, as needed, newly fertilized embryos from a laboratory stock culture unit, or (4) buying disease-free embryos from a commercial hatchery or another laboratory. The species to be tested will in many cases dictate the method used to obtain embryos.

The ELS test is started by impartially placing 50–100 newly fertilized (<8 h old) embryos into

Table 4 Recommended Species, Test Temperature, Salinity, Duration, and Photoperiod for the Testing of Fish Early Life Stages[a]

Species[b]	Temperature (°C)	Salinity (ppt)	Duration (d)
Freshwater:			
Coho salmon, *Oncorhynchus kisutch*	10, 12[c]	NA[d]	60[e]
Chinook salmon, *O. tshawytscha*	10, 12	NA	60
Rainbow trout, *Salmo gairdneri*	10, 12	NA	60
Brown trout, *S. trutta*	10	NA	60
Brook trout, *Salvelinus fontinalis*	10	NA	60
Lake trout, *S. namaycush*	7	NA	60
Northern pike, *Esox lucius*	12, 18	NA	28
Fathead minnow, *Pimephales promelas*	25	NA	28
White sucker, *Catostomus commersoni*	15	NA	28
Bluegill, *Lepomis macrochirus*	28	NA	28
Channel catfish, *Ictalurus punctatus*	26	NA	28
Saltwater:			
Sheepshead minnow, *Cyprinodon variegatus*	30	15-30	28[f]
Silversides, *Menidia* spp.	25	15-35	28

[a]Modified from ASTM (1981); copyright American Society for Testing and Materials; adapted with permission.
[b]Scientific name must be verified.
[c]10°C refers to embryo test temperature, 12°C to larval test temperature.
[d]Not applicable.
[e]Posthatch exposure duration.
[f]Includes incubation.

the incubation chambers in each exposure chamber. An alternative procedure involves mixing the gametes together and fertilizing the eggs in the exposure water before placing the embryos in the incubation chambers. The latter method is preferred since effects on fertilization and water hardening of the eggs are included in the same test. If the first method is used, the exact stage of embryo development should be recorded (see Figure 2 and Lagler et al., 1977, pp. 302–323). The embryos are incubated in vertically oscillating 120-ml glass jars with 40-mesh nylon screen bottoms similar to those of Mount (1968) or in stationary incubation containers with fluctuating water levels controlled by self-starting siphons. The embryo incubation chambers are randomly placed in the exposure chambers. Salmonid embryos are sensitive to light and must be incubated in the dark from fertilization to 1 wk after hatching. During incubation, dead embryos must be removed daily to prevent the spread of fungal infections. Dead embryos usually appear opaque and are easily distinguished from living ones.

Larval Period

This phase begins with hatching. An accurate daily record of abnormal larvae, with descriptions of abnormalities, should be kept. Except for salmonids, release from the incubation cup should occur when hatching is 90% complete or 48 h after the first hatch in that incubation cup. Unhatched embryos should remain in the cup, and larvae should be allowed to swim out of the cup into the exposure chamber. If more embryos were used for hatchability than are to be used for the larval portion of the test, the fish retained to complete the test should be randomly selected from all larvae hatched, both normal and abnormal. Salmonids are released from the incubation cups 21 d after hatching or at swim-up and are handled as described above.

Food should be provided to the growing larvae as the larvae absorb the yolk and begin to swim up off the bottom of the exposure tank. The interval between hatching and swim-up varies greatly. For example, it is 2-3 d for fathead minnows at 25°C and 21-28 d for brook trout at 10°C. Newly hatched brine shrimp nauplii are the food of choice for the fish listed in Table 4 except channel catfish and salmonids, which require a commercially prepared starter food, and silversides, which require a rotifer (*Brachionus* sp.). Feeding should take place at least three times a day and be spaced so that food is present in the water continously during daylight. A weight or volume system should be used to quantitate the food placed in the exposure chambers so that the amounts fed are equal. The amounts should be reduced on a percentage basis as larvae die in the higher concentrations.

Survival is determined daily for larval fish in each exposure chamber. This is best done by counting live larvae in each exposure chamber, because during the first few weeks many newly hatched species deteriorate rapidly after death. Behavior (feeding, sluggish movement, loss of equilibrium, and so on) should be observed. The development and number of morphological abnormalities should also be recorded.

Early Juvenile Period

At the end of the ELS test the number of surviving juveniles at each exposure is recorded, weights and lengths are recorded for growth, and all deformities are described (with photographs if possible) and quantitated.

Biological Data

The following biological data should be collected during a typical ELS toxicity test for each developmental period.

Embryonic period:

1 Incubation period, the time from fertilization to complete hatching.
2 Hatching time, the time required for all embryos to hatch at a given exposure concentration once the first one hatches.
3 Time from complete hatch to first feeding.
4 Percent of normal and abnormal larvae at complete hatch.
5 Weight of larvae at complete hatch.

Larval and early juvenile periods:

1 Abnormal behavior.
2 Total percent of deformed fish at the end of the test.
3 Percent survival of fish at the end of the test.
4 Weights and lengths of all fish at the end of the test.
5 Occurrence of normal metamorphosis from the metalarval phase to the juvenile phase as compared to controls.

Special Examinations

This category can always be expanded and can result in the generation of useful end points for future ELS tests. Examples of the kinds of measurements that can be made are shown in Table 1. These include embryological, physiological, biochemical, histological, and behavioral observations.

Conclusions and Interpretation of Data

Once the data have been collected, evaluated, and categorized (for example, see Table 5), they must be analyzed statistically to determine whether there are significant effects of the toxicant.

If an organic carrier is used to introduce the toxicant, a *t*-test (Steel and Torrie, 1960) must be performed to determine the differences (if any) between the regular control and the carrier control. If differences exist, the results are questionable. If both controls are the same, then all the control data can be combined for the remaining statistical analyses.

Once the controls have been analyzed, the question is whether significant differences exist in each data set (items in Table 5) collected during the ELS

Table 5 Hatchability of Embryos, Normal Larvae at Hatch, and Survival and Weight of 28-d-old Fathead Minnows Exposed to Various 1,2-Dichloropropane Concentrations[a]

Item	1,2-Dichloropropane (mg/l)					Control[b]
	110 ± 3[c]	51 ± 2	25 ± 1	11 ± 2	6 ± 1	(0.1 ± 0.02)
Hatchability of embryos (%)[d]	96 (90–100)	96 (93–100)	98 (93–100)	98 (93–100)	96 (93–100)	97 (97)
Normal larvae at hatch (%)	0	67[f] (53–80)	100	100	100	100
Survival of 28-d-old fish (%)[e]	0	27[f] (0–40)	58[f] (0–93)	95 (87–100)	92 (87–100)	95 (93–100)
Weight of 28-d-old fish (mg)	—	18[f] ±6	79[f] ±20	126[f] ±33	140 ±29	145 ±45

[a]Each entry represents the mean of four replicates per treatment and numbers in parentheses denote replicate range. From Benoit et al. (1981b).
[b]Slight contamination of control water was from air inside vented enclosure.
[c]Mean and standard deviation.
[d]Thirty 2–5-h-old embryos exposed per replicate.
[e]Fifteen fish exposed per replicate.
[f]Significantly different from control, $P = 0.05$ (analysis of variance with Dunnett's test).

test. Therefore an analysis of variance (ANOVA) F-test is run on each data set (Steel and Torrie, 1960; McClave et al., 1981; Sokal and Rohlf, 1969; Box et al., 1978). For data sets or items showing significant differences ($P = 0.05$), all exposure concentration measurements are compared with the control measurements by an appropriate multiple comparison test, such as Dunnett's procedure (Dunnett, 1955, 1964) and Williams' procedure (Williams, 1971, 1972). Significant differences from the controls ($P = 0.05$) are then used to establish the *lowest observed effect concentration* (LOEC) and the highest *no observed effect concentration* (NOEC), or the MATC range. In the example in Table 5, the LOEC and NOEC were 11 and 6 mg/l, respectively, based on the most sensitive effects in the test. With 1,2-dichloropropane that effect was 28-d growth (Benoit et al., 1981b). Therefore, the estimated MATC of 1,2-dichloropropane for fathead minnows can be reported as $>6 < 11$ mg/l based on 28-d growth.

The sublethal effects used to establish the NOEC (growth in the example above) are assumed to be important for the individuals' well-being and the maintenance of a healthy population. Decisions about the importance of specific effects must be made by the researcher. The validity of such decisions in most cases depends on the experience and biological expertise of the individual. The importance of selected sublethal effects was discussed in the earlier section on end points.

INVERTEBRATE EARLY LIFE STAGE TOXICITY TESTS

Freshwater and saltwater aquatic invertebrates have become very important in aquatic toxicology in recent years because of (1) the speed with which an invertebrate life cycle test can be performed (1–5 mo compared to 12–24 mo for a fish life cycle test) and (2) the similar sensitivities of the two groups. The latter fact has enhanced the development of fish partial life cycle and ELS toxicity tests, while little need existed for the development of shorter predictive invertebrate tests.

Daphnia magna is the best known and most used invertebrate test species in freshwater toxicology. Life cycle toxicity tests with this species range from 21-28 d to lifetime studies requiring several months. The 21-28 d studies have been more or less standardized for use in long-term toxicity testing. There is some evidence, however, that lifetime studies with *Daphnia* are more sensitive than the 21-28 d exposures (Winner and Farrell, 1976; Buikema et al., 1980). Buikema et al. (1980) reported that with the water-soluble fractions of hydrocarbons the 21-d daphnid test measuring growth and reproduction was predictive for lifetime studies only 56% of the time. Some work on the development of short predictive tests for the 21-28 d daphnid life cycle tests (e.g., Geiger et al., 1980) indicated that the size of daphnid preadults after 7 d of exposure to the water-soluble fraction of hydrocarbons gave the same toxic effect values as the 21-d exposures. There are enough chronic toxicity data for daphnids to evaluate new predictive methods, but a standardized short predictor of chronic toxicity does not yet exist and the standard daphnid 21-28 d chronic toxicity test continues to be the major freshwater invertebrate test (see Chapter 4 for details on chronic daphnid tests). At present, the only short predictive test with an invertebrate on a par with the fish ELS toxicity test is that with *Daphnia magna.*

Review articles on two other less tested freshwater invertebrates, the chironomids (Anderson, 1980) and the amphipods (Arthur, 1980), show considerable progress in life cycle and partial life cycle test development. However, the already short duration of the life cycle tests and lack of life cycle data for evaluating the predictive capabilities of shorter tests inhibit the development of short predictive tests. Therefore, as with *Daphnia,* no standardized short tests for predicting chronic toxicity have been established and life cycle tests with these species continue to be important.

Essentially, the same arguments hold for saltwater invertebrates. Mysids (Nimmo et al., 1977), grass shrimp (Tyler-Schoeder, 1979) and polychaete annelids (Reish, 1980) are used in life cycle tests that require 1-2 mo. Until more data on life cycle tests with these animals are available for evaluating new shorter predictive tests, and until more is known about the biology and sensitivity of the developmental stages of each life cycle, few useful predictive tests will be developed with these saltwater invertebrates. Other saltwater invertebrates used in toxicity tests are larvae of the oyster (Calabrese et al., 1973; Roberts, 1980) and larvae of the crab and lobster (APHA et al., 1975). The time intervals and culturing difficulties with these species make routine life cycle tests nearly impossible at this time.

Invertebrates have an excellent potential as rapid (a few days or weeks), sensitive predictors of long-term toxicity because of their diversity and short, sensitive life cycles. However, the biology of each species must be better known and a chronic data base made available in order to evaluate the usefulness of new short-term predictive tests.

EXTRAPOLATION FROM EARLY LIFE STAGE TOXICITY TESTS TO LIFE CYCLE TOXICITY TESTS

Evaluation of the usefulness of ELS toxicity tests with fish has been made possible by the large data base from life cycle toxicity tests generated over the past decade. In addition, several investigators have shown that the ELS portions of life cycle tests are the most sensitive to single toxicants. However, acceptance of the ELS toxicity test approach and its widespread use came only after comprehensive reviews of all existing data from freshwater life cycle toxicity tests (Macek and Sleight, 1977; McKim, 1977). Table 6 gives a summary of these data, as well as recent life cycle test data from freshwater and saltwater laboratories. The new data presented here were subjected to the same rigorous criteria for inclusion that were used previously. These data confirm that the embryo-larval, early juvenile portion of the life cycle is the most, or one of the most, sensitive in the life cycle of both freshwater and saltwater fish. This updated evaluation is based on 72 life cycle

Table 6 Results of Partial (P) and Complete (C) Life Cycle Toxicity Tests with Fish Including the MATC Established by Each Life Cycle Test and an MATC Estimated by Embryo, Larval, and Early Juvenile (ELEJ) Exposures Conducted as Part of Each Life Cycle Test[a]

Toxicant	Species	Type of life cycle test	Critical life stage end points
	Freshwater		
Metals			
1. Cadmium	Fathead minnow	P	ELEJ–J
2. Cadmium	Bluegill	P	ELEJ–M, G
3. Cadmium	Flagfish	C	A–S
4. Cadmium	Flagfish	C	ELEJ–M
5. Cadmium	Flagfish	C	ELEJ–M
6. Cadmium	Flagfish	C	ELEJ–M
7. Cadmium	Brook trout	C	ELEJ–G
8. Copper	Fathead minnow	P	ELEJ–M; A–S
9. Copper	Brook trout	P	ELEJ–G
10. Copper	Brook trout	C	ELEJ–G
11. Copper	Bluegill	P	ELEJ–M
12. Chromium	Fathead minnow	P	ELEJ–M, G
13. Chromium	Brook trout	C	ELEJ–M
14. Lead	Brook trout	C	ELEJ–M
15. Lead	Flagfish	C	ELEJ–D
16. Mercury	Fathead minnow	C	A–S
17. Nickel	Fathead minnow	C	ELEJ–G
18. Zinc	Fathead minnow	P	ELEJ–H
19. Zinc	Fathead minnow	P	ELEJ–F
20. Zinc	Flagfish	C	A–G
21. Zinc	Flagfish	C	A–G
22. Zinc	Brook trout	P	ELEJ–F
	Fathead minnow	C	ELEJ–F
Trimetallic			
23. Mixture: zinc + cadmium + copper	Fathead minnow	P	ELEJ–F
Organometallics			
24. Methylmercury	Fathead minnow	C	A–M, S
25. Methylmercury	Flagfish	P	ELEJ–G; A–S
26. Methylmercury	Brook trout	C	ELEJ–M, G
Halogens			
27. Chlorine	Fathead minnow	P	ELEJ–M
28. Chloramine	Fathead minnow	P	A–S
29. Fluoride	Fathead minnow	P	ELEJ–M
Polychlorinated biphenyls			
30. PCB-1254	Fathead minnow	C	ELEJ–M
31. PCB-1242	Fathead minnow	C	ELEJ–M
32. PCB-1248	Fathead minnow	C	ELEJ–M
33. PCB-1260	Fathead minnow	C	ELEJ–M
Pesticides			
34. Acrolein	Fathead minnow	P	ELEJ–M
35. Atrazine	Brook trout	P	A–G
36. Captan	Fathead minnow	P	ELEJ–M, G

Note: See page 84 for footnotes.

MATC (μg/l)	Estimated MATC (μg/l)	Type of embryolarval test (d)	Reference
		Freshwater	
37.0–57.0	37.0–57.0	60 (Pue) (Pe)[b]	Pickering and Gast (1972)
31.0–80.0	31.0–80.0	60 (Pe)	Eaton (1974)
4.1–8.1	8.1–16.0	60 (Pue)	Spehar (1976)
7.4–16.9	7.4–16.9	30 (Pue) (Pe)[b]	A. R. Carlson and J. H. Tucker[c]
3.0–6.5	3.0–6.5	30 (Pue) (Pe)	A. R. Carlson and J. H. Tucker[c]
3.4–7.3	3.4–7.3	30 (Pue) (Pe)	A. R. Carlson and J. H. Tucker[c]
1.7–3.4	1.7–3.4	90 (Pe) (Pe_1)	Benoit et al. (1976)
10.6–18.4	10.6–18.4	120 (Pue) (Pe)[b]	Mount and Stephan (1969)
9.5–17.4	9.5–17.4	90 (Pe)	McKim and Benoit (1971)
9.5–17.4	9.5–17.4	90 (Pue) (Pe) (Pe_1)[b]	McKim and Benoit (1974)
21.0–40.0	21.0–40.0	90 (Pue)	Benoit (1975)
1000.0–3950.0	1000.0–3950.0	30 (Pue)	Q. H. Pickering[c]
200.0–350.0	200.0–350.0	90 (Pue) (Pe)[b]	Benoit (1976)
58.0–119.0	58.0–119.0	120 (Pe) (Pe_1)	Holcombe et al. (1976)
31.3–62.5	62.5–125.0	120 (Pue) (Pe)[b]	R. L. Anderson[c]
<0.26	<0.26	30 (Pe)	V. M. Snarski and G. F. Olson (1981)[c]
380.0–730.0	380.0–730.0	30 (Pe)	Pickering (1974)
30.0–180.0	30.0–180.0	30 (Pe)	Brungs (1969)
26.0–51.0	51.0–85.0	30 (Pue)	Spehar (1976)
75.0–1239.0	139.0–267.0	30 (Pue)	Spehar (1976)
532.0–1368.0	532.0–1368.0	90 (Pue) (Pe) (Pe_1)[b]	Holcombe et al. (1979)
78.0–145.0	78.0–145.0	90 (Pue) (Pe)[b]	Benoit and Holcombe (1978)
27.3–42.3	27.3–42.3	30 (Pe)	Eaton (1973)
3.9–7.1	3.9–7.1		
5.3–6.7	5.3–6.7		
0.07–0.13	0.13–0.23	60 (Pue)	D. I. Mount and G. F. Olson[c]
0.17–0.33	0.17–0.33	30 (Pue)	W. E. Smith[c]
0.29–0.93	0.29–0.93	90 (Pe) (Pe_1)	McKim et al. (1976)
14.0–42.0	14.0–42.0	60 (Pe)	Arthur et al. (1975)
16.0–43.0	43.0–108.0	30 (Pe)	Arthur and Eaton (1971)
6800.0–13,600.0	6800.0–13,600.0	30 (Pe)	D. Olson[c]
1.8–4.6	1.8–4.6	90 (Pue) (Pe)	Nebeker et al. (1974)
5.4–15.0	5.4–15.0	90 (Pue) (Pe)	Nebeker et al. (1974)
1.1–3.0	1.1–4.4	60 (Pue) (Pe)	DeFoe et al. (1978)
2.1–4.0	2.1–4.0	60 (Pue) (Pe)	DeFoe et al. (1978)
11.4–41.7	11.4–41.7	30 (Pe)	Macek et al. (1976c)
60.0–120.0	120.0–240.0	90 (Pue) (Pe)[b]	Macek et al. (1976b)
16.5–39.5	16.5–39.5	30 (Pe)	Hermanutz et al. (1973)

Table 6 Results of Partial (P) and Complete (C) Life Cycle Toxicity Tests with Fish Including the MATC Established by Each Life Cycle Test and an MATC Estimated by Embryo, Larval, and Early Juvenile (ELEJ) Exposures Conducted as Part of Each Life Cycle Test[a] *(Continued)*

Toxicant	Species	Type of life cycle test	Critical life stage end points
37. Carbaryl	Fathead minnow	C	ELEJ–M
38. DDT	Fathead minnow	C	ELEJ–H, M
39. Diazinon	Flagfish	C	ELEJ–H
40. Diazinon	Fathead minnow	P	ELEJ–M, G
41. Diazinon	Brook trout	P	ELEJ–G
42. Endosulfan	Fathead minnow	P	ELEJ–H
43. Endrin	Flagfish	C	ELEJ–G
44. Endrin	Fathead minnow	C	ELEJ–M
45. Guthion®	Fathead minnow	C	A–S
46. Heptachlor	Fathead minnow	C	ELEJ–M
47. Lindane	Fathead minnow	C	A–G
48. Malathion	Flagfish	C	ELEJ–G
49. Trifluralin	Fathead minnow	P	A–M
50. Toxaphene	Brook trout	P	ELEJ–M, G
51. Toxaphene	Fathead minnow	C	ELEJ–G
52. Toxaphene	Channel catfish	P	ELEJ–G, M
Sewage			
53. Nondisinfected	Fathead minnow	C	ELEJ–M, G
54. Chlorinated	Fathead minnow	C	ELEJ–G
55. Dechlorinated	Fathead minnow	C	ELEJ–M
56. Chlorobrominated	Fathead minnow	C	ELEJ–M
Other			
57. Linear alkylate sulfonate	Fathead minnow	P	ELEJ–M
58. pH	Fathead minnow	C	ELEJ–H
59. pH	Brook trout	P	ELEJ–M, G
60. NTA	Fathead minnow	P	A–M
61. Oxygen	Fathead minnow	P	ELEJ–M
	Saltwater		
1. Carbofuran	Sheepshead minnow	C	ELEJ–M
2. Caustic waste	Sheepshead minnow	C	ELEJ–M
3. Chlordane	Sheepshead minnow	C	ELEJ–H
4. Endrin	Sheepshead minnow	C	ELEJ–H, M
5. EPN	Sheepshead minnow	C	ELEJ–G
6. Kepone	Sheepshead minnow	C	ELEJ–G
7. Malathion	Sheepshead minnow	C	ELEJ–M
8. Methoxychlor	Sheepshead minnow	C	ELEJ–H, M
9. Pentachlorophenol	Sheepshead minnow	C	A–M
10. Tributyltin oxide	Sheepshead minnow	C	ELEJ–M
11. Trifluralin	Sheepshead minnow	C	A–S, ELEJ–M

[a]Pue, embryos, and larvae from previously unexposed parents; Pe, from previously exposed parents; Pe_1, from previously exposed first-generation parents; A, adult; M, mortality; G, growth; F, egg fragility; H, hatchability; D, deformities; S, spawning.

[b]No difference in sensitivity between embryos and larvae from previously exposed (Pe) and unexposed (Pue) parents *[modified from McKim (1977); reproduced by permission of the Ministry of Supply and Services Canada]*.

[c]Personal communication, Environmental Research Laboratory, Duluth, Minn.

[d]Author states value may be high by 50%.

MATC (μg/l)	Estimated MATC (μg/l)	Type of embryolarval test (d)	Reference
210.0–680.0	210.0–680.0	60 (Pue) (Pe)[b]	Carlson (1971)
0.36–1.5	0.36–1.5	30 (Pe)	Jarvinen et al. (1977)
54.0–88.0	54.0–88.0	60 (Pue) (Pe)[b]	Allison (1977)
6.8–13.5	6.8–13.5	60 (Pue) (Pe)	Allison and Hermanutz (1977)
<0.80	<0.80	120 (Pue) (Pe)	Allison and Hermanutz (1977)
0.20–0.40	0.20–0.40	30 (Pue)	Macek et al. (1976c)
0.22–30	0.22–0.30	30 (Pue) (Pe)	Hermanutz (1978)
<0.14	<0.14	30 (Pe)	Jarvinen and Tyo (1978)
0.33–0.51	0.7–1.8[d]	50 (Pue) (Pe)[b]	Adelman et al. (1976)
0.86–1.84	0.86–1.84	60 (Pue) (Pe)[b]	Macek et al. (1976c)
9.1–23.5	No effect	60 (Pe)	Macek et al. (1976a)
8.6–10.9	8.6–10.9	30 (Pue) (Pe)	Hermanutz (1978)
1.95–5.1	5.1–8.2	60 (Pe)	Macek et al. (1976c)
<0.039	<0.039	60 (Pue)	Mayer et al. (1975)
0.025–0.054	0.025–0.054	30 (Pe)	Mayer et al. (1977)
0.129–0.299	0.129–0.299	30 (Pe)	Mayer et al. (1977)
25–50%	25–50%	30 (Pue) (Pe)[b]	Ward et al. (1976)
7–14%	7–14%	30 (Pue) (Pe)[b]	Ward et al. (1976)
50–100%	50–100%	30 (Pue) (Pe)[b]	Ward et al. (1976)
50–100%	50–100%	30 (Pue) (Pe)[b]	Ward et al. (1976)
630.0–1200.0	630.0–1200.0	30 (Pe)	Pickering and Thatcher (1970)
6.6–5.9 (pH)	6.6–5.9 (pH)	30 (Pue) (Pe)[b]	Mount (1973)
6.5–6.1 (pH)	6.5–6.1 (pH)	90 (Pue) (Pe)[b]	Menendez (1976)
54,000.0–114,000.0	54,000.0–114,000.0	–	Arthur et al. (1974)
5.01–4.02 (mg/l)	5.01–4.02 (mg/l)	30 (Pe)	Brungs (1971)
		Saltwater	
15.0–23.0	15.0–23.0	30 (Pe) (Pue)	Parrish et al. (1977)
1500.0–3000.0	1500.0–3000.0	30 (Pue) (Pe)	Hollister et al. (1980)
0.5–0.8	0.5–0.8	30 (Pe) (Pue)	Parrish et al. (1976, 1978)
0.12–0.31	0.12–0.31	30 (Pe) (Pue)[b]	Hansen et al. (1977)
4.1–7.9	4.1–7.9	30 (Pue) (Pe)	U.S. EPA (1981a)
0.074–0.12	0.074–0.12	30 (Pe) (Pue)	Goodman et al. (1981)
4.0–9.0	4.0–9.0	30 (Pe) (Pue)	Parrish et al. (1977)
12.0–23.0	12.0–23.0	30 (Pe) (Pue)[b]	Parrish et al. (1977)
47.0–88.0	88.0–195.0	30 (Pe) (Pue)	Parrish et al. (1978)
1.0–4.8	1.0–4.8	30 (Pe) (Pue)[b]	Ward and Parrish (1980)
1.3–4.8	1.3–4.8	30 (Pe) (Pue)[b]	Parrish et al. (1978)

tests with 4 freshwater species and 37 organic and inorganic chemicals plus 1 saltwater species and 11 organic chemicals. In 83% of these tests the ELS portion of the life cycle gave the same MATC as the full life cycle, while the remaining 17% of the ELS tests showed more or less sensitivity than the life cycle test by a factor of 2. However, a factor of 2 variation is almost meaningless, since the MATCs for specific toxicants, species, and water combinations can easily vary by a factor of 2. Data for flagfish exposed to cadmium in soft water (Table 6, tests 3, 4, 5, and 6) and to zinc in soft water (Table 6, tests 19 and 20) illustrate this.

McKim (1977) showed that the sensitivity of embryo-larval, early juvenile stages was usually the same when embryos from exposed and unexposed parents were compared. There still seems to be considerable variation in the data (Table 6), but in general the determination of freshwater MATCs does not seem to be drastically affected by parental exposure. This is due in part to the variability and insensitivity of the ELS and life cycle test designs as well as the mode of action of the chemical. Saltwater ELS exposures started with embryos from exposed (Pe) or unexposed (Pue) parents show essentially the same variability as the freshwater studies. Specifically, 4 of the 11 saltwater studies showed no differences in sensitivity between Pue and Pe embryos (tests 4, 8, 10, and 11), 2 studies showed Pue embryos to be more sensitive (tests 2 and 5), and 5 studies showed Pe embryos to be more sensitive (tests 1, 3, 6, 7, and 9). Again, a combination of variations in experimental design and toxicant mode of action seems to make this difficult to evaluate. These variations usually make at most a twofold difference in the MATC.

The critical life stage effects listed for each toxicant in Table 6 are representative of the toxic effects used to determine the MATC in each case. In cases where the MATC was determined by the ELS portion of the life cycle test, growth and mortality of larval and early juvenile fish were the major effects observed. Embryo hatchability was next, followed by chorion fragility and deformities. Other symptomatologies of toxic effect were noted in most of these studies and are described in the references cited. These symptomatologies (ELS effects), shown in column 3 of Table 1, are manifested later in development through growth and mortality. These later manifestations are the major effects used to establish the statistically significant effect levels (MATCs) listed in Table 6. Ward and Parrish (1980) found that in 90% of their ELS studies with sheepshead minnows (saltwater fish), mortality was the most sensitive indicator of effect. In only one case was growth the most sensitive indicator. This was not the case in freshwater fish or several saltwater studies listed in Table 6, although most of the sheepshead minnow effects were mortality.

Data on the sensitivity of various species to single toxicants in the same water are summarized in Table 7. With some toxicants, species differences in sensitivity were not very great, while with other toxicants sensitivity varied considerably among species. Thus the most important species or the most sensitive, if known, should be tested in a particular situation. For screening large numbers of chemicals, one or two consistently sensitive species (e.g., fathead minnows and sheepshead minnows) will suffice until more is known about species sensitivity to specific types or classes of compounds. The ELS toxicity test should prove effective for rapidly testing the sensitivity of selected species to specific classes of chemicals in order to help determine the minimum number of species required in developing criteria for safe environmental concentrations.

CONCLUSIONS

Early life stage toxicity tests with fish have developed rapidly and can be used to estimate the MATC within a factor of 2 for most freshwater and saltwater fish. Eighty-three percent of the time the MATC estimated from ELS effects was identical to that established by the longer, more involved, and more costly partial or complete life cycle toxicity test. Water quality, species, type of

Table 7 Estimated MATC for 12 Species of Fish Established By Embryo, Larval, and Early Juvenile Tests and Grouped According to Quality of Diluent Water Used[a]

Water type and test location	Species	Estimated MATC (μg/l)	Type of embryo, larval test	Reference
		Cadmium		
Soft water (45 mg/l as $CaCO_3$) Duluth, Minn.	Flagfish	8.1–16.0	60-d (Pue) 25°C	Spehar (1976)
	Flagfish	7.4–16.9	30-d (Pue) (Pe) 25°C	A. R. Carlson and J. H. Tucker[b]
	Flagfish	3.0–6.5	30-d (Pue) (Pe) 25°C	A. R. Carlson and J. H. Tucker[b]
	Flagfish	3.4–7.3	30-d (Pue) (Pe) 25°C	A. R. Carlson and J. H. Tucker[b]
	Brook trout	1.7–3.4	90-d (Pe) 9°C	Benoit et al. (1976)
	Brook trout	1.1–3.8	90-d (Pue) 10°C	Eaton et al. (1978)
	Brown trout	3.8–11.7	60-d (Pue) 10°C	Eaton et al. (1978)
	Lake trout	4.4–12.3	60-d (Pue) 10°C	Eaton et al. (1978)
	Coho salmon	4.1–12.5	60-d (Pue) 10°C	Eaton et al. (1978)
	Northern pike	4.0–12.9	30-d (Pue) 16°C	Eaton et al. (1978)
	Smallmouth bass	4.3–12.7	30-d (Pue) 20°C	Eaton et al. (1978)
	White sucker	4.2–12.0	30-d (Pue) 18°C	Eaton et al. (1978)
Soft water (36 mg/l as $CaCO_3$) Wareham, Mass.	Brook trout	1.0–3.0	60-d (Pue) 10°C	Sauter et al. (1978)
	Channel catfish	11.0–17.0	60-d (Pue) 22°C	Sauter et al. (1978)
Hard water (200 mg/l as $CaCO_3$) Newtown, Ohio	Fathead minnow	37.0–57.0	60-d (Pue) (Pe) 24°C	Pickering and Gast (1972)
	Bluegill	31.0–80.0	60-d (Pe) 26°C	Eaton (1974)
Hard water (187 mg/l as $CaCO_3$) Wareham, Mass.	Brook trout	7.0–12.0	60-d (Pue) 10°C	Sauter et al. (1976)
	Channel catfish	12.0–17.0	60-d (Pue) 22°C	Sauter et al. (1976)
		Chromium		
Soft water (36 mg/l as $CaCO_3$) Wareham, Mass.	Rainbow trout	51.0–105.0	60-d (Pue) 10°C	Sauter et al. (1976)
	Lake trout	105.0–194.0	60-d (Pue) 10°C	Sauter et al. (1976)
	Channel catfish	150.0–305.0	60-d (Pue) 22°C	Sauter et al. (1976)
	Bluegill	522.0–1122.0	60-d (Pue) 25°C	Sauter et al. (1976)
	White sucker	290.0–538.0	60-d (Pue) 17°C	Sauter et al. (1976)
Soft water (45 mg/l as $CaCO_3$) Duluth, Minn.	Rainbow trout	200.0–350.0	90-d (Pue) 10°C	Benoit (1976)
	Brook trout	200.0–350.0	90-d (Pue) 9°C	Benoit (1976)
		Copper		
Soft water (40 mg/l as $CaCO_3$) Newtown, Ohio	Fathead minnow	10.6–18.4	90-d (Pue) (Pe) 24°C	Mount and Stephan (1969)
Soft water (45 mg/l as $CaCO_3$) Duluth, Minn.	Brook trout	9.5–17.4	90-d (Pe) 9°C	McKim and Benoit (1971)
	Brook trout	9.5–17.4	90-d (Pue) (Pe) 9°C	McKim and Benoit (1974)
	Bluegill	21.0–40.0	90-d (Pue) 27°C	Benoit (1975)
	Brook trout	22.3–43.5	60-d (Pue) 6°C	McKim et al. (1978)
	Lake trout	22.0–42.3	60-d (Pue) 6°C	McKim et al. (1978)
	Brown trout	22.0–43.2	60-d (Pue) 6°C	McKim et al. (1978)

Note: See page 88 for footnotes.

Table 7 Estimated MATC for 12 Species of Fish Established by Embryo, Larval, and Early Juvenile Tests and Grouped According to Quality of Diluent Water Used[a] (*Continued*)

Water type and test location	Species	Estimated MATC (μg/l)	Type of embryo, larval test	Reference
	Rainbow trout	11.4–31.7	30-d (Pue) 11°C	McKim et al. (1978)
	White sucker	12.9–33.8	30-d (Pue) 15°C	McKim et al. (1978)
	Northern pike	34.9–104.4	30-d (Pue) 16°C	McKim et al. (1978)
Soft water	Brook trout	5.0–7.0	60-d (Pue) 10°C	Sauter et al. (1976)
(36 mg/l as $CaCO_3$)	Channel catfish	12.0–18.0	60-d (Pue) 22°C	Sauter et al. (1976)
Wareham, Mass.				
Hard water	Brook trout	5.0–8.0	60-d (Pue) 10°C	Sauter et al. (1976)
(187 mg/l as $CaCO_3$)	Channel catfish	13.0–19.0	60-d (Pue) 22°C	Sauter et al. (1976)
Wareham, Mass.				
		Lead		
Soft water	Brook trout	58.0–119.0	(Pe) 9°C	Holcombe et al. (1976)
(45 mg/l as $CaCO_3$)	Flagfish	60.0–125.0	(Pue) 27°C	L. E. Anderson et al.[b]
Duluth, Minn.				
Soft water	Rainbow trout	71.0–146.0	60-d (Pue) 10°C	Sauter et al. (1976)
(36 mg/l as $CaCO_3$)	Lake trout	48.0–83.0	60-d (Pue) 10°C	Sauter et al. (1976)
Wareham, Mass.	Channel catfish	75.0–136.0	60-d (Pue) 22°C	Sauter et al. (1976)
	Bluegill	70.0–120.0	60-d (Pue) 25°C	Sauter et al. (1976)
	White sucker	119.0–253.0	60-d (Pue) 17°C	Sauter et al. (1976)
		Zinc		
Soft water	Flagfish	51.0–85.0	30-d (Pue) 25°C	Spehar (1976)
(45 mg/l as $CaCO_3$)	Flagfish	139.0–267.0	30-d (Pue) 25°C	Spehar (1976)
Duluth, Minn.	Brook trout	532.0–1368.0	90-d (Pue) (Pe) 9°C	Holcombe and Benoit (1979)
	Fathead minnow	78.0–145.0	90-d (Pue) (Pe) 23°C	Benoit and Holcombe (1978)
		Organometallics methylmercuric chloride		
Soft water	Fathead minnow	0.13–0.23	60-d (Pue) 23°C	D. I. Mount and G. F. Olson[b]
(45 mg/l as $CaCO_3$)	Flagfish	0.17–0.33	30-d (Pue) 25°C	W. E. Smith[b]
Duluth, Minn.	Brook trout	0.29–0.93	90-d (Pe) 9°C	McKim et al. (1976)
		Linear alkylate sulfate (LAS)		
Soft water	Fathead minnow	1100.0	30-d (Pue) 23°C	McKim et al. (1975)
(45 mg/l as $CaCO_3$)	Northern pike	1200.0	30-d (Pue) 15°C	McKim et al. (1975)
Duluth, Minn.	Smallmouth bass	5800.0	30-d (Pue) 15°C	McKim et al. (1975)
	White sucker	500.0	30-d (Pue) 15°C	McKim et al. (1975)

[a]From McKim (1977); reproduced by permission of the Ministry of Supply and Services Canada.
[b]Personal communication, Environmental Research Laboratory, Duluth, Minn.

toxicant, and parental exposure had little effect on these conclusions.

These conclusions for individual toxicants are probably also true for toxicant mixtures or complex effluents. Little information is available on life cycle toxicity tests with complex effluents or mixtures, but the work cited in Table 6 on sewage (tests 53 to 56) and metal mixtures (test 23) supports this assumption to some degree. Laboratory and field methods now available for ELS tests allow relatively rapid screening of single chemicals and complex effluents with selected species both in the laboratory and on site (field testing). Further development of these tests is under way; the major emphasis is on reducing the duration of exposure through the development of new and more sophisticated toxicity end points by embryological, physiological, and biochemical approaches.

Muul et al. (1976) described the need in mammalian toxicology for faster, less costly tests to screen out potentially carcinogenic and toxic chemicals from the ever-expanding list of potentially useful chemicals. Faster, less costly testing procedures are also of extreme importance in aquatic toxicology. In the United States these procedures are especially desirable for the implementation of the Federal Insecticide, Fungicide, and Rodenticide Act of 1972 and the Toxic Substances Control Act of 1976. These laws require prescreening by industry of some chemicals before commercial production. The use of ELS fish toxicity tests permits testing with species which, because of their extensive life cycle requirements, could not be used in a life cycle test. In addition, these tests provide the opportunity to screen large numbers of chemicals rapidly at a far lower cost per test than a life cycle toxicity test.

Life cycle tests with aquatic invertebrates will continue to be important for evaluating chemical impacts on the aquatic environment. However, the development of standardized predictive tests is less important than with fish because of the short time period required for definitive invertebrate life cycle tests.

LITERATURE CITED

Adelman IR, Smith LL, Jr, Siesennop GD: Chronic toxicity of guthion to the fathead minnow (*Pimephales promelas* Rafinesque). Bull Environ Contam Toxicol 15:726–733, 1976.

Allison DT: Use of exposure units for estimating aquatic toxicity of organophosphate pesticides. EPA-600/3-77-077. Duluth, Minn.: U.S. EPA, 1977.

Allison DT, Hermanutz RO: Toxicity of diazinon to brook trout and fathead minnows. EPA-600/3-77-060. Duluth, Minn.: U.S. EPA, 1977.

American Public Health Association, American Water Works Association, and Water Pollution Control Federation: Standard Methods for the Examination of Water and Wastewater, 14th ed. Washington, D.C.: American Public Health Association, 1975.

American Society for Testing and Materials: Standard practice for conducting acute toxicity tests with fishes, macroinvertebrates, and amphibians. ASTM E729-80. Philadelphia: ASTM, 1980a.

American Society for Testing and Materials: Annual Book of ASTM Standards. Part 31: Water. Philadelphia: ASTM, 1980b.

American Society for Testing and Materials: Standard practice for conducting chronic toxicity tests with the early life stages of fish. ASTM E-47.01. Philadelphia: ASTM, 1981 (draft).

Anderson RL: Chironomidae toxicity tests—Biological background and procedures. In: Aquatic Invertebrates Bioassays, edited by AL Buikema Jr., J Cairns Jr., pp. 70–80. ASTM STP 715. Philadelphia: ASTM, 1980.

Arthur JW: Chronic effects of linear alkylate sulfonate detergent on *Gammarus pseudolimnaeus, Campeloma decisum,* and *Physa integra.* Water Res 4:251–257, 1970.

Arthur JW: Review of freshwater bioassay procedure for selected amphipods. In: Aquatic Invertebrates Bioassays, edited by AL Buikema Jr., J Cairns Jr., pp. 98–108. ASTM STP 715. Philadelphia: ASTM, 1980.

Arthur JW, Eaton JG: Chloramine toxicity to the amphipod *Gammarus pseudolimnaeus* and the fathead minnow (*Pimephales promelas*). J Fish Res Bd Can 28:1841–1845, 1971.

Arthur JW, Lemke AE, Mattson VR, Halligan BJ: Toxicity of sodium nitrilotriacetate (NTA) to the fathead minnow and an amphipod in soft water. Water Res 8:187–193, 1974.

Arthur JW, Andrew RW, Mattson VR, Olson DT, Glass GE, Halligan BJ, Walbridge CT: Comparative toxicity of sewage-effluent disinfectants to freshwater aquatic life. EPA-600/3-75-012. U.S. EPA, 1975.

Balon EK: Terminology of intervals in fish development. J Fish Res Bd Can 32:1663–1670, 1975.

Benoit DA: Chronic effects of copper on survival, growth, and reproduction of the bluegill (*Lepomis macrochirus*). Trans Am Fish Soc 104:353–358, 1975.

Benoit DA: Toxic effects of hexavalent chromium on brook trout (*Salvelinus fontinalis*) and rainbow trout (*Salmo gairdneri*). Water Res 10: 497–500, 1976.

Benoit DA, Holcombe GW: Toxic effects of zinc on fathead minnows (*Pimephales promelas*) in soft water. J Fish Biol 13:701–708, 1978.

Benoit DA, Leonard EN, Christensen GM, Hunt EP: Toxic effects of cadmium on three generations of brook trout (*Salvelinus fontinalis*). Trans Am Fish Soc 105:550–560, 1976.

Benoit DA, Mattson VR, Olson DL: A compact continuous flow mini-diluter exposure system for testing early life stages of fish and invertebrates in single chemicals and complex effluents. Water Res, 1981a, in press.

Benoit DA, Puglisi FA, Olson DL: A fathead minnow (*Pimephales promelas*) early life stage toxicity test method evaluation and exposure to four organic chemicals. Environ Pollut, 1981b, in press.

Benoit DA, Syrett RF, Freeman FB: Design manual for construction of a continuous flow mini-diluter exposure system. Duluth, Minn.: U.S. EPA, 1981c (unpublished report).

Biesinger KE, Christensen GM: Effects of various metals on survival, growth, reproduction, and metabolism of *Daphnia magna*. J Fish Res Bd Can 29:1691–1700, 1972.

Box GEP, Hunter WG, Hunter JS: Statistics for Experimenters. New York: Wiley-Interscience, 1978.

Brungs WA: Chronic toxicity of zinc to the fathead minnow, *Pimephales promelas* Rafinesque. Trans Am Fish Soc 98:272–279, 1969.

Brungs WA: Chronic effects of low dissolved oxygen concentrations on the fathead minnow (*Pimephales promelas*). J Fish Res Bd Can 28: 1119–1123, 1971.

Brungs WA, Jones BR: Temperature criteria for freshwater fish: Protocol and procedures. EPA-600/3-77-061. Duluth, Minn.: U.S. EPA, 1977.

Buikema AL Jr., Cairns J Jr. (eds): Aquatic invertebrate bioassays, pp. 1–209. ASTM STP 715. Philadelphia: ASTM, 1980.

Buikema AL Jr., Geiger JG, Lee DR: *Daphnia* toxicity tests. In: Aquatic Invertebrate Bioassays, edited by AL Buikema Jr., J Cairns Jr., pp. 48–69. ASTM STP 715. Philadelphia: ASTM, 1980.

Calabrese A, Collier RS, Nelson DA, MacInnes JR: The toxicity of heavy metals to embryos of the American oyster (*Crassostrea virginica*). Mar Biol 18:162–166, 1973.

Carlson AR: Effects of long-term exposure of carbaryl (sevin) on survival, growth, and reproduction of the fathead minnow (*Pimephales promelas*). J Fish Res Bd Can 29:583–587, 1971.

Chapman GA, Shumway DL: Effects of sodium pentachlorophenate on survival and energy metabolism of embryonic and larval steelhead trout. In: Pentachlorophenol, edited by RK Rao, pp. 285–299. New York: Plenum, 1977.

Derr SK, Zabik MJ: Biologically active compounds in the aquatic environment: The effect of DDE on the egg viability of *Chironomus tentans*. Bull Environ Contam Toxicol 7:366–368, 1972a.

Derr SK, Zabik MJ: Biologically active compounds in the aquatic environment: The uptake and distribution of [1,1-dichloro-2,2-bis(*p*-chlorophenyl) ethylene], DDE by *Chironomus tentans* Fabricius (Diptera: Chiromonidae). Trans Am Fish Soc 101:323–329, 1972b.

DeFoe DL, Veith GD, Carlson RW: Effects of Aroclor 1248 and 1260 on the fathead minnow (*Pimephales promelas*). J Fish Res Bd Can 35: 997–1002, 1978.

Dunnett CW: A multiple comparisons procedure for comparing several treatments with a control. J Am Stat Assoc 50:1096–1121, 1955.

Dunnett CW: New tables for multiple comparisons with a control. Biometrics 20:482–491, 1964.

Eaton JG: Chronic malathion toxicity to the bluegill (*Lepomis macrochirus* Rafinesque). Water Res 4:673–684, 1970.

Eaton JG: Chronic toxicity of copper, cadmium, and zinc mixture to the fathead minnow (*Pimephales promelas*). Water Res 7:1723–1736, 1973.

Eaton JG: Chronic cadmium toxicity to the bluegill (*Lepomis macrochirus* Rafinesque). Trans Am Fish Soc 103:729–735, 1974.

Eaton JG, McKim JM, Holcombe GW: Metal toxicity to embryos and larvae of seven freshwater fish species. I: Cadmium. Bull Environ Contam Toxicol 19:95–103, 1978.

Geiger JG, Buikema AL Jr., Cairns J Jr.: A tentative seven-day test for predicting effects of stress on populations of *Daphnia pulex*. In: Aquatic Toxicology, edited by JG Eaton, PR Parrish, AC Hendricks, pp. 13–26. ASTM STP 707. Philadelphia: ASTM, 1980.

Gingerich WH, Seim WK, Schonbrod RD: An apparatus for the continuous generation of stock solutions of hydrophobic chemicals. Bull Environ Contam Toxicol 23:685–689, 1979.

Goodman LR, Hansen DJ, Couch JA, Forester J: Effects of heptachlor toxaphene on laboratory-reared embryos and fry of the sheepshead minnow. 13th Annu Conf Southeast Assoc Game Fish Comm, pp. 192–202, 1976.

Goodman LR, Hansen DJ, Manning CS, Faas LF: Effects of Kepone on the sheepshead minnow in an entire life-cycle toxicity test. Arch Environ Contam Toxicol 11:335–342, 1982.

Hansen DJ, Parrish PR: Suitability of sheepshead minnows (*Cyprindon variegatus*) for life-cycle toxicity tests. In: Aquatic Toxicology and Hazard Evaluation, edited by FL Mayer, JL Hamelink, pp. 117–126. ASTM STP 634. Philadelphia: ASTM, 1977a.

Hansen DJ, Schimmel SC, Forester J: Endrin: Effects on the entire life-cycle of saltwater fish, *Cyprinodon variegatus*. J Toxicol Environ Health 3:721–726, 1977b.

Helder T: Effects of 2,3,7,8-tetrachlorodibenzo-*p*-dioxin (TCDD) on early life stages of the pike (*Esox lucius* L.). Sci Total Environ 14:255–264, 1980.

Hermanutz R: Endrin and malathion toxicity to flagfish (*Jordanella floridae*). Arch Environ Contam Toxicol 7:159–168, 1978.

Hermanutz RO, Mueller LH, Kempfert KD: Captan toxicity to fathead minnows (*Pimephales promelas*), bluegills (*Lepomis macrochirus*), and brook trout (*Salvelinus fontinalis*). J Fish Res Bd Can 30:1811–1817, 1973.

Hokanson KEF, McCormick JH, Jones BR, Tucker JH: Thermal requirements for maturation, spawning, and embryo survival of the brook trout (*Salvelinus fontinalis*). J Fish Res Bd Can 30:975–984, 1973a.

Hokanson KEF, McCormick JH, Jones BR: Temperature requirements for embryos and larvae of the northern pike, *Esox lucius* (Linnaeus). Trans Am Fish Soc 102:89–100, 1973b.

Holcombe GW, Benoit DA, Leonard EN, McKim JM: Long-term effects of lead exposure on three generations of brook trout (*Salvelinus fontinalis*). J Fish Res Bd Can 33:1731–1741, 1976.

Holcombe GW, Benoit DA, Leonard EN: Long-term effects of zinc exposures on brook trout (*Salvelinus fontinalis*). Trans Am Fish Soc 108: 76–87, 1979.

Hollister TA, Heitmuller PT, Parrish PR, and Dyar EE: Studies to determine relationships between time and toxicity of an acidic effluent and an alkaline effluent to two estuarine species. In: Aquatic Toxicology, edited by JG Eaton, PR Parrish, AC Hendricks, pp. 251–265. ASTM STP 707. Philadelphia: ASTM, 1980.

Ishida J, Tagushi S, Maruyama K: ATP content and ATPase activity in developing embryos of the teleost (*Oryzias latipes*). Annot Zool Jpn 32:1–5, 1959.

Jarvinen AW, Tyo RM: Toxicity to fathead minnows of endrin in food and water. Arch Environ Contam Toxicol 7:409–421, 1978.

Jarvinen AW, Hoffman MJ, Thorslund TW: Long-term toxic effects of DDT food and water exposure on fathead minnows (*Pimephales promelas*). J Fish Res Bd Can 34:2089–2103, 1977.

Jones PW, Martin FD, Hardy JD Jr.: Development

of fishes of the mid-Atlantic bight: An atlas of egg, larval and juvenile stages, vol. 1. Biol Serv Prog US Dept Inter Publ FWS/OBS-78/12, 1978.

Lagler KF, Bardach JE, Miller RM, Passino DM: Ichthyology, 2d ed. New York: Wiley, 1977.

Lemke AE, Brungs WA, Halligan BJ: Manual for construction and operation of toxicity-testing proportional diluters. EPA-600/3-78-072. Duluth, Minn.: U.S. EPA, 1978.

Macek KJ, Sleight BH: Utility of toxicity tests with embryos and fry of fish in evaluating hazards associated with the chronic toxicity of chemicals to fishes. In: Aquatic Toxicology and Hazard Evaluation, edited by FL Mayer, JL Hamelink, pp. 137–146. ASTM STP 634. Philadelphia: ASTM, 1977.

Macek KJ, Buxton KS, Derr SK, Dean JW, Sauter S: Chronic toxicity of lindane to selected aquatic invertebrates and fishes. EPA-600/3-76-046. U.S. EPA, 1976a.

Macek KJ, Buxton KS, Sauter S, Gnilka S, Dean JW: Chronic toxicity of atrazine to selected invertebrates and fishes. EPA-600/3-76-047. U.S. EPA, 1976b.

Macek KJ, Lindberg MA, Sauter S, Buxton KS, Costa PA: Toxicity of four pesticides to water fleas and fathead minnows. Acute and chronic toxicity to acrolein, heptachlor, endosulfan, and trifluralin to the water flea (*Daphnia magna*) and the fathead minnow (*Pimephales promelas*), EPA-600/3-76-099. U.S. EPA, 1976c.

Mayer FL Jr., Mehrle PM Jr., Dwyer WP: Toxaphene effects on reproduction, growth, and mortality of brook trout. EPA-600/3-075-013. Duluth, Minn.: U.S. EPA, 1975.

Mayer FL Jr., Mehrle PM Jr., Dwyer WP: Toxaphene: Chronic toxicity to fathead minnows and channel catfish. EPA-600/3-77-069. Duluth, Minn.: U.S. EPA, 1977.

McClave JT, Sullivan JH, Pearson JG: Statistical analysis of fish chronic toxicity. In: Aquatic Toxicology and Hazard Evaluation, edited by DR Branson and KL Dickson, pp. 359–376. ASTM STP 737. Philadelphia: ASTM, 1981.

McCormick JH, Jones BR, and Hokanson KEF: White sucker (*Catostomus commersoni*) embryo development, and early growth and survival at different temperatures. J Fish Res Bd Can 34:1019–1025, 1977.

McKim JM: Evaluation of tests with early life stages of fish for predicting long-term toxicity. J Fish Res Bd Can 34:1148–1154, 1977.

McKim JM, Benoit DA: Effects of long-term exposures to copper on the survival, growth, and reproduction of brook trout. J Fish Res Bd Can 28:655–662, 1971.

McKim JM, Benoit DA: Duration of toxicity tests for establishing "no effect" concentrations for copper with brook trout. J Fish Res Bd Can 31: 449–452, 1974.

McKim JM, Arthur JW, Thorslund TW: Toxicity of a linear alkylate sulfonate detergent to larvae of four species of freshwater fish. Bull Environ Contam Toxicol 14:1–7, 1975.

McKim JM, Holcombe GW, Olson GF, Hunt EP: Long-term effects of methylmercuric chloride on three generations of brook trout (*Salvelinus fontinalis*): Toxicity, accumulation, distribution, and elimination. J Fish Res Bd Can 33: 2726–2739, 1976.

McKim JM, Eaton JG, Holcombe GW: Metal toxicity to embryos and larvae of eight species of freshwater fish. II: Copper. Bull Environ Contam Toxicol 19:608–616, 1978.

Mehrle PM, Mayer FL Jr.: Toxaphene effects on growth and bone composition of fathead minnows, *Pimephales promelas*. J Fish Res Bd Can 32:593–598, 1975.

Menendez R: Chronic effects of reduced pH on brook trout (*Salvelinus fontinalis*). J Fish Res Bd Can 33:118–123, 1976.

Meyer W, Sauerbier I: The development of catecholamines in embryos and larvae of the rainbow trout (*Salmo gairdneri* Rich.). J Fish Biol 10:431–435, 1977.

Middaugh DP, Dean JM: Comparative sensitivity of eggs, larvae, and adults of the estuarine teleosts, *Fundulus heteroclitus* and *Menidia menidia* to cadmium. Bull Environ Contam Toxicol 17:645–651, 1977.

Middaugh DP, Lempesis PW: Laboratory spawning and rearing of a marine fish, the silverside *Menidia menidia menidia*. Mar Biol 35:295–300, 1976.

Mount DI: Chronic toxicity of copper to fathead minnows. Water Res 2:215–223, 1968.

Mount DI: Chronic effect of low pH on fathead minnow survival, growth, and reproduction. Water Res 7:987–993, 1973.

Mount DI: An assessment of application factors in aquatic toxicology. In: Recent Advances in Fish Toxicology—A Symposium, edited by RA Tubb, pp. 183–189. EPA-600/3-77-085. U.S. EPA, 1977.

Mount DI, Brungs WA: A simplified dosing apparatus for fish toxicology studies. Water Res 1:21–29, 1967.

Mount DI, Stephan CE: A method for establishing acceptable limits for fish—Malathion and the butoxyethanol ester of 2,4-D. Trans Am Fish Soc 96:185–193, 1967.

Mount DI, Stephan CE: Chronic toxicity of copper to the fathead minnow (*Pimephales promelas*) in soft water. J Fish Res Bd Can 26: 2449–2457, 1969.

Muul I, Hegyeli AF, Dacre J, Woodard G: Toxicological testing dilemma. Science 193:834, 1976.

Nebeker AV: Temperature requirements and life cycle of the midge *Tanytarsus dissimilis*. J Kans Entomol Soc 46:160–165, 1973.

Nebeker AV, Puglisi FA, DeFoe DL: Effect of polychlorinated biphenyl compounds on survival and reproduction of the fathead minnow and flagfish. Trans Am Fish Soc 103:562–568, 1974.

Nebeker AV, Andros JD, McCrady JK, Stevens DG: Survival of steelhead trout (*Salmo gairdneri*) eggs, embryos, and fry in air-supersaturated water. J Fish Res Bd Can 35:261–264, 1977.

Nimmo DR, Bahner LH, Rigby RA, Sheppard JM, Wilson AJ Jr.: *Mysidopsis bahia:* An estuarine species suitable for life-cycle toxicity tests to determine the effects of a pollutant. In: Aquatic Toxicology and Hazard Evaluation, edited by FL Mayer, JL Hamelink, pp. 109–116. ASTM STP 634. Philadelphia: ASTM, 1977.

Parrish PR, Schimmel SC, Hansen DJ, Patrick JM Jr., Forester J: Chlordane: Effects on several estuarine organisms. J Toxicol Environ Health 1:485–494, 1976.

Parrish PR, Dyar EE, Lindberg MA, Shanike CM, Enos JM: Chronic toxicity of methoxychlor, malathion, and carbofuran to sheepshead minnows (*Cyprinodon variegatus*). EPA-600/3-77-059. U.S. EPA, 1977.

Parrish PR, Dyar EE, Enos JM, Wilson WG: Chronic toxicity of chlordane, trifluralin, and pentachlorophenol to sheepshead minnows (*Cyprinodon variegatus*). EPA-600/3-78-010. U.S. EPA, 1978.

Pickering QH: Chronic toxicity of nickel to the fathead minnow. J Water Pollut Control Fed 46:760–765, 1974.

Pickering QH, Gast MH: Acute and chronic toxicity of cadmium to the fathead minnow *Pimephales promelas.* J Fish Res Bd Can 29:1099–1106, 1972.

Pickering QH, Thatcher TO: The chronic toxicity of linear alkylate sulfonate (LAS) to *Pimephales promelas* Rafinesque. J Water Pollut Control Fed 42:243–254, 1970.

Pickering QH, Brungs W, Gast M: Effect of exposure time and copper concentration on reproduction of the fathead minnow (*Pimephales promelas*). Water Res 11:1079–1083, 1977.

Reish DJ: Use of polychaetous annelids as test organisms for marine bioassay experiments. In: Aquatic Invertebrate Bioassays, edited by AL Buikema Jr., J Cairns Jr., pp. 140–154. ASTM STP 715. Philadelphia: ASTM, 1980.

Roberts MH Jr.: Flow-through toxicity testing system for molluscan larvae as applied to halogen toxicity in estuarine water. In: Aquatic Invertebrate Bioassays, edited by AL Buikema Jr., J Cairns Jr., pp. 131–139. ASTM STP 715. Philadelphia: ASTM, 1980.

Rosenthal H, Alderdice DF: Sublethal effects of environmental stressors, natural and pollutional, on marine fish eggs and larvae. J Fish Res Bd Can 33:2047–2065, 1976.

Sauter S, Buxton KS, Macek KJ, Petrocelli SR: Effects of exposure to heavy metals on selected freshwater fish. EPA-600/3-76-105. U.S. EPA, 1976.

Schimmel SC, Hansen DJ: Sheepshead minnow (*Cyprinodon variegatus*): An estuarine fish suitable for chronic (entire life-cycle) bioassays.

Proc 28th Am Conf Southeast Game Fish Comm, pp. 392–398, 1974.

Smith, WE: A cyprinodontid fish, *Jordanella floridae,* as reference animals for rapid chronic bioassays. J Fish Res Bd Can 39:329–330, 1973.

Snyder DE, Snyder MBM, Douglas SC: Identification of golden shiner, *Notemigonus crysoleucas,* spotfin shiner, *Notropis spilopterus,* and fathead minnow, *Pimephales promelas,* larvae. J Fish Res Bd Can 34:1397–1409, 1977.

Sokal RR, Rohlf FJ: Biometry. San Francisco: Freeman, 1969.

Spehar RL: Cadmium and zinc toxicity to *Jordanella floridae.* J Fish Res Bd Can 33:1939–1945, 1976.

Steel RG, Torrie JH: Principles and Procedures of Statistics with Special Reference to Biological Sciences. New York: McGraw-Hill, 1960.

Strickland JDG, Parsons TR: A practical handbook of seawater analysis. Fish Res Bd Can Bull 167:21–26, 1968.

Tyler-Schroeder DB: Use of the grass shrimp (*Palaemonetes pugio*) in a life-cycle toxicity test. In: Aquatic Toxicology, edited by LL Marking, AA Kimerle, pp. 159–170. ASTM STP 667. Philadelphia: ASTM, 1979.

Uesugi S, Yamazoe S: Acetyl cholinesterase activity during the development of rainbow trout eggs. Gunma J Med Sci 13:91–93, 1964.

U.S. Environmental Protection Agency: Methods for chemical analysis of water and wastes. EPA-600/4-79-020. U.S. EPA, 1979.

U.S. Environmental Protection Agency: Manual of analytical methods for the analysis of pesticides in humans and environmental samples. EPA-600/8-80-038. U.S. EPA, 1980.

U.S. Environmental Protection Agency: Acephate, aldicarb, carbophenothion, DEF, EPN, ethoprop, methyl parathion, and phorate: Their acute and chronic toxicity, bioconcentration potential, and persistence as related to marine environments. EPA-600/4-81-023. Gulf Breeze, Fla.: U.S. EPA, 1981a.

U.S. Environmental Protection Agency: Test standard for conducting an early life stage toxicity test using the fathead minnow, sheepshead minnow, and brook trout or rainbow trout. Washington, D.C.: U.S. EPA, Health and Environmental Review Division (TS-792), 1981b (draft).

Veith GD, Comstock VM: Apparatus for continuously saturating water with hydrophobic organic chemicals. J Fish Res Bd Can 32:1849–1851, 1975.

Von Westernhagen H, Dethlefsen V, Rosenthal H: Combined effects of cadmium and salinity on development and survival of garpike eggs. Helgol Wiss Meeresunters 27:268–282, 1975.

Ward GS, Parrish PR: Evaluation of early life-stage toxicity tests with embryos and juveniles of sheepshead minnows (*Cyprinodon variegatus*). In: Aquatic Toxicology, edited by JG Eaton, PR Parrish, AC Hendricks, pp. 243–247. ASTM STP 707. Philadelphia: ASTM, 1980.

Ward RW, Griffin RD, DeGraeve GM, Stone RA: Disinfection efficiency and residual toxicity of several wastewater disinfectants, vol. 1. EPA-600/2-76-156. Grandville, Mich.: U.S. EPA, 1976.

Warner RE: Toxicant-induced behavioral and histological pathology: A quantitative study of sublethal toxication in the aquatic environment. Contract PH 66-63-72. U.S. Public Health Service, 1964.

Weis P, Weis JS: Methylmercury teratogenesis in the killifish, *Fundulus heteroclitus.* Teratology 16:321–324, 1977.

Williams DA: A test for differences between treatment means when several dose levels are compared with a zero dose control. Biometrics 27: 103–117, 1971.

Williams DA: The comparison of several dose levels with a zero dose control. Biometrics 28: 519–531, 1972.

Winner RW, Farell MP: Acute and chronic toxicity of copper to four species of *Daphnia.* J Fish Res Bd Can 33:1685–1691, 1976.

SUPPLEMENTAL READING

Blaxter JHS: Fish Physiology, chapter 4, Development: Eggs and Larvae. New York: Academic, 1969.

Blaxter JHS: The Early Life History of Fish, Proceedings of an International Symposium held at

Dunstaffnage Marine Research Laboratory of the Scottish Marine Biological Association at Oban, Scotland, March 17–23, 1973. New York: Springer-Verlag, 1974.

Buikema AL Jr., Cairns J Jr. (eds): Aquatic Invertebrate Bioassays, pp. 1–209, ASTM STP 715. Philadelphia: ASTM, 1980.

Lagler KF, Bardach JE, Miller RM, Passino DM: Icthyology, 2d ed. New York: Wiley, 1977.

Mansueti A, Hardy JD Jr.: Development of Fishes of the Chesapeake Bay Region: An Atlas of Egg, Larval and Juvenile Stages. College Park, Md.: Natural Resources Institute, University of Maryland, 1967.

New DAT: The Culture of Vertebrate Embryos, chapter 6, Fish. New York: Academic, 1966.

Yamamoto T: Medaka (Killifish): Biology and Strains. Tokyo: Yugaku-sha, 1975.

Appendix C

FACTORS THAT MODIFY TOXICITY

J. B. Sprague

INTRODUCTION

Various characteristics of water and of organisms that may change the toxicity of water pollutants are considered in this chapter. The preceding chapters on test methods placed considerable emphasis on standard procedures and control of test conditions, and the reader may occasionally have wondered whether it was necessary to be so careful about them. Here the reasons for the recommended procedures will become evident—they are largely an attempt to eliminate extraneous factors that may affect results of the toxicity test. For example, it will be seen below that poor control of pH during a test could have an enormous influence on the estimated toxicity of some pollutants. Thus, control of conditions in toxicity tests is simply an application of a basic scientific procedure: eliminating all the variables in an experiment except the one of interest. In this case, the series of concentrations of the toxicant is the variable of interest.

It may seem a little surprising that a chapter of this book emphasizes the variation in toxicity that is sometimes encountered. At times the reader may even be discouraged about the accuracy and value of toxicity testing, but this is not the intention. In some cases the modifying effects are surprisingly small, and the reader may then assimilate the interpretations and principles of other chapters. On the other hand, some major changes in toxicant potency are caused by modifying characteristics of the dilution water, and it is important to keep these in mind when considering the information in the rest of the book. Awareness of possible variations may help one to ignore some of the

Originally appeared as Chapter 6 in Rand and Petrocelli: *Fundamentals of Aquatic Toxicology*, © 1985 Taylor & Franicis.

less important minor differences in toxicity data and instead look for broader and more useful generalizations.

A section on unanticipated variation has been placed near the beginning of the chapter. Some examples are reassuringly precise, and others show rather shocking scatter of data. Some of this unexplained variation must make us uncomfortable as scientists, but it must not be ignored. Awareness of the background variation explains why a twofold difference in toxicity may sometimes be described here as a modest one. Perhaps it should be repeated that the intention is not to cast doubt on aquatic toxicology as a field of science, but to focus in a realistic way on the more important major effects of modifying factors.

Both biotic and abiotic characteristics may act as modifying factors. The biotic ones include all the features that are within the organism. Primary features are the type of organism–alga, insect, or fish–and beyond that the particular species, since one species may respond to pollutants in a different way than others in the same general group. Other biotic features are stage of life (larva, juvenile, adult), size of individual, nutritional status and health, seasonal changes in physiological state, and degree of acclimation to natural environmental conditions or to a pollutant. The abiotic conditions that can act as modifying factors are the multitude of physicochemical characteristics of the water surrounding an organism. Obvious ones are water temperature, pH, dissolved oxygen content, salinity or hardness, and the extent of fluctuation of each of these in a particular body of water. Other abiotic conditions could include suspended materials, organic or inorganic; dissolved salts or nutrients and their relative proportions; dissolved CO_2 and other gases; intensity of light and photoperiod; water movements and their velocity; and binding or chelating action of substances in the water.

A complete review of the literature on modifying factors is not attempted; that would expand this chapter to a book. However, the major effects are covered and illustrated by specific examples. Some items have been left out; an example is the sex of test organisms, since it usually has only a slight effect on reactions to pollutants. In any case, tests are often performed with sexually immature individuals, which are usually more sensitive than adults. The presence of another toxicant could obviously affect the action of a given pollutant, and this topic is covered in Chapter 7 by Marking. Information on how survival times are affected by modifying factors has been largely omitted. It often happens that there are differences in short-term survival, say in the first 24 h of an experiment, and yet the threshold of lethal effect is much the same if the experiment is continued long enough to determine that parameter.

TERMINOLOGY

Modifying factor. Any characteristic of the organism or the surrounding water that affects toxicity of a pollutant is considered to act as a modifying factor. Thus the size or species of the organism reacting to the toxicant could be a biotic modifying factor. Physicochemical entities such as pH or temperature of the water could be abiotic modifying factors. The environmental or abiotic entities should probably be called *masking factors,* following F. E. J. Fry's well-known classification of the environment according to the ways in which such entities act on an organism. By his definition, "A masking factor is an identity which modifies the operation of a second identity on the organism" (Fry, 1971). However, the term is not used here, because Fry's examples of the action of masking factors concern interactive effects within the organism, that is, physiological actions and the regulatory systems of the organism. In aquatic toxicology we are often still at the stage of observing a changed response without understanding the mechanism. Indeed, in many cases the modifying entity has a purely chemical effect on the toxic material, not within the organism but in the water. For example, pH can affect the chemical forms of metals, ammonia, and cyanide present in water. Use of the more general term, modifying factor, also allows inclusion of biotic characteristics that affect toxicity.

Resistance. Ability of an organism or species to resist a condition (e.g., a toxicant) which is at a level that will ultimately be lethal. Thus the description should ordinarily be accompanied by an indication of the time of exposure, and *median resistance times* (*median lethal times* or *median survival times*) are often used.

Tolerance. Ability of an organism to tolerate a condition for an indefinitely long exposure without dying. For toxicants, an organism is tolerant of concentrations at or below the incipient lethal level.

Isosmotic. Of equal osmotic pressure. Applicable when the dissolved materials in the body fluids of an organism exert an osmotic pressure equal to that of substances dissolved in the surrounding water. Thus there would be no tendency for the fish to gain or lose water.

Hyperosmotic. Having greater osmotic pressure than another fluid system. For example, a fish in fresh water would have body fluids with a greater concentration of dissolved materials than the concentration in the water, and the fish would tend to take in water. A *hypoosmotic* organism (most fish in the sea) would tend to lose water to the environment. In each case, the organisms would need regulatory systems to control the flux of water.

Euryhaline. Having a wide range of tolerance for salinity. For example, some estuarine organisms can exist in full-strength seawater or in low salinities approaching those of fresh water.

Stenohaline. Restricted to a narrow and constant range of salinity.

UNANTICIPATED VARIATION

When toxicity tests are repeated, results may not coincide exactly. A small part of such variation is due to truly random factors such as individual differences in resistance within any group of organisms. At present, however, large portions of the variation often cannot be explained; they could result from unrecognized changes in the organisms or in the test conditions.

The extremes of such variation may be illustrated by the experiences of three researchers in my laboratory. The least variation, comparing favorably with anything in the literature, was in the following series of 25 tests on copper lethality, carried out periodically during 2 yr as controls in an acclimation experiment with rainbow trout in very hard water (Dixon and Sprague, 1981b). The LC50s (μg/l) are listed in order of magnitude, not time of occurrence.

381	367	329	320	302
374	365	328	316	299
371	360	328	311	298
371	335	324	305	294
369	333	320	304	274

The geometric mean value was 330 μg/l and the highest and lowest values differed by only a factor of 1.39.

At the opposite extreme were two sets of results obtained about 3 yr earlier. These LC50s were obtained in adjacent parts of the laboratory with rainbow trout from the same stock, the same hard well-water mentioned above, measured concentrations of dissolved copper, and a flow-through apparatus but somewhat different flow rates. The LC50s were 94 μg/l (Fogels and Sprague, 1977) and 520 μg/l (Howarth and Sprague, 1978), different by a factor of 5.5 and different from the results given above. No explanation was evident to any of the researchers, and each result appears valid.

The situation is not unique to one laboratory. Brown (1968) reported that "the 48-hour LC50 values obtained for ammonia, phenol, and zinc in successive tests on a stock of rainbow trout over a period of 9 months varied by a factor of 2.5. The causes of this variation have not been identified." Adelman and Smith (1976) conducted eight sets of acute toxicity tests over 2 yr, using four different toxicants. They reported a coefficient of variation for NaCl of 6, a very low value which suggests that the ratio of the highest and lowest LC50s was about 1.3. Coefficients for the organophosphate insecticide Guthion were 29 and 39 for two species of fish, suggesting that the high and low LC50s differed by factors of about 5 or more.

Aquatic toxicologists may be uncomfortable about revealing such information, but it should not be hidden. Fogels and Sprague (1977) review the subject and conclude that between-laboratory differences in LC50 may approach a factor of 10.

An example of between-laboratory variability is a "ring test" or "round-robin" test, using the insecticide endosulfan against fish with a standard test procedure (EPA, 1979). LC50s in duplicate from four laboratories were as follows:

(Lab 1)	(Lab 2)	(Lab 3)	(Lab 4)
3.30	1.88	1.08	0.96
2.10	1.25	0.89	0.68

The geometric mean was 1.34, and the lowest and highest values differed by a factor of 4.9.

Some extremely variable examples were chosen above in order to make a point, but often results will be much closer. For example, there was only a twofold variation in an interlaboratory comparison of the lethality of sodium pentachlorophenate (Davis and Hoos, 1975). In a European ring test of 11 detergents, each laboratory used its particular standard method and species of fish, which varied from brown trout to the tropical harlequin fish. Eleven laboratories conducted either 2-d or 4-d tests. For the detergent showing the least interlaboratory variation, the factor between high and low LC50s was only 2.2--remarkably close considering the variety of approaches. For the most variable detergent the factor was 12, and the median of the factors was 6.8. The authors concluded that "a reliable order of magnitude for the acute toxicity to fish of the group of detergents tested can be obtained by a variety of test fish and by a variety of test procedures" (Reiff et al., 1979).

This section was begun with an example of good agreement among toxicity tests, and perhaps it would be well to end it with a similar example. Lloyd (1961b) published a guide for estimating the lethal level of ammonia, based on his research with rainbow trout and the chemical behavior of ammonia in water. In using the guide, an LC50 is read from a graph according to the pH and alkalinity of the water. Correction factors are applied for water temperature and for amounts of dissolved oxygen and carbon dioxide. The system may be used to predict LC50s for the water characteristics used in a recent study of ammonia toxicity to rainbow trout (Thurston et al., 1981). Over the usable ranges given by Lloyd the agreement with actual LC50s (total ammonia as milligrams of N per liter) is very good, as shown in Table 1. The greatest difference is a factor of 1.14, rather satisfactory since the two pieces of research span 20 yr and the Atlantic Ocean, and involve five chemical characteristics of the water which act as modifying factors.

TEST CONDITIONS

Here we consider more subtle things than test temperature or basic type of water. Obvious conditions such as having enough test solution per gram of fish fall in the category of good practice; these are fairly well established and are covered in other parts of this book.

An assortment of biotic and abiotic components of test procedure may affect results, although usually not in a major way. Seasonal

Table 1 Predicted and Measured LC50s for Ammonia in Rainbow Trout

Oxygen in test (% saturation)	LC50 (total ammonia as mg N/l)	
	Predicted (Lloyd, 1961b)	Actual (Thurston et al., 1981)
81	42	45
62	34	38
41	25	24
30	21	24

variation is a possibility, and may have contributed to the 2.5-fold variation in response to ammonia, phenol, and zinc over 9 mo mentioned above (Brown, 1968). However, trout showed the same LC50s for pulp mill effluent whether acclimated to summer or winter photoperiod (McLeay and Gordon, 1978). In another test with zinc, a significant change in resistance time was associated with photoperiod acclimation and the time of day for starting a test of acute lethality, but the changes were not great, all values being in the range 160-230 min (McLeay and Munro, 1979). Both of the papers cited above review other work on these topics.

The activity of a fish in a test tank would not seem to bear much relation to its behavior in a natural habitat. However, one of the most obvious differences, the physical activity of fish, has less effect on resistance than might be thought. Trout forced to swim at 55% of their maximum sustainable velocity showed no increase in susceptibility to ammonia or zinc (Herbert and Shurben, 1963). Even at 85% of maximum cruising speed, lethal concentrations dropped only to 0.8 and 0.7 of the values for still water. The differences were a little less than might be expected from the increased irrigation of gills (see the section on dissolved oxygen, below).

The artificial nature of the exposure tank is usually ignored as a factor, but it may have small effects on estimates of tolerance, or at least on the variability of fish response. Sparks et al. (1972) are among the few people to assess this. They kept pairs of bluegills in bare aquaria and found that one fish became dominant over the other. When exposed to zinc, the dominant fish survived significantly longer (e.g., 20 compared to 9.5 h). Most interestingly, when the experiment was done with a flowerpot in the tank for shelter, there were no significant differences in survival time between dominant and submissive fish. Some unusual results related to aggression were obtained by Sprague et al. (1978). During a 2-yr project on sublethal effects, 90 screening tests were conducted to assess the lethality of batches of effluent. Most of the batches gave standard and satisfactory results, but during two periods there was extreme aggression between fish after they were removed from the holding tank and put into the test tanks. There were odd results for four batches of effluent in particular; the numbers of dead fish are shown in Table 2. It was obvious from looking into the tanks that fish were killing each other, and the results were interpreted as follows. In tests 38 and 39, 100% effluent was lethal to some or all fish, while in the control and low concentration, fish were being killed by aggression. In the middle concentrations fish were pacified by the effluent, which was apparently not lethal at concentrations of 50% or less. In tests 15 and 41 the effluent had little or no lethality, but high concentrations dampened the aggression. Test 41 must be some sort of classic for reversal of results in an acute toxicity test. These results demonstrate that biotic factors may sometimes be of importance in toxicology.

If invertebrate animals are being tested, some, such as *Daphnia,* may be quite at home in an ordinary container of water, but others have special habitat requirements that may be important for applying laboratory results to the real world. Pesch and Morgan (1978) demonstrated this for a marine polychaete worm. When the worms were provided with sand and made burrows, as in nature, the long-term LC50 for copper was four times the value for worms in bare tanks. Those in sand had lower body levels of copper, and it ap-

Table 2 Effects of Aggression between Fish on Mortality in Different Batches of Refinery Effluent

Batch and test number	Control	Fish mortality out of 10 (% effluent)			
		12	25	50	100
15	5	5	8	6	1
38	5	2	0	0	6
39	4	3	0	0	10
41	4	2	3	1	0

peared that they had reduced their exposure by circulating only small amounts of water through their burrows.

The keys to reducing such variation from miscellaneous causes would be to follow good laboratory practice and to standardize everything except the variable being studied, that is, concentration of toxicant. If it is suspected that an aspect of the test procedure, such as the habitat provided for an invertebrate, is important and of interest, then this aspect should also be singled out as a variable in a suitably designed experiment. It is therefore important that the investigator be familiar with the life history and general requirements of organisms used in testing. The alternative is to use the standard test species, whose requirements are well known.

BIOTIC CHARACTERISTICS AS MODIFYING FACTORS

This topic has received considerably less attention than the abiotic modifying conditions described below, but recent work shows that it should not be ignored. Such variation as is described here should be kept in mind when considering the overall significance of differences caused by all the abiotic modifying factors.

Test Species

There are real differences among different groups of organisms, and many authors have suggested that pollutants should be tested against an alga and one or more invertebrates, as well as fish, to obtain a more complete picture of toxicity. For example, crustaceans seem to be particularly susceptible to metals, pesticides, and many other pollutants. This is usually taken into consideration in modern criteria for natural waters.

A general idea of species variability is provided by the comparison of data in a literature review by Klapow and Lewis (1979). Some impressions are provided by Fig. 3, discussed in the salinity section below, but Klapow and Lewis made a more precise comparison by collecting results for various kinds of organisms and life stages tested under similar conditions. One kind of comparison involved tests by different investigators with the same species and toxicants. In 21 of 36 cases, the highest LC50 was within an order of magnitude of the lowest LC50, and the remainder were within two orders of magnitude. Differences were much greater in the second kind of comparison, between results for different kinds of organisms tested against a given toxicant in single studies. When the highest LC50 was divided by the lowest LC50 in the same study, there were 47 cases in which the spread was an order of magnitude or less (i.e., ratio of 10 or less), 16 cases with the spread ranging up to two orders of magnitude, and 11 cases with the spread mostly within three orders of magnitude. From the two comparisons it is evident that much greater variability is associated with the kind of test organism used than with test conditions and techniques. It should be kept in mind, however, that Klapow and Lewis placed no restrictions on the data used in their comparisons. For a given toxicant, they could have accepted results for fish, algae, sensitive microcrustacean larvae, or notoriously resistant invertebrates.

There are some differences even within one group, fishes. For example, the careful review of data by Spear and Pierce (1979) indicates that sunfishes (Centrarchidae) are about 15 times more tolerant of copper than are salmonids and minnows. Organophosphate insecticides also provide an extreme example of differences between species. Thirteen insecticides produced differences that ranged from 4- to 900-fold between bluegills and guppies, which were sensitive, and goldfish and fathead minnows, which were generally tolerant of the chemicals (Pickering et al., 1962). Results were also compared for different species with chlorinated hydrocarbons, cyanides, organic solvents, metals, detergents, and other chemicals, and the authors concluded, for their organophosphate results, that "In no other group of compounds that have been tested in our laboratories have the differences in species sensitivity been so great . . . the range . . . has generally been less than 10" (Pickering et al., 1962).

Organophosphates may be unusual in this respect; there often seems to be more concern than is warranted about possible differences between species of fish. For example, the small tropical flagfish and zebra fish averaged only 4.2 and 2.6 times the tolerance of rainbow trout for five representative toxicants, and it was concluded that this was not a major difference compared to within- and between-laboratory variation (Fogels and Sprague, 1977). Comparing trout with three species of "coarse" fish, Ball (1967) found that the lethal thresholds were almost identical for ammonia. When oil refinery waste was tested against 57 species of fish, there was only a fivefold difference between gizzard shad, the most sensitive species, and the guppy, which was the most resistant (Irwin, 1965). Similarly, the variation may not be large for closely related invertebrates. When 40 insecticides and herbicides were tested with two daphnids of different genera, 23 pairs of LC50s were not significantly different. The overall ratio of LC50s of the more resistant species to those of the more sensitive one was 1.8, a fairly modest difference (Sanders and Cope, 1966).

Life Stage and Size

For aquatic arthropods, the time of molting may be particularly susceptible and could appreciably affect results, even in acute toxicity tests. Lee and Buikema (1979) demonstrated this for *Daphnia pulex,* and recommended 48-h tests so that all or most of the animals will have molted during the exposure. They pointed out that there may be near-synchrony of molting in some populations, and 24-h tests might involve an all-or-none aspect of molting that would complicate comparisons.

For fish, the most sensitive life stages seem to be the embryo-larval and early juvenile stages, and advantage has been taken of this in designing chronic tests. McKim (1977) reviewed 56 life cycle tests and found that the estimated "no-effect" concentrations were virtually identical, in 80% of the cases, for the short embryo-larval-juvenile phase and the lengthy life cycle test. In the rest of the cases the differences were factors of 2 or less.

Concerning size, larger fish might be expected to be more tolerant of a toxicant, and this is often the case. However, there is no overall relation that can be applied across species, toxicants, and size ranges. This is summed up succinctly by Anderson and Spear (1980): "The results of this study demonstrate that lethal tolerance to copper may or may not vary with fish size." They found that pumpkinseed sunfish weighing 1.2 g (wet) had an LC50 of 1.24 mg/l for copper, but resistance gradually increased with weight so that fish of 7.6 g had an LC50 of 1.94 mg/l. On the other hand, there was no change in the LC50 of rainbow trout over the size range 3.9-176 g. To add more confusion, Howarth and Sprague (1978) reported that copper resistance of rainbow trout gradually increased with size–in fact, by a factor of 2.5 in LC50 for 10-g fish over those weighing 0.7 g. It may be that there is a change with weight in the smaller trout which disappears when they reach the larger weight range studied by Anderson and Spear. Such a picture was recently presented for the insecticide Permethrin (Kumaraguru and Beamish, 1981). Tolerance of rainbow trout doubled between 1 and 5 g, increased regularly by a factor of 5 between 5 and 50 g, and then did not show any change between 50 and 200 g. The results seemed to fit a fairly standard explanation, inasmuch as the weight-specific metabolic rate of fish decreases in larger fish. It seems reasonable that the smaller fish, with higher metabolic rates, would take up more of the poison. In addition, hydrolytic breakdown of the poison by microsomal esterases in the liver would be more effective in the larger fish, because of the relatively larger ratio of enzyme to insecticide.

It is apparent that if size is a factor in toxicity studies, the effect must be determined empirically in each case. For tests of acute lethality, the standard relation used is

$$\text{LC50} = aW^b$$

where W is the weight of the organism and a and b must be determined from the experiment. The value of b is the slope of the relation between LC50 and weight, more clearly seen if the equation above is put in the form $\log LC50 = \log a + b \log W$. Although values of b are frequently around 0.7 (Anderson and Weber, 1975), b may have any value, such as 0.27 for the sunfish of Anderson and Spear, mentioned above, and 0.0 for their trout.

Nutrition, Health, and Parasitism

Most descriptions of standard methods for toxicity tests caution that only healthy stocks of fish must be used and that mortality during holding must be low. It would seem likely that any stress of disease would increase or enhance the stress from the toxicant and affect the results. However there is only slim evidence for this, and effects seem to be small.

Adelman and Smith (1976) compared "normal" goldfish with a stock that had been infested with skin flukes and probably with bacterial disease. The latter group had been heavily treated and had suffered 50% mortality during holding. The unhealthy fish were more sensitive to NaCl, with LC50s about 84% of the "normal" LC50 and outside the expected range of variation. They also seemed much more sensitive to Guthion (an organophosphate insecticide), but this could not be proved because of extreme variation in normal results. For chromium and pentachlorophenol, unhealthy fish showed LC50s within the range of normal values. It is possible, of course, that "weaker" diseased fish died during holding and that testing of "stronger" fish tended to balance the effects of disease. Small decreases in survival time have been reported for parasitized fish and snails exposed to zinc or cadmium, but effects on threshold tolerance, if any, were not included in the studies (Boyce and Yamada, 1977; Guth et al., 1977; Pascoe and Cram, 1977).

Nutrition of test animals has almost certainly received too little attention as a variable affecting toxicity results. In some cases the effect could be major. Wild populations of the marine copepod showed a sixfold variation in lethal levels of copper, being more resistant when food was more abundant (Sosnowski et al., 1979). That could have resulted, partly or wholly, from chelation by products of those food organisms, but copper resistance of trout is definitely affected by diet. As carbohydrate formed an increasing proportion of an equicaloric diet, trout became less tolerant of the metal. The LC50 increased by a factor of 1.7 in direct proportion to the percentage of protein in the liver (Dixon and Hilton, 1981). Apparently the high-carbohydrate diet brought about glycogen-filled livers, which were less effective in detoxifying copper. The finding is relevant inasmuch as wild carnivorous fish would receive a low-carbohydrate diet.

Susceptibility to pesticides seems particularly affected by diet. Increased protein in the food was associated with a six times greater tolerance of rainbow trout to lethal levels of chlordane. There was much more variability of response in the low-protein fish, and the type of protein also played a role (Mehrle et al., 1977). Diet (pellet versus worms) can also affect results of chronic tests, judged by growth of fish exposed to dieldrin (Phillips and Buhler, 1979). Fish exposed to toxaphene use vitamin C in its detoxification, resulting in deficiency symptoms in bones. Large increases of vitamin C in the diet can counteract this and raise the threshold for chronic effects of toxaphene by a factor of 2 to 4 (Mayer et al., 1978). However, a lack of vitamin C, or an excess, does not appear to change symptoms of lead toxicity, despite their similarity to the symptoms of vitamin C deficiency (Hodson et al., 1980).

Acclimation

It has long been known that aquatic animals can acclimate to temperature and to low oxygen. Fry (1947, 1971) provided comprehensive reviews showing that higher acclimation temperatures could raise the upper and lower lethal temperatures of fish by 3-10°C, and change other phenomena, such as temperature for maximum swimming speed, by similar amounts. Fish also acclimate to low levels of dissolved oxygen. Shepard (1955) demonstrated that as the acclimation level in brook trout decreased from an oxygen saturation of 11 mg/l to 2.5 mg/l, the lethal level of oxygen also decreased from 1.8 to 1.0 mg/l. However, young salmonids apparently do not acclimate to low pH (Daye, 1980).

Many hints and incidental observations in the literature suggest that aquatic organisms may acclimate to some toxicants. Much of this information is based on changes in survival times, not threshold lethal concentrations, which would be more meaningful. It might be hypothesized that animals exposed to a sublethal level of a pollutant could become more tolerant or become weakened, depending on the mode of action of the poison and the type of detoxifying mechanism, if any, available to the animal. A recent series of papers show that both situations occur (Dixon and Sprague, 1981a,b,c). Trout were acclimated for 3 wk to constant sublethal levels of various pollutants, and their response to lethal levels of the same pollutant was used as a measure of acclimation. For acclimation to about 0.22 of the lethal level of arsenic, the threshold LC50 of arsenic increased by a factor of 1.5, and acclimation to copper could double the tolerance for that metal. For acclimation at one-third of the lethal level of cyanide, fish became more sensitive by about one-third in the first week, then, with continued acclimation, their tolerance climbed back to the original level by the end of 3 wk.

A thorough study of the copper response showed that there was a greater increase in tolerance with higher preexposure concentrations up to about 0.6 of the lethal level, and also with time of preexposure over a 3-wk period (Fig. 1). When returned to clean water, trout lost their increased tolerance. Acclimation stimulated production of a liver protein, probably a metallothionein, the apparent defense mechanism. Acclimated fish were able to maintain their body burden of copper at a steady level, with no increase during subsequent exposures that should have been quickly lethal. Acclimation at 0.18 of the lethal level was about the threshold for triggering a strengthened defense mechanism, the ill effects of preexposure being just balanced by increased resistance to the subsequent lethal exposure. At a lower level of acclimation (0.09 of the lethal level), fish had an increased body burden of copper but apparently were not

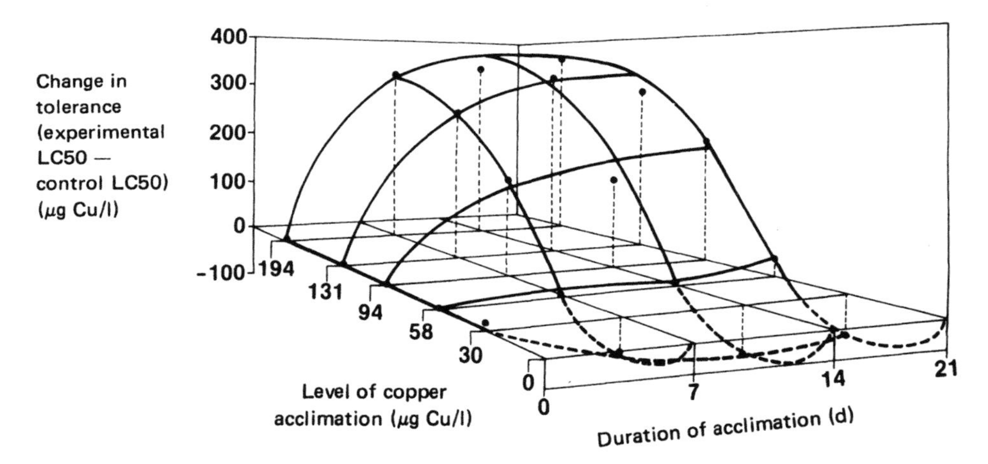

Figure 1 Change in incipient lethal levels of copper for rainbow trout following acclimation to different sublethal concentrations of copper for periods up to 3 wk. Points are observed LC50s; lines represent a fitted response surface. Average control LC50 was 330 μg/l; thus acclimation could double tolerance. Following acclimation to 30 μg/l (0.09 of the incipient lethal level), fish were more sensitive in the subsequent lethal exposures. *[From Dixon and Sprague (1981b).]*

"physiologically aware" of the copper and did not acclimate upward. They were, in fact, more sensitive than control fish to subsequent lethal exposure. The threshold of 0.18 of the LC50 for triggering the defense mechanism was apparently a meaningful one; above that there was reduced growth, and below it there was no effect on growth.

The effects of acclimation described above are not major, being twofold or less for these three toxicants, but they could be of some importance for aquatic organisms in waters that are mildly polluted over long periods. Also important could be genetic selection for increased tolerance. This appears to have occurred in polychaete worms from sediments that were high in metals (Bryan and Hummerstone, 1971, 1973) and in an isopod from rivers with a long history of pollution from metal-mining waste (Brown, 1976).

BACKGROUND ON ABIOTIC CHARACTERISTICS OF WATER

These are natural characteristics of water that may be influenced by human activity. Although this chapter emphasizes the ways in which they modify the potency of toxic substances, it must not be forgotten that each abiotic condition can be of fundamental importance by virtue of its direct action on aquatic organisms; for example, high temperature may act as a lethal factor. Because of this direct importance, the major physicochemical characteristics of water, and their actions on organisms, are briefly described below. There is a vast literature on effects of these natural characteristics, especially for fish. For more details, the reader may consult reviews of environmental factors by Fry (1971), of temperature, oxygen, pH, and suspended solids by Alabaster and Lloyd (1980), and of dissolved oxygen by Doudoroff and Shumway (1970) and Davis (1975).

For each of these natural characteristics, there will be a certain optimal level or zone for a given species; the organism functions most efficiently at that level. For example, there will be an optimum for temperature, and the "preferred temperature" of a fish will usually coincide. If the environmental condition changes to a less favorable level, the organism usually has physiological processes that allow it to partially compensate, but at some metabolic cost. The change within the organism may be semipermanent, in which case the animal has "acclimated" to the new condition. For example, fish held at an elevated temperature show an increase in upper lethal temperature and may, for a while, prefer a temperature above that normally selected. Near the outer limits of the biokinetic range, the metabolic loading on the organism becomes greater, and it will have less capacity to carry out normal activities such as feeding and reproduction. Such a situation is analogous to sublethal effects of a toxic substance. Eventually a limit is reached, beyond which the organism will die from the environmental condition, be it high temperature or pH or low dissolved oxygen.

The implications of the direct actions of many natural environmental entities for modifying pollutant toxicity are obvious. If an organism is already under stress or metabolic loading from such an environmental condition, then faced with the added complication of a toxic pollutant, it would probably be less capable of dealing with that pollutant, and indeed might show greatly increased susceptibility. The environmental entity would have acted as a factor modifying the toxicity of the pollutant.

Temperature

Temperatures in fresh surface waters of the temperate zone are usually within a range from the freezing point to 30°C; values to 35°C or even higher have been recorded–for example, in rivers near the Gulf of Mexico (Fry, 1960; Alabaster and Lloyd, 1980). Of course, there are streams in mountains or wooded regions where temperatures seldom exceed 20°C. Latitude governs not so much the maximum temperature attained as the length of time for which it prevails. The species present in a given aquatic community are largely determined by the temperature regime.

Most aquatic organisms are at the same temperature as the water, and there are upper and lower lethal temperatures beyond which they cannot survive. In various species of fish, these lethal levels may change by 3-10°C, depending on the temperature to which they are acclimated. Many fish seem to have an upper incipient lethal temperature that is at least 5°C higher than the maximum usually encountered in their habitat.

Within the lethal zone, temperature has profound effects on metabolic processes of poikilotherms, and indeed the metabolic rate may double for every 10° rise in temperature. There are optimal temperatures for such things as growth efficiency and performance of various activities. However, the favorable range can be modified to some extent; salmonids can give similar performance in spawning and growth over a 3 or 4° range if given a chance to acclimate, and goldfish can adapt over about 15° to show similar swimming speeds (Fry, 1971; NAS/NAE, 1974).

Water temperatures can be raised by discharge of cooling water, especially from power plants; diversion of watercourses for irrigation; and clearing of forest cover. Effects are greater in streams and rivers than in lakes or bays. To protect aquatic life against such changes, the all-pervading influence of temperature requires complex criteria. Modern ones (EPA, 1976) are individually designed for each species, and include at least: (1) a temperature that will not cause lethality in short exposures, (2) a maximum during the cold season that protects against lethality from sudden drops in temperature, (3) a weekly average in the warm season that protects against sublethal metabolic loading, and (4) specific requirements for successful reproduction.

Increased water temperature also increases the solubility of many substances, influences the chemical form of some, and governs the amount of oxygen that dissolves in water. Such changes can interact with the direct deleterious effects of elevated temperature.

Dissolved Oxygen Content

The dissolved oxygen content of fresh water is about 14.6 mg/l for saturation at 0°C and decreases gradually with temperature to 9.1 mg/l at 20°C and 7.5 mg/l at 30°C. In full-strength seawater the corresponding values range from 11.4 to 6.2 mg/l. Natural oxygen levels are often below saturation, particularly in the bottom layer of lakes. They may fluctuate because of algal activity, being below saturation at night and near saturation or even supersaturated during the day, when photosynthesis takes place.

One of the most studied effects of pollution is the lowering of dissolved oxygen caused by organic wastes; this occurs because oxygen is used in the decomposition process. From the literature of the first half of the century, one would almost gather that this was the only kind of pollution. Much fieldwork was done; systems of "indicator species" were developed to classify the degree of pollution, and these continue to be useful. There is a dramatic decrease and change of species in waters deoxygenated by heavy sewage pollution, the normal mixed assemblages of bottom-living invertebrates being replaced in extreme cases by vast numbers of red (hemoglobin-containing) midge larvae or red oligochaete worms.

Since oxygen is required for respiration, decreased levels can be effective as a limiting factor for aquatic organisms. Lethal levels are surprisingly low, most fish showing incipient values in the vicinity of 2 mg/l (Doudoroff and Shumway, 1970). However, these values are mostly from the laboratory, where fish are not required to pursue the activities that would be normal in a natural habitat. Such activities are hampered at much higher levels of oxygen. For example, the swimming speed of fish is directly related to the concentration of dissolved oxygen, and they can improve their performance above expected levels if the water is supersaturated (Fry, 1971). Modern water quality criteria recognize that any reduction of oxygen below natural levels will have some deleterious effect on organisms (Doudoroff and Shumway, 1970; Davis, 1975), and there is no single "magic number" that can be used as a criterion of acceptable levels. Rather, a series of levels of protection have been established for various degrees of impairment of the fauna, and a selection must be made of the desired level—that is, socioeconomic factors enter into the decision. For example, if the natural oxygen minimum of a stream was 9.1 mg/l in a particular season, the criterion of Doudoroff and Shumway (1970) for protection of important fisheries (an estimated 3% impairment) would be a minimum of 7.6 mg/l dissolved oxygen.

Hydrogen Ion Concentration

The hydrogen ion concentration is usually measured as pH, the logarithm of the reciprocal of hydrogen ion activity (the negative logarithm of $[H^+]$). Thus pH 7 represents 1×10^{-7} mole of H^+ per liter and, this being the point of neutrality, it is balanced by the same concentration of OH^- (hydroxyl ions). Most natural fresh waters are in the range of pH 5-9, soft waters often being pH 6-7 and hard waters 8-8.3. The oceans are relatively well buffered, usually in the pH range 8.0-8.2, although sometimes a little lower.

Related to pH is the buffering capacity of a water. *Acidity* is the ability of a water to resist a change in pH when a base is added. The acidity is largely caused by carbon dioxide, the salts of strong acids and weak bases, and other factors; above pH 8.3 there is considered to be no measurable acidity. *Alkalinity* in fresh water results chiefly from the carbonate-bicarbonate buffering system, and below pH 4.5 the alkalinity is zero. Fresh waters with high levels of calcium and magnesium (hard waters) have relatively high alkalinity, often in the vicinity of 60% of the total hardness when each is expressed as milligrams of $CaCO_3$ per liter.

Direct effects of pH on organisms are graded, becoming more severe as pH deviates either way from the neutral zone. Alabaster and Lloyd (1980; see also NAS/NAE, 1974) give a useful table summarizing these effects, with a middle range of pH 6.5-9.0 being considered "harmless to fish." A few fish such as perch and pike can acclimate to the acid pH range of 4.0-4.5, but most could not reproduce and only a few kinds of invertebrates would live in such waters. On the alkaline side, at pH 9.0-9.5 most invertebrates are not affected, but this range approaches the tolerance of many fish; pH 9.5-10.0 may be lethal to salmonids over a prolonged period.

Historically, major effects on pH due to human activity have come from acid runoff in coal and metal mining areas. This is chiefly caused by exposure of sulfide-containing minerals. In the past two decades, however, acid rain has dwarfed other types of pollution as a factor in pH. There has been catastrophic acidification of soft-water lakes and rivers by fallout of combustion products, chiefly oxides of sulfur and nitrogen. This has resulted in hundreds of fishless lakes in Scandinavia, Canada, and the northeast United States.

Some modern water quality criteria recognize the graded nature of pH effects. NAS/NAE proposed levels of protection with increasingly wider limits and a wider allowance of fluctuation. The highest level of protection for freshwater organisms specified pH in the range 6.5-8.5, with no change greater than 0.5 unit beyond the estimated natural limits.

Dissolved Salts

Calcium and, to a lesser extent, magnesium are the predominant dissolved cations in fresh water and are chiefly responsible for water hardness. A commonly used classification is: soft water, 0-75 mg/l expressed as $CaCO_3$; moderately hard, 75-150; hard, 150-300; and very hard, 300 mg/l and above (EPA, 1976). The mean hardness of the rivers of the world is about 50 mg/l, the median hardness of major U.S. municipal supplies about 90 mg/l, but all levels may be found in nature. Streams may consistently run at 20 mg/l or less in some forested, mountainous, or igneous areas. Generally, biological productivity is low in such waters, probably because of a concomitant lack of nutrient minerals for primary production. There is also some evidence that fish are under appreciable osmotic stress in such "pure" water. Ground water from wells in limestone areas often seems to emerge at 350–380 mg/l total hardness, about the maximum that is usually encountered, and a level that causes severe encrustation in water pipes of houses. Hard-water streams and lakes are productive ones; certainly these natural dissolved salts are not harmful to organisms.

Along with the cations go higher levels of carbonates, bicarbonates, and some sulfates. Their important role in complexing toxic metals is described in a later section.

Inert Suspended Solids

Here are included suspended particles such as silt which do not have a toxic action, but do have other effects on aquatic systems. Streams in forested areas have low levels of suspended solids, usually in the range 0-20 or 0-50 mg/l at all times. Some waters are naturally high in solids; the Mississippi is famous for this and may approach 300 mg/l in Louisiana, although present levels are partly due to human activity. However, even the Mackenzie River, which flows through pristine areas of subarctic Canada, may run at 200 mg/l of nonfilterable residue (suspended solids), reducing underwater vision to a meter or less. Anthropogenic causes of high levels in surface waters are erosion and direct discharge of wastes such as pulp mill effluent and tailings from mining and milling.

These particulate solids are commonly measured in milligrams per liter by drying those that are retained on a filter. However, they may also be assessed by their light-scattering effect in water (turbidity) or as the percent transmission of light at depth. There is no general way of converting from one measurement to another, since this will depend on the specific gravity of particles and their size, shape, and angularity.

Lack of light penetration into turbid lakes will reduce primary production and hence the production of the whole system. Solids that settle to the bottom can blanket the interstices and reduce numbers of benthic algae and invertebrates, and can drastically reduce survival of buried fish eggs by impeding flow through bottom gravel. Alabaster and Lloyd (1980) give the general conclusions of European experts concerning damaging levels of solids for aquatic life, and these have been widely adopted as water quality criteria. Again, there is a graded response. Less than 25 mg/l finely divided suspended solids seems to have no harmful effect on fisheries. From 25 to 80 mg/l may reduce fish production, waters of 80–400 mg/l are unlikely to support good fisheries, and poor fisheries are likely at levels above 400 mg/l. Extremely high levels of suspended solids can be acutely lethal to fish, presumably by clogging or abrading the gills. No single level can be specified as lethal, since it depends on the type of particle, but most reported values are in the tens or hundreds of thousands of milligrams per liter.

Some suspended materials, especially organic ones, can have a major detoxifying action on certain pollutants by sorbing or chelating them.

ABIOTIC CONDITIONS AS MODIFYING FACTORS

Water Temperature

There is no single pattern for effects of temperature on toxicity of pollutants to aquatic organisms. Temperature change in a given direction may increase, decrease, or cause no change in toxicity, depending on the toxicant, the species, and in many cases the experimenter who has selected a particular procedure and response. Limited evidence suggests that temperature may not have much effect on the chronic "no-effect" thresholds of pollutants, which are ultimately of most importance to aquatic organisms.

One statement that can be made is that the short resistance times caused by a severely lethal level of toxicant are changed by temperature roughly in accordance with van't Hoff's equation;

that is, survival time decreases by a factor of 2 or 3 for each rise of 10°C (Lloyd and Herbert, 1962; Hodson and Sprague, 1975; MacLeod and Pessah, 1973). This is interesting, since most physiological processes of poikilotherms show similar increases in rate with rising temperature, but it is not emphasized here because of the greater interest in *tolerance,* that is, the threshold LC50. Of even more importance is the effect on chronic sublethal toxicity, and it is unfortunate that there has been relatively little research on that topic.

Temperature effects on zinc toxicity are described first as an example of diversity in response; then some other examples are considered. It might be expected that zinc would be more lethal to a poikilothermic animal at high temperature, but this is far from true as a generalization. A good example is Atlantic salmon acclimated and tested at one of three temperatures: 3, 11, or 19°C (Hodson and Sprague, 1975). The lethal concentrations for 24-h exposures indicate that fish died more slowly of zinc poisoning in cold water. That is, they were more resistant, and the lethal concentration was higher by a factor of about 4 in cold water than in warm water, as shown in Table 3. At 96 h there was some evidence of a similar effect, but the overall picture was not clear. However, at 2 wk the threshold LC50 was higher in warm water; that is, salmon were more tolerant in warm water than in cold, the opposite of the usual expectation.

To explain this, it might be hypothesized that the "warm" fish, with their higher metabolic activity, were able to detoxify zinc more rapidly, perhaps by moving it out of the gills, the primary site of acutely lethal damage (Skidmore and Tovell, 1972; Burton et al., 1972). Hodson (1975) found the opposite when he pursued the mechanism of the temperature effect. Salmonids that died of zinc at 19°C had about 2.5 times *more* zinc in their gill tissue than those killed at 3 and 11°C.

Table 3 Effect of Temperature on Zinc Toxicity

	LC50 of zinc (mg/l) at different temperatures		
Exposure	3°C	11°C	19°C
24 h	20	8.4	5.4
96 h	7.2	3.6	5.4
2 wk (threshold)	3.5	3.6	5.4

This surprising result was not related to temperature-induced changes in lactate-glycogen metabolism (Hodson, 1976), which might have remedied tissue hypoxia, the secondary lethal effect of gill destruction (Skidmore, 1970). In fact, the warm-temperature salmonid gill tissue could accommodate more zinc before the cells broke down in the usual syndrome of zinc poisoning. The best-fit lines in Fig. 2 show that, in colder water, a given concentration of zinc caused more histological damage to the gills (Hodson, 1974). The damage was measured by four criteria: distance of delamination of epithelium from its basement membrane and from the gill filament, angle of bending of lamellae, and amount of cellular debris.

This example shows that temperature can indeed affect the actions of toxicants. It also shows that one should not prejudge the direction of temperature effects. An even stronger example of this unpredictability was shown in the same series of experiments with zinc and salmon (Hodson and Sprague, 1975). Fish tested at a temperature lower than their acclimation temperature (i.e., a cold stress) became less tolerant of zinc than they would have been at their acclimation temperature. The opposite might have been expected if one considered the slowing of metabolic rate when fish are moved to a lower temperature, with slower gill irrigation and a probable decrease in zinc uptake.

Results of other work add to the complexity. Rainbow trout have been reported to show no change in zinc tolerance between 13.5 and 21.5°C (Lloyd and Herbert, 1962) and also an increase of about 65% in tolerance at 3 compared to 20°C (Brown, 1968). Thus in the same genus we may quote all three possible results for the effect of temperature on zinc lethality.

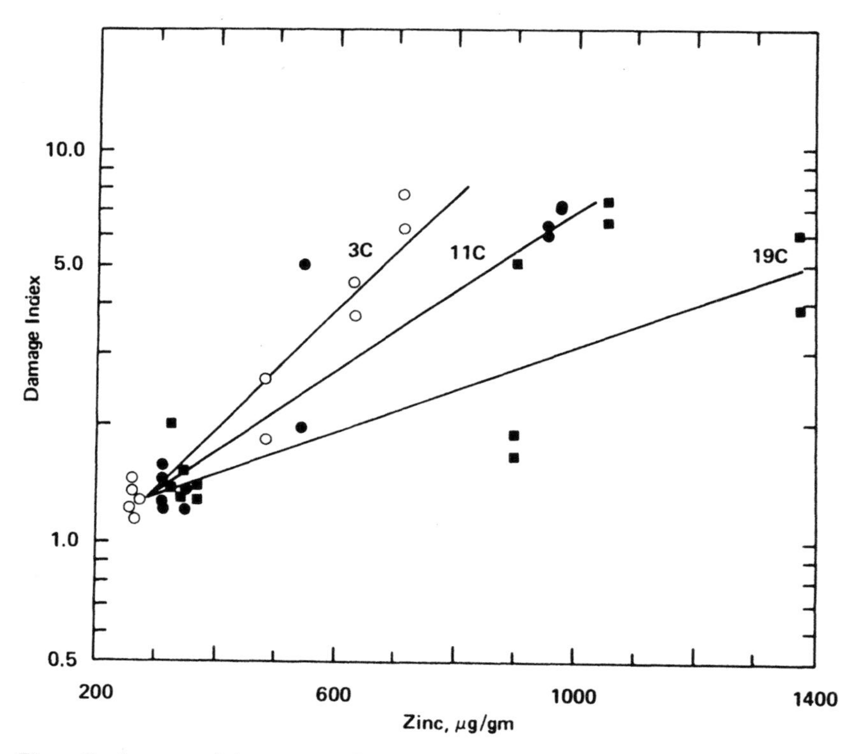

Figure 2 Amount of damage to gill tissue of salmonid fish related to amounts of zinc in that tissue. Fish were sampled at various stages of zinc intoxication, following acclimation and testing at 3, 11, or 19°C. Histological damage was assessed by a combination of four criteria from photographs of gill preparations, with a higher index for greater damage. *[From Hodson (1974).]*

Among other species, fathead minnows were three times more tolerant of zinc at 15 than at 25°C, yet bluegills showed no difference in threshold LC50 (Pickering and Henderson, 1966b). Several Hudson River fishes displayed no difference in tolerance to zinc at 15 and 28°C (Rehwoldt et al., 1972). (Nor were there differences for four other metals. Only for mercury was there decreased tolerance at the high temperature.) Various other reviewers also concluded that temperature has diverse effects on zinc toxicity to aquatic organisms (Alabaster and Lloyd, 1980; Cairns et al., 1975; Eedy, 1974).

Similar diverse effects on lethality are found for other pollutants. Some show a negligible temperature effect or become more toxic in cold water. Temperature has a profound but purely chemical influence on the proportion of ammonia present in the toxic unionized state, as discussed in the section on pH. However, when allowance is made for that chemical change, the threshold lethality of unionized ammonia per se does not change above 10°C, while at lower temperatures toxicity may increase by a factor of 2 or more (EIFAC, 1973b). Phenol is twice as lethal to trout at 6 as at 18°C (Brown et al., 1967). Similarly, naphthenic acids are twice as toxic to snails at 20 than at 30°C in soft water. Oddly enough, no difference with temperature has been reported for snails in hard water, and no differences for bluegills in either kind of water (Cairns and Scheier, 1962). Linear alkyl sulfonate (LAS), a component of detergents, is not appreciably different in lethality to bluegills at 15 and 25°C (Hokanson and Smith, 1971).

In contrast, some pollutants become more toxic at higher temperatures. An increase of about 40% was noted for cyanide between 18 and 30°C (Cairns and Sheier, 1963). Whole effluent from a pulp mill became more toxic over the range 8-19°C (Loch and MacLeod, 1974), but inspection of the data suggests a modest change in threshold LC50, perhaps only by a factor of 2. One of the largest temperature effects is for hydrogen sulfide. Adelman and Smith (1972) found that goldfish tested at 25°C were an order of magnitude less tolerant than those at 6.5°C (530 μg/l compared to 44 μg/l, the logarithm of LC50 showing a linear relation to temperature). These results were for 4-d exposures; 11-d LC50s were closer in warm and cool waters, but still different. The findings were based on toxic undissociated H_2S only, and so variation in proportions of ionic species was not a factor (see later section).

As a lethal agent, pH can be affected in either direction by temperature. Fingerling trout were more resistant to acid in cold water (LC50 at 10°C = pH 4.1, LC50 at 20°C = pH 4.3, representing a 60% increase in actual concentration of hydrogen ions resisted at the cooler temperature). However, trout eggs were less resistant to acid in cold water (LC50 at 5°C = pH 5.5, LC50 at 10°C = pH 4.8, a fivefold decrease in concentration resisted in colder water) (Kwain, 1975).

There have been many toxicity tests with pesticides at different temperatures, and in this group there is also a diversity of modifying effects. Many are more toxic in warm water, including endrin to five species of fish (Johnson, 1968), 14 of 15 pesticides tested against bluegills and rainbow trout (Macek et al., 1969), two herbicides (Woodward, 1976), and pentachlorophenol (Asano et al., 1969). Other pesticides showed stronger lethal action at low temperatures. This was true of several of the 22 chemicals tested against cladocerans by Sanders and Cope (1966); in particular, DDT was three times as toxic at 10 as at 27°C. A very large change occurred with Permethrin, a synthetic pyrethroid, which was an order of magnitude more toxic to rainbow trout at low temperatures. The LC50 dropped from 6.4 μg/l at 20°C to 3.2 μg/l at 15°C, and 0.7 and 0.6 μg/l at 10 and 5°C (Kumaraguru and Beamish, 1981). Still other pesticides are little affected by temperature, including Zectran at 7 and 17°C (Mauck et al., 1977) and rotenone at 7 and 22°C (Marking and Bills, 1976). Two studies on toxaphene show the same picture—that is, differences in survival times but no effect on eventual lethal concentrations in long exposures (Johnson, 1961; Koeppe, 1961).

The preceding review shows that modifying effects of temperature on acute lethality are diverse, but for the most part they are only small or moderate. Whether temperature plays a major role in sublethal or chronic toxicity is a more important question for the well-being of organisms in their environment. On this topic there has been much less relevant research. One perceptive study of the toxicity of chromate and arsenate to a rotifer is particularly interesting. In the first part of their research, Schaefer and Pipes (1973) showed what was generalized above—that in short lethal exposures there was more apparent toxicity at high temperatures, but in long exposures the difference decreased to a factor of only about 2, between 5 and 30°C. The work then proceeded to a masterful study of life spans, which were, of course, much longer at cold temperatures in clean water (Table 4). Tests were then run to determine the median effective concentrations (EC50s) of added toxicants that allowed the same life spans as in clean water. One might attempt to see trends in their results, but in fact the variation in EC50 at any temperature is well within the confidence limits of EC50s at other temperatures (Table 4). The authors stated that at higher temperatures the EC50 for life span "may either increase or decrease; that is, no general pattern is evident . . . . The conclusion . . . is that temperature has no effect on the life span [EC50]."

This finding is an important one. If further studies resulted in similar findings, it would mean that we could use the same water quality criteria from the tropics to the Arctic. There would be no need for expensive repeated testing in warm, tem-

Table 4 Lack of Influence of Temperature on Effective Concentrations of Chromate and Arsenate that Resulted in No Change in Life Span, Compared to Controls, in the Rotifer *Philodina roseola*[a]

Temperature (°C)	Median life span, control conditions (d)	Concentration (and confidence limits) resulting in same median life span (mg/l)	
		Chromate	Arsenate
15	28	3.5 (2.2–5.6)	6.0 (3.8–9.5)
20	10	4.6 (2.8–7.7)	9.0 (5.0–16)
25	5.9	4.5 (3.1–6.4)	10 (7.8–13)
30	3.7	3.8 (2.5–5.9)	13 (7.1–23)
35	3.0	3.7 (2.7–5.0)	11 (7.2–18)

[a]Data reproduced with permission from Schaefer and Pipes (1973); copyright 1973, Pergamon Press, Ltd.

perate, and cold conditions. An important consideration is that the results are for one species of rotifer over a range of temperatures that includes the extremes of its spectrum (the species did not reproduce at 5°C). If there was no major effect of temperature in this case, there seems little reason to expect great differences in no-effect levels of pollutants between an arctic organism tested in cold water, to which it is adapted, and a similar tropical species tested in its native warm water.

In some studies temperature was a modifying factor in avoidance reactions of fish. Goldfish avoided low levels of copper in a particular experimental configuration with steep gradients (Kleerekoper et al., 1973). However, if the copper-containing water mass was warmer, fish were attracted to it more than to warm water alone. The situation is not simplified if we look at another side of it; fathead minnows selected a lower temperature if preexposed to a sublethal level of copper (Opuszyński, 1971). Like most avoidance work, this is difficult to extrapolate into field conditions. A simpler finding is lack of change in levels of hypochlorous acid avoided by fish at 6 and 30°C (Cherry et al., 1977).

A within-organism study showed a temperature difference in apparent physiological defense mechanism. *Fundulus* exposed to an aromatic hydrocarbon in cold water (6.5°C) showed no induction of cytochrome P-450 (Stegeman, 1979). Induction was found, however, after exposure at 16.5°C. The study did not actually show whether lack of this metabolic pathway would result in deleterious effects on the whole organism. Finally, there is an indirect but important effect of temperature on the activity of DDT within organisms, an effect that may exist for other toxicants which are stored in body fat. Salmonids accumulate more of the pesticide at a warm temperature (Reinert et al., 1974), but following such accumulation there can be near-complete mortality when the fish are subjected to a dropping temperature and their food intake decreases. The latter aspect has been demonstrated in a laboratory experiment (Macek, 1968) and also in wild fish from a sprayed stream (Elson and Kerswill, 1966). Presumably DDT is released from storage and becomes acutely toxic when body fats are mobilized by the fish at times of low food intake. Such delayed mortality could be of major significance in temperate climates, although it would not occur with pollutants that were not stored in body fat.

Dissolved Oxygen

It might be expected that stresses on aquatic organisms caused by a reduction in ambient dissolved oxygen would greatly increase the toxicity of a pollutant in the water. Effects of this kind exist, but often they are not major ones. Most findings show a change in toxicity by only a factor of 2 or less for tests conducted at low and high levels of dissolved oxygen.

The most general examination of the question goes back two decades to Lloyd (1961a). He showed very similar increases in lethality of copper, lead, zinc, and phenols in relation to degree of deoxygenation, and fitted a curve with values approximating those in Table 5. The values labeled "toxicity" were essentially ratios of near-threshold lethal concentrations, that is, the LC50 at full oxygen saturation divided by the LC50 at specified partial saturation. Lloyd advanced a convincing explanation of the increase in lethality, relating it to increased irrigation of the gills at reduced oxygen levels. The increased water flow was thought to bring about speedier laminar flow, causing higher concentrations of pollutants in the immediate vicinity of gill membranes and thus a higher rate of diffusion of toxicants into the gills. The explanation was supported by calculations based on known values for gill irrigation and gradients with such laminar flow. Lloyd concluded that the explanation

> . . . implies that any environmental or physiological change which affects the rate of respiratory flow of a fish will also affect the concentration of poison at the surface of the gill epithelium, and that a known relation exists between these two factors. It also implies that the relation between the increase in toxicity of poisons to fish and a reduced dissolved oxygen concentration of the water will be the same for all poisons except those whose toxicities are affected by the pH value of the water.

Table 5 Effect of Dissolved Oxygen on Toxicity

Oxygen (% saturation)	Toxicity
100	1.0
80	1.06
60	1.18
50	1.28
40	1.4
30	1.56

Table 6 Toxicity of Pulp Mill Effluent at Reduced Oxygen Levels

Oxygen (% saturation)	Toxicity
55	1.23
47	1.34
38	1.53

If toxicity of a substance depends on pH, it may change to a greater extent at reduced levels of oxygen. Ammonia behaved in this way, with up to 2.5-fold increases in toxicity (Lloyd 1961a). Briefly, the explanation was that lower dissolved oxygen meant greater irrigation and lower respiratory CO_2 near the gills, causing higher pH locally, an increase in the concentration of unionized ammonia, and greater toxicity (also see the section on pH). Recent work on ammonia toxicity is not in conflict with the hypothesis. Thurston et al. (1981) found that the toxicity of ammonia to rainbow trout increased by a factor of 1.9 when dissolved oxygen dropped from 80 to 30% saturation. This is close to the factor of 2.0 predicted by Lloyd (1961a), and greater than the factor of about 1.5 for other toxicants, as calculated in Table 5 for these oxygen saturations.

There has not been a great deal of additional research on the topic. Hicks and DeWitt (1971) measured 24-h LC50s of whole pulp mill effluent at reduced oxygen levels. The highest oxygen value was 79% saturation, and compared to this, toxicity factors at various oxygen levels were those shown in Table 6. These values do not appear to be greatly different in magnitude from those of Lloyd.

The 20-d LC50s of zinc to bluegills were similarly reduced by low oxygen (Pickering, 1968).

Average lethal concentrations of zinc at the various oxygen saturations are shown in Table 7. The LC50 at the lowest oxygen level was lower than the other two (statistically significant), and toxicity was increased by a factor of 1.5 over that at 67% saturation. Unfortunately, no test was done at full saturation, but the modifying effect of oxygen is similar to those described above.

The surfactant LAS appeared to be an order of magnitude more lethal to bluegills at low oxygen (22% saturation) than at near-saturation (Hokanson and Smith, 1971). However, the difference decreased to a factor of 1.7 if the fish were first given some acclimation to low oxygen and LAS. Further, tolerance to LAS by the most sensitive life stage (feeding sac fry) did not vary between 40% and full oxygen saturation. Hokanson and Smith concluded that the "acclimated" findings would be more relevant to field conditions, and that the lack of modifying effect on the most sensitive life stage was more important for establishing water quality criteria than were the variations found in other parts of the work.

Adelman and Smith (1972) found H_2S to be about 1.4 times as toxic to goldfish at 10% oxygen saturation as at 63% saturation. Cairns and Scheier (1958) cycled oxygen levels, with daily 8-h periods during which oxygen dropped to either 53 or 21% saturation. For the remainder of each day, the dissolved oxygen was at 95% saturation. For the more extreme depletion, toxicity was increased by factors of 4.5 for KCN, 2.8 for naphthenic acid, 1.6 for zinc, and 1.0 for potassium permanganate. The variation among toxicants is considerable and cannot be related precisely to the results of experiments with constant levels of dissolved oxygen.

Table 7 LC50s of Zinc to Bluegills at Reduced Oxygen Levels

Oxygen (% saturation)	LC50 (mg/l)
67	11.3
38	10.6
21	7.3

Some lethality results show no modifying effect of dissolved oxygen. No difference in 4-d LC50s of cadmium for a marine fish were found between 4 mg/l oxygen and saturation (Voyer, 1975). Oseid and Smith (1974) noted differences in survival times of invertebrates exposed to H_2S at various levels of oxygen, but the final LC50s of H_2S were not greatly different for 35 and 60% oxygen saturation.

Unfortunately, there appears to have been very little research on *sublethal* effects of toxicants, as influenced by reduced oxygen. Pickering (1968) assessed the growth of bluegills in a well-designed experiment, with three levels of zinc at each of three levels of dissolved oxygen (21, 38, and 67% saturation). Growth was reduced in the lower levels of oxygen, but zinc did not have a statistically significant effect on growth, nor was there a significant interactive effect of oxygen and zinc on growth (confirmed by reanalysis of Pickering's data). Even at the lowest oxygen level, and a very high measured concentration of 4.7 mg/l zinc (equivalent to 0.64 of the 20-d LC50), the reduced growth was statistically ascribable to low oxygen, not to zinc or the interaction of the two. This is an interesting experiment inasmuch as it is the same one mentioned four paragraphs above, in which 21% saturation of oxygen made the fish less tolerant to lethal levels of zinc, yet the same level of oxygen did not modify sublethal toxicity. Although such a modifying effect could have been hidden by random variability in the experimental design, it is clear that low oxygen was less important than might be expected as a modifier of sublethal toxicity. Similarly, eggs and fry of walleye and suckers showed little difference in deformities and mortality caused by H_2S at 28 and 56% oxygen saturation (Smith and Oseid, 1972). It might have been predicted that the sensitivity of these stages to low oxygen would result in major effects when a toxic compound was also present.

A major modifying effect of oxygen has, however, been documented for the respiratory behavior of young Pacific salmon exposed to pulp mill effluent in about half-strength seawater. Alderdice and Brett (1957) estimated that at 72% oxygen saturation, respiratory distress (gasping at the surface) was caused in the salmon by 18% effluent. When oxygen was lowered, less effluent was required to cause the effect, so that at 36% oxygen saturation the critical level was only 2% pulp mill effluent. The change is almost an order of magnitude, a sharp contrast between this behavioral response and the other sublethal effects documented above which were not modified by oxygen level.

It is evident that there is not yet a comprehensive picture of the influence of reduced oxygen on chronic or sublethal actions of toxicants. However, the effects may be as small as, or even smaller than, the modest effects on acute lethality.

Hydrogen Ion Concentration

Major modifying effects of pH might be expected for toxicants that ionize under the influence of pH, and this is the case for some. Undissociated molecules may be more toxic since they penetrate cell membranes more easily.

A classic example is ammonia. Its toxic activity is now fairly well understood and can be predicted from characteristics of the water, as reviewed in Alabaster and Lloyd (1980), EIFAC (1973b), and NAS/NAE (1974). Ionized ammonia (NH_4^+) has little or no toxicity, but the unionized form (NH_3) is quite toxic, the LC50 for salmonids being in the range 0.2–0.7 mg/l as N. A rise of one pH unit, within the usual middle range of surface waters, increases the proportion of NH_3 about sixfold, with a concomitant increase in toxicity. Factors other than pH also affect the toxicity of ammonia. A second modifying factor is temperature, which also affects ionization. Lloyd (1961b) showed a 2.5-fold increase in toxicity for an increase in temperature from 7 to 20°C, and that factor has been confirmed by the dissociation constants of Emerson et al. (1975). For a temperature change from 0 to 30°C, the dissociation constants indicate an approximately ninefold increase in the proportion of NH_3 at a representative pH of 8.0. In most waters alkalinity per se would not be a major factor in ammonia toxicity, but in unusual waters the interrelation of pH, alkalinity, and CO_2 could have an influence. Lowered alkalinity would be associated with higher CO_2 and lower pH, hence a lower proportion of NH_3 and reduced toxicity. In waters with an extremely low CO_2 content, respiratory CO_2 could reduce pH in the immediate vicinity of the fish's gill and thus diminish the apparent toxicity of ammonia [see the oxygen section above (Lloyd 1961a) or the review by Alabaster and Lloyd (1980)]. Thus the toxicity of ammonia is complex, but is primarily governed by pH-caused ionization and is thus predictable. Since only total ammonia can be measured in the water, the amount of NH_3 must be calculated from known dissociation constants (Emerson et al., 1975; NAS/NAE, 1974) or interpolated from graphs (Lloyd, 1961b). Sublethal effects of ammonia seem to follow the same pattern of toxicity, related to unionized ammonia.

Cyanide is another toxicant affected by pH. Molecular hydrogen cyanide (HCN) predominates in acid and middle pH values, but above pH 8.5 an appreciable porportion of CN^- is present. Apparent toxicity drops somewhat if calculated on the basis of total free cyanide, since the undissociated molecule (HCN) seems to be about twice as toxic as the ion (Broderius et al., 1977). Even more striking examples are provided by changes in toxicity of various complex metal cyanides. For instance, a change from pH 7.8 to 7.5 causes an apparent tenfold increase in toxicity of nickelocyanide. This has been convincingly explained by Doudoroff et al. (1966). The toxicity is related to the free cyanide, not to the metal complex, and dissociation is governed primarily by pH. The situation is similar for hydrogen sulfide, whose toxicity is chiefly due to H_2S, not dissociated ions. The effect of pH can be major, since undissociated H_2S forms only 4% of the total at pH 8.4 but more than 90% at pH 6 (NAS/NAE, 1974). The observed LC50 of total dissolved sulfide (i.e., $H_2S + SH^- + S^{2-}$) changed from 64 μg/l at pH 6.5 to 800 μg/l at pH 8.7, apparently because HCN is about 15 times as toxic as the ionized forms (Broderius et al., 1977).

Sulfite becomes more toxic at low pH because the predominant form is HSO_3^-, which is about an order of magnitude more toxic than the SO_3^{2-} ion (Sano, 1976). Nitrite can be 20 times as toxic at high pH, as HNO_2 becomes less predominant over NO_2^-, but both are toxic and whether one is more so is not evident (Russo et al., 1981).

In most of the cases above, the undissociated molecule, or the "less dissociated" ion, had greater toxicity. The situation is usually the opposite for metals; that is, the free or ionic form is thought to have relatively high toxicity (see separate section below).

For some biocides toxicity changes with pH and for some it does not. The lethality of a dinitrophenol herbicide was about five times as great to fish at pH 6.9 as at pH 8 (Lipschuetz and Cooper, 1961). Holcombe et al. (1980) found that 2,4-dichlorophenol had somewhat less lethality at high pH and also caused much less disturbance of schooling behavior of fathead minnows. This was explained in terms of lower proportions of the undissociated form at higher pH. The same general result and explanation apply to the piscicide antimycin, for which there is much less toxicity above pH 8.5 (Marking, 1975). In fact, the toxicity results indicated that the previously estimated dissociation constant was erroneous. Similarly, the insecticide Zectran was 6-20 times as lethal to various fish at pH 9 as at near-neutrality (Mauck et al., 1977). However, the same explanation did not apply since the material was unionized at all tested pH values. Hydrolysis in alkaline water to more toxic products was the apparent explanation for Zectran and also for malathion (Bender, 1969). Among the biocides that are little affected by pH are rotenone (Marking and Bills, 1976) and three esters of 2,4-Dichlorophenoxyacetic acid (2,4-D) (Woodward and Mayer, 1978), which, unlike the amine form of 2,4-D, did not behave as weak acids and change in toxicity.

Some other substances do not change much in toxicity with pH—for example, phenols (EIFAC, 1973a) and the surfactant alkyl benzenesulfonate (ABS) (Marchetti, 1965). Probably this is the case for many other common pollutants. However, whole effluents may be expected to vary in toxicity with pH if an active constituent is subject to ionization. Pulp mill effluents were about three times as lethal to rainbow trout at pH 9 as at pH 5 in both fresh and seawater dilutions (McLeay et al., 1979b). The explanation appeared to be increased toxicity of the resin acid component, which was much more toxic in the unionized form dominant at high pH (McLeay et al., 1979a).

Salinity

The greatest differences in chemical characteristics of surface waters are those between sea water and fresh waters. It might be expected that this could cause enormous differences in toxicity of pollutants to organisms, but this is not the case. The general picture is that marine organisms are similar in tolerance to their freshwater cousins, *when both are tested in their own environments.* Euryhaline organisms seem to be most resistant in about one-third seawater, close to the isosmotic level. Marine animals probably become less tolerant of other stresses if the salinity decreases appreciably.

An overall view of this similarity is presented by Klapow and Lewis (1979), who made a nonselective comparison of published acute lethality data for nine metals and six other pollutants to marine, estuarine, and freshwater organisms. In their figures (Fig. 3), results for a given toxicant are usually clustered over one or two orders of magnitude, with outlying values spreading the range to about four orders of magnitude. However, what is *not* seen in most of the figures is a spread of marine data at one end of the distribution and freshwater values at the other end. Instead the scatters seem similar, considering the larger number of freshwater LC50s (e.g., chromium; Fig. 3). Klapow and Lewis tested median values for statistically significant differences, and found no differences between marine and freshwater organisms for 14 of the pollutants, the only significant difference being for cadmium. Inspection of their cadmium data (Fig. 3) shows a median LC50 of about 8 mg/l for marine and estuarine organisms and about 2 mg/l for freshwater ones. The similarity for most pollutants is perhaps surprising, particularly since there were so many metals included in the comparison and their toxicity may be greatly influenced by dissolved salts (see later section). From their overall comparisons, Klapow and Lewis concluded that, if there are influences of salinity on toxicity, they are obscured by "more important factors producing variation in bioassay results."

A similar indication of little difference between marine and freshwater species arises from tabulations by the U.S. EPA, presented in their proposed guidelines for processing diverse sets of toxicity data into unified water quality criteria (EPA, 1978). The document indicates that chronically toxic concentrations for saltwater fish and invertebrates were "in the same range" as those for freshwater fish and invertebrates. One of their tabulations compares lethal concentrations of many pollutants to a variety of fish. From the generalized presentation, a toxicant-by-toxicant comparison of saltwater and freshwater species cannot be made, but an overall comparison emerges. Tabulations of LC50s of 62 toxicants for various freshwater fish yielded an average value of 0.83 for the logarithm to base e ($\log_e$) of standard deviations (SDs) among species. For 38 toxicants acting on various marine fish, the average of $\log_e$ SD was 0.79. This in itself indicates only that the degree of variability was similar in the two groups. However, when marine and freshwater fish were combined for the individual toxicants, the average $\log_e$ SD remained almost identical at 0.83. This indicates that the variability among all species was no greater than the variability within freshwater species or within marine species. If either group of fish had been more sensitive overall, there would have been increased variability in the combined data, that is, a larger value of average $\log_e$ SD.

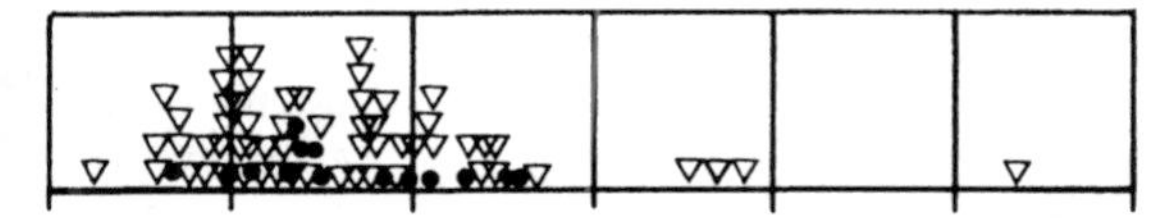

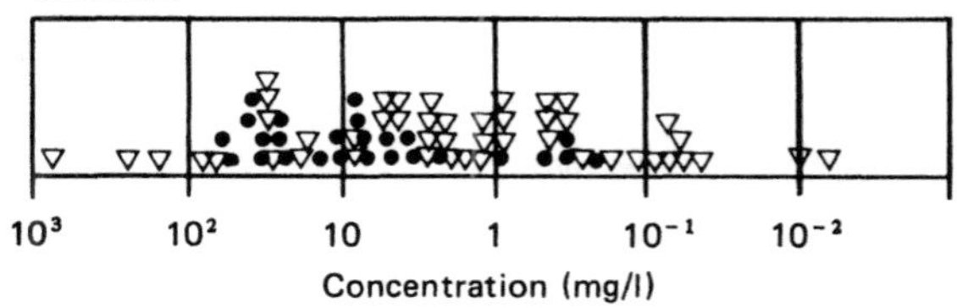

Figure 3 Range of reported lethal concentrations of two toxicants for (•) marine and estuarine organisms and (▽) freshwater organisms. For chromium there was no significant difference in median LC50 between the two groups of organisms, but the median for marine organisms was higher for cadmium. *[Modified from Klapow and Lewis (1979).]*

From other chapters, it will become evident why logarithms were used by the EPA in the calculations above. Briefly, a logarithmic scale of concentration deals with proportional changes in concentration instead of absolute changes and thus gives a truer representation of fish response in different parts of a concentration range. It follows that averaging of concentrations is best done as the geometric mean, that is, the antilog of the average of logarithmic concentrations (or the nth root of the product of n concentrations). In the example above, basing calculations on the logarithms of individual LC50s and using the logarithms of SD meant that comparisons of variability were always made as proportions of the LC50s. This allowed comparison of variability among species, even for LC50s that had large differences in the absolute arithmetic values.

The two general comparisons above indicate that there is little difference in overall tolerance to pollutants between freshwater organisms in fresh water and marine organisms in the sea. However, a particular species of fish may show differences in tolerance if the salinity of the water is changed. This might be expected, since all fish in fresh water are hyperosmotic with respect to their environment and most fish in the sea are hypoosmotic. Different strategies are required to maintain osmotic balance, but in either case there is an imposed physiological load that could be considered a form of stress. If the fish is euryhaline, it would be in overall osmotic balance at an intermediate salinity, and it could be hypothesized that at this balance point the fish would be physiologically more effective. For example, *Cyprinodon macularius* shows better food conversion efficiency in half-strength sea water than in either fresh water or full-strength sea water (Kinne, 1966). Euryhaline fish may also be more effective in dealing with pollutants at their isosmotic point.

This was clearly demonstrated by Herbert and Wakeford (1964) and Herbert and Shurben (1965), who tested the tolerance of rainbow trout and Atlantic salmon smolts to various pollutants at different salinities. For zinc and ammonium chloride, tolerance increased from a minimum in fresh water to a peak in the isosmotic range of 30–40% seawater, then dropped off in higher salinities. For zinc there was a large change in tolerance, by a factor of about 14, and for ammonia a smaller change of 3. The authors advanced the explanation that there was decreased osmotic stress as salinity was increased toward the isosmotic point, with a decreased inward flow of water, which would presumably be accompanied by a reduced intake of toxic ions.

A similar relation was found for naphthalene toxicity to the euryhaline mummichog, with least mortality at isosmotic salinities (Levitan and Taylor, 1979). Less mortality occurred in low salinities than in high salinities, which were apparently associated with osmoregulatory dysfunction and increased uptake of naphthalene. Near the isosmotic point, mummichogs were twice as tolerant of cadmium as in 5% seawater (Voyer, 1975) and mummichogs and eels experienced minimum toxicity for a detergent (Eisler, 1965). Even lethal temperatures seem to follow the same pattern; Garside and Jordan (1968) found greatest tolerance of the mummichog and another cyprinodontid in isosmotic water, with decreases of 4.6° in lethal temperature in both fresh- and seawater. They again advanced the explanation of osmoregulatory stress and cited other work showing similar patterns.

Brackish-water clams (*Rangia cuneata*) were about equally tolerant to copper, chromium, and mercury in salinities equivalent to about 70 and 17% seawater (Olson and Harrel, 1973). They were an order of magnitude less tolerant to copper and cadmium at or somewhat below 3% seawater, the lower limit of their range, and slightly less tolerant of mercury. Work on whole effluents is less complete, but pulp mill effluent was equally toxic to coho salmon in fresh and salt water if the pH was the same and the fish were acclimated to the water (McLeay et al., 1979a).

Truly marine organisms seem fully adapted to seawater, and we might expect them to become less tolerant of other stresses if salinity were reduced. This was illustrated by the classic work of McLeese (1956) on the American lobster, an inhabitant of cool marine waters. In full seawater they could withstand very low levels of dissolved oxygen (about 1 mg/l) and high temperatures (around 30°C, depending on acclimation). Resistance to those variables dropped precipitously with any lowering of salinity, and osmotic stress became lethal at about 25% seawater. That pattern may or may not hold for toxic materials. There was no difference in survival times of lobsters exposed to copper in two-thirds and full-strength seawater (McLeese, 1974), nor did uptake of mercury by fiddler crabs differ at 17% and full-strength seawater (Vernberg and O'Hara, 1972).

From these examples it is clear that salinity per se does not greatly modify the toxicity of pollutants. The important aspect is the genetic nature of the organism, whether it is a marine, euryhaline, or freshwater organism. This determines whether it can adapt to a given salinity and thus affects tolerance to pollutants. Also important, although not stressed in the discussion, would be the immediate prehistory or acclimation of the organism with respect to salinity. Most other parts of this chapter will deal with modifying effects in fresh water, partly because most of the research has been done in fresh water but also because its variations are much greater than those in relatively uniform oceanic waters.

Water Hardness

The major effects of hardness on metal toxicity are covered in a separate section below. Other pollutants are covered in this section, and for them the hardness effect is small or negligible with few exceptions. For example, there are no differences in ammonia toxicity with hardness (EIFAC, 1973b) and little difference for phenols (EIFAC, 1973a).

Surfactant toxicity is less affected by water hardness than might be predicted. Alkyl benzene sulfonate was, on the average, 1.5 times more toxic in hard water than in soft water when tested with fathead minnows (Henderson et al., 1959a). A similar result was obtained for commercial detergent formulations. LAS was also 1.5 times as toxic to bluegills in hard water as in soft water (Hokanson and Smith, 1971). However, hardness caused no difference in ABS lethality to bluegills or to the snail *Physa;* an apparent difference for diatoms was confounded with variation in test species (Cairns et al., 1964). Lethal levels of sodium lauryl sulfate were the same in hard and soft water (Henderson et al., 1959a), and survival times did not change greatly in exposures of guppies to another nonionic detergent (Tovell et al., 1975). As might be expected, commercial soaps were one or two orders of magnitude more toxic in soft water, since they are deactivated by calcium and magnesium (Henderson et al., 1959a).

A comparison of the toxicity of 14 petrochemicals in hard and soft water showed significant differences for only 4 of them (Pickering and Henderson, 1966a). All were more toxic to fathead minnows in soft water, but only by a factor of 2 or less. A study of four hydrazines showed no differences for two of them in toxicity to guppies (Slonin, 1977). Hydrazine itself was six times more toxic in soft water, while dimethylhydrazine was 2.5 times more toxic in hard water. Part of the explanation for hydrazine was probably its rapid decomposition in hard water but not in soft water.

Pesticide toxicity seems little affected by water hardness. This was the case for lethality of 16 organophosphates and 12 chlorinated hydrocarbons tested with various fish (Henderson and Pickering, 1958; Henderson et al., 1959b; Pickering and Henderson, 1966c). It was also the case for four of six carbamates; the other two showed about a threefold difference, but in opposite directions, with respect to water hardness (Mauck et al., 1977; Pickering and Henderson, 1966c). For rotenone (Marking and Bills, 1976) and three herbicides (Woodward and Mayer, 1978), there was no difference in toxicity to fish in hard and soft water.

HARDNESS AND ASSOCIATED QUALITIES THAT MODIFY METAL TOXICITY

For a long time, it has been known that most heavy metals become less toxic in harder water. A general relation was developed by the British Water Pollution Research Laboratory (Fig. 4; from Brown, 1968) and this has often been reproduced and utilized (e.g., NAS/NAE, 1974). This empirical relation is useful, although it is now clear that hardness alone does not govern the potency of the metals. Copper will be used to illustrate the complexity of the situation, and similar pictures probably apply to most of the metals. For complete assessments of the wealth of information about copper, readers should consult the superb reviews of Alabaster and Lloyd (1980) or Spear and Pierce (1979). Much of the following is a condensed version of material covered in the latter review.

Spear and Pierce (1979) summarized the lethal levels of copper for fish, as determined by many investigators, and found a reassuringly consistent pattern with hardness. Results were divided according to four taxa of fish, with bluegills and their relatives being most tolerant of copper. The relation for most salmonid fish is shown in Fig. 5. This embodies results from 19 separate papers, with only a few estimates excluded as being atypic. For salmon and trout, the lethal level of copper could be predicted by the equation shown

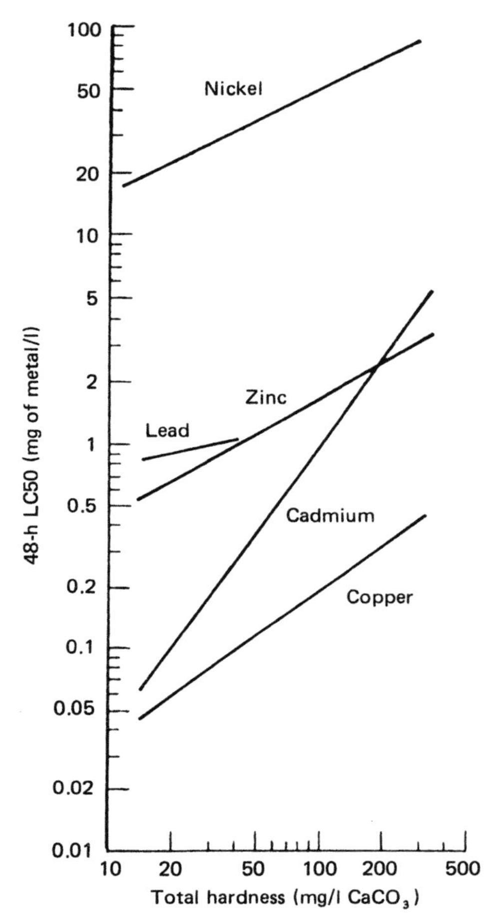

Figure 4 Relation between total hardness of water and 48-h LC50 to rainbow trout of nickel, lead, zinc, cadmium, and copper. *[Reprinted from Brown (1968) with permission from the Controller of Her Majesty's Stationery Office.]*

in Fig. 5. Fathead minnows and certain other members of the minnow family showed a lethality relation very close to the one in Fig. 5.

Even more important with respect to water quality criteria, the sublethal toxicity of copper shows a similar change with water hardness. Some results for growth are shown in Fig. 5. Other sublethal responses such as critical swimming speed and food conversion efficiency are also more affected by copper in soft water (Waiwood and Beamish, 1978a,b). The generally consistent patterns in Fig. 5 enable an easy prediction of the effects of copper, at least in laboratory situations. There is an apparently straightforward relation with water hardness, and sublethal thresholds are in the vicinity of 0.1-0.3 of the lethal concentrations of copper.

It is generally considered that this sparing effect of increased hardness is a phenomenon within the fish. The usual explanation is that higher levels of calcium in the fish tissues make the cell membranes in the gills less permeable, so that less metal enters the fish. The calcium content of fish tissue does increase with that of the water (Houston, 1959), and fish reared in hard water apparently require several days to lose calcium before they become as sensitive to heavy metal intoxication as fish reared in soft water (Lloyd, 1965). Simply adding calcium or magnesium to test water has little effect on copper resistance of fish (Zitko and Carson, 1976). However, there do not appear to be any published measurements of the relative amounts of the metal entering gills in hard and soft water.

The simple empirical relation with hardness disappears if tests are done in unusual waters, in which pH and alkalinity are not associated with hardness in the usual way. The two factors, pH and alkalinity, in fact govern the forms of copper present in water (Spear and Pierce, 1979) and a typical array of "ionic species" is shown in Fig. 6. The picture would change slightly for different alkalinities. At pH 5, for example, a large proportion of the copper is present as Cu^{2+}, the free copper ion, but at higher pH there is a variety of hydroxides and carbonates.* It would stretch the imagination a little too far to assume that all these forms were equally toxic, and indeed they are not.

A number of investigations on the toxicity of "ionic species" have been carried out in the past

*Actually, free copper is thought to be present as the aquo ion $[Cu(H_2O)_6]^{2+}$ (Spear and Pierce, 1979). Some other hydroxides and carbonates precipitate, including $Cu(OH)_2$, malachite $Cu_2(OH)_2CO_3$, and azurite $Cu_3(OH)_2(CO_3)_2$ (Stiff, 1971b; Silva, 1976).

decade, and a toxicity picture is shown in Fig. 7 for diverse combinations of hardness and pH, with alkalinity allowed to find its own level. Some general effect of hardness may be seen. Several groups of investigators have shown that for fixed combinations of pH and alkalinity, increasing hardness causes a reduction in toxicity (Anderson et al., 1979; Chakoumakos et al., 1979; Inglis and Davis, 1972; Miller and MacKay, 1980). But it is clear from Fig. 7 that great changes in toxicity are also associated with pH. Here, again, there is ample documentation that at a fixed hardness, toxicity can be changed by alkalinity or pH changes (Andrew, 1976; Chakoumakos et al., 1979; Chapman and McCrady, 1977; Mancy and Allen, 1977; Pagenkopf et al., 1974; Sunda and Guillard, 1976). The general conclusion of these workers is that the toxicity to fish and phytoplankton is

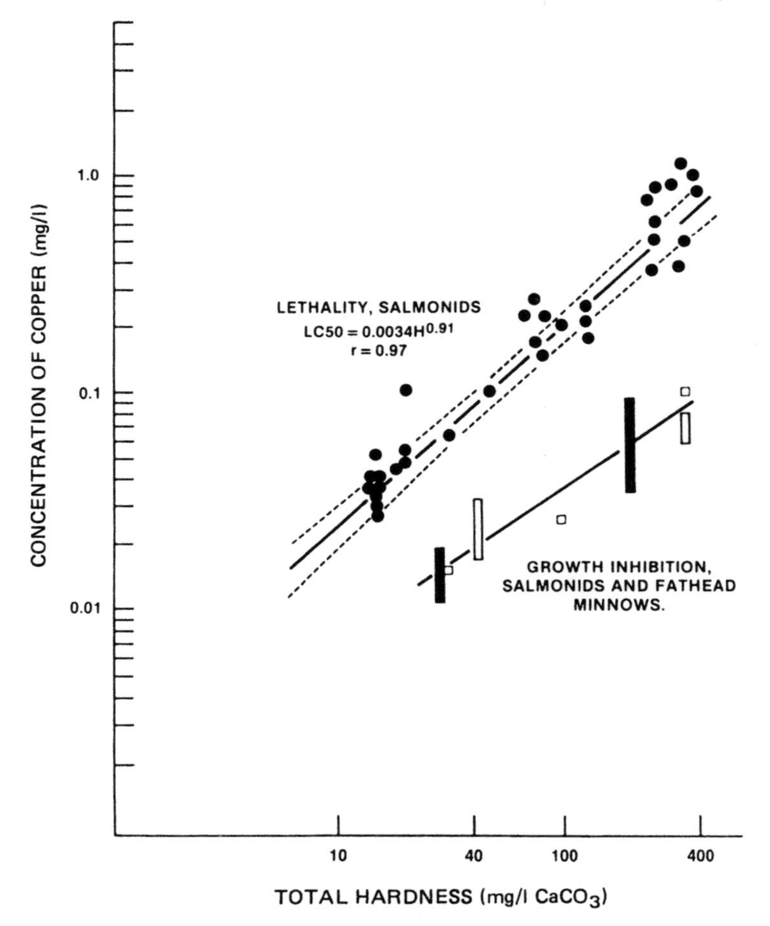

Figure 5 Empirical relation between copper toxicity and total hardness of fresh water. The upper line represents median lethal concentrations for most salmon and trout. The lower line represents thresholds for inhibition of growth. Growth results shown in darkened bars are for fathead minnows; those in open bars are for brook and rainbow trout in neutral or alkaline water. Modified from figures in Spear and Pierce (1979). Lethal data are from many reports in the literature. Growth results are from Lett et al. (1976), McKim and Benoit (1971), Mount (1968), Mount and Stephan (1969), and Waiwood and Beamish (1978b). In the formula for predicting lethal concentration, *H* is the total hardness of the water.

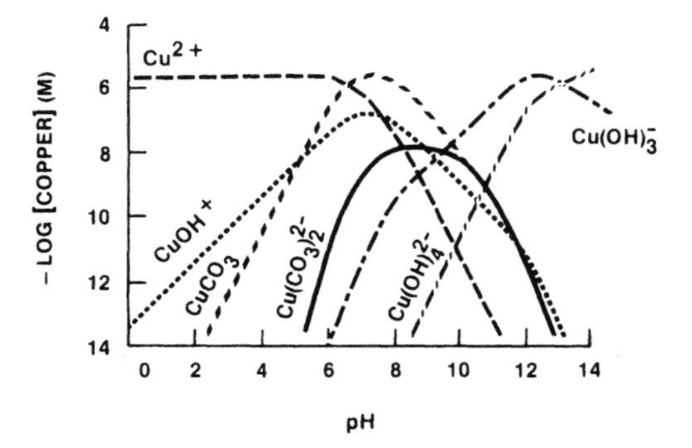

Figure 6 Comparison of log concentration of various forms of copper with pH for a total copper content of 2.4×10^{-6} M and a total carbonate content of 10^{-3} M. *[Reproduced with permission from Ernst et al. (1975); copyright 1975, Pergamon Press, Ltd.]*

due either to ionic copper or to that ion plus ionized hydroxides. The large hump at pH 8 in Fig. 7 represents high concentrations of copper carbonates and unionized hydroxides, which are not thought to be toxic. This suggestion was made a decade ago by Stiff (1971a) and is agreed with by most of the investigators listed above. When Fig. 7 was replotted, and the vertical axis displayed the concentration of supposedly toxic forms (Cu^{2+} and two ionized hydroxides) instead of total dissolved copper, the response surface lost its wild undulations and became fairly smooth.

However, the replotted response surface was far from flat, leading to a final complexity. The lethal concentrations of "toxic" forms were 200–2000

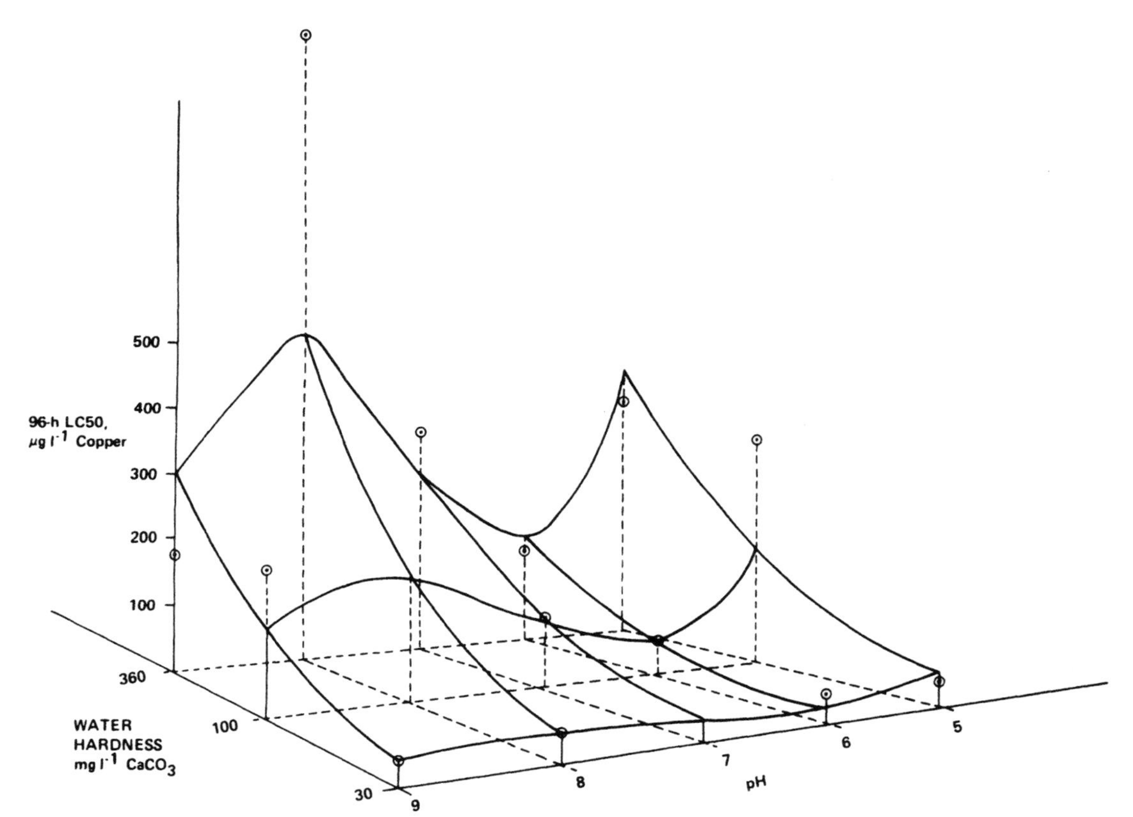

Figure 7 Lethal concentrations of total dissolved copper to 10-gram rainbow trout, at different combinations of water hardness and pH. Points show the LC50s obtained, the response surface was mathematically fitted. *[Reproduced with permission from Howarth and Sprague (1978); copyright 1978, Pergamon Press, Ltd.]*

times higher at pH 5 than at pH 9, depending on the hardness in a given set of tests. In other words, the active forms of copper were much more toxic at high pH. It is difficult to give credence to this variation, but it may exist.* Andrew (1976) suggests that the increasing toxicity of the cupric ion at high pH may result from interactions with sulfhydryl-containing proteins or enzymes. Other authors confirmed this change in toxicity with pH. When the LC50 of copper for guppies was calculated on the basis of labile copper (i.e., that measured by pulse polarography), it changed from 240 μg/l at pH 5.7 to 15 μg/l at pH 8.1, with the same hardness in each test–a 16-fold increase in toxicity at high pH (Mancy and Allen, 1977).

This research indicates that to predict copper lethality in unusual fresh waters, it will be necessary to take into account three modifying factors: hardness, alkalinity, and pH. Unfortunately, it requires fairly sophisticated chemical calculations to estimate the proportions of the various forms of copper, or else access to a good computer program to make these predictions. Even after that, there does not yet exist a single value for the toxicity of the apparently toxic forms of copper–it varies with water characteristics. Perhaps the working biologist will be able to use the empirical relation with hardness (Fig. 4), which will fit the usual conditions in surface waters. As partial justification, it may be considered that unusual combinations of pH, alkalinity, and hardness, perhaps brought about by acid or alkaline pollution, will change to the more usual combinations, given some time and the natural aeration that takes place in surface waters.

There are similar quandaries about the forms of other metals that should be measured in order to evaluate toxicity. For example, Davies et al. (1976) found great differences in the toxicity of lead between soft and hard water when the metal was measured as total concentration. The differences were minor when only dissolved lead was considered. Their results are simplified in Table 8 by using geometric averages of their toxicity ranges. In this case it seems clear that dissolved lead is the active toxic agent. The chronic thresholds of dissolved lead almost exactly fit findings of Hodson et al. (1978), and it appears that the difference between 0.010 and 0.024 mg/l is not an effect of hardness but of pH. Hodson and colleagues showed that the level of lead in the blood of trout, which is directly related to sublethal toxicity, increases by a factor of 2.1 for a drop of one unit of pH. The soft water of Davies et al. was just about one unit lower in pH than their hard water, and the toxicity increase of 2.4 agrees rather well with the factor of Hodson et al.

Thus for some metals we are coming close to understanding which forms are the toxic ones, and the information should be utilized in evaluating pollution of natural waters. One note of caution might be added for such applied work. The total content of metal, such as the large amount of nontoxic lead in the hard water of Table 2, should not be ignored. If water conditions changed (say from acid pollution), the ionic species of a metal could change and previously nontoxic forms could become toxic.

*Reexamination of the data of Howarth and Sprague (1978) suggests a way of decreasing the large variation in apparent toxicity. If one form of ionized carbonate $Cu(CO_3)_2^{2-}$ were also considered toxic and added in to the LC50 with the other three forms, the variation with pH would become only three- to eightfold. There would still be more toxicity in alkaline pH.

Table 8 Differences in Lead Toxicity to Rainbow Trout in Soft and Hard Water (28 and 350 mg/l $CaCO_3$)[a]

Measurement	Soft (mg/l)	Hard (mg/l)
LC50		
Total lead	1.2	510
Dissolved lead	1.2	1.4
Chronic threshold		
Total lead	0.010	0.21
Dissolved lead	0.010	0.024

[a]Modified from Davies et al. (1976); copyright 1976, Pergamon Press, Ltd.

Most of the discussion above concerns lethality, but the influence of pH, alkalinity, and so on, on the sublethal action of metals has become much more important in recent years. Severe acidification of soft-water lakes is occurring in many parts of the world, and is accompanied by increased metal content of the waters. There is some evidence that the metals may play a primary role in the observed reproductive failure of fish in such lakes, which leads eventually to fishless lakes. Considerable research on this is under way, and any attempt to review it would be quickly out of date. However, it can be said that the knowledge about toxic forms of metal that has been gained from work on lethality should be most useful for designing the lengthier experiments needed to assess the modifying factors of sublethal metal toxicity.

Binding and Sorption

Suspended and dissolved materials in surface waters may partly detoxify some pollutants. This is true for many metals, particularly copper, and for some other toxicants. Metals in water are commonly analyzed by atomic absorption spectroscopy (see chapter 16), which includes all configurations of metal in the measurement. Such measurements may not agree with responses in biological tests, which may show toxicity that is equivalent to much lower concentrations. As mentioned above, there is some justification for thinking that all forms of metal should be included in pollution assessment or in setting water quality criteria, since nontoxic forms can change to toxic forms (e.g., NAS/NAE, 1974). However, at least for some metals, understanding has progressed to the point where more realistic assessments can be based on the active forms. The following discussion is based mainly on complexing of metals, particularly copper, since there has been considerable research on that metal.

One form of complexation is with natural inorganic ions in surface water such as carbonates and hydroxides. This was outlined in the section immediately above. Excellent discussions of copper complexation have been provided by Pagenkopf et al. (1974) and Andrew et al. (1977), who demonstrated that carbonate, orthophosphate, and pyrophosphate complexes of copper were nontoxic. Other types of complexation will be considered in this section.

Many field studies have shown that aquatic organisms may be unaffected by supposedly toxic levels of metals. An early example was a trial of lake poisoning which showed persistence of copper in the water, but in a nontoxic form (Smith, 1939). In upwelling deep oceanic water, copper seems to be about an order of magnitude more toxic to algae than in surface waters, because the deep waters apparently lack natural chelating agents (Steeman Neilsen and Wium-Andersen, 1970). In the water of a Newfoundland river, 125 μg/l copper was required for lethality to young salmon, instead of the expected level of about 30 μg/l (Wilson, 1972). Addition of spent sulfite liquor from a pulp mill raised the LC50 to more than 250 μg/l. In certain Norwegian rivers and lakes, healthy populations of salmonids existed in copper-zinc concentrations that exceeded the carefully established European criteria for "safe" levels by factors of 3-9 (EIFAC, 1977). Chemical examinations indicated that as much as 86% of the copper might be present as organometal complexes that were not toxic.

A massive field experiment in Ohio was designed to assess the validity of laboratory research on copper, when applied to surface waters (Geckler et al., 1976). An important finding was that lethal levels of total copper for fathead minnows averaged 12 times higher in the stream than in nearby laboratory water. The no observed effect level for reproduction was three to four times higher in stream water. Clearly, there was considerable complexing or other detoxification of copper in the stream water, which contained final products from an upstream sewage treatment plant. LC50s in the stream fluctuated greatly over 1.5 yr, from 1.6 to 21 mg/l based on a measure of *total* copper, but values were relatively constant at 0.60–0.98 mg/l when based on "dissolved" copper that passed a 0.45-μm membrane filter. It was concluded that predictive ability was much improved for several of the heavy metals if assessments were based on dissolved metal (Brungs et al., 1976).

Laboratory studies on complexing have further added to the understanding of these effects. As might be expected, certain organic chemicals designed to be chelators of cations are very good at making metals nontoxic to aquatic organisms. Nitrilotriacetic acid (NTA) effectively eliminates copper and zinc lethality to brook trout even when the metals are present at 33 times their usual threshold LC50 (Sprague, 1968). Similar findings are reported for guppies (Gudernatsch, 1970) and the crustacean *Daphnia magna* (Biesinger et al., 1974). Cadmium is also detoxified for *Daphnia* by EDTA, a comparable chelating agent (Teulin, 1978). The actions of the agents above are quite predictable; for example, NTA binds cations in a one-to-one molecular ratio, with a known order of priority for various cations.

Laboratory tests have documented similar actions for other, more natural organic molecules. Algae exude complexing ligands that can detoxify metals such as copper or bind essential micronutrients (Van den Berg et al., 1979). The amino acid glycine detoxified copper by an order of magnitude (Brown et al., 1974). Other amino acids and polypeptides can be expected to do so (Stiff, 1971b). Sewage plant effluents and suspended solids have a smaller effect (Brown et al., 1974), but colloids, clays, and soil particles can adsorb major proportions of copper in water, with largely unknown effects on toxicity. Between 10 and 80% of various metals were bound to suspended particles in Lake Ontario (Nriagu et al., 1981; Spear and Pierce, 1979). Finally, humic and fulvic acids, which darken streams in many forested areas, have a well-documented effect. Humic acid can cause an order-of-magnitude decrease in copper toxicity in very soft water, but its binding action falls off in harder water (Brown et al., 1974; Zitko et al., 1973). Fulvic acid is somewhat less effective, and neither of these agents has an appreciable effect in detoxifying zinc and cadmium (Zitko et al., 1973; Giesy et al., 1977). All the above findings concern lethality, but Carson and Carson (1973) made the significant finding that avoidance reactions of Atlantic salmon to copper could be decreased by a factor of about 5 with respect to copper concentration if 20 mg/l humic acid was added to soft water. Such levels of humic acid could be found in natural waters.

It is clear that both inorganic and organic complexing can lower toxicity of metals, particularly copper. The problem is how to assess concentrations of the toxic fractions, and recent work brings us closer to satisfactory methods. Atomic absorption spectrometry, commonly used at present, measures all forms of metal in the sample—that is, total metal. A relevant comment is that of Andrew (1976) about the "futility of basing the results of toxicity bioassays on total metal analyses. . . . Such data are not useful to any extent." Earlier in this chapter, reference was made to the use of chemical equilibrium calculations to distinguish between toxic forms such as ions and apparently nontoxic inorganic complexes. These calculations can be useful; however, they are not a direct measurement but a prediction based on equilibrium constants and the chemical characteristics of the dilution water. McCrady and Chapman (1979) urged that such predictions "be used with caution, realizing that at best, quantitative results might be obtained at higher concentrations."

It was also mentioned above that measurements of so-called dissolved metal, which passes a 0.45-μm filter, can give better estimates of concentrations of the toxic forms than measurements of total metal. This, too, can be deceptive, since many organic ligands would pass such a filter (Giesy et al., 1977). Most "nontoxic" suspended precipitates would be expected to stay on the filter, but some may pass, and some dissolved molecules may be taken up by the filter.

A closer approach to measuring concentrations of toxic forms of metals would be provided by differential pulse polarography, which is said to measure "labile" forms. These include the free ion and other ions that are easily dissociated; obviously, this is another operational division of forms of metals rather than an absolute one. Polarography has the advantage of detecting very low concentrations—for example, 0.2–0.7 μg/l for some common metals in water (Chau and Lum-Shue-Chan, 1974).

Finally, a technique that is receiving a great deal of attention in recent literature involves ion-selective electrodes, which are more or less specific for measurement of one metallic ion (or its hydrated form). Detection of low concentrations is not as satisfactory as with polarography. McCrady and Chapman (1979) reported reasonable agreement between cupric ion concentrations measured with such an electrode and concentrations predicted from a chemical equilibrium model. However, measured concentrations in natural waters were lower than predicted, presumably because of complexing by organic ligands and other materials in the water. Specific ion electrodes have been used with success to measure proportions of metals not bound by humic acid (Zitko et al., 1973) and to assess which species of algae are primarily responsible for binding heavy metals in natural waters (Briand et al., 1978).

Thus, for the complicated subject of toxic species of metals and their chelation in natural waters, a good deal of progress has been made in the past decade. By applying recent technology, admittedly rather sophisticated technology, a reasonable picture can be obtained of the true toxicity of metals in water, and fairly accurate predictions can be made.

The effect of binding and sorption on other, nonmetallic pollutants appears to have been studied very little and, for the most part, strong modifying effects have not been noted. Toxicity of the insecticide endrin to fathead minnows was not reduced when it was added to suspensions of clay (Brungs and Bailey, 1966). In contrast, addition of a soil suspension to endrin solutions seemed to moderate its toxicity somewhat, and endrin already sorbed onto runoff sediments was not released into water in sufficient amounts to be lethal to fish (Ferguson et al., 1965). Clay suspensions at levels that might be found in nature had little influence on the toxicity of the surfactant LAS (Hokanson and Smith, 1971). Very strong suspensions (about 20,000 mg/l) of drilling fluid, mostly a mixture of inert barium sulfate and clay, reduced the toxicity of three surfactants but not of two others, and also reduced toxicity of paraformaldehyde and capryl alcohol (Sprague and Logan, 1979). Reductions were only by factors of 1.5-2.0, not very large considering the large amount of suspended matter, including the clay bentonite, which has a relatively high cation exchange capacity. It may be that the lack of such research on complexing of nonmetallic pollutants is an indication that effects are not major.

SUMMARY

1 Factors that modify the toxicity of pollutants to aquatic organisms, including biotic variables such as species and abiotic variables such as water temperature, have been considered in this chapter.

2 Effects of modifying factors must be cautiously interpreted within experiments and not between them, bearing in mind that even repeated toxicity tests in the same laboratory will not give identical results. Some within-laboratory variation is difficult or impossible to explain. A careful investigator may obtain a series of LC50s with a range of only ±20% of the median value. On the other hand, LC50s from the same laboratory have been known to vary by a factor of 5. Interlaboratory comparisons may show variation by a factor of only 1.2 or differences as great as 10-fold.

3 If good basic procedures are followed in acute lethality tests, other differences in test procedure do not affect results greatly. For example, when trout were tested at their maximum cruising speed, their LC50s dropped to only 0.7 or 0.8 of the values for unexercised fish.

4 The kind of test organism has a major influence on toxicity data, but perhaps less than might be expected. A literature review with indi-vidiual comparisons of 15 pollutants tested under similar conditions, with results accepted for all types of organisms, showed that well over half of the lethal levels fell within one order of magnitude. The range was spread over three orders of magnitude because of variation in only 15% of the observed LC50s. Within one group such as fish, lethal levels for a given pollutant would generally fall within an order of magnitude. For example, oil refinery waste showed only a fivefold variation in LC50s between the most and least sensitive of 57 species. Some heavy metals and organophosphate insecticides are exceptions, with extreme variation between species of fish. However, bluegills are tolerant of the former and sensitive to the latter, which suggests that caution should be used in generalizing about "tolerant species."

5 In the life cycle of fish, the egg, larval, and early juvenile stages are most sensitive in chronic exposures. Large fish are not necessarily more tolerant than small juveniles; the effect of size on LC50 may go either way, or may be absent.

6 There is only slim evidence that diseased stocks of fish are less tolerant of toxicants. On the other hand, nutrition of test organisms can be quite important. Fish on a low-protein diet may be two to six times less tolerant of some pollutants. Acclimation of fish to a toxicant may result in a doubling of tolerance or in no increase, depending on the substance involved and whether a defense mechanism is stimulated.

7 There is no general effect of temperature on toxicity. Depending on the species and pollutant, fish in warmer water may be more, less, or equally tolerant. Limited evidence suggests that thresholds of sublethal effect may be about the same at all water temperatures.

8 Dissolved oxygen concentrations in the vicinity of 20–30% of saturation seem to cause an increase of only about 1.5-fold in lethality of toxicants to fish. Some experimental results indicate that sublethal effects of a toxicant are not greatly modified by low oxygen, even when the reduced oxygen itself has such sublethal action.

9 Some pollutants show large changes in toxicity with pH of the water if they are ionized under the influence of that variable. Usually it is the undissociated or less dissociated form that is most toxic, and the change may be an order of magnitude for a difference of one pH unit or less. Ammonia, cyanide, and hydrogen sulfide show such changes, and their lethality can be predicted accurately on the basis of the chemical transformations. Most metals act in the opposite way, with the free ionic species being more toxic than the undissociated compounds. Changes in sublethal toxicity appear to parallel changes in chemical behavior and lethality.

10 Salinity is a major variable among the characteristics of surface water, but its influence on toxicity is not so important. Marine organisms in seawater and freshwater organisms in their environment seems to be about equally tolerant of a given pollutant. Euryhaline organisms, which are less common, attain maximum tolerance in isosmotic water—that is, about 30–40% seawater—and the increase in tolerance may be an order of magnitude. Modification of sublethal effects by salinity has been less studied.

11 Total hardness of fresh water has little effect on the potency of most pollutants (except metals). A few substances change their toxicity up to two- or threefold, but the direction of change with hardness varies.

12 Many heavy metals become an order of magnitude more lethal in very soft water than in very hard water, and sublethal changes are roughly parallel. This is thought to result from changes in gill permeability caused by calcium content of the fish. For peculiar waters in which pH and alkalinity do not follow the usual relations with hardness, the importance of those two variables becomes more obvious. They may partially or wholly govern the proportions of various ionic species of metal that are present, and these differ in toxicity. To generalize from knowledge of copper, it appears that the free or ionic metal is very toxic. Ionized hydroxides of metal may also be toxic, but unionized carbonates are probably not. Furthermore, for copper at least, toxicity changes when alkalinity is varied but hardness kept constant, and there is much higher potency of the toxic forms of metal at high pH than at low pH, again for the same hardness.

13 Natural waters often contain suspended and dissolved matter including organic ligands or chelators. Metals are the chief examples of pollutants that may be detoxified because of sorption or binding by these materials; most other pollutants seem much less affected. Among the metals copper is particularly affected, and binding and sorption may decrease its lethal and sublethal toxic effects by an order of magnitude.

14 Because of the extreme effects of modifying factors on metal toxicity described in items 12 and 13, a number of approaches to chemical measurement of metallic pollutants should be considered. Measurement of toxic forms of metal is better approximated by the "dissolved" metal (i.e., the metal able to pass through a 0.45-μm filter) than by the total metal (atomic absorption spectrometry). The empirical relation between toxicity of total dissolved metal and water hardness may suffice for routine use in many natural waters. It is conceivable that the nontoxic ionic forms included with total dissolved metal could change to toxic forms following an event such as acid pollution. The proportions of supposedly toxic ionic forms, present as parts of the total dissolved metal, could be predicted from models using chemical equilibrium constants, although this requires effective use of computer programs. None of the methods above adequately allows for metals detoxified by binding to dissolved or suspended materials. A closer approximation of toxic species can be made by polarography, which detects ionic or free metals plus easily dissociating forms. Specific ion electrodes can measure the free forms only for certain metals, but with less sensitivity to low concentrations than is obtained by polarography.

15 It is difficult to sift out "concepts of aquatic toxicology" from this review of modifying factors. They will have to be particular relations rather than sweeping principles. First, we must expect an order-of-magnitude variation in toxicity data gathered from a number of laboratories. Second, biotic variables seem to have generally more important modifying effects than do abiotic ones, and the principal one is the kind of organism. Although most kinds would show tolerances within an order-of-magnitude range, some groups are sensitive and total variation is often about three orders of magnitude. Nutritional status of test organisms appears to be another important biotic modifier. Third, the pollutants most affected by abiotic modification are those that ionize, and the most important modifying entities are pH and alkalinity. Fourth, metals are subject to large changes in toxicity, not only through ionization but through precipitation, sorption, and binding to suspended and dissolved matter. Finally, temperature, low dissolved oxygen, saltwater versus freshwater medium, and hardness of fresh water are not generally *major* modifying entities. Their effects are often only two- or threefold or much less, with some exceptions, including a hardness effect on metals.

LITERATURE CITED

Adelman IR, Smith LL, Jr: Toxicity of hydrogen sulfide to goldfish (*Carassius auratus*) as influenced by temperature, oxygen, and bioassay techniques. J Fish Res Bd Can 29:1309–1317, 1972.

Adelman IR, Smith LL, Jr: Fathead minnows (*Pimephales promelas*) and goldfish (*Carassius auratus*) as standard fish in bioassays and their reaction to potential reference toxicants. J Fish Res Bd Can 33:209–214, 1976.

Alabaster JS, Lloyd R: Water Quality Criteria for Freshwater Fish. London: Butterworths, 1980.

Alderdice DF, Brett JR: Some effects of kraft mill effluent on young Pacific salmon. J Fish Res Bd Can 14:783–795, 1957.

Anderson PD, Spear PA: Copper pharmacokinetics in fish gills. II. Body size relationships for accumulation and tolerance. Water Res 14:1107–1111, 1980.

Anderson PD, Weber LJ: Toxic response as a quantitative function of body size. Toxicol Appl Pharmacol 33:471–483, 1975.

Anderson PD, Horovitch H, Weinstein NL: Pollutant mixtures in the aquatic environment: A complex problem in toxic hazard assessment. Proc 5th Ann Aquatic Toxicity Workshop, Hamilton, Ont, Nov 7–9, 1978. Can Fish Mar Serv Tech Rept 862, pp. 100–114, 1979.

Andrew RW: Toxicity relationships to copper forms in natural waters. In: Toxicity to Biota of Metal Forms in Natural Water. Proc Workshop, Duluth, Minn, Oct 7–8, 1975, edited by RW Andrew, PV Hodson, DE Konasewich, pp. 127–143. Windsor, Ont: International Joint Commission, 1976.

Andrew RW, Biesinger KE, Glass GE: Effects of inorganic complexing on the toxicity of copper to *Daphnia magna.* Water Res 11:309–315, 1977.

Asano S, Nagasawa S, Fushimi S: Biological trials of chemicals on fish. VI. Relation between temperature and the toxicity of pentachlorophenol (PCP) to carp. Bochu-Kagaku 34:13–21, 1969; Chem Abstr 71(11):260, 1969.

Ball IR: The relative susceptibilities of some species of fresh-water fish to poisons. I. Ammonia. Water Res 1:767–775, 1967.

Bender ME: The toxicity of the hydrolysis and breakdown products of malathion to the fathead minnow (*Pimephales promelas* Rafinesque). Water Res 3:571–582, 1969.

Biesinger KE, Andrew RW, Arthur JW: Chronic toxicity of NTA (nitrilotriacetate) and metal-NTA complexes to *Daphnia magna*. J Fish Res Bd Can 31:486–490, 1974.

Boyce NP, Yamada SB: Effects of a parasite, *Eubothrium salvelini* (Cestoda; Pseudophyllidea), on the resistance of juvenile sockeye salmon (*Oncorhynchus nerka*) to zinc. J Fish Res Bd Can 34:706–709, 1977.

Briand F, Trucco R, Ramamoorthy S: Correlations between specific algae and heavy metal binding in lakes. J Fish Res Bd Can 35:1482–1485, 1978.

Broderius SJ, Smith LL, Jr, Lind DT: Relative toxicity of free cyanide and dissolved sulfide forms to the fathead minnow (*Pimephales promelas*). J Fish Res Bd Can 34:2323–2332, 1977.

Brown BE: Observations on the tolerance of the isopod *Asellus meridianus* Rac. to copper and lead. Water Res 10:555–559, 1976.

Brown VM: The calculation of the acute toxicity of mixtures of poisons to rainbow trout. Water Res 2:723–733, 1968.

Brown VM, Jordan DHM, Tiller BA: The effect of temperature on the acute toxicity of phenol to rainbow trout in hard water. Water Res 1:587–594, 1967.

Brown VM, Shaw TL, Shurben DG: Aspects of water quality and the toxicity of copper to rainbow trout. Water Res 8:797–803, 1974.

Brungs WA, Bailey GW: Influence of suspended solids on the acute toxicity of endrin to fathead minnows. Proc 21st Ind Waste Conf. Purdue Univ Eng Extn Ser No 121, part 1, 50(2):4–12, 1967.

Brungs WA, Geckler JR, Gast M: Acute and chronic toxicity of copper to the fathead minnow in a surface water of variable quality. Water Res 10:37–43, 1976.

Bryan GW, Hummerstone LG: Adaptation of the polychaete *Nereis diversicolor* to estuarine sediments containing high concentrations of heavy metals. 1. General observation and adaptations to copper. J Mar Biol Assoc UK 51:207–217, 1971.

Bryan GW, Hummerstone LG: Adaptation of the polychaete *Nereis diversicolor* to estuarine sediments containing high concentrations of zinc and cadmium. J Mar Biol Assoc UK 53:839–859, 1973.

Burton DT, Jones AH, Cairns J, Jr: Acute zinc toxicity to rainbow trout (*Salmo gairdneri*): Confirmation of the hypothesis that death is related to tissue hypoxia. J Fish Res Bd Can 29:1463–1466, 1972.

Cairns J, Scheier A: The effects of periodic low oxygen upon the toxicity of various chemicals to aquatic organisms. Proc 12th Ind Waste Conf Purdue Univ Eng Ext Ser 94:165–176, 1958.

Cairns J, Scheier A: The effects of temperature and water hardness upon the toxicity of naphthenic acids to the common bluegill sunfish, *Lepomis macrochirus* Raf., and the pond snail, *Physa heterostropha* Say. Not Nat Acad Nat Sci Philadelphia No 353, 1962.

Cairns J, Jr, Scheier A: Environmental effects upon cyanide toxicity to fish. Not Nat Acad Nat Sci Philadelphia No 361, 1963.

Cairns J, Jr, Scheier A, Hess NE: The effects of alkyl benzene sulfonate on aquatic organisms. Ind Water Wastes 9:22–28, 1964.

Cairns J, Jr, Heath AG, Parker BC: Temperature influence on chemical toxicity to aquatic organisms. J Water Pollut Control Fed 47:267–280, 1975.

Carson WG, Carson WV: Avoidance of copper in the presence of humic acid by juvenile Atlantic salmon. Fish Res Bd Can Ms Rept Ser No 1237, 1973.

Chakoumakos C, Russo RC, Thurston RV: Toxicity of copper to cutthroat trout (*Salmo clarki*) under different conditions of alkalinity, pH, and hardness. Environ Sci Technol 13:213–219, 1979.

Chapman GA, McCrady JK: Copper toxicity: A question of form. In: Recent Advances in Fish Toxicology: A Symposium, edited by RA Tubb, pp. 132–151. U.S. EPA Rept EPA 660/3-77/085, 1977.

Chau YK, Lum-Shue-Chan K: Determination of labile and strongly bound metals in lake water. Water Res 8:383–388, 1974.

Cherry DS, Larrick SR, Dickson KL, Hoehn RC, Cairns J, Jr: Significance of hypochlorous acid in free residual chlorine to the avoidance response of spotted bass (*Micropterus punctulatus*) and rosyface shiner (*Notropis rubellus*). J Fish Res Bd Can 34:1365–1372, 1977.

Davies PH, Goettl JP, Jr, Sinley JR, Smith NF: Acute and chronic toxicity of lead to rainbow trout *Salmo gairdneri,* in hard and soft water. Water Res 10:199–206, 1976.

Davis JC: Minimal dissolved oxygen requirements of aquatic life with emphasis on Canadian species: A review. J Fish Res Bd Can 32:2295–2332, 1975.

Davis JS, Hoos RAW: Use of sodium pentachlorophenate and dehydroabietic acid as reference toxicants for salmonid bioassays. J Fish Res Bd Can 32:411–416, 1975.

Daye PG: Attempts to acclimate embryos and alevins of Atlantic salmon, *Salmo salar,* and rainbow trout, *S. gairdneri,* to low pH. Can J Fish Aquat Sci 37:1035–1038, 1980.

Dixon DG, Hilton JW: Influence of available dietary carbohydrate content on tolerance of waterborne copper by rainbow trout. J Fish Biol 19:509–517, 1981.

Dixon DG, Sprague JB: Acclimation-induced changes in toxicity of arsenic and cyanide to rainbow trout. J Fish Biol 18:579–589, 1981a.

Dixon DG, Sprague JB: Acclimation to copper by rainbow trout (*Salmo gairdneri*)—a modifying factor in toxicity. Can J Fish Aquat Sci 38: 880–888, 1981b.

Dixon DG, Sprague JB: Copper bioaccumulation and hepatoprotein synthesis during acclimation to copper by juvenile rainbow trout. Aquat Toxicol 1:69–81, 1981c.

Doudoroff P, Shumway DL: Dissolved oxygen requirements of freshwater fishes. FAO Fish Tech Pap 86, 1970.

Doudoroff P, Leduc G, Schneider CR: Acute toxicity to fish of solutions containing complex metal cyanides, in relation to concentrations of molecular hydrocyanic acid. Trans Am Fish Soc 95:6–22, 1966.

Eedy W: Environmental cause/effect phenomena relating to technological development in the Canadian Arctic. Natl Res Counc Can Publ NRCC 13688, 1974.

EIFAC (European Inland Fisheries Advisory Commission): Water quality criteria for European freshwater fish. Report on monohydric phenols and inland fisheries. Water Res 7:929–941, 1973a.

EIFAC: Water quality criteria for European freshwater fish. Report on ammonia and inland fisheries. Water Res 7:1011–1022, 1973b.

EIFAC: Report on the effect of zinc and copper pollution on the salmonid fisheries in a river and lake system in central Norway. FAO Eur Inland Fish Adv Comm Tech Pap 29, 1977.

Eisler R: Some effects of a synthetic detergent on estuarine fishes. Trans Am Fish Soc 94:26–31, 1965.

Elson PF, Kerswill CJ: Impact on salmon of spraying insecticide over forests. In: Advances in Water Pollution Research, vol. 1, Proceedings of a Conference of International Associations on Water Pollution Research, Munich, pp. 55–74. Washington, D.C.: Water Pollution Control Federation, 1966.

Emerson K, Russo RC, Lund RE, Thurston RV: Aqueous ammonia equilibrium calculations: Effect of pH and temperature. J Fish Res Bd Can 32:2379–2383, 1975.

EPA: Quality criteria for water. U.S. Environmental Protection Agency, Washington, D.C., 256 p, 1976.

EPA: Water quality criteria. Fed Regist 43(97): 21506–21518, 1978.

EPA: Interlaboratory comparisons of toxicity tests. U.S. EPA Environ Res Lab Duluth Q Rep, pp. 2–3, April–June 1979.

Ernst R, Allen HE, Mancy KH: Characterization of trace metal species and measurement of trace metal stability constants by electrochemical techniques. Water Res 9:969–979, 1975.

Ferguson DE, Ludke JL, Wood JP, and Prather JW: The effects of mud on the bioactivity of pesticides on fishes. J Miss Acad Sci 11:219–228, 1965.

Fogels A, Sprague JB: Comparative short-term tolerance of zebrafish, flagfish, and rainbow trout to five poisons including potential reference toxicants. Water Res 11:811–817, 1977.

Fry FEJ: Effects of the environment on animal activity. Univ Toronto Stud Biol Ser 55; Publ Ont Fish Res Lab 68, 1947.

Fry FEJ: Requirements for the aquatic habitat. Pulp Pap Mag Can 61:T61–T66, 1960.

Fry FEJ: The effect of environmental factors on the physiology of fish. In: Fish Physiology, vol. 6, Environmental Relations and Behavior, edited by WS Hoar, DJ Randall, pp. 1–98. New York: Academic, 1971.

Garside ET, Jordan CM: Upper lethal temperatures at various levels of salinity in the euryhaline cyprinodontids *Fundulus heteroclitus* and *F. diaphanus* after isosmotic acclimation. J Fish Res Bd Can 25:2717–2720, 1968.

Geckler JR, Horning WB, Nieheisel TM, Pickering QH, Robinson, EL, Stephan CE: Validity of laboratory tests for predicting copper toxicity in streams. U.S. EPA Ecol Res Ser EPA-600/3-76-116, 1976.

Giesy JP, Jr, Leversee GJ, Williams DR: Effects of naturally occurring aquatic organic fractions on cadmium toxicity to *Simocephalus serrulatus* (Daphnidae) and *Gambusia affinis* (Poeciliidae). Water Res 11:1013–1020, 1977.

Gudernatsch H: Verhalten von Nitrilotriessigsäure im Klärprozess und im Abwasser. Gas Wasserfach 111:511–516, 1970.

Guth DJ, Blankespoor HD, Cairns J, Jr: Potentiation of zinc stress caused by parasitic infection of snails. Hydrobiologia 55:225–229, 1977.

Henderson C, Pickering QH: Toxicity of organic phosphorus insecticides to fish. Trans Am Fish Soc 87:39–51, 1958.

Henderson C, Pickering QH, Cohen JM: The toxicity of synthetic detergents and soaps to fish. Sewage Ind Wastes 31:295–306, 1959a.

Henderson C, Pickering, QH, Tarzwell CM: Relative toxicity of ten chlorinated hydrocarbon insecticides to four species of fish. Trans Am Fish Soc 88:23–32, 1959b.

Herbert DWM, Shurben DS: A preliminary study of the effect of physical activity on the resistance of rainbow trout (*Salmo gairdneri* Richardson) to two poisons. Ann Appl Biol 52:321–326, 1963.

Herbert DWM, Shurben DS: The susceptibility of salmonid fish to poisons under estuarine conditions. II. Ammonium chloride. Int J Air Water Pollut 9:89–91, 1965.

Herbert DWM, Wakeford AC: The susceptibility of salmonid fish to poisons under estuarine conditions. I. Zinc sulphate. Int J Air Water Pollut 8: 251–256, 1964.

Hicks DB, DeWitt JW: Effects of dissolved oxygen on kraft pulp mill effluent toxicity. Water Res 5:693–701, 1971.

Hodson PV: The effect of temperature on the toxicity of zinc to fish of the genus *Salmo*. PhD thesis, Univ of Guelph, Guelph, Ontario, Canada, 1974.

Hodson PV: Zinc uptake by Atlantic salmon (*Salmo salar*) exposed to a lethal concentration of zinc at 3, 11, and 19°C. J Fish Res Bd Can 32:2552–2556, 1975.

Hodson PV: Temperature effects on lactate-glycogen metabolism in zinc-intoxicated rainbow trout (*Salmo gairdneri*). J Fish Res Bd Can 33:1393–1397, 1976.

Hodson PV, Sprague JB: Temperature-induced changes in acute toxicity of zinc to Atlantic salmon (*Salmo salar*). J Fish Res Bd Can 32:1–10, 1975.

Hodson PV, Blunt BR, Spry DJ: pH-induced changes in blood lead of lead-exposed rainbow trout. J Fish Res Bd Can 35:437–445, 1978.

Hodson PV, Hilton JW, Blunt BR, Slinger SJ: Effects of dietary ascorbic acid on chronic lead toxicity to young rainbow trout (*Salmo gairdneri*). Can J Fish Aquat Sci 37:170–176, 1980.

Hokanson KEF, Smith LL, Jr: Some factors influencing toxicity of linear alkylate sulfonate (LAS) to the bluegill. Trans Am Fish Soc 100: 1–12, 1971.

Holcombe GW, Fiandt JT, Phipps GL: Effects of pH increases and sodium chloride additions on the acute toxicity of 2,4-dichlorophenol to the fathead minnow. Water Res 14:1073–1077, 1980.

Houston AH: Osmoregulatory adaptation of steelhead trout (*Salmo gairdneri* Richardson) to sea water. Can J Zool 37:729–743, 1959.

Howarth RS, Sprague JB: Copper lethality to rainbow trout in waters of various hardness and pH. Water Res 12:455–462, 1978.

Inglis A, Davis EL: Effects of water hardness on the toxicity of several organic and inorganic herbicides to fish. U.S. Bur Sport Fish Wildl Tech Pap 67, 1972.

Irwin WH: Fifty-seven species of fish in oil-refinery waste bioassay. Trans 30th North Am Wildl Nat Resour Conf, pp. 89–99, 1965.

Johnson DW: Pesticides and fishes–A review of selected literature. Trans Am Fish Soc 97:398–424, 1968.

Johnson MW: The effectiveness of toxaphene at low temperatures. Minn Fish Game Invest Fish Ser 2:52–54, 1961.

Kinne O: Physiological aspects of animal life in estuaries with special reference to salinity. Neth J Sea Res 3:222–244, 1966.

Klapow LA, Lewis RH: Analysis of toxicity data for California marine water quality standards. J Water Pollut Control Fed 51:2054–2070, 1979.

Kleerekoper H, Waxman JB, Matis J: Interaction of temperature and copper ions as orienting stimuli in the locomotor behavior of the goldfish (*Carassius auratus*). J Fish Res Bd Can 30: 725–728, 1973.

Koeppe R: The toxicology and toxicity of toxaphene with respect to fish and aquatic food animals. Z Fisch Deren Hilfswiss 9:771–794, 1961.

Kumaraguru AK, Beamish FWH: Lethal toxicity of Permethrin (NRDC-143) to rainbow trout, *Salmo gairdneri*, in relation to body weight and water temperature. Water Res 15:503–505, 1981.

Kwain W: Effects of temperature on development and survival of rainbow trout, *Salmo gairdneri*, in acid waters. J Fish Res Bd Can 32:493–497, 1975.

Lee DR, Buikema AR, Jr: Molt-related sensitivity of *Daphnia pulex* in toxicity testing. J Fish Res Bd Can 36:1129–1133, 1979.

Lett PF, Farmer GJ, Beamish FWH: Effect of copper on some aspects of the bioenergetics of rainbow trout (*Salmo gairdneri*). J Fish Res Bd Can 33:1335–1342, 1976.

Levitan WM, Taylor MH: Physiology of salinity-dependent naphthalene toxicity in *Fundulus heteroclitus*. J Fish Res Bd Can 36:615–620, 1979.

Lipschuetz M, Cooper AL: Toxicity of 2-secondary-butyl-4,6-dinitrophenol to blacknose dace and rainbow trout. NY Fish Game J 8: 110–121, 1961.

Lloyd R: Effect of dissolved oxygen concentrations on the toxicity of several poisons to rainbow trout (*Salmo gairdnerii* Richardson). J Exp Biol 38:447–455, 1961a.

Lloyd R: The toxicity of ammonia to rainbow trout (*Salmo gairdnerii* Richardson). Water Waste Treatm J 8:278–279, 1961b.

Lloyd R: Factors that affect the tolerance of fish to heavy metal poisoning. Biological problems in water pollution, third seminar, 1962. U.S. Public Health Serv Publ 999-WP-25, pp. 181–187, 1965.

Lloyd R, Herbert DWM: The effect of the environment on the toxicity of poisons to fish. J Inst Public Health Eng 61:132–145, 1962.

Loch JS, MacLeod JC: Factors affecting acute toxicity bioassays with pulp mill effluent. Can Dept Environ Fish Mar Serv Tech Rept Ser No CEN/T-74-2, 1974.

Macek KJ: Growth and resistance to stress in brook trout fed sublethal levels of DDT. J Fish Res Bd Can 25:2443–2451, 1968.

Macek KJ, Hutchinson C, Cope OB: The effects of temperature on the susceptibility of bluegills and rainbow trout to selected pesticides. Bull Environ Contam Toxicol 4:174–183, 1969.

MacLeod JC, Pessah E: Temperature effects on mercury accumulation, toxicity, and metabolic rate in rainbow trout (*Salmo gairdneri*). J Fish Res Bd Can 30:485–492, 1973.

Mancy KH, Allen HE: A controlled bioassay system for measuring toxicity of heavy metals. U.S. EPA Rept EPA-600/3-77-037, 1977.

Marchetti R: Critical review of the effects of synthetic detergents on aquatic life. Stud Rev FAO Gen Fish Counc Mediterranean No. 26, 1965.

Marking LL: Effects of pH on toxicity of antimycin to fish. J Fish Res Bd Can 32:769–773, 1975.

Marking LL, Bills TD: Toxicity of rotenone to fish in standardized laboratory tests. U.S. Dept

Inter Fish Wildl Serv Invest Fish Control No. 72, 1976.

Mauck WL, Olson LE, Hogan JW: Effects of water quality on deactivation and toxicity of mexacarbate (Zectran) to fish. Arch Environ Contam Toxicol 6:385–393, 1977.

Mayer FL, Mehrle PM, Crutcher PL: Interactions of toxaphene and vitamin C in channel catfish. Trans Am Fish Soc 107:326–333, 1978.

McCrady JK, Chapman GA: Determination of copper complexing capacity of natural river water, well water and artificially reconstituted water. Water Res 13:143–150, 1979.

McKim JM: Evaluation of tests with early life stages of fish for predicting long-term toxicity. J Fish Res Bd Can 34:1148–1154, 1977.

McKim JM, Benoit DA: Effects of long-term exposures to copper on survival, growth, and reproduction of brook trout (*Salvelinus fontinalis*). J Fish Res Bd Can 28:655–662, 1971.

McLeay DJ, Gordon MR: Effect of seasonal photoperiod on acute toxic responses of juvenile rainbow trout (*Salmo gairdneri*) to pulp mill effluent. J Fish Res Bd Can 35:1388–1392, 1978.

McLeay DJ, Munro JR: Photoperiodic acclimation and circadian variations in tolerance of juvenile rainbow trout (*Salmo gairdneri*) to zinc. Bull Environ Contam Toxicol 23:552–557, 1979.

McLeay DJ, Walden CC, Munro JR: Influence of dilution water on the toxicity of kraft pulp and paper mill effluent, including mechanisms of effect. Water Res 13:151–158, 1979a.

McLeay DJ, Walden CC, Munro JR: Effect of pH on toxicity of kraft pulp and paper mill effluent to salmonid fish in fresh and seawater. Water Res 13:249–254, 1979b.

McLeese DW: Effects of temperature, salinity and oxygen on the survival of the American lobster. J Fish Res Bd Can 13:247–272, 1956.

McLeese DW: Toxicity of copper at two temperatures and three salinities to the American lobster (*Homarus americanus*). J Fish Res Bd Can 31:1949–1952, 1974.

Mehrle PM, Mayer FL, Johnson WW: Diet quality in fish toxicology: Effects on acute and chronic toxicity. In: Aquatic Toxicology and Hazard Evaluation, edited by FL Mayer and JL Hamelink, pp. 269–280. ASTM Spec Tech Rept 634, 1977.

Miller TG, Mackay WC: The effects of hardness, alkalinity and pH of test water on the toxicity of copper to rainbow trout (*Salmo gairdneri*). Water Res 14:129–133, 1980.

Mount DI: Chronic toxicity of copper to fathead minnows (*Pimephales promelas,* Rafinesque). Water Res 2:215–223, 1968.

Mount DI, Stephan CE: Chronic toxicity of copper to the fathead minnow (*Pimephales promelas*) in soft water. J Fish Res Bd Can 26: 2449–2457, 1969.

NAS/NAE (National Academy of Sciences, National Academy of Engineering): Water Quality Criteria 1972. U.S. EPA Ecol Res Ser EPA-R-73-033, p. 594, 1974.

Nriagu JO, Wong HKT, Coker RD: Particulate and dissolved trace metals in Lake Ontario. Water Res 15:91–96, 1981.

Olson R, Harrel RC: Effect of salinity on acute toxicity of mercury, copper, and chromium for *Rangia cuneata* (Pelecypoda, Mactridae). Contrib Mar Sci 17:9–13, 1973.

Opuszyński K: Temperature preference of fathead minnow *Pimephales promelas* (Rafinesque) and its changes induced by copper salt $CuSO_4$. Pol Arch Hydrobiol 18:401–408, 1971.

Oseid DM, Smith LL, Jr: Factors influencing acute toxicity estimates of hydrogen sulfide to freshwater invertebrates. Water Res 8:739–746, 1974.

Pagenkopf GK, Russo RC, Thurston RV: Effect of complexation on toxicity of copper to fishes. J Fish Res Bd Can 31:461–465, 1974.

Pascoe D, Cram P: The effect of parasitism on the toxicity of cadmium to the three-spined stickleback, *Gasterosteus aculeatus* L. J Fish Biol 10: 467–472, 1977.

Pesch CE, Morgan D: Influence of sediment in copper toxicity tests with the polychaeta *Neanthes arenaceodentata.* Water Res 12:747–751, 1978.

Phillips GR, Buhler DR: Influences of dieldrin on the growth and body composition of fingerling rainbow trout (*Salmo gairdneri*) fed Oregon moist pellets or tubificid worms (*Tubifex* sp.). J Fish Res Bd Can 36:77–80, 1979.

Pickering QH: Some effects of dissolved oxygen concentrations upon the toxicity of zinc to the bluegill, *Lepomis macrochirus,* Raf. Water Res 2:187–194, 1968.

Pickering QH, Henderson C: Acute toxicity of some important petrochemicals to fish. J Water Pollut Control Fed 38:1419–1429, 1966a.

Pickering QH, Henderson C: The acute toxicity of some heavy metals to different species of warmwater fishes. Int J Air Water Pollut 10:453–463, 1966b.

Pickering QH, Henderson C: The acute toxicity of some pesticides to fish. Ohio J Sci 66:508–513, 1966c.

Pickering QH, Henderson C, Lemke AE: The toxicity of organic phosphorus insecticides to different species of warmwater fishes. Trans Am Fish Soc 91:175–184, 1962.

Rehwoldt R, Menapace LW, Nerrie B, Alessandrello D: The effect of increased temperature upon the acute toxicity of some heavy metal ions. Bull Environ Contam Toxicol 8:91–96, 1972.

Reiff B, Lloyd R, How MJ, Brown D, Alabaster JS: The acute toxicity of eleven detergents to fish: Results of an interlaboratory exercise. Water Res 13:207–210, 1979.

Reinert RE, Stone LJ, Willford WA: Effect of temperature on accumulation of methylmercuric chloride and *p,p'*-DDT by rainbow trout (*Salmo gairdneri*). J Fish Res Bd Can 31:1649–1652, 1974.

Russo RC, Thurston RV, Emerson K: Acute toxicity of nitrite to rainbow trout (*Salmo gairdneri*): Effects of pH, nitrite species, and anion species. Can J Fish Aquat Sci 38:387–393, 1981.

Sanders HO, Cope OB: Toxicities of several pesticides to two species of cladocerans. Trans Am Fish Soc 95:165–169, 1966.

Sano H: The role of pH on the acute toxicity of sulfite in water. Water Res 10:139–142, 1976.

Schaefer ED, Pipes WO: Temperature and the toxicity of chromate and arsenate to the rotifer, *Philodina roseola.* Water Res 7:1781–1790, 1973.

Shepard MP: Resistance and tolerance of young speckled trout (*Salvelinus fontinalis*) to oxygen lack, with special reference to low oxygen acclimation. J Fish Res Bd Can 12:387–446, 1955.

Skidmore JF: Respiration and osmoregulation in rainbow trout with gills damaged by zinc sulphate. J Exp Biol 52:481–494, 1970.

Skidmore JF, Tovell PWA: Toxic effects of zinc sulphate on the gills of rainbow trout. Water Res 6:217–230, 1972.

Slonin AR: Acute toxicity of selected hydrazines to the common guppy. Water Res 11:889–895, 1977.

Smith LL, Oseid DM: Effects of hydrogen sulfide on fish eggs and fry. Water Res 6:711–720, 1972.

Smith MW: Copper sulphate and rotenone as fish poisons. Trans Am Fish Soc 69:141–157, 1939.

Sosnowski SL, Germond DJ, Gentile JH: The effect of nutrition on the response of field populations of the calanoid copepod *Acartia tonsa* to copper. Water Res 13:449–452, 1979.

Sparks RE, Walker WT, Cairns J, Jr: Effect of shelters on the resistance of dominant and submissive bluegills (*Lepomis macrochirus*) to a lethal concentration of zinc. J Fish Res Bd Can 29:1356–1358, 1972.

Spear PA, Pierce RC: Copper in the aquatic environment: Chemistry, distribution and toxicology. Natl Res Counc Can Environ Secretariat NRCC No. 16454, 1979.

Sprague JB: Promising anti-pollutant: Chelating agent NTA protects fish from copper and zinc. Nature (London) 220:1345–1346, 1968.

Sprague JB, Logan WJ: Separate and joint toxicity to rainbow trout of substances used in drilling fluids for oil exploration. Environ Pollut 19: 269–281, 1979.

Sprague JB, Rowe DW, Westlake GF, Heming TA, Brown IT: Sublethal effects of treated liquid effluent from a petroleum refinery on freshwater organisms. Ottawa: Petroleum Association for Conservation of the Canadian Environment, and Fisheries and Environment Canada, Environmental Protection Service, 1978.

Steemann Nielsen E, Wium-Andersen S: Copper ions as poison in the sea and in freshwater. Mar Biol 6:93–97, 1970.

Stegeman JJ: Temperature influence on basal activity and induction of mixed function oxygenase activity in *Fundulus heteroclitus.* J Fish Res Bd Can 36:1400–1405, 1979.

Stiff MJ: Copper/bicarbonate equilibria in solutions of bicarbonate ion at concentrations similar to those found in natural water. Water Res 5:171–176, 1971a.

Stiff MJ: The chemical states of copper in polluted freshwater and a scheme to differentiate them. Water Res 5:585–599, 1971b.

Sunda WG, Guillard RR: Relationship between cupric ion activity and toxicity of copper to phytoplankton. J Mar Res 34:511–529, 1976.

Sylva RN: The environmental chemistry of copper (II) in aquatic systems. Water Res 10:789–792, 1976.

Teulin MP: An improved experimental medium for freshwater toxicity studies using *Daphnia magna.* Water Res 12:1027–1034, 1978.

Thurston RV, Phillips GR, Russo RC: Increased toxicity of ammonia to rainbow trout (*Salmo gairdneri*) resulting from reduced concentrations of dissolved oxygen. Can J Fish Aquat Sci 38:983–988, 1981.

Tovell PWA, Newsome C, Howes D: Effect of water hardness on the toxicity of a nonionic detergent to fish. Water Res 9:31–36, 1975.

Van den Berg CMG, Wong PTS, Chau YK: Measurement of complexing materials excreted from algae and their ability to ameliorate copper toxicity. J Fish Res Bd Can 36:901–905, 1979.

Vernberg WB, O'Hara J: Temperature-salinity stress and mercury uptake in the fiddler crab, *Uca pugilator.* J Fish Res Bd Can 29:1491–1494, 1972.

Voyer RA: Effect of dissolved oxygen concentrations on the acute toxicity of cadmium to the mummichog *Fundulus heteroclitus* (L.) at various salinities. Trans Am Fish Soc 104:129–134, 1975.

Waiwood KG, Beamish FWH: Effects of copper, pH and hardness on the critical swimming performance of rainbow trout (*Salmo gairdneri* Richardson). Water Res 12:611–619, 1978a.

Waiwood KG, Beamish FWH: The effect of copper, hardness and pH on the growth of rainbow trout *Salmo gairdneri.* J Fish Biol 13:591–598, 1978b.

Wilson RCH: Prediction of copper toxicity in receiving waters. J Fish Res Bd Can 29:1500–1502, 1972.

Woodward DF: Toxicity of the herbicides Dinoseb and Picloram to cutthroat (*Salmo clarkii*) and lake trout (*Salvelinus namaycush*). J Fish Res Bd Can 33:1671–1676, 1976.

Woodward DF, Mayer FL, Jr: Toxicity of three herbicides (butyl, isoocytl, and propylene glycol butyl ether esters of 2,4-D) to cutthroat trout and lake trout. U.S. Dept Inter Fish Wildl Serv Tech Pap 97, 1978.

Zitko V, Carson WG: A mechanism of the effects of water hardness on the lethality of heavy metals to fish. Chemosphere 5:299–303, 1976.

Zitko V, Carson WV, Carson WG: Prediction of incipient lethal levels of copper to juvenile Atlantic salmon in the presence of humic acid by cupric electrode. Bull Environ Contam Toxicol 10:265–271, 1973.

SUPPLEMENTAL READING

Alabaster JS, Lloyd R: Water quality criteria for freshwater fish. London: Butterworths, 1980.

Alderdice DF: Factor combinations. Responses of marine poikioltherms to environmental factors acting in concert. In: Marine Ecology, vol. 1, Environmental Factors, part 3, edited by O Kinne, pp. 1659–1722. London: Wiley-Interscience, 1972.

Brown VM: The calculation of the acute toxicity of mixtures of poisons to rainbow trout. Water Res 2:723–733, 1968.

Cairns J, Jr, Heath AG, Parker BC: Temperature influence on chemical toxicity to aquatic organisms. J Water Pollut Control Fed 47:267–280, 1975.

Fry FEJ: The effect of environmental factors on the physiology of fish. In: Fish Physiology, vol. 6, Environmental Relations and Behaviour, edited by WS Hoar, DJ Randall, pp. 1–98. New York: Academic, 1971.

Appendix D

BIOACCUMULATION

A. Spacie and *J. L. Hamelink*

INTRODUCTION

Bioaccumulation of pollutants first gained public attention in the 1960s with the discovery of DDT, DDD, and methyl mercury residues in fish and wildlife. Mortalities and reproductive failures in fish and fish-eating birds were linked to unusually high concentrations of DDT or its metabolites in the fat of these animals. Since the top-level carnivores, especially birds, had higher residue concentrations of these chemicals than the food they consumed, it was logical to postulate that accumulation occurred primarily by transfer through the food chain. This idea was supported indirectly by the observation that DDT residues increased in a stepwise fashion from one trophic level to the next (Woodwell et al., 1967). The net efficiency of energy transfer between trophic levels is only about 10%. If the transfer efficiency for a chemical contaminant from food to consumer were greater, say 50–100%, and if there were no significant losses from the organism, then residues would continue to accumulate throughout the life of the consumer. The higher the trophic level, the greater would be the body burden of residues. Although the actual mechanism for such a process was not clear, the concept of "biomagnification" or "transfer up the food chain" became popular.

Later, when laboratory and microcosm experiments failed to show the same degree of magnification, debates arose over the relative importance of food and water in the accumulation process. This was the situation until quite recently, when detailed dynamic models were used to distinguish the various routes of uptake. It is now clear that the degree of accumulation in aquatic organisms depends on the type of food chain, on the availability and persistence of the contaminant in water, and especially on the physical-chemical properties of the contaminant.

Originally appeared as Chapter 17 in Rand and Petrocelli: *Fundamentals of Aquatic Toxicology*, © 1985 Taylor & Franicis.

Persistent contaminants such as DDT and the highly chlorinated PCBs share several characteristics, including poor biodegradability, low water solubility, and high lipid solubility. When such nonpolar chemicals are allowed to partition or distribute between aqueous and organic solvents, nearly all of the molecules diffuse into the organic phase. For example, about 562,000 times more DDT will accumulate in octanol than in water in the same system. Nonpolar chemicals also partition into the fat of aquatic animals exposed to them in water. The degree of bioaccumulation is generally correlated with the partition coefficient (P) measured in an octanol-water mixture. Chemicals with very large partition coefficients are more likely to be bioaccumulated than those with low P values.

Nevertheless, the distribution of a chemical in aquatic organisms is not quite as simple to model or measure as it is in organic solvents. Since animals move about, feed, grow, and actively transport and transform chemicals within their tissues, it is more difficult to describe the factors that affect chemical uptake and distribution. Mass balance or kinetic models are required to describe the various steps in accumulation, distribution, and elimination of residues. Models similar to those used by pharmacologists to describe the time course of drugs in the human body are transferable to aquatic systems. They are especially useful for describing residue dynamics over time and for distinguishing the routes of uptake from food and water.

This chapter introduces the use of kinetic models in describing bioaccumulation. It emphasizes the application of models and physical-chemical correlations to the design of appropriate test methods for measuring bioaccumulation potential.

TERMINOLOGY

Uptake. Transfer of a chemical into or onto an aquatic organism. The uptake phase of an accumulation test is the period during which test organisms are exposed to the chemical.

Depuration. Elimination of a chemical from an organism by desorption, diffusion, excretion, egestion, biotransformation, or another route. The depuration phase of a test is the period during which previously exposed organisms are held in uncontaminated water.

Half-life or half-time. Time required for an organism held in clean water to eliminate 50% of the total body burden or tissue concentration of a chemical.

Bioavailable. Term used for the fraction of the total chemical in the surrounding environment which is available for uptake by organisms. The environment may include water, sediment, suspended particles, and food items.

Partitioning. Distribution of a chemical between two immiscible solvents. The partition coefficient (P or K_{ow}) is the ratio of the chemical concentrations in the two solvents at equilibrium. Partition coefficients are commonly measured between *n*-octanol and water.

Steady state or dynamic equilibrium. The state at which the competing rates of uptake and elimination of a chemical within an organism or tissue are equal. An apparent steady state is reached when the concentration of a chemical in tissue remains essentially constant during a continuous exposure. Bioconcentration factors are usually measured at steady state.

Compartment. Quantity of chemical that displays uniform rates of uptake and elimination in a biological system and whose kinetics can be distinguished from those of other compartments (Moriarty, 1975).

THE PROCESS OF ACCUMULATION

Bioaccumulation can occur only if the rate of uptake of a chemical by an organism exceeds its rate of elimination. It is important to consider all the possible routes of chemical uptake so that appropriate models and test methods can be designed. Although many of the biochemical processes involved in xenobiotic transport are poorly understood, the general patterns of uptake from water, sediment, and food are clear enough to be modeled in a reasonably realistic way. This section describes the major features of the uptake process.

Uptake from Water

Direct uptake of chemicals from water has been shown for many aquatic organisms including algae, annelids, arthropods, mollusks, and fish. Considering the tremendous size range and structural diversity of aquatic organisms, a variety of uptake sites and mechanisms would be expected. Nevertheless, the discussion can be simplified by concentrating on the three most significant transport processes: diffusion, special transport, and adsorption.

Diffusion

Most xenobiotics are taken into the body by passive diffusion through semipermeable membranes such as the gill, lining of the mouth, or gastrointestinal tract. Fish gills are especially vulnerable to foreign chemicals because their design maximizes diffusion. Gill membranes are thin (2–4 μm) and typically represent 2–10 times the surface area of the body. In contrast, the fish skin, crustacean carapace, and insect cuticle are relatively impermeable because of their dense, nonliving layers composed of hydrophobic and hydrophilic regions. Factors governing the transport of toxicants through insect cuticles have been studied intensively but are still not well understood (Matsumura, 1975).

Passive diffusion can occur across any barrier that is permeable to the chemical and across which a concentration gradient (ΔC) exists. It is a physical process that requires no energy expenditure by the organism. The lipid bilayer of simple biological membranes permits rapid diffusion of nonpolar (lipophilic) organic molecules, but little diffusion of water, ions, or polar molecules. Weak acids and bases cross the membrane primarily in their unionized form. Proteinaceous pores in the membrane, however, allow passage of water, small ions such as chloride, and small molecules up to a molecular weight (MW) of approximately 100. The permeability of these membrane pores can vary with salinity and with the physiological state of the organism.

According to Fick's law, the rate of diffusion is proportional to

$$\frac{\Delta C \times \text{area} \times \text{temperature}}{\text{distance}} \tag{1}$$

Isolated membranes would be quickly saturated and the rate of diffusion would decline if the chemical were allowed to accumulate locally. The countercurrent blood flow pattern at the gill, which transports substances away from the site of diffusion, is especially effective in maintaining a large concentration gradient. The rate of passive uptake at the gill tissue (dC/dt) is directly proportional to the difference between the chemical concentration in water (C_w) and that in blood (C_b). Since the initial value of C_b for a xenobiotic substance is usually zero at the beginning of an exposure,

$$\frac{dC}{dt} = k_u C_w \tag{2}$$

initially. Here the rate coefficient k_u incorporates all the physical factors that affect the rate of diffusion. Not only surface area and membrane thickness, but also characteristics such as the respiration rate and the size and lipid solubility of the chemical molecule, are involved. Assuming that the chemical concentration in water is held constant, the initial rate of diffusion into the organism is also constant. Later we will see that Eq. (2) forms the basis for a simple bioconcentration model.

Experiments with isolated perfused trout gills showed diffusion of the insecticide dieldrin from water to blood (Fromm and Hunter, 1969). The transport into blood apparently requires the presence of plasma proteins to help bind the dieldrin. Passive diffusion of cadmium was also shown with isolated gills of the mussel *Mytilus edulis* (Carpene and George, 1981). The uptake was linear over time and directly proportional to the external cadmium concentration.

In vivo uptake at the gill has been observed for DDT in salmonids (Premdas and Anderson, 1963; Holden, 1962), methyl mercury and mercuric chloride in rainbow trout (Olson et al., 1973), surfactants in carp (Kikuchi et al., 1980), copper in sunfish (Anderson and Spear, 1980), and fish anesthetics (Hunn and Allen, 1974). In general, measured uptake rates are greater for living than for dead animals because of the role of the blood circulation in transport.

Arthropods can apparently absorb some chemicals through the general body surface. Derr and Zabik (1974) found passive diffusion of DDE through the cuticle of midge larvae. Uptake was related to the surface area and was similar for both living and dead animals. DDT also accumulates through the integument of dragonfly nymphs (Wilkes and Weiss, 1971) and *Daphnia* (Crosby and Tucker, 1971), where uptake is proportional to exposure concentration. Dead *Daphnia* accumulate about half as much DDT in 24 h as live *Daphnia.*

The molting process can affect body burdens, at least for metals. Jennings and Rainbow (1979) found that the gills of the crab *Carcinus maenas* absorbed cadmium, but 59–80% of the total body burden was sorbed to the exoskeleton. Zinc may also be taken up by the gills of *C. maenas* (Bryan, 1976), but 61% is lost at each molt (Renfro et al., 1975). About half of the cadmium concentration in shrimp *Lysmata seticaudata* was also lost at each molt (Fowler and Benayoun, 1974). PCBs were taken up across the general integument of *Gammarus oceanicus,* but in this case no effect of molting was noted (Wildish and Zitko, 1971).

Special Transport

Special transport of chemicals into the body includes both active and facilitated transport. Active transport can occur against a concentration gradient, while facilitated transport cannot. In both processes the chemical forms a reversible complex with a macromolecular carrier in the membrane. Thus the kinetics of special transport can be saturated at high chemical concentrations. Competitive inhibition of the uptake of a chemical in the presence of chemically similar substances is also possible. There are numerous examples of metals interacting to affect uptake. For example, cadmium tends to reduce the uptake of zinc and copper. Cobalt and manganese can compete for uptake via the iron transport system (Klaassen, 1980).

Since active transport requires energy to move the chemical against a gradient, it is a true concentrating mechanism. Metabolic blocking agents can be used experimentally to prevent active transport. In contrast, facilitated transport requires no cellular energy input and cannot concentrate chemicals against a gradient (Klaassen, 1980). The two special transport processes are intended to regulate biologically important materials such as essential metals, sugars, and amino acids. Metals can be accumulated by both active and passive transport processes and it is not clear which is more common.

Regardless of mechanism, metal uptake in aquatic organisms is usually directly proportional to the metal concentration in water (C_w), at least initially. Bryan (1976) showed linear rates of uptake for mercury, copper, silver, manganese, zinc, cadmium, and arsenic in the marine worm *Nereis diversicolor.* Direct proportionality between uptake and C_w for lead was also found with aquatic insects (Spehar et al., 1978). In the same study, cadmium residues in insects after 28 d were proportional to exposure concentration at low C_w but approached a constant concentration at the upper end of the range tested. This suggests saturation of binding sites on or in the insects. Cadmium uptake in perfused trout gills also decreased at high exposure concentrations (Part and Svanberg, 1981).

Low salinity can increase the rate of uptake of cadmium by marine organisms that actively accumulate salt as salinity declines (O'Hara, 1973; Jackim et al., 1977). Low salinity also enhances the uptake of cesium in aquatic organisms because of reduced competition from the similar ion, potassium (Thomann, 1981). Thus it may be difficult to generalize about bioaccumulation of metals in aquatic organisms because of the importance of water chemistry and the active role played by organisms in regulating normal body constituents.

Adsorption

Adsorption is the binding of a chemical to a surface by covalent, electrostatic, or molecular forces. Since it is a surface phenomenon, it is important mainly as the initial step in the accumulation process. Substances that bind to the integument of an animal contribute to the total body burden and may affect vulnerable functions of the epithelium, but they do not generally contribute to the toxic effect level within the body.

Adsorption is especially important for microorganisms because of their extremely high surface-to-volume ratio. Sodergren (1968) found initial rapid adsorption of DDT onto *Chlorella* cells followed by absorption across the cell membrane. Similarly, zinc was adsorbed on diatoms and immediately transported by diffusion into the cytoplasm (Davies, 1973). Diffusion rather than adsorption controlled the overall rate of zinc uptake. Adsorption mechanisms have been cited for bacteria with certain PCBs and toxaphene (Paris et al., 1977, 1978) and for algae with TCDD (Isensee and Jones, 1975) and DDT (Hamelink et al., 1971; Cox, 1970). Because adsorption is a physical process, it should be equally effective in living and dead organisms. Paris et al. (1978) found very little difference in the degree of accumulation of PCB between bacteria, nonliving seston, and sediment. Södergren (1968) reported similar rates of DDT uptake for living and dead algal cells.

A common method of expressing adsorption to a substrate is the Freundlich isotherm:

$$\frac{X}{M} = kC_w^{1/n} \quad \text{or} \quad \log\frac{X}{M} = \log k + \frac{\log C_w}{n} \tag{3}$$

where X/M is the mass of chemical sorbed per gram of sorbent at equilibrium, k is the Freundlich adsorption constant, C_w is the equilibrium solution concentration of the chemical, and $1/n$ represents the slope of the isotherm. The slope may be found by plotting log (X/M) against log C_w for various concentrations. The higher the intercept (log k), the greater the degree of adsorption, and the larger the slope, the greater the efficiency of adsorption.

It is important to note that this relationship assumes a constant ratio of sorbent to water. Higher sorbent concentrations tend to produce proportionately less adsorption. If the constant k is thought of as a type of concentration factor, then that factor may vary with the quantity of sorbent in the system tested. A good example of this mass effect is the uptake of 2,4,5,2′,5′-pentachlorobiphenyl by marine phytoplankton, which follows a Freundlich-type relationship (Harding and Phillips, 1978). The quantity of PCB adsorbed (X/M) decreases linearly with increasing density of the algal suspension when plotted logarithmically (Fig. 1). Ellgehausen et al. (1980) also found a Freundlich relationship for the uptake of *p,p*′-DDT by algae,

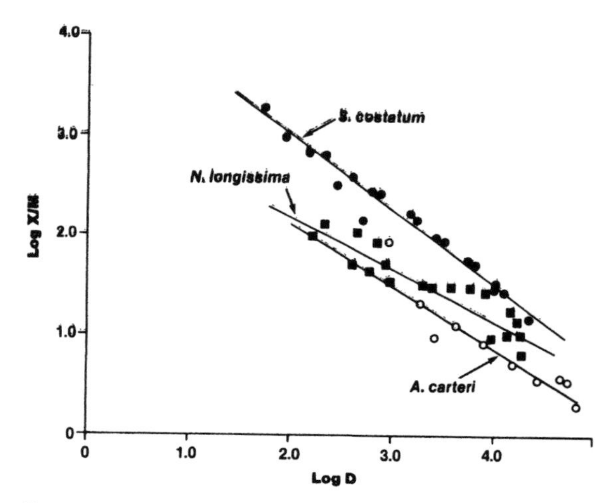

Figure 1 Relationships between algal density as log *D* (micrograms of cell carbon per liter) and equilibrium concentration of 2,4,5,2′,5′-pentachlorobiphenyl (PCB) in phytoplankton as log *X/M* (micrograms of PCB per gram of cell carbon). Suspensions of *Skeletonema costatum, Nitzschia longissima,* and *Amphidinium carteri* were incubated at 12°C with an initial PCB concentration of 0.31 μg/l. *[From Harding and Phillips (1978).]*

Daphnia, and fish (Table 1). The algae appeared to have a greater binding capacity (k) than the other organisms because they were tested at a lower biomass concentration (0.2 mg/ml). When expressed on the basis of equivalent weight (1 mg/ml), all three taxa showed approximately the same degree of accumulation. Thus the relative proportions of chemical, water, and biomass can affect the outcome of bioaccumulation experiments. Field studies may also show this mass effect. For example, lower DDT concentrations in marine phytoplankton were associated with higher densities of the phytoplankton standing crop (Cox, 1970).

The slope of the isotherm ($1/n$) is often close to 1.0 under equilibrium conditions (Table 1). Harding and Phillips (1978) found slopes ranging from 0.8 to 2.5 for various algal species, with most values near 1.0. A value of 1.0 indicates a direct proportionality between the concentration of a chemical in or on the organism (C) and its concentration in water (C_w):

$$C = \frac{X}{M} = kC_w \tag{4}$$

so that k becomes a simple equilibrium distribution coefficient that cannot be distinguished mathematically from a partition coefficient or a steady-state bioconcentration factor:

Table 1 Freundlich Sorption Constants (k) and Slopes ($1/n$) for p,p'-DDT Accumulation in Aquatic Organisms[a]

Organism	log k	$1/n$	Biomass (mg/ml)
Algae	4.200	1.07	0.2
Algae	3.322[b]	–	1.0
Daphnia	3.350	1.09	1.0
Catfish	3.539	1.21	1.0

[a]Adapted from Ellgehausen et al. (1980).
[b]Calculated from an experimentally determined regression relating DDT residue in algae to algal biomass.

$$\text{Partition coefficient} = \frac{\text{concentration in solvent 1}}{\text{concentration in solvent 2}} \tag{5}$$

Bioconcentration factor

$$= \frac{\text{concentration in organism}}{\text{concentration in water}} \tag{6}$$

Thus adsorption isotherms with slopes near unity do not necessarily represent simple physical adsorption, but rather may describe a more general distribution or partitioning phenomenon. Although PCB uptake by marine plankton can be described by a Freundlich relationship, the accumulation is certainly not limited to surface adsorption (Pavlou and Dexter, 1979).

Bioavailability in Water

Transport into biological membranes requires that the xenobiotic in the surrounding water be available in a dissolved form. Environmental factors that reduce the amount of chemical in true solution (C_w) will also reduce the rate of uptake [Eq. (2)]. Among the most important processes that reduce bioavailability are adsorption to suspended solids, adsorption to sediments, adsorption to humic acids and other macromolecules, formation of colloidal suspensions, chelation, complexation, and ionization.

Lipophilic chemicals with a strong tendency to be bioconcentrated are also likely to partition into the organic fraction of sediment or suspended solids (Baughman and Lassiter, 1978; Kenaga and Goring, 1980; Hamelink, 1980; Kenaga, 1980). The concentration factor (K) for a chemical adsorbed to sediments or soil (C_a) at equilibrium can be defined as

$$K = \frac{C_a}{C_w}$$

or, normalized for the organic content of the soil,

$$K_{oc} = \frac{K}{\text{fraction of organic content}}$$

The adsorption-desorption process for lipophilic organics is rapid (equilibrium in minutes or a few hours) and usually reversible (Baughman and Lassiter, 1978). Kenaga and Goring (1980) measured the concentration factors (K_{oc}) for a number of pollutants, mostly pesticides, as a function of water solubility (S, ppm) or octanol-water partition coefficient (P):

$$\log K_{oc} = 3.64 - 0.55 \log S \quad (7)$$

$$\log K_{oc} = 1.377 + 0.544 \log P \quad (8)$$

Thus association with soil or sediments is greatest for chemicals with low water solubility and high P values. Bioconcentration factors (BCFs) were also expressed as functions of S or P, giving correlations such as

$$\log \mathrm{BCF} = 2.791 - 0.564 \log S \quad (9)$$

The slopes of Eqs. (7) and (9) are remarkably similar, illustrating the tendency of aquatic organisms to "compete" with sediments for the same dissolved chemicals.

Sorption of organics and cations increases as the particle size of the soil becomes smaller (Baughman and Lassiter, 1978). For example, the concentration factor (K_{oc}) for methoxychlor increases from 22,000 for sand to 93,000 for fine silt. Thus the nature of the sediments may profoundly influence the availability of a chemical in the environment. For example, Lynch (1979) found that amphipods exposed to hexachlorobiphenyl in the sediments of a stream microcosm accumulated substrate-desorbed residues directly from water. Bioaccumulation was greatest in amphipods exposed to sediments with the least organic content and the largest particle sizes. Suspended particulates and adsorbents such as humic acids should have a similar effect of reducing uptake of lipophilic chemicals in the environment by reducing C_w. For example, the presence of humic acids reduces the rate of uptake of benzo-[*a*]pyrene from water by sunfish but does not affect the uptake of anthracene, which is less lipophilic (Spacie et al., 1983). Lower total DDT residues were measured in fish taken from hypereutrophic lakes than from the same species taken from nearby oligotrophic lakes (Vanderford and Hamelink, 1977). Differences in plankton standing crop and organic content of the water and sediments may explain this result.

The importance of complexation in reducing the uptake and toxicity of trace metal ions is well documented (Sprague, 1968; Sunda et al., 1979; Zitko et al., 1973; Pagenkopf et al., 1974). Greater accumulation of metals usually occurs in soft water and at lower pH values, as reported for cadmium (Kinkade and Erdman, 1975), lead (Merlini and Pozzi, 1977), and mercury (Tsai et al., 1975). Part and Svanberg (1981) found that an organic complexing agent EDNTA (Na_2-(ethylenedinitrilo)tetraacetate) reduced the rate of cadmium diffusion through perfused trout gills. While it is generally correct to assume that the free metal ion is the available form, some metal-ligand complexes may be sufficiently lipophilic to accumulate as well. For example, uptake of cadmium by *Daphnia* was increased by the formation of a nonpolar complex with diethyldithiocarbamate (Poldoski, 1979).

Uptake from Food

Chemicals that are taken in at the gill or outer membranes should also be readily absorbed in the gastrointestinal tract. The same models of uptake by diffusion and special transport function in the gut. In addition, certain macromolecules such as carrageenans (MW ~ 40,000) and polystyrene particles (2200 Å) can apparently be absorbed in the intestine by a process similar to pinocytosis (Guthrie, 1980). Lipophilic organics in food are expected to transfer efficiently because of the relatively long time of contact between food and membranes. However, efficiency of assimilation from food has been measured for very few xenobiotics in aquatic species.

Weak organic acids and bases are absorbed mainly in the un-ionized form, according to the pH of the gut environment. The acidic conditions of the stomach favor diffusion of weak acids that are protonated at pH 1–3. The intestinal pH is higher, although variable, and tends to favor absorption of neutral or weakly basic chemicals.

Absorption of metals from food is highly variable because of the variety of free and bound forms of the ions that are possible in food. Competition between related elements for active transport sites is a further variable. The type of food and method of its contamination can have a very large effect on the amount of uptake measured experimentally. Plaice, *Pleuronectes platessa,* fed zinc in gelatin or starch pellets absorbed greater amounts of zinc than those fed zinc-contaminated *Nereis* worms (Pentreath, 1973). Sunfish, *Lepomis gibbosus,* also accumulated more zinc from an artificial diet than from a natural diet of snails contaminated with the same levels (Merlini et al., 1976). Apparently, the organically bound fraction of metal in the food is relatively unavailable for uptake in the gut.

THE PROCESS OF ELIMINATION

Both the toxicity and the bioaccumulation potential of a foreign compound are greatly affected by the rate of elimination from the organism. If an unaltered chemical can be eliminated rapidly, residues will not accumulate and tissue damage is less likely. In vertebrates, elimination can proceed by several routes, including transport across the integument or respiratory surfaces, secretion in gallbladder bile, and excretion from the kidney in urine. Biotransformation to more polar metabolites is also an important means of eliminating foreign chemicals (see Chapter 18). The metabolites may have rates and routes of elimination different from those of the parent material. Two other elimination processes are important in certain situations. Arthropods may lose residues in exuviae through the process of molting (Renfro et al., 1975; van Weers, 1975). Furthermore, fish and invertebrates eliminate lipid-soluble chemicals through egg deposition (Derr and Zabik, 1974; Guiney et al., 1979). Surprisingly few attempts have been made to distinguish between the various routes of elimination in aquatic organisms, perhaps because of experimental difficulties. At least a few chemicals are known to proceed by each of the principal routes in certain species. But considering the importance of elimination processes in both toxicity and bioaccumulation, it is clear that more work needs to be done in this area.

Gill

Passive elimination of lipid-soluble chemicals can occur across the skin and gills by the same partitioning process involved in uptake. The physical characteristics of the gill make it the principal organ for elimination of this type. Gill elimination appears to be most important for nonpolar compounds that are not biotransformed rapidly. Their rates of elimination are generally a function of the partition coefficient (Maren et al., 1968). In the case of weak electrolytes, gill elimination depends on the relative proportions of protonated and unprotonated forms, as well as the difference between blood and external pH (Lo and Hayton, 1981). The nonionized form of the weak base tricane methanesulfonate (MS-222) (pK_a = 3.5) diffuses rapidly across the gills of the dogfish shark (*Squalus acanthias*) (Maren et al., 1968), rainbow trout (*Salmo gairdneri*), and channel catfish (*Ictalurus punctatus*) (Hunn and Allen, 1974). Quinaldine (pK_a = 5.4) is another base that is eliminated at the gill (Hunn and Allen, 1974; Allen and Hunn, 1977). Studies of elimination of phenol, DDT, and di-2-ethylhexylphthalate (DEHP) from dogfish shark suggest significant elimination through gill or other body surfaces (Guarino and Arnold, 1979). Approximately half of the pentachlorophenol residue in goldfish (*Carassius auratus*) is eliminated through the gill, and the remainder is eliminated equally in bile and urine (Kobayashi, 1979). Both free and conjugated forms may be excreted.

Liver and Gallbladder

Polar chemicals and metabolites of nonpolar substances are usually excreted more effectively by the liver or the kidney. Metabolites formed in the vertebrate liver are transported to the gallbladder, where they are discharged with bile into the small intestine. Although some metabolites in bile are eventually eliminated with the feces, a significant portion may be reabsorbed in the intestine and returned to the blood. This short-circuiting of the biliary excretion process, termed enterohepatic cycling, complicates the study of elimination routes.

Statham et al. (1976) recommended analysis of fish bile as an aid to water quality monitoring because a wide variety of pollutants tend to accumulate in bile. Metals such as mercury, lead, and arsenic are actively transported to the gallbladder. Furthermore, the liver is capable of conjugating many organic chemicals that are bound to plasma proteins and are thus unavailable for excretion by other routes. These conjugates are secreted into the bile. Benzo[*a*]pyrene is a nonpolar organic that is effectively biotransformed and eliminated in bile. Chemicals of moderate to high molecular weight (>400) tend to be eliminated in bile, whereas those of lower weight are often excreted more readily in urine. Despite this general pattern, there are significant differences among species in their ability to biotransform and eliminate foreign chemicals. Aquatic invertebrates seem especially variable in transformation abilities, and the role of the hepatopancreas in elimination is still not clear.

Kidney

The kidney eliminates chemicals in the course of urine formation, either by glomerular filtration or by diffusion or secretory processes in the kidney tubules. Most xenobiotics (MW <60,000) dissolved in the blood are small enough to be removed by glomerular filtration. However, binding of chemicals to plasma proteins reduces elimination by this route, because plasma proteins are retained by the glomeruli. Thus the urinary excretion of many lipid-soluble organics and metals is impeded by their affinity for proteins. Furthermore, nonpolar chemicals in the glomerular filtrate may be reabsorbed by passive diffusion across tubular membranes. Weak organic acids and bases are excreted or reabsorbed, depending on urinary pH. Renal tubules are also capable of actively secreting certain organic acids and bases into the urine. Protein binding does not prevent this type of active transport. However, competition between similar chemicals for the same transport sites may affect the overall rate of elimination (Klaassen, 1980).

The importance of physical-chemical properties is shown by a study comparing the distribution and excretion of DDT and its polar metabolite DDA (Pritchard et al., 1977). After administration of equal doses to the winter flounder (*Pseudopleuronectes americanus*), DDA was excreted into the urine 250 times as quickly as DDT. The difference could not be explained by plasma binding, which was similar for both chemicals (97%). *In vitro* studies with isolated renal tubules showed that DDA was actively accumulated by the renal organic acid transport system. Some of the accumulation could be blocked by metabolic inhibitors. The more lipid-soluble DDT was not actively transported and was probably reabsorbed from tubular fluids.

It is important to distinguish between the primary routes of elimination in aquatic organisms because the routes can be affected differently by temperature, water chemistry, tissue damage, preexposure to toxicants, or the presence of competing chemicals. Such complications are common for exposed populations in the field. Changes in elimination pathways can affect the success of laboratory tests as well as the ability of models to predict the degree of bioaccumulation found in nature.

COMPARTMENT MODELS

A compartment model is a mathematical description of the quantity of a chemical within a uniform system, as determined by competing rates of chemical input and output. Compartment models are commonly used in pharmacokinetic studies and the properties of such models are discussed elsewhere (Levy and Gibaldi, 1975; Moriarty, 1975; Roland, 1977; Tuey, 1980).

This section outlines the compartment models

that are most frequently used in bioaccumulation studies. Throughout this discussion the reader should note the data requirements and assumptions incorporated in each model, since these will apply directly to the design of bioaccumulation experiments. Compartment models are not descriptions of biological processes, but rather descriptions of the data set collected by the experimenter. They can be used successfully as predictive tools provided they are used in situations that do not violate the original constraints on the data. Many environmental and physiological factors, including those discussed in the previous sections, can modify the results predicted by such models.

One-Compartment Bioconcentration Model

The simplest technique for modeling residue dynamics is to treat each organism as a single compartment (Fig. 2). Uptake of the chemical from water is the only input considered, and it is assumed to be directly proportional to the exposure concentration (C_w). All residues in the organisms are assumed to belong to a common pool, where they are equally available for depuration. The rate of depuration, regardless of mechanism, is assumed to be first-order, that is, directly proportional to the concentration in the organism (C). The volume of the compartment must be constant (no growth). In this case the rate of change in residues over time is given by

$$\frac{dC}{dt} = \text{uptake} - \text{loss} = k_u C_w - k_d C \qquad (10)$$

where C_w = concentration of chemical in water (μg/ml)

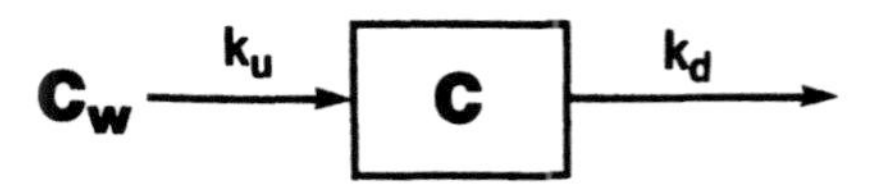

Figure 2 Representation of a one-compartment model for accumulation of a chemical concentration (C) in a whole organism exposed to a constant concentration in water (C_w).

C = concentration of chemical in animal (μg/g)

t = time (h)

k_u = uptake rate constant (ml/g·h or, assuming a tissue density of 1 g/ml, h^{-1}).

k_d = first-order rate constant for depuration (h^{-1}).

During initial uptake, when C is very small, the increase in residues over time should be linear and proportional to C_w. As C becomes larger, the uptake rate (dC/dt) declines and eventually reaches zero. Steady state occurs when the rate of uptake balances the rate of loss, so that

$$\frac{dC}{dt} = 0 = k_u C_w - k_d C_{ss}$$

Therefore, after a sufficient duration of exposure,

$$C_{ss} = \frac{k_u}{k_d} C_w \qquad (11)$$

where C_{ss} is the residue concentration in the body at steady state. The steady-state bioconcentration factor can then be conveniently defined as

$$\text{BCF} = \frac{C_{ss}}{C_w} = \frac{k_u}{k_d} \qquad (12)$$

To integrate Eq. (10), it is necessary to assume that the coefficients k_u and k_d as well as the concentration in water remain constant. The integrated form is

$$C = \frac{k_u}{k_d} C_w [1 - \exp(-k_d t)] \qquad (13)$$

where $\exp(-k_d t)$ represents the base of natural logarithms (e) raised to the power $-k_d t$. Equivalent forms combining Eqs. (12) and (13) are

$$C = C_{ss}[1 - \exp(-k_d t)] \qquad (14)$$

$$C = \text{BCF}\ C_w [1 - \exp(k_d t)] \qquad (15)$$

Equations (13)-(15) describe a curvilinear increase in C, with an asymptotic approach to a "plateau" at later times (Fig. 3).

The time required to approach steady state is determined solely by the depuration rate constant (k_d). For example, 90% of steady state will be reached when $C = 0.9C_{ss}$. Substituting this into Eq. (14), the time required may be found from

$$t_{90} = -\frac{\ln 0.1}{k_d} = \frac{2.303}{k_d} \qquad (16)$$

Figure 3 illustrates the relationship between k_d, the shape of the uptake curve, and the time to steady state. A useful graph for estimating time to steady state is given in Fig. 4. If an apparent steady state is reached during a continuous exposure experiment, the value of k_d may be estimated by finding the time (t_{50}) required to reach 50% of C_{ss}:

$$k_d = -\frac{\ln 0.5}{t_{50}} = \frac{0.693}{t_{50}}$$

provided there are sufficient observations during uptake to fit a smooth curve with confidence. Unfortunately, uptake curves resembling Fig. 3 may be produced by mechanisms other than simple exchange processes, and such cases do not fit the one-compartment model.

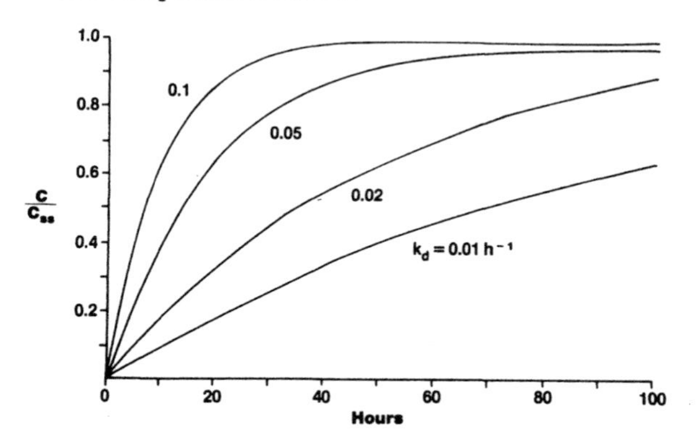

Figure 3 Influence of k_d on time required to approach a steady-state concentration of chemical (C_{ss}) in a one-compartment model.

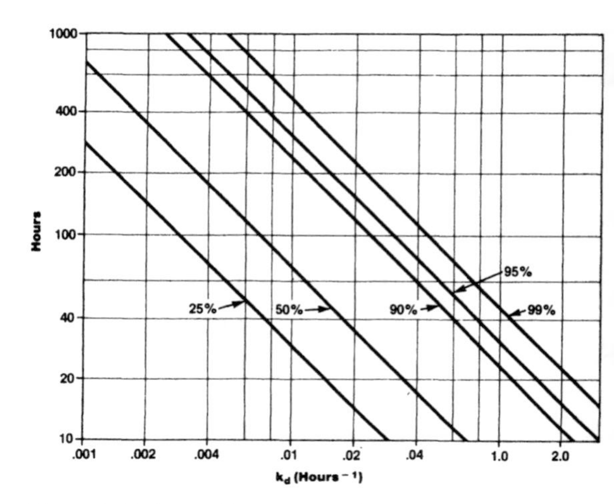

Figure 4 Relationship between depuration rate constant (k_d) and time required to reach a given percentage of steady state within an organism.

Graphical Method

For this reason it is usually better to verify k_d directly by transferring exposed animals to uncontaminated water. The organisms do not need to be at steady state when this depuration phase is initiated. A plot of the decline in C, on a logarithmic scale, against depuration time can be used to calculate k_d from the slope of the straight line:

$$\ln C = \ln C_0 - k_d t \qquad (17)$$

where C_0 is the residue concentration at the start of the depuration phase (Fig. 5). If logarithms to the base 10 are plotted instead of natural logarithms, the slope must be multiplied by 2.303 to determine k_d. A depuration half-life ($t_{1/2}$) can be found from

$$t_{1/2} = \frac{\ln 0.5}{k_d} = \frac{0.693}{k_d} \qquad (18)$$

This direct method has several advantages. First, a line can be fitted to the data points by using a simple linear least-squares regression that provides confidence limits for the slope. Second, if the depuration phase does not fit a straight line with a

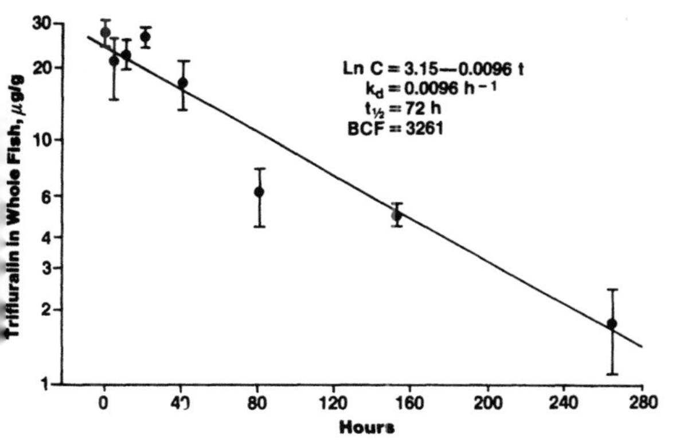

Figure 5 Depuration of trifluralin from fathead minnows (*Pimephales promelas*) transferred to clean water after a 40-h exposure to 20 μg/l trifluralin. Bars represent ±1 standard error. *[From Spacie (1975).]*

measurable slope, the simple one-compartment model is in question and other models should be sought (see the following section).

Once k_d is determined, the uptake coefficient k_u can be calculated manually by measuring the tangent to the uptake curve ($\Delta C/\Delta t$) at several points (Blanchard et al., 1977). Each tangent is an approximation of dC/dt that can be used to calculate a value of k_u from Eq. (10), as shown in Table 2. The k_u estimates for each sample time are then averaged. Small variations in C_w during the experiment can be accommodated by this method of calculating k_u.

Nonlinear Method

It is also possible to find both k_u and k_d directly from the uptake phase of the test without using a separate depuration experiment. In this approach the quantity C/C_w is followed over time:

$$\frac{C}{C_w} = \text{BCF}[1 - \exp(-k_d t)] \tag{19}$$

The two parameters k_d and BCF must be evaluated by a nonlinear least-squares regression procedure such as those available in the Biomedical Computer Programs (Dixon and Brown, 1979), SPSS (Nie et al., 1975), or similar programs written specifically for kinetic models (Metzler et al., 1974; Blau and Agin, 1978). Such iterative programs require the input of reasonably good initial estimates for k_d and BCF (or k_u). An initial estimate of k_d can be obtained graphically from the depuration curve. Alternatively, it can be calculated by choosing two separate data points C_1, t_1 and C_2, t_2 on the uptake curve. The parameter k_d that appears in

Table 2 Sample Calculation of k_u Based on Hypothetical Data for Uptake Phase of an Experiment[a]

Exposure time (h)	Tangent ($\Delta C/\Delta t$)	k_d (h^{-1})	C (mg/g)	C_w (mg/ml)	Estimated k_u[b]
0	–	–	0	0.76	–
12	2.8	0.015	52	0.44	8.14
24	2.0	0.015	82	0.78	4.14
48	1.7	0.015	124	0.66	5.39
72	1.6	0.015	164	0.76	5.34
96	1.0	0.015	195	0.64	6.13
Means				0.67	5.83

[a]The tangent to the uptake curve was determined graphically at each observation time. The rate constant k_d was found previously from the depuration phase.

[b]Estimated $k_u = (\Delta C/\Delta t + k_d C)/C_w$; BCF $= k_u/k_d = 5.83/0.015 = 388$; $C_{ss} = \text{BCF} \times C_w =$ 260 μg/g. Time to 90% of steady state $= 2.303/k_d = 154$ h.

$$\frac{C_1}{C_2} = \frac{1 - \exp(-k_d t_1)}{1 - \exp(-k_d t_2)} \quad (20)$$

can be determined numerically by trial and error on a hand calculator. A rough estimate for k_d is substituted into the right-hand side of Eq. (20) and the expression calculated. The process is repeated with better estimates of k_d until the expression approaches C_1/C_2. For example, a k_d estimate of 0.0175 h^{-1} was obtained with the 24- and 96-h data in Table 2. Once k_d is determined, an initial estimate for k_u or BCF can be made by using the method of Table 2.

One drawback of nonlinear estimation procedures is that they do not provide a unique solution of the equation representing the model. There is always a chance that the iterative program will converge on an unrealistic answer, particularly if the initial parameter estimates are poor. Thus the researcher should check that the original model is justified and that the numerical results make sense biologically.

Growth

Chemicals such as DDT and PCB with large partition coefficients approach steady state slowly. Results of long-term bioconcentration tests designed for these materials must be corrected for the growth of the test animals, since growth has the effect of diluting the accumulated residues. Growth can be incorporated as a loss term in the simple model:

$$\frac{dC}{dt} = k_u C_w - k_d C - gC \quad (21)$$

In this relationship, growth (dW/dt) is assumed to be an exponential function:

$$\frac{dW}{dt} = W_0 \exp(gt) \quad (22)$$

where W_0 is the weight at the start of the growth period or test and g (d^{-1}) is the growth rate constant (other growth functions could be substituted for the exponential one). For convenience, the two loss coefficients in Eq. (21) may be combined as

$$k' = k_d + g \quad (23)$$

so that the simple compartment model with growth can be integrated to yield

$$C = \frac{k_u}{k'} C_w [1 - \exp(-k't)] \quad (24)$$

Since growth contributes to the overall decline of residue concentration, rapid growth of test organisms may cause a reduction in the time required to approach an apparent steady state and also reduce the depuration half-life.

Examples of the application of one-compartment models to the study of bioconcentration are given by Harding and Vass (1979); Kimerle et al. (1981); McLeese et al. (1979); Spacie and Hamelink (1979); and Wong et al. (1981).

Two-Compartment Bioconcentration Model

Tissue residues do not always behave as if they belong to a common pool. Often a portion of the chemical is eliminated rapidly while the remainder takes considerably longer to depart. Metals may bind strongly to proteins or other cell components (Miettinen, 1974); organics in poorly perfused adipose tissue may require longer to be eliminated than those circulating in the blood. Biphasic or biexponential kinetics are recognized by a noticeable change in slope of the depuration curve, representing loss from "fast" and "slow" compartments (Fig. 6). The transport process can be visualized as a two-compartment design such as Fig. 7, where uptake and elimination by the body occur only through the fast or central compartment. Residues in the slower peripheral compartment accumulate as a function of the concentration in compartment 1. It is important to realize that a compartment does not necessarily represent an anatomic unit

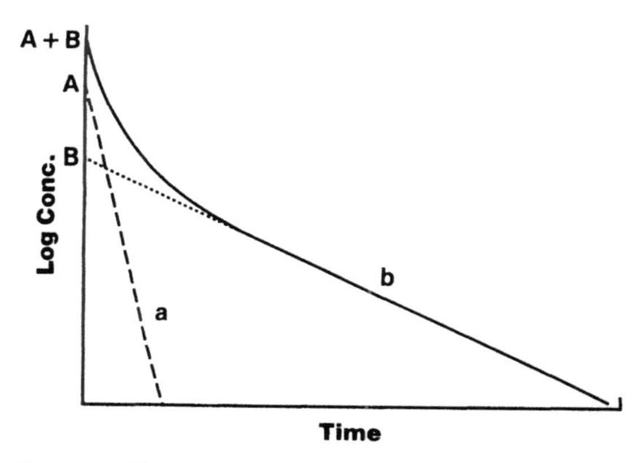

Figure 6 Illustration of a depuration curve showing biphasic kinetics. The slopes of the "fast" and "slow" phases are *a* and *b*, respectively.

(although it is convenient to think in terms of blood, fat, etc.). It represents a quantity of chemical with identifiable kinetics.

The total residue concentration in the animal at any time during depuration (Fig. 6) is described by Eq. (25):

$$C = A\ \exp(-at) + B\ \exp(-bt) \qquad (25)$$

where A and B are the intercepts of the lines with slopes a and b, respectively, plotted logarithmically. The quantity $A + B$ is equivalent to C_0, the initial body concentration of chemical at the start

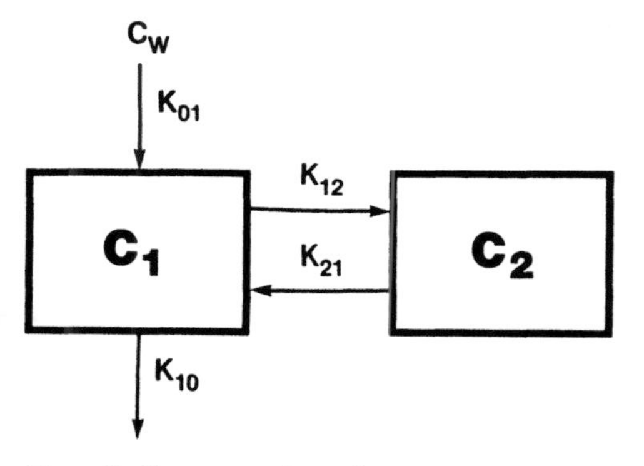

Figure 7 Representation of a two-compartment model for accumulation of residues in an organism exposed to a chemical in water. It is assumed that uptake from water and elimination can occur only through compartment 1.

of depuration. The two lines can be resolved from the experimental points by first using a linear regression on the later portion of the slow phase. Once B and b are found, the remaining line can be found by subtracting the calculated values of C at each observation time from the experimental points during the early period. A plot of the differences will yield a straight line with slope a and intercept A. The graphically determined values are related to the rate constants for the model shown in Fig. 6 according to the following equations:

$$k_{21} = \frac{Ab + Ba}{A + B} \qquad (26)$$

$$k_{10} = \frac{ab}{k_{21}} \qquad (27)$$

$$k_{12} = a + B - k_{21} - k_{10} \qquad (28)$$

The uptake rate constant k_{01} must be measured independently by using the initial portion of the uptake curve. If it is known, the BCF can be calculated as

$$\mathrm{BCF} = \frac{k_{01}}{ab}(a + b - k_{10}) \qquad (29)$$

or

$$\mathrm{BCF} = \frac{k_{01}}{ab}(k_{12} + k_{21}) \qquad (30)$$

In the special case where the peripheral compartment clears rapidly ($k_{21} \gg k_{12}$), the model reduces to a one-compartment type.

One practical drawback of the two-compartment model is the requirement for many data points during the later phases of depuration. The added time and expense of providing these later samples, though, are often justified by the added realism of the model. The work of Könemann and van Leeuwen (1980) on the bioconcentration of chlorobenzenes is a good illustration of the use of the two-compartment model. Nagel and Urich

(1980) describe the use of a slightly different type of two-compartment analysis to study the accumulation of substituted phenols in goldfish.

Verifying Kinetics

The compartment models presented in this section assume constant rate coefficients (k_u and k_d) over the range of experimental conditions. This assumption should be tested by measuring k_u and k_d at several exposure levels and, perhaps, temperatures. One method of verifying a constant rate of uptake is to rearrange Eq. (13) as

$$C = k_u C_w \left[\frac{1 - \exp(-k_d t)}{k_d} \right] \tag{31}$$

A plot of the function in brackets against C during the experiment should yield a linear relationship if k_u and C_w are both constant. An example of this approach for the uptake of methyl mercury at several temperatures is shown in Fig. 8.

First-order elimination kinetics should be verified by plotting the logarithm of C/C_0, the fraction of the initial tissue concentration for the depuration phase, against time at several levels of exposure. The depuration lines for the various exposure concentrations should all coincide and should have the same slope and half-life if depuration is first-order.

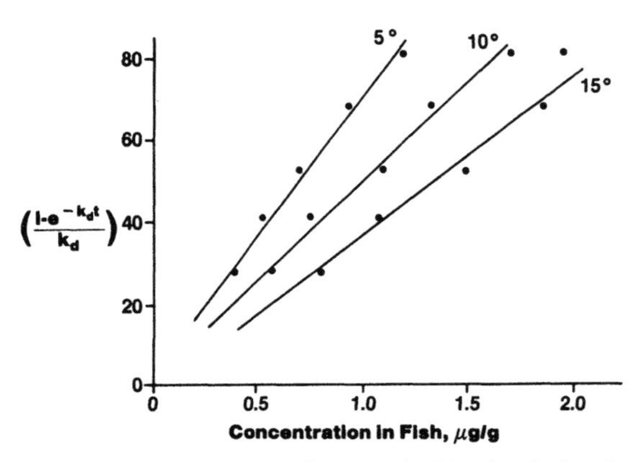

Figure 8 Demonstration of zero-order kinetics during the uptake of methyl mercuric chloride in whole rainbow trout (*Salmo gairdneri*) at three temperatures. The estimated k_d is 0.00099 h^{-1}. *[From Hartung (1976), using data of Reinert et al. (1974).]*

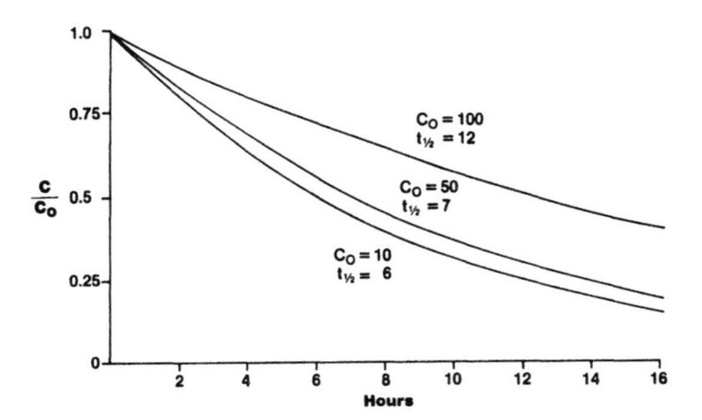

Figure 9 Hypothetical depuration of residues at several initial residue concentrations in the organism (C_0). The long half-life at the highest initial concentration suggests Michaelis-Menten or saturation kinetics during elimination.

Elimination often involves special transport or a series of enzyme-mediated biotransformation steps. Unlike simple diffusion, these processes can become saturated at high residue concentrations. As a result, elimination at high doses may be slower than elimination at lower doses (Fig. 9). In such cases the depuration half-life appears to increase at the upper concentrations. For example, Mayer (1976) found somewhat slower elimination of DEHP from fathead minnows at the highest exposure concentration tested.

Under such circumstances, elimination can often be described by Michaelis-Menten kinetics:

$$-\frac{dC}{dt} = \frac{V_m C}{K_m + C} \tag{32}$$

where V_m = maximum or "saturated" rate of elimination

K_m = concentration of contaminant that can be eliminated at half the maximum rate

First-order elimination can be viewed as a special case of Eq. (32) for which $C \ll K_m$, so that elimination is directly proportional to C:

$$\frac{dC}{dt} = -\frac{V_m}{K_m} C = -k_u C \tag{33}$$

At large values of C, when $C \gg K_m$, the rate of elimination decreases to the point where saturation of the transport or transformation systems occurs and the rate of loss approaches V_m (Fig. 9).

Most xenobiotics occur at such low concentrations in the environment that saturation kinetics are not often encountered. Laboratory or field situations involving acutely toxic exposures are more likely to require a Michaelis-Menten type of model.

Rates of Metabolite Elimination

Bioaccumulation tests are commonly performed with chemicals labeled with radioactive carbon (^{14}C) or another radioisotope (e.g., ^{3}H, ^{36}Cl, ^{35}S) because it is easy to analyze for these atoms. If the labeled material is biotransformed to a labeled metabolite during the experiment, results based on total isotope activity may lead to an incorrect estimate of k_d for the parent compound. Metabolites carrying the ^{14}C label may be eliminated at rates (k_m) that are greater or less than that of the parent (k_d). Figure 10 shows elimination patterns for two extreme cases. When $k_m \gg k_d$, the overall elimination process will be controlled by loss of parent material. Only a small proportion of the ^{14}C body burden will consist of metabolite and the slope of the depuration line will yield an appropriate value of k_d (Fig. 10*a*). However, if $k_m \ll k_d$, metabolite residues will persist longer than parent residues and a plot of total ^{14}C activity will produce a slope that underestimates k_d for the parent material (Fig. 10*b*). The latter case can be resolved only by direct analysis of either the parent compound or total activity plus metabolite.

The elimination of the phthalate DEHP and the hydrocarbon benz[*a*]acridine from fathead minnow (*Pimephales promelas*) illustrates the two situations. Mayer (1976) found that the half-life for DEHP elimination was approximately the same whether it was based on total ^{14}C activity or analysis of parent material. In contrast, the biotransformation of benz[*a*]acridine to metabolites is so rapid that the overall loss rate of ^{14}C activity is controlled by metabolite elimination, giving the appearance of much slower loss of benz[*a*]acridine than actually occurs (Southworth et al., 1981). Chemicals such as benz[*a*]acridine can give variable results in bioconcentration tests, depending on the rate of biotransformation in the test organisms.

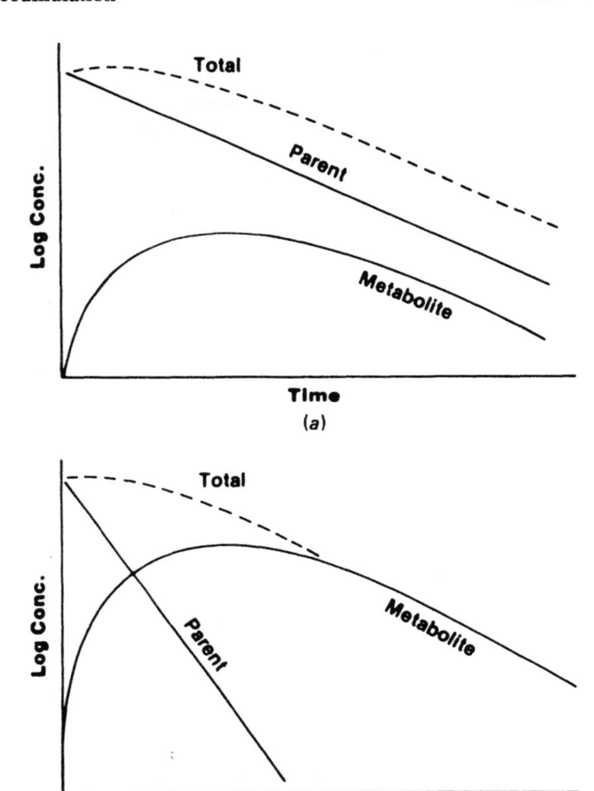

Figure 10 Elimination of a parent chemical and its metabolite from an organism following transfer to clean water. (*a*) The rate constant for metabolite elimination is greater than the rate constant for parent elimination. (*b*) The rate constant for metabolite elimination is less than that of the parent chemical. In both cases the dashed line shows quantity of total material, such as the result obtained by using a radioisotope label. *[Adapted from Levy and Gibaldi (1975).]*

Bioaccumulation Model

A term for the assimilation of contaminant from food can be incorporated into the simple one-compartment model with growth [Eqs. (21)–(23)] according to the analysis of Thomann (1981):

$$\frac{dC}{dt} = k_u C_w + \alpha R C_f - k'C \tag{34}$$

where α = assimilation efficiency of contaminant from food (micrograms absorbed divided by micrograms ingested)

R = weight-specific ration for feeding rate (grams ingested per gram of body weight per day)

C_f = concentration of contaminant in food (μg/g)

A similar expression can be written for the rate of accumulation of the total body burden:

$$\frac{dX}{dt} = k_u C_w W + \alpha R C_f W - k_d X \tag{35}$$

where X is the total weight of residue (μg) in the body and W is body weight. The integrated forms of the two expressions are

$$C = \frac{k_u C_w + \alpha R C_f}{k'} [1 - \exp(-k't)] + C_0 \exp(-k't) \tag{36}$$

$$X = \frac{(k_u C_w + \alpha R C_f) W_0}{k'} [\exp(gt) - \exp(-k_d t)] + X_0 \exp(-k_d t) \tag{37}$$

where X_0, W_0, and C_0 represent the total residue, body weight, and residue concentration, respectively, at the beginning of the experiment or observation period.

Equations (36) and (37) are helpful for evaluating the relative contributions of food, water, and growth to the overall bioaccumulation of a chemical. At long exposure times, as $t \to \infty$, the term $\exp(-k_d t)$ of Eq. (37) approaches zero. Thus the total body burden (X) can continue to increase as long as growth [$\exp(gt)$] continues. A steady state of X cannot be reached unless growth is zero. Even with growth, however, the residue concentration (C) can reach a steady state as $t \to \infty$:

$$C_{ss} = \frac{k_u C_w + \alpha R C_f}{k'} \tag{38}$$

Referring to the bioconcentration factor defined in Eq. (12),

$$\text{BCF} = N_w = \frac{k_u}{k'} \tag{39}$$

where N_w represents the bioconcentration factor for uptake from water only. An analogous bioaccumulation factor (N) for total uptake from food and water can be derived from Eqs. (38) and (39):

$$N = \frac{C_{ss}}{C_w} = \frac{k_u}{k'} + \frac{\alpha R}{k'} \frac{C_f}{C_w} \tag{40}$$

or

$$N = N_w + \frac{\alpha R}{k'} \frac{C_f}{C_w} \tag{41}$$

Equation (41) shows that bioaccumulation will exceed bioconcentration (according to this model) whenever the food term ($\alpha R C_f$) is nonzero. In practice, though, the contribution from food will be insignificant if the loss term (k') is large—that is, whenever the chemical has a short half-life of depuration. For substances with long half-lives, k' will be very small, causing the expression for food intake ($\alpha R/k'$) to increase in importance.

This model assumes that the food and water sources of a chemical are additive. Macek et al. (1979) concluded, on the basis of experimental evidence, that additivity is a good assumption for most residues. Nevertheless, there may be special situations where chemicals from the two sources compete or interfere with each other's uptake.

The model also assumes that elimination rates of food- and water-derived residues are first-order and identical. Experiments on the bioaccumulation of metals (Giesy et al., 1980; Vighi, 1981) suggest that metals derived from food may be eliminated more slowly than those derived from water.

In actual field situations, the uptake, assimilation, and elimination rates are also affected by the metabolic rate of the animal, which may vary (Roberts et al., 1979). Thus, water temperature influences respiration and caloric intake, which in turn control contaminant uptake, growth, and lipid content of trout (Spigarelli et al., 1983). Elimination rates may also slow as an animal grows and accumulates fat, so that it may be necessary to normalize chemical concentrations and rate constants according to size or lipid content (Bruggeman et al., 1981).

Food Chain Transfer

The bioaccumulation model can be expanded to represent the transfer of a chemical through an aquatic food chain with trophic levels such as (1) phytoplankton, (2) zooplankton, (3) small fish, and (4) large fish. The concentration of chemical at each trophic level (i) is determined by the bioconcentration factor for that level (N_{iw}) plus the chemical contributed by feeding on the next lower level ($i-1$). In general,

$$C_i = \frac{k_{ui}}{k_i'} C_w + \frac{\alpha_{i,i-1} R_{i,i-1}}{k_i'} C_{i-1} \qquad (42)$$

The bioaccumulation factor for level i can also be derived:

$$N_i = \frac{C_i}{C_w} = N_{iw} + \frac{\alpha_{i,i-1} R_{i,i-1}}{k_i'} \frac{C_{i-1}}{C_w}$$

or

$$N_i = N_{iw} + \frac{\alpha_{i,i-1} R_{i,i-1}}{k_i'} N_{i-1,w} \qquad (43)$$

where $N_{i-1,w}$ = bioconcentration factor for trophic level $i-1$

N_{iw} = bioconcentration factor for trophic level i

The notation can be simplified by substituting

$$f_i = \frac{\alpha_{i,i-1} R_{i,i-1}}{k_i'} \qquad (44)$$

into Eq. (43) to give the general expression

$$N_i = N_{iw} + f_i N_{i-1} \qquad (45)$$

Expressions for the overall accumulation at each trophic level, derived by Thomann (1981), show the cumulative effect of food sources to successive trophic levels:

Level 1: $$N_{1w} = \frac{k_{u1}}{k_1'} \quad \text{(no feeding)} \qquad (46)$$

Level 2: $$N_2 = N_{2w} + f_2 N_{1w} \qquad (47)$$

Level 3: $$N_3 = N_{3w} + f_3 N_{2w} + f_3 f_2 N_{1w} \qquad (48)$$

Level 4: $$N_4 = N_{4w} + f_4 N_{3w} + f_4 f_3 N_{2w} + f_4 f_3 f_2 N_{1w} \qquad (49)$$

Thus the bioaccumulation at any trophic level can be calculated from a knowledge of the bioconcentration factors (N_{iw}) at each level and the food chain transfer numbers (f_i) for the consumers. Values of f_i greater than 1.0 increase the food contribution at successive trophic levels, causing biomagnification. The longer the food chain, the greater the magnification can be. However, if $f_i < 1$, the contribution of phytoplankton residues to the upper trophic levels becomes insignificant.

A numerical example can be used to illustrate the biomagnification effect. Thomann (1981) employed rate constants from the literature to calculate PCB residues in a four-step food chain (Table 3). The rate constants k_u, k_d, and α were not measured directly. Nevertheless, the resulting bioaccumulation factors agreed within an order of magnitude with residues measured in the field. Trophic levels 2, 3, and 4 showed stepwise magnification of residue concentrations caused by the important contributions from food. In contrast, calculations for plutonium (^{239}Pu) showed that transfer through food did not contribute significantly to the bioaccumulation of plutonium in upper trophic levels.

One cautionary note should be mentioned in discussing food chain transfer. The model is highly sensitive to the relative values of α and k'. A small error in estimating either coefficient will affect the outcome significantly, especially at the higher trophic levels. Rates of depuration and assimilation from food have been measured for few organic pollutants in fewer species. The larger predaceous fish are especially inconvenient for laboratory testing. Thus calculations such as those in Table 3 should be viewed as a helpful guide in analyzing food chain transfer, not as proof of a particular mechanism.

Macek et al. (1979) reviewed data on bioaccumulation in simple experimental food chains and concluded that most chemicals are not biomagnified. Most organic compounds have depuration rate constants of 0.1 d^{-1} or more, with half-lives of hours or a few days (Zitko, 1979). Even assuming efficient assimilation from food (e.g., $\alpha = 0.8$ or 0.9), the rate of loss from the body will exceed the rate of intake from food, giving $f_i \ll 1$. Organochlorines such as DDT and PCB are unusual because of their large bioconcentration factors and half-lives of 1 wk or more in fish. Hence, slow-growing, long-lived species have the potential to accumulate such residues from their food at a rate exceeding that of depuration. Bruggeman et al. (1981) evaluated the relative importance of bioaccumulation from food and water, concluding that food chain biomagnification is likely only for organics with octanol-water partition coefficients of approximately 10^5 or greater.

PREDICTING BIOCONCENTRATION FACTORS

Correlations between the partition coefficients (P) of toxicants such as alcohol and ether and their

Table 3 Sample Calculation of Food Chain Transfer from Parameters for PCB Estimated by Thomann (1981)[a]

Parameter	Trophic level 1	2	3	4
k_d, 1/d	–	0.01	0.004	0.001
g, 1/d	–	0.0092	0.0048	0.0018
R, g/g·d	–	0.105	0.017	0.009
α	–	0.9	0.9	0.9
N_W	$10^{5.5}$	$10^{5.29}$	$10^{4.59}$	$10^{4.9}$
N	$10^{5.5}$	$10^{6.24}$	$10^{6.49}$	$10^{6.95}$
Magnification (N/N_W)	–	8.9X	79X	112X

[a]Sample calculation for level 3: $N_3 = N_{3W} + f_3 N_{2W} + f_3 f_2 N_{1W}$; $N_3 = 10^{4.59} + (1.74)10^{5.29} + (1.74)(4.92)10^{5.5}$; $N_3 = 10^{6.49}$; where $f_3 = \alpha_3 R_3/(k_d + g) = (0.9)(0.017)/(0.004 + 0.0048) = 1.74$.

effective concentrations in organisms have been recognized for many years. Partitioning is also the dominant process governing bioconcentration of trace organics in algae, fish and invertebrates (Hamelink et al., 1971; Neely et al., 1974; Pavlou and Dexter, 1977; Southworth et al., 1978; Voice et al., 1983). Veith et al. (1979a) reported a good correlation between the bioconcentration factor and the partition coefficient for a group of 84 organics in fish:

$$\log \mathrm{BCF} = 0.76 \log P - 0.23 \qquad r = 0.907 \quad (50)$$

This relationship is illustrated in Fig. 11. More recently, Mackay (1982) reexamined the available data on bioconcentraton and concluded that BCF is approximately 5% of the octanol-water partition coefficient. Such relationships are useful because the expected BCF of a new chemical can be estimated before laboratory tests are designed. Detailed methods for using partition coefficients to predict bioconcentration potentials are given by Bysshe (1982). The partition coefficients used in these estimates must be either measured directly (Chiou et al., 1977; Veith and Morris, 1978) or calculated from structural constants. In general, the calculated values are more reliable because of practical difficulties in measuring equilibrium concentrations of the chemical in the partitioning phases. Methods for calculating P values are described by Leo et al. (1971), Lyman (1982), and Rekker (1977). Compilations of P values are also available (Leo et al., 1971; Veith et al., 1979b; Kenaga and Goring, 1980).

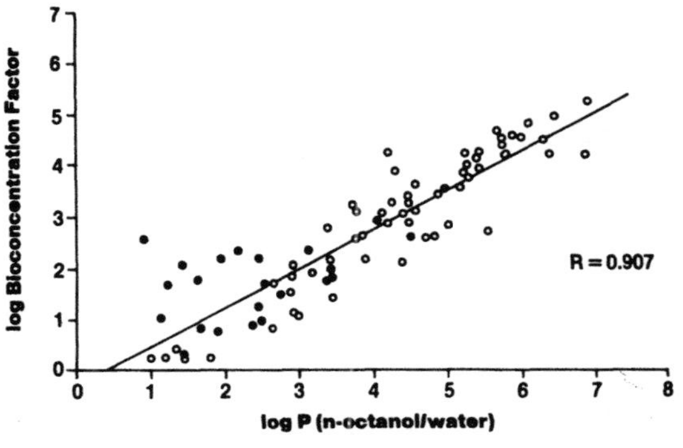

Figure 11 Correlation between bioconcentration factor and n-octanol-water partition coefficient (P) for a variety of organic chemicals in whole fish. *[From Veith et al. (1979a).]*

Partition coefficients can be estimated indirectly from a knowledge of water solubility, since the two properties are directly related. Regression equations such as the one published by Chiou et al. (1977) can be used to calculate P:

$$\log P = 5.00 - 0.670 \log S \qquad (51)$$

where S is the water solubility (μmol/l). Solubilities of a number of organics are tabulated by Kenaga and Goring (1980), Kenaga (1980), Chiou et al. (1977), and Martin and Worthington (1977).

Equation (50) is most appropriate for relatively inert compounds that do not undergo rapid biotransformation in the body. Degradable chemicals usually have a lower BCF because elimination is enhanced. For example, Southworth et al. (1980) found a lower than predicted BCF for azaarenes in fish because of rapid biotransformation and elimination.

The rate constants k_u and k_d can also be predicted for certain organics. Neely (1979) analyzed uptake rates in fish as a function of log P and respiration rate. A direct correlation between k_u and log P for organochlorines and several other chemicals is also available (Fig. 12). Based on the relationship provided by these few substances, it appears that k_u is relatively constant over a wide range of partition coefficients. By comparison, the depuration rate constant (k_d) is highly dependent on log P (Fig. 13). A regression such as the one shown in Fig. 13 is especially useful for designing bioconcentration tests because the predicted k_d can be used to estimate the time required to approach steady state [Eq. (16)] and to reach the depuration half-life [Eq. (18)]. The length of the uptake and depuration phases of the test, sampling intervals, and number of organisms required can be judged from the predicted rate constants.

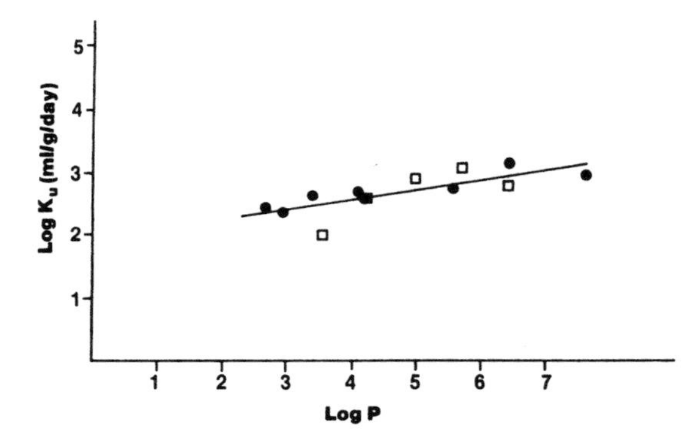

Figure 12 Correlation between uptake rate constant (k_u) and *n*-octanol-water partition coefficient (*P*) for several organic chemicals in fish. Data are for (•) rainbow trout (Neely et al., 1974) and (□) guppies (Könemann and van Leeuwen, 1980) converted to the basis of whole fish weight. *[From Spacie and Hamelink (1982).]*

TEST METHODS

Plateau Method for Bioconcentration

The most direct way to measure bioconcentration is to expose a group of test organisms to a constant concentration of a chemical in water until accumulation reaches an apparent steady state. The test usually consists of an uptake phase followed by a depuration phase. During uptake, randomized groups of test organisms are exposed to one or more chemical concentrations in a system

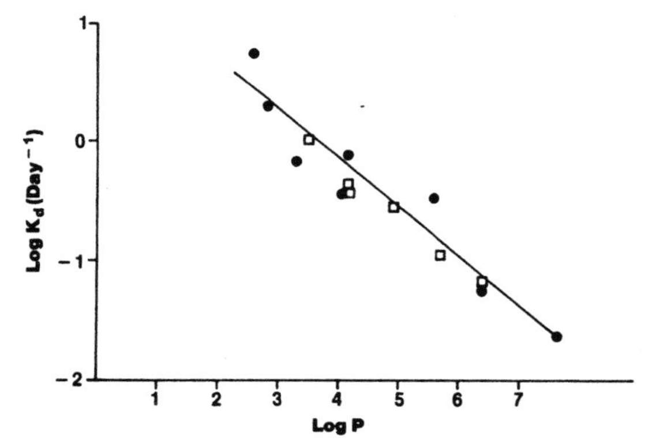

Figure 13 Correlation between depuration rate constant (k_d) and *n*-octanol-water partition coefficient (*P*) for several organic chemicals in fish. Symbols as in Fig. 12. *[From Spacie and Hamelink (1982).]*

with continuous or intermittent flow, as described for the acute toxicity tests (see Chapters 1 and 2). Basic procedures for handling and exposing the organisms are also similar to those used in toxicity tests (see Chapters 2, 3, and 4). The flow of water provided should be sufficient that the organisms do not significantly deplete either the test chemical or the oxygen in their surroundings. The test water should be analyzed before and during the uptake phase to verify the exposure concentration. A control group consisting of untreated animals should be maintained throughout the experiment to provide uncontaminated samples for analysis and to demonstrate the acceptability of the living conditions.

In situations where a flow-through aquarium system is impractical, a static exposure is sometimes adequate for measuring the accumulation of stable materials. Recommendations for conducting static bioconcentration tests with bivalves are discussed by Ernst (1977, 1979). Since the concentration of test chemical in a static system usually declines exponentially during the test, the actual analytical data for chemical concentrations over time must be fitted to an exponential equation. The bioconcentration results must then be interpreted according to the changing value of C_w. Static exposures are not recommended for long-term tests or for materials that degrade rapidly in water.

Choice of exposure levels in a bioconcentration test depends on several practical constraints. First, the chemical must be tested at a concentration high enough to permit residue analyses in both water and tissue. Nevertheless, the concentration should not be so high that it exceeds the solubility limit of the chemical in water. Concentrations that are toxic to the organisms during the test period should also be avoided, since organisms that are stressed or damaged by the material may exhibit unusual rates of accumulation. Finally, whenever possible, the test concentration should be reasonably close to those anticipated or observed in the environment, so that the bioconcentration results provide some degree of realism.

The bioconcentration test can best be designed if a preliminary estimate of k_d is available. If k_d is unknown, a pretest may be performed by exposing a small number of organisms to the chemical for several hours under static or flow-through conditions. All organisms are then transferred to uncontaminated water, where they are sampled periodically for analysis. A preliminary estimate of k_d may be found from Eq. (17).

The length of the uptake phase of the plateau test is determined by the time required to reach an apparent steady state. In practice, it is the time (t_{95}) to reach approximately 95% of C_{ss}, where $t_{95} = 3.0/k_d$. Organisms are usually sampled at least five times during the uptake phase, with the fifth sample near t_{95}. In addition, samples at two or more times after t_{95} should be taken to establish the plateau or steady-state concentration (Fig. 14). In general, if t_{95} is 12 d or less, the additional samples may be taken at 1-d intervals; if t_{95} is greater than 12 d, the interval can be two or more days. Steady state is demonstrated when replicate samples at three or more sampling intervals do not differ significantly, according to an analysis of variance test.

The depuration phase of the test, following transfer of exposed organisms to uncontaminated water, is usually continued for approximately three half-lives, or until the concentration of residues in the remaining organisms is about 10% of C_{ss}. At least four sampling times are usually used to establish k_d for the one-compartment model (the last sample time during uptake can be used as $t = 0$ for depuration if the animals are immediately transferred to uncontaminated water). Test animals may be fed daily during both phases of the test but they should not be fed on the day that they are removed for analysis.

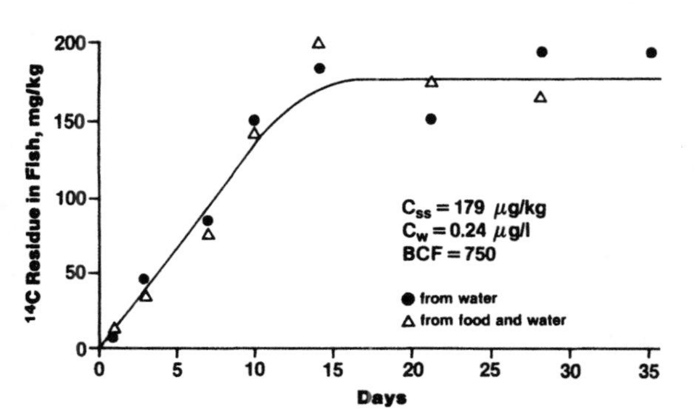

Figure 14 Accumulation of ^{14}C-labeled leptophos in bluegill sunfish (*Lepomis macrochirus*) during continuous exposure through water or food plus water. The steady-state residue concentration (C_{ss}) and bioconcentration factor are for the aqueous exposure. *[Drawn from data of Macek et al. (1979); copyright ASTM, Philadelphia, Pa. Adapted with permission.]*

Duplicate water samples and triplicate residue analyses of the organisms are typically taken at each sampling time. Each residue analysis can represent an individual, if the organisms are sufficiently large, or it can consist of a group of pooled organisms. The animals should be removed from the tanks, rinsed, killed, weighed, and preserved by an acceptable method. Residues are ordinarily reported on a whole-body, wet-weight basis. However, for special studies, residues in separate tissues (e.g., muscle, fat, blood, and fillet) or concentrations expressed on the basis of dry weight or lipid weight may be desirable. If other bases are used, a conversion factor for whole body, wet weight should be reported.

Results of the plateau test are easily calculated, since the BCF is the mean of the residue concentrations during steady state, C_{ss}, divided by the mean C_w for the uptake phase. The standard deviation should be calculated as well. The k_d for a one-compartment model is found from a regression of ln C versus time during depuration [Eq. (17)]. Multiple-compartment models may also be applied if the data warrant such treatment. An estimate of k_u can be calculated from BCF $\times$ k_d.

The plateau test has the advantage of measuring a steady-state BCF directly. It requires no assumptions about the type of compartment model or rates of metabolism. On the other hand, it is an expensive test, lasting up to 1 mo or more for chemicals with high bioconcentration potentials. The information gained in a long exposure may not justify the time and expense. Long tests may also require a correction for growth of the organisms.

Accelerated Bioconcentration Test

Branson et al. (1975) described a shorter method for measuring BCF based on the one-compartment model. The uptake phase is terminated after 4 d whether or not steady state has been reached. The depuration phase is performed as for the plateau test. After calculation of k_d, the value of k_u is calculated as illustrated in Table 2 or by a nonlinear least-squares regression. The projected BCF is k_u/k_d.

The accelerated method is more economical than the plateau test and is especially good for biologically stable materials with large BCF values. It can also be used as a range-finding test for more elaborate studies. Its limitations are those of the one-compartment model, which fails to account for "fast" and "slow" tissues, changes in biotransformation rate, irreversible binding, saturation kinetics, or other nonideal processes.

Uptake from Water and Sediment

Tests can be designed to measure the bioavailability of contaminants incorporated in sediments. It is often possible to use a simple static system containing sediments or soil equilibrated with water. Organisms are added in quantities that do not seriously deplete the total supply of test chemical or the available dissolved oxygen. The organisms can either be given direct access to the sediments or separated from the sediments by screening (Halter and Johnson, 1977).

Aquarium-scale tests typically employ about 1 kg soil for 10–15 kg water. In contrast, lakes usually have sediment-to-water proportions closer to 1:100 or 1:1000. For chemicals with a moderate to high bioconcentration potential (log P = 4–6), a 1:10 soil-water system in an aquarium leaves very little of the total chemical in solution (Table 4). The resulting low values of C_w may cause analytical problems and slow uptake by the test organisms. To avoid such problems, the equilibrium water concentration (C_w) may be estimated before designing the test, according to:

Table 4 Effect of Partition Coefficient and Soil: Water Ratio on the Proportion of Test Chemical Remaining in Water[a]

		Percent of chemical in water for soil/water ratio		
log P	K^b	1:10	1:100	1:1000
1	4.2	70	95	99
2	14.6	40	86	98
4	179	5	36	84
6	2190	0.5	4	31

[a]Each system contains enough fish to accumulate 1% of total chemical added.
[b]Calculated from Eq. (8) with 5% organic matter.

$$C_w = \frac{X_T}{G_w + G_s K} \tag{52}$$

where X_T = total grams of chemical added to the system
G_w = weight of water (g)
G_s = weight of sediment or soil (g)
K = distribution coefficient for that soil, calculated from log P or water solubility [Eqs. (7) and (8)]

The desirable biomass loading in a static uptake experiment is also influenced by log P and the soil-water ratio (Table 5). If the values of log P and ex-

Table 5 Effect of Partition Coefficient and Soil: Water Ratio on Biomass of Fish Used in Model Ecosystems[a]

		Grams of fish per kilogram of water for soil:water ratio		
log P	BCF[b]	1:10	1:100	1:1000
1	3.4	4.1	3.06	2.95
2	19.5	1.3	0.59	0.52
4	646	0.3	0.04	0.018
6	21,380	0.1	0.01	0.0015

[a]The fish accumulate 1% of the total chemical added to each test, as for Table 4.
[b]Calculated from Eq. (50).

pected BCF are high, fewer animals can be added to the system without seriously depleting the dissolved chemical. For example, Table 5 shows the weight of fish (G_f) that will accumulate 1% of the total chemical in a static system, according to:

$$G_f = \frac{0.01}{\text{BCF}} (G_w + G_s K) \tag{53}$$

At a log P of 6 (similar to that of DDT) and a soil-water ratio of 1:10, a loading of only 0.1 g of organisms per kilogram of water will accumulate approximately twice as much chemical as the amount dissolved in water. This limitation presents a serious problem in the physical design of tests to show accumulation rates over time. Isensee and Jones (1975) found a wide range of "bioaccumulation factors," depending on the soil-water ratio of static tests.

Soil-water tests can be done in flow-through systems to maintain concentrations of unstable materials or to simulate conditions in rivers or estuaries. Unless the influent test water contains a level of contaminant equal to C_w at equilibrium with the sediments, the system may continually release contaminant from the sediment. Thus the concentration in both water and sediment may change over time, making it difficult to calculate the rate constants for bioconcentration.

Aquarium-scale microcosms containing soil, water, plants, and animals of several trophic levels have been widely used to screen environmental chemicals (Metcalf et al., 1971; Isensee and Jones, 1975; Metcalf, 1977a,b). Generally they are not recommended for measuring bioaccumulation potential in aquatic animals because of their very high ratios of soil to water and of biomass to water. Larger organisms such as fish often do not reach a steady-state body burden because they are exposed to a continually changing C_w following the single or multiple applications of chemical. Such aquarium-scale microcosms are not steady-state and are difficult to replicate, it is usually more informative to use a simpler, larger system where the soil-water and biomass-water ratios can be controlled and rate constants for uptake and depuration can be measured (Ribeyre et al., 1979).

Large-scale microcosms or field enclosures designed to simulate specific portions of lake, stream, or oceanic habitats have more representative proportions of water, sediment, and biomass. Some are able to reproduce physical characteristics (such as thermal stratification, light extinction, and eddy diffusion) that are impossible to mimic in aquaria (see, for example, Hamelink and Waybrant, 1976). Large systems may be used to measure the bioconcentration and fate of a chemical, provided precautions are taken to prevent release of hazardous materials into the surrounding environment. Advantages and disadvantages of working with microcosms for chemical testing have been reviewed by Giesy (1980) and others (National Research Council, 1981). While large-scale microcosms offer a high degree of realism for chemical fate testing and for measuring responses of aquatic communities. they are often too cumbersome and costly to be used simply for bioconcentration testing.

Regardless of the physical design of the system, soil-water bioaccumulation tests should be based on a careful analysis of the dissolved contaminant concentration. The analyst must distinguish between dissolved material and material that is bound to sediments and suspended particulates. The substrate should also be characterized according to percent organic matter and particle size.

Bioaccumulation Tests

Bioaccumulation or food chain transfer tests should be undertaken only for chemicals showing a high degree of bioconcentration or for especially hazardous and persistent environmental contaminants. The relative contributions of food and water to overall bioaccumulation for a particular species can be measured by a plateau test with three groups of test organisms. Two of the groups, fed uncontaminated food, serve as the control group and the treatment group for aqueous exposure. The third group is exposed to both contaminated water and food. The water concentration (C_w) should be the same for both treated groups. The food usually consists of living organisms such as algae or microcrustaceans that have been exposed to the same C_w and are at steady state. They may be exposed in an individual chamber of the same flow-through system or, if they equilibrate rapidly, they may be exposed separately in a static system. The contaminated food can be fed alive or frozen. However, the feeding rates and type of food should be identical for all three treatment groups. The food organisms should be characterized in terms of percentages of lipid and protein. The terms R and C_f in Eq. (41) are measured directly, while the term α is found by comparing the steady-state body burdens in the "water-only" and "food-plus-water" treatments [Eq. (38)].

It is important to perform the depuration phases for both the water-only and food-plus-water groups, because the depuration rate constants may not be the same for both sources. For example, Vighi (1981) found an overall half-life of 27 d for lead in *Poecilia* fish. The lead acquired from water was eliminated with a half-life of only 9 d, while the food-derived lead required 40 d.

Experimental designs for feeding studies have been described by Johnson (1975), Petrocelli (1975), Bryan (1976), Jarvinen et al. (1977), Jarvinen and Tyo (1978), Canton et al., (1978), Macek et al. (1979), Giesy et al. (1980), and Skaar et al. (1981).

Food Chain Transfer

The bioaccumulation test may be expanded to a simple three-step food chain by using the same three treatment groups for the highest trophic level. The organisms are typically exposed in a flow-through system of tanks connected in series, with the lowest trophic level receiving the test water first. A good example of such a design is given by Vighi (1981). Other designs are described by Canton et al. (1975), Reinert (1972), Chadwick and Brocksen (1969), Ribeyre et al. (1979), and Klumpp (1980).

Feeding rates (R_i) and contaminant concentrations (C_{i-1}) at each level must be monitored throughout the test. Since the top-level organisms usually require the longest time to reach steady state, it may be a good idea to equilibrate the first two levels with the chemical first. The third-level organisms can then be brought to steady state more rapidly, saving time and the number of samples used for analysis.

If the parameters R, k_d, g, C_{i-1}, and α_i have been determined over the course of the experiment at each level, the food chain transfer numbers (f_i) can be determined for each trophic level [Eq. (44)].

CONCLUSIONS

A small number of the trace contaminants in aquatic ecosystems pose an environmental hazard because of their great toxicity or persistence. Nevertheless, it is now possible to measure bioaccumulation potential and, in many cases, to predict potential residue problems before they develop. New organic chemicals can be evaluated according to their partition coefficients and rates of biodegradation. Those with moderate to high partition coefficients can be tested for bioconcentration and, if necessary, bioaccumulation potential. Depuration rate constants can also be used to predict loss of residues from contaminated organisms following pollution abatement.

The type of kinetic model and testing scheme chosen for measuring bioaccumulation depends largely on the level of detail required. One-compartment models, log P correlations, and short-term tests are adequate for prescreening the great majority of chemicals. Those that "look suspicious" in preliminary tests can be scrutinized more carefully in bioaccumulation or food chain transfer tests. The internal dynamics of residues within the body can also be studied for especially hazardous materials. Whatever the level of detail, the primary goal in designing a bioaccumulation test should be to produce data that can be interpreted and evaluated in terms of a clearly defined conceptual model.

LITERATURE CITED

Allen JL, Hunn JB: Renal excretion in channel catfish following injection of quinaldine sulphate or 3-trifluoromethyl-4-nitrophenol. J Fish Biol 10:473–479, 1977.

Anderson PD, Spear PA: Copper pharmacokinetics in fish gills. I. Kinetics in pumpkinseed sunfish, *Lepomis gibbosus,* of different body sizes. Water Res 14:1101–1105, 1980.

Baughman GL, Lassiter RR: Prediction of environmental pollutant concentration. In: Estimating the Hazard of Chemical Substances to Aquatic Life, edited by J Cairns, Jr, KL Dickson, AW Maki, ASTM STP 657, pp. 35–54. Philadelphia, Pa.: ASTM, 1978.

Blanchard FA, Takahashi IT, Alexander HC, Bartlett EA: Uptake, clearance, and bioconcentration of ^{14}C-*sec*-butyl-4-chlorodiphenyl oxide in rainbow trout. In: Aquatic Toxicology and Hazard Evaluation, edited by FL Mayer, JL Hamelink, pp. 162–177. ASTM STP 634. Philadelphia, Pa.: ASTM, 1977.

Blau GE, Agin GL: A User's Manual for BIOFAC: A Computer Program for Characterizing the Rates of Uptake and Clearance of Chemicals in Aquatic Organisms. Midland, Mich.: Dow Chemical, 1978.

Branson DR, Blau GE, Alexander HC, Neely WB: Bioconcentration of 2,2′,4,4′-tetrachlorobiphenyl in rainbow trout as measured by an accelerated test. Trans Am Fish Soc 104:785–792, 1975.

Bruggeman WA, Martron LBJM, Kooiman D, Hutzinger, O: Accumulation and elimination kinetics of di-, tri- and tetra chlorobiphenyls by goldfish after dietary and aqueous exposure. Chemosphere 10:811–832, 1981.

Bryan GW: Some aspects of heavy metal tolerance in aquatic organism. In: Effects of Pollutants on Aquatic Organisms, edited by APM Lockwood, pp. 7–34. London: Cambridge, 1976.

Bysshe SE: Bioconcentration factor in aquatic organisms. In: Handbook of Chemical Property Estimation Methods, edited by WJ Lyman, WF Reehl, DH Rosenblatt, pp. 5-1–5-30, New York: McGraw-Hill, 1982.

Canton JH, Greve PA, Slooff W, van Esch GJ: Toxicity, accumulation and elimination studies of γ-hexachlorocyclohexane (γ-HCH) with freshwater organisms of different trophic levels. Water Res 9:1163–1169, 1975.

Canton JH, Wegman RCC, Vulto TJA, Verhoef CH, van Esch GJ: Toxicity, accumulation, and elimination studies of γ-hexachlorocyclohexane (γ-HCH) with saltwater organisms of different trophic levels. Water Res 12:687–690, 1978.

Carpene E, George SG: Absorption of cadmium by gills of *Mytilus edulis* (L.). Mol Physiol 1:23–34, 1981.

Chadwick GG, Brocksen RW: Accumulation of dieldrin by fish and selected fish-food organisms. J Wildl Manage 33(3):693–700, 1969.

Chiou CT, Freed VH, Schmedding DW, Kohnert RL: Partition coefficient and bioaccumulation of selected organic chemicals. Environ Sci Technol 11:475–478, 1977.

Cox JL: DDT residues in marine phytoplankton: Increase from 1955 to 1969. Science 170:71–73, 1970.

Crosby DG, Tucker RK: Accumulation of DDT by *Daphnia magna.* Environ Sci Technol 5:714–716, 1971.

Davies AG: In: Radioactive Contamination of the Marine Environment, pp. 403–420. Vienna: International Atomic Energy Agency, 1973.

Derr SK, Zabik MJ: Bioactive compounds in the aquatic environment: Studies on the mode of uptake of DDE by the aquatic midge, *Chironomus tentans* (Diptera: Chironomidae). Arch Environ Contam Toxicol 2(2):152–164, 1974.

Dixon WJ, Brown MB (eds.): BMDP Biomedical computer programs, P-series. Berkeley: Univ. of California Press, 1979.

Ellgehausen H, Guth JA, Essner HO: Factors determining the bioaccumulation potential of pesticides in the individual compartments of aquatic food chains. Ecotoxicol Environ Saf 4: 134–157, 1980.

Ernst W: Determination of the bioconcentration potential of marine organisms—a steady state approach. I. Bioconcentration data for seven chlorinated pesticides in mussels (*Mytilus edulis*) and their relation to solubility data. Chemosphere 6:731–740, 1977.

Ernst W: Factors affecting the evaluation of chemicals in laboratory experiments using marine organisms. Ecotoxicol Environ Saf 3: 90–98, 1979.

Fowler SW, Benayoun G: Experimental studies on cadmium flux through marine biota. In: Comparative Studies of Food and Environmental Contamination, pp. 159–178. Vienna: International Atomic Energy Agency, 1974.

Fromm PO, Hunter RC: Uptake of dieldrin by isolated perfused gills of rainbow trout. J Fish Res Board Can 26:1939–1942, 1969.

Giesy JP, Jr (ed.): Microcosms in Ecological Research. DOE CONF-781101. Springfield, Va.: National Technical Information Service, 1980.

Giesy JP, Jr, Bowling JW, Kania HJ: Cadmium and zinc accumulation and elimination by freshwater crayfish. Arch Environ Contam Toxicol 9:685–699, 1980.

Guarino AM, Arnold ST: Xenobiotic transport mechanisms and pharmacokinetics in the dogfish shark. In: Pesticide and Xenobiotic Metabolism in Aquatic Organisms, edited by MAQ Khan, JJ Lech, JJ Menn, ACS Symp Ser 99: 131–143, 1979.

Guiney PD, Melancon MJ, Jr, Lech JJ, Peterson RE: Effects of egg and sperm maturation and spawning on the distribution and elimination of a polychlorinated biphenyl in rainbow trout (*Salmo gairdneri*). Toxicol Appl Pharmacol 47: 261–272, 1979.

Guthrie FE: Absorption and distribution. In: Introduction to Biochemical Toxicology, edited by E Hodgson, FE Guthrie, pp. 10–39. New York: Elsevier, 1980.

Halter MT, Johnson HE: A model system to study the desorption and biological availability of PCB in hydrosoils. In: Aquatic Toxicology and Hazard Evaluation, edited by FL Mayer, JL Hamelink, pp. 178–195. ASTM STP 634. Philadelphia, Pa.: ASTM, 1977.

Hamelink JL: Bioavailability of chemicals in aquatic environments. In: Biotransformation and Fate of Chemicals in the Aquatic Environment, edited by AW Maki, KL Dickson, J Cairns, Jr, pp. 56–62. Washington, D.C.: American Society of Microbiology, 1980.

Hamelink JL, Waybrant RC: DDE and lindane in a large-scale model lentic ecosystem. Trans Am Fish Soc 105:124–134, 1976.

Hamelink JL, Waybrant RC, Ball RC: A proposal: Exchange equilibria control the degree chlorinated hydrocarbons are biologically magnified in lentic environments. Trans Am Fish Soc 100: 207–214, 1971.

Harding GCH, Vass WP: Uptake from seawater and clearance of *p,p'*-DDT by marine planktonic crustacea. J Fish Res Board Can 36:247–254, 1979.

Harding LW, Jr, Phillips JH, Jr: Polychlorinated biphenyl (PCB) uptake by marine phytoplankton. Mar Biol 49:103–111, 1978.

Hartung R: Pharmacokinetic approaches to the evaluation of methylmercury in fish. In: Toxicity to Biota of Metal Forms in Natural Water, edited by RW Andrew, PV Hodson, DE Konasewich, pp. 233–248. Windsor, Ontario: International Joint Commission, 1976.

Holden AV: A study of the absorption of ^{14}C-labeled DDT from water by fish. Ann Appl Biol 50:467–477, 1962.

Hunn JB, Allen JL: Movement of drugs across the gills of fishes. Annu Rev Pharmacol 14:47–55, 1974.

Isensee AR, Jones GE: Distribution of 2,3,7,8-tetrachlorodibenzo-*p*-dioxin (TCDD) in aquatic model ecosystem. Environ Sci Technol 9:668–672, 1975.

Jackim E, Morrison G, Steele R: Effects of environmental factors on radiocadmium uptake by four species of marine bivalves. Mar Biol 40: 303–308, 1977.

Jarvinen AW, Hoffman MJ, Thorslund TW: Long-term toxic effects of DDT food and water exposure on fathead minnows (*Pimephales promelas*). J Fish Res Board Can 34:2089–2103, 1977.

Jarvinen AW, Tyo RM: Toxicity to fathead minnows of endrin in food and water. Arch Environ Contam Toxicol 7:409–421, 1978.

Jennings JR, Rainbow PS: Studies on the uptake of cadmium by the crab *Carcinus maenas* in the laboratory. I. Accumulation from seawater and food source. Mar Biol 50:131–139, 1979.

Johnson BT: Aquatic food chain models for estimating bioaccumulation and biodegradation of xenobiotics. Columbia, Mo.: U.S. Dept. of Interior, Fish and Wildlife Service, Fish-Pesticide Research Laboratory, 1975.

Kenaga EE: Predicted bioconcentration factors

and soil sorption coefficients of pesticides and other chemicals. Ecotoxicol Environ Saf 4:26–38, 1980.

Kenaga EE, Goring CAI: Relationship between water solubility, soil sorption, octanol-water partitioning, and concentration of chemicals in biota. In: Aquatic Toxicology, edited by JG Eaton, PR Parrish, AC Hendricks, pp. 78–115. ASTM STP 707. Philadelphia, Pa.: ASTM, 1980.

Kikuchi M, Wakabayashi M, Kojima H, Yoshida T: Bioaccumulation profiles of ^{35}S-labeled sodium alkylpoly(oxyethylene) sulfates in carp (*Cyprinus carpio*). Water Res 14:1541–1548, 1980.

Kimerle RA, Macek KJ, Sleight BH, III, Burrows ME: Bioconcentration of linear alkylbenzene sulfonate (LAS) in bluegill (*Lepomis macrochirus*). Water Res 15:251–256, 1981.

Kinkade ML, Erdman HE: The influence of hardness components (Ca^{2+} and Mg^{2+}) in water on the uptake and concentration of cadmium in a simulated freshwater ecosystem. Environ Res 10(2):308–313, 1975.

Klaassen CD: Absorption, distribution, and excretion of toxicants. In: Toxicology: The Basic Science of Poisons, 2nd ed, edited by J Doull, CD Klaassen, MO Amdur, pp. 28–55. New York: Macmillan, 1980.

Klumpp DW: Accumulation of arsenic from water and food by *Littorina littoralis* and *Nucella lapillus*. Mar Biol 58:265–274, 1980.

Kobayashi K: Metabolism of pentachlorophenol in fish. In: Pesticide and Xenobiotic Metabolism in Aquatic Organisms, edited by MAQ Khan, JJ Lech, JJ Menn, ACS Symp Ser 99:131–143, 1979.

Könemann H, van Leeuwen K: Toxicokinetics in fish: Accumulation and elimination of six chlorobenzenes by guppies. Chemosphere 9:3–19, 1980.

Leo A, Hansch C, Elkins D: Partition coefficients and their uses. Chem Rev 71:525, 1971.

Levy G, Gibaldi M: Pharmacokinetics. In: Concepts in Biochemical Pharmacology: Handbook of Experimental Pharmacology, edited by JR Gillette, JR Mitchell, pp. 1–34. Berlin: Springer Verlag, 1975.

Lo IH, Hayton WL: Effects of pH on the accumulation of sulfonamides by fish. J Pharmacokinet Biopharmacol 9:443–459, 1981.

Lyman WJ: Octanol/water partition coefficient. In: Handbook of Chemical Property Estimation Methods, edited by WJ Lyman, WF Reehl, DH Rosenblatt, pp. 1-1–1-54. New York: McGraw-Hill, 1982.

Lynch TR: Residue dynamics and availability of 2,4,5,2′,4′,5′-hexachlorobiphenyl in aquatic model ecosystems. Ph.D. thesis, Michigan State Univ., Ann Arbor, Mich., 1979.

Macek KJ, Petrocelli SR, Sleight BH, III: Considerations in assessing the potential for, and significance of, biomagnification of chemical residues in aquatic food chains. In: Aquatic Toxicology, edited by LL Marking, RA Kimerle, pp. 251–268. ASTM STP 667. Philadelphia, Pa.: ASTM, 1979.

MacKay, D: Correlation of bioconcentration factors. Environ Sci Technol 16:274–278, 1982.

Maren TH, Embry R, Broder LE: The excretion of drugs across the gill of the dogfish, *Squalus acanthias*. Comp Biochem Physiol 26:853–864, 1968.

Martin H, Worthing CR: Pesticide Manual, 5th ed. Worcestershire, England: British Crop Protection Council, 1977.

Matsumura F: Toxicology of Insecticides. New York: Plenum, 1975.

Mayer FL: Residue dynamics of di-2-ethylhexylphthalate in fathead minnows (*Pimephales promelas*). J Fish Res Board Can 33:2610–2613, 1976.

McLeese DW, Zitko V, Sergeant DB: Uptake and excretion of fenitrothion by clams and mussels. Bull Environ Contam Toxicol 22:800–806, 1979.

Merlini M, Pozzi G: Lead and freshwater fishes. Part 1. Lead accumulation and water pH. Environ Pollut 12:168–172, 1977.

Merlini M, Pozzi G, Brazzelli A, Berg A: The transfer of ^{65}Zn from natural and synthetic foods to a freshwater fish. In: Radioecology and Energy Resources. Ecol Soc Am Spec Publ 1:226–229, 1976.

Metcalf RL: Model ecosystem approach to insecticide degradation: A critique. Annu Rev Entomol 22:241–261, 1977a.

Metcalf RL: Model ecosystem studies of bioconcentration and biodegradation of pesticides. In: Pesticides in Aquatic Environments, edited by

MAQ Khan, pp. 127–144. New York: Plenum, 1977b.

Metcalf RL, Sangha GK, Kapoor IP: Ecosystem for the evaluation of pesticide biodegradability and ecological magnification. Environ Sci Technol 5:709–713, 1971.

Metzler CM, Elfring GL, McEwen AJ: A package of computer programs for pharmacokinetic modeling. Biometrics 30:562–563, 1974.

Miettinen JK: The accumulation and excretion of heavy metals in organisms. In: Ecological Toxicology Research, edited by AD McIntyre, CF Mills, pp. 215–229. New York: Plenum, 1974.

Nagel R, Urich K: Kinetic studies on the elimination of different substituted phenols by goldfish (*Carassius auratus*). Bull Environ Contam Toxicol 24:374–378, 1980.

National Research Council: Testing for Effects of Chemicals on Ecosystems. Washington, D.C.: National Academy Press, 1981.

Neely WB: Estimating rate constants for the uptake and clearance of chemicals by fish. Environ Sci Technol 13:1506–1510, 1979.

Neely WB, Branson DR, Blau GE: Partition coefficient to measure bioconcentration potential of organic chemicals in fish. Environ Sci Technol 8:1113–1115, 1974.

Nie NH, Hull CH, Jenkins JG, Steinbrenner K, Bent DH: SPSS statistical package for the social sciences, 2nd ed., New York: McGraw-Hill, 1975.

O'Hara J: Cadmium uptake by fiddler crabs exposed to temperature and salinity stress. J Fish Res Board Can 30(6):846–848, 1973.

Olson KR, Bergman HL, Fromm PO: Uptake of methyl mercuric chloride and mercuric chloride by trout: A study of uptake pathways into the whole animal and uptake by erythrocytes in vitro. J Fish Res Board Can 30:1293–1299, 1973.

Pagenkopf GK, Russo RC, Thurston RV: Effect of complexation on toxicity of copper to fishes. J Fish Res Board Can 31:462–465, 1974.

Paris DF, Lewis DL, Barnett JT: Bioconcentration of toxaphene by microorganisms. Bull Environ Contam Toxicol 17:564–572, 1977.

Paris DF, Steen WC, Baughman GL: Rate of physico-chemical properties of Aroclors 1016 and 1242 in determining their fate and transport in aquatic environments. Chemosphere 7: 319–325, 1978.

Part P, Svanberg O: Uptake of cadmium in perfused rainbow trout (*Salmo gairdneri*) gills. Can J Fish Aquat Sci 38:917–924, 1981.

Pavlou SP, Dexter RN: Distribution of polychlorinated biphenyls (PCB) in estuarine ecosystems. Testing the concept of equilibrium partitioning in the marine environment. Environ Sci Technol 13:65–70, 1979.

Pentreath RJ: The accumulation and retention of ^{65}Zn and ^{54}Mn by the plaice, *Pleuronectes platessa* L. J Exp Mar Biol Ecol 12:1–18, 1973.

Petrocelli SR: Biomagnification of dieldrin residues by food-chain transfer from clams to blue crabs under controlled conditions. Bull Environ Contam Toxicol 13:108–116, 1975.

Poldoski JE: Cadmium bioaccumulation assays. Their relationship to various ionic equilibria in Lake Superior water. Environ Sci Technol 13: 701–706, 1979.

Premdas FH, Anderson JM: The uptake and detoxicification of C^{14}-labeled DDT in Atlantic salmon, *Salmo salar*. J Fish Res Board Can 20: 827, 1963.

Pritchard JB, Karnaky KJ, Jr, Guarino AM, Kinter WB: Renal handling of the polar DDT metabolite DDA (2,2-bis[*p*-chlorophenyl] acetic acid) by marine fish. Am J Physiol 233:F126–F132, 1977.

Reinert RE: Accumulation of dieldrin in an alga (*Scenedesmus obliquus*), *Daphnia magna*, and the guppy (*Poecilia reticulata*). J Fish Res Board Can 29:1413–1418, 1972.

Reinert RE, Stone LJ, Willford WA: Effect of temperature on accumulation of methylmercuric chloride and *p,p'*-DDT by rainbow trout (*Salmo gairdneri*). J Fish Res Board Can 31:1649–1652, 1974.

Rekker RF: The Hydrophobic Fragmental Constant. Amsterdam: Elsevier, 1977.

Renfro WC, Fowler SW, Heyraud M, La Rosa J: Relative importance of food and water in long-term zinc^{65} accumulation by marine biota. J Fish Res Board Can 32:1339–1345, 1975.

Ribeyre F, Boudou A, Delarche A: Interest of the experimental trophic chains as ecotoxicological models for the study of the ecosystem contaminations. Ecotoxicol Environ Saf 3:411–427, 1979.

Roberts JR, deFreitas ASW, Gidney, MAJ: Control factors on uptake and clearance of xenobiotic chemicals by fish. In: Animals as monitors of environmental pollutants, pp. 3–14. Washington, D.C.: National Academy of Sciences, 1979.

Rowland M: Pharmacokinetics. In: Drug Metabolism—From Microbe to Man, edited by DV Parke, RL Smith, pp. 123–145. London: Taylor and Francis, 1977.

Skaar DR, Johnson BT, Jones JR, Huckins JN: Fate of Kepone and mirex in a model aquatic environment: Sediment, fish, and diet. Can J Fish Aquat Sci 38:931–938, 1981.

Sodergren A: Uptake and accumulation of C^{14}-DDT by *Chlorella* sp. (Chlorophyceae). Oikos 19:126–138, 1968.

Southworth GR, Beauchamp JJ, Schmieder PK: Bioaccumulation potential of polycyclic aromatic hydrocarbons in *Daphnia pulex.* Water Res 12:973–977, 1978.

Southworth GR, Keffer CC, Beauchamp JJ: Potential and realized bioconcentration. A comparison of observed and predicted bioconcentration of azaarenes in the fathead minnow (*Pimephales promelas*). Environ Sci Technol 14:1529–1531, 1980.

Southworth GR, Keffer CC, Beauchamp JJ: The accumulation and disposition of benz(a)acridine in the fathead minnow, *Pimephales promelas.* Arch Environ Contam Toxicol 10:561–570, 1981.

Spacie A: The bioconcentration of trifluralin from a manufacturing effluent by fish in the Wabash River. Ph.D. thesis, Purdue Univ., West Lafayette, Ind., 1975.

Spacie A, Hamelink JL: Dynamics of trifluralin accumulation in river fishes. Environ Sci Technol 13:817–822, 1979.

Spacie A, Hamelink JL: Alternative models for describing the bioconcentration of organics in fish. Environ Toxicol Chem 1:309–320, 1982.

Spacie, A, Landrum PF, Leversee GJ: Uptake, depuration, and biotransformation of anthracene and benzo(a)pyrene in bluegill sunfish. Ecotoxicol Environ Safety 7:330–341, 1983.

Spehar RL, Anderson RL, Fiandt JT: Toxicity and bioaccumulation of cadmium and lead in aquatic invertebrates. Environ Pollut 15:195–208, 1978.

Spigarelli SA, Thommes MM, Prepejchal W: Thermal and metabolic factors affecting PCB uptake by adult brown trout. Environ Sci Technol 17: 88–94, 1983.

Sprague JB: Promising anti-pollutant: Chelating agent NTA protects fish from copper and zinc. Nature (London) 220:1345–1346, 1968.

Statham CN, Melancon MJ, Jr, Lech JJ: Bioconcentration of xenobiotics in trout bile: A proposed monitoring aid for some waterborne chemicals. Science 193:680–681, 1976.

Sunda WG, Engel DW, Thuotte RM: Effects of chemical speciation on toxicity of cadmium to grass shrimp *Palaemonetes pugio:* Importance of free cadmium ion. Environ Sci Technol 12: 409–413, 1979.

Thomann RV: Equilibrium model of fate of microcontaminants in diverse aquatic food chains. Can J Fish Aquat Sci 38:280–296, 1981.

Tsai SC, Boush GM, Matsumura F: Importance of water pH in accumulation of inorganic mercury in fish. Bull Environ Contam Toxicol 13:188–193, 1975.

Tuey DB: Toxicokinetics. In: Introduction to Biochemical Toxicology, edited by E Hodgson, FE Guthrie, pp. 40–66. New York: Elsevier, 1980.

Vanderford MJ, Hamelink JL: Influence of environmental factors on pesticide levels in sport fish. Pestic Monit J 11:138–145, 1977.

Van Weers AW: The effect of temperature on the uptake and retention of ^{60}Co and ^{65}Zn by the common shrimp (*Crangon crangon* L.). In: Combined Effects of Radioactive, Chemical and Thermal Releases into the Environment, pp. 35–49. Vienna: International Atomic Energy Agency, 1975.

Veith GD, Morris RT: A rapid method for estimating log P for organic chemicals. Ecol Res Series EPA-600/3-78-049. Duluth, Minn.: U.S. Environmental Protection Agency, 1978.

Veith GD, Macek KJ, Petrocelli SR, Carroll J: An evaluation of using partition coefficients and water solubility to estimate bioconcentration factors for organic chemicals in fish. Fed Regist 44:15926–15981, 1979a.

Veith GD, DeFoe DL, Bergstedt BV: Measuring and estimating the bioconcentration factor of chemicals in fish. J Fish Res Board Can 36: 1040–1048, 1979b.

Vighi M: Lead uptake and release in an experimental trophic chain. Ecotoxicol Environ Saf 5: 177–193, 1981.

Voice TC, Rice CP, Weber WJ, Jr: Effect of solids concentration on the sorptive partitioning of hydrophobic pollutants in aquatic systems. Environ Sci Technol 17:513–518, 1983.

Wildish DJ, Zitko V: Uptake of polychlorinated biphenyls from seawater by *Gammarus oceanicus*. Mar Biol 9:213–218, 1971.

Wilkes FG, Weiss CM: The accumulation of DDT by the dragonfly nymph, *Tetragoneuria*. Trans Am Fish Soc 100:222–236, 1971.

Wong PTS, Chau YK, Kramar O, Bengert GA: Accumulation and depuration of tetramethyllead by rainbow trout. Water Res 15:621–625, 1981.

Woodwell GM, Wurster CF, Jr, Isaacson PA: DDT residues in an East Coast estuary: A case of biological concentration of a persistent insecticide. Science 156:821–824, 1967.

Zitko V: Relationships governing the behavior of pollutants in aquatic ecosystems and their use in risk assessment. Can Tech Rep Aquat Sci, 1979.

Zitko V, Carson WV, Carson WG: Prediction of incipient lethal levels of copper to juvenile Atlantic salmon in the presence of humic acid by cupric electrode. Bull Environ Contam Toxicol 10:265–271, 1973.

SUPPLEMENTAL READING

Clark B, Smith DA: An introduction to pharmacokinetics. Oxford: Blackwell Scientific Publications, 1981.

Hodson PV, Blunt BR, Borgmann U, Minns CK, McGaw S: Effect of fluctuating lead exposures on lead accumulation by rainbow trout (*Salmo gairdneri*). Environ Toxicol Chem 2:225–238, 1983.

Oliver BG, Niimi AJ: Bioconcentration of chlorobenzenes from water by rainbow trout: Correlations with partition coefficients and environmental residues. Environ Sci Technol 17:287–291, 1983.

Phillips GR, Russo RC: Metal bioaccumulation in fishes and aquatic invertebrates: A literature review. EPA-600/3-78-103. Duluth, Minn.: Environmental Research Laboratory, U.S. Environmental Protection Agency, 1978.

Pritchard PH: Model ecosystems. In: Environmental Risk Analysis for Chemicals, edited by RA Conway, pp. 257–353. New York: Van Nostrand Reinhold, 1982.

INDEX

t indicates tables and *f* indicates figures.